Mathematical Formulas*

Quadratic Formula

If $ax^2 + bx + c = 0$, then $x = \dfrac{-b \pm \sqrt{b^2 - 4ac}}{2a}$

Binomial Theorem

$$(1 + x)^n = 1 + \frac{nx}{1!} + \frac{n(n-1)x^2}{2!} + \ldots \qquad (x^2 < 1)$$

Products of Vectors

Let θ be the smaller of the two angles between $\vec{a}$ and $\vec{b}$. Then

$$\vec{a} \cdot \vec{b} = \vec{b} \cdot \vec{a} = a_x b_x + a_y b_y + a_z b_z = ab \cos \theta$$

$$\vec{a} \times \vec{b} = -\vec{b} \times \vec{a} = \begin{vmatrix} \hat{\mathrm{i}} & \hat{\mathrm{j}} & \hat{\mathrm{k}} \\ a_x & a_y & a_z \\ b_x & b_y & b_z \end{vmatrix}$$

$$= \hat{\mathrm{i}} \begin{vmatrix} a_y & a_z \\ b_y & b_z \end{vmatrix} - \hat{\mathrm{j}} \begin{vmatrix} a_x & a_z \\ b_x & b_z \end{vmatrix} + \hat{\mathrm{k}} \begin{vmatrix} a_x & a_y \\ b_x & b_y \end{vmatrix}$$

$$= (a_y b_z - b_y a_z)\hat{\mathrm{i}} + (a_z b_x - b_z a_x)\hat{\mathrm{j}} + (a_x b_y - b_x a_y)\hat{\mathrm{k}}$$

$$|\vec{a} \times \vec{b}| = ab \sin \theta$$

Trigonometric Identities

$$\sin \alpha \pm \sin \beta = 2 \sin \tfrac{1}{2}(\alpha \pm \beta) \cos \tfrac{1}{2}(\alpha \mp \beta)$$
$$\cos \alpha + \cos \beta = 2 \cos \tfrac{1}{2}(\alpha + \beta) \cos \tfrac{1}{2}(\alpha - \beta)$$

* See Appendix E for a more complete list.

Derivatives and Integrals

$$\frac{d}{dx} \sin x = \cos x \qquad \int \sin x \, dx = -\cos x$$

$$\frac{d}{dx} \cos x = -\sin x \qquad \int \cos x \, dx = \sin x$$

$$\frac{d}{dx} e^x = e^x \qquad \int e^x \, dx = e^x$$

$$\int \frac{dx}{\sqrt{x^2 + a^2}} = \ln(x + \sqrt{x^2 + a^2})$$

$$\int \frac{x\,dx}{(x^2 + a^2)^{3/2}} = -\frac{1}{(x^2 + a^2)^{1/2}}$$

$$\int \frac{dx}{(x^2 + a^2)^{3/2}} = \frac{x}{a^2(x^2 + a^2)^{1/2}}$$

Cramer's Rule

Two simultaneous equations in unknowns x and y,

$$a_1 x + b_1 y = c_1 \qquad \text{and} \qquad a_2 x + b_2 y = c_2,$$

have the solutions

$$x = \frac{\begin{vmatrix} c_1 & b_1 \\ c_2 & b_2 \end{vmatrix}}{\begin{vmatrix} a_1 & b_1 \\ a_2 & b_2 \end{vmatrix}} = \frac{c_1 b_2 - c_2 b_1}{a_1 b_2 - a_2 b_1}$$

and

$$y = \frac{\begin{vmatrix} a_1 & c_1 \\ a_2 & c_2 \end{vmatrix}}{\begin{vmatrix} a_1 & b_1 \\ a_2 & b_2 \end{vmatrix}} = \frac{a_1 c_2 - a_2 c_1}{a_1 b_2 - a_2 b_1}.$$

The Greek Alphabet

Alpha	A	α	Iota	I	ι	Rho	P	ρ
Beta	B	β	Kappa	K	κ	Sigma	Σ	σ
Gamma	Γ	γ	Lambda	Λ	λ	Tau	T	τ
Delta	Δ	δ	Mu	M	μ	Upsilon	Y	υ
Epsilon	E	ϵ	Nu	N	ν	Phi	Φ	ϕ, φ
Zeta	Z	ζ	Xi	Ξ	ξ	Chi	X	χ
Eta	H	η	Omicron	O	o	Psi	Ψ	ψ
Theta	Θ	θ	Pi	Π	π	Omega	Ω	ω

SIXTH EDITION

Fundamentals of Physics

VOLUME 2 / EXTENDED

ENHANCED PROBLEMS VERSION

SIXTH EDITION

Fundamentals of Physics

VOLUME 2/EXTENDED

ENHANCED PROBLEMS VERSION

David Halliday
University of Pittsburgh

Robert Resnick
Rensselaer Polytechnic Institute

Jearl Walker
Cleveland State University

John Wiley & Sons, Inc.

ACQUISITIONS EDITOR Stuart Johnson
DEVELOPMENTAL EDITOR Ellen Ford
MARKETING MANAGER Bob Smith
ASSOCIATE PRODUCTION DIRECTOR Lucille Buonocore
SENIOR PRODUCTION EDITORS Monique Calello, Elizabeth Swain
TEXT/COVER DESIGNER Madelyn Lesure
COVER PHOTO Tsuyoshi Nishiinoue/Orion Press
PHOTO MANAGER Hilary Newman
PHOTO RESEARCHER Jennifer Atkins
DUMMY DESIGNER Lee Goldstein
ILLUSTRATION EDITORS Edward Starr and Anna Melhorn
ILLUSTRATION Radiant/Precision Graphics
COPYEDITOR Helen Walden
PROOFREADER Lilian Brady
TECHNICAL PROOFREADER Georgia Kamvosoulis Mederer
INDEXER Dorothy M. Jahoda

This book was set in 10/12 Times Roman by Progressive Information Technologies and printed and bound by Von Hoffmann Press, Inc. The cover was printed by Von Hoffman Press.

This book is printed on acid-free paper. 

The paper in this book was manufactured by a mill whose forest management programs include sustained yield harvesting of its timberlands. Sustained yield harvesting principles ensure that the number of trees cut each year does not exceed the amount of new growth.

ISBN 0-471-22858-3

Printed in the United States of America

10 9 8 7 6 5 4 3 2

BRIEF CONTENTS

VOLUME 1

PART 1

PART 2

VOLUME 2

PART 3

PART 4

PART 5

TABLES

CONTENTS

PART 3

CHAPTER 22

CHAPTER 23

CHAPTER 26

CHAPTER 27

CHAPTER 28

CHAPTER 29

CHAPTER 30

CHAPTER 31

PART 4

CHAPTER 34

CHAPTER 35

CHAPTER 38

PART 5

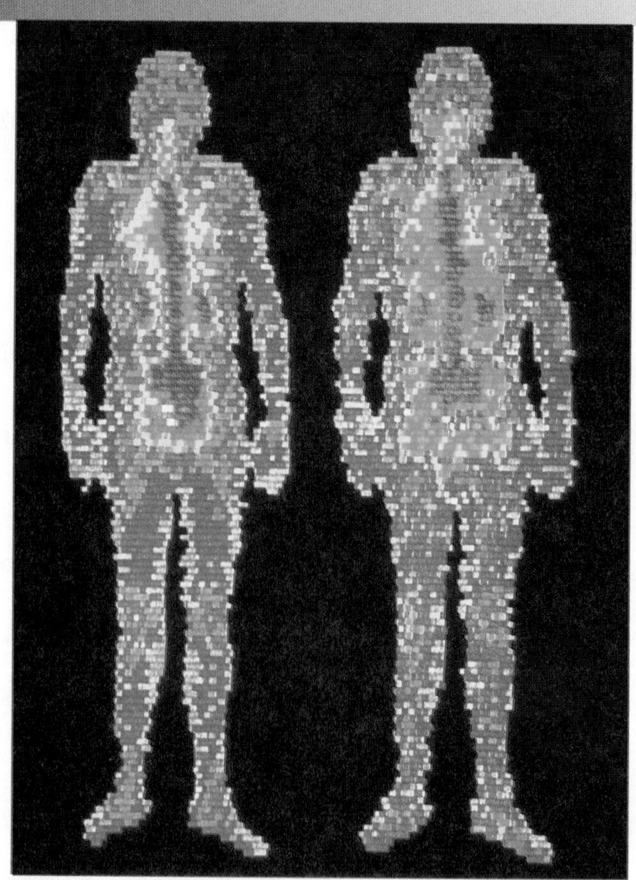

CHAPTER 39

Photons and Matter Waves 953

How can a particle such as an electron be a wave?

CHAPTER 40

More About Matter Waves 979

How can you corral an electron?

CHAPTER 41

All About Atoms 1006

What is so different about light from a laser?

CHAPTER 44

CHAPTER 45

APPENDICES

ANSWERS TO CHECKPOINTS AND ODD-NUMBERED QUESTIONS, EXERCISES, AND PROBLEMS

INDEX

PREFACE

ENHANCED PROBLEMS VERSION

The Goal

The principle goal of this book is to provide instructors with a tool by which they can teach students how to effectively read scientific material and successfully reason through scientific questions. To sharpen this tool, this enhanced version of the sixth edition of *Fundamentals of Physics* contains over 1000 new, high-quality problems that require thought and reasoning rather than simplistic plugging of data into formulas.

The new problems are printed at the end of each chapter. They are not presorted in any way for students but, pedagogically, most of them can be grouped into three types:

➤ ***Graphs as puzzles.*** These are problems that give a graph and ask for a result that requires much more than just reading off a data point from the graph. Rather, the solution requires an understanding of the physical arrangement in a problem and the principles behind the associated equations. These problems are more like puzzles in that a student must decide what data are important, like Sherlock Holmes must decide what evidence is important. Here are three examples from over 130 similar problems:

- Chapter 9, Problem N14
- Chapter 22, Problem N3
- Chapter 30, Problem N6

➤ ***Novel situations.*** Here is one example:

- Chapter 5, problem N1: A true story of how Air Canada flight 143 ran out of fuel at an altitude of 7.9 km because the crew and airport personnel did not consider the units for the fuel. An important lesson for students who tend to "blow off" units.

➤ ***Sorting-out problems.*** These are problems that focus a student on a derivation, a commonly confusing point, or the limitation of an approximation. Here are three examples:

- Chapter 11, Problem N9
- Chapter 18, Problem N14
- Chapter 26, Problem N12

Sixth Edition

The sixth edition of *Fundamentals of Physics* contains a redesign and major rewrites of the widely used fifth edition, while maintaining many elements of the classic text first written by David Halliday and Robert Resnick. Nearly all the changes are based on suggestions from instructors and students using the fifth edition, from reviewers of the manuscripts for the sixth edition, and from research done on the process of learning. You can send suggestions, corrections, and positive or negative comments to John Wiley & Sons (http://www.wiley.com/college/halliday) or Jearl Walker (mail address: Physics Department, Cleveland State University, Cleveland OH 44115 USA; fax number: (USA) (216) 687-2424; or email address: physics@wiley.com). We may not be able to respond to all suggestions, but we keep and study each of them.

Design Changes

➤ ***More open format.*** Previous editions have been printed in a double-column format, which many students and instructors have found cluttered and distracting. In this edition, the narrative is presented in a single-column format with a wide margin for note-taking.

➤ ***Streamlined presentation.*** It is a common complaint of all texts that they cover too much material. As a response to this criticism, the sixth edition has been shortened in two ways.

1. Material regarding special relativity and quantum physics has been moved from the early chapters to the later chapters devoted to those subjects.
2. Only the essential sample problems have been retained in the book.

➤ ***Vector notation.*** Vectors are now presented with an overhead arrow (such as $\vec{F}$) instead of as a bold symbol (such as **F**).

➤ ***Emphasis on metric units.*** Except in Chapter 1 (in which various systems of units are employed) and certain problems involving baseball (in which English units are traditional), metric units are used almost exclusively.

➤ ***Structured versus unstructured order of problems.*** The homework problems in this book are still ordered approximately according to their difficulty and grouped under section titles corresponding to the narrative of the chapter. However, the new problems that are now at the end of each chapter are not ordered or grouped in any way.

➤ ***Icons for additional help.*** When worked-out solutions are provided either in print or electronically for certain of the odd-numbered homework problems, the statements for those problems include a trailing icon to alert both student and instructor as to where the solutions are located. An icon guide is provided here and at the beginning of each set of homework problems:

ssm Solution is in the Student Solutions Manual.
www Solution is available on the World Wide Web at: http://www.wiley.com/college/halliday
ilw Solution is available on the Interactive LearningWare

These resources are described later in this preface.

Pedagogy Changes

➤ ***Reasoning versus plug-and-chug.*** A primary goal of this book is to teach students to reason through challenging situations, from basic principles to a solution. Although some plug-and-chug homework problems remain in this book, most homework problems emphasize reasoning.

➤ ***Key Ideas in the sample problems.*** The solutions to all the sample problems have been rewritten to begin with one or more Key Ideas based on basic principles.

➤ ***Lengthened solutions to sample problems.*** Most of the solutions to the sample problems are now longer because they build step by step from the beginning Key Ideas to an answer, often repeating some of the important reasoning of the narrative preceding the sample problems. For example, see Sample Problem 8-3 on page 148 and Sample Problem 10-2 on pages 200–201.

➤ ***Use of vector-capable calculators.*** When vector calculations in a sample problem can be performed directly on-screen with a vector-capable calculator, the solution of the sample problem indicates that fact but still carries through the traditional component analysis. When vector calculations cannot be performed directly on-screen, the solution explains why.

➤ ***Problems with applied physics,*** based on published research, have been added in many places, either as sample problems or homework problems. For example, see Sample Problem 11-6 on page 229, homework problem 64 on page 71, and homework problem 56 on page 214. For an example of homework problems that build with a continuing story, see problems 4, 32, and 48 (on pages 112, 114, and 115, respectively) in Chapter 6.

Content Changes

➤ ***Chapter 5 on force and motion*** contains clearer explanations of the gravitational force, weight, and normal force (pages 80–82).

➤ ***Chapter 7 on kinetic energy and work*** begins with a rough definition of energy. It then defines kinetic energy, work, and the work–kinetic energy theorem in ways that are more closely tied to Newton's second law than in the fifth edition, while keeping those definitions consistent with thermodynamics (pages 117–120).

➤ ***Chapter 8 on the conservation of energy*** avoids the much criticized definition of work done by a nonconservative force by explaining, instead, the energy transfers that occur due to a nonconservative force (page 153). (The wording still allows an instructor to superimpose a definition of work done by a nonconservative force.)

➤ ***Chapter 10 on collisions*** now presents the general situation of inelastic one-dimensional collisions (pages 198–200) before the special situation of elastic one-dimensional collisions (pages 202–204).

➤ ***Chapters 16, 17, and 18 on SHM and waves*** have been rewritten to better ease a student into these difficult subjects.

➤ ***Chapter 21 on entropy*** now presents a Carnot engine as the ideal heat engine with the greatest efficiency.

Chapter Features

➤ ***Opening puzzlers.*** A curious puzzling situation opens each chapter and is explained somewhere within the chapter, to entice a student to read the chapter.

➤ ***Checkpoints*** are stopping points that effectively ask the student, "Can you answer this question with some reasoning based on the narrative or sample problem that you just read?" If not, then the student should go back over that previous material before traveling deeper into the chapter. For example, see Checkpoint 3 on page 78 and Checkpoint 1 on page 101. **Answers to all checkpoints are in the back of the book.**

➤ ***Sample problems*** have been chosen to help the student organize the basic concepts of the narrative and to develop problem-solving skills. Each sample problem builds step by step from one or more Key Ideas to a solution.

➤ ***Problem-solving tactics*** contain helpful instructions to guide the beginning physics student as to how to solve problems and to avoid common errors.

➤ ***Review & Summary*** is a brief outline of the chapter contents that contains the essential concepts but which is not a substitute for reading the chapter.

➤ ***Questions*** are like the checkpoints and require reasoning and understanding rather than calculations. **Answers to the odd-number questions are in the back of the book.**

➤ ***Exercises & Problems*** are ordered approximately according to difficulty and grouped under section titles. **The odd-numbered ones are answered in the back of the book.** Worked-out solutions to the odd-numbered problems with trailing icons are available either in print or electronically. (See the icon guide at the beginning of the Exercises & Problems.) A problem number with a star indicates an especially challenging problem.

➤ ***Additional Problems*** appear at the end of the Exercises & Problems in certain chapters. They are not sorted according to section titles and many involve applied physics.

➤ ***New Problems*** appear at the ends of Chapters 1 through 38. They are not ordered or sorted in any way. Many use graphs where students must decide what data are important.

Versions of the Text

The Enhanced Problems Version of the sixth edition of *Fundamentals of Physics* is available in a number of different versions, to accommodate the individual needs of instructors and students. The Regular Edition consists of Chapters 1 through 38 (ISBN 0-471-22863-X). The Extended Edition contains seven additional chapters on quantum physics and cosmology (Chapters 1–45) (ISBN 0-471-22858-3). Both editions are available as single, hardcover books, or in the following alternative versions:

➤ ***Volume 1—Chapters 1–21 (Mechanics/Thermodynamics), hardcover, 0-471-22861-3***

➤ ***Volume 2—Chapters 22–45 (E&M and Modern Physics), hardcover, 0-471-22858-3***

➤ ***Part 1—Chapters 1–12, paperback, 0-471-22860-5***

➤ ***Part 2—Chapters 13–21, paperback, 0-471-22859-1***

➤ ***Part 3—Chapters 22–33, paperback, 0-471-22857-5***

➤ ***Part 4—Chapters 34–38, paperback, 0-471-22856-7***

➤ ***Part 5—Chapters 39–45, paperback, 0-471-36038-4***

Supplements

The sixth edition of *Fundamentals of Physics* is supplemented by a comprehensive ancillary package carefully developed to help teachers teach and students learn.

Instructor's Supplements

➤ ***Instructor's Manual*** by J. RICHARD CHRISTMAN, U.S. Coast Guard Academy. This manual contains lecture notes outlining the most important topics of each chapter, demonstration experiments, laboratory and computer projects, film and video sources, answers to all Questions, Exercises & Problems, and Checkpoints, and a correlation guide to the Questions and Exercises & Problems in the previous edition.

➤ ***Instructor's Solutions Manual*** by JAMES WHITENTON, Southern Polytechnic University. This manual provides worked-out solutions for all the exercises and problems found at the end of each chapter within the text and in the Problem Supplement #1. *This supplement is available only to instructors.*

➤ ***Test Bank*** by J. RICHARD CHRISTMAN, U.S. Coast Guard Academy. More than 2200 multiple-choice questions are included in this manual. These items are also available in the Computerized Test Bank (see below).

➤ ***Instructor's Resource CD.*** This CD contains:

- All of the Instructor's Solutions Manual in both LaTex and PDF files.
- Computerized Test Bank in both IBM and Macintosh versions, with full editing features to help instructors customize tests.
- All text illustrations suitable for both classroom presentation and printing.

➤ ***Transparencies.*** More than 200 four-color illustrations from the text are provided in a form suitable for projection in the classroom.

➤ ***On-line Course Management.***

- WebAssign, CAPA, and WebTest are on-line homework and quizzing programs that give instructors the ability to deliver and grade homework and quizzes over the Internet.
- Instructors will also have access to WebCT course materials. WebCT is a powerful Web site program that allows instructors to set up complete on-line courses with chat rooms, bulletin boards, quizzing, student tracking, etc. Please contact your local Wiley representative for more information.

Student's Supplements

➤ ***A Student Companion*** by J. RICHARD CHRISTMAN, U.S. Coast Guard Academy. This student study guide consists of a traditional print component and an accompanying Web site, which together provide a rich, interactive environment for review and study. The Student Companion Web site includes self-quizzes, simulation exercises, hints for solving end-of-chapter problems, the *Interactive LearningWare* program (see the next page), and links to other Web sites that offer physics tutorial help.

➤ ***Student Solutions Manual*** by J. RICHARD CHRISTMAN, U.S. Coast Guard Academy and EDWARD DERRINGH, Wentworth Institute. This manual provides students with complete worked-out solutions to 30 percent of the exercises and problems found at the end of each chapter within the text. These problems are indicated with an ssm icon in the text.

➤ ***Interactive LearningWare.*** This software guides students through solutions to 200 of the end-of-chapter problems. The solutions process is developed interactively, with appropriate feedback and access to error-specific help for the most common mistakes. These problems are indicated with an ilw icon in the text.

➤ ***CD-Physics, 3.0.*** This CD-ROM based version of *Fundamentals of Physics,* Sixth Edition, contains the complete, extended version of the text, *A Student's Companion,* the *Student's Solutions Manual,* the *Interactive LearningWare,* and numerous simulations all connected with extensive hyperlinking.

➤ ***Take Note!*** This bound notebook lets students take notes directly onto large, black-and-white versions of textbook illustrations. All of the illustrations from the transparency set are included. In-class time spent copying illustrations is substantially reduced by this supplement.

➤ ***Physics Web Site.*** This Web site, **http://www.wiley.com/college/halliday**, was developed specifically for *Fundamentals of Physics,* Sixth Edition, and is designed to further assist students in the study of physics and offers additional physics resources. The site also includes solutions to selected end-of-chapter problems. These problems are identified with a www icon in the text.

ACKNOWLEDGMENTS

A textbook contains far more contributions to the elucidation of a subject than those made by the authors alone. J. Richard Christman, of the U.S. Coast Guard Academy, has once again created many fine supplements for us; his knowledge of our book and his recommendations to students and faculty are invaluable. James Tanner, of Georgia Institute of Technology, and Gary Lewis, of Kennesaw State College, have provided us with innovative software, closely tied to the text exercises and problems. James Whitenton, of Southern Polytechnic State University, and Jerry Shi, of Pasadena City College, performed the Herculean task of working out solutions for every one of the Exercises & Problems in the text. We thank John Merrill, of Brigham Young University, and Edward Derringh, of the Wentworth Institute of Technology, for their many contributions in the past. We also thank George W. Hukle of Oxnard, California, for his check of the "old problems" and Renee M. Goertzen of Cleveland State University for her check of the "new problems."

At John Wiley publishers, we have been fortunate to receive strong coordination and support from our former editor, Cliff Mills. Cliff guided our efforts and encouraged us along the way. When Cliff moved on to other responsibilities at Wiley, we were ably guided to completion by his successor, Stuart Johnson. Ellen Ford has coordinated the developmental editing and multilayered preproduction process. Sue Lyons, our marketing manager, has been tireless in her efforts on behalf of this edition. Joan Kalkut has built a fine supporting package of ancillary materials. Thomas Hempstead managed the reviews of manuscript and the multiple administrative duties admirably.

We thank Lucille Buonocore and Pam Kennedy, our production directors, and Monique Calello and Elizabeth Swain, our production editors, for pulling all the pieces together and guiding us through the complex production process. We also thank Maddy Lesure, for her design; Helen Walden for her copyediting; Edward Starr and Anna Melhorn, for managing the illustration program; Georgia Kamvosoulis Mederer, Katrina Avery, and Lilian Brady, for their proofreading; and all other members of the production team.

Hilary Newman and her team of photo researchers were inspired in their search for unusual and interesting photographs that communicate physics principles beautifully. We also owe a debt of gratitude for the line art to the late John Balbalis, whose careful hand and understanding of physics can still be seen in every diagram.

We especially thank Edward Millman for his developmental work on the manuscript. With us, he has read every word, asking many questions from the point of view of a student. Many of his questions and suggested changes have added to the clarity of this volume.

We owe a particular debt of gratitude to the numerous students who used the previous editions of *Fundamentals of Physics* and took the time to fill out the response cards and return them to us. As the ultimate consumers of this text, students are extremely important to us. By sharing their opinions with us, your students help us ensure that we are providing the best possible product and the most value for their textbook dollars. We encourage the users of this book to contact us with their thoughts and concerns so that we can continue to improve this text in the years to come.

Finally, our external reviewers have been outstanding and we acknowledge here our debt to each member of that team:

Edward Adelson
Ohio State University

Mark Arnett
Kirkwood Community College

Arun Bansil
Northeastern University

J. Richard Christman
U.S. Coast Guard Academy

Robert N. Davie, Jr.
St. Petersburg Junior College

Cheryl K. Dellai
Glendale Community College

Eric R. Dietz
California State University at Chico

N. John DiNardo
Drexel University

Harold B. Hart
Western Illinois University

Rebecca Hartzler
Edmonds Community College

Joey Huston
Michigan State University

Shawn Jackson
University of Tulsa

Hector Jimenez
University of Puerto Rico

Sudhakar B. Joshi
York University

Leonard M. Kahn
University of Rhode Island

Yuichi Kubota
Cornell University

Priscilla Laws
Dickinson College

Edbertho Leal
Polytechnic University of Puerto Rico

Dale Long
Virginia Tech

Andreas Mandelis
University of Toronto

Paul Marquard
Caspar College

James Napolitano
Rensselaer Polytechnic Institute

Des Penny
Southern Utah University

Joe Redish
University of Maryland

Timothy M. Ritter
University of North Carolina at Pembroke

Gerardo A. Rodriguez
Skidmore College

John Rosendahl
University of California at Irvine

Michael Schatz
Georgia Institute of Technology

Michael G. Strauss
University of Oklahoma

Dan Styer
Oberlin College

Marshall Thomsen
Eastern Michigan University

Fred F. Tomblin
New Jersey Institute of Technology

B. R. Weinberger
Trinity College

William M. Whelan
Ryerson Polytechnic University

William Zimmerman, Jr.
University of Minnesota

Reviewers of the Fifth and Previous Editions

Maris A. Abolins
Michigan State University

Barbara Andereck
Ohio Wesleyan University

Albert Bartlett
University of Colorado

Michael E. Browne
University of Idaho

Timothy J. Burns
Leeward Community College

Joseph Buschi
Manhattan College

Philip A. Casabella
Rensselaer Polytechnic Institute

Randall Caton
Christopher Newport College

J. Richard Christman
U.S. Coast Guard Academy

Roger Clapp
University of South Florida

W. R. Conkie
Queen's University

Peter Crooker
University of Hawaii at Manoa

William P. Crummett
Montana College of Mineral Science and Technology

Eugene Dunnam
University of Florida

Robert Endorf
University of Cincinnati

F. Paul Esposito
University of Cincinnati

Jerry Finkelstein
San Jose State University

Alexander Firestone
Iowa State University

Alexander Gardner
Howard University

Andrew L. Gardner
Brigham Young University

John Gieniec
Central Missouri State University

John B. Gruber
San Jose State University

Ann Hanks
American River College

Samuel Harris
Purdue University

Emily Haught
Georgia Institute of Technology

Laurent Hodges
Iowa State University

John Hubisz
North Carolina State University

Joey Huston
Michigan State University

Darrell Huwe
Ohio University

Claude Kacser
University of Maryland

Leonard Kleinman
University of Texas at Austin

Earl Koller
Stevens Institute of Technology

Arthur Z. Kovacs
Rochester Institute of Technology

Kenneth Krane
Oregon State University

Sol Krasner
University of Illinois at Chicago

Peter Loly
University of Manitoba

Robert R. Marchini
Memphis State University

David Markowitz
University of Connecticut

Howard C. McAllister
University of Hawaii at Manoa

W. Scott McCullough
Oklahoma State University

James H. McGuire
Tulane University

David M. McKinstry
Eastern Washington University

Joe P. Meyer
Georgia Institute of Technology

Roy Middleton
University of Pennsylvania

Irvin A. Miller
Drexel University

Eugene Mosca
United States Naval Academy

Michael O'Shea
Kansas State University

Patrick Papin
San Diego State University

George Parker
North Carolina State University

Robert Pelcovits
Brown University

Oren P. Quist
South Dakota State University

Jonathan Reichart
SUNY—Buffalo

Manuel Schwartz
University of Louisville

Darrell Seeley
Milwaukee School of Engineering

Bruce Arne Sherwood
Carnegie Mellon University

John Spangler
St. Norbert College

Ross L. Spencer
Brigham Young University

Harold Stokes
Brigham Young University

Jay D. Strieb
Villanova University

David Toot
Alfred University

J. S. Turner
University of Texas at Austin

T. S. Venkataraman
Drexel University

Gianfranco Vidali
Syracuse University

Fred Wang
Prairie View A & M

Robert C. Webb
Texas A & M University

George Williams
University of Utah

David Wolfe
University of New Mexico

22 Electric Charge

If you adapt your eyes to darkness for about 15 minutes and then have a friend chew a wintergreen LifeSaver, you will see a faint flash of blue light from your friend's mouth with each chomp. (To avoid wear on the teeth, you might crush the candy with pliers, as in the photograph.)

What causes this display of light, commonly called "sparking"?

The answer is in this chapter.

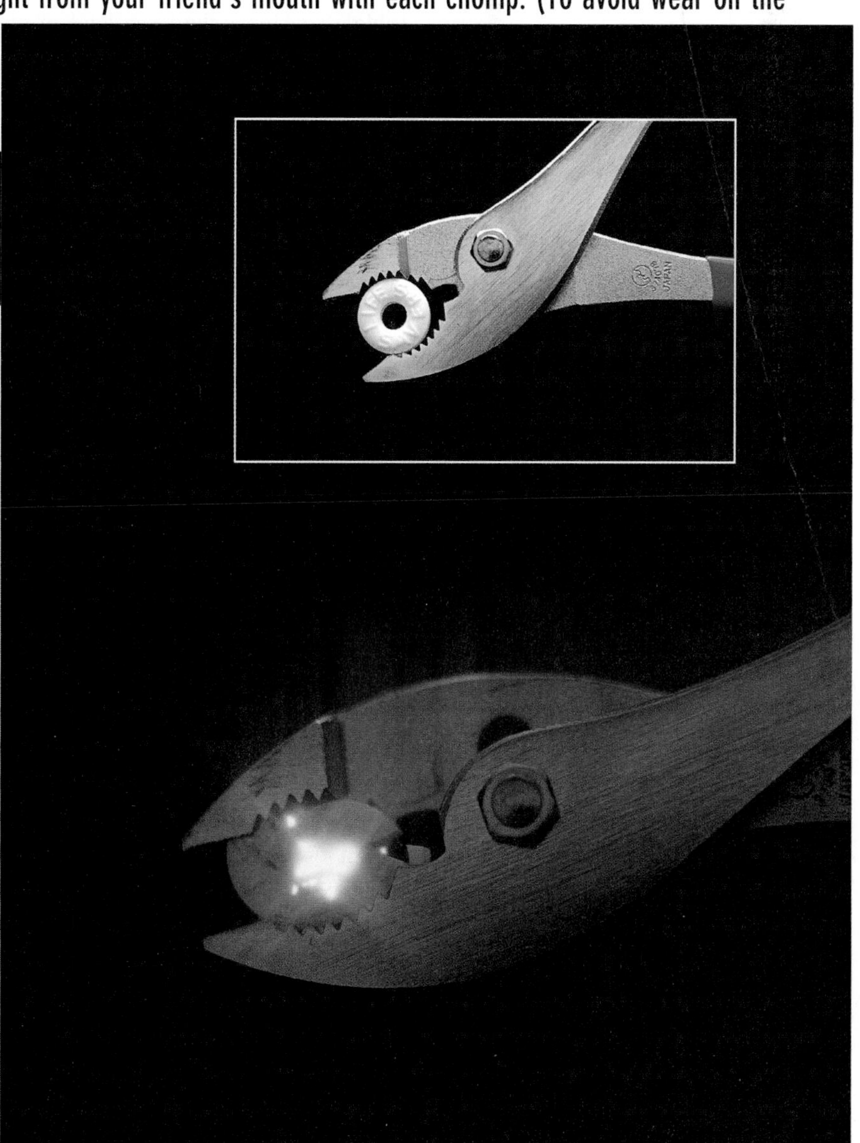

22-1 Electromagnetism

The early Greek philosophers knew that if you rubbed a piece of amber, it would attract bits of straw. This ancient observation is a direct ancestor of the electronic age in which we live. (The strength of the connection is indicated by our word *electron*, which is derived from the Greek word for amber.) The Greeks also recorded the observation that some naturally occurring "stones," known today as the mineral magnetite, would attract iron.

From these modest origins, the sciences of electricity and magnetism developed separately for centuries—until 1820, in fact, when Hans Christian Oersted found a connection between them: an electric current in a wire can deflect a magnetic compass needle. Interestingly enough, Oersted made this discovery while preparing a lecture demonstration for his physics students.

The new science of *electromagnetism* (the combination of electrical and magnetic phenomena) was developed further by workers in many countries. One of the best was Michael Faraday, a truly gifted experimenter with a talent for physical intuition and visualization. That talent is attested to by the fact that his collected laboratory notebooks do not contain a single equation. In the mid-19th century, James Clerk Maxwell put Faraday's ideas into mathematical form, introduced many new ideas of his own, and put electromagnetism on a sound theoretical basis.

Table 32-1 shows the basic laws of electromagnetism, now called Maxwell's equations. We plan to work our way through them in the chapters between here and there, but you might want to glance at them now, to see our goal.

22-2 Electric Charge

If you walk across a carpet in dry weather, you can produce a spark by bringing your finger close to a metal doorknob. Television advertising has alerted us to the problem of "static cling" in clothing (Fig. 22-1). On a grander scale, lightning is familiar to everyone. Each of these phenomena represents a tiny glimpse of the vast amount of *electric charge* that is stored in the familiar objects that surround us and—indeed—in our own bodies. **Electric charge** is an intrinsic characteristic of the fundamental particles making up those objects; that is, it is a characteristic that automatically accompanies those particles wherever they exist.

The vast amount of charge in an everyday object is usually hidden because the object contains equal amounts of the two kinds of charge: *positive charge* and *negative charge*. With such an equality—or *balance*—of charge, the object is said to be *electrically neutral*; that is, it contains no *net* charge. If the two types of charge are not in balance, then there *is* a net charge. We say that an object is *charged* to indicate that it has a charge imbalance, or net charge. The imbalance is always very small compared to the total amounts of positive charge and negative charge contained in the object.

Fig. 22-1 Static cling, an electrical phenomenon that accompanies dry weather, causes these pieces of paper to stick to one another and to the plastic comb, and your clothing to stick to your body.

Charged objects interact by exerting forces on one another. To show this, we first charge a glass rod by rubbing one end with silk. At points of contact between the rod and the silk, tiny amounts of charge are transferred from one to the other, slightly upsetting the electrical neutrality of each. (We *rub* the silk over the rod to increase the number of contact points and thus the amount, still tiny, of transferred charge.)

Suppose we now suspend the charged rod from a thread to *electrically isolate* it from its surroundings so that its charge cannot change. If we bring a second, similarly charged, glass rod nearby (Fig. 22-2*a*), the two rods *repel* each other; that is, each rod experiences a force directed away from the other rod. However, if we rub a plastic rod with fur and bring it near the suspended glass rod (Fig. 22-2*b*), the

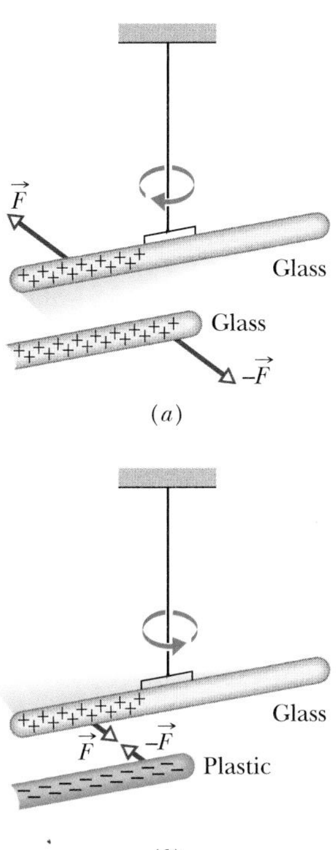

Fig. 22-2 (*a*) Two charged rods of the same signs repel each other. (*b*) Two charged rods of opposite signs attract each other. Plus signs indicate a positive net charge, and minus signs a negative net charge.

two rods *attract* each other; that is, each rod experiences a force directed toward the other rod.

We can understand these two demonstrations in terms of positive and negative charges. When a glass rod is rubbed with silk, the glass loses some of its negative charge and then has a small unbalanced positive charge (represented by the plus signs in Fig. 22-2*a*). When the plastic rod is rubbed with fur, the plastic gains a small unbalanced negative charge (represented by the minus signs in Fig. 22-2*b*). Our two demonstrations reveal the following:

> Charges with the same electrical sign repel each other, and charges with opposite electrical signs attract each other.

In Section 22-4, we shall put this rule into quantitative form as Coulomb's law of *electrostatic force* (or *electric force*) between charges. The term *electrostatic* is used to emphasize that, relative to each other, the charges are either stationary or moving only very slowly.

The "positive" and "negative" labels and signs for electric charge were chosen arbitrarily by Benjamin Franklin. He could easily have interchanged the labels or used some other pair of opposites to distinguish the two kinds of charge. (Franklin was a scientist of international reputation. It has even been said that Franklin's triumphs in diplomacy in France during the American War of Independence were facilitated, and perhaps even made possible, because he was so highly regarded as a scientist.)

The attraction and repulsion between charged bodies have many industrial applications, including electrostatic paint spraying and powder coating, fly-ash collection in chimneys, nonimpact ink-jet printing, and photocopying. Figure 22-3 shows a tiny carrier bead in a Xerox copying machine, covered with particles of black powder called *toner*, which stick to it by means of electrostatic forces. The negatively charged toner particles are eventually attracted from the carrier bead to a rotating drum, where a positively charged image of the document being copied has formed. A charged sheet of paper then attracts the toner particles from the drum to itself, after which they are heat-fused in place to produce the copy.

Fig. 22-3 A carrier bead from a Xerox copying machine; it is covered with toner particles that cling to it by electrostatic attraction. The diameter of the bead is about 0.3 mm.

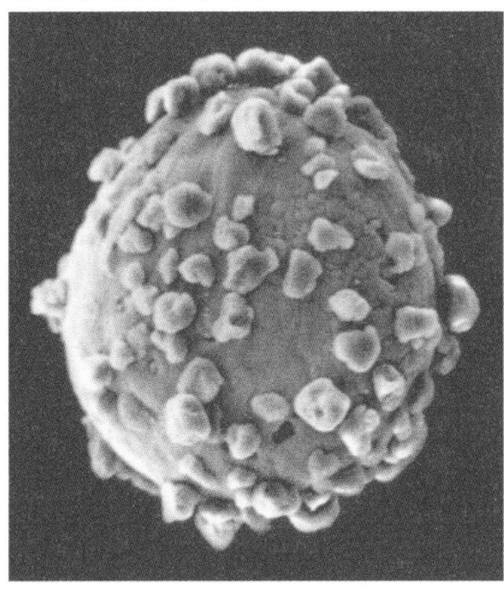

22-3 Conductors and Insulators

In some materials, such as metals, tap water, and the human body, some of the negative charge can move rather freely. We call such materials **conductors.** In other materials, such as glass, chemically pure water, and plastic, none of the charge can move freely. We call these materials **nonconductors** or **insulators.**

If you rub a copper rod with wool while holding the rod in your hand, you will not be able to charge the rod, because both you and the rod are conductors. The rubbing will cause a charge imbalance on the rod, but the excess charge will immediately move from the rod through you to the floor (which is connected to Earth's surface), and the rod will quickly be neutralized.

In thus setting up a pathway of conductors between an object and Earth's surface, we are said to *ground* the object, and in neutralizing the object (by eliminating an unbalanced positive or negative charge), we are said to *discharge* the object. If instead of holding the copper rod in your hand, you hold it by an insulating handle, you eliminate the conducting path to Earth, and the rod can then be charged by rubbing, as long as you do not touch it directly with your hand.

The properties of conductors and insulators are due to the structure and electrical nature of atoms. Atoms consist of positively charged *protons*, negatively charged

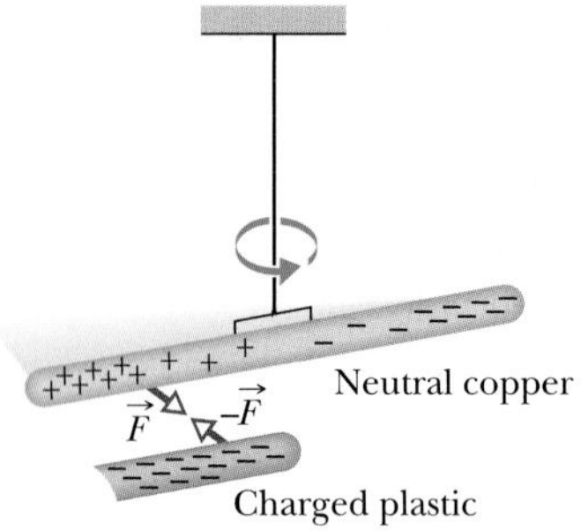

Fig. 22-4 A neutral copper rod is electrically isolated from its surroundings by being suspended on a nonconducting thread. Either end of the copper rod will be attracted by a charged rod. Here, conduction electrons in the copper rod are repelled to the far end of that rod by the negative charge on the plastic rod. Then that negative charge attracts the remaining positive charge on the near end of the copper rod, rotating the copper rod to bring that near end closer to the plastic rod.

Fig. 22-5 This is not a parlor stunt but a serious experiment carried out in 1774 to prove that the human body is a conductor of electricity. The etching shows a person suspended by nonconducting ropes while being charged by a charged rod (which probably touched flesh instead of the trousers). When the person brought his face, left hand, or the conducting ball and rod in his right hand near bits of paper on the plates, charge was induced on the paper, which then flew through the intermediate air to him.

electrons, and electrically neutral *neutrons*. The protons and neutrons are packed tightly together in a central *nucleus*.

The charge of a single electron and that of a single proton have the same magnitude but are opposite in sign. Hence, an electrically neutral atom contains equal numbers of electrons and protons. Electrons are held near the nucleus because they have the electrical sign opposite that of the protons in the nucleus and thus are attracted to the nucleus.

When atoms of a conductor like copper come together to form the solid, some of their outermost (and so most loosely held) electrons do not remain attached to the individual atoms but become free to wander about within the solid, leaving behind positively charged atoms (*positive ions*). We call the mobile electrons *conduction electrons*. There are few (if any) free electrons in a nonconductor.

The experiment of Fig. 22-4 demonstrates the mobility of charge in a conductor. A negatively charged plastic rod will attract either end of an isolated neutral copper rod. What happens is that many of the conduction electrons in the closer end of the copper rod are repelled by the negative charge on the plastic rod. They move to the far end of the copper rod, leaving the near end depleted in electrons and thus with an unbalanced positive charge. This positive charge is attracted to the negative charge in the plastic rod. Although the copper rod is still neutral, it is said to have an *induced charge*, which means that some of its positive and negative charges have been separated owing to the presence of a nearby charge.

Similarly, if a positively charged glass rod is brought near one end of a neutral copper rod, conduction electrons in the copper rod are attracted to that end. That end becomes negatively charged and the other end positively charged, so again an induced charge is set up in the copper rod. Although the copper rod is still neutral, it and the glass rod attract each other. (Figure 22-5 shows another demonstration of induced charge.)

Note that only conduction electrons, with their negative charges, can move; positive ions are fixed in place. Thus, an object becomes positively charged only through the *removal of negative charges*.

Semiconductors, such as silicon and germanium, are materials that are intermediate between conductors and insulators. The microelectronic revolution that has transformed our lives in so many ways is due to devices constructed of semiconducting materials.

Finally, there are **superconductors,** so called because they present no resistance to the movement of electric charge through them. When charge moves through a material, we say that an **electric current** exists in the material. Ordinary materials, even good conductors, tend to resist the flow of charge through them. In a superconductor, however, the resistance is not just small; it is precisely zero. If you set up a current in a superconducting ring, it lasts "forever," with no battery or other source of energy needed to maintain it.

✔CHECKPOINT 1: The figure shows five pairs of plates: A, B, and D are charged plastic plates and C is an electrically neutral copper plate. The electrostatic forces between the pairs of plates are shown for three of the pairs. For the remaining two pairs, do the plates repel or attract each other?

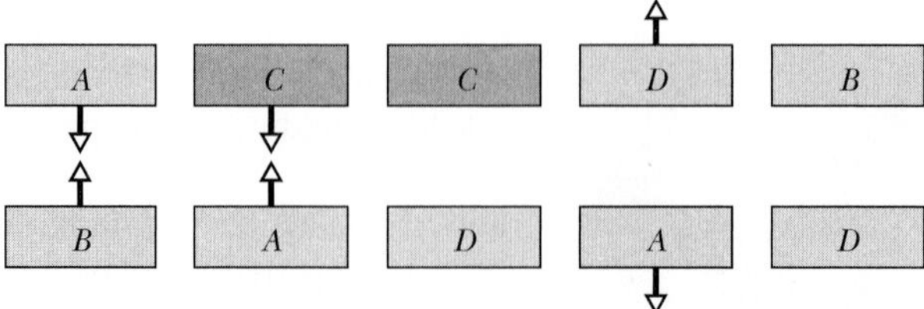

22-4 Coulomb's Law

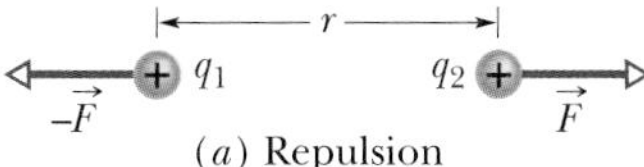

(a) Repulsion

(b) Repulsion

q_1 q_2 $\vec{F}$ $-\vec{F}$

(c) Attraction

Fig. 22-6 Two charged particles, separated by distance r, repel each other if their charges are (a) both positive and (b) both negative. (c) They attract each other if their charges are of opposite signs. In each of the three situations, the force acting on one particle is equal in magnitude to the force acting on the other particle but has the opposite direction.

Let two charged particles (also called *point charges*) have charge magnitudes q_1 and q_2 and be separated by a distance r. The **electrostatic force** of attraction or repulsion between them has the magnitude

$$F = k\frac{|q_1||q_2|}{r^2} \quad \text{(Coulomb's law)}, \tag{22-1}$$

in which k is a constant. Each particle exerts a force of this magnitude on the other particle; the two forces form a third-law force pair. If the particles *repel* each other, the force on each particle is directed *away from* the other particle (as in Figs. 22-6*a* and *b*). If the particles *attract* each other, the force on each particle is directed *toward* the other particle (as in Fig. 22-6*c*).

Equation 22-1 is called **Coulomb's law** after Charles Augustin Coulomb, whose experiments in 1785 led him to it. Curiously, the form of Eq. 22-1 is the same as that of Newton's equation for the gravitational force between two particles with masses m_1 and m_2 that are separated by a distance r:

$$F = G\frac{m_1 m_2}{r^2}, \tag{22-2}$$

in which G is the gravitational constant.

The constant k in Eq. 22-1, by analogy with the gravitational constant G in Eq. 22-2, may be called the *electrostatic constant.* Both equations describe inverse square laws that involve a property of the interacting particles—the mass in one case and the charge in the other. The laws differ in that gravitational forces are always attractive but electrostatic forces may be either attractive or repulsive, depending on the signs of the two charges. This difference arises from the fact that, although there is only one kind of mass, there are two kinds of charge (and that is why absolute value signs are needed in Eq. 22-1 but not in Eq. 22-2).

Coulomb's law has survived every experimental test; no exceptions to it have ever been found. It holds even within the atom, correctly describing the force between the positively charged nucleus and each of the negatively charged electrons, even though classical Newtonian mechanics fails in that realm and is replaced there by quantum physics. This simple law also correctly accounts for the forces that bind atoms together to form molecules, and for the forces that bind atoms and molecules together to form solids and liquids.

For practical reasons having to do with the accuracy of measurements, the SI unit of charge is derived from the SI unit of electric current, the ampere (A). The SI unit of charge is the **coulomb** (C): *One coulomb is the amount of charge that is transferred through the cross section of a wire in 1 second when there is a current of 1 ampere in the wire.* In Section 30-2 we shall describe how the ampere is defined experimentally. In general, we can write

$$dq = i\,dt, \tag{22-3}$$

in which dq (in coulombs) is the charge transferred by a current i (in amperes) during the time interval dt (in seconds).

For historical reasons (and because doing so simplifies many other formulas), the electrostatic constant k of Eq. 22-1 is usually written $1/4\pi\varepsilon_0$. Then Coulomb's law becomes

$$F = \frac{1}{4\pi\varepsilon_0}\frac{|q_1||q_2|}{r^2} \quad \text{(Coulomb's law)}. \tag{22-4}$$

The constants in Eqs. 22-1 and 22-4 have the value

$$k = \frac{1}{4\pi\varepsilon_0} = 8.99 \times 10^9\ \mathrm{N \cdot m^2/C^2}. \tag{22-5}$$

The quantity ε_0, called the **permittivity constant,** sometimes appears separately in equations and is

$$\varepsilon_0 = 8.85 \times 10^{-12}\ \mathrm{C^2/N \cdot m^2}. \tag{22-6}$$

Still another parallel between the gravitational force and the electrostatic force is that both obey the principle of superposition. If we have n charged particles, they interact independently in pairs, and the force on any one of them, let us say particle 1, is given by the vector sum

$$\vec{F}_{1,\mathrm{net}} = \vec{F}_{12} + \vec{F}_{13} + \vec{F}_{14} + \vec{F}_{15} + \cdots + \vec{F}_{1n}, \tag{22-7}$$

in which, for example, $\vec{F}_{14}$ is the force acting on particle 1 owing to the presence of particle 4. An identical formula holds for the gravitational force.

Finally, the two shell theorems that we found so useful in our study of gravitation have analogs in electrostatics:

> A shell of uniform charge attracts or repels a charged particle that is outside the shell as if all the shell's charge were concentrated at its center.

> If a charged particle is located inside a shell of uniform charge, there is no net electrostatic force on the particle from the shell.

(In the first theorem, we assume that the charge on the shell is much greater than that of the particle. Then any redistribution of the charge on the shell due to the presence of the particle's charge can be neglected.)

Spherical Conductors

If excess charge is placed on a spherical shell that is made of conducting material, the excess charge spreads uniformly over the (external) surface. For example, if we place excess electrons on a spherical metal shell, those electrons repel one another and tend to move apart, spreading over the available surface until they are uniformly distributed. That arrangement maximizes the distances between all pairs of the excess electrons. According to the first shell theorem, the shell then will attract or repel an external charge as if all the excess charge on the shell were concentrated at its center.

If we remove negative charge from a spherical metal shell, the resulting positive charge of the shell is also spread uniformly over the surface of the shell. For example, if we remove n electrons, there are then n sites of positive charge (sites missing an electron) that are spread uniformly over the shell. According to the first shell theorem, the shell will again attract or repel an external charge as if all the shell's excess charge were concentrated at its center.

CHECKPOINT 2: The figure shows two protons (symbol p) and one electron (symbol e) on an axis. What are the directions of (a) the electrostatic force on the central proton due to the electron, (b) the electrostatic force on the central proton due to the other proton, and (c) the net electrostatic force on the central proton?

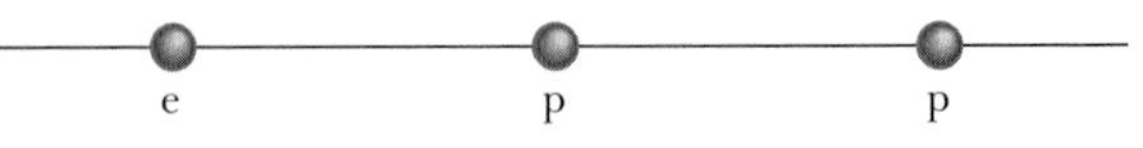

Sample Problem 22-1

(a) Figure 22-7*a* shows two positively charged particles fixed in place on an x axis. The charges are $q_1 = 1.60 \times 10^{-19}$ C and $q_2 = 3.20 \times 10^{-19}$ C, and the particle separation is $R = 0.0200$ m. What are the magnitude and direction of the electrostatic force $\vec{F}_{12}$ on particle 1 from particle 2?

SOLUTION: The **Key Idea** here is that, because both particles are positively charged, particle 1 is repelled by particle 2, with a force magnitude given by Eq. 22-4. Thus, the direction of force $\vec{F}_{12}$ on particle 1 is *away from* particle 2, in the negative direction of the x axis, as indicated in the free-body diagram of Fig. 22-7*b*. Using Eq. 22-4 with separation R substituted for r, we can write the magnitude F_{12} of this force as

$$\begin{aligned} F_{12} &= \frac{1}{4\pi\varepsilon_0}\frac{|q_1||q_2|}{R^2} \\ &= (8.99 \times 10^9\ \text{N}\cdot\text{m}^2/\text{C}^2) \\ &\quad \times \frac{(1.60 \times 10^{-19}\ \text{C})(3.20 \times 10^{-19}\ \text{C})}{(0.0200\ \text{m})^2} \\ &= 1.15 \times 10^{-24}\ \text{N}. \end{aligned}$$

Thus, force $\vec{F}_{12}$ has the following magnitude and direction (relative to the positive direction of the x axis):

$$1.15 \times 10^{-24}\ \text{N} \quad \text{and} \quad 180°. \qquad \text{(Answer)}$$

We can also write $\vec{F}_{12}$ in unit-vector notation as

$$\vec{F}_{12} = -(1.15 \times 10^{-24}\ \text{N})\hat{\text{i}}. \qquad \text{(Answer)}$$

(b) Figure 22-7*c* is identical to Fig. 22-7*a* except that particle 3 now lies on the x axis between particles 1 and 2. Particle 3 has charge $q_3 = -3.20 \times 10^{-19}$ C and is at a distance $\frac{3}{4}R$ from particle 1. What is the net electrostatic force $\vec{F}_{1,\text{net}}$ on particle 1 due to particles 2 and 3?

SOLUTION: One **Key Idea** here is that the presence of particle 3 does not alter the electrostatic force on particle 1 from particle 2. Thus, force $\vec{F}_{12}$ still acts on particle 1. Similarly, the force $\vec{F}_{13}$ that acts on particle 1 due to particle 3 is not affected by the presence of particle 2. Because particles 1 and 3 have charge of opposite sign, particle 1 is attracted to particle 3. Thus, force $\vec{F}_{13}$ is directed *toward* particle 3, as indicated in the free-body diagram of Fig. 22-7*d*.

To find the magnitude of $\vec{F}_{13}$, we can rewrite Eq. 22-4 as

$$\begin{aligned} F_{13} &= \frac{1}{4\pi\varepsilon_0}\frac{|q_1||q_3|}{(\frac{3}{4}R)^2} \\ &= (8.99 \times 10^9\ \text{N}\cdot\text{m}^2/\text{C}^2) \\ &\quad \times \frac{(1.60 \times 10^{-19}\ \text{C})(3.20 \times 10^{-19}\ \text{C})}{(\frac{3}{4})^2(0.0200\ \text{m})^2} \\ &= 2.05 \times 10^{-24}\ \text{N}. \end{aligned}$$

We can also write $\vec{F}_{13}$ in unit-vector notation:

$$\vec{F}_{13} = (2.05 \times 10^{-24}\ \text{N})\hat{\text{i}}.$$

A second **Key Idea** here is that the net force $\vec{F}_{1,\text{net}}$ on particle 1 is the vector sum of $\vec{F}_{12}$ and $\vec{F}_{13}$; that is, from Eq. 22-7, we can

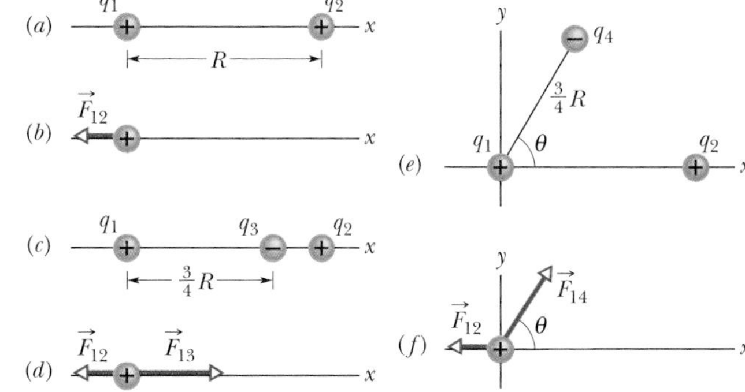

Fig. 22-7 Sample Problem 22-1. (*a*) Two charged particles of charges q_1 and q_2 are fixed in place on an x axis, with separation R. (*b*) The free-body diagram for particle 1, showing the electrostatic force on it from particle 2. (*c*) Particle 3 is now fixed in place on the x axis, along with particles 1 and 2. (*d*) The free-body diagram for particle 1. (*e*) Particle 4 is fixed in place on a line at angle θ to the x axis, again with particles 1 and 2. (*f*) The free-body diagram for particle 1.

write the net force $\vec{F}_{1,\text{net}}$ on particle 1 in unit-vector notation as

$$\begin{aligned} \vec{F}_{1,\text{net}} &= \vec{F}_{12} + \vec{F}_{13} \\ &= -(1.15 \times 10^{-24}\ \text{N})\hat{\text{i}} + (2.05 \times 10^{-24}\ \text{N})\hat{\text{i}} \\ &= (9.00 \times 10^{-25}\ \text{N})\hat{\text{i}}. \qquad \text{(Answer)} \end{aligned}$$

Thus, $\vec{F}_{1,\text{net}}$ has the following magnitude and direction (relative to the positive direction of the x axis):

$$9.00 \times 10^{-25}\ \text{N} \quad \text{and} \quad 0°. \qquad \text{(Answer)}$$

(c) Figure 22-7*e* is identical to Fig. 22-7*a* except that particle 4 is now positioned as shown. Particle 4 has charge $q_4 = -3.20 \times 10^{-19}$ C, is at a distance $\frac{3}{4}R$ from particle 1, and lies on a line that makes an angle $\theta = 60°$ with the x axis. What is the net electrostatic force $\vec{F}_{1,\text{net}}$ on particle 1 due to particles 2 and 4?

SOLUTION: The **Key Idea** here is that the net force $\vec{F}_{1,\text{net}}$ is the vector sum of $\vec{F}_{12}$ and a new force $\vec{F}_{14}$ acting on particle 1 due to particle 4. Because particles 1 and 4 have charges of opposite sign, particle 1 is attracted to particle 4. Thus, force $\vec{F}_{14}$ on particle 1 is directed *toward* particle 4, at angle $\theta = 60°$, as indicated in the free-body diagram of Fig. 22-7*f*.

To find the magnitude of $\vec{F}_{14}$, we can rewrite Eq. 22-4 as

$$\begin{aligned} F_{14} &= \frac{1}{4\pi\varepsilon_0}\frac{|q_1||q_4|}{(\frac{3}{4}R)^2} \\ &= (8.99 \times 10^9\ \text{N}\cdot\text{m}^2/\text{C}^2) \\ &\quad \times \frac{(1.60 \times 10^{-19}\ \text{C})(3.20 \times 10^{-19}\ \text{C})}{(\frac{3}{4})^2(0.0200\ \text{m})^2} \\ &= 2.05 \times 10^{-24}\ \text{N}. \end{aligned}$$

Then from Eq. 22-7, we can write the net force $\vec{F}_{1,\text{net}}$ on particle 1 as

$$\vec{F}_{1,\text{net}} = \vec{F}_{12} + \vec{F}_{14}.$$

To evaluate the right side of this equation, we need another Key Idea: Because the forces $\vec{F}_{12}$ and $\vec{F}_{14}$ are not directed along the same axis, we *cannot* sum by simply combining their magnitudes. Instead, we must add them as vectors, using one of the following methods.

Method 1. *Summing directly on a vector-capable calculator*. For $\vec{F}_{12}$, we enter the magnitude 1.15×10^{-24} and the angle 180°. For $\vec{F}_{14}$, we enter the magnitude 2.05×10^{-24} and the angle 60°. Then we add the vectors.

Method 2. *Summing in unit-vector notation*. First we rewrite $\vec{F}_{14}$ as

$$\vec{F}_{14} = (F_{14} \cos \theta)\hat{\mathrm{i}} + (F_{14} \sin \theta)\hat{\mathrm{j}}.$$

Substituting 2.05×10^{-24} N for F_{14} and 60° for θ, this becomes

$$\vec{F}_{14} = (1.025 \times 10^{-24}\ \mathrm{N})\hat{\mathrm{i}} + (1.775 \times 10^{-24}\ \mathrm{N})\hat{\mathrm{j}}.$$

Then we sum:

$$\begin{aligned}\vec{F}_{1,\mathrm{net}} &= \vec{F}_{12} + \vec{F}_{14} \\ &= -(1.15 \times 10^{-24}\ \mathrm{N})\hat{\mathrm{i}} \\ &\quad + (1.025 \times 10^{-24}\ \mathrm{N})\hat{\mathrm{i}} + (1.775 \times 10^{-24}\ \mathrm{N})\hat{\mathrm{j}} \\ &\approx (-1.25 \times 10^{-25}\ \mathrm{N})\hat{\mathrm{i}} + (1.78 \times 10^{-24}\ \mathrm{N})\hat{\mathrm{j}}.\end{aligned}$$

(Answer)

Method 3. *Summing components axis by axis*. The sum of the x components gives us

$$\begin{aligned}F_{1,\mathrm{net},x} &= F_{12,x} + F_{14,x} = F_{12} + F_{14} \cos 60° \\ &= -1.15 \times 10^{-24}\ \mathrm{N} + (2.05 \times 10^{-24}\ \mathrm{N})(\cos 60°) \\ &= -1.25 \times 10^{-25}\ \mathrm{N}.\end{aligned}$$

The sum of the y components gives us

$$\begin{aligned}F_{1,\mathrm{net},y} &= F_{12,y} + F_{14,y} = 0 + F_{14} \sin 60° \\ &= (2.05 \times 10^{-24}\ \mathrm{N})(\sin 60°) \\ &= 1.78 \times 10^{-24}\ \mathrm{N}.\end{aligned}$$

The net force $\vec{F}_{1,\mathrm{net}}$ has the magnitude

$$F_{1,\mathrm{net}} = \sqrt{F_{1,\mathrm{net},x}^2 + F_{1,\mathrm{net},y}^2} = 1.78 \times 10^{-24}\ \mathrm{N}. \quad \text{(Answer)}$$

To find the direction of $\vec{F}_{1,\mathrm{net}}$, we take

$$\theta = \tan^{-1} \frac{F_{1,\mathrm{net},y}}{F_{1,\mathrm{net},x}} = -86.0°.$$

However, this is an unreasonable result because $\vec{F}_{1,\mathrm{net}}$ must have a direction between the directions of $\vec{F}_{12}$ and $\vec{F}_{14}$. To correct θ, we add 180°, obtaining

$$-86.0° + 180° = 94.0°. \quad \text{(Answer)}$$

CHECKPOINT 3: The figure here shows three arrangements of an electron e and two protons p. (a) Rank the arrangements according to the magnitude of the net electrostatic force on the electron due to the protons, largest first. (b) In situation c, is the angle between the net force on the electron and the line labeled d less than or more than 45°?

PROBLEM-SOLVING TACTICS

Tactic 1: *Symbols Representing Charge*

Here is a general guide to the symbols representing charge. If the symbol q, with or without a subscript, is used in a sentence when no electrical sign has been specified, the charge can be either positive or negative. Sometimes the sign is explicitly shown, as in the notation $+q$ or $-q$.

When more than one charged object is being considered, their charges might be given as multiples of a charge magnitude. As examples, the notation $+2q$ means a positive charge with magnitude twice that of some reference charge magnitude q, and $-3q$ means a negative charge with magnitude three times that of the reference charge magnitude q.

Sample Problem 22-2

Figure 22-8a shows two particles fixed in place: a particle of charge $q_1 = +8q$ at the origin and a particle of charge $q_2 = -2q$ at $x = L$. At what point (other than infinitely far away) can a proton be placed so that it is in *equilibrium* (meaning that the net force on it is zero)? Is that equilibrium *stable* or *unstable*?

SOLUTION: The Key Idea here is that, if $\vec{F}_1$ is the force on the proton due to charge q_1 and $\vec{F}_2$ is the force on the proton due to charge q_2, then the point we seek is where $\vec{F}_1 + \vec{F}_2 = 0$. This condition requires that

$$\vec{F}_1 = -\vec{F}_2. \tag{22-8}$$

This tells us that at the point we seek, the forces acting on the

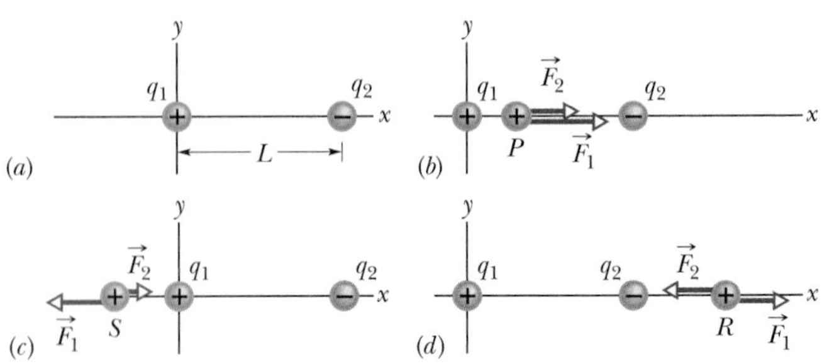

Fig. 22-8 Sample Problem 22-2. (a) Two particles of charges q_1 and q_2 are fixed in place on an x axis, with separation L. (b)–(d) Three possible locations P, S, and R for a proton. At each location, $\vec{F}_1$ is the force on the proton from particle 1 and $\vec{F}_2$ is the force on the proton from particle 2.

proton due to the other two particles must be of equal magnitudes,

$$F_1 = F_2, \tag{22-9}$$

and that the forces must have opposite directions.

A proton has a positive charge. Thus, the proton and the particle of charge q_1 are of the same sign, and force $\vec{F}_1$ on the proton must point away from q_1. Also, the proton and the particle of charge q_2 are of opposite signs, so force $\vec{F}_2$ on the proton must point toward q_2. "Away from q_1" and "toward q_2" can be in opposite directions only if the proton is located on the x axis.

If the proton is on the x axis at any point between q_1 and q_2, such as P in Fig. 22-8*b*, then $\vec{F}_1$ and $\vec{F}_2$ are in the same direction and not in opposite directions as required. If the proton is at any point on the x axis to the left of q_1, such as point S in Fig. 22-8*c*, then $\vec{F}_1$ and $\vec{F}_2$ are in opposite directions. However, Eq. 22-4 tells us that $\vec{F}_1$ and $\vec{F}_2$ cannot have equal magnitudes there: F_1 must be greater than F_2, because F_1 is produced by a closer charge (with lesser r) of greater magnitude ($8q$ versus $2q$).

Finally, if the proton is at any point on the x axis to the right of q_2, such as point R in Fig. 22-8*d*, then $\vec{F}_1$ and $\vec{F}_2$ are again in opposite directions. However, because now the charge of greater magnitude (q_1) is *farther* away from the proton than the charge of lesser magnitude, there is a point at which F_1 is equal to F_2. Let x be the coordinate of this point, and let q_p be the charge of the proton. Then with the aid of Eq. 22-4, we can rewrite Eq. 22-9 as

$$\frac{1}{4\pi\varepsilon_0}\frac{8qq_p}{x^2} = \frac{1}{4\pi\varepsilon_0}\frac{2qq_p}{(x-L)^2}. \tag{22-10}$$

(Note that only the charge magnitudes appear in Eq. 22-10.) Rearranging Eq. 22-10 gives us

$$\left(\frac{x-L}{x}\right)^2 = \frac{1}{4}.$$

After taking the square roots of both sides, we have

$$\frac{x-L}{x} = \frac{1}{2},$$

which gives us

$$x = 2L. \qquad \text{(Answer)}$$

The equilibrium at $x = 2L$ is unstable; that is, if the proton is displaced leftward from point R, then F_1 and F_2 both increase but F_2 increases more (because q_2 is closer than q_1), and a net force will drive the proton farther leftward. If the proton is displaced rightward, both F_1 and F_2 decrease but F_2 decreases more, and a net force will then drive the proton farther rightward. In a stable equilibrium, each time the proton is displaced slightly, it would return to the equilibrium position.

PROBLEM-SOLVING TACTICS

Tactic 2: *Drawing Electrostatic Force Vectors*
When you are given a diagram of charged particles, such as Fig. 22-7*a*, and are asked to find the net electrostatic force on one of them, you should usually draw a free-body diagram showing only the particle of concern and the forces *it* experiences, as in Fig. 22-7*b*. If, instead, you choose to superimpose those forces on the given diagram showing all the particles, be sure to draw the force vectors with either their tails (preferably) or their heads on the particle of concern. If you draw the vectors elsewhere in the diagram, you invite confusion—and confusion is guaranteed if you draw the vectors on the particles *causing* the forces on the particle of concern.

Sample Problem 22-3

In Fig. 22-9*a*, two identical, electrically isolated conducting spheres A and B are separated by a (center-to-center) distance a that is large compared to the spheres. Sphere A has a positive charge of $+Q$, and sphere B is electrically neutral. Initially, there is no electrostatic force between the spheres. (Assume that there is no induced charge on the spheres because of their large separation.)

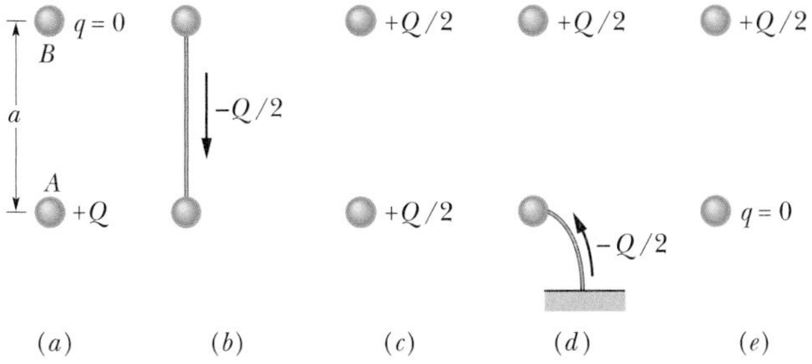

Fig. 22-9 Sample Problem 22-3. Two small conducting spheres A and B. (*a*) To start, sphere A is charged positively. (*b*) Negative charge is transferred between the spheres through a connecting wire. (*c*) Both spheres are then charged positively. (*d*) Negative charge is transferred through a grounding wire to sphere A. (*e*) Sphere A is then neutral.

(a) Suppose the spheres are connected for a moment by a conducting wire. The wire is thin enough so that any net charge on it is negligible. What is the electrostatic force between the spheres after the wire is removed?

SOLUTION: A **Key Idea** here is that when the spheres are wired together, the (negative) conduction electrons on sphere B, which always repel one another, have a way to move farther away from one another (along the wire to positively charged sphere A, which attracts them). (See Fig. 22-9*b*.) As sphere B loses negative charge, it becomes positively charged, and as A gains negative charge, it becomes *less* positively charged. A second **Key Idea** is that the spheres must end up with the same charge because they are identical. Thus, the transfer of charge stops when the excess charge on B has increased to $+Q/2$ and the excess charge on A has decreased to $+Q/2$. This condition occurs when a charge of $-Q/2$ has been transferred.

After the wire has been removed (Fig. 22-9*c*), we can assume that the charge on either sphere does not disturb the uniformity of the charge distribution on the other sphere, because the spheres are small relative to their separation. Thus, we can apply the first shell theorem to each sphere. By Eq. 22-4 with $q_1 = q_2 = Q/2$ and $r = a$, the electrostatic force between the spheres has a magnitude of

$$F = \frac{1}{4\pi\varepsilon_0}\frac{(Q/2)(Q/2)}{a^2} = \frac{1}{16\pi\varepsilon_0}\left(\frac{Q}{a}\right)^2. \quad \text{(Answer)}$$

The spheres, now positively charged, repel each other.

(b) Next, suppose sphere A is grounded momentarily, and then the ground connection is removed. What now is the electrostatic force between the spheres?

SOLUTION: The **Key Idea** here is that the ground connection allows electrons, with a total charge of $-Q/2$, to move from the ground to sphere A (Fig. 22-9*d*), neutralizing that sphere (Fig. 22-9*e*). With no charge on sphere A, there is no electrostatic force between the two spheres (just as initially, in Fig. 22-9*a*).

22-5 Charge Is Quantized

In Benjamin Franklin's day, electric charge was thought to be a continuous fluid—an idea that was useful for many purposes. However, we now know that fluids themselves, such as air and water, are not continuous but are made up of atoms and molecules; matter is discrete. Experiment shows that "electrical fluid" is also not continuous but is made up of multiples of a certain elementary charge. Any positive or negative charge q that can be detected can be written as

$$q = ne, \qquad n = \pm 1, \pm 2, \pm 3, \ldots, \tag{22-11}$$

in which e, the **elementary charge,** has the value

$$e = 1.60 \times 10^{-19}\ \text{C}. \tag{22-12}$$

The elementary charge e is one of the important constants of nature. The electron and proton both have a charge of magnitude e (Table 22-1). (Quarks, the constituent particles of protons and neutrons, have charges of $\pm e/3$ or $\pm 2e/3$, but they apparently cannot be detected individually. For this and for historical reasons, we do not take their charges to be the elementary charge.)

You often see phrases—such as "the charge on a sphere," "the amount of charge transferred," and "the charge carried by the electron"—that suggest that charge is a substance. (Indeed, such statements have already appeared in this chapter.) You should, however, keep in mind what is intended: *Particles* are the substance and charge happens to be one of their properties, just as mass is.

When a physical quantity such as charge can have only discrete values rather than any value, we say that the quantity is **quantized.** It is possible, for example, to find a particle that has no charge at all or a charge of $+10e$ or $-6e$, but not a particle with a charge of, say, $3.57e$.

The quantum of charge is small. In an ordinary 100 W lightbulb, for example, about 10^{19} elementary charges enter the bulb every second and just as many leave. However, the graininess of electricity does not show up in such large-scale phenomena (the bulb does not flicker with each electron), just as you cannot feel the individual molecules of water with your hand.

The graininess of electricity is responsible for the blue glow that is emitted by a wintergreen LifeSaver while it is being crushed. When the sugar (sucrose) crystals in the candy rupture, one part of each ruptured crystal has excess electrons while the other part has excess positive ions. Almost immediately, electrons and ions jump across the gap of the rupture to neutralize the two sides. During the jumps, the electrons and positive ions collide with nitrogen molecules in the air that is then flowing into the gap.

The collisions cause the nitrogen to emit ultraviolet light that you cannot see, as well as blue light (from the visible region of the spectrum) that is, however, too

TABLE 22-1 The Charges of Three Particles

Particle	Symbol	Charge
Electron	e or e^-	$-e$
Proton	p	$+e$
Neutron	n	0

dim to see. Oil of wintergreen in the crystals absorbs the ultraviolet light and immediately emits enough blue light to light up a mouth or a pair of pliers. However, if the candy is wet with saliva, the demonstration fails, because the conducting saliva neutralizes the two parts of a fractured crystal before sparking can occur.

CHECKPOINT 4: Initially, sphere A has a charge of $-50e$ and sphere B has a charge of $+20e$. The spheres are made of conducting material and are identical in size. If the spheres then touch, what is the resulting charge on sphere A?

Sample Problem 22-4

The nucleus in an iron atom has a radius of about 4.0×10^{-15} m and contains 26 protons.

(a) What is the magnitude of the repulsive electrostatic force between two of the protons that are separated by 4.0×10^{-15} m?

SOLUTION: The Key Idea here is that the protons can be treated as charged particles, so the magnitude of the electrostatic force on one from the other is given by Coulomb's law. Table 22-1 tells us that their charge is $+e$. Thus, Eq. 22-4 gives us

$$F = \frac{1}{4\pi\varepsilon_0}\frac{e^2}{r^2} = \frac{(8.99 \times 10^9\ \text{N}\cdot\text{m}^2/\text{C}^2)(1.60 \times 10^{-19}\ \text{C})^2}{(4.0 \times 10^{-15}\ \text{m})^2} = 14\ \text{N}. \qquad \text{(Answer)}$$

This is a small force to be acting on a macroscopic object like a cantaloupe, but an enormous force to be acting on a proton. Such forces should blow apart the nucleus of any element but hydrogen (which has only one proton in its nucleus). However, they don't, not even in nuclei with a great many protons. Therefore, there must be some enormous attractive force to counter this enormous repulsive electrostatic force.

(b) What is the magnitude of the gravitational force between those same two protons?

SOLUTION: The Key Idea here is like that in part (a): Because the protons are particles, the magnitude of the gravitational force on one from the other is given by Newton's equation for the gravitational force (Eq. 22-2). With m_p ($= 1.67 \times 10^{-27}$ kg) representing the mass of a proton, Eq. 22-2 gives us

$$F = G\frac{m_p^2}{r^2} = \frac{(6.67 \times 10^{-11}\ \text{N}\cdot\text{m}^2/\text{kg}^2)(1.67 \times 10^{-27}\ \text{kg})^2}{(4.0 \times 10^{-15}\ \text{m})^2} = 1.2 \times 10^{-35}\ \text{N}. \qquad \text{(Answer)}$$

This result tells us that the (attractive) gravitational force is far too weak to counter the repulsive electrostatic forces between protons in a nucleus. Instead, the protons are bound together by an enormous force called (aptly) the *strong nuclear force*—a force that acts between protons (and neutrons) when they are close together, as in a nucleus.

Although the gravitational force is many times weaker than the electrostatic force, it is more important in large-scale situations because it is always attractive. This means that it can collect many small bodies into huge bodies with huge masses, such as planets and stars, that then exert large gravitational forces. The electrostatic force, on the other hand, is repulsive for charges of the same sign, so it is unable to collect either positive charge or negative charge into large concentrations that would then exert large electrostatic forces.

22-6 Charge Is Conserved

If you rub a glass rod with silk, a positive charge appears on the rod. Measurement shows that a negative charge of equal magnitude appears on the silk. This suggests that rubbing does not create charge but only transfers it from one body to another, upsetting the electrical neutrality of each body during the process. This hypothesis of **conservation of charge,** first put forward by Benjamin Franklin, has stood up under close examination, both for large-scale charged bodies and for atoms, nuclei, and elementary particles. No exceptions have ever been found. Thus, we add electric charge to our list of quantities—including energy and both linear and angular momentum—that obey a conservation law.

Radioactive decay of nuclei, in which a nucleus spontaneously transforms into a different type of nucleus, gives us many instances of charge conservation at the nuclear level. For example, uranium-238, or ^{238}U, which is found in common uranium ore, can decay by emitting an alpha particle (which is a helium nucleus, ^{4}He)

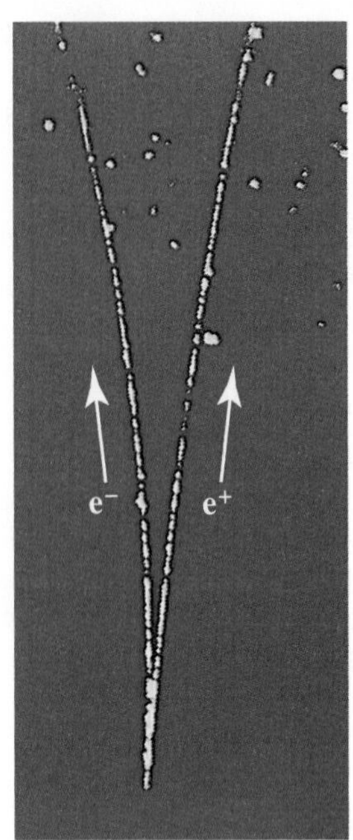

Fig. 22-10 A photograph of trails of bubbles left in a bubble chamber by an electron and a positron. The pair of particles was produced by a gamma ray that entered the chamber from the bottom. Being electrically neutral, the gamma ray did not generate a telltale trail of bubbles along its path, as the electron and positron did.

and transforming to thorium, ^{234}Th:

$$^{238}\text{U} \rightarrow {}^{234}\text{Th} + {}^{4}\text{He} \qquad \text{(radioactive decay).} \qquad (22\text{-}13)$$

The atomic number Z of the radioactive *parent* nucleus ^{238}U is 92, which tells us that this nucleus contains 92 protons and has a charge of $92e$. The emitted alpha particle has $Z = 2$, and the *daughter* nucleus ^{234}Th has $Z = 90$. Thus, the amount of charge present before the decay, $92e$, is equal to the total amount present after the decay, $90e + 2e$. Charge is conserved.

Another example of charge conservation occurs when an electron e^- (whose charge is $-e$) and its antiparticle, the *positron* e^+ (whose charge is $+e$), undergo an *annihilation process* in which they transform into two *gamma rays* (high-energy light):

$$e^- + e^+ \rightarrow \gamma + \gamma \qquad \text{(annihilation).} \qquad (22\text{-}14)$$

In applying the conservation-of-charge principle, we must add the charges algebraically, with due regard for their signs. In the annihilation process of Eq. 22-14 then, the net charge of the system is zero both before and after the event. Charge is conserved.

In *pair production*, the converse of annihilation, charge is also conserved. In this process a gamma ray transforms into an electron and a positron:

$$\gamma \rightarrow e^- + e^+ \qquad \text{(pair production).} \qquad (22\text{-}15)$$

Figure 22-10 shows such a pair-production event that occurred in a bubble chamber. A gamma ray entered the chamber from the bottom and at one point transformed into an electron and a positron. Because those new particles were charged and moving, each left a trail of tiny bubbles. (The trails were curved because a magnetic field had been set up in the chamber.) The gamma ray, being electrically neutral, left no trail. Still, you can tell exactly where it underwent pair production—at the tip of the curved **V**, where the trails of the electron and positron begin.

REVIEW & SUMMARY

Electric Charge The strength of a particle's electric interaction with objects around it depends on its **electric charge,** which can be either positive or negative. Charges with the same sign repel each other and charges with opposite signs attract each other. An object with equal amounts of the two kinds of charge is electrically neutral, whereas one with an imbalance is electrically charged.

Conductors are materials in which a significant number of charged particles (electrons in metals) are free to move. The charged particles in **nonconductors,** or **insulators,** are not free to move. When charge moves through a material, we say that an **electric current** exists in the material.

The Coulomb and Ampere The SI unit of charge is the **coulomb** (C). It is defined in terms of the unit of current, the ampere (A), as the charge passing a particular point in 1 second when there is a current of 1 ampere at that point.

Coulomb's Law *Coulomb's law* describes the **electrostatic force** between small (point) electric charges q_1 and q_2 at rest (or nearly at rest) and separated by a distance r:

$$F = \frac{1}{4\pi\varepsilon_0}\frac{|q_1||q_2|}{r^2} \qquad \text{(Coulomb's law).} \qquad (22\text{-}4)$$

Here $\varepsilon_0 = 8.85 \times 10^{-12}\ \text{C}^2/\text{N}\cdot\text{m}^2$ is the **permittivity constant,** and $1/4\pi\varepsilon_0 = k = 8.99 \times 10^9\ \text{N}\cdot\text{m}^2/\text{C}^2$.

The force of attraction or repulsion between point charges at rest acts along the line joining the two charges. If more than two charges are present, Eq. 22-4 holds for each pair of charges. The net force on each charge is then found, using the superposition principle, as the vector sum of the forces exerted on the charge by all the others.

The two shell theorems for electrostatics are

A shell of uniform charge attracts or repels a charged particle that is outside the shell as if all the shell's charge were concentrated at its center.

If a charged particle is located inside a shell of uniform charge, there is no net electrostatic force on the particle from the shell.

The Elementary Charge Electric charge is **quantized:** any charge can be written as ne, where n is a positive or negative integer, and e is a constant of nature called the **elementary charge** (approximately 1.60×10^{-19} C). Electric charge is **conserved:** the (algebraic) net charge of any isolated system cannot change.

QUESTIONS

1. Does Coulomb's law hold for all charged objects?

2. A particle of charge q is to be placed, in turn, outside four metal objects, each of uniform charge Q: (1) a large solid sphere, (2) a large spherical shell, (3) a small solid sphere, and (4) a small spherical shell. The distance between the particle and the center of the object is the same, and q is small enough not to alter significantly the uniform distribution of Q. Rank the objects according to the electrostatic force they exert on the particle, greatest first.

3. Figure 22-11 shows four situations in which charged particles are fixed in place on an axis. In which situations is there a point to the left of the particles where an electron will be in equilibrium?

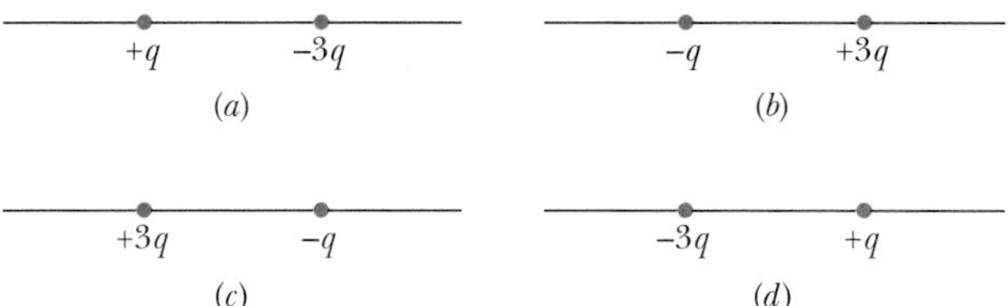

Fig. 22-11 Question 3.

4. Figure 22-12 shows two charged particles on an axis. The charges are free to move. At one point, however, a third charged particle can be placed such that all three particles are in equilibrium. (a) Is that point to the left of the first two particles, to their right, or between them? (b) Should the third particle be positively or negatively charged? (c) Is the equilibrium stable or unstable?

Fig. 22-12 Question 4.

5. In Fig. 22-13, a central particle of charge $-q$ is surrounded by two circular rings of charged particles, of radii r and R, with $R > r$. What are the magnitude and direction of the net electrostatic force on the central particle due to the other particles?

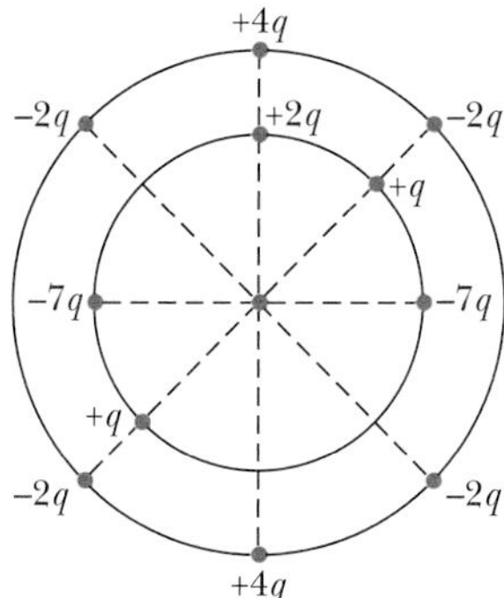

Fig. 22-13 Question 5.

6. Figure 22-14 shows four arrangements of charged particles. Rank the arrangements according to the magnitude of the net electrostatic force on the particle with charge $+Q$, greatest first.

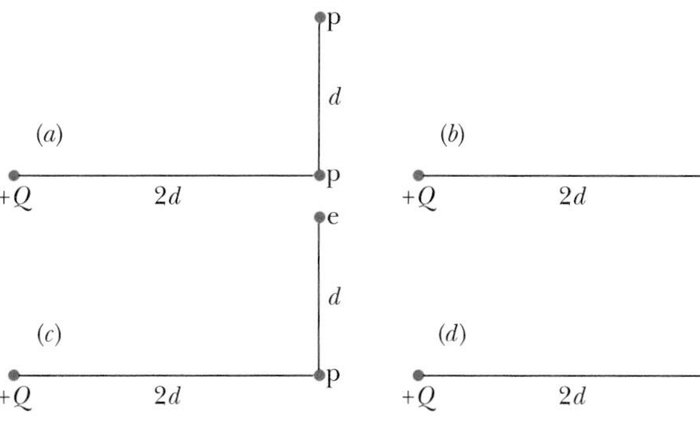

Fig. 22-14 Question 6.

7. Figure 22-15 shows four situations in which particles of charge $+q$ or $-q$ are fixed in place. In each, the particles on the x axis are equidistant from the y axis. First, consider the middle particle in situation 1; the middle particle experiences an electrostatic force from each of the other two particles. (a) Are the magnitudes F of those forces the same or different? (b) Is the magnitude of the net force on the middle particle equal to, greater than, or less than $2F$? (c) Do the x components of the two forces add or cancel? (d) Do their y components add or cancel? (e) Is the direction of the net force on the middle particle that of the canceling components or the adding components? (f) What is the direction of that net force? Now consider the remaining situations: What is the direction of the net force on the middle particle in (g) situation 2, (h) situation 3, and (i) situation 4? (In each, consider the symmetry of the charge distribution and determine the canceling components and the adding components.)

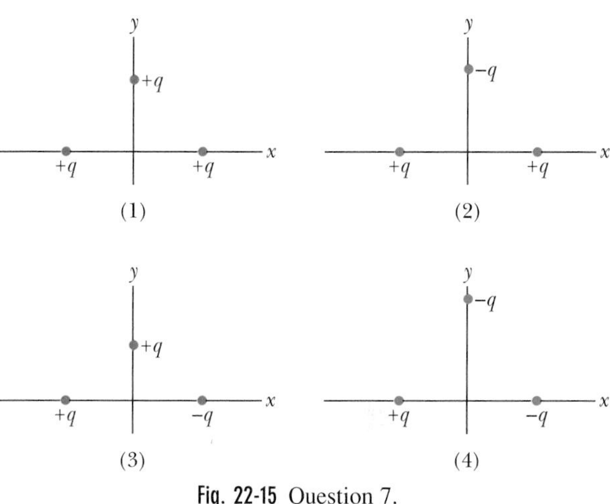

Fig. 22-15 Question 7.

8. A positively charged ball is brought close to a neutral isolated conductor. The conductor is then grounded while the ball is kept close. Is the conductor charged positively or negatively, or is it neutral, if (a) the ball is first taken away and then the ground connection is removed and (b) the ground connection is first removed and then the ball is taken away?

9. (a) A positively charged glass rod attracts an object suspended by a nonconducting thread. Is the object definitely negatively charged or only possibly negatively charged? (b) A positively charged glass rod repels a similarly suspended object. Is the object definitely positively charged or only possibly?

10. In Fig. 22-4, the nearby (negatively charged) plastic rod causes some of the conduction electrons in the copper rod to move to the far end of the copper rod. Why does the flow of the conduction electrons quickly cease? After all, a huge number of them are free to move to that far end.

11. A person standing on an electrically insulated platform touches a charged, electrically isolated conductor. Does this discharge the conductor completely?

EXERCISES & PROBLEMS

ssm Solution is in the Student Solutions Manual.
www Solution is available on the World Wide Web at: http://www.wiley.com/college/hrw
ilw Solution is available on the Interactive LearningWare.

SEC. 22-4 Coulomb's Law

1E. What must be the distance between point charge $q_1 = 26.0\ \mu C$ and point charge $q_2 = -47.0\ \mu C$ for the electrostatic force between them to have a magnitude of 5.70 N? ssm

2E. A point charge of $+3.00 \times 10^{-6}$ C is 12.0 cm distant from a second point charge of -1.50×10^{-6} C. Calculate the magnitude of the force on each charge.

3E. Two equally charged particles, held 3.2×10^{-3} m apart, are released from rest. The initial acceleration of the first particle is observed to be 7.0 m/s^2 and that of the second to be 9.0 m/s^2. If the mass of the first particle is 6.3×10^{-7} kg, what are (a) the mass of the second particle and (b) the magnitude of the charge of each particle? ilw

4E. Identical isolated conducting spheres 1 and 2 have equal charges and are separated by a distance that is large compared with their diameters (Fig. 22-16*a*). The electrostatic force acting on sphere 2 due to sphere 1 is $\vec{F}$. Suppose now that a third identical sphere 3, having an insulating handle and initially neutral, is touched first to sphere 1 (Fig. 22-16*b*), then to sphere 2 (Fig. 22-16*c*), and finally removed (Fig. 22-16*d*). In terms of magnitude F, what is the magnitude of the electrostatic force $\vec{F}'$ that now acts on sphere 2?

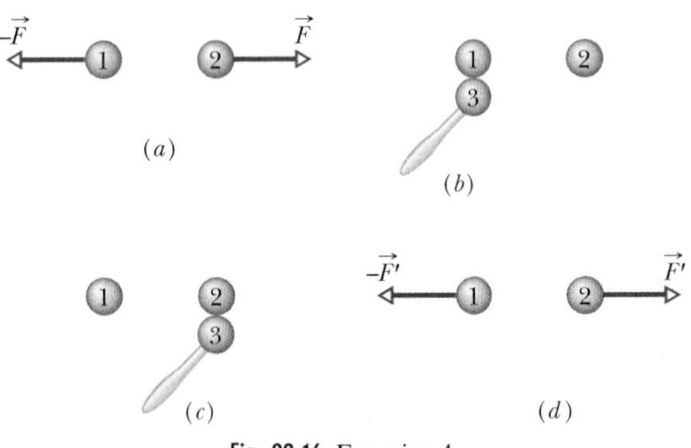

Fig. 22-16 Exercise 4.

5P. In Fig. 22-17, what are the (a) horizontal and (b) vertical components of the net electrostatic force on the charged particle in the lower left corner of the square if $q = 1.0 \times 10^{-7}$ C and $a = 5.0$ cm? ilw

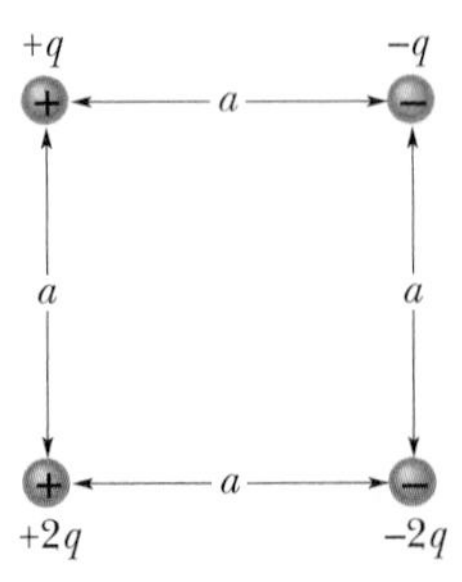

Fig. 22-17 Problem 5.

6P. Point charges q_1 and q_2 lie on the x axis at points $x = -a$ and $x = +a$, respectively. (a) How must q_1 and q_2 be related for the net electrostatic force on point charge $+Q$, placed at $x = +a/2$, to be zero? (b) Repeat (a) but with point charge $+Q$ now placed at $x = +3a/2$.

7P. Two identical conducting spheres, fixed in place, attract each other with an electrostatic force of 0.108 N when separated by 50.0 cm, center-to-center. The spheres are then connected by a thin conducting wire. When the wire is removed, the spheres repel each other with an electrostatic force of 0.0360 N. What were the initial charges on the spheres? ssm

8P. In Fig. 22-18, three charged particles lie on a straight line and are separated by distances d. Charges q_1 and q_2 are held fixed. Charge q_3 is free to move but happens to be in equilibrium (no net electrostatic force acts on it). Find q_1 in terms of q_2.

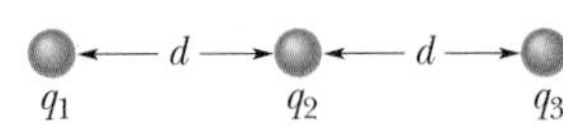

Fig. 22-18 Problem 8.

9P. Two *free* particles (that is, free to move) with charges $+q$ and $+4q$ are a distance L apart. A third charge is placed so that the entire system is in equilibrium. (a) Find the location, magnitude, and sign of the third charge. (b) Show that the equilibrium is unstable. ssm www

10P. Two fixed particles, of charges $q_1 = +1.0\ \mu C$ and $q_2 = -3.0\ \mu C$, are 10 cm apart. How far from each should a third charge be located so that no net electrostatic force acts on it?

11P. (a) What equal positive charges would have to be placed on Earth and on the Moon to neutralize their gravitational attraction? Do you need to know the lunar distance to solve this problem? Why or why not? (b) How many kilograms of hydrogen would be needed to provide the positive charge calculated in (a)? ssm

12P. The charges and coordinates of two charged particles held fixed in the xy plane are $q_1 = +3.0\ \mu C$, $x_1 = 3.5$ cm, $y_1 = 0.50$ cm, and $q_2 = -4.0\ \mu C$, $x_2 = -2.0$ cm, $y_2 = 1.5$ cm. (a) Find the magnitude and direction of the electrostatic force on q_2. (b) Where could you locate a third charge $q_3 = +4.0\ \mu C$ such that the net electrostatic force on q_2 is zero?

13P. A certain charge Q is divided into two parts q and $Q - q$, which are then separated by a certain distance. What must q be in terms of Q to maximize the electrostatic repulsion between the two charges? ssm ilw

14P. A particle with charge Q is fixed at each of two opposite corners of a square, and a particle with charge q is placed at each of the other two corners. (a) If the net electrostatic force on each particle with charge Q is zero, what is Q in terms of q? (b) Is there any value of q that makes the net electrostatic force on each of the four particles zero? Explain.

15P. In Fig. 22-19, two tiny conducting balls of identical mass m and identical charge q hang from nonconducting threads of length L. Assume that θ is so small that

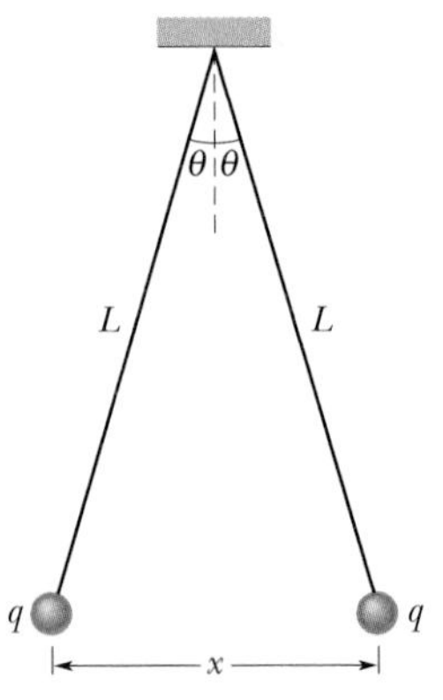

Fig. 22-19 Problem 15.

tan θ can be replaced by its approximate equal, sin θ. (a) Show that, for equilibrium,

$$x = \left(\frac{q^2L}{2\pi\varepsilon_0 mg}\right)^{1/3},$$

where x is the separation between the balls. (b) If $L = 120$ cm, $m = 10$ g, and $x = 5.0$ cm, what is q? ssm

16P. Explain what happens to the balls of Problem 15b if one of them is discharged (loses it charge q to, say, the ground), and find the new equilibrium separation x, using the given values of L and m and the computed value of q.

17P. Figure 22-20 shows a long, nonconducting, massless rod of length L, pivoted at its center and balanced with a block of weight W at a distance x from the left end. At the left and right ends of the rod are attached small conducting spheres with positive charges q and $2q$, respectively. A distance h directly beneath each of these spheres is a fixed sphere with positive charge Q. (a) Find the distance x when the rod is horizontal and balanced. (b) What value should h have so that the rod exerts no vertical force on the bearing when the rod is horizontal and balanced? ssm

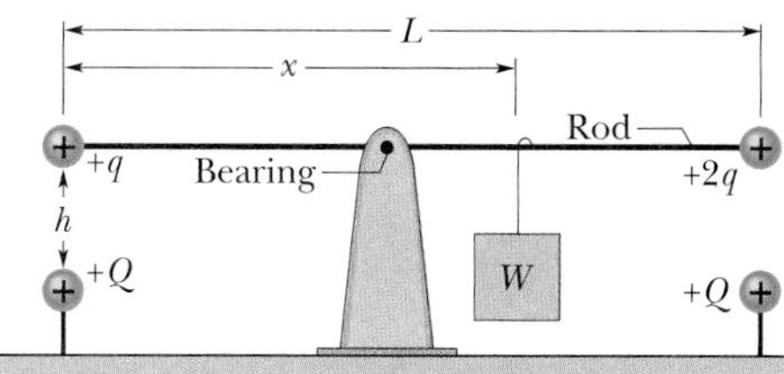

Fig. 22-20 Problem 17.

SEC. 22-5 Charge Is Quantized

18E. What is the magnitude of the electrostatic force between a singly charged sodium ion (Na^+, of charge $+e$) and an adjacent singly charged chlorine ion (Cl^-, of charge $-e$) in a salt crystal if their separation is 2.82×10^{-10} m?

19E. What is the total charge in coulombs of 75.0 kg of electrons? ssm

20E. How many megacoulombs of positive (or negative) charge are in 1.00 mol of neutral molecular-hydrogen gas (H_2)?

21E. The magnitude of the electrostatic force between two identical ions that are separated by a distance of 5.0×10^{-10} m is 3.7×10^{-9} N. (a) What is the charge of each ion? (b) How many electrons are "missing" from each ion (thus giving the ion its charge imbalance)? ssm

22E. Two tiny, spherical water drops, with identical charges of -1.00×10^{-16} C, have a center-to-center separation of 1.00 cm. (a) What is the magnitude of the electrostatic force acting between them? (b) How many excess electrons are on each drop, giving it its charge imbalance?

23E. How many electrons would have to be removed from a coin to leave it with a charge of $+1.0 \times 10^{-7}$ C? ilw

24E. An electron is in a vacuum near the surface of Earth. Where should a second electron be placed so that the electrostatic force it exerts on the first electron balances the gravitational force on the first electron due to Earth?

25P. Earth's atmosphere is constantly bombarded by *cosmic ray protons* that originate somewhere in space. If the protons all passed through the atmosphere, each square meter of Earth's surface would intercept protons at the average rate of 1500 protons per second. What would be the corresponding electric current intercepted by the total surface area of the planet? ilw

26P. Calculate the number of coulombs of positive charge in 250 cm^3 of (neutral) water (about a glassful).

27P. In the basic CsCl (cesium chloride) crystal structure, Cs^+ ions form the corners of a cube and a Cl^- ion is at the cube's center (Fig. 22-21). The edge length of the cube is 0.40 nm. The Cs^+ ions are each deficient by one electron (and thus each has a charge of $+e$), and the Cl^- ion has one excess electron (and thus has a charge of $-e$). (a) What is the magnitude of the net electrostatic force exerted on the Cl^- ion by the eight Cs^+ ions at the corners of the cube? (b) If one of the Cs^+ ions is missing, the crystal is said to have a *defect*; what is the magnitude of the net electrostatic force exerted on the Cl^- ion by the seven remaining Cs^+ ions? ssm www

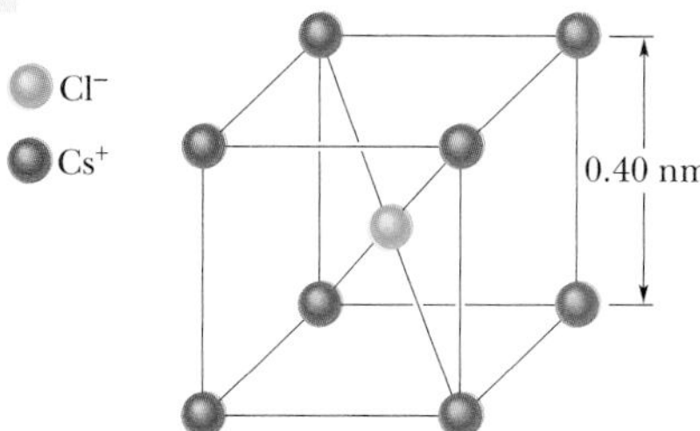

Fig. 22-21 Problem 27.

28P. We know that the negative charge on the electron and the positive charge on the proton are equal. Suppose, however, that these magnitudes differ from each other by 0.00010%. With what force would two copper coins, placed 1.0 m apart, repel each other? Assume that each coin contains 3×10^{22} copper atoms. (*Hint:* A neutral copper atom contains 29 protons and 29 electrons.) What do you conclude?

SEC. 22-6 Charge Is Conserved

29E. Identify X in the following nuclear reactions (in the first, n represents a neutron): (a) $^1H + {}^9Be \rightarrow X + n$; (b) $^{12}C + {}^1H \rightarrow X$; (c) $^{15}N + {}^1H \rightarrow {}^4He + X$. Appendix F will help. ssm

Additional Problem

30. In Problem 13, let $q = \alpha Q$. (a) Write an expression for the magnitude F of the force between the charges in terms of α, Q, and the charge separation d. (b) Graph F as a function of α. Graphically find the values of α that give (c) the maximum value of F and (d) half the maximum value of F.

NEW PROBLEMS

N1. Figure 22N-1 shows four tiny charged beads that can be slid or fixed in place on wires that stretch along x and y axes. A central bead at the crossing point of the wires (the origin) has a charge of $+e$. The other beads each have a charge of $-e$. Initially beads 1, 2, and 3 are at distance $d = 10.0$ cm from the central bead, and bead 4 is at a distance of $d/2$. (a) How far from the central bead must you position bead 1 so that the direction of the net electrostatic force $\vec{F}_{net}$ on the central bead rotates counterclockwise by 30°? (b) With bead 1 still in its new position, where must you slide bead 3 so that the direction of $\vec{F}_{net}$ rotates back by 30°?

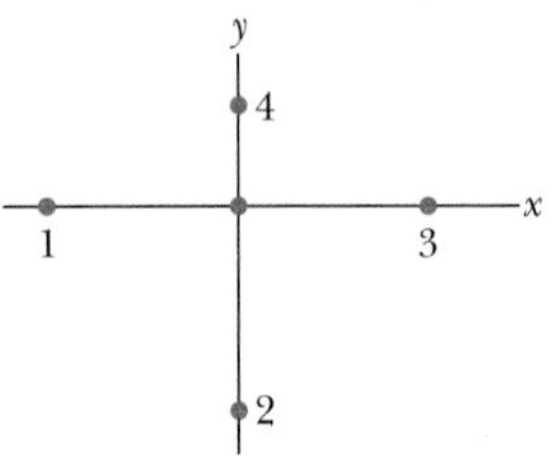

Fig. 22N-1 Problem N1.

N2. In Fig. 22N-2, a particle of charge $+4e$ is above a floor by distance $d_1 = 2.0$ mm and a particle of charge $+6e$ is on the floor, at horizontal distance $d_2 = 6.0$ mm from the first particle. What is the x component of the electrostatic force on the second particle due to the first particle?

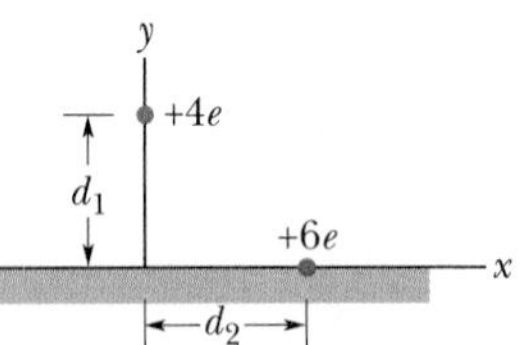

Fig. 22N-2 Problem N2.

N3. Figure 22N-3a shows charged particles 1 and 2 that are fixed in place on an x axis. Particle 1 has a charge with a magnitude of $|q_1| = 8.00e$. Particle 3, with a charge of $q_3 = +8.00e$, is initially on the x axis near particle 2. Then particle 3 is gradually moved in the positive direction of the x axis. As a result, the magnitude of the net electrostatic force $\vec{F}_{2,net}$ on particle 2 due to particles 1 and 3 changes. Figure 22N-3b gives the x component of that net force as a function of the position x of particle 3. The plot has an asymptote of $F_{2,net} = 1.5 \times 10^{-25}$ N as $x \to \infty$. As a multiple of e, what is the charge q_2 of particle 2?

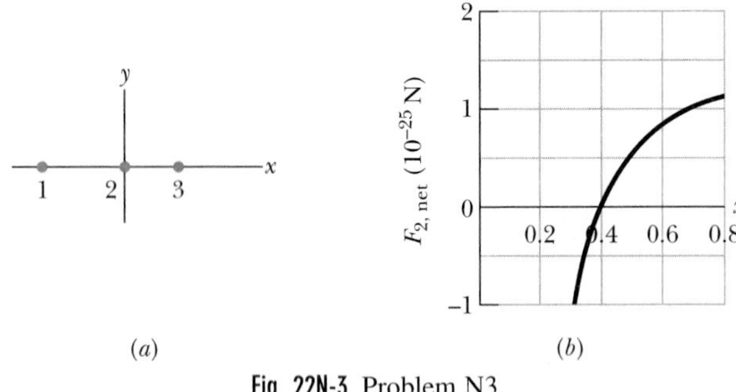

Fig. 22N-3 Problem N3.

N4. Figure 22N-4 shows four charged particles that are fixed along an axis, separated by distance $d = 2.00$ cm. The charges are indicated. Find the magnitude and direction of the net electrostatic force on (a) the particle with charge $+2e$ and (b) the particle with charge $-e$, due to the other particles.

d d d
+2e −e +e +4e

Fig. 22N-4 Problem N4.

N5. In Fig. 22N-5a, particle 1 (with charge q_1) and particle 2 (with charge q_2) are fixed in place on an x axis, 8.00 cm apart. Particle 3 with a charge $q_3 = +5e$ is to be placed on the line between particles 1 and 2, so that they produce a net electrostatic force $\vec{F}_{3,net}$ on it. Figure 22N-5b gives the x component of that force versus the coordinate x at which particle 3 is placed. What are (a) the sign of charge q_1 and (b) the ratio q_2/q_1?

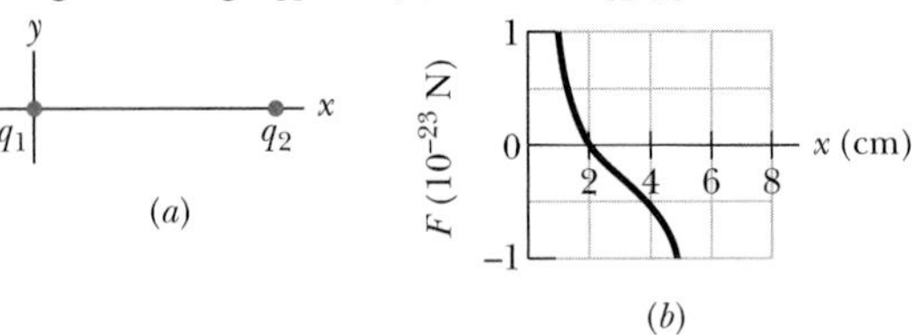

Fig. 22N-5 Problem N5.

N6. Figure 22N-6 shows two particles, each of charge $+2e$, that are fixed on a y axis, each at a distance $d = 17$ cm from the x axis. A third particle, of charge $+4e$, is moved slowly along the x axis, from $x = 0$ to $x = +5.0$ m. At what values of x will the magnitude of the electrostatic force on the third particle from the other two particles be (a) minimum and (b) maximum? What are (c) the minimum magnitude and (d) the maximum magnitude?

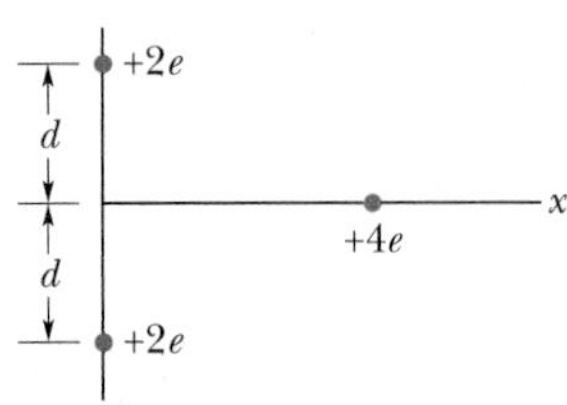

Fig. 22N-6 Problem N6.

N7. A charge of 6.0 μC is to be split into two parts that are then separated by 3.0 mm. What is the maximum possible magnitude of the electrostatic force between those two parts?

N8. Figure 22N-7 shows four identical conducting spheres that are actually well separated from one another. Sphere W (with an initial charge of zero) is touched to sphere A and then they are separated. Next, sphere W is touched to sphere B (with an initial charge of $-32e$) and then they are separated. Finally, sphere W is touched to sphere C (with an initial charge of $+48e$) and then they are separated. The final charge on sphere W is $+18e$. What was the initial charge on sphere A?

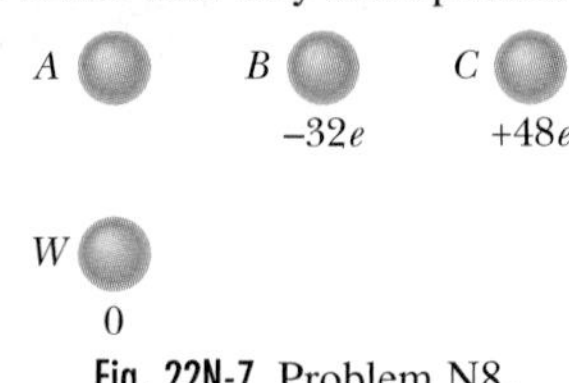

Fig. 22N-7 Problem N8.

N9. In Fig. 22N-8, how far from the charged particle on the right and in what direction is there a point where a third charged particle will be in balance?

a
−5.00q +2.00q

Fig. 22N-8 Problem N9.

N10. In Fig. 22N-9a, three positively charged particles are fixed on an x axis. Particles B and C are so close to each other that they can be considered to be at the same distance from particle A. The net force on particle A due to particles B and C is 2.014 ×

x
A BC
(a)
x
B A C
(b)

Fig. 22N-9 Problem N10.

10^{-23} N in the negative direction of the x axis. In Fig. 22N-9b, particle B has been moved to the opposite side of A but is still at the same distance from it. The net force on A is now 2.877×10^{-24} N in the negative direction of the x axis. What is the ratio of the charge of particle C to that of particle B?

N11. A particle of charge Q is fixed at the origin of an xy coordinate system. At $t = 0$ a particle ($m = 0.800$ g, $q = 4.00\ \mu$C) is located on the x axis at $x = 20.0$ cm, moving with a speed of 50.0 m/s in the positive y direction. For what value of Q will the moving particle execute circular motion? (Assume that the gravitational force on the particle may be neglected.)

N12. Figure 22N-10 shows an arrangement of seven positively charged particles that are separated from the central particle by distances of either d (= 1.0 cm) or $2d$, as drawn. The charges are indicated. What are the magnitude and direction of the net electrostatic force on the central particle due to the other six particles?

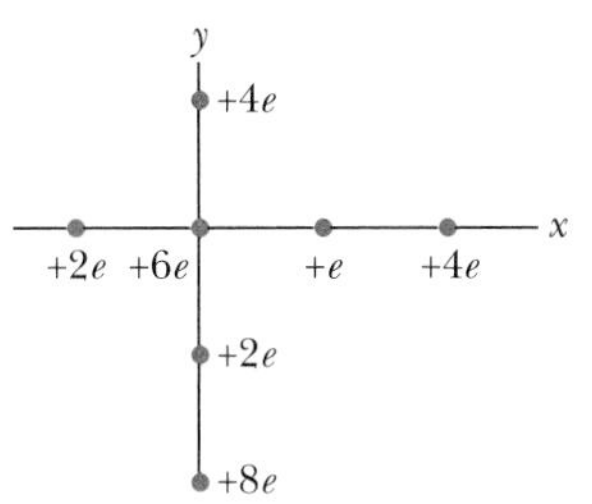

Fig. 22N-10 Problem N12.

N13. In Fig. 22N-11, what is q in terms of Q if the net electrostatic force on the charged particle at the upper left corner of the square array is to be zero?

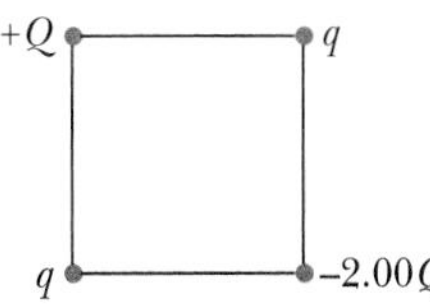

Fig. 22N-11 Problem N13.

N14. Figure 22N-12a shows an arrangement of three charged particles separated by distance d. Particles A and C are fixed on the x axis, but particle B can be moved along a circle centered on particle A. During the movement, a radial line between A and B makes an angle θ relative to the positive direction of the x axis (Fig. 22N-12b). The curves in Fig. 22N-12c give, for two situations, the magnitude F_{net} of the net electrostatic force on particle A due to the other particles. That net force is given as a function of angle θ and as a multiple of a basic amount F_0. For example on curve 1, at $\theta = 180°$, we see that $F_{net} = 2F_0$. (a) For the situation corresponding to curve 1, what is the ratio of the charge of particle C to that of particle B (including sign)? (b) For the situation corresponding to curve 2, what is that ratio?

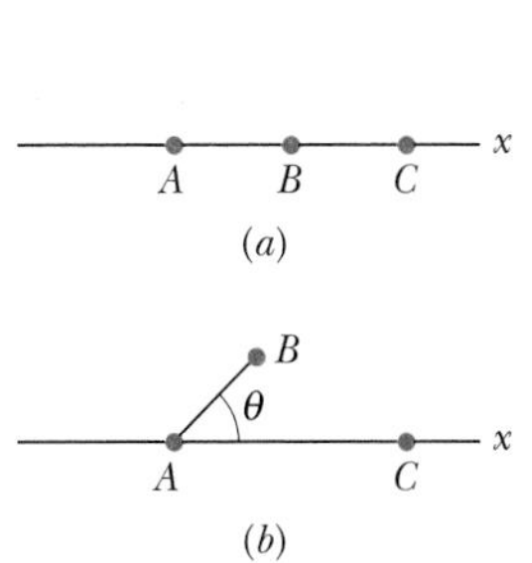

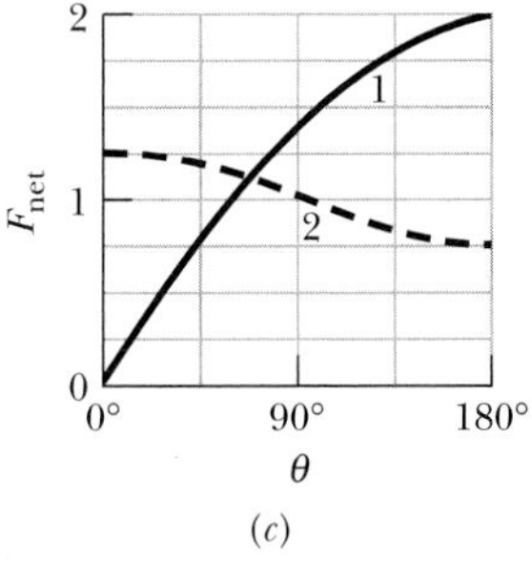

Fig. 22N-12 Problem N14.

N15. Figure 22N-13 shows two electrons (charge $-e$) on an x axis and two charged ions of identical charges $-q$ and identical angles θ. The central electron is free to move; the other particles are fixed in place at horizontal distances R and are intended to hold the free electron in place. (a) Plot the required magnitude of q versus angle θ if this is to happen. (b) From the plot, determine which values of θ will be needed for physically possible values of $q \leq 5e$.

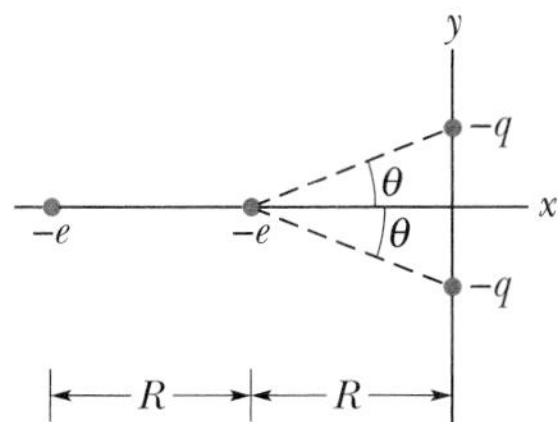

Fig. 22N-13 Problem N15.

N16. Figure 22N-14 shows an arrangement of four charged particles, with angle $\theta = 30°$ and distance $d = 2.00$ cm. The two negatively charged particles on the y axis are electrons that are fixed in place; the particle at the right has a charge $q_2 = +5e$. (a) Find distance D such that the net force on the particle at the left, due to the three other particles, is zero. (b) If the two electrons were moved closer to the x axis, would the required value of D be greater than, less than, or the same as in part (a)?

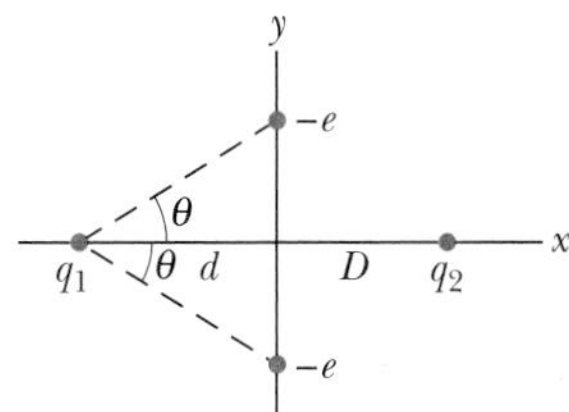

Fig. 22N-14 Problem N16.

N17. In *beta decay* a massive fundamental particle changes to another massive particle, and either an electron or a positron is emitted. (a) If a proton undergoes beta decay to become a neutron, which particle is emitted? (b) If a neutron undergoes beta decay to become a proton, which particle is emitted?

N18. In Fig. 22N-15, particles 1 and 2 are fixed in place on an x axis, at a separation of $L = 8.00$ cm. Their charges are $q_1 = +e$ and $q_2 = -27e$. Particle 3 with charge $q_3 = +4e$ is to be placed on the line between particles 1 and 2, so that they produce a net electrostatic force $\vec{F}_{3,net}$ on it. (a) At what coordinate should particle 3 be placed to minimize the magnitude of that force? (b) What is that minimum magnitude?

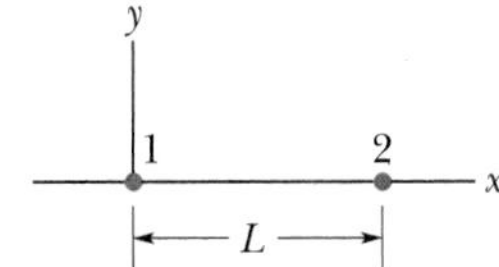

Fig. 22N-15 Problem N18.

N19. Two particles, each of positive charge q, are fixed in place on an x axis, one at $x = 0$ and the other at $x = d$. A particle of positive charge Q is to be placed along that axis at locations given by $x = \alpha d$. (a) Write expressions, in terms of α, that give the net electrostatic force $\vec{F}$ acting on the third particle when it is in the three regions $x < 0$, $0 < x < d$, and $d < x$. The expressions should give a positive result when $\vec{F}$ is in the positive direction of the x axis and a negative result when $\vec{F}$ is in the negative direction. (b) Graph $\vec{F}$ versus α for the range $-2 < \alpha < 3$.

N20. A current of 0.300 A through your chest can send your heart into fibrillation, disrupting the flow of blood (and thus oxygen) to your brain. If that current persists for 2.00 min, how many conduction electrons pass through your chest?

23 Electric Fields

During the frequent eruptions of the Sakurajima volcano in Japan, multiple electrical discharges (sparks) flash over the volcano's crater, lighting up the sky and sending out sound waves that resemble thunder. However, this is not a lightning display in a thunderstorm, with electrified clouds of water drops discharging to the ground. This is something different.

How does the region above the volcano become electrified, and is there any way to tell whether the sparks travel up from the crater or down to it?

The answer is in this chapter.

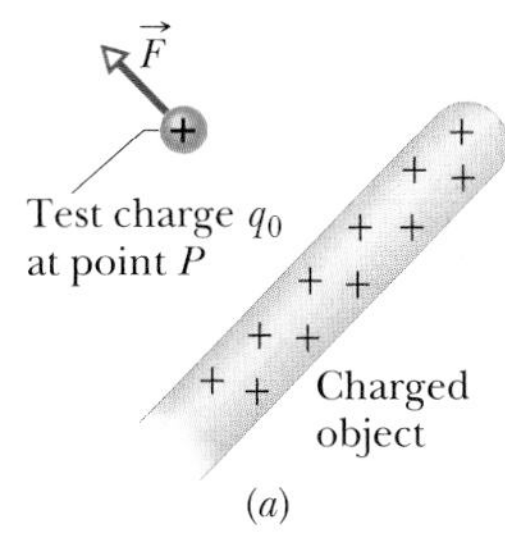

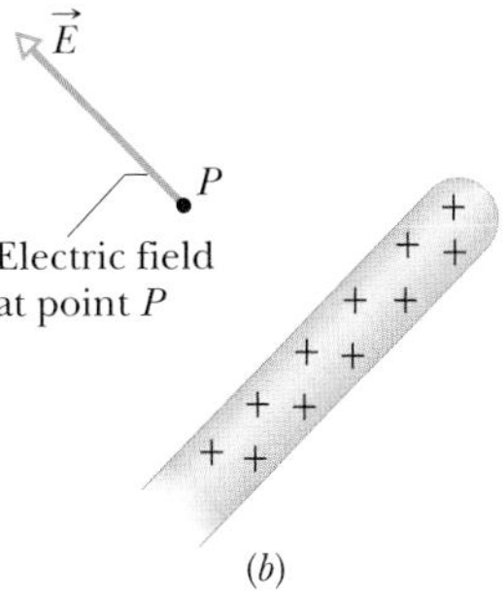

Fig. 23-1 (*a*) A positive test charge q_0 placed at point P near a charged object. An electrostatic force $\vec{F}$ acts on the test charge. (*b*) The electric field $\vec{E}$ at point P produced by the charged object.

23-1 Charges and Forces: A Closer Look

Suppose we fix a positive point charge q_1 in place and then put a second positive point charge q_2 near it. From Coulomb's law we know that q_1 exerts a repulsive electrostatic force on q_2 and, given enough data, we could determine the magnitude and direction of that force. Still, a nagging question remains: How does q_1 "know" of the presence of q_2? Since the charges do not touch, how can q_1 exert a force on q_2?

This question about *action at a distance* can be answered by saying that q_1 sets up an **electric field** in the space surrounding it. At any given point P in that space, the field has both magnitude and direction. The magnitude depends on the magnitude of q_1 and the distance between P and q_1. The direction depends on the direction from q_1 to P and the electrical sign of q_1. Thus when we place q_2 at P, q_1 interacts with q_2 through the electric field at P. The magnitude and direction of that electric field determine the magnitude and direction of the force acting on q_2.

Another action-at-a-distance problem arises if we move q_1, say, toward q_2. Coulomb's law tells us that when q_1 is closer to q_2, the repulsive electrostatic force acting on q_2 must be greater—and it is. However, here the nagging question is: Does the electric field at q_2, and thus the force acting on q_2, change immediately?

The answer is no. Instead, the information about the move by q_1 travels outward from q_1 (in all directions) as an electromagnetic wave at the speed of light c. The change in the electric field at q_2, and thus the change in the force acting on q_2, occurs when the wave finally reaches q_2.

23-2 The Electric Field

The temperature at every point in a room has a definite value. You can measure the temperature at any given point or combination of points by putting a thermometer there. We call the resulting distribution of temperatures a *temperature field*. In much the same way, you can imagine a *pressure field* in the atmosphere; it consists of the distribution of air pressure values, one for each point in the atmosphere. These two examples are of *scalar fields*, because temperature and air pressure are scalar quantities.

The electric field is a *vector field*; it consists of a distribution of *vectors*, one for each point in the region around a charged object, such as a charged rod. In principle, we can define the electric field at some point near the charged object, such as point P in Fig. 23-1*a*, as follows: We first place a *positive* charge q_0, called a *test charge*, at the point. We then measure the electrostatic force $\vec{F}$ that acts on the test charge. Finally, we define the electric field $\vec{E}$ at point P due to the charged object as

$$\vec{E} = \frac{\vec{F}}{q_0} \quad \text{(electric field).} \tag{23-1}$$

Thus, the magnitude of the electric field $\vec{E}$ at point P is $E = F/q_0$, and the direction of $\vec{E}$ is that of the force $\vec{F}$ that acts on the *positive* test charge. As shown in Fig. 23-1*b*, we represent the electric field at P with a vector whose tail is at P. To define the electric field within some region, we must similarly define it at all points in the region.

The SI unit for the electric field is the newton per coulomb (N/C). Table 23-1 shows the electric fields that occur in a few physical situations.

TABLE 23-1 Some Electric Fields

Field Location or Situation	Value (N/C)
At the surface of a uranium nucleus	3×10^{21}
Within a hydrogen atom, at a radius of 5.29×10^{-11} m	5×10^{11}
Electric breakdown occurs in air	3×10^{6}
Near the charged drum of a photocopier	10^{5}
Near a charged comb	10^{3}
In the lower atmosphere	10^{2}
Inside the copper wire of household circuits	10^{-2}

Although we use a positive test charge to define the electric field of a charged object, that field exists independently of the test charge. The field at point P in Figure 23-1*b* existed both before and after the test charge of Fig. 23-1*a* was put there. (We

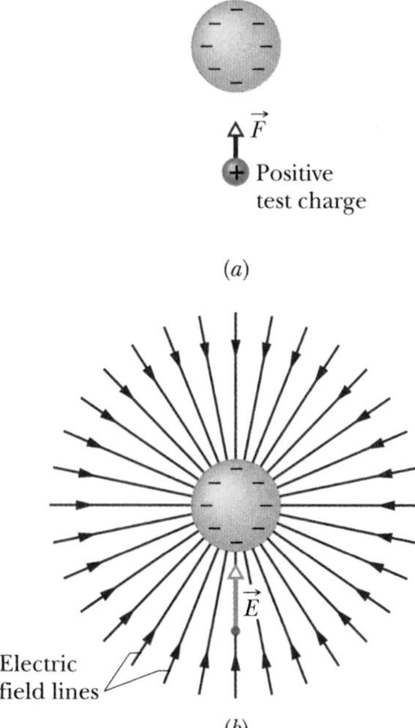

Fig. 23-2 (*a*) The electrostatic force $\vec{F}$ acting on a positive test charge near a sphere of uniform negative charge. (*b*) The electric field vector $\vec{E}$ at the location of the test charge, and the electric field lines in the space near the sphere. The field lines extend *toward* the negatively charged sphere. (They originate on distant positive charges.)

assume that in our defining procedure, the presence of the test charge does not affect the charge distribution on the charged object, and thus does not alter the electric field we are defining.)

To examine the role of an electric field in the interaction between charged objects, we have two tasks: (1) calculating the electric field produced by a given distribution of charge, and (2) calculating the force that a given field exerts on a charge placed in it. We perform the first task in Sections 23-4 through 23-7 for several charge distributions. We perform the second task in Sections 23-8 and 23-9 by considering a point charge and a pair of point charges in an electric field. First, however, we discuss a way to visualize electric fields.

23-3 Electric Field Lines

Michael Faraday, who introduced the idea of electric fields in the 19th century, thought of the space around a charged body as filled with *lines of force*. Although we no longer attach much reality to these lines, now usually called **electric field lines,** they still provide a nice way to visualize patterns in electric fields.

The relation between the field lines and electric field vectors is this: (1) At any point, the direction of a straight field line or the direction of the tangent to a curved field line gives the direction of $\vec{E}$ at that point, and (2) the field lines are drawn so that the number of lines per unit area, measured in a plane that is perpendicular to the lines, is proportional to the *magnitude* of $\vec{E}$. This second relation means that where the field lines are close together, E is large; and where they are far apart, E is small.

Figure 23-2*a* shows a sphere of uniform negative charge. If we place a *positive* test charge anywhere near the sphere, an electrostatic force pointing *toward* the center of the sphere will act on the test charge as shown. In other words, the electric field vectors at all points near the sphere are directed radially toward the sphere. This pattern of vectors is neatly displayed by the field lines in Fig. 23-2*b*, which point in the same directions as the force and field vectors. Moreover, the spreading of the field lines with distance from the sphere tells us that the magnitude of the electric field decreases with distance from the sphere.

If the sphere of Fig. 23-2 were of uniform *positive* charge, the electric field vectors at all points near the sphere would be directed radially *away from* the sphere. Thus, the electric field lines would also extend radially away from the sphere. We then have the following rule:

> Electric field lines extend away from positive charge (where they originate) and toward negative charge (where they terminate).

Figure 23-3*a* shows part of an infinitely large, nonconducting *sheet* (or plane) with a uniform distribution of positive charge on one side. If we were to place a positive test charge at any point near the sheet of Fig. 23-3*a*, the net electrostatic force acting on the test charge would be perpendicular to the sheet, because forces acting in all other directions would cancel one another as a result of the symmetry. Moreover, the net force on the test charge would point away from the sheet as shown. Thus, the electric field vector at any point in the space on either side of the sheet is also perpendicular to the sheet and directed away from it (Figs. 23-3*b* and *c*). Since the charge is uniformly distributed along the sheet, all the field vectors have the same magnitude. Such an electric field, with the same magnitude and direction at every point, is a *uniform electric field.*

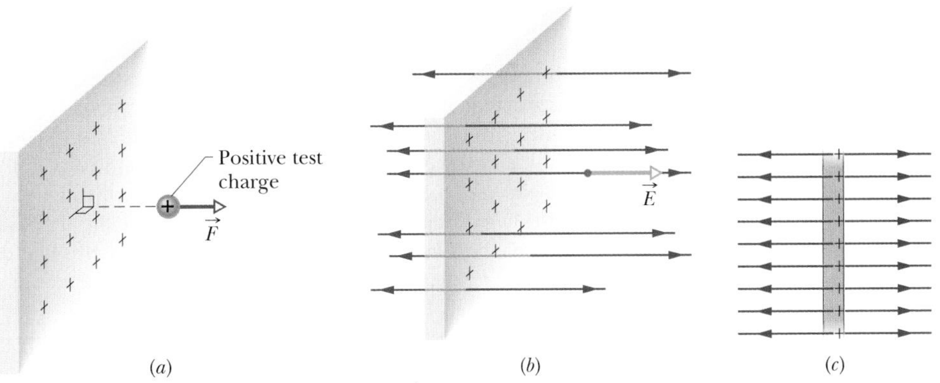

Fig. 23-3 (*a*) The electrostatic force $\vec{F}$ on a positive test charge near a very large, nonconducting sheet with uniformly distributed positive charge on one side. (*b*) The electric field vector $\vec{E}$ at the location of the test charge, and the electric field lines in the space near the sheet. The field lines extend *away from* the positively charged sheet. (*c*) Side view of (*b*).

Of course, no real nonconducting sheet (such as a flat expanse of plastic) is infinitely large, but if we consider a region that is near the middle of a real sheet and not near its edges, the field lines through that region are arranged as in Figs. 23-3*b* and *c*.

Figure 23-4 shows the field lines for two equal positive charges. Figure 23-5 shows the pattern for two charges that are equal in magnitude but of opposite sign, a configuration that we call an **electric dipole.** Although we do not often use field lines quantitatively, they are very useful to visualize what is going on. Can you not almost "see" the charges being pushed apart in Fig. 23-4 and pulled together in Fig. 23-5?

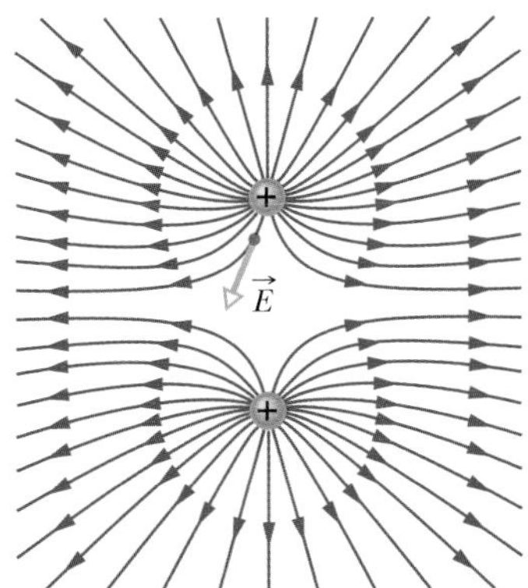

Fig. 23-4 Field lines for two equal positive point charges. The charges repel each other. (The lines terminate on distant negative charges.) To "see" the actual three-dimensional pattern of field lines, mentally rotate the pattern shown here about an axis passing through both charges in the plane of the page. The three-dimensional pattern and the electric field it represents are said to have *rotational symmetry* about that axis. The electric field vector at one point is shown; note that it is tangent to the field line through that point.

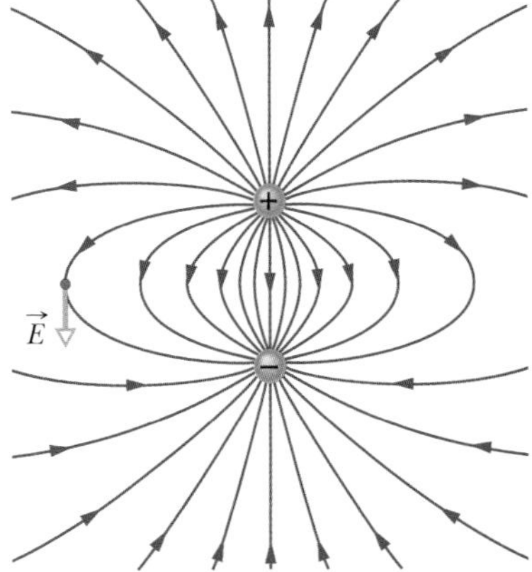

Fig. 23-5 Field lines for a positive and a nearby negative point charge that are equal in magnitude. The charges attract each other. The pattern of field lines and the electric field it represents have rotational symmetry about an axis passing through both charges in the plane of the page. The electric field vector at one point is shown; the vector is tangent to the field line through the point.

Sample Problem 23-1

How does the magnitude of the electric field vary with distance from the center of the uniformly charged sphere in Fig. 23-2? Use an argument based on the electric field lines.

SOLUTION: One Key Idea here is that the field lines are uniformly distributed around the sphere and extend outward from it without interruption. Thus, if we place a concentric spherical shell of radius r around the charged sphere, all the field lines terminating on the charged sphere must pass through the concentric shell. Let the number of field lines be N. Then, because the shell has surface area $4\pi r^2$, the number of field lines per unit area passing through the shell is $N/4\pi r^2$.

A second Key Idea is that the magnitude E of the electric field is proportional to the number of lines per unit area perpendicular to the lines. Since the shell is perpendicular to the field lines, E is proportional to $N/4\pi r^2$. Because r is the only variable in that term, E varies as the inverse square of the distance from the center of the charged sphere.

23-4 The Electric Field Due to a Point Charge

To find the electric field due to a point charge q (or charged particle) at any point a distance r from the point charge, we put a positive test charge q_0 at that point. From Coulomb's law (Eq. 22-4), the magnitude of the electrostatic force acting on q_0 is

$$F = \frac{1}{4\pi\varepsilon_0}\frac{|q||q_0|}{r^2}. \qquad (23\text{-}2)$$

The direction of $\vec{F}$ is directly away from the point charge if q is positive, and directly toward the point charge if q is negative. The magnitude of the electric field vector is, from Eq. 23-1,

$$E = \frac{F}{q_0} = \frac{1}{4\pi\varepsilon_0}\frac{|q|}{r^2} \quad \text{(point charge).} \qquad (23\text{-}3)$$

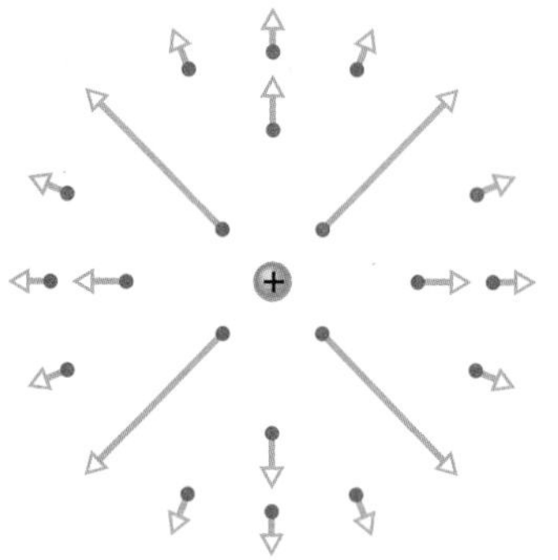

Fig. 23-6 The electric field vectors at several points around a positive point charge.

The direction of $\vec{E}$ is the same as that of the force on the positive test charge: directly away from the point charge if q is positive, and toward it if q is negative.

Because there is nothing special about the point we chose for q_0, Eq. 23-3 gives the field at every point around the point charge q. The field for a positive point charge is shown in Fig. 23-6 in vector form (not as field lines).

We can quickly find the net, or resultant, electric field due to more than one point charge. If we place a positive test charge q_0 near n point charges $q_1, q_2, \ldots, q_n$, then, from Eq. 22-7, the net force $\vec{F}_0$ from the n point charges acting on the test charge is

$$\vec{F}_0 = \vec{F}_{01} + \vec{F}_{02} + \cdots + \vec{F}_{0n}.$$

Therefore, from Eq. 23-1, the net electric field at the position of the test charge is

$$\begin{aligned}\vec{E} &= \frac{\vec{F}_0}{q_0} = \frac{\vec{F}_{01}}{q_0} + \frac{\vec{F}_{02}}{q_0} + \cdots + \frac{\vec{F}_{0n}}{q_0} \\ &= \vec{E}_1 + \vec{E}_2 + \cdots + \vec{E}_n. \qquad (23\text{-}4)\end{aligned}$$

Here $\vec{E}_i$ is the electric field that would be set up by point charge i acting alone. Equation 23-4 shows us that the principle of superposition applies to electric fields as well as to electrostatic forces.

CHECKPOINT 1: The figure here shows a proton p and an electron e on an x axis. What is the direction of the electric field due to the electron at (a) point S and (b) point R? What is the direction of the net electric field at (c) point R and (d) point S?

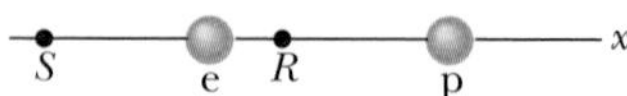

Sample Problem 23-2

Figure 23-7*a* shows three particles with charges $q_1 = +2Q$, $q_2 = -2Q$, and $q_3 = -4Q$, each a distance d from the origin. What net electric field $\vec{E}$ is produced at the origin?

SOLUTION: The **Key Idea** is that charges q_1, q_2, and q_3 produce electric field vectors $\vec{E}_1$, $\vec{E}_2$, and $\vec{E}_3$, respectively, at the origin, and the net electric field is the vector sum $\vec{E} = \vec{E}_1 + \vec{E}_2 + \vec{E}_3$. To find this sum, we first must find the magnitudes and orientations of the three field vectors. To find the magnitude of $\vec{E}_1$, which is due to q_1, we use Eq. 23-3, substituting d for r and $2Q$ for $|q|$ and obtaining

$$E_1 = \frac{1}{4\pi\varepsilon_0}\frac{2Q}{d^2}.$$

Similarly, we find the magnitudes of the fields $\vec{E}_2$ and $\vec{E}_3$ to be

$$E_2 = \frac{1}{4\pi\varepsilon_0}\frac{2Q}{d^2} \quad \text{and} \quad E_3 = \frac{1}{4\pi\varepsilon_0}\frac{4Q}{d^2}.$$

We next must find the orientations of the three electric field vectors at the origin. Because q_1 is a positive charge, the field vector it produces points directly *away* from it, and because q_2 and q_3 are both negative, the field vectors they produce point directly *toward* each of them. Thus, the three electric fields produced at the origin by the three charged particles are oriented as in Fig. 23-7*b*. (*Caution:* Note that we have placed the tails of the vectors at the point where the fields are to be evaluated; doing so decreases the chance of error.)

We can now add the fields vectorially as outlined for forces in Sample Problem 22-1c. However, here we can use symmetry to simplify the procedure. From Fig. 23-7*b*, we see that $\vec{E}_1$ and $\vec{E}_2$ have the same direction. Hence, their vector sum has that direction and has the magnitude

$$E_1 + E_2 = \frac{1}{4\pi\varepsilon_0}\frac{2Q}{d^2} + \frac{1}{4\pi\varepsilon_0}\frac{2Q}{d^2} = \frac{1}{4\pi\varepsilon_0}\frac{4Q}{d^2},$$

which happens to equal the magnitude of field $\vec{E}_3$.

We must now combine two vectors, $\vec{E}_3$ and the vector sum $\vec{E}_1 + \vec{E}_2$, that have the same magnitude and that are oriented symmetrically about the x axis, as shown in Fig. 23-7*c*. From the symmetry of Fig. 23-7*c*, we realize that the equal y components of our two vectors cancel and the equal x components add. Thus, the net electric field $\vec{E}$ at the origin is in the positive direction of the x axis and has the magnitude

$$E = 2E_{3x} = 2E_3 \cos 30^\circ = (2)\frac{1}{4\pi\varepsilon_0}\frac{4Q}{d^2}(0.866) = \frac{6.93Q}{4\pi\varepsilon_0 d^2}. \quad \text{(Answer)}$$

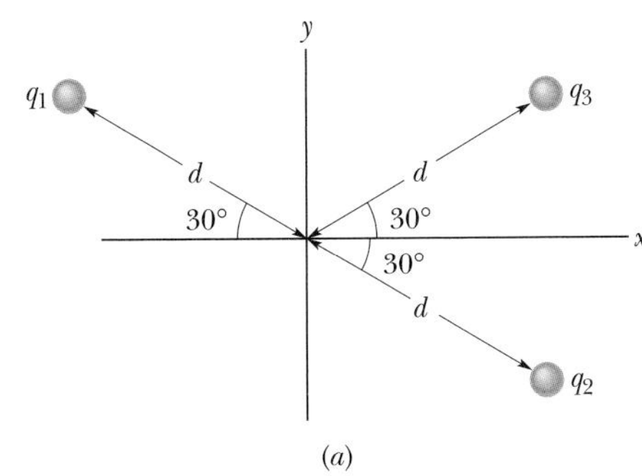

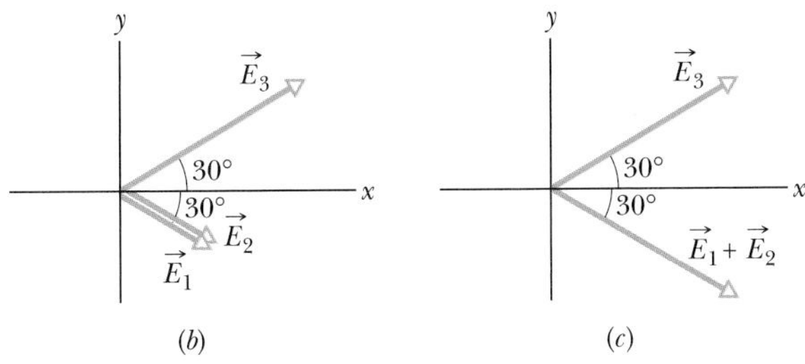

Fig. 23-7 Sample Problem 23-2. (*a*) Three particles with charges q_1, q_2, and q_3 are at the same distance d from the origin. (*b*) The electric field vectors $\vec{E}_1$, $\vec{E}_2$, and $\vec{E}_3$ at the origin due to the three particles. (*c*) The electric field vector $\vec{E}_3$ and the vector sum $\vec{E}_1 + \vec{E}_2$ at the origin.

CHECKPOINT 2: The figure here shows four situations in which charged particles are at equal distances from the origin. Rank the situations according to the magnitude of the net electric field at the origin, greatest first.

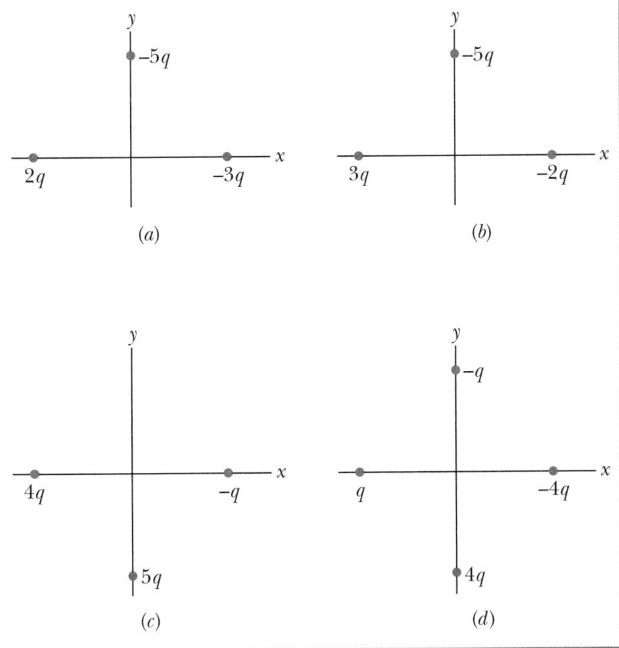

23-5 The Electric Field Due to an Electric Dipole

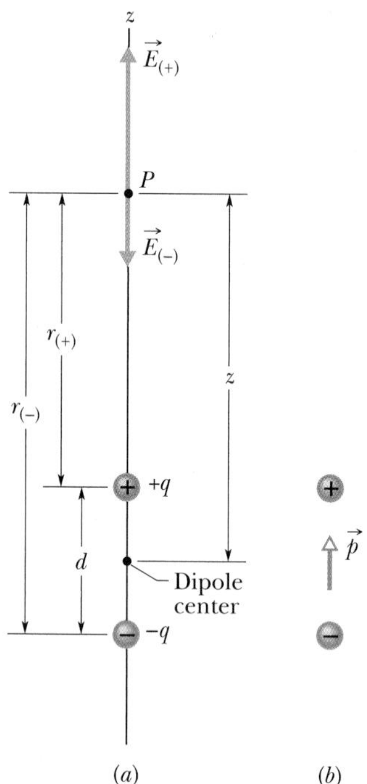

Fig. 23-8 (*a*) An electric dipole. The electric field vectors $\vec{E}_{(+)}$ and $\vec{E}_{(-)}$ at point P on the dipole axis result from the dipole's two charges. P is at distances $r_{(+)}$ and $r_{(-)}$ from the individual charges that make up the dipole. (*b*) The dipole moment $\vec{p}$ of the dipole points from the negative charge to the positive charge.

Figure 23-8*a* shows two charged particles of magnitude q but of opposite sign, separated by a distance d. As was noted in connection with Fig. 23-5, we call this configuration an *electric dipole*. Let us find the electric field due to the dipole of Fig. 23-8*a* at a point P, a distance z from the midpoint of the dipole and on the axis through the particles, which is called the *dipole axis*.

From symmetry, the electric field $\vec{E}$ at point P—and also the fields $\vec{E}_{(+)}$ and $\vec{E}_{(-)}$ due to the separate charges that make up the dipole—must lie along the dipole axis, which we have taken to be a z axis. Applying the superposition principle for electric fields, we find that the magnitude E of the electric field at P is

$$\begin{aligned} E &= E_{(+)} - E_{(-)} \\ &= \frac{1}{4\pi\varepsilon_0}\frac{q}{r_{(+)}^2} - \frac{1}{4\pi\varepsilon_0}\frac{q}{r_{(-)}^2} \\ &= \frac{q}{4\pi\varepsilon_0(z - \frac{1}{2}d)^2} - \frac{q}{4\pi\varepsilon_0(z + \frac{1}{2}d)^2}. \end{aligned} \tag{23-5}$$

After a little algebra, we can rewrite this equation as

$$E = \frac{q}{4\pi\varepsilon_0 z^2}\left[\left(1 - \frac{d}{2z}\right)^{-2} - \left(1 + \frac{d}{2z}\right)^{-2}\right]. \tag{23-6}$$

We are usually interested in the electrical effect of a dipole only at distances that are large compared with the dimensions of the dipole—that is, at distances such that $z \gg d$. At such large distances, we have $d/2z \ll 1$ in Eq. 23-6. We can then expand the two quantities in the brackets in that equation by the binomial theorem (Appendix E), obtaining for those quantities

$$\left[\left(1 + \frac{2d}{2z(1!)} + \cdots\right) - \left(1 - \frac{2d}{2z(1!)} + \cdots\right)\right].$$

Thus,

$$E = \frac{q}{4\pi\varepsilon_0 z^2}\left[\left(1 + \frac{d}{z} + \cdots\right) - \left(1 - \frac{d}{z} + \cdots\right)\right]. \tag{23-7}$$

The unwritten terms in the two expansions in Eq. 23-7 involve d/z raised to progressively higher powers. Since $d/z \ll 1$, the contributions of those terms are progressively less, and to approximate E at large distances, we can neglect them. Then, in our approximation, we can rewrite Eq. 23-7 as

$$E = \frac{q}{4\pi\varepsilon_0 z^2}\frac{2d}{z} = \frac{1}{2\pi\varepsilon_0}\frac{qd}{z^3}. \tag{23-8}$$

The product qd, which involves the two intrinsic properties q and d of the dipole, is the magnitude p of a vector quantity known as the **electric dipole moment $\vec{p}$** of the dipole. (The unit of $\vec{p}$ is the Coulomb–meter.) Thus, we can write Eq. 23-8 as

$$E = \frac{1}{2\pi\varepsilon_0}\frac{p}{z^3} \quad \text{(electric dipole).} \tag{23-9}$$

The direction of $\vec{p}$ is taken to be from the negative to the positive end of the dipole, as indicated in Fig. 23-8*b*. We can use $\vec{p}$ to specify the orientation of a dipole.

Equation 23-9 shows that, if we measure the electric field of a dipole only at distant points, we can never find q and d separately, only their product. The field at

TABLE 23-2 **Some Measures of Electric Charge**

Name	Symbol	SI Unit
Charge	q	C
Linear charge density	λ	C/m
Surface charge density	σ	C/m^2
Volume charge density	ρ	C/m^3

distant points would be unchanged if, for example, q were doubled and d simultaneously halved. Thus, the dipole moment is a basic property of a dipole.

Although Eq. 23-9 holds only for distant points along the dipole axis, it turns out that E for a dipole varies as $1/r^3$ for *all* distant points, regardless of whether they lie on the dipole axis; here r is the distance between the point in question and the dipole center.

Inspection of Fig. 23-8 and of the field lines in Fig. 23-5 shows that the direction of $\vec{E}$ for distant points on the dipole axis is always the direction of the dipole moment vector $\vec{p}$. This is true whether point P in Fig. 23-8a is on the upper or the lower part of the dipole axis.

Inspection of Eq. 23-9 shows that if you double the distance of a point from a dipole, the electric field at the point drops by a factor of 8. If you double the distance from a single point charge, however (see Eq. 23-3), the electric field drops only by a factor of 4. Thus the electric field of a dipole decreases more rapidly with distance than does the electric field of a single charge. The physical reason for this rapid decrease in electric field for a dipole is that from distant points a dipole looks like two equal but opposite charges that almost—but not quite—coincide. Thus, their electric fields at distant points almost—but not quite—cancel each other.

23-6 The Electric Field Due to a Line of Charge

So far we have considered the electric field that is produced by one or, at most, a few point charges. We now consider charge distributions that consist of a great many closely spaced point charges (perhaps billions) that are spread along a line, over a surface, or within a volume. Such distributions are said to be **continuous** rather than discrete. Since these distributions can include an enormous number of point charges, we find the electric fields that they produce by means of calculus rather than by considering the point charges one by one. In this section we discuss the electric field caused by a line of charge. We consider a charged surface in the next section. In the next chapter, we shall find the field inside a uniformly charged sphere.

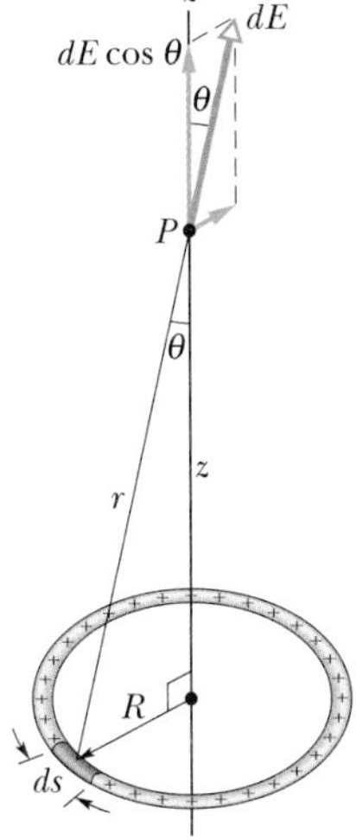

Fig. 23-9 A ring of uniform positive charge. A differential element of charge occupies a length ds (greatly exaggerated for clarity). This element sets up an electric field $d\vec{E}$ at point P. The component of $d\vec{E}$ along the central axis of the ring is $dE\cos\theta$.

When we deal with continuous charge distributions, it is most convenient to express the charge on an object as a *charge density* rather than as a total charge. For a line of charge, for example, we would report the linear charge density (or charge per unit length) λ, whose SI unit is the coulomb per meter. Table 23-2 shows the other charge densities we shall be using.

Figure 23-9 shows a thin ring of radius R with a uniform positive linear charge density λ around its circumference. We may imagine the ring to be made of plastic or some other insulator, so that the charges can be regarded as fixed in place. What is the electric field $\vec{E}$ at point P, a distance z from the plane of the ring along its central axis?

To answer, we cannot just apply Eq. 23-3, which gives the electric field set up by a point charge, because the ring is obviously not a point charge. However, we can mentally divide the ring into differential elements of charge that are so small that they are like point charges, and then we can apply Eq. 23-3 to each of them. Next, we can add the electric fields set up at P by all the differential elements. The vector sum of all those fields gives us the field set up at P by the ring.

Let ds be the (arc) length of any differential element of the ring. Since λ is the charge per unit length, the element has a charge of magnitude

$$dq = \lambda\, ds. \tag{23-10}$$

This differential charge sets up a differential electric field $d\vec{E}$ at point P, which is a distance r from the element. Treating the element as a point charge, and using Eq. 23-10, we can rewrite Eq. 23-3 to express the magnitude of $d\vec{E}$ as

$$dE = \frac{1}{4\pi\varepsilon_0}\frac{dq}{r^2} = \frac{1}{4\pi\varepsilon_0}\frac{\lambda\, ds}{r^2}. \tag{23-11}$$

From Fig. 23-9, we can rewrite Eq. 23-11 as

$$dE = \frac{1}{4\pi\varepsilon_0}\frac{\lambda\, ds}{(z^2 + R^2)}. \tag{23-12}$$

Figure 23-9 shows us that $d\vec{E}$ is at an angle θ to the central axis (which we have taken to be a z axis) and has components perpendicular to and parallel to that axis.

Every charge element in the ring sets up a differential field $d\vec{E}$ at P, with magnitude given by Eq. 23-12. All the $d\vec{E}$ vectors have identical components parallel to the central axis, in both magnitude and direction. All these $d\vec{E}$ vectors have components perpendicular to the central axis as well; these perpendicular components are identical in magnitude but point in different directions. In fact, for any perpendicular component that points in a given direction, there is another one that points in the opposite direction. The sum of this pair of components, like the sum of all other pairs of oppositely directed components, is zero.

Thus, the perpendicular components cancel and we need not consider them further. This leaves the parallel components; they all have the same direction, so the net electric field at P is their sum.

The parallel component of $d\vec{E}$ shown in Fig. 23-9 has magnitude $dE \cos\theta$. The figure also shows us that

$$\cos\theta = \frac{z}{r} = \frac{z}{(z^2 + R^2)^{1/2}}. \tag{23-13}$$

Then Eqs. 23-13 and 23-12 give us, for the parallel component of $d\vec{E}$,

$$dE\cos\theta = \frac{z\lambda}{4\pi\varepsilon_0(z^2 + R^2)^{3/2}}\, ds. \tag{23-14}$$

To add the parallel components $dE \cos\theta$ produced by all the elements, we integrate Eq. 23-14 around the circumference of the ring, from $s = 0$ to $s = 2\pi R$. Since the only quantity in Eq. 23-14 that varies during the integration is s, the other quantities can be moved outside the integral sign. The integration then gives us

$$E = \int dE\cos\theta = \frac{z\lambda}{4\pi\varepsilon_0(z^2 + R^2)^{3/2}}\int_0^{2\pi R} ds$$

$$= \frac{z\lambda(2\pi R)}{4\pi\varepsilon_0(z^2 + R^2)^{3/2}}. \tag{23-15}$$

Since λ is the charge per length of the ring, the term $\lambda(2\pi R)$ in Eq. 23-15 is q, the total charge on the ring. We then can rewrite Eq. 23-15 as

$$E = \frac{qz}{4\pi\varepsilon_0(z^2 + R^2)^{3/2}} \quad \text{(charged ring)}. \tag{23-16}$$

If the charge on the ring is negative, instead of positive as we have assumed, the magnitude of the field at P is still given by Eq. 23-16. However, the electric field vector then points toward the ring instead of away from it.

Let us check Eq. 23-16 for a point on the central axis that is so far away that $z \gg R$. For such a point, the expression $z^2 + R^2$ in Eq. 23-16 can be approximated as z^2, and Eq. 23-16 becomes

$$E = \frac{1}{4\pi\varepsilon_0}\frac{q}{z^2} \quad \text{(charged ring at large distance).} \tag{23-17}$$

This is a reasonable result, because from a large distance, the ring "looks like" a point charge. If we replace z with r in Eq. 23-17, we indeed do have Eq. 23-3, the magnitude of the electric field due to a point charge.

Let us next check Eq. 23-16 for a point at the center of the ring—that is, for $z = 0$. At that point, Eq. 23-16 tells us that $E = 0$. This is a reasonable result, because if we were to place a test charge at the center of the ring, there would be no net electrostatic force acting on it; the force due to any element of the ring would be canceled by the force due to the element on the opposite side of the ring. By Eq. 23-1, if the force at the center of the ring were zero, the electric field there would also have to be zero.

Sample Problem 23-3

Figure 23-10a shows a plastic rod having a uniformly distributed charge $-Q$. The rod has been bent in a 120° circular arc of radius r. We place coordinate axes such that the axis of symmetry of the rod lies along the x axis and the origin is at the center of curvature P of the rod. In terms of Q and r, what is the electric field $\vec{E}$ due to the rod at point P?

SOLUTION: The **Key Idea** here is that, because the rod has a continuous charge distribution, we must find an expression for the electric fields due to differential elements of the rod and then sum those fields via calculus. Consider a differential element having arc length ds and located at an angle θ above the x axis (Fig. 23-10b). If we let λ represent the linear charge density of the rod, our element ds has a differential charge of magnitude

$$dq = \lambda\, ds. \tag{23-18}$$

Our element produces a differential electric field $d\vec{E}$ at point P, which is a distance r from the element. Treating the element as a point charge, we can rewrite Eq. 23-3 to express the magnitude of $d\vec{E}$ as

$$dE = \frac{1}{4\pi\varepsilon_0}\frac{dq}{r^2} = \frac{1}{4\pi\varepsilon_0}\frac{\lambda\, ds}{r^2}. \tag{23-19}$$

The direction of $d\vec{E}$ is toward ds, because charge dq is negative.

Our element has a symmetrically located (mirror image) element ds' in the bottom half of the rod. The electric field $d\vec{E}'$ set up at P by ds' also has the magnitude given by Eq. 23-19, but the field vector points toward ds' as shown in Fig. 23-10b. If we resolve the electric field vectors of ds and ds' into x and y components as shown in Fig. 23-10b, we see that their y components cancel (because they have equal magnitudes and are in opposite directions). We also see that their x components have equal magnitudes and are in the same direction.

Thus, to find the electric field set up by the rod, we need sum (via integration) only the x components of the differential electric fields set up by all the differential elements of the rod. From

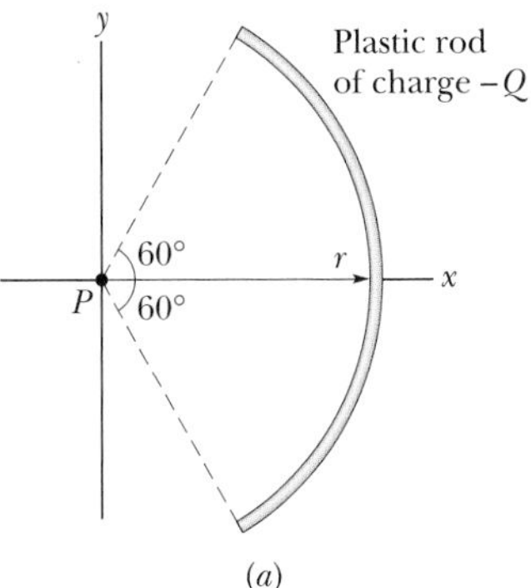

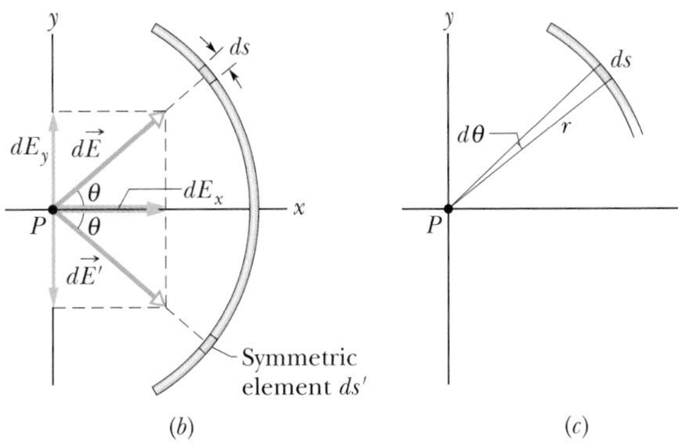

Fig. 23-10 Sample Problem 23-3. (a) A plastic rod of charge $-Q$ is a circular section of radius r and central angle 120°; point P is the center of curvature of the rod. (b) A differential element in the top half of the rod, at an angle θ to the x axis and of arc length ds, sets up a differential electric field $d\vec{E}$ at P. An element ds', symmetric to ds about the x axis, sets up a field $d\vec{E}'$ at P with the same magnitude. (c) Arc length ds makes an angle $d\theta$ about point P.

Fig. 23-10b and Eq. 23-19, we can write the component dE_x set up by ds as

$$dE_x = dE\cos\theta = \frac{1}{4\pi\varepsilon_0}\frac{\lambda}{r^2}\cos\theta\, ds. \qquad (23\text{-}20)$$

Equation 23-20 has two variables, θ and s. Before we can integrate it, we must eliminate one variable. We do so by replacing ds, using the relation

$$ds = r\,d\theta,$$

in which $d\theta$ is the angle at P that includes arc length ds (Fig. 23-10c). With this replacement, we can integrate Eq. 23-20 over the angle made by the rod at P, from $\theta = -60°$ to $\theta = 60°$; that will give us the magnitude of the electric field at P due to the rod:

$$\begin{aligned} E &= \int dE_x = \int_{-60°}^{60°} \frac{1}{4\pi\varepsilon_0}\frac{\lambda}{r^2}\cos\theta\, r\,d\theta \\ &= \frac{\lambda}{4\pi\varepsilon_0 r}\int_{-60°}^{60°}\cos\theta\,d\theta = \frac{\lambda}{4\pi\varepsilon_0 r}\Big[\sin\theta\Big]_{-60°}^{60°} \\ &= \frac{\lambda}{4\pi\varepsilon_0 r}[\sin 60° - \sin(-60°)] \\ &= \frac{1.73\lambda}{4\pi\varepsilon_0 r}. \end{aligned} \qquad (23\text{-}21)$$

(If we had reversed the limits on the integration, we would have gotten the same result but with a minus sign. Since the integration gives only the magnitude of $\vec{E}$, we would then have discarded the minus sign.)

To evaluate λ, we note that the rod has an angle of 120° and so is one-third of a full circle. Its arc length is then $2\pi r/3$, and its linear charge density must be

$$\lambda = \frac{\text{charge}}{\text{length}} = \frac{Q}{2\pi r/3} = \frac{0.477Q}{r}.$$

Substituting this into Eq. 23-21 and simplifying give us

$$\begin{aligned} E &= \frac{(1.73)(0.477Q)}{4\pi\varepsilon_0 r^2} \\ &= \frac{0.83Q}{4\pi\varepsilon_0 r^2}. \end{aligned} \qquad \text{(Answer)}$$

The direction of $\vec{E}$ is toward the rod, along the axis of symmetry of the charge distribution. We can write $\vec{E}$ in unit-vector notation as

$$\vec{E} = \frac{0.83Q}{4\pi\varepsilon_0 r^2}\,\hat{\text{i}}.$$

PROBLEM-SOLVING TACTICS

Tactic 1: *A Field Guide for Lines of Charge*

Here is a generic guide for finding the electric field $\vec{E}$ produced at a point P by a line of uniform charge, either circular or straight. The general strategy is to pick out an element dq of the charge, find $d\vec{E}$ due to that element, and integrate $d\vec{E}$ over the entire line of charge.

Step 1. If the line of charge is circular, let ds be the arc length of an element of the distribution. If the line is straight, run an x axis along it and let dx be the length of an element. Mark the element on a sketch.

Step 2. Relate the charge dq of the element to the length of the element with either $dq = \lambda\, ds$ or $dq = \lambda\, dx$. Consider dq and λ to be positive, even if the charge is actually negative. (The sign of the charge is used in the next step.)

Step 3. Express the field $d\vec{E}$ produced at P by dq with Eq. 23-3, replacing q in that equation with either $\lambda\, ds$ or $\lambda\, dx$. If the charge on the line is positive, then at P draw a vector $d\vec{E}$ that points directly away from dq. If the charge is negative, draw the vector pointing directly toward dq.

Step 4. Always look for any symmetry in the situation. If P is on an axis of symmetry of the charge distribution, resolve the field $d\vec{E}$ produced by dq into components that are perpendicular and parallel to the axis of symmetry. Then consider a second element dq' that is located symmetrically to dq about the line of symmetry. At P draw the vector $d\vec{E}'$ that this symmetrical element produces, and resolve it into components. One of the components produced by dq is a *canceling component*; it is canceled by the corresponding component produced by dq' and needs no further attention. The other component produced by dq is an *adding component*; it adds to the corresponding component produced by dq'. Add the adding components of all the elements via integration.

Step 5. Here are four general types of uniform charge distributions, with strategies for simplifying the integral of step 4.

Ring, with point P on (central) axis of symmetry, as in Fig. 23-9. In the expression for dE, replace r^2 with $z^2 + R^2$, as in Eq. 23-12. Express the adding component of $d\vec{E}$ in terms of θ. That introduces $\cos\theta$, but θ is identical for all elements and thus is not a variable. Replace $\cos\theta$ as in Eq. 23-13. Integrate over s, around the circumference of the ring.

Circular arc, with point P at the center of curvature, as in Fig. 23-10. Express the adding component of $d\vec{E}$ in terms of θ. That introduces either $\sin\theta$ or $\cos\theta$. Reduce the resulting two variables s and θ to one, θ, by replacing ds with $r\,d\theta$. Integrate over θ, as in Sample Problem 23-3, from one end of the arc to the other end.

Straight line, with point P on an extension of the line, as in Fig. 23-11a. In the expression for dE, replace r with x. Integrate over x, from end to end of the line of charge.

Straight line, with point P at perpendicular distance y from the line of charge, as in Fig. 23-11b. In the expression for dE, replace r with an expression involving x and y. If P is on the perpendicular bisector of the line of charge, find an expression for the adding component of $d\vec{E}$. That will introduce either $\sin\theta$ or $\cos\theta$. Reduce the resulting two variables x and θ to one, x, by replacing the trigonometric function with an expression (its definition) involving x and y. Integrate over x from end to end of the line of charge. If P is not on a line

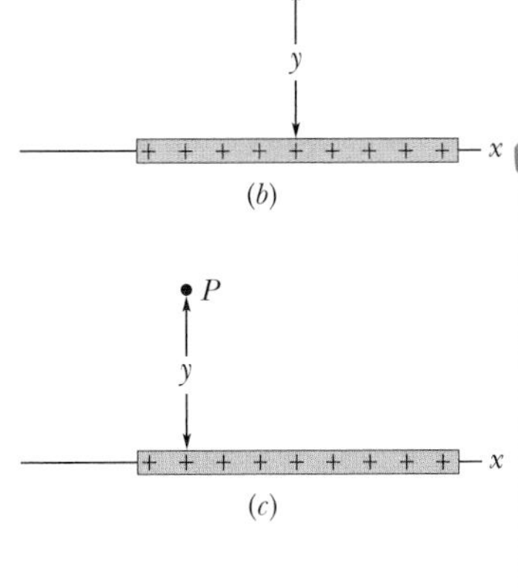

Fig. 23-11 (*a*) Point P is on an extension of the line of charge. (*b*) P is on a line of symmetry of the line of charge, at perpendicular distance y from that line. (*c*) Same as (*b*) except that P is not on a line of symmetry.

of symmetry, as in Fig. 23-11*c*, set up an integral to sum the components dE_x, and integrate over x to find E_x. Also set up an integral to sum the components dE_y, and integrate over x again to find E_y. Use the components E_x and E_y in the usual way to find the magnitude E and the orientation of $\vec{E}$.

Step 6. One arrangement of the integration limits gives a positive result. The reverse arrangement gives the same result with a minus sign; discard the minus sign. If the result is to be stated in terms of the total charge Q of the distribution, replace λ with Q/L, in which L is the length of the distribution. For a ring, L is the ring's circumference.

✔CHECKPOINT 3: The figure here shows three nonconducting rods, one circular and two straight. Each has a uniform charge of magnitude Q along its top half and another along its bottom half. For each rod, what is the direction of the net electric field at point P?

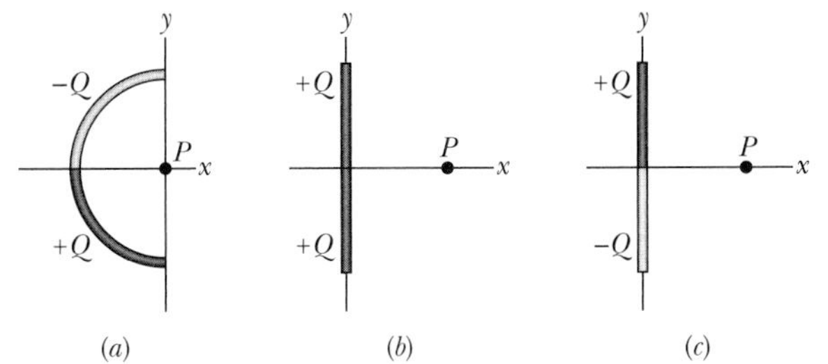

23-7 The Electric Field Due to a Charged Disk

Figure 23-12 shows a circular plastic disk of radius R that has a positive surface charge of uniform density σ on its upper surface (see Table 23-2). What is the electric field at point P, a distance z from the disk along its central axis?

Our plan is to divide the disk into concentric flat rings and then to calculate the electric field at point P by adding up (that is, by integrating) the contributions of all the rings. Figure 23-12 shows one such ring, with radius r and radial width dr. Since σ is the charge per unit area, the charge on the ring is

$$dq = \sigma\, dA = \sigma(2\pi r\, dr), \tag{23-22}$$

where dA is the differential area of the ring.

We have already solved the problem of the electric field due to a ring of charge. Substituting dq from Eq. 23-22 for q in Eq. 23-16, and replacing R in Eq. 23-16 with r, we obtain an expression for the electric field dE at P due to our flat ring:

$$dE = \frac{z\sigma 2\pi r\, dr}{4\pi\varepsilon_0(z^2 + r^2)^{3/2}},$$

which we may write as

$$dE = \frac{\sigma z}{4\varepsilon_0} \frac{2r\, dr}{(z^2 + r^2)^{3/2}}. \tag{23-23}$$

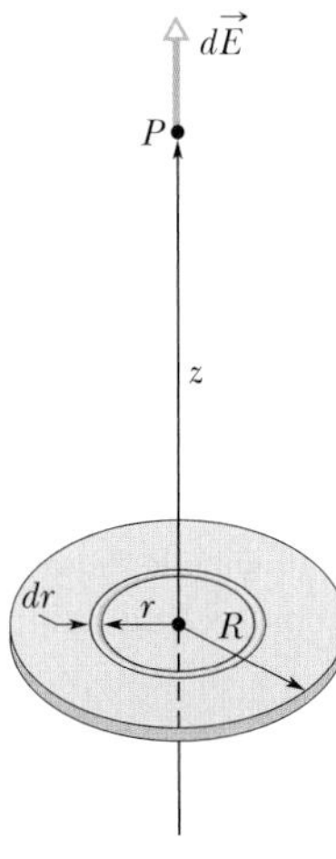

Fig. 23-12 A disk of radius R and uniform positive charge. The ring shown has radius r and radial width dr. It sets up a differential electric field $d\vec{E}$ at point P on its central axis.

We can now find E by integrating Eq. 23-23 over the surface of the disk—that is, by integrating with respect to the variable r from $r = 0$ to $r = R$. Note that z remains constant during this process. We get

$$E = \int dE = \frac{\sigma z}{4\varepsilon_0} \int_0^R (z^2 + r^2)^{-3/2}(2r)\, dr. \tag{23-24}$$

To solve this integral, we cast it in the form $\int X^m\,dX$ by setting $X = (z^2 + r^2)$, $m = -\frac{3}{2}$, and $dX = (2r)\,dr$. For the recast integral we have

$$\int X^m\,dX = \frac{X^{m+1}}{m+1},$$

so Eq. 23-24 becomes

$$E = \frac{\sigma z}{4\varepsilon_0}\left[\frac{(z^2 + r^2)^{-1/2}}{-\frac{1}{2}}\right]_0^R. \tag{23-25}$$

Taking the limits in Eq. 23-25 and rearranging, we find

$$E = \frac{\sigma}{2\varepsilon_0}\left(1 - \frac{z}{\sqrt{z^2 + R^2}}\right) \quad \text{(charged disk)} \tag{23-26}$$

as the magnitude of the electric field produced by a flat, circular, charged disk at points on its central axis. (In carrying out the integration, we assumed that $z \geq 0$.)

If we let $R \rightarrow \infty$ while keeping z finite, the second term in the parentheses in Eq. 23-26 approaches zero, and this equation reduces to

$$E = \frac{\sigma}{2\varepsilon_0} \quad \text{(infinite sheet).} \tag{23-27}$$

This is the electric field produced by an infinite sheet of uniform charge located on one side of a nonconductor such as plastic. The electric field lines for such a situation are shown in Fig. 23-3.

We also get Eq. 23-27 if we let $z \rightarrow 0$ in Eq. 23-26 while keeping R finite. This shows that at points very close to the disk, the electric field set up by the disk is the same as if the disk were infinite in extent.

23-8 A Point Charge in an Electric Field

In the preceding four sections we worked at the first of our two tasks: given a charge distribution, to find the electric field it produces in the surrounding space. Here we begin the second task: to determine what happens to a charged particle when it is in an electric field that is produced by other stationary or slowly moving charges.

What happens is that an electrostatic force acts on the particle, as given by

$$\vec{F} = q\vec{E}, \tag{23-28}$$

in which q is the charge of the particle (including its sign) and $\vec{E}$ is the electric field that other charges have produced at the location of the particle. (The field is *not* the field set up by the particle itself; to distinguish the two fields, the field acting on the particle in Eq. 23-28 is often called the *external field.* A charged particle (or object) is not affected by its own electric field.) Equation 23-28 tells us

> The electrostatic force $\vec{F}$ acting on a charged particle located in an external electric field $\vec{E}$ has the direction of $\vec{E}$ if the charge q of the particle is positive and has the opposite direction if q is negative.

CHECKPOINT 4: (a) In the figure, what is the direction of the electrostatic force on the electron due to the electric field shown? (b) In which direction will the electron accelerate if it is moving parallel to the y axis before it encounters the electric field? (c) If, instead, the electron is initially moving rightward, will its speed increase, decrease, or remain constant?

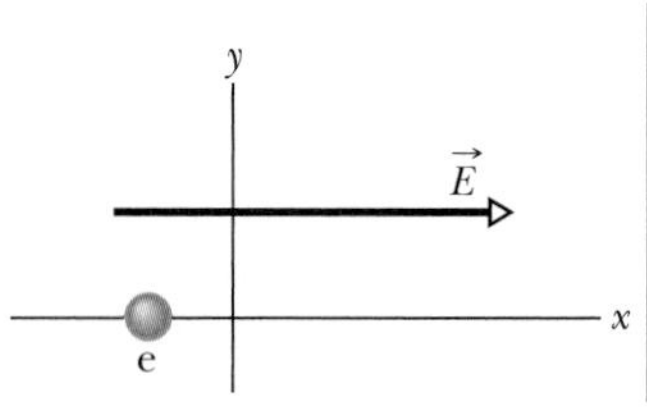

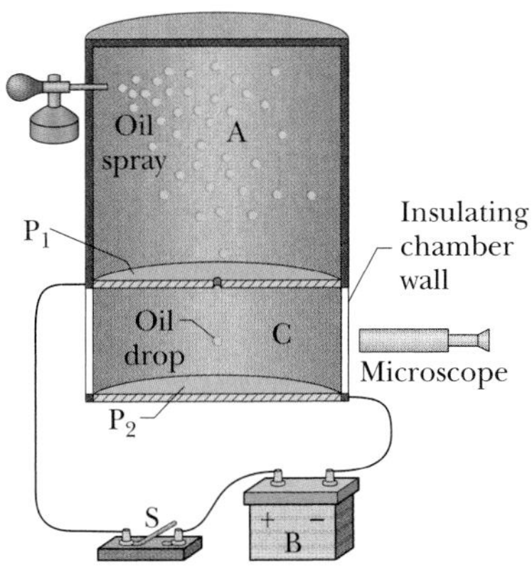

Fig. 23-13 The Millikan oil-drop apparatus for measuring the elementary charge e. When a charged oil drop drifted into chamber C through the hole in plate P_1, its motion could be controlled by closing and opening switch S and thereby setting up or eliminating an electric field in chamber C. The microscope was used to view the drop, to permit timing of its motion.

Measuring the Elementary Charge

Equation 23-28 played a role in the measurement of the elementary charge e by American physicist Robert A. Millikan in 1910–1913. Figure 23-13 is a representation of his apparatus. When tiny oil drops are sprayed into chamber A, some of them become charged, either positively or negatively, in the process. Consider a drop that drifts downward through the small hole in plate P_1 and into chamber C. Let us assume that this drop has a negative charge q.

If switch S in Fig. 23-13 is open as shown, battery B has no electrical effect on chamber C. If the switch is closed (the connection between chamber C and the positive terminal of the battery is then complete), the battery causes an excess positive charge on conducting plate P_1 and an excess negative charge on conducting plate P_2. The charged plates set up a downward-directed electric field $\vec{E}$ in chamber C. According to Eq. 23-28, this field exerts an electrostatic force on any charged drop that happens to be in the chamber and affects its motion. In particular, our negatively charged drop will tend to drift upward.

By timing the motion of oil drops with the switch opened and with it closed and thus determining the effect of the charge q, Millikan discovered that the values of q were always given by

$$q = ne, \qquad \text{for } n = 0, \pm 1, \pm 2, \pm 3, \ldots, \tag{23-29}$$

in which e turned out to be the fundamental constant we call the *elementary charge*, 1.60×10^{-19} C. Millikan's experiment is convincing proof that charge is quantized, and he earned the 1923 Nobel prize in physics in part for this work. Modern measurements of the elementary charge rely on a variety of interlocking experiments, all more precise than the pioneering experiment of Millikan.

Ink-Jet Printing

The need for high-quality, high-speed printing has caused a search for an alternative to impact printing, such as occurs in a standard typewriter. Building up letters by squirting tiny drops of ink at the paper is one such alternative.

Figure 23-14 shows a negatively charged drop moving between two conducting deflecting plates, between which a uniform, downward-directed electric field $\vec{E}$ has been set up. The drop is deflected upward according to Eq. 23-28 and then strikes the paper at a position that is determined by the magnitudes of $\vec{E}$ and the charge q of the drop.

In practice, E is held constant and the position of the drop is determined by the charge q delivered to the drop in the charging unit, through which the drop must pass before entering the deflecting system. The charging unit, in turn, is activated by electronic signals that encode the material to be printed.

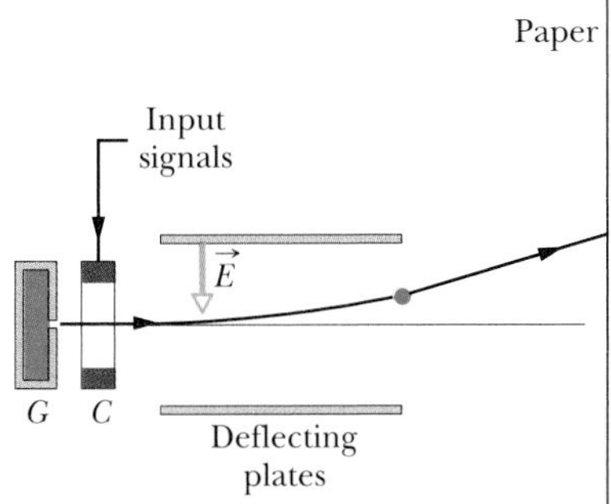

Fig. 23-14 The essential features of an ink-jet printer. Drops are shot out from generator G and receive a charge in charging unit C. An input signal from a computer controls the charge given to each drop and thus the effect of field $\vec{E}$ on the drop and the position on the paper at which the drop lands. About 100 tiny drops are needed to form a single character.

Volcanic Lightning

When the Sakurajima volcano erupts, as seen in this chapter's opening photograph, it spews ash into the air. That ash results when liquid water within the volcano, suddenly converted to steam by the flow of hot lava, shatters rock, which is then burnt. The liquid-to-steam conversion and the explosion of rock cause positive and negative charges to separate. Then, as the steam and ash are spewed into the air, they form a cloud that contains pockets of positive charge and pockets of negative charge.

As these pockets grow, the electric fields between adjacent pockets and between pockets and the volcano crater increase in magnitude. Whenever the magnitude of about 3×10^6 N/C is reached, the air undergoes *electric breakdown* and begins to conduct current. These momentary conducting paths appear in the air where the electric field has ionized air molecules, freeing some of their electrons. These electrons, propelled by the field, collide with air molecules in their way, which causes those molecules to emit light. We can see these brief paths, commonly called *sparks,* because of the light they emit. (A small-scale example of *sparking* can be seen around the charged metal cap in Fig. 23-15.)

The sparks above the volcano snake their way either down from a charge pocket to the crater wall or vice versa. You can tell the direction of a spark by how any dead-end branches on it are forked. If the branches fork downward, then the spark snaked its way downward. (See the bright spark extending from the right side of the photograph to the crater wall.) If the branches fork upward, then the spark snaked its way upward. (See the lower part of the central bright spark on the crater wall.) Sometimes a downward-snaking spark and an upward-snaking spark meet each other. Can you find an example in the photograph?

Fig. 23-15 The metal cap is so charged that the electric field it produces in the surrounding space causes the air there to undergo electric breakdown. The visible sparks reveal where momentary conducting paths are set up in the air, along which the electric field has removed electrons from their molecules and then accelerated them into collisions with the molecules.

Sample Problem 23-4

Figure 23-16 shows the deflecting plates of an ink-jet printer, with superimposed coordinate axes. An ink drop with a mass m of 1.3×10^{-10} kg and a negative charge of magnitude $Q = 1.5 \times 10^{-13}$ C enters the region between the plates, initially moving along the x axis with speed $v_x = 18$ m/s. The length L of the plates is 1.6 cm. The plates are charged and thus produce an electric field at all points between them. Assume that field $\vec{E}$ is downward directed, uniform, and has a magnitude of 1.4×10^6 N/C. What is the vertical deflection of the drop at the far edge of the plates? (The gravitational force on the drop is small relative to the electrostatic force acting on the drop and can be neglected.)

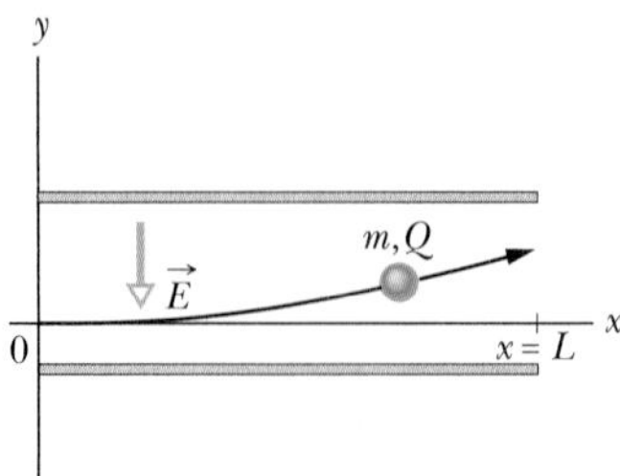

Fig. 23-16 Sample Problem 23-4. An ink drop of mass m and charge magnitude Q is deflected in the electric field of an ink-jet printer.

SOLUTION: The drop is negatively charged and the electric field is directed *downward.* The **Key Idea** here is that, from Eq. 23-28, a constant electrostatic force of magnitude QE acts *upward* on the charged drop. Thus, as the drop travels parallel to the x axis at constant speed v_x, it accelerates upward with some constant acceleration a_y. Applying Newton's second law ($F = ma$) for components along the y axis, we find that

$$a_y = \frac{F}{m} = \frac{QE}{m}. \tag{23-30}$$

Let t represent the time required for the drop to pass through the region between the plates. During t the vertical and horizontal displacements of the drop are

$$y = \tfrac{1}{2}a_y t^2 \quad \text{and} \quad L = v_x t, \tag{23-31}$$

respectively. Eliminating t between these two equations and substituting Eq. 23-30 for a_y, we find

$$\begin{aligned} y &= \frac{QEL^2}{2mv_x^2} \\ &= \frac{(1.5 \times 10^{-13}\ \text{C})(1.4 \times 10^6\ \text{N/C})(1.6 \times 10^{-2}\ \text{m})^2}{(2)(1.3 \times 10^{-10}\ \text{kg})(18\ \text{m/s})^2} \\ &= 6.4 \times 10^{-4}\ \text{m} \\ &= 0.64\ \text{mm}. \end{aligned}$$

(Answer)

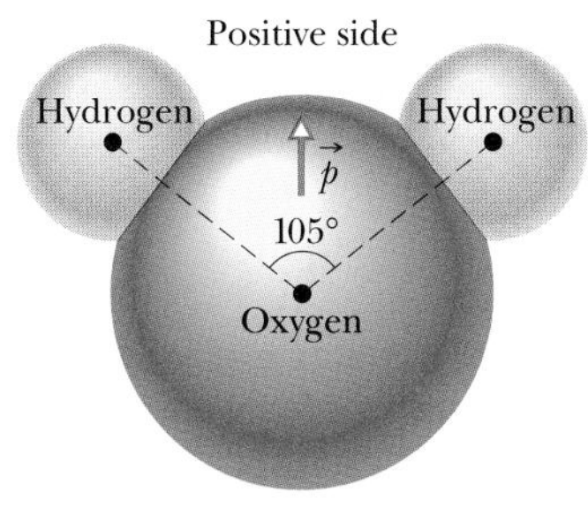

Fig. 23-17 A molecule of H_2O, showing the three nuclei (represented by dots) and the regions in which the electrons can be located. The electric dipole moment $\vec{p}$ points from the (negative) oxygen side to the (positive) hydrogen side of the molecule.

23-9 A Dipole in an Electric Field

We have defined the electric dipole moment $\vec{p}$ of an electric dipole to be a vector that points from the negative to the positive end of the dipole. As you will see, the behavior of a dipole in a uniform external electric field $\vec{E}$ can be described completely in terms of the two vectors $\vec{E}$ and $\vec{p}$, with no need of any details about the dipole's structure.

A molecule of water (H_2O) is an electric dipole; Fig. 23-17 shows why. There the black dots represent the oxygen nucleus (having eight protons) and the two hydrogen nuclei (having one proton each). The colored enclosed areas represent the regions in which electrons can be located around the nuclei.

In a water molecule, the two hydrogen atoms and the oxygen atom do not lie on a straight line but form an angle of about 105°, as shown in Fig. 23-17. As a result, the molecule has a definite "oxygen side" and "hydrogen side." Moreover, the 10 electrons of the molecule tend to remain closer to the oxygen nucleus than to the hydrogen nuclei. This makes the oxygen side of the molecule slightly more negative than the hydrogen side and creates an electric dipole moment $\vec{p}$ that points along the symmetry axis of the molecule as shown. If the water molecule is placed in an external electric field, it behaves as would be expected of the more abstract electric dipole of Fig. 23-8.

To examine this behavior, we now consider such an abstract dipole in a uniform external electric field $\vec{E}$, as shown in Fig. 23-18*a*. We assume that the dipole is a rigid structure that consists of two centers of opposite charge, each of magnitude q, separated by a distance d. The dipole moment $\vec{p}$ makes an angle θ with field $\vec{E}$.

Electrostatic forces act on the charged ends of the dipole. Because the electric field is uniform, those forces act in opposite directions (as shown in Fig. 23-18) and with the same magnitude $F = qE$. Thus, *because the field is uniform,* the net force on the dipole from the field is zero and the center of mass of the dipole does not move. However, the forces on the charged ends do produce a net torque $\vec{\tau}$ on the dipole about its center of mass. The center of mass lies on the line connecting the charged ends, at some distance x from one end and thus a distance $d - x$ from the other end. From Eq. 11-31 ($\tau = rF \sin \phi$), we can write the magnitude of the net torque $\vec{\tau}$ as

$$\tau = Fx \sin \theta + F(d - x) \sin \theta = Fd \sin \theta. \tag{23-32}$$

We can also write the magnitude of $\vec{\tau}$ in terms of the magnitudes of the electric field E and the dipole moment $p = qd$. To do so, we substitute qE for F and p/q for d in Eq. 23-32, finding that the magnitude of $\vec{\tau}$ is

$$\tau = pE \sin \theta. \tag{23-33}$$

We can generalize this equation to vector form as

$$\vec{\tau} = \vec{p} \times \vec{E} \quad \text{(torque on a dipole)}. \tag{23-34}$$

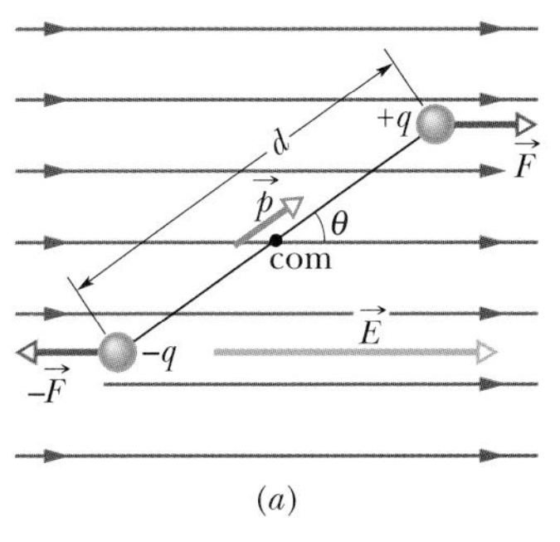

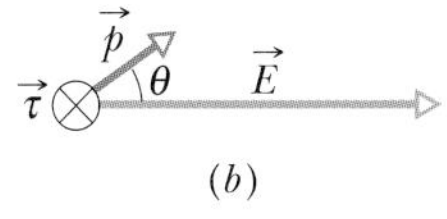

Fig. 23-18 (*a*) An electric dipole in a uniform electric field $\vec{E}$. Two centers of equal but opposite charge are separated by distance d. The line between them represents their rigid connection. (*b*) Field $\vec{E}$ causes a torque τ on the dipole. The direction of $\vec{\tau}$ is into the page, as represented by the symbol ⊗.

Vectors $\vec{p}$ and $\vec{E}$ are shown in Fig. 23-18*b*. The torque acting on a dipole tends to rotate $\vec{p}$ (hence the dipole) into the direction of field $\vec{E}$, thereby reducing θ. In Fig. 23-18, such rotation is clockwise. As we discussed in Chapter 11, we can represent a torque that gives rise to a clockwise rotation by including a minus sign with the magnitude of the torque. With that notation, the torque of Fig. 23-18 is

$$\tau = -pE \sin \theta. \tag{23-35}$$

Potential Energy of an Electric Dipole

Potential energy can be associated with the orientation of an electric dipole in an electric field. The dipole has its least potential energy when it is in its equilibrium orientation, which is when its moment $\vec{p}$ is lined up with the field $\vec{E}$ (then $\vec{\tau} = \vec{p} \times \vec{E} = 0$). It has greater potential energy in all other orientations. Thus the dipole is like a pendulum, which has *its* least gravitational potential energy in *its* equilibrium orientation—at its lowest point. To rotate the dipole or the pendulum to any other orientation requires work by some external agent.

In any situation involving potential energy, we are free to define the zero-potential-energy configuration in a perfectly arbitrary way, because only differences in potential energy have physical meaning. It turns out that the expression for the potential energy of an electric dipole in an external electric field is simplest if we choose the potential energy to be zero when the angle θ in Fig. 23-18 is 90°. We then can find the potential energy U of the dipole at any other value of θ with Eq. 8-1 ($\Delta U = -W$) by calculating the work W done by the field on the dipole when the dipole is rotated to that value of θ from 90°. With the aid of Eq. 11-45 ($W = \int \tau \, d\theta$) and Eq. 23-35, we find that the potential energy U at any angle θ is

$$U = -W = -\int_{90^\circ}^{\theta} \tau \, d\theta = \int_{90^\circ}^{\theta} pE \sin \theta \, d\theta. \tag{23-36}$$

Evaluating the integral leads to

$$U = -pE \cos \theta. \tag{23-37}$$

We can generalize this equation to vector form as

$$U = -\vec{p} \cdot \vec{E} \quad \text{(potential energy of a dipole).} \tag{23-38}$$

Equations 23-37 and 23-38 show us that the potential energy of the dipole is least ($U = -pE$) when $\theta = 0$, which is when $\vec{p}$ and $\vec{E}$ are in the same direction; the potential energy is greatest ($U = pE$) when $\theta = 180°$, which is when $\vec{p}$ and $\vec{E}$ are in opposite directions.

When a dipole rotates from an initial orientation θ_i to another orientation θ_f, the work W done on the dipole by the electric field is

$$W = -\Delta U = -(U_f - U_i), \tag{23-39}$$

where U_f and U_i are calculated with Eq. 23-38. If the change in orientation is caused by an applied torque (commonly said to be due to an external agent), then the work W_a done on the dipole by the applied torque is the negative of the work done on the dipole by the field; that is,

$$W_a = -W = (U_f - U_i). \tag{23-40}$$

CHECKPOINT 5: The figure shows four orientations of an electric dipole in an external electric field. Rank the orientations according to (a) the magnitude of the torque on the dipole and (b) the potential energy of the dipole, greatest first.

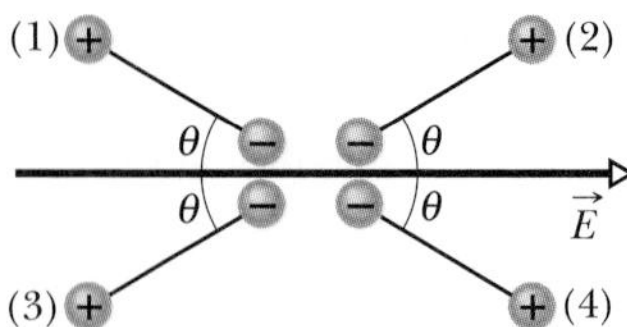

Sample Problem 23-5

A neutral water molecule (H_2O) in its vapor state has an electric dipole moment of magnitude 6.2×10^{-30} C · m.

(a) How far apart are the molecule's centers of positive and negative charge?

SOLUTION: The Key Idea here is that a molecule's dipole moment depends on the magnitude q of the molecule's positive or negative charge and the charge separation d. There are 10 electrons and 10 protons in a neutral water molecule, so the magnitude of its dipole moment is

$$p = qd = (10e)(d),$$

in which d is the separation we are seeking and e is the elementary charge. Thus,

$$d = \frac{p}{10e} = \frac{6.2 \times 10^{-30}\ \mathrm{C \cdot m}}{(10)(1.60 \times 10^{-19}\ \mathrm{C})} = 3.9 \times 10^{-12}\ \mathrm{m} = 3.9\ \mathrm{pm}. \quad \text{(Answer)}$$

This distance is not only small, but it is actually smaller than the radius of a hydrogen atom.

(b) If the molecule is placed in an electric field of 1.5×10^4 N/C, what maximum torque can the field exert on it? (Such a field can easily be set up in the laboratory.)

SOLUTION: The Key Idea here is that the torque on a dipole is maximum when the angle θ between $\vec{p}$ and $\vec{E}$ is 90°. Substituting this value in Eq. 23-33 yields

$$\begin{aligned}\tau &= pE \sin \theta \\ &= (6.2 \times 10^{-30}\ \mathrm{C \cdot m})(1.5 \times 10^4\ \mathrm{N/C})(\sin 90^\circ) \\ &= 9.3 \times 10^{-26}\ \mathrm{N \cdot m}. \quad \text{(Answer)}\end{aligned}$$

(c) How much work must an *external agent* do to turn this molecule end for end in this field, starting from its fully aligned position, for which $\theta = 0$?

SOLUTION: The Key Idea here is that the work done by an external agent (by means of a torque applied to the molecule) is equal to the change in the molecule's potential energy due to the change in orientation. From Eq. 23-40, we find

$$\begin{aligned}W_a &= U(180^\circ) - U(0) \\ &= (-pE \cos 180^\circ) - (-pE \cos 0) \\ &= 2pE = (2)(6.2 \times 10^{-30}\ \mathrm{C \cdot m})(1.5 \times 10^4\ \mathrm{N/C}) \\ &= 1.9 \times 10^{-25}\ \mathrm{J}. \quad \text{(Answer)}\end{aligned}$$

REVIEW & SUMMARY

Electric Field One way to explain the electrostatic force between two charges is to assume that each charge sets up an electric field in the space around it. The electrostatic force acting on any one charge is then due to the electric field set up at its location by the other charge.

Definition of Electric Field The *electric field* $\vec{E}$ at any point is defined in terms of the electrostatic force $\vec{F}$ that would be exerted on a positive test charge q_0 placed there:

$$\vec{E} = \frac{\vec{F}}{q_0}. \quad (23\text{-}1)$$

Electric Field Lines *Electric field lines* provide a means for visualizing the direction and magnitude of electric fields. The electric field vector at any point is tangent to a field line through that point. The density of field lines in any region is proportional to the magnitude of the electric field in that region. Field lines originate on positive charges and terminate on negative charges.

Field Due to a Point Charge The magnitude of the electric field $\vec{E}$ set up by a point charge q at a distance r from the charge is

$$E = \frac{1}{4\pi\varepsilon_0}\frac{|q|}{r^2}. \quad (23\text{-}3)$$

The direction of $\vec{E}$ is away from the point charge if the charge is positive and toward the point charge if the charge is negative.

Field Due to an Electric Dipole An *electric dipole* consists of two particles with charges of equal magnitude q but opposite sign, separated by a small distance d. Their **dipole moment** $\vec{p}$ has magnitude qd and points from the negative charge to the positive charge. The magnitude of the electric field set up by the dipole at a distant point on the dipole axis (which runs through both charges) is

$$E = \frac{1}{2\pi\varepsilon_0}\frac{p}{z^3}, \quad (23\text{-}9)$$

where z is the distance between the point and the dipole center.

Field Due to a Continuous Charge Distribution The electric field due to a *continuous charge distribution* is found by treating charge elements as point charges and then summing, via integration, the electric field vectors produced by all the charge elements.

Force on a Point Charge in an Electric Field When a point charge q is placed in an electric field $\vec{E}$ set up by other charges, the electrostatic force $\vec{F}$ that acts on the point charge is

$$\vec{F} = q\vec{E}. \quad (23\text{-}28)$$

Force $\vec{F}$ has the same direction as $\vec{E}$ if q is positive and the opposite direction if q is negative.

Dipole in an Electric Field When an electric dipole of dipole moment $\vec{p}$ is placed in an electric field $\vec{E}$, the field exerts a torque $\vec{\tau}$ on the dipole:

$$\vec{\tau} = \vec{p} \times \vec{E}. \tag{23-34}$$

The dipole has a potential energy U associated with its orientation in the field:

$$U = -\vec{p} \cdot \vec{E}. \tag{23-38}$$

This potential energy is defined to be zero when $\vec{p}$ is perpendicular to $\vec{E}$; it is least ($U = -pE$) when $\vec{p}$ is aligned with $\vec{E}$, and most ($U = pE$) when $\vec{p}$ is directed opposite $\vec{E}$.

QUESTIONS

1. Figure 23-19 shows three electric field lines. What is the direction of the electrostatic force on a positive test charge placed at (a) point A and (b) point B? (c) At which point, A or B, will the acceleration of the test charge be greater if the charge is released?

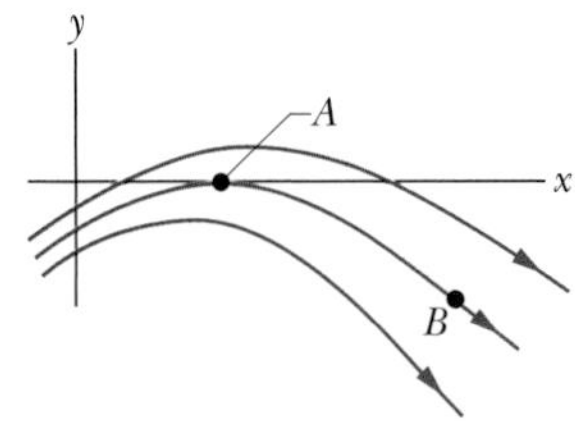

Fig. 23-19 Question 1.

2. Figure 23-20a shows two charged particles on an axis. (a) Where on the axis (other than at an infinite distance) is there a point at which their net electric field is zero: between the charges, to their left, or to their right? (b) Is there a point of zero electric field off the axis (other than at an infinite distance)?

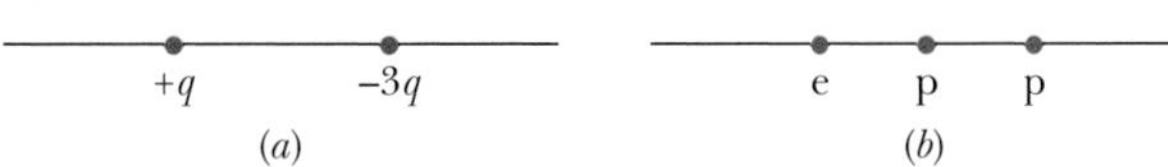

Fig. 23-20 Questions 2 and 3.

3. Figure 23-20b shows two protons and an electron that are evenly spaced on an axis. Where on the axis (other than at an infinite distance) is there a point at which their net electric field is zero: to the left of the particles, to their right, between the two protons, or between the electron and the nearer proton?

4. Figure 23-21 shows two square arrays of charged particles. The squares, which are centered on point P, are misaligned. The particles are separated by either d or $d/2$ along the perimeters of the squares. What are the magnitude and direction of the net electric field at P?

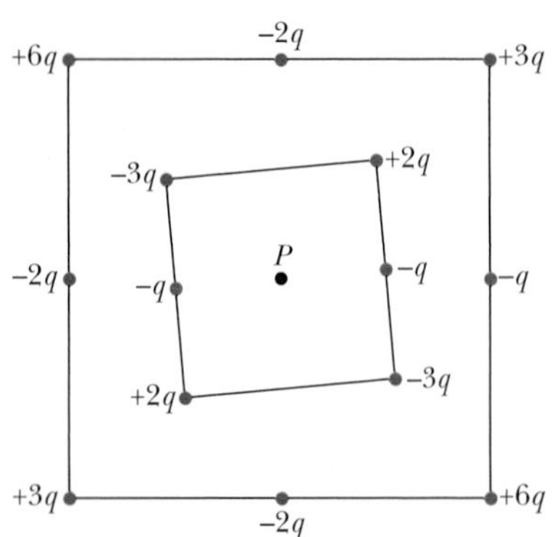

Fig. 23-21 Question 4.

5. In Fig. 23-22, two particles of charge $-q$ are arranged symmetrically about the y axis; each produces an electric field at point P on that axis. (a) Are the magnitudes of the fields at P equal? (b) Is each electric field directed toward or away from the charge producing it? (c) Is the magnitude of the net electric field at P equal to the sum of the magnitudes E of the two field vectors (is it equal to $2E$)? (d) Do the x components of those two field vectors add or cancel? (e) Do their y components add or cancel? (f) Is the direction of the net field at P that of the canceling components or the adding components? (g) What is the direction of the net field?

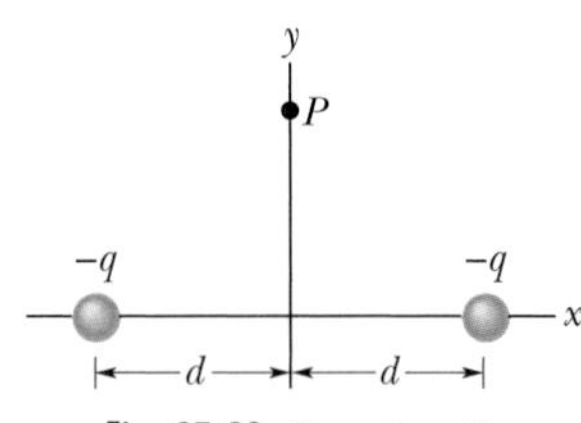

Fig. 23-22 Question 5.

6. Three circular nonconducting rods of the same radius of curvature have uniform charges. Rod A has charge $+2Q$ and subtends an arc of 30°, rod B has charge $+6Q$ and subtends 90°, and rod C has charge $+4Q$ and subtends 60°. Rank the rods according to their linear charge density, greatest first.

7. In Fig. 23-23a, a circular plastic rod with uniform charge $+Q$ produces an electric field of magnitude E at the center of curvature (at the origin). In Figs. 23-23b, c, and d, more circular rods with

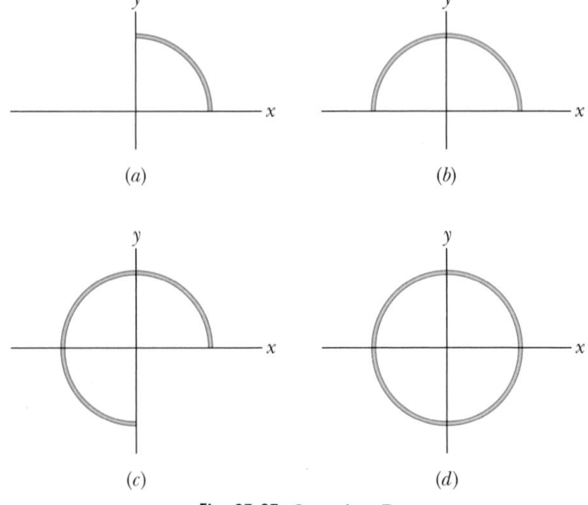

Fig. 23-23 Question 7.

identical uniform charges $+Q$ are added until the circle is complete. A fifth arrangement (which would be labeled *e*) is like that in *d* except that the rod in the fourth quadrant has charge $-Q$. Rank the five arrangements according to the magnitude of the electric field at the center of curvature, greatest first.

8. In Fig. 23-24, an electron e travels through a small hole in plate *A* and then toward plate *B*. A uniform electric field in the region between the plates then slows the electron without deflecting it. (a) What is the direction of the field? (b) Four other particles similarly travel through small holes in either plate *A* or plate *B* and then into the region between the plates. Three have charges $+q_1$, $+q_2$, and $-q_3$. The fourth (labeled n) is a neutron, which is electrically neutral. Does the speed of each of those four other particles increase, decrease, or remain the same in the region between the plates?

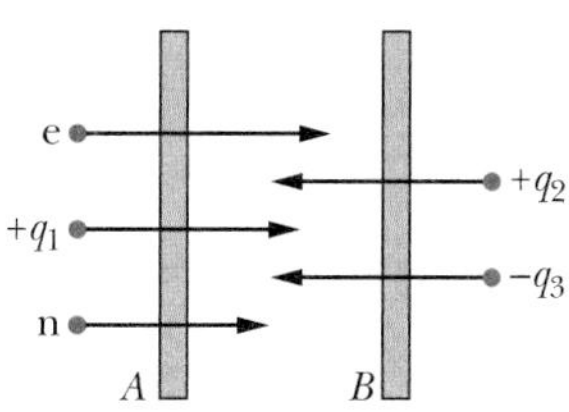

Fig. 23-24 Question 8.

9. Figure 23-25 shows the path of negatively charged particle 1 through a rectangular region of uniform electric field; the particle is deflected toward the top of the page. (a) Is the field directed leftward, rightward, toward the top of the page, or toward the bottom? (b) Three other charged particles are shown approaching the region of electric field. Which are deflected toward the top of the page and which toward the bottom?

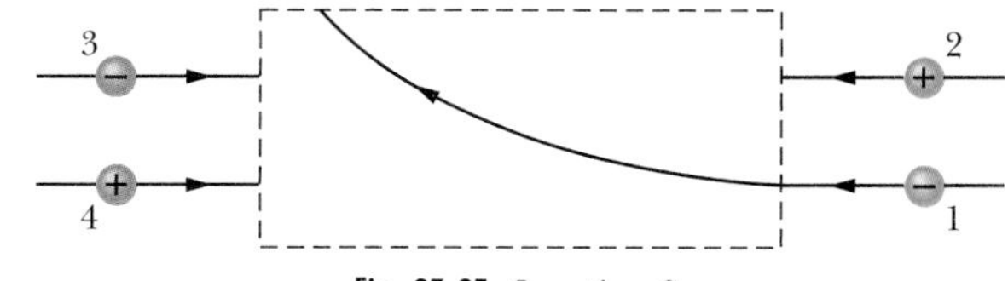

Fig. 23-25 Question 9.

10. (a) In Checkpoint 5, if the dipole rotates from orientation 1 to orientation 2, is the work done on the dipole by the field positive, negative, or zero? (b) If, instead, the dipole rotates from orientation 1 to orientation 4, is the work done by the field more than, less than, or the same as in (a)?

11. The potential energies associated with four orientations of an electric dipole in an electric field are (1) $-5U_0$, (2) $-7U_0$, (3) $3U_0$, and (4) $5U_0$, where U_0 is positive. Rank the orientations according to (a) the angle between the electric dipole moment $\vec{p}$ and the electric field $\vec{E}$, and (b) the magnitude of the torque on the electric dipole, greatest first.

12. If you walk across some types of carpet on a dry day and then reach for a metal doorknob or (for more fun) the back of someone's neck, you might produce a spark. Why does the spark occur? (You can increase the brightness and noise of the spark if you reach with a pointed finger or, even better, a metal key with the pointed end forward.)

EXERCISES & PROBLEMS

ssm Solution is in the Student Solutions Manual.
www Solution is available on the World Wide Web at: http://www.wiley.com/college/hrw
ilw Solution is available on the Interactive LearningWare.

SEC. 23-3 Electric Field Lines

1E. In Fig. 23-26 the electric field lines on the left have twice the separation of those on the right. (a) If the magnitude of the field at *A* is 40 N/C, what force acts on a proton at *A*? (b) What is the magnitude of the field at *B*?

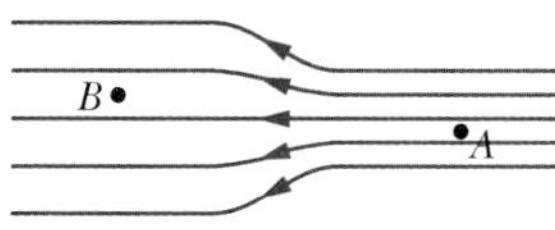

Fig. 23-26 Exercise 1.

2E. Sketch qualitatively the electric field lines both between and outside two concentric conducting spherical shells when a uniform positive charge q_1 is on the inner shell and a uniform negative charge $-q_2$ is on the outer. Consider the cases $q_1 > q_2$, $q_1 = q_2$, and $q_1 < q_2$.

3E. Sketch qualitatively the electric field lines for a thin, circular, uniformly charged disk of radius *R*. (*Hint:* Consider as limiting cases points very close to the disk, where the electric field is directed perpendicular to the surface, and points very far from it, where the electric field is like that of a point charge.) ssm

SEC. 23-4 The Electric Field Due to a Point Charge

4E. What is the magnitude of a point charge that would create an electric field of 1.00 N/C at points 1.00 m away?

5E. What is the magnitude of a point charge whose electric field 50 cm away has the magnitude 2.0 N/C? ssm

6E. Two particles with equal charge magnitudes 2.0×10^{-7} C but opposite signs are held 15 cm apart. What are the magnitude and direction of $\vec{E}$ at the point midway between the charges?

7E. An atom of plutonium-239 has a nuclear radius of 6.64 fm and the atomic number $Z = 94$. Assuming that the positive charge is distributed uniformly within the nucleus, what are the magnitude and direction of the electric field at the surface of the nucleus due to the positive charge? ssm

8P. In Fig. 23-27, two fixed point charges $q_1 = +1.0 \times 10^{-6}$ C and $q_2 = +3.0 \times 10^{-6}$ C are separated by a distance $d = 10$ cm. Plot their net electric field $E(x)$ as a function of x for both positive and negative values of x, taking E to be positive when the vector $\vec{E}$ points to the right and negative when $\vec{E}$ points to the left.

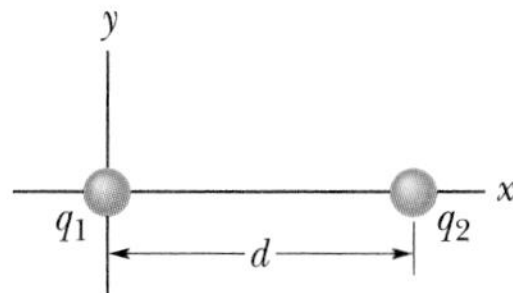

Fig. 23-27 Problems 8 and 10.

9P. Two point charges $q_1 = 2.1 \times 10^{-8}$ C and $q_2 = -4.0q_1$ are fixed in place 50 cm apart. Find the point along the straight line passing through the two charges at which the electric field is zero. ssm www

10P. (a) In Fig. 23-27, two fixed point charges $q_1 = -5q$ and $q_2 = +2q$ are separated by distance d. Locate the point (or points) at which the net electric field due to the two charges is zero. (b) Sketch the net electric field lines qualitatively.

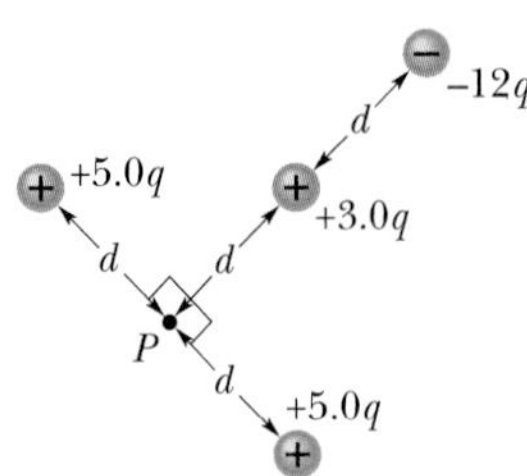

Fig. 23-28 Problem 11.

11P. In Fig. 23-28, what is the magnitude of the electric field at point P due to the four point charges shown?

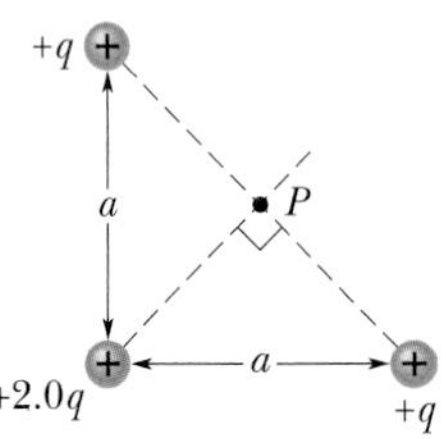

Fig. 23-29 Problem 12.

12P. Calculate the direction and magnitude of the electric field at point P in Fig. 23-29, due to the three point charges.

13P. What are the magnitude and direction of the electric field at the center of the square of Fig. 23-30 if $q = 1.0 \times 10^{-8}$ C and $a = 5.0$ cm? ssm ilw

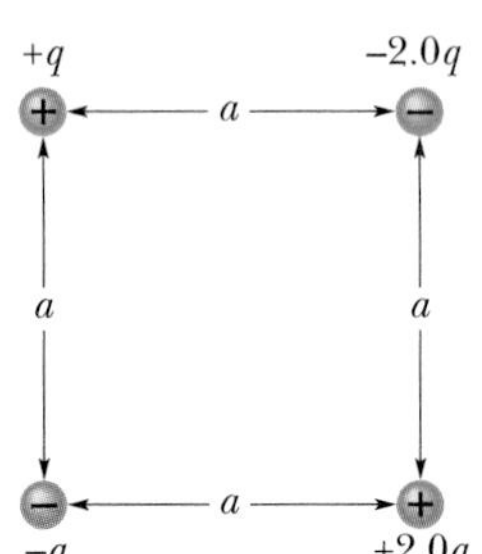

Fig. 23-30 Problem 13.

SEC. 23-5 The Electric Field Due to an Electric Dipole

14E. In Fig. 23-8, let both charges be positive. Assuming $z \gg d$, show that E at point P in that figure is then given by

$$E = \frac{1}{4\pi\varepsilon_0}\frac{2q}{z^2}.$$

15E. Calculate the electric dipole moment of an electron and a proton 4.30 nm apart. ssm

16P. Find the magnitude and direction of the electric field at point P due to the electric dipole in Fig. 23-31. P is located at a distance $r \gg d$ along the perpendicular bisector of the line joining the charges. Express your answer in terms of the magnitude and direction of the electric dipole moment $\vec{p}$.

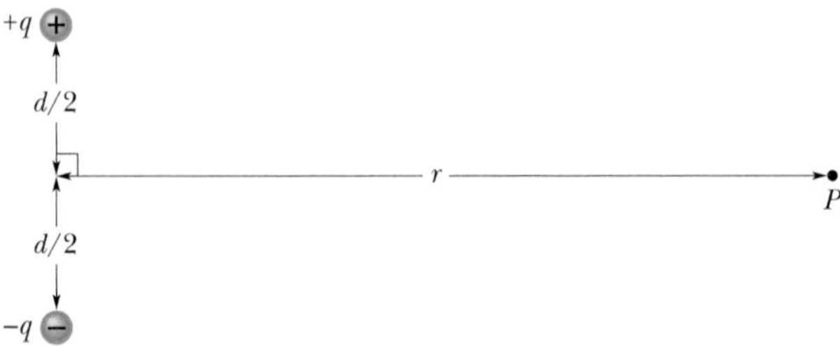

Fig. 23-31 Problem 16.

17P*. *Electric quadrupole.* Figure 23-32 shows an electric quadrupole. It consists of two dipoles with dipole moments that are equal in magnitude but opposite in direction. Show that the value of E on the axis of the quadrupole for a point P a distance z from its center (assume $z \gg d$) is given by

$$E = \frac{3Q}{4\pi\varepsilon_0 z^4},$$

in which Q $(= 2qd^2)$ is known as the *quadrupole moment* of the charge distribution. ssm

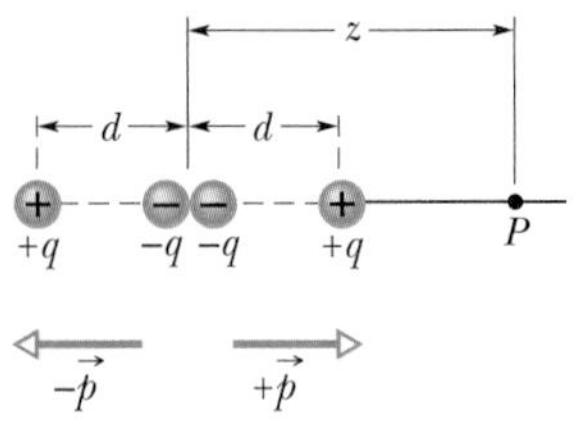

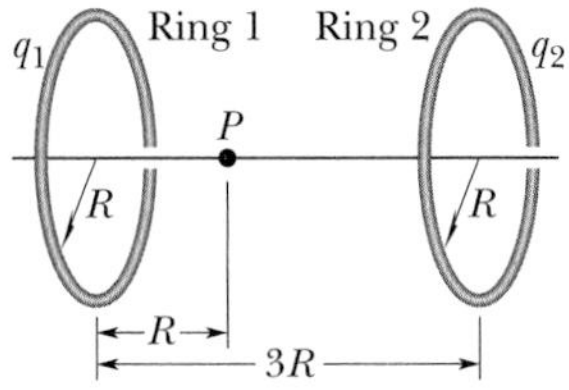

Fig. 23-32 Problem 17.

SEC. 23-6 The Electric Field Due to a Line of Charge

18E. Figure 23-33 shows two parallel nonconducting rings arranged with their central axes along a common line. Ring 1 has uniform charge q_1 and radius R; ring 2 has uniform charge q_2 and the same radius R. The rings are separated by a distance $3R$. The net electric field at point P on the common line, at distance R from ring 1, is zero. What is the ratio q_1/q_2?

Fig. 23-33 Exercise 18.

19P. An electron is constrained to the central axis of the ring of charge of radius R discussed in Section 23-6. Show that the electrostatic force on the electron can cause it to oscillate through the center of the ring with an angular frequency

$$\omega = \sqrt{\frac{eq}{4\pi\varepsilon_0 mR^3}},$$

where q is the ring's charge and m is the electron's mass. ssm

20P. In Fig. 23-34a, two curved plastic rods, one of charge $+q$ and the other of charge $-q$, form a circle of radius R in an xy plane. The x axis passes through their connecting points, and the charge is distributed uniformly on both rods. What are the magnitude and direction of the electric field $\vec{E}$ produced at P, the center of the circle?

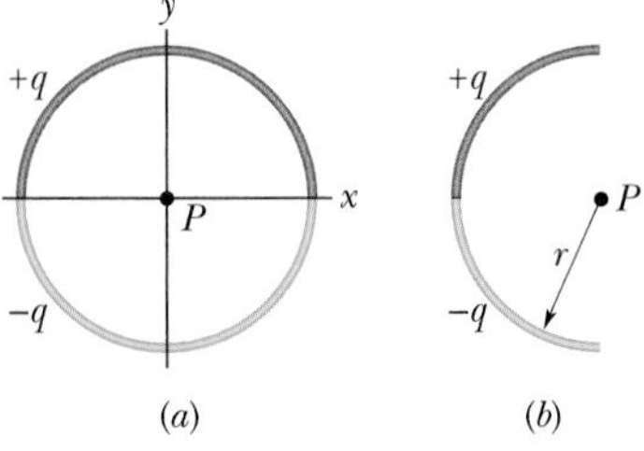

Fig. 23-34 Problems 20 and 21.

21P. A thin glass rod is bent into a semicircle of radius r. A charge $+q$ is uniformly distributed along the upper half, and a charge $-q$ is uniformly distributed along the lower half, as shown in Fig. 23-34b. Find the magnitude and direction of the electric field $\vec{E}$ at P, the center of the semicircle. ilw

22P. At what distance along the central axis of a ring of radius R and uniform charge is the magnitude of the electric field due to the ring's charge maximum?

23P. In Fig. 23-35, a nonconducting rod of length L has charge $-q$ uniformly distributed along its length. (a) What is the linear charge density of the rod? (b) What is the electric field at point P, a distance a from the end of the rod? (c) If P were very far from the rod compared to L, the rod would look like a point charge. Show that your answer to (b) reduces to the electric field of a point charge for $a \gg L$. ssm ilw www

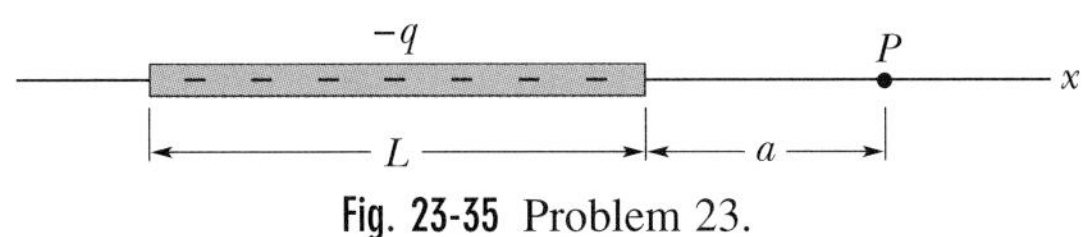

Fig. 23-35 Problem 23.

24P. A thin nonconducting rod of finite length L has a charge q spread uniformly along it. Show that

$$E = \frac{q}{2\pi\varepsilon_0 y} \frac{1}{(L^2 + 4y^2)^{1/2}}$$

gives the magnitude E of the electric field at point P on the perpendicular bisector of the rod (Fig. 23-36).

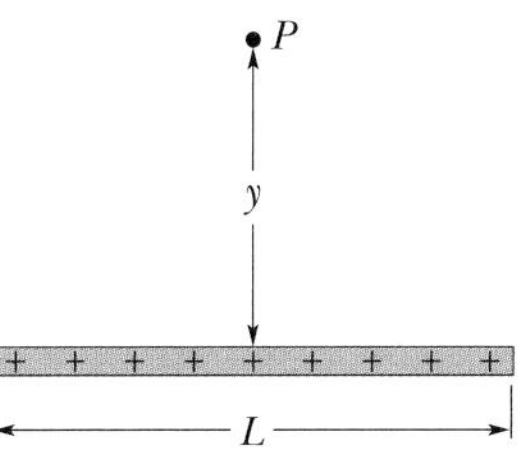

Fig. 23-36 Problem 24.

25P*. In Fig. 23-37, a "semi-infinite" nonconducting rod (that is, infinite in one direction only) has uniform linear charge density λ. Show that the electric field at point P makes an angle of 45° with the rod and that this result is independent of the distance R. (*Hint:* Separately find the parallel and perpendicular (to the rod) components of the electric field at P, and then compare those components.) ssm

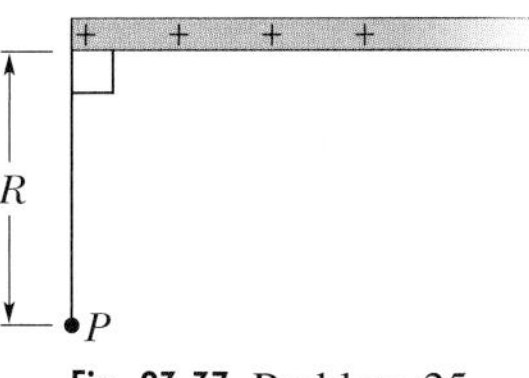

Fig. 23-37 Problem 25.

SEC. 23-7 The Electric Field Due to a Charged Disk

26E. A disk of radius 2.5 cm has a surface charge density of 5.3 μC/m^2 on its upper face. What is the magnitude of the electric field produced by the disk at a point on its central axis at distance z = 12 cm from the disk?

27P. At what distance along the central axis of a uniformly charged plastic disk of radius R is the magnitude of the electric field equal to one-half the magnitude of the field at the center of the surface of the disk? ssm

SEC. 23-8 A Point Charge in an Electric Field

28E. An electron is accelerated eastward at 1.80×10^9 m/s^2 by an electric field. Determine the magnitude and direction of the electric field.

29E. An electron is released from rest in a uniform electric field of magnitude 2.00×10^4 N/C. Calculate the acceleration of the electron. (Ignore gravitation.) ssm

30E. An alpha particle (the nucleus of a helium atom) has a mass of 6.64×10^{-27} kg and a charge of $+2e$. What are the magnitude and direction of the electric field that will balance the gravitational force on it?

31E. Calculate the magnitude of the force, due to an electric dipole of dipole moment 3.6×10^{-29} C·m, on an electron 25 nm from the center of the dipole, along the dipole axis. Assume that this distance is large relative to the dipole's charge separation. ilw

32E. Humid air breaks down (its molecules become ionized) in an electric field of 3.0×10^6 N/C. In that field, what is the magnitude of the electrostatic force on (a) an electron and (b) an ion with a single electron missing?

33E. A charged cloud system produces an electric field in the air near Earth's surface. A particle of charge -2.0×10^{-9} C is acted on by a downward electrostatic force of 3.0×10^{-6} N when placed in this field. (a) What is the magnitude of the electric field? (b) What are the magnitude and direction of the electrostatic force exerted on a proton placed in this field? (c) What is the gravitational force on the proton? (d) What is the ratio of the magnitude of the electrostatic force to the magnitude of the gravitational force in this case? ssm

34E. An electric field $\vec{E}$ with an average magnitude of about 150 N/C points downward in the atmosphere near Earth's surface. We wish to "float" a sulfur sphere weighing 4.4 N in this field by charging the sphere. (a) What charge (both sign and magnitude) must be used? (b) Why is the experiment impractical?

35E. Beams of high-speed protons can be produced in "guns" using electric fields to accelerate the protons. (a) What acceleration would a proton experience if the gun's electric field were 2.00×10^4 N/C? (b) What speed would the proton attain if the field accelerated the proton through a distance of 1.00 cm? ssm

36E. An electron with a speed of 5.00×10^8 cm/s enters an electric field of magnitude 1.00×10^3 N/C, traveling along the field lines in the direction that retards its motion. (a) How far will the electron travel in the field before stopping momentarily and (b) how much time will have elapsed? (c) If the region with the electric field is only 8.00 mm long (too short for the electron to stop within it), what fraction of the electron's initial kinetic energy will be lost in that region?

37E. In Millikan's experiment, an oil drop of radius 1.64 μm and density 0.851 g/cm^3 is suspended in chamber C (Fig. 23-13) when a downward-directed electric field of 1.92×10^5 N/C is applied. Find the charge on the drop, in terms of e. ssm

38P. In one of his experiments, Millikan observed that the following measured charges, among others, appeared at different times on a single drop:

6.563×10^{-19} C	13.13×10^{-19} C	19.71×10^{-19} C
8.204×10^{-19} C	16.48×10^{-19} C	22.89×10^{-19} C
11.50×10^{-19} C	18.08×10^{-19} C	26.13×10^{-19} C

What value for the elementary charge e can be deduced from these data?

39P. A uniform electric field exists in a region between two oppositely charged plates. An electron is released from rest at the surface of the negatively charged plate and strikes the surface of the opposite plate, 2.0 cm away, in a time 1.5×10^{-8} s. (a) What is the speed of the electron as it strikes the second plate? (b) What is the magnitude of the electric field $\vec{E}$? ilw

40P. At some instant the velocity components of an electron moving between two charged parallel plates are $v_x = 1.5 \times 10^5$ m/s and $v_y = 3.0 \times 10^3$ m/s. Suppose that the electric field between the plates is given by $\vec{E} = (120\text{ N/C})\hat{j}$. (a) What is the acceleration of the electron? (b) What will be the velocity of the electron after its x coordinate has changed by 2.0 cm?

41P. Two large parallel copper plates are 5.0 cm apart and have a uniform electric field between them as depicted in Fig. 23-38. An electron is released from the negative plate at the same time that a proton is released from the positive plate. Neglect the force of the particles on each other and find their distance from the positive plate when they pass each other. (Does it surprise you that you need not know the electric field to solve this problem?) ssm www

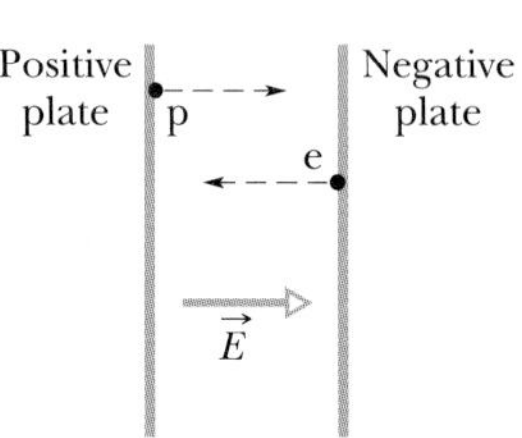

Fig. 23-38 Problem 41.

42P. A 10.0 g block with a charge of $+8.00 \times 10^{-5}$ C is placed in electric field $\vec{E} = (3.00 \times 10^3)\hat{i} - 600\hat{j}$, where $\vec{E}$ is in newtons per coulomb. (a) What are the magnitude and direction of the force on the block? (b) If the block is released from rest at the origin at $t = 0$, what will be its coordinates at $t = 3.00$ s?

43P. In Fig. 23-39, a uniform, upward-directed electric field $\vec{E}$ of magnitude 2.00×10^3 N/C has been set up between two horizontal plates by charging the lower plate positively and the upper plate negatively. The plates have length $L = 10.0$ cm and separation $d = 2.00$ cm. An electron is then shot between the plates from the left edge of the lower plate. The initial velocity $\vec{v}_0$ of the electron makes an angle $\theta = 45.0°$ with the lower plate and has a magnitude of 6.00×10^6 m/s. (a) Will the electron strike one of the plates? (b) If so, which plate and how far horizontally from the left edge will the electron strike? ssm

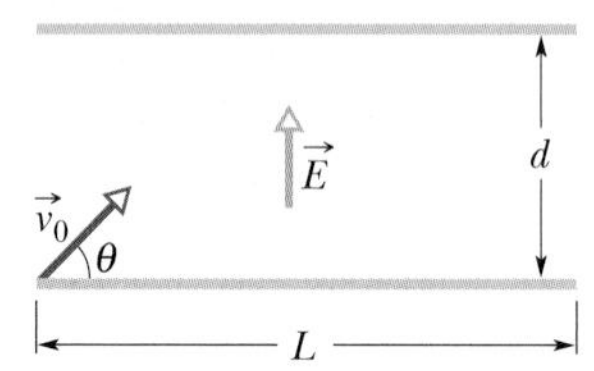

Fig. 23-39 Problem 43.

SEC. 23-9 A Dipole in an Electric Field

44E. An electric dipole, consisting of charges of magnitude 1.50 nC separated by 6.20 μm, is in an electric field of strength 1100 N/C. (a) What is the magnitude of the electric dipole moment? (b) What is the difference between the potential energies corresponding to dipole orientations parallel to and antiparallel to the field?

45E. An electric dipole consists of charges $+2e$ and $-2e$ separated by 0.78 nm. It is in an electric field of strength 3.4×10^6 N/C. Calculate the magnitude of the torque on the dipole when the dipole moment is (a) parallel to, (b) perpendicular to, and (c) antiparallel to the electric field.

46P. Find the work required to turn an electric dipole end for end in a uniform electric field $\vec{E}$, in terms of the magnitude p of the dipole moment, the magnitude E of the field, and the initial angle θ_0 between $\vec{p}$ and $\vec{E}$.

47P. Find the frequency of oscillation of an electric dipole, of dipole moment $\vec{p}$ and rotational inertia I, for small amplitudes of oscillation about its equilibrium position in a uniform electric field of magnitude E. ssm

Additional Problem

48. The reproduction of flowers depends on insects carrying pollen grains from one flower to another. One way in which honeybees can do this is by collecting the grains electrically, because the bees are usually positively charged. When a bee hovers near a flower's anther (Fig. 23-40), which is electrically insulated, the pollen grains (which are moderately conducting) jump to the bee, where they cling during the flight to the next flower. As the bee nears that flower's stigma, which is electrically connected to ground through the flower's interior, the pollen grains jump from the bee to the stigma, fertilizing the flower.

(a) Assuming that a bee with a typical charge of 45 pC is a spherical conductor, find the magnitude of the bee's electric field at the location of a pollen grain 2.0 cm from the bee's center. (b) Is that field uniform or nonuniform? (c) Give a plausible explanation of why the pollen grains jump to the bee, cling to the bee during the flight, and then jump away from the bee to the grounded stigma. (*Hint:* Consider Fig. 22-5.) When a pollen grain reaches the bee, does it make electrical contact with it, so that the charge on the grain changes?

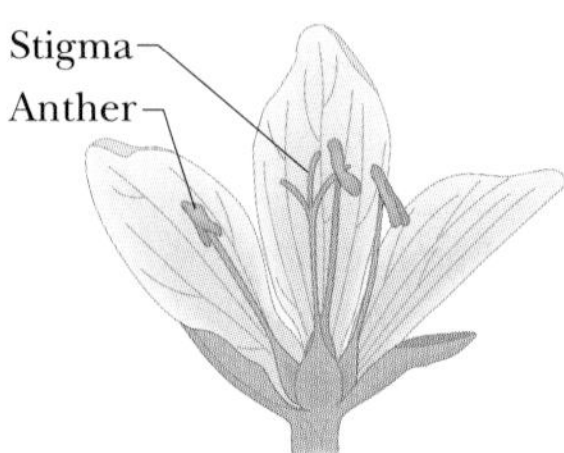

Fig. 23-40 Problem 48.

NEW PROBLEMS

N1. Suppose that you design an apparatus in which a uniformly charged disk of radius R is to produce an electric field. The field magnitude is most important along the central axis of the disk, at a point P at distance $2R$ from the disk (Fig. 23N-1*a*). Cost analysis suggests that you switch to a ring of the same outer radius R but with an inner radius of $R/2$ (Fig. 23N-1*b*). Assume that the ring will have the same surface charge density as the original disk. If you switch to the ring, by what percentage will you decrease the electric field magnitude at point P?

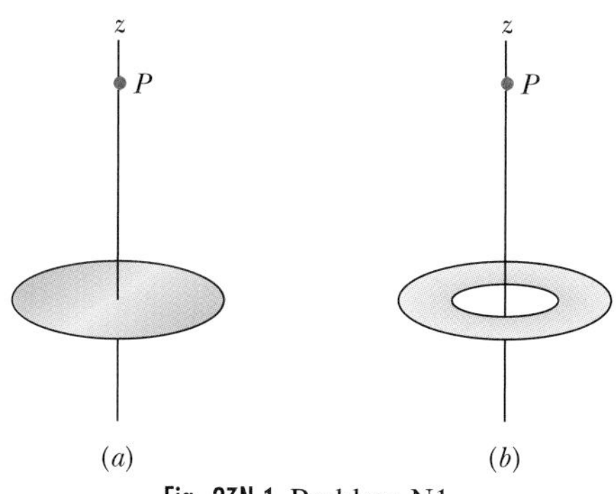

Fig. 23N-1 Problem N1.

N2. Two charged beads are on the plastic ring in Fig. 23N-2*a*. Bead 2, which is not shown, is fixed in place on the ring, which has radius $R = 60.0$ cm. Bead 1 is initially at the right side of the ring, at angle $\theta = 0°$. It is then moved to the left side, at angle $\theta = 180°$, through the first and second quadrants of the xy coordinate system. Figure 23N-2*b* gives the x component of the net electric field produced at the origin by the two beads as a function of θ. Similarly, Fig. 23N-2*c* gives the y component. (a) At what angle θ is bead 2 located? What are the charges of (b) bead 1 and (c) bead 2?

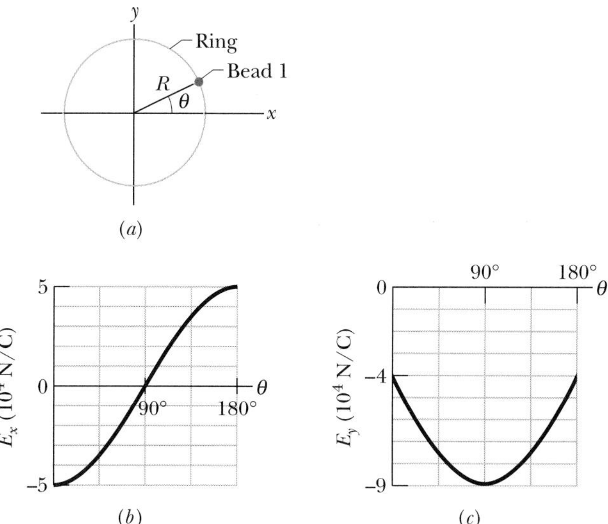

Fig. 23N-2 Problem N2.

N3. *Density, density, density.* (a) A charge of $-300e$ is uniformly distributed along a circular arc of radius 4.00 cm, which subtends an angle of 40°. What is the linear charge density along the arc? (b) A charge of $-300e$ is uniformly distributed over one face of a circular disk of radius 2.00 cm. What is the surface charge density over that face? (c) A charge of $-300e$ is uniformly distributed over the surface of a sphere of radius 2.00 cm. What is the surface charge density over that surface? (d) A charge of $-300e$ is uniformly spread through the volume of a sphere of radius 2.00 cm. What is the volume charge density in that sphere?

N4. Figure 23N-3*a* shows two charged particles fixed in place on an x axis with separation L. The ratio q_1/q_2 of their charge magnitudes is 4.00. Figure 23N-3*b* shows the x component $E_{net,x}$ of their net electric field along the x axis just to the right of particle 2. (a) At what value of $x > 0$ is $E_{net,x}$ maximum? (b) If particle 2 has charge $-q_2 = -3e$, what is the value of that maximum?

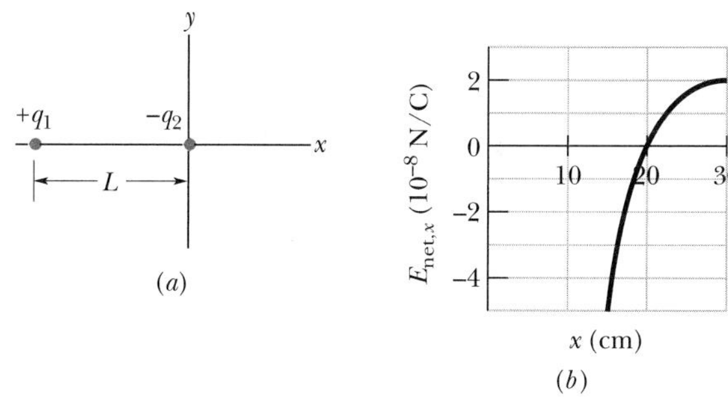

Fig. 23N-3 Problem N4.

N5. A circular plastic disk with radius $R = 2.00$ cm has a uniformly distributed charge of $Q = +(2.00 \times 10^6)e$ on one face. A circular ring of width 30 μm is centered on that face, with the center of the ring at radius $r = 0.50$ cm. In coulombs, what charge is contained within the width of the ring?

N6. A thin nonconducting rod with a uniform distribution of positive charge Q is bent into a circle of radius R (Fig. 23N-4). The central axis through the ring is a z axis, with the origin at the center of the ring. What is the magnitude of the electric field due to the rod at (a) $z = 0$ and (b) $z = \pm\infty$? (c) In terms of R, at what values of z is that magnitude maximum? (d) If radius $R = 2.00$ cm and charge $Q = 4.00$ μC, what is the maximum magnitude?

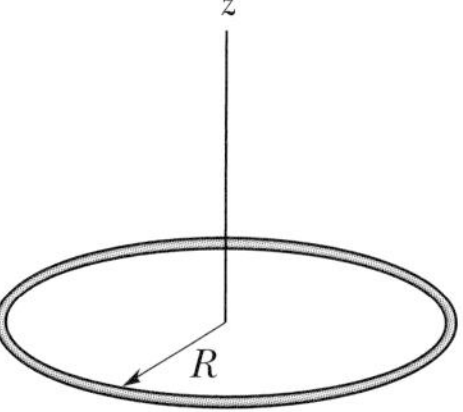

Fig. 23N-4 Problem N6.

N7. A circular rod has a radius of curvature R and a uniformly distributed charge Q and it subtends an angle θ (in radians). What is the magnitude of the electric field it produces at the center of curvature?

N8. Figure 23N-5 shows two concentric rings, of radii R and $R' = 3.00R$, that lie on the same plane. Point P lies on the central z axis, at distance $D = 2.00R$ from the center of the rings. The smaller ring has uniformly distributed charge $+Q$. What must be the uniformly distributed charge on the larger ring if the net electric field at point P due to the two rings is to be zero?

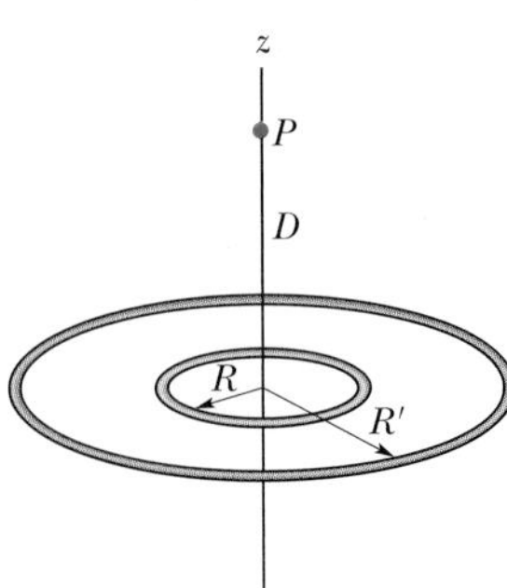

Fig. 23N-5 Problem N8.

N9. In Fig. 23N-6*a*, a particle of charge $+Q$ produces an electric field with a magnitude E_{part} at point P, at distance R from it. In Fig. 23N-6*b*, that same amount of charge is spread uniformly along a circular arc that has radius R and subtends an angle θ. The charge on the arc produces an electric field with a magnitude E_{arc} at its center of curvature P. For what value of θ does $E_{arc} = 0.500E_{part}$? (*Hint:* You will probably resort to a graphical solution.)

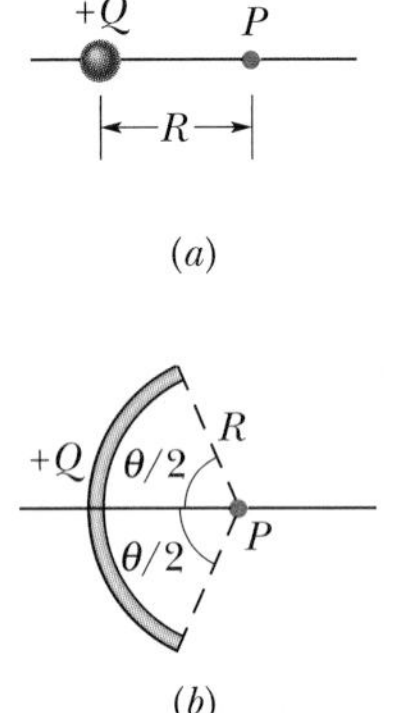

Fig. 23N-6 Problem N9.

N10. Figure 23N-7*a* shows a nonconducting rod with a uniformly distributed charge $+Q$. The rod forms a half circle with radius R and produces an electric field of magnitude E_{arc} at its center of curvature P. If the arc is collapsed to a point at distance R from P (Fig. 23N-7*b*), by what factor is the magnitude of the electric field at P multiplied?

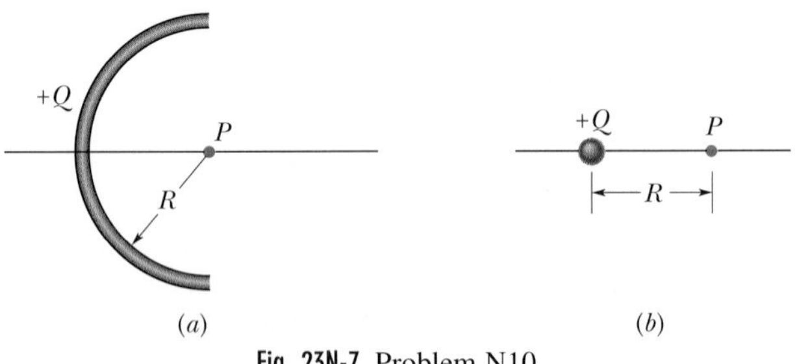

Fig. 23N-7 Problem N10.

N11. For the data of Problem 38, assume that the charge q on the drop is given by $q = ne$, where n is an integer and e is the elementary charge. (a) Find n for each measurement of q. (b) Do a linear regression fit of the values of q versus the values of n; from it find e.

N12. Equations 23-8 and 23-9 are approximations of the magnitude of the electric field of an electric dipole, at points along the dipole axis. Consider a point P on that axis at distance $z = 5d$ from the dipole center (d is the separation distance between the particles of the dipole). Let E_{appr} be the magnitude of the field at point P as approximated by Eqs. 23-8 and 23-9. Let E_{act} be the actual magnitude. What is the ratio E_{appr}/E_{act}?

N13. In Fig. 23N-8, particles with charges $+1.0q$ and $-2.0q$ are fixed a distance d apart. Find the magnitude and direction of the net electric field at points (a) A, (b) B, and (c) C. (d) Sketch the electric field lines.

$$A \quad \xleftarrow{d} \quad +1.0q \quad \frac{d}{2} \quad B \quad \frac{d}{2} \quad -2.0q \quad \xleftarrow{d} \quad C$$

Fig. 23N-8 Problem N13.

N14. Figure 23N-9 shows a plastic ring of radius $R = 50.0$ cm. Two small charged beads are on the ring: Bead 1 of charge $+2.00\ \mu$C is fixed in place at the left side; bead 2 of charge $+6.00\ \mu$C can be moved along the ring. The two beads produce a net electric field of magnitude E at the center of the ring. At what angle θ should bead 2 be positioned such that $E = 2.00 \times 10^5$ N/C?

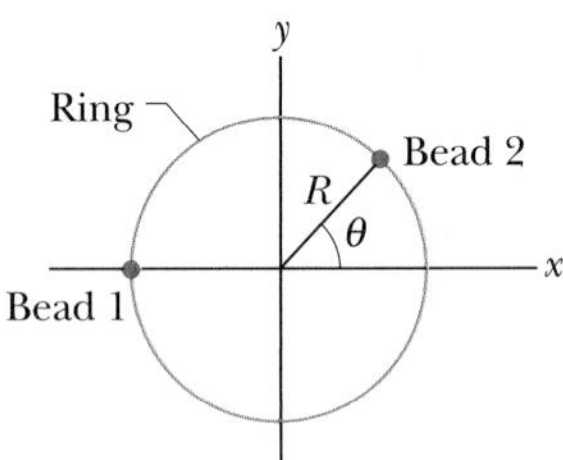

Fig. 23N-9 Problem N14.

N15. In Fig. 23N-10, three point charges are arranged in an equilateral triangle. (a) Sketch the field lines due to $+Q$ and $-Q$, and from them determine the direction of the force that acts on $+q$ because of the presence of the other two charges. (*Hint:* See Fig. 23-5.) (b) What is the magnitude of that net electric force on $+q$?

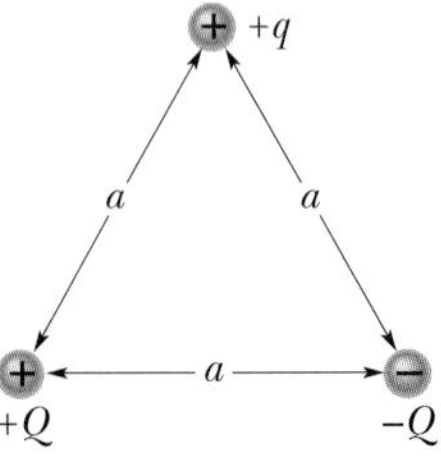

Fig. 23N-10 Problem N15.

N16. The electric field of an electric dipole along the dipole axis is approximated by Eqs. 23-8 and 23-9, which result from a termination of two binomial expansions as shown just before those equations. If the expansions were carried out further, what would be the next nonzero term in the expression for the dipole's electric field along the dipole axis? That is, if

$$E = \frac{1}{2\pi\varepsilon_0}\frac{qd}{z^3} + E_{\text{next}},$$

what is E_{next}?

N17. Two particles, each with a charge of magnitude 12 nC, are placed at two of the vertices of an equilateral triangle. The length of each side of the triangle is 2.0 m. What is the magnitude of the electric field at the third vertex of the triangle if (a) both of the charges are positive and (b) one of the charges is positive and the other is negative?

N18. A certain electric dipole is placed in a uniform electric field $\vec{E}$ of magnitude 40 N/C. Figure 23N-11 gives the magnitude τ of the torque on the dipole versus the angle θ between $\vec{E}$ and the dipole moment $\vec{p}$. What is the magnitude of $\vec{p}$?

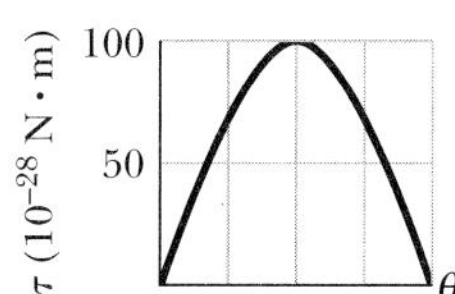

Fig. 23N-11 Problem N18.

N19. Three particles, each with positive charge Q, form an equilateral triangle, with each side of length d. What is the magnitude of the electric field produced by the particles at the midpoint of any side?

N20. In Fig. 23N-12, an electron (e) is to be released from rest on the central axis of a uniformly charged disk of radius R. The surface charge density on the disk is $+4.00\ \mu\text{C/m}^2$. What is the magnitude of the electron's initial acceleration if it is released at a distance (a) R, (b) $R/100$, and (c) $R/1000$ from the center of the disk? (d) Why does the acceleration magnitude increase only slightly as the release point is moved closer to the disk?

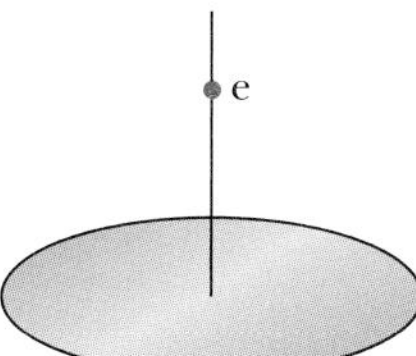

Fig. 23N-12 Problem N20.

N21. An electron enters a region of uniform electric field with an initial velocity of 40 km/s in the same direction as the electric field, which has magnitude $E = 50$ N/C. (a) What is the speed of the electron 1.5 ns after entering this region? (b) How far does the electron travel during the 1.5 ns interval?

N22. In Fig. 23N-13, an electron is shot at an initial speed of $v_0 = 2.00 \times 10^6$ m/s, at angle $\theta_0 = 40°$ from an x axis. It moves in a region with uniform electric field $\vec{E} = (5.00\ \text{N/C})\hat{\text{j}}$. A screen for detecting electrons is positioned parallel to the y axis, at distance $x = 3.00$ m. In unit-vector notation, what is the velocity of the electron when it hits the screen?

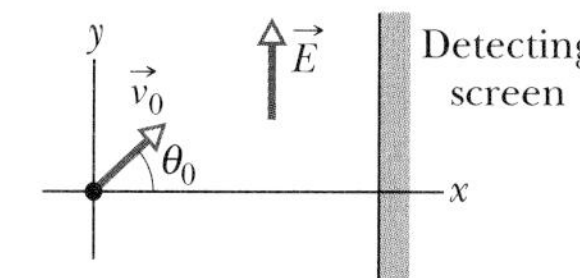

Fig. 23N-13 Problem N22.

N23. Figure 23N-14 shows the deflection-plate system of a conventional TV tube. The length of the plates is 3.0 cm and the electric field between the two plates is 10^6 N/C (vertically up). If the electron enters the plates with a horizontal velocity of 3.9×10^7 m/s, what is the vertical deflection Δy at the end of the plates?

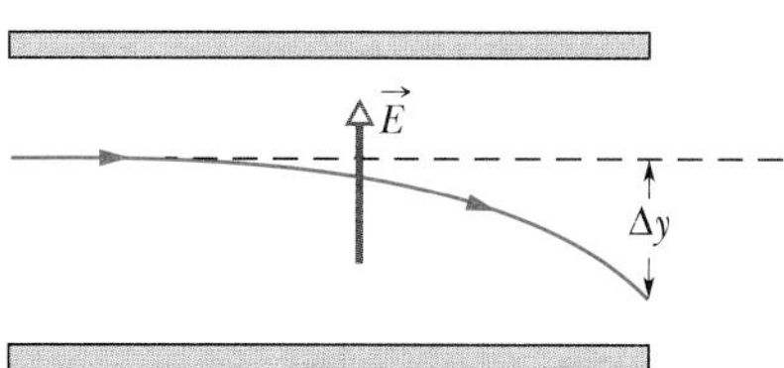

Fig. 23N-14 Problem N23.

N24. Figure 23N-15 shows a proton (p) on the central axis through a disk with a uniform charge density due to excess electrons. Three of those electrons are shown: Electron e_c is at the center of the disk and electrons e_s are at opposite sides of the disk, at radius R from the center of the disk. The proton is initially at a distance $z = R = 2.0$ cm from the disk. At the proton's location, what are the magnitudes of (a) the electric field $\vec{E}_c$ due to electron e_c and (b) the *net* electric field $\vec{E}_{s,\text{net}}$ due to electrons e_s? The proton is then moved to a distance $z = R/10$. What then are the magnitudes of (c) $\vec{E}_c$ and (d) $\vec{E}_{s,\text{net}}$ at the proton's location? From (a) and (c) we see that as the proton moves nearer to the disk, the magnitude of $\vec{E}_c$ increases. (e) Why does the magnitude of $\vec{E}_{s,\text{net}}$ decrease, as we see from (b) and (d)?

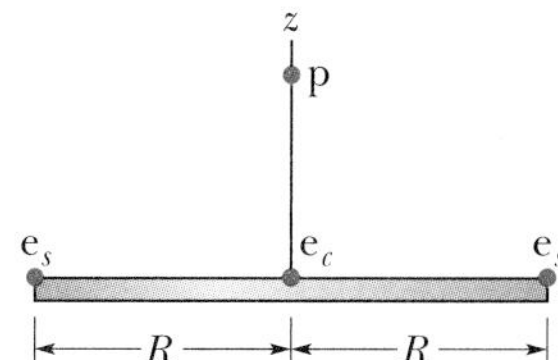

Fig. 23N-15 Problem N24.

N25. Figure 23N-16 shows two charged particles on an x axis: $-q = -3.20 \times 10^{-19}$ C at $x = -3.00$ m and $q = 3.20 \times 10^{-19}$ C at $x = +3.00$ m. What are the magnitude and direction of the net electric field they produce at point P at $y = 4.00$ m?

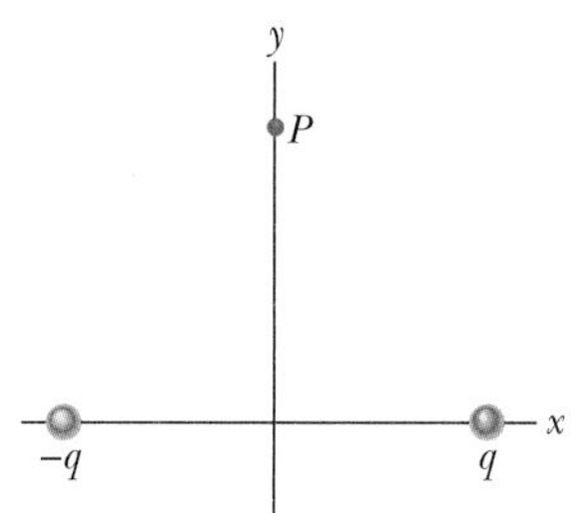

Fig. 23N-16 Problem N25.

N26. Figure 23N-17 shows three circular arcs centered on the origin of a coordinate system. The uniformly distributed charge on each arc is given in terms of $Q = 2.00\ \mu C$. The radii are given in terms of $R = 10.0$ cm. What are the magnitude and direction of the net electric field at the origin due to the arcs?

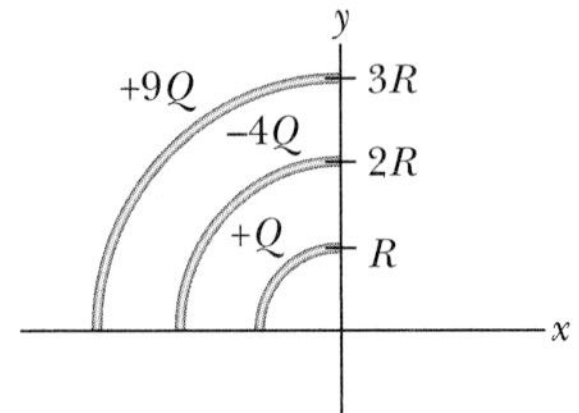

Fig. 23N-17 Problem N26.

N27. In Fig. 23N-18, eight charged particles form a square array; charge $q = e$ and distance $d = 2.0$ cm. What are the magnitude and direction of the net electric field at the center?

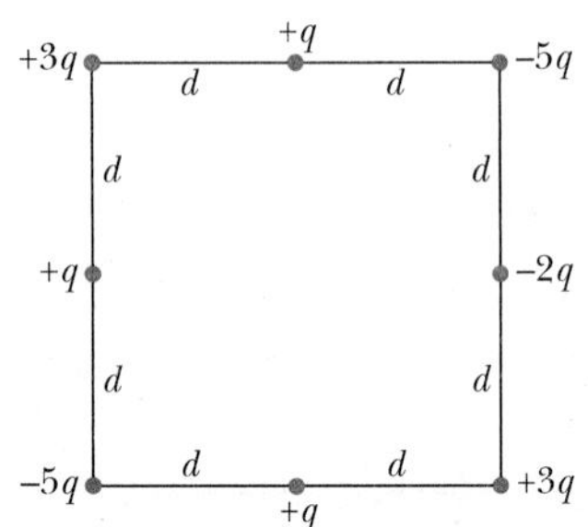

Fig. 23N-18 Problem N27.

N28. Figure 23N-19*a* shows a circular disk that is uniformly charged. The central z axis is perpendicular to the disk face, with the origin at the disk. Figure 23N-19*b* gives the magnitude of the electric field along that axis in terms of the maximum magnitude E_m at the disk surface. What is the radius of the disk?

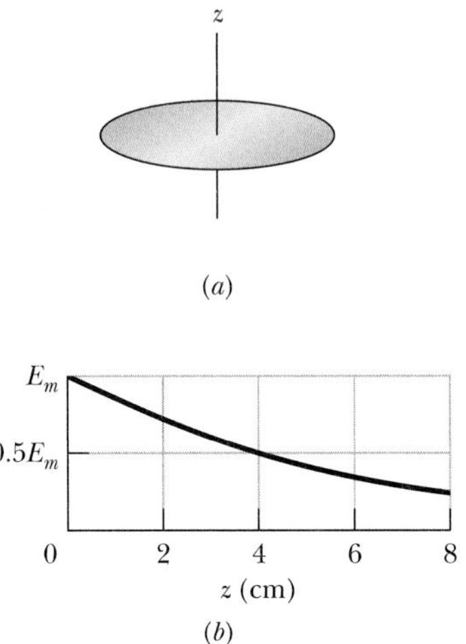

Fig. 23N-19 Problem N28.

N29. How much energy is needed to flip an electric dipole from being lined up with a uniform external electric field to being lined up opposite the field? The dipole consists of an electron and a proton at a separation of 2.00 nm, and it is in a uniform field of magnitude 3.00×10^6 N/C.

N30. A certain electric dipole is placed in a uniform electric field $\vec{E}$ of magnitude 20 N/C. Figure 23N-20 gives the potential energy U of the dipole versus the angle θ between $\vec{E}$ and the dipole moment $\vec{p}$. What is the magnitude of $\vec{p}$?

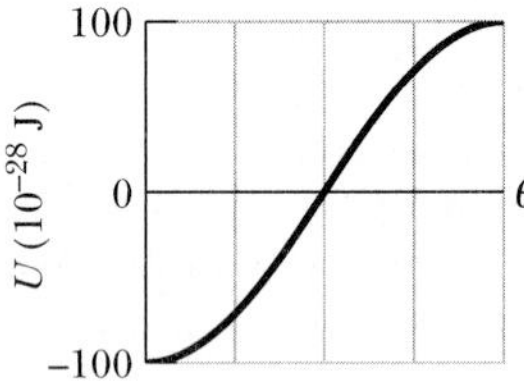

Fig. 23N-20 Problem N30.

24 Gauss' Law

Lightning strikes Tucson in a brilliant display, each strike delivering about 10^{20} electrons from the cloud base to the ground.

How wide is a lightning strike? Since a strike can be seen from kilometers away, is it as wide as, say, a car?

The answer is in this chapter.

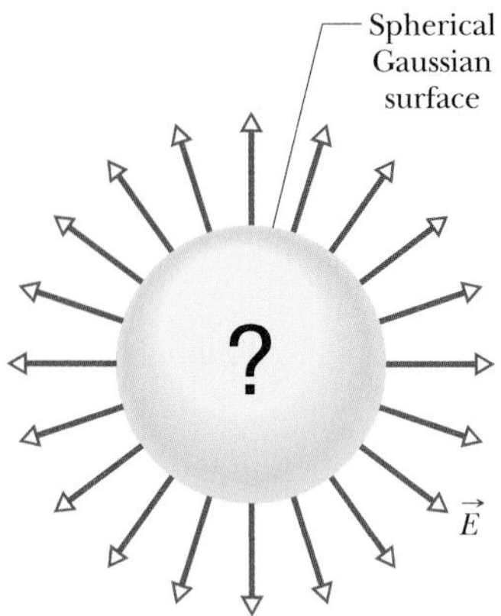

Fig. 24-1 A spherical Gaussian surface. If the electric field vectors are of uniform magnitude and point radially outward at all surface points, you can conclude that a net positive distribution of charge must lie within the surface and have spherical symmetry.

24-1 A New Look at Coulomb's Law

If you want to find the center of mass of a potato, you can do so by experiment or by laborious calculation, involving the numerical evaluation of a triple integral. However, if the potato happens to be a uniform ellipsoid, you know from its symmetry exactly where the center of mass is without calculation. Such are the advantages of symmetry. Symmetrical situations arise in all areas of physics; when possible, it makes sense to cast the laws of physics in forms that take full advantage of this fact.

Coulomb's law is the governing law in electrostatics, but it is not cast in a form that particularly simplifies the work in situations involving symmetry. In this chapter we introduce a new formulation of Coulomb's law, derived by German mathematician and physicist Carl Friedrich Gauss (1777–1855). This law, called **Gauss' law,** *can* be used to take advantage of special symmetry situations. For electrostatics problems, it is the full equivalent of Coulomb's law; which of them we choose to use depends only on the problem at hand.

Central to Gauss' law is a hypothetical closed surface called a **Gaussian surface.** The Gaussian surface can be of any shape you wish to make it, but the most useful surface is one that mimics the symmetry of the problem at hand. Thus, the Gaussian surface will often be a sphere, a cylinder, or some other symmetrical form. It must always be a *closed* surface, so that a clear distinction can be made between points that are inside the surface, on the surface, and outside the surface.

Imagine that you have established a Gaussian surface around a distribution of charges. Then Gauss' law comes into play:

> Gauss' law relates the electric fields at points on a (closed) Gaussian surface and the net charge enclosed by that surface.

Figure 24-1 shows a simple situation in which the Gaussian surface is a sphere. Suppose you know that there is an electric field at every point on the surface and that all the fields have the same magnitude and point radially outward. Without knowing anything about Gauss' law, you can guess that some net positive charge must be enclosed by the Gaussian surface. If you *do* know Gauss' law, you can calculate just how much net positive charge is enclosed. To make the calculation, you need know only "how much" electric field is intercepted by the surface—this "how much" involves the *flux* of the electric field through the surface.

24-2 Flux

Suppose that, as in Fig. 24-2*a*, you aim a wide airstream of uniform velocity $\vec{v}$ at a small square loop of area A. Let Φ represent the *volume flow rate* (volume per unit time) at which air flows through the loop. This rate depends on the angle between $\vec{v}$ and the plane of the loop. If $\vec{v}$ is perpendicular to the plane, the rate Φ is equal to vA.

If $\vec{v}$ is parallel to the plane of the loop, no air moves through the loop, so Φ is zero. For an intermediate angle θ, the rate Φ depends on the component of $\vec{v}$ that is normal to the plane (Fig. 24-2*b*). Since that component is $v \cos \theta$, the rate of volume flow through the loop is

$$\Phi = (v \cos \theta)A. \tag{24-1}$$

This rate of flow through an area is an example of a **flux**—a *volume flux* in this situation. Before we discuss a flux that is involved in electrostatics, we need to rewrite Eq. 24-1 in terms of vectors.

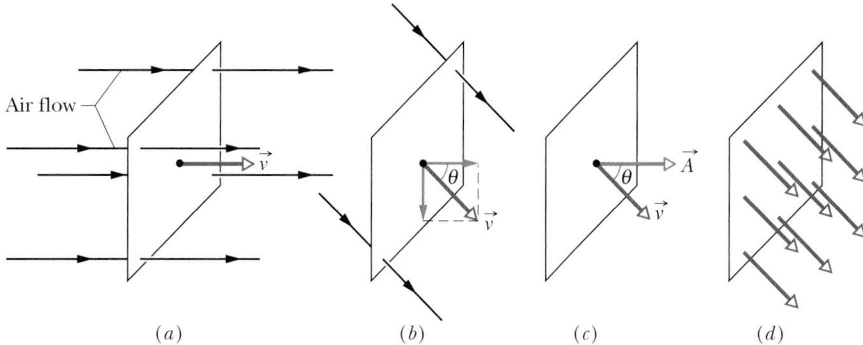

Fig. 24-2 (*a*) A uniform airstream of velocity $\vec{v}$ is perpendicular to the plane of a square loop of area A. (*b*) The component of $\vec{v}$ perpendicular to the plane of the loop is $v \cos \theta$, where θ is the angle between $\vec{v}$ and a normal to the plane. (*c*) The area vector $\vec{A}$ is perpendicular to the plane of the loop and makes an angle θ with $\vec{v}$. (*d*) The velocity field intercepted by the area of the loop.

To do this, we first define an *area vector* $\vec{A}$ as being a vector whose magnitude is equal to an area (here the area of the loop) and whose direction is normal to the plane of the area (Fig. 24-2*c*). We then rewrite Eq. 24-1 as the scalar (or dot) product of the velocity vector $\vec{v}$ of the airstream and the area vector $\vec{A}$ of the loop:

$$\Phi = vA \cos \theta = \vec{v} \cdot \vec{A}, \tag{24-2}$$

where θ is the angle between $\vec{v}$ and $\vec{A}$.

The word "flux" comes from the Latin word meaning "to flow." That meaning makes sense if we talk about the flow of air volume through the loop. However, Eq. 24-2 can be regarded in a more abstract way. To see it, note that we can assign a velocity vector to each point in the airstream passing through the loop (Fig. 24-2*d*). The composite of all those vectors is a *velocity field*, so we can interpret Eq. 24-2 as giving the *flux of the velocity field through the loop.* With this interpretation, flux no longer means the actual flow of something through an area—rather it means the product of an area and the field across that area.

24-3 Flux of an Electric Field

To define the flux of an electric field, consider Fig. 24-3, which shows an arbitrary (asymmetric) Gaussian surface immersed in a nonuniform electric field. Let us divide the surface into small squares of area ΔA, each square being small enough to permit us to neglect any curvature and to consider the individual square to be flat. We represent each such element of area with an area vector $\Delta\vec{A}$, whose magnitude is the area ΔA. Each vector $\Delta\vec{A}$ is perpendicular to the Gaussian surface and directed away from the interior of the surface.

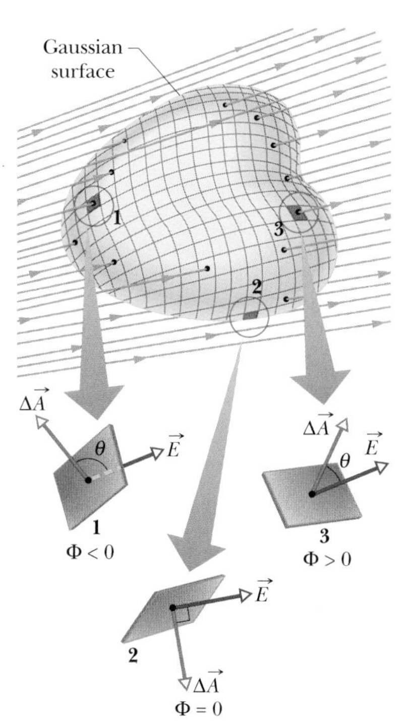

Fig. 24-3 A Gaussian surface of arbitrary shape immersed in an electric field. The surface is divided into small squares of area ΔA. The electric field vectors $\vec{E}$ and the area vectors $\Delta\vec{A}$ for three representative squares, marked 1, 2, and 3, are shown.

Because the squares have been taken to be arbitrarily small, the electric field $\vec{E}$ may be taken as constant over any given square. The vectors $\Delta\vec{A}$ and $\vec{E}$ for each square then make some angle θ with each other. Figure 24-3 shows an enlarged view of three squares (1, 2, and 3) on the Gaussian surface, and the angle θ for each.

A provisional definition for the flux of the electric field for the Gaussian surface of Fig. 24-3 is

$$\Phi = \sum \vec{E} \cdot \Delta\vec{A}. \tag{24-3}$$

This equation instructs us to visit each square on the Gaussian surface, to evaluate the scalar product $\vec{E} \cdot \Delta\vec{A}$ for the two vectors $\vec{E}$ and $\Delta\vec{A}$ that we find there, and to sum the results algebraically (that is, with signs included) for all the squares that make up the surface. The sign or a zero resulting from each scalar product determines whether the flux through its square is positive, negative, or zero. Squares like 1, in which $\vec{E}$ points inward, make a negative contribution to the sum of Eq. 24-3. Squares like 2, in which $\vec{E}$ lies in the surface, make zero contribution. Squares like 3, in which $\vec{E}$ points outward, make a positive contribution.

The exact definition of the flux of the electric field through a closed surface is found by allowing the area of the squares shown in Fig. 24-3 to become smaller and smaller, approaching a differential limit dA. The area vectors then approach a differential limit $d\vec{A}$. The sum of Eq. 24-3 then becomes an integral and we have, for the definition of electric flux,

$$\Phi = \oint \vec{E} \cdot d\vec{A} \quad \text{(electric flux through a Gaussian surface).} \tag{24-4}$$

The circle on the integral sign indicates that the integration is to be taken over the entire (closed) surface. The flux of the electric field is a scalar, and its SI unit is the newton–square-meter per coulomb ($\text{N} \cdot \text{m}^2/\text{C}$).

We can interpret Eq. 24-4 in the following way: First recall that we can use the density of electric field lines passing through an area as a proportional measure of an electric field $\vec{E}$ there. Specifically, the magnitude E is proportional to the number of electric field lines per unit area. Thus, the scalar product $\vec{E} \cdot d\vec{A}$ in Eq. 24-4 is proportional to the number of electric field lines passing through area $d\vec{A}$. Then, because the integration in Eq. 24-4 is carried out over a Gaussian surface, which is closed, we see that

> The electric flux Φ through a Gaussian surface is proportional to the net number of electric field lines passing through that surface.

Sample Problem 24-1

Figure 24-4 shows a Gaussian surface in the form of a cylinder of radius R immersed in a uniform electric field $\vec{E}$, with the cylinder axis parallel to the field. What is the flux Φ of the electric field through this closed surface?

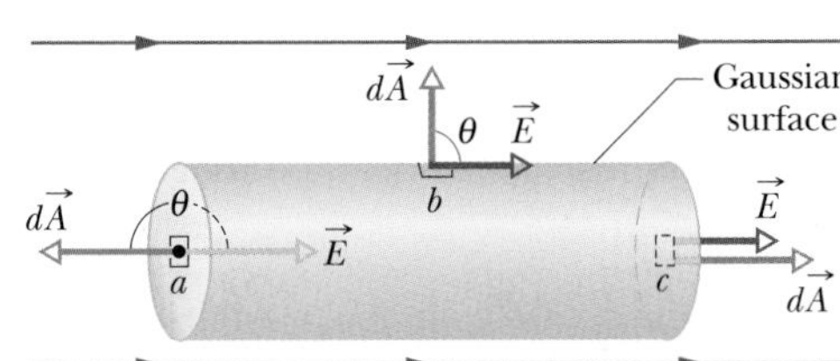

Fig. 24-4 Sample Problem 24-1. A cylindrical Gaussian surface, closed by end caps, is immersed in a uniform electric field. The cylinder axis is parallel to the field direction.

SOLUTION: The **Key Idea** here is that we can find the flux Φ through the surface by integrating the scalar product $\vec{E} \cdot d\vec{A}$ over the Gaussian surface. We can do this by writing the flux as the sum of three terms: integrals over the left cylinder cap a, the cylindrical surface b, and the right cap c. Thus, from Eq. 24-4,

$$\begin{aligned}\Phi &= \oint \vec{E} \cdot d\vec{A} \\ &= \int_a \vec{E} \cdot d\vec{A} + \int_b \vec{E} \cdot d\vec{A} + \int_c \vec{E} \cdot d\vec{A}.\end{aligned} \tag{24-5}$$

For all points on the left cap, the angle θ between $\vec{E}$ and $d\vec{A}$ is 180° and the magnitude E of the field is constant. Thus,

$$\int_a \vec{E} \cdot d\vec{A} = \int E(\cos 180°)\, dA = -E \int dA = -EA,$$

where $\int dA$ gives the cap's area, $A\ (= \pi R^2)$. Similarly, for the right cap, where $\theta = 0$ for all points,

$$\int_c \vec{E} \cdot d\vec{A} = \int E(\cos 0)\, dA = EA.$$

Finally, for the cylindrical surface, where the angle θ is 90° at all points,

$$\int_b \vec{E} \cdot d\vec{A} = \int E(\cos 90°)\, dA = 0.$$

Substituting these results into Eq. 24-5 leads us to

$$\Phi = -EA + 0 + EA = 0. \quad \text{(Answer)}$$

This result is perhaps not surprising because the field lines that represent the electric field all pass entirely through the Gaussian surface, entering through the left end cap, leaving through the right end cap, and giving a net flux of zero.

CHECKPOINT 1: The figure here shows a Gaussian cube of face area A immersed in a uniform electric field $\vec{E}$ that has the positive direction of the z axis. In terms of E and A, what is the flux through (a) the front face (which is in the xy plane), (b) the rear face, (c) the top face, and (d) the whole cube?

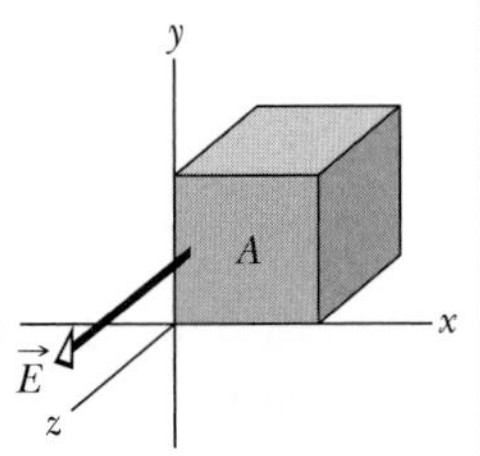

Sample Problem 24-2

A *nonuniform* electric field given by $\vec{E} = 3.0x\hat{i} + 4.0\hat{j}$ pierces the Gaussian cube shown in Fig. 24-5. (E is in newtons per coulomb and x is in meters.) What is the electric flux through the right face, the left face, and the top face?

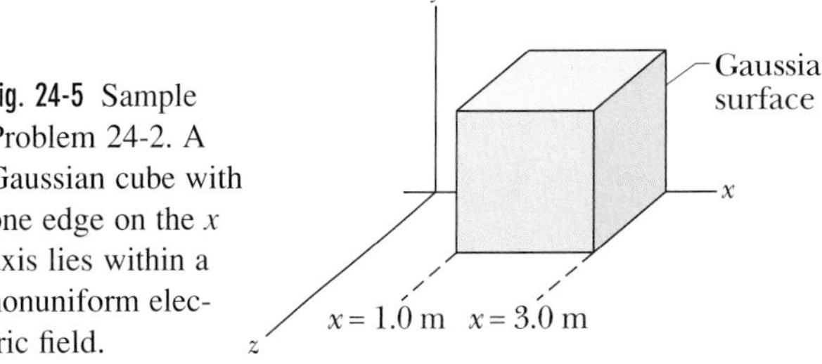

Fig. 24-5 Sample Problem 24-2. A Gaussian cube with one edge on the x axis lies within a nonuniform electric field.

SOLUTION: The **Key Idea** here is that we can find the flux Φ through the surface by integrating the scalar product $\vec{E} \cdot d\vec{A}$ over each face.

Right face: An area vector $\vec{A}$ is always perpendicular to its surface and always points away from the interior of a Gaussian surface. Thus, the vector $d\vec{A}$ for the right face of the cube must point in the positive x direction. In unit vector notation, then,

$$d\vec{A} = dA\hat{i}.$$

From Eq. 24-4, the flux Φ_r through the right face is then

$$\begin{aligned}\Phi_r &= \int \vec{E} \cdot d\vec{A} = \int (3.0x\hat{i} + 4.0\hat{j}) \cdot (dA\hat{i}) \\ &= \int [(3.0x)(dA)\hat{i} \cdot \hat{i} + (4.0)(dA)\hat{j} \cdot \hat{i})] \\ &= \int (3.0x\, dA + 0) = 3.0 \int x\, dA.\end{aligned}$$

We are about to integrate over the right face, but we note that x has the same value everywhere on that face—namely, $x = 3.0$ m. This means we can substitute that constant value for x. Then

$$\Phi_r = 3.0 \int (3.0)\, dA = 9.0 \int dA.$$

Now the integral merely gives us the area $A = 4.0\ \text{m}^2$ of the right face, so

$$\Phi_r = (9.0\ \text{N/C})(4.0\ \text{m}^2) = 36\ \text{N} \cdot \text{m}^2/\text{C}. \qquad \text{(Answer)}$$

Left face: The procedure for finding the flux through the left face is the same as that for the right face. However, two factors change. (1) The differential area vector $d\vec{A}$ points in the negative x direction and thus $d\vec{A} = -dA\hat{i}$. (2) The term x again appears in our integration, and it is again constant over the face being considered. However, on the left face, $x = 1.0$ m. With these two changes, we find that the flux Φ_l through the left face is

$$\Phi_l = -12\text{N} \cdot \text{m}^2/\text{C}. \qquad \text{(Answer)}$$

Top face: The differential area vector $d\vec{A}$ points in the positive y direction and thus $d\vec{A} = dA\hat{j}$. The flux Φ_t through the top face is then

$$\begin{aligned}\Phi_t &= \int (3.0x\hat{i} + 4.0\hat{j}) \cdot (dA\hat{j}) \\ &= \int [(3.0x)(dA)\hat{i} \cdot \hat{j} + (4.0)(dA)\hat{j} \cdot \hat{j})] \\ &= \int (0 + 4.0\, dA) = 4.0 \int dA \\ &= 16\ \text{N} \cdot \text{m}^2/\text{C}. \qquad \text{(Answer)}\end{aligned}$$

24-4 Gauss' Law

Gauss' law relates the net flux Φ of an electric field through a closed surface (a Gaussian surface) to the *net* charge q_{enc} that is *enclosed* by that surface. It tells us that

$$\varepsilon_0 \Phi = q_{enc} \qquad \text{(Gauss' law)}. \qquad (24\text{-}6)$$

By substituting Eq. 24-4, the definition of flux, we can also write Gauss' law as

$$\varepsilon_0 \oint \vec{E} \cdot d\vec{A} = q_{enc} \qquad \text{(Gauss' law)}. \qquad (24\text{-}7)$$

Equations 24-6 and 24-7 hold only when the net charge is located in a vacuum or (what is the same for most practical purposes) in air. In Section 26-8, we modify Gauss' law to include situations in which a material such as mica, oil, or glass is present.

In Eqs. 24-6 and 24-7, the net charge q_{enc} is the algebraic sum of all the *enclosed* positive and negative charges, and it can be positive, negative, or zero. We include the sign, rather than just use the magnitude of the enclosed charge, because the sign

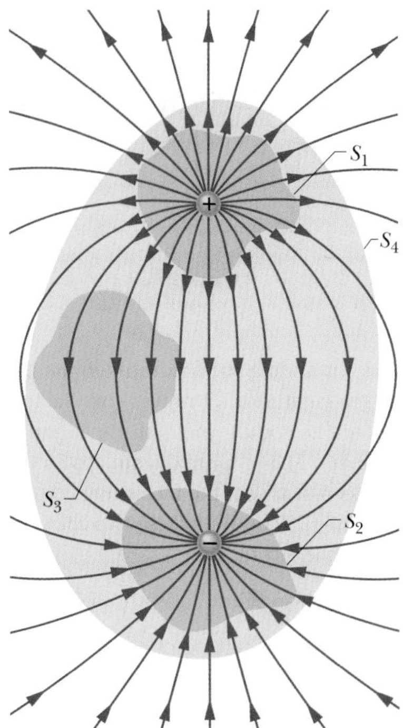

Fig. 24-6 Two point charges, equal in magnitude but opposite in sign, and the field lines that represent their net electric field. Four Gaussian surfaces are shown in cross section. Surface S_1 encloses the positive charge. Surface S_2 encloses the negative charge. Surface S_3 encloses no charge. Surface S_4 encloses both charges, and thus no net charge.

tells us something about the net flux through the Gaussian surface: If q_{enc} is positive, the net flux is *outward*; if q_{enc} is negative, the net flux is *inward*.

Charge outside the surface, no matter how large or how close it may be, is not included in the term q_{enc} in Gauss' law. The exact form or location of the charges inside the Gaussian surface is also of no concern; the only things that matter, on the right side of Eq. 24-7, are the magnitude and sign of the net enclosed charge. The quantity $\vec{E}$ on the left side of Eq. 24-7, however, is the electric field resulting from *all* charges, both those inside and those outside the Gaussian surface. This may seem to be inconsistent, but keep in mind what we saw in Sample Problem 24-1: The electric field due to a charge outside the Gaussian surface contributes zero net flux *through* the surface, because as many field lines due to that charge enter the surface as leave it.

Let us apply these ideas to Fig. 24-6, which shows two point charges, equal in magnitude but opposite in sign, and the field lines describing the electric fields that they set up in the surrounding space. Four Gaussian surfaces are also shown, in cross section. Let us consider each in turn.

Surface S_1. The electric field is outward for all points on this surface. Thus, the flux of the electric field through this surface is positive, and so is the net charge within the surface, as Gauss' law requires. (That is, in Eq. 24-6, if Φ is positive, q_{enc} must be also.)

Surface S_2. The electric field is inward for all points on this surface. Thus, the flux of the electric field is negative and so is the enclosed charge, as Gauss' law requires.

Surface S_3. This surface encloses no charge, and thus $q_{enc} = 0$. Gauss' law (Eq. 24-6) requires that the net flux of the electric field through this surface be zero. That is reasonable because all the field lines pass entirely through the surface, entering it at the top and leaving at the bottom.

Surface S_4. This surface encloses no *net* charge, because the enclosed positive and negative charges have equal magnitudes. Gauss' law requires that the net flux of the electric field through this surface be zero. That is reasonable because there are as many field lines leaving surface S_4 as entering it.

What would happen if we were to bring an enormous charge Q up close to surface S_4 in Fig. 24-6? The pattern of the field lines would certainly change, but the net flux for each of the four Gaussian surfaces would not change. We can understand this because the field lines associated with the added Q would pass entirely through each of the four Gaussian surfaces, making no contribution to the net flux through any of them. The value of Q would not enter Gauss' law in any way, because Q lies outside all four of the Gaussian surfaces that we are considering.

CHECKPOINT 2: The figure shows three situations in which a Gaussian cube sits in an electric field. The arrows and the values indicate the directions of the field lines and the magnitudes (in $N \cdot m^2/C$) of the flux through the six sides of each cube. (The lighter arrows are for the hidden faces.) In which situations does the cube enclose (a) a positive net charge, (b) a negative net charge, and (c) zero net charge?

(1) (2) (3)

Sample Problem 24-3

Figure 24-7 shows five charged lumps of plastic and an electrically neutral coin. The cross section of a Gaussian surface S is indicated. What is the net electric flux through the surface if $q_1 = q_4 = +3.1$ nC, $q_2 = q_5 = -5.9$ nC, and $q_3 = -3.1$ nC?

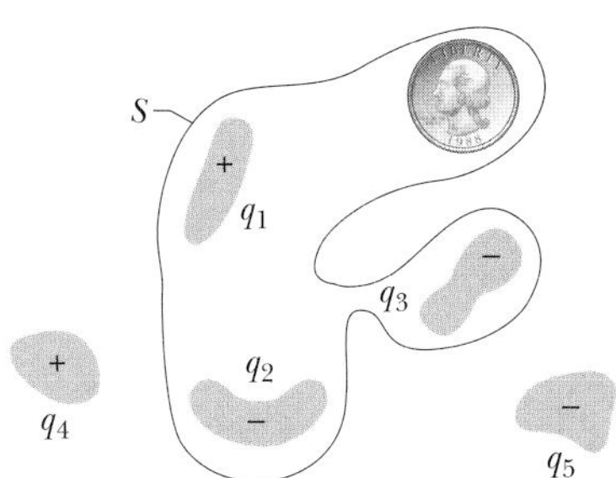

Fig. 24-7 Sample Problem 24-3. Five plastic objects, each with an electric charge, and a coin, which has no net charge. A Gaussian surface, shown in cross section, encloses three of the plastic objects and the coin.

SOLUTION: The **Key Idea** here is that the *net* flux Φ through the surface depends on the *net* charge q_{enc} enclosed by surface S. This means that the coin and charges q_4 and q_5 do not contribute to Φ. The coin does not contribute because it is neutral and thus contains equal amounts of positive and negative charge. Charges q_4 and q_5 do not contribute because they are outside surface S. Thus, q_{enc} is $q_1 + q_2 + q_3$ and Eq. 24-6 gives us

$$\Phi = \frac{q_{enc}}{\varepsilon_0} = \frac{q_1 + q_2 + q_3}{\varepsilon_0}$$

$$= \frac{+3.1 \times 10^{-9}\text{ C} - 5.9 \times 10^{-9}\text{ C} - 3.1 \times 10^{-9}\text{ C}}{8.85 \times 10^{-12}\text{ C}^2/\text{N}\cdot\text{m}^2}$$

$$= -670\text{ N}\cdot\text{m}^2/\text{C}. \qquad \text{(Answer)}$$

The minus sign shows that the net flux through the surface is inward and thus that the net charge within the surface is negative.

24-5 Gauss' Law and Coulomb's Law

If Gauss' law and Coulomb's law are equivalent, we should be able to derive each from the other. Here we derive Coulomb's law from Gauss' law and some symmetry considerations.

Figure 24-8 shows a positive point charge q, around which we have drawn a concentric spherical Gaussian surface of radius r. Let us divide this surface into differential areas dA. By definition, the area vector $d\vec{A}$ at any point is perpendicular to the surface and directed outward from the interior. From the symmetry of the situation, we know that at any point the electric field $\vec{E}$ is also perpendicular to the surface and directed outward from the interior. Thus, since the angle θ between $\vec{E}$ and $d\vec{A}$ is zero, we can rewrite Eq. 24-7 for Gauss' law as

$$\varepsilon_0 \oint \vec{E}\cdot d\vec{A} = \varepsilon_0 \oint E\,dA = q_{enc}. \qquad (24\text{-}8)$$

Here $q_{enc} = q$. Although E varies radially with the distance from q, it has the same value everywhere on the spherical surface. Since the integral in Eq. 24-8 is taken over that surface, E is a constant in the integration and can be brought out in front of the integral sign. That gives us

$$\varepsilon_0 E \oint dA = q. \qquad (24\text{-}9)$$

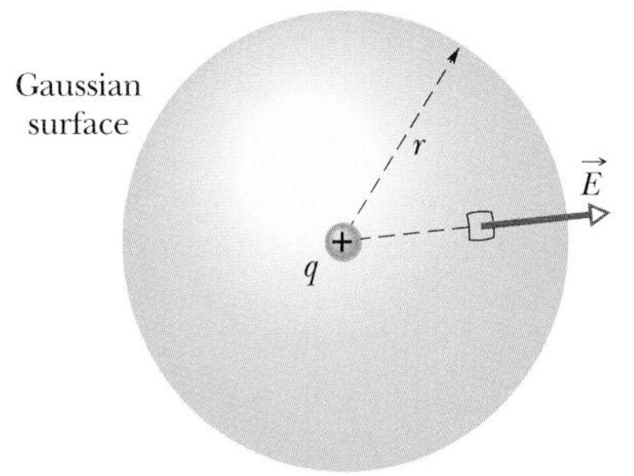

Fig. 24-8 A spherical Gaussian surface centered on a point charge q.

The integral is now merely the sum of all the differential areas dA on the sphere and thus is just the surface area, $4\pi r^2$. Substituting this, we have

$$\varepsilon_0 E(4\pi r^2) = q$$

or

$$E = \frac{1}{4\pi\varepsilon_0}\frac{q}{r^2}. \qquad (24\text{-}10)$$

This is exactly the electric field due to a point charge (Eq. 23-3), which we found using Coulomb's law. Thus, Gauss' law is equivalent to Coulomb's law.

CHECKPOINT 3: There is a certain net flux Φ_i through a Gaussian sphere of radius r enclosing an isolated charged particle. Suppose the enclosing Gaussian surface is changed to (a) a larger Gaussian sphere, (b) a Gaussian cube with edge length equal to r, and (c) a Gaussian cube with edge length equal to $2r$. In each case, is the net flux through the new Gaussian surface greater than, less than, or equal to Φ_i?

PROBLEM-SOLVING TACTICS

Tactic 1: *Choosing a Gaussian Surface*
The derivation of Eq. 24-10 using Gauss' law is a warm-up for derivations of electric fields produced by other charge configurations, so let us go back over the steps involved. We started with a given positive point charge q; we know that electric field lines extend radially outward from q in a spherically symmetric pattern.

To find the magnitude E of the electric field at a distance r by Gauss' law (Eq. 24-7), we had to place a hypothetical closed Gaussian surface around q, through a point that is a distance r from q. Then we had to sum via integration the values of $\vec{E} \cdot d\vec{A}$ over the full Gaussian surface. To make this integration as simple as possible, we chose a spherical Gaussian surface (to mimic the spherical symmetry of the electric field). That choice produced three simplifying features. (1) The dot product $\vec{E} \cdot d\vec{A}$ became simple, because at all points on the Gaussian surface the angle between $\vec{E}$ and $d\vec{A}$ is zero, and so at all points we have $\vec{E} \cdot d\vec{A} = E\,dA$. (2) The electric field magnitude E is the same at all points on the spherical Gaussian surface, so E was a constant in the integration and could be brought out in front of the integral sign. (3) The result was a very simple integration—a summation of the differential areas of the sphere, which we could immediately write as $4\pi r^2$.

Note that Gauss' law holds regardless of the shape of the Gaussian surface we choose to place around charge q_{enc}. However, if we had chosen, say, a cubical Gaussian surface, our three simplifying features would have disappeared and the integration of $\vec{E} \cdot d\vec{A}$ over the cubical surface would have been very difficult. The moral here is to choose the Gaussian surface that most simplifies the integration in Gauss' law.

24-6 A Charged Isolated Conductor

Gauss' law permits us to prove an important theorem about isolated conductors:

> If an excess charge is placed on an isolated conductor, that amount of charge will move entirely to the surface of the conductor. None of the excess charge will be found within the body of the conductor.

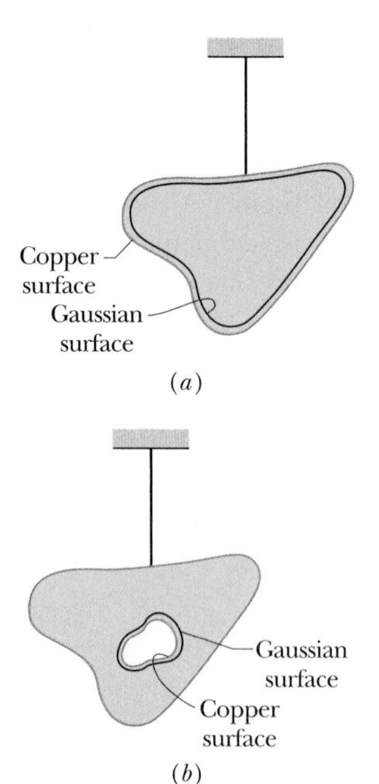

Fig. 24-9 (*a*) A lump of copper with a charge q hangs from an insulating thread. A Gaussian surface is placed within the metal, just inside the actual surface. (*b*) The lump of copper now has a cavity within it. A Gaussian surface lies within the metal, close to the cavity surface.

This might seem reasonable, considering that charges with the same sign repel each other. You might imagine that, by moving to the surface, the added charges are getting as far away from each other as they can. We turn to Gauss' law for verification of this speculation.

Figure 24-9*a* shows, in cross section, an isolated lump of copper hanging from an insulating thread and having an excess charge q. We place a Gaussian surface just inside the actual surface of the conductor.

The electric field inside this conductor must be zero. If this were not so, the field would exert forces on the conduction (free) electrons, which are always present in a conductor, and thus current would always exist within a conductor. (That is, charge would flow from place to place within the conductor.) Of course, there are no such perpetual currents in an isolated conductor, and so the internal electric field is zero.

(An internal electric field *does* appear as a conductor is being charged. However, the added charge quickly distributes itself in such a way that the net internal electric field—the vector sum of the electric fields due to all the charges, both inside and outside—is zero. The movement of charge then ceases, because the net force on each charge is zero; the charges are then in *electrostatic equilibrium.*)

If $\vec{E}$ is zero everywhere inside our copper conductor, it must be zero for all points on the Gaussian surface because that surface, though close to the surface of the conductor, is definitely inside the conductor. This means that the flux through the Gaussian surface must be zero. Gauss' law then tells us that the net charge inside the Gaussian surface must also be zero. Then because the excess charge is not inside the Gaussian surface, it must be outside that surface, which means it must lie on the actual surface of the conductor.

An Isolated Conductor with a Cavity

Figure 24-9*b* shows the same hanging conductor, but now with a cavity that is totally within the conductor. It is perhaps reasonable to suppose that when we scoop out the electrically neutral material to form the cavity, we do not change the distribution of charge or the pattern of the electric field that exists in Fig. 24-9*a*. Again, we must turn to Gauss' law for a quantitative proof.

We draw a Gaussian surface surrounding the cavity, close to its surface but inside the conducting body. Because $\vec{E} = 0$ inside the conductor, there can be no flux through this new Gaussian surface. Therefore, from Gauss' law, that surface can enclose no net charge. We conclude that there is no net charge on the cavity walls; all the excess charge remains on the outer surface of the conductor, as in Fig. 24-9*a*.

The Conductor Removed

Suppose that, by some magic, the excess charges could be "frozen" into position on the conductor's surface, perhaps by embedding them in a thin plastic coating, and suppose that then the conductor could be removed completely. This is equivalent to enlarging the cavity of Fig. 24-9*b* until it consumes the entire conductor, leaving only the charges. The electric field would not change at all; it would remain zero inside the thin shell of charge and would remain unchanged for all external points. This shows us that the electric field is set up by the charges and not by the conductor. The conductor simply provides an initial pathway for the charges to take up their positions.

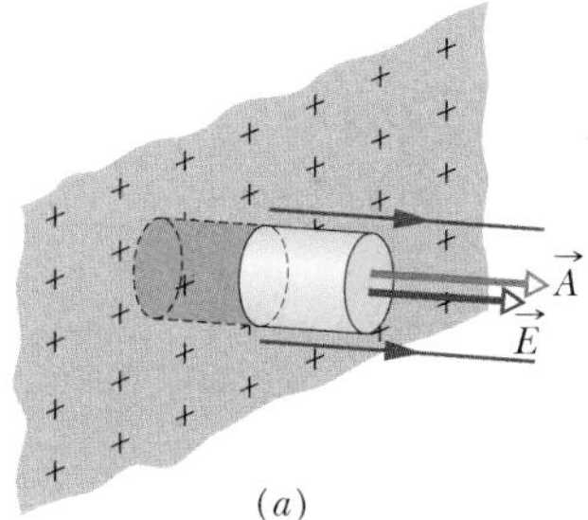

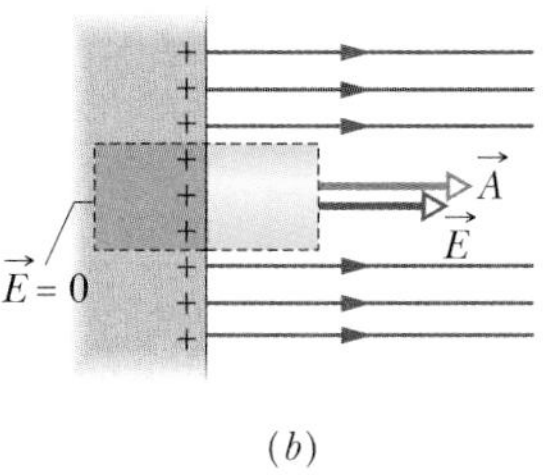

Fig. 24-10 Perspective view (*a*) and side view (*b*) of a tiny portion of a large, isolated conductor with excess positive charge on its surface. A (closed) cylindrical Gaussian surface, embedded perpendicularly in the conductor, encloses some of the charge. Electric field lines pierce the external end cap of the cylinder, but not the internal end cap. The external end cap has area A and area vector $\vec{A}$.

The External Electric Field

You have seen that the excess charge on an isolated conductor moves entirely to the conductor's surface. However, unless the conductor is spherical, the charge does not distribute itself uniformly. Put another way, the surface charge density σ (charge per unit area) varies over the surface of any nonspherical conductor. Generally, this variation makes the determination of the electric field set up by the surface charges very difficult.

However, the electric field just outside the surface of a conductor is easy to determine using Gauss' law. To do this, we consider a section of the surface that is small enough to permit us to neglect any curvature and thus to take the section to be flat. We then imagine a tiny cylindrical Gaussian surface to be embedded in the section as in Fig. 24-10: One end cap is fully inside the conductor, the other is fully outside, and the cylinder is perpendicular to the conductor's surface.

The electric field $\vec{E}$ at and just outside the conductor's surface must also be perpendicular to that surface. If it were not, then it would have a component along the conductor's surface that would exert forces on the surface charges, causing them to move. However, such motion would violate our implicit assumption that we are dealing with electrostatic equilibrium. Therefore, $\vec{E}$ is perpendicular to the conductor's surface.

We now sum the flux through the Gaussian surface. There is no flux through the internal end cap, because the electric field within the conductor is zero. There is no flux through the curved surface of the cylinder, because internally (in the conductor) there is no electric field and externally the electric field is parallel to the curved portion of the Gaussian surface. The only flux through the Gaussian surface is that through the external end cap, where $\vec{E}$ is perpendicular to the plane of the cap. We assume that the cap area A is small enough that the field magnitude E is constant over the cap. Then the flux through the cap is EA, and that is the net flux Φ through the Gaussian surface.

The charge q_{enc} enclosed by the Gaussian surface lies on the conductor's surface in an area A. If σ is the charge per unit area, then q_{enc} is equal to σA. When we substitute σA for q_{enc} and EA for Φ, Gauss' law (Eq. 24-6) becomes

$$\varepsilon_0 EA = \sigma A,$$

from which we find

$$E = \frac{\sigma}{\varepsilon_0} \quad \text{(conducting surface).} \tag{24-11}$$

Thus, the magnitude of the electric field at a location just outside a conductor is proportional to the surface charge density at that location on the conductor. If the charge on the conductor is positive, the electric field is directed away from the conductor as in Fig. 24-10. It is directed toward the conductor if the charge is negative.

The field lines in Fig. 24-10 must terminate on negative charges somewhere in the environment. If we bring those charges near the conductor, the charge density at any given location on the conductor's surface changes, and so does the magnitude of the electric field. However, the relation between σ and E is still given by Eq. 24-11.

Sample Problem 24-4

Figure 24-11*a* shows a cross section of a spherical metal shell of inner radius R. A point charge of $-5.0\ \mu C$ is located at a distance $R/2$ from the center of the shell. If the shell is electrically neutral, what are the (induced) charges on its inner and outer surfaces? Are those charges uniformly distributed? What is the field pattern inside and outside the shell?

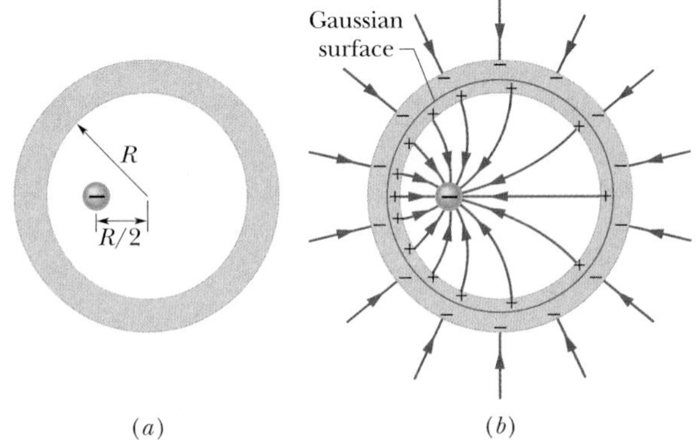

Fig. 24-11 Sample Problem 24-4. (*a*) A negative point charge is located within a spherical metal shell that is electrically neutral. (*b*) As a result, positive charge is nonuniformly distributed on the inner wall of the shell, and an equal amount of negative charge is uniformly distributed on the outer wall.

SOLUTION: Figure 24-11*b* shows a cross section of a spherical Gaussian surface within the metal, just outside the inner wall of the shell. One **Key Idea** here is that the electric field must be zero inside the metal (and thus on the Gaussian surface inside the metal). This means that the electric flux through the Gaussian surface must also be zero. Gauss' law then tells us that the *net* charge enclosed by the Gaussian surface must be zero. With a point charge of $-5.0\ \mu C$ within the shell, a charge of $+5.0\ \mu C$ must lie on the inner wall of the shell.

If the point charge were centered, this positive charge would be uniformly distributed along the inner wall. However, since the point charge is off-center, the distribution of positive charge is skewed, as suggested by Fig. 24-11*b*, because the positive charge tends to collect on the section of the inner wall nearest the (negative) point charge.

A second **Key Idea** is that because the shell is electrically neutral, its inner wall can have a charge of $+5.0\ \mu C$ only if electrons, with a total charge of $-5.0\ \mu C$, leave the inner wall and move to the outer wall. There they spread out uniformly, as is also suggested by Fig. 24-11*b*. This distribution of negative charge is uniform because the shell is spherical and because the skewed distribution of positive charge on the inner wall cannot produce an electric field in the shell to affect the distribution of charge on the outer wall.

The field lines inside and outside the shell are shown approximately in Fig. 24-11*b*. All the field lines intersect the shell and the point charge perpendicularly. Inside the shell the pattern of field lines is skewed owing to the skew of the positive charge distribution. Outside the shell the pattern is the same as if the point charge were centered and the shell were missing. In fact, this would be true no matter where inside the shell the point charge happened to be located.

CHECKPOINT 4: A ball of charge $-50e$ lies at the center of a hollow spherical metal shell that has a net charge of $-100e$. What is the charge on (a) the shell's inner surface and (b) its outer surface?

24-7 Applying Gauss' Law: Cylindrical Symmetry

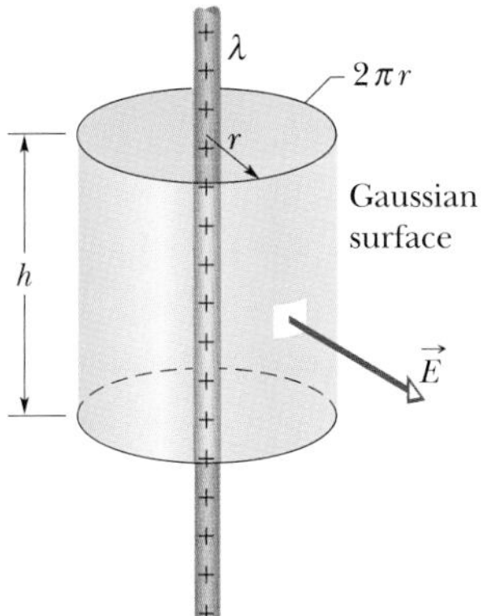

Fig. 24-12 A Gaussian surface in the form of a closed cylinder surrounds a section of a very long, uniformly charged, cylindrical plastic rod.

Figure 24-12 shows a section of an infinitely long cylindrical plastic rod with a uniform positive linear charge density λ. Let us find an expression for the magnitude of the electric field $\vec{E}$ at a distance r from the axis of the rod.

Our Gaussian surface should match the symmetry of the problem, which is cylindrical. We choose a circular cylinder of radius r and length h, coaxial with the rod. The Gaussian surface must be closed, so we include two end caps as part of the surface.

Imagine now that, while you are not watching, someone rotates the plastic rod around its longitudinal axis or turns it end for end. When you look again at the rod, you will not be able to detect any change. We conclude from this symmetry that the only uniquely specified direction in this problem is along a radial line. Thus, at every point on the cylindrical part of the Gaussian surface, $\vec{E}$ must have the same magnitude E and (for a positively charged rod) must be directed radially outward.

Since $2\pi r$ is the cylinder's circumference and h is its height, the area A of the cylindrical surface is $2\pi rh$. The flux of $\vec{E}$ through this cylindrical surface is then

$$\Phi = EA\cos\theta = E(2\pi rh)\cos 0 = E(2\pi rh).$$

There is no flux through the end caps because $\vec{E}$, being radially directed, is parallel to the end caps at every point.

The charge enclosed by the surface is λh, so Gauss' law,

$$\varepsilon_0\Phi = q_{enc},$$

reduces to

$$\varepsilon_0 E(2\pi rh) = \lambda h,$$

yielding

$$E = \frac{\lambda}{2\pi\varepsilon_0 r} \quad \text{(line of charge).} \tag{24-12}$$

This is the electric field due to an infinitely long, straight line of charge, at a point that is a radial distance r from the line. The direction of $\vec{E}$ is radially outward from the line of charge if the charge is positive, and radially inward if it is negative. Equation 24-12 also approximates the field of a *finite* line of charge, at points that are not too near the ends (compared with the distance from the line).

Sample Problem 24-5

The visible portion of a lightning strike is preceded by an invisible stage in which a column of electrons extends downward from a cloud to the ground. These electrons come from the cloud and from air molecules that are ionized within the column. The linear charge density λ along the column is typically -1×10^{-3} C/m. Once the column reaches the ground, electrons within it are rapidly dumped to the ground. During the dumping, collisions between the moving electrons and the air within the column result in a brilliant flash of light. If air molecules break down (ionize) in an electric field exceeding 3×10^6 N/C, what is the radius of the column?

SOLUTION: One **Key Idea** here is that, although the column is not straight or infinitely long, we can approximate it as being a line of charge as in Fig. 24-12. (Since it contains a net negative charge,

Fig. 24-13 Lightning strikes a 20-m-high sycamore. Because the tree was wet, most of the charge traveled through the water on it and the tree was unharmed.

its electric field $\vec{E}$ points radially inward.) Then, according to Eq. 24-12, the field's magnitude E decreases with distance from the axis of the column of charge.

A second **Key Idea** is that the surface of the column of charge must be at the radius r where the magnitude of $\vec{E}$ is 3×10^6 N/C, because air molecules within that radius ionize while those farther out do not. Solving Eq. 24-12 for r and inserting the known data, we find the radius of the column to be

$$r = \frac{\lambda}{2\pi\varepsilon_0 E}$$

$$= \frac{1 \times 10^{-3}\ \text{C/m}}{(2\pi)(8.85 \times 10^{-12}\ \text{C}^2/\text{N}\cdot\text{m}^2)(3 \times 10^6\ \text{N/C})}$$

$$= 6\ \text{m}. \qquad \text{(Answer)}$$

(The radius of the luminous portion of a lightning strike is smaller, perhaps only 0.5 m. You can get an idea of the width from Fig. 24-13.) Although the radius of the column may be only 6 m, do not assume that you are safe if you are at a somewhat greater distance from the strike point, because the electrons dumped by the strike travel along the ground. Such *ground currents* are lethal. Figure 24-14 shows evidence of ground currents.

Fig. 24-14 Ground currents from a lightning strike have burned grass off this golf green, exposing the soil.

24-8 Applying Gauss' Law: Planar Symmetry

Nonconducting Sheet

Figure 24-15 shows a portion of a thin, infinite, nonconducting sheet with a uniform (positive) surface charge density σ. A sheet of thin plastic wrap, uniformly charged on one side, can serve as a simple model. Let us find the electric field $\vec{E}$ a distance r in front of the sheet.

A useful Gaussian surface is a closed cylinder with end caps of area A, arranged to pierce the sheet perpendicularly as shown. From symmetry, $\vec{E}$ must be perpendicular to the sheet and hence to the end caps. Furthermore, since the charge is positive, $\vec{E}$ is directed *away* from the sheet, and thus the electric field lines pierce the two Gaussian end caps in an outward direction. Because the field lines do not pierce the curved surface, there is no flux through this portion of the Gaussian

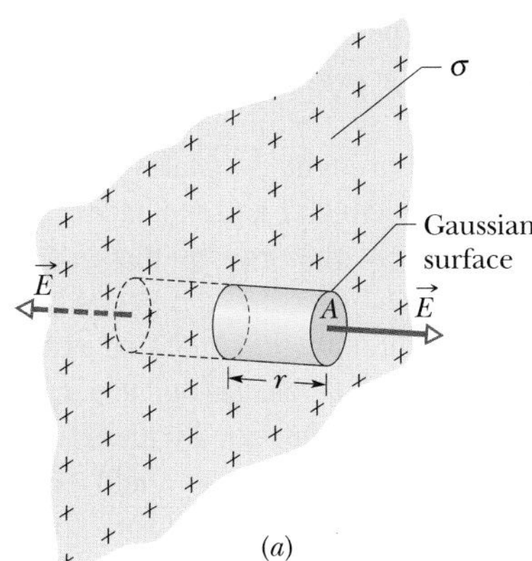

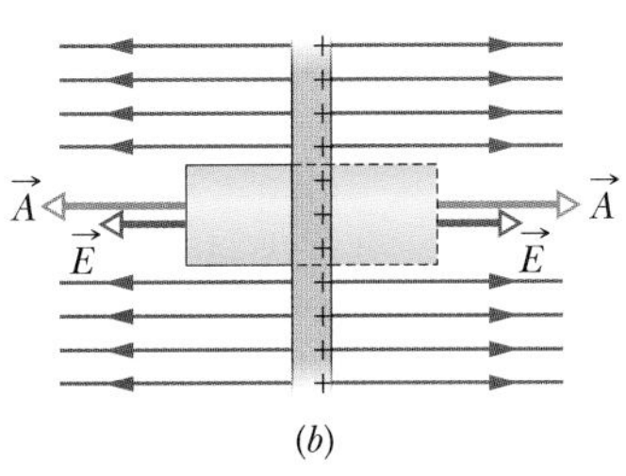

Fig. 24-15 Perspective view (*a*) and side view (*b*) of a portion of a very large, thin plastic sheet, uniformly charged on one side to surface charge density σ. A closed cylindrical Gaussian surface passes through the sheet and is perpendicular to it.

surface. Thus $\vec{E} \cdot d\vec{A}$ is simply $E\,dA$; then Gauss' law,

$$\varepsilon_0 \oint \vec{E} \cdot d\vec{A} = q_{\text{enc}},$$

becomes

$$\varepsilon_0(EA + EA) = \sigma A,$$

where σA is the charge enclosed by the Gaussian surface. This gives

$$E = \frac{\sigma}{2\varepsilon_0} \quad \text{(sheet of charge).} \tag{24-13}$$

Since we are considering an infinite sheet with uniform charge density, this result holds for any point at a finite distance from the sheet. Equation 24-13 agrees with Eq. 23-27, which we found by integration of the electric field components that are produced by individual charges. (Look back to that time-consuming and challenging integration, and note how much more easily we obtain the result with Gauss' law. That is one reason for devoting a whole chapter to that law: for certain symmetric arrangements of charge, it is very much easier to use than integration of field components.)

Two Conducting Plates

Figure 24-16*a* shows a cross section of a thin, infinite conducting plate with excess positive charge. From Section 24-6 we know that this excess charge lies on the surface of the plate. Since the plate is thin and very large, we can assume that essentially all the excess charge is on the two large faces of the plate.

If there is no external electric field to force the positive charge into some particular distribution, it will spread out on the two faces with a uniform surface charge density of magnitude σ_1. From Eq. 24-11 we know that just outside the plate this charge sets up an electric field of magnitude $E = \sigma_1/\varepsilon_0$. Because the excess charge is positive, the field is directed away from the plate.

Figure 24-16*b* shows an identical plate with excess negative charge having the same magnitude of surface charge density σ_1. The only difference is that now the electric field is directed toward the plate.

Suppose we arrange for the plates of Figs. 24-16*a* and *b* to be close to each other and parallel (Fig. 24-16*c*). Since the plates are conductors, when we bring them into this arrangement, the excess charge on one plate attracts the excess charge on the other plate, and all the excess charge moves onto the inner faces of the plates as in Fig. 24-16*c*. With twice as much charge now on each inner face, the new surface charge density (call it σ) on each inner face is twice σ_1. Thus, the electric field at any point between the plates has the magnitude

$$E = \frac{2\sigma_1}{\varepsilon_0} = \frac{\sigma}{\varepsilon_0}. \tag{24-14}$$

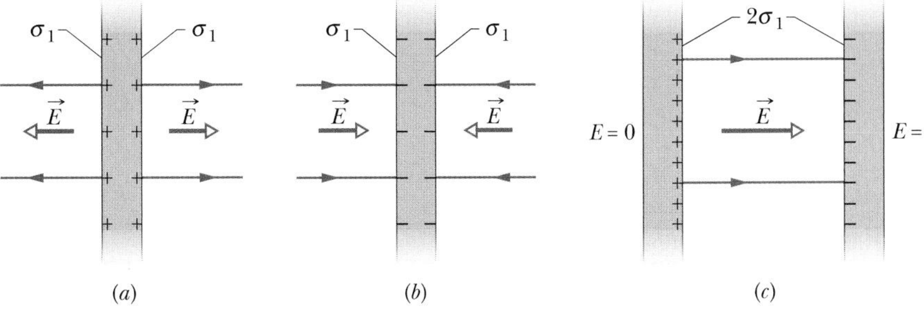

Fig. 24-16 (*a*) A thin, very large conducting plate with excess positive charge. (*b*) An identical plate with excess negative charge. (*c*) The two plates arranged so they are parallel and close.

This field is directed away from the positively charged plate and toward the negatively charged plate. Since no excess charge is left on the outer faces, the electric field to the left and right of the plates is zero.

Because the charges on the plates moved when we brought the plates close to each other, Fig. 24-16*c* is *not* the superposition of Figs. 24-16*a* and *b*; that is, the charge distribution of the two-plate system is not merely the sum of the charge distributions of the individual plates.

You may wonder why we discuss such seemingly unrealistic situations as the field set up by an infinite line of charge, an infinite sheet of charge, or a pair of infinite plates of charge. One reason is that analyzing such situations with Gauss' law is easy. More important is that analyses for "infinite" situations yield good approximations to many real-world problems. Thus, Eq. 24-13 holds well for a finite nonconducting sheet as long as we are dealing with points close to the sheet and not too near its edges. Equation 24-14 holds well for a pair of finite conducting plates as long as we consider points that are not too close to their edges.

The trouble with the edges of a sheet or a plate, and the reason we take care not to deal with them, is that near an edge we can no longer use planar symmetry to find expressions for the fields. In fact, the field lines there are curved (said to be an *edge effect* or *fringing*), and the fields can be very difficult to express algebraically.

Sample Problem 24-6

Figure 24-17*a* shows portions of two large, parallel, nonconducting sheets, each with a fixed uniform charge on one side. The magnitudes of the surface charge densities are $\sigma_{(+)} = 6.8\ \mu\text{C/m}^2$ for the positively charged sheet and $\sigma_{(-)} = 4.3\ \mu\text{C/m}^2$ for the negatively charged sheet.

Find the electric field $\vec{E}$ (a) to the left of the sheets, (b) between the sheets, and (c) to the right of the sheets.

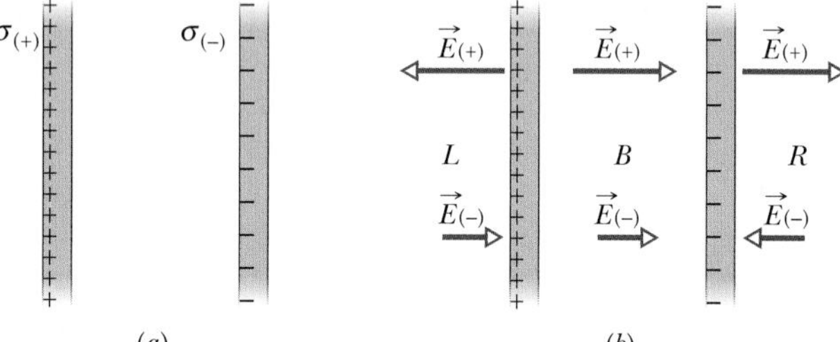

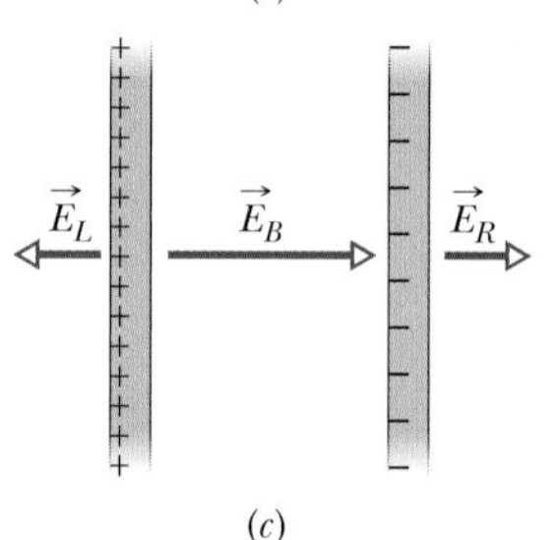

Fig. 24-17 Sample Problem 24-6. (*a*) Two large, parallel sheets, uniformly charged on one side. (*b*) The individual electric fields resulting from the two charged sheets. (*c*) The net field due to both charged sheets, found by superposition.

SOLUTION: The **Key Idea** here is that with the charges fixed in place, we can find the electric field of the sheets in Fig. 24-17*a* by (1) finding the field of each sheet as if that sheet were isolated and (2) algebraically adding the fields of the isolated sheets via the superposition principle. (We can add the fields algebraically because they are parallel to each other.) From Eq. 24-13, the magnitude $E_{(+)}$ of the electric field due to the positive sheet at any point is

$$E_{(+)} = \frac{\sigma_{(+)}}{2\varepsilon_0} = \frac{6.8 \times 10^{-6}\ \text{C/m}^2}{(2)(8.85 \times 10^{-12}\ \text{C}^2/\text{N}\cdot\text{m}^2)}$$
$$= 3.84 \times 10^5\ \text{N/C}.$$

Similarly, the magnitude $E_{(-)}$ of the electric field at any point due to the negative sheet is

$$E_{(-)} = \frac{\sigma_{(-)}}{2\varepsilon_0} = \frac{4.3 \times 10^{-6}\ \text{C/m}^2}{(2)(8.85 \times 10^{-12}\ \text{C}^2/\text{N}\cdot\text{m}^2)}$$
$$= 2.43 \times 10^5\ \text{N/C}.$$

Figure 24-17*b* shows the fields set up by the sheets to the left of the sheets (*L*), between them (*B*), and to their right (*R*).

The resultant fields in these three regions follow from the superposition principle. To the left, the field magnitude is

$$\begin{aligned} E_L &= E_{(+)} - E_{(-)} \\ &= 3.84 \times 10^5\ \text{N/C} - 2.43 \times 10^5\ \text{N/C} \\ &= 1.4 \times 10^5\ \text{N/C}. \qquad \text{(Answer)} \end{aligned}$$

Because $E_{(+)}$ is larger than $E_{(-)}$, the net electric field $\vec{E}_L$ in this region is directed to the left, as Fig. 24-17*c* shows. To the right of the sheets, the electric field $\vec{E}_R$ has the same magnitude but is directed to the right, as Fig. 24-17*c* shows.

Between the sheets, the two fields add and we have

$$\begin{aligned} E_B &= E_{(+)} + E_{(-)} \\ &= 3.84 \times 10^5\ \text{N/C} + 2.43 \times 10^5\ \text{N/C} \\ &= 6.3 \times 10^5\ \text{N/C}. \qquad \text{(Answer)} \end{aligned}$$

The electric field $\vec{E}_B$ is directed to the right.

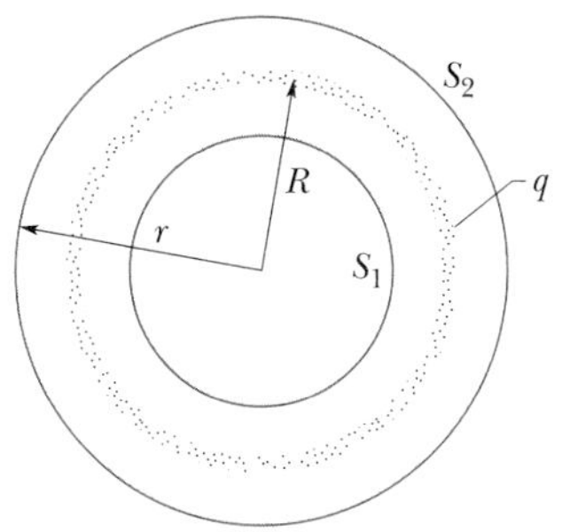

Fig. 24-18 A thin, uniformly charged, spherical shell with total charge q, in cross section. Two Gaussian surfaces S_1 and S_2 are also shown in cross section. Surface S_2 encloses the shell, and S_1 encloses only the empty interior of the shell.

24-9 Applying Gauss' Law: Spherical Symmetry

Here we use Gauss' law to prove the two shell theorems presented without proof in Section 22-4:

> A shell of uniform charge attracts or repels a charged particle that is outside the shell as if all the shell's charge were concentrated at the center of the shell.

> A shell of uniform charge exerts no electrostatic force on a charged particle that is located inside the shell.

Figure 24-18 shows a charged spherical shell of total charge q and radius R and two concentric spherical Gaussian surfaces, S_1 and S_2. If we followed the procedure of Section 24-5 as we applied Gauss' law to surface S_2, for which $r \geq R$, we would find that

$$E = \frac{1}{4\pi\varepsilon_0}\frac{q}{r^2} \quad \text{(spherical shell, field at } r \geq R\text{).} \tag{24-15}$$

This is the same field that would be set up by a point charge q at the center of the shell of charge. Thus, a shell of charge q would produce the same force on a charged particle placed outside the shell as would a point charge q located at the center of the shell. This proves the first shell theorem.

Applying Gauss' law to surface S_1, for which $r < R$, leads directly to

$$E = 0 \quad \text{(spherical shell, field at } r < R\text{),} \tag{24-16}$$

because this Gaussian surface encloses no charge. Thus, if a charged particle were enclosed by the shell, the shell would exert no net electrostatic force on it. This proves the second shell theorem.

Any spherically symmetric charge distribution, such as that of Fig. 24-19, can be constructed with a nest of concentric spherical shells. For purposes of applying the two shell theorems, the volume charge density ρ should have a single value for each shell but need not be the same from shell to shell. Thus, for the charge distribution as a whole, ρ can vary, but only with r, the radial distance from the center. We can then examine the effect of the charge distribution "shell by shell."

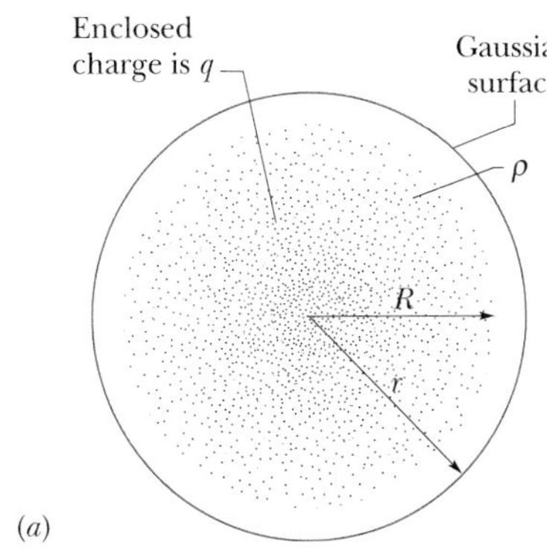

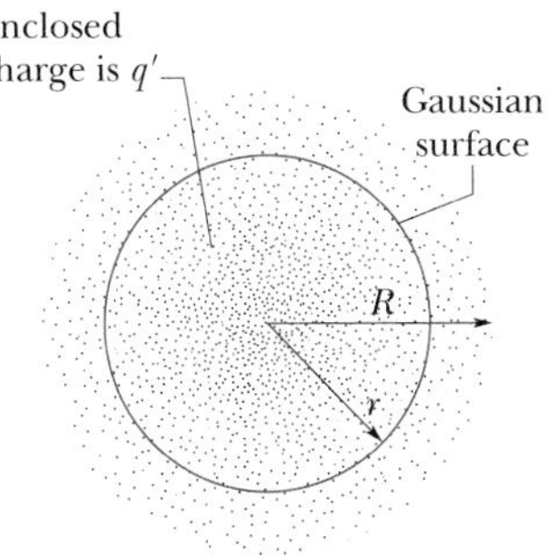

Fig. 24-19 The dots represent a spherically symmetric distribution of charge of radius R, whose volume charge density ρ is a function only of distance from the center. The charged object is not a conductor, and the charge is assumed to be fixed in position. A concentric spherical Gaussian surface with $r > R$ is shown in (*a*). A similar Gaussian surface with $r < R$ is shown in (*b*).

In Fig. 24-19*a* the entire charge lies within a Gaussian surface with $r > R$. The charge produces an electric field on the Gaussian surface as if the charge were a point charge located at the center, and Eq. 24-15 holds.

Figure 24-19*b* shows a Gaussian surface with $r < R$. To find the electric field at points on this Gaussian surface, we consider two sets of charged shells—one set inside the Gaussian surface and one set outside. Equation 24-16 says that the charge lying *outside* the Gaussian surface does not set up a net electric field on the Gaussian surface. Equation 24-15 says that the charge *enclosed* by the surface sets up an electric field as if that enclosed charge were concentrated at the center. Letting q' represent that enclosed charge, we can then rewrite Eq. 24-15 as

$$E = \frac{1}{4\pi\varepsilon_0}\frac{q'}{r^2} \quad \text{(spherical distribution, field at } r \leq R\text{).} \tag{24-17}$$

If the full charge q enclosed within radius R is uniform, then q' enclosed within radius r in Fig. 24-19*b* is proportional to q:

$$\frac{\text{charge enclosed by } r}{\text{volume enclosed by } r} = \frac{\text{full charge}}{\text{full volume}}$$

or

$$\frac{q'}{\frac{4}{3}\pi r^3} = \frac{q}{\frac{4}{3}\pi R^3}. \tag{24-18}$$

This gives us

$$q' = q\frac{r^3}{R^3}. \tag{24-19}$$

Substituting this into Eq. 24-17 yields

$$E = \left(\frac{q}{4\pi\varepsilon_0 R^3}\right)r \quad \text{(uniform charge, field at } r \leq R\text{)}. \tag{24-20}$$

CHECKPOINT 5: The figure shows two large, parallel, nonconducting sheets with identical (positive) uniform surface charge densities, and a sphere with a uniform (positive) volume charge density. Rank the four numbered points according to the magnitude of the net electric field there, greatest first.

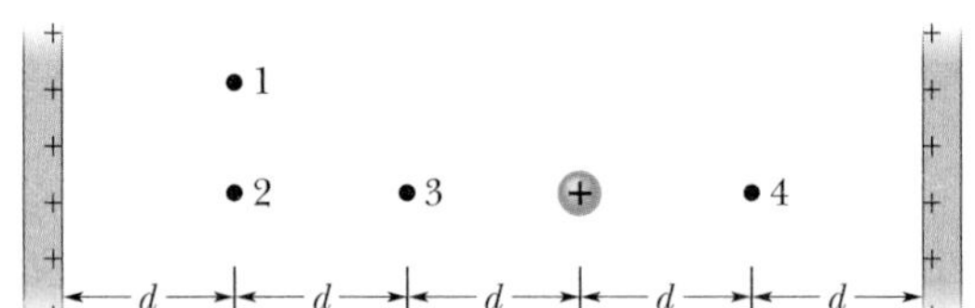

REVIEW & SUMMARY

Gauss' Law *Gauss' law* and Coulomb's law, although expressed in different forms, are equivalent ways of describing the relation between charge and electric field in static situations. Gauss' law is

$$\varepsilon_0\Phi = q_{enc} \quad \text{(Gauss' law)}, \tag{24-6}$$

in which q_{enc} is the net charge inside an imaginary closed surface (a **Gaussian surface**) and Φ is the net **flux** of the electric field through the surface:

$$\Phi = \oint \vec{E}\cdot d\vec{A} \quad \text{(electric flux through a Gaussian surface)}. \tag{24-4}$$

Coulomb's law can readily be derived from Gauss' law.

Applications of Gauss' Law Using Gauss' law and, in some cases, symmetry arguments, we can derive several important results in electrostatic situations. Among these are:

1. An excess charge on a *conductor* is located entirely on the outer surface of the conductor.
2. The external electric field near the *surface of a charged conductor* is perpendicular to the surface and has magnitude

$$E = \frac{\sigma}{\varepsilon_0} \quad \text{(conducting surface)}. \tag{24-11}$$

Within the conductor, $E = 0$.

3. The electric field at any point due to an infinite *line of charge* with uniform linear charge density λ is perpendicular to the line of charge and has magnitude

$$E = \frac{\lambda}{2\pi\varepsilon_0 r} \quad \text{(line of charge)}, \tag{24-12}$$

where r is the perpendicular distance from the line of charge to the point.

4. The electric field due to an *infinite nonconducting sheet* with uniform surface charge density σ is perpendicular to the plane of the sheet and has magnitude

$$E = \frac{\sigma}{2\varepsilon_0} \quad \text{(sheet of charge)}. \tag{24-13}$$

5. The electric field *outside a spherical shell of charge* with radius R and total charge q is directed radially and has magnitude

$$E = \frac{1}{4\pi\varepsilon_0}\frac{q}{r^2} \quad \text{(spherical shell, for } r \geq R\text{)}. \tag{24-15}$$

Here r is the distance from the center of the shell to the point at which E is measured. (The charge behaves, for external points, as if it were all located at the center of the sphere.) The field *inside* a uniform spherical shell of charge is exactly zero:

$$E = 0 \quad \text{(spherical shell, for } r < R\text{)}. \tag{24-16}$$

6. The electric field *inside a uniform sphere of charge* is directed radially and has magnitude

$$E = \left(\frac{q}{4\pi\varepsilon_0 R^3}\right)r. \tag{24-20}$$

QUESTIONS

1. A surface has the area vector $\vec{A} = (2\hat{i} + 3\hat{j})\ m^2$. What is the flux of an electric field through it if the field is (a) $\vec{E} = 4\hat{i}$ N/C and (b) $\vec{E} = 4\hat{k}$ N/C?

2. What is $\int dA$ for (a) a square of edge length a, (b) a circle of radius r, and (c) the curved surface of a cylinder of length h and radius r?

3. In Fig. 24-20, a full Gaussian surface encloses two of the four positively charged particles. (a) Which of the particles contribute to the electric field at point P on the surface? (b) Which net flux of electric field through the surface is greater (if either): that due to q_1 and q_2 or that due to all four charges?

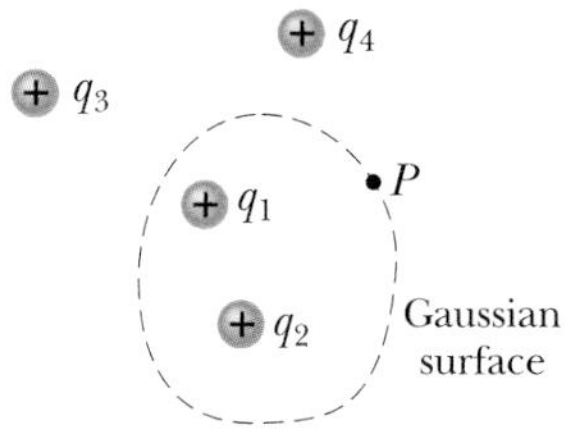

Fig. 24-20 Question 3.

4. Figure 24-21 shows, in cross section, a central metal ball, two spherical metal shells, and three spherical Gaussian surfaces of radii R, $2R$, and $3R$, all with the same center. The uniform charges on the three objects are: ball, Q; smaller shell, $3Q$; larger shell, $5Q$. Rank the Gaussian surfaces according to the magnitude of the electric field at any point on the surface, greatest first.

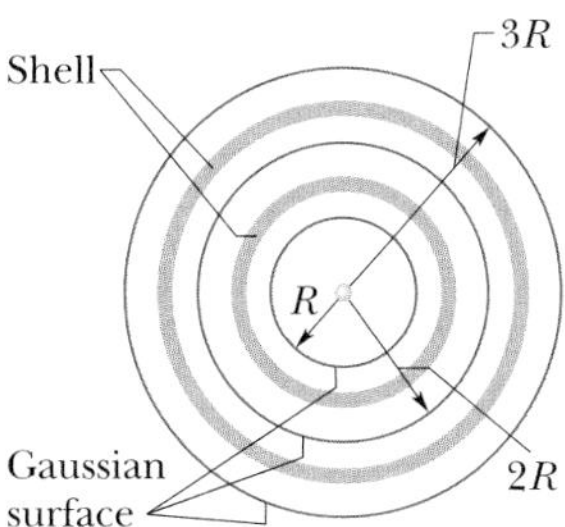

Fig. 24-21 Question 4.

5. Figure 24-22 shows three Gaussian surfaces, each half-submerged in a large, thick metal plate with a uniform surface charge density. Gaussian surface S_1 is the tallest and has the smallest square end caps; surface S_3 is shortest and has the largest square end caps; and S_2 has intermediate values. Rank the surfaces according to (a) the charge they enclose, (b) the magnitude of the electric field at points on their top end cap, (c) the net electric flux through that top end cap, and (d) the net electric flux through their bottom end cap, greatest first.

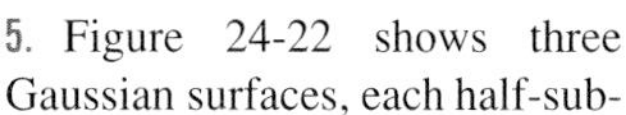

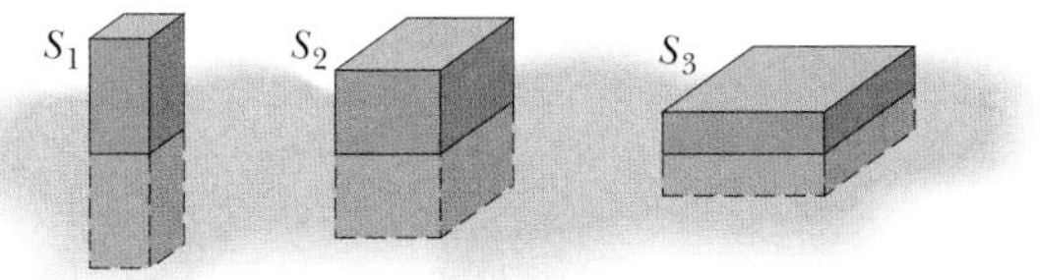

Fig. 24-22 Question 5.

6. Figure 24-23 shows, in cross section, three cylinders, each of uniform charge Q. Concentric with each cylinder is a cylindrical Gaussian surface, all three with the same radius. Rank the Gaussian surfaces according to the electric field at any point on the surface, greatest first.

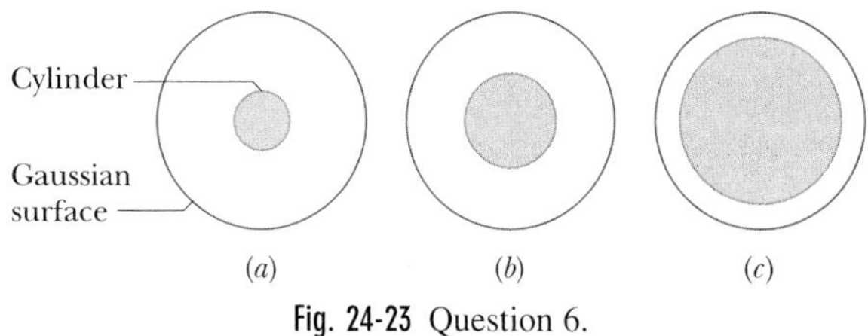

Fig. 24-23 Question 6.

7. Three infinite nonconducting sheets, with uniform surface charge densities σ, 2σ, and 3σ, are arranged to be parallel like the two sheets in Fig. 24-17*a*. What is their order, from left to right, if the electric field $\vec{E}$ produced by the arrangement has magnitude $E = 0$ in one region and $E = 2\sigma/\varepsilon_0$ in another region?

8. A small charged ball lies within the hollow of a metallic spherical shell of radius R. Here, for three situations, are the net charges on the ball and shell, respectively: (1) $+4q$, 0; (2) $-6q$, $+10q$; (3) $+16q$, $-12q$. Rank the situations according to the charge on (a) the inner surface of the shell and (b) the outer surface, most positive first.

9. Rank the situations of Question 8 according to the magnitude of the electric field (a) halfway through the shell and (b) at a point $2R$ from the center of the shell, greatest first.

10. Figure 24-24 shows four spheres, each with charge Q uniformly distributed through its volume. (a) Rank the spheres according to their volume charge density, greatest first. The figure also shows a point P for each sphere, all at the same distance from the center of the sphere. (b) Rank the spheres according to the magnitude of the electric field they produce at point P, greatest first.

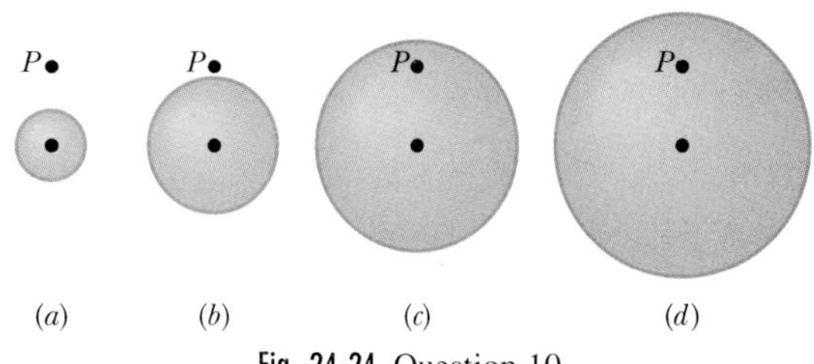

Fig. 24-24 Question 10.

EXERCISES & PROBLEMS

ssm Solution is in the Student Solutions Manual.
www Solution is available on the World Wide Web at: http://www.wiley.com/college/hrw
ilw Solution is available on the Interactive LearningWare.

SEC. 24-2 Flux

1E. Water in an irrigation ditch of width $w = 3.22$ m and depth $d = 1.04$ m flows with a speed of 0.207 m/s. The *mass flux* of the flowing water through an imaginary surface is the product of the water's density (1000 kg/m^3) and its volume flux through that surface. Find the mass flux through the following imaginary surfaces: (a) a surface of area wd, entirely in the water, perpendicular to the flow; (b) a surface with area $3wd/2$, of which wd is in the water, perpendicular to the flow; (c) a surface of area $wd/2$, entirely in the water, perpendicular to the flow; (d) a surface of area wd, half in the water and half out, perpendicular to the flow; (e) a surface of area wd, entirely in the water, with its normal 34° from the direction of flow.

SEC. 24-3 Flux of an Electric Field

2E. The square surface shown in Fig. 24-25 measures 3.2 mm on each side. It is immersed in a uniform electric field with magnitude E = 1800 N/C. The field lines make an angle of 35° with a normal to the surface, as shown. Take that normal to be directed "outward," as though the surface were one face of a box. Calculate the electric flux through the surface. ssm

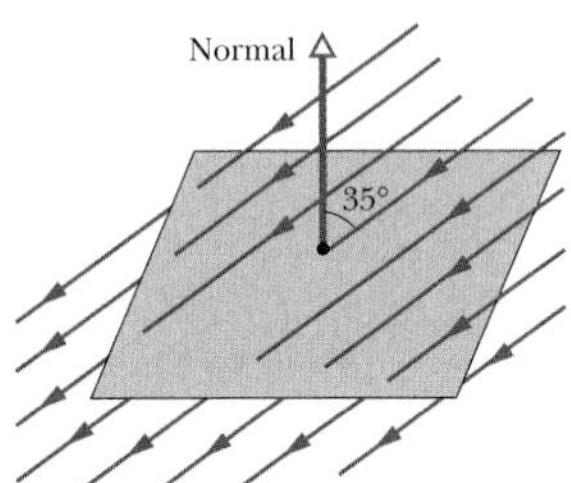

Fig. 24-25 Exercise 2.

3E. The cube in Fig. 24-26 has edge lengths of 1.40 m and is oriented as shown in a region of uniform electric field. Find the electric flux through the right face if the electric field, in newtons per coulomb, is given by (a) $6.00\hat{\mathrm{i}}$, (b) $-2.00\hat{\mathrm{j}}$, and (c) $-3.00\hat{\mathrm{i}} + 4.00\hat{\mathrm{k}}$. (d) What is the total flux through the cube for each of these fields?

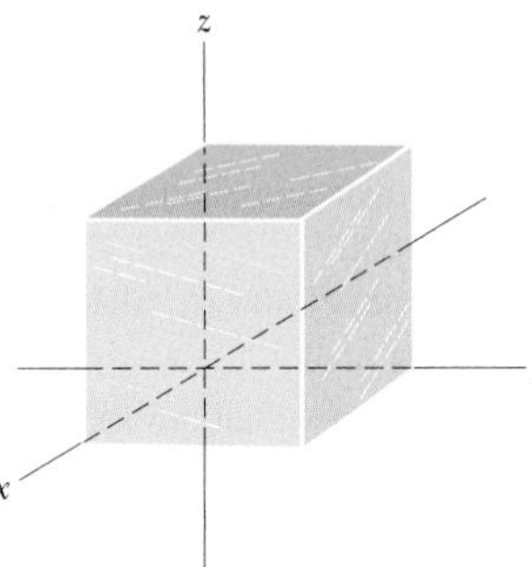

Fig. 24-26 Exercise 3 and Problems 7 and 10.

SEC. 24-4 Gauss' Law

4E. You have four point charges, $2q$, q, $-q$, and $-2q$. If possible, describe how you would place a closed surface that encloses at least the charge $2q$ (and perhaps other charges) and through which the net electric flux is (a) 0, (b) $+3q/\varepsilon_0$, and (c) $-2q/\varepsilon_0$.

5E. A point charge of 1.8 μC is at the center of a cubical Gaussian surface 55 cm on edge. What is the net electric flux through the surface? ssm

6E. In Fig. 24-27, a butterfly net is in a uniform electric field of magnitude E. The rim, a circle of radius a, is aligned perpendicular to the field. Find the electric flux through the netting.

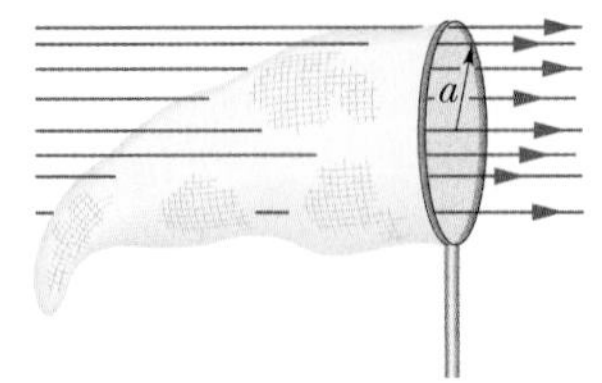

Fig. 24-27 Exercise 6.

7P. Find the net flux through the cube given in Exercise 3 and Fig. 24-26 if the electric field is given by (a) $\vec{E} = 3.00y\hat{\mathrm{j}}$ and (b) $\vec{E} = -4.00\hat{\mathrm{i}} + (6.00 + 3.00y)\hat{\mathrm{j}}$. E is in newtons per coulomb, and y is in meters. (c) In each case, how much charge is enclosed by the cube? ilw

8P. When a shower is turned on in a closed bathroom, the splashing of the water on the bare tub can fill the room's air with negatively charged ions and produce an electric field in the air as great as 1000 N/C. Consider a bathroom with dimensions of 2.5 m × 3.0 m × 2.0 m. Along the ceiling, floor, and four walls, approximate the electric field in the air as being directed perpendicular to the surface and as having a uniform magnitude of 600 N/C. Also, treat those surfaces as forming a closed Gaussian surface around the room's air. What are (a) the volume charge density ρ and (b) the number of excess elementary charges e per cubic meter in the room's air?

9P. It is found experimentally that the electric field in a certain region of Earth's atmosphere is directed vertically down. At an altitude of 300 m the field has magnitude 60.0 N/C; at an altitude of 200 m, the magnitude is 100 N/C. Find the net amount of charge contained in a cube 100 m on edge, with horizontal faces at altitudes of 200 and 300 m. Neglect the curvature of Earth. ssm

10P. At each point on the surface of the cube shown in Fig. 24-26, the electric field is in the positive direction of z. The length of each edge of the cube is 3.0 m. On the top surface of the cube $\vec{E} = -34\hat{\mathrm{k}}$ N/C, and on the bottom face of the cube $\vec{E} = +20\hat{\mathrm{k}}$ N/C. Determine the net charge contained within the cube.

11P. A point charge q is placed at one corner of a cube of edge a. What is the flux through each of the cube faces? (*Hint:* Use Gauss' law and symmetry arguments.) ssm

SEC. 24-6 A Charged Isolated Conductor

12E. The electric field just above the surface of the charged drum of a photocopying machine has a magnitude E of 2.3×10^5 N/C. What is the surface charge density on the drum, assuming that the drum is a conductor?

13E. A uniformly charged conducting sphere of 1.2 m diameter has a surface charge density of 8.1 μC/m^2. (a) Find the net charge on the sphere. (b) What is the total electric flux leaving the surface of the sphere? ssm

14E. Space vehicles traveling through Earth's radiation belts can intercept a significant number of electrons. The resulting charge buildup can damage electronic components and disrupt operations. Suppose a spherical metallic satellite 1.3 m in diameter accumulates 2.4 μC of charge in one orbital revolution. (a) Find the resulting surface charge density. (b) Calculate the magnitude of the electric field just outside the surface of the satellite, due to the surface charge.

15P. An isolated conductor of arbitrary shape has a net charge of $+10 \times 10^{-6}$ C. Inside the conductor is a cavity within which is a point charge $q = +3.0 \times 10^{-6}$ C. What is the charge (a) on the cavity wall and (b) on the outer surface of the conductor? ssm www

SEC. 24-7 Applying Gauss' Law: Cylindrical Symmetry

16E. (a) The drum of the photocopying machine in Exercise 12 has a length of 42 cm and a diameter of 12 cm. What is the total charge on the drum? (b) The manufacturer wishes to produce a desktop version of the machine. This requires reducing the size of the drum to a length of 28 cm and a diameter of 8.0 cm. The electric field at the drum surface must remain unchanged. What must be the charge on this new drum?

17E. An infinite line of charge produces a field of 4.5×10^4 N/C at a distance of 2.0 m. Calculate the linear charge density. ssm

18P. Figure 24-28 shows a section of a long, thin-walled metal tube of radius R, with a charge per unit length λ on its surface. Derive expressions for E in terms of the distance r from the tube axis, considering both (a) $r > R$ and (b) $r < R$. Plot your results for the range $r = 0$ to $r = 5.0$ cm, assuming that $\lambda = 2.0 \times 10^{-8}$ C/m and $R = 3.0$ cm. (*Hint:* Use cylindrical Gaussian surfaces, coaxial with the metal tube.)

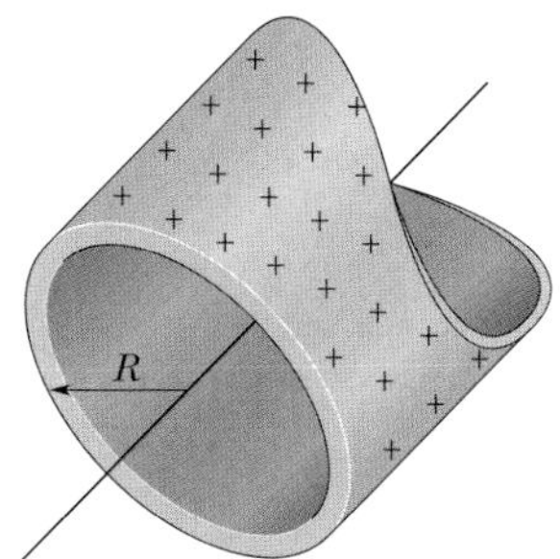

Fig. 24-28 Problem 18.

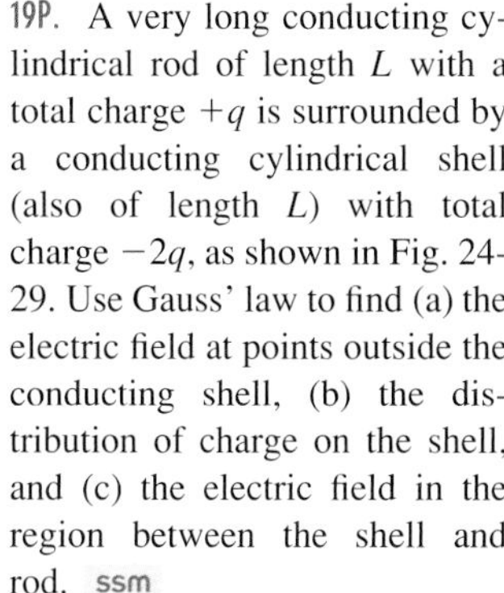

19P. A very long conducting cylindrical rod of length L with a total charge $+q$ is surrounded by a conducting cylindrical shell (also of length L) with total charge $-2q$, as shown in Fig. 24-29. Use Gauss' law to find (a) the electric field at points outside the conducting shell, (b) the distribution of charge on the shell, and (c) the electric field in the region between the shell and rod. ssm

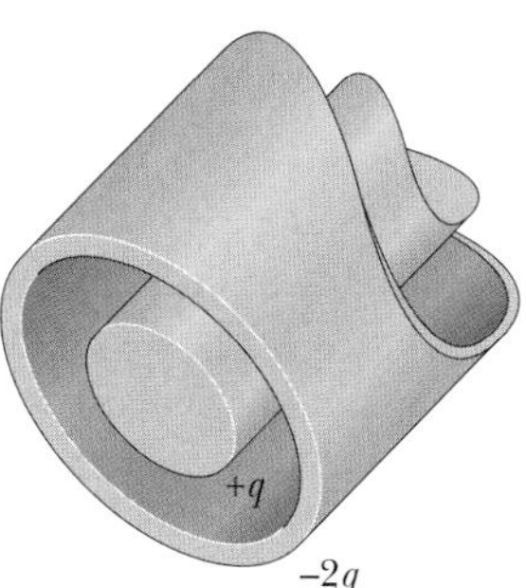

Fig. 24-29 Problem 19.

20P. A long, straight wire has fixed negative charge with a linear charge density of magnitude 3.6 nC/m. The wire is to be enclosed by a thin, nonconducting cylinder of outside radius 1.5 cm, coaxial with the wire. The cylinder is to have positive charge on its outside surface with a surface charge density σ such that the net external electric field is zero. Calculate the required σ.

21P. Two long, charged, concentric cylinders have radii of 3.0 and 6.0 cm. The charge per unit length is 5.0×10^{-6} C/m on the inner cylinder and -7.0×10^{-6} C/m on the outer cylinder. Find the electric field at (a) $r = 4.0$ cm and (b) $r = 8.0$ cm, where r is the radial distance from the common central axis. ilw

22P. A long, nonconducting, solid cylinder of radius 4.0 cm has a nonuniform volume charge density ρ that is a function of the radial distance r from the axis of the cylinder, as given by $\rho = Ar^2$, with $A = 2.5\ \mu\text{C/m}^5$. What is the magnitude of the electric field at a radial distance of (a) 3.0 cm and (b) 5.0 cm from the axis of the cylinder?

23P. Figure 24-30 shows a Geiger counter, a device used to detect ionizing radiation (radiation that causes ionization of atoms). The counter consists of a thin, positively charged central wire surrounded by a concentric, circular, conducting cylinder with an equal negative charge. Thus, a strong radial electric field is set up inside the cylinder. The cylinder contains a low-pressure inert gas. When a particle of radiation enters the device through the cylinder wall, it ionizes a few of the gas atoms. The resulting free electrons (label e) are drawn to the positive wire. However, the electric field is so intense that, between collisions with other gas atoms, the free electrons gain energy sufficient to ionize these atoms also. More free electrons are thereby created, and the process is repeated until the electrons reach the wire. The resulting "avalanche" of electrons is collected by the wire, generating a signal that is used to record the passage of the original particle of radiation. Suppose that the radius of the central wire is 25 μm, the radius of the cylinder 1.4 cm, and the length of the tube 16 cm. If the electric field at the cylinder's inner wall is 2.9×10^4 N/C, what is the total positive charge on the central wire? ssm

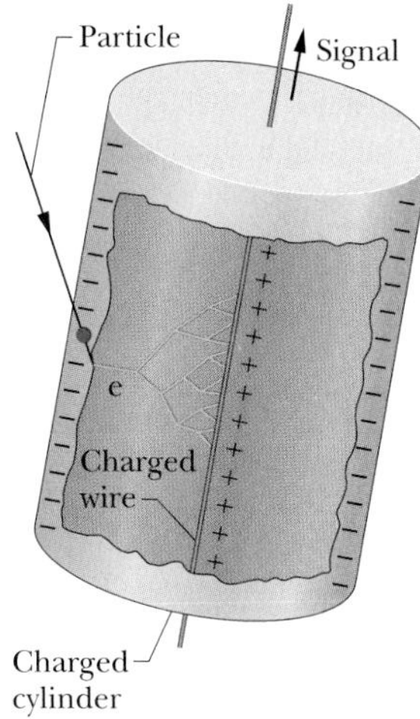

Fig. 24-30 Problem 23.

24P. A charge of uniform linear density 2.0 nC/m is distributed along a long, thin, nonconducting rod. The rod is coaxial with a long, hollow, conducting cylinder (inner radius = 5.0 cm, outer radius = 10 cm). The net charge on the conductor is zero. (a) What is the magnitude of the electric field 15 cm from the axis of the cylinder? What is the surface charge density on (b) the inner surface and (c) the outer surface of the conductor?

25P. Charge is distributed uniformly throughout the volume of an infinitely long cylinder of radius R. (a) Show that, at a distance r from the cylinder axis (for $r < R$),

$$E = \frac{\rho r}{2\varepsilon_0},$$

where ρ is the volume charge density. (b) Write an expression for E when $r > R$. ssm www

SEC. 24-8 Applying Gauss' Law: Planar Symmetry

26E. Figure 24-31 shows cross-sections through two large, parallel, nonconducting sheets with identical distributions of positive charge with surface charge density σ. What is $\vec{E}$ at points (a) above the sheets, (b) between them, and (c) below them?

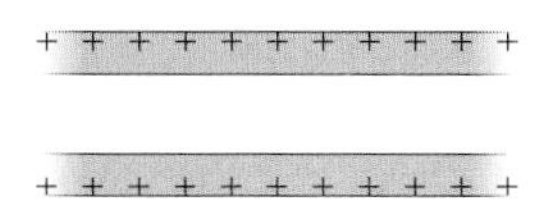

Fig. 24-31 Exercise 26.

27E. A square metal plate of edge length 8.0 cm and negligible thickness has a total charge of 6.0×10^{-6} C. (a) Estimate the magnitude E of the electric field just off the center of the plate (at, say, a distance of 0.50 mm) by assuming that the charge is spread uniformly over the two faces of the plate. (b) Estimate E at a distance of 30 m (large relative to the plate size) by assuming that the plate is a point charge. ssm

28E. A large, flat, nonconducting surface has a uniform charge density σ. A small circular hole of radius R has been cut in the middle of the surface, as shown in Fig. 24-32. Ignore fringing of the field lines around all edges, and calculate the electric field at point P, a distance z from the center of the hole along its axis. (*Hint:* See Eq. 23-26 and use superposition.)

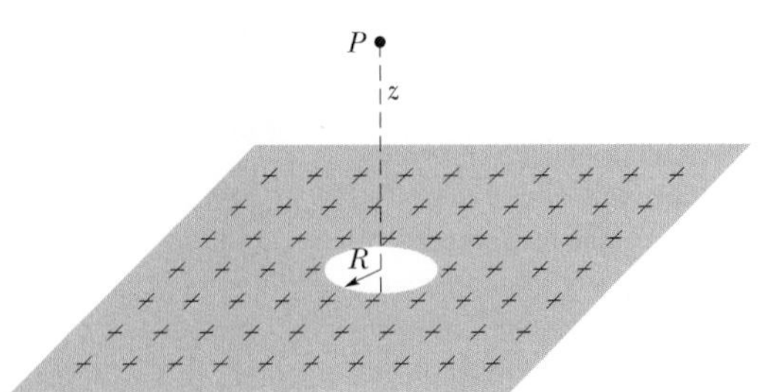

Fig. 24-32 Exercise 28.

29P. In Fig. 24-33, a small, nonconducting ball of mass m = 1.0 mg and charge q = 2.0×10^{-8} C (distributed uniformly through its volume) hangs from an insulating thread that makes an angle $\theta = 30°$ with a vertical, uniformly charged nonconducting sheet (shown in cross section). Considering the gravitational force on the ball and assuming that the sheet extends far vertically and into and out of the page, calculate the surface charge density σ of the sheet. ssm

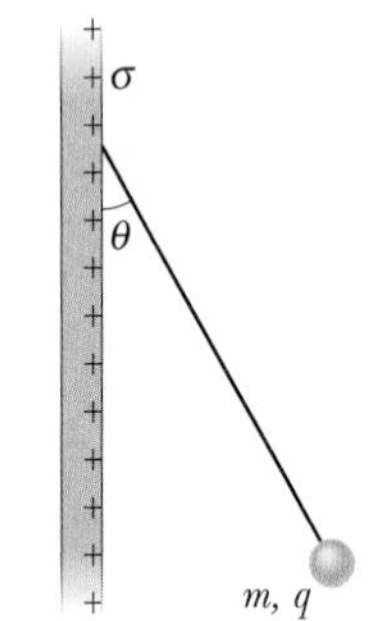

Fig. 24-33 Problem 29.

30P. Two large, thin metal plates are parallel and close to each other, as in Fig. 24-16*c*. On their inner faces, the plates have excess surface charge densities of opposite signs and with a magnitude 7.0×10^{-22} C/m^2, with the negatively charged plate on the left. What are the magnitude and direction of the electric field $\vec{E}$ (a) to the left of the plates, (b) to the right of the plates, and (c) between the plates?

31P. An electron is shot directly toward the center of a large metal plate that has excess negative charge with surface charge density 2.0×10^{-6} C/m^2. If the initial kinetic energy of the electron is 100 eV and if the electron is to stop (owing to electrostatic repulsion from the plate) just as it reaches the plate, how far from the plate must it be shot? ssm

32P. Two large metal plates of area 1.0 m^2 face each other. They are 5.0 cm apart and have equal but opposite charges on their inner surfaces. If the magnitude E of the electric field between the plates is 55 N/C, what is the magnitude of the charge on each plate? Neglect edge effects.

33P*. A planar slab of thickness d has a uniform volume charge density ρ. Find the magnitude of the electric field at all points in space both (a) inside and (b) outside the slab, in terms of x, the distance measured from the central plane of the slab. ssm

SEC. 24-9 Applying Gauss' Law: Spherical Symmetry

34E. A point charge causes an electric flux of -750 N·m^2/C to pass through a spherical Gaussian surface of 10.0 cm radius centered on the charge. (a) If the radius of the Gaussian surface were doubled, how much flux would pass through the surface? (b) What is the value of the point charge?

35E. A conducting sphere of radius 10 cm has an unknown charge. If the electric field 15 cm from the center of the sphere has the magnitude 3.0×10^3 N/C and is directed radially inward, what is the net charge on the sphere? ssm

36E. Two charged concentric spheres have radii of 10.0 cm and 15.0 cm. The charge on the inner sphere is 4.00×10^{-8} C, and that on the outer sphere is 2.00×10^{-8} C. Find the electric field (a) at r = 12.0 cm and (b) at r = 20.0 cm.

37E. In a 1911 paper, Ernest Rutherford said: "In order to form some idea of the forces required to deflect an α particle through a large angle, consider an atom [as] containing a point positive charge Ze at its centre and surrounded by a distribution of negative electricity $-Ze$ uniformly distributed within a sphere of radius R. The electric field E . . . at a distance r from the center for a point *inside* the atom [is]

$$E = \frac{Ze}{4\pi\varepsilon_0}\left(\frac{1}{r^2} - \frac{r}{R^3}\right)."$$

Verify this equation. ssm

38E. Equation 24-11 ($E = \sigma/\varepsilon_0$) gives the electric field at points near a charged conducting surface. Apply this equation to a conducting sphere of radius r and charge q, and show that the electric field outside the sphere is the same as the field of a point charge located at the center of the sphere.

39P. A proton with speed $v = 3.00 \times 10^5$ m/s orbits just outside a charged sphere of radius r = 1.00 cm. What is the charge on the sphere? ssm www

40P. A point charge $+q$ is placed at the center of an electrically neutral, spherical conducting shell with inner radius a and outer radius b. What charge appears on (a) the inner surface of the shell and (b) the outer surface? What is the net electric field at a distance r from the center of the shell if (c) $r < a$, (d) $b > r > a$, and (e) $r > b$? Sketch field lines for those three regions. For $r > b$, what is the net electric field due to (f) the central point charge plus the inner surface charge and (g) the outer surface charge? A point charge $-q$ is now placed outside the shell. Does this point charge change the charge distribution on (h) the outer surface and (i) the inner surface? Sketch the field lines now. (j) Is there an electrostatic force on the second point charge? (k) Is there a net electrostatic force on the first point charge? (l) Does this situation violate Newton's third law?

41P. A solid nonconducting sphere of radius R has a nonuniform charge distribution of volume charge density $\rho = \rho_s r/R$, where ρ_s is a constant and r is the distance from the center of the sphere. Show (a) that the total charge on the sphere is $Q = \pi\rho_s R^3$ and (b) that

$$E = \frac{1}{4\pi\varepsilon_0}\frac{Q}{R^4}r^2$$

gives the magnitude of the electric field inside the sphere. ilw

42P. A hydrogen atom can be considered as having a central point-like proton of positive charge $+e$ and an electron of negative charge $-e$ that is distributed about the proton according to the volume charge density $\rho = A \exp(-2r/a_0)$. Here A is a constant, $a_0 = 0.53 \times 10^{-10}$ m is the *Bohr radius*, and r is the distance from the center of the atom. (a) Using the fact that hydrogen is electrically neutral, find A. (b) Then find the electric field produced by the atom at the Bohr radius.

43P. In Fig. 24-34 a sphere, of radius a and charge $+q$ uniformly distributed throughout its volume, is concentric with a spherical conducting shell of inner radius b and outer radius c. This shell has a net charge of $-q$. Find expressions for the electric field, as a function of the radius r, (a) within the sphere ($r < a$), (b) between the sphere and the shell ($a < r < b$), (c) inside the shell ($b < r < c$), and (d) outside the shell ($r > c$). (e) What are the charges on the inner and outer surfaces of the shell? ssm

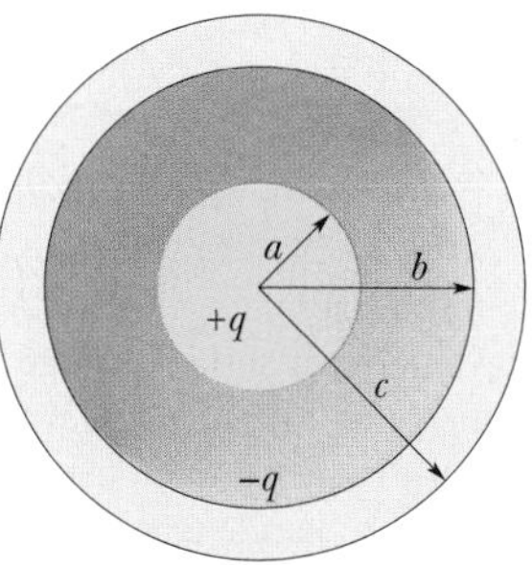

Fig. 24-34 Problem 43.

44P. Figure 24-35*a* shows a spherical shell of charge with uniform volume charge density ρ. Plot E due to the shell for distances r from the center of the shell ranging from zero to 30 cm. Assume that $\rho = 1.0 \times 10^{-6}$ C/m³, $a = 10$ cm, and $b = 20$ cm.

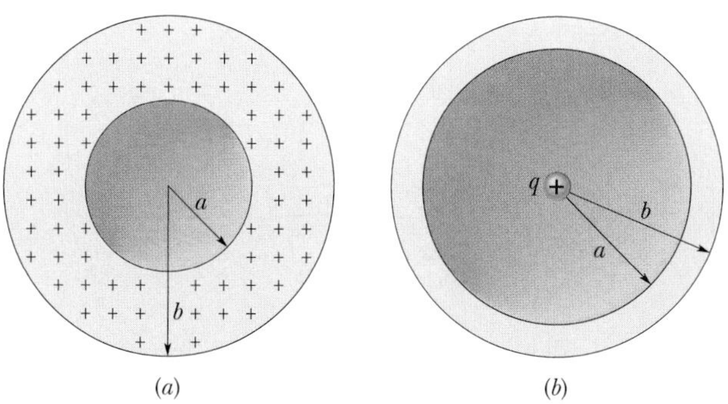

Fig. 24-35 Problems 44 and 45.

45P. In Fig. 24-35*b*, a nonconducting spherical shell, of inner radius a and outer radius b, has a positive volume charge density $\rho = A/r$ (within its thickness), where A is a constant and r is the distance from the center of the shell. In addition, a positive point charge q is located at that center. What value should A have if the electric field in the shell ($a \leq r \leq b$) is to be uniform? (*Hint:* The constant A depends on a but not on b.) ssm

46P*. A nonconducting sphere has a uniform volume charge density ρ. Let $\vec{r}$ be the vector from the center of the sphere to a general point P within the sphere. (a) Show that the electric field at P is given by $\vec{E} = \rho\vec{r}/3\varepsilon_0$. (Note that the result is independent of the radius of the sphere.) (b) A spherical cavity is hollowed out of the sphere, as shown in Fig. 24-36. Using superposition concepts, show that the electric field at all points within the cavity is uniform and equal to $\vec{E} = \rho\vec{a}/3\varepsilon_0$, where $\vec{a}$ is the position vector from the center of the sphere to the center of the cavity. (Note that this result is independent of the radius of the sphere and the radius of the cavity.)

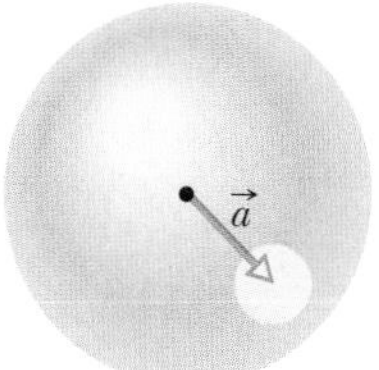

Fig. 24-36 Problem 46.

47P*. A spherically symmetrical but nonuniform volume distribution of charge produces an electric field of magnitude $E = Kr^4$, directed radially outward from the center of the sphere. Here r is the radial distance from that center, and K is a constant. What is the volume density ρ of the charge distribution?

Additional Problem

48. *The chocolate crumb mystery.* Explosions ignited by electrostatic discharges (sparks) constitute a serious danger in facilities handling grain or powder. Such an explosion occurred in chocolate crumb powder at a biscuit factory in the 1970s. At the factory, workers usually emptied newly delivered sacks of the powder into a loading bin, from which it was blown through grounded PVC pipes to a silo for storage. Somewhere along this route, two conditions for an explosion were met: (1) The magnitude of an electric field became 3.0×10^6 N/C or greater, so that electric breakdown and thus sparking could occur. (2) The energy of a spark was 150 mJ or greater so that it could ignite the powder explosively. Let us check for the first condition in the powder flow through the PVC pipes.

Suppose a stream of *negatively* charged chocolate crumb powder is blown through a cylindrical PVC pipe with radius $R = 5.0$ cm. Assume that the powder and its charge are spread uniformly through the pipe with a volume charge density ρ. (a) Using Gauss' law, find an expression for the magnitude of the electric field $\vec{E}$ in the pipe as a function of the radial distance r from the center of the pipe. (b) Does the magnitude increase or decrease with r? (c) Is the electric field $\vec{E}$ directed radially inward or outward? (d) Assuming a volume charge density ρ of magnitude 1.1×10^{-3} C/m³ (which was typical at the biscuit factory), find the maximum magnitude of the electric field and determine where that maximum field occurs. (e) Could sparking occur, and if so, where? (The story continues with Problem 57 in Chapter 25.)

NEW PROBLEMS

N1. *Flux and nonconducting shells.* A charged particle is suspended at the center of two concentric spherical shells that are very thin and made of nonconducting material. Figure 24N-1a shows a cross section. Figure 24N-1b gives the net flux Φ through a Gaussian sphere centered on the particle, as a function of the radius r of the sphere. (a) What is the charge of the central particle? What are the net charges of (b) shell A and (c) shell B?

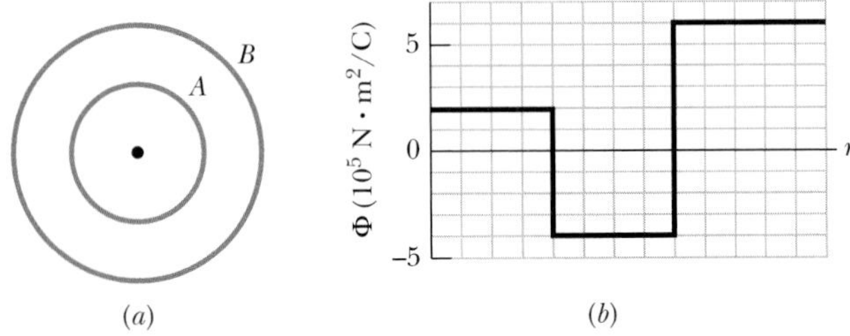

Fig. 24N-1 Problem N1.

N2. *Flux and conducting shells.* A charged particle is held at the center of two concentric conducting spherical shells. Figure 24N-2a shows a cross section. Figure 24N-2b gives the net flux Φ through a Gaussian sphere centered on the particle, as a function of the radius r of the sphere. What are (a) the charge of the central particle and the net charges of (b) shell A and (c) shell B?

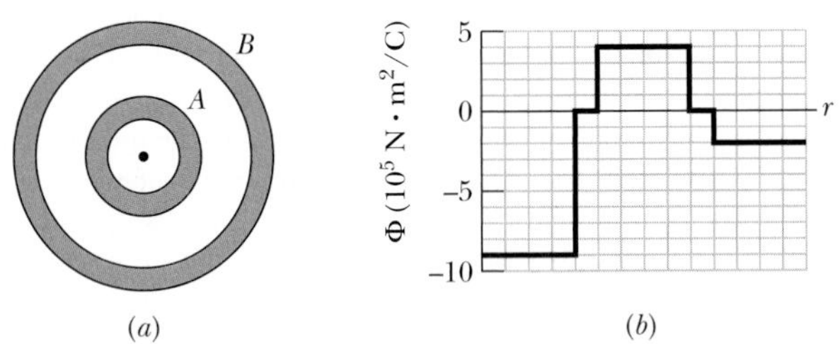

Fig. 24N-2 Problem N2.

N3. Figure 24N-3 shows a very large nonconducting sheet that has a uniform surface charge density of $\sigma = -2.00\ \mu\text{C/m}^2$; it also shows a particle of charge $Q = 6.00\ \mu\text{C}$, at distance d from the sheet. Both are fixed in place. Other than at infinite distances, where on the x axis is the net electric field from the sheet and particle zero if (a) $d = 0.20$ m and (b) $d = 0.80$ m?

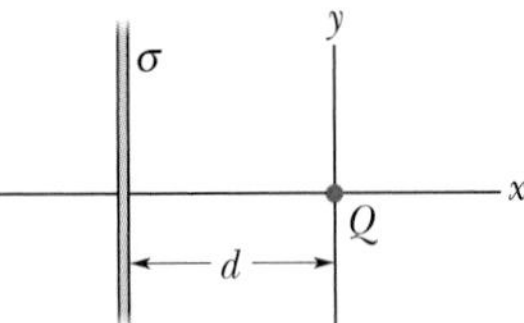

Fig. 24N-3 Problem N3.

N4. Figure 24N-4 shows two nonconducting spherical shells, fixed in place. Shell 1 has uniform surface charge density $+6.0\ \mu\text{C/m}^2$ on its outer surface and radius 3.0 cm; shell 2 has uniform surface charge density $+4.0\ \mu\text{C/m}^2$ on its outer surface and radius 2.0 cm; the shell centers are separated by $L = 10$ cm. What are the magnitude and direction of the net electric field at $x = 2.0$ cm?

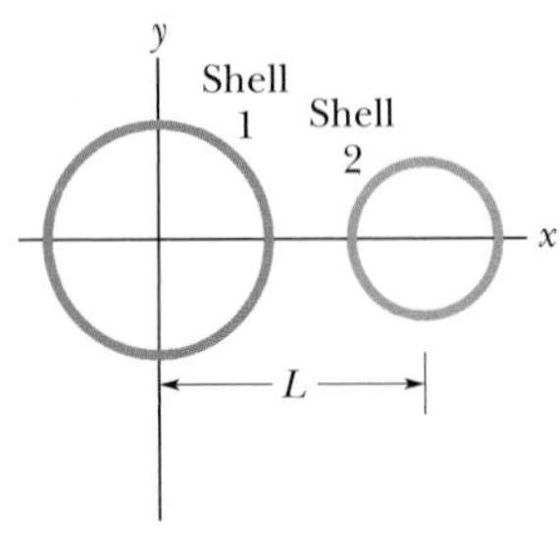

Fig. 24N-4 Problem N4.

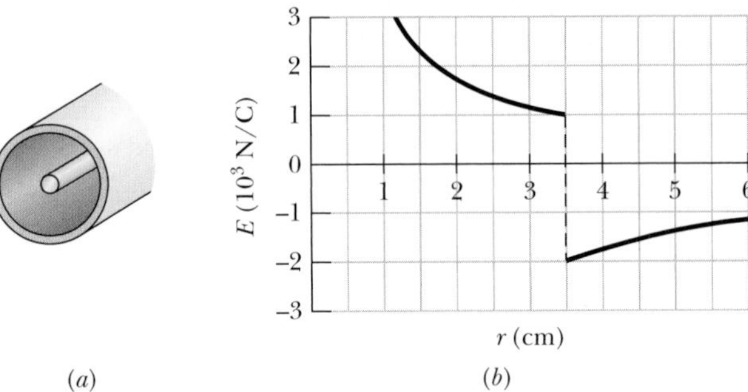

Fig. 24N-5 Problem N5.

N5. Figure 24N-5a shows a narrow charged cylinder that is coaxial with a larger charged cylinder. Both are nonconducting, thin, and have uniform surface charge densities on their outer surfaces. Figure 24N-5b gives the radial component E of the electric field versus radial distance r from the common axis of the cylinders. What is the linear charge density of the larger cylinder?

N6. A charged particle is held at the center of a spherical shell. Figure 24N-6 gives the magnitude E of the electric field versus radial distance r. Approximately, what is the net charge on the shell?

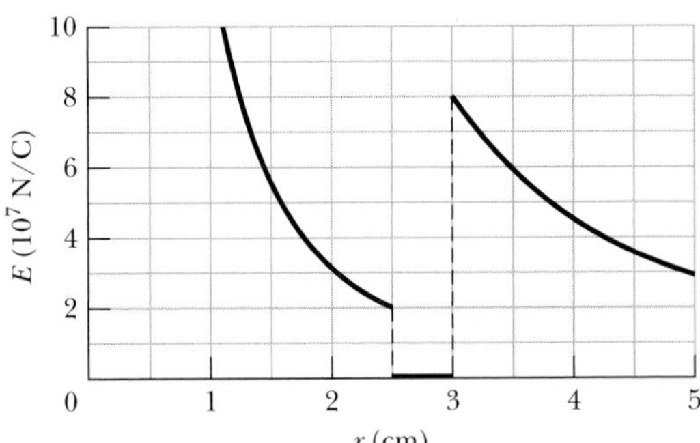

Fig. 24N-6 Problem N6.

N7. The electric field just outside the outer surface of a hollow spherical conductor of inner radius = 10 cm and outer radius = 20 cm has a magnitude of 450 N/C and is directed outward. When an unknown point charge Q is introduced into the center of the sphere, the electric field is still directed outward but has decreased to 180 N/C. (a) What is the net charge enclosed within the outer spherical surface of the conductor before Q is introduced? (b) What is charge Q? After Q is introduced, what are the charges on (c) the inner surface and (d) the outer surface of the conductor?

N8. A spherical ball of charged particles has a uniform charge density. In terms of the ball's radius R, at what radial distances from the center of the ball is the magnitude of the ball's electrical field equal to $\frac{1}{4}$ of the maximum magnitude of that field?

N9. Charge of uniform density $\rho = 3.2\ \mu\text{C/m}^3$ fills a nonconducting sphere of radius 5.0 cm. What are the magnitudes of the electric field (a) 3.5 cm and (b) 8.0 cm from the center of the sphere?

N10. Figure 24N-7a shows a closed Gaussian surface in the shape of a cube of edge length 2.00 m. It lies in a region where the electric field is given by $\vec{E} = (3.00x + 4.00)\hat{\text{i}} + 6.00\hat{\text{j}} + 7.00\hat{\text{k}}$ N/C. What is the net charge contained by the cube?

Fig. 24N-7 Problems N10 and N11.

N11. Figure 24N-7*b* shows a closed Gaussian surface in the shape of a cube of edge length 2.00 m. It lies in a region where the electric field is given by $\vec{E} = -3.00\hat{\mathrm{i}} - 4.00y^2\hat{\mathrm{j}} + 3.00\hat{\mathrm{k}}$ N/C. What is the net charge contained by the cube?

N12. Figure 24N-8 shows two nonconducting spherical shells that are fixed in place on an x axis. Shell 1 has uniform surface charge density $+4.0\ \mu\text{C/m}^2$ on its outer surface and radius 0.50 cm; shell 2 has uniform surface charge density $-2.0\ \mu\text{C/m}^2$ on its outer surface and radius 2.0 cm; the centers are separated by $L = 6.0$ cm. Other than at $x = \infty$, where on the x axis is the net electric field from the two shells equal to zero?

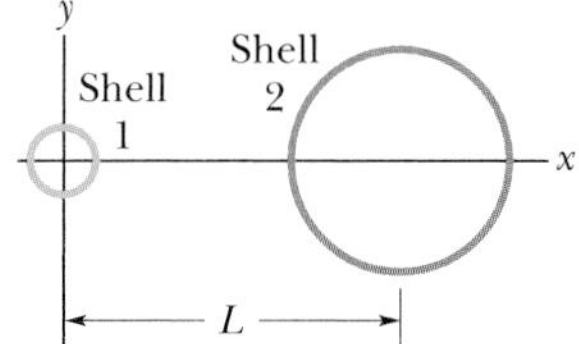

Fig. 24N-8 Problem N12.

N13. Charge Q is uniformly distributed in a sphere of radius R. (a) How much charge is contained within radius $r = R/2$? (b) What is the ratio of the electric field magnitude at $r = R/2$ to that on the surface of the sphere?

N14. Figure 24N-9 shows short sections of two very long parallel lines of charge that are fixed in place, separated by $L = 8.0$ cm. The uniform linear charge densities are $+6.0\ \mu\text{C/m}$ for line 1 and $-2.0\ \mu\text{C/m}$ for line 2. Where along the x axis shown is the net electric field from the two lines zero?

N15. A uniform surface charge of $8.0\ \text{nC/m}^2$ is distributed over the entire xy plane. Consider a spherical Gaussian surface centered on the origin and having a radius of 5.0 cm. Determine the electric flux for this surface.

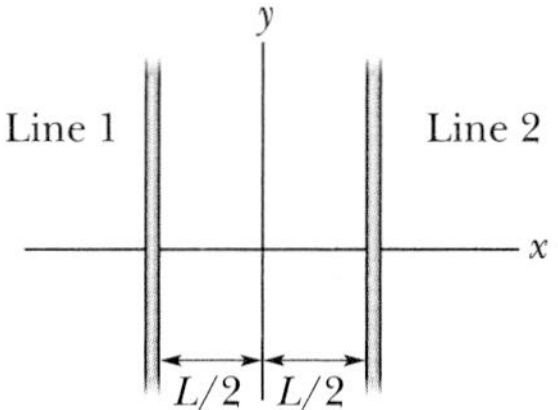

Fig. 24N-9 Problem N14.

N16. Figure 24N-10 gives the magnitude of the electric field inside and outside a sphere with a uniformly distributed positive charge. What is the charge on the sphere?

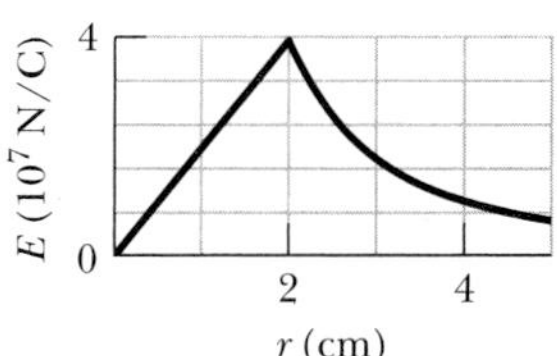

Fig. 24N-10 Problem N16.

N17. A solid ball of radius R has a uniform volume charge density and produces a certain electric field magnitude E_1 at point P, at a distance of $2R$ from the ball's center. If, instead, the ball had a hollow core of radius $R/2$, what would have been the electric field magnitude at point P?

N18. Figure 24N-11*a* shows three plastic sheets that are large, parallel, and uniformly charged. Figure 24N-11*b* gives the component of the net electric field along an x axis through the sheets. What is the ratio of the charge density on sheet 3 to that on sheet 2?

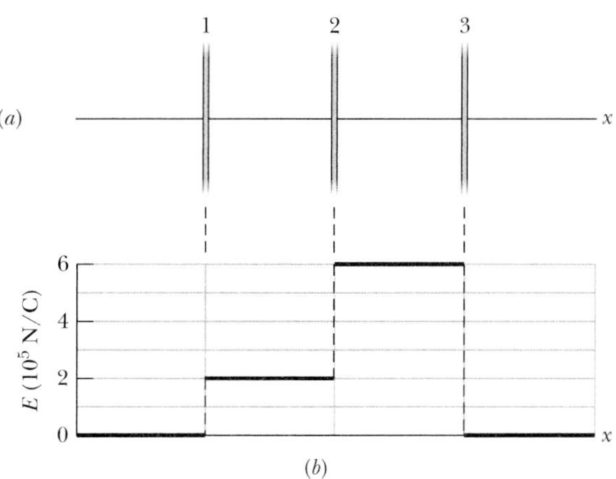

Fig. 24N-11 Problem N18.

N19. Figure 24N-12 shows, in cross section, two spheres with uniformly distributed charges throughout their volumes. Each has radius R. Point P lies on a line connecting the centers of the spheres, at radial distance $R/2$ from the center of sphere 1. If the net electric field at point P is zero, what is the ratio q_2/q_1 of the total charge q_2 in sphere 2 to the total charge q_1 in sphere 1?

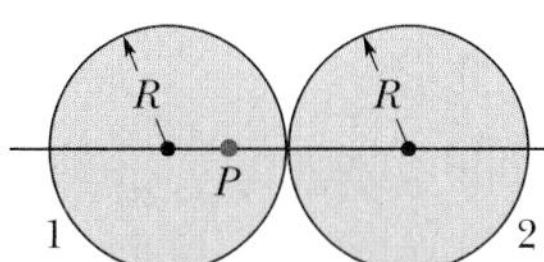

Fig. 24N-12 Problem N19.

N20. Assume that a ball of charged particles has a uniformly distributed negative charge density except for a narrow radial tunnel through its center, from the surface on one side to the surface on the opposite side. Also assume that we can position a proton anywhere along the tunnel or outside the ball. Let F_R be the magnitude of the electrostatic force on the proton when it is located at the ball's surface, at radius R. How far from the surface is there a point where the force magnitude is $\frac{1}{2}F_R$ if we move the proton (a) away from the ball and (b) into the tunnel?

N21. An electron is released from rest at a perpendicular distance of 9.0 cm from a line of charge on a very long nonconducting rod. That charge is uniformly distributed, with 6.0 μC per meter. What is the magnitude of the electron's initial acceleration?

N22. In Fig. 24N-13*a*, an electron is shot directly away from a uniformly charged plastic sheet, at speed 2.0×10^5 m/s. The sheet is nonconducting, flat, and very large. Figure 24N-13*b* gives the electron's vertical velocity component v versus time t until the return to the launch point. What is the sheet's surface charge density?

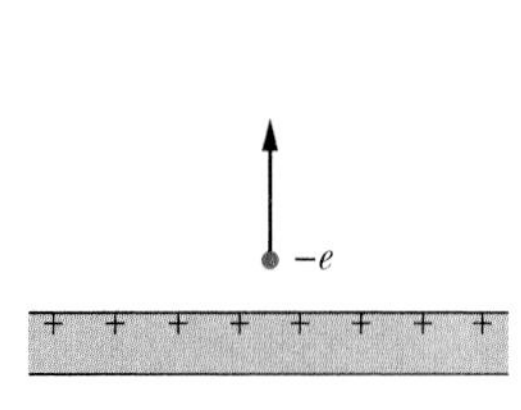

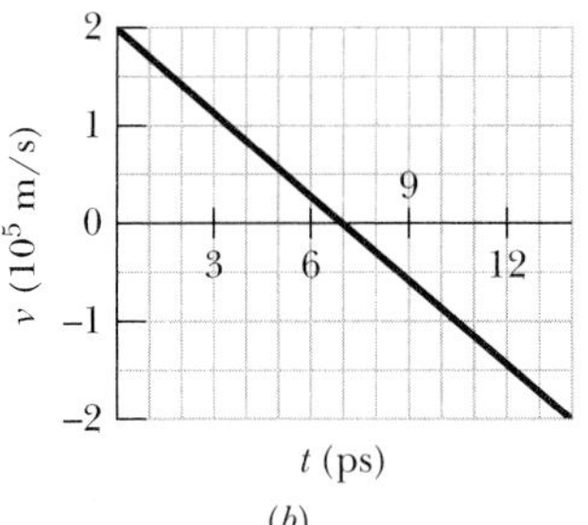

Fig. 24N-13 Problem N22.

25 Electric Potential

While enjoying the Sequoia National Park from a lookout platform, this woman found her hair rising from her head. Amused, her brother took her photograph. Five minutes after they left, lightning struck the platform, killing one person and injuring seven.

What had caused the woman's hair to rise?

The answer is in this chapter.

25-1 Electric Potential Energy

Newton's law for the gravitational force and Coulomb's law for the electrostatic force are mathematically identical. Thus, the general features we have discussed for the gravitational force should apply to the electrostatic force.

In particular, we can infer, correctly, that the electrostatic force is a *conservative force*. Thus, when an electrostatic force acts between two or more charged particles within a system of particles, we can assign an **electric potential energy** U to the system. Moreover, if the system changes its configuration from an initial state i to a different final state f, the electrostatic force does work W on the particles. From Eq. 8-1, we then know that the resulting change ΔU in the potential energy of the system is

$$\Delta U = U_f - U_i = -W. \tag{25-1}$$

As with other conservative forces, the work done by the electrostatic force is *path independent*. Suppose a charged particle within the system moves from point i to point f while an electrostatic force between it and the rest of the system acts on it. Provided the rest of the system does not change, the work W done by the force is the same for *all* paths between points i and f.

For convenience, we usually take the *reference configuration* of a system of charged particles to be that in which the particles are all infinitely separated from each other. Also, we usually set the corresponding *reference potential energy* to be zero. Suppose that several charged particles come together from initially infinite separations (state i) to form a system of nearby particles (state f). Let the initial potential energy U_i be zero, and let W_∞ represent the work done by the electrostatic forces between the particles during the move in from infinity. Then from Eq. 25-1, the final potential energy U of the system is

$$U = -W_\infty. \tag{25-2}$$

As is true of other kinds of potential energy, electric potential energy is considered to be a type of mechanical energy. Recall from Chapter 8 that if only conservative forces act within a (closed) system, the mechanical energy of the system is conserved. We shall use this fact extensively in the rest of this chapter.

PROBLEM-SOLVING TACTICS

Tactic 1: *Electric Potential Energy; Work Done by a Field*
An electric potential energy is associated with a system of particles as a whole. However, you will see statements (starting with Sample Problem 25-1) that associate it with only one particle within a system. For example, you might read, "An electron in an electric field has a potential energy of 10^{-7} J." Such statements are often acceptable, but you should always keep in mind that the potential energy is actually associated with a system—here the electron plus the charged particles that set up the electric field. Also keep in mind that it makes sense to assign a particular potential energy value, such as 10^{-7} J here, to a particle or even a system *only* if the reference potential energy value is known.

When the potential energy is associated with only one particle within a system, you often will read that work is done on the particle *by the electric field*. What is meant is that work is done by the force on the particle due to the charges that set up the field.

Sample Problem 25-1

Electrons are continually being knocked out of air molecules in the atmosphere by cosmic-ray particles coming in from space. Once released, each electron experiences an electrostatic force $\vec{F}$ due to the electric field $\vec{E}$ that is produced in the atmosphere by charged particles already on Earth. Near Earth's surface the electric field has the magnitude $E = 150$ N/C and is directed downward. What is the change ΔU in the electric potential energy of a released electron when the electrostatic force causes it to move vertically upward through a distance $d = 520$ m (Fig. 25-1)?

SOLUTION: We need three **Key Ideas** here. One is that the change ΔU in the electric potential energy of the electron is related to the work

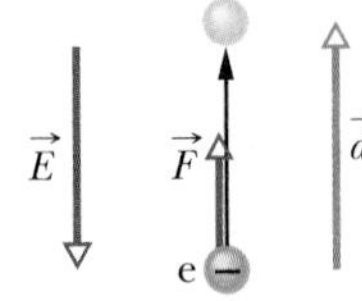

Fig. 25-1 Sample Problem 25-1. An electron in the atmosphere is moved upward through displacement $\vec{d}$ by an electrostatic force $\vec{F}$ due to an electric field $\vec{E}$.

W done on the electron by the electric field. Equation 25-1 ($\Delta U = -W$) gives the relation. A second Key Idea is that the work done by a constant force $\vec{F}$ on a particle undergoing a displacement $\vec{d}$ is

$$W = \vec{F} \cdot \vec{d}. \tag{25-3}$$

Finally, the third Key Idea is that the electrostatic force and the electric field are related by $\vec{F} = q\vec{E}$, where here q is the charge of an electron ($= -1.6 \times 10^{-19}$ C). Substituting for $\vec{F}$ in Eq. 25-3 and taking the dot product yield

$$W = q\vec{E} \cdot \vec{d} = qEd \cos \theta, \tag{25-4}$$

where θ is the angle between the directions of $\vec{E}$ and $\vec{d}$. The field $\vec{E}$ is directed downward and the displacement $\vec{d}$ is directed upward, so $\theta = 180°$. Substituting this and other data into Eq. 25-4, we find

$$W = (-1.6 \times 10^{-19}\ \text{C})(150\ \text{N/C})(520\ \text{m}) \cos 180°$$
$$= 1.2 \times 10^{-14}\ \text{J}.$$

Equation 25-1 then yields

$$\Delta U = -W = -1.2 \times 10^{-14}\ \text{J}. \qquad \text{(Answer)}$$

This result tells us that during the 520 m ascent, the electric potential energy of the electron *decreases* by 1.2×10^{-14} J.

CHECKPOINT 1: In the figure, a proton moves from point i to point f in a uniform electric field directed as shown. (a) Does the electric field do positive or negative work on the proton? (b) Does the electric potential energy of the proton increase or decrease?

$\vec{E}$ f i

25-2 Electric Potential

As you can infer from Sample Problem 25-1, the potential energy of a charged particle in an electric field depends on the magnitude of the charge. However, the potential energy *per unit charge* has a unique value at any point in an electric field.

For an example of this, suppose we place a test particle of positive charge 1.60×10^{-19} C at a point in an electric field where the particle has an electric potential energy of 2.40×10^{-17} J. Then the potential energy per unit charge is

$$\frac{2.40 \times 10^{-17}\ \text{J}}{1.60 \times 10^{-19}\ \text{C}} = 150\ \text{J/C}.$$

Next, suppose we replace that test particle with one having twice as much positive charge, 3.20×10^{-19} C. We would find that the second particle has an electric potential energy of 4.80×10^{-17} J, twice that of the first particle. However, the potential energy per unit charge would be the same, still 150 J/C.

Thus, the potential energy per unit charge, which can be symbolized as U/q, is independent of the charge q of the particle we happen to use and is *characteristic only of the electric field* we are investigating. The potential energy per unit charge at a point in an electric field is called the **electric potential** V (or simply the **potential**) at that point. Thus,

$$V = \frac{U}{q}. \tag{25-5}$$

Note that electric potential is a scalar, not a vector.

The *electric potential difference* ΔV between any two points i and f in an electric field is equal to the difference in potential energy per unit charge between the two points:

$$\Delta V = V_f - V_i = \frac{U_f}{q} - \frac{U_i}{q} = \frac{\Delta U}{q}. \tag{25-6}$$

Using Eq. 25-1 to substitute $-W$ for ΔU in Eq. 25-6, we can define the potential difference between points i and f as

$$\Delta V = V_f - V_i = -\frac{W}{q} \quad \text{(potential difference defined).} \tag{25-7}$$

The potential difference between two points is thus the negative of the work done by the electrostatic force to move a unit charge from one point to the other. A potential difference can be positive, negative, or zero, depending on the signs and magnitudes of q and W.

If we set $U_i = 0$ at infinity as our reference potential energy, then by Eq. 25-5, the electric potential must also be zero there. Then from Eq. 25-7, we can define the electric potential V at any point in an electric field to be

$$V = -\frac{W_\infty}{q} \quad \text{(potential defined),} \tag{25-8}$$

where W_∞ is the work done by the electric field on a charged particle as that particle moves in from infinity to point f. A potential V can be positive, negative, or zero, depending on the signs and magnitudes of q and W_∞.

The SI unit for potential that follows from Eq. 25-8 is the joule per coulomb. This combination occurs so often that a special unit, the *volt* (abbreviated V) is used to represent it. Thus,

$$1 \text{ volt} = 1 \text{ joule per coulomb.} \tag{25-9}$$

This new unit allows us to adopt a more conventional unit for the electric field $\vec{E}$, which we have measured up to now in newtons per coulomb. With two unit conversions, we obtain

$$\begin{aligned} 1 \text{ N/C} &= \left(1\frac{\text{N}}{\text{C}}\right)\left(\frac{1 \text{ V}\cdot\text{C}}{1 \text{ J}}\right)\left(\frac{1 \text{ J}}{1 \text{ N}\cdot\text{m}}\right) \\ &= 1 \text{ V/m.} \end{aligned} \tag{25-10}$$

The conversion factor in the second set of parentheses comes from Eq. 25-9; that in the third set of parentheses is derived from the definition of the joule. From now on, we shall express values of the electric field in volts per meter rather than in newtons per coulomb.

Finally, we can now define an energy unit that is a convenient one for energy measurements in the atomic and subatomic domain: One *electron-volt* (eV) is the energy equal to the work required to move a single elementary charge e, such as that of the electron or the proton, through a potential difference of exactly one volt. Equation 25-7 tells us that the magnitude of this work is $q\,\Delta V$, so

$$\begin{aligned} 1 \text{ eV} &= e(1 \text{ V}) \\ &= (1.60 \times 10^{-19} \text{ C})(1 \text{ J/C}) = 1.60 \times 10^{-19} \text{ J.} \end{aligned}$$

PROBLEM-SOLVING TACTICS

Tactic 2: *Electric Potential and Electric Potential Energy*
Electric potential V and electric potential energy U are quite different quantities and should not be confused.

- *Electric potential* is a property of an electric field, regardless of whether a charged object has been placed in that field; it is measured in joules per coulomb, or volts.

- *Electric potential energy* is an energy of a charged object in an external electric field (or more precisely, an energy of the system consisting of the object and the external electric field); it is measured in joules.

Work Done by an Applied Force

Suppose we move a particle of charge q from point i to point f in an electric field by applying a force to it. During the move, our applied force does work W_{app} on the charge while the electric field does work W on it. By the work–kinetic energy theorem of Eq. 7-10, the change ΔK in the kinetic energy of the particle is

$$\Delta K = K_f - K_i = W_{app} + W. \tag{25-11}$$

Now suppose the particle is stationary before and after the move. Then K_f and K_i are both zero, and Eq. 25-11 reduces to

$$W_{app} = -W. \tag{25-12}$$

In words, the work W_{app} done by our applied force during the move is equal to the negative of the work W done by the electric field—provided there is no change in kinetic energy.

By using Eq. 25-12 to substitute W_{app} into Eq. 25-1, we can relate the work done by our applied force to the change in the potential energy of the particle during the move. We find

$$\Delta U = U_f - U_i = W_{app}. \tag{25-13}$$

By similarly using Eq. 25-12 to substitute W_{app} into Eq. 25-7, we can relate our work W_{app} to the electric potential difference ΔV between the initial and final locations of the particle. We find

$$W_{app} = q\,\Delta V. \tag{25-14}$$

W_{app} can be positive, negative, or zero depending on the signs and magnitudes of q and ΔV. It is the work we must do to move a particle of charge q through a potential difference ΔV with no change in the particle's kinetic energy.

CHECKPOINT 2: In the figure of Checkpoint 1, we move a proton from point i to point f in a uniform electric field directed as shown. (a) Does our force do positive or negative work? (b) Does the proton move to a point of higher or lower potential?

25-3 Equipotential Surfaces

Adjacent points that have the same electric potential form an **equipotential surface,** which can be either an imaginary surface or a real, physical surface. No net work W is done on a charged particle by an electric field when the particle moves between two points i and f on the same equipotential surface. This follows from Eq. 25-7, which tells us that W must be zero if $V_f = V_i$. Because of the path independence of work (and thus of potential energy and potential), $W = 0$ for *any* path connecting points i and f, regardless of whether that path lies entirely on the equipotential surface.

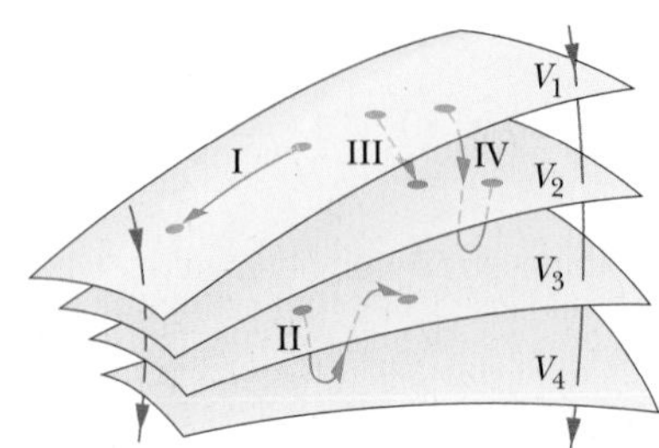

Fig. 25-2 Portions of four equipotential surfaces at electric potentials $V_1 = 100$ V, $V_2 = 80$ V, $V_3 = 60$ V, and $V_4 = 40$ V. Four paths along which a test charge may move are shown. Two electric field lines are also indicated.

Figure 25-2 shows a *family* of equipotential surfaces, associated with the electric field due to some distribution of charges. The work done by the electric field on a charged particle as the particle moves from one end to the other of paths I and II is zero because each of these paths begins and ends on the same equipotential surface. The work done as the charged particle moves from one end to the other of paths III and IV is not zero but has the same value for both these paths because the initial and final potentials are identical for the two paths; that is, paths III and IV connect the same pair of equipotential surfaces.

From symmetry, the equipotential surfaces produced by a point charge or a spherically symmetrical charge distribution are a family of concentric spheres. For a uniform electric field, the surfaces are a family of planes perpendicular to the field

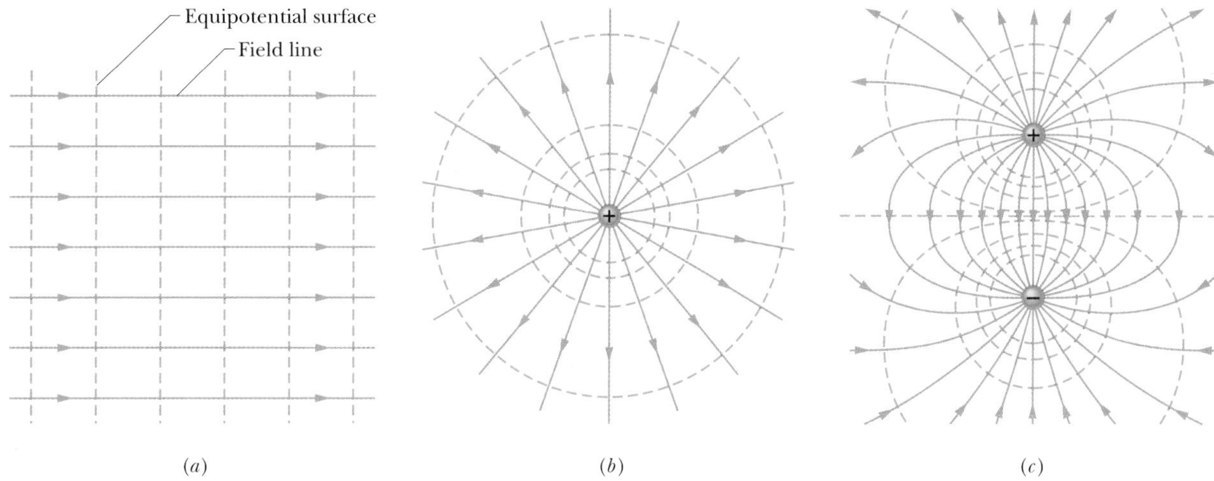

Fig. 25-3 Electric field lines (purple) and cross sections of equipotential surfaces (gold) for (*a*) a uniform field, (*b*) the field of a point charge, and (*c*) the field of an electric dipole.

lines. In fact, equipotential surfaces are always perpendicular to electric field lines and thus to $\vec{E}$, which is always tangent to these lines. If $\vec{E}$ were *not* perpendicular to an equipotential surface, it would have a component lying along that surface. This component would then do work on a charged particle as it moved along the surface. However, by Eq. 25-7 work cannot be done if the surface is truly an equipotential surface; the only possible conclusion is that $\vec{E}$ must be everywhere perpendicular to the surface. Figure 25-3 shows electric field lines and cross sections of the equipotential surfaces for a uniform electric field and for the field associated with a point charge and with an electric dipole.

Fig. 25-4 This enhancement of the chapter's opening photograph shows the result of overhead clouds creating a strong electric field $\vec{E}$ near a woman's head. Many of the hair strands extended along the field, which was perpendicular to the equipotential surfaces and greatest where those surfaces were closest, near the top of her head.

We now return to the woman in the opening photograph for this chapter. Because she was standing on a platform that was connected to the mountainside, she was at about the same potential as the mountainside. Overhead, a highly charged cloud system had moved in and created a strong electric field around her and the mountainside, with $\vec{E}$ pointing outward from her and the mountain. Electrostatic forces due to this field drove some of the conduction electrons in the woman downward through her body, leaving strands of her hair positively charged. The magnitude of $\vec{E}$ was apparently large, but less than the value of about 3×10^6 V/m that would have caused electrical breakdown of the air molecules. (That value was exceeded shortly later when lightning struck the platform.)

The equipotential surfaces surrounding the woman on the mountainside platform can be inferred from her hair; the strands are extended along the direction of $\vec{E}$ and thus are perpendicular to the equipotential surfaces, so the surfaces must be as drawn in Fig. 25-4. The field magnitude E was apparently greatest (the equipotential surfaces were apparently most closely spaced) just above her head, because the hair there extended farther than the hair around the sides.

The lesson here is simple. If an electric field causes the hairs on your head to stand up, you would do better to run for shelter than to pose for a snapshot.

25-4 Calculating the Potential from the Field

We can calculate the potential difference between any two points i and f in an electric field if we know the electric field vector $\vec{E}$ all along any path connecting those points. To make the calculation, we find the work done on a positive test charge by the field as the charge moves from i to f, and then use Eq. 25-7.

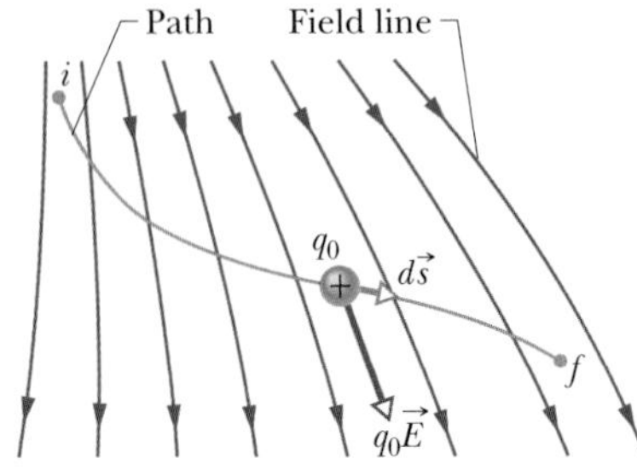

Fig. 25-5 A test charge q_0 moves from point i to point f along the path shown in a nonuniform electric field. During a displacement $d\vec{s}$, an electrostatic force $q_0\vec{E}$ acts on the test charge. This force points in the direction of the field line at the location of the test charge.

Consider an arbitrary electric field, represented by the field lines in Fig. 25-5, and a positive test charge q_0 that moves along the path shown from point i to point f. At any point on the path, an electrostatic force $q_0\vec{E}$ acts on the charge as it moves through a differential displacement $d\vec{s}$. From Chapter 7, we know that the differential work dW done on a particle by a force $\vec{F}$ during a displacement $d\vec{s}$ is

$$dW = \vec{F} \cdot d\vec{s}. \tag{25-15}$$

For the situation of Fig. 25-5, $\vec{F} = q_0\vec{E}$ and Eq. 25-15 becomes

$$dW = q_0\vec{E} \cdot d\vec{s}. \tag{25-16}$$

To find the total work W done on the particle by the field as the particle moves from point i to point f, we sum—via integration—the differential works done on the charge as it moves through all the differential displacements $d\vec{s}$ along the path:

$$W = q_0 \int_i^f \vec{E} \cdot d\vec{s}. \tag{25-17}$$

If we substitute the total work W from Eq. 25-17 into Eq. 25-7, we find

$$V_f - V_i = -\int_i^f \vec{E} \cdot d\vec{s}. \tag{25-18}$$

Thus, the potential difference $V_f - V_i$ between any two points i and f in an electric field is equal to the negative of the *line integral* (meaning the integral along a particular path) of $\vec{E} \cdot d\vec{s}$ from i to f. However, because the electrostatic force is conservative, all paths (whether easy or difficult to use) yield the same result.

If the electric field is known throughout a certain region, Eq. 25-18 allows us to calculate the difference in potential between any two points in the field. If we choose the potential V_i at point i to be zero, then Eq. 25-18 becomes

$$V = -\int_i^f \vec{E} \cdot d\vec{s}, \tag{25-19}$$

in which we have dropped the subscript f on V_f. Equation 25-19 gives us the potential V at any point f in the electric field *relative to the zero potential* at point i. If we let point i be at infinity, then Eq. 25-19 gives us the potential V at any point f relative to the zero potential at infinity.

CHECKPOINT 3: The figure shows a family of parallel equipotential surfaces (in cross section) and five paths along which we shall move an electron from one surface to another. (a) What is the direction of the electric field associated with the surfaces? (b) For each path, is the work we do positive, negative, or zero? (c) Rank the paths according to the work we do, greatest first.

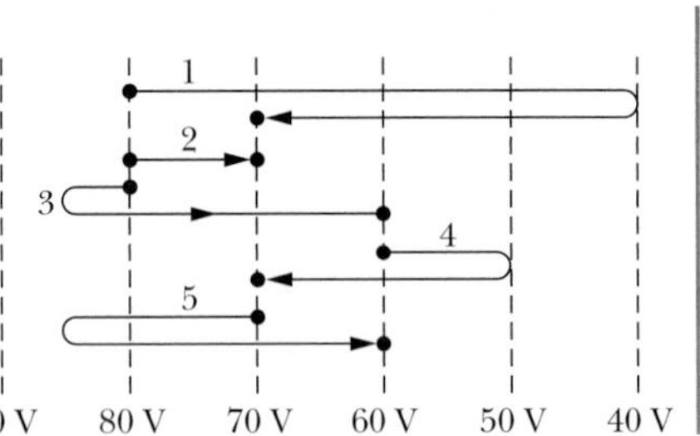

Sample Problem 25-2

(a) Figure 25-6*a* shows two points i and f in a uniform electric field $\vec{E}$. The points lie on the same electric field line (not shown) and are separated by a distance d. Find the potential difference $V_f - V_i$ by moving a positive test charge q_0 from i to f along the path shown, which is parallel to the field direction.

SOLUTION: The **Key Idea** here is that we can find the potential difference between any two points in an electric field by integrating $\vec{E} \cdot d\vec{s}$ along a path connecting those two points according to Eq. 25-18. We do this by mentally moving a test charge q_0 along that path, from initial point i to final point f. As we move such a test charge along the path in Fig. 25-6*a*, its differential displacement $d\vec{s}$ always has the same direction as $\vec{E}$. Thus, the angle θ between $\vec{E}$ and $d\vec{s}$ is zero and the dot product in Eq. 25-18 is

$$\vec{E} \cdot d\vec{s} = E\, ds \cos \theta = E\, ds. \qquad (25\text{-}20)$$

Equations 25-18 and 25-20 then give us

$$V_f - V_i = -\int_i^f \vec{E} \cdot d\vec{s} = -\int_i^f E\, ds. \qquad (25\text{-}21)$$

Since the field is uniform, E is constant over the path and can be moved outside the integral, giving us

$$V_f - V_i = -E \int_i^f ds = -Ed, \qquad \text{(Answer)}$$

in which the integral is simply the length d of the path. The minus sign in the result shows that the potential at point f in Fig. 25-6*a* is lower than the potential at point i. This is a general result: The potential always decreases along a path that extends in the direction of the electric field lines.

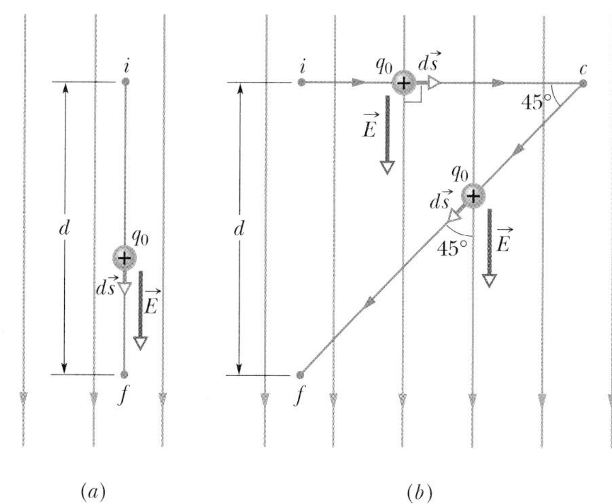

Fig. 25-6 Sample Problem 25-2. (*a*) A test charge q_0 moves in a straight line from point i to point f, along the direction of a uniform electric field. (*b*) Charge q_0 moves along path icf in the same electric field.

(b) Now find the potential difference $V_f - V_i$ by moving the positive test charge q_0 from i to f along the path icf shown in Fig. 25-6*b*.

SOLUTION: The Key Idea of (a) applies here too, except now we move the test charge along a path that consists of two lines: ic and cf. At all points along line ic, the displacement $d\vec{s}$ of the test charge is perpendicular to $\vec{E}$. Thus, the angle θ between $\vec{E}$ and $d\vec{s}$ is 90°, and the dot product $\vec{E} \cdot d\vec{s}$ is 0. Equation 25-18 then tells us that points i and c are at the same potential: $V_c - V_i = 0$.

For line cf we have $\theta = 45°$ and, from Eq. 25-18,

$$V_f - V_i = -\int_c^f \vec{E} \cdot d\vec{s} = -\int_c^f E(\cos 45°)\, ds$$
$$= -E(\cos 45°) \int_c^f ds.$$

The integral in this equation is just the length of line cf; from Fig. 25-6*b*, that length is $d/\sin 45°$. Thus,

$$V_f - V_i = -E(\cos 45°) \frac{d}{\sin 45°} = -Ed. \qquad \text{(Answer)}$$

This is the same result we obtained in (a), as it must be; the potential difference between two points does not depend on the path connecting them. Moral: When you want to find the potential difference between two points by moving a test charge between them, you can save time and work by choosing a path that simplifies the use of Eq. 25-18.

25-5 Potential Due to a Point Charge

We will now use Eq. 25-18 to derive an expression for the electric potential V in the space around a charged particle, relative to the zero potential at infinity. Consider a point P at a distance R from a fixed particle of positive charge q (Fig. 25-7). To use Eq. 25-18, we imagine that we move a positive test charge q_0 from point P to infinity. Because the path we take does not matter, let us choose the simplest one—a line that extends radially from the fixed particle through P to infinity.

To use Eq. 25-18, we must evaluate the dot product

$$\vec{E} \cdot d\vec{s} = E \cos \theta\, ds. \qquad (25\text{-}22)$$

The electric field $\vec{E}$ in Fig. 25-7 is directed radially outward from the fixed particle. Thus, the differential displacement $d\vec{s}$ of the test particle along its path has the same direction as $\vec{E}$. That means that in Eq. 25-22, angle $\theta = 0$ and $\cos \theta = 1$. Because

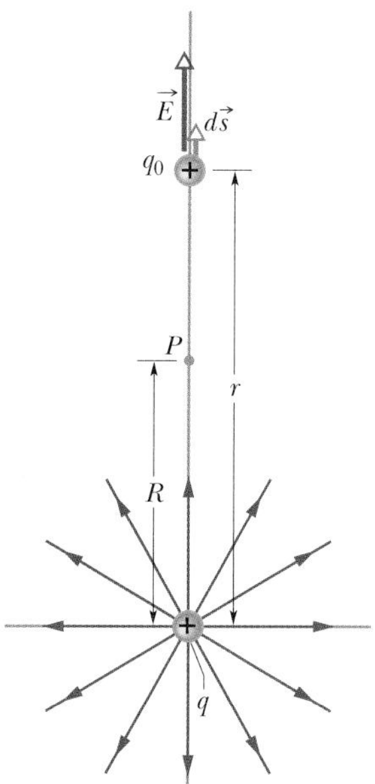

Fig. 25-7 The positive point charge q produces an electric field $\vec{E}$ and an electric potential V at point P. We find the potential by moving a test charge q_0 from P to infinity. The test charge is shown at distance r from the point charge, during differential displacement $d\vec{s}$.

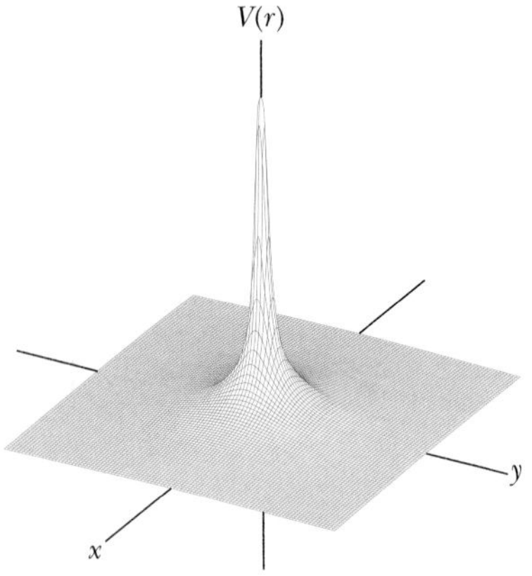

Fig. 25-8 A computer-generated plot of the electric potential $V(r)$ due to a positive point charge located at the origin of an xy plane. The potentials at points in that plane are plotted vertically. (Curved lines have been added to help you visualize the plot.) The infinite value of V predicted by Eq. 25-26 for $r = 0$ is not plotted.

the path is radial, let us write ds as dr. Then, substituting the limits R and ∞, we can write Eq. 25-18 as

$$V_f - V_i = -\int_R^\infty E\,dr. \tag{25-23}$$

Next, we set $V_f = 0$ (at ∞) and $V_i = V$ (at R). Then, for the magnitude of the electric field at the site of the test charge, we substitute from Eq. 23-3:

$$E = \frac{1}{4\pi\varepsilon_0}\frac{q}{r^2}. \tag{25-24}$$

With these changes, Eq. 25-23 then gives us

$$0 - V = -\frac{q}{4\pi\varepsilon_0}\int_R^\infty \frac{1}{r^2}\,dr = \frac{q}{4\pi\varepsilon_0}\left[\frac{1}{r}\right]_R^\infty$$

$$= -\frac{1}{4\pi\varepsilon_0}\frac{q}{R}. \tag{25-25}$$

Solving for V and switching R to r, we then have

$$V = \frac{1}{4\pi\varepsilon_0}\frac{q}{r} \tag{25-26}$$

as the electric potential V due to a particle of charge q, at any radial distance r from the particle.

Although we have derived Eq. 25-26 for a positively charged particle, the derivation holds also for a negatively charged particle, in which case, q is a negative quantity. Note that the sign of V is the same as the sign of q:

> A positively charged particle produces a positive electric potential. A negatively charged particle produces a negative electric potential.

Figure 25-8 shows a computer-generated plot of Eq. 25-26 for a positively charged particle; the magnitude of V is plotted vertically. Note that the magnitude increases as $r \to 0$. In fact, according to Eq. 25-26, V is infinite at $r = 0$, although Fig. 25-8 shows a finite, smoothed-off value there.

Equation 25-26 also gives the electric potential *outside or on the external surface of* a spherically symmetric charge distribution. We can prove this by using one of the shell theorems of Sections 22-4 and 24-9 to replace the actual spherical charge distribution with an equal charge concentrated at its center. Then the derivation leading to Eq. 25-26 follows, provided we do not consider a point within the actual distribution.

PROBLEM-SOLVING TACTICS

Tactic 3: *Finding a Potential Difference*
To find the potential difference ΔV between any two points in the field of an isolated point charge, we can evaluate Eq. 25-26 at each point and then subtract the results. The value of ΔV will be the same for any choice of reference potential energy because that choice is eliminated by the subtraction.

25-6 Potential Due to a Group of Point Charges

We can find the net potential at a point due to a group of point charges with the help of the superposition principle. We calculate the potential resulting from each charge at the given point separately, using Eq. 25-26 with the sign of the charge included. Then we sum the potentials. For n charges, the net potential is

$$V = \sum_{i=1}^{n} V_i = \frac{1}{4\pi\varepsilon_0} \sum_{i=1}^{n} \frac{q_i}{r_i} \quad (n \text{ point charges}). \tag{25-27}$$

Here q_i is the value of the ith charge, and r_i is the radial distance of the given point from the ith charge. The sum in Eq. 25-27 is an *algebraic sum,* not a vector sum like the sum that would be used to calculate the electric field resulting from a group of point charges. Herein lies an important computational advantage of potential over electric field: It is a lot easier to sum several scalar quantities than to sum several vector quantities whose directions and components must be considered.

CHECKPOINT 4: The figure here shows three arrangements of two protons. Rank the arrangements according to the net electric potential produced at point P by the protons, greatest first.

Sample Problem 25-3

What is the electric potential at point P, located at the center of the square of point charges shown in Fig. 25-9*a*? The distance d is 1.3 m, and the charges are

$$q_1 = +12 \text{ nC}, \qquad q_3 = +31 \text{ nC},$$
$$q_2 = -24 \text{ nC}, \qquad q_4 = +17 \text{ nC}.$$

SOLUTION: The **Key Idea** here is that the electric potential V at P is the algebraic sum of the electric potentials contributed by the four point charges. (Because electric potential is a scalar, the orientations of the point charges do not matter.) Thus, from Eq. 25-27, we have

$$V = \sum_{i=1}^{4} V_i = \frac{1}{4\pi\varepsilon_0}\left(\frac{q_1}{r} + \frac{q_2}{r} + \frac{q_3}{r} + \frac{q_4}{r}\right).$$

The distance r is $d/\sqrt{2}$, which is 0.919 m, and the sum of the charges is

$$q_1 + q_2 + q_3 + q_4 = (12 - 24 + 31 + 17) \times 10^{-9} \text{ C}$$
$$= 36 \times 10^{-9} \text{ C}.$$

Thus, $$V = \frac{(8.99 \times 10^9 \text{ N}\cdot\text{m}^2/\text{C}^2)(36 \times 10^{-9} \text{ C})}{0.919 \text{ m}}$$
$$\approx 350 \text{ V}. \qquad \text{(Answer)}$$

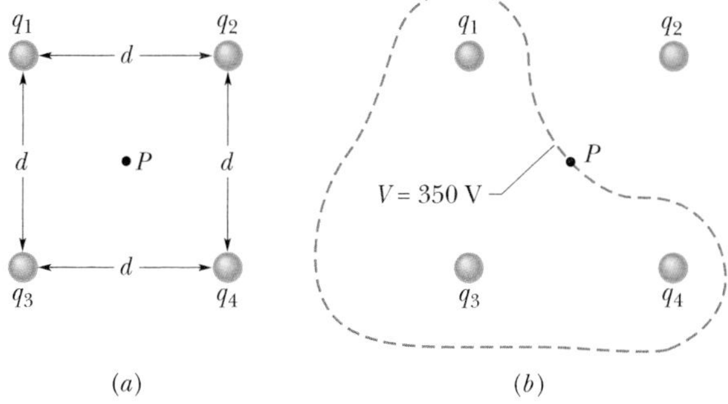

Fig. 25-9 Sample Problem 25-3. (*a*) Four point charges are held fixed at the corners of a square. (*b*) The closed curve is a cross section, in the plane of the figure, of the equipotential surface that contains point P. (The curve is only roughly drawn.)

Close to any of the three positive charges in Fig. 25-9*a*, the potential has very large positive values. Close to the single negative charge, the potential has very large negative values. Therefore, there must be points within the square that have the same intermediate potential as that at point P. The curve in Fig. 25-9*b* shows the intersection of the plane of the figure with the equipotential surface that contains point P. Any point along that curve has the same potential as point P.

Sample Problem 25-4

(a) In Fig. 25-10*a*, 12 electrons (of charge $-e$) are equally spaced and fixed around a circle of radius R. Relative to $V = 0$ at infinity, what are the electric potential and electric field at the center C of the circle due to these electrons?

SOLUTION: The **Key Idea** here is that the electric potential V at C is the algebraic sum of the electric potentials contributed by all the electrons. (Because electric potential is a scalar, the orientations of the electrons do not matter.) Because the electrons all have the same negative charge $-e$ and are all the same distance R from C, Eq. 25-27 gives us

$$V = -12 \frac{1}{4\pi\varepsilon_0} \frac{e}{R}. \quad \text{(Answer)} \quad (25\text{-}28)$$

For the electric field at C, the **Key Idea** is that electric field is a vector quantity and thus the orientation of the electrons *is* important. Because of the symmetry of the arrangement in Fig. 25-10*a*, the electric field vector at C due to any given electron is canceled by the field vector due to the electron that is diametrically opposite it. Thus, at C,

$$\vec{E} = 0. \quad \text{(Answer)}$$

(b) If the electrons are moved along the circle until they are nonuniformly spaced over a 120° arc (Fig. 25-10*b*), what then is the potential at C? How does the electric field at C change (if at all)?

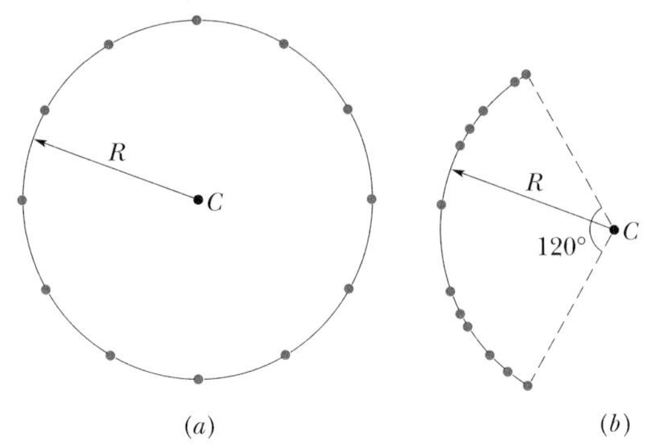

Fig. 25-10 Sample Problem 25-4. (*a*) Twelve electrons uniformly spaced around a circle. (*b*) Those electrons are now nonuniformly spaced along an arc of the original circle.

SOLUTION: The potential is still given by Eq. 25-28, because the distance between C and each electron is unchanged and orientation is irrelevant. The electric field is no longer zero, because the arrangement is no longer symmetric. There is now a net field that is directed toward the charge distribution.

25-7 Potential Due to an Electric Dipole

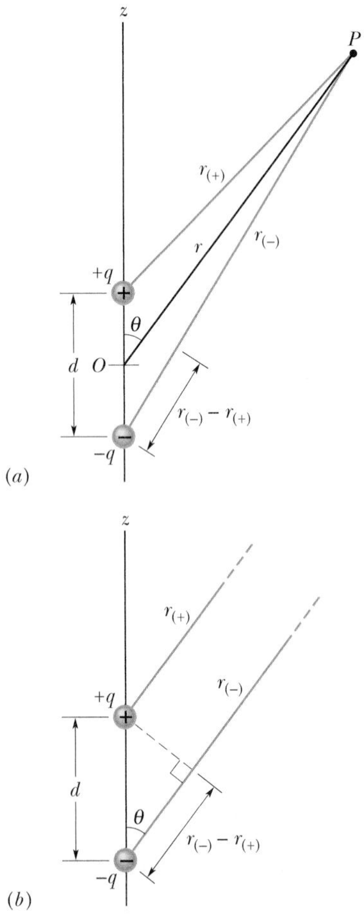

Now let us apply Eq. 25-27 to an electric dipole to find the potential at an arbitrary point P in Fig. 25-11*a*. At P, the positive point charge (at distance $r_{(+)}$) sets up potential $V_{(+)}$ and the negative point charge (at distance $r_{(-)}$) sets up potential $V_{(-)}$. Then the net potential at P is given by Eq. 25-27 as

$$V = \sum_{i=1}^{2} V_i = V_{(+)} + V_{(-)} = \frac{1}{4\pi\varepsilon_0}\left(\frac{q}{r_{(+)}} + \frac{-q}{r_{(-)}}\right)$$
$$= \frac{q}{4\pi\varepsilon_0} \frac{r_{(-)} - r_{(+)}}{r_{(-)} r_{(+)}}. \quad (25\text{-}29)$$

Naturally occurring dipoles—such as those possessed by many molecules—are quite small, so we are usually interested only in points that are relatively far from the dipole, such that $r \gg d$, where d is the distance between the charges. Under those conditions, the approximations that follow from Fig. 25-11*b* are

$$r_{(-)} - r_{(+)} \approx d \cos\theta \quad \text{and} \quad r_{(-)} r_{(+)} \approx r^2.$$

If we substitute these quantities into Eq. 25-29, we can approximate V to be

$$V = \frac{q}{4\pi\varepsilon_0} \frac{d\cos\theta}{r^2},$$

where θ is measured from the dipole axis as shown in Fig. 25-11*a*. We can now

Fig. 25-11 (*a*) Point P is a distance r from the midpoint O of a dipole. The line OP makes an angle θ with the dipole axis. (*b*) If P is far from the dipole, the lines of lengths $r_{(+)}$ and $r_{(-)}$ are approximately parallel to the line of length r, and the dashed black line is approximately perpendicular to the line of length $r_{(-)}$.

(a)

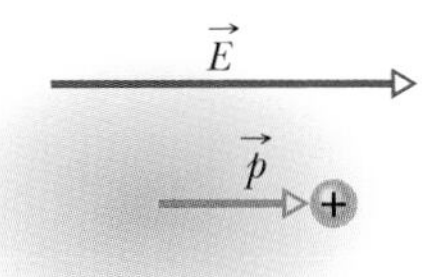

(b)

Fig. 25-12 (a) An atom, showing the positively charged nucleus (green) and the negatively charged electrons (gold shading). The centers of positive and negative charge coincide. (b) If the atom is placed in an external electric field $\vec{E}$, the electron orbits are distorted so that the centers of positive and negative charge no longer coincide. An induced dipole moment $\vec{p}$ appears. The distortion is greatly exaggerated here.

write V as

$$V = \frac{1}{4\pi\varepsilon_0}\frac{p\cos\theta}{r^2} \quad \text{(electric dipole)}, \tag{25-30}$$

in which p (= qd) is the magnitude of the electric dipole moment $\vec{p}$ defined in Section 23-5. The vector $\vec{p}$ is directed along the dipole axis, from the negative to the positive charge. (Thus, θ is measured from the direction of $\vec{p}$.)

CHECKPOINT 5: Suppose that three points are set at equal (large) distances r from the center of the dipole in Fig. 25-11: Point a is on the dipole axis above the positive charge, point b is on the axis below the negative charge, and point c is on a perpendicular bisector through the line connecting the two charges. Rank the points according to the electric potential of the dipole there, greatest (most positive) first.

Induced Dipole Moment

Many molecules such as water have *permanent* electric dipole moments. In other molecules (called *nonpolar molecules*) and in every isolated atom, the centers of the positive and negative charges coincide (Fig. 25-12*a*) and thus no dipole moment is set up. However, if we place an atom or a nonpolar molecule in an external electric field, the field distorts the electron orbits and separates the centers of positive and negative charge (Fig. 25-12*b*). Because the electrons are negatively charged, they tend to be shifted in a direction opposite the field. This shift sets up a dipole moment $\vec{p}$ that points in the direction of the field. This dipole moment is said to be *induced* by the field, and the atom or molecule is then said to be *polarized* by the field (it has a positive side and a negative side). When the field is removed, the induced dipole moment and the polarization disappear.

25-8 Potential Due to a Continuous Charge Distribution

When a charge distribution q is continuous (as on a uniformly charged thin rod or disk), we cannot use the summation of Eq. 25-27 to find the potential V at a point P. Instead, we must choose a differential element of charge dq, determine the potential dV at P due to dq, and then integrate over the entire charge distribution.

Let us again take the zero of potential to be at infinity. If we treat the element of charge dq as a point charge, then we can use Eq. 25-26 to express the potential dV at point P due to dq:

$$dV = \frac{1}{4\pi\varepsilon_0}\frac{dq}{r} \quad \text{(positive or negative } dq\text{)}. \tag{25-31}$$

Here r is the distance between P and dq. To find the total potential V at P, we integrate to sum the potentials due to all the charge elements:

$$V = \int dV = \frac{1}{4\pi\varepsilon_0}\int \frac{dq}{r}. \tag{25-32}$$

The integral must be taken over the entire charge distribution. Note that because the electric potential is a scalar, there are *no vector components* to consider in Eq. 25-32.

We now examine two continuous charge distributions, a line of charge and a charged disk.

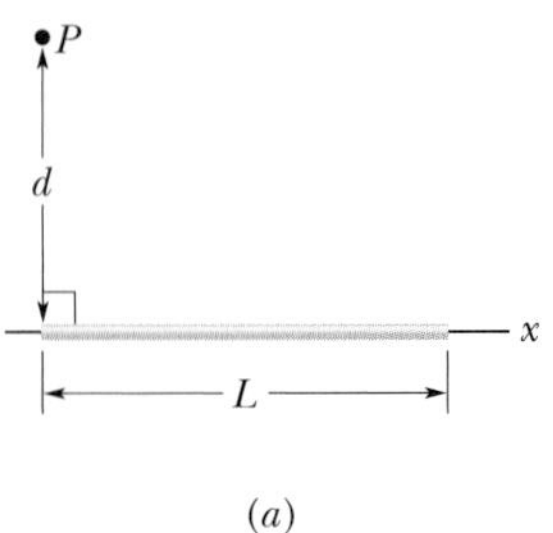

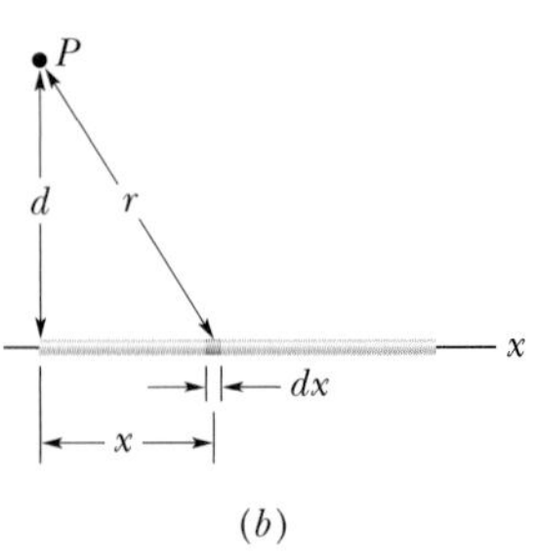

Fig. 25-13 (*a*) A thin, uniformly charged rod produces an electric potential V at point P. (*b*) An element of charge produces a differential potential dV at P.

Line of Charge

In Fig. 25-13*a*, a thin nonconducting rod of length L has a positive charge of uniform linear density λ. Let us determine the electric potential V due to the rod at point P, a perpendicular distance d from the left end of the rod.

We consider a differential element dx of the rod as shown in Fig. 25-13*b*. This (or any other) element of the rod has a differential charge of

$$dq = \lambda\, dx. \tag{25-33}$$

This element produces an electric potential dV at point P, which is a distance $r = (x^2 + d^2)^{1/2}$ from the element. Treating the element as a point charge, we can use Eq. 25-31 to write the potential dV as

$$dV = \frac{1}{4\pi\varepsilon_0}\frac{dq}{r} = \frac{1}{4\pi\varepsilon_0}\frac{\lambda\, dx}{(x^2 + d^2)^{1/2}}. \tag{25-34}$$

Since the charge on the rod is positive and we have taken $V = 0$ at infinity, we know from Section 25-5 that dV in Eq. 25-34 must be positive.

We now find the total potential V produced by the rod at point P by integrating Eq. 25-34 along the length of the rod, from $x = 0$ to $x = L$, using integral 17 in Appendix E. We find

$$\begin{aligned} V &= \int dV = \int_0^L \frac{1}{4\pi\varepsilon_0}\frac{\lambda}{(x^2 + d^2)^{1/2}}\, dx \\ &= \frac{\lambda}{4\pi\varepsilon_0}\int_0^L \frac{dx}{(x^2 + d^2)^{1/2}} \\ &= \frac{\lambda}{4\pi\varepsilon_0}\left[\ln\left(x + (x^2 + d^2)^{1/2}\right)\right]_0^L \\ &= \frac{\lambda}{4\pi\varepsilon_0}\left[\ln\left(L + (L^2 + d^2)^{1/2}\right) - \ln d\right]. \end{aligned}$$

We can simplify this result by using the general relation $\ln A - \ln B = \ln(A/B)$. We then find

$$V = \frac{\lambda}{4\pi\varepsilon_0}\ln\left[\frac{L + (L^2 + d^2)^{1/2}}{d}\right]. \tag{25-35}$$

Because V is the sum of positive values of dV, it should be positive—but does Eq. 25-35 give a positive V? Since the argument of the logarithm is greater than one, the logarithm is a positive number and V is indeed positive.

Charged Disk

In Section 23-7, we calculated the magnitude of the electric field at points on the central axis of a plastic disk of radius R that has a uniform charge density σ on one surface. Here we derive an expression for $V(z)$, the electric potential at any point on the central axis.

In Fig. 25-14, consider a differential element consisting of a flat ring of radius R' and radial width dR'. Its charge has magnitude

$$dq = \sigma(2\pi R')(dR'),$$

in which $(2\pi R')(dR')$ is the upper surface area of the ring. All parts of this charged element are the same distance r from point P on the disk's axis. With the aid of Fig. 25-14, we can use Eq. 25-31 to write the contribution of this ring to the electric

Fig. 25-14 A plastic disk of radius R, charged on its top surface to a uniform surface charge density σ. We wish to find the potential V at point P on the central axis of the disk.

potential at P as

$$dV = \frac{1}{4\pi\varepsilon_0}\frac{dq}{r} = \frac{1}{4\pi\varepsilon_0}\frac{\sigma(2\pi R')(dR')}{\sqrt{z^2 + R'^2}}. \tag{25-36}$$

We find the net potential at P by adding (via integration) the contributions of all the strips from $R' = 0$ to $R' = R$:

$$V = \int dV = \frac{\sigma}{2\varepsilon_0}\int_0^R \frac{R'\,dR'}{\sqrt{z^2 + R'^2}} = \frac{\sigma}{2\varepsilon_0}(\sqrt{z^2 + R^2} - z). \tag{25-37}$$

Note that the variable in the second integral of Eq. 25-37 is R' and not z, which remains constant while the integration over the surface of the disk is carried out. (Note also that, in evaluating the integral, we have assumed that $z \geq 0$.)

PROBLEM-SOLVING TACTICS

Tactic 4: *Signs of Trouble with Electric Potential*
When you calculate the potential V at some point P due to a line of charge or any other continuous charge configuration, the signs can cause you trouble. Here is a generic guide to sort out the signs.

If the charge is negative, should the symbols dq and λ represent negative quantities, or should you explicitly show the signs, using $-dq$ and $-\lambda$? You can do either as long as you remember what your notation means, so that when you get to the final step, you can correctly interpret the sign of V.

Another approach, which can be used when the entire charge distribution is of a single sign, is to let the symbols dq and λ represent magnitudes only. The result of the calculation will give you the magnitude of V at P. Then add a sign to V based on the sign of the charge. (If the zero potential is at infinity, positive charge gives a positive potential and negative charge gives a negative potential.)

If you happen to reverse the limits on the integral used to calculate a potential, you will obtain a negative value for V. The magnitude will be correct, but discard the minus sign. Then determine the proper sign for V from the sign of the charge. As an example, we would have obtained a minus sign in Eq. 25-35 if we had reversed the limits in the integral above that equation. We would then have discarded that minus sign and noted that the potential is positive because the charge producing it is positive.

25-9 Calculating the Field from the Potential

In Section 25-4, you saw how to find the potential at a point f if you know the electric field along a path from a reference point to point f. In this section, we propose to go the other way—that is, to find the electric field when we know the potential. As Fig. 25-3 shows, solving this problem graphically is easy: If we know the potential V at all points near an assembly of charges, we can draw in a family of equipotential surfaces. The electric field lines, sketched perpendicular to those surfaces, reveal the variation of $\vec{E}$. What we are seeking here is the mathematical equivalent of this graphical procedure.

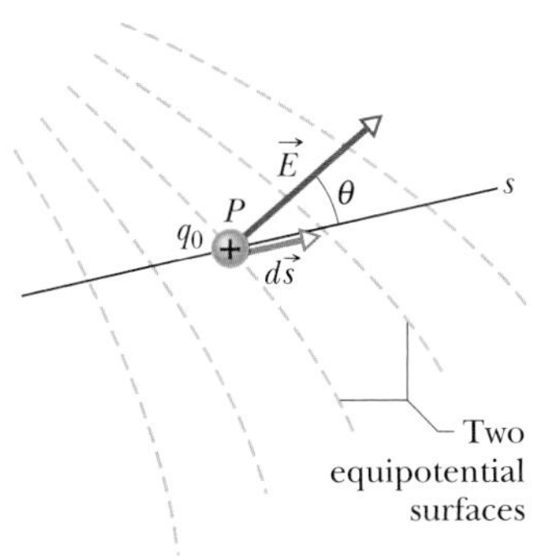

Fig. 25-15 A test charge q_0 moves a distance $d\vec{s}$ from one equipotential surface to another. (The separation between the surfaces has been exaggerated for clarity.) The displacement $d\vec{s}$ makes an angle θ with the direction of the electric field $\vec{E}$.

Figure 25-15 shows cross sections of a family of closely spaced equipotential surfaces, the potential difference between each pair of adjacent surfaces being dV. As the figure suggests, the field $\vec{E}$ at any point P is perpendicular to the equipotential surface through P.

Suppose that a positive test charge q_0 moves through a displacement $d\vec{s}$ from one equipotential surface to the adjacent surface. From Eq. 25-7, we see that the work the electric field does on the test charge during the move is $-q_0\,dV$. From Eq. 25-16 and Fig. 25-15, we see that the work done by the electric field may also be written as the scalar product $(q_0\vec{E}) \cdot d\vec{s}$, or $q_0E(\cos\theta)\,ds$. Equating these two expressions for the work yields

$$-q_0\,dV = q_0E(\cos\theta)\,ds, \tag{25-38}$$

or

$$E\cos\theta = -\frac{dV}{ds}. \tag{25-39}$$

Since $E \cos \theta$ is the component of $\vec{E}$ in the direction of $d\vec{s}$, Eq. 25-39 becomes

$$E_s = -\frac{\partial V}{\partial s}. \qquad (25\text{-}40)$$

We have added a subscript to E and switched to the partial derivative symbols to emphasize that Eq. 25-40 involves only the variation of V along a specified axis (here called the s axis) and only the component of $\vec{E}$ along that axis. In words, Eq. 25-40 (which is essentially the inverse of Eq. 25-18) states:

> The component of $\vec{E}$ in any direction is the negative of the rate of change of the electric potential with distance in that direction.

If we take the s axis to be, in turn, the x, y, and z axes, we find that the x, y, and z components of $\vec{E}$ at any point are

$$E_x = -\frac{\partial V}{\partial x}; \qquad E_y = -\frac{\partial V}{\partial y}; \qquad E_z = -\frac{\partial V}{\partial z}. \qquad (25\text{-}41)$$

Thus, if we know V for all points in the region around a charge distribution—that is, if we know the function $V(x, y, z)$—we can find the components of $\vec{E}$, and thus $\vec{E}$ itself, at any point by taking partial derivatives.

For the simple situation in which the electric field $\vec{E}$ is uniform, Eq. 25-40 becomes

$$E = -\frac{\Delta V}{\Delta s}, \qquad (25\text{-}42)$$

where s is perpendicular to the equipotential surfaces. The component of the electric field is zero in any direction parallel to the equipotential surfaces.

CHECKPOINT 6: The figure shows three pairs of parallel plates with the same separation, and the electric potential of each plate. The electric field between the plates is uniform and perpendicular to the plates. (a) Rank the pairs according to the magnitude of the electric field between the plates, greatest first. (b) For which pair is the electric field pointing rightward? (c) If an electron is released midway between the third pair of plates, does it remain there, move rightward at constant speed, move leftward at constant speed, accelerate rightward, or accelerate leftward?

−50 V +150 V
(1)

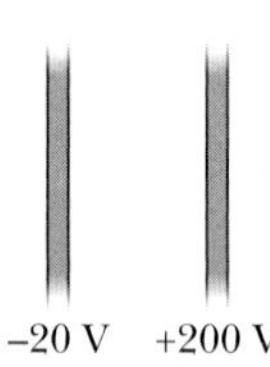

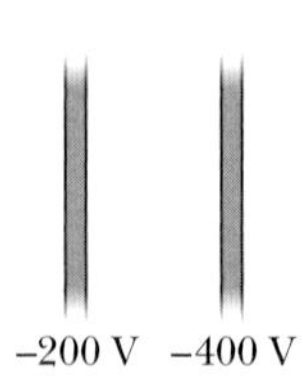

Sample Problem 25-5

The electric potential at any point on the central axis of a uniformly charged disk is given by Eq. 25-37,

$$V = \frac{\sigma}{2\varepsilon_0}\left(\sqrt{z^2 + R^2} - z\right).$$

Starting with this expression, derive an expression for the electric field at any point on the axis of the disk.

SOLUTION: We want the electric field $\vec{E}$ as a function of distance z along the axis of the disk. For any value of z, the direction of $\vec{E}$ must be along that axis because the disk has circular symmetry about that axis. Thus, we want the component E_z of $\vec{E}$ in the direction of z. Then the **Key Idea** is that this component is the negative of the rate of change of the electric potential with distance z. Thus, from the last of Eqs. 25-41, we can write

$$E_z = -\frac{\partial V}{\partial z} = -\frac{\sigma}{2\varepsilon_0}\frac{d}{dz}\left(\sqrt{z^2 + R^2} - z\right)$$
$$= \frac{\sigma}{2\varepsilon_0}\left(1 - \frac{z}{\sqrt{z^2 + R^2}}\right). \qquad \text{(Answer)}$$

This is the same expression that we derived in Section 23-7 by integration, using Coulomb's law.

25-10 Electric Potential Energy of a System of Point Charges

In Section 25-1, we discussed the electric potential energy of a charged particle as an electrostatic force does work on it. In that section, we assumed that the charges that produced the force were fixed in place, so that neither the force nor the corresponding electric field could be influenced by the presence of the test charge. In this section we can take a broader view, to find the electric potential energy of a *system* of charges due to the electric field produced *by* those same charges.

For a simple example, suppose you push together two bodies that have charges of the same electrical sign. The work that you must do is stored as electric potential energy in the two-body system (provided the kinetic energy of the bodies does not change). If you later release the charges, you can recover this stored energy, in whole or in part, as kinetic energy of the charged bodies as they rush away from each other.

We define the electric potential energy *of a system of point charges,* held in fixed positions by forces not specified, as follows:

> The electric potential energy of a system of fixed point charges is equal to the work that must be done by an external agent to assemble the system, bringing each charge in from an infinite distance.

We assume that the charges are stationary both in their initial infinitely distant positions and in their final assembled configuration.

Figure 25-16 shows two point charges q_1 and q_2, separated by a distance r. To find the electric potential energy of this two-charge system, we must mentally build the system, starting with both charges infinitely far away and at rest. When we bring q_1 in from infinity and put it in place we do no work, because no electrostatic force acts on q_1. However, when we next bring q_2 in from infinity and put it in place, we must do work, because q_1 exerts an electrostatic force on q_2 during the move.

We can calculate that work with Eq. 25-8 by dropping the minus sign (so that the equation gives the work *we* do rather than the field's work) and substituting q_2 for the general charge q. Our work is then equal to q_2V, where V is the potential that has been set up by q_1 at the point where we put q_2. From Eq. 25-26, that potential is

$$V = \frac{1}{4\pi\varepsilon_0}\frac{q_1}{r}.$$

Thus, from our definition, the electric potential energy of the pair of point charges of Fig. 25-16 is

$$U = W = q_2V = \frac{1}{4\pi\varepsilon_0}\frac{q_1q_2}{r}. \tag{25-43}$$

If the charges have the same sign, we have to do positive work to push them together against their mutual repulsion. Hence, as Eq. 25-43 shows, the potential energy of the system is then positive. If the charges have opposite signs, we have to do negative work against their mutual attraction to bring them together if they are to be stationary. The potential energy of the system is then negative. Sample Problem 25-6 shows how to extend this process to more than two charges.

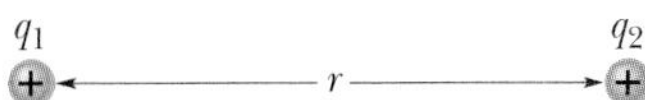

Fig. 25-16 Two charges held a fixed distance r apart.

Sample Problem 25-6

Figure 25-17 shows three point charges held in fixed positions by forces that are not shown. What is the electric potential energy U of this system of charges? Assume that $d = 12$ cm and that

$$q_1 = +q, \quad q_2 = -4q, \quad \text{and} \quad q_3 = +2q,$$

in which $q = 150$ nC.

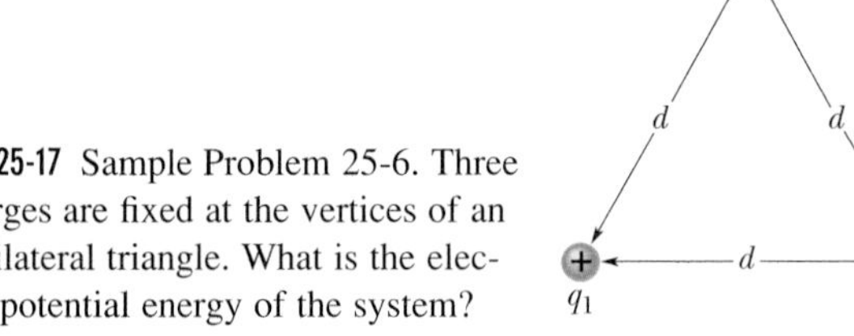

Fig. 25-17 Sample Problem 25-6. Three charges are fixed at the vertices of an equilateral triangle. What is the electric potential energy of the system?

SOLUTION: The **Key Idea** here is that the potential energy U of the system is equal to the work we must do to assemble the system, bringing in each charge from an infinite distance. Therefore, let's mentally build the system of Fig. 25-17, starting with one of the point charges, say q_1, in place and the others at infinity. Then we bring another one, say q_2, in from infinity and put it in place. From Eq. 25-43 with d substituted for r, the potential energy U_{12} associated with the pair of point charges q_1 and q_2 is

$$U_{12} = \frac{1}{4\pi\varepsilon_0}\frac{q_1 q_2}{d}.$$

We then bring the last point charge q_3 in from infinity and put it in place. The work that we must do in this last step is equal to the sum of the work we must do to bring q_3 near q_1 and the work we must do to bring it near q_2. From Eq. 25-43, with d substituted for r, that sum is

$$W_{13} + W_{23} = U_{13} + U_{23} = \frac{1}{4\pi\varepsilon_0}\frac{q_1 q_3}{d} + \frac{1}{4\pi\varepsilon_0}\frac{q_2 q_3}{d}.$$

The total potential energy U of the three-charge system is the sum of the potential energies associated with the three pairs of charges. This sum (which is actually independent of the order in which the charges are brought together) is

$$\begin{aligned} U &= U_{12} + U_{13} + U_{23} \\ &= \frac{1}{4\pi\varepsilon_0}\left(\frac{(+q)(-4q)}{d} + \frac{(+q)(+2q)}{d} + \frac{(-4q)(+2q)}{d}\right) \\ &= -\frac{10q^2}{4\pi\varepsilon_0 d} \\ &= -\frac{(8.99 \times 10^9\ \text{N}\cdot\text{m}^2/\text{C}^2)(10)(150 \times 10^{-9}\ \text{C})^2}{0.12\ \text{m}} \\ &= -1.7 \times 10^{-2}\ \text{J} = -17\ \text{mJ}. \end{aligned} \qquad \text{(Answer)}$$

The negative potential energy means that negative work would have to be done to assemble this structure, starting with the three charges infinitely separated and at rest. Put another way, an external agent would have to do 17 mJ of work to disassemble the structure completely, ending with the three charges infinitely far apart.

25-11 Potential of a Charged Isolated Conductor

In Section 24-6, we concluded that $\vec{E} = 0$ for all points inside an isolated conductor. We then used Gauss' law to prove that an excess charge placed on an isolated conductor lies entirely on its surface. (This is true even if the conductor has an empty internal cavity.) Here we use the first of these facts to prove an extension of the second:

> An excess charge placed on an isolated conductor will distribute itself on the surface of that conductor so that all points of the conductor—whether on the surface or inside—come to the same potential. This is true even if the conductor has an internal cavity and even if that cavity contains a net charge.

Our proof follows directly from Eq. 25-18, which is

$$V_f - V_i = -\int_i^f \vec{E} \cdot d\vec{s}.$$

Since $\vec{E} = 0$ for all points within a conductor, it follows directly that $V_f = V_i$ for all possible pairs of points i and f in the conductor.

Figure 25-18a is a plot of potential against radial distance r from the center for an isolated spherical conducting shell of 1.0 m radius, having a charge of 1.0 μC.

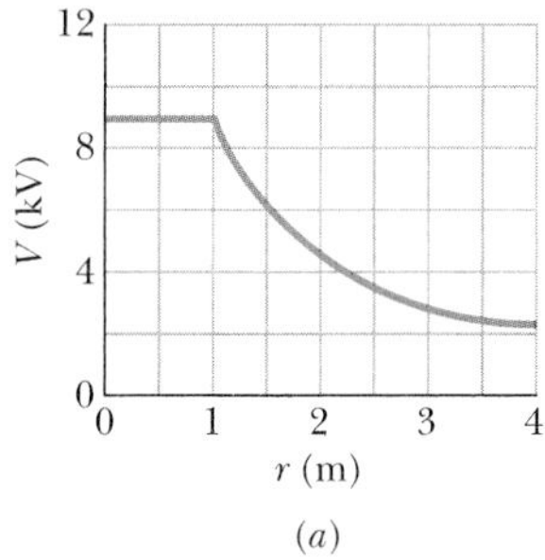

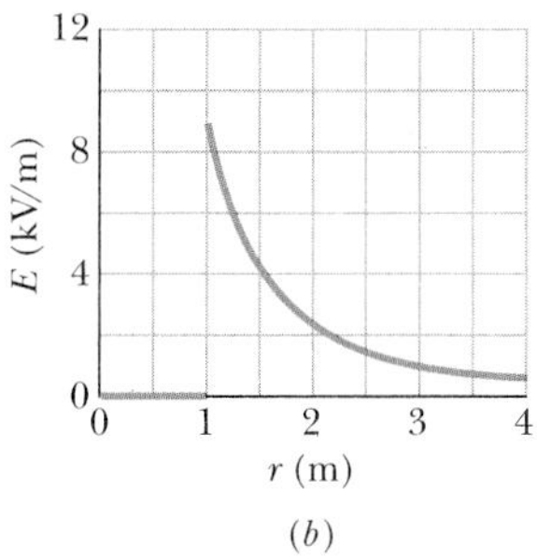

Fig. 25-18 (a) A plot of $V(r)$ both inside and outside a charged spherical shell of radius 1.0 m. (b) A plot of $E(r)$ for the same shell.

For points outside the shell, we can calculate $V(r)$ from Eq. 25-26 because the charge q behaves for such external points as if it were concentrated at the center of the shell. That equation holds right up to the surface of the shell. Now let us push a small test charge through the shell—assuming a small hole exists—to its center. No extra work is needed to do this because no net electric force acts on the test charge once it is inside the shell. Thus, the potential at all points inside the shell has the same value as that on the surface, as Fig. 25-18a shows.

Figure 25-18b shows the variation of electric field with radial distance for the same shell. Note that $E = 0$ everywhere inside the shell. The curves of Fig. 25-18b can be derived from the curve of Fig. 25-18a by differentiating with respect to r, using Eq. 25-40 (recall that the derivative of any constant is zero). The curve of Fig. 25-18a can be derived from the curves of Fig. 25-18b by integrating with respect to r, using Eq. 25-19.

On nonspherical conductors, a surface charge does not distribute itself uniformly over the surface of the conductor. At sharp points or edges, the surface charge density—and thus the external electric field, which is proportional to it—may reach very high values. The air around such sharp points may become ionized, producing the corona discharge that golfers and mountaineers see on the tips of bushes, golf clubs, and rock hammers when thunderstorms threaten. Such corona discharges, like hair that stands on end, are often the precursors of lightning strikes. In such circumstances, it is wise to enclose yourself in a cavity inside a conducting shell, where the electric field is guaranteed to be zero. A car (unless it is a convertible or made with a plastic body) is almost ideal (Fig. 25-19).

If an isolated conductor is placed in an *external electric field,* as in Fig. 25-20, all points of the conductor still come to a single potential regardless of whether the conductor has an excess charge. The free conduction electrons distribute themselves on the surface in such a way that the electric field they produce at interior points cancels the external electric field that would otherwise be there. Furthermore, the electron distribution causes the net electric field at all points on the surface to be perpendicular to the surface. If the conductor in Fig. 25-20 could be somehow removed, leaving the surface charges frozen in place, the pattern of the electric field would remain absolutely unchanged, for both exterior and interior points.

Fig. 25-19 A large spark jumps to a car's body and then exits by moving across the insulating left front tire (note the flash there), leaving the person inside unharmed.

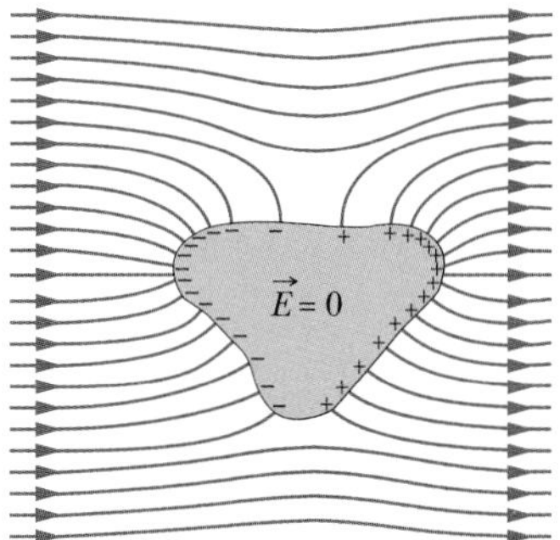

Fig. 25-20 An uncharged conductor is suspended in an external electric field. The free electrons in the conductor distribute themselves on the surface as shown, so as to reduce the net electric field inside the conductor to zero and make the net field at the surface perpendicular to the surface.

REVIEW & SUMMARY

Electric Potential Energy The change ΔU in the electric potential energy U of a point charge as the charge moves from an initial point i to a final point f in an electric field is

$$\Delta U = U_f - U_i = -W, \tag{25-1}$$

where W is the work done by the electrostatic force (due to the electric field) on the point charge during the move from i to f. If the potential energy is defined to be zero at infinity, the **electric potential energy** U of the point charge at a particular point is

$$U = -W_\infty. \tag{25-2}$$

Here W_∞ is the work done by the electrostatic force on the point charge as the charge moves from infinity to the particular point.

Electric Potential Difference and Electric Potential We define the **potential difference** ΔV between two points i and f in an electric field as

$$\Delta V = V_f - V_i = -\frac{W}{q}, \tag{25-7}$$

where q is the charge of a particle on which work is done by the field. The **potential** at a point is

$$V = -\frac{W_\infty}{q}. \tag{25-8}$$

The SI unit of potential is the *volt:* 1 volt = 1 joule per coulomb.

Potential and potential difference can also be written in terms of the electric potential energy U of a particle of charge q in an electric field:

$$V = \frac{U}{q}, \tag{25-5}$$

$$\Delta V = V_f - V_i = \frac{U_f}{q} - \frac{U_i}{q} = \frac{\Delta U}{q}. \tag{25-6}$$

Equipotential Surfaces The points on an **equipotential surface** all have the same electric potential. The work done on a test charge in moving it from one such surface to another is independent of the locations of the initial and final points on these surfaces and of the path that joins the points. The electric field $\vec{E}$ is always directed perpendicularly to corresponding equipotential surfaces.

Finding V from $\vec{E}$ The electric potential difference between two points i and f is

$$V_f - V_i = -\int_i^f \vec{E} \cdot d\vec{s}, \tag{25-18}$$

where the integral is taken over any path connecting the points. If we choose $V_i = 0$ we have, for the potential at a particular point,

$$V = -\int_i^f \vec{E} \cdot d\vec{s}. \tag{25-19}$$

Potential Due to Point Charges The electric potential due to a single point charge at a distance r from that point charge is

$$V = \frac{1}{4\pi\varepsilon_0}\frac{q}{r}. \tag{25-26}$$

V has the same sign as q. The potential due to a collection of point charges is

$$V = \sum_{i=1}^{n} V_i = \frac{1}{4\pi\varepsilon_0}\sum_{i=1}^{n}\frac{q_i}{r_i}. \tag{25-27}$$

Potential Due to an Electric Dipole At a distance r from an electric dipole with dipole moment magnitude $p = qd$, the electric potential of the dipole is

$$V = \frac{1}{4\pi\varepsilon_0}\frac{p\cos\theta}{r^2} \tag{25-30}$$

for $r \gg d$; the angle θ is defined in Fig. 25-11.

Potential Due to a Continuous Charge Distribution For a continuous distribution of charge, Eq. 25-27 becomes

$$V = \frac{1}{4\pi\varepsilon_0}\int\frac{dq}{r}, \tag{25-32}$$

in which the integral is taken over the entire distribution.

Calculating $\vec{E}$ from V The component of $\vec{E}$ in any direction is the negative of the rate of change of the potential with distance in that direction:

$$E_s = -\frac{\partial V}{\partial s}. \tag{25-40}$$

The x, y, and z components of $\vec{E}$ may be found from

$$E_x = -\frac{\partial V}{\partial x}; \quad E_y = -\frac{\partial V}{\partial y}; \quad E_z = -\frac{\partial V}{\partial z}. \tag{25-41}$$

When $\vec{E}$ is uniform, Eq. 25-40 reduces to

$$E = -\frac{\Delta V}{\Delta s}, \tag{25-42}$$

where s is perpendicular to the equipotential surfaces. The electric field is zero in the direction parallel to an equipotential surface.

Electric Potential Energy of a System of Point Charges The electric potential energy of a system of point charges is equal to the work needed to assemble the system with the charges initially at rest and infinitely distant from each other. For two charges at separation r,

$$U = W = \frac{1}{4\pi\varepsilon_0}\frac{q_1 q_2}{r}. \tag{25-43}$$

Potential of a Charged Conductor An excess charge placed on a conductor will, in the equilibrium state, be located entirely on the outer surface of the conductor. The charge will distribute itself so that the entire conductor, including interior points, is at a uniform potential.

QUESTIONS

1. Figure 25-21 shows three paths along which we can move positively charged sphere A closer to positively charged sphere B, which is fixed in place. (a) Would sphere A be moved to a higher or lower electric potential? Is the work done (b) by our force and (c) by the electric field (due to the second sphere) positive, negative, or zero? (d) Rank the paths according to the work our force does, greatest first.

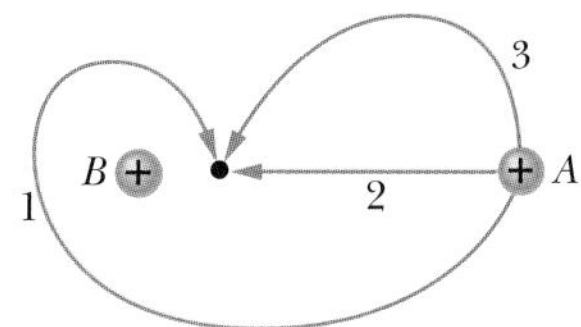

Fig. 25-21 Question 1.

2. Figure 25-22 shows four pairs of charged particles. Let $V = 0$ at infinity. For which pairs is there another point of zero net electric potential *on the axis shown,* (a) between the particles and (b) to their right? (c) Where such a zero potential point exists, is the net electric field $\vec{E}$ due to the particles equal to zero? (d) For each pair, are there points off the axis (other than at infinity, of course) where $V = 0$?

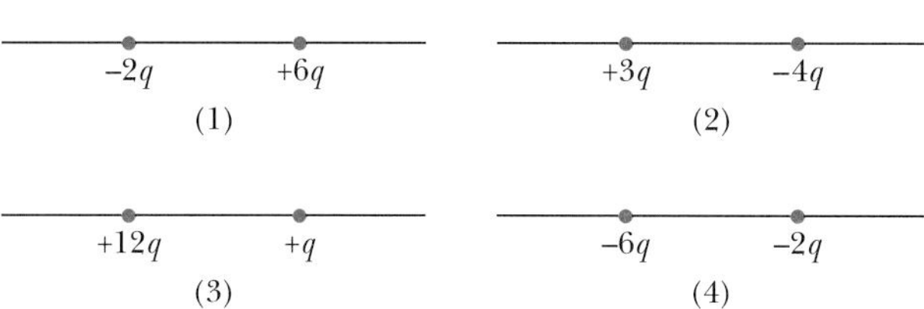

Fig. 25-22 Questions 2 and 8.

3. Figure 25-23 shows a square array of charged particles, with distance d between adjacent particles. What is the electric potential at point P at the center of the square if the electric potential is zero at infinity?

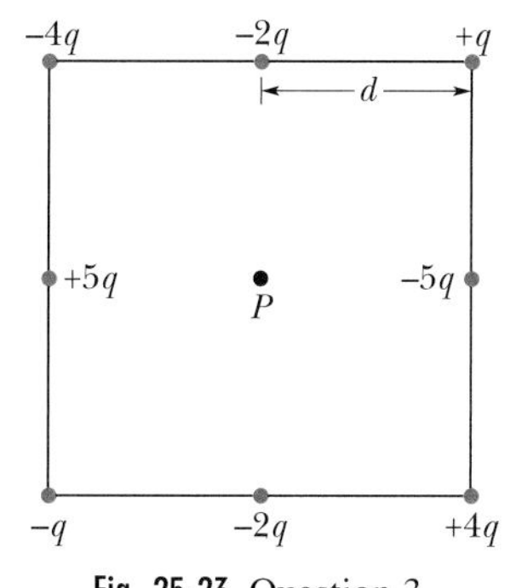

Fig. 25-23 Question 3.

4. Figure 25-24 shows four arrangements of charged particles, all the same distance from the origin. Rank the situations according to the net electric potential at the origin, most positive first. Take the potential to be zero at infinity.

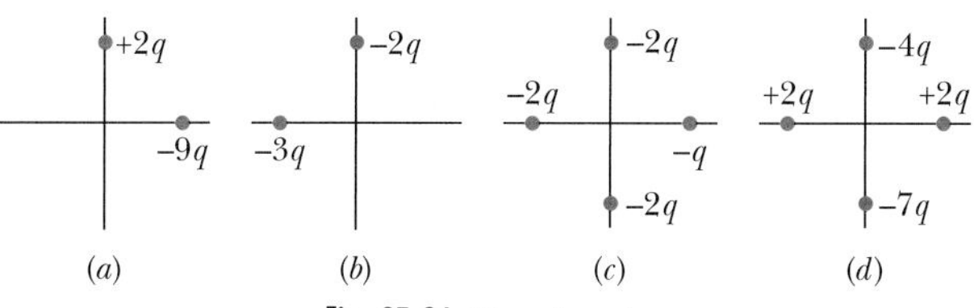

Fig. 25-24 Question 4.

5. (a) In Fig. 25-25a, what is the potential at point P due to charge Q at distance R from P? Set $V = 0$ at infinity. (b) In Fig. 25-25b, the same charge Q has been spread uniformly over a circular arc of radius R and central angle 40°. What is the potential at point P, the center of curvature of the arc? (c) In Fig. 25-25c, the same charge Q has been spread uniformly over a circle of radius R. What is the potential at point P, the center of the circle? (d) Rank the three situations according to the magnitude of the electric field that is set up at P, greatest first.

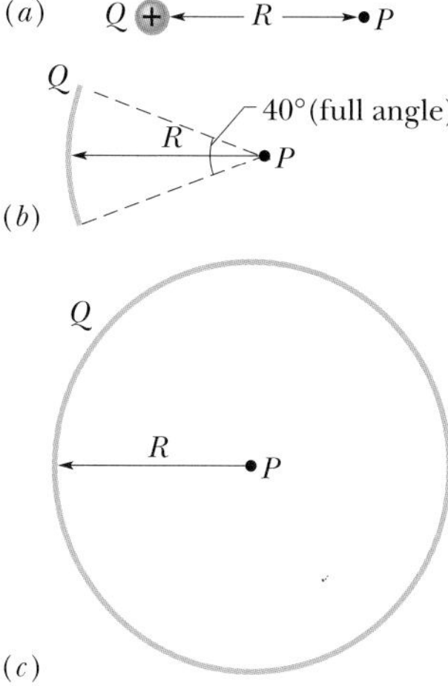

Fig. 25-25 Question 5.

6. Figure 25-26 shows three sets of cross sections of equipotential surfaces; all three cover the same size region of space. (a) Rank the arrangements according to the magnitude of the electric field present in the region, greatest first. (b) In which is the electric field directed down the page?

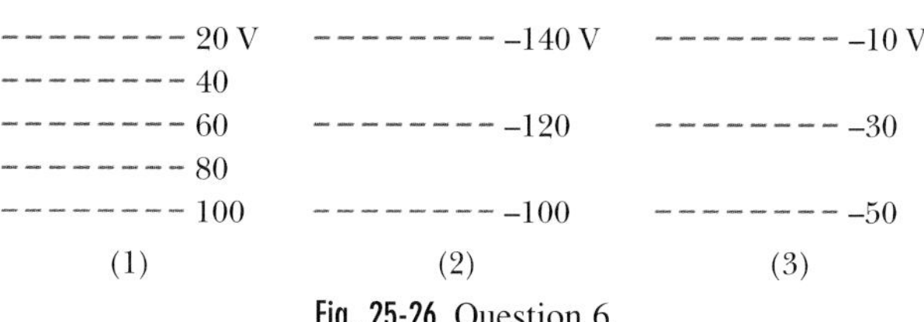

Fig. 25-26 Question 6.

7. Figure 25-27 gives the electric potential V as a function of x. (a) Rank the five regions according to the magnitude of the x component of the electric field within them, greatest first. What is the direction of the field along the x axis in (b) region 2 and (c) region 4?

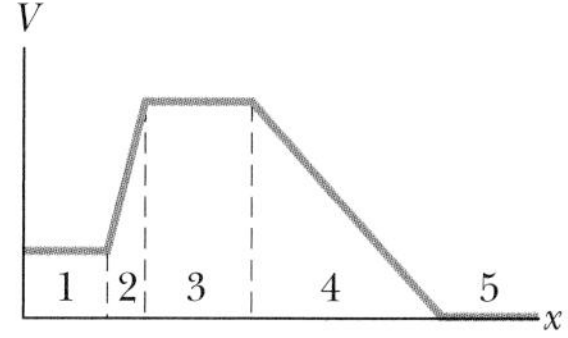

Fig. 25-27 Question 7.

8. Figure 25-22 shows four pairs of charged particles with identical separations. (a) Rank the pairs according to their electric potential energy, greatest (most positive) first. (b) For each pair, if the separation between the particles is increased, does the potential energy of the pair increase or decrease?

9. Figure 25-28 shows a system of three charged particles. If you move the particle of charge $+q$ from point A to point D, are the following positive, negative, or zero: (a) the change in the electric potential energy of the three-particle system, (b) the work done by the net electrostatic force on the particle you moved, and (c) the work done by your force? (d) What are the answers to (a) through (c) if, instead, the move is from point B to point C?

10. In the situation of Question 9, is the work done by your force positive, negative, or zero if the move is (a) from A to B, (b) from A to C, and (c) from B to D? (d) Rank those moves according to the magnitude of the work done by your force, greatest first.

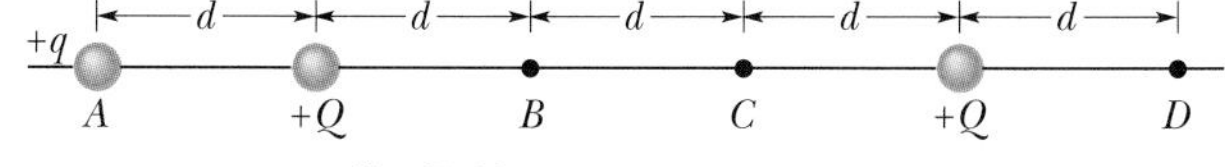

Fig. 25-28 Questions 9 and 10.

EXERCISES & PROBLEMS

ssm Solution is in the Student Solutions Manual.
www Solution is available on the World Wide Web at: http://www.wiley.com/college/hrw
ilw Solution is available on the Interactive LearningWare.

SEC. 25-2 Electric Potential

1E. A particular 12 V car battery can send a total charge of 84 A · h (ampere-hours) through a circuit, from one terminal to the other. (a) How many coulombs of charge does this represent? (*Hint:* See Eq. 22-3.) (b) If this entire charge undergoes a potential difference of 12 V, how much energy is involved? ssm

2E. The electric potential difference between the ground and a cloud in a particular thunderstorm is 1.2×10^9 V. What is the magnitude of the change in the electric potential energy (in multiples of the electron-volt) of an electron that moves between the ground and the cloud?

3P. In a given lightning flash, the potential difference between a cloud and the ground is 1.0×10^9 V and the quantity of charge transferred is 30 C. (a) What is the decrease in energy of that transferred charge? (b) If all that energy could be used to accelerate a 1000 kg automobile from rest, what would be the automobile's final speed? (c) If the energy could be used to melt ice, how much ice would it melt at 0°C? The heat of fusion of ice is 3.33×10^5 J/kg. ssm

SEC. 25-4 Calculating the Potential from the Field

4E. When an electron moves from A to B along an electric field line in Fig. 25-29, the electric field does 3.94×10^{-19} J of work on it. What are the electric potential differences (a) $V_B - V_A$, (b) $V_C - V_A$, and (c) $V_C - V_B$?

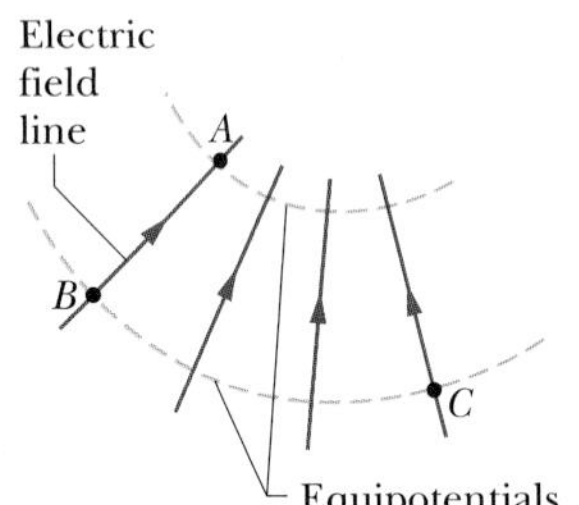

Fig. 25-29 Exercise 4.

5E. An infinite nonconducting sheet has a surface charge density $\sigma = 0.10\ \mu\text{C/m}^2$ on one side. How far apart are equipotential surfaces whose potentials differ by 50 V? ssm

6E. Two large, parallel, conducting plates are 12 cm apart and have charges of equal magnitude and opposite sign on their facing surfaces. An electrostatic force of 3.9×10^{-15} N acts on an electron placed anywhere between the two plates. (Neglect fringing.) (a) Find the electric field at the position of the electron. (b) What is the potential difference between the plates?

7P. A Geiger counter has a metal cylinder 2.00 cm in diameter along whose axis is stretched a wire 1.30×10^{-4} cm in diameter. If the potential difference between the wire and the cylinder is 850 V, what is the electric field at the surface of (a) the wire and (b) the cylinder? (*Hint:* Use the result of Problem 23 of Chapter 24.) ssm

8P. The electric field inside a nonconducting sphere of radius R, with charge spread uniformly throughout its volume, is radially directed and has magnitude

$$E(r) = \frac{qr}{4\pi\varepsilon_0 R^3}.$$

Here q (positive or negative) is the total charge within the sphere, and r is the distance from the sphere's center. (a) Taking $V = 0$ at the center of the sphere, find the electric potential $V(r)$ inside the sphere. (b) What is the difference in electric potential between a point on the surface and the sphere's center? (c) If q is positive, which of those two points is at the higher potential?

9P*. A charge q is distributed uniformly throughout a spherical volume of radius R. (a) Setting $V = 0$ at infinity, show that the potential at a distance r from the center, where $r < R$, is given by

$$V = \frac{q(3R^2 - r^2)}{8\pi\varepsilon_0 R^3}.$$

(*Hint:* See Section 24-9.) (b) Why does this result differ from that in (a) of Problem 8? (c) What is the potential difference between a point on the surface and the sphere's center? (d) Why doesn't this result differ from that of (b) of Problem 8? ssm

10P. Figure 25-30 shows, edge-on, an infinite nonconducting sheet with positive surface charge density σ on one side. (a) Use Eq. 25-18 and Eq. 24-13 to show that the electric potential of an infinite sheet of charge can be written $V = V_0 - (\sigma/2\varepsilon_0)z$, where V_0 is the electric potential at the surface of the sheet and z is the perpendicular distance from the sheet. (b) How much work is done by the electric field of the sheet as a small positive test charge q_0 is moved from an initial position on the sheet to a final position located a distance z from the sheet?

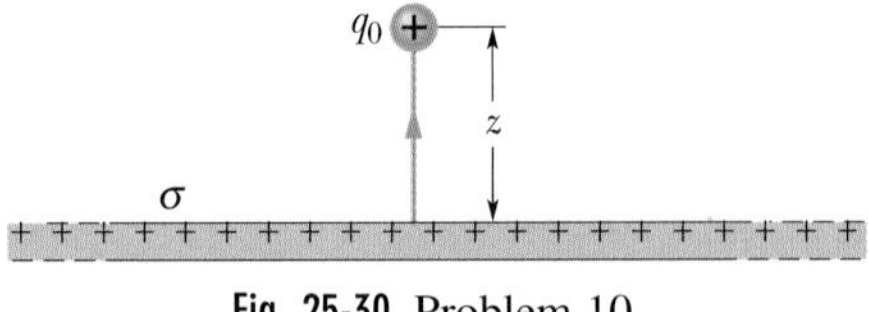

Fig. 25-30 Problem 10.

11P*. A thick spherical shell of charge Q and uniform volume charge density ρ is bounded by radii r_1 and r_2, where $r_2 > r_1$. With $V = 0$ at infinity, find the electric potential V as a function of the distance r from the center of the distribution, considering the regions (a) $r > r_2$, (b) $r_2 > r > r_1$, and (c) $r < r_1$. (d) Do these solutions agree at $r = r_2$ and $r = r_1$? (*Hint:* See Section 24-9.) ssm

SEC. 25-6 Potential Due to a Group of Point Charges

12E. As a space shuttle moves through the dilute ionized gas of Earth's ionosphere, its potential is typically changed by -1.0 V during one revolution. By assuming that the shuttle is a sphere of radius 10 m, estimate the amount of charge it collects.

13E. Consider a point charge $q = 1.0\ \mu$C, point A at distance $d_1 = 2.0$ m from q, and point B at distance $d_2 = 1.0$ m. (a) If these points are diametrically opposite each other, as in Fig. 25-31a, what is the electric potential difference $V_A - V_B$? (b) What is that electric potential difference if points A and B are located as in Fig. 25-31b?

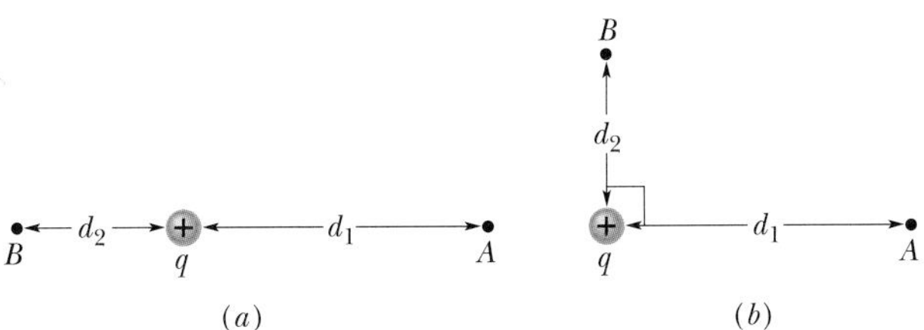

Fig. 25-31 Exercise 13.

14E. Figure 25-32 shows two charged particles on an axis. Sketch the electric field lines and the equipotential surfaces in the plane of the page for (a) $q_1 = +q$ and $q_2 = +2q$ and (b) $q_1 = +q$ and $q_2 = -3q$.

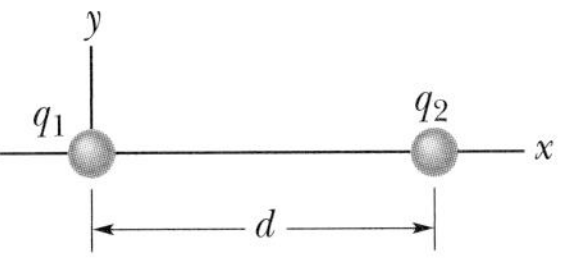

Fig. 25-32 Exercises 14, 15, and 16.

15E. In Fig. 25-32, set $V = 0$ at infinity and let the particles have charges $q_1 = +q$ and $q_2 = -3q$. Then locate (in terms of the separation distance d) any point on the x axis (other than at infinity) at which the net potential due to the two particles is zero.

16E. Two particles, of charges q_1 and q_2, are separated by distance d in Fig. 25-32. The net electric field of the particles is zero at $x = d/4$. With $V = 0$ at infinity, locate (in terms of d) any point on the x axis (other than at infinity) at which the electric potential due to the two particles is zero.

17P. A spherical drop of water carrying a charge of 30 pC has a potential of 500 V at its surface (with $V = 0$ at infinity). (a) What is the radius of the drop? (b) If two such drops of the same charge and radius combine to form a single spherical drop, what is the potential at the surface of the new drop? ssm ilw www

18P. What are (a) the charge and (b) the charge density on the surface of a conducting sphere of radius 0.15 m whose potential is 200 V (with $V = 0$ at infinity)?

19P. An electric field of approximately 100 V/m is often observed near the surface of Earth. If this were the field over the entire surface, what would be the electric potential of a point on the surface? (Set $V = 0$ at infinity.) ssm

20P. In Fig. 25-33, point P is at the center of the rectangle. With $V = 0$ at infinity, what is the net electric potential at P due to the six charged particles?

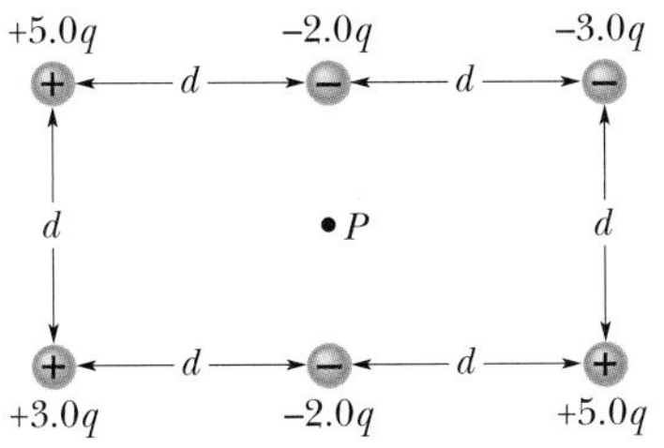

Fig. 25-33 Problem 20.

21P. In Fig. 25-34, what is the net potential at point P due to the four point charges, if $V = 0$ at infinity? ssm

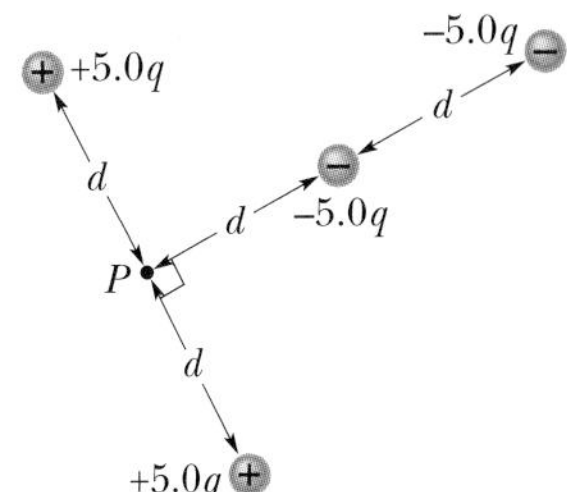

Fig. 25-34 Problem 21.

SEC. 25-7 Potential Due to an Electric Dipole

22E. The ammonia molecule NH_3 has a permanent electric dipole moment equal to 1.47 D, where 1 D = 1 debye unit = 3.34×10^{-30} C·m. Calculate the electric potential due to an ammonia molecule at a point 52.0 nm away along the axis of the dipole. (Set $V = 0$ at infinity.) ilw

23P. Figure 25-35 shows three charged particles located on a horizontal axis. For points (such as P) on the axis with $r \gg d$, show that the electric potential $V(r)$ is given by

$$V = \frac{1}{4\pi\varepsilon_0}\frac{q}{r}\left(1 + \frac{2d}{r}\right).$$

(*Hint:* The charge configuration can be viewed as the sum of an isolated charge and a dipole.) ssm www

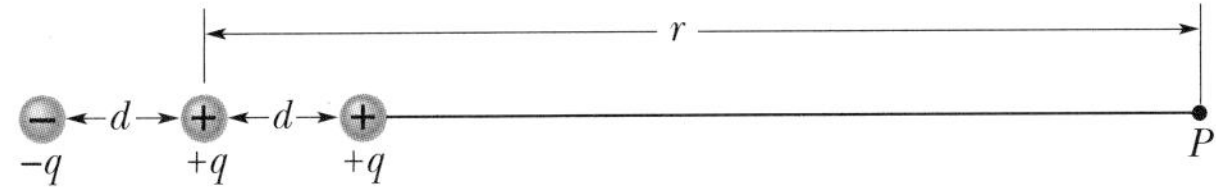

Fig. 25-35 Problem 23.

SEC. 25-8 Potential Due to a Continuous Charge Distribution

24E. (a) Figure 25-36a shows a positively charged plastic rod of length L and uniform linear charge density λ. Setting $V = 0$ at infinity and considering Fig. 25-13 and Eq. 25-35, find the electric potential at point P without written calculation. (b) Figure 25-36b shows an identical rod, except that it is split in half and the right half is negatively charged; the left and right halves have the same magnitude λ of uniform linear charge density. With V still zero at infinity, what is the electric potential at point P in Fig. 25-36b?

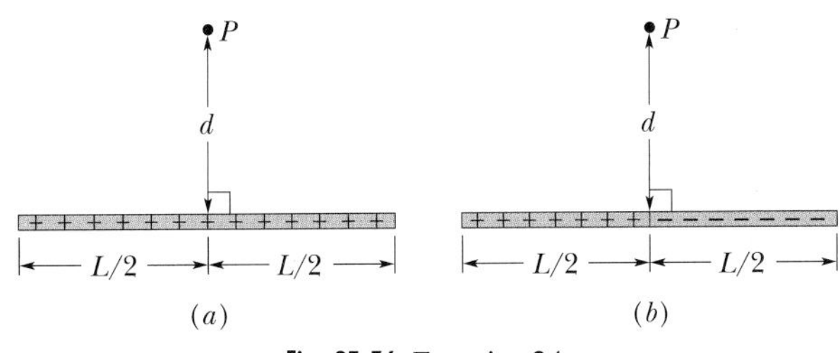

Fig. 25-36 Exercise 24.

25E. A plastic rod has been formed into a circle of radius R. It has a positive charge $+Q$ uniformly distributed along one-quarter of its circumference and a negative charge of $-6Q$ uniformly distributed along the rest of the circumference (Fig. 25-37). With $V = 0$ at infinity, what is the electric potential (a) at the center C of the circle and (b) at point P, which is on the central axis of the circle at distance z from the center? ssm

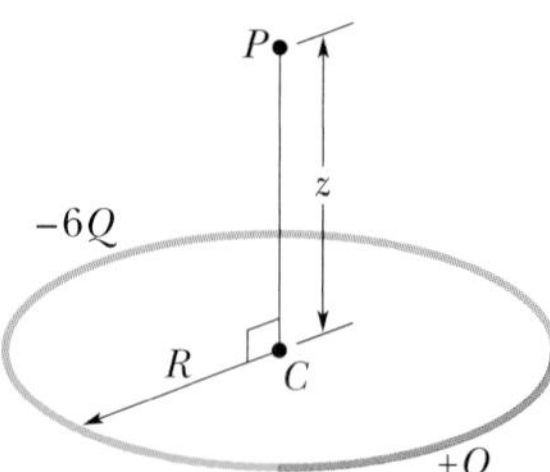

Fig. 25-37 Exercise 25.

26E. In Fig. 25-38, a plastic rod having a uniformly distributed charge $-Q$ has been bent into a circular arc of radius R and central angle 120°. With $V = 0$ at infinity, what is the electric potential at P, the center of curvature of the rod?

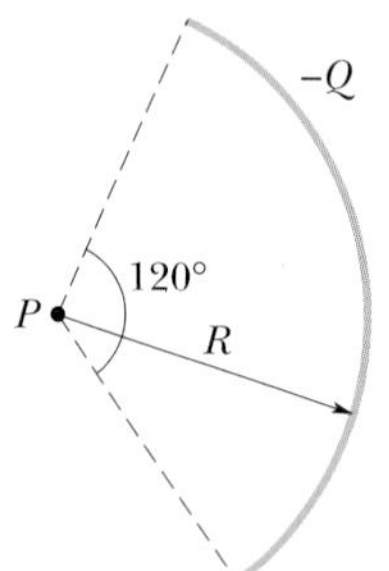

Fig. 25-38 Exercise 26.

27E. A plastic disk is charged on one side with a uniform surface charge density σ, and then three quadrants of the disk are removed. The remaining quadrant is shown in Fig. 25-39. With $V = 0$ at infinity, what is the potential due to the remaining quadrant at point P, which is on the central axis of the original disk at a distance z from the original center? ssm

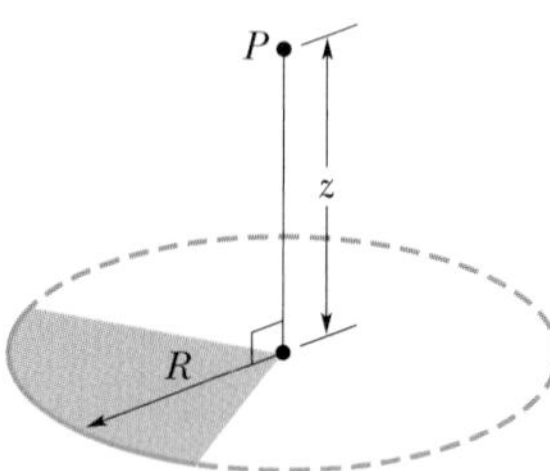

Fig. 25-39 Exercise 27.

28P. Figure 25-40 shows a plastic rod of length L and uniform positive charge Q lying on an x axis. With $V = 0$ at infinity, find the electric potential at point P_1 on the axis, at distance d from one end of the rod.

29P. The plastic rod shown in Fig. 25-40 has length L and a nonuniform linear charge density $\lambda = cx$, where c is a positive constant. With $V = 0$ at infinity, find the electric potential at point P_1 on the axis, at distance d from one end.

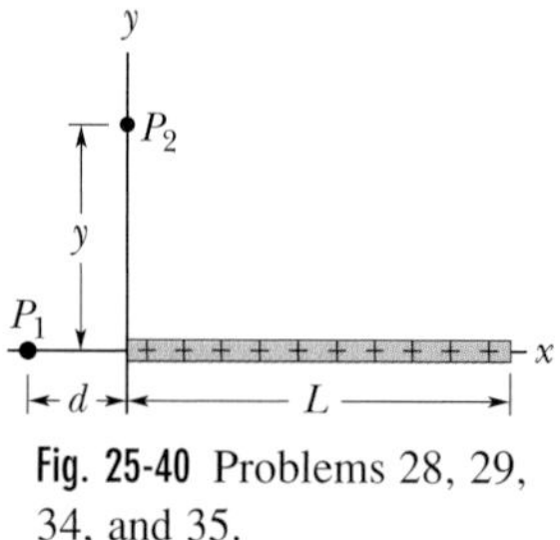

Fig. 25-40 Problems 28, 29, 34, and 35.

SEC. 25-9 Calculating the Field from the Potential

30E. Two large parallel metal plates are 1.5 cm apart and have equal but opposite charges on their facing surfaces. Take the potential of the negative plate to be zero. If the potential halfway between the plates is then +5.0 V, what is the electric field in the region between the plates?

31E. The electric potential at points in an xy plane is given by $V = (2.0\ \text{V/m}^2)x^2 - (3.0\ \text{V/m}^2)y^2$. What are the magnitude and direction of the electric field at the point (3.0 m, 2.0 m)?

32E. The electric potential V in the space between two flat parallel plates is given by $V = 1500x^2$, where V is in volts if x, the distance from one of the plates, is in meters. Calculate the magnitude and direction of the electric field at $x = 1.3$ cm.

33P. (a) Using Eq. 25-32, show that the electric potential at a point on the central axis of a thin ring of charge of radius R and a distance z from the ring is

$$V = \frac{1}{4\pi\varepsilon_0}\frac{q}{\sqrt{z^2 + R^2}}.$$

(b) From this result, derive an expression for E at points on the ring's axis; compare your result with the calculation of E in Section 23-6. ssm www

34P. The plastic rod of length L in Fig. 25-40 has the nonuniform linear charge density $\lambda = cx$, where c is a positive constant. (a) With $V = 0$ at infinity, find the electric potential at point P_2 on the y axis, a distance y from one end of the rod. (b) From that result, find the electric field component E_y at P_2. (c) Why cannot the field component E_x at P_2 be found using the result of (a)?

35P. (a) Use the result of Problem 28 to find the electric field component E_x at point P_1 in Fig. 25-40. (*Hint:* First substitute the variable x for the distance d in the result.) (b) Use symmetry to determine the electric field component E_y at P_1. ssm

SEC. 25-10 Electric Potential Energy of a System of Point Charges

36E. (a) What is the electric potential energy of two electrons separated by 2.00 nm? (b) If the separation increases, does the potential energy increase or decrease?

37E. Derive an expression for the work required to set up the four-charge configuration of Fig. 25-41, assuming the charges are initially infinitely far apart. ilw

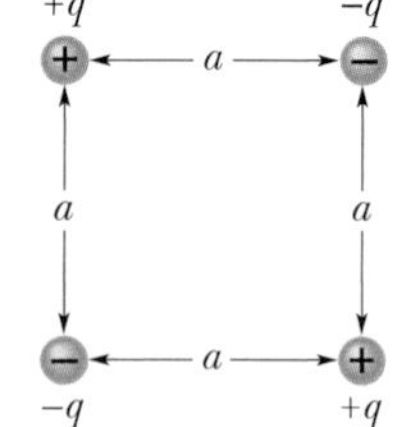

Fig. 25-41 Exercise 37.

38E. What is the electric potential energy of the charge configuration of Fig. 25-9a? Use the numerical values provided in Sample Problem 25-3.

39P. In the rectangle of Fig. 25-42, the sides have lengths 5.0 cm and 15 cm, $q_1 = -5.0\ \mu\text{C}$, and $q_2 = +2.0\ \mu\text{C}$. With $V = 0$ at infinity, what are the electric potentials (a) at corner A and (b) at corner B? (c) How much work is required to move a third charge $q_3 = +3.0\ \mu\text{C}$ from B to A along a diagonal of the rectangle? (d) Does this work increase or decrease the electric energy of the three-charge system? Is more, less, or the same work required if q_3 is moved along paths that are (e) inside the rectangle but not on a diagonal and (f) outside the rectangle? ssm

Fig. 25-42 Problem 39.

40P. In Fig. 25-43, how much work is required to bring the charge of $+5q$ in from infinity along the dashed line and place it

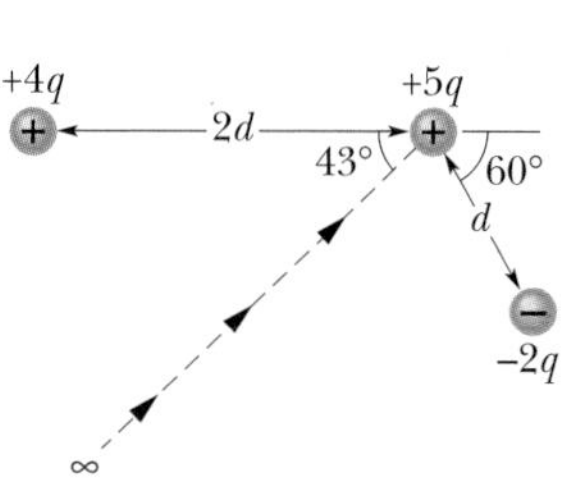

Fig. 25-43 Problem 40.

as shown near the two fixed charges $+4q$ and $-2q$? Take distance $d = 1.40$ cm and charge $q = 1.6 \times 10^{-19}$ C.

41P. A particle of positive charge Q is fixed at point P. A second particle of mass m and negative charge $-q$ moves at constant speed in a circle of radius r_1, centered at P. Derive an expression for the work W that must be done by an external agent on the second particle to increase the radius of the circle of motion to r_2. ssm www

42P. Calculate (a) the electric potential established by the nucleus of a hydrogen atom at the average distance ($r = 5.29 \times 10^{-11}$ m) of the atom's electron (take $V = 0$ at infinite distance), (b) the electric potential energy of the atom when the electron is at this radius, and (c) the kinetic energy of the electron, assuming it to be moving in a circular orbit of this radius centered on the nucleus. (d) How much energy is required to ionize the hydrogen atom (that is, to remove the electron from the nucleus so that the separation is effectively infinite)? Express all energies in electron-volts.

43P. A particle of charge q is fixed at point P, and a second particle of mass m and the same charge q is initially held a distance r_1 from P. The second particle is then released. Determine its speed when it is a distance r_2 from P. Let $q = 3.1\ \mu$C, $m = 20$ mg, $r_1 = 0.90$ mm, and $r_2 = 2.5$ mm. ilw

44P. A charge of -9.0 nC is uniformly distributed around a thin plastic ring of radius 1.5 m that lies in the yz plane with its center at the origin. A point charge of -6.0 pC is located on the x axis at $x = 3.0$ m. Calculate the work done on the point charge by an external force to move the point charge to the origin.

45P. Two tiny metal spheres A and B of mass $m_A = 5.00$ g and $m_B = 10.0$ g have equal positive charges $q = 5.00\ \mu$C. The spheres are connected by a massless nonconducting string of length $d = 1.00$ m, which is much greater than the radii of the spheres. (a) What is the electric potential energy of the system? (b) Suppose you cut the string. At that instant, what is the acceleration of each sphere? (c) A long time after you cut the string, what is the speed of each sphere? ssm

46P. A thin, spherical, conducting shell of radius R is mounted on an isolating support and charged to a potential of $-V$. An electron is then fired from point P at distance r from the center of the shell ($r \gg R$) with initial speed v_0 and directly toward the shell's center. What value of v_0 is needed for the electron to just reach the shell before reversing direction?

47P. Two electrons are fixed 2.0 cm apart. Another electron is shot from infinity and stops midway between the two. What is its initial speed? ssm

48P. Two charged, parallel, flat conducting surfaces are spaced $d = 1.00$ cm apart and produce a potential difference $\Delta V = 625$ V between them. An electron is projected from one surface directly toward the second. What is the initial speed of the electron if it stops just at the second surface?

49P. An electron is projected with an initial speed of 3.2×10^5 m/s directly toward a proton that is fixed in place. If the electron is initially a great distance from the proton, at what distance from the proton is the speed of the electron instantaneously equal to twice the initial value?

SEC. 25-11 Potential of a Charged Isolated Conductor

50E. An empty hollow metal sphere has a potential of +400 V with respect to ground (defined to be at $V = 0$) and has a charge of 5.0×10^{-9} C. Find the electric potential at the center of the sphere.

51E. What is the excess charge on a conducting sphere of radius $r = 0.15$ m if the potential of the sphere is 1500 V and $V = 0$ at infinity? ssm

52E. Consider two widely separated conducting spheres, 1 and 2, the second having twice the diameter of the first. The smaller sphere initially has a positive charge q, and the larger one is initially uncharged. You now connect the spheres with a long thin wire. (a) How are the final potentials V_1 and V_2 of the spheres related? (b) What are the final charges q_1 and q_2 on the spheres, in terms of q? (c) What is the ratio of the final surface charge density of sphere 1 to that of sphere 2?

53P. Two metal spheres, each of radius 3.0 cm, have a center-to-center separation of 2.0 m. One has a charge of $+1.0 \times 10^{-8}$ C; the other has a charge of -3.0×10^{-8} C. Assume that the separation is large enough relative to the size of the spheres to permit us to consider the charge on each to be uniformly distributed (the spheres do not affect each other). With $V = 0$ at infinity, calculate (a) the potential at the point halfway between their centers and (b) the potential of each sphere. ssm

54P. A charged metal sphere of radius 15 cm has a net charge of 3.0×10^{-8} C. (a) What is the electric field at the sphere's surface? (b) If $V = 0$ at infinity, what is the electric potential at the sphere's surface? (c) At what distance from the sphere's surface has the electric potential decreased by 500 V?

55P. (a) If Earth had a net surface charge density of 1.0 electron per square meter (a very artificial assumption), what would its potential be? (Set $V = 0$ at infinity.) (b) What would be the electric field due to Earth just outside its surface?

56P. Two thin, isolated, concentric conducting spheres of radii R_1 and R_2 (with $R_1 < R_2$) have charges q_1 and q_2. With $V = 0$ at infinity, derive expressions for $E(r)$ and $V(r)$, where r is distance from the center of the spheres. Plot $E(r)$ and $V(r)$ from $r = 0$ to $r = 4.0$ m for $R_1 = 0.50$ m, $R_2 = 1.0$ m, $q_1 = +2.0\ \mu$C, and $q_2 = +1.0\ \mu$C.

Additional Problem

57. *The chocolate crumb mystery.* This story begins with Problem 48 in Chapter 24. (a) From the answer to part (a) of that problem, find an expression for the electric potential as a function of the radial distance r from the center of the pipe. (The electric potential is zero on the grounded pipe wall.) (b) For the typical volume charge density $\rho = -1.1 \times 10^{-3}$ C/m^3, what is the difference in the electric potential between the pipe's center and its inside wall? (The story continues with Problem 48 in Chapter 26.)

NEW PROBLEMS

N1. Identical 50 μC charges are fixed on an x axis at $x = \pm 3.0$ m. A particle of charge $q = -15$ μC is then released from rest at a point on the positive part of the y axis. Due to the symmetry of the situation, the particle moves along the y axis and has a kinetic energy of 1.2 J as it passes through the point $x = 0$, $y = 4.0$ m. (a) What is the kinetic energy of the particle as it passes through the origin? (b) At what negative value of y will the particle momentarily stop?

N2. Two charged particles are shown in Fig. 25N-1a. Particle 1, with charge q_1, is fixed in place, at distance $d = 2.0$ cm from the origin. Particle 2, with charge q_2, can be moved along the x axis. Figure 25N-1b gives the net electric potential V at the origin due to the two particles as a function of the x coordinate of particle 2. The plot has an asymptote of $V = 5.76 \times 10^{-7}$ V as $x \to \infty$. What is q_2 in terms of e?

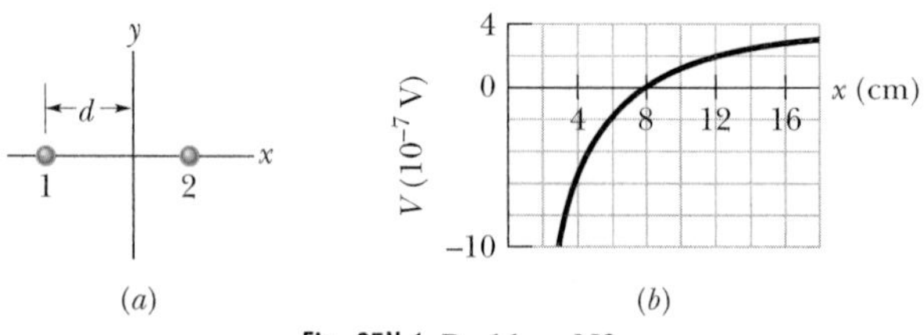

Fig. 25N-1 Problem N2.

N3. Figure 25N-2 shows a rectangular array of charged particles fixed in place. What is the net electric potential at the center of the array? (*Hint:* First consider just the corner particles.)

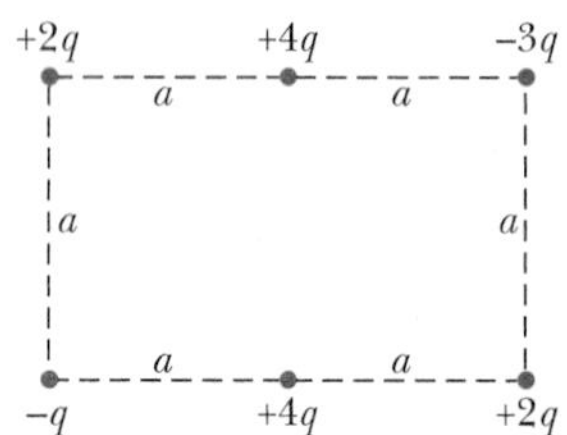

Fig. 25N-2 Problem N3.

N4. Figure 25N-3a shows three particles on an x axis. Particle 1 (with a charge of $+5.0$ μC) and particle 2 (with a charge of $+3.0$ μC) are fixed in place with separation $d = 4.0$ cm. Particle 3 can be moved along the x axis to the right of particle 2. Figure 25N-3b gives the electric potential energy U of the three-particle system as a function of the x coordinate of particle 3. What is the charge of particle 3?

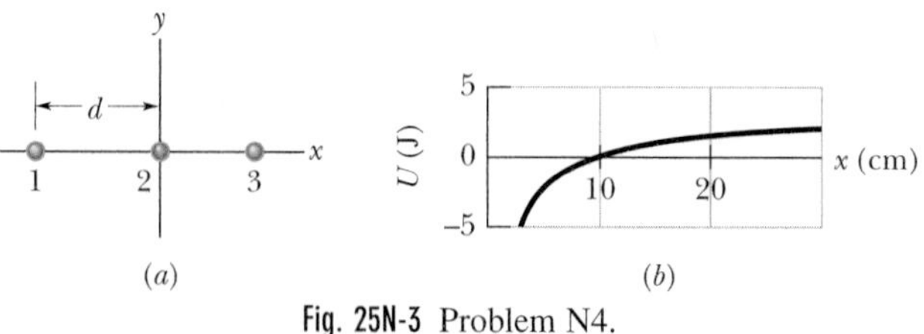

Fig. 25N-3 Problem N4.

N5. A solid conducting sphere of radius 3.0 cm has a charge of 30 nC distributed over its surface. Let A be a point 1.0 cm from the center of the sphere, S be a point on the surface of the sphere, and B be a point 5.0 cm from the center of the sphere. What are the electric potential differences (a) $V_S - V_B$ and (b) $V_A - V_B$?

N6. In Fig. 25N-4a, a particle of charge $+e$ is initially at coordinate $z = 20$ nm on the dipole axis through an electric dipole, on the positive side of the dipole. (The origin of z is at the dipole center.) The particle is then moved along a circular path around the dipole center until it is at coordinate $z = -20$ nm. Figure 25N-4b gives the work W_a done by the force moving the particle versus the angle θ that locates the particle. What is the magnitude of the dipole moment?

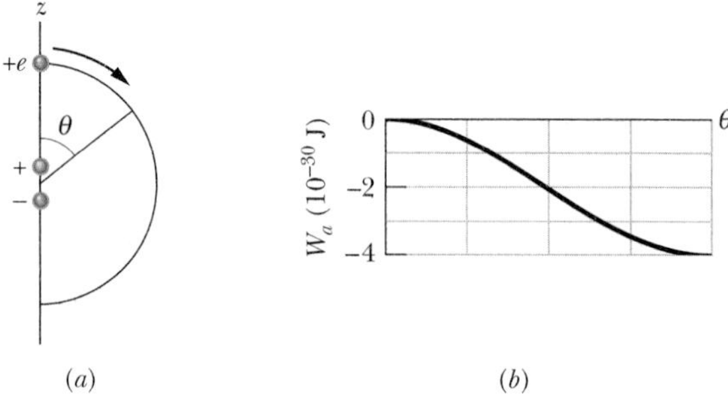

Fig. 25N-4 Problem N6.

N7. A particle of charge = 7.5 μC is released from rest at a point on an x axis at $x = 60$ cm. It begins to move due to the presence of a charge Q that remains fixed at the origin. What is the kinetic energy of the particle at the instant it has moved 40 cm if (a) $Q = +20$ μC and (b) $Q = -20$ μC?

N8. Figure 25N-5a shows an electron moving along an electric dipole axis toward the negative side of the dipole. The dipole is fixed in place. The electron was initially very far from the dipole, with a kinetic energy of 100 eV. Figure 25N-5b gives the kinetic energy K of the electron versus its distance r from the dipole center. What is the magnitude of the dipole moment?

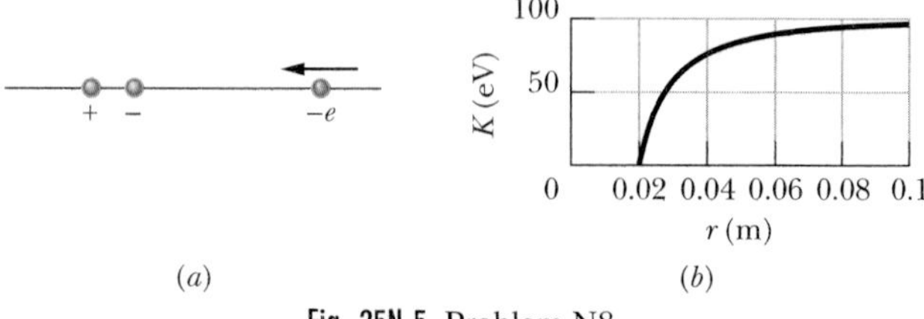

Fig. 25N-5 Problem N8.

N9. Consider an electron on the surface of a uniformly charged sphere of radius 1.0 cm and total charge 1.6×10^{-15} C. What is the *escape speed* for this electron? That is, what initial speed must it have to reach an infinite distance from the sphere and there have zero kinetic energy? (This escape speed is defined similarly to that in Chapter 14 for escaping the gravitational force, but here neglect that force.)

N10. The smiling face of Fig. 25N-6 consists of three items:

1. a thin rod of charge $-3.0\ \mu C$ that forms a full circle of radius 6.0 cm,
2. a second thin rod of charge $2.0\ \mu C$ that forms a circular arc of radius 4.0 cm, subtending an angle of 90° about the center of the full circle,
3. an electric dipole with a dipole moment that is perpendicular to a radial line and that has magnitude $1.28 \times 10^{-21}\ C \cdot m$.

What is the net electric potential at the center?

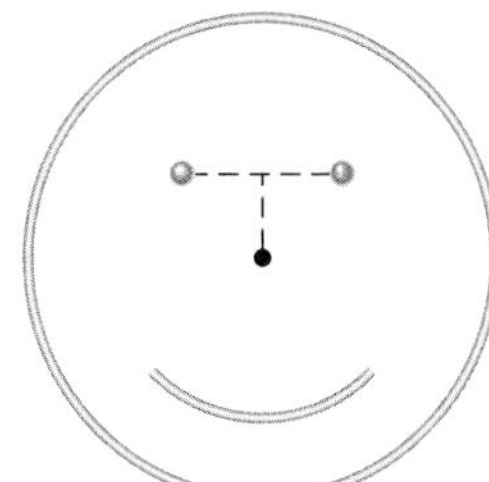

Fig. 25N-6 Problem N10.

N11. (a) In Fig. 25N-7, what is the net electric potential at point P due to the two charged particles, which are fixed in place? We bring a particle of charge $+2e$ from infinity to point P. (b) How much work do we do? (c) What is the electric potential energy of the three-particle system once the third particle is in place?

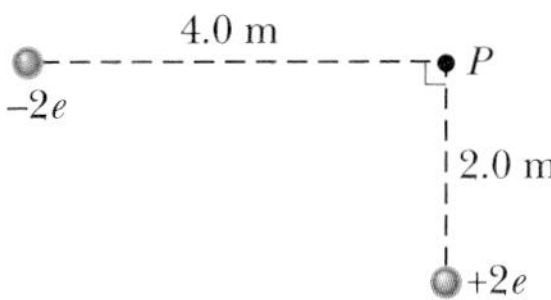

Fig. 25N-7 Problem N11.

N12. Figure 25N-8 shows a thin rod with a uniform charge density of $2.00\ \mu C/m$. Evaluate the electric potential at point P if $d = D = L/4$.

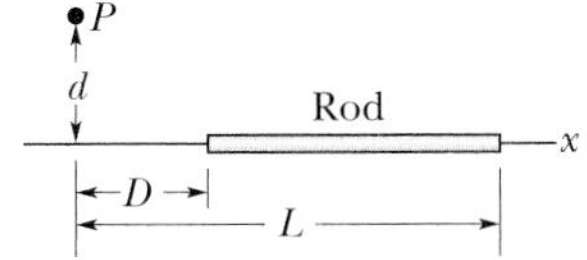

Fig. 25N-8 Problem N12.

N13. In Fig. 25N-9, we move a particle of charge $+2e$ in from infinity to the x axis. How much work do we do? Distance D is 4.00 m.

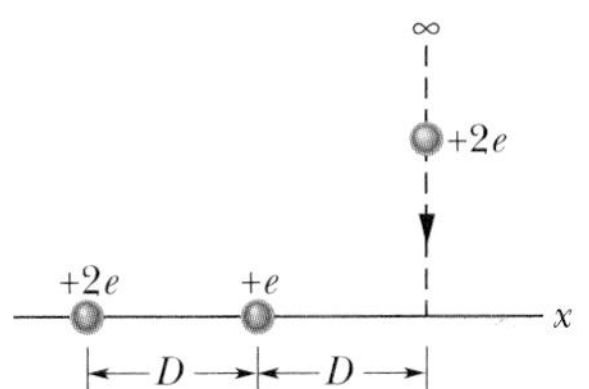

Fig. 25N-9 Problem N13.

N14. Figure 25N-10 shows a hemisphere with a charge of $4.00\ \mu C$ distributed uniformly through its volume. The hemisphere lies on an xy plane like half a grapefruit might lie face down on a kitchen table. Point P is located on that plane, along a radial line from the hemisphere's center of curvature, at radial distance 15 cm. What is the electric potential at point P due to the hemisphere?

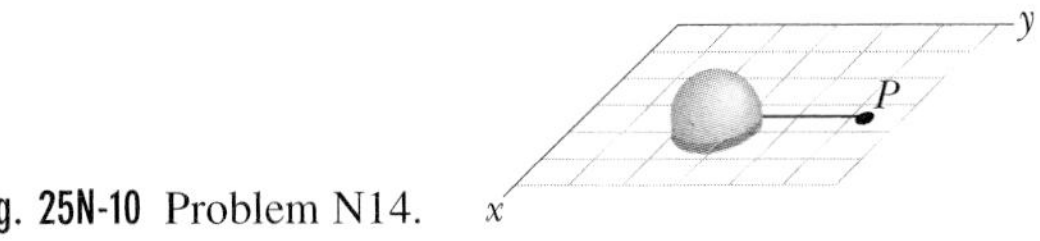

Fig. 25N-10 Problem N14.

N15. Initially two electrons are fixed in place with a separation of $2.00\ \mu m$. How much work must we do to bring a third electron in from infinity to complete an equilateral triangle?

N16. In Fig. 25N-11, seven charged particles are fixed in place to form a square with an edge length of 4.0 cm. How much work must we do to bring a particle of charge $+6e$ from an infinite distance to the center of the square?

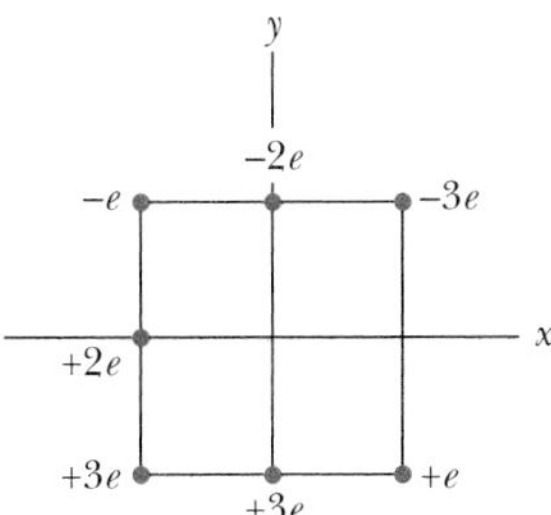

Fig. 25N-11 Problem N16.

N17. In Fig. 25N-12, a charged particle (which is either an electron or a proton) is moving rightward between two parallel charged plates separated by distance $d = 2.00$ mm. The particle is slowing from an initial speed of 90.0 km/s at the left plate. (a) Is the particle an electron or a proton? (b) What is its speed just as it reaches the plate at the right?

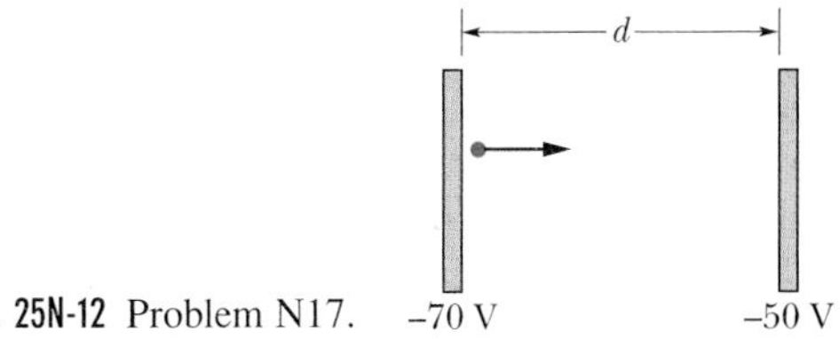

Fig. 25N-12 Problem N17.

N18. A Gaussian sphere of radius 4.00 cm is centered on a ball of radius 1.00 cm, with a uniform charge distribution. The total (net) electric flux through the surface of the Gaussian sphere is $+5.60 \times 10^4\ N \cdot m^2/C$. What is the electric potential at a radial distance of 12.0 cm from the center of the ball?

N19. A decade before Einstein published his theory of relativity, J. J. Thomson proposed that the electron might consist of small parts and attributed its mass m to the electrical interaction of the parts. Furthermore, he suggested that the energy equals mc^2, where c is the speed of light. Make a rough estimate of the electron mass in the following way: Assume that the electron is composed of three identical parts that are brought in from infinity and placed at the vertices of an equilateral triangle having sides equal to the *classical radius* of the electron, 2.82×10^{-15} m. (a) Find the total electric potential energy of this arrangement. (b) Divide by c^2 and compare your result to the accepted electron mass. (The result improves if more parts are assumed.)

N20. *Proton in well.* Figure 25N-13 shows the electric potential along an x axis. A proton is to be released at $x = 3.5$ cm with an initial kinetic energy of 4.00 eV. (a) If it is initially moving in the negative direction of the axis, does it reach a turning point (if so, where is that point) or does it escape from the plotted region (if so, what is its speed at $x = 0$)? (b) If it is initially moving in the positive direction of the axis, does it reach a turning point (if so, where is that point) or does it escape from the plotted region (if so, what is its speed at $x = 6.0$ cm)? What would be the magnitude and direction of the electric force acting on the proton if it moves (c) just to the left of $x = 3.0$ cm and (d) just to the right of $x = 5.0$ cm?

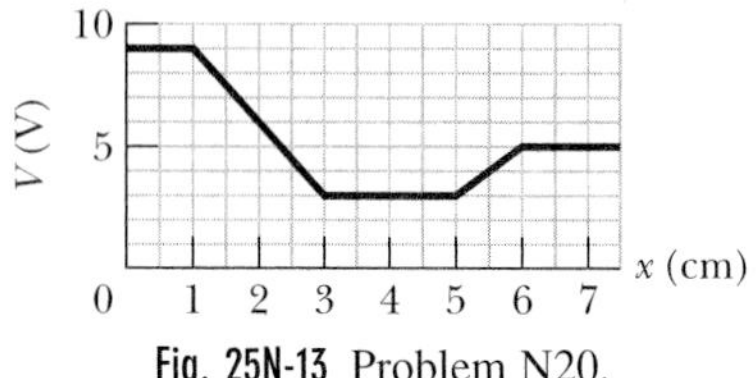

Fig. 25N-13 Problem N20.

N21. Charge $q_1 = -1.2 \times 10^{-9}$ C is at the origin, and charge $q_2 = 2.5 \times 10^{-9}$ C is on the y axis at $y = 0.50$ m. Take the electric potential to be zero far from both charges. (a) Plot the intersection of the $V = 5.0$ V equipotential surface with the xy plane. It encloses one of the charges. (b) There are two equipotential surfaces corresponding to $V = 3.0$ V. One encloses one of the charges and the other encloses both charges. Plot their intersections with the xy plane. (c) Find the value of the potential for which the pattern of the electric potential switches from one to two equipotential surfaces.

N22. *Electron in well.* Figure 25N-14 shows the electric potential V along an x axis. An electron is to be released at $x = 4.5$ cm with an initial kinetic energy of 3.00 eV. (a) If it is initially moving in the negative direction of the axis, does it reach a turning point (if so, where is that point) or does it escape from the plotted region (if so, what is its speed at $x = 0$)? (b) If it is initially moving in the positive direction of the axis, does it reach a turning point (if so, where is that point) or does it escape from the plotted region (if so, what is its speed at $x = 7.0$ cm)? What would be the magnitude and direction of the electric force acting on the electron if it moves (c) just to the left of $x = 4.0$ cm and (d) just to the right of $x = 5.0$ cm?

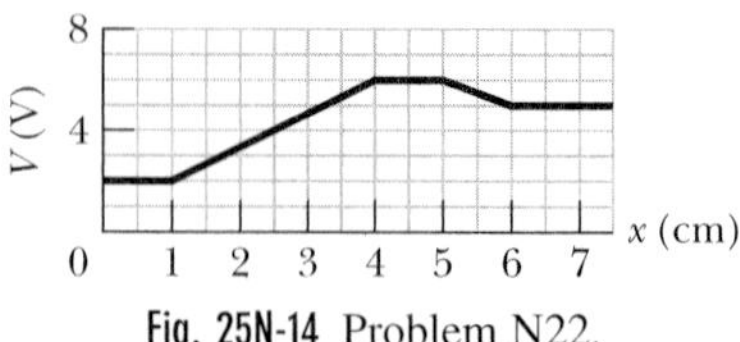

Fig. 25N-14 Problem N22.

N23. An alpha particle (which has two protons) is sent directly toward a target nucleus with 92 protons. The alpha particle has an initial kinetic energy of 0.48 pJ. What is the least center-to-center distance that the alpha particle will be from the target nucleus, assuming that the nucleus does not move?

N24. An electron is placed in an xy plane. The electric potential in the region changes with the electron's displacement parallel to the x and y axis according to Fig. 25N-15. (It does not change with displacement parallel to the z axis.) What are the magnitude and direction of the electric force on the electron?

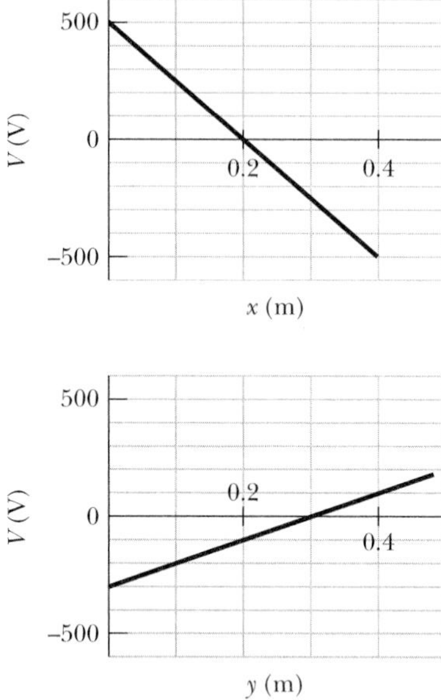

Fig. 25N-15 Problem N24.

N25. The magnitude E of an electric field depends on the radial distance r according to $E = A/r^4$, where A is a constant with the unit volt-cubic meter. What is the magnitude of the electric potential difference between $r = 2.00$ m and $r = 3.00$ m?

N26. In Fig. 25N-16, two particles of charges q_1 and q_2 are fixed to an x axis. If a third particle, of charge $+6.0\ \mu$C, is brought from an infinite distance to point P, the three-particle system has the same electric potential energy as the original two-particle system. What is the charge ratio q_1/q_2?

$$1 \quad 2 \quad P \quad x \qquad \leftarrow d \rightarrow \leftarrow 1.5d \rightarrow$$

Fig. 25N-16 Problem N26.

N27. Three +0.12 C charges form an equilateral triangle, 1.7 m on a side. Using energy that is supplied at the rate of 0.83 kW, how many days would be required to move one of the charges to the midpoint of the line joining the other two charges?

N28. A positron (charge $= +e$ and mass equal to the electron mass) is initially moving at 1.0×10^7 m/s in the positive direction of an x axis when, at $x = 0$, it encounters an electric field that is directed along the x axis. The electric potential V associated with that field is given in Fig. 25N-17. (a) Does the positron emerge from the field at $x = 0$ (its motion is reversed) or at $x = 0.50$ m (its motion is not reversed)? (b) What is its speed when it emerges?

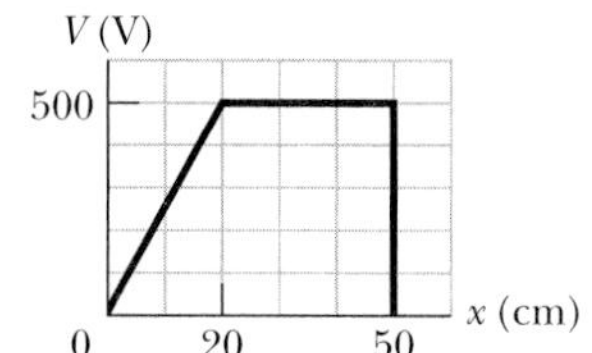

Fig. 25N-17 Problem N28.

N29. A point charge $q_1 = +6.0e$ is fixed at the origin of a rectangular coordinate system, and a second point charge $q_2 = -10e$ is fixed at $x = 8.6$ nm, $y = 0$. The locus of all points in the xy plane with $V = 0$ (other than at infinity) is a circle centered on the x axis, as shown in Fig. 25N-18. Find (a) the location x_c of the center of the circle and (b) the radius R of the circle. (c) Is the xy cross section of the 5 V equipotential surface also a circle?

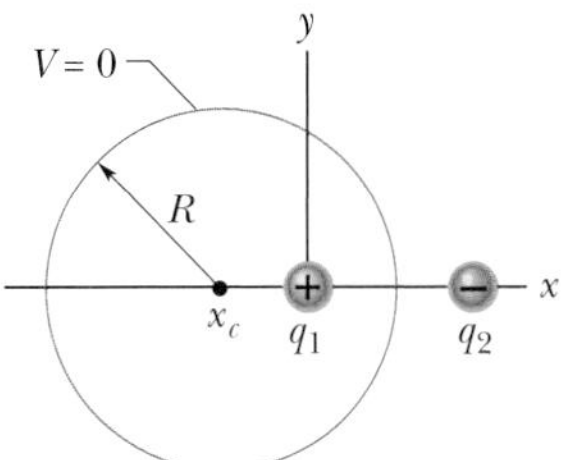

Fig. 25N-18 Problem N29.

N30. In Fig. 25N-19a, we move an electron from an infinite distance to a point at distance $R = 8.00$ cm from a tiny charged ball. The move requires work $W = 2.16 \times 10^{-13}$ J by us. (a) What is the charge Q on the ball? In Fig. 25N-19b, the ball has been sliced up and the slices spread out so that an equal amount of charge is at the hour positions on a circle of radius $R = 8.00$ cm. Now the electron is brought from an infinite distance to the center of the circle. (b) With that addition of the electron to the system of charged particles, what is the change in the electric potential energy of the system?

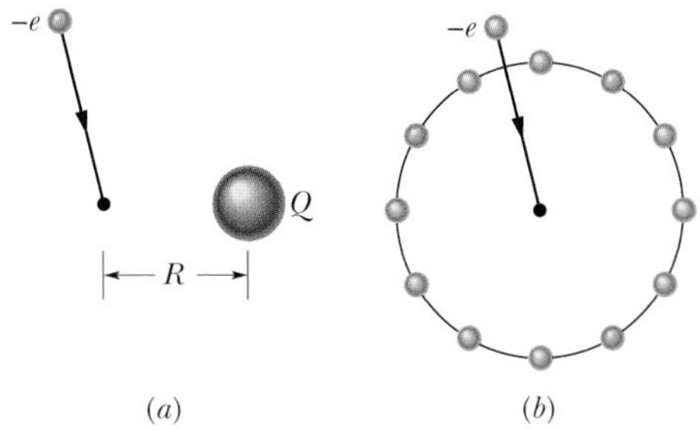

Fig. 25N-19 Problem N30.

N31. Three particles with charges $q_1 = +10\ \mu$C, $q_2 = -20\ \mu$C, and $q_3 = +30\ \mu$C are positioned at the vertices of an isosceles triangle as shown in Fig. 25N-20. If $a = 10$ cm and $b = 6.0$ cm, how much work must an external agent do to exchange the positions of (a) q_1 and q_3 and, instead, (b) q_1 and q_2?

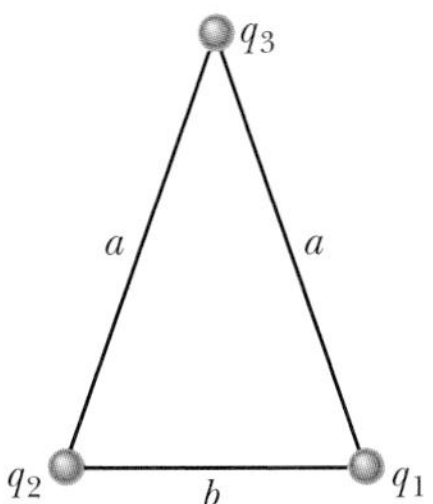

Fig. 25N-20 Problem N31.

N32. Three thin plastic rods are bent to form quarter circles and then arranged in the xy plane so that they have a common center of curvature at the origin. Figure 25N-21 gives the charge on each rod in terms of $Q = 30$ nC. What is the net electric potential due to the three rods at their common center of curvature?

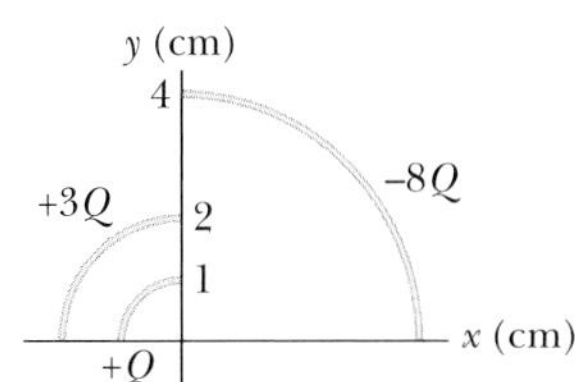

Fig. 25N-21 Problem N32.

N33. Let the separation d between the particles in Fig. 25N-22 be 1.0 m; let their charges be $q_1 = +q$ and $q_2 = +2q$; and let $V = 0$ at infinity. Then locate any point on the x axis (other than at infinity) at which (a) the net electric potential due to the two particles is zero and (b) the net electric field due to them is zero.

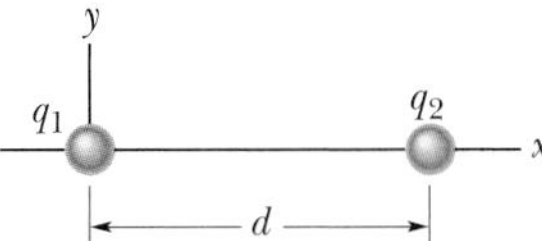

Fig. 25N-22 Problem N33.

N34. An electron is released from rest on the axis of an electric dipole that has charge e and charge separation $d = 20$ pm and that is fixed in place. The release point is on the positive side of the dipole, at a distance of $7.0d$ from the dipole center. What is the electron's speed when it reaches a point $5.0d$ from the dipole center?

N35. A solid copper sphere whose radius is 1.0 cm has a very thin surface coating of nickel. Some of the nickel atoms are radioactive, each atom emitting an electron as it decays. Half of these electrons enter the copper sphere, each depositing 100 keV of energy there. The other half of the electrons escape, each carrying away a charge of $-e$. The nickel coating has an activity of 10 mCi (= 10 millicuries = 3.70×10^8 radioactive decays per second). The sphere is hung from a long, nonconducting string and isolated from its surroundings. (a) How long will it take for the potential of the sphere to increase by 1000 V? (b) How long will it take for the temperature of the sphere to increase by 5.0 K due to the energy deposited by the electrons? The heat capacity of the sphere is 14 J/K.

N36. A particle of mass m, positive charge q, and initial kinetic energy K is projected (from a large distance) toward a heavy nucleus of charge Q that is fixed in place. Assuming that the particle approaches head-on, how close to the center of the nucleus is the particle when it momentarily comes to rest?

N37. Exercise 37 in Chapter 24 deals with Rutherford's calculation of the electric field at a distance r from the center of an atom and inside the atom. He also gave the electric potential as

$$V = \frac{Ze}{4\pi\varepsilon_0}\left(\frac{1}{r} - \frac{3}{2R} + \frac{r^2}{2R^3}\right).$$

(a) Show how the expression for the electric field given in Exercise 37 of Chapter 24 follows from the above expression for V. (b) Why does this expression for V not go to zero as $r \to \infty$?

26 Capacitance

During ventricular fibrillation, a common type of heart attack, the chambers of the heart fail to pump blood because their muscle fibers randomly contract and relax. To save a victim of ventricular fibrillation, the heart muscle must be shocked to reestablish its normal rhythm. For that, 20 A of current must be sent through the chest cavity to transfer 200 J of electrical energy in about 2.0 ms. This requires about 100 kW of electric power. Such a requirement may easily be met in a hospital, but not by, say, the electrical system of an ambulance arriving to help the victim.

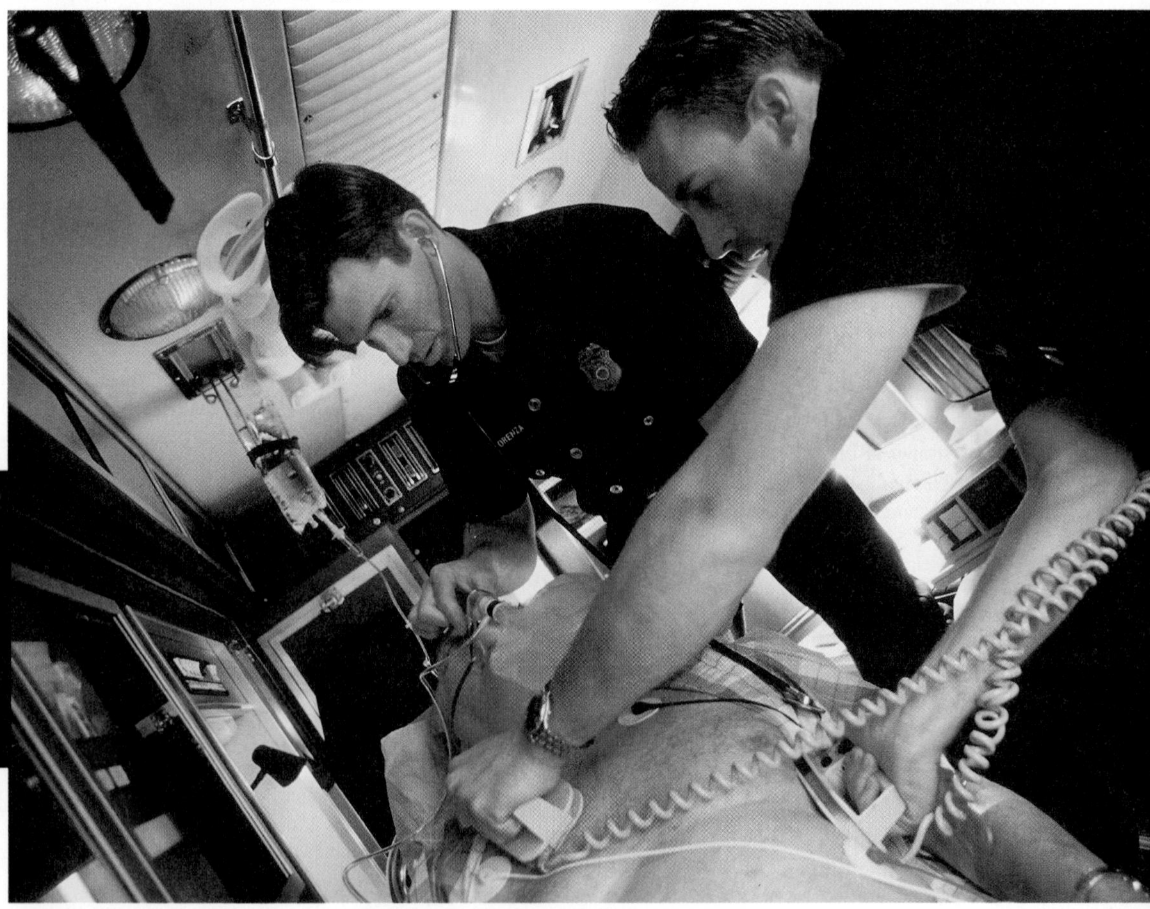

What, then, can provide the power needed for defibrillation at remote locations?

The answer is in this chapter.

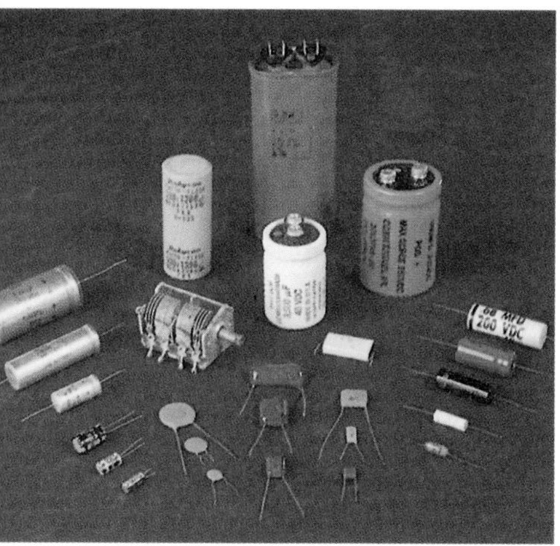

Fig. 26-1 An assortment of capacitors.

26-1 The Uses of Capacitors

You can store energy as potential energy by pulling a bowstring, stretching a spring, compressing a gas, or lifting a book. You can also store energy as potential energy in an electric field, and a **capacitor** is a device you can use to do exactly that.

There is a capacitor in a portable battery-operated photoflash unit, for example. It accumulates charge relatively slowly during the readying process between flashes, building up an electric field as it does so. It holds this field and the associated energy until the energy is rapidly released to initiate the flash.

Capacitors have many uses in our electronic and microelectronic age beyond serving as storehouses for potential energy. As one example, they are vital elements in the circuits with which we tune radio and television transmitters and receivers. As another example, microscopic capacitors form the memory banks of computers. These tiny devices are not as important for their stored energy as for the ON–OFF information that the presence or absence of their electric fields provides.

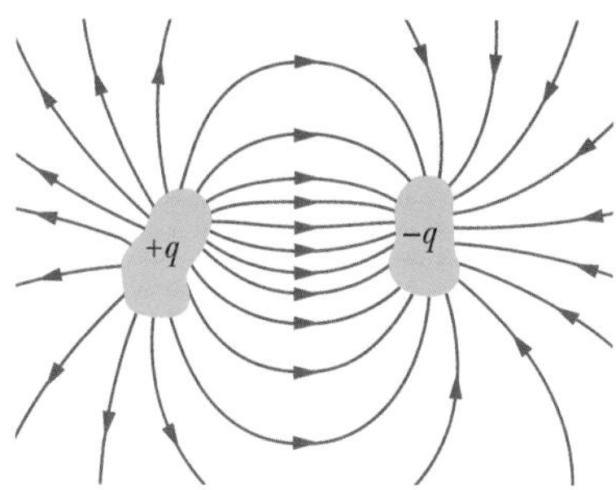

Fig. 26-2 Two conductors, isolated electrically from each other and from their surroundings, form a *capacitor*. When the capacitor is charged, the conductors, or *plates* as they are called, have equal but opposite charges of magnitude q.

26-2 Capacitance

Figure 26-1 shows some of the many sizes and shapes of capacitors. Figure 26-2 shows the basic elements of *any* capacitor—two isolated conductors of any shape. No matter what their geometry, flat or not, we call these conductors *plates.*

Figure 26-3*a* shows a less general but more conventional arrangement, called a *parallel-plate capacitor*, consisting of two parallel conducting plates of area A separated by a distance d. The symbol that we use to represent a capacitor (⊣⊢) is based on the structure of a parallel-plate capacitor but is used for capacitors of all geometries. We assume for the time being that no material medium (such as glass or plastic) is present in the region between the plates. In Section 26-6, we shall remove this restriction.

When a capacitor is *charged,* its plates have equal but opposite charges of $+q$ and $-q$. However, we refer to the *charge of a capacitor* as being q, the absolute value of these charges on the plates. (Note that q is not the net charge on the capacitor, which is zero.)

Because the plates are conductors, they are equipotential surfaces; all points on a plate are at the same electric potential. Moreover, there is a potential difference between the two plates. For historical reasons, we represent the absolute value of this potential difference with V rather than with ΔV as we would with previous notation.

The charge q and the potential difference V for a capacitor are proportional to each other; that is,

$$q = CV. \tag{26-1}$$

The proportionality constant C is called the **capacitance** of the capacitor. Its value depends only on the geometry of the plates and *not* on their charge or potential difference. The capacitance is a measure of how much charge must be put on the

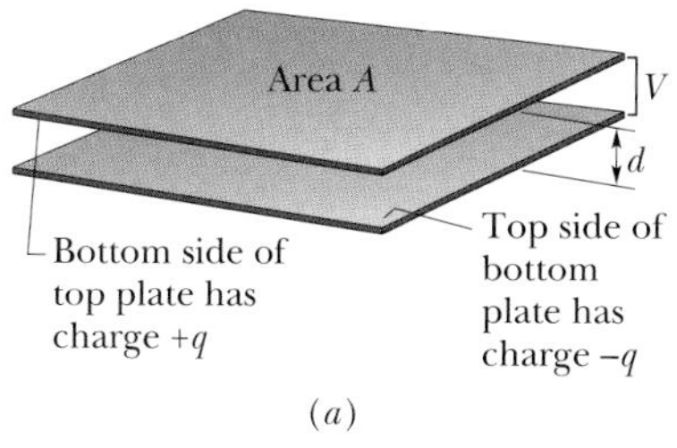

Fig. 26-3 (*a*) A parallel-plate capacitor, made up of two plates of area A separated by a distance d. The plates have equal and opposite charges of magnitude q on their facing surfaces. (*b*) As the field lines show, the electric field due to the charged plates is uniform in the central region between the plates. The field is not uniform at the edges of the plates, as indicated by the 'fringing' of the field lines there.

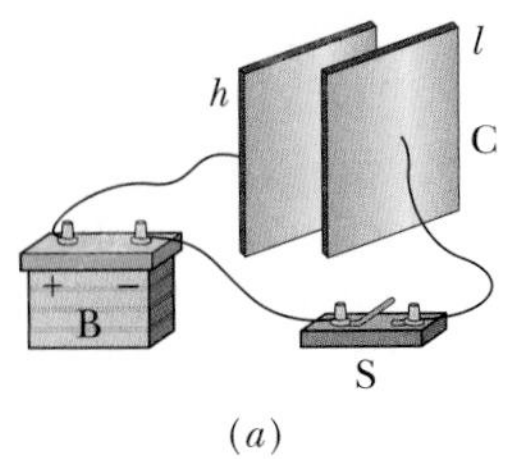

(a)

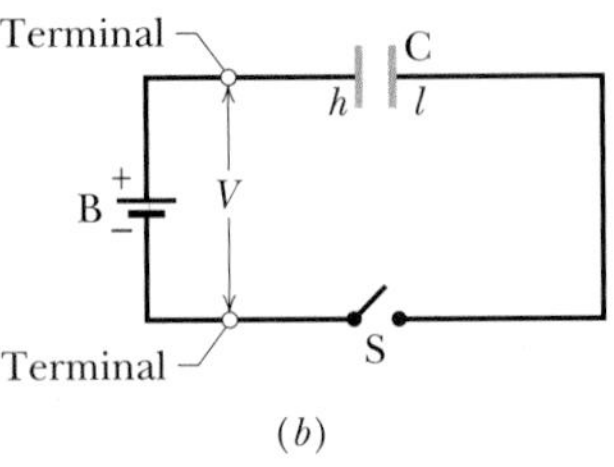

(b)

Fig. 26-4 (a) Battery B, switch S, and plates h and l of capacitor C, connected in a circuit. (b) A schematic diagram with the *circuit elements* represented by their symbols.

plates to produce a certain potential difference between them: The *greater the capacitance, the more charge is required.*

The SI unit of capacitance that follows from Eq. 26-1 is the coulomb per volt. This unit occurs so often that it is given a special name, the *farad* (F):

$$1 \text{ farad} = 1 \text{ F} = 1 \text{ coulomb per volt} = 1 \text{ C/V}. \qquad (26\text{-}2)$$

As you will see, the farad is a very large unit. Submultiples of the farad, such as the microfarad ($1\ \mu\text{F} = 10^{-6}$ F) and the picofarad ($1 \text{ pF} = 10^{-12}$ F), are more convenient units in practice.

Charging a Capacitor

One way to charge a capacitor is to place it in an electric circuit with a battery. An *electric circuit* is a path through which charge can flow. A *battery* is a device that maintains a certain potential difference between its *terminals* (points at which charge can enter or leave the battery) by means of internal electrochemical reactions in which electric forces can move internal charge.

In Fig. 26-4*a*, a battery B, a switch S, an uncharged capacitor C, and interconnecting wires form a circuit. The same circuit is shown in the *schematic diagram* of Fig. 26-4*b*, in which the symbols for a battery, a switch, and a capacitor represent those devices. The battery maintains potential difference V between its terminals. The terminal of higher potential is labeled + and is often called the *positive* terminal; the terminal of lower potential is labeled − and is often called the *negative* terminal.

The circuit shown in Figs. 26-4*a* and *b* is said to be *incomplete* because switch S is *open*; that is, it does not electrically connect the wires attached to it. When the switch is *closed,* electrically connecting those wires, the circuit is complete and charge can then flow through the switch and the wires. As we discussed in Chapter 22, the charge that can flow through a conductor, such as a wire, is that of electrons. When the circuit of Fig. 26-4 is completed, electrons are driven through the wires by an electric field that the battery sets up in the wires. The field drives electrons from capacitor plate h to the positive terminal of the battery; thus, plate h, losing electrons, becomes positively charged. The field drives just as many electrons from the negative terminal of the battery to capacitor plate l; thus, plate l, gaining electrons, becomes negatively charged *just as much* as plate h, losing electrons, becomes positively charged.

Initially, when the plates are uncharged, the potential difference between them is zero. As the plates become oppositely charged, that potential difference increases until it equals the potential difference V between the terminals of the battery. Then plate h and the positive terminal of the battery are at the same potential, and there is no longer an electric field in the wire between them. Similarly, plate l and the negative terminal reach the same potential and there is then no electric field in the wire between them. Thus, with the field zero, there is no further drive of electrons. The capacitor is then said to be *fully charged,* with a potential difference V and charge q that are related by Eq. 26-1.

In this book we assume that during the charging of a capacitor and afterward, charge cannot pass from one plate to the other across the gap separating them. Also, we assume that a capacitor can retain (or *store*) charge indefinitely, until it is put into a circuit where it can be *discharged.*

✔CHECKPOINT 1: Does the capacitance C of a capacitor increase, decrease, or remain the same (a) when the charge q on it is doubled and (b) when the potential difference V across it is tripled?

PROBLEM-SOLVING TACTICS

Tactic 1: *The Symbol V and Potential Difference*

In previous chapters, the symbol V represents an electric potential at a point or along an equipotential surface. However, in matters concerning electrical devices, V often represents a *potential difference* between two points or two equipotential surfaces. Equation 26-1 is an example of this second use of the symbol. In Section 26-3, you will see a mixture of the two meanings of V. There and in later chapters, you need to be alert as to the intent of this symbol.

You will also be seeing, in this book and elsewhere, a variety of phrases regarding potential difference. A potential difference or a "potential" or a "voltage" may be *applied* to a device, or it may be *across* a device. A capacitor can be charged to a potential difference, as in "a capacitor is charged to 12 V." Also, a battery can be characterized by the potential difference across it, as in "a 12 V battery." Always keep in mind what is meant by such phrases: There is a potential difference between two points, such as two points in a circuit or at the terminals of a device such as a battery.

26-3 Calculating the Capacitance

Our task here is to calculate the capacitance of a capacitor once we know its geometry. Because we will consider a number of different geometries, it seems wise to develop a general plan to simplify the work. In brief our plan is as follows: (1) Assume a charge q on the plates; (2) calculate the electric field $\vec{E}$ between the plates in terms of this charge, using Gauss' law; (3) knowing $\vec{E}$, calculate the potential difference V between the plates from Eq. 25-18; (4) calculate C from Eq. 26-1.

Before we start, we can simplify the calculation of both the electric field and the potential difference by making certain assumptions. We discuss each in turn.

Calculating the Electric Field

To relate the electric field $\vec{E}$ between the plates of a capacitor to the charge q on either plate, we shall use Gauss' law:

$$\varepsilon_0 \oint \vec{E} \cdot d\vec{A} = q. \tag{26-3}$$

Here q is the charge enclosed by a Gaussian surface, and $\oint \vec{E} \cdot d\vec{A}$ is the net electric flux through that surface. In all cases that we shall consider, the Gaussian surface will be such that whenever electric flux passes through it, $\vec{E}$ will have a uniform magnitude E and the vectors $\vec{E}$ and $d\vec{A}$ will be parallel. Equation 26-3 then will reduce to

$$q = \varepsilon_0 EA \quad \text{(special case of Eq. 26-3)}, \tag{26-4}$$

in which A is the area of that part of the Gaussian surface through which flux passes. For convenience, we shall always draw the Gaussian surface in such a way that it completely encloses the charge on the positive plate; see Fig. 26-5 for an example.

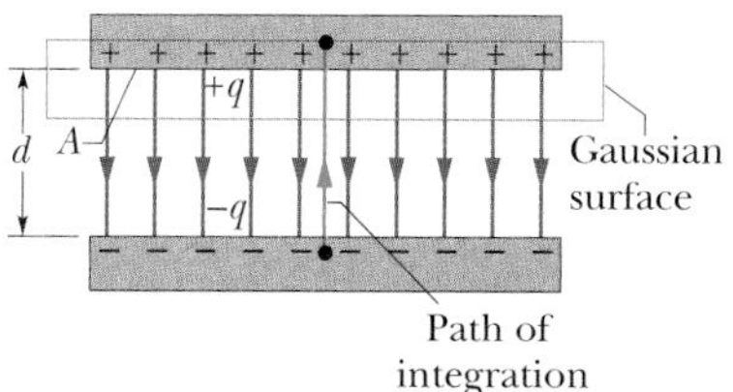

Fig. 26-5 A charged parallel-plate capacitor. A Gaussian surface encloses the charge on the positive plate. The integration of Eq. 26-6 is taken along a path extending directly from the negative plate to the positive plate.

Calculating the Potential Difference

In the notation of Chapter 25 (Eq. 25-18), the potential difference between the plates of a capacitor is related to the field $\vec{E}$ by

$$V_f - V_i = -\int_i^f \vec{E} \cdot d\vec{s}, \tag{26-5}$$

in which the integral is to be evaluated along any path that starts on one plate and ends on the other. We shall always choose a path that follows an electric field line, from the negative plate to the positive plate. For this path, the vectors $\vec{E}$ and $d\vec{s}$ will have opposite directions, so the dot product $\vec{E} \cdot d\vec{s}$ will be equal to $-E\,ds$. Thus, the right side of Eq. 26-5 will then be positive. Letting V represent the difference $V_f - V_i$, we can then recast Eq. 26-5 as

$$V = \int_{-}^{+} E\,ds \qquad \text{(special case of Eq. 26-5),} \tag{26-6}$$

in which the − and + remind us that our path of integration starts on the negative plate and ends on the positive plate.

We are now ready to apply Eqs. 26-4 and 26-6 to some particular cases.

A Parallel-Plate Capacitor

We assume, as Fig. 26-5 suggests, that the plates of our parallel-plate capacitor are so large and so close together that we can neglect the fringing of the electric field at the edges of the plates, taking $\vec{E}$ to be constant throughout the region between the plates.

We draw a Gaussian surface that encloses just the charge q on the positive plate, as in Fig. 26-5. From Eq. 26-4 we can then write

$$q = \varepsilon_0 EA, \tag{26-7}$$

where A is the area of the plate.

Equation 26-6 yields

$$V = \int_{-}^{+} E\,ds = E\int_{0}^{d} ds = Ed. \tag{26-8}$$

In Eq. 26-8, E can be placed outside the integral because it is a constant; the second integral then is simply the plate separation d.

If we now substitute q from Eq. 26-7 and V from Eq. 26-8 into the relation $q = CV$ (Eq. 26-1), we find

$$C = \frac{\varepsilon_0 A}{d} \qquad \text{(parallel-plate capacitor).} \tag{26-9}$$

Thus, the capacitance does indeed depend only on geometrical factors—namely, the plate area A and the plate separation d. Note that C increases as we increase the plate area A or decrease the separation d.

As an aside, we point out that Eq. 26-9 suggests one of our reasons for writing the electrostatic constant in Coulomb's law in the form $1/4\pi\varepsilon_0$. If we had not done so, Eq. 26-9—which is used more often in engineering practice than Coulomb's law—would have been less simple in form. We note further that Eq. 26-9 permits us to express the permittivity constant ε_0 in a unit more appropriate for use in problems involving capacitors; namely,

$$\varepsilon_0 = 8.85 \times 10^{-12}\ \text{F/m} = 8.85\ \text{pF/m}. \tag{26-10}$$

We have previously expressed this constant as

$$\varepsilon_0 = 8.85 \times 10^{-12}\ \text{C}^2/\text{N}\cdot\text{m}^2. \tag{26-11}$$

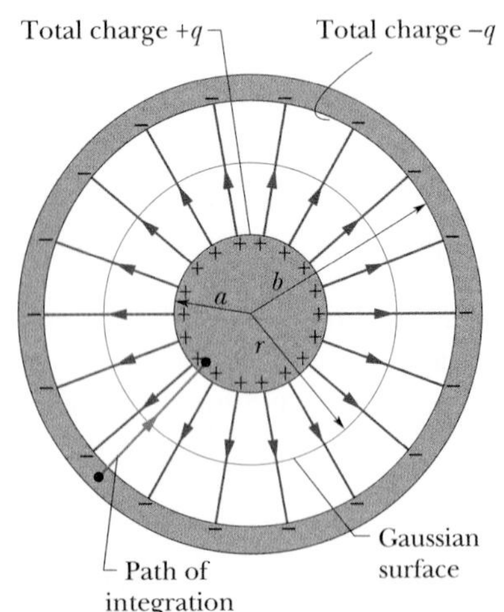

Fig. 26-6 A cross section of a long cylindrical capacitor, showing a cylindrical Gaussian surface of radius r (that encloses the positive plate) and the radial path of integration along which Eq. 26-6 is to be applied. This figure also serves to illustrate a spherical capacitor in a cross section through its center.

A Cylindrical Capacitor

Figure 26-6 shows, in cross section, a cylindrical capacitor of length L formed by two coaxial cylinders of radii a and b. We assume that $L \gg b$ so that we can neglect the fringing of the electric field that occurs at the ends of the cylinders. Each plate contains a charge of magnitude q.

As a Gaussian surface, we choose a cylinder of length L and radius r, closed by end caps and placed as is shown in Fig. 26-6. Equation 26-4 then yields

$$q = \varepsilon_0 EA = \varepsilon_0 E(2\pi rL),$$

in which $2\pi rL$ is the area of the curved part of the Gaussian surface. There is no flux through the end caps. Solving for E yields

$$E = \frac{q}{2\pi\varepsilon_0 Lr}. \tag{26-12}$$

Substitution of this result into Eq. 26-6 yields

$$V = \int_{-}^{+} E\,ds = -\frac{q}{2\pi\varepsilon_0 L}\int_b^a \frac{dr}{r} = \frac{q}{2\pi\varepsilon_0 L}\ln\left(\frac{b}{a}\right), \tag{26-13}$$

where we have used the fact that here $ds = -dr$ (we integrated radially inward). From the relation $C = q/V$, we then have

$$C = 2\pi\varepsilon_0 \frac{L}{\ln(b/a)} \quad \text{(cylindrical capacitor).} \tag{26-14}$$

We see that the capacitance of a cylindrical capacitor, like that of a parallel-plate capacitor, depends only on geometrical factors, in this case L, b, and a.

A Spherical Capacitor

Figure 26-6 can also serve as a central cross section of a capacitor that consists of two concentric spherical shells, of radii a and b. As a Gaussian surface we draw a sphere of radius r concentric with the two shells; then Eq. 26-4 yields

$$q = \varepsilon_0 EA = \varepsilon_0 E(4\pi r^2),$$

in which $4\pi r^2$ is the area of the spherical Gaussian surface. We solve this equation for E, obtaining

$$E = \frac{1}{4\pi\varepsilon_0}\frac{q}{r^2}, \tag{26-15}$$

which we recognize as the expression for the electric field due to a uniform spherical charge distribution (Eq. 24-15).

If we substitute this expression into Eq. 26-6, we find

$$V = \int_{-}^{+} E\,ds = -\frac{q}{4\pi\varepsilon_0}\int_b^a \frac{dr}{r^2} = \frac{q}{4\pi\varepsilon_0}\left(\frac{1}{a} - \frac{1}{b}\right) = \frac{q}{4\pi\varepsilon_0}\frac{b-a}{ab}, \tag{26-16}$$

where again we have substituted $-dr$ for ds. If we now substitute Eq. 26-16 into Eq. 26-1 and solve for C, we find

$$C = 4\pi\varepsilon_0 \frac{ab}{b-a} \quad \text{(spherical capacitor).} \tag{26-17}$$

An Isolated Sphere

We can assign a capacitance to a *single* isolated spherical conductor of radius R by assuming that the "missing plate" is a conducting sphere of infinite radius. After all, the field lines that leave the surface of a positively charged isolated conductor must end somewhere; the walls of the room in which the conductor is housed can serve effectively as our sphere of infinite radius.

To find the capacitance of the isolated conductor, we first rewrite Eq. 26-17 as

$$C = 4\pi\varepsilon_0 \frac{a}{1 - a/b}.$$

If we then let $b \to \infty$ and substitute R for a, we find

$$C = 4\pi\varepsilon_0 R \qquad \text{(isolated sphere).} \qquad (26\text{-}18)$$

Note that this formula and the others we have derived for capacitance (Eqs. 26-9, 26-14, and 26-17) involve the constant ε_0 multiplied by a quantity that has the dimensions of a length.

CHECKPOINT 2: For capacitors charged by the same battery, does the charge stored by the capacitor increase, decrease, or remain the same in each of the following situations? (a) The plate separation of a parallel-plate capacitor is increased. (b) The radius of the inner cylinder of a cylindrical capacitor is increased. (c) The radius of the outer spherical shell of a spherical capacitor is increased.

Sample Problem 26-1

A storage capacitor on a random access memory (RAM) chip has a capacitance of 55 fF. If the capacitor is charged to 5.3 V, how many excess electrons are on its negative plate?

SOLUTION: One **Key Idea** here is that we can find the number n of excess electrons on the negative plate if we know q, the total *amount* of excess charge on that plate. Then $n = q/e$, where e is the magnitude of the charge on each electron. A second **Key Idea** is that q is related to the potential difference V to which the capacitor is charged, according to Eq. 26-1 ($q = CV$). Combining these two ideas then gives us

$$n = \frac{q}{e} = \frac{CV}{e} = \frac{(55 \times 10^{-15}\ \text{F})(5.3\ \text{V})}{1.60 \times 10^{-19}\ \text{C}}$$

$$= 1.8 \times 10^6 \text{ electrons.} \qquad \text{(Answer)}$$

For electrons, this is a very small number. A speck of household dust, so tiny that it essentially never settles, contains about 10^{17} electrons (and the same number of protons).

26-4 Capacitors in Parallel and in Series

When there is a combination of capacitors in a circuit, we can sometimes replace that combination with an **equivalent capacitor**—that is, a single capacitor that has the same capacitance as the actual combination of capacitors. With such a replacement, we can simplify the circuit, affording easier solutions for unknown quantities of the circuit. Here we discuss two basic combinations of capacitors that allow such a replacement.

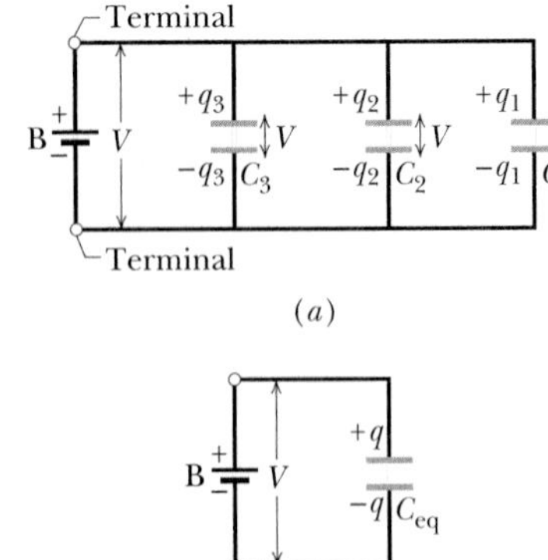

Fig. 26-7 (*a*) Three capacitors connected in parallel to battery B. The battery maintains potential difference V across its terminals and thus across *each* capacitor. (*b*) The equivalent capacitor, with capacitance C_{eq}, replaces the parallel combination.

Capacitors in Parallel

Figure 26-7*a* shows an electric circuit in which three capacitors are connected *in parallel* to battery B. This description has little to do with how the capacitor plates are drawn. Rather, "in parallel" means that the capacitors are directly wired together at one plate and directly wired together at the other plate, and that a potential difference V is applied across the two groups of wired-together plates. Thus, each capacitor has the same potential difference V, which produces charge on the capacitor. (In Fig. 26-7*a*, the applied potential V is maintained by the battery.) In general,

> When a potential difference V is applied across several capacitors connected in parallel, that potential difference V is applied across each capacitor. The total charge q stored on the capacitors is the sum of the charges stored on all the capacitors.

When we analyze a circuit of capacitors in parallel, we can simplify it with this mental replacement:

Capacitors connected in parallel can be replaced with an equivalent capacitor that has the same *total* charge q and the same potential difference V as the actual capacitors.

(You might remember this result with the nonsense word "par-V," which is close to "party.") Figure 26-7*b* shows the equivalent capacitor (with equivalent capacitance C_{eq}) that has replaced the three capacitors (with actual capacitances C_1, C_2, and C_3) of Fig. 26-7*a*.

To derive an expression for C_{eq} in Fig. 26-7*b*, we first use Eq. 26-1 to find the charge on each actual capacitor:

$$q_1 = C_1V, \quad q_2 = C_2V, \quad \text{and} \quad q_3 = C_3V.$$

The total charge on the parallel combination of Fig. 26-7*a* is then

$$q = q_1 + q_2 + q_3 = (C_1 + C_2 + C_3)V.$$

The equivalent capacitance, with the same total charge q and applied potential difference V as the combination, is then

$$C_{eq} = \frac{q}{V} = C_1 + C_2 + C_3,$$

a result that we can easily extend to any number n of capacitors, as

$$C_{eq} = \sum_{j=1}^{n} C_j \quad (n \text{ capacitors in parallel}). \tag{26-19}$$

Thus, to find the equivalent capacitance of a parallel combination, we simply add the individual capacitances.

Capacitors in Series

Figure 26-8*a* shows three capacitors connected *in series* to battery B. This description has little to do with how the capacitors are drawn. Rather, "in series" means that the capacitors are wired serially, one after the other, and that a potential difference V is applied across the two ends of the series. (In Fig. 26-8*a*, this potential difference V is maintained by battery B.) The potential differences that then exist across the capacitors in the series produce identical charges q on them.

When a potential difference V is applied across several capacitors connected in series, the capacitors have identical charges q. The sum of the potential differences across all the capacitors is equal to the applied potential difference V.

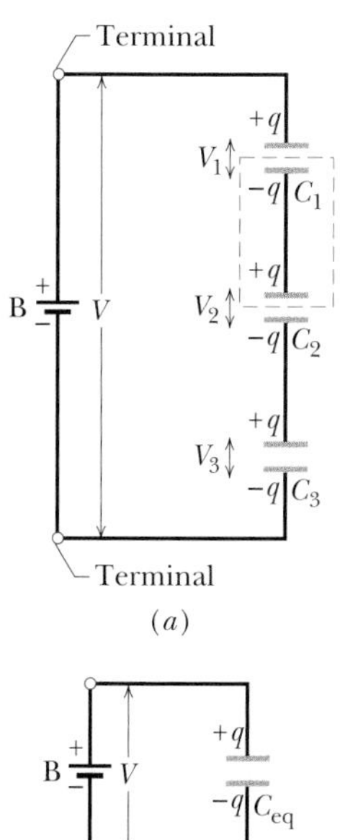

Fig. 26-8 (*a*) Three capacitors connected in series to battery B. The battery maintains potential difference V between the top and bottom plates of the series combination. (*b*) The equivalent capacitor, with capacitance C_{eq}, replaces the series combination.

We can explain how the capacitors end up with identical charges by following a *chain reaction* of events, in which the charging of each capacitor causes the charging of the next capacitor. We start with capacitor 3 and work upward to capacitor 1. When the battery is first connected to the series of capacitors, it produces charge $-q$ on the bottom plate of capacitor 3. That charge then repels negative charge from the top plate of capacitor 3 (leaving it with charge $+q$). The repelled negative charge moves to the bottom plate of capacitor 2 (giving it charge $-q$). That charge on the bottom plate of capacitor 2 then repels negative charge from the top plate of capacitor 2 (leaving it with charge $+q$) to the bottom plate of capacitor 1 (giving it charge $-q$). Finally the charge on the bottom plate of capacitor 1 helps move negative charge from the top plate of capacitor 1 to the battery, leaving that top plate with charge $+q$.

Here are two important points about capacitors in series:

1. When charge is shifted from one capacitor to another in a series of capacitors, it can move along only one route, such as from capacitor 3 to capacitor 2 in Fig. 26-8*a*. If there are additional routes, the capacitors are not in series. An example is given in Sample Problem 26-2.
2. The battery directly produces charges on only the two plates to which it is connected (the bottom plate of capacitor 3 and the top plate of capacitor 1 in Fig. 26-8*a*). Charges that are produced on the other plates are due merely to the shifting of charge already there. For example, in Fig. 26-8, the part of the circuit enclosed by dashed lines is electrically isolated from the rest of the circuit. Thus, the net charge of that part cannot be changed by the battery—its charge can only be redistributed.

When we analyze a circuit of capacitors in series, we can simplify it with this mental replacement:

> Capacitors that are connected in series can be replaced with an equivalent capacitor that has the same charge q and the same *total* potential difference V as the actual series capacitors.

(You might remember this with the nonsense word "seri-q.") Figure 26-8*b* shows the equivalent capacitor (with equivalent capacitance C_{eq}) that has replaced the three actual capacitors (with actual capacitances C_1, C_2, and C_3) of Fig. 26-8*a*.

To derive an expression for C_{eq} in Fig. 26-8*b*, we first use Eq. 26-1 to find the potential difference of each actual capacitor:

$$V_1 = \frac{q}{C_1}, \quad V_2 = \frac{q}{C_2}, \quad \text{and} \quad V_3 = \frac{q}{C_3}.$$

The total potential difference V due to the battery is the sum of these three potential differences. Thus,

$$V = V_1 + V_2 + V_3 = q\left(\frac{1}{C_1} + \frac{1}{C_2} + \frac{1}{C_3}\right).$$

The equivalent capacitance is then

$$C_{eq} = \frac{q}{V} = \frac{1}{1/C_1 + 1/C_2 + 1/C_3},$$

or

$$\frac{1}{C_{eq}} = \frac{1}{C_1} + \frac{1}{C_2} + \frac{1}{C_3}.$$

We can easily extend this to any number n of capacitors as

$$\frac{1}{C_{eq}} = \sum_{j=1}^{n} \frac{1}{C_j} \quad \text{(n capacitors in series).} \tag{26-20}$$

Using Eq. 26-20 you can show that the equivalent of a series of capacitances is always *less* than the least capacitance in the series.

CHECKPOINT 3: A battery of potential V stores charge q on a combination of two identical capacitors. What are the potential difference across and the charge on either capacitor if the capacitors are (a) in parallel and (b) in series?

Sample Problem 26-2

(a) Find the equivalent capacitance for the combination of capacitances shown in Fig. 26-9*a*, across which potential difference V is applied. Assume

$$C_1 = 12.0\ \mu\text{F}, \quad C_2 = 5.30\ \mu\text{F}, \quad \text{and} \quad C_3 = 4.50\ \mu\text{F}.$$

SOLUTION: The **Key Idea** here is that any capacitors connected in series can be replaced with their equivalent capacitor, and any capacitors connected in parallel can be replaced with their equivalent capacitor. Therefore, we should first check whether any of the capacitors in Fig. 26-9*a* are in parallel or series.

Capacitors 1 and 3 are connected one after the other, but are they in series? No. The potential V that is applied to the capacitors produces charge on the bottom plate of capacitor 3. That charge causes charge to shift from the top plate of capacitor 3. However, note that the shifting charge can move to the bottom plates of both capacitor 1 and capacitor 2. Because there is more than one route for the shifting charge, capacitor 3 is *not* in series with capacitor 1 (or capacitor 2).

Are capacitor 1 and capacitor 2 in parallel? Yes. Their top plates are directly wired together and their bottom plates are directly wired together, and electric potential is applied between the top-plate pair and the bottom-plate pair. Thus, capacitor 1 and capacitor 2 are in parallel, and Eq. 26-19 tells us that their equivalent capacitance C_{12} is

$$C_{12} = C_1 + C_2 = 12.0\ \mu\text{F} + 5.30\ \mu\text{F} = 17.3\ \mu\text{F}.$$

In Fig. 26-9*b*, we have replaced capacitors 1 and 2 with their equivalent capacitor, call it capacitor 12 (say "one two"). (The connections at points A and B are exactly the same in Figs. 26-9*a* and *b*.)

Is capacitor 12 in series with capacitor 3? Again applying the test for series capacitances, we see that the charge that shifts from the top plate of capacitor 3 must entirely go to the bottom plate of capacitor 12. Thus, capacitor 12 and capacitor 3 are in series, and we can replace them with their equivalent C_{123}, as shown in Fig. 26-9*c*. From Eq. 26-20, we have

$$\frac{1}{C_{123}} = \frac{1}{C_{12}} + \frac{1}{C_3} = \frac{1}{17.3\ \mu\text{F}} + \frac{1}{4.50\ \mu\text{F}} = 0.280\ \mu\text{F}^{-1},$$

from which

$$C_{123} = \frac{1}{0.280\ \mu\text{F}^{-1}} = 3.57\ \mu\text{F}. \qquad \text{(Answer)}$$

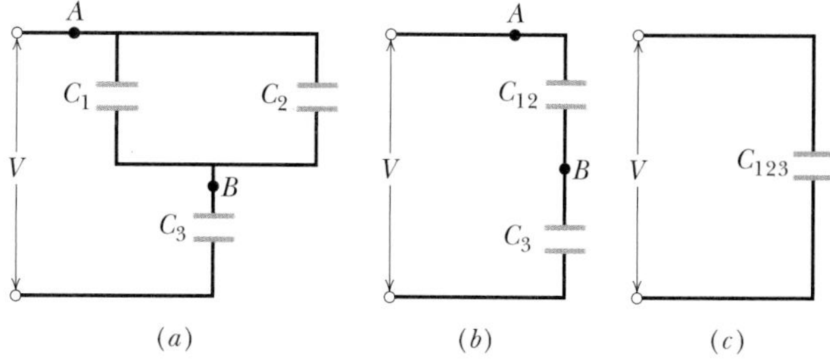

Fig. 26-9 Sample Problem 26-2. (*a*) Three capacitors. (*b*) C_1 and C_2, a parallel combination, are replaced by C_{12}. (*c*) C_{12} and C_3, a series combination, are replaced by the equivalent capacitance C_{123}.

(b) The potential difference that is applied to the input terminals in Fig. 26-9*a* is $V = 12.5$ V. What is the charge on C_1?

SOLUTION: One **Key Idea** here is that, to get the charge q_1 on capacitor 1, we now have to work backward to that capacitor, starting with the equivalent capacitor 123. Since the given potential difference V (= 12.5 V) is applied across the actual combination of three capacitors in Fig. 26-9*a*, it is also applied across capacitor 123 in Fig. 26-9*c*. Thus, Eq. 26-1 ($q = CV$) gives us

$$q_{123} = C_{123}V = (3.57\ \mu\text{F})(12.5\ \text{V}) = 44.6\ \mu\text{C}.$$

A second **Key Idea** is that the series capacitors 12 and 3 in Fig. 26-9*b* have the same charge as their equivalent capacitor 123 (recall "seri-q"). Thus, capacitor 12 has charge $q_{12} = q_{123} = 44.6\ \mu$C. From Eq. 26-1, the potential difference across capacitor 12 must be

$$V_{12} = \frac{q_{12}}{C_{12}} = \frac{44.6\ \mu\text{C}}{17.3\ \mu\text{F}} = 2.58\ \text{V}.$$

A third **Key Idea** is that the parallel capacitors 1 and 2 both have the same potential difference as their equivalent capacitor 12 (recall "par-V"). Thus, capacitor 1 has the potential difference $V_1 = V_{12} = 2.58$ V. Thus, from Eq. 26-1, the charge on capacitor 1 must be

$$q_1 = C_1V_1 = (12.0\ \mu\text{F})(2.58\ \text{V})$$
$$= 31.0\ \mu\text{C}. \qquad \text{(Answer)}$$

Sample Problem 26-3

Capacitor 1, with $C_1 = 3.55\ \mu$F, is charged to a potential difference $V_0 = 6.30$ V, using a 6.30 V battery. The battery is then removed and the capacitor is connected as in Fig. 26-10 to an uncharged capacitor 2, with $C_2 = 8.95\ \mu$F. When switch S is closed, charge flows between the capacitors until they have the same potential difference V. Find V.

SOLUTION: The situation here differs from the previous example because an applied electric potential is *not* maintained across a combination of capacitors by a battery or some other source. Here, just

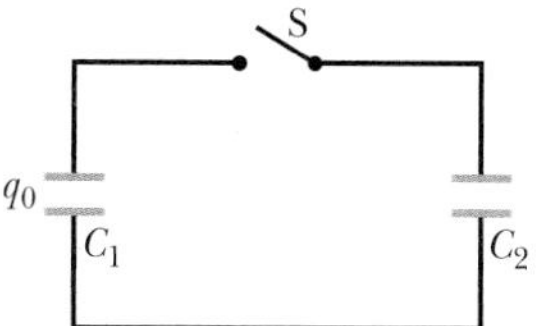

Fig. 26-10 Sample Problem 26-3. A potential difference V_0 is applied to capacitor 1 and the charging battery is removed. Switch S is then closed so that the charge on capacitor 1 is shared with capacitor 2.

after switch S is closed, the only applied electric potential is that of capacitor 1 on capacitor 2, and that potential is decreasing. Thus, although the capacitors in Fig. 26-10 are connected end to end, in this situation they are not *in series*; and although they are drawn parallel, in this situation they are not *in parallel*.

To find the final electric potential (when the system comes to equilibrium and charge stops flowing), we use this Key Idea: After the switch is closed, the original charge q_0 on capacitor 1 is redistributed (shared) between capacitor 1 and capacitor 2. When equilibrium is reached, we can relate the original charge q_0 with the final charges q_1 and q_2 by writing

$$q_0 = q_1 + q_2.$$

Applying the relation $q = CV$ to each term of this equation yields

$$C_1V_0 = C_1V + C_2V,$$

from which

$$V = V_0\frac{C_1}{C_1 + C_2} = \frac{(6.30\text{ V})(3.55\ \mu\text{F})}{3.55\ \mu\text{F} + 8.95\ \mu\text{F}}$$
$$= 1.79\text{ V}. \qquad \text{(Answer)}$$

When the capacitors reach this value of electric potential difference, the charge flow stops.

CHECKPOINT 4: In this sample problem, suppose capacitor 2 is replaced by a series combination of capacitors 3 and 4. (a) After the switch is closed and charge has stopped flowing, what is the relation between the initial charge q_0, the charge q_1 then on capacitor 1, and the charge q_{34} then on the equivalent capacitor 34? (b) If $C_3 > C_4$, is the charge q_3 on capacitor 3 more than, less than, or equal to the charge q_4 on capacitor 4?

PROBLEM-SOLVING TACTICS

Tactic 2: *Multiple-Capacitor Circuits*

Let us review the procedure used in the solution of Sample Problem 26-2, in which several capacitors are connected to a battery. To find a single equivalent capacitance, we simplify the given arrangement of capacitances by replacing them, in steps, with equivalent capacitances, using Eq. 26-19 when we find capacitances in parallel and Eq. 26-20 when we find capacitances in series. Then, to find the charge stored by that single equivalent capacitance, we use Eq. 26-1 and the potential difference V applied by the battery.

That result tells us the net charge stored on the actual arrangement of capacitors. However, to find the charge on, or the potential difference across, any particular capacitor in the actual arrangement, we need to reverse our steps of simplification. With each reversed step, we use these two rules: When capacitances are in parallel, they have the same potential difference as their equivalent capacitance, and we use Eq. 26-1 to find the charge on each capacitance; when they are in series, they have the same charge as their equivalent capacitance, and we use Eq. 26-1 to find the potential difference across each capacitance.

Tactic 3: *Batteries and Capacitors*

A battery maintains a certain potential difference across its terminals. Thus, when capacitor 1 of Sample Problem 26-3 is connected to the 6.30 V battery, charge flows between the capacitor and the battery until the capacitor has the same potential difference across it as the battery.

A capacitor differs from a battery in that a capacitor lacks the internal electrochemical reactions needed to release charged particles (electrons) from internal atoms and molecules. Thus, when the charged capacitor 1 of Sample Problem 26-3 is disconnected from the battery and then connected to the uncharged capacitor 2 with switch S closed, the potential difference across capacitor 1 is not maintained. The quantity that *is* maintained is the charge q_0 of the two-capacitor system; that is, charge obeys a conservation law, *not* electric potential.

26-5 Energy Stored in an Electric Field

Work must be done by an external agent to charge a capacitor. Starting with an uncharged capacitor, for example, imagine that—using "magic tweezers"—you remove electrons from one plate and transfer them one at a time to the other plate. The electric field that builds up in the space between the plates has a direction that tends to oppose further transfer. Thus, as charge accumulates on the capacitor plates, you have to do increasingly larger amounts of work to transfer additional electrons. In practice, this work is done not by "magic tweezers" but by a battery, at the expense of its store of chemical energy.

We visualize the work required to charge a capacitor as being stored in the form of **electric potential energy** U in the electric field between the plates. You can recover this energy at will, by discharging the capacitor in a circuit, just as you can recover the potential energy stored in a stretched bow by releasing the bowstring to transfer the energy to the kinetic energy of an arrow.

Suppose that, at a given instant, a charge q' has been transferred from one plate of a capacitor to the other. The potential difference V' between the plates at that instant will be q'/C. If an extra increment of charge dq' is then transferred, the increment of work required will be, from Eq. 25-7,

$$dW = V'\,dq' = \frac{q'}{C}\,dq'.$$

The work required to bring the total capacitor charge up to a final value q is

$$W = \int dW = \frac{1}{C}\int_0^q q'\,dq' = \frac{q^2}{2C}.$$

This work is stored as potential energy U in the capacitor, so that

$$U = \frac{q^2}{2C} \quad \text{(potential energy).} \tag{26-21}$$

From Eq. 26-1, we can also write this as

$$U = \tfrac{1}{2}CV^2 \quad \text{(potential energy).} \tag{26-22}$$

Equations 26-21 and 26-22 hold no matter what the geometry of the capacitor is.

To gain some physical insight into energy storage, consider two parallel-plate capacitors that are identical except that capacitor 1 has twice the plate separation of capacitor 2. Then capacitor 1 has twice the volume between its plates and also, from Eq. 26-9, half the capacitance of capacitor 2. Equation 26-4 tells us that if both capacitors have the same charge q, the electric fields between their plates are identical. And Eq. 26-21 tells us that capacitor 1 has twice the stored potential energy of capacitor 2. Thus, of two otherwise identical capacitors with the same charge and same electric field, the one with twice the volume between its plates has twice the stored potential energy. Arguments like this tend to verify our earlier assumption:

➤ The potential energy of a charged capacitor may be viewed as being stored in the electric field between its plates.

The Medical Defibrillator

The ability of a capacitor to store potential energy is the basis of *defibrillator* devices, which are used by emergency medical teams to stop the fibrillation of heart attack victims. In the portable version, a battery charges a capacitor to a high potential difference, storing a large amount of energy in less than a minute. The battery maintains only a modest potential difference; an electronic circuit repeatedly uses that potential difference to greatly increase the potential difference of the capacitor. The power, or rate of energy transfer, during this process is also modest.

Conducting leads ("paddles") are placed on the victim's chest. When a control switch is closed, the capacitor sends a portion of its stored energy from paddle to paddle through the victim. As an example, when a 70 μF capacitor in a defibrillator is charged to 5000 V, Eq. 26-22 gives the energy stored in the capacitor as

$$U = \tfrac{1}{2}CV^2 = \tfrac{1}{2}(70 \times 10^{-6}\text{ F})(5000\text{ V})^2 = 875\text{ J}.$$

About 200 J of this energy is sent through the victim during a pulse of about 2.0 ms. The power of the pulse is

$$P = \frac{U}{t} = \frac{200\text{ J}}{2.0 \times 10^{-3}\text{ s}} = 100\text{ kW},$$

Fig. 26-11 To photograph a bullet blowing apart a banana, Harold Edgerton, the inventor of the stroboscope, used a capacitor to dump electrical energy into one of his stroboscopic lamps, which then brightly illuminated the banana for only 0.3 μs.

which is much greater than the power of the battery itself. This same technique of slowly charging a capacitor with a battery and then discharging the capacitor at a much higher power is commonly used in flash photography and stroboscopic photography (Fig. 26-11).

Energy Density

In a parallel-plate capacitor, neglecting fringing, the electric field has the same value at all points between the plates. Thus, the **energy density** u—that is, the potential energy per unit volume between the plates—should also be uniform. We can find u by dividing the total potential energy by the volume Ad of the space between the plates. Using Eq. 26-22, we obtain

$$u = \frac{U}{Ad} = \frac{CV^2}{2Ad}.$$

With Eq. 26-9 ($C = \varepsilon_0 A/d$), this result becomes

$$u = \tfrac{1}{2}\varepsilon_0\left(\frac{V}{d}\right)^2.$$

However, from Eq. 25-42, V/d equals the electric field magnitude E, so

$$u = \tfrac{1}{2}\varepsilon_0 E^2 \quad \text{(energy density).} \tag{26-23}$$

Although we derived this result for the special case of a parallel-plate capacitor, it holds generally, whatever may be the source of the electric field. If an electric field $\vec{E}$ exists at any point in space, we can think of that point as a site of electric potential energy whose amount per unit volume is given by Eq. 26-23.

Sample Problem 26-4

An isolated conducting sphere whose radius R is 6.85 cm has a charge $q = 1.25$ nC.

(a) How much potential energy is stored in the electric field of this charged conductor?

SOLUTION: The Key Idea here is that the energy U stored in a capacitor depends on the charge q on the capacitor and the capacitance C of the capacitor, according to Eq. 26-21. Substituting from Eq. 26-18 for C, Eq. 26-21 gives us

$$U = \frac{q^2}{2C} = \frac{q^2}{8\pi\varepsilon_0 R}$$
$$= \frac{(1.25 \times 10^{-9}\ \text{C})^2}{(8\pi)(8.85 \times 10^{-12}\ \text{F/m})(0.0685\ \text{m})}$$
$$= 1.03 \times 10^{-7}\ \text{J} = 103\ \text{nJ}. \quad \text{(Answer)}$$

(b) What is the energy density at the surface of the sphere?

SOLUTION: The Key Idea here is that the density u of the energy stored in an electric field depends on the magnitude E of the field, according to Eq. 26-23 ($u = \frac{1}{2}\varepsilon_0 E^2$), so we must first find E at the surface of the sphere. This is given by Eq. 24-15:

$$E = \frac{1}{4\pi\varepsilon_0}\frac{q}{R^2}.$$

The energy density is then

$$u = \tfrac{1}{2}\varepsilon_0 E^2 = \frac{q^2}{32\pi^2\varepsilon_0 R^4}$$
$$= \frac{(1.25 \times 10^{-9}\ \text{C})^2}{(32\pi^2)(8.85 \times 10^{-12}\ \text{C}^2/\text{N}\cdot\text{m}^2)(0.0685\ \text{m})^4}$$
$$= 2.54 \times 10^{-5}\ \text{J/m}^3 = 25.4\ \mu\text{J/m}^3. \quad \text{(Answer)}$$

26-6 Capacitor with a Dielectric

If you fill the space between the plates of a capacitor with a *dielectric*, which is an insulating material such as mineral oil or plastic, what happens to the capacitance? Michael Faraday—to whom the whole concept of capacitance is largely due and for whom the SI unit of capacitance is named—first looked into this matter in 1837. Using simple equipment much like that shown in Fig. 26-12, he found that the

Fig. 26-12 The simple electrostatic apparatus used by Faraday. An assembled apparatus (second from left) forms a spherical capacitor consisting of a central brass ball and a concentric brass shell. Faraday placed dielectric materials in the space between the ball and the shell.

TABLE 26-1 **Some Properties of Dielectrics**[a]

Material	Dielectric Constant κ	Dielectric Strength (kV/mm)
Air (1 atm)	1.00054	3
Polystyrene	2.6	24
Paper	3.5	16
Transformer oil	4.5	
Pyrex	4.7	14
Ruby mica	5.4	
Porcelain	6.5	
Silicon	12	
Germanium	16	
Ethanol	25	
Water (20°C)	80.4	
Water (25°C)	78.5	
Titania ceramic	130	
Strontium titanate	310	8

For a vacuum, κ = unity.

[a]Measured at room temperature, except for the water.

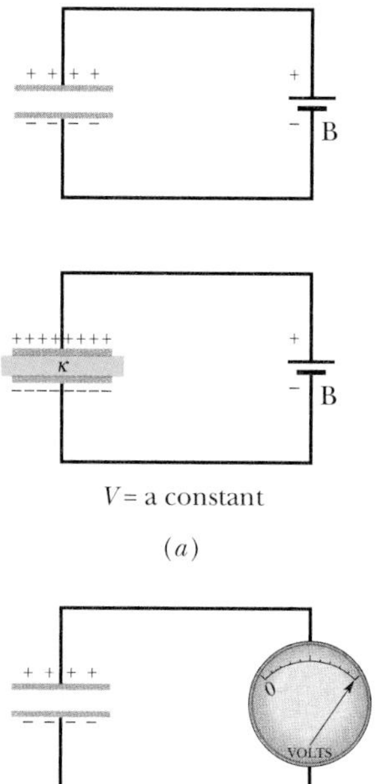

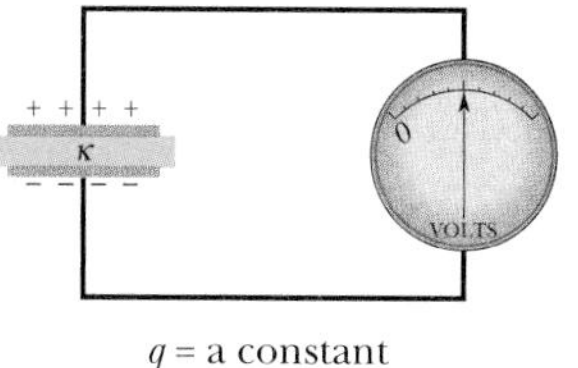

capacitance *increased* by a numerical factor κ, which he called the **dielectric constant** of the insulating material. Table 26-1 shows some dielectric materials and their dielectric constants. The dielectric constant of a vacuum is unity by definition. Because air is mostly empty space, its measured dielectric constant is only slightly greater than unity.

Another effect of the introduction of a dielectric is to limit the potential difference that can be applied between the plates to a certain value V_{max}, called the *breakdown potential.* If this value is substantially exceeded, the dielectric material will break down and form a conducting path between the plates. Every dielectric material has a characteristic *dielectric strength,* which is the maximum value of the electric field that it can tolerate without breakdown. A few such values are listed in Table 26-1.

As we discussed in connection with Eq. 26-18, the capacitance of any capacitor can be written in the form

$$C = \varepsilon_0 \mathcal{L}, \tag{26-24}$$

in which $\mathcal{L}$ has the dimensions of a length. For example, $\mathcal{L} = A/d$ for a parallel-plate capacitor. Faraday's discovery was that, with a dielectric *completely* filling the space between the plates, Eq. 26-24 becomes

$$C = \kappa\varepsilon_0 \mathcal{L} = \kappa C_{air}, \tag{26-25}$$

where C_{air} is the value of the capacitance with only air between the plates.

Figure 26-13 provides some insight into Faraday's experiments. In Fig. 26-13*a* the battery ensures that the potential difference V between the plates will remain

Fig. 26-13 (*a*) If the potential difference between the plates of a capacitor is maintained, as by battery B, the effect of a dielectric is to increase the charge on the plates. (*b*) If the charge on the capacitor plates is maintained, as in this case, the effect of a dielectric is to reduce the potential difference between the plates. The scale shown is that of a *potentiometer*, a device used to measure potential difference (here, between the plates). A capacitor cannot discharge through a potentiometer.

constant. When a dielectric slab is inserted between the plates, the charge q on the plates increases by a factor of κ; the additional charge is delivered to the capacitor plates by the battery. In Fig. 26-13*b* there is no battery and therefore the charge q must remain constant when the dielectric slab is inserted; then the potential difference V between the plates decreases by a factor of κ. Both these observations are consistent (through the relation $q = CV$) with the increase in capacitance caused by the dielectric.

Comparison of Eqs. 26-24 and 26-25 suggests that the effect of a dielectric can be summed up in more general terms:

> In a region completely filled by a dielectric material of dielectric constant κ, all electrostatic equations containing the permittivity constant ε_0 are to be modified by replacing ε_0 with $\kappa\varepsilon_0$.

Thus, a point charge inside a dielectric produces an electric field that, by Coulomb's law, has the magnitude

$$E = \frac{1}{4\pi\kappa\varepsilon_0}\frac{q}{r^2}. \tag{26-26}$$

Also, the expression for the electric field just outside an isolated conductor immersed in a dielectric (see Eq. 24-11) becomes

$$E = \frac{\sigma}{\kappa\varepsilon_0}. \tag{26-27}$$

Both these equations show that *for a fixed distribution of charges, the effect of a dielectric is to weaken the electric field* that would otherwise be present.

Sample Problem 26-5

A parallel-plate capacitor whose capacitance C is 13.5 pF is charged by a battery to a potential difference $V = 12.5$ V between its plates. The charging battery is now disconnected and a porcelain slab ($\kappa = 6.50$) is slipped between the plates. What is the potential energy of the capacitor–slab device, both before and after the slab is put into place?

SOLUTION: The Key Idea here is that we can relate the potential energy U of the capacitor to the capacitance C and either the potential V (with Eq. 26-22) or the charge q (with Eq. 26-21):

$$U_i = \tfrac{1}{2}CV^2 = \frac{q^2}{2C}.$$

Because we are given the initial potential V (= 12.5 V), we use Eq. 26-22 to find the initial stored energy:

$$U_i = \tfrac{1}{2}CV^2 = \tfrac{1}{2}(13.5 \times 10^{-12}\ \text{F})(12.5\ \text{V})^2$$
$$= 1.055 \times 10^{-9}\ \text{J} = 1055\ \text{pJ} \approx 1100\ \text{pJ}. \quad \text{(Answer)}$$

To find the final potential energy U_f (after the slab is introduced), we need another Key Idea: Because the battery has been disconnected, the charge on the capacitor cannot change when the dielectric is inserted. However, the potential *does* change. Thus, we must now use Eq. 26-21 (based on q) to write the final potential energy U_f, but now that the slab is within the capacitor, the capacitance is κC. We then have

$$U_f = \frac{q^2}{2\kappa C} = \frac{U_i}{\kappa} = \frac{1055\ \text{pJ}}{6.50} = 162\ \text{pJ} \approx 160\ \text{pJ}. \quad \text{(Answer)}$$

When the slab is introduced, the potential energy decreases by a factor of κ.

The "missing" energy, in principle, would be apparent to the person who introduced the slab. The capacitor would exert a tiny tug on the slab and would do work on it, in amount

$$W = U_i - U_f = (1055 - 162)\ \text{pJ} = 893\ \text{pJ}.$$

If the slab were allowed to slide between the plates with no restraint and if there were no friction, the slab would oscillate back and forth between the plates with a (constant) mechanical energy of 893 pJ, and this system energy would transfer back and forth between kinetic energy of the moving slab and potential energy stored in the electric field.

CHECKPOINT 5: If the battery in this sample problem remains connected, do the following increase, decrease, or remain the same when the slab is introduced: (a) the potential difference between the capacitor plates, (b) the capacitance, (c) the charge on the capacitor, (d) the potential energy of the device, (e) the electric field between the plates? (*Hint:* For (e), note that the charge is not fixed.)

26-7 Dielectrics: An Atomic View

What happens, in atomic and molecular terms, when we put a dielectric in an electric field? There are two possibilities, depending on the nature of the molecules:

1. *Polar dielectrics.* The molecules of some dielectrics, like water, have permanent electric dipole moments. In such materials (called *polar dielectrics*), the electric dipoles tend to line up with an external electric field as in Fig. 26-14. Because the molecules are continuously jostling each other as a result of their random thermal motion, this alignment is not complete, but it becomes more complete as the magnitude of the applied field is increased (or as the temperature, and thus the jostling, is decreased). The alignment of the electric dipoles produces an electric field that is directed opposite the applied field and smaller in magnitude.
2. *Nonpolar dielectrics.* Regardless of whether they have permanent electric dipole moments, molecules acquire dipole moments by induction when placed in an external electric field. In Section 25-7 (see Fig. 25-12), we saw that this occurs because the external field tends to "stretch" the molecules, slightly separating the centers of negative and positive charge.

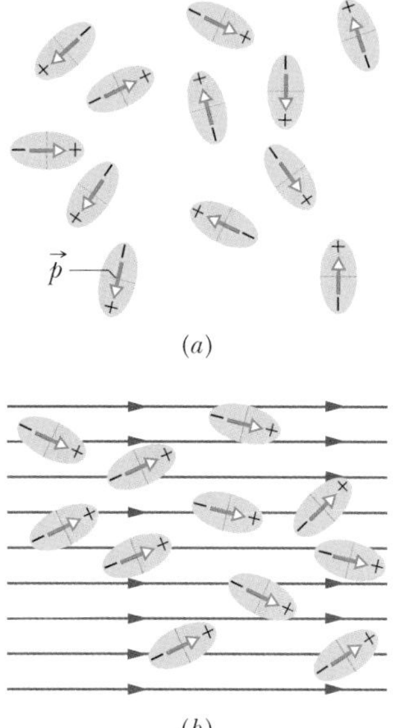

Fig. 26-14 (*a*) Molecules with a permanent electric dipole moment, showing their random orientation in the absence of an external electric field. (*b*) An electric field is applied, producing partial alignment of the dipoles. Thermal agitation prevents complete alignment.

Figure 26-15*a* shows a nonpolar dielectric slab with no external electric field applied. In Fig. 26-15*b*, an electric field $\vec{E}_0$ is applied via a capacitor, whose plates are charged as shown. The result is a slight separation of the centers of the positive and negative charge distributions within the slab, producing positive charge on one face of the slab (due to the positive ends of dipoles there) and negative charge on the opposite face (due to the negative ends of dipoles there). The slab as a whole remains electrically neutral and—within the slab—there is no excess charge in any volume element.

Figure 26-15*c* shows that the induced surface charges on the faces produce an electric field $\vec{E}'$ in the direction opposite that of the applied electric field $\vec{E}_0$. The resultant field $\vec{E}$ inside the dielectric (the vector sum of fields $\vec{E}_0$ and $\vec{E}'$) has the direction of $\vec{E}_0$ but is smaller in magnitude.

Both the field $\vec{E}'$ produced by the surface charges in Fig. 26-15*c* and the electric field produced by the permanent electric dipoles in Fig. 26-14 act in the same way—they oppose the applied field $\vec{E}$. Thus, the effect of both polar and nonpolar dielectrics is to weaken any applied field within them, as between the plates of a capacitor.

We can now see why the dielectric porcelain slab in Sample Problem 26-5 is pulled into the capacitor: As it enters the space between the plates, the surface charge that appears on each slab face has the sign opposite that of the charge on the nearby capacitor plate. Thus, slab and plates attract each other.

(*a*)
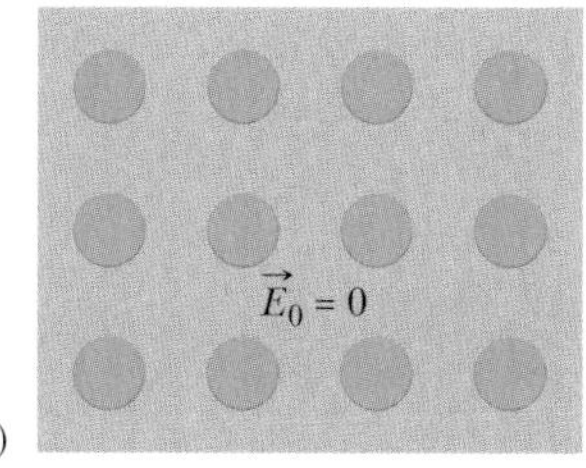

(*b*)
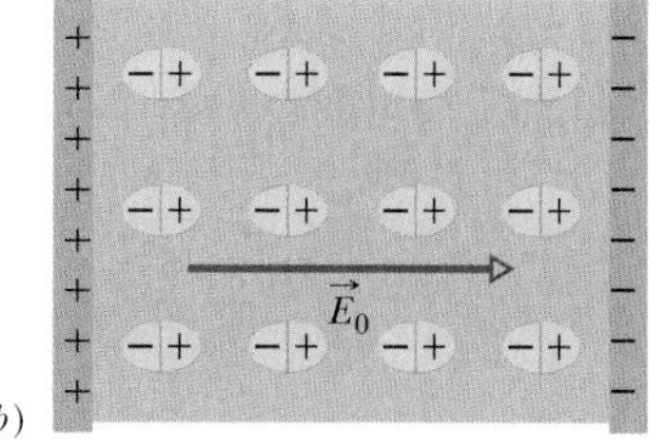

(*c*)
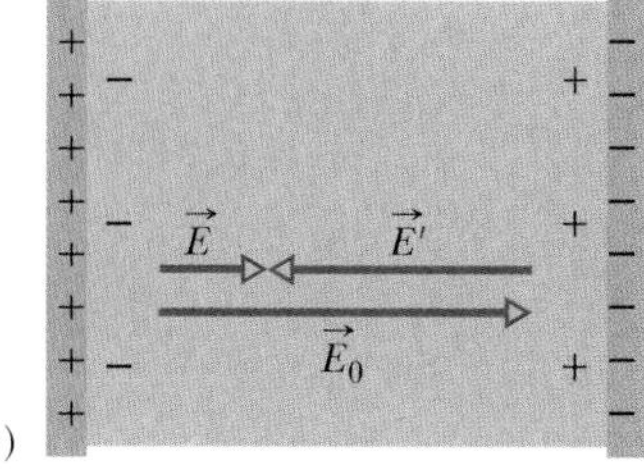

Fig. 26-15 (*a*) A nonpolar dielectric slab. The circles represent the electrically neutral atoms within the slab. (*b*) An electric field is applied via charged capacitor plates; the field slightly stretches the atoms, separating the centers of positive and negative charge. (*c*) The separation produces surface charges on the slab faces. These charges set up a field $\vec{E}'$, which opposes the applied field $\vec{E}_0$. The resultant field $\vec{E}$ inside the dielectric (the vector sum of $\vec{E}_0$ and $\vec{E}'$) has the same direction as $\vec{E}_0$ but less magnitude.

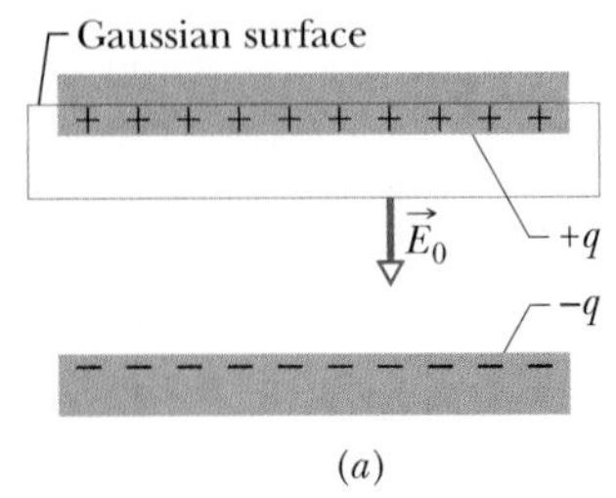

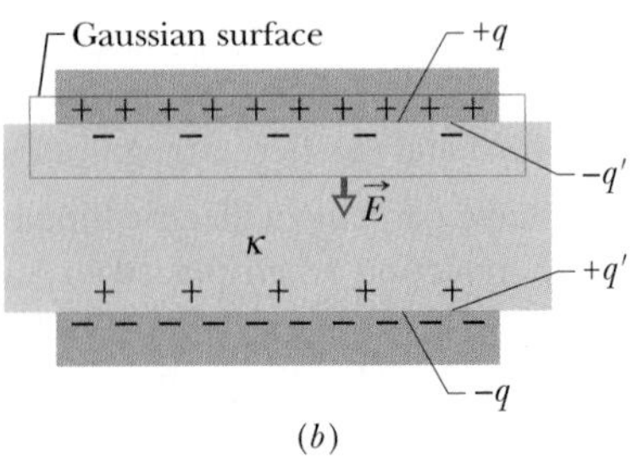

Fig. 26-16 A parallel-plate capacitor (*a*) without and (*b*) with a dielectric slab inserted. The charge q on the plates is assumed to be the same in both cases.

26-8 Dielectrics and Gauss' Law

In our discussion of Gauss' law in Chapter 24, we assumed that the charges existed in a vacuum. Here we shall see how to modify and generalize that law if dielectric materials, such as those listed in Table 26-1, are present. Figure 26-16 shows a parallel-plate capacitor of plate area A, both with and without a dielectric. We assume that the charge q on the plates is the same in both situations. Note that the field between the plates induces charges on the faces of the dielectric by one of the methods of Section 26-7.

For the situation of Fig. 26-16*a*, without a dielectric, we can find the electric field $\vec{E}_0$ between the plates as we did in Fig. 26-5: We enclose the charge $+q$ on the top plate with a Gaussian surface and then apply Gauss' law. Letting E_0 represent the magnitude of the field, we find

$$\varepsilon_0 \oint \vec{E} \cdot d\vec{A} = \varepsilon_0 E_0 A = q, \tag{26-28}$$

or

$$E_0 = \frac{q}{\varepsilon_0 A}. \tag{26-29}$$

In Fig. 26-16*b*, with the dielectric in place, we can find the electric field between the plates (and within the dielectric) by using the same Gaussian surface. However, now the surface encloses two types of charge: It still encloses charge $+q$ on the top plate, but it now also encloses the induced charge $-q'$ on the top face of the dielectric. The charge on the conducting plate is said to be *free charge* because it can move if we change the electric potential of the plate; the induced charge on the surface of the dielectric is not free charge because it cannot move from that surface.

The net charge enclosed by the Gaussian surface in Fig. 26-16*b* is $q - q'$, so Gauss' law now gives

$$\varepsilon_0 \oint \vec{E} \cdot d\vec{A} = \varepsilon_0 E A = q - q', \tag{26-30}$$

or

$$E = \frac{q - q'}{\varepsilon_0 A}. \tag{26-31}$$

The effect of the dielectric is to weaken the original field E_0 by a factor of κ, so we may write

$$E = \frac{E_0}{\kappa} = \frac{q}{\kappa \varepsilon_0 A}. \tag{26-32}$$

Comparison of Eqs. 26-31 and 26-32 shows that

$$q - q' = \frac{q}{\kappa}. \tag{26-33}$$

Equation 26-33 shows correctly that the magnitude q' of the induced surface charge is less than that of the free charge q and is zero if no dielectric is present (then, $\kappa = 1$ in Eq. 26-33).

By substituting for $q - q'$ from Eq. 26-33 in Eq. 26-30, we can write Gauss' law in the form

$$\varepsilon_0 \oint \kappa \vec{E} \cdot d\vec{A} = q \quad \text{(Gauss' law with dielectric).} \tag{26-34}$$

This important equation, although derived for a parallel-plate capacitor, is true generally and is the most general form in which Gauss' law can be written. Note the following:

1. The flux integral now involves $\kappa\vec{E}$, not just $\vec{E}$. (The vector $\varepsilon_0\kappa\vec{E}$ is sometimes called the *electric displacement* $\vec{D}$, so that Eq. 26-34 can be written in the form $\oint \vec{D} \cdot d\vec{A} = q$.)
2. The charge q enclosed by the Gaussian surface is now taken to be the *free charge only*. The induced surface charge is deliberately ignored on the right side of Eq. 26-34, having been taken fully into account by introducing the dielectric constant κ on the left side.
3. Equation 26-34 differs from Eq. 24-7, our original statement of Gauss' law, only in that ε_0 in the latter equation has been replaced by $\kappa\varepsilon_0$. We keep κ inside the integral of Eq. 26-34 to allow for cases in which κ is not constant over the entire Gaussian surface.

Sample Problem 26-6

Figure 26-17 shows a parallel-plate capacitor of plate area A and plate separation d. A potential difference V_0 is applied between the plates. The battery is then disconnected, and a dielectric slab of thickness b and dielectric constant κ is placed between the plates as shown. Assume

$$A = 115 \text{ cm}^2, \quad d = 1.24 \text{ cm}, \quad V_0 = 85.5 \text{ V},$$
$$b = 0.780 \text{ cm}, \quad \kappa = 2.61.$$

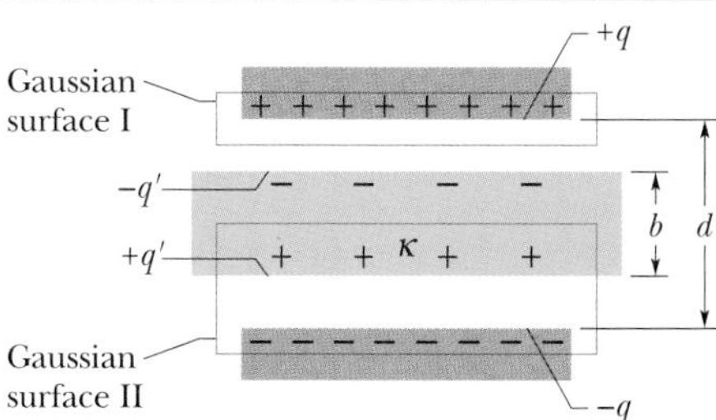

Fig. 26-17 Sample Problem 26-6. A parallel-plate capacitor containing a dielectric slab that only partially fills the space between the plates.

(a) What is the capacitance C_0 before the dielectric slab is inserted?

SOLUTION: From Eq. 26-9 we have

$$C_0 = \frac{\varepsilon_0 A}{d} = \frac{(8.85 \times 10^{-12} \text{ F/m})(115 \times 10^{-4} \text{ m}^2)}{1.24 \times 10^{-2} \text{ m}}$$
$$= 8.21 \times 10^{-12} \text{ F} = 8.21 \text{ pF}. \quad \text{(Answer)}$$

(b) What free charge appears on the plates?

SOLUTION: From Eq. 26-1,

$$q = C_0V_0 = (8.21 \times 10^{-12} \text{ F})(85.5 \text{ V})$$
$$= 7.02 \times 10^{-10} \text{ C} = 702 \text{ pC}. \quad \text{(Answer)}$$

Because the charging battery was disconnected before the slab was introduced, the free charge remains unchanged as the slab is put into place.

(c) What is the electric field E_0 in the gaps between the plates and the dielectric slab?

SOLUTION: A Key Idea here is to apply Gauss' law, in the form of Eq. 26-34, to Gaussian surface I in Fig. 26-17—that surface passes through the gap, and so it encloses *only* the free charge on the upper capacitor plate. Because the area vector $d\vec{A}$ and the field vector $\vec{E}_0$ are both directed downward, the dot product in Eq. 26-34 becomes

$$\vec{E}_0 \cdot d\vec{A} = E_0 \, dA \cos 0° = E_0 \, dA.$$

Equation 26-34 then becomes

$$\varepsilon_0\kappa E_0 \oint dA = q.$$

The integration now simply gives the surface area A of the plate. Thus, we obtain

$$\varepsilon_0\kappa E_0 A = q,$$

or

$$E_0 = \frac{q}{\varepsilon_0\kappa A}.$$

One more Key Idea is needed before we evaluate E_0; that is, we must put $\kappa = 1$ here because Gaussian surface I does not pass through the dielectric. Thus, we have

$$E_0 = \frac{q}{\varepsilon_0\kappa A} = \frac{7.02 \times 10^{-10} \text{ C}}{(8.85 \times 10^{-12} \text{ F/m})(1)(115 \times 10^{-4} \text{ m}^2)}$$
$$= 6900 \text{ V/m} = 6.90 \text{ kV/m}. \quad \text{(Answer)}$$

Note that the value of E_0 does not change when the slab is introduced because the amount of charge enclosed by Gaussian surface I in Fig. 26-17 does not change.

(d) What is the electric field E_1 in the dielectric slab?

SOLUTION: The Key Idea here is to apply Eq. 26-34 to Gaussian surface II in Fig. 26-17. That surface encloses free charge $-q$ and induced charge $+q'$, but we ignore the latter when we use Eq. 26-34. We find

$$\varepsilon_0 \oint \kappa\vec{E}_1 \cdot d\vec{A} = -\varepsilon_0\kappa E_1 A = -q. \quad (26\text{-}35)$$

(The first minus sign in this equation comes from the dot product $\vec{E}_1 \cdot d\vec{A}$, because now the field vector $\vec{E}_1$ is directed downward and the area vector $d\vec{A}$ is directed upward.) Equation 26-35 gives us

$$E_1 = \frac{q}{\varepsilon_0\kappa A} = \frac{E_0}{\kappa} = \frac{6.90 \text{ kV/m}}{2.61} = 2.64 \text{ kV/m}. \quad \text{(Answer)}$$

(e) What is the potential difference V between the plates after the slab has been introduced?

SOLUTION: The Key Idea here is to find V by integrating along a straight-line path extending directly from the bottom plate to the top plate. Within the dielectric, the path length is b and the electric field is E_1. Within the two gaps above and below the dielectric, the total path length is $d - b$ and the electric field is E_0. Equation 26-6 then yields

$$\begin{aligned} V &= \int_-^+ E\,ds = E_0(d-b) + E_1 b \\ &= (6900\ \text{V/m})(0.0124\ \text{m} - 0.00780\ \text{m}) \\ &\quad + (2640\ \text{V/m})(0.00780\ \text{m}) \\ &= 52.3\ \text{V}. \qquad \text{(Answer)} \end{aligned}$$

This is less than the original potential difference of 85.5 V.

(f) What is the capacitance with the slab in place?

SOLUTION: The Key Idea now is that the capacitance C is related to the free charge q and the potential difference V via Eq. 26-1, just as when a dielectric is not in place. Taking q from (b) and V from (e), we have

$$\begin{aligned} C &= \frac{q}{V} = \frac{7.02 \times 10^{-10}\ \text{C}}{52.3\ \text{V}} \\ &= 1.34 \times 10^{-11}\ \text{F} = 13.4\ \text{pF}. \qquad \text{(Answer)} \end{aligned}$$

This is greater than the original capacitance of 8.21 pF.

CHECKPOINT 6: In this sample problem, if the thickness b of the slab increases, do the following increase, decrease, or remain the same: (a) the electric field E_1, (b) the potential difference between the plates, and (c) the capacitance of the capacitor?

REVIEW & SUMMARY

Capacitor; Capacitance A **capacitor** consists of two isolated conductors (the *plates*) with equal and opposite charges $+q$ and $-q$. Its **capacitance** C is defined from

$$q = CV, \qquad (26\text{-}1)$$

where V is the potential difference between the plates. The SI unit of capacitance is the farad (1 farad = 1 F = 1 coulomb per volt).

Determining Capacitance We generally determine the capacitance of a particular capacitor configuration by (1) assuming a charge q to have been placed on the plates, (2) finding the electric field $\vec{E}$ due to this charge, (3) evaluating the potential difference V, and (4) calculating C from Eq. 26-1. Some specific results are the following:

A *parallel-plate capacitor* with flat parallel plates of area A and spacing d has capacitance

$$C = \frac{\varepsilon_0 A}{d}. \qquad (26\text{-}9)$$

A *cylindrical capacitor* (two long coaxial cylinders) of length L and radii a and b has capacitance

$$C = 2\pi\varepsilon_0 \frac{L}{\ln(b/a)}. \qquad (26\text{-}14)$$

A *spherical capacitor* with concentric spherical plates of radii a and b has capacitance

$$C = 4\pi\varepsilon_0 \frac{ab}{b-a}. \qquad (26\text{-}17)$$

If we let $b \to \infty$ and $a = R$ in Eq. 26-17, we obtain the capacitance of an *isolated sphere* of radius R:

$$C = 4\pi\varepsilon_0 R. \qquad (26\text{-}18)$$

Capacitors in Parallel and in Series The **equivalent capacitances** C_{eq} of combinations of individual capacitors connected in **parallel** and in **series** can be found from

$$C_{eq} = \sum_{j=1}^{n} C_j \qquad (n \text{ capacitors in parallel}) \qquad (26\text{-}19)$$

and

$$\frac{1}{C_{eq}} = \sum_{j=1}^{n} \frac{1}{C_j} \qquad (n \text{ capacitors in series}). \qquad (26\text{-}20)$$

Equivalent capacitances can be used to calculate the capacitances of more complicated series-parallel combinations.

Potential Energy and Energy Density The **electric potential energy** U of a charged capacitor,

$$U = \frac{q^2}{2C} = \tfrac{1}{2}CV^2, \qquad (26\text{-}21, 26\text{-}22)$$

is equal to the work required to charge it. This energy can be associated with the capacitor's electric field $\vec{E}$. By extension we can associate stored energy with an electric field. In vacuum, the **energy density** u, or potential energy per unit volume, within an electric field of magnitude E is given by

$$u = \tfrac{1}{2}\varepsilon_0 E^2. \qquad (26\text{-}23)$$

Capacitance with a Dielectric If the space between the plates of a capacitor is completely filled with a dielectric material, the capacitance C is increased by a factor κ, called the **dielectric constant,** which is characteristic of the material. In a region that is completely filled by a dielectric, all electrostatic equations containing ε_0 must be modified by replacing ε_0 with $\kappa\varepsilon_0$.

The effects of adding a dielectric can be understood physically in terms of the action of an electric field on the permanent or induced electric dipoles in the dielectric slab. The result is the formation of induced charges on the surfaces of the dielectric, which results in a weakening of the field within the dielectric for the same free charge on the plates.

Gauss' Law with a Dielectric When a dielectric is present, Gauss' law may be generalized to

$$\varepsilon_0 \oint \kappa \vec{E} \cdot d\vec{A} = q. \tag{26-34}$$

Here q is the free charge; any induced surface charge is accounted for by including the dielectric constant κ inside the integral.

QUESTIONS

1. Figure 26-18 shows plots of charge versus potential difference for three parallel-plate capacitors, which have the plate areas and separations given in the table. Which of the plots goes with which of the capacitors?

Capacitor	Area	Separation
1	A	d
2	$2A$	d
3	A	$2d$

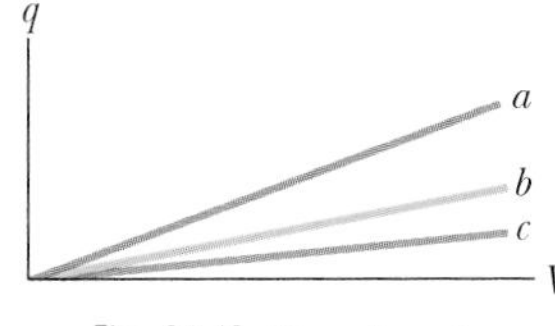

Fig. 26-18 Question 1.

2. Figure 26-19 shows an open switch, a battery of potential difference V, a current-measuring meter A, and three identical uncharged capacitors of capacitance C. When the switch is closed and the circuit reaches equilibrium, what are (a) the potential difference across each capacitor and (b) the charge on the left-hand plate of each capacitor? (c) During the charging process, what is the net charge that passes through the meter?

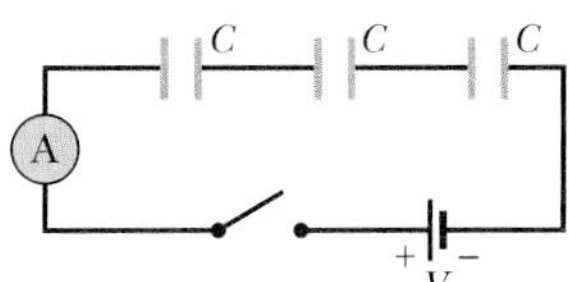

Fig. 26-19 Question 2.

3. For each circuit in Fig. 26-20, are the capacitors connected in series, in parallel, or in neither mode?

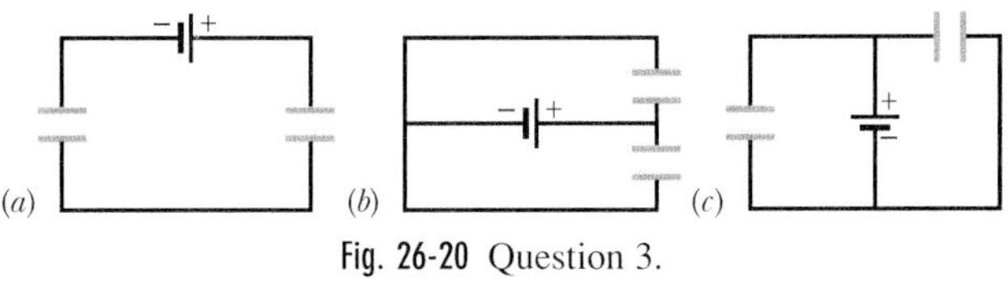

Fig. 26-20 Question 3.

4. (a) In Fig. 26-21a, are capacitors C_1 and C_3 in series? (b) In the same figure, are capacitors C_1 and C_2 in parallel? (c) Rank the equivalent capacitances of the four circuits shown in Fig. 26-21, greatest first.

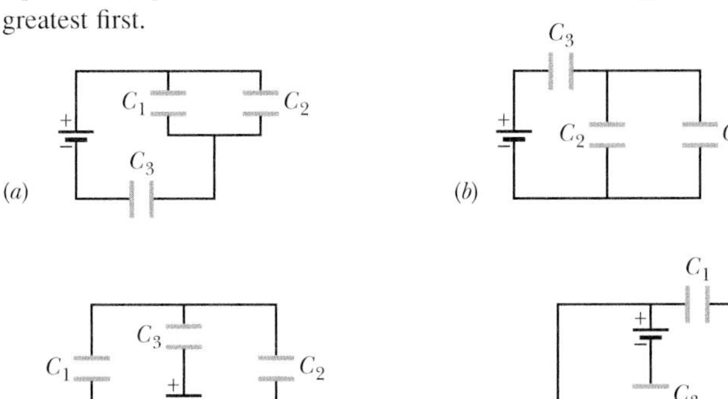

Fig. 26-21 Question 4.

5. What is the equivalent capacitance of three capacitors, each of capacitance C, if they are connected to a battery (a) in series with one another and (b) in parallel? (c) In which arrangement is there more charge on the equivalent capacitance?

6. You are to connect capacitances C_1 and C_2, with $C_1 > C_2$, to a battery, first individually, then in series, and then in parallel. Rank those arrangements according to the amount of charge stored, greatest first.

7. Initially, a single capacitance C_1 is wired to a battery. Then capacitance C_2 is added in parallel. Are (a) the potential difference across C_1 and (b) the charge q_1 on C_1 now more than, less than, or the same as previously? (c) Is the equivalent capacitance C_{12} of C_1 and C_2 more than, less than, or equal to C_1? (d) Is the total charge stored on C_1 and C_2 together more than, less than, or equal to the charge stored previously on C_1?

8. Repeat Question 7 for C_2 added in series, not in parallel.

9. Figure 26-22 shows three circuits, each consisting of a switch and two capacitors, initially charged as indicated. After the switches have been closed, in which circuit (if any) will the charge on the left-hand capacitor (a) increase, (b) decrease, and (c) remain the same?

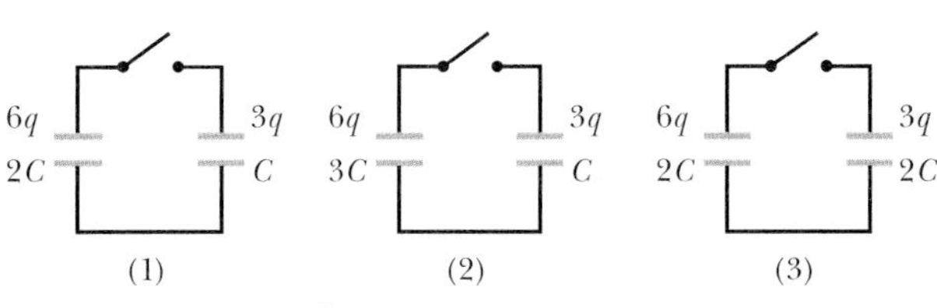

Fig. 26-22 Question 9.

10. Two isolated metal spheres A and B have radii R and $2R$, respectively, and the same charge q. (a) Is the capacitance of A more than, less than, or equal to that of B? (b) Is the energy density just outside the surface of A more than, less than, or equal to that of B? (c) Is the energy density at distance $3R$ from the center of A more than, less than, or equal to that at the same distance from the center of B? (d) Is the total energy of the electric field due to A more than, less than, or equal to that of B?

11. When a dielectric slab is inserted between the plates of one of the two identical capacitors in Fig. 26-23, do the following properties of that capacitor increase, decrease, or remain the same: (a) capacitance, (b) charge, (c) potential difference, and (d) potential energy? (e) How about the same properties of the other capacitor?

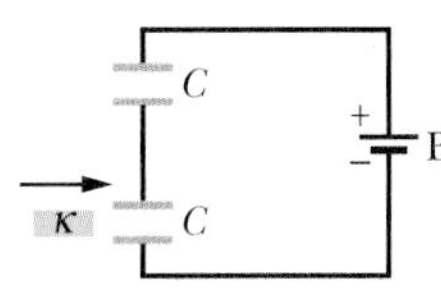

Fig. 26-23 Question 11.

EXERCISES & PROBLEMS

ssm Solution is in the Student Solutions Manual.
www Solution is available on the World Wide Web at: http://www.wiley.com/college/hrw
ilw Solution is available on the Interactive LearningWare.

SEC. 26-2 Capacitance

1E. An electrometer is a device used to measure static charge—an unknown charge is placed on the plates of the meter's capacitor, and the potential difference is measured. What minimum charge can be measured by an electrometer with a capacitance of 50 pF and a voltage sensitivity of 0.15 V?

2E. The two metal objects in Fig. 26-24 have net charges of +70 pC and −70 pC, which result in a 20 V potential difference between them. (a) What is the capacitance of the system? (b) If the charges are changed to +200 pC and −200 pC, what does the capacitance become? (c) What does the potential difference become?

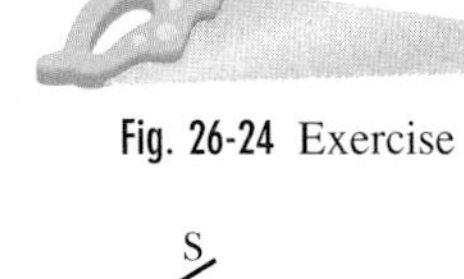

Fig. 26-24 Exercise 2.

3E. The capacitor in Fig. 26-25 has a capacitance of 25 μF and is initially uncharged. The battery provides a potential difference of 120 V. After switch S is closed, how much charge will pass through it? ssm

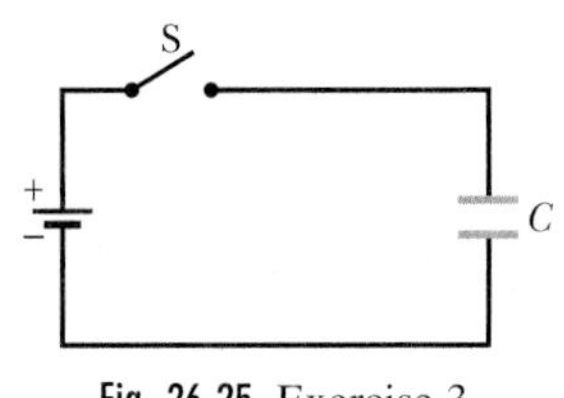

Fig. 26-25 Exercise 3.

SEC. 26-3 Calculating the Capacitance

4E. If we solve Eq. 26-9 for ε_0, we see that its SI unit is the farad per meter. Show that this unit is equivalent to that obtained earlier for ε_0—namely, the coulomb squared per newton-meter squared ($C^2/N \cdot m^2$).

5E. A parallel-plate capacitor has circular plates of 8.2 cm radius and 1.3 mm separation. (a) Calculate the capacitance. (b) What charge will appear on the plates if a potential difference of 120 V is applied? ssm

6E. You have two flat metal plates, each of area 1.00 m^2, with which to construct a parallel-plate capacitor. If the capacitance of the device is to be 1.00 F, what must be the separation between the plates? Could this capacitor actually be constructed?

7E. A spherical drop of mercury of radius R has a capacitance given by $C = 4\pi\varepsilon_0 R$. If two such drops combine to form a single larger drop, what is its capacitance? ssm

8E. The plates of a spherical capacitor have radii 38.0 mm and 40.0 mm. (a) Calculate the capacitance. (b) What must be the plate area of a parallel-plate capacitor with the same plate separation and capacitance?

9P. Suppose that the two spherical shells of a spherical capacitor have approximately equal radii. Under these conditions the device approximates a parallel-plate capacitor with $b - a = d$. Show that Eq. 26-17 does indeed reduce to Eq. 26-9 in this case. ssm

SEC. 26-4 Capacitors in Parallel and in Series

10E. In Fig. 26-26, find the equivalent capacitance of the combination. Assume that $C_1 = 10.0$ μF, $C_2 = 5.00$ μF, and $C_3 = 4.00$ μF.

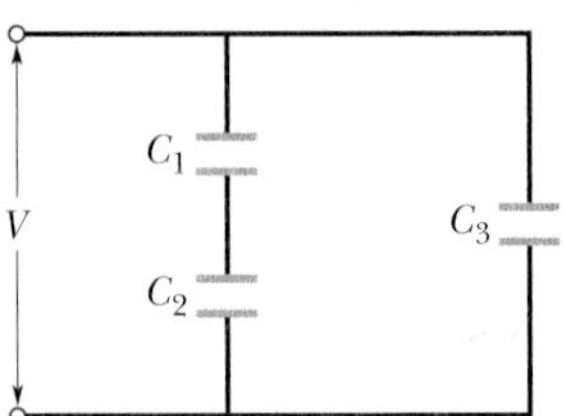

Fig. 26-26 Exercise 10 and Problem 30.

11E. How many 1.00 μF capacitors must be connected in parallel to store a charge of 1.00 C with a potential of 110 V across the capacitors? ssm

12E. Each of the uncharged capacitors in Fig. 26-27 has a capacitance of 25.0 μF. A potential difference of 4200 V is established when the switch is closed. How many coulombs of charge then pass through meter A?

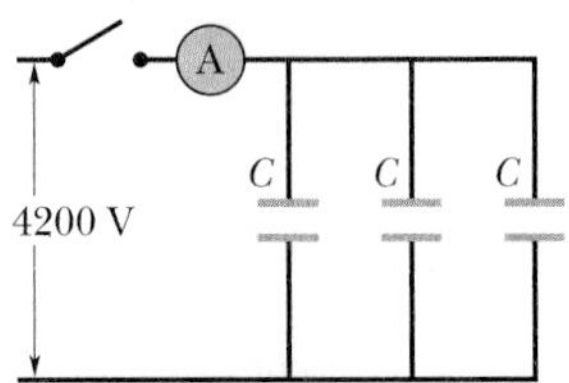

Fig. 26-27 Exercise 12.

13E. In Fig. 26-28 find the equivalent capacitance of the combination. Assume that $C_1 = 10.0$ μF, $C_2 = 5.00$ μF, and $C_3 = 4.00$ μF. ilw

14P. In Fig. 26-28 suppose that capacitor 3 breaks down electrically, becoming equivalent to a conducting path. What *changes* in (a) the charge and (b) the potential difference occur for capacitor 1? Assume that $V = 100$ V.

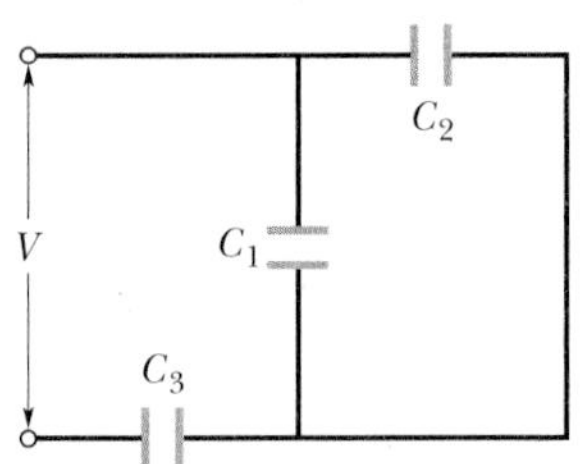

Fig. 26-28 Exercise 13, Problems 14 and 28.

15P. Figure 26-29 shows two capacitors in series; the center section of length b is movable vertically. Show that the equivalent capacitance of this series combination is independent of the position of the center section and is given by $C = \varepsilon_0 A/(a - b)$, where A is the plate area. ssm www

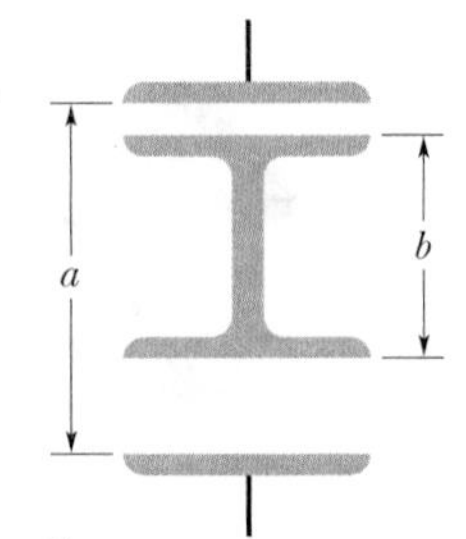

Fig. 26-29 Problem 15.

16P. In Fig. 26-30, the battery has a potential difference of 10 V and the five capacitors each have a capacitance of 10 μF. What is the charge on (a) capacitor 1 and (b) capacitor 2?

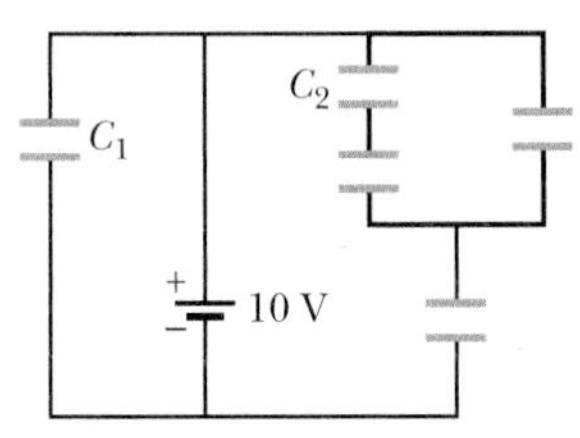

Fig. 26-30 Problem 16.

17P. A 100 pF capacitor is charged to a potential difference of 50 V, and the charging battery is disconnected. The capacitor is then connected in parallel with a second (initially uncharged) capacitor. If the potential difference across the first capacitor drops to 35 V, what is the capacitance of this second capacitor? ssm ilw

18P. In Fig. 26-31, the battery has a potential difference of 20 V. Find (a) the equivalent capacitance of all the capacitors and (b) the charge stored on that equivalent capacitance. Find the potential across and charge on (c) capacitor 1, (d) capacitor 2, and (e) capacitor 3.

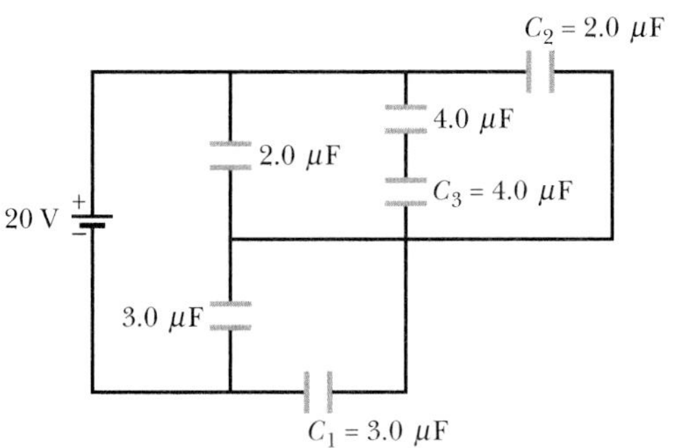

Fig. 26-31 Problem 18.

19P. In Fig. 26-32, the capacitances are $C_1 = 1.0\ \mu F$ and $C_2 = 3.0\ \mu F$ and both capacitors are charged to a potential difference of $V = 100$ V but with opposite polarity as shown. Switches S_1 and S_2 are now closed. (a) What is now the potential difference between points a and b? What are now the charges on capacitors (b) 1 and (c) 2? ssm www

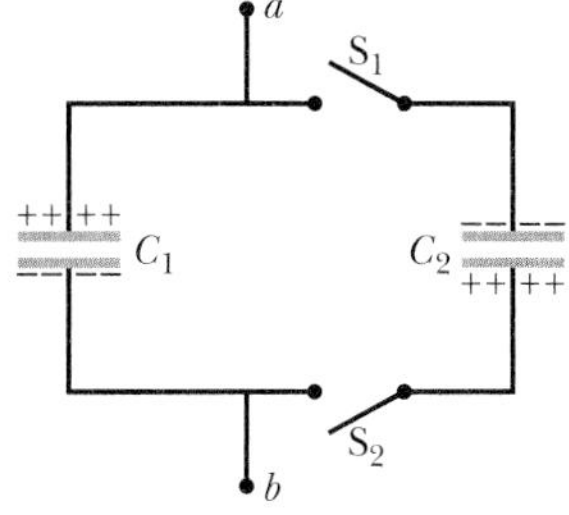

Fig. 26-32 Problem 19.

20P. In Fig. 26-33, battery B supplies 12 V. Find the charge on each capacitor (a) first when only switch S_1 is closed and (b) later when switch S_2 is also closed. Take $C_1 = 1.0\ \mu F$, $C_2 = 2.0\ \mu F$, $C_3 = 3.0\ \mu F$, and $C_4 = 4.0\ \mu F$.

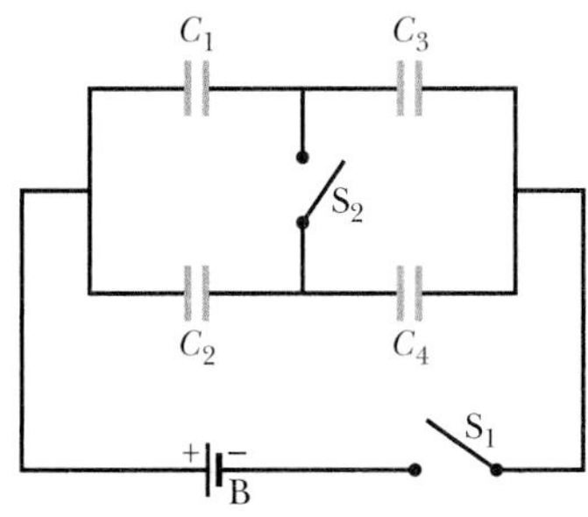

Fig. 26-33 Problem 20.

21P. When switch S is thrown to the left in Fig. 26-34, the plates of capacitor 1 acquire a potential difference V_0. Capacitors 2 and 3 are initially uncharged. The switch is now thrown to the right. What are the final charges q_1, q_2, and q_3 on the capacitors? ssm

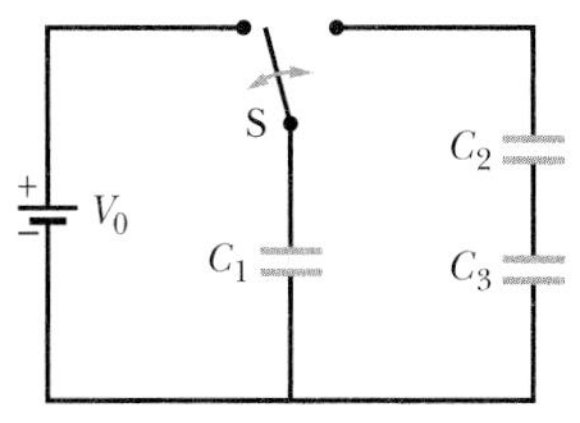

Fig. 26-34 Problem 21.

SEC. 26-5 Energy Stored in an Electric Field

22E. How much energy is stored in one cubic meter of air due to the "fair weather" electric field of magnitude 150 V/m?

23E. What capacitance is required to store an energy of 10 kW · h at a potential difference of 1000 V? ssm

24E. A parallel-plate air-filled capacitor having area 40 cm^2 and plate spacing 1.0 mm is charged to a potential difference of 600 V. Find (a) the capacitance, (b) the magnitude of the charge on each plate, (c) the stored energy, (d) the electric field between the plates, and (e) the energy density between the plates.

25E. Two capacitors, of 2.0 and 4.0 μF capacitance, are connected in parallel across a 300 V potential difference. Calculate the total energy stored in the capacitors. ssm

26P. A parallel-connected bank of 5.00 μF capacitors is used to store electric energy. What does it cost to charge the 2000 capacitors of the bank to 50 000 V, assuming 3.0¢/kW · h?

27P. One capacitor is charged until its stored energy is 4.0 J. A second uncharged capacitor is then connected to it in parallel. (a) If the charge distributes equally, what is now the total energy stored in the electric fields? (b) Where did the excess energy go? ilw

28P. In Fig. 26-28 find (a) the charge, (b) the potential difference, and (c) the stored energy for each capacitor. Assume the numerical values of Exercise 13, with $V = 100$ V.

29P. A parallel-plate capacitor has plates of area A and separation d and is charged to a potential difference V. The charging battery is then disconnected, and the plates are pulled apart until their separation is $2d$. Derive expressions in terms of A, d, and V for (a) the new potential difference; (b) the initial and final stored energies, U_i and U_f; and (c) the work required to separate the plates. ssm ilw

30P. In Fig. 26-26, find (a) the charge, (b) the potential difference, and (c) the stored energy for each capacitor. Assume the numerical values of Exercise 10, with $V = 100$ V.

31P. A cylindrical capacitor has radii a and b as in Fig. 26-6. Show that half the stored electric potential energy lies within a cylinder whose radius is $r = \sqrt{ab}$. ssm www

32P. A charged isolated metal sphere of diameter 10 cm has a potential of 8000 V relative to $V = 0$ at infinity. Calculate the energy density in the electric field near the surface of the sphere.

33P. (a) Show that the plates of a parallel-plate capacitor attract each other with a force given by $F = q^2/2\varepsilon_0 A$. Do so by calculating the work needed to increase the plate separation from x to $x + dx$, with the charge q remaining constant. (b) Next show that the force per unit area (the *electrostatic stress*) acting on either capacitor plate is given by $\frac{1}{2}\varepsilon_0 E^2$. (Actually, this is the force per unit area on *any* conductor of *any* shape with an electric field $\vec{E}$ at its surface.) ssm

SEC. 26-6 Capacitor with a Dielectric

34E. An air-filled parallel-plate capacitor has a capacitance of 1.3 pF. The separation of the plates is doubled and wax is inserted between them. The new capacitance is 2.6 pF. Find the dielectric constant of the wax.

35E. Given a 7.4 pF air-filled capacitor, you are asked to convert it to a capacitor that can store up to 7.4 μJ with a maximum potential difference of 652 V. What dielectric in Table 26-1 should you use to fill the gap in the air capacitor if you do not allow for a margin of error? ssm

36E. A parallel-plate air-filled capacitor has a capacitance of 50 pF. (a) If each of its plates has an area of 0.35 m^2, what is the separation? (b) If the region between the plates is now filled with material having $\kappa = 5.6$, what is the capacitance?

37E. A coaxial cable used in a transmission line has an inner radius of 0.10 mm and an outer radius of 0.60 mm. Calculate the capacitance per meter for the cable. Assume that the space between the conductors is filled with polystyrene. ssm

38P. You are asked to construct a capacitor having a capacitance near 1 nF and a breakdown potential in excess of 10 000 V. You think of using the sides of a tall Pyrex drinking glass as a dielectric, lining the inside and outside curved surfaces with aluminum foil to act as the plates. The glass is 15 cm tall with an inner radius of 3.6 cm and an outer radius of 3.8 cm. What are the (a) capacitance and (b) breakdown potential of this capacitor?

39P. A certain substance has a dielectric constant of 2.8 and a dielectric strength of 18 MV/m. If it is used as the dielectric material in a parallel-plate capacitor, what minimum area should the plates of the capacitor have to obtain a capacitance of 7.0×10^{-2} μF and to ensure that the capacitor will be able to withstand a potential difference of 4.0 kV? ssm ilw

40P. A parallel-plate capacitor of plate area A is filled with two dielectrics as in Fig. 26-35*a*. Show that the capacitance is

$$C = \frac{\varepsilon_0 A}{d} \frac{\kappa_1 + \kappa_2}{2}.$$

Check this formula for limiting cases. (*Hint:* Can you justify this arrangement as being two capacitors in parallel?)

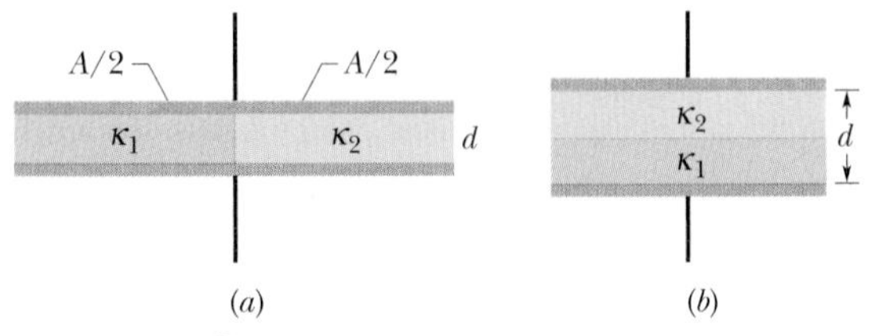

Fig. 26-35 Problems 40 and 41.

41P. A parallel-plate capacitor of plate area A is filled with two dielectrics as in Fig. 26-35*b*. Show that the capacitance is

$$C = \frac{2\varepsilon_0 A}{d} \frac{\kappa_1 \kappa_2}{\kappa_1 + \kappa_2}.$$

Check this formula for limiting cases. (*Hint:* Can you justify this arrangement as being two capacitors in series?) ssm

42P. What is the capacitance of the capacitor, of plate area A, shown in Fig. 26-36? (*Hint:* See Problems 40 and 41.)

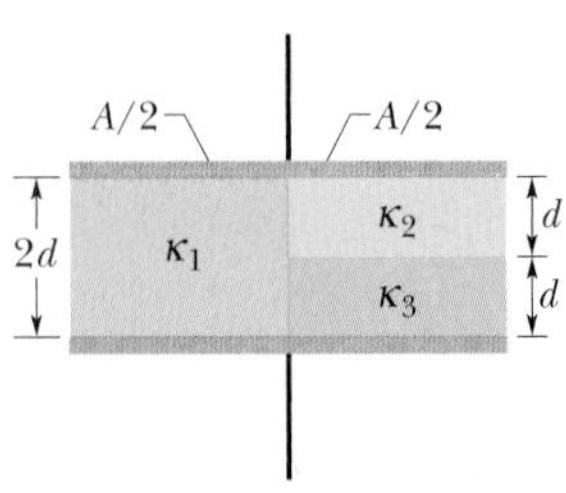

Fig. 26-36 Problem 42.

SEC. 26-8 Dielectrics and Gauss' Law

43E. A parallel-plate capacitor has a capacitance of 100 pF, a plate area of 100 cm^2, and a mica dielectric ($\kappa = 5.4$) completely filling the space between the plates. At 50 V potential difference, calculate (a) the electric field magnitude E in the mica, (b) the magnitude of the free charge on the plates, and (c) the magnitude of the induced surface charge on the mica. ssm

44E. In Sample Problem 26-6, suppose that the battery remains connected while the dielectric slab is being introduced. Calculate (a) the capacitance, (b) the charge on the capacitor plates, (c) the electric field in the gap, and (d) the electric field in the slab, after the slab is in place.

45P. The space between two concentric conducting spherical shells of radii b and a (where $b > a$) is filled with a substance of dielectric constant κ. A potential difference V exists between the inner and outer shells. Determine (a) the capacitance of the device, (b) the free charge q on the inner shell, and (c) the charge q' induced along the surface of the inner shell. ssm www

46P. Two parallel plates of area 100 cm^2 are given charges of equal magnitudes 8.9×10^{-7} C but opposite signs. The electric field within the dielectric material filling the space between the plates is 1.4×10^6 V/m. (a) Calculate the dielectric constant of the material. (b) Determine the magnitude of the charge induced on each dielectric surface.

47P. A dielectric slab of thickness b is inserted between the plates of a parallel-plate capacitor of plate separation d. Show that the capacitance is then given by

$$C = \frac{\kappa \varepsilon_0 A}{\kappa d - b(\kappa - 1)}.$$

(*Hint:* You can derive the formula following the procedure outlined in Sample Problem 26-6.) Does this formula predict the correct numerical result of Sample Problem 26-6? Verify that the formula gives reasonable results for the special cases of $b = 0$, $\kappa = 1$, and $b = d$. ssm

Additional Problem

48. *The chocolate crumb mystery.* This story begins with Problem 48 in Chapter 24 and Problem 57 in Chapter 25. As part of the investigation of the biscuit factory explosion, the electric potentials of the workers were measured as they emptied sacks of chocolate crumb powder into the loading bin, stirring up a cloud of the powder around themselves. Each worker had an electric potential of about 7.0 kV relative to the ground, which was taken as zero potential. (a) Assuming that each worker was effectively a capacitor with a typical capacitance of 200 pF, find the energy stored in that effective capacitor. If a single spark between the worker and any conducting object connected to the ground neutralized the worker, that energy would be transferred to the spark. According to measurements, a spark that could ignite a cloud of chocolate crumb powder, and thus set off an explosion, had to have an energy of at least 150 mJ. (b) Could a spark from a worker have set off an explosion in the cloud of powder in the loading bin? (The story continues with Problem 44 in Chapter 27.)

NEW PROBLEMS

N1. A 10 V battery is connected to a series of n capacitors, each of capacitance 2.0 μF. If the total energy stored in the capacitors is 25 μJ, what is n?

N2. Figure 26N-1 shows capacitor 1 ($C_1 = 8.00\ \mu$F), capacitor 2 ($C_2 = 6.00\ \mu$F), and capacitor 3 ($C_3 = 8.00\ \mu$F) connected to a 12 V battery. When switch S is closed so as to connect uncharged capacitor 4 ($C_4 = 6.00\ \mu$F) to that initial circuit, (a) how much charge passes through point P from the battery and (b) how much charge shows up on capacitor 4? (c) Explain the discrepancy in those two results.

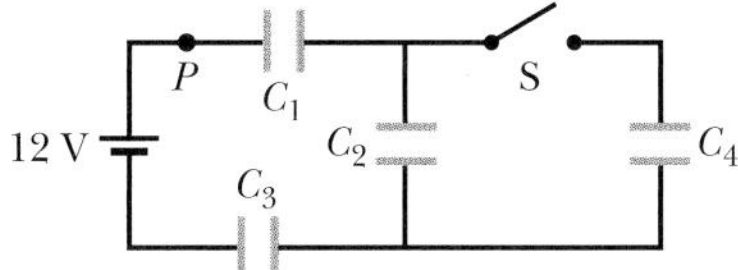

Fig. 26N-1 Problem N2.

N3. Two parallel-plate capacitors, 6.0 μF each, are connected in parallel to a 10 V battery. One of the capacitors is then squeezed so that its plate separation is halved. Because of the squeezing, (a) how much additional charge is transferred to the capacitors by the battery and (b) what is the change in the total charge stored on the capacitors?

N4. Capacitor 3 in Fig. 26N-2a is a *variable capacitor* (its capacitance C_3 can be varied). Figure 26N-2b gives the electric potential V_1 across capacitor 1 versus C_3. Electric potential V_1 approaches an asymptote of 10 V as $C_3 \to \infty$. What are (a) the electric potential V across the battery, (b) capacitance C_1 of capacitor 1, and (c) capacitance C_2 of capacitor 2?

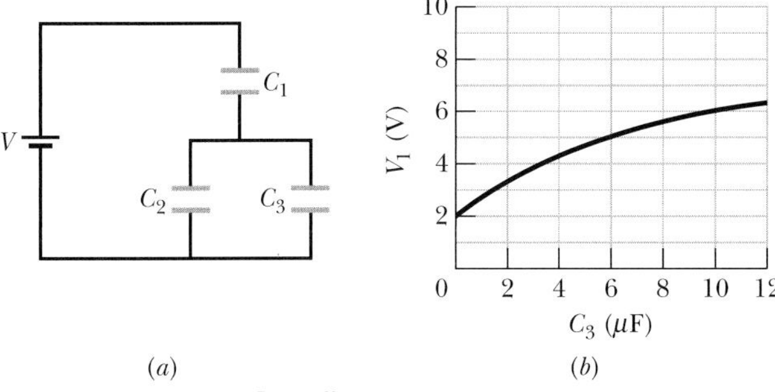

Fig. 26N-2 Problem N4.

N5. In Fig. 26N-3, two parallel-plate capacitors (with air between the plates) are connected to a battery. Capacitor 1 has a plate area of 1.5 cm^2 and an electric field (between its plates) of magnitude 2000 V/m. Capacitor 2 has a plate area of 0.70 cm^2 and an electric field (between its plates) of magnitude 1500 V/m. What is the total charge on the two capacitors?

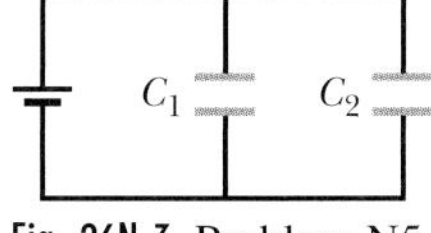

Fig. 26N-3 Problem N5.

N6. In Fig. 26N-4, capacitor 2 has capacitance $C_2 = 3.0\ \mu$F and capacitor 4 has capacitance $C_4 = 4.0\ \mu$F, and all the capacitors are initially uncharged. When switch S is closed, a total charge of 12 μC passes through point a and a total charge of 8.0 μC passes through point b. What are capacitances (a) C_1 and (b) C_3?

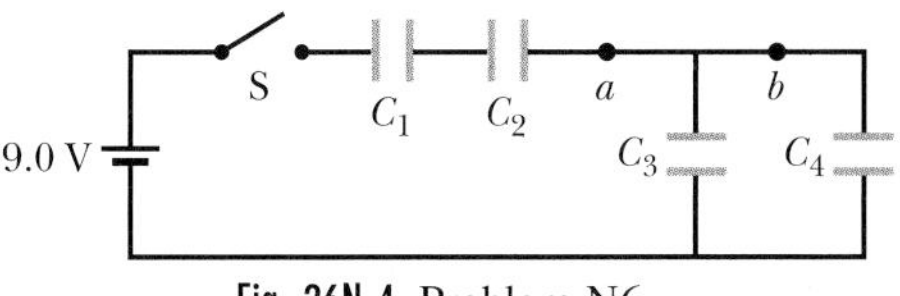

Fig. 26N-4 Problem N6.

N7. Two capacitors consist of parallel plates with air between the plates. They are to be connected to a 10 V battery, first individually, then in series, and then in parallel. In those arrangements, the energy stored in the capacitors turns out to be the following, listed least to greatest: 75 μJ, 100 μJ, 300 μJ, and 400 μJ. What are the capacitances of the capacitors?

N8. The capacitors in Fig. 26N-5 are initially uncharged. The capacitances are $C_1 = 4.0\ \mu$F, $C_2 = 8.0\ \mu$F, and $C_3 = 12\ \mu$F. When switch S is closed, how many electrons travel through (a) point a, (b) point b, (c) point c, and (d) point d, and in which direction in the figure do they travel?

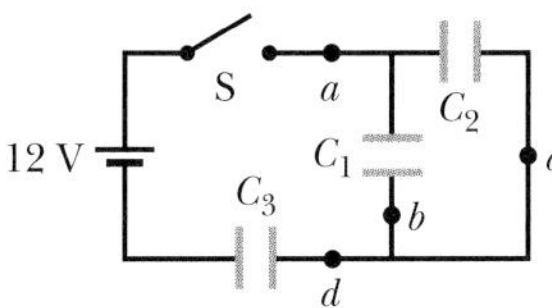

Fig. 26N-5 Problem N8.

N9. In Fig. 26N-6, how much charge is stored on the parallel plate capacitors? One is filled with air, and the other has a dielectric with $\kappa = 3.00$; both have a plate area of 5.00×10^{-3} m^2 and a plate separation of 2.00 mm.

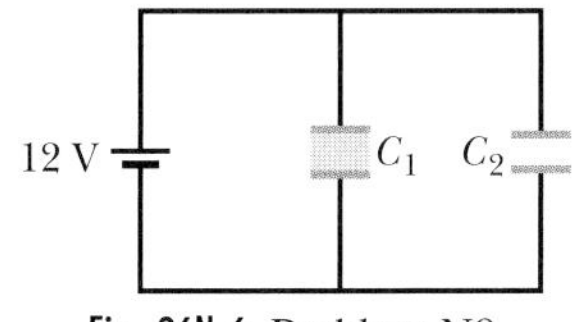

Fig. 26N-6 Problem N9.

N10. Figure 26N-7 shows a circuit of four air-filled capacitors that are connected to a larger circuit. The graph below the circuit shows the electric potential $V(x)$ as a function of position x along the lower part of the circuit, through capacitor 4. Similarly, the graph above the circuit shows the electric potential $V(x)$ as a function of position x along the upper part of the circuit, through capacitors 1, 2, and 3. Capacitor 3 has a capacitance of 0.80 μF. What are the capacitances of (a) capacitor 1 and (b) capacitor 2?

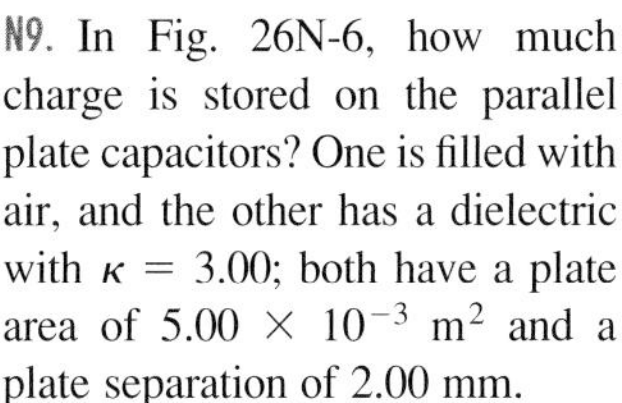

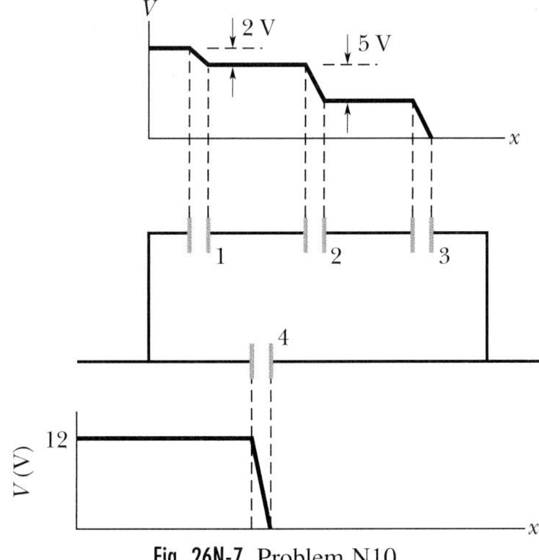

Fig. 26N-7 Problem N10.

N11. In Fig. 26N-8, what are (a) the charge on the bottom capacitor and (b) the potential across it?

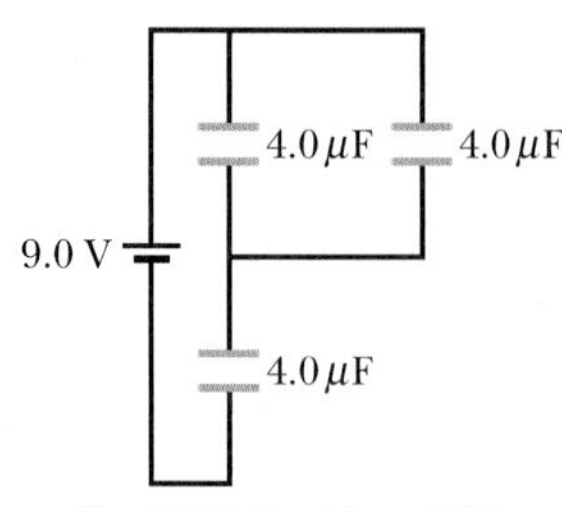

Fig. 26N-8 Problem N11.

N12. In Fig. 26N-9a, when switch S is closed to connect the (uncharged) capacitor to the battery, we can imagine that a negative charge moves *from* the battery to the lower capacitor plate and that an equal amount of negative charge moves *to* the battery from the upper capacitor plate. Let us examine this motion for the lower plate. Suppose the plate has a thickness of $L = 0.50$ cm and a face area of $A = 2.0 \times 10^{-4}\ \text{m}^2$. If the plate is made of copper, the electrons that are free to move to the plate face have a density (number of electrons per unit volume) of $8.49 \times 10^{28}\ \text{m}^{-3}$. From what depth d within the plate (Fig. 26N-9b) must electrons move to the face if the face is to gain a charge of $q = -3.0\ \mu\text{C}$? (Do the electrons actually come from the battery?)

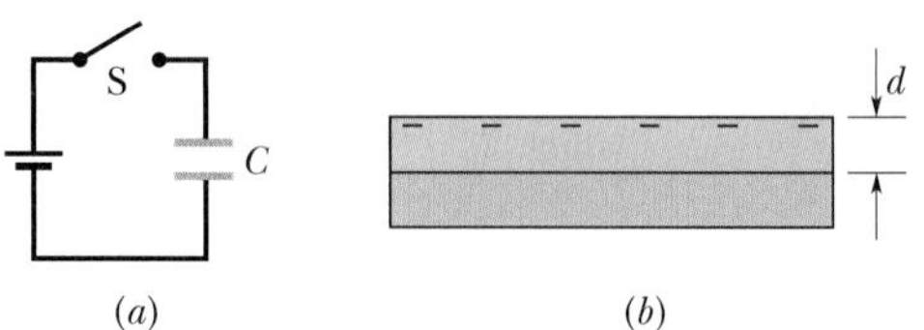

Fig. 26N-9 Problem N12.

N13. In Fig. 26N-10, $C_1 = 10\ \mu\text{F}$ and $C_2 = C_3 = 20\ \mu\text{F}$. Switch S is first thrown to the left until C_1 reaches equilibrium. Then the switch is thrown to the right. When equilibrium is again reached, how much charge is on C_1?

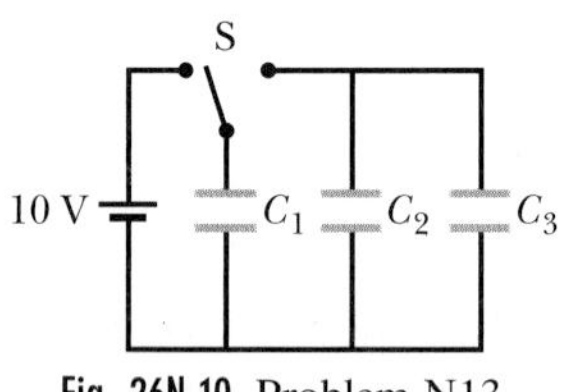

Fig. 26N-10 Problem N13.

N14. Plot 1 in Fig. 26N-11a gives the charge q that can be stored on capacitor 1 versus the electric potential V set up across it. Plots 2 and 3 are similar plots for capacitors 2 and 3, respectively. Figure 26N-11b shows a circuit with those three capacitors and a 6.0 V battery. What is the charge stored on capacitor 2 in that circuit?

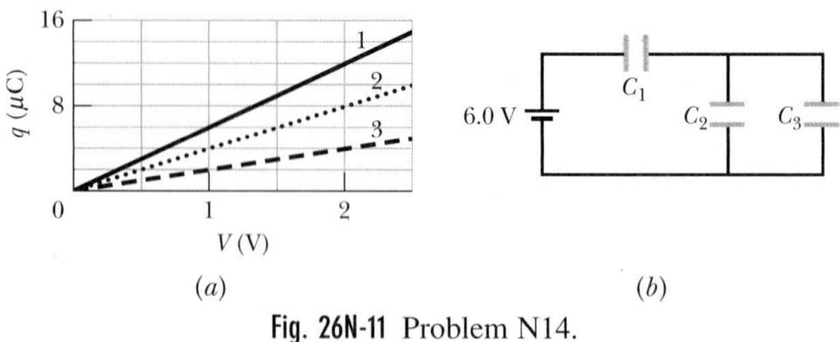

Fig. 26N-11 Problem N14.

N15. In Fig. 26N-12, $C_1 = C_2 = 30\ \mu\text{F}$ and $C_3 = C_4 = 15\ \mu\text{F}$. What is the charge on C_4?

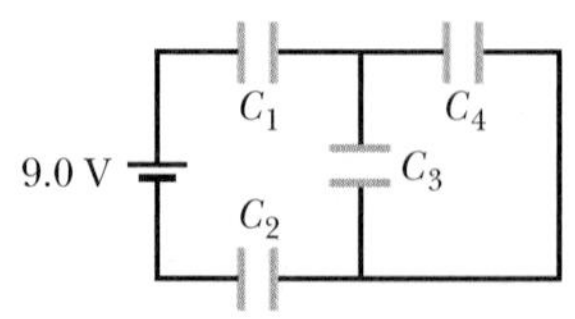

Fig. 26N-12 Problem N15.

N16. Two parallel-plate capacitors, 6.0 μF each, are connected in series to a 10 V battery. One of the capacitors is then squeezed so that its plate separation is halved. Because of the squeezing, (a) how much additional charge is transferred to the capacitors by the battery and (b) what is the change in the *total* charge stored on the capacitors (the charge on the positive plate of one capacitor plus the charge on the positive plate of the other capacitor)?

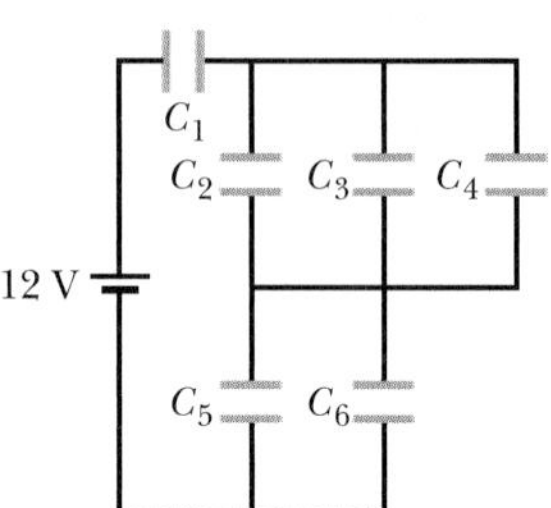

Fig. 26N-13 Problem N17.

N17. In Fig. 26N-13, $C_1 = C_5 = C_6 = 6.0\ \mu\text{F}$ and $C_2 = C_3 = C_4 = 4.0\ \mu\text{F}$. What are (a) the net charge stored on the capacitors and (b) the charge on C_4?

N18. Figure 26N-14 shows two air-filled cylindrical capacitors connected in series across a battery with potential $V = 10$ V. Capacitor 1 has an inner plate radius of 5.0 mm, an outer plate radius of 1.5 cm, and a length of 5.0 cm. Capacitor 2 has an inner plate radius of 2.5 mm, an outer plate radius of 1.0 cm, and a length of 9.0 cm. The outer plate of capacitor 2 is a conducting organic membrane that can be stretched, and the capacitor can be inflated to increase the plate separation. If the outer plate radius is increased to 2.5 cm by inflation, (a) how many electrons move through point P and (b) do they move toward or away from the battery?

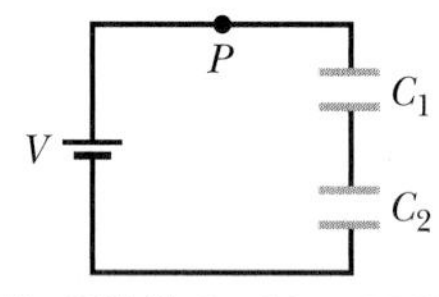

Fig. 26N-14 Problem N18.

N19. In Fig. 26N-15, the parallel-plate capacitor of plate area $2.00 \times 10^{-2}\ \text{m}^2$ is filled with two dielectric slabs, each with a thickness of 2.00 mm. One slab has dielectric constant 3.00; the other has dielectric constant 4.00. How much charge is on the capacitor?

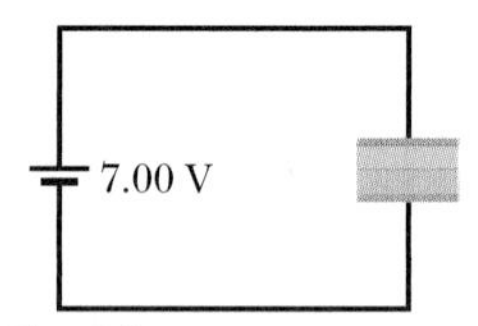

Fig. 26N-15 Problem N19.

N20. The capacitors in Fig. 26N-16a each have capacitance 10 μF. What are the charges on (a) C_1 and (b) C_2?

N21. Using the approximation that $\ln(1 + x) \approx x$ when $x \ll 1$ (see Appendix E), show that the capacitance of a cylindrical capacitor approaches that of a parallel-plate capacitor when the spacing between the two cylinders is small.

N22. In Fig. 26N-16b, $C_1 = C_4 = 2.0\ \mu\text{F}$, $C_2 = 4.0\ \mu\text{F}$, and $C_3 = 1.0\ \mu\text{F}$. What is the charge on capacitor C_4?

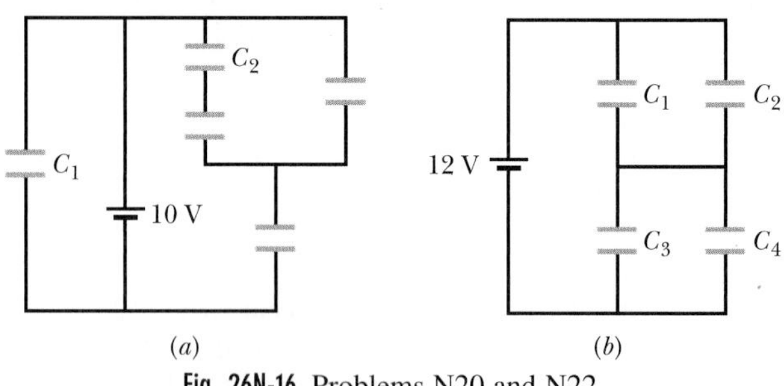

Fig. 26N-16 Problems N20 and N22.

27 Current and Resistance

The pride of Germany and a wonder of its time, the zeppelin *Hindenburg*—almost the length of three football fields—was the largest flying machine that had ever been built. Although it was kept aloft by 16 cells of highly flammable hydrogen gas, it made many trans-Atlantic trips without incident. In fact, German zeppelins, which all depended on hydrogen, had never suffered an accident due to the hydrogen. However, shortly after 7:21 p.m. on May 6, 1937, as the *Hindenburg* was ready to land at the U.S. Naval Air Station at Lakehurst, New Jersey, the ship burst into flames. Its crew had been waiting for a rainstorm to diminish, and handling ropes had just been let down to a navy ground crew, when ripples were sighted on the outer fabric of the ship about one-third of the way forward from the stern. Seconds later a flame erupted from that region, and a red glow illuminated the interior of the ship. Within 32 seconds the burning ship fell to the ground.

After so many successful flights of hydrogen-floated zeppelins, why did this zeppelin burst into flames?

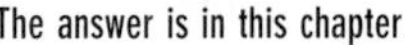

The answer is in this chapter.

27-1 Moving Charges and Electric Currents

Chapters 22 through 26 deal largely with *electrostatics*—that is, with charges at rest. With this chapter we begin to focus on **electric currents**—that is, charges in motion.

Examples of electric currents abound, ranging from the large currents that constitute lightning strokes to the tiny nerve currents that regulate our muscular activity. The currents in household wiring, in lightbulbs, and in electrical appliances are familiar to all. A beam of electrons—a current—moves through an evacuated space in the picture tube of a common television set. Charged particles of *both* signs flow in the ionized gases of fluorescent lamps, in the batteries of radios, and in car batteries. Electric currents can also be found in the semiconductors in calculators and in the chips that control microwave ovens and electric dishwashers.

On a global scale, charged particles trapped in the Van Allen radiation belts surge back and forth above the atmosphere between Earth's north and south magnetic poles. On the scale of the solar system, enormous currents of protons, electrons, and ions fly radially outward from the Sun as the *solar wind.* On the galactic scale, cosmic rays, which are largely energetic protons, stream through our Milky Way galaxy, some reaching Earth.

Although an electric current is a stream of moving charges, not all moving charges constitute an electric current. If there is to be an electric current through a given surface, there must be a net flow of charge through that surface. Two examples clarify our meaning.

1. The free electrons (conduction electrons) in an isolated length of copper wire are in random motion at speeds of the order of 10^6 m/s. If you pass a hypothetical plane through such a wire, conduction electrons pass through it *in both directions* at the rate of many billions per second—but there is *no net transport* of charge and thus *no current* through the wire. However, if you connect the ends of the wire to a battery, you slightly bias the flow in one direction, with the result that there now is a net transport of charge and thus an electric current through the wire.
2. The flow of water through a garden hose represents the directed flow of positive charge (the protons in the water molecules) at a rate of perhaps several million coulombs per second. There is no net transport of charge, however, because there is a parallel flow of negative charge (the electrons in the water molecules) of exactly the same amount moving in exactly the same direction.

In this chapter we restrict ourselves largely to the study—within the framework of classical physics—of *steady* currents of *conduction electrons* moving through *metallic conductors* such as copper wires.

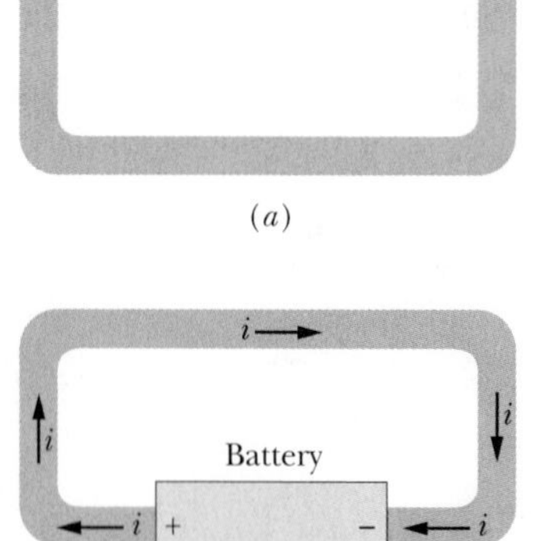

Fig. 27-1 (*a*) A loop of copper in electrostatic equilibrium. The entire loop is at a single potential, and the electric field is zero at all points inside the copper. (*b*) Adding a battery imposes an electric potential difference between the ends of the loop that are connected to the terminals of the battery. The battery thus produces an electric field within the loop, from terminal to terminal, and the field causes charges to move around the loop. This movement of charges is a current *i*.

27-2 Electric Current

As Fig. 27-1*a* reminds us, an isolated conducting loop—regardless of whether it has an excess charge—is all at the same potential. No electric field can exist within it or along its surface. Although conduction electrons are available, no net electric force acts on them and thus there is no current.

If, as in Fig. 27-1*b*, we insert a battery in the loop, the conducting loop is no longer at a single potential. Electric fields act inside the material making up the loop, exerting forces on the conduction electrons, causing them to move, and thus establishing a current. After a very short time, the electron flow reaches a constant value and the current is in its *steady state* (it does not vary with time).

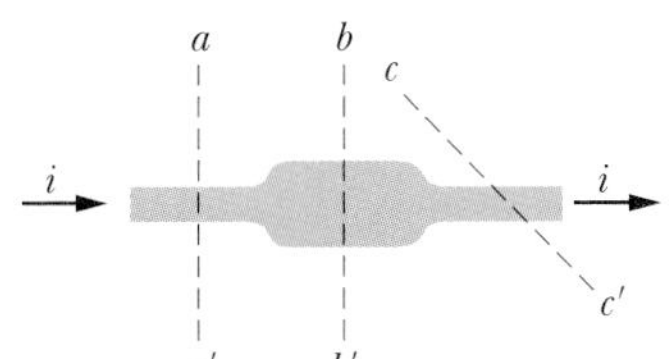

Fig. 27-2 The current i through the conductor has the same value at planes aa', bb', and cc'.

Figure 27-2 shows a section of a conductor, part of a conducting loop in which current has been established. If charge dq passes through a hypothetical plane (such as aa') in time dt, then the current i through that plane is defined as

$$i = \frac{dq}{dt} \quad \text{(definition of current).} \tag{27-1}$$

We can find the charge that passes through the plane in a time interval extending from 0 to t by integration:

$$q = \int dq = \int_0^t i\,dt, \tag{27-2}$$

in which the current i may vary with time.

Under steady-state conditions, the current is the same for planes aa', bb', and cc' and indeed for all planes that pass completely through the conductor, no matter what their location or orientation. This follows from the fact that charge is conserved. Under the steady-state conditions assumed here, an electron must pass through plane aa' for every electron that passes through plane cc'. In the same way, if we have a steady flow of water through a garden hose, a drop of water must leave the nozzle for every drop that enters the hose at the other end. The amount of water in the hose is a conserved quantity.

The SI unit for current is the coulomb per second, also called the *ampere* (A):

$$1 \text{ ampere} = 1 \text{ A} = 1 \text{ coulomb per second} = 1 \text{ C/s}.$$

The ampere is an SI base unit; the coulomb is defined in terms of the ampere, as we discussed in Chapter 22. The formal definition of the ampere is discussed in Chapter 30.

Current, as defined by Eq. 27-1, is a scalar because both charge and time in that equation are scalars. Yet, as in Fig. 27-1*b*, we often represent a current with an arrow to indicate that charge is moving. Such arrows are not vectors, however, and they do not require vector addition. Figure 27-3*a* shows a conductor with current i_0 splitting at a junction into two branches. Because charge is conserved, the magnitudes of the currents in the branches must add to yield the magnitude of the current in the original conductor, so that

$$i_0 = i_1 + i_2. \tag{27-3}$$

As Fig. 27-3*b* suggests, bending or reorienting the wires in space does not change the validity of Eq. 27-3. Current arrows show only a direction (or sense) of flow along a conductor, not a direction in space.

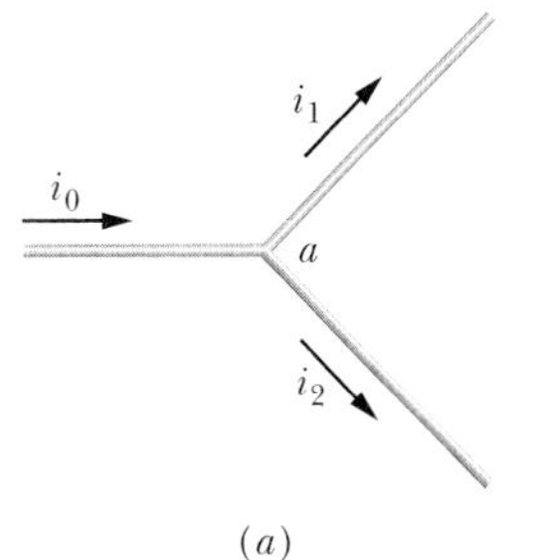

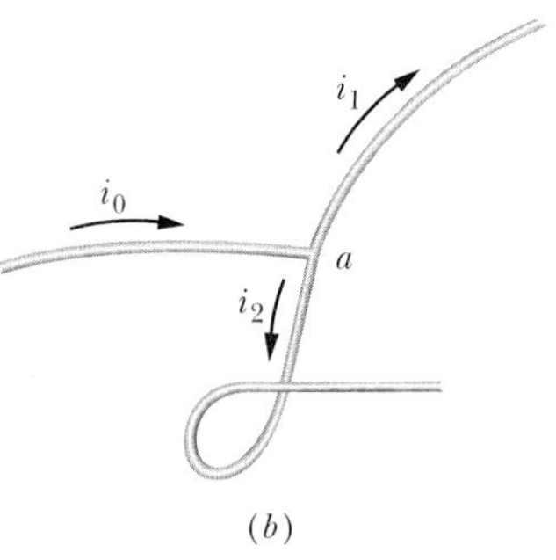

Fig. 27-3 The relation $i_0 = i_1 + i_2$ is true at junction a no matter what the orientation in space of the three wires. Currents are scalars, not vectors.

The Directions of Currents

In Fig. 27-1*b* we drew the current arrows in the direction in which positively charged particles would be forced to move through the loop by the electric field. Such positive *charge carriers,* as they are often called, would move away from the positive battery terminal and toward the negative terminal. Actually, the charge carriers in the copper loop of Fig. 27-1*b* are electrons and thus are negatively charged. The electric field forces them to move in the direction opposite the current arrows, from the negative terminal to the positive terminal. For historical reasons, however, we use the following convention:

A current arrow is drawn in the direction in which positive charge carriers would move, even if the actual charge carriers are negative and move in the opposite direction.

We can use this convention because in *most* situations, the assumed motion of positive charge carriers in one direction has the same effect as the actual motion of negative charge carriers in the opposite direction. (When the effect is not the same, we shall, of course, drop the convention and describe the actual motion.)

✔CHECKPOINT 1: The figure here shows a portion of a circuit. What are the magnitude and direction of the current i in the lower right-hand wire?

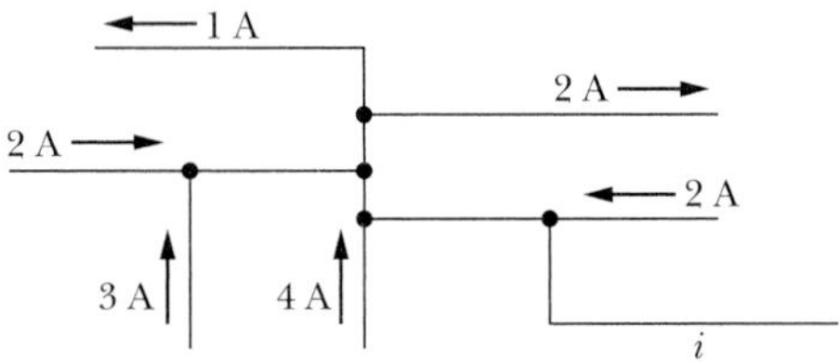

Sample Problem 27-1

Water flows through a garden hose at a volume flow rate dV/dt of 450 cm^3/s. What is the current of negative charge?

SOLUTION: The current i of negative charge is due to the electrons in the water molecules moving through the hose. The current is the rate at which that negative charge passes through any plane that cuts completely across the hose. Thus, the **Key Idea** here is that we can write the current in terms of the number of molecules that pass through such a plane per second as

$$i = \begin{pmatrix}\text{charge}\\ \text{per}\\ \text{electron}\end{pmatrix}\begin{pmatrix}\text{electrons}\\ \text{per}\\ \text{molecule}\end{pmatrix}\begin{pmatrix}\text{molecules}\\ \text{per}\\ \text{second}\end{pmatrix}$$

or

$$i = (e)(10)\frac{dN}{dt}.$$

We substitute 10 electrons per molecule because a water (H_2O) molecule contains 8 electrons in the single oxygen atom and 1 electron in each of the two hydrogen atoms.

We can express the rate dN/dt in terms of the given volume flow rate dV/dt by first writing

$$\begin{pmatrix}\text{molecules}\\ \text{per}\\ \text{second}\end{pmatrix} = \begin{pmatrix}\text{molecules}\\ \text{per}\\ \text{mole}\end{pmatrix}\begin{pmatrix}\text{moles}\\ \text{per unit}\\ \text{mass}\end{pmatrix}\begin{pmatrix}\text{mass}\\ \text{per unit}\\ \text{volume}\end{pmatrix}\begin{pmatrix}\text{volume}\\ \text{per}\\ \text{second}\end{pmatrix}.$$

"Molecules per mole" is Avogadro's number N_A. "Moles per unit mass" is the inverse of the mass per mole, which is the molar mass M of water. "Mass per unit volume" is the (mass) density ρ_{mass} of water. The volume per second is the volume flow rate dV/dt. Thus, we have

$$\frac{dN}{dt} = N_A\left(\frac{1}{M}\right)\rho_{mass}\left(\frac{dV}{dt}\right) = \frac{N_A\rho_{mass}}{M}\frac{dV}{dt}.$$

Substituting this into the equation for i, we find

$$i = 10eN_AM^{-1}\rho_{mass}\frac{dV}{dt}.$$

N_A is 6.02×10^{23} molecules/mol, or 6.02×10^{23} mol^{-1}, and ρ_{mass} is 1000 kg/m^3. We can get the molar mass of water from the molar masses listed in Appendix F: We add the molar mass of oxygen (16 g/mol) to twice the molar mass of hydrogen (1 g/mol), obtaining 18 g/mol = 0.018 kg/mol. Then

$$\begin{aligned} i &= (10)(1.6 \times 10^{-19}\text{ C})(6.02 \times 10^{23}\text{ mol}^{-1}) \\ &\quad \times (0.018\text{ kg/mol})^{-1}(1000\text{ kg/m}^3)(450 \times 10^{-6}\text{ m}^3\text{/s}) \\ &= 2.41 \times 10^7\text{ C/s} = 2.41 \times 10^7\text{ A} \\ &= 24.1\text{ MA}. \end{aligned}$$ (Answer)

This current of negative charge is exactly compensated by a current of positive charge associated with the nuclei of the three atoms that make up the water molecule. Thus, there is no net flow of charge through the hose.

27-3 Current Density

Sometimes we are interested in the current i in a particular conductor. At other times we take a localized view and study the flow of charge through a cross section of the conductor at a particular point. To describe this flow, we can use the **current density** $\vec{J}$, which has the same direction as the velocity of the moving charges if they are positive and the opposite direction if they are negative. For each element of the cross section, the magnitude J is equal to the current per unit area through that element. We can write the amount of current through the element as $\vec{J} \cdot d\vec{A}$, where $d\vec{A}$ is the

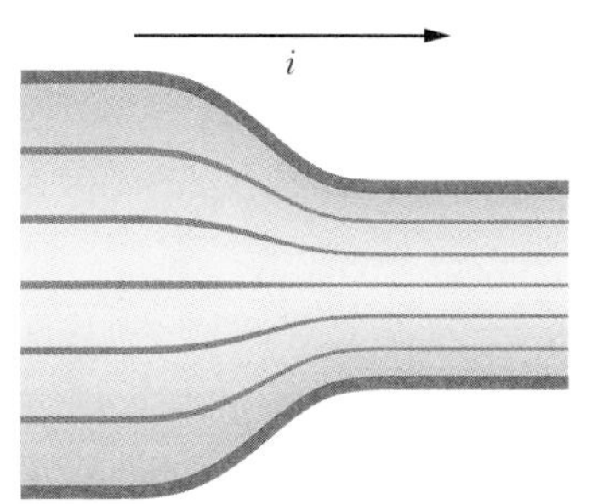

Fig. 27-4 Streamlines representing current density in the flow of charge through a constricted conductor.

area vector of the element, perpendicular to the element. The total current through the surface is then

$$i = \int \vec{J} \cdot d\vec{A}. \tag{27-4}$$

If the current is uniform across the surface and parallel to $d\vec{A}$, then $\vec{J}$ is also uniform and parallel to $d\vec{A}$. Then Eq. 27-4 becomes

$$i = \int J\, dA = J \int dA = JA,$$

so

$$J = \frac{i}{A}, \tag{27-5}$$

where A is the total area of the surface. From Eq. 27-4 or 27-5 we see that the SI unit for current density is the ampere per square meter (A/m^2).

In Chapter 23 we saw that we can represent an electric field with electric field lines. Figure 27-4 shows how current density can be represented with a similar set of lines, which we can call *streamlines*. The current, which is toward the right in Fig. 27-4, makes a transition from the wider conductor at the left to the narrower conductor at the right. Because charge is conserved during the transition, the amount of charge and thus the amount of current cannot change. However, the current density does change—it is greater in the narrower conductor. The spacing of the streamlines suggests this increase in current density; streamlines that are closer together imply greater current density.

Drift Speed

When a conductor does not have a current through it, its conduction electrons move randomly, with no net motion in any direction. When the conductor does have a current through it, these electrons actually still move randomly, but now they tend to *drift* with a **drift speed** v_d in the direction opposite that of the applied electric field that causes the current. The drift speed is tiny compared to the speeds in the random motion. For example, in the copper conductors of household wiring, electron drift speeds are perhaps 10^{-5} or 10^{-4} m/s, whereas the random-motion speeds are around 10^6 m/s.

We can use Fig. 27-5 to relate the drift speed v_d of the conduction electrons in a current through a wire to the magnitude J of the current density in the wire. For convenience, Fig. 27-5 shows the equivalent drift of *positive* charge carriers in the direction of the applied electric field $\vec{E}$. Let us assume that these charge carriers all move with the same drift speed v_d and that the current density J is uniform across the wire's cross-sectional area A. The number of charge carriers in a length L of the wire is nAL, where n is the number of carriers per unit volume. The total charge of the carriers in the length L, each with charge e, is then

$$q = (nAL)e.$$

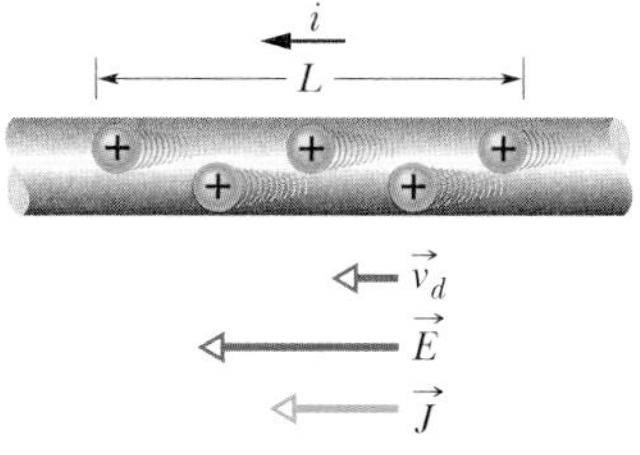

Fig. 27-5 Positive charge carriers drift at speed v_d in the direction of the applied electric field $\vec{E}$. By convention, the direction of the current density $\vec{J}$ and the sense of the current arrow are drawn in that same direction.

Because the carriers all move along the wire with speed v_d, this total charge moves through any cross section of the wire in the time interval

$$t = \frac{L}{v_d}.$$

Equation 27-1 tells us that the current i is the time rate of transfer of charge across a cross section, so here we have

$$i = \frac{q}{t} = \frac{nALe}{L/v_d} = nAev_d. \qquad (27\text{-}6)$$

Solving for v_d and recalling Eq. 27-5 ($J = i/A$), we obtain

$$v_d = \frac{i}{nAe} = \frac{J}{ne}$$

or, extended to vector form,

$$\vec{J} = (ne)\vec{v}_d. \qquad (27\text{-}7)$$

Here the product ne, whose SI unit is the coulomb per cubic meter (C/m^3), is the *carrier charge density*. For positive carriers, ne is positive and Eq. 27-7 predicts that $\vec{J}$ and $\vec{v}_d$ have the same direction. For negative carriers, ne is negative and $\vec{J}$ and $\vec{v}_d$ have opposite directions.

CHECKPOINT 2: The figure here shows conduction electrons moving leftward through a wire. Are the following leftward or rightward: (a) the current i, (b) the current density $\vec{J}$, (c) the electric field $\vec{E}$ in the wire?

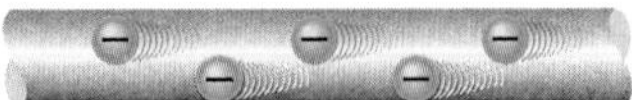

Sample Problem 27-2

(a) The current density in a cylindrical wire of radius $R = 2.0$ mm is uniform across a cross section of the wire and is $J = 2.0 \times 10^5$ A/m^2. What is the current through the outer portion of the wire between radial distances $R/2$ and R (Fig. 27-6*a*)?

SOLUTION: The **Key Idea** here is that, because the current density is uniform across the cross section, the current density J, the current i, and the cross-sectional area A are related by Eq. 27-5 ($J = i/A$). However, we want only the current through a reduced cross-sectional area A' of the wire (rather than the entire area), where

$$A' = \pi R^2 - \pi\left(\frac{R}{2}\right)^2 = \pi\left(\frac{3R^2}{4}\right)$$
$$= \frac{3\pi}{4}(0.002\text{ m})^2 = 9.424 \times 10^{-6}\text{ m}^2.$$

We now rewrite Eq. 27-5 as

$$i = JA'$$

and then substitute the data to find

$$i = (2.0 \times 10^5\text{ A/m}^2)(9.424 \times 10^{-6}\text{ m}^2)$$
$$= 1.9\text{ A}. \qquad \text{(Answer)}$$

(b) Suppose, instead, that the current density through a cross section varies with radial distance r as $J = ar^2$, in which $a = 3.0 \times 10^{11}$ A/m^4 and r is in meters. What now is the current through the same outer portion of the wire?

SOLUTION: The **Key Idea** here is that, because the current density is not uniform across a cross section of the wire, we must resort to Eq. 27-4 ($i = \int \vec{J} \cdot d\vec{A}$) and integrate the current density over the portion of the wire from $r = R/2$ to $r = R$. The current density vector $\vec{J}$ (along the wire's length) and the differential area vector $d\vec{A}$ (perpendicular to a cross section of the wire) have the same direction. Thus,

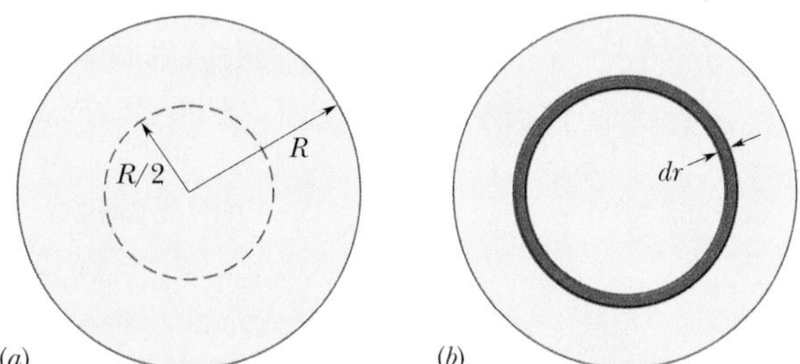

Fig. 27-6 Sample Problem 27-2. (*a*) Cross section of a wire of radius R. (*b*) A thin ring has width dr and circumference $2\pi r$, and thus a differential area $dA = 2\pi r\, dr$.

$$\vec{J} \cdot d\vec{A} = J\, dA \cos 0 = J\, dA.$$

We need to replace the differential area dA with something we can actually integrate between the limits $r = R/2$ and $r = R$. The simplest replacement (because J is given as a function of r) is the area $2\pi r\, dr$ of a thin ring of circumference $2\pi r$ and width dr (Fig. 27-6*b*). We can then integrate with r as the variable of integration. Equation 27-4 then gives us

$$i = \int \vec{J} \cdot d\vec{A} = \int J\, dA$$
$$= \int_{R/2}^{R} ar^2\, 2\pi r\, dr = 2\pi a \int_{R/2}^{R} r^3\, dr$$
$$= 2\pi a\left[\frac{r^4}{4}\right]_{R/2}^{R} = \frac{\pi a}{2}\left[R^4 - \frac{R^4}{16}\right] = \frac{15}{32}\pi a R^4$$
$$= \frac{15}{32}\pi(3.0 \times 10^{11}\text{ A/m}^4)(0.002\text{ m})^4 = 7.1\text{ A}. \qquad \text{(Answer)}$$

Sample Problem 27-3

What is the drift speed of the conduction electrons in a copper wire with radius $r = 900\ \mu\text{m}$ when it has a uniform current $i = 17$ mA? Assume that each copper atom contributes one conduction electron to the current and the current density is uniform across the wire's cross section.

SOLUTION: We need three Key Ideas here:

1. The drift speed v_d is related to the current density $\vec{J}$ and the number n of conduction electrons per unit volume according to Eq. 27-7, which we can write in magnitude form as $J = nev_d$.
2. Because the current density is uniform, its magnitude J is related to the given current i and wire size by Eq. 27-5 ($J = i/A$, where A is the cross-sectional area of the wire).
3. Because we assume one conduction electron per atom, the number n of conduction electrons per unit volume is the same as the number of atoms per unit volume.

Let us start with the third idea by writing

$$n = \begin{pmatrix}\text{atoms}\\\text{per unit}\\\text{volume}\end{pmatrix} = \begin{pmatrix}\text{atoms}\\\text{per}\\\text{mole}\end{pmatrix}\begin{pmatrix}\text{moles}\\\text{per unit}\\\text{mass}\end{pmatrix}\begin{pmatrix}\text{mass}\\\text{per unit}\\\text{volume}\end{pmatrix}.$$

The number of atoms per mole is just Avogadro's number N_A ($= 6.02 \times 10^{23}\ \text{mol}^{-1}$). Moles per unit mass is the inverse of the mass per mole, which here is the molar mass M of copper. The mass per unit volume is the (mass) density ρ_{mass} of copper. Thus,

$$n = N_A\left(\frac{1}{M}\right)\rho_{\text{mass}} = \frac{N_A\rho_{\text{mass}}}{M}.$$

Taking copper's molar mass M and density ρ_{mass} from Appendix F, we then have (with some conversions of units)

$$n = \frac{(6.02 \times 10^{23}\ \text{mol}^{-1})(8.96 \times 10^3\ \text{kg/m}^3)}{63.54 \times 10^{-3}\ \text{kg/mol}}$$
$$= 8.49 \times 10^{28}\ \text{electrons/m}^3$$

or

$$n = 8.49 \times 10^{28}\ \text{m}^{-3}.$$

Next let us combine the first two key ideas by writing

$$\frac{i}{A} = nev_d.$$

Substituting for A with πr^2 ($= 2.54 \times 10^{-6}\ \text{m}^2$), and solving for v_d, we then find

$$v_d = \frac{i}{ne(\pi r^2)}$$
$$= \frac{17 \times 10^{-3}\ \text{A}}{(8.49 \times 10^{28}\ \text{m}^{-3})(1.6 \times 10^{-19}\ \text{C})(2.54 \times 10^{-6}\ \text{m}^2)}$$
$$= 4.9 \times 10^{-7}\ \text{m/s}, \qquad \text{(Answer)}$$

which is only 1.8 mm/h, slower than a sluggish snail.

You may well ask: "If the electrons drift so slowly, why do the room lights turn on so quickly when I throw the switch?" Confusion on this point results from not distinguishing between the drift speed of the electrons and the speed at which *changes* in the electric field configuration travel along wires. This latter speed is nearly that of light; electrons everywhere in the wire begin drifting almost at once, including into the lightbulbs. Similarly, when you open the valve on your garden hose, with the hose full of water, a pressure wave travels along the hose at the speed of sound in water. The speed at which the water itself moves through the hose—measured perhaps with a dye marker—is much slower.

27-4 Resistance and Resistivity

If we apply the same potential difference between the ends of geometrically similar rods of copper and of glass, very different currents result. The characteristic of the conductor that enters here is its electrical **resistance.** We determine the resistance between any two points of a conductor by applying a potential difference V between those points and measuring the current i that results. The resistance R is then

$$R = \frac{V}{i} \quad \text{(definition of } R\text{)}. \tag{27-8}$$

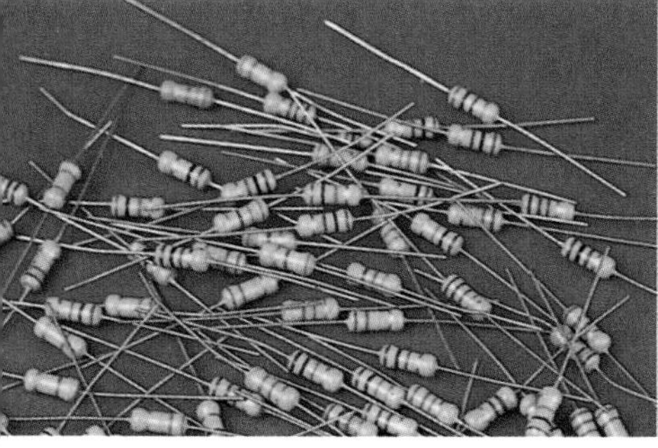

Fig. 27-7 An assortment of resistors. The circular bands are color-coding marks that identify the value of the resistance.

The SI unit for resistance that follows from Eq. 27-8 is the volt per ampere. This combination occurs so often that we give it a special name, the **ohm** (symbol Ω); that is,

$$1\ \text{ohm} = 1\ \Omega = 1\ \text{volt per ampere}$$
$$= 1\ \text{V/A}. \tag{27-9}$$

A conductor whose function in a circuit is to provide a specified resistance is called a **resistor** (see Fig. 27-7). In a circuit diagram, we represent a resistor and a resistance

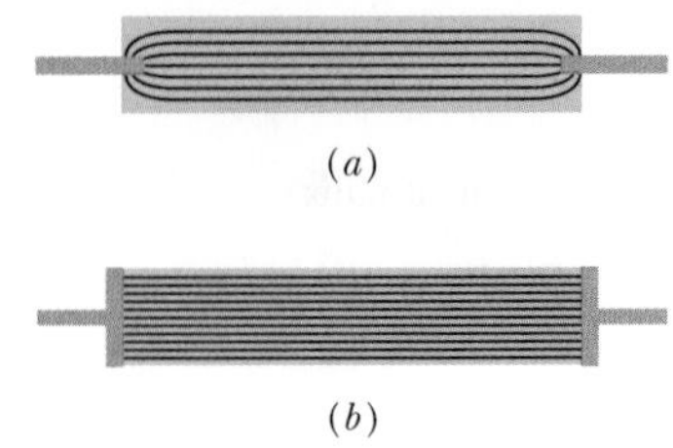

Fig. 27-8 Two ways of applying a potential difference to a conducting rod. The heavy gray connectors are assumed to have negligible resistance. When they are arranged as in (*a*), the measured resistance is larger than when they are arranged as in (*b*).

with the symbol -ΛΛΛ-. If we write Eq. 27-8 as

$$i = \frac{V}{R},$$

we see that "resistance" is aptly named. For a given potential difference, the greater the resistance (to current), the smaller the current.

The resistance of a conductor depends on the manner in which the potential difference is applied to it. Figure 27-8, for example, shows a given potential difference applied in two different ways to the same conductor. As the current density streamlines suggest, the currents in the two cases—hence the measured resistances—will be different. Unless otherwise stated, we shall assume that any given potential difference is applied as in Fig. 27-8*b*.

As we have done several times in other connections, we often wish to take a general view and deal not with particular objects but with materials. Here we do so by focusing not on the potential difference V across a particular resistor but on the electric field $\vec{E}$ at a point in a resistive material. Instead of dealing with the current i through the resistor, we deal with the current density $\vec{J}$ at the point in question. Instead of the resistance R of an object, we deal with the **resistivity** ρ of the *material*:

$$\rho = \frac{E}{J} \quad \text{(definition of } \rho\text{).} \tag{27-10}$$

(Compare this equation with Eq. 27-8.)

If we combine the SI units of E and J according to Eq. 27-10, we get, for the unit of ρ, the ohm-meter ($\Omega \cdot \text{m}$):

$$\frac{\text{unit } (E)}{\text{unit } (J)} = \frac{\text{V/m}}{\text{A/m}^2} = \frac{\text{V}}{\text{A}}\,\text{m} = \Omega \cdot \text{m}.$$

(Do not confuse the *ohm-meter,* the unit of resistivity, with the *ohmmeter,* which is an instrument that measures resistance.) Table 27-1 lists the resistivities of some materials.

We can write Eq. 27-10 in vector form as

$$\vec{E} = \rho\vec{J}. \tag{27-11}$$

Equations 27-10 and 27-11 hold only for *isotropic* materials—materials whose electrical properties are the same in all directions.

We often speak of the **conductivity** σ of a material. This is simply the reciprocal of its resistivity, so

$$\sigma = \frac{1}{\rho} \quad \text{(definition of } \sigma\text{).} \tag{27-12}$$

The SI unit of conductivity is the reciprocal ohm-meter, $(\Omega \cdot \text{m})^{-1}$. The unit name mhos per meter is sometimes used (mho is ohm backwards). The definition of σ allows us to write Eq. 27-11 in the alternative form

$$\vec{J} = \sigma\vec{E}. \tag{27-13}$$

Calculating Resistance from Resistivity

We have just made an important distinction:

➤ Resistance is a property of an object. Resistivity is a property of a material.

TABLE 27-1 **Resistivities of Some Materials at Room Temperature (20°C)**

Material	Resistivity, ρ ($\Omega \cdot m$)	Temperature Coefficient of Resistivity, α (K^{-1})
	Typical Metals	
Silver	1.62×10^{-8}	4.1×10^{-3}
Copper	1.69×10^{-8}	4.3×10^{-3}
Aluminum	2.75×10^{-8}	4.4×10^{-3}
Tungsten	5.25×10^{-8}	4.5×10^{-3}
Iron	9.68×10^{-8}	6.5×10^{-3}
Platinum	10.6×10^{-8}	3.9×10^{-3}
Manganin[a]	4.82×10^{-8}	0.002×10^{-3}
	Typical Semiconductors	
Silicon, pure	2.5×10^{3}	-70×10^{-3}
Silicon, *n*-type[b]	8.7×10^{-4}	
Silicon, *p*-type[c]	2.8×10^{-3}	
	Typical Insulators	
Glass	10^{10}–10^{14}	
Fused quartz	$\sim 10^{16}$	

[a]An alloy specifically designed to have a small value of α.

[b]Pure silicon doped with phosphorus impurities to a charge carrier density of 10^{23} m^{-3}.

[c]Pure silicon doped with aluminum impurities to a charge carrier density of 10^{23} m^{-3}.

If we know the resistivity of a substance such as copper, we can calculate the resistance of a length of wire made of that substance. Let A be the cross-sectional area of the wire, let L be its length, and let a potential difference V exist between its ends (Fig. 27-9). If the streamlines representing the current density are uniform throughout the wire, the electric field and the current density will be constant for all points within the wire and, from Eqs. 25-42 and 27-5, will have the values

$$E = V/L \quad \text{and} \quad J = i/A. \tag{27-14}$$

We can then combine Eqs. 27-10 and 27-14 to write

$$\rho = \frac{E}{J} = \frac{V/L}{i/A}. \tag{27-15}$$

However, V/i is the resistance R, which allows us to recast Eq. 27-15 as

$$R = \rho \frac{L}{A}. \tag{27-16}$$

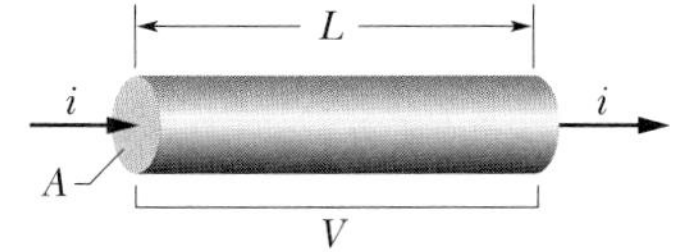

Fig. 27-9 A potential difference V is applied between the ends of a wire of length L and cross section A, establishing a current i.

Equation 27-16 can be applied only to a homogeneous isotropic conductor of uniform cross section, with the potential difference applied as in Fig. 27-8*b*.

The macroscopic quantities V, i, and R are of greatest interest when we are making electrical measurements on specific conductors. They are the quantities that we read directly on meters. We turn to the microscopic quantities E, J, and ρ when we are interested in the fundamental electrical properties of materials.

CHECKPOINT 3: The figure here shows three cylindrical copper conductors along with their face areas and lengths. Rank them according to the current through them, greatest first, when the same potential difference V is placed across their lengths.

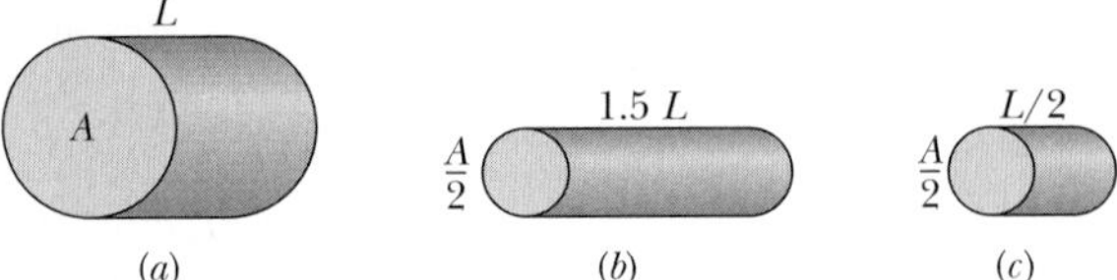

Variation with Temperature

The values of most physical properties vary with temperature, and resistivity is no exception. Figure 27-10, for example, shows the variation of this property for copper over a wide temperature range. The relation between temperature and resistivity for copper—and for metals in general—is fairly linear over a rather broad temperature range. For such linear relations we can write an empirical approximation that is good enough for most engineering purposes:

$$\rho - \rho_0 = \rho_0\alpha(T - T_0). \quad (27\text{-}17)$$

Here T_0 is a selected reference temperature and ρ_0 is the resistivity at that temperature. Usually $T_0 = 293$ K (room temperature), for which $\rho_0 = 1.69 \times 10^{-8}\ \Omega \cdot \text{m}$ for copper.

Because temperature enters Eq. 27-17 only as a difference, it does not matter whether you use the Celsius or Kelvin scale in that equation because the sizes of degrees on these scales are identical. The quantity α in Eq. 27-17, called the *temperature coefficient of resistivity,* is chosen so that the equation gives good agreement with experiment for temperatures in the chosen range. Some values of α for metals are listed in Table 27-1.

The *Hindenburg*

When the zeppelin *Hindenburg* was preparing to land, the handling ropes were let down to the ground crew. Exposed to the rain, the ropes became wet (and thus were able to conduct a current). In this condition, the ropes "grounded" the metal framework of the zeppelin to which they were attached; that is, the wet ropes formed a conducting path between the framework and the ground, making the electric potential of the framework the same as that of the ground. This should have also grounded the outer fabric of the zeppelin. The *Hindenburg*, however, had been the first zeppelin to have its outer fabric painted with a sealant of large electrical resistivity. Thus, the fabric remained at the electric potential of the atmosphere at the zeppelin's altitude of about 43 m. Due to the rainstorm, that potential was large relative to the potential at ground level.

The handling of the ropes apparently ruptured one of the hydrogen cells and released hydrogen between that cell and the zeppelin's outer fabric, causing the reported rippling of the fabric. There was then a dangerous situation: The fabric was wet with conducting rainwater and was at a potential much different from that of the framework of the zeppelin. Apparently, charge flowed along the wet fabric and then sparked through the released hydrogen to reach the metal framework of the zeppelin, igniting the hydrogen in the process. The burning rapidly ignited the cells of hydrogen in the zeppelin and brought the ship down. If the sealant on the outer fabric of the *Hindenburg* had been of less resistivity (like that of earlier and later zeppelins), the *Hindenburg* disaster probably would not have occurred.

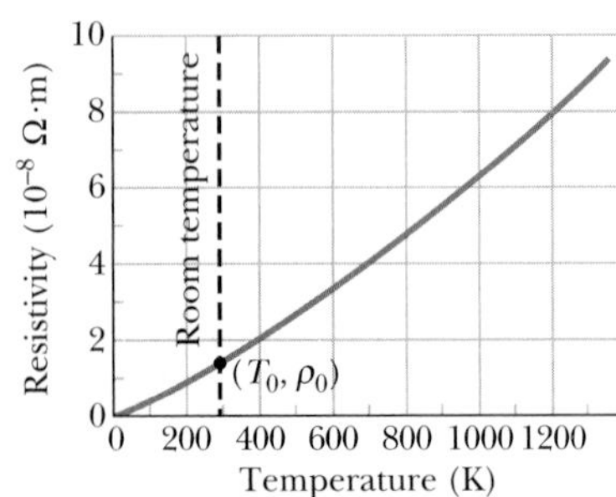

Fig. 27-10 The resistivity of copper as a function of temperature. The dot on the curve marks a convenient reference point at temperature $T_0 = 293$ K and resistivity $\rho_0 = 1.69 \times 10^{-8}\ \Omega \cdot \text{m}$.

Sample Problem 27-4

A rectangular block of iron has dimensions 1.2 cm × 1.2 cm × 15 cm. A potential difference is to be applied to the block between parallel sides and in such a way that those sides are equipotential surfaces (as in Fig. 27-8*b*). What is the resistance of the block if the two parallel sides are (1) the square ends (with dimensions 1.2 cm × 1.2 cm) and (2) two rectangular sides (with dimensions 1.2 cm × 15 cm)?

SOLUTION: The **Key Idea** here is that the resistance R of an object depends on how the electric potential is applied to the object. In particular, it depends on the ratio L/A, according to Eq. 27-16 ($R = \rho L/A$), where A is the area of the surfaces to which the potential difference is applied and L is the distance between those surfaces. For arrangement 1, L = 15 cm = 0.15 m and

$$A = (1.2 \text{ cm})^2 = 1.44 \times 10^{-4} \text{ m}^2.$$

Substituting into Eq. 27-16 with the resistivity ρ from Table 27-1, we then find that for arrangement 1,

$$R = \frac{\rho L}{A} = \frac{(9.68 \times 10^{-8}\ \Omega \cdot \text{m})(0.15 \text{ m})}{1.44 \times 10^{-4} \text{ m}^2}$$
$$= 1.0 \times 10^{-4}\ \Omega = 100\ \mu\Omega. \quad \text{(Answer)}$$

Similarly, for arrangement 2, with distance L = 1.2 cm and area A = (1.2 cm)(15 cm), we obtain

$$R = \frac{\rho L}{A} = \frac{(9.68 \times 10^{-8}\ \Omega \cdot \text{m})(1.2 \times 10^{-2} \text{ m})}{1.80 \times 10^{-3} \text{ m}^2}$$
$$= 6.5 \times 10^{-7}\ \Omega = 0.65\ \mu\Omega. \quad \text{(Answer)}$$

27-5 Ohm's Law

As we just discussed in Section 27-4, a resistor is a conductor with a specified resistance. It has that same resistance no matter what the magnitude and direction (*polarity*) of the applied potential difference. Other conducting devices, however, might have resistances that change with the applied potential difference.

Figure 27-11*a* shows how to distinguish such devices. A potential difference V is applied across the device being tested, and the resulting current i through the device is measured as V is varied in both magnitude and polarity. The polarity of V is arbitrarily taken to be positive when the left terminal of the device is at a higher potential than the right terminal. The direction of the resulting current (from left to right) is arbitrarily assigned a plus sign. The reverse polarity of V (with the right terminal at a higher potential) is then negative; the current it causes is assigned a minus sign.

Figure 27-11*b* is a plot of i versus V for one device. This plot is a straight line passing through the origin, so the ratio i/V (which is the slope of the straight line) is the same for all values of V. This means that the resistance $R = V/i$ of the device is independent of the magnitude and polarity of the applied potential difference V.

Figure 27-11*c* is a plot for another conducting device. Current can exist in this device only when the polarity of V is positive and the applied potential difference is more than about 1.5 V. When current does exist, the relation between i and V is not linear; it depends on the value of the applied potential difference V.

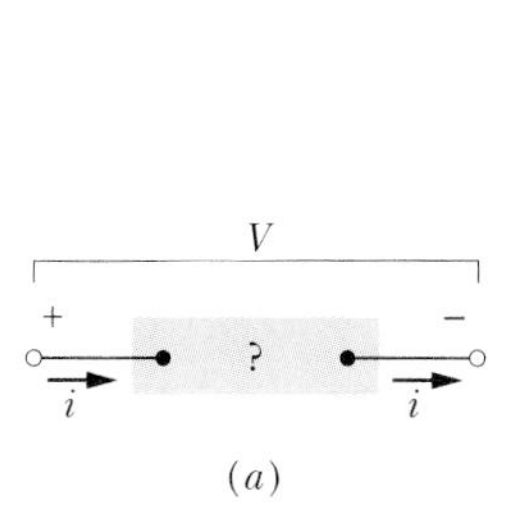

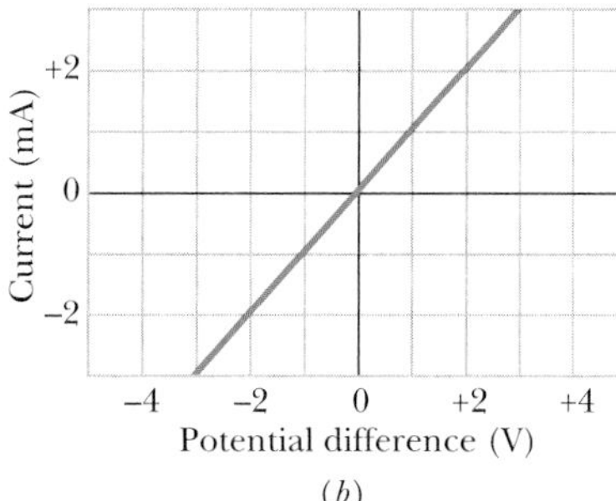

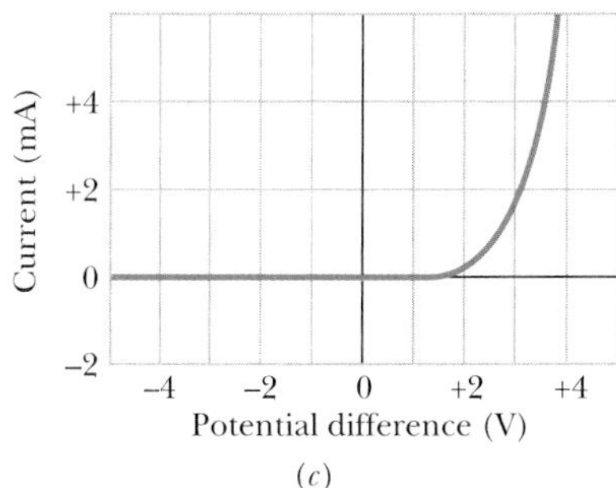

Fig. 27-11 (*a*) A potential difference V is applied to the terminals of a device, establishing a current i. (*b*) A plot of current i versus applied potential difference V when the device is a 1000 Ω resistor. (*c*) A plot when the device is a semiconducting *pn* junction diode.

We distinguish between the two types of device by saying that one obeys Ohm's law and the other does not.

> **Ohm's law** is an assertion that the current through a device is *always* directly proportional to the potential difference applied to the device.

(This assertion is correct only in certain situations; still, for historical reasons, the term "law" is used.) The device of Fig. 27-11*b*—which turns out to be a 1000 Ω resistor—obeys Ohm's law. The device of Fig. 27-11*c*—which turns out to be a so-called *pn* junction diode—does not.

> A conducting device obeys Ohm's law when the resistance of the device is independent of the magnitude and polarity of the applied potential difference.

Modern microelectronics—and therefore much of the character of our present technological civilization—depends almost totally on devices that do *not* obey Ohm's law. Your calculator, for example, is full of them.

It is often contended that $V = iR$ is a statement of Ohm's law. That is not true! This equation is the defining equation for resistance, and it applies to all conducting devices, whether they obey Ohm's law or not. If we measure the potential difference V across, and the current i through, any device, even a *pn* junction diode, we can find its resistance *at that value of* V as $R = V/i$. The essence of Ohm's law, however, is that a plot of i versus V is linear; that is, R is independent of V.

We can express Ohm's law in a more general way if we focus on conducting *materials* rather than on conducting *devices*. The relevant relation is then Eq. 27-11 ($\vec{E} = \rho\vec{J}$), which is the analog of $V = iR$.

> A conducting material obeys Ohm's law when the resistivity of the material is independent of the magnitude and direction of the applied electric field.

All homogeneous materials, whether they are conductors like copper or semiconductors like pure silicon or silicon containing special impurities, obey Ohm's law within some range of values of the electric field. If the field is too strong, however, there are departures from Ohm's law in all cases.

CHECKPOINT 4: The following table gives the current i (in amperes) through two devices for several values of potential difference V (in volts). From these data, determine which device does not obey Ohm's law.

Device 1		Device 2	
V	i	V	i
2.00	4.50	2.00	1.50
3.00	6.75	3.00	2.20
4.00	9.00	4.00	2.80

27-6 A Microscopic View of Ohm's Law

To find out *why* particular materials obey Ohm's law, we must look into the details of the conduction process at the atomic level. Here we consider only conduction in metals, such as copper. We base our analysis on the *free-electron model,* in which we assume that the conduction electrons in the metal are free to move throughout the volume of a sample, like the molecules of a gas in a closed container. We

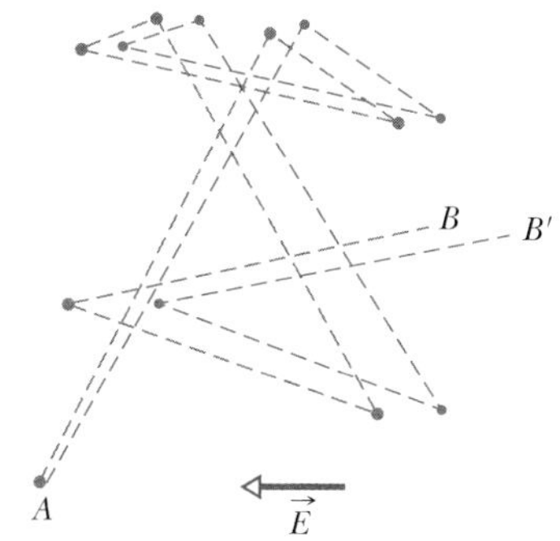

Fig. 27-12 The gray lines show an electron moving from A to B, making six collisions en route. The green lines show what its path might be in the presence of an applied electric field $\vec{E}$. Note the steady drift in the direction of $-\vec{E}$. (Actually, the green lines should be slightly curved, to represent the parabolic paths followed by the electrons between collisions, under the influence of an electric field.)

also assume that the electrons collide not with one another but only with atoms of the metal.

According to classical physics, the electrons should have a Maxwellian speed distribution somewhat like that of the molecules in a gas. In such a distribution (see Section 20-7), the average electron speed would be proportional to the square root of the absolute temperature. The motions of electrons are, however, governed not by the laws of classical physics but by those of quantum physics. As it turns out, an assumption that is much closer to the quantum reality is that conduction electrons in a metal move with a single effective speed v_{eff}, and this speed is essentially independent of the temperature. For copper, $v_{eff} \approx 1.6 \times 10^6$ m/s.

When we apply an electric field to a metal sample, the electrons modify their random motions slightly and drift very slowly—in a direction opposite that of the field—with an average drift speed v_d. As we saw in Sample Problem 27-3, the drift speed in a typical metallic conductor is about 5×10^{-7} m/s, less than the effective speed (1.6×10^6 m/s) by many orders of magnitude. Figure 27-12 suggests the relation between these two speeds. The gray lines show a possible random path for an electron in the absence of an applied field; the electron proceeds from A to B, making six collisions along the way. The green lines show how the same events *might* occur when an electric field $\vec{E}$ is applied. We see that the electron drifts steadily to the right, ending at B' rather than at B. Figure 27-12 was drawn with the assumption that $v_d \approx 0.02v_{eff}$. However, because the actual value is more like $v_d \approx (10^{-13})v_{eff}$, the drift displayed in the figure is greatly exaggerated.

The motion of the conduction electrons in an electric field $\vec{E}$ is thus a combination of the motion due to random collisions and that due to $\vec{E}$. When we consider all the free electrons, their random motions average to zero and make no contribution to the drift speed. Thus, the drift speed is due only to the effect of the electric field on the electrons.

If an electron of mass m is placed in an electric field of magnitude E, the electron will experience an acceleration given by Newton's second law:

$$a = \frac{F}{m} = \frac{eE}{m}. \tag{27-18}$$

The nature of the collisions experienced by conduction electrons is such that, after a typical collision, each electron will—so to speak—completely lose its memory of its previous drift velocity. Each electron will then start off fresh after every encounter, moving off in a random direction. In the average time τ between collisions, the average electron will acquire a drift speed of $v_d = a\tau$. Moreover, if we measure the drift speeds of all the electrons at any instant, we will find that their average drift speed is also $a\tau$. Thus, at any instant, on average, the electrons will have drift speed $v_d = a\tau$. Then Eq. 27-18 gives us

$$v_d = a\tau = \frac{eE\tau}{m}. \tag{27-19}$$

Combining this result with Eq. 27-7 ($\vec{J} = ne\vec{v}_d$), in magnitude form, yields

$$v_d = \frac{J}{ne} = \frac{eE\tau}{m},$$

which we can write as

$$E = \left(\frac{m}{e^2 n\tau}\right) J.$$

Comparing this with Eq. 27-11 ($\vec{E} = \rho\vec{J}$), in magnitude form, leads to

$$\rho = \frac{m}{e^2 n\tau}. \tag{27-20}$$

Equation 27-20 may be taken as a statement that metals obey Ohm's law if we can show that, for metals, their resistivity ρ is a constant, independent of the strength of the applied electric field $\vec{E}$. Because n, m, and e are constant, this reduces to convincing ourselves that τ, the average time (or *mean free time*) between collisions, is a constant, independent of the strength of the applied electric field. Indeed, τ can be considered to be a constant because the drift speed v_d caused by the field is so much smaller than the effective speed v_{eff} that the electron speed—and thus τ—is hardly affected by the field.

Sample Problem 27-5

(a) What is the mean free time τ between collisions for the conduction electrons in copper?

SOLUTION: The Key Idea here is that the mean free time τ of copper is approximately constant, and in particular does not depend on any electric field that might be applied to a sample of the copper. Thus, we need not consider any particular value of applied electric field. However, because the resistivity ρ displayed by copper under an electric field depends on τ, we can find the mean free time τ from Eq. 27-20 ($\rho = m/e^2 n\tau$). That equation gives us

$$\tau = \frac{m}{ne^2\rho}.$$

We take the value of n, the number of conduction electrons per unit volume in copper, from Sample Problem 27-3. We take the value of ρ from Table 27-1. The denominator then becomes

$$(8.49 \times 10^{28}\ \text{m}^{-3})(1.6 \times 10^{-19}\ \text{C})^2(1.69 \times 10^{-8}\ \Omega\cdot\text{m})$$
$$= 3.67 \times 10^{-17}\ \text{C}^2\cdot\Omega/\text{m}^2 = 3.67 \times 10^{-17}\ \text{kg/s},$$

where we converted units as

$$\frac{\text{C}^2\cdot\Omega}{\text{m}^2} = \frac{\text{C}^2\cdot\text{V}}{\text{m}^2\cdot\text{A}} = \frac{\text{C}^2\cdot\text{J/C}}{\text{m}^2\cdot\text{C/s}} = \frac{\text{kg}\cdot\text{m}^2/\text{s}^2}{\text{m}^2/\text{s}} = \frac{\text{kg}}{\text{s}}.$$

Using these results and substituting for the electron mass m, we then have

$$\tau = \frac{9.1 \times 10^{-31}\ \text{kg}}{3.67 \times 10^{-17}\ \text{kg/s}} = 2.5 \times 10^{-14}\ \text{s}. \quad \text{(Answer)}$$

(b) The mean free path λ of the conduction electrons in a conductor is the average distance traveled by an electron between collisions. (This definition parallels that in Section 20-6 for the mean free path of molecules in a gas.) What is λ for the conduction electrons in copper, assuming that their effective speed v_{eff} is 1.6×10^6 m/s?

SOLUTION: The Key Idea here is that the distance d any particle travels in a certain time t at a constant speed v is $d = vt$. For the electrons in copper, this gives us

$$\lambda = v_{eff}\tau = (1.6 \times 10^6\ \text{m/s})(2.5 \times 10^{-14}\ \text{s})$$
$$= 4.0 \times 10^{-8}\ \text{m} = 40\ \text{nm}. \quad \text{(Answer)}$$

This is about 150 times the distance between nearest-neighbor atoms in a copper lattice. Thus, on the average, each conduction electron passes many copper atoms before finally hitting one.

27-7 Power in Electric Circuits

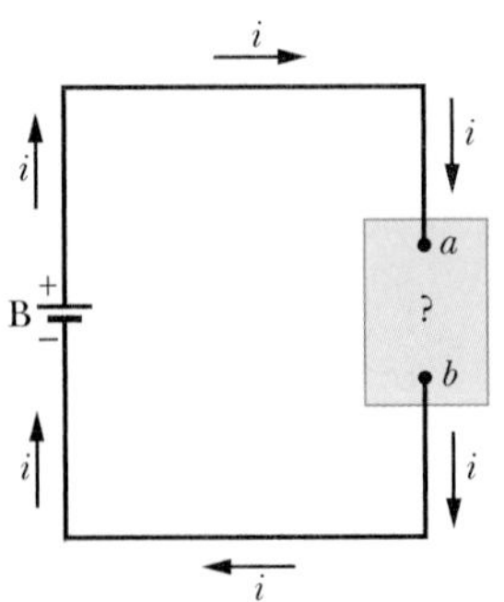

Fig. 27-13 A battery B sets up a current i in a circuit containing an unspecified conducting device.

Figure 27-13 shows a circuit consisting of a battery B that is connected by wires, which we assume have negligible resistance, to an unspecified conducting device. The device might be a resistor, a storage battery (a rechargeable battery), a motor, or some other electrical device. The battery maintains a potential difference of magnitude V across its own terminals, and thus (because of the wires) across the terminals of the unspecified device, with a greater potential at terminal a of the device than at terminal b.

Since there is an external conducting path between the two terminals of the battery, and since the potential differences set up by the battery are maintained, a steady current i is produced in the circuit, directed from terminal a to terminal b. The amount of charge dq that moves between those terminals in time interval dt is equal to $i\,dt$. This charge dq moves through a decrease in potential of magnitude V, and thus its electric potential energy decreases in magnitude by the amount

$$dU = dq\,V = i\,dt\,V.$$

The principle of conservation of energy tells us that the decrease in electric potential energy from a to b is accompanied by a transfer of energy to some other

The wire coils within a toaster have appreciable resistance. When there is a current through them, electric energy is transferred to thermal energy of the coils, increasing their temperature. The coils then emit infrared radiation and visible light that will toast (or burn) bread.

form. The power P associated with that transfer is the rate of transfer dU/dt, which is

$$P = iV \qquad \text{(rate of electric energy transfer).} \tag{27-21}$$

Moreover, this power P is also the rate that energy is transferred from the battery to the unspecified device. If that device is a motor connected to a mechanical load, the energy is transferred as work done on the load. If the device is a storage battery that is being charged, the energy is transferred to stored chemical energy in the storage battery. If the device is a resistor, the energy is transferred to internal thermal energy, tending to increase the resistor's temperature.

The unit of power that follows from Eq. 27-21 is the volt-ampere (V · A). We can write it as

$$1\ \text{V} \cdot \text{A} = \left(1\,\frac{\text{J}}{\text{C}}\right)\left(1\,\frac{\text{C}}{\text{s}}\right) = 1\,\frac{\text{J}}{\text{s}} = 1\ \text{W}.$$

The course of an electron moving through a resistor at constant drift speed is much like that of a stone falling through water at constant terminal speed. The average kinetic energy of the electron remains constant, and its lost electric potential energy appears as thermal energy in the resistor and the surroundings. On a microscopic scale this energy transfer is due to collisions between the electron and the molecules of the resistor, which leads to an increase in the temperature of the resistor lattice. The mechanical energy thus transferred to thermal energy is *dissipated* (lost), because the transfer cannot be reversed.

For a resistor or some other device with resistance R, we can combine Eqs. 27-8 ($R = V/i$) and 27-21 to obtain, for the rate of electric energy dissipation due to a resistance, either

$$P = i^2R \qquad \text{(resistive dissipation)} \tag{27-22}$$

or

$$P = \frac{V^2}{R} \qquad \text{(resistive dissipation).} \tag{27-23}$$

Caution: We must be careful to distinguish these two equations from Eq. 27-21: $P = iV$ applies to electric energy transfers of all kinds; $P = i^2R$ and $P = V^2/R$ apply only to the transfer of electric potential energy to thermal energy in a device with resistance.

CHECKPOINT 5: A potential difference V is connected across a device with resistance R, causing current i through the device. Rank the following variations according to the change in the rate at which electric energy is converted to thermal energy due to the resistance, greatest change first: (a) V is doubled with R unchanged, (b) i is doubled with R unchanged, (c) R is doubled with V unchanged, (d) R is doubled with i unchanged.

Sample Problem 27-6

You are given a length of uniform heating wire made of a nickel–chromium–iron alloy called Nichrome; it has a resistance R of 72 Ω. At what rate is energy dissipated in each of the following situations? (1) A potential difference of 120 V is applied across the full length of the wire. (2) The wire is cut in half, and a potential difference of 120 V is applied across the length of each half.

SOLUTION: The Key Idea is that a current in a resistive material produces a transfer of mechanical energy to thermal energy; the rate of transfer (dissipation) is given by Eqs. 27-21 to 27-23. Because we know the potential V and resistance R, we use Eq. 27-23, which yields, for situation 1,

$$P = \frac{V^2}{R} = \frac{(120\ \text{V})^2}{72\ \Omega} = 200\ \text{W}. \qquad \text{(Answer)}$$

In situation 2, the resistance of each half of the wire is (72 Ω)/2, or 36 Ω. Thus, the dissipation rate for each half is

$$P' = \frac{(120 \text{ V})^2}{36 \ \Omega} = 400 \text{ W},$$

and that for the two halves is

$$P = 2P' = 800 \text{ W}. \qquad \text{(Answer)}$$

This is four times the dissipation rate of the full length of wire. Thus, you might conclude that you could buy a heating coil, cut it in half, and reconnect it to obtain four times the heat output. Why is this unwise? (What would happen to the amount of current in the coil?)

27-8 Semiconductors

Semiconducting devices are at the heart of the microelectronic revolution that ushered in the information age. Table 27-2 compares the properties of silicon—a typical semiconductor—and copper—a typical metallic conductor. We see that silicon has many fewer charge carriers, a much higher resistivity, and a temperature coefficient of resistivity that is both large and negative. Thus, although the resistivity of copper increases with temperature, that of pure silicon decreases.

Pure silicon has such a high resistivity that it is effectively an insulator and thus not of much direct use in microelectronic circuits. However, its resistivity can be greatly reduced in a controlled way by adding minute amounts of specific "impurity" atoms in a process called *doping*. Table 27-1 gives typical values of resistivity for silicon before and after doping with two different impurities.

We can roughly explain the differences in resistivity (and thus in conductivity) between semiconductors, insulators, and metallic conductors in terms of the energies of their electrons. (We need quantum physics to explain in more detail.) In a metallic conductor such as copper wire, most of the electrons are firmly locked into place within the molecules; much energy would be required to free them so they could move and participate in an electric current. However, there are also some electrons that, roughly speaking, are only loosely held in place and that require only little energy to become free. Thermal energy can supply that energy, as can an electric field applied across the conductor. The field would not only free these loosely held electrons but would also propel them along the wire; thus, the field would drive a current through the conductor.

In an insulator, significantly greater energy is required to free electrons so they can move through the material. Thermal energy cannot supply enough energy, and neither can any reasonable electric field applied to the insulator. Thus, no electrons are available to move through the insulator, and hence no current occurs even with an applied electric field.

A semiconductor is like an insulator *except* that the energy required to free some electrons is not quite so great. More important, doping can supply electrons or positive charge carriers that are very loosely held within the material and thus are easy to get moving. Moreover, by controlling the doping of a semiconductor, we can

TABLE 27-2 Some Electrical Properties of Copper and Silicon[a]

Property	Copper	Silicon
Type of material	Metal	Semiconductor
Charge carrier density, m^{-3}	9×10^{28}	1×10^{16}
Resistivity, $\Omega \cdot m$	2×10^{-8}	3×10^{3}
Temperature coefficient of resistivity, K^{-1}	$+4 \times 10^{-3}$	-70×10^{-3}

[a]Rounded to one significant figure for easy comparison.

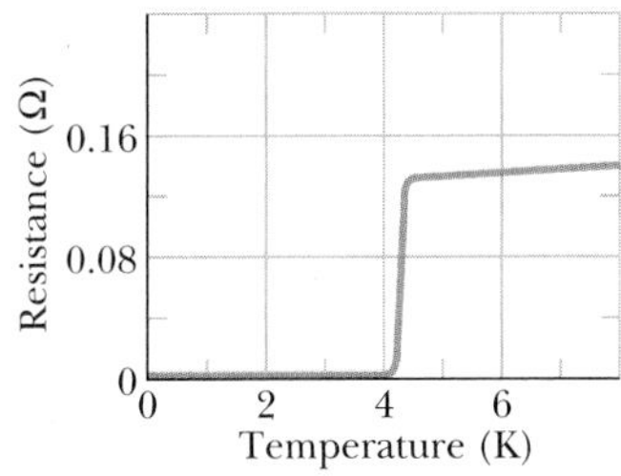

Fig. 27-14 The resistance of mercury drops to zero at a temperature of about 4 K.

control the density of charge carriers that can participate in a current, and thereby can control some of its electrical properties. Most semiconducting devices, such as transistors and junction diodes, are fabricated by the selective doping of different regions of the silicon with impurity atoms of different kinds.

Let us now look again at Eq. 27-20 for the resistivity of a conductor:

$$\rho = \frac{m}{e^2 n \tau}, \tag{27-24}$$

where n is the number of charge carriers per unit volume and τ is the mean time between collisions of the charge carriers. (We derived this equation for conductors, but it also applies to semiconductors.) Let us consider how the variables n and τ change as the temperature is increased.

In a conductor, n is large but very nearly constant with any change in temperature. The increase of resistivity with temperature for metals (Fig. 27-10) is due to an increase in the collision rate of the charge carriers, which shows up in Eq. 27-24 as a decrease in τ, the mean time between collisions.

In a semiconductor, n is small but increases very rapidly with temperature as the increased thermal agitation makes more charge carriers available. This causes a *decrease* of resistivity with increasing temperature, as indicated by the negative temperature coefficient of resistivity for silicon in Table 27-2. The same increase in collision rate that we noted for metals also occurs for semiconductors, but its effect is swamped by the rapid increase in the number of charge carriers.

27-9 Superconductors

In 1911, Dutch physicist Kamerlingh Onnes discovered that the resistivity of mercury absolutely disappears at temperatures below about 4 K (Fig. 27-14). This phenomenon of **superconductivity** is of vast potential importance in technology because it means that charge can flow through a superconducting conductor without producing thermal energy losses. Currents created in a superconducting ring, for example, have persisted for several years without diminution; the electrons making up the current require a force and a source of energy at start-up time, but not thereafter.

Prior to 1986, the technological development of superconductivity was throttled by the cost of producing the extremely low temperatures that were required to achieve the effect. In 1986, however, new ceramic materials were discovered that become superconducting at considerably higher (and thus cheaper to produce) temperatures. Practical application of superconducting devices at room temperature may eventually become feasible.

Superconductivity is a much different phenomenon from conductivity. In fact, the best of the normal conductors, such as silver and copper, cannot become superconducting at any temperature, and the new ceramic superconductors are actually good insulators when they are not at low enough temperatures to be in a superconducting state.

One explanation for superconductivity is that the electrons which make up the current move in coordinated pairs. One of the electrons in a pair may electrically distort the molecular structure of the superconducting material as it moves through, creating nearby a short-lived concentration of positive charge. The other electron in the pair may then be attracted toward this positive charge. According to the theory, such coordination between electrons would prevent them from colliding with the molecules of the material and thus would eliminate electrical resistance. The theory worked well to explain the pre-1986, lower temperature superconductors, but new theories appear to be needed for the newer, higher temperature superconductors.

A disk-shaped magnet is levitated above a superconducting material that has been cooled by liquid nitrogen. The goldfish is along for the ride.

REVIEW & SUMMARY

Current An **electric current** i in a conductor is defined by

$$i = \frac{dq}{dt}. \tag{27-1}$$

Here dq is the amount of (positive) charge that passes in time dt through a hypothetical surface that cuts across the conductor. By convention, the direction of electric current is taken as the direction in which positive charge carriers would move. The SI unit of electric current is the **ampere** (A): 1 A = 1 C/s.

Current Density Current (a scalar) is related to **current density** $\vec{J}$ (a vector) by

$$i = \int \vec{J} \cdot d\vec{A}, \tag{27-4}$$

where $d\vec{A}$ is a vector perpendicular to a surface element of area dA, and the integral is taken over any surface cutting across the conductor. $\vec{J}$ has the same direction as the velocity of the moving charges if they are positive and the opposite direction if they are negative.

Drift Speed of the Charge Carriers When an electric field $\vec{E}$ is established in a conductor, the charge carriers (assumed positive) acquire a **drift speed** v_d in the direction of $\vec{E}$; the velocity $\vec{v}_d$ is related to the current density by

$$\vec{J} = (ne)\vec{v}_d, \tag{27-7}$$

where ne is the carrier charge density.

Resistance of a Conductor The **resistance** R of a conductor is defined as

$$R = \frac{V}{i} \quad \text{(definition of } R\text{)}, \tag{27-8}$$

where V is the potential difference across the conductor and i is the current. The SI unit of resistance is the **ohm** (Ω): 1 Ω = 1 V/A. Similar equations define the **resistivity** ρ and **conductivity** σ of a material:

$$\rho = \frac{1}{\sigma} = \frac{E}{J} \quad \text{(definitions of } \rho \text{ and } \sigma\text{)}, \tag{27-12, 27-10}$$

where E is the magnitude of the applied electric field. The SI unit of resistivity is the ohm-meter ($\Omega \cdot$ m). Equation 27-10 corresponds to the vector equation

$$\vec{E} = \rho\vec{J}. \tag{27-11}$$

The resistance R of a conducting wire of length L and uniform cross section is

$$R = \rho\frac{L}{A}, \tag{27-16}$$

where A is the cross-sectional area.

Change of ρ with Temperature The resistivity ρ for most materials changes with temperature. For many materials, including metals, the relation between ρ and temperature T is approximated by the equation

$$\rho - \rho_0 = \rho_0\alpha(T - T_0). \tag{27-17}$$

Here T_0 is a reference temperature, ρ_0 is the resistivity at T_0, and α is the temperature coefficient of resistivity for the material.

Ohm's Law A given device (conductor, resistor, or any other electrical device) obeys *Ohm's law* if its resistance R, defined by Eq. 27-8 as V/i, is independent of the applied potential difference V. A given *material* obeys Ohm's law if its resistivity, defined by Eq. 27-10, is independent of the magnitude and direction of the applied electric field $\vec{E}$.

Resistivity of a Metal By assuming that the conduction electrons in a metal are free to move like the molecules of a gas, it is possible to derive an expression for the resistivity of a metal:

$$\rho = \frac{m}{e^2 n\tau}. \tag{27-20}$$

Here n is the number of free electrons per unit volume, and τ is the mean time between the collisions of an electron with the atoms of the metal. We can explain why metals obey Ohm's law by pointing out that τ is essentially independent of the magnitude E of any electric field applied to a metal.

Power The power P, or rate of energy transfer, in an electrical device across which a potential difference V is maintained is

$$P = iV \quad \text{(rate of electric energy transfer)}. \tag{27-21}$$

Resistive Dissipation If the device is a resistor, we can write Eq. 27-21 as

$$P = i^2R = \frac{V^2}{R} \quad \text{(resistive dissipation)}. \tag{27-22, 27-23}$$

In a resistor, electric potential energy is converted to internal thermal energy via collisions between charge carriers and atoms.

Semiconductors *Semiconductors* are materials with few conduction electrons but can become conductors when they are *doped* with other atoms that contribute free electrons.

Superconductors *Superconductors* are materials that lose all electrical resistance at low temperatures. Recent research has discovered materials that are superconducting at surprisingly high temperatures.

QUESTIONS

1. Figure 27-15 shows plots of the current i through a certain cross section of a wire over four different time periods. Rank the periods according to the net charge that passes through the cross section during each, greatest first.

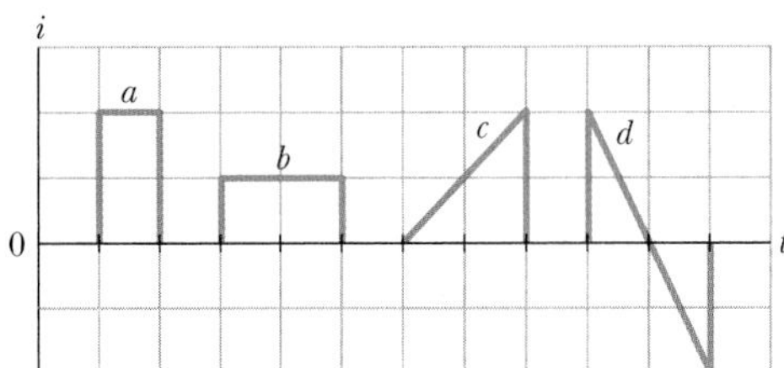

Fig. 27-15 Question 1.

2. Figure 27-16 shows four situations in which positive and negative charges move horizontally through a region and gives the rate at which each charge moves. Rank the situations according to the effective current through the regions, greatest first.

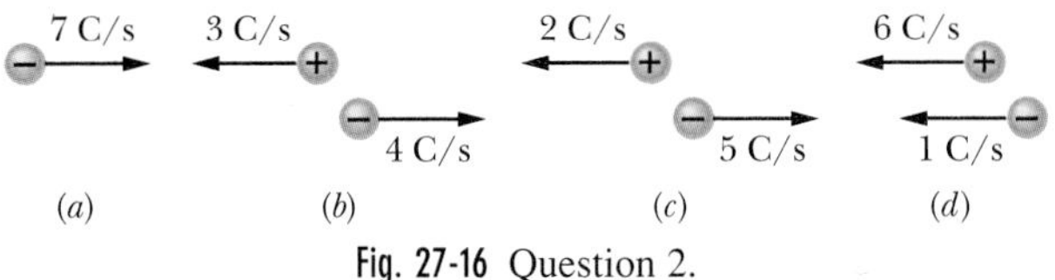

Fig. 27-16 Question 2.

3. Figure 27-17 shows cross sections through three wires of equal length and of the same material. The figure also gives the length of each side in millimeters. Rank the wires according to their resistances (measured end to end along each wire's length), greatest first.

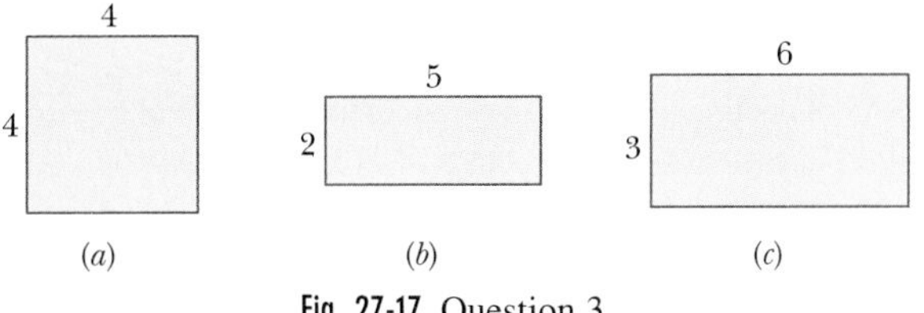

Fig. 27-17 Question 3.

4. If you stretch a cylindrical wire and it remains cylindrical, does the resistance of the wire (measured end to end along its length) increase, decrease, or remain the same?

5. Figure 27-18 shows cross sections through three long conductors of the same length and material, with square cross sections of edge lengths as shown. Conductor B will fit snugly within conductor A, and conductor C will fit snugly within conductor B. Rank the following according to their end-to-end resistances, greatest first: the individual conductors and the combinations of $A + B$, $B + C$, and $A + B + C$.

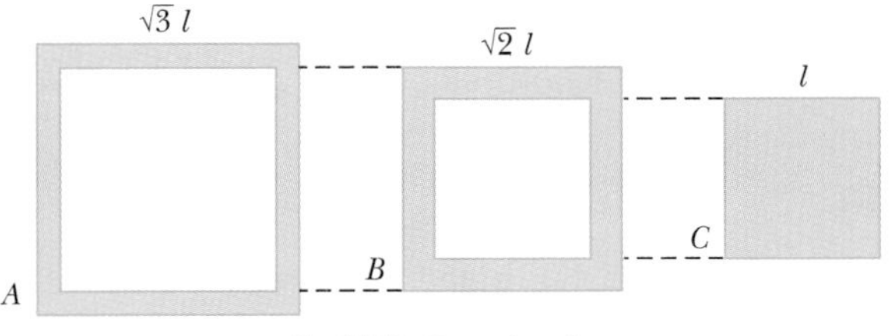

Fig. 27-18 Question 5.

6. Figure 27-19 shows a rectangular solid conductor of edge lengths L, $2L$, and $3L$. A certain potential difference V is to be applied between pairs of opposite faces of the conductor as in Fig. 27-8b: left–right, top–bottom, and front–back. Rank those pairs according to (a) the magnitude of the electric field within the conductor, (b) the current density within the conductor, (c) the current through the conductor, and (d) the drift speed of the electrons through the conductor, greatest first.

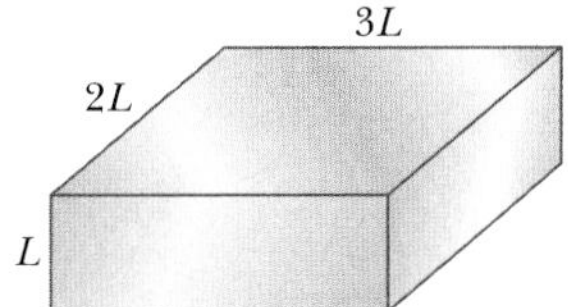

Fig. 27-19 Question 6.

7. The following table gives the lengths of three copper rods, their diameters, and the potential differences between their ends. Rank the rods according to (a) the magnitude of the electric field within them, (b) the current density within them, and (c) the drift speed of electrons through them, greatest first.

Rod	Length	Diameter	Potential Difference
1	L	$3d$	V
2	$2L$	d	$2V$
3	$3L$	$2d$	$2V$

8. The following table gives the conductivity and the density of conduction electrons for materials A, B, C, and D. Rank the materials according to the average time between collisions of the conduction electrons in the materials, greatest first.

	A	B	C	D
Conductivity	σ	2σ	2σ	σ
Electrons/m^3	n	$2n$	n	$2n$

9. Three wires, of the same diameter, are connected in turn between two points maintained at a constant potential difference. Their resistivities and lengths are ρ and L (wire A), 1.2ρ and $1.2L$ (wire B), and 0.9ρ and L (wire C). Rank the wires according to the rate at which energy is transferred to thermal energy within them, greatest first.

10. Figure 27-20 gives the resistivities of four materials as a function of temperature. (a) Which materials are conductors, and which are semiconductors? In which materials does an increase in temperature result in (b) an increase in the number of conduction electrons per unit volume and (c) an increase in the collision rate of conduction electrons?

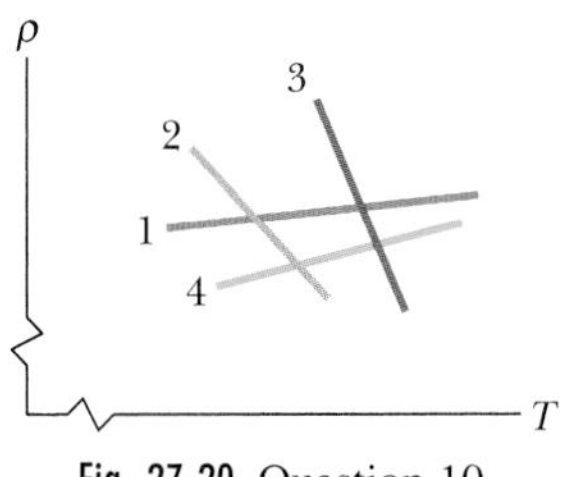

Fig. 27-20 Question 10.

EXERCISES & PROBLEMS

ssm Solution is in the Student Solutions Manual.
www Solution is available on the World Wide Web at: http://www.wiley.com/college/hrw
ilw Solution is available on the Interactive LearningWare.

SEC. 27-2 Electric Current

1E. A current of 5.0 A exists in a 10 Ω resistor for 4.0 min. How many (a) coulombs and (b) electrons pass through any cross section of the resistor in this time? ssm

2P. A charged belt, 50 cm wide, travels at 30 m/s between a source of charge and a sphere. The belt carries charge into the sphere at a rate corresponding to 100 μA. Compute the surface charge density on the belt.

3P. An isolated conducting sphere has a 10 cm radius. One wire carries a current of 1.000 002 0 A into it. Another wire carries a current of 1.000 000 0 A out of it. How long would it take for the sphere to increase in potential by 1000 V? ssm

SEC. 27-3 Current Density

4E. A small but measurable current of 1.2×10^{-10} A exists in a copper wire whose diameter is 2.5 mm. Assuming the current is uniform, calculate (a) the current density and (b) the electron drift speed. (See Sample Problem 27-3.)

5E. A beam contains 2.0×10^8 doubly charged positive ions per cubic centimeter, all of which are moving north with a speed of 1.0×10^5 m/s. (a) What are the magnitude and direction of the current density $\vec{J}$? (b) Can you calculate the total current i in this ion beam? If not, what additional information is needed? ssm

6E. The (United States) National Electric Code, which sets maximum safe currents for insulated copper wires of various diameters, is given (in part) in the table. Plot the safe current density as a function of diameter. Which wire gauge has the maximum safe current density? ("Gauge" is a way of identifying wire diameters, and 1 mil = 10^{-3} in.)

Gauge	4	6	8	10	12	14	16	18
Diameter, mils	204	162	129	102	81	64	51	40
Safe current, A	70	50	35	25	20	15	6	3

7E. A fuse in an electric circuit is a wire that is designed to melt, and thereby open the circuit, if the current exceeds a predetermined value. Suppose that the material to be used in a fuse melts when the current density rises to 440 A/cm^2. What diameter of cylindrical wire should be used to make a fuse that will limit the current to 0.50 A? ssm

8P. Near Earth, the density of protons in the solar wind (a stream of particles from the Sun) is 8.70 cm^{-3}, and their speed is 470 km/s. (a) Find the current density of these protons. (b) If Earth's magnetic field did not deflect them, the protons would strike the planet. What total current would Earth then receive?

9P. A steady beam of alpha particles ($q = +2e$) traveling with constant kinetic energy 20 MeV carries a current of 0.25 μA. (a) If the beam is directed perpendicular to a plane surface, how many alpha particles strike the surface in 3.0 s? (b) At any instant, how many alpha particles are there in a given 20 cm length of the beam? (c) Through what potential difference is it necessary to accelerate each alpha particle from rest to bring it to an energy of 20 MeV? ssm

10P. (a) The current density across a cylindrical conductor of radius R varies in magnitude according to the equation

$$J = J_0\left(1 - \frac{r}{R}\right),$$

where r is the distance from the central axis. Thus, the current density is a maximum J_0 at that axis ($r = 0$) and decreases linearly to zero at the surface ($r = R$). Calculate the current in terms of J_0 and the conductor's cross-sectional area $A = \pi R^2$. (b) Suppose that, instead, the current density is a maximum J_0 at the cylinder's surface and decreases linearly to zero at the axis: $J = J_0 r/R$. Calculate the current. Why is the result different from that in (a)?

11P. How long does it take electrons to get from a car battery to the starting motor? Assume the current is 300 A and the electrons travel through a copper wire with cross-sectional area 0.21 cm^2 and length 0.85 m. (See Sample Problem 27-3.) ilw

SEC. 27-4 Resistance and Resistivity

12E. A wire of Nichrome (a nickel–chromium–iron alloy commonly used in heating elements) is 1.0 m long and 1.0 mm^2 in cross-sectional area. It carries a current of 4.0 A when a 2.0 V potential difference is applied between its ends. Calculate the conductivity σ of Nichrome.

13E. A conducting wire has a 1.0 mm diameter, a 2.0 m length, and a 50 mΩ resistance. What is the resistivity of the material? ssm

14E. A steel trolley-car rail has a cross-sectional area of 56.0 cm^2. What is the resistance of 10.0 km of rail? The resistivity of the steel is $3.00 \times 10^{-7}\ \Omega \cdot$m.

15E. A human being can be electrocuted if a current as small as 50 mA passes near the heart. An electrician working with sweaty hands makes good contact with the two conductors he is holding, one in each hand. If his resistance is 2000 Ω, what might the fatal voltage be? ssm

16E. A wire 4.00 m long and 6.00 mm in diameter has a resistance of 15.0 mΩ. A potential difference of 23.0 V is applied between the ends. (a) What is the current in the wire? (b) What is the current density? (c) Calculate the resistivity of the wire material. Identify the material. (Use Table 27-1.)

17E. A coil is formed by winding 250 turns of insulated 16-gauge copper wire (diameter = 1.3 mm) in a single layer on a cylindrical form of radius 12 cm. What is the resistance of the coil? Neglect the thickness of the insulation. (Use Table 27-1.) ssm

18E. (a) At what temperature would the resistance of a copper conductor be double its resistance at 20.0°C? (Use 20.0°C as the reference point in Eq. 27-17; compare your answer with Fig. 27-10.)

(b) Does this same "doubling temperature" hold for all copper conductors, regardless of shape or size?

19E. A wire with a resistance of 6.0 Ω is drawn out through a die so that its new length is three times its original length. Find the resistance of the longer wire, assuming that the resistivity and density of the material are unchanged. ssm ilw

20E. A certain wire has a resistance R. What is the resistance of a second wire, made of the same material, that is half as long and has half the diameter?

21P. Two conductors are made of the same material and have the same length. Conductor A is a solid wire of diameter 1.0 mm. Conductor B is a hollow tube of outside diameter 2.0 mm and inside diameter 1.0 mm. What is the resistance ratio R_A/R_B, measured between their ends? ssm www

22P. An electrical cable consists of 125 strands of fine wire, each having 2.65 $\mu\Omega$ resistance. The same potential difference is applied between the ends of all the strands and results in a total current of 0.750 A. (a) What is the current in each strand? (b) What is the applied potential difference? (c) What is the resistance of the cable?

23P. When 115 V is applied across a wire that is 10 m long and has a 0.30 mm radius, the current density is 1.4×10^4 A/m^2. Find the resistivity of the wire. ssm

24P. A block in the shape of a rectangular solid has a cross-sectional area of 3.50 cm^2 across its width, a front-to-rear length of 15.8 cm, and a resistance of 935 Ω. The material of which the block is made has 5.33×10^{22} conduction electrons/m^3. A potential difference of 35.8 V is maintained between its front and rear faces. (a) What is the current in the block? (b) If the current density is uniform, what is its value? (c) What is the drift velocity of the conduction electrons? (d) What is the magnitude of the electric field in the block?

25P. A common flashlight bulb is rated at 0.30 A and 2.9 V (the values of the current and voltage under operating conditions). If the resistance of the bulb filament at room temperature (20°C) is 1.1 Ω, what is the temperature of the filament when the bulb is on? The filament is made of tungsten. ilw

26P. Earth's lower atmosphere contains negative and positive ions that are produced by radioactive elements in the soil and cosmic rays from space. In a certain region, the atmospheric electric field strength is 120 V/m, directed vertically down. This field causes singly charged positive ions, at a density of 620/cm^3, to drift downward and singly charged negative ions, at a density of 550/cm^3, to drift upward (Fig. 27-21). The measured conductivity of the air in that region is $2.70 \times 10^{-14}/\Omega \cdot$ m. Calculate (a) the ion drift speed, assumed to be the same for positive and negative ions, and (b) the current density.

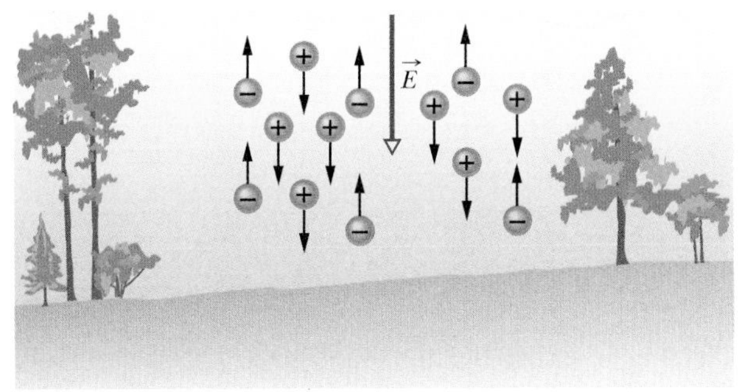

Fig. 27-21 Problem 26.

27P. When a metal rod is heated, not only its resistance but also its length and its cross-sectional area change. The relation $R = \rho L/A$ suggests that all three factors should be taken into account in measuring ρ at various temperatures. (a) If the temperature changes by 1.0 C°, what percentage changes in R, L, and A occur for a copper conductor? (b) The coefficient of linear expansion for copper is 1.7×10^{-5}/K. What conclusion do you draw? ssm www

28P. If the gauge number of a wire is increased by 6, the diameter is halved; if a gauge number is increased by 1, the diameter decreases by the factor $2^{1/6}$ (see the table in Exercise 6). Knowing this, and knowing that 1000 ft of 10-gauge copper wire has a resistance of approximately 1.00 Ω, estimate the resistance of 25 ft of 22-gauge copper wire.

29P. A resistor has the shape of a truncated right-circular cone (Fig. 27-22). The end radii are a and b, and the altitude is L. If the taper is small, we may assume that the current density is uniform across any cross section. (a) Calculate the resistance of this object. (b) Show that your answer reduces to $\rho(L/A)$ for the special case of zero taper (that is, for $a = b$). ssm

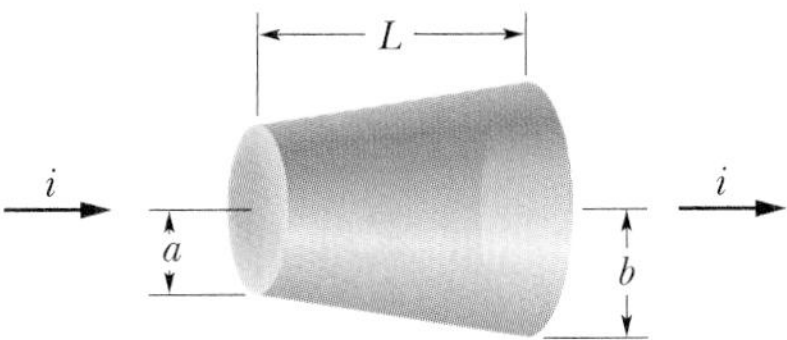

Fig. 27-22 Problem 29.

SEC. 27-6 A Microscopic View of Ohm's Law

30P. Show that, according to the free-electron model of electrical conduction in metals and classical physics, the resistivity of metals should be proportional to $\sqrt{T}$, where T is the temperature in kelvins. (See Eq. 20-31.)

SEC. 27-7 Power in Electric Circuits

31E. A certain x-ray tube operates at a current of 7.0 mA and a potential difference of 80 kV. What is its power in watts? ssm

32E. A student kept his 9.0 V, 7.0 W radio turned on at full volume from 9:00 p.m. until 2:00 a.m. How much charge went through it?

33E. A 120 V potential difference is applied to a space heater whose resistance is 14 Ω when hot. (a) At what rate is electric energy transferred to heat? (b) At 5.0¢/kW · h, what does it cost to operate the device for 5.0 h? ssm

34E. Thermal energy is produced in a resistor at a rate of 100 W when the current is 3.00 A. What is the resistance?

35E. An unknown resistor is connected between the terminals of a 3.00 V battery. Energy is dissipated in the resistor at the rate of 0.540 W. The same resistor is then connected between the terminals of a 1.50 V battery. At what rate is energy now dissipated? ilw

36E. A 120 V potential difference is applied to a space heater that dissipates 500 W during operation. (a) What is its resistance during operation? (b) At what rate do electrons flow through any cross section of the heater element?

37P. A 1250 W radiant heater is constructed to operate at 115 V. (a) What will be the current in the heater? (b) What is the resistance of the heating coil? (c) How much thermal energy is produced in 1.0 h by the heater? ssm ilw www

38P. A heating element is made by maintaining a potential difference of 75.0 V across the length of a Nichrome wire that has a 2.60×10^{-6} m^2 cross section. Nichrome has a resistivity of $5.00 \times 10^{-7}\ \Omega \cdot$m. (a) If the element dissipates 5000 W, what is its length? (b) If a potential difference of 100 V is used to obtain the same dissipation rate, what should the length be?

39P. A Nichrome heater dissipates 500 W when the applied potential difference is 110 V and the wire temperature is 800°C. What would be the dissipation rate if the wire temperature were held at 200°C by immersing the wire in a bath of cooling oil? The applied potential difference remains the same, and α for Nichrome at 800°C is 4.0×10^{-4}/K. ssm

40P. A 100 W lightbulb is plugged into a standard 120 V outlet. (a) How much does it cost per month to leave the light turned on continuously? Assume electric energy costs 6¢/kW · h. (b) What is the resistance of the bulb? (c) What is the current in the bulb? (d) Is the resistance different when the bulb is turned off?

41P. A linear accelerator produces a pulsed beam of electrons. The pulse current is 0.50 A, and each pulse has a duration of 0.10 μs. (a) How many electrons are accelerated per pulse? (b) What is the average current for an accelerator operating at 500 pulses/s? (c) If the electrons are accelerated to an energy of 50 MeV, what are the average and peak powers of the accelerator? ssm

42P. A cylindrical resistor of radius 5.0 mm and length 2.0 cm is made of material that has a resistivity of $3.5 \times 10^{-5}\ \Omega \cdot$m. What are (a) the current density and (b) the potential difference when the energy dissipation rate in the resistor is 1.0 W?

43P. A copper wire of cross-sectional area 2.0×10^{-6} m^2 and length 4.0 m has a current of 2.0 A uniformly distributed across that area. (a) What is the magnitude of the electric field along the wire? (b) How much electric energy is transferred to thermal energy in 30 min?

Additional Problems

44. *The chocolate crumb mystery.* This story begins with Problem 48 in Chapter 24 and continues through Chapters 25 and 26. The chocolate crumb powder moved to the silo through a pipe of radius R with uniform speed v and uniform charge density ρ. (a) Find an expression for the current i (the rate at which charge on the powder moved) through a perpendicular cross section of the pipe. (b) Evaluate i for the conditions at the factory: pipe radius R = 5.0 cm, speed v = 2.0 m/s, and charge density $\rho = 1.1 \times 10^{-3}$ C/m^3.

If the powder were to flow through a change V in electric potential, its energy could be transferred to a spark at the rate $P = iV$. (c) Could there be such a transfer within the pipe due to the radial potential difference discussed in Problem 57 of Chapter 25?

As the powder flowed from the pipe into the silo, the electric potential of the powder changed. The magnitude of that change was at least equal to the radial potential difference within the pipe (as evaluated in Problem 57 of Chapter 25). (d) Assuming that value for the potential difference and using the current found in (b) above, find the rate at which energy could have been transferred from the powder to a spark as the powder exited the pipe. (e) If a spark did occur at the exit and lasted for 0.20 s (a reasonable expectation), how much energy would have been transferred to the spark?

Recall from Problem 48 in Chapter 24 that a minimum energy transfer of 150 mJ is needed to cause an explosion. (f) Where did the powder explosion most likely occur: in the powder cloud at the unloading bin (considered in Problem 48 of Chapter 26), within the pipe, or at the exit of the pipe into the silo?

45. *Heart attack or electrocution?* One morning a man walks barefooted away from a picnic onto moist ground near a tower that supports electric transmission lines. Suddenly he collapses. His relatives at the picnic table see him fall and, on reaching him a few seconds later, find that he is in ventricular fibrillation. The man dies before an emergency team can reach him with defibrillation equipment. Later the family files a lawsuit against the power company, claiming that the victim was electrocuted because of accidental current leakage from the tower. You are hired as part of the forensic team to investigate the death—was it due to heart attack or electrocution?

Investigation of the power company records reveals that there was indeed an *electrical fault* at the tower that morning—for about 1.0 s a current I leaked from a rod into the ground. Assume that the current spread uniformly (hemispherically) into the ground (Fig. 27-23). Let ρ be the resistivity of the ground and r be the distance from the rod. Find expressions for (a) the current density and (b) the electric field magnitude, both as functions of r. The lower end of the rod was spherical with a radius of b. (c) From your expression for the electric field magnitude, find an expression for the potential difference ΔV between the lower end of the rod and a point at distance r. Your investigation finds that I = 100 A, $\rho = 100\ \Omega \cdot$m, and b = 1.0 cm, and that the victim was located at r = 10 m. At the victim's location, what were (d) the current density, (e) the electric field magnitude, and (f) the potential difference ΔV? (This story continues with Problem 56 in Chapter 28.)

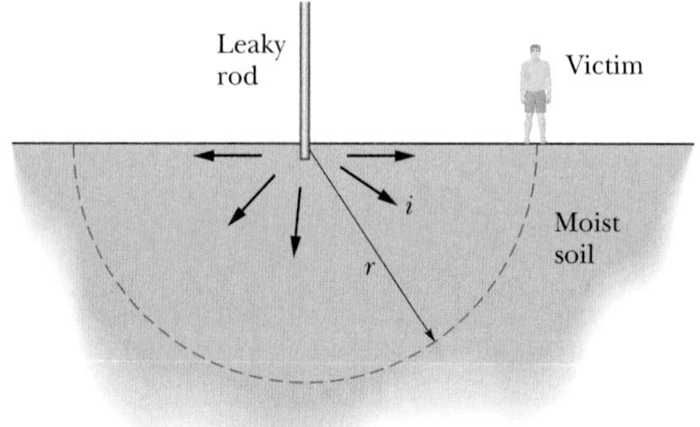

Fig. 27-23 Problem 45.

NEW PROBLEMS

N1. When a metal rod is heated, not only its resistance but also its length and its cross-sectional area change. The relation $R = \rho L/A$ suggests that all three factors should be taken into account in measuring ρ at various temperatures. (a) If the temperature changes by 1.0 C°, what percentage changes in R, L, and A occur for a copper conductor? The coefficient of linear expansion is 1.7×10^{-5}/K. (b) What conclusion do you draw?

N2. Figure 27N-1 shows wire section 1 of radius $2R$ and wire section 2 of radius R, connected by a tapered section. The wire is copper and carries a current. Assume that the current is uniformly distributed across any cross-sectional area through the wire's width. The electric potential change V along the length $L = 2.00$ m shown in section 2 is 10.0 μV. What is the drift speed of the conduction electrons in section 1? (*Hint:* A result of Sample Problem 27-3 is helpful.)

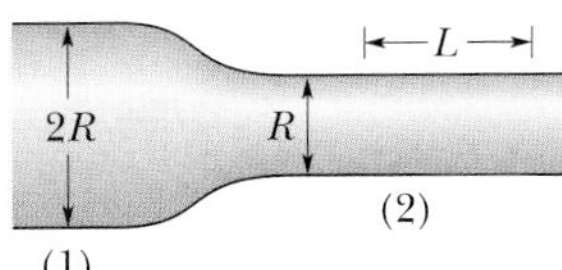

Fig. 27N-1 Problem N2.

N3. A cylindrical resistor of radius 5.0 mm and length 2.0 cm is made of material that has a resistivity of $3.5 \times 10^{-5}\ \Omega \cdot$ m. What are (a) the current density and (b) the potential difference when the energy dissipation rate in the resistor is 1.0 W?

N4. Figure 27N-2 gives the electric potential $V(x)$ along a copper wire carrying uniform current, from a point of higher potential at $x = 0$ to a point of lower potential at $x = 3.0$ m. The wire has a radius of 2.00 mm. What is the current in the wire?

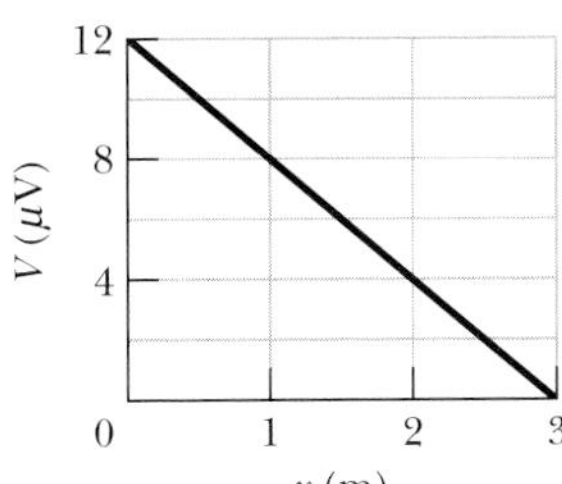

Fig. 27N-2 Problem N4.

N5. A 2.0 kW heater element from a dryer has a length of 80 cm. If a 10 cm section is removed, what power is used by the now shortened element at 120 V?

N6. In Fig. 27N-3, a 12 V battery is connected to a resistive strip of resistance $R = 6.0\ \Omega$. When an electron moves through the strip from one end to the other, (a) in which direction in the figure does the electron move, (b) how much work is done on the electron by the electric field in the strip, and (c) how much energy is transferred to the thermal energy of the strip by the electron?

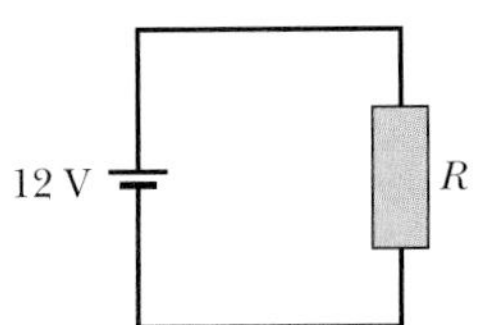

Fig. 27N-3 Problem N6.

N7. A current is established in a gas discharge tube when a sufficiently high potential difference is applied across the two electrodes in the tube. The gas ionizes; electrons move toward the positive terminal and singly charged positive ions toward the negative terminal. (a) What is the magnitude of the current in a hydrogen discharge tube in which 3.1×10^{18} electrons and 1.1×10^{18} protons move past a cross-sectional area of the tube each second? (b) What is the direction of the current density $\vec{J}$?

N8. Figure 27N-4a gives the electric fields $E(x)$ that have been set up by a battery along a resistive rod of length 9.00 mm (Fig. 27N-4b). The rod consists of three sections of the same material but with different radii. (The schematic diagram of Fig. 27N-4b does not indicate the different radii.) The radius of section 3 is 2.00 mm. What are the radii of (a) section 1 and (b) section 2?

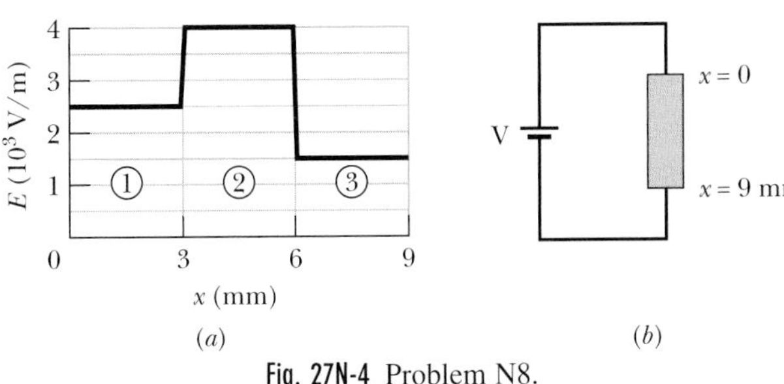

Fig. 27N-4 Problem N8.

N9. A coil of current-carrying Nichrome wire is immersed in a liquid contained in a calorimeter. (Nichrome is a nickel–chromium–iron alloy commonly used in heating elements.) When the potential difference across the coil is 12 V and the current through the coil is 5.2 A, the liquid boils at a steady rate, evaporating at the rate of 21 mg/s. Calculate the heat of vaporization of the liquid, in joules per kilogram (see Section 19-7).

N10. In Fig. 27N-5a, a 9.0 V battery is connected to a resistive strip that consists of three sections with the same cross-sectional areas but with different conductivities. Figure 27N-5b gives the electric potential $V(x)$ versus position x along the strip. The conductivity of section 3 is $3.00 \times 10^7\ (\Omega \cdot \text{m})^{-1}$. What are the conductivities of (a) section 1 and (b) section 2?

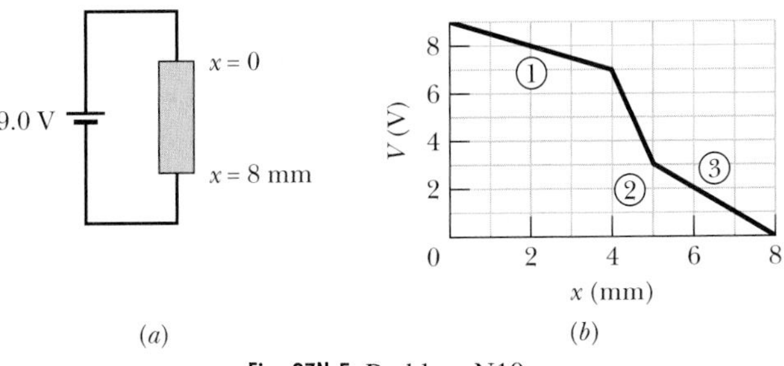

Fig. 27N-5 Problem N10.

N11. A copper wire of cross-sectional area 2.0×10^{-6} m² and length 4.0 m has a current of 2.0 A uniformly distributed across that area. (a) What is the magnitude of the electric field along the wire? (b) How much energy is converted to thermal energy in 30 min?

N12. Figure 27N-6a shows a rod of resistive material. The resistance per unit length of the rod increases in the positive direction of the x axis. At any position x along the rod, the resistance dR of a narrow (differential) section of width dx is given by $dR = 5.00x\,dx$, where dR is in ohms and x is in meters. Figure 27N-6b shows such a narrow section. You are to slice off a length of the rod between $x = 0$ and some position $x = L$ and then connect that length to a 5.0 V battery (Fig. 27N-6c). You want the current in the length to transfer energy to thermal energy at the rate of 200 W. At what position $x = L$ should you cut the rod?

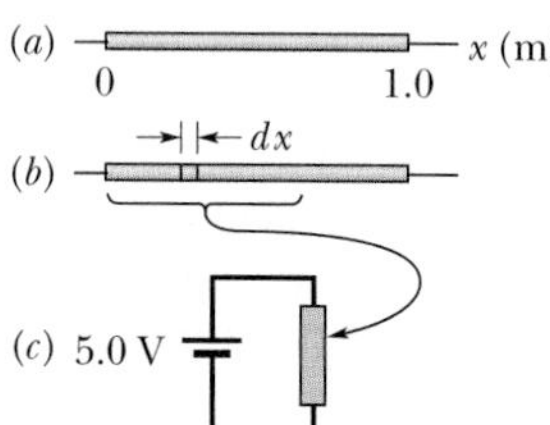

Fig. 27N-6 Problem N12.

N13. A conducting wire has a 1.0 mm diameter, a 2.0 m length, and a 50 mΩ resistance. What is the resistivity of the material?

N14. In Fig. 27N-7a, a 20 Ω resistor is connected to a battery. Figure 27N-7b shows the increase of thermal energy E_{th} in the resistor as a function of time t. What is the electric potential across the battery?

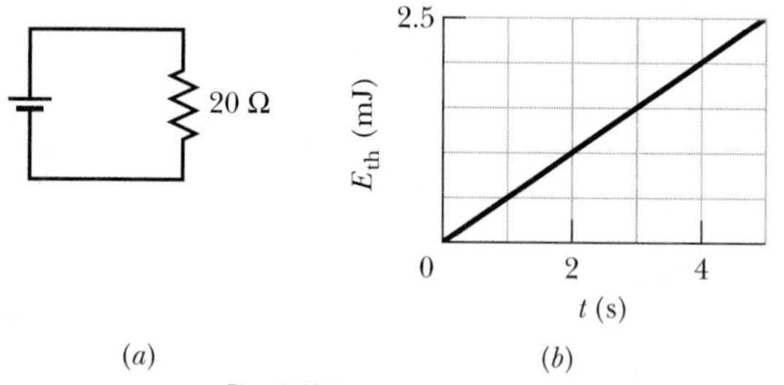

Fig. 27N-7 Problem N14.

N15. A wire of Nichrome (a nickel–chromium–iron alloy commonly used in heating elements) is 1.0 m long and 1.0 mm^2 in cross-sectional area. It carries a current of 4.0 A when a 2.0 V potential difference is applied between its ends. Calculate the conductivity σ of Nichrome.

N16. The current density $J(r)$ in a certain cylindrical wire is given as a function of radial distance from the center of the wire's cross section as $J(r) = Br$, where r is in meters, J is in amperes per square meter, and $B = 2.00 \times 10^5$ A/m^3. This function applies out to the wire's radius of 2.00 mm. How much current is contained within the width of a thin ring concentric with the wire if the ring has a radial width of 10.0 μm and is at a radial distance of 1.20 mm?

N17. A cylindrical rod is reformed so that its length is 4.00 times its original length (with no change in its volume). What is the ratio of its resistance (end to end) to its original resistance?

N18. A potential difference of 3.00 nV is set up across a 2.00 cm length of copper wire which has a radius of 2.00 mm. How much charge drifts through a cross section of the wire in 3.00 ms?

N19. A resistor, with a potential difference of 200 V across it, transfers electrical energy to thermal energy at the rate of 3000 W. What is the resistance of the resistor?

N20. The current through the battery and resistors 1 and 2 in Fig. 27N-8a is 2.00 A. Energy is transferred from the current to thermal energy E_{th} in both resistors. Curves 1 and 2 in Fig. 27N-8b give that thermal energy E_{th} for resistors 1 and 2, respectively, as a function of time t. What is the power of the battery?

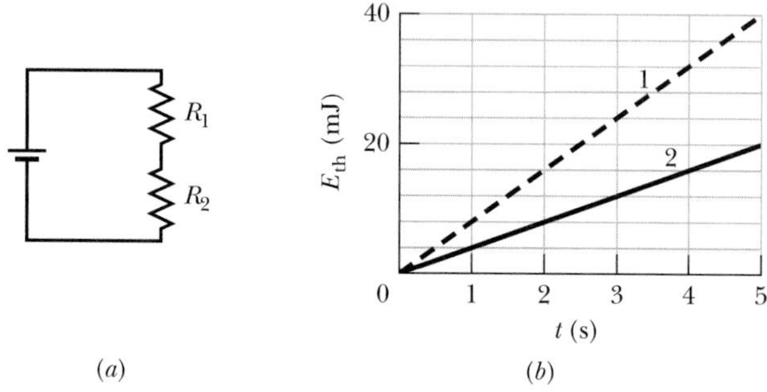

Fig. 27N-8 Problem N20.

N21. The current density in a certain circular wire of radius 3.00 mm is given by $J = (2.75 \times 10^{10}\ \text{A/m}^4)r^2$, where r is the radial distance. The potential applied to the wire (end to end) is 60.0 V. How much energy is converted to thermal energy within the wire in 1.00 h?

N22. A certain cylindrical wire carries current. We draw a circle of radius r around its central axis in Fig. 27N-9a to determine the current i within the circle. Figure 27N-9b shows current i as a function of r^2. (a) Is the current density uniform? (b) If so, what is the current density?

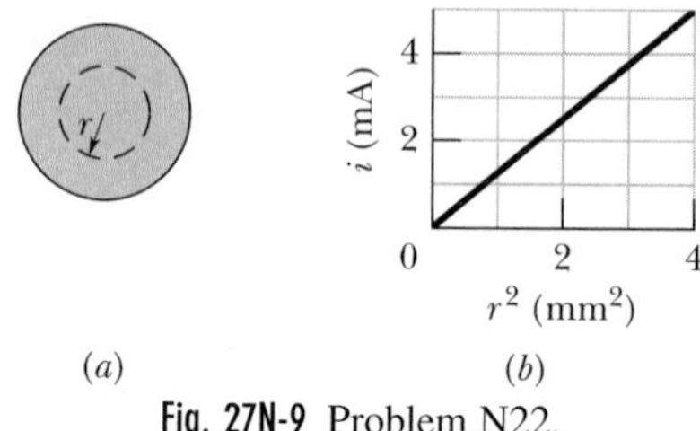

Fig. 27N-9 Problem N22.

N23. How much energy is consumed in 2.0 h by an electrical resistance of 400 Ω when the potential applied across it is 90.0 V?

N24. A potential difference of 12 V is applied to (circular) copper wire that has a length of 45 m and a radius of 2.0 mm. How much thermal energy is produced in the wire by the current in 40 s?

N25. Aluminum wire has been used to replace copper wire during times of high copper prices. (See Table 27-1 for the resistivities of aluminum and copper.) (a) If 20-gauge copper wire has a resistance of 33 Ω per kilometer of length, what is its diameter? (b) What is the diameter of an aluminum wire with the same resistance per unit length as 20-gauge copper wire?

N26. An 18.0 W device has 9.00 V across it. How much charge goes through the device in 4.00 h?

28 Circuits

The electric eel (*Electrophorus*) lurks in rivers of South America, killing the fish on which it preys with pulses of current. It does so by producing a potential difference of several hundred volts along its length; the resulting current in the surrounding water, from near the eel's head to the tail region, can be as much as one ampere. If you were to brush up against this eel while swimming, you might wonder (after recovering from the very painful stun):

How can the creature manage to produce a current that large without shocking itself?

The answer is in this chapter.

The world's largest battery, housed in Chino, California, has a power capability of 10 MW, which is put to use during peak power demands on the electric system served by Southern California Edison. Because the battery does work on charge carriers, it is an emf device.

28-1 "Pumping" Charges

If you want to make charge carriers flow through a resistor, you must establish a potential difference between the ends of the device. One way to do this is to connect each end of the resistor to one plate of a charged capacitor. The trouble with this scheme is that the flow of charge acts to discharge the capacitor, quickly bringing the plates to the same potential. When that happens, there is no longer an electric field in the resistor and thus the flow of charge stops.

To produce a steady flow of charge, you need a "charge pump," a device that—by doing work on the charge carriers—maintains a potential difference between a pair of terminals. We call such a device an **emf device,** and the device is said to provide an **emf** $\mathscr{E}$, which means that it does work on charge carriers. An emf device is sometimes called a *seat of emf.* The term *emf* comes from the outdated phrase *electromotive force,* which was adopted before scientists clearly understood the function of an emf device.

In Chapter 27, we discussed the motion of charge carriers through a circuit in terms of the electric field set up in the circuit—the field produces forces that move the charge carriers. In this chapter we take a different approach: We discuss the motion of the charge carriers in terms of the required energy—an emf device supplies the energy for the motion via the work it does.

A common emf device is the *battery,* used to power a wide variety of machines from wristwatches to submarines. The emf device that most influences our daily lives, however, is the *electric generator,* which, by means of electrical connections (wires) from a generating plant, creates a potential difference in our homes and workplaces. The emf devices known as *solar cells,* long familiar as the winglike panels on spacecraft, also dot the countryside for domestic applications. Less familiar emf devices are the *fuel cells* that power the space shuttles and the *thermopiles* that provide onboard electrical power for some spacecraft and for remote stations in Antarctica and elsewhere. An emf device does not have to be an instrument—living systems, ranging from electric eels and human beings to plants, have physiological emf devices.

Although the devices we have listed differ widely in their modes of operation, they all perform the same basic function—they do work on charge carriers and thus maintain a potential difference between their terminals.

28-2 Work, Energy, and Emf

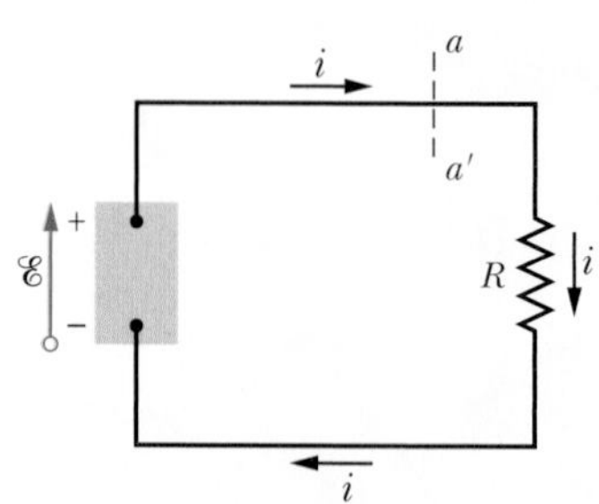

Fig. 28-1 A simple electric circuit, in which a device of emf $\mathscr{E}$ does work on the charge carriers and maintains a steady current i in a resistor of resistance R.

Figure 28-1 shows an emf device (consider it to be a battery) that is part of a simple circuit containing a single resistance R (the symbol for resistance and a resistor is -ᴡᴡ-). The emf device keeps one of its terminals (called the positive terminal and often labeled +) at a higher electric potential than the other terminal (called the negative terminal and labeled −). We can represent the emf of the device with an arrow that points from the negative terminal toward the positive terminal as in Fig. 28-1. A small circle on the tail of the emf arrow distinguishes it from the arrows that indicate current direction.

When an emf device is not connected to a circuit, its internal chemistry does not cause any net flow of charge carriers within it. However, when it is connected to a circuit as in Fig. 28-1, its internal chemistry causes a net flow of positive charge carriers from the negative terminal to the positive terminal, in the direction of the emf arrow. This flow is part of the current that is set up around the circuit in that same direction (clockwise in Fig. 28-1).

Within the emf device, positive charge carriers move from a region of low electric potential and thus low electric potential energy (at the negative terminal) to a region of higher electric potential and higher electric potential energy (at the positive terminal). This motion is just the opposite of what the electric field between the terminals (which is directed from the positive terminal toward the negative terminal) would cause the charge carriers to do.

Thus, there must be some source of energy within the device, enabling it to do work on the charges by forcing them to move as they do. The energy source may be chemical, as in a battery or a fuel cell. It may involve mechanical forces, as in an electric generator. Temperature differences may supply the energy, as in a thermopile; or the Sun may supply it, as in a solar cell.

Let us now analyze the circuit of Fig. 28-1 from the point of view of work and energy transfers. In any time interval dt, a charge dq passes through any cross section of this circuit, such as aa'. This same amount of charge must enter the emf device at its low-potential end and leave at its high-potential end. The device must do an amount of work dW on the charge dq to force it to move in this way. We define the emf of the emf device in terms of this work:

$$\mathscr{E} = \frac{dW}{dq} \quad \text{(definition of } \mathscr{E}\text{)}. \tag{28-1}$$

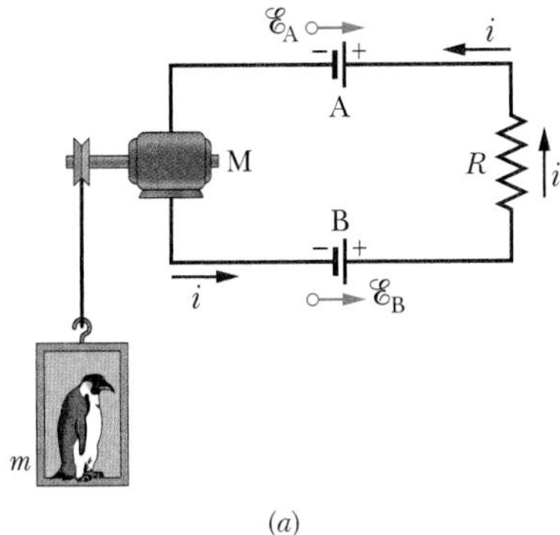

(a)

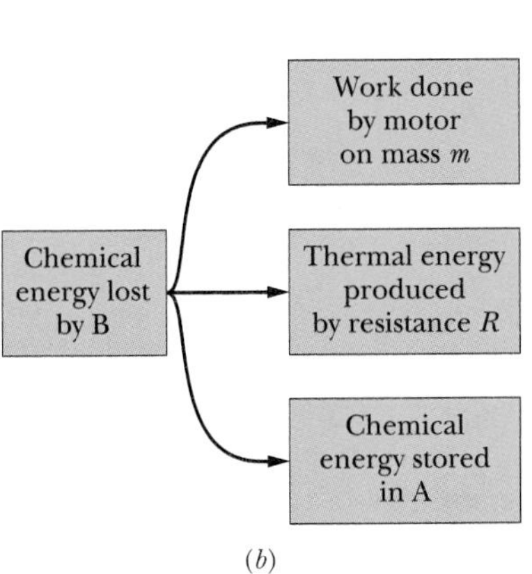

(b)

Fig. 28-2 (a) In the circuit, $\mathscr{E}_B > \mathscr{E}_A$; so battery B determines the direction of the current. (b) The energy transfers in the circuit, assuming that no dissipation occurs in the motor.

In words, the emf of an emf device is the work per unit charge that the device does in moving charge from its low-potential terminal to its high-potential terminal. The SI unit for emf is the joule per coulomb; in Chapter 25 we defined that unit as the *volt.*

An **ideal emf device** is one that lacks any internal resistance to the internal movement of charge from terminal to terminal. The potential difference between the terminals of an ideal emf device is equal to the emf of the device. For example, an ideal battery with an emf of 12.0 V always has a potential difference of 12.0 V between its terminals.

A **real emf device,** such as any real battery, has internal resistance to the internal movement of charge. When a real emf device is not connected to a circuit, and thus does not have current through it, the potential difference between its terminals is equal to its emf. However, when that device has current through it, the potential difference between its terminals differs from its emf. We will discuss such real batteries in Section 28-4.

When an emf device is connected to a circuit, the device transfers energy to the charge carriers passing through it. This energy can then be transferred from the charge carriers to other devices in the circuit, for example, to light a bulb. Figure 28-2*a* shows a circuit containing two ideal rechargeable (*storage*) batteries A and B, a resistance R, and an electric motor M that can lift an object by using energy it obtains from charge carriers in the circuit. Note that the batteries are connected so that they tend to send charges around the circuit in opposite directions. The actual direction of the current in the circuit is determined by the battery with the larger emf, which happens to be battery B, so the chemical energy within battery B is decreasing as energy is transferred to the charge carriers passing through it. However, the chemical energy within battery A is increasing because the current in it is directed from the positive terminal to the negative terminal. Thus, battery B is charging battery A. Battery B is also providing energy to motor M and energy that is being dissipated by resistance R. Figure 28-2*b* shows all three energy transfers from battery B; each decreases that battery's chemical energy.

28-3 Calculating the Current in a Single-Loop Circuit

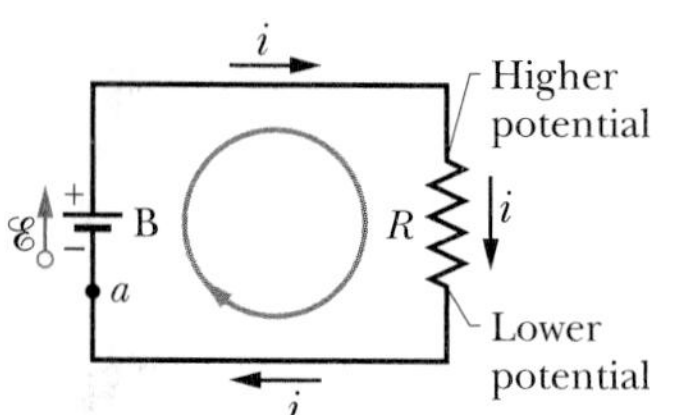

Fig. 28-3 A single-loop circuit in which a resistance R is connected across an ideal battery B with emf $\mathscr{E}$. The resulting current i is the same throughout the circuit.

We discuss here two equivalent ways to calculate the current in the simple *single-loop* circuit of Fig. 28-3; one method is based on energy conservation considerations, and the other on the concept of potential. The circuit consists of an ideal battery B with emf $\mathscr{E}$, a resistor of resistance R, and two connecting wires. (Unless otherwise indicated, we assume that wires in circuits have negligible resistance. Their function, then, is merely to provide pathways along which charge carriers can move.)

Energy Method

Equation 27-22 ($P = i^2R$) tells us that in a time interval dt an amount of energy given by $i^2R\,dt$ will appear in the resistor of Fig. 28-3 as thermal energy. (Since we assume the wires to have negligible resistance, no thermal energy will appear in them.) During the same interval, a charge $dq = i\,dt$ will have moved through battery B, and the work that the battery will have done on this charge, according to Eq. 28-1, is

$$dW = \mathscr{E}\,dq = \mathscr{E}i\,dt.$$

From the principle of conservation of energy, the work done by the (ideal) battery must equal the thermal energy that appears in the resistor:

$$\mathscr{E}i\,dt = i^2R\,dt.$$

This gives us

$$\mathscr{E} = iR.$$

The emf $\mathscr{E}$ is the energy per unit charge transferred to the moving charges by the battery. The quantity iR is the energy per unit charge transferred *from* the moving charges to thermal energy within the resistor. Therefore, this equation means that the energy per unit charge transferred to the moving charges is equal to the energy per unit charge transferred from them. Solving for i, we find

$$i = \frac{\mathscr{E}}{R}. \tag{28-2}$$

Potential Method

Suppose we start at any point in the circuit of Fig. 28-3 and mentally proceed around the circuit in either direction, adding algebraically the potential differences that we encounter. Then when we return to our starting point, we must also have returned to our starting potential. Before actually doing so, we shall formalize this idea in a statement that holds not only for single-loop circuits such as that of Fig. 28-3 but also for any complete loop in a *multiloop* circuit, as we shall discuss in Section 28-6:

LOOP RULE: The algebraic sum of the changes in potential encountered in a complete traversal of any loop of a circuit must be zero.

This is often referred to as *Kirchhoff's loop rule* (or *Kirchhoff's voltage law*), after German physicist Gustav Robert Kirchhoff. This rule is equivalent to saying that each point on a mountain has only one elevation above sea level. If you start from any point and return to it after walking around the mountain, the algebraic sum of the changes in elevation that you encounter must be zero.

In Fig. 28-3, let us start at point a, whose potential is V_a, and mentally walk clockwise around the circuit until we are back at a, keeping track of potential changes as we move. Our starting point is at the low-potential terminal of the battery. Since the battery is ideal, the potential difference between its terminals is equal to $\mathscr{E}$. When we pass through the battery to the high-potential terminal, the change in potential is $+\mathscr{E}$.

As we walk along the top wire to the top end of the resistor, there is no potential change because the wire has negligible resistance; it is at the same potential as the high-potential terminal of the battery. So too is the top end of the resistor. When we pass through the resistor, however, the potential changes according to Eq. 27-8 (which we can rewrite as $V = iR$). Moreover, the potential must decrease because we are moving from the higher potential side of the resistor. Thus, the change in potential is $-iR$.

We return to point a by moving along the bottom wire. Since this wire also has negligible resistance, we again find no potential change. Back at point a, the potential is again V_a. Because we traversed a complete loop, our initial potential, as modified for potential changes along the way, must be equal to our final potential; that is,

$$V_a + \mathscr{E} - iR = V_a.$$

The value of V_a cancels from this equation, which becomes

$$\mathscr{E} - iR = 0.$$

Solving this equation for i gives us the same result, $i = \mathscr{E}/R$, as the energy method (Eq. 28-2).

If we apply the loop rule to a complete *counterclockwise* walk around the circuit, the rule gives us

$$-\mathscr{E} + iR = 0$$

and we again find that $i = \mathscr{E}/R$. Thus, you may mentally circle a loop in either direction to apply the loop rule.

To prepare for circuits more complex than that of Fig. 28-3, let us set down two rules for finding potential differences as we move around a loop:

RESISTANCE RULE: For a move through a resistance in the direction of the current, the change in potential is $-iR$; in the opposite direction it is $+iR$.

EMF RULE: For a move through an ideal emf device in the direction of the emf arrow, the change in potential is $+\mathscr{E}$; in the opposite direction it is $-\mathscr{E}$.

CHECKPOINT 1: The figure shows the current i in a single-loop circuit with a battery B and a resistance R (and wires of negligible resistance). (a) Should the emf arrow at B be drawn pointing leftward or rightward? At points a, b, and c, rank (b) the magnitude of the current, (c) the electric potential, and (d) the electric potential energy of the charge carriers, greatest first.

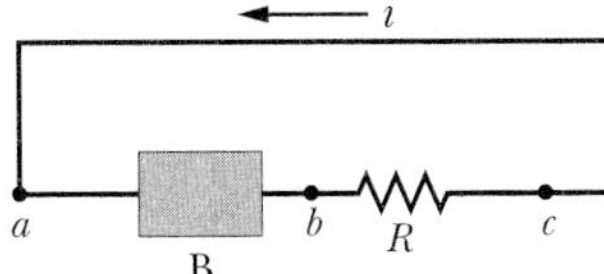

28-4 Other Single-Loop Circuits

In this section we extend the simple circuit of Fig. 28-3 in two ways.

Internal Resistance

Figure 28-4*a* shows a real battery, with internal resistance r, wired to an external resistor of resistance R. The internal resistance of the battery is the electrical resistance of the conducting materials of the battery and thus is an unremovable feature of the battery. In Fig. 28-4*a*, however, the battery is drawn as if it could be separated into an ideal battery with emf $\mathscr{E}$ and a resistor of resistance r. The order in which the symbols for these separated parts are drawn does not matter.

If we apply the loop rule clockwise beginning at point a, the *changes* in potential give us

$$\mathscr{E} - ir - iR = 0. \tag{28-3}$$

Solving for the current, we find

$$i = \frac{\mathscr{E}}{R + r}. \tag{28-4}$$

Note that this equation reduces to Eq. 28-2 if the battery is ideal—that is, if $r = 0$.

Figure 28-4*b* shows graphically the changes in electric potential around the circuit. (To better link Fig. 28-4*b* with the *closed circuit* in Fig. 28-4*a*, imagine curling the graph into a cylinder with point a at the left overlapping point a at the right.) Note how traversing the circuit is like walking around a (potential) mountain and returning to your starting point—you also return to the starting elevation.

In this book, when a battery is not described as real or if no internal resistance is indicated, you can generally assume that it is ideal—but, of course, in the real world batteries are always real and have internal resistance.

Resistances in Series

Figure 28-5*a* shows three resistances connected **in series** to an ideal battery with emf $\mathscr{E}$. This description has little to do with how the resistances are drawn. Rather, "in series" means that the resistances are wired one after another and that a potential difference V is applied across the two ends of the series. In Fig. 28-5*a*, the resistances are connected one after another between a and b, and a potential difference is main-

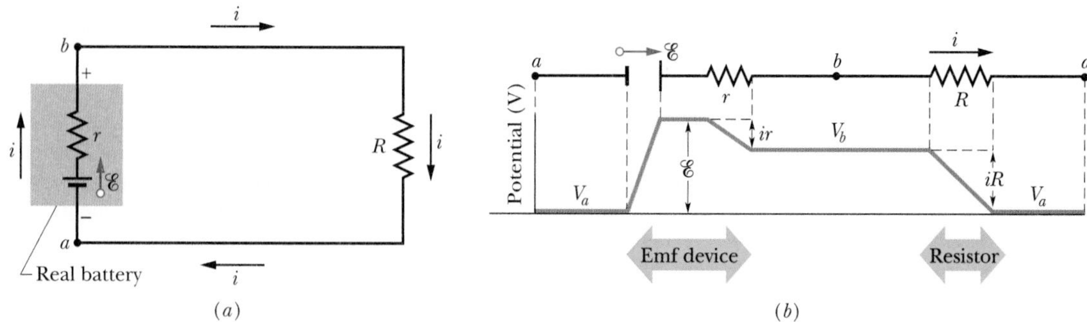

Fig. 28-4 (*a*) A single-loop circuit containing a real battery having internal resistance r and emf $\mathscr{E}$. (*b*) The same circuit, now spread out in a line. The potentials encountered in traversing the circuit clockwise from a are also shown. The potential V_a is arbitrarily assigned a value of zero, and other potentials in the circuit are graphed relative to V_a.

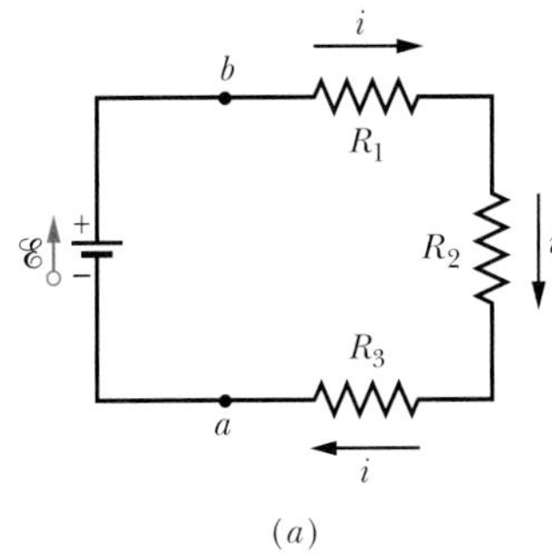

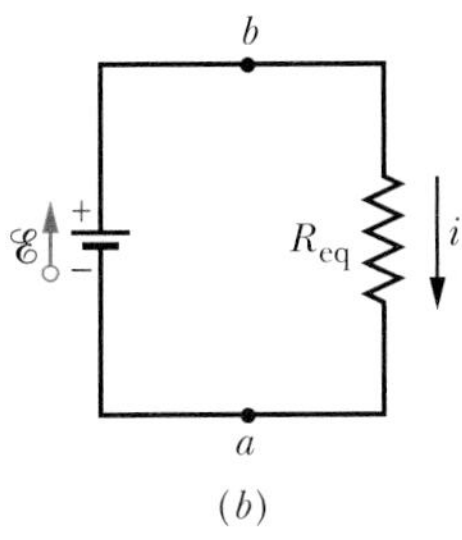

Fig. 28-5 (*a*) Three resistors are connected in series between points *a* and *b*. (*b*) An equivalent circuit, with the three resistors replaced with their equivalent resistance R_{eq}.

tained across *a* and *b* by the battery. The potential differences that then exist across the resistances in the series produce identical currents *i* in them. In general,

➤ When a potential difference *V* is applied across resistances connected in series, the resistances have identical currents *i*. The sum of the potential differences across the resistances is equal to the applied potential difference *V*.

Note that charge moving through the series resistances can move along only a single route. If there are additional routes, so that the currents in different resistances are different, the resistances are not connected in series.

➤ Resistances connected in series can be replaced with an equivalent resistance R_{eq} that has the same current *i* and the same *total* potential difference *V* as the actual resistances.

You might remember that R_{eq} and all the actual series resistances have the same current *i* with the nonsense word "ser-i." Figure 28-5*b* shows the equivalent resistance R_{eq} that can replace the three resistances of Fig. 28-5*a*.

To derive an expression for R_{eq} in Fig. 28-5*b*, we apply the loop rule to both circuits. For Fig. 28-5*a*, starting at terminal *a* and going clockwise around the circuit, we find

$$\mathscr{E} - iR_1 - iR_2 - iR_3 = 0,$$

or

$$i = \frac{\mathscr{E}}{R_1 + R_2 + R_3}. \tag{28-5}$$

For Fig. 28-5*b*, with the three resistances replaced with a single equivalent resistance R_{eq}, we find

$$\mathscr{E} - iR_{eq} = 0,$$

or

$$i = \frac{\mathscr{E}}{R_{eq}}. \tag{28-6}$$

Comparison of Eqs. 28-5 and 28-6 shows that

$$R_{eq} = R_1 + R_2 + R_3.$$

The extension to *n* resistances is straightforward and is

$$R_{eq} = \sum_{j=1}^{n} R_j \qquad (n \text{ resistances in series}). \tag{28-7}$$

Note that when resistances are in series, their equivalent resistance is greater than any of the individual resistances.

✔CHECKPOINT 2: In Fig. 28-5*a*, if $R_1 > R_2 > R_3$, rank the three resistances according to (a) the current through them and (b) the potential difference across them, greatest first.

28-5 Potential Differences

We often want to find the potential difference between two points in a circuit. In Fig. 28-4*a*, for example, what is the potential difference between points *b* and *a*? To find out, let us start at point *b* and traverse the circuit clockwise to point *a*, passing through resistor *R*. If V_a and V_b are the potentials at *a* and *b*, respectively, we have

$$V_b - iR = V_a$$

because (according to our resistance rule) we experience a decrease in potential in going through a resistance in the direction of the current. We rewrite this as

$$V_b - V_a = +iR, \tag{28-8}$$

which tells us that point b is at greater potential than point a. Combining Eq. 28-8 with Eq. 28-4, we have

$$V_b - V_a = \mathscr{E}\frac{R}{R+r}, \tag{28-9}$$

where again r is the internal resistance of the emf device.

> To find the potential difference between any two points in a circuit, start at one point and traverse the circuit to the other, following any path, and add algebraically the changes in potential that you encounter.

Let us again calculate $V_b - V_a$, starting again at point b but this time proceeding counterclockwise to a through the battery. We have

$$V_b + ir - \mathscr{E} = V_a,$$

or

$$V_b - V_a = \mathscr{E} - ir. \tag{28-10}$$

Combining this with Eq. 28-4 again leads to Eq. 28-9.

The quantity $V_b - V_a$ in Fig. 28-4 is the potential difference that the battery sets up across the battery terminals. As noted earlier, $V_b - V_a$ is equal to the emf $\mathscr{E}$ of the battery only if the battery has no internal resistance ($r = 0$ in Eq. 28-9) or if the circuit is open ($i = 0$ in Eq. 28-10).

Suppose that in Fig. 28-4, $\mathscr{E} = 12$ V, $R = 10\ \Omega$, and $r = 2.0\ \Omega$. Then Eq. 28-9 tells us that the potential across the battery's terminals is

$$V_b - V_a = (12\text{ V})\frac{10\ \Omega}{10\ \Omega + 2.0\ \Omega} = 10\text{ V}.$$

In "pumping" charge through itself, the battery does work per unit charge of $\mathscr{E} =$ 12 J/C, or 12 V. However, because of the internal resistance of the battery, it produces a potential difference of only 10 J/C, or 10 V, across its terminals.

Power, Potential, and Emf

When a battery or some other type of emf device does work on the charge carriers to establish a current i, it transfers energy from its source of energy (such as the chemical source in a battery) to the charge carriers. Because a real emf device has an internal resistance r, it also transfers energy to internal thermal energy via resistive dissipation, discussed in Section 27-7. Let us relate these transfers.

The net rate P of energy transfer from the emf device to the charge carriers is given by Eq. 27-21:

$$P = iV, \tag{28-11}$$

where V is the potential across the terminals of the emf device. From Eq. 28-10, we can substitute $V = \mathscr{E} - ir$ into Eq. 28-11 to find

$$P = i(\mathscr{E} - ir) = i\mathscr{E} - i^2r. \tag{28-12}$$

We see that the term i^2r in Eq. 28-12 is the rate P_r of energy transfer to thermal energy within the emf device:

$$P_r = i^2r \quad \text{(internal dissipation rate).} \tag{28-13}$$

Then the term $i\mathscr{E}$ in Eq. 28-12 must be the rate P_{emf} at which the emf device transfers energy to *both* the charge carriers and to internal thermal energy. Thus,

$$P_{\text{emf}} = i\mathscr{E} \quad \text{(power of emf device).} \tag{28-14}$$

If a battery is being *recharged*, with a "wrong way" current through it, the energy transfer is then *from* the charge carriers *to* the battery—both to the battery's chemical energy and to the energy dissipated in the internal resistance r. The rate of change of the chemical energy is given by Eq. 28-14, the rate of dissipation is given by Eq. 28-13, and the rate at which the carriers supply energy is given by Eq. 28-11.

Sample Problem 28-1

The emfs and resistances in the circuit of Fig. 28-6a have the following values:

$$\mathscr{E}_1 = 4.4 \text{ V}, \quad \mathscr{E}_2 = 2.1 \text{ V},$$

$$r_1 = 2.3\ \Omega, \quad r_2 = 1.8\ \Omega, \quad R = 5.5\ \Omega.$$

(a) What is the current i in the circuit?

SOLUTION: The Key Idea here is that we can get an expression involving the current i in this single-loop circuit by applying the loop rule. Although knowing the direction of i is not necessary, we can easily determine it from the emfs of the two batteries. Because $\mathscr{E}_1$ is greater than $\mathscr{E}_2$, battery 1 controls the direction of i, so the direction is clockwise. Let us then apply the loop rule by going counterclockwise—against the current—and starting at point a. We find

$$-\mathscr{E}_1 + ir_1 + iR + ir_2 + \mathscr{E}_2 = 0.$$

Check that this equation also results if we apply the loop rule clockwise or start at some point other than a. Also, take the time to compare this equation term by term with Fig. 28-6b, which shows the potential changes graphically (with the potential at point a arbitrarily taken to be zero).

Solving the above loop equation for the current i, we obtain

$$i = \frac{\mathscr{E}_1 - \mathscr{E}_2}{R + r_1 + r_2} = \frac{4.4 \text{ V} - 2.1 \text{ V}}{5.5\ \Omega + 2.3\ \Omega + 1.8\ \Omega}$$

$$= 0.2396 \text{ A} \approx 240 \text{ mA}. \quad \text{(Answer)}$$

(b) What is the potential difference between the terminals of battery 1 in Fig. 28-6a?

SOLUTION: The Key Idea is to sum the potential differences between points a and b. Let us start at point b (effectively the negative terminal of battery 1) and travel clockwise through battery 1 to point a (effectively the positive terminal), keeping track of potential changes. We find that

$$V_b - ir_1 + \mathscr{E}_1 = V_a,$$

which gives us

$$V_a - V_b = -ir_1 + \mathscr{E}_1$$
$$= -(0.2396 \text{ A})(2.3\ \Omega) + 4.4 \text{ V}$$
$$= +3.84 \text{ V} \approx 3.8 \text{ V}, \quad \text{(Answer)}$$

which is less than the emf of the battery. You can verify this result by starting at point b in Fig. 28-6a and traversing the circuit counterclockwise to point a.

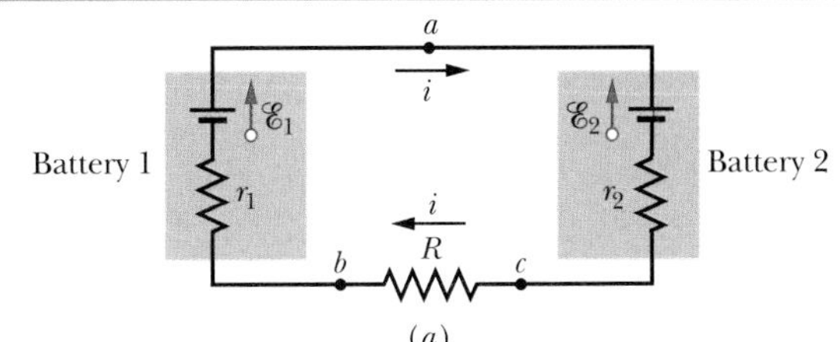

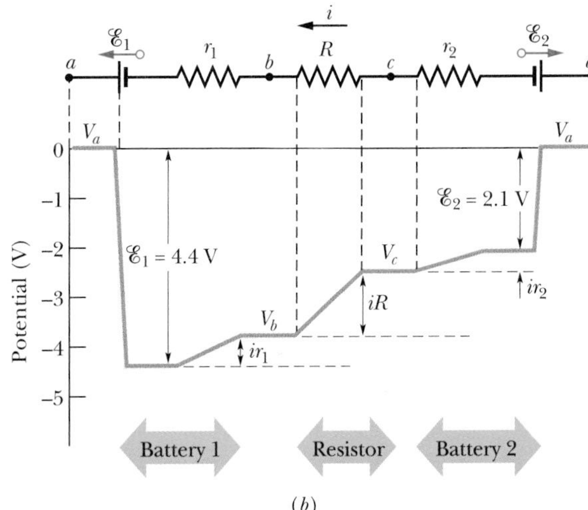

Fig. 28-6 Sample Problem 28-1. (a) A single-loop circuit containing two real batteries and a resistor. The batteries oppose each other; that is, they tend to send current in opposite directions through the resistor. (b) A graph of the potentials, counterclockwise from point a, with the potential at a arbitrarily taken to be zero. (To better link the circuit with the graph, mentally cut the circuit at a and then unfold the left side of the circuit toward the left and the right side of the circuit toward the right.)

CHECKPOINT 3: A battery has an emf of 12 V and an internal resistance of 2 Ω. Is the terminal-to-terminal potential difference greater than, less than, or equal to 12 V if the current in the battery is (a) from the negative to the positive terminal, (b) from the positive to the negative terminal, and (c) zero?

PROBLEM-SOLVING TACTICS

Tactic 1: *Assuming the Direction of a Current*

In solving circuit problems, you do not need to know the direction of a current in advance. Instead, you can just assume its direction, although that may take some physics courage. To show this, assume the current in Fig. 28-6*a* is counterclockwise; that is, reverse the direction of the current arrows shown. Applying the loop rule counterclockwise from point a now yields

$$-\mathscr{E}_1 - ir_1 - iR - ir_2 + \mathscr{E}_2 = 0,$$

or

$$i = -\frac{\mathscr{E}_1 - \mathscr{E}_2}{R + r_1 + r_2}.$$

Substituting the numerical values of Sample Problem 28-1 yields $i = -240$ mA for the current. The minus sign is a signal that the current is opposite the direction we initially assumed.

28-6 Multiloop Circuits

Figure 28-7 shows a circuit containing more than one loop. For simplicity, we assume the batteries are ideal. There are two *junctions* in this circuit, at b and d, and there are three *branches* connecting these junctions. The branches are the left branch (*bad*), the right branch (*bcd*), and the central branch (*bd*). What are the currents in the three branches?

We arbitrarily label the currents, using a different subscript for each branch. Current i_1 has the same value everywhere in branch *bad*, i_2 has the same value everywhere in branch *bcd*, and i_3 is the current through branch *bd*. The directions of the currents are assumed arbitrarily.

Consider junction d for a moment: Charge comes into that junction via incoming currents i_1 and i_3, and it leaves via outgoing current i_2. Because there is no variation in the charge at the junction, the total incoming current must equal the total outgoing current:

$$i_1 + i_3 = i_2. \tag{28-15}$$

You can easily check that applying this condition to junction b leads to exactly the same equation. Equation 28-15 thus suggests a general principle:

JUNCTION RULE: The sum of the currents entering any junction must be equal to the sum of the currents leaving that junction.

This rule is often called *Kirchhoff's junction rule* (or *Kirchhoff's current law*). It is simply a statement of the conservation of charge for a steady flow of charge—there is neither a build-up nor a depletion of charge at a junction. Thus, our basic tools for solving complex circuits are the *loop rule* (based on the conservation of energy) and the *junction rule* (based on the conservation of charge).

Equation 28-15 is a single equation involving three unknowns. To solve the circuit completely (that is, to find all three currents), we need two more equations involving those same unknowns. We obtain them by applying the loop rule twice. In the circuit of Fig. 28-7, we have three loops from which to choose: the left-hand loop (*badb*), the right-hand loop (*bcdb*), and the big loop (*badcb*). Which two loops we choose does not matter—let's choose the left-hand loop and the right-hand loop.

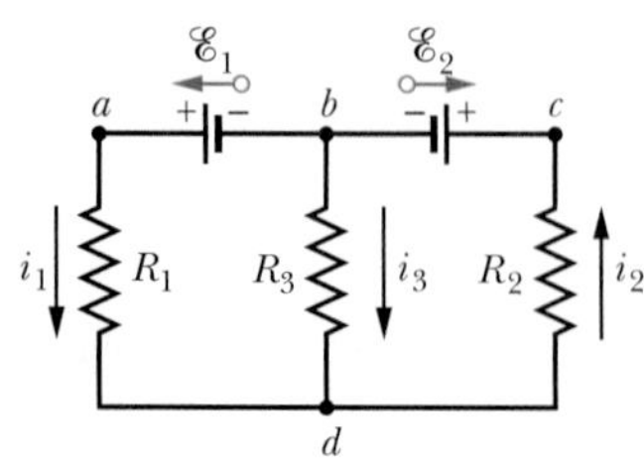

Fig. 28-7 A multiloop circuit consisting of three branches: left-hand branch *bad*, right-hand branch *bcd*, and central branch *bd*. The circuit also consists of three loops: left-hand loop *badb*, right-hand loop *bcdb*, and big loop *badcb*.

If we traverse the left-hand loop in a counterclockwise direction from point b, the loop rule gives us

$$\mathscr{E}_1 - i_1R_1 + i_3R_3 = 0. \tag{28-16}$$

If we traverse the right-hand loop in a counterclockwise direction from point b, the loop rule gives us

$$-i_3R_3 - i_2R_2 - \mathscr{E}_2 = 0. \tag{28-17}$$

We now have three equations (Eqs. 28-15, 28-16, and 28-17) in the three unknown currents, and they can be solved by a variety of techniques.

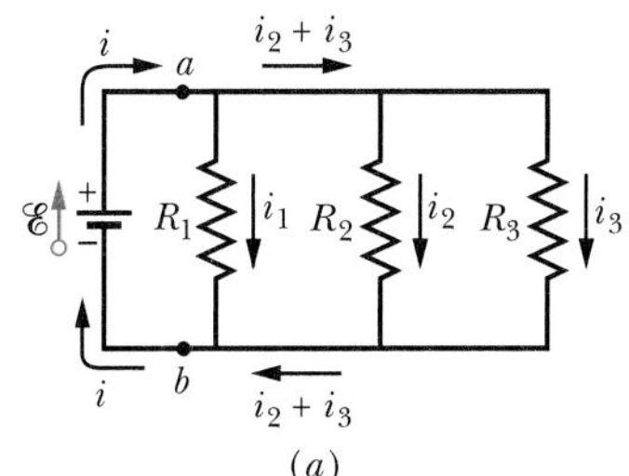

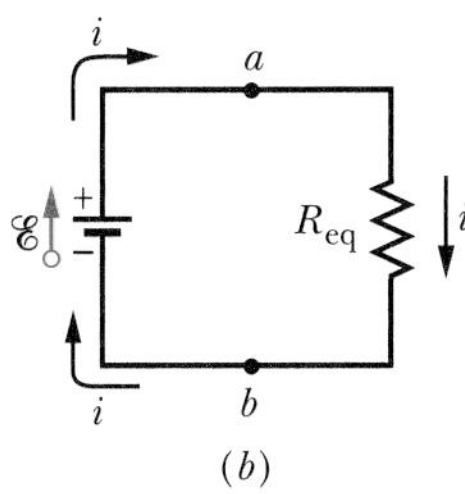

Fig. 28-8 (*a*) Three resistors connected in parallel across points *a* and *b*. (*b*) An equivalent circuit, with the three resistors replaced with their equivalent resistance R_{eq}.

If we had applied the loop rule to the big loop, we would have obtained (moving counterclockwise from b) the equation

$$\mathscr{E}_1 - i_1R_1 - i_2R_2 - \mathscr{E}_2 = 0.$$

This equation may look like fresh information, but in fact it is only the sum of Eqs. 28-16 and 28-17. (It would, however, yield the proper results when used with Eq. 28-15 and either 28-16 or 28-17.)

Resistances in Parallel

Figure 28-8*a* shows three resistances connected *in parallel* to an ideal battery of emf $\mathscr{E}$. The term "in parallel" means that the resistances are directly wired together on one side and directly wired together on the other side, and that a potential difference V is applied across the pair of connected sides. Thus, all three resistances have the same potential difference V across them, producing a current through each. In general,

> When a potential difference V is applied across resistances connected in parallel, the resistances all have that same potential difference V.

In Fig. 28-8*a*, the applied potential difference V is maintained by the battery. In Fig. 28-8*b*, the three parallel resistances have been replaced with an equivalent resistance R_{eq}.

> Resistances connected in parallel can be replaced with an equivalent resistance R_{eq} that has the same potential difference V and the same *total* current i as the actual resistances.

You might remember that R_{eq} and all the actual parallel resistances have the same potential difference V with the nonsense word "par-V."

To derive an expression for R_{eq} in Fig. 28-8*b*, we first write the current in each actual resistance in Fig. 28-8*a* as

$$i_1 = \frac{V}{R_1}, \quad i_2 = \frac{V}{R_2}, \quad \text{and} \quad i_3 = \frac{V}{R_3},$$

where V is the potential difference between a and b. If we apply the junction rule at point a in Fig. 28-8*a* and then substitute these values, we find

$$i = i_1 + i_2 + i_3 = V\left(\frac{1}{R_1} + \frac{1}{R_2} + \frac{1}{R_3}\right). \tag{28-18}$$

If we replaced the parallel combination with the equivalent resistance R_{eq} (Fig. 28-8*b*), we would have

$$i = \frac{V}{R_{eq}}. \tag{28-19}$$

Comparing Eqs. 28-18 and 28-19 leads to

$$\frac{1}{R_{eq}} = \frac{1}{R_1} + \frac{1}{R_2} + \frac{1}{R_3}. \tag{28-20}$$

Extending this result to the case of n resistances, we have

$$\frac{1}{R_{eq}} = \sum_{j=1}^{n} \frac{1}{R_j} \quad (n \text{ resistances in parallel}). \tag{28-21}$$

For the case of two resistances, the equivalent resistance is their product divided by

TABLE 28-1 Series and Parallel Resistors and Capacitors

Series	Parallel	Series	Parallel
Resistors		Capacitors	
$R_{eq} = \sum_{j=1}^{n} R_j$ Eq. 28-7	$\frac{1}{R_{eq}} = \sum_{j=1}^{n} \frac{1}{R_j}$ Eq. 28-21	$\frac{1}{C_{eq}} = \sum_{j=1}^{n} \frac{1}{C_j}$ Eq. 26-20	$C_{eq} = \sum_{j=1}^{n} C_j$ Eq. 26-19
Same current through all resistors	Same potential difference across all resistors	Same charge on all capacitors	Same potential difference across all capacitors

their sum; that is,

$$R_{eq} = \frac{R_1 R_2}{R_1 + R_2}. \tag{28-22}$$

If you accidentally took the equivalent resistance to be the sum divided by the product, you would notice at once that this result would be dimensionally incorrect.

Note that when two or more resistances are connected in parallel, the equivalent resistance is smaller than any of the combining resistances. Table 28-1 summarizes the equivalence relations for resistors and capacitors in series and in parallel.

CHECKPOINT 4: A battery, with potential V across it, is connected to a combination of two identical resistors and then has current i through it. What are the potential difference across and the current through either resistor if the resistors are (a) in series and (b) in parallel?

Sample Problem 28-2

Figure 28-9*a* shows a multiloop circuit containing one ideal battery and four resistances with the following values:

$$R_1 = 20\ \Omega, \quad R_2 = 20\ \Omega, \quad \mathscr{E} = 12\ \text{V},$$
$$R_3 = 30\ \Omega, \quad R_4 = 8.0\ \Omega.$$

(a) What is the current through the battery?

SOLUTION: First note that the current through the battery must also be the current through R_1. Thus, one **Key Idea** here is that we might find that current by applying the loop rule to a loop that includes R_1 because the current would be included in the potential difference across R_1. Either the left-hand loop or the big loop will do. Noting that the emf arrow of the battery points upward, so the current the battery supplies is clockwise, we might apply the loop rule to the left-hand loop, clockwise from point a. With i being the current through the battery, we would get

$$+\mathscr{E} - iR_1 - iR_2 - iR_4 = 0 \quad \text{(incorrect).}$$

However, this equation is incorrect because it assumes that R_1, R_2, and R_4 all have the same current i. Resistances R_1 and R_4 do have the same current, because the current passing through R_4 must pass through the battery and then through R_1 with no change in value. However, that current splits at junction point b—only part passes through R_2, the rest through R_3.

To distinguish the several currents in the circuit, we must label them individually as in Fig. 28-9*b*. Then, circling clockwise from a, we can write the loop rule for the left-hand loop as

$$+\mathscr{E} - i_1R_1 - i_2R_2 - i_1R_4 = 0.$$

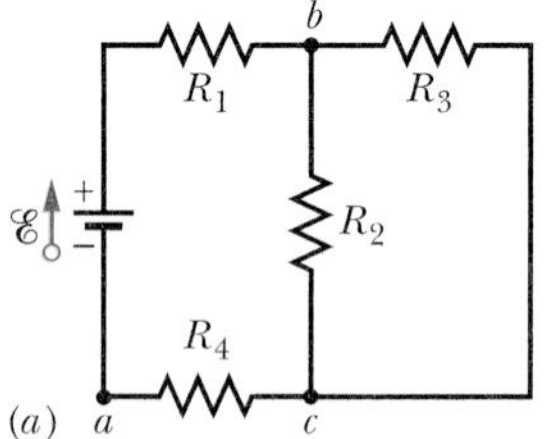

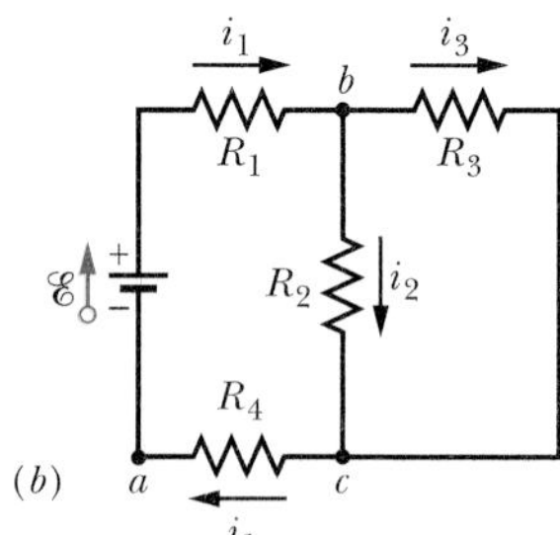

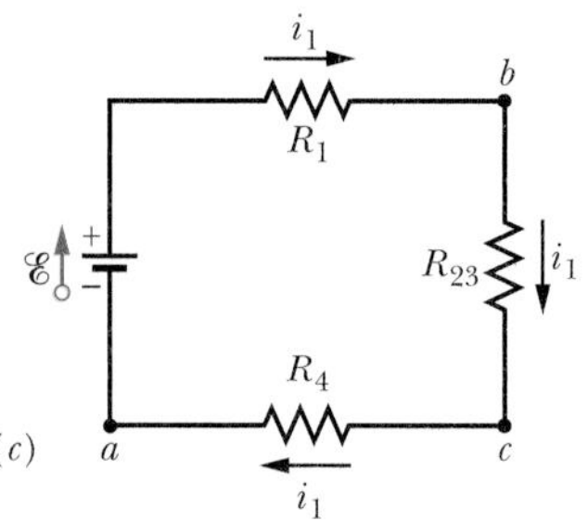

Fig. 28-9 Sample Problem 28-2. (*a*) A multiloop circuit with an ideal battery of emf $\mathscr{E}$ and four resistances. (*b*) Assumed currents through the resistances. (*c*) A simplification of the circuit, with resistances R_2 and R_3 replaced with their equivalent resistance R_{23}. The current through R_{23} is equal to that through R_1 and R_4.

Unfortunately, this equation contains two unknowns, i_1 and i_2; we would need at least one more equation to find them.

A second **Key Idea** is that a much easier option is to simplify the circuit of Fig. 28-9*b* by finding equivalent resistances. Note carefully that R_1 and R_2 are *not* in series and thus cannot be replaced with an equivalent resistance. However, R_2 and R_3 are in parallel, so we can use either Eq. 28-21 or Eq. 28-22 to find their equivalent resistance R_{23}. From the latter,

$$R_{23} = \frac{R_2R_3}{R_2 + R_3} = \frac{(20\ \Omega)(30\ \Omega)}{50\ \Omega} = 12\ \Omega.$$

We can now redraw the circuit as in Fig. 28-9*c*; note that the current through R_{23} must be i_1 because charge that moves through R_1 and R_4 must also move through R_{23}. For this simple one-loop circuit, the loop rule (applied clockwise from point *a*) yields

$$+\mathscr{E} - i_1R_1 - i_1R_{23} - i_1R_4 = 0.$$

Substituting the given data, we find

$$12\ \text{V} - i_1(20\ \Omega) - i_1(12\ \Omega) - i_1(8.0\ \Omega) = 0,$$

which gives us

$$i_1 = \frac{12\ \text{V}}{40\ \Omega} = 0.30\ \text{A}. \qquad \text{(Answer)}$$

(b) What is the current i_2 through R_2?

SOLUTION: One **Key Idea** here is that we must work backward from the equivalent circuit of Fig. 28-9*c*, where R_{23} has replaced the parallel resistances R_2 and R_3. A second **Key Idea** is that, because R_2 and R_3 are in parallel, they both have the same potential difference across them as their equivalent R_{23}. We know the current through R_{23} is $i_1 = 0.30$ A. Thus, we can use Eq. 27-8 ($R = V/i$) to find the potential difference V_{23} across R_{23}:

$$V_{23} = i_1R_{23} = (0.30\ \text{A})(12\ \Omega) = 3.6\ \text{V}.$$

The potential difference across R_2 is thus 3.6 V, so the current i_2 in R_2 must be, by Eq. 27-8,

$$i_2 = \frac{V_2}{R_2} = \frac{3.6\ \text{V}}{20\ \Omega} = 0.18\ \text{A}. \qquad \text{(Answer)}$$

(c) What is the current i_3 through R_3?

SOLUTION: We can answer by using the same technique as in (b), or we can use this **Key Idea**: The junction rule tells us that at point *b* in Fig. 28-9*b*, the incoming current i_1 and the outgoing currents i_2 and i_3 are related by

$$i_1 = i_2 + i_3.$$

This gives us

$$i_3 = i_1 - i_2 = 0.30\ \text{A} - 0.18\ \text{A} = 0.12\ \text{A}. \qquad \text{(Answer)}$$

Sample Problem 28-3

Figure 28-10 shows a circuit whose elements have the following values:

$$\mathscr{E}_1 = 3.0\ \text{V}, \quad \mathscr{E}_2 = 6.0\ \text{V},$$

$$R_1 = 2.0\ \Omega, \quad R_2 = 4.0\ \Omega.$$

The three batteries are ideal batteries. Find the magnitude and direction of the current in each of the three branches.

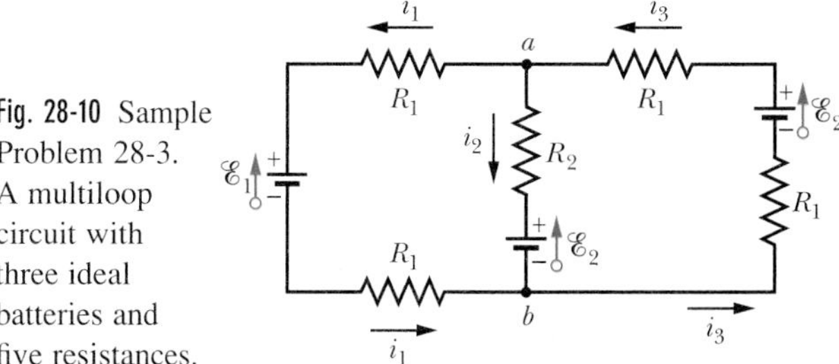

Fig. 28-10 Sample Problem 28-3. A multiloop circuit with three ideal batteries and five resistances.

SOLUTION: It is not worthwhile to try to simplify this circuit, because no two resistors are in parallel, and the resistors that are in series (those in the right branch or those in the left branch) present no problem. So, our **Key Idea** is to apply the junction and loop rules.

Using arbitrarily chosen directions for the currents as shown in Fig. 28-10, we apply the junction rule at point *a* by writing

$$i_3 = i_1 + i_2. \qquad (28\text{-}23)$$

An application of the junction rule at junction *b* gives only the same equation, so we next apply the loop rule to any two of the three loops of the circuit. We first arbitrarily choose the left-hand loop, arbitrarily start at point *a*, and arbitrarily traverse the loop in the counterclockwise direction, obtaining

$$-i_1R_1 - \mathscr{E}_1 - i_1R_1 + \mathscr{E}_2 + i_2R_2 = 0.$$

Substituting the given data and simplifying yield

$$i_1(4.0\ \Omega) - i_2(4.0\ \Omega) = 3.0\ \text{V}. \qquad (28\text{-}24)$$

For our second application of the loop rule, we arbitrarily choose to traverse the right-hand loop clockwise from point *a*, finding

$$+i_3R_1 - \mathscr{E}_2 + i_3R_1 + \mathscr{E}_2 + i_2R_2 = 0.$$

Substituting the given data and simplifying yield

$$i_2(4.0\ \Omega) + i_3(4.0\ \Omega) = 0. \qquad (28\text{-}25)$$

Using Eq. 28-23 to eliminate i_3 from Eq. 28-25 and simplifying give us

$$i_1(4.0\ \Omega) + i_2(8.0\ \Omega) = 0. \qquad (28\text{-}26)$$

We now have a system of two equations (Eqs. 28-24 and 28-26) in two unknowns (i_1 and i_2) to solve either "by hand" (which is easy enough here) or with a "math package." (One solution technique is Cramer's rule, given in Appendix E.) We find

$$i_2 = -0.25\ \text{A}.$$

(The minus sign signals that our arbitrary choice of direction for i_2 in Fig. 28-10 is wrong; i_2 should point up through $\mathscr{E}_2$ and R_2.) Substituting $i_2 = -0.25$ A into Eq. 28-26 and solving for i_1 then give us

$$i_1 = 0.50\ \text{A}. \qquad \text{(Answer)}$$

With Eq. 28-23 we then find that

$$i_3 = i_1 + i_2 = 0.25\ \text{A}. \qquad \text{(Answer)}$$

The positive answers we obtained for i_1 and i_3 signal that our choices of directions for these currents are correct. We can now correct the direction for i_2 and write its magnitude as

$$i_2 = 0.25\ \text{A}. \qquad \text{(Answer)}$$

Sample Problem 28-4

Electric fish are able to generate current with biological cells called *electroplaques,* which are physiological emf devices. The electroplaques in the South American eel shown in the photograph that opens this chapter are arranged in 140 rows, each row stretching horizontally along the body and each containing 5000 electroplaques. The arrangement is suggested in Fig. 28-11*a*; each electroplaque has an emf $\mathscr{E}$ of 0.15 V and an internal resistance r of 0.25 Ω. The water surrounding the eel completes a circuit between the two ends of the electroplaque array, one end at the animal's head and the other near its tail.

(a) If the water surrounding the eel has resistance $R_w = 800\ \Omega$, how much current can the eel produce in the water?

SOLUTION: The **Key Idea** here is that we can simplify the circuit of Fig. 28-11*a* by replacing combinations of emfs and internal resistances with equivalent emfs and resistances. We first consider a single row. The total emf $\mathscr{E}_{\text{row}}$ along a row of 5000 electroplaques is the sum of the emfs:

$$\mathscr{E}_{\text{row}} = 5000\mathscr{E} = (5000)(0.15\ \text{V}) = 750\ \text{V}.$$

The total resistance R_{row} along a row is the sum of the internal resistances of the 5000 electroplaques:

$$R_{\text{row}} = 5000r = (5000)(0.25\ \Omega) = 1250\ \Omega.$$

We can now represent each of the 140 identical rows as having a single emf $\mathscr{E}_{\text{row}}$ and a single resistance R_{row}, as shown in Fig. 28-11*b*.

In Fig. 28-11*b*, the emf between point a and point b on any row is $\mathscr{E}_{\text{row}} = 750$ V. Because the rows are identical and because they are all connected together at the left in Fig. 28-11*b*, all points b in that figure are at the same electric potential. Thus, we can consider them to be connected so that there is only a single point b. The emf between point a and this single point b is $\mathscr{E}_{\text{row}} = 750$ V, so we can draw the circuit as shown in Fig. 28-11*c*.

Between points b and c in Fig. 28-11*c* are 140 resistances $R_{\text{row}} = 1250\ \Omega$, all in parallel. The equivalent resistance R_{eq} of this combination is given by Eq. 28-21 as

$$\frac{1}{R_{\text{eq}}} = \sum_{j=1}^{140} \frac{1}{R_j} = 140\,\frac{1}{R_{\text{row}}},$$

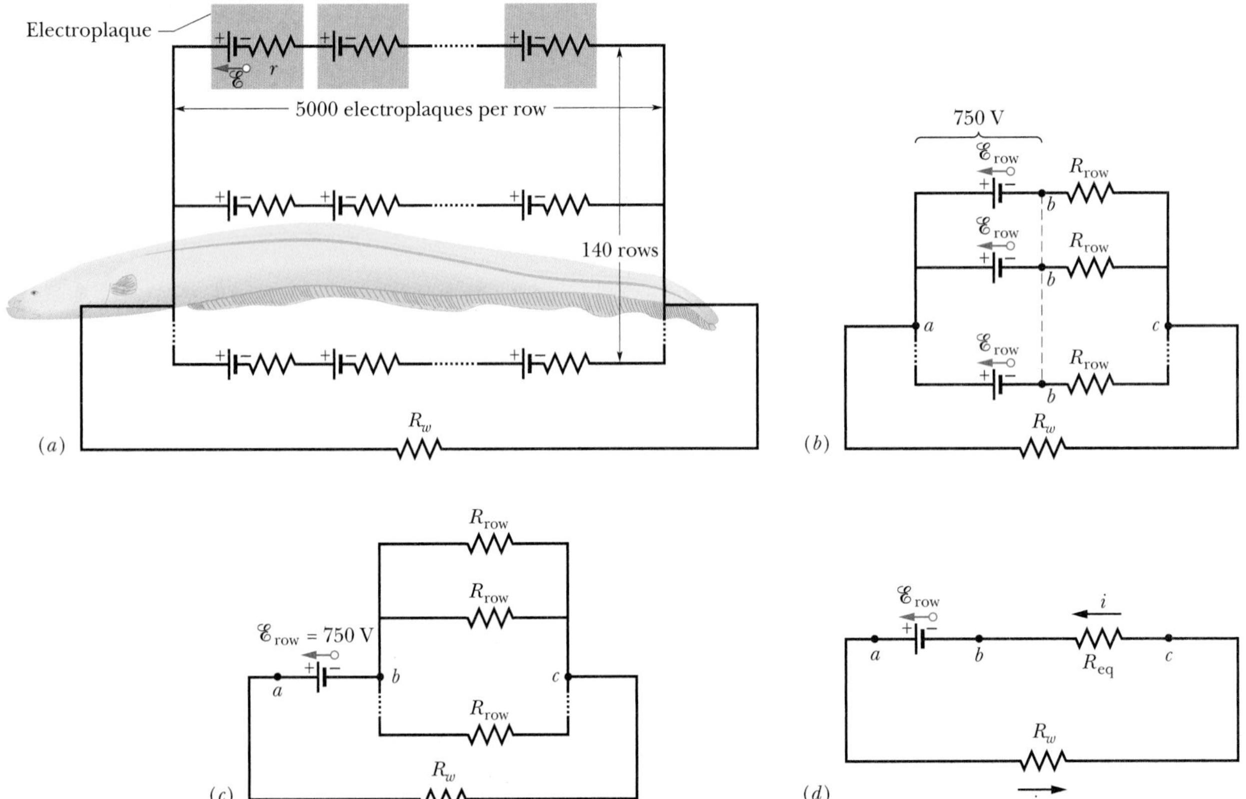

Fig. 28-11 Sample Problem 28-4. (*a*) A model of the electric circuit of an eel in water. Each electroplaque of the eel has an emf $\mathscr{E}$ and internal resistance r. Along each of 140 rows extending from the head to the tail of the eel, there are 5000 electroplaques. The surrounding water has resistance R_w. (*b*) The emf $\mathscr{E}_{\text{row}}$ and resistance R_{row} of each row. (*c*) The emf between points a and b is $\mathscr{E}_{\text{row}}$. Between points b and c are 140 parallel resistances R_{row}. (*d*) The simplified circuit, with R_{eq} replacing the parallel combination.

or
$$R_{eq} = \frac{R_{row}}{140} = \frac{1250\ \Omega}{140} = 8.93\ \Omega.$$

Replacing the parallel combination with R_{eq}, we obtain the simplified circuit of Fig. 28-11*d*. Applying the loop rule to this circuit counterclockwise from point *b*, we have

$$\mathscr{E}_{row} - iR_w - iR_{eq} = 0.$$

Solving for *i* and substituting the known data, we find

$$i = \frac{\mathscr{E}_{row}}{R_w + R_{eq}} = \frac{750\ \text{V}}{800\ \Omega + 8.93\ \Omega}$$
$$= 0.927\ \text{A} \approx 0.93\ \text{A}. \qquad \text{(Answer)}$$

If the head or tail of the eel is near a fish, some of this current could pass along a narrow path through the fish, stunning or killing it.

(b) How much current i_{row} travels through each row of Fig. 28-11*a*?

SOLUTION: The **Key Idea** here is that since the rows are identical, the current into and out of the eel is evenly divided among them:

$$i_{row} = \frac{i}{140} = \frac{0.927\ \text{A}}{140} = 6.6 \times 10^{-3}\ \text{A}. \qquad \text{(Answer)}$$

Thus, the current through each row is small, about two orders of magnitude smaller than the current through the water. This tends to spread the current through the eel's body, so that it need not stun or kill itself when it stuns or kills a fish.

PROBLEM-SOLVING TACTICS

Tactic 2: *Solving Circuits of Batteries and Resistors*
Here are two general techniques for solving circuits for unknown currents or potential differences.

1. If a circuit can be simplified by replacing resistors in series or in parallel with their equivalents, do so. If you can reduce the circuit to a single loop, then you can find the current through the battery with that loop, as in Sample Problem 28-2a. You may then have to "work backward," undoing the resistor simplification process, to find the current or potential difference for any particular resistor, as in Sample Problem 28-2b.
2. If a circuit cannot be simplified to a single loop, use the junction rule and the loop rule to write a set of simultaneous equations, as in Sample Problem 28-3. You need have only as many independent equations as there are unknowns in those equations. If you have to find the current or potential difference for a particular resistor, you can ensure that its current or potential difference appears in the equations by having at least one of the loops pass through it.

Tactic 3: *Arbitrary Choices in Solving Circuit Problems*

In Sample Problem 28-3, we made several arbitrary choices. (1) We assumed directions for the currents in Fig. 28-10 arbitrarily. (2) We chose which of the three possible loops to write equations for arbitrarily. (3) We chose the direction in which to traverse each loop arbitrarily. (4) We chose the starting and ending point for each traversal arbitrarily.

Such arbitrariness often worries a beginning circuit solver, but an experienced circuit solver knows that it does not matter. Just keep two rules firmly in mind. First, make sure you traverse each chosen loop completely. Second, once you have chosen a direction for a current, stick with it until you get numerical values for all the currents. If you were wrong about a direction, the algebra will signal you with a minus sign. Then you can make a correction by simply erasing the minus sign and reversing the arrow representing that current in the circuit diagram. However, *you should not make this correction until you have completed all the required calculations for the circuit,* as we did in Sample Problem 28-3.

28-7 The Ammeter and the Voltmeter

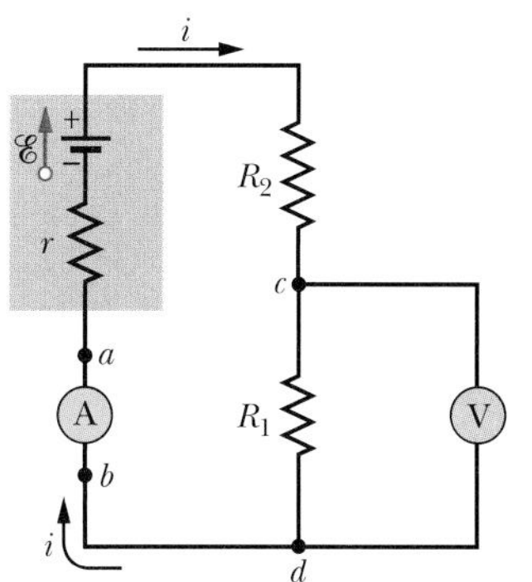

Fig. 28-12 A single-loop circuit, showing how to connect an ammeter (A) and a voltmeter (V).

An instrument used to measure currents is called an *ammeter*. To measure the current in a wire, you usually have to break or cut the wire and insert the ammeter so that the current to be measured passes through the meter. (In Fig. 28-12, ammeter A is set up to measure current *i*.)

It is essential that the resistance R_A of the ammeter be very small compared to other resistances in the circuit. Otherwise, the very presence of the meter will change the current to be measured.

A meter used to measure potential differences is called a *voltmeter*. To find the potential difference between any two points in the circuit, the voltmeter terminals are connected between those points, without breaking or cutting the wire. (In Fig. 28-12, voltmeter V is set up to measure the voltage across R_1.)

It is essential that the resistance R_V of a voltmeter be very large compared to the resistance of any circuit element across which the voltmeter is connected. Otherwise, the meter itself becomes an important circuit element and alters the potential difference that is to be measured.

Often a single meter is packaged so that, by means of a switch, it can be made to serve as either an ammeter or a voltmeter—and usually also as an *ohmmeter,* designed to measure the resistance of any element connected between its terminals. Such a versatile unit is called a *multimeter*.

28-8 *RC* Circuits

In preceding sections we dealt only with circuits in which the currents did not vary with time. Here we begin a discussion of time-varying currents.

Charging a Capacitor

The capacitor of capacitance C in Fig. 28-13 is initially uncharged. To charge it, we close switch S on point a. This completes an *RC series circuit* consisting of the capacitor, an ideal battery of emf $\mathscr{E}$, and a resistance R.

From Section 26-2, we already know that as soon as the circuit is complete, charge begins to flow (current exists) between a capacitor plate and a battery terminal on each side of the capacitor. This current increases the charge q on the plates and the potential difference V_C $(= q/C)$ across the capacitor. When that potential difference equals the potential difference across the battery (which here is equal to the emf $\mathscr{E}$), the current is zero. From Eq. 26-1 $(q = CV)$, the *equilibrium* (final) *charge* on the then fully charged capacitor is equal to $C\mathscr{E}$.

Here we want to examine the charging process. In particular we want to know how the charge $q(t)$ on the capacitor plates, the potential difference $V_C(t)$ across the capacitor, and the current $i(t)$ in the circuit vary with time during the charging process. We begin by applying the loop rule to the circuit, traversing it clockwise from the negative terminal of the battery. We find

$$\mathscr{E} - iR - \frac{q}{C} = 0. \tag{28-27}$$

The last term on the left side represents the potential difference across the capacitor. The term is negative because the capacitor's top plate, which is connected to the battery's positive terminal, is at a higher potential than the lower plate. Thus, there is a drop in potential as we move down through the capacitor.

We cannot immediately solve Eq. 28-27 because it contains two variables, i and q. However, those variables are not independent but are related by

$$i = \frac{dq}{dt}. \tag{28-28}$$

Substituting this for i in Eq. 28-27 and rearranging, we find

$$R\frac{dq}{dt} + \frac{q}{C} = \mathscr{E} \qquad \text{(charging equation).} \tag{28-29}$$

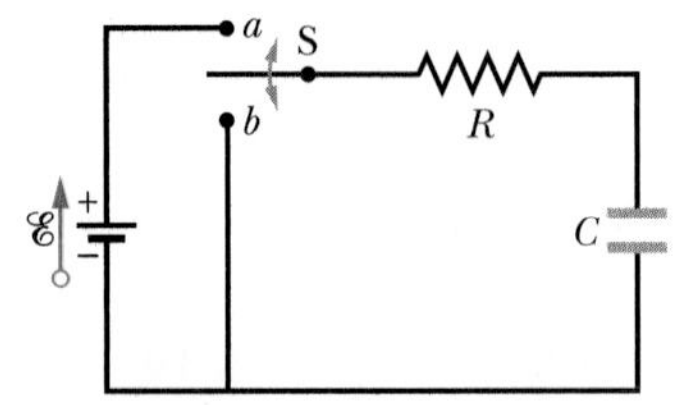

Fig. 28-13 When switch S is closed on a, the capacitor is *charged* through the resistor. When the switch is afterward closed on b, the capacitor *discharges* through the resistor.

This differential equation describes the time variation of the charge q on the capacitor in Fig. 28-13. To solve it, we need to find the function $q(t)$ that satisfies this equation and also satisfies the condition that the capacitor be initially uncharged; that is, $q = 0$ at $t = 0$.

We shall soon show that the solution to Eq. 28-29 is

$$q = C\mathscr{E}(1 - e^{-t/RC}) \qquad \text{(charging a capacitor).} \tag{28-30}$$

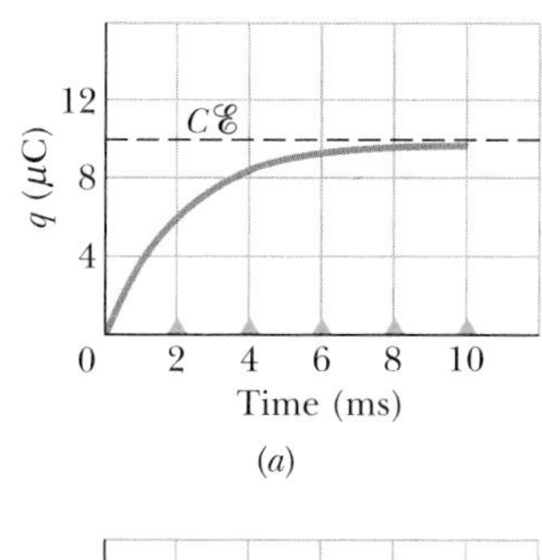

(*a*)

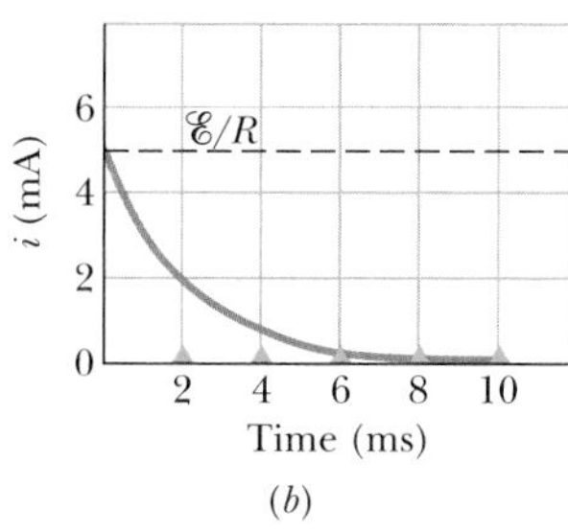

(*b*)

Fig. 28-14 (*a*) A plot of Eq. 28-30, which shows the buildup of charge on the capacitor of Fig. 28-13. (*b*) A plot of Eq. 28-31, which shows the decline of the charging current in the circuit of Fig. 28-13. The curves are plotted for $R = 2000\ \Omega$, $C = 1\ \mu\text{F}$, and $\mathscr{E} = 10$ V; the small triangles represent successive intervals of one time constant τ.

(Here e is the exponential base, 2.718 . . . , and not the elementary charge.) Note that Eq. 28-30 does indeed satisfy our required initial condition, because at $t = 0$ the term $e^{-t/RC}$ is unity; so the equation gives $q = 0$. Note also that as t goes to infinity (that is, a long time later), the term $e^{-t/RC}$ goes to zero; so the equation gives the proper value for the full (equilibrium) charge on the capacitor—namely, $q = C\mathscr{E}$. A plot of $q(t)$ for the charging process is given in Fig. 28-14*a*.

The derivative of $q(t)$ is the current $i(t)$ charging the capacitor:

$$i = \frac{dq}{dt} = \left(\frac{\mathscr{E}}{R}\right)e^{-t/RC} \quad \text{(charging a capacitor).} \tag{28-31}$$

A plot of $i(t)$ for the charging process is given in Fig. 28-14*b*. Note that the current has the initial value $\mathscr{E}/R$ and that it decreases to zero as the capacitor becomes fully charged.

> A capacitor that is being charged initially acts like ordinary connecting wire relative to the charging current. A long time later, it acts like a broken wire.

By combining Eq. 26-1 ($q = CV$) and Eq. 28-30, we find that the potential difference $V_C(t)$ across the capacitor during the charging process is

$$V_C = \frac{q}{C} = \mathscr{E}(1 - e^{-t/RC}) \quad \text{(charging a capacitor).} \tag{28-32}$$

This tells us that $V_C = 0$ at $t = 0$ and that $V_C = \mathscr{E}$ when the capacitor becomes fully charged as $t \to \infty$.

The Time Constant

The product RC that appears in Eqs. 28-30, 28-31, and 28-32 has the dimensions of time (both because the argument of an exponential must be dimensionless and because, in fact, $1.0\ \Omega \times 1.0\ \text{F} = 1.0$ s). RC is called the **capacitive time constant** of the circuit and is represented with the symbol τ:

$$\tau = RC \quad \text{(time constant).} \tag{28-33}$$

From Eq. 28-30, we can now see that at time $t = \tau\ (= RC)$, the charge on the initially uncharged capacitor of Fig. 28-13 has increased from zero to

$$q = C\mathscr{E}(1 - e^{-1}) = 0.63C\mathscr{E}. \tag{28-34}$$

In words, during the first time constant τ the charge has increased from zero to 63% of its final value $C\mathscr{E}$. In Fig. 28-14, the small triangles along the time axes mark successive intervals of one time constant during the charging of the capacitor. The charging times for RC circuits are often stated in terms of τ; the greater τ is, the greater the charging time.

Discharging a Capacitor

Assume now that the capacitor of Fig. 28-13 is fully charged to a potential V_0 equal to the emf $\mathscr{E}$ of the battery. At a new time $t = 0$, switch S is thrown from a to b so that the capacitor can *discharge* through resistance R. How do the charge $q(t)$ on the capacitor and the current $i(t)$ through the discharge loop of capacitor and resistance now vary with time?

The differential equation describing $q(t)$ is like Eq. 28-29 except that now, with no battery in the discharge loop, $\mathscr{E} = 0$. Thus,

$$R\frac{dq}{dt} + \frac{q}{C} = 0 \quad \text{(discharging equation).} \tag{28-35}$$

The solution to this differential equation is

$$q = q_0 e^{-t/RC} \quad \text{(discharging a capacitor),} \tag{28-36}$$

where $q_0\,(= CV_0)$ is the initial charge on the capacitor. You can verify by substitution that Eq. 28-36 is indeed a solution of Eq. 28-35.

Equation 28-36 tells us that q decreases exponentially with time, at a rate that is set by the capacitive time constant $\tau = RC$. At time $t = \tau$, the capacitor's charge has been reduced to $q_0 e^{-1}$, or about 37% of the initial value. Note that a greater τ means a greater discharge time.

Differentiating Eq. 28-36 gives us the current $i(t)$:

$$i = \frac{dq}{dt} = -\left(\frac{q_0}{RC}\right)e^{-t/RC} \quad \text{(discharging a capacitor).} \tag{28-37}$$

This tells us that the current also decreases exponentially with time, at a rate set by τ. The initial current i_0 is equal to q_0/RC. Note that you can find i_0 by simply applying the loop rule to the circuit at $t = 0$; just then the capacitor's initial potential V_0 is connected across the resistance R, so the current must be $i_0 = V_0/R = (q_0/C)/R = q_0/RC$. The minus sign in Eq. 28-37 can be ignored; it merely means that the capacitor's charge q is decreasing.

Derivation of Eq. 28-30

To solve Eq. 28-29, we first rewrite it as

$$\frac{dq}{dt} + \frac{q}{RC} = \frac{\mathscr{E}}{R}. \tag{28-38}$$

The general solution to this differential equation is of the form

$$q = q_p + Ke^{-at}, \tag{28-39}$$

where q_p is a *particular solution* of the differential equation, K is a constant to be evaluated from the initial conditions, and $a = 1/RC$ is the coefficient of q in Eq. 28-38. To find q_p, we set $dq/dt = 0$ in Eq. 28-38 (corresponding to the final condition of no further charging), let $q = q_p$, and solve, obtaining

$$q_p = C\mathscr{E}. \tag{28-40}$$

To evaluate K, we first substitute this into Eq. 28-39 to get

$$q = C\mathscr{E} + Ke^{-at}.$$

Then substituting the initial conditions $q = 0$ and $t = 0$ yields

$$0 = C\mathscr{E} + K,$$

or $K = -C\mathscr{E}$. Finally, with the values of q_p, a, and K inserted, Eq. 28-39 becomes

$$q = C\mathscr{E} - C\mathscr{E}e^{-t/RC},$$

which, with a slight modification, is Eq. 28-30.

CHECKPOINT 5: The table gives four sets of values for the circuit elements in Fig. 28-13. Rank the sets according to (a) the initial current (as the switch is closed on a) and (b) the time required for the current to decrease to half its initial value, greatest first.

	1	2	3	4
$\mathscr{E}$ (V)	12	12	10	10
R (Ω)	2	3	10	5
C (μF)	3	2	0.5	2

Sample Problem 28-5

A capacitor of capacitance C is discharging through a resistor of resistance R.

(a) In terms of the time constant $\tau = RC$, when will the charge on the capacitor be half its initial value?

SOLUTION: The Key Idea here is that the charge on the capacitor varies according to Eq. 28-36,

$$q = q_0 e^{-t/RC},$$

in which q_0 is the initial charge. We are asked to find the time t at which $q = \frac{1}{2}q_0$, or at which

$$\tfrac{1}{2}q_0 = q_0 e^{-t/RC}. \tag{28-41}$$

After canceling q_0, we realize that the time t we seek is "buried" inside an exponential function. To expose the symbol t in Eq. 28-41, we take the natural logarithms of both sides of the equation. (The natural logarithm is the inverse function of the exponential function.) We find

$$\ln \tfrac{1}{2} = \ln(e^{-t/RC}) = -\frac{t}{RC},$$

or

$$t = (-\ln \tfrac{1}{2})RC = 0.69RC = 0.69\tau. \qquad \text{(Answer)}$$

(b) When will the energy stored in the capacitor be half its initial value?

SOLUTION: There are two Key Ideas here. First, the energy U stored in a capacitor is related to the charge q on the capacitor according to Eq. 26-21 ($U = Q^2/2C$). Second, that charge is decreasing according to Eq. 28-36. Combining these two ideas gives us

$$U = \frac{q^2}{2C} = \frac{q_0^2}{2C} e^{-2t/RC} = U_0 e^{-2t/RC},$$

in which U_0 is the initial stored energy. We are asked to find the time at which $U = \frac{1}{2}U_0$, or at which

$$\tfrac{1}{2}U_0 = U_0 e^{-2t/RC}.$$

Canceling U_0 and taking the natural logarithms of both sides, we obtain

$$\ln \tfrac{1}{2} = -\frac{2t}{RC},$$

or

$$t = -RC\,\frac{\ln \frac{1}{2}}{2} = 0.35RC = 0.35\tau. \qquad \text{(Answer)}$$

It takes longer (0.69τ versus 0.35τ) for the *charge* to fall to half its initial value than for the *stored energy* to fall to half its initial value. Doesn't this result surprise you?

REVIEW & SUMMARY

Emf An **emf device** does work on charges to maintain a potential difference between its output terminals. If dW is the work the device does to force positive charge dq from the negative to the positive terminal, then the **emf** (work per unit charge) of the device is

$$\mathscr{E} = \frac{dW}{dq} \quad \text{(definition of } \mathscr{E}\text{)}. \tag{28-1}$$

The volt is the SI unit of emf as well as of potential difference. An **ideal emf device** is one that lacks any internal resistance. The potential difference between its terminals is equal to the emf. A **real emf device** has internal resistance. The potential difference between its terminals is equal to the emf only if there is no current through the device.

Analyzing Circuits The change in potential in traversing a resistance R in the direction of the current is $-iR$; in the opposite direction it is $+iR$. The change in potential in traversing an ideal emf device in the direction of the emf arrow is $+\mathscr{E}$; in the opposite direction it is $-\mathscr{E}$. Conservation of energy leads to the loop rule:

Loop Rule. *The algebraic sum of the changes in potential encountered in a complete traversal of any loop of a circuit must be zero.*

Conservation of charge gives us the junction rule:

Junction Rule. *The sum of the currents entering any junction must be equal to the sum of the currents leaving that junction.*

Single-Loop Circuits The current in a single-loop circuit containing a single resistance R and an emf device with emf $\mathscr{E}$ and internal resistance r is

$$i = \frac{\mathscr{E}}{R + r}, \quad (28\text{-}4)$$

which reduces to $i = \mathscr{E}/R$ for an ideal emf device with $r = 0$.

Power When a real battery of emf $\mathscr{E}$ and internal resistance r does work on the charge carriers in a current i through it, the rate P of energy transfer to the charge carriers is

$$P = iV, \quad (28\text{-}11)$$

where V is the potential across the terminals of the battery. The rate P_r of energy transfer to thermal energy within the battery is

$$P_r = i^2 r. \quad (28\text{-}13)$$

The rate P_{emf} at which the chemical energy within the battery changes is

$$P_{emf} = i\mathscr{E}. \quad (28\text{-}14)$$

Series Resistances When resistances are in **series,** they have the same current. The equivalent resistance that can replace a series combination of resistances is

$$R_{eq} = \sum_{j=1}^{n} R_j \quad (n \text{ resistances in series}). \quad (28\text{-}7)$$

Parallel Resistances When resistances are in **parallel,** they have the same potential difference. The equivalent resistance that can replace a parallel combination of resistances is given by

$$\frac{1}{R_{eq}} = \sum_{j=1}^{n} \frac{1}{R_j} \quad (n \text{ resistances in parallel}). \quad (28\text{-}21)$$

RC Circuits When an emf $\mathscr{E}$ is applied to a resistance R and capacitance C in series, as in Fig. 28-13 with the switch at a, the charge on the capacitor increases according to

$$q = C\mathscr{E}(1 - e^{-t/RC}) \quad \text{(charging a capacitor)}, \quad (28\text{-}30)$$

in which $C\mathscr{E} = q_0$ is the equilibrium (final) charge and $RC = \tau$ is the **capacitive time constant** of the circuit. During the charging, the current is

$$i = \frac{dq}{dt} = \left(\frac{\mathscr{E}}{R}\right)e^{-t/RC} \quad \text{(charging a capacitor)}. \quad (28\text{-}31)$$

When a capacitor discharges through a resistance R, the charge on the capacitor decays according to

$$q = q_0 e^{-t/RC} \quad \text{(discharging a capacitor)}. \quad (28\text{-}36)$$

During the discharging, the current is

$$i = \frac{dq}{dt} = -\left(\frac{q_0}{RC}\right)e^{-t/RC} \quad \text{(discharging a capacitor)}. \quad (28\text{-}37)$$

QUESTIONS

1. Figure 28-15 shows current i passing through a battery. The following table gives four sets of values for i and the battery's emf $\mathscr{E}$ and internal resistance r; it also gives the *polarity* (orientation of the terminals) of the battery. Rank the sets according to the rate at which energy is transferred between the battery and the charge carriers, greatest transfer *to* the carriers first and greatest transfer *from* the carriers last.

	$\mathscr{E}$	r	i	Polarity
(1)	$15\mathscr{E}_1$	0	i_1	+ at left
(2)	$10\mathscr{E}_1$	0	$2i_1$	+ at left
(3)	$10\mathscr{E}_1$	0	$2i_1$	− at left
(4)	$10\mathscr{E}_1$	r_1	$2i_1$	− at left

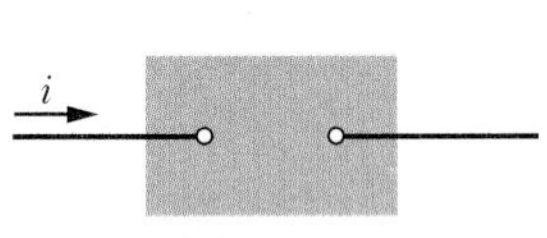

Fig. 28-15 Question 1.

2. For each circuit in Fig. 28-16, are the resistors connected in series, in parallel, or neither?

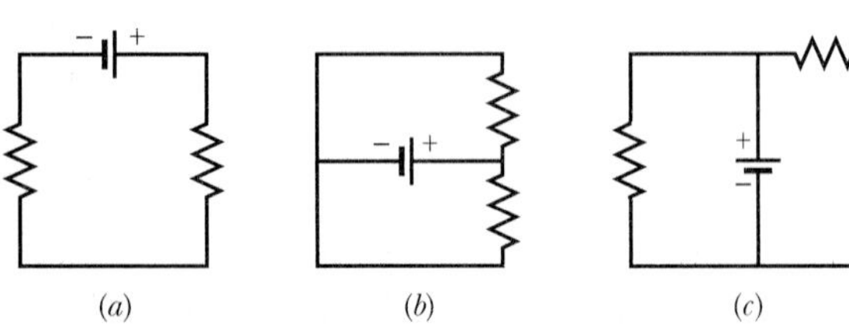

Fig. 28-16 Question 2.

3. (a) In Fig. 28-17a, are resistors R_1 and R_3 in series? (b) Are resistors R_1 and R_2 in parallel? (c) Rank the equivalent resistances of the four circuits shown in Fig. 28-17, greatest first.

4. (a) In Fig. 28-17a, with $R_1 > R_2$, is the potential difference across R_2 more than, less than, or equal to that across R_1? (b) Is the current through resistor R_2 more than, less than, or equal to that through resistor R_1?

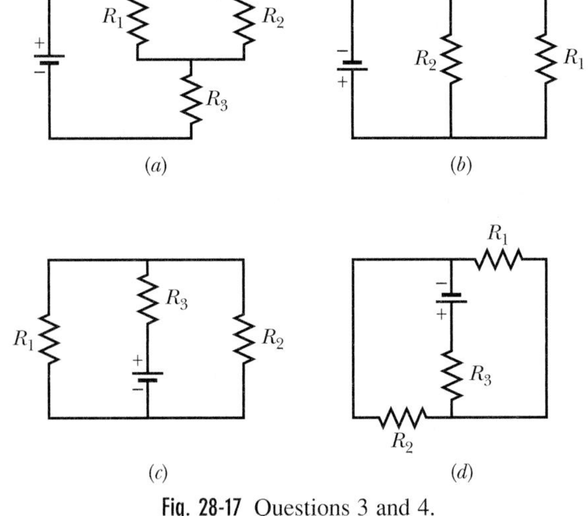

Fig. 28-17 Questions 3 and 4.

5. You are to connect resistors R_1 and R_2, with $R_1 > R_2$, to a battery, first individually, then in series, and then in parallel. Rank those arrangements according to the amount of current through the battery, greatest first.

6. *Res-monster maze.* In Fig. 28-18, all the resistors have a resistance of 4.0 Ω and all the (ideal) batteries have an emf of 4.0 V. What is the current through resistor R? (If you can find the proper loop through this maze, you can answer the question with a few seconds of mental calculation.)

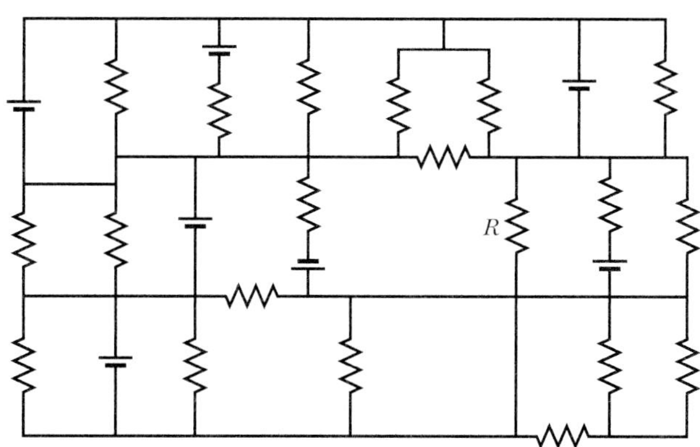

Fig. 28-18 Question 6.

7. Initially, a single resistor R_1 is wired to a battery. Then resistor R_2 is added in parallel. Are (a) the potential difference across R_1 and (b) the current i_1 through R_1 now more than, less than, or the same as previously? (c) Is the equivalent resistance R_{12} of R_1 and R_2 more than, less than, or equal to R_1? (d) Is the total current through R_1 and R_2 together more than, less than, or equal to the current through R_1 previously?

8. *Cap-monster maze.* In Fig. 28-19, all the capacitors have a capacitance of 6.0 μF, and all the batteries have an emf of 10 V. What is the charge on capacitor C? (If you can find the proper loop through this maze, you can answer the question with a few seconds of mental calculation.)

9. A resistor R_1 is wired to a battery, then resistor R_2 is added in series. Are (a) the potential difference across R_1 and (b) the current i_1 through R_1 now more than, less than, or the same as previously? (c) Is the equivalent resistance R_{12} of R_1 and R_2 more than, less than, or equal to R_1?

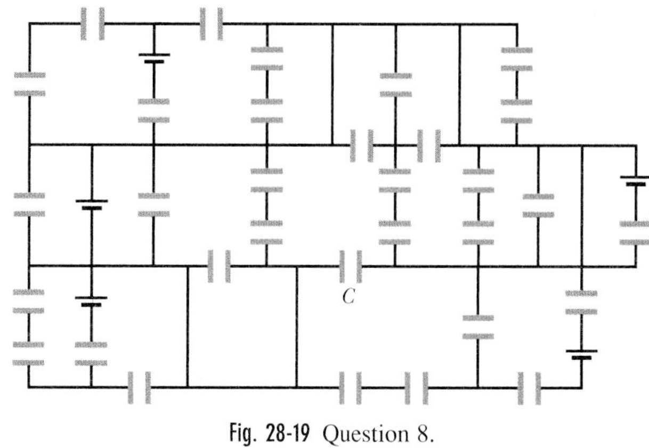

Fig. 28-19 Question 8.

10. Figure 28-20 shows three sections of circuit that are to be connected in turn to the same battery via a switch as in Fig. 28-13. The resistors are all identical, as are the capacitors. Rank the sections according to (a) the final (equilibrium) charge on the capacitor and (b) the time required for the capacitor to reach 50% of its final charge, greatest first.

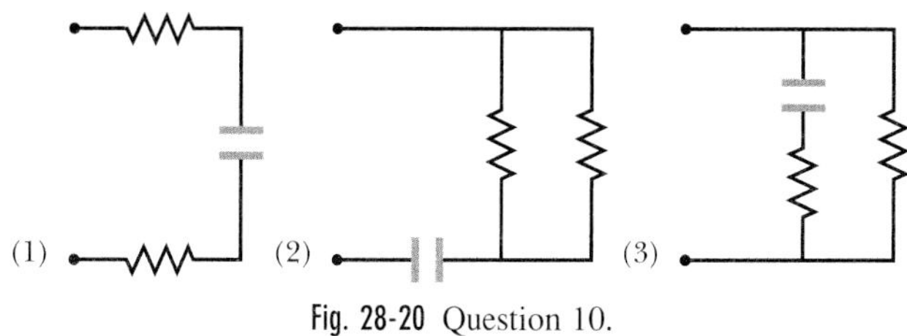

Fig. 28-20 Question 10.

11. Figure 28-21 shows plots of $V(t)$ for three capacitors that discharge (separately) through the same resistor. Rank the plots according to the capacitances of the capacitors, greatest first.

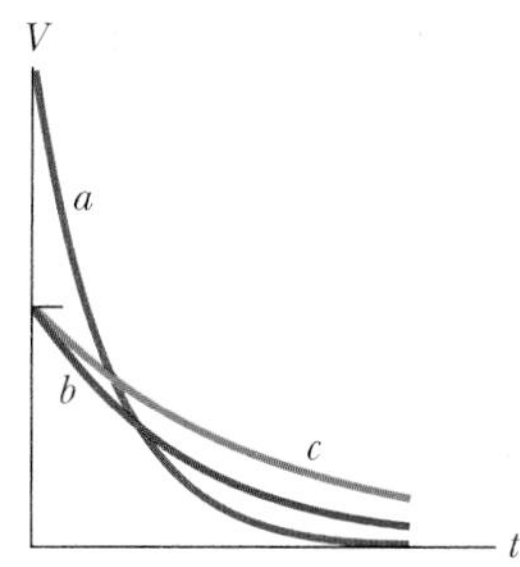

Fig. 28-21 Question 11.

EXERCISES & PROBLEMS

ssm Solution is in the Student Solutions Manual.
www Solution is available on the World Wide Web at: http://www.wiley.com/college/hrw
ilw Solution is available on the Interactive LearningWare.

SEC. 28-5 Potential Differences

1E. A standard flashlight battery can deliver about 2.0 W · h of energy before it runs down. (a) If a battery costs 80¢, what is the cost of operating a 100 W lamp for 8.0 h using batteries? (b) What is the cost if energy is provided at 6¢ per kilowatt-hour? ssm

2E. A 5.0 A current is set up in a circuit for 6.0 min by a rechargeable battery with a 6.0 V emf. By how much is the chemical energy of the battery reduced?

3E. A certain car battery with a 12 V emf has an initial charge of 120 A · h. Assuming that the potential across the terminals stays constant until the battery is completely discharged, for how long can it deliver energy at the rate of 100 W? ssm

4E. In Fig. 28-22, $\mathscr{E}_1 = 12$ V and $\mathscr{E}_2 = 8$ V. (a) What is the direction of the current in the resistor? (b) Which battery is doing positive work? (c) Which point, A or B, is at the higher potential?

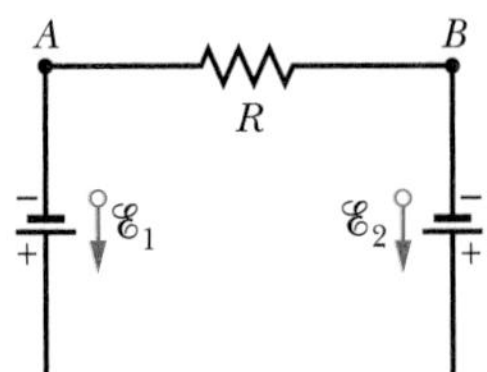

Fig. 28-22 Exercise 4.

5E. Assume that the batteries in Fig. 28-23 have negligible internal resistance. Find (a) the current in the circuit, (b) the power dissipated in each resistor, and (c) the power of each battery, stating whether energy is supplied by or absorbed by it. ssm

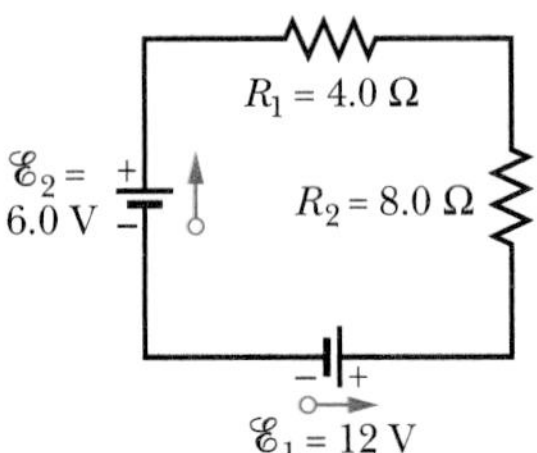

Fig. 28-23 Exercise 5.

6E. A wire of resistance 5.0 Ω is connected to a battery whose emf $\mathscr{E}$ is 2.0 V and whose internal resistance is 1.0 Ω. In 2.0 min, (a) how much energy is transferred from chemical to electrical form? (b) How much energy appears in the wire as thermal energy? (c) Account for the difference between (a) and (b).

7E. A car battery with a 12 V emf and an internal resistance of 0.040 Ω is being charged with a current of 50 A. (a) What is the potential difference across its terminals? (b) At what rate is energy being dissipated as thermal energy in the battery? (c) At what rate is electric energy being converted to chemical energy? (d) What are the answers to (a) and (b) when the battery is used to supply 50 A to the starter motor? ilw

8E. In Fig. 28-4a, put $\mathscr{E} = 2.0$ V and $r = 100\ \Omega$. Plot (a) the current and (b) the potential difference across R, as functions of R over the range 0 to 500 Ω. Make both plots on the same graph. (c) Make a third plot by multiplying together, for various values of R, the corresponding values on the two plotted curves. What is the physical significance of this third plot?

9E. In Fig. 28-24, circuit section AB absorbs energy at a rate of 50 W when a current $i = 1.0$ A passes through it in the indicated direction. (a) What is the potential difference between A and B? (b) Emf device X does not have internal resistance. What is its emf? (c) What is its *polarity* (the orientation of its positive and negative terminals)? ssm

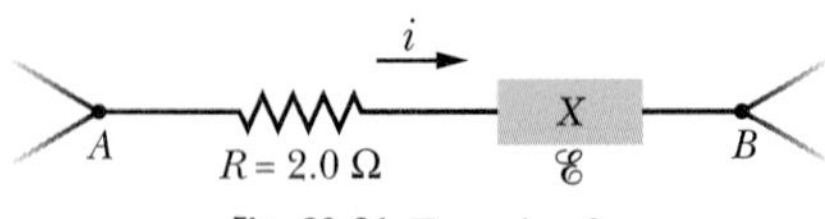

Fig. 28-24 Exercise 9.

10E. In Fig. 28-25, if the potential at point P is 100 V, what is the potential at point Q?

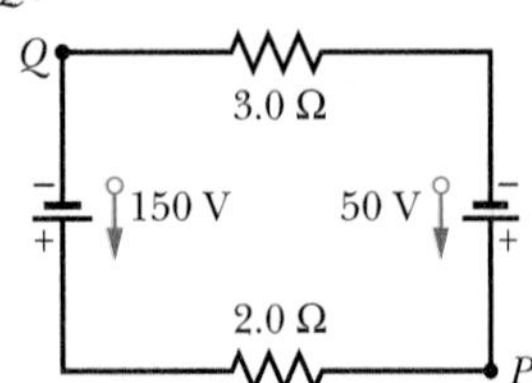

Fig. 28-25 Exercise 10.

11E. In Fig. 28-6a, calculate the potential difference between a and c by considering a path that contains R, r_1, and $\mathscr{E}_1$.

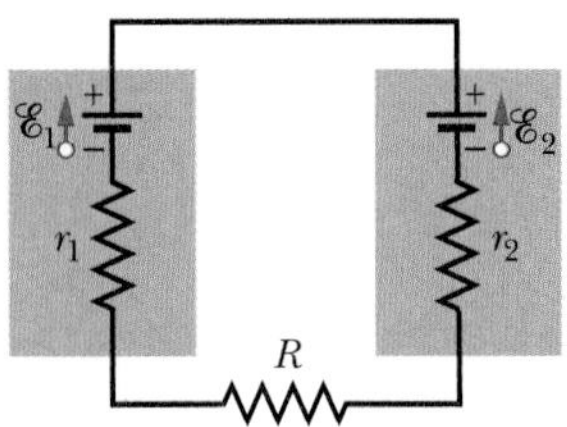

Fig. 28-26 Problem 12.

12P. (a) In Fig. 28-26, what value must R have if the current in the circuit is to be 1.0 mA? Take $\mathscr{E}_1 = 2.0$ V, $\mathscr{E}_2 = 3.0$ V, and $r_1 = r_2 = 3.0\ \Omega$. (b) What is the rate at which thermal energy appears in R?

13P. The current in a single-loop circuit with one resistance R is 5.0 A. When an additional resistance of 2.0 Ω is inserted in series with R, the current drops to 4.0 A. What is R? ilw

14P. The starting motor of an automobile is turning too slowly, and the mechanic has to decide whether to replace the motor, the cable, or the battery. The manufacturer's manual says that the 12 V battery should have no more than 0.020 Ω internal resistance, the motor no more than 0.200 Ω resistance, and the cable no more than 0.040 Ω resistance. The mechanic turns on the motor and measures 11.4 V across the battery, 3.0 V across the cable, and a current of 50 A. Which part is defective?

15P. Two batteries having the same emf $\mathscr{E}$ but different internal resistances r_1 and r_2 ($r_1 > r_2$) are connected in series to an external resistance R. (a) Find the value of R that makes the potential difference zero between the terminals of one battery. (b) Which battery is it? ssm

16P. A solar cell generates a potential difference of 0.10 V when a 500 Ω resistor is connected across it, and a potential difference of 0.15 V when a 1000 Ω resistor is substituted. What are (a) the internal resistance and (b) the emf of the solar cell? (c) The area of the cell is 5.0 cm^2, and the rate per unit area at which it receives energy from light is 2.0 mW/cm^2. What is the efficiency of the cell for converting light energy to thermal energy in the 1000 Ω external resistor?

17P. (a) In Fig. 28-4a, show that the rate at which energy is dissipated in R as thermal energy is a maximum when $R = r$. (b) Show that this maximum power is $P = \mathscr{E}^2/4r$. ssm www

SEC. 28-6 Multiloop Circuits

18E. By using only two resistors—singly, in series, or in parallel—you are able to obtain resistances of 3.0, 4.0, 12, and 16 Ω. What are the two resistances?

19E. Four 18.0 Ω resistors are connected in parallel across a 25.0 V ideal battery. What is the current through the battery? ssm

20E. In Fig. 28-27, find the equivalent resistance between points D and E. (*Hint:* Imagine that a battery is connected between points D and E.)

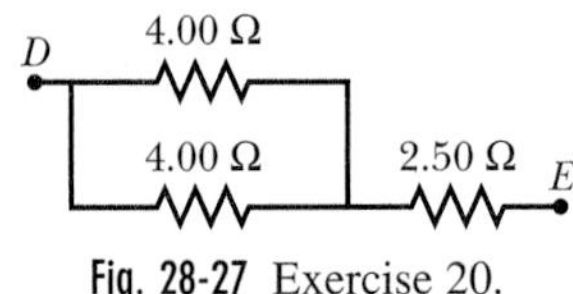

Fig. 28-27 Exercise 20.

21E. In Fig. 28-28 find the current in each resistor and the potential difference between points a and b. Put $\mathscr{E}_1 = 6.0$ V, $\mathscr{E}_2 = 5.0$ V, $\mathscr{E}_3 = 4.0$ V, $R_1 = 100\ \Omega$, and $R_2 = 50\ \Omega$. ssm

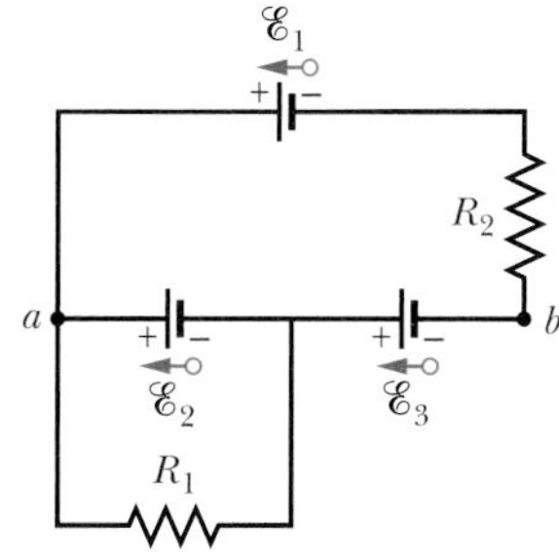

Fig. 28-28 Exercise 21.

22E. Figure 28-29 shows a circuit containing three switches, labeled S_1, S_2, and S_3. Find the current at a for all possible combinations of switch settings. Put $\mathscr{E} = 120$ V, $R_1 = 20.0\ \Omega$, and $R_2 = 10.0\ \Omega$. Assume that the battery has no resistance.

23E. Two lightbulbs, one of resistance R_1 and the other of resistance R_2, where $R_1 > R_2$, are connected to a battery (a) in parallel and (b) in series. Which bulb is brighter (dissipates more energy) in each case? ssm

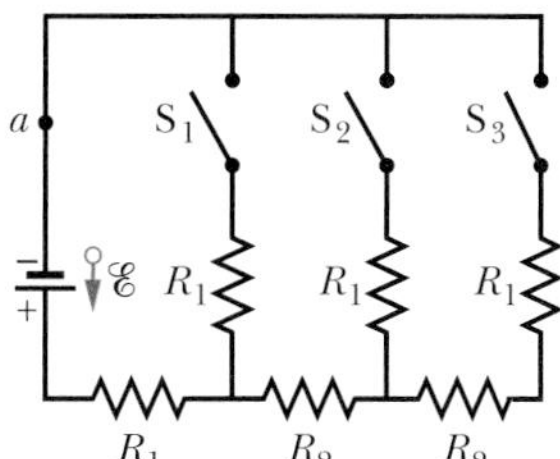

Fig. 28-29 Exercise 22.

24E. In Fig. 28-7, calculate the potential difference between points c and d by as many paths as possible. Assume that $\mathscr{E}_1 = 4.0$ V, $\mathscr{E}_2 = 1.0$ V, $R_1 = R_2 = 10\ \Omega$, and $R_3 = 5.0\ \Omega$.

25E. Nine copper wires of length l and diameter d are connected in parallel to form a single composite conductor of resistance R. What must be the diameter D of a single copper wire of length l if it is to have the same resistance? ssm

26P. In Fig. 28-30, find the equivalent resistance between points (a) F and H and (b) F and G. (*Hint:* For each pair of points, imagine that a battery is connected across the pair.)

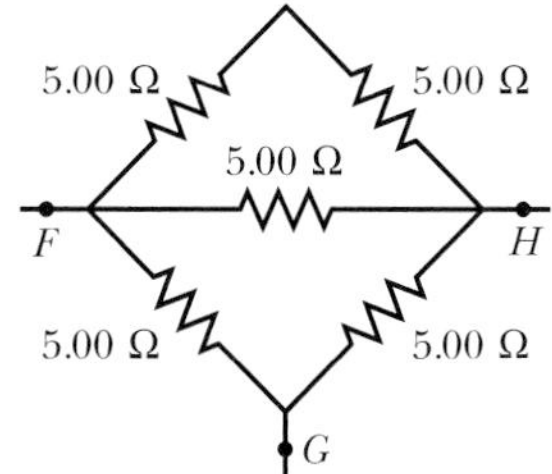

Fig. 28-30 Problem 26.

27P. You are given a number of 10 Ω resistors, each capable of dissipating only 1.0 W without being destroyed. What is the minimum number of such resistors that you need to combine in series or in parallel to make a 10 Ω resistance that is capable of dissipating at least 5.0 W? ssm

28P. (a) In Fig. 28-31, what is the equivalent resistance of the network shown? (b) What is the current in each resistor? Put $R_1 = 100\ \Omega$, $R_2 = R_3 = 50\ \Omega$, $R_4 = 75\ \Omega$, and $\mathscr{E} = 6.0$ V; assume the battery is ideal.

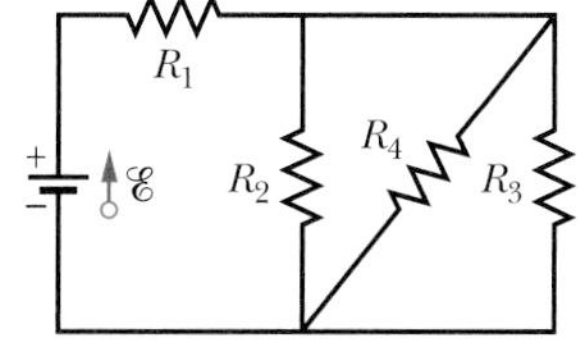

Fig. 28-31 Problem 28.

29P. Two batteries of emf $\mathscr{E}$ and internal resistance r are connected in parallel across a resistor R, as in Fig. 28-32a. (a) For what value of R is the rate of electrical energy dissipation by the resistor a maximum? (b) What is the maximum energy dissipation rate? ssm

30P. You are given two batteries of emf $\mathscr{E}$ and internal resistance r. They may be connected either in parallel (Fig. 28-32a) or in series (Fig. 28-32b) and are to be used to establish a current in a resistor R. (a) Derive expressions for the current in R for both arrangements. Which will yield the larger current (b) when $R > r$ and (c) when $R < r$?

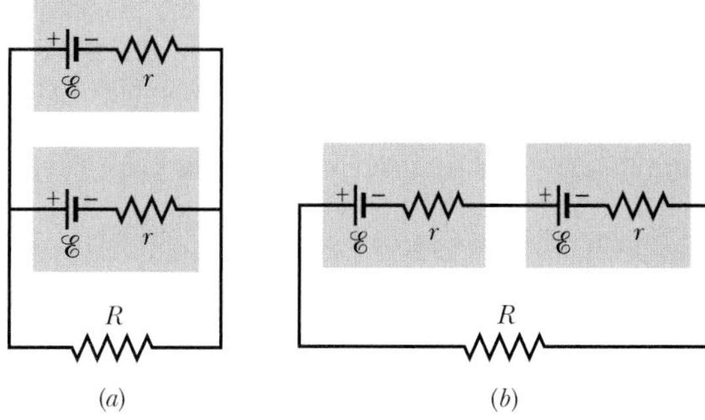

Fig. 28-32 Problems 29 and 30.

31P. In Fig. 28-33, $\mathscr{E}_1 = 3.00$ V, $\mathscr{E}_2 = 1.00$ V, $R_1 = 5.00\ \Omega$, $R_2 = 2.00\ \Omega$, $R_3 = 4.00\ \Omega$, and both batteries are ideal. What is the rate at which energy is dissipated in (a) R_1, (b) R_2, and (c) R_3? What is the power of (d) battery 1 and (e) battery 2? ssm www

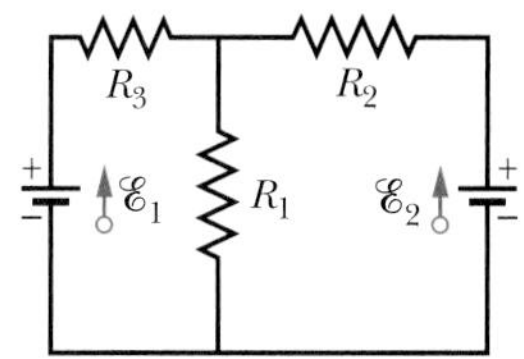

Fig. 28-33 Problem 31.

32P. In the circuit of Fig. 28-34, for what value of R will the ideal battery transfer energy to the resistors (a) at a rate of 60.0 W, (b) at the maximum possible rate, and (c) at the minimum possible rate? (d) What are those rates?

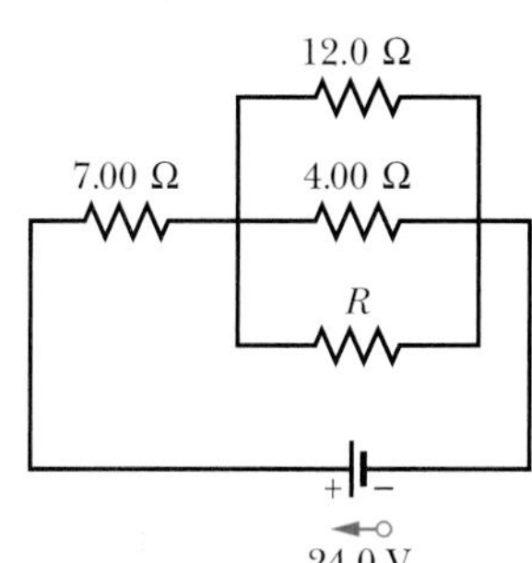

Fig. 28-34 Problem 32.

33P. (a) Calculate the current through each ideal battery in Fig. 28-35. Assume that $R_1 = 1.0\ \Omega$, $R_2 = 2.0\ \Omega$, $\mathscr{E}_1 = 2.0$ V, and $\mathscr{E}_2 = \mathscr{E}_3 = 4.0$ V. (b) Calculate $V_a - V_b$. ilw

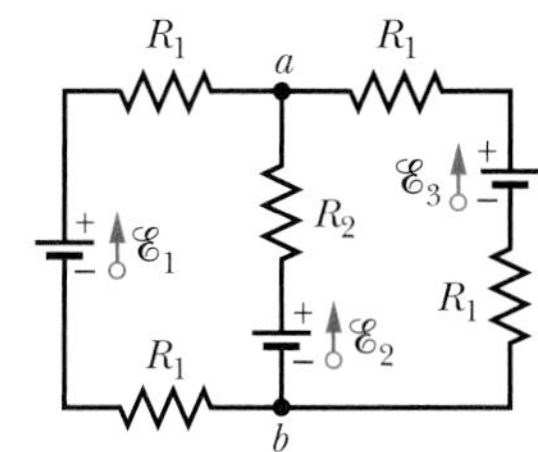

Fig. 28-35 Problem 33.

34P. In the circuit of Fig. 28-36, $\mathscr{E}$ has a constant value but R can be varied. Find the value of R that results in the maximum heating in that resistor. The battery is ideal.

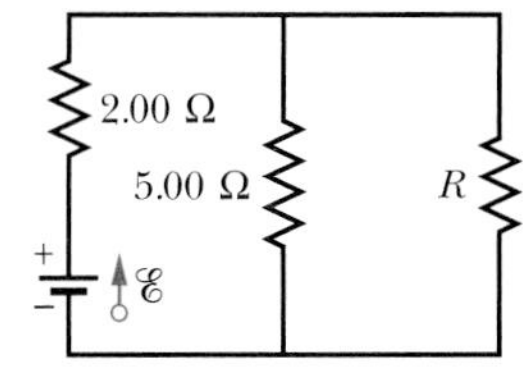

Fig. 28-36 Problem 34.

35P. A copper wire of radius $a = 0.250$ mm has an aluminum jacket of outer radius $b = 0.380$ mm. (a) There is a current $i = 2.00$ A in the composite wire. Using Table 27-1, calculate the current in each material. (b) If a potential difference $V = 12.0$ V between the ends maintains the current, what is the length of the composite wire? ssm

SEC. 28-7 The Ammeter and the Voltmeter

36E. A simple ohmmeter is made by connecting a 1.50 V flashlight battery in series with a resistance R and an ammeter that reads from 0 to 1.00 mA, as shown in Fig. 28-37. Resistance R is adjusted so that when the clip leads are shorted together, the meter deflects to its full-scale value of 1.00 mA. What external resistance across the leads results in a deflection of (a) 10%, (b) 50%, and (c) 90% of full scale? (d) If the ammeter has a resistance of 20.0 Ω and the internal resistance of the battery is negligible, what is the value of R?

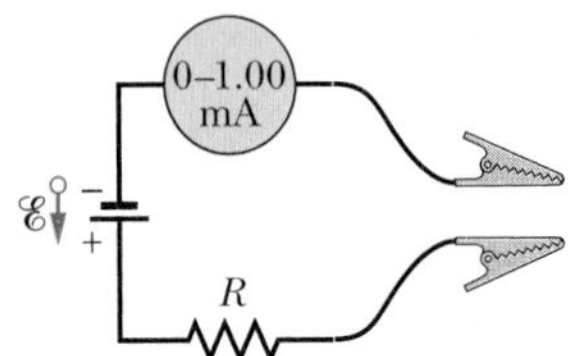

Fig. 28-37 Exercise 36.

37P. (a) In Fig. 28-38, determine what the ammeter will read, assuming $\mathscr{E}$ = 5.0 V (for the ideal battery), R_1 = 2.0 Ω, R_2 = 4.0 Ω, and R_3 = 6.0 Ω. (b) The ammeter and the source of emf are now physically interchanged. Show that the ammeter reading remains unchanged. ilw

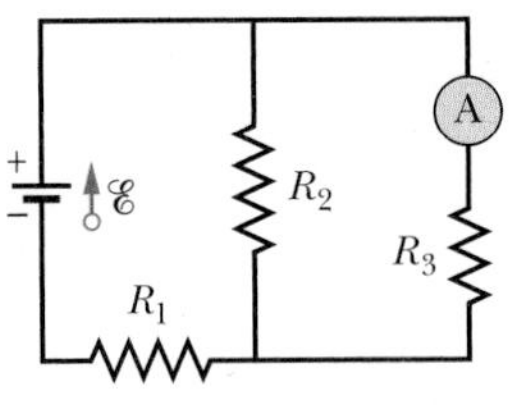

Fig. 28-38 Problem 37.

38P. When the lights of an automobile are switched on, an ammeter in series with them reads 10 A and a voltmeter connected across them reads 12 V. See Fig. 28-39. When the electric starting motor is turned on, the ammeter reading drops to 8.0 A and the lights dim somewhat. If the internal resistance of the battery is 0.050 Ω and that of the ammeter is negligible, what are (a) the emf of the battery and (b) the current through the starting motor when the lights are on?

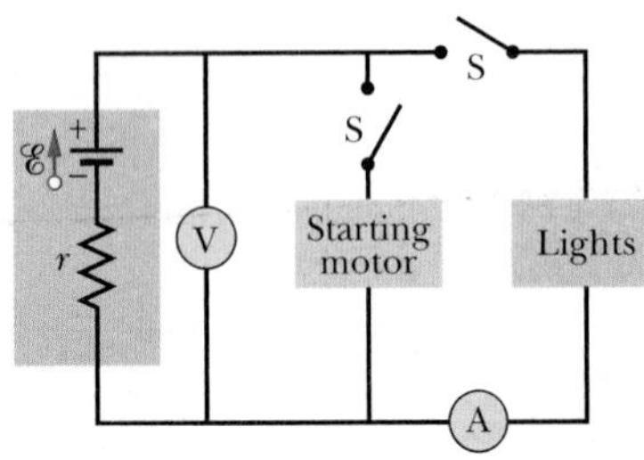

Fig. 28-39 Problem 38.

39P. In Fig. 28-12, assume that $\mathscr{E}$ = 3.0 V, r = 100 Ω, R_1 = 250 Ω, and R_2 = 300 Ω. If the voltmeter resistance R_V is 5.0 kΩ, what percent error does it introduce into the measurement of the potential difference across R_1? Ignore the presence of the ammeter. ssm

40P. A voltmeter (of resistance R_V) and an ammeter (of resistance R_A) are connected to measure a resistance R, as in Fig. 28-40*a*. The resistance is given by $R = V/i$, where V is the voltmeter reading and i is the current in the resistance R. Some of the current i' registered by the ammeter goes through the voltmeter, so that the ratio of the meter readings ($= V/i'$) gives only an *apparent* resistance reading R'. Show that R and R' are related by

$$\frac{1}{R} = \frac{1}{R'} - \frac{1}{R_V}.$$

Note that as $R_V \to \infty$, $R' \to R$.

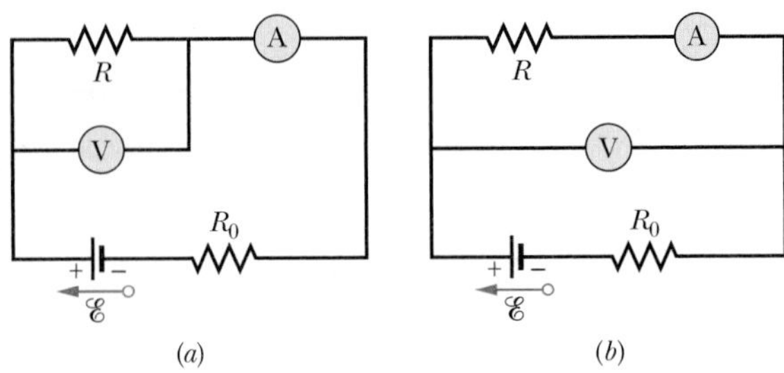

Fig. 28-40 Problems 40 to 42.

41P. (See Problem 40.) If an ammeter and a voltmeter are used to measure resistance, they may also be connected as in Fig. 28-40*b*. Again the ratio of the meter readings gives only an apparent resistance R'. Show that now R' is related to R by

$$R = R' - R_A,$$

in which R_A is the ammeter resistance. Note that as $R_A \to 0$, $R' \to R$.

42P. (See Problems 40 and 41.) In Fig. 28-40, the ammeter and voltmeter resistances are 3.00 and 300 Ω, respectively. Take $\mathscr{E}$ = 12.0 V for the ideal battery and R_0 = 100 Ω. If R = 85.0 Ω, (a) what will the meters read for the two different connections (Figs. 28-40*a* and *b*)? (b) What apparent resistance R' will be computed in each case?

43P. In Fig. 28-41, R_s is to be adjusted in value by moving the sliding contact across it until points a and b are brought to the same potential. (One tests for this condition by momentarily connecting a sensitive ammeter between a and b; if these points are at the same potential, the ammeter will not deflect.) Show that when this adjustment is made, the following relation holds:

$$R_x = R_s\left(\frac{R_2}{R_1}\right).$$

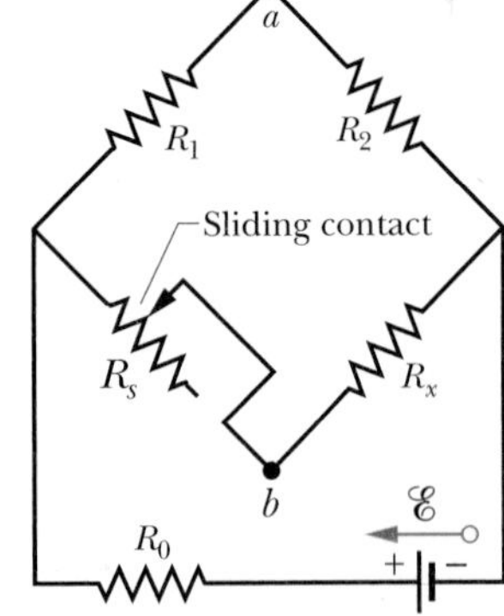

Fig. 28-41 Problem 43.

An unknown resistance (R_x) can be measured in terms of a standard (R_s) using this device, which is called a Wheatstone bridge. ssm

SEC. 28-8 *RC* Circuits

44E. A capacitor with initial charge q_0 is discharged through a resistor. In terms of the time constant τ, how long is required for the capacitor to lose (a) the first one-third of its charge and (b) two-thirds of its charge?

45E. How many time constants must elapse for an initially uncharged capacitor in an *RC* series circuit to be charged to 99.0% of its equilibrium charge? ssm

46E. In an *RC* series circuit, $\mathscr{E}$ = 12.0 V, R = 1.40 MΩ, and C = 1.80 μF. (a) Calculate the time constant. (b) Find the maximum charge that will appear on the capacitor during charging. (c) How long does it take for the charge to build up to 16.0 μC?

47E. A 15.0 kΩ resistor and a capacitor are connected in series and then a 12.0 V potential difference is suddenly applied across them. The potential difference across the capacitor rises to 5.00 V in 1.30 μs. (a) Calculate the time constant of the circuit. (b) Find the capacitance of the capacitor. ilw

48P. The potential difference between the plates of a leaky (meaning that charge leaks from one plate to the other) 2.0 μF capacitor drops to one-fourth its initial value in 2.0 s. What is the equivalent resistance between the capacitor plates?

49P. A 3.00 MΩ resistor and a 1.00 μF capacitor are connected in series with an ideal battery of emf $\mathscr{E}$ = 4.00 V. At 1.00 s after the connection is made, what are the rates at which (a) the charge of the capacitor is increasing, (b) energy is being stored in the capacitor, (c) thermal energy is appearing in the resistor, and (d) energy is being delivered by the battery? ssm

50P. An initially uncharged capacitor C is fully charged by a device of constant emf $\mathscr{E}$ connected in series with a resistor R. (a) Show that the final energy stored in the capacitor is half the energy supplied by the emf device. (b) By direct integration of i^2R over the charging time, show that the thermal energy dissipated by the resistor is also half the energy supplied by the emf device.

51P. A capacitor with an initial potential difference of 100 V is discharged through a resistor when a switch between them is closed at $t = 0$. At $t = 10.0$ s, the potential difference across the capacitor is 1.00 V. (a) What is the time constant of the circuit? (b) What is the potential difference across the capacitor at $t = 17.0$ s? ssm

52P. Figure 28-42 shows the circuit of a flashing lamp, like those attached to barrels at highway construction sites. The fluorescent lamp L (of negligible capacitance) is connected in parallel across the capacitor C of an RC circuit. There is a current through the lamp only when the potential difference across it reaches the breakdown voltage V_L; in this event, the capacitor discharges completely through the lamp and the lamp flashes briefly. Suppose that two flashes per second are needed. For a lamp with breakdown voltage $V_L = 72.0$ V, wired to a 95.0 V ideal battery and a 0.150 μF capacitor, what should be the resistance R?

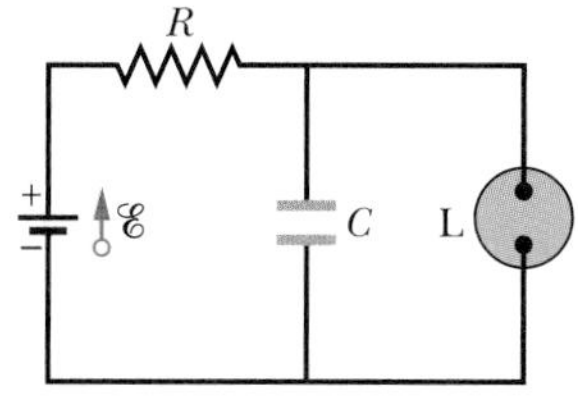

Fig. 28-42 Problem 52.

53P. A 1.0 μF capacitor with an initial stored energy of 0.50 J is discharged through a 1.0 MΩ resistor. (a) What is the initial charge on the capacitor? (b) What is the current through the resistor when the discharge starts? (c) Determine V_C, the potential difference across the capacitor, and V_R, the potential difference across the resistor, as functions of time. (d) Express the production rate of thermal energy in the resistor as a function of time. ssm www

54P. A controller on an electronic arcade game consists of a variable resistor connected across the plates of a 0.220 μF capacitor. The capacitor is charged to 5.00 V, then discharged through the resistor. The time for the potential difference across the plates to decrease to 0.800 V is measured by a clock inside the game. If the range of discharge times that can be handled effectively is from 10.0 μs to 6.00 ms, what should be the resistance range of the resistor?

55P*. In the circuit of Fig. 28-43, $\mathscr{E}$ = 1.2 kV, C = 6.5 μF, $R_1 = R_2 = R_3 = 0.73$ MΩ. With C completely uncharged, switch S is suddenly closed (at $t = 0$). (a) Determine the current through each resistor at $t = 0$ and as $t \to \infty$. (b) Draw qualitatively a graph of the potential difference V_2 across R_2 from $t = 0$ to $t = \infty$. (c) What are the numerical values of V_2 at $t = 0$ and as $t \to \infty$? (d) What is the physical meaning of "$t \to \infty$" in this case? ssm

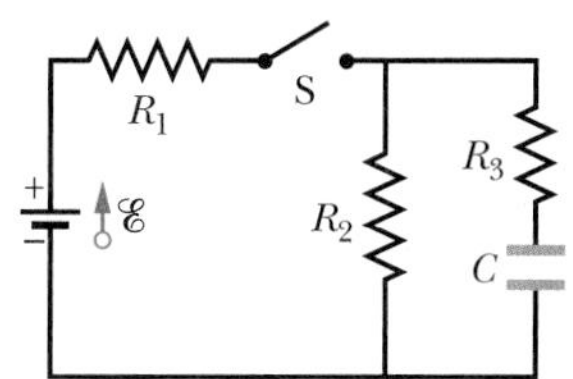

Fig. 28-43 Problem 55.

Additional Problem

56. *Heart attack or electrocution?* This story begins with Problem 45 in Chapter 27. Figure 28-44 shows the electrical pathway of the current up through one foot of the victim, across the torso (including the heart), and down through the other foot. (a) From the given data, find the potential difference between the man's feet, assuming that one foot was 0.50 m closer to the leaky rod than the other. (b) Assume that the resistance of a foot on wet soil is the typical value of 300 Ω and the resistance of the torso interior is the commonly accepted value of 1000 Ω. What then was the current across the victim's torso? (c) The human heart can be put into fibrillation by a current of 0.10 A to 1.0 A through the torso. Was the victim's fibrillation due to current leakage from the rod?

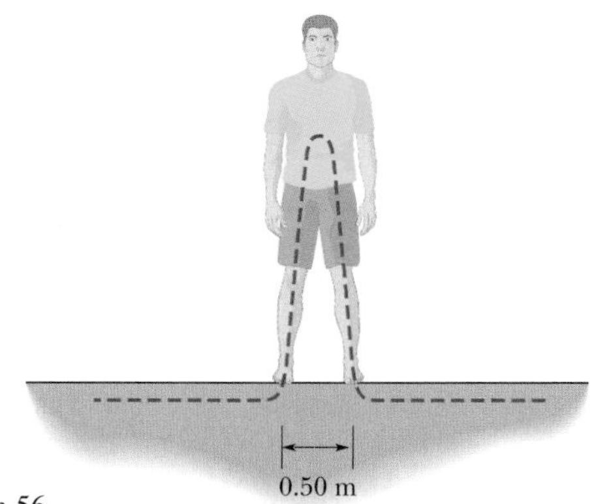

Fig. 28-44 Problem 56.

NEW PROBLEMS

N1. What are the sizes and directions of the currents through resistors (a) R_2 and (b) R_3 in Fig. 28N-1, where each of the three resistances is 4.0 Ω?

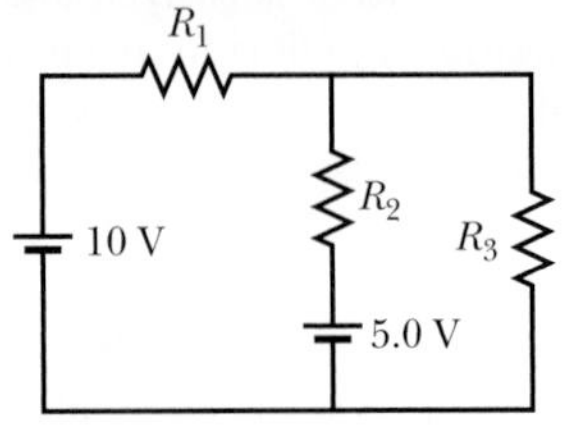

Fig. 28N-1 Problem N1.

N2. In Fig. 28N-2*a*, resistor 3 is a variable resistor and the battery is an ideal 12 V battery. Figure 28N-2*b* gives the current i through the battery as a function of R_3. The curve has an asymptote of 2.0 mA as $R_3 \to \infty$. What are (a) resistance R_1 and (b) resistance R_2?

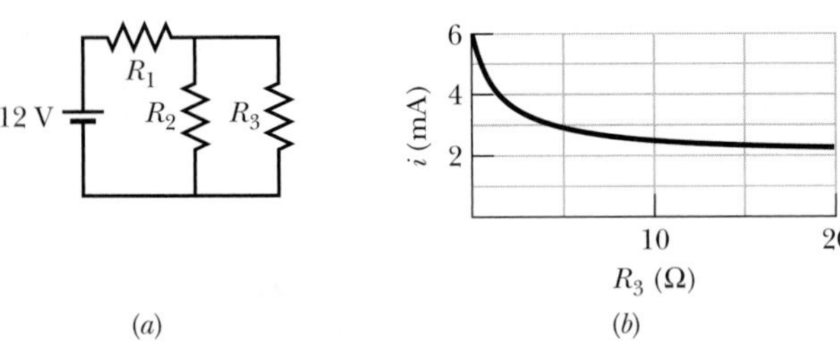

Fig. 28N-2 Problem N2.

N3. Power is supplied by a device of emf $\mathscr{E}$ to a transmission line with resistance R. Find the ratio of the power dissipated in the line for $\mathscr{E} = 110\,000$ V to that dissipated for $\mathscr{E} = 110$ V, assuming the power supplied is the same for the two cases.

N4. The resistances in Figs. 28N-3*a* and *b* are all 6.0 Ω, and the batteries are ideal 12 V batteries. (a) When switch S in Fig. 28N-3*a* is closed, what is the change in the electric potential V_1 across resistor 1, or does V_1 remain the same? (b) When switch S in Fig. 28N-3*b* is closed, what is the change in the electric potential V_1 across resistor 1, or does V_1 remain the same?

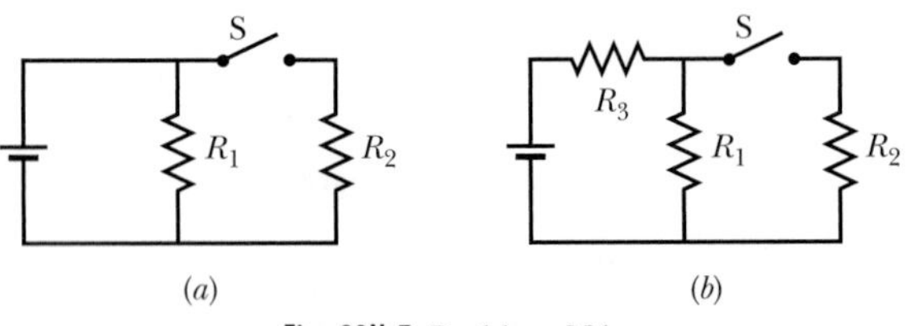

Fig. 28N-3 Problem N4.

N5. (a) How much work does an ideal battery with a 12.0 V emf do on an electron that passes through the battery from the positive to the negative terminal? (b) If 3.4×10^{18} electrons pass through each second, what is the power of the battery?

N6. Figure 28N-4 shows a 6.00 Ω resistor connected to an ideal 12.0 V battery by means of two copper wires. The wires each have length 20.0 cm and radius 1.00 mm. In such circuits in Chapter 28, we generally neglect the potential differences along wires and the transfer of energy to thermal energy in them. Check the validity of this neglect for the circuit of Fig. 28N-4: What are the potential differences across (a) the resistor and (b) each of the two sections of wire? At what rate is energy lost to thermal energy in (c) the resistor and (d) each of the two sections of wire?

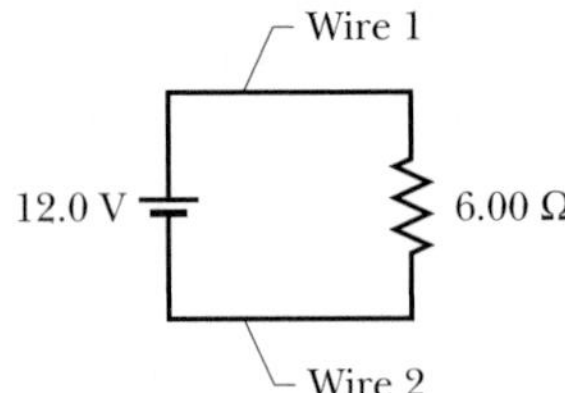

Fig. 28N-4 Problem N6.

N7. (a) In Fig. 28N-5, what is the equivalent resistance of the network shown? (b) What is the current in each resistor? Put $R_1 = 100$ Ω, $R_2 = R_3 = 50$ Ω, $R_4 = 75$ Ω, and $\mathscr{E} = 6.0$ V; assume the battery is ideal.

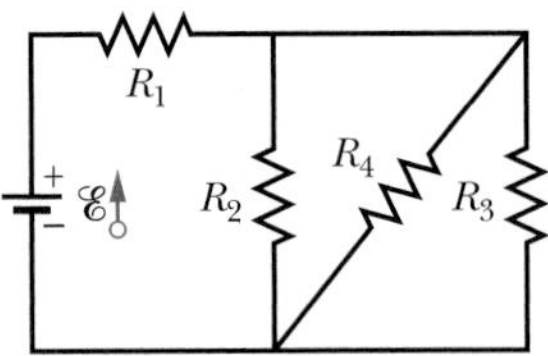

Fig. 28N-5 Problem N7.

N8. Switch S in Fig. 28N-6 is closed at time $t = 0$, to begin charging an initially uncharged capacitor of capacitance $C = 15.0$ μF through a resistor of resistance $R = 20.0$ Ω. At what time is the electric potential across the capacitor equal to that across the resistor?

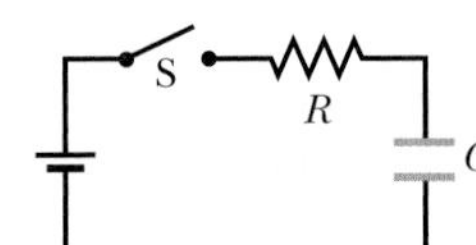

Fig. 28N-6 Problems N8 and N10.

N9. Figure 28N-7 shows a portion of a circuit. The rest of the circuit draws current i at the connections A and B, as indicated. Take $\mathscr{E}_1 = 10$ V, $\mathscr{E}_2 = 15$ V, $R_1 = R_2 = 5.0$ Ω, $R_3 = R_4 = 8.0$ Ω, and $R_5 = 12$ Ω. (a) For each of four values of i—0, 4.0, 8.0, and 12 A—find the current through each ideal battery and state whether the battery is charging or discharging. Also find the potential difference V_{AB}. (b) The portion of the circuit not shown consists of an emf and a resistor in series. What are their values?

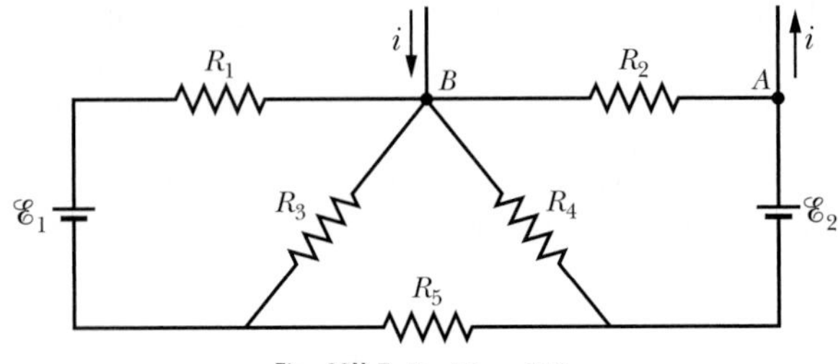

Fig. 28N-7 Problem N9.

N10. Figure 28N-6 shows an ideal battery of emf $\mathscr{E} = 12$ V, a resistor of resistance $R = 4.0$ Ω, and an uncharged capacitor of capacitance $C = 4.0$ μF. After switch S is closed, what is the current through the resistor when the charge on the capacitor is 8.0 μC?

N11. The following table gives the electric potential difference V_T across the terminals of a battery as a function of current i being drawn from the battery. (a) Write an equation that represents the relationship between the terminal potential difference V_T and the current i. Enter the data into your graphing calculator and perform a linear regression fit of V_T versus i. From the parameters of the fit, find (b) the battery's emf and (c) its internal resistance.

i (A):	50	75	100	125	150	175	200
V_T (V):	10.7	9.0	7.7	6.0	4.8	3.0	1.7

N12. In Fig. 28N-8*a*, both batteries have emf $\mathscr{E} = 1.20$ V and the external resistance R is a variable resistor. Figure 28N-8*b* gives the electric potentials V between the terminals of each battery as functions of R: Curve 1 corresponds to battery 1 and curve 2 corresponds to battery 2. What are the internal resistances of (a) battery 1 and (b) battery 2?

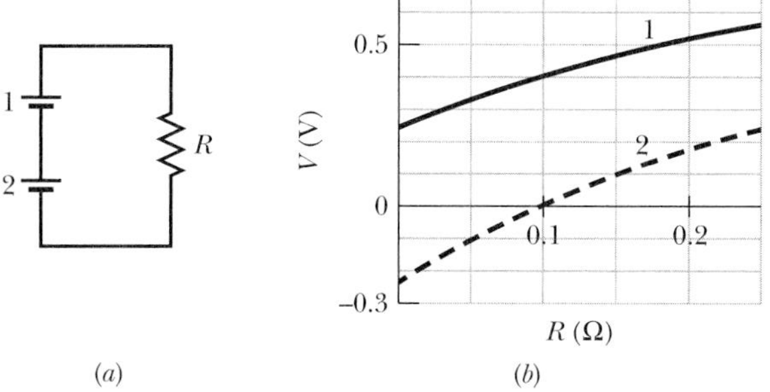

Fig. 28N-8 Problem N12.

N13. What are the sizes and directions of (a) current i_1 and (b) current i_2 in Fig. 28N-9, where each resistance is 2.00 Ω? (Can you answer this making only mental calculations?) (c) At what rate is energy being transferred in the 5.00 V battery at the left, and is the energy being supplied or absorbed by the battery?

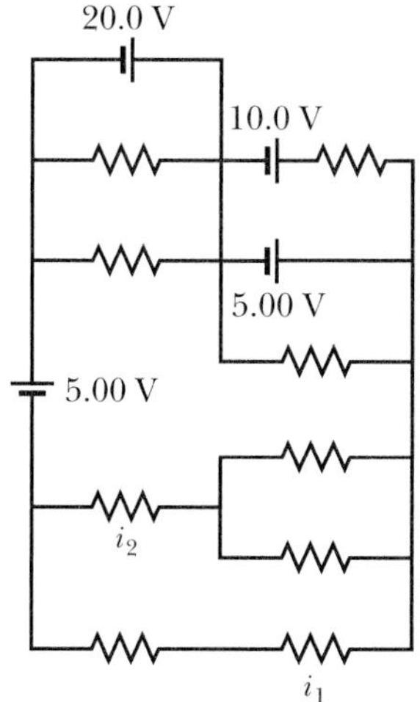

Fig. 28N-9 Problem N13.

N14. In Fig. 28N-10, a resistor and an arrangement of n resistors in parallel are connected in series with an ideal battery. All the resistors have the same resistance. If one more identical resistor were added in parallel to the n resistors already in parallel, the current through the battery would change by 1.25%. What is the value of n?

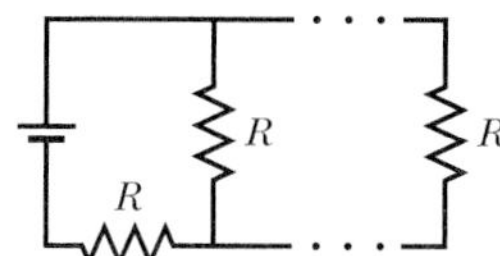

Fig. 28N-10 Problem N14.

N15. In Fig. 28N-11, $R = 10$ Ω. What is the equivalent resistance between points A and B? (*Hint:* This circuit section might look simpler if you first assume that points A and B are connected to a battery.)

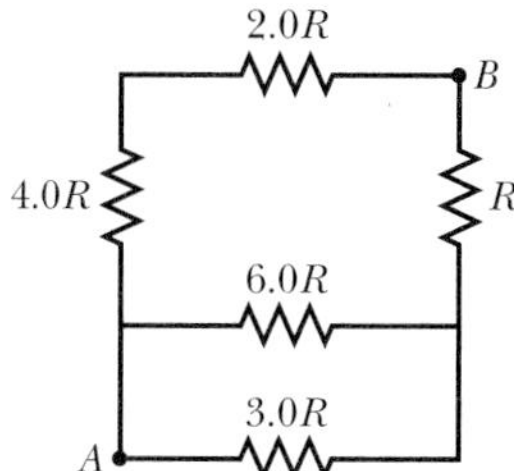

Fig. 28N-11 Problem N15.

N16. Figure 28N-12 shows a circuit of four resistors that are connected to a larger circuit. The graph below the circuit shows the electric potential $V(x)$ as a function of position x along the lower branch of the circuit, through resistor 4. Similarly, the graph above the circuit shows the electric potential $V(x)$ as a function of position x along the upper branch of the circuit, through resistors 1, 2, and 3. Resistor 3 has a resistance of 200 Ω. What are the resistances of (a) resistor 1 and (b) resistor 2?

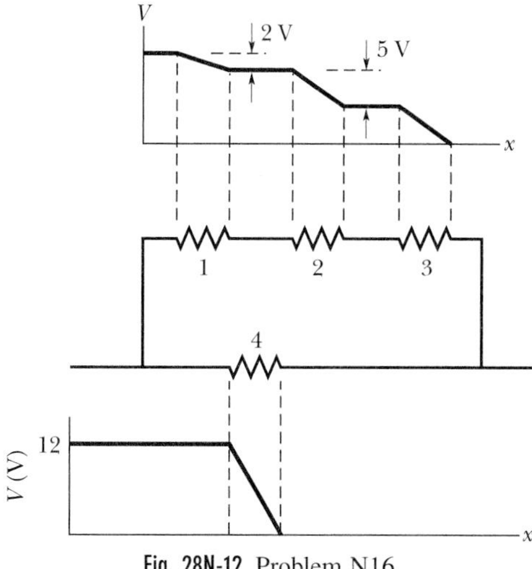

Fig. 28N-12 Problem N16.

N17. In Fig. 28N-13, what are currents (a) i_2, (b) i_4, (c) i_1, (d) i_3, and (e) i_5?

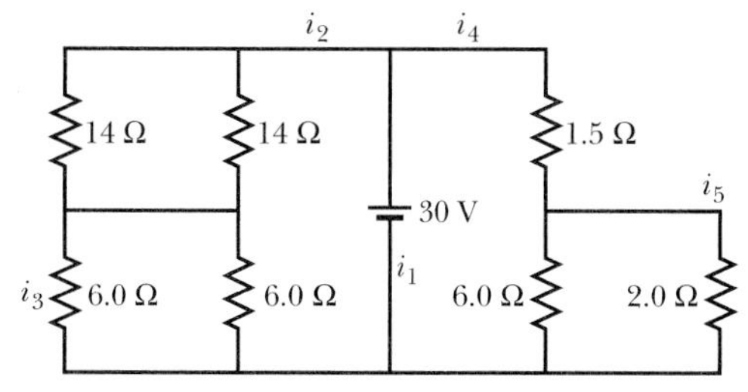

Fig. 28N-13 Problem N17.

N18. Both batteries in Fig. 28N-14*a* are ideal. Emf $\mathscr{E}_1$ of battery 1 has a fixed value but emf $\mathscr{E}_2$ of battery 2 can be varied between 1.0 V and 10 V. The plots in Fig. 28N-14*b* give the currents through the two batteries as a function of $\mathscr{E}_2$. You must decide which plot corresponds to which battery, but for both plots, a negative current occurs when the direction of the current through the battery is opposite the direction of that battery's emf. What are (a) emf $\mathscr{E}_1$, (b) resistance R_1, and (c) resistance R_2?

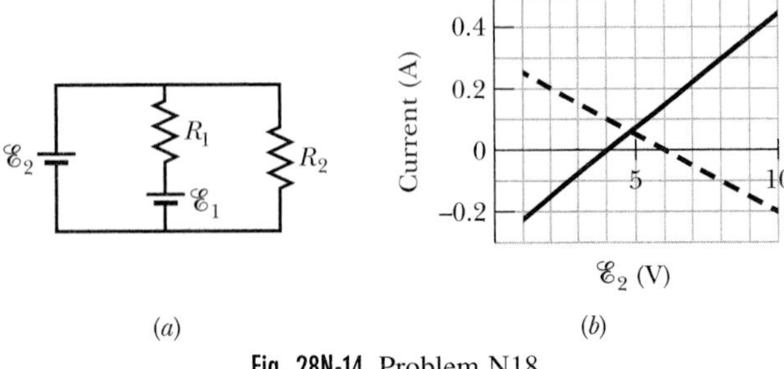

Fig. 28N-14 Problem N18.

N19. In Fig. 28N-15, where each resistance is 4.00 Ω, what are the sizes and directions of currents (a) i_1 and (b) i_2? At what rates is energy being transferred at (c) the 4.00 V battery and (d) the 12.0 V battery, and for each, is the battery supplying or absorbing energy?

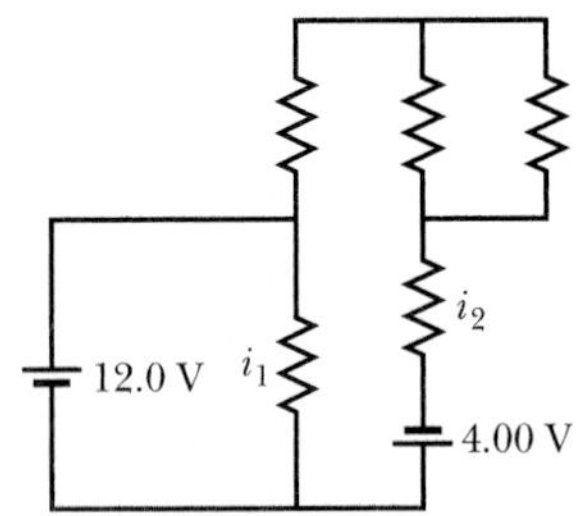

Fig. 28N-15 Problem N19.

N20. (a) What are the size and direction of current i_1 in Fig. 28N-16, where each resistance is 2.0 Ω? What are the powers of (b) the 20 V battery, (c) the 10 V battery, and (d) the 5.0 V battery, and for each, is energy being supplied or absorbed?

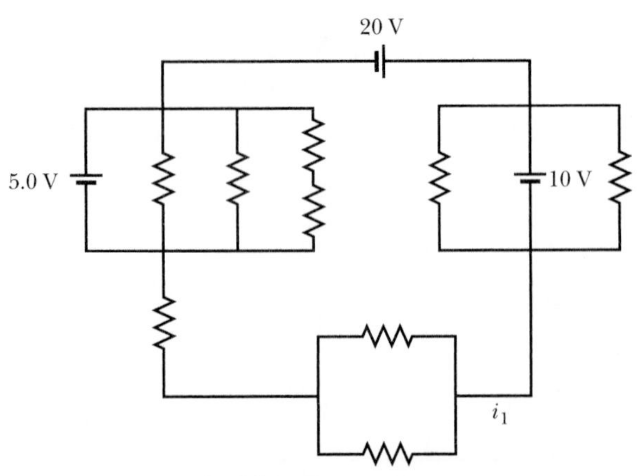

Fig. 28N-16 Problem N20.

N21. (a) What are the size and direction of current i_1 in Fig. 28N-17? (b) How much energy is dissipated by all four resistors in 1.0 min?

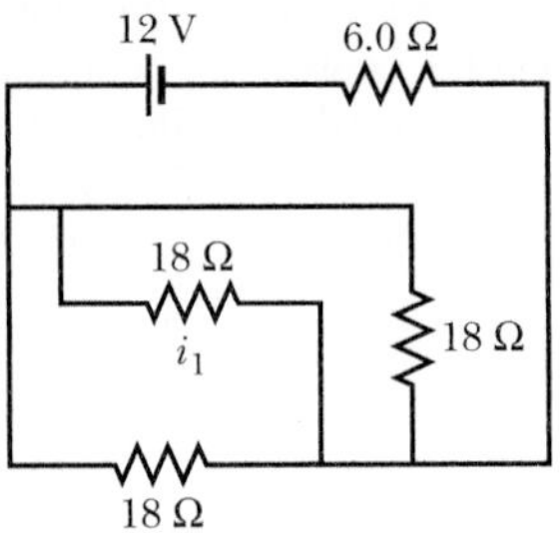

Fig. 28N-17 Problem N21.

N22. Plot 1 in Fig. 28N-18*a* gives the current *i* that can appear in resistor 1 versus the electric potential *V* set up across it. Plots 2 and 3 are similar plots for resistors 2 and 3, respectively. Figure 28N-18*b* shows a circuit with those three resistors and a 6.0 V battery. What is the current in resistor 2 in that circuit?

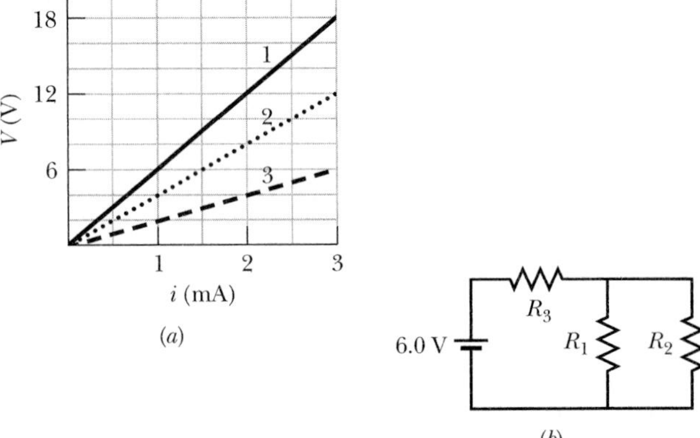

Fig. 28N-18 Problem N22.

N23. Figure 28N-19 shows a section of a circuit. The electric potential difference between points *A* and *B* that connect the section to the rest of the circuit is $V_A - V_B = 78$ V, and the current through the 6.0 Ω resistor is 6.0 A. Is the device represented by "Box" absorbing or providing energy to the circuit and at what rate?

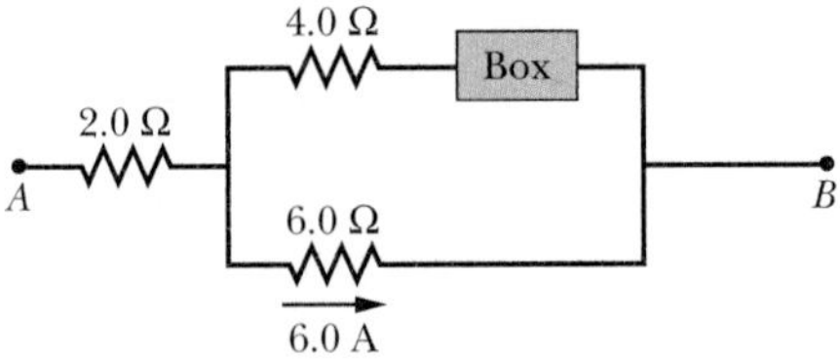

Fig. 28N-19 Problem N23.

N24. Consider the circuit in Fig. 28N-20. (a) Apply the junction rule to junctions *d* and *a* and the loop rule to the three loops to produce five simultaneous, linearly independent equations. (b) Represent the five linear equations by the matrix equation $[A][B] = [C]$, where

$$[B] = \begin{bmatrix} i_1 \\ i_2 \\ i_3 \\ i_4 \\ i_5 \end{bmatrix}.$$

What are the matrices [A] and [C]? (c) Have the calculator perform $[A]^{-1}[C]$ to find the values of i_1, i_2, i_3, i_4, and i_5.

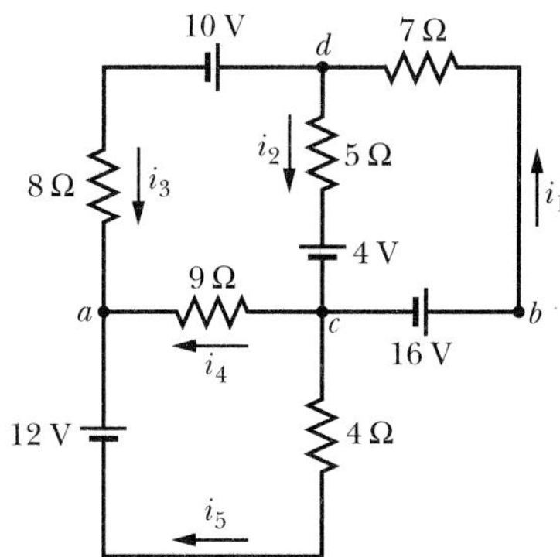

Fig. 28N-20 Problems N24 and N27.

N25. When steady-state conditions are reached in the circuit of Fig. 28N-21, what is the total energy stored in the two capacitors?

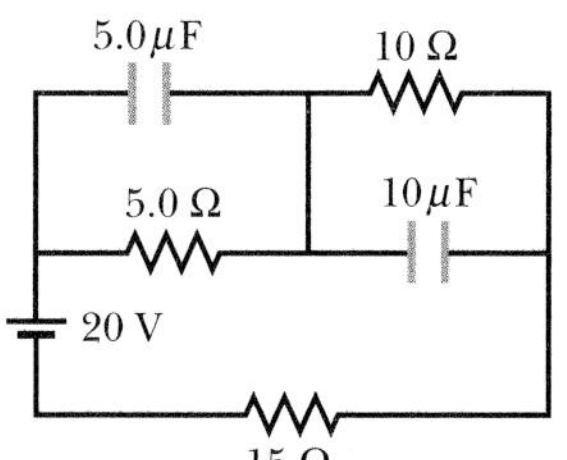

Fig. 28N-21 Problem N25.

N26. Figure 28N-22 shows two circuits with a charged capacitor that is to be discharged through a resistor when a switch is closed. In Fig. 28N-22*a*, resistance $R_1 = 20.0\ \Omega$ and capacitance $C_1 = 5.00\ \mu\text{F}$. In Fig. 28N-22*b*, resistance $R_2 = 10.0\ \Omega$ and capacitance $C_2 = 8.00\ \mu\text{F}$. The ratio of the initial charges on the two capacitors is $q_{02}/q_{01} = 1.50$. At time $t = 0$, both switches in the two circuits are closed. At what time t do the two capacitors have the same charge?

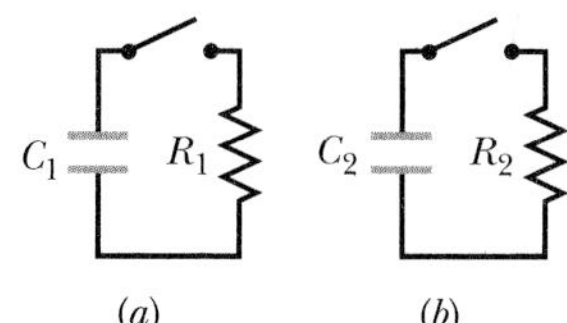

Fig. 28N-22 Problem N26.

N27. For the same situation as in Problem N24 and having already solved for the five unknown currents, do the following. (a) Find the electric potential difference across the 9 Ω resistor. (b) Find the rate at which work is being done on the 7 Ω resistor. (c) Find the rate at which the 12 V battery is doing work on the circuit. (d) Find the rate at which the 4 V battery is doing work on the circuit. (e) Of the points in the circuit labeled *a* and *c*, which is at the higher electric potential?

N28. A capacitor with capacitance C_0, after having been connected to a battery with emf $\mathscr{E}_0$ for a long time, is discharged through a 200 000 Ω resistor at time $t = 0$. The potential difference across the capacitor is then measured as a function of time for a brief time interval; the results are recorded below. (a) Write an equation that describes the potential difference across the capacitor as a function of time. Enter the data into your calculator and have the calculator perform a linear regression fit of $\ln V_C$ versus t. From the parameters of the fit, determine (b) the emf $\mathscr{E}_0$ of the battery and (c) the time constant τ for the circuit. (d) Finally, determine the value of capacitance C_0.

V_C (V):	9.9	7.2	5.7	4.4	3.4	2.7	2.0
t (s):	0.2	0.4	0.6	0.8	1.0	1.2	1.4

N29. In Fig. 28N-23, the symbol at the left indicates that the circuit is grounded there, which means that the potential V is defined to be zero there. What are the potentials (a) V_1, (b) V_2, and (c) V_3 at the points indicated? (*Hint:* The whole circuit need not be solved. Only two independent loop equations need to be solved.)

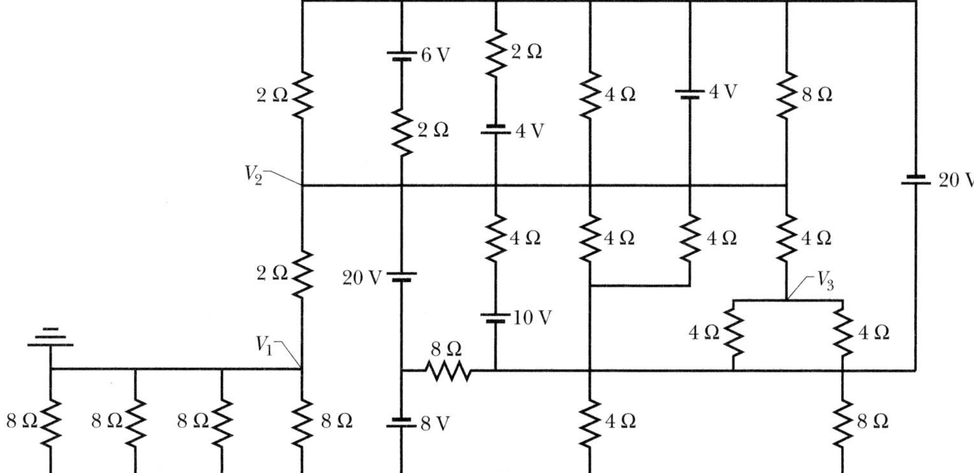

Fig. 28N-23 Problem N29.

29 Magnetic Fields

If you are outside on a dark night in the middle to high latitudes, you might be able to see an aurora, a ghostly "curtain" of light that hangs down from the sky. This curtain is not only local; it may be several hundred kilometers high and several thousand kilometers long, stretching around Earth in an arc. However, it is less than 1 km thick.

What produces this huge display, and what makes it so thin?

The answer is in this chapter.

Fig. 29-1 Using an electromagnet to collect and transport scrap metal at a steel mill.

29-1 The Magnetic Field

We have discussed how a charged plastic rod produces a vector field—the electric field $\vec{E}$—at all points in the space around it. Similarly, a magnet produces a vector field—the **magnetic field $\vec{B}$**—at all points in the space around it. You get a hint of that magnetic field whenever you attach a note to a refrigerator door with a small magnet, or accidentally erase a computer disk by bringing it near a magnet. The magnet acts on the door or disk *by means of* its magnetic field.

In a familiar type of magnet, a wire coil is wound around an iron core and a current is sent through the coil; the strength of the magnetic field is determined by the size of the current. In industry, such **electromagnets** are used for sorting scrap metal (Fig. 29-1) among many other things. You are probably more familiar with **permanent magnets**—magnets, like the refrigerator-door type, that do not need current to have a magnetic field.

In Chapter 23 we saw that an *electric charge* sets up an electric field that can then affect other electric charges. Here, we might reasonably expect that a *magnetic charge* sets up a magnetic field that can then affect other magnetic charges. Although such magnetic charges, called *magnetic monopoles,* are predicted by certain theories, their existence has not been confirmed.

How then are magnetic fields set up? There are two ways. (1) Moving electrically charged particles, such as a current in a wire, create magnetic fields. (2) Elementary particles such as electrons have an *intrinsic* magnetic field around them; that is, this field is a basic characteristic of the particles, just as are their mass and electric charge (or lack of charge). As we shall discuss in Chapter 32, the magnetic fields of the electrons in certain materials add together to give a net magnetic field around the material. This is true for the material in permanent magnets (which is good, because they can then hold notes to a refrigerator door). In other materials, the magnetic fields of all the electrons cancel out, giving no net magnetic field surrounding the material. This is true for the material in your body (which is also good, because otherwise you might be slammed up against a refrigerator door every time you passed one).

Experimentally we find that when a charged particle (either alone or as part of a current) moves through a magnetic field, a force due to the field can act on the particle. In this chapter we focus on the relation between the magnetic field and this force.

29-2 The Definition of $\vec{B}$

We determined the electric field $\vec{E}$ at a point by putting a test particle of charge q at rest at that point and measuring the electric force $\vec{F}_E$ acting on the particle. We then defined $\vec{E}$ as

$$\vec{E} = \frac{\vec{F}_E}{q}. \tag{29-1}$$

If a magnetic monopole were available, we could define $\vec{B}$ in a similar way. Because such particles have not been found, we must define $\vec{B}$ in another way, in terms of the magnetic force $\vec{F}_B$ exerted on a moving electrically charged test particle.

In principle, we do this by firing a charged particle through the point at which $\vec{B}$ is to be defined, using various directions and speeds for the particle and determining the force $\vec{F}_B$ that acts on the particle at that point. After many such trials we would find that when the particle's velocity $\vec{v}$ is along a particular axis through the point, force $\vec{F}_B$ is zero. For all other directions of $\vec{v}$, the magnitude of $\vec{F}_B$ is always proportional to $v \sin \phi$, where ϕ is the angle between the zero-force axis and the direction of $\vec{v}$. Furthermore, the direction of $\vec{F}_B$ is always perpendicular to the direction of $\vec{v}$. (These results suggest that a cross product is involved.)

We can then define a magnetic field $\vec{B}$ to be a vector quantity that is directed along the zero-force axis. We can next measure the magnitude of $\vec{F}_B$ when $\vec{v}$ is directed perpendicular to that axis and then define the magnitude of $\vec{B}$ in terms of that force magnitude:

$$B = \frac{F_B}{|q|v},$$

where q is the charge of the particle.

We can summarize all these results with the following vector equation:

$$\vec{F}_B = q\vec{v} \times \vec{B}; \tag{29-2}$$

that is, the force $\vec{F}_B$ on the particle is equal to the charge q times the cross product of its velocity $\vec{v}$ and the field $\vec{B}$ (all measured in the same reference frame). Using Eq. 3-20 for the cross product, we can write the magnitude of $\vec{F}_B$ as

$$F_B = |q|vB \sin \phi, \tag{29-3}$$

where ϕ is the angle between the directions of velocity $\vec{v}$ and magnetic field $\vec{B}$.

Finding the Magnetic Force on a Particle

Equation 29-3 tells us that the magnitude of the force $\vec{F}_B$ acting on a particle in a magnetic field is proportional to the charge q and speed v of the particle. Thus, the force is equal to zero if the charge is zero or if the particle is stationary. Equation 29-3 also tells us that the magnitude of the force is zero if $\vec{v}$ and $\vec{B}$ are either parallel ($\phi = 0°$) or antiparallel ($\phi = 180°$), and the force is at its maximum when $\vec{v}$ and $\vec{B}$ are perpendicular to each other.

Equation 29-2 tells us all this plus the direction of $\vec{F}_B$. From Section 3-7, we know that the cross product $\vec{v} \times \vec{B}$ in Eq. 29-2 is a vector that is perpendicular to the two vectors $\vec{v}$ and $\vec{B}$. The right-hand rule (Fig. 29-2*a*) tells us that the thumb of the right hand points in the direction of $\vec{v} \times \vec{B}$ when the fingers sweep $\vec{v}$ into $\vec{B}$. If q is positive, then (by Eq. 29-2) the force $\vec{F}_B$ has the same sign as $\vec{v} \times \vec{B}$ and thus must be in the same direction; that is, for positive q, $\vec{F}_B$ is directed along the thumb (Fig. 29-2*b*). If q is negative, then the force $\vec{F}_B$ and cross product

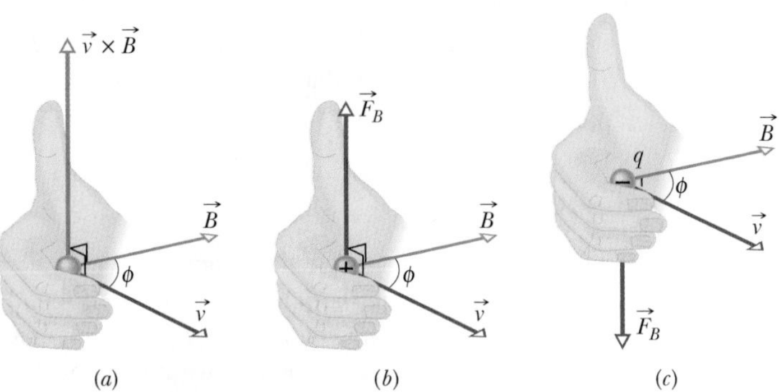

Fig. 29-2 (*a*) The right-hand rule (in which $\vec{v}$ is swept into $\vec{B}$ through the smaller angle ϕ between them) gives the direction of $\vec{v} \times \vec{B}$ as the direction of the thumb. (*b*) If q is positive, then the direction of $\vec{F}_B = q\vec{v} \times \vec{B}$ is in the direction of $\vec{v} \times \vec{B}$. (*c*) If q is negative, then the direction of $\vec{F}_B$ is opposite that of $\vec{v} \times \vec{B}$.

$\vec{v} \times \vec{B}$ have opposite signs and thus must be in opposite directions. For negative q, $\vec{F}_B$ is directed opposite the thumb (Fig. 29-2c).

Regardless of the sign of the charge, however,

> The force $\vec{F}_B$ acting on a charged particle moving with velocity $\vec{v}$ through a magnetic field $\vec{B}$ is *always* perpendicular to $\vec{v}$ and $\vec{B}$.

Thus, $\vec{F}_B$ *never* has a component parallel to $\vec{v}$. This means that $\vec{F}_B$ cannot change the particle's speed v (and thus it cannot change the particle's kinetic energy). The force can change only the direction of $\vec{v}$ (and thus the direction of travel); only in this sense can $\vec{F}_B$ accelerate the particle.

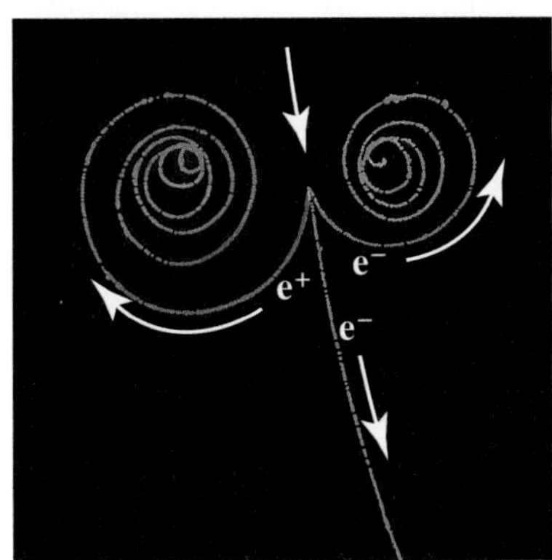

Fig. 29-3 The tracks of two electrons (e^-) and a positron (e^+) in a bubble chamber that is immersed in a uniform magnetic field that is directed out of the plane of the page.

To develop a feeling for Eq. 29-2, consider Fig. 29-3, which shows some tracks left by charged particles moving rapidly through a *bubble chamber* at the Lawrence Berkeley Laboratory. The chamber, which is filled with liquid hydrogen, is immersed in a strong uniform magnetic field that is directed out of the plane of the figure. An incoming gamma ray particle—which leaves no track because it is uncharged—transforms into an electron (spiral track marked e^-) and a positron (track marked e^+) while it knocks an electron out of a hydrogen atom (long track marked e^-). Check with Eq. 29-2 and Fig. 29-2 that the three tracks made by these two negative particles and one positive particle curve in the proper directions.

The SI unit for $\vec{B}$ that follows from Eqs. 29-2 and 29-3 is the newton per coulomb-meter per second. For convenience, this is called the **tesla** (T):

$$1 \text{ tesla} = 1 \text{ T} = 1 \frac{\text{newton}}{(\text{coulomb})(\text{meter/second})}.$$

Recalling that a coulomb per second is an ampere, we have

$$1 \text{ T} = 1 \frac{\text{newton}}{(\text{coulomb/second})(\text{meter})} = 1 \frac{\text{N}}{\text{A} \cdot \text{m}}. \tag{29-4}$$

An earlier (non-SI) unit for $\vec{B}$, still in common use, is the *gauss* (G), and

$$1 \text{ tesla} = 10^4 \text{ gauss}. \tag{29-5}$$

Table 29-1 lists the magnetic fields that occur in a few situations. Note that Earth's magnetic field near the planet's surface is about 10^{-4} T ($= 100\ \mu$T or 1 gauss).

TABLE 29-1 Some Approximate Magnetic Fields

At the surface of a neutron star	10^8 T
Near a big electromagnet	1.5 T
Near a small bar magnet	10^{-2} T
At Earth's surface	10^{-4} T
In interstellar space	10^{-10} T
Smallest value in a magnetically shielded room	10^{-14} T

CHECKPOINT 1: The figure shows three situations in which a charged particle with velocity $\vec{v}$ travels through a uniform magnetic field $\vec{B}$. In each situation, what is the direction of the magnetic force $\vec{F}_B$ on the particle?

(a) (b) (c)

Magnetic Field Lines

We can represent magnetic fields with field lines, as we did for electric fields. Similar rules apply; that is, (1) the direction of the tangent to a magnetic field line at any point gives the direction of $\vec{B}$ at that point, and (2) the spacing of the lines represents the magnitude of $\vec{B}$—the magnetic field is stronger where the lines are closer together, and conversely.

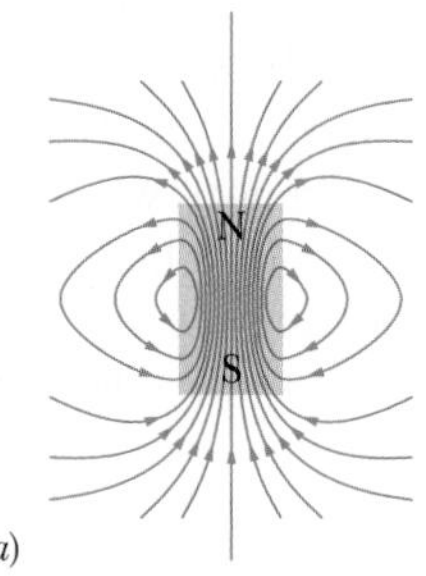

Fig. 29-4 (*a*) The magnetic field lines for a bar magnet. (*b*) A "cow magnet"—a bar magnet that is intended to be slipped down into the rumen of a cow to prevent accidentally ingested bits of scrap iron from reaching the cow's intestines. The iron filings at its ends reveal the magnetic field lines.

Figure 29-4*a* shows how the magnetic field near a *bar magnet* (a permanent magnet in the shape of a bar) can be represented by magnetic field lines. The lines all pass through the magnet, and they all form closed loops (even those that are not shown closed in the figure). The external magnetic effects of a bar magnet are strongest near its ends, where the field lines are most closely spaced. Thus, the magnetic field of the bar magnet in Fig. 29-4*b* collects the iron filings mainly near the two ends of the magnet.

The (closed) field lines enter one end of a magnet and exit the other end. The end of a magnet from which the field lines emerge is called the *north pole* of the magnet; the other end, where field lines enter the magnet, is called the *south pole*. The magnets we use to fix notes on refrigerators are short bar magnets. Figure 29-5 shows two other common shapes for magnets: a *horseshoe magnet* and a magnet that has been bent around into the shape of a **C** so that the *pole faces* are facing each other. (The magnetic field between the pole faces can then be approximately uniform.) Regardless of the shape of the magnets, if we place two of them near each other we find:

> Opposite magnetic poles attract each other, and like magnetic poles repel each other.

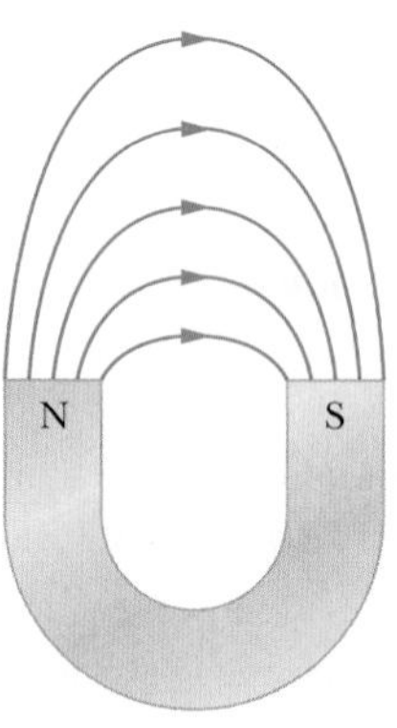

(*a*)

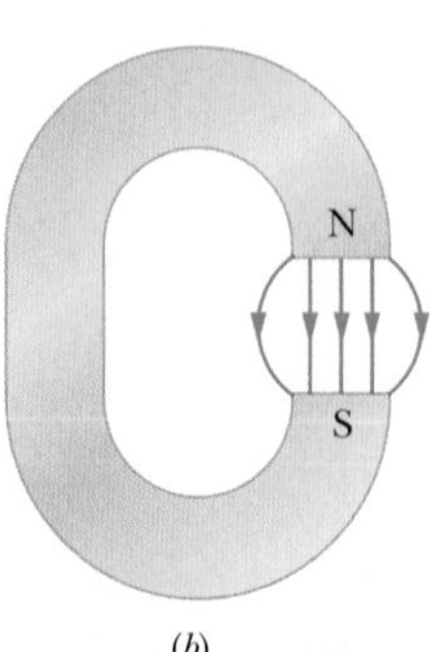

(*b*)

Fig. 29-5 (*a*) A horseshoe magnet and (*b*) a **C**-shaped magnet. (Only some of the external field lines are shown.)

Earth has a magnetic field that is produced in its core by still unknown mechanisms. On Earth's surface, we can detect this magnetic field with a compass, which is essentially a slender bar magnet on a low-friction pivot. This bar magnet, or this needle, turns because its north-pole end is attracted toward the Arctic region of Earth. Thus, the *south* pole of Earth's magnetic field must be located toward the Arctic. Logically, we then should call the pole there a south pole. However, because we call that direction north, we are trapped into the statement that Earth has a *geomagnetic north pole* in that direction.

With more careful measurement we would find that in the northern hemisphere, the magnetic field lines of Earth generally point down into Earth and toward the Arctic. In the southern hemisphere, they generally point up out of Earth and away from the Antarctic—that is, away from Earth's *geomagnetic south pole*.

Sample Problem 29-1

A uniform magnetic field $\vec{B}$, with magnitude 1.2 mT, is directed vertically upward throughout the volume of a laboratory chamber. A proton with kinetic energy 5.3 MeV enters the chamber, moving horizontally from south to north. What magnetic deflecting force acts on the proton as it enters the chamber? The proton mass is 1.67×10^{-27} kg. (Neglect Earth's magnetic field.)

SOLUTION: Because the proton is charged and moving through a magnetic field, a magnetic force $\vec{F}_B$ can act on it. The **Key Idea** here is that, because the initial direction of the proton's velocity is not along a magnetic field line, $\vec{F}_B$ is not simply zero. To find the magnitude of $\vec{F}_B$, we can use Eq. 29-3 provided we first find the proton's speed v. We can find v from the given kinetic energy, since $K = \frac{1}{2}mv^2$. Solving for v, we obtain

$$v = \sqrt{\frac{2K}{m}} = \sqrt{\frac{(2)(5.3\text{ MeV})(1.60 \times 10^{-13}\text{ J/MeV})}{1.67 \times 10^{-27}\text{ kg}}}$$
$$= 3.2 \times 10^7\text{ m/s}.$$

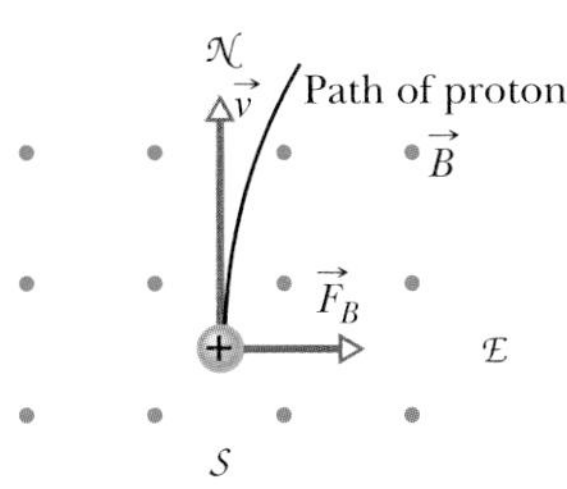

Fig. 29-6 Sample Problem 29-1. An overhead view of a proton moving from south to north with velocity $\vec{v}$ in a chamber. A magnetic field is directed vertically upward in the chamber, as represented by the array of dots (which resemble the tips of arrows). The proton is deflected toward the east.

Equation 29-3 then yields

$$F_B = |q|vB \sin \phi$$
$$= (1.60 \times 10^{-19}\text{ C})(3.2 \times 10^7\text{ m/s})$$
$$\times (1.2 \times 10^{-3}\text{ T})(\sin 90°)$$
$$= 6.1 \times 10^{-15}\text{ N}. \qquad \text{(Answer)}$$

This may seem like a small force, but it acts on a particle of small mass, producing a large acceleration; namely,

$$a = \frac{F_B}{m} = \frac{6.1 \times 10^{-15}\text{ N}}{1.67 \times 10^{-27}\text{ kg}} = 3.7 \times 10^{12}\text{ m/s}^2.$$

To find the direction of $\vec{F}_B$, we use the **Key Idea** that $\vec{F}_B$ has the direction of the cross product $q\vec{v} \times \vec{B}$. Because the charge q is positive, $\vec{F}_B$ must have the same direction as $\vec{v} \times \vec{B}$, which can be determined with the right-hand rule for cross products (as in Fig. 29-2*b*). We know that $\vec{v}$ is directed horizontally from south to north and $\vec{B}$ is directed vertically up. The right-hand rule shows us that the deflecting force $\vec{F}_B$ must be directed horizontally from west to east, as Fig. 29-6 shows. (The array of dots in the figure represents a magnetic field directed out of the plane of the figure. An array of **X**s would have represented a magnetic field directed into that plane.)

If the charge of the particle were negative, the magnetic deflecting force would be directed in the opposite direction—that is, horizontally from east to west. This is predicted automatically by Eq. 29-2 if we substitute a negative value for q.

29-3 Crossed Fields: Discovery of the Electron

Both an electric field $\vec{E}$ and a magnetic field $\vec{B}$ can produce a force on a charged particle. When the two fields are perpendicular to each other, they are said to be *crossed fields*. Here we shall examine what happens to charged particles—namely, electrons—as they move through crossed fields. We use as our example the experiment that led to the discovery of the electron in 1897 by J. J. Thomson at Cambridge University.

Figure 29-7 shows a modern, simplified version of Thomson's experimental apparatus—a *cathode ray tube* (which is like the picture tube in a standard television

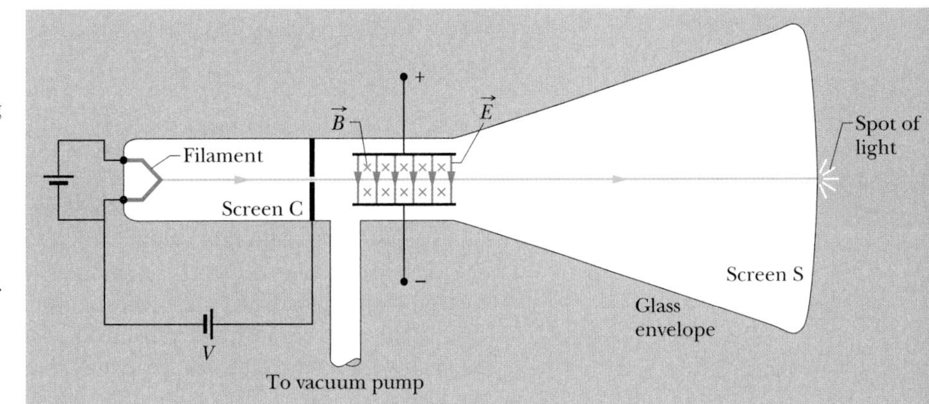

Fig. 29-7 A modern version of J. J. Thomson's apparatus for measuring the ratio of mass to charge for the electron. An electric field $\vec{E}$ is established by connecting a battery across the deflecting-plate terminals. A magnetic field $\vec{B}$ is set up by means of a current in a system of coils (not shown). The magnetic field shown is into the plane of the figure, as represented by the array of **X**s (which resemble the feathered ends of arrows).

set). Charged particles (which we now know as electrons) are emitted by a hot filament at the rear of the evacuated tube and are accelerated by an applied potential difference V. After they pass through a slit in screen C, they form a narrow beam. They then pass through a region of crossed $\vec{E}$ and $\vec{B}$ fields, headed toward a fluorescent screen S, where they produce a spot of light (on a television screen the spot is part of the picture). The forces on the charged particles in the crossed-fields region can deflect them from the center of the screen. By controlling the magnitudes and directions of the fields, Thomson could thus control where the spot of light appeared on the screen. Recall that the force on a negatively charged particle due to an electric field is directed opposite the field. Thus, for the particular field arrangement of Fig. 29-7, electrons are forced up the page by the electric field $\vec{E}$ and down the page by the magnetic field $\vec{B}$; that is, the forces are *in opposition.* Thomson's procedure was equivalent to the following series of steps.

1. Set $E = 0$ and $B = 0$ and note the position of the spot on screen S due to the undeflected beam.
2. Turn on $\vec{E}$ and measure the resulting beam deflection.
3. Maintaining $\vec{E}$, now turn on $\vec{B}$ and adjust its value until the beam returns to the undeflected position. (With the forces in opposition, they can be made to cancel.)

We discussed the deflection of a charged particle moving through an electric field $\vec{E}$ between two plates (step 2 here) in Sample Problem 23-4. We found that the deflection of the particle at the far end of the plates is

$$y = \frac{qEL^2}{2mv^2}, \tag{29-6}$$

where v is the particle's speed, m its mass, and q its charge, and L is the length of the plates. We can apply this same equation to the beam of electrons in Fig. 29-7; if need be, we can calculate the deflection by measuring the deflection of the beam on screen S and then working back to calculate the deflection y at the end of the plates. (Because the direction of the deflection is set by the sign of the particle's charge, Thomson was able to show that the particles that were lighting up his screen were negatively charged.)

When the two fields in Fig. 29-7 are adjusted so that the two deflecting forces cancel (step 3), we have from Eqs. 29-1 and 29-3a

$$|q|E = |q|vB \sin(90°) = |q|vB$$

or

$$v = \frac{E}{B}. \tag{29-7}$$

Thus, the crossed fields allow us to measure the speed of the charged particles passing through them. Substituting Eq. 29-7 for v in Eq. 29-6 and rearranging yield

$$\frac{m}{q} = \frac{B^2L^2}{2yE}, \tag{29-8}$$

in which all quantities on the right can be measured. Thus, the crossed fields allow us to measure the ratio m/q of the particles moving through Thomson's apparatus.

Thomson claimed that these particles are found in all matter. He also claimed that they are lighter than the lightest known atom (hydrogen) by a factor of more than 1000. (The exact ratio proved later to be 1836.15.) His m/q measurement, coupled with the boldness of his two claims, is considered to be the "discovery of the electron."

CHECKPOINT 2: The figure shows four directions for the velocity vector $\vec{v}$ of a positively charged particle moving through a uniform electric field $\vec{E}$ (directed out of the page and represented with an encircled dot) and a uniform magnetic field $\vec{B}$. (a) Rank directions 1, 2, and 3 according to the magnitude of the net force on the particle, greatest first. (b) Of all four directions, which might result in a net force of zero?

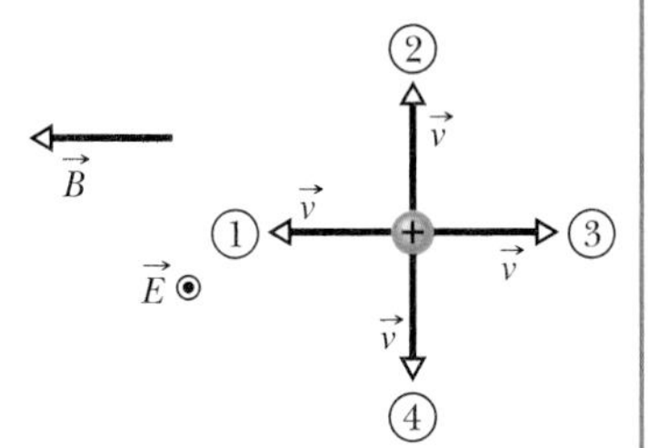

29-4 Crossed Fields: The Hall Effect

As we just discussed, a beam of electrons in a vacuum can be deflected by a magnetic field. Can the drifting conduction electrons in a copper wire also be deflected by a magnetic field? In 1879, Edwin H. Hall, then a 24-year-old graduate student at the Johns Hopkins University, showed that they can. This **Hall effect** allows us to find out whether the charge carriers in a conductor are positively or negatively charged. Beyond that, we can measure the number of such carriers per unit volume of the conductor.

Figure 29-8*a* shows a copper strip of width d, carrying a current i whose conventional direction is from the top of the figure to the bottom. The charge carriers are electrons and, as we know, they drift (with drift speed v_d) in the opposite direction, from bottom to top. At the instant shown in Fig. 29-8*a*, an external magnetic field $\vec{B}$, pointing into the plane of the figure, has just been turned on. From Eq. 29-2 we see that a magnetic deflecting force $\vec{F}_B$ will act on each drifting electron, pushing it toward the right edge of the strip.

As time goes on, electrons move to the right, mostly piling up on the right edge of the strip, leaving uncompensated positive charges in fixed positions at the left edge. The separation of positive and negative charges produces an electric field $\vec{E}$ within the strip, pointing from left to right in Fig. 29-8*b*. This field exerts an electric force $\vec{F}_E$ on each electron, tending to push it to the left.

An equilibrium quickly develops in which the electric force on each electron builds up until it just cancels the magnetic force. When this happens, as Fig. 29-8*b* shows, the force due to $\vec{B}$ and the force due to $\vec{E}$ are in balance. The drifting electrons then move along the strip toward the top of the page at velocity $\vec{v}_d$, with no further collection of electrons on the right edge of the strip and thus no further increase in the electric field $\vec{E}$.

A *Hall potential difference* V is associated with the electric field across strip width d. From Eq. 25-42, the magnitude of that potential difference is

$$V = Ed. \tag{29-9}$$

By connecting a voltmeter across the width, we can measure the potential difference between the two edges of the strip. Moreover, the voltmeter can tell us which edge is at higher potential. For the situation of Fig. 29-8*a*, we would find that the left

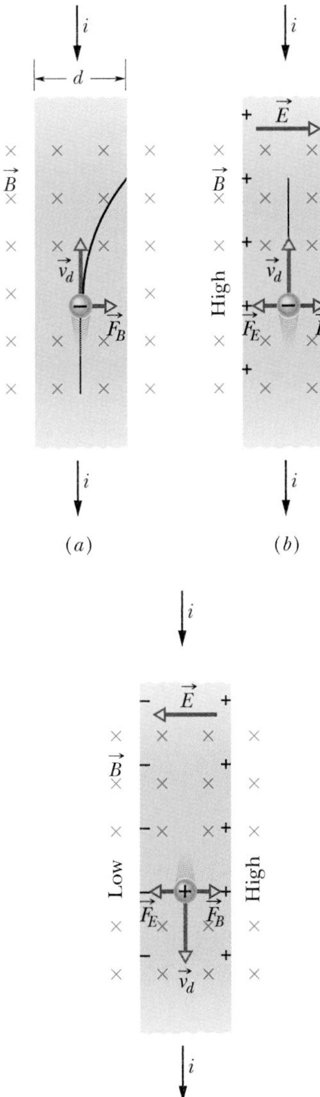

Fig. 29-8 A strip of copper carrying a current i is immersed in a magnetic field $\vec{B}$. (*a*) The situation immediately after the magnetic field is turned on. The curved path that will then be taken by an electron is shown. (*b*) The situation at equilibrium, which quickly follows. Note that negative charges pile up on the right side of the strip, leaving uncompensated positive charges on the left. Thus, the left side is at a higher potential than the right side. (*c*) For the same current direction, if the charge carriers were positively charged, *they* would pile up on the right side, and the right side would be at the higher potential.

edge is at higher potential, which is consistent with our assumption that the charge carriers are negatively charged.

For a moment, let us make the opposite assumption, that the charge carriers in current i are positively charged (Fig. 29-8c). Convince yourself that as these charge carriers move from top to bottom in the strip, they are pushed to the right edge by $\vec{F}_B$ and thus that the *right* edge is at higher potential. Because that last statement is contradicted by our voltmeter reading, the charge carriers must be negatively charged.

Now for the quantitative part. When the electric and magnetic forces are in balance (Fig. 29-8b), Eqs. 29-1 and 29-3 give us

$$eE = ev_dB. \tag{29-10}$$

From Eq. 27-7, the drift speed v_d is

$$v_d = \frac{J}{ne} = \frac{i}{neA}, \tag{29-11}$$

in which J (= i/A) is the current density in the strip, A is the cross-sectional area of the strip, and n is the *number density* of charge carriers (their number per unit volume).

In Eq. 29-10, substituting for E with Eq. 29-9 and substituting for v_d with Eq. 29-11, we obtain

$$n = \frac{Bi}{Vle}, \tag{29-12}$$

in which l (= A/d) is the thickness of the strip. With this equation we can find n from measurable quantities.

It is also possible to use the Hall effect to measure directly the drift speed v_d of the charge carriers, which you may recall is of the order of centimeters per hour. In this clever experiment, the metal strip is moved mechanically through the magnetic field in a direction opposite that of the drift velocity of the charge carriers. The speed of the moving strip is then adjusted until the Hall potential difference vanishes. At this condition, with no Hall effect, the velocity of the charge carriers *with respect to the laboratory frame* must be zero, so the velocity of the strip must be equal in magnitude but opposite the direction of the velocity of the negative charge carriers.

Sample Problem 29-2

Figure 29-9 shows a solid metal cube, of edge length d = 1.5 cm, moving in the positive y direction at a constant velocity $\vec{v}$ of magnitude 4.0 m/s. The cube moves through a uniform magnetic field $\vec{B}$ of magnitude 0.050 T directed toward positive z.

(a) Which cube face is at a lower electric potential and which is at a higher electric potential because of the motion through the field?

SOLUTION: One **Key Idea** here is that, because the cube is moving through a magnetic field $\vec{B}$, a magnetic force $\vec{F}_B$ acts on its charged particles, including its conduction electrons. A second **Key Idea** is how $\vec{F}_B$ causes an electric potential difference between certain faces of the cube. When the cube first begins to move through the magnetic field, its electrons do also. Because each electron has charge q and is moving through a magnetic field with velocity $\vec{v}$, the magnetic force $\vec{F}_B$ acting on it is given by Eq. 29-2. Because q is negative, the direction of $\vec{F}_B$ is opposite the cross product $\vec{v} \times \vec{B}$, which is in the positive direction of the x axis in Fig. 29-9. Thus, $\vec{F}_B$ acts in the negative direction of the x axis, toward the left face of the cube (which is hidden from view in Fig. 29-9).

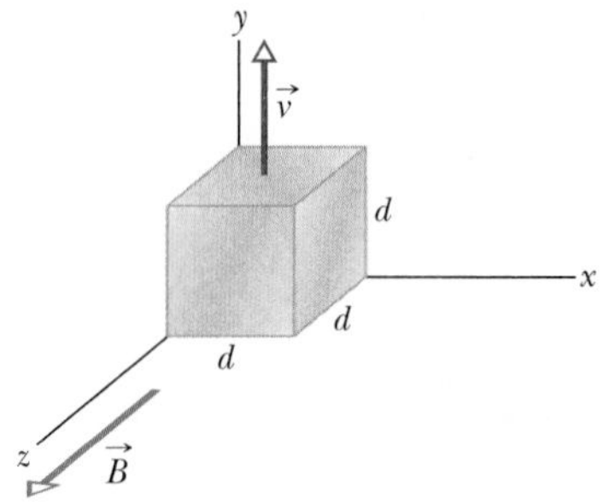

Fig. 29-9 Sample Problem 29-2. A solid metal cube of edge length d, at constant velocity $\vec{v}$ through a uniform magnetic field $\vec{B}$.

Most of the electrons are fixed in place in the molecules of the cube. However, because the cube is a metal, it contains conduction electrons that are free to move. Some of those conduction electrons are deflected by $\vec{F}_B$ to the left cube face, making that face negatively charged and leaving the right face positively charged. This charge separation produces an electric field $\vec{E}$ directed from the positively charged right face to the negatively charged left face. Thus, the left face is at a lower electric potential, and the right face is at a higher electric potential.

(b) What is the potential difference between the faces of higher and lower electric potential?

SOLUTION: The Key Ideas here are these:

1. The electric field $\vec{E}$ created by the charge separation produces an electric force $\vec{F}_E = q\vec{E}$ on each electron. Because q is negative, this force is directed opposite the field $\vec{E}$—that is, toward the right. Thus on each electron, $\vec{F}_E$ acts toward the right and $\vec{F}_B$ acts toward the left.
2. When the cube had just begun to move through the magnetic field and the charge separation had just begun, the magnitude of $\vec{E}$ began to increase from zero. Thus, the magnitude of $\vec{F}_E$ also began to increase from zero and was initially smaller than the magnitude $\vec{F}_B$. During this early stage, the net force on any electron was dominated by $\vec{F}_B$, which continuously moved additional electrons to the left cube face, increasing the charge separation.
3. However, as the charge separation increased, eventually magnitude F_E became equal to magnitude F_B. The net force on any electron was then zero, and no additional electrons were moved to the left cube face. Thus, the magnitude of $\vec{F}_E$ could not increase further, and the electrons were then in equilibrium.

We seek the potential difference V between the left and right cube faces after equilibrium was reached (which occurred quickly). We can obtain V with Eq. 29-9 ($V = Ed$) provided we first find the magnitude E of the electric field at equilibrium. We can do so with the equation for the balance of forces ($F_E = F_B$).

For F_E, we substitute $|q|E$. For F_B, we substitute $|q|vB \sin\phi$ from Eq. 29-3. From Fig. 29-9, we see that the angle ϕ between vectors $\vec{v}$ and $\vec{B}$ is 90°; so $\sin\phi = 1$ and $F_E = F_B$ yields

$$|q|E = |q|vB \sin 90° = |q|vB.$$

This gives us $E = vB$, so Eq. 29-9 ($V = Ed$) becomes

$$V = vBd. \qquad (29\text{-}13)$$

Substituting known values gives us

$$V = (4.0 \text{ m/s})(0.050 \text{ T})(0.015 \text{ m})$$
$$= 0.0030 \text{ V} = 3.0 \text{ mV}. \qquad \text{(Answer)}$$

✔CHECKPOINT 3: The figure shows a metallic, rectangular solid that is to move at a certain speed v through the uniform magnetic field $\vec{B}$. Its dimensions are multiples of d, as shown. You have six choices for the direction of the velocity of the solid: it can be parallel to x, y, or z, in either the positive or negative direction. (a) Rank the six choices according to the potential difference that would be set up across the solid, greatest first. (b) For which choice is the front face at lower potential?

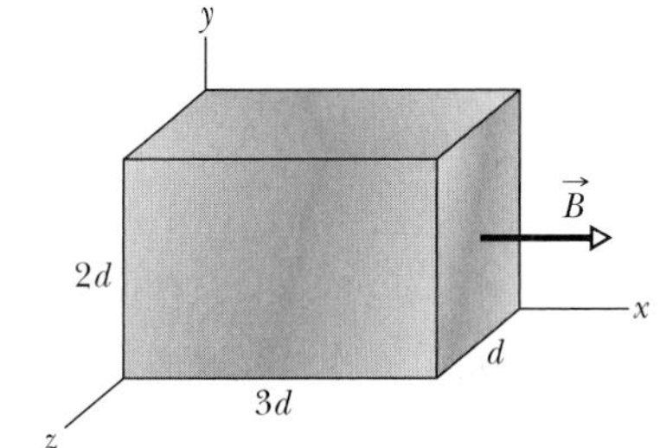

29-5 A Circulating Charged Particle

If a particle moves in a circle at constant speed, we can be sure that the net force acting on the particle is constant in magnitude and points toward the center of the circle, always perpendicular to the particle's velocity. Think of a stone tied to a string and whirled in a circle on a smooth horizontal surface, or of a satellite moving in a circular orbit around Earth. In the first case, the tension in the string provides the necessary force and centripetal acceleration. In the second case, Earth's gravitational attraction provides the force and acceleration.

Figure 29-10 shows another example: A beam of electrons is projected into a chamber by an *electron gun* G. The electrons enter in the plane of the page with speed v and then move in a region of uniform magnetic field $\vec{B}$ directed out of that plane. As a result, a magnetic force $\vec{F}_B = q\vec{v} \times \vec{B}$ continually deflects the electrons, and because $\vec{v}$ and $\vec{B}$ are always perpendicular to each other, this deflection causes the electrons to follow a circular path. The path is visible in the photo because atoms of gas in the chamber emit light when some of the circulating electrons collide with them.

Fig. 29-10 Electrons circulating in a chamber containing gas at low pressure (their path is the glowing circle). A uniform magnetic field $\vec{B}$, pointing directly out of the plane of the page, fills the chamber. Note the radially directed magnetic force $\vec{F}_B$; for circular motion to occur, $\vec{F}_B$ *must* point toward the center of the circle. Use the right-hand rule for cross products to confirm that $\vec{F}_B = q\vec{v} \times \vec{B}$ gives $\vec{F}_B$ the proper direction. (Don't forget the sign of q.)

We would like to determine the parameters that characterize the circular motion of these electrons, or of any particle of charge magnitude q and mass m moving perpendicular to a uniform magnetic field $\vec{B}$ at speed v. From Eq. 29-3, the force acting on the particle has a magnitude of qvB. From Newton's second law ($\vec{F} = m\vec{a}$) applied to uniform circular motion (Eq. 6-18),

$$F = m\frac{v^2}{r}, \tag{29-14}$$

we have

$$qvB = \frac{mv^2}{r}. \tag{29-15}$$

Solving for r, we find the radius of the circular path as

$$r = \frac{mv}{qB} \quad \text{(radius)}. \tag{29-16}$$

The period T (the time for one full revolution) is equal to the circumference divided by the speed:

$$T = \frac{2\pi r}{v} = \frac{2\pi}{v}\frac{mv}{qB} = \frac{2\pi m}{qB} \quad \text{(period)}. \tag{29-17}$$

The frequency f (the number of revolutions per unit time) is

$$f = \frac{1}{T} = \frac{qB}{2\pi m} \quad \text{(frequency)}. \tag{29-18}$$

The angular frequency ω of the motion is then

$$\omega = 2\pi f = \frac{qB}{m} \quad \text{(angular frequency)}. \tag{29-19}$$

The quantities T, f, and ω do not depend on the speed of the particle (provided that speed is much less than the speed of light). Fast particles move in large circles and slow ones in small circles, but all particles with the same charge-to-mass ratio q/m

take the same time T (the period) to complete one round trip. Using Eq. 29-2, you can show that if you are looking in the direction of $\vec{B}$, the direction of rotation for a positive particle is always counterclockwise, and the direction for a negative particle is always clockwise.

Helical Paths

If the velocity of a charged particle has a component parallel to the (uniform) magnetic field, the particle will move in a helical path about the direction of the field vector. Figure 29-11*a*, for example, shows the velocity vector $\vec{v}$ of such a particle resolved into two components, one parallel to $\vec{B}$ and one perpendicular to it:

$$v_{\parallel} = v \cos \phi \quad \text{and} \quad v_{\perp} = v \sin \phi. \tag{29-20}$$

The parallel component determines the *pitch p* of the helix—that is, the distance between adjacent turns (Fig. 29-11*b*). The perpendicular component determines the radius of the helix and is the quantity to be substituted for v in Eq. 29-16.

Figure 29-11*c* shows a charged particle spiraling in a nonuniform magnetic field. The more closely spaced field lines at the left and right sides indicate that the magnetic field is stronger there. When the field at an end is strong enough, the particle "reflects" from that end. If the particle reflects from both ends, it is said to be trapped in a *magnetic bottle*.

Electrons and protons are trapped in this way by the terrestrial magnetic field; the trapped particles form the *Van Allen radiation belts,* which loop well above Earth's atmosphere between Earth's north and south geomagnetic poles. These particles bounce back and forth, from one end of this magnetic bottle to the other, within a few seconds.

When a large solar flare shoots additional energetic electrons and protons into the radiation belts, an electric field is produced in the region where electrons normally reflect. This field eliminates the reflection and instead drives electrons down into the atmosphere, where they collide with atoms and molecules of air, causing

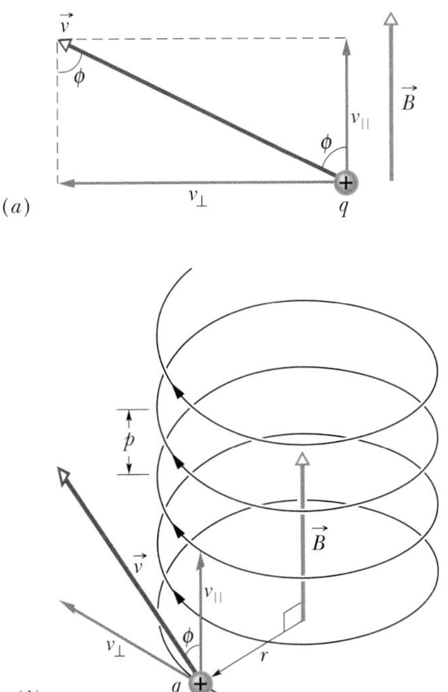

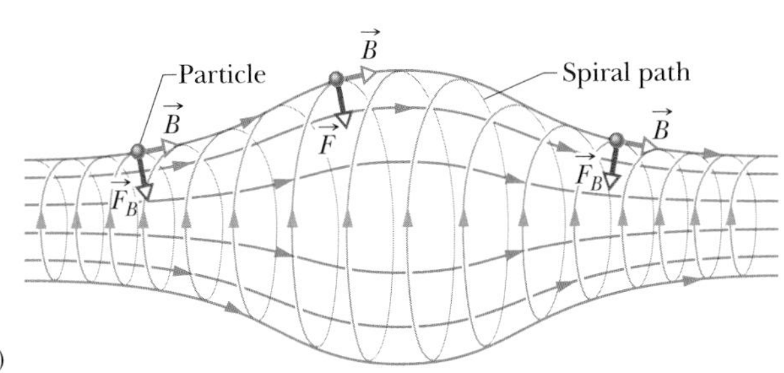

Fig. 29-11 (*a*) A charged particle moves in a uniform magnetic field $\vec{B}$, its velocity $\vec{v}$ making an angle ϕ with the field direction. (*b*) The particle follows a helical path, of radius r and pitch p. (*c*) A charged particle spiraling in a nonuniform magnetic field. (The particle can become trapped, spiraling back and forth between the strong field regions at either end.) Note that the magnetic force vectors at the left and right sides have a component pointing toward the center of the figure.

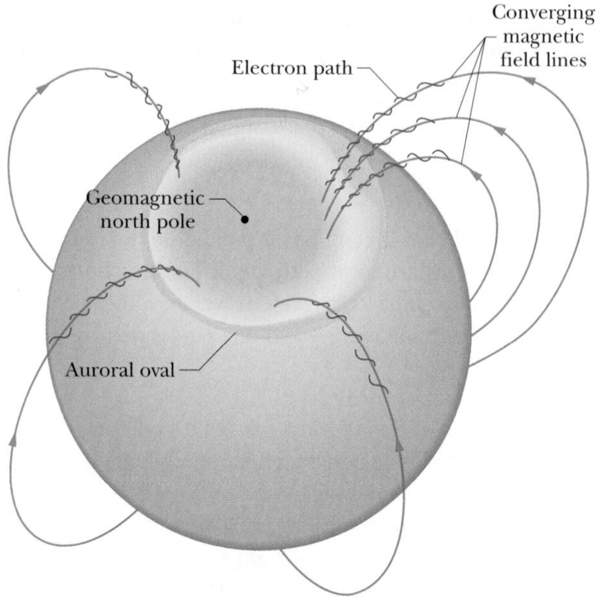

Converging magnetic field lines
Electron path
Geomagnetic north pole
Auroral oval

Fig. 29-12 The auroral oval surrounding Earth's geomagnetic north pole (in northwestern Greenland). Magnetic field lines converge toward that pole. Electrons moving toward Earth are "caught by" and spiral around these field lines, entering the terrestrial atmosphere at high latitudes and producing aurora within the oval.

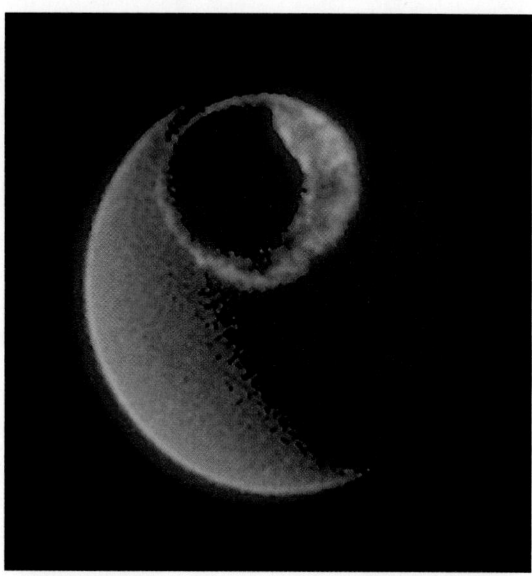

Fig. 29-13 A false-color image of aurora inside the north auroral oval, recorded by the satellite *Dynamic Explorer*, using ultraviolet light emitted by oxygen atoms excited in the aurora. The sun-lit portion of Earth is the crescent at the left.

that air to emit light. This light forms the aurora—a curtain of light that hangs down to an altitude of about 100 km. Green light is emitted by oxygen atoms, and pink light is emitted by nitrogen molecules, but often the light is so dim that we perceive only white light.

Aurora extend in arcs above Earth and can occur in a region called the *auroral oval* that is shown in Figs. 29-12 and 29-13 as seen from space. Although an aurora is long, it is less than 1 km thick (north to south) because the paths of the electrons producing it converge as the electrons spiral down the converging magnetic field lines (Fig. 29-12).

CHECKPOINT 4: The figure here shows the circular paths of two particles that travel at the same speed in a uniform magnetic field $\vec{B}$, which is directed into the page. One particle is a proton; the other is an electron (which is less massive). (a) Which particle follows the smaller circle, and (b) does that particle travel clockwise or counterclockwise?

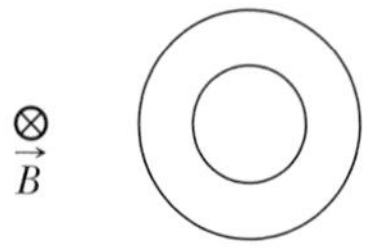

$\otimes \vec{B}$

Sample Problem 29-3

Figure 29-14 shows the essentials of a *mass spectrometer*, which can be used to measure the mass of an ion; an ion of mass m (to be measured) and charge q is produced in source S. The initially stationary ion is accelerated by the electric field due to a potential difference V. The ion leaves S and enters a separator chamber in which a uniform magnetic field $\vec{B}$ is perpendicular to the path of the ion. The magnetic field causes the ion to move in a semicircle, striking (and thus altering) a photographic plate at distance x from the entry slit. Suppose that in a certain trial $B = 80.000$ mT and $V = 1000.0$ V, and ions of charge $q = +1.6022 \times 10^{-19}$ C strike the plate at $x = 1.6254$ m. What is the mass m of the individual ions, in unified atomic mass units ($1 \text{ u} = 1.6605 \times 10^{-27}$ kg)?

SOLUTION: One **Key Idea** here is that, because the (uniform) magnetic field causes the (charged) ion to follow a circular path, we can relate the ion's mass m to the path's radius r with Eq. 29-16 ($r = mv/qB$). From Fig. 29-14 we see that $r = x/2$, and we are given the magnitude B of the magnetic field. However, we lack the ion's speed v in the magnetic field, after it has been accelerated due to the potential difference V.

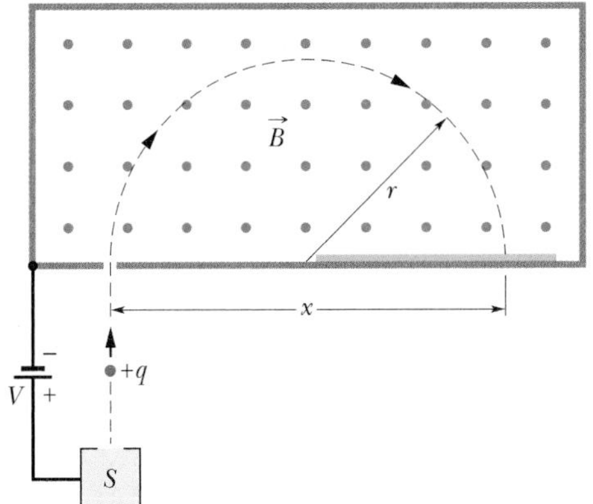

Fig. 29-14 Sample Problem 29-3. Essentials of a mass spectrometer. A positive ion, after being accelerated from its source S by potential difference V, enters a chamber of uniform magnetic field $\vec{B}$. There it travels through a semicircle of radius r and strikes a photographic plate at a distance x from where it entered the chamber.

To relate v and V, we use the Key Idea that mechanical energy ($E_{mec} = K + U$) is conserved during the acceleration. When the ion emerges from the source, its kinetic energy is approximately zero. At the end of the acceleration, its kinetic energy is $\frac{1}{2}mv^2$. Also, during the acceleration, the positive ion moves through a change in potential of $-V$. Thus, because the ion has positive charge q, its potential energy changes by $-qV$. If we now write the conservation of mechanical energy as

$$\Delta K + \Delta U = 0,$$

we get

$$\tfrac{1}{2}mv^2 - qV = 0$$

or

$$v = \sqrt{\frac{2qV}{m}}. \tag{29-21}$$

Substituting this into Eq. 29-16 gives us

$$r = \frac{mv}{qB} = \frac{m}{qB}\sqrt{\frac{2qV}{m}} = \frac{1}{B}\sqrt{\frac{2mV}{q}}.$$

Thus,

$$x = 2r = \frac{2}{B}\sqrt{\frac{2mV}{q}}.$$

Solving this for m and substituting the given data yield

$$m = \frac{B^2qx^2}{8V}$$
$$= \frac{(0.080000\text{ T})^2(1.6022 \times 10^{-19}\text{ C})(1.6254\text{ m})^2}{8(1000.0\text{ V})}$$
$$= 3.3863 \times 10^{-25}\text{ kg} = 203.93\text{ u}. \qquad \text{(Answer)}$$

Sample Problem 29-4

An electron with a kinetic energy of 22.5 eV moves into a region of uniform magnetic field $\vec{B}$ of magnitude 4.55×10^{-4} T. The angle between the directions of $\vec{B}$ and the electron's velocity $\vec{v}$ is 65.5°. What is the pitch of the helical path taken by the electron?

SOLUTION: One Key Idea here is that the pitch p is the distance the electron travels parallel to the magnetic field $\vec{B}$ during one period T of circulation. A second Key Idea is that the period T is given by Eq. 29-17, regardless of the angle between the directions of $\vec{v}$ and $\vec{B}$ (provided the angle is not zero, for which there is no circulation of the electron). Thus, using Eqs. 29-20 and 29-17, we find

$$p = v_{\parallel}T = (v\cos\phi)\,\frac{2\pi m}{qB}. \tag{29-22}$$

We can calculate the electron's speed v from its kinetic energy as we did for the proton in Sample Problem 29-1. We find that $v = 2.81 \times 10^6$ m/s. Substituting this and known data in Eq. 29-22 gives us

$$p = (2.81 \times 10^6\text{ m/s})(\cos 65.5°)$$
$$\times \frac{2\pi(9.11 \times 10^{-31}\text{ kg})}{(1.60 \times 10^{-19}\text{ C})(4.55 \times 10^{-4}\text{ T})}$$
$$= 9.16\text{ cm}. \qquad \text{(Answer)}$$

29-6 Cyclotrons and Synchrotrons

What is the structure of matter on the smallest scale? This question has always intrigued physicists. One way of getting at the answer is to allow an energetic charged particle (a proton, for example) to slam into a solid target. Better yet, allow two such energetic protons to collide head-on. Then analyze the debris from many such collisions to learn the nature of the subatomic particles of matter. The Nobel Prizes in physics for 1976 and 1984 were awarded for just such studies.

How can we give a proton enough kinetic energy for such an experiment? The direct approach is to allow the proton to "fall" through a potential difference V, thereby increasing its kinetic energy by eV. As we want higher and higher energies, however, it becomes more and more difficult to establish the necessary potential difference.

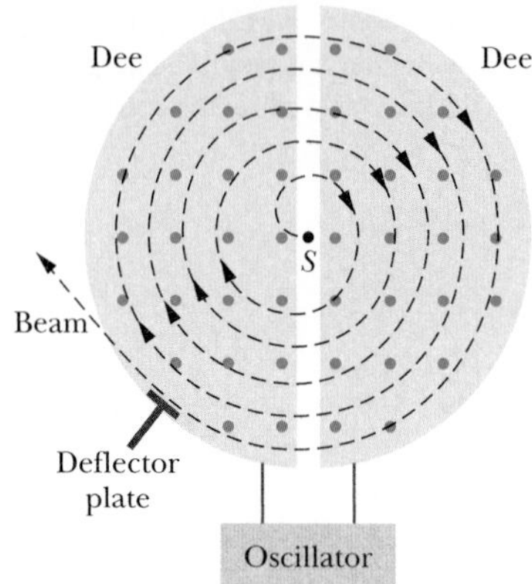

Fig. 29-15 The elements of a cyclotron, showing the particle source S and the dees. A uniform magnetic field is directed up from the plane of the page. Circulating protons spiral outward within the hollow dees, gaining energy every time they cross the gap between the dees.

A better way is to arrange for the proton to circulate in a magnetic field, and to give it a modest electrical "kick" once per revolution. For example, if a proton circulates 100 times in a magnetic field and receives an energy boost of 100 keV every time it completes an orbit, it will end up with a kinetic energy of (100)(100 keV) or 10 MeV. Two very useful accelerating devices are based on this principle.

The Cyclotron

Figure 29-15 is a top view of the region of a *cyclotron* in which the particles (protons, say) circulate. The two hollow **D**-shaped objects (open on their straight edges) are made of sheet copper. These *dees,* as they are called, are part of an electrical oscillator that alternates the electric potential difference across the gap between the dees. The electrical signs of the dees are alternated so that the electric field in the gap alternates in direction, first toward one dee and then the other dee, back and forth. The dees are immersed in a magnetic field ($B = 1.5$ T) whose direction is out of the plane of the page and that is set up by a large electromagnet.

Suppose that a proton, injected by source S at the center of the cyclotron in Fig. 29-15, initially moves toward a negatively charged dee. It will accelerate toward this dee and enter it. Once inside, it is shielded from electric fields by the copper walls of the dee; that is, the electric field does not enter the dee. The magnetic field, however, is not screened by the (nonmagnetic) copper dee, so the proton moves in a circular path whose radius, which depends on its speed, is given by Eq. 29-16 ($r = mv/qB$).

Let us assume that at the instant the proton emerges into the center gap from the first dee, the potential difference between the dees is reversed. Thus, the proton *again* faces a negatively charged dee and is *again* accelerated. This process continues, the circulating proton always being in step with the oscillations of the dee potential, until the proton has spiraled out to the edge of the dee system. There a deflector plate sends it out through a portal.

The key to the operation of the cyclotron is that the frequency f at which the proton circulates in the field (and that does *not* depend on its speed) must be equal to the fixed frequency f_{osc} of the electrical oscillator, or

$$f = f_{osc} \quad \text{(resonance condition).} \tag{29-23}$$

This *resonance condition* says that, if the energy of the circulating proton is to increase, energy must be fed to it at a frequency f_{osc} that is equal to the natural frequency f at which the proton circulates in the magnetic field.

Combining Eqs. 29-18 and 29-23 allows us to write the resonance condition as

$$qB = 2\pi m f_{osc}. \tag{29-24}$$

For the proton, q and m are fixed. The oscillator (we assume) is designed to work at a single fixed frequency f_{osc}. We then "tune" the cyclotron by varying B until Eq. 29-24 is satisfied and then many protons circulate through the magnetic field, to emerge as a beam.

The Proton Synchrotron

At proton energies above 50 MeV, the conventional cyclotron begins to fail because one of the assumptions of its design—that the frequency of revolution of a charged particle circulating in a magnetic field is independent of the particle's speed—is true only for speeds that are much less than the speed of light. At greater proton speeds (above about 10% of the speed of light), we must treat the problem relativistically. According to relativity theory, as the speed of a circulating proton approaches that of light, the proton's frequency of revolution decreases steadily. Thus, the protons get out of step with the cyclotron's oscillator—whose frequency remains fixed at f_{osc}—and eventually the energy of the circulating proton stops increasing.

There is another problem. For a 500 GeV proton in a magnetic field of 1.5 T, the path radius is 1.1 km. The corresponding magnet for a conventional cyclotron of the proper size would be impossibly expensive, the area of its pole faces being about 4×10^6 m^2.

The *proton synchrotron* is designed to meet these two difficulties. The magnetic field B and the oscillator frequency f_{osc}, instead of having fixed values as in the conventional cyclotron, are made to vary with time during the accelerating cycle. When this is done properly, (1) the frequency of the circulating protons remains in step with the oscillator at all times, and (2) the protons follow a circular—not a spiral—path. Thus, the magnet need extend only along that circular path, not over some 4×10^6 m^2. The circular path, however, still must be large if high energies are to be achieved. The proton synchrotron at the Fermi National Accelerator Laboratory (Fermilab) in Illinois has a circumference of 6.3 km and can produce protons with energies of about 1 TeV ($= 10^{12}$ eV).

Sample Problem 29-5

Suppose a cyclotron is operated at an oscillator frequency of 12 MHz and has a dee radius $R = 53$ cm.

(a) What is the magnitude of the magnetic field needed for deuterons to be accelerated in the cyclotron? A deuteron is the nucleus of deuterium, an isotope of hydrogen. It consists of a proton and a neutron and thus has the same charge as a proton. Its mass is $m = 3.34 \times 10^{-27}$ kg.

SOLUTION: The **Key Idea** here is that, for a given oscillator frequency f_{osc}, the magnetic field magnitude B required to accelerate any particle in a cyclotron depends on the ratio m/q of mass to charge for the particle, according to Eq. 29-24. For deuterons and the oscillator frequency $f_{osc} = 12$ MHz, we find

$$B = \frac{2\pi m f_{osc}}{q} = \frac{(2\pi)(3.34 \times 10^{-27}\ \text{kg})(12 \times 10^6\ \text{s}^{-1})}{1.60 \times 10^{-19}\ \text{C}}$$
$$= 1.57\ \text{T} \approx 1.6\ \text{T}. \qquad \text{(Answer)}$$

Note that, to accelerate protons, B would have to be reduced by a factor of 2, provided that the oscillator frequency remained fixed at 12 MHz.

(b) What is the resulting kinetic energy of the deuterons?

SOLUTION: One **Key Idea** here is that the kinetic energy ($\frac{1}{2}mv^2$) of a deuteron exiting the cyclotron is equal to the kinetic energy it had just before exiting, when it was traveling in a circular path with a radius approximately equal to the radius R of the cyclotron dees. A second **Key Idea** is that we can find the speed v of the deuteron in that circular path with Eq. 29-16 ($r = mv/qB$). Solving that equation for v, substituting R for r, and then substituting known data, we find

$$v = \frac{RqB}{m} = \frac{(0.53\ \text{m})(1.60 \times 10^{-19}\ \text{C})(1.57\ \text{T})}{3.34 \times 10^{-27}\ \text{kg}}$$
$$= 3.99 \times 10^7\ \text{m/s}.$$

This speed corresponds to a kinetic energy of

$$K = \tfrac{1}{2}mv^2$$
$$= \tfrac{1}{2}(3.34 \times 10^{-27}\ \text{kg})(3.99 \times 10^7\ \text{m/s})^2$$
$$= 2.7 \times 10^{-12}\ \text{J}, \qquad \text{(Answer)}$$

or about 17 MeV.

29-7 Magnetic Force on a Current-Carrying Wire

We have already seen (in connection with the Hall effect) that a magnetic field exerts a sideways force on electrons moving in a wire. This force must then be transmitted to the wire itself, because the conduction electrons cannot escape sideways out of the wire.

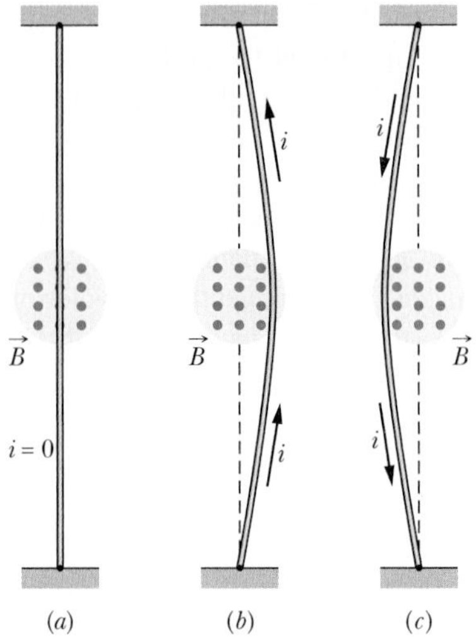

Fig. 29-16 A flexible wire passes between the pole faces of a magnet (only the farther pole face is shown). (*a*) Without current in the wire, the wire is straight. (*b*) With upward current, the wire is deflected rightward. (*c*) With downward current, the deflection is leftward. The connections for getting the current into the wire at one end and out of it at the other end are not shown.

In Fig. 29-16*a*, a vertical wire, carrying no current and fixed in place at both ends, extends through the gap between the vertical pole faces of a magnet. The magnetic field between the faces is directed outward from the page. In Fig. 29-16*b*, a current is sent upward through the wire; the wire deflects to the right. In Fig. 29-16*c*, we reverse the direction of the current and the wire deflects to the left.

Figure 29-17 shows what happens inside the wire of Fig. 29-16. We see one of the conduction electrons, drifting downward with an assumed drift speed v_d. Equation 29-3, in which we must put $\phi = 90°$, tells us that a force $\vec{F}_B$ of magnitude ev_dB must act on each such electron. From Eq. 29-2 we see that this force must be directed to the right. We expect then that the wire as a whole will experience a force to the right, in agreement with Fig. 29-16*b*.

If, in Fig. 29-17, we were to reverse *either* the direction of the magnetic field *or* the direction of the current, the force on the wire would reverse, being directed now to the left. Note too that it does not matter whether we consider negative charges drifting downward in the wire (the actual case) or positive charges drifting upward. The direction of the deflecting force on the wire is the same. We are safe then in dealing with a current of positive charge.

Consider a length L of the wire in Fig. 29-17. All the conduction electrons in this section of wire will drift past plane xx in Fig. 29-17 in a time $t = L/v_d$. Thus, in that time a charge given by

$$q = it = i\frac{L}{v_d}$$

will pass through that plane. Substituting this into Eq. 29-3 yields

$$F_B = qv_dB \sin \phi = \frac{iL}{v_d} v_dB \sin 90°$$

or

$$F_B = iLB. \tag{29-25}$$

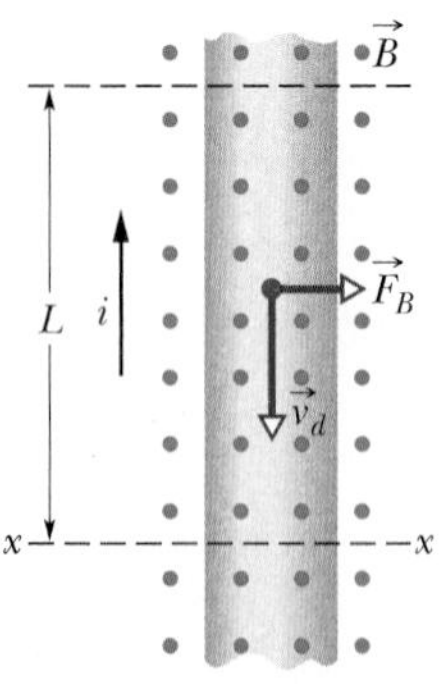

Fig. 29-17 A close-up view of a section of the wire of Fig. 29-16*b*. The current direction is upward, which means that electrons drift downward. A magnetic field that emerges from the plane of the page causes the electrons and the wire to be deflected to the right.

This equation gives the magnetic force that acts on a length L of straight wire carrying a current i and immersed in a magnetic field $\vec{B}$ that is perpendicular to the wire.

If the magnetic field is *not* perpendicular to the wire, as in Fig. 29-18, the magnetic force is given by a generalization of Eq. 29-25:

$$\vec{F}_B = i\vec{L} \times \vec{B} \quad \text{(force on a current).} \tag{29-26}$$

Here $\vec{L}$ is a *length vector* that has magnitude L and is directed along the wire segment in the direction of the (conventional) current. The force magnitude F_B is

$$F_B = iLB \sin \phi, \tag{29-27}$$

where ϕ is the angle between the directions of $\vec{L}$ and $\vec{B}$. The direction of $\vec{F}_B$ is that of the cross product $\vec{L} \times \vec{B}$, because we take current i to be a positive quantity. Equation 29-26 tells us that $\vec{F}_B$ is always perpendicular to the plane defined by vectors $\vec{L}$ and $\vec{B}$, as indicated in Fig. 29-18.

Equation 29-26 is equivalent to Eq. 29-2 in that either can be taken as the

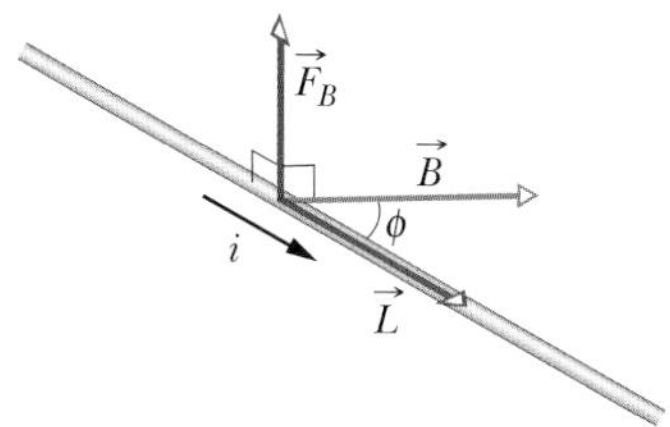

Fig. 29-18 A wire carrying current i makes an angle ϕ with magnetic field $\vec{B}$. The wire has length L in the field and length vector $\vec{L}$ (in the direction of the current). A magnetic force $\vec{F}_B = i\vec{L} \times \vec{B}$ acts on the wire.

defining equation for $\vec{B}$. In practice, we define $\vec{B}$ from Eq. 29-26. It is much easier to measure the magnetic force acting on a wire than that on a single moving charge.

If a wire is not straight or the field is not uniform, we can imagine it broken up into small straight segments and apply Eq. 29-26 to each segment. The force on the wire as a whole is then the vector sum of all the forces on the segments that make it up. In the differential limit, we can write

$$d\vec{F}_B = i\, d\vec{L} \times \vec{B}, \tag{29-28}$$

and we can find the resultant force on any given arrangement of currents by integrating Eq. 29-28 over that arrangement.

In using Eq. 29-28, bear in mind that there is no such thing as an isolated current-carrying wire segment of length dL. There must always be a way to introduce the current into the segment at one end and take it out at the other end.

CHECKPOINT 5: The figure shows a current i through a wire in a uniform magnetic field $\vec{B}$, as well as the magnetic force $\vec{F}_B$ acting on the wire. The field is oriented so that the force is maximum. In what direction is the field?

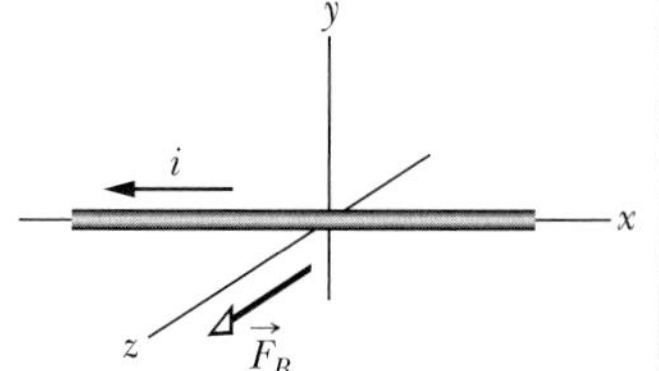

Sample Problem 29-6

A straight, horizontal length of copper wire has a current $i = 28$ A through it. What are the magnitude and direction of the minimum magnetic field $\vec{B}$ needed to suspend the wire—that is, to balance the gravitational force on it? The linear density (mass per unit length) of the wire is 46.6 g/m.

SOLUTION: One **Key Idea** is that, because the wire carries a current, a magnetic force $\vec{F}_B$ can act on the wire if we place it in a magnetic field $\vec{B}$. To balance the downward gravitational force $\vec{F}_g$ on the wire, we want $\vec{F}_B$ to be directed upward (Fig. 29-19).

A second **Key Idea** is that the direction of $\vec{F}_B$ is related to the directions of $\vec{B}$ and the wire's length vector $\vec{L}$ by Eq. 29-26. Because $\vec{L}$ is directed horizontally (and the current is taken to be positive), Eq. 29-26 and the right-hand rule for cross products tell us that $\vec{B}$ must be horizontal and rightward (in Fig. 29-19) to give the required upward $\vec{F}_B$.

The magnitude of $\vec{F}_B$ is given by Eq. 29-27 ($F_B = iLB \sin \phi$). Because we want $\vec{F}_B$ to balance $\vec{F}_g$, we want

$$iLB \sin \phi = mg, \tag{29-29}$$

where mg is the magnitude of $\vec{F}_g$ and m is the mass of the wire. We also want the minimal field magnitude B for $\vec{F}_B$ to balance $\vec{F}_g$. Thus, we need to maximize $\sin \phi$ in Eq. 29-29. To do so, we set $\phi = 90°$, thereby arranging for $\vec{B}$ to be perpendicular to the wire. We then have $\sin \phi = 1$, so Eq. 29-29 yields

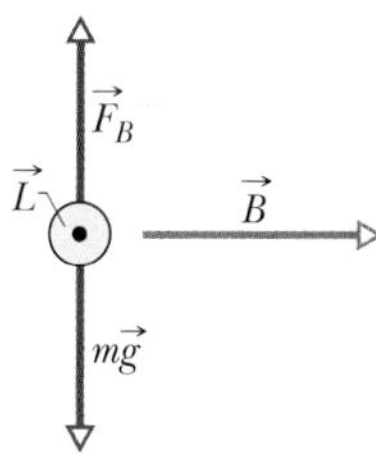

Fig. 29-19 Sample Problem 29-6. A current-carrying wire (shown in cross section) can be made to "float" in a magnetic field. The current in the wire emerges from the plane of the page, and the magnetic field is directed to the right.

$$B = \frac{mg}{iL \sin \phi} = \frac{(m/L)g}{i}. \tag{29-30}$$

We write the result this way because we know m/L, the linear density of the wire. Substituting known data then gives us

$$B = \frac{(46.6 \times 10^{-3}\ \text{kg/m})(9.8\ \text{m/s}^2)}{28\ \text{A}}$$

$$= 1.6 \times 10^{-2}\ \text{T}. \quad \text{(Answer)}$$

This is about 160 times the strength of Earth's magnetic field.

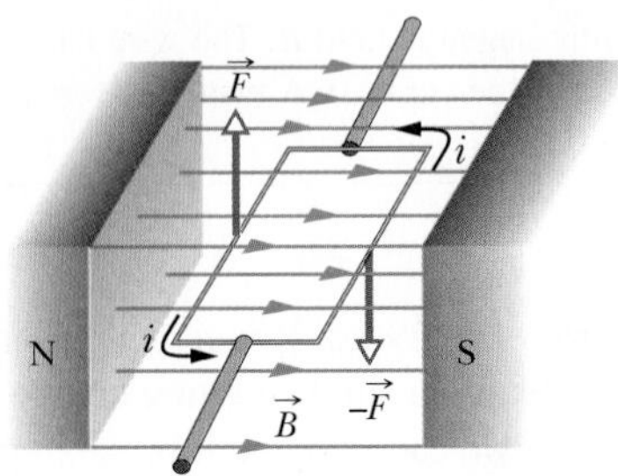

Fig. 29-20 The elements of an electric motor. A rectangular loop of wire, carrying a current and free to rotate about a fixed axis, is placed in a magnetic field. Magnetic forces on the wire produce a torque that rotates it. A commutator (not shown) reverses the direction of the current every half-revolution so that the torque always acts in the same direction.

29-8 Torque on a Current Loop

Much of the world's work is done by electric motors. The forces behind this work are the magnetic forces that we studied in the preceding section—that is, the forces that a magnetic field exerts on a wire that carries a current.

Figure 29-20 shows a simple motor, consisting of a single current-carrying loop immersed in a magnetic field $\vec{B}$. The two magnetic forces $\vec{F}$ and $-\vec{F}$ produce a torque on the loop, tending to rotate it about its central axis. Although many essential details have been omitted, the figure does suggest how the action of a magnetic field on a current loop produces rotary motion. Let us analyze that action.

Figure 29-21*a* shows a rectangular loop of sides *a* and *b*, carrying current *i* through uniform magnetic field $\vec{B}$. We place it in the field so that its long sides, labeled 1 and 3, are perpendicular to the field direction (which is into the page), but its short sides, labeled 2 and 4, are not. Wires to lead the current into and out of the loop are needed but, for simplicity, they are not shown.

To define the orientation of the loop in the magnetic field, we use a normal vector $\vec{n}$ that is perpendicular to the plane of the loop. Figure 29-21*b* shows a right-hand rule for finding the direction of $\vec{n}$. Point or curl the fingers of your right hand in the direction of the current at any point on the loop. Your extended thumb then points in the direction of the normal vector $\vec{n}$.

In Fig. 29-21*c*, the normal vector of the loop is shown at an arbitrary angle θ to the direction of the magnetic field $\vec{B}$. We wish to find the net force and net torque acting on the loop in this orientation.

The net force on the loop is the vector sum of the forces acting on its four sides. For side 2 the vector $\vec{L}$ in Eq. 29-26 points in the direction of the current and has magnitude *b*. The angle between $\vec{L}$ and $\vec{B}$ for side 2 (see Fig. 29-21*c*) is $90° - \theta$. Thus, the magnitude of the force acting on this side is

$$F_2 = ibB \sin(90° - \theta) = ibB \cos \theta. \tag{29-31}$$

You can show that the force $\vec{F}_4$ acting on side 4 has the same magnitude as $\vec{F}_2$ but the opposite direction. Thus, $\vec{F}_2$ and $\vec{F}_4$ cancel out exactly. Their net force is zero and, because their common line of action is through the center of the loop, their net torque is also zero.

The situation is different for sides 1 and 3. For them, $\vec{L}$ is perpendicular to $\vec{B}$,

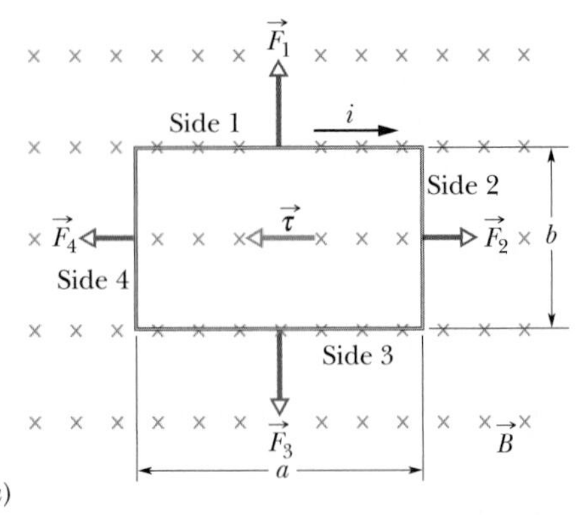

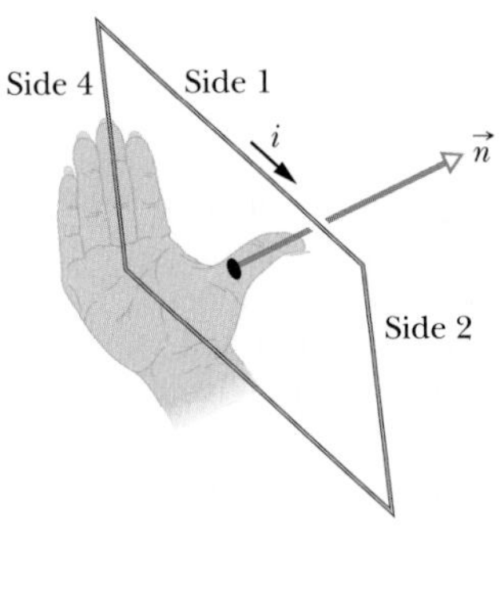

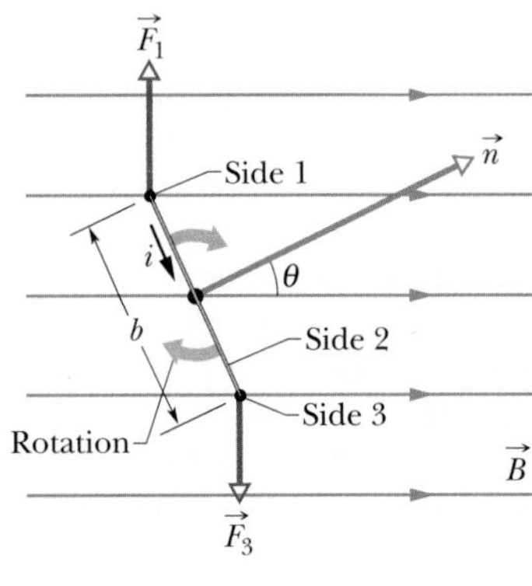

Fig. 29-21 A rectangular loop, of length *a* and width *b* and carrying a current *i*, is located in a uniform magnetic field. A torque $\vec{\tau}$ acts to align the normal vector $\vec{n}$ with the direction of the field. (*a*) The loop as seen by looking in the direction of the magnetic field. (*b*) A perspective of the loop showing how the right-hand rule gives the direction of $\vec{n}$, which is perpendicular to the plane of the loop. (*c*) A side view of the loop, from side 2. The loop rotates as indicated.

so the forces $\vec{F}_1$ and $\vec{F}_3$ have the common magnitude iaB. Because these two forces have opposite directions, they do not tend to move the loop up or down. However, as Fig. 29-21c shows, these two forces do *not* share the same line of action so they *do* produce a net torque. The torque tends to rotate the loop so as to align its normal vector $\vec{n}$ with the direction of the magnetic field $\vec{B}$. That torque has moment arm $(b/2)\sin\theta$ about the central axis of the loop. The magnitude τ' of the torque due to forces $\vec{F}_1$ and $\vec{F}_3$ is then (see Fig. 29-21c)

$$\tau' = \left(iaB\frac{b}{2}\sin\theta\right) + \left(iaB\frac{b}{2}\sin\theta\right) = iabB\sin\theta. \tag{29-32}$$

Suppose we replace the single loop of current with a *coil* of N loops, or *turns*. Further, suppose that the turns are wound tightly enough that they can be approximated as all having the same dimensions and lying in a plane. Then the turns form a *flat coil* and a torque τ' with the magnitude given in Eq. 29-32 acts on each of them. The total torque on the coil then has magnitude

$$\tau = N\tau' = NiabB\sin\theta = (NiA)B\sin\theta, \tag{29-33}$$

in which A $(= ab)$ is the area enclosed by the coil. The quantities in parentheses (NiA) are grouped together because they are all properties of the coil: its number of turns, its area, and the current it carries. Equation 29-33 holds for all flat coils, no matter what their shape, provided the magnetic field is uniform.

Instead of focusing on the motion of the coil, it is simpler to keep track of the vector $\vec{n}$, which is normal to the plane of the coil. Equation 29-33 tells us that a current-carrying flat coil placed in a magnetic field will tend to rotate so that $\vec{n}$ has the same direction as the field.

In a motor, the current in the coil is reversed as $\vec{n}$ begins to line up with the field direction, so that a torque continues to rotate the coil. This automatic reversal of the current is done via a commutator that electrically connects the rotating coil with the stationary contacts on the wires that supply the current from some source.

Sample Problem 29-7

Analog voltmeters and ammeters work by measuring the torque exerted by a magnetic field on a current-carrying coil. The reading is displayed by means of the deflection of a pointer over a scale. Figure 29-22 shows the essentials of a *galvanometer*, on which both analog ammeters and analog voltmeters are based. Assume the coil is 2.1 cm high and 1.2 cm wide, has 250 turns, and is mounted so that it can rotate about an axis (into the page) in a uniform *radial* magnetic field with $B = 0.23$ T. For any orientation of the coil, the net magnetic field through the coil is perpendicular to the normal vector of the coil (and thus parallel to the plane of the coil). A spring Sp provides a countertorque that balances the magnetic torque, so that a given steady current i in the coil results in a steady angular deflection ϕ. The greater the current is, the greater the deflection is, and thus the greater the torque required of the spring is. If a current of 100 μA produces an angular deflection of 28°, what must be the torsional constant κ of the spring, as used in Eq. 16-22 ($\tau = -\kappa\phi$)?

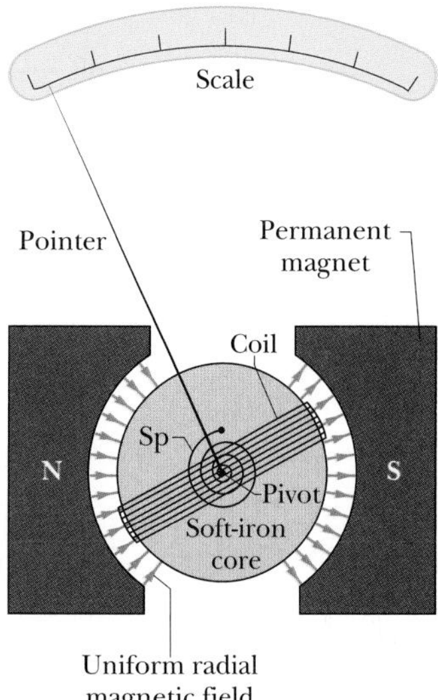

Fig. 29-22 Sample Problem 29-7. The elements of a galvanometer. Depending on the external circuit, this device can be wired up as either a voltmeter or an ammeter.

SOLUTION: The **Key Idea** here is that, with a constant current through the device, the resulting magnetic torque (Eq. 29-33) is balanced by the spring torque. Thus, the magnitudes of those torques are equal:

$$NiAB \sin \theta = \kappa\phi. \tag{29-34}$$

Here ϕ is the angular deflection of the coil and pointer, and A ($= 2.52 \times 10^{-4}$ m^2) is the area encircled by the coil. Since the net magnetic field through the coil is always perpendicular to the normal vector of the coil, $\theta = 90°$ for any orientation of the pointer.

Solving Eq. 29-34 for κ, we find

$$\begin{aligned}\kappa &= \frac{NiAB \sin \theta}{\phi} \\ &= (250)(100 \times 10^{-6}\ \text{A})(2.52 \times 10^{-4}\ \text{m}^2) \\ &\quad \times \frac{(0.23\ \text{T})(\sin 90°)}{28°} \\ &= 5.2 \times 10^{-8}\ \text{N} \cdot \text{m/degree}. \qquad \text{(Answer)}\end{aligned}$$

Many modern ammeters and voltmeters are of the digital, direct-reading type and do not use a moving coil.

29-9 The Magnetic Dipole Moment

We can describe the current-carrying coil of the preceding section with a single vector $\vec{\mu}$, its **magnetic dipole moment.** We take the direction of $\vec{\mu}$ to be that of the normal vector $\vec{n}$ to the plane of the coil, as in Fig. 29-21*c*. We define the magnitude of $\vec{\mu}$ as

$$\mu = NiA \qquad \text{(magnetic moment)}, \tag{29-35}$$

in which N is the number of turns in the coil, i is the current through the coil, and A is the area enclosed by each turn of the coil. (Equation 29-35 tells us that the unit of $\vec{\mu}$ is the ampere-square meter.) Using $\vec{\mu}$, we can rewrite Eq. 29-33 for the torque on the coil due to a magnetic field as

$$\tau = \mu B \sin \theta, \tag{29-36}$$

in which θ is the angle between the vectors $\vec{\mu}$ and $\vec{B}$.

We can generalize this to the vector relation

$$\vec{\tau} = \vec{\mu} \times \vec{B}, \tag{29-37}$$

which reminds us very much of the corresponding equation for the torque exerted by an *electric* field on an *electric* dipole—namely, Eq. 23-34:

$$\vec{\tau} = \vec{p} \times \vec{E}.$$

In each case the torque due to the field—either magnetic or electric—is equal to the vector product of the corresponding dipole moment and the field vector.

A magnetic dipole in an external magnetic field has a **magnetic potential energy** that depends on the dipole's orientation in the field. For electric dipoles we have shown (Eq. 23-38) that

$$U(\theta) = -\vec{p} \cdot \vec{E}.$$

In strict analogy, we can write for the magnetic case

$$U(\theta) = -\vec{\mu} \cdot \vec{B}. \tag{29-38}$$

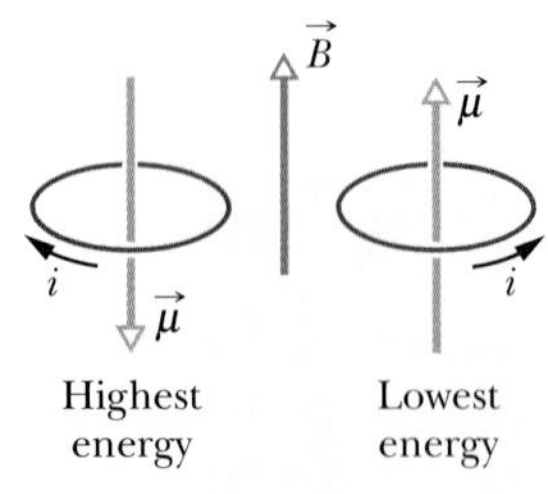

Fig. 29-23 The orientations of highest and lowest energy of a magnetic dipole (here a coil carrying current) in an external magnetic field $\vec{B}$. The direction of the current i gives the direction of the magnetic dipole moment $\vec{\mu}$ via the right-hand rule shown for $\vec{n}$ in Fig. 29-21*b*.

A magnetic dipole has its lowest energy ($= -\mu B \cos 0 = -\mu B$) when its dipole moment $\vec{\mu}$ is lined up with the magnetic field (Fig. 29-23). It has its highest energy ($= -\mu B \cos 180° = +\mu B$) when $\vec{\mu}$ is directed opposite the field.

When a magnetic dipole rotates from an initial orientation θ_i to another orientation θ_f, the work W done on the dipole by the magnetic field is

$$W = -\Delta U = -(U_f - U_i), \tag{29-39}$$

where U_f and U_i are calculated with Eq. 29-38. If an applied torque (due to "an

TABLE 29-2 Some Magnetic Dipole Moments

A small bar magnet	5 J/T
Earth	8.0×10^{22} J/T
A proton	1.4×10^{-26} J/T
An electron	9.3×10^{-24} J/T

external agent") acts on the dipole during the change in its orientation, then work W_a is done on the dipole by the applied torque. *If the dipole is stationary* before and after the change in its orientation, then work W_a is the negative of the work done on the dipole by the field. Thus,

$$W_a = -W = U_f - U_i. \tag{29-40}$$

So far, we have identified only a current-carrying coil as a magnetic dipole. However, a simple bar magnet is also a magnetic dipole, as is a rotating sphere of charge. Earth itself is (approximately) a magnetic dipole. Finally, most subatomic particles, including the electron, the proton, and the neutron, have magnetic dipole moments. As you will see in Chapter 32, all these quantities can be viewed as current loops. For comparison, some approximate magnetic dipole moments are shown in Table 29-2.

CHECKPOINT 6: The figure shows four orientations, at angle θ, of a magnetic dipole moment $\vec{\mu}$ in a magnetic field. Rank the orientations according to (a) the magnitude of the torque on the dipole and (b) the potential energy of the dipole, greatest first.

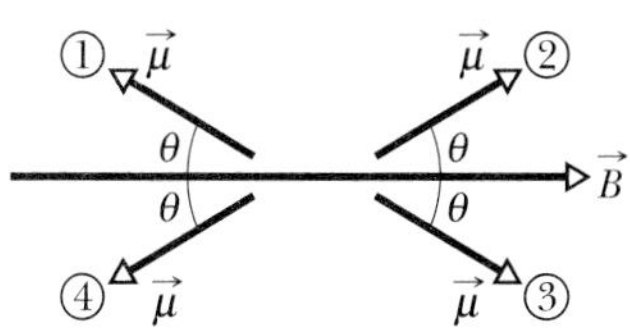

Sample Problem 29-8

Figure 29-24 shows a circular coil with 250 turns, an area A of 2.52×10^{-4} m^2, and a current of 100 μA. The coil is at rest in a uniform magnetic field of magnitude $B = 0.85$ T, with its magnetic dipole moment $\vec{\mu}$ initially aligned with $\vec{B}$.

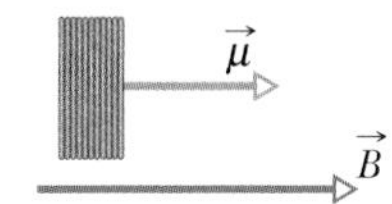

Fig. 29-24 Sample Problem 29-8. A side view of a circular coil carrying a current and oriented so that its magnetic dipole moment $\vec{\mu}$ is aligned with magnetic field $\vec{B}$.

(a) In Fig. 29-24, what is the direction of the current in the coil?

SOLUTION: The **Key Idea** here is to apply the following right-hand rule to the coil: Imagine cupping the coil with your right hand so that your right thumb is outstretched in the direction of $\vec{\mu}$. The direction in which your fingers curl around the coil is the direction of the current in the coil. Thus, in the wires on the near side of the coil—those we see in Fig. 29-24—the current is from top to bottom.

(b) How much work would the torque applied by an external agent have to do on the coil to rotate it 90° from its initial orientation, so that $\vec{\mu}$ is perpendicular to $\vec{B}$ and the coil is again at rest?

SOLUTION: The **Key Idea** here is that the work W_a done by the applied torque would be equal to the change in the coil's potential energy due to its change in orientation. From Eq. 29-40 ($W_a = U_f - U_i$), we find

$$\begin{aligned} W_a &= U(90°) - U(0°) \\ &= -\mu B \cos 90° - (-\mu B \cos 0°) = 0 + \mu B \\ &= \mu B. \end{aligned}$$

Substituting for μ from Eq. 29-35 ($\mu = NiA$), we find that

$$\begin{aligned} W_a &= (NiA)B \\ &= (250)(100 \times 10^{-6}\ \text{A})(2.52 \times 10^{-4}\ \text{m}^2)(0.85\ \text{T}) \\ &= 5.356 \times 10^{-6}\ \text{J} \approx 5.4\ \mu\text{J}. \end{aligned} \quad \text{(Answer)}$$

REVIEW & SUMMARY

Magnetic Field $\vec{B}$ A **magnetic field** $\vec{B}$ is defined in terms of the force $\vec{F}_B$ acting on a test particle with charge q moving through the field with velocity $\vec{v}$:

$$\vec{F}_B = q\vec{v} \times \vec{B}. \tag{29-2}$$

The SI unit for $\vec{B}$ is the **tesla** (T): 1 T = 1 N/(A · m) = 10^4 gauss.

The Hall Effect When a conducting strip of thickness l carrying a current i is placed in a uniform magnetic field $\vec{B}$, some charge carriers (with charge e) build up on the sides of the conductor, creating a potential difference V across the strip. The polarities of the sides indicate the sign of the charge carriers; the number density n of charge carriers can be calculated with

$$n = \frac{Bi}{Vle}. \tag{29-12}$$

A Charged Particle Circulating in a Magnetic Field A charged particle with mass m and charge magnitude q moving with velocity $\vec{v}$ perpendicular to a uniform magnetic field $\vec{B}$ will travel in a circle. Applying Newton's second law to the circular motion yields

$$qvB = \frac{mv^2}{r}, \tag{29-15}$$

from which we find the radius r of the circle to be

$$r = \frac{mv}{qB}. \tag{29-16}$$

The frequency of revolution f, the angular frequency ω, and the period of the motion T are given by

$$f = \frac{\omega}{2\pi} = \frac{1}{T} = \frac{qB}{2\pi m}. \qquad \text{(29-19, 29-18, 29-17)}$$

Cyclotrons and Synchrotrons A cyclotron is a particle accelerator that uses a magnetic field to hold a charged particle in a circular orbit of increasing radius so that a modest accelerating potential may act on the particle repeatedly, providing it with high energy. Because the moving particle gets out of step with the oscillator as its speed approaches that of light, there is an upper limit to the energy attainable with the cyclotron. A synchrotron avoids this difficulty. Here both B and the oscillator frequency f_{osc} are programmed to change cyclically so that the particle not only can go to high energies but can do so at a constant orbital radius.

Magnetic Force on a Current-Carrying Wire A straight wire carrying a current i in a uniform magnetic field experiences a sideways force

$$\vec{F}_B = i\vec{L} \times \vec{B}. \tag{29-26}$$

The force acting on a current element $i\,d\vec{L}$ in a magnetic field is

$$d\vec{F}_B = i\,d\vec{L} \times \vec{B}. \tag{29-28}$$

The direction of the length vector $\vec{L}$ or $d\vec{L}$ is that of the current i.

Torque on a Current-Carrying Coil A coil (of area A and carrying current i, with N turns) in a uniform magnetic field $\vec{B}$ will experience a torque $\vec{\tau}$ given by

$$\vec{\tau} = \vec{\mu} \times \vec{B}. \tag{29-37}$$

Here $\vec{\mu}$ is the **magnetic dipole moment** of the coil, with magnitude $\mu = NiA$ and direction given by the right-hand rule.

Orientation Energy of a Magnetic Dipole The **magnetic potential energy** of a magnetic dipole in a magnetic field is

$$U(\theta) = -\vec{\mu}\cdot\vec{B}. \tag{29-38}$$

If a magnetic dipole rotates from an initial orientation θ_i to another orientation θ_f, the work W done on the dipole by the magnetic field is

$$W = -\Delta U = -(U_f - U_i). \tag{29-39}$$

QUESTIONS

1. Figure 29-25 shows three situations in which a positive particle of velocity $\vec{v}$ moves through a uniform magnetic field $\vec{B}$ and experiences a magnetic force $\vec{F}_B$. In each situation, determine whether the orientations of the vectors are physically reasonable.

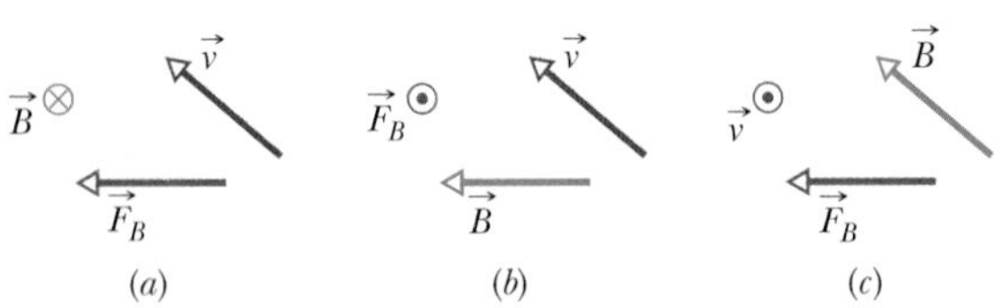

Fig. 29-25 Question 1.

2. For four situations, here is the velocity $\vec{v}$ of a proton at a certain instant as it moves through a uniform magnetic field $\vec{B}$:

(a) $\vec{v} = 2\hat{i} - 3\hat{j}$ and $\vec{B} = 4\hat{k}$
(b) $\vec{v} = 3\hat{i} + 2\hat{j}$ and $\vec{B} = -4\hat{k}$
(c) $\vec{v} = 3\hat{j} - 2\hat{k}$ and $\vec{B} = 4\hat{i}$
(d) $\vec{v} = 20\hat{i}$ and $\vec{B} = -4\hat{i}$.

Without written calculation, rank the situations according to the magnitude of the magnetic force on the proton, greatest first.

3. In Section 29-3, we discussed a charged particle moving through crossed fields with the forces $\vec{F}_E$ and $\vec{F}_B$ in opposition. We found that the particle moves in a straight line (that is, neither force dominates the motion) if its speed is given by Eq. 29-7 ($v = E/B$). Which of the two forces dominates if the speed of the particle is, instead, (a) $v < E/B$ and (b) $v > E/B$?

4. Figure 29-26 shows crossed and uniform electric and magnetic fields $\vec{E}$ and $\vec{B}$ and, at a certain instant, the velocity vectors of the 10 charged particles listed in Table 29-3. (The vectors are not drawn to scale.) The table gives the signs of the charges and the speeds of the particles; the speeds are given as either less than or greater than E/B (see Question 3). Which particles will move out of the page toward you after the instant of Fig. 29-26?

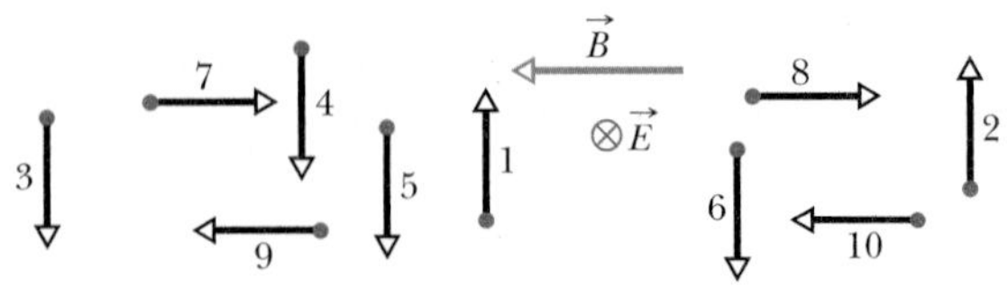

Fig. 29-26 Question 4.

TABLE 29-3 Question 4

Particle	Charge	Speed	Particle	Charge	Speed
1	+	Less	6	−	Greater
2	+	Greater	7	+	Less
3	+	Less	8	+	Greater
4	+	Greater	9	−	Less
5	−	Less	10	−	Greater

5. In Fig. 29-27, a charged particle enters a uniform magnetic field $\vec{B}$ with speed v_0, moves through a half-circle in time T_0, and then leaves the field. (a) Is the charge positive or negative? (b) Is the final speed of the particle greater than, less than, or equal to v_0? (c) If the initial speed had been $0.5v_0$, would the time spent in field $\vec{B}$ have been greater than, less than, or equal to T_0? (d) Would the path have been a half-circle, more than a half-circle, or less than a half-circle?

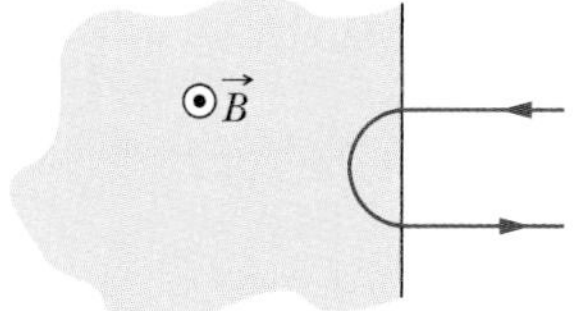

Fig. 29-27 Question 5.

6. Figure 29-28 shows the path of a particle through six regions of uniform magnetic field, where the path is either a half-circle or a quarter-circle. Upon leaving the last region, the particle travels between two charged, parallel plates and is deflected toward the plate of higher potential. What are the directions of the magnetic fields in the six regions?

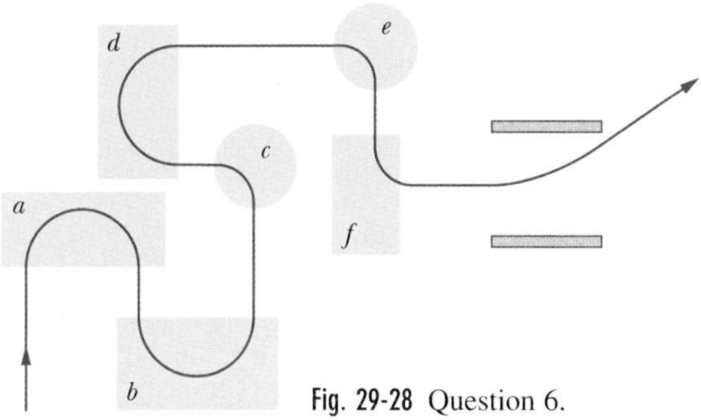

Fig. 29-28 Question 6.

7. Figure 29-29 shows the path of an electron that passes through two regions containing uniform magnetic fields of magnitudes B_1 and B_2. Its path in each region is a half-circle. (a) Which field is stronger? (b) What are the directions of the two fields? (c) Is the time spent by the electron in the $\vec{B}_1$ region greater than, less than, or the same as the time spent in the $\vec{B}_2$ region?

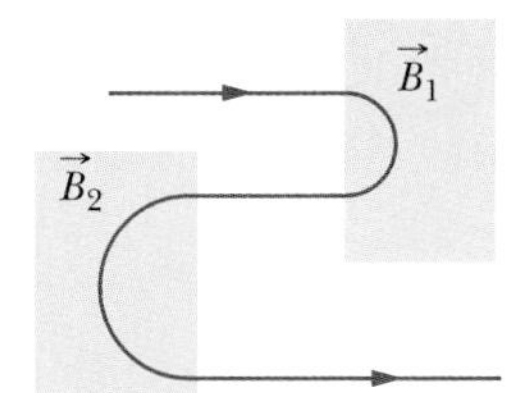

Fig. 29-29 Question 7.

8. *Particle Roundabout.* Figure 29-30 shows 11 paths through a region of uniform magnetic field. One path is a straight line; the rest are half-circles. Table 29-4 gives the masses, charges, and speeds of 11 particles that take these paths through the field in the directions shown. Which path in the figure corresponds to which particle in the table?

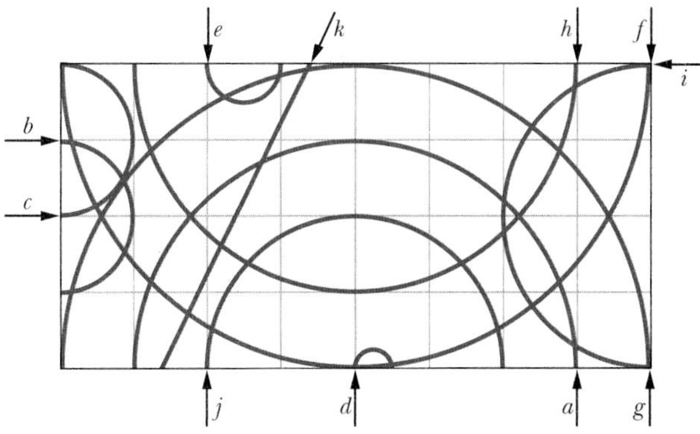

Fig. 29-30 Question 8.

TABLE 29-4 Question 8

Particle	Mass	Charge	Speed
1	$2m$	q	v
2	m	$2q$	v
3	$m/2$	q	$2v$
4	$3m$	$3q$	$3v$
5	$2m$	q	$2v$
6	m	$-q$	$2v$
7	m	$-4q$	v
8	m	$-q$	v
9	$2m$	$-2q$	$3v$
10	m	$-2q$	$8v$
11	$3m$	0	$3v$

9. Figure 29-31 shows eight wires that carry identical currents through the same uniform magnetic field (directed into the page) in eight separate experiments. Each wire consists of two straight sections (each of length L and either parallel or perpendicular to the x and y axes shown) and one curved section (with radius of curvature R). The directions of the currents through the wires are indicated by the arrows next to the wires. (a) Give the direction of the net magnetic force on each wire in terms of an angle measured counterclockwise from the positive direction of the x axis. (b) Rank wires 1 through 4 according to the magnitude of the net magnetic force on them, greatest first. (c) Do the same type of ranking for wires 5 through 8.

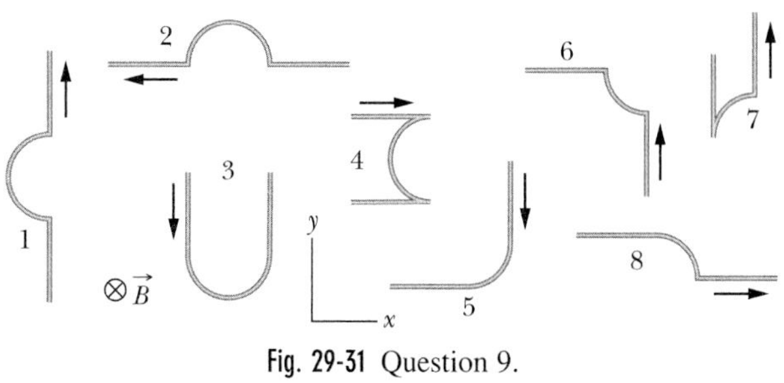

Fig. 29-31 Question 9.

10. (a) In Checkpoint 6, if the dipole moment $\vec{\mu}$ rotates from orientation 1 to orientation 2, is the work done on the dipole *by the magnetic field* positive, negative, or zero? (b) Rank the work done on the dipole by the magnetic field for rotations from orientation 1 to (1) orientation 2, (2) orientation 3, and (3) orientation 4, greatest first.

EXERCISES & PROBLEMS

ssm Solution is in the Student Solutions Manual.
www Solution is available on the World Wide Web at: http://www.wiley.com/college/hrw
ilw Solution is available on the Interactive LearningWare.

SEC. 29-2 The Definition of $\vec{B}$

1E. An alpha particle travels at a velocity $\vec{v}$ of magnitude 550 m/s through a uniform magnetic field $\vec{B}$ of magnitude 0.045 T. (An alpha particle has a charge of $+3.2 \times 10^{-19}$ C and a mass of 6.6×10^{-27} kg.) The angle between $\vec{v}$ and $\vec{B}$ is 52°. What are the magnitudes of (a) the force $\vec{F}_B$ acting on the particle due to the field and (b) the acceleration of the particle due to $\vec{F}_B$? (c) Does the speed of the particle increase, decrease, or remain equal to 550 m/s?

2E. An electron in a TV camera tube is moving at 7.20×10^6 m/s in a magnetic field of strength 83.0 mT. (a) Without knowing the direction of the field, what can you say about the greatest and least magnitudes of the force acting on the electron due to the field? (b) At one point the electron has an acceleration of magnitude 4.90×10^{14} m/s^2. What is the angle between the electron's velocity and the magnetic field?

3E. A proton traveling at 23.0° with respect to the direction of a magnetic field of strength 2.60 mT experiences a magnetic force of 6.50×10^{-17} N. Calculate (a) the proton's speed and (b) its kinetic energy in electron-volts. ssm ilw

4P. An electron that has velocity

$$\vec{v} = (2.0 \times 10^6 \text{ m/s})\hat{\imath} + (3.0 \times 10^6 \text{ m/s})\hat{\jmath}$$

moves through the magnetic field $\vec{B} = (0.030 \text{ T})\hat{\imath} - (0.15 \text{ T})\hat{\jmath}$. (a) Find the force on the electron. (b) Repeat your calculation for a proton having the same velocity.

5P. Each of the electrons in the beam of a television tube has a kinetic energy of 12.0 keV. The tube is oriented so that the electrons move horizontally from geomagnetic south to geomagnetic north. The vertical component of Earth's magnetic field points down and has a magnitude of 55.0 μT. (a) In what direction will the beam deflect? (b) What is the acceleration of a single electron due to the magnetic field? (c) How far will the beam deflect in moving 20.0 cm through the television tube? ssm www

SEC. 29-3 Crossed Fields: Discovery of the Electron

6E. A proton travels through uniform magnetic and electric fields. The magnetic field is $\vec{B} = -2.5\hat{\imath}$ mT. At one instant the velocity of the proton is $\vec{v} = 2000\hat{\jmath}$ m/s. At that instant, what is the magnitude of the net force acting on the proton if the electric field is (a) $4.0\hat{k}$ V/m, (b) $-4.0\hat{k}$ V/m, and (c) $4.0\hat{\imath}$ V/m?

7E. An electron with kinetic energy 2.5 keV moves horizontally into a region of space in which there is a downward-directed uniform electric field of magnitude 10 kV/m. (a) What are the magnitude and direction of the (smallest) uniform magnetic field that will cause the electron to continue to move horizontally? Ignore the gravitational force, which is rather small. (b) Is it possible for a proton to pass through this combination of fields undeflected? If so, under what circumstances? ssm

8E. An electric field of 1.50 kV/m and a magnetic field of 0.400 T act on a moving electron to produce no net force. (a) Calculate the minimum speed v of the electron. (b) Draw the vectors $\vec{E}$, $\vec{B}$, and $\vec{v}$.

9P. An electron is accelerated through a potential difference of 1.0 kV and directed into a region between two parallel plates separated by 20 mm with a potential difference of 100 V between them. The electron is moving perpendicular to the electric field of the plates when it enters the region between the plates. What uniform magnetic field, applied perpendicular to both the electron path and the electric field, will allow the electron to travel in a straight line? ilw

10P. An electron has an initial velocity of $(12.0\hat{\jmath} + 15.0\hat{k})$ km/s and a constant acceleration of $(2.00 \times 10^{12} \text{ m/s}^2)\hat{\imath}$ in a region in which uniform electric and magnetic fields are present. If $\vec{B} = (400\ \mu\text{T})\hat{\imath}$, find the electric field $\vec{E}$.

11P. An ion source is producing ions of ^{6}Li (mass = 6.0 u), each with a charge of $+e$. The ions are accelerated by a potential difference of 10 kV and pass horizontally into a region in which there is a uniform vertical magnetic field of magnitude $B = 1.2$ T. Calculate the strength of the smallest electric field, to be set up over the same region, that will allow the ^{6}Li ions to pass through undeflected. ssm

SEC. 29-4 Crossed Fields: The Hall Effect

12E. A strip of copper 150 μm wide is placed in a uniform magnetic field $\vec{B}$ of magnitude 0.65 T, with $\vec{B}$ perpendicular to the strip. A current $i = 23$ A is then sent through the strip such that a Hall potential difference V appears across the width of the strip. Calculate V. (The number of charge carriers per unit volume for copper is 8.47×10^{28} electrons/m^3.)

13P. (a) In Fig. 29-8, show that the ratio of the Hall electric field E to the electric field E_C responsible for moving charge (the current) along the length of the strip is

$$\frac{E}{E_C} = \frac{B}{ne\rho},$$

where ρ is the resistivity of the material and n is the number density of the charge carriers. (b) Compute this ratio numerically for Exercise 12. (See Table 27-1.)

14P. A metal strip 6.50 cm long, 0.850 cm wide, and 0.760 mm thick moves with constant velocity $\vec{v}$ through a uniform magnetic field $B = 1.20$ mT directed perpendicular to the strip, as shown in Fig. 29-32. A potential difference of 3.90 μV is measured between points x and y across the strip. Calculate the speed v.

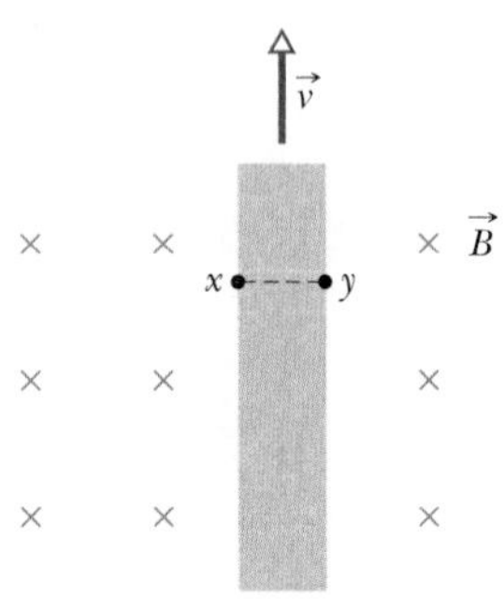

Fig. 29-32 Problem 14.

SEC. 29-5 A Circulating Charged Particle

15E. What uniform magnetic field, applied perpendicular to a beam of electrons moving at 1.3×10^6 m/s, is required to make the electrons travel in a circular arc of radius 0.35 m? ssm

16E. An electron is accelerated from rest by a potential difference of 350 V. It then enters a uniform magnetic field of magnitude 200 mT with its velocity perpendicular to the field. Calculate (a) the speed of the electron and (b) the radius of its path in the magnetic field.

17E. An electron with kinetic energy 1.20 keV circles in a plane perpendicular to a uniform magnetic field. The orbit radius is 25.0 cm. Find (a) the speed of the electron, (b) the magnetic field, (c) the frequency of circling, and (d) the period of the motion. ssm

18E. Physicist S. A. Goudsmit devised a method for measuring the masses of heavy ions by timing their periods of revolution in a known magnetic field. A singly charged ion of iodine makes 7.00 rev in a field of 45.0 mT in 1.29 ms. Calculate its mass, in unified atomic mass units. (Actually, the method allows mass measurements to be carried out to much greater accuracy than these approximate data suggest.)

19E. (a) Find the frequency of revolution of an electron with an energy of 100 eV in a uniform magnetic field of 35.0 μT. (b) Calculate the radius of the path of this electron if its velocity is perpendicular to the magnetic field. ilw

20E. An alpha particle ($q = +2e$, $m = 4.00$ u) travels in a circular path of radius 4.50 cm in a uniform magnetic field with $B =$ 1.20 T. Calculate (a) its speed, (b) its period of revolution, (c) its kinetic energy in electron-volts, and (d) the potential difference through which it would have to be accelerated to achieve this energy.

21E. A beam of electrons whose kinetic energy is K emerges from a thin-foil "window" at the end of an accelerator tube. There is a metal plate a distance d from this window and perpendicular to the direction of the emerging beam (Fig. 29-33). Show that we can prevent the beam from hitting the plate if we apply a uniform magnetic field $\vec{B}$ such that

$$B \geq \sqrt{\frac{2mK}{e^2d^2}},$$

in which m and e are the electron mass and charge. How should $\vec{B}$ be oriented? ssm

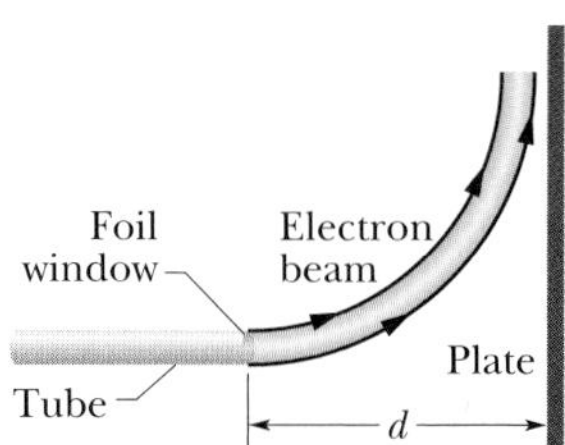

Fig. 29-33 Exercise 21.

22P. A source injects an electron of speed $v = 1.5 \times 10^7$ m/s into a uniform magnetic field of magnitude $B = 1.0 \times 10^{-3}$ T. The velocity of the electron makes an angle $\theta = 10°$ with the direction of the magnetic field. Find the distance d from the point of injection at which the electron next crosses the field line that passes through the injection point.

23P. In a nuclear experiment a proton with kinetic energy 1.0 MeV moves in a circular path in a uniform magnetic field. What energy must (a) an alpha particle ($q = +2e$, $m = 4.0$ u) and (b) a deuteron ($q = +e$, $m = 2.0$ u) have if they are to circulate in the same circular path?

24P. A proton, a deuteron ($q = +e$, $m = 2.0$ u), and an alpha particle ($q = +2e$, $m = 4.0$ u) with the same kinetic energies enter a region of uniform magnetic field $\vec{B}$, moving perpendicular to $\vec{B}$. Compare the radii of their circular paths.

25P. A certain commercial mass spectrometer (see Sample Problem 29-3) is used to separate uranium ions of mass 3.92×10^{-25} kg and charge 3.20×10^{-19} C from related species. The ions are accelerated through a potential difference of 100 kV and then pass into a uniform magnetic field, where they are bent in a path of radius 1.00 m. After traveling through 180° and passing through a slit of width 1.00 mm and height 1.00 cm, they are collected in a cup. (a) What is the magnitude of the (perpendicular) magnetic field in the separator? If the machine is used to separate out 100 mg of material per hour, calculate (b) the current of the desired ions in the machine and (c) the thermal energy produced in the cup in 1.00 h. ssm

26P. A proton of charge $+e$ and mass m enters a uniform magnetic field $\vec{B} = B\hat{i}$ with an initial velocity $\vec{v} = v_{0x}\hat{i} + v_{0y}\hat{j}$. Find an expression in unit-vector notation for its velocity $\vec{v}$ at any later time t.

27P. A positron with kinetic energy 2.0 keV is projected into a uniform magnetic field $\vec{B}$ of magnitude 0.10 T, with its velocity vector making an angle of 89° with $\vec{B}$. Find (a) the period, (b) the pitch p, and (c) the radius r of its helical path. ssm www

28P. In Fig. 29-34, a charged particle moves into a region of uniform magnetic field $\vec{B}$, goes through half a circle, and then exits that region. The particle is either a proton or an electron (you must decide which). It spends 130 ns within the region. (a) What is the magnitude of $\vec{B}$? (b) If the particle is sent back through the magnetic field (along the same initial path) but with 2.00 times its previous kinetic energy, how much time does it spend within the field?

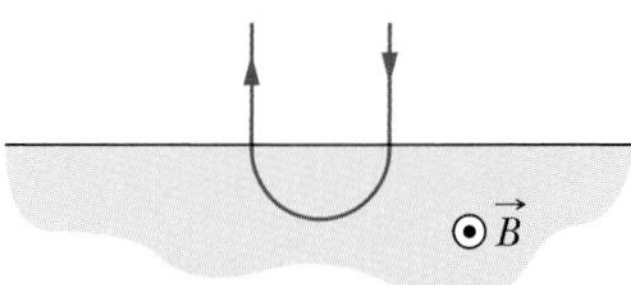

Fig. 29-34 Problem 28.

29P. A neutral particle is at rest in a uniform magnetic field $\vec{B}$. At time $t = 0$ it decays into two charged particles, each of mass m. (a) If the charge of one of the particles is $+q$, what is the charge of the other? (b) The two particles move off in separate paths, both of which lie in the plane perpendicular to $\vec{B}$. At a later time the particles collide. Express the time from decay until collision in terms of m, B, and q. ssm

SEC. 29-6 Cyclotrons and Synchrotrons

30E. In a certain cyclotron a proton moves in a circle of radius 0.50 m. The magnitude of the magnetic field is 1.2 T. (a) What is the oscillator frequency? (b) What is the kinetic energy of the proton, in electron-volts?

31P. Estimate the total path length traveled by a deuteron in the cyclotron of Sample Problem 29-5 during the (entire) acceleration process. Assume that the accelerating potential between the dees is 80 kV. ssm www

32P. The oscillator frequency of the cyclotron in Sample Problem 29-5 has been adjusted to accelerate deuterons ($q = +e$, $m =$ 2.0 u). (a) If protons are injected instead of deuterons, to what kinetic energy can the protons be accelerated, using the same oscillator frequency? (b) What magnetic field would be required? (c) What kinetic energy could be produced for protons if the magnetic field were left at the value used for deuterons? (d) What oscillator frequency would then be required? (e) Answer the same questions for alpha particles ($q = +2e$, $m = 4.0$ u).

SEC. 29-7 Magnetic Force on a Current-Carrying Wire

33E. A horizontal conductor that is part of a power line carries a current of 5000 A from south to north. Earth's magnetic field (60.0 μT) is directed toward the north and is inclined downward at 70° to the horizontal. Find the magnitude and direction of the magnetic force on 100 m of the conductor due to Earth's field. ssm

34E. A wire 1.80 m long carries a current of 13.0 A and makes an angle of 35.0° with a uniform magnetic field $B = 1.50$ T. Calculate the magnetic force on the wire.

35E. A wire of 62.0 cm length and 13.0 g mass is suspended by a pair of flexible leads in a uniform magnetic field of magnitude 0.440 T (Fig. 29-35). What are the magnitude and direction of the current required to remove the tension in the supporting leads? ssm ilw

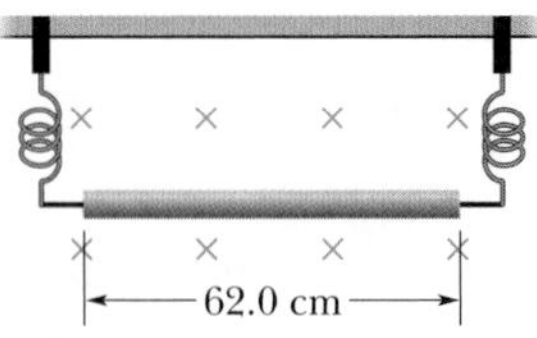

Fig. 29-35 Exercise 35.

36P. A wire 50 cm long lying along the x axis carries a current of 0.50 A in the positive x direction, through a magnetic field $\vec{B} = (0.0030\text{ T})\hat{j} + (0.010\text{ T})\hat{k}$. Find the magnetic force on the wire.

37P. A 1.0 kg copper rod rests on two horizontal rails 1.0 m apart and carries a current of 50 A from one rail to the other. The coefficient of static friction between rod and rails is 0.60. What is the smallest magnetic field (not necessarily vertical) that would cause the rod to slide? ssm

38P. Consider the possibility of a new design for an electric train. The engine is driven by the force on a conducting axle due to the vertical component of Earth's magnetic field. To produce the force, current is maintained down one rail, through a conducting wheel, through the axle, through another conducting wheel, and then back to the source via the other rail. (a) What current is needed to provide a modest 10 kN force? Take the vertical component of Earth's field to be 10 μT and the length of the axle to be 3.0 m. (b) At what rate would electric energy be lost for each ohm of resistance in the rails? (c) Is such a train totally or just marginally unrealistic?

SEC. 29-8 Torque on a Current Loop

39E. Figure 29-36 shows a rectangular 20-turn coil of wire, of dimensions 10 cm by 5.0 cm. It carries a current of 0.10 A and is hinged along one long side. It is mounted in the xy plane, at 30° to the direction of a uniform magnetic field of magnitude 0.50 T. Find the magnitude and direction of the torque acting on the coil about the hinge line. ssm

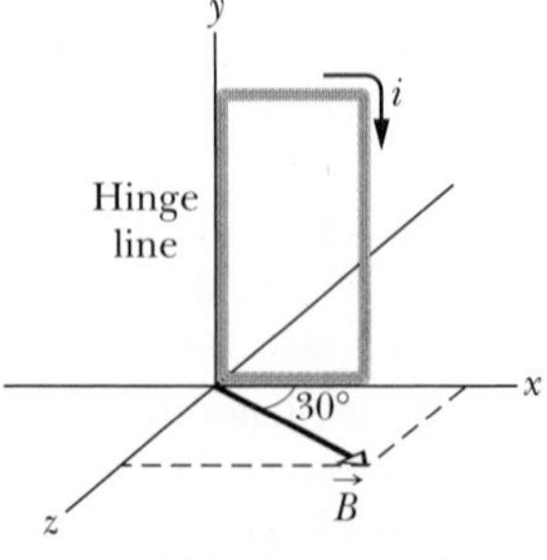

Fig. 29-36 Exercise 39.

40E. A single-turn current loop, carrying a current of 4.00 A, is in the shape of a right triangle with sides 50.0, 120, and 130 cm. The loop is in a uniform magnetic field of magnitude 75.0 mT whose direction is parallel to the current in the 130 cm side of the loop. (a) Find the magnitude of the magnetic force on each of the three sides of the loop. (b) Show that the total magnetic force on the loop is zero.

41E. A length L of wire carries a current i. Show that if the wire is formed into a circular coil, then the maximum torque in a given magnetic field is developed when the coil has one turn only, and that maximum torque has the magnitude $\tau = L^2iB/4\pi$. ssm ilw

42P. Prove that the relation $\tau = NiAB \sin \theta$ holds for closed loops of arbitrary shape and not only for rectangular loops as in Fig. 29-21. (*Hint:* Replace the loop of arbitrary shape with an assembly of adjacent long, thin, approximately rectangular loops that are nearly equivalent to the loop of arbitrary shape as far as the distribution of current is concerned.)

43P. Figure 29-37 shows a wire ring of radius a that is perpendicular to the general direction of a radially symmetric, diverging magnetic field. The magnetic field at the ring is everywhere of the same magnitude B, and its direction at the ring everywhere makes an angle θ with a normal to the plane of the ring. The twisted lead wires have no effect on the problem. Find the magnitude and direction of the force the field exerts on the ring if the ring carries a current i. ssm www

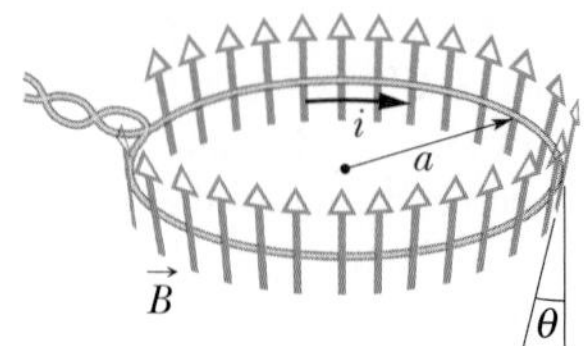

Fig. 29-37 Problem 43.

44P. A closed wire loop with current i is in a uniform magnetic field $\vec{B}$, with the plane of the loop at angle θ to the direction of $\vec{B}$. Show that the total magnetic force on the loop is zero. Does your proof also hold for a nonuniform magnetic field?

45P. The coil of a certain galvanometer (see Sample Problem 29-7) has a resistance of 75.3 Ω; its needle shows a full-scale deflection when a current of 1.62 mA passes through the coil. (a) Determine the value of the auxiliary resistance required to convert the galvanometer to a voltmeter that reads 1.00 V at full-scale deflection. How should this resistance be connected? (b) Determine the value of the auxiliary resistance required to convert the galvanometer to an ammeter that reads 50.0 mA at full-scale deflection. How should this resistance be connected? ssm

46P. A particle of charge q moves in a circle of radius a with speed v. Treating the circular path as a current loop with constant current equal to its average current, find the maximum torque exerted on the loop by a uniform magnetic field of magnitude B.

47P. Figure 29-38 shows a wood cylinder of mass $m = 0.250$ kg and length $L = 0.100$ m, with $N = 10.0$ turns of wire wrapped around it longitudinally, so that the plane of the wire coil contains the axis of the cylinder. What is the least current i through the coil

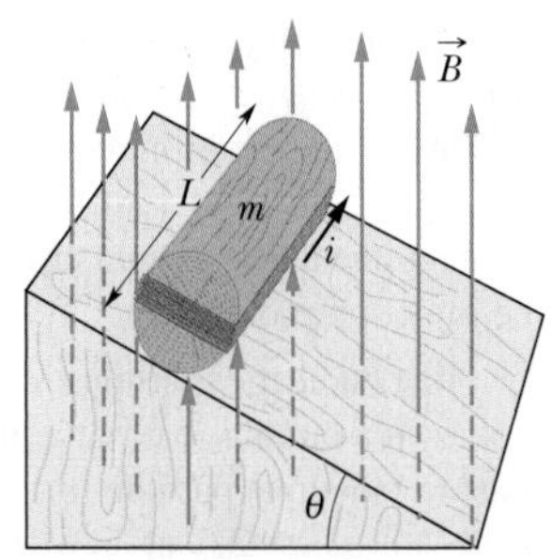

Fig. 29-38 Problem 47.

that will prevent the cylinder from rolling down a plane inclined at an angle θ to the horizontal, in the presence of a vertical, uniform magnetic field of magnitude 0.500 T, if the plane of the coil is parallel to the inclined plane? ssm

SEC. 29-9 The Magnetic Dipole Moment

48E. The magnetic dipole moment of Earth is 8.00×10^{22} J/T. Assume that this is produced by charges flowing in Earth's molten outer core. If the radius of their circular path is 3500 km, calculate the current they produce.

49E. A circular coil of 160 turns has a radius of 1.90 cm. (a) Calculate the current that results in a magnetic dipole moment of $2.30\ \text{A}\cdot\text{m}^2$. (b) Find the maximum torque that the coil, carrying this current, can experience in a uniform 35.0 mT magnetic field. ssm

50E. A circular wire loop whose radius is 15.0 cm carries a current of 2.60 A. It is placed so that the normal to its plane makes an angle of 41.0° with a uniform magnetic field of 12.0 T. (a) Calculate the magnetic dipole moment of the loop. (b) What torque acts on the loop?

51E. A current loop, carrying a current of 5.0 A, is in the shape of a right triangle with sides 30, 40, and 50 cm. The loop is in a uniform magnetic field of magnitude 80 mT whose direction is parallel to the current in the 50 cm side of the loop. Find the magnitude of (a) the magnetic dipole moment of the loop and (b) the torque on the loop. ssm

52E. A stationary circular wall clock has a face with a radius of 15 cm. Six turns of wire are wound around its perimeter; the wire carries a current of 2.0 A in the clockwise direction. The clock is located where there is a constant, uniform external magnetic field of magnitude 70 mT (but the clock still keeps perfect time). At exactly 1:00 p.m., the hour hand of the clock points in the direction of the external magnetic field. (a) After how many minutes will the minute hand point in the direction of the torque on the winding due to the magnetic field? (b) Find the torque magnitude.

53E. Two concentric, circular wire loops, of radii 20.0 and 30.0 cm, are located in the xy plane; each carries a clockwise current of 7.00 A (Fig. 29-39). (a) Find the net magnetic dipole moment of this system. (b) Repeat for reversed current in the inner loop. ssm

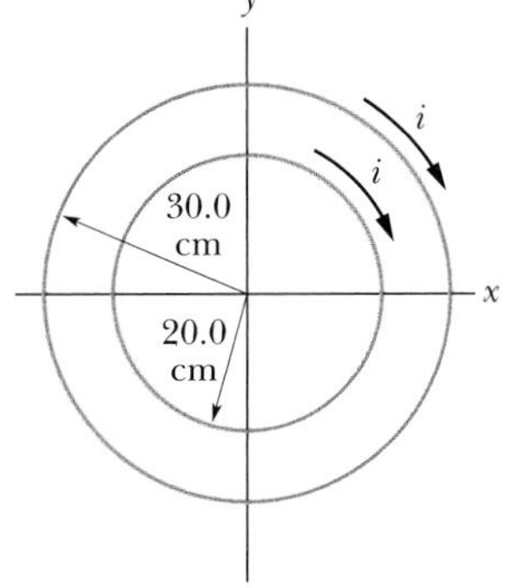

Fig. 29-39 Exercise 53.

54P. Figure 29-40 shows a current loop $ABCDEFA$ carrying a current $i = 5.00$ A. The sides of the loop are parallel to the coordinate axes, with $AB = 20.0$ cm, $BC = 30.0$ cm, and $FA = 10.0$ cm. Calculate the magnitude and direction of the magnetic dipole moment of this loop. (*Hint:* Imagine equal and opposite currents i in the line segment AD; then treat the two rectangular loops $ABCDA$ and $ADEFA$.)

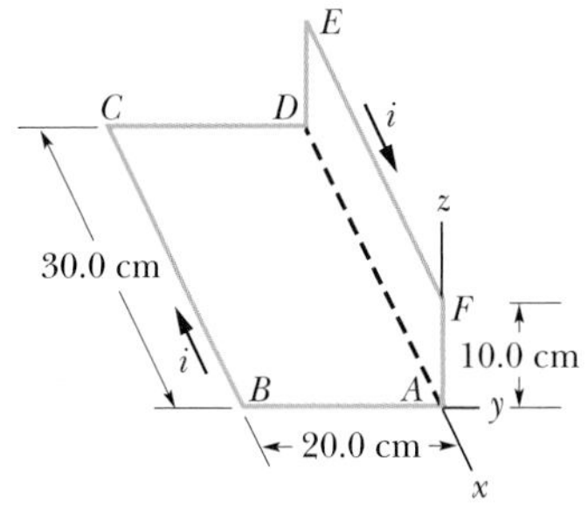

Fig. 29-40 Problem 54.

55P. A circular loop of wire having a radius of 8.0 cm carries a current of 0.20 A. A vector of unit length and parallel to the dipole moment $\vec{\mu}$ of the loop is given by $0.60\hat{\text{i}} - 0.80\hat{\text{j}}$. If the loop is located in a uniform magnetic field given by $\vec{B} = (0.25\ \text{T})\hat{\text{i}} + (0.30\ \text{T})\hat{\text{k}}$, find (a) the torque on the loop (in unit-vector notation) and (b) the magnetic potential energy of the loop. ssm

Additional Problems

56. A wire lying along a y axis from $y = 0$ to $y = 0.250$ m carries a current of 2.00 mA in the negative y direction. The wire lies in a nonuniform magnetic field given by

$$\vec{B} = (0.300\ \text{T/m})y\hat{\text{i}} + (0.400\ \text{T/m})y\hat{\text{j}}.$$

In unit-vector notation, what is the magnetic force on (a) an element dy of the wire at position y and (b) the entire wire?

57. A proton moves at a constant velocity of +50 m/s along an x axis, through crossed electric and magnetic fields. The magnetic field is $\vec{B} = (2.0\ \text{mT})\hat{\text{j}}$. What is the electric field?

58. A magnetic dipole with a dipole moment of magnitude 0.020 J/T is released from rest in a uniform magnetic field of magnitude 52 mT. The rotation of the dipole due to the magnetic force on it is unimpeded. When the dipole rotates through the orientation where its dipole moment is aligned with the magnetic field, its kinetic energy is 0.80 mJ. (a) What is the initial angle between the dipole moment and the magnetic field? (b) What is the angle when the dipole is next (momentarily) at rest?

59. An electron moves through a uniform magnetic field given by $\vec{B} = B_x\hat{\text{i}} + (3B_x)\hat{\text{j}}$. At a particular instant, the electron has the velocity $\vec{v} = (2.0\hat{\text{i}} + 4.0\hat{\text{j}})$ m/s and the magnetic force acting on it is $(6.4 \times 10^{-19}\ \text{N})\hat{\text{k}}$. Find B_x.

NEW PROBLEMS

N1. A proton moves through a uniform magnetic field given by $\vec{B} = (10\hat{i} - 20\hat{j} + 30\hat{k})$ mT. At time t_1, the proton has a velocity given by $\vec{v} = v_x\hat{i} + v_y\hat{j} + (2000\text{ m/s})\hat{k}$ and the magnetic force on the proton is $\vec{F}_B = (4.0 \times 10^{-17}\text{ N})\hat{i} + (2.0 \times 10^{-17}\text{ N})\hat{j}$. At that instant, what are (a) v_x and (b) v_y?

N2. At time t_1, an electron is sent along the positive direction of an x axis, through an electric field and a magnetic field. The electric field is directed parallel to the y axis. Figure 29N-1 gives the y component $F_{\text{net},y}$ of the net force on the electron due to the two fields, as a function of the speed v of the electron at time t_1. The x and z components of the net force are zero at t_1. Assuming $B_x = 0$, what are (a) the magnitude of the electric field, (b) the magnitude of the magnetic field, and (c) the direction of the magnetic field?

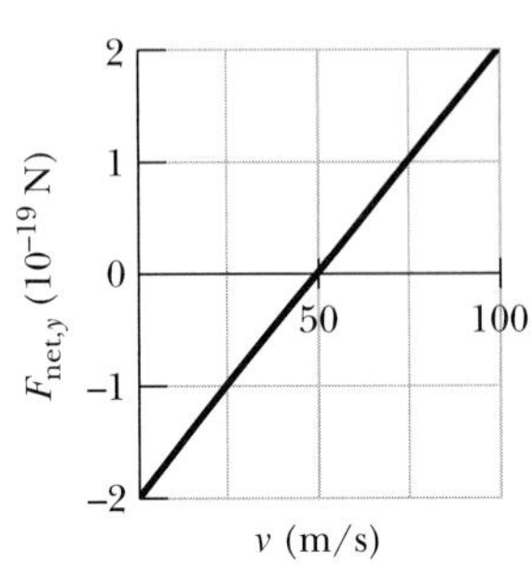

Fig. 29N-1 Problem N2.

N3. In Fig. 29N-2a, two concentric coils, lying in the same plane, carry currents in opposite directions. The current in the larger coil 1 is fixed. Current i_2 in coil 2 can be varied. Figure 29N-2b gives the net magnetic moment of the two-coil system as a function of i_2. If the current in coil 2 is then reversed, what is the magnitude of the net magnetic moment of the two-coil system when i_2 = 7.0 mA?

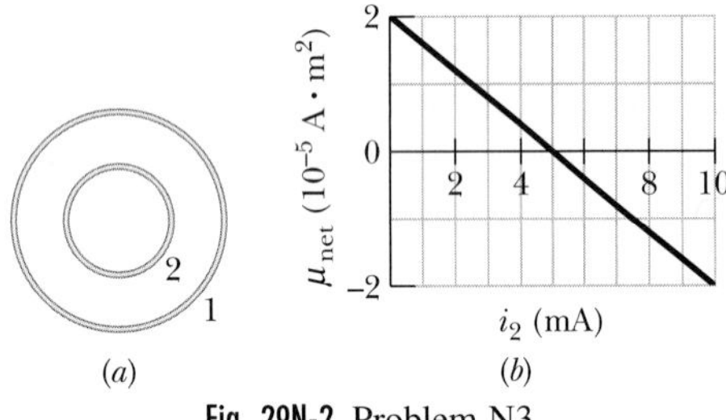

Fig. 29N-2 Problem N3.

N4. In Fig. 29N-3, a rectangular loop carrying current lies in the plane of a uniform magnetic field of magnitude 0.040 T. The loop consists of a single turn of flexible conducting wire that is wrapped around a flexible mount such that the dimensions of the rectangle can be changed. (The total length of the wire is not changed.) As edge length x is varied from approximately zero to its maximum value of approximately 4.0 cm, the magnitude τ of the torque on the loop changes. The maximum value of τ is 4.80×10^{-8} N · m. What is the current in the loop?

Fig. 29N-3 Problem N4.

N5. Bainbridge's mass spectrometer, shown in Fig. 29N-4, separates ions having the same velocity. The ions, after entering through slits, S_1 and S_2, pass through a velocity selector composed of an electric field produced by the charged plates P and P′, and a magnetic field $\vec{B}$ perpendicular to the electric field and the ion path. The ions that then pass undeviated through the crossed $\vec{E}$ and $\vec{B}$ fields enter into a region where a second magnetic field $\vec{B}'$ exists, where they are made to follow circular paths. A photographic plate registers their arrival. Show that, for the ions, $q/m = E/rBB'$, where r is the radius of the circular orbit.

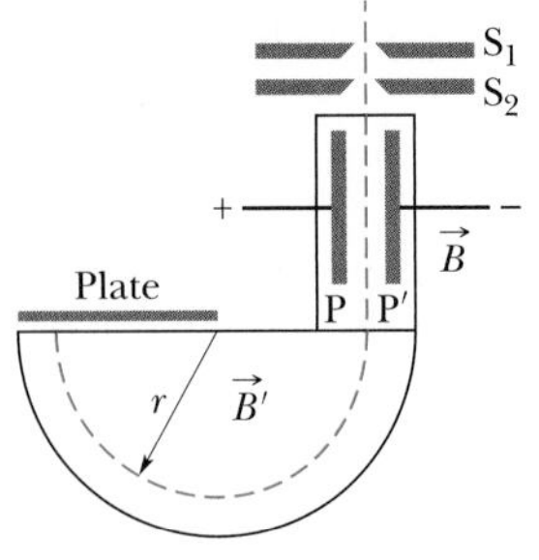

Fig. 29N-4 Problem N5.

N6. A certain particle is sent into a uniform magnetic field, with its velocity vector perpendicular to the direction of the field. Figure 29N-5 gives the period T of the particle's motion versus the inverse B^{-1} of the field's magnitude. What is the ratio m/q of the particle's mass to the magnitude of its charge?

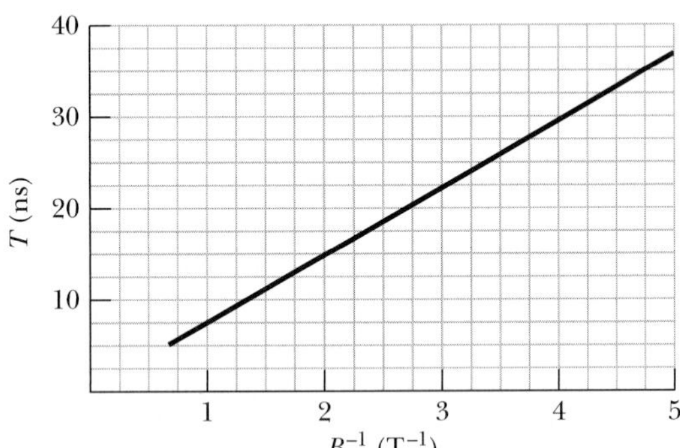

Fig. 29N-5 Problem N6.

N7. A deuteron in a cyclotron is moving in a magnetic field with B = 1.5 T and an orbit radius of 50 cm. Because of a grazing collision with a target, the deuteron breaks up, with negligible loss of kinetic energy, into a proton and a neutron. Discuss the subsequent motion of each. Assume that the deuteron energy is shared equally by the proton and neutron at breakup.

N8. A particle undergoes uniform circular motion of radius 26.1 μm in a uniform magnetic field. The magnetic force on the particle has a magnitude of 1.60×10^{-17} N. What is the kinetic energy of the particle?

N9. An electron that is moving through a uniform magnetic field has a velocity $\vec{v} = (40\text{ km/s})\hat{i} + (35\text{ km/s})\hat{j}$ when it experiences a force $\vec{F} = -(4.2\text{ fN})\hat{i} + (4.8\text{ fN})\hat{j}$ due to the magnetic field. If $B_x = 0$, calculate the magnetic field $\vec{B}$.

N10. An electron follows a helical path in a uniform magnetic field of magnitude 0.300 T. The pitch of the path is 6.00 μm, and the magnitude of the magnetic force on the electron is 2.00×10^{-15} N. What is the electron's speed?

N11. What uniform magnetic field must be set up in space to permit a proton of speed 1.0×10^7 m/s to move in a circle the size of Earth's equator?

N12. An electron is accelerated from rest through potential difference V and then enters a region of uniform magnetic field, where it undergoes uniform circular motion. Figure 29N-6 gives the radius r of that motion versus $V^{1/2}$. What is the magnitude of the magnetic field?

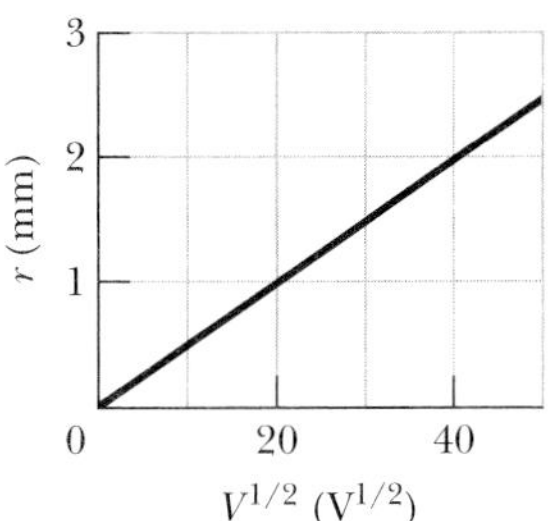

Fig. 29N-6 Problem N12.

N13. Show that, in terms of the Hall electric field E and the current density J, the number of charge carriers per unit volume is given by $n = JB/eE$.

N14. A proton circulates in a cyclotron, beginning approximately at rest at the center. Whenever it passes through the gap between dees, the electric potential difference between the dees is 200 V. (a) By how much does its kinetic energy increase with each passage through the gap? (b) What is its kinetic energy as it completes 100 passes through the gap? Let r_{100} be the radius of its circular path in the dee that it enters as it completes those 100 passes, and let r_{101} be its next radius, as it enters a dee the next time. (c) By what percentage does the radius increase when it changes from r_{100} to r_{101}? That is, what is

$$\text{percentage increase} = \frac{r_{101} - r_{100}}{r_{100}} 100\%?$$

N15. The coil in Fig. 29N-7 carries a current of 2.00 A in the direction indicated, is parallel to an xz plane, has 3.00 turns and an area of 4.00×10^{-3} m^2, and lies within a uniform magnetic field $\vec{B} = (2.00\hat{i} - 3.00\hat{j} - 4.00\hat{k})$ mT. What are (a) the magnetic potential energy of the coil–magnetic field system and (b) the magnetic torque (in unit-vector notation) on the coil?

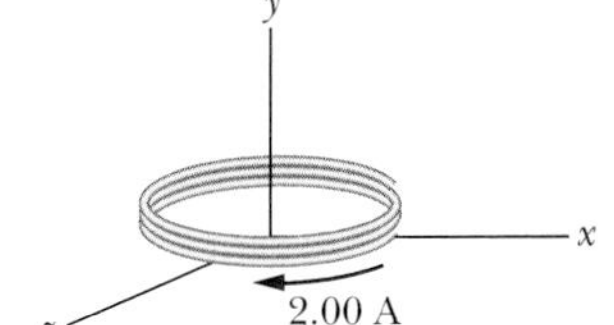

Fig. 29N-7 Problem N15.

N16. An electron follows a helical path in a uniform magnetic field given by $\vec{B} = (20\hat{i} - 50\hat{j} - 30\hat{k})$ mT. At time $t = 0$, the electron's velocity is given by $\vec{v} = (20\hat{i} - 30\hat{j} + 50\hat{k})$ m/s. (a) What is the angle ϕ between $\vec{v}$ and $\vec{B}$? The electron's velocity changes with time. Do (b) its speed and (c) the angle ϕ change with time? (d) What is the radius of the helical path?

N17. In Fig. 29N-8, an electron with an initial kinetic energy of 4.0 keV enters region 1 at time $t = 0$. That region contains a uniform magnetic field directed into the page, with magnitude 0.010 T. The electron goes through a half-circle and then exits region 1, headed toward region 2 across a gap of 25.0 cm. There is an electric potential difference of $\Delta V = 2000$ V across the gap, with a polarity such that the electron's speed increases uniformly (its acceleration is constant). Region 2 contains a uniform magnetic field directed out of the page, with magnitude 0.020 T. The electron goes through a half circle and then leaves region 2. At what time t does it leave?

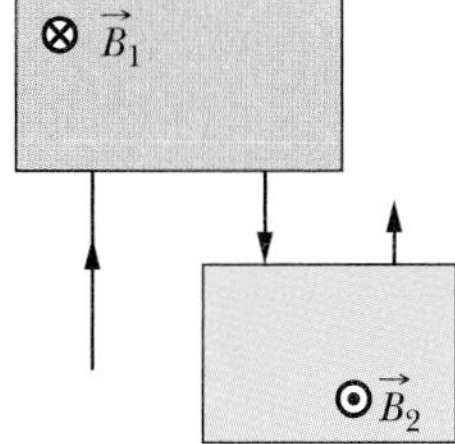

Fig. 29N-8 Problem N17.

N18. Figure 29N-9 gives the potential energy U of a magnetic dipole in an external magnetic field $\vec{B}$, as a function of angle ϕ between the directions of $\vec{B}$ and the dipole moment. The dipole can be rotated about an axle with negligible friction so as to change ϕ. Counterclockwise rotation from $\phi = 0$ yields positive values of ϕ, and clockwise rotations yield negative values. The dipole is to be released at angle $\phi = 0$ with a rotational kinetic energy of 6.7×10^{-4} J, so that it rotates counterclockwise. To what maximum value of ϕ will it rotate? (In the language of Section 8-5, what value ϕ is the turning point in the potential well of Fig. 29N-9?)

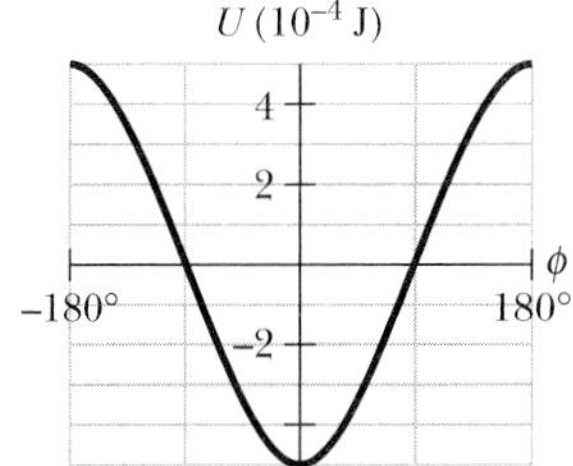

Fig. 29N-9 Problem N18.

N19. Express the unit of a magnetic field B in terms of the dimensions M, L, T, and Q (mass, length, time, and charge).

N20. Figure 29N-10 shows a metallic block, with its faces parallel to coordinate axes. The block is in a uniform magnetic field of magnitude 0.020 T. One edge length of the block is 25 cm; the block is *not* drawn to scale. The block is moved at 3.0 m/s parallel to each axis, in turn, and the resulting potential difference V that appears across the block is measured. With the motion parallel to the y axis, $V = 12$ mV; with the motion parallel to the z axis, $V = 18$ mV; with the motion parallel to the x axis, $V = 0$. What are the block lengths (a) d_x, (b) d_y, and (c) d_z?

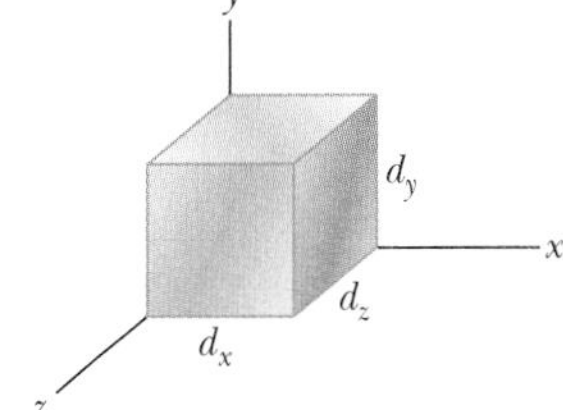

Fig. 29N-10 Problem N20.

30 Magnetic Fields Due to Currents

This is the way we presently launch materials into space. However, when we begin mining the Moon and the asteroids, where we will not have a source of fuel for such conventional rockets, we shall need a more effective way. Electromagnetic launchers may be the answer. A small prototype, the *electromagnetic rail gun,* can presently accelerate a projectile from rest to a speed of 10 km/s (36 000 km/h) within 1 ms.

How can such rapid acceleration possibly be accomplished?

The answer is in this chapter.

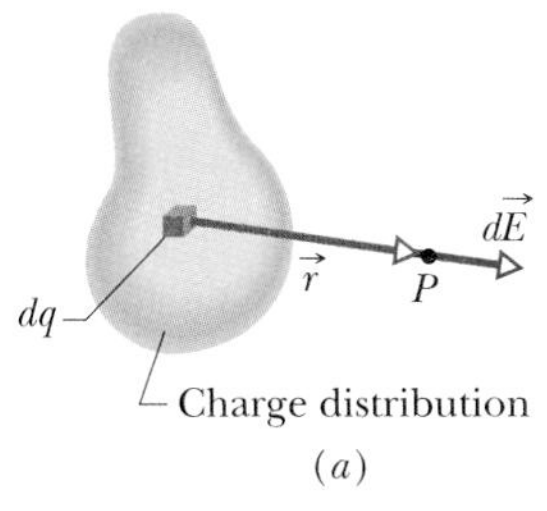

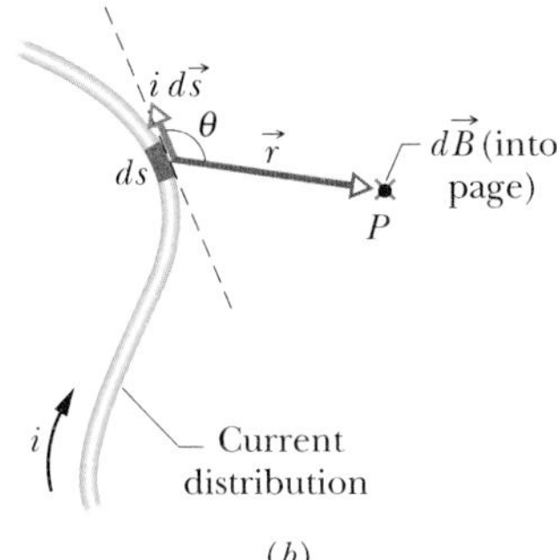

Fig. 30-1 (*a*) A charge element dq produces a differential electric field $d\vec{E}$ at point P. (*b*) A current-length element $i\,d\vec{s}$ produces a differential magnetic field $d\vec{B}$ at point P. The green $\times$ (the tail of an arrow) at the dot for point P indicates that $d\vec{B}$ is directed *into* the page there.

30-1 Calculating the Magnetic Field Due to a Current

As we discussed in Section 29-1, one way to produce a magnetic field is with moving charges—that is, with a current. Our goal in this chapter is to calculate the magnetic field that is produced by a given distribution of currents. We shall use the same basic procedure we used in Chapter 23 to calculate the electric field produced by a given distribution of charged particles.

Let us quickly review that basic procedure. We first mentally divide the charge distribution into charge elements dq, as is done for a charge distribution of arbitrary shape in Fig. 30-1a. We then calculate the field $d\vec{E}$ produced at some point P by a typical charge element. Because the electric fields contributed by different elements can be superimposed, we calculate the net field $\vec{E}$ at P by summing, via integration, the contributions $d\vec{E}$ from all the elements.

Recall that we express the magnitude of $d\vec{E}$ as

$$dE = \frac{1}{4\pi\varepsilon_0}\frac{dq}{r^2}, \tag{30-1}$$

in which r is the distance from the charge element dq to point P. For a positively charged element, the direction of $d\vec{E}$ is that of $\vec{r}$, where $\vec{r}$ is the vector that extends from the charge element dq to the point P. Using $\vec{r}$, we can rewrite Eq. 30-1 in vector form as

$$d\vec{E} = \frac{1}{4\pi\varepsilon_0}\frac{dq}{r^3}\vec{r}, \tag{30-2}$$

which indicates that the direction of the vector $d\vec{E}$ produced by a positively charged element is the direction of the vector $\vec{r}$. Note that Eq. 30-2 is an inverse-square law ($d\vec{E}$ depends on inverse r^2) in spite of the exponent 3 in the denominator. That exponent is in the equation only because we added a factor of magnitude r in the numerator.

Now let us use the same basic procedure to calculate the magnetic field due to a current. Figure 30-1b shows a wire of arbitrary shape carrying a current i. We want to find the magnetic field $\vec{B}$ at a nearby point P. We first mentally divide the wire into differential elements ds and then define for each element a length vector $d\vec{s}$ that has length ds and whose direction is the direction of the current in ds. We can then define a differential *current-length element* to be $i\,d\vec{s}$; we wish to calculate the field $d\vec{B}$ produced at P by a typical current-length element. From experiment we find that magnetic fields, like electric fields, can be superimposed to find a net field. Thus, we can calculate the net field $\vec{B}$ at P by summing, via integration, the contributions $d\vec{B}$ from all the current-length elements. However, this summation is more challenging than the process associated with electric fields because of a complexity; whereas a charge element dq producing an electric field is a scalar, a current-length element $i\,d\vec{s}$ producing a magnetic field is the product of a scalar and a vector.

The magnitude of the field $d\vec{B}$ produced at point P by a current-length element $i\,d\vec{s}$ turns out to be

$$dB = \frac{\mu_0}{4\pi}\frac{i\,ds\sin\theta}{r^2}, \tag{30-3}$$

where θ is the angle between the directions of $d\vec{s}$ and $\vec{r}$, the vector that extends from ds to P. Symbol μ_0 is a constant, called the *permeability constant,* whose value is defined to be exactly

$$\mu_0 = 4\pi \times 10^{-7}\ \mathrm{T\cdot m/A} \approx 1.26 \times 10^{-6}\ \mathrm{T\cdot m/A}. \tag{30-4}$$

The direction of $d\vec{B}$, shown as being into the page in Fig. 30-1*b*, is that of the cross product $d\vec{s} \times \vec{r}$. We can therefore write Eq. 30-3 in vector form as

$$d\vec{B} = \frac{\mu_0}{4\pi} \frac{i\, d\vec{s} \times \vec{r}}{r^3} \quad \text{(Biot–Savart law).} \tag{30-5}$$

This vector equation and its scalar form, Eq. 30-3, are known as the **law of Biot and Savart** (rhymes with "Leo and bazaar"). The law, which is experimentally deduced, is an inverse-square law (the exponent in the denominator of Eq. 30-5 is 3 only because of the factor $\vec{r}$ in the numerator). We shall use this law to calculate the net magnetic field $\vec{B}$ produced at a point by various distributions of current.

Magnetic Field Due to a Current in a Long Straight Wire

Shortly we shall use the law of Biot and Savart to prove that the magnitude of the magnetic field at a perpendicular distance R from a long (infinite) straight wire carrying a current i is given by

$$B = \frac{\mu_0 i}{2\pi R} \quad \text{(long straight wire).} \tag{30-6}$$

The field magnitude B in Eq. 30-6 depends only on the current and the perpendicular distance R of the point from the wire. We shall show in our derivation that the field lines of $\vec{B}$ form concentric circles around the wire, as Fig. 30-2 shows and as the iron filings in Fig. 30-3 suggest. The increase in the spacing of the lines in Fig. 30-2 with increasing distance from the wire represents the $1/R$ decrease in the magnitude of $\vec{B}$ predicted by Eq. 30-6. The lengths of the two vectors $\vec{B}$ in the figure also show the $1/R$ decrease.

Here is a simple right-hand rule for finding the direction of the magnetic field set up by a current-length element, such as a section of a long wire:

> *Right-hand rule:* Grasp the element in your right hand with your extended thumb pointing in the direction of the current. Your fingers will then naturally curl around in the direction of the magnetic field lines due to that element.

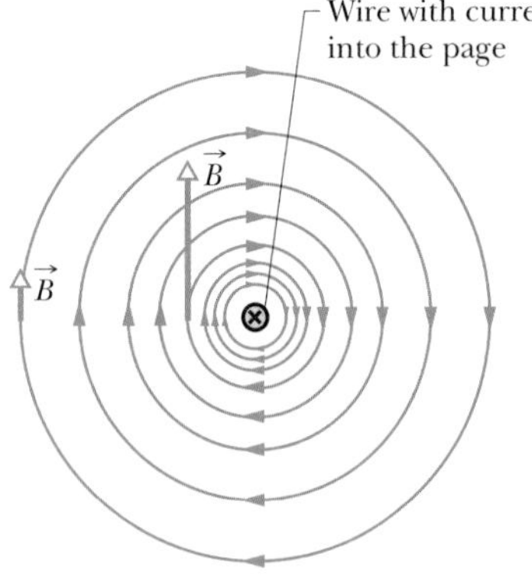

Fig. 30-2 The magnetic field lines produced by a current in a long straight wire form concentric circles around the wire. Here the current is into the page, as indicated by the ×.

Fig. 30-3 Iron filings that have been sprinkled onto cardboard collect in concentric circles when current is sent through the central wire. The alignment, which is along magnetic field lines, is caused by the magnetic field produced by the current.

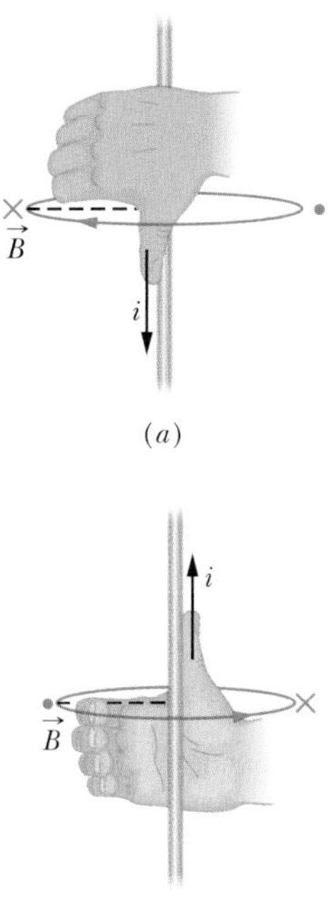

Fig. 30-4 A right-hand rule gives the direction of the magnetic field due to a current in a wire. (*a*) The situation of Fig. 30-2, seen from the side. The magnetic field $\vec{B}$ at any point to the left of the wire is perpendicular to the dashed radial line and directed into the page, in the direction of the fingertips, as indicated by the $\times$. (*b*) If the current is reversed, $\vec{B}$ at any point to the left is still perpendicular to the dashed radial line but now is directed out of the page, as indicated by the dot.

The result of applying this right-hand rule to the current in the straight wire of Fig. 30-2 is shown in a side view in Fig. 30-4*a*. To determine the direction of the magnetic field $\vec{B}$ set up at any particular point by this current, mentally wrap your right hand around the wire with your thumb in the direction of the current. Let your fingertips pass through the point; their direction is then the direction of the magnetic field at that point. In the view of Fig. 30-2, $\vec{B}$ at any point is *tangent to a magnetic field line;* in the view of Fig. 30-4, it is *perpendicular to a dashed radial line connecting the point and the current.*

Proof of Equation 30-6

Figure 30-5, which is just like Fig. 30-1*b* except that now the wire is straight and of infinite length, illustrates the task at hand; we seek the field $\vec{B}$ at point P, a perpendicular distance R from the wire. The magnitude of the differential magnetic field produced at P by the current-length element $i\,d\vec{s}$ located a distance r from P is given by Eq. 30-3:

$$dB = \frac{\mu_0}{4\pi}\frac{i\,ds\,\sin\theta}{r^2}.$$

The direction of $d\vec{B}$ in Fig. 30-5 is that of the vector $d\vec{s} \times \vec{r}$—namely, directly into the page.

Note that $d\vec{B}$ at point P has this same direction for all the current-length elements into which the wire can be divided. Thus, we can find the magnitude of the magnetic field produced at P by the current-length elements in the upper half of the infinitely long wire by integrating dB in Eq. 30-3 from 0 to ∞.

Now consider a current-length element in the lower half of the wire, one that is as far below P as $d\vec{s}$ is above P. By Eq. 30-5, the magnetic field produced at P by this current-length element has the same magnitude and direction as that from element $i\,d\vec{s}$ in Fig. 30-5. Further, the magnetic field produced by the lower half of the wire is exactly the same as that produced by the upper half. To find the magnitude of the *total* magnetic field $\vec{B}$ at P, we need only multiply the result of our integration by 2. We get

$$B = 2\int_0^\infty dB = \frac{\mu_0 i}{2\pi}\int_0^\infty \frac{\sin\theta\,ds}{r^2}. \qquad (30\text{-}7)$$

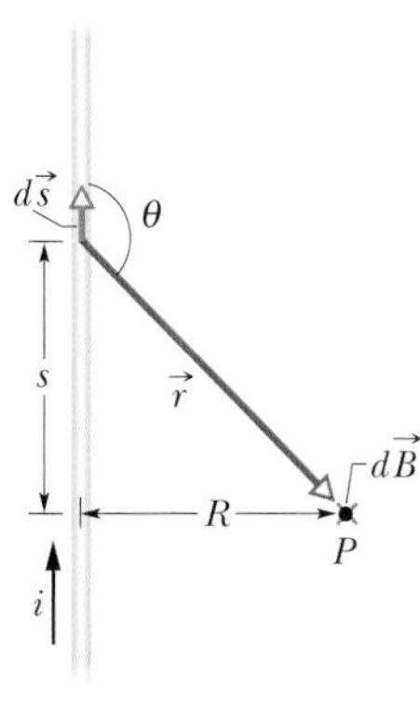

Fig. 30-5 Calculating the magnetic field produced by a current i in a long straight wire. The field $d\vec{B}$ at P associated with the current-length element $i\,d\vec{s}$ is into the page, as shown.

The variables θ, s, and r in this equation are not independent but (see Fig. 30-5) are related by

$$r = \sqrt{s^2 + R^2}$$

and

$$\sin\theta = \sin(\pi - \theta) = \frac{R}{\sqrt{s^2 + R^2}}.$$

With these substitutions and integral 19 in Appendix E, Eq. 30-7 becomes

$$B = \frac{\mu_0 i}{2\pi} \int_0^\infty \frac{R\,ds}{(s^2 + R^2)^{3/2}}$$

$$= \frac{\mu_0 i}{2\pi R} \left[\frac{s}{(s^2 + R^2)^{1/2}} \right]_0^\infty = \frac{\mu_0 i}{2\pi R}, \qquad (30\text{-}8)$$

which is the relation we set out to prove. Note that the magnetic field at P due to either the lower half or the upper half of the infinite wire in Fig. 30-5 is half this value; that is,

$$B = \frac{\mu_0 i}{4\pi R} \quad \text{(semi-infinite straight wire).} \qquad (30\text{-}9)$$

Magnetic Field Due to a Current in a Circular Arc of Wire

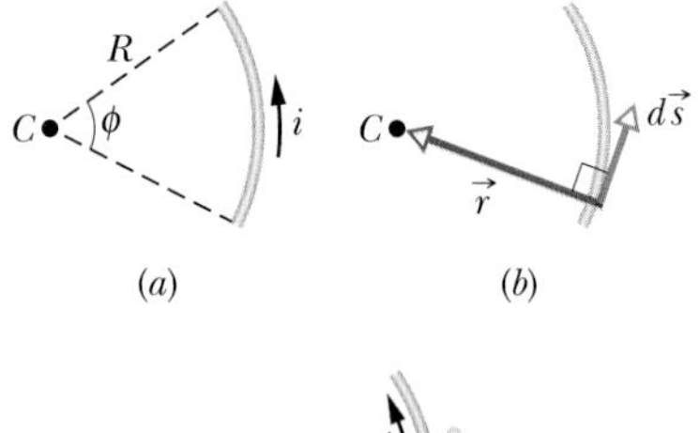

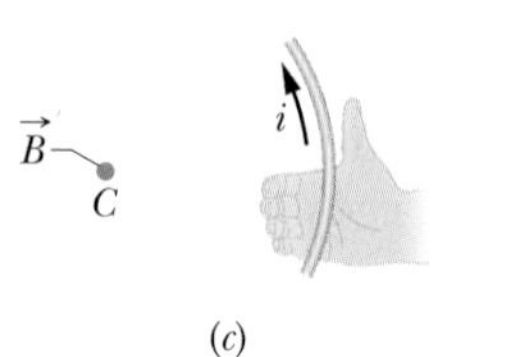

Fig. 30-6 (*a*) A wire in the shape of a circular arc with center C carries current i. (*b*) For any element of wire along the arc, the angle between the directions of $d\vec{s}$ and $\vec{r}$ is 90°. (*c*) Determining the direction of the magnetic field at the center C due to the current in the wire; the field is out of the page, in the direction of the fingertips, as indicated by the colored dot at C.

To find the magnetic field produced at a point by a current in a curved wire, we would again use Eq. 30-3 to write the magnitude of the field produced by a single current-length element, and we would again integrate to find the net field produced by all the current-length elements. That integration can be difficult, depending on the shape of the wire; it is fairly straightforward, however, when the wire is a circular arc and the point is the center of curvature.

Figure 30-6*a* shows such an arc-shaped wire with central angle ϕ, radius R, and center C, carrying current i. At C, each current-length element $i\,d\vec{s}$ of the wire produces a magnetic field of magnitude dB given by Eq. 30-3. Moreover, as Fig. 30-6*b* shows, no matter where the element is located on the wire, the angle θ between the vectors $d\vec{s}$ and $\vec{r}$ is 90°; also, $r = R$. Thus, by substituting R for r and 90° for θ, we obtain from Eq. 30-3,

$$dB = \frac{\mu_0}{4\pi} \frac{i\,ds \sin 90^\circ}{R^2} = \frac{\mu_0}{4\pi} \frac{i\,ds}{R^2}. \qquad (30\text{-}10)$$

The field at C due to each current-length element in the arc has this magnitude.

An application of the right-hand rule anywhere along the wire (as in Fig. 30-6*c*) will show that all the differential fields $d\vec{B}$ have the same direction at C—directly out of the page. Thus, the total field at C is simply the sum (via integration) of all the differential fields $d\vec{B}$. We use the identity $ds = R\,d\phi$ to change the variable of integration from ds to $d\phi$ and obtain, from Eq. 30-10,

$$B = \int dB = \int_0^\phi \frac{\mu_0}{4\pi} \frac{iR\,d\phi}{R^2} = \frac{\mu_0 i}{4\pi R} \int_0^\phi d\phi.$$

Integrating, we find that

$$B = \frac{\mu_0 i \phi}{4\pi R} \quad \text{(at center of circular arc).} \qquad (30\text{-}11)$$

Note that this equation gives us the magnetic field *only* at the center of curvature of a circular arc of current. When you insert data into the equation, you must be careful to express ϕ in radians rather than degrees. For example, to find the magnitude of the magnetic field at the center of a full circle of current, you would substitute 2π rad for ϕ in Eq. 30-11, finding

$$B = \frac{\mu_0 i (2\pi)}{4\pi R} = \frac{\mu_0 i}{2R} \quad \text{(at center of full circle).} \qquad (30\text{-}12)$$

Sample Problem 30-1

The wire in Fig. 30-7a carries a current i and consists of a circular arc of radius R and central angle $\pi/2$ rad, and two straight sections whose extensions intersect the center C of the arc. What magnetic field $\vec{B}$ does the current produce at C?

SOLUTION: One Key Idea here is that we can find the magnetic field $\vec{B}$ at point C by applying the Biot–Savart law of Eq. 30-5 to the wire. A second Key Idea is that the application of Eq. 30-5 can be simplified by evaluating $\vec{B}$ separately for the three distinguishable sections of the wire—namely, (1) the straight section at the left, (2) the straight section at the right, and (3) the circular arc.

Straight sections. For any current-length element in section 1, the angle θ between $d\vec{s}$ and $\vec{r}$ is zero (Fig. 30-7b), so Eq. 30-3 gives us

$$dB_1 = \frac{\mu_0}{4\pi}\frac{i\,ds\sin\theta}{r^2} = \frac{\mu_0}{4\pi}\frac{i\,ds\sin 0}{r^2} = 0.$$

Thus, the current along the entire length of wire in straight section 1 contributes no magnetic field at C:

$$B_1 = 0.$$

The same situation prevails in straight section 2, where the angle θ between $d\vec{s}$ and $\vec{r}$ for any current-length element is 180°. Thus,

$$B_2 = 0.$$

Circular arc. The Key Idea here is that application of the Biot–Savart law to evaluate the magnetic field at the center of a circular arc leads to Eq. 30-11 ($B = \mu_0 i\phi/4\pi R$). Here the central angle ϕ of the arc is $\pi/2$ rad. Thus from Eq. 30-11, the magnitude of the magnetic field $\vec{B}_3$ at the arc's center C is

$$B_3 = \frac{\mu_0 i(\pi/2)}{4\pi R} = \frac{\mu_0 i}{8R}.$$

To find the direction of $\vec{B}_3$, we apply the right-hand rule displayed in Fig. 30-4. Mentally grasp the circular arc with your right hand as in Fig. 30-7c, with your thumb in the direction of the current. The direction in which your fingers curl around the wire indicates the direction of the magnetic field lines around the wire. In the region of point C (inside the circular arc), your fingertips point *into the plane* of the page. Thus, $\vec{B}_3$ is directed into that plane.

Net field. Generally, when we must combine two or more magnetic fields to find the net magnetic field, we must combine the fields as vectors and not simply add their magnitudes. Here, however, only the circular arc produces a magnetic field at point C.

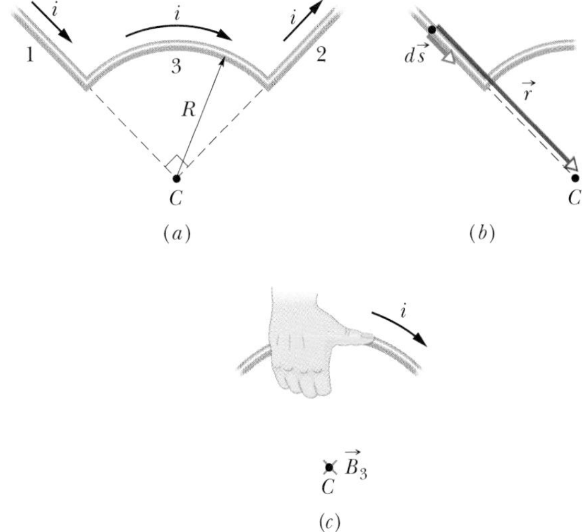

Fig. 30-7 Sample Problem 30-1. (a) A wire consists of two straight sections (1 and 2) and a circular arc (3), and carries current i. (b) For a current-length element in section 1, the angle between $d\vec{s}$ and $\vec{r}$ is zero. (c) Determining the direction of magnetic field $\vec{B}_3$ at C due to the current in the circular arc; the field is into the page there.

Thus, we can write the magnitude of the net field $\vec{B}$ as

$$B = B_1 + B_2 + B_3 = 0 + 0 + \frac{\mu_0 i}{8R} = \frac{\mu_0 i}{8R}. \quad \text{(Answer)}$$

The direction of $\vec{B}$ is the direction of $\vec{B}_3$—namely, into the plane of Fig. 30-7.

CHECKPOINT 1: The figure here shows three circuits consisting of concentric circular arcs (either half- or quarter-circles of radii r, $2r$, and $3r$) and radial lengths. The circuits carry the same current. Rank them according to the magnitude of the magnetic field produced at the center of curvature (the dot), greatest first.

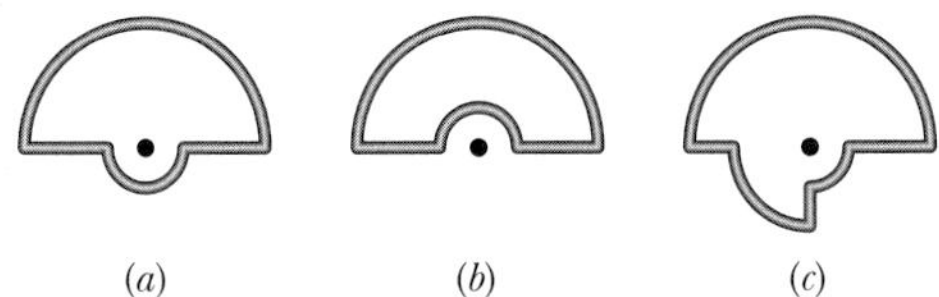

Sample Problem 30-2

Figure 30-8a shows two long parallel wires carrying currents i_1 and i_2 in opposite directions. What are the magnitude and direction of the net magnetic field at point P? Assume the following values: $i_1 = 15$ A, $i_2 = 32$ A, and $d = 5.3$ cm.

SOLUTION: One Key Idea here is that the net magnetic field $\vec{B}$ at point P is the vector sum of the magnetic fields due to the currents in the two wires. A second Key Idea is that we can find the magnetic field due to any current by applying the Biot–Savart law to the

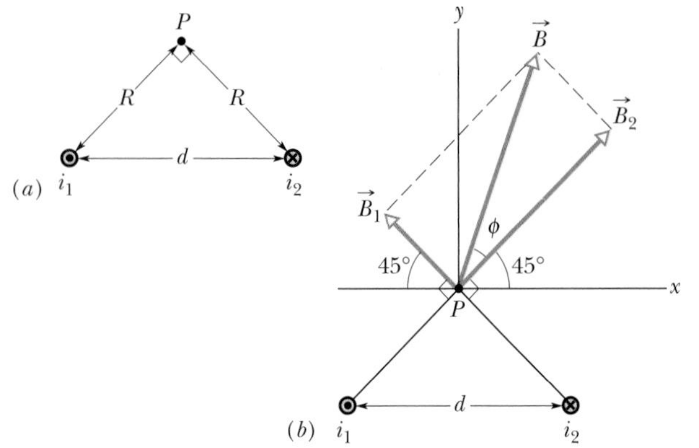

Fig. 30-8 Sample Problem 30-2. (*a*) Two wires carry currents i_1 and i_2 in opposite directions (out of and into the page). Note the right angle at P. (*b*) The separate fields $\vec{B}_1$ and $\vec{B}_2$ are combined vectorially to yield the net field $\vec{B}$.

current. For points near the current in a long straight wire, that law leads to Eq. 30-6.

In Fig. 30-8*a*, point P is distance R from both currents i_1 and i_2. Thus, Eq. 30-6 tells us that at point P those currents produce magnetic fields $\vec{B}_1$ and $\vec{B}_2$ with magnitudes

$$B_1 = \frac{\mu_0 i_1}{2\pi R} \quad \text{and} \quad B_2 = \frac{\mu_0 i_2}{2\pi R}.$$

In the right triangle of Fig. 30-8*a*, note that the base angles (between sides R and d) are both 45°. Thus, we may write $\cos 45° = R/d$ and replace R with $d \cos 45°$. Then the field magnitudes B_1 and B_2 become

$$B_1 = \frac{\mu_0 i_1}{2\pi d \cos 45°} \quad \text{and} \quad B_2 = \frac{\mu_0 i_2}{2\pi d \cos 45°}.$$

We want to combine $\vec{B}_1$ and $\vec{B}_2$ to find their vector sum, which is the net field $\vec{B}$ at P. To find the directions of $\vec{B}_1$ and $\vec{B}_2$, we apply the right-hand rule of Fig. 30-4 to each current in Fig. 30-8*a*. For wire 1, with current out of the page, we mentally grasp the wire with the right hand, with the thumb pointing out of the page. Then the curled fingers indicate that the field lines run counterclockwise. In particular, in the region of point P, they are directed upward to the left. Recall that the magnetic field at a point near a long, straight current-carrying wire must be directed perpendicular to a radial line between the point and the current. Thus, $\vec{B}_1$ must be directed upward to the left as drawn in Fig. 30-8*b*. (Note carefully the perpendicular symbol between vector $\vec{B}_1$ and the line connecting point P and wire 1.)

Repeating this analysis for the current in wire 2, we find that $\vec{B}_2$ is directed upward to the right as drawn in Fig. 30-8*b*. (Note the perpendicular symbol between vector $\vec{B}_2$ and the line connecting point P and wire 2.)

We can now vectorially add $\vec{B}_1$ and $\vec{B}_2$ to find the net magnetic field $\vec{B}$ at point P, either by using a vector-capable calculator or by resolving the vectors into components and then combining the components of $\vec{B}$. However, in Fig. 30-8*b*, there is a third method: Because $\vec{B}_1$ and $\vec{B}_2$ are perpendicular to each other, they form the legs of a right triangle, with $\vec{B}$ as the hypotenuse. The Pythagorean theorem then gives us

$$B = \sqrt{B_1^2 + B_2^2} = \frac{\mu_0}{2\pi d(\cos 45°)}\sqrt{i_1^2 + i_2^2}$$

$$= \frac{(4\pi \times 10^{-7}\ \text{T}\cdot\text{m/A})\sqrt{(15\ \text{A})^2 + (32\ \text{A})^2}}{(2\pi)(5.3 \times 10^{-2}\ \text{m})(\cos 45°)}$$

$$= 1.89 \times 10^{-4}\ \text{T} \approx 190\ \mu\text{T}. \qquad \text{(Answer)}$$

The angle ϕ between the directions of $\vec{B}$ and $\vec{B}_2$ in Fig. 30-8*b* follows from

$$\phi = \tan^{-1}\frac{B_1}{B_2},$$

which, with B_1 and B_2 as given above, yields

$$\phi = \tan^{-1}\frac{i_1}{i_2} = \tan^{-1}\frac{15\ \text{A}}{32\ \text{A}} = 25°.$$

The angle between the direction of $\vec{B}$ and the x axis shown in Fig. 30-8*b* is then

$$\phi + 45° = 25° + 45° = 70°. \qquad \text{(Answer)}$$

PROBLEM-SOLVING TACTICS

Tactic 1: *Right-Hand Rules*

To help you sort out the right-hand rules you have now seen (and the ones coming up), here is a review.

Right-Hand Rule for Cross Products. Introduced in Section 3-7, this is a way to determine the direction of the vector that results from a cross product. You point the fingers of your right hand so as to sweep the first vector expressed in the product into the second vector, through the smaller angle between the two vectors. Your outstretched thumb gives you the direction of the vector resulting from the cross product. In Chapter 12, we used this right-hand rule to find the directions of torque and angular momentum vectors; in Chapter 29, we used it to find the direction of the force on a current-carrying wire in a magnetic field.

Curled–Straight Right-Hand Rules for Magnetism. In many situations involving magnetism, you need to relate a "curled" element and a "straight" element. You can do so with the (curled) fingers and the (straight) thumb on your right hand. You have already seen an example in Section 29-8, in which we related the current around a loop (curled element) to the normal vector $\vec{n}$ (straight element) of the loop: You curl the fingers of your right hand around in the direction of the current along the loop; your outstretched thumb then gives the direction of $\vec{n}$. This is also the direction of the magnetic dipole moment $\vec{\mu}$ of the loop.

In this section you were introduced to a second curled–straight right-hand rule. To determine the direction of the magnetic field lines around a current-length element, you point the outstretched thumb of your right hand in the direction of the current. The fingers then curl around the current-length element in the direction of the field lines.

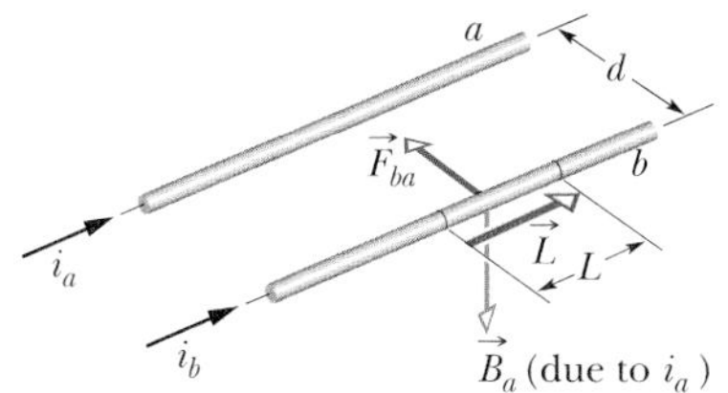

Fig. 30-9 Two parallel wires carrying currents in the same direction attract each other. $\vec{B}_a$ is the magnetic field at wire b produced by the current in wire a. $\vec{F}_{ba}$ is the resulting force acting on wire b because it carries current in $\vec{B}_a$.

30-2 Force Between Two Parallel Currents

Two long parallel wires carrying currents exert forces on each other. Figure 30-9 shows two such wires, separated by a distance d and carrying currents i_a and i_b. Let us analyze the forces on these wires due to each other.

We seek first the force on wire b in Fig. 30-9 due to the current in wire a. That current produces a magnetic field $\vec{B}_a$, and it is this magnetic field that actually causes the force we seek. To find the force, then, we need the magnitude and direction of the field $\vec{B}_a$ *at the site of wire b*. The magnitude of $\vec{B}_a$ at every point of wire b is, from Eq. 30-6,

$$B_a = \frac{\mu_0 i_a}{2\pi d}. \tag{30-13}$$

A (curled–straight) right-hand rule tells us that the direction of $\vec{B}_a$ at wire b is down, as Fig. 30-9 shows.

Now that we have the field, we can find the force it produces on wire b. Equation 29-27 tells us that the force $\vec{F}_{ba}$ on a length L of wire b due to the external magnetic field $\vec{B}_a$ is

$$\vec{F}_{ba} = i_b \vec{L} \times \vec{B}_a, \tag{30-14}$$

where $\vec{L}$ is the length vector of the wire. In Fig. 30-9, vectors $\vec{L}$ and $\vec{B}_a$ are perpendicular, so with Eq. 30-13, we can write

$$F_{ba} = i_b L B_a \sin 90° = \frac{\mu_0 L i_a i_b}{2\pi d}. \tag{30-15}$$

The direction of $\vec{F}_{ba}$ is the direction of the cross product $\vec{L} \times \vec{B}_a$. Applying the right-hand rule for cross products to $\vec{L}$ and $\vec{B}_a$ in Fig. 30-9, we see that $\vec{F}_{ba}$ is directly toward wire a, as shown.

The general procedure for finding the force on a current-carrying wire is this:

> To find the force on a current-carrying wire due to a second current-carrying wire, first find the field due to the second wire at the site of the first wire. Then find the force on the first wire due to that field.

We could now use this procedure to compute the force on wire a due to the current in wire b. We would find that the force is directly toward wire b; hence, the two wires with parallel currents attract each other. Similarly, if the two currents were antiparallel, we could show that the two wires repel each other. Thus,

> Parallel currents attract, and antiparallel currents repel.

The force acting between currents in parallel wires is the basis for the definition of the ampere, which is one of the seven SI base units. The definition, adopted in 1946, is this: The ampere is that constant current which, if maintained in two straight, parallel conductors of infinite length, of negligible circular cross section, and placed 1 m apart in vacuum, would produce on each of these conductors a force of magnitude 2×10^{-7} newton per meter of length.

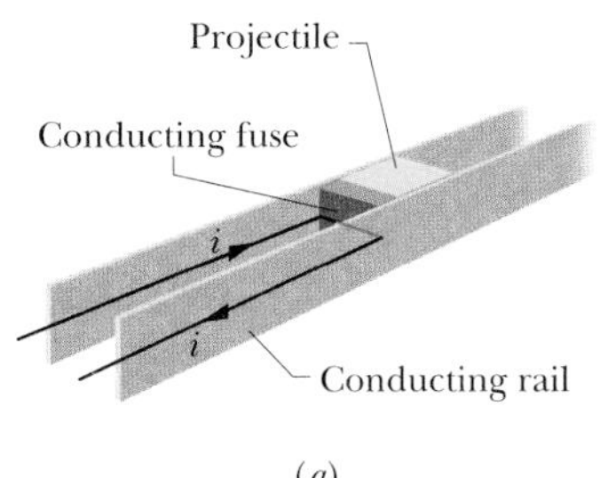

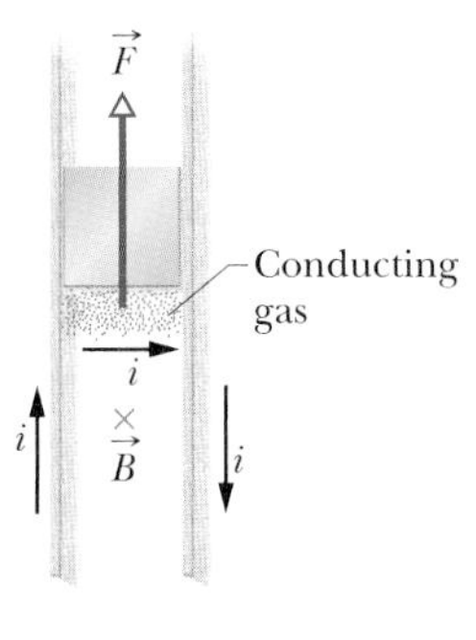

Fig. 30-10 (*a*) A rail gun, as a current i is set up in it. The current rapidly causes the conducting fuse to vaporize. (*b*) The current produces a magnetic field $\vec{B}$ between the rails, and the field causes a force $\vec{F}$ to act on the conducting gas, which is part of the current path. The gas propels the projectile along the rails, launching it.

Rail Gun

A rail gun is a device in which a magnetic force can accelerate a projectile to a high speed in a short time. The basics of a rail gun are shown in Fig. 30-10*a*. A large current is sent out along one of two parallel conducting rails, across a conducting "fuse" (such as a narrow piece of copper) between the rails, and then back to the current source along the second rail. The projectile to be fired lies on the far side of the fuse and fits loosely between the rails. Immediately after the current begins, the fuse element melts and vaporizes, creating a conducting gas between the rails where the fuse had been.

The curled–straight right-hand rule of Fig. 30-4 reveals that the currents in the rails of Fig. 30-10*a* produce magnetic fields that are directed downward between the rails. The net magnetic field $\vec{B}$ exerts a force $\vec{F}$ on the gas due to the current i through the gas (Fig. 30-10*b*). With Eq. 30-14 and the right-hand rule for cross products, we find that $\vec{F}$ points outward along the rails. As the gas is forced outward along the rails, it pushes the projectile, accelerating it by as much as $5 \times 10^6 g$, and then launches it with a speed of 10 km/s, all within 1 ms.

CHECKPOINT 2: The figure here shows three long, straight, parallel, equally spaced wires with identical currents either into or out of the page. Rank the wires according to the magnitude of the force on each due to the currents in the other two wires, greatest first.

a b c

30-3 Ampere's Law

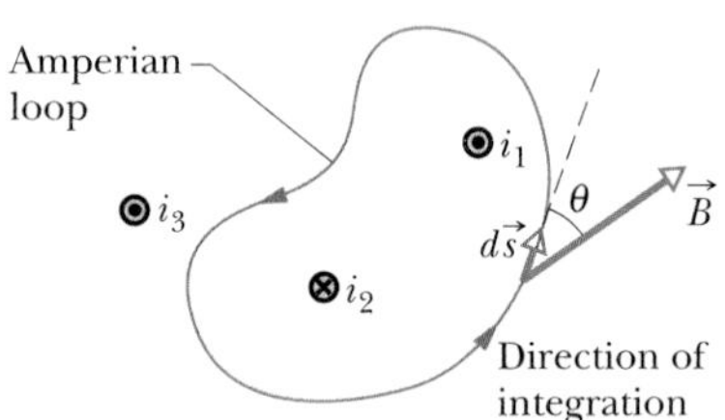

Fig. 30-11 Ampere's law applied to an arbitrary Amperian loop that encircles two long straight wires but excludes a third wire. Note the directions of the currents.

We can find the net electric field due to *any* distribution of charges with the inverse-square law for the differential field $d\vec{E}$ (Eq. 30-2), but if the distribution is complicated, we may have to use a computer. Recall, however, that if the distribution has planar, cylindrical, or spherical symmetry, we can apply Gauss' law to find the net electric field with considerably less effort.

Similarly, we can find the net magnetic field due to *any* distribution of currents with the inverse-square law for the differential field $d\vec{B}$ (Eq. 30-5), but again we may have to use a computer for a complicated distribution. However, if the distribution has some symmetry, we may be able to apply **Ampere's law** to find the magnetic field with considerably less effort. This law, which can be derived from the Biot–Savart law, has traditionally been credited to André Marie Ampère (1775–1836), for whom the SI unit of current is named. However, the law actually was advanced by English physicist James Clerk Maxwell.

Ampere's law is

$$\oint \vec{B} \cdot d\vec{s} = \mu_0 i_{\text{enc}} \quad \text{(Ampere's law)}. \tag{30-16}$$

The circle on the integral sign means that the scalar (or dot) product $\vec{B} \cdot d\vec{s}$ is to be integrated around a *closed* loop, called an *Amperian loop*. The current i_{enc} on the right is the *net* current encircled by that loop.

To see the meaning of the scalar product $\vec{B} \cdot d\vec{s}$ and its integral, let us first apply Ampere's law to the general situation of Fig. 30-11. The figure shows cross sections of three long straight wires that carry currents i_1, i_2, and i_3 either directly into or directly out of the page. An arbitrary Amperian loop lying in the plane of the page encircles two of the currents but not the third. The counterclockwise direction marked on the loop indicates the arbitrarily chosen direction of integration for Eq. 30-16.

To apply Ampere's law, we mentally divide the loop into differential vector elements $d\vec{s}$ that are everywhere directed along the tangent to the loop in the direction of integration. Assume that at the location of the element $d\vec{s}$ shown in Fig. 30-11, the net magnetic field due to the three currents is $\vec{B}$. Because the wires are perpendicular to the page, we know that the magnetic field at $d\vec{s}$ due to each current is in the plane of Fig. 30-11; thus, their net magnetic field $\vec{B}$ at $d\vec{s}$ must also be in that plane. However, we do not know the orientation of $\vec{B}$ within the plane. In Fig. 30-11, $\vec{B}$ is arbitrarily drawn at an angle θ to the direction of $d\vec{s}$.

The scalar product $\vec{B} \cdot d\vec{s}$ on the left side of Eq. 30-16 is then equal to $B \cos \theta \, ds$. Thus, Ampere's law can be written as

$$\oint \vec{B} \cdot d\vec{s} = \oint B \cos \theta \, ds = \mu_0 i_{\text{enc}}. \tag{30-17}$$

We can now interpret the scalar product $\vec{B} \cdot d\vec{s}$ as being the product of a length ds of the Amperian loop and the field component $B \cos \theta$ that is tangent to the loop. Then we can interpret the integration as being the summation of all such products around the entire loop.

When we can actually perform this integration, we do not need to know the direction of $\vec{B}$ before integrating. Instead, we arbitrarily assume $\vec{B}$ to be generally in the direction of integration (as in Fig. 30-11). Then we use the following curled–straight right-hand rule to assign a plus sign or a minus sign to each of the currents that make up the net encircled current i_{enc}:

Curl your right hand around the Amperian loop, with the fingers pointing in the direction of integration. A current through the loop in the general direction of your outstretched thumb is assigned a plus sign, and a current generally in the opposite direction is assigned a minus sign.

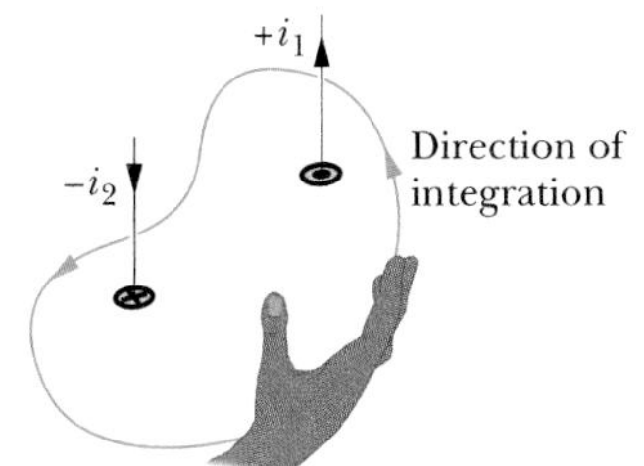

Fig. 30-12 A right-hand rule for Ampere's law, to determine the signs for currents encircled by an Amperian loop. The situation is that of Fig. 30-11.

Finally, we solve Eq. 30-17 for the magnitude of $\vec{B}$. If B turns out positive, then the direction we assumed for $\vec{B}$ is correct. If it turns out negative, we neglect the minus sign and redraw $\vec{B}$ in the opposite direction.

In Fig. 30-12 we apply the curled–straight rule for Ampere's law to the situation of Fig. 30-11. With the indicated counterclockwise direction of integration, the net current encircled by the loop is

$$i_{\text{enc}} = i_1 - i_2.$$

(Current i_3 is not encircled by the loop.) We can then rewrite Eq. 30-17 as

$$\oint B \cos \theta \, ds = \mu_0(i_1 - i_2). \tag{30-18}$$

You might wonder why, since current i_3 contributes to the magnetic-field magnitude B on the left side of Eq. 30-18, it is not needed on the right side. The answer is that the contributions of current i_3 to the magnetic field cancel out because the integration in Eq. 30-18 is made around the full loop. In contrast, the contributions of an encircled current to the magnetic field do not cancel out.

We cannot solve Eq. 30-18 for the magnitude B of the magnetic field, because for the situation of Fig. 30-11 we do not have enough information to simplify and solve the integral. However, we do know the outcome of the integration; it must be equal to the value of $\mu_0(i_1 - i_2)$, which is set by the net current passing through the loop.

We shall now apply Ampere's law to two situations in which symmetry does allow us to simplify and solve the integral, hence to find the magnetic field.

The Magnetic Field Outside a Long Straight Wire with Current

Figure 30-13 shows a long straight wire that carries current i directly out of the page. Equation 30-6 tells us that the magnetic field $\vec{B}$ produced by the current has the same

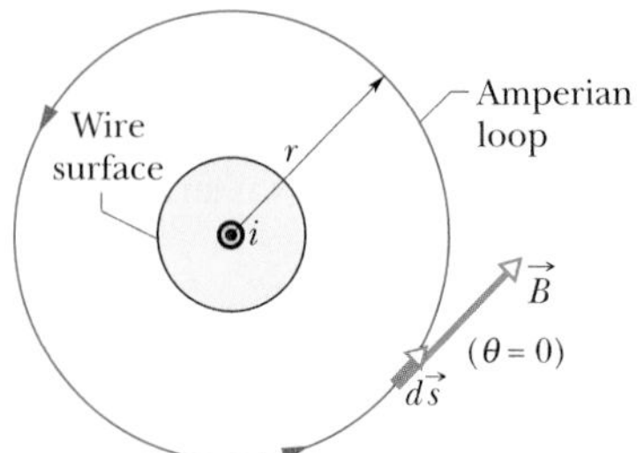

Fig. 30-13 Using Ampere's law to find the magnetic field produced by a current i in a long straight wire. The Amperian loop is a concentric circle that lies outside the wire.

magnitude at all points that are the same distance r from the wire; that is, the field $\vec{B}$ has cylindrical symmetry about the wire. We can take advantage of that symmetry to simplify the integral in Ampere's law (Eqs. 30-16 and 30-17) if we encircle the wire with a concentric circular Amperian loop of radius r, as in Fig. 30-13. The magnetic field $\vec{B}$ then has the same magnitude B at every point on the loop. We shall integrate counterclockwise, so that $d\vec{s}$ has the direction shown in Fig. 30-13.

We can further simplify the quantity $B \cos \theta$ in Eq. 30-17 by noting that $\vec{B}$ is tangent to the loop at every point along the loop, as is $d\vec{s}$. Thus, $\vec{B}$ and $d\vec{s}$ are either parallel or antiparallel at each point of the loop, and we shall arbitrarily assume the former. Then at every point the angle θ between $d\vec{s}$ and $\vec{B}$ is 0°, so $\cos \theta = \cos 0° = 1$. The integral in Eq. 30-17 then becomes

$$\oint \vec{B} \cdot d\vec{s} = \oint B \cos \theta \, ds = B \oint ds = B(2\pi r).$$

Note that $\oint ds$ above is the summation of all the line segment lengths ds around the circular loop; that is, it simply gives the circumference $2\pi r$ of the loop.

Our right-hand rule gives us a plus sign for the current of Fig. 30-13. The right side of Ampere's law becomes $+\mu_0 i$ and we then have

$$B(2\pi r) = \mu_0 i$$

or

$$B = \frac{\mu_0 i}{2\pi r}. \tag{30-19}$$

With a slight change in notation, this is Eq. 30-6, which we derived earlier—with considerably more effort—using the law of Biot and Savart. In addition, because the magnitude B turned out positive, we know that the correct direction of $\vec{B}$ must be the one shown in Fig. 30-13.

The Magnetic Field Inside a Long Straight Wire with Current

Figure 30-14 shows the cross section of a long straight wire of radius R that carries a uniformly distributed current i directly out of the page. Because the current is uniformly distributed over a cross section of the wire, the magnetic field $\vec{B}$ that it produces must be cylindrically symmetrical. Thus, to find the magnetic field at points inside the wire, we can again use an Amperian loop of radius r, as shown in Fig. 30-14, where now $r < R$. Symmetry again suggests that $\vec{B}$ is tangent to the loop, as shown, so the left side of Ampere's law again yields

$$\oint \vec{B} \cdot d\vec{s} = B \oint ds = B(2\pi r). \tag{30-20}$$

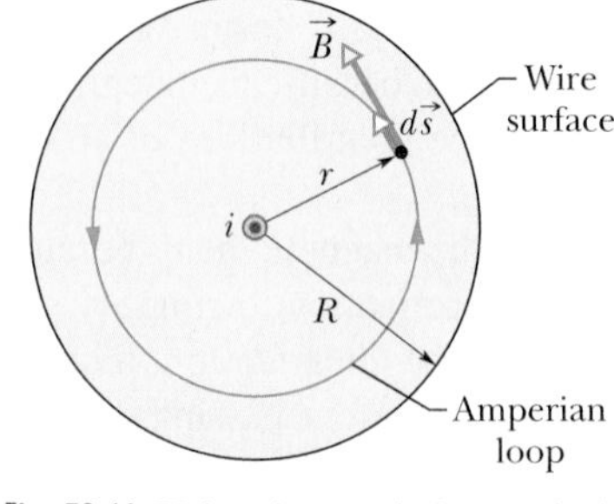

Fig. 30-14 Using Ampere's law to find the magnetic field that a current i produces inside a long straight wire of circular cross section. The current is uniformly distributed over the cross section of the wire and emerges from the page. An Amperian loop is drawn inside the wire.

To find the right side of Ampere's law, we note that because the current is uniformly distributed, the current i_{enc} encircled by the loop is proportional to the area encircled by the loop; that is,

$$i_{enc} = i \frac{\pi r^2}{\pi R^2}. \tag{30-21}$$

Our right-hand rule tells us that i_{enc} gets a plus sign. Then Ampere's law gives us

$$B(2\pi r) = \mu_0 i \frac{\pi r^2}{\pi R^2}$$

or

$$B = \left(\frac{\mu_0 i}{2\pi R^2}\right) r. \tag{30-22}$$

Thus, inside the wire, the magnitude B of the magnetic field is proportional to r; that magnitude is zero at the center and a maximum at the surface, where $r = R$. Note that Eqs. 30-19 and 30-22 give the same value for B at $r = R$; that is, the expressions for the magnetic field outside the wire and inside the wire yield the same result at the surface of the wire.

CHECKPOINT 3: The figure here shows three equal currents i (two parallel and one antiparallel) and four Amperian loops. Rank the loops according to the magnitude of $\oint \vec{B} \cdot d\vec{s}$ along each, greatest first.

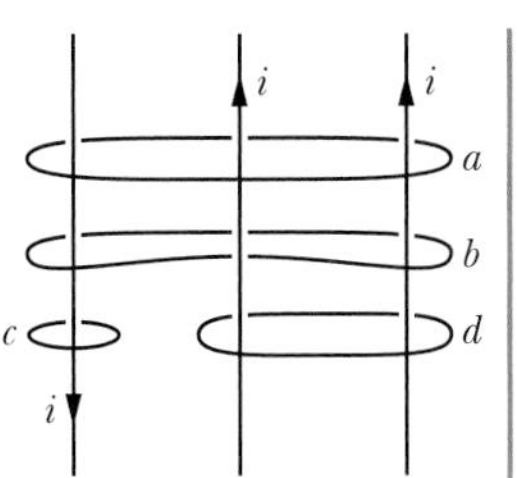

Sample Problem 30-3

Figure 30-15a shows the cross section of a long conducting cylinder with inner radius a = 2.0 cm and outer radius b = 4.0 cm. The cylinder carries a current out of the page, and the current density in the cross section is given by $J = cr^2$, with $c = 3.0 \times 10^6$ A/m^4 and r in meters. What is the magnetic field $\vec{B}$ at a point that is 3.0 cm from the central axis of the cylinder?

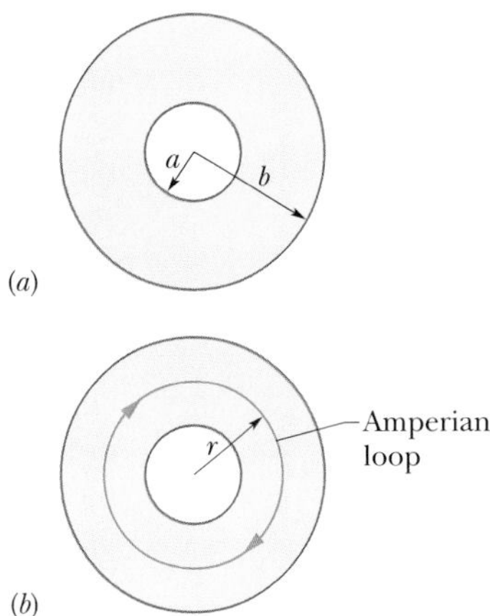

Fig. 30-15 Sample Problem 30-3. (a) Cross section of a conducting cylinder of inner radius a and outer radius b. (b) An Amperian loop of radius r is added to compute the magnetic field at points that are a distance r from the central axis.

SOLUTION: The point at which we want to evaluate $\vec{B}$ is inside the material of the conducting cylinder, between its inner and outer radii. We note that the current distribution has cylindrical symmetry (it is the same all around the cross section for any given radius). Thus, the **Key Idea** here is that the symmetry allows us to use Ampere's law to find $\vec{B}$ at the point. We first draw the Amperian loop shown in Fig. 30-15b. The loop is concentric with the cylinder and has radius r = 3.0 cm, because we want to evaluate $\vec{B}$ at that distance from the cylinder's central axis.

Next, we must compute the current i_{enc} that is encircled by the Amperian loop. However, a second **Key Idea** is that we *cannot* set up a proportionality as in Eq. 30-21, because here the current is not uniformly distributed. Instead, following the procedure of Sample Problem 27-2b, we must integrate the current density from the cylinder's inner radius a to the loop radius r:

$$\begin{aligned} i_{enc} &= \int J\,dA = \int_a^r cr^2\,(2\pi r\,dr) \\ &= 2\pi c \int_a^r r^3\,dr = 2\pi c\left[\frac{r^4}{4}\right]_a^r \\ &= \frac{\pi c(r^4 - a^4)}{2}. \end{aligned}$$

The direction of integration indicated in Fig. 30-15b is (arbitrarily) clockwise. Applying the right-hand rule for Ampere's law to that loop, we find that we should take i_{enc} as negative because the current is directed out of the page but our thumb is directed into the page.

We next evaluate the left side of Ampere's law exactly as we did in Fig. 30-14, and we again obtain Eq. 30-20. Then Ampere's law,

$$\oint \vec{B} \cdot d\vec{s} = \mu_0 i_{enc},$$

gives us

$$B(2\pi r) = -\frac{\mu_0 \pi c}{2}(r^4 - a^4).$$

Solving for B and substituting known data yield

$$\begin{aligned} B &= -\frac{\mu_0 c}{4r}(r^4 - a^4) \\ &= -\frac{(4\pi \times 10^{-7}\ \text{T}\cdot\text{m/A})(3.0 \times 10^6\ \text{A/m}^4)}{4(0.030\ \text{m})} \\ &\quad \times [(0.030\ \text{m})^4 - (0.020\ \text{m})^4] \\ &= -2.0 \times 10^{-5}\ \text{T}. \end{aligned}$$

Thus, the magnetic field $\vec{B}$ at a point 3.0 cm from the central axis has the magnitude

$$B = 2.0 \times 10^{-5}\ \text{T} \qquad \text{(Answer)}$$

and forms magnetic field lines that are directed opposite our direction of integration, hence counterclockwise in Fig. 30-15b.

30-4 Solenoids and Toroids

Magnetic Field of a Solenoid

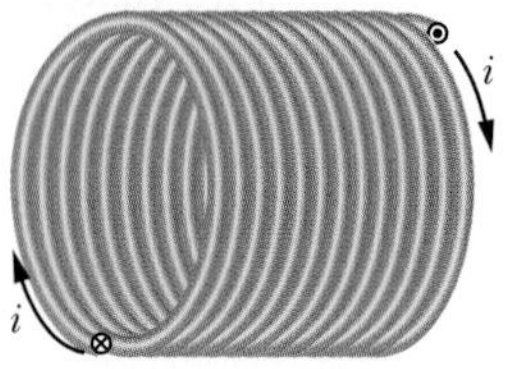

Fig. 30-16 A solenoid carrying current i.

We now turn our attention to another situation in which Ampere's law proves useful. It concerns the magnetic field produced by the current in a long, tightly wound helical coil of wire. Such a coil is called a **solenoid** (Fig. 30-16). We assume that the length of the solenoid is much greater than the diameter.

Figure 30-17 shows a section through a portion of a "stretched-out" solenoid. The solenoid's magnetic field is the vector sum of the fields produced by the individual turns (loops) that make up the solenoid. For points very close to each turn, the wire behaves magnetically almost like a long straight wire, and the lines of $\vec{B}$ there are almost concentric circles. Figure 30-17 suggests that the field tends to cancel between adjacent turns. It also suggests that, at points inside the solenoid and reasonably far from the wire, $\vec{B}$ is approximately parallel to the (central) solenoid axis. In the limiting case of an *ideal solenoid,* which is infinitely long and consists of tightly packed (*close-packed*) turns of square wire, the field inside the coil is uniform and parallel to the solenoid axis.

At points above the solenoid, such as P in Fig. 30-17, the field set up by the upper parts of the solenoid turns (marked ⊙) is directed to the left (as drawn near P) and tends to cancel the field set up by the lower parts of the turns (marked ⊗), which is directed to the right (not drawn). In the limiting case of an ideal solenoid, the magnetic field outside the solenoid is zero. Taking the external field to be zero is an excellent assumption for a real solenoid if its length is much greater than its diameter and if we consider external points such as point P that are not at either end of the solenoid. The direction of the magnetic field along the solenoid axis is given by a curled–straight right-hand rule: Grasp the solenoid with your right hand so that your fingers follow the direction of the current in the windings; your extended right thumb then points in the direction of the axial magnetic field.

Figure 30-18 shows the lines of $\vec{B}$ for a real solenoid. The spacing of the lines of $\vec{B}$ in the central region shows that the field inside the coil is fairly strong and uniform over the cross section of the coil. The external field, however, is relatively weak.

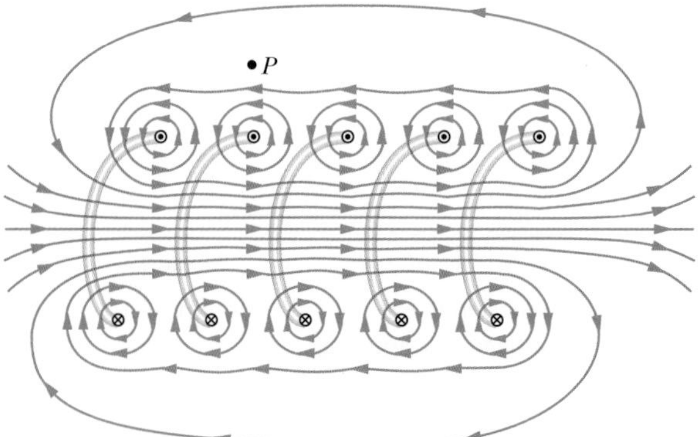

Fig. 30-17 A vertical cross section through the central axis of a "stretched-out" solenoid. The back portions of five turns are shown, as are the magnetic field lines due to a current through the solenoid. Each turn produces circular magnetic field lines near it. Near the solenoid's axis, the field lines combine into a net magnetic field that is directed along the axis. The closely spaced field lines there indicate a strong magnetic field. Outside the solenoid the field lines are widely spaced; the field there is very weak.

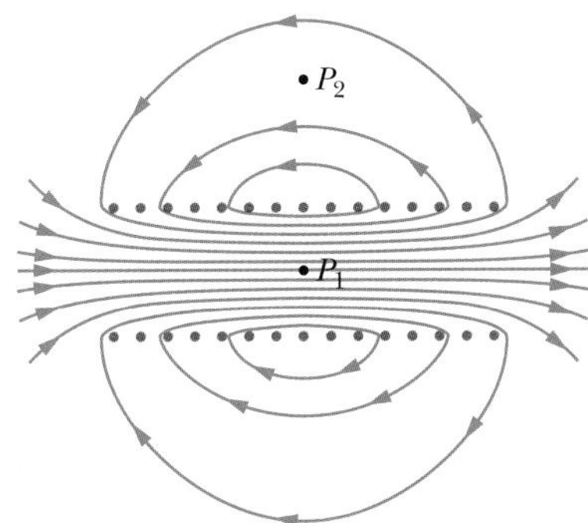

Fig. 30-18 Magnetic field lines for a real solenoid of finite length. The field is strong and uniform at interior points such as P_1 but relatively weak at external points such as P_2.

Let us now apply Ampere's law,

$$\oint \vec{B} \cdot d\vec{s} = \mu_0 i_{\text{enc}}, \tag{30-23}$$

to the ideal solenoid of Fig. 30-19, where $\vec{B}$ is uniform within the solenoid and zero outside it, using the rectangular Amperian loop *abcda*. We write $\oint \vec{B} \cdot d\vec{s}$ as the sum of four integrals, one for each loop segment:

$$\oint \vec{B} \cdot d\vec{s} = \int_a^b \vec{B} \cdot d\vec{s} + \int_b^c \vec{B} \cdot d\vec{s} + \int_c^d \vec{B} \cdot d\vec{s} + \int_d^a \vec{B} \cdot d\vec{s}. \tag{30-24}$$

The first integral on the right of Eq. 30-24 is Bh, where B is the magnitude of the uniform field $\vec{B}$ inside the solenoid and h is the (arbitrary) length of the segment from a to b. The second and fourth integrals are zero because for every element ds of these segments, $\vec{B}$ either is perpendicular to ds or is zero, and thus the product $\vec{B} \cdot d\vec{s}$ is zero. The third integral, which is taken along a segment that lies outside the solenoid, is zero because $B = 0$ at all external points. Thus, $\oint \vec{B} \cdot d\vec{s}$ for the entire rectangular loop has the value Bh.

The net current i_{enc} encircled by the rectangular Amperian loop in Fig. 30-19 is not the same as the current i in the solenoid windings because the windings pass more than once through this loop. Let n be the number of turns per unit length of the solenoid; then the loop encloses nh turns and

$$i_{\text{enc}} = i(nh).$$

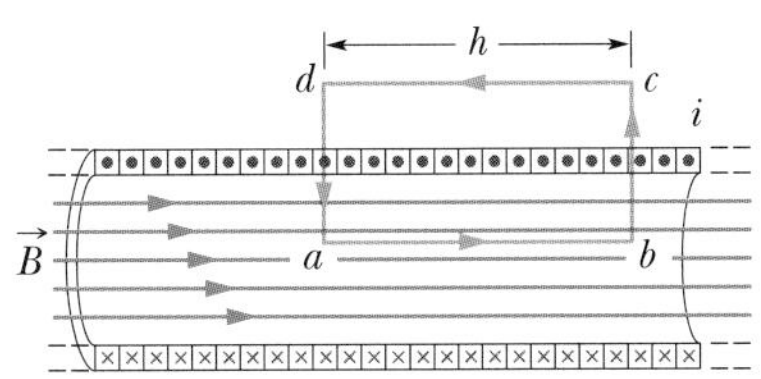

Fig. 30-19 Application of Ampere's law to a section of a long ideal solenoid carrying a current i. The Amperian loop is the rectangle *abcd*.

Ampere's law then gives us

$$Bh = \mu_0 inh$$

or

$$B = \mu_0 in \quad \text{(ideal solenoid).} \tag{30-25}$$

Although we derived Eq. 30-25 for an infinitely long ideal solenoid, it holds quite well for actual solenoids if we apply it only at interior points, well away from the solenoid ends. Equation 30-25 is consistent with the experimental fact that the magnetic field magnitude B within a solenoid does not depend on the diameter or the length of the solenoid and that B is uniform over the solenoidal cross section. A solenoid thus provides a practical way to set up a known uniform magnetic field for experimentation, just as a parallel-plate capacitor provides a practical way to set up a known uniform electric field.

Magnetic Field of a Toroid

Figure 30-20*a* shows a **toroid**, which we may describe as a solenoid bent into the shape of a hollow doughnut. What magnetic field $\vec{B}$ is set up at its interior points (within the hollow of the doughnut)? We can find out from Ampere's law and the symmetry of the doughnut.

From the symmetry, we see that the lines of $\vec{B}$ form concentric circles inside the toroid, directed as shown in Fig. 30-20*b*. Let us choose a concentric circle of

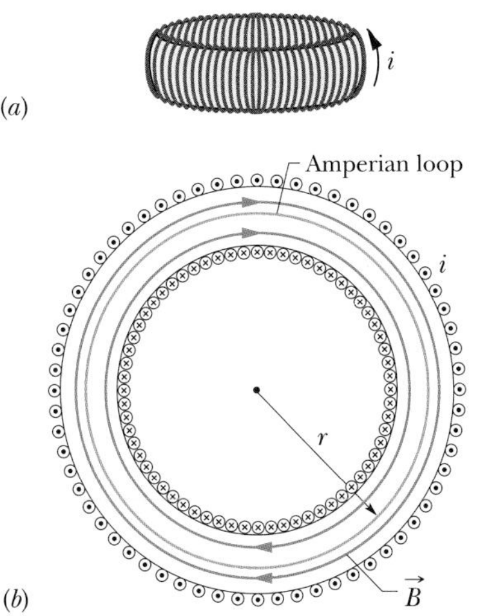

Fig. 30-20 (*a*) A toroid carrying a current i. (*b*) A horizontal cross section of the toroid. The interior magnetic field (inside the doughnut-shaped tube) can be found by applying Ampere's law with the Amperian loop shown.

radius r as an Amperian loop and traverse it in the clockwise direction. Ampere's law (Eq. 30-16) yields

$$(B)(2\pi r) = \mu_0 iN,$$

where i is the current in the toroid windings (and is positive for those windings enclosed by the Amperian loop) and N is the total number of turns. This gives

$$B = \frac{\mu_0 iN}{2\pi}\frac{1}{r} \quad \text{(toroid).} \tag{30-26}$$

In contrast to the situation for a solenoid, B is not constant over the cross section of a toroid. It is easy to show, with Ampere's law, that $B = 0$ for points outside an ideal toroid (as if the toroid were made from an ideal solenoid).

The direction of the magnetic field within a toroid follows from our curled–straight right-hand rule: Grasp the toroid with the fingers of your right hand curled in the direction of the current in the windings; your extended right thumb points in the direction of the magnetic field.

Sample Problem 30-4

A solenoid has length $L = 1.23$ m and inner diameter $d = 3.55$ cm, and it carries a current $i = 5.57$ A. It consists of five close-packed layers, each with 850 turns along length L. What is B at its center?

SOLUTION: One **Key Idea** here is that the magnitude B of the magnetic field along the solenoid's center is related to the solenoid's current i and number of turns per unit length n by Eq. 30-25. A second **Key Idea** is that B does not depend on the diameter of the windings, so the value of n for five identical layers is simply five times the value for each layer. Equation 30-25 then tells us

$$B = \mu_0 in = (4\pi \times 10^{-7}\ \text{T}\cdot\text{m/A})(5.57\ \text{A})\frac{5 \times 850\ \text{turns}}{1.23\ \text{m}}$$

$$= 2.42 \times 10^{-2}\ \text{T} = 24.2\ \text{mT}. \quad \text{(Answer)}$$

30-5 A Current-Carrying Coil as a Magnetic Dipole

So far we have examined the magnetic fields produced by current in a long straight wire, a solenoid, and a toroid. We turn our attention here to the field produced by a coil carrying a current. You saw in Section 29-9 that such a coil behaves as a magnetic dipole in that, if we place it in an external magnetic field $\vec{B}$, a torque $\vec{\tau}$ given by

$$\vec{\tau} = \vec{\mu} \times \vec{B} \tag{30-27}$$

acts on it. Here $\vec{\mu}$ is the magnetic dipole moment of the coil and has the magnitude NiA, where N is the number of turns (or loops), i is the current in each turn, and A is the area enclosed by each turn.

Recall that the direction of $\vec{\mu}$ is given by a curled–straight right-hand rule: Grasp the coil so that the fingers of your right hand curl around it in the direction of the current; your extended thumb then points in the direction of the dipole moment $\vec{\mu}$.

Magnetic Field of a Coil

We turn now to the other aspect of a current-carrying coil as a magnetic dipole. What magnetic field does *it* produce at a point in the surrounding space? The problem does not have enough symmetry to make Ampere's law useful, so we must turn to

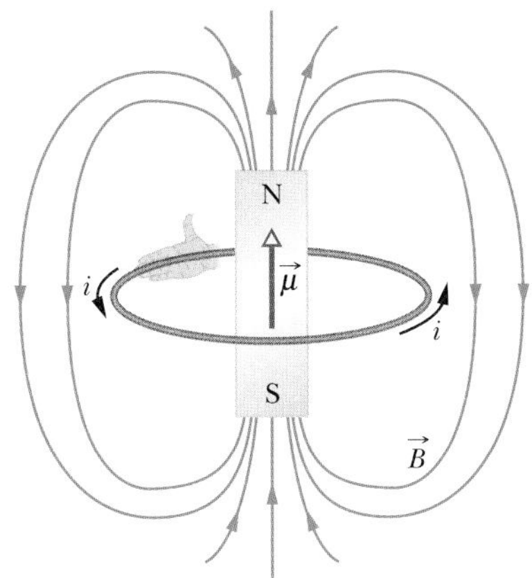

Fig. 30-21 A current loop produces a magnetic field like that of a bar magnet and thus has associated north and south poles. The magnetic dipole moment $\vec{\mu}$ of the loop, given by a curled–straight right-hand rule, points from the south pole to the north pole, in the direction of the field $\vec{B}$ within the loop.

the law of Biot and Savart. For simplicity, we first consider only a coil with a single circular loop and only points on its central axis, which we take to be a z axis. We shall show that the magnitude of the magnetic field at such points is

$$B(z) = \frac{\mu_0 i R^2}{2(R^2 + z^2)^{3/2}}, \tag{30-28}$$

in which R is the radius of the circular loop and z is the distance of the point in question from the center of the loop. Furthermore, the direction of the magnetic field $\vec{B}$ is the same as the direction of the magnetic dipole moment $\vec{\mu}$ of the loop.

For axial points far from the loop, we have $z \gg R$ in Eq. 30-28. With that approximation, the equation reduces to

$$B(z) \approx \frac{\mu_0 i R^2}{2z^3}.$$

Recalling that πR^2 is the area A of the loop and extending our result to include a coil of N turns, we can write this equation as

$$B(z) = \frac{\mu_0}{2\pi}\frac{NiA}{z^3}.$$

Further, since $\vec{B}$ and $\vec{\mu}$ have the same direction, we can write the equation in vector form, substituting from the identity $\mu = NiA$:

$$\vec{B}(z) = \frac{\mu_0}{2\pi}\frac{\vec{\mu}}{z^3} \quad \text{(current-carrying coil).} \tag{30-29}$$

Thus, we have two ways in which we can regard a current-carrying coil as a magnetic dipole: (1) it experiences a torque when we place it in an external magnetic field; (2) it generates its own intrinsic magnetic field, given, for distant points along its axis, by Eq. 30-29. Figure 30-21 shows the magnetic field of a current loop; one side of the loop acts as a north pole (in the direction of $\vec{\mu}$) and the other side as a south pole, as suggested by the lightly drawn magnet in the figure.

CHECKPOINT 4: The figure here shows four arrangements of circular loops of radius r or $2r$, centered on vertical axes (perpendicular to the loops) and carrying identical currents in the directions indicated. Rank the arrangements according to the magnitude of the net magnetic field at the dot, midway between the loops on the central axis, greatest first.

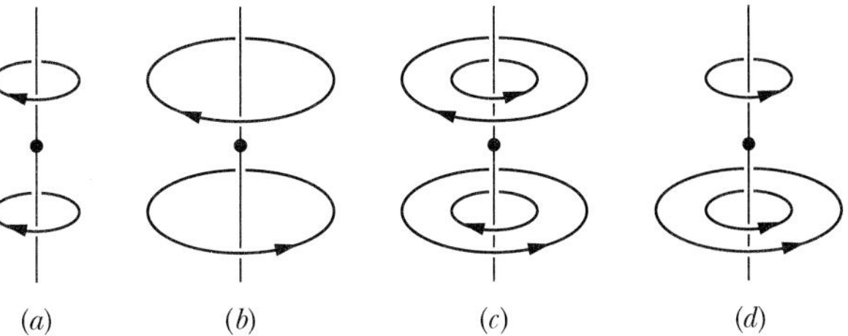

Proof of Equation 30-28

Figure 30-22 shows the back half of a circular loop of radius R carrying a current i. Consider a point P on the axis of the loop, a distance z from its plane. Let us apply the law of Biot and Savart to a differential element ds of the loop, located at the left side of the loop. The length vector $d\vec{s}$ for this element points perpendicularly out of

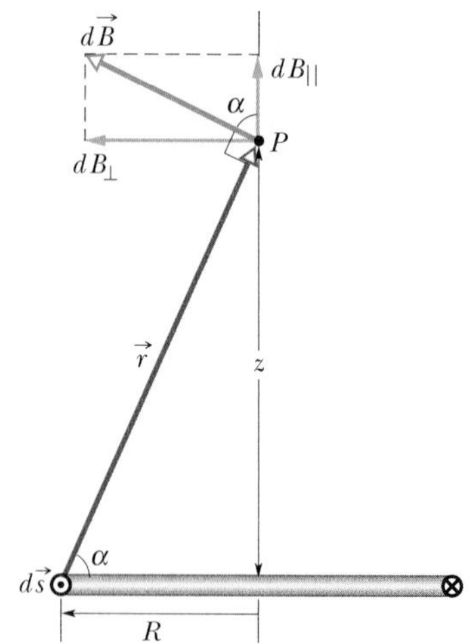

Fig. 30-22 A current loop of radius R. The plane of the loop is perpendicular to the page and only the back half of the loop is shown. We use the law of Biot and Savart to find the magnetic field at point P on the central axis of the loop.

the page. The angle θ between $d\vec{s}$ and $\vec{r}$ in Fig. 30-22 is 90°; the plane formed by these two vectors is perpendicular to the plane of the figure and contains both $\vec{r}$ and $d\vec{s}$. From the law of Biot and Savart (and the right-hand rule), the differential field $d\vec{B}$ produced at point P by the current in this element is perpendicular to this plane and thus is directed in the plane of the figure, perpendicular to $\vec{r}$, as indicated in Fig. 30-22.

Let us resolve $d\vec{B}$ into two components: $dB_{||}$ along the axis of the loop and $dB_{\perp}$ perpendicular to this axis. From the symmetry, the vector sum of all the perpendicular components $dB_{\perp}$ due to all the loop elements ds is zero. This leaves only the axial components $dB_{||}$ and we have

$$B = \int dB_{||}.$$

For the element $d\vec{s}$ in Fig. 30-22, the law of Biot and Savart (Eq. 30-3) tells us that the magnetic field at distance r is

$$dB = \frac{\mu_0}{4\pi} \frac{i\,ds \sin 90^\circ}{r^2}.$$

We also have

$$dB_{||} = dB \cos \alpha.$$

Combining these two relations, we obtain

$$dB_{||} = \frac{\mu_0 i \cos \alpha\, ds}{4\pi r^2}. \tag{30-30}$$

Figure 30-22 shows that r and α are not independent but are related to each other. Let us express each in terms of the variable z, the distance between point P and the center of the loop. The relations are

$$r = \sqrt{R^2 + z^2} \tag{30-31}$$

and

$$\cos \alpha = \frac{R}{r} = \frac{R}{\sqrt{R^2 + z^2}}. \tag{30-32}$$

Substituting Eqs. 30-31 and 30-32 into Eq. 30-30, we find

$$dB_{||} = \frac{\mu_0 iR}{4\pi(R^2 + z^2)^{3/2}}\, ds.$$

Note that i, R, and z have the same values for all elements ds around the loop, so when we integrate this equation, we find that

$$B = \int dB_{||}$$

$$= \frac{\mu_0 iR}{4\pi(R^2 + z^2)^{3/2}} \int ds$$

or, since $\int ds$ is simply the circumference $2\pi R$ of the loop,

$$B(z) = \frac{\mu_0 iR^2}{2(R^2 + z^2)^{3/2}}.$$

This is Eq. 30-28, the relation we sought to prove.

REVIEW & SUMMARY

The Biot–Savart Law The magnetic field set up by a current-carrying conductor can be found from the *Biot–Savart law*. This law asserts that the contribution $d\vec{B}$ to the field produced by a current-length element $i\,d\vec{s}$ at a point P, a distance r from the current element, is

$$d\vec{B} = \frac{\mu_0}{4\pi}\frac{i\,d\vec{s} \times \vec{r}}{r^3} \quad \text{(Biot–Savart law).} \qquad (30\text{-}5)$$

Here $\vec{r}$ is a vector that points from the element to P. The quantity μ_0, called the permeability constant, has the value $4\pi \times 10^{-7}\ \text{T}\cdot\text{m/A} \approx 1.26 \times 10^{-6}\ \text{T}\cdot\text{m/A}$.

Magnetic Field of a Long Straight Wire For a long straight wire carrying a current i, the Biot–Savart law gives, for the magnitude of the magnetic field at a perpendicular distance R from the wire,

$$B = \frac{\mu_0 i}{2\pi R} \quad \text{(long straight wire).} \qquad (30\text{-}6)$$

Magnetic Field of a Circular Arc The magnitude of the magnetic field at the center of a circular arc, of radius R and central angle ϕ (in radians), carrying current i, is

$$B = \frac{\mu_0 i \phi}{4\pi R} \quad \text{(at center of circular arc).} \qquad (30\text{-}11)$$

The Force Between Parallel Wires Carrying Currents Parallel wires carrying currents in the same direction attract each other, whereas parallel wires carrying currents in opposite directions repel each other. The magnitude of the force on a length L of either wire is

$$F_{ba} = i_b L B_a \sin 90° = \frac{\mu_0 L i_a i_b}{2\pi d}, \qquad (30\text{-}15)$$

where d is the wire separation, and i_a and i_b are the currents in the wires.

Ampere's Law **Ampere's law** states that

$$\oint \vec{B}\cdot d\vec{s} = \mu_0 i_{\text{enc}} \quad \text{(Ampere's law).} \qquad (30\text{-}16)$$

The line integral in this equation is evaluated around a closed loop called an *Amperian loop*. The current i is the *net* current encircled by the loop. For some current distributions, Eq. 30-16 is easier to use than Eq. 30-5 to calculate the magnetic field due to the currents.

Fields of a Solenoid and a Toroid Inside a *long solenoid* carrying current i, at points not near its ends, the magnitude B of the magnetic field is

$$B = \mu_0 i n \quad \text{(ideal solenoid),} \qquad (30\text{-}25)$$

where n is the number of turns per unit length. At a point inside a *toroid*, the magnitude B of the magnetic field is

$$B = \frac{\mu_0 i N}{2\pi}\frac{1}{r} \quad \text{(toroid),} \qquad (30\text{-}26)$$

where r is the distance from the center of the toroid to the point.

Field of a Magnetic Dipole The magnetic field produced by a current-carrying coil, which is a *magnetic dipole*, at a point P located a distance z along the coil's central axis is parallel to the axis and is given by

$$\vec{B}(z) = \frac{\mu_0}{2\pi}\frac{\vec{\mu}}{z^3}, \qquad (30\text{-}29)$$

where $\vec{\mu}$ is the dipole moment of the coil. This equation applies only when z is much greater than the dimensions of the coil.

QUESTIONS

1. Figure 30-23 shows four arrangements in which long parallel wires carry equal currents directly into or out of the page at the corners of identical squares. Rank the arrangements according to the magnitude of the net magnetic field at the center of the square, greatest first.

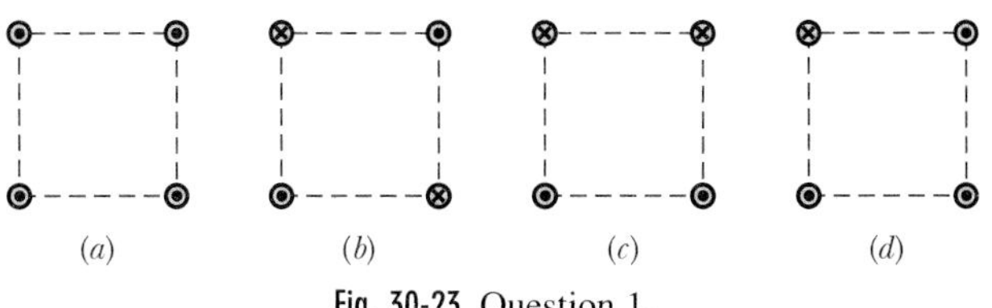

Fig. 30-23 Question 1.

2. Figure 30-24 shows cross sections of two long straight wires; the left-hand wire carries current i_1 directly out of the page. If the net magnetic field due to the two currents is to be zero at point P, (a) should the direction of current i_2 in the right-hand wire be directly into or out of the page and (b) should i_2 be greater than, less than, or equal to i_1?

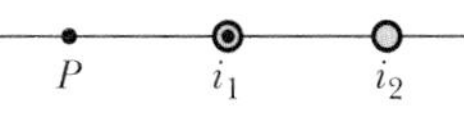

Fig. 30-24 Question 2.

3. Figure 30-25 shows three circuits, each consisting of two concentric circular arcs, one of radius r and the other of a larger radius

Fig. 30-25 Question 3.

R, and two radial lengths. The circuits have the same current through them, and the radial lengths have the same angle between them. Rank the circuits according to the magnitude of the net magnetic field at the center, greatest first.

4. Figure 30-26 shows four arrangements in which long, parallel, equally spaced wires carry equal currents directly into or out of the page. Rank the arrangements according to the magnitude of the net force on the central wire due to the currents in the other wires, greatest first.

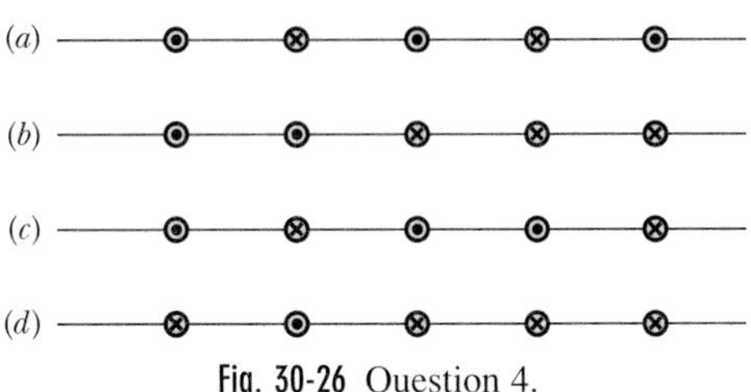

Fig. 30-26 Question 4.

5. Figure 30-27 shows three arrangements of three long straight wires, carrying equal currents directly into or out of the page. (a) Rank the arrangements according to the magnitude of the net force on the wire with the current directed out of the page due to the currents in the other wires, greatest first. (b) In arrangement 3, is the angle between the net force on that wire and the dashed line equal to, less than, or more than 45°?

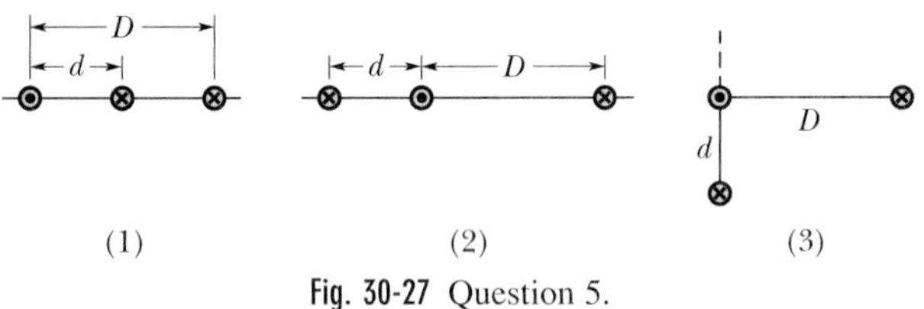

Fig. 30-27 Question 5.

6. Figure 30-28 shows a uniform magnetic field $\vec{B}$ and four straight-line paths of equal lengths. Rank the paths according to the magnitude of $\int \vec{B} \cdot d\vec{s}$ taken along the paths, greatest first.

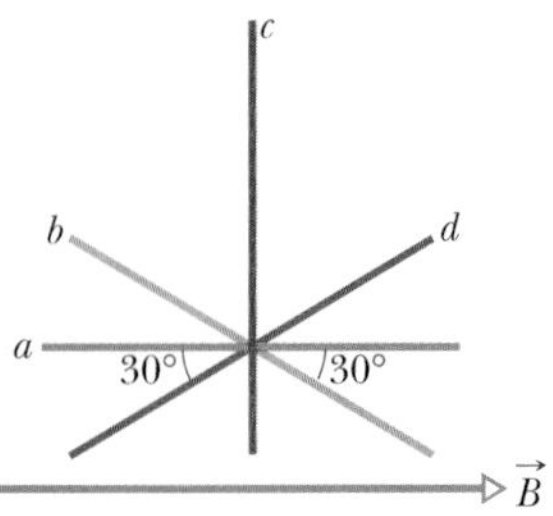

Fig. 30-28 Question 6.

7. Figure 30-29a shows four circular Amperian loops concentric with a wire whose current is directed out of the page. The current is uniform across the wire's circular cross section. Rank the loops according to the magnitude of $\oint \vec{B} \cdot d\vec{s}$ around each, greatest first.

8. Figure 30-29b shows four circular Amperian loops (red) and, in cross section, four long circular conductors (blue), all of which are concentric. Three of the conductors are hollow cylinders; the central conductor is a solid cylinder. The currents in the conductors are, from smallest radius to largest radius, 4 A out of the page, 9 A into the page, 5 A out of the page, and 3 A into the page. Rank the loops according to the magnitude of $\oint \vec{B} \cdot d\vec{s}$ around each, greatest first.

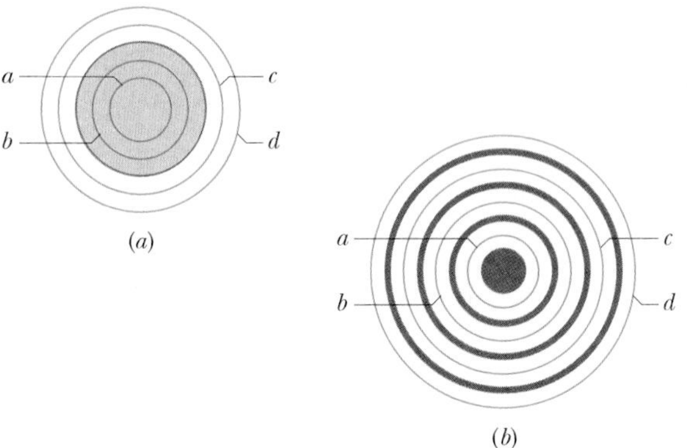

Fig. 30-29 Questions 7 and 8.

9. Figure 30-30 shows four identical currents i and five Amperian paths encircling them. Rank the paths according to the value of $\oint \vec{B} \cdot d\vec{s}$ taken in the directions shown, most positive first and most negative last.

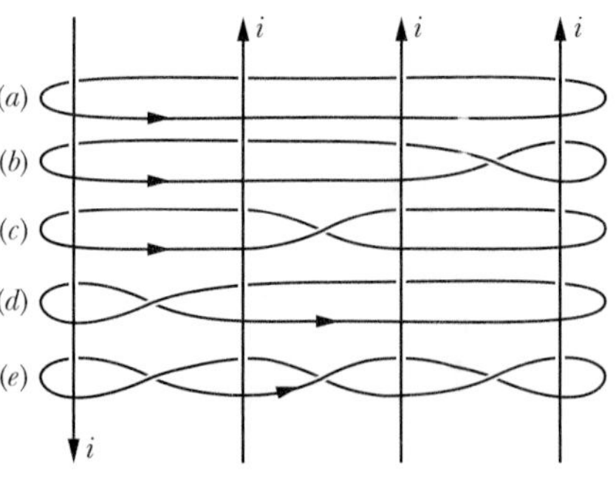

Fig. 30-30 Question 9.

10. The following table gives the number of turns per unit length n and the current i through six ideal solenoids of different radii. You want to combine several of them concentrically to produce a net magnetic field of zero along the central axis. Can this be done with (a) two of them, (b) three of them, (c) four of them, and (d) five of them? If so, answer by listing which solenoids are to be used and indicate the directions of the currents.

Solenoid:	1	2	3	4	5	6
n:	5	4	3	2	10	8
i:	5	3	7	6	2	3

EXERCISES & PROBLEMS

ssm Solution is in the Student Solutions Manual.
www Solution is available on the World Wide Web at: http://www.wiley.com/college/hrw
ilw Solution is available on the Interactive LearningWare.

SEC. 30-1 Calculating the Magnetic Field Due to a Current

1E. A surveyor is using a magnetic compass 6.1 m below a power line in which there is a steady current of 100 A. (a) What is the magnetic field at the site of the compass due to the power line? (b) Will this interfere seriously with the compass reading? The horizontal component of Earth's magnetic field at the site is 20 μT. ssm

2E. The electron gun in a traditional television tube fires electrons of kinetic energy 25 keV at the screen in a circular beam 0.22 mm in diameter; 5.6×10^{14} electrons arrive each second. Calculate the magnetic field produced by the beam at a point 1.5 mm from the beam axis.

3E. At a certain position in the Philippines, Earth's magnetic field of 39 μT is horizontal and directed due north. Suppose the net field is zero exactly 8.0 cm above a long, straight, horizontal wire that carries a constant current. What are (a) the magnitude and (b) the direction of the current? ssm

4E. A long wire carrying a current of 100 A is placed in a uniform external magnetic field of 5.0 mT. The wire is perpendicular to this magnetic field. Locate the points at which the net magnetic field is zero.

5E. A particle with positive charge q is a distance d from a long straight wire that carries a current i; the particle is traveling with speed v perpendicular to the wire. What are the direction and magnitude of the force on the particle if it is moving (a) toward and (b) away from the wire? ilw

6E. A straight conductor carrying a current i splits into identical semicircular arcs as shown in Fig. 30-31. What is the magnetic field at the center C of the resulting circular loop?

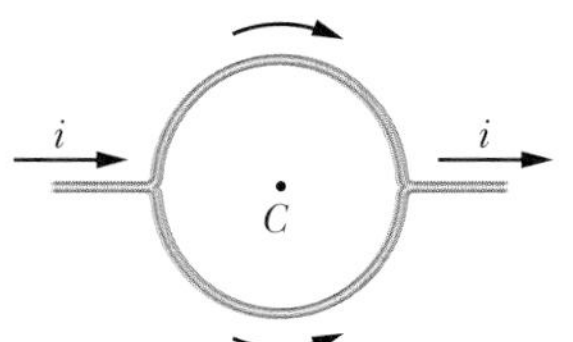

Fig. 30-31 Exercise 6.

7P. A wire carrying current i has the configuration shown in Fig. 30-32. Two semi-infinite straight sections, both tangent to the same circle, are connected by a circular arc, of central angle θ, along the circumference of the circle, with all sections lying in the same plane. What must θ be in order for B to be zero at the center of the circle? ssm

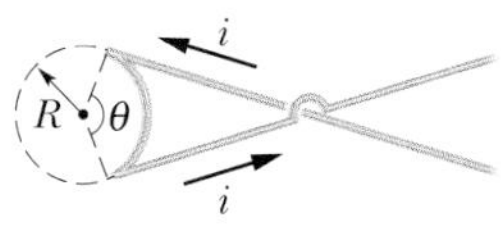

Fig. 30-32 Problem 7.

8P. Use the Biot–Savart law to calculate the magnetic field $\vec{B}$ at C, the common center of the semicircular arcs AD and HJ in Fig. 30-33a. The two arcs, of radii R_2 and R_1, respectively, form part of the circuit $ADJHA$ carrying current i.

9P. In the circuit of Fig. 30-33b, the curved segments are arcs of circles of radii a and b with common center P. The straight segments are along radii. Find the magnetic field $\vec{B}$ at point P, assuming a current i in the circuit. ssm

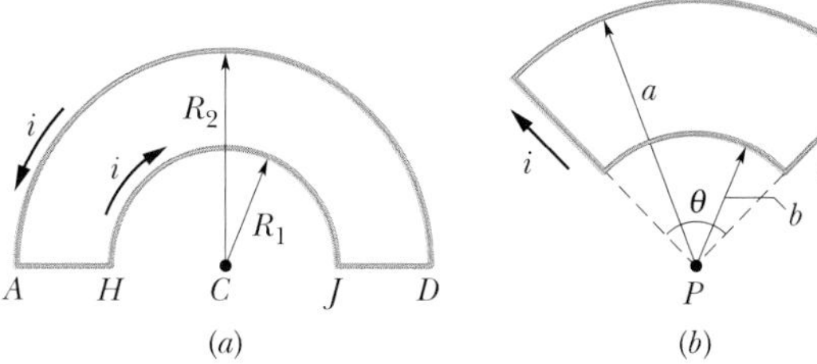

Fig. 30-33 Problems 8 and 9.

10P. The wire shown in Fig. 30-34 carries current i. What magnetic field $\vec{B}$ is produced at the center C of the semicircle by (a) each straight segment of length L, (b) the semicircular segment of radius R, and (c) the entire wire?

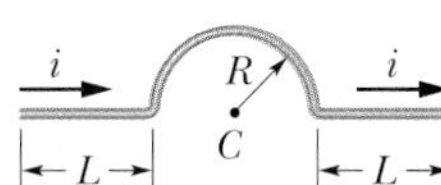

Fig. 30-34 Problem 10.

11P. In Fig. 30-35, a straight wire of length L carries current i. Show that the magnitude of the magnetic field $\vec{B}$ produced by this segment at P_1, a distance R from the segment along a perpendicular bisector, is

$$B = \frac{\mu_0 i}{2\pi R} \frac{L}{(L^2 + 4R^2)^{1/2}}.$$

Show that this expression for B reduces to an expected result as $L \to \infty$. ssm www

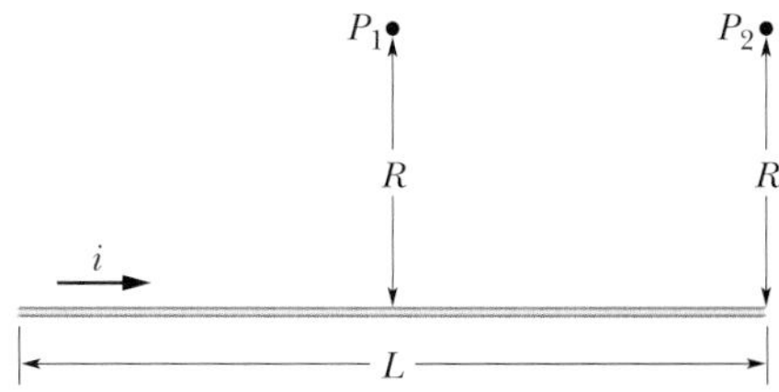

Fig. 30-35 Problems 11 and 13.

12P. A square loop of wire of edge length a carries current i. Using Problem 11, show that, at the center of the loop, the magnitude of the magnetic field produced by the current is

$$B = \frac{2\sqrt{2}\mu_0 i}{\pi a}.$$

13P. In Fig. 30-35, a straight wire of length L carries current i. Show that

$$B = \frac{\mu_0 i}{4\pi R} \frac{L}{(L^2 + R^2)^{1/2}}$$

gives the magnitude of the magnetic field $\vec{B}$ produced by the wire at P_2, a perpendicular distance R from one end of the wire. ssm

14P. Using Problem 11, show that the magnitude of the magnetic field produced at the center of a rectangular loop of wire of length L and width W, carrying a current i, is

$$B = \frac{2\mu_0 i}{\pi}\frac{(L^2 + W^2)^{1/2}}{LW}.$$

15P. A square loop of wire of edge length a carries current i. Using Problem 11, show that the magnitude of the magnetic field produced at a point on the axis of the loop and a distance x from its center is

$$B(x) = \frac{4\mu_0 i a^2}{\pi(4x^2 + a^2)(4x^2 + 2a^2)^{1/2}}.$$

Prove that this result is consistent with the result of Problem 12. ssm

16P. In Fig. 30-36, a straight wire of length a carries a current i. Show that the magnitude of the magnetic field produced by the current at point P is $B = \sqrt{2}\mu_0 i/8\pi a$.

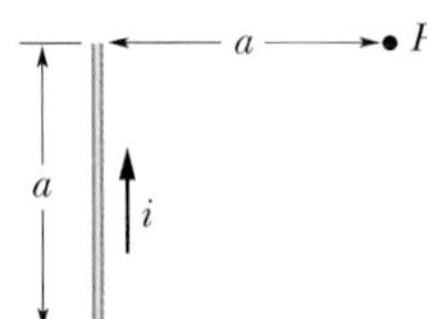

Fig. 30-36 Problem 16.

17P. Two wires, both of length L, are formed into a circle and a square, and each carries current i. Show that the square produces a greater magnetic field at its center than the circle produces at its center. (See Problem 12.) ssm

18P. Find the magnetic field $\vec{B}$ at point P in Fig. 30-37. (See Problem 16.)

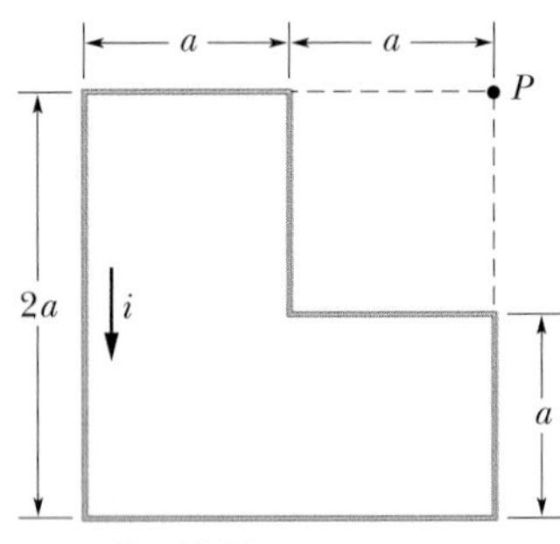

Fig. 30-37 Problem 18.

19P. Figure 30-38 shows a cross section of a long thin ribbon of width w that is carrying a uniformly distributed total current i into the page. Calculate the magnitude and direction of the magnetic field $\vec{B}$ at a point P in the plane of the ribbon at a distance d from its edge. (*Hint:* Imagine the ribbon to be constructed from many long, thin, parallel wires.) ilw

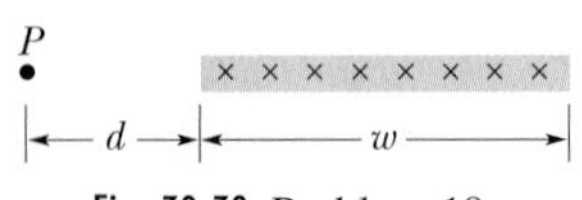

Fig. 30-38 Problem 19.

20P. Find the magnetic field $\vec{B}$ at point P in Fig. 30-39 for $i = 10$ A and $a = 8.0$ cm. (See Problems 13 and 16.)

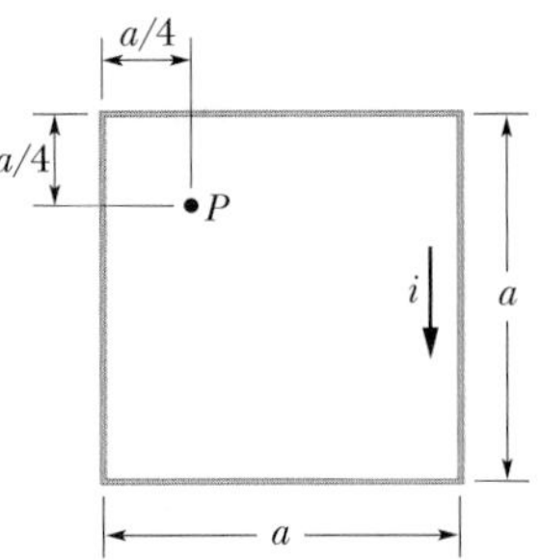

Fig. 30-39 Problem 20.

SEC. 30-2 Force Between Two Parallel Currents

21E. Two long parallel wires are 8.0 cm apart. What equal currents must be in the wires if the magnetic field halfway between them is to have a magnitude of 300 μT? Answer for both (a) parallel and (b) antiparallel currents. ssm

22E. Two long parallel wires a distance d apart carry currents of i and $3i$ in the same direction. Locate the point or points at which their magnetic fields cancel.

23E. Two long, straight, parallel wires, separated by 0.75 cm, are perpendicular to the plane of the page as shown in Fig. 30-40. Wire 1 carries a current of 6.5 A into the page. What must be the current (magnitude and direction) in wire 2 for the resultant magnetic field at point P to be zero?

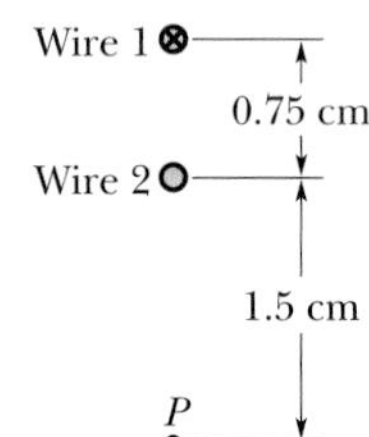

Fig. 30-40 Exercise 23.

24E. Figure 30-41 shows five long parallel wires in the xy plane. Each wire carries a current $i = 3.00$ A in the positive x direction. The separation between adjacent wires is $d = 8.00$ cm. In unit-vector notation, what is the magnetic force per meter exerted on each of these five wires by the other wires?

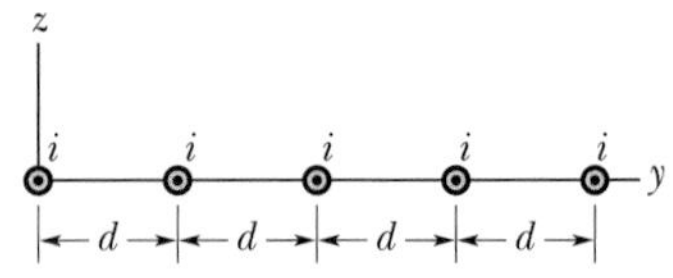

Fig. 30-41 Exercise 24.

25P. Four long copper wires are parallel to each other, their cross sections forming the corners of a square with sides $a = 20$ cm. A 20 A current exists in each wire in the direction shown in Fig. 30-42. What are the magnitude and direction of $\vec{B}$ at the center of the square? ssm www

26P. Four identical parallel currents i are arranged to form a square of edge length a as in Fig. 30-42, *except* that they are *all* out of the page. What is the force per unit length (magnitude and direction) on any one wire?

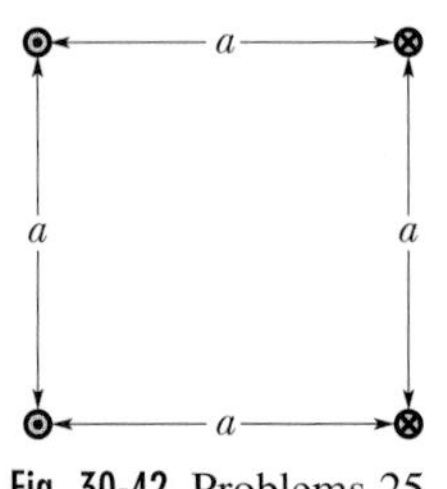

Fig. 30-42 Problems 25, 26, and 27.

27P. In Fig. 30-42, what is the force per unit length acting on the lower left wire, in magnitude and direction, with the current directions as shown? The currents are i.

28P. Figure 30-43 is an idealized schematic drawing of a rail gun. Projectile P sits between two wide rails of circular cross section; a source of current sends current through the rails and through the (conducting) projectile itself (a fuse is not used). (a) Let w be the distance between the rails, R the radius of the rails, and i the current. Show that the force on the projectile is directed to the right along the rails and is given approximately by

$$F = \frac{i^2\mu_0}{2\pi} \ln \frac{w + R}{R}.$$

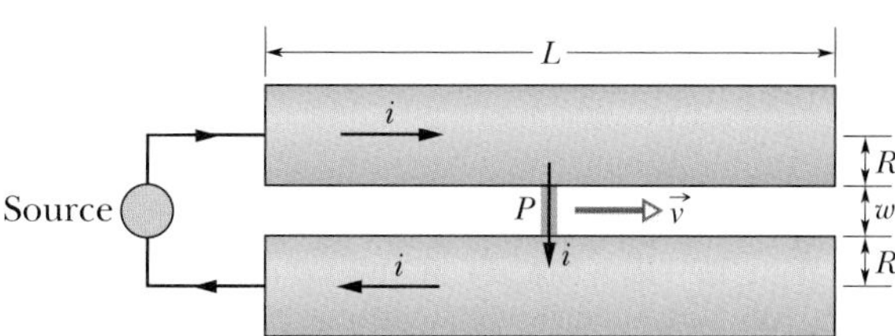

Fig. 30-43 Problem 28.

(b) If the projectile starts from the left end of the rails at rest, find the speed v at which it is expelled at the right. Assume that $i = 450$ kA, $w = 12$ mm, $R = 6.7$ cm, $L = 4.0$ m, and the mass of the projectile is $m = 10$ g.

29P. In Fig. 30-44, the long straight wire carries a current of 30 A and the rectangular loop carries a current of 20 A. Calculate the resultant force acting on the loop. Assume that $a = 1.0$ cm, $b = 8.0$ cm, and $L = 30$ cm. ilw

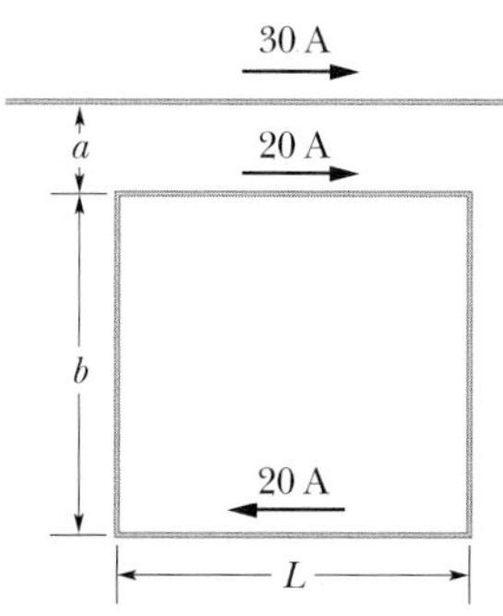

Fig. 30-44 Problem 29.

SEC. 30-3 Ampere's Law

30E. Eight wires cut the page perpendicularly at the points shown in Fig. 30-45. A wire labeled with the integer k ($k = 1, 2, \ldots, 8$) carries the current ki. For those with odd k, the current is out of the page; for those with even k, it is into the page. Evaluate $\oint \vec{B} \cdot d\vec{s}$ along the closed path in the direction shown.

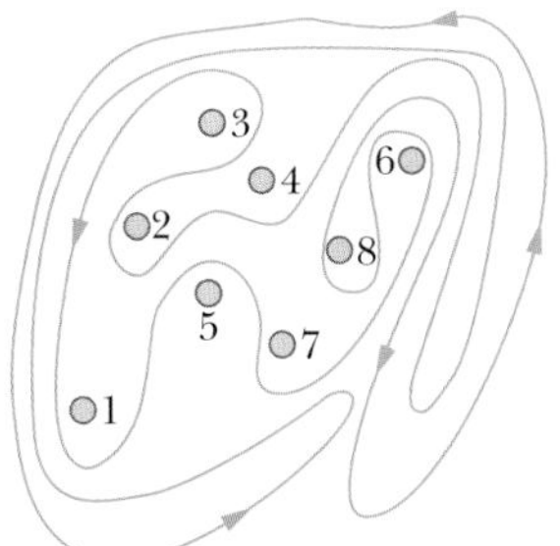

Fig. 30-45 Exercise 30.

31E. Each of the eight conductors in Fig. 30-46 carries 2.0 A of current into or out of the page. Two paths are indicated for the line integral $\oint \vec{B} \cdot d\vec{s}$. What is the value of the integral for the path (a) at the left and (b) at the right? ssm

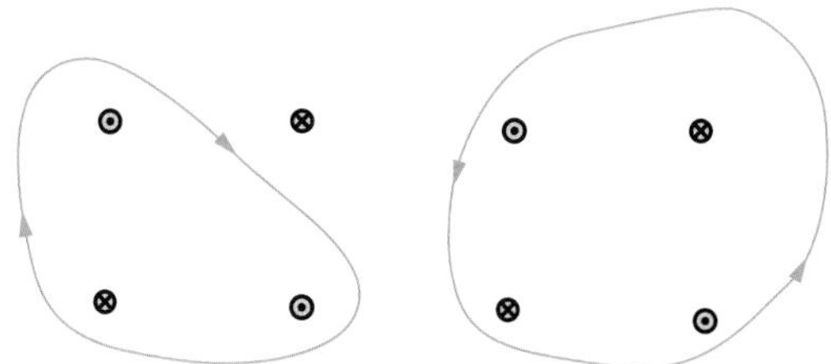

Fig. 30-46 Exercise 31.

32E. Figure 30-47 shows a cross section of a long cylindrical conductor of radius a, carrying a uniformly distributed current i. Assume that $a = 2.0$ cm and $i = 100$ A, and plot $B(r)$ over the range $0 < r < 6.0$ cm.

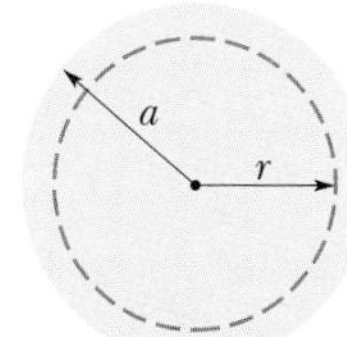

Fig. 30-47 Exercise 32.

33P. Show that a uniform magnetic field $\vec{B}$ cannot drop abruptly to zero (as is suggested by the lack of field lines to the right of point a in Fig. 30-48) as one moves perpendicular to $\vec{B}$, say along the horizontal arrow in the figure. (*Hint:* Apply Ampere's law to the rectangular path shown by the dashed lines.) In actual magnets "fringing" of the magnetic field lines always occurs, which means that $\vec{B}$ approaches zero in a gradual manner. Modify the field lines in the figure to indicate a more realistic situation. ssm

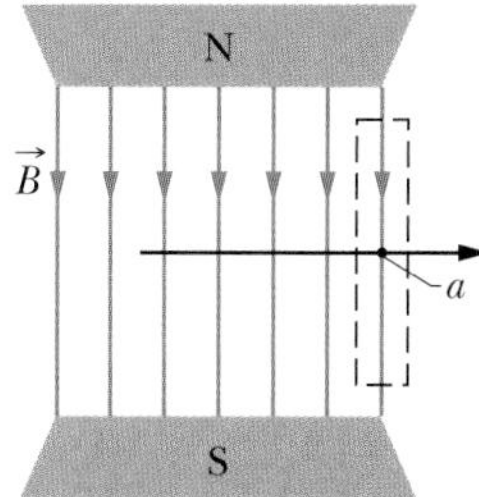

Fig. 30-48 Problem 33.

34P. Two square conducting loops carry currents of 5.0 and 3.0 A as shown in Fig. 30-49. What is the value of $\oint \vec{B} \cdot d\vec{s}$ for each of the two closed paths shown?

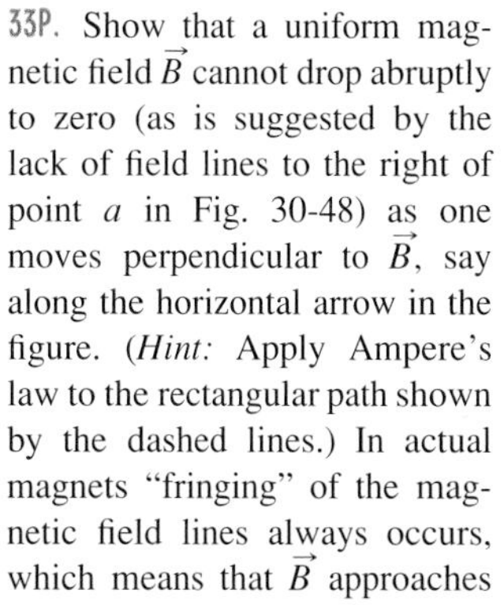

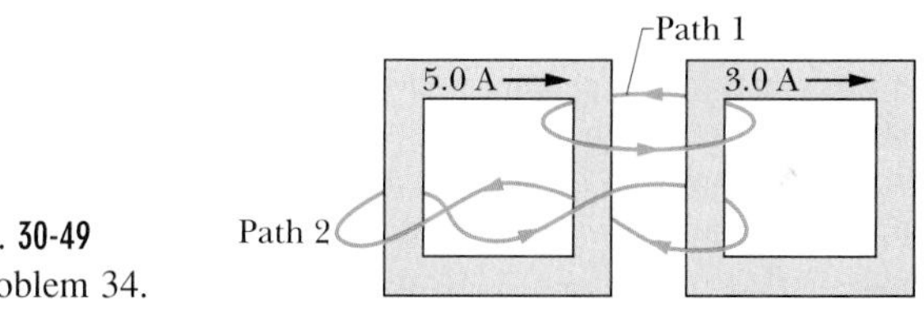

Fig. 30-49 Problem 34.

35P. The current density inside a long, solid, cylindrical wire of radius a is in the direction of the central axis and varies linearly with radial distance r from the axis according to $J = J_0 r/a$. Find the magnetic field inside the wire. ilw

36P. A long straight wire (radius = 3.0 mm) carries a constant current distributed uniformly over a cross section perpendicular to the axis of the wire. If the current density is 100 A/m^2, what are the magnitudes of the magnetic fields (a) 2.0 mm from the axis of the wire and (b) 4.0 mm from the axis of the wire?

37P. Figure 30-50 shows a cross section of a long cylindrical conductor of radius a containing a long cylindrical hole of radius b. The axes of the cylinder and hole are parallel and are a distance d apart; a current i is uniformly distributed over the tinted area. (a) Use superposition to show that the magnetic field at the center of the hole is

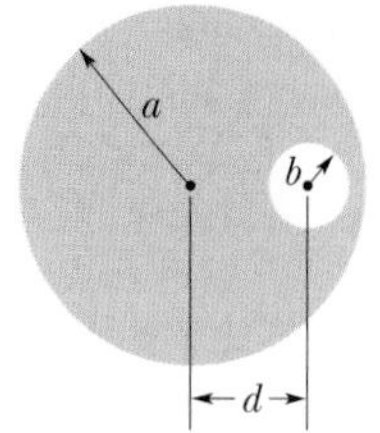

Fig. 30-50 Problem 37.

$$B = \frac{\mu_0 i d}{2\pi(a^2 - b^2)}.$$

(b) Discuss the two special cases $b = 0$ and $d = 0$. (c) Use Ampere's law to show that the magnetic field in the hole is uniform. (*Hint:* Regard the cylindrical hole as resulting from the superposition of a complete cylinder (no hole) carrying a current in one direction and a cylinder of radius b carrying a current in the opposite direction, both cylinders having the same current density.) ssm www

38P. A long circular pipe with outside radius R carries a (uniformly distributed) current i into the page as shown in Fig. 30-51. A wire runs parallel to the pipe at a distance of $3R$ from center to center. Find the magnitude and direction of the current in the wire such that the net magnetic field at point P has the same magnitude as the net magnetic field at the center of the pipe but is in the opposite direction.

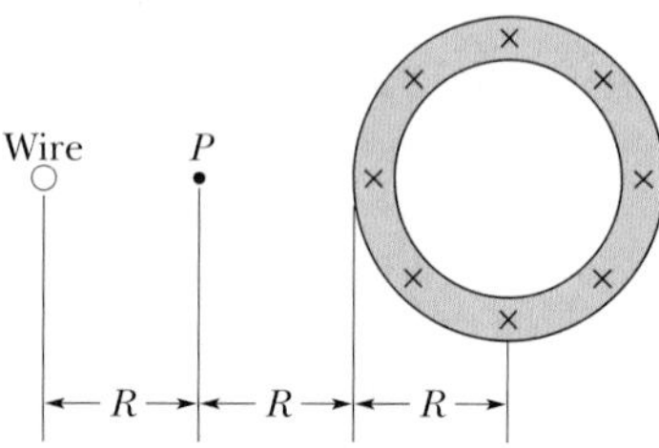

Fig. 30-51 Problem 38.

39P. Figure 30-52 shows a cross section of an infinite conducting sheet carrying a current per unit x-length of λ; the current emerges perpendicularly out of the page. (a) Use the Biot–Savart law and symmetry to show that for all points P above the sheet, and all points P' below it, the magnetic field $\vec{B}$ is parallel to the sheet and directed as shown. (b) Use Ampere's law to prove that $B = \frac{1}{2}\mu_0\lambda$ at all points P and P'. ssm

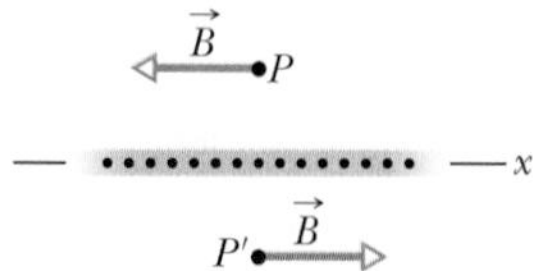

Fig. 30-52 Problems 39 and 44.

SEC. 30-4 Solenoids and Toroids

40E. A solenoid that is 95.0 cm long has a radius of 2.00 cm and a winding of 1200 turns; it carries a current of 3.60 A. Calculate the magnitude of the magnetic field inside the solenoid.

41E. A 200-turn solenoid having a length of 25 cm and a diameter of 10 cm carries a current of 0.30 A. Calculate the magnitude of the magnetic field $\vec{B}$ inside the solenoid. ssm

42E. A solenoid 1.30 m long and 2.60 cm in diameter carries a current of 18.0 A. The magnetic field inside the solenoid is 23.0 mT. Find the length of the wire forming the solenoid.

43E. A toroid having a square cross section, 5.00 cm on a side, and an inner radius of 15.0 cm has 500 turns and carries a current of 0.800 A. (It is made up of a square solenoid—instead of a round one as in Fig. 30-16—bent into a doughnut shape.) What is the magnetic field inside the toroid at (a) the inner radius and (b) the outer radius of the toroid? ssm

44P. Treat an ideal solenoid as a thin cylindrical conductor whose current per unit length, measured parallel to the cylinder axis, is λ. (a) By doing so, show that the magnitude of the magnetic field inside an ideal solenoid can be written as $B = \mu_0\lambda$. This is the value of the *change* in $\vec{B}$ that you encounter as you move from inside the solenoid to outside, through the solenoid wall. (b) Show that the same change occurs as you move through an infinite flat current sheet such as that of Fig. 30-52 (see Problem 39). Does this equality surprise you?

45P. In Section 30-4, we showed that the magnetic field at any radius r *inside* a toroid is given by

$$B = \frac{\mu_0 i N}{2\pi r}.$$

Show that as you move from any point just inside a toroid to a point just outside, the magnitude of the *change* in $\vec{B}$ that you encounter is just $\mu_0\lambda$. Here λ is the current per unit length along a circumference of radius r within the toroid. Compare this with the similar result found in Problem 44. Isn't the equality surprising? ssm

46P. A long solenoid has 100 turns/cm and carries current i. An electron moves within the solenoid in a circle of radius 2.30 cm perpendicular to the solenoid axis. The speed of the electron is $0.0460c$ (c = speed of light). Find the current i in the solenoid.

47P. A long solenoid with 10.0 turns/cm and a radius of 7.00 cm carries a current of 20.0 mA. A current of 6.00 A exists in a straight conductor located along the central axis of the solenoid. (a) At what radial distance from the axis will the direction of the resulting magnetic field be at 45.0° to the axial direction? (b) What is the magnitude of the magnetic field there? ssm ilw www

SEC. 30-5 A Current-Carrying Coil as a Magnetic Dipole

48E. Figure 30-53a shows a length of wire carrying a current i and bent into a circular coil of one turn. In Fig. 30-53b the same length of wire has been bent more sharply, to give a coil of two

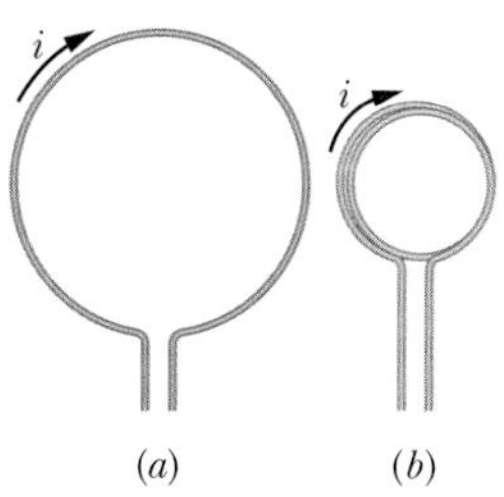

Fig. 30-53 Exercise 48.

turns, each of half the original radius. (a) If B_a and B_b are the magnitudes of the magnetic fields at the centers of the two coils, what is the ratio B_b/B_a? (b) What is the ratio of the dipole moments, μ_b/μ_a, of the coils?

49E. What is the magnetic dipole moment $\vec{\mu}$ of the solenoid described in Exercise 41? ssm

50E. Figure 30-54 shows an arrangement known as a Helmholtz coil. It consists of two circular coaxial coils, each of N turns and radius R, separated by a distance R. The two coils carry equal currents i in the same direction. Find the magnitude of the net magnetic field at P, midway between the coils.

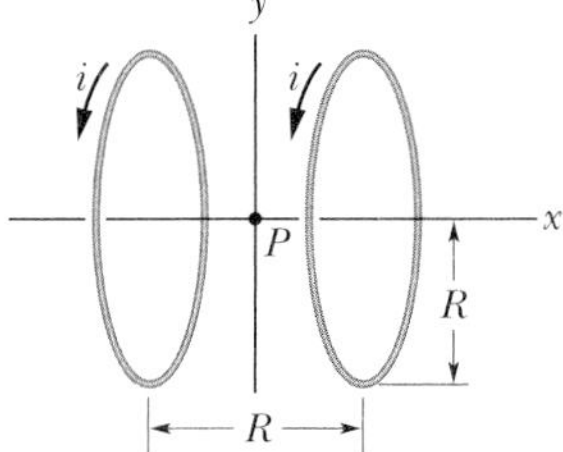

Fig. 30-54 Exercise 50; Problems 53 and 55.

51E. A student makes a short electromagnet by winding 300 turns of wire around a wooden cylinder of diameter $d = 5.0$ cm. The coil is connected to a battery producing a current of 4.0 A in the wire. (a) What is the magnetic moment of this device? (b) At what axial distance $z \gg d$ will the magnetic field of this dipole have the magnitude 5.0 μT (approximately one-tenth that of Earth's magnetic field)? ssm

52E. The magnitude $B(x)$ of the magnetic field at points on the axis of a square current loop of side a is given in Problem 15. (a) Show that the axial magnetic field of this loop, for $x \gg a$, is that of a magnetic dipole (see Eq. 30-29). (b) What is the magnetic dipole moment of this loop?

53P. Two 300-turn coils of radius R each carry a current i. They are arranged a distance R apart, as in Fig. 30-54. For $R = 5.0$ cm and $i = 50$ A, plot the magnitude B of the net magnetic field as a function of distance x along the common x axis over the range $x = -5$ cm to $x = +5$ cm, taking $x = 0$ at the midpoint P. (Such coils provide an especially uniform field $\vec{B}$ near point P.) (*Hint:* See Eq. 30-28.)

54P. A conductor carries a current of 6.0 A along the closed path *abcdefgha* involving 8 of the 12 edges of a cube of side 10 cm as shown in Fig. 30-55. (a) Why can one regard this as the superposition of three square loops: *bcfgb*, *abgha*, and *cdefc*? (*Hint:* Draw currents around those square loops.) (b) Use this superposition to find the magnetic dipole moment $\vec{\mu}$ (magnitude and direction) of the closed path. (c) Calculate $\vec{B}$ at the points $(x, y, z) =$ (0, 5.0 m, 0) and (5.0 m, 0, 0).

55P. In Exercise 50 (Fig. 30-54), let the separation of the coils be a variable s (not necessarily equal to the coil radius R). (a) Show that the first derivative of the magnitude of the net magnetic field of the coils (dB/dx) vanishes at the midpoint P regardless of the value of s. Why would you expect this to be true from symmetry? (b) Show that the second derivative (d^2B/dx^2) also vanishes at P, provided $s = R$. This accounts for the uniformity of B near P for this particular coil separation. ssm

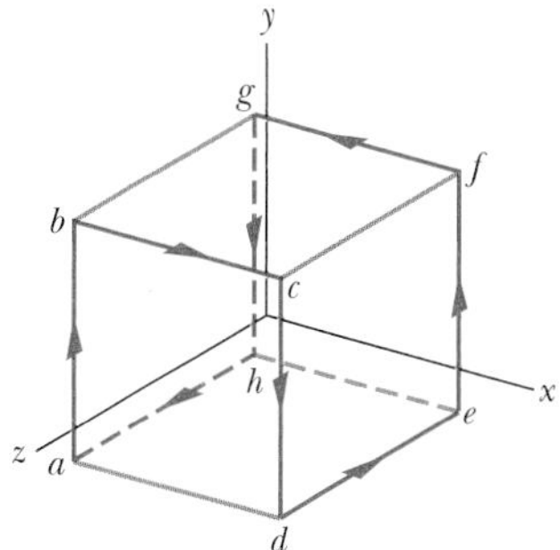

Fig. 30-55 Problem 54.

56P. A length of wire is formed into a closed circuit with radii a and b, as shown in Fig. 30-56, and carries a current i. (a) What are the magnitude and direction of $\vec{B}$ at point P? (b) Find the magnetic dipole moment of the circuit.

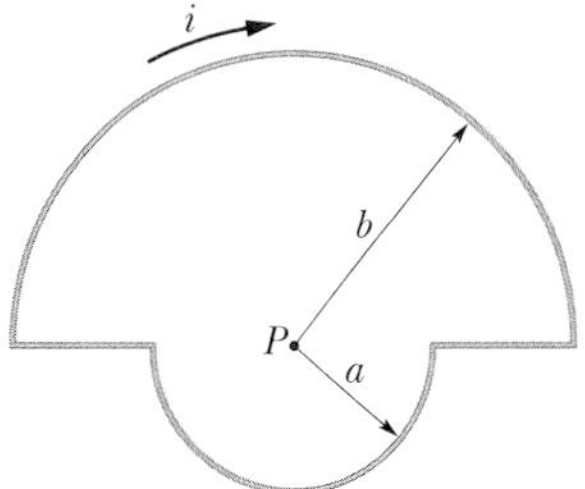

Fig. 30-56 Problem 56.

57P. A circular loop of radius 12 cm carries a current of 15 A. A flat coil of radius 0.82 cm, having 50 turns and a current of 1.3 A, is concentric with the loop. (a) What magnetic field $\vec{B}$ does the loop produce at its center? (b) What torque acts on the coil? Assume that the planes of the loop and coil are perpendicular and that the magnetic field due to the loop is essentially uniform throughout the volume occupied by the coil.

58P. (a) A long wire is bent into the shape shown in Fig. 30-57, without the wire actually touching itself at P. The radius of the circular section is R. Determine the magnitude and direction of $\vec{B}$ at the center C of the circular section when the current i is as indicated. (b) Suppose the circular section of the wire is rotated without distortion about the indicated diameter, until the plane of the circle is perpendicular to the straight sections of the wire. The magnetic dipole moment associated with the circular section is now in the direction of the current in the straight section of the wire. Determine $\vec{B}$ at C in this case.

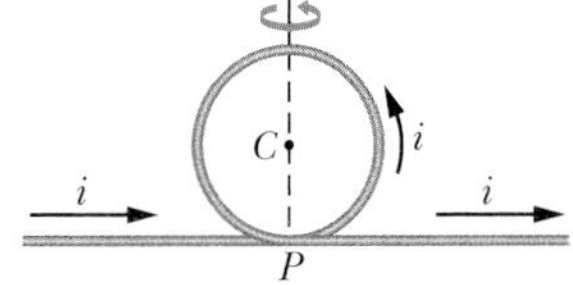

Fig. 30-57 Problem 58.

NEW PROBLEMS

N1. Equation 30-6 gives the magnitude B of the magnetic field set up by a current in an *infinitely long* straight wire, at a point P with perpendicular distance R from the wire. Suppose that point P is actually at perpendicular distance R from the midpoint of a wire with a *finite* length L. Using Eq. 30-6 to calculate B then results in a certain percentage error. In terms of R, what value must L exceed if the percentage error is to be less than 1.00%? That is, what L gives

$$\frac{(B \text{ from Eq. 30-6}) - (B \text{ actual})}{(B \text{ actual})}(100\%) = 1.00\%?$$

N2. In Fig. 30N-1, point P is at perpendicular distance R = 2.00 cm from a very long straight wire carrying a current. The magnetic field $\vec{B}$ set up at point P is due to contributions from all the identical current-length elements $i\,d\vec{s}$ along the wire. What is the distance s to the current-length element that makes (a) the greatest contribution to field $\vec{B}$ and (b) 10% of the greatest contribution?

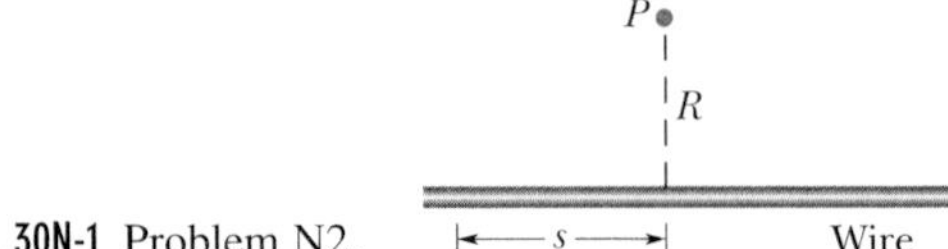

Fig. 30N-1 Problem N2.

N3. Figure 30N-2a shows an element of length ds = 1.00 μm in a very long straight wire carrying current. The current in that element sets up a differential magnetic field $d\vec{B}$ at points in the surrounding space. Figure 30N-2b gives the magnitude dB of the field for points 2.5 cm from the element, as a function of angle θ between the wire and a straight line to the point. What is the magnitude of the magnetic field set up by the entire wire at perpendicular distance 2.5 cm from the wire?

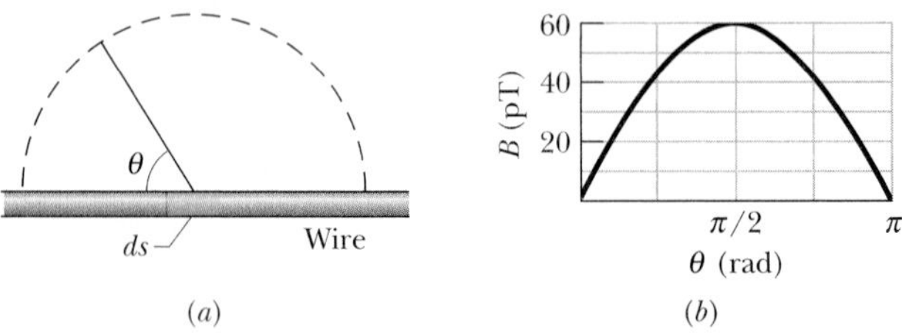

Fig. 30N-2 Problem N3.

N4. Figure 30N-3 shows, in cross section, two long straight wires held against a plastic cylinder of radius 20.0 cm. Wire 1 carries current i_1 = 60.0 mA out of the page and is fixed in place at the left side of the cylinder. Wire 2 carries current i_2 = 40.0 mA out of the page and can be moved around the cylinder. At what angle θ_2 should wire 2 be positioned such that the net magnetic field at the origin from the two currents has a magnitude of 80.0 nT?

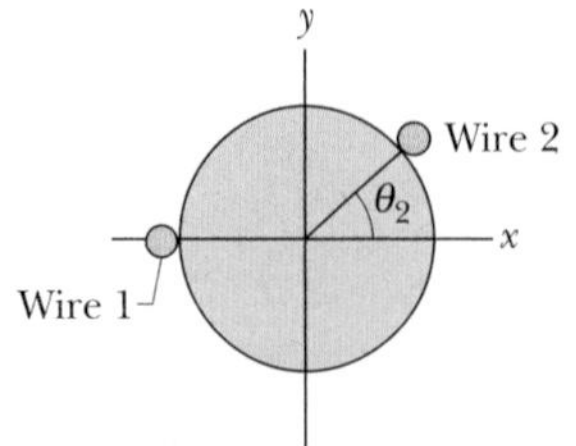

Fig. 30N-3 Problem N4.

N5. Figure 30N-4a shows two wires carrying currents. Wire 1 consists of a circular arc of radius R and two radial lengths; it carries current i_1 = 2.0 A in the direction indicated. Wire 2 is long and straight; it carries a current i_2 that can be varied; and it is at distance $R/2$ from the center of the arc. The net magnetic field $\vec{B}$ due to the two currents is measured at the center of curvature of the arc. Figure 30N-4b is a plot of the component of $\vec{B}$ in the direction perpendicular to the figure versus current i_2. What is the angle subtended by the arc?

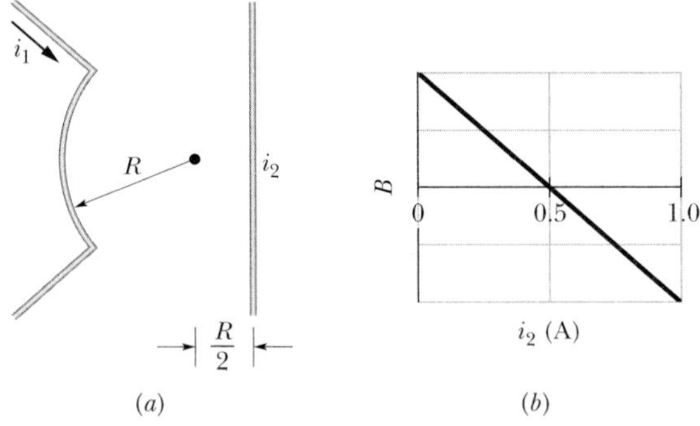

Fig. 30N-4 Problem N5.

N6. Two long straight thin wires with current lie against an equally long plastic cylinder, at radius R = 20.0 cm from the cylinder's central axis. Figure 30N-5a shows, in cross section, the cylinder and wire 1 but not wire 2. With wire 2 fixed in place, wire 1 is moved around the cylinder, from angle $\theta_1 = 0°$ to angle $\theta_1 = 180°$, through the first and second quadrants of the xy coordinate system. The net magnetic field $\vec{B}$ at the center of the cylinder is measured as a function of θ_1. Figure 30N-5b gives the x component B_x of that field and Fig. 30N-5c gives the y component B_y, both as functions of θ_1. (a) At what angle θ_2 is wire 2 located? What are the size and direction of the currents in (b) wire 1 and (c) wire 2?

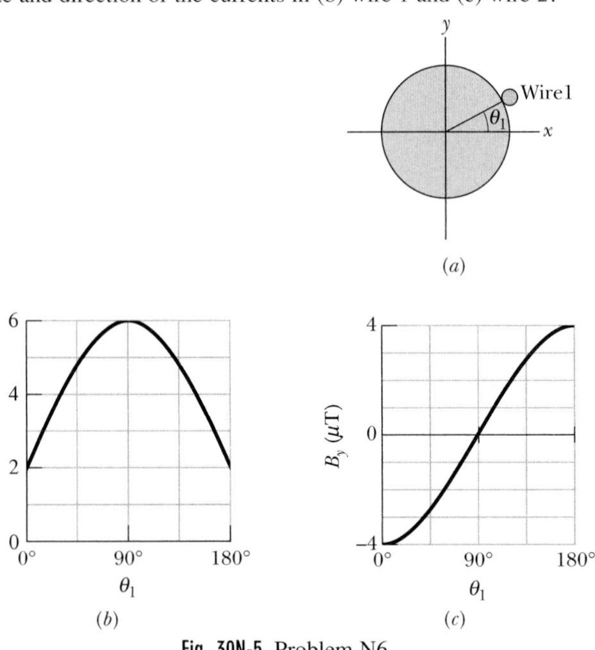

Fig. 30N-5 Problem N6.

N7. Figure 30N-6a shows, in cross section, two long, parallel wires carrying current and separated by distance L. The ratio i_1/i_2 of their currents is 4.00; the directions of the currents are not indicated. Figure 30N-6b shows the y component B_y of their net magnetic field along the x axis to the right of wire 2. (a) At what value of $x > 0$ is B_y maximum? (b) If $i_2 = 3$ mA, what is the value of that maximum? What are the directions of (c) current i_1 and (d) current i_2?

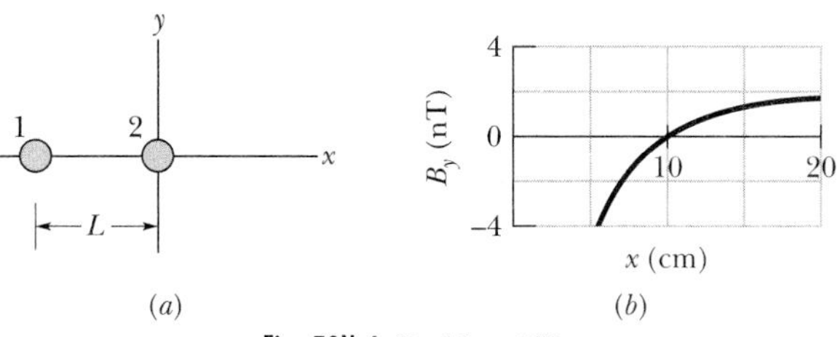

Fig. 30N-6 Problem N7.

N8. In Fig. 30N-7a, wire 1 consists of a circular arc and two radial lengths; it carries current $i_1 = 0.50$ A in the direction indicated. Wire 2, shown in cross section, is long, straight, and perpendicular to the plane of the figure. Its distance from the center of the arc is equal to the radius R of the arc, and it carries a current i_2 that can be varied. The two currents set up a net magnetic field $\vec{B}$ at the center of the arc. Figure 30N-7b gives the square of the field's magnitude B^2 versus the square of the current i_2^2. What angle is subtended by the arc?

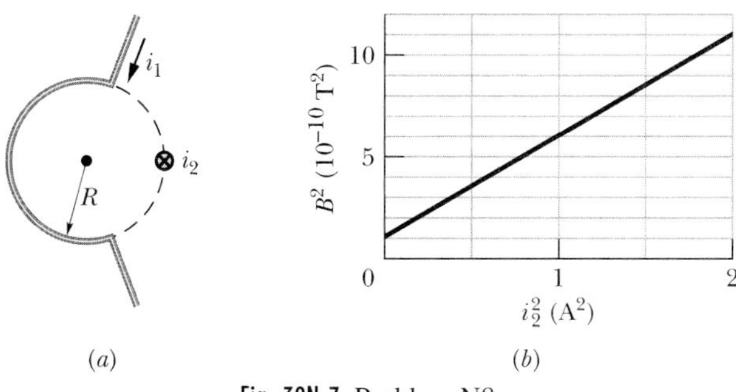

Fig. 30N-7 Problem N8.

N9. A cylindrical cable, with radius 8.00 mm, carries a current of 25.0 A, uniformly spread over its cross-sectional area. At what distance from the center of the wire is there a point within the wire where the magnetic field is 0.100 mT?

N10. In Fig. 30N-8, two concentric circular loops of wire carrying current in the same direction lie in the same plane. Loop 1 has radius 1.50 cm and carries 4.00 mA. Loop 2 has radius 2.50 cm and carries 6.00 mA. Loop 2 is be rotated about a diameter while the net magnetic field $\vec{B}$ set up by the two loops at their common center is measured. Through what angle must loop 2 be rotated so that the magnitude of that net field is 100 nT?

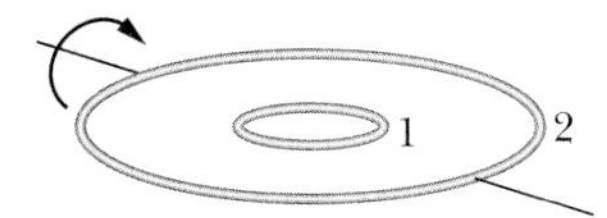

Fig. 30N-8 Problem N10.

N11. One long wire lies along an entire x axis and carries a current of 30 A in the positive x direction. A second long wire is perpendicular to the xy plane, passes through the point (0, 4 m, 0), and carries a current of 40 A in the positive z direction. What is the magnitude of the resulting magnetic field at the point $y = 2.0$ m on the y axis?

N12. In Fig. 30N-9a, two circular loops, with different currents but the same radius of 4.0 cm, are centered on a y axis. They are initially separated by distance $L = 3.0$ cm, with loop 2 positioned at the origin of the axis. The currents in the two loops produce a net magnetic field at the origin, with y component B_y. That component is to be measured as loop 2 is gradually moved in the positive direction of the y axis. Figure 30N-9b gives B_y as a function of the position y of loop 2. The curve approaches an asymptote of $B_y = 7.20$ μT as $y \to \infty$. What are (a) current i_1 in loop 1 and (b) current i_2 in loop 2?

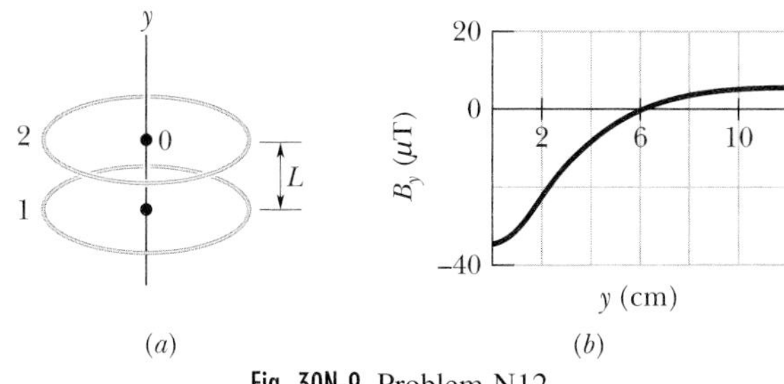

Fig. 30N-9 Problem N12.

N13. Figure 30N-10 shows two very long straight wires (in cross section) that carry currents of 4.00 A directly out of the page. Distance $d_1 = 6.00$ m and distance $d_2 = 4.00$ m. What is the magnitude of the net magnetic field at point P, which lies on a perpendicular bisector to the wires?

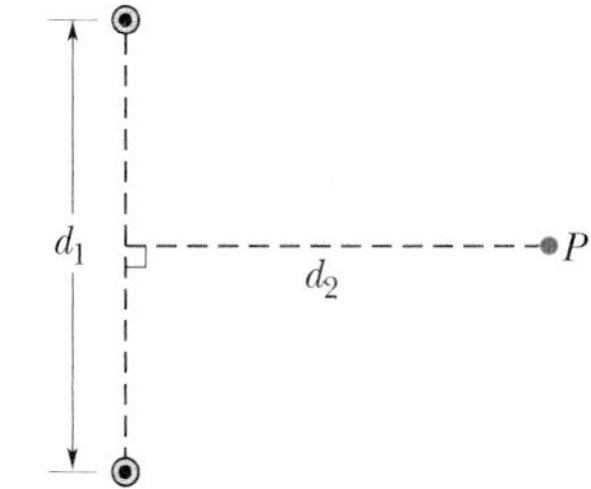

Fig. 30N-10 Problem N13.

N14. Figure 30N-11 shows, in cross section, four thin wires that are parallel, straight, and very long. They carry identical currents, in the directions indicated. Initially all four wires are at distance $d = 15.0$ cm from the origin of the coordinate system, where they create a net magnetic field $\vec{B}$. (a) To what value of x must you move wire 1 along the x axis in order to rotate $\vec{B}$ counterclockwise by 30°? (b) With wire 1 in that new position, to what value of x must you move wire 3 along the x axis to rotate $\vec{B}$ by 30° back to its initial orientation?

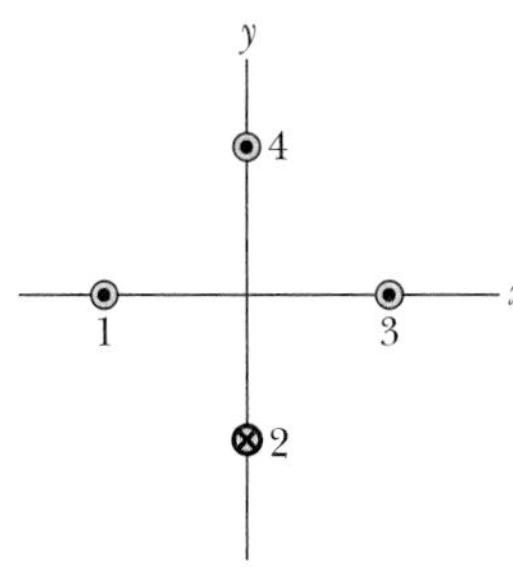

Fig. 30N-11 Problem N14.

N15. In Fig. 30N-12, a closed loop carries a current of 200 mA. The loop consists of two radial straight wires and two concentric circular arcs of radii 2.00 m and 4.00 m. The angle θ is $\pi/4$ rad. What are the magnitude and direction of the net magnetic field at point P, which is at the center of curvature?

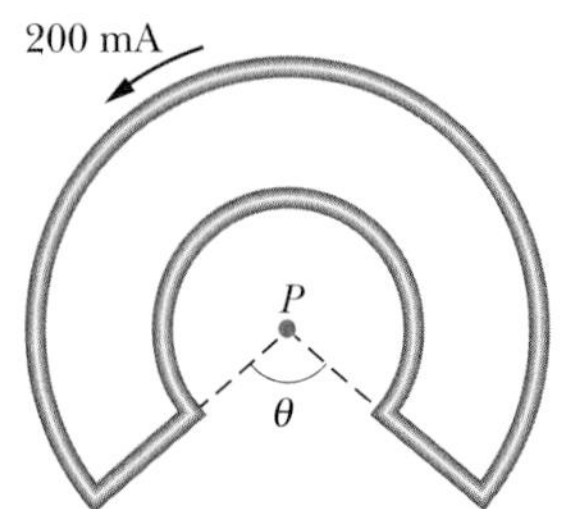

Fig. 30N-12 Problem N15.

N16. The wire loop in Fig. 30N-13a lies in a plane and consists of a semicircle of radius 10.0 cm, a smaller semicircle with the same center, and two radial lengths. The smaller semicircle is rotated out of that plane by angle θ, until it is perpendicular to the plane (Fig. 30N-13b). Figure 30N-13c gives the magnitude of the net magnetic field at the center of curvature versus angle θ. What is the radius of the smaller semicircle?

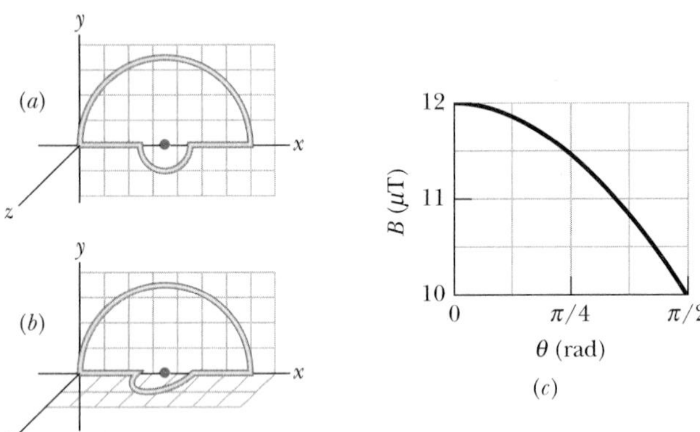

Fig. 30N-13 Problem N16.

N17. Figure 30N-14 shows a closed loop carrying a current of 2.00 A. The loop consists of a half circle of radius 4.00 m, two quarter circles of radii 2.00 m, and three radial straight wires. What is the magnitude of the net magnetic field at the common center of the circular sections?

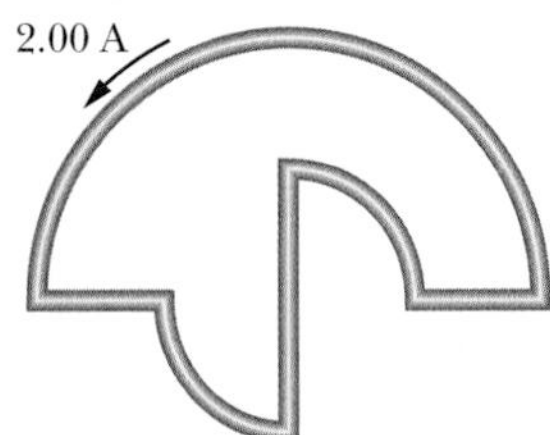

Fig. 30N-14 Problem N17.

N18. Figure 30N-15 shows wire 1 in cross section; the wire is long and straight, carries a current of 4.00 mA out of the page, and is at distance $d_1 = 2.4$ cm from a surface. Wire 2, which is parallel to wire 1 and also long, is at horizontal distance $d_2 = 5.0$ cm from wire 1 and carries a current of 6.80 mA into the page. What is the x component of the magnetic force per unit length on wire 2 due to the current in wire 1?

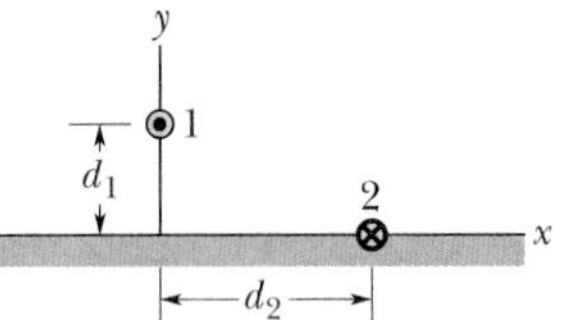

Fig. 30N-15 Problem N18.

N19. The magnetic field in a certain region is given by

$$\vec{B} = 3.0\hat{\text{i}} + 8.0(x^2/d^2)\hat{\text{j}},$$

with $\vec{B}$ in milliteslas and x and length d in meters. The field is due to current that is parallel to the z axis. (a) Evaluate $\oint \vec{B} \cdot d\vec{s}$ around the square of edge length d shown in Fig. 30N-16. (b) Assuming $d = 0.50$ m, evaluate the current through the square. (c) Is the current in the direction of $+\hat{\text{k}}$ or $-\hat{\text{k}}$?

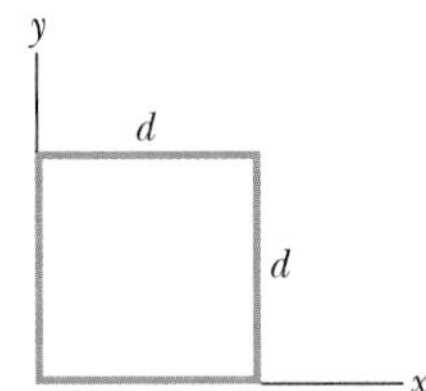

Fig. 30N-16 Problem N19.

N20. Figure 30N-17a shows, in cross section, three current-carrying wires that are long, straight, and parallel to one another. Wires 1 and 2 are fixed in place on an x axis, with separation d. Wire 1 has a current of 0.750 A, but the direction of the current is not given. Wire 3, with a current of 0.250 A out of the page, can be moved along the x axis to the right of wire 2. As it is moved, the magnitude of the net magnetic force $\vec{F}_2$ on wire 2 due to the currents in wires 1 and 3 changes. The y component of that force is F_{2y} and the value per unit length of wire 2 is F_{2y}/L_2. Figure 30N-17b gives F_{2y}/L_2 versus the position x of wire 3. The plot has an asymptote of $F_{2y}/L_2 = -0.627\ \mu\text{N/m}$ as $x \to \infty$. What is the size and direction of the current in wire 2?

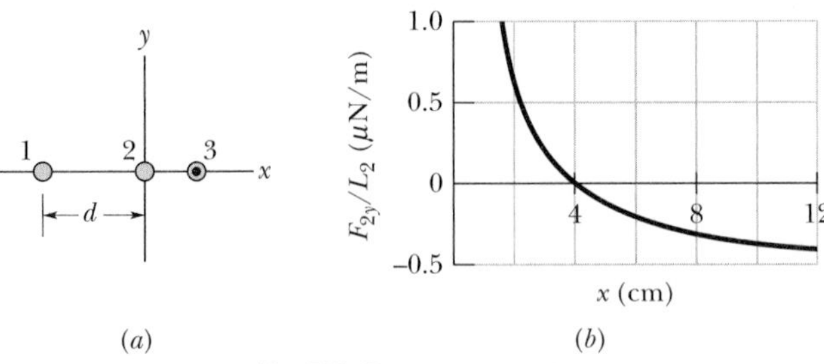

Fig. 30N-17 Problem N20.

N21. Figure 30N-18 shows, in cross section, two long straight wires; the 3.0 A current in the right-hand wire is out of the page. What are the size and direction of the current in the left-hand wire if the net magnetic field at point P is to be zero?

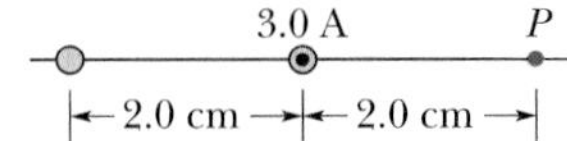

Fig. 30N-18 Problem N21.

N22. A current is set up in a wire loop consisting of a semicircle of radius 4.00 cm, a smaller, concentric semicircle, and two radial straight lengths, all in a plane. Figure 30N-19*a* shows the arrangement but is not drawn to scale. The magnitude of the magnetic field produced at the center of curvature is 47.25 μT. The smaller semicircle is then flipped over (rotated) until the loop is again entirely in the plane (Fig. 30N-19*b*). The magnetic field produced at the (same) center of curvature now has the magnitude of 15.75 μT, and its direction is now reversed. What is the radius of the smaller semicircle?

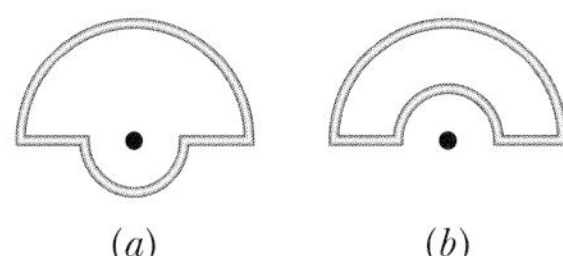

Fig. 30N-19 Problem N22.

N23. The magnetic field of a circular loop with current i, at a point on the central axis through the loop, is parallel to that axis. The magnitude of the field is given by

$$B = \frac{\mu_0 i R^2}{2(R^2 + z^2)^{3/2}},$$

where R is the radius of the loop and z is the distance from the center of the loop. A solenoid can be constructed mathematically by using many such circular loops that are identical in radius and current, coaxial, and closely spaced. Suppose the solenoid has a length of 25.0 cm and a radius of 1.00 cm and consists of N equally spaced loops, each with a current of 1.00 A. For (a) $N = 11$, (b) $N = 21$, and (c) $N = 51$, compute the magnitude of the magnetic field at the center of the solenoid by summing the fields produced by the individual loops. For each value of N, compare the result with the value found using Eq. 30-25, which holds for a long solenoid with a large number of tightly spaced loops.

N24. An electron is shot into one end of a solenoid. As it enters the uniform magnetic field within the solenoid, its speed is 800 m/s and its velocity vector makes an angle of 30° with the central axis of the solenoid. The solenoid carries 4.0 A and has 8000 turns along its length. How many revolutions does the electron make along its helical path within the solenoid by the time it emerges from the solenoid's opposite end? (In a real solenoid, where the field is not uniform at the two ends, the number of revolutions would be slightly less than the answer here.)

N25. In Fig. 30N-20, two long straight wires (shown in cross section) carry currents $i_1 = 30.0$ mA and $i_2 = 40.0$ mA directly out of the page. They are equal distances from the origin, where they set up a certain magnetic field $\vec{B}$. To what value must current i_1 be changed in order to rotate $\vec{B}$ by an angle of 20.0° clockwise?

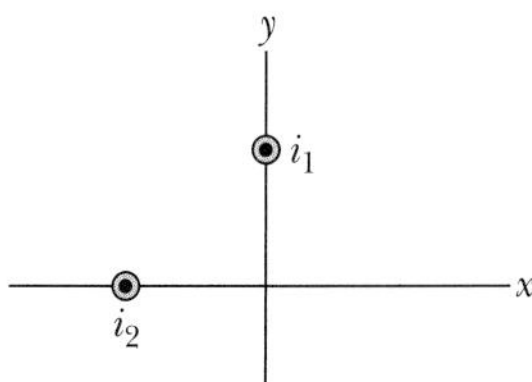

Fig. 30N-20 Problem N25.

N26. A computer can be used to demonstrate Ampere's law for a situation in which the Amperian loop does not coincide with a magnetic field line. Suppose that a square with edges of length a is centered at the origin of a coordinate system whose x and y axes are parallel to sides of the square. A long straight wire carrying current i is perpendicular to the plane of the square and crosses the x axis at $x = x'$. Evaluate $\oint \vec{B} \cdot d\vec{s}$ numerically. Divide a side of the square into N segments of equal length Δs and for each segment evaluate $\vec{B} \cdot \vec{u}\,\Delta s$. Here $\vec{B}$ is the magnetic field at the center of the segment and $\vec{u}$ is a unit vector that is parallel to the segment and is in the direction of integration. For different segments, $\vec{u}$ might be $\hat{\mathrm{i}}$, $\hat{\mathrm{j}}$, $-\hat{\mathrm{i}}$, or $-\hat{\mathrm{j}}$. The magnetic field at a point with coordinates x and y is given by

$$\vec{B} = \frac{\mu_0 i[-y\hat{\mathrm{i}} + (x - x')\hat{\mathrm{j}}]}{2\pi[(x - x')^2 + y^2]}.$$

For sides that are parallel to the x axis, take $y = a/2$ or $-a/2$; for sides that are parallel to the y axis, take $x = a/2$ or $-a/2$. Suppose that the length of a side is 1.00 m and the current is 1.00 A. Then for each of the following cases, calculate the sum over segments for each side of the square separately; next add the results to find the total for the square. Compare the total to $\mu_0 i_{enc}$. The value $N = 50$ should give you three-significant-figure accuracy: (a) $x' = 0$ (the wire is at the center of the square), (b) $x' = 0.200$ m (the wire passes inside the square at an off-center point), (c) $x' = 0.400$ m (the wire passes through the square near the center of a side), and (d) $x' = 0.600$ m (the wire is outside the square).

N27. Two long parallel conductors carry currents parallel to the z axis. The conductors intersect the xy plane at points along the x axis: one intersects at $x = a$ and carries a current i_1 in the $+z$ direction; the other conductor intersects at $x = 0$ and carries a current i_2 that can be varied in both magnitude and direction. The current is considered to be positive if directed in the positive z direction and negative if directed in the negative z direction. (a) Write an equation that gives the net magnetic field $\vec{B}$ along the x axis for $x > a$. (b) Rewrite the equation for $x = 2a$ after substituting for i_2 with $i_2 = bi_1$, where b is a variable. (c) With this rewritten equation, graph $\vec{B}$ versus b for the range $3 > b > -3$. Positive $\vec{B}$ corresponds to the magnetic field being directed in the positive y direction, negative in the negative y direction.

31 Induction and Inductance

Soon after rock began in the mid-1950s, guitarists switched from acoustic guitars to electric guitars—but it was Jimi Hendrix who first understood the electric guitar as an electronic instrument. He exploded on the scene in the 1960s, ripping his pick along the strings, positioning himself and his guitar in front of a speaker to sustain feedback, and then laying down chords on top of the feedback. He shoved rock forward from the melodies of Buddy Holly into the psychedelia of the late 1960s and into the early heavy metal of Led Zeppelin and the raw energy of Joy Division in the 1970s, and his ideas continue to influence rock today.

What is it about an electric guitar that distinguishes it from an acoustic guitar and enabled Hendrix to make so much broader use of this electronic instrument?

The answer is in this chapter.

31-1 Two Symmetric Situations

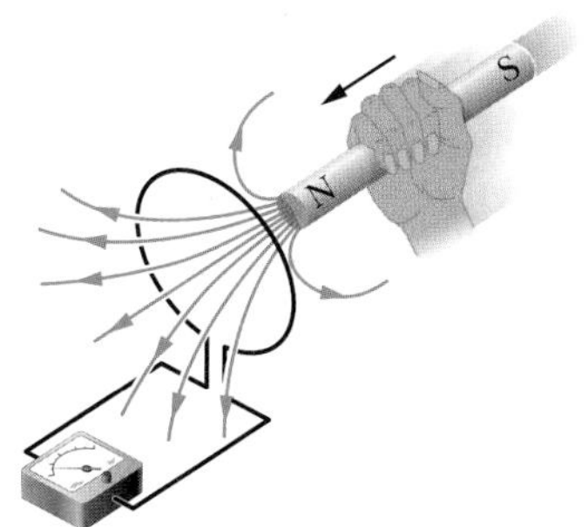

Fig. 31-1 A current meter registers a current in the wire loop when the magnet is moving with respect to the loop.

In Section 29-8, we saw that if we put a closed conducting loop in a magnetic field and then send current through the loop, forces due to the magnetic field create a torque to turn the loop:

$$\text{current loop} + \text{magnetic field} \Rightarrow \text{torque}. \tag{31-1}$$

Suppose that, instead, with the current off, we turn the loop by hand. Will the opposite of Eq. 31-1 occur? That is, will a current now appear in the loop:

$$\text{torque} + \text{magnetic field} \Rightarrow \text{current?} \tag{31-2}$$

The answer is yes—a current does appear. The situations of Eqs. 31-1 and 31-2 are symmetric. The physical law on which Eq. 31-2 depends is called *Faraday's law of induction.* Whereas Eq. 31-1 is the basis for the electric motor, Eq. 31-2 and Faraday's law are the basis for the electric generator. This chapter is concerned with that law and the process it describes.

31-2 Two Experiments

Let us examine two simple experiments to prepare for our discussion of Faraday's law of induction.

First Experiment. Figure 31-1 shows a conducting loop connected to a sensitive current meter. Since there is no battery or other source of emf included, there is no current in the circuit. However, if we move a bar magnet toward the loop, a current suddenly appears in the circuit. The current disappears when the magnet stops. If we then move the magnet away, a current again suddenly appears, but now in the opposite direction. If we experimented for a while, we would discover the following:

1. A current appears only if there is relative motion between the loop and the magnet (one must move relative to the other); the current disappears when the relative motion between them ceases.
2. Faster motion produces a greater current.
3. If moving the magnet's north pole toward the loop causes, say, clockwise current, then moving the north pole away causes counterclockwise current. Moving the south pole toward or away from the loop also causes currents, but in the reversed directions.

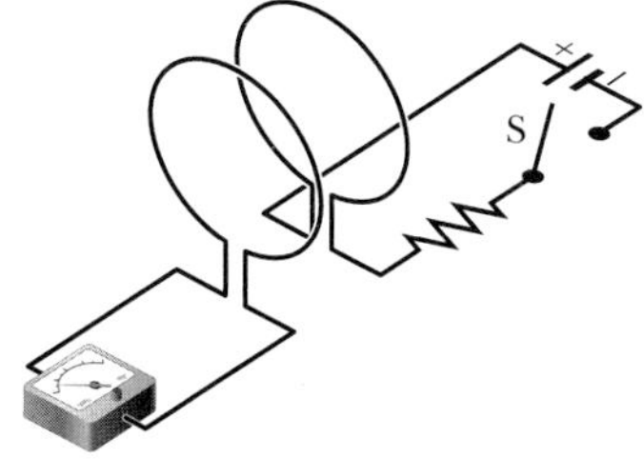

Fig. 31-2 The current meter registers a current in the left-hand wire loop just as switch S is closed (to turn on the current in the right-hand wire loop) or opened (to turn off the current in the right-hand loop). No motion of the coils is involved.

The current produced in the loop is called an **induced current;** the work done per unit charge to produce that current (to move the conduction electrons that constitute the current) is called an **induced emf;** and the process of producing the current and emf is called **induction.**

Second Experiment. For this experiment we use the apparatus of Fig. 31-2, with the two conducting loops close to each other but not touching. If we close switch S, to turn on a current in the right-hand loop, the meter suddenly and briefly registers a current—an induced current—in the left-hand loop. If we then open the switch, another sudden and brief induced current appears in the left-hand loop, but in the opposite direction. We get an induced current (and thus an induced emf) only when the current in the right-hand loop is changing (either turning on or turning off) and not when it is constant (even if it is large).

The induced emf and induced current in these experiments are apparently caused when something changes—but what is that "something"? Faraday knew.

31-3 Faraday's Law of Induction

Faraday realized that an emf and a current can be induced in a loop, as in our two experiments, by changing the *amount of magnetic field* passing through the loop. He further realized that the "amount of magnetic field" can be visualized in terms of the magnetic field lines passing through the loop. **Faraday's law of induction,** stated in terms of our experiments, is this:

> An emf is induced in the loop at the left in Figs. 31-1 and 31-2 when the number of magnetic field lines that pass through the loop is changing.

The actual number of field lines passing through the loop does not matter; the values of the induced emf and induced current are determined by the *rate* at which that number changes.

In our first experiment (Fig. 31-1), the magnetic field lines spread out from the north pole of the magnet. Thus, as we move the north pole closer to the loop, the number of field lines passing through the loop increases. That increase apparently causes conduction electrons in the loop to move (the induced current) and provides energy (the induced emf) for their motion. When the magnet stops moving, the number of field lines through the loop no longer changes and the induced current and induced emf disappear.

In our second experiment (Fig. 31-2), when the switch is open (no current), there are no field lines. However, when we turn on the current in the right-hand loop, the increasing current builds up a magnetic field around that loop and at the left-hand loop. While the field builds, the number of magnetic field lines through the left-hand loop increases. As in the first experiment, the increase in field lines through that loop apparently induces a current and an emf there. When the current in the right-hand loop reaches a final, steady value, the number of field lines through the left-hand loop no longer changes, and the induced current and induced emf disappear.

Faraday's law does not explain *why* a current and an emf are induced in either experiment; it is just a statement that helps us visualize the induction.

A Quantitative Treatment

To put Faraday's law to work, we need a way to calculate the *amount of magnetic field* that passes through a loop. In Chapter 24, in a similar situation, we needed to calculate the amount of an electric field that passes through a surface. There we defined an electric flux $\Phi_E = \int \vec{E} \cdot d\vec{A}$. Here we define a *magnetic flux:* Suppose a loop enclosing an area A is placed in a magnetic field $\vec{B}$. Then the magnetic flux through the loop is

$$\Phi_B = \int \vec{B} \cdot d\vec{A} \quad \text{(magnetic flux through area } A\text{).} \tag{31-3}$$

As in Chapter 24, $d\vec{A}$ is a vector of magnitude dA that is perpendicular to a differential area dA.

As a special case of Eq. 31-3, suppose that the loop lies in a plane and that the magnetic field is perpendicular to the plane of the loop. Then we can write the dot product in Eq. 31-3 as $B\,dA \cos 0° = B\,dA$. If the magnetic field is also uniform, then B can be brought out in front of the integral sign. The remaining $\int dA$ then gives just the area A of the loop. Thus, Eq. 31-3 reduces to

$$\Phi_B = BA \qquad (\vec{B} \perp \text{area } A, \vec{B} \text{ uniform}). \tag{31-4}$$

From Eqs. 31-3 and 31-4, we see that the SI unit for magnetic flux is the tesla–square meter, which is called the *weber* (abbreviated Wb):

$$1 \text{ weber} = 1 \text{ Wb} = 1 \text{ T} \cdot \text{m}^2. \tag{31-5}$$

With the notion of magnetic flux, we can state Faraday's law in a more quantitative and useful way:

> The magnitude of the emf $\mathscr{E}$ induced in a conducting loop is equal to the rate at which the magnetic flux Φ_B through that loop changes with time.

As you will see in the next section, the induced emf $\mathscr{E}$ tends to oppose the flux change, so Faraday's law is formally written as

$$\mathscr{E} = -\frac{d\Phi_B}{dt} \qquad \text{(Faraday's law)}, \tag{31-6}$$

with the minus sign indicating that opposition. We often neglect the minus sign in Eq. 31-6, seeking only the magnitude of the induced emf.

If we change the magnetic flux through a coil of N turns, an induced emf appears in every turn and the total emf induced in the coil is the sum of these individual induced emfs. If the coil is tightly wound (*closely packed*), so that the same magnetic flux Φ_B passes through all the turns, the total emf induced in the coil is

$$\mathscr{E} = -N\frac{d\Phi_B}{dt} \qquad \text{(coil of } N \text{ turns)}. \tag{31-7}$$

Here are the general means by which we can change the magnetic flux through a coil:

1. Change the magnitude B of the magnetic field within the coil.
2. Change the area of the coil, or the portion of that area that happens to lie within the magnetic field (for example, by expanding the coil or sliding it in or out of the field).
3. Change the angle between the direction of the magnetic field $\vec{B}$ and the area of the coil (for example, by rotating the coil so that field $\vec{B}$ is first perpendicular to the plane of the coil and then is along that plane).

CHECKPOINT 1: The graph gives the magnitude $B(t)$ of a uniform magnetic field that exists throughout a conducting loop, perpendicular to the plane of the loop. Rank the five regions of the graph according to the magnitude of the emf induced in the loop, greatest first.

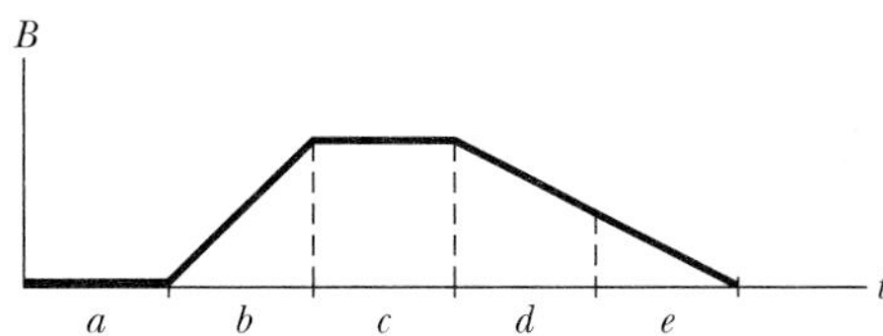

Sample Problem 31-1

The long solenoid S shown (in cross section) in Fig. 31-3 has 220 turns/cm and carries a current $i = 1.5$ A; its diameter D is 3.2 cm. At its center we place a 130-turn closely packed coil C of diameter $d = 2.1$ cm. The current in the solenoid is reduced to zero at a steady rate in 25 ms. What is the magnitude of the emf that is induced in coil C while the current in the solenoid is changing?

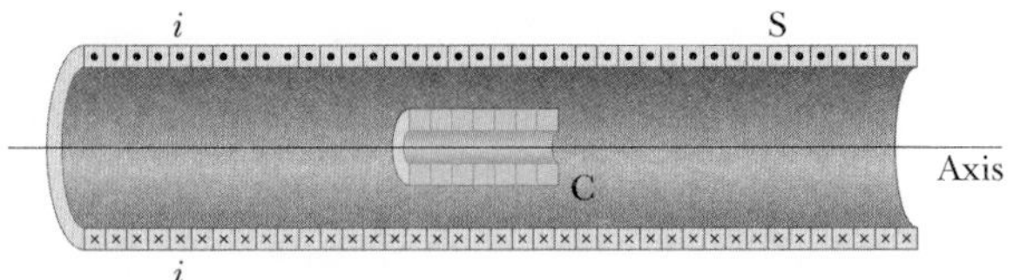

Fig. 31-3 Sample Problem 31-1. A coil C is located inside a solenoid S, which carries current i.

SOLUTION: The Key Ideas here are these:

1. Because coil C is located in the interior of the solenoid, it lies within the magnetic field produced by current i in the solenoid; thus, there is a magnetic flux Φ_B through coil C.
2. Because current i decreases, flux Φ_B also decreases.
3. As Φ_B decreases, emf $\mathscr{E}$ is induced in coil C, according to Faraday's law.

Because coil C consists of more than one turn, we apply Faraday's law in the form of Eq. 31-7 ($\mathscr{E} = -N\, d\Phi_B/dt$), where the number of turns N is 130 and $d\Phi_B/dt$ is the rate at which the flux in each turn changes.

Because the current in the solenoid decreases at a steady rate, flux Φ_B also decreases at a steady rate and we can write $d\Phi_B/dt$ as $\Delta\Phi_B/\Delta t$. Then, to evaluate $\Delta\Phi_B$, we need the final and initial flux. The final flux $\Phi_{B,f}$ is zero because the final current in the solenoid is zero. To find the initial flux $\Phi_{B,i}$, we need two more Key Ideas:

4. The flux through each turn of coil C depends on the area A and orientation of that turn in the solenoid's magnetic field $\vec{B}$. Because $\vec{B}$ is uniform and directed perpendicular to area A, the flux is given by Eq. 31-4 ($\Phi_B = BA$).
5. The magnitude B of the magnetic field in the interior of a solenoid depends on the solenoid's current i and its number n of turns per unit length, according to Eq. 30-25 ($B = \mu_0 in$).

For the situation of Fig. 31-3, A is $\frac{1}{4}\pi d^2$ ($= 3.46 \times 10^{-4}\ \text{m}^2$) and n is 220 turns/cm, or 22 000 turns/m. Substituting Eq. 30-25 into Eq. 31-4 then leads to

$$\begin{aligned}\Phi_{B,i} &= BA = (\mu_0 in)A \\ &= (4\pi \times 10^{-7}\ \text{T}\cdot\text{m/A})(1.5\ \text{A})(22\ 000\ \text{turns/m}) \\ &\quad \times (3.46 \times 10^{-4}\ \text{m}^2) \\ &= 1.44 \times 10^{-5}\ \text{Wb}.\end{aligned}$$

Now we can write

$$\begin{aligned}\frac{d\Phi_B}{dt} &= \frac{\Delta\Phi_B}{\Delta t} = \frac{\Phi_{B,f} - \Phi_{B,i}}{\Delta t} \\ &= \frac{(0 - 1.44 \times 10^{-5}\ \text{Wb})}{25 \times 10^{-3}\ \text{s}} \\ &= -5.76 \times 10^{-4}\ \text{Wb/s} = -5.76 \times 10^{-4}\ \text{V}.\end{aligned}$$

We are interested only in magnitudes, so we ignore the minus signs here and in Eq. 31-7, writing

$$\begin{aligned}\mathscr{E} &= N\frac{d\Phi_B}{dt} = (130\ \text{turns})(5.76 \times 10^{-4}\ \text{V}) \\ &= 7.5 \times 10^{-2}\ \text{V} = 75\ \text{mV}. \qquad \text{(Answer)}\end{aligned}$$

31-4 Lenz's Law

Soon after Faraday proposed his law of induction, Heinrich Friedrich Lenz devised a rule—now known as **Lenz's law**—for determining the direction of an induced current in a loop:

> An induced current has a direction such that the magnetic field due to *the current* opposes the change in the magnetic flux that induces the current.

Furthermore, the direction of an induced emf is that of the induced current. To get a feel for Lenz's law, let us apply it in two different but equivalent ways to Fig. 31-4, where the north pole of a magnet is being moved toward a conducting loop.

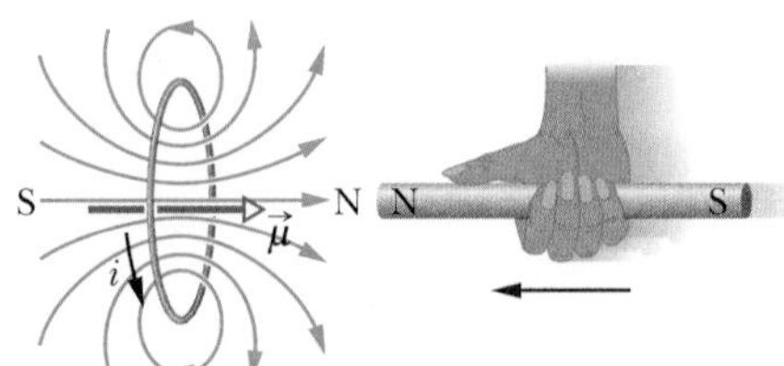

Fig. 31-4 Lenz's law at work. As the magnet is moved toward the loop, a current is induced in the loop. The current produces its own magnetic field, with magnetic dipole moment $\vec{\mu}$ oriented so as to oppose the motion of the magnet. Thus, the induced current must be counterclockwise as shown.

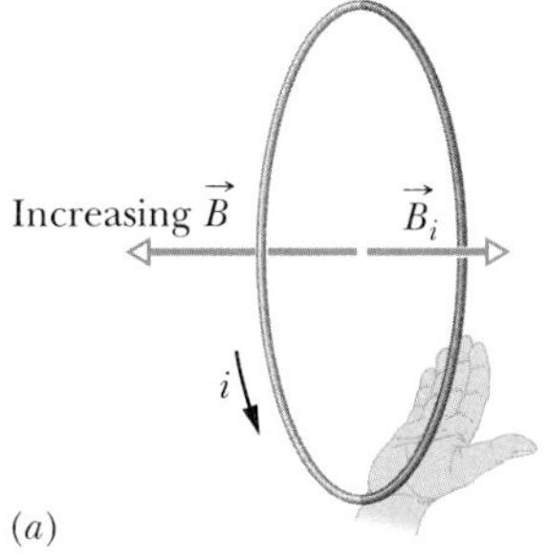

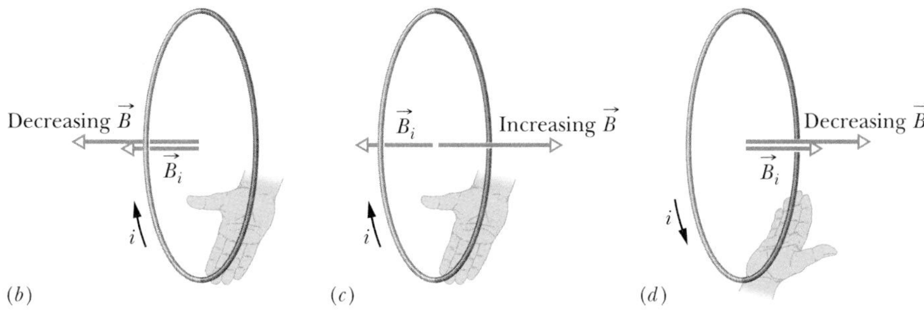

Fig. 31-5 The current i induced in a loop has the direction such that the current's magnetic field $\vec{B}_i$ opposes the *change* in the magnetic field $\vec{B}$ inducing i. The field $\vec{B}_i$ is always directed opposite an increasing field $\vec{B}$ (a, c) and in the same direction as a decreasing field $\vec{B}$ (b, d). The curled–straight right-hand rule gives the direction of the induced current based on the direction of the induced field.

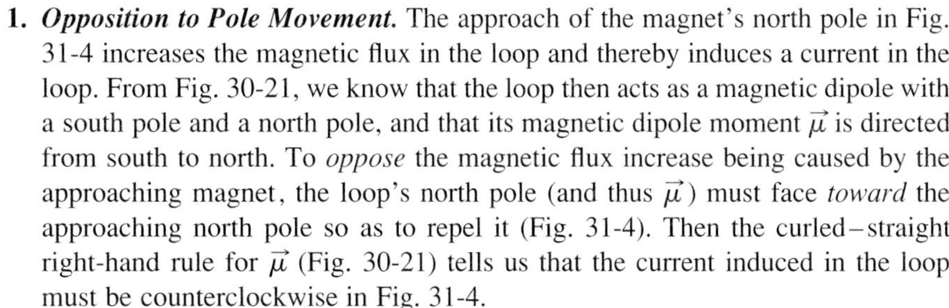

1. ***Opposition to Pole Movement.*** The approach of the magnet's north pole in Fig. 31-4 increases the magnetic flux in the loop and thereby induces a current in the loop. From Fig. 30-21, we know that the loop then acts as a magnetic dipole with a south pole and a north pole, and that its magnetic dipole moment $\vec{\mu}$ is directed from south to north. To *oppose* the magnetic flux increase being caused by the approaching magnet, the loop's north pole (and thus $\vec{\mu}$) must face *toward* the approaching north pole so as to repel it (Fig. 31-4). Then the curled–straight right-hand rule for $\vec{\mu}$ (Fig. 30-21) tells us that the current induced in the loop must be counterclockwise in Fig. 31-4.

   If we next pull the magnet away from the loop, a current will again be induced in the loop. Now, however, the loop will have a south pole facing the retreating north pole of the magnet, so as to oppose the retreat. Thus, the induced current will be clockwise.

2. ***Opposition to Flux Change.*** In Fig. 31-4, with the magnet initially distant, no magnetic flux passes through the loop. As the north pole of the magnet then nears the loop with its magnetic field $\vec{B}$ directed *toward the left*, the flux through the loop increases. To oppose this increase in flux, the induced current i must set up its own field $\vec{B}_i$ *directed toward the right* inside the loop, as shown in Fig. 31-5a; then the rightward flux of field $\vec{B}_i$ opposes the increasing leftward flux of field $\vec{B}$. The curled–straight right-hand rule of Fig. 30-21 then tells us that i must be counterclockwise in Fig. 31-5a.

   Note carefully that the flux of $\vec{B}_i$ always opposes the *change* in the flux of $\vec{B}$, but that does not always mean that $\vec{B}_i$ points opposite $\vec{B}$. For example, if we next pull the magnet away from the loop, the flux Φ_B from the magnet is still directed to the left through the loop, but it is now decreasing. The flux of $\vec{B}_i$ must now be to the left inside the loop, to oppose the *decrease* in Φ_B, as shown in Fig. 31-5b. Thus, $\vec{B}_i$ and $\vec{B}$ are now in the same direction.

   Figures 31-5c and d show the situations in which the south pole of the magnet approaches and retreats from the loop, respectively.

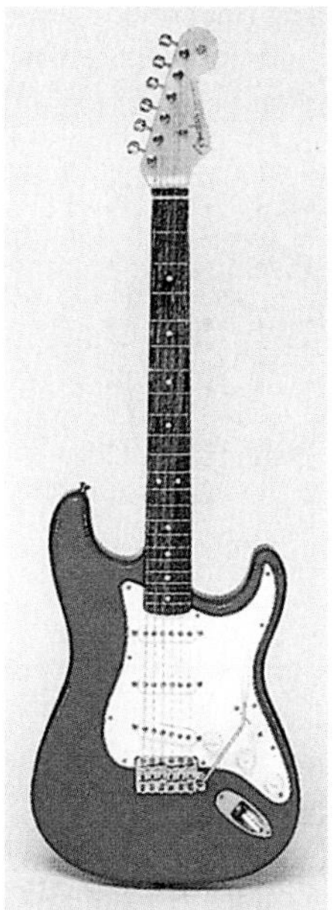

Fig. 31-6 A Fender Stratocaster has three groups of six electric pickups each (within the wide part of the body). A toggle switch (at the bottom of the guitar) allows the musician to determine which group of pickups sends signals to an amplifier and thus to a speaker system.

Electric Guitars

Figure 31-6 shows a Fender Stratocaster, the type of electric guitar that was used by Jimi Hendrix and by many other musicians. Whereas an acoustic guitar depends for its sound on the acoustic resonance produced in the hollow body of the instrument

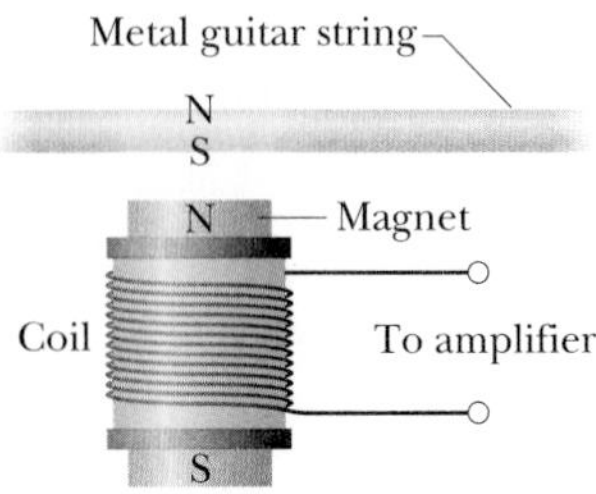

Fig. 31-7 A side view of an electric guitar pickup. When the metal string (which acts like a magnet) is made to oscillate, it causes a variation in magnetic flux that induces a current in the coil.

by the oscillations of the strings, an electric guitar is a solid instrument, so there is no body resonance. Instead, the oscillations of the metal strings are sensed by electric "pickups" that send signals to an amplifier and a set of speakers.

The basic construction of a pickup is shown in Fig. 31-7. Wire connecting the instrument to the amplifier is coiled around a small magnet. The magnetic field of the magnet produces a north and south pole in the section of the metal string just above the magnet. That section of string then has its own magnetic field. When the string is plucked and thus made to oscillate, its motion relative to the coil changes the flux of its magnetic field through the coil, inducing a current in the coil. As the string oscillates toward and away from the coil, the induced current changes direction at the same frequency as the string's oscillations, thus relaying the frequency of oscillation to the amplifier and speaker.

On a Stratocaster, there are three groups of pickups, placed at the near end of the strings (on the wide part of the body). The group closest to the near end better detects the high-frequency oscillations of the strings; the group farthest from the near end better detects the low-frequency oscillations. By throwing a toggle switch on the guitar, the musician can select which group or which pair of groups will send signals to the amplifier and speakers.

To gain further control over his music, Hendrix sometimes rewrapped the wire in the pickup coils of his guitar to change the number of turns. In this way, he altered the amount of emf induced in the coils and thus their relative sensitivity to string oscillations. Even without this additional measure, you can see that the electric guitar offers far more control over the sound that is produced than can be obtained with an acoustic guitar.

CHECKPOINT 2: The figure shows three situations in which identical circular conducting loops are in uniform magnetic fields that are either increasing (Inc) or decreasing (Dec) in magnitude at identical rates. In each, the dashed line coincides with a diameter. Rank the situations according to the magnitude of the current induced in the loops, greatest first.

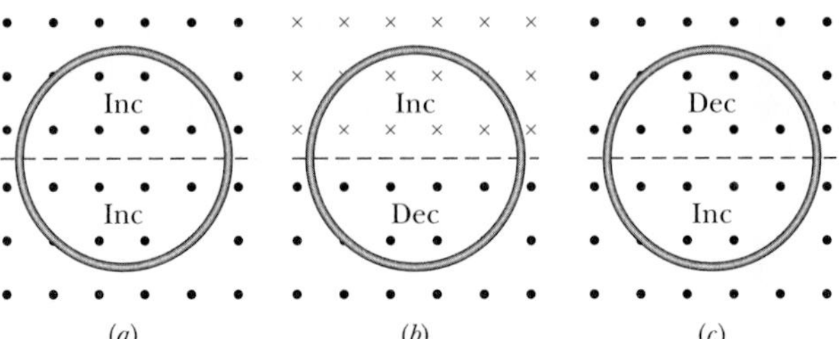

Sample Problem 31-2

Figure 31-8 shows a conducting loop consisting of a half-circle of radius $r = 0.20$ m and three straight sections. The half-circle lies in a uniform magnetic field $\vec{B}$ that is directed out of the page; the field magnitude is given by $B = 4.0t^2 + 2.0t + 3.0$, with B in teslas and t in seconds. An ideal battery with emf $\mathscr{E}_{bat} = 2.0$ V is connected to the loop. The resistance of the loop is 2.0 Ω.

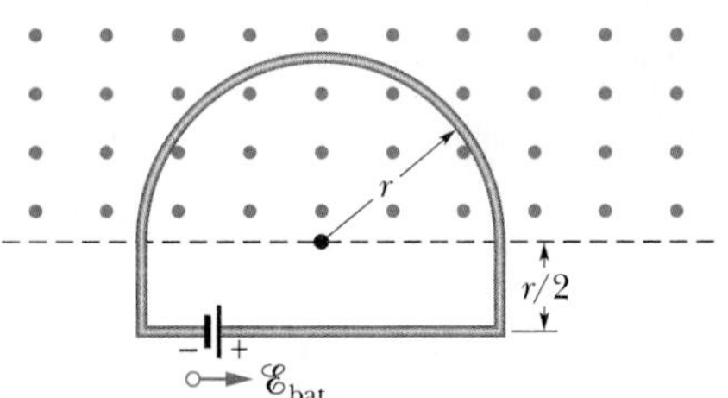

Fig. 31-8 Sample Problem 31-2. A battery is connected to a conducting loop consisting of a half-circle of radius r that lies in a uniform magnetic field. The field is directed out of the page; its magnitude is changing.

(a) What are the magnitude and direction of the emf $\mathscr{E}_{ind}$ induced around the loop by field $\vec{B}$ at $t = 10$ s?

SOLUTION: One **Key Idea** here is that, according to Faraday's law, the magnitude of $\mathscr{E}_{ind}$ is equal to the rate $d\Phi_B/dt$ at which the magnetic flux through the loop changes. A second Key Idea is that the flux through the loop depends on the loop's area A and its orientation in the magnetic field $\vec{B}$. Because $\vec{B}$ is uniform and is perpendicular to the plane of the loop, the flux is given by Eq. 31-4 ($\Phi_B = BA$). Using this equation and realizing that only the field magnitude B changes in time (not the area A), we rewrite Faraday's law, Eq. 31-6, as

$$\mathscr{E}_{\text{ind}} = \frac{d\Phi_B}{dt} = \frac{d(BA)}{dt} = A\frac{dB}{dt}.$$

A third Key Idea is that, because the flux penetrates the loop only within the half-circle, the area A in this equation is $\frac{1}{2}\pi r^2$. Substituting this and the given expression for B yields

$$\mathscr{E}_{\text{ind}} = A\frac{dB}{dt} = \frac{\pi r^2}{2}\frac{d}{dt}(4.0t^2 + 2.0t + 3.0)$$
$$= \frac{\pi r^2}{2}(8.0t + 2.0).$$

At $t = 10$ s, then,

$$\mathscr{E}_{\text{ind}} = \frac{\pi(0.20\text{ m})^2}{2}[8.0(10) + 2.0]$$
$$= 5.152\text{ V} \approx 5.2\text{ V}. \quad \text{(Answer)}$$

To find the direction of $\mathscr{E}_{\text{ind}}$, we first note that in Fig. 31-8 the flux through the loop is out of the page and increasing. Then the Key Idea here is that the induced field B_i (due to the induced current) must oppose that increase, and thus be *into* the page. Using the curled–straight right-hand rule (Fig. 30-7*c*), we find that the induced current must be clockwise around the loop. The induced emf $\mathscr{E}_{\text{ind}}$ must then also be clockwise.

(b) What is the current in the loop at $t = 10$ s?

SOLUTION: The Key Idea here is that two emfs tend to move charges around the loop. The induced emf $\mathscr{E}_{\text{ind}}$ tends to drive a current clockwise around the loop; the battery's emf $\mathscr{E}_{\text{bat}}$ tends to drive a current counterclockwise. Because $\mathscr{E}_{\text{ind}}$ is greater than $\mathscr{E}_{\text{bat}}$, the net emf $\mathscr{E}_{\text{net}}$ is clockwise, and thus so is the current. To find the current at $t = 10$ s, we use Eq. 28-2 ($i = \mathscr{E}/R$):

$$i = \frac{\mathscr{E}_{\text{net}}}{R} = \frac{\mathscr{E}_{\text{ind}} - \mathscr{E}_{\text{bat}}}{R}$$
$$= \frac{5.152\text{ V} - 2.0\text{ V}}{2.0\ \Omega} = 1.58\text{ A} \approx 1.6\text{ A}. \quad \text{(Answer)}$$

Sample Problem 31-3

Figure 31-9 shows a rectangular loop of wire immersed in a nonuniform and varying magnetic field $\vec{B}$ that is perpendicular to and directed into the page. The field's magnitude is given by $B = 4t^2x^2$, with B in teslas, t in seconds, and x in meters. The loop has width $W = 3.0$ m and height $H = 2.0$ m. What are the magnitude and direction of the induced emf $\mathscr{E}$ around the loop at $t = 0.10$ s?

SOLUTION: One Key Idea here is that because the magnitude of the magnetic field $\vec{B}$ is changing with time, the magnetic flux Φ_B through the loop is also changing. A second Key Idea is that the changing flux induces an emf $\mathscr{E}$ in the loop according to Faraday's law, which we can write as $\mathscr{E} = d\Phi_B/dt$.

To use that law, we need an expression for the flux Φ_B at any time t. However, a third Key Idea is that because B is *not* uniform over the area enclosed by the loop, we *cannot* use Eq. 31-4 ($\Phi_B = BA$) to find that expression; instead we must use Eq. 31-3 ($\Phi_B = \int \vec{B}\cdot d\vec{A}$).

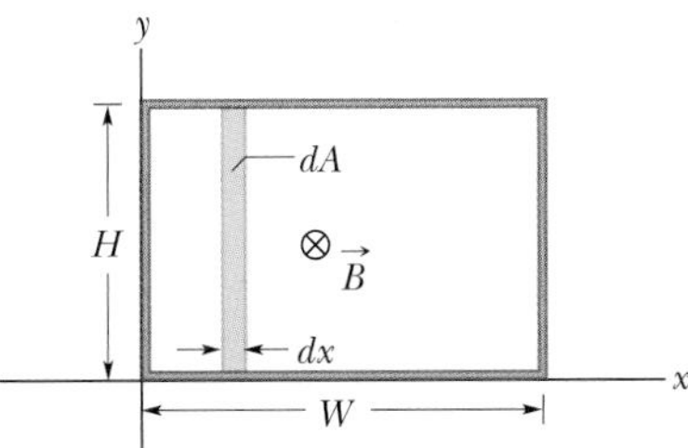

Fig. 31-9 Sample Problem 31-3. A closed conducting loop, of width W and height H, lies in a nonuniform, varying magnetic field that points directly into the page. To apply Faraday's law, we use the vertical strip of height H, width dx, and area dA.

In Fig. 31-9, $\vec{B}$ is perpendicular to the plane of the loop (and hence parallel to the differential area vector $d\vec{A}$), so the dot product in Eq. 31-3 gives $B\,dA$. Because the magnetic field varies with the coordinate x but not with the coordinate y, we can take the differential area dA to be the area of a vertical strip of height H and width dx (as shown in Fig. 31-9). Then $dA = H\,dx$, and the flux through the loop is

$$\Phi_B = \int \vec{B}\cdot d\vec{A} = \int B\,dA = \int BH\,dx = \int 4t^2x^2H\,dx.$$

Treating t as a constant for this integration and inserting the integration limits $x = 0$ and $x = 3.0$ m, we obtain

$$\Phi_B = 4t^2H\int_0^{3.0} x^2\,dx = 4t^2H\left[\frac{x^3}{3}\right]_0^{3.0} = 72t^2,$$

where we have substituted $H = 2.0$ m and Φ_B is in webers. Now we can use Faraday's law to find the magnitude of $\mathscr{E}$ at any time t:

$$\mathscr{E} = \frac{d\Phi_B}{dt} = \frac{d(72t^2)}{dt} = 144t,$$

in which $\mathscr{E}$ is in volts. At $t = 0.10$ s,

$$\mathscr{E} = (144\text{ V/s})(0.10\text{ s}) \approx 14\text{ V}. \quad \text{(Answer)}$$

The flux of $\vec{B}$ through the loop is into the page in Fig. 31-9 and is increasing in magnitude because B is increasing in magnitude with time. According to Lenz's law, the field B_i of the induced current must oppose this increase and so is directed out of the page. The curled–straight right-hand rule of Fig. 31-5 then tells us that the induced current is counterclockwise around the loop, and thus so is the induced emf $\mathscr{E}$.

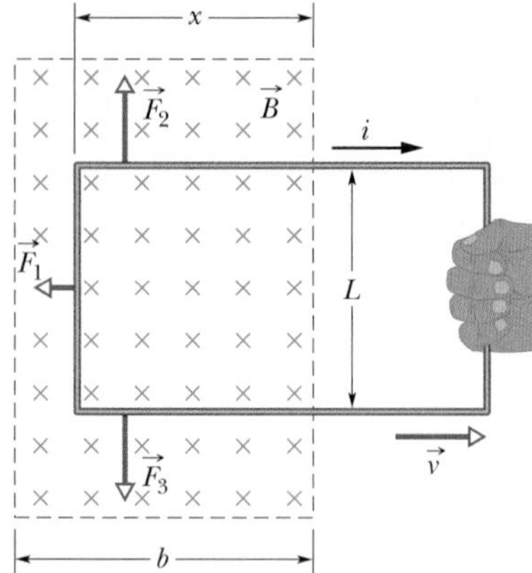

Fig. 31-10 You pull a closed conducting loop out of a magnetic field at constant velocity $\vec{v}$. While the loop is moving, a clockwise current i is induced in the loop, and the loop segments still within the magnetic field experience forces $\vec{F}_1$, $\vec{F}_2$, and $\vec{F}_3$.

31-5 Induction and Energy Transfers

By Lenz's law, whether you move the magnet toward or away from the loop in Fig. 31-1, a magnetic force resists the motion, requiring your applied force to do positive work. At the same time, thermal energy is produced in the material of the loop because of the material's electrical resistance to the current that is induced by the motion. The energy you transfer to the closed *loop* + *magnet* system via your applied force ends up in this thermal energy. (For now, we neglect energy that is radiated away from the loop as electromagnetic waves during the induction.) The faster you move the magnet, the more rapidly your applied force does work, and the greater the rate at which your energy is transferred to thermal energy in the loop; that is, the power of the transfer is greater.

Regardless of how current is induced in a loop, energy is always transferred to thermal energy during the process because of the electrical resistance of the loop (unless the loop is superconducting). For example, in Fig. 31-2, when switch S is closed and a current is briefly induced in the left-hand loop, energy is transferred from the battery to thermal energy in that loop.

Figure 31-10 shows another situation involving induced current. A rectangular loop of wire of width L has one end in a uniform external magnetic field that is directed perpendicularly into the plane of the loop. This field may be produced, for example, by a large electromagnet. The dashed lines in Fig. 31-10 show the assumed limits of the magnetic field; the fringing of the field at its edges is neglected. You are asked to pull this loop to the right at a constant velocity $\vec{v}$.

The situation of Fig. 31-10 does not differ in any essential way from that of Fig. 31-1. In each case a magnetic field and a conducting loop are in relative motion; in each case the flux of the field through the loop is changing with time. It is true that in Fig. 31-1 the flux is changing because $\vec{B}$ is changing and in Fig. 31-10 the flux is changing because the area of the loop still in the magnetic field is changing, but that difference is not important. The important difference between the two arrangements is that the arrangement of Fig. 31-10 makes calculations easier. Let us now calculate the rate at which you do mechanical work as you pull steadily on the loop in Fig. 31-10.

As you will see, to pull the loop at a constant velocity $\vec{v}$, you must apply a constant force $\vec{F}$ to the loop because a magnetic force of equal magnitude but opposite direction acts on the loop to oppose you. From Eq. 7-48, the rate at which you do work is then

$$P = Fv, \tag{31-8}$$

where F is the magnitude of your force. We wish to find an expression for P in terms of the magnitude B of the magnetic field and the characteristics of the loop—namely, its resistance R to current and its dimension L.

As you move the loop to the right in Fig. 31-10, the portion of its area within the magnetic field decreases. Thus, the flux through the loop also decreases and, according to Lenz's law, a current is produced in the loop. It is the presence of this current that causes the force that opposes your pull.

To find the current, we first apply Faraday's law. When x is the length of the loop still in the magnetic field, the area of the loop still in the field is Lx. Then from Eq. 31-4, the magnitude of the flux through the loop is

$$\Phi_B = BA = BLx. \tag{31-9}$$

As x decreases, the flux decreases. Faraday's law tells us that with this flux decrease, an emf is induced in the loop. Dropping the minus sign in Eq. 31-6 and using Eq.

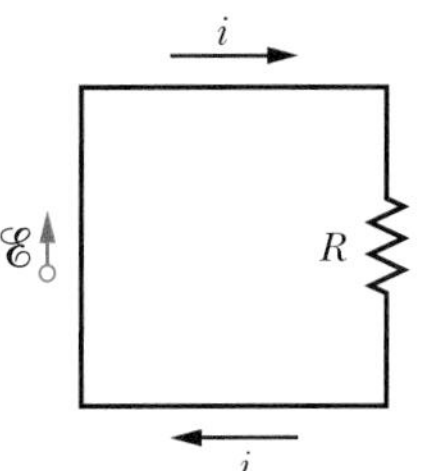

Fig. 31-11 A circuit diagram for the loop of Fig. 31-10 while the loop is moving.

31-9, we can write the magnitude of this emf as

$$\mathscr{E} = \frac{d\Phi_B}{dt} = \frac{d}{dt} BLx = BL\frac{dx}{dt} = BLv, \tag{31-10}$$

in which we have replaced dx/dt with v, the speed at which the loop moves.

Figure 31-11 shows the loop as a circuit: induced emf $\mathscr{E}$ is represented on the left, and the collective resistance R of the loop is represented on the right. The direction of the induced current i is obtained with a right-hand rule as in Fig. 31-5*b*; $\mathscr{E}$ must have the same direction.

To find the magnitude of the induced current, we cannot apply the loop rule for potential differences in a circuit because, as you will see in Section 31-6, we cannot define a potential difference for an induced emf. However, we can apply the equation $i = \mathscr{E}/R$, as we did in Sample Problem 31-2. With Eq. 31-10, this becomes

$$i = \frac{BLv}{R}. \tag{31-11}$$

Because three segments of the loop in Fig. 31-10 carry this current through the magnetic field, sideways deflecting forces act on those segments. From Eq. 29-26 we know that such a deflecting force is, in general notation,

$$\vec{F}_d = i\vec{L} \times \vec{B}. \tag{31-12}$$

In Fig. 31-10, the deflecting forces acting on the three segments of the loop are marked $\vec{F}_1$, $\vec{F}_2$, and $\vec{F}_3$. Note, however, that from the symmetry, forces $\vec{F}_2$ and $\vec{F}_3$ are equal in magnitude and cancel. This leaves only $\vec{F}_1$, which is directed opposite your force $\vec{F}$ on the loop and thus is the force opposing you. So, $\vec{F} = -\vec{F}_1$.

Using Eq. 31-12 to obtain the magnitude of $\vec{F}_1$ and noting that the angle between $\vec{B}$ and the length vector $\vec{L}$ for the left segment is 90°, we write

$$F = F_1 = iLB \sin 90° = iLB. \tag{31-13}$$

Substituting Eq. 31-11 for i in Eq. 31-13 then gives us

$$F = \frac{B^2L^2v}{R}. \tag{31-14}$$

Since B, L, and R are constants, the speed v at which you move the loop is constant if the magnitude F of the force you apply to the loop is also constant.

By substituting Eq. 31-14 into Eq. 31-8, we find the rate at which you do work on the loop as you pull it from the magnetic field:

$$P = Fv = \frac{B^2L^2v^2}{R} \quad \text{(rate of doing work).} \tag{31-15}$$

To complete our analysis, let us find the rate at which thermal energy appears in the loop as you pull it along at constant speed. We calculate it from Eq. 27-22,

$$P = i^2R. \tag{31-16}$$

Substituting for i from Eq. 31-11, we find

$$P = \left(\frac{BLv}{R}\right)^2 R = \frac{B^2L^2v^2}{R} \quad \text{(thermal energy rate),} \tag{31-17}$$

which is exactly equal to the rate at which you are doing work on the loop (Eq. 31-15). Thus, the work that you do in pulling the loop through the magnetic field appears as thermal energy in the loop.

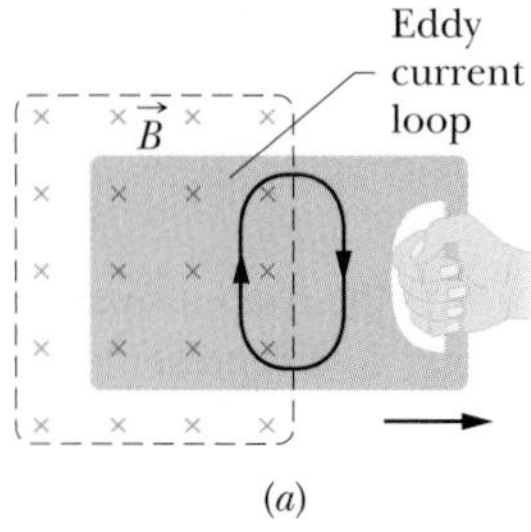

(a)

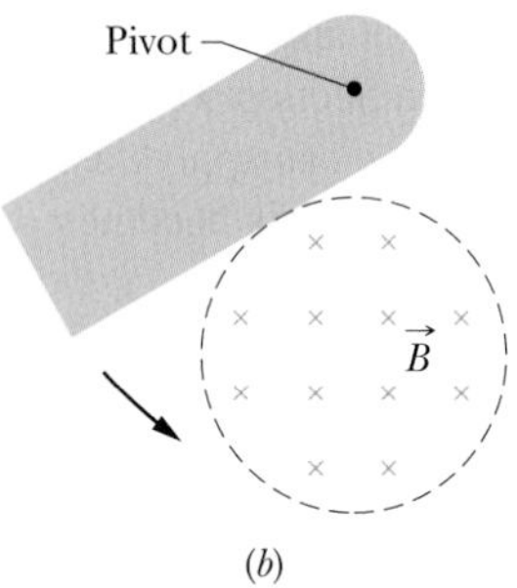

(b)

Fig. 31-12 (a) As you pull a solid conducting plate out of a magnetic field, *eddy currents* are induced in the plate. A typical loop of eddy current is shown. (b) A conducting plate is allowed to swing like a pendulum about a pivot and into a region of magnetic field. As it enters and leaves the field, eddy currents are induced in the plate.

Eddy Currents

Suppose we replace the conducting loop of Fig. 31-10 with a solid conducting plate. If we then move the plate out of the magnetic field as we did the loop (Fig. 31-12*a*), the relative motion of the field and the conductor again induces a current in the conductor. Thus, we again encounter an opposing force and must do work because of the induced current. With the plate, however, the conduction electrons making up the induced current do not follow one path as they do with the loop. Instead, the electrons swirl about within the plate as if they were caught in an eddy (or whirlpool) of water. Such a current is called an *eddy current* and can be represented as in Fig. 31-12*a as if* it followed a single path.

As with the conducting loop of Fig. 31-10, the current induced in the plate results in mechanical energy being dissipated as thermal energy. The dissipation is more apparent in the arrangement of Fig. 31-12*b*; a conducting plate, free to rotate about a pivot, is allowed to swing down through a magnetic field like a pendulum. Each time the plate enters and leaves the field, a portion of its mechanical energy is transferred to its thermal energy. After several swings, no mechanical energy remains and the warmed-up plate just hangs from its pivot.

CHECKPOINT 3: The figure shows four wire loops, with edge lengths of either L or $2L$. All four loops will move through a region of uniform magnetic field $\vec{B}$ (directed out of the page) at the same constant velocity. Rank the four loops according to the maximum magnitude of the emf induced as they move through the field, greatest first.

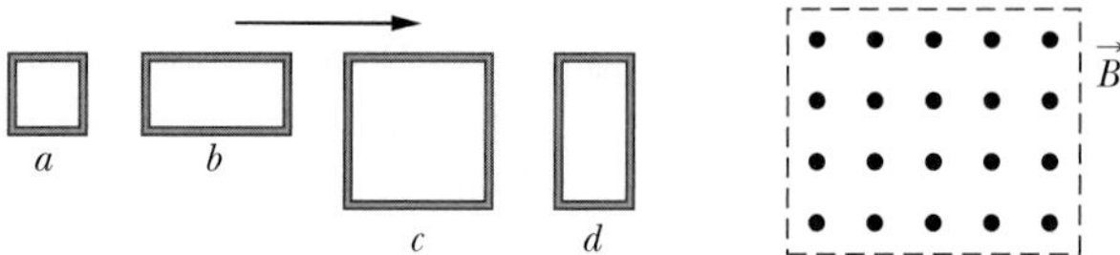

31-6 Induced Electric Fields

Let us place a copper ring of radius r in a uniform external magnetic field, as in Fig. 31-13*a*. The field—neglecting fringing—fills a cylindrical volume of radius R. Suppose that we increase the strength of this field at a steady rate, perhaps by increasing—in an appropriate way—the current in the windings of the electromagnet that produces the field. The magnetic flux through the ring will then change at a steady rate and—by Faraday's law—an induced emf and thus an induced current will appear in the ring. From Lenz's law we can deduce that the direction of the induced current is counterclockwise in Fig. 31-13*a*.

If there is a current in the copper ring, an electric field must be present along the ring; an electric field is needed to do the work of moving the conduction electrons. Moreover, the electric field must have been produced by the changing magnetic flux. This **induced electric field** $\vec{E}$ is just as real as an electric field produced by static charges; either field will exert a force $q_0\vec{E}$ on a particle of charge q_0.

By this line of reasoning, we are led to a useful and informative restatement of Faraday's law of induction:

> A changing magnetic field produces an electric field.

The striking feature of this statement is that the electric field is induced even if there is no copper ring.

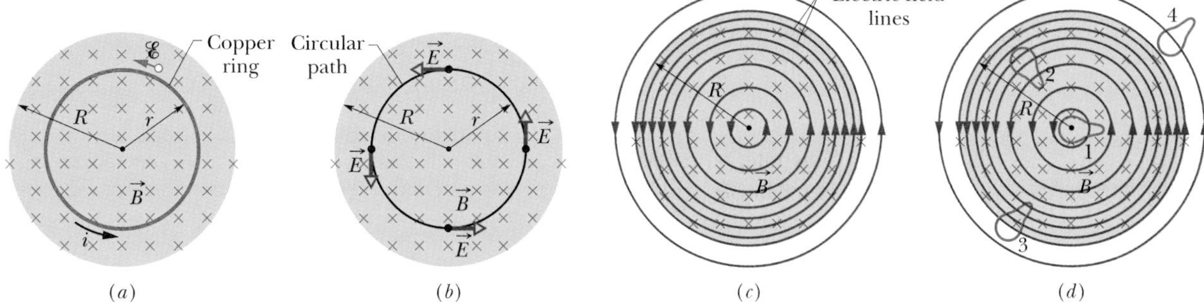

Fig. 31-13 (*a*) If the magnetic field increases at a steady rate, a constant induced current appears, as shown, in the copper ring of radius r. (*b*) An induced electric field exists even when the ring is removed; the electric field is shown at four points. (*c*) The complete picture of the induced electric field, displayed as field lines. (*d*) Four similar closed paths that enclose identical areas. Equal emfs are induced around paths 1 and 2, which lie entirely within the region of changing magnetic field. A smaller emf is induced around path 3, which only partially lies in that region. No emf is induced around path 4, which lies entirely outside the magnetic field.

To fix these ideas, consider Fig. 31-13*b*, which is just like Fig. 31-13*a* except the copper ring has been replaced by a hypothetical circular path of radius r. We assume, as previously, that the magnetic field $\vec{B}$ is increasing in magnitude at a constant rate dB/dt. The electric field induced at various points around the circular path must—from the symmetry—be tangent to the circle, as Fig. 31-13*b* shows.* Hence, the circular path is an electric field line. There is nothing special about the circle of radius r, so the electric field lines produced by the changing magnetic field must be a set of concentric circles, as in Fig. 31-13*c*.

As long as the magnetic field is *increasing* with time, the electric field represented by the circular field lines in Fig. 31-13*c* will be present. If the magnetic field remains *constant* with time, there will be no induced electric field and thus no electric field lines. If the magnetic field is *decreasing* with time (at a constant rate), the electric field lines will still be concentric circles as in Fig. 31-13*c*, but they will now have the opposite direction. All this is what we have in mind when we say: "A changing magnetic field produces an electric field."

A Reformulation of Faraday's Law

Consider a particle of charge q_0 moving around the circular path of Fig. 31-13*b*. The work W done on it in one revolution by the induced electric field is $\mathscr{E}q_0$, where $\mathscr{E}$ is the induced emf—that is, the work done per unit charge in moving the test charge around the path. From another point of view, the work is

$$\int \vec{F} \cdot d\vec{s} = (q_0 E)(2\pi r), \tag{31-18}$$

where $q_0 E$ is the magnitude of the force acting on the test charge and $2\pi r$ is the distance over which that force acts. Setting these two expressions for W equal to each other and canceling q_0, we find that

$$\mathscr{E} = 2\pi r E. \tag{31-19}$$

More generally, we can rewrite Eq. 31-18 to give the work done on a particle of charge q_0 moving along any closed path:

$$W = \oint \vec{F} \cdot d\vec{s} = q_0 \oint \vec{E} \cdot d\vec{s}. \tag{31-20}$$

*Arguments of symmetry would also permit the lines of $\vec{E}$ around the circular path to be *radial*, rather than tangential. However, such radial lines would imply that there are free charges, distributed symmetrically about the axis of symmetry, on which the electric field lines could begin or end; there are no such charges.

(The circle indicates that the integral is to be taken around the closed path.) Substituting $\mathscr{E}q_0$ for W, we find that

$$\mathscr{E} = \oint \vec{E} \cdot d\vec{s}. \tag{31-21}$$

This integral reduces at once to Eq. 31-19 if we evaluate it for the special case of Fig. 31-13*b*.

With Eq. 31-21, we can expand the meaning of induced emf. Previously, induced emf has meant the work per unit charge done in maintaining current due to a changing magnetic flux, or it has meant the work done per unit charge on a charged particle that moves around a closed path in a changing magnetic flux. However, with Fig. 31-13*b* and Eq. 31-21, an induced emf can exist without the need of a current or particle: An induced emf is the sum—via integration—of quantities $\vec{E} \cdot d\vec{s}$ around a closed path, where $\vec{E}$ is the electric field induced by a changing magnetic flux and $d\vec{s}$ is a differential length vector along the closed path.

If we combine Eq. 31-21 with Faraday's law in Eq. 31-6 ($\mathscr{E} = -d\Phi_B/dt$), we can rewrite Faraday's law as

$$\oint \vec{E} \cdot d\vec{s} = -\frac{d\Phi_B}{dt} \quad \text{(Faraday's law).} \tag{31-22}$$

This equation says simply that a changing magnetic field induces an electric field. The changing magnetic field appears on the right side of this equation, the electric field on the left.

Faraday's law in the form of Eq. 31-22 can be applied to *any* closed path that can be drawn in a changing magnetic field. Figure 31-13*d*, for example, shows four such paths, all having the same shape and area but located in different positions in the changing field. For paths 1 and 2, the induced emfs $\mathscr{E}$ ($= \oint \vec{E} \cdot d\vec{s}$) are equal because these paths lie entirely in the magnetic field and thus have the same value of $d\Phi_B/dt$. This is true even though the electric field vectors at points along these paths are different, as indicated by the patterns of electric field lines in the figure. For path 3 the induced emf is smaller because the enclosed flux Φ_B (hence $d\Phi_B/dt$) is smaller, and for path 4 the induced emf is zero, even though the electric field is not zero at any point on the path.

A New Look at Electric Potential

Induced electric fields are produced not by static charges but by a changing magnetic flux. Although electric fields produced in either way exert forces on charged particles, there is an important difference between them. The simplest evidence of this difference is that the field lines of induced electric fields form closed loops, as in Fig. 31-13*c*. Field lines produced by static charges never do so but must start on positive charges and end on negative charges.

In a more formal sense, we can state the difference between electric fields produced by induction and those produced by static charges in these words:

> Electric potential has meaning only for electric fields that are produced by static charges; it has no meaning for electric fields that are produced by induction.

You can understand this statement qualitatively by considering what happens to a charged particle that makes a single journey around the circular path in Fig. 31-13*b*. It starts at a certain point and, on its return to that same point, has experienced an emf $\mathscr{E}$ of, let us say, 5 V; that is, work of 5 J/C has been done on the particle, and thus the particle should then be at a point that is 5 V greater in potential. However, that is impossible because the particle is back at the same point, which cannot have two different values of potential. We must conclude that potential has no meaning for electric fields that are set up by changing magnetic fields.

We can take a more formal look by recalling Eq. 25-18, which defines the potential difference between two points i and f in an electric field $\vec{E}$:

$$V_f - V_i = -\int_i^f \vec{E} \cdot d\vec{s}. \tag{31-23}$$

In Chapter 25 we had not yet encountered Faraday's law of induction, so the electric fields involved in the derivation of Eq. 25-18 were those due to static charges. If i and f in Eq. 31-23 are the same point, the path connecting them is a closed loop, V_i and V_f are identical, and Eq. 31-23 reduces to

$$\oint \vec{E} \cdot d\vec{s} = 0. \tag{31-24}$$

However, when a changing magnetic flux is present, this integral is *not* zero but is $-d\Phi_B/dt$, as Eq. 31-22 asserts. Thus, assigning electric potential to an induced electric field leads us to a contradiction. We must conclude that electric potential has no meaning for electric fields associated with induction.

CHECKPOINT 4: The figure shows five lettered regions in which a uniform magnetic field extends either directly out of the page (as in region a) or into the page. The field is increasing in magnitude at the same steady rate in all five regions; the regions are identical in area. Also shown are four numbered paths along which $\oint \vec{E} \cdot d\vec{s}$ has the magnitudes given below in terms of a quantity mag. Determine whether the magnetic fields in regions b through e are directed into or out of the page.

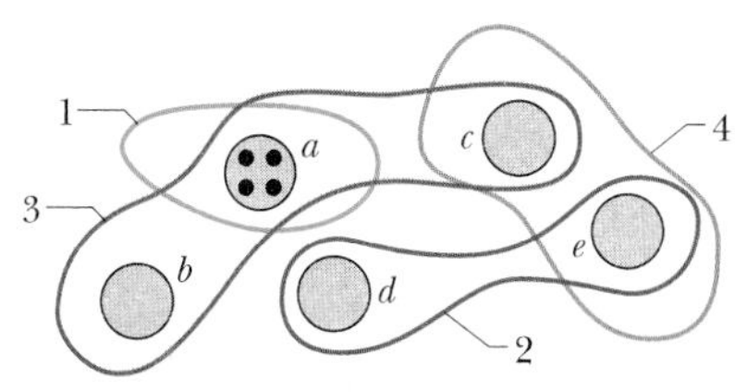

Path:	1	2	3	4
$\oint \vec{E} \cdot d\vec{s}$:	mag	2(mag)	3(mag)	0

Sample Problem 31-4

In Fig. 31-13*b*, take $R = 8.5$ cm and $dB/dt = 0.13$ T/s.

(a) Find an expression for the magnitude E of the induced electric field at points within the magnetic field, at radius r from the center of the magnetic field. Evaluate the expression for $r = 5.2$ cm.

SOLUTION: The Key Idea here is that an electric field is induced by the changing magnetic field, according to Faraday's law. To calculate the field magnitude E, we apply Faraday's law in the form of Eq. 31-22. We use a circular path of integration with radius $r \leq R$ because we want E for points within the magnetic field. We assume from the symmetry that $\vec{E}$ in Fig. 31-13*b* is tangent to the circular path at all points. The path vector $d\vec{s}$ is also always tangent to the circular path, so the dot product $\vec{E} \cdot d\vec{s}$ in Eq. 31-22 must have the magnitude $E\,ds$ at all points on the path. We can also assume from the symmetry that E has the same value at all points along the circular path. Then the left side of Eq. 31-22 becomes

$$\oint \vec{E} \cdot d\vec{s} = \oint E\,ds = E \oint ds = E(2\pi r). \tag{31-25}$$

(The integral $\oint ds$ is the circumference $2\pi r$ of the circular path.)

Next, we need to evaluate the right side of Eq. 31-22. Because $\vec{B}$ is uniform over the area A encircled by the path of integration and is directed perpendicular to that area, the magnetic flux is given by Eq. 31-4:

$$\Phi_B = BA = B(\pi r^2). \tag{31-26}$$

Substituting this and Eq. 31-25 into Eq. 31-22 and dropping the minus sign, we find that

$$E(2\pi r) = (\pi r^2)\frac{dB}{dt}$$

or

$$E = \frac{r}{2}\frac{dB}{dt}. \qquad \text{(Answer)} \tag{31-27}$$

Equation 31-27 gives the magnitude of the electric field at any point for which $r \leq R$ (that is, within the magnetic field). Substituting given values yields, for the magnitude of $\vec{E}$ at $r = 5.2$ cm,

$$E = \frac{(5.2 \times 10^{-2}\text{ m})}{2}(0.13\text{ T/s})$$

$$= 0.0034\text{ V/m} = 3.4\text{ mV/m}. \qquad \text{(Answer)}$$

(b) Find an expression for the magnitude E of the induced electric field at points that are outside the magnetic field, at radius r. Evaluate the expression for $r = 12.5$ cm.

SOLUTION: The **Key Idea** of part (a) applies here also, except that we use a circular path of integration with radius $r \geq R$, because we want to evaluate E for points outside the magnetic field. Proceeding as in (a), we again obtain Eq. 31-25. However, we do not then obtain Eq. 31-26, because the new path of integration is now outside the magnetic field, and we need this **Key Idea**: The magnetic flux encircled by the new path is only that in the area πR^2 of the magnetic field region. Therefore,

$$\Phi_B = BA = B(\pi R^2). \tag{31-28}$$

Substituting this and Eq. 31-25 into Eq. 31-22 (without the minus sign) and solving for E yield

$$E = \frac{R^2}{2r}\frac{dB}{dt}. \qquad \text{(Answer)} \tag{31-29}$$

Since E is not zero here, we know that an electric field is induced even at points that are outside the changing magnetic field, an important result that (as you shall see in Section 33-11) makes transformers possible. With the given data, Eq. 31-29 yields the magnitude of $\vec{E}$ at $r = 12.5$ cm:

$$E = \frac{(8.5 \times 10^{-2}\text{ m})^2}{(2)(12.5 \times 10^{-2}\text{ m})}(0.13\text{ T/s})$$

$$= 3.8 \times 10^{-3}\text{ V/m} = 3.8\text{ mV/m}. \qquad \text{(Answer)}$$

Equations 31-27 and 31-29 give the same result, as they must, for $r = R$. Figure 31-14 shows a plot of $E(r)$ based on these two equations.

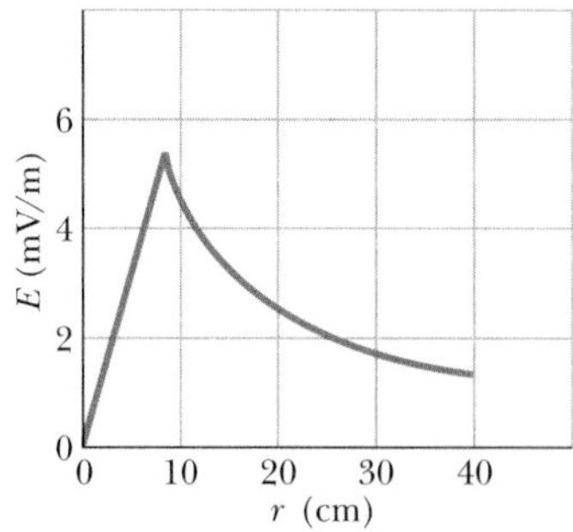

Fig. 31-14 A plot of the induced electric field $E(r)$ for the conditions of Sample Problem 31-4.

31-7 Inductors and Inductance

We found in Chapter 26 that a capacitor can be used to produce a desired electric field. We considered the parallel-plate arrangement as a basic type of capacitor. Similarly, an **inductor** (symbol ꝏꝏ) can be used to produce a desired magnetic field. We shall consider a long solenoid (more specifically, a short length near the middle of a long solenoid) as our basic type of inductor.

If we establish a current i in the windings (or turns) of an inductor (a solenoid), the current produces a magnetic flux Φ_B through the central region of the inductor. The **inductance** of the inductor is then

$$L = \frac{N\Phi_B}{i} \quad \text{(inductance defined)}, \tag{31-30}$$

in which N is the number of turns. The windings of the inductor are said to be *linked* by the shared flux, and the product $N\Phi_B$ is called the *magnetic flux linkage*. The inductance L is thus a measure of the flux linkage produced by the inductor per unit of current.

Because the SI unit of magnetic flux is the tesla–square meter, the SI unit of inductance is the tesla–square meter per ampere ($\text{T} \cdot \text{m}^2/\text{A}$). We call this the **henry** (H), after American physicist Joseph Henry, the codiscoverer of the law of induction and a contemporary of Faraday. Thus,

$$1\text{ henry} = 1\text{ H} = 1\text{ T} \cdot \text{m}^2/\text{A}. \tag{31-31}$$

Through the rest of this chapter we assume that all inductors, no matter what their geometric arrangement, have no magnetic materials such as iron in their vicinity. Such materials would distort the magnetic field of an inductor.

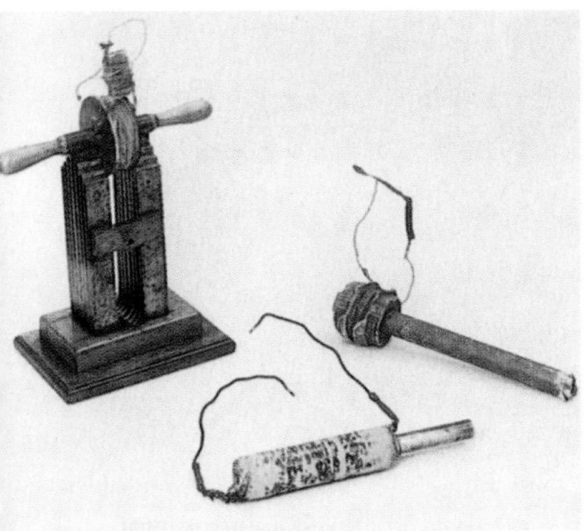

The crude inductors with which Michael Faraday discovered the law of induction. In those days amenities such as insulated wire were not commercially available. It is said that Faraday insulated his wires by wrapping them with strips cut from one of his wife's petticoats.

Inductance of a Solenoid

Consider a long solenoid of cross-sectional area A. What is the inductance per unit length near its middle?

To use the defining equation for inductance (Eq. 31-30), we must calculate the flux linkage set up by a given current in the solenoid windings. Consider a length l near the middle of this solenoid. The flux linkage for this section of the solenoid is

$$N\Phi_B = (nl)(BA)$$

in which n is the number of turns per unit length of the solenoid and B is the magnitude of the magnetic field within the solenoid.

The magnitude B is given by Eq. 30-25,

$$B = \mu_0 in,$$

so from Eq. 31-30,

$$L = \frac{N\Phi_B}{i} = \frac{(nl)(BA)}{i} = \frac{(nl)(\mu_0 in)(A)}{i}$$
$$= \mu_0 n^2 lA. \qquad (31\text{-}32)$$

Thus, the inductance per unit length for a long solenoid near its center is

$$\frac{L}{l} = \mu_0 n^2 A \qquad \text{(solenoid)}. \qquad (31\text{-}33)$$

Inductance—like capacitance—depends only on the geometry of the device. The dependence on the square of the number of turns per unit length is to be expected. If you, say, triple n, you not only triple the number of turns (N) but you also triple the flux ($\Phi_B = BA = \mu_0 inA$) through each turn, multiplying the flux linkage $N\Phi_B$ and thus the inductance L by a factor of 9.

If the solenoid is very much longer than its radius, then Eq. 31-32 gives its inductance to a good approximation. This approximation neglects the spreading of the magnetic field lines near the ends of the solenoid, just as the parallel-plate capacitor formula ($C = \varepsilon_0 A/d$) neglects the fringing of the electric field lines near the edges of the capacitor plates.

From Eq. 31-32, and recalling that n is a number per unit length, we can see that an inductance can be written as a product of the permeability constant μ_0 and a quantity with the dimensions of a length. This means that μ_0 can be expressed in the unit henry per meter:

$$\mu_0 = 4\pi \times 10^{-7}\ \text{T}\cdot\text{m/A}$$
$$= 4\pi \times 10^{-7}\ \text{H/m}. \qquad (31\text{-}34)$$

31-8 Self-Induction

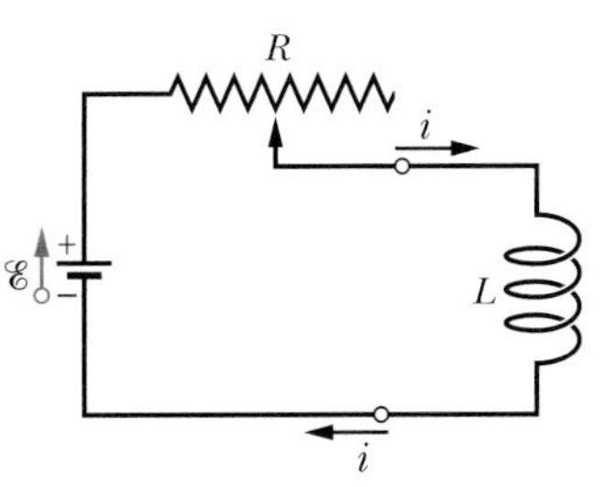

Fig. 31-15 If the current in a coil is changed by varying the contact position on a variable resistor, a self-induced emf $\mathscr{E}_L$ will appear in the coil *while the current is changing.*

If two coils—which we can now call inductors—are near each other, a current i in one coil produces a magnetic flux Φ_B through the second coil. We have seen that if we change this flux by changing the current, an induced emf appears in the second coil according to Faraday's law. An induced emf appears in the first coil as well.

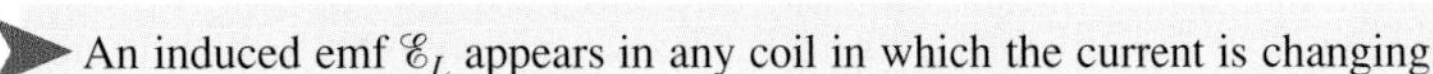
An induced emf $\mathscr{E}_L$ appears in any coil in which the current is changing.

This process (see Fig. 31-15) is called **self-induction,** and the emf that appears is called a **self-induced emf.** It obeys Faraday's law of induction just as other induced emfs do.

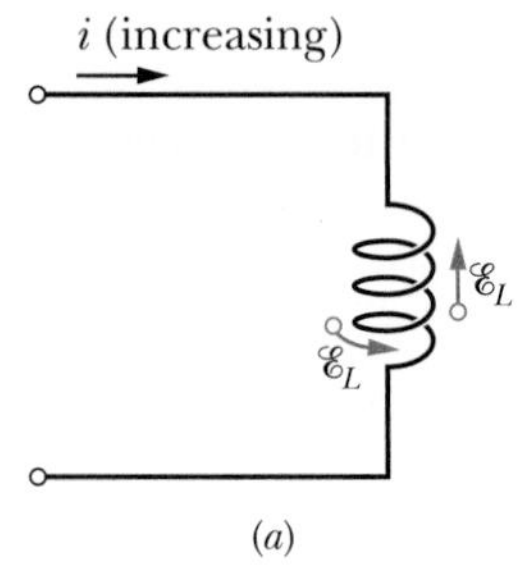

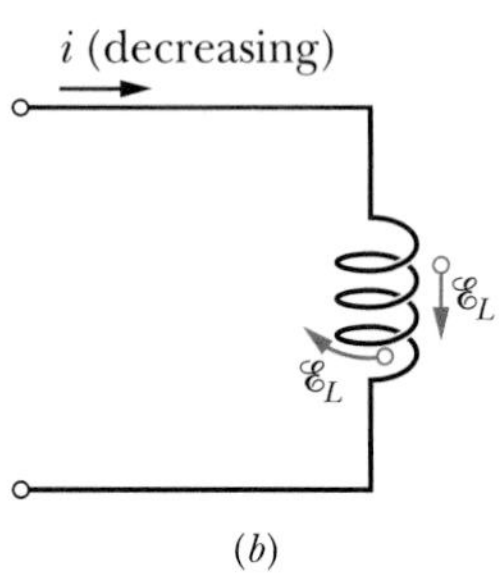

Fig. 31-16 (*a*) The current i is increasing and the self-induced emf $\mathscr{E}_L$ appears along the coil in a direction such that it opposes the increase. The arrow representing $\mathscr{E}_L$ can be drawn along a turn of the coil or alongside the coil. Both are shown. (*b*) The current i is decreasing and the self-induced emf appears in a direction such that it opposes the decrease.

For any inductor, Eq. 31-30 tells us that

$$N\Phi_B = Li. \tag{31-35}$$

Faraday's law tells us that

$$\mathscr{E}_L = -\frac{d(N\Phi_B)}{dt}. \tag{31-36}$$

By combining Eqs. 31-35 and 31-36 we can write:

$$\mathscr{E}_L = -L\frac{di}{dt} \quad \text{(self-induced emf).} \tag{31-37}$$

Thus, in any inductor (such as a coil, a solenoid, or a toroid) a self-induced emf appears whenever the current changes with time. The magnitude of the current has no influence on the magnitude of the induced emf; only the rate of change of the current counts.

You can find the *direction* of a self-induced emf from Lenz's law. The minus sign in Eq. 31-37 indicates that—as the law states—the self-induced emf $\mathscr{E}_L$ has the orientation such that it opposes the change in current i. We can drop the minus sign when we want only the magnitude of $\mathscr{E}_L$.

Suppose that, as in Fig. 31-16*a*, you set up a current i in a coil and arrange to have it increase with time at a rate di/dt. In the language of Lenz's law, this increase in the current is the "change" that the self-induction must oppose. For such opposition to occur, a self-induced emf must appear in the coil, pointing—as the figure shows—so as to oppose the increase in the current. If you cause the current to decrease with time, as in Fig. 31-16*b*, the self-induced emf must point in a direction that tends to oppose the decrease in the current, as the figure shows.

In Section 31-6 we saw that we cannot define an electric potential for an electric field (and thus for an emf) that is induced by a changing magnetic flux. This means that when a self-induced emf is produced in the inductor of Fig. 31-15, we cannot define an electric potential within the inductor itself, where the flux is changing. However, potentials can still be defined at points of the circuit that are not within the inductor—points where the electric fields are due to charge distributions and their associated electric potentials.

Moreover, we can define a self-induced potential difference V_L *across an inductor* (between its terminals, which we assume to be outside the region of changing flux). If the inductor is an *ideal inductor* (its wire has negligible resistance), the magnitude of V_L is equal to the magnitude of the self-induced emf $\mathscr{E}_L$.

If, instead, the wire in the inductor has resistance r, we mentally separate the inductor into a resistance r (which we take to be outside the region of changing flux) and an ideal inductor of self-induced emf $\mathscr{E}_L$. As with a real battery of emf $\mathscr{E}$ and internal resistance r, the potential difference across the terminals of a real inductor then differs from the emf. Unless otherwise indicated, we assume here that inductors are ideal.

CHECKPOINT 5: The figure shows an emf $\mathscr{E}_L$ induced in a coil. Which of the following can describe the current through the coil: (a) constant and rightward, (b) constant and leftward, (c) increasing and rightward, (d) decreasing and rightward, (e) increasing and leftward, (f) decreasing and leftward?

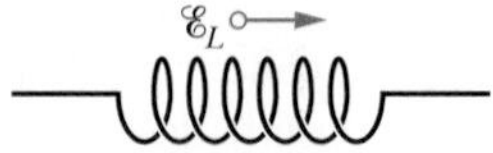

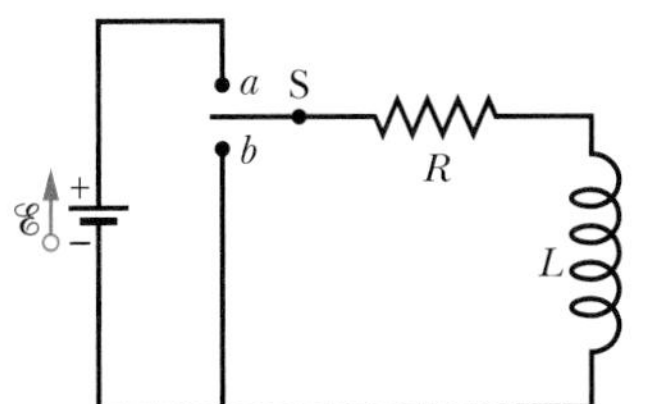

Fig. 31-17 An *RL* circuit. When switch S is closed on *a*, the current rises and approaches a limiting value of $\mathscr{E}/R$.

31-9 *RL* Circuits

In Section 28-8 we saw that if we suddenly introduce an emf $\mathscr{E}$ into a single-loop circuit containing a resistor R and a capacitor C, the charge on the capacitor does not build up immediately to its final equilibrium value $C\mathscr{E}$ but approaches it in an exponential fashion:

$$q = C\mathscr{E}(1 - e^{-t/\tau_C}). \tag{31-38}$$

The rate at which the charge builds up is determined by the capacitive time constant τ_C, defined in Eq. 28-33 as

$$\tau_C = RC. \tag{31-39}$$

If we suddenly remove the emf from this same circuit, the charge does not immediately fall to zero but approaches zero in an exponential fashion:

$$q = q_0 e^{-t/\tau_C}. \tag{31-40}$$

The time constant τ_C describes the fall of the charge as well as its rise.

An analogous slowing of the rise (or fall) of the current occurs if we introduce an emf $\mathscr{E}$ into (or remove it from) a single-loop circuit containing a resistor R and an inductor L. When the switch S in Fig. 31-17 is closed on a, for example, the current in the resistor starts to rise. If the inductor were not present, the current would rise rapidly to a steady value $\mathscr{E}/R$. Because of the inductor, however, a self-induced emf $\mathscr{E}_L$ appears in the circuit; from Lenz's law, this emf opposes the rise of the current, which means that it opposes the battery emf $\mathscr{E}$ in polarity. Thus, the current in the resistor responds to the difference between two emfs, a constant one $\mathscr{E}$ due to the battery and a variable one $\mathscr{E}_L$ ($= -L\, di/dt$) due to self-induction. As long as $\mathscr{E}_L$ is present, the current in the resistor will be less than $\mathscr{E}/R$.

As time goes on, the rate at which the current increases becomes less rapid and the magnitude of the self-induced emf, which is proportional to di/dt, becomes smaller. Thus, the current in the circuit approaches $\mathscr{E}/R$ asymptotically.

We can generalize these results as follows:

> Initially, an inductor acts to oppose changes in the current through it. A long time later, it acts like ordinary connecting wire.

Now let us analyze the situation quantitatively. With the switch S in Fig. 31-17 thrown to a, the circuit is equivalent to that of Fig. 31-18. Let us apply the loop rule, starting at point x in this figure and moving clockwise around the loop along with current i.

1. *Resistor.* Because we move through the resistor in the direction of current i, the electric potential decreases by iR. Thus, as we move from point x to point y, we encounter a potential change of $-iR$.
2. *Inductor.* Because current i is changing, there is a self-induced emf $\mathscr{E}_L$ in the inductor. The magnitude of $\mathscr{E}_L$ is given by Eq. 31-37 as $L\, di/dt$. The direction of $\mathscr{E}_L$ is upward in Fig. 31-18 because current i is downward through the inductor *and* increasing. Thus, as we move from point y to point z, opposite the direction of $\mathscr{E}_L$, we encounter a potential change of $-L\, di/dt$.
3. *Battery.* As we move from point z back to starting point x, we encounter a potential change of $+\mathscr{E}$ due to the battery's emf.

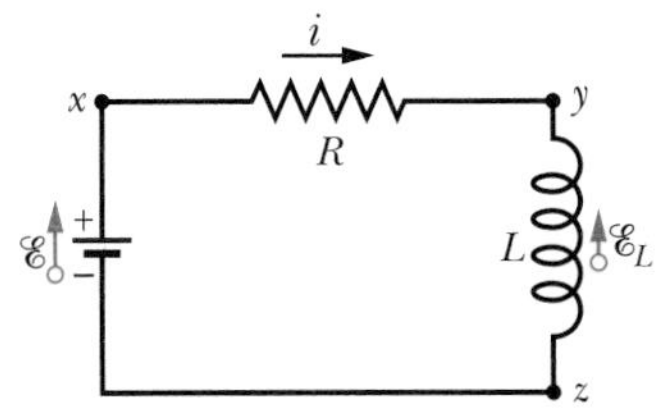

Fig. 31-18 The circuit of Fig. 31-17 with the switch closed on *a*. We apply the loop rule for circuits clockwise, starting at *x*.

Thus, the loop rule gives us

$$-iR - L\frac{di}{dt} + \mathscr{E} = 0$$

or $$L\frac{di}{dt} + Ri = \mathscr{E} \quad \text{(RL circuit)}. \tag{31-41}$$

Equation 31-41 is a differential equation involving the variable i and its first derivative di/dt. To solve it, we seek the function $i(t)$ such that when $i(t)$ and its first derivative are substituted in Eq. 31-41, the equation is satisfied and the initial condition $i(0) = 0$ is satisfied.

Equation 31-41 and its initial condition are of exactly the form of Eq. 28-29 for an RC circuit, with i replacing q, L replacing R, and R replacing $1/C$. The solution of Eq. 31-41 must then be of exactly the form of Eq. 28-30 with the same replacements. That solution is

$$i = \frac{\mathscr{E}}{R}(1 - e^{-Rt/L}), \tag{31-42}$$

which we can rewrite as

$$i = \frac{\mathscr{E}}{R}(1 - e^{-t/\tau_L}) \quad \text{(rise of current)}. \tag{31-43}$$

Here τ_L, the **inductive time constant,** is given by

$$\tau_L = \frac{L}{R} \quad \text{(time constant)}. \tag{31-44}$$

Let's examine Eq. 31-43 for when the switch is closed (at time $t = 0$) and for a time long after the switch is closed ($t \to \infty$). If we substitute $t = 0$ into Eq. 31-43, the exponential becomes $e^{-0} = 1$. Thus, Eq. 31-43 tells us that the current is initially $i = 0$, as we expected. Next, if we let t go to ∞, then the exponential goes to $e^{-\infty} = 0$. Thus, Eq. 31-43 tells us that the current goes to its equilibrium value of $\mathscr{E}/R$.

We can also examine the potential differences in the circuit. Figure 31-19 shows how the potential differences V_R ($= iR$) across the resistor and V_L ($= L\,di/dt$) across the inductor vary with time for particular values of $\mathscr{E}$, L, and R. Compare this figure carefully with the corresponding figure for an RC circuit (Fig. 28-14).

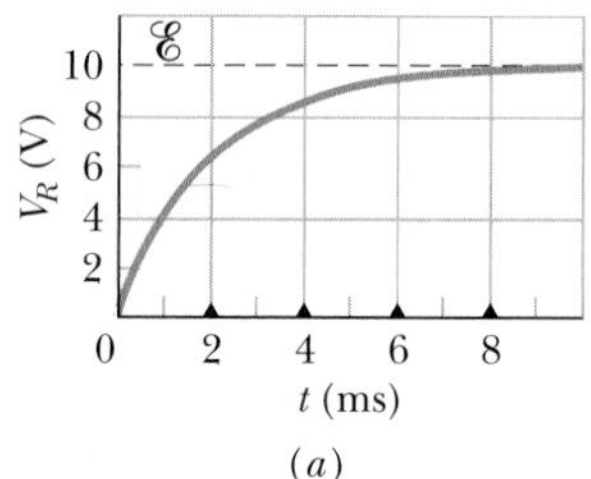

(a)

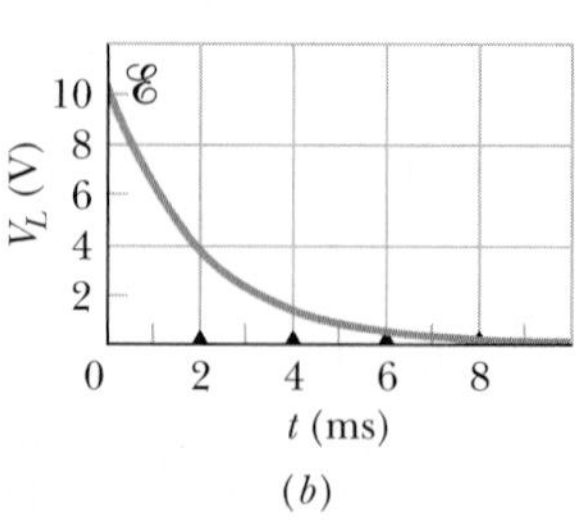

(b)

Fig. 31-19 The variation with time of (a) V_R, the potential difference across the resistor in the circuit of Fig. 31-18, and (b) V_L, the potential difference across the inductor in that circuit. The small triangles represent successive intervals of one inductive time constant $\tau_L = L/R$. The figure is plotted for $R = 2000\ \Omega$, $L = 4.0$ H, and $\mathscr{E} = 10$ V.

To show that the quantity τ_L ($= L/R$) has the dimension of time, we convert from Henries per ohm as follows:

$$1\frac{\text{H}}{\Omega} = 1\frac{\text{H}}{\Omega}\left(\frac{1\ \text{V}\cdot\text{s}}{1\ \text{H}\cdot\text{A}}\right)\left(\frac{1\ \Omega\cdot\text{A}}{1\ \text{V}}\right) = 1\ \text{s}.$$

The first quantity in parentheses is a conversion factor based on Eq. 31-37, and the second one is a conversion factor based on the relation $V = iR$.

The physical significance of the time constant follows from Eq. 31-43. If we put $t = \tau_L = L/R$ in this equation, it reduces to

$$i = \frac{\mathscr{E}}{R}(1 - e^{-1}) = 0.63\frac{\mathscr{E}}{R}. \tag{31-45}$$

Thus, the time constant τ_L is the time it takes the current in the circuit to reach about 63% of its final equilibrium value $\mathscr{E}/R$. Since the potential difference V_R across the resistor is proportional to the current i, a graph of the increasing current versus time has the same shape as that of V_R in Fig. 31-19a.

If the switch S in Fig. 31-17 is closed on a long enough for the equilibrium current $\mathscr{E}/R$ to be established, and then is thrown to b, the effect will be to remove the battery from the circuit. (The connection to b must actually be made an instant before the connection to a is broken. A switch that does this is called a *make-before-break* switch.)

With the battery gone, the current through the resistor will decrease. However, it cannot drop immediately to zero but must decay to zero over time. The differential equation that governs the decay can be found by putting $\mathscr{E} = 0$ in Eq. 31-41:

$$L\frac{di}{dt} + iR = 0. \qquad (31\text{-}46)$$

By analogy with Eqs. 28-35 and 28-36, the solution of this differential equation that satisfies the initial condition $i(0) = i_0 = \mathscr{E}/R$ is

$$i = \frac{\mathscr{E}}{R}e^{-t/\tau_L} = i_0 e^{-t/\tau_L} \quad \text{(decay of current).} \qquad (31\text{-}47)$$

We see that both current rise (Eq. 31-43) and current decay (Eq. 31-47) in an RL circuit are governed by the same inductive time constant, τ_L.

We have used i_0 in Eq. 31-47 to represent the current at time $t = 0$. In our case that happened to be $\mathscr{E}/R$, but it could be any other initial value.

Sample Problem 31-5

Figure 31-20*a* shows a circuit that contains three identical resistors with resistance $R = 9.0\ \Omega$, two identical inductors with inductance $L = 2.0$ mH, and an ideal battery with emf $\mathscr{E} = 18$ V.

(a) What is the current i through the battery just after the switch is closed?

SOLUTION: The **Key Idea** here is that just after the switch is closed, the inductor acts to oppose a change in the current through it. Because the current through each inductor is zero before the switch is closed, it will also be zero just afterward. Thus, immediately after the switch is closed, the inductors act as broken wires, as indicated in Fig. 31-20*b*. We then have a single-loop circuit for which the loop rule gives us

$$\mathscr{E} - iR = 0.$$

Substituting given data, we find that

$$i = \frac{\mathscr{E}}{R} = \frac{18\ \text{V}}{9.0\ \Omega} = 2.0\ \text{A}. \qquad \text{(Answer)}$$

(b) What is the current i through the battery long after the switch has been closed?

SOLUTION: The **Key Idea** here is that long after the switch has been closed, the currents in the circuit have reached their equilibrium values, and the inductors act as simple connecting wires, as indicated in Fig. 31-20*c*. We then have a circuit with three identical resistors in parallel; from Eq. 28-20, their equivalent resistance is $R_{eq} = R/3 = (9.0\ \Omega)/3 = 3.0\ \Omega$. The equivalent circuit shown in Fig. 31-20*d* then yields the loop equation $\mathscr{E} - iR_{eq} = 0$, or

$$i = \frac{\mathscr{E}}{R_{eq}} = \frac{18\ \text{V}}{3.0\ \Omega} = 6.0\ \text{A}. \qquad \text{(Answer)}$$

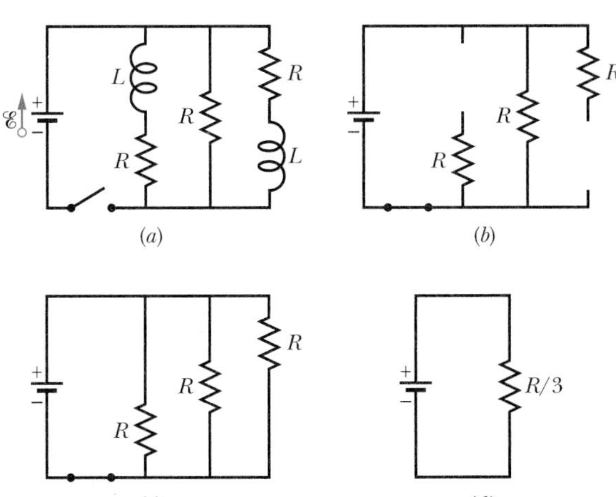

Fig. 31-20 Sample Problem 31-5. (*a*) A multiloop *RL* circuit with an open switch. (*b*) The equivalent circuit just after the switch has been closed. (*c*) The equivalent circuit a long time later. (*d*) The single-loop circuit that is equivalent to circuit (*c*).

CHECKPOINT 6: The figure shows three circuits with identical batteries, inductors, and resistors. Rank the circuits according to the current through the battery (a) just after the switch is closed and (b) a long time later, greatest first.

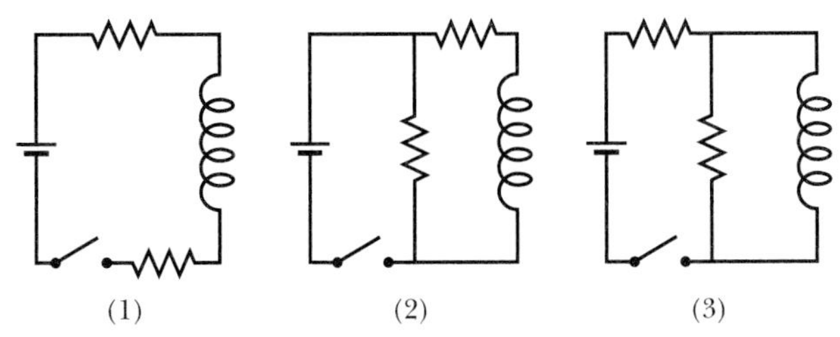

Sample Problem 31-6

A solenoid has an inductance of 53 mH and a resistance of 0.37 Ω. If it is connected to a battery, how long will the current take to reach half its final equilibrium value?

SOLUTION: One **Key Idea** here is that we can mentally separate the solenoid into a resistance and an inductance that are wired in series with a battery, as in Fig. 31-18. Then application of the loop rule leads to Eq. 31-41, which has the solution of Eq. 31-43 for the current i in the circuit.

The second **Key Idea** is that, according to that solution, current i increases exponentially from zero to its final equilibrium value of $\mathscr{E}/R$. Let t_0 be the time that current i takes to reach half its equilibrium value. Then Eq. 31-43 gives us

$$\frac{1}{2}\frac{\mathscr{E}}{R} = \frac{\mathscr{E}}{R}(1 - e^{-t_0/\tau_L}).$$

We solve for t_0 by canceling $\mathscr{E}/R$, isolating the exponential, and taking the natural logarithm of each side. We find

$$t_0 = \tau_L \ln 2 = \frac{L}{R}\ln 2 = \frac{53 \times 10^{-3}\ \text{H}}{0.37\ \Omega}\ln 2$$
$$= 0.10\ \text{s}. \qquad \text{(Answer)}$$

31-10 Energy Stored in a Magnetic Field

When we pull two particles with opposite signs of charge away from each other, we say that the resulting electric potential energy is stored in the electric field of the particles. We get it back from the field by letting the particles move closer together again. In the same way we can consider energy to be stored in a magnetic field.

To derive a quantitative expression for that stored energy, consider again Fig. 31-18, which shows a source of emf $\mathscr{E}$ connected to a resistor R and an inductor L. Equation 31-41, restated here for convenience,

$$\mathscr{E} = L\frac{di}{dt} + iR, \qquad (31\text{-}48)$$

is the differential equation that describes the growth of current in this circuit. Recall that this equation follows immediately from the loop rule and that the loop rule in turn is an expression of the principle of conservation of energy for single-loop circuits. If we multiply each side of Eq. 31-48 by i, we obtain

$$\mathscr{E}i = Li\frac{di}{dt} + i^2R, \qquad (31\text{-}49)$$

which has the following physical interpretation in terms of work and energy:

1. If a charge dq passes through the battery of emf $\mathscr{E}$ in Fig. 31-18 in time dt, the battery does work on it in the amount $\mathscr{E}\,dq$. The rate at which the battery does work is $(\mathscr{E}\,dq)/dt$, or $\mathscr{E}i$. Thus, the left side of Eq. 31-49 represents the rate at which the emf device delivers energy to the rest of the circuit.
2. The rightmost term in Eq. 31-49 represents the rate at which energy appears as thermal energy in the resistor.
3. Energy that is delivered to the circuit but does not appear as thermal energy must, by the conservation-of-energy hypothesis, be stored in the magnetic field of the inductor. Since Eq. 31-49 represents the principle of conservation of energy for RL circuits, the middle term must represent the rate dU_B/dt at which energy is stored in the magnetic field.

Thus

$$\frac{dU_B}{dt} = Li\frac{di}{dt}. \qquad (31\text{-}50)$$

We can write this as

$$dU_B = Li\,di.$$

Integrating yields

$$\int_0^{U_B} dU_B = \int_0^i Li\,di$$

or

$$U_B = \tfrac{1}{2}Li^2 \quad \text{(magnetic energy)}, \tag{31-51}$$

which represents the total energy stored by an inductor L carrying a current i. Note the similarity in form between this expression and the expression for the energy stored by a capacitor with capacitance C and charge q; namely,

$$U_E = \frac{q^2}{2C}. \tag{31-52}$$

(The variable i^2 corresponds to q^2, and the constant L corresponds to $1/C$.)

Sample Problem 31-7

A coil has an inductance of 53 mH and a resistance of 0.35 Ω.

(a) If a 12 V emf is applied across the coil, how much energy is stored in the magnetic field after the current has built up to its equilibrium value?

SOLUTION: The **Key Idea** here is that the energy stored in the magnetic field of a coil at any time depends on the current through the coil at that time, according to Eq. 31-51 ($U_B = \frac{1}{2}Li^2$). Thus, to find the energy $U_{B\infty}$ stored at equilibrium, we must first find the equilibrium current. From Eq. 31-43, the equilibrium current is

$$i_\infty = \frac{\mathscr{E}}{R} = \frac{12\text{ V}}{0.35\ \Omega} = 34.3\text{ A}. \tag{31-53}$$

Then substitution yields

$$U_{B\infty} = \tfrac{1}{2}Li_\infty^2 = (\tfrac{1}{2})(53 \times 10^{-3}\text{ H})(34.3\text{ A})^2$$
$$= 31\text{ J}. \quad \text{(Answer)}$$

(b) After how many time constants will half this equilibrium energy be stored in the magnetic field?

SOLUTION: The **Key Idea** of part (a) applies here also. Now we are being asked: At what time t will the relation

$$U_B = \tfrac{1}{2}U_{B\infty}$$

be satisfied? Using Eq. 31-51 twice allows us to rewrite this energy condition as

$$\tfrac{1}{2}Li^2 = (\tfrac{1}{2})\tfrac{1}{2}Li_\infty^2$$

or

$$i = \left(\frac{1}{\sqrt{2}}\right)i_\infty. \tag{31-54}$$

However, i is given by Eq. 31-43 and i_∞ (see Eq. 31-53) is $\mathscr{E}/R$, so Eq. 31-54 becomes

$$\frac{\mathscr{E}}{R}(1 - e^{-t/\tau_L}) = \frac{\mathscr{E}}{\sqrt{2}R}.$$

By canceling $\mathscr{E}/R$ and rearranging, this can be written as

$$e^{-t/\tau_L} = 1 - \frac{1}{\sqrt{2}} = 0.293,$$

which yields

$$\frac{t}{\tau_L} = -\ln 0.293 = 1.23$$

or

$$t \approx 1.2\tau_L. \quad \text{(Answer)}$$

Thus, the stored energy will reach half its equilibrium value 1.2 time constants after the emf is applied.

31-11 Energy Density of a Magnetic Field

Consider a length l near the middle of a long solenoid of cross-sectional area A carrying current i; the volume associated with this length is Al. The energy U_B stored by the length l of the solenoid must lie entirely within this volume because the magnetic field outside such a solenoid is approximately zero. Moreover, the stored energy must be uniformly distributed within the solenoid because the magnetic field is (approximately) uniform everywhere inside.

Thus, the energy stored per unit volume of the field is

$$u_B = \frac{U_B}{Al}$$

or, since

$$U_B = \tfrac{1}{2}Li^2,$$

we have

$$u_B = \frac{Li^2}{2Al} = \frac{L}{l}\frac{i^2}{2A}.$$

Here L is the inductance of length l of the solenoid.

Substituting for L/l from Eq. 31-33, we find

$$u_B = \tfrac{1}{2}\mu_0 n^2 i^2, \tag{31-55}$$

where n is the number of turns per unit length. From Eq. 30-25 ($B = \mu_0 in$) we can write this *energy density* as

$$u_B = \frac{B^2}{2\mu_0} \quad \text{(magnetic energy density).} \tag{31-56}$$

This equation gives the density of stored energy at any point where the magnetic field is B. Even though we derived it by considering the special case of a solenoid, Eq. 31-56 holds for all magnetic fields, no matter how they are generated. The equation is comparable to Eq. 26-23; namely,

$$u_E = \tfrac{1}{2}\varepsilon_0 E^2, \tag{31-57}$$

which gives the energy density (in a vacuum) at any point in an electric field. Note that both u_B and u_E are proportional to the square of the appropriate field magnitude, B or E.

CHECKPOINT 7: The table lists the number of turns per unit length, current, and cross-sectional area for three solenoids. Rank the solenoids according to the magnetic energy density within them, greatest first.

Solenoid	Turns per Unit Length	Current	Area
a	$2n_1$	i_1	$2A_1$
b	n_1	$2i_1$	A_1
c	n_1	i_1	$6A_1$

Sample Problem 31-8

A long coaxial cable (Fig. 31-21) consists of two thin-walled concentric conducting cylinders with radii a and b. The inner cylinder carries a steady current i, the outer cylinder providing the return path for that current. The current sets up a magnetic field between the two cylinders.

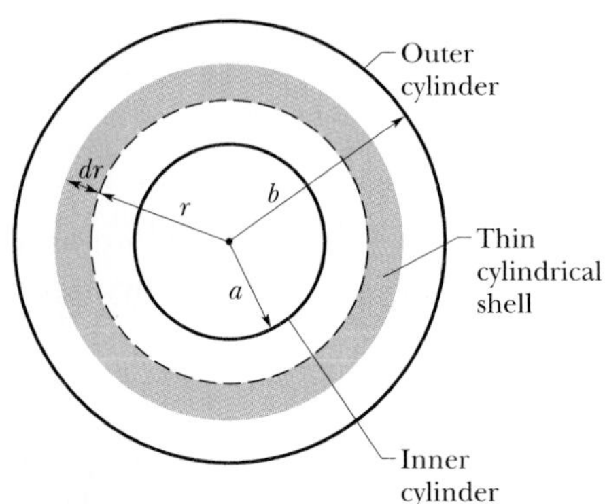

Fig. 31-21 Sample Problem 31-8. A cross section of a long coaxial cable consisting of two thin-walled conducting cylinders, the inner cylinder of radius a and the outer cylinder of radius b.

(a) Calculate the energy stored in the magnetic field for a length ℓ of the cable.

SOLUTION: The **Key Ideas** for this challenging problem are these:

1. We can calculate the (total) energy U_B stored in the magnetic field from the energy density u_B of the field.
2. That energy density depends on the magnitude B of the magnetic field according to Eq. 31-56 ($u_B = B^2/2\mu_0$).

3. Because of the circular symmetry of the cable, we can find B by using Ampere's law with the given current i.

Finding B: To apply these ideas, we begin with Ampere's law, using a circular path of integration with radius r such that $a < r < b$ (between the two cylinders, as indicated by the dashed line in Fig. 31-21). The only current enclosed by this path is current i on the inner cylinder. Thus, we can write Ampere's law as

$$\oint \vec{B} \cdot d\vec{s} = \mu_0 i. \tag{31-58}$$

Next, we simplify the integral: Because of the circular symmetry, at all points along the circular path, $\vec{B}$ is tangent to the path and has the same magnitude B. Let us take the direction of integration along the path as the direction of the magnetic field around the path. Then we can replace $\vec{B} \cdot d\vec{s}$ with $B\,ds \cos 0 = B\,ds$, and then move magnitude B in front of the integration. The integral that remains is $\oint ds$, which just gives the circumference $2\pi r$ of the path. Thus, Eq. 31-58 simplifies to

$$B(2\pi r) = \mu_0 i$$

or

$$B = \frac{\mu_0 i}{2\pi r}. \tag{31-59}$$

Finding u_B: Next, to obtain the energy density, we substitute Eq. 31-59 into Eq. 31-56:

$$u_B = \frac{B^2}{2\mu_0} = \frac{\mu_0 i^2}{8\pi^2 r^2}. \tag{31-60}$$

Finding U_B: Note that u_B is not uniform in the volume between the two cylinders, but instead depends on the radial distance r. Thus, to find the total energy U_B stored between the cylinders, we must integrate u_B over that volume.

Because the volume between the cylinders has circular symmetry about the cable's central axis, we consider the volume dV of a cylindrical shell located between the cylinders; the shell has inner radius r, outer radius $r + dr$ (Fig. 31-21), and length ℓ. The shell's cross-sectional area (or face area) is the product of its circumference $2\pi r$ and thickness dr. Thus, the shell's volume dV is $(2\pi r)(dr)(\ell)$; that is, $dV = 2\pi r\ell\,dr$.

Because points within this shell are all at approximately the same radial distance r, they all have approximately the same energy density u_B. Thus, the total energy dU_B contained in the shell of volume dV is given by

$$\text{energy} = \left(\begin{array}{c}\text{energy per}\\ \text{unit volume}\end{array}\right)(\text{volume})$$

or

$$dU_B = u_B\,dV.$$

Substituting Eq. 31-60 for u_B and $2\pi r\ell\,dr$ for dV, we obtain

$$dU_B = \frac{\mu_0 i^2}{8\pi^2 r^2}(2\pi r\ell)\,dr = \frac{\mu_0 i^2 \ell}{4\pi}\frac{dr}{r}.$$

To find the total energy contained between the two cylinders, we integrate this equation over the volume between the two cylinders:

$$U_B = \int dU_B = \frac{\mu_0 i^2 \ell}{4\pi}\int_a^b \frac{dr}{r}$$

$$= \frac{\mu_0 i^2 \ell}{4\pi}\ln\frac{b}{a}. \quad \text{(Answer)} \tag{31-61}$$

No energy is stored outside the outer cylinder or inside the inner cylinder because the magnetic field is zero in both locations, as you can show with Ampere's law.

(b) What is the stored energy per unit length of the cable if $a = 1.2$ mm, $b = 3.5$ mm, and $i = 2.7$ A?

SOLUTION: From Eq. 31-61 we have

$$\frac{U_B}{\ell} = \frac{\mu_0 i^2}{4\pi}\ln\frac{b}{a}$$

$$= \frac{(4\pi \times 10^{-7}\text{ H/m})(2.7\text{ A})^2}{4\pi}\ln\frac{3.5\text{ mm}}{1.2\text{ mm}}$$

$$= 7.8 \times 10^{-7}\text{ J/m} = 780\text{ nJ/m}. \quad \text{(Answer)}$$

31-12 Mutual Induction

In this section we return to the case of two interacting coils, which we first discussed in Section 31-2, and we treat it in a somewhat more formal manner. We saw earlier that if two coils are close together as in Fig. 31-2, a steady current i in one coil will set up a magnetic flux Φ through the other coil (*linking* the other coil). If we change i with time, an emf $\mathscr{E}$ given by Faraday's law appears in the second coil; we called this process *induction*. We could better have called it **mutual induction,** to suggest the mutual interaction of the two coils and to distinguish it from *self-induction,* in which only one coil is involved.

Let us look a little more quantitatively at mutual induction. Figure 31-22*a* shows two circular close-packed coils near each other and sharing a common central axis. There is a steady current i_1 in coil 1, produced by the battery in the external circuit. This current creates a magnetic field represented by the lines of $\vec{B}_1$ in the figure. Coil 2 is connected to a sensitive meter but contains no battery; a magnetic flux Φ_{21} (the flux through coil 2 associated with the current in coil 1) links the N_2 turns of coil 2.

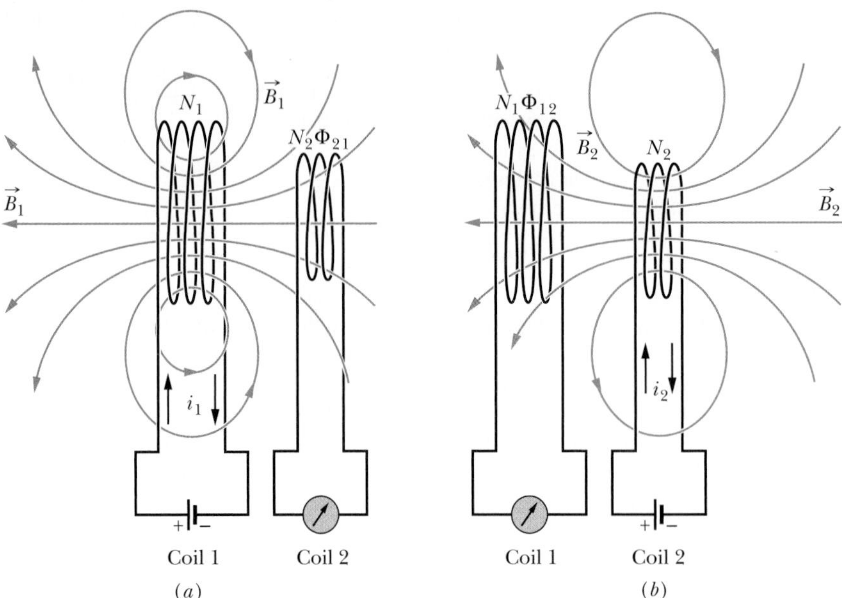

Fig. 31-22 Mutual induction. (*a*) If the current in coil 1 changes, an emf will be induced in coil 2. (*b*) If the current in coil 2 changes, an emf will be induced in coil 1.

We define the mutual inductance M_{21} of coil 2 with respect to coil 1 as

$$M_{21} = \frac{N_2\Phi_{21}}{i_1}, \tag{31-62}$$

which has the same form as Eq. 31-30 ($L = N\Phi/i$), the definition of (self) inductance. We can recast Eq. 31-62 as

$$M_{21}i_1 = N_2\Phi_{21}.$$

If, by external means, we cause i_1 to vary with time, we have

$$M_{21}\frac{di_1}{dt} = N_2\frac{d\Phi_{21}}{dt}.$$

The right side of this equation is, according to Faraday's law, just the magnitude of the emf $\mathscr{E}_2$ appearing in coil 2 due to the changing current in coil 1. Thus, with a minus sign to indicate direction,

$$\mathscr{E}_2 = -M_{21}\frac{di_1}{dt}, \tag{31-63}$$

which you should compare with Eq. 31-37 for self-induction ($\mathscr{E} = -L\,di/dt$).

Let us now interchange the roles of coils 1 and 2, as in Fig. 31-22*b*; that is, we set up a current i_2 in coil 2 by means of a battery, and this produces a magnetic flux Φ_{12} that links coil 1. If we change i_2 with time, we have, by the argument given above,

$$\mathscr{E}_1 = -M_{12}\frac{di_2}{dt}. \tag{31-64}$$

Thus, we see that the emf induced in either coil is proportional to the rate of change of current in the other coil. The proportionality constants M_{21} and M_{12} seem to be different. We assert, without proof, that they are in fact the same so that no subscripts are needed. (This conclusion is true but is in no way obvious.) Thus, we have

$$M_{21} = M_{12} = M, \tag{31-65}$$

and we can rewrite Eqs. 31-63 and 31-64 as

$$\mathscr{E}_2 = -M\frac{di_1}{dt} \qquad (31\text{-}66)$$

and

$$\mathscr{E}_1 = -M\frac{di_2}{dt}. \qquad (31\text{-}67)$$

The induction is indeed mutual. The SI unit for M (as for L) is the henry.

Sample Problem 31-9

Figure 31-23 shows two circular close-packed coils, the smaller (radius R_2, with N_2 turns) being coaxial with the larger (radius R_1, with N_1 turns) and in the same plane.

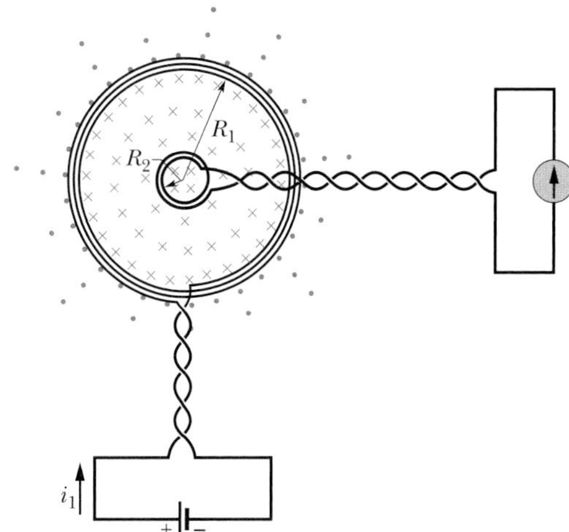

Fig. 31-23 Sample Problem 31-9. A small coil is located at the center of a large coil. The mutual inductance of the coils can be determined by sending current i_1 through the large coil.

(a) Derive an expression for the mutual inductance M for this arrangement of these two coils, assuming that $R_1 \gg R_2$.

SOLUTION: The Key Idea here is that the mutual inductance M for these coils is the ratio of the flux linkage ($N\Phi$) through one coil to the current i in the other coil, which produces that flux linkage. Thus, we need to assume that currents exist in the coils; then we need to calculate the flux linkage in one of the coils.

The magnetic field through the larger coil due to the smaller coil is nonuniform in both magnitude and direction, so the flux through the larger coil due to the smaller coil is nonuniform and difficult to calculate. However, the smaller coil is small enough for us to assume that the magnetic field through it due to the larger coil is approximately uniform. Thus, the flux through it due to the larger coil is also approximately uniform. Hence, to find M we shall assume a current i_1 in the larger coil and calculate the flux linkage $N_2\Phi_{21}$ in the smaller coil:

$$M = \frac{N_2\Phi_{21}}{i_1}. \qquad (31\text{-}68)$$

A second Key Idea is that the flux Φ_{21} through each turn of the smaller coil is, from Eq. 31-4,

$$\Phi_{21} = B_1A_2,$$

where B_1 is the magnitude of the magnetic field at points within the small coil due to the larger coil, and A_2 ($= \pi R_2^2$) is the area enclosed by the turn. Thus, the flux linkage in the smaller coil (with its N_2 turns) is

$$N_2\Phi_{21} = N_2B_1A_2. \qquad (31\text{-}69)$$

A third Key Idea is that to find B_1 at points within the smaller coil, we can use Eq. 30-28, with z set to 0 because the smaller coil is in the plane of the larger coil. That equation tells us that each turn of the larger coil produces a magnetic field of magnitude $\mu_0 i_1/2R_1$ at points within the smaller coil. Thus, the larger coil (with its N_1 turns) produces a total magnetic field of magnitude

$$B_1 = N_1\frac{\mu_0 i_1}{2R_1} \qquad (31\text{-}70)$$

at points within the smaller coil.

Substituting Eq. 31-70 for B_1 and πR_2^2 for A_2 in Eq. 31-69 yields

$$N_2\Phi_{21} = \frac{\pi\mu_0 N_1 N_2 R_2^2 i_1}{2R_1}.$$

Substituting this result into Eq. 31-68, we find

$$M = \frac{N_2\Phi_{21}}{i_1} = \frac{\pi\mu_0 N_1 N_2 R_2^2}{2R_1}. \qquad \text{(Answer)} \quad (31\text{-}71)$$

(b) What is the value of M for $N_1 = N_2 = 1200$ turns, $R_2 = 1.1$ cm, and $R_1 = 15$ cm?

SOLUTION: Equation 31-71 yields

$$M = \frac{(\pi)(4\pi \times 10^{-7}\ \text{H/m})(1200)(1200)(0.011\ \text{m})^2}{(2)(0.15\ \text{m})}$$
$$= 2.29 \times 10^{-3}\ \text{H} \approx 2.3\ \text{mH}. \qquad \text{(Answer)}$$

Consider the situation if we reverse the roles of the two coils—that is, if we produce a current i_2 in the smaller coil and try to calculate M from Eq. 31-62 in the form

$$M = \frac{N_1\Phi_{12}}{i_2}.$$

The calculation of Φ_{12} (the nonuniform flux of the smaller coil's magnetic field encompassed by the larger coil) is not simple. If we were to do the calculation numerically using a computer, we would find M to be 2.3 mH, as above! This emphasizes that Eq. 31-65 ($M_{21} = M_{12} = M$) is not obvious.

REVIEW & SUMMARY

Magnetic Flux The *magnetic flux* Φ_B through an area A in a magnetic field $\vec{B}$ is defined as

$$\Phi_B = \int \vec{B} \cdot d\vec{A}, \qquad (31\text{-}3)$$

where the integral is taken over the area. The SI unit of magnetic flux is the weber, where 1 Wb = 1 T · m². If $\vec{B}$ is perpendicular to the area and uniform over it, Eq. 31-3 becomes

$$\Phi_B = BA \qquad (\vec{B} \perp A, \vec{B} \text{ uniform}). \qquad (31\text{-}4)$$

Faraday's Law of Induction If the magnetic flux Φ_B through an area bounded by a closed conducting loop changes with time, a current and an emf are produced in the loop; this process is called *induction*. The induced emf is

$$\mathscr{E} = -\frac{d\Phi_B}{dt} \qquad \text{(Faraday's law)}. \qquad (31\text{-}6)$$

If the loop is replaced by a closely packed coil of N turns, the induced emf is

$$\mathscr{E} = -N\frac{d\Phi_B}{dt}. \qquad (31\text{-}7)$$

Lenz's Law An induced current has a direction such that the magnetic field *of the current* opposes the change in the magnetic flux that produces the current. The induced emf has the same direction as the induced current.

Emf and the Induced Electric Field An emf is induced by a changing magnetic flux even if the loop through which the flux is changing is not a physical conductor but an imaginary line. The changing magnetic field induces an electric field $\vec{E}$ at every point of such a loop; the induced emf is related to $\vec{E}$ by

$$\mathscr{E} = \oint \vec{E} \cdot d\vec{s}, \qquad (31\text{-}21)$$

where the integration is taken around the loop. From Eq. 31-21 we can write Faraday's law in its most general form,

$$\oint \vec{E} \cdot d\vec{s} = -\frac{d\Phi_B}{dt} \qquad \text{(Faraday's law)}. \qquad (31\text{-}22)$$

The essence of this law is that *a changing magnetic field induces an electric field* $\vec{E}$.

Inductors An **inductor** is a device that can be used to produce a known magnetic field in a specified region. If a current i is established through each of the N windings of an inductor, a magnetic flux Φ_B links those windings. The **inductance** L of the inductor is

$$L = \frac{N\Phi_B}{i} \qquad \text{(inductance defined)}. \qquad (31\text{-}30)$$

The SI unit of inductance is the **henry** (H), with

$$1 \text{ henry} = 1 \text{ H} = 1 \text{ T} \cdot \text{m}^2/\text{A}. \qquad (31\text{-}31)$$

The inductance per unit length near the middle of a long solenoid of cross-sectional area A and n turns per unit length is

$$\frac{L}{l} = \mu_0 n^2 A \qquad \text{(solenoid)}. \qquad (31\text{-}33)$$

Self-Induction If a current i in a coil changes with time, an emf is induced in the coil. This self-induced emf is

$$\mathscr{E}_L = -L\frac{di}{dt}. \qquad (31\text{-}37)$$

The direction of $\mathscr{E}_L$ is found from Lenz's law: The self-induced emf acts to oppose the change that produces it.

Series RL Circuits If a constant emf $\mathscr{E}$ is introduced into a single-loop circuit containing a resistance R and an inductance L, the current rises to an equilibrium value of $\mathscr{E}/R$ according to

$$i = \frac{\mathscr{E}}{R}(1 - e^{-t/\tau_L}) \qquad \text{(rise of current)}. \qquad (31\text{-}43)$$

Here τ_L (= L/R) governs the rate of rise of the current and is called the **inductive time constant** of the circuit. When the source of constant emf is removed, the current decays from a value i_0 according to

$$i = i_0 e^{-t/\tau_L} \qquad \text{(decay of current)}. \qquad (31\text{-}47)$$

Magnetic Energy If an inductor L carries a current i, the inductor's magnetic field stores an energy given by

$$U_B = \tfrac{1}{2}Li^2 \qquad \text{(magnetic energy)}. \qquad (31\text{-}51)$$

If B is the magnitude of a magnetic field at any point (in an inductor or anywhere else), the density of stored magnetic energy at that point is

$$u_B = \frac{B^2}{2\mu_0} \qquad \text{(magnetic energy density)}. \qquad (31\text{-}56)$$

Mutual Induction If two coils (labeled 1 and 2) are near each other, a changing current in either coil can induce an emf in the other. This mutual induction is described by

$$\mathscr{E}_2 = -M\frac{di_1}{dt} \qquad (31\text{-}66)$$

and

$$\mathscr{E}_1 = -M\frac{di_2}{dt}, \qquad (31\text{-}67)$$

where M (measured in henries) is the mutual inductance for the coil arrangement.

QUESTIONS

1. In Fig. 31-24, a long straight wire with current i passes (without touching) three rectangular wire loops with edge lengths L, $1.5L$, and $2L$. The loops are widely spaced (so as to not affect one another). Loops 1 and 3 are symmetric about the long wire. Rank the loops according to the size of the current induced in them if current i is (a) constant and (b) increasing, greatest first.

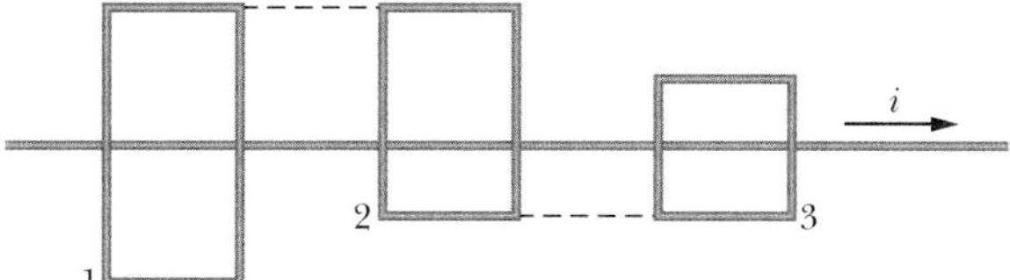

Fig. 31-24 Question 1.

2. If the circular conductor in Fig. 31-25 undergoes thermal expansion while it is in a uniform magnetic field, a current will be induced clockwise around it. Is the magnetic field directed into the page or out of it?

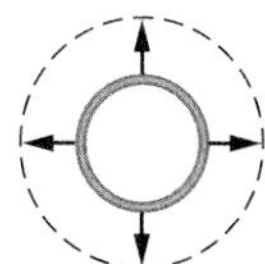

Fig. 31-25 Question 2.

3. Figure 31-26 shows two circuits in which a conducting bar is slid at the same speed v through the same uniform magnetic field and along a U-shaped wire. The parallel lengths of the wire are separated by $2L$ in circuit 1 and by L in circuit 2. The current induced in circuit 1 is counterclockwise. (a) Is the direction of the magnetic field into or out of the page? (b) Is the direction of the current induced in circuit 2 clockwise or counterclockwise? (c) Is the emf induced in circuit 1 larger than, smaller than, or the same as that in circuit 2?

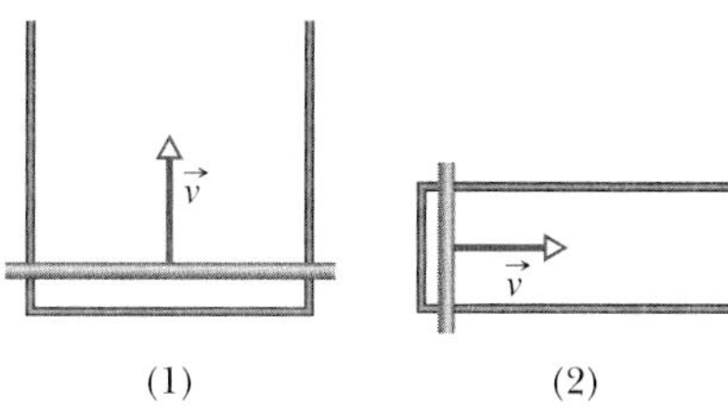

Fig. 31-26 Question 3.

4. Figure 31-27 shows two coils wrapped around nonconducting rods. Coil X is connected to a battery and a variable resistance. What is the direction of the induced current through the current meter connected to coil Y (a) when coil Y is moved toward coil X and (b) when the current in coil X is decreased without any change in the relative positions of the coils?

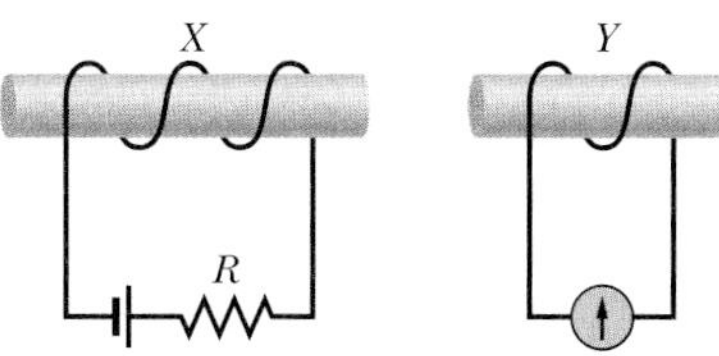

Fig. 31-27 Question 4.

5. Figure 31-28a shows a circular region in which an increasing uniform magnetic field is directed out of the page, as well as a concentric circular path along which $\oint \vec{E} \cdot d\vec{s}$ is to be evaluated. The table gives the initial magnitude of the magnetic field, the increase in that magnitude, and the time interval for the increase, in three situations. Rank the situations according to the magnitude of the electric field induced along the path, greatest first.

Situation	Initial Field	Increase	Time
a	B_1	ΔB_1	Δt_1
b	$2B_1$	$\Delta B_1/2$	Δt_1
c	$B_1/4$	ΔB_1	$\Delta t_1/2$

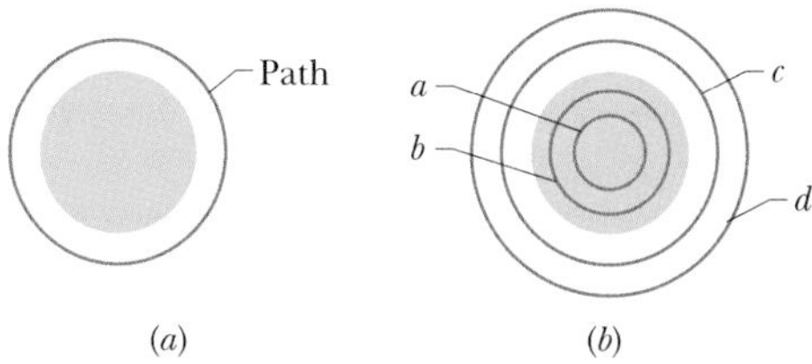

Fig. 31-28 Questions 5 and 6.

6. Figure 31-28b shows a circular region in which a decreasing uniform magnetic field is directed out of the page, as well as four concentric circular paths. Rank the paths according to the magnitude of $\oint \vec{E} \cdot d\vec{s}$ evaluated along them, greatest first.

7. Figure 31-29 gives the variation with time of the potential difference V_R across a resistor in three circuits wired as shown in Fig. 31-18. The circuits contain the same resistance R and emf $\mathscr{E}$ but differ in the inductance L. Rank the circuits according to the value of L, greatest first.

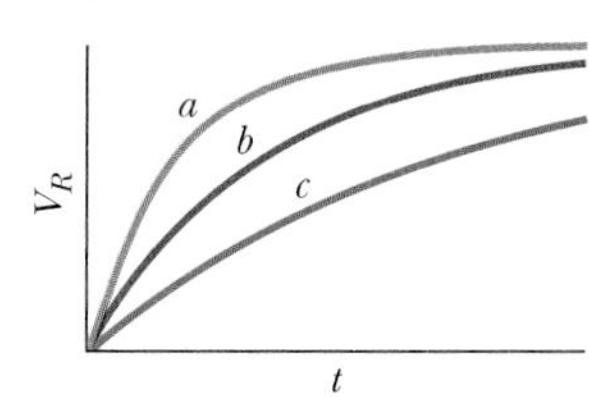

Fig. 31-29 Question 7.

8. Figure 31-30 shows three circuits with identical batteries, inductors, and resistors. Rank the circuits according to the time for the current to reach 50% of its equilibrium value after the switches are closed, greatest first.

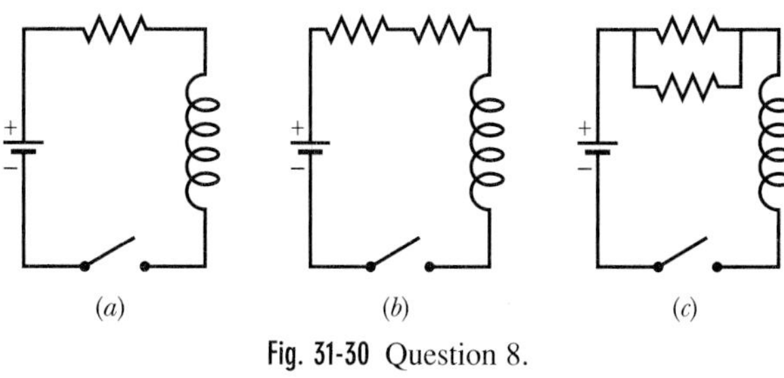

Fig. 31-30 Question 8.

9. Figure 31-31 shows a circuit with two identical resistors and an ideal inductor. Is the current through the central resistor more than, less than, or the same as that through the other resistor (a) just after the closing of switch S, (b) a long time after the closing of S, (c) just after S is reopened, a long time later, and (d) a long time after the reopening of S?

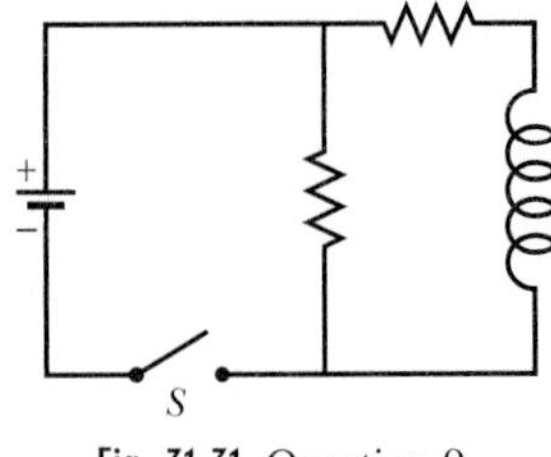

Fig. 31-31 Question 9.

10. The switch in the circuit of Fig. 31-17 has been closed on a for a very long time when it is then thrown to b. The resulting current through the inductor is indicated in Fig. 31-32 for four sets of values for the resistance R and inductance L: (1) R_0 and L_0, (2) $2R_0$ and L_0, (3) R_0 and $2L_0$, (4) $2R_0$ and $2L_0$. Which set goes with which curve?

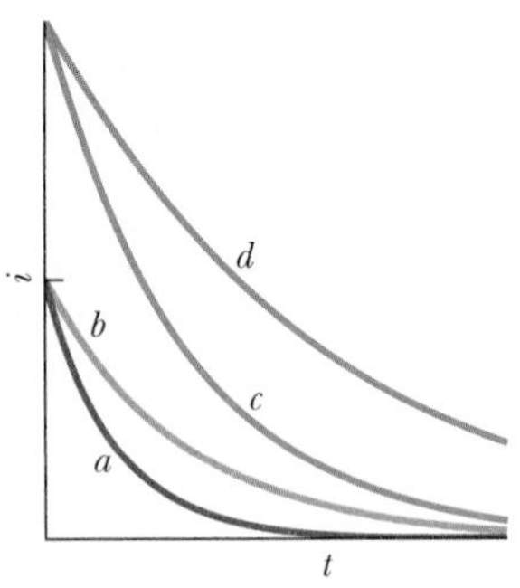

Fig. 31-32 Question 10.

EXERCISES & PROBLEMS

ssm Solution is in the Student Solutions Manual.
www Solution is available on the World Wide Web at: http://www.wiley.com/college/hrw
ilw Solution is available on the Interactive LearningWare.

SEC. 31-4 Lenz's Law

1E. A UHF television loop antenna has a diameter of 11 cm. The magnetic field of a TV signal is normal to the plane of the loop and, at one instant of time, its magnitude is changing at the rate 0.16 T/s. The magnetic field is uniform. What emf is induced in the antenna? ssm

2E. A small loop of area A is inside of, and has its axis in the same direction as, a long solenoid of n turns per unit length and current i. If $i = i_0 \sin \omega t$, find the emf induced in the loop.

3E. The magnetic flux through the loop shown in Fig. 31-33 increases according to the relation $\Phi_B = 6.0t^2 + 7.0t$, where Φ_B is in milliwebers and t is in seconds. (a) What is the magnitude of the emf induced in the loop when $t = 2.0$ s? (b) What is the direction of the current through R?

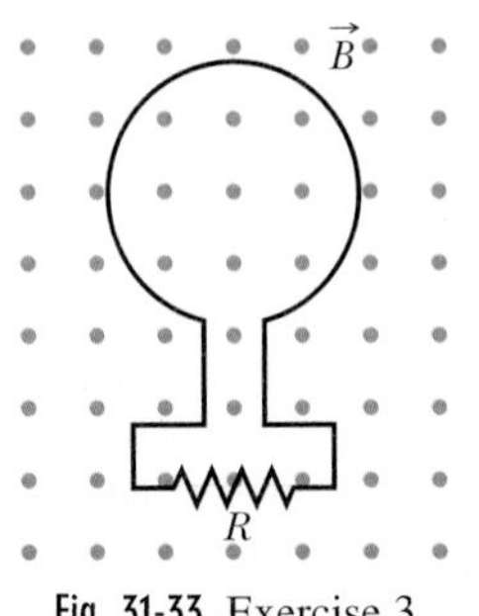

Fig. 31-33 Exercise 3 and Problem 11.

4E. The magnetic field through a single loop of wire, 12 cm in radius and of 8.5 Ω resistance, changes with time as shown in Fig. 31-34. Calculate the emf in the loop as a function of time. Consider the time intervals (a) $t = 0$ to $t = 2.0$ s, (b) $t = 2.0$ s to $t = 4.0$ s, (c) $t = 4.0$ s to $t = 6.0$ s. The (uniform) magnetic field is perpendicular to the plane of the loop.

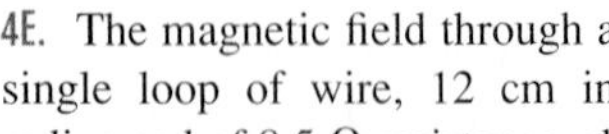
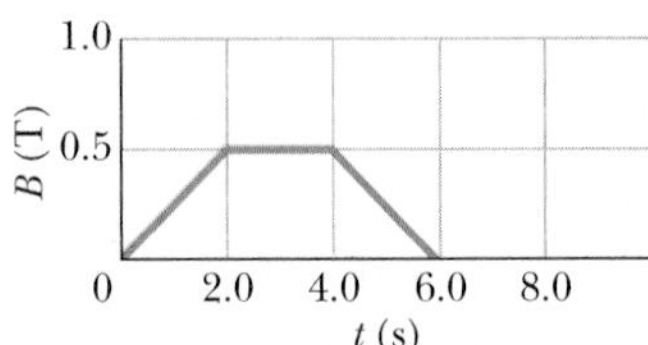

Fig. 31-34 Exercise 4.

5E. A uniform magnetic field is normal to the plane of a circular loop 10 cm in diameter and made of copper wire (of diameter 2.5 mm). (a) Calculate the resistance of the wire. (See Table 27-1.) (b) At what rate must the magnetic field change with time if an induced current of 10 A is to appear in the loop?

6P. The current in the solenoid of Sample Problem 31-1 changes, not as stated there, but according to $i = 3.0t + 1.0t^2$, where i is in amperes and t is in seconds. (a) Plot the induced emf in the coil from $t = 0$ to $t = 4.0$ s. (b) The resistance of the coil is 0.15 Ω. What is the current in the coil at $t = 2.0$ s?

7P. In Fig. 31-35 a 120-turn coil of radius 1.8 cm and resistance 5.3 Ω is placed *outside* a solenoid like that of Sample Problem 31-1. If the current in the solenoid is changed as in that sample problem, what current appears in the coil while the solenoid current is being changed? ssm

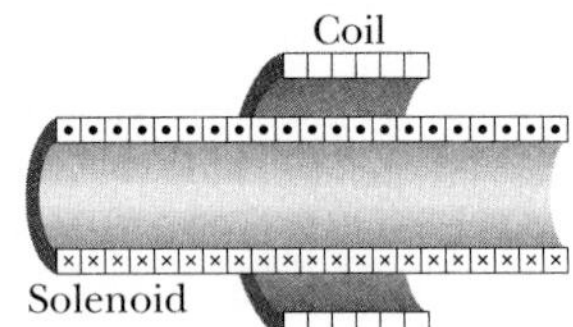

Fig. 31-35 Problem 7.

8P. An elastic conducting material is stretched into a circular loop of 12.0 cm radius. It is placed with its plane perpendicular to a uniform 0.800 T magnetic field. When released, the radius of the loop starts to shrink at an instantaneous rate of 75.0 cm/s. What emf is induced in the loop at that instant?

9P. Figure 31-36 shows two parallel loops of wire having a common axis. The smaller loop (radius r) is above the larger loop (radius R) by a distance $x \gg R$. Consequently, the magnetic field due to the current i in the larger loop is nearly constant throughout the smaller loop. Suppose that x is increasing at the constant rate of $dx/dt = v$. (a) Determine the magnetic flux through the area bounded by the smaller loop as a function of x. (*Hint:* See Eq. 30-29.) In the smaller loop, find (b) the induced emf and (c) the direction of the induced current. ssm www

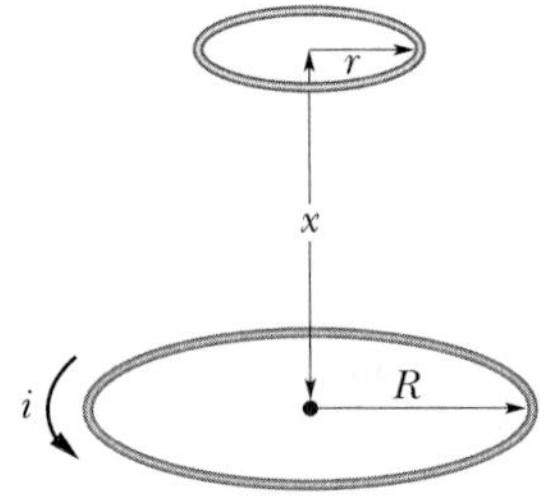

Fig. 31-36 Problem 9.

10P. In Fig. 31-37, a circular loop of wire 10 cm in diameter (seen

edge-on) is placed with its normal $\vec{N}$ at an angle $\theta = 30°$ with the direction of a uniform magnetic field $\vec{B}$ of magnitude 0.50 T. The loop is then rotated such that $\vec{N}$ rotates in a cone about the field direction at the constant rate of 100 rev/min; the angle θ remains unchanged during the process. What is the emf induced in the loop?

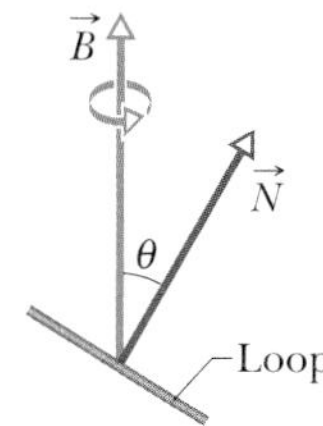

Fig. 31-37 Problem 10.

11P. In Fig. 31-33 let the flux through the loop be $\Phi_B(0)$ at time $t = 0$. Then let the magnetic field $\vec{B}$ vary in a continuous but unspecified way, in both magnitude and direction, so that at time t the flux is represented by $\Phi_B(t)$. (a) Show that the net charge $q(t)$ that has passed through resistor R in time t is

$$q(t) = \frac{1}{R}[\Phi_B(0) - \Phi_B(t)]$$

and is independent of the way $\vec{B}$ has changed. (b) If $\Phi_B(t) = \Phi_B(0)$ in a particular case, we have $q(t) = 0$. Is the induced current necessarily zero throughout the interval from 0 to t? ssm

12P. A small circular loop of area 2.00 cm^2 is placed in the plane of, and concentric with, a large circular loop of radius 1.00 m. The current in the large loop is changed uniformly from 200 A to −200 A (a change in direction) in a time of 1.00 s, beginning at $t = 0$. (a) What is the magnetic field at the center of the small circular loop due to the current in the large loop at $t = 0$, $t = 0.500$ s, and $t = 1.00$ s? (b) What emf is induced in the small loop at $t = 0.500$ s? (Since the inner loop is small, assume the field $\vec{B}$ due to the outer loop is uniform over the area of the smaller loop.)

13P. One hundred turns of insulated copper wire are wrapped around a wooden cylindrical core of cross-sectional area 1.20×10^{-3} m^2. The two terminals are connected to a resistor. The total resistance in the circuit is 13.0 Ω. If an externally applied uniform longitudinal magnetic field in the core changes from 1.60 T in one direction to 1.60 T in the opposite direction, how much charge flows through the circuit? (*Hint:* See Problem 11.) ilw

14P. At a certain place, Earth's magnetic field has magnitude $B = 0.590$ gauss and is inclined downward at an angle of 70.0° to the horizontal. A flat horizontal circular coil of wire with a radius of 10.0 cm has 1000 turns and a total resistance of 85.0 Ω. It is connected to a meter with 140 Ω resistance. The coil is flipped through a half-revolution about a diameter, so that it is again horizontal. How much charge flows through the meter during the flip? (*Hint:* See Problem 11.)

15P. A square wire loop with 2.00 m sides is perpendicular to a uniform magnetic field, with half the area of the loop in the field as shown in Fig. 31-38. The loop contains a 20.0 V battery with negligible internal resistance. If the magnitude of the field varies with time according to $B = 0.0420 - 0.870t$, with B in teslas and t in seconds, what are (a) the net emf in the circuit and (b) the direction of the current through the battery? ssm

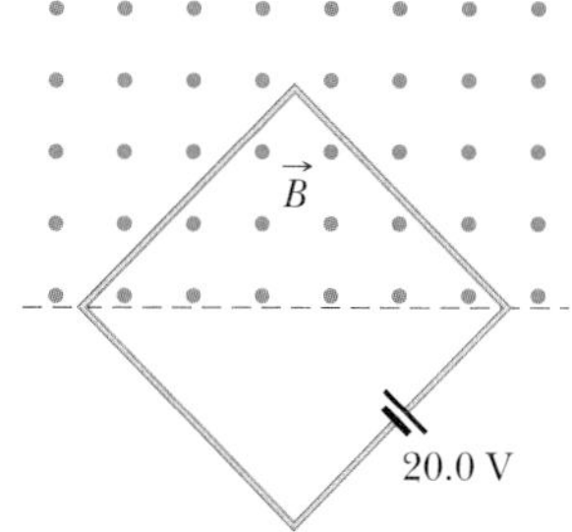

Fig. 31-38 Problem 15.

16P. A wire is bent into three circular segments, each of radius $r = 10$ cm, as shown in Fig. 31-39. Each segment is a quadrant of a circle, ab lying in the xy plane, bc lying in the yz plane, and ca lying in the zx plane. (a) If a uniform magnetic field $\vec{B}$ points in the positive x direction, what is the magnitude of the emf developed in the wire when B increases at the rate of 3.0 mT/s? (b) What is the direction of the current in segment bc?

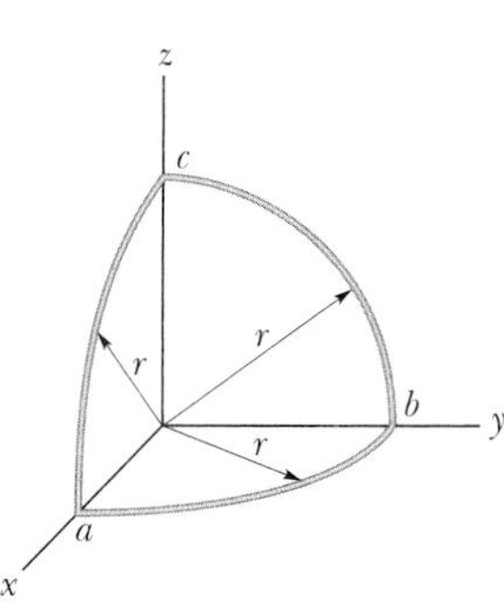

Fig. 31-39 Problem 16.

17P. A rectangular coil of N turns and of length a and width b is rotated at frequency f in a uniform magnetic field $\vec{B}$, as indicated in Fig. 31-40. The coil is connected to co-rotating cylinders, against which metal brushes slide to make contact. (a) Show that the emf induced in the coil is given (as a function of time t) by

$$\mathscr{E} = 2\pi fNabB \sin(2\pi ft) = \mathscr{E}_0 \sin(2\pi ft).$$

This is the principle of the commercial alternating-current generator. (b) Design a loop that will produce an emf with $\mathscr{E}_0 = 150$ V when rotated at 60.0 rev/s in a uniform magnetic field of 0.500 T. ssm www

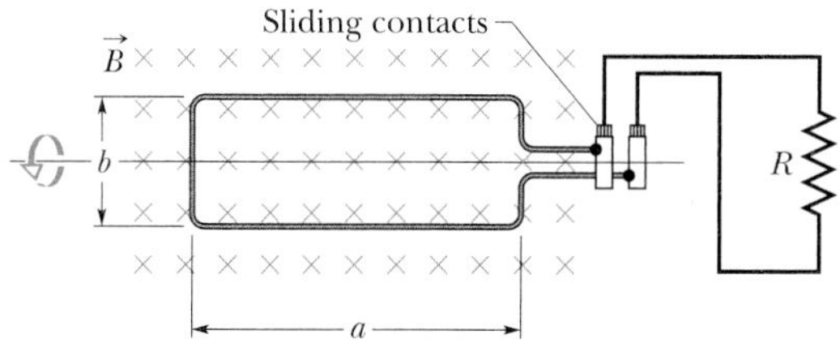

Fig. 31-40 Problem 17.

18P. A stiff wire bent into a semicircle of radius a is rotated with frequency f in a uniform magnetic field, as suggested in Fig. 31-41. What are (a) the frequency and (b) the amplitude of the varying emf induced in the loop?

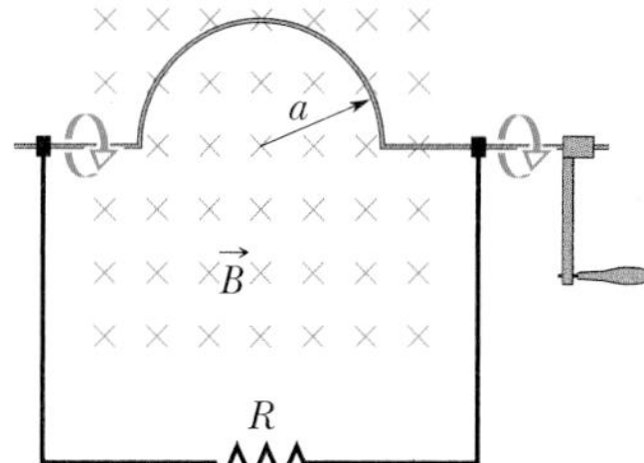

Fig. 31-41 Problem 18

19P. An electric generator consists of 100 turns of wire formed into a rectangular loop 50.0 cm by 30.0 cm, placed entirely in a uniform magnetic field with magnitude $B = 3.50$ T. What is the maximum value of the emf produced when the loop is spun at 1000 rev/min about an axis perpendicular to $\vec{B}$? ilw

20P. In Fig. 31-42, a wire forms a closed circular loop, with radius $R = 2.0$ m and resistance 4.0 Ω. The circle is centered on a long straight wire; at time $t = 0$, the current in the long straight wire is 5.0 A rightward. Thereafter, the current changes according to $i = 5.0\text{ A} - (2.0\text{ A/s}^2)t^2$. (The straight wire is insulated, so there is no electrical contact between it and the wire of the loop.) What are the magnitude and direction of the current induced in the loop at times $t > 0$?

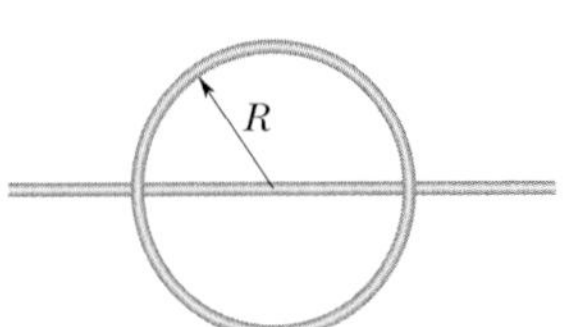

Fig. 31-42 Problem 20.

21P. In Fig. 31-43, the square loop of wire has sides of length 2.0 cm. A magnetic field is directed out of the page; its magnitude is given by $B = 4.0t^2y$, where B is in teslas, t is in seconds, and y is in meters. Determine the emf around the square at $t = 2.5$ s and give its direction. ssm ilw

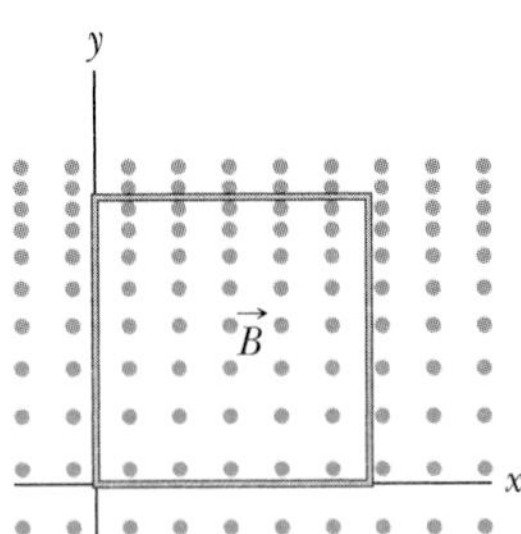

Fig. 31-43 Problem 21.

22P. For the situation shown in Fig. 31-44, $a = 12.0$ cm and $b = 16.0$ cm. The current in the long straight wire is given by $i = 4.50t^2 - 10.0t$, where i is in amperes and t is in seconds. (a) Find the emf in the square loop at $t = 3.00$ s. (b) What is the direction of the induced current in the loop?

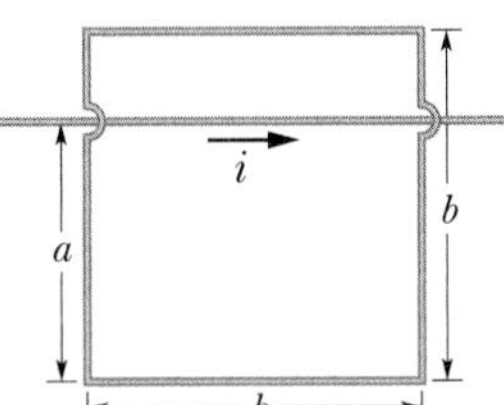

Fig. 31-44 Problem 22.

23P*. Two long, parallel copper wires of diameter 2.5 mm carry currents of 10 A in opposite directions. (a) Assuming that their central axes are 20 mm apart, calculate the magnetic flux per meter of wire that exists in the space between those axes. (b) What fraction of this flux lies inside the wires? (c) Repeat part (a) for parallel currents. ssm

24P. A rectangular loop of wire with length a, width b, and resistance R is placed near an infinitely long wire carrying current i, as shown in Fig. 31-45. The distance from the long wire to the center of the loop is r. Find (a) the magnitude of the magnetic flux through the loop and (b) the current in the loop as it moves away from the long wire with speed v.

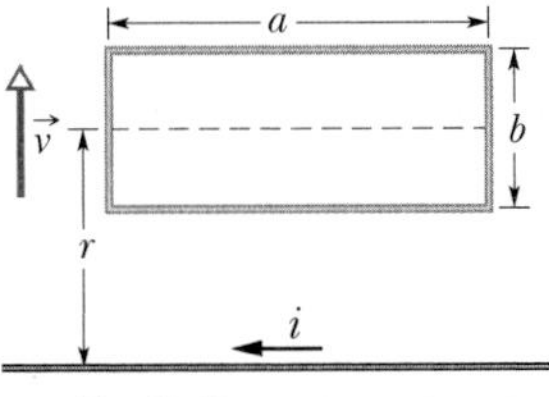

Fig. 31-45 Problem 24.

SEC. 31-5 Induction and Energy Transfers

25E. If 50.0 cm of copper wire (diameter = 1.00 mm) is formed into a circular loop and placed perpendicular to a uniform magnetic field that is increasing at the constant rate of 10.0 mT/s, at what rate is thermal energy generated in the loop? ssm ilw

26E. A loop antenna of area A and resistance R is perpendicular to a uniform magnetic field $\vec{B}$. The field drops linearly to zero in a time interval Δt. Find an expression for the total thermal energy dissipated in the loop.

27E. A metal rod is forced to move with constant velocity $\vec{v}$ along two parallel metal rails, connected with a strip of metal at one end, as shown in Fig. 31-46. A magnetic field $B = 0.350$ T points out of the page. (a) If the rails are separated by 25.0 cm and the speed of the rod is 55.0 cm/s, what emf is generated? (b) If the rod has a resistance of 18.0 Ω and the rails and connector have negligible resistance, what is the current in the rod? (c) At what rate is energy being transferred to thermal energy? ssm

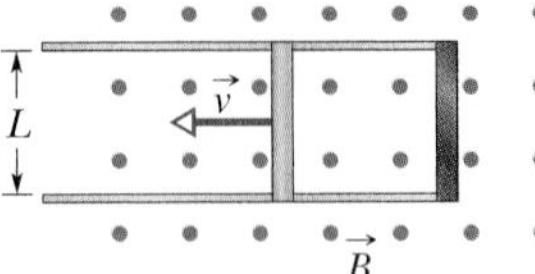

Fig. 31-46 Exercise 27 and Problem 29.

28P. In Fig. 31-47, a long rectangular conducting loop, of width L, resistance R, and mass m, is hung in a horizontal, uniform magnetic field $\vec{B}$ that is directed into the page and that exists only above line aa. The loop is then dropped; during its fall, it accelerates until it reaches a certain terminal speed v_t. Ignoring air drag, find that terminal speed.

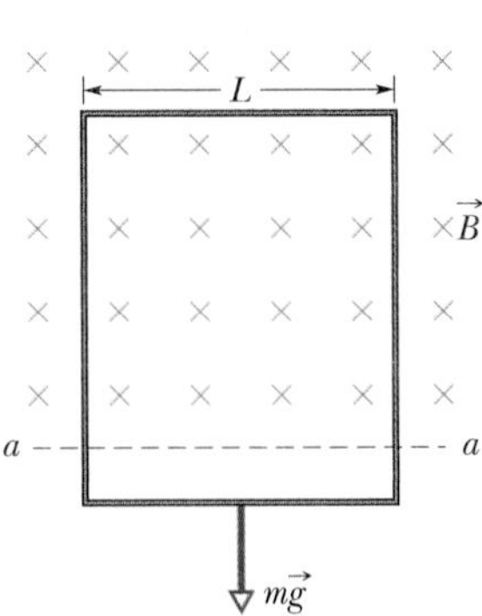

Fig. 31-47 Problem 28.

29P. The conducting rod shown in Fig. 31-46 has length L and is being pulled along horizontal, frictionless conducting rails at a constant velocity $\vec{v}$. The rails are connected at one end with a metal strip. A uniform magnetic field $\vec{B}$, directed out of the page, fills the region in which the rod moves. Assume that $L = 10$ cm, $v = 5.0$ m/s, and $B = 1.2$ T. (a) What are the magnitude and direction of the emf induced in the rod? (b) What is the current in the conducting loop? Assume that the resistance of the rod is 0.40 Ω and that the resistance of the rails and metal strip is negligibly small. (c) At what rate is thermal energy being generated in the rod? (d) What force must be applied to the rod by an external agent to maintain its motion? (e) At what rate does this external agent do work on the rod? Compare this answer with the answer to (c). ssm

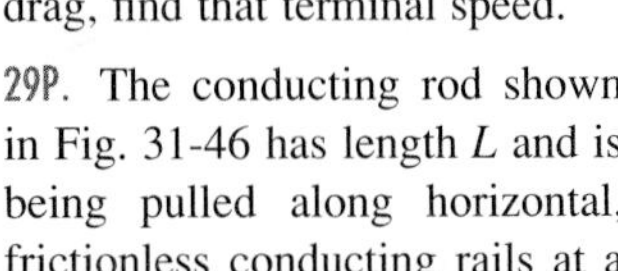

30P. Two straight conducting rails form a right angle where their ends are joined. A conducting bar in contact with the rails starts at the vertex at time $t = 0$ and moves with a constant velocity of 5.20 m/s along them, as shown in Fig. 31-48. A magnetic field with $B = 0.350$ T is directed out of the page. Calculate (a) the flux through the triangle formed by the rails and bar at $t = 3.00$ s and (b) the emf around the triangle at that time. (c) If we write the emf as $\mathscr{E} = at^n$, where a and n are constants, what is the value of n?

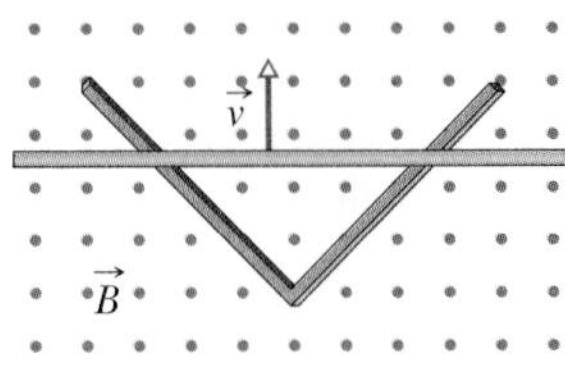

Fig. 31-48 Problem 30.

31P. Figure 31-49 shows a rod of length L caused to move at constant speed v along horizontal conducting rails. The magnetic field in which the rod moves is *not uniform* but is provided by a current

i in a long wire parallel to the rails. Assume that $v = 5.00$ m/s, $a = 10.0$ mm, $L = 10.0$ cm, and $i = 100$ A. (a) Calculate the emf induced in the rod. (b) What is the current in the conducting loop? Assume that the resistance of the rod is 0.400 Ω and that the resistance of the rails and the strip that connects them at the right is negligible. (c) At what rate is thermal energy being generated in the rod? (d) What force must be applied to the rod by an external agent to maintain its motion? (e) At what rate does this external agent do work on the rod? Compare this answer to that for (c). ssm

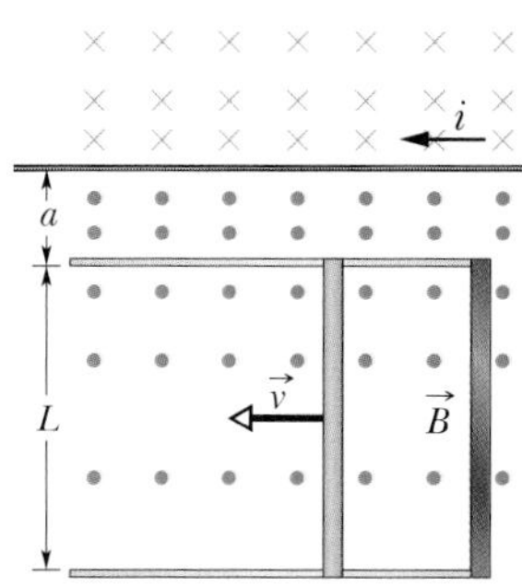

Fig. 31-49 Problem 31.

SEC. 31-6 Induced Electric Fields

32E. Figure 31-50 shows two circular regions R_1 and R_2 with radii $r_1 = 20.0$ cm and $r_2 = 30.0$ cm. In R_1 there is a uniform magnetic field $B_1 = 50.0$ mT into the page, and in R_2 there is a uniform magnetic field $B_2 = 75.0$ mT out of the page (ignore any fringing of these fields). Both fields are decreasing at the rate of 8.50 mT/s. Calculate the integral $\oint \vec{E} \cdot d\vec{s}$ for each of the three dashed paths.

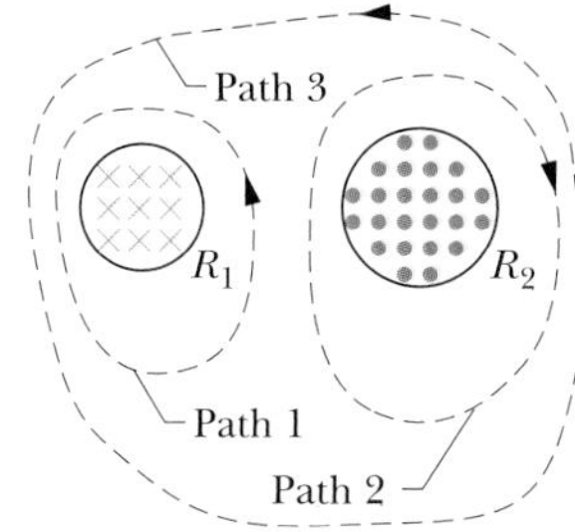

Fig. 31-50 Exercise 32.

33E. A long solenoid has a diameter of 12.0 cm. When a current i exists in its windings, a uniform magnetic field $B = 30.0$ mT is produced in its interior. By decreasing i, the field is caused to decrease at the rate of 6.50 mT/s. Calculate the magnitude of the induced electric field (a) 2.20 cm and (b) 8.20 cm from the axis of the solenoid. ssm ilw

34P. Early in 1981 the Francis Bitter National Magnet Laboratory at M.I.T. commenced operation of a 3.3-cm-diameter cylindrical magnet, which produces a 30 T field, then the world's largest steady-state field. The field can be varied sinusoidally between the limits of 29.6 and 30.0 T at a frequency of 15 Hz. When this is done, what is the maximum value of the induced electric field at a radial distance of 1.6 cm from the axis? (*Hint:* See Sample Problem 31-4.)

35P. Prove that the electric field $\vec{E}$ in a charged parallel-plate capacitor cannot drop abruptly to zero (as is suggested at point a in Fig. 31-51), as one moves perpendicular to the field, say, along the horizontal arrow in the figure. Fringing of the field lines always occurs in actual capacitors, which means that $\vec{E}$ approaches zero in a continuous and gradual way (see Problem 33 in Chapter 30). (*Hint:* Apply Faraday's law to the rectangular path shown by the dashed lines.) ssm

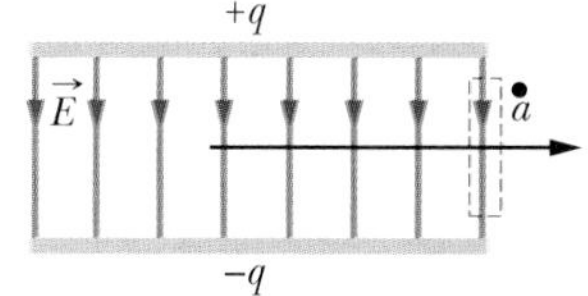

Fig. 31-51 Problem 35.

SEC. 31-7 Inductors and Inductance

36E. A circular coil has a 10.0 cm radius and consists of 30.0 closely wound turns of wire. An externally produced magnetic field of 2.60 mT is perpendicular to the coil. (a) If no current is in the coil, what magnetic flux links its turns? (b) When the current in the coil is 3.80 A in a certain direction, the net flux through the coil is found to vanish. What is the inductance of the coil?

37E. The inductance of a close-packed coil of 400 turns is 8.0 mH. Calculate the magnetic flux through the coil when the current is 5.0 mA. ssm

38P. A wide copper strip of width W is bent to form a tube of radius R with two parallel planar extensions, as shown in Fig. 31-52. There is a current i through the strip, distributed uniformly over its width. In this way a "one-turn solenoid" is formed. (a) Derive an expression for the magnitude of the magnetic field $\vec{B}$ in the tubular part (far away from the edges). (*Hint:* Assume that the magnetic field outside this one-turn solenoid is negligibly small.) (b) Find the inductance of this one-turn solenoid, neglecting the two planar extensions.

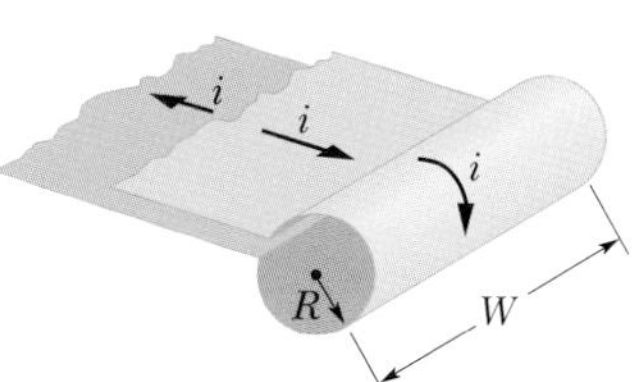

Fig. 31-52 Problem 38.

39P. Two long parallel wires, both of radius a and whose centers are a distance d apart, carry equal currents in opposite directions. Show that, neglecting the flux within the wires, the inductance of a length l of such a pair of wires is given by

$$L = \frac{\mu_0 l}{\pi} \ln \frac{d - a}{a}.$$

(*Hint:* Calculate the flux through a rectangle of which the wires form two opposite sides.) ssm www

SEC. 31-8 Self-Induction

40E. At a given instant the current and self-induced emf in an inductor are directed as indicated in Fig. 31-53. (a) Is the current increasing or decreasing? (b) The induced emf is 17 V and the rate of change of the current is 25 kA/s; find the inductance.

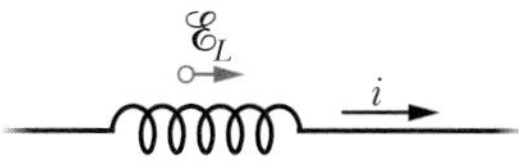

Fig. 31-53 Exercise 40.

41E. A 12 H inductor carries a steady current of 2.0 A. How can a 60 V self-induced emf be made to appear in the inductor? ssm

42P. The current i through a 4.6 H inductor varies with time t as shown by the graph of Fig. 31-54. The inductor has a resistance of

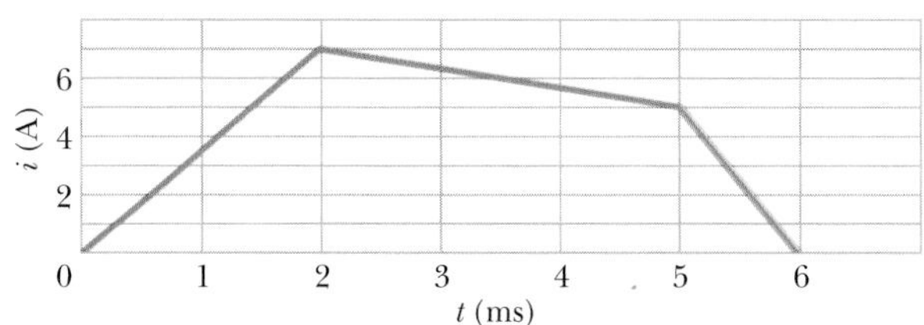

Fig. 31-54 Problem 42.

12 Ω. Find the magnitude of the induced emf $\mathscr{E}$ during the time intervals (a) $t = 0$ to $t = 2$ ms, (b) $t = 2$ ms to $t = 5$ ms, (c) $t = 5$ ms to $t = 6$ ms. (Ignore the behavior at the ends of the intervals.)

43P. *Inductors in series.* Two inductors L_1 and L_2 are connected in series and are separated by a large distance. (a) Show that the equivalent inductance is given by

$$L_{eq} = L_1 + L_2.$$

(*Hint:* Review the derivations for resistors in series and capacitors in series. Which is similar here?) (b) Why must their separation be large for this relationship to hold? (c) What is the generalization of (a) for N inductors in series?

44P. *Inductors in parallel.* Two inductors L_1 and L_2 are connected in parallel and separated by a large distance. (a) Show that the equivalent inductance is given by

$$\frac{1}{L_{eq}} = \frac{1}{L_1} + \frac{1}{L_2}.$$

(*Hint:* Review the derivations for resistors in parallel and capacitors in parallel. Which is similar here?) (b) Why must their separation be large for this relationship to hold? (c) What is the generalization of (a) for N inductors in parallel?

SEC. 31-9 *RL* Circuits

45E. In terms of τ_L, how long must we wait for the current in an *RL* circuit to build up to within 0.100% of its equilibrium value? ssm

46E. The current in an *RL* circuit builds up to one-third of its steady-state value in 5.00 s. Find the inductive time constant.

47E. The current in an *RL* circuit drops from 1.0 A to 10 mA in the first second following removal of the battery from the circuit. If L is 10 H, find the resistance R in the circuit. ilw

48E. Consider the *RL* circuit of Fig. 31-17. In terms of the battery emf $\mathscr{E}$, (a) what is the self-induced emf $\mathscr{E}_L$ when the switch has just been closed on a, and (b) what is $\mathscr{E}_L$ when $t = 2.0\tau_L$? (c) In terms of τ_L, when will $\mathscr{E}_L$ be just one-half the battery emf $\mathscr{E}$?

49E. A solenoid having an inductance of 6.30 μH is connected in series with a 1.20 kΩ resistor. (a) If a 14.0 V battery is switched across the pair, how long will it take for the current through the resistor to reach 80.0% of its final value? (b) What is the current through the resistor at time $t = 1.0\tau_L$? ssm

50P. Suppose the emf of the battery in the circuit of Fig. 31-18 varies with time t so that the current is given by $i(t) = 3.0 + 5.0t$, where i is in amperes and t is in seconds. Take $R = 4.0$ Ω and $L = 6.0$ H, and find an expression for the battery emf as a function of time. (*Hint:* Apply the loop rule.)

51P. At time $t = 0$, a 45.0 V potential difference is suddenly applied to a coil with $L = 50.0$ mH and $R = 180$ Ω. At what rate is the current increasing at $t = 1.20$ ms? ilw

52P. A wooden toroidal core with a square cross section has an inner radius of 10 cm and an outer radius of 12 cm. It is wound with one layer of wire (of diameter 1.0 mm and resistance per meter 0.020 Ω/m). What are (a) the inductance and (b) the inductive time constant of the resulting toroid? Ignore the thickness of the insulation on the wire.

53P. In Fig. 31-55, $\mathscr{E} = 100$ V, $R_1 = 10.0$ Ω, $R_2 = 20.0$ Ω, $R_3 = 30.0$ Ω, and $L = 2.00$ H. Find the values of i_1 and i_2 (a) immediately after the closing of switch S, (b) a long time later, (c) immediately after the reopening of switch S, and (d) a long time after the reopening. ssm

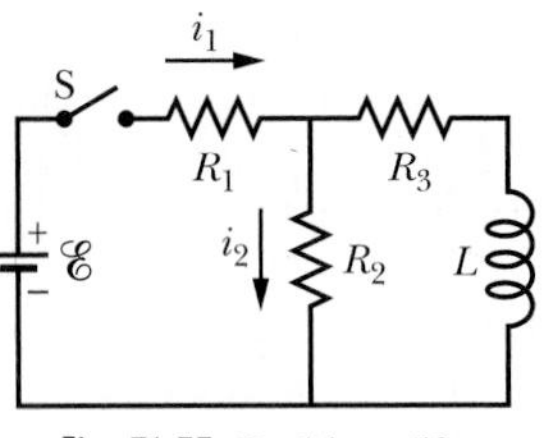

Fig. 31-55 Problem 53.

54P. In the circuit of Fig. 31-56, $\mathscr{E} = 10$ V, $R_1 = 5.0$ Ω, $R_2 = 10$ Ω, and $L = 5.0$ H. For the two separate conditions (I) switch S just closed and (II) switch S closed for a long time, calculate (a) the current i_1 through R_1, (b) the current i_2 through R_2, (c) the current i through the switch, (d) the potential difference across R_2, (e) the potential difference across L, and (f) the rate of change di_2/dt.

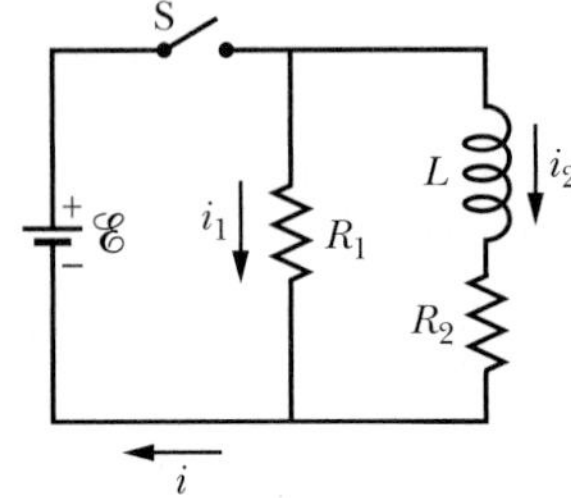

Fig. 31-56 Problem 54.

55P*. In the circuit shown in Fig. 31-57, switch S is closed at time $t = 0$. Thereafter, the constant current source, by varying its emf, maintains a constant current i out of its upper terminal. (a) Derive an expression for the current through the inductor as a function of time. (b) Show that the current through the resistor equals the current through the inductor at time $t = (L/R) \ln 2$. ssm www

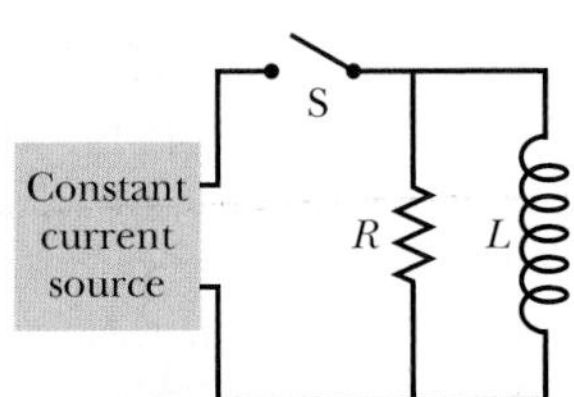

Fig. 31-57 Problem 55.

SEC. 31-10 Energy Stored in a Magnetic Field

56E. Consider the circuit of Fig. 31-18. In terms of the inductive time constant, at what instant after the battery is connected will the energy stored in the magnetic field of the inductor be half its steady-state value?

57E. Suppose that the inductive time constant for the circuit of Fig. 31-18 is 37.0 ms and the current in the circuit is zero at time $t = 0$. At what time does the rate at which energy is dissipated in the resistor equal the rate at which energy is being stored in the inductor? ilw

58E. A coil with an inductance of 2.0 H and a resistance of 10 Ω is suddenly connected to a resistanceless battery with $\mathscr{E} = 100$ V. At 0.10 s after the connection is made, what are the rates at which (a) energy is being stored in the magnetic field, (b) thermal energy is appearing in the resistance, and (c) energy is being delivered by the battery?

59P. A coil is connected in series with a 10.0 kΩ resistor. A 50.0 V battery is applied across the two devices, and the current reaches a value of 2.00 mA after 5.00 ms. (a) Find the inductance of the coil. (b) How much energy is stored in the coil at this same moment? ssm

60P. For the circuit of Fig. 31-18, assume that $\mathscr{E} = 10.0$ V, $R =$

6.70 Ω, and $L = 5.50$ H. The battery is connected at time $t = 0$. (a) How much energy is delivered by the battery during the first 2.00 s? (b) How much of this energy is stored in the magnetic field of the inductor? (c) How much of this energy is dissipated in the resistor?

61P. Prove that, after switch S in Fig. 31-17 has been thrown from a to b, all the energy stored in the inductor will ultimately appear as thermal energy in the resistor. ssm

SEC. 31-11 Energy Density of a Magnetic Field

62E. A toroidal inductor with an inductance of 90.0 mH encloses a volume of 0.0200 m^3. If the average energy density in the toroid is 70.0 J/m^3, what is the current through the inductor?

63E. A solenoid that is 85.0 cm long has a cross-sectional area of 17.0 cm^2. There are 950 turns of wire carrying a current of 6.60 A. (a) Calculate the energy density of the magnetic field inside the solenoid. (b) Find the total energy stored in the magnetic field there (neglect end effects). ssm

64E. The magnetic field in the interstellar space of our galaxy has a magnitude of about 10^{-10} T. How much energy is stored in this field in a cube 10 light-years on edge? (For scale, note that the nearest star is 4.3 light-years distant and the radius of our galaxy is about 8×10^4 light-years.)

65E. What must be the magnitude of a uniform electric field if it is to have the same energy density as that possessed by a 0.50 T magnetic field? ilw

66E. A circular loop of wire 50 mm in radius carries a current of 100 A. (a) Find the magnetic field strength at the center of the loop. (b) Calculate the energy density at the center of the loop.

67P. A length of copper wire carries a current of 10 A, uniformly distributed through its cross section. Calculate the energy density of (a) the magnetic field and (b) the electric field at the surface of the wire. The wire diameter is 2.5 mm, and its resistance per unit length is 3.3 Ω/km. ssm www

SEC. 31-12 Mutual Induction

68E. Coil 1 has $L_1 = 25$ mH and $N_1 = 100$ turns. Coil 2 has $L_2 = 40$ mH and $N_2 = 200$ turns. The coils are rigidly positioned with respect to each other; their mutual inductance M is 3.0 mH. A 6.0 mA current in coil 1 is changing at the rate of 4.0 A/s. (a) What magnetic flux Φ_{12} links coil 1, and what self-induced emf appears there? (b) What magnetic flux Φ_{21} links coil 2, and what mutually induced emf appears there?

69E. Two coils are at fixed locations. When coil 1 has no current and the current in coil 2 increases at the rate 15.0 A/s, the emf in coil 1 is 25.0 mV. (a) What is their mutual inductance? (b) When coil 2 has no current and coil 1 has a current of 3.60 A, what is the flux linkage in coil 2? ssm

70E. Two solenoids are part of the spark coil of an automobile. When the current in one solenoid falls from 6.0 A to zero in 2.5 ms, an emf of 30 kV is induced in the other solenoid. What is the mutual inductance M of the solenoids?

71P. Two coils, connected as shown in Fig. 31-58, separately have inductances L_1 and L_2. Their mutual inductance is M. (a) Show that this combination can be replaced by a single coil of equivalent inductance given by

$$L_{eq} = L_1 + L_2 + 2M.$$

(b) How could the coils in Fig. 31-58 be reconnected to yield an equivalent inductance of

$$L_{eq} = L_1 + L_2 - 2M?$$

(This problem is an extension of Problem 43, but the requirement that the coils be far apart has been removed.) ssm

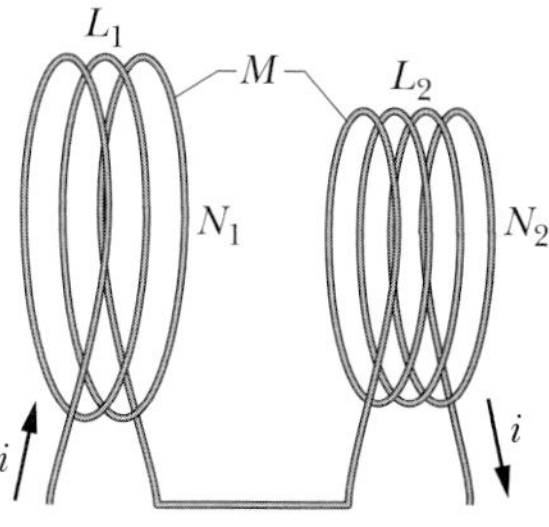

Fig. 31-58 Problem 71.

72P. A coil C of N turns is placed around a long solenoid S of radius R and n turns per unit length, as in Fig. 31-59. Show that the mutual inductance for the coil–solenoid combination is given by $M = \mu_0 \pi R^2 n N$. Explain why M does not depend on the shape, size, or possible lack of close-packing of the coil.

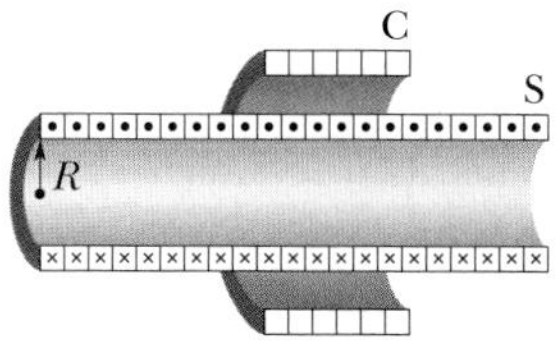

Fig. 31-59 Problem 72.

73P. Figure 31-60 shows, in cross section, two coaxial solenoids. Show that the mutual inductance M for a length l of this solenoid–solenoid combination is given by $M = \pi R_1^2 l \mu_0 n_1 n_2$, in which n_1 and n_2 are the respective numbers of turns per unit length and R_1 is the radius of the inner solenoid. Why does M depend on R_1 and not on R_2? ssm

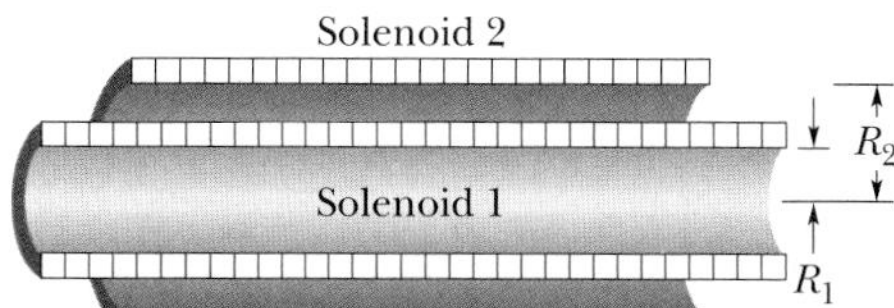

Fig. 31-60 Problem 73.

74P. Figure 31-61 shows a coil of N_2 turns wound as shown around part of a toroid of N_1 turns. The toroid's inner radius is a, its outer radius is b, and its height is h. Show that the mutual inductance M for the toroid–coil combination is

$$M = \frac{\mu_0 N_1 N_2 h}{2\pi} \ln \frac{b}{a}.$$

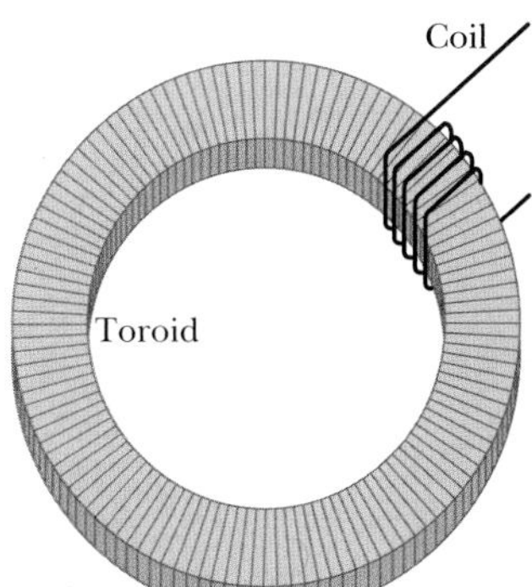

Fig. 31-61 Problem 74.

75P. A rectangular loop of N close-packed turns is positioned near a long straight wire as shown in Fig. 31-62. (a) What is the mutual inductance M for the loop–wire combination? (b) Evaluate M for $N = 100$, $a = 1.0$ cm, $b = 8.0$ cm, and $l = 30$ cm. ilw

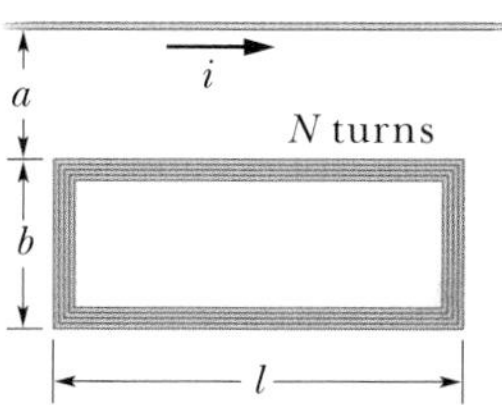

Fig. 31-62 Problem 75.

NEW PROBLEMS

N1. Once the switch S is closed in Fig. 31N-1, the time required for the current to reach any obtainable value depends, in part, on the value of resistance R. Suppose the emf $\mathscr{E}$ of the ideal battery is 12 V and the inductance of the ideal (resistanceless) inductor is 18 mH. How much time is needed for the current to reach 2.00 A if R is (a) 1.00 Ω, (b) 5.00 Ω, and (c) 6.00 Ω? (d) Why is there such a huge jump between the answers to (b) and (c)? (e) For what value of R is the time required for the current to reach 2.00 A least? (f) What is that least time? (*Hint:* Rethink Eq. 31-41.)

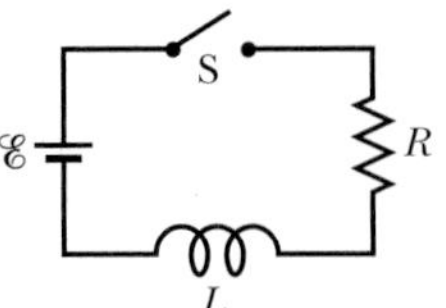

Fig. 31N-1 Problems N1, N4, N6, N18.

N2. Figure 31N-2a shows a circuit consisting of an ideal battery with emf $\mathscr{E}$ = 6.00 μV, a resistance R, and a small wire loop of area 5.0 cm^2. For the time interval t = 10 to t = 20 s, an external magnetic field is set up throughout the loop. The field is uniform, its direction is into the page in Fig. 31N-2a, and the field magnitude is given by $B = at$, where B is in teslas, a is a constant, and t is in seconds. Figure 31N-2b gives the current i in the circuit before, during, and after the external field is set up. Find a.

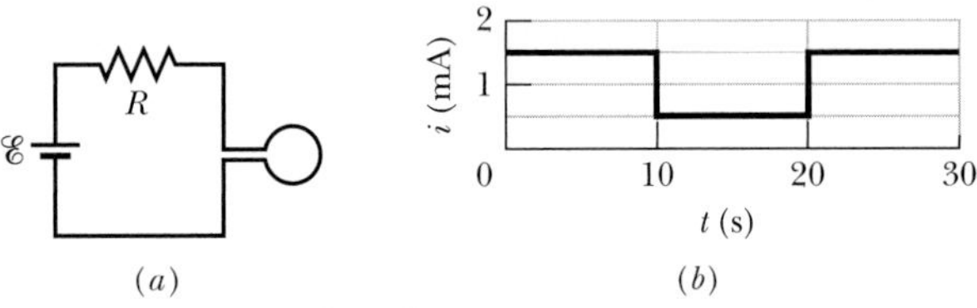

Fig. 31N-2 Problem N2.

N3. In Fig. 31N-3a, a uniform magnetic field $\vec{B}$ increases in magnitude with time t as given by Fig. 31N-3b. A circular conducting loop of area 8.0 × 10^{-4} m^2 lies in the field, in the plane of the page. The amount of charge q passing point A on the loop is given in Fig. 31N-3c as a function of t. What is the loop's resistance?

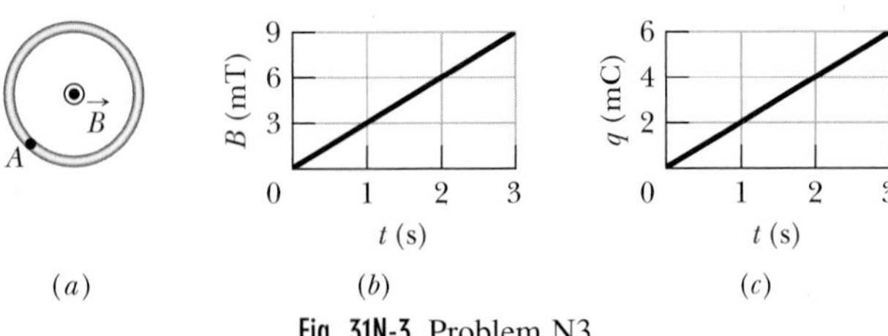

Fig. 31N-3 Problem N3.

N4. In Fig. 31N-1, a 12.0 V ideal battery, a 20 Ω resistor, and an inductor are connected by a switch at time t = 0. At what rate is the battery transferring energy to the inductor's field at t = 1.61τ_L?

N5. A rectangular loop (area = 0.15 m^2) turns in a uniform magnetic field, B = 0.20 T. When the angle between the field and the normal to the plane of the loop is $\pi/2$ rad and increasing at 0.60 rad/s, what emf is induced in the loop?

N6. In Fig. 31N-1, the inductor has 25 turns and the ideal battery has an emf of 16 V. Figure 31N-4 gives the magnetic flux Φ through each turn versus the current i through the inductor. If switch S is closed at time t = 0, at what rate di/dt will the current be changing at t = 1.5τ_L?

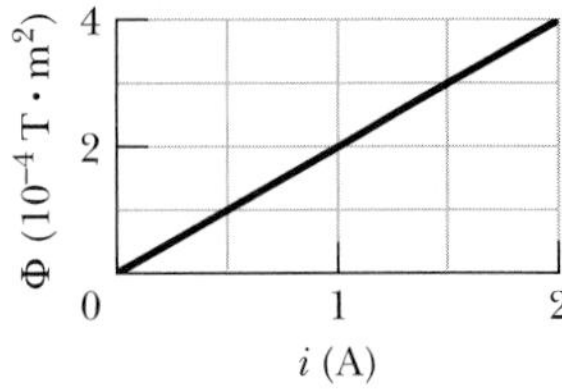

Fig. 31N-4 Problem N6.

N7. A square loop of wire is held in a uniform, magnetic field 0.24 T directed perpendicularly to the plane of the loop. The length of each side of the square is decreasing at a constant rate of 5.0 cm/s. What emf is induced in the loop when the length is 12 cm?

N8. In Fig. 31N-5a, a circular loop of wire is concentric with a solenoid and lies in a plane that is perpendicular to the solenoid's central axis. The loop has radius 6.00 cm. The solenoid has radius 2.00 cm, consists of 8000 turns per meter, and has a current i_{sol} that varies with time t as given in Fig. 31N-5b. Figure 31N-5c shows, as a function of time, the energy E_{th} that is transferred to thermal energy of the loop. What is the loop's resistance?

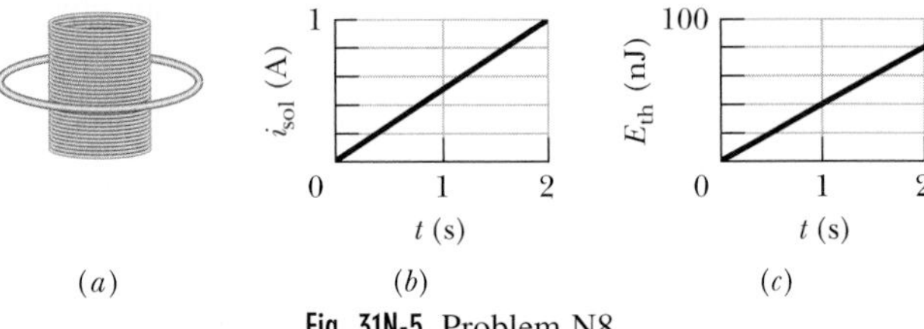

Fig. 31N-5 Problem N8.

N9. Figure 31N-6a shows a wire that forms a rectangle and which has a resistance of 5.0 mΩ. Its interior is split into three equal areas with different magnetic fields $\vec{B}_1$, $\vec{B}_2$, and $\vec{B}_3$ that are either directly out of or into the page, as indicated. The fields are uniform within each region. Figure 31N-6b gives the change in the z components B_z of the three fields with time t. What are the magnitude and direction of the current induced in the wire?

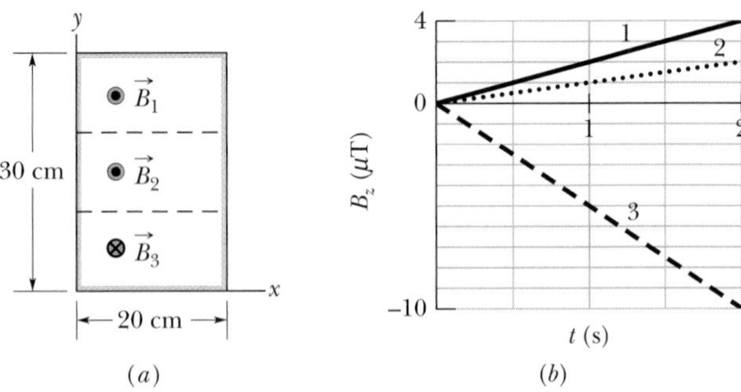

Fig. 31N-6 Problem N9.

N10. Figure 31N-7a shows, in cross section, two wires that are straight, parallel, and very long. The ratio i_1/i_2 of the current carried by wire 1 to that carried by wire 2 is 1/3. Wire 1 is fixed in place to the left of the origin. Wire 2 can be moved along the positive side of the x axis so as to change the magnetic energy density u_B set up by the two currents at the origin. Figure 31N-7b gives that energy density u_B as a function of the position x of wire 2. The curve has an asymptote of u_B = 1.96 nJ/m^3 as $x \to \infty$. What are the values of (a) current i_1 and (b) current i_2?

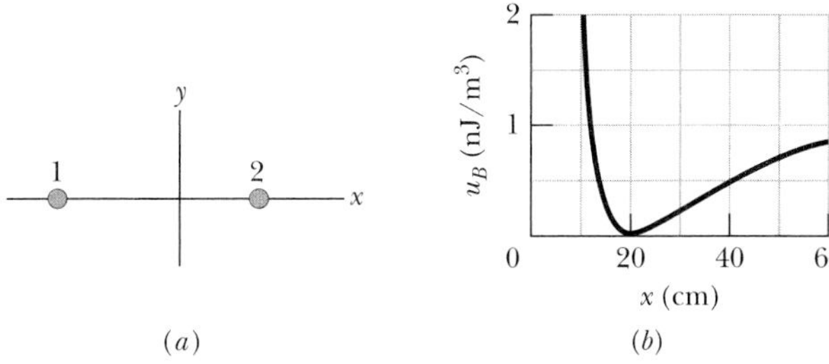

Fig. 31N-7 Problem N10.

N11. The inductance of a closely wound coil is such that an emf of 3.0 mV is induced when the current changes at the rate of 5.0 A/s. A steady current of 8.0 A produces a magnetic flux of 40 μWb through each turn. (a) Calculate the inductance of the coil. (b) How many turns does the coil have?

N12. Figure 31N-8a shows two concentric circular regions in which uniform magnetic fields can change. Region 1, with radius $r_1 =$ 1.0 cm, has an outward magnetic field $\vec{B}_1$ that is increasing in magnitude. Region 2, with radius $r_2 = 2.0$ cm, has an outward magnetic field $\vec{B}_2$ that may also be changing. Imagine that a conducting ring of radius R is centered on the two regions and then the emf $\mathscr{E}$ around the ring is determined. Figure 31N-8b gives emf $\mathscr{E}$ as a function of the square R^2 of the ring's radius, to the outer edge of region 2. What are the rates (a) dB_1/dt and (b) dB_2/dt? (c) Is the magnitude of $\vec{B}_2$ increasing, decreasing, or remaining constant?

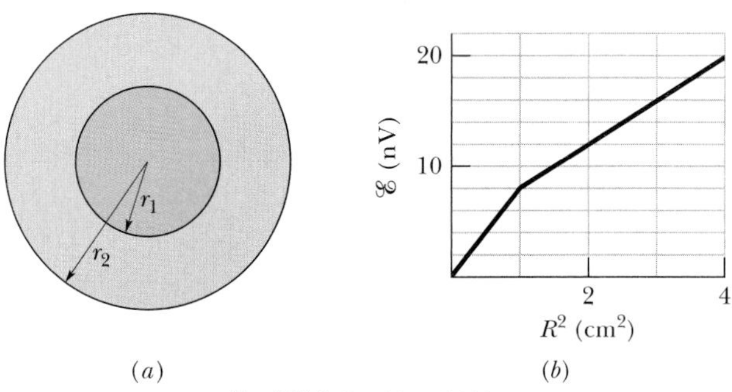

Fig. 31N-8 Problem N12.

N13. At time $t = 0$, a 45 V potential difference is suddenly applied to the leads of a coil with inductance $L = 50$ mH and resistance $R = 180\ \Omega$. At what rate is the current through the coil increasing at $t = 1.2$ ms?

N14. A coil with 150 turns has a flux of 50.0 nT · m^2 through each turn when the current is 2.00 mA. (a) What is the inductance of the coil? What are (b) the inductance and (c) the flux through each turn when the current is increased to 4.00 mA? (d) What is the maximum emf $\mathscr{E}$ across the coil when the current through it is given by $i = (3.0 \text{ mA}) \cos(377t)$, with t in seconds?

N15. In Fig. 31N-9a, switch S has been closed on A long enough to establish a steady current in the inductor of inductance $L_1 =$ 5.00 mH and the resistor of resistance $R_1 = 25\ \Omega$. Similarly, in Fig. 31N-9b, switch S has been closed on A long enough to establish a steady current in the inductor of inductance $L_2 = 3.00$ mH and the resistor of resistance $R_2 = 30\ \Omega$. The ratio Φ_{02}/Φ_{01} of the magnetic flux through a turn in inductor 2 to that in inductor 1 is 1.5. At time $t = 0$, the two switches are closed on B. At what time t is the flux through a turn in the two inductors equal?

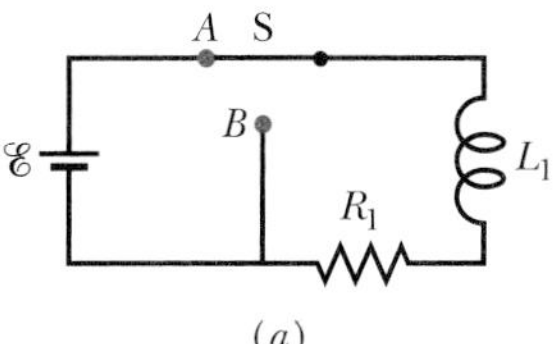

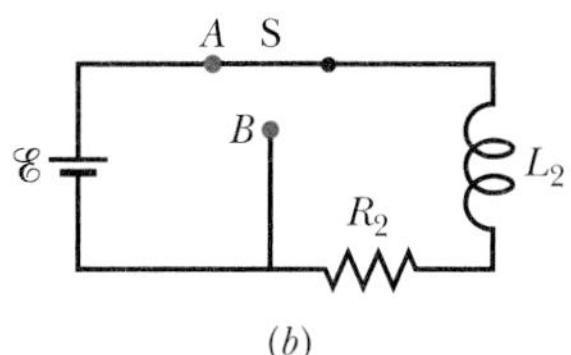

Fig. 31N-9 Problem N15.

N16. A circular region in the xy plane is penetrated by a uniform magnetic field in the positive direction of the z axis. The field's magnitude B (in teslas) increases with time t (in seconds) according to $B = at$, where a is a constant. The magnitude E of the electric field set up by that increase in the magnetic field is given by Fig. 31N-10 as a function of the distance r from the center of the region. Find a.

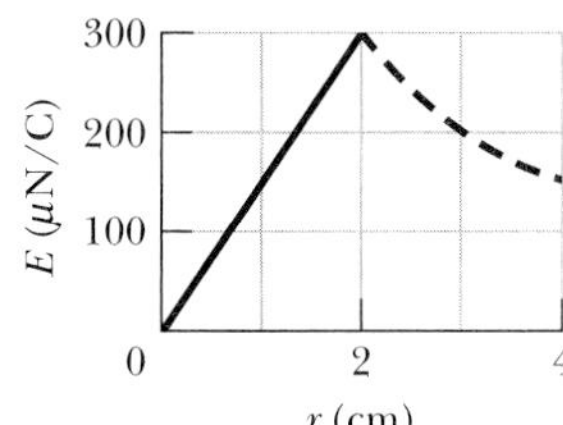

Fig. 31N-10 Problem N16.

N17. At time $t = 0$, a 12 V potential difference is suddenly applied to the leads of a coil of inductance 23.0 mH and a certain resistance R. At time $t = 0.150$ ms, the current through the inductor is changing at the rate of 280 A/s. Evaluate R.

N18. Switch S in Fig. 31N-1 is closed at time $t = 0$, initiating the buildup of current in the 15.0 mH inductor and the 20.0 Ω resistor. At what time is the emf across the inductor equal to the potential difference across the resistor?

N19. A uniform magnetic field $\vec{B}$ is perpendicular to the plane of a circular wire loop of radius r. The magnitude of the field varies with time according to $B = B_0 e^{-t/\tau}$, where B_0 and τ are constants. Find the emf in the loop as a function of time.

N20. A wire rectangle lies in an xy plane: $x = 0$ to 40.0 cm, $y = 0$ to 25.0 cm. What are the magnitude and direction of the emfs $\mathscr{E}$ induced around the wire for the following magnetic fields, with $\vec{B}$ in teslas, x and y in meters, and t in seconds:

(a) $\vec{B} = (4.00 \times 10^{-2})y\hat{k}$, (b) $\vec{B} = (6.00 \times 10^{-2})t\hat{k}$,
(c) $\vec{B} = (8.00 \times 10^{-2})yt\hat{k}$, (d) $\vec{B} = (3.00 \times 10^{-2})xt\hat{j}$,
(e) $\vec{B} = (5.00 \times 10^{-2})yt\hat{i}$.

N21. Figure 31N-11a shows a rectangular conducting loop of resistance $R = 0.020\ \Omega$, height $H = 1.5$ cm, and length $D = 2.5$ cm being pulled at constant speed $v = 40$ cm/s through two regions of uniform magnetic field. Figure 31N-11b gives the current i induced in the loop as a function of the position x of the right side of the loop. For example, a current of 3.0 μA is induced clockwise as the loop enters region 1. What are the magnitudes and directions of the magnetic field in (a) region 1 and (b) region 2?

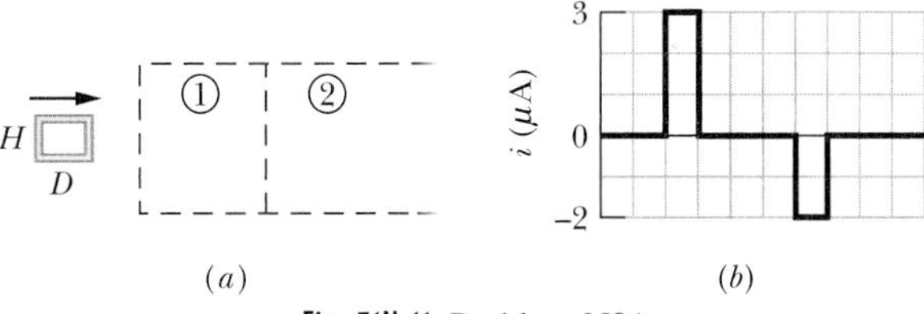

Fig. 31N-11 Problem N21.

32 Magnetism of Matter: Maxwell's Equations

This is an overhead view of a frog that is being levitated in a magnetic field produced by current in a vertical solenoid below the frog. The solenoid's upward magnetic force on the frog balances the downward gravitational force on the frog. (The frog is not in discomfort; the sensation is like floating in water, which frogs like very much.) However, a frog is not magnetic (it would not, for example, stick to a refrigerator door).

How, then, can there be a magnetic force on the frog?

The answer is in this chapter.

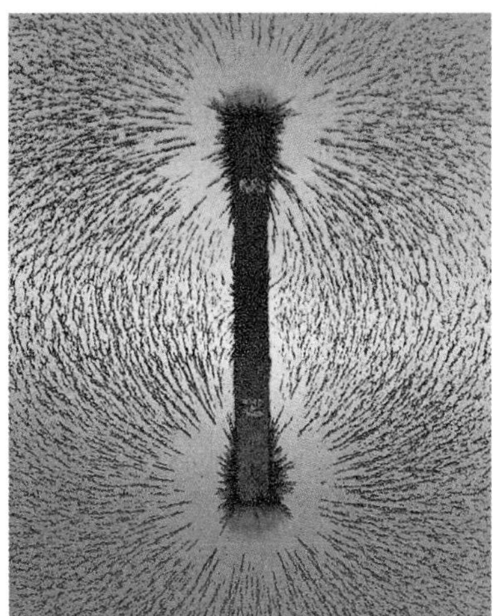

Fig. 32-1 A bar magnet is a magnetic dipole. The iron filings suggest the magnetic field lines. (The background is illuminated with colored light.)

32-1 Magnets

The first known magnets were *lodestones*, which are stones that have been *magnetized* (made magnetic) naturally. When the ancient Greeks and ancient Chinese discovered these rare stones, they were amused by the stones' ability to attract metal over a short distance, as if by magic. Only much later did they learn to use lodestones (and artificially magnetized pieces of iron) in compasses to determine direction.

Today, magnets and magnetic materials are ubiquitous. We find them in VCRs, audio cassettes, ATM and credit cards, audio headsets, and even in the inks for paper money. In fact, some breakfast cereals that are "iron fortified" contain small bits of magnetic materials (you can collect them from a slurry of cereal and water with a magnet). More important, the modern electronics industry as we know it (including the music and information sectors) would not exist without magnetic materials.

The magnetic properties of materials can be traced back to their atoms and electrons. We begin here, however, with the bar magnet in Fig. 32-1. As you have seen, iron filings sprinkled around such a magnet tend to align with the magnetic field of the magnet, and their pattern reveals the magnetic field lines. The clustering of the lines at the ends of the magnet suggests that one end is a *source* of the lines (the field diverges from it) and the other end is a *sink* of the lines (the field converges toward it). By convention, we call the source the *north pole* of the magnet and the opposite end the *south pole,* and we say that the magnet, with its two poles, is an example of a **magnetic dipole.**

Suppose we break apart a bar magnet the way we break a piece of chalk (Fig. 32-2). We should, it seems, be able to isolate a single pole, or *monopole*. However, we cannot—not even if we break the magnet down to its individual atoms and then to its electrons and nuclei. Each fragment has a north pole and a south pole. Thus:

> The simplest magnetic structure that can exist is a magnetic dipole. Magnetic monopoles do not exist (as far as we know).

32-2 Gauss' Law for Magnetic Fields

Gauss' law for magnetic fields is a formal way of saying that magnetic monopoles do not exist. The law asserts that the net magnetic flux Φ_B through any closed Gaussian surface is zero:

$$\Phi_B = \oint \vec{B} \cdot d\vec{A} = 0 \quad \text{(Gauss' law for magnetic fields).} \tag{32-1}$$

Contrast this with Gauss' law for electric fields,

$$\Phi_E = \oint \vec{E} \cdot d\vec{A} = \frac{q_{\text{enc}}}{\varepsilon_0} \quad \text{(Gauss' law for electric fields).}$$

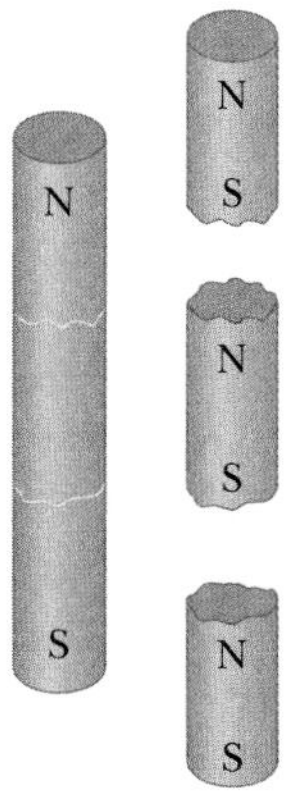

Fig. 32-2 If you break a magnet, each fragment becomes a separate magnet, with its own north and south poles.

In both equations, the integral is taken over a *closed* Gaussian surface. Gauss' law for electric fields says that this integral (the net electric flux through the surface) is proportional to the net electric charge q_{enc} enclosed by the surface. Gauss' law for magnetic fields says that there can be no net magnetic flux through the surface because there can be no net "magnetic charge" (individual magnetic poles) enclosed by the surface. The simplest magnetic structure that can exist and thus be enclosed by a Gaussian surface is a dipole, which consists of both a source and a sink for the field lines. Thus, there must always be as much magnetic flux into the surface as out of it, and the net magnetic flux must always be zero.

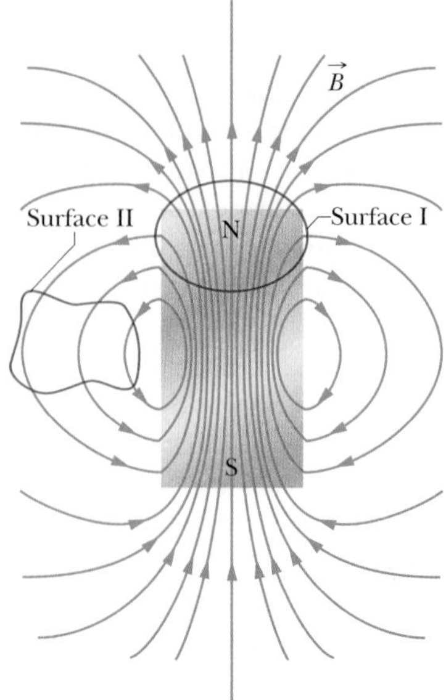

Fig. 32-3 The field lines for the magnetic field $\vec{B}$ of a short bar magnet. The red curves represent cross sections of closed, three-dimensional Gaussian surfaces.

Gauss' law for magnetic fields holds for more complicated structures than a magnetic dipole, and it holds even if the Gaussian surface does not enclose the entire structure. Gaussian surface II near the bar magnet of Fig. 32-3 encloses no poles, and we can easily conclude that the net magnetic flux through it is zero. Gaussian surface I is more difficult. It may seem to enclose only the north pole of the magnet because it encloses the label N and not the label S. However, a south pole must be associated with the lower boundary of the surface, because magnetic field lines enter the surface there. (The enclosed section is like one piece of the broken bar magnet in Fig. 32-2.) Thus, Gaussian surface I encloses a magnetic dipole and the net flux through the surface is zero.

CHECKPOINT 1: The figure here shows four closed surfaces with flat top and bottom faces and curved sides. The table gives the areas A of the faces and the magnitudes B of the uniform and perpendicular magnetic fields through those faces; the units of A and B are arbitrary but consistent. Rank the surfaces according to the magnitudes of the magnetic flux through their curved sides, greatest first.

Surface	A_{top}	B_{top}	A_{bot}	B_{bot}
a	2	6, outward	4	3, inward
b	2	1, inward	4	2, inward
c	2	6, inward	2	8, outward
d	2	3, outward	3	2, outward

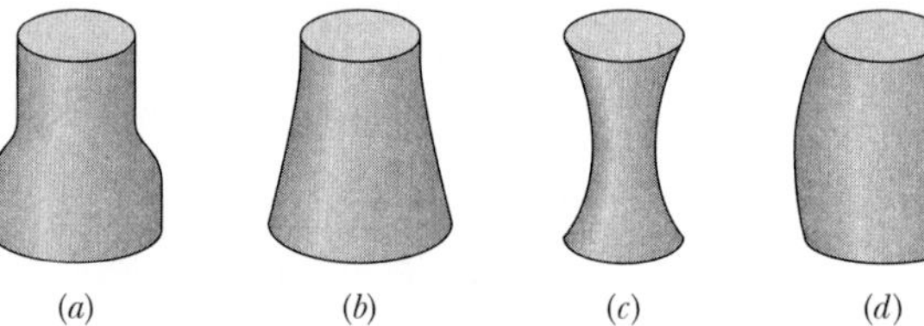

32-3 The Magnetism of Earth

Earth is a huge magnet; for points near Earth's surface, its magnetic field can be approximated as the field of a huge bar magnet—a magnetic dipole—that straddles the center of the planet. Figure 32-4 is an idealized symmetric depiction of the dipole field, without the distortion caused by passing charged particles from the Sun.

Because Earth's magnetic field is that of a magnetic dipole, a magnetic dipole moment $\vec{\mu}$ is associated with the field. For the idealized field of Fig. 32-4, the magnitude of $\vec{\mu}$ is 8.0×10^{22} J/T and the direction of $\vec{\mu}$ makes an angle of 11.5° with the rotation axis (RR) of Earth. The *dipole axis* (MM in Fig. 32-4) lies along $\vec{\mu}$ and intersects Earth's surface at the *geomagnetic north pole* in northwest Greenland and the *geomagnetic south pole* in Antarctica. The lines of the magnetic field $\vec{B}$ generally emerge in the southern hemisphere and reenter Earth in the northern hemisphere. Thus, the magnetic pole that is in Earth's northern hemisphere and known as a "north magnetic pole" *is really the south pole of Earth's magnetic dipole.*

The direction of the magnetic field at any location on Earth's surface is commonly specified in terms of two angles. The **field declination** is the angle (left or right) between geographic north (which is toward 90° latitude) and the horizontal component of the field. The **field inclination** is the angle (up or down) between a horizontal plane and the field's direction.

Magnetometers measure these angles and determine the field with much precision. However, you can do reasonably well with just a *compass* and a *dip meter*. A

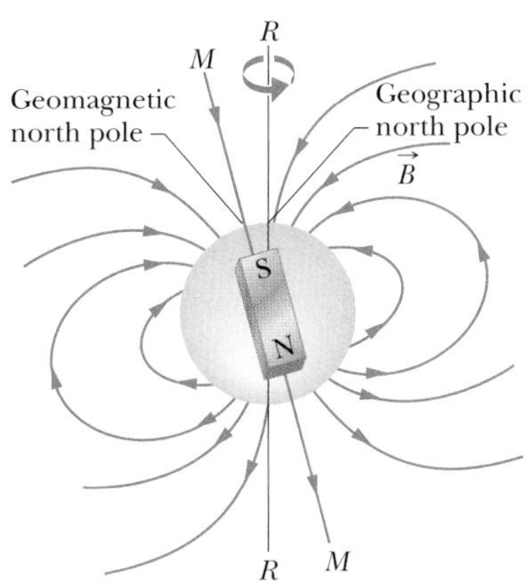

Fig. 32-4 Earth's magnetic field represented as a dipole field. The dipole axis *MM* makes an angle of 11.5° with Earth's rotational axis *RR*. The south pole of the dipole is in Earth's northern hemisphere.

compass is simply a needle-shaped magnet that is mounted so it can rotate freely about a vertical axis. When it is held in a horizontal plane, the north-pole end of the needle points, generally, toward the geomagnetic north pole (really a south magnetic pole, remember). The angle between the needle and geographic north is the field declination. A dip meter is a similar magnet that can rotate freely about a horizontal axis. When its vertical plane of rotation is aligned with the direction of the compass, the angle between the meter's needle and the horizontal is the field inclination.

At any point on Earth's surface, the measured magnetic field may differ appreciably, in both magnitude and direction, from the idealized dipole field of Fig. 32-4. In fact, the point where the field is actually perpendicular to Earth's surface and inward is not located at the geomagnetic north pole in Greenland as we would expect; instead, this so-called *dip north pole* is located in the Queen Elizabeth Islands in northern Canada, far from Greenland.

In addition, the field observed at any location on the surface of Earth varies with time, by measurable amounts over a period of a few years and by substantial amounts over, say, 100 years. For example, between 1580 and 1820 the direction indicated by compass needles in London changed by 35°.

In spite of these local variations, the average dipole field changes only slowly over such relatively short time periods. Variations over longer periods can be studied by measuring the weak magnetism of the ocean floor on either side of the Mid-Atlantic Ridge (Fig. 32-5). This floor has been formed by molten magma that oozed up through the ridge from Earth's interior, solidified, and was pulled away from the ridge (by the drift of tectonic plates) at the rate of a few centimeters per year. As the magma solidified, it became weakly magnetized with its magnetic field in the direction of Earth's magnetic field at the time of solidification. Study of this solidified magma across the ocean floor reveals that Earth's field has reversed its *polarity* (directions of the north pole and south pole) about every million years. The reason for the reversals is not known. In fact, the mechanism that produces Earth's magnetic field is only vaguely understood.

32-4 Magnetism and Electrons

Magnetic materials, from lodestones to videotapes, are magnetic because of the electrons within them. We have already seen one way in which electrons can generate a magnetic field: Send them through a wire as an electric current, and their motion produces a magnetic field around the wire. There are two more ways, each involving a magnetic dipole moment that produces a magnetic field in the surrounding space. However, their explanation requires quantum physics that is beyond the physics presented in this book, so here we shall only outline the results.

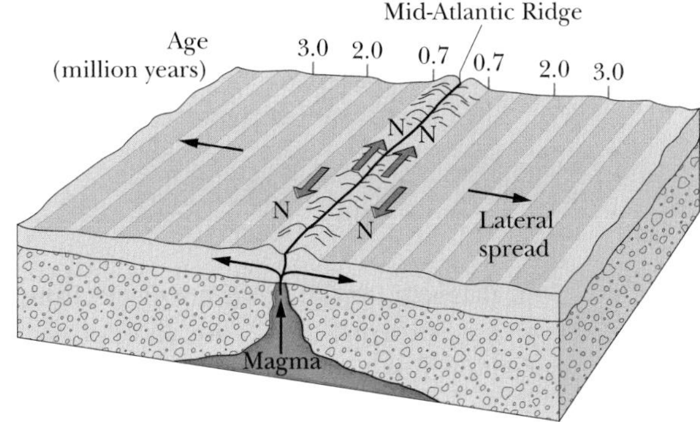

Fig. 32-5 A magnetic profile of the seafloor on either side of the Mid-Atlantic Ridge. The seafloor, extruded through the ridge and spreading out as part of the tectonic drift system, displays a record of the past magnetic history of Earth's core. The direction of the magnetic field produced by the core reverses about every million years.

Spin Magnetic Dipole Moment

An electron has an intrinsic angular momentum called its **spin angular momentum** (or just **spin**) $\vec{S}$; associated with this spin is an intrinsic **spin magnetic dipole moment** $\vec{\mu}_s$. (By *intrinsic*, we mean that $\vec{S}$ and $\vec{\mu}_s$ are basic characteristics of an electron, like its mass and electric charge.) $\vec{S}$ and $\vec{\mu}_s$ are related by

$$\vec{\mu}_s = -\frac{e}{m}\vec{S}, \tag{32-2}$$

in which e is the elementary charge (1.60×10^{-19} C) and m is the mass of an electron (9.11×10^{-31} kg). The minus sign means that $\vec{\mu}_s$ and $\vec{S}$ are oppositely directed.

Spin $\vec{S}$ is different from the angular momenta of Chapter 12 in two respects:

1. $\vec{S}$ itself cannot be measured. However, its component along any axis can be measured.
2. A measured component of $\vec{S}$ is *quantized*, which is a general term that means it is restricted to certain values. A measured component of $\vec{S}$ can have only two values, which differ only in sign.

Let us assume that the component of spin $\vec{S}$ is measured along the z axis of a coordinate system. Then the measured component S_z can have only the two values given by

$$S_z = m_s \frac{h}{2\pi}, \quad \text{for } m_s = \pm\tfrac{1}{2}, \tag{32-3}$$

where m_s is called the *spin magnetic quantum number* and h ($= 6.63 \times 10^{-34}$ J·s) is the Planck constant, the ubiquitous constant of quantum physics. The signs given in Eq. 32-3 have to do with the direction of S_z along the z axis. When S_z is parallel to the z axis, m_s is $+\frac{1}{2}$ and the electron is said to be *spin up*. When S_z is antiparallel to the z axis, m_s is $-\frac{1}{2}$ and the electron is said to be *spin down*.

The spin magnetic dipole moment $\vec{\mu}_s$ of an electron also cannot be measured; only its component along any axis can be measured, and that component too is quantized, with two possible values of the same magnitude but different signs. We can relate the component $\mu_{s,z}$ measured on the z axis to S_z by rewriting Eq. 32-2 in component form for the z axis as

$$\mu_{s,z} = -\frac{e}{m}S_z.$$

Substituting for S_z from Eq. 32-3 then gives us

$$\mu_{s,z} = \pm\frac{eh}{4\pi m}, \tag{32-4}$$

where the plus and minus signs correspond to $\mu_{s,z}$ being parallel and antiparallel to the z axis, respectively.

The quantity on the right side of Eq. 32-4 is called the *Bohr magneton* μ_B:

$$\mu_B = \frac{eh}{4\pi m} = 9.27 \times 10^{-24}\ \text{J/T} \quad \text{(Bohr magneton)}. \tag{32-5}$$

Spin magnetic dipole moments of electrons and other elementary particles can be expressed in terms of μ_B. For an electron, the magnitude of the measured z component of $\vec{\mu}_s$ is

$$\mu_{s,z} = 1\mu_B. \tag{32-6}$$

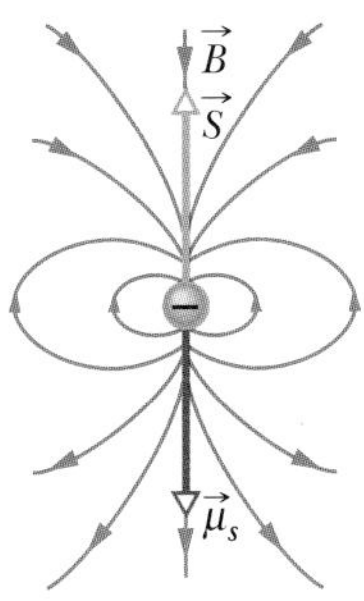

Fig. 32-6 The spin $\vec{S}$, spin magnetic dipole moment $\vec{\mu}_s$, and magnetic dipole field $\vec{B}$ of an electron represented as a microscopic sphere.

(The quantum physics of the electron, called *quantum electrodynamics,* or QED, reveals that $\mu_{s,z}$ is actually slightly greater than $1\mu_B$, but we shall neglect that fact.)

When an electron is placed in an external magnetic field $\vec{B}_{ext}$, a potential energy U can be associated with the orientation of the electron's spin magnetic dipole moment $\vec{\mu}_s$ just as a potential energy can be associated with the orientation of the magnetic dipole moment $\vec{\mu}$ of a current loop placed in $\vec{B}_{ext}$. From Eq. 29-38, the potential energy for the electron is

$$U = -\vec{\mu}_s \cdot \vec{B}_{ext} = -\mu_{s,z} B_{ext}, \tag{32-7}$$

where the z axis is taken to be in the direction of $\vec{B}_{ext}$.

If we imagine an electron to be a microscopic sphere (which it is not), we can represent the spin $\vec{S}$, the spin magnetic dipole moment $\vec{\mu}_s$, and the associated magnetic dipole field as in Fig. 32-6. Although we use the word "spin" here, electrons do not spin like tops. How, then, can something have angular momentum without actually rotating? Again, we would need quantum physics to provide the answer.

Protons and neutrons also have an intrinsic angular momentum called spin and an associated intrinsic spin magnetic dipole moment. For a proton those two vectors have the same direction, and for a neutron they have opposite directions. We shall not examine the contributions of these dipole moments to the magnetic fields of atoms because they are about a thousand times smaller than that due to an electron.

CHECKPOINT 2: The figure here shows the spin orientations of two particles in an external magnetic field $\vec{B}_{ext}$. (a) If the particles are electrons, which spin orientation is at lower potential energy? (b) If, instead, the particles are protons, which spin orientation is at lower potential energy?

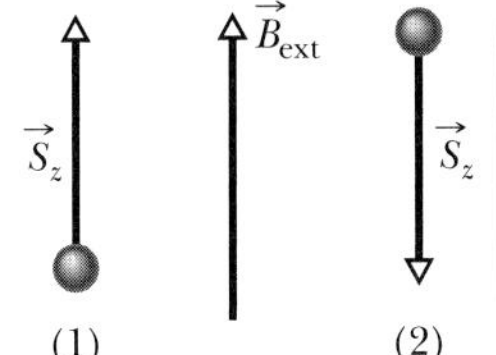

Orbital Magnetic Dipole Moment

When it is in an atom, an electron has an additional angular momentum called its **orbital angular momentum** $\vec{L}_{orb}$. Associated with $\vec{L}_{orb}$ is an **orbital magnetic dipole moment** $\vec{\mu}_{orb}$; the two are related by

$$\vec{\mu}_{orb} = -\frac{e}{2m}\vec{L}_{orb}. \tag{32-8}$$

The minus sign means that $\vec{\mu}_{orb}$ and $\vec{L}_{orb}$ have opposite directions.

Orbital angular momentum $\vec{L}_{orb}$ cannot be measured; only its component along any axis can be measured, and that component is quantized. The component along, say, a z axis can have only the values given by

$$L_{orb,z} = m_l \frac{h}{2\pi}, \qquad \text{for } m_l = 0, \pm 1, \pm 2, \ldots, \pm(\text{limit}), \tag{32-9}$$

in which m_l is called the *orbital magnetic quantum number* and "limit" refers to some largest allowed integer value for m_l. The signs in Eq. 32-9 have to do with the direction of $L_{orb,z}$ along the z axis.

The orbital magnetic dipole moment $\vec{\mu}_{orb}$ of an electron also cannot itself be measured; only its component along an axis can be measured, and that component is again quantized. By writing Eq. 32-8 for a component along the same z axis as above and then substituting for $L_{orb,z}$ from Eq. 32-9, we can write the z component $\mu_{orb,z}$ of the orbital magnetic dipole moment as

$$\mu_{orb,z} = -m_l \frac{eh}{4\pi m} \tag{32-10}$$

and, in terms of the Bohr magneton, as

$$\mu_{\text{orb},z} = -m_l \mu_B. \tag{32-11}$$

When an atom is placed in an external magnetic field $\vec{B}_{\text{ext}}$, a potential energy U can be associated with the orientation of the orbital magnetic dipole moment of each electron in the atom. Its value is

$$U = -\vec{\mu}_{\text{orb}} \cdot \vec{B}_{\text{ext}} = -\mu_{\text{orb},z} B_{\text{ext}}, \tag{32-12}$$

where the z axis is taken in the direction of $\vec{B}_{\text{ext}}$.

Although we have used the words "orbit" and "orbital" here, electrons do not orbit the nucleus of an atom like planets orbiting the Sun. How can an electron have an orbital angular momentum without orbiting in the common meaning of the term? Once again, this can be explained only with quantum physics.

Loop Model for Electron Orbits

We can obtain Eq. 32-8 with the nonquantum derivation that follows, in which we assume that an electron moves along a circular path with a radius that is much larger than an atomic radius (hence the name "loop model"). However, the derivation does not apply to an electron within an atom (for which we need quantum physics).

We imagine an electron moving at constant speed v in a circular path of radius r, counterclockwise as shown in Fig. 32-7. The motion of the negative charge of the electron is equivalent to a conventional current i (of positive charge) that is clockwise, as also shown in Fig. 32-7. The magnitude of the orbital magnetic dipole moment of such a *current loop* is obtained from Eq. 29-35 with $N = 1$:

$$\mu_{\text{orb}} = iA, \tag{32-13}$$

where A is the area enclosed by the loop. The direction of this magnetic dipole moment is, from the right-hand rule of Fig. 30-21, downward in Fig. 32-7.

To evaluate Eq. 32-13, we need the current i. Current is, generally, the rate at which charge passes some point in a circuit. Here, the charge of magnitude e takes a time $T = 2\pi r/v$ to circle from any point back through that point, so

$$i = \frac{\text{charge}}{\text{time}} = \frac{e}{2\pi r/v}. \tag{32-14}$$

Substituting this and the area $A = \pi r^2$ of the loop into Eq. 32-13 gives us

$$\mu_{\text{orb}} = \frac{e}{2\pi r/v}\,\pi r^2 = \frac{evr}{2}. \tag{32-15}$$

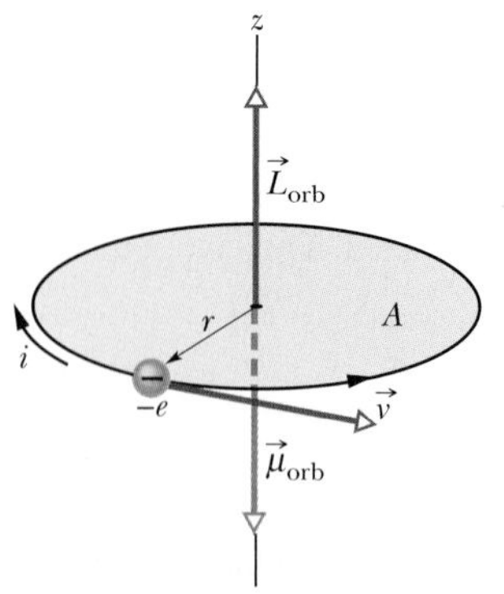

Fig. 32-7 An electron moving at constant speed v in a circular path of radius r that encloses an area A. The electron has an orbital angular momentum $\vec{L}_{\text{orb}}$ and an associated orbital magnetic dipole moment $\vec{\mu}_{\text{orb}}$. A clockwise current i (of positive charge) is equivalent to the counterclockwise circulation of the negatively charged electron.

To find the electron's orbital angular momentum $\vec{L}_{\text{orb}}$, we use Eq. 12-18, $\vec{\ell} = m(\vec{r} \times \vec{v})$. Because $\vec{r}$ and $\vec{v}$ are perpendicular, $\vec{L}_{\text{orb}}$ has the magnitude

$$L_{\text{orb}} = mrv \sin 90° = mrv. \tag{32-16}$$

$\vec{L}_{\text{orb}}$ is directed upward in Fig. 32-7 (see Fig. 12-11). Combining Eqs. 32-15 and 32-16, generalizing to a vector formulation, and indicating the opposite directions of the vectors with a minus sign yield

$$\vec{\mu}_{\text{orb}} = -\frac{e}{2m}\vec{L}_{\text{orb}},$$

which is Eq. 32-8. Thus, by "classical" (nonquantum) analysis we have obtained the same result, in both magnitude and direction, given by quantum physics. You might

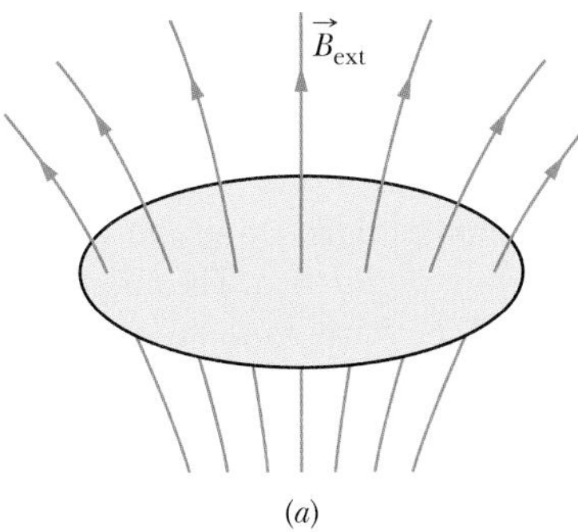

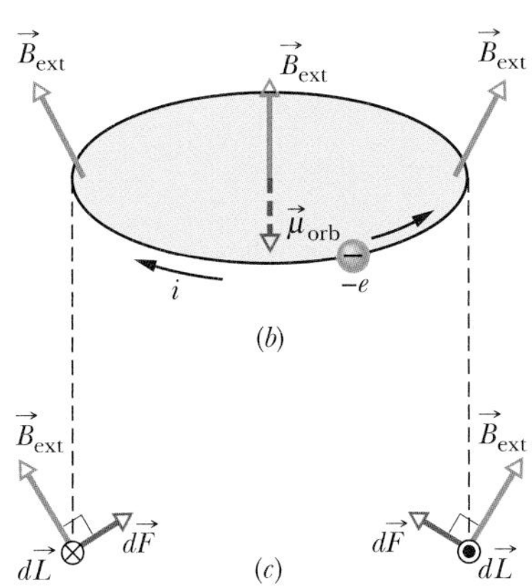

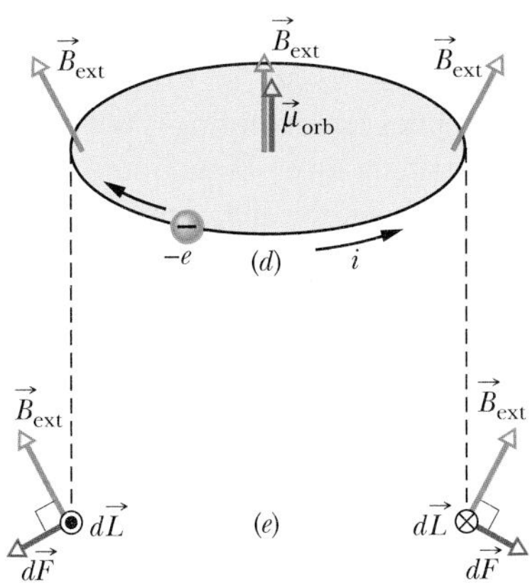

Fig. 32-8 (*a*) A loop model for an electron orbiting in an atom while in a nonuniform magnetic field $\vec{B}_{ext}$. (*b*) Charge $-e$ moves counterclockwise; the associated conventional current i is clockwise. (*c*) The magnetic forces $d\vec{F}$ on the left and right sides of the loop, as seen from the plane of the loop. The net force on the loop is upward. (*d*) Charge $-e$ now moves clockwise. (*e*) The net force on the loop is now downward.

wonder, since this derivation gives the correct result for an electron within an atom, why the derivation is invalid for that situation. The answer is that this line of reasoning yields other results that are contradicted by experiments.

Loop Model in a Nonuniform Field

We continue to consider an electron orbit as a current loop, as we did in Fig. 32-7. Now, however, we draw the loop in a nonuniform magnetic field $\vec{B}_{ext}$ as shown in Fig. 32-8*a*. (This field could be the diverging field near the north pole of the magnet in Fig. 32-3.) We make this change to prepare for the next several sections, in which we shall discuss the forces that act on magnetic materials when the materials are placed in a nonuniform magnetic field. We shall discuss these forces by assuming that the electron orbits in the materials are tiny current loops like that in Fig. 32-8*a*.

Here we assume that the magnetic field vectors all around the electron's circular path have the same magnitude and form the same angle with the vertical, as shown in Figs. 32-8*b* and *d*. We also assume that all the electrons in an atom move either counterclockwise (Fig. 32-8*b*) or clockwise (Fig. 32-8*d*). The associated conventional current i around the current loop and the orbital magnetic dipole moment $\vec{\mu}_{orb}$ produced by i are shown for each of these directions of motion.

Figures 32-8*c* and *e* show diametrically opposite views of a length element $d\vec{L}$ of the loop with the same direction as i, as seen from the plane of the orbit. Also shown are the field $\vec{B}_{ext}$ and the resulting magnetic force $d\vec{F}$ on $d\vec{L}$. Recall that a current along an element $d\vec{L}$ in a magnetic field $\vec{B}_{ext}$ experiences a magnetic force $d\vec{F}$ as given by Eq. 29-28:

$$d\vec{F} = i\, d\vec{L} \times \vec{B}_{ext}. \tag{32-17}$$

On the left side of Fig. 32-8*c*, Eq. 32-17 tells us that the force $d\vec{F}$ is directed upward and rightward. On the right side, the force $d\vec{F}$ is just as large and is directed upward and leftward. Because their angles are the same, the horizontal components of these two forces cancel and the vertical components add. The same is true at any other two symmetric points on the loop. Thus, the net force on the current loop of Fig. 32-8*b* must be upward. The same reasoning leads to a downward net force on the loop in Fig. 32-8*d*. We shall use these two results shortly when we examine the behavior of magnetic materials in nonuniform magnetic fields.

32-5 Magnetic Materials

Each electron in an atom has an orbital magnetic dipole moment and a spin magnetic dipole moment that combine vectorially. The resultant of these two vector quantities combines vectorially with similar resultants for all other electrons in the atom, and the resultant for each atom combines with those for all the other atoms in a sample of a material. If the combination of all these magnetic dipole moments produces a magnetic field, then the material is magnetic. There are three general types of magnetism: diamagnetism, paramagnetism, and ferromagnetism.

1. ***Diamagnetism*** is exhibited by all common materials but is so feeble that it is masked if the material also exhibits magnetism of either of the other two types. In diamagnetism, weak magnetic dipole moments are produced in the atoms of the material when the material is placed in an external magnetic field $\vec{B}_{ext}$; the combination of all those induced dipole moments gives the material as a whole only a feeble net magnetic field. The dipole moments and thus their net field disappear when $\vec{B}_{ext}$ is removed. The term *diamagnetic material* usually refers to materials that exhibit only diamagnetism.

2. ***Paramagnetism*** is exhibited by materials containing transition elements, rare earth elements, and actinide elements (see Appendix G). Each atom of such a material has a permanent resultant magnetic dipole moment, but the moments are randomly oriented in the material and the material as a whole lacks a net magnetic field. However, an external magnetic field $\vec{B}_{ext}$ can partially align the atomic magnetic dipole moments to give the material a net magnetic field. The alignment and thus its field disappear when $\vec{B}_{ext}$ is removed. The term *paramagnetic material* usually refers to materials that exhibit primarily paramagnetism.
3. ***Ferromagnetism*** is a property of iron, nickel, and certain other elements (and of compounds and alloys of these elements). Some of the electrons in these materials have their resultant magnetic dipole moments aligned, which produces regions with strong magnetic dipole moments. An external field $\vec{B}_{ext}$ can then align the magnetic moments of such regions, producing a strong magnetic field for a sample of the material; the field partially persists when $\vec{B}_{ext}$ is removed. We usually use the term *ferromagnetic material,* and even the common term *magnetic material,* to refer to materials that exhibit primarily ferromagnetism.

The next three sections explore these three types of magnetism.

32-6 Diamagnetism

We cannot yet discuss the quantum physical explanation of diamagnetism, but we can provide a classical explanation with the loop model of Figs. 32-7 and 32-8. To begin, we assume that in an atom of a diamagnetic material each electron can orbit only clockwise as in Fig. 32-8*d* or counterclockwise as in Fig. 32-8*b*. To account for the lack of magnetism in the absence of an external magnetic field $\vec{B}_{ext}$, we assume the atom lacks a net magnetic dipole moment. This implies that before $\vec{B}_{ext}$ is applied, as many electrons orbit one way as orbit the other, with the result that the net upward magnetic dipole moment of the atom equals the net downward magnetic dipole moment.

Now let's turn on the nonuniform field $\vec{B}_{ext}$ of Fig. 32-8*a*, in which $\vec{B}_{ext}$ is directed upward but is diverging (the magnetic field lines are diverging). We could do this by increasing the current through an electromagnet or by moving the north pole of a bar magnet closer to, and below, the orbits. As the magnitude of $\vec{B}_{ext}$ increases from zero to its final maximum, steady-state value, a clockwise electric field is induced around each electron's orbital loop according to Faraday's law and Lenz's law. Let us see how this induced electric field affects the orbiting electrons in Figs. 32-8*b* and *d*.

In Fig. 32-8*b*, the counterclockwise electron is accelerated by the clockwise electric field. Thus, as the magnetic field $\vec{B}_{ext}$ increases to its maximum value, the electron speed increases to a maximum value. This means that the associated conventional current i and the downward magnetic dipole moment $\vec{\mu}$ due to i also *increase*.

In Fig. 32-8*d*, the clockwise electron is decelerated by the clockwise electric field. Thus, here, the electron speed, the associated current i, and the upward magnetic dipole moment $\vec{\mu}$ due to i all *decrease*. By turning on field $\vec{B}_{ext}$, we have given the atom a *net* magnetic dipole moment that is upward. This would also be so if the magnetic field were uniform.

The nonuniformity of field $\vec{B}_{ext}$ also affects the atom. Because the current i in Fig. 32-8*b* increases, the upward magnetic forces $d\vec{F}$ in Fig. 32-8*c* also increase, as does the net upward force on the current loop. Because current i in Fig. 32-8*d* decreases, the downward magnetic forces $d\vec{F}$ in Fig. 32-8*e* also decrease, as does the net downward force on the current loop. Thus, by turning on the *nonuniform* field $\vec{B}_{ext}$, we have produced a net force on the atom; moreover, that force is directed *away* from the region of greater magnetic field.

We have argued with fictitious electron orbits (current loops), but we have ended up with exactly what happens to a diamagnetic material: If we apply the magnetic field of Fig. 32-8, the material develops a downward magnetic dipole moment and experiences an upward force. When the field is removed, both the dipole moment and the force disappear. The external field need not be positioned as shown; similar arguments can be made for other orientations of $\vec{B}_{ext}$. In general,

> A diamagnetic material placed in an external magnetic field $\vec{B}_{ext}$ develops a magnetic dipole moment directed opposite $\vec{B}_{ext}$. If the field is nonuniform, the diamagnetic material is repelled *from* a region of greater magnetic field *toward* a region of lesser field.

The frog in the photograph opening this chapter is diamagnetic (as is any other animal). When the frog was placed in the diverging magnetic field near the top end of a vertical current-carrying solenoid, every atom in the frog was repelled upward, away from the region of stronger magnetic field at that end of the solenoid. The frog moved upward into weaker and weaker magnetic field until the upward magnetic force balanced the gravitational force on it, and there it hung in midair. If we built a solenoid that was large enough, we could similarly levitate a person in midair owing to the person's diamagnetism.

CHECKPOINT 3: The figure shows two diamagnetic spheres located near the south pole of a bar magnet. Are (a) the magnetic forces on the spheres and (b) the magnetic dipole moments of the spheres directed toward or away from the bar magnet? (c) Is the magnetic force on sphere 1 greater than, less than, or equal to that on sphere 2?

1 2 S N

32-7 Paramagnetism

In paramagnetic materials, the spin and orbital magnetic dipole moments of the electrons in each atom do not cancel but add vectorially to give the atom a net (and permanent) magnetic dipole moment $\vec{\mu}$. In the absence of an external magnetic field, these atomic dipole moments are randomly oriented, and the net magnetic dipole moment of the material is zero. However, if a sample of the material is placed in an external magnetic field $\vec{B}_{ext}$, the magnetic dipole moments tend to line up with the field, which gives the sample a net magnetic dipole moment. This alignment with the external field is the opposite of what we saw with diamagnetic materials.

Liquid oxygen is suspended between the two pole faces of a magnet because the liquid is paramagnetic and is magnetically attracted to the magnet.

> A paramagnetic material placed in an external magnetic field $\vec{B}_{ext}$ develops a magnetic dipole moment in the direction of $\vec{B}_{ext}$. If the field is nonuniform, the paramagnetic material is attracted *toward* a region of greater magnetic field *from* a region of lesser field.

A paramagnetic sample with N atoms would have a magnetic dipole moment of magnitude $N\mu$ if alignment of its atomic dipoles were complete. However, random collisions of atoms due to thermal agitation transfer energy among them, disrupting their alignment and thus reducing the sample's magnetic dipole moment.

The importance of thermal agitation may be measured by comparing two energies. One, from Eq. 20-24, is the mean translational kinetic energy K $(=\frac{3}{2}kT)$ of an atom at temperature T, where k is the Boltzmann constant $(1.38 \times 10^{-23}$ J/K) and T is in kelvins (not Celsius degrees). The other, from Eq. 29-38, is the difference in energy ΔU_B $(= 2\mu B_{ext})$ between parallel alignment and antiparallel alignment of

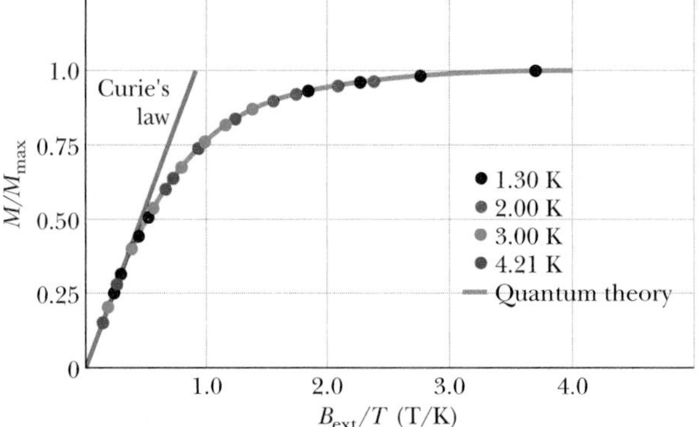

Fig. 32-9 A *magnetization curve* for potassium chromium sulfate, a paramagnetic salt. The ratio of magnetization M of the salt to the maximum possible magnetization M_{max} is plotted versus the ratio of the applied magnetic field B_{ext} to the temperature T. Curie's law fits the data at the left; quantum theory fits all the data. After W. E. Henry.

the magnetic dipole moment of an atom and the external field. As we shall show below, $K \gg \Delta U_B$, even for ordinary temperatures and field magnitudes. Thus, energy transfers during collisions among atoms can significantly disrupt the alignment of the atomic dipole moments, keeping the magnetic dipole moment of a sample much less than $N\mu$.

We can express the extent to which a given paramagnetic sample is magnetized by finding the ratio of its magnetic dipole moment to its volume V. This vector quantity, the magnetic dipole moment per unit volume, is the **magnetization** $\vec{M}$ of the sample, and its magnitude is

$$M = \frac{\text{measured magnetic moment}}{V}. \tag{32-18}$$

The unit of $\vec{M}$ is the ampere–square meter per cubic meter, or ampere per meter (A/m). Complete alignment of the atomic dipole moments, called *saturation* of the sample, corresponds to the maximum value $M_{max} = N\mu/V$.

In 1895 Pierre Curie discovered experimentally that the magnetization of a paramagnetic sample is directly proportional to the external magnetic field $\vec{B}_{ext}$ and inversely proportional to the temperature T in kelvins; that is,

$$M = C\frac{B_{ext}}{T}. \tag{32-19}$$

Equation 32-19 is known as *Curie's law,* and C is called the *Curie constant.* Curie's law is reasonable in that increasing B_{ext} tends to align the atomic dipole moments in a sample and thus to increase M, whereas increasing T tends to disrupt the alignment via thermal agitation and thus to decrease M. However, the law is actually an approximation that is valid only when the ratio B_{ext}/T is not too large.

Figure 32-9 shows the ratio M/M_{max} as a function of B_{ext}/T for a sample of the salt potassium chromium sulfate, in which chromium ions are the paramagnetic substance. The plot is called a *magnetization curve.* The straight line for Curie's law fits the experimental data at the left, for B_{ext}/T below about 0.5 T/K. The curve that fits all the data points is based on quantum physics. The data on the right side, near saturation, are very difficult to obtain because they require very strong magnetic fields (about 100 000 times Earth's field), even at very low temperatures.

CHECKPOINT 4: The figure here shows two paramagnetic spheres located near the south pole of a bar magnet. Are (a) the magnetic forces on the spheres and (b) the magnetic dipole moments of the spheres directed toward or away from the bar magnet? (c) Is the magnetic force on sphere 1 greater than, less than, or equal to that on sphere 2?

1 2 S N

Sample Problem 32-1

A paramagnetic gas at room temperature ($T = 300$ K) is placed in an external uniform magnetic field of magnitude $B = 1.5$ T; the atoms of the gas have magnetic dipole moment $\mu = 1.0\mu_B$. Calculate the mean translational kinetic energy K of an atom of the gas and the energy difference ΔU_B between parallel alignment and antiparallel alignment of the atom's magnetic dipole moment with the external field.

SOLUTION: The first **Key Idea** here is that the mean translational kinetic energy K of an atom in a gas depends on the temperature of the gas. From Eq. 20-24, we have

$$K = \tfrac{3}{2}kT = \tfrac{3}{2}(1.38 \times 10^{-23}\ \text{J/K})(300\ \text{K})$$
$$= 6.2 \times 10^{-21}\ \text{J} = 0.039\ \text{eV}. \qquad \text{(Answer)}$$

The second **Key Idea** is that the potential energy U_B of a magnetic dipole $\vec{\mu}$ in an external magnetic field $\vec{B}$ depends on the angle θ between the directions of $\vec{\mu}$ and $\vec{B}$. From Eq. 29-38 ($U_B = -\vec{\mu}\cdot\vec{B}$), we can write the difference ΔU_B between parallel alignment ($\theta = 0°$) and antiparallel alignment ($\theta = 180°$) as

$$\Delta U_B = -\mu B \cos 180° - (-\mu B \cos 0°) = 2\mu B$$
$$= 2\mu_B B = 2(9.27 \times 10^{-24}\ \text{J/T})(1.5\ \text{T})$$
$$= 2.8 \times 10^{-23}\ \text{J} = 0.000\ 17\ \text{eV}. \qquad \text{(Answer)}$$

Here K is about 230 times ΔU_B, so energy exchanges among the atoms during their collisions with one another can easily reorient any magnetic dipole moments that might be aligned with the external magnetic field. The magnetic dipole moment exhibited by the paramagnetic gas must then be due to fleeting partial alignments of the atomic dipole moments.

32-8 Ferromagnetism

When we speak of magnetism in everyday conversation, we almost always have a mental picture of a bar magnet or a disk magnet (probably clinging to a refrigerator door). That is, we picture a ferromagnetic material having strong, permanent magnetism, and not a diamagnetic or paramagnetic material having weak, temporary magnetism.

Iron, cobalt, nickel, gadolinium, dysprosium, and alloys containing these elements exhibit ferromagnetism because of a quantum physical effect called *exchange coupling* in which the electron spins of one atom interact with those of neighboring atoms. The result is alignment of the magnetic dipole moments of the atoms, in spite of the randomizing tendency of atomic collisions. This persistent alignment is what gives ferromagnetic materials their permanent magnetism.

If the temperature of a ferromagnetic material is raised above a certain critical value, called the *Curie temperature,* the exchange coupling ceases to be effective. Most such materials then become simply paramagnetic; that is, the dipoles still tend to align with an external field but much more weakly, and thermal agitation can now more easily disrupt the alignment. The Curie temperature for iron is 1043 K (= 770°C).

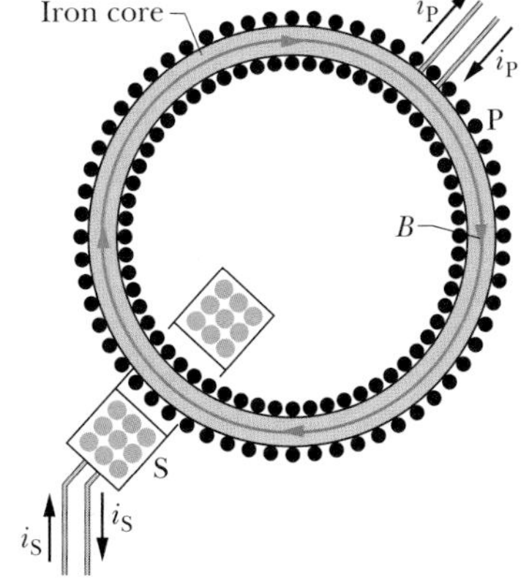

Fig. 32-10 A Rowland ring. Current i_P is sent through a primary coil P whose core is the ferromagnetic material to be studied (here iron) and that is magnetized by the current. (The turns of the coil are represented by dots.) The extent of magnetization of the core determines the total magnetic field $\vec{B}$ within coil P. Field $\vec{B}$ can be measured by means of a secondary coil S.

The magnetization of a ferromagnetic material such as iron can be studied with an arrangement called a *Rowland ring* (Fig. 32-10). The material is formed into a thin toroidal core of circular cross section. A primary coil P having n turns per unit length is wrapped around the core and carries current i_P. (The coil is essentially a long solenoid bent into a circle.) If the iron core were not present, the magnitude of the magnetic field inside the coil would be, from Eq. 30-25,

$$B_0 = \mu_0 i_P n. \qquad (32\text{-}20)$$

However, with the iron core present, the magnetic field $\vec{B}$ inside the coil is greater than $\vec{B}_0$, usually by a large amount. We can write the magnitude of this field as

$$B = B_0 + B_M, \qquad (32\text{-}21)$$

where B_M is the magnitude of the magnetic field contributed by the iron core. This contribution results from the alignment of the atomic dipole moments within the iron, due to exchange coupling and to the applied magnetic field B_0, and is proportional to the magnetization M of the iron. That is, the contribution B_M is proportional

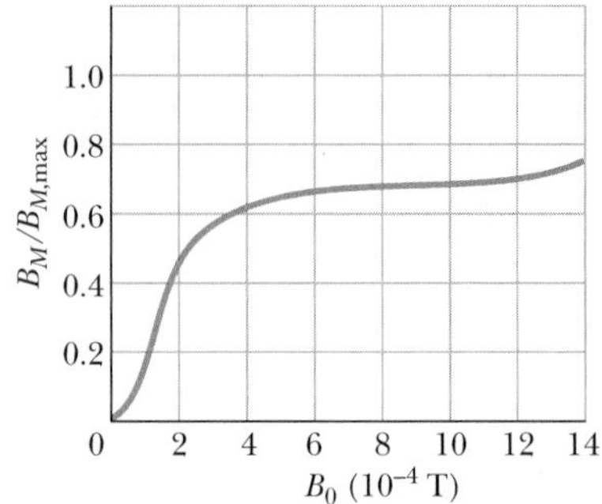

Fig. 32-11 A magnetization curve for a ferromagnetic core material in the Rowland ring of Fig. 32-10. On the vertical axis, 1.0 corresponds to complete alignment (saturation) of the atomic dipoles within the material.

to the magnetic dipole moment per unit volume of the iron. To determine B_M we use a secondary coil S to measure B, compute B_0 with Eq. 32-20, and subtract as suggested by Eq. 32-21.

Figure 32-11 shows a magnetization curve for a ferromagnetic material in a Rowland ring: the ratio $B_M/B_{M,\max}$, where $B_{M,\max}$ is the maximum possible value of B_M, corresponding to saturation, is plotted versus B_0. The curve is like Fig. 32-9, the magnetization curve for a paramagnetic substance: both curves show the extent to which an applied magnetic field can align the atomic dipole moments of a material.

For the ferromagnetic core yielding Fig. 32-11, the alignment of the dipole moments is about 70% complete for $B_0 \approx 1 \times 10^{-3}$ T. If B_0 were increased to 1 T, the alignment would be almost complete (but $B_0 = 1$ T, and thus almost complete saturation, is quite difficult to obtain).

Magnetic Domains

Exchange coupling produces strong alignment of adjacent atomic dipoles in a ferromagnetic material at a temperature below the Curie temperature. Why, then, isn't the material naturally at saturation even when there is no applied magnetic field B_0? That is, why isn't every piece of iron, such as an iron nail, a naturally strong magnet?

To understand this, consider a specimen of a ferromagnetic material such as iron that is in the form of a single crystal; that is, the arrangement of the atoms that make it up—its crystal lattice—extends with unbroken regularity throughout the volume of the specimen. Such a crystal will, in its normal state, be made up of a number of *magnetic domains*. These are regions of the crystal throughout which the alignment of the atomic dipoles is essentially perfect. The domains, however, are not all aligned. For the crystal as a whole, the domains are so oriented that they largely cancel each other as far as their external magnetic effects are concerned.

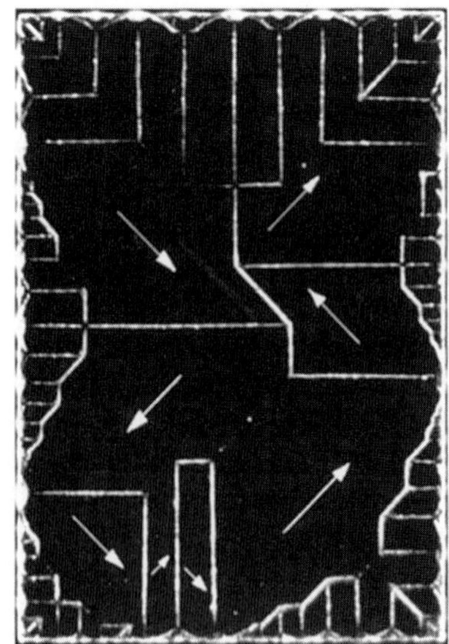

Fig. 32-12 A photograph of domain patterns within a single crystal of nickel; white lines reveal the boundaries of the domains. The white arrows superimposed on the photograph show the orientations of the magnetic dipoles within the domains and thus the orientations of the net magnetic dipoles of the domains. The crystal as a whole is unmagnetized if the net magnetic field (the vector sum over all the domains) is zero.

Figure 32-12 is a magnified photograph of such an assembly of domains in a single crystal of nickel. It was made by sprinkling a colloidal suspension of finely powdered iron oxide on the surface of the crystal. The domain boundaries, which are thin regions in which the alignment of the elementary dipoles changes from a certain orientation in one domain to a different orientation in the other, are the sites of intense, but highly localized and nonuniform, magnetic fields. The suspended colloidal particles are attracted to these boundaries and show up as the white lines (not all the domain boundaries are apparent in Fig. 32-12). Although the atomic dipoles in each domain are completely aligned as shown by the arrows, the crystal as a whole may have a very small resultant magnetic moment.

Actually, a piece of iron as we ordinarily find it is not a single crystal but an assembly of many tiny crystals, randomly arranged; we call it a *polycrystalline solid*. Each tiny crystal, however, has its array of variously oriented domains, just as in Fig. 32-12. If we magnetize such a specimen by placing it in an external magnetic field of gradually increasing strength, we produce two effects; together they produce a magnetization curve of the shape shown in Fig. 32-11. One effect is a growth in size of the domains that are oriented along the external field at the expense of those that are not. The second effect is a shift of the orientation of the dipoles within a domain, as a unit, to become closer to the field direction.

Exchange coupling and domain shifting give us the following result:

> A ferromagnetic material placed in an external magnetic field $\vec{B}_{ext}$ develops a strong magnetic dipole moment in the direction of $\vec{B}_{ext}$. If the field is nonuniform, the ferromagnetic material is attracted *toward* a region of greater magnetic field *from* a region of lesser field.

You can actually hear sound produced by shifting domains: Put an audio cassette player into its play mode without a cassette in place (or with a blank cassette) and turn the volume control to maximum. Then bring a strong magnet up to the play head (which is ferromagnetic). The magnetic field causes the domains in the play head to shift abruptly, which abruptly alters the magnetic field through a coil wrapped around the play head. The resulting suddenly induced currents in the coil are amplified and fed to the speaker, producing a fizzing sound.

Sample Problem 32-2

A compass needle made of pure iron (with density 7900 kg/m^3) has a length L of 3.0 cm, a width of 1.0 mm, and a thickness of 0.50 mm. The magnitude of the magnetic dipole moment of an iron atom is $\mu_{Fe} = 2.1 \times 10^{-23}$ J/T. If the magnetization of the needle is equivalent to the alignment of 10% of the atoms in the needle, what is the magnitude of the needle's magnetic dipole moment $\vec{\mu}$?

SOLUTION: One Key Idea here is that alignment of all N atoms in the needle would give a magnitude of $N\mu_{Fe}$ for the needle's magnetic dipole moment $\vec{\mu}$. However, the needle has only 10% alignment (the random orientation of the rest does not give any net contribution to $\vec{\mu}$). Thus,

$$\mu = 0.10N\mu_{Fe}. \tag{32-22}$$

A second Key Idea is that we can find the number of atoms N in the needle from the needle's mass:

$$N = \frac{\text{needle's mass}}{\text{iron's atomic mass}}. \tag{32-23}$$

Iron's atomic mass is not listed in Appendix F, but its molar mass M is. Thus, we write

$$\text{iron's atomic mass} = \frac{\text{iron's molar mass } M}{\text{Avogadro's number } N_A}. \tag{32-24}$$

Equation 32-23 then becomes

$$N = \frac{mN_A}{M}. \tag{32-25}$$

The needle's mass m is the product of its density and its volume. The volume works out to be 1.5×10^{-8} m^3, so we can write

$$\begin{aligned}\text{needle's mass } m &= (\text{needle's density})(\text{needle's volume})\\ &= (7900 \text{ kg/m}^3)(1.5 \times 10^{-8} \text{ m}^3)\\ &= 1.185 \times 10^{-4} \text{ kg}.\end{aligned}$$

Substituting into Eq. 32-25 with this value for m, and also 55.847 g/mole (= 0.055 847 kg/mole) for M and 6.02×10^{23} for N_A, we find

$$\begin{aligned}N &= \frac{(1.185 \times 10^{-4} \text{ kg})(6.02 \times 10^{23})}{0.055\,847 \text{ kg/mole}}\\ &= 1.2774 \times 10^{21}.\end{aligned}$$

Substituting this and the value of μ_{Fe} into Eq. 32-22 then yields

$$\begin{aligned}\mu &= (0.10)(1.2774 \times 10^{21})(2.1 \times 10^{-23} \text{ J/T})\\ &= 2.682 \times 10^{-3} \text{ J/T} \approx 2.7 \times 10^{-3} \text{ J/T}. \quad \text{(Answer)}\end{aligned}$$

Hysteresis

Magnetization curves for ferromagnetic materials are not retraced as we increase and then decrease the external magnetic field B_0. Figure 32-13 is a plot of B_M versus B_0 during the following operations with a Rowland ring: (1) Starting with the iron unmagnetized (point a), increase the current in the toroid until B_0 (= $\mu_0 in$) has the value corresponding to point b; (2) reduce the current in the toroid winding (and thus B_0) back to zero (point c); (3) reverse the toroid current and increase it in magnitude until B_0 has the value corresponding to point d; (4) reduce the current to zero again (point e); (5) reverse the current once more until point b is reached again.

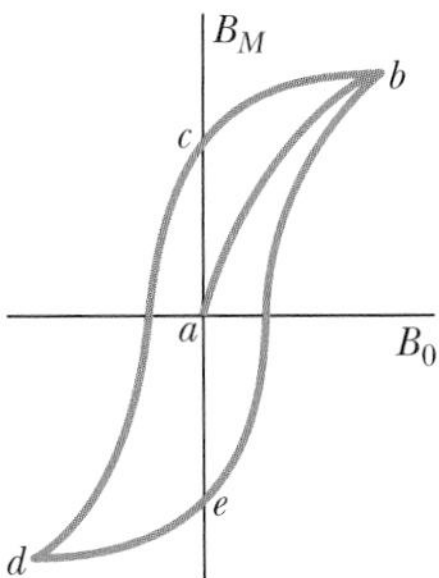

Fig. 32-13 A magnetization curve (ab) for a ferromagnetic specimen and an associated hysteresis loop ($bcdeb$).

The lack of retraceability shown in Fig. 32-13 is called **hysteresis,** and the curve $bcdeb$ is called a *hysteresis loop*. Note that at points c and e the iron core is magnetized, even though there is no current in the toroid windings; this is the familiar phenomenon of permanent magnetism.

Hysteresis can be understood through the concept of magnetic domains. Evidently the motions of the domain boundaries and the reorientations of the domain directions are not totally reversible. When the applied magnetic field B_0 is increased and then decreased back to its initial value, the domains do not return completely to their original configuration but retain some "memory" of their alignment after the initial increase. This memory of magnetic materials is essential for the magnetic storage of information, as on cassette tapes and computer disks.

This memory of the alignment of domains can also occur naturally. When lightning sends currents along multiple tortuous paths through the ground, the currents produce intense magnetic fields that can suddenly magnetize any ferromagnetic material in nearby rock. Because of hysteresis, such rock material retains some of that magnetization after the lightning strike (after the currents disappear). Pieces of the rock, later exposed, broken, and loosened by weathering, are then lodestones.

32-9 Induced Magnetic Fields

In Chapter 31 you saw that a changing magnetic flux induces an electric field, and we ended up with Faraday's law of induction in the form

$$\oint \vec{E} \cdot d\vec{s} = -\frac{d\Phi_B}{dt} \quad \text{(Faraday's law of induction).} \tag{32-26}$$

Here $\vec{E}$ is the electric field induced along a closed loop by the changing magnetic flux Φ_B encircled by that loop. Because symmetry is often so powerful in physics, we should be tempted to ask whether induction can occur in the opposite sense; that is, can a changing electric flux induce a magnetic field?

The answer is that it can; furthermore, the equation governing the induction of a magnetic field is almost symmetric with Eq. 32-26. We often call it Maxwell's law of induction after James Clerk Maxwell, and we write it as

$$\oint \vec{B} \cdot d\vec{s} = \mu_0 \varepsilon_0 \frac{d\Phi_E}{dt} \quad \text{(Maxwell's law of induction).} \tag{32-27}$$

Here $\vec{B}$ is the magnetic field induced along a closed loop by the changing electric flux Φ_E in the region encircled by that loop.

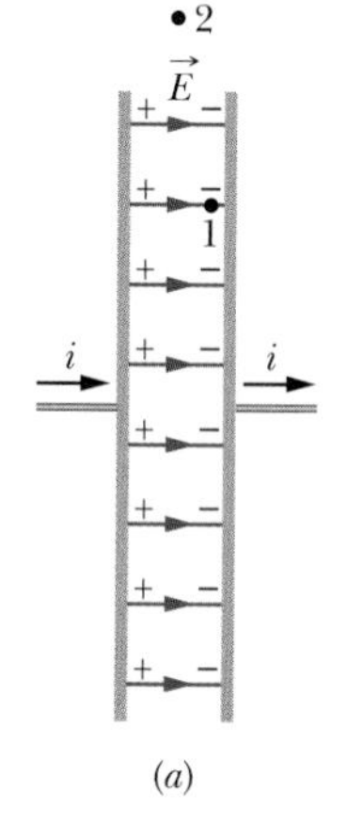

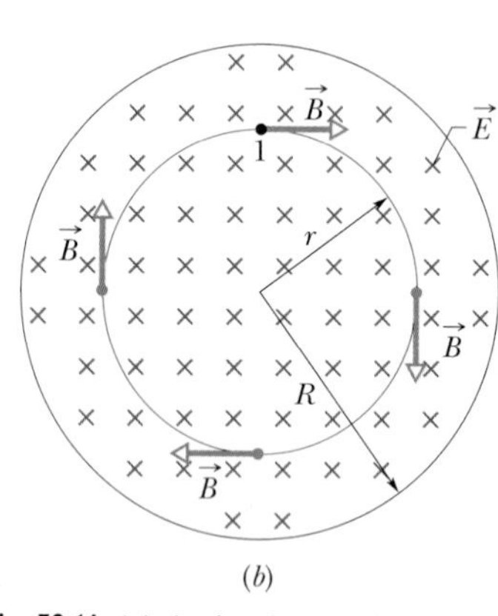

Fig. 32-14 (*a*) A circular parallel-plate capacitor, shown in side view, is being charged by a constant current *i*. (*b*) A view from within the capacitor, toward the plate at the right. The electric field $\vec{E}$ is uniform, is directed into the page (toward the plate), and grows in magnitude as the charge on the capacitor increases. The magnetic field $\vec{B}$ induced by this changing electric field is shown at four points on a circle with a radius *r* less than the plate radius *R*.

As an example of this sort of induction, we consider the charging of a parallel-plate capacitor with circular plates (Fig. 32-14*a*). (Although we shall focus on this particular arrangement, a changing electric flux will always induce a magnetic field whenever it occurs.) We assume that the charge on the capacitor is being increased at a steady rate by a constant current *i* in the connecting wires. Then the electric field magnitude between the plates must also be increasing at a steady rate.

Figure 32-14*b* is a view of the right-hand plate of Fig. 32-14*a* from between the plates. The electric field is directed into the page. Let us consider a circular loop through point 1 in Figs. 32-14*a* and *b*, concentric with the capacitor plates and with a radius smaller than that of the plates. Because the electric field through the loop is changing, the electric flux through the loop must also be changing. According to Eq. 32-27, this changing electric flux induces a magnetic field around the loop.

Experiment proves that a magnetic field $\vec{B}$ *is* indeed induced around such a loop, directed as shown. This magnetic field has the same magnitude at every point around the loop and thus has circular symmetry about the central axis of the capacitor plates.

If we now consider a larger loop, say, through point 2 outside the plates in Figs. 32-14*a* and *b*, we find that a magnetic field is induced around that loop as well. Thus, while the electric field is changing, magnetic fields are induced between the plates, both inside and outside the gap. When the electric field stops changing, these induced magnetic fields disappear.

Although Eq. 32-27 is similar to Eq. 32-26, the equations differ in two ways. First, Eq. 32-27 has the two extra symbols, μ_0 and ε_0, but they appear only because

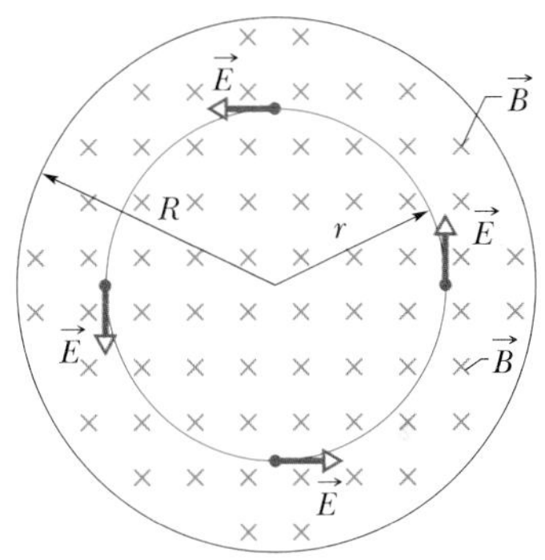

Fig. 32-15 A uniform magnetic field $\vec{B}$ in a circular region. The field, directed into the page, is increasing in magnitude. The electric field $\vec{E}$ induced by the changing magnetic field is shown at four points on a circle concentric with the circular region. Compare this situation with that of Fig. 32-14*b*.

we employ SI units. Second, Eq. 32-27 lacks the minus sign of Eq. 32-26, meaning that the induced electric field $\vec{E}$ and the induced magnetic field $\vec{B}$ have opposite directions when they are produced in otherwise similar situations. To see this opposition, examine Fig. 32-15, in which an increasing magnetic field $\vec{B}$, directed into the page, induces an electric field $\vec{E}$. The induced field $\vec{E}$ is counterclockwise, opposite the induced magnetic field $\vec{B}$ in Fig. 32-14*b*.

Ampere–Maxwell Law

Now recall that the left side of Eq. 32-27, the integral of the dot product $\vec{B}\cdot d\vec{s}$ around a closed loop, appears in another equation—namely, Ampere's law:

$$\oint \vec{B}\cdot d\vec{s} = \mu_0 i_{\text{enc}} \quad \text{(Ampere's law)}, \tag{32-28}$$

where i_{enc} is the current encircled by the closed loop. Thus, our two equations that specify the magnetic field $\vec{B}$ produced by means other than a magnetic material (that is, by a current and by a changing electric field) give the field in exactly the same form. We can combine the two equations into the single equation

$$\oint \vec{B}\cdot d\vec{s} = \mu_0\varepsilon_0 \frac{d\Phi_E}{dt} + \mu_0 i_{\text{enc}} \quad \text{(Ampere–Maxwell law)}. \tag{32-29}$$

When there is a current but no change in electric flux (such as with a wire carrying a constant current), the first term on the right side of Eq. 32-29 is zero, and so Eq. 32-29 reduces to Eq. 32-28, Ampere's law. When there is a change in electric flux but no current (such as inside or outside the gap of a charging capacitor), the second term on the right side of Eq. 32-29 is zero, and so Eq. 32-29 reduces to Eq. 32-27, Maxwell's law of induction.

Sample Problem 32-3

A parallel-plate capacitor with circular plates of radius R is being charged as in Fig. 32-14*a*.

(a) Derive an expression for the magnetic field at radii r for the case $r \le R$.

SOLUTION: The **Key Idea** here is that a magnetic field can be set up by a current and by induction due to a changing electric flux; both effects are included in Eq. 32-29. There is no current between the capacitor plates of Fig. 32-14, but the electric flux there is changing. Thus, Eq. 32-29 reduces to

$$\oint \vec{B}\cdot d\vec{s} = \mu_0\varepsilon_0 \frac{d\Phi_E}{dt}. \tag{32-30}$$

We shall separately evaluate the left and right sides of this equation.

Left side of Eq. 32-30: We choose a circular Amperian loop with a radius $r \le R$ as shown in Fig. 32-14, because we want to evaluate the magnetic field for $r \le R$—that is, inside the capacitor. The magnetic field $\vec{B}$ at all points along the loop is tangent to the loop, as is the path element $d\vec{s}$. Thus, $\vec{B}$ and $d\vec{s}$ are either parallel or antiparallel at each point of the loop. For simplicity, assume they are parallel (the choice does not alter our outcome here). Then

$$\oint \vec{B}\cdot d\vec{s} = \oint B\, ds \cos 0^\circ = \oint B\, ds.$$

Due to the circular symmetry of the plates, we can also assume that $\vec{B}$ has the same magnitude at every point around the loop. Thus, B can be taken outside the integral on the right side of the above equation. The integral that remains is $\oint ds$, which simply gives the circumference $2\pi r$ of the loop. The left side of Eq. 32-30 is then $(B)(2\pi r)$.

Right side of Eq. 32-30: We assume that the electric field $\vec{E}$ is uniform between the capacitor plates and directed perpendicular to the plates. Then the electric flux Φ_E through the Amperian loop is EA, where A is the area encircled by the loop within the electric field. Thus, the right side of Eq. 32-30 is $\mu_0\varepsilon_0\, d(EA)/dt$.

Substituting our results for the left and right sides into Eq. 32-30, we get

$$(B)(2\pi r) = \mu_0\varepsilon_0 \frac{d(EA)}{dt}.$$

Because A is a constant, we write $d(EA)$ as $A\, dE$, so we have

$$(B)(2\pi r) = \mu_0\varepsilon_0 A \frac{dE}{dt}. \tag{32-31}$$

We next use this **Key Idea**: The area A that is encircled by the Amperian loop within the electric field is the full area πr^2 of the loop, because the loop's radius r is less than (or equal to) the plate radius R. Substituting πr^2 for A in Eq. 32-31 and solving the result for B give us, for $r \le R$,

$$B = \frac{\mu_0 \varepsilon_0 r}{2} \frac{dE}{dt}. \quad \text{(Answer)} \quad (32\text{-}32)$$

This equation tells us that, inside the capacitor, B increases linearly with increased radial distance r, from zero at the center of the plates to a maximum value at the plate edges (where $r = R$).

(b) Evaluate the field magnitude B for $r = R/5 = 11.0$ mm and $dE/dt = 1.50 \times 10^{12}$ V/m · s.

SOLUTION: From the answer to (a), we have

$$\begin{aligned} B &= \frac{1}{2} \mu_0 \varepsilon_0 r \frac{dE}{dt} \\ &= \tfrac{1}{2}(4\pi \times 10^{-7}\ \text{T}\cdot\text{m/A})(8.85 \times 10^{-12}\ \text{C}^2/\text{N}\cdot\text{m}^2) \\ &\quad \times (11.0 \times 10^{-3}\ \text{m})(1.50 \times 10^{12}\ \text{V/m}\cdot\text{s}) \\ &= 9.18 \times 10^{-8}\ \text{T}. \quad \text{(Answer)} \end{aligned}$$

(c) Derive an expression for the induced magnetic field for the case $r \ge R$.

SOLUTION: Our procedure is the same as in (a) except we now use an Amperian loop with a radius r that is greater than the plate radius R, to evaluate B outside the capacitor. Evaluating the left and right sides of Eq. 32-30 again leads to Eq. 32-31. However, we then need this subtle **Key Idea**: The electric field exists only between the plates, not outside the plates. Thus, the area A that is encircled by the Amperian loop in the electric field is *not* the full area πr^2 of the loop. Rather, A is only the plate area πR^2.

Substituting πR^2 for A in Eq. 32-31 and solving the result for B give us, for $r \ge R$,

$$B = \frac{\mu_0 \varepsilon_0 R^2}{2r} \frac{dE}{dt}. \quad \text{(Answer)} \quad (32\text{-}33)$$

This equation tells us that, outside the capacitor, B decreases with increased radial distance r, from a maximum value at the plate edges (where $r = R$). By substituting $r = R$ into Eqs. 32-32 and 32-33, you can show that these equations are consistent; that is, they give the same maximum value of B at the plate radius.

The magnitude of the induced magnetic field calculated in (b) is so small that it can scarcely be measured with simple apparatus. This is in sharp contrast to the magnitudes of induced electric fields (Faraday's law), which can be measured easily. This experimental difference exists partly because induced emfs can easily be multiplied by using a coil of many turns. No technique of comparable simplicity exists for multiplying induced magnetic fields. In any case, the experiment suggested by this sample problem has been done, and the presence of the induced magnetic fields has been verified quantitatively.

CHECKPOINT 5: The figure shows graphs of the electric field magnitude E versus time t for four uniform electric fields, all contained within identical circular regions as in Fig. 32-14*b*. Rank the fields according to the magnitudes of the magnetic fields they induce at the edge of the region, greatest first.

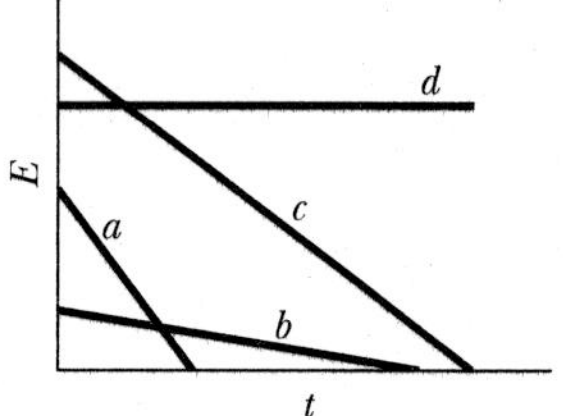

32-10 Displacement Current

If you compare the two terms on the right side of Eq. 32-29, you will see that the product $\varepsilon_0(d\Phi_E/dt)$ must have the dimension of a current. In fact, that product has been treated as being a fictitious current called the **displacement current** i_d:

$$i_d = \varepsilon_0 \frac{d\Phi_E}{dt} \quad \text{(displacement current).} \quad (32\text{-}34)$$

"Displacement" is poorly chosen in that nothing is being displaced, but we are stuck with the word. Nevertheless, we can now rewrite Eq. 32-29 as

$$\oint \vec{B} \cdot d\vec{s} = \mu_0 i_{d,\text{enc}} + \mu_0 i_{\text{enc}} \quad \text{(Ampere–Maxwell law),} \quad (32\text{-}35)$$

in which $i_{d,\text{enc}}$ is the displacement current that is encircled by the integration loop.

Let us again focus on a charging capacitor with circular plates, as in Fig. 32-16*a*. The real current i that is charging the plates changes the electric field $\vec{E}$ between the plates. The fictitious displacement current i_d between the plates is associated with that changing field $\vec{E}$. Let us relate these two currents.

The charge q on the plates at any time is related to the magnitude E of the field between the plates at that time by Eq. 26-4:

$$q = \varepsilon_0 AE, \quad (32\text{-}36)$$

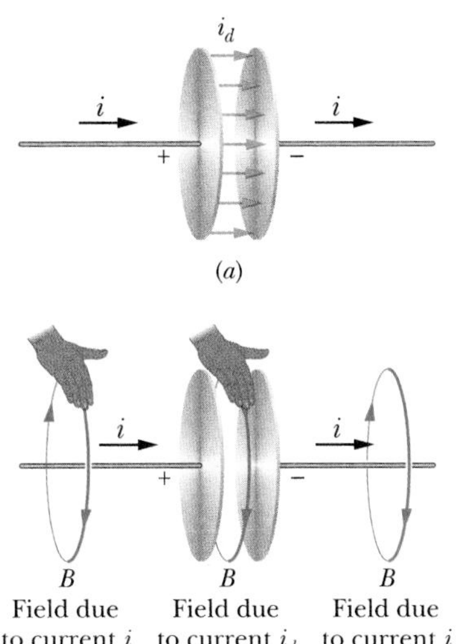

Fig. 32-16 (a) The displacement current i_d between the plates of a capacitor that is being charged by a current i. (b) The right-hand rule for finding the direction of the magnetic field around a wire with a real current (as at the left) also gives the direction of the magnetic field around a displacement current (as in the center).

in which A is the plate area. To get the real current i, we differentiate Eq. 32-36 with respect to time, finding

$$\frac{dq}{dt} = i = \varepsilon_0 A \frac{dE}{dt}. \tag{32-37}$$

To get the displacement current i_d, we can use Eq. 32-34. Assuming that the electric field $\vec{E}$ between the two plates is uniform (we neglect any fringing), we can replace the electric flux Φ_E in that equation with EA. Then Eq. 32-34 becomes

$$i_d = \varepsilon_0 \frac{d\Phi_E}{dt} = \varepsilon_0 \frac{d(EA)}{dt} = \varepsilon_0 A \frac{dE}{dt}. \tag{32-38}$$

Comparing Eqs. 32-37 and 32-38, we see that the real current i charging the capacitor and the fictitious displacement current i_d between the plates have the same magnitude:

$$i_d = i \qquad \text{(displacement current in a capacitor).} \tag{32-39}$$

Thus, we can consider the fictitious displacement current i_d to be simply a continuation of the real current i from one plate, across the capacitor gap, to the other plate. Because the electric field is uniformly spread over the plates, the same is true of this fictitious displacement current i_d, as suggested by the spread of current arrows in Fig. 32-16*a*. Although no charge actually moves across the gap between the plates, the idea of the fictitious current i_d can help us to quickly find the direction and magnitude of an induced magnetic field, as follows.

Finding the Induced Magnetic Field

In Chapter 30 we found the direction of the magnetic field produced by a real current i by using the right-hand rule of Fig. 30-4. We can apply the same rule to find the direction of an induced magnetic field produced by a fictitious displacement current i_d, as is shown in the center of Fig. 32-16*b* for a capacitor.

We can also use i_d to find the magnitude of the magnetic field induced by a charging capacitor with parallel circular plates of radius R. We simply consider the space between the plates to be an imaginary circular wire of radius R carrying the imaginary current i_d. Then, from Eq. 30-22, the magnitude of the magnetic field at a point inside the capacitor at radius r from the center is

$$B = \left(\frac{\mu_0 i_d}{2\pi R^2}\right) r \qquad \text{(inside a circular capacitor).} \tag{32-40}$$

Similarly, from Eq. 30-19, the magnitude of the magnetic field at a point outside the capacitor at radius r is

$$B = \frac{\mu_0 i_d}{2\pi r} \qquad \text{(outside a circular capacitor).} \tag{32-41}$$

Sample Problem 32-4

The circular parallel-plate capacitor in Sample Problem 32-3 is being charged with a current i.

(a) Between the plates, what is the magnitude of $\oint \vec{B} \cdot d\vec{s}$, in terms of μ_0 and i, at a radius $r = R/5$ from their center?

SOLUTION: The first **Key Idea** of Sample Problem 32-3a holds here too. However, now we can replace the product $\varepsilon_0\, d\Phi_E/dt$ in Eq. 32-29 with a fictitious displacement current i_d. Then integral $\oint \vec{B} \cdot d\vec{s}$ is given by Eq. 32-35, but because there is no real current i between the capacitor plates, the equation reduces to

$$\oint \vec{B} \cdot d\vec{s} = \mu_0 i_{d,\text{enc}}. \tag{32-42}$$

Because we want to evaluate $\oint \vec{B} \cdot d\vec{s}$ at radius $r = R/5$ (within the capacitor), the integration loop encircles only a portion $i_{d,\text{enc}}$ of the total displacement current i_d. A second Key Idea is to assume that i_d is uniformly spread over the full plate area. Then the portion of the displacement current encircled by the loop is proportional to the area encircled by the loop:

$$\frac{\begin{pmatrix}\text{encircled displacement} \\ \text{current } i_{d,\text{enc}}\end{pmatrix}}{\begin{pmatrix}\text{total displacement} \\ \text{current } i_d\end{pmatrix}} = \frac{\text{encircled area } \pi r^2}{\text{full plate area } \pi R^2}.$$

This gives us

$$i_{d,\text{enc}} = i_d \frac{\pi r^2}{\pi R^2}.$$

Substituting this into Eq. 32-42, we obtain

$$\oint \vec{B} \cdot d\vec{s} = \mu_0 i_d \frac{\pi r^2}{\pi R^2}. \tag{32-43}$$

Now substituting $i_d = i$ (from Eq. 32-39) and $r = R/5$ into Eq. 32-43 leads to

$$\oint \vec{B} \cdot d\vec{s} = \mu_0 i \frac{(R/5)^2}{R^2} = \frac{\mu_0 i}{25}. \quad \text{(Answer)}$$

(b) In terms of the maximum induced magnetic field, what is the magnitude of the magnetic field induced at $r = R/5$, inside the capacitor?

SOLUTION: The Key Idea here is that, because the capacitor has parallel circular plates, we can treat the space between the plates as an imaginary wire of radius R carrying the imaginary current i_d. Then we can use Eq. 32-40 to find the induced magnetic field magnitude B at any point inside the capacitor. At $r = R/5$, that equation yields

$$B = \left(\frac{\mu_0 i_d}{2\pi R^2}\right) r = \frac{\mu_0 i_d (R/5)}{2\pi R^2} = \frac{\mu_0 i_d}{10\pi R}. \tag{32-44}$$

The maximum field magnitude B_{max} within the capacitor occurs at $r = R$. It is

$$B_{\text{max}} = \left(\frac{\mu_0 i_d}{2\pi R^2}\right) R = \frac{\mu_0 i_d}{2\pi R}. \tag{32-45}$$

Dividing Eq. 32-44 by Eq. 32-45 and rearranging the result, we find

$$B = \frac{B_{\text{max}}}{5}. \quad \text{(Answer)}$$

We should be able to obtain this result with a little reasoning and less work. Equation 32-40 tells us that inside the capacitor, B increases linearly with r. Therefore, a point $\frac{1}{5}$ the distance out to the full radius R of the plates, where B_{max} occurs, should have a field B that is $\frac{1}{5}B_{\text{max}}$.

CHECKPOINT 6: The figure is a view of one plate of a parallel-plate capacitor from within the capacitor. The dashed lines show four integration paths (path b follows the edge of the plate). Rank the paths according to the magnitude of $\oint \vec{B} \cdot d\vec{s}$ along the paths during the discharging of the capacitor, greatest first.

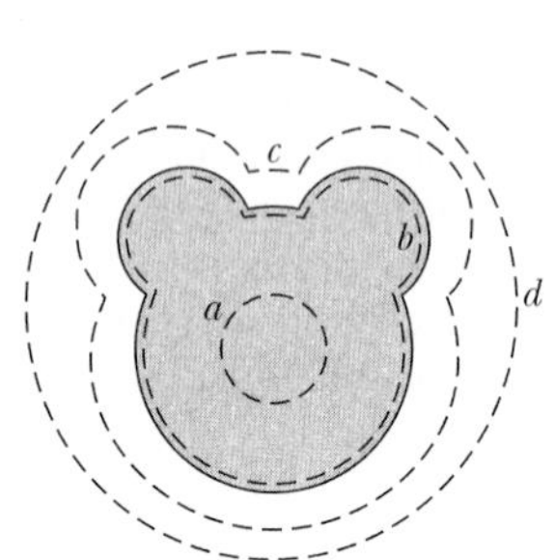

32-11 Maxwell's Equations

Equation 32-29 is the last of the four fundamental equations of electromagnetism, called *Maxwell's equations* and displayed in Table 32-1. These four equations explain a diverse range of phenomena, from why a compass needle points north to why a car starts when you turn the ignition key. They are the basis for the functioning of such electromagnetic devices as electric motors, cyclotrons, television transmitters and receivers, telephones, fax machines, radar, and microwave ovens.

Maxwell's equations are the basis from which many of the equations you have seen since Chapter 22 can be derived. They are also the basis of many of the equations you will see in Chapters 34 through 37, which introduce you to optics.

TABLE 32-1 **Maxwell's Equations**[a]

Name	Equation	
Gauss' law for electricity	$\oint \vec{E} \cdot d\vec{A} = q_{\text{enc}}/\varepsilon_0$	Relates net electric flux to net enclosed electric charge
Gauss' law for magnetism	$\oint \vec{B} \cdot d\vec{A} = 0$	Relates net magnetic flux to net enclosed magnetic charge
Faraday's law	$\oint \vec{E} \cdot d\vec{s} = -\frac{d\Phi_B}{dt}$	Relates induced electric field to changing magnetic flux
Ampere–Maxwell law	$\oint \vec{B} \cdot d\vec{s} = \mu_0 \varepsilon_0 \frac{d\Phi_E}{dt} + \mu_0 i_{\text{enc}}$	Relates induced magnetic field to changing electric flux and to current

[a] Written on the assumption that no dielectric or magnetic materials are present.

REVIEW & SUMMARY

Gauss' Law for Magnetic Fields The simplest magnetic structures are magnetic dipoles. Magnetic monopoles do not exist (as far as we know). **Gauss' law** for magnetic fields,

$$\Phi_B = \oint \vec{B} \cdot d\vec{A} = 0, \quad (32\text{-}1)$$

states that the net magnetic flux through any (closed) Gaussian surface is zero. It implies that magnetic monopoles don't exist.

Earth's Magnetic Field Earth's magnetic field can be approximated as being that of a magnetic dipole whose dipole moment makes an angle of 11.5° with Earth's rotation axis, and with the south pole of the dipole in the northern hemisphere. The direction of the local magnetic field at any point on Earth's surface is given by the *field declination* (the angle left or right from geographic north) and the *field inclination* (the angle up or down from the horizontal).

Spin Magnetic Dipole Moment An electron has an intrinsic angular momentum called *spin angular momentum* (or *spin*) $\vec{S}$, with which an intrinsic *spin magnetic dipole moment* $\vec{\mu}_s$ is associated:

$$\vec{\mu}_s = -\frac{e}{m}\vec{S}. \quad (32\text{-}2)$$

Spin $\vec{S}$ cannot itself be measured, but any component can be measured. Assuming that the measurement is along the z axis of a coordinate system, the component S_z can have only the values given by

$$S_z = m_s \frac{h}{2\pi}, \quad \text{for } m_s = \pm\tfrac{1}{2}, \quad (32\text{-}3)$$

where h ($= 6.63 \times 10^{-34}$ J·s) is the Planck constant. Similarly, the electron's spin magnetic dipole moment $\vec{\mu}_s$ cannot itself be measured but its component can be measured. Along a z axis, the component is

$$\mu_{s,z} = \pm\frac{eh}{4\pi m} = \pm\mu_B, \quad (32\text{-}4, 32\text{-}6)$$

where μ_B is the *Bohr magneton:*

$$\mu_B = \frac{eh}{4\pi m} = 9.27 \times 10^{-24} \text{ J/T}. \quad (32\text{-}5)$$

The potential energy U associated with the orientation of the spin magnetic dipole moment in an external magnetic field $\vec{B}_{ext}$ is

$$U = -\vec{\mu}_s \cdot \vec{B}_{ext} = -\mu_{s,z}B. \quad (32\text{-}7)$$

Orbital Magnetic Dipole Moment An electron in an atom has an additional angular momentum called its *orbital angular momentum* $\vec{L}_{orb}$, with which an *orbital magnetic dipole moment* $\vec{\mu}_{orb}$ is associated:

$$\vec{\mu}_{orb} = -\frac{e}{2m}\vec{L}_{orb}. \quad (32\text{-}8)$$

Orbital angular momentum is quantized and can have only values given by

$$L_{orb,z} = m_l \frac{h}{2\pi}, \quad \text{for } m_l = 0, \pm 1, \pm 2, \ldots, \pm(\text{limit}). \quad (32\text{-}9)$$

Thus, the magnitude of the orbital angular momentum is

$$\mu_{orb,z} = -m_l \frac{eh}{4\pi m} = -m_l \mu_B. \quad (32\text{-}10, 32\text{-}11)$$

The potential energy U associated with the orientation of the orbital magnetic dipole moment in an external magnetic field $\vec{B}_{ext}$ is

$$U = -\vec{\mu}_{orb} \cdot \vec{B}_{ext} = -\mu_{orb,z}B_{ext}. \quad (32\text{-}12)$$

Diamagnetism *Diamagnetic materials* do not exhibit magnetism until they are placed in an external magnetic field $\vec{B}_{ext}$. They then develop a magnetic dipole moment directed opposite $\vec{B}_{ext}$. If the field is nonuniform, the diamagnetic material is repelled from regions of greater magnetic field. This property is called *diamagnetism*.

Paramagnetism In a *paramagnetic material,* each atom has a permanent magnetic dipole moment $\vec{\mu}$, but the dipole moments are randomly oriented and the material as a whole lacks a magnetic field. However, an external magnetic field $\vec{B}_{ext}$ can partially align the atomic dipole moments to give the material a net magnetic dipole moment in the direction of $\vec{B}_{ext}$. If $\vec{B}_{ext}$ is nonuniform, the material is attracted to regions of greater magnetic field. These properties are called *paramagnetism*.

The alignment of the atomic dipole moments increases with an increase in $\vec{B}_{ext}$ and decreases with an increase in temperature T. The extent to which a sample of volume V is magnetized is given by its *magnetization* $\vec{M}$, whose magnitude is

$$M = \frac{\text{measured magnetic moment}}{V}. \quad (32\text{-}18)$$

Complete alignment of all N atomic magnetic dipoles in a sample, called *saturation* of the sample, corresponds to the maximum magnetization value $M_{max} = N\mu/V$. For low values of the ratio B_{ext}/T, we have the approximation

$$M = C\frac{B_{ext}}{T} \quad \text{(Curie's law)}, \quad (32\text{-}19)$$

where C is called the *Curie constant*.

Ferromagnetism In the absence of an external magnetic field, some of the electrons in a ferromagnetic material have their magnetic dipole moments aligned by means of a quantum physical interaction called *exchange coupling*, producing regions (domains) within the material with strong magnetic dipole moments. An external field $\vec{B}_{ext}$ can align the magnetic dipole moments of those regions, producing a strong net magnetic dipole moment for the material as a whole, in the direction of $\vec{B}_{ext}$. This net magnetic dipole moment can partially persist when field $\vec{B}_{ext}$ is removed. If $\vec{B}_{ext}$ is nonuniform, the ferromagnetic material is attracted to regions of greater magnetic field. These properties are called *ferromagnetism*. Exchange coupling disappears when a sample's temperature exceeds its *Curie temperature*, and then the sample has only paramagnetism.

Maxwell's Extension of Ampere's Law A changing electric flux induces a magnetic field $\vec{B}$. Maxwell's law,

$$\oint \vec{B} \cdot d\vec{s} = \mu_0 \varepsilon_0 \frac{d\Phi_E}{dt} \quad \text{(Maxwell's law of induction)}, \qquad (32\text{-}27)$$

relates the magnetic field induced along a closed loop to the changing electric flux Φ_E through the loop. Ampere's law, $\oint \vec{B} \cdot d\vec{s} = \mu_0 i_{enc}$ (Eq. 32-28), gives the magnetic field generated by a current i_{enc} encircled by a closed loop. Maxwell's law and Ampere's law can be written as the single equation

$$\oint \vec{B} \cdot d\vec{s} = \mu_0 \varepsilon_0 \frac{d\Phi_E}{dt} + \mu_0 i_{enc} \quad \text{(Ampere–Maxwell law)}. \qquad (32\text{-}29)$$

Displacement Current We define the fictitious *displacement current* due to a changing electric field as

$$i_d = \varepsilon_0 \frac{d\Phi_E}{dt}. \qquad (32\text{-}34)$$

Equation 32-29 then becomes

$$\oint \vec{B} \cdot d\vec{s} = \mu_0 i_{d,enc} + \mu_0 i_{enc} \quad \text{(Ampere–Maxwell law)}, \qquad (32\text{-}35)$$

where $i_{d,enc}$ is the displacement current encircled by the integration loop. The idea of a displacement current allows us to retain the notion of continuity of current through a capacitor. However, displacement current is *not* a transfer of charge.

Maxwell's Equations Maxwell's equations, displayed in Table 32-1, summarize electromagnetism and form its foundation.

QUESTIONS

1. An electron in an external magnetic field $\vec{B}_{ext}$ has its spin angular momentum S_z antiparallel to $\vec{B}_{ext}$. If the electron undergoes a *spin-flip* so that S_z is then parallel with $\vec{B}_{ext}$, must energy be supplied to or lost by the electron?

2. Figure 32-17*a* shows a pair of opposite spin orientations for an electron in an external magnetic field $\vec{B}_{ext}$. Figure 32-17*b* gives three choices for the graph of the potential energies associated with those orientations as a function of the magnitude of $\vec{B}_{ext}$. Choices *b* and *c* consist of intersecting lines, choice *a* of parallel lines. Which is the correct choice?

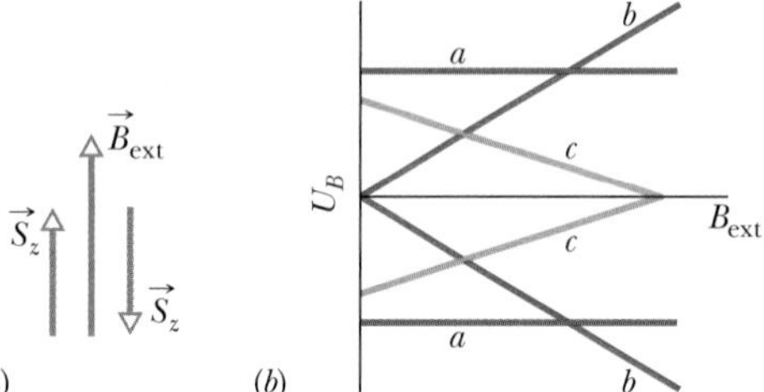

Fig. 32-17 Question 2.

3. Figure 32-18 shows three loop models of an electron orbiting counterclockwise within a magnetic field. The fields are nonuniform for models 1 and 2 and uniform for model 3. For each model, are (a) the magnetic dipole moment of the loop and (b) the magnetic force on the loop directed up, directed down, or zero?

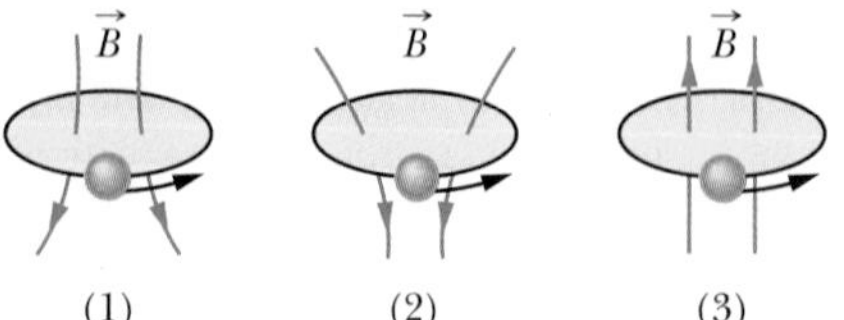

Fig. 32-18 Questions 3, 5, and 6.

4. Does the magnitude of the net force on the loop of Figs. 32-8*a* and *b* increase, decrease, or remain the same if we increase (a) the magnitude of $\vec{B}_{ext}$ and (b) the divergence of $\vec{B}_{ext}$?

5. Replace the current loops of Question 3 and Fig. 32-18 with diamagnetic spheres. For each field, are (a) the magnetic dipole moment of the sphere and (b) the magnetic force on the sphere directed up, directed down, or zero?

6. Replace the current loops of Question 3 and Fig. 32-18 with paramagnetic spheres. For each field, are (a) the magnetic dipole moment of the sphere and (b) the magnetic force on the sphere up, down, or zero?

7. The magnetic dipoles in a diamagnetic material are represented, for three situations, in Fig. 32-19. (For simplicity, the dipole moments are assumed to be directed only up or down the page.) The three situations differ in the magnitude of a magnetic field applied to the material. (a) For each situation, is the applied field directed up or down the page? Rank the three situations according to (b) the magnitude of the applied field and (c) the magnetization of the material, greatest first.

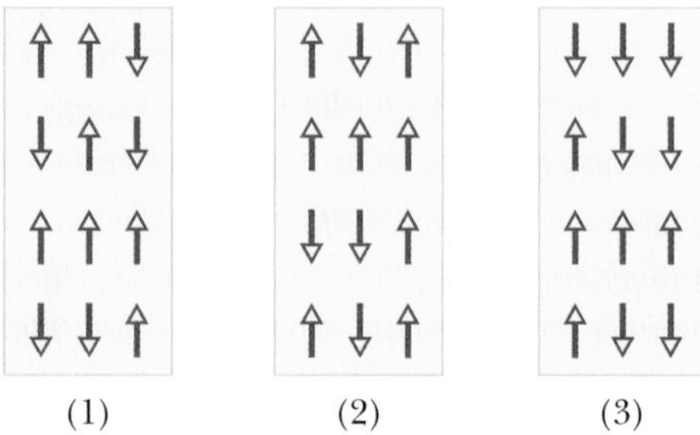

Fig. 32-19 Question 7.

8. Figure 32-20 shows, in two situations, an electric field vector and an induced magnetic field line. In each, is the magnitude of $\vec{E}$ increasing or decreasing?

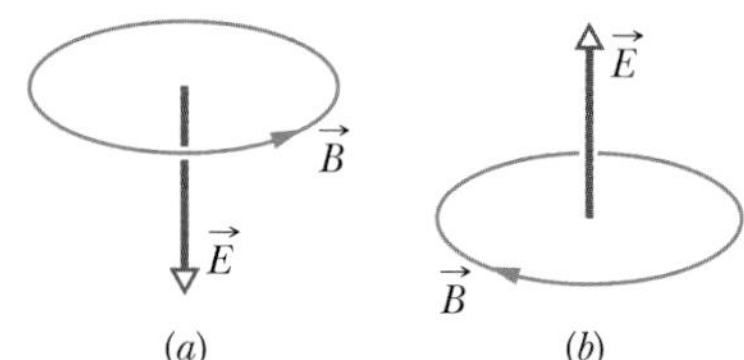

Fig. 32-20 Question 8.

9. Figure 32-21 shows a parallel-plate capacitor and the current in the connecting wires that is discharging the capacitor. Are the directions of (a) electric field $\vec{E}$ and (b) displacement current i_d leftward or rightward between the plates? (c) Is the magnetic field at point P into the page or out of the page?

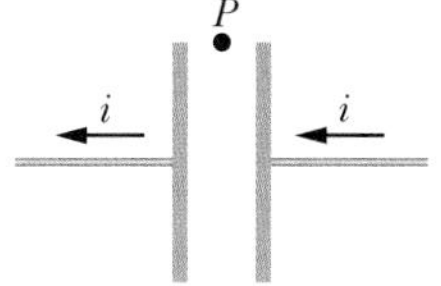

Fig. 32-21 Question 9.

10. A parallel-plate capacitor with rectangular plates is being discharged. A rectangular loop, centered on the plates and between them, measures L by $2L$; the plates measure $2L$ by $4L$. What fraction of the displacement current is encircled by the loop if that current is uniform?

11. Figure 32-22a shows a capacitor, with circular plates, that is being charged. Point a (near one of the connecting wires) and point b (inside the capacitor gap) are equidistant from the central axis, as are point c (not so near the wire) and point d (between the plates but outside the gap). In Fig. 32-22b, one curve gives the variation with distance r of the magnitude of the magnetic field inside and outside the wire. The other curve gives the variation with distance r of the magnitude of the magnetic field inside and outside the gap. The two curves partially overlap. Which of the three points on the curves correspond to which of the four points of Fig. 32-22a?

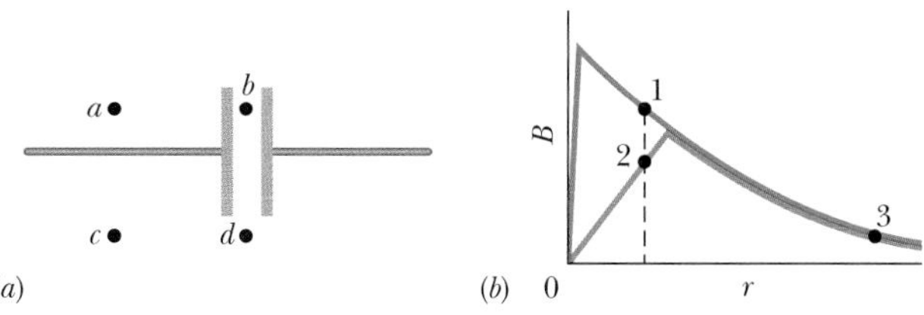

Fig. 32-22 Question 11.

EXERCISES & PROBLEMS

ssm Solution is in the Student Solutions Manual.
www Solution is available on the World Wide Web at: http://www.wiley.com/college/hrw
ilw Solution is available on the Interactive LearningWare.

SEC. 32-2 Gauss' Law for Magnetic Fields

1E. Imagine rolling a sheet of paper into a cylinder and placing a bar magnet near its end as shown in Fig. 32-23. (a) Sketch the magnetic field lines that pass through the surface of the cylinder. (b) What can you say about the sign of $\vec{B} \cdot d\vec{A}$ for every area $d\vec{A}$ on the surface? (c) Does this contradict Gauss' law for magnetism? Explain.

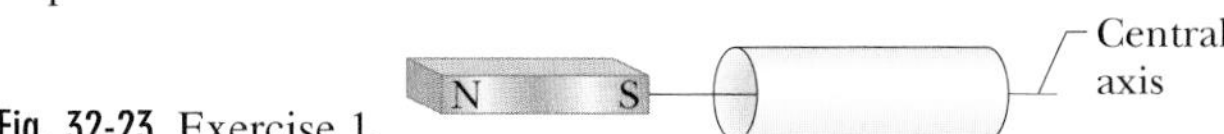

Fig. 32-23 Exercise 1.

2E. The magnetic flux through each of five faces of a die (singular of "dice") is given by $\Phi_B = \pm N$ Wb, where N (= 1 to 5) is the number of spots on the face. The flux is positive (outward) for N even and negative (inward) for N odd. What is the flux through the sixth face of the die?

3P. A Gaussian surface in the shape of a right circular cylinder with end caps has a radius of 12.0 cm and a length of 80.0 cm. Through one end there is an inward magnetic flux of 25.0 μWb. At the other end there is a uniform magnetic field of 1.60 mT, normal to the surface and directed outward. What is the net magnetic flux through the curved surface? ssm ilw www

SEC. 32-3 The Magnetism of Earth

4E. Assume the average value of the vertical component of Earth's magnetic field is 43 μT (downward) for all of Arizona, which has an area of 2.95×10^5 km^2, and calculate the net magnetic flux through the rest of Earth's surface (the entire surface excluding Arizona). Is that net magnetic flux outward or inward?

5E. In New Hampshire the average horizontal component of Earth's magnetic field in 1912 was 16 μT and the average inclination or "dip" was 73°. What was the corresponding magnitude of Earth's magnetic field? ssm

6P. The magnetic field of Earth can be approximated as the magnetic field of a dipole, with horizontal and vertical components, at a point a distance r from Earth's center, given by

$$B_h = \frac{\mu_0 \mu}{4\pi r^3} \cos \lambda_m, \qquad B_v = \frac{\mu_0 \mu}{2\pi r^3} \sin \lambda_m,$$

where λ_m is the *magnetic latitude* (this type of latitude is measured from the geomagnetic equator toward the north or south geomagnetic pole). Assume that Earth's magnetic dipole moment is $\mu = 8.00 \times 10^{22}$ A $\cdot$ m^2. (a) Show that the magnitude of Earth's field at latitude λ_m is given by

$$B = \frac{\mu_0 \mu}{4\pi r^3} \sqrt{1 + 3 \sin^2 \lambda_m}.$$

(b) Show that the inclination ϕ_i of the magnetic field is related to the magnetic latitude λ_m by

$$\tan \phi_i = 2 \tan \lambda_m.$$

7P. Use the results displayed in Problem 6 to predict Earth's magnetic field (both magnitude and inclination) at (a) the geomagnetic equator, (b) a point at geomagnetic latitude 60°, and (c) the north geomagnetic pole. ssm

8P. Using the approximations given in Problem 6, find (a) the altitude above Earth's surface where the magnitude of its magnetic field is 50% of the surface value at the same latitude; (b) the maximum magnitude of the magnetic field at the core–mantle boundary, 2900 km below Earth's surface; and (c) the magnitude and inclination of Earth's magnetic field at the north geographic pole. Suggest why the values you calculated for (c) differ from measured values.

SEC. 32-4 Magnetism and Electrons

9E. What is the measured component of the orbital magnetic dipole moment of an electron with (a) $m_l = 1$ and (b) $m_l = -2$? ssm

10E. What is the energy difference between parallel and antiparallel alignment of the z component of an electron's spin magnetic dipole moment with an external magnetic field of magnitude 0.25 T, directed parallel to the z axis?

11E. If an electron in an atom has an orbital angular momentum with $m_l = 0$, what are the components (a) $L_{orb,z}$ and (b) $\mu_{orb,z}$? If the atom is in an external magnetic field $\vec{B}$ of magnitude 35 mT and directed along the z axis, what are the potential energies associated with the orientations of (c) the electron's orbital magnetic dipole moment and (d) the electron's spin magnetic dipole moment? (e) Repeat (a) through (d) for $m_l = -3$. ssm

SEC. 32-6 Diamagnetism

12E. Figure 32-24 shows a loop model (loop L) for a diamagnetic material. (a) Sketch the magnetic field lines through and about the material due to the bar magnet. (b) What are the directions of the loop's net magnetic dipole moment $\vec{\mu}$ and the conventional current i in the loop? (c) What is the direction of the magnetic force on the loop?

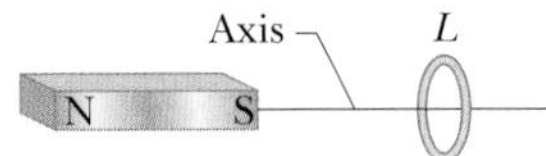

Fig. 32-24 Exercises 12 and 16.

13P*. Assume that an electron of mass m and charge magnitude e moves in a circular orbit of radius r about a nucleus. A uniform magnetic field $\vec{B}$ is then established perpendicular to the plane of the orbit. Assuming also that the radius of the orbit does not change and that the change in the speed of the electron due to field $\vec{B}$ is small, find an expression for the change in the orbital magnetic dipole moment of the electron due to the field. ssm

SEC. 32-7 Paramagnetism

14E. A 0.50 T magnetic field is applied to a paramagnetic gas whose atoms have an intrinsic magnetic dipole moment of 1.0×10^{-23} J/T. At what temperature will the mean kinetic energy of translation of the gas atoms be equal to the energy required to reverse such a dipole end for end in this magnetic field?

15E. A magnet in the form of a cylindrical rod has a length of 5.00 cm and a diameter of 1.00 cm. It has a uniform magnetization of 5.30×10^3 A/m. What is its magnetic dipole moment? ssm ilw

16E. Repeat Exercise 12 for the case in which loop L is the model for a paramagnetic material.

17E. A sample of the paramagnetic salt to which the magnetization curve of Fig. 32-9 applies is to be tested to see whether it obeys Curie's law. The sample is placed in a uniform 0.50 T magnetic field that remains constant throughout the experiment. The magnetization M is then measured at temperatures ranging from 10 to 300 K. Will it be found that Curie's law is valid under these conditions? ssm

18E. A sample of the paramagnetic salt to which the magnetization curve of Fig. 32-9 applies is held at room temperature (300 K). At what applied magnetic field will the degree of magnetic saturation of the sample be (a) 50% and (b) 90%? (c) Are these fields attainable in the laboratory?

19P. An electron with kinetic energy K_e travels in a circular path that is perpendicular to a uniform magnetic field, the electron's motion subject only to the force due to the field. (a) Show that the magnetic dipole moment of the electron due to its orbital motion has magnitude $\mu = K_e/B$ and that it is in the direction opposite that of $\vec{B}$. (b) What are the magnitude and direction of the magnetic dipole moment of a positive ion with kinetic energy K_i under the same circumstances? (c) An ionized gas consists of 5.3×10^{21} electrons/m^3 and the same number density of ions. Take the average electron kinetic energy to be 6.2×10^{-20} J and the average ion kinetic energy to be 7.6×10^{-21} J. Calculate the magnetization of the gas when it is in a magnetic field of 1.2 T. ssm www

SEC. 32-8 Ferromagnetism

20E. Measurements in mines and boreholes indicate that Earth's interior temperature increases with depth at the average rate of 30 C°/km. Assuming a surface temperature of 10°C, at what depth does iron cease to be ferromagnetic? (The Curie temperature of iron varies very little with pressure.)

21E. The exchange coupling mentioned in Section 32-8 as being responsible for ferromagnetism is *not* the mutual magnetic interaction between two elementary magnetic dipoles. To show this, calculate (a) the magnitude of the magnetic field a distance of 10 nm away, along the dipole axis, from an atom with magnetic dipole moment 1.5×10^{-23} J/T (cobalt), and (b) the minimum energy required to turn a second identical dipole end for end in this field. By comparing the latter with the results of Sample Problem 32-1, what can you conclude? ssm

22E. The dipole moment associated with an atom of iron in an iron bar is 2.1×10^{-23} J/T. Assume that all the atoms in the bar, which is 5.0 cm long and has a cross-sectional area of 1.0 cm^2, have their dipole moments aligned. (a) What is the dipole moment of the bar? (b) What torque must be exerted to hold this magnet perpendicular to an external field of 1.5 T? (The density of iron is 7.9 g/cm^3.)

23E. The saturation magnetization M_{max} of the ferromagnetic metal nickel is 4.70×10^5 A/m. Calculate the magnetic moment of a single nickel atom. (The density of nickel is 8.90 g/cm^3 and its molar mass is 58.71 g/mol.) ssm

24P. Figure 32-25 shows the apparatus used in a lecture demonstration of para- and diamagnetism. A sample of the magnetic material is suspended by a long string in the nonuniform field (d = 2 cm) between the poles of a powerful electromagnet. Pole P_1 is sharply pointed and pole P_2 is rounded as indicated. Any deflection of the string from the vertical is visible to the audience by means of an optical projection system (not shown). (a) First a bismuth (highly diamagnetic) sample is used. When the electromagnet is turned on, the sample is observed to deflect slightly (about 1 mm) toward one of the poles. What is the direction of this deflection? (b) Next an aluminum (paramagnetic, conducting) sample is used. When the electromagnet is turned on, the sample is observed to de-

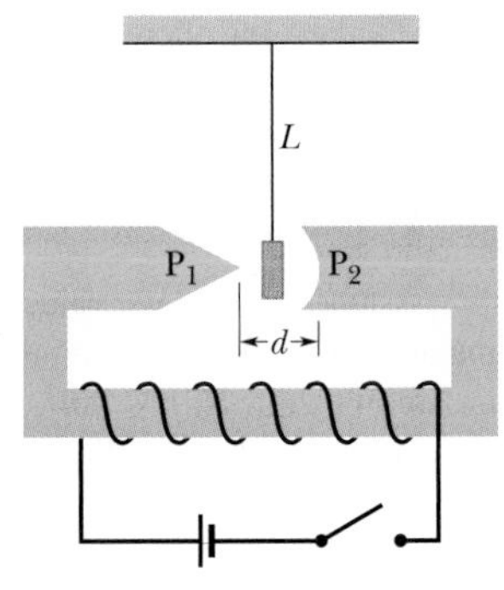

Fig. 32-25 Problem 24.

flect strongly (about 1 cm) toward one pole for about a second and then deflect moderately (a few millimeters) toward the other pole. Explain and indicate the direction of these deflections. (*Hint:* The aluminum sample is a conductor, for which Lenz's law applies.) (c) What would happen if a ferromagnetic sample were used?

25P. The magnetic dipole moment of Earth is 8.0×10^{22} J/T. (a) If the origin of this magnetism were a magnetized iron sphere at the center of Earth, what would be its radius? (b) What fraction of the volume of Earth would such a sphere occupy? Assume complete alignment of the dipoles. The density of Earth's inner core is 14 g/cm^3. The magnetic dipole moment of an iron atom is 2.1×10^{-23} J/T. (*Note:* Earth's inner core is in fact thought to be in both liquid and solid forms and partly iron, but a permanent magnet as the source of Earth's magnetism has been ruled out by several considerations. For one, the temperature is certainly above the Curie point.) ssm ilw

SEC. 32-9 Induced Magnetic Fields

26E. Sample Problem 32-3 describes the charging of a parallel-plate capacitor with circular plates of radius 55.0 mm. At what two radii r from the central axis of the capacitor is the magnitude of the induced magnetic field equal to 50% of its maximum value?

27E. The induced magnetic field 6.0 mm from the central axis of a circular parallel-plate capacitor and between the plates is 2.0×10^{-7} T. The plates have radius 3.0 mm. At what rate $d\vec{E}/dt$ is the electric field between the plates changing? ssm

28P. Suppose that a parallel-plate capacitor has circular plates with radius $R = 30$ mm and a plate separation of 5.0 mm. Suppose also that a sinusoidal potential difference with a maximum value of 150 V and a frequency of 60 Hz is applied across the plates; that is,

$$V = (150 \text{ V}) \sin[2\pi(60 \text{ Hz})t].$$

(a) Find $B_{max}(R)$, the maximum value of the induced magnetic field that occurs at $r = R$. (b) Plot $B_{max}(r)$ for $0 < r < 10$ cm.

SEC. 32-10 Displacement Current

29E. Prove that the displacement current in a parallel-plate capacitor of capacitance C can be written as $i_d = C(dV/dt)$, where V is the potential difference between the plates. ssm

30E. At what rate must the potential difference between the plates of a parallel-plate capacitor with a 2.0 μF capacitance be changed to produce a displacement current of 1.5 A?

31E. For the situation of Sample Problem 32-3, show that the current density of the displacement current is $J_d = \varepsilon_0(dE/dt)$ for $r \leq R$. ssm

32E. A parallel-plate capacitor with circular plates of radius 0.10 m is being discharged. A circular loop of radius 0.20 m is concentric with the capacitor and halfway between the plates. The displacement current through the loop is 2.0 A. At what rate is the electric field between the plates changing?

33P. As a parallel-plate capacitor with circular plates 20 cm in diameter is being charged, the current density of the displacement current in the region between the plates is uniform and has a magnitude of 20 A/m^2. (a) Calculate the magnitude B of the magnetic field at a distance $r = 50$ mm from the axis of symmetry of this region. (b) Calculate dE/dt in this region. ssm ilw

34P. The magnitude of the electric field between the two circular parallel plates in Fig. 32-26 is $E = (4.0 \times 10^5) - (6.0 \times 10^4 t)$, with E in volts per meter and t in seconds. At $t = 0$, the field is upward as shown. The plate area is 4.0×10^{-2} m^2. For $t \geq 0$, (a) what are the magnitude and direction of the displacement current between the plates and (b) is the direction of the induced magnetic field clockwise or counterclockwise around the plates?

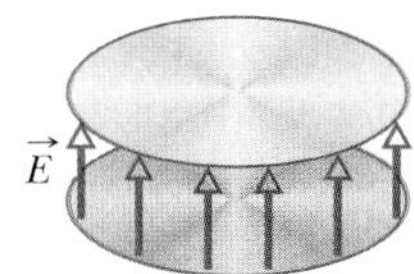

Fig. 32-26 Problem 34.

35P. A uniform electric field collapses to zero from an initial strength of 6.0×10^5 N/C in a time of 15 μs in the manner shown in Fig. 32-27. Calculate the magnitude of the displacement current, through a 1.6 m^2 area perpendicular to the field, during each of the time intervals a, b, and c shown on the graph. (Ignore the behavior at the ends of the intervals.) ssm ilw

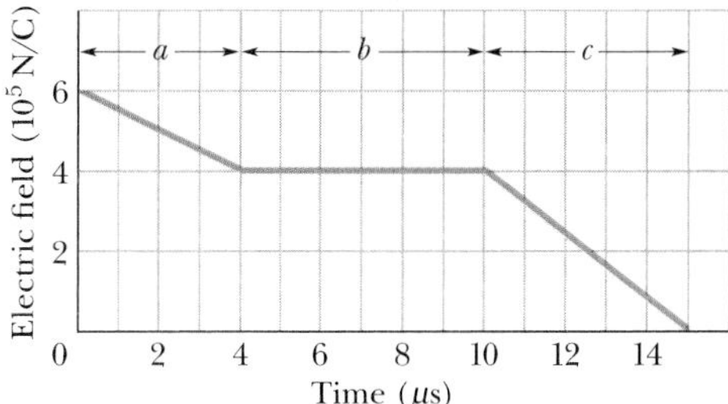

Fig. 32-27 Problem 35.

36P. A parallel-plate capacitor with circular plates is being charged. Consider a circular loop centered on the central axis between the plates. The loop radius is 0.20 m; the plate radius is 0.10 m; and the displacement current through the loop is 2.0 A. What is the rate at which the electric field between the plates is changing?

37P. A parallel-plate capacitor has square plates 1.0 m on a side as shown in Fig. 32-28. A current of 2.0 A charges the capacitor, producing a uniform electric field $\vec{E}$ between the plates, with $\vec{E}$ perpendicular to the plates. (a) What is the displacement current i_d through the region between the plates? (b) What is dE/dt in this region? (c) What is the displacement current through the square dashed path between the plates? (d) What is $\oint \vec{B} \cdot d\vec{s}$ around this square dashed path? ssm ilw www

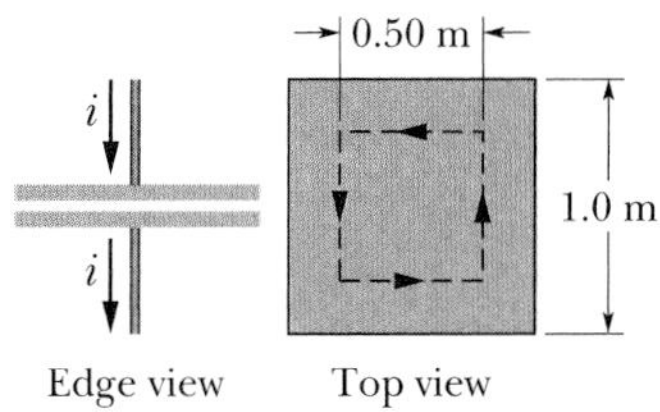

Fig. 32-28 Problem 37.

38P. A capacitor with parallel circular plates of radius R is discharging via a current of 12.0 A. Consider a loop of radius $R/3$ that is centered on the central axis between the plates. (a) How much displacement current is encircled by the loop? The maximum induced magnetic field has a magnitude of 12.0 mT. (b) At what radial distance from the central axis of the plate is the magnitude of the induced magnetic field 3.00 mT?

NEW PROBLEMS

N1. Figure 32N-1 shows a closed surface. Along the flat top face, which has a radius of 2.0 cm, a magnetic field $\vec{B}$ of magnitude 0.30 T is directed outward. Along the flat bottom face, a magnetic flux of 0.70 mWb is directed outward. What are (a) the magnitude and (b) the direction of the magnetic flux through the curved part of the surface?

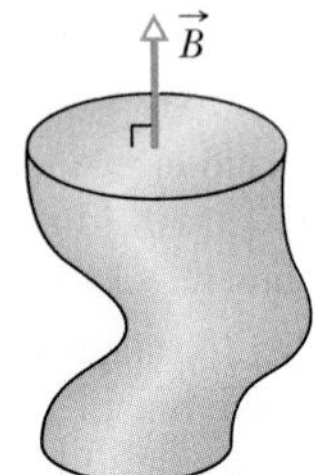

Fig. 32N-1 Problem N1.

N2. Figure 32N-2*a* shows current i that is produced in a wire of resistivity $1.62 \times 10^{-8}\ \Omega \cdot m$ in the direction indicated. The magnitude of the current versus time t is shown in Fig. 32N-2*b*. Point P is at radius 9.00 mm from the wire's center. Determine the magnitude of the magnetic field at point P due to the actual current i in the wire at (a) $t = 20$ ms, (b) $t = 40$ ms, (c) $t = 60$ ms, and (d) $t = 70$ ms. Next, assume that the electric field driving the current is confined to the wire. Then determine the magnitude of the magnetic field at point P due to the displacement current i_d in the wire at (e) $t = 20$ ms, (f) $t = 40$ ms, (g) $t = 60$ ms, and (h) $t = 70$ ms. (i) When both magnetic fields are present at point P, what are their directions in Fig. 32N-2*a*?

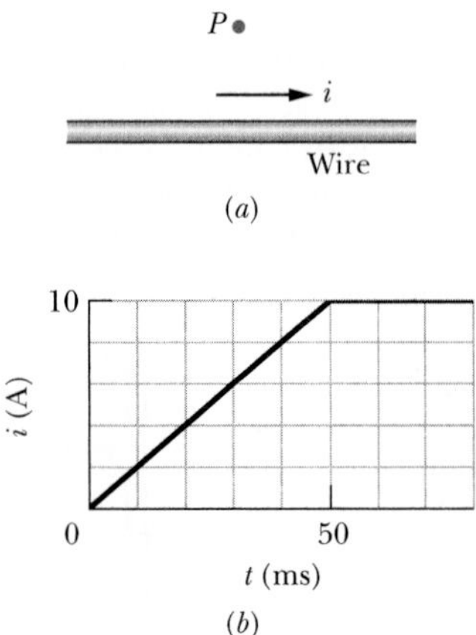

Fig. 32N-2 Problem N2.

N3. An electron is placed in a magnetic field $\vec{B}$ that is directed along a z axis. The energy difference between parallel and antiparallel alignments of the z component of the electron's spin magnetic moment with $\vec{B}$ is 6.00×10^{-25} J. What is the magnitude of $\vec{B}$?

N4. Two plates (as in Fig. 32-16) are being discharged by a constant current. Each plate has a radius of 4.00 cm. During the discharging, at a point between the plates at radius 2.00 cm, the magnetic field has a magnitude of 12.5 nT. (a) What is the magnitude of the magnetic field at radius 6.00 cm? (b) What is the current in the wires attached to the plates?

N5. A capacitor with square plates of edge length L is being discharged by a current of 0.75 A. Figure 32N-3 is a head-on view of one of the plates from inside the capacitor. A dashed rectangular path is shown. If $L = 12$ cm, $W = 4.0$ cm, and $H = 2.0$ cm, what is the value of $\oint \vec{B} \cdot d\vec{s}$ around the dashed path?

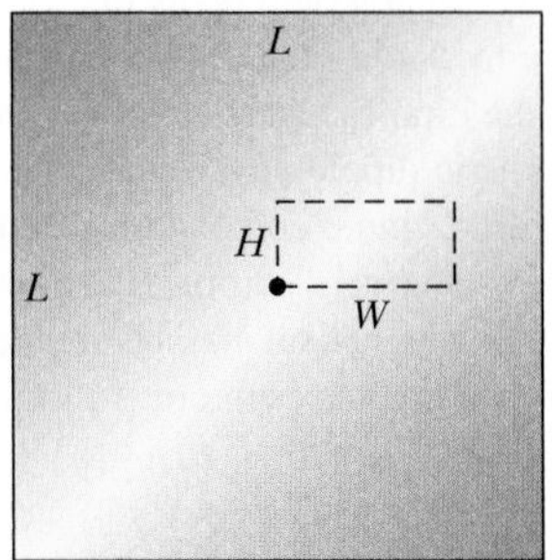

Fig. 32N-3 Problem N5.

N6. You place a magnetic compass on a horizontal surface, allow the needle to settle into its equilibrium position, and then give the compass a gentle wiggle to cause the needle to oscillate about that equilibrium position. The frequency of oscillation is 0.312 Hz. Earth's magnetic field at the location of the compass has a horizontal component of 18.0 μT. The needle has a magnetic moment of 0.680 mJ/T. What is the needle's rotational inertia about its (vertical) axis of rotation?

N7. *Uniform electric flux.* Figure 32N-4 shows a circular region of radius $R = 3.00$ cm in which a uniform electric flux is directed out of the page. The total electric flux through the region is given by $\Phi_E = (3.00\ \text{mV} \cdot \text{m/s})t$, where t is time. What is the magnitude of the magnetic field that is induced at radial distances (a) 2.00 cm and (b) 5.00 cm?

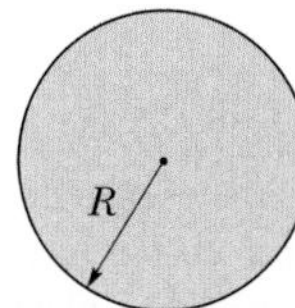

Fig. 32N-4 Problems N7 through N10 and N17 through N20.

N8. *Nonuniform electric flux.* Figure 32N-4 shows a circular region of radius $R = 3.00$ cm in which an electric flux is directed out of the page. The flux encircled by a concentric circle of radius r is given by $\Phi_{E,\text{enc}} = (0.600\ \text{V} \cdot \text{m/s})(r/R)t$, where $r \leq R$ and t is time. What is the magnitude of the induced magnetic field at radial distances (a) 2.00 cm and (b) 5.00 cm?

N9. *Uniform electric field.* In Fig. 32N-4, a uniform electric field is directed out of the page within a circular region of radius $R =$ 3.00 cm. The magnitude of the electric field is given by $E = (4.5 \times 10^{-3}\ \text{V/m} \cdot \text{s})t$, where t is time. What is the magnitude of the induced magnetic field at radial distances (a) 2.00 cm and (b) 5.00 cm?

N10. *Nonuniform electric field.* In Fig. 32N-4, an electric field is directed out of the page within a circular region of radius R = 3.00 cm. The magnitude of the electric field is given by $E = (0.500 \text{ V/m} \cdot \text{s})(1 - r/R)t$, where t is the time and r is the radial distance ($r \leq R$). What is the magnitude of the induced magnetic field at radial distances (a) 2.00 cm and (b) 5.00 cm?

N11. The circuit in Fig. 32N-5 consists of switch S, a 12.0 V ideal battery, a 20.0 MΩ resistor, and an air-filled capacitor. The capacitor has parallel circular plates of radius 5.00 cm, separated by 3.00 mm. At time $t = 0$, switch S is closed to begin charging the capacitor. The electric field between the plates is uniform. At t = 250 μs, what is the magnitude of the magnetic field within the capacitor, at radial distance 3.00 cm?

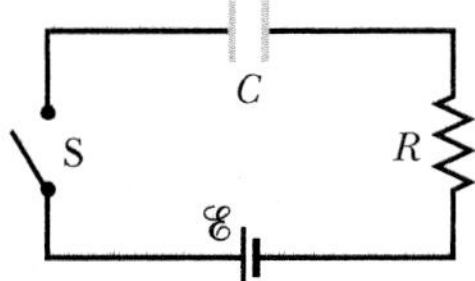

Fig. 32N-5 Problem N11.

N12. A magnetic rod with length 6.00 cm, radius 3.00 mm, and (uniform) magnetization of 2.70×10^3 A/m can turn about its center like a compass needle. It is placed in a uniform magnetic field $\vec{B}$ of magnitude 35.0 mT, such that the directions of its dipole moment and $\vec{B}$ make an angle of 68.0°. (a) What is the magnitude of the torque on the rod due to $\vec{B}$? (b) What is the change in the magnetic potential energy of the rod if the angle changes to 34.0°?

N13. The capacitor in Fig. 32-16 is being charged with a 2.50 A current. The wire radius is 1.50 mm, and the plate radius is 2.00 cm. Assume that the current in the wire and the displacement current in the capacitor gap are both uniformly distributed. What are the magnitudes of the magnetic field due to the current in the wire at the following radial distances: (a) 1.00 mm (inside the wire), (b) 3.00 mm (outside the wire), and (c) 2.20 cm (outside the wire)? What are the magnitudes of the magnetic field due to the displacement current in the capacitor gap at those same radial distances: (d) 1.00 mm (inside the gap), (e) 3.00 mm (inside the gap), and (f) 2.20 cm (outside the gap)? (g) Explain why the fields at the two smaller radii are so different for the wire and the gap, while the fields at the largest radius are not.

N14. The energy difference between parallel and antiparallel alignments of the z component of an electron's spin magnetic dipole moment with an external magnetic field $\vec{B}$ directed along the z component is 6.0×10^{-25} J. What is the magnitude of $\vec{B}$?

N15. Suppose that ±4 are the limits to the values of m_l for an electron in an atom. (a) How many different values of the z component $\mu_{\text{orb},z}$ of the electron's orbital magnetic dipole moment are possible? (b) What is the greatest magnitude of those possible values? Next, suppose that the atom is in a magnetic field of magnitude 0.250 T, in the positive direction of the z axis. What are (c) the maximum potential energy and (d) the minimum potential energy associated with those possible values of $\mu_{\text{orb},z}$?

N16. Figure 32N-6 gives the magnetization curve for a paramagnetic material. Let μ_{sam} be the measured net magnetic moment of a sample of the material and μ_{max} be the maximum possible net magnetic moment of that sample. According to Curie's law, what would be the ratio $\mu_{\text{sam}}/\mu_{\text{max}}$ were the sample placed in a uniform magnetic field of magnitude 0.800 T, at a temperature of 2.00 K?

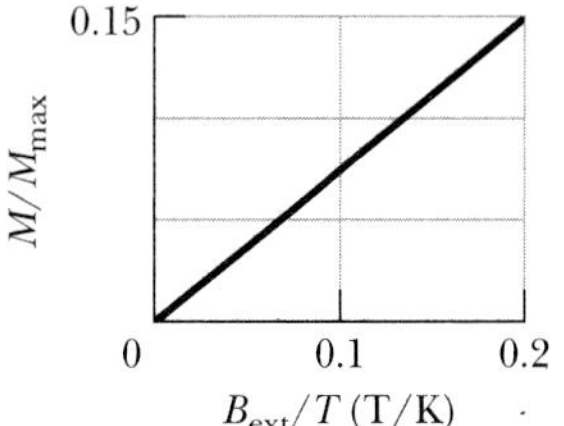

Fig. 32N-6 Problem N16.

N17. *Uniform displacement-current density.* Figure 32N-4 shows a circular region of radius R = 3.00 cm in which a displacement current is directed out of the page. The displacement current has a uniform density J_d = 6.00 A/m². What is the magnitude of the magnetic field due to the displacement current at radial distances (a) 2.00 cm and (b) 5.00 cm?

N18. *Uniform displacement current.* Figure 32N-4 shows a circular region of radius R = 3.00 cm in which a uniform displacement current i_d = 0.500 A is directed out of the page. What is the magnitude of the magnetic field due to the displacement current at radial distances (a) 2.00 cm and (b) 5.00 cm?

N19. *Nonuniform displacement-current density.* Figure 32N-4 shows a circular region of radius R = 3.00 cm in which a displacement current is directed out of the page. The displacement current has a density given by $J_d = (4.00 \text{ A/m}^2)(1 - r/R)$, where r is the radial distance $r \leq R$. What is the magnitude of the magnetic field due to the displacement current at radial distances (a) 2.00 cm and (b) 5.00 cm?

N20. *Nonuniform displacement current.* Figure 32N-4 shows a circular region of radius R = 3.00 cm in which a displacement current i_d is directed out of the page. The magnitude of the displacement current is given by $i_d = (3.00 \text{ A})(r/R)$, where r is the radial distance $r \leq R$. What is the magnitude of the magnetic field due to the displacement current at radial distances (a) 2.00 cm and (b) 5.00 cm?

33 Electromagnetic Oscillations and Alternating Current

When a high-voltage power transmission line requires repair, a utility company cannot just shut it down, perhaps blacking out an entire city. Repairs must be made while the lines are electrically "hot." The man outside the helicopter in this photograph has just replaced a spacer between 500 kV lines *by hand*, a procedure that requires considerable expertise.

How does he manage this repair without being electrocuted?

The answer is in this chapter.

33-1 New Physics—Old Mathematics

In this chapter you will see how the electric charge q varies with time in a circuit made up of an inductor L, a capacitor C, and a resistor R. From another point of view, we shall discuss how energy shuttles back and forth between the magnetic field of the inductor and the electric field of the capacitor, while it is being gradually dissipated as thermal energy in the resistor.

We have discussed oscillations before, in another context. In Chapter 16 we saw how displacement x varies with time in a mechanical oscillating system made up of a block of mass m, a spring of spring constant k, and a viscous or frictional element such as oil; Fig. 16-15 shows such a system. We also saw how energy shuttles back and forth between the kinetic energy of the oscillating mass and the potential energy of the spring, being gradually dissipated as thermal energy.

The parallel between these two idealized systems is exact, and the controlling differential equations are identical. Thus, there is no new mathematics to be learned; we can simply change the symbols and give our full attention to the physics of the situation.

33-2 *LC* Oscillations, Qualitatively

Of the three circuit elements, resistance R, capacitance C, and inductance L, we have so far discussed the series combinations RC (in Section 28-8) and RL (in Section 31-9). In these two kinds of circuit we found that the charge, current, and potential difference grow and decay exponentially. The time scale of the growth or decay is given by a *time constant* τ, which is either capacitive or inductive.

We now examine the remaining two-element circuit combination LC. You will see that in this case the charge, current, and potential difference do not decay exponentially with time but vary sinusoidally (with period T and angular frequency ω). The resulting oscillations of the capacitor's electric field and the inductor's magnetic field are said to be **electromagnetic oscillations.** Such a circuit is said to oscillate.

Parts a through h of Fig. 33-1 show succeeding stages of the oscillations in a simple LC circuit. From Eq. 26-21, the energy stored in the electric field of the capacitor at any time is

$$U_E = \frac{q^2}{2C}, \tag{33-1}$$

where q is the charge on the capacitor at that time. From Eq. 31-51, the energy stored in the magnetic field of the inductor at any time is

$$U_B = \frac{Li^2}{2}, \tag{33-2}$$

where i is the current through the inductor at that time.

The method of repairing high-voltage lines shown in the opening photograph is patented by Scott H. Yenzer and is licensed exclusively to Haverfield Corporation of Gettysburg, PA. As the lineman approaches a hot line, the electric field surrounding the line brings his body to nearly the potential of the line. To match the two potentials, he then extends a conducting "wand" to the line. To avoid being electrocuted, he must be isolated from anything electrically connected to the ground. To ensure that his body is always at a single potential—that of the line he is working on—he wears a conducting suit, hood, and gloves, all of which are electrically connected to the line via the wand.

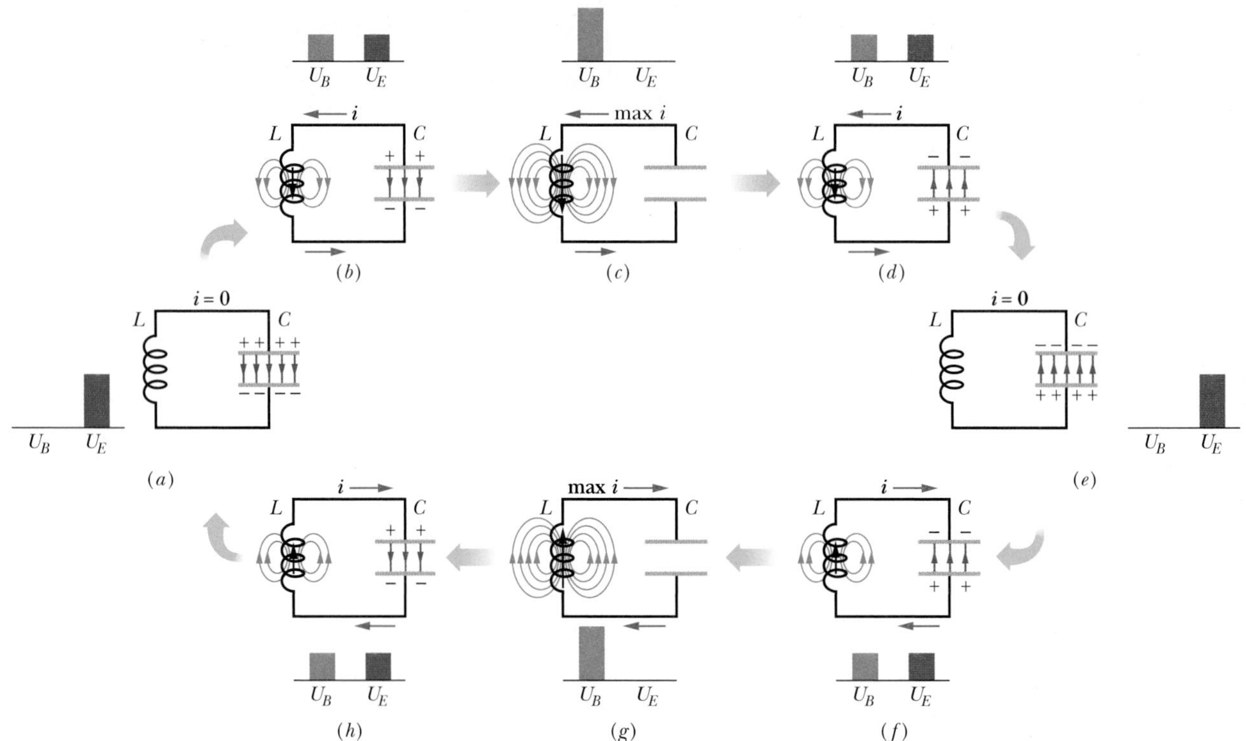

Fig. 33-1 Eight stages in a single cycle of oscillation of a resistanceless *LC* circuit. The bar graphs by each figure show the stored magnetic and electric energies. The magnetic field lines of the inductor and the electric field lines of the capacitor are shown. (*a*) Capacitor with maximum charge, no current. (*b*) Capacitor discharging, current increasing. (*c*) Capacitor fully discharged, current maximum. (*d*) Capacitor charging but with polarity opposite that in (*a*), current decreasing. (*e*) Capacitor with maximum charge having polarity opposite that in (*a*), no current. (*f*) Capacitor discharging, current increasing with direction opposite that in (*b*). (*g*) Capacitor fully discharged, current maximum. (*h*) Capacitor charging, current decreasing.

We now adopt the convention of representing *instantaneous values* of the electrical quantities of a sinusoidally oscillating circuit with small letters, such as q, and the *amplitudes* of those quantities with capital letters, such as Q. With this convention in mind, let us assume that initially the charge q on the capacitor in Fig. 33-1 is at its maximum value Q and that the current i through the inductor is zero. This initial state of the circuit is shown in Fig. 33-1*a*. The bar graphs for energy included there indicate that at this instant, with zero current through the inductor and maximum charge on the capacitor, the energy U_B of the magnetic field is zero and the energy U_E of the electric field is a maximum.

The capacitor now starts to discharge through the inductor, positive charge carriers moving counterclockwise, as shown in Fig. 33-1*b*. This means that a current i, given by dq/dt and pointing down in the inductor, is established. As the capacitor's charge decreases, the energy stored in the electric field within the capacitor also decreases. This energy is transferred to the magnetic field that appears around the inductor because of the current i that is building up there. Thus, the electric field decreases and the magnetic field builds up as energy is transferred from the electric field to the magnetic field.

The capacitor eventually loses all its charge (Fig. 33-1*c*) and thus also loses its electric field and the energy stored in that field. The energy has then been fully transferred to the magnetic field of the inductor. The magnetic field is then at its maximum magnitude, and the current through the inductor is then at its maximum value I.

Although the charge on the capacitor is now zero, the counterclockwise current must continue because the inductor does not allow it to change suddenly to zero. The current continues to transfer positive charge from the top plate to the bottom plate through the circuit (Fig. 33-1*d*). Energy now flows from the inductor back to the capacitor as the electric field within the capacitor builds up again. The current

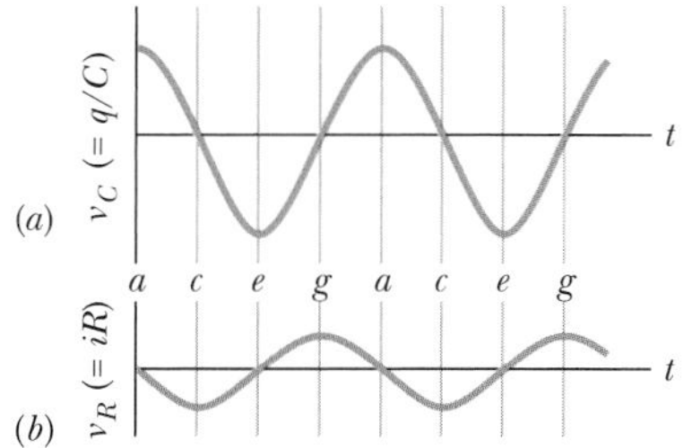

Fig. 33-2 (*a*) The potential difference across the capacitor of the circuit of Fig. 33-1 as a function of time. This quantity is proportional to the charge on the capacitor. (*b*) A potential proportional to the current in the circuit of Fig. 33-1. The letters refer to the correspondingly labeled oscillation stages in Fig. 33-1.

gradually decreases during this energy transfer. When, eventually, the energy has been transferred completely back to the capacitor (Fig. 33-1*e*), the current has decreased to zero (momentarily). The situation of Fig. 33-1*e* is like the initial situation, except that the capacitor is now charged oppositely.

The capacitor then starts to discharge again but now with a clockwise current (Fig. 33-1*f*). Reasoning as before, we see that the clockwise current builds to a maximum (Fig. 33-1*g*) and then decreases (Fig. 33-1*h*), until the circuit eventually returns to its initial situation (Fig. 33-1*a*). The process then repeats at some frequency f and thus at an angular frequency $\omega = 2\pi f$. In the ideal *LC* circuit with no resistance, there are no energy transfers other than that between the electric field of the capacitor and the magnetic field of the inductor. Owing to the conservation of energy, the oscillations continue indefinitely. The oscillations need not begin with the energy all in the electric field; the initial situation could be any other stage of the oscillation.

To determine the charge q on the capacitor as a function of time, we can put in a voltmeter to measure the time-varying potential difference (or *voltage*) v_C that exists across the capacitor C. From Eq. 26-1 we can write

$$v_C = \left(\frac{1}{C}\right)q,$$

which allows us to find q. To measure the current, we can connect a small resistance R in series with the capacitor and inductor and measure the time-varying potential difference v_R across it; v_R is proportional to i through the relation

$$v_R = iR.$$

We assume here that R is so small that its effect on the behavior of the circuit is negligible. The variations in time of v_C and v_R, and thus of q and i, are shown in Fig. 33-2. All four quantities vary sinusoidally.

In an actual *LC* circuit, the oscillations will not continue indefinitely because there is always some resistance present that will drain energy from the electric and magnetic fields and dissipate it as thermal energy (the circuit may become warmer). The oscillations, once started, will die away as Fig. 33-3 suggests. Compare this figure with Fig. 16-16, which shows the decay of mechanical oscillations caused by frictional damping in a block–spring system.

CHECKPOINT 1: A charged capacitor and an inductor are connected in series at time $t = 0$. In terms of the period T of the resulting oscillations, determine how much later the following reach their maximums: (a) the charge on the capacitor; (b) the voltage across the capacitor, with its original polarity; (c) the energy stored in the electric field; and (d) the current.

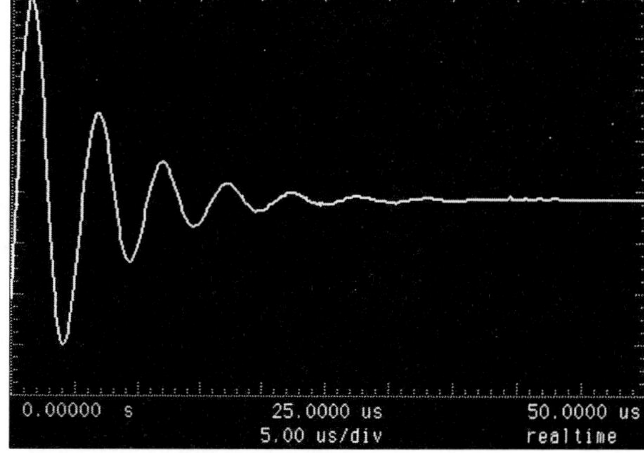

Fig. 33-3 An oscilloscope trace showing how the oscillations in an *RLC* circuit actually die away because energy is dissipated in the resistor as thermal energy.

Sample Problem 33-1

A 1.5 μF capacitor is charged to 57 V. The charging battery is then disconnected, and a 12 mH coil is connected in series with the capacitor so that *LC* oscillations occur. What is the maximum current in the coil? Assume that the circuit contains no resistance.

SOLUTION: The **Key Ideas** here are these:

1. Because the circuit contains no resistance, the electromagnetic energy of the circuit is conserved as the energy is transferred back and forth between the electric field of the capacitor and the magnetic field of the coil (inductor).
2. At any time t, the energy $U_B(t)$ of the magnetic field is related to the current $i(t)$ through the coil by Eq. 33-2 ($U_B = Li^2/2$). When all the energy is stored as magnetic energy, the current is at its maximum value I and that energy is $U_{B,\max} = LI^2/2$.
3. At any time t, the energy $U_E(t)$ of the electric field is related to the charge $q(t)$ on the capacitor by Eq. 33-1 ($U_E = q^2/2C$). When all the energy is stored as electric energy, the charge is at its maximum value Q and that energy is $U_{E,\max} = Q^2/2C$.

With these ideas, we can now write the conservation of energy as

$$U_{B,\max} = U_{E,\max}$$

or

$$\frac{LI^2}{2} = \frac{Q^2}{2C}.$$

Solving for I gives us

$$I = \sqrt{\frac{Q^2}{LC}}.$$

We know L and C, but not Q. However, with Eq. 26-1 ($q = CV$) we can relate Q to the maximum potential difference V across the capacitor, which is the initial potential difference of 57 V. Thus, substituting $Q = CV$ leads to

$$I = V\sqrt{\frac{C}{L}} = (57\text{ V})\sqrt{\frac{1.5\times 10^{-6}\text{ F}}{12\times 10^{-3}\text{ H}}}$$
$$= 0.637\text{ A} \approx 640\text{ mA}. \quad \text{(Answer)}$$

33-3 The Electrical–Mechanical Analogy

Let us look a little closer at the analogy between the oscillating *LC* system of Fig. 33-1 and an oscillating block–spring system. Two kinds of energy are involved in the block–spring system. One is potential energy of the compressed or extended spring; the other is kinetic energy of the moving block. These two energies are given by the familiar formulas in the left energy column in Table 33-1.

The table also shows, in the right energy column, the two kinds of energy involved in *LC* oscillations. By looking across the table, we can see an analogy between the forms of the two pairs of energies—the mechanical energies of the block–spring system and the electromagnetic energies of the *LC* oscillator. The equations for v and i at the bottom of the table help us see the details of the analogy. They tell us that q corresponds to x, and i corresponds to v (in both equations, the former is differentiated to obtain the latter). These correspondences then suggest that, in the energy expressions, $1/C$ corresponds to k and L corresponds to m. Thus,

$$q \text{ corresponds to } x, \qquad 1/C \text{ corresponds to } k,$$
$$i \text{ corresponds to } v, \quad \text{and} \quad L \text{ corresponds to } m.$$

These correspondences suggest that in an *LC* oscillator, the capacitor is mathematically like the spring in a block–spring system, and the inductor is like the block.

In Section 16-3 we saw that the angular frequency of oscillation of a (friction-

TABLE 33-1 The Energy in Two Oscillating Systems Compared

Block–Spring System		*LC* Oscillator	
Element	Energy	Element	Energy
Spring	Potential, $\frac{1}{2}kx^2$	Capacitor	Electric, $\frac{1}{2}(1/C)q^2$
Block	Kinetic, $\frac{1}{2}mv^2$	Inductor	Magnetic, $\frac{1}{2}Li^2$
$v = dx/dt$		$i = dq/dt$	

less) block–spring system is

$$\omega = \sqrt{\frac{k}{m}} \quad \text{(block–spring system).} \tag{33-3}$$

The correspondences listed above suggest that to find the angular frequency of oscillation for a (resistanceless) *LC* circuit, k should be replaced by $1/C$ and m by L, yielding

$$\omega = \frac{1}{\sqrt{LC}} \quad (LC \text{ circuit}). \tag{33-4}$$

We derive this result in the next section.

33-4 *LC* Oscillations, Quantitatively

Here we want to show explicitly that Eq. 33-4 for the angular frequency of *LC* oscillations is correct. At the same time, we want to examine even more closely the analogy between *LC* oscillations and block–spring oscillations. We start by extending somewhat our earlier treatment of the mechanical block–spring oscillator.

The Block–Spring Oscillator

We analyzed block–spring oscillations in Chapter 16 in terms of energy transfers and did not—at that early stage—derive the fundamental differential equation that governs those oscillations. We do so now.

We can write, for the total energy U of a block–spring oscillator at any instant,

$$U = U_b + U_s = \tfrac{1}{2}mv^2 + \tfrac{1}{2}kx^2, \tag{33-5}$$

where U_b and U_s are, respectively, the kinetic energy of the moving block and the potential energy of the stretched or compressed spring. If there is no friction—which we assume—the total energy U remains constant with time, even though v and x vary. In more formal language, $dU/dt = 0$. This leads to

$$\frac{dU}{dt} = \frac{d}{dt}(\tfrac{1}{2}mv^2 + \tfrac{1}{2}kx^2) = mv\frac{dv}{dt} + kx\frac{dx}{dt} = 0. \tag{33-6}$$

However, $v = dx/dt$ and $dv/dt = d^2x/dt^2$. With these substitutions, Eq. 33-6 becomes

$$m\frac{d^2x}{dt^2} + kx = 0 \quad \text{(block–spring oscillations).} \tag{33-7}$$

Equation 33-7 is the fundamental *differential equation* that governs the frictionless block–spring oscillations.

The general solution to Eq. 33-7—that is, the function $x(t)$ that describes the block–spring oscillations—is (as we saw in Eq. 16-3)

$$x = X\cos(\omega t + \phi) \quad \text{(displacement),} \tag{33-8}$$

in which X is the amplitude of the mechanical oscillations (represented by x_m in Chapter 16), ω is the angular frequency of the oscillations, and ϕ is a phase constant.

The *LC* Oscillator

Now let us analyze the oscillations of a resistanceless *LC* circuit, proceeding exactly as we just did for the block–spring oscillator. The total energy U present at any instant in an oscillating LC circuit is given by

$$U = U_B + U_E = \frac{Li^2}{2} + \frac{q^2}{2C}, \tag{33-9}$$

in which U_B is the energy stored in the magnetic field of the inductor and U_E is the energy stored in the electric field of the capacitor. Since we have assumed the circuit resistance to be zero, no energy is transferred to thermal energy and U remains constant with time. In more formal language, dU/dt must be zero. This leads to

$$\frac{dU}{dt} = \frac{d}{dt}\left(\frac{Li^2}{2} + \frac{q^2}{2C}\right) = Li\frac{di}{dt} + \frac{q}{C}\frac{dq}{dt} = 0. \tag{33-10}$$

However, $i = dq/dt$ and $di/dt = d^2q/dt^2$. With these substitutions, Eq. 33-10 becomes

$$L\frac{d^2q}{dt^2} + \frac{1}{C}q = 0 \quad \text{(}LC\text{ oscillations).} \tag{33-11}$$

This is the *differential equation* that describes the oscillations of a resistanceless LC circuit. Equations 33-11 and 33-7 are exactly of the same mathematical form.

Charge and Current Oscillations

Since the differential equations are mathematically identical, their solutions must also be mathematically identical. Because q corresponds to x, we can write the general solution of Eq. 33-11, by analogy to Eq. 33-8, as

$$q = Q\cos(\omega t + \phi) \quad \text{(charge),} \tag{33-12}$$

where Q is the amplitude of the charge variations, ω is the angular frequency of the electromagnetic oscillations, and ϕ is the phase constant.

Taking the first derivative of Eq. 33-12 with respect to time gives us the current of the LC oscillator:

$$i = \frac{dq}{dt} = -\omega Q\sin(\omega t + \phi) \quad \text{(current).} \tag{33-13}$$

The amplitude I of this sinusoidally varying current is

$$I = \omega Q, \tag{33-14}$$

so we can rewrite Eq. 33-13 as

$$i = -I\sin(\omega t + \phi). \tag{33-15}$$

Angular Frequencies

We can test whether Eq. 33-12 is a solution of Eq. 33-11 by substituting it and its second derivative with respect to time into Eq. 33-11. The first derivative of Eq. 33-12 is Eq. 33-13. The second derivative is then

$$\frac{d^2q}{dt^2} = -\omega^2 Q\cos(\omega t + \phi).$$

Substituting for q and d^2q/dt^2 into Eq. 33-11, we obtain

$$-L\omega^2 Q\cos(\omega t + \phi) + \frac{1}{C}Q\cos(\omega t + \phi) = 0.$$

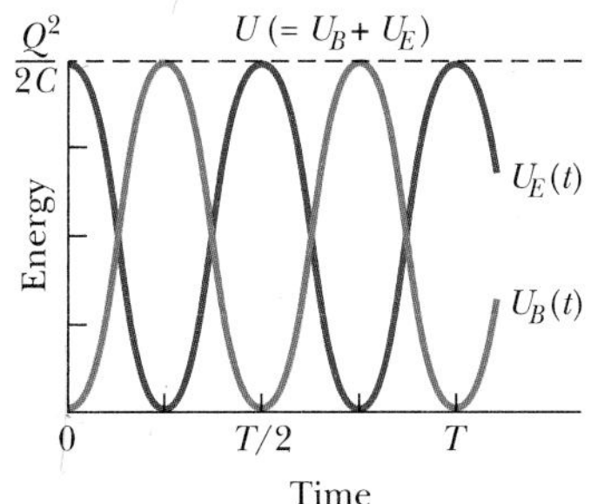

Fig. 33-4 The stored magnetic energy and electric energy in the circuit of Fig. 33-1 as a function of time. Note that their sum remains constant. T is the period of oscillation.

Canceling $Q \cos(\omega t + \phi)$ and rearranging lead to

$$\omega = \frac{1}{\sqrt{LC}}.$$

Thus, Eq. 33-12 is indeed a solution of Eq. 33-11 if ω has the constant value $1/\sqrt{LC}$. Note that this expression for ω is exactly that given by Eq. 33-4, which we arrived at by examining correspondences.

The phase constant ϕ in Eq. 33-12 is determined by the conditions that exist at any certain time, say, $t = 0$. If the conditions yield $\phi = 0$ at $t = 0$, Eq. 33-12 requires that $q = Q$ and Eq. 33-13 requires that $i = 0$; these are the initial conditions represented by Fig. 33-1*a*.

Electric and Magnetic Energy Oscillations

The electric energy stored in the *LC* circuit at any time t is, from Eqs. 33-1 and 33-12,

$$U_E = \frac{q^2}{2C} = \frac{Q^2}{2C}\cos^2(\omega t + \phi). \qquad (33\text{-}16)$$

The magnetic energy is, from Eqs. 33-2 and 33-13,

$$U_B = \tfrac{1}{2}Li^2 = \tfrac{1}{2}L\omega^2Q^2\sin^2(\omega t + \phi).$$

Substituting for ω from Eq. 33-4 then gives us

$$U_B = \frac{Q^2}{2C}\sin^2(\omega t + \phi). \qquad (33\text{-}17)$$

Figure 33-4 shows plots of $U_E(t)$ and $U_B(t)$ for the case of $\phi = 0$. Note that

1. The maximum values of U_E and U_B are both $Q^2/2C$.
2. At any instant the sum of U_E and U_B is equal to $Q^2/2C$, a constant.
3. When U_E is maximum, U_B is zero, and conversely.

CHECKPOINT 2: A capacitor in an *LC* oscillator has a maximum potential difference of 17 V and a maximum energy of 160 μJ. When the capacitor has a potential difference of 5 V and an energy of 10 μJ, what are (a) the emf across the inductor and (b) the energy stored in the magnetic field?

Sample Problem 33-2

For the situation described in Sample Problem 33-1, let the coil (inductor) be connected to the charged capacitor at time $t = 0$. The result is an *LC* circuit like that in Fig. 33-1.

(a) What is the potential difference $v_L(t)$ across the inductor as a function of time?

SOLUTION: One **Key Idea** here is that the current and potential differences of the circuit undergo sinusoidal oscillations. Another **Key Idea** is that we can still apply the loop rule to this oscillating circuit—just as we did for the nonoscillating circuits of Chapter 28. At any time t during the oscillations, the loop rule and Fig. 33-1 give us

$$v_L(t) = v_C(t); \qquad (33\text{-}18)$$

that is, the potential difference v_L across the inductor must always be equal to the potential difference v_C across the capacitor, so that the net potential difference around the circuit is zero. Thus, we will find $v_L(t)$ if we can find $v_C(t)$, and we can find $v_C(t)$ from $q(t)$ with Eq. 26-1 ($q = CV$).

Because the potential difference $v_C(t)$ is maximum when the oscillations begin at time $t = 0$, the charge q on the capacitor must also be maximum then. Thus, phase constant ϕ must be zero, so that Eq. 33-12 gives us

$$q = Q \cos \omega t. \tag{33-19}$$

(Note that this cosine function does indeed yield maximum q (= Q) when $t = 0$.) To get the potential difference $v_C(t)$, we divide both sides of Eq. 33-19 by C to write

$$\frac{q}{C} = \frac{Q}{C} \cos \omega t,$$

and then use Eq. 26-1 to write

$$v_C = V_C \cos \omega t. \tag{33-20}$$

Here, V_C is the amplitude of the oscillations in the potential difference v_C across the capacitor.

Next, substituting $v_C = v_L$ from Eq. 33-18, we find

$$v_L = V_C \cos \omega t. \tag{33-21}$$

We can evaluate the right side of this equation by first noting that the amplitude V_C is equal to the initial (maximum) potential difference of 57 V across the capacitor. Then, using the values of L and C from Sample Problem 33-1, we find ω with Eq. 33-4:

$$\omega = \frac{1}{\sqrt{LC}} = \frac{1}{[(0.012\text{ H})(1.5 \times 10^{-6}\text{ F})]^{0.5}}$$
$$= 7454\text{ rad/s} \approx 7500\text{ rad/s}.$$

Thus, Eq. 33-21 becomes

$$v_L = (57\text{ V}) \cos(7500\text{ rad/s})t. \quad \text{(Answer)}$$

(b) What is the maximum rate $(di/dt)_{max}$ at which the current i changes in the circuit?

SOLUTION: The **Key Idea** here is that, with the charge on the capacitor oscillating as in Eq. 33-12, the current is in the form of Eq. 33-13. Because $\phi = 0$, that equation gives us

$$i = -\omega Q \sin \omega t.$$

Then $$\frac{di}{dt} = \frac{d}{dt}(-\omega Q \sin \omega t) = -\omega^2 Q \cos \omega t.$$

We can simplify this equation by substituting CV_C for Q (because we know C and V_C but not Q) and $1/\sqrt{LC}$ for ω according to Eq. 33-4. We get

$$\frac{di}{dt} = -\frac{1}{LC} CV_C \cos \omega t = -\frac{V_C}{L} \cos \omega t.$$

This tells us that the current changes at a varying (sinusoidal) rate, with its maximum rate of change being

$$\frac{V_C}{L} = \frac{57\text{ V}}{0.012\text{ H}} = 4750\text{ A/s} \approx 4800\text{ A/s}. \quad \text{(Answer)}$$

33-5 Damped Oscillations in an *RLC* Circuit

A circuit containing resistance, inductance, and capacitance is called an *RLC circuit.* We shall here discuss only *series RLC circuits* like that shown in Fig. 33-5. With a resistance R present, the total *electromagnetic energy U* of the circuit (the sum of the electric energy and magnetic energy) is no longer constant; instead, it decreases with time as energy is transferred to thermal energy in the resistance. Because of this loss of energy, the oscillations of charge, current, and potential difference continuously decrease in amplitude, and the oscillations are said to be *damped.* As you will see, they are damped in exactly the same way as those of the damped block–spring oscillator of Section 16-8.

To analyze the oscillations of this circuit, we write an equation for the total electromagnetic energy U in the circuit at any instant. Because the resistance does not store electromagnetic energy, we can use Eq. 33-9:

$$U = U_B + U_E = \frac{Li^2}{2} + \frac{q^2}{2C}. \tag{33-22}$$

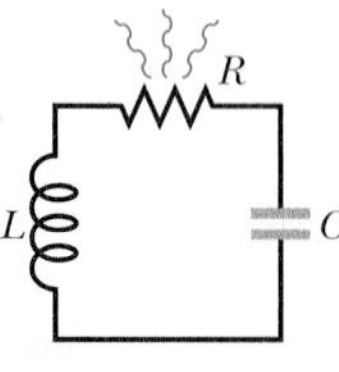

Fig. 33-5 A series *RLC* circuit. As the charge contained in the circuit oscillates back and forth through the resistance, electromagnetic energy is dissipated as thermal energy, damping (decreasing the amplitude of) the oscillations.

Now, however, this total energy decreases as energy is transferred to thermal energy. The rate of that transfer is, from Eq. 27-22,

$$\frac{dU}{dt} = -i^2R, \tag{33-23}$$

where the minus sign indicates that U decreases. By differentiating Eq. 33-22 with respect to time and then substituting the result in Eq. 33-23, we obtain

$$\frac{dU}{dt} = Li\frac{di}{dt} + \frac{q}{C}\frac{dq}{dt} = -i^2R.$$

Substituting dq/dt for i and d^2q/dt^2 for di/dt, we obtain

$$L\frac{d^2q}{dt^2} + R\frac{dq}{dt} + \frac{1}{C}q = 0 \qquad (RLC \text{ circuit}), \tag{33-24}$$

which is the differential equation for damped oscillations in an *RLC* circuit.

The solution to Eq. 33-24 is

$$q = Qe^{-Rt/2L}\cos(\omega' t + \phi), \tag{33-25}$$

in which

$$\omega' = \sqrt{\omega^2 - (R/2L)^2}, \tag{33-26}$$

where $\omega = 1/\sqrt{LC}$, as with an undamped oscillator. Equation 33-25 tells us how the charge on the capacitor oscillates in a damped *RLC* circuit; that equation is the electromagnetic counterpart of Eq. 16-40, which gives the displacement of a damped block–spring oscillator.

Equation 33-25 describes a sinusoidal oscillation (the cosine function) with an *exponentially decaying amplitude* $Qe^{-Rt/2L}$ (the factor that multiplies the cosine). The angular frequency ω' of the damped oscillations is always less than the angular frequency ω of the undamped oscillations; however, we shall here consider only situations in which R is small enough for us to replace ω' with ω.

Let us next find an expression for the total electromagnetic energy U of the circuit as a function of time. One way to do so is to monitor the energy of the electric field in the capacitor, which is given by Eq. 33-1 ($U_E = q^2/2C$). By substituting Eq. 33-25 into Eq. 33-1, we obtain

$$U_E = \frac{q^2}{2C} = \frac{[Qe^{-Rt/2L}\cos(\omega' t + \phi)]^2}{2C} = \frac{Q^2}{2C}e^{-Rt/L}\cos^2(\omega' t + \phi). \tag{33-27}$$

Thus, the energy of the electric field oscillates according to a cosine-squared term and the amplitude of that oscillation decreases exponentially with time.

Sample Problem 33-3

A series *RLC* circuit has inductance $L = 12$ mH, capacitance $C = 1.6\ \mu$F, and resistance $R = 1.5\ \Omega$.

(a) At what time t will the amplitude of the charge oscillations in the circuit be 50% of its initial value?

SOLUTION: The Key Idea here is that the amplitude of the charge oscillations decreases exponentially with time t: According to Eq. 33-25, the charge amplitude at any time t is $Qe^{-Rt/2L}$, in which Q is the amplitude at time $t = 0$. We want the time when the charge amplitude has decreased to $0.500Q$, that is, when

$$Qe^{-Rt/2L} = 0.50Q.$$

Canceling Q and taking the natural logarithms of both sides, we have

$$-\frac{Rt}{2L} = \ln 0.50.$$

Solving for t and then substituting given data yield

$$t = -\frac{2L}{R}\ln 0.50 = -\frac{(2)(12\times 10^{-3}\text{ H})(\ln 0.50)}{1.5\ \Omega}$$
$$= 0.0111\text{ s} \approx 11\text{ ms}. \qquad \text{(Answer)}$$

(b) How many oscillations are completed within this time?

SOLUTION: The Key Idea here is that the time for one complete oscillation is the period $T = 2\pi/\omega$, where the angular frequency for *LC* oscillations is given by Eq. 33-4 ($\omega = 1/\sqrt{LC}$). Thus, in the time interval $\Delta t = 0.0111$ s, the number of complete oscillations is

$$\frac{\Delta t}{T} = \frac{\Delta t}{2\pi\sqrt{LC}}$$
$$= \frac{0.0111\text{ s}}{2\pi[(12\times 10^{-3}\text{ H})(1.6\times 10^{-6}\text{ F})]^{1/2}} \approx 13. \qquad \text{(Answer)}$$

Thus, the amplitude decays by 50% in about 13 complete oscillations. This damping is less severe than that shown in Fig. 33-3, where the amplitude decays by a little more than 50% in one oscillation.

33-6 Alternating Current

The oscillations in an *RLC* circuit will not damp out if an external emf device supplies enough energy to make up for the energy dissipated as thermal energy in the resistance R. Circuits in homes, offices, and factories, including countless *RLC* circuits, receive such energy from local power companies. In most countries the energy is supplied via oscillating emfs and currents—the current is said to be an **alternating current,** or **ac** for short. (The nonoscillating current from a battery is said to be a **direct current,** or **dc**.) These oscillating emfs and currents vary sinusoidally with time, reversing direction (in North America) 120 times per second and thus having frequency $f = 60$ Hz.

At first sight this may seem to be a strange arrangement. We have seen that the drift speed of the conduction electrons in household wiring may typically be 4×10^{-5} m/s. If we now reverse their direction every $\frac{1}{120}$ s, such electrons can move only about 3×10^{-7} m in a half-cycle. At this rate, a typical electron can drift past no more than about 10 atoms in the wiring before it is required to reverse its direction. How, you may wonder, can the electron ever get anywhere?

Although this question may be worrisome, it is a needless concern. The conduction electrons do not have to "get anywhere." When we say that the current in a wire is one ampere, we mean that charge passes through any plane cutting across that wire at the rate of one coulomb per second. The speed at which the charge carriers cross that plane does not matter directly; one ampere may correspond to many charge carriers moving very slowly or to a few moving very rapidly. Furthermore, the signal to the electrons to reverse directions—which originates in the alternating emf provided by the power company's generator—is propagated along the conductor at a speed close to that of light. All electrons, no matter where they are located, get their reversal instructions at about the same instant. Finally, we note that for many devices, such as lightbulbs and toasters, the direction of motion is unimportant as long as the electrons do move so as to transfer energy to the device via collisions with atoms in the device.

The basic advantage of alternating current is this: *As the current alternates, so does the magnetic field that surrounds the conductor.* This makes possible the use of Faraday's law of induction, which, among other things, means that we can step up (increase) or step down (decrease) the magnitude of an alternating potential difference at will, using a device called a transformer, as you will see later in this chapter. Moreover, alternating current is more readily adaptable to rotating machinery such as generators and motors than is (nonalternating) direct current.

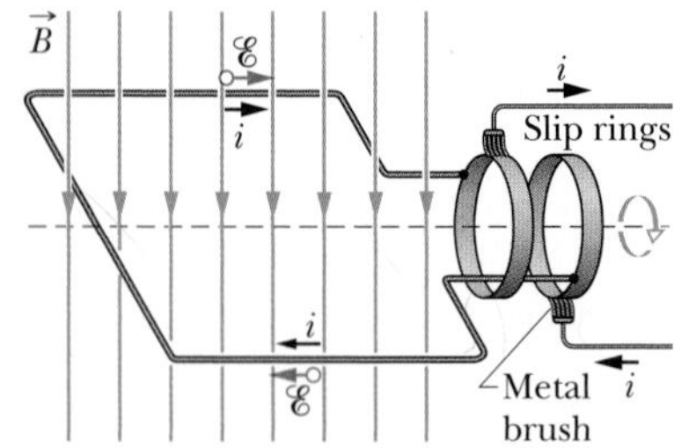

Fig. 33-6 The basic mechanism of an alternating-current generator is a conducting loop rotated in an external magnetic field. In practice, the alternating emf induced in a coil of many turns of wire is made accessible by means of slip rings attached to the rotating loop. Each ring is connected to one end of the loop wire and is electrically connected to the rest of the generator circuit by a conducting brush against which it slips as the loop (and it) rotates.

Figure 33-6 shows a simple model of an ac generator. As the conducting loop is forced to rotate through the external magnetic field $\vec{B}$, a sinusoidally oscillating emf $\mathscr{E}$ is induced in the loop:

$$\mathscr{E} = \mathscr{E}_m \sin \omega_d t. \tag{33-28}$$

The *angular frequency* ω_d of the emf is equal to the angular speed with which the loop rotates in the magnetic field; the *phase* of the emf is $\omega_d t$; and the *amplitude* of the emf is $\mathscr{E}_m$ (where the subscript stands for maximum). When the rotating loop is part of a closed conducting path, this emf produces (*drives*) a sinusoidal (alternating) current along the path with the same angular frequency ω_d, which then is called the **driving angular frequency.** We can write the current as

$$i = I \sin(\omega_d t - \phi), \tag{33-29}$$

in which I is the amplitude of the driven current. (The phase $\omega_d t - \phi$ of the current is traditionally written with a minus sign instead of as $\omega_d t + \phi$.) We include a phase

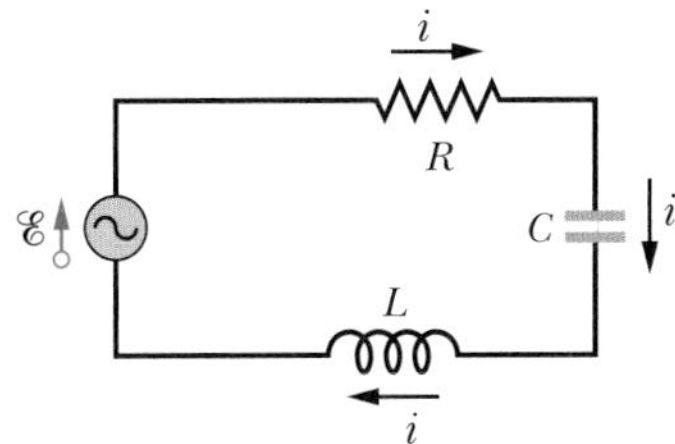

Fig. 33-7 A single-loop circuit containing a resistor, a capacitor, and an inductor. A generator, represented by a sine wave in a circle, produces an alternating emf that establishes an alternating current; the directions of the emf and current are indicated here at only one instant.

constant ϕ in Eq. 33-29 because the current i may not be in phase with the emf $\mathscr{E}$. (As you will see, the phase constant depends on the circuit to which the generator is connected.) We can also write the current i in terms of the **driving frequency** f_d of the emf, by substituting $2\pi f_d$ for ω_d in Eq. 33-29.

33-7 Forced Oscillations

We have seen that once started, the charge, potential difference, and current in both undamped LC circuits and damped RLC circuits (with small enough R) oscillate at angular frequency $\omega = 1/\sqrt{LC}$. Such oscillations are said to be *free oscillations* (free of any external emf), and the angular frequency ω is said to be the circuit's **natural angular frequency.**

When the external alternating emf of Eq. 33-28 is connected to an RLC circuit, the oscillations of charge, potential difference, and current are said to be *driven oscillations* or *forced oscillations.* These oscillations always occur at the driving angular frequency ω_d:

> Whatever the natural angular frequency ω of a circuit may be, forced oscillations of charge, current, and potential difference in the circuit always occur at the driving angular frequency ω_d.

However, as you will see in Section 33-9, the amplitudes of the oscillations very much depend on how close ω_d is to ω. When the two angular frequencies match—a condition known as **resonance**—the amplitude I of the current in the circuit is maximum.

33-8 Three Simple Circuits

Later in this chapter, we shall connect an external alternating emf device to a series RLC circuit as in Fig. 33-7. We shall then find expressions for the amplitude I and phase constant ϕ of the sinusoidally oscillating current in terms of the amplitude $\mathscr{E}_m$ and angular frequency ω_d of the external emf. First, however, let us consider three simpler circuits, each having an external emf and only one other circuit element: R, C, or L. We start with a resistive element (a purely *resistive load*).

A Resistive Load

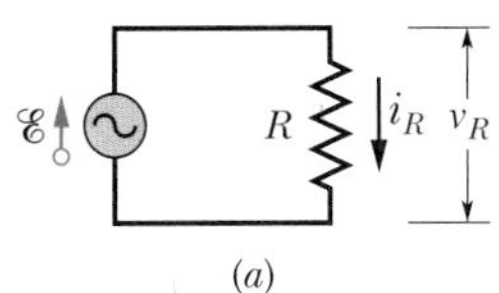

(a)

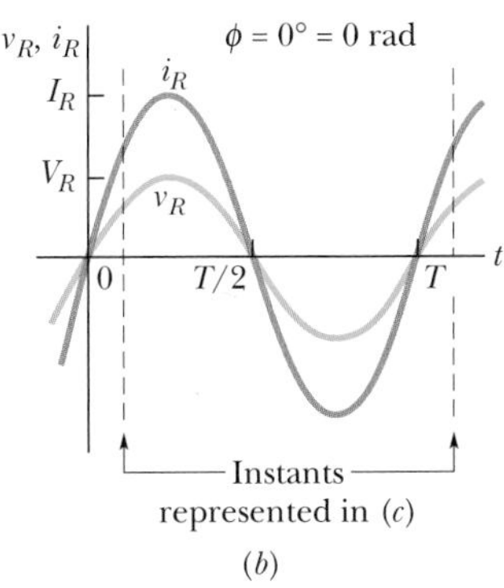

(b)

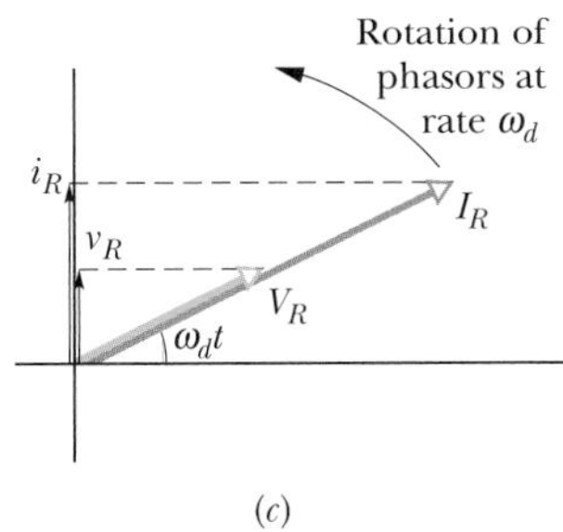

(c)

Fig. 33-8 (a) A resistor is connected across an alternating-current generator. (b) The current i_R and the potential difference v_R across the resistor are plotted on the same graph, both versus time t. They are in phase and complete one cycle in one period T. (c) A phasor diagram shows the same thing as (b).

Figure 33-8*a* shows a circuit containing a resistance element of value R and an ac generator with the alternating emf of Eq. 33-28. By the loop rule, we have

$$\mathscr{E} - v_R = 0.$$

With Eq. 33-28, this gives us

$$v_R = \mathscr{E}_m \sin \omega_d t.$$

Because the amplitude V_R of the alternating potential difference (or voltage) across the resistance is equal to the amplitude $\mathscr{E}_m$ of the alternating emf, we can write this as

$$v_R = V_R \sin \omega_d t. \tag{33-30}$$

From the definition of resistance ($R = V/i$), we can now write the current i_R in the resistance as

$$i_R = \frac{v_R}{R} = \frac{V_R}{R} \sin \omega_d t. \tag{33-31}$$

From Eq. 33-29, we can also write this current as

$$i_R = I_R \sin(\omega_d t - \phi), \tag{33-32}$$

where I_R is the amplitude of the current i_R in the resistance. Comparing Eqs. 33-31 and 33-32, we see that for a purely resistive load the phase constant $\phi = 0°$. We also see that the voltage amplitude and current amplitude are related by

$$V_R = I_R R \qquad \text{(resistor)}. \tag{33-33}$$

Although we found this relation for the circuit of Fig. 33-8*a*, it applies to any resistance in any ac circuit.

By comparing Eqs. 33-30 and 33-31, we see that the time-varying quantities v_R and i_R are both functions of $\sin \omega_d t$ with $\phi = 0°$. Thus, these two quantities are *in phase,* which means that their corresponding maxima (and minima) occur at the same times. Figure 33-8*b*, which is a plot of $v_R(t)$ and $i_R(t)$, illustrates this fact. Note that v_R and i_R do not decay here, because the generator supplies energy to the circuit to make up for the energy dissipated in R.

The time-varying quantities v_R and i_R can also be represented geometrically by *phasors.* Recall from Section 17-10 that phasors are vectors that rotate around an origin. Those that represent the voltage across and current in the resistor of Fig. 33-8*a* are shown in Fig. 33-8*c* at an arbitrary time t. Such phasors have the following properties:

Angular speed: Both phasors rotate counterclockwise about the origin with an angular speed equal to the angular frequency ω_d of v_R and i_R.

Length: The length of each phasor represents the amplitude of the alternating quantity: V_R for the voltage and I_R for the current.

Projection: The projection of each phasor on the *vertical* axis represents the value of the alternating quantity at time t: v_R for the voltage and i_R for the current.

Rotation angle: The rotation angle of each phasor is equal to the phase of the alternating quantity at time t. In Fig. 33-8*c*, the voltage and current are in phase, so their phasors always have the same phase $\omega_d t$ and the same rotation angle, and thus they rotate together.

Mentally follow the rotation. Can you see that when the phasors have rotated so that $\omega_d t = 90°$ (they point vertically upward), they indicate that just then $v_R = V_R$ and $i_R = I_R$? Equations 33-30 and 33-32 give the same results.

Sample Problem 33-4

Purely resistive load. In Fig. 33-8*a*, resistance R is 200 Ω and the sinusoidal alternating emf device operates at amplitude $\mathscr{E}_m = 36.0$ V and frequency $f_d = 60.0$ Hz.

(a) What is the potential difference $v_R(t)$ across the resistance as a function of time t, and what is the amplitude V_R of $v_R(t)$?

SOLUTION: When we applied the loop rule to the circuit of Fig. 33-8*a*, we found this **Key Idea**: In a circuit with a purely resistive load, the potential difference $v_R(t)$ across the resistance is always equal to the potential difference $\mathscr{E}(t)$ across the emf device. Thus, $v_R(t) = \mathscr{E}(t)$ and $V_R = \mathscr{E}_m$. Since $\mathscr{E}_m$ is given, we can write

$$V_R = \mathscr{E}_m = 36.0 \text{ V}. \qquad \text{(Answer)}$$

To find $v_R(t)$, we use Eq. 33-28 to write

$$v_R(t) = \mathscr{E}(t) = \mathscr{E}_m \sin \omega_d t, \tag{33-34}$$

and then substitute $\mathscr{E}_m = 36.0$ V and

$$\omega_d = 2\pi f_d = 2\pi(60 \text{ Hz}) = 120\pi$$

to obtain

$$v_R = (36.0 \text{ V}) \sin(120\pi t). \qquad \text{(Answer)}$$

We can leave the argument of the sine in this form for convenience, or we can write it as (377 rad/s)t or as (377 s^{-1})t.

(b) What are the current $i_R(t)$ in the resistance and the amplitude I_R of $i_R(t)$?

SOLUTION: The **Key Idea** here is that in an ac circuit with a purely resistive load, the alternating current $i_R(t)$ in the resistance is *in phase* with the alternating potential difference $v_R(t)$ across the resistance; that is, the phase constant ϕ for the current is zero. Thus, we can write Eq. 33-29 as

$$i_R = I_R \sin(\omega_d t - \phi) = I_R \sin \omega_d t. \tag{33-35}$$

From Eq. 33-33, the amplitude I_R is

$$I_R = \frac{V_R}{R} = \frac{36.0 \text{ V}}{200 \ \Omega} = 0.180 \text{ A}. \qquad \text{(Answer)}$$

Substituting this and $\omega_d = 2\pi f_d = 120\pi$ into Eq. 33-35, we have

$$i_R = (0.180 \text{ A}) \sin(120\pi t). \qquad \text{(Answer)}$$

CHECKPOINT 3: If we increase the driving frequency in a circuit with a purely resistive load, do (a) amplitude V_R and (b) amplitude I_R increase, decrease, or remain the same?

A Capacitive Load

Figure 33-9*a* shows a circuit containing a capacitance and a generator with the alternating emf of Eq. 33-28. Using the loop rule and proceeding as we did when we obtained Eq. 33-30, we find that the potential difference across the capacitor is

$$v_C = V_C \sin \omega_d t, \tag{33-36}$$

where V_C is the amplitude of the alternating voltage across the capacitor. From the definition of capacitance we can also write

$$q_C = C v_C = C V_C \sin \omega_d t. \tag{33-37}$$

Our concern, however, is with the current rather than the charge. Thus, we differentiate Eq. 33-37 to find

$$i_C = \frac{dq_C}{dt} = \omega_d C V_C \cos \omega_d t. \tag{33-38}$$

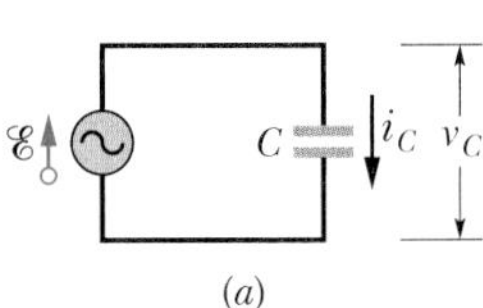

(*a*)

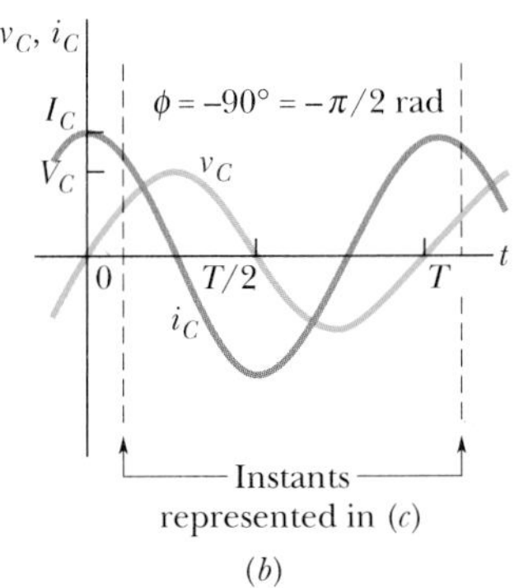

(*b*)

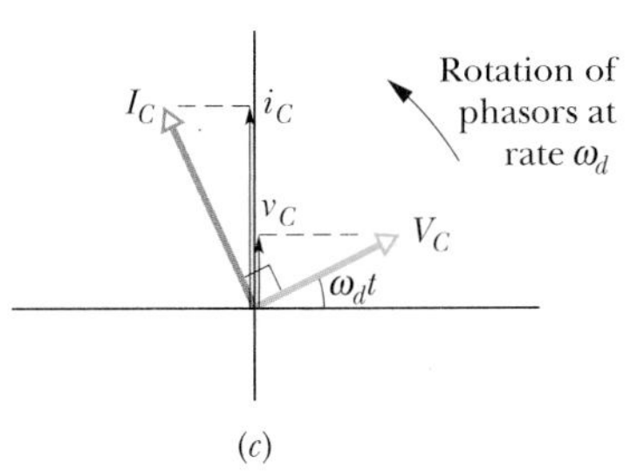

(*c*)

Fig. 33-9 (*a*) A capacitor is connected across an alternating-current generator. (*b*) The current in the capacitor leads the voltage by 90° (= $\pi/2$ rad). (*c*) A phasor diagram shows the same thing.

We now modify Eq. 33-38 in two ways. First, for reasons of symmetry of notation, we introduce the quantity X_C, called the **capacitive reactance** of a capacitor, defined as

$$X_C = \frac{1}{\omega_d C} \quad \text{(capacitive reactance).} \tag{33-39}$$

Its value depends not only on the capacitance but also on the driving angular frequency ω_d. We know from the definition of the capacitive time constant ($\tau = RC$) that the SI unit for C can be expressed as seconds per ohm. Applying this to Eq. 33-39 shows that the SI unit of X_C is the *ohm*, just as for resistance R.

Second, we replace $\cos \omega_d t$ in Eq. 33-38 with a phase-shifted sine:

$$\cos \omega_d t = \sin(\omega_d t + 90°).$$

You can verify this identity by shifting a sine curve in the negative direction by 90°.

With these two modifications, Eq. 33-38 becomes

$$i_C = \left(\frac{V_C}{X_C}\right) \sin(\omega_d t + 90°). \tag{33-40}$$

From Eq. 33-29, we can also write the current i_C in C as

$$i_C = I_C \sin(\omega_d t - \phi),$$

where I_C is the amplitude of i_C. Comparing Eqs. 33-40 and 33-41, w… purely capacitive load the phase constant ϕ for the current is -90… the voltage amplitude and current amplitude are related by

$$V_C = I_C X_C \quad \text{(capacitor).}$$

Although we found this relation for the circuit of Fig. 33-9*a*, it applies to any capacitance in any ac circuit.

Comparison of Eqs. 33-36 and 33-40, or inspection of Fig. 33-9*b*, shows that the quantities v_C and i_C are 90°, or one-quarter cycle, out of phase. Furthermore, we see that i_C *leads* v_C, which means that, if you monitored the current i_C and the potential difference v_C in the circuit of Fig. 33-9*a*, you would find that i_C reaches its maximum *before* v_C does, by one-quarter cycle.

This relation between i_C and v_C is illustrated by the phasor diagram of Fig. 33-9*c*. As the phasors representing these two quantities rotate counterclockwise together, the phasor labeled I_C does indeed lead that labeled V_C, and by an angle of 90°; that is, the phasor I_C coincides with the vertical axis one-quarter cycle before the phasor V_C does. Be sure to convince yourself that the phasor diagram of Fig. 33-9*c* is consistent with Eqs. 33-36 and 33-40.

CHECKPOINT 4: The figure shows, in (*a*), a sine curve $S(t) = \sin(\omega_d t)$ and three other sinusoidal curves $A(t)$, $B(t)$, and $C(t)$, each of the form $\sin(\omega_d t - \phi)$. (a) Rank the three other curves according to the value of ϕ, most positive first and most negative last. (b) Which curve corresponds to which phasor in (*b*) of the figure? (c) Which curve leads the others?

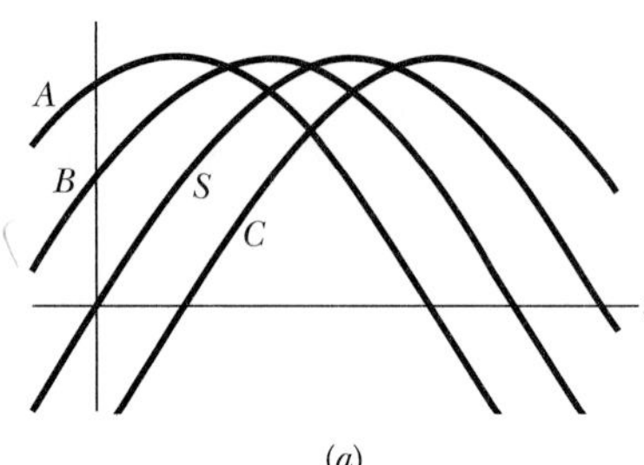

(*a*)

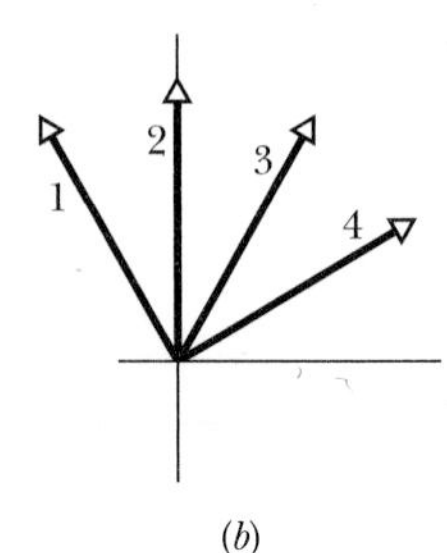

(*b*)

Sample Problem 33-5

Purely capacitive load. In Fig. 33-9*a*, capacitance C is 15.0 μF and the sinusoidal alternating emf device operates at amplitude $\mathscr{E}_m = 36.0$ V and frequency $f_d = 60.0$ Hz.

(a) What are the potential difference $v_C(t)$ across the capacitance and the amplitude V_C of $v_C(t)$?

SOLUTION: If we apply the loop rule to the circuit of Fig. 33-9*a*, we find this **Key Idea**: In a circuit with a purely capacitive load, the potential difference $v_C(t)$ across the capacitance is always equal to the potential difference $\mathscr{E}(t)$ across the emf device. Thus, $v_C(t) = \mathscr{E}(t)$ and $V_C = \mathscr{E}_m$. Since $\mathscr{E}_m$ is given, we have

$$V_C = \mathscr{E}_m = 36.0 \text{ V}. \qquad \text{(Answer)}$$

To find $v_C(t)$, we use Eq. 33-28 to write

$$v_C(t) = \mathscr{E}(t) = \mathscr{E}_m \sin \omega_d t. \qquad (33\text{-}43)$$

Then, substituting $\mathscr{E}_m = 36.0$ V and $\omega_d = 2\pi f_d = 120\pi$ into Eq. 33-43, we have

$$v_C = (36.0 \text{ V}) \sin(120\pi t). \qquad \text{(Answer)}$$

What are the current $i_C(t)$ in the circuit as a function of time the amplitude I_C of $i_C(t)$?

N: The **Key Idea** here is that in an ac circuit with a purely capacitive load, the alternating current $i_C(t)$ in the capacitance leads the alternating potential difference $v_C(t)$ by 90°; that is, the phase constant ϕ for the current is $-90°$ or $-\pi/2$ rad. Thus, we can write Eq. 33-29 as

$$i_C = I_C \sin(\omega_d t - \phi) = I_C \sin(\omega_d t + \pi/2). \qquad (33\text{-}44)$$

A second **Key Idea** is that we can find the amplitude I_C from Eq. 33-42 ($V_C = I_C X_C$) if we first find the capacitive reactance X_C. From Eq. 33-39 ($X_C = 1/\omega_d C$), with $\omega_d = 2\pi f_d$, we can write

$$X_C = \frac{1}{2\pi f_d C} = \frac{1}{(2\pi)(60.0 \text{ Hz})(15.0 \times 10^{-6} \text{ F})} = 177\ \Omega.$$

Then Eq. 33-42 tells us that the current amplitude is

$$I_C = \frac{V_C}{X_C} = \frac{36.0 \text{ V}}{177\ \Omega} = 0.203 \text{ A}. \qquad \text{(Answer)}$$

Substituting this and $\omega_d = 2\pi f_d = 120\pi$ into Eq. 33-44, we have

$$i_C = (0.203 \text{ A}) \sin(120\pi t + \pi/2). \qquad \text{(Answer)}$$

CHECKPOINT 5: If we increase the driving frequency in a circuit with a purely capacitive load, do (a) amplitude V_C and (b) amplitude I_C increase, decrease, or remain the same?

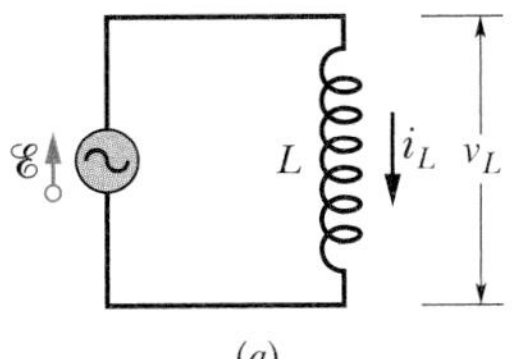

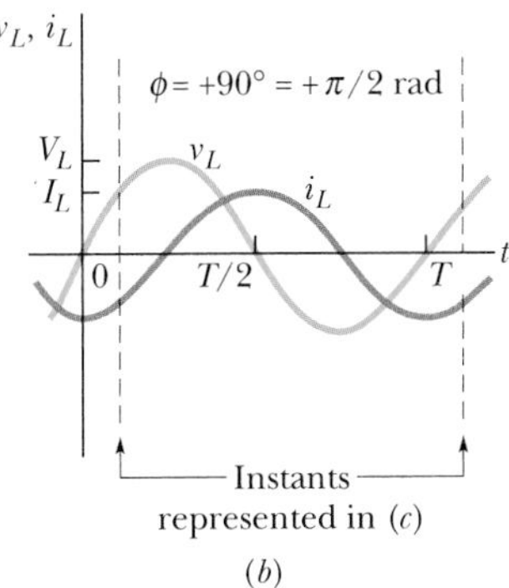

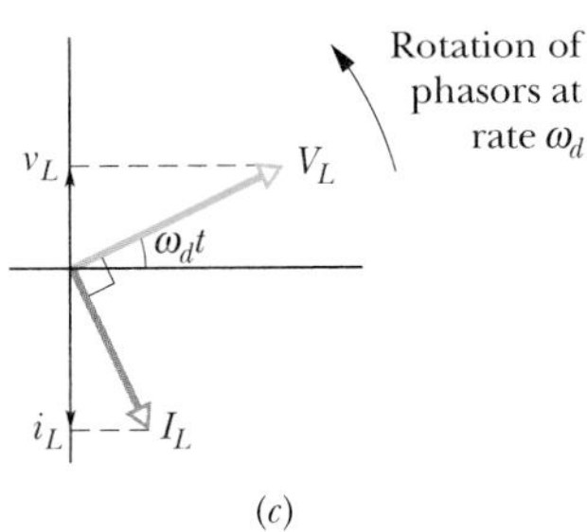

Fig. 33-10 (*a*) An inductor is connected across an alternating-current generator. (*b*) The current in the inductor lags the voltage by 90° (= $\pi/2$ rad). (*c*) A phasor diagram shows the same thing.

An Inductive Load

Figure 33-10*a* shows a circuit containing an inductance and a generator with the alternating emf of Eq. 33-28. Using the loop rule and proceeding as we did to obtain Eq. 33-30, we find that the potential difference across the inductance is

$$v_L = V_L \sin \omega_d t, \tag{33-45}$$

where V_L is the amplitude of v_L. From Eq. 31-37, we can write the potential difference across an inductance L, in which the current is changing at the rate di_L/dt, as

$$v_L = L\frac{di_L}{dt}. \tag{33-46}$$

If we combine Eqs. 33-45 and 33-46, we have

$$\frac{di_L}{dt} = \frac{V_L}{L} \sin \omega_d t. \tag{33-47}$$

Our concern, however, is with the current rather than with its time derivative. We find the former by integrating Eq. 33-47, obtaining

$$i_L = \int di_L = \frac{V_L}{L}\int \sin \omega_d t\, dt = -\left(\frac{V_L}{\omega_d L}\right) \cos \omega_d t. \tag{33-48}$$

We now modify this equation in two ways. First, for reasons of symmetry of notation, we introduce the quantity X_L, called the **inductive reactance** of an inductor, which is defined as

$$X_L = \omega_d L \quad \text{(inductive reactance).} \tag{33-49}$$

The value of X_L depends on the driving angular frequency ω_d. The unit of the inductive time constant τ_L indicates that the SI unit of X_L is the *ohm*, just as it is for X_C and for R.

Second, we replace $-\cos \omega_d t$ in Eq. 33-48 with a phase-shifted sine:

$$-\cos \omega_d t = \sin(\omega_d t - 90°).$$

You can verify this identity by shifting a sine curve in the positive direction by 90°.

With these two changes, Eq. 33-48 becomes

$$i_L = \left(\frac{V_L}{X_L}\right) \sin(\omega_d t - 90°). \tag{33-50}$$

From Eq. 33-29, we can also write this current in the inductance as

$$i_L = I_L \sin(\omega_d t - \phi), \tag{33-51}$$

where I_L is the amplitude of the current i_L. Comparing Eqs. 33-50 and 33-51, we see that for a purely inductive load the phase constant ϕ for the current is +90°. We also see that the voltage amplitude and current amplitude are related by

$$V_L = I_L X_L \quad \text{(inductor).} \tag{33-52}$$

Although we found this relation for the circuit of Fig. 33-10*a*, it applies to any inductance in any ac circuit.

Comparison of Eqs. 33-45 and 33-50, or inspection of Fig. 33-10*b*, shows that the quantities i_L and v_L are 90° out of phase. In this case, however, i_L *lags* v_L; that is, if you monitored the current i_L and the potential difference v_L in the circuit of Fig. 33-10*a*, you would find that i_L reaches its maximum value *after* v_L does, by one-quarter cycle.

The phasor diagram of Fig. 33-10*c* also contains this information. As the phasors rotate counterclockwise in the figure, the phasor labeled I_L does indeed lag that labeled V_L, and by an angle of 90°. Be sure to convince yourself that Fig. 33-10*c* represents Eqs. 33-45 and 33-50.

PROBLEM-SOLVING TACTICS

Tactic 1: *Leading and Lagging in AC Circuits*

Table 33-2 summarizes the relations between the current i and the voltage v for each of the three kinds of circuit elements we have considered. When an applied alternating voltage produces an alternating current in them, the current is in phase with the voltage across a resistor, leads the voltage across a capacitor, and lags the voltage across an inductor.

Many students remember these results with the mnemonic "*ELI* the *ICE* man." *ELI* contains the letter L (for inductor), and in it the letter I (for current) comes after the letter E (for emf or voltage). Thus, for an inductor, the current *lags* the voltage. Similarly *ICE* (which contains a C for capacitor) means that the current *leads* the voltage. You might also use the modified mnemonic "*ELI positively* is the *ICE* man" to remember that the phase constant ϕ is positive for an inductor.

If you have difficulty in remembering whether X_C is equal to $\omega_d C$ (wrong) or $1/\omega_d C$ (right), try remembering that C is in the "cellar"—that is, in the denominator.

TABLE 33-2 Phase and Amplitude Relations for Alternating Currents and Voltages

Circuit Element	Symbol	Resistance or Reactance	Phase of the Current	Phase Constant (or Angle) ϕ	Amplitude Relation
Resistor	R	R	In phase with v_R	0° (= 0 rad)	$V_R = I_R R$
Capacitor	C	$X_C = 1/\omega_d C$	Leads v_C by 90° (= $\pi/2$ rad)	−90° (= $-\pi/2$ rad)	$V_C = I_C X_C$
Inductor	L	$X_L = \omega_d L$	Lags v_L by 90° (= $\pi/2$ rad)	+90° (= $+\pi/2$ rad)	$V_L = I_L X_L$

Sample Problem 33-6

Purely inductive load. In Fig. 33-10*a*, inductance L is 230 mH and the sinusoidal alternating emf device operates at amplitude $\mathscr{E}_m =$ 36.0 V and frequency $f_d = 60.0$ Hz.

(a) What are the potential difference $v_L(t)$ across the inductance and the amplitude V_L of $v_L(t)$?

SOLUTION: If we apply the loop rule to the circuit in Fig. 33-10*a*, we find this **Key Idea**: In a circuit with a purely inductive load, the potential difference $v_L(t)$ across the inductance is always equal to the potential difference $\mathscr{E}(t)$ across the emf device. Thus, $v_L(t) = \mathscr{E}(t)$ and $V_L = \mathscr{E}_m$. Since $\mathscr{E}_m$ is given, we have

$$V_L = \mathscr{E}_m = 36.0 \text{ V}. \qquad \text{(Answer)}$$

To find $v_L(t)$, we use Eq. 33-28 to write

$$v_L(t) = \mathscr{E}(t) = \mathscr{E}_m \sin \omega_d t. \qquad (33\text{-}53)$$

Then, substituting $\mathscr{E}_m = 36.0$ V and $\omega_d = 2\pi f_d = 120\pi$ into Eq. 33-53, we have

$$v_L = (36.0 \text{ V}) \sin(120\pi t). \qquad \text{(Answer)}$$

(b) What are the current $i_L(t)$ in the circuit as a function of time and the amplitude I_L of $i_L(t)$?

SOLUTION: The **Key Idea** here is that in an ac circuit with a purely inductive load, the alternating current $i_L(t)$ in the inductance lags the alternating potential difference $v_L(t)$ by 90°. (In the mnemonic of Problem-Solving Tactic 1, this circuit is "positively an *ELI* circuit," which tells us that the emf E leads the current I and that ϕ is *positive*.) Thus, the phase constant ϕ for the current is +90° or $+\pi/2$ rad, and we can write Eq. 33-29 as

$$i_L = I_L \sin(\omega_d t - \phi) = I_L \sin(\omega_d t - \pi/2). \qquad (33\text{-}54)$$

A second **Key Idea** is that we can find the amplitude I_L from Eq. 33-52 ($V_L = I_L X_L$) if we first find the inductive reactance X_L. From Eq. 33-49 ($X_L = \omega_d L$), with $\omega_d = 2\pi f_d$, we can write

$$X_L = 2\pi f_d L = (2\pi)(60.0 \text{ Hz})(230 \times 10^{-3} \text{ H}) = 86.7\ \Omega.$$

Then Eq. 33-52 tells us that the current amplitude is

$$I_L = \frac{V_L}{X_L} = \frac{36.0 \text{ V}}{86.7\ \Omega} = 0.415 \text{ A}. \qquad \text{(Answer)}$$

Substituting this and $\omega_d = 2\pi f_d = 120\pi$ into Eq. 33-54, we have

$$i_L = (0.415 \text{ A}) \sin(120\pi t - \pi/2). \qquad \text{(Answer)}$$

CHECKPOINT 6: If we increase the driving frequency in a circuit with a purely inductive load, do (a) amplitude V_L and (b) amplitude I_L increase, decrease, or remain the same?

33-9 The Series *RLC* Circuit

We are now ready to apply the alternating emf of Eq. 33-28,

$$\mathscr{E} = \mathscr{E}_m \sin \omega_d t \qquad \text{(applied emf)}, \tag{33-55}$$

to the full *RLC* circuit of Fig. 33-7. Because R, L, and C are in series, the same current

$$i = I \sin(\omega_d t - \phi) \tag{33-56}$$

is driven in all three of them. We wish to find the current amplitude I and the phase constant ϕ. The solution is simplified by the use of phasor diagrams.

The Current Amplitude

We start with Fig. 33-11*a*, which shows the phasor representing the current of Eq. 33-56 at an arbitrary time t. The length of the phasor is the current amplitude I, the projection of the phasor on the vertical axis is the current i at time t, and the angle of rotation of the phasor is the phase $\omega_d t - \phi$ of the current at time t.

Figure 33-11*b* shows the phasors representing the voltages across R, L, and C at the same time t. Each phasor is oriented relative to the angle of rotation of current phasor I in Fig. 33-11*a*, based on the information in Table 33-2:

> ***Resistor:*** Here current and voltage are in phase, so the angle of rotation of voltage phasor V_R is the same as that of phasor I.
>
> ***Capacitor:*** Here current leads voltage by 90°, so the angle of rotation of voltage phasor V_C is 90° less than that of phasor I.
>
> ***Inductor:*** Here current lags voltage by 90°, so the angle of rotation of voltage phasor v_L is 90° greater than that of phasor I.

Figure 33-11*b* also shows the instantaneous voltages v_R, v_C, and v_L across R, C, and L at time t; those voltages are the projections of the corresponding phasors on the vertical axis of the figure.

Figure 33-11*c* shows the phasor representing the applied emf of Eq. 33-55. The length of the phasor is the emf amplitude $\mathscr{E}_m$, the projection of the phasor on the vertical axis is the emf $\mathscr{E}$ at time t, and the angle of rotation of the phasor is the phase $\omega_d t$ of the emf at time t.

From the loop rule we know that at any instant the sum of the voltages v_R, v_C, and v_L is equal to the applied emf $\mathscr{E}$:

$$\mathscr{E} = v_R + v_C + v_L. \tag{33-57}$$

Thus, at time t the projection $\mathscr{E}$ in Fig. 33-11*c* is equal to the algebraic sum of the projections v_R, v_C, and v_L in Fig. 33-11*b*. In fact, as the phasors rotate together, this equality always holds. This means that phasor $\mathscr{E}_m$ in Fig. 33-11*c* must be equal to the vector sum of the three voltage phasors V_R, V_C, and V_L in Fig. 33-11*b*.

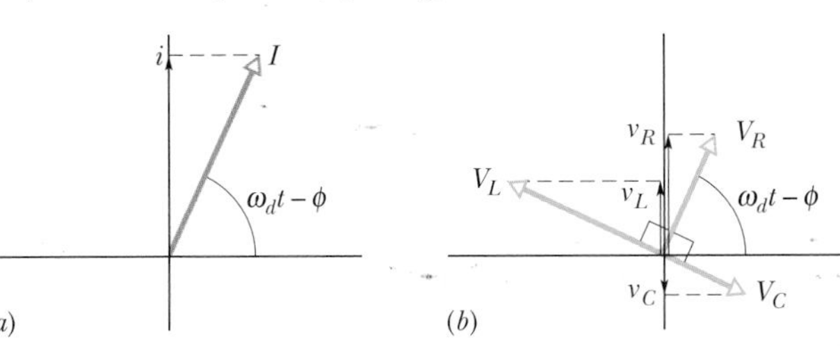

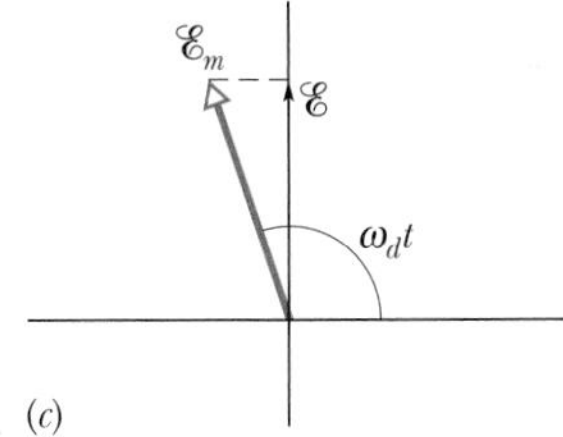

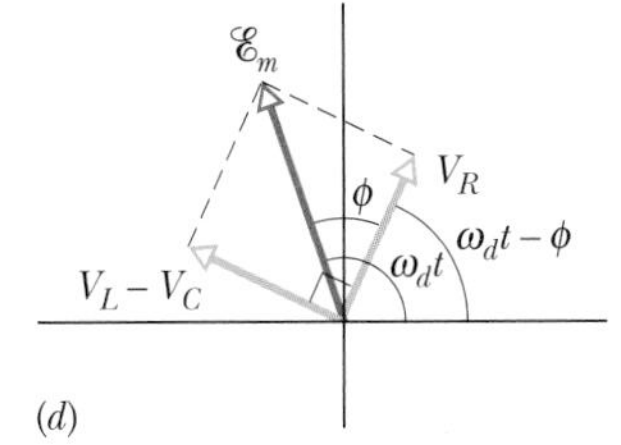

Fig. 33-11 (*a*) A phasor representing the alternating current in the driven *RLC* circuit of Fig. 33-7 at time t. The amplitude I, the instantaneous value i, and the phase $(\omega_d t - \phi)$ are shown. (*b*) Phasors representing the voltages across the inductor, resistor, and capacitor, oriented with respect to the current phasor in (*a*). (*c*) A phasor representing the alternating emf that drives the current of (*a*). (*d*) The emf phasor is equal to the vector sum of the three voltage phasors of (*b*). Here, voltage phasors V_L and V_C have been added to yield their net phasor $(V_L - V_C)$.

That requirement is indicated in Fig. 33-11*d*, where phasor $\mathscr{E}_m$ is drawn as the sum of phasors V_R, V_L, and V_C. Because phasors V_L and V_C have opposite directions in the figure, we simplify the vector sum by first combining V_L and V_C to form the single phasor $V_L - V_C$. Then we combine that single phasor with V_R to find the net phasor. Again, the net phasor must coincide with phasor $\mathscr{E}_m$, as shown.

Both triangles in Fig. 33-11*d* are right triangles. Applying the Pythagorean theorem to either one yields

$$\mathscr{E}_m^2 = V_R^2 + (V_L - V_C)^2. \tag{33-58}$$

From the amplitude information displayed in Table 33-2 we can rewrite this as

$$\mathscr{E}_m^2 = (IR)^2 + (IX_L - IX_C)^2, \tag{33-59}$$

and then rearrange it to the form

$$I = \frac{\mathscr{E}_m}{\sqrt{R^2 + (X_L - X_C)^2}}. \tag{33-60}$$

The denominator in Eq. 33-60 is called the **impedance** Z of the circuit for the driving angular frequency ω_d:

$$Z = \sqrt{R^2 + (X_L - X_C)^2} \quad \text{(impedance defined)}. \tag{33-61}$$

We can then write Eq. 33-60 as

$$I = \frac{\mathscr{E}_m}{Z}. \tag{33-62}$$

If we substitute for X_C and X_L from Eqs. 33-39 and 33-49, we can write Eq. 33-60 more explicitly as

$$I = \frac{\mathscr{E}_m}{\sqrt{R^2 + (\omega_d L - 1/\omega_d C)^2}} \quad \text{(current amplitude)}. \tag{33-63}$$

We have now accomplished half our goal: We have obtained an expression for the current amplitude I in terms of the sinusoidal driving emf and the circuit elements in a series *RLC* circuit.

The value of I depends on the difference between $\omega_d L$ and $1/\omega_d C$ in Eq. 33-63 or, equivalently, the difference between X_L and X_C in Eq. 33-60. In either equation, it does not matter which of the two quantities is greater because the difference is always squared.

The current that we have been describing in this section is the *steady-state current* that occurs after the alternating emf has been applied for some time. When the emf is first applied to a circuit, a brief *transient current* occurs. Its duration (before settling down into the steady-state current) is determined by the time constants $\tau_L = L/R$ and $\tau_C = RC$ as the inductive and capacitive elements "turn on." This transient current can be large and can, for example, destroy a motor on start-up if it is not properly taken into account in the motor's circuit design.

The Phase Constant

From the right-hand phasor triangle in Fig. 33-11*d* and from Table 33-2 we can write

$$\tan\phi = \frac{V_L - V_C}{V_R} = \frac{IX_L - IX_C}{IR}, \tag{33-64}$$

which gives us

$$\tan \phi = \frac{X_L - X_C}{R} \quad \text{(phase constant).} \tag{33-65}$$

This is the other half of our goal: an equation for the phase constant ϕ in a sinusoidally driven series *RLC* circuit. In essence, it gives us three different results for the phase constant, depending on the relative values of X_L and X_C:

$X_L > X_C$**:** The circuit is said to be *more inductive than capacitive*. Equation 33-65 tells us that ϕ is positive for such a circuit, which means that phasor I rotates behind phasor $\mathscr{E}_m$ (Fig. 33-12*a*). A plot of $\mathscr{E}$ and i versus time is like that in Fig. 33-12*b*. (Figures 33-11*c* and *d* were drawn assuming $X_L > X_C$.)

$X_C > X_L$**:** The circuit is said to be *more capacitive than inductive*. Equation 33-65 tells us that ϕ is negative for such a circuit, which means that phasor I rotates ahead of phasor $\mathscr{E}_m$ (Fig. 33-12*c*). A plot of $\mathscr{E}$ and i versus time is like that in Fig. 33-12*d*.

$X_C = X_L$**:** The circuit is said to be in *resonance*, a state that is discussed next. Equation 33-65 tells us that $\phi = 0°$ for such a circuit, which means that phasors $\mathscr{E}_m$ and I rotate together (Fig. 33-12*e*). A plot of $\mathscr{E}$ and i versus time is like that in Fig. 33-12*f*.

As illustration, let us reconsider two extreme circuits: In the *purely inductive circuit* of Fig. 33-10*a*, where X_L is nonzero and $X_C = R = 0$, Eq. 33-65 tells us that $\phi = +90°$ (the greatest value of ϕ), consistent with Fig. 33-10*c*. In the *purely capacitive circuit* of Fig. 33-9*a*, where X_C is nonzero and $X_L = R = 0$, Eq. 33-65 tells us that $\phi = -90°$ (the least value of ϕ), consistent with Fig. 33-9*c*.

Resonance

Equation 33-63 gives the current amplitude I in an *RLC* circuit as a function of the driving angular frequency ω_d of the external alternating emf. For a given resistance R, that amplitude is a maximum when the quantity $\omega_d L - 1/\omega_d C$ in the denominator

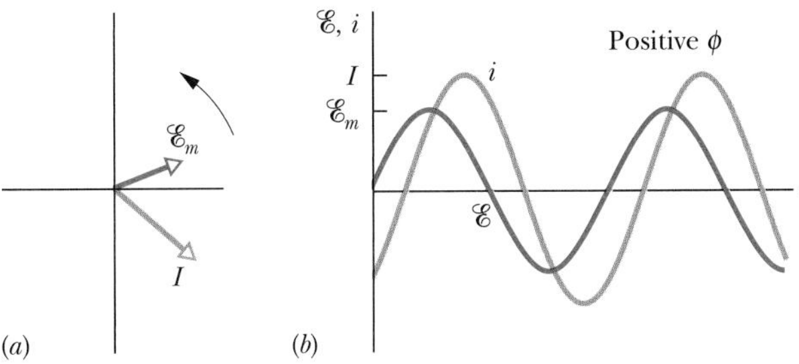

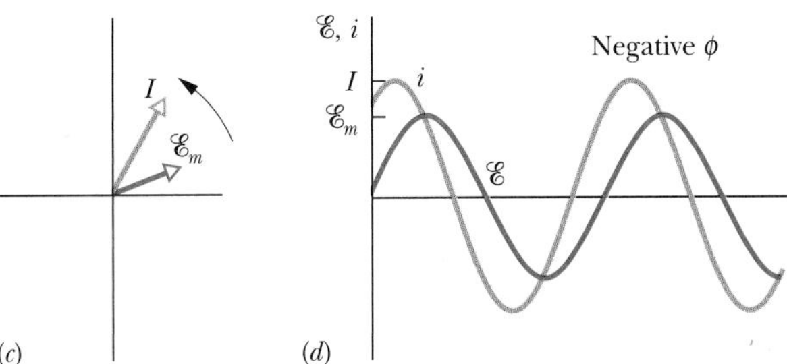

Fig. 33-12 Phasor diagrams and graphs of the alternating emf $\mathscr{E}$ and current i for the driven *RLC* circuit of Fig. 33-7. In the phasor diagram of (*a*) and the graph of (*b*), the current i lags the driving emf $\mathscr{E}$ and the current's phase constant ϕ is positive. In (*c*) and (*d*), the current i leads the driving emf $\mathscr{E}$ and its phase constant ϕ is negative. In (*e*) and (*f*), the current i is in phase with the driving emf $\mathscr{E}$ and its phase constant ϕ is zero.

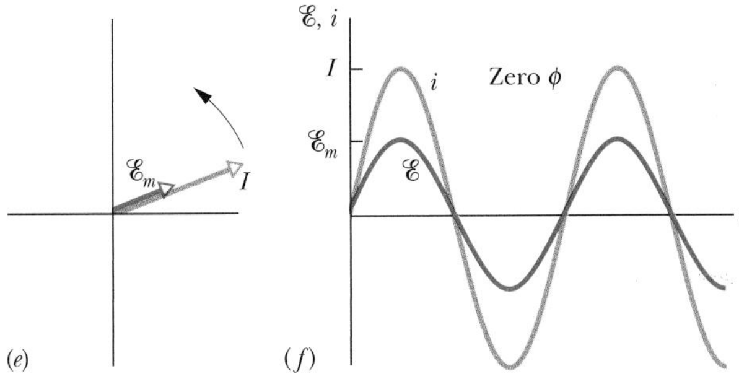

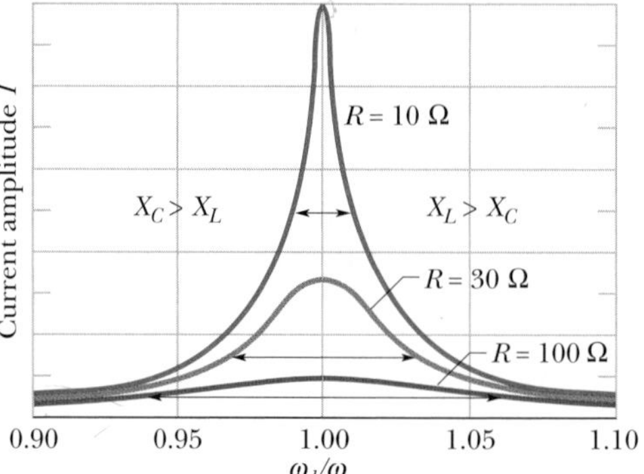

Fig. 33-13 *Resonance curves* for the driven *RLC* circuit of Fig. 33-7 with $L = 100\ \mu\text{H}$, $C = 100$ pF, and three values of R. The current amplitude I of the alternating current depends on how close the driving angular frequency ω_d is to the natural angular frequency ω. The horizontal arrow on each curve measures the curve's width at the half-maximum level, a measure of the sharpness of the resonance. To the left of $\omega_d/\omega = 1.00$, the circuit is mainly capacitive, with $X_C > X_L$; to the right, it is mainly inductive, with $X_L > X_C$.

is zero—that is, when

$$\omega_d L = \frac{1}{\omega_d C}$$

or

$$\omega_d = \frac{1}{\sqrt{LC}} \quad \text{(maximum } I\text{)}. \tag{33-66}$$

Because the natural angular frequency ω of the *RLC* circuit is also equal to $1/\sqrt{LC}$, the maximum value of I occurs when the driving angular frequency matches the natural angular frequency—that is, at resonance. Thus, in an *RLC* circuit, resonance and maximum current amplitude I occur when

$$\omega_d = \omega = \frac{1}{\sqrt{LC}} \quad \text{(resonance)}. \tag{33-67}$$

Figure 33-13 shows three *resonance curves* for sinusoidally driven oscillations in three series *RLC* circuits differing only in R. Each curve peaks at its maximum current amplitude I when the ratio ω_d/ω is 1.00, but the maximum value of I decreases with increasing R. (The maximum I is always $\mathscr{E}_m/R$; to see why, combine Eqs. 33-61 and 33-62.) In addition, the curves increase in width (measured in Fig. 33-13 at half the maximum value of I) with increasing R.

To make physical sense of Fig. 33-13, consider how the reactances X_L and X_C change as we increase the driving angular frequency ω_d, starting with a value much less than the natural frequency ω. For small ω_d, reactance X_L ($= \omega_d L$) is small and reactance X_C ($= 1/\omega_d C$) is large. Thus, the circuit is mainly capacitive and the impedance is dominated by the large X_C, which keeps the current low.

As we increase ω_d, reactance X_C remains dominant but decreases while reactance X_L increases. The decrease in X_C decreases the impedance, allowing the current to increase, as we see on the left side of any resonance curve in Fig. 33-13. When the increasing X_L and the decreasing X_C reach equal values, the current is greatest and the circuit is in resonance, with $\omega_d = \omega$.

As we continue to increase ω_d, the increasing reactance X_L becomes progressively more dominant over the decreasing reactance X_C. The impedance increases because of X_L and the current decreases, as on the right side of any resonance curve in Fig. 33-13. In summary, then: The low-angular-frequency side of a resonance curve is dominated by the capacitor's reactance, the high-angular-frequency side is dominated by the inductor's reactance, and resonance occurs in the middle.

CHECKPOINT 7: Here are the capacitive reactance and inductive reactance, respectively, for three sinusoidally driven series *RLC* circuits: (1) 50 Ω, 100 Ω; (2) 100 Ω, 50 Ω; (3) 50 Ω, 50 Ω. (a) For each, does the current lead or lag the applied emf, or are the two in phase? (b) Which circuit is in resonance?

Sample Problem 33-7

In Fig. 33-7 let $R = 200\ \Omega$, $C = 15.0\ \mu F$, $L = 230$ mH, $f_d = 60.0$ Hz, and $\mathscr{E}_m = 36.0$ V. (These parameters are those used in Sample Problems 33-4, 33-5, and 33-6.)

(a) What is the current amplitude I?

SOLUTION: The **Key Idea** here is that current amplitude I depends on the amplitude $\mathscr{E}_m$ of the driving emf and on the impedance Z of the circuit, according to Eq. 33-62 ($I = \mathscr{E}_m/Z$). Thus, we need to find Z, which depends on the circuit's resistance R, capacitive reactance X_C, and inductive reactance X_L.

The circuit's only resistance is the given resistance R. Its only capacitive reactance is due to the given capacitance and, from Sample Problem 33-5, $X_C = 177\ \Omega$. Its only inductive reactance is due to the given inductance and, from Sample Problem 33-6, $X_L = 86.7\ \Omega$. Thus, the circuit's impedance is

$$Z = \sqrt{R^2 + (X_L - X_C)^2} = \sqrt{(200\ \Omega)^2 + (86.7\ \Omega - 177\ \Omega)^2} = 219\ \Omega.$$

We then find

$$I = \frac{\mathscr{E}_m}{Z} = \frac{36.0\ \text{V}}{219\ \Omega} = 0.164\ \text{A}. \qquad \text{(Answer)}$$

(b) What is the phase constant ϕ of the current in the circuit relative to the driving emf?

SOLUTION: The **Key Idea** here is that the phase constant depends on the inductive reactance, the capacitive reactance, and the resistance of the circuit, according to Eq. 33-65. Solving that equation for ϕ leads to

$$\phi = \tan^{-1}\frac{X_L - X_C}{R} = \tan^{-1}\frac{86.7\ \Omega - 177\ \Omega}{200\ \Omega} = -24.3^\circ = -0.424\ \text{rad}. \qquad \text{(Answer)}$$

The negative phase constant is consistent with the fact that the load is mainly capacitive; that is, $X_C > X_L$. In the mnemonic of Problem-Solving Tactic 1, this circuit is an *ICE* circuit—the current *leads* the driving emf.

33-10 Power in Alternating-Current Circuits

In the *RLC* circuit of Fig. 33-7, the source of energy is the alternating-current generator. Some of the energy that it provides is stored in the electric field in the capacitor, some is stored in the magnetic field in the inductor, and some is dissipated as thermal energy in the resistor. In steady-state operation—which we assume—the average energy stored in the capacitor and inductor together remains constant. The net transfer of energy is thus from the generator to the resistor, where electromagnetic energy is dissipated as thermal energy.

The instantaneous rate at which energy is dissipated in the resistor can be written, with the help of Eqs. 27-22 and 33-29, as

$$P = i^2R = [I\sin(\omega_d t - \phi)]^2 R = I^2R\sin^2(\omega_d t - \phi). \qquad (33\text{-}68)$$

The *average* rate at which energy is dissipated in the resistor, however, is the average of Eq. 33-68 over time. Over one complete cycle, the average value of $\sin\theta$, where θ is any variable, is zero (Fig. 33-14*a*) but the average value of $\sin^2\theta$ is $\frac{1}{2}$ (Fig. 33-14*b*). (Note in Fig. 33-14*b* how the shaded areas under the curve but above the horizontal line marked $+\frac{1}{2}$ exactly fill in the unshaded spaces below that line.) Thus, we can write, from Eq. 33-68,

$$P_{avg} = \frac{I^2R}{2} = \left(\frac{I}{\sqrt{2}}\right)^2 R. \qquad (33\text{-}69)$$

The quantity $I/\sqrt{2}$ is called the **root-mean-square,** or **rms,** value of the current i:

$$I_{rms} = \frac{I}{\sqrt{2}} \quad \text{(rms current)}. \qquad (33\text{-}70)$$

We can now rewrite Eq. 33-69 as

$$P_{avg} = I_{rms}^2 R \quad \text{(average power)}. \qquad (33\text{-}71)$$

Equation 33-71 looks much like Eq. 27-22 ($P = i^2R$); the message is that if we

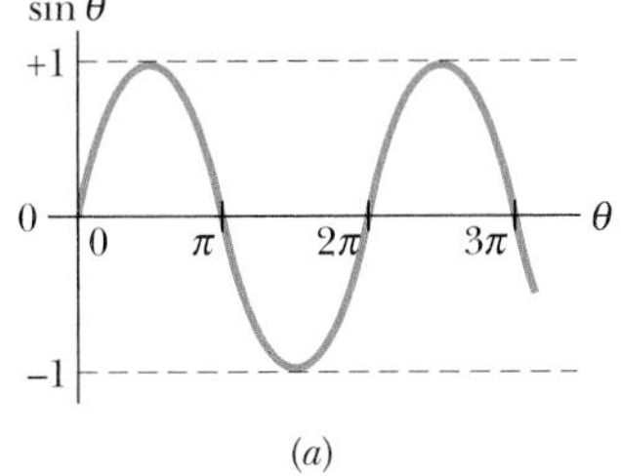

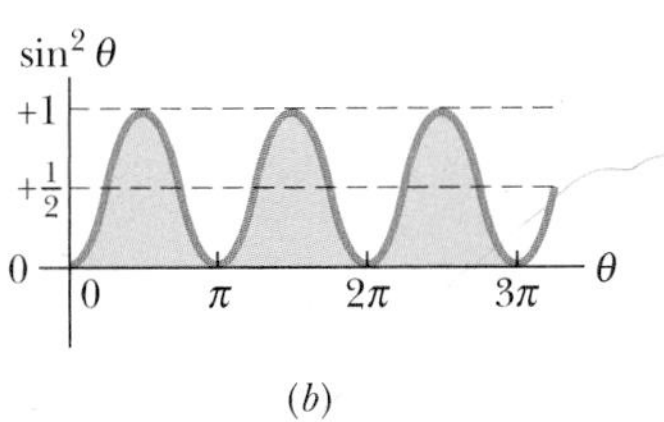

Fig. 33-14 (*a*) A plot of $\sin\theta$ versus θ. The average value over one cycle is zero. (*b*) A plot of $\sin^2\theta$ versus θ. The average value over one cycle is $\frac{1}{2}$.

switch to the rms current, we can compute the average rate of energy dissipation for alternating-current circuits just as for direct-current circuits.

We can also define rms values of voltages and emfs for alternating-current circuits:

$$V_{rms} = \frac{V}{\sqrt{2}} \quad \text{and} \quad \mathscr{E}_{rms} = \frac{\mathscr{E}_m}{\sqrt{2}} \quad \text{(rms voltage; rms emf).} \tag{33-72}$$

Alternating-current instruments, such as ammeters and voltmeters, are usually calibrated to read I_{rms}, V_{rms}, and $\mathscr{E}_{rms}$. Thus, if you plug an alternating-current voltmeter into a household electric outlet and it reads 120 V, that is an rms voltage. The *maximum* value of the potential difference at the outlet is $\sqrt{2} \times (120\text{ V})$, or 170 V.

Because the proportionality factor $1/\sqrt{2}$ in Eqs. 33-70 and 33-72 is the same for all three variables, we can write Eqs. 33-62 and 33-60 as

$$I_{rms} = \frac{\mathscr{E}_{rms}}{Z} = \frac{\mathscr{E}_{rms}}{\sqrt{R^2 + (X_L - X_C)^2}}, \tag{33-73}$$

and, indeed, this is the form that we almost always use.

We can use the relationship $I_{rms} = \mathscr{E}_{rms}/Z$ to recast Eq. 33-71 in a useful equivalent way. We write

$$P_{avg} = \frac{\mathscr{E}_{rms}}{Z} I_{rms} R = \mathscr{E}_{rms} I_{rms} \frac{R}{Z}. \tag{33-74}$$

From Fig. 33-11*d*, Table 33-2, and Eq. 33-62, however, we see that R/Z is just the cosine of the phase constant ϕ:

$$\cos\phi = \frac{V_R}{\mathscr{E}_m} = \frac{IR}{IZ} = \frac{R}{Z}. \tag{33-75}$$

Equation 33-74 then becomes

$$P_{avg} = \mathscr{E}_{rms} I_{rms} \cos\phi \quad \text{(average power),} \tag{33-76}$$

in which the term $\cos\phi$ is called the **power factor.** Because $\cos\phi = \cos(-\phi)$, Eq. 33-76 is independent of the sign of the phase constant ϕ.

To maximize the rate at which energy is supplied to a resistive load in an *RLC* circuit, we should keep the power factor $\cos\phi$ as close to unity as possible. This is equivalent to keeping the phase constant ϕ in Eq. 33-29 as close to zero as possible. If, for example, the circuit is highly inductive, it can be made less so by putting more capacitance in the circuit, connected in series. (Recall that putting an additional capacitance into a series of capacitances decreases the equivalent capacitance C_{eq} of the series.) Thus, the resulting decrease in C_{eq} in the circuit reduces the phase constant and increases the power factor in Eq. 33-76. Power companies place series-connected capacitors throughout their transmission systems to get these results.

CHECKPOINT 8: (a) If the current in a sinusoidally driven series *RLC* circuit leads the emf, would we increase or decrease the capacitance to increase the rate at which energy is supplied to the resistance? (b) Will this change bring the resonant angular frequency of the circuit closer to the angular frequency of the emf or put it farther away?

Sample Problem 33-8

A series *RLC* circuit, driven with $\mathscr{E}_{rms} = 120$ V at frequency $f_d = 60.0$ Hz, contains a resistance $R = 200\ \Omega$, an inductance with $X_L = 80.0\ \Omega$, and a capacitance with $X_C = 150\ \Omega$.

(a) What are the power factor $\cos\phi$ and phase constant ϕ of the circuit?

SOLUTION: The **Key Idea** here is that the power factor $\cos\phi$ can be found from the resistance R and impedance Z via Eq. 33-75

($\cos\phi = R/Z$). To calculate Z, we use Eq. 33-61:

$$Z = \sqrt{R^2 + (X_L - X_C)^2}$$
$$= \sqrt{(200\ \Omega)^2 + (80.0\ \Omega - 150\ \Omega)^2} = 211.90\ \Omega.$$

Equation 33-75 then gives us

$$\cos\phi = \frac{R}{Z} = \frac{200\ \Omega}{211.90\ \Omega} = 0.9438 \approx 0.944. \quad \text{(Answer)}$$

Taking the inverse cosine then yields

$$\phi = \cos^{-1} 0.944 = \pm 19.3°.$$

Both $+19.3°$ and $-19.3°$ have a cosine of 0.944. To determine which sign is correct, we must consider whether the current leads or lags the driving emf. Because $X_C > X_L$, this circuit is mainly capacitive, with the current leading the emf. Thus, ϕ must be negative:

$$\phi = -19.3°. \quad \text{(Answer)}$$

We could, instead, have found ϕ with Eq. 33-65. A calculator would then have given us the complete answer, with the minus sign.

(b) What is the average rate P_{avg} at which energy is dissipated in the resistance?

SOLUTION: One way to answer this question is to use this **Key Idea**: Because the circuit is assumed to be in steady-state operation, the rate at which energy is dissipated in the resistance is equal to the rate at which energy is supplied to the circuit, as given by Eq. 33-76 ($P_{avg} = \mathscr{E}_{rms} I_{rms} \cos\phi$).

We are given the rms driving emf $\mathscr{E}_{rms}$ and we know $\cos\phi$ from part (a). To find I_{rms} we use the **Key Idea** that the rms current is determined by the rms value of the driving emf and the circuit's impedance Z (which we know), according to Eq. 33-73:

$$I_{rms} = \frac{\mathscr{E}_{rms}}{Z}.$$

Substituting this into Eq. 33-76 then leads to

$$P_{avg} = \mathscr{E}_{rms} I_{rms} \cos\phi = \frac{\mathscr{E}_{rms}^2}{Z} \cos\phi$$
$$= \frac{(120\ \text{V})^2}{211.90\ \Omega}(0.9438) = 64.1\ \text{W}. \quad \text{(Answer)}$$

A second way to answer the question is to use the **Key Idea** that the rate at which energy is dissipated in a resistance R depends on the square of the rms current I_{rms} through it, according to Eq. 33-71. We then find

$$P_{avg} = I_{rms}^2 R = \frac{\mathscr{E}_{rms}^2}{Z^2} R$$
$$= \frac{(120\ \text{V})^2}{(211.90\ \Omega)^2}(200\ \Omega) = 64.1\ \text{W}. \quad \text{(Answer)}$$

(c) What new capacitance C_{new} is needed to maximize P_{avg} if the other parameters of the circuit are not changed?

SOLUTION: One **Key Idea** here is that the average rate P_{avg} at which energy is supplied and dissipated is maximized if the circuit is brought into resonance with the driving emf. A second **Key Idea** is that resonance occurs when $X_C = X_L$. From the given data, we have $X_C > X_L$. Thus, we must decrease X_C to reach resonance. From Eq. 33-39 ($X_C = 1/\omega_d C$), we see that this means we must increase C to the new value C_{new}.

Using Eq. 33-39, we can write the condition $X_C = X_L$ as

$$\frac{1}{\omega_d C_{new}} = X_L.$$

Substituting $2\pi f_d$ for ω_d (because we are given f_d and not ω_d) and then solving for C_{new}, we find

$$C_{new} = \frac{1}{2\pi f_d X_L} = \frac{1}{(2\pi)(60\ \text{Hz})(80.0\ \Omega)}$$
$$= 3.32 \times 10^{-5}\ \text{F} = 33.2\ \mu\text{F}. \quad \text{(Answer)}$$

Following the procedure of part (b), you can show that with C_{new}, P_{avg} would then be at its maximum value of 72.0 W.

33-11 Transformers

Energy Transmission Requirements

When an ac circuit has only a resistive load, the power factor in Eq. 33-76 is $\cos 0° = 1$ and the applied rms emf $\mathscr{E}_{rms}$ is equal to the rms voltage V_{rms} across the load. Thus, with an rms current I_{rms} in the load, energy is supplied and dissipated at the average rate of

$$P_{avg} = \mathscr{E}I = IV. \quad (33\text{-}77)$$

(In Eq. 33-77 and the rest of this section, we follow conventional practice and drop the subscripts identifying rms quantities. Engineers and scientists assume that all time-varying currents and voltages are reported as rms values; that is what the meters read.) Equation 33-77 tells us that, to satisfy a given power requirement, we have a range of choices, from a relatively large current I and a relatively small voltage V to just the reverse, provided only that the product IV is as required.

At 5:17 p.m. on November 9, 1965, a faulty relay in the power system near Niagara Falls opened a circuit breaker on a transmission line, automatically causing the current to switch to other lines, which overloaded those lines and made other circuit breakers open. Within minutes a runaway shutdown had blacked out much of New York, New England, and Ontario.

In electric power distribution systems it is desirable for reasons of safety and for efficient equipment design to deal with relatively low voltages at both the generating end (the electric power plant) and the receiving end (the home or factory). Nobody wants an electric toaster or a child's electric train to operate at, say, 10 kV. On the other hand, in the transmission of electric energy from the generating plant to the consumer, we want the lowest practical current (hence the largest practical voltage) to minimize I^2R losses (often called *ohmic losses*) in the transmission line.

As an example, consider the 735 kV line used to transmit electric energy from the La Grande 2 hydroelectric plant in Quebec to Montreal, 1000 km away. Suppose that the current is 500 A and the power factor is close to unity. Then from Eq. 33-77, energy is supplied at the average rate

$$P_{avg} = \mathscr{E}I = (7.35 \times 10^5 \text{ V})(500 \text{ A}) = 368 \text{ MW}.$$

The resistance of the transmission line is about 0.220 Ω/km; thus, there is a total resistance of about 220 Ω for the 1000 km stretch. Energy is dissipated owing to that resistance at a rate of about

$$P_{avg} = I^2R = (500 \text{ A})^2(220 \ \Omega) = 55.0 \text{ MW},$$

which is nearly 15% of the supply rate.

Imagine what would happen if we doubled the current and halved the voltage. Energy would be supplied by the plant at the same average rate of 368 MW as previously, but now energy would be dissipated at the rate of about

$$P_{avg} = I^2R = (1000 \text{ A})^2(220 \ \Omega) = 220 \text{ MW},$$

which is *almost 60% of the supply rate*. Hence the general energy transmission rule: Transmit at the highest possible voltage and the lowest possible current.

The Ideal Transformer

The transmission rule leads to a fundamental mismatch between the requirement for efficient high-voltage transmission and the need for safe low-voltage generation and consumption. We need a device with which we can raise (for transmission) and lower (for use) the ac voltage in a circuit, keeping the product current × voltage essentially constant. The **transformer** is such a device. It has no moving parts, operates by Faraday's law of induction, and has no simple direct-current counterpart.

The *ideal transformer* in Fig. 33-15 consists of two coils, with different numbers of turns, wound around an iron core. (The coils are insulated from the core.) In use, the primary winding, of N_p turns, is connected to an alternating-current generator whose emf $\mathscr{E}$ at any time t is given by

$$\mathscr{E} = \mathscr{E}_m \sin \omega t. \tag{33-78}$$

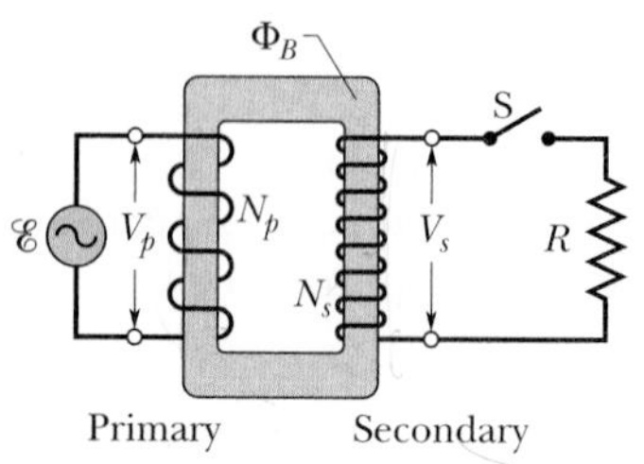

Fig. 33-15 An ideal transformer (two coils wound on an iron core) in a basic transformer circuit. An ac generator produces current in the coil at the left (the *primary*). The coil at the right (the *secondary*) is connected to the resistive load R when switch S is closed.

The secondary winding, of N_s turns, is connected to load resistance R, but its circuit is an open circuit as long as switch S is open (which we assume for the present). Thus, there can be no current through the secondary coil. We assume further for this ideal transformer that the resistances of the primary and secondary windings are negligible, as are energy losses due to magnetic hysteresis in the iron core. Well-designed, high-capacity transformers can have energy losses as low as 1%, so our assumptions are reasonable.

For the assumed conditions, the primary winding (or *primary*) is a pure inductance, and the primary circuit is like that in Fig. 33-10*a*. Thus, the (very small) primary current, also called the *magnetizing current* I_{mag}, lags the primary voltage V_p by 90°; the primary's power factor (= cos ϕ in Eq. 33-76) is zero, so no power is delivered from the generator to the transformer.

However, the small alternating primary current I_{mag} induces an alternating magnetic flux Φ_B in the iron core. Because the core extends through the secondary winding (or *secondary*), this induced flux also extends through the turns of the secondary. From Faraday's law of induction (Eq. 31-6), the induced emf per turn $\mathscr{E}_{turn}$ is the same for both the primary and the secondary. Also, the voltage V_p across the primary is equal to the emf induced in the primary, and the voltage V_s across the secondary is equal to the emf induced in the secondary. Thus, we can write

$$\mathscr{E}_{turn} = \frac{d\Phi_B}{dt} = \frac{V_p}{N_p} = \frac{V_s}{N_s}$$

and thus,

$$V_s = V_p \frac{N_s}{N_p} \quad \text{(transformation of voltage).} \tag{33-79}$$

If $N_s > N_p$, the transformer is called a *step-up transformer* because it steps the primary's voltage V_p *up* to a higher voltage V_s. Similarly, if $N_s < N_p$, the device is a *step-down transformer*.

So far, with switch S open, no energy is transferred from the generator to the rest of the circuit. Now let us close S to connect the secondary to the resistive load R. (In general, the load would also contain inductive and capacitive elements, but here we consider just resistance R.) We find that now energy *is* transferred from the generator. Let us see why.

Several things happen when we close switch S.

1. An alternating current I_s appears in the secondary circuit, with corresponding energy dissipation rate I_s^2R $(= V_s^2/R)$ in the resistive load.
2. This current produces its own alternating magnetic flux in the iron core, and this flux induces (from Faraday's law and Lenz's law) an opposing emf in the primary windings.
3. The voltage V_p of the primary, however, cannot change in response to this opposing emf because it must always be equal to the emf $\mathscr{E}$ that is provided by the generator; closing switch S cannot change this fact.
4. To maintain V_p, the generator now produces (in addition to I_{mag}) an alternating current I_p in the primary circuit; the magnitude and phase constant of I_p are just those required for the emf induced by I_p in the primary to exactly cancel the emf induced there by I_s. Because the phase constant of I_p is not 90° like that of I_{mag}, this current I_p can transfer energy to the primary.

We want to relate I_s to I_p. However, rather than analyze the foregoing complex process in detail, let us just apply the principle of conservation of energy. The rate at which the generator transfers energy to the primary is equal to I_pV_p. The rate at which the primary then transfers energy to the secondary (via the alternating magnetic field linking the two coils) is I_sV_s. Because we assume that no energy is lost along the way, conservation of energy requires that

$$I_pV_p = I_sV_s.$$

Substituting for V_s from Eq. 33-79, we find that

$$I_s = I_p \frac{N_p}{N_s} \quad \text{(transformation of currents).} \tag{33-80}$$

This equation tells us that the current I_s in the secondary can differ from the current I_p in the primary, depending on the *turns ratio* N_p/N_s.

Current I_p appears in the primary circuit because of the resistive load R in the secondary circuit. To find I_p, we substitute $I_s = V_s/R$ into Eq. 33-80 and then we substitute for V_s from Eq. 33-79. We find

$$I_p = \frac{1}{R}\left(\frac{N_s}{N_p}\right)^2 V_p. \tag{33-81}$$

This equation has the form $I_p = V_p/R_{eq}$, where equivalent resistance R_{eq} is

$$R_{eq} = \left(\frac{N_p}{N_s}\right)^2 R. \tag{33-82}$$

This R_{eq} is the value of the load resistance as "seen" by the generator; the generator produces the current I_p and voltage V_p as if it were connected to a resistance R_{eq}.

Impedance Matching

Equation 33-82 suggests still another function for the transformer. For maximum transfer of energy from an emf device to a resistive load, the resistance of the emf device and the resistance of the load must be equal. The same relation holds for ac circuits except that the *impedance* (rather than just the resistance) of the generator must be matched to that of the load. Often this condition is not met. For example, in a music-playing system, the amplifier has high impedance and the speaker set has low impedance. We can match the impedances of the two devices by coupling them through a transformer with a suitable turns ratio N_p/N_s.

CHECKPOINT 9: An alternating-current emf device has a smaller resistance than that of the resistive load; to increase the transfer of energy from the device to the load, a transformer will be connected between the two. (a) Should N_s be greater than or less than N_p? (b) Will that make it a step-up or step-down transformer?

Sample Problem 33-9

A transformer on a utility pole operates at $V_p = 8.5$ kV on the primary side and supplies electric energy to a number of nearby houses at $V_s = 120$ V, both quantities being rms values. Assume an ideal step-down transformer, a purely resistive load, and a power factor of unity.

(a) What is the turns ratio N_p/N_s of the transformer?

SOLUTION: The **Key Idea** here is that the turns ratio N_p/N_s is related to the (given) rms primary and secondary voltages via Eq. 33-79, which we can write as

$$\frac{V_s}{V_p} = \frac{N_s}{N_p}.$$

(Note that the right side of this equation is the *inverse* of the turns ratio.) Inverting both sides then gives us

$$\frac{N_p}{N_s} = \frac{V_p}{V_s} = \frac{8.5 \times 10^3 \text{ V}}{120 \text{ V}} = 70.83 \approx 71. \quad \text{(Answer)}$$

(b) The average rate of energy consumption (or dissipation) in the houses served by the transformer is 78 kW. What are the rms currents in the primary and secondary of the transformer?

SOLUTION: The **Key Idea** here is that for a purely resistive load, the power factor $\cos\phi$ is unity; thus, the average rate at which energy is supplied and dissipated is given by Eq. 33-77. In the primary circuit, with $V_p = 8.5$ kV, Eq. 33-77 yields

$$I_p = \frac{P_{avg}}{V_p} = \frac{78 \times 10^3 \text{ W}}{8.5 \times 10^3 \text{ V}} = 9.176 \text{ A} \approx 9.2 \text{ A}. \quad \text{(Answer)}$$

Similarly, in the secondary circuit,

$$I_s = \frac{P_{avg}}{V_s} = \frac{78 \times 10^3 \text{ W}}{120 \text{ V}} = 650 \text{ A}. \quad \text{(Answer)}$$

You can check that $I_s = I_p(N_p/N_s)$ as required by Eq. 33-80.

(c) What is the resistive load R_s in the secondary circuit? What is the corresponding resistive load R_p in the primary circuit?

SOLUTION: For both circuits, the **Key Idea** here is that we can relate the resistive load to the rms voltage and current with $V = IR$. For the secondary circuit, we find

$$R_s = \frac{V_s}{I_s} = \frac{120 \text{ V}}{650 \text{ A}} = 0.1846\ \Omega \approx 0.18\ \Omega. \quad \text{(Answer)}$$

Similarly, for the primary circuit we find

$$R_p = \frac{V_p}{I_p} = \frac{8.5 \times 10^3\ \text{V}}{9.176\ \text{A}} = 926\ \Omega \approx 930\ \Omega. \quad \text{(Answer)}$$

Another Key Idea that we can use to find R_p is that R_p is the equivalent resistive load "seen" from the primary side of the transformer, as given by Eq. 33-82. If we substitute R_p for R_{eq} and R_s for R, that equation yields

$$R_p = \left(\frac{N_p}{N_s}\right)^2 R_s = (70.83)^2(0.1846\ \Omega)$$

$$= 926\ \Omega \approx 930\ \Omega. \quad \text{(Answer)}$$

REVIEW & SUMMARY

***LC* Energy Transfers** In an oscillating LC circuit, energy is shuttled periodically between the electric field of the capacitor and the magnetic field of the inductor; instantaneous values of the two forms of energy are

$$U_E = \frac{q^2}{2C} \quad \text{and} \quad U_B = \frac{Li^2}{2}, \qquad (33\text{-}1, 33\text{-}2)$$

where q is the instantaneous charge on the capacitor and i is the instantaneous current through the inductor. The total energy $U\ (= U_E + U_B)$ remains constant.

***LC* Charge and Current Oscillations** The principle of conservation of energy leads to

$$L\frac{d^2q}{dt^2} + \frac{1}{C}q = 0 \quad (LC \text{ oscillations}) \qquad (33\text{-}11)$$

as the differential equation of LC oscillations (with no resistance). The solution of Eq. 33-11 is

$$q = Q\cos(\omega t + \phi) \quad \text{(charge)}, \qquad (33\text{-}12)$$

in which Q is the *charge amplitude* (maximum charge on the capacitor) and the angular frequency ω of the oscillations is

$$\omega = \frac{1}{\sqrt{LC}}. \qquad (33\text{-}4)$$

The phase constant ϕ in Eq. 33-12 is determined by the initial conditions (at $t = 0$) of the system.

The current i in the system at any time t is

$$i = -\omega Q\sin(\omega t + \phi) \quad \text{(current)}, \qquad (33\text{-}13)$$

in which ωQ is the *current amplitude* I.

Damped Oscillations Oscillations in an LC circuit are damped when a dissipative element R is also present in the circuit. Then

$$L\frac{d^2q}{dt^2} + R\frac{dq}{dt} + \frac{1}{C}q = 0 \quad (RLC \text{ circuit}). \qquad (33\text{-}24)$$

The solution of this differential equation is

$$q = Qe^{-Rt/2L}\cos(\omega' t + \phi), \qquad (33\text{-}25)$$

where

$$\omega' = \sqrt{\omega^2 - (R/2L)^2}. \qquad (33\text{-}26)$$

We consider only situations with small R and thus small damping; then $\omega' \approx \omega$.

Alternating Currents; Forced Oscillations A series RLC circuit may be set into *forced oscillation* at a *driving angular frequency* ω_d by an external alternating emf

$$\mathscr{E} = \mathscr{E}_m \sin \omega_d t. \qquad (33\text{-}28)$$

The current driven in the circuit by the emf is

$$i = I\sin(\omega_d t - \phi), \qquad (33\text{-}29)$$

where ϕ is the phase constant of the current.

Resonance The current amplitude I in a series RLC circuit driven by a sinusoidal external emf is a maximum ($I = \mathscr{E}_m/R$) when the driving angular frequency ω_d equals the natural angular frequency ω of the circuit (that is, at *resonance*). Then $X_C = X_L$, $\phi = 0$, and the current is in phase with the emf.

Single Circuit Elements The alternating potential difference across a resistor has amplitude $V_R = IR$; the current is in phase with the potential difference.

For a *capacitor*, $V_C = IX_C$, in which $X_C = 1/\omega_d C$ is the **capacitive reactance;** the current here leads the potential difference by 90° ($\phi = -90° = -\pi/2$ rad).

For an *inductor*, $V_L = IX_L$, in which $X_L = \omega_d L$ is the **inductive reactance;** the current here lags the potential difference by 90° ($\phi = +90° = +\pi/2$ rad).

Series *RLC* Circuits For a series RLC circuit with external emf given by Eq. 33-28 and current given by Eq. 33-29,

$$I = \frac{\mathscr{E}_m}{\sqrt{R^2 + (X_L - X_C)^2}}$$

$$= \frac{\mathscr{E}_m}{\sqrt{R^2 + (\omega_d L - 1/\omega_d C)^2}} \quad \text{(current amplitude)} \qquad (33\text{-}60, 33\text{-}63)$$

and

$$\tan\phi = \frac{X_L - X_C}{R} \quad \text{(phase constant)}. \qquad (33\text{-}65)$$

Defining the impedance Z of the circuit as

$$Z = \sqrt{R^2 + (X_L - X_C)^2} \quad \text{(impedance)} \qquad (33\text{-}61)$$

allows us to write Eq. 33-60 as $I = \mathscr{E}_m/Z$.

Power In a series RLC circuit, the **average power** P_{avg} of the generator is equal to the production rate of thermal energy in the resistor:

$$P_{avg} = I_{rms}^2 R = \mathscr{E}_{rms} I_{rms}\cos\phi. \qquad (33\text{-}71, 33\text{-}76)$$

Here rms stands for **root-mean-square;** the rms quantities are related to the maximum quantities by $I_{rms} = I/\sqrt{2}$, $V_{rms} = V_m/\sqrt{2}$, and $\mathscr{E}_{rms} = \mathscr{E}_m/\sqrt{2}$. The term cos ϕ is called the **power factor** of the circuit.

Transformers A *transformer* (assumed to be ideal) is an iron core on which are wound a primary coil of N_p turns and a secondary coil of N_s turns. If the primary coil is connected across an alternating-current generator, the primary and secondary voltages are related by

$$V_s = V_p \frac{N_s}{N_p} \quad \text{(transformation of voltage).} \qquad (33\text{-}79)$$

The currents through the coils are related by

$$I_s = I_p \frac{N_p}{N_s} \quad \text{(transformation of currents),} \qquad (33\text{-}80)$$

and the equivalent resistance of the secondary circuit, as seen by the generator, is

$$R_{eq} = \left(\frac{N_p}{N_s}\right)^2 R, \qquad (33\text{-}82)$$

where R is the resistive load in the secondary circuit. The ratio N_p/N_s is called the transformer's *turns ratio*.

QUESTIONS

1. A charged capacitor and an inductor are connected at time $t = 0$. In terms of the period T of the resulting oscillations, what is the first later time at which the following reach a maximum: (a) U_B, (b) the magnetic flux through the inductor, (c) di/dt, and (d) the emf of the inductor?

2. What values of phase constant ϕ in Eq. 33-12 allow situations (*a*), (*c*), (*e*), and (*g*) of Fig. 33-1 to occur at $t = 0$?

3. Figure 33-16 shows three oscillating LC circuits with identical inductors and capacitors. Rank the circuits according to the time taken to fully discharge the capacitors during the oscillations, greatest first.

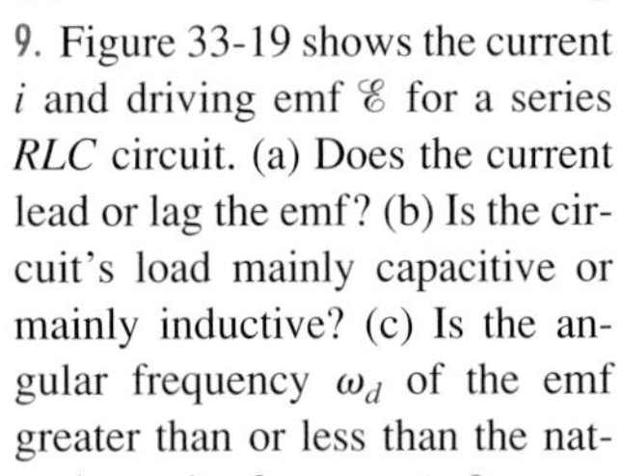

Fig. 33-16 Question 3.

4. Figure 33-17 shows graphs of capacitor voltage v_C for LC circuits 1 and 2, which contain identical capacitances and have the same maximum charge Q. Are (a) the inductance L and (b) the maximum current I in circuit 1 greater than, less than, or the same as those in circuit 2?

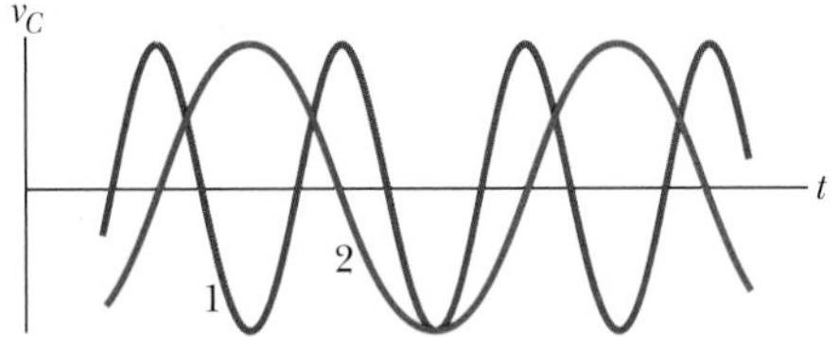

Fig. 33-17 Question 4.

5. Charges on the capacitors in three oscillating LC circuits vary as follows: (1) $q = 2 \cos 4t$, (2) $q = 4 \cos t$, (3) $q = 3 \cos 4t$ (with q in coulombs and t in seconds). Rank the circuits according to (a) the current amplitude and (b) the period, greatest first.

6. If you increase the inductance L in an oscillating LC circuit having a given maximum charge Q, do (a) the current magnitude I and (b) the maximum magnetic energy U_B increase, decrease, or stay the same?

7. An alternating emf source with a certain emf amplitude is connected, in turn, to a resistor, a capacitor, and then an inductor. Once connected to one of the devices, the driving frequency f_d is varied and the amplitude I of the resulting current through the device is measured and plotted. Which of the three plots in Fig. 33-18 corresponds to which of the three devices?

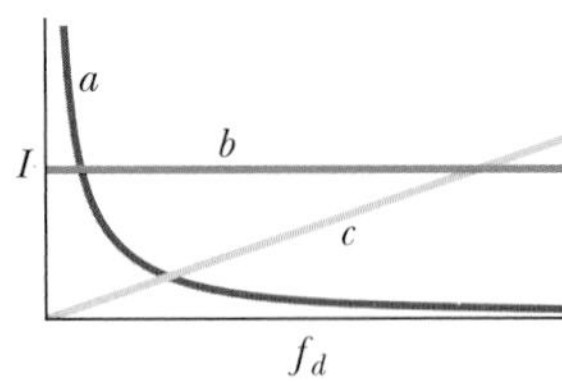

Fig. 33-18 Question 7.

8. The values of the phase constant ϕ for four sinusoidally driven series RLC circuits are (1) $-15°$, (2) $+35°$, (3) $\pi/3$ rad, and (4) $-\pi/6$ rad. (a) In which is the load primarily capacitive? (b) In which does the current lag the alternating emf?

9. Figure 33-19 shows the current i and driving emf $\mathscr{E}$ for a series RLC circuit. (a) Does the current lead or lag the emf? (b) Is the circuit's load mainly capacitive or mainly inductive? (c) Is the angular frequency ω_d of the emf greater than or less than the natural angular frequency ω?

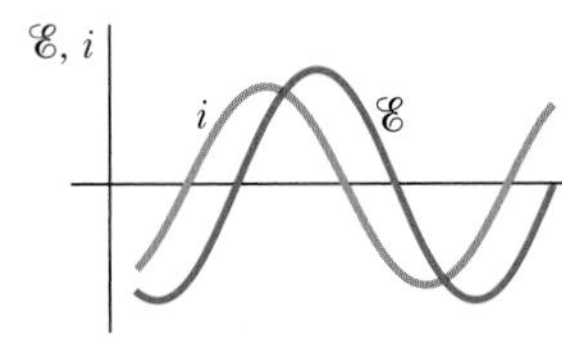

Fig. 33-19 Questions 9 and 11.

10. Figure 33-20 shows three situations like those of Fig. 33-12. For each situation, is the driving angular frequency greater than, less than, or equal to the resonant angular frequency of the circuit?

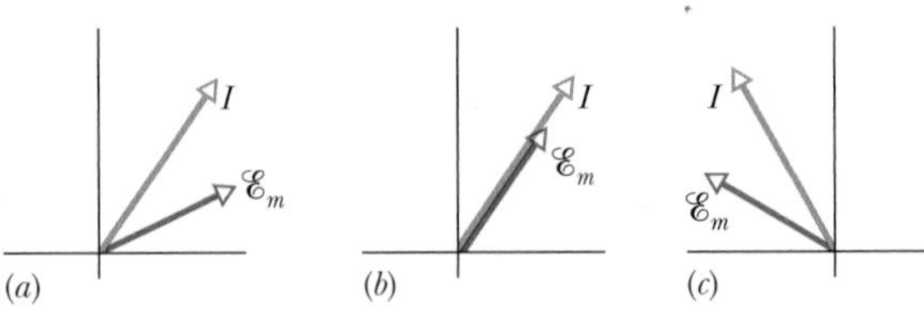

Fig. 33-20 Question 10.

11. Figure 33-19 shows the current i and driving emf $\mathscr{E}$ for a series RLC circuit. Relative to the emf curve, does the current curve shift leftward or rightward and does the amplitude of that curve increase or decrease if we slightly increase (a) L, (b) C, and (c) ω_d?

12. Figure 33-21 shows the current i and driving emf $\mathscr{E}$ for a series RLC circuit. (a) Is the phase constant positive or negative? (b) To increase the rate at which energy is transferred to the resistive load, should L be increased or decreased? (c) Should, instead, C be increased or decreased?

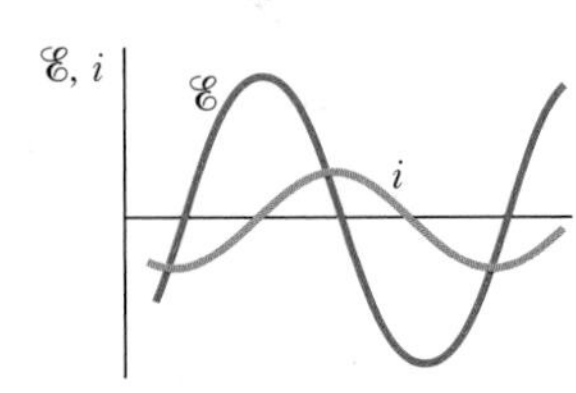

Fig. 33-21 Question 12.

EXERCISES & PROBLEMS

ssm Solution is in the Student Solutions Manual.
www Solution is available on the World Wide Web at: http://www.wiley.com/college/hrw
ilw Solution is available on the Interactive LearningWare.

SEC. 33-2 *LC* Oscillations, Qualitatively

1E. What is the capacitance of an oscillating *LC* circuit if the maximum charge on the capacitor is 1.60 μC and the total energy is 140 μJ?

2E. In an oscillating *LC* circuit, $L = 1.10$ mH and $C = 4.00$ μF. The maximum charge on the capacitor is 3.00 μC. Find the maximum current.

3E. An oscillating *LC* circuit consists of a 75.0 mH inductor and a 3.60 μF capacitor. If the maximum charge on the capacitor is 2.90 μC, (a) what is the total energy in the circuit and (b) what is the maximum current? ssm

4E. In a certain oscillating *LC* circuit the total energy is converted from electric energy in the capacitor to magnetic energy in the inductor in 1.50 μs. (a) What is the period of oscillation? (b) What is the frequency of oscillation? (c) How long after the magnetic energy is a maximum will it be a maximum again?

5P. The frequency of oscillation of a certain *LC* circuit is 200 kHz. At time $t = 0$, plate *A* of the capacitor has maximum positive charge. At what times $t > 0$ will (a) plate *A* again have maximum positive charge, (b) the other plate of the capacitor have maximum positive charge, and (c) the inductor have maximum magnetic field?

SEC. 33-3 The Electrical–Mechanical Analogy

6E. A 0.50 kg body oscillates in SHM on a spring that, when extended 2.0 mm from its equilibrium, has an 8.0 N restoring force. (a) What is the angular frequency of oscillation? (b) What is the period of oscillation? (c) What is the capacitance of an *LC* circuit with the same period if *L* is chosen to be 5.0 H?

7P. The energy in an oscillating *LC* circuit containing a 1.25 H inductor is 5.70 μJ. The maximum charge on the capacitor is 175 μC. Find (a) the mass, (b) the spring constant, (c) the maximum displacement, and (d) the maximum speed for a mechanical system with the same period. ssm

SEC. 33-4 *LC* Oscillations, Quantitatively

8E. *LC* oscillators have been used in circuits connected to loudspeakers to create some of the sounds of electronic music. What inductance must be used with a 6.7 μF capacitor to produce a frequency of 10 kHz, which is near the middle of the audible range of frequencies?

9E. In an oscillating *LC* circuit with $L = 50$ mH and $C = 4.0$ μF, the current is initially a maximum. How long will it take before the capacitor is fully charged for the first time? ssm ilw

10E. A single loop consists of inductors ($L_1, L_2, \ldots$), capacitors ($C_1, C_2, \ldots$), and resistors ($R_1, R_2, \ldots$) connected in series as shown, for example, in Fig. 33-22*a*. Show that regardless of the sequence of these circuit elements in the loop, the behavior of this circuit is identical to that of the simple *LC* circuit shown in Fig. 33-22*b*. (*Hint:* Consider the loop rule and see Problem 43 in Chapter 31.)

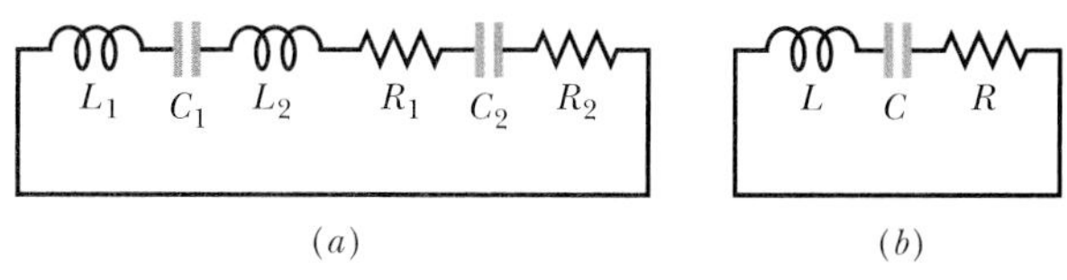

Fig. 33-22 Exercise 10.

11P. An oscillating *LC* circuit consisting of a 1.0 nF capacitor and a 3.0 mH coil has a maximum voltage of 3.0 V. (a) What is the maximum charge on the capacitor? (b) What is the maximum current through the circuit? (c) What is the maximum energy stored in the magnetic field of the coil? ilw

12P. In an oscillating *LC* circuit in which $C = 4.00$ μF, the maximum potential difference across the capacitor during the oscillations is 1.50 V and the maximum current through the inductor is 50.0 mA. (a) What is the inductance *L*? (b) What is the frequency of the oscillations? (c) How much time is required for the charge on the capacitor to rise from zero to its maximum value?

13P. In the circuit shown in Fig. 33-23 the switch is kept in position *a* for a long time. It is then thrown to position *b*. (a) Calculate the frequency of the resulting oscillating current. (b) What is the amplitude of the current oscillations? ssm ilw

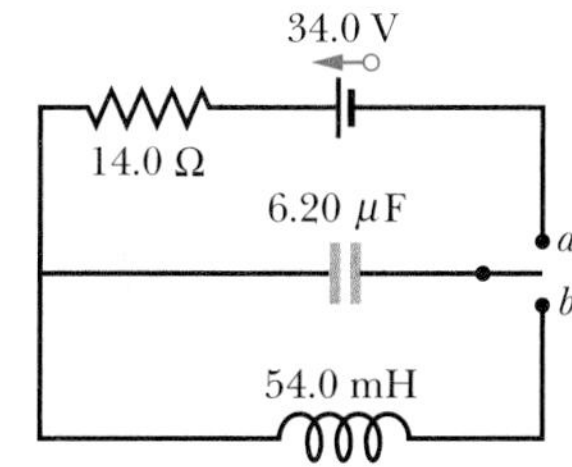

Fig. 33-23 Problem 13.

14P. You are given a 10 mH inductor and two capacitors, of 5.0 μF and 2.0 μF capacitance. List the oscillation frequencies that can be generated by connecting these elements in various combinations.

15P. A variable capacitor with a range from 10 to 365 pF is used with a coil to form a variable-frequency *LC* circuit to tune the input to a radio. (a) What ratio of maximum to minimum frequencies may be obtained with such a capacitor? (b) If this circuit is to obtain frequencies from 0.54 MHz to 1.60 MHz, the ratio computed in (a) is too large. By adding a capacitor in parallel to the variable capacitor, this range may be adjusted. What should be the capacitance of this added capacitor, and what inductance should be used to obtain the desired range of frequencies? ssm www

16P. In an oscillating *LC* circuit, 75.0% of the total energy is stored in the magnetic field of the inductor at a certain instant. (a) In terms of the maximum charge on the capacitor, what is the charge there at that instant? (b) In terms of the maximum current in the inductor, what is the current there at that instant?

17P. In an oscillating *LC* circuit, $L = 25.0$ mH and $C = 7.80$ μF. At time $t = 0$ the current is 9.20 mA, the charge on the capacitor is 3.80 μC, and the capacitor is charging. (a) What is the total energy in the circuit? (b) What is the maximum charge on the capacitor? (c) What is the maximum current? (d) If the charge on the capacitor is given by $q = Q\cos(\omega t + \phi)$, what is the phase angle ϕ? (e) Suppose the data are the same, except that the capacitor is discharging at $t = 0$. What then is ϕ? ssm

18P. An inductor is connected across a capacitor whose capacitance can be varied by turning a knob. We wish to make the frequency of oscillation of this *LC* circuit vary linearly with the angle of rotation of the knob, going from 2×10^5 to 4×10^5 Hz as the knob turns through 180°. If $L = 1.0$ mH, plot the required capacitance C as a function of the angle of rotation of the knob.

19P. In an oscillating *LC* circuit, $L = 3.00$ mH and $C = 2.70\ \mu$F. At $t = 0$ the charge on the capacitor is zero and the current is 2.00 A. (a) What is the maximum charge that will appear on the capacitor? (b) In terms of the period T of oscillation, how much time will elapse after $t = 0$ until the energy stored in the capacitor will be increasing at its greatest rate? (c) What is this greatest rate at which energy is transferred to the capacitor? ssm

20P. A series circuit containing inductance L_1 and capacitance C_1 oscillates at angular frequency ω. A second series circuit, containing inductance L_2 and capacitance C_2, oscillates at the same angular frequency. In terms of ω, what is the angular frequency of oscillation of a series circuit containing all four of these elements? Neglect resistance. (*Hint:* Use the formulas for equivalent capacitance and equivalent inductance; see Section 26-4 and Problem 43 in Chapter 31.)

21P. In an oscillating *LC* circuit with $C = 64.0\ \mu$F, the current as a function of time is given by $i = (1.60)\sin(2500t + 0.680)$, where t is in seconds, i in amperes, and the phase constant in radians. (a) How soon after $t = 0$ will the current reach its maximum value? What are (b) the inductance L and (c) the total energy? ilw

22P. Three identical inductors L and two identical capacitors C are connected in a two-loop circuit as shown in Fig. 33-24. (a) Suppose the currents are as shown in Fig. 33-24*a*. What is the current in the middle inductor? Write the loop equations and show that they are satisfied if the current oscillates with angular frequency $\omega = 1/\sqrt{LC}$. (b) Now suppose the currents are as shown in Fig. 33-24*b*. What is the current in the middle inductor? Write the loop equations and show that they are satisfied if the current oscillates with angular frequency $\omega = 1/\sqrt{3LC}$. Because the circuit can oscillate at two different frequencies, we cannot find an equivalent single-loop *LC* circuit to replace it.

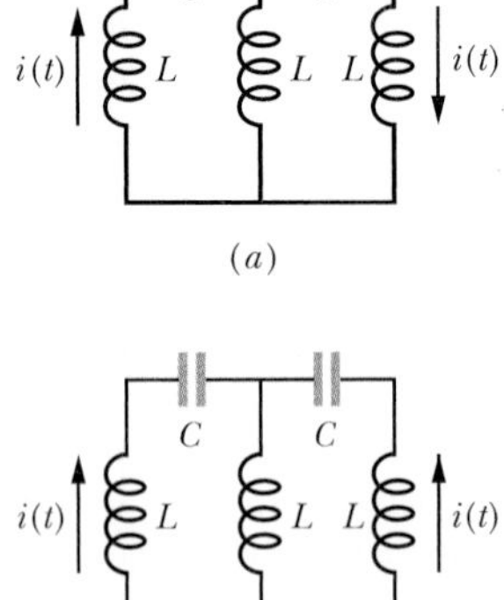

Fig. 33-24 Problem 22.

23P*. In Fig. 33-25, capacitor 1 with $C_1 = 900\ \mu$F is initially charged to 100 V and capacitor 2 with $C_2 = 100\ \mu$F is uncharged. The inductor has an inductance of 10.0 H. Describe in detail how one might charge capacitor 2 to 300 V by manipulating switches S_1 and S_2. ssm www

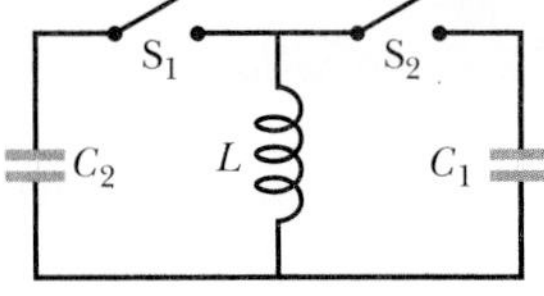

Fig. 33-25 Problem 23.

SEC. 33-5 Damped Oscillations in an *RLC* Circuit

24E. Consider a damped *LC* circuit. (a) Show that the damping term $e^{-Rt/2L}$ (which involves L but not C) can be rewritten in a more symmetric manner (involving L and C) as $e^{-\pi R(\sqrt{C/L})t/T}$. Here T is the period of oscillation (neglecting resistance). (b) Using (a), show that the SI unit of $\sqrt{L/C}$ is the ohm. (c) Using (a), show that the condition that the fractional energy loss per cycle be small is $R \ll \sqrt{L/C}$.

25E. What resistance R should be connected in series with an inductance $L = 220$ mH and capacitance $C = 12.0\ \mu$F for the maximum charge on the capacitor to decay to 99.0% of its initial value in 50.0 cycles? (Assume $\omega' \approx \omega$.) ssm ilw

26P. A single-loop circuit consists of a 7.20 Ω resistor, a 12.0 H inductor, and a 3.20 μF capacitor. Initially the capacitor has a charge of 6.20 μC and the current is zero. Calculate the charge on the capacitor N complete cycles later for $N = 5$, 10, and 100.

27P. In an oscillating series *RLC* circuit, find the time required for the maximum energy present in the capacitor during an oscillation to fall to half its initial value. Assume $q = Q$ at $t = 0$. ssm

28P. At time $t = 0$ there is no charge on the capacitor of a series *RLC* circuit but there is current I through the inductor. (a) Find the phase constant ϕ in Eq. 33-25 for the circuit. (b) Write an expression for the charge q on the capacitor as a function of time t and in terms of the current amplitude and angular frequency ω' of the oscillations.

29P*. In an oscillating series *RLC* circuit, show that the fraction of the energy lost per cycle of oscillation, $\Delta U/U$, is given to a close approximation by $2\pi R/\omega L$. The quantity $\omega L/R$ is often called the Q of the circuit (for *quality*). A high-Q circuit has low resistance and a low fractional energy loss ($= 2\pi/Q$) per cycle. ssm www

SEC. 33-8 Three Simple Circuits

30E. A 1.50 μF capacitor is connected as in Fig. 33-9*a* to an ac generator with $\mathscr{E}_m = 30.0$ V. What is the amplitude of the resulting alternating current if the frequency of the emf is (a) 1.00 kHz and (b) 8.00 kHz?

31E. A 50.0 mH inductor is connected as in Fig. 33-10*a* to an ac generator with $\mathscr{E}_m = 30.0$ V. What is the amplitude of the resulting alternating current if the frequency of the emf is (a) 1.00 kHz and (b) 8.00 kHz? ssm ilw

32E. A 50 Ω resistor is connected as in Fig. 33-8*a* to an ac generator with $\mathscr{E}_m = 30.0$ V. What is the amplitude of the resulting alternating current if the frequency of the emf is (a) 1.00 kHz and (b) 8.00 kHz?

33E. (a) At what frequency would a 6.0 mH inductor and a 10 μF capacitor have the same reactance? (b) What would the reactance be? (c) Show that this frequency would be the natural frequency of an oscillating circuit with the same L and C. ssm

34P. An ac generator has emf $\mathscr{E} = \mathscr{E}_m \sin\omega_d t$, with $\mathscr{E}_m = 25.0$ V and $\omega_d = 377$ rad/s. It is connected to a 12.7 H inductor. (a) What is the maximum value of the current? (b) When the current is a maximum, what is the emf of the generator? (c) When the emf of the generator is -12.5 V and increasing in magnitude, what is the current?

35P. An ac generator has emf $\mathscr{E} = \mathscr{E}_m \sin(\omega_d t - \pi/4)$, where $\mathscr{E}_m = 30.0$ V and $\omega_d = 350$ rad/s. The current produced in a connected circuit is $i(t) = I \sin(\omega_d t - 3\pi/4)$, where $I = 620$ mA. (a) At what time after $t = 0$ does the generator emf first reach a maximum? (b) At what time after $t = 0$ does the current first reach a maximum? (c) The circuit contains a single element other than the generator. Is it a capacitor, an inductor, or a resistor? Justify your answer. (d) What is the value of the capacitance, inductance, or resistance, as the case may be? ssm

36P. The ac generator of Problem 34 is connected to a 4.15 μF capacitor. (a) What is the maximum value of the current? (b) When the current is a maximum, what is the emf of the generator? (c) When the emf of the generator is -12.5 V and increasing in magnitude, what is the current?

SEC. 33-9 The Series *RLC* Circuit

37E. (a) Find Z, ϕ, and I for the situation of Sample Problem 33-7 with the capacitor removed from the circuit, all other parameters remaining unchanged. (b) Draw to scale a phasor diagram like that of Fig. 33-11*d* for this new situation.

38E. (a) Find Z, ϕ, and I for the situation of Sample Problem 33-7 with the inductor removed from the circuit, all other parameters remaining unchanged. (b) Draw to scale a phasor diagram like that of Fig. 33-11*d* for this new situation.

39E. (a) Find Z, ϕ, and I for the situation of Sample Problem 33-7 with $C = 70.0$ μF, the other parameters remaining unchanged. (b) Draw a phasor diagram like that of Fig. 33-11*d* for this new situation and compare the two diagrams closely. ssm www

40P. In Fig. 33-26, a generator with an adjustable frequency of oscillation is connected to a variable resistance R, a capacitor of $C = 5.50$ μF, and an inductor of inductance L. The amplitude of the current produced in the circuit by the generator is at half-maximum level when the generator's frequency is 1.30 or 1.50 kHz. (a) What is L? (b) If R is increased, what happens to the frequencies at which the current amplitude is at half-maximum level?

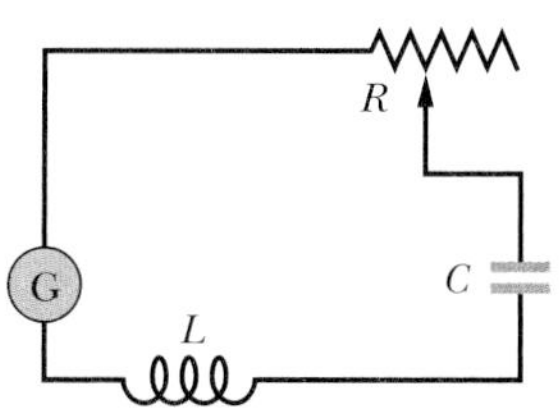

Fig. 33-26 Problem 40.

41P. In an *RLC* circuit, can the amplitude of the voltage across an inductor be greater than the amplitude of the generator emf? Consider an *RLC* circuit with $\mathscr{E}_m = 10$ V, $R = 10$ Ω, $L = 1.0$ H, and $C = 1.0$ μF. Find the amplitude of the voltage across the inductor at resonance. ssm ilw

42P. When the generator emf in Sample Problem 33-7 is a maximum, what is the voltage across (a) the generator, (b) the resistance, (c) the capacitance, and (d) the inductance? (e) By summing these with appropriate signs, verify that the loop rule is satisfied.

43P. A coil of inductance 88 mH and unknown resistance and a 0.94 μF capacitor are connected in series with an alternating emf of frequency 930 Hz. If the phase constant between the applied voltage and the current is 75°, what is the resistance of the coil? ssm ilw

44P. An ac generator with $\mathscr{E}_m = 220$ V and operating at 400 Hz causes oscillations in a series *RLC* circuit having $R = 220$ Ω, $L = 150$ mH, and $C = 24.0$ μF. Find (a) the capacitive reactance X_C, (b) the impedance Z, and (c) the current amplitude I. A second capacitor of the same capacitance is then connected in series with the other components. Determine whether the values of (d) X_C, (e) Z, and (f) I increase, decrease, or remain the same.

45P. An *RLC* circuit such as that of Fig. 33-7 has $R = 5.00$ Ω, $C = 20.0$ μF, $L = 1.00$ H, and $\mathscr{E}_m = 30.0$ V. (a) At what angular frequency ω_d will the current amplitude have its maximum value, as in the resonance curves of Fig. 33-13? (b) What is this maximum value? (c) At what two angular frequencies ω_{d1} and ω_{d2} will the current amplitude be half this maximum value? (d) What is the fractional half-width $[= (\omega_{d1} - \omega_{d2})/\omega]$ of the resonance curve for this circuit? ssm

46P. An ac generator is to be connected in series with an inductor of $L = 2.00$ mH and a capacitance C. You are to produce C by using capacitors of capacitances $C_1 = 4.00$ μF and $C_2 = 6.00$ μF, either singly or together. What resonant frequencies can the circuit have, depending on how you use C_1 and C_2?

47P. Show that the fractional half-width (see Problem 45) of a resonance curve is given by

$$\frac{\Delta\omega_d}{\omega} = \sqrt{\frac{3C}{L}}\,R,$$

in which ω is the angular frequency at resonance and $\Delta\omega_d$ is the width of the resonance curve at half-amplitude. Note that $\Delta\omega_d/\omega$ increases with R, as Fig. 33-13 shows. Use this formula to check the answer to Problem 45d. ssm

48P. In Fig. 33-27, a generator with an adjustable frequency of oscillation is connected to resistance $R = 100$ Ω, inductances $L_1 = 1.70$ mH and $L_2 = 2.30$ mH, and capacitances $C_1 = 4.00$ μF, $C_2 = 2.50$ μF, and $C_3 = 3.50$ μF. (a) What is the resonant frequency of the circuit? (*Hint:* See Problem 43 in Chapter 31.) What happens to the resonant frequency if (b) the value of R is increased, (c) the value of L_1 is increased, and (d) capacitance C_3 is removed from the circuit?

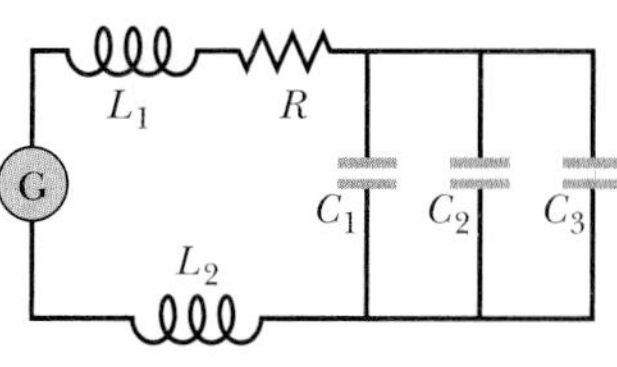

Fig. 33-27 Problem 48.

SEC. 33-10 Power in Alternating-Current Circuits

49E. What direct current will produce the same amount of thermal energy, in a particular resistor, as an alternating current that has a maximum value of 2.60 A? ssm

50E. An ac voltmeter with large impedance is connected in turn across the inductor, the capacitor, and the resistor in a series circuit having an alternating emf of 100 V (rms); it gives the same reading in volts in each case. What is this reading?

51E. What is the maximum value of an ac voltage whose rms value is 100 V?

52E. (a) For the conditions in Problem 34c, is the generator supplying energy to or taking energy from the rest of the circuit? (b) Repeat for the conditions of Problem 36c.

53E. Calculate the average rate of energy dissipation in the circuits of Exercises 31, 32, 37, and 38.

54E. Show that the average rate at which energy is supplied to the circuit of Fig. 33-7 can also be written as $P_{avg} = \mathscr{E}^2_{rms}R/Z^2$. Show that this expression for average power gives reasonable results for a purely resistive circuit, for an *RLC* circuit at resonance, for a purely capacitive circuit, and for a purely inductive circuit.

55E. An air conditioner connected to a 120 V rms ac line is equivalent to a 12.0 Ω resistance and a 1.30 Ω inductive reactance in series. (a) Calculate the impedance of the air conditioner. (b) Find the average rate at which energy is supplied to the appliance. ssm

56P. In a series oscillating *RLC* circuit, $R = 16.0\ \Omega$, $C = 31.2\ \mu F$, $L = 9.20$ mH, and $\mathscr{E} = \mathscr{E}_m \sin \omega_d t$ with $\mathscr{E}_m = 45.0$ V and $\omega_d = 3000$ rad/s. For time $t = 0.442$ ms find (a) the rate at which energy is being supplied by the generator, (b) the rate at which the energy in the capacitor is changing, (c) the rate at which the energy in the inductor is changing, and (d) the rate at which energy is being dissipated in the resistor. (e) What is the meaning of a negative result for any of (a), (b), and (c)? (f) Show that the results of (b), (c), and (d) sum to the result of (a).

57P. Figure 33-28 shows an ac generator connected to a "black box" through a pair of terminals. The box contains an *RLC* circuit, possibly even a multiloop circuit, whose elements and connections we do not know. Measurements outside the box reveal that

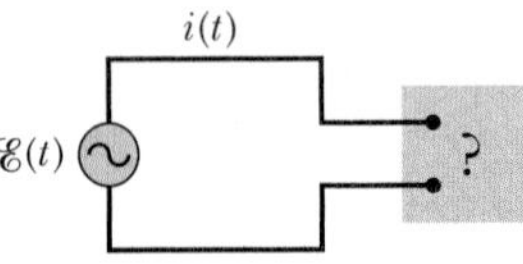

Fig. 33-28 Problem 57.

$$\mathscr{E}(t) = (75.0 \text{ V}) \sin \omega_d t$$

and

$$i(t) = (1.20 \text{ A}) \sin(\omega_d t + 42.0°).$$

(a) What is the power factor? (b) Does the current lead or lag the emf? (c) Is the circuit in the box largely inductive or largely capacitive? (d) Is the circuit in the box in resonance? (e) Must there be a capacitor in the box? An inductor? A resistor? (f) At what average rate is energy delivered to the box by the generator? (g) Why don't you need to know the angular frequency ω_d to answer all these questions? ssm www

58P. In Fig. 33-29 show that the average rate at which energy is dissipated in resistance R is a maximum when R is equal to the internal resistance r of the ac generator. (In the text discussion we have tacitly assumed that $r = 0$.)

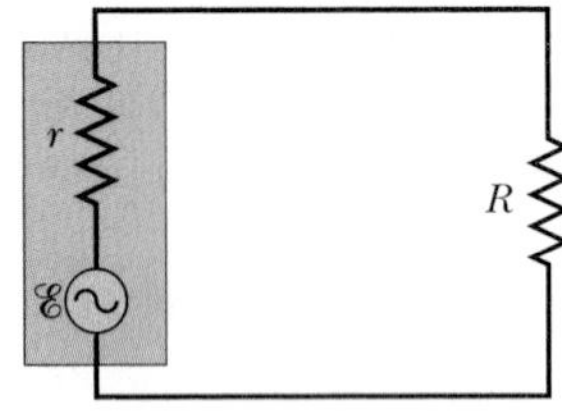

Fig. 33-29 Problems 58 and 65.

59P. In an *RLC* circuit such as that of Fig. 33-7 assume that $R = 5.00\ \Omega$, $L = 60.0$ mH, $f_d = 60.0$ Hz, and $\mathscr{E}_m = 30.0$ V. For what values of the capacitance would the average rate at which energy is dissipated in the resistance be (a) a maximum and (b) a minimum? (c) What are these maximum and minimum energy dissipation rates? What are (d) the corresponding phase angles and (e) the corresponding power factors? ssm

60P. A typical "light dimmer" used to dim the stage lights in a theater consists of a variable inductor L (whose inductance is adjustable between zero and L_{max}) connected in series with the lightbulb B as shown in Fig. 33-30. The electrical supply is 120 V (rms) at 60.0 Hz; the lightbulb is rated as "120 V, 1000 W." (a) What L_{max} is required if the rate of energy dissipation in the lightbulb is to be varied by a factor of 5 from its upper limit of 1000 W? Assume that the resistance of the lightbulb is independent of its temperature. (b) Could one use a variable resistor (adjustable between zero and R_{max}) instead of an inductor? If so, what R_{max} is required? Why isn't this done?

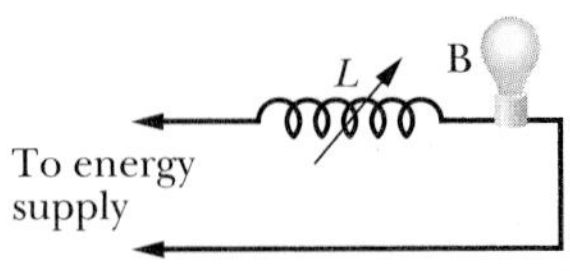

Fig. 33-30 Problem 60.

61P. In Fig. 33-31, $R = 15.0\ \Omega$, $C = 4.70\ \mu F$, and $L = 25.0$ mH. The generator provides a sinusoidal voltage of 75.0 V (rms) and frequency $f = 550$ Hz. (a) Calculate the rms current. (b) Find the rms voltages V_{ab}, V_{bc}, V_{cd}, V_{bd}, V_{ad}. (c) At what average rate is energy dissipated by each of the three circuit elements? ilw

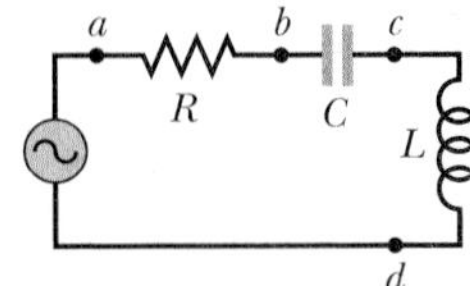

Fig. 33-31 Problem 61.

SEC. 33-11 Transformers

62E. A generator supplies 100 V to the primary coil of a transformer of 50 turns. If the secondary coil has 500 turns, what is the secondary voltage?

63E. A transformer has 500 primary turns and 10 secondary turns. (a) If V_p is 120 V (rms), what is V_s with an open circuit? (b) If the secondary now has a resistive load of 15 Ω, what are the currents in the primary and secondary? ssm ilw

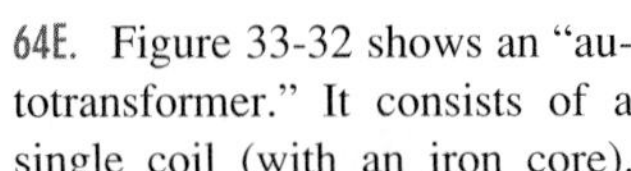

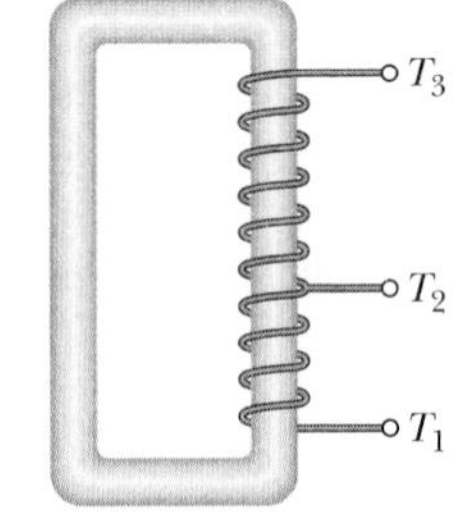

Fig. 33-32 Exercise 64.

64E. Figure 33-32 shows an "autotransformer." It consists of a single coil (with an iron core). Three taps T_i are provided. Between taps T_1 and T_2 there are 200 turns, and between taps T_2 and T_3 there are 800 turns. Any two taps can be considered the "primary terminals" and any two taps can be considered the "secondary terminals." List all the ratios by which the primary voltage may be changed to a secondary voltage.

65P. In Fig. 33-29 let the rectangular box on the left represent the (high-impedance) output of an audio amplifier, with $r = 1000\ \Omega$. Let $R = 10\ \Omega$ represent the (low-impedance) coil of a loudspeaker. For maximum transfer of energy to the load R we must have $R = r$, and that is not true in this case. However, a transformer can be used to "transform" resistances, making them behave electrically as if they were larger or smaller than they actually are. Sketch the primary and secondary coils of a transformer that can be introduced between the amplifier and the speaker in Fig. 33-29 to match the impedances. What must be the turns ratio? ssm

NEW PROBLEMS

N1. What capacitance would you connect across a 1.30 mH inductor to make the resulting oscillator resonate at 3.50 kHz?

N2. The current amplitude I versus driving angular frequency ω_d for a driven RLC circuit is given in Fig. 33N-1. The inductance is 200 μH, and the emf amplitude is 8.0 V. What are the values of (a) C and (b) R?

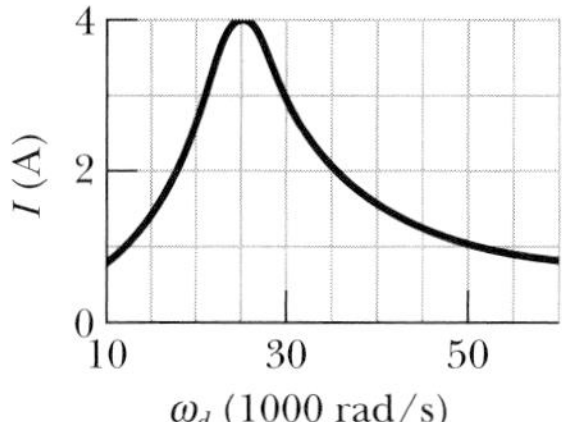

Fig. 33N-1 Problem N2.

N3. An oscillating LC circuit has an inductance of 3.00 mH and a capacitance of 10.0 μF. Calculate (a) the angular frequency and (b) the period of the oscillation. (c) At time $t = 0$, the capacitor is charged to 200 μC and the current is zero. Sketch roughly the charge on the capacitor as a function of time.

N4. An alternating source with a variable frequency, a capacitor with capacitance C, and a resistor with resistance R are connected in series. Figure 33N-2 gives the impedance Z of the circuit versus the driving angular frequency ω_d; the curve reaches an asymptote of 500 Ω. The figure also gives the reactance X_C for the capacitor versus ω_d. What are (a) R and (b) C?

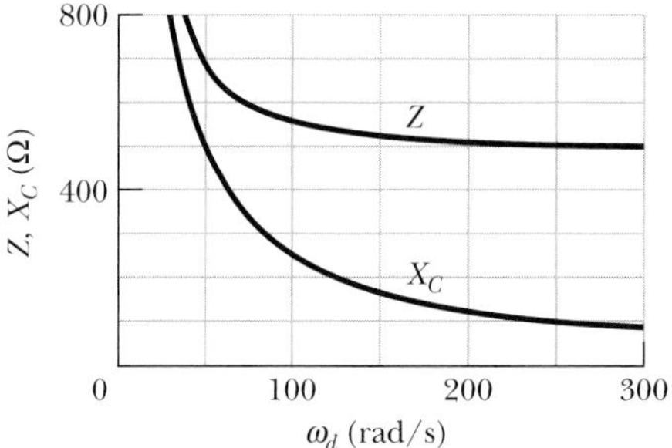

Fig. 33N-2 Problem N4.

N5. A 1.50 μF capacitor has a capacitive reactance of 12.0 Ω. (a) What must be its operating frequency? (b) What will be the capacitive reactance if the frequency is doubled?

N6. An alternating source with a variable frequency f_d, a 50.0 Ω resistor, and a 20.0 μF capacitor are connected in series. The emf amplitude is 12.0 V. (a) Draw a phasor diagram for phasor V_R (for the potential across the resistor) and phasor V_C (for the potential across the capacitor). (b) At what driving frequency f_d do the two phasors have the same length? At that driving frequency, what are (c) the phase angle in degrees, (d) the angular speed at which the phasors rotate, and (e) the current amplitude?

N7. A generator with an adjustable frequency of oscillation is wired in series to an inductor of L = 2.50 mH and a capacitor of C = 3.00 μF. At what frequency does the generator produce the largest possible current amplitude in the circuit?

N8. An alternating source with a variable frequency f_d, a 80.0 Ω resistor, and a 40.0 mH inductor are connected in series. The emf amplitude is 6.00 V. (a) Draw a phasor diagram for phasor V_R (for the potential across the resistor) and phasor V_L (for the potential across the inductor). (b) At what driving frequency f_d do the two phasors have the same length? At that driving frequency, what are (c) the phase angle in degrees, (d) the angular speed at which the phasors rotate, and (e) the current amplitude?

N9. A 40.0 mH inductor and a 200 Ω resistor are connected in series with an ac source with an emf amplitude $\mathscr{E}_m$ of 100 V. The frequency f_d of the source can be varied from 0 to 2500 Hz. (a) Write an equation for the inductive reactance X_L. (b) Simultaneously plot the resistance R, the inductive reactance X_L, and the impedance Z versus f_d for the range $0 < f_d < 2500$ Hz. (c) Determine the value of f_d for which $X_L = R$.

N10. An alternating source with a variable frequency, an inductor with inductance L, and a resistor with resistance R are connected in series. Figure 33N-3 gives the impedance Z of the circuit versus the driving angular frequency ω_d. The figure also gives the reactance X_L for the inductor versus ω_d. What are (a) R and (b) L?

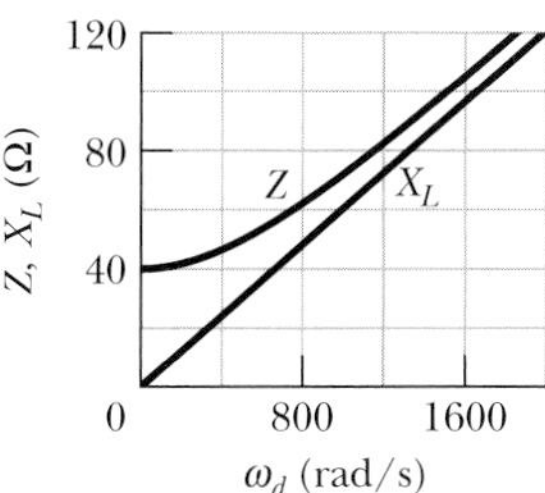

Fig. 33N-3 Problem N10.

N11. A series RLC circuit has a resonant frequency of 6.00 kHz. When it is driven at 8.00 kHz, it has an impedance of 1.00 kΩ and a phase constant of 45°. What are the values of (a) R, (b) L, and (c) C for this circuit?

N12. An alternating source drives a series RLC circuit with an emf amplitude of 6.00 V, at a phase angle of +30.0°. When the potential difference across the capacitor reaches its maximum positive value of +5.00 V, what is the potential difference across the inductor (sign included)?

N13. A series circuit with resistor–inductor–capacitor combination R_1, L_1, C_1 has the same resonant frequency as a second circuit with a different combination R_2, L_2, C_2. You now connect the two combinations in series. Show that this new circuit has the same resonant frequency as the separate circuits.

N14. An oscillating LC circuit has a current amplitude of 7.50 mA, a potential amplitude of 250 mV, and a capacitance of 220 nF. What are (a) the period of oscillation, (b) the maximum energy stored in the capacitor, (c) the maximum energy stored in the inductor, (d) the maximum rate at which the current changes, and (e) the maximum rate at which the inductor gains energy?

N15. Derive the differential equation for an LC circuit (Eq. 33-11) using the loop rule.

N16. A capacitor of capacitance 158 μF and an inductor form an LC circuit that oscillates at 8.15 kHz, with a current amplitude of 4.21 mA. What are (a) the inductance, (b) the total energy in the circuit, and (c) the maximum charge on the capacitor?

N17. Verify mathematically that the following geometric construction correctly gives both the impedance Z and the phase constant ϕ. Referring to Fig. 33N-4, first draw an arrow in the positive y direction of magnitude X_L; next draw a second arrow in the negative y direction of magnitude X_C; then draw a third arrow of magnitude R in the positive x direction. Then the magnitude of the "resultant" of these arrows is Z, and the angle of this resultant is ϕ.

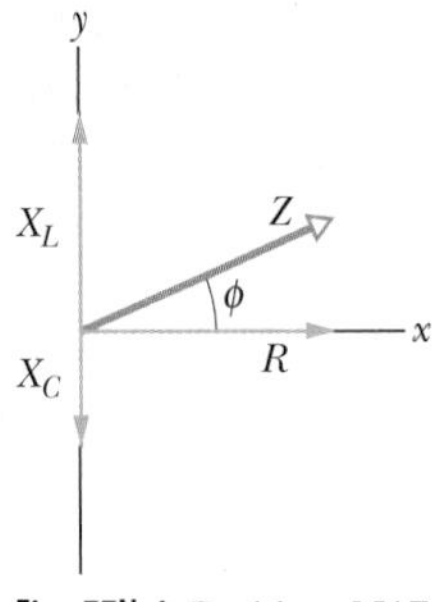

Fig. 33N-4 Problem N17.

N18. A generator drives a series *RLC* circuit in which the potential amplitude across the inductance is 2.4 times the potential amplitude across the capacitance and 1.75 times the potential amplitude across the resistance. (a) What is the phase angle in radians? (b) Is the circuit capacitive, inductive, or in resonance? The resistance is 120 Ω, and the current amplitude is 150 mA. (c) What is the amplitude of the emf?

N19. (a) For the situation of Problem N9, simultaneously plot the voltage V_L across the inductor, the voltage V_R across the resistor, and the (constant) emf amplitude $\mathscr{E}_m$ across the source versus f_d for the range $0 < f_d < 2500$ Hz. (b) Determine the value of f_d for which $V_L = V_R$. (c) What is V_R at that frequency? (d) Determine the value of f_d for which $V_R = \mathscr{E}_m/3$. (e) What is V_L at that frequency? (f) Determine the value of f_d for which $V_L = \mathscr{E}_m/3$. (g) What is V_R at that frequency?

N20. An *RLC* circuit is driven by a generator with an emf amplitude of 80.0 V and a current amplitude of 1.25 A. The current leads the emf by 0.650 rad. What are (a) the impedance and (b) the resistance of the circuit? (c) Is the circuit inductive, capacitive, or in resonance?

N21. A 150.0 mH inductor, a 45.0 μF capacitor, and a 90.0 Ω resistor are connected in series with an ac source with an emf amplitude $\mathscr{E}_m$ of 100 V. The frequency f_d of the source can be varied from 0 to 1000 Hz. (a) Simultaneously plot the capacitive reactance X_C and the inductive reactance X_L versus f_d for the range $0 < f_d < 200$ Hz. (b) Determine f_d at which $X_C = X_L$. Plot the impedance Z of the circuit versus f_d for the range $0 < f_d < 188$ Hz and determine (c) the minimum value of Z and (d) the value of f_d at which it occurs.

N22. A generator of frequency 3000 Hz drives a series *RLC* circuit with an emf amplitude of 120 V. The resistance is 40.0 Ω, the capacitance is 1.60 μF, and the inductance is 850 μH. What are (a) the phase constant in radians and (b) the current amplitude? (c) Is the circuit capacitive, inductive, or in resonance?

N23. Figure 33N-5 shows an *RLC* circuit that is driven by an emf source of fixed amplitude $\mathscr{E}_m$. Initially the circuit consists of one resistor of resistance R, one inductor of inductance L, and one capacitor of capacitance C, and the driving frequency matches the natural frequency. Then switches S_1, S_2, S_3, and S_4 are closed, in that order. The closings bring in capacitors identical to the first one or resistors identical to the first one.

Let $\mathscr{E}_m = 12.0$ V, $C = 2.00$ μF, $L = 2.00$ mH, and $R = 12.0$ Ω. (a) Fill in the first blank column of the following table with the values of the equivalent capacitance C_{eq} of the circuit after each switch closing. Similarly, fill in the other columns with values for (b) the natural (or resonance) frequency f, (c) the equivalent resistance R_{eq}, (d) the impedance Z, and (e) the current amplitude I. (Avoid rounding off the numbers until all calculations are finished.)

Closing	C_{eq}	f	R_{eq}	Z	I
S_1					
S_2					
S_3					
S_4					

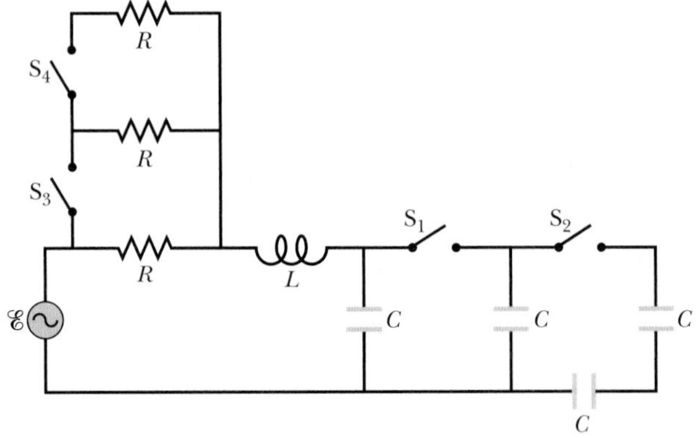

Fig. 33N-5 Problem N23.

N24. Figure 33N-6 shows a driven *RLC* circuit that contains two identical capacitors and two switches. The emf amplitude is set at 12.0 V, and the driving frequency is set at 60.0 Hz. With both switches open, the current leads the emf by 30.9°. With switch S_1 closed and switch S_2 still open, the emf leads the current by 15.0°. With both switches closed, the current amplitude is 447 mA. What are the values of (a) R, (b) C, and (c) L?

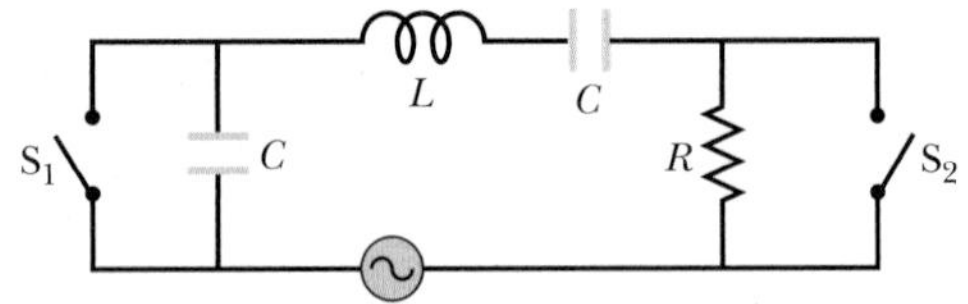

Fig. 33N-6 Problem N24.

N25. A 1.50 mH inductor in an oscillating *LC* circuit stores a maximum energy of 10.0 μJ. What is the maximum current?

N26. (a) List at least two ways you can tell that the circuit of Sample Problem 33-7 is not in resonance. (b) What capacitance could you combine *in parallel* with the capacitance already in the circuit to bring the circuit to resonance? (c) If it were at resonance with that change, what would be the current amplitude?

N27. A series *RLC* circuit is driven by an alternating source at a frequency of 400 Hz and an emf amplitude of 90.0 V. The resistance is 20.0 Ω, the capacitance is 12.1 μF, and the inductance is 24.2 mH. What are the rms potential differences across (a) the resistor, (b) the capacitor, and (c) the inductor? (d) What is the average rate at which energy is dissipated in the circuit?

34 Electromagnetic Waves

As a comet swings around the Sun, ice on its surface vaporizes, releasing trapped dust and charged particles. The electrically charged "solar wind" forces the charged particles into a straight "tail" that points radially away from the Sun. However, the dust is unaffected by the solar wind and seemingly should continue to travel along the comet's orbit.

Why, instead, does much of the dust fashion the curved lower tail seen in the photograph?

The answer is in this chapter.

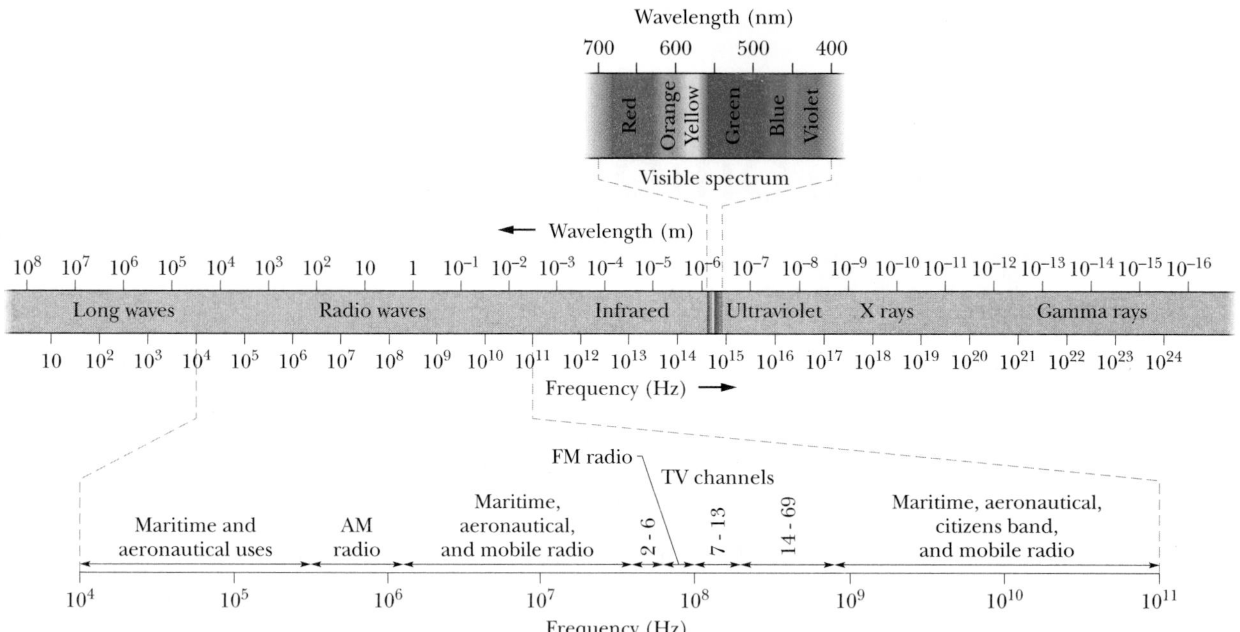

Fig. 34-1 The electromagnetic spectrum.

34-1 Maxwell's Rainbow

James Clerk Maxwell's crowning achievement was to show that a beam of light is a traveling wave of electric and magnetic fields—an **electromagnetic wave**—and thus that optics, the study of visible light, is a branch of electromagnetism. In this chapter we move from one to the other: we conclude our discussion of strictly electric and magnetic phenomena, and we build a foundation for optics.

In Maxwell's time (the mid 1800s), the visible, infrared, and ultraviolet forms of light were the only electromagnetic waves known. Spurred on by Maxwell's work, however, Heinrich Hertz discovered what we now call radio waves and verified that they move through the laboratory at the same speed as visible light.

As Fig. 34-1 shows, we now know a wide *spectrum* (or range) of electromagnetic waves, referred to by one imaginative writer as "Maxwell's rainbow." Consider the extent to which we are bathed in electromagnetic waves throughout this spectrum. The Sun, whose radiations define the environment in which we as a species have evolved and adapted, is the dominant source. We are also crisscrossed by radio and television signals. Microwaves from radar systems and from telephone relay systems may reach us. There are electromagnetic waves from lightbulbs, from the heated engine blocks of automobiles, from x-ray machines, from lightning flashes, and from buried radioactive materials. Beyond this, radiation reaches us from stars and other objects in our galaxy and from other galaxies. Electromagnetic waves also travel in the other direction. Television signals, transmitted from Earth since about 1950, have now taken news about us (along with episodes of *I Love Lucy,* albeit *very* faintly) to whatever technically sophisticated inhabitants there may be on whatever planets may encircle the nearest 400 or so stars.

In the wavelength scale in Fig. 34-1 (and similarly the corresponding frequency scale), each scale marker represents a change in wavelength (and correspondingly in frequency) by a factor of 10. The scale is open-ended; the wavelengths of electromagnetic waves have no inherent upper or lower bounds.

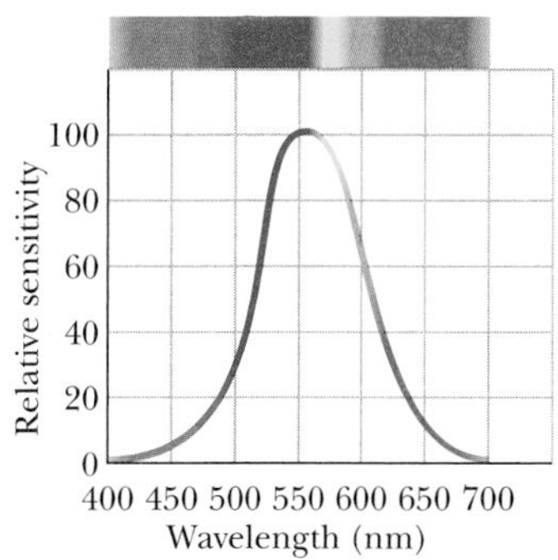

Fig. 34-2 The relative sensitivity of the average human eye to electromagnetic waves at different wavelengths. This portion of the electromagnetic spectrum to which the eye is sensitive is called *visible light.*

Certain regions of the electromagnetic spectrum in Fig. 34-1 are identified by familiar labels, such as *x rays* and *radio waves.* These labels denote roughly defined wavelength ranges within which certain kinds of sources and detectors of electromagnetic waves are in common use. Other regions of Fig. 34-1, such as those labeled television and AM radio, represent specific wavelength bands assigned by law for certain commercial or other purposes. There are no gaps in the electromagnetic spectrum—and all electromagnetic waves, no matter where they lie in the spectrum, travel through *free space* (vacuum) with the same speed c.

The visible region of the spectrum is of course of particular interest to us. Figure 34-2 shows the relative sensitivity of the human eye to light of various wavelengths. The center of the visible region is about 555 nm, which produces the sensation that we call yellow-green.

The limits of this visible spectrum are not well defined because the eye sensitivity curve approaches the zero-sensitivity line asymptotically at both long and short wavelengths. If we take the limits, arbitrarily, as the wavelengths at which eye sensitivity has dropped to 1% of its maximum value, these limits are about 430 and 690 nm; however, the eye can detect electromagnetic waves somewhat beyond these limits if they are intense enough.

34-2 The Traveling Electromagnetic Wave, Qualitatively

Some electromagnetic waves, including x rays, gamma rays, and visible light, are *radiated* (emitted) from sources that are of atomic or nuclear size, where quantum physics rules. Here we discuss how other electromagnetic waves are generated. To simplify matters, we restrict ourselves to that region of the spectrum (wavelength $\lambda \approx 1$ m) in which the source of the *radiation* (the emitted waves) is both macroscopic and of manageable dimensions.

Figure 34-3 shows, in broad outline, the generation of such waves. At its heart is an *LC oscillator,* which establishes an angular frequency $\omega\ (= 1/\sqrt{LC})$. Charges and currents in this circuit vary sinusoidally at this frequency, as depicted in Fig. 33-1. An external source—possibly an ac generator—must be included to supply energy to compensate both for thermal losses in the circuit and for energy carried away by the radiated electromagnetic wave.

The LC oscillator of Fig. 34-3 is coupled by a transformer and a transmission line to an *antenna,* which consists essentially of two thin, solid, conducting rods. Through this coupling, the sinusoidally varying current in the oscillator causes charge to oscillate sinusoidally along the rods of the antenna at the angular frequency ω of the LC oscillator. The current in the rods associated with this movement of

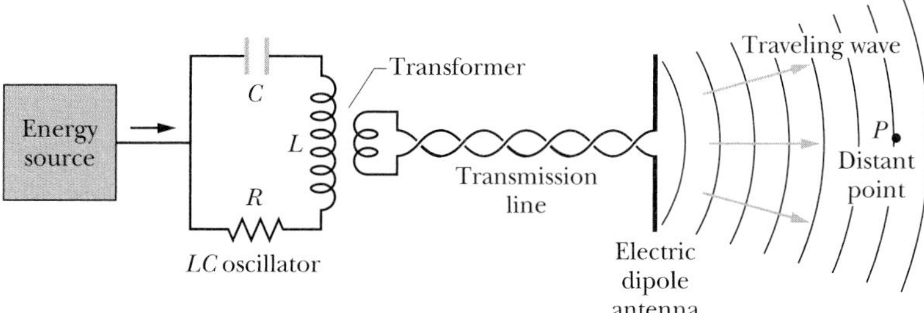

Fig. 34-3 An arrangement for generating a traveling electromagnetic wave in the shortwave radio region of the spectrum: an LC oscillator produces a sinusoidal current in the antenna, which generates the wave. P is a distant point at which a detector can monitor the wave traveling past it.

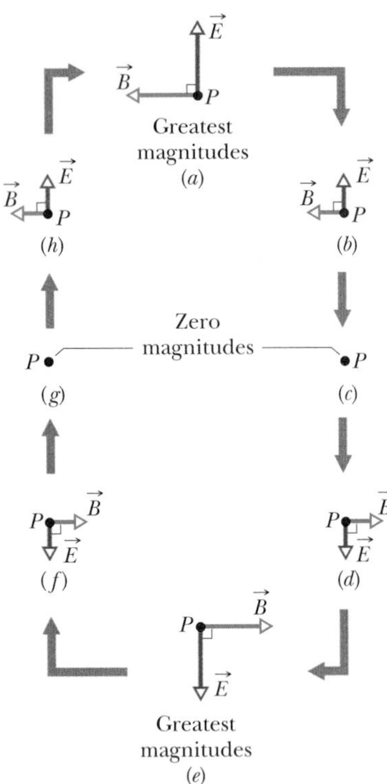

Fig. 34-4 (*a*)–(*h*) The variation in the electric field $\vec{E}$ and the magnetic field $\vec{B}$ at the distant point P of Fig. 34-3 as one wavelength of the electromagnetic wave travels past it. In this perspective, the wave is traveling directly out of the page. The two fields vary sinusoidally in magnitude and direction. Note that they are always perpendicular to each other and to the direction of travel of the wave.

charge also varies sinusoidally, in magnitude and direction, at angular frequency ω. The antenna has the effect of an electric dipole whose electric dipole moment varies sinusoidally in magnitude and direction along the length of the antenna.

Because the dipole moment varies in magnitude and direction, the electric field produced by the dipole varies in magnitude and direction. Also, because the current varies, the magnetic field produced by the current varies in magnitude and direction. However, the changes in the electric and magnetic fields do not happen everywhere instantaneously; rather, the changes travel outward from the antenna at the speed of light c. Together the changing fields form an electromagnetic wave that travels away from the antenna at speed c. The angular frequency of this wave is ω, the same as that of the LC oscillator.

Figure 34-4 shows how the electric field $\vec{E}$ and the magnetic field $\vec{B}$ change with time as one wavelength of the wave sweeps past the distant point P of Fig. 34-3; in each part of Fig. 34-4, the wave is traveling directly out of the page. (We choose a distant point so that the curvature of the waves suggested in Fig. 34-3 is small enough to neglect. At such points, the wave is said to be a *plane wave,* and discussion of the wave is much simplified.) Note several key features in Fig. 34-4; they are present regardless of how the wave is created:

1. The electric and magnetic fields $\vec{E}$ and $\vec{B}$ are always perpendicular to the direction of travel of the wave. Thus, the wave is a *transverse wave,* as discussed in Chapter 17.
2. The electric field is always perpendicular to the magnetic field.
3. The cross product $\vec{E} \times \vec{B}$ always gives the direction of travel of the wave.
4. The fields always vary sinusoidally, just like the transverse waves discussed in Chapter 17. Moreover, the fields vary with the same frequency and *in phase* (in step) with each other.

In keeping with these features, we can assume that the electromagnetic wave is traveling toward P in the positive direction of an x axis, that the electric field in Fig. 34-4 is oscillating parallel to the y axis, and that the magnetic field is then oscillating parallel to the z axis (using a right-handed coordinate system, of course). Then we can write the electric and magnetic fields as sinusoidal functions of position x (along the path of the wave) and time t:

$$E = E_m \sin(kx - \omega t), \tag{34-1}$$

$$B = B_m \sin(kx - \omega t), \tag{34-2}$$

in which E_m and B_m are the amplitudes of the fields and, as in Chapter 17, ω and k are the angular frequency and angular wave number of the wave, respectively. From these equations, we note that not only do the two fields form the electromagnetic wave but each forms its own wave. Equation 34-1 gives the *electric wave component* of the electromagnetic wave, and Eq. 34-2 gives the *magnetic wave component.* As we shall discuss below, these two wave components cannot exist independently.

From Eq. 17-12, we know that the speed of the wave is ω/k. However, since this is an electromagnetic wave, its speed (in vacuum) is given the symbol c rather than v. In the next section you will see that c has the value

$$c = \frac{1}{\sqrt{\mu_0 \varepsilon_0}} \quad \text{(wave speed)}, \tag{34-3}$$

which is about 3.0×10^8 m/s. In other words,

All electromagnetic waves, including visible light, have the same speed c in vacuum.

You will also see that the wave speed c and the amplitudes of the electric and magnetic fields are related by

$$\frac{E_m}{B_m} = c \quad \text{(amplitude ratio).} \tag{34-4}$$

If we divide Eq. 34-1 by Eq. 34-2 and then substitute with Eq. 34-4, we find that the magnitudes of the fields at every instant and at any point are related by

$$\frac{E}{B} = c \quad \text{(magnitude ratio).} \tag{34-5}$$

We can represent the electromagnetic wave as in Fig. 34-5*a*, with a *ray* (a directed line showing the wave's direction of travel) or with *wavefronts* (imaginary surfaces over which the wave has the same magnitude of electric field), or both. The two wavefronts shown in Fig. 34-5*a* are separated by one wavelength λ ($= 2\pi/k$) of the wave. (Waves traveling in approximately the same direction form a *beam*, such as a laser beam, which can also be represented with a ray.)

We can also represent the wave as in Fig. 34-5*b*, which shows the electric and magnetic field vectors in a "snapshot" of the wave at a certain instant. The curves through the tips of the vectors represent the sinusoidal oscillations given by Eqs. 34-1 and 34-2; the wave components $\vec{E}$ and $\vec{B}$ are in phase, perpendicular to each other, and perpendicular to the wave's direction of travel.

Interpretation of Fig. 34-5*b* requires some care. The similar drawings for a transverse wave on a taut string that we discussed in Chapter 17 represented the up and down displacement of sections of the string as the wave passed (*something actually moved*). Figure 34-5*b* is more abstract. At the instant shown, the electric and magnetic fields each have a certain magnitude and direction (but always perpendicular to the x axis) at each point along the x axis. We choose to represent these vector quantities with a pair of arrows for each point, so we must draw arrows of different lengths for different points, all directed away from the x axis, like thorns on a rose stem. However, the arrows represent only field values at points that are on the x axis. Neither the arrows nor the sinusoidal curves represent a sideways motion of anything, nor do the arrows connect points on the x axis with points off the axis.

Drawings like Fig. 34-5 help us visualize what is actually a very complicated situation. First consider the magnetic field. Because it varies sinusoidally, it induces (via Faraday's law of induction) a perpendicular electric field that also varies sinusoidally. However, because that electric field is varying sinusoidally, it induces (via

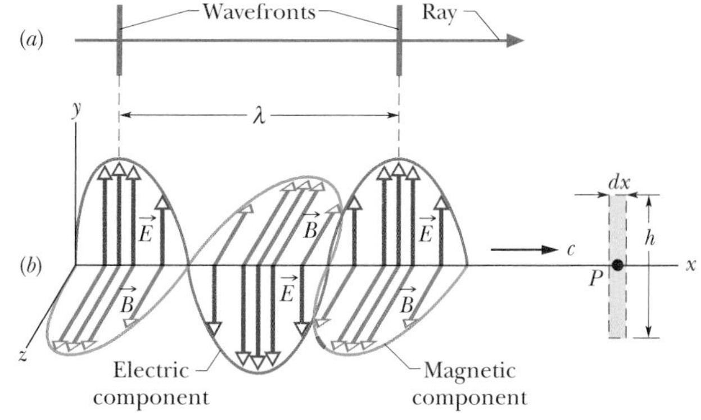

Fig. 34-5 (*a*) An electromagnetic wave represented with a ray and two wavefronts; the wavefronts are separated by one wavelength λ. (*b*) The same wave represented in a "snapshot" of its electric field $\vec{E}$ and magnetic field $\vec{B}$ at points on the x axis, along which the wave travels at speed c. As it travels past point P, the fields vary as shown in Fig. 34-4. The electric component of the wave consists of only the electric fields; the magnetic component consists of only the magnetic fields. The dashed rectangle at P is used in Fig. 34-6.

Maxwell's law of induction) a perpendicular magnetic field that also varies sinusoidally. And so on. The two fields continuously create each other via induction, and the resulting sinusoidal variations in the fields travel as a wave—the electromagnetic wave. Without this amazing result, we could not see; indeed, because we need electromagnetic waves from the Sun to maintain Earth's temperature, without this result we could not even exist.

A Most Curious Wave

The waves we discussed in Chapters 17 and 18 require a *medium* (some material) through which or along which to travel. We had waves traveling along a string, through Earth, and through the air. However, an electromagnetic wave (let's use the term *light wave* or *light*) is curiously different in that it requires no medium for its travel. It can, indeed, travel through a medium such as air or glass, but it can also travel through the vacuum of space between a star and us.

Once the special theory of relativity became accepted, long after Einstein published it in 1905, the speed of light waves was realized to be special. One reason is that light has the same speed regardless of the frame of reference from which it is measured. If you send a beam of light along an axis and ask several observers to measure its speed while they move at different speeds along that axis, either in the direction of the light or opposite it, they will all measure the *same speed* for the light. This result is an amazing one and quite different from what would have been found if those observers had measured the speed of any other type of wave; for other waves, the speed of the observers relative to the wave would have affected their measurements.

The meter has now been defined so that the speed of light (any electromagnetic wave) in vacuum has the exact value

$$c = 299\,792\,458 \text{ m/s},$$

which can be used as a standard. In fact, if you now measure the travel time of a pulse of light from one point to another, you are not really measuring the speed of the light but rather the distance between those two points.

34-3 The Traveling Electromagnetic Wave, Quantitatively

We shall now derive Eqs. 34-3 and 34-4 and, even more important, explore the dual induction of electric and magnetic fields that gives us light.

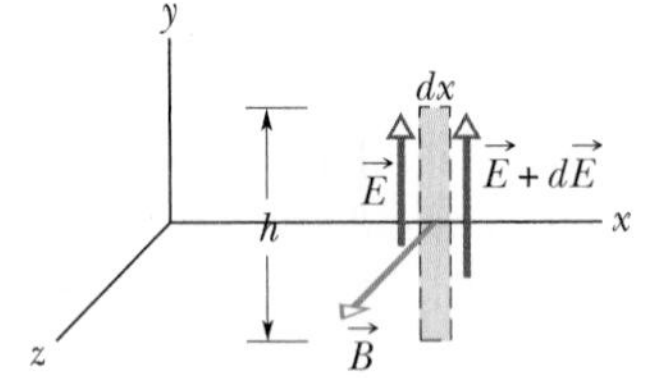

Fig. 34-6 As the electromagnetic wave travels rightward past point P in Fig. 34-5, the sinusoidal variation of the magnetic field $\vec{B}$ through a rectangle centered at P induces electric fields along the rectangle. At the instant shown, $\vec{B}$ is decreasing in magnitude and the induced electric field is therefore greater in magnitude on the right side of the rectangle than on the left.

Equation 34-4 and the Induced Electric Field

The dashed rectangle of dimensions dx and h in Fig. 34-6 is fixed at point P on the x axis and in the xy plane (it is shown on the right in Fig. 34-5b). As the electromagnetic wave moves rightward past the rectangle, the magnetic flux Φ_B through the rectangle changes and—according to Faraday's law of induction—induced electric fields appear throughout the region of the rectangle. We take $\vec{E}$ and $\vec{E} + d\vec{E}$ to be the induced fields along the two long sides of the rectangle. These induced electric fields are, in fact, the electric component of the electromagnetic wave.

Let us consider these fields at the instant when the magnetic wave component passing through the rectangle is the small section marked with red in Fig. 34-5b. Just then, the magnetic field through the rectangle points in the positive z direction and is decreasing in magnitude (the magnitude was greater just before the red section arrived). Because the magnetic field is decreasing, the magnetic flux Φ_B through the rectangle is also decreasing. According to Faraday's law, this change in flux is opposed by induced electric fields, which produce a magnetic field $\vec{B}$ in the positive z direction.

According to Lenz's law, this in turn means that if we imagine the boundary of the rectangle to be a conducting loop, a counterclockwise induced current would have to appear in it. There is, of course, no conducting loop; but this analysis shows that the induced electric field vectors $\vec{E}$ and $\vec{E} + d\vec{E}$ are indeed oriented as shown in Fig. 34-6, with the magnitude of $\vec{E} + d\vec{E}$ greater than that of $\vec{E}$. Otherwise, the net induced electric field would not act counterclockwise around the rectangle.

Let us now apply Faraday's law of induction,

$$\oint \vec{E} \cdot d\vec{s} = -\frac{d\Phi_B}{dt}, \tag{34-6}$$

counterclockwise around the rectangle of Fig. 34-6. There is no contribution to the integral from the top or bottom of the rectangle because $\vec{E}$ and $d\vec{s}$ are perpendicular there. The integral then has the value

$$\oint \vec{E} \cdot d\vec{s} = (E + dE)h - Eh = h\,dE. \tag{34-7}$$

The flux Φ_B through this rectangle is

$$\Phi_B = (B)(h\,dx), \tag{34-8}$$

where B is the magnitude of $\vec{B}$ within the rectangle and $h\,dx$ is the area of the rectangle. Differentiating Eq. 34-8 with respect to t gives

$$\frac{d\Phi_B}{dt} = h\,dx\,\frac{dB}{dt}. \tag{34-9}$$

If we substitute Eqs. 34-7 and 34-9 into Eq. 34-6, we find

$$h\,dE = -h\,dx\,\frac{dB}{dt}$$

or

$$\frac{dE}{dx} = -\frac{dB}{dt}. \tag{34-10}$$

Actually, both B and E are functions of *two* variables, x and t, as Eqs. 34-1 and 34-2 imply. However, in evaluating dE/dx, we must assume that t is constant because Fig. 34-6 is an "instantaneous snapshot." Also, in evaluating dB/dt we must assume that x is constant because we are dealing with the time rate of change of B at a particular place, the point P in Fig. 34-5*b*. The derivatives under these circumstances are *partial derivatives*, and Eq. 34-10 must be written

$$\frac{\partial E}{\partial x} = -\frac{\partial B}{\partial t}. \tag{34-11}$$

The minus sign in this equation is appropriate and necessary because, although E is increasing with x at the site of the rectangle in Fig. 34-6, B is decreasing with t.

From Eq. 34-1 we have

$$\frac{\partial E}{\partial x} = kE_m \cos(kx - \omega t)$$

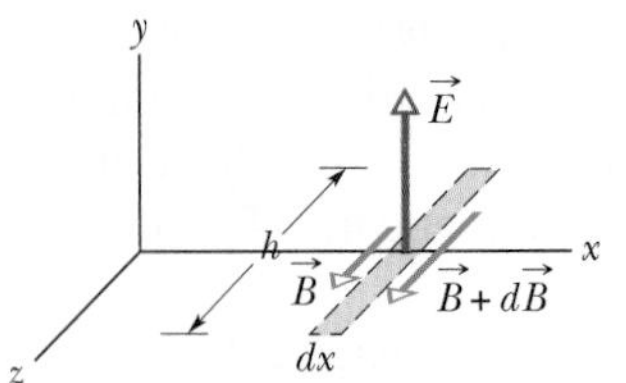

Fig. 34-7 The sinusoidal variation of the electric field through this rectangle, located (but not shown) at point P in Fig. 34-5, induces magnetic fields along the rectangle. The instant shown is that of Fig. 34-6: $\vec{E}$ is decreasing in magnitude and the induced magnetic field is greater in magnitude on the right side of the rectangle than on the left.

and from Eq. 34-2

$$\frac{\partial B}{\partial t} = -\omega B_m \cos(kx - \omega t).$$

Then Eq. 34-11 reduces to

$$kE_m \cos(kx - \omega t) = \omega B_m \cos(kx - \omega t). \tag{34-12}$$

The ratio ω/k for a traveling wave is its speed, which we are calling c. Equation 34-12 then becomes

$$\frac{E_m}{B_m} = c \qquad \text{(amplitude ratio)}, \tag{34-13}$$

which is just Eq. 34-4.

Equation 34-3 and the Induced Magnetic Field

Figure 34-7 shows another dashed rectangle at point P of Fig. 34-5; this one is in the xz plane. As the electromagnetic wave moves rightward past this new rectangle, the electric flux Φ_E through the rectangle changes and—according to Maxwell's law of induction—induced magnetic fields appear throughout the region of the rectangle. These induced magnetic fields are, in fact, the magnetic component of the electromagnetic wave.

We see from Fig. 34-5 that at the instant chosen for the magnetic field in Fig. 34-6, the electric field through the rectangle of Fig. 34-7 is directed as shown. Recall that at the chosen instant, the magnetic field in Fig. 34-6 is decreasing. Because the two fields are in phase, the electric field in Fig. 34-7 must also be decreasing, and so must the electric flux Φ_E through the rectangle. By applying the same reasoning we applied to Fig. 34-6, we see that the changing flux Φ_E will induce a magnetic field with vectors $\vec{B}$ and $\vec{B} + d\vec{B}$ oriented as shown in Fig. 34-7, where $\vec{B} + d\vec{B}$ is greater than $\vec{B}$.

Let us apply Maxwell's law of induction,

$$\oint \vec{B} \cdot d\vec{s} = \mu_0 \varepsilon_0 \frac{d\Phi_E}{dt}, \tag{34-14}$$

by proceeding counterclockwise around the dashed rectangle of Fig. 34-7. Only the long sides of the rectangle contribute to the integral, whose value is

$$\oint \vec{B} \cdot d\vec{s} = -(B + dB)h + Bh = -h\,dB. \tag{34-15}$$

The flux Φ_E through the rectangle is

$$\Phi_E = (E)(h\,dx), \tag{34-16}$$

where E is the average magnitude of $\vec{E}$ within the rectangle. Differentiating Eq. 34-16 with respect to t gives

$$\frac{d\Phi_E}{dt} = h\,dx\,\frac{dE}{dt}.$$

If we substitute this and Eq. 34-15 into Eq. 34-14, we find

$$-h\,dB = \mu_0 \varepsilon_0 \left(h\,dx\,\frac{dE}{dt} \right)$$

or, changing to partial-derivative notation as we did before (Eq. 34-11),

$$-\frac{\partial B}{\partial x} = \mu_0 \varepsilon_0 \frac{\partial E}{\partial t}. \tag{34-17}$$

Again, the minus sign in this equation is necessary because, although B is increasing with x at point P in the rectangle in Fig. 34-7, E is decreasing with t.

Evaluating Eq. 34-17 by using Eqs. 34-1 and 34-2 leads to

$$-kB_m \cos(kx - \omega t) = -\mu_0 \varepsilon_0 \omega E_m \cos(kx - \omega t),$$

which we can write as

$$\frac{E_m}{B_m} = \frac{1}{\mu_0 \varepsilon_0 (\omega/k)} = \frac{1}{\mu_0 \varepsilon_0 c}.$$

Combining this with Eq. 34-13 leads at once to

$$c = \frac{1}{\sqrt{\mu_0 \varepsilon_0}} \quad \text{(wave speed),} \tag{34-18}$$

which is exactly Eq. 34-3.

CHECKPOINT 1: The magnetic field $\vec{B}$ through the rectangle of Fig. 34-6 is shown at a different instant in part 1 of the figure here; $\vec{B}$ is directed in the xz plane, parallel to the z axis, and its magnitude is increasing. (a) Complete part 1 by drawing the induced electric fields, indicating both directions and relative magnitudes (as in Fig. 34-6). (b) For the same instant, complete part 2 of the figure by drawing the electric field of the electromagnetic wave. Also draw the induced magnetic fields, indicating both directions and relative magnitudes (as in Fig. 34-7).

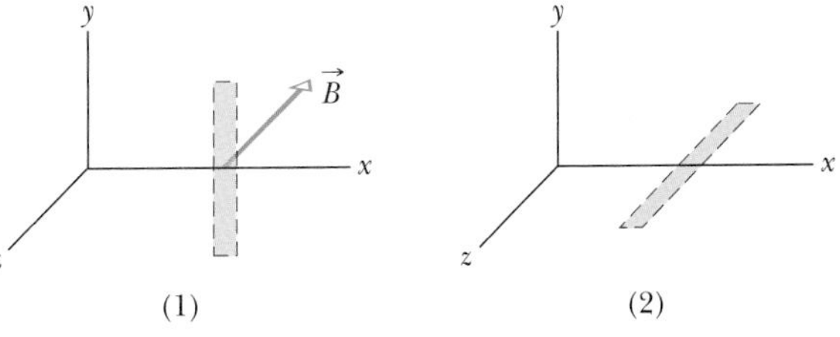

34-4 Energy Transport and the Poynting Vector

All sunbathers know that an electromagnetic wave can transport energy and deliver it to a body on which it falls. The rate of energy transport per unit area in such a wave is described by a vector $\vec{S}$, called the **Poynting vector** after physicist John Henry Poynting (1852–1914), who first discussed its properties. $\vec{S}$ is defined as

$$\vec{S} = \frac{1}{\mu_0} \vec{E} \times \vec{B} \quad \text{(Poynting vector).} \tag{34-19}$$

Its magnitude S is related to the rate at which energy is transported by a wave across a unit area at any instant (inst):

$$S = \left(\frac{\text{energy/time}}{\text{area}}\right)_{\text{inst}} = \left(\frac{\text{power}}{\text{area}}\right)_{\text{inst}}. \tag{34-20}$$

From this we can see that the SI unit for $\vec{S}$ is the watt per square meter (W/m^2).

The direction of the Poynting vector $\vec{S}$ of an electromagnetic wave at any point gives the wave's direction of travel and the direction of energy transport at that point.

Because $\vec{E}$ and $\vec{B}$ are perpendicular to each other in an electromagnetic wave, the magnitude of $\vec{E} \times \vec{B}$ is EB. Then the magnitude of $\vec{S}$ is

$$S = \frac{1}{\mu_0} EB, \tag{34-21}$$

in which S, E, and B are instantaneous values. E and B are so closely coupled to each other that we need to deal with only one of them; we choose E, largely because most instruments for detecting electromagnetic waves deal with the electric component of the wave rather than the magnetic component. Using $B = E/c$ from Eq. 34-5, we can rewrite Eq. 34-21 as

$$S = \frac{1}{c\mu_0} E^2 \quad \text{(instantaneous energy flow rate).} \tag{34-22}$$

By substituting $E = E_m \sin(kx - \omega t)$ into Eq. 34-22, we could obtain an equation for the energy transport rate as a function of time. More useful in practice, however, is the average energy transported over time; for that, we need to find the time-averaged value of S, written S_{avg} and also called the **intensity** I of the wave. Thus from Eq. 34-20, the intensity I is

$$I = S_{avg} = \left(\frac{\text{energy/time}}{\text{area}}\right)_{avg} = \left(\frac{\text{power}}{\text{area}}\right)_{avg}. \tag{34-23}$$

From Eq. 34-22, we find

$$I = S_{avg} = \frac{1}{c\mu_0} [E^2]_{avg} = \frac{1}{c\mu_0} [E_m^2 \sin^2(kx - \omega t)]_{avg}. \tag{34-24}$$

Over a full cycle, the average value of $\sin^2 \theta$, for any angular variable θ, is $\frac{1}{2}$ (see Fig. 33-14). In addition, we define a new quantity E_{rms}, the *root-mean-square* value of the electric field, as

$$E_{rms} = \frac{E_m}{\sqrt{2}}. \tag{34-25}$$

We can then rewrite Eq. 34-24 as

$$I = \frac{1}{c\mu_0} E_{rms}^2. \tag{34-26}$$

Because $E = cB$ and c is such a very large number, you might conclude that the energy associated with the electric field is much greater than that associated with the magnetic field. That conclusion is incorrect; the two energies are exactly equal. To show this, we start with Eq. 26-23, which gives the energy density u $(= \frac{1}{2}\varepsilon_0 E^2)$ within an electric field, and substitute cB for E; then we can write

$$u_E = \tfrac{1}{2}\varepsilon_0 E^2 = \tfrac{1}{2}\varepsilon_0 (cB)^2.$$

If we now substitute for c with Eq. 34-3, we get

$$u_E = \tfrac{1}{2}\varepsilon_0 \frac{1}{\mu_0 \varepsilon_0} B^2 = \frac{B^2}{2\mu_0}.$$

However, Eq. 31-56 tells us that $B^2/2\mu_0$ is the energy density u_B of a magnetic field $\vec{B}$, so we see that $u_E = u_B$ everywhere along an electromagnetic wave.

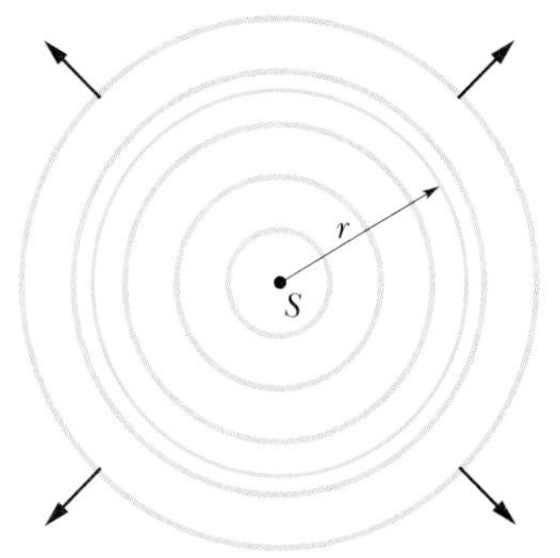

Fig. 34-8 A point source S emits electromagnetic waves uniformly in all directions. The spherical wavefronts pass through an imaginary sphere of radius r that is centered on S.

Variation of Intensity with Distance

How intensity varies with distance from a real source of electromagnetic radiation is often complex—especially when the source (like a searchlight at a movie premier) beams the radiation in a particular direction. However, in some situations we can assume that the source is a *point source* that emits the light *isotropically*—that is, with equal intensity in all directions. The spherical wavefronts spreading from such an isotropic point source S at a particular instant are shown in cross section in Fig. 34-8.

Let us assume that the energy of the waves is conserved as they spread from this source. Let us also center an imaginary sphere of radius r on the source, as shown in Fig. 34-8. All the energy emitted by the source must pass through the sphere. Thus, the rate at which energy is transferred through the sphere by the radiation must equal the rate at which energy is emitted by the source—that is, the power P_s of the source. The intensity I at the sphere must then be

$$I = \frac{P_s}{4\pi r^2}, \tag{34-27}$$

where $4\pi r^2$ is the area of the sphere. Equation 34-27 tells us that the intensity of the electromagnetic radiation from an isotropic point source decreases with the square of the distance r from the source.

CHECKPOINT 2: The figure here gives the electric field of an electromagnetic wave at a certain point and a certain instant. The wave is transporting energy in the negative z direction. What is the direction of the magnetic field of the wave at that point and instant?

Sample Problem 34-1

An observer is 1.8 m from an isotropic point light source whose power P_s is 250 W. Calculate the rms values of the electric and magnetic fields due to the source at the position of the observer.

SOLUTION: The first two Key Ideas here are these:

1. The rms value E_{rms} of the electric field in light is related to the intensity I of the light via Eq. 34-26 ($I = E_{\text{rms}}^2/c\mu_0$).
2. Because the source is a point source emitting light with equal intensity in all directions, the intensity I at any distance r from the source is related to the source's power P_s via Eq. 34-27 ($I = P_s/4\pi r^2$).

Putting these two ideas together gives us

$$I = \frac{P_s}{4\pi r^2} = \frac{E_{\text{rms}}^2}{c\mu_0}$$

which leads to

$$E_{\text{rms}} = \sqrt{\frac{P_s c\mu_0}{4\pi r^2}}$$

$$= \sqrt{\frac{(250\text{ W})(3.00 \times 10^8\text{ m/s})(4\pi \times 10^{-7}\text{ H/m})}{(4\pi)(1.8\text{ m})^2}}$$

$$= 48.1\text{ V/m} \approx 48\text{ V/m}. \qquad \text{(Answer)}$$

The third Key Idea here is that magnitudes of the electric field and magnetic field of an electromagnetic wave at any instant and at any point in the wave are related by the speed of light c according to Eq. 34-5 ($E/B = c$). Thus, the rms values of those fields are also related by Eq. 34-5 and we can write

$$B_{\text{rms}} = \frac{E_{\text{rms}}}{c} = \frac{48.1\text{ V/m}}{3.00 \times 10^8\text{ m/s}} = 1.6 \times 10^{-7}\text{ T}. \qquad \text{(Answer)}$$

Note that E_{rms} ($= 48$ V/m) is appreciable as judged by ordinary laboratory standards, but B_{rms} ($= 1.6 \times 10^{-7}$ T) is quite small. This difference helps to explain why most instruments used for the detection and measurement of electromagnetic waves are designed to respond to the electric component of the wave. It is wrong, however, to say that the electric component of an electromagnetic wave is "stronger" than the magnetic component. You cannot compare quantities that are measured in different units. As we have seen, the electric and magnetic components are on an equal basis as far as the propagation of the wave is concerned, because their average energies, which *can* be compared, are exactly equal.

34-5 Radiation Pressure

Electromagnetic waves have linear momentum as well as energy. This means that we can exert a pressure—a **radiation pressure**—on an object by shining light on it. However, the pressure must be very small because, for example, you do not feel a camera flash when it is used to take your photograph.

To find an expression for the pressure, let us shine a beam of electromagnetic radiation—light, for example—on an object for a time interval Δt. Further, let us assume that the object is free to move and that the radiation is entirely **absorbed** (taken up) by the object. This means that during the interval Δt, the object gains an energy ΔU from the radiation. Maxwell showed that the object also gains linear momentum. The magnitude Δp of the momentum change of the object is related to the energy change ΔU by

$$\Delta p = \frac{\Delta U}{c} \quad \text{(total absorption)}, \tag{34-28}$$

where c is the speed of light. The direction of the momentum change of the object is the direction of the *incident* (incoming) beam that the object absorbs.

Instead of being absorbed, the radiation can be **reflected** by the object; that is, the radiation can be sent off in a new direction as if it bounced off the object. If the radiation is entirely reflected back along its original path, the magnitude of the momentum change of the object is twice that given above, or

$$\Delta p = \frac{2\,\Delta U}{c} \quad \text{(total reflection back along path)}. \tag{34-29}$$

In the same way, an object undergoes twice as much momentum change when a perfectly elastic tennis ball is bounced from it as when it is struck by a perfectly inelastic ball (a lump of wet putty, say) of the same mass and velocity. If the incident radiation is partly absorbed and partly reflected, the momentum change of the object is between $\Delta U/c$ and $2\,\Delta U/c$.

From Newton's second law, we know that a change in momentum is related to a force by

$$F = \frac{\Delta p}{\Delta t}. \tag{34-30}$$

To find expressions for the force exerted by radiation in terms of the intensity I of the radiation, suppose that a flat surface of area A, perpendicular to the path of the radiation, intercepts the radiation. In time interval Δt, the energy intercepted by area A is

$$\Delta U = IA\,\Delta t. \tag{34-31}$$

If the energy is completely absorbed, then Eq. 34-28 tells us that $\Delta p = IA\,\Delta t/c$ and, from Eq. 34-30, the magnitude of the force on the area A is

$$F = \frac{IA}{c} \quad \text{(total absorption)}. \tag{34-32}$$

Similarly, if the radiation is totally reflected back along its original path, Eq. 34-29 tells us that $\Delta p = 2IA\,\Delta t/c$ and, from Eq. 34-30,

$$F = \frac{2IA}{c} \quad \text{(total reflection back along path)}. \tag{34-33}$$

If the radiation is partly absorbed and partly reflected, the magnitude of the force on area A is between the values of IA/c and $2IA/c$.

The force per unit area on an object due to radiation is the radiation pressure p_r. We can find it for the situations of Eqs. 34-32 and 34-33 by dividing both sides of each equation by A. We obtain

$$p_r = \frac{I}{c} \quad \text{(total absorption)} \tag{34-34}$$

and

$$p_r = \frac{2I}{c} \quad \text{(total reflection back along path).} \tag{34-35}$$

Be careful not to confuse the symbol p_r for radiation pressure with the symbol p for momentum. Just as with fluid pressure in Chapter 15, the SI unit of radiation pressure is the newton per square meter (N/m^2), which is called the pascal (Pa).

The development of laser technology has permitted researchers to achieve radiation pressures much greater than, say, that due to a camera flashlamp. This comes about because a beam of laser light—unlike a beam of light from a small lamp filament—can be focused to a tiny spot only a few wavelengths in diameter. This permits the delivery of great amounts of energy to small objects placed at that spot.

CHECKPOINT 3: Light of uniform intensity shines perpendicularly on a totally absorbing surface, fully illuminating the surface. If the area of the surface is decreased, do (a) the radiation pressure and (b) the radiation force on the surface increase, decrease, or stay the same?

Sample Problem 34-2

When dust is released by a comet, it does not continue along the comet's orbit because radiation pressure from sunlight pushes it radially outward from the Sun. Assume that a dust particle is spherical with radius R, has density $\rho = 3.5 \times 10^3$ kg/m^3, and totally absorbs the sunlight it intercepts. For what value of R does the gravitational force $\vec{F}_g$ on the dust particle due to the Sun just balance the radiation force $\vec{F}_r$ on it from the sunlight?

SOLUTION: We can assume that the Sun is far enough from the particle to act as an isotropic point source of light. Then because we are told that the radiation pressure pushes the particle radially outward from the Sun, we know that the radiation force $\vec{F}_r$ on the particle is directed radially outward from the center of the Sun. At the same time, the gravitational force $\vec{F}_g$ on the particle is directed radially inward *toward* the center of the Sun. Thus, we can write the balance of these two forces as

$$F_r = F_g. \tag{34-36}$$

Let us consider these forces separately.

Radiation force: To evaluate the left side of Eq. 34-36, we use these three **Key Ideas**.

1. Because the particle is totally absorbing, the force magnitude F_r can be found from the intensity I of sunlight at the particle's location and the particle's cross-sectional area A, via Eq. 34-32 ($F = IA/c$).
2. Because we assume that the Sun is an isotropic point source of light, we can use Eq. 34-27 ($I = P_s/4\pi r^2$) to relate the Sun's power P_S to the intensity I of the sunlight at the particle's distance r from the Sun.
3. Because the particle is spherical, its cross-sectional area A is πR^2 (*not* half its surface area).

Putting these three ideas together gives us

$$F_r = \frac{IA}{c} = \frac{P_S \pi R^2}{4\pi r^2 c} = \frac{P_S R^2}{4r^2 c}. \tag{34-37}$$

Gravitational force: The **Key Idea** here is Newton's law of gravitation (Eq. 14-1), which gives us the magnitude of the gravitational force on the particle as

$$F_g = \frac{GM_S m}{r^2}, \tag{34-38}$$

where M_S is the Sun's mass and m is the particle's mass. Next, the particle's mass is related to its density ρ and volume V ($= \frac{4}{3}\pi R^3$, for a sphere) by

$$\rho = \frac{m}{V} = \frac{m}{\frac{4}{3}\pi R^3}.$$

Solving this for m and substituting the result into Eq. 34-38 give us

$$F_g = \frac{GM_S\rho(\frac{4}{3}\pi R^3)}{r^2}. \tag{34-39}$$

Then substituting Eqs. 34-37 and 34-39 into Eq. 34-36 and solving for R yield

$$R = \frac{3P_S}{16\pi c\rho G M_S}.$$

Using the given value of ρ and the known values of G (Appendix B) and M_S (Appendix C), we can evaluate the denominator:

$$(16\pi)(3 \times 10^8 \text{ m/s})(3.5 \times 10^3 \text{ kg/m}^3) \times (6.67 \times 10^{-11} \text{ N}\cdot\text{m}^2/\text{kg}^2)(1.99 \times 10^{30} \text{ kg}) = 7.0 \times 10^{33} \text{ N/s}.$$

Using P_S from Appendix C, we then have

$$R = \frac{(3)(3.9 \times 10^{26} \text{ W})}{7.0 \times 10^{33} \text{ N/s}} = 1.7 \times 10^{-7} \text{ m}. \quad \text{(Answer)}$$

Note that this result is independent of the particle's distance r from the Sun.

Dust particles with radius $R \approx 1.7 \times 10^{-7}$ m follow an approximately straight path like path b in Fig. 34-9. For larger values of R, comparison of Eqs. 34-37 and 34-39 shows that, because F_g varies with R^3 and F_r varies with R^2, the gravitational force F_g dominates the radiation force F_r. Thus, such particles follow a path that is curved toward the Sun like path c in Fig. 34-9. Similarly, for smaller values of R, the radiation force dominates, and the dust follows a path that is curved away from the Sun like path a. The composite of these dust particles is the dust tail of the comet.

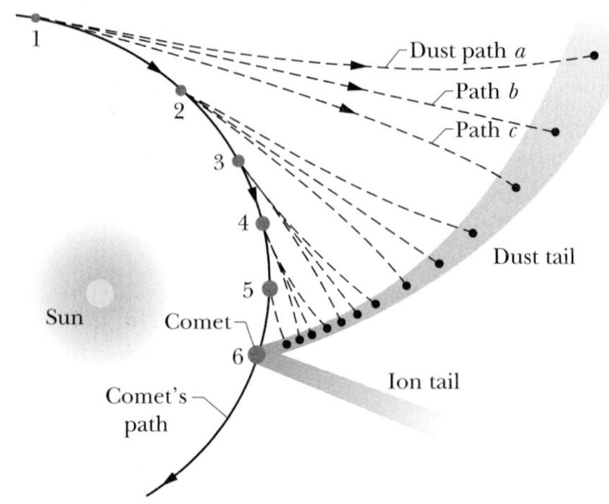

Fig. 34-9 Sample Problem 34-2. A comet is now at position 6. Dust it has released at five previous positions has been pushed outward by radiation pressure from sunlight, has taken the dashed paths, and now forms the comet's curved dust tail.

34-6 Polarization

VHF (very high frequency) television antennas in England are oriented vertically, but those in North America are horizontal. The difference is due to the direction of oscillation of the electromagnetic waves carrying the TV signal. In England, the transmitting equipment is designed to produce waves that are **polarized** vertically; that is, their electric field oscillates vertically. Thus, for the electric field of the incident television waves to drive a current along an antenna (and provide a signal to a television set), the antenna must be vertical. In North America, the waves are polarized horizontally .

Figure 34-10a shows an electromagnetic wave with its electric field oscillating parallel to the vertical y axis. The plane containing the $\vec{E}$ vectors is called the **plane of oscillation** of the wave (hence, the wave is said to be *plane-polarized* in the y direction). We can represent the wave's *polarization* (state of being polarized) by showing the directions of the electric field oscillations in a "head-on" view of the plane of oscillation, as in Fig. 34-10b. The vertical "double arrow" in that figure indicates that as the wave travels past us, its electric field oscillates vertically—it continuously changes between being directed up and down along the y axis.

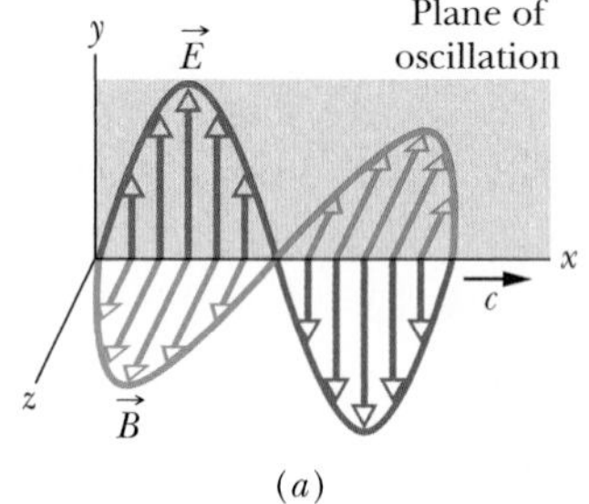

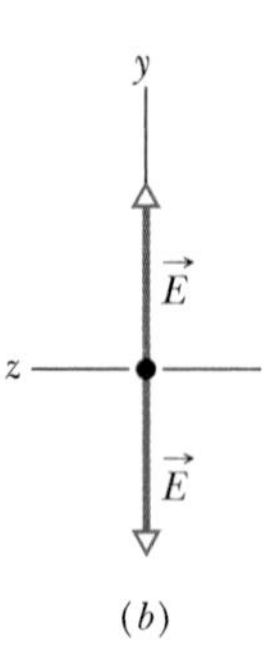

Fig. 34-10 (a) The plane of oscillation of a polarized electromagnetic wave. (b) To represent the polarization, we view the plane of oscillation "head-on" and indicate the directions of the oscillating electric field with a double arrow.

Polarized Light

The electromagnetic waves emitted by a television station all have the same polarization, but the electromagnetic waves emitted by any common source of light (such as the Sun or a bulb) are **polarized randomly** or **unpolarized;** that is, the electric field at any given point is always perpendicular to the direction of travel of the waves but changes directions randomly. Thus, if we try to represent a head-on view of the oscillations over some time period, we do not have a simple drawing with a single double arrow like that of Fig. 34-10b; instead we have a mess of double arrows like that in Fig. 34-11a.

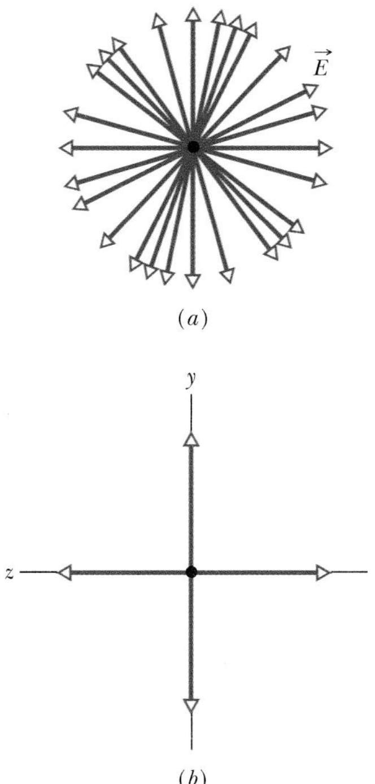

Fig. 34-11 (*a*) Unpolarized light consists of waves with randomly directed electric fields. Here the waves are all traveling along the same axis, directly out of the page, and all have the same amplitude *E*. (*b*) A second way of representing unpolarized light—the light is the superposition of two polarized waves whose planes of oscillation are perpendicular to each other.

In principle, we can simplify the mess by resolving each electric field of Fig. 34-11*a* into *y* and *z* components. Then as the wave travels past us, the net *y* component oscillates parallel to the *y* axis and the net *z* component oscillates parallel to the *z* axis. We can then represent the unpolarized light with a pair of double arrows as shown in Fig. 34-11*b*. The double arrow along the *y* axis represents the oscillations of the net *y* component of the electric field. The double arrow along the *z* axis represents the oscillations of the net *z* component of the electric field. In doing all this, we effectively change unpolarized light into the superposition of two polarized waves whose planes of oscillation are perpendicular to each other—one plane contains the *y* axis and the other contains the *z* axis. One reason to make this change is that drawing Fig. 34-11*b* is a lot easier than drawing Fig. 34-11*a*.

We can draw similar figures to represent light that is **partially polarized** (its field oscillations are not completely random as in Fig. 34-11*a* nor are they parallel to a single axis as in Fig. 34-10*b*). For this situation, we can draw one of the double arrows in a perpendicular pair of double arrows longer than the other one.

We can transform unpolarized visible light into polarized light by sending it through a *polarizing sheet,* as is shown in Fig. 34-12. Such sheets, commercially known as Polaroids or Polaroid filters, were invented in 1932 by Edwin Land while he was an undergraduate student. A polarizing sheet consists of certain long molecules embedded in plastic. When the sheet is manufactured, it is stretched to align the molecules in parallel rows, like rows in a plowed field. When light is then sent through the sheet, electric field components along one direction pass through the sheet, while components perpendicular to that direction are absorbed by the molecules and disappear.

We shall not dwell on the molecules but, instead, shall assign to the sheet a *polarizing direction,* along which electric field components are passed:

> An electric field component parallel to the polarizing direction is passed (*transmitted*) by a polarizing sheet; a component perpendicular to it is absorbed.

Thus, the electric field of the light emerging from the sheet consists of only the components that are parallel to the polarizing direction of the sheet; hence the light is polarized in that direction. In Fig. 34-12, the vertical electric field components are transmitted by the sheet; the horizontal components are absorbed. The transmitted waves are then vertically polarized.

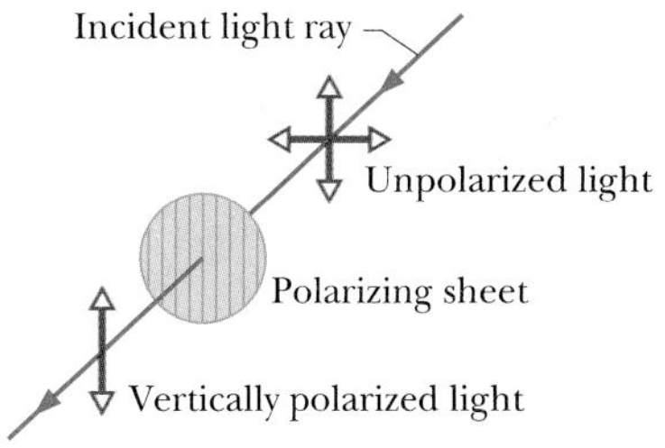

Fig. 34-12 Unpolarized light becomes polarized when it is sent through a polarizing sheet. Its direction of polarization is then parallel to the polarizing direction of the sheet, which is represented here by the vertical lines drawn in the sheet.

Intensity of Transmitted Polarized Light

We now consider the intensity of light transmitted by a polarizing sheet. We start with unpolarized light, whose electric field oscillations we can resolve into *y* and *z* components as represented in Fig. 34-11*b*. Further, we can arrange for the *y* axis to be parallel to the polarizing direction of the sheet. Then only the *y* components of the light's electric field are passed by the sheet; the *z* components are absorbed. As suggested by Fig. 34-11*b*, if the original waves are randomly oriented, the sum of the *y* components and the sum of the *z* components are equal. When the *z* components are absorbed, half the intensity I_0 of the original light is lost. The intensity *I* of the emerging polarized light is then

$$I = \tfrac{1}{2}I_0. \tag{34-40}$$

Let us call this the *one-half rule;* we can use it *only* when the light reaching a polarizing sheet is unpolarized.

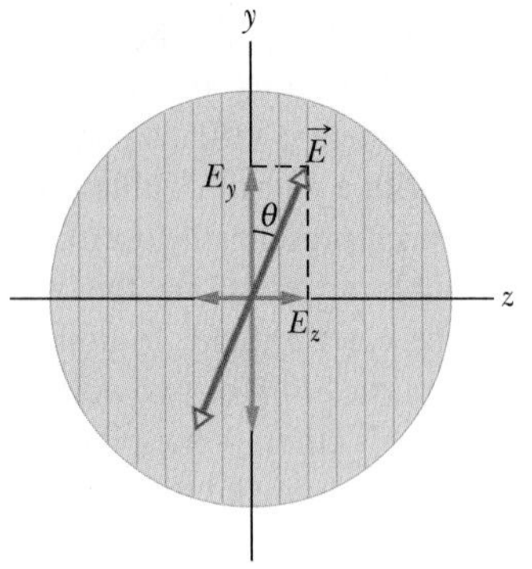

Fig. 34-13 Polarized light approaching a polarizing sheet. The electric field $\vec{E}$ of the light can be resolved into components E_y (parallel to the polarizing direction of the sheet) and E_z (perpendicular to that direction). Component E_y will be transmitted by the sheet; component E_z will be absorbed.

Suppose now that the light reaching a polarizing sheet is already polarized. Figure 34-13 shows a polarizing sheet in the plane of the page and the electric field $\vec{E}$ of such a polarized light wave traveling toward the sheet (and thus prior to any absorption). We can resolve $\vec{E}$ into two components relative to the polarizing direction of the sheet: parallel component E_y is transmitted by the sheet, and perpendicular component E_z is absorbed. Since θ is the angle between $\vec{E}$ and the polarizing direction of the sheet, the transmitted parallel component is

$$E_y = E \cos \theta. \tag{34-41}$$

Recall that the intensity of an electromagnetic wave (such as our light wave) is proportional to the square of the electric field's magnitude (Eq. 34-26). In our present case then, the intensity I of the emerging wave is proportional to E_y^2 and the intensity I_0 of the original wave is proportional to E^2. Hence, from Eq. 34-41 we can write $I/I_0 = \cos^2 \theta$, or

$$I = I_0 \cos^2 \theta. \tag{34-42}$$

Let us call this the *cosine-squared rule;* we can use it *only* when the light reaching a polarizing sheet is already polarized. Then the transmitted intensity I is a maximum and is equal to the original intensity I_0 when the original wave is polarized parallel to the polarizing direction of the sheet (when θ in Eq. 34-42 is 0° or 180°). I is zero when the original wave is polarized perpendicular to the polarizing direction of the sheet (when θ is 90°).

Figure 34-14 shows an arrangement in which initially unpolarized light is sent through two polarizing sheets P_1 and P_2. (Often, the first sheet is called the *polarizer,* and the second the *analyzer.*) Because the polarizing direction of P_1 is vertical, the light transmitted by P_1 to P_2 is polarized vertically. If the polarizing direction of P_2 is also vertical, then all the light transmitted by P_1 is transmitted by P_2. If the polarizing direction of P_2 is horizontal, none of the light transmitted by P_1 is transmitted by P_2. We reach the same conclusions by considering only the *relative* orientations of the two sheets: If their polarizing directions are parallel, all the light passed by the first sheet is passed by the second sheet. If those directions are perpendicular (the sheets are said to be *crossed*), no light is passed by the second sheet. These two extremes are displayed with polarized sunglasses in Fig. 34-15.

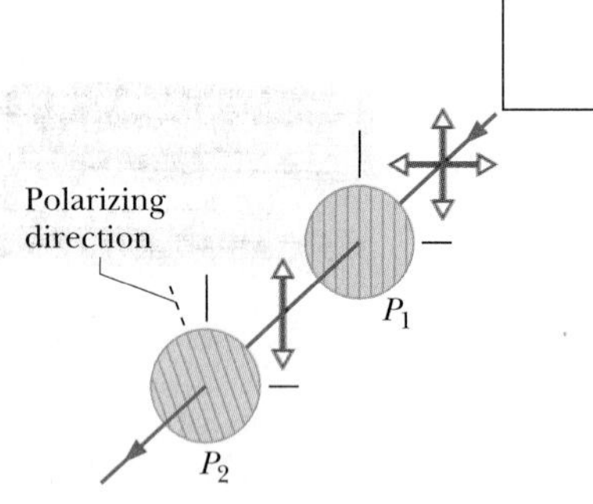

Fig. 34-14 The light transmitted by polarizing sheet P_1 is vertically polarized, as represented by the vertical double arrow. The amount of that light that is then transmitted by polarizing sheet P_2 depends on the angle between the polarization direction of that light and the polarizing direction of P_2 (indicated by the lines drawn in the sheet and by the dashed line).

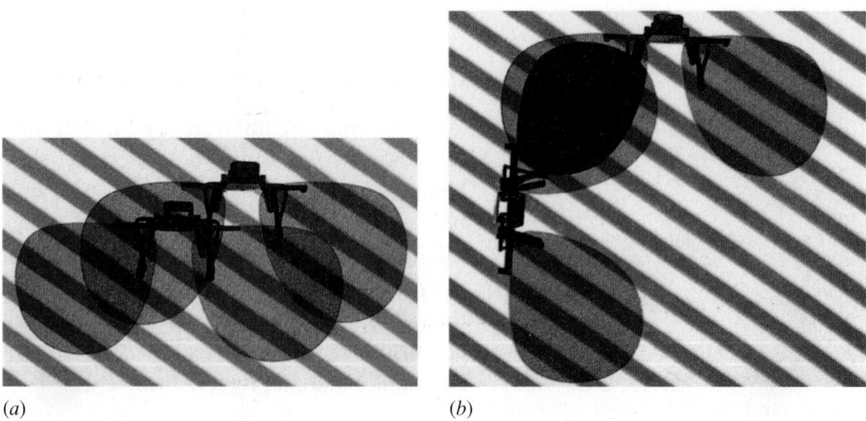

Fig. 34-15 Polarizing sunglasses consist of sheets whose polarizing directions are vertical when the sunglasses are worn. (*a*) Overlapping sunglasses transmit light fairly well when their polarizing directions have the same orientation, but (*b*) they block most of the light when they are crossed.

Finally, if the two polarizing directions of Fig. 34-14 make an angle between 0° and 90°, some of the light transmitted by P_1 will be transmitted by P_2. The intensity of that light is determined by Eq. 34-42.

Light can be polarized by means other than polarizing sheets, such as by reflection (discussed in Section 34-9) and by scattering from atoms or molecules. In *scattering,* light that is intercepted by an object, such as a molecule, is sent off in many, perhaps random, directions. An example is the scattering of sunlight by molecules in the atmosphere, which gives the sky its general glow.

Although direct sunlight is unpolarized, light from much of the sky is at least partially polarized by such scattering. Bees use the polarization of sky light in navigating to and from their hives. Similarly, the Vikings used it to navigate across the North Sea when the daytime Sun was below the horizon (because of the high latitude of the North Sea). These early seafarers had discovered certain crystals (now called cordierite) that changed color when rotated in polarized light. By looking at the sky through such a crystal while rotating it about their line of sight, they could locate the hidden Sun and thus determine which way was south.

Sample Problem 34-3

Figure 34-16*a* shows a system of three polarizing sheets in the path of initially unpolarized light. The polarizing direction of the first sheet is parallel to the y axis, that of the second sheet is 60° counterclockwise from the y axis, and that of the third sheet is parallel to the x axis. What fraction of the initial intensity I_0 of the light emerges from the system, and how is that light polarized?

SOLUTION: The **Key Ideas** here are these:

1. We work through the system sheet by sheet, from the first one encountered by the light to the last one.
2. To find the intensity transmitted by any sheet, we apply either the one-half rule or the cosine-squared rule, depending on whether the light reaching the sheet is unpolarized or already polarized.
3. The light that is transmitted by a polarizing sheet is always polarized parallel to the polarizing direction of the sheet.

First sheet: The original light wave is represented in Fig. 34-16*b*, using the head-on, double-arrow representation of Fig. 34-11*b*. Because the light is initially unpolarized, the intensity I_1 of the light transmitted by the first sheet is given by the one-half rule (Eq. 34-40):

$$I_1 = \tfrac{1}{2}I_0.$$

Because the polarizing direction of the first sheet is parallel to the y axis, the polarization of the light transmitted by it is also, as shown in the head-on view of Fig. 34-16*c*.

Second sheet: Since the light reaching the second sheet is polarized, the intensity I_2 of the light transmitted by that sheet is given by the cosine-squared rule (Eq. 34-42). The angle θ in the rule is the angle between the polarization direction of the entering light (parallel to the y axis) and the polarizing direction of the second sheet (60° counterclockwise from the y axis), and so θ is 60°. Then

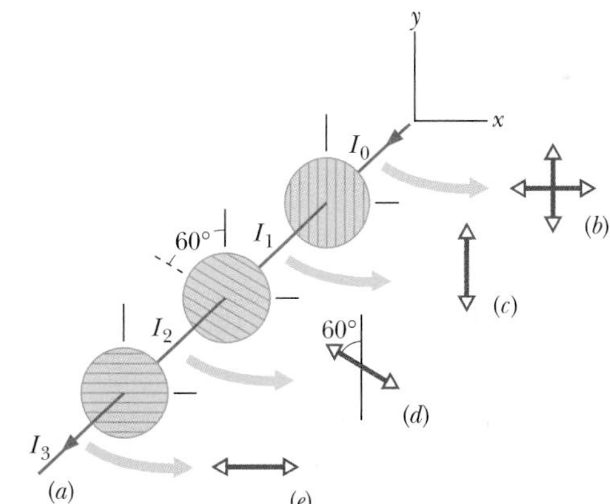

Fig. 34-16 Sample Problem 34-3. (*a*) Initially unpolarized light of intensity I_0 is sent into a system of three polarizing sheets. The intensities I_1, I_2, and I_3 of the light transmitted by the sheets are labeled. Shown also are the polarizations, from head-on views, of (*b*) the initial light and the light transmitted by (*c*) the first sheet, (*d*) the second sheet, and (*e*) the third sheet.

$$I_2 = I_1 \cos^2 60°.$$

The polarization of this transmitted light is parallel to the polarizing direction of the sheet transmitting it—that is, 60° counterclockwise from the y axis, as shown in the head-on view of Fig. 34-16*d*.

Third sheet: Because the light reaching the third sheet is polarized, the intensity I_3 of the light transmitted by that sheet is given by the cosine-squared rule. The angle θ is now the angle between the polarization direction of the entering light (Fig. 34-16*d*) and the polarizing direction of the third sheet (parallel to the x axis), and so $\theta = 30°$. Thus,

$$I_3 = I_2 \cos^2 30°.$$

This final transmitted light is polarized parallel to the x axis (Fig. 34-16e). We find its intensity by substituting first for I_2 and then for I_1 in the equation above:

$$I_3 = I_2 \cos^2 30° = (I_1 \cos^2 60°) \cos^2 30°$$
$$= (\tfrac{1}{2}I_0) \cos^2 60° \cos^2 30° = 0.094 I_0.$$

Thus, $$\frac{I_3}{I_0} = 0.094. \qquad \text{(Answer)}$$

That is to say, 9.4% of the initial intensity emerges from the three-sheet system. (If we now remove the second sheet, what fraction of the initial intensity emerges from the system?)

CHECKPOINT 4: The figure shows four pairs of polarizing sheets, seen face-on. Each pair is mounted in the path of initially unpolarized light (like the three sheets in Fig. 34-16a). The polarizing direction of each sheet (indicated by the dashed line) is referenced to either a horizontal x axis or a vertical y axis. Rank the pairs according to the fraction of the initial intensity that they pass, greatest first.

30° 30° 30° 30°
60° 60° 60° 60°
(a) (b) (c) (d)

34-7 Reflection and Refraction

Although a light wave spreads as it moves away from its source, we can often approximate its travel as being in a straight line; we did so for the light wave in Fig. 34-5a. The study of the properties of light waves under that approximation is called *geometrical optics*. For the rest of this chapter and all of Chapter 35, we shall discuss the geometrical optics of visible light.

The black-and-white photograph in Fig. 34-17a shows an example of light waves traveling in approximately straight lines. A narrow beam of light (the *incident* beam), angled downward from the left and traveling through air, encounters a *plane* (flat) glass surface. Part of the light is **reflected** by the surface, forming a beam directed upward toward the right, traveling as if the original beam had bounced from the surface. The rest of the light travels through the surface and into the glass, forming a beam directed downward to the right. Because light can travel through the glass like this, the glass is said to be *transparent;* that is, we can see through it. (In this chapter we shall consider only transparent materials.)

The travel of light through a surface (or *interface*) that separates two media is called **refraction,** and the light is said to be *refracted.* Unless an incident beam of light is perpendicular to a surface, refraction by the surface changes the light's direction of travel. For this reason, the beam is said to be "bent" by the refraction. Note in Fig. 34-17a that the bending occurs only at the surface; within the glass, the light travels in a straight line.

In Figure 34-17b, the beams of light in the photograph are represented with an *incident ray,* a *reflected ray,* and a *refracted ray* (and wavefronts). Each ray is oriented with respect to a line, called the *normal,* that is perpendicular to the surface at the point of reflection and refraction. In Fig. 34-17b, the **angle of incidence** is θ_1, the **angle of reflection** is θ'_1, and the **angle of refraction** is θ_2, all measured *relative to the normal* as shown. The plane containing the incident ray and the normal is the *plane of incidence,* which is in the plane of the page in Fig. 34-17b.

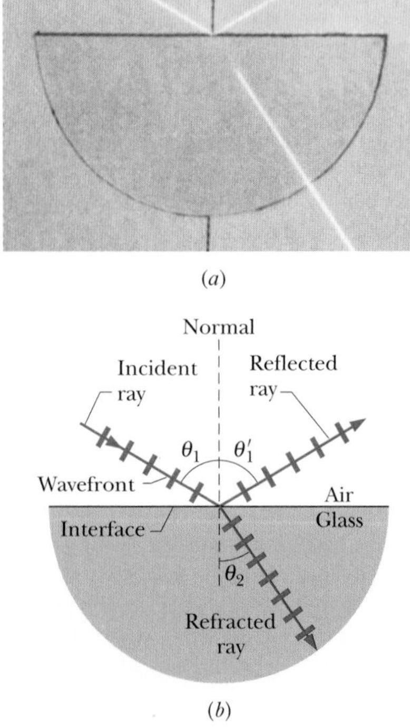

Fig. 34-17 (a) A black-and-white photograph showing the reflection and refraction of an incident beam of light by a horizontal plane glass surface. (A portion of the refracted beam within the glass was not well photographed.) At the bottom surface, which is curved, the beam is perpendicular to the surface; so the refraction there does not bend the beam. (b) A representation of (a) using rays. The angles of incidence (θ_1), of reflection (θ'_1), and of refraction (θ_2) are marked.

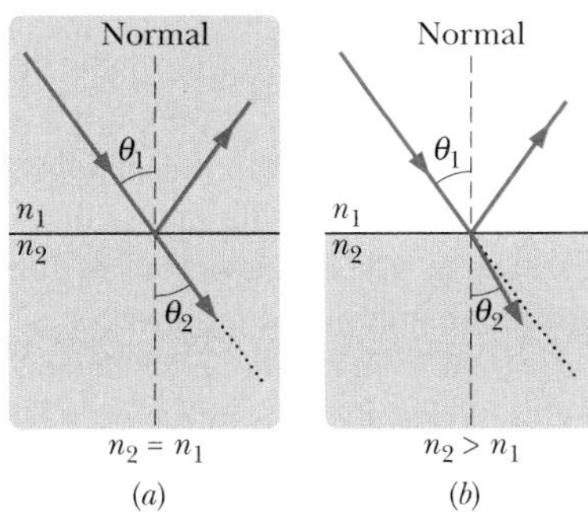

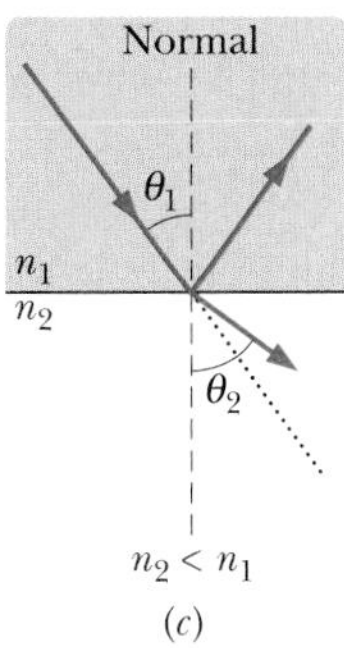

Fig. 34-18 Light refracting from a medium with an index of refraction n_1 and into a medium with an index of refraction n_2. (*a*) The beam does not bend when $n_2 = n_1$; the refracted light then travels in the *undeflected direction* (the dotted line), which is the same as the direction of the incident beam. The beam bends (*b*) toward the normal when $n_2 > n_1$ and (*c*) away from the normal when $n_2 < n_1$.

Experiment shows that reflection and refraction are governed by two laws:

Law of reflection: A reflected ray lies in the plane of incidence and has an angle of reflection equal to the angle of incidence. In Fig. 34-17*b*, this means that

$$\theta_1' = \theta_1 \qquad \text{(reflection).} \tag{34-43}$$

(We shall now usually drop the prime on the angle of reflection.)

Law of refraction: A refracted ray lies in the plane of incidence and has an angle of refraction θ_2 that is related to the angle of incidence θ_1 by

$$n_2 \sin \theta_2 = n_1 \sin \theta_1 \qquad \text{(refraction).} \tag{34-44}$$

Here each of the symbols n_1 and n_2 is a dimensionless constant, called the **index of refraction,** that is associated with a medium involved in the refraction. We derive this equation, called Snell's law, in Chapter 36. As we shall discuss there, the index of refraction of a medium is equal to c/v, where v is the speed of light in that medium and c is its speed in vacuum.

Table 34-1 gives the indexes of refraction of vacuum and some common substances. For vacuum, n is defined to be exactly 1; for air, n is very close to 1.0 (an approximation we shall often make). Nothing has an index of refraction below 1.

We can rearrange Eq. 34-44 as

$$\sin \theta_2 = \frac{n_1}{n_2} \sin \theta_1 \tag{34-45}$$

to compare the angle of refraction θ_2 with the angle of incidence θ_1. We can then see that the relative value of θ_2 depends on the relative values of n_2 and n_1. In fact, we can have three basic results:

1. If n_2 is equal to n_1, then θ_2 is equal to θ_1. In this case, refraction does not bend the light beam, which continues in the *undeflected direction,* as in Fig. 34-18*a*.
2. If n_2 is greater than n_1, then θ_2 is less than θ_1. In this case, refraction bends the light beam away from the undeflected direction and toward the normal, as in Fig. 34-18*b*.
3. If n_2 is less than n_1, then θ_2 is greater than θ_1. In this case, refraction bends the light beam away from the undeflected direction and away from the normal, as in Fig. 34-18*c*.

Refraction *cannot* bend a beam so much that the refracted ray is on the same side of the normal as the incident ray.

Table 34-1 Some Indexes of Refraction[a]

Medium	Index	Medium	Index
Vacuum	Exactly 1	Typical crown glass	1.52
Air (STP)[b]	1.00029	Sodium chloride	1.54
Water (20°C)	1.33	Polystyrene	1.55
Acetone	1.36	Carbon disulfide	1.63
Ethyl alcohol	1.36	Heavy flint glass	1.65
Sugar solution (30%)	1.38	Sapphire	1.77
Fused quartz	1.46	Heaviest flint glass	1.89
Sugar solution (80%)	1.49	Diamond	2.42

[a]For a wavelength of 589 nm (yellow sodium light).

[b]STP means "standard temperature (0°C) and pressure (1 atm)."

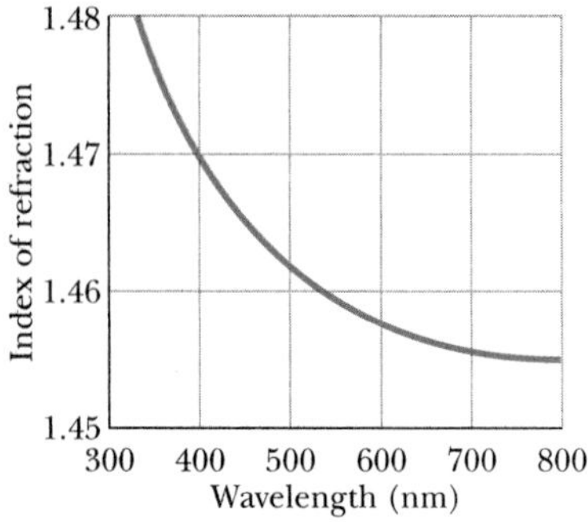

Fig. 34-19 The index of refraction as a function of wavelength for fused quartz. The graph indicates that a beam of short-wavelength light, for which the index of refraction is higher, is bent more upon entering or leaving quartz than a beam of long-wavelength light.

Chromatic Dispersion

The index of refraction n encountered by light in any medium except vacuum depends on the wavelength of the light. The dependence of n on wavelength implies that when a light beam consists of rays of different wavelengths, the rays will be refracted at different angles by a surface; that is, the light will be spread out by the refraction. This spreading of light is called **chromatic dispersion,** in which "chromatic" refers to the colors associated with the individual wavelengths and "dispersion" refers to the spreading of the light according to its wavelengths or colors. The refractions of Figs. 34-17 and 34-18 do not show chromatic dispersion because the beams are *monochromatic* (of a single wavelength or color).

Generally, the index of refraction of a given medium is *greater* for a shorter wavelength (corresponding to, say, blue light) than for a longer wavelength (say, red light). As an example, Fig. 34-19 shows how the index of refraction of fused quartz depends on the wavelength of light. Such dependence means that when a beam with waves of both blue and red light is refracted through a surface, such as from air into quartz or vice versa, the blue *component* (the ray corresponding to the wave of blue light) bends more than the red component.

A beam of *white light* consists of components of all (or nearly all) the colors in the visible spectrum with approximately uniform intensities. When you see such a beam, you perceive white rather than the individual colors. In Fig. 34-20*a*, a beam of white light in air is incident on a glass surface. (Because the pages of this book are white, a beam of white light is represented with a gray ray here. Also, a beam of monochromatic light is generally represented with a red ray.) Of the refracted light in Fig. 34-20*a*, only the red and blue components are shown. Because the blue component is bent more than the red component, the angle of refraction θ_{2b} for the blue component is *smaller* than the angle of refraction θ_{2r} for the red component. (Remember, angles are measured relative to the normal.) In Fig. 34-20*b*, a ray of white light in glass is incident on a glass–air interface. Again, the blue component is bent more than the red component, but now θ_{2b} is greater than θ_{2r}.

To increase the color separation, we can use a solid glass prism with a triangular cross section, as in Fig. 34-21*a*. The dispersion at the first surface (on the left in Fig. 34-21*a*, *b*) is then enhanced by that at the second surface.

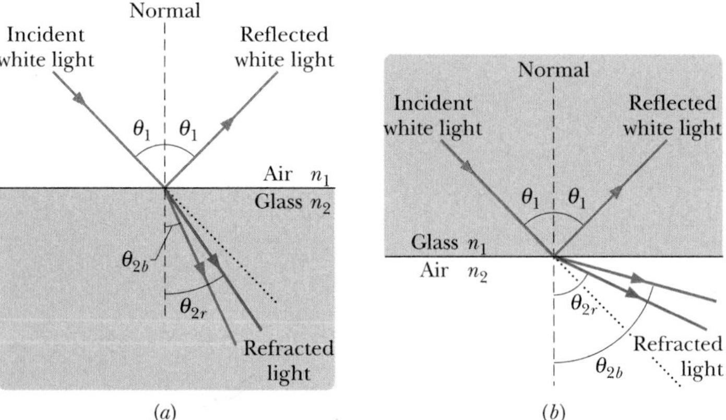

Fig. 34-20 Chromatic dispersion of white light. The blue component is bent more than the red component. (*a*) Passing from air to glass, the blue component ends up with the smaller angle of refraction. (*b*) Passing from glass to air, the blue component ends up with the greater angle of refraction.

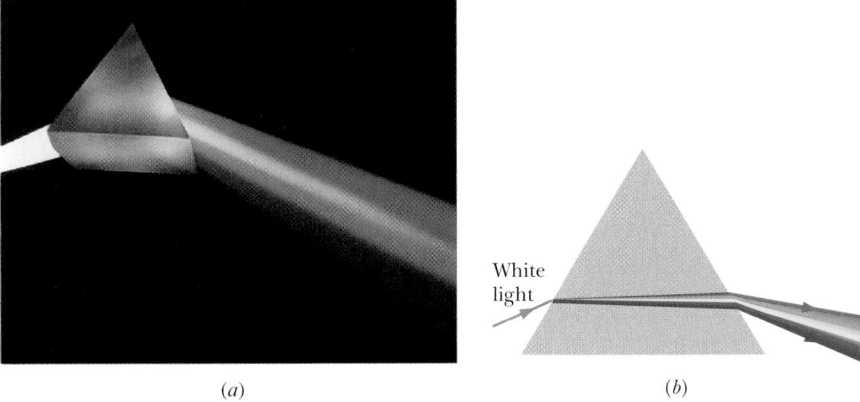

Fig. 34-21 (*a*) A triangular prism separating white light into its component colors. (*b*) Chromatic dispersion occurs at the first surface and is increased at the second surface.

The most charming example of chromatic dispersion is a rainbow. When white sunlight is intercepted by a falling raindrop, some of the light refracts into the drop, reflects from the drop's inner surface, and then refracts out of the drop (Fig. 34-22). As with a prism, the first refraction separates the sunlight into its component colors, and the second refraction increases the separation.

The rainbow you see is formed by light refracted by many such drops; the red comes from drops angled slightly higher in the sky, the blue from drops angled slightly lower, and the intermediate colors from drops at intermediate angles. All the drops sending separated colors to you are angled at about 42° from a point that is directly opposite the Sun in your view. If the rainfall is extensive and brightly lit, you see a circular arc of color, with red on top and blue on bottom. Your rainbow is a personal one, because another observer intercepts light from other drops.

Fig. 34-22 (*a*) A rainbow is always a circular arc that is centered on the direction you would look if you looked directly away from the Sun. Under normal conditions, you are lucky if you see a long arc, but if you are looking downward from an elevated position, you might actually see a full circle. (*b*) The separation of colors when sunlight refracts into and out of falling raindrops leads to a rainbow. The figure shows the situation for the Sun on the horizon (the rays of sunlight are then horizontal). The paths of red and blue rays from two drops are indicated. Many other drops also contribute red and blue rays, as well as the intermediate colors of the visible spectrum.

✔CHECKPOINT 5: Which of the three drawings here (if any) show physically possible refraction?

$n = 1.6$, $n = 1.4$ — (a)

$n = 1.8$, $n = 1.6$ — (b)

$n = 1.5$, $n = 1.6$ — (c)

Sample Problem 34-4

(a) In Fig. 34-23*a*, a beam of monochromatic light reflects and refracts at point *A* on the interface between material 1 with index of refraction $n_1 = 1.33$ and material 2 with index of refraction $n_2 = 1.77$. The incident beam makes an angle of 50° with the interface. What is the angle of reflection at point *A*? What is the angle of refraction there?

SOLUTION: The **Key Idea** of any reflection is that the angle of reflection is equal to the angle of incidence. Further, both angles are measured between the corresponding light ray and a normal to the interface at the point of reflection. In Fig. 34-23*a*, the normal at point *A* is drawn as a dashed line through the point. Note that the angle of incidence θ_1 is not the given 50° but rather is $90° - 50° = 40°$. Thus, the angle of reflection is

$$\theta_1' = \theta_1 = 40°. \quad \text{(Answer)}$$

The light that passes from material 1 into material 2 undergoes refraction at point *A* on the interface between the two materials. The **Key Idea** of any refraction is that we can relate the angle of incidence, the angle of refraction, and the indexes of refraction of the two materials via Eq. 34-44:

$$n_2 \sin \theta_2 = n_1 \sin \theta_1. \quad (34\text{-}46)$$

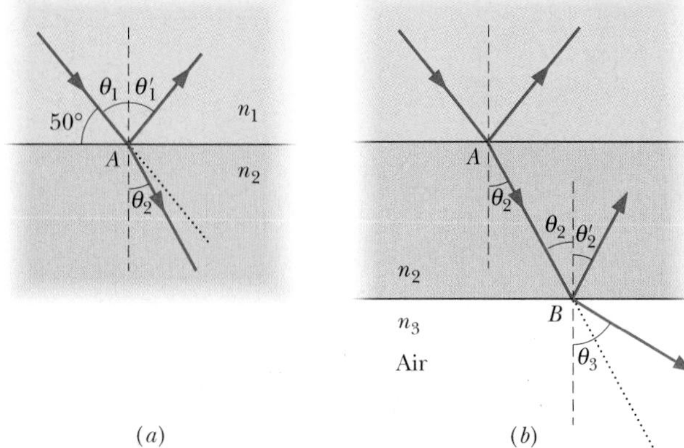

Fig. 34-23 Sample Problem 34-4. (*a*) Light reflects and refracts at point *A* on the interface between materials 1 and 2. (*b*) The light that passes through material 2 reflects and refracts at point *B* on the interface between materials 2 and 3 (air).

Again we measure angles between light rays and a normal, here at the point of refraction. Thus, in Fig. 34-23*a*, the angle of refraction is the angle marked θ_2. Solving Eq. 34-46 for θ_2 gives us

$$\theta_2 = \sin^{-1}\left(\frac{n_1}{n_2}\sin\theta_1\right) = \sin^{-1}\left(\frac{1.33}{1.77}\sin 40°\right)$$
$$= 28.88° \approx 29°. \quad \text{(Answer)}$$

This result means that the beam swings toward the normal (it was at 40° to the normal and is now at 29°). The reason is that when the light travels across the interface, it moves into a material with a greater index of refraction.

(b) The light that enters material 2 at point *A* then reaches point *B* on the interface between material 2 and material 3, which is air, as shown in Fig. 34-23*b*. The interface through *B* is parallel to that through *A*. At *B*, some of the light reflects and the rest enters the air. What is the angle of reflection? What is the angle of refraction into the air?

SOLUTION: We first need to relate one of the angles at point *B* with a known angle at point *A*. Because the interface through point *B* is parallel to that through point *A*, the incident angle at *B* must be equal to the angle of refraction θ_2, as shown in Fig. 34-23*b*. Then for reflection, we use the same **Key Idea** as in (a): the law of reflection. Thus, the angle of reflection at *B* is

$$\theta_2' = \theta_2 = 28.88° \approx 29°. \quad \text{(Answer)}$$

Next, the light that passes from material 2 into the air undergoes refraction at point *B*, with refraction angle θ_3. Thus, the **Key Idea** here is again to apply the law of refraction, but this time by writing Eq. 34-46 as

$$n_3 \sin \theta_3 = n_2 \sin \theta_2.$$

Solving for θ_3 then leads to

$$\theta_3 = \sin^{-1}\left(\frac{n_2}{n_3}\sin\theta_2\right) = \sin^{-1}\left(\frac{1.77}{1.00}\sin 28.88°\right)$$
$$= 58.75° \approx 59°. \quad \text{(Answer)}$$

This result means that the beam swings away from the normal (it was at 29° to the normal and is now at 59°). The reason is that when the light travels across the interface, it moves into a material (air) with a lower index of refraction.

34-8 Total Internal Reflection

Figure 34-24 shows rays of monochromatic light from a point source S in glass incident on the interface between the glass and air. For ray a, which is perpendicular to the interface, part of the light reflects at the interface and the rest travels through it with no change in direction.

For rays b through e, which have progressively larger angles of incidence at the interface, there are also both reflection and refraction at the interface. As the angle of incidence increases, the angle of refraction increases; for ray e it is 90°, which means that the refracted ray points directly along the interface. The angle of incidence giving this situation is called the **critical angle** θ_c. For angles of incidence larger than θ_c, such as for rays f and g, there is no refracted ray and *all* the light is reflected; this effect is called **total internal reflection.**

To find θ_c, we use Eq. 34-44; we arbitrarily associate subscript 1 with the glass and subscript 2 with the air, and then we substitute θ_c for θ_1 and 90° for θ_2, finding

$$n_1 \sin \theta_c = n_2 \sin 90°,$$

which gives us

$$\theta_c = \sin^{-1} \frac{n_2}{n_1} \quad \text{(critical angle).} \tag{34-47}$$

Because the sine of an angle cannot exceed unity, n_2 cannot exceed n_1 in this equation. This restriction tells us that total internal reflection cannot occur when the incident light is in the medium of lower index of refraction. If source S were in the air in Fig. 34-24, all its rays that are incident on the air–glass interface (including f and g) would be both reflected *and* refracted at the interface.

Total internal reflection has found many applications in medical technology. For example, a physician can search for an ulcer in the stomach of a patient by running two thin bundles of *optical fibers* (Fig. 34-25) down the patient's throat. Light introduced at the outer end of one bundle undergoes repeated total internal reflection within the fibers so that, even though the bundle provides a curved path, most of the light ends up exiting the other end and illuminating the interior of the stomach. Some of the light reflected from the interior then comes back up the second bundle in a similar way, to be detected and converted to an image on a monitor's screen for the physician to view.

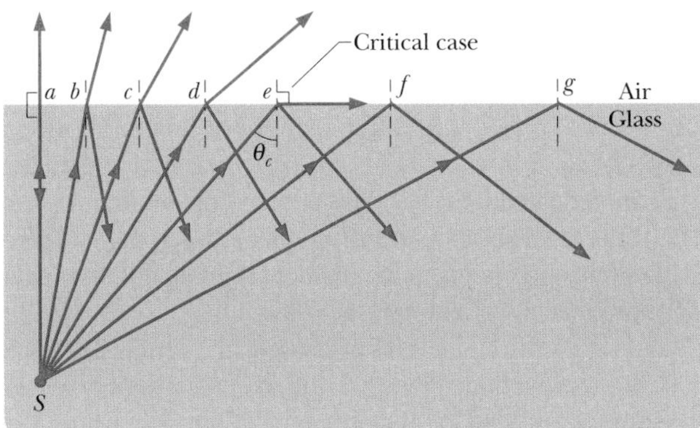

Fig. 34-24 Total internal reflection of light from a point source S in glass occurs for all angles of incidence greater than the critical angle θ_c. At the critical angle, the refracted ray points along the air–glass interface.

Fig. 34-25 Light sent into one end of an optical fiber like those shown here is transmitted to the opposite end with little loss of light through the sides of the fiber.

Sample Problem 34-5

Figure 34-26 shows a triangular prism of glass in air; an incident ray enters the glass perpendicular to one face and is totally reflected at the far glass–air interface as indicated. If θ_1 is 45°, what can you say about the index of refraction n of the glass?

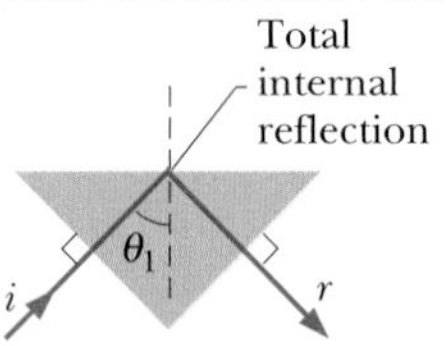

Fig. 34-26 Sample Problem 34-5. The incident ray i is totally internally reflected at the glass–air interface, becoming the reflected ray r.

SOLUTION: One Key Idea here is that because the light ray is totally reflected at the interface, the critical angle θ_c for that interface must be less than the incident angle of 45°. A second Key Idea is that we can relate the index of refraction n of the glass to θ_c with the law of refraction, which leads to Eq. 34-47. Substituting $n_2 = 1$ (for the air) and $n_1 = n$ (for the glass) into that equation yields

$$\theta_c = \sin^{-1}\frac{n_2}{n_1} = \sin^{-1}\frac{1}{n}.$$

Because θ_c must be less than the incident angle of 45°, we have

$$\sin^{-1}\frac{1}{n} < 45°,$$

which gives us

$$\frac{1}{n} < \sin 45°$$

or

$$n > \frac{1}{\sin 45°} = 1.4. \quad \text{(Answer)}$$

The index of refraction of the glass must be greater than 1.4; otherwise, total internal reflection would not occur for the incident ray shown.

CHECKPOINT 6: Suppose the prism in this sample problem has the index of refraction $n = 1.4$. Does the light still totally internally reflect if we keep the incident ray horizontal but rotate the prism (a) 10° clockwise and (b) 10° counterclockwise in Fig. 34-26?

34-9 Polarization by Reflection

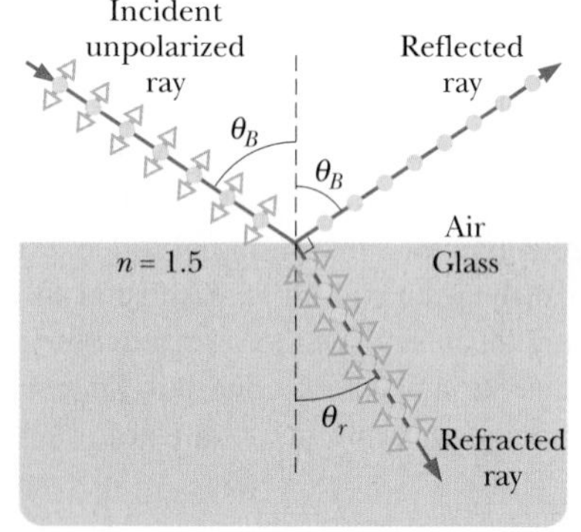

Fig. 34-27 A ray of unpolarized light in air is incident on a glass surface at the Brewster angle θ_B. The electric fields along that ray have been resolved into components perpendicular to the page (the plane of incidence, reflection, and refraction) and components parallel to the page. The reflected light consists only of components perpendicular to the page and is thus polarized in that direction. The refracted light consists of the original components parallel to the page and weaker components perpendicular to the page; this light is partially polarized.

You can vary the glare you see in sunlight that has been reflected from, say, water by looking through a polarizing sheet (such as a polarizing sunglass lens) and then rotating the sheet's polarizing axis around your line of sight. You can do so because reflected light is fully or partially polarized by the reflection from a surface.

Figure 34-27 shows a ray of unpolarized light incident on a glass surface. Let us resolve the electric field vectors of the light into two components. The *perpendicular components* are perpendicular to the plane of incidence and thus also to the page in Fig. 34-27; these components are represented with dots (as if we see the tips of the vectors). The *parallel components* are parallel to the plane of incidence and the page; they are represented with double-headed arrows. Because the light is unpolarized, these two components are of equal magnitude.

In general, the reflected light also has both components but with unequal magnitudes. This means that the reflected light is partially polarized—the electric fields oscillating along one direction have greater amplitudes than those oscillating along other directions. However, when the light is incident at a particular incident angle, called the *Brewster angle* θ_B, the reflected light has only perpendicular components, as shown in Fig. 34-27. The reflected light is then fully polarized perpendicular to the plane of incidence. The parallel components of the incident light do not disappear but (with the perpendicular components) refract into the glass.

Glass, water, and the other dielectric materials discussed in Section 26-7 can partially and fully polarize light by reflection. When you intercept sunlight reflected from such a surface, you see a bright spot (the glare) on the surface where the reflection takes place. If the surface is horizontal as in Fig. 34-27, the reflected light is partially or fully polarized horizontally. To eliminate such glare from horizontal surfaces, the lenses in polarizing sunglasses are mounted with their polarizing direction vertical.

Brewster's Law

For light incident at the Brewster angle θ_B, we find experimentally that the reflected and refracted rays are perpendicular to each other. Because the reflected ray is reflected at the angle θ_B in Fig. 34-27 and the refracted ray is at an angle θ_r, we have

$$\theta_B + \theta_r = 90°. \tag{34-48}$$

These two angles can also be related with Eq. 34-44. Arbitrarily assigning subscript 1 in Eq. 34-44 to the material through which the incident and reflected rays travel, we have, from that equation,

$$n_1 \sin \theta_B = n_2 \sin \theta_r.$$

Combining these equations leads to

$$n_1 \sin \theta_B = n_2 \sin(90° - \theta_B) = n_2 \cos \theta_B,$$

which gives us

$$\theta_B = \tan^{-1} \frac{n_2}{n_1} \quad \text{(Brewster angle).} \tag{34-49}$$

(Note carefully that the subscripts in Eq. 34-49 are *not* arbitrary because of our decision as to their meanings.) If the incident and reflected rays travel *in air,* we can approximate n_1 as unity and let n represent n_2 in order to write Eq. 34-49 as

$$\theta_B = \tan^{-1} n \quad \text{(Brewster's law).} \tag{34-50}$$

This simplified version of Eq. 34-49 is known as **Brewster's law.** Like θ_B, it is named after Sir David Brewster, who found both experimentally in 1812.

REVIEW & SUMMARY

Electromagnetic Waves An electromagnetic wave consists of oscillating electric and magnetic fields. The various possible frequencies of electromagnetic waves form a *spectrum,* a small part of which is visible light. An electromagnetic wave traveling along an x axis has an electric field $\vec{E}$ and a magnetic field $\vec{B}$ with magnitudes that depend on x and t:

$$E = E_m \sin(kx - \omega t)$$

and

$$B = B_m \sin(kx - \omega t), \tag{34-1, 34-2}$$

where E_m and B_m are the amplitudes of $\vec{E}$ and $\vec{B}$. The electric field induces the magnetic field and vice versa. The speed of any electromagnetic wave in vacuum is c, which can be written as

$$c = \frac{E}{B} = \frac{1}{\sqrt{\mu_0 \varepsilon_0}}, \tag{34-5, 34-3}$$

where E and B are the simultaneous magnitudes of the fields.

Energy Flow The rate per unit area at which energy is transported via an electromagnetic wave is given by the Poynting vector $\vec{S}$:

$$\vec{S} = \frac{1}{\mu_0} \vec{E} \times \vec{B}. \tag{34-19}$$

The direction of $\vec{S}$ (and thus of the wave's travel and the energy transport) is perpendicular to the directions of both $\vec{E}$ and $\vec{B}$. The time-averaged rate per unit area at which energy is transported is S_{avg}, which is called the *intensity* I of the wave:

$$I = \frac{1}{c\mu_0} E_{rms}^2, \tag{34-26}$$

in which $E_{rms} = E_m/\sqrt{2}$. A *point source* of electromagnetic waves emits the waves *isotropically*—that is, with equal intensity in all directions. The intensity of the waves at distance r from a point source of power P_s is

$$I = \frac{P_s}{4\pi r^2}. \tag{34-27}$$

Radiation Pressure When a surface intercepts electromagnetic radiation, a force and a pressure are exerted on the surface. If the radiation is totally absorbed by the surface, the force is

$$F = \frac{IA}{c} \quad \text{(total absorption),} \tag{34-32}$$

in which I is the intensity of the radiation and A is the area of the surface perpendicular to the path of the radiation. If the radiation is totally reflected back along its original path, the force is

$$F = \frac{2IA}{c} \quad \text{(total reflection back along path).} \qquad (34\text{-}33)$$

The radiation pressure p_r is the force per unit area:

$$p_r = \frac{I}{c} \quad \text{(total absorption)} \qquad (34\text{-}34)$$

and

$$p_r = \frac{2I}{c} \quad \text{(total reflection back along path).} \qquad (34\text{-}35)$$

Polarization Electromagnetic waves are **polarized** if their electric field vectors are all in a single plane, called the *plane of oscillation*. Light waves from common sources are not polarized; that is, they are **unpolarized** or **randomly polarized.**

Polarizing Sheets When a polarizing sheet is placed in the path of light, only electric field components of the light parallel to the sheet's **polarizing direction** are *transmitted* by the sheet; components perpendicular to the polarizing direction are absorbed. The light that emerges from a polarizing sheet is polarized parallel to the polarizing direction of the sheet.

If the original light is initially unpolarized, the transmitted intensity I is half the original intensity I_0:

$$I = \tfrac{1}{2}I_0. \qquad (34\text{-}40)$$

If the original light is initially polarized, the transmitted intensity depends on the angle θ between the polarization direction of the original light and the polarizing direction of the sheet:

$$I = I_0 \cos^2 \theta. \qquad (34\text{-}42)$$

Geometrical Optics *Geometrical optics* is an approximate treatment of light in which light waves are represented as straight-line rays.

Reflection and Refraction When a light ray encounters a boundary between two transparent media, a **reflected** ray and a **refracted** ray generally appear. Both rays remain in the plane of incidence. The **angle of reflection** is equal to the angle of incidence, and the **angle of refraction** is related to the angle of incidence by

$$n_2 \sin \theta_2 = n_1 \sin \theta_1 \quad \text{(refraction),} \qquad (34\text{-}44)$$

where n_1 and n_2 are the indexes of refraction of the media in which the incident and refracted rays travel.

Total Internal Reflection A wave encountering a boundary across which the index of refraction decreases will experience **total internal reflection** if the angle of incidence exceeds a **critical angle** θ_c, where

$$\theta_c = \sin^{-1}\frac{n_2}{n_1} \quad \text{(critical angle).} \qquad (34\text{-}47)$$

Polarization by Reflection A reflected wave will be fully **polarized,** with its $\vec{E}$ vectors perpendicular to the plane of incidence, if it strikes a boundary at the **Brewster angle** θ_B, where

$$\theta_B = \tan^{-1}\frac{n_2}{n_1} \quad \text{(Brewster angle).} \qquad (34\text{-}49)$$

QUESTIONS

1. If the magnetic field of a light wave oscillates parallel to a y axis and is given by $B_y = B_m \sin(kz - \omega t)$, (a) in what direction does the wave travel and (b) parallel to which axis does the associated electric field oscillate?

2. Figure 34-28 shows the electric and magnetic fields of an electromagnetic wave at a certain instant. Is the wave traveling into the page or out of it?

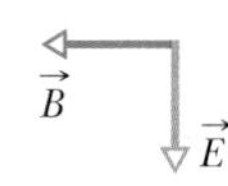

Fig. 34-28 Question 2.

3. (a) Figure 34-29 shows light reaching a polarizing sheet whose polarizing direction is parallel to a y axis. We shall rotate the sheet 40° clockwise about the light's indicated line of travel. During this rotation, does the fraction of the initial light intensity passed by the sheet increase, decrease, or remain the same if the light is (a) initially unpolarized, (b) initially polarized parallel to the x axis, and (c) initially polarized parallel to the y axis?

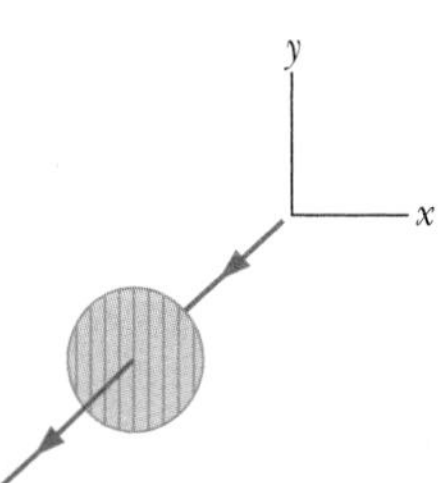

Fig. 34-29 Question 3.

4. In Fig. 34-16a, start with light that is initially polarized parallel to the x axis, and write the ratio of its final intensity I_3 to its initial intensity I_0 as $I_3/I_0 = A \cos^n \theta$. What are A, n, and θ if we rotate the polarizing direction of the first sheet (a) 60° counterclockwise and (b) 90° clockwise from what is shown?

5. Suppose we rotate the second sheet in Fig. 34-16a, starting with its polarization direction aligned with the y axis ($\theta = 0$) and ending with its polarization direction aligned with the x axis ($\theta = 90°$). Which of the three curves in Fig. 34-30 best shows the intensity of the light through the three-sheet system during this 90° rotation?

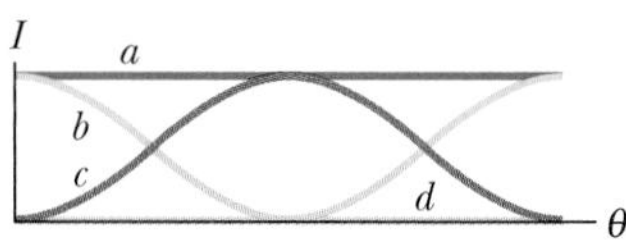

Fig. 34-30 Question 5.

6. Figure 34-31 shows the multiple reflections of a light ray along a glass corridor where the walls are either parallel or perpendicular to one another. If the angle of incidence at point a is 30°, what are the angles of reflection of the light ray at points b, c, d, e, and f?

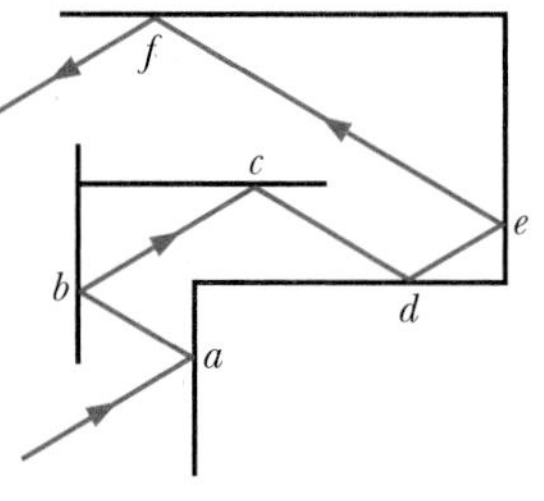

Fig. 34-31 Question 6.

7. Figure 34-32 shows rays of monochromatic light passing through three materials a, b, and c. Rank the materials according to their indexes of refraction, greatest first.

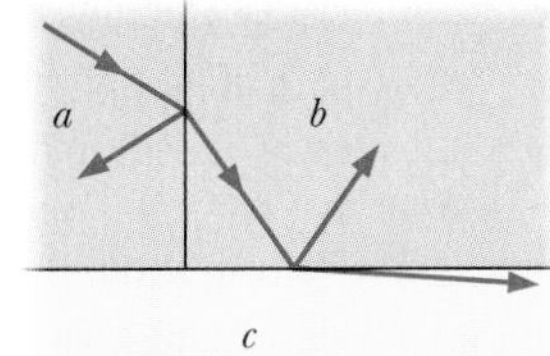

Fig. 34-32 Question 7.

8. In Fig. 34-33, light travels from material a, through three layers of other materials with surfaces parallel to one another, and then back into another layer of material a. The refractions (but not the associated reflections) at the surfaces are shown. Rank the materials according to their indexes of refraction, greatest first.

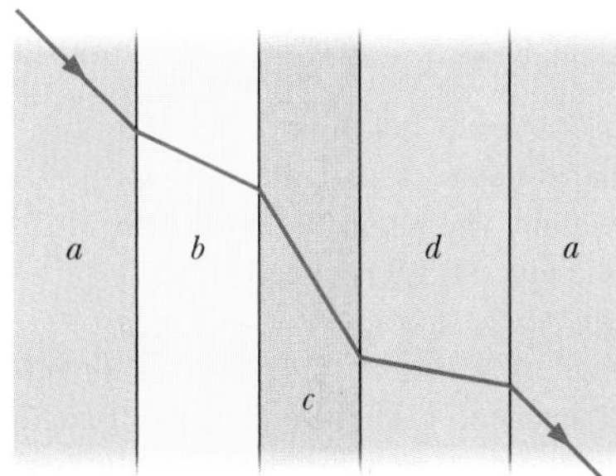

Fig. 34-33 Question 8.

9. Each part of Fig. 34-34 shows light that refracts through an interface between two materials. The incident ray (shown gray in the figure) consists of red and blue light. The approximate index of refraction for visible light is indicated for each material. Which of the three parts show physically possible refraction?

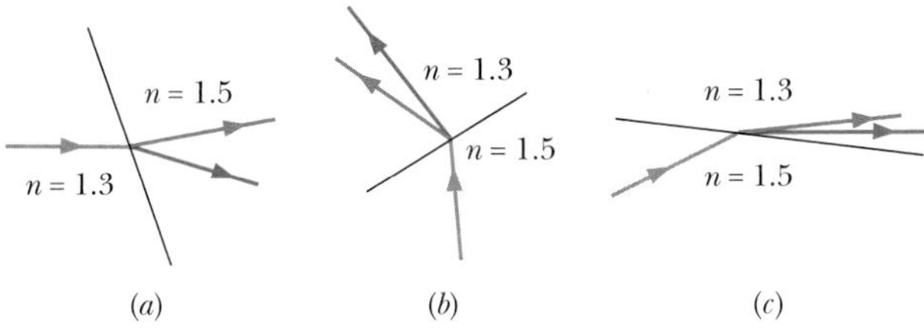

Fig. 34-34 Question 9.

10. (a) Figure 34-35a shows a ray of sunlight that just barely passes a vertical stick in a pool of water. Does that ray end in the general region of point a or point b? (b) Does the red or the blue component of the ray end up closer to the stick? (c) Figure 34-35b shows a flat object (such as a double-edged razor blade) floating in shallow water and illuminated vertically. The gravitational pull on the object and the cohesion of the water cause the water surface to curve as shown. In which general region (a, b, or c) is the edge of the object's shadow? (To the right of the edge of the shadow, many rays of sunlight are concentrated and produce an especially bright region, said to be a *caustic*.)

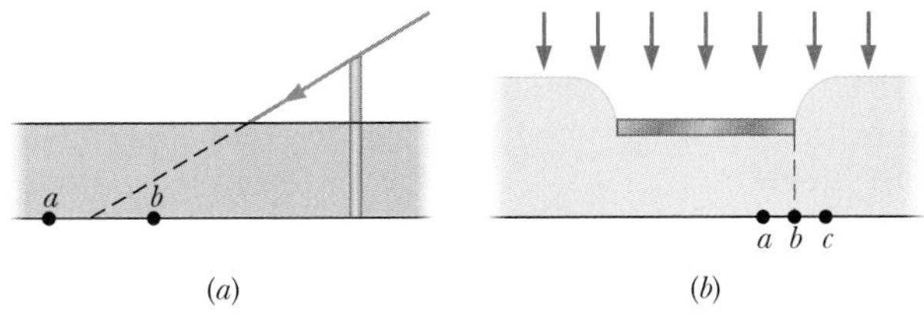

Fig. 34-35 Question 10.

11. Figure 34-22 shows some of the rays of sunlight responsible for the *primary rainbow* (which involves one reflection inside each water drop). A fainter, less frequent *secondary rainbow* (involving two reflections inside each water drop) can appear above a primary rainbow, formed by rays entering and exiting water drops as shown in Fig. 34-36 (without color indicated). Which ray, a or b, corresponds to red light?

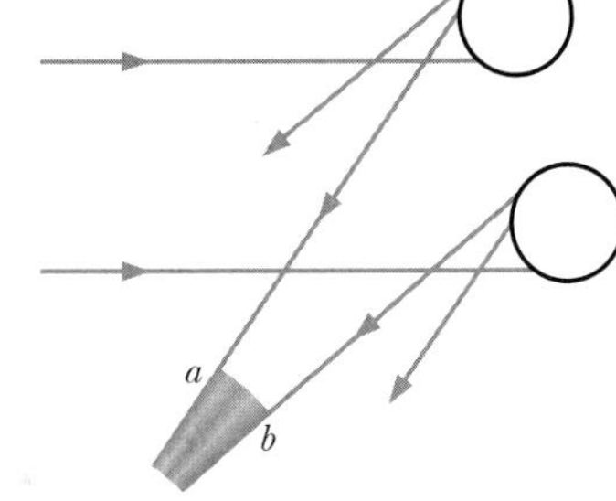

Fig. 34-36 Question 11.

12. Figure 34-37 shows four long horizontal layers of different materials, with air above and below them. The index of refraction of each material is given. Rays of light are sent into the left ends of each layer as shown. In which layer (give the index of refraction) is there the possibility of totally trapping the light in that layer so that, after many reflections, all the light reaches the right end of the layer?

Air
1.3
1.5
1.4
1.3
Air

Fig. 34-37 Question 12.

EXERCISES & PROBLEMS

ssm Solution is in the Student Solutions Manual.
www Solution is available on the World Wide Web at: http://www.wiley.com/college/hrw
ilw Solution is available on the Interactive LearningWare.

SEC. 34-1 Maxwell's Rainbow

1E. (a) How long does it take a radio signal to travel 150 km from a transmitter to a receiving antenna? (b) We see a full Moon by reflected sunlight. How much earlier did the light that enters our eye leave the Sun? The Earth–Moon and Earth–Sun distances are 3.8×10^5 km and 1.5×10^8 km. (c) What is the round-trip travel time for light between Earth and a spaceship orbiting Saturn, 1.3×10^9 km distant? (d) The Crab nebula, which is about 6500 light-years (ly) distant, is thought to be the result of a supernova explosion recorded by Chinese astronomers in A.D. 1054. In approximately what year did the explosion actually occur? ssm

2E. Project Seafarer was an ambitious program to construct an enormous antenna, buried underground on a site about 10 000 km^2 in area. Its purpose was to transmit signals to submarines while they were deeply submerged. If the effective wavelength were 1.0×10^4 Earth radii, what would be (a) the frequency and (b) the period of the radiations emitted? Ordinarily, electromagnetic radiations do not penetrate very far into conductors such as seawater.

3E. (a) At what wavelengths does the eye of a standard observer have half its maximum sensitivity? (b) What are the wavelength, the frequency, and the period of the light for which the eye is the most sensitive?

4E. A certain helium–neon laser emits red light in a narrow band of wavelengths centered at 632.8 nm and with a "wavelength width" (such as on the scale of Fig. 34-1) of 0.0100 nm. What is the corresponding "frequency width" for the emission?

5P. One method for measuring the speed of light, based on observations by Roemer in 1676, consisted of observing the apparent times of revolution of one of the moons of Jupiter. The true period of revolution is 42.5 h. (a) Taking into account the finite speed of light, how would you expect the apparent time for one revolution to change as Earth moves in its orbit from point x to point y in Fig. 34-38? (b) What observations would be needed to compute the speed of light? Neglect the motion of Jupiter in its orbit. Figure 34-38 is not drawn to scale. ssm

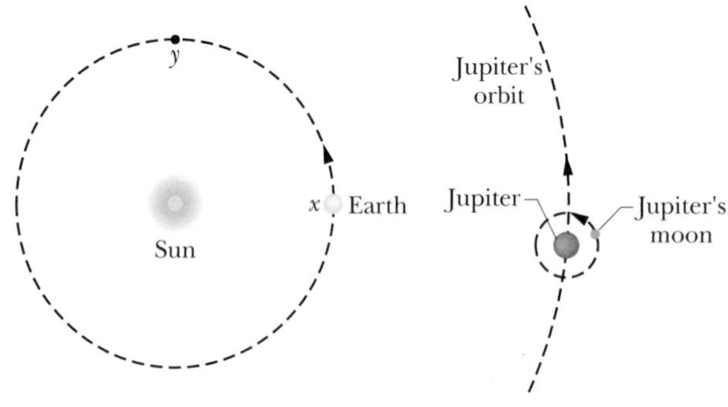

Fig. 34-38 Problem 5.

SEC. 34-2 The Traveling Electromagnetic Wave, Qualitatively

6E. What is the wavelength of the electromagnetic wave emitted by the oscillator–antenna system of Fig. 34-3 if $L = 0.253\ \mu$H and $C = 25.0$ pF?

7E. What inductance must be connected to a 17 pF capacitor in an oscillator capable of generating 550 nm (i.e., visible) electromagnetic waves? Comment on your answer. ssm

SEC. 34-3 The Traveling Electromagnetic Wave, Quantitatively

8E. A plane electromagnetic wave has a maximum electric field of 3.20×10^{-4} V/m. Find the maximum magnetic field.

9E. The electric field of a certain plane electromagnetic wave is given by $E_x = 0$; $E_y = 0$; $E_z = 2.0 \cos[\pi \times 10^{15}(t - x/c)]$, with $c = 3.0 \times 10^8$ m/s and all quantities in SI units. The wave is propagating in the positive x direction. Write expressions for the components of the magnetic field of the wave. ilw

SEC. 34-4 Energy Transport and the Poynting Vector

10E. Show, by finding the direction of the Poynting vector $\vec{S}$, that the directions of the electric and magnetic fields at all points in Figs. 34-4 to 34-7 are consistent at all times with the assumed directions of propagation.

11E. Some neodymium–glass lasers can provide 100 TW of power in 1.0 ns pulses at a wavelength of 0.26 μm. How much energy is contained in a single pulse? ssm ilw

12E. Our closest stellar neighbor, Proxima Centauri, is 4.3 ly away. It has been suggested that TV programs from our planet have reached this star and may have been viewed by the hypothetical inhabitants of a hypothetical planet orbiting it. Suppose a television station on Earth has a power of 1.0 MW. What is the intensity of its signal at Proxima Centauri?

13E. The radiation emitted by a laser spreads out in the form of a narrow cone with circular cross section. The angle θ of the cone (see Fig. 34-39) is called the *full-angle beam divergence*. An argon laser, radiating at 514.5 nm, is aimed at the Moon in a ranging experiment. If the beam has a full-angle beam divergence of 0.880 μrad, what area on the Moon's surface is illuminated by the laser? ssm

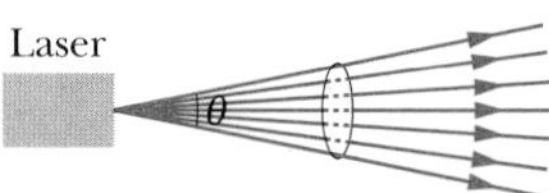

Fig. 34-39 Exercise 13.

14E. What is the intensity of a plane traveling electromagnetic wave if B_m is 1.0×10^{-4} T?

15E. In a plane radio wave the maximum value of the electric field component is 5.00 V/m. Calculate (a) the maximum value of the magnetic field component and (b) the wave intensity.

16P. Sunlight just outside Earth's atmosphere has an intensity of 1.40 kW/m^2. Calculate E_m and B_m for sunlight there, assuming it to be a plane wave.

17P. The maximum electric field at a distance of 10 m from an isotropic point light source is 2.0 V/m. What are (a) the maximum value of the magnetic field and (b) the average intensity of the light there? (c) What is the power of the source? ilw

18P. Figure 34-40: Frank D. Drake, an investigator in the SETI (Search for Extra-Terrestrial Intelligence) program, once said that

Fig. 34-40 Problem 18. Radio telescope at Arecibo.

the large radio telescope in Arecibo, Puerto Rico, "can detect a signal which lays down on the entire surface of the earth a power of only one picowatt." (a) What is the power that would be received by the Arecibo antenna for such a signal? The antenna diameter is 300 m. (b) What would be the power of a source at the center of our galaxy that could provide such a signal? The galactic center is 2.2×10^4 ly away. Take the source as radiating uniformly in all directions.

19P. An airplane flying at a distance of 10 km from a radio transmitter receives a signal of intensity 10 μW/m^2. Calculate (a) the amplitude of the electric field at the airplane due to this signal, (b) the amplitude of the magnetic field at the airplane, and (c) the total power of the transmitter, assuming the transmitter to radiate uniformly in all directions. ssm www

SEC. 34-5 Radiation Pressure

20E. A black, totally absorbing piece of cardboard of area $A = 2.0$ cm^2 intercepts light with an intensity of 10 W/m^2 from a camera strobe light. What radiation pressure is produced on the cardboard by the light?

21E. High-power lasers are used to compress a plasma (a gas of charged particles) by radiation pressure. A laser generating pulses of radiation of peak power 1.5×10^3 MW is focused onto 1.0 mm^2 of high-electron-density plasma. Find the pressure exerted on the plasma if the plasma reflects all the light beams directly back along their paths. ssm

22E. Radiation from the Sun reaching Earth (just outside the atmosphere) has an intensity of 1.4 kW/m^2. (a) Assuming that Earth (and its atmosphere) behaves like a flat disk perpendicular to the Sun's rays and that all the incident energy is absorbed, calculate the force on Earth due to radiation pressure. (b) Compare it with the force due to the Sun's gravitational attraction.

23E. What is the radiation pressure 1.5 m away from a 500 W lightbulb? Assume that the surface on which the pressure is exerted faces the bulb and is perfectly absorbing and that the bulb radiates uniformly in all directions. ssm ilw

24P. A helium–neon laser of the type often found in physics laboratories has a beam power of 5.00 mW at a wavelength of 633 nm. The beam is focused by a lens to a circular spot whose effective diameter may be taken to be equal to 2.00 wavelengths. Calculate (a) the intensity of the focused beam, (b) the radiation pressure exerted on a tiny perfectly absorbing sphere whose diameter is that of the focal spot, (c) the force exerted on this sphere, and (d) the magnitude of the acceleration imparted to it. Assume a sphere density of 5.00×10^3 kg/m^3.

25P. A plane electromagnetic wave, with wavelength 3.0 m, travels in vacuum in the positive x direction with its electric field $\vec{E}$, of amplitude 300 V/m, directed along the y axis. (a) What is the frequency f of the wave? (b) What are the direction and amplitude of the magnetic field associated with the wave? (c) What are the values of k and ω if $E = E_m \sin(kx - \omega t)$? (d) What is the time-averaged rate of energy flow in watts per square meter associated with this wave? (e) If the wave falls on a perfectly absorbing sheet of area 2.0 m^2, at what rate is momentum delivered to the sheet and what is the radiation pressure exerted on the sheet? ssm www

26P. In Fig. 34-41, a laser beam of power 4.60 W and diameter 2.60 mm is directed upward at one circular face (of diameter $d <$ 2.60 mm) of a perfectly reflecting cylinder, which is made to "hover" by the beam's radiation pressure. The cylinder's density is 1.20 g/cm^3. What is the cylinder's height H?

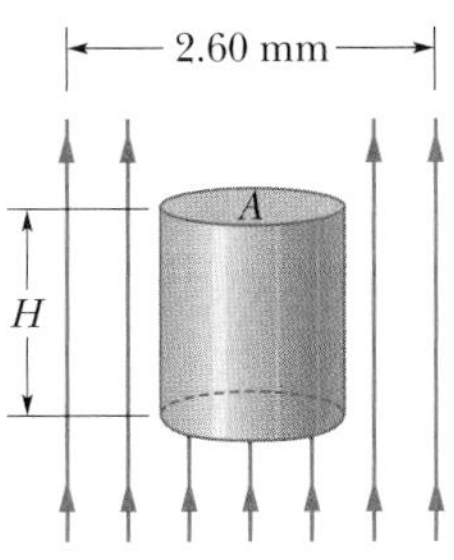

Fig. 34-41 Problem 26.

27P. Prove, for a plane electromagnetic wave that is normally incident on a plane surface, that the radiation pressure on the surface is equal to the energy density in the incident beam. (This relation between pressure and energy density holds no matter what fraction of the incident energy is reflected.) ssm

28P. Prove that the average pressure of a stream of bullets striking a plane surface perpendicularly is twice the kinetic energy density in the stream outside the surface. Assume that the bullets are completely absorbed by the surface. Contrast this with Problem 27.

29P. A small spaceship whose mass is 1.5×10^3 kg (including an astronaut) is drifting in outer space with negligible gravitational forces acting on it. If the astronaut turns on a 10 kW laser beam, what speed will the ship attain in 1.0 day because of the momentum carried away by the beam? ssm

30P. It has been proposed that a spaceship might be propelled in the solar system by radiation pressure, using a large sail made of foil. How large must the sail be if the radiation force is to be equal in magnitude in the Sun's gravitational attraction? Assume that the mass of the ship + sail is 1500 kg, that the sail is perfectly reflecting, and that the sail is oriented perpendicular to the Sun's rays. See Appendix C for needed data. (With a larger sail, the ship is continually driven away from the Sun.)

31P. A particle in the solar system is under the combined influence of the Sun's gravitational attraction and the radiation force due to the Sun's rays. Assume that the particle is a sphere of density 1.0×10^3 kg/m^3 and that all the incident light is absorbed. (a) Show that, if its radius is less than some critical radius R, the particle will be blown out of the solar system. (b) Calculate the critical radius. ssm

SEC. 34-6 Polarization

32E. The magnetic field equations for an electromagnetic wave in vacuum are $B_x = B \sin(ky + \omega t)$, $B_y = B_z = 0$. (a) What is the direction of propagation? (b) Write the electric field equations. (c) Is the wave polarized? If so, in what direction?

33E. A beam of unpolarized light of intensity 10 mW/m^2 is sent through a polarizing sheet as in Fig. 34-12. (a) Find the maximum value of the electric field of the transmitted beam. (b) What radiation pressure is exerted on the polarizing sheet? ssm

34E. In Fig. 34-42, initially unpolarized light is sent through three polarizing sheets whose polarizing directions make angles of $\theta_1 = \theta_2 = \theta_3 = 50°$ with the direction of the y axis. What percentage of the initial intensity is transmitted by the system of the three sheets? (*Hint:* Be careful with the angles.)

35E. In Fig. 34-42, initially unpolarized light is sent through three polarizing sheets whose polarizing directions make angles of $\theta_1 = 40°$, $\theta_2 = 20°$, and $\theta_3 = 40°$ with the direction of the y axis. What percentage of the light's initial intensity is transmitted by the system? (*Hint:* Be careful with the angles.) ssm

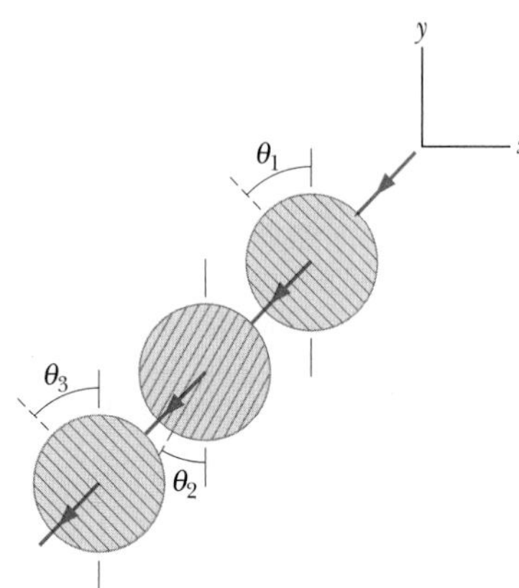

Fig. 34-42 Exercises 34 and 35.

36P. A beam of polarized light is sent through a system of two polarizing sheets. Relative to the polarization direction of that incident light, the polarizing directions of the sheets are at angles θ for the first sheet and 90° for the second sheet. If 0.10 of the incident intensity is transmitted by the two sheets, what is θ?

37P. A horizontal beam of vertically polarized light of intensity 43 W/m^2 is sent through two polarizing sheets. The polarizing direction of the first is at 70° to the vertical, and that of the second is horizontal. What is the intensity of the light transmitted by the pair of sheets? ssm ilw

38P. Suppose that in Problem 37 the initial beam is unpolarized. What then is the intensity of the transmitted light?

39P. A beam of partially polarized light can be considered to be a mixture of polarized and unpolarized light. Suppose we send such a beam through a polarizing filter and then rotate the filter through 360° while keeping it perpendicular to the beam. If the transmitted intensity varies by a factor of 5.0 during the rotation, what fraction of the intensity of the original beam is associated with the beam's polarized light? ssm www

40P. At a beach the light is generally partially polarized owing to reflections off sand and water. At a particular beach on a particular day near sundown, the horizontal component of the electric field vector is 2.3 times the vertical component. A standing sunbather puts on polarizing sunglasses; the glasses eliminate the horizontal field component. (a) What fraction of the light intensity received before the glasses were put on now reaches the sunbather's eyes? (b) The sunbather, still wearing the glasses, lies on his side. What fraction of the light intensity received before the glasses were put on now reaches his eyes?

41P. We want to rotate the direction of polarization of a beam of polarized light through 90° by sending the beam through one or more polarizing sheets. (a) What is the minimum number of sheets required? (b) What is the minimum number of sheets required if the transmitted intensity is to be more than 60% of the original intensity? ssm

SEC. 34-7 Reflection and Refraction

42E. Figure 34-43 shows light reflecting from two perpendicular reflecting surfaces A and B. Find the angle between the incoming ray i and the outgoing ray r'.

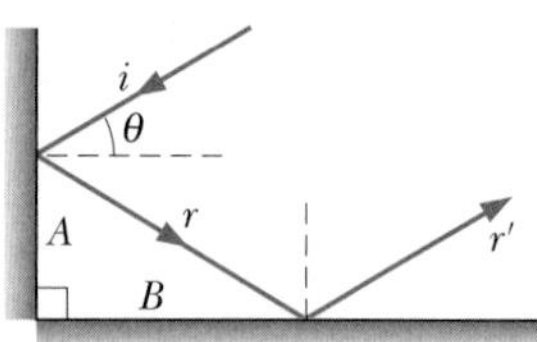

Fig. 34-43 Exercise 42.

43E. Light in vacuum is incident on the surface of a glass slab. In the vacuum the beam makes an angle of 32.0° with the normal to the surface, while in the glass it makes an angle of 21.0° with the normal. What is the index of refraction of the glass? ssm

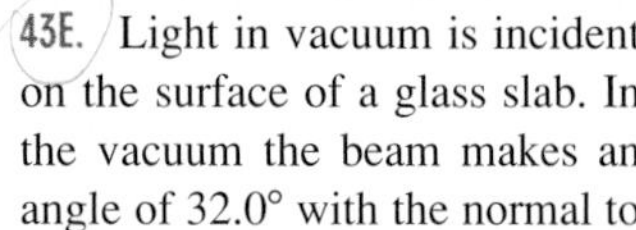

44E. In about A.D. 150, Claudius Ptolemy gave the following measured values for the angle of incidence θ_1 and the angle of refraction θ_2 for a light beam passing from air to water:

θ_1	θ_2	θ_1	θ_2
10°	8°	50°	35°
20°	15°30′	60°	40°30′
30°	22°30′	70°	45°30′
40°	29°	80°	50°

(a) Are these data consistent with the law of refraction? (b) If so, what index of refraction results? These data are interesting as perhaps the oldest recorded physical measurements.

45E. When the rectangular metal tank in Fig. 34-44 is filled to the top with an unknown liquid, an observer with eyes level with the top of the tank can just see the corner E; a ray that refracts toward the observer at the top surface of the liquid is shown. Find the index of refraction of the liquid. ssm

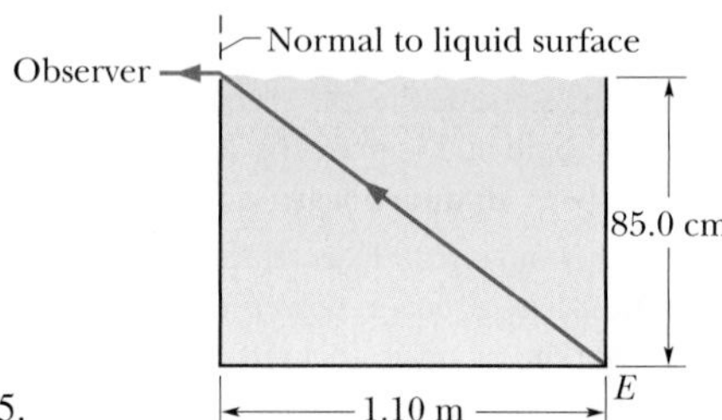

Fig. 34-44 Exercise 45.

46P. In Fig. 34-45, light is incident at angle $\theta_1 = 40.1°$ on a boundary between two transparent materials. Some of the light then travels down through the next three layers of transparent materials, while some of it reflects upward and then escapes into the air. What are the values of (a) θ_5 and (b) θ_4?

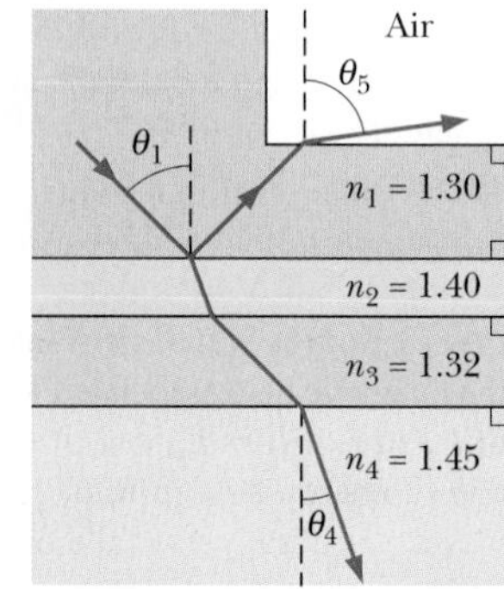

Fig. 34-45 Problem 46.

47P. In Fig. 34-46, a 2.00-m-long vertical pole extends from the bottom of a swimming pool to a point 50.0 cm above the water. Sunlight is incident at 55.0° above the horizon. What is the length of the shadow of the pole on the level bottom of the pool? ssm

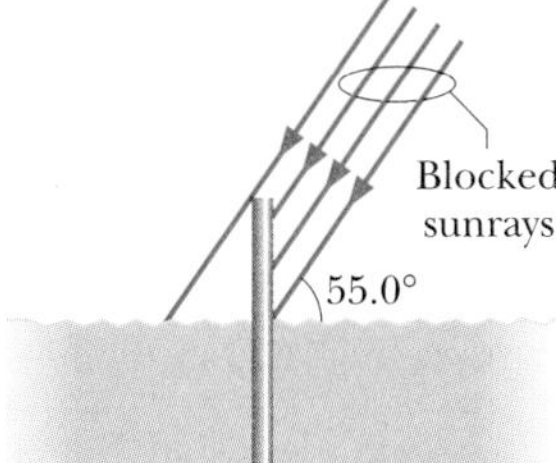

Fig. 34-46 Problem 47.

48P. A ray of white light makes an angle of incidence of 35° on one face of a prism of fused quartz; the prism's cross section is an equilateral triangle. Sketch the light as it passes through the prism, showing the paths traveled by rays representing (a) blue light, (b) yellow-green light, and (c) red light.

49P. Prove that a ray of light incident on the surface of a sheet of plate glass of thickness t emerges from the opposite face parallel to its initial direction but displaced sideways, as in Fig. 34-47. Show that, for small angles of incidence θ, this displacement is given by

$$x = t\theta \frac{n-1}{n},$$

where n is the index of refraction of the glass and θ is measured in radians. ssm www

Fig. 34-47 Problem 49.

50P. In Fig. 34-48, two perpendicular mirrors form the sides of a vessel filled with water. (a) A light ray is incident from above, normal to the water surface. Show that the emerging ray is parallel to the incident ray. Assume that there are two reflections at the mirror surfaces. (b) Repeat the analysis for the case of oblique incidence, with the incident ray in the plane of the figure.

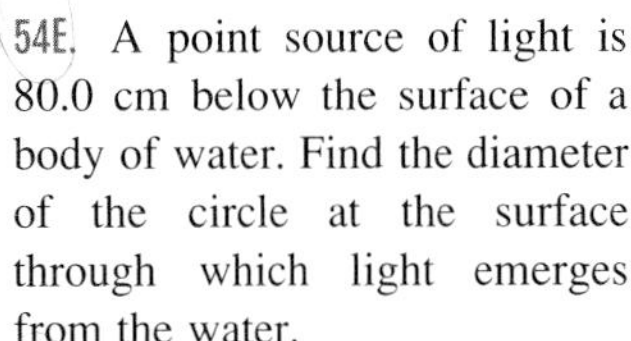

Fig. 34-48 Problem 50.

51P. In Fig. 34-49, a ray is incident on one face of a triangular glass prism in air. The angle of incidence θ is chosen so that the emerging ray also makes the same angle θ with the normal to the other face. Show that the index of refraction n of the glass prism is given by

$$n = \frac{\sin \frac{1}{2}(\psi + \phi)}{\sin \frac{1}{2}\phi},$$

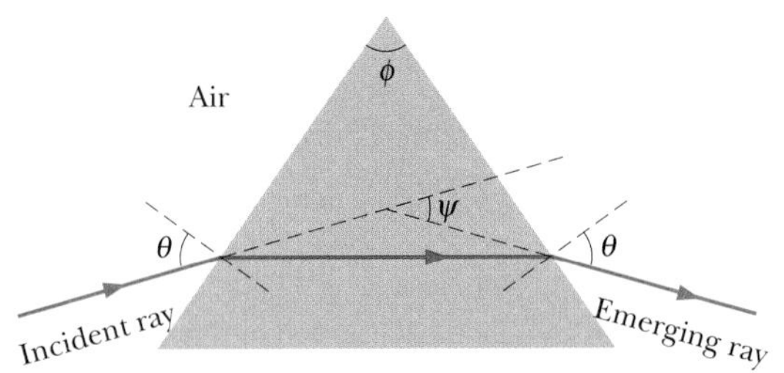

Fig. 34-49 Problems 51 and 58.

where ϕ is the vertex angle of the prism and ψ is the *deviation angle,* the total angle through which the beam is turned in passing through the prism. (Under these conditions the deviation angle ψ has the smallest possible value, which is called the *angle of minimum deviation.*) ilw

SEC. 34-8 Total Internal Reflection

52E. The index of refraction of benzene is 1.8. What is the critical angle for a light ray traveling in benzene toward a plane layer of air above the benzene?

53E. In Fig. 34-50, a light ray enters a glass slab at point A and then undergoes total internal reflection at point B. What minimum value for the index of refraction of the glass can be inferred from this information? ssm

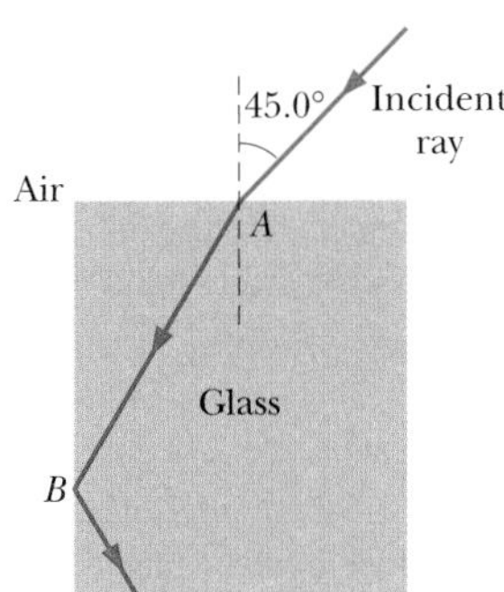

Fig. 34-50 Exercise 53.

54E. A point source of light is 80.0 cm below the surface of a body of water. Find the diameter of the circle at the surface through which light emerges from the water.

55E. In Fig. 34-51, a ray of light is perpendicular to the face ab of a glass prism ($n = 1.52$). Find the largest value for the angle ϕ so that the ray is totally reflected at face ac if the prism is immersed (a) in air and (b) in water. ssm ilw

Fig. 34-51 Exercise 55.

56P. A ray of white light travels through fused quartz that is surrounded by air. If all the color components of the light undergo total internal reflection at the surface, then the reflected light forms a reflected ray of white light. However, if the color component at one end of the visible range (either blue or red) partially refracts through the surface into the air, there is less of that component in the reflected light. Then the reflected light is not white but has the tint of the opposite end of the visible range. (If blue were partially lost to refraction, then the reflected beam would be reddish, and vice versa.) Is it possible for the reflected light to be (a) bluish or (b) reddish? (c) If so, what must be the angle of incidence of the original white light on the quartz surface? (See Fig. 34-19.)

57P. A solid glass cube, of edge length 10 mm and index of refraction 1.5, has a small spot at its center. (a) What parts of each cube face must be covered to prevent the spot from being seen, no matter what the direction of viewing? (Neglect light that reflects inside the cube and then refracts out into the air.) (b) What fraction of the cube surface must be so covered? ssm

58P. Suppose the prism of Fig. 34-49 has apex angle $\phi = 60.0°$ and index of refraction $n = 1.60$. (a) What is the smallest angle of incidence θ for which a ray can enter the left face of the prism and exit the right face? (b) What angle of incidence θ is required for the ray to exit the prism with an identical angle θ for its refraction, as it does in Fig. 34-49? (See Problem 51.)

59P. In Fig. 34-52, light enters a 90° triangular prism at point P with incident angle θ and then some of it refracts at point Q with an angle of refraction of 90°. (a) What is the index of refraction of the prism in terms of θ? (b) What, numerically, is the maximum value that the index of refraction can have? Explain what happens to the light at Q if the incident angle at Q is (c) increased slightly and (d) decreased slightly. ssm

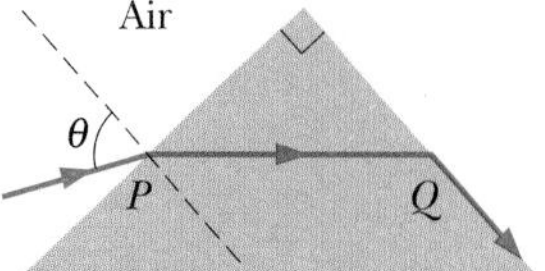

Fig. 34-52 Problem 59.

SEC. 34-9 Polarization by Reflection

60E. (a) At what angle of incidence will the light reflected from water be completely polarized? (b) Does this angle depend on the wavelength of the light?

61E. Light traveling in water of refractive index 1.33 is incident on a plate of glass with index of refraction 1.53. At what angle of incidence is the reflected light fully polarized? ssm

62E. Calculate the upper and lower limits of the Brewster angles for white light incident on fused quartz. Assume that the wavelength limits of the light are 400 and 700 nm.

Additional Problems

63. In Fig. 34-53, an albatross glides at a constant 15 m/s horizontally above level ground, moving in a vertical plane that contains the Sun. It glides toward a solid wall of height h = 2.0 m, which it will just barely clear. At that time of day, the angle of the Sun relative to the ground is θ = 30°. At what speed does the shadow of the albatross move (a) across the level ground and then (b) up the wall? Suppose that later a hawk happens to glide along the same path, also at 15 m/s. You see that when its shadow reaches the wall, the speed of the shadow noticeably increases. (c) Is the Sun now higher or lower in the sky than when the albatross flew by earlier? (d) If the speed of the hawk's shadow on the wall is 45 m/s, what is the angle θ of the Sun just then?

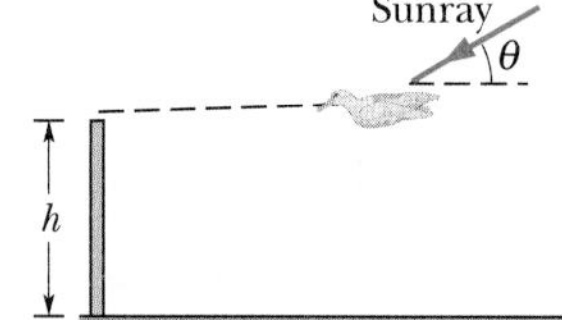

Fig. 34-53 Problem 63.

64. *Searching for graves.* In an archaeological investigation, unmarked graves and underground tombs can be located and mapped by *ground-penetrating radar* without disruption of the grave site. The radar unit emits a wave pulse directly downward into the ground; the pulse is then partially reflected upward by any underground interface. That is to say, a pulse is reflected upward by any horizontal boundary across which the speed of the pulse changes. The equipment detects the reflection and records the time interval between emission and detection. If this procedure is repeated at several locations along the ground, an archaeologist can determine the shape of underground structures.

A ground-penetrating radar unit was used at eight locations lying along a straight line on level ground and numbered west to east, as indicated in Fig. 34-54. Adjacent locations are separated by 2.0 m. An empty tomb formed with equally thick horizontal and vertical slabs of stone lies below these locations; the horizontal slabs form the ceiling and floor of the tomb, the vertical slabs form walls. The following table gives the time intervals Δt (in nanoseconds) recorded for the pulses at the eight locations. For example, at location 4, the original pulse emitted into the ground resulted in four reflected pulses, the first one returning 63.00 ns after the emission and the last one returning 86.54 ns after the emission.

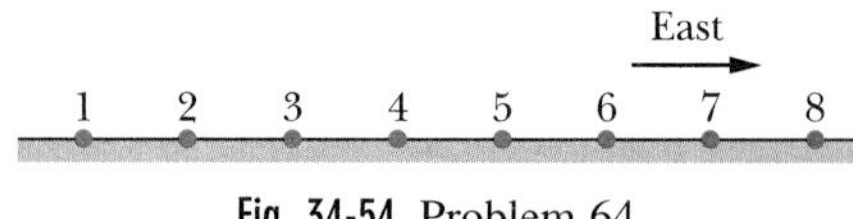

Fig. 34-54 Problem 64.

Assume that the pulses have a wave speed of 10.0 cm/ns in the soil above, below, and to the sides of the tomb; 10.6 cm/ns in the stone slabs; and 30 cm/ns in the air within the tomb. Find (a) the depth of the top surface of the tomb's ceiling, (b) the horizontal length of the tomb along the west–east line containing the eight locations, and (c) the vertical dimensions of the tomb's interior.

Location	1	2	3	4	5	6	7	8
Δt	None	63.00	63.00	63.00	63.00	63.00	63.00	None
		115.8	66.77	66.77	66.77	66.77	93.19	
			82.77	82.77	74.77	74.77		
			86.54	86.54	101.2	78.54		

65. In Fig. 34-55, a light ray in air is incident on a flat layer of material 2 that has an index of refraction n_2 = 1.5. Beneath material 2 is material 3 with an index of refraction n_3. The ray is incident on the air–material 2 interface at the Brewster angle for that interface. The ray of light refracted into material 3 happens to be incident on the material 2–material 3 interface at the Brewster angle for that interface. What is the value of n_3?

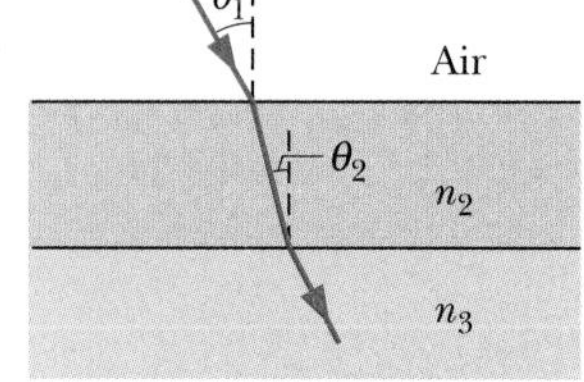

Fig. 34-55 Problem 65.

66. About how far apart must you hold your hands for them to be separated by 1.0 nano-light-second?

NEW PROBLEMS

N1. The intensity of direct solar radiation that is not absorbed by the atmosphere on a particular summer day is 100 W/m^2. How close would you have to stand to a 1.0 kW electric heater to feel the same intensity? Assume that the heater radiates uniformly in all directions.

N2. In Fig. 34N-1*a*, a light ray in water ($n = 1.33$) is incident on a boundary with a second material. Figure 34N-1*b* gives two plots of the resulting angle of refraction θ_2 versus the incident angle θ_1, for two choices of that second material. (a) Without calculation, determine whether the indexes of refraction of those second materials are greater than or less than that of water. What is the index of refraction of the second material corresponding to (b) plot 1 and (c) plot 2?

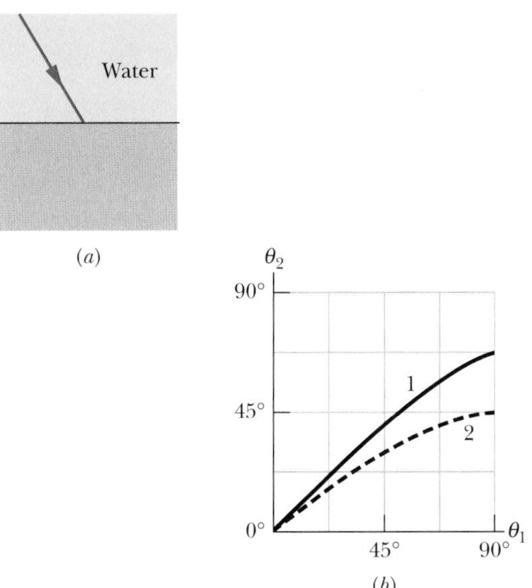

Fig. 34N-1 Problem N2.

N3. During a test, a NATO surveillance radar system, operating at 12 GHz at 180 kW of power, attempts to detect an incoming stealth aircraft at 90 km. Assume that the radar beam is emitted uniformly over a hemisphere. (a) What is the intensity of the beam when the beam reaches the aircraft's location? The aircraft reflects radar waves as though it has a cross-sectional area of only 0.22 m^2. (b) What is the power of the aircraft's reflection? Assume that the beam is reflected uniformly over a hemisphere. Back at the radar site, what are (c) the intensity, (d) the maximum value of the electric field vector, and (e) the rms value of the magnetic field of the reflected (and now detected) radar beam?

N4. In Fig. 34N-2*a*, a light ray in a solid material is incident on a boundary with water ($n = 1.33$). Figure 34N-2*b* gives two plots of the resulting angle of refraction θ_2 versus the incident angle θ_1, corresponding to two choices of the solid material. (a) Without calculation, determine whether the indexes of refraction of those solid materials are greater than or less than that of water. What is the index of refraction of the second material corresponding to (b) plot 1 and (c) plot 2?

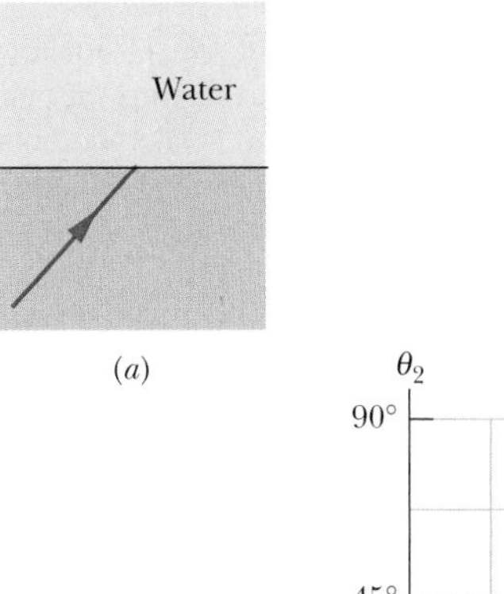

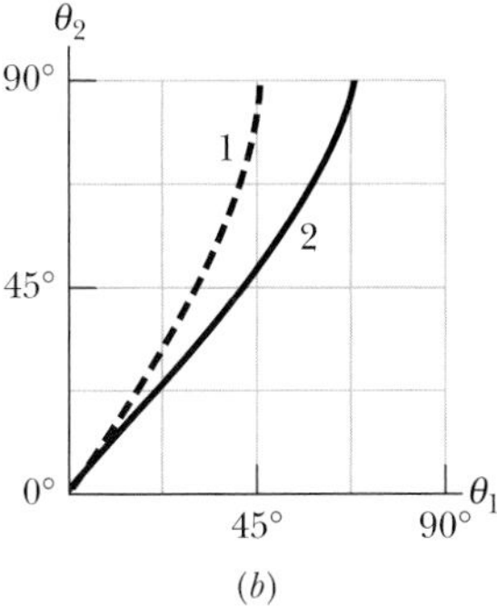

(*b*)

Fig. 34N-2 Problem N4.

N5. Show that in a plane traveling electromagnetic wave the intensity—that is, the average rate of energy transport per unit area—is given by

$$S_{avg} = \frac{E_m^2}{2\mu_0 c} = \frac{cB_m^2}{2\mu_0}.$$

N6. In Fig. 34N-3*a*, a beam of light in material 1 is incident on a boundary at an angle of 30°. The extent of refraction of the light into material 2 depends, in part, on the index of refraction n_2 of material 2. Figure 34N-3*b* gives the angle of refraction θ_2 versus n_2 for a range of possible n_2 values. (a) What is the index of refraction of material 1? (b) If the incident angle is changed to 60° and the second material has an index of refraction of 2.4, then what is angle θ_2?

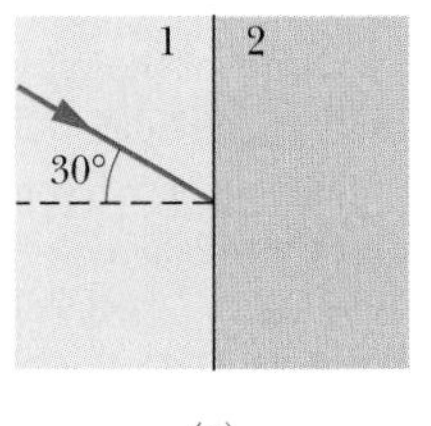

(*a*)

θ_2: 40°, 30°, 20°; n_2: 1.3, 1.5, 1.7, 1.9

(*b*)

Fig. 34N-3 Problem N6.

N7. Radiation of intensity I is normally incident on an object that absorbs a fraction *frac* of it and reflects the rest back along the original path. What is the radiation pressure on the object?

N8. In Fig. 34N-4*a*, a beam of light in material 1 is incident on a boundary at an angle of 40°, some of it travels through material 2, and then some of it emerges into material 3. The two boundaries between the three materials are parallel. The final direction of the beam depends, in part, on the index of refraction n_3 of the third material. Figure 34N-4*b* gives the angle of refraction θ_3 in that material versus n_3 for a range of possible n_3 values. (a) What is the index of refraction of material 1, or is the index impossible to calculate without more information? (b) What is the index of refraction of material 2, or is the index impossible to calculate without more information? (c) If the incident angle is changed to 70° and the index of refraction of material 3 is 2.4, what is angle θ_3?

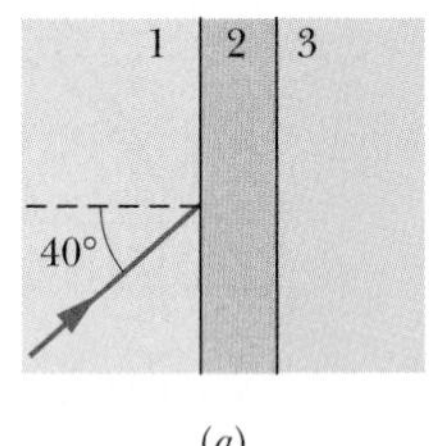

(*a*)

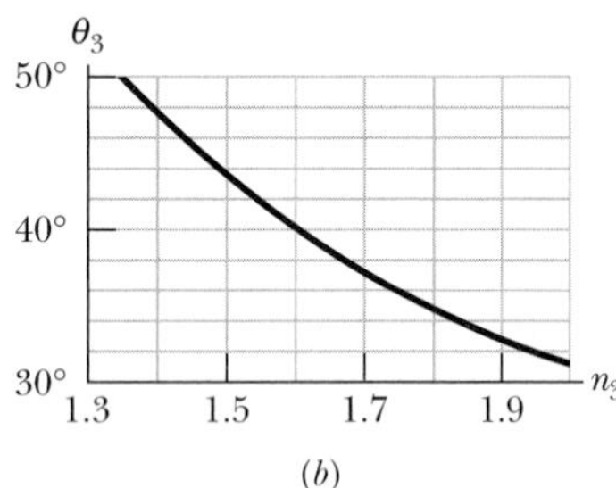

(*b*)

Fig. 34N-4 Problem N8.

N9. A laser beam of intensity I reflects from a flat, totally reflecting surface of area A whose normal makes an angle θ with the direction of the beam. Write an expression for the radiation pressure $p_r(\theta)$ exerted on the surface, in terms of the pressure $p_{r\perp}$ that would be exerted if the beam were perpendicular to the surface.

N10. The intensity I of light from an isotropic point light source is determined as a function of the distance r from the source. Figure 34N-5 gives intensity I versus the inverse square r^{-2} of that distance. What is the power of the source?

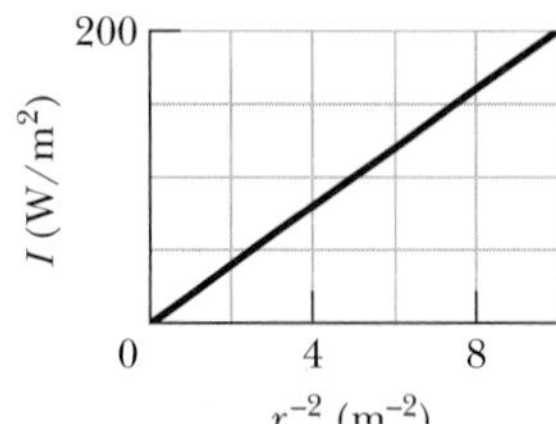

Fig. 34N-5 Problem N10.

N11. A beam of unpolarized light is sent through two polarizing sheets placed one on top of the other. What must be the angle between the polarizing directions of the sheets if the intensity of the transmitted light is to be one-third the incident intensity?

N12. In Fig. 34N-6*a*, unpolarized light is sent through a system of two polarizing sheets. The angles θ_1 and θ_2 of the polarizing axes of the sheets are measured counterclockwise from the positive direction of the y axis (they are not drawn to scale in the figure). Angle θ_1 is fixed but angle θ_2 can be varied. Figure 34N-6*b* gives the intensity of the light emerging from sheet 2 as a function of θ_2. (The scale of the intensity axis is not indicated.) What percentage of the light's initial intensity is transmitted by the two-sheet system when $\theta_2 = 90°$?

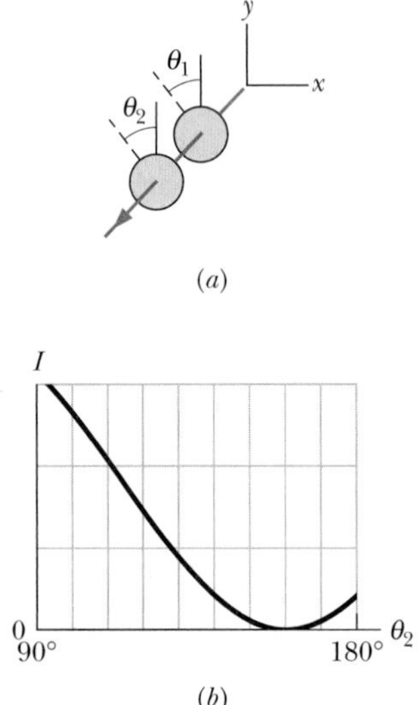

Fig. 34N-6 Problem N12.

N13. An unpolarized beam of light is sent through a stack of four polarizing sheets, oriented so that the angle between the polarizing directions of adjacent sheets is 30°. What fraction of the incident intensity is transmitted by the system?

N14. In Fig. 34N-7, a light ray in air is incident at angle θ_1 on a block of transparent plastic with an index of refraction of 1.56. The dimensions indicated are $H = 2.00$ cm and $W = 3.00$ cm. The light passes through the block to one of its sides and there undergoes reflection (inside the block) and possibly refraction (out into the air). This is the point of *first reflection*. The reflected light then passes through the block to another of its sides — a point of *second reflection*. If $\theta_1 = 40°$, on which sides are the points of (a) first reflection and (b) second reflection? What are the angles of refraction at (c) the point of first reflection and (d) the point of second reflection? If $\theta_1 = 70°$, on which sides are the points of (e) first reflection and (f) second reflection? What are the angles of refraction at (g) the point of first reflection and (h) the point of second reflection?

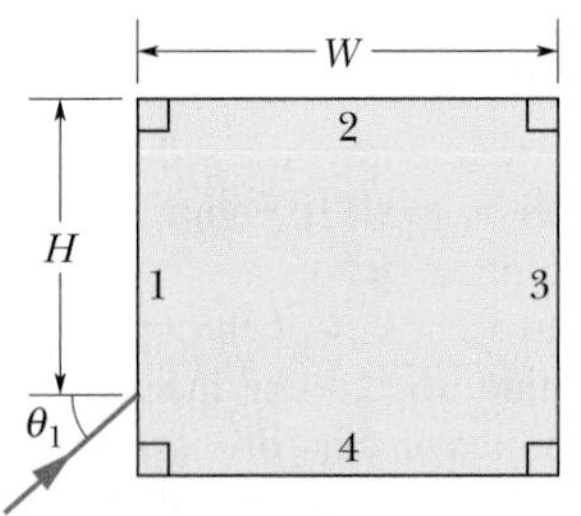

Fig. 34N-7 Problem N14.

N15. At Earth's surface, what intensity of light is needed to suspend a totally absorbing spherical particle against its own weight if the mass of the particle is 2.0×10^{-13} kg and its radius is 2.0 μm?

N16. In Fig. 34N-8*a*, unpolarized light is sent through a system of three polarizing sheets. The angles θ_1, θ_2, and θ_3 of the polarizing axes of the sheets are measured counterclockwise from the positive direction of the y axis (they are not drawn to scale). Angles θ_1 and θ_3 are fixed but angle θ_2 can be varied. Figure 34N-8*b* gives the intensity of the light emerging from sheet 3 as a function of θ_2. (The scale of the intensity axis is not indicated.) What percentage of the light's initial intensity is transmitted by the three-sheet system when $\theta_2 = 30°$?

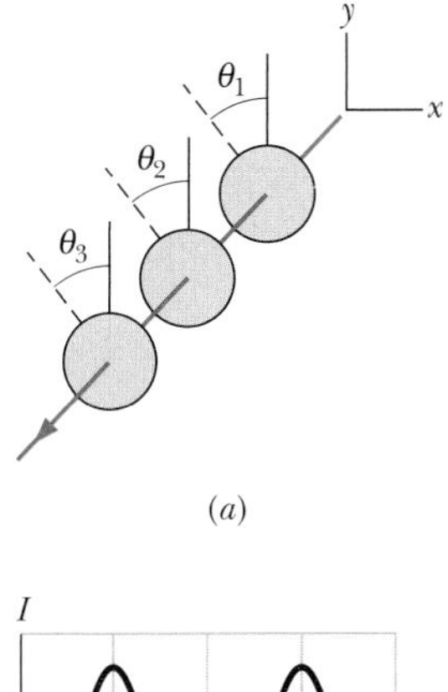

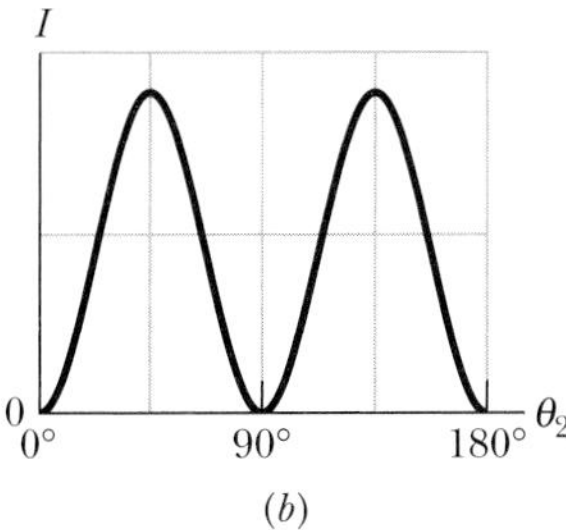

Fig. 34N-8 Problems N16 and N18.

N17. In Fig. 34N-9, light that is initially unpolarized is sent into a system of three polarizing sheets. What fraction of the initial light intensity emerges from the system?

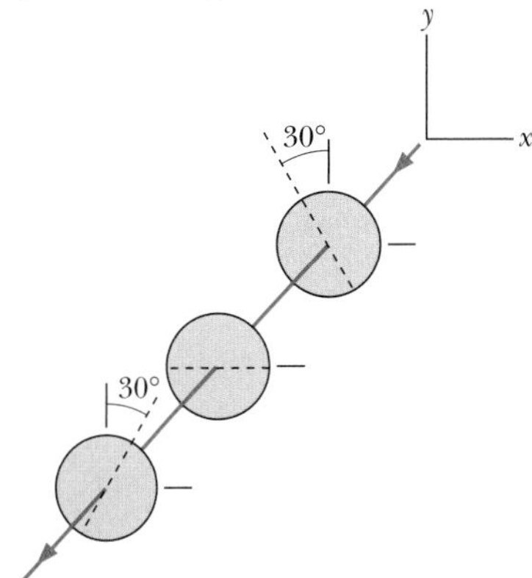

Fig. 34N-9 Problem N17.

N18. In Fig. 34N-8*a*, unpolarized light is sent through a system of three polarizing sheets. The angles θ_1, θ_2, and θ_3 of the polarizing axes of the sheets are measured counterclockwise from the positive direction of the y axis (they are not drawn to scale). Angles θ_1 and θ_3 are fixed but angle θ_2 can be varied. Figure 34N-10 gives the intensity of the light emerging from sheet 3 as a function of θ_2. (The scale of the intensity axis is not indicated.) What percentage of the light's initial intensity is transmitted by the three-sheet system when $\theta_2 = 90°$?

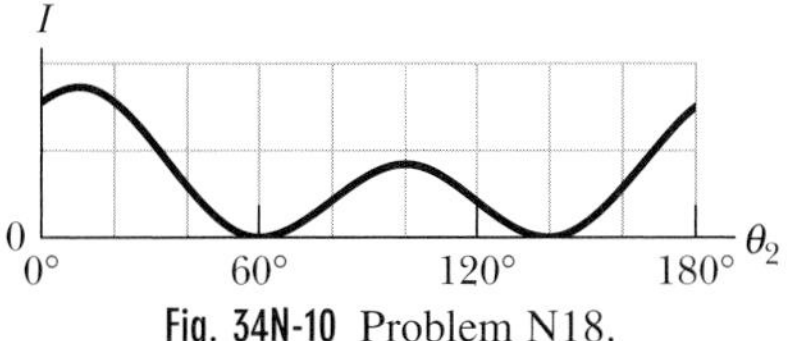

Fig. 34N-10 Problem N18.

N19. A system of three polarizing sheets is shown in Fig. 34N-11. When initially unpolarized light is sent into the system, the intensity of the transmitted light is 5.0×10^{-2} of the initial intensity. What is the value of θ?

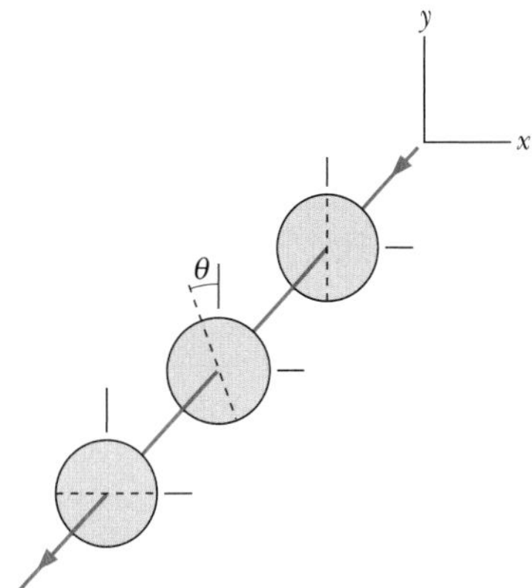

Fig. 34N-11 Problem N19.

N20. An electromagnetic wave with frequency 4.00×10^{14} Hz travels through vacuum in the positive direction of an x axis. The wave is polarized, with its electric field directed parallel to the y axis, with amplitude E_m. At time $t = 0$, the electric field at point P on the x axis has a value of $+E_m/4$ and is decreasing with time. What is the distance along the x axis from point P to the first point with $E = 0$ if we search in (a) the negative direction and (b) the positive direction of the x axis?

N21. In Fig. 34N-12, light refracts into material 2, crosses that material, and is then incident at the critical angle on the interface between materials 2 and 3. (a) What is angle θ? (b) If θ is increased, is there refraction of light into material 3?

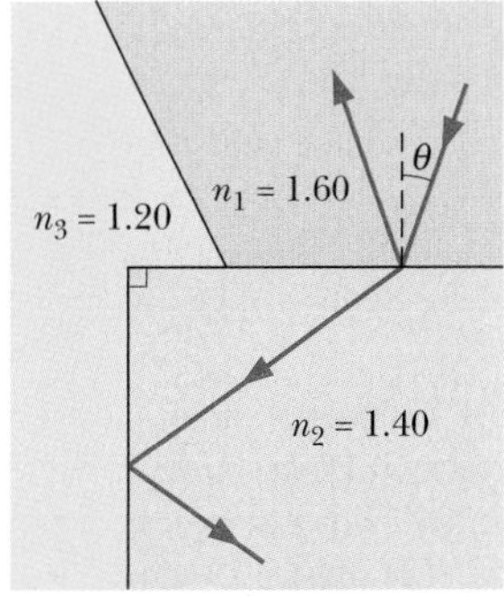

Fig. 34N-12 Problem N21.

N22. *Rainbows from square drops.* Suppose that, on some surreal world, raindrops fell with square cross sections, with one face horizontal. Figure 34N-13 shows such a falling drop, with a white beam of sunlight incident at $\theta = 70.0°$ at point P. The part of the light that enters the drop then travels to point A, where some of it refracts out into the air and the rest reflects. That reflected light then travels to point B, where again some of the light refracts out into the air and the rest reflects. What are the differences in the angles of the red light ($n = 1.331$) and the blue light ($n = 1.343$) that emerge at (a) point A and (b) point B? (This angular difference in the light emerging at, say, point A would be the angular width of the rainbow you would see were you to intercept the light emerging there.)

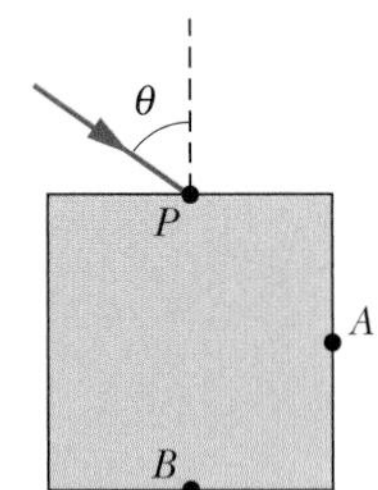

Fig. 34N-13 Problem N22.

N23. In Fig. 34N-14, light refracts from material 1 into material 2. If it is incident at point A at the critical angle for the interface between materials 2 and 3, what are (a) the angle of refraction at point B and (b) the initial angle θ? If, instead, it is incident at point B at the critical angle for the interface between materials 2 and 3, what are (c) the angle of refraction at point A and (d) the initial angle θ? If, instead of all that, it is incident at point A at Brewster's angle for the interface between materials 2 and 3, what are (e) the angle of refraction at point B and (f) the initial angle θ?

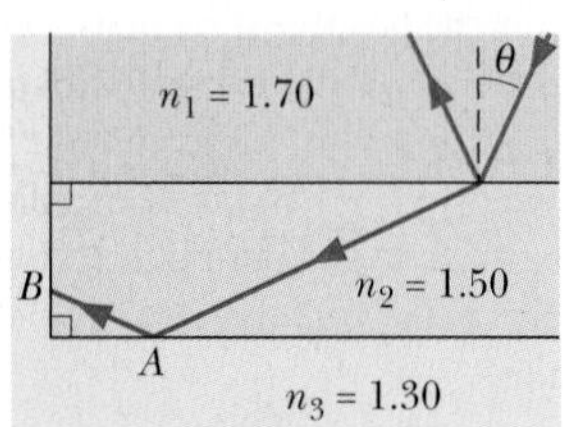

Fig. 34N-14 Problem N23.

N24. The magnetic component of an electromagnetic wave in vacuum has an amplitude of 85.8 nT and an angular wave number of $4.00\ m^{-1}$. What are (a) the frequency of the wave, (b) the rms value of the electric component, and (c) the intensity of the light?

N25. An isotropic point source emits light at wavelength 500 nm, at the rate of 200 W. A light detector is positioned 400 m from the source. What is the maximum rate $\partial B/\partial t$ at which the magnetic component of the light changes with time at the detector's location?

N26. Someone plans to float a small, totally absorbing sphere 0.500 m above an isotropic point source of light, so that the upward radiation force from the light matches the downward gravitational force on the sphere. The sphere's density is $19.0\ g/cm^3$ and its radius is 2.00 mm. (a) What power would be required of the light source? (b) Even if such a source were made, why would the support of the sphere be unstable?

N27. *Dispersion in a window pane.* In Fig. 34N-15, a beam of white light is incident at angle $\theta = 50°$ on a common window pane (shown in cross section). For the pane's type of glass, the index of refraction for visible light ranges from 1.524 at the blue end of the spectrum to 1.509 at the red end. The two sides of the pane are parallel. What is the angular spread of the colors in the beam when (a) the light enters the pane and (b) it emerges from the opposite side? (*Hint:* When you look at an object through a window pane, are the colors in the light from the object dispersed as shown in, say, Fig. 34-21?)

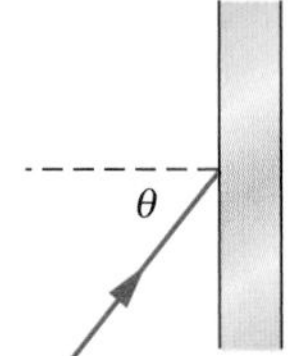

Fig. 34N-15 Problem N27.

N28. An electromagnetic wave with a wavelength of 450 nm travels through vacuum in the negative direction of a y axis with its electric component directed parallel to the x axis. The rms value of the electric component is 5.31×10^{-6} V/m. Write an equation for the magnetic component in the form of Eq. 34-2, but complete with numbers.

N29. In Fig. 34N-16, unpolarized light with an intensity of $25\ W/m^2$ is sent into a system of four polarizing sheets. What is the intensity of the light that emerges from the system?

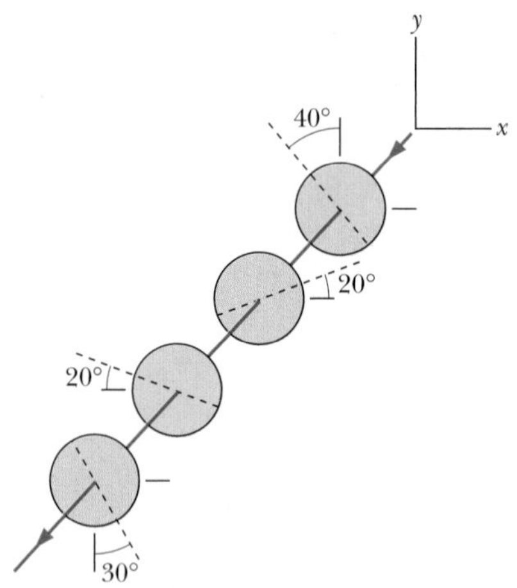

Fig. 34N-16 Problem N29.

N30. A square, perfectly reflecting surface is oriented in space to be perpendicular to the light rays from the Sun. The surface has an edge length of 2.0 m and is located 3.0×10^{11} m from the Sun's center. What is the radiation force on the surface from the light rays?

N31. The magnetic component of a polarized wave of light is

$$B_x = (4.0 \times 10^{-6}\ \text{T}) \sin[(1.57 \times 10^7\ \text{m}^{-1})y + \omega t]$$

(a) Parallel to which axis is the light polarized? What are (b) the frequency and (c) the intensity of the light?

35 Images

Edouard Manet's *A Bar at the Folies-Bergère* has enchanted viewers ever since it was painted in 1882. Part of its appeal lies in the contrast between an audience ready for entertainment and a bartender whose eyes betray her fatigue. Its appeal also depends on subtle distortions of reality that Manet hid in the painting—distortions that give an eerie feel to the scene even before you recognize what is "wrong."

Can you find those subtle distortions of reality?

The answer is in this chapter.

35-1 Two Types of Image

For you to see, say, a penguin, your eye must intercept some of the light rays spreading from the penguin and then redirect them onto the retina at the rear of the eye. Your visual system, starting with the retina and ending with the visual cortex at the rear of your brain, automatically and subconsciously processes the information provided by the light. That system identifies edges, orientations, textures, shapes, and colors and then rapidly brings to your consciousness an **image** (a reproduction derived from light) of the penguin; you perceive and recognize the penguin as being in the direction from which the light rays came and at the proper distance.

Your visual system goes through this processing and recognition even if the light rays do not come directly from the penguin, but instead reflect toward you from a mirror or refract through the lenses in a pair of binoculars. However, you now see the penguin in the direction from which the light rays came after they reflected or refracted, and the distance you perceive may be quite different from the penguin's true distance.

For example, if the light rays have been reflected toward you from a standard flat mirror, the penguin appears to be behind the mirror because the rays you intercept come from that direction. Of course, the penguin is not back there. This type of image, which is called a **virtual image,** truly exists only within the brain but nevertheless is *said* to exist at the perceived location.

A **real image** differs in that it can be formed on a surface, such as a card or a movie screen. You can see a real image (otherwise movie theaters would be empty), but the existence of the image does not depend on your seeing it and it is present even if you are not.

In this chapter we explore several ways in which virtual and real images are formed by reflection (as with mirrors) and refraction (as with lenses). We also distinguish between the two types of image more clearly, but here first is an example of a natural virtual image.

A Common Mirage

A common example of a virtual image is a pool of water that appears to lie on the road some distance ahead of you on a sunny day, but that you can never reach. The pool is a *mirage* (a type of illusion), formed by light rays coming from the low section of the sky in front of you (Fig. 35-1*a*). As the rays approach the road, they travel through progressively warmer air that has been heated by the road, which is usually relatively warm. With an increase in air temperature, the speed of light in air increases slightly and, correspondingly, the index of refraction of the air decreases

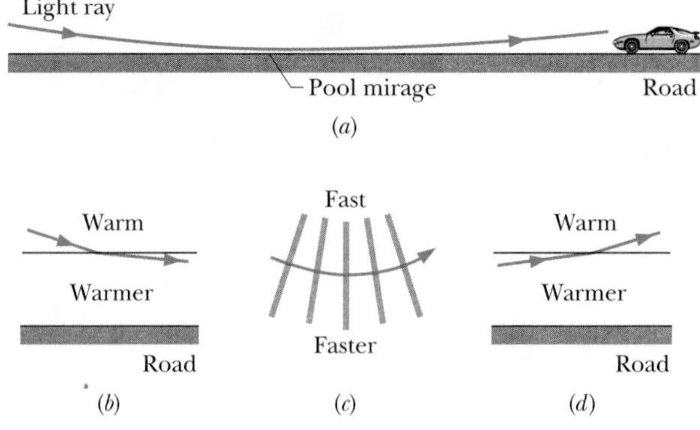

Fig. 35-1 (*a*) A ray from a low section of the sky refracts through air that is heated by a road (without reaching the road). An observer who intercepts the light perceives it to be from a pool of water on the road. (*b*) Bending (exaggerated) of a light ray descending across an imaginary boundary from warm air to warmer air. (*c*) Shifting of wavefronts and associated bending of a ray, which occurs because the lower ends of wavefronts move faster in warmer air. (*d*) Bending of a ray ascending across an imaginary boundary to warm air from warmer air.

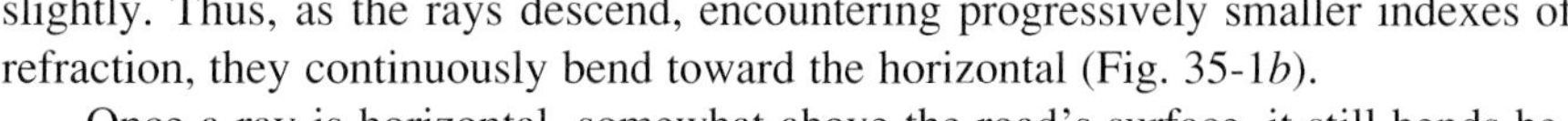

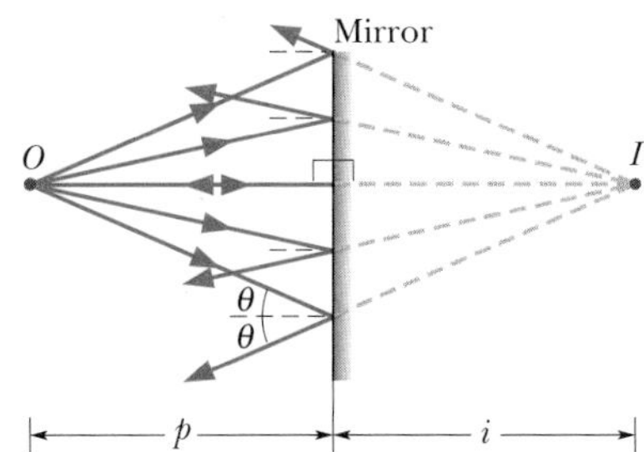

Fig. 35-2 A point source of light O, called the *object,* is a perpendicular distance p in front of a plane mirror. Light rays reaching the mirror from O reflect from the mirror. If your eye intercepts some of the reflected rays, you perceive a point source of light I to be behind the mirror, at a perpendicular distance i. The perceived source I is a virtual image of object O.

slightly. Thus, as the rays descend, encountering progressively smaller indexes of refraction, they continuously bend toward the horizontal (Fig. 35-1*b*).

Once a ray is horizontal, somewhat above the road's surface, it still bends because the lower portion of each associated wavefront is in slightly warmer air and is moving slightly faster than the upper portion of the wavefront (Fig. 35-1*c*). This nonuniform motion of the wavefronts bends the ray upward. As the ray then ascends, it continues to bend upward through progressively greater indexes of refraction (Fig. 35-1*d*).

If you intercept some of this light, your visual system automatically infers that it originated along a backward extension of the rays you have intercepted and, to make sense of the light, assumes that it came from the road surface. If the light happens to be bluish from blue sky, the mirage appears bluish, like water. Because the air is probably turbulent due to the heating, the mirage shimmies, as if water waves were present. The bluish coloring and the shimmy enhance the illusion of a pool of water, but you are actually seeing a virtual image of a low section of the sky.

35-2 Plane Mirrors

A **mirror** is a surface that can reflect a beam of light in one direction instead of either scattering it widely in many directions or absorbing it. A shiny metal surface acts as a mirror; a concrete wall does not. In this section we examine the images that a **plane mirror** (a flat reflecting surface) can produce.

Figure 35-2 shows a point source of light O, which we shall call the *object*, at a perpendicular distance p in front of a plane mirror. The light that is incident on the mirror is represented with rays spreading from O. The reflection of that light is represented with reflected rays spreading from the mirror. If we extend the reflected rays backward (behind the mirror), we find that the extensions intersect at a point that is a perpendicular distance i behind the mirror.

If you look into the mirror of Fig. 35-2, your eyes intercept some of the reflected light. To make sense of what you see, you perceive a point source of light located at the point of intersection of the extensions. This point source is the image I of object O. It is called a *point image* because it is a point, and it is a virtual image because the rays do not actually pass through it. (As you will see, rays *do* pass through a point of intersection for a real image.)

Figure 35-3 shows two rays selected from the many rays in Fig. 35-2. One reaches the mirror at point b, perpendicularly. The other reaches it at an arbitrary point a, with an angle of incidence θ. The extensions of the two reflected rays are also shown. The right triangles $aOba$ and $aIba$ have a common side and three equal angles and are thus congruent (equal in size), so their horizontal sides have the same length. That is,

$$Ib = Ob, \tag{35-1}$$

where Ib and Ob are the distances from the mirror to the image and the object, respectively. Equation 35-1 tells us that the image is as far behind the mirror as the object is in front of it. By convention (that is, to get our equations to work out), *object distances p* are taken to be positive quantities, and *image distances i* for virtual images (as here) are taken to be negative quantities. Thus, Eq. 35-1 can be written as $|i| = p$, or as

$$i = -p \quad \text{(plane mirror).} \tag{35-2}$$

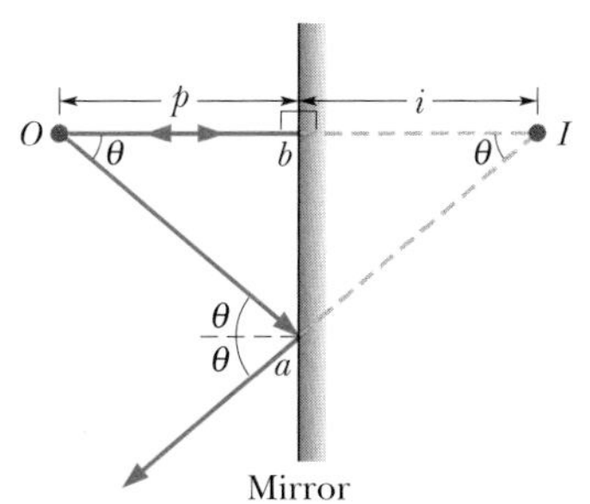

Fig. 35-3 Two rays from Fig. 35-2. Ray Oa makes an arbitrary angle θ with the normal to the mirror surface. Ray Ob is perpendicular to the mirror.

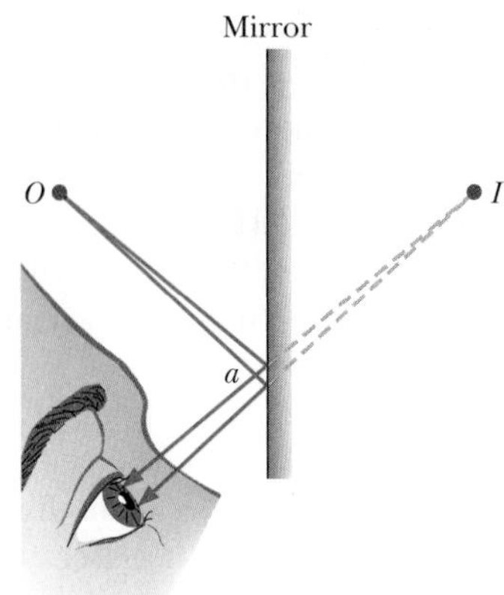

Fig. 35-4 A "pencil" of rays from O enters the eye after reflection at the mirror. Only a small portion of the mirror near a is involved in this reflection. The light appears to originate at point I behind the mirror.

Only rays that are fairly close together can enter the eye after reflection at a mirror. For the eye position shown in Fig. 35-4, only a small portion of the mirror near point a (a portion smaller than the pupil of the eye) is useful in forming the image. To find this portion, close one eye and look at the mirror image of a small object such as the tip of a pencil. Then move your fingertip over the mirror surface until you cannot see the image. Only that small portion of the mirror under your fingertip produced the image.

Extended Objects

In Fig. 35-5, an extended object O, represented by an upright arrow, is at perpendicular distance p in front of a plane mirror. Each small portion of the object that faces the mirror acts like the point source O of Figs. 35-2 and 35-3. If you intercept the light reflected by the mirror, you perceive a virtual image I that is a composite of the virtual point images of all those portions of the object and seems to be at distance i behind the mirror. Distances i and p are related by Eq. 35-2.

We can also locate the image of an extended object as we did for a point object in Fig. 35-2: we draw some of the rays that reach the mirror from the top of the object, draw the corresponding reflected rays, and then extend those reflected rays behind the mirror until they intersect to form an image of the top of the object. We then do the same for rays from the bottom of the object. As shown in Fig. 35-5, we find that virtual image I has the same orientation and *height* (measured parallel to the mirror) as object O.

Manet's "Folies-Bergère"

In *A Bar at the Folies-Bergère* you see the barroom via reflection by a large mirror on the wall behind the woman tending bar, but the reflection is subtly wrong in three ways. First note the bottles at the left. Manet painted their reflections in the mirror but misplaced them, painting them farther toward the front of the bar than they should be.

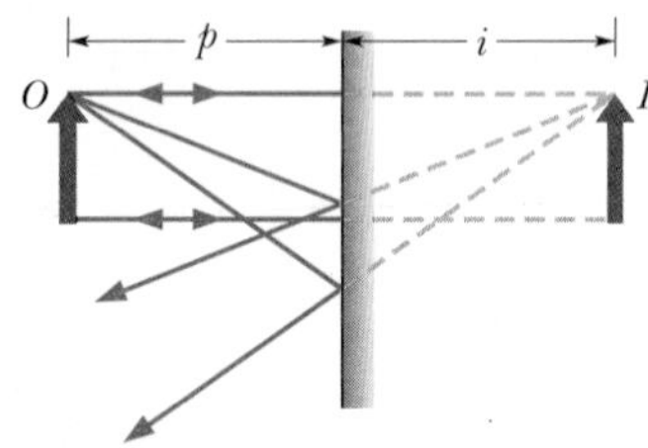

Fig. 35-5 An extended object O and its virtual image I in a plane mirror.

Now note the reflection of the woman. Since your view is from directly in front of the woman, her reflection should be behind her, with only a little of it (if any) visible to you; yet Manet painted her reflection well off to the right. Finally, note the reflection of the man facing her. He must be you, because the reflection shows that he is directly in front of the woman, and thus he must be the viewer of the painting. You are looking into Manet's work and seeing your reflection well off to your right. The effect is eerie because it is not what we expect from a painting or from a mirror.

CHECKPOINT 1: In the figure you look into a system of two vertical parallel mirrors A and B separated by distance d. A grinning gargoyle is perched at point O, a distance $0.2d$ from mirror A. Each mirror produces a *first* (least deep) image of the gargoyle. Then each mirror produces a *second* image with the object being the first image in the opposite mirror. Then each mirror produces a *third* image with the object being the second image in the opposite mirror, and so on—you might see hundreds of grinning gargoyle images. How deep behind mirror A are the first, second, and third images in mirror A?

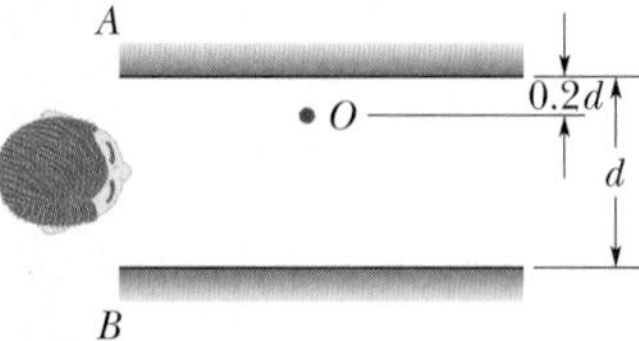

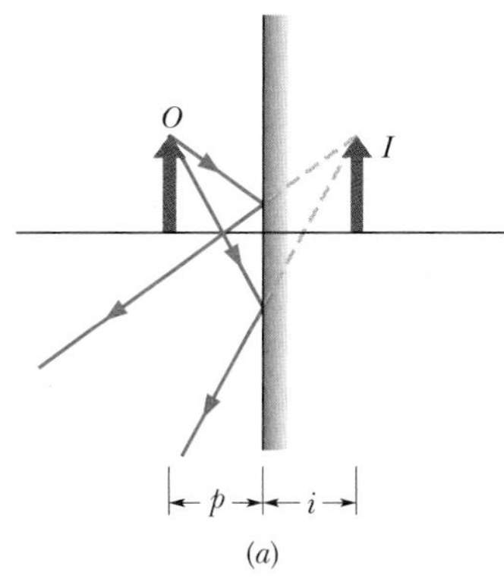

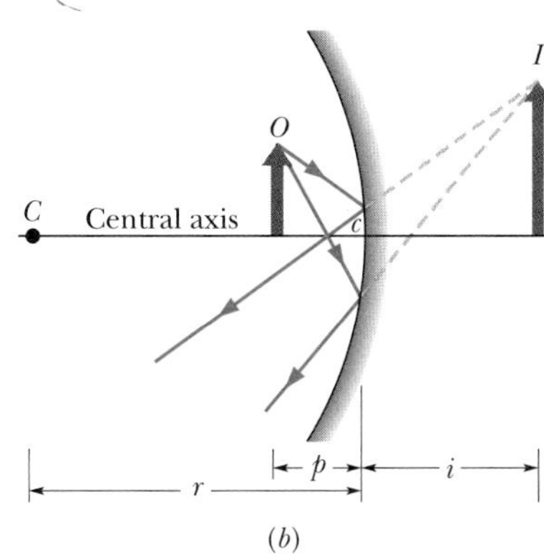

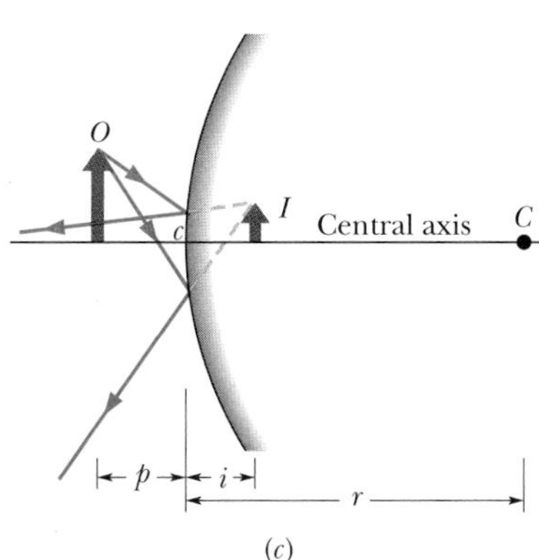

Fig. 35-6 (*a*) An object O forms a virtual image I in a plane mirror. (*b*) If the mirror is bent so that it becomes *concave,* the image moves farther away and becomes larger. (*c*) If the plane mirror is bent so that it becomes *convex,* the image moves closer and becomes smaller.

35-3 Spherical Mirrors

We turn now from images produced by plane mirrors to images produced by mirrors with curved surfaces. In particular, we shall consider spherical mirrors, which are simply mirrors in the shape of a small section of the surface of a sphere. A plane mirror is in fact a spherical mirror with an infinitely large *radius of curvature*.

Making a Spherical Mirror

We start with the plane mirror of Fig. 35-6*a*, which faces leftward toward an object O that is shown and an observer that is not shown. We make a **concave mirror** by curving the mirror's surface so it is *concave* ("caved in") as in Fig. 35-6*b*. Curving the surface in this way changes several characteristics of the mirror and the image it produces of the object:

1. The *center of curvature* C (the center of the sphere of which the mirror's surface is part) was infinitely far from the plane mirror; it is now closer but still in front of the concave mirror.
2. The *field of view*—the extent of the scene that is reflected to the observer—was wide; it is now smaller.
3. The image of the object was as far behind the plane mirror as the object was in front; the image is farther behind the concave mirror; that is, $|i|$ is greater.
4. The height of the image was equal to the height of the object; the height of the image is now greater. This feature is why many makeup mirrors and shaving mirrors are concave—they produce a larger image of a face.

We can make a **convex mirror** by curving a plane mirror so its surface is *convex* ("flexed out") as in Fig. 35-6*c*. Curving the surface in this way (1) moves the center of curvature C to *behind* the mirror and (2) *increases* the field of view. It also (3) moves the image of the object *closer* to the mirror and (4) *shrinks* it. Store surveillance mirrors are usually convex to take advantage of the increase in the field of view—more of the store can then be monitored with a single mirror.

Focal Points of Spherical Mirrors

For a plane mirror, the magnitude of the image distance i is always equal to the object distance p. Before we can determine how these two distances are related for a spherical mirror, we must consider the reflection of light from an object O located an effectively infinite distance in front of a spherical mirror, on the mirror's *central axis*. That axis extends through the center of curvature C and the center c of the mirror. Because of the great distance between the object and the mirror, the light waves spreading from the object are plane waves when they reach the mirror along the central axis. This means that the rays representing the light waves are all parallel to the central axis when they reach the mirror.

When these parallel rays reach a concave mirror like that of Fig. 35-7*a*, those near the central axis are reflected through a common point F; two of these reflected rays are shown in the figure. If we placed a (small) card at F, a point image of the infinitely distant object O would appear on the card. (This would occur for any infinitely distant object.) Point F is called the **focal point** (or **focus**) of the mirror, and its distance from the center of the mirror is the **focal length** f of the mirror.

If we now substitute a convex mirror for the concave mirror, we find that the parallel rays are no longer reflected through a common point. Instead, they diverge

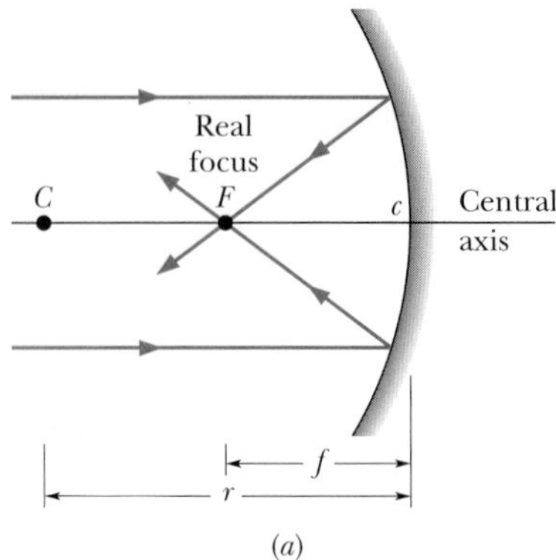

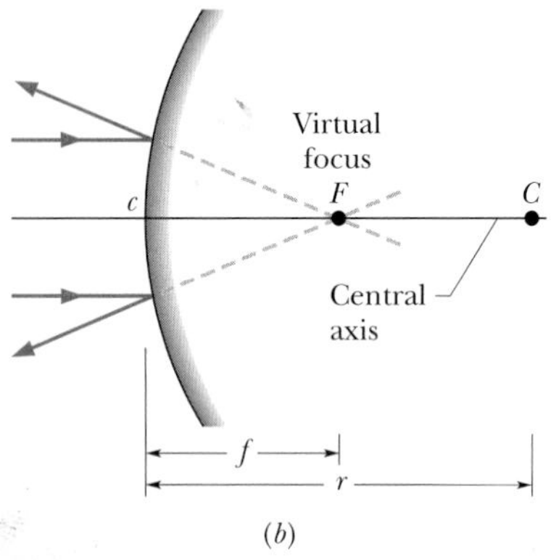

Fig. 35-7 (*a*) In a concave mirror, incident parallel light rays are brought to a real focus at F, on the same side of the mirror as the light rays. (*b*) In a convex mirror, incident parallel light rays seem to diverge from a virtual focus at F, on the side of the mirror opposite the light rays.

as shown in Fig. 35-7*b*. However, if your eye intercepts some of the reflected light, you perceive the light as originating from a point source behind the mirror. This perceived source is located where extensions of the reflected rays pass through a common point (F in Fig. 35-7*b*). That point is the focal point (or focus) F of the convex mirror, and its distance from the mirror surface is the focal length f of the mirror. If we placed a card at this focal point, an image of object O would *not* appear on the card, so this focal point is not like that of a concave mirror.

To distinguish the actual focal point of a concave mirror from the perceived focal point of a convex mirror, the former is said to be a *real focal point* and the latter is said to be a *virtual focal point*. Moreover, the focal length f of a concave mirror is taken to be a positive quantity, and that of a convex mirror a negative quantity. For mirrors of both types, the focal length f is related to the radius of curvature r of the mirror by

$$f = \tfrac{1}{2}r \qquad \text{(spherical mirror)}, \tag{35-3}$$

where, consistent with the signs for the focal length, r is a positive quantity for a concave mirror and a negative quantity for a convex mirror.

35-4 Images from Spherical Mirrors

With the focal point of a spherical mirror defined, we can find the relation between image distance i and object distance p for concave and convex spherical mirrors. We begin by placing the object O *inside the focal point* of the concave mirror—that is, between the mirror and its focal point F (Fig. 35-8*a*). An observer can then see a virtual image of O in the mirror: The image appears to be behind the mirror, and it has the same orientation as the object.

If we now move the object away from the mirror until it is at the focal point, the image moves farther back from the mirror until it is at infinity (Fig. 35-8*b*). The image is then ambiguous and imperceptible because neither the rays reflected by the mirror nor the ray extensions behind the mirror cross to form an image of O.

If we next move the object *outside the focal point*—that is, farther away from the mirror than the focal point—the rays reflected by the mirror converge to form an *inverted* image of object O (Fig. 35-8*c*) in front of the mirror. That image moves in from infinity as we move the object farther outside F. If you were to hold a card at the position of the image, the image would show up on the card—the image is said to be *focused* on the card by the mirror. (The verb "focus," which in this context means to produce an image, differs from the noun "focus," which is another name for the focal point.) Because this image can actually appear on a surface, it is a real image—the rays actually intersect to create the image, regardless of whether an observer is present. The image distance i of a real image is a positive quantity, in contrast to that for a virtual image. We also see that

Fig. 35-8 (*a*) An object O inside the focal point of a concave mirror, and its virtual image I. (*b*) The object at the focal point F. (*c*) The object outside the focal point, and its real image I.

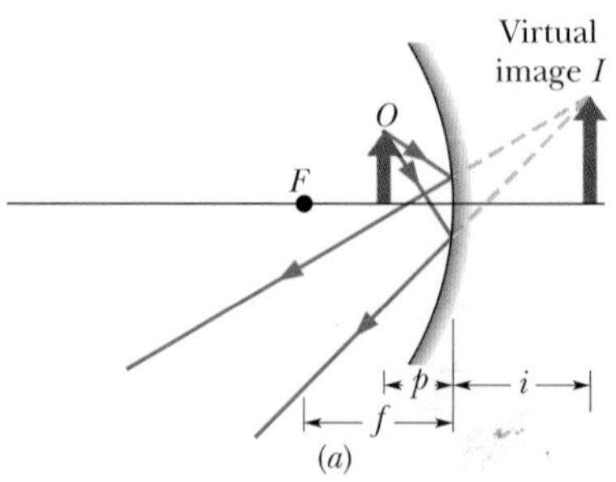

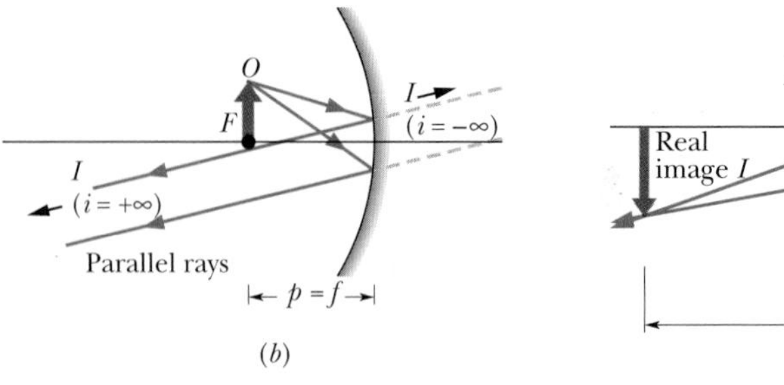

> Real images form on the side of a mirror where the object is, and virtual images form on the opposite side.

As we shall prove in Section 35-8, when light rays from an object make only small angles with the central axis of a spherical mirror, a simple equation relates the object distance p, the image distance i, and the focal length f:

$$\frac{1}{p} + \frac{1}{i} = \frac{1}{f} \quad \text{(spherical mirror).} \tag{35-4}$$

We assume such small angles in figures such as Fig. 35-8, but for clarity the rays are drawn with exaggerated angles. With that assumption, Eq. 35-4 applies to any concave, convex, or plane mirror. For a convex or plane mirror, only a virtual image can be formed, regardless of the object's location on the central axis. As shown in the example of a convex mirror in Fig. 35-6*c*, the image is always on the opposite side of the mirror from the object and has the same orientation as the object.

The size of an object or image, as measured *perpendicular* to the mirror's central axis, is called the object or image *height.* Let h represent the height of the object, and h' the height of the image. Then the ratio h'/h is called the **lateral magnification** m produced by the mirror. However, by convention, the lateral magnification always includes a plus sign when the image orientation is that of the object and a minus sign when the image orientation is opposite that of the object. For this reason, we write the formula for m as

$$|m| = \frac{h'}{h} \quad \text{(lateral magnification).} \tag{35-5}$$

We shall soon prove that the lateral magnification can also be written as

$$m = -\frac{i}{p} \quad \text{(lateral magnification).} \tag{35-6}$$

For a plane mirror, for which $i = -p$, we have $m = +1$. The magnification of 1 means that the image is the same size as the object. The plus sign means that the image and the object have the same orientation. For the concave mirror of Fig. 35-8*c*, $m \approx -1.5$.

Equations 35-3 through 35-6 hold for all plane mirrors, concave spherical mirrors, and convex spherical mirrors. In addition to those equations, you have been asked to absorb a lot of information about these mirrors, and you should organize it for yourself by filling in Table 35-1. Under Image Location, note whether the image is on the *same* side of the mirror as the object or on the *opposite* side. Under Image Type, note whether the image is *real* or *virtual.* Under Image Orientation, note whether the image has the *same* orientation as the object or is *inverted.* Under Sign,

Table 35-1 Your Organizing Table for Mirrors

Mirror	Object	Image			Sign		
Type	Location	Location	Type	Orientation	of f	of r	of m
Plane	Anywhere	opposite	virtual	non-inverted	+	+	+
Concave	Inside F	opposite	virtual	non-inverted	+	+	+
	Outside F	same	real	inverted	+	+	−
Convex	Anywhere	opposite	virtual	non-inverted	−	−	+

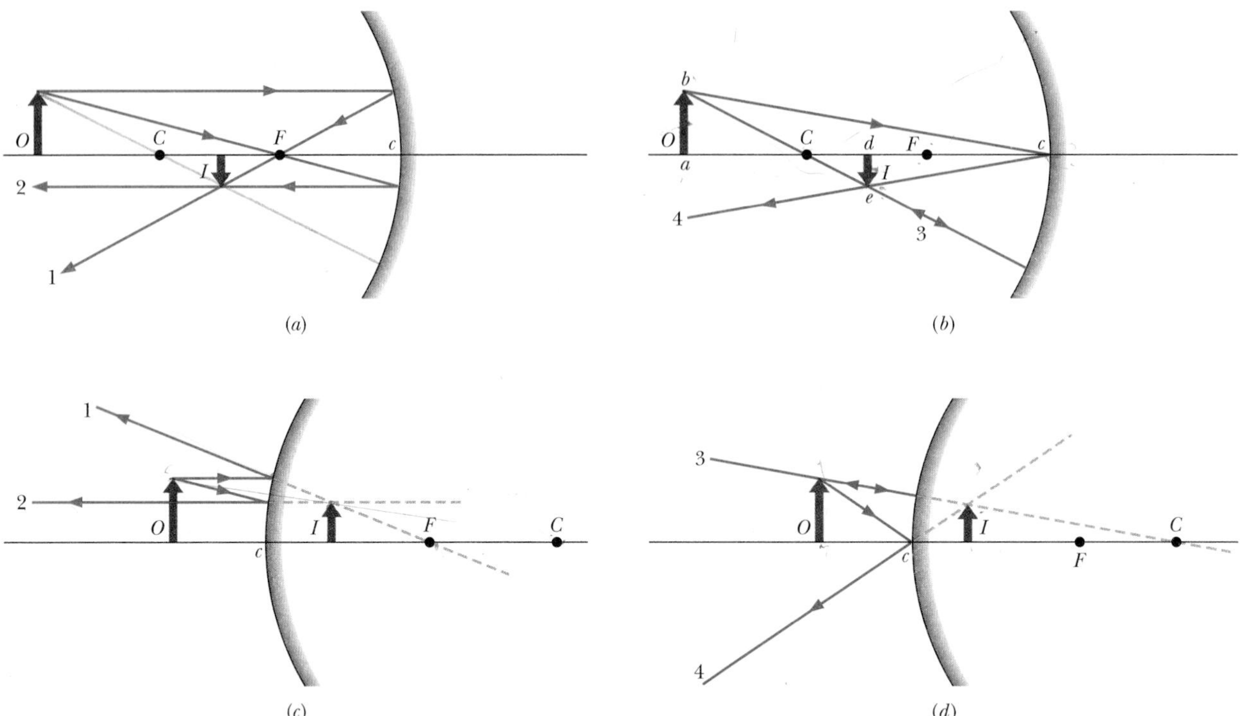

Fig. 35-9 (*a*, *b*) Four rays that may be drawn to find the image of an object in a concave mirror. For the object position shown, the image is real, inverted, and smaller than the object. (*c*, *d*) Four similar rays for the case of a convex mirror. For a convex mirror, the image is always virtual, oriented like the object, and smaller than the object. [In (*c*), ray 2 is initially directed toward focal point *F*. In (*d*), ray 3 is initially directed toward center of curvature *C*.]

give the sign of the quantity or fill in ± if the sign is ambiguous. You will need this organization to tackle homework or a test.

Locating Images by Drawing Rays

Figures 35-9*a* and *b* show an object *O* in front of a concave mirror. We can graphically locate the image of any off-axis point of the object by drawing a *ray diagram* with any two of four special rays through the point:

1. A ray that is initially parallel to the central axis reflects through the focal point *F* (ray 1 in Fig. 35-9*a*).
2. A ray that reflects from the mirror after passing through the focal point emerges parallel to the central axis (ray 2 in Fig. 35-9*a*).
3. A ray that reflects from the mirror after passing through the center of curvature *C* returns along itself (ray 3 in Fig. 35-9*b*).
4. A ray that reflects from the mirror at its intersection *c* with the central axis is reflected symmetrically about that axis (ray 4 in Fig. 35-9*b*).

The image of the point is at the intersection of the two special rays you choose. The image of the object can then be found by locating the images of two or more of its off-axis points. You need to modify the descriptions of the rays slightly to apply them to convex mirrors, as in Figs. 35-9*c* and *d*.

Proof of Equation 35-6

We are now in a position to derive Eq. 35-6 ($m = -i/p$), the equation for the lateral magnification of an object reflected in a mirror. Consider ray 4 in Fig. 35-9*b*. It is reflected at point *c* so that the incident and reflected rays make equal angles with the axis of the mirror at that point.

The two right triangles *abc* and *dec* in the figure are similar (have the same set of angles), so we can write

$$\frac{de}{ab} = \frac{cd}{ca}.$$

The quantity on the left (apart from the question of sign) is the lateral magnification m produced by the mirror. Since we indicate an inverted image as a *negative* magnification, we symbolize this as $-m$. However, $cd = i$ and $ca = p$, so we have

$$m = -\frac{i}{p} \quad \text{(magnification)}, \tag{35-7}$$

which is the relation we set out to prove.

Sample Problem 35-1

A tarantula of height h sits cautiously before a spherical mirror whose focal length has absolute value $|f| = 40$ cm. The image of the tarantula produced by the mirror has the same orientation as the tarantula and has height $h' = 0.20h$.

(a) Is the image real or virtual, and is it on the same side of the mirror as the tarantula or the opposite side?

SOLUTION: The **Key Idea** here is that because the image has the same orientation as the tarantula (the object), it must be virtual and on the opposite side of the mirror. (You can easily see this result if you have filled out Table 35-1.)

(b) Is the mirror concave or convex, and what is its focal length f, sign included?

SOLUTION: We *cannot* tell the type of mirror from the type of image, because both types of mirror can produce virtual images. Similarly, we cannot tell the type of mirror from the sign of the focal length f, as obtained from Eq. 35-3 or Eq. 35-4, because we lack enough information to use either equation. However—and this is the **Key Idea** here—we can make use of the magnification information. We know that the ratio of image height h' to object height h is 0.20. Thus, from Eq. 35-5 we have

$$|m| = \frac{h'}{h} = 0.20.$$

Because the object and image have the same orientation, we know that m must be positive: $m = +0.20$. Substituting this into Eq. 35-6 and solving for, say, i gives us

$$i = -0.20p,$$

which does not appear to be of help in finding f. However, it is helpful if we substitute it into Eq. 35-4. That equation gives us

$$\frac{1}{f} = \frac{1}{i} + \frac{1}{p} = \frac{1}{-0.20p} + \frac{1}{p} = \frac{1}{p}(-5 + 1),$$

from which we find

$$f = -p/4.$$

Now we have it: Because p is positive, f must be negative, which means that the mirror is convex with

$$f = -40 \text{ cm}. \quad \text{(Answer)}$$

CHECKPOINT 2: A Central American vampire bat, dozing on the central axis of a spherical mirror, is magnified by $m = -4$. Is its image (a) real or virtual, (b) inverted or of the same orientation as the bat, and (c) on the same side of the mirror as the bat or on the opposite side?

35-5 Spherical Refracting Surfaces

We now turn from images formed by reflections to images formed by refraction through surfaces of transparent materials, such as glass. We shall consider only spherical surfaces, with radius of curvature r and center of curvature C. The light will be emitted by a point object O in a medium with index of refraction n_1; it will refract through a spherical surface into a medium of index of refraction n_2.

Our concern is whether the light rays, after refracting through the surface, form a real image (no observer necessary) or a virtual image (assuming that an observer intercepts the rays). The answer depends on the relative values of n_1 and n_2 and on the geometry of the situation.

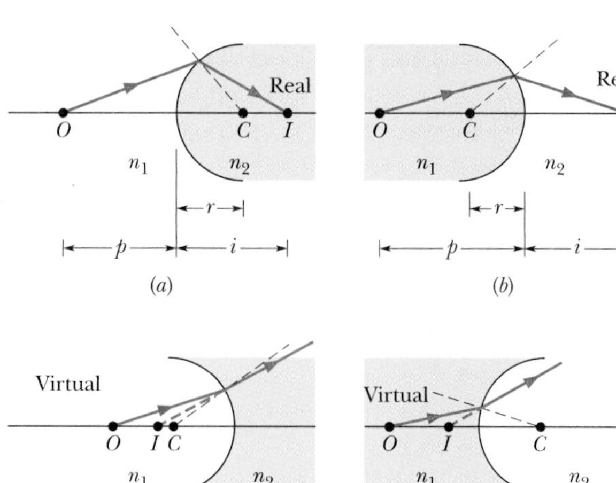

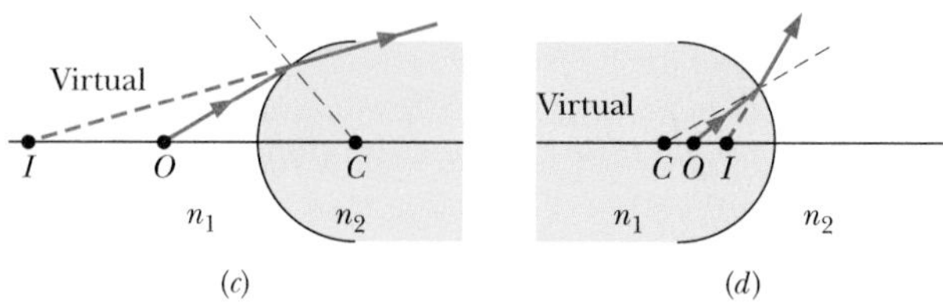

Fig. 35-10 Six possible ways in which an image can be formed by refraction through a spherical surface of radius r and center of curvature C. The surface separates a medium with index of refraction n_1 from a medium with index of refraction n_2. The point object O is always in the medium with n_1, to the left of the surface. The material with the lesser index of refraction is unshaded (think of it as being air, and the other material as being glass). Real images are formed in (*a*) and (*b*); virtual images are formed in the other four situations.

Six possible results are shown in Fig. 35-10. In each part of the figure, the medium with the greater index of refraction is shaded, and object O is always in the medium with index of refraction n_1, to the left of the refracting surface. In each part, a representative ray is shown refracting through the surface. (That ray and a ray along the central axis suffice to determine the position of the image in each case.)

At the point of refraction of each ray, the normal to the refracting surface is a radial line through the center of curvature C. Because of the refraction, the ray bends toward the normal if it is entering a medium of greater index of refraction, and away from the normal if it is entering a medium of lesser index of refraction. If the refracted ray is then directed toward the central axis, it and other (undrawn) rays will form a real image on that axis. If it is directed away from the central axis, it cannot form a real image; however, backward extensions of it and other refracted rays can form a virtual image, provided (as with mirrors) some of those rays are intercepted by an observer.

This insect has been entombed in amber for about 25 million years. Because we view the insect through a curved refracting surface, the image we see does not coincide with the insect.

Real images I are formed (at image distance i) in parts a and b of Fig. 35-10, where the refraction directs the ray *toward* the central axis. Virtual images are formed in parts c and d, where the refraction directs the ray *away* from the central axis. Note, in these four parts, that real images are formed when the object is relatively far from the refracting surface, and virtual images are formed when the object is nearer the refracting surface. In the final situations (Figs. 35-10e and f), refraction always directs the ray away from the central axis and virtual images are always formed, regardless of the object distance.

Note the following major difference from reflected images:

➤ Real images form on the side of a refracting surface that is opposite the object, and virtual images form on the same side as the object.

In Section 35-8, we shall show that (for light rays making only small angles with the central axis)

$$\frac{n_1}{p} + \frac{n_2}{i} = \frac{n_2 - n_1}{r}. \tag{35-8}$$

Just as with mirrors, the object distance p is positive, and the image distance i is positive for a real image and negative for a virtual image. However, to keep all the signs correct in Eq. 35-8, we must use the following rule for the sign of the radius of curvature r:

> When the object faces a convex refracting surface, the radius of curvature r is positive. When it faces a concave surface, r is negative.

Be careful: This is just the reverse of the sign convention we have for mirrors.

CHECKPOINT 3: A bee is hovering in front of the concave spherical refracting surface of a glass sculpture. (a) Which of the general situations of Fig. 35-10 is like this situation? (b) Is the image produced by the surface real or virtual, and is it on the same side as the bee or the opposite side?

Sample Problem 35-2

A Jurassic mosquito is discovered embedded in a chunk of amber, which has index of refraction 1.6. One surface of the amber is spherically convex with radius of curvature 3.0 mm (Fig. 35-11). The mosquito head happens to be on the central axis of that surface and, when viewed along the axis, appears to be buried 5.0 mm into the amber. How deep is it really?

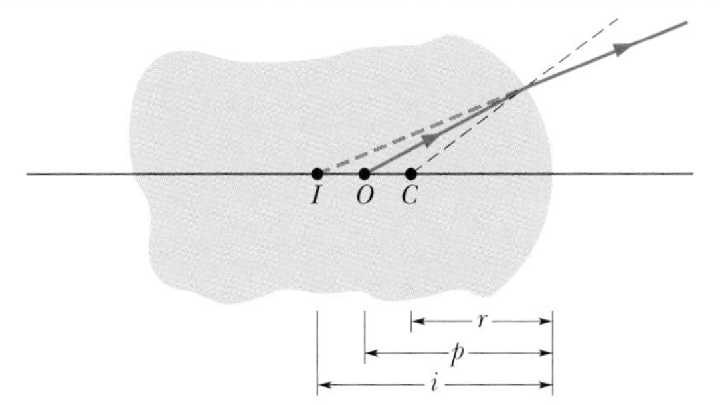

Fig. 35-11 Sample Problem 35-2. A piece of amber with a mosquito from the Jurassic period, with the head buried at point O. The spherical refracting surface at the right end, with center of curvature C, provides an image I to an observer intercepting rays from the object at O.

SOLUTION: The Key Idea here is that the head only appears to be 5.0 mm into the amber because the light rays that the observer intercepts are bent by refraction at the convex amber surface. The image distance i differs from the actual object distance p according to Eq. 35-8. To use that equation to find the actual object distance, we first note:

1. Because the object (the head) and its image are on the same side of the refracting surface, the image must be virtual and so $i = -5.0$ mm.
2. Because the object is always taken to be in the medium of index of refraction n_1, we must have $n_1 = 1.6$ and $n_2 = 1.0$.
3. Because the object faces a concave refracting surface, the radius of curvature r is negative and so $r = -3.0$ mm.

Making these substitutions in Eq. 35-8,

$$\frac{n_1}{p} + \frac{n_2}{i} = \frac{n_2 - n_1}{r},$$

yields

$$\frac{1.6}{p} + \frac{1.0}{-5.0 \text{ mm}} = \frac{1.0 - 1.6}{-3.0 \text{ mm}}$$

and

$$p = 4.0 \text{ mm}. \quad \text{(Answer)}$$

35-6 Thin Lenses

A **lens** is a transparent object with two refracting surfaces whose central axes coincide. The common central axis is the central axis of the lens. When a lens is surrounded by air, light refracts from the air into the lens, crosses through the lens, and then refracts back into the air. Each refraction can change the direction of travel of the light.

A lens that causes light rays initially parallel to the central axis to converge is (reasonably) called a **converging lens.** If, instead, it causes such rays to diverge, the lens is a **diverging lens.** When an object is placed in front of a lens of either type, refraction by the lens's surface of light rays from the object can produce an image of the object.

We shall consider only the special case of a **thin lens**—that is, a lens in which the thickest part is thin compared to the object distance p, the image distance i, and the radii of curvature r_1 and r_2 of the two surfaces of the lens. We shall also consider

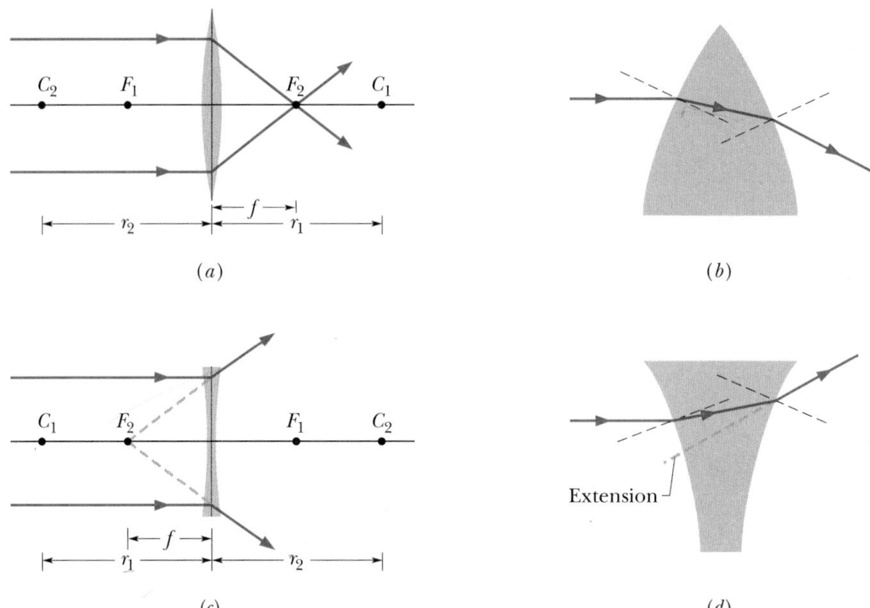

Fig. 35-12 (*a*) Rays initially parallel to the central axis of a converging lens are made to converge to a real focal point F_2 by the lens. The lens is thinner than drawn, with a width like that of the vertical line through it, where we shall consider all the bending of rays to occur. (*b*) An enlargement of the top part of the lens of (*a*); normals to the surfaces are shown dashed. Note that both refractions of the ray at the surfaces bend the ray downward, toward the central axis. (*c*) The same initially parallel rays are made to diverge by a diverging lens. Extensions of the diverging rays pass through a virtual focal point F_2. (*d*) An enlargement of the top part of the lens of (*c*). Note that both refractions of the ray at the surfaces bend the ray upward, away from the central axis.

only light rays that make small angles with the central axis (they are exaggerated in the figures here). In Section 35-8 we shall prove that for such rays, a thin lens has a focal length *f*. Moreover, *i* and *p* are related to each other by

$$\frac{1}{f} = \frac{1}{p} + \frac{1}{i} \quad \text{(thin lens)}, \tag{35-9}$$

which is the same as we had for mirrors. We shall also prove that when a thin lens with index of refraction *n* is surrounded by air, this focal length *f* is given by

$$\frac{1}{f} = (n - 1)\left(\frac{1}{r_1} - \frac{1}{r_2}\right) \quad \text{(thin lens in air)}, \tag{35-10}$$

which is often called the *lens maker's equation.* Here r_1 is the radius of curvature of the lens surface nearer the object, and r_2 is that of the other surface. The signs of these radii are found with the rules in Section 35-5 for the radii of spherical refracting surfaces. If the lens is surrounded by some medium other than air (say, corn oil) with index of refraction n_{medium}, we replace *n* in Eq. 35-10 with n/n_{medium}. Keep in mind the basis of Eqs. 35-9 and 35-10:

> A lens can produce an image of an object only because it can bend light rays; but it can bend light rays only if its index of refraction differs from that of the surrounding medium.

Figure 35-12*a* shows a thin lens with convex refracting surfaces, or *sides*. When rays that are parallel to the central axis of the lens are sent through the lens, they refract twice, as is shown enlarged in Fig. 35-12*b*. This double refraction causes the rays to converge and pass through a common point F_2 at a distance *f* from the center of the lens. Hence, this lens is a converging lens; further, a *real* focal point (or focus) exists at F_2 (because the rays really do pass through it), and the associated focal length is *f*. When rays parallel to the central axis are sent in the opposite direction through the lens, we find another real focal point at F_1 on the other side of the lens. For a thin lens, these two focal points are equidistant from the lens.

A fire is being started by focusing sunlight onto newspaper by means of a converging lens made of clear ice. The lens was made by melting both sides of a flat piece of ice into a convex shape in the shallow vessel (which has a curved bottom).

Because the focal points of a converging lens are real, we take the associated focal lengths f to be positive, just as we do with a real focus of a concave mirror. However, signs in optics can be tricky, so we had better check this in Eq. 35-10. The left side of that equation is positive if f is positive; how about the right side? We examine it term by term. Because the index of refraction n of glass or any other material is greater than 1, the term $(n - 1)$ must be positive. Because the source of the light (which is the object) is at the left and faces the convex left side of the lens, the radius of curvature r_1 of that side must be positive according to the sign rule for refracting surfaces. Similarly, because the object faces a concave right side of the lens, the radius of curvature r_2 of that side must be negative according to that rule. Thus, the term $(1/r_1 - 1/r_2)$ is positive, the whole right side of Eq. 35-10 is positive, and all the signs are consistent.

Figure 35-12*c* shows a thin lens with concave sides. When rays that are parallel to the central axis of the lens are sent through this lens, they refract twice, as is shown enlarged in Fig. 35-12*d*; these rays *diverge,* never passing through any common point, and so this lens is a diverging lens. However, extensions of the rays do pass through a common point F_2 at a distance f from the center of the lens. Hence, the lens has a *virtual* focal point at F_2. (If your eye intercepts some of the diverging rays, you perceive a bright spot to be at F_2, as if it is the source of the light.) Another virtual focus exists on the opposite side of the lens at F_1, symmetrically placed if the lens is thin. Because the focal points of a diverging lens are virtual, we take the focal length f to be negative.

Images from Thin Lenses

We now consider the types of image formed by converging and diverging lenses. Figure 35-13*a* shows an object O outside the focal point F_1 of a converging lens. The two rays drawn in the figure show that the lens forms a real, inverted image I of the object on the side of the lens opposite the object.

When the object is placed inside the focal point F_1, as in Fig. 35-13*b*, the lens forms a virtual image I on the same side of the lens as the object and with the same orientation. Hence, a converging lens can form either a real image or a virtual image, depending on whether the object is outside or inside the focal point, respectively.

Figure 35-13*c* shows an object O in front of a diverging lens. Regardless of the object distance (regardless of whether O is inside or outside the virtual focal point), this lens produces a virtual image that is on the same side of the lens as the object and has the same orientation.

As with mirrors, we take the image distance i to be positive when the image is real and negative when the image is virtual. However, the locations of real and virtual images from lenses are the reverse of those from mirrors:

> Real images form on the side of a lens that is opposite the object, and virtual images form on the side where the object is.

Fig. 35-13 (*a*) A real, inverted image I is formed by a converging lens when the object O is outside the focal point F_1. (*b*) The image I is virtual and has the same orientation as O when O is inside the focal point. (*c*) A diverging lens forms a virtual image I, with the same orientation as the object O, whether O is inside or outside the focal point of the lens.

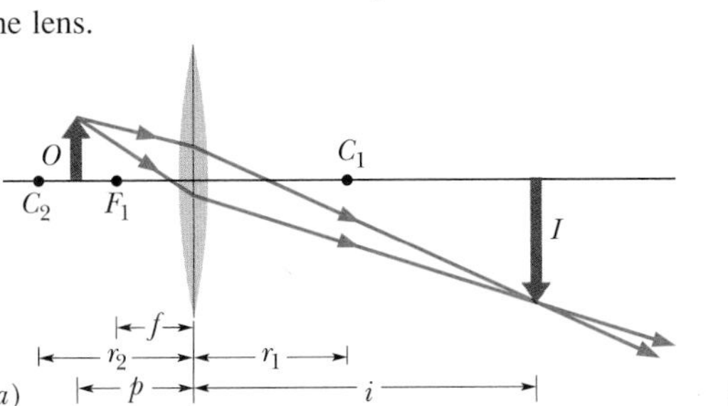

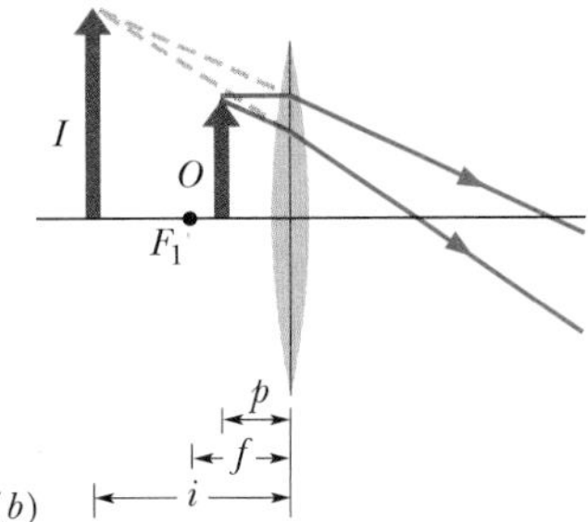

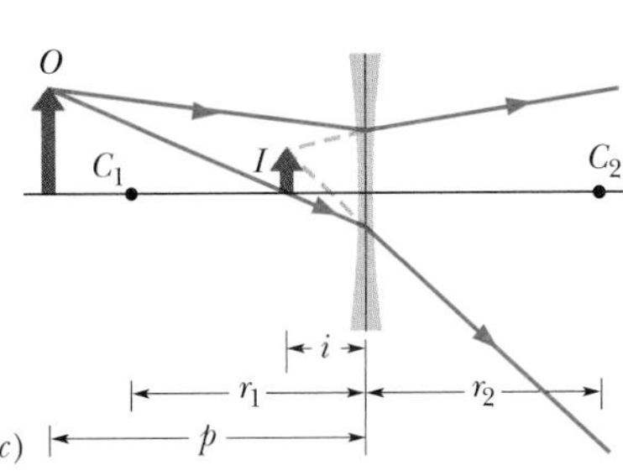

TABLE 35-2 **Your Organizing Table for Thin Lenses**

Lens Type	Object Location	Image			Sign		
		Location	Type	Orientation	of f	of i	of m
Converging	Inside F						
	Outside F						
Diverging	Anywhere						

The lateral magnification m produced by converging and diverging lenses is given by Eqs. 35-5 and 35-6, the same as for mirrors.

You have been asked to absorb a lot of information in this section, and you should organize it for yourself by filling in Table 35-2 for thin *symmetric lenses* (both sides are convex or both sides are concave). Under Image Location note whether the image is on the *same* side of the lens as the object or on the *opposite* side. Under Image Type note whether the image is *real* or *virtual*. Under Image Orientation note whether the image has the *same* orientation as the object or is *inverted*.

PROBLEM-SOLVING TACTICS

Tactic 1: *Signs of Trouble with Mirrors and Lenses*
Be careful: A mirror with a convex surface has a negative focal length f, just the opposite of a lens with convex surfaces. A mirror with a concave surface has a positive focal length f, just the opposite of a lens with concave surfaces. Confusing lens properties with mirror properties is a common mistake.

Locating Images of Extended Objects by Drawing Rays

Figure 35-14*a* shows an object O outside focal point F_1 of a converging lens. We can graphically locate the image of any off-axis point on such an object (such as the tip of the arrow in Fig. 35-14*a*) by drawing a ray diagram with any two of three special rays through the point. These special rays, chosen from all those that pass through the lens to form the image, are the following:

1. A ray that is initially parallel to the central axis of the lens will pass through focal point F_2 (ray 1 in Fig. 35-14*a*).
2. A ray that initially passes through focal point F_1 will emerge from the lens parallel to the central axis (ray 2 in Fig. 35-14*a*).
3. A ray that is initially directed toward the center of the lens will emerge from the lens with no change in its direction (ray 3 in Fig. 35-14*a*) because the ray encounters the two sides of the lens where they are almost parallel.

The image of the point is located where the rays intersect on the far side of the lens. The image of the object is found by locating the images of two or more of its points.

Figure 35-14*b* shows how the extensions of the three special rays can be used to locate the image of an object placed inside focal point F_1 of a converging lens. Note that the description of ray 2 requires modification (it is now a ray whose backward extension passes through F_1).

You need to modify the descriptions of rays 1 and 2 to use them to locate an image placed (anywhere) in front of a diverging lens. In Fig. 35-14*c*, for example, we find the intersection of ray 3 and the backward extensions of rays 1 and 2.

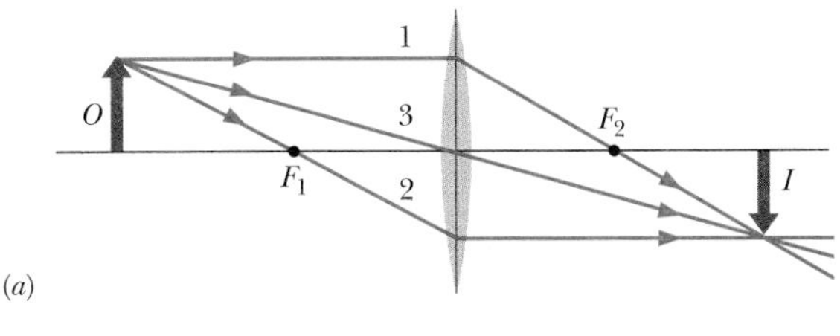

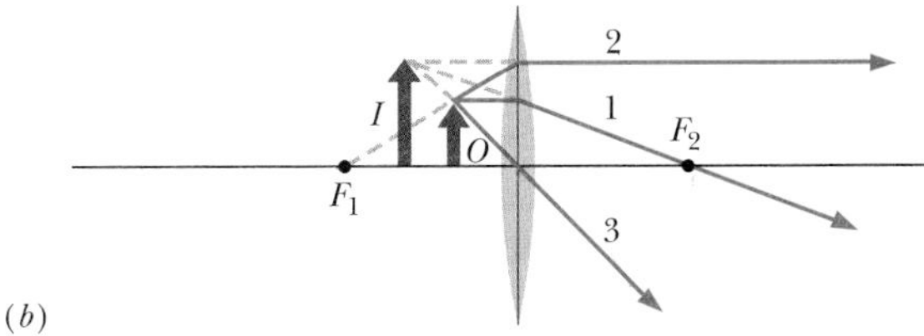

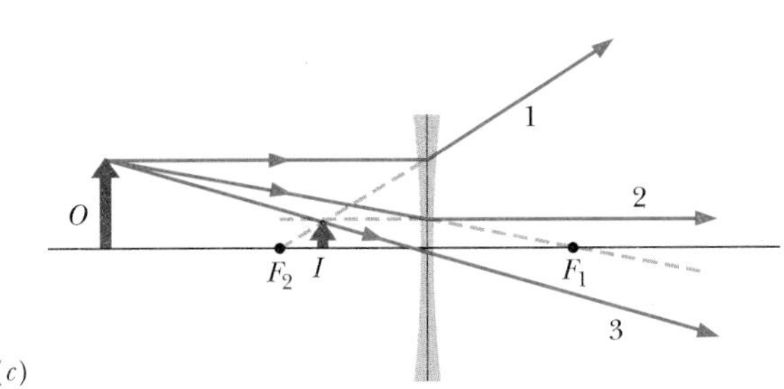

Fig. 35-14 Three special rays allow us to locate an image formed by a thin lens whether the object O is (*a*) outside or (*b*) inside the focal point of a converging lens, or (*c*) anywhere in front of a diverging lens.

Two-Lens Systems

When an object O is placed in front of a system of two lenses whose central axes coincide, we can locate the final image of the system (that is, the image produced by the lens farther from the object) by working in steps. Let lens 1 be the nearer lens and lens 2 the farther lens.

Step 1. We let p_1 represent the distance of object O from lens 1. We then find the distance i_1 of the image produced by lens 1, either by use of Eq. 35-9 or by drawing rays.

Step 2. Now, ignoring the presence of lens 1, we treat the image found in step 1 *as the object* for lens 2. If this new object is located beyond lens 2, the object distance p_2 for lens 2 is taken to be negative. (Note this exception to the rule that says the object distance is positive; the exception occurs because the object here is on the side opposite the source of light.) Otherwise, p_2 is taken to be positive as usual. We then find the distance i_2 of the (final) image produced by lens 2 by use of Eq. 35-9 or by drawing rays.

A similar step-by-step solution can be used for any number of lenses or if a mirror is substituted for lens 2.

The overall lateral magnification M produced by a system of two lenses is the product of the lateral magnifications m_1 and m_2 produced by the two lenses:

$$M = m_1 m_2. \tag{35-11}$$

Sample Problem 35-3

A praying mantis preys along the central axis of a thin symmetric lens, 20 cm from the lens. The lateral magnification of the mantis provided by the lens is $m = -0.25$, and the index of refraction of the lens material is 1.65.

(a) Determine the type of image produced by the lens; the type of lens; whether the object (mantis) is inside or outside the focal point; on which side of the lens the image appears; and whether the image is inverted.

SOLUTION: The Key Idea here is that we can tell a lot about the lens and the image from the given value of m. From it and Eq. 35-6 ($m = -i/p$), we see that

$$i = -mp = 0.25p.$$

Even without finishing the calculation, we can answer the questions. Because p is positive, i here must be positive. That means we have a real image, which means we have a converging lens (the only lens that can produce a real image). The object must be outside the focal point (the only way a real image can be produced). Also, the image is inverted and on the side of the lens opposite the object. (That is how a converging lens makes a real image.)

(b) What are the two radii of curvature of the lens?

SOLUTION: The Key Ideas here are these:

1. Because the lens is symmetric, r_1 (for the surface nearer the object) and r_2 have the same magnitude r.
2. Because the lens is a converging lens, the object faces a convex surface on the nearer side and so $r_1 = +r$. Similarly, it faces a concave surface on the farther side and so $r_2 = -r$.
3. We can relate these radii of curvature to the focal length f via the lens maker's equation, Eq. 35-10 (our only equation involving the radii of curvature of a lens).
4. We can relate f to the object distance p and image distance i via Eq. 35-9.

We know p but we do not know i. Thus, our starting point is to finish the calculation for i in part (a); we obtain

$$i = (0.25)(20 \text{ cm}) = 5.0 \text{ cm}.$$

Now Eq. 35-9 gives us

$$\frac{1}{f} = \frac{1}{p} + \frac{1}{i} = \frac{1}{20 \text{ cm}} + \frac{1}{5.0 \text{ cm}},$$

from which we find $f = 4.0$ cm.

Equation 35-10 then gives us

$$\frac{1}{f} = (n-1)\left(\frac{1}{r_1} - \frac{1}{r_2}\right) = (n-1)\left(\frac{1}{+r} - \frac{1}{-r}\right)$$

or, with known values inserted,

$$\frac{1}{4.0 \text{ cm}} = (1.65 - 1)\frac{2}{r},$$

which yields

$$r = (0.65)(2)(4.0 \text{ cm}) = 5.2 \text{ cm}. \qquad \text{(Answer)}$$

CHECKPOINT 4: A thin symmetric lens provides an image of a fingerprint with a magnification of +0.2 when the fingerprint is 1.0 cm farther from the lens than the focal point of the lens. What are the type and orientation of the image, and what is the type of lens?

Sample Problem 35-4

Figure 35-15*a* shows a jalapeño seed O_1 that is placed in front of two thin symmetrical coaxial lenses 1 and 2, with focal lengths $f_1 = +24$ cm and $f_2 = +9.0$ cm, respectively, and with lens separation $L = 10$ cm. The seed is 6.0 cm from lens 1. Where does the system of two lenses produce an image of the seed?

SOLUTION: We could locate the image produced by the system of lenses by tracing light rays from the seed through the two lenses. However, the Key Idea here is that we can, instead, calculate the location of that image by working through the system in steps, lens by lens. We begin with the lens closer to the seed. The image we seek is the final one—that is, image I_2 produced by lens 2.

Lens 1. Ignoring lens 2, we locate the image I_1 produced by lens 1 by applying Eq. 35-9 to lens 1 alone:

$$\frac{1}{p_1} + \frac{1}{i_1} = \frac{1}{f_1}.$$

The object O_1 for lens 1 is the seed, which is 6.0 cm from the lens; thus, we substitute $p_1 = +6.0$ cm. Also substituting the given value of f_1, we then have

$$\frac{1}{+6.0 \text{ cm}} + \frac{1}{i_1} = \frac{1}{+24 \text{ cm}},$$

which yields $i_1 = -8.0$ cm.

This tells us that image I_1 is 8.0 cm from lens 1 and virtual. (We could have guessed that it is virtual by noting that the seed is inside the focal point of lens 1.) Since I_1 is virtual, it is on the same

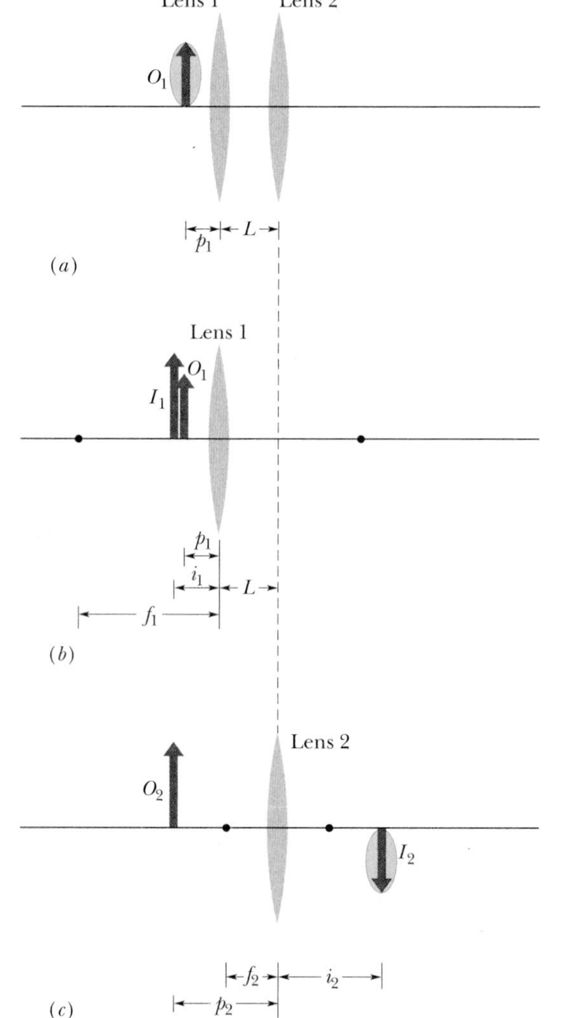

Fig. 35-15 Sample Problem 35-4. (*a*) Seed O_1 is distance p_1 from a two-lens system with lens separation L. We use the arrow to orient the seed. (*b*) The image I_1 produced by lens 1 alone. (*c*) Image I_1 acts as object O_2 for lens 2 alone, which produces the final image I_2.

side of the lens as object O_1 and has the same orientation as the seed, as shown in Fig. 35-15*b*.

Lens 2. In the second step of our solution, the **Key Idea** is that we can treat image I_1 as an object O_2 for the second lens and now ignore lens 1. We first note that this object O_2 is outside the focal point of lens 2. So the image I_2 produced by lens 2 must be real, inverted, and on the side of the lens opposite O_2. Let us see.

The distance p_2 between this object O_2 and lens 2 is, from Fig. 35-15*c*,

$$p_2 = L + |i_1| = 10 \text{ cm} + 8.0 \text{ cm} = 18 \text{ cm}.$$

Then Eq. 35-9, now written for lens 2, yields

$$\frac{1}{+18 \text{ cm}} + \frac{1}{i_2} = \frac{1}{+9.0 \text{ cm}}.$$

Hence,

$$i_2 = +18 \text{ cm}. \quad \text{(Answer)}$$

The plus sign confirms our guess: Image I_2 produced by lens 2 is real, inverted, and on the side of lens 2 opposite O_2, as shown in Fig. 35-15*c*.

35-7 Optical Instruments

The human eye is a remarkably effective organ, but its range can be extended in many ways by optical instruments such as eyeglasses, simple magnifying lenses, motion picture projectors, cameras (including TV cameras), microscopes, and telescopes. Many such devices extend the scope of our vision beyond the visible range; satellite-borne infrared cameras and x-ray microscopes are just two examples.

The mirror and thin-lens formulas can be applied only as approximations to most sophisticated optical instruments. The lenses in typical laboratory microscopes are by no means "thin." In most optical instruments the lenses are compound lenses; that is, they are made of several components, the interfaces rarely being exactly spherical. Now we discuss three optical instruments, assuming, for simplicity, that the thin-lens formulas apply.

Simple Magnifying Lens

The normal human eye can focus a sharp image of an object on the retina (at the rear of the eye) if the object is located anywhere from infinity to a certain point called the *near point* P_n. If you move the object closer to the eye than the near point, the perceived retinal image becomes fuzzy. The location of the near point normally varies with age. We have all heard about people who claim not to need glasses but

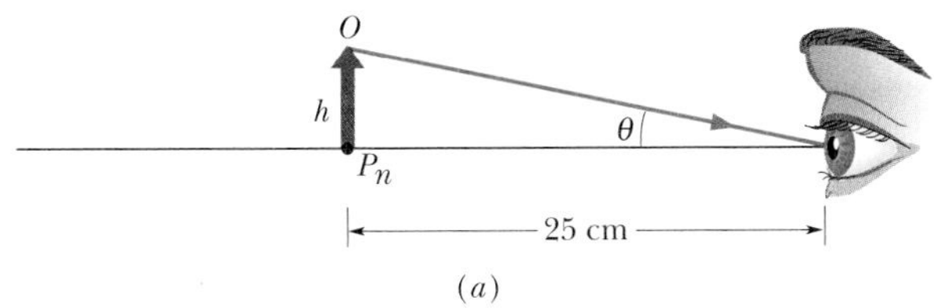

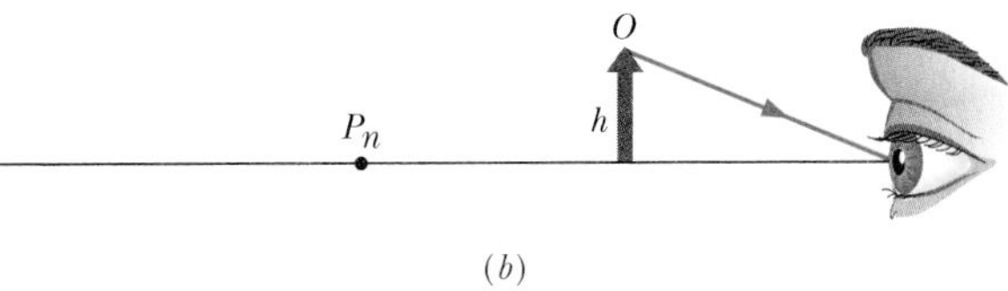

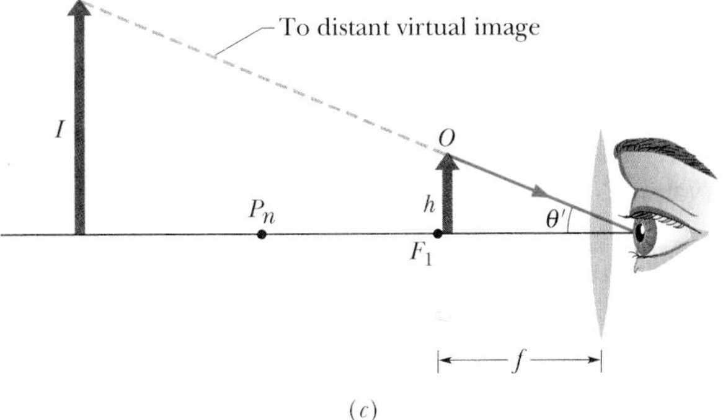

Fig. 35-16 (*a*) An object O of height h, placed at the near point of a human eye, occupies angle θ in the eye's view. (*b*) The object is moved closer to increase the angle, but now the observer cannot bring the object into focus. (*c*) A converging lens is placed between the object and the eye, with the object just inside the focal point F_1 of the lens. The image produced by the lens is then far enough away to be focused by the eye, and the image occupies a larger angle θ' than object O does in (*a*).

read their newspapers at arm's length; their near points are receding. To find your own near point, remove your glasses or contacts if you wear any, close one eye, and then bring this page closer to your open eye until it becomes indistinct. In what follows, we take the near point to be 25 cm from the eye, a bit more than the typical value for 20-year-olds.

Figure 35-16*a* shows an object O placed at the near point P_n of an eye. The size of the image of the object produced on the retina depends on the angle θ that the object occupies in the field of view from that eye. By moving the object closer to the eye, as in Fig. 35-16*b*, you can increase the angle and, hence, the possibility of distinguishing details of the object. However, because the object is then closer than the near point, it is no longer *in focus*; that is, the image is no longer clear.

You can restore the clarity by looking at O through a converging lens, placed so that O is just inside the focal point F_1 of the lens, which is at focal length f (Fig. 35-16*c*). What you then see is the virtual image of O produced by the lens. That image is farther away than the near point; thus, the eye can see it clearly.

Moreover, the angle θ' occupied by the virtual image is larger than the largest angle θ that the object alone can occupy and still be seen clearly. The *angular magnification* m_θ (not to be confused with lateral magnification m) of what is seen is

$$m_\theta = \theta'/\theta.$$

In words, the angular magnification of a simple magnifying lens is a comparison of the angle occupied by the image the lens produces with the angle occupied by the object when the object is moved to the near point of the viewer.

From Fig. 35-16, assuming that O is at the focal point of the lens, and approximating $\tan\theta$ as θ and $\tan\theta'$ as θ' for small angles, we have

$$\theta \approx h/25 \text{ cm} \quad \text{and} \quad \theta' \approx h/f.$$

We then find that

$$m_\theta \approx \frac{25 \text{ cm}}{f} \quad \text{(simple magnifier).} \tag{35-12}$$

Compound Microscope

Figure 35-17 shows a thin-lens version of a compound microscope. The instrument consists of an *objective* (the front lens) of focal length f_{ob} and an *eyepiece* (the lens near the eye) of focal length f_{ey}. It is used for viewing small objects that are very close to the objective.

The object O to be viewed is placed just outside the first focal point F_1 of the objective, close enough to F_1 that we can approximate its distance p from the lens

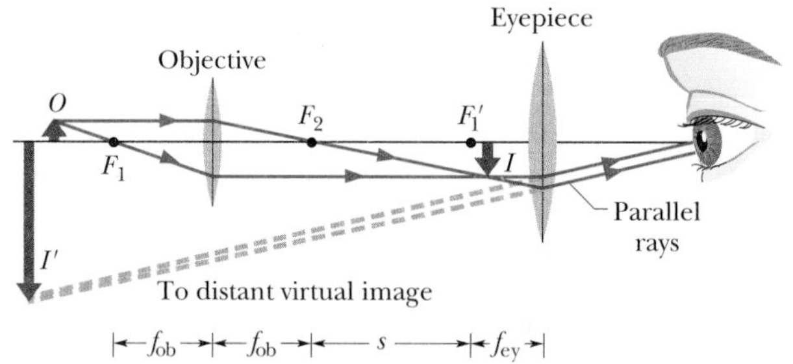

Fig. 35-17 A thin-lens representation of a compound microscope (not to scale). The objective produces a real image I of object O just inside the focal point F_1' of the eyepiece. Image I then acts as an object for the eyepiece, which produces a virtual final image I' that is seen by the observer. The objective has focal length f_{ob}; the eyepiece has focal length f_{ey}; and s is the tube length.

as being f_{ob}. The separation between the lenses is then adjusted so that the enlarged, inverted, real image I produced by the objective is located just inside the first focal point F_1' of the eyepiece. The *tube length* s shown in Fig. 35-17 is actually large relative to f_{ob}, and we can approximate the distance i between the objective and the image I as being length s.

From Eq. 35-6, and using our approximations for p and i, we can write the lateral magnification produced by the objective as

$$m = -\frac{i}{p} = -\frac{s}{f_{ob}}. \tag{35-13}$$

Since the image I is located just inside the focal point F_1' of the eyepiece, the eyepiece acts as a simple magnifying lens, and an observer sees a final (virtual, inverted) image I' through it. The overall magnification of the instrument is the product of the lateral magnification m produced by the objective, given by Eq. 35-13, and the angular magnification m_θ produced by the eyepiece, given by Eq. 35-12; that is,

$$M = mm_\theta = -\frac{s}{f_{ob}}\frac{25\text{ cm}}{f_{ey}} \quad \text{(microscope)}. \tag{35-14}$$

Refracting Telescope

Telescopes come in a variety of forms. The form we describe here is the simple refracting telescope that consists of an objective and an eyepiece; both are represented in Fig. 35-18 with simple lenses, although in practice, as is also true for most microscopes, each lens is actually a compound lens system.

The lens arrangements for telescopes and for microscopes are similar, but telescopes are designed to view large objects, such as galaxies, stars, and planets, at large distances, whereas microscopes are designed for just the opposite purpose. This difference requires that in the telescope of Fig. 35-18 the second focal point of the objective F_2 coincide with the first focal point of the eyepiece F_1', whereas in the microscope of Fig. 35-17 these points are separated by the tube length s.

In Fig. 35-18*a*, parallel rays from a distant object strike the objective, making an angle θ_{ob} with the telescope axis and forming a real, inverted image at the common focal point F_2, F_1'. This image I acts as an object for the eyepiece, through which

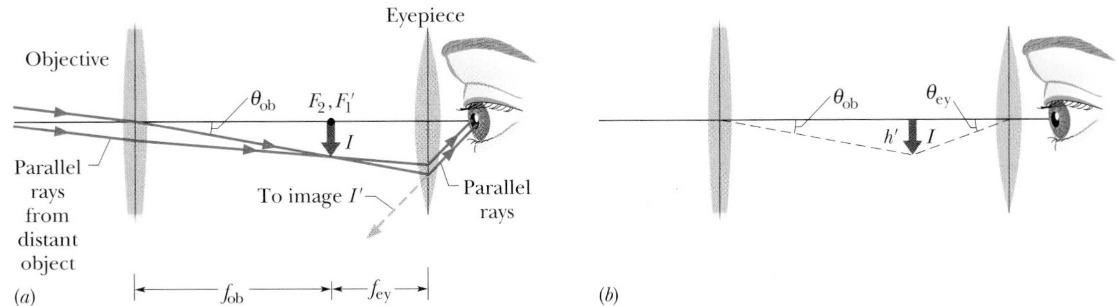

Fig. 35-18 (*a*) A thin-lens representation of a refracting telescope. The objective produces a real image I of a distant source of light (the object), with approximately parallel light rays at the objective. (One end of the object is assumed to lie on the central axis.) Image I, formed at the common focal points F_2 and F_1', acts as an object for the eyepiece, which produces a virtual final image I' at a great distance from the observer. The objective has focal length f_{ob}; the eyepiece has focal length f_{ey}. (*b*) Image I has height h' and takes up angle θ_{ob} measured from the objective and angle θ_{ey} measured from the eyepiece.

an observer sees a distant (still inverted) virtual image I'. The rays defining the image make an angle θ_{ey} with the telescope axis.

The angular magnification m_θ of the telescope is θ_{ey}/θ_{ob}. From Fig. 35-18*b*, for rays close to the central axis, we can write $\theta_{ob} = h'/f_{ob}$ and $\theta_{ey} \approx h'/f_{ey}$, which gives us

$$m_\theta = -\frac{f_{ob}}{f_{ey}} \quad \text{(telescope)}, \tag{35-15}$$

where the minus sign indicates that I' is inverted. In words, the angular magnification of a telescope is a comparison of the angle occupied by the image the telescope produces with the angle occupied by the distant object as seen without the telescope.

Magnification is only one of the design factors for an astronomical telescope and is indeed easily achieved. A good telescope needs *light-gathering power*, which determines how bright the image is. This is important for viewing faint objects such as distant galaxies and is accomplished by making the objective diameter as large as possible. A telescope also needs *resolving power*, which is the ability to distinguish between two distant objects (stars, say) whose angular separation is small. *Field of view* is another important design parameter. A telescope designed to look at galaxies (which occupy a tiny field of view) is much different from one designed to track meteors (which move over a wide field of view).

The telescope designer must also take into account the difference between real lenses and the ideal thin lenses we have discussed. A real lens with spherical surfaces does not form sharp images, a flaw called *spherical aberration*. Also, because refraction by the two surfaces of a real lens depends on wavelength, a real lens does not focus light of different wavelengths to the same point, a flaw called *chromatic aberration*.

This brief discussion by no means exhausts the design parameters of astronomical telescopes—many others are involved. We could make a similar listing for any other high-performance optical instrument.

35-8 Three Proofs

The Spherical Mirror Formula (Eq. 35-4)

Figure 35-19 shows a point object O placed on the central axis of a concave spherical mirror, outside its center of curvature C. A ray from O that makes an angle α with the axis intersects the axis at I after reflection from the mirror at a. A ray that leaves O along the axis is reflected back along itself at c and also passes through I. Thus, I is the image of O; it is a *real* image because light actually passes through it. Let us find the image distance i.

A trigonometry theorem that is useful here tells us that an exterior angle of a triangle is equal to the sum of the two opposite interior angles. Applying this to triangles OaC and OaI in Fig. 35-19 yields

$$\beta = \alpha + \theta \quad \text{and} \quad \gamma = \alpha + 2\theta.$$

If we eliminate θ between these two equations, we find

$$\alpha + \gamma = 2\beta. \tag{35-16}$$

We can write angles α, β, and γ, in radian measure, as

$$\alpha \approx \frac{\widehat{ac}}{cO} = \frac{\widehat{ac}}{p}, \qquad \beta = \frac{\widehat{ac}}{cC} = \frac{\widehat{ac}}{r},$$

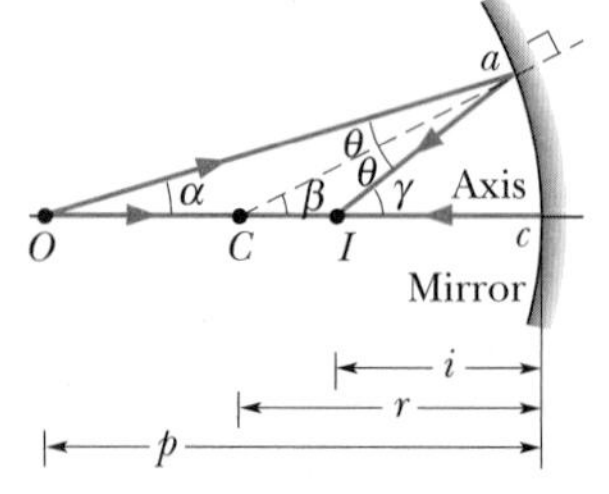

Fig. 35-19 A concave spherical mirror forms a real point image I by reflecting light rays from a point object O.

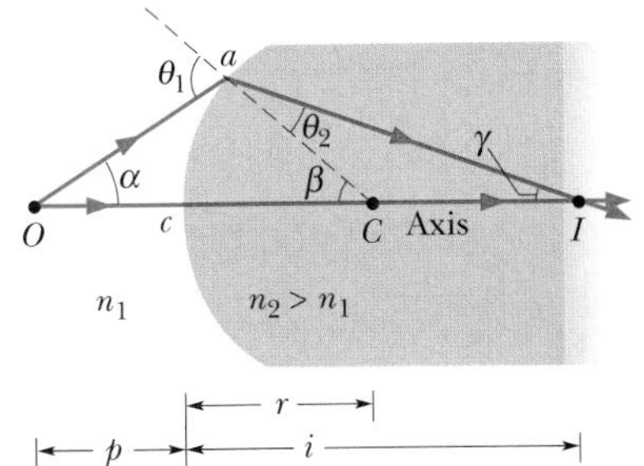

Fig. 35-20 A real point image I of a point object O is formed by refraction at a spherical convex surface between two media.

and

$$\gamma \approx \frac{\widehat{ac}}{cI} = \frac{\widehat{ac}}{i}. \tag{35-17}$$

Only the equation for β is exact, because the center of curvature of arc ac is at C. However, the equations for α and γ are approximately correct if these angles are small enough (that is, for rays close to the central axis). Substituting Eqs. 35-17 into Eq. 35-16, using Eq. 35-3 to replace r with $2f$, and canceling $\widehat{ac}$ lead exactly to Eq. 35-4, the relation that we set out to prove.

The Refracting Surface Formula (Eq. 35-8)

The incident ray from point object O in Fig. 35-20 that falls on point a of a spherical refracting surface is refracted there according to Eq. 34-44,

$$n_1 \sin \theta_1 = n_2 \sin \theta_2.$$

If α is small, θ_1 and θ_2 will also be small and we can replace the sines of these angles with the angles themselves. Thus, the equation above becomes

$$n_1\theta_1 \approx n_2\theta_2. \tag{35-18}$$

We again use the fact that an exterior angle of a triangle is equal to the sum of the two opposite interior angles. Applying this to triangles COa and ICa yields

$$\theta_1 = \alpha + \beta \quad \text{and} \quad \beta = \theta_2 + \gamma. \tag{35-19}$$

If we use Eqs. 35-19 to eliminate θ_1 and θ_2 from Eq. 35-18, we find

$$n_1\alpha + n_2\gamma = (n_2 - n_1)\beta. \tag{35-20}$$

In radian measure the angles α, β, and γ are

$$\alpha \approx \frac{\widehat{ac}}{p}; \qquad \beta = \frac{\widehat{ac}}{r}; \qquad \gamma \approx \frac{\widehat{ac}}{i}. \tag{35-21}$$

Only the second of these equations is exact. The other two are approximate because I and O are not the centers of circles of which $\widehat{ac}$ is a part. However, for α small enough (for rays close to the axis), the inaccuracies in Eqs. 35-21 are small. Substituting Eqs. 35-21 into Eq. 35-20 leads directly to Eq. 35-8, as we wanted.

The Thin-Lens Formulas (Eqs. 35-9 and 35-10)

Our plan is to consider each lens surface as a separate refracting surface, and to use the image formed by the first surface as the object for the second.

We start with the thick glass "lens" of length L in Fig. 35-21a whose left and right refracting surfaces are ground to radii r' and r''. A point object O' is placed near the left surface as shown. A ray leaving O' along the central axis is not deflected on entering or leaving the lens.

A second ray leaving O' at an angle α with the central axis intersects the left surface at point a', is refracted, and intersects the second (right) surface at point a''. The ray is again refracted and crosses the axis at I'', which, being the intersection of two rays from O', is the image of point O', formed after refraction at two surfaces.

Figure 35-21b shows that the first (left) surface also forms a virtual image of O' at I'. To locate I', we use Eq. 35-8,

$$\frac{n_1}{p} + \frac{n_2}{i} = \frac{n_2 - n_1}{r}.$$

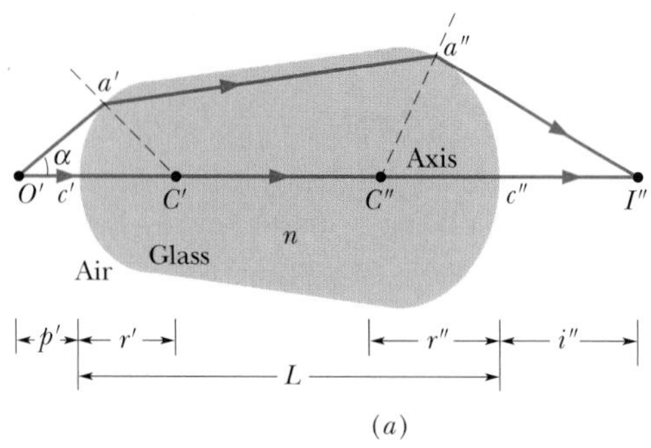

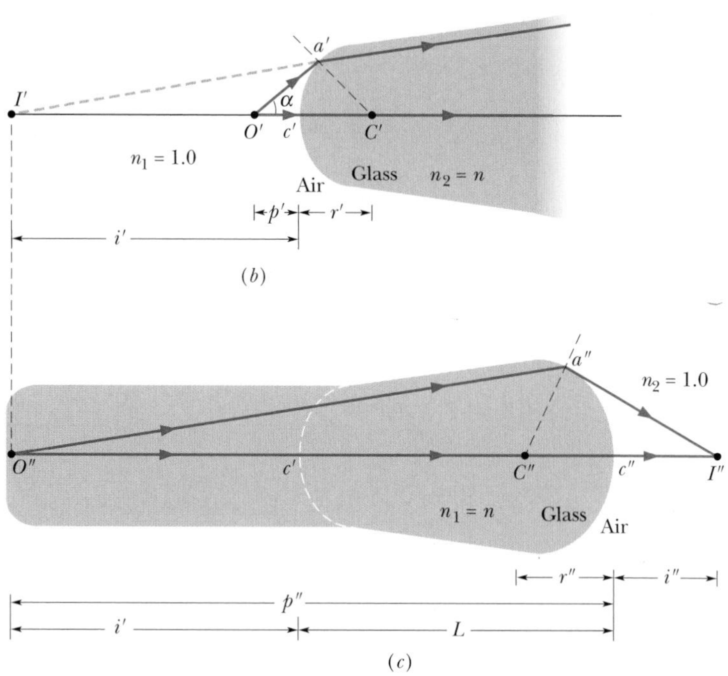

Fig. 35-21 (*a*) Two rays from point object O' form a real image I'' after refracting through two spherical surfaces of a "lens." The object faces a convex surface at the left side of the lens and a concave surface at the right side. The ray traveling through points a' and a'' is actually close to the central axis through the lens. (*b*) The left side and (*c*) the right side of the "lens" in (*a*), shown separately.

Putting $n_1 = 1$ for air and $n_2 = n$ for lens glass and bearing in mind that the image distance is negative (that is, $i = -i'$ in Fig. 35-21*b*), we obtain

$$\frac{1}{p'} - \frac{n}{i'} = \frac{n-1}{r'}. \tag{35-22}$$

In this equation i' will be a positive number because we have already introduced the minus sign appropriate to a virtual image.

Figure 35-21*c* shows the second surface again. Unless an observer at point a'' were aware of the existence of the first surface, the observer would think that the light striking that point originated at point I' in Fig. 35-21*b* and that the region to the left of the surface was filled with glass as indicated. Thus, the (virtual) image I' formed by the first surface serves as a real object O'' for the second surface. The distance of this object from the second surface is

$$p'' = i' + L. \tag{35-23}$$

To apply Eq. 35-8 to the second surface, we must insert $n_1 = n$ and $n_2 = 1$ because the object now is effectively imbedded in glass. If we substitute with Eq. 35-23, then Eq. 35-8 becomes

$$\frac{n}{i' + L} + \frac{1}{i''} = \frac{1-n}{r''}. \tag{35-24}$$

Let us now assume that the thickness L of the "lens" in Fig. 35-21*a* is so small that we can neglect it in comparison with our other linear quantities (such as p', i', p'', i'', r', and r''). In all that follows we make this *thin-lens approximation.* Putting $L = 0$ in Eq. 35-24 and rearranging the right side lead to

$$\frac{n}{i'} + \frac{1}{i''} = -\frac{n-1}{r''}. \tag{35-25}$$

Adding Eqs. 35-22 and 35-25 leads to

$$\frac{1}{p'} + \frac{1}{i''} = (n - 1)\left(\frac{1}{r'} - \frac{1}{r''}\right).$$

⁻inally, calling the original object distance simply p and the final image distance ·ds to

$$\frac{1}{p} + \frac{1}{i} = (n - 1)\left(\frac{1}{r'} - \frac{1}{r''}\right), \qquad (35\text{-}26)$$

·hange in notation, is Eqs. 35-9 and 35-10, the relations we set out to ·

REVIEW & SUMMARY

Real and Virtual Images An *image* is a reproduction of an object via light. If the image can form on a surface, it is a *real image* and can exist even if no observer is present. If the image requires the visual system of an observer, it is a *virtual image*.

Image Formation *Spherical mirrors, spherical refracting surfaces,* and *thin lenses* can form images of a source of light—the object—by redirecting rays emerging from the source. The image occurs where the redirected rays cross (forming a real image) or where backward extensions of those rays cross (forming a virtual image). If the rays are sufficiently close to the *central axis* through the spherical mirror, refracting surface, or thin lens, we have the following relations between the *object distance p* (which is positive) and the *image distance i* (which is positive for real images and negative for virtual images):

1. Spherical Mirror:

$$\frac{1}{p} + \frac{1}{i} = \frac{1}{f} = \frac{2}{r}, \qquad (35\text{-}4, 35\text{-}3)$$

where f is the mirror's focal length and r is the mirror's radius of curvature. A *plane mirror* is a special case for which $r \to \infty$, so that $p = -i$. Real images form on the side of a mirror where the object is located, and virtual images form on the opposite side.

2. Spherical Refracting Surface:

$$\frac{n_1}{p} + \frac{n_2}{i} = \frac{n_2 - n_1}{r} \quad \text{(single surface)}, \qquad (35\text{-}8)$$

where n_1 is the index of refraction of the material where the object is located, n_2 is the index of refraction of the material on the other side of the refracting surface, and r is the radius of curvature of the surface. When the object faces a convex refracting surface, the radius r is positive. When it faces a concave surface, r is negative. Real images form on the side of a refracting surface that is opposite the object, and virtual images form on the same side as the object.

3. Thin Lens:

$$\frac{1}{p} + \frac{1}{i} = \frac{1}{f} = (n - 1)\left(\frac{1}{r_1} - \frac{1}{r_2}\right), \qquad (35\text{-}9, 35\text{-}10)$$

where f is the lens's focal length, n is the index of refraction of the lens material, and r_1 and r_2 are the radii of curvature of the two sides of the lens, which are spherical surfaces. A convex lens surface that faces the object has a positive radius of curvature; a concave lens surface that faces the object has a negative radius of curvature. Real images form on the side of a lens that is opposite the object, and virtual images form on the same side as the object.

Lateral Magnification The *lateral magnification m* produced by a spherical mirror or a thin lens is

$$m = -\frac{i}{p}. \qquad (35\text{-}6)$$

The magnitude of m is given by

$$|m| = \frac{h'}{h}, \qquad (35\text{-}5)$$

where h and h' are the heights (measured perpendicular to the central axis) of the object and image, respectively.

Optical Instruments Three optical instruments that extend human vision are:

1. The *simple magnifying lens,* which produces an *angular magnification* m_θ given by

$$m_\theta = \frac{25\text{ cm}}{f}, \qquad (35\text{-}12)$$

where f is the focal length of the magnifying lens.

2. The *compound microscope,* which produces an *overall magnification M* given by

$$M = mm_\theta = -\frac{s}{f_{ob}}\frac{25\text{ cm}}{f_{ey}}, \qquad (35\text{-}14)$$

where m is the lateral magnification produced by the objective, m_θ is the angular magnification produced by the eyepiece, s is the tube length, and f_{ob} and f_{ey} are the focal lengths of the objective and eyepiece, respectively.

3. The *refracting telescope,* which produces an *angular magnification* m_θ given by

$$m_\theta = -\frac{f_{ob}}{f_{ey}}. \qquad (35\text{-}15)$$

QUESTIONS

1. Lake monsters, mermen, and mermaids have long been "sighted" by observers located either on a shore or on a low deck of a ship. From such a low point, an observer can intercept rays of light that leave a floating object (say, a log or a porpoise) and bend slightly back downward toward the observer (one such refracted ray, exaggerated, is shown in Fig. 35-22*a*). The observer then perceives the object as being elongated upward from the water (and probably oscillating because of air turbulence) in a mirage that might easily resemble one of the fabled creatures. Figure 35-22*b* gives several plots of height from the water surface versus air temperature. Which one of the plots best illustrates the air-temperature conditions that can bend the light rays so as to create this mirage?

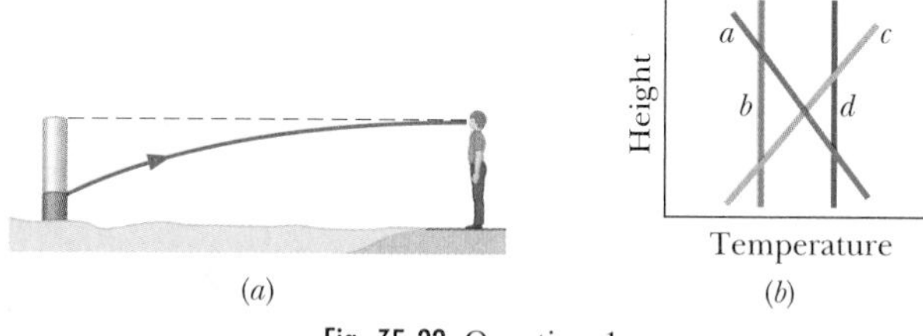

Fig. 35-22 Question 1.

2. Figure 35-23 shows a fish and a fish stalker in water. (a) Does the stalker see the fish in the general region of point *a* or point *b*? (b) Does the fish see the (wild) eyes of the stalker in the general region of point *c* or point *d*?

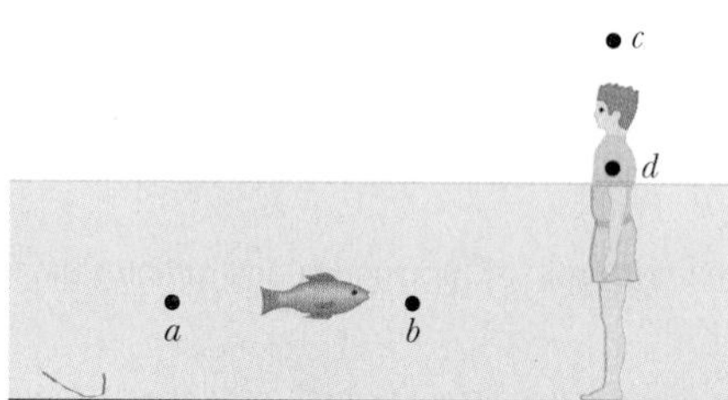

Fig. 35-23 Question 2.

3. In the mirror maze of Fig. 35-24*a*, many "virtual hallways" seem to extend away from you because you see multiple reflections from the mirrors that form the walls of the maze. Those mirrors are placed along some sides of repeated equilateral triangles on the floor. The floor plan for a similar but different maze is shown in Fig. 35-24*b*; every wall section within this maze is mirrored. If you stand at entrance *x*, (a) which of the maze monsters *a*, *b*, and *c* hiding in the maze can you see along the virtual hallways extending from entrance *x*; (b) how many times does each visible monster appear in a hallway; and (c) what is at the far end of a hallway? (*Hint:* The two rays shown are coming down virtual hallways; follow them back into the maze, using the law of reflection at each mirror along the path of each ray. Do they pass through a triangle with a monster? If so, how many times? For additional analysis, see J. Walker, "The Amateur Scientist," *Scientific American*, Vol. 254, pages 120–126, June 1986.)

(*a*)

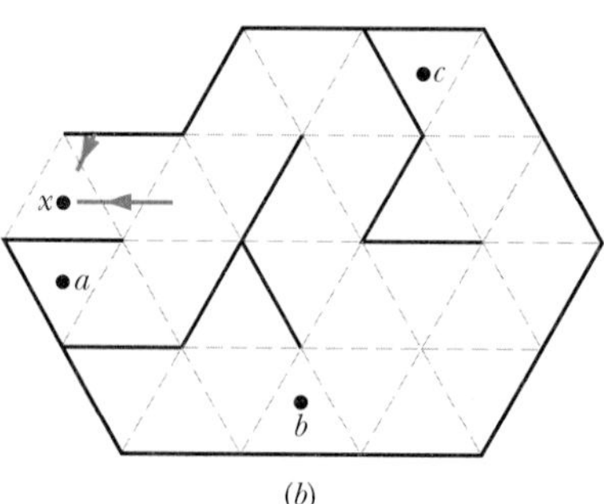

(*b*)

Fig. 35-24 Question 3.

4. A penguin waddles along the central axis of a concave mirror, from the focal point to an effectively infinite distance. (a) How does its image move? (b) Does the height of its image increase continually, decrease continually, or change in some more complicated manner?

5. When a *T. rex* pursues a jeep in the movie *Jurassic Park*, we see a reflected image of the *T. rex* via a side-view mirror, on which is printed the (then darkly humorous) warning: "Objects in mirror are closer than they appear." Is the mirror flat, convex, or concave?

6. Figure 35-25 shows four thin lenses, all of the same material, ... are flat or

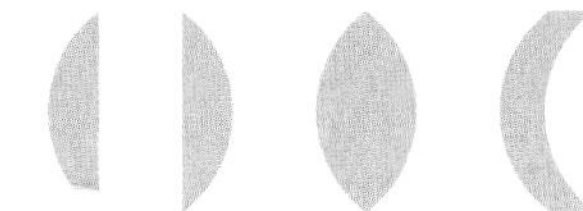

magnitude ... ten calculation, rank the ... according to the magnitude of the focal length, greatest first.

7. An object lies before a thin, symmetric, converging lens. Does the image distance increase, decrease, or remain the same if we increase (a) the index of refraction n of the lens, (b) the magnitude of the radius of curvature of the two sides, and (c) the index of refraction n_{med} of the surrounding medium, keeping n_{med} less than n?

8. A concave mirror and a converging lens (glass with $n = 1.5$) both have a focal length of 3 cm when in air. When they are in water ($n = 1.33$), are their focal lengths greater than, less than, or equal to 3 cm?

9. The table details six variations of the basic arrangement of two thin lenses represented in Fig. 35-26. (The points labeled F_1 and F_2 are the focal points of lenses 1 and 2.) An object is distance p_1 to the left of lens 1, as in Fig. 35-15. (a) For which variations can we tell, *without calculation,* whether the final image (that due to lens 2) is to the left or right of lens 2 and whether it has the same orientation as the object? (b) For those "easy" variations, give the image location as "left" or "right" and the orientation as "same" or "inverted."

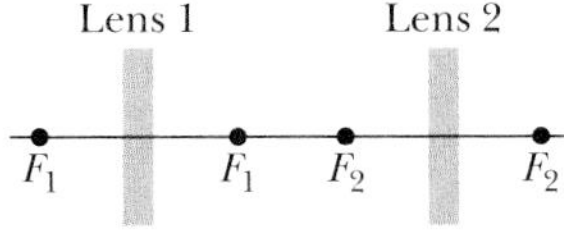

Fig. 35-26 Question 9.

2	Converging	Converging	$p_1 > f_1$
3	Diverging	Converging	$p_1 < f_1$
4	Diverging	Converging	$p_1 > f_1$
5	Diverging	Diverging	$p_1 < f_1$
6	Diverging	Diverging	$p_1 > f_1$

10. Much of the bending of light rays necessary for human vision occurs at the cornea (at the air–eye interface). The cornea has an index of refraction somewhat greater than that of water. (a) When your eye is submerged in a swimming pool, is the bending of light rays at the cornea greater than, less than, or the same as in air? (b) The Central American fish *Anableps anableps* can see simultaneously above and below water because it swims with its eyes partially extending above the water surface. To provide clear sight in both media, is the radius of curvature of the submerged portion of the cornea greater than, less than, or equal to that of the exposed portion?

EXERCISES & PROBLEMS

ssm Solution is in the Student Solutions Manual.
www Solution is available on the World Wide Web at: http://www.wiley.com/college/hrw
ilw Solution is available on the Interactive LearningWare.

SEC. 35-2 Plane Mirrors

1E. A moth at about eye level is 10 cm in front of a plane mirror; you are behind the moth, 30 cm from the mirror. What is the distance between your eyes and the apparent position of the moth's image in the mirror? ssm ilw

2E. You look through a camera toward an image of a hummingbird in a plane mirror. The camera is 4.30 m in front of the mirror. The bird is at camera level, 5.00 m to your right and 3.30 m from the mirror. What is the distance between the camera and the apparent position of the bird's image in the mirror?

3E. Figure 35-27*a* is an overhead view of two vertical plane mirrors with an object O placed between them. If you look into the mirrors, you see multiple images of O. You can find them by drawing the reflection in each mirror of the angular region between the mirrors, as is done for the left-hand mirror in Fig. 35-27*b*. Then draw the reflection of the reflection. Continue this on the left and on the right until the reflections meet or overlap at the rear of the mirrors. Then you can count the number of images of O. (a) If $\theta = 90°$, how many images of O would you see? (b) Draw their locations and orientations (as in Fig. 35-27*b*). ssm

4P. Repeat Exercise 3 for the mirror angle θ equal to (a) 45°, (b) 60°, and (c) 120°. (d) Explain why there are several possible answers for (c).

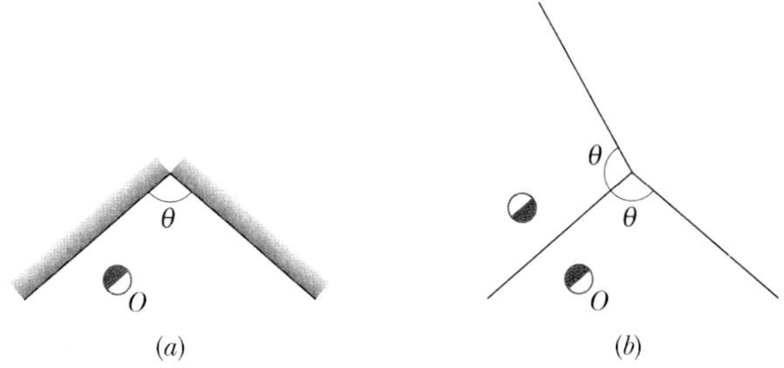

Fig. 35-27 Exercise 3 and Problem 4.

5P. Prove that if a plane mirror is rotated through an angle α, the reflected beam is rotated through an angle 2α. Show that this result is reasonable for $\alpha = 45°$. ssm

6P. Figure 35-28 shows an overhead view of a corridor with a

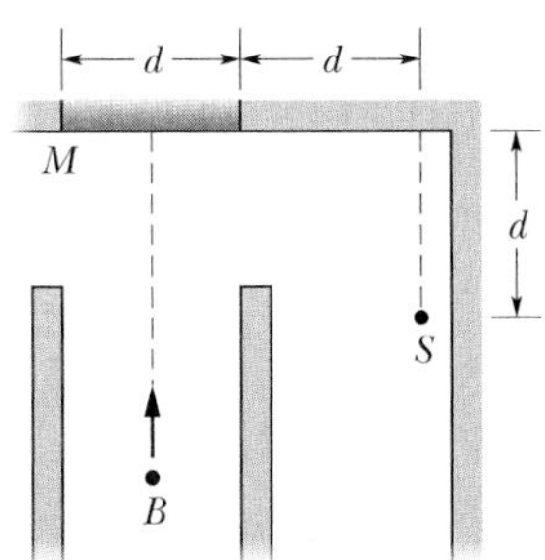

Fig. 35-28 Problem 6.

plane mirror M mounted at one end. A burglar B sneaks along the corridor directly toward the center of the mirror. If $d = 3.0$ m, how far from the mirror will she be when the security guard S can first see her in the mirror?

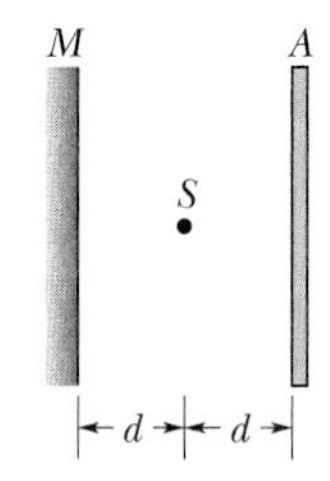

Fig. 35-29 Problem 7.

7P. You put a point source of light S a distance d in front of a screen A. How is the light intensity at the center of the screen changed if you put a completely reflecting mirror M a distance d behind the source, as in Fig. 35-29? (*Hint:* Use Eq. 34-27.) ssm

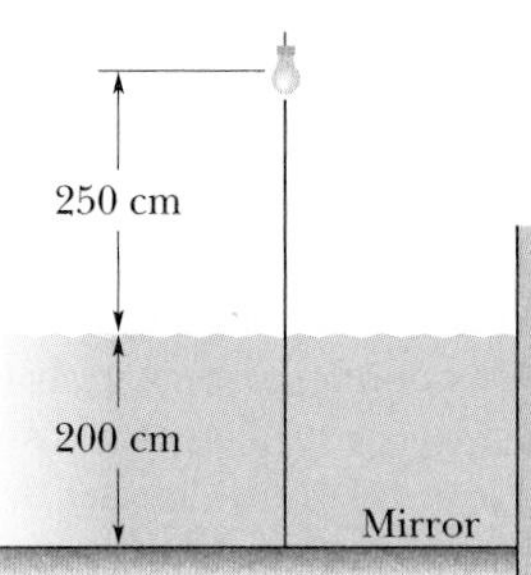

Fig. 35-30 Problem 8.

8P. Figure 35-30 shows a small lightbulb suspended above the surface of the water in a swimming pool. The bottom of the pool is a large mirror. How far below the mirror's surface is the image of the bulb? (*Hint:* Construct a diagram of two rays like that of Fig. 35-3, but take into account the bending of light rays by refraction. Assume that the rays are close to a vertical axis through the bulb, and use the small-angle approximation that $\sin\theta \approx \tan\theta \approx \theta$.)

SEC. 35-4 Images from Spherical Mirrors

9E. A concave shaving mirror has a radius of curvature of 35.0 cm. It is positioned so that the (upright) image of a man's face is 2.50 times the size of the face. How far is the mirror from the face?

10P. Fill in Table 35-3, each row of which refers to a different combination of an object and either a plane mirror, a spherical convex mirror, or a spherical concave mirror. Distances are in centimeters. If a number lacks a sign, find the sign. Sketch each combination and draw in enough rays to locate the object and its image.

11P. A short straight object of length L lies along the central axis of a spherical mirror, a distance p from the mirror. (a) Show that its image in the mirror has a length L' where

$$L' = L\left(\frac{f}{p-f}\right)^2.$$

(*Hint:* Locate the two ends of the object.) (b) Show that the *longitudinal magnification* m' $(= L'/L)$ is equal to m^2, where m is the lateral magnification. ssm www

12P. (a) A luminous point is moving at speed v_O toward a spherical mirror with radius of curvature r, along the central axis of the mirror. Show that the image of this point is moving at speed

$$v_I = -\left(\frac{r}{2p-r}\right)^2 v_O,$$

where p is the distance of the luminous point from the mirror at any given time. (*Hint:* Start with Eq. 35-4.) Now assume that the mirror is concave, with $r = 15$ cm, and let $v_O = 5.0$ cm/s. Find the speed of the image when (b) $p = 30$ cm (far outside the focal point), (c) $p = 8.0$ cm (just outside the focal point), and (d) $p = 10$ mm (very near the mirror).

SEC. 35-5 Spherical Refracting Surfaces

13P. A beam of parallel light rays from a laser is incident on a solid transparent sphere of index of refraction n (Fig. 35-31). (a) If a point image is produced at the back of the sphere, what is the index of refraction of the sphere? (b) What index of refraction, if any, will produce a point image at the center of the sphere?

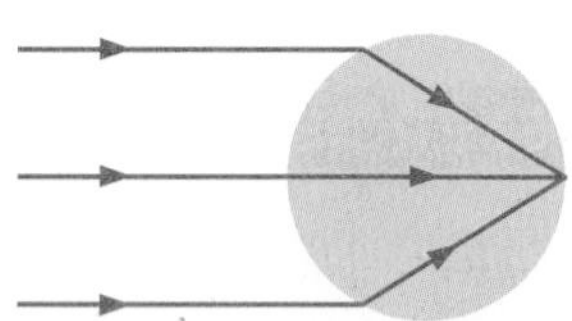
Fig. 35-31 Problem 13.

14P. Fill in Table 35-4, each row of which refers to a different combination of a point object and a spherical refracting surface separating two media with different indexes of refraction. Distances are in centimeters. If a number lacks a sign, find the sign. Sketch

TABLE 35-3 Problem 10: Mirrors

Type	f	r	i	p	m	Real Image?	Inverted Image?
(a) Concave	20			+10			
(b)				+10	+1.0	No	
(c)	+20			+30			
(d)				+60	−0.50		
(e)		−40	−10				
(f)	20				+0.10		
(g) Convex		40	4.0				
(h)				+24	0.50		Yes

TABLE 35-4 Problem 14: Spherical Refracting Surfaces

	n_1	n_2	p	i	r	Inverted Image?
(a)	1.0	1.5	+10		+30	
(b)	1.0	1.5	+10	−13		
(c)	1.0	1.5		+600	+30	
(d)	1.0		+20	−20	−20	
(e)	1.5	1.0	+10	−6.0		
(f)	1.5	1.0		−7.5	−30	
(g)	1.5	1.0	+70		+30	
(h)	1.5		+100	+600	−30	

TABLE 35-5 Problem 24: Thin Lenses

	Type	f	r_1	r_2	i	p	n	m	Real Image?	Inverted Image?
(a)	C	10				+20				
(b)		+10				+5.0				
(c)		10				+5.0		>1.0		
(d)		10				+5.0		<1.0		
(e)			+30	−30		+10	1.5			
(f)			−30	+30		+10	1.5			
(g)			−30	−60		+10	1.5			
(h)						+10		0.50		No
(i)						+10		−0.50		

each combination and draw in enough rays to locate the object and image.

15P. You look downward at a coin that lies at the bottom of a pool of liquid with depth d and index of refraction n (Fig. 35-32). Because you view with two eyes, which intercept different rays of light from the coin, you perceive the coin to be where extensions of the intercepted rays cross, at depth d_a instead of d. Assuming that the intercepted rays in Fig. 35-32 are close to a vertical axis through the coin, show that $d_a = d/n$. (*Hint:* Use the small-angle approximation that $\sin\theta \approx \tan\theta \approx \theta$.)

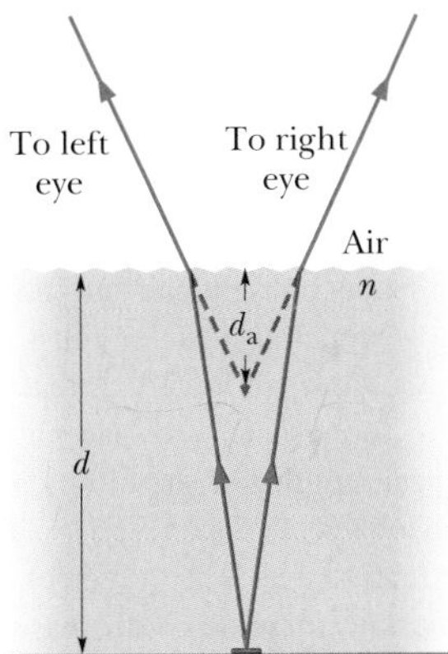

Fig. 35-32 Problem 15.

16P. A 20-mm-thick layer of water ($n = 1.33$) floats on a 40-mm-thick layer of carbon tetrachloride ($n = 1.46$) in a tank. A coin lies at the bottom of the tank. At what depth below the top water surface do you perceive the coin? (*Hint:* Use the result and assumptions of Problem 15 and work with a ray diagram of the situation.)

SEC. 35-6 Thin Lenses

17E. An object is 20 cm to the left of a thin diverging lens having a 30 cm focal length. What is the image distance i? Find the image position with a ray diagram. ssm ilw

18E. You produce an image of the Sun on a screen, using a thin lens whose focal length is 20.0 cm. What is the diameter of the image? (See Appendix C for needed data on the Sun.)

19E. A double-convex lens is to be made of glass with an index of refraction of 1.5. One surface is to have twice the radius of curvature of the other and the focal length is to be 60 mm. What are the radii? ssm

20E. A lens is made of glass having an index of refraction of 1.5. One side of the lens is flat, and the other is convex with a radius of curvature of 20 cm. (a) Find the focal length of the lens. (b) If an object is placed 40 cm in front of the lens, where will the image be located?

21E. The formula
$$\frac{1}{p} + \frac{1}{i} = \frac{1}{f}$$
is called the *Gaussian* form of the thin-lens formula. Another form of this formula, the *Newtonian* form, is obtained by considering the distance x from the object to the first focal point and the distance x' from the second focal point to the image. Show that
$$xx' = f^2$$
is the Newtonian form of the thin-lens formula. ssm

22E. A movie camera with a (single) lens of focal length 75 mm takes a picture of a 180-cm-high person standing 27 m away. What is the height of the image of the person on the film?

23P. An illuminated slide is held 44 cm from a screen. How far from the slide must a lens of focal length 11 cm be placed to form an image of the slide's picture on the screen? ilw

24P. To the extent possible, fill in Table 35-5, each row of which refers to a different combination of an object and a thin lens. Distances are in centimeters. For the type of lens, use C for converging and D for diverging. If a number (except for the index of refraction) lacks a sign, find the sign. Sketch each combination and draw in enough rays to locate the object and image.

25P. Show that the distance between an object and its real image formed by a thin converging lens is always greater than or equal to four times the focal length of the lens. ssm www

26P. A diverging lens with a focal length of −15 cm and a converging lens with a focal length of 12 cm have a common central axis. Their separation is 12 cm. An object of height 1.0 cm is 10 cm in front of the diverging lens, on the common central axis. (a) Where does the lens combination produce the final image of the object (the one produced by the second, converging lens)? (b) What is the height of that image? (c) Is the image real or virtual? (d) Does the image have the same orientation as the object or is it inverted?

27P. A converging lens with a focal length of +20 cm is located 10 cm to the left of a diverging lens having a focal length of

−15 cm. If an object is located 40 cm to the left of the converging lens, locate and describe completely the final image formed by the diverging lens.

28P. An object is 20 cm to the left of a lens with a focal length of +10 cm. A second lens of focal length +12.5 cm is 30 cm to the right of the first lens. (a) Find the location and relative size of the final image. (b) Verify your conclusions by drawing the lens system to scale and constructing a ray diagram. (c) Is the final image real or virtual? (d) Is it inverted?

29P. Two thin lenses of focal lengths f_1 and f_2 are in contact. Show that they are equivalent to a single thin lens with

$$f = \frac{f_1 f_2}{f_1 + f_2}$$

as its focal length. ssm

30P. In Fig. 35-33, a real inverted image I of an object O is formed by a certain lens (not shown); the object–image separation is $d = 40.0$ cm, measured along the central axis of the lens. The image is just half the size of the object. (a) What kind of lens must be used to produce this image? (b) How far from the object must the lens be placed? (c) What is the focal length of the lens?

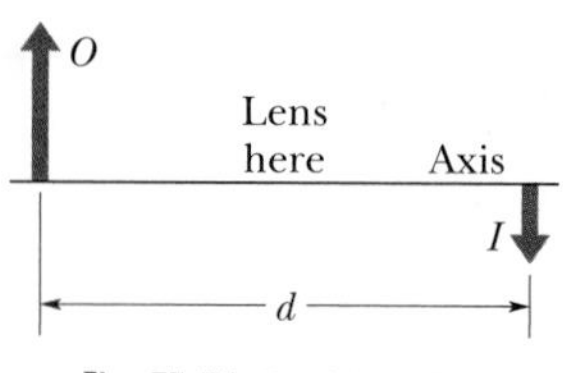

Fig. 35-33 Problem 30.

31P. A luminous object and a screen are a fixed distance D apart. (a) Show that a converging lens of focal length f, placed between object and screen, will form a real image on the screen for two lens positions that are separated by a distance

$$d = \sqrt{D(D - 4f)}.$$

(b) Show that

$$\left(\frac{D - d}{D + d}\right)^2$$

gives the ratio of the two image sizes for these two positions of the lens. ssm www

SEC. 35-7 Optical Instruments

32E. If the angular magnification of an astronomical telescope is 36 and the diameter of the objective is 75 mm, what is the minimum diameter of the eyepiece required to collect all the light entering the objective from a distant point source on the telescope axis?

33E. In a microscope of the type shown in Fig. 35-17, the focal length of the objective is 4.00 cm, and that of the eyepiece is 8.00 cm. The distance between the lenses is 25.0 cm. (a) What is the tube length s? (b) If image I in Fig. 35-17 is to be just inside focal point F_1', how far from the objective should the object be? What then are (c) the lateral magnification m of the objective, (d) the angular magnification m_θ of the eyepiece, and (e) the overall magnification M of the microscope? ssm

34P. A simple magnifying lens of focal length f is placed near the eye of someone whose near point P_n is 25 cm from the eye. An object is positioned so that its image in the magnifying lens appears at P_n. (a) What is the lens's angular magnification? (b) What is the angular magnification if the object is moved so that its image appears at infinity? (c) Evaluate the angular magnifications of (a) and (b) for $f = 10$ cm. (Viewing an image at P_n requires effort by muscles in the eye, whereas for many people viewing an image at infinity requires no effort.)

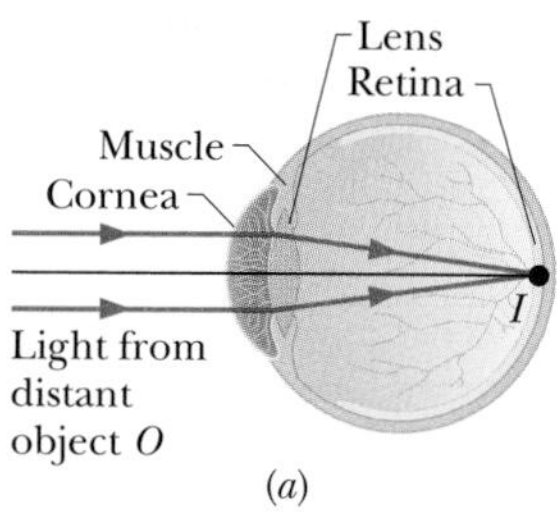

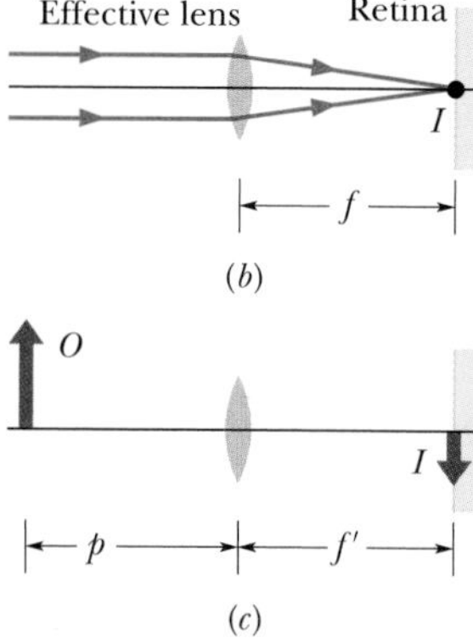

Fig. 35-34 Problem 35.

35P. Figure 35-34a shows the basic structure of a human eye. Light refracts into the eye through the cornea and is then further redirected by a lens whose shape (and thus ability to focus the light) is controlled by muscles. We can treat the cornea and eye lens as a single effective thin lens (Fig. 35-34b). A "normal" eye can focus parallel light rays from a distant object O to a point on the retina at the back of the eye, where processing of the visual information begins. As an object is brought close to the eye, however, the muscles must change the shape of the lens so that rays form an inverted real image on the retina (Fig. 35-34c). (a) Suppose that for the parallel rays of Figs. 35-34a and b, the focal length f of the effective thin lens of the eye is 2.50 cm. For an object at distance $p = 40.0$ cm, what focal length f' of the effective lens is required for it to be seen clearly? (b) Must the eye muscles increase or decrease the radii of curvature of the eye lens to give focal length f'? ssm

36P. An object is 10.0 mm from the objective of a certain compound microscope. The lenses are 300 mm apart and the intermediate image is 50.0 mm from the eyepiece. What overall magnification is produced by the instrument?

37P. Figure 35-35a shows the basic structure of a camera. A lens can be moved forward or back to produce an image on film at the back of the camera. For a certain camera, with the distance i between the lens and the film set at $f = 5.0$ cm, parallel light rays from a very distant object O converge to a point image on the film, as shown. The object is now brought closer, to a distance of $p = 100$ cm, and the lens–film distance is adjusted so that an inverted real image forms on the film (Fig. 35-35b). (a) What is the lens–film distance i now? (b) By how much was i changed?

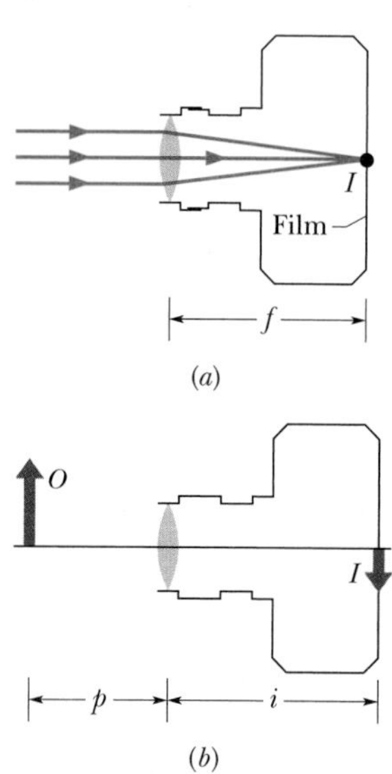

Fig. 35-35 Problem 37.

NEW PROBLEMS

N1. An object is moved along the central axis of a spherical mirror while the lateral magnification m of it is measured. Figure 35N-1 gives m versus object distance p for a range of p. What is the magnification of the object when the object is 14.0 cm from the mirror?

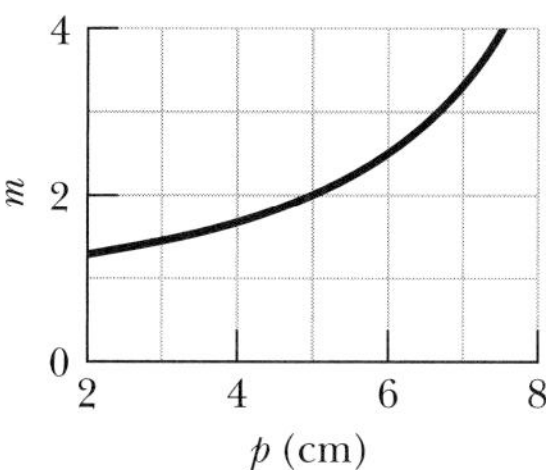

Fig. 35N-1 Problem N1.

N2. An object is placed against the center of a spherical mirror and then moved 70 cm from it along the central axis as the image distance i is measured. Figure 35N-2 gives i versus object distance p out to $p = 40$ cm. What is the image distance when the object is on the central axis and 70 cm from the mirror?

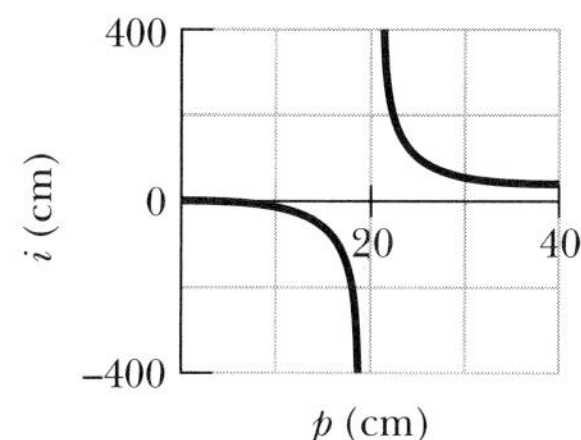

Fig. 35N-2 Problem N2.

N3 through N11. *Spherical mirrors.* Object O stands on the central axis of a spherical mirror. For this situation, each problem in Table 35-6 gives object distance p (centimeters), the type of mirror, and then the distance (centimeters, without proper sign) between the focal point and the mirror. Find the image distance i and the lateral magnification m of the object, including signs. Also, determine whether the image is real (R) or virtual (V), inverted (I) from object O or noninverted (NI), and on the *same* side of the mirror as object O or on the *opposite* side.

TABLE 35-6 Problems N3 through N11: Spherical Mirrors
See the setup for these problems.

	p	Mirror	i	m	R/V	I/NI	Side
N3.	+18	concave, 12					
N4.	+12	concave, 18					
N5.	+8.0	convex, 10					
N6.	+10	convex, 8.0					
N7.	+24	concave, 36					
N8.	+17	convex, 14					
N9.	+22	convex, 35					
N10.	+15	concave, 10					
N11.	+19	concave, 19					

N12. An object is placed against the center of a spherical mirror and then moved 70 cm from it along the central axis as the image distance i is measured. Figure 35N-3 gives i versus object distance p out to $p = 40$ cm. What is the image distance when the object is 70 cm from the mirror?

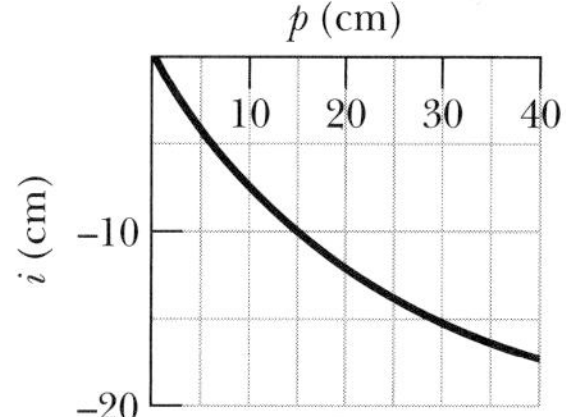

Fig. 35N-3 Problem N12.

N13 through N17. *More mirrors.* Object O stands on the central axis of a spherical mirror. For this situation, each problem in Table 35-7 refers to the type of mirror, the focal distance f, the object distance p, the image distance i, and the lateral magnification m. (All distances are in centimeters.) Complete a line by filling in the mirror type and the missing numbers and signs. Also, determine whether the image is real (R) or virtual (V), inverted (I) from object O or noninverted (NI), and on the *same* side of the mirror as object O or on the *opposite* side.

TABLE 35-7 Problems N13 through N17: More Mirrors
See the setup for these problems.

	Type	f	p	i	m	R/V	I/NI	Side
N13.			+30		0.40		I	
N14.		30			+0.20			
N15.		−30		−15				
N16.			+40		−0.70			
N17.		20	+60					same

N18. Figure 35N-4 gives the lateral magnification m of an object versus the object distance p from a spherical mirror as the object is moved along the mirror's central axis through a range of values for p. What is the magnification of the object when the object is 21 cm from the mirror?

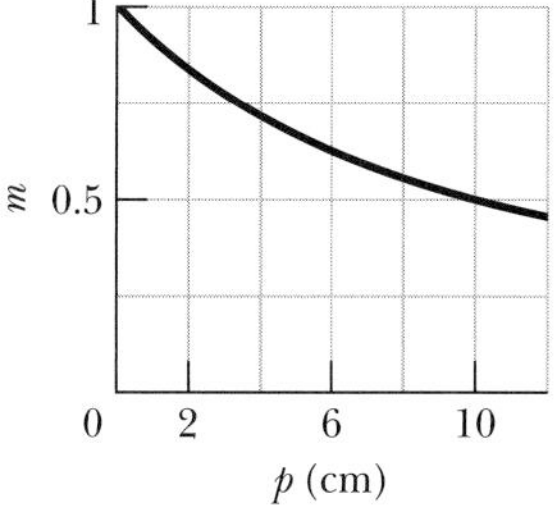

Fig. 35N-4 Problem N18.

N19 through N23. *Spherical refracting surfaces.* An object O stands on the central axis of a spherical refracting surface. For this situation, each problem in Table 35-8 gives the index of refraction n_1 where the object is located, the index of refraction n_2 on the other side of the refracting surface, object distance p (centimeters), and radius of curvature r of the surface (centimeters). Find image distance i. Also, determine whether the image is real (R) or virtual (V), inverted (I) from object O or noninverted (NI), and on the *same* side of the mirror as object O or on the *opposite* side.

TABLE 35-8 Problems N19 through N23: Spherical Refracting Surfaces
See the setup for these problems.

	n_1	n_2	p	r	i	R/V	I/NI	Side
N19.	1.0	1.5	+12	+9.0				
N20.	1.0	1.5	+4.0	+9.0				
N21.	1.0	1.5	+20	+4.0				
N22.	1.0	1.5	+14	−4.0				
N23.	1.5	1.0	+30	−8.0				

N24. An object is placed against the center of a thin lens and then moved away from it along the central axis as the image distance i is measured. Figure 35N-5 gives i versus object distance p out to $p = 60$ cm. What is the image distance when the object is 100 cm from the lens?

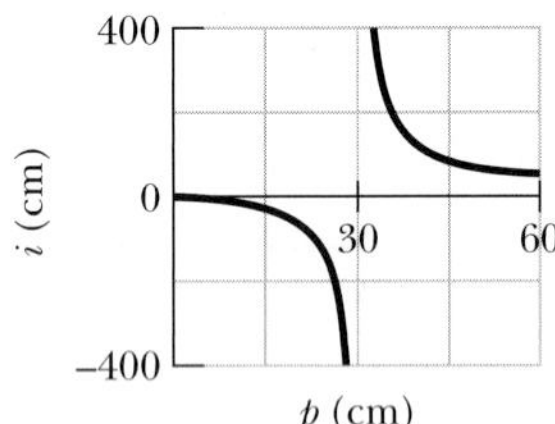

Fig. 35N-5 Problem N24.

N25 through N32. *Thin lenses.* Object O stands on the central axis of a thin, symmetric lens. For this situation, each problem in Table 35-9 gives object distance p (centimeters), the type of lens (C stands for converging and D for diverging), and then the distance (centimeters, without proper sign) between a focal point and the lens. Find the image distance i and the lateral magnification m of the object, including signs. Also, determine whether the image is real (R) or virtual (V), inverted (I) from object O or noninverted (NI), and on the *same* side of the lens as object O or on the *opposite* side.

TABLE 35-9 Problems N25 through N32: Thin Lenses
See the setup for these problems.

	p	Lens	i	m	R/V	I/NI	Side
N25.	+16	C, 4.0					
N26.	+12	C, 16					
N27.	+10	D, 6.0					
N28.	+8.0	D, 12					
N29.	+25	C, 35					
N30.	+22	D, 14					
N31.	+12	D, 31					
N32.	+45	C, 20					

N33 through N39. *Lenses with given radii.* Object O stands in front of a lens, on the central axis. For this situation, each problem in Table 35-10 gives object distance p, index of refraction n of the lens, radius r_1 of the nearer lens surface, and radius r_2 of the farther lens surface. (The distances are in centimeters.) Find the image distance i and the lateral magnification m of the object, including signs. Also, determine whether the image is real (R) or virtual (V), inverted (I) from object O or noninverted (NI), and on the *same* side of the lens as object O or on the *opposite* side.

TABLE 35-10 Problems N33 through N39: Lenses with Given Radii
See the setup for these problems.

	p	n	r_1	r_2	i	m	R/V	I/NI	Side
N33.	+60	1.50	+35	−35					
N34.	+6.0	1.70	+10	−12					
N35.	+24	1.50	−15	−25					
N36.	+18	1.60	−27	+24					
N37.	+35	1.70	+42	+33					
N38.	+29	1.65	+35	∞					
N39.	+75	1.55	+30	−42					

N40. An object is placed against the center of a thin lens and then moved 70 cm from it along the central axis as the image distance i is measured. Figure 35N-6 gives i versus object distance p out to $p = 40$ cm. What is the image distance when the object is 70 cm from the lens?

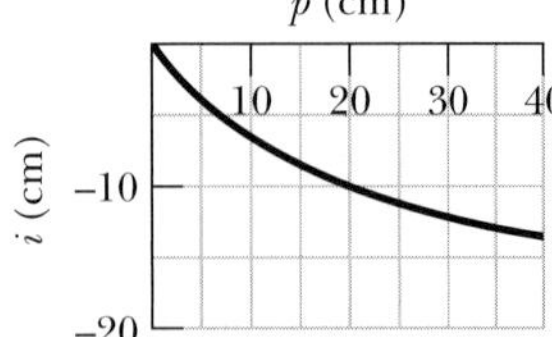

Fig. 35N-6 Problem N40.

N41. An object is moved along the central axis of a thin lens while the lateral magnification m of it is measured. Figure 35N-7 gives m versus object distance p for a range of p. What is the magnification of the object when the object is 14.0 cm from the lens?

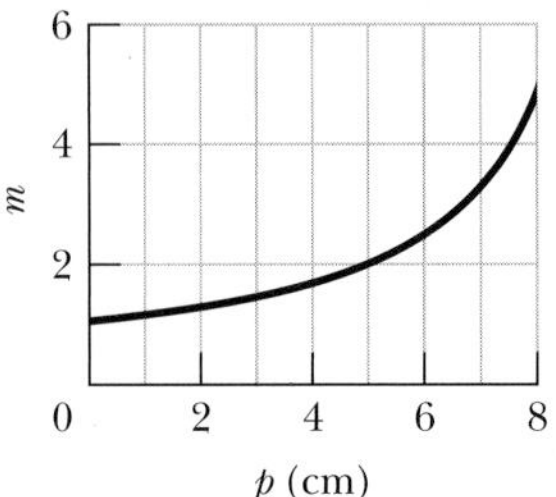

Fig. 35N-7 Problem N41.

N42. Figure 35N-8 gives the lateral magnification m of an object versus the object distance p from a lens as the object is moved along the central axis of the lens through a range of values for p. What is the magnification of the object when the object is 35 cm from the lens?

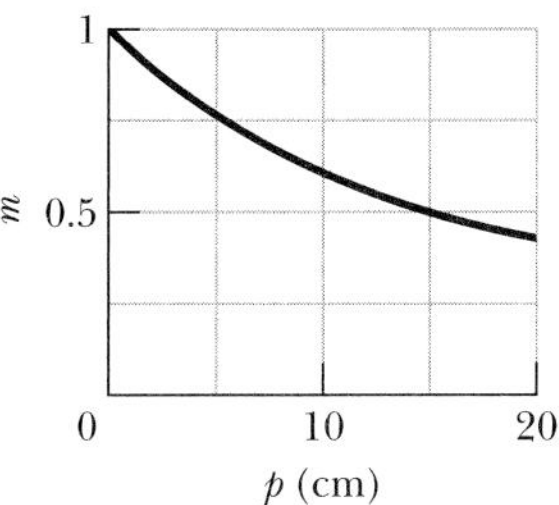

Fig. 35N-8 Problem N42.

N43 through N47. *More lenses*. Object O stands at distance p on the central axis of a thin symmetric lens. Each problem in Table 35-11 refers to different lenses and distances (in centimeters) for that arrangement. Complete a line by filling in the type of lens (converging or diverging) and the missing numbers and signs, including the values for m in the first two problems. Also, determine whether the image is real (R) or virtual (V), inverted (I) from object O or noninverted (NI), and on the *same* side of the lens as object O or on the *opposite* side.

TABLE 35-11 Problems N43 through N47: More Lenses
See the setup for these problems.

	Type	f	p	i	m	R/V	I/NI	Side
N43.		20	+8.0		>1.0			
N44.		20	+8.0		<1.0		NI	
N45.			+16		+0.25			
N46.			+16		−0.25			
N47.			+16		+1.25			

N48 through N55. *Two-lens systems*. In Fig. 35N-9, stick figure O (the object) stands on the common central axis of two thin, symmetric lenses, which are mounted in the boxed regions. Lens 1 is mounted within the boxed region closer to O, which is at object distance p_1. Lens 2 is mounted within the farther boxed region, at distance d. Each problem in Table 35-12 refers to a different combination of lenses and different values for distances, which are given in centimeters. The type of lens is indicated by C for converging and D for diverging; the number after C or D is the distance between a lens and either of its focal points (the proper sign of the focal distance is not indicated).

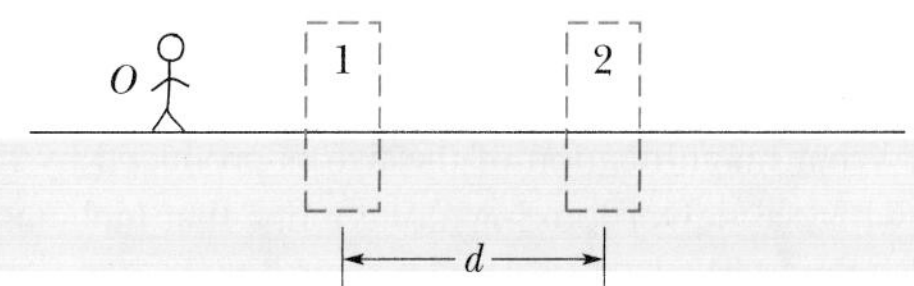

Fig. 35N-9 Problems N48 through N55.

Find the image distance i_2 for the (final) image produced by lens 2 (the final image produced by the system) and the overall lateral magnification m for the system, including signs. Also, determine whether the final image is real (R) or virtual (V), inverted (I) from object O or noninverted (NI), and on the *same* side of lens 2 as object O or on the *opposite* side.

TABLE 35-12 Problems N48 through N55: Two-lens Systems
See the setup for these problems.

	p_1	Lens 1	d	Lens 2	i_2	m	R/V	I/NI	Side
N48.	+10	C, 15	10	C, 8.0					
N49.	+12	C, 8.0	32	C, 6.0					
N50.	+15	C, 12	67	C, 10					
N51.	+20	C, 9.0	8.0	C, 5.0					
N52.	+8.0	D, 6.0	12	C, 6.0					
N53.	+4.0	C, 6.0	8.0	D, 6.0					
N54.	+12	C, 8.0	30	D, 8.0					
N55.	+20	D, 12	10	D, 8.0					

N56. (a) Show that if the object O in Fig. 35-16c is moved from focal point F_1 toward the observer's eye, the image moves in from infinity and the angle θ' (and thus the angular magnification m_θ) increases. (b) If you continue this process, at what image location will m_θ have its maximum usable value? (You can then still increase m_θ, but the image will no longer be clear.) (c) Show that the maximum usable value of m_θ is $1 + (25 \text{ cm})/f$. (d) Show that in this situation the angular magnification is equal to the lateral magnification.

N57. A narrow beam of parallel light rays is incident on a glass sphere from the left, directed toward the center of the sphere. (The sphere is a lens but certainly not a *thin* lens.) Approximate the angle of incidence of the rays as 0°, and assume that the index of refraction of the glass is $n < 2.0$. Find the image distance i (from the right side of the sphere) in terms of n and the radius r of the sphere. (*Hint:* Apply Eq. 35-8 to locate the image that is produced by refraction at the left side of the sphere; then use that image as the object for refraction at the right side of the sphere to locate the final image. In the second refraction, is the object distance p positive or negative?)

N58*. A goldfish in a spherical fish bowl of radius R is at the level of the center C of the bowl and at distance $R/2$ from the glass (Fig. 35N-10). What magnification of the fish is produced by the water of the bowl for a viewer looking along a line that includes the fish and the center, with the fish on the near side of the center? The index of refraction of the water in the bowl is 1.33. Neglect the glass wall of the bowl. Assume the viewer looks with one eye. (*Hint:* Equation 35-5 holds, but Eq. 35-6 does not. You need to work with a ray diagram of the situation and assume that the rays are close to the observer's line of sight—that is, they deviate from that line by only small angles.)

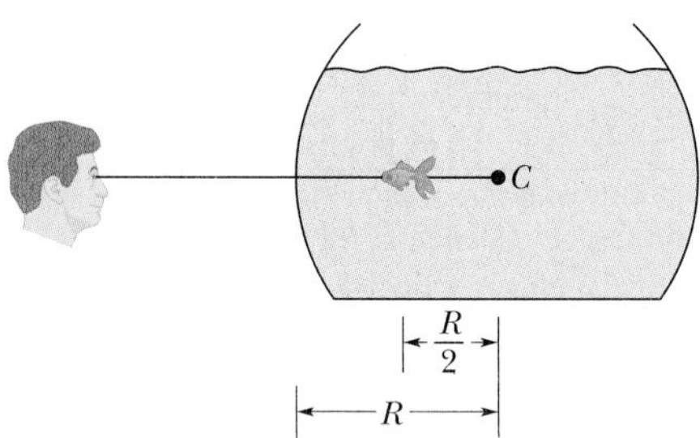

Fig. 35N-10 Problem N58.

N59*. A *corner reflector,* much used in optical, microwave, and other applications, consists of three plane mirrors fastened together to form the corner of a cube. The device has the following property: after three reflections, an incident ray is returned with its direction exactly reversed. Prove this result.

N60. The *power* P of a lens is defined as $P = 1/f$, where f is the focal length. The unit of power is the *diopter,* where 1 diopter = 1 m^{-1}. (a) Why is this a reasonable definition to use for lens power? (b) Show that the net power of two lenses in contact is given by $P = P_1 + P_2$, where P_1 and P_2 are the powers of the two lenses.

N61. The equation $1/p + 1/i = 2/r$ for spherical mirrors is an approximation that is valid if the image is formed by rays that make only small angles with the central axis. In reality, many of the angles are large, which smears out the image a little. You can determine how much. Refer to Fig. 35-19 and consider a ray that leaves a point source (the object) on the central axis and that makes an angle α with that axis.

First, find the point of intersection of the ray with the mirror. If the coordinates of this point are x and y and the origin is placed at the center of curvature, then $y = (x + p - r)\tan\alpha$ and $x^2 + y^2 = r^2$, where p is the object distance and r is the mirror's radius of curvature. Next, use $\tan\beta = y/x$ to find the angle β at the point of intersection, and then use $\alpha + \gamma = 2\beta$ to find the value of γ. Finally, use $\tan\gamma = y/(x + i - r)$ to find the distance i of the image.

(a) Suppose $r = 12$ cm and $p = 20$ cm. For each of the following values of α, find the position of the image—that is, the position of the point where the reflected ray crosses the central axis: 0.500, 0.100, 0.0100 rad. Compare the results with those obtained with the equation $1/p + 1/i = 2/r$. (b) Repeat the calculations for $p = 4.00$ cm.

N62 through N67. *Three-lens systems.* In Fig. 35N-11, stick figure O (the object) stands on the common central axis of three thin, symmetric lenses, which are mounted in the boxed regions. Lens 1 is mounted within the boxed region closest to O, which is at object distance p_1. Lens 2 is mounted within the middle boxed region, at distance d_{12} from lens 1. Lens 3 is mounted in the farther boxed region, at distance d_{23} from lens 2. Each problem in Table 35-13 refers to a different combination of lenses and different values for distances, which are given in centimeters. The type of lens is indicated by C for converging and D for diverging; the number after C or D is the distance between a lens and either of the focal points (the proper sign of the focal distance is not indicated).

Find the image distance i_3 for the (final) image produced by lens 3 (the final image produced by the system) and the overall lateral magnification m for the system, including signs. Also, determine whether the final image is real (R) or virtual (V), inverted (I) from object O or noninverted (NI), and on the same side of lens 3 as object O or on the opposite side.

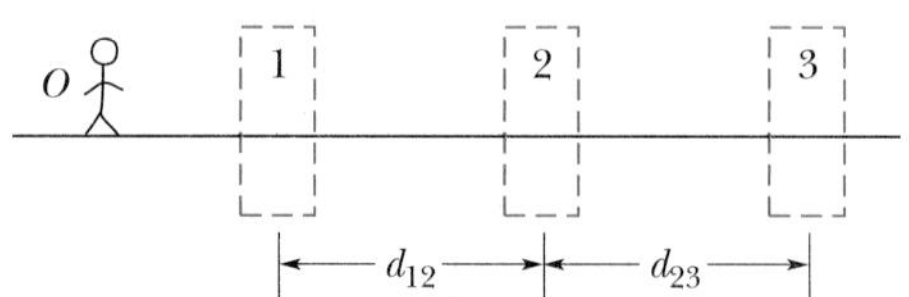

Fig. 35N-11 Problems N62 through N67.

TABLE 35-13 Problems N62 through N67: Three-Lens Systems
See the setup for these problems.

	p_1	Lens 1	d_{12}	Lens 2	d_{23}	Lens 3	i_3	m	R/V	I/NI	Side
N62.	+18	C, 6.0	15	C, 3.0	11	C, 3.0					
N63.	+2.0	C, 6.0	15	C, 6.0	19	C, 5.0					
N64.	+12	C, 8.0	28	C, 6.0	8.0	C, 6.0					
N65.	+4.0	D, 6.0	9.6	C, 6.0	14	C, 4.0					
N66.	+8.0	D, 8.0	8.0	D, 16	5.1	C, 8.0					
N67.	+4.0	C, 6.0	8.0	D, 4.0	5.7	D, 12					

36 Interference

At first glance, the top surface of the *Morpho* butterfly's wing is simply a beautiful blue-green. There is something strange about the color, however, for it almost glimmers, unlike the colors of most objects—and if you change your perspective, or if the wing moves, the tint of the color changes. The wing is said to be iridescent, and the blue-green we see hides the wing's "true" dull brown color that appears on the bottom surface.

What, then, is so different about the top surface that gives us this arresting display?

The answer is in this chapter.

36-1 Interference

Sunlight, as the rainbow shows us, is a composite of all the colors of the visible spectrum. The colors reveal themselves in the rainbow because the incident wavelengths are bent through different angles as they pass through raindrops that produce the bow. However, soap bubbles and oil slicks can also show striking colors, produced not by refraction but by constructive and destructive **interference** of light. The interfering waves combine either to enhance or to suppress certain colors in the spectrum of the incident sunlight. Interference of light waves is thus a superposition phenomenon like those we discussed in Chapter 17.

This selective enhancement or suppression of wavelengths has many applications. When light encounters an ordinary glass surface, for example, about 4% of the incident energy is reflected, thus weakening the transmitted beam by that amount. This unwanted loss of light can be a real problem in optical systems with many components. A thin, transparent "interference film," deposited on the glass surface, can reduce the amount of reflected light (and thus enhance the transmitted light) by destructive interference. The bluish cast of a camera lens reveals the presence of such a coating. Interference coatings can also be used to enhance—rather than reduce—the ability of a surface to reflect light.

To understand interference, we must go beyond the restrictions of geometrical optics and employ the full power of wave optics. In fact, as you will see, the existence of interference phenomena is perhaps our most convincing evidence that light is a wave—because interference cannot be explained other than with waves.

36-2 Light as a Wave

The first person to advance a convincing wave theory for light was Dutch physicist Christian Huygens, in 1678. Although much less comprehensive than the later electromagnetic theory of Maxwell, Huygens' theory was simpler mathematically and remains useful today. Its great advantages are that it accounts for the laws of reflection and refraction in terms of waves and gives physical meaning to the index of refraction.

Huygens' wave theory is based on a geometrical construction that allows us to tell where a given wavefront will be at any time in the future if we know its present position. This construction is based on **Huygens' principle,** which is:

> All points on a wavefront serve as point sources of spherical secondary wavelets. After a time t, the new position of the wavefront will be that of a surface tangent to these secondary wavelets.

Here is a simple example. At the left in Fig. 36-1, the present location of a wavefront of a plane wave traveling to the right in vacuum is represented by plane ab, perpendicular to the page. Where will the wavefront be at time Δt later? We let several points on plane ab (the dots) serve as sources of spherical secondary wavelets that are emitted at $t = 0$. At time Δt, the radius of all these spherical wavelets will have grown to $c\,\Delta t$, where c is the speed of light in vacuum. We draw plane de tangent to these wavelets at time Δt. This plane represents the wavefront of the plane wave at time Δt; it is parallel to plane ab and a perpendicular distance $c\,\Delta t$ from it.

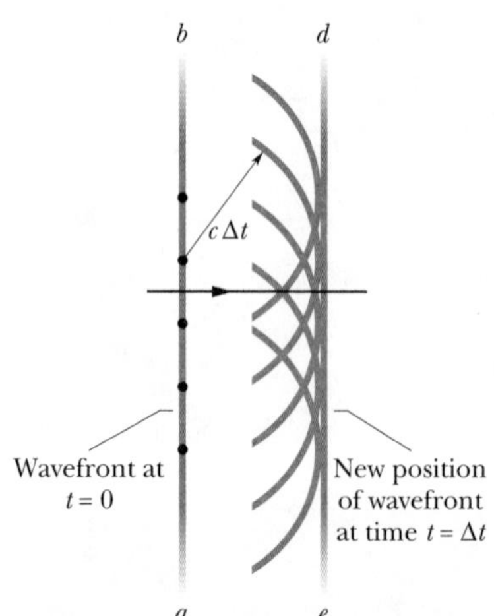

Fig. 36-1 The propagation of a plane wave in vacuum, as portrayed by Huygens' principle.

The Law of Refraction

We now use Huygens' principle to derive the law of refraction, Eq. 34-44 (Snell's law). Figure 36-2 shows three stages in the refraction of several wavefronts at a plane interface between air (medium 1) and glass (medium 2). We arbitrarily choose the wavefronts in the incident light beam to be separated by λ_1, the wavelength in medium 1. Let the speed of light in air be v_1 and that in glass be v_2. We assume that $v_2 < v_1$, which happens to be true.

Angle θ_1 in Fig. 36-2*a* is the angle between the wavefront and the interface; it has the same value as the angle between the *normal* to the wavefront (that is, the incident ray) and the *normal* to the interface. Thus, θ_1 is the angle of incidence.

As the wave moves into the glass, a Huygens wavelet at point *e* will expand to pass through point *c*, at a distance of λ_1 from point *e*. The time interval required for this expansion is that distance divided by the speed of the wavelet, or λ_1/v_1. Now note that in this same time interval, a Huygens wavelet at point *h* will expand to pass through point *g*, at the reduced speed v_2 and with wavelength λ_2. Thus, this time interval must also be equal to λ_2/v_2. By equating these times of travel, we obtain the relation

$$\frac{\lambda_1}{\lambda_2} = \frac{v_1}{v_2}, \tag{36-1}$$

which shows that the wavelengths of light in two media are proportional to the speeds of light in those media.

By Huygens' principle, the refracted wavefront must be tangent to an arc of radius λ_2 centered on *h*, say at point *g*. The refracted wavefront must also be tangent to an arc of radius λ_1 centered on *e*, say at *c*. Then the refracted wavefront must be oriented as shown. Note that θ_2, the angle between the refracted wavefront and the interface, is actually the angle of refraction.

For the right triangles *hce* and *hcg* in Fig. 36-2*b* we may write

$$\sin\theta_1 = \frac{\lambda_1}{hc} \qquad \text{(for triangle } hce\text{)}$$

and

$$\sin\theta_2 = \frac{\lambda_2}{hc} \qquad \text{(for triangle } hcg\text{).}$$

Dividing the first of these two equations by the second and using Eq. 36-1, we find

$$\frac{\sin\theta_1}{\sin\theta_2} = \frac{\lambda_1}{\lambda_2} = \frac{v_1}{v_2}. \tag{36-2}$$

We can define an **index of refraction** *n* for each medium as the ratio of the speed of light in vacuum to the speed of light *v* in the medium. Thus,

$$n = \frac{c}{v} \qquad \text{(index of refraction).} \tag{36-3}$$

In particular, for our two media, we have

$$n_1 = \frac{c}{v_1} \quad \text{and} \quad n_2 = \frac{c}{v_2}. \tag{36-4}$$

If we combine Eqs. 36-2 and 36-4, we find

$$\frac{\sin\theta_1}{\sin\theta_2} = \frac{c/n_1}{c/n_2} = \frac{n_2}{n_1} \tag{36-5}$$

or

$$n_1 \sin\theta_1 = n_2 \sin\theta_2 \qquad \text{(law of refraction),} \tag{36-6}$$

as introduced in Chapter 34.

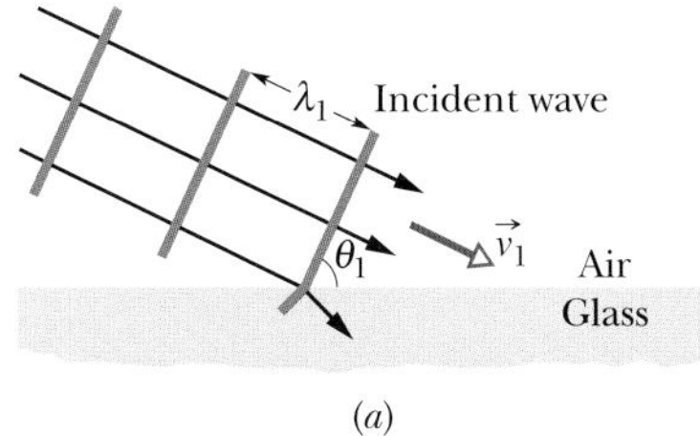

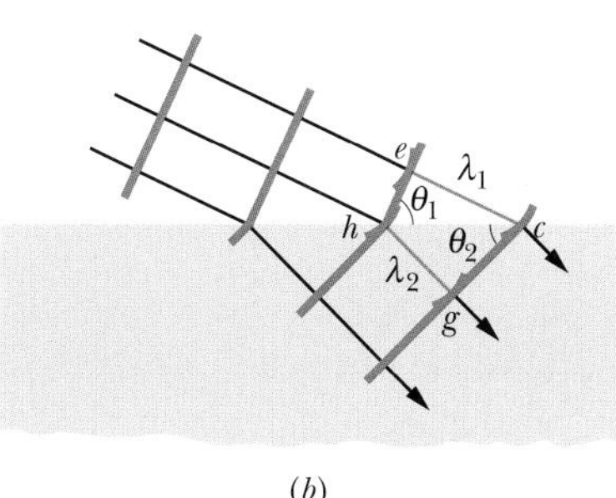

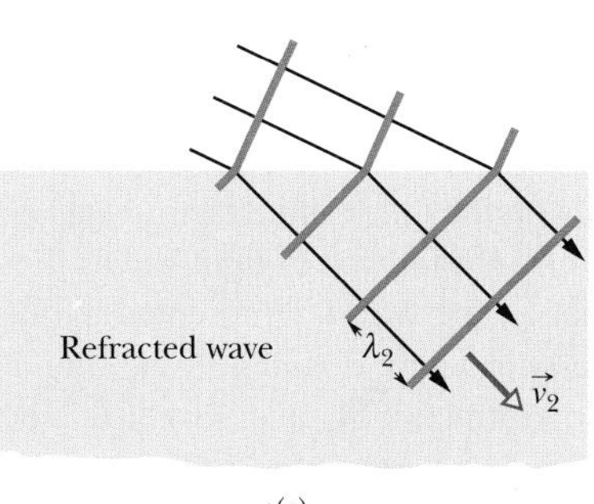

Fig. 36-2 The refraction of a plane wave at an air–glass interface, as portrayed by Huygens' principle. The wavelength in glass is smaller than that in air. For simplicity, the reflected wave is not shown. Parts (*a*) through (*c*) represent three successive stages of the refraction.

✔CHECKPOINT 1: The figure shows a monochromatic ray of light traveling across parallel interfaces, from an original material a, through layers of materials b and c, and then back into material a. Rank the materials according to the speed of light in them, greatest first.

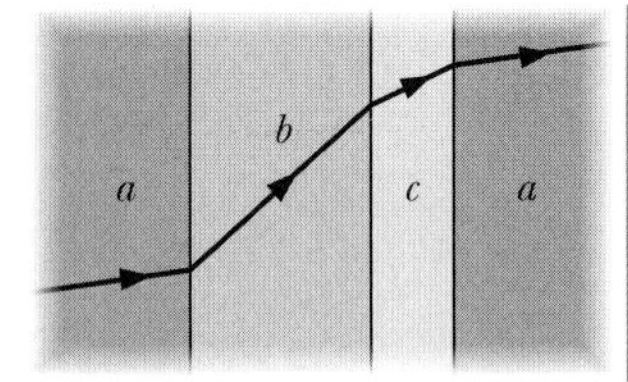

Wavelength and Index of Refraction

We have now seen that the wavelength of light changes when the speed of the light changes, as happens when light crosses an interface from one medium into another. Further, the speed of light in any medium depends on the index of refraction of the medium, according to Eq. 36-3. Thus, the wavelength of light in any medium depends on the index of refraction of the medium. Let a certain monochromatic light have wavelength λ and speed c in vacuum and wavelength λ_n and speed v in a medium with an index of refraction n. Now we can rewrite Eq. 36-1 as

$$\lambda_n = \lambda \frac{v}{c}. \tag{36-7}$$

Using Eq. 36-3 to substitute $1/n$ for v/c then yields

$$\lambda_n = \frac{\lambda}{n}. \tag{36-8}$$

This equation relates the wavelength of light in any medium to its wavelength in vacuum. It tells us that the greater the index of refraction of a medium, the smaller the wavelength of light in that medium.

What about the frequency of the light? Let f_n represent the frequency of the light in a medium with index of refraction n. Then from the general relation of Eq. 17-12 ($v = \lambda f$), we can write

$$f_n = \frac{v}{\lambda_n}.$$

Substituting Eqs. 36-3 and 36-8 then gives us

$$f_n = \frac{c/n}{\lambda/n} = \frac{c}{\lambda} = f,$$

where f is the frequency of the light in vacuum. Thus, although the speed and wavelength of light are different in the medium than in vacuum, *the frequency of the light in the medium is the same as it is in vacuum.*

The fact that the wavelength of light depends on the index of refraction via Eq. 36-8 is important in certain situations involving the interference of light waves. For example, in Fig. 36-3, the *waves of the rays* (that is, the waves represented by the rays) have identical wavelengths λ and are initially in phase in air ($n \approx 1$). One of the waves travels through medium 1 of index of refraction n_1 and length L. The other travels through medium 2 of index of refraction n_2 and the same length L. When the waves leave the two media, they will have the same wavelength—their wavelength λ in air. However, because their wavelengths differed in the two media, the two waves may no longer be in phase.

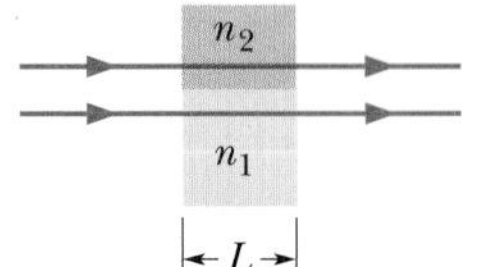

Fig. 36-3 Two light rays travel through two media having different indexes of refraction.

➤ The phase difference between two light waves can change if the waves travel through different materials having different indexes of refraction.

As we shall discuss soon, this change in the phase difference can determine how the light waves will interfere if they reach some common point.

To find their new phase difference in terms of wavelengths, we first count the number N_1 of wavelengths there are in the length L of medium 1. From Eq. 36-8, the wavelength in medium 1 is $\lambda_{n1} = \lambda/n_1$, so

$$N_1 = \frac{L}{\lambda_{n1}} = \frac{Ln_1}{\lambda}. \tag{36-9}$$

Similarly, we count the number N_2 of wavelengths there are in the length L of medium 2, where the wavelength is $\lambda_{n2} = \lambda/n_2$:

$$N_2 = \frac{L}{\lambda_{n2}} = \frac{Ln_2}{\lambda}. \tag{36-10}$$

To find the new phase difference between the waves, we subtract the smaller of N_1 and N_2 from the larger. Assuming $n_2 > n_1$, we obtain

$$N_2 - N_1 = \frac{Ln_2}{\lambda} - \frac{Ln_1}{\lambda} = \frac{L}{\lambda}(n_2 - n_1). \tag{36-11}$$

Suppose Eq. 36-11 tells us that the waves now have a phase difference of 45.6 wavelengths. That is equivalent to taking the initially in-phase waves and shifting one of them by 45.6 wavelengths. However, a shift of an integer number of wavelengths (such as 45) would put the waves back in phase, so it is only the decimal fraction (here, 0.6) that is important. A phase difference of 45.6 wavelengths is equivalent to an *effective phase difference* of 0.6 wavelength.

A phase difference of 0.5 wavelength puts two waves exactly out of phase. If the waves had equal amplitudes and were to reach some common point, they would then undergo fully destructive interference, producing darkness at that point. With a phase difference of 0.0 or 1.0 wavelength, they would, instead, undergo fully constructive interference, resulting in brightness at the common point. Our phase difference of 0.6 wavelength is an intermediate situation, but closer to destructive interference, and the waves would produce a dimly illuminated common point.

We can also express phase difference in terms of radians and degrees, as we have done already. A phase difference of one wavelength is equivalent to phase differences of 2π rad and 360°.

Sample Problem 36-1

In Fig. 36-3, the two light waves that are represented by the rays have wavelength 550.0 nm before entering media 1 and 2. They also have equal amplitudes and are in phase. Medium 1 is now just air, and medium 2 is a transparent plastic layer of index of refraction 1.600 and thickness 2.600 μm.

(a) What is the phase difference of the emerging waves in wavelengths, radians, and degrees? What is their effective phase difference (in wavelengths)?

SOLUTION: One **Key Idea** here is that the phase difference of two light waves can change if they travel through different media, with different indexes of refraction. The reason is that their wavelengths are different in the different media. We can calculate the change in phase difference by counting the number of wavelengths that fits into each medium and then subtracting those numbers. When the path lengths of the waves in the two media are identical, Eq. 36-11 gives the result. Here we have $n_1 = 1.000$ (for the air), $n_2 = 1.600$, $L = 2.600$ μm, and $\lambda = 550.0$ nm. Thus, Eq. 36-11 yields

$$\begin{aligned} N_2 - N_1 &= \frac{L}{\lambda}(n_2 - n_1) \\ &= \frac{2.600 \times 10^{-6}\text{ m}}{5.500 \times 10^{-7}\text{ m}}(1.600 - 1.000) \\ &= 2.84. \qquad \text{(Answer)} \end{aligned}$$

Thus, the phase difference of the emerging waves is 2.84 wavelengths. Because 1.0 wavelength is equivalent to 2π rad and 360°, you can show that this phase difference is equivalent to

$$\text{phase difference} = 17.8 \text{ rad} \approx 1020°. \qquad \text{(Answer)}$$

A second Key Idea is that the effective phase difference is the decimal part of the actual phase difference *expressed in wavelengths*. Thus, we have

effective phase difference = 0.84 wavelength. (Answer)

You can show that this is equivalent to 5.3 rad and about 300°. *Caution:* We do *not* find the effective phase difference by taking the decimal part of the actual phase difference as expressed in radians or degrees. For example, we do *not* take 0.8 rad from the actual phase difference of 17.8 rad.

(b) If the rays of the waves were angled slightly so that the waves reached the same point on a distant viewing screen, what type of interference would the waves produce at that point?

SOLUTION: The Key Idea here is to compare the effective phase difference of the waves with the phase differences that give the extreme types of interference. Here the effective phase difference of 0.84 wavelength is between 0.5 wavelength (for fully destructive interference, or the darkest possible result) and 1.0 wavelength (for fully constructive interference, or the brightest possible result), but closer to 1.0 wavelength. Thus, the waves would produce intermediate interference that is closer to fully constructive interference—they would produce a relatively bright spot.

CHECKPOINT 2: The light waves of the rays in Fig. 36-3 have the same wavelength and amplitude and are initially in phase. (a) If 7.60 wavelengths fit within the length of the top material and 5.50 wavelengths fit within that of the bottom material, which material has the greater index of refraction? (b) If the rays are angled slightly so that they meet at the same point on a distant screen, will the interference there result in the brightest possible illumination, bright intermediate illumination, dark intermediate illumination, or darkness?

36-3 Diffraction

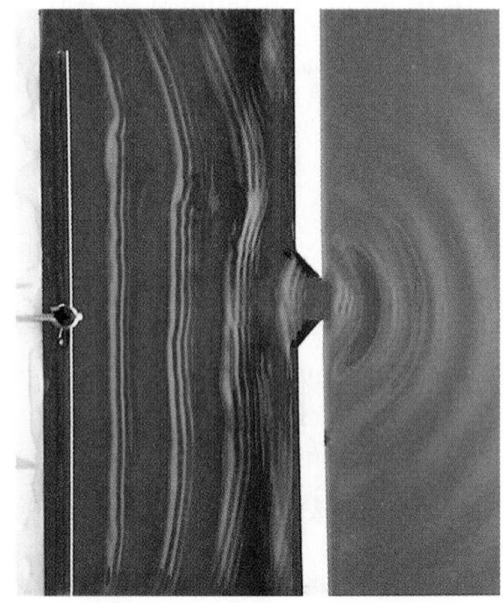

Fig. 36-4 The diffraction of water waves in a ripple tank. The waves are produced by an oscillating paddle at the left. As they move from left to right, they flare out through an opening in a barrier along the water surface.

In the next section we shall discuss the experiment that first proved that light is a wave. To prepare for that discussion, we must introduce the idea of **diffraction** of waves, a phenomenon that we explore much more fully in Chapter 37. Its essence is this: If a wave encounters a barrier that has an opening of dimensions similar to the wavelength, the part of the wave that passes through the opening will flare (spread) out —will *diffract*—into the region beyond the barrier. The flaring is consistent with the spreading of wavelets in the Huygens construction of Fig. 36-1. Diffraction occurs for waves of all types, not just light waves; Fig. 36-4 shows the diffraction of water waves traveling across the surface of water in a shallow tank.

Figure 36-5*a* shows the situation schematically for an incident plane wave of wavelength λ encountering a slit that has width $a = 6.0\lambda$ and extends into and out of the page. The wave flares out on the far side of the slit. Figures 36-5*b* (with $a = 3.0\lambda$) and 36-5*c* ($a = 1.5\lambda$) illustrate the main feature of diffraction: the narrower the slit, the greater the diffraction.

Diffraction limits geometrical optics, in which we represent an electromagnetic wave with a ray. If we actually try to form a ray by sending light through a narrow slit, or through a series of narrow slits, diffraction will always defeat our effort because it always causes the light to spread. Indeed, the narrower we make the slits (in the hope of producing a narrower beam), the greater the spreading is. Thus, geometrical optics holds only when slits or other apertures that might be located in the path of light do not have dimensions comparable to or smaller than the wavelength of the light.

36-4 Young's Interference Experiment

In 1801, Thomas Young experimentally proved that light is a wave, contrary to what most other scientists then thought. He did so by demonstrating that light undergoes interference, as do water waves, sound waves, and waves of all other types. In addition, he was able to measure the average wavelength of sunlight; his value, 570 nm, is impressively close to the modern accepted value of 555 nm. We shall here examine Young's experiment as an example of the interference of light waves.

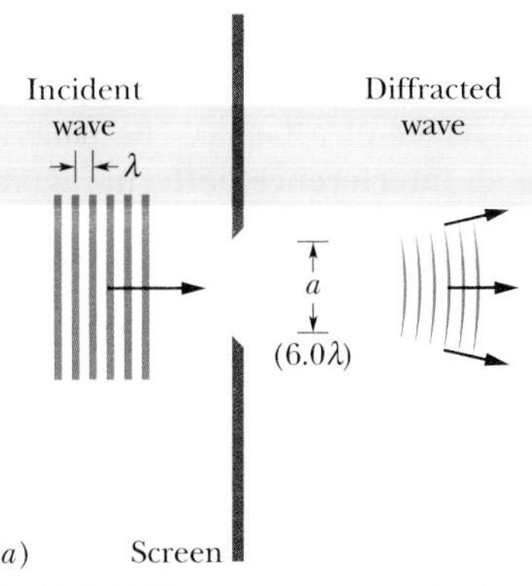

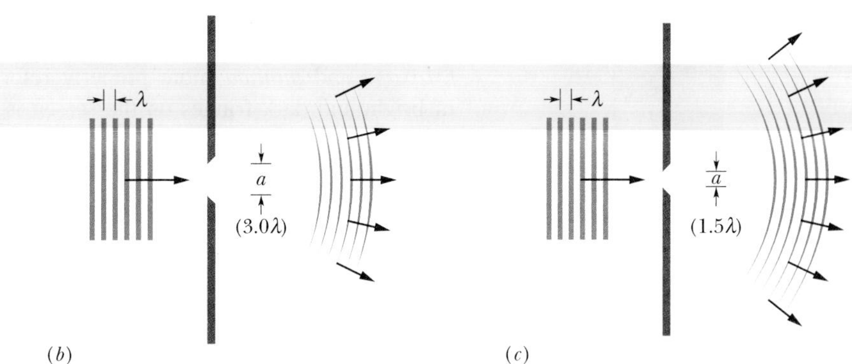

Fig. 36-5 Diffraction represented schematically. For a given wavelength λ, the diffraction is more pronounced the smaller the slit width a. The figures show the cases for (*a*) slit width $a = 6.0\lambda$, (*b*) slit width $a = 3.0\lambda$, and (*c*) slit width $a = 1.5\lambda$. In all three cases, the screen and the length of the slit extend well into and out of the page, perpendicular to it.

Figure 36-6 gives the basic arrangement of Young's experiment. Light from a distant monochromatic source illuminates slit S_0 in screen A. The emerging light then spreads via diffraction to illuminate two slits S_1 and S_2 in screen B. Diffraction of the light by these two slits sends overlapping circular waves into the region beyond screen B, where the waves from one slit interfere with the waves from the other slit.

The "snapshot" of Fig. 36-6 depicts the interference of the ovelapping waves. However, we cannot see evidence for the interference except where a viewing screen C intercepts the light. Where it does so, points of interference maxima form visible bright rows—called *bright bands, bright fringes*, or (loosely speaking) *maxima*—that extend across the screen (into and out of the page in Fig. 36-6). Dark regions—

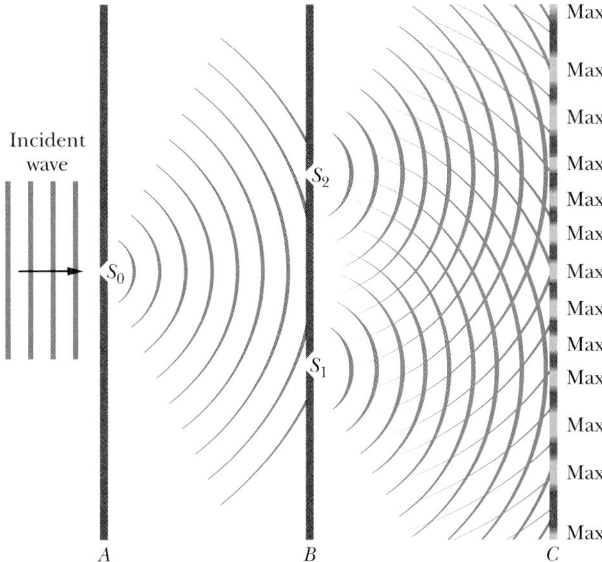

Fig. 36-6 In Young's interference experiment, incident monochromatic light is diffracted by slit S_0, which then acts as a point source of light that emits semicircular wavefronts. As that light reaches screen B, it is diffracted by slits S_1 and S_2, which then act as two point sources of light. The light waves traveling from slits S_1 and S_2 overlap and undergo interference, forming an interference pattern of maxima and minima on viewing screen C. This figure is a cross section; the screens, slits, and interference pattern extend into and out of the page. Between screens B and C, the semicircular wavefronts centered on S_2 depict the waves that would be there if only S_2 were open. Similarly, those centered on S_1 depict waves that would be there if only S_1 were open.

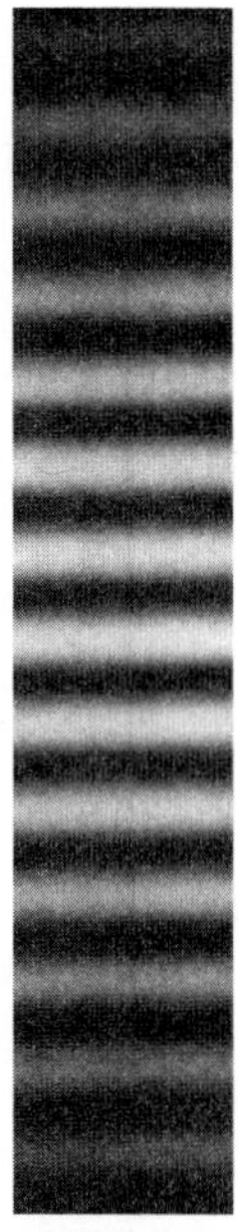

Fig. 36-7 A photograph of the interference pattern produced by the arrangement shown in Fig. 36-6. (The photograph is a front view of part of screen C.) The alternating maxima and minima are called *interference fringes* (because they resemble the decorative fringe sometimes used on clothing and rugs).

called *dark bands, dark fringes,* or (loosely speaking) *minima*—result from fully destructive interference and are visible between adjacent pairs of bright fringes. (*Maxima* and *minima* more properly refer to the center of a band.) The pattern of bright and dark fringes on the screen is called an **interference pattern.** Figure 36-7 is a photograph of part of the interference pattern as seen from the left in Fig. 36-6.

Locating the Fringes

Light waves produce fringes in a *Young's double-slit interference experiment,* as it is called, but what exactly determines the locations of the fringes? To answer, we shall use the arrangement in Fig. 36-8a. There, a plane wave of monochromatic light is incident on two slits S_1 and S_2 in screen B; the light diffracts through the slits and produces an interference pattern on screen C. We draw a central axis from the point halfway between the slits to screen C as a reference. We then pick, for discussion, an arbitrary point P on the screen, at angle θ to the central axis. This point intercepts the wave of ray r_1 from the bottom slit and the wave of ray r_2 from the top slit.

These waves are in phase when they pass through the two slits because there they are just portions of the same incident wave. However, once they have passed the slits, the two waves must travel different distances to reach P. We saw a similar situation in Section 18-4 with sound waves and concluded that

> The phase difference between two waves can change if the waves travel paths of different lengths.

The change in phase difference is due to the *path length difference* ΔL in the paths taken by the waves. Consider two waves initially exactly in phase, traveling along paths with a path length difference ΔL, and then passing through some common point. When ΔL is zero or an integer number of wavelengths, the waves arrive at the common point exactly in phase and they interfere fully constructively there. If that is true for the waves of rays r_1 and r_2 in Fig. 36-8, then point P is part of a bright fringe. When, instead, ΔL is an odd multiple of half a wavelength, the waves arrive at the common point exactly out of phase and they interfere fully destructively there. If that is true for the waves of rays r_1 and r_2, then point P is part of a dark

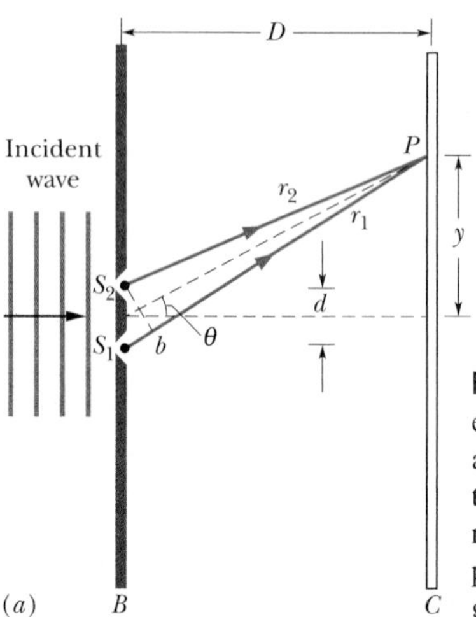

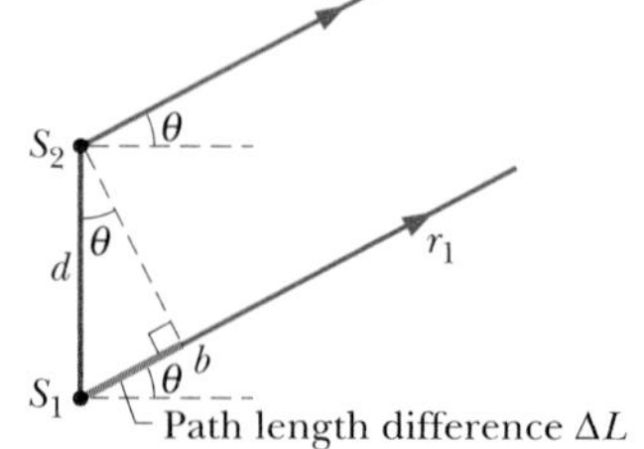

Fig. 36-8 (a) Waves from slits S_1 and S_2 (which extend into and out of the page) combine at P, an arbitrary point on screen C at distance y from the central axis. The angle θ serves as a convenient locator for P. (b) For $D \gg d$, we can approximate rays r_1 and r_2 as being parallel, at angle θ to the central axis.

fringe. (And, of course, we can have intermediate situations of interference and thus intermediate illumination at P.) Thus,

> What appears at each point on the viewing screen in a Young's double-slit interference experiment is determined by the path length difference ΔL of the rays reaching that point.

We can specify where each bright or dark fringe is located on the screen by giving the angle θ from the central axis to that fringe. To find θ, we must relate it to ΔL. We start with Fig. 36-8*a* by finding a point b along ray r_1 such that the path length from b to P equals the path length from S_2 to P. Then the path length difference ΔL between the two rays is the distance from S_1 to b.

The relation between this S_1-to-b distance and θ is complicated, but we can simplify it considerably if we arrange for the distance D from the slits to the screen to be much greater than the slit separation d. Then we can approximate rays r_1 and r_2 as being parallel to each other and at angle θ to the central axis (Fig. 36-8*b*). We can also approximate the triangle formed by S_1, S_2, and b as being a right triangle, and approximate the angle inside that triangle at S_2 as being θ. Then, for that triangle, $\sin\theta = \Delta L/d$ and thus

$$\Delta L = d\sin\theta \qquad \text{(path length difference).} \tag{36-12}$$

For a bright fringe, we saw that ΔL must be zero or an integer number of wavelengths. Using Eq. 36-12, we can write this requirement as

$$\Delta L = d\sin\theta = (\text{integer})(\lambda), \tag{36-13}$$

or as

$$d\sin\theta = m\lambda, \quad \text{for } m = 0, 1, 2, \ldots \qquad \text{(maxima—bright fringes).} \tag{36-14}$$

For a dark fringe, ΔL must be an odd multiple of half a wavelength. Again using Eq. 36-12, we can write this requirement as

$$\Delta L = d\sin\theta = (\text{odd number})(\tfrac{1}{2}\lambda), \tag{36-15}$$

or as

$$d\sin\theta = (m + \tfrac{1}{2})\lambda, \quad \text{for } m = 0, 1, 2, \ldots \qquad \text{(minima—dark fringes).} \tag{36-16}$$

With Eqs. 36-14 and 36-16, we can find the angle θ to any fringe and thus locate that fringe; further, we can use the values of m to label the fringes. For the value and label $m = 0$, Eq. 36-14 tells us that a bright fringe is at $\theta = 0$—that is, on the central axis. This *central maximum* is the point at which waves arriving from the two slits have a path length difference $\Delta L = 0$, hence zero phase difference.

For, say, $m = 2$, Eq. 36-14 tells us that *bright* fringes are at the angle

$$\theta = \sin^{-1}\left(\frac{2\lambda}{d}\right)$$

above and below the central axis. Waves from the two slits arrive at these two fringes with $\Delta L = 2\lambda$ and with a phase difference of two wavelengths. These fringes are said to be the *second-order fringes* (meaning $m = 2$) or the *second side maxima* (the second maxima to the side of the central maximum), or they are described as being the second fringes from the central maximum.

For $m = 1$, Eq. 36-16 tells us that *dark* fringes are at the angle

$$\theta = \sin^{-1}\left(\frac{1.5\lambda}{d}\right)$$

above and below the central axis. Waves from the two slits arrive at these two fringes with $\Delta L = 1.5\lambda$ and with a phase difference, in wavelengths, of 1.5. These fringes are called the *second dark fringes* or *second minima* because they are the second dark fringes from the central axis. (The first dark fringes, or first minima, are at locations for which $m = 0$ in Eq. 36-16.)

We derived Eqs. 36-14 and 36-16 for the situation $D \gg d$. However, they also apply if we place a converging lens between the slits and the viewing screen and then move the viewing screen closer to the slits, to the focal point of the lens. (The screen is then said to be in the *focal plane* of the lens; that is, it is in the plane perpendicular to the central axis at the focal point.) One property of a converging lens is that it focuses all rays that are parallel to one another to the same point on its focal plane. Thus, the rays that now arrive at any point on the screen (in the focal plane) were exactly parallel (rather than approximately) when they left the slits. They are like the initially parallel rays in Fig. 35-12*a* that are directed to a point (the focal point) by a lens.

CHECKPOINT 3: In Fig. 36-8*a*, what are ΔL (as a multiple of the wavelength) and the phase difference (in wavelengths) for the two rays if point P is (a) a third side maximum and (b) a third minimum?

Sample Problem 36-2

What is the distance on screen C in Fig. 36-8*a* between adjacent maxima near the center of the interference pattern? The wavelength λ of the light is 546 nm, the slit separation d is 0.12 mm, and the slit–screen separation D is 55 cm. Assume that θ in Fig. 36-8 is small enough to permit use of the approximations $\sin\theta \approx \tan\theta \approx \theta$, in which θ is expressed in radian measure.

SOLUTION: First, let us pick a maximum with a low value of m to ensure that it is near the center of the pattern. Then one **Key Idea** is that, from the geometry of Fig. 36-8*a*, the maximum's vertical distance y_m from the center of the pattern is related to its angle θ from the central axis by

$$\tan\theta \approx \theta = \frac{y_m}{D}.$$

A second **Key Idea** is that, from Eq. 36-14, this angle θ for the mth maximum is given by

$$\sin\theta \approx \theta = \frac{m\lambda}{d}.$$

If we equate these two expressions for θ and solve for y_m, we find

$$y_m = \frac{m\lambda D}{d}. \qquad (36\text{-}17)$$

For the next farther out maximum, we have

$$y_{m+1} = \frac{(m+1)\lambda D}{d}. \qquad (36\text{-}18)$$

We find the distance between these adjacent maxima by subtracting Eq. 36-17 from Eq. 36-18:

$$\Delta y = y_{m+1} - y_m = \frac{\lambda D}{d}$$
$$= \frac{(546 \times 10^{-9}\text{ m})(55 \times 10^{-2}\text{ m})}{0.12 \times 10^{-3}\text{ m}}$$
$$= 2.50 \times 10^{-3}\text{ m} \approx 2.5\text{ mm}. \qquad \text{(Answer)}$$

As long as d and θ in Fig. 36-8*a* are small, the separation of the interference fringes is independent of m; that is, the fringes are evenly spaced.

36-5 Coherence

For the interference pattern to appear on viewing screen C in Fig. 36-6, the light waves reaching any point P on the screen must have a phase difference that does not vary in time. That is the case in Fig. 36-6, because the waves passing through slits S_1 and S_2 are portions of the single light wave that illuminates the slits. Because the phase difference remains constant, the light from slits S_1 and S_2 is said to be completely **coherent.**

Direct sunlight is partially coherent; that is, sunlight waves intercepted at two points have a constant phase difference only if the points are very close. If you look closely at your fingernail in bright sunlight, you can see a faint interference pattern called *speckle* that causes the nail to appear to be covered with specks. You see this effect because light waves scattering from very close points on the nail are sufficiently coherent to interfere with one another at your eye. The slits in a double-slit experiment, however, are not close enough, and in direct sunlight, the light at the slits would be **incoherent.** To get coherent light, we would have to send the sunlight through a single slit as in Fig. 36-6; because that single slit is small, light that passes through it is coherent. In addition, the smallness of the slit causes the coherent light to spread via diffraction to illuminate both slits in the double-slit experiment.

If we replace the double slits with two similar but independent monochromatic light sources, such as two fine incandescent wires, the phase difference between the waves emitted by the sources varies rapidly and randomly. (This occurs because the light is emitted by vast numbers of atoms in the wires, acting randomly and independently for extremely short times—of the order of nanoseconds.) As a result, at any given point on the viewing screen, the interference between the waves from the two sources varies rapidly and randomly between fully constructive and fully destructive. The eye (and most common optical detectors) cannot follow such changes, and no interference pattern can be seen. The fringes disappear, and the screen is seen as being uniformly illuminated.

A *laser* differs from common light sources in that its atoms emit light in a cooperative manner, thereby making the light coherent. Moreover, the light is almost monochromatic, is emitted in a thin beam with little spreading, and can be focused to a width that almost matches the wavelength of the light.

36-6 Intensity in Double-Slit Interference

Equations 36-14 and 36-16 tell us how to locate the maxima and minima of the double-slit interference pattern on screen C of Fig. 36-8 as a function of the angle θ in that figure. Here we wish to derive an expression for the intensity I of the fringes as a function of θ.

The light leaving the slits is in phase. However, let us assume that the light waves from the two slits are not in phase when they arrive at point P. Instead, the electric field components of those waves at point P are not in phase and vary with time as

$$E_1 = E_0 \sin \omega t \tag{36-19}$$

and

$$E_2 = E_0 \sin(\omega t + \phi), \tag{36-20}$$

where ω is the angular frequency of the waves and ϕ is the phase constant of wave E_2. Note that the two waves have the same amplitude E_0 and a phase difference of ϕ. Because that phase difference does not vary, the waves are coherent. We shall show that these two waves will combine at P to produce an intensity I given by

$$I = 4I_0 \cos^2 \tfrac{1}{2}\phi, \tag{36-21}$$

and that

$$\phi = \frac{2\pi d}{\lambda} \sin \theta. \tag{36-22}$$

In Eq. 36-21, I_0 is the intensity of the light that arrives on the screen from one slit when the other slit is temporarily covered. We assume that the slits are so narrow in comparison to the wavelength that this single-slit intensity is essentially uniform over the region of the screen in which we wish to examine the fringes.

Equations 36-21 and 36-22, which together tell us how the intensity I of the fringe pattern varies with the angle θ in Fig. 36-8, necessarily contain information about the location of the maxima and minima. Let us see if we can extract that information, to find equations about those locations.

Study of Eq. 36-21 shows that intensity maxima will occur when

$$\tfrac{1}{2}\phi = m\pi, \qquad \text{for } m = 0, 1, 2, \ldots. \tag{36-23}$$

If we put this result into Eq. 36-22, we find

$$2m\pi = \frac{2\pi d}{\lambda}\sin\theta, \qquad \text{for } m = 0, 1, 2, \ldots$$

or

$$d\sin\theta = m\lambda, \qquad \text{for } m = 0, 1, 2, \ldots \quad \text{(maxima)}, \tag{36-24}$$

which is exactly Eq. 36-14, the expression that we derived earlier for the locations of the maxima.

The minima in the fringe pattern occur when

$$\tfrac{1}{2}\phi = (m + \tfrac{1}{2})\pi, \qquad \text{for } m = 0, 1, 2, \ldots.$$

If we combine this relation with Eq. 36-22, we are led at once to

$$d\sin\theta = (m + \tfrac{1}{2})\lambda \qquad \text{for } m = 0, 1, 2, \ldots \quad \text{(minima)}, \tag{36-25}$$

which is just Eq. 36-16, the expression we derived earlier for the locations of the fringe minima.

Figure 36-9, which is a plot of Eq. 36-21, shows the intensity of double-slit interference patterns as a function of the phase difference ϕ between the waves at the screen. The horizontal solid line is I_0, the (uniform) intensity on the screen when one of the slits is covered up. Note in Eq. 36-21 and the graph that the intensity I varies from zero at the fringe minima to $4I_0$ at the fringe maxima.

If the waves from the two sources (slits) were *incoherent*, so that no enduring phase relation existed between them, there would be no fringe pattern and the intensity would have the uniform value $2I_0$ for all points on the screen; the horizontal dashed line in Fig. 36-9 shows this uniform value.

Interference cannot create or destroy energy but merely redistributes it over the screen. Thus, the *average* intensity on the screen must be the same $2I_0$ regardless of whether the sources are coherent. This follows at once from Eq. 36-21; if we substitute $\frac{1}{2}$, the average value of the cosine-squared function, this equation reduces to $I_{avg} = 2I_0$.

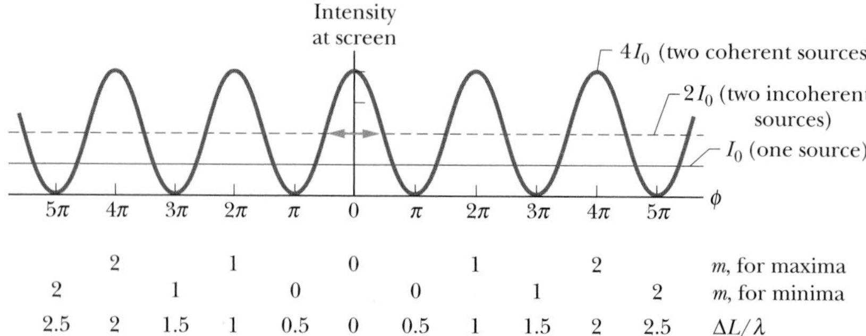

Fig. 36-9 A plot of Eq. 36-21, showing the intensity of a double-slit interference pattern as a function of the phase difference between the waves when they arrive from the two slits. I_0 is the (uniform) intensity that would appear on the screen if one slit were covered. The average intensity of the fringe pattern is $2I_0$, and the *maximum* intensity (for coherent light) is $4I_0$.

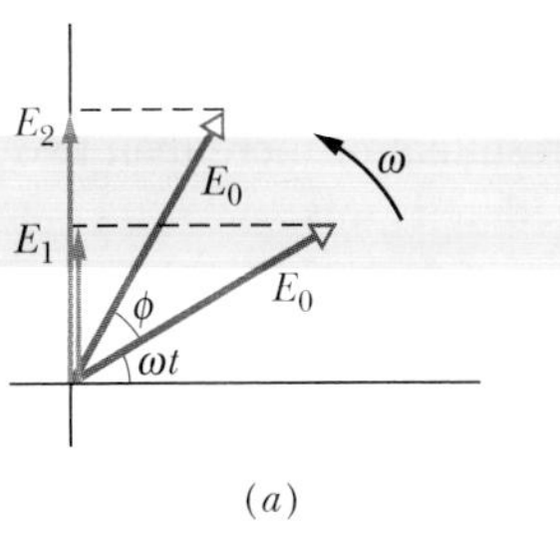

(a)

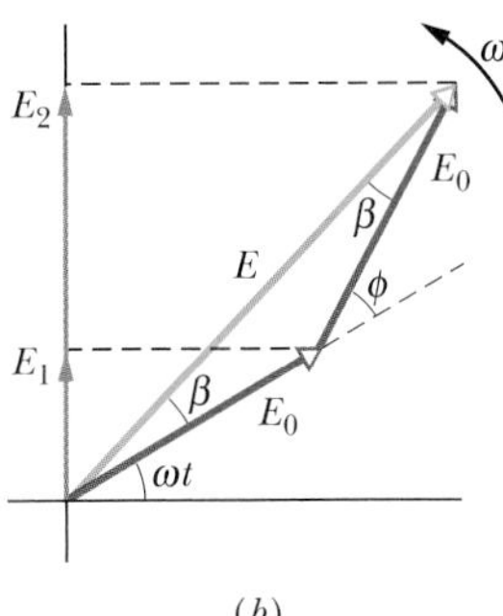

(b)

Fig. 36-10 (a) Phasors representing, at time t, the electric field components given by Eqs. 36-19 and 36-20. Both phasors have magnitude E_0 and rotate with angular speed ω. Their phase difference is ϕ. (b) Vector addition of the two phasors gives the phasor representing the resultant wave, with amplitude E and phase constant β.

Proof of Eqs. 36-21 and 36-22

We shall combine the electric field components E_1 and E_2, given by Eqs. 36-19 and 36-20, respectively, by the method of phasors as is discussed in Section 17-10. In Fig. 36-10a, the waves with components E_1 and E_2 are represented by phasors of magnitude E_0 that rotate around the origin at angular speed ω. The values of E_1 and E_2 at any time are the projections of the corresponding phasors on the vertical axis. Figure 36-10a shows the phasors and their projections at an arbitrary time t. Consistent with Eqs. 36-19 and 36-20, the phasor for E_1 has a rotation angle ωt and the phasor for E_2 has a rotation angle $\omega t + \phi$.

To combine the field components E_1 and E_2 at any point P in Fig. 36-8, we add their phasors vectorially, as shown in Fig. 36-10b. The magnitude of the vector sum is the amplitude E of the resultant wave at point P, and that wave has a certain phase constant β. To find the amplitude E in Fig. 36-10b, we first note that the two angles marked β are equal because they are opposite equal-length sides of a triangle. From the theorem (for triangles) that an exterior angle (here ϕ, as shown in Fig. 36-10b) is equal to the sum of the two opposite interior angles (here that sum is $\beta + \beta$), we see that $\beta = \frac{1}{2}\phi$. Thus, we have

$$\begin{aligned} E &= 2(E_0 \cos \beta) \\ &= 2E_0 \cos \tfrac{1}{2}\phi. \end{aligned} \tag{36-26}$$

If we square each side of this relation, we obtain

$$E^2 = 4E_0^2 \cos^2 \tfrac{1}{2}\phi. \tag{36-27}$$

Now, from Eq. 34-24, we know that the intensity of an electromagnetic wave is proportional to the square of its amplitude. Therefore, the waves we are combining in Fig. 36-10b, whose amplitudes are E_0, each has an intensity I_0 that is proportional to E_0^2, and the resultant wave, with amplitude E, has an intensity I that is proportional to E^2. Thus,

$$\frac{I}{I_0} = \frac{E^2}{E_0^2}.$$

Substituting Eq. 36-27 into this equation and rearranging then yield

$$I = 4I_0 \cos^2 \tfrac{1}{2}\phi,$$

which is Eq. 36-21, which we set out to prove.

It remains to prove Eq. 36-22, which relates the phase difference ϕ between the waves arriving at any point P on the screen of Fig. 36-8 to the angle θ that serves as a locator of that point.

The phase difference ϕ in Eq. 36-20 is associated with the path length difference S_1b in Fig. 36-8b. If S_1b is $\frac{1}{2}\lambda$, then ϕ is π; if S_1b is λ, then ϕ is 2π, and so on. This suggests

$$\begin{pmatrix}\text{phase}\\ \text{difference}\end{pmatrix} = \frac{2\pi}{\lambda}\begin{pmatrix}\text{path length}\\ \text{difference}\end{pmatrix}. \tag{36-28}$$

The path length difference S_1b in Fig. 36-8b is $d \sin \theta$, so Eq. 36-28 becomes

$$\phi = \frac{2\pi d}{\lambda} \sin \theta,$$

which is Eq. 36-22, the other equation that we set out to prove.

Combining More Than Two Waves

In a more general case, we might want to find the resultant of more than two sinusoidally varying waves at a point. The general procedure is this:

1. Construct a series of phasors representing the waves to be combined. Draw them end to end, maintaining the proper phase relations between adjacent phasors.
2. Construct the vector sum of this array. The length of this vector sum gives the amplitude of the resultant phasor. The angle between the vector sum and the first phasor is the phase of the resultant with respect to this first phasor. The projection of this vector-sum phasor on the vertical axis gives the time variation of the resultant wave.

Sample Problem 36-3

Three light waves combine at a certain point where their electric field components are

$$E_1 = E_0 \sin \omega t,$$
$$E_2 = E_0 \sin(\omega t + 60°),$$
$$E_3 = E_0 \sin(\omega t - 30°).$$

Find their resultant component $E(t)$ at that point.

SOLUTION: The resultant wave is

$$E(t) = E_1(t) + E_2(t) + E_3(t).$$

The Key Idea here is a two-fold idea: We can use the method of phasors to find this sum and we are free to evaluate the phasors at any time t. To simplify the solution we choose $t = 0$, for which the phasors representing the three waves are shown in Fig. 36-11. We can add these three phasors either directly on a vector-capable calculator or by components. For the component approach, we first write the sum of their horizontal components as

$$\sum E_h = E_0 \cos 0 + E_0 \cos 60° + E_0 \cos(-30°) = 2.37E_0.$$

The sum of their vertical components, which is the value of E at $t = 0$, is

$$\sum E_v = E_0 \sin 0 + E_0 \sin 60° + E_0 \sin(-30°) = 0.366E_0.$$

The resultant wave $E(t)$ thus has an amplitude E_R of

$$E_R = \sqrt{(2.37E_0)^2 + (0.366E_0)^2} = 2.4E_0,$$

and a phase angle β relative to the phasor representing E_1 of

$$\beta = \tan^{-1}\left(\frac{0.366E_0}{2.37E_0}\right) = 8.8°.$$

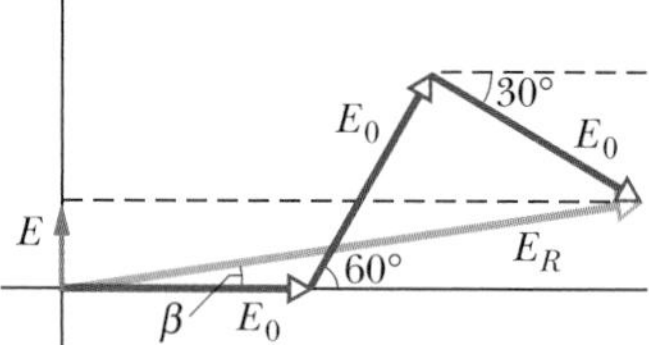

Fig. 36-11 Sample Problem 36-3. Three phasors, representing waves with equal amplitudes E_0 and with phase constants 0°, 60°, and −30°, shown at time $t = 0$. The phasors combine to give a resultant phasor with magnitude E_R, at angle β.

We can now write, for the resultant wave $E(t)$,

$$E = E_R \sin(\omega t + \beta)$$
$$= 2.4E_0 \sin(\omega t + 8.8°). \qquad \text{(Answer)}$$

Be careful to interpret the angle β correctly in Fig. 36-11: It is the constant angle between E_R and the phasor representing E_1 as the four phasors rotate as a single unit around the origin. The angle between E_R and the horizontal axis in Fig. 36-11 does not remain equal to β.

CHECKPOINT 4: Each of four pairs of light waves arrives at a certain point on a screen. The waves have the same wavelength. At the arrival point, their amplitudes and phase differences are (a) $2E_0$, $6E_0$, and π rad; (b) $3E_0$, $5E_0$, and π rad; (c) $9E_0$, $7E_0$, and 3π rad; (d) $2E_0$, $2E_0$, and 0 rad. Rank the four pairs according to the intensity of the light at those points, greatest first. (*Hint:* Draw phasors.)

36-7 Interference from Thin Films

The colors we see when sunlight illuminates a soap bubble or an oil slick are caused by the interference of light waves reflected from the front and back surfaces of a thin transparent film. The thickness of the soap or oil film is typically of the order of magnitude of the wavelength of the (visible) light involved. (Greater thicknesses spoil the coherence of the light needed to produce the colors.)

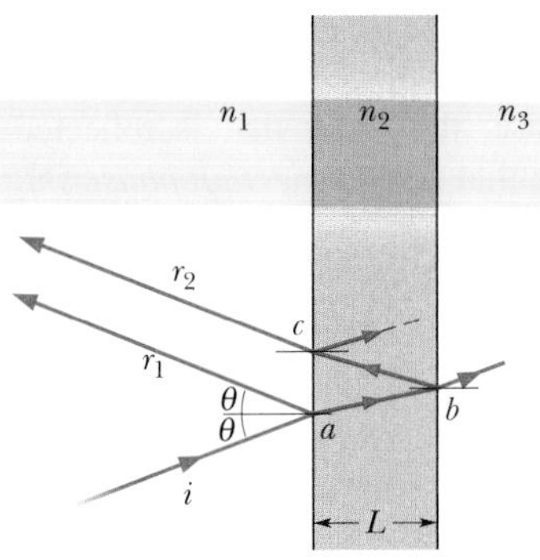

Fig. 36-12 Light waves, represented with ray i, are incident on a thin film of thickness L and index of refraction n_2. Rays r_1 and r_2 represent light waves that have been reflected by the front and back surfaces of the film, respectively. (All three rays are actually nearly perpendicular to the film.) The interference of the waves of r_1 and r_2 with each other depends on their phase difference. The index of refraction n_1 of the medium at the left can differ from the index of refraction n_3 of the medium at the right, but for now we assume that both media are air, with $n_1 = n_3 = 1.0$, which is less than n_2.

Figure 36-12 shows a thin transparent film of uniform thickness L and index of refraction n_2, illuminated by bright light of wavelength λ from a distant point source. For now, we assume that air lies on both sides of the film and thus that $n_1 = n_3$ in Fig. 36-12. For simplicity, we also assume that the light rays are almost perpendicular to the film ($\theta \approx 0$). We are interested in whether the film is bright or dark to an observer viewing it almost perpendicularly. (Since the film is brightly illuminated, how could it possibly be dark? You will see.)

The incident light, represented by ray i, intercepts the front (left) surface of the film at point a and undergoes both reflection and refraction there. The reflected ray r_1 is intercepted by the observer's eye. The refracted light crosses the film to point b on the back surface, where it undergoes both reflection and refraction. The light reflected at b crosses back through the film to point c, where it undergoes both reflection and refraction. The light refracted at c, represented by ray r_2, is intercepted by the observer's eye.

If the light waves of rays r_1 and r_2 are exactly in phase at the eye, they produce an interference maximum, and region ac on the film is bright to the observer. If they are exactly out of phase, they produce an interference minimum, and region ac is dark to the observer, *even though it is illuminated.* If there is some intermediate phase difference, there are intermediate interference and intermediate brightness.

Thus, the key to what the observer sees is the phase difference between the waves of rays r_1 and r_2. Both rays are derived from the same ray i, but the path involved in producing r_2 involves light traveling twice across the film (a to b, and then b to c), whereas the path involved in producing r_1 involves no travel through the film. Because θ is about zero, we approximate the path length difference between the waves of r_1 and r_2 as $2L$. However, to find the phase difference between the waves, we cannot just find the number of wavelengths λ that is equivalent to a path length difference of $2L$. This simple approach is impossible for two reasons: (1) the path length difference occurs in a medium other than air, and (2) reflections are involved, which can change the phase.

▶ The phase difference between two waves can change if one or both are reflected.

Before we continue our discussion of interference from thin films, we must discuss changes in phase that are caused by reflections.

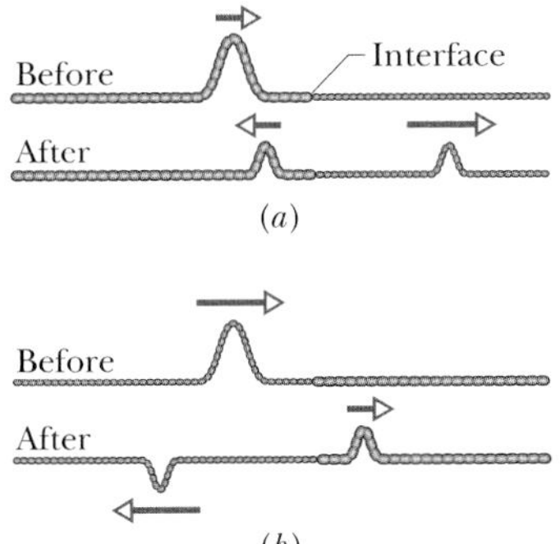

Fig. 36-13 Phase changes when a pulse is reflected at the interface between two stretched strings of different linear densities. The wave speed is greater in the lighter string. (*a*) The incident pulse is in the denser string. (*b*) The incident pulse is in the lighter string. Only here is there a phase change, and only in the reflected wave.

Reflection Phase Shifts

Refraction at an interface never causes a phase change—but reflection can, depending on the indexes of refraction on the two sides of the interface. Figure 36-13 shows what happens when reflection causes a phase change, using as an example pulses on a denser string (along which pulse travel is relatively slow) and a lighter string (along which pulse travel is relatively fast).

When a pulse traveling relatively slowly along the denser string in Fig. 36-13*a* reaches the interface with the lighter string, the pulse is partially transmitted and partially reflected, with no change in orientation. For light, this situation corresponds to the incident wave traveling in the medium of greater index of refraction n (recall that greater n means slower speed). In that case, the wave that is reflected at the interface does not undergo a change in phase; that is, its *reflection phase shift* is zero.

When a pulse traveling more quickly along the lighter string in Fig. 36-13*b* reaches the interface with the denser string, the pulse is again partially transmitted and partially reflected. The transmitted pulse again has the same orientation as the incident pulse, but now the reflected pulse is inverted. For a sinusoidal wave, such an inversion involves a phase change of π rad, or half a wavelength. For light, this situation corresponds to the incident wave traveling in the medium of lesser index of refraction (with greater speed). In that case, the wave that is reflected at the interface undergoes a phase shift of π rad, or half a wavelength.

We can summarize these results for light in terms of the index of refraction of the medium off which (or from which) the light reflects:

Reflection	Reflection phase shift
Off lower index	0
Off higher index	0.5 wavelength

This might be remembered as "higher means half."

Equations for Thin-Film Interference

In this chapter we have now seen three ways in which the phase difference between two waves can change:

1. by reflection
2. by the waves traveling along paths of different lengths
3. by the waves traveling through media of different indexes of refraction

When light reflects from a thin film, producing the waves of rays r_1 and r_2 shown in Fig. 36-12, all three ways are involved. Let us consider them one by one.

We first reexamine the two reflections in Fig. 36-12. At point a on the front interface, the incident wave (in air) reflects from the medium having the higher of the two indexes of refraction, so the wave of reflected ray r_1 has its phase shifted by 0.5 wavelength. At point b on the back interface, the incident wave reflects from the medium (air) having the lower of the two indexes of refraction, so the wave reflected there is not shifted in phase by the reflection, and thus neither is the portion of it that exits the film as ray r_2. We can organize this information with the first line in Table 36-1. It tells us that, so far, as a result of the reflection phase shifts, the waves of r_1 and r_2 have a phase difference of 0.5 wavelength and thus are exactly out of phase.

Now we must consider the path length difference $2L$ that occurs because the wave of ray r_2 crosses the film twice. (This difference $2L$ is shown on the second line in Table 36-1.) If the waves of r_1 and r_2 are to be exactly in phase so that they produce fully constructive interference, the path length $2L$ must cause an additional phase difference of 0.5, 1.5, 2.5, . . . wavelengths. Only then will the net phase difference be an integer number of wavelengths. Thus, for a bright film, we must have

$$2L = \frac{\text{odd number}}{2} \times \text{wavelength} \quad \text{(in-phase waves).} \tag{36-29}$$

TABLE 36-1 **An Organizing Table for Thin-Film Interference in Air**[a]

	r_1	r_2
Reflection phase shifts	0.5 wavelength	0
Path length difference	$2L$	
Index in which path length difference occurs	n_2	
In phase[a]:	$2L = \frac{\text{odd number}}{2} \times \frac{\lambda}{n_2}$	
Out of phase[a]:	$2L = \text{integer} \times \frac{\lambda}{n_2}$	

[a]Valid for $n_2 > n_1$ and $n_2 > n_3$.

The wavelength we need here is the wavelength λ_{n2} of the light in the medium containing path length $2L$—that is, in the medium with index of refraction n_2. Thus, we can rewrite Eq. 36-29 as

$$2L = \frac{\text{odd number}}{2} \times \lambda_{n2} \quad \text{(in-phase waves).} \tag{36-30}$$

If, instead, the waves are to be exactly out of phase so that there is fully destructive interference, the path length $2L$ must cause either no additional phase difference or a phase difference of 1, 2, 3, . . . wavelengths. Only then will the net phase difference be an odd number of half-wavelengths. For a dark film, we must have

$$2L = \text{integer} \times \text{wavelength}, \tag{36-31}$$

where, again, the wavelength is the wavelength λ_{n2} in the medium containing $2L$. Thus, this time we have

$$2L = \text{integer} \times \lambda_{n2} \quad \text{(out-of-phase waves).} \tag{36-32}$$

Now we can use Eq. 36-8 ($\lambda_n = \lambda/n$) to write the wavelength of the wave of ray r_2 inside the film as

$$\lambda_{n2} = \frac{\lambda}{n_2}, \tag{36-33}$$

where λ is the wavelength of the incident light in vacuum (and approximately also in air). Substituting Eq. 36-33 into Eq. 36-30 and replacing "odd number/2" with $(m + \frac{1}{2})$ give us

$$2L = (m + \tfrac{1}{2})\frac{\lambda}{n_2}, \quad \text{for } m = 0, 1, 2, \ldots \quad \text{(maxima—bright film in air).} \tag{36-34}$$

Similarly, with m replacing "integer," Eq. 36-32 yields

$$2L = m\frac{\lambda}{n_2}, \quad \text{for } m = 0, 1, 2, \ldots \quad \text{(minima—dark film in air).} \tag{36-35}$$

For a given film thickness L, Eqs. 36-34 and 36-35 tell us the wavelengths of light for which the film appears bright and dark, respectively, one wavelength for each value of m. Intermediate wavelengths give intermediate brightnesses. For a given wavelength λ, Eqs. 36-34 and 36-35 tell us the thicknesses of the films that appear bright and dark in that light, respectively, one thickness for each value of m. Intermediate thicknesses give intermediate brightnesses.

A special situation arises when a film is so thin that L is much less than λ, say, $L < 0.1\lambda$. Then the path length difference $2L$ can be neglected, and the phase difference between r_1 and r_2 is due *only* to reflection phase shifts. If the film of Fig. 36-12, where the reflections cause a phase difference of 0.5 wavelength, has thickness $L < 0.1\lambda$, then r_1 and r_2 are exactly out of phase, and thus the film is dark, regardless of the wavelength and even the intensity of the light that illuminates it. This special situation corresponds to $m = 0$ in Eq. 36-35. We shall count any thickness $L < 0.1\lambda$ as being the least thickness specified by Eq. 36-35 to make the film of Fig. 36-12 dark. (Every such thickness will correspond to $m = 0$.) The next greater thickness that will make the film dark is that corresponding to $m = 1$.

Figure 36-14 shows a vertical soap film whose thickness increases from top to bottom because gravitation has caused the film to slump. Bright white light illumi-

Fig. 36-14 The reflection of light from a soapy water film spanning a vertical loop. The top portion is so thin that the light reflected there undergoes destructive interference, making that portion dark. Colored interference fringes, or bands, decorate the rest of the film but are marred by circulation of liquid within the film as the liquid is gradually pulled downward by gravitation.

nates the film. However, the top portion is so thin that it is dark. In the (somewhat thicker) middle we see fringes, or bands, whose color depends primarily on the wavelength at which reflected light undergoes fully constructive interference for a particular thickness. Toward the (thickest) bottom of the film the fringes become progressively narrower and the colors begin to overlap and fade.

Iridescence of a *Morpho* Butterfly Wing

A surface that displays colors due to thin-film interference is said to be *iridescent* because the tints of the colors change as you change your view of the surface. The iridescence of the top surface of a *Morpho* butterfly wing is due to thin-film interference of light reflected by thin terraces of transparent cuticle-like material on the wing. These terraces are arranged like wide, flat branches on a tree-like structure that extends perpendicular to the wing.

Suppose you look directly down on these terraces as white light shines directly down on the wing. Then the light reflected back up to you from the terraces undergoes fully constructive interference in the blue-green region of the visible spectrum. Light in the yellow and red regions, at the opposite end of the spectrum, is weaker because it undergoes only intermediate interference. Thus, the top surface of the wing looks blue-green to you.

If you intercept light that reflects from the wing in some other direction, the light has traveled along a slanted path through the terraces. Then the wavelength at which there is fully constructive interference is somewhat different from that for light reflected directly upward. Thus, if the wing moves in your view so that the angle at which you view it changes, the color at which the wing is brightest changes somewhat, producing the iridescence of the wing.

PROBLEM-SOLVING TACTICS

Tactic 1: *Thin-Film Equations*
Some students believe that Eq. 36-34 gives the maxima and Eq. 36-35 gives the minima for *all* thin-film situations. This is not true. These relations were derived only for the situation in which $n_2 > n_1$ and $n_2 > n_3$ in Fig. 36-12.

The appropriate equations for other relative values of the indexes of refraction can be derived by following the reasoning of this section and constructing new versions of Table 36-1. In each case you will end up with Eqs. 36-34 and 36-35, but sometimes Eq. 36-34 will give the minima and Eq. 36-35 will give the maxima—the opposite of what we found here. Which equation gives which depends on whether the reflections at the two interfaces give the same reflection phase shift.

CHECKPOINT 5: The figure shows four situations in which light reflects perpendicularly from a thin film of thickness L (as in Fig. 36-12), with indexes of refraction as given. (a) For which situations does reflection at the film interfaces cause a zero phase difference for the two reflected rays? (b) For which situations will the film be dark if the path length difference $2L$ causes a phase difference of 0.5 wavelength?

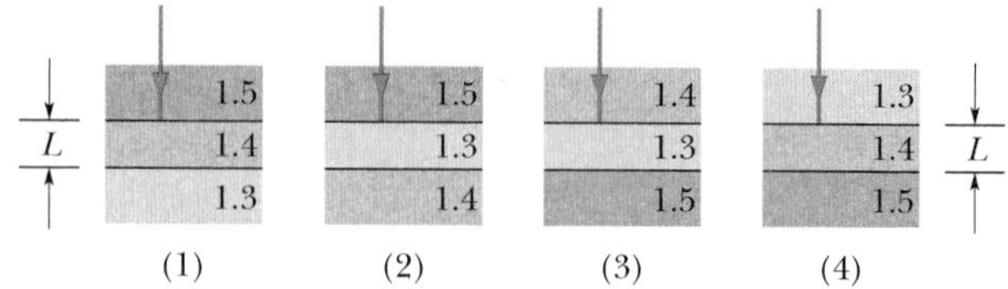

Sample Problem 36-4

White light, with a uniform intensity across the visible wavelength range of 400 to 690 nm, is perpendicularly incident on a water film, of index of refraction $n_2 = 1.33$ and thickness $L = 320$ nm, that is suspended in air. At what wavelength λ is the light reflected by the film brightest to an observer?

SOLUTION: The **Key Idea** here is that the reflected light from the film is brightest at the wavelengths λ for which the reflected rays are in phase with one another. The equation relating these wavelengths λ to the given film thickness L and film index of refraction n_2 is either Eq. 36-34 or Eq. 36-35, depending on the reflection phase shifts for this particular film.

To determine which equation is needed, we should fill out an organizing table like Table 36-1. However, because there is air on both sides of the water film, the situation here is exactly like that in Fig. 36-12, and thus the table would be exactly like Table 36-1. Then from Table 36-1, we see that the reflected rays are in phase (and thus the film is brightest) when

$$2L = \frac{\text{odd number}}{2} \times \frac{\lambda}{n_2},$$

which leads to Eq. 36-34:

$$2L = (m + \tfrac{1}{2})\frac{\lambda}{n_2}.$$

Solving for λ and substituting for L and n_2, we find

$$\lambda = \frac{2n_2L}{m + \frac{1}{2}} = \frac{(2)(1.33)(320 \text{ nm})}{m + \frac{1}{2}} = \frac{851 \text{ nm}}{m + \frac{1}{2}}.$$

For $m = 0$, this gives us $\lambda = 1700$ nm, which is in the infrared region. For $m = 1$, we find $\lambda = 567$ nm, which is yellow-green light, near the middle of the visible spectrum. For $m = 2$, $\lambda = 340$ nm, which is in the ultraviolet region. Thus, the wavelength at which the light seen by the observer is brightest is

$$\lambda = 567 \text{ nm}. \qquad \text{(Answer)}$$

Sample Problem 36-5

In Fig. 36-15, a glass lens is coated on one side with a thin film of magnesium fluoride (MgF_2) to reduce reflection from the lens surface. The index of refraction of MgF_2 is 1.38; that of the glass is 1.50. What is the least coating thickness that eliminates (via interference) the reflections at the middle of the visible spectrum ($\lambda = 550$ nm)? Assume that the light is approximately perpendicular to the lens surface.

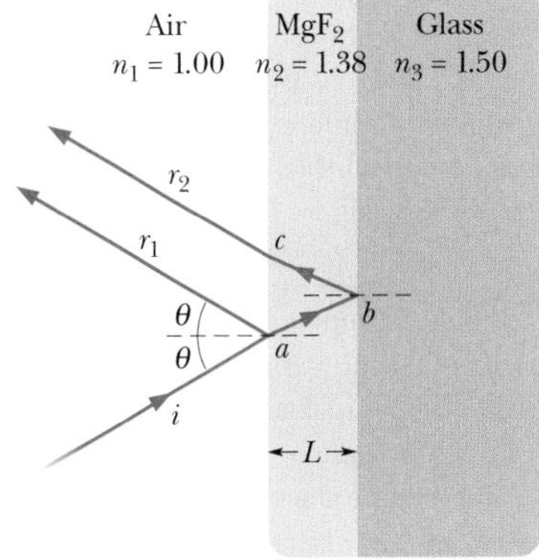

Fig. 36-15 Sample Problem 36-5. Unwanted reflections from glass can be suppressed (at a chosen wavelength) by coating the glass with a thin transparent film of magnesium fluoride of the properly chosen thickness.

SOLUTION: The Key Idea here is that reflection is eliminated if the film thickness L is such that light waves reflected from the two film interfaces are exactly out of phase. The equation relating L to the given wavelength λ and the index of refraction n_2 of the thin film is either Eq. 36-34 or Eq. 36-35, depending on the reflection phase shifts at the interfaces.

To determine which equation is needed, we fill out an organizing table like Table 36-1. At the first interface, the incident light is in air, which has a lesser index of refraction than the MgF_2 (the thin film). Thus, we fill in 0.5 wavelength under r_1 in our organizing table (meaning that the waves of ray r_1 are shifted by 0.5λ at the first interface). At the second interface, the incident light is in the MgF_2, which has a lesser index of refraction than the glass on the other side of the interface. Thus, we fill in 0.5 wavelength under r_2 in our table.

Because both reflections cause the same phase shift, they tend to put the waves of r_1 and r_2 in phase. Since we want those waves to be *out of phase*, their path length difference $2L$ must be an odd number of half-wavelengths:

$$2L = \frac{\text{odd number}}{2} \times \frac{\lambda}{n_2}.$$

This leads to Eq. 36-34. Solving that equation for L then gives us the film thicknesses that will eliminate reflection from the lens and coating:

$$L = (m + \tfrac{1}{2})\frac{\lambda}{2n_2}, \quad \text{for } m = 0, 1, 2, \ldots. \qquad (36\text{-}36)$$

We want the least thickness for the coating—that is, the least L. Thus, we choose $m = 0$, the least possible value of m. Substituting it and the given data in Eq. 36-36, we obtain

$$L = \frac{\lambda}{4n_2} = \frac{550 \text{ nm}}{(4)(1.38)} = 99.6 \text{ nm}. \qquad \text{(Answer)}$$

Sample Problem 36-6

Figure 36-16*a* shows a transparent plastic block with a thin wedge of air at the right. (The wedge thickness is exaggerated in the figure.) A broad beam of red light, with wavelength $\lambda = 632.8$ nm, is directed downward through the top of the block (at an incidence angle of 0°). Some of the light is reflected back up from the top and bottom surfaces of the wedge, which acts as a thin film (of air) with a thickness that varies uniformly and gradually from L_L at the left-hand end to L_R at the right-hand end. (The plastic layers above and below the wedge of air are too thick to act as thin films.) An observer looking down on the block sees an interference pattern consisting of six dark fringes and five bright red fringes along the wedge. What is the change in thickness ΔL $(= L_R - L_L)$ along the wedge?

SOLUTION: One Key Idea here is that the brightness at any point along the left–right length of the air wedge is due to the interference of

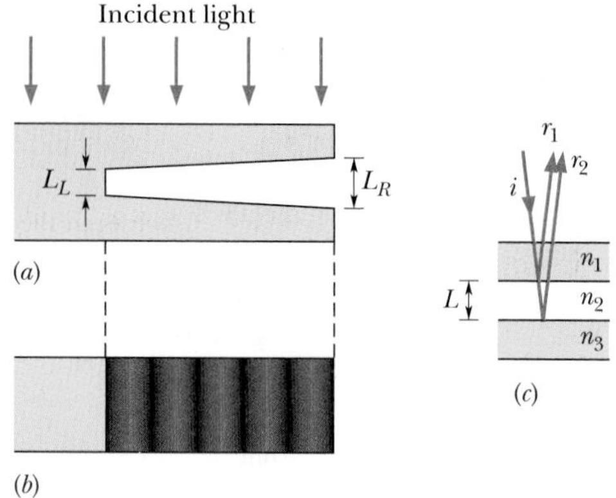

Fig. 36-16 Sample Problem 36-6. (*a*) Red light is incident on a thin, air-filled wedge in the side of a transparent plastic block. The thickness of the wedge is L_L at the left end and L_R at the right end. (*b*) The view from above the block: an interference pattern of six dark fringes and five bright red fringes lies over the region of the wedge. (*c*) A representation of the incident ray i, reflected rays r_1 and r_2, and thickness L of the wedge anywhere along the length of the wedge.

the waves reflected at the top and bottom interfaces of the wedge. A second Key Idea is that the variation of brightness in the pattern of bright and dark fringes is due to the variation in the thickness of the wedge. In some regions, the thickness puts the reflected waves in phase and thus produces a bright reflection (a bright red fringe). In other regions, the thickness puts the reflected waves out of phase and thus produces no reflection (a dark fringe).

Because the observer sees more dark fringes than bright fringes, we can assume that a dark fringe is produced at both the left and right ends of the wedge. Thus, the interference pattern is that shown in Fig. 36-16*b*, which we can use to determine the change in thickness ΔL of the wedge.

Another Key Idea is that we can represent the reflection of light at the top and bottom interfaces of the wedge, at any point along its length, with Fig. 36-16*c*, in which L is the wedge thickness at that point. Let us apply this figure to the left end of the wedge, where the reflections give a dark fringe.

We know that, for a dark fringe, the waves of rays r_1 and r_2 in Fig. 36-16*c* must be out of phase. We also know that the equation relating the film thickness L to the light's wavelength λ and the film's index of refraction n_2 is either Eq. 36-34 or Eq. 36-35, depending on the reflection phase shifts. To determine which equation gives a dark fringe at the left end of the wedge, we should fill out an organizing table like Table 36-1.

At the top interface of the wedge, the incident light is in the plastic, which has a greater index of refraction than the air beneath that interface. Thus, we fill in 0 under r_1 in our organizing table. At the bottom interface of the wedge, the incident light is in air, which has a lesser index of refraction than the plastic beneath that interface. Thus, we fill in 0.5 wavelength under r_2 in our organizing table. Therefore, the reflections alone tend to put the waves of r_1 and r_2 out of phase.

Since the waves are, in fact, out of phase at the left end of the air wedge, the path length difference $2L$ at that end of the wedge must be given by

$$2L = \text{integer} \times \frac{\lambda}{n_2},$$

which leads to Eq. 36-35:

$$2L = m\frac{\lambda}{n_2}, \qquad \text{for } m = 0, 1, 2, \ldots . \tag{36-37}$$

Here is another Key Idea: Eq. 36-37 holds not only for the left end of the wedge but also at any point along the wedge where a dark fringe is observed, including the right end—with a different integer value of m for each fringe. The least value of m is associated with the least thickness of the wedge where a dark fringe is observed. Progressively greater values of m are associated with progressively greater thicknesses of the wedge where a dark fringe is observed. Let m_L be the value at the left end. Then the value at the right end must be $m_L + 5$ because, from Fig. 36-16*b*, the right end is located at the fifth dark fringe from the left end.

We want the change ΔL in thickness, from the left end to the right end of the wedge. To find it we first solve Eq. 36-37 twice—once for the thickness L_L at the left end and once for the thickness L_R at the right end:

$$L_L = (m_L)\frac{\lambda}{2n_2}, \qquad L_R = (m_L + 5)\frac{\lambda}{2n_2}. \tag{36-38}$$

To find the change in thickness ΔL, we can now subtract L_L from L_R and substitute known data, including $n_2 = 1.00$ for the air within the wedge:

$$\Delta L = L_R - L_L = \frac{(m_L + 5)\lambda}{2n_2} - \frac{m_L\lambda}{2n_2} = \frac{5}{2}\frac{\lambda}{n_2}$$

$$= \frac{5}{2}\frac{632.8 \times 10^{-9}\text{ m}}{1.00}$$

$$= 1.58 \times 10^{-6}\text{ m}. \qquad \text{(Answer)}$$

36-8 Michelson's Interferometer

An **interferometer** is a device that can be used to measure lengths or changes in length with great accuracy by means of interference fringes. We describe the form originally devised and built by A. A. Michelson in 1881.

Consider light that leaves point P on extended source S in Fig. 36-17 and encounters *beam splitter M*. A beam splitter is a mirror that transmits half the incident light and reflects the other half. In the figure we have assumed, for convenience, that

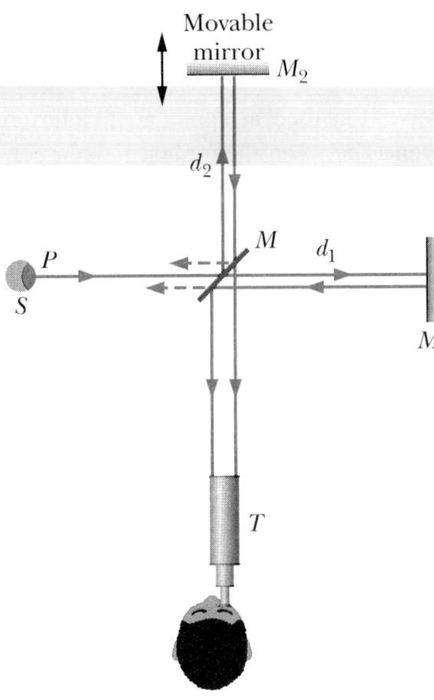

Fig. 36-17 Michelson's interferometer, showing the path of light originating at point P of an extended source S. Mirror M splits the light into two beams, which reflect from mirrors M_1 and M_2 back to M and then to telescope T. In the telescope an observer sees a pattern of interference fringes.

this mirror possesses negligible thickness. At M the light thus divides into two waves. One proceeds by transmission toward mirror M_1; the other proceeds by reflection toward mirror M_2. The waves are entirely reflected at these mirrors and are sent back along their directions of incidence, each wave eventually entering telescope T. What the observer sees is a pattern of curved or approximately straight interference fringes; in the latter case the fringes resemble the stripes on a zebra.

The path length difference for the two waves when they recombine at the telescope is $2d_2 - 2d_1$, and anything that changes this path length difference will cause a change in the phase difference between these two waves at the eye. As an example, if mirror M_2 is moved by a distance $\frac{1}{2}\lambda$, the path length difference is changed by λ and the fringe pattern is shifted by one fringe (as if each dark stripe on a zebra had moved to where the adjacent dark stripe had been). Similarly, moving mirror M_2 by $\frac{1}{4}\lambda$ causes a shift by half a fringe (each dark zebra stripe shifts to where the adjacent white stripe was).

A shift in the fringe pattern can also be caused by the insertion of a thin transparent material into the optical path of one of the mirrors, say, M_1. If the material has thickness L and index of refraction n, then the number of wavelengths along the light's to-and-fro path through the material is, from Eq. 36-9,

$$N_m = \frac{2L}{\lambda_n} = \frac{2Ln}{\lambda}. \tag{36-39}$$

The number of wavelengths in the same thickness $2L$ of air before the insertion of the material is

$$N_a = \frac{2L}{\lambda}. \tag{36-40}$$

When the material is inserted, the light returned by mirror M_1 undergoes a phase change (in terms of wavelengths) of

$$N_m - N_a = \frac{2Ln}{\lambda} - \frac{2L}{\lambda} = \frac{2L}{\lambda}(n - 1). \tag{36-41}$$

For each phase change of one wavelength, the fringe pattern is shifted by one fringe. Thus, by counting the number of fringes through which the material causes the pattern to shift, and substituting that number for $N_m - N_a$ in Eq. 36-41, you can determine the thickness L of the material in terms of λ.

By such techniques the lengths of objects can be expressed in terms of the wavelengths of light. In Michelson's day, the standard of length—the meter —was chosen by international agreement to be the distance between two fine scratches on a certain metal bar preserved at Sèvres, near Paris. Michelson was able to show, using his interferometer, that the standard meter was equivalent to 1 553 163.5 wavelengths of a certain monochromatic red light emitted from a light source containing cadmium. For this careful measurement, Michelson received the 1907 Nobel prize in physics. His work laid the foundation for the eventual abandonment (in 1961) of the meter bar as a standard of length and for the redefinition of the meter in terms of the wavelength of light. By 1983, even this wavelength standard was not precise enough to meet the growing requirements of science and technology, and it was replaced with a new standard based on a defined value for the speed of light.

REVIEW & SUMMARY

Huygens' Principle The three-dimensional transmission of waves, including light, may often be predicted by *Huygens' principle,* which states that all points on a wavefront serve as point sources of spherical secondary wavelets. After a time t, the new position of the wavefront will be that of a surface tangent to these secondary wavelets.

The law of refraction can be derived from Huygens' principle by assuming that the index of refraction of any medium is $n = c/v$, in which v is the speed of light in the medium and c is the speed of light in vacuum.

Wavelength and Index of Refraction The wavelength λ_n of light in a medium depends on the index of refraction n of the medium:

$$\lambda_n = \frac{\lambda}{n}, \qquad (36\text{-}8)$$

in which λ is the wavelength of the light in vacuum. Because of this dependency, the phase difference between two waves can change if they pass through different materials with different indexes of refraction.

Young's Experiment In **Young's interference experiment,** light passing through a single slit falls on two slits in a screen. The light leaving these slits flares out (by diffraction), and interference occurs in the region beyond the screen. A fringe pattern, due to the interference, forms on a viewing screen.

The light intensity at any point on the viewing screen depends in part on the difference in the path lengths from the slits to that point. If this difference is an integer number of wavelengths, the waves interfere constructively and an intensity maximum results. If it is an odd number of half-wavelengths, there is destructive interference and an intensity minimum occurs. The conditions for maximum and minimum intensity are

$$d \sin\theta = m\lambda, \quad \text{for } m = 0, 1, 2, \ldots \quad \text{(maxima—bright fringes)}, \qquad (36\text{-}14)$$

$$d \sin\theta = (m + \tfrac{1}{2})\lambda, \quad \text{for } m = 0, 1, 2, \ldots \quad \text{(minima—dark fringes)}, \qquad (36\text{-}16)$$

where θ is the angle the light path makes with a central axis and d is the slit separation.

Coherence If two light waves that meet at a point are to interfere perceptibly, the phase difference between them must remain constant with time; that is, the waves must be **coherent.** When two coherent waves meet, the resulting intensity may be found by using phasors.

Intensity in Two-Slit Interference In Young's interference experiment, two waves, each with intensity I_0, yield a resultant wave of intensity I at the viewing screen, with

$$I = 4I_0 \cos^2 \tfrac{1}{2}\phi, \qquad \text{where } \phi = \frac{2\pi d}{\lambda}\sin\theta. \qquad (36\text{-}21, 36\text{-}22)$$

Equations 36-14 and 36-16, which identify the positions of the fringe maxima and minima, are contained within this relation.

Thin-Film Interference When light is incident on a thin transparent film, the light waves reflected from the front and back surfaces interfere. For near-normal incidence the wavelength conditions for maximum and minimum intensity of the light reflected from a *film in air* are

$$2L = (m + \tfrac{1}{2})\frac{\lambda}{n_2}, \quad \text{for } m = 0, 1, 2, \ldots \quad \text{(maxima—bright film in air)}, \qquad (36\text{-}34)$$

$$2L = m\frac{\lambda}{n_2}, \quad \text{for } m = 0, 1, 2, \ldots \quad \text{(minima—dark film in air)}, \qquad (36\text{-}35)$$

where n_2 is the index of refraction of the film, L is its thickness, and λ is the wavelength of the light in air.

If the light incident at an interface between media with different indexes of refraction is in the medium with the smaller index of refraction, the reflection causes a phase change of π rad, or half a wavelength, in the reflected wave. Otherwise, there is no phase change due to the reflection. Refraction at an interface does not cause a phase shift.

The Michelson Interferometer In *Michelson's interferometer* a light wave is split into two beams that, after traversing paths of different lengths, are recombined so they interfere and form a fringe pattern. Varying the path length of one of the beams allows distances to be accurately expressed in terms of wavelengths of light, by counting the number of fringes through which the pattern shifts because of the change.

QUESTIONS

1. In Fig. 36-18, three pulses of light—a, b, and c—of the same wavelength are sent through layers of plastic whose indexes of refraction are given. Rank the pulses according to their travel time through the plastic, greatest first.

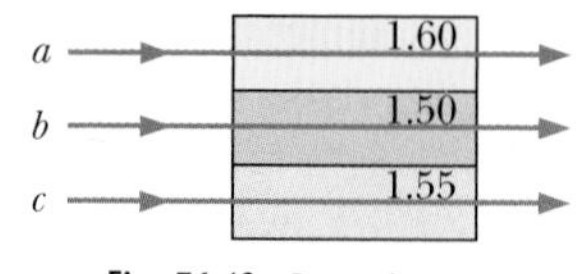

Fig. 36-18 Question 1.

2. Light travels along the length of a 1500-nm-long nanostructure. When a peak of the wave is at one end of the nanostructure, is there a peak or a valley at the other end if the wavelength is (a) 500 nm and (b) 1000 nm?

3. Figure 36-19 shows two rays of light, of wavelength 600 nm, that reflect from glass surfaces separated by 150 nm. The rays are initially in phase. (a) What is the path length difference of the rays?

(b) When they have cleared the reflection region, are the rays exactly in phase, exactly out of phase, or in some intermediate state?

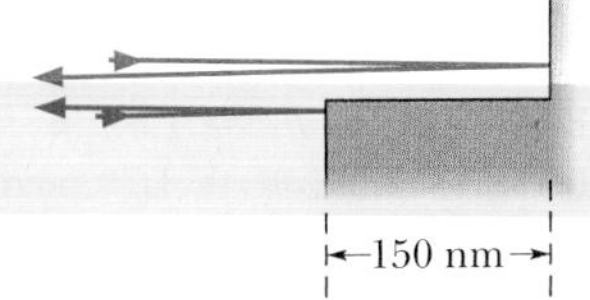

Fig. 36-19 Question 3.

4. Figure 36-20 shows two light rays that are initially exactly in phase and that reflect from several glass surfaces. Neglect the slight slant in the path of the light in the second arrangement. (a) What is the path length difference of the rays? In wavelengths λ, (b) what should that path length difference equal if the rays are to be exactly out of phase when they emerge, and (c) what is the smallest value of d that will allow that final phase difference?

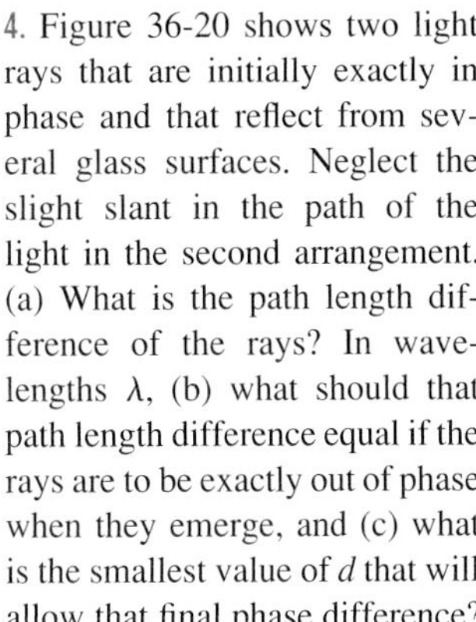

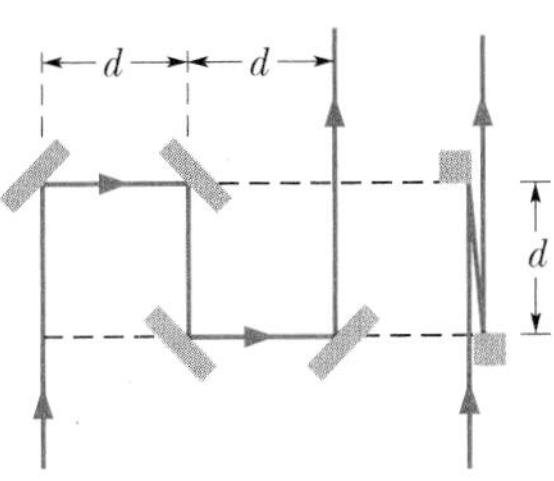

Fig. 36-20 Question 4.

5. Is there an interference maximum, a minimum, an intermediate state closer to a maximum, or an intermediate state closer to a minimum at point P in Fig. 36-8 if the path length difference of the two rays is (a) 2.2λ, (b) 3.5λ, (c) 1.8λ, and (d) 1.0λ? For each situation, give the value of m associated with the maximum or minimum involved.

6. (a) If you move from one bright fringe in a two-slit interference pattern to the next one farther out, (a) does the path length difference ΔL increase or decrease and (b) by how much does it change, in wavelengths λ?

7. Does the spacing between fringes in a two-slit interference pattern increase, decrease, or stay the same if (a) the slit separation is increased, (b) the color of the light is switched from red to blue, and (c) the whole apparatus is submerged in cooking sherry? (d) If the slits are illuminated with white light, then at any side maximum, does the blue component or the red component peak closer to the central maximum?

8. Each part of Fig. 36-21 shows phasors representing the two light waves in a double-slit interference experiment. Further, each part represents a different point on the viewing screen, at a different time. Assuming all eight phasors have the same length, rank the points according to the intensity of the light there, greatest first.

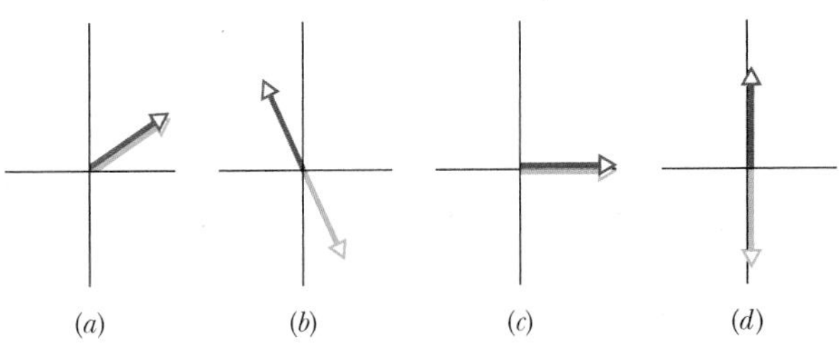

Fig. 36-21 Question 8.

9. Figure 36-22 shows two sources S_1 and S_2 that emit radio waves of wavelength λ in all directions. The sources are exactly in phase and are separated by a distance equal to 1.5λ. The vertical broken line is the perpendicular bisector of the distance between the sources. (a) If we start at the indicated start point and travel along path 1, does the interference produce a maximum all along the path, a minimum all along the path, or alternating maxima and minima? Repeat for (b) path 2 and (c) path 3.

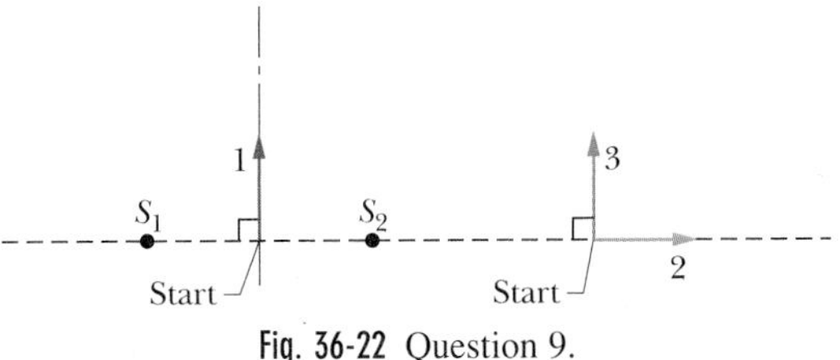

Fig. 36-22 Question 9.

10. Figure 36-23 shows two rays of light encountering interfaces, where they reflect and refract. Which of the resulting waves are shifted in phase at the interface?

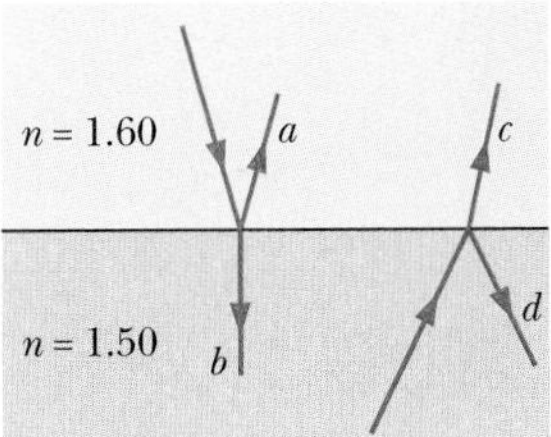

Fig. 36-23 Question 10.

11. Figure 36-24a shows the cross section of a vertical thin film whose width increases downward because gravitation causes slumping. Figure 36-24b is a face-on view of the film, showing four bright interference fringes that result when the film is illuminated with a perpendicular beam of red light. Points in the cross section corresponding to the bright fringes are labeled. In terms of the wavelength of the light inside the film, what is the difference in film thickness between (a) points a and b and (b) points b and d?

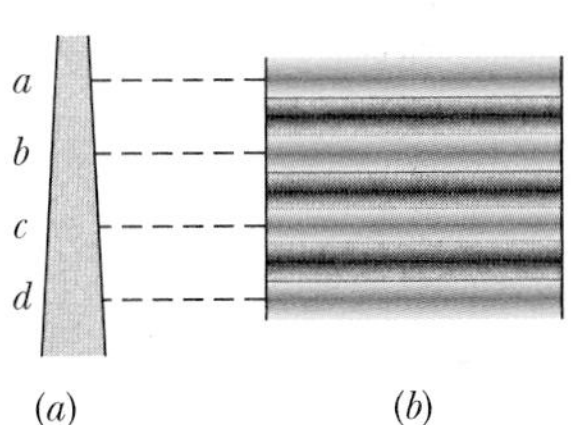

Fig. 36-24 Question 11.

12. Figure 36-25 shows the transmission of light through a thin film in air by a perpendicular beam (tilted in the figure for clarity). (a) Did ray r_3 undergo a phase shift due to reflection? (b) In wavelengths, what is the reflection phase shift for ray r_4? (c) If the film thickness is L, what is the path length difference between rays r_3 and r_4?

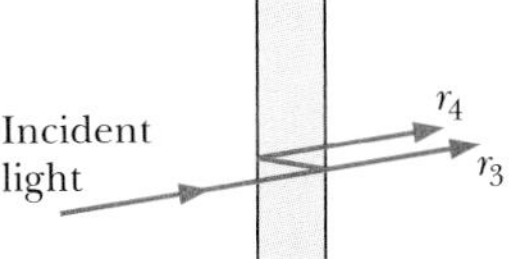

Fig. 36-25 Question 12.

EXERCISES & PROBLEMS

ssm Solution is in the Student Solutions Manual.
www Solution is available on the World Wide Web at: http://www.wiley.com/college/hrw
ilw Solution is available on the Interactive LearningWare.

SEC. 36-2 Light as a Wave

1E. The wavelength of yellow sodium light in air is 589 nm. (a) What is its frequency? (b) What is its wavelength in glass whose index of refraction is 1.52? (c) From the results of (a) and (b) find its speed in this glass.

2E. How much faster, in meters per second, does light travel in sapphire than in diamond? See Table 34-1.

3E. The speed of yellow light (from a sodium lamp) in a certain liquid is measured to be 1.92×10^8 m/s. What is the index of refraction of this liquid for the light?

4E. What is the speed in fused quartz of light of wavelength 550 nm? (See Fig. 34-19.)

5P. Ocean waves moving at a speed of 4.0 m/s are approaching a beach at an angle of 30° to the normal, as shown from above in Fig. 36-26. Suppose the water depth changes abruptly at a certain distance from the beach and the wave speed there drops to 3.0 m/s. Close to the beach, what is the angle θ between the direction of wave motion and the normal? (Assume the same law of refraction as for light.) Explain why most waves come in normal to a shore even though at large distances they approach at a variety of angles.

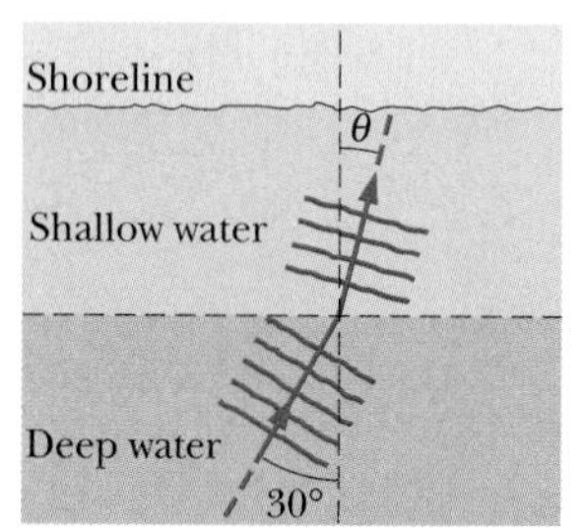

Fig. 36-26 Problem 5.

6P. In Fig. 36-27, two pulses of light are sent through layers of plastic with the indexes of refraction indicated and with thicknesses of either L or $2L$ as shown. (a) Which pulse travels through the plastic in less time? (b) In terms of L/c, what is the difference in the traversal times of the pulses?

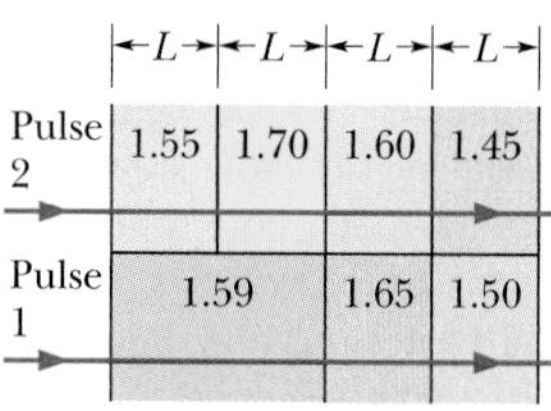

Fig. 36-27 Problem 6.

7P. In Fig. 36-3, assume that two waves of light in air, of wavelength 400 nm, are initially in phase. One travels through a glass layer of index of refraction $n_1 = 1.60$ and thickness L. The other travels through an equally thick plastic layer of index of refraction $n_2 = 1.50$. (a) What is the least value L should have if the waves are to end up with a phase difference of 5.65 rad? (b) If the waves arrive at some common point after emerging, what type of interference do they undergo? ssm

8P. Suppose that the two waves in Fig. 36-3 have wavelength 500 nm in air. In wavelengths, what is their phase difference after traversing media 1 and 2 if (a) $n_1 = 1.50$, $n_2 = 1.60$, and $L = 8.50\ \mu$m; (b) $n_1 = 1.62$, $n_2 = 1.72$, and $L = 8.50\ \mu$m; and (c) $n_1 = 1.59$, $n_2 = 1.79$, and $L = 3.25\ \mu$m? (d) Suppose that in each of these three situations the waves arrive at a common point after emerging. Rank the situations according to the brightness the waves produce at the common point.

9P. Two waves of light in air, of wavelength 600.0 nm, are initially in phase. They then travel through plastic layers as shown in Fig. 36-28, with $L_1 = 4.00\ \mu$m, $L_2 = 3.50\ \mu$m, $n_1 = 1.40$, and $n_2 = 1.60$. (a) In wavelengths, what is their phase difference after they both have emerged from the layers? (b) If the waves later arrive at some common point, what type of interference do they undergo? ilw

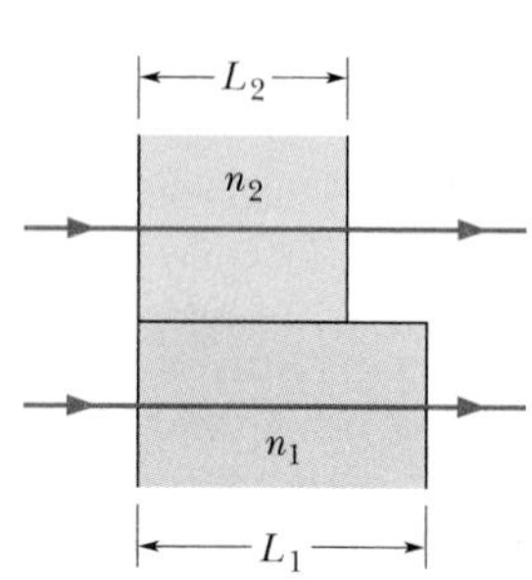

Fig. 36-28 Problem 9.

10P. In Fig. 36-3, assume that the two light waves, of wavelength 620 nm in air, are initially out of phase by π rad. The indexes of refraction of the media are $n_1 = 1.45$ and $n_2 = 1.65$. (a) What is the least thickness L that will put the waves exactly in phase once they pass through the two media? (b) What is the next greater L that will do this?

SEC. 36-4 Young's Interference Experiment

11E. Monochromatic green light, of wavelength 550 nm, illuminates two parallel narrow slits 7.70 μm apart. Calculate the angular deviation (θ in Fig. 36-8) of the third-order (for $m = 3$) bright fringe (a) in radians and (b) in degrees.

12E. What is the phase difference of the waves from the two slits when they arrive at the mth dark fringe in a Young's double-slit experiment?

13E. Suppose that Young's experiment is performed with blue-green light of wavelength 500 nm. The slits are 1.20 mm apart, and the viewing screen is 5.40 m from the slits. How far apart are the bright fringes? ssm ilw

14E. In a double-slit arrangement the slits are separated by a distance equal to 100 times the wavelength of the light passing through the slits. (a) What is the angular separation in radians between the central maximum and an adjacent maximum? (b) What is the distance between these maxima on a screen 50.0 cm from the slits?

15E. A double-slit arrangement produces interference fringes for sodium light ($\lambda = 589$ nm) that have an angular separation of 3.50×10^{-3} rad. For what wavelength would the angular separation be 10.0% greater? ssm

16E. A double-slit arrangement produces interference fringes for sodium light ($\lambda = 589$ nm) that are 0.20° apart. What is the angular fringe separation if the entire arrangement is immersed in water ($n = 1.33$)?

17E. Two radio-frequency point sources separated by 2.0 m are radiating in phase with $\lambda = 0.50$ m. A detector moves in a circular path around the two sources in a plane containing them. Without written calculation, find how many maxima it detects. ssm

18E. Sources A and B emit long-range radio waves of wavelength 400 m, with the phase of the emission from A ahead of that from source B by 90°. The distance r_A from A to a detector is greater than the corresponding distance r_B by 100 m. What is the phase difference at the detector?

19P. In a double-slit experiment the distance between slits is 5.0 mm and the slits are 1.0 m from the screen. Two interference patterns can be seen on the screen: one due to light with wavelength 480 nm, and the other due to light with wavelength 600 nm. What is the separation on the screen between the third-order ($m = 3$) bright fringes of the two interference patterns? ssm

20P. In Fig. 36-29, S_1 and S_2 are identical radiators of waves that are in phase and of the same wavelength λ. The radiators are separated by distance $d = 3.00\lambda$. Find the greatest distance from S_1, along the x axis, for which fully destructive interference occurs. Express this distance in wavelengths.

Fig. 36-29 Problems 20, 27, and 59.

21P. A thin flake of mica ($n = 1.58$) is used to cover one slit of a double-slit interference arrangement. The central point on the viewing screen is now occupied by what had been the seventh bright side fringe ($m = 7$) before the mica was used. If $\lambda = 550$ nm, what is the thickness of the mica? (*Hint:* Consider the wavelength of the light within the mica.) ssm www

22P. Laser light of wavelength 632.8 nm passes through a double-slit arrangement at the front of a lecture room, reflects off a mirror 20.0 m away at the back of the room, and then produces an interference pattern on a screen at the front of the room. The distance between adjacent bright fringes is 10.0 cm. (a) What is the slit separation? (b) What happens to the pattern when the lecturer places a thin cellophane sheet over one slit, thereby increasing by 2.50 the number of wavelengths along the path that includes the cellophane?

SEC. 36-6 Intensity in Double-Slit Interference

23E. Two waves of the same frequency have amplitudes 1.00 and 2.00. They interfere at a point where their phase difference is 60.0°. What is the resultant amplitude? ssm

24E. Find the sum y of the following quantities:

$$y_1 = 10 \sin \omega t \quad \text{and} \quad y_2 = 8.0 \sin(\omega t + 30°).$$

25E. Add the quantities

$$y_1 = 10 \sin \omega t$$
$$y_2 = 15 \sin(\omega t + 30°)$$
$$y_3 = 5.0 \sin(\omega t - 45°)$$

using the phasor method. ilw

26E. Light of wavelength 600 nm is incident normally on two parallel narrow slits separated by 0.60 mm. Sketch the intensity pattern observed on a distant screen as a function of angle θ from the pattern's center for the range of values $0 \le \theta \le 0.0040$ rad.

27P. S_1 and S_2 in Fig. 36-29 are point sources of electromagnetic waves of wavelength 1.00 m. They are in phase and separated by $d = 4.00$ m, and they emit at the same power. (a) If a detector is moved to the right along the x axis from source S_1, at what distances from S_1 are the first three interference maxima detected? (b) Is the intensity of the nearest minimum exactly zero? (*Hint:* Does the intensity of a wave from a point source remain constant with an increase in distance from the source?) ssm

28P. The double horizontal arrow in Fig. 36-9 marks the points on the intensity curve where the intensity of the central fringe is half the maximum intensity. Show that the angular separation $\Delta\theta$ between the corresponding points on the viewing screen is

$$\Delta\theta = \frac{\lambda}{2d}$$

if θ in Fig. 36-8 is small enough so that $\sin\theta \approx \theta$.

29P*. Suppose that one of the slits of a double-slit interference experiment is wider than the other, so the amplitude of the light reaching the central part of the screen from one slit, acting alone, is twice that from the other slit, acting alone. Derive an expression for the light intensity I at the screen as a function of θ, corresponding to Eqs. 36-21 and 36-22. ssm www

SEC. 36-7 Interference from Thin Films

30E. In Fig. 36-30, light wave W_1 reflects once from a reflecting surface while light wave W_2 reflects twice from that surface and once from a reflecting sliver at distance L from the mirror. The waves are initially in phase and have a wavelength of 620 nm. Neglect the slight tilt of the rays. (a) For what least value of L are the reflected waves exactly out of phase? (b) How far must the sliver be moved to put the waves exactly out of phase again?

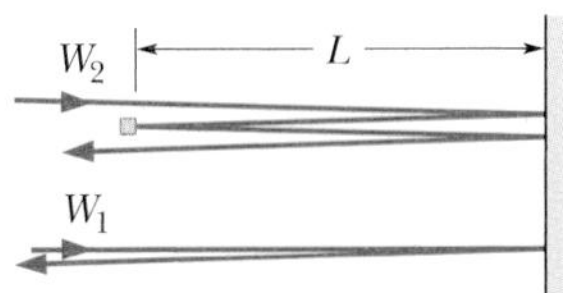

Fig. 36-30 Exercises 30 and 32.

31E. Bright light of wavelength 585 nm is incident perpendicularly on a soap film ($n = 1.33$) of thickness 1.21 μm, suspended in air. Is the light reflected by the two surfaces of the film closer to interfering fully destructively or fully constructively? ssm

32E. Suppose the light waves of Exercise 30 are initially exactly out of phase. Find an expression for the values of L (in terms of the wavelength λ) that put the reflected waves exactly in phase.

33E. Light of wavelength 624 nm is incident perpendicularly on a soap film (with $n = 1.33$) suspended in air. What are the least two thicknesses of the film for which the reflections from the film undergo fully constructive interference? ilw

34E. A camera lens with index of refraction greater than 1.30 is coated with a thin transparent film of index of refraction 1.25 to eliminate by interference the reflection of light at wavelength λ that is incident perpendicularly on the lens. In terms of λ, what minimum film thickness is needed?

35E. The rhinestones in costume jewelry are glass with index of refraction 1.50. To make them more reflective, they are often coated with a layer of silicon monoxide of index of refraction 2.00. What is the minimum coating thickness needed to ensure that light of wavelength 560 nm and of perpendicular incidence will be reflected from the two surfaces of the coating with fully constructive interference? ssm

36E. In Fig. 36-31, light of wavelength 600 nm is incident perpendicularly on five sections of a transparent structure suspended in air. The structure has index of refraction 1.50. The thickness of each section is given in terms of $L = 4.00\ \mu m$. For which sections will the light that is reflected from the top and bottom surfaces of that section undergo fully constructive interference?

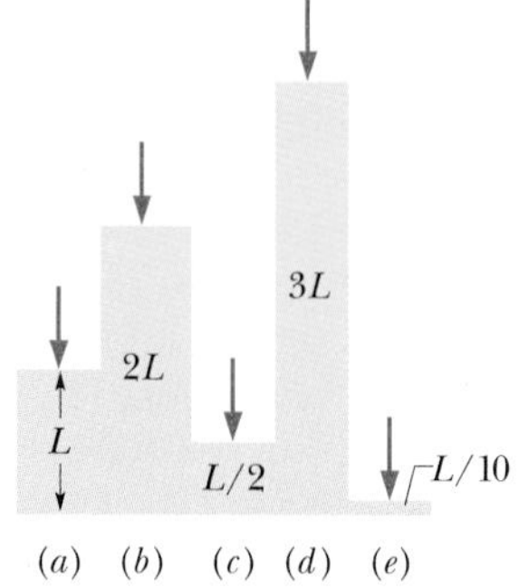

Fig. 36-31 Exercise 36.

37E. We wish to coat flat glass ($n = 1.50$) with a transparent material ($n = 1.25$) so that reflection of light at wavelength 600 nm is eliminated by interference. What minimum thickness can the coating have to do this? ssm www

38P. In Fig. 36-32, light is incident perpendicularly on four thin layers of thickness L. The indexes of refraction of the thin layers and of the media above and below these layers are given. Let λ represent the wavelength of the light in air, and n_2 represent the index of refraction of the thin layer in each situation. Consider only the transmission of light that undergoes no reflection or two reflections, as in Fig. 36-32*a*. For which of the situations does the expression

$$\lambda = \frac{2Ln_2}{m}, \quad \text{for } m = 1, 2, 3, \ldots,$$

give the wavelengths of the transmitted light that undergoes fully constructive interference?

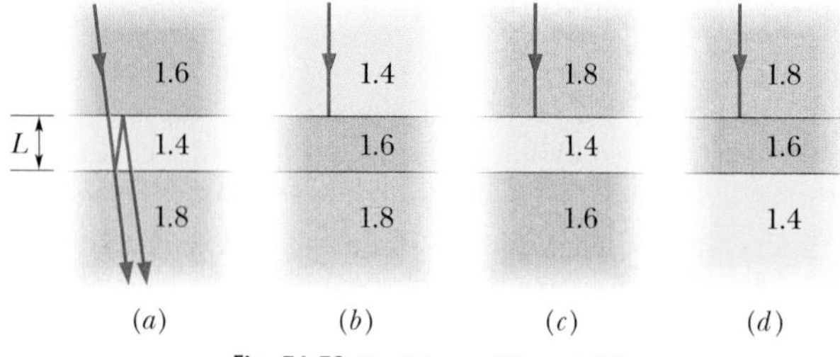

Fig. 36-32 Problems 38 and 39.

39P. A disabled tanker leaks kerosene ($n = 1.20$) into the Persian Gulf, creating a large slick on top of the water ($n = 1.30$). (a) If you are looking straight down from an airplane, while the Sun is overhead, at a region of the slick where its thickness is 460 nm, for which wavelength(s) of visible light is the reflection brightest because of constructive interference? (b) If you are scuba diving directly under this same region of the slick, for which wavelength(s) of visible light is the transmitted intensity strongest? (*Hint:* Use Fig. 36-32*a* with appropriate indexes of refraction.)

40P. A plane wave of monochromatic light is incident normally on a uniform thin film of oil that covers a glass plate. The wavelength of the source can be varied continuously. Fully destructive interference of the reflected light is observed for wavelengths of 500 and 700 nm and for no wavelengths in between. If the index of refraction of the oil is 1.30 and that of the glass is 1.50, find the thickness of the oil film.

41P. A plane monochromatic light wave in air is perpendicularly incident on a thin film of oil that covers a glass plate. The wavelength of the source may be varied continuously. Fully destructive interference of the reflected light is observed for wavelengths of 500 and 700 nm and for no wavelength in between. The index of refraction of the glass is 1.50. Show that the index of refraction of the oil must be less than 1.50. ssm

42P. The reflection of perpendicularly incident white light by a soap film in air has an interference maximum at 600 nm and a minimum at 450 nm, with no minimum in between. If $n = 1.33$ for the film, what is the film thickness, assumed uniform?

43P. In Fig. 36-33, a broad beam of light of wavelength 683 nm is sent directly downward through the top plate of a pair of glass plates. The plates are 120 mm long, touch at the left end, and are separated by a wire of diameter 0.048 mm at the right end. The air between the plates acts as a thin film. How many bright fringes will be seen by an observer looking down through the top plate? ssm

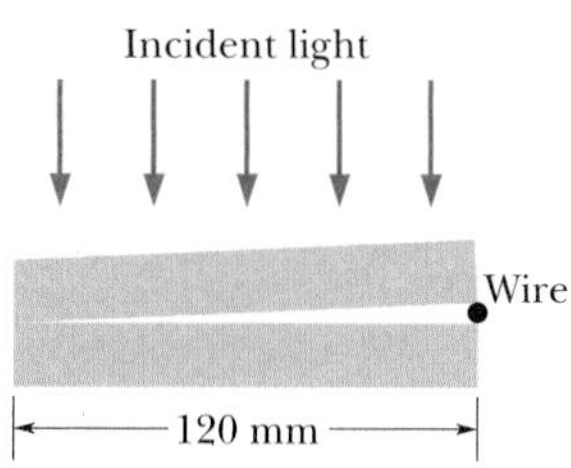

Fig. 36-33 Problems 43 and 44.

44P. In Fig. 36-33, white light is sent directly downward through the top plate of a pair of glass plates. The plates touch at the left end and are separated by a wire of diameter 0.048 mm at the right end; the air between the plates acts as a thin film. An observer looking down through the top plate sees bright and dark fringes due to that film. (a) Is a dark fringe or a bright fringe seen at the left end? (b) To the right of that end, fully destructive interference occurs at different locations for different wavelengths of the light. Does it occur first for the red end or the blue end of the visible spectrum?

45P. A broad beam of light of wavelength 630 nm is incident at 90° on a thin, wedge-shaped film with index of refraction 1.50. An observer intercepting the light transmitted by the film sees 10 bright and 9 dark fringes along the length of the film. By how much does the film thickness change over this length? ssm www

46P. A thin film of acetone ($n = 1.25$) coats a thick glass plate ($n = 1.50$). White light is incident normal to the film. In the reflections, fully destructive interference occurs at 600 nm and fully constructive interference at 700 nm. Calculate the thickness of the acetone film.

47P. Two glass plates are held together at one end to form a wedge of air that acts as a thin film. A broad beam of light of wavelength 480 nm is directed through the plates, perpendicular to the first plate. An observer intercepting light reflected from the plates sees on the plates an interference pattern that is due to the wedge of air. How much thicker is the wedge at the sixteenth bright fringe than it is at the sixth bright fringe, counting from where the plates touch?

48P. A broad beam of monochromatic light is directed perpendicularly through two glass plates that are held together at one end to create a wedge of air between them. An observer intercepting light reflected from the wedge of air, which acts as a thin film, sees 4001 dark fringes along the length of the wedge. When the air between the plates is evacuated, only 4000 dark fringes are seen. Calculate the index of refraction of air from these data.

49P. Figure 36-34a shows a lens with radius of curvature R lying on a plane glass plate and illuminated from above by light with wavelength λ. Figure 36-34b (a photograph taken from above the lens) shows that circular interference fringes (called *Newton's rings*) appear, associated with the variable thickness d of the air film between the lens and the plate. Find the radii r of the interference maxima assuming $r/R \ll 1$. ssm ilw

(a)

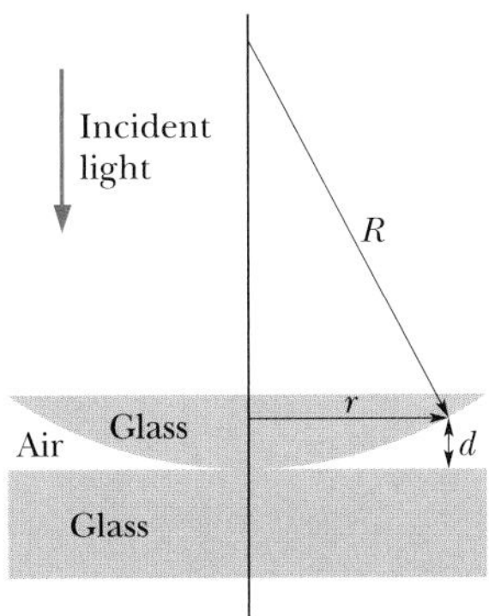

(b)

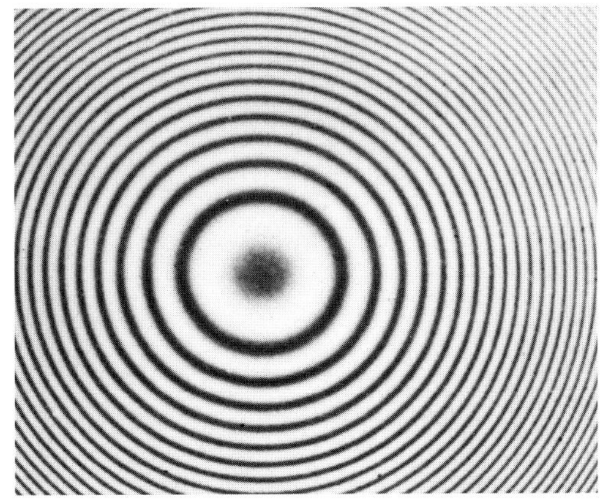

Fig. 36-34 Problems 49 through 52.

50P. In a Newton's rings experiment (see Problem 49), the radius of curvature R of the lens is 5.0 m and the lens diameter is 20 mm. (a) How many bright rings are produced? Assume that $\lambda = 589$ nm. (b) How many bright rings would be produced if the arrangement were immersed in water ($n = 1.33$)?

51P. A Newton's rings apparatus is to be used to determine the radius of curvature of a lens (see Fig. 36-34 and Problem 49). The radii of the nth and $(n + 20)$th bright rings are measured and found to be 0.162 and 0.368 cm, respectively, in light of wavelength 546 nm. Calculate the radius of curvature of the lower surface of the lens.

52P. (a) Use the result of Problem 49 to show that, in a Newton's rings experiment, the difference in radius between adjacent bright rings (maxima) is given by

$$\Delta r = r_{m+1} - r_m \approx \tfrac{1}{2}\sqrt{\lambda R/m},$$

assuming $m \gg 1$. (b) Now show that the *area* between adjacent bright rings is given by

$$A = \pi\lambda R,$$

assuming $m \gg 1$. Note that this area is independent of m.

53P. In Fig. 36-35, a microwave transmitter at height a above the water level of a wide lake transmits microwaves of wavelength λ toward a receiver on the opposite shore, a distance x above the water level. The microwaves reflecting from the water interfere with the microwaves arriving directly from the transmitter. Assuming that the lake width D is much greater than a and x, and that $\lambda \geq a$, at what values of x is the signal at the receiver maximum? (*Hint:* Does the reflection cause a phase change?) ssm

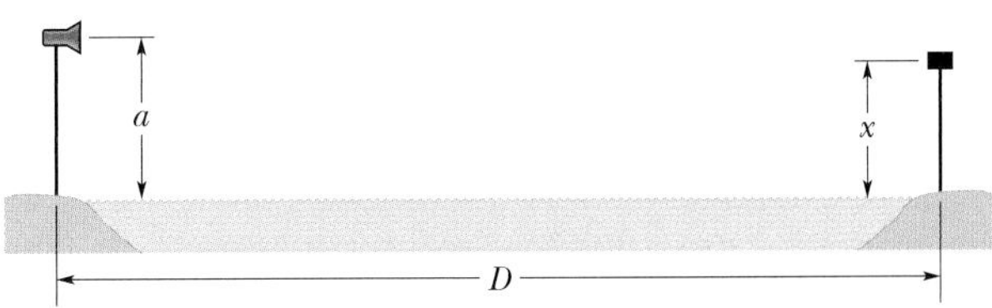

Fig. 36-35 Problem 53.

SEC. 36-8 Michelson's Interferometer

54E. A thin film with index of refraction $n = 1.40$ is placed in one arm of a Michelson interferometer, perpendicular to the optical path. If this causes a shift of 7.0 fringes of the pattern produced by light of wavelength 589 nm, what is the film thickness?

55E. If mirror M_2 in a Michelson interferometer (Fig. 36-17) is moved through 0.233 mm, a shift of 792 fringes occurs. What is the wavelength of the light producing the fringe pattern? ssm

56P. The element sodium can emit light at two wavelengths, $\lambda_1 = 589.10$ nm and $\lambda_2 = 589.59$ nm. Light from sodium is being used in a Michelson interferometer (Fig. 36-17). Through what distance must mirror M_2 be moved to shift the fringe pattern for one wavelength by 1.00 fringe more than the fringe pattern for the other wavelength?

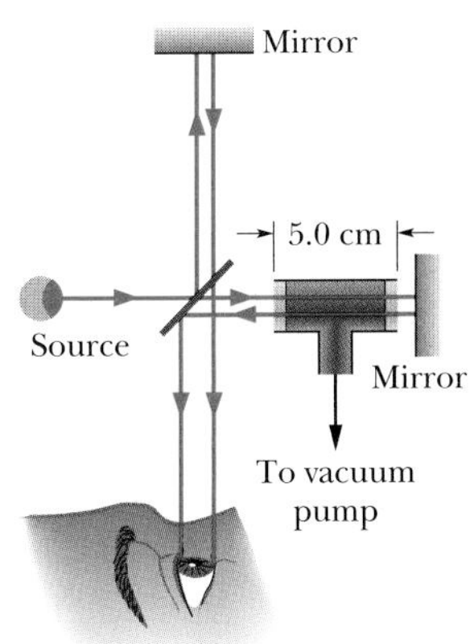

Fig. 36-36 Problem 57.

57P. In Fig. 36-36, an airtight chamber 5.0 cm long with glass windows is placed in one arm of a Michelson interferometer. Light of wavelength $\lambda = 500$ nm is used. Evacuating the air from the chamber causes a shift of 60 fringes. From these data, find the index of refraction of air at atmospheric pressure. ssm

58P. Write an expression for the intensity observed in a Michelson interferometer (Fig. 36-17) as a function of the position of the movable mirror. Measure the position of the mirror from the point at which $d_2 = d_1$.

Additional Problems

59. Figure 36-29 shows two point sources S_1 and S_2 that emit light of wavelength $\lambda = 500$ nm. The emissions are isotropic and in phase, and the separation between the sources is $d = 2.00\ \mu$m. At any point P on the x axis, the wave from S_1 and the wave from S_2 interfere. When P is very far away ($x \approx \infty$), what are (a) the phase difference between the waves arriving from S_1 and S_2 and (b) the type of interference they produce (approximately fully constructive or fully destructive)? (c) As we then move P along the x axis toward S_1, does the phase difference between the waves from S_1 and S_2 increase or decrease? (d) Produce a table that gives the positions x at which the phase differences are 0, 0.50λ, 1.00λ, . . . , 2.50λ, and for each indicate the corresponding type of interference—either fully destructive (fd) or fully constructive (fc).

60. By the late 1800s, most scientists believed that light (any electromagnetic wave) required a medium in which to travel, that it could not travel through vacuum. One reason for this belief was that any other type of wave known to scientists requires a medium. For example, sound waves can travel through air, water, or ground but not through vacuum. Thus, reasoned the scientists, when light travels from the Sun or any other star to Earth, it cannot be traveling through vacuum; instead, it must be traveling through a medium that fills all of space and through which Earth slips. Presumably, light has a certain speed c through this medium, which was called *aether* (or *ether*).

In 1887, Michelson and Edward Morely used a version of Michelson's interferometer to test for the effects of aether on the travel of light within the device. Specifically, the motion of the device through aether as Earth moves around the Sun should affect the interference pattern produced by the device. Scientists assumed that the Sun is approximately stationary in aether; hence the speed of the interferometer through aether should be Earth's speed v about the Sun.

Figure 36-37*a* shows the basic arrangement of mirrors in the 1887 experiment. The mirrors were mounted on a heavy slab that was suspended on a pool of mercury so that the slab could be rotated smoothly about a vertical axis. Michelson and Morely wanted to monitor the interference pattern as they rotated the slab, thus changing the orientation of the interferometer arms relative to the motion through aether. A fringe shift in the interference pattern during the rotation would clearly signal the presence of aether.

Figure 36-37*b*, an overhead view of the equipment, shows the path of the light. To improve the possibility of fringe shift, the light was reflected several times along the arms of the interferometer, instead of only once along each arm as indicated in the basic interferometer of Fig. 36-17. This repeated reflection increased the effective length of each arm to about 10 m. In spite of the added complexity, the interferometer of Figs. 36-37*a* and *b* functions just like the simpler interferometer of Fig. 36-17; so we can use Fig. 36-17 in our discussion here by merely taking the arm lengths d_1 and d_2 to be 10 m each.

Let us assume that there is aether through which light has speed c. Figure 36-37*c* shows a side view of the arm of length d_1 from the aether reference frame as the interferometer moves rightward through it with velocity $\vec{v}$. (For simplicity, the beam splitter M of Fig. 36-17 is drawn parallel to the mirror M_1 at the far end of the arm.) Figure 36-37*d* shows the arm just as a particular portion of the light (represented by a dot) begins its travel along the arm. We shall follow this light to find the path length along the arm, from the beam splitter to M_1 and then back to the beam splitter.

As the light moves at speed c rightward through aether and toward mirror M_1, that mirror moves rightward at speed v. Figure 36-37*e* shows the positions of M and M_1 when the light reaches M_1, reflecting there. The light now moves leftward through aether at speed c while M moves rightward. Figure 36-37*f* shows the positions of M and M_1 when the light has returned to M. (a) Show that the total time of travel for this light, from M to M_1 and then back to M, is

$$t_1 = \frac{2cd_1}{c^2 - v^2}$$

and thus that the path length L_1 traveled by the light along this

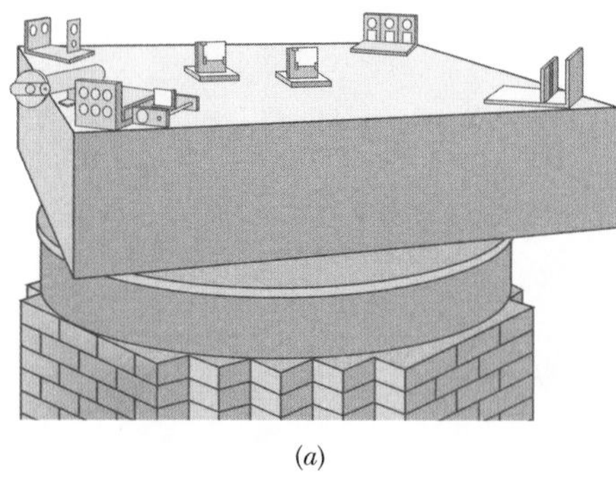
(*a*)

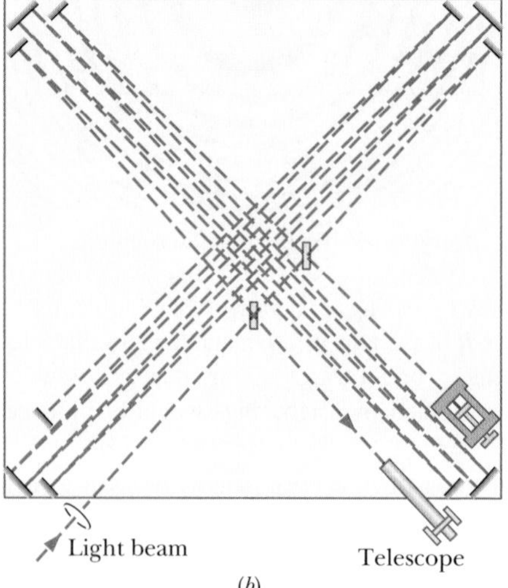

(*b*)

Fig. 36-37 Problem 60.

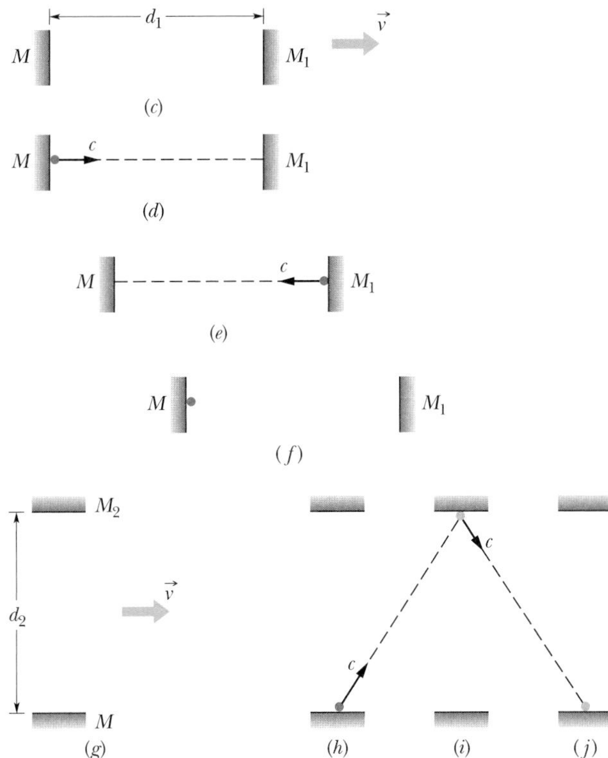

Fig. 36-37 Problem 60 (*continued*).

arm is

$$L_1 = ct_1 = \frac{2c^2 d_1}{c^2 - v^2}.$$

Figure 36-37*g* shows a view of the arm of length d_2; that arm also moves rightward with velocity $\vec{v}$ through the aether. For simplicity, the beam splitter M of Fig. 36-17 is now drawn parallel to the mirror M_2 at the far end of this arm. Figure 36-37*h* shows the arm just as a particular portion of the light (the dot) begins its travel along the arm. Because the arm moves rightward during the flight of the light, the path of the light is angled rightward toward the position that M_2 will have when the light reaches that mirror (Fig. 36-37*i*). The reflection of the light from M_2 sends the light angled rightward toward the position that M will have when the light returns to it (Fig. 36-37*j*). (b) Show that the total time of travel for the light, from M to M_2 and then back to M, is

$$t_2 = \frac{2d_2}{\sqrt{c^2 - v^2}}$$

and thus that the path length L_2 traveled by the light along this arm is

$$L_2 = ct_2 = \frac{2cd_2}{\sqrt{c^2 - v^2}}.$$

Substitute d for d_1 and d_2 in the expressions for L_1 and L_2. Then expand the two expressions by using the binomial expansion (given in Appendix E); retain the first two terms in each expansion. (c) Show that path length L_1 is greater than path length L_2 and that their difference ΔL is

$$\Delta L = \frac{dv^2}{c^2}.$$

(d) Next show that, at the telescope, the phase difference (in terms of wavelengths) between the light traveling along L_1 and that along L_2 is

$$\frac{\Delta L}{\lambda} = \frac{dv^2}{\lambda c^2},$$

where λ is the wavelength of the light. This phase difference determines the fringe pattern produced by the light arriving at the telescope in the interferometer.

Now rotate the interferometer by 90° so that the arm of length d_2 is along the direction of motion through the aether and the arm of length d_1 is perpendicular to that direction. (e) Show that the shift in the fringe pattern due to this rotation is

$$\text{shift} = \frac{2dv^2}{\lambda c^2}.$$

(f) Evaluate the shift, setting $c = 3.0 \times 10^8$ m/s, $d = 10$ m, and $\lambda = 500$ nm and using data about Earth given in Appendix C.

This expected fringe shift would have been easily observable. However, Michelson and Morely observed no fringe shift, which cast grave doubt on the existence of aether. In fact, the idea of aether soon disappeared. Moreover, the null result of Michelson and Morely led, at least indirectly, to Einstein's special theory of relativity.

NEW PROBLEMS

N1. A 600-nm-thick soap film ($n = 1.40$) in air is illuminated with white light in a direction perpendicular to the film. For how many different wavelengths in the 300 to 700 nm range is there (a) fully constructive interference and (b) fully destructive interference in the reflected light?

N2. A thin film of liquid is held in a horizontal circular ring like the soap film is held in the vertical ring in Fig. 36-14. Air lies on both sides of the film. A beam of light at wavelength 550 nm is directed perpendicularly onto the film, and the intensity I of its reflection is monitored. Figure 36N-1 gives intensity I as a function of time t. The intensity changes because of evaporation from the two sides of the film. Assume that the film is flat and has parallel sides, a radius of 1.80 cm, and an index of refraction of 1.40. Also assume that the film's volume decreases at a constant rate. Find that rate.

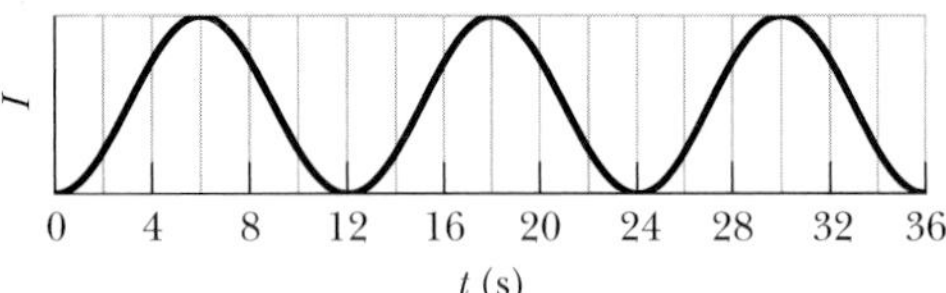

Fig. 36N-1 Problem N2.

N3. Derive the law of reflection using Huygens' principle.

N4. Figure 36N-2a shows two light rays that are initially in phase as they travel upward through a block of plastic, with wavelength 400 nm as measured in air. Light ray r_1 exits directly into air. However, before light ray r_2 exits into air, it travels through a liquid in a hollow cylinder within the plastic. Initially the height L_{liq} of the liquid is 40.0 μm, but then the liquid begins to evaporate. Let ϕ be the phase difference between rays r_1 and r_2 once they both exit into the air. Figure 36N-2b shows ϕ versus the liquid's height L_{liq} until the liquid disappears, with ϕ given in terms of wavelength. What are (a) the index of refraction of the plastic and (b) the index of refraction of the liquid?

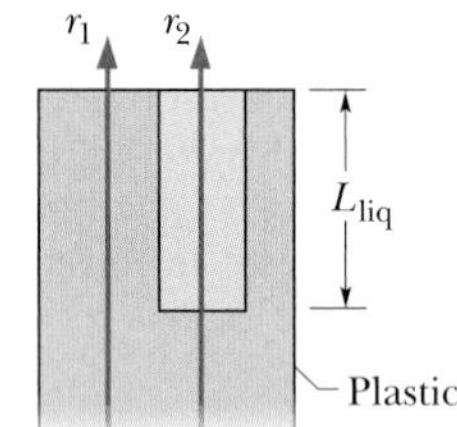

(a)

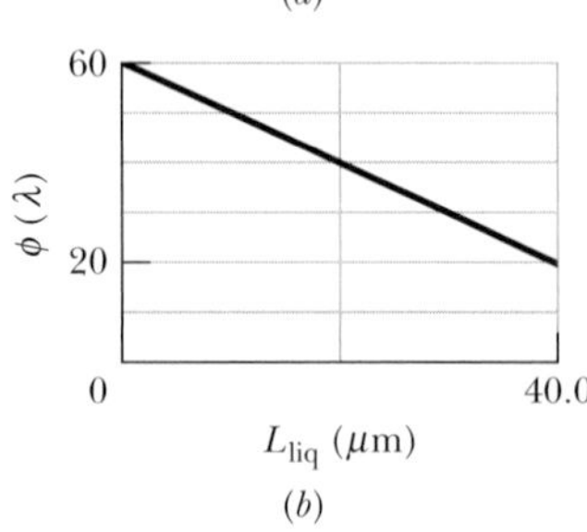

(b)

Fig. 36N-2 Problem N4.

N5. A laser beam travels along the axis of a straight section of pipeline, 1 mi long. The pipe normally contains air at standard temperature and pressure (see Table 34-1), but it may also be evacuated. In which case would the travel time for the beam be greater, and by how much?

N6. In Figure 36N-3, two isotropic point sources S_1 and S_2 emit light in phase at wavelength λ. The sources are separated by distance $2d = 6.00\lambda$. They lie on an axis that is parallel to an x axis, which runs along a viewing screen at distance $D = 20\lambda$. The origin lies on the perpendicular bisector between the sources. The figure shows two rays reaching point P on the screen, at position x_P. (a) At what value of x_P do the rays have the minimum possible phase difference? (b) What is that minimum phase difference in terms of wavelength? (c) At what value of x_P do the rays have the maximum possible phase difference? (d) What is that maximum phase difference in terms of wavelength? (e) In terms of wavelength, what is their phase difference when $x_P = 6.00\lambda$? (f) Is the resulting intensity at point P maximum, minimum, intermediate but closer to maximum, or intermediate but closer to minimum?

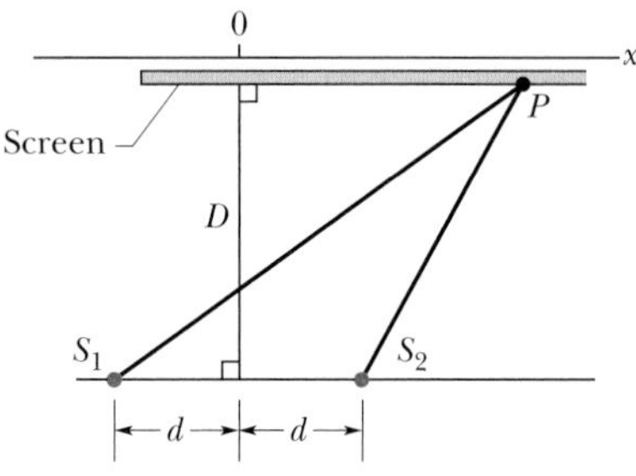

Fig. 36N-3 Problem N6.

N7. In Fig. 36N-4, light travels from point A to point B, through two regions having indexes of refraction n_1 and n_2. Show that the path that requires the least travel time from A to B is the path for which θ_1 and θ_2 in the figure satisfy Eq. 36-6.

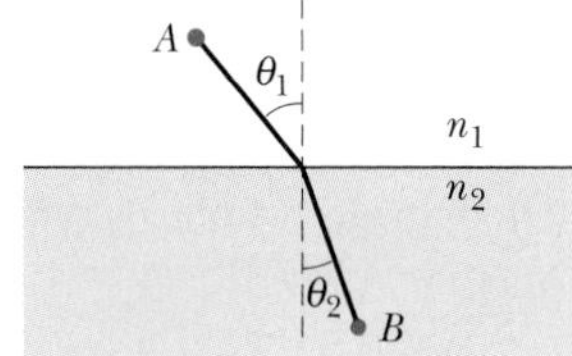

Fig. 36N-4 Problem N7.

N8. In Figure 36N-5, two isotropic point sources S_1 and S_2 emit light in phase at wavelength λ. The sources are separated by distance $d = 6.00\lambda$ on an x axis. A viewing screen is at distance $D = 20\lambda$ from S_2 and parallel to the y axis. The figure shows two rays reaching point P on the screen, at height y_P. (a) At what value of y_P do the rays have the minimum possible phase difference? (b) What is that minimum phase difference in terms of wavelength? (c) At what value of y_P do the rays have the maximum possible phase difference? (d) What is that maximum phase difference in terms of wavelength? (e) In terms of wavelength, what is their phase difference when $y_P = d$? (f) Is the resulting intensity at point

P maximum, minimum, intermediate but closer to maximum, or intermediate but closer to minimum?

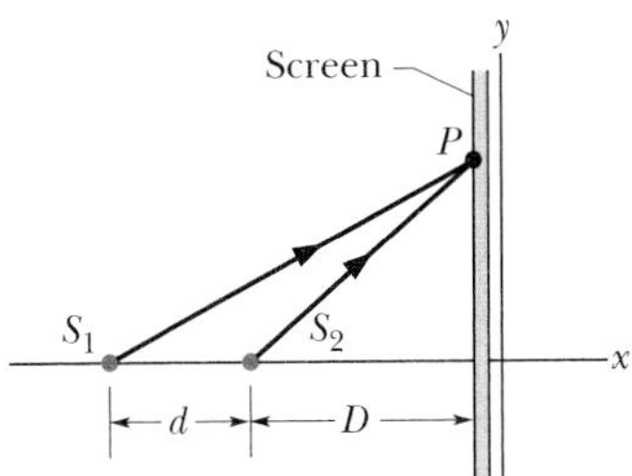

Fig. 36N-5 Problem N8.

N9. Two point sources, S_1 and S_2 in Fig. 36N-6, emit waves in phase and at the same frequency. The sources are separated by distance d. Show that all curves (such as that given) over which the phase difference for rays r_1 and r_2 is a constant are hyperbolas. (*Hint:* A constant phase difference implies a constant difference in length between r_1 and r_2.)

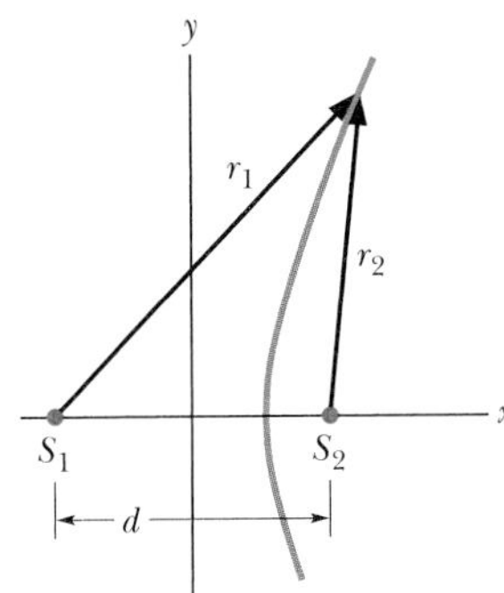

Fig. 36N-6 Problem N9.

N10. In the double-slit experiment of Fig. 36-8, the electric fields of the waves arriving at point P are given by

$$E_1 = (2.00\ \mu\text{V/m})\sin[(1.26 \times 10^{15})t]$$
$$E_2 = (2.00\ \mu\text{V/m})\sin[(1.26 \times 10^{15})t + 39.6\ \text{rad}],$$

where time t is in seconds. (a) What is the amplitude of the resultant electric field at point P? (b) What is the ratio of the intensity I_P at point P to the intensity I_{cen} at the center of the interference pattern? (c) Describe where point P is in the interference pattern by giving the maximum or minimum on which it lies, or the maximum and minimum between which it lies. In a phasor diagram of the electric fields, (d) at what rate would the phasors rotate around the origin and (e) what is the angle between the phasors?

N11. A lens with index of refraction greater than 1.30 is coated with a thin transparent film of index of refraction 1.30. The film is to eliminate by interference the reflection of red light at wavelength 680 nm that is incident perpendicularly on the lens in air. What minimum film thickness is needed?

N12. In the double-slit experiment of Fig. 36-8, the viewing screen is at distance $D = 4.00$ m, point P lies at distance $y = 20.5$ cm from the center of the pattern, the slit separation d is 4.50 μm, and the wavelength λ is 580 nm. (a) Determine where point P is in the interference pattern by giving the maximum or minimum on which it lies, or the maximum and minimum between which it lies. (b) What is the ratio of the intensity I_P at point P to the intensity I_{cen} at the center of the pattern?

N13. A broad beam of light of wavelength 630 nm is incident at 90° on a thin, wedge-shaped film with index of refraction 1.50. An observer intercepting the light transmitted by the film sees 10 bright and 9 dark fringes along the length of the film. By how much does the film thickness change over this length?

N14. In Fig. 36N-7a, a beam of light in material 1 is incident on a boundary at an angle of 30°. The extent of refraction of the light depends, in part, on the index of refraction n_2 of material 2. Figure 36N-7b gives the angle of refraction θ_2 versus n_2 for a range of possible n_2 values. What is the speed of light in material 1?

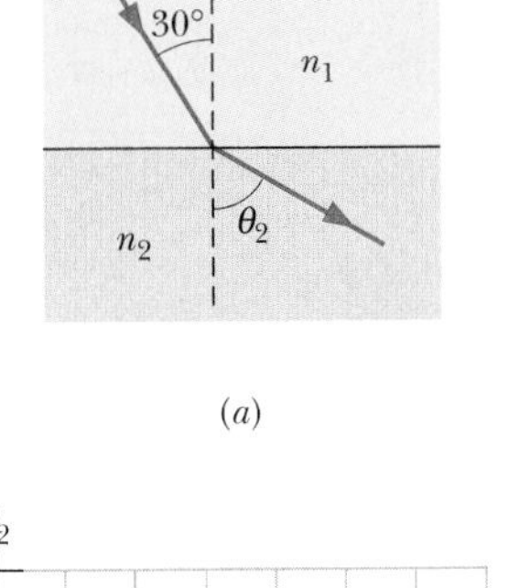

(*a*)

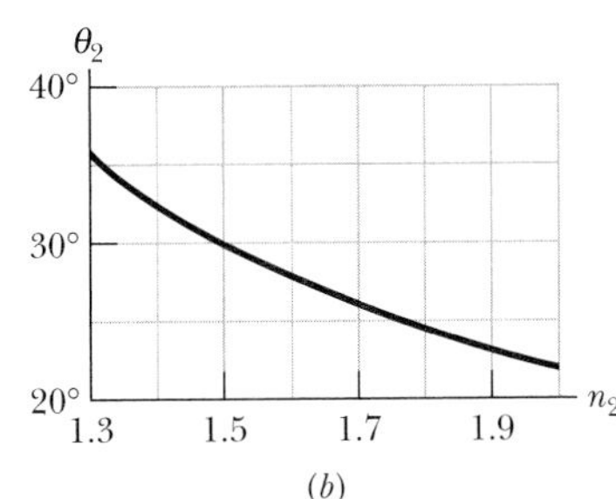

(*b*)

Fig. 36N-7 Problem N14.

N15. Light of wavelength 700.0 nm is sent along a route of length 2000 nm. The route is then filled with a medium having an index of refraction of 1.400. In degrees, by how much does the medium phase shift the light? Give the full shift and then the equivalent shift of less than 360°.

N16. Sunlight is used in a double-slit interference experiment. The fourth-order maximum for a wavelength of 450 nm occurs at an angle of $\theta = 90°$. Thus, it is on the verge of being eliminated from the pattern because θ cannot exceed 90° in Eq. 36-14. (a) What range of wavelengths in the visible range (400 nm to 700 nm) are not present in the third-order maxima? (b) By what least amount must the slit separation be changed to eliminate all of the visible light in the fourth-order maxima?

N17. Two rectangular, optically flat, glass plates ($n = 1.60$) are in contact along one edge and are separated along the opposite edge by a thin foil of unknown thickness. Light with a wavelength of 600 nm is incident perpendicularly onto the top plate. Nine dark fringes and eight bright fringes are observed across the top plate. If the distance between the two plates along the separated edges is increased by 600 nm, how many dark fringes will there then be across the top plate?

N18. In Fig. 36N-8, a broad beam of light of wavelength 620 nm is sent directly downward through the top plate of a pair of glass plates, which are shown in cross section. The plates touch at the left end. The air between the plates acts as a thin film, and an interference pattern can be seen from above the plates. Initially, a dark fringe lies at the left end, a bright fringe lies at the right end, and nine dark fringes lie between those two end fringes. The plates are then very gradually squeezed together at a constant rate to decrease the angle between them. As a result the fringe at the right side changes from being bright to being dark, and vice versa, every 15.0 s. (a) At what rate is the spacing between the plates at the right end being changed? (b) By how much has the spacing there changed when both left and right ends have a dark fringe and there are five dark fringes between them?

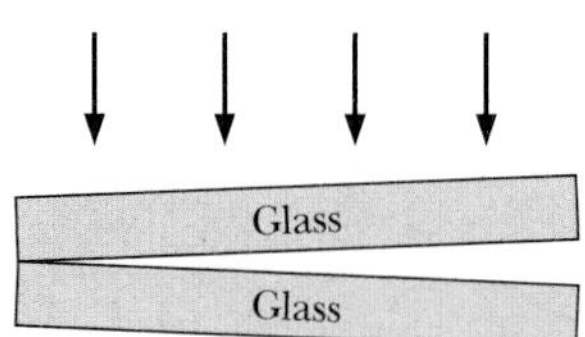

Fig. 36N-8 Problem N18.

N19. White light is sent downward onto a horizontal thin film that has been formed between two materials. The indexes of refraction are 1.80 for the top material, 1.70 for the thin film, and 1.50 for the bottom material. The film thickness is 5.00×10^{-7} m. (a) Which visible wavelengths (400 to 700 nm) result in fully constructive interference at an observer above the film? The materials and film are then heated so that the film thickness increases. (b) Does the light resulting in fully constructive interference shift toward longer or shorter wavelengths?

N20. In the two-slit experiment of Fig. 36-8, let angle θ be 20°, the slit separation be 4.24 μm, and the wavelength be 500 nm. In terms of (a) wavelengths and (b) radians, what is the phase difference between the waves of rays r_1 and r_2 when they arrive at point P on the distant screen? (c) Determine where in the interference pattern point P lies by giving the maximum or minimum on which it lies, or the maximum and minimum between which it lies.

N21. In Fig. 36N-9, two glass plates ($n = 1.60$) form a wedge, and a fluid ($n = 1.50$) fills the interior. At the left end the plates touch; at the right, they are separated by 580 nm. Light with a wavelength (in air) of 580 nm shines downward on the assembly, and an observer intercepts light sent back upward. Do dark or bright bands lie at (a) the left end and (b) the right end? (c) How many dark bands are along the plates?

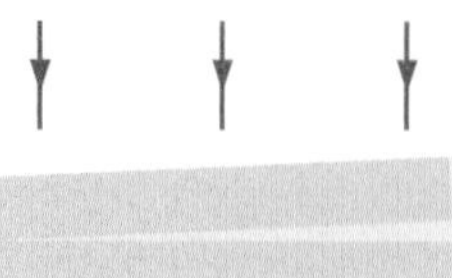

Fig. 36N-9 Problem N21.

N22. In Fig. 36N-10, two isotropic point sources S_1 and S_2 emit light in phase at wavelength λ. The sources lie on an x axis, and a light detector is moved in a circle of large radius around the midpoint between them. It detects 30 points of zero intensity, including two on the x axis, one of them to the left of the sources and the other to the right of the sources. In terms of λ, what is the separation between the two sources?

Fig. 36N-10 Problem N22.

N23. In Fig. 36N-11, two light rays go through different paths by reflecting from the various flat surfaces shown. The light waves have a wavelength of 420.0 nm and are initially in phase. What are (a) the least value and (b) the second least value of distance L that will put the waves exactly out of phase as they emerge from the region?

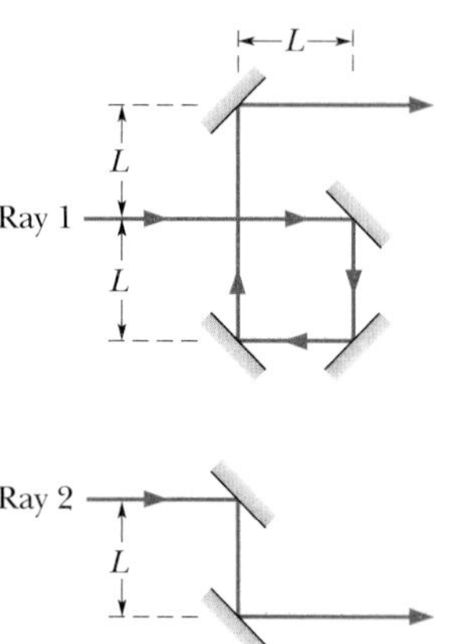

Fig. 36N-11 Problems N23 and N24.

N24. In two experiments, light is to be sent along the two paths shown in Fig. 36N-11, by reflecting it from the various flat surfaces shown. In the first experiment, rays 1 and 2 are initially in phase and have a wavelength of 620.0 nm. In the second experiment, rays 1 and 2 are initially in phase and have a wavelength of 496.0 nm. What least value of distance L is required such that the 620.0 nm waves emerge from the region exactly in phase but the 496.0 nm waves emerge exactly out of phase?

N25. In a phasor diagram for the waves at any point on the viewing screen for the two-slit experiment in Fig. 36-8, the phasor of the resultant wave rotates 60.0° in 2.50×10^{-16} s. What is the wavelength of the light?

N26. A thin film, with a thickness of 272.7 nm and with air on both sides, is illuminated with a beam of white light. The beam is perpendicular to the film and consists of the full range of wavelengths for the visible spectrum. In the light reflected by the film, light with a wavelength of 600.0 nm undergoes fully constructive interference. At what wavelength does the reflected light undergo fully destructive interference? (*Hint*: You must make a reasonable assumption about the index of refraction.)

N27. In Fig. 36N-12*a*, the waves along rays 1 and 2 are initially in phase, with the same wavelength λ in air. Ray 2 goes through a material with length L and index of refraction n. The rays are then reflected by mirrors to a common point P on a screen. Suppose that we can vary n from $n = 1.0$ to $n = 2.5$. Suppose also that, from $n = 1.0$ to $n = 1.5$, the intensity I of the light at point P varies with n as given in Fig. 36N-12*b*. At what values of n greater than 1.4 is intensity I (a) maximum and (b) zero? (c) In terms of wavelength, what is the phase difference between the rays at point P when $n = 2.0$?

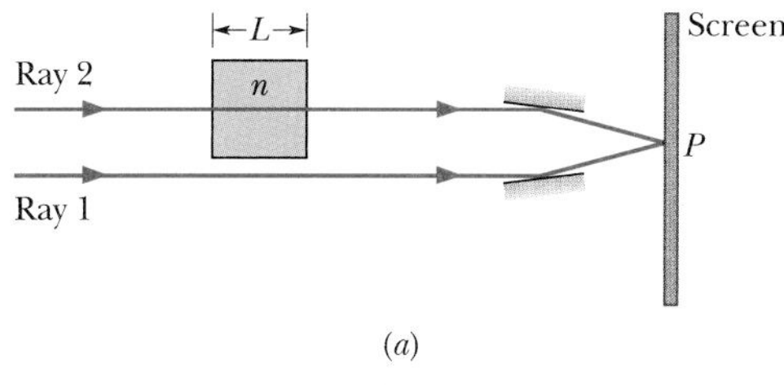

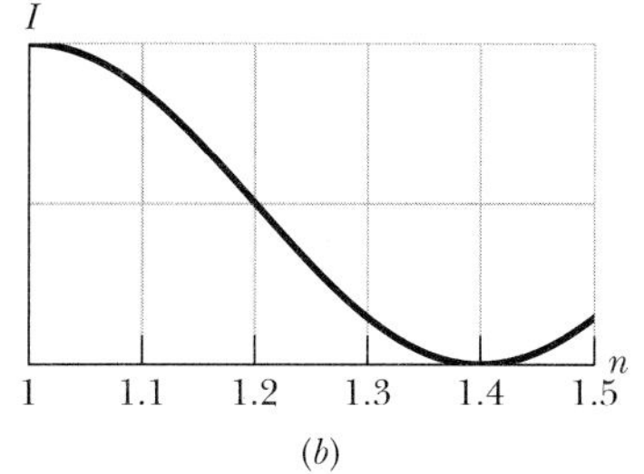

Fig. 36N-12 Problems N27 and N28.

N28. In Fig. 36N-12a, the waves along rays 1 and 2 are initially in phase, with the same wavelength λ in air. Ray 2 goes through a material with length L and index of refraction n. The rays are then reflected by mirrors to a common point P on a screen. Suppose that we can vary L from 0 to 2400 nm. Suppose also that, from $L = 0$ to $L = 900$ nm, the intensity I of the light at point P varies with L as given in Fig. 36N-13. At what values of L greater than 900 nm is intensity I (a) maximum and (b) zero? (c) In terms of wavelength, what is the phase difference between the rays at point P when $L = 1200$ nm?

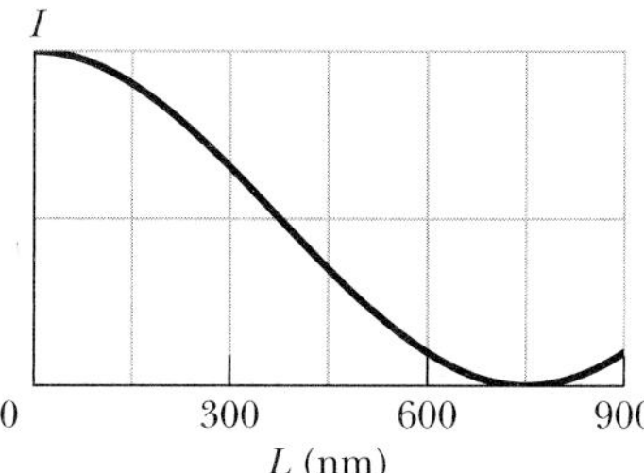

Fig. 36N-13 Problem N28.

N29. In Fig. 36N-14, a light ray passes from air through five transparent layers of different materials whose boundaries are parallel. The thickness and index of refraction are given for two of the layers. (a) At what angle does the light emerge back into air at the right? (b) How long does the light take to travel through the layer with an index of refraction of 1.45?

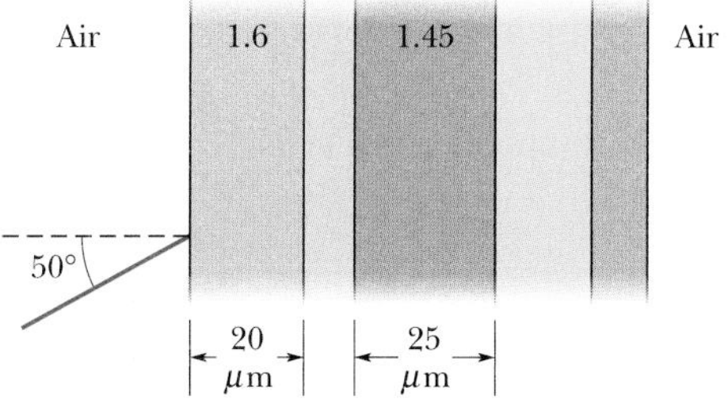

Fig. 36N-14 Problem N29.

N30. In Figure 36N-15, two isotropic point sources S_1 and S_2 emit light at wavelength $\lambda = 400$ nm. Source S_1 is located at $y = 640$ nm; source S_2 is located at $y = -640$ nm. At point P_1 (at $x = 720$ nm), the wave from S_2 arrives ahead of the wave from S_1 by a phase difference of 0.60π rad. (a) In terms of wavelength, what is the phase difference between the waves from the two sources as the waves arrive at point P_2, which is located at $y = 720$ nm. (The figure is not drawn to scale.) (b) Is the interference at point P_2 fully constructive, fully destructive, intermediate but closer to fully constructive, or intermediate but closer to fully destructive?

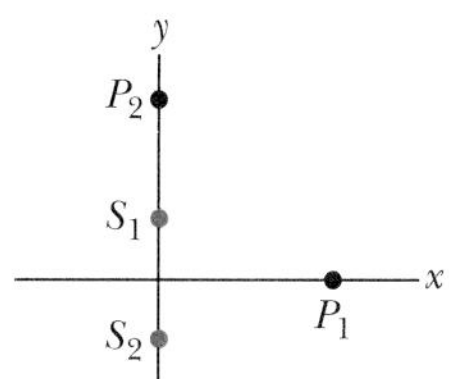

Fig. 36N-15 Problem N30.

N31. Figure 36N-16a shows two isotropic point sources of light (S_1 and S_2) that emit in phase at wavelength 400 nm. A detection point P is shown on an x axis that extends through source S_1. The phase difference ϕ between the light arriving at point P from the two sources is to be measured as P is moved along the x axis from $x = 0$ out to $x = +\infty$. The results out to $x = 10 \times 10^{-7}$ m are given in Fig. 36N-16b. On the way out to $+\infty$, what is the greatest value of x at which the light arriving at P from S_1 is exactly out of phase with the light arriving at P from S_2?

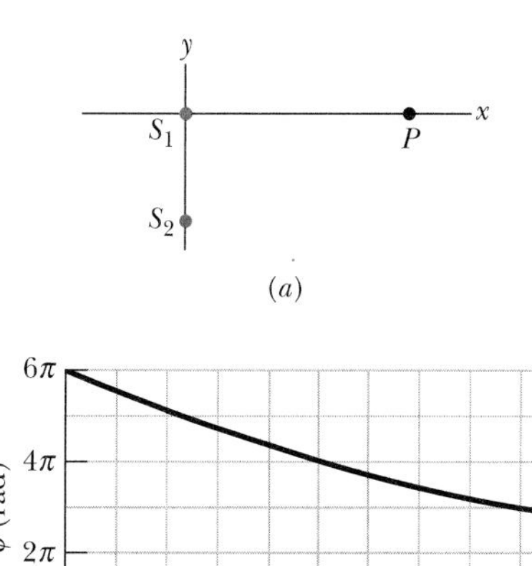

Fig. 36N-16 Problem N31.

37 Diffraction

Georges Seurat painted *Sunday Afternoon on the Island of La Grande Jatte* using not brush strokes in the usual sense, but rather a myriad of small colored dots, in a style of painting now known as pointillism. You can see the dots if you stand close enough to the painting, but as you move away from it, they eventually blend and cannot be distinguished. Moreover, the color that you see at any given place on the painting changes as you move away—which is why Seurat painted with the dots.

What causes this change in color?

The answer is in this chapter.

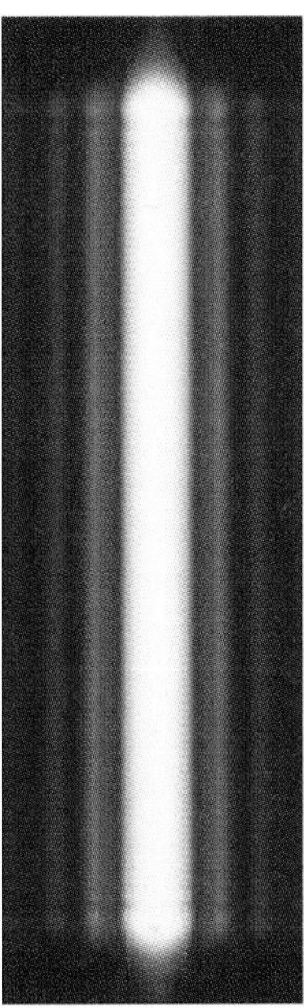

Fig. 37-1 This diffraction pattern appeared on a viewing screen when light that had passed through a narrow vertical slit reached the screen. Diffraction causes light to flare out perpendicular to the long sides of the slit. That produces an interference pattern consisting of a broad central maximum and less intense and narrower secondary (or side) maxima, with minima between them.

37-1 Diffraction and the Wave Theory of Light

In Chapter 36 we defined diffraction rather loosely as the flaring of light as it emerges from a narrow slit. More than just flaring occurs, however, because the light produces an interference pattern called a **diffraction pattern.** For example, when monochromatic light from a distant source (or a laser) passes through a narrow slit and is then intercepted by a viewing screen, the light produces on the screen a diffraction pattern like that in Fig. 37-1. This pattern consists of a broad and intense (very bright) central maximum and a number of narrower and less intense maxima (called **secondary** or **side** maxima) to both sides. In between the maxima are minima.

Such a pattern would be totally unexpected in geometrical optics: If light traveled in straight lines as rays, then the slit would allow some of those rays through and they would form a sharp, bright rendition of the slit on the viewing screen. As in Chapter 36, we must conclude that geometrical optics is only an approximation.

Diffraction of light is not limited to situations of light passing through a narrow opening (such as a slit or pinhole). It also occurs when light passes an edge, such as the edges of the razor blade whose diffraction pattern is shown in Fig. 37-2. Note the lines of maxima and minima that run approximately parallel to the edges, at both the inside edges of the blade and the outside edges. As the light passes, say, the vertical edge at the left, it flares left and right and undergoes interference, producing the pattern along the left edge. The rightmost portion of that pattern actually lies within what would have been the shadow of the blade if geometrical optics prevailed.

You encounter a common example of diffraction when you look at a clear blue sky and see tiny specks and hairlike structures floating in your view. These *floaters*, as they are called, are produced when light passes the edges of tiny deposits in the vitreous humor, the transparent material filling most of the eyeball. What you are seeing when a floater is in your field of vision is the diffraction pattern produced on the retina by one of these deposits. If you sight through a pinhole in an otherwise opaque sheet so as to make the light entering your eye approximately a plane wave, you can distinguish individual maxima and minima in the patterns.

The Fresnel Bright Spot

Diffraction finds a ready explanation in the wave theory of light. However, this theory, originally advanced in the late 1600s by Huygens and used 123 years later by Young to explain double-slit interference, was very slow in being adopted, largely because it ran counter to Newton's theory that light was a stream of particles.

Newton's view was the prevailing view in French scientific circles of the early nineteenth century, when Augustin Fresnel was a young military engineer. Fresnel, who believed in the wave theory of light, submitted a paper to the French Academy of Sciences describing his experiments with light and his wave-theory explanations of them.

In 1819, the Academy, dominated by supporters of Newton and thinking to challenge the wave point of view, organized a prize competition for an essay on the subject of diffraction. Fresnel won. The Newtonians, however, were neither converted nor silenced. One of them, S. D. Poisson, pointed out the "strange result" that if Fresnel's theories were correct, then light waves should flare into the shadow region of a sphere as they pass the edge of the sphere, producing a bright spot at the center of the shadow. The prize committee arranged a test of the famous mathe-

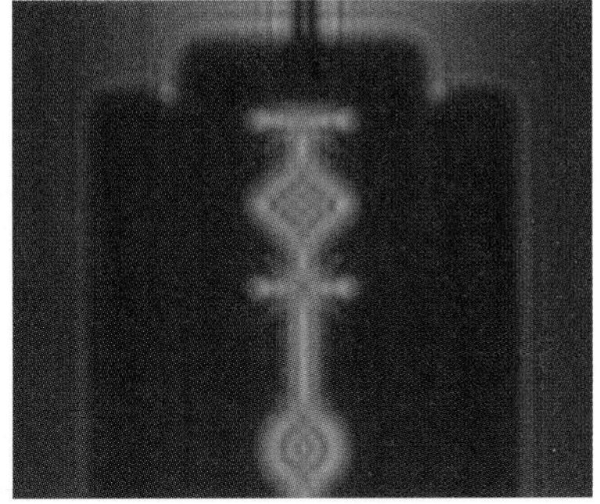

Fig. 37-2 The diffraction pattern produced by a razor blade in monochromatic light. Note the lines of alternating maximum and minimum intensity.

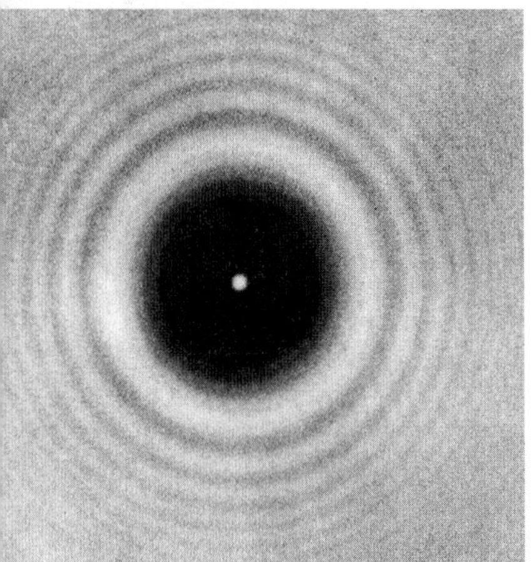

Fig. 37-3 A photograph of the diffraction pattern of a disk. Note the concentric diffraction rings and the Fresnel bright spot at the center of the pattern. This experiment is essentially identical to that arranged by the committee testing Fresnel's theories, because both the sphere they used and the disk used here have a cross section with a circular edge.

matician's prediction and discovered (see Fig. 37-3) that the predicted *Fresnel bright spot,* as we call it today, was indeed there! Nothing builds confidence in a theory so much as having one of its unexpected and counterintuitive predictions verified by experiment.

37-2 Diffraction by a Single Slit: Locating the Minima

Let us now examine the diffraction pattern of plane waves of light of wavelength λ that are diffracted by a single, long, narrow slit of width a in an otherwise opaque screen B, as shown in cross section in Fig. 37-4a. (In that figure, the slit's length extends into and out of the page, and the incoming wavefronts are parallel to screen B.) When the diffracted light reaches viewing screen C, waves from different points within the slit undergo interference and produce a diffraction pattern of bright and dark fringes (interference maxima and minima) on the screen. To locate the fringes, we shall use a procedure somewhat similar to the one we used to locate the fringes in a two-slit interference pattern. However, diffraction is more mathematically challenging, and here we shall be able to find equations for only the dark fringes.

Before we do that, however, we can justify the central bright fringe seen in Fig. 37-1 by noting that the Huygens wavelets from all points in the slit travel about the same distance to reach the center of the pattern and thus are in phase there. As for the other bright fringes, we can say only that they are approximately halfway between adjacent dark fringes.

To find the dark fringes, we shall use a clever (and simplifying) strategy that involves pairing up all the rays coming through the slit and then finding what conditions cause the wavelets of the rays in each pair to cancel each other. We apply this strategy in Fig. 37-4a to locate the first dark fringe, at point P_1. First, we mentally divide the slit into two *zones* of equal widths $a/2$. Then we extend to P_1 a light ray r_1 from the top point of the top zone and a light ray r_2 from the top point of the bottom zone. A central axis is drawn from the center of the slit to screen C, and P_1 is located at an angle θ to that axis.

The wavelets of the pair of rays r_1 and r_2 are in phase within the slit because they originate from the same wavefront passing through the slit, along the width of the slit. However, to produce the first dark fringe they must be out of phase by $\lambda/2$ when they reach P_1; this phase difference is due to their path length difference, with the wavelet of r_2 traveling a longer path to reach P_1 than the wavelet of r_1. To display this path length difference, we find a point b on ray r_2 such that the path length from b to P_1 matches the path length of ray r_1. Then the path length difference between the two rays is the distance from the center of the slit to b.

When viewing screen C is near screen B, as in Fig. 37-4a, the diffraction pattern on C is difficult to describe mathematically. However, we can simplify the mathematics considerably if we arrange for the screen separation D to be much larger than the slit width a. Then we can approximate rays r_1 and r_2 as being parallel, at angle θ to the central axis (Fig. 37-4b). We can also approximate the triangle formed by point b, the top point of the slit, and the center point of the slit as being a right triangle, and one of the angles inside that triangle as being θ. The path length difference between rays r_1 and r_2 (which is still the distance from the center of the slit to point b) is then equal to $(a/2)\sin\theta$.

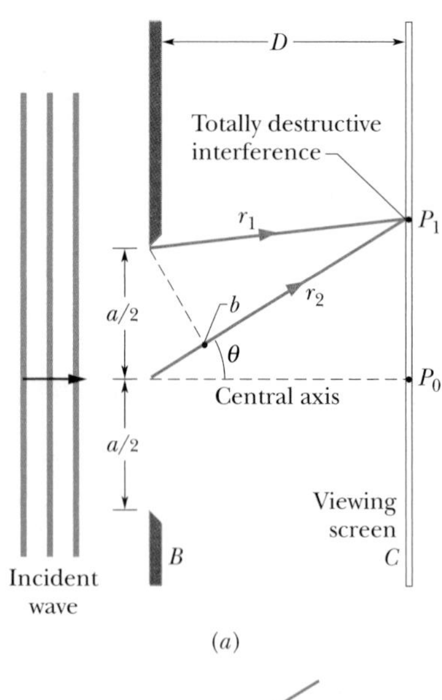

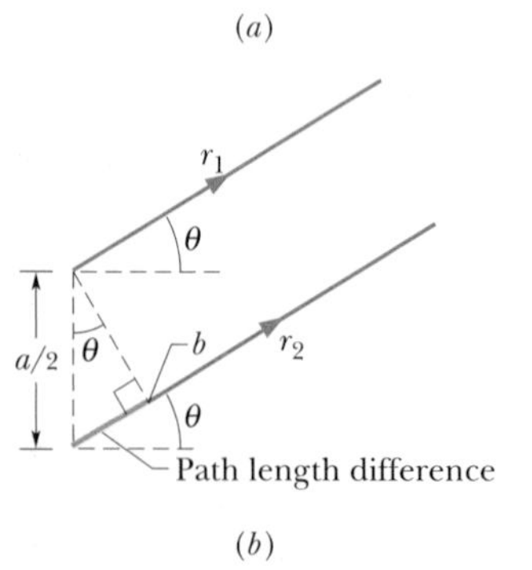

Fig. 37-4 (a) Waves from the top points of two zones of width $a/2$ undergo totally destructive interference at point P_1 on viewing screen C. (b) For $D \gg a$, we can approximate rays r_1 and r_2 as being parallel, at angle θ to the central axis.

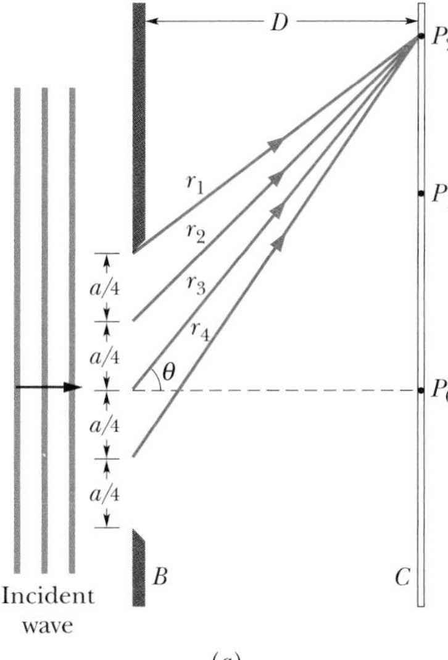

(a)

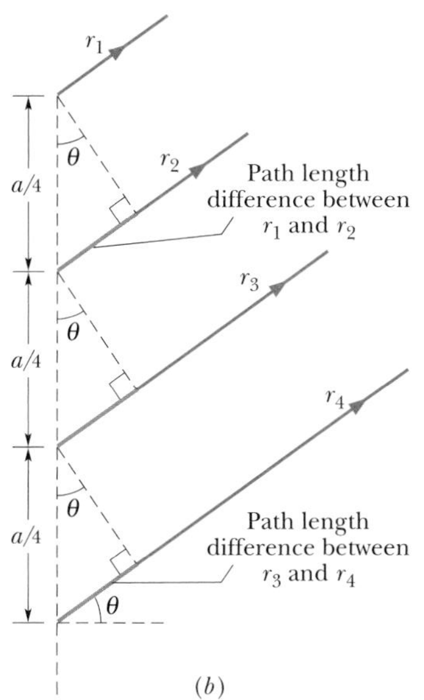

(b)

Fig. 37-5 (a) Waves from the top points of four zones of width $a/4$ undergo totally destructive interference at point P_2. (b) For $D \gg a$, we can approximate rays r_1, r_2, r_3, and r_4 as being parallel, at angle θ to the central axis.

We can repeat this analysis for any other pair of rays originating at corresponding points in the two zones (say, at the midpoints of the zones) and extending to point P_1. Each such pair of rays has the same path length difference $(a/2)\sin\theta$. Setting this common path length difference equal to $\lambda/2$ (our condition for the first dark fringe), we have

$$\frac{a}{2}\sin\theta = \frac{\lambda}{2},$$

which gives us

$$a\sin\theta = \lambda \qquad \text{(first minimum).} \qquad (37\text{-}1)$$

Given slit width a and wavelength λ, Eq. 37-1 tells us the angle θ of the first dark fringe above and (by symmetry) below the central axis.

Note that if we begin with $a > \lambda$ and then narrow the slit while holding the wavelength constant, we increase the angle at which the first dark fringes appear; that is, the extent of the diffraction (the extent of the flaring and the width of the pattern) is *greater* for a *narrower* slit. When we have reduced the slit width to the wavelength (that is, $a = \lambda$), the angle of the first dark fringes is 90°. Since the first dark fringes mark the two edges of the central bright fringe, that bright fringe must then cover the entire viewing screen.

We find the second dark fringes above and below the central axis as we found the first dark fringes, except that we now divide the slit into *four* zones of equal widths $a/4$, as shown in Fig. 37-5*a*. We then extend rays r_1, r_2, r_3, and r_4 from the top points of the zones to point P_2, the location of the second dark fringe above the central axis. To produce that fringe, the path length difference between r_1 and r_2, that between r_2 and r_3, and that between r_3 and r_4 must all be equal to $\lambda/2$.

For $D \gg a$, we can approximate these four rays as being parallel, at angle θ to the central axis. To display their path length differences, we extend a perpendicular line through each adjacent pair of rays, as shown in Fig. 37-5*b*, to form a series of right triangles, each of which has a path length difference as one side. We see from the top triangle that the path length difference between r_1 and r_2 is $(a/4)\sin\theta$. Similarly, from the bottom triangle, the path length difference between r_3 and r_4 is also $(a/4)\sin\theta$. In fact, the path length difference for any two rays that originate at corresponding points in two adjacent zones is $(a/4)\sin\theta$. Since in each such case the path length difference is equal to $\lambda/2$, we have

$$\frac{a}{4}\sin\theta = \frac{\lambda}{2},$$

which gives us

$$a\sin\theta = 2\lambda \qquad \text{(second minimum).} \qquad (37\text{-}2)$$

We could now continue to locate dark fringes in the diffraction pattern by splitting up the slit into more zones of equal width. We would always choose an even number of zones so that the zones (and their waves) could be paired as we have been doing. We would find that the dark fringes above and below the central axis can be located with the following general equation:

$$a\sin\theta = m\lambda, \qquad \text{for } m = 1, 2, 3, \ldots \qquad \text{(minima—dark fringes).} \qquad (37\text{-}3)$$

You can remember this result in the following way. Draw a triangle like the one in Fig. 37-4*b*, but for the full slit width a, and note that the path length difference between the top and bottom rays from the slit equals $a\sin\theta$. Thus, Eq. 37-3 says:

> In a single-slit diffraction experiment, dark fringes are produced where the path length differences ($a\sin\theta$) between the top and bottom rays are equal to $\lambda, 2\lambda, 3\lambda, \ldots$.

This may seem to be wrong, because the waves of those two particular rays will be exactly in phase with each other when their path length difference is an integer number of wavelengths. However, they each will still be part of a pair of waves that are exactly out of phase with each other; thus, *each* wave will be canceled by some other wave, resulting in darkness.

Equations 37-1, 37-2, and 37-3 are derived for the case of $D \gg a$. However, they also apply if we place a converging lens between the slit and the viewing screen and then move the screen in so that it coincides with the focal plane of the lens. The lens ensures that rays which now reach any point on the screen are *exactly* parallel (rather than approximately) back at the slit. They are like the initially parallel rays of Fig. 35-12*a* that are directed to the focal point by a converging lens.

CHECKPOINT 1: We produce a diffraction pattern on a viewing screen by means of a long narrow slit illuminated by blue light. Does the pattern expand away from the bright center (the maxima and minima shift away from the center) or contract toward it if we (a) switch to yellow light or (b) decrease the slit width?

Sample Problem 37-1

A slit of width a is illuminated by white light (which consists of all the wavelengths in the visible range).

(a) For what value of a will the first minimum for red light of wavelength $\lambda = 650$ nm appear at $\theta = 15°$?

SOLUTION: The Key Idea here is that diffraction occurs separately for each wavelength in the range of wavelengths passing through the slit, with the locations of the minima for each wavelength given by Eq. 37-3 ($a \sin \theta = m\lambda$). When we set $m = 1$ (for the first minimum) and substitute the given values of θ and λ, Eq. 37-3 yields

$$a = \frac{m\lambda}{\sin \theta} = \frac{(1)(650 \text{ nm})}{\sin 15°} = 2511 \text{ nm} \approx 2.5\ \mu\text{m}. \qquad \text{(Answer)}$$

For the incident light to flare out that much ($\pm 15°$ to the first minima) the slit has to be very fine indeed—about four times the wavelength. For comparison, note that a fine human hair may be about 100 μm in diameter.

(b) What is the wavelength λ' of the light whose first side diffraction maximum is at 15°, thus coinciding with the first minimum for the red light?

SOLUTION: The Key Idea here is that the first side maximum for any wavelength is about halfway between the first and second minima for that wavelength. Those first and second minima can be located with Eq. 37-3 by setting $m = 1$ and $m = 2$, respectively. Thus, the first side maximum can be located *approximately* by setting $m = 1.5$. Then Eq. 37-3 becomes

$$a \sin \theta = 1.5\lambda'.$$

Solving for λ' and substituting known data yield

$$\lambda' = \frac{a \sin \theta}{1.5} = \frac{(2511 \text{ nm})(\sin 15°)}{1.5} = 430 \text{ nm}. \qquad \text{(Answer)}$$

Light of this wavelength is violet. The first side maximum for light of wavelength 430 nm will always coincide with the first minimum for light of wavelength 650 nm, no matter what the slit width is. If the slit is relatively narrow, the angle θ at which this overlap occurs will be relatively large, and conversely.

37-3 Intensity in Single-Slit Diffraction, Qualitatively

In Section 37-2 we saw how to find the positions of the minima and the maxima in a single-slit diffraction pattern. Now we turn to a more general problem: find an expression for the intensity I of the pattern as a function of θ, the angular position of a point on a viewing screen.

To do this, we divide the slit of Fig. 37-4*a* into N zones of equal widths Δx small enough that we can assume each zone acts as a source of Huygens wavelets. We wish to superimpose the wavelets arriving at an arbitrary point P on the viewing screen, at angle θ to the central axis, so that we can determine the amplitude E_θ of

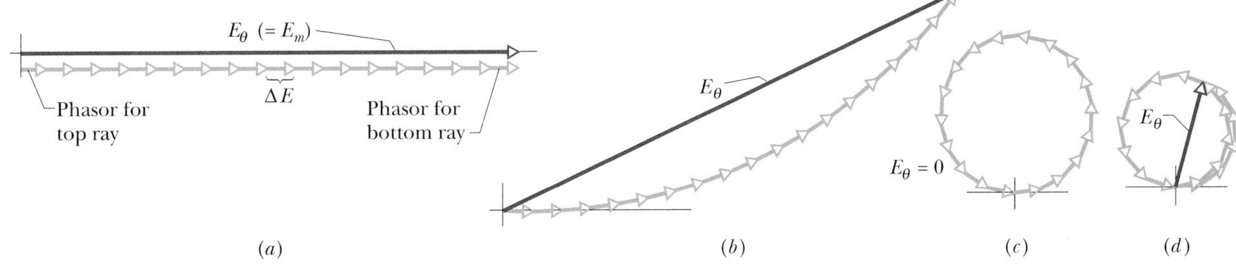

Fig. 37-6 Phasor diagrams for $N = 18$ phasors, corresponding to the division of a single slit into 18 zones. Resultant amplitudes E_θ are shown for (*a*) the central maximum at $\theta = 0$, (*b*) a point on the screen lying at a small angle θ to the central axis, (*c*) the first minimum, and (*d*) the first side maximum.

the electric component of the resultant wave at P. The intensity of the light at P is then proportional to the square of that amplitude.

To find E_θ, we need the phase relationships among the arriving wavelets. The phase difference between wavelets from adjacent zones is given by

$$\begin{pmatrix}\text{phase}\\ \text{difference}\end{pmatrix} = \left(\frac{2\pi}{\lambda}\right)\begin{pmatrix}\text{path length}\\ \text{difference}\end{pmatrix}.$$

For point P at angle θ, the path length difference between wavelets from adjacent zones is $\Delta x \sin\theta$, so the phase difference $\Delta\phi$ between wavelets from adjacent zones is

$$\Delta\phi = \left(\frac{2\pi}{\lambda}\right)(\Delta x \sin\theta). \tag{37-4}$$

We assume that the wavelets arriving at P all have the same amplitude ΔE. To find the amplitude E_θ of the resultant wave at P, we add the amplitudes ΔE via phasors. To do this, we construct a diagram of N phasors, one corresponding to the wavelet from each zone in the slit.

For point P_0 at $\theta = 0$ on the central axis of Fig. 37-4*a*, Eq. 37-4 tells us that the phase difference $\Delta\phi$ between the wavelets is zero; that is, the wavelets all arrive in phase. Figure 37-6*a* is the corresponding phasor diagram; adjacent phasors represent wavelets from adjacent zones and are arranged head to tail. Because there is zero phase difference between the wavelets, there is zero angle between each pair of adjacent phasors. The amplitude E_θ of the net wave at P_0 is the vector sum of these phasors. This arrangement of the phasors turns out to be the one that gives the greatest value for the amplitude E_θ. We call this value E_m; that is, E_m is the value of E_θ for $\theta = 0$.

We next consider a point P that is at a small angle θ to the central axis. Equation 37-4 now tells us that the phase difference $\Delta\phi$ between wavelets from adjacent zones is no longer zero. Figure 37-6*b* shows the corresponding phasor diagram; as before, the phasors are arranged head to tail, but now there is an angle $\Delta\phi$ between adjacent phasors. The amplitude E_θ at this new point is still the vector sum of the phasors, but it is smaller than that in Fig. 37-6*a*, which means that the intensity of the light is less at this new point P than at P_0.

If we continue to increase θ, the angle $\Delta\phi$ between adjacent phasors increases, and eventually the chain of phasors curls completely around so that the head of the last phasor just reaches the tail of the first phasor (Fig. 37-6*c*). The amplitude E_θ is now zero, which means that the intensity of the light is also zero. We have reached the first minimum, or dark fringe, in the diffraction pattern. The first and last phasors now have a phase difference of 2π rad, which means that the path length difference between the top and bottom rays through the slit equals one wavelength. Recall that this is the condition we determined for the first diffraction minimum.

As we continue to increase θ, the angle $\Delta\phi$ between adjacent phasors continues to increase, the chain of phasors begins to wrap back on itself, and the resulting coil begins to shrink. Amplitude E_θ now increases until it reaches a maximum value in the arrangement shown in Fig. 37-6*d*. This arrangement corresponds to the first side maximum in the diffraction pattern.

If we increase θ a bit more, the resulting shrinkage of the coil decreases E_θ, which means that the intensity also decreases. When θ is increased enough, the head of the last phasor again meets the tail of the first phasor. We have then reached the second minimum.

We could continue this qualitative method of determining the maxima and minima of the diffraction pattern but, instead, we shall now turn to a quantitative method.

CHECKPOINT 2: The figures represent, in smoother form (with more phasors) than Fig. 37-6, the phasor diagrams for two points of a diffraction pattern that are on opposite sides of a certain diffraction maximum. (a) Which maximum is it? (b) What is the approximate value of m (in Eq. 37-3) that corresponds to this maximum?

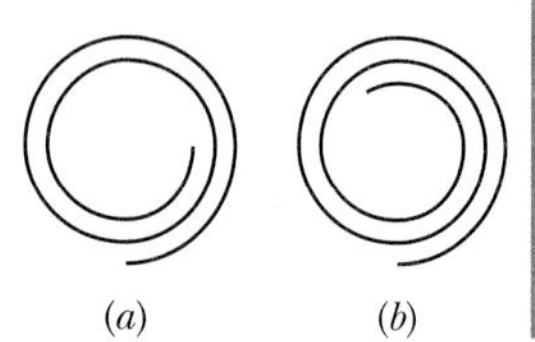

37-4 Intensity in Single-Slit Diffraction, Quantitatively

Equation 37-3 tells us how to locate the minima of the single-slit diffraction pattern on screen C of Fig. 37-4*a* as a function of the angle θ in that figure. Here we wish to derive an expression for the intensity $I(\theta)$ of the pattern as a function of θ. We state, and shall prove below, that the intensity is given by

$$I(\theta) = I_m\left(\frac{\sin\alpha}{\alpha}\right)^2, \qquad (37\text{-}5)$$

where

$$\alpha = \tfrac{1}{2}\phi = \frac{\pi a}{\lambda}\sin\theta. \qquad (37\text{-}6)$$

The symbol α is just a convenient connection between the angle θ that locates a point on the viewing screen and the light intensity $I(\theta)$ at that point. I_m is the greatest value of the intensities $I(\theta)$ in the pattern and occurs at the central maximum (where $\theta = 0$), and ϕ is the phase difference (in radians) between the top and bottom rays from the slit width a.

Study of Eq. 37-5 shows that intensity minima will occur where

$$\alpha = m\pi, \qquad \text{for } m = 1, 2, 3, \ldots. \qquad (37\text{-}7)$$

If we put this result into Eq. 37-6, we find

$$m\pi = \frac{\pi a}{\lambda}\sin\theta, \qquad \text{for } m = 1, 2, 3, \ldots$$

or

$$a\sin\theta = m\lambda, \qquad \text{for } m = 1, 2, 3, \ldots \qquad \text{(minima—dark fringes)}, \qquad (37\text{-}8)$$

which is exactly Eq. 37-3, the expression that we derived earlier for the location of the minima.

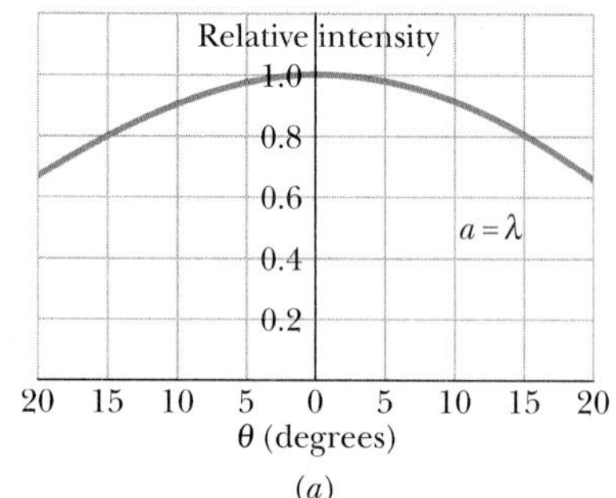

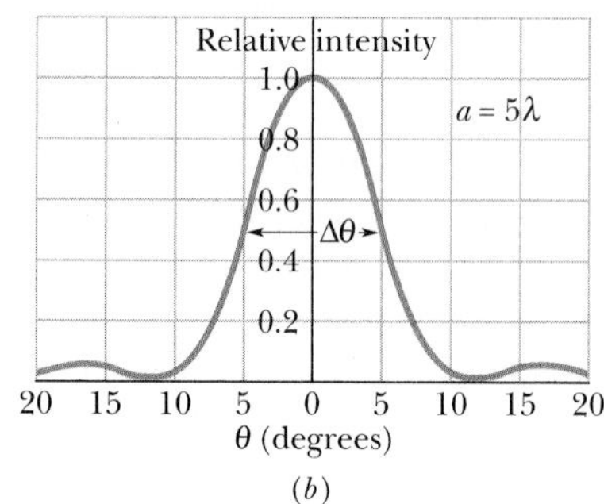

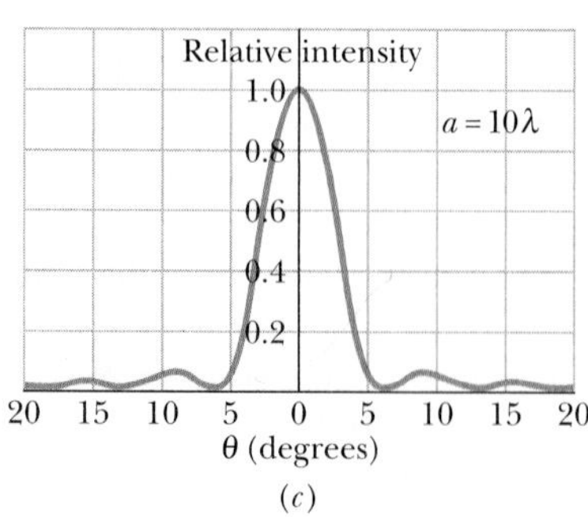

Fig. 37-7 The relative intensity in single-slit diffraction for three values of the ratio a/λ. The wider the slit is, the narrower is the central diffraction maximum.

Figure 37-7 shows plots of the intensity of a single-slit diffraction pattern, calculated with Eqs. 37-5 and 37-6 for three slit widths: $a = \lambda$, $a = 5\lambda$, and $a = 10\lambda$. Note that as the slit width increases (relative to the wavelength), the width of the *central diffraction maximum* (the central hill-like region of the graphs) decreases;

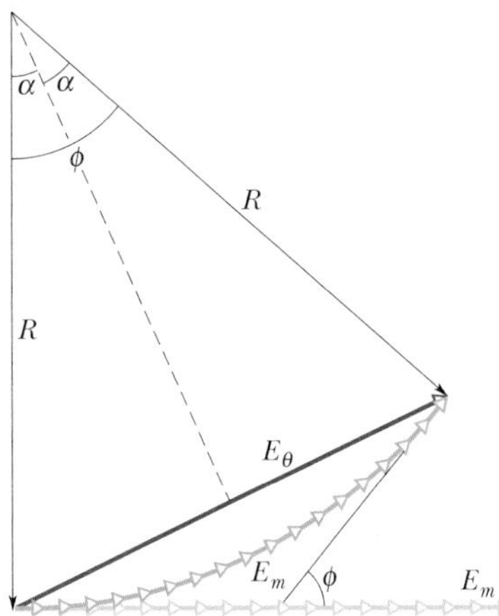

Fig. 37-8 A construction used to calculate the intensity in single-slit diffraction. The situation shown corresponds to that of Fig. 37-6*b*.

that is, the light undergoes less flaring by the slit. The secondary maxima also decrease in width (and become weaker). In the limit of slit width a being much greater than wavelength λ, the secondary maxima due to the slit disappear; we then no longer have single-slit diffraction (but we still have diffraction due to the edges of the wide slit, like that produced by the edges of the razor blade in Fig. 37-2).

Proof of Eqs. 37-5 and 37-6

The arc of phasors in Fig. 37-8 represents the wavelets that reach an arbitrary point P on the viewing screen of Fig. 37-4, corresponding to a particular small angle θ. The amplitude E_θ of the resultant wave at P is the vector sum of these phasors. If we divide the slit of Fig. 37-4 into infinitesimal zones of width Δx, the arc of phasors in Fig. 37-8 approaches the arc of a circle; we call its radius R as indicated in that figure. The length of the arc must be E_m, the amplitude at the center of the diffraction pattern, because if we straightened out the arc we would have the phasor arrangement of Fig. 37-6*a* (shown lightly in Fig. 37-8).

The angle ϕ in the lower part of Fig. 37-8 is the difference in phase between the infinitesimal vectors at the left and right ends of arc E_m. From the geometry, ϕ is also the angle between the two radii marked R in Fig. 37-8. The dashed line in that figure, which bisects ϕ, then forms two congruent right triangles. From either triangle we can write

$$\sin \tfrac{1}{2}\phi = \frac{E_\theta}{2R}. \tag{37-9}$$

In radian measure, ϕ is (with E_m considered to be a circular arc)

$$\phi = \frac{E_m}{R}.$$

Solving this equation for R and substituting in Eq. 37-9 lead to

$$E_\theta = \frac{E_m}{\frac{1}{2}\phi} \sin \tfrac{1}{2}\phi. \tag{37-10}$$

In Section 34-4 we saw that the intensity of an electromagnetic wave is proportional to the square of the amplitude of its electric field. Here, this means that the maximum intensity I_m (which occurs at the center of the diffraction pattern) is proportional to E_m^2 and the intensity $I(\theta)$ at angle θ is proportional to E_θ^2. Thus, we may write

$$\frac{I(\theta)}{I_m} = \frac{E_\theta^2}{E_m^2}. \tag{37-11}$$

Substituting for E_θ with Eq. 37-10 and then substituting $\alpha = \frac{1}{2}\phi$, we are led to the following expression for the intensity as a function of θ:

$$I(\theta) = I_m \left(\frac{\sin \alpha}{\alpha}\right)^2.$$

This is exactly Eq. 37-5, one of the two equations we set out to prove.

The second equation we wish to prove relates α to θ. The phase difference ϕ between the rays from the top and bottom of the entire slit may be related to a path length difference with Eq. 37-4; it tells us that

$$\phi = \left(\frac{2\pi}{\lambda}\right)(a \sin \theta),$$

where a is the sum of the widths Δx of the infinitesimal zones. However, $\phi = 2\alpha$, so this equation reduces to Eq. 37-6.

CHECKPOINT 3: Two wavelengths, 650 and 430 nm, are used separately in a single-slit diffraction experiment. The figure shows the results as graphs of intensity I versus angle θ for the two diffraction patterns. If both wavelengths are then used simultaneously, what color will be seen in the combined diffraction pattern at (a) angle A and (b) angle B?

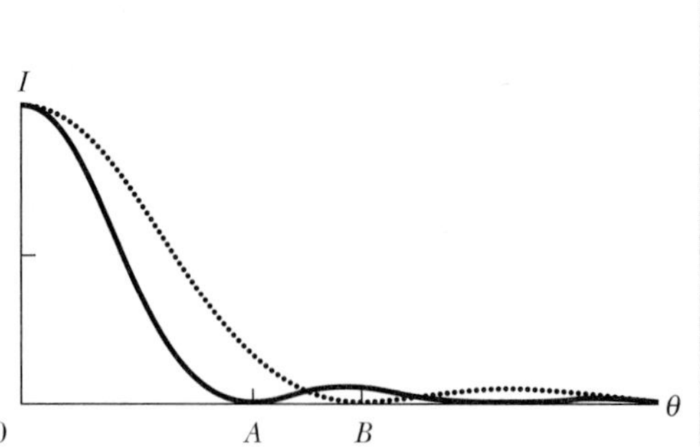

Sample Problem 37-2

Find the intensities of the first three secondary maxima (side maxima) in the single-slit diffraction pattern of Fig. 37-1, measured relative to the intensity of the central maximum.

SOLUTION: One **Key Idea** here is that the secondary maxima lie approximately halfway between the minima, whose angular locations are given by Eq. 37-7 ($\alpha = m\pi$). The locations of the secondary maxima are then given (approximately) by

$$\alpha = (m + \tfrac{1}{2})\pi, \qquad \text{for } m = 1, 2, 3, \ldots,$$

with α in radian measure.

A second **Key Idea** is that we can relate the intensity I at any point in the diffraction pattern to the intensity I_m of the central maximum via Eq. 37-5. Thus, we can substitute the approximate values of α for the secondary maxima into Eq. 37-5 to obtain the relative intensities at those maxima. We get

$$\frac{I}{I_m} = \left(\frac{\sin\alpha}{\alpha}\right)^2 = \left(\frac{\sin(m + \frac{1}{2})\pi}{(m + \frac{1}{2})\pi}\right)^2, \qquad \text{for } m = 1, 2, 3, \ldots.$$

The first of the secondary maxima occurs for $m = 1$, and its relative intensity is

$$\frac{I_1}{I_m} = \left(\frac{\sin(1 + \frac{1}{2})\pi}{(1 + \frac{1}{2})\pi}\right)^2 = \left(\frac{\sin 1.5\pi}{1.5\pi}\right)^2$$
$$= 4.50 \times 10^{-2} \approx 4.5\%. \qquad \text{(Answer)}$$

For $m = 2$ and $m = 3$ we find that

$$\frac{I_2}{I_m} = 1.6\% \quad \text{and} \quad \frac{I_3}{I_m} = 0.83\%. \qquad \text{(Answer)}$$

Successive secondary maxima decrease rapidly in intensity. Figure 37-1 was deliberately overexposed to reveal them.

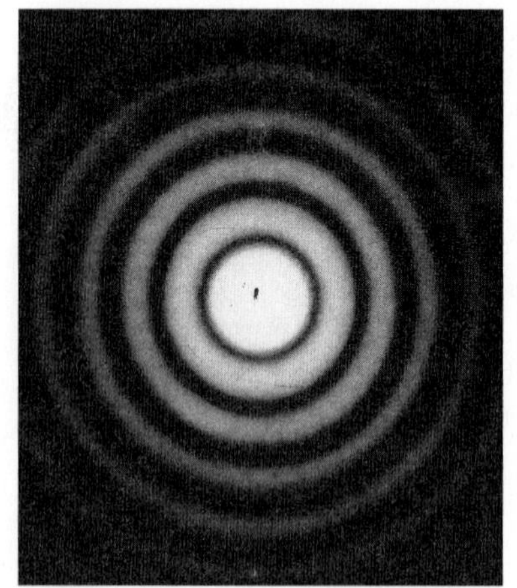

Fig. 37-9 The diffraction pattern of a circular aperture. Note the central maximum and the circular secondary maxima. The figure has been overexposed to bring out these secondary maxima, which are much less intense than the central maximum.

37-5 Diffraction by a Circular Aperture

Here we consider diffraction by a circular aperture—that is, a circular opening such as a circular lens, through which light can pass. Figure 37-9 shows the image of a distant point source of light (a star, for instance) formed on photographic film placed in the focal plane of a converging lens. This image is not a point, as geometrical optics would suggest, but a circular disk surrounded by several progressively fainter secondary rings. Comparison with Fig. 37-1 leaves little doubt that we are dealing with a diffraction phenomenon. Here, however, the aperture is a circle of diameter d rather than a rectangular slit.

The analysis of such patterns is complex. It shows, however, that the first minimum for the diffraction pattern of a circular aperture of diameter d is located by

$$\sin\theta = 1.22\frac{\lambda}{d} \qquad \text{(first minimum; circular aperture).} \qquad (37\text{-}12)$$

The angle θ here is the angle from the central axis to any point on that (circular) minimum. Compare this with Eq. 37-1,

$$\sin\theta = \frac{\lambda}{a} \qquad \text{(first minimum; single slit),} \qquad (37\text{-}13)$$

which locates the first minimum for a long narrow slit of width a. The main difference is the factor 1.22, which enters because of the circular shape of the aperture.

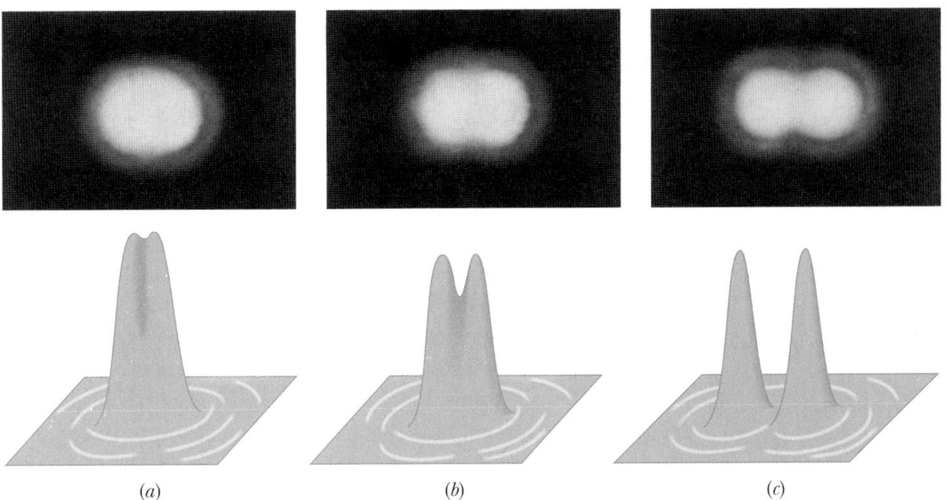

Fig. 37-10 At the top, the images of two point sources (stars), formed by a converging lens. At the bottom, representations of the image intensities. In (*a*) the angular separation of the sources is too small for them to be distinguished, in (*b*) they can be marginally distinguished, and in (*c*) they are clearly distinguished. Rayleigh's criterion is satisfied in (*b*), with the central maximum of one diffraction pattern coinciding with the first minimum of the other.

Resolvability

The fact that lens images are diffraction patterns is important when we wish to *resolve* (distinguish) two distant point objects whose angular separation is small. Figure 37-10 shows, in three different cases, the visual appearance and corresponding intensity pattern for two distant point objects (stars, say) with small angular separation. In Figure 37-10*a*, the objects are not resolved because of diffraction; that is, their diffraction patterns (mainly their central maxima) overlap so much that the two objects cannot be distinguished from a single point object. In Fig. 37-10*b* the objects are barely resolved, and in Fig. 37-10*c* they are fully resolved.

In Fig. 37-10*b* the angular separation of the two point sources is such that the central maximum of the diffraction pattern of one source is centered on the first minimum of the diffraction pattern of the other, a condition called **Rayleigh's criterion** for resolvability. From Eq. 37-12, two objects that are barely resolvable by this criterion must have an angular separation θ_R of

$$\theta_R = \sin^{-1}\frac{1.22\lambda}{d}.$$

Since the angles are small, we can replace $\sin\theta_R$ with θ_R expressed in radians:

$$\theta_R = 1.22\,\frac{\lambda}{d} \qquad \text{(Rayleigh's criterion).} \qquad (37\text{-}14)$$

Rayleigh's criterion for resolvability is only an approximation, because resolvability depends on many factors, such as the relative brightness of the sources and their surroundings, turbulence in the air between the sources and the observer, and the functioning of the observer's visual system. Experimental results show that the least angular separation that can actually be resolved by a person is generally somewhat greater than the value given by Eq. 37-14. However, for the sake of calculations here, we shall take Eq. 37-14 as being a precise criterion: If the angular separation θ between the sources is greater than θ_R, we can resolve the sources; if it is less, we cannot.

Rayleigh's criterion can explain the colors in Seurat's *Sunday Afternoon on the Island of La Grande Jatte* (or any other pointillistic painting). When you stand close enough to the painting, the angular separations θ of adjacent dots are greater than θ_R and thus the dots can be seen individually. Their colors are the colors of the

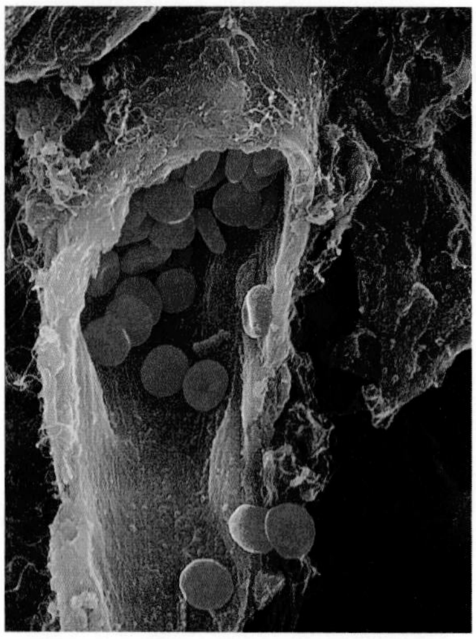

Fig. 37-11 A false-color scanning electron micrograph of a vein containing red blood cells.

paints Seurat used. However, when you stand far enough from the painting, the angular separations θ are less than θ_R and the dots cannot be seen individually. The resulting blend of colors coming into your eye from any group of dots can then cause your brain to "make up" a color for that group—a color that may not actually exist in the group. In this way, Seurat uses your visual system to create the colors of his art.

When we wish to use a lens instead of our visual system to resolve objects of small angular separation, it is desirable to make the diffraction pattern as small as possible. According to Eq. 37-14, this can be done either by increasing the lens diameter or by using light of a shorter wavelength.

For this reason ultraviolet light is often used with microscopes; because of its shorter wavelength, it permits finer detail to be examined than would be possible for the same microscope operated with visible light. In Chapter 39 of the extended version of this text, we show that beams of electrons behave like waves under some circumstances. In an *electron microscope* such beams may have an effective wavelength that is 10^{-5} of the wavelength of visible light. They permit the detailed examination of tiny structures, like that in Fig. 37-11, that would be blurred by diffraction if viewed with an optical microscope.

CHECKPOINT 4: Suppose that you can barely resolve two red dots, owing to diffraction by the pupil of your eye. If we increase the general illumination around you so that the pupil decreases in diameter, does the resolvability of the dots improve or diminish? Consider only diffraction. (You might experiment to check your answer.)

Sample Problem 37-3

A circular converging lens, with diameter $d = 32$ mm and focal length $f = 24$ cm, forms images of distant point objects in the focal plane of the lens. Light of wavelength $\lambda = 550$ nm is used.

(a) Considering diffraction by the lens, what angular separation must two distant point objects have to satisfy Rayleigh's criterion?

SOLUTION: Figure 37-12 shows two distant point objects P_1 and P_2, the lens, and a viewing screen in the focal plane of the lens. It also shows, on the right, plots of light intensity I versus position on the screen for the central maxima of the images formed by the lens. Note that the angular separation θ_o of the objects equals the angular separation θ_i of the images. Thus, the **Key Idea** here is that if the images are to satisfy Rayleigh's criterion for resolvability, the angular separations on both sides of the lens must be given by Eq. 37-14 (assuming small angles). Substituting the given data, we obtain from Eq. 37-14

$$\theta_o = \theta_i = \theta_R = 1.22\frac{\lambda}{d}$$

$$= \frac{(1.22)(550 \times 10^{-9}\ \text{m})}{32 \times 10^{-3}\ \text{m}} = 2.1 \times 10^{-5}\ \text{rad.} \qquad \text{(Answer)}$$

At this angular separation, each central maximum in the two intensity curves of Fig. 37-12 is centered on the first minimum of the other curve.

(b) What is the separation Δx of the centers of the *images* in the focal plane? (That is, what is the separation of the *central* peaks in the two curves?)

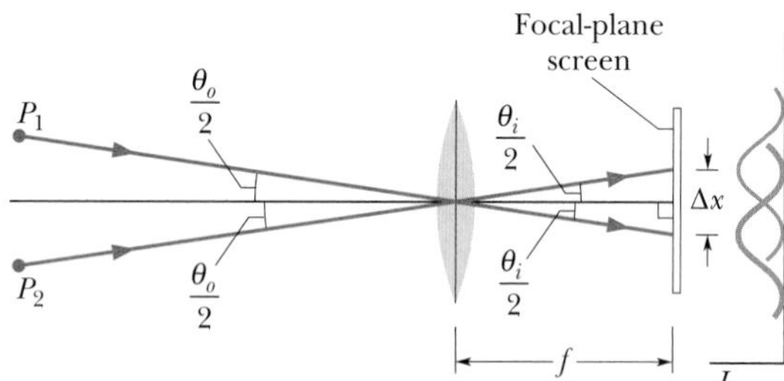

Fig. 37-12 Sample Problem 37-3. Light from two distant point objects P_1 and P_2 passes through a converging lens and forms images on a viewing screen in the focal plane of the lens. Only one representative ray from each object is shown. The images are not points but diffraction patterns, with intensities approximately as plotted at the right. The angular separation of the objects is θ_o and that of the images is θ_i; the central maxima of the images have a separation Δx.

SOLUTION: The **Key Idea** here is to relate the separation Δx to the angle θ_i, which we now know. From either triangle between the lens and the screen in Fig. 37-12, we see that $\tan \theta_i/2 = \Delta x/2f$. Rearranging this and making the approximation $\tan \theta \approx \theta$, we find

$$\Delta x = f\theta_i, \qquad (37\text{-}15)$$

where θ_i is in radian measure. Substituting known data then yields

$$\Delta x = (0.24\ \text{m})(2.1 \times 10^{-5}\ \text{rad}) = 5.0\ \mu\text{m}. \qquad \text{(Answer)}$$

37-6 Diffraction by a Double Slit

In the double-slit experiments of Chapter 36, we implicitly assumed that the slits were narrow compared to the wavelength of the light illuminating them; that is, $a \ll \lambda$. For such narrow slits, the central maximum of the diffraction pattern of either slit covers the entire viewing screen. Moreover, the interference of light from the two slits produces bright fringes with approximately the same intensity (Fig. 36-9).

In practice with visible light, however, the condition $a \ll \lambda$ is often not met. For relatively wide slits, the interference of light from two slits produces bright fringes that do not all have the same intensity. That is, the intensities of the fringes produced by double-slit interference (as discussed in Chapter 36) are modified by diffraction of the light passing through each slit (as discussed in this chapter).

As an example, the intensity plot of Fig. 37-13*a* suggests the double-slit interference pattern that would occur if the slits were infinitely narrow (and thus $a \ll \lambda$); all the bright interference fringes would have the same intensity. The intensity plot of Fig. 37-13*b* is that for diffraction by a single actual slit; the diffraction pattern has a broad central maximum and weaker secondary maxima at $\pm 17°$. The plot of Fig. 37-13*c* suggests the interference pattern for two actual slits. That plot was constructed by using the curve of Fig. 37-13*b* as an *envelope* on the intensity plot in Fig. 37-13*a*. The positions of the fringes are not changed; only the intensities are affected.

Figure 37-14*a* shows an actual pattern in which both double-slit interference and diffraction are evident. If one slit is covered, the single-slit diffraction pattern of Fig. 37-14*b* results. Note the correspondence between Figs. 37-14*a* and 37-13*c*, and between Figs. 37-14*b* and 37-13*b*. In comparing these figures, bear in mind that Fig. 37-14 has been deliberately overexposed to bring out the faint secondary maxima and that two secondary maxima (rather than one) are shown.

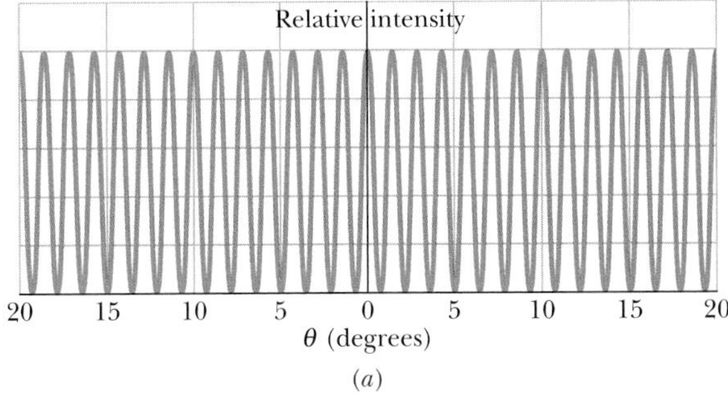

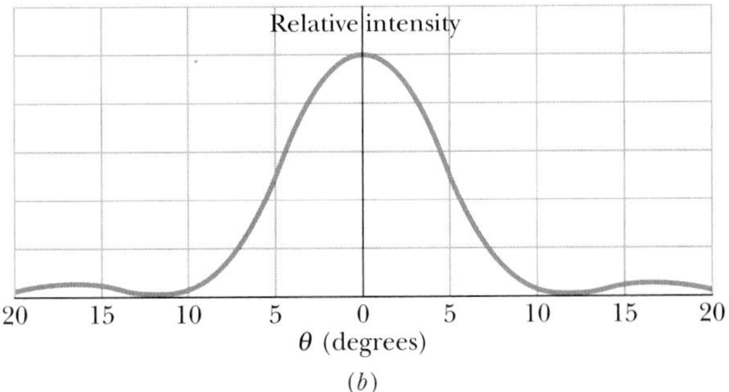

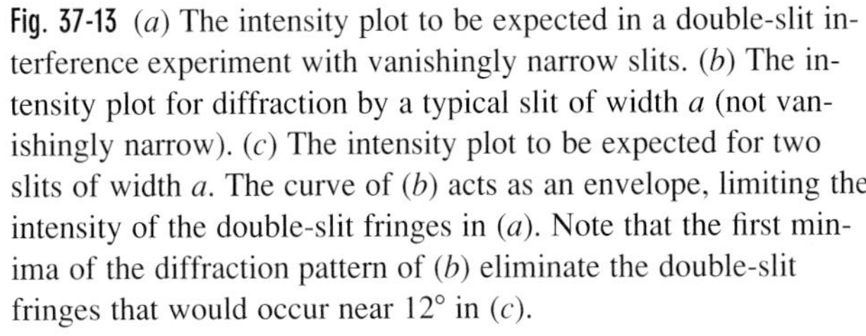
Fig. 37-13 (*a*) The intensity plot to be expected in a double-slit interference experiment with vanishingly narrow slits. (*b*) The intensity plot for diffraction by a typical slit of width a (not vanishingly narrow). (*c*) The intensity plot to be expected for two slits of width a. The curve of (*b*) acts as an envelope, limiting the intensity of the double-slit fringes in (*a*). Note that the first minima of the diffraction pattern of (*b*) eliminate the double-slit fringes that would occur near 12° in (*c*).

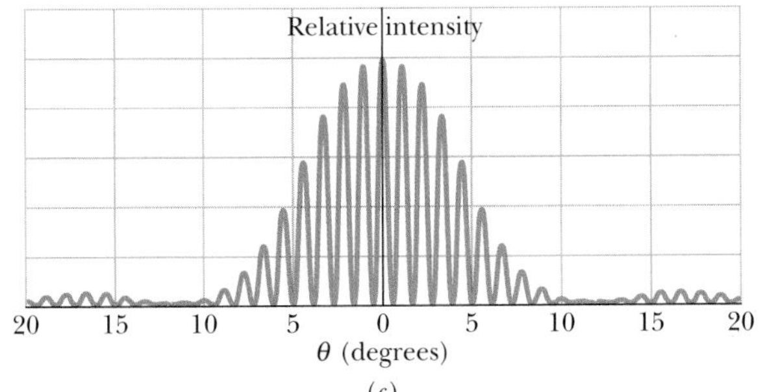

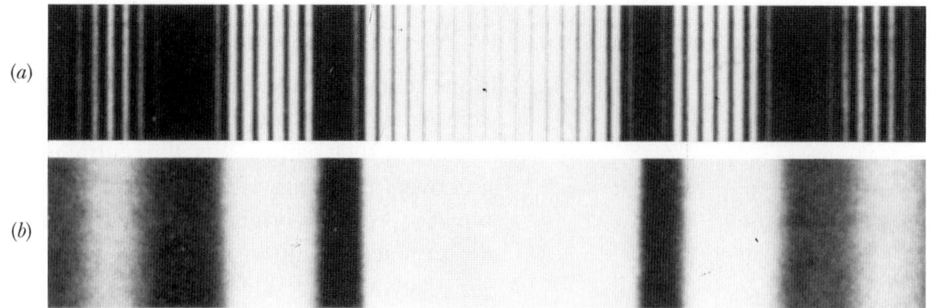

Fig. 37-14 (*a*) Interference fringes for an actual double-slit system; compare with Fig. 37-13*c*. (*b*) The diffraction pattern of a single slit; compare with Fig. 37-13*b*.

With diffraction effects taken into account, the intensity of a double-slit interference pattern is given by

$$I(\theta) = I_m(\cos^2 \beta)\left(\frac{\sin \alpha}{\alpha}\right)^2 \quad \text{(double slit)}, \tag{37-16}$$

in which

$$\beta = \frac{\pi d}{\lambda} \sin \theta \tag{37-17}$$

and

$$\alpha = \frac{\pi a}{\lambda} \sin \theta. \tag{37-18}$$

Here d is the distance between the centers of the slits, and a is the slit width. Note carefully that the right side of Eq. 37-16 is the product of I_m and two factors. (1) The *interference factor* $\cos^2 \beta$ is due to the interference between two slits with slit separation d (as given by Eqs. 36-17 and 36-18). (2) The *diffraction factor* $[(\sin \alpha)/\alpha]^2$ is due to diffraction by a single slit of width a (as given by Eqs. 37-5 and 37-6).

Let us check these factors. If we let $a \to 0$ in Eq. 37-18, for example, then $\alpha \to 0$ and $(\sin \alpha)/\alpha \to 1$. Equation 37-16 then reduces, as it must, to an equation describing the interference pattern for a pair of vanishingly narrow slits with slit separation d. Similarly, putting $d = 0$ in Eq. 37-17 is equivalent physically to causing the two slits to merge into a single slit of width a. Then Eq. 37-17 yields $\beta = 0$ and $\cos^2 \beta = 1$. In this case Eq. 37-16 reduces, as it must, to an equation describing the diffraction pattern for a single slit of width a.

The double-slit pattern described by Eq. 37-16 and displayed in Fig. 37-14*a* combines interference and diffraction in an intimate way. Both are superposition effects, in that they result from the combining of waves with different phases at a given point. If the combining waves originate from a small number of elementary coherent sources—as in a double-slit experiment with $a \ll \lambda$—we call the process *interference*. If the combining waves originate in a single wavefront—as in a single-slit experiment—we call the process *diffraction*. This distinction between interference and diffraction (which is somewhat arbitrary and not always adhered to) is a convenient one, but we should not forget that both are superposition effects and usually both are present simultaneously (as in Fig. 37-14*a*).

Sample Problem 37-4

In a double-slit experiment, the wavelength λ of the light source is 405 nm, the slit separation d is 19.44 μm, and the slit width a is 4.050 μm. Consider the interference of the light from the two slits and also the diffraction of the light through each slit.

(a) How many bright interference fringes are within the central peak of the diffraction envelope?

SOLUTION: Let us first analyze the two basic mechanisms responsible for the optical pattern produced in the experiment:

Single-slit diffraction: The Key Idea here is that the limits of the central peak are the first minima in the diffraction pattern due to either slit, individually. (See Fig. 37-13.) The angular locations of those minima are given by Eq. 37-3 ($a \sin \theta = m\lambda$). Let us write this equation as $a \sin \theta = m_1\lambda$, with the subscript 1 referring to the one-slit diffraction. For the first minima in the diffraction pattern, we substitute $m_1 = 1$, obtaining

$$a \sin \theta = \lambda. \tag{37-19}$$

Double-slit interference: The Key Idea here is that the angular locations of the bright fringes of the double-slit interference pattern are given by Eq. 36-14, which we can write as

$$d \sin \theta = m_2\lambda, \quad \text{for } m_2 = 0, 1, 2, \ldots. \tag{37-20}$$

Here the subscript 2 refers to the double-slit interference.

We can locate the first diffraction minimum within the double-slit fringe pattern by dividing Eq. 37-20 by Eq. 37-19 and solving for m_2. By doing so and then substituting the given data, we obtain

$$m_2 = \frac{d}{a} = \frac{19.44\ \mu\text{m}}{4.050\ \mu\text{m}} = 4.8.$$

This tells us that the bright interference fringe for $m_2 = 4$ fits into the central peak of the one-slit diffraction pattern, but the fringe for $m_2 = 5$ does not fit. Within the central diffraction peak we have the central bright fringe ($m_2 = 0$), and four bright fringes (up to $m_2 = 4$) on each side of it. Thus, a total of nine bright fringes of the double-slit interference pattern are within the central peak of the diffraction envelope. The bright fringes to one side of the central bright fringe are shown in Fig. 37-15.

(b) How many bright fringes are within either of the first side peaks of the diffraction envelope?

SOLUTION: The Key Idea here is that the outer limits of the first side diffraction peaks are the second diffraction minima, each of which is at the angle θ given by $a \sin \theta = m_1\lambda$ with $m_1 = 2$:

$$a \sin \theta = 2\lambda. \tag{37-21}$$

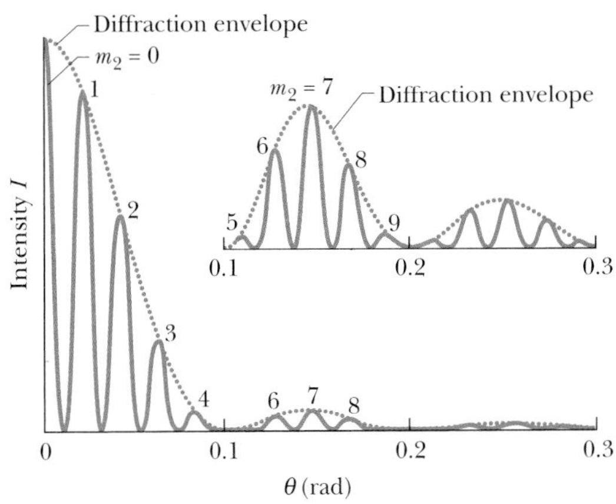

Fig. 37-15 Sample Problem 37-4. One side of the intensity plot for a two-slit interference experiment; the diffraction envelope is indicated by the dotted curve. The smaller inset shows (vertically expanded) the intensity plot within the first and second side peaks of the diffraction envelope.

Dividing Eq. 37-20 by Eq. 37-21, we find

$$m_2 = \frac{2d}{a} = \frac{(2)(19.44\ \mu\text{m})}{4.050\ \mu\text{m}} = 9.6.$$

This tells us that the second diffraction minimum occurs just before the bright interference fringe for $m_2 = 10$ in Eq. 37-20. Within either first side diffraction peak we have the fringes from $m_2 = 5$ to $m_2 = 9$ for a total of five bright fringes of the double-slit interference pattern (shown in the inset of Fig. 37-15). However, if the $m_2 = 5$ bright fringe, which is almost eliminated by the first diffraction minimum, is considered too dim to count, then only four bright fringes are in the first side diffraction peak.

✔CHECKPOINT 5: If we increase the wavelength of the light source in this sample problem to 550 nm, do (a) the width of the central diffraction peak and (b) the number of bright interference fringes within that peak increase, decrease, or remain the same?

37-7 Diffraction Gratings

One of the most useful tools in the study of light and of objects that emit and absorb light is the **diffraction grating.** This device is somewhat like the double-slit arrangement of Fig. 36-8 but has a much greater number N of slits, often called *rulings*, perhaps as many as several thousand per millimeter. An idealized grating consisting of only five slits is represented in Fig. 37-16. When monochromatic light is sent through the slits, it forms narrow interference fringes that can be analyzed to determine the wavelength of the light. (Diffraction gratings can also be opaque surfaces with narrow parallel grooves arranged like the slits in Fig. 37-16. Light then scatters

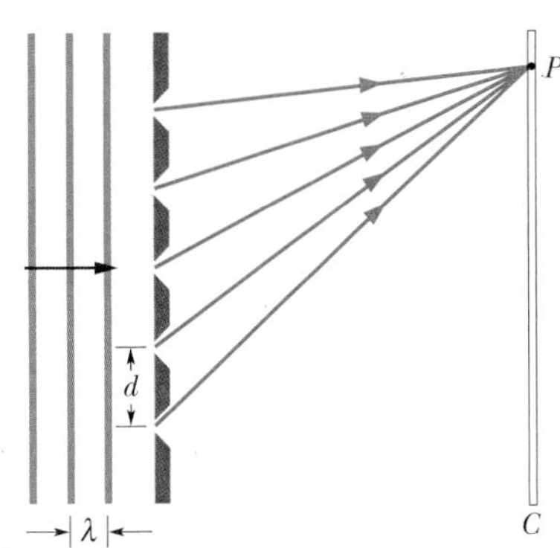

Fig. 37-16 An idealized diffraction grating, consisting of only five rulings, that produces an interference pattern on a distant viewing screen C.

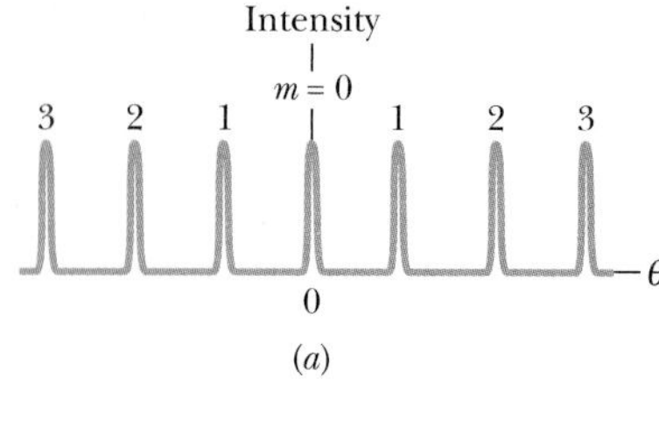

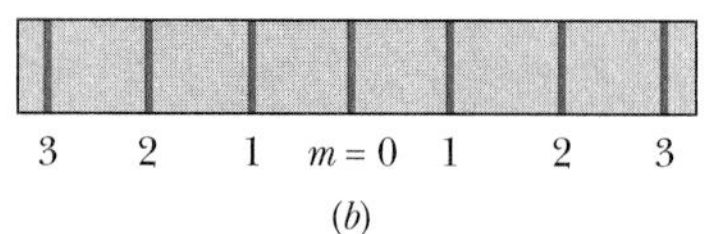

Fig. 37-17 (*a*) The intensity plot produced by a diffraction grating with a great many rulings consists of narrow peaks, here labeled with their order numbers m. (*b*) The corresponding bright fringes seen on the screen are called lines and are here also labeled with order numbers m. Lines of the zeroth, first, second, and third orders are shown.

back from the grooves to form interference fringes rather than being transmitted through open slits.)

With monochromatic light incident on a diffraction grating, if we gradually increase the number of slits from two to a large number N, the intensity plot changes from the typical double-slit plot of Fig. 37-13*c* to a much more complicated one and then eventually to a simple graph like that shown in Fig. 37-17*a*. The pattern you would see on a viewing screen using monochromatic red light from, say, a helium–neon laser, is shown in Fig. 37-17*b*. The maxima are now very narrow (and so are called *lines*); they are separated by relatively wide dark regions.

We use a familiar procedure to find the locations of the bright lines on the viewing screen. We first assume that the screen is far enough from the grating so that the rays reaching a particular point P on the screen are approximately parallel when they leave the grating (Fig. 37-18). Then we apply to each pair of adjacent rulings the same reasoning we used for double-slit interference. The separation d between rulings is called the *grating spacing*. (If N rulings occupy a total width w, then $d = w/N$.) The path length difference between adjacent rays is again $d \sin \theta$ (Fig. 37-18), where θ is the angle from the central axis of the grating (and of the diffraction pattern) to point P. A line will be located at P if the path length difference between adjacent rays is an integer number of wavelengths—that is, if

$$d \sin \theta = m\lambda, \quad \text{for } m = 0, 1, 2, \ldots \quad \text{(maxima—lines)}, \qquad (37\text{-}22)$$

where λ is the wavelength of the light. Each integer m represents a different line; hence these integers can be used to label the lines, as in Fig. 37-17. The integers are then called the *order numbers*, and the lines are called the zeroth-order line (the central line, with $m = 0$), the first-order line, the second-order line, and so on.

If we rewrite Eq. 37-22 as $\theta = \sin^{-1}(m\lambda/d)$, we see that, for a given diffraction grating, the angle from the central axis to any line (say, the third-order line) depends on the wavelength of the light being used. Thus, when light of an unknown wavelength is sent through a diffraction grating, measurements of the angles to the higher-order lines can be used in Eq. 37-22 to determine the wavelength. Even light of several unknown wavelengths can be distinguished and identified in this way. We cannot do that with the double-slit arrangement of Section 36-4, even though the same equation and wavelength dependence apply there. In double-slit interference, the bright fringes due to different wavelengths overlap too much to be distinguished.

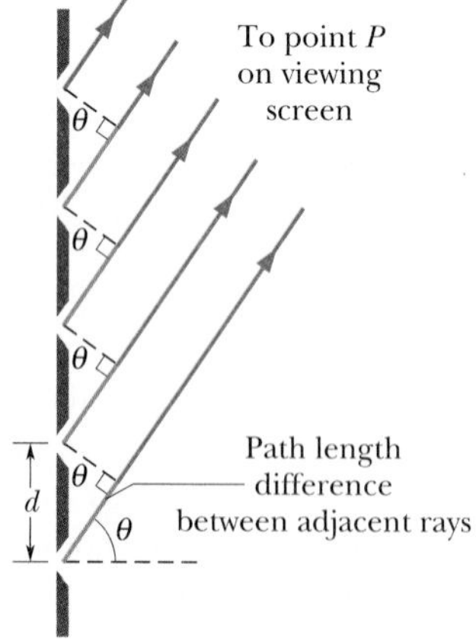

Fig. 37-18 The rays from the rulings in a diffraction grating to a distant point P are approximately parallel. The path length difference between each two adjacent rays is $d \sin \theta$, where θ is measured as shown. (The rulings extend into and out of the page.)

Width of the Lines

A grating's ability to resolve (separate) lines of different wavelengths depends on the width of the lines. We shall here derive an expression for the *half-width* of the central line (the line for which $m = 0$) and then state an expression for the half-widths of the higher-order lines. We measure the half-width of the central line as the angle $\Delta\theta_{hw}$ from the center of the line at $\theta = 0$ outward to where the line effectively ends and darkness effectively begins with the first minimum (Fig. 37-19). At such a minimum, the N rays from the N slits of the grating cancel one another. (The actual width of the central line is, of course $2(\Delta\theta_{hw})$, but line widths are usually compared via half-widths.)

In Section 37-2 we were also concerned with the cancellation of a great many rays, there due to diffraction through a single slit. We obtained Eq. 37-3, which, owing to the similarity of the two situations, we can use to find the first minimum here. It tells us that the first minimum occurs where the path length difference between the top and bottom rays equals λ. For single-slit diffraction, this difference is $a \sin \theta$. For a grating of N rulings, each separated from the next by distance d, the

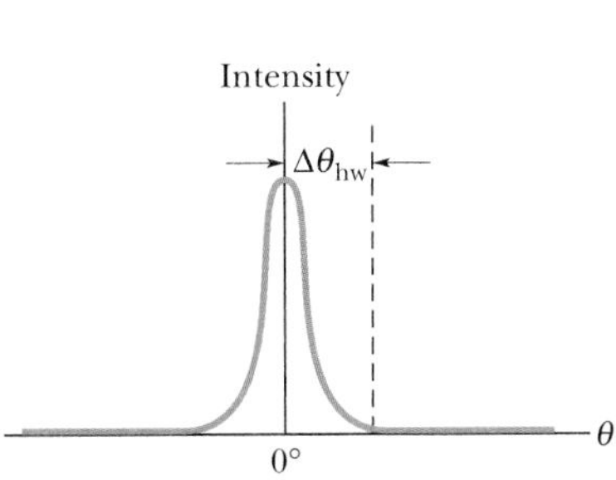

Fig. 37-19 The half-width $\Delta\theta_{hw}$ of the central line is measured from the center of that line to the adjacent minimum on a plot of I versus θ like Fig. 37-17*a*.

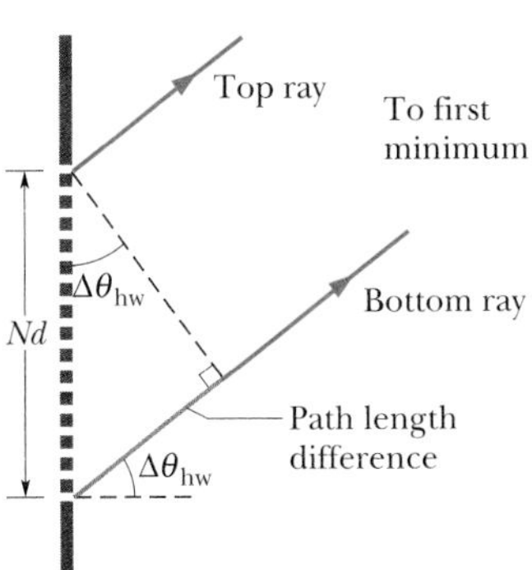

Fig. 37-20 The top and bottom rulings of a diffraction grating of N rulings are separated by distance Nd. The top and bottom rays passing through these rulings have a path length difference of $Nd \sin \Delta\theta_{hw}$, where $\Delta\theta_{hw}$ is the angle to the first minimum. (The angle is here greatly exaggerated for clarity.)

distance between the top and bottom rulings is Nd (Fig. 37-20), so the path length difference between the top and bottom rays here is $Nd \sin \Delta\theta_{hw}$. Thus, the first minimum occurs where

$$Nd \sin \Delta\theta_{hw} = \lambda. \tag{37-23}$$

Because $\Delta\theta_{hw}$ is small, $\sin \Delta\theta_{hw} = \Delta\theta_{hw}$ (in radian measure). Substituting this in Eq. 37-23 gives the half-width of the central line as

$$\Delta\theta_{hw} = \frac{\lambda}{Nd} \quad \text{(half-width of central line).} \tag{37-24}$$

We state without proof that the half-width of any other line depends on its location relative to the central axis and is

$$\Delta\theta_{hw} = \frac{\lambda}{Nd \cos \theta} \quad \text{(half-width of line at } \theta\text{).} \tag{37-25}$$

Note that for light of a given wavelength λ and a given ruling separation d, the widths of the lines decrease with an increase in the number N of rulings. Thus, of two diffraction gratings, the grating with the larger value of N is better able to distinguish between wavelengths because its diffraction lines are narrower and so produce less overlap.

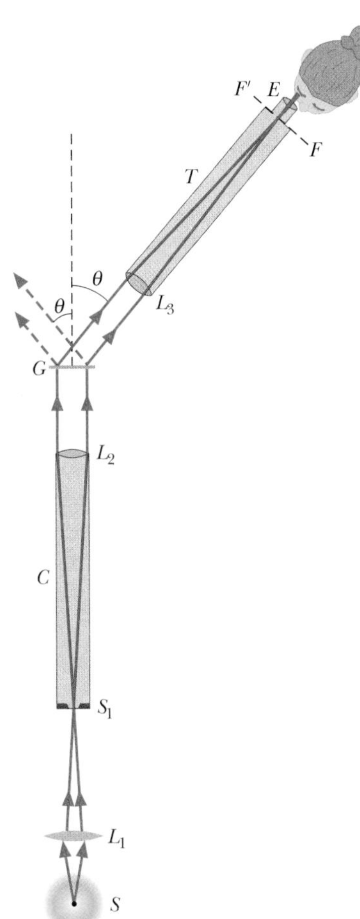

Fig. 37-21 A simple type of grating spectroscope used to analyze the wavelengths of light emitted by source S.

An Application of Diffraction Gratings

Diffraction gratings are widely used to determine the wavelengths that are emitted by sources of light ranging from lamps to stars. Figure 37-21 shows a simple *grating spectroscope* in which a grating is used for this purpose. Light from source S is focused by lens L_1 on a vertical slit S_1 placed in the focal plane of lens L_2. The light emerging from tube C (called a *collimator*) is a plane wave and is incident perpendicularly on grating G, where it is diffracted into a diffraction pattern, with the $m = 0$ order diffracted at angle $\theta = 0$ along the central axis of the grating.

We can view the diffraction pattern that would appear on a viewing screen at any angle θ simply by orienting telescope T in Fig. 37-21 to that angle. Lens L_3 of the telescope then focuses the light diffracted at angle θ (and at slightly smaller and

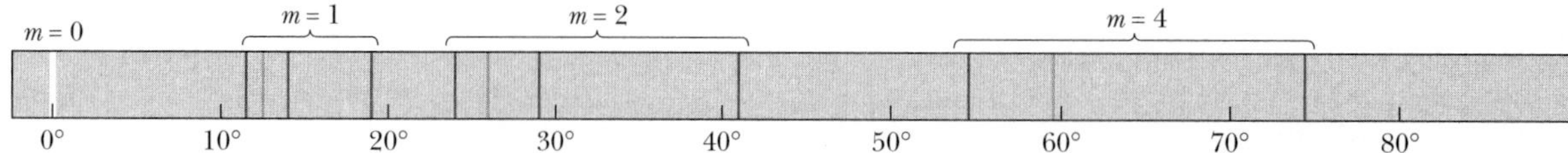

Fig. 37-22 The zeroth, first, second, and fourth orders of the visible emission lines from hydrogen. Note that the lines are farther apart at greater angles. (They are also dimmer and wider, although that is not shown here.)

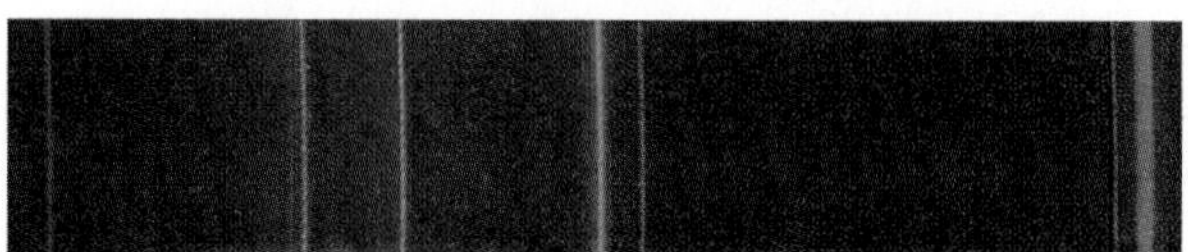

Fig. 37-23 The visible emission lines of cadmium, as seen through a grating spectroscope.

larger angles) onto a focal plane FF' within the telescope. When we look through eyepiece E, we see a magnified view of this focused image.

By changing the angle θ of the telescope, we can examine the entire diffraction pattern. For any order number other than $m = 0$, the original light is spread out according to wavelength (or color) so that we can determine, with Eq. 37-22, just what wavelengths are being emitted by the source. If the source emits discrete wavelengths, what we see as we rotate the telescope horizontally through the angles corresponding to an order m is a vertical line of color for each wavelength, with the shorter-wavelength line at a smaller angle θ than the longer-wavelength line.

For example, the light emitted by a hydrogen lamp, which contains hydrogen gas, has four discrete wavelengths in the visible range. If our eyes intercept this light directly, it appears to be white. If, instead, we view it through a grating spectroscope, we can distinguish, in several orders, the lines of the four colors corresponding to these visible wavelengths. (Such lines are called *emission lines.*) Four orders are represented in Fig. 37-22. In the central order ($m = 0$), the lines corresponding to all four wavelengths are superimposed, giving a single white line at $\theta = 0$. The colors are separated in the higher orders.

The third order is not shown in Fig. 37-22 for the sake of clarity; it actually overlaps the second and fourth orders. The fourth-order red line is missing because it is not formed by the grating used here. That is, when we attempt to solve Eq. 37-22 for the angle θ for the red wavelength when $m = 4$, we find that $\sin\theta$ is greater than unity, which is not possible. The fourth order is then said to be *incomplete* for this grating; it might not be incomplete for a grating with greater spacing d, which will spread the lines less than in Fig. 37-22. Figure 37-23 is a photograph of the visible emission lines produced by cadmium.

CHECKPOINT 6: The figure shows lines of different orders produced by a diffraction grating in monochromatic red light. (a) Is the center of the pattern to the left or right? (b) If we switch to monochromatic green light, will the half-widths of the lines then produced in the same orders be greater than, less than, or the same as the half-widths of the lines shown?

The fine rulings, each 0.5 μm wide, on a compact disc function as a diffraction grating. When a small source of white light illuminates a disc, the diffracted light forms colored "lanes" that are the composite of the diffraction patterns from the rulings.

37-8 Gratings: Dispersion and Resolving Power

Dispersion

To be useful in distinguishing wavelengths that are close to each other (as in a grating spectroscope), a grating must spread apart the diffraction lines associated with the various wavelengths. This spreading, called **dispersion,** is defined as

$$D = \frac{\Delta\theta}{\Delta\lambda} \quad \text{(dispersion defined).} \tag{37-26}$$

Here $\Delta\theta$ is the angular separation of two lines whose wavelengths differ by $\Delta\lambda$. The greater D is, the greater is the distance between two emission lines whose wavelengths differ by $\Delta\lambda$. We show below that the dispersion of a grating at angle θ is given by

$$D = \frac{m}{d\cos\theta} \quad \text{(dispersion of a grating).} \tag{37-27}$$

Thus, to achieve higher dispersion we must use a grating of smaller grating spacing d and work in a higher order m. Note that the dispersion does not depend on the number of rulings N in the grating. The SI unit for D is the degree per meter or the radian per meter.

Resolving Power

To *resolve* lines whose wavelengths are close together (that is, to make the lines distinguishable), the line should also be as narrow as possible. Expressed otherwise, the grating should have a high **resolving power** R, defined as

$$R = \frac{\lambda_{avg}}{\Delta\lambda} \quad \text{(resolving power defined).} \tag{37-28}$$

Here λ_{avg} is the mean wavelength of two emission lines that can barely be recognized as separate, and $\Delta\lambda$ is the wavelength difference between them. The greater R is, the closer two emission lines can be and still be resolved. We shall show below that the resolving power of a grating is given by the simple expression

$$R = Nm \quad \text{(resolving power of a grating).} \tag{37-29}$$

To achieve high resolving power, we must use many rulings (large N in Eq. 37-29).

Proof of Eq. 37-27

Let us start with Eq. 37-22, the expression for the locations of the lines in the diffraction pattern of a grating:

$$d\sin\theta = m\lambda.$$

Let us regard θ and λ as variables and take differentials of this equation. We find

$$d\cos\theta\, d\theta = m\, d\lambda.$$

For small enough angles, we can write these differentials as small differences, obtaining

$$d \cos \theta \, \Delta\theta = m \, \Delta\lambda \tag{37-30}$$

or

$$\frac{\Delta\theta}{\Delta\lambda} = \frac{m}{d \cos \theta}.$$

The ratio on the left is simply D (see Eq. 37-26), so we have indeed derived Eq. 37-27.

Proof of Eq. 37-29

We start with Eq. 37-30, which was derived from Eq. 37-22, the expression for the locations of the lines in the diffraction pattern formed by a grating. Here $\Delta\lambda$ is the small wavelength difference between two waves that are diffracted by the grating, and $\Delta\theta$ is the angular separation between them in the diffraction pattern. If $\Delta\theta$ is to be the smallest angle that will permit the two lines to be resolved, it must (by Rayleigh's criterion) be equal to the half-width of each line, which is given by Eq. 37-25:

$$\Delta\theta_{\text{hw}} = \frac{\lambda}{Nd \cos \theta}.$$

If we substitute $\Delta\theta_{\text{hw}}$ as given here for $\Delta\theta$ in Eq. 37-30, we find that

$$\frac{\lambda}{N} = m \, \Delta\lambda,$$

from which it readily follows that

$$R = \frac{\lambda}{\Delta\lambda} = Nm.$$

This is Eq. 37-29, which we set out to derive.

Dispersion and Resolving Power Compared

The resolving power of a grating must not be confused with its dispersion. Table 37-1 shows the characteristics of three gratings, all illuminated with light of wavelength $\lambda = 589$ nm, whose diffracted light is viewed in the first order ($m = 1$ in Eq. 37-22). You should verify that the values of D and R as given in the table can be calculated with Eqs. 37-27 and 37-29, respectively. (In the calculations for D, you will need to convert radians per meter to degrees per micrometer.)

For the conditions noted in Table 37-1, gratings A and B have the same *dispersion* and A and C have the same *resolving power*.

Figure 37-24 shows the intensity patterns (also called *line shapes*) that would be produced by these gratings for two lines of wavelengths λ_1 and λ_2, in the vicinity of $\lambda = 589$ nm. Grating B, with the higher resolving power, produces narrower lines and thus is capable of distinguishing lines that are much closer together in wavelength than those in the figure. Grating C, with the higher dispersion, produces the greater angular separation between the lines.

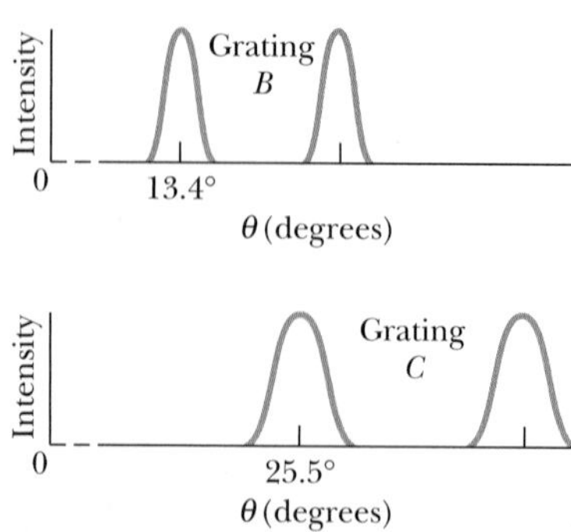

Fig. 37-24 The intensity patterns for light of two wavelengths sent through the gratings of Table 37-1. Grating B has the highest resolving power, and grating C the highest dispersion.

TABLE 37-1 Three Gratings[a]

Grating	N	d (nm)	θ	D (°/μm)	R
A	10 000	2540	13.4°	23.2	10 000
B	20 000	2540	13.4°	23.2	20 000
C	10 000	1370	25.5°	46.3	10 000

[a]Data are for $\lambda = 589$ nm and $m = 1$.

Sample Problem 37-5

A diffraction grating has 1.26×10^4 rulings uniformly spaced over width $w = 25.4$ mm. It is illuminated at normal incidence by yellow light from a sodium vapor lamp. This light contains two closely spaced emission lines (known as the sodium doublet) of wavelengths 589.00 nm and 589.59 nm.

(a) At what angle does the first-order maximum occur (on either side of the center of the diffraction pattern) for the wavelength of 589.00 nm?

SOLUTION: The Key Idea here is that the maxima produced by the diffraction grating can be located with Eq. 37-22 ($d \sin \theta = m\lambda$). The grating spacing d for this diffraction grating is

$$d = \frac{w}{N} = \frac{25.4 \times 10^{-3} \text{ m}}{1.26 \times 10^4}$$
$$= 2.016 \times 10^{-6} \text{ m} = 2016 \text{ nm}.$$

The first-order maximum corresponds to $m = 1$. Substituting these values for d and m into Eq. 37-22 leads to

$$\theta = \sin^{-1} \frac{m\lambda}{d} = \sin^{-1} \frac{(1)(589.00 \text{ nm})}{2016 \text{ nm}}$$
$$= 16.99^\circ \approx 17.0^\circ. \quad \text{(Answer)}$$

(b) Using the dispersion of the grating, calculate the angular separation between the two lines in the first order.

SOLUTION: One Key Idea here is that the angular separation $\Delta\theta$ between the two lines in the first order depends on their wavelength difference $\Delta\lambda$ and the dispersion D of the grating, according to Eq. 37-26 ($D = \Delta\theta/\Delta\lambda$). A second Key Idea is that the dispersion D depends on the angle θ at which it is to be evaluated. We can assume that, in the first order, the two sodium lines occur close enough to each other for us to evaluate D at the angle $\theta = 16.99^\circ$ we found in part (a) for one of those lines. Then Eq. 37-27 gives the dispersion as

$$D = \frac{m}{d \cos \theta} = \frac{1}{(2016 \text{ nm})(\cos 16.99^\circ)}$$
$$= 5.187 \times 10^{-4} \text{ rad/nm}.$$

From Eq. 37-26, we then have

$$\Delta\theta = D\,\Delta\lambda = (5.187 \times 10^{-4} \text{ rad/nm})(589.59 \text{ nm} - 589.00 \text{ nm})$$
$$= 3.06 \times 10^{-4} \text{ rad} = 0.0175^\circ. \quad \text{(Answer)}$$

You can show that this result depends on the grating spacing d but not on the number of rulings there are in the grating.

(c) What is the least number of rulings a grating can have and still be able to resolve the sodium doublet in the first order?

SOLUTION: One Key Idea here is that the resolving power of a grating in any order m is physically set by the number of rulings N in the grating according to Eq. 37-29 ($R = Nm$). A second Key Idea is that the least wavelength difference $\Delta\lambda$ that can be resolved depends on the average wavelength involved and the resolving power R of the grating, according to Eq. 37-28 ($R = \lambda_{avg}/\Delta\lambda$). For the sodium doublet to be barely resolved, $\Delta\lambda$ must be their wavelength separation of 0.59 nm, and λ_{avg} must be their average wavelength of 589.30 nm.

Putting these ideas together, we find that the least number of rulings for a grating to resolve the sodium doublet is

$$N = \frac{R}{m} = \frac{\lambda_{avg}}{m\,\Delta\lambda}$$
$$= \frac{589.30 \text{ nm}}{(1)(0.59 \text{ nm})} = 999 \text{ rulings}. \quad \text{(Answer)}$$

37-9 X-Ray Diffraction

X rays are electromagnetic radiation whose wavelengths are of the order of 1 Å ($= 10^{-10}$ m). Compare this with a wavelength of 550 nm ($= 5.5 \times 10^{-7}$ m) at the center of the visible spectrum. Figure 37-25 shows that x rays are produced when electrons escaping from a heated filament F are accelerated by a potential difference V and strike a metal target T.

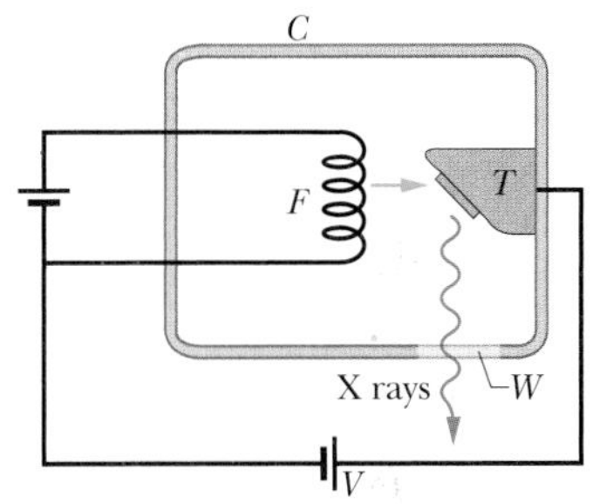

Fig. 37-25 X rays are generated when electrons leaving heated filament F are accelerated through a potential difference V and strike a metal target T. The "window" W in the evacuated chamber C is transparent to x rays.

A standard optical diffraction grating cannot be used to discriminate between different wavelengths in the x-ray wavelength range. For $\lambda = 1$ Å ($= 0.1$ nm) and $d = 3000$ nm, for example, Eq. 37-22 shows that the first-order maximum occurs at

$$\theta = \sin^{-1} \frac{m\lambda}{d} = \sin^{-1} \frac{(1)(0.1 \text{ nm})}{3000 \text{ nm}} = 0.0019^\circ.$$

This is too close to the central maximum to be practical. A grating with $d \approx \lambda$ is desirable, but, since x-ray wavelengths are about equal to atomic diameters, such gratings cannot be constructed mechanically.

In 1912, it occurred to German physicist Max von Laue that a crystalline solid, which consists of a regular array of atoms, might form a natural three-dimensional "diffraction grating" for x rays. The idea is that, in a crystal such as sodium chloride (NaCl), a basic unit of atoms (called the *unit cell*) repeats itself throughout the array. In NaCl four sodium ions and four chlorine ions are associated with each unit cell. Figure 37-26*a* represents a section through a crystal of NaCl and identifies this basic unit. The unit cell is a cube measuring a_0 on each side.

When an x-ray beam enters a crystal such as NaCl, x rays are *scattered*—that is, redirected—in all directions by the crystal structure. In some directions the scattered waves undergo destructive interference, resulting in intensity minima; in other directions the interference is constructive, resulting in intensity maxima. This process of scattering and interference is a form of diffraction, although it is unlike the diffraction of light traveling through a slit or past an edge as we discussed earlier.

Although the process of diffraction of x rays by a crystal is complicated, the maxima turn out to be in directions *as if* the x rays were reflected by a family of parallel *reflecting planes* (or *crystal planes*) that extend through the atoms within the crystal and that contain regular arrays of the atoms. (The x rays are not actually reflected; we use these fictional planes only to simplify the analysis of the actual diffraction process.)

Figure 37-26*b* shows three of the family of planes, with *interplanar spacing d*, from which the incident rays shown are said to reflect. Rays 1, 2, and 3 reflect from the first, second, and third planes, respectively. At each reflection the angle of incidence and the angle of reflection are represented with θ. Contrary to the custom

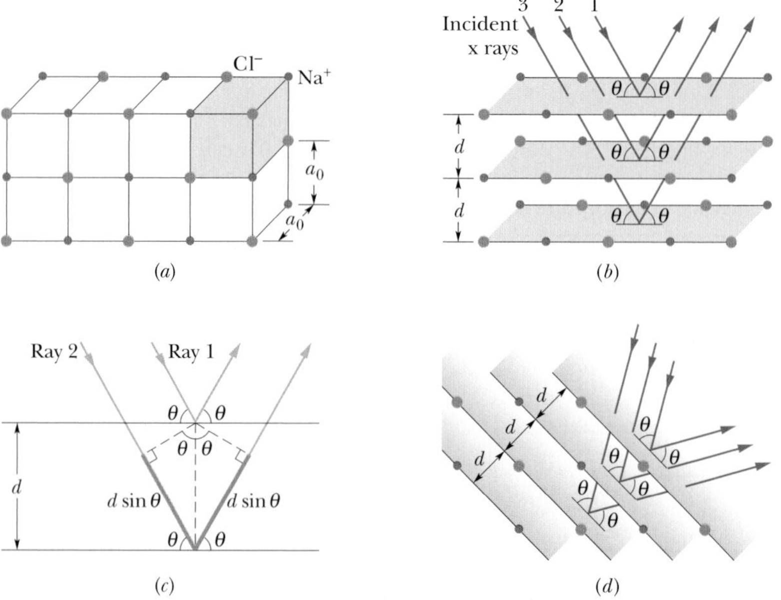

Fig. 37-26 (*a*) The cubic structure of NaCl, showing the sodium and chlorine ions and a unit cell (shaded). (*b*) Incident x rays undergo diffraction by the structure of (*a*). The x rays are diffracted as if they were reflected by a family of parallel planes, with the angle of reflection equal to the angle of incidence, both angles measured relative to the planes (not relative to a normal as in optics). (*c*) The path length difference between waves effectively reflected by two adjacent planes is $2d \sin \theta$. (*d*) A different orientation of the incident x rays relative to the structure. A different family of parallel planes now effectively reflects the x rays.

in optics, these angles are defined relative to the *surface* of the reflecting plane rather than a normal to that surface. For the situation of Fig. 37-26*b*, the interplanar spacing happens to be equal to the unit cell dimension a_0.

Figure 37-26*c* shows an edge-on view of reflection from an adjacent pair of planes. The waves of rays 1 and 2 arrive at the crystal in phase. After they are reflected, they must again be in phase, because the reflections and the reflecting planes have been defined solely to explain the intensity maxima in the diffraction of x rays by a crystal. Unlike light rays, the x rays do not refract upon entering the crystal; moreover, we do not define an index of refraction for this situation. Thus, the relative phase between the waves of rays 1 and 2 as they leave the crystal is set solely by their path length difference. For these rays to be in phase, the path length difference must be equal to an integer multiple of the wavelength λ of the x rays.

By drawing the dashed perpendiculars in Fig. 37-26*c*, we find that the path length difference is $2d \sin \theta$. In fact, this is true for any pair of adjacent planes in the family of planes represented in Fig. 37-26*b*. Thus, we have, as the criterion for intensity maxima for x-ray diffraction,

$$2d \sin \theta = m\lambda, \qquad \text{for } m = 1, 2, 3, \ldots \qquad \text{(Bragg's law)}, \tag{37-31}$$

where m is the order number of an intensity maximum. Equation 37-31 is called **Bragg's law** after British physicist W. L. Bragg, who first derived it. (He and his father shared the 1915 Nobel prize for their use of x rays to study the structures of crystals.) The angle of incidence and reflection in Eq. 37-31 is called a *Bragg angle*.

Regardless of the angle at which x rays enter a crystal, there is always a family of planes from which they can be said to reflect so that we can apply Bragg's law. In Fig. 37-26*d*, notice that the crystal structure has the same orientation as it does in Fig. 37-26*a*, but the angle at which the beam enters the structure differs from that shown in Fig. 37-26*b*. This new angle requires a new family of reflecting planes, with a different interplanar spacing d and different Bragg angle θ, in order to explain the x-ray diffraction via Bragg's law.

Figure 37-27 shows how the interplanar spacing d can be related to the unit cell dimension a_0. For the particular family of planes shown there, the Pythagorean theorem gives

$$5d = \sqrt{5}a_0,$$

or

$$d = \frac{a_0}{\sqrt{5}}. \tag{37-32}$$

Figure 37-27 suggests how the dimensions of the unit cell can be found once the interplanar spacing has been measured by means of x-ray diffraction.

X-ray diffraction is a powerful tool for studying both x-ray spectra and the arrangement of atoms in crystals. To study spectra, a particular set of crystal planes, having a known spacing d, is chosen. These planes effectively reflect different wavelengths at different angles. A detector that can discriminate one angle from another can then be used to determine the wavelength of radiation reaching it. The crystal itself can be studied with a monochromatic x-ray beam, to determine not only the spacing of various crystal planes but also the structure of the unit cell.

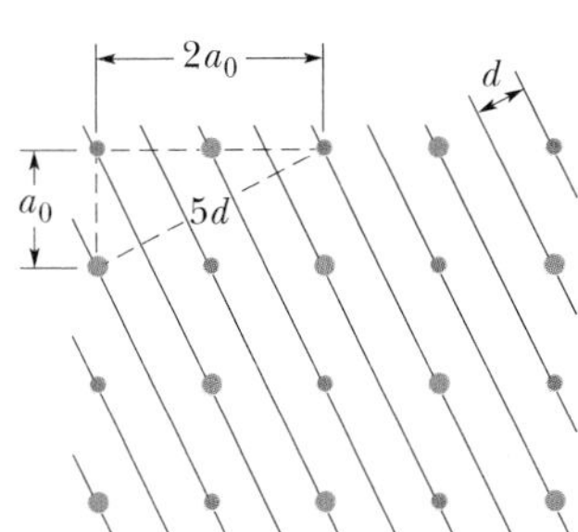

Fig. 37-27 A family of planes through the structure of Fig. 37-26*a*, and a way to relate the edge length a_0 of a unit cell to the interplanar spacing d.

REVIEW & SUMMARY

Diffraction When waves encounter an edge or an obstacle or an aperture with a size comparable to the wavelength of the waves, those waves spread out as they travel and, as a result, undergo interference. This is called **diffraction.**

Single-Slit Diffraction Waves passing through a long narrow slit of width a produce, on a viewing screen, a **single-slit diffraction pattern** that includes a central maximum and other maxima, separated by minima located at angles θ to the central axis that satisfy

$$a \sin \theta = m\lambda, \quad \text{for } m = 1, 2, 3, \ldots \quad \text{(minima)}. \quad (37\text{-}3)$$

The intensity of the diffraction pattern at any given angle θ is

$$I(\theta) = I_m \left(\frac{\sin \alpha}{\alpha} \right)^2, \quad \text{where} \quad \alpha = \frac{\pi a}{\lambda} \sin \theta \quad (37\text{-}5, 37\text{-}6)$$

and I_m is the intensity at the center of the pattern.

Circular-Aperture Diffraction Diffraction by a circular aperture or a lens with diameter d produces a central maximum and concentric maxima and minima, with the first minimum at an angle θ given by

$$\sin \theta = 1.22 \frac{\lambda}{d} \quad \text{(first minimum; circular aperture)}. \quad (37\text{-}12)$$

Rayleigh's Criterion *Rayleigh's criterion* suggests that two objects are on the verge of resolvability if the central diffraction maximum of one is at the first minimum of the other. Their angular separation must then be at least

$$\theta_R = 1.22 \frac{\lambda}{d} \quad \text{(Rayleigh's criterion)}, \quad (37\text{-}14)$$

in which d is the diameter of the aperture through which the light passes.

Double-Slit Diffraction Waves passing through two slits, each of width a, whose centers are a distance d apart, display diffraction patterns whose intensity I at angle θ is

$$I(\theta) = I_m (\cos^2 \beta) \left(\frac{\sin \alpha}{\alpha} \right)^2 \quad \text{(double slit)}, \quad (37\text{-}16)$$

with $\beta = (\pi d/\lambda) \sin \theta$ and α the same as for the case of single-slit diffraction.

Multiple-Slit Diffraction Diffraction by N (multiple) slits results in maxima (lines) at angles θ such that

$$d \sin \theta = m\lambda, \quad \text{for } m = 0, 1, 2, \ldots \quad \text{(maxima)}, \quad (37\text{-}22)$$

with the half-widths of the lines given by

$$\Delta\theta_{hw} = \frac{\lambda}{Nd \cos \theta} \quad \text{(half-widths)}. \quad (37\text{-}25)$$

Diffraction Gratings A *diffraction grating* is a series of "slits" used to separate an incident wave into its component wavelengths by separating and displaying their diffraction maxima. A grating is characterized by its dispersion D and resolving power R:

$$D = \frac{\Delta\theta}{\Delta\lambda} = \frac{m}{d \cos \theta} \quad (37\text{-}26, 37\text{-}27)$$

$$R = \frac{\lambda_{avg}}{\Delta\lambda} = Nm. \quad (37\text{-}28, 37\text{-}29)$$

X-Ray Diffraction The regular array of atoms in a crystal is a three-dimensional diffraction grating for short-wavelength waves such as x rays. For analysis purposes, the atoms can be visualized as being arranged in planes with characteristic interplanar spacing d. Diffraction maxima (due to constructive interference) occur if the incident direction of the wave, measured from the surfaces of these planes, and the wavelength λ of the radiation satisfy **Bragg's law:**

$$2d \sin \theta = m\lambda, \quad \text{for } m = 1, 2, 3, \ldots \quad \text{(Bragg's law)}. \quad (37\text{-}31)$$

QUESTIONS

1. Light of frequency f illuminating a long narrow slit produces a diffraction pattern. (a) If we switch to light of frequency $1.3f$, does the pattern expand away from the center or contract toward the center? (b) Does the pattern expand or contract if, instead, we submerge the equipment in clear corn syrup?

2. You are conducting a single-slit diffraction experiment with light of wavelength λ. What appears, on a distant viewing screen, at a point at which the top and bottom rays through the slit have a path length difference equal to (a) 5λ and (b) 4.5λ?

3. If you speak with the same intensity with and without a megaphone in front of your mouth, in which situation do you sound louder to someone directly in front of you?

4. Figure 37-28 shows four choices for the rectangular opening of a source of either sound waves or light waves. The sides have lengths of either L or $2L$, with L being 3.0 times the wavelength of the waves. Rank the openings according to the extent of (a) left–right spreading and (b) up–down spreading of the waves due to diffraction, greatest first.

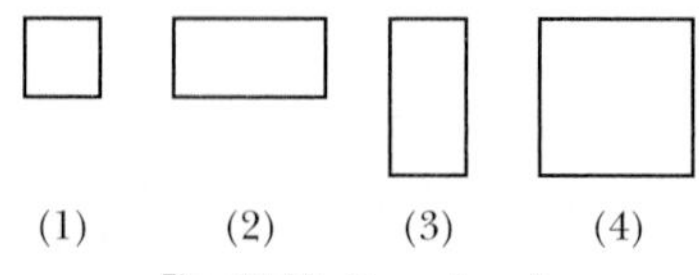

Fig. 37-28 Question 4.

5. In a single-slit diffraction experiment, the top and bottom rays through the slit arrive at a certain point on the viewing screen with a path length difference of 4.0 wavelengths. In a phasor representation like those in Fig 37-6, how many overlapping circles does the chain of phasors make?

6. At night many people see rings (called *entoptic halos*) surrounding bright outdoor lamps in otherwise dark surroundings. The rings are the first of the side maxima in diffraction patterns produced by structures that are thought to be within the cornea (or possibly the lens) of the observer's eye. (The central maxima of such patterns overlap the lamp.) (a) Would a particular ring become smaller or larger if the lamp were switched from blue to red light? (b) If a lamp emits white light, is blue or red on the outside edge of the ring?

7. Figure 37-29 shows the bright fringes that lie within the central diffraction envelope in two double-slit diffraction experiments using the same wavelength of light. Are (a) the slit width a, (b) the slit separation d, and (c) the ratio d/a in experiment B greater than, less than, or the same as those in experiment A?

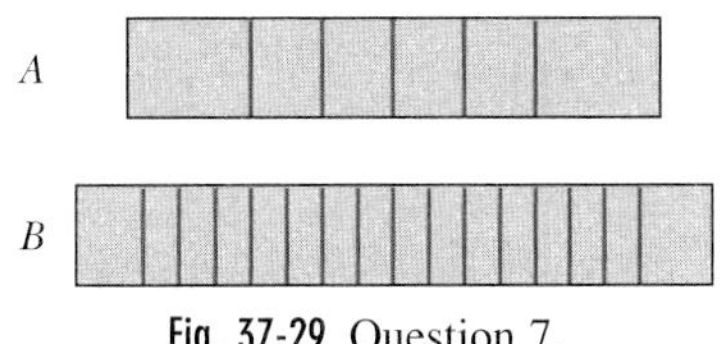

Fig. 37-29 Question 7.

8. Figure 37-30 shows a red line and a green line of the same order in the pattern produced by a diffraction grating. If we increased the number of rulings in the grating, say, by removing tape that had covered half the rulings, would (a) the half-widths of the lines and (b) the separation of the lines increase, decrease, or remain the same? (c) Would the lines shift to the right, shift to the left, or remain in place?

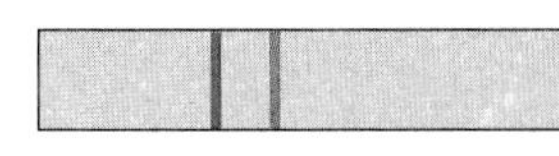
Fig. 37-30 Questions 8 and 9.

9. For the situation of Question 8 and Fig. 37-30, if instead we increased the grating spacing, would (a) the half-widths of the lines and (b) the separation of the lines increase, decrease, or remain the same? (c) Would the lines shift to the right, shift to the left, or remain in place?

10. (a) Figure 37-31a shows the lines produced by diffraction gratings A and B using light of the same wavelength; the lines are of the same order and appear at the same angles θ. Which grating has the greater number of rulings? (b) Figure 37-31b shows lines of two orders produced by a single diffraction grating using light of two wavelengths, both in the red region of the spectrum. Which lines, the left pair or right pair, are in the order with greater m? Is the center of the diffraction pattern to the left or to the right in (c) Fig. 37-31a and (d) Fig. 37-31b?

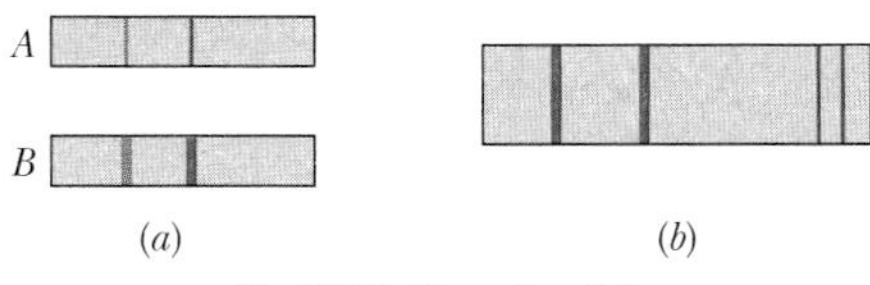

Fig. 37-31 Question 10.

11. (a) For a given diffraction grating, does the least difference $\Delta\lambda$ in two wavelengths that can be resolved increase, decrease, or remain the same as the wavelength increases? (b) For a given wavelength region (say, around 500 nm), is $\Delta\lambda$ greater in the first order or in the third order?

EXERCISES & PROBLEMS

ssm Solution is in the Student Solutions Manual.
www Solution is available on the World Wide Web at: http://www.wiley.com/college/hrw
ilw Solution is available on the Interactive LearningWare.

SEC. 37-2 Diffraction by a Single Slit: Locating the Minima

1E. Light of wavelength 633 nm is incident on a narrow slit. The angle between the first diffraction minimum on one side of the central maximum and the first minimum on the other side is 1.20°. What is the width of the slit? ssm

2E. Monochromatic light of wavelength 441 nm is incident on a narrow slit. On a screen 2.00 m away, the distance between the second diffraction minimum and the central maximum is 1.50 cm. (a) Calculate the angle of diffraction θ of the second minimum. (b) Find the width of the slit.

3E. A single slit is illuminated by light of wavelengths λ_a and λ_b, chosen so the first diffraction minimum of the λ_a component coincides with the second minimum of the λ_b component. (a) What relationship exists between the two wavelengths? (b) Do any other minima in the two diffraction patterns coincide? ssm

4E. The distance between the first and fifth minima of a single-slit diffraction pattern is 0.35 mm with the screen 40 cm away from the slit, when light of wavelength 550 nm is used. (a) Find the slit width. (b) Calculate the angle θ of the first diffraction minimum.

5E. A plane wave of wavelength 590 nm is incident on a slit with a width of $a = 0.40$ mm. A thin converging lens of focal length +70 cm is placed between the slit and a viewing screen and focuses the light on the screen. (a) How far is the screen from the lens? (b) What is the distance on the screen from the center of the diffraction pattern to the first minimum? ssm

6P. Sound waves with frequency 3000 Hz and speed 343 m/s diffract through the rectangular opening of a speaker cabinet and into a large auditorium. The opening, which has a horizontal width of 30.0 cm, faces a wall 100 m away (Fig. 37-32). Where along that wall will a listener be at the first diffraction minimum and thus have difficulty hearing the sound? (Neglect reflections.)

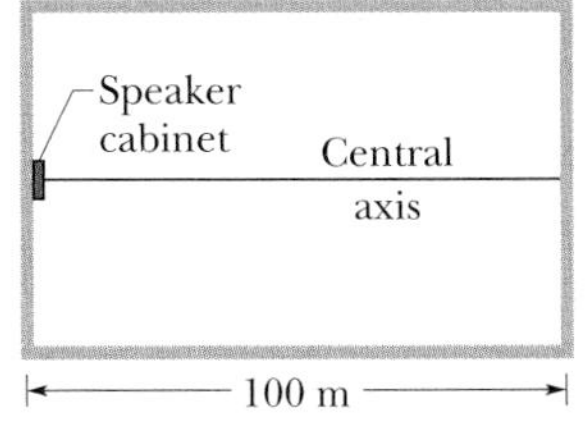

Fig. 37-32 Problem 6.

7P. A slit 1.00 mm wide is illuminated by light of wavelength 589 nm. We see a diffraction pattern on a screen 3.00 m away. What is the distance between the first two diffraction minima on the same side of the central diffraction maximum? ssm ilw

SEC. 37-4 Intensity in Single-Slit Diffraction, Quantitatively

8E. A 0.10-mm-wide slit is illuminated by light of wavelength 589 nm. Consider a point P on a viewing screen on which the diffraction pattern of the slit is viewed; the point is at 30° from the central axis of the slit. What is the phase difference between the Huygens wavelets arriving at point P from the top and midpoint of the slit? (*Hint:* See Eq. 37-4.)

9E. If you double the width of a single slit, the intensity of the central maximum of the diffraction pattern increases by a factor of 4, even though the energy passing through the slit only doubles. Explain this quantitatively. ssm

10E. Monochromatic light with wavelength 538 nm is incident on a slit with width 0.025 mm. The distance from the slit to a screen is 3.5 m. Consider a point on the screen 1.1 cm from the central maximum. (a) Calculate θ for that point. (b) Calculate α. (c) Calculate the ratio of the intensity at this point to the intensity at the central maximum.

11P. The full width at half-maximum (FWHM) of a central diffraction maximum is defined as the angle between the two points in the pattern where the intensity is one-half that at the center of the pattern. (See Fig. 37-7*b*.) (a) Show that the intensity drops to one-half the maximum value when $\sin^2 \alpha = \alpha^2/2$. (b) Verify that $\alpha =$ 1.39 rad (about 80°) is a solution to the transcendental equation of (a). (c) Show that the FWHM is $\Delta\theta = 2\sin^{-1}(0.443\lambda/a)$, where a is the slit width. (d) Calculate the FWHM of the central maximum for slits whose widths are 1.0, 5.0, and 10 wavelengths. ssm www

12P. *Babinet's Principle.* A monochromatic beam of parallel light is incident on a "collimating" hole of diameter $x \gg \lambda$. Point P lies in the geometrical shadow region on a *distant* screen (Fig. 37-33*a*). Two diffracting objects, shown in Fig. 37-33*b*, are placed in turn over the collimating hole. A is an opaque circle with a hole in it and B is the "photographic negative" of A. Using superposition concepts, show that the intensity at P is identical for the two diffracting objects A and B.

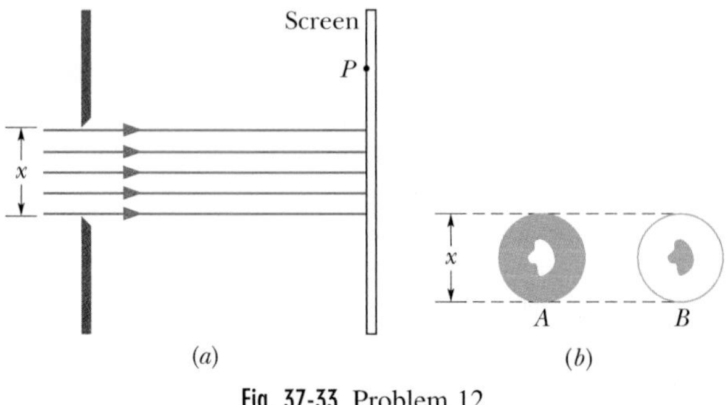

Fig. 37-33 Problem 12.

13P. (a) Show that the values of α at which intensity maxima for single-slit diffraction occur can be found exactly by differentiating Eq. 37-5 with respect to α and equating the result to zero, obtaining the condition $\tan \alpha = \alpha$. (b) Find the values of α satisfying this relation by plotting the curve $y = \tan \alpha$ and the straight line $y = \alpha$ and finding their intersections or by using a calculator to find an appropriate value of α by trial and error. (c) Find the (noninteger) values of m corresponding to successive maxima in the single-slit pattern. Note that the secondary maxima do not lie exactly halfway between minima. ssm

SEC. 37-5 Diffraction by a Circular Aperture

14E. Assume that the lamp in Question 6 emits light at wavelength 550 nm. If a ring has an angular diameter of 2.5°, approximately what is the (linear) diameter of the structure in the eye that causes the ring?

15E. The two headlights of an approaching automobile are 1.4 m apart. At what (a) angular separation and (b) maximum distance will the eye resolve them? Assume that the pupil diameter is 5.0 mm, and use a wavelength of 550 nm for the light. Also assume that diffraction effects alone limit the resolution so that Rayleigh's criterion can be applied. ssm

16E. An astronaut in a space shuttle claims she can just barely resolve two point sources on Earth's surface, 160 km below. Calculate their (a) angular and (b) linear separation, assuming ideal conditions. Take $\lambda = 540$ nm and the pupil diameter of the astronaut's eye to be 5.0 mm.

17E. Find the separation of two points on the Moon's surface that can just be resolved by the 200 in. (= 5.1 m) telescope at Mount Palomar, assuming that this separation is determined by diffraction effects. The distance from Earth to the Moon is 3.8×10^5 km. Assume a wavelength of 550 nm for the light. ilw

18E. The wall of a large room is covered with acoustic tile in which small holes are drilled 5.0 mm from center to center. How far can a person be from such a tile and still distinguish the individual holes, assuming ideal conditions, the pupil diameter of the observer's eye to be 4.0 mm, and the wavelength of the room light to be 550 nm?

19E. Estimate the linear separation of two objects on the planet Mars that can just be resolved under ideal conditions by an observer on Earth (a) using the naked eye and (b) using the 200 in. (= 5.1 m) Mount Palomar telescope. Use the following data: distance to Mars $= 8.0 \times 10^7$ km, diameter of pupil $= 5.0$ mm, wavelength of light $= 550$ nm. ssm

20E. The radar system of a navy cruiser transmits at a wavelength of 1.6 cm, from a circular antenna with a diameter of 2.3 m. At a range of 6.2 km, what is the smallest distance that two speedboats can be from each other and still be resolved as two separate objects by the radar system?

21P. The wings of tiger beetles (Fig. 37-34) are colored by interference due to thin cuticle-like layers. In addition, these layers are arranged in patches that are 60 μm across and produce different colors. The color you see is a pointillistic mixture of thin-film interference colors that varies with perspective. Approximately what viewing distance from a wing puts you at the limit of resolving the different colored patches according to Rayleigh's criterion? Use

Fig. 37-34 Problem 21. Tiger beetles are colored by pointillistic mixtures of thin-film interference colors.

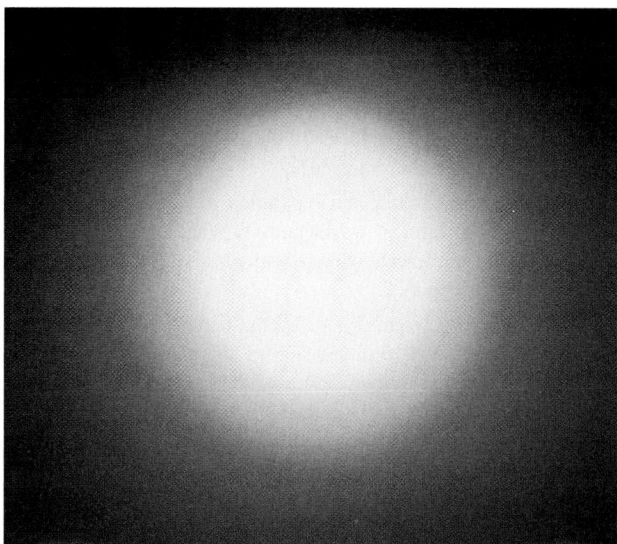

Fig. 37-35 Problem 24. The corona around the Moon is a composite of the diffraction patterns of airborne water drops.

550 nm as the wavelength of light and 3.00 mm as the diameter of your pupil.

22P. In June 1985, a laser beam was sent out from the Air Force Optical Station on Maui, Hawaii, and reflected back from the shuttle *Discovery* as it sped by, 354 km overhead. The diameter of the central maximum of the beam at the shuttle position was said to be 9.1 m, and the beam wavelength was 500 nm. What is the effective diameter of the laser aperture at the Maui ground station? (*Hint:* A laser beam spreads only because of diffraction; assume a circular exit aperture.)

23P. Millimeter-wave radar generates a narrower beam than conventional microwave radar, making it less vulnerable to antiradar missiles. (a) Calculate the angular width of the central maximum, from first minimum to first minimum, produced by a 220 GHz radar beam emitted by a 55.0-cm-diameter circular antenna. (The frequency is chosen to coincide with a low-absorption atmospheric "window.") (b) Calculate the same quantity for the ship's radar described in Exercise 20. ssm www

24P. A circular obstacle produces the same diffraction pattern as a circular hole of the same diameter (except very near $\theta = 0$). Airborne water drops are examples of such obstacles. When you see the Moon through suspended water drops, such as in a fog, you intercept the diffraction pattern from many drops. The composite of the central diffraction maxima of those drops forms a white region that surrounds the Moon and may obscure it. Figure 37-35 is a photograph in which the Moon is obscured. There are two, faint, colored rings around the Moon (the larger one may be too faint to be seen in your copy of the photograph). The smaller ring is on the outer edge of the central maxima from the drops; the somewhat larger ring is on the outer edge of the smallest of the secondary maxima from the drops (see Fig. 37-9). The color is visible because the rings are adjacent to the diffraction minima (dark rings) in the patterns. (Colors in other parts of the pattern overlap too much to be visible.)

(a) What is the color of these rings on the outer edges of the diffraction maxima? (b) The colored ring around the central maxima in Fig. 37-35 has an angular diameter that is 1.35 times the angular diameter of the Moon, which is 0.50°. Assume that the drops all have about the same diameter. Approximately what is that diameter?

25P. (a) What is the angular separation of two stars if their images are barely resolved by the Thaw refracting telescope at the Allegheny Observatory in Pittsburgh? The lens diameter is 76 cm and its focal length is 14 m. Assume $\lambda = 550$ nm. (b) Find the distance between these barely resolved stars if each of them is 10 light-years distant from Earth. (c) For the image of a single star in this telescope, find the diameter of the first dark ring in the diffraction pattern, as measured on a photographic plate placed at the focal plane of the telescope lens. Assume that the structure of the image is associated entirely with diffraction at the lens aperture and not with lens "errors."

26P. In a joint Soviet–French experiment to monitor the Moon's surface with a light beam, pulsed radiation from a ruby laser ($\lambda = 0.69\ \mu$m) was directed to the Moon through a reflecting telescope with a mirror radius of 1.3 m. A reflector on the Moon behaved like a circular plane mirror with radius 10 cm, reflecting the light directly back toward the telescope on Earth. The reflected light was then detected after being brought to a focus by this telescope. What fraction of the original light energy was picked up by the detector? Assume that for each direction of travel all the energy is in the central diffraction peak.

SEC. 37-6 Diffraction by a Double Slit

27E. Suppose that the central diffraction envelope of a double-slit diffraction pattern contains 11 bright fringes and the first diffraction minima eliminate (are coincident with) bright fringes. How many bright fringes lie between the first and second minima of the diffraction envelope? ssm

28E. In a double-slit experiment, the slit separation d is 2.00 times the slit width w. How many bright interference fringes are in the central diffraction envelope?

29P. (a) In a double-slit experiment, what ratio of d to a causes diffraction to eliminate the fourth bright side fringe? (b) What other bright fringes are also eliminated?

30P. Two slits of width a and separation d are illuminated by a coherent beam of light of wavelength λ. What is the linear separation of the bright interference fringes observed on a screen that is at a distance D away?

31P. (a) How many bright fringes appear between the first diffraction-envelope minima to either side of the central maximum in a double-slit pattern if $\lambda = 550$ nm, $d = 0.150$ mm, and $a = 30.0$ μm? (b) What is the ratio of the intensity of the third bright fringe to the intensity of the central fringe? ssm

32P. Light of wavelength 440 nm passes through a double slit, yielding a diffraction pattern whose graph of intensity I versus angular position θ is shown in Fig. 37-36. Calculate (a) the slit width and (b) the slit separation. (c) Verify the displayed intensities of the $m = 1$ and $m = 2$ interference fringes.

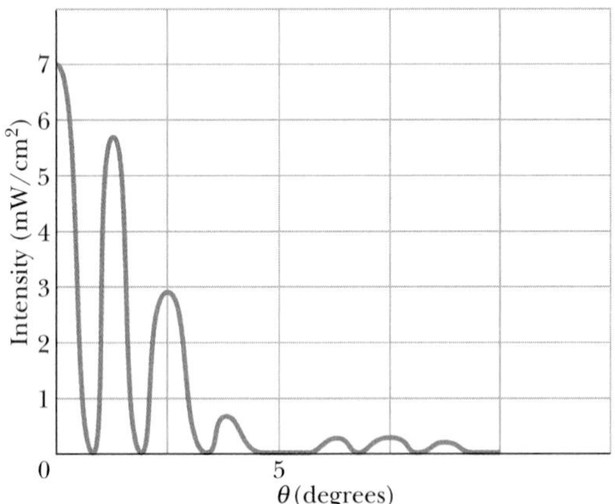

Fig. 37-36 Problem 32.

SEC. 37-7 Diffraction Gratings

33E. A diffraction grating 20.0 mm wide has 6000 rulings. (a) Calculate the distance d between adjacent rulings. (b) At what angles θ will intensity maxima occur on a viewing screen if the radiation incident on the grating has a wavelength of 589 nm?

34E. A grating has 315 rulings/mm. For what wavelengths in the visible spectrum can fifth-order diffraction be observed when this grating is used in a diffraction experiment?

35E. A grating has 400 lines/mm. How many orders of the entire visible spectrum (400–700 nm) can it produce in a diffraction experiment, in addition to the $m = 0$ order? ssm ilw

36E. Perhaps to confuse a predator, some tropical gyrinid beetles (whirligig beetles) are colored by optical interference that is due to scales whose alignment forms a diffraction grating (which scatters light instead of transmiting it). When the incident light rays are perpendicular to the grating, the angle between the first-order maxima (on opposite sides of the zeroth-order maximum) is about 26° in light with a wavelength of 550 nm. What is the grating spacing of the beetle?

37P. Light of wavelength 600 nm is incident normally on a diffraction grating. Two adjacent maxima occur at angles given by $\sin\theta = 0.2$ and $\sin\theta = 0.3$. The fourth-order maxima are missing. (a) What is the separation between adjacent slits? (b) What is the smallest slit width this grating can have? (c) Which orders of intensity maxima are produced by the grating, assuming the values derived in (a) and (b)? ssm

38P. A diffraction grating is made up of slits of width 300 nm with separation 900 nm. The grating is illuminated by monochromatic plane waves of wavelength $\lambda = 600$ nm at normal incidence. (a) How many maxima are there in the full diffraction pattern? (b) What is the width of a spectral line observed in the first order if the grating has 1000 slits?

39P. Assume that the limits of the visible spectrum are arbitrarily chosen as 430 and 680 nm. Calculate the number of rulings per millimeter of a grating that will spread the first-order spectrum through an angle of 20°. ssm www

40P. With light from a gaseous discharge tube incident normally on a grating with slit separation 1.73 μm, sharp maxima of green light are produced at angles $\theta = \pm 17.6°, 37.3°, -37.1°, 65.2°$, and $-65.0°$. Compute the wavelength of the green light that best fits these data.

41P. Light is incident on a grating at an angle ψ as shown in Fig. 37-37. Show that bright fringes occur at angles θ that satisfy the equation

$$d(\sin\psi + \sin\theta) = m\lambda, \quad \text{for } m = 0, 1, 2, \ldots.$$

(Compare this equation with Eq. 37-22.) Only the special case $\psi = 0$ has been treated in this chapter.

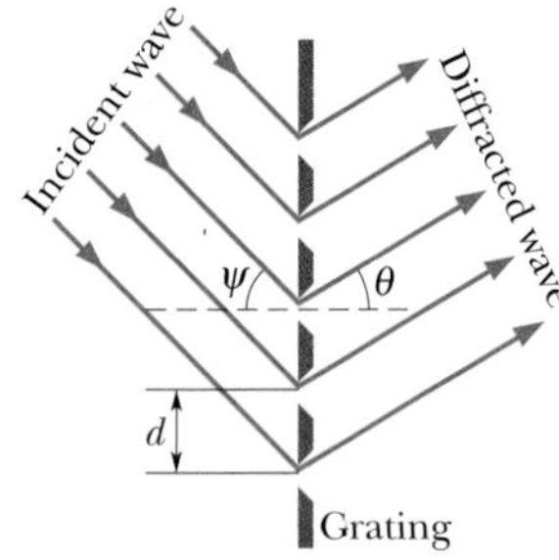

Fig. 37-37 Problem 41.

42P. A grating with $d = 1.50$ μm is illuminated at various angles of incidence by light of wavelength 600 nm. Plot, as a function of the angle of incidence (0 to 90°), the angular deviation of the first-order maximum from the incident direction. (See Problem 41.)

43P. Derive Eq. 37-25, the expression for the half-widths of lines in a grating's diffraction pattern. ssm

44P. A grating has 350 rulings per millimeter and is illuminated at normal incidence by white light. A spectrum is formed on a screen 30 cm from the grating. If a hole 10 mm square is cut in the screen, its inner edge being 50 mm from the central maximum and parallel to it, what is the range in the wavelengths of the light that passes through the hole?

45P*. Derive this expression for the intensity pattern for a three-slit "grating":

$$I = \tfrac{1}{9}I_m(1 + 4\cos\phi + 4\cos^2\phi),$$

where $\phi = (2\pi d \sin\theta)/\lambda$. Assume that $a \ll \lambda$; be guided by the derivation of the corresponding double-slit formula (Eq. 36-21). ssm

SEC. 37-8 Gratings: Dispersion and Resolving Power

46E. The D line in the spectrum of sodium is a doublet with wavelengths 589.0 and 589.6 nm. Calculate the minimum number of lines needed in a grating that will resolve this doublet in the second-order spectrum. See Sample Problem 37-5.

47E. A source containing a mixture of hydrogen and deuterium atoms emits red light at two wavelengths whose mean is 656.3 nm and whose separation is 0.180 nm. Find the minimum number of lines needed in a diffraction grating that can resolve these lines in the first order. ssm ilw

48E. A grating has 600 rulings/mm and is 5.0 mm wide. (a) What is the smallest wavelength interval it can resolve in the third order at $\lambda = 500$ nm? (b) How many higher orders of maxima can be seen?

49E. Show that the dispersion of a grating is $D = (\tan\theta)/\lambda$. ssm

50E. With a particular grating the sodium doublet (see Sample Problem 37-5) is viewed in the third order at 10° to the normal and is barely resolved. Find (a) the grating spacing and (b) the total width of the rulings.

51P. A diffraction grating has resolving power $R = \lambda_{avg}/\Delta\lambda = Nm$. (a) Show that the corresponding frequency range Δf that can just be resolved is given by $\Delta f = c/Nm\lambda$. (b) From Fig. 37-18, show that the times required for light to travel along the ray at the bottom of the figure and the ray at the top differ by an amount $\Delta t = (Nd/c)\sin\theta$. (c) Show that $(\Delta f)(\Delta t) = 1$, this relation being independent of the various grating parameters. Assume $N \gg 1$. ssm

52P. (a) In terms of the angle θ locating a line produced by a grating, find the product of that line's half-width and the resolving power of the grating. (b) Evaluate that product for the grating of Problem 38, for the first order.

SEC. 37-9 X-Ray Diffraction

53E. X rays of wavelength 0.12 nm are found to undergo second-order reflection at a Bragg angle of 28° from a lithium fluoride crystal. What is the interplanar spacing of the reflecting planes in the crystal? ssm

54E. Figure 37-38 is a graph of intensity versus angular position θ for the diffraction of an x-ray beam by a crystal. The beam consists of two wavelengths, and the spacing between the reflecting planes is 0.94 nm. What are the two wavelengths?

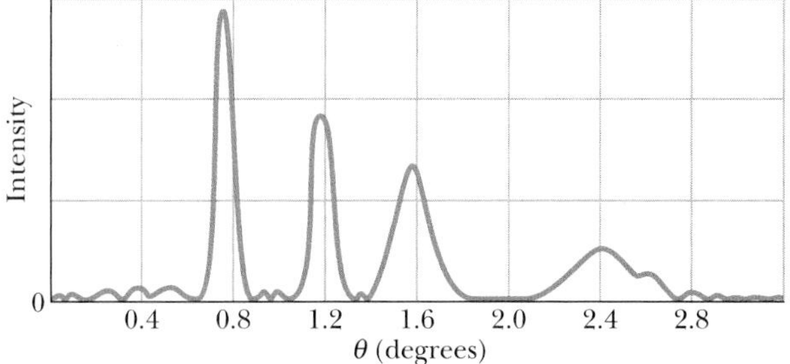

Fig. 37-38 Exercise 54.

55E. An x-ray beam of a certain wavelength is incident on a NaCl crystal, at 30.0° to a certain family of reflecting planes of spacing 39.8 pm. If the reflection from those planes is of the first order, what is the wavelength of the x rays?

56E. An x-ray beam of wavelength A undergoes first-order reflection from a crystal when its angle of incidence to a crystal face is 23°, and an x-ray beam of wavelength 97 pm undergoes third-order reflection when its angle of incidence to that face is 60°. Assuming that the two beams reflect from the same family of reflecting planes, find (a) the interplanar spacing and (b) the wavelength A.

57P. Prove that it is not possible to determine both wavelength of incident radiation and spacing of reflecting planes in a crystal by measuring the Bragg angles for several orders. ssm

58P. In Fig. 37-39, first-order reflection from the reflection planes shown occurs when an x-ray beam of wavelength 0.260 nm makes an angle of 63.8° with the top face of the crystal. What is the unit cell size a_0?

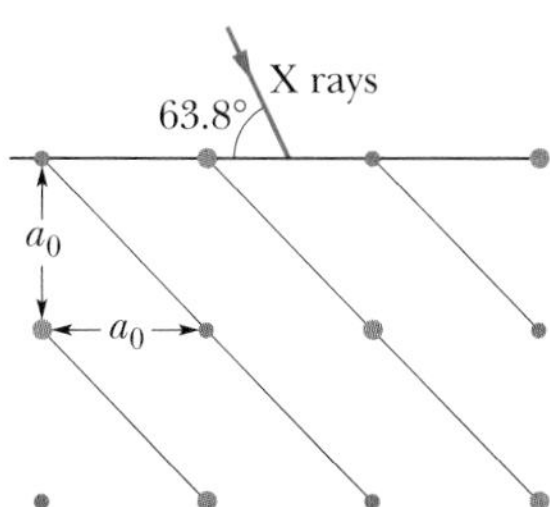

Fig. 37-39 Problem 58.

59P. Consider a two-dimensional square crystal structure, such as one side of the structure shown in Fig. 37-26a. One interplanar spacing of reflecting planes is the unit cell size a_0. (a) Calculate and sketch the next five smaller interplanar spacings. (b) Show that your results in (a) are consistent with the general formula

$$d = \frac{a_0}{\sqrt{h^2 + k^2}},$$

where h and k are relatively prime integers (they have no common factor other than unity). ssm www

60P. In Fig. 37-40, an x-ray beam of wavelengths from 95.0 pm to 140 pm is incident at 45° to a family of reflecting planes with spacing $d = 275$ pm. At which wavelengths will these planes produce intensity maxima in their reflections?

61P. In Fig. 37-40, let a beam of x rays of wavelength 0.125 nm be incident on an NaCl crystal at an angle of 45.0° to the top face of the crystal and a family of reflecting planes. Let the reflecting planes have separation $d = 0.252$ nm. Through what angles must the crystal be turned about an axis that is perpendicular to the plane of the page for these reflecting planes to give intensity maxima in their reflections? ssm

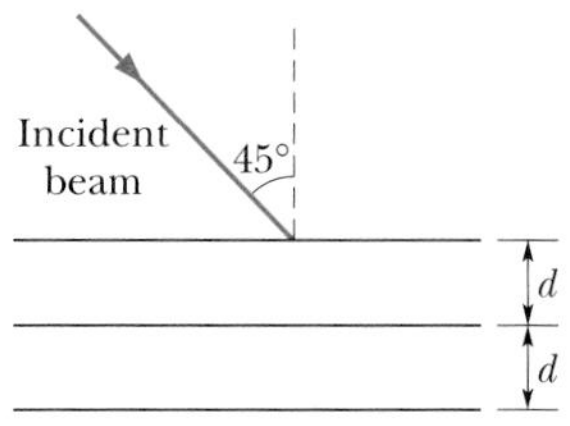

Fig. 37-40 Problems 60 and 61.

Additional Problems

62. In conventional television, signals are broadcast from towers to home receivers. Even when a receiver is not in direct view of a tower because of a hill or building, it can still intercept a signal if the signal diffracts enough around the obstacle, into the obstacle's "shadow region." Current television signals have a wavelength of about 50 cm, but future digital television signals that are to be transmitted from towers will have a wavelength of about 10 mm. (a) Will this change in wavelength increase or decrease the diffraction of the signals into the shadow regions of obstacles? Assume that a signal passes through an opening of 5.0 m width between two adjacent buildings. What is the angular spread of the central diffraction maximum (out to the first minima) for wavelengths of (b) 50 cm and (c) 10 mm?

63. Assume that Rayleigh's criterion gives the limit of resolution of an astronaut's eye looking down on Earth's surface from a typical space shuttle altitude of 400 km. (a) Under that idealized assumption, estimate the least linear width on Earth's surface that the astronaut can resolve. Take the astronaut's pupil diameter to be 5 mm and the wavelength of visible light to be 550 nm. (b) Can the astronaut resolve the Great Wall of China (Fig. 37-41), which is over 3000 km long, 5 to 10 m thick at its base, 4 m thick at its top, and 8 m in height? (c) Would the astronaut be able to resolve any unmistakable sign of intelligent life on Earth's surface?

Fig. 37-41 Problem 63. The Great Wall of China.

64. *Floaters.* As described in Section 37-1, the specks and hairlike structures that you sometimes see floating in your field of view are actually diffraction patterns cast on your retina. They are always present but are noticeable only if you view a featureless background, such as the sky or a brightly lit wall. The patterns are produced when light passes deposits in the gel (*vitreous humor*) that fills most of your eye. The light diffracts around these deposits and into their "shadow" region, much as light did in the Fresnel experiment of Section 37-1. You perceive not the deposits themselves, but their diffraction patterns on your retina. The patterns are called "floaters" because when you move your eye, the gel shimmies (somewhat like a gelatin dessert when shook), causing the diffraction patterns to move around on your retina. As you age, the gel can shimmy more because its attachment to the interior wall of the eye weakens; thus, with age, your floaters will be more noticeable (and a frequent reminder of diffraction physics).

To study the patterns, you can make them more distinct by looking through a pinhole, because the pinhole acts as a single point source of light (as in Fig. 36-5*c*). Then you can tell that floaters can be circular with a bright center and one or more dark rings (Fig. 37-42*a*); they can also be in the shape of a hair, with a bright interior and one or more dark bands running along the sides (Fig. 37-42*b*).

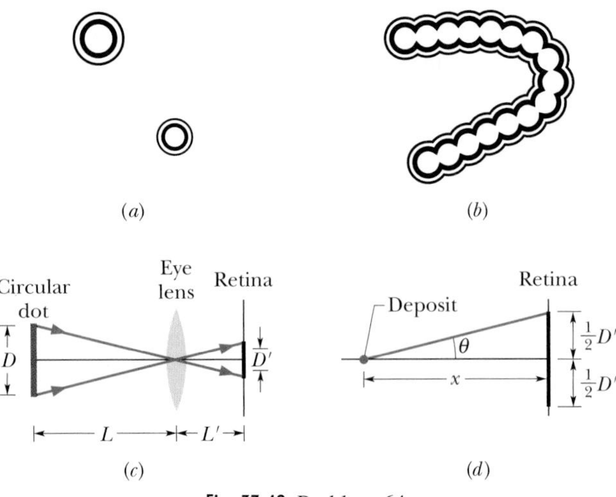

Fig. 37-42 Problem 64.

Estimate the size of the deposits in your eye's gel with the following procedure. Punch a pinhole through an opaque sheet of cardboard, about as distant from one edge as your nose is distant from the center of your eye. Draw a circular dot of diameter D = 2 mm on another sheet of cardboard. Position the pinhole immediately in front of your right eye and hold the circular dot in front of your left eye. Simultaneously look at the sky through the pinhole with your right eye and at the dot with your left eye. With a little practice, you can mentally merge the two views so that the dot mentally appears among the floaters.

Adjust the distance of the dot from your left eye until the dot's size approximates the size of one of the circular floaters. Have someone measure the distance L between the dot and your left eye (an estimate will do). Figure 37-42*c* shows a simplified schematic of your view of the dot: Rays pass straight through an eye lens to form the image of the dot on the retina, at a distance L' = 2.0 cm behind the lens. From this view of the dot and the value of L, find the diameter D' of the dot's image (and the floater's pattern) on the retina.

Let us approximate the deposit as spherical. Then its diffraction pattern is identical (except at the very center) to that of a circular aperture of the same diameter; that is, what you see from the deposit is identical (except at the very center) to the pattern shown in Fig. 37-9. Moreover, the location of the first minimum in the deposit's diffraction pattern is given by Eq. 37-12 ($\sin\theta = 1.22\lambda/d$). Assume the wavelength of the light is 550 nm. Use Fig. 37-42*d* to relate the angle θ to the radius $\frac{1}{2}D'$ of the dot's image on the retina and the distance x between the deposit and the retina. Let us assume that x ranges from about 1 mm to about 1.5 cm. What, then, is the approximate diameter of the deposits in the gel of your eye?

NEW PROBLEMS

N1. The pupil of a person's eye has a diameter of 5.00 mm. According to Rayleigh's criterion, what distance apart must two small objects be if their images are just barely resolved when they are 250 mm from the eye. Assume that they are illuminated with light of wavelength 500 nm?

N2. In the single-slit diffraction experiment of Fig. 37-4, let the wavelength of the light be 500 nm, the slit width be 6.00 μm, and the viewing screen be at distance 3.00 m. Let a y axis extend upward along the viewing screen, with its origin at the center of the diffraction pattern. Also let I_P represent the intensity of the diffracted light at point P at $y = 15.0$ cm. (a) What is the ratio of I_P to the intensity I_m at the center of the pattern? (b) Determine where point P is in the diffraction pattern by giving the maximum and minimum between which it lies, or the two minima between which it lies.

N3. Nuclear-pumped x-ray lasers are seen as a possible weapon to destroy ICBM booster rockets at ranges up to 2000 km. One limitation on such a device is the spreading of the beam due to diffraction, with resulting dilution of beam intensity. Consider such a laser operating at a wavelength of 1.40 nm. The element that emits light is the end of a wire with diameter 0.200 mm. (a) Calculate the diameter of the central beam at a target 2000 km away from the beam source. (b) By what factor is the beam intensity reduced in transit to the target? (The laser is fired from space, so that atmospheric absorption can be ignored.)

N4. Figure 37N-1 gives α versus the sine of the angle θ in a single-slit diffraction experiment using light of wavelength 610 nm. What are (a) the slit width, (b) the total number of diffraction minima in the pattern (count them on both sides of the center of the diffraction pattern), (c) the least angle for a minimum, and (d) the greatest angle for a minimum?

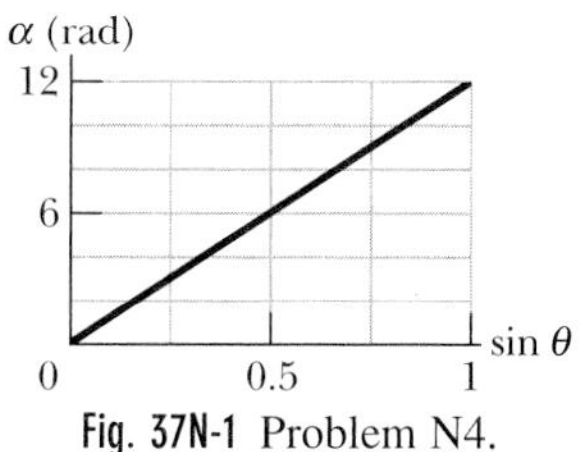

Fig. 37N-1 Problem N4.

N5. A spy satellite orbiting at 160 km above Earth's surface has a lens with a focal length of 3.6 m and can resolve objects on the ground as small as 30 cm. For example, it can easily measure the size of an aircraft's air intake port. What is the effective diameter of the lens as determined by diffraction consideration alone? Assume $\lambda = 550$ nm.

N6. In the two-slit interference experiment of Fig. 36-8, the slit widths are each 12.0 μm, their separation is 24 μm, the wavelength is 600 nm, and the viewing screen is at a distance of 4.00 m. Let I_P represent the intensity at point P on the screen, at height $y = 70.0$ cm. (a) What is the ratio of I_P to the intensity I_m at the center of the pattern? (b) Determine where P is in the two-slit interference pattern by giving the maximum or minimum on which it lies or the maximum and minimum between which it lies. (c) Next, for the diffraction that occurs, determine where point P is in the diffraction pattern by giving the minimum on which it lies or the two minima between which it lies.

N7. How many orders to one side of the central fringe can be produced of the entire visible spectrum (400–700 nm) by a grating of 500 lines/mm?

N8. Figure 37N-2 gives the parameter β of Eq. 37–17 versus the sine of the angle θ in a two-slit interference experiment using light of wavelength 435 nm. What are (a) the slit separation, (b) the total number of interference maxima (count them on both sides of the center of the interference pattern), (c) the least angle for a maxima, and (d) the greatest angle for a minimum? Assume that none of the interference maxima are completely eliminated by a diffraction minimum.

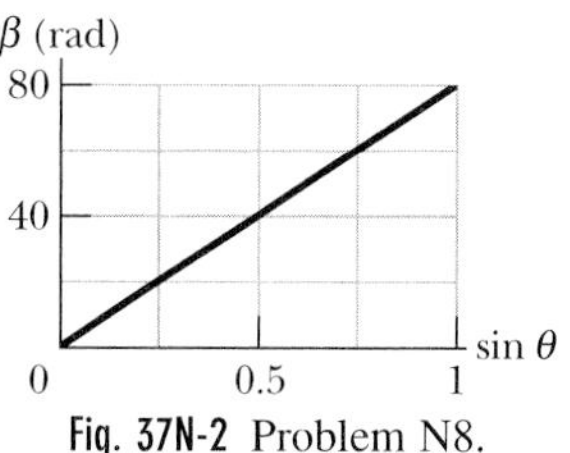

Fig. 37N-2 Problem N8.

N9. Show that a grating made up of alternately transparent and opaque strips of equal width eliminates all the even orders of maxima (except $m = 0$).

N10. A beam of light consists of two wavelengths, 590.159 nm and 590.220 nm, that are to be resolved with a diffraction grating. If the grating has lines across a width of 3.80 cm, what is the minimum number of lines required for the two wavelengths to be resolved in the second order?

N11. A diffraction grating 3.00 cm wide produces the second order at 33.0° with light of wavelength 600 nm. What is the total number of lines on the grating?

N12. A beam of light consisting of wavelengths from 460.0 nm to 640.0 nm is directed perpendicularly onto a diffraction grating with 160 lines/mm. (a) What is the lowest order that is overlapped by another order? (b) What is the highest order for which the complete wavelength range of the beam is present? In that highest order, at what angles does the light at wavelengths (c) 460.0 nm and (d) 640.0 nm appear? (e) What is the greatest angle at which light at wavelength 460.0 nm appears?

N13. Two emission lines have wavelengths λ and $\lambda + \Delta\lambda$, respectively, where $\Delta\lambda \ll \lambda$. Show that their angular separation $\Delta\theta$ in a grating spectrometer is given approximately by

$$\Delta\theta = \frac{\Delta\lambda}{\sqrt{(d/m)^2 - \lambda^2}}$$

where d is the slit separation and m is the order at which the lines are observed. Note that the angular separation is greater in the higher orders than in the lower orders.

N14. In a single-slit diffraction experiment, what must be the ratio of the slit width to the wavelength if the second diffraction minima are to occur at an angle of 37.0° from the center of the diffraction pattern on a viewing screen?

N15. An acoustic double-slit system (of slit separation d and slit width a) is driven by two loudspeakers as shown in Fig. 37N-3. By use of a variable delay line, the phase of one of the speakers may be varied relative to the other speaker. Describe in detail what changes occur in the double-slit diffraction pattern at large distances as the phase difference between the speakers is varied from zero to 2π. Take both interference and diffraction effects into account.

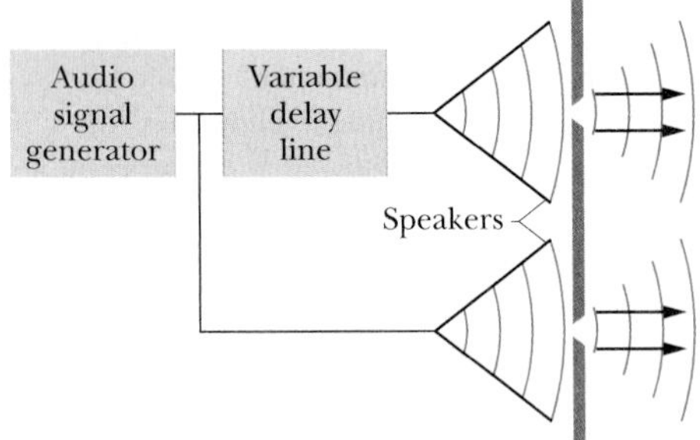

Fig. 37N-3 Problem N15.

N16. Two yellow flowers are separated by 60 cm along a line perpendicular to your line of sight of the flowers. How far are you from the flowers when they are at the limit of resolution according to the Rayleigh criterion? Assume the light from the flowers has a single wavelength of 550 nm and that your pupil has a diameter of 5.5 mm.

N17. White light (consisting of wavelengths from 400 nm to 700 nm) is normally incident on a grating. Show that, no matter what the value of the grating spacing d, the second order and third order overlap.

N18. In a two-slit interference experiment, what is the ratio of slit separation to slit width if there are 21 bright fringes within the central diffraction envelope and the diffraction minima at the edges of that envelope coincide with two-slit interference maxima?

N19. If first-order reflection occurs in a crystal at Bragg angle 3.4°, at what Bragg angle does second-order reflection occur from the same family of reflecting planes?

N20. If you look at something 40 m from you, what is the smallest length (perpendicular to your line of sight) that you can resolve, according to Rayleigh's criterion? Assume your pupil (opening) has a diameter of 4.00 mm, and use 500 nm as the wavelength of the light reaching you.

N21. In two-slit interference, if the slit separation is 14 μm and the slit widths are each 2.0 μm, then (a) how many two-slit maxima (bright bands of the two-slit pattern) are in the central diffraction envelope and (b) how many are in the first diffraction envelope to either side?

N22. A beam of light of a single wavelength is incident perpendicularly on a double-slit arrangement, as in Fig. 36-8. The slit widths are each 46 μm and the slit separation is 0.30 mm. How many complete bright fringes appear between the two first-order minima of the diffraction pattern?

N23. A beam of light with a narrow wavelength range centered on 450 nm is incident perpendicularly on a diffraction grating with a width of 1.80 cm and a line density of 1400 lines/cm across that width. What is the smallest wavelength difference that this grating can resolve in the light in the third order?

N24. If there are 17 bright fringes within the central diffraction envelope in a two-slit interference pattern, then what is the ratio of slit separation to the slit width?

N25. If we put $d = a$ in Fig. 37N-4, the two slits coalesce into a single slit of width $2a$. Show that Eq. 37-16 reduces to give the diffraction pattern for such a slit.

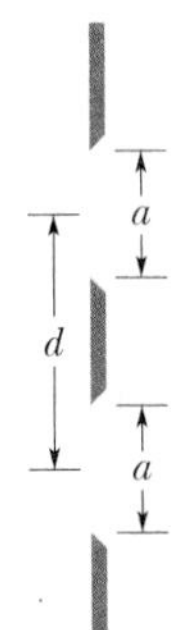

Fig. 37N-4 Problem N25.

N26. A diffraction grating has 8900 slits across 1.20 cm. If light with a wavelength of 500 nm is sent through it, how many orders (maxima) lie to one side of the central maximum?

N27. A single-slit diffraction experiment is set up with light of wavelength 420 nm, incident perpendicularly on a slit of width 5.10 μm. The viewing screen is 3.20 m distant. On the screen, what is the distance between the center of the diffraction pattern and the second diffraction minimum?

N28. A computer can be used to sum the phasors corresponding to Huygens' wavelets and so find a diffraction pattern. Suppose light with a wavelength of 500 nm is incident normally on a single slit with a width of 5.00×10^{-6} m. To approximate the diffraction pattern, sum the phasors corresponding to $N = 200$ wavelets spreading from uniformly distributed sources within the slit. The horizontal and vertical components of the resultant are proportional to

$$E_h = \sum_{i=1}^{N} \cos \phi_i \quad \text{and} \quad E_v = \sum_{i=1}^{N} \sin \phi_i,$$

respectively, where ϕ_i is the phase of wavelet i. The intensity ratio is $I/I_m = (E_h^2 + E_v^2)/N^2$. The factor $1/N^2$ assures that $I/I_m = 1$ when all the wavelets have the same phase. If you consider light that is diffracted at the angle θ to the straight-ahead direction, then you may take the phase of the first wavelet to be zero and the phase of each successive wavelet to be $(2\pi/\lambda)\Delta x \sin \theta$ greater than that of the preceding wavelet. Here Δx is the distance between wavelet sources; that is, $\Delta x = a/(N - 1)$, where a is the slit width. Use this technique to search for the diffraction angles corresponding to the first three secondary maxima and find the intensity ratios for those maxima.

38 Relativity

In modern long-range navigation, the precise location and speed of moving craft are continuously monitored and updated. A system of navigation satellites called NAVSTAR permits locations and speeds anywhere on Earth to be determined to within about 16 m and 2 cm/s. However, if relativity effects were not taken into account, speeds could not be determined any closer than about 20 cm/s, which is unacceptable for modern navigation systems.

How can something as abstract as Einstein's special theory of relativity be involved in something as practical as navigation?

The answer is in this chapter.

38-1 What Is Relativity All About?

One principal focus of **relativity** has to do with measurements of events (things that happen): where and when they happen, and by how much any two events are separated in space and in time. In addition, relativity has to do with transforming such measurements and others between reference frames that move relative to each other. (Hence the name *relativity.*) We discussed such matters in Sections 4-8 and 4-9.

Transformations and moving reference frames were well understood and quite routine to physicists in 1905. Then Albert Einstein (Fig. 38-1) published his **special theory of relativity.** The adjective *special* means that the theory deals only with **inertial reference frames,** which are frames in which Newton's laws are valid. This means that the frames do not accelerate; instead they can move only at constant velocities relative to one another. (Einstein's *general theory of relativity* treats the more challenging situation in which reference frames accelerate; in this chapter the term *relativity* implies only inertial reference frames.)

Starting with two deceivingly simple postulates, Einstein stunned the scientific world by showing that the old ideas about relativity were wrong, even though everyone was so accustomed to them that they seemed to be unquestionable common sense. This supposed common sense, however, was derived from experience only with things that move rather slowly. Einstein's relativity, which turns out to be correct for all possible speeds, predicted many effects that were, at first study, bizarre because no one had experienced them.

In particular, Einstein demonstrated that space and time are entangled; that is, the time between two events depends on how far apart they occur, and vice versa. Also, the entanglement is different for observers who move relative to each other. One result is that time does not pass at a fixed rate, as if it were ticked off with mechanical regularity on some master grandfather clock that controls the universe. Rather, that rate is adjustable: Relative motion can change the rate at which time passes. Prior to 1905, no one but a few daydreamers would have thought that. Now,

Fig. 38-1 Einstein in the early 1900s, at his desk at the patent office in Bern, Switzerland, where he was employed when he published his special theory of relativity.

engineers and scientists take it for granted because their experience with special relativity has reshaped their common sense.

Special relativity has the reputation of being difficult. It is not difficult mathematically, at least not here. However, it is difficult in that we must be very careful about *who* measures *what* about an event and just *how* that measurement is made—and it can be difficult because it can contradict experience.

38-2 The Postulates

We now examine the two postulates of relativity, on which Einstein's theory is based:

1. The Relativity Postulate: The laws of physics are the same for observers in all inertial reference frames. No frame is preferred.

Galileo assumed that the laws of *mechanics* were the same in all inertial reference frames. (Newton's first law of motion is one important consequence.) Einstein extended that idea to include *all* the laws of physics, especially electromagnetism and optics. This postulate does *not* say that the measured values of all physical quantities are the same for all inertial observers; most are not the same. It is the *laws of physics,* which relate these measurements to each other, that are the same.

2. The Speed of Light Postulate: The speed of light in vacuum has the same value c in all directions and in all inertial reference frames.

We can also phrase this postulate to say that there is in nature an *ultimate speed* c, the same in all directions and in all inertial reference frames. Light happens to travel at this ultimate speed, as do any massless particles (neutrinos might be an example). However, no entity that carries energy or information can exceed this limit. Moreover, no particle that does have mass can actually reach speed c, no matter how much or how long it is accelerated.

Both postulates have been exhaustively tested, and no exceptions have ever been found.

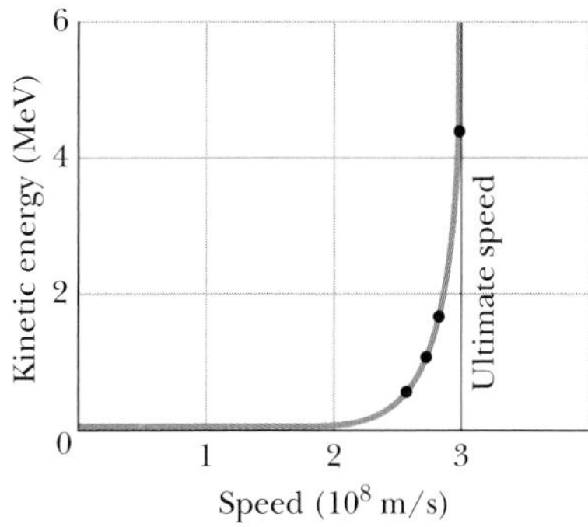

Fig. 38-2 The dots show measured values of the kinetic energy of an electron plotted against its measured speed. No matter how much energy is given to an electron (or to any other particle having mass), its speed can never equal or exceed the ultimate limiting speed c. (The plotted curve through the dots shows the predictions of Einstein's special theory of relativity.)

The Ultimate Speed

The existence of a limit to the speed of accelerated electrons was shown in a 1964 experiment of W. Bertozzi. He accelerated electrons to various measured speeds (see Fig. 38-2) and—by an independent method—also measured their kinetic energies. He found that as the force on a very fast electron is increased, the electron's measured kinetic energy increases toward very large values but its speed does not increase appreciably. Electrons have been accelerated to at least 0.999 999 999 95 times the speed of light but—close though it may be—that speed is still less than the ultimate speed c.

This ultimate speed has been defined to be exactly

$$c = 299\,792\,458 \text{ m/s}. \tag{38-1}$$

So far in this book we have (appropriately) approximated c as 3.0×10^8 m/s, but in this chapter we shall approximate it as 2.998×10^8 m/s. You might want to store the exact value in your calculator's memory (if it is not there already), to be called up when needed.

Testing the Speed of Light Postulate

If the speed of light is the same in all inertial reference frames, then the speed of light that is emitted by a moving source should be the same as the speed of light that is emitted by a source at rest in the laboratory. This claim has been tested directly, in an experiment of high precision. The "light source" was the *neutral pion* (symbol π^0), an unstable, short-lived particle that can be produced by collisions in a particle accelerator. It decays into two gamma rays by the process

$$\pi^0 \rightarrow \gamma + \gamma. \tag{38-2}$$

Gamma rays are part of the electromagnetic spectrum (at very high frequencies) and so obey the speed of light postulate, just as visible light does.

In a 1964 experiment, physicists at CERN, the European particle-physics laboratory near Geneva, generated a beam of pions moving at a speed of $0.999\,75c$ with respect to the laboratory. The experimenters then measured the speed of the gamma rays emitted from these very rapidly moving sources. They found that the speed of the light emitted by the pions was the same as would be measured if the pions were at rest in the laboratory.

38-3 Measuring an Event

An **event** is something that happens, to which an observer can assign three space coordinates and one time coordinate. Among many possible events are (1) the turning on or off of a tiny lightbulb, (2) the collision of two particles, (3) the passage of a pulse of light through a specified point, (4) an explosion, and (5) the coincidence of the hand of a clock with a marker on the rim of the clock. A certain observer, fixed in a certain inertial reference frame, might, for example, assign to an event A the coordinates given in Table 38-1. Because space and time are entangled with each other in relativity, we can describe these coordinates collectively as *spacetime* coordinates. The coordinate system itself is part of the reference frame of the observer.

A given event may be recorded by any number of observers, each in a different inertial reference frame. In general, different observers will assign different spacetime coordinates to the same event. Note that an event does not "belong" to a particular inertial reference frame. An event is just something that happens, and anyone in any reference frame may detect it and assign spacetime coordinates to it.

Making such an assignment can be complicated by a practical problem. For example, suppose a balloon bursts 1 km to your right while a firecracker pops 2 km to your left, both at 9:00 a.m. However, you do not detect either event precisely at 9:00 a.m. because light from the events has not yet reached you. Because light from the pop has farther to go, it arrives at your eyes later than does light from the balloon burst, and thus the pop will seem to have occurred later than the burst. To sort out the actual times and to assign 9:00 a.m. to both events, you must calculate the travel times of the light and then subtract them from the arrival times.

This procedure can be very messy in more challenging situations, and we need an easier procedure that automatically eliminates any concern about the travel time from an event to an observer. To set up such a procedure, we shall construct an imaginary array of measuring rods and clocks throughout the observer's inertial frame (the array moves rigidly with the observer). This construction may seem contrived, but it spares us much confusion and calculation and allows us to find the space coordinates, the time coordinate, and the spacetime coordinates, as follows.

TABLE 38-1 Record of Event A

Coordinate	Value
x	3.58 m
y	1.29 m
z	0 m
t	34.5 s

1. **The Space Coordinates.** We imagine the observer's coordinate system fitted with a close-packed, three-dimensional array of measuring rods, one set of rods parallel

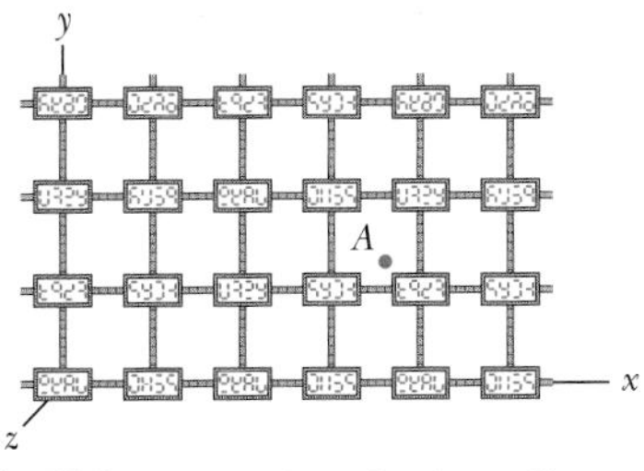

Fig. 38-3 One section of a three-dimensional array of clocks and measuring rods by which an observer can assign spacetime coordinates to an event, such as a flash of light at point A. The event's space coordinates are approximately $x = 3.7$ rod lengths, $y = 1.2$ rod lengths, and $z = 0$. The time coordinate is whatever time appears on the clock closest to A at the instant of the flash.

to each of the three coordinate axes. These rods provide a way to determine coordinates along the axes. Thus, if the event is, say, the turning on of a small lightbulb, the observer, in order to locate the position of the event, need only read the three space coordinates at the bulb's location.

2. **The Time Coordinate.** For the time coordinate, we imagine that every point of intersection in the array of measuring rods includes a tiny clock, which the observer can read by the light generated by the event. Figure 38-3 suggests one plane in the "jungle gym" of clocks and measuring rods that we have described.

   The array of clocks must be synchronized properly. It is not enough to assemble a set of identical clocks, set them all to the same time, and then move them to their assigned positions. We do not know, for example, whether moving the clocks will change their rates. (Actually, it will.) We must put the clocks in place and *then* synchronize them.

   If we had a method of transmitting signals at infinite speed, synchronization would be a simple matter. However, no known signal has this property. We therefore choose light (interpreted broadly to include the entire electromagnetic spectrum) to send out our synchronizing signals because, in vacuum, light travels at the greatest possible speed, the limiting speed c.

   Here is one of many ways in which an observer might synchronize an array of clocks using light signals: The observer enlists the help of a great number of temporary helpers, one for each clock. The observer then stands at a point selected as the origin and sends out a pulse of light when the origin clock reads $t = 0$. When the light pulse reaches the location of a helper, that helper sets the clock there to read $t = r/c$, where r is the distance between the helper and the origin. The clocks are then synchronized.

3. **The Spacetime Coordinates.** The observer can now assign spacetime coordinates to an event by simply recording the time on the clock nearest the event and the position as measured on the nearest measuring rods. If there are two events, the observer computes their separation in time as the difference of the times on clocks near each, and their separation in space from the differences of coordinates on rods near each. We thus avoid the practical problem of calculating the travel times of signals that reach the observer from the events.

38-4 The Relativity of Simultaneity

Suppose that one observer (Sam) notes that two independent events (event Red and event Blue) occur at the same time. Suppose also that another observer (Sally), who is moving at a constant velocity $\vec{v}$ with respect to Sam, also records these same two events. Will Sally also find that they occur at the same time?

The answer is that in general she will not:

> If two observers are in relative motion, they will not, in general, agree as to whether two events are simultaneous. If one observer finds them to be simultaneous, the other generally will not.

We cannot say that one observer is right and the other wrong. Their observations are equally valid, and there is no reason to favor one over the other.

The realization that two contradictory statements about the same natural event can be correct is a seemingly strange outcome of Einstein's theory. However, in Chapter 18 we discussed another way in which motion can affect measurement, without balking at the contradictory results: In the Doppler effect, the frequency an observer measures for a sound wave depends on the relative motion of the observer and the source. Thus, two observers moving relative to one another can measure different frequencies for the same wave—and both measurements are correct.

We conclude the following:

> Simultaneity is not an absolute concept but a relative one, depending on the motion of the observer.

If the relative speed of the observers is very much less than the speed of light, then measured departures from simultaneity are so small that they are not noticeable. Such is the case for all our experiences of daily living; that is why the relativity of simultaneity is unfamiliar.

A Closer Look at Simultaneity

Let us clarify the relativity of simultaneity with an example based on the postulates of relativity, no clocks or measuring rods being directly involved. Figure 38-4 shows two long spaceships (the SS *Sally* and the SS *Sam*), which can serve as inertial reference frames for observers Sally and Sam. The two observers are stationed at the midpoints of their ships. The ships are separating along a common x axis, the relative velocity of *Sally* with respect to *Sam* being $\vec{v}$. Figure 38-4*a* shows the ships with the two observer stations momentarily aligned opposite each other.

Two large meteorites strike the ships, one setting off a red flare (event Red) and the other a blue flare (event Blue), not necessarily simultaneously. Each event leaves a permanent mark on each ship, at positions R,R' and B,B'.

Let us suppose that the expanding wavefronts from the two events happen to reach Sam at the same time, as Fig. 38-4*c* shows. Let us further suppose that, after the episode, Sam finds, by measurements of the marks on his spaceship, that he was indeed stationed exactly halfway between the markers B and R on his ship when the two events occurred. He will say:

Sam: Light from event Red and light from event Blue reached me at the same time. From the marks on my spaceship, I find that I was standing halfway between the two sources when the light from them reached me. Therefore, event Red and event Blue were simultaneous events.

As study of Fig. 38-4 shows, Sally and the expanding wavefront from event Red are moving *toward* each other, while she and the expanding wavefront from event Blue

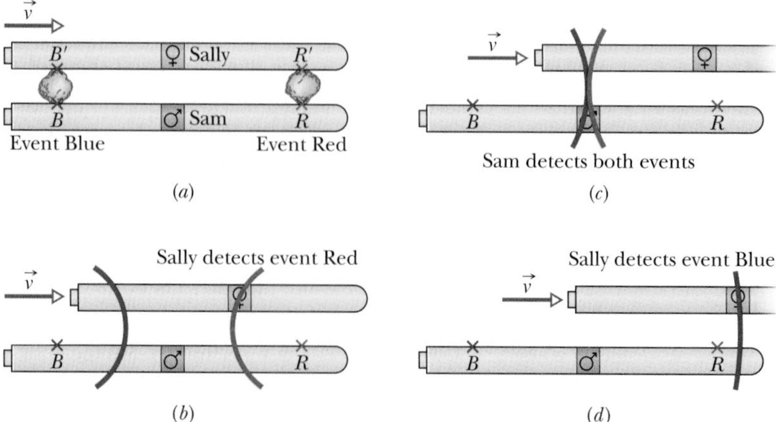

Fig. 38-4 The spaceships of Sally and Sam and the occurrences of events from Sam's view. Sally's ship moves rightward with velocity $\vec{v}$. (*a*) Event Red occurs at positions R,R' and event Blue occurs at positions B,B'; each event sends out a wave of light. (*b*) Sally detects the wave from event Red. (*c*) Sam simultaneously detects the waves from event Red and event Blue. (*d*) Sally detects the wave from event Blue.

are moving in the *same direction*. Thus, the wavefront from event Red will reach Sally *before* the wavefront from event Blue does. She will say:

Sally: Light from event Red reached me before light from event Blue did. From the marks on my spaceship, I found that I too was standing halfway between the two sources. Therefore, the events were *not* simultaneous; event Red occurred first, followed by event Blue.

These reports do not agree. Nevertheless, *both* observers are correct.

Note carefully that there is only one wavefront expanding from the site of each event and that *this wavefront travels with the same speed c in both reference frames,* exactly as the speed of light postulate requires.

It *might* have happened that the meteorites struck the ships in such a way that the two hits appeared to Sally to be simultaneous. If that had been the case, then Sam would have declared them not to be simultaneous.

38-5 The Relativity of Time

If observers who move relative to each other measure the time interval (or *temporal separation*) between two events, they generally will find different results. Why? Because the spatial separation of the events can affect the time intervals measured by the observers.

> The time interval between two events depends on how far apart they occur, in both space and time; that is, their spatial and temporal separations are entangled.

In this section we discuss this entanglement by means of an example; however, the example is restricted in a crucial way: *To one of two observers, the two events occur at the same location.* We shall not get to more general examples until Section 38-7.

Figure 38-5a shows the basics of an experiment Sally conducts while she and her equipment ride in a train moving with constant velocity $\vec{v}$ relative to a station.

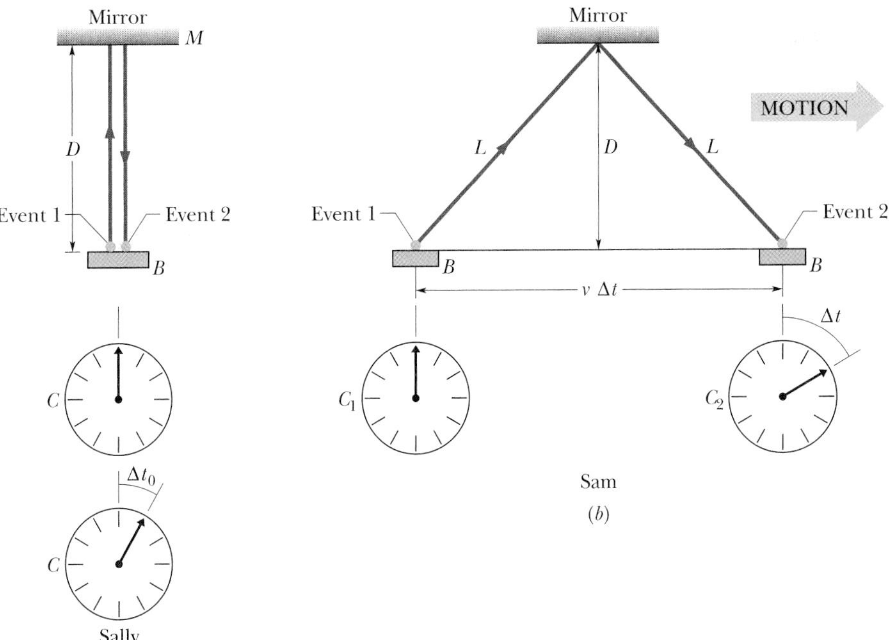

Fig. 38-5 (a) Sally, on the train, measures the time interval Δt_0 between events 1 and 2 using a single clock C on the train. That clock is shown twice: first for event 1 and then for event 2. (b) Sam, watching from the station as the events occur, requires two synchronized clocks, C_1 at event 1 and C_2 at event 2, to measure the time interval between the two events; his measured time interval is Δt.

A pulse of light leaves a light source B (event 1), travels vertically upward, is reflected vertically downward by a mirror, and then is detected back at the source (event 2). Sally measures a certain time interval Δt_0 between the two events, related to the distance D from source to mirror by

$$\Delta t_0 = \frac{2D}{c} \quad \text{(Sally).} \tag{38-3}$$

The two events occur at the same location in Sally's reference frame, and she needs only one clock C at that location to measure the time interval. Clock C is shown twice in Fig. 38-5*a*, at the beginning and end of the interval.

Consider now how these same two events are measured by Sam, who is standing on the station platform as the train passes. Because the equipment moves with the train during the travel time of the light, Sam sees the path of the light as shown in Fig. 38-5*b*. For him, the two events occur at different places in his reference frame, so to measure the time interval between events, Sam must use *two* synchronized clocks, C_1 and C_2, one at each event. According to Einstein's speed of light postulate, the light travels at the same speed c for Sam as for Sally. Now, however, the light travels distance $2L$ between events 1 and 2. The time interval measured by Sam between the two events is

$$\Delta t = \frac{2L}{c} \quad \text{(Sam),} \tag{38-4}$$

in which

$$L = \sqrt{(\tfrac{1}{2}v\,\Delta t)^2 + D^2}\,. \tag{38-5}$$

From Eq. 38-3, we can write this as

$$L = \sqrt{(\tfrac{1}{2}v\,\Delta t)^2 + (\tfrac{1}{2}c\,\Delta t_0)^2}\,. \tag{38-6}$$

If we eliminate L between Eqs. 38-4 and 38-6 and solve for Δt, we find

$$\Delta t = \frac{\Delta t_0}{\sqrt{1 - (v/c)^2}}\,. \tag{38-7}$$

Equation 38-7 tells us how Sam's measured interval Δt between the events compares with Sally's interval Δt_0. Because v must be less than c, the denominator in Eq. 38-7 must be less than unity. Thus, Δt must be greater than Δt_0: Sam measures a *greater* time interval between the two events than does Sally. Sam and Sally have measured the time interval between the *same* two events, but the relative motion between Sam and Sally made their measurements *different*. We conclude that relative motion can change the *rate* at which time passes between two events; the key to this effect is the fact that the speed of light is the same for both observers.

We distinguish between the measurements of Sam and Sally with the following terminology:

When two events occur at the same location in an inertial reference frame, the time interval between them, measured in that frame, is called the **proper time interval** or the **proper time.** Measurements of the same time interval from any other inertial reference frame are always greater.

Thus, Sally measures a proper time interval, and Sam measures a greater time interval. (The term *proper* is unfortunate in that it implies that any other measurement is improper or nonreal. That is just not so.) The amount by which a measured time interval is greater than the corresponding proper time interval is called **time dilation.** (To dilate is to expand or stretch; here the time interval is expanded or stretched.)

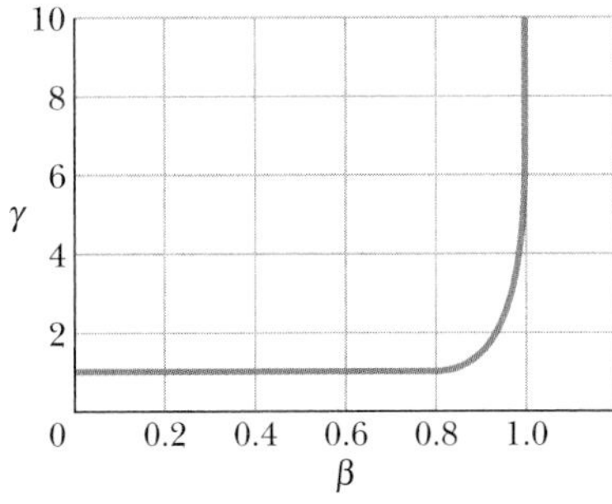

Fig. 38-6 A plot of the Lorentz factor γ as a function of the speed parameter $\beta\ (= v/c)$.

Often the dimensionless ratio v/c in Eq. 38-7 is replaced with β, called the **speed parameter,** and the dimensionless inverse square root in Eq. 38-7 is often replaced with γ, called the **Lorentz factor:**

$$\gamma = \frac{1}{\sqrt{1-\beta^2}} = \frac{1}{\sqrt{1-(v/c)^2}}. \tag{38-8}$$

With these replacements, we can rewrite Eq. 38-7 as

$$\Delta t = \gamma\,\Delta t_0 \qquad \text{(time dilation).} \tag{38-9}$$

The speed parameter β is always less than unity and, provided v is not zero, γ is always greater than unity. However, the difference between γ and 1 is not significant unless $v > 0.1c$. Thus, in general, "old relativity" works well enough for $v < 0.1c$, but we must use special relativity for greater values of v. As shown in Fig. 38-6, γ increases rapidly in magnitude as β approaches 1 (as v approaches c). Therefore, the greater the relative speed between Sally and Sam is, the greater will be the time interval measured by Sam, until at a great enough speed, the interval takes "forever."

You might wonder what Sally says about Sam's having measured a greater time interval than she did. His measurement comes as no surprise to her, because to her, he failed to synchronize his clocks C_1 and C_2 in spite of his insistence that he did. Recall that observers in relative motion generally do not agree about simultaneity. Here, Sam insists that his two clocks simultaneously read the same time when event 1 occurred. To Sally, however, Sam's clock C_2 was erroneously set ahead during the synchronization process. Thus, when Sam read the time of event 2 on it, to Sally he was reading off a time that was too large, and that is why the time interval he measured between the two events was greater than the interval she measured.

Two Tests of Time Dilation

1. **Microscopic Clocks.** Subatomic particles called *muons* are unstable; that is, when a muon is produced, it lasts for only a short time before it *decays* (transforms into particles of other types). The *lifetime* of a muon is the time interval between its production (event 1) and its decay (event 2). When muons are stationary and their lifetimes are measured with stationary clocks (say, in a laboratory), their average lifetime is 2.200 μs. This is a proper time interval because, for each muon, events 1 and 2 occur at the same location in the reference frame of the muon, namely at the muon itself. We can represent this proper time interval with Δt_0; moreover, we can call the reference frame in which it is measured the *rest frame* of the muon.

   If, instead, the muons are moving, say, through a laboratory, then measurements of their lifetimes made with the laboratory clocks should yield a greater average lifetime (a dilated average lifetime). To check this conclusion, measurements were made of the average lifetime of muons moving with a speed of $0.9994c$ relative to laboratory clocks. From Eq. 38-8, with $\beta = 0.9994$, the Lorentz factor for this speed is

$$\gamma = \frac{1}{\sqrt{1-\beta^2}} = \frac{1}{\sqrt{1-(0.9994)^2}} = 28.87.$$

   Equation 38-9 then yields, for the average dilated lifetime,

$$\Delta t = \gamma\,\Delta t_0 = (28.87)(2.200\ \mu\text{s}) = 63.51\ \mu\text{s}.$$

   The actual measured value matched this result within experimental error.

2. **Macroscopic Clocks.** In October 1977, Joseph Hafele and Richard Keating carried out what must have been a grueling experiment. They flew four portable atomic clocks twice around the world on commercial airlines, once in each direction. Their purpose was "to test Einstein's theory of relativity with macroscopic clocks." As we have just seen, the time dilation predictions of Einstein's theory have been confirmed on a microscopic scale, but there is great comfort in seeing a confirmation made with an actual clock. Such macroscopic measurements became possible only because of the very high precision of modern atomic clocks. Hafele and Keating verified the predictions of the theory to within 10%. (Einstein's *general* theory of relativity, which predicts that the rate at which time passes on a clock is influenced by the gravitational force on the clock, also plays a role in this experiment.)

A few years later, physicists at the University of Maryland carried out a similar experiment with improved precision. They flew an atomic clock round and round over Chesapeake Bay for flights lasting 15 h and succeeded in checking the time dilation prediction to better than 1%. Today, when atomic clocks are transported from one place to another for calibration or other purposes, the time dilation caused by their motion is always taken into account.

CHECKPOINT 1: Standing beside railroad tracks, we are suddenly startled by a relativistic boxcar traveling past us as shown in the figure. Inside, a well-equipped hobo fires a laser pulse from the front of the boxcar to its rear. (a) Is our measurement of the speed of the pulse greater than, less than, or the same as that measured by the hobo? (b) Is his measurement of the flight time of the pulse a proper time? (c) Are his measurement and our measurement of the flight time related by Eq. 38-9?

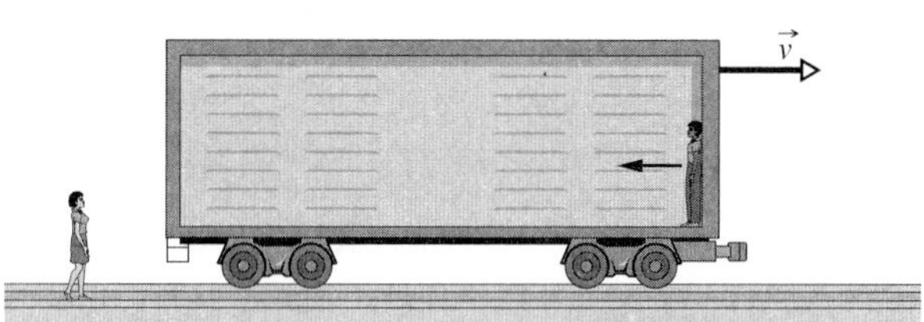

Sample Problem 38-1

Your starship passes Earth with a relative speed of $0.9990c$. After traveling 10.0 y (your time), you stop at lookout post LP13, turn, and then travel back to Earth with the same relative speed. The trip back takes another 10.0 y (your time). How long does the round trip take according to measurements made on Earth? (Neglect any effects due to the accelerations involved with stopping, turning, and getting back up to speed.)

SOLUTION: We begin by analyzing the outward trip only, with these **Key Ideas:**

1. This problem involves measurements made from two (inertial) reference frames, one attached to Earth and the other (your reference frame) attached to your ship.
2. The outward trip involves two events: the start of the trip at Earth and the end of the trip at LP13.
3. Your measurement of 10 y for the outward trip is the proper time Δt_0 between those two events, because the events occur at the same location in your reference frame, namely on your ship.
4. The Earth-frame measurement of the time interval Δt for the outward trip must be greater than Δt_0, according to Eq. 38-9 ($\Delta t = \gamma\, \Delta t_0$) for time dilation.

Using Eq. 38-8 to substitute for γ in Eq. 38-9, we find

$$\Delta t = \frac{\Delta t_0}{\sqrt{1 - (v/c)^2}}$$

$$= \frac{10.0 \text{ y}}{\sqrt{1 - (0.9990c/c)^2}} = (22.37)(10.0 \text{ y}) = 224 \text{ y}.$$

On the return trip, we have the same situation and the same data. Thus, the round trip requires 20 y of your time but

$$\Delta t_{\text{total}} = (2)(224 \text{ y}) = 448 \text{ y} \qquad \text{(Answer)}$$

of Earth time. In other words, you have aged 20 y while the Earth has aged 448 y. Although you cannot travel into the past (as far as we know), you can travel into the future of, say, Earth, by using high-speed relative motion to adjust the rate at which time passes.

Sample Problem 38-2

The elementary particle known as the *positive kaon* (K^+) has, on average, a lifetime of 0.1237 μs when stationary—that is, when the lifetime is measured in the rest frame of the kaon. If a positive kaon has a speed of $0.990c$ relative to a laboratory reference frame when it is produced, how far can it travel in that frame during its lifetime according to *classical physics* (which is a reasonable approximation for speeds much less than c) and according to special relativity (which is correct for all physically possible speeds)?

SOLUTION: We begin with these Key Ideas:

1. This problem involves measurements made from two (inertial) reference frames, one attached to the kaon and the other attached to the laboratory.
2. This problem also involves two events: the start of the kaon's travel (when the kaon is produced) and the end of that travel (at the end of the kaon's lifetime).
3. The distance traveled by the kaon between those two events is related to its speed v and the time interval for the travel by

$$v = \frac{\text{distance}}{\text{time interval}}. \tag{38-10}$$

With these ideas in mind, let us solve for the distance first with classical physics and then with special relativity.

Classical physics: In classical physics, we have this Key Idea: We would find the same distance and time interval (in Eq. 38-10) whether we measured them from the kaon frame or from the laboratory frame. Thus, we need not be careful about the frame in which the measurements are made. To find the kaon's travel distance d_{cp} according to classical physics, we first rewrite Eq. 38-10 as

$$d_{cp} = v\,\Delta t, \tag{38-11}$$

where Δt is the time interval between the two events in either frame. Then, substituting $0.990c$ for v and 0.1237 μs for Δt in Eq. 38-11, we find

$$\begin{aligned} d_{cp} &= (0.990c)\,\Delta t \\ &= (0.990)(2.998 \times 10^8 \text{ m/s})(0.1237 \times 10^{-6} \text{ s}) \\ &= 36.7 \text{ m.} \qquad \text{(Answer)} \end{aligned}$$

This is how far the kaon would travel if classical physics were correct at speeds close to c.

Special relativity: In special relativity we have this Key Idea: We must be very careful that both the distance and the time interval in Eq. 38-10 are measured in the *same* reference frame—especially when the speed is close to c, as here. Thus, to find the actual travel distance d_{sr} of the kaon *as measured from the laboratory frame* and according to special relativity, we rewrite Eq. 38-10 as

$$d_{sr} = v\,\Delta t, \tag{38-12}$$

where Δt is the time interval between the two events *as measured from the laboratory frame*.

Before we can evaluate d_{sr} in Eq. 38-12, we must find Δt by using this Key Idea: the 0.1237 μs time interval is a proper time, because the two events occur at the same location in the kaon frame—namely, at the kaon itself. Therefore, let Δt_0 represent this proper time interval. Then we can use Eq. 38-9 ($\Delta t = \gamma\,\Delta t_0$) for time dilation to find the time interval Δt as measured from the laboratory frame. Using Eq. 38-8 to substitute for γ in Eq. 38-9 leads to

$$\Delta t = \frac{\Delta t_0}{\sqrt{1-(v/c)^2}} = \frac{0.1237 \times 10^{-6} \text{ s}}{\sqrt{1-(0.990c/c)^2}} = 8.769 \times 10^{-7} \text{ s.}$$

This is about seven times longer than the kaon's proper lifetime. That is, the kaon's lifetime is about seven times longer in the laboratory frame than in its own frame—the kaon's lifetime is dilated. We can now evaluate Eq. 38-12 for the travel distance d_{sr} in the laboratory frame as

$$\begin{aligned} d_{sr} &= v\,\Delta t = (0.990c)\,\Delta t \\ &= (0.990)(2.998 \times 10^8 \text{ m/s})(8.769 \times 10^{-7} \text{ s}) \\ &= 260 \text{ m.} \qquad \text{(Answer)} \end{aligned}$$

This is about seven times d_{cp}. Experiments like the one outlined here, which verify special relativity, became routine in physics laboratories decades ago. The engineering design and the construction of any scientific or medical facility that employs high-speed particles must take relativity into account.

38-6 The Relativity of Length

If you want to measure the length of a rod that is at rest with respect to you, you can—at your leisure—note the positions of its end points on a long stationary scale and subtract one reading from the other. If the rod is moving, however, you must note the positions of the end points *simultaneously* (in your reference frame) or your measurement cannot be called a length. Figure 38-7 suggests the difficulty of trying to measure the length of a moving penguin by locating its front and back at different times. Because simultaneity is relative and it enters into length measurements, length should also be a relative quantity. It is.

Let L_0 be the length of a rod that you measure when the rod is stationary (meaning you and it are in the same reference frame, the rod's rest frame). If, instead,

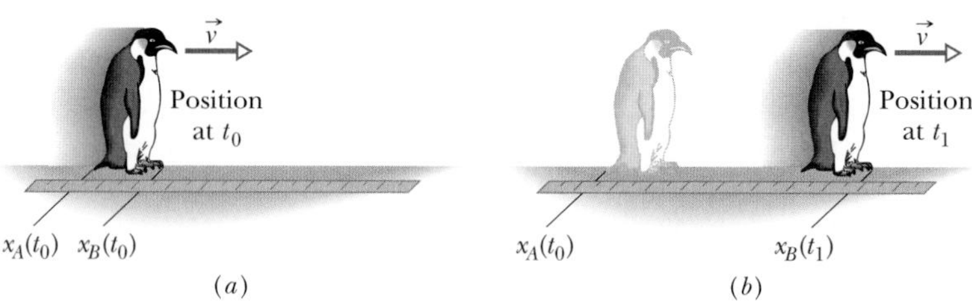

Fig. 38-7 If you want to measure the front-to-back length of a penguin while it is moving, you must mark the positions of its front and back simultaneously (in your reference frame), as in (*a*), rather than at different times, as in (*b*).

there is relative motion at speed v between you and the rod *along the length of the rod,* then with simultaneous measurement you obtain a length L given by

$$L = L_0\sqrt{1-\beta^2} = \frac{L_0}{\gamma} \quad \text{(length contraction).} \tag{38-13}$$

Because the Lorentz factor γ is always greater than unity if there is relative motion, L is less than L_0. The relative motion causes a *length contraction,* and L is called a *contracted length.* Because γ increases with speed v, the length contraction also increases with v.

The length L_0 of an object measured in the rest frame of the object is its **proper length** or **rest length.** Measurements of the length from any reference frame that is in relative motion parallel to that length are always less than the proper length.

Be careful: Length contraction occurs only along the direction of relative motion. Also, the length that is measured does not have to be that of an object like a rod or a circle. Instead, it can be the length (or distance) between two objects in the same rest frame—for example, the Sun and a nearby star (which are, at least approximately, at rest relative to each other).

Does a moving object *really* shrink? Reality is based on observations and measurements; if the results are always consistent and if no error can be determined, then what is observed and measured is real. In that sense, the object really does shrink. However, a more precise statement is that the object *is really measured* to shrink—motion affects that measurement and thus reality.

When you measure a contracted length for, say, a rod, what does an observer moving with the rod say of your measurement? To that observer, you did not locate the two ends of the rod simultaneously. (Recall that observers in motion relative to each other do not agree about simultaneity.) To the observer, you first located the rod's front end and then, slightly later, its rear end, and that is why you measured a length that is less than the proper length.

Proof of Eq. 38-13

Length contraction is a direct consequence of time dilation. Consider once more our two observers. This time, both Sally, seated on a train moving through a station, and Sam, again on the station platform, want to measure the length of the platform. Sam, using a tape measure, finds the length to be L_0, a proper length because the platform is at rest with respect to him. Sam also notes that Sally, on the train, moves through this length in a time $\Delta t = L_0/v$, where v is the speed of the train; that is,

$$L_0 = v\,\Delta t \quad \text{(Sam).} \tag{38-14}$$

This time interval Δt is not a proper time interval because the two events that define it (Sally passes the back of the platform and Sally passes the front of the platform) occur at two different places and Sam must use two synchronized clocks to measure the time interval Δt.

For Sally, however, the platform is moving past her. She finds that the two events measured by Sam occur *at the same place* in her reference frame. She can time them with a single stationary clock, so the interval Δt_0 that she measures is a proper time interval. To her, the length L of the platform is given by

$$L = v\,\Delta t_0 \qquad \text{(Sally)}. \tag{38-15}$$

If we divide Eq. 38-15 by Eq. 38-14 and apply Eq. 38-9, the time dilation equation, we have

$$\frac{L}{L_0} = \frac{v\,\Delta t_0}{v\,\Delta t} = \frac{1}{\gamma},$$

or

$$L = \frac{L_0}{\gamma}, \tag{38-16}$$

which is Eq. 38-13, the length contraction equation.

Sample Problem 38-3

In Fig. 38-8, Sally (at point A) and Sam's spaceship (of proper length $L_0 = 230$ m) pass each other with constant relative speed v. Sally measures a time interval of 3.57 μs for the ship to pass her (from the passage of point B to the passage of point C). In terms of c, what is the relative speed v between Sally and the ship?

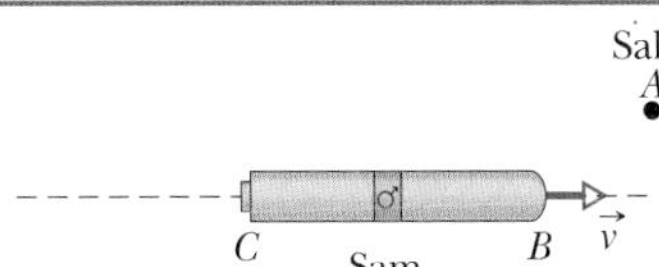

Fig. 38-8 Sample Problem 38-3. Sally, at point A, measures the time a spaceship takes to pass her.

SOLUTION: Let us assume that speed v is near c. Then we begin with these Key Ideas:

1. This problem involves measurements made from two (inertial) reference frames, one attached to Sally and the other attached to Sam and his spaceship.
2. This problem also involves two events: the first is the passage of point B and the second is the passage of point C.
3. From either reference frame, the other reference frame passes at speed v and moves a certain distance in the time interval between the two events:

$$v = \frac{\text{distance}}{\text{time interval}}. \tag{38-17}$$

Because speed v is assumed to be near the speed of light, we must be careful that the distance and the time interval in Eq. 38-17 are measured in the *same* reference frame.

We are free to use either frame for the measurements. Because we know that the time interval Δt between the two events measured from Sally's frame is 3.57 μs, let us also use the distance L between the two events measured from her frame. Equation 38-17 then becomes

$$v = \frac{L}{\Delta t}. \tag{38-18}$$

We do not know L but we can relate it to the given L_0 with this additional Key Idea: The distance between the two events as measured from Sam's frame is the ship's proper length L_0. Thus, the distance L measured from Sally's frame must be less than L_0, as given by Eq. 38-13 ($L = L_0/\gamma$) for length contraction. Substituting L_0/γ for L in Eq. 38-18 and then substituting Eq. 38-8 for γ, we find

$$v = \frac{L_0/\gamma}{\Delta t} = \frac{L_0\sqrt{(1-(v/c)^2}}{\Delta t}.$$

Solving this equation for v leads us to

$$v = \frac{L_0 c}{\sqrt{(c\,\Delta t)^2 + L_0^2}}$$

$$= \frac{(230\text{ m})c}{\sqrt{(2.998\times10^8\text{ m/s})^2(3.57\times10^{-6}\text{ s})^2 + (230\text{ m})^2}}$$

$$= 0.210c. \qquad \text{(Answer)}$$

Thus, the relative speed between Sally and the ship is 21% of the speed of light. Note that only the relative motion of Sally and Sam matters here; whether either is stationary relative to, say, a space station is irrelevant. In Fig. 38-8 we took Sally to be stationary, but we could instead have taken the ship to be stationary, with Sally moving past it. Nothing would have changed in our result.

✔CHECKPOINT 2: In this sample problem, Sally measures the passage time of the ship. If Sam does also, (a) which measurement, if either, is a proper time and (b) which measurement is smaller?

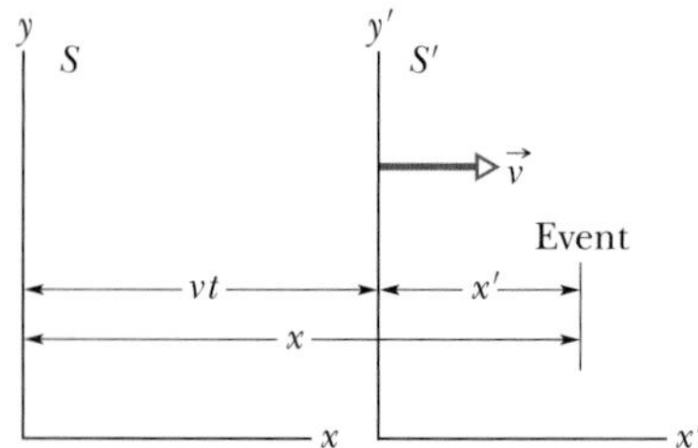

Fig. 38-9 Two inertial reference frames: frame S' has velocity $\vec{v}$ relative to frame S.

38-7 The Lorentz Transformation

Figure 38-9 shows inertial reference frame S' moving with speed v relative to frame S, in the common positive direction of their horizontal axes (marked x and x'). An observer in S reports spacetime coordinates x, y, z, t for an event, and an observer in S' reports x', y', z', t' for the same event. How are these sets of numbers related?

We claim at once (although it requires proof) that the y and z coordinates, which are perpendicular to the motion, are not affected by the motion; that is, $y = y'$ and $z = z'$. Our interest then reduces to the relation between x and x' and between t and t'.

The Galilean Transformation Equations

Prior to Einstein's publication of his special theory of relativity, the four coordinates of interest were assumed to be related by the *Galilean transformation equations:*

$$\begin{aligned} x' &= x - vt \\ t' &= t \end{aligned} \qquad \text{(Galilean transformation equations; approximately valid at low speeds).} \qquad (38\text{-}19)$$

(These equations are written with the assumption that $t = t' = 0$ when the origins of S and S' coincide.) You can verify the first equation with Fig. 38-9. The second equation effectively claims that time passes at the same rate for observers in both reference frames. That would have been so obviously true to a scientist prior to Einstein that it would not even have been mentioned. When speed v is small compared to c, Eqs. 38-19 generally work well.

The Lorentz Transformation Equations

We state without proof that the correct transformation equations, which remain valid for all speeds up to the speed of light, can be derived from the postulates of relativity. The results, called the **Lorentz transformation equations*** or sometimes (more loosely) just the Lorentz transformations, are

$$\begin{aligned} x' &= \gamma(x - vt), \\ y' &= y, \\ z' &= z, \\ t' &= \gamma(t - vx/c^2) \end{aligned} \qquad \text{(Lorentz transformation equations; valid at all physically possible speeds).} \qquad (38\text{-}20)$$

(The equations are written with the assumption that $t = t' = 0$ when the origins of S and S' coincide.) Note that the spatial values x and the temporal values t are bound together in the first and last equations. This entanglement of space and time was a prime message of Einstein's theory, a message that was long rejected by many of his contemporaries.

It is a formal requirement of relativistic equations that they should reduce to familiar classical equations if we let c approach infinity. That is, if the speed of light were infinitely great, *all* finite speeds would be "low" and classical equations would

*You may wonder why we do not call these the *Einstein transformation equations* (and why not the *Einstein factor* for γ). H. A. Lorentz actually derived these equations before Einstein did, but as the great Dutch physicist graciously conceded, he did not take the further bold step of interpreting these equations as describing the true nature of space and time. It is this interpretation, first made by Einstein, that is at the heart of relativity.

TABLE 38-2 **The Lorentz Transformation Equations for Pairs of Events**

1. $\Delta x = \gamma(\Delta x' + v\,\Delta t')$	1′. $\Delta x' = \gamma(\Delta x - v\,\Delta t)$
2. $\Delta t = \gamma(\Delta t' + v\,\Delta x'/c^2)$	2′. $\Delta t' = \gamma(\Delta t - v\,\Delta x/c^2)$

$$\gamma = \frac{1}{\sqrt{1-(v/c)^2}} = \frac{1}{\sqrt{1-\beta^2}}$$

Frame S' moves at velocity v relative to frame S.

never fail. If we let $c \to \infty$ in Eqs. 38-20, $\gamma \to 1$ and these equations reduce—as we expect—to the Galilean equations (Eqs. 38-19). You should check this.

Equations 38-20 are written in a form that is useful if we are given x and t and wish to find x' and t'. We may wish to go the other way, however. In that case we simply solve Eqs. 38-20 for x and t, obtaining

$$x = \gamma(x' + vt') \quad \text{and} \quad t = \gamma(t' + vx'/c^2). \tag{38-21}$$

Comparison shows that, starting from either Eqs. 38-20 or Eqs. 38-21, you can find the other set by interchanging primed and unprimed quantities and reversing the sign of the relative velocity v.

Equations 38-20 and 38-21 relate the coordinates of a single event as seen by two observers. Sometimes we want to know not the coordinates of a single event but the differences between coordinates for a pair of events. That is, if we label our events 1 and 2, we may want to relate

$$\Delta x = x_2 - x_1 \quad \text{and} \quad \Delta t = t_2 - t_1,$$

as measured by an observer in S, and

$$\Delta x' = x_2' - x_1' \quad \text{and} \quad \Delta t' = t_2' - t_1',$$

as measured by an observer in S'.

Table 38-2 displays the Lorentz equations in difference form, suitable for analyzing pairs of events. The equations in the table were derived by simply substituting differences (such as Δx and $\Delta x'$) for the four variables in Eqs. 38-20 and 38-21.

Be careful: When substituting values for these differences, you must be consistent and not mix the values for the first event with those for the second event. Also, if, say, Δx is a negative quantity, you must be certain to include the minus sign in a substitution.

CHECKPOINT 3: The figure here shows three situations in which a blue reference frame and a green reference frame are in relative motion along the common direction of their x and x' axes, as indicated by the velocity vector attached to one of the frames. For each situation, if we choose the blue frame to be stationary, then is v in the equations of Table 38-2 a positive or negative quantity?

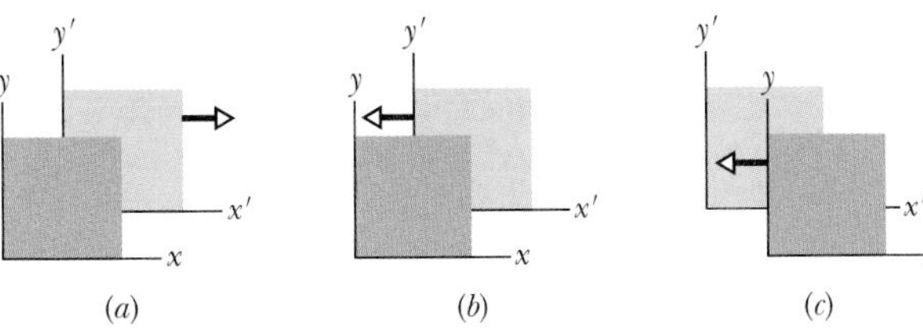

38-8 Some Consequences of the Lorentz Equations

Here we use the transformation equations of Table 38-2 to affirm some of the conclusions that we reached earlier by arguments based directly on the postulates.

Simultaneity

Consider Eq. 2 of Table 38-2,

$$\Delta t = \gamma\left(\Delta t' + \frac{v\,\Delta x'}{c^2}\right). \tag{38-22}$$

If two events occur at different places in reference frame S' of Fig. 38-9, then $\Delta x'$ in this equation is not zero. It follows that even if the events are simultaneous in S' (so $\Delta t' = 0$), they will not be simultaneous in frame S. (This is in accord with our conclusion in Section 38-4.) The time interval between the events in S will be

$$\Delta t = \gamma \frac{v\,\Delta x'}{c^2} \qquad \text{(simultaneous events in } S'\text{)}.$$

Time Dilation

Suppose now that two events occur at the same place in S' (so $\Delta x' = 0$) but at different times (so $\Delta t' \neq 0$). Equation 38-22 then reduces to

$$\Delta t = \gamma\,\Delta t' \qquad \text{(events in same place in } S'\text{)}. \tag{38-23}$$

This confirms time dilation. Because the two events occur at the same place in S', the time interval $\Delta t'$ between them can be measured with a single clock, located at that place. Under these conditions, the measured interval is a proper time interval, and we can label it Δt_0. Thus, Eq. 38-23 becomes

$$\Delta t = \gamma\,\Delta t_0 \qquad \text{(time dilation)},$$

which is exactly Eq. 38-9, the time dilation equation.

Length Contraction

Consider Eq. 1$'$ of Table 38-2,

$$\Delta x' = \gamma(\Delta x - v\,\Delta t). \tag{38-24}$$

If a rod lies parallel to the x and x' axes of Fig. 38-9 and is at rest in reference frame S', an observer in S' can measure its length at leisure. One way to do so is by subtracting the coordinates of the end points of the rod. The value of $\Delta x'$ that is obtained will be the proper length L_0 of the rod.

Suppose the rod is moving in frame S. This means that Δx can be identified as the length L of the rod in frame S only if the coordinates of the rod's end points are measured *simultaneously*—that is, if $\Delta t = 0$. If we put $\Delta x' = L_0$, $\Delta x = L$, and $\Delta t = 0$ in Eq. 38-24, we find

$$L = \frac{L_0}{\gamma} \qquad \text{(length contraction)}, \tag{38-25}$$

which is exactly Eq. 38-13, the length contraction equation.

Sample Problem 38-4

An Earth starship has been sent to check an Earth outpost on the planet P1407, whose moon houses a battle group of the often hostile Reptulians. As the ship follows a straight-line course first past the planet and then past the moon, it detects a high-energy microwave burst at the Reptulian moon base and then, 1.10 s later, an explosion at the Earth outpost, which is 4.00×10^8 m from the Reptulian base as measured from the ship's reference frame. The Reptulians have obviously attacked the Earth outpost, so the starship begins to prepare for a confrontation with them.

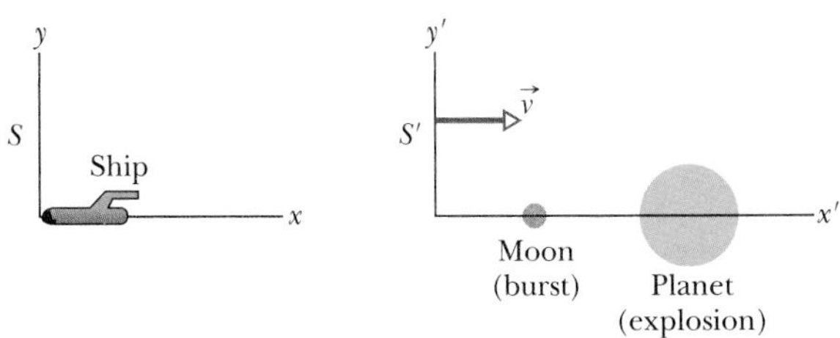

Fig. 38-10 Sample Problem 38-4. A planet and its moon in reference frame S' move rightward with speed v relative to a starship in reference frame S.

(a) The speed of the ship relative to the planet and its moon is $0.980c$. What are the distance and time interval between the burst and the explosion as measured in the planet–moon inertial frame (and thus according to the occupants of the stations)?

SOLUTION: We begin with these Key Ideas:

1. This problem involves measurements made from two reference frames, the planet–moon frame and the starship frame.
2. This problem involves two events: the burst and the explosion.
3. We need to transform the given data about the time and distance between the two events as measured in the starship frame to the corresponding data as measured in the planet–moon frame.

Before we get to the transformation, we need to carefully choose our notation. We begin with a sketch of the situation as shown in Fig. 38-10. There, we have chosen the ship's frame S to be stationary and the planet–moon frame S' to be moving with positive velocity (rightward). (This is an arbitrary choice; we could, instead, have chosen the planet–moon frame to be stationary. Then we would redraw $\vec{v}$ in Fig. 38-10 as being attached to the S frame and indicating leftward motion; v would then be a negative quantity. The results would be the same.) Let subscripts e and b represent the explosion and burst, respectively. Then the given data, all in the unprimed (starship) reference frame, are

$$\Delta x = x_e - x_b = +4.00 \times 10^8 \text{ m}$$

and

$$\Delta t = t_e - t_b = +1.10 \text{ s}.$$

Here, Δx is a positive quantity because in Fig. 38-10, the coordinate x_e for the explosion is greater than the coordinate x_b for the burst; Δt is also a positive quantity because the time t_e of the explosion is greater (later) than the time t_b of the burst.

We seek $\Delta x'$ and $\Delta t'$, which we shall get by transforming the given S-frame data to the planet–moon frame S'. Because we are considering a pair of events, we choose transformation equations from Table 38-2, namely Eqs. 1′ and 2′:

$$\Delta x' = \gamma(\Delta x - v\,\Delta t) \quad (38\text{-}26)$$

and

$$\Delta t' = \gamma\left(\Delta t - \frac{v\,\Delta x}{c^2}\right). \quad (38\text{-}27)$$

Here, $v = +0.980c$, and the Lorentz factor is

$$\gamma = \frac{1}{\sqrt{1-(v/c)^2}} = \frac{1}{\sqrt{1-(+0.980c/c)^2}} = 5.0252.$$

Equation 38-26 then becomes

$$\begin{aligned}\Delta x' &= (5.0252)\\ &\quad \times [4.00 \times 10^8 \text{ m} - (+0.980)(2.998 \times 10^8 \text{ m/s})(1.10 \text{ s})]\\ &= 3.86 \times 10^8 \text{ m}, \qquad \text{(Answer)}\end{aligned}$$

and Eq. 38-27 becomes

$$\begin{aligned}\Delta t' &= (5.0252)\\ &\quad \times \left[(1.10 \text{ s}) - \frac{(+0.980)(2.998 \times 10^8 \text{ m/s})(4.00 \times 10^8 \text{ m})}{(2.998 \times 10^8 \text{ m/s})^2}\right]\\ &= -1.04 \text{ s}. \qquad \text{(Answer)}\end{aligned}$$

(b) What is the meaning of the minus sign in the value for $\Delta t'$?

SOLUTION: The Key Idea here is to be consistent with the notation we set up in part (a). Recall how we originally defined the time interval between burst and explosion: $\Delta t = t_e - t_b = +1.10$ s. To be consistent with that choice of notation, our definition of $\Delta t'$ must be $t'_e - t'_b$; thus, we have found that

$$\Delta t' = t'_e - t'_b = -1.04 \text{ s}.$$

The minus sign here tells us that $t'_b > t'_e$; that is, in the planet–moon reference frame, the burst occurred 1.04 s *after* the explosion, not 1.10 s *before* the explosion as detected in the ship frame.

(c) Did the burst cause the explosion, or vice versa?

SOLUTION: The sequence of events measured in the planet–moon reference frame is the reverse of that measured in the ship frame. The Key Idea here is that in either situation, if there is a causal relationship between the two events, information must travel from the location of one event to the location of the other to cause it. Let us check the required speed of the information. In the ship frame, this speed is

$$v_{\text{info}} = \frac{\Delta x}{\Delta t} = \frac{4.00 \times 10^8 \text{ m}}{1.10 \text{ s}} = 3.64 \times 10^8 \text{ m/s},$$

but that speed is impossible because it exceeds c. In the planet–moon frame, the speed comes out to be 3.70×10^8 m/s, also impossible. Therefore, neither event could possibly have caused the other event; that is, they are *unrelated* events. Thus, the starship should not confront the Reptulians.

38-9 The Relativity of Velocities

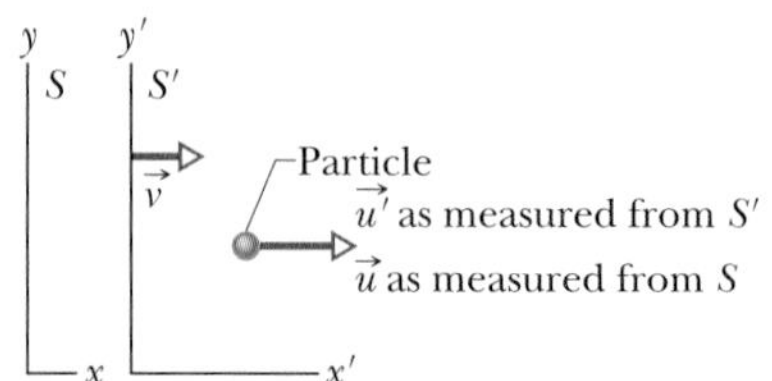

Fig. 38-11 Reference frame S' moves with velocity $\vec{v}$ relative to frame S. A particle has velocity $\vec{u}'$ relative to reference frame S' and velocity $\vec{u}$ relative to reference frame S.

Here we wish to use the Lorentz transformation equations to compare the velocities that two observers in different inertial reference frames S and S' would measure for the same moving particle. Let S' move with velocity v relative to S.

Suppose that the particle, moving with constant velocity parallel to the x and x' axes in Fig. 38-11, sends out two signals as it moves. Each observer measures the space interval and the time interval between these two events. These four measurements are related by Eqs. 1 and 2 of Table 38-2,

$$\Delta x = \gamma(\Delta x' + v\,\Delta t')$$

and

$$\Delta t = \gamma\left(\Delta t' + \frac{v\,\Delta x'}{c^2}\right).$$

If we divide the first of these equations by the second, we find

$$\frac{\Delta x}{\Delta t} = \frac{\Delta x' + v\,\Delta t'}{\Delta t' + v\,\Delta x'/c^2}.$$

Dividing the numerator and denominator of the right side by $\Delta t'$, we find

$$\frac{\Delta x}{\Delta t} = \frac{\Delta x'/\Delta t' + v}{1 + v(\Delta x'/\Delta t')/c^2}.$$

However, in the differential limit, $\Delta x/\Delta t$ is u, the velocity of the particle as measured in S, and $\Delta x'/\Delta t'$ is u', the velocity of the particle as measured in S'. Then we have, finally,

$$u = \frac{u' + v}{1 + u'v/c^2} \quad \text{(relativistic velocity transformation)} \tag{38-28}$$

as the relativistic velocity transformation equation. This equation reduces to the classical, or Galilean, velocity transformation equation,

$$u = u' + v \quad \text{(classical velocity transformation)}, \tag{38-29}$$

when we apply the formal test of letting $c \rightarrow \infty$. In other words, Eq. 38-28 is correct for all physically possible speeds while Eq. 38-29 is approximately correct for speeds much less than c.

38-10 Doppler Effect for Light

In Section 18-8 we discussed the Doppler effect (a shift in detected frequency) for sound waves traveling in air. For such waves, the Doppler effect depends on two velocities—namely, the velocities of the source and detector with respect to the air. (Air is the medium that transmits the waves.)

That is not the situation with light waves, for they (and other electromagnetic waves) require no medium, being able to travel even through vacuum. The Doppler effect for light waves depends on only one velocity, the relative velocity $\vec{v}$ between source and detector, as measured from the reference frame of either. Let f_0 represent the **proper frequency** of the source—that is, the frequency that is measured by an observer in the rest frame of the source. Let f represent the frequency detected by an observer moving with velocity $\vec{v}$ relative to that rest frame. Then, when the direction of $\vec{v}$ is directly away from the source,

$$f = f_0\sqrt{\frac{1-\beta}{1+\beta}} \quad \text{(source and detector separating)}, \tag{38-30}$$

where $\beta = v/c$. When the direction of $\vec{v}$ is directly toward the source, we must change the signs in front of both β symbols in Eq. 38-30.

Low-Speed Doppler Effect

For low speeds ($\beta \ll 1$), Eq. 38-30 can be expanded in a power series in β and approximated as

$$f = f_0(1 - \beta + \tfrac{1}{2}\beta^2) \quad \text{(source and detector separating, } \beta \ll 1\text{)}. \tag{38-31}$$

The corresponding low-speed equation for the Doppler effect with sound waves (or any waves except light waves) has the same first two terms but a different coefficient in the third term. Thus, the relativistic effect for low-speed light sources and detectors shows up only with the β^2 term.

A police radar unit employs the Doppler effect with microwaves to measure the speed v of a car. A source in the radar unit emits a microwave beam at a certain (proper) frequency f_0 along the road. A car that is moving toward the unit intercepts that beam but at a frequency that is shifted upward by the Doppler effect, due to the car's motion toward the radar unit. The car reflects the beam back toward the radar unit. Because the car is moving toward the radar unit, the detector in the unit intercepts a reflected beam that is further shifted up in frequency. The unit compares that detected frequency with f_0 and computes the speed v of the car.

Astronomical Doppler Effect

In astronomical observations of stars, galaxies, and other sources of light, we can determine how fast the sources are moving, either directly away from us or directly toward us, by measuring the *Doppler shift* of the light that reaches us. If a certain star were at rest relative to us, we would detect light from it with a certain proper frequency f_0. However, if the star is moving either directly away from us or directly toward us, the light we detect has a frequency f that is shifted from f_0 by the Doppler effect. This Doppler shift is due only to the *radial* motion of the star (its motion directly toward us or away from us), and the speed we can determine by measuring this Doppler shift is only the *radial speed* v of the star—that is, only the radial component of the star's velocity relative to us.

Let us assume that the radial speed v of a certain light source is low enough (β is small enough) for us to neglect the β^2 term in Eq. 38-31. Also, let us explicitly show a $\pm$ option in front of the β term—the minus sign corresponding to radial motion away from us and the plus sign corresponding to radial motion toward us. Then Eq. 38-31 becomes

$$f = f_0(1 \pm \beta). \tag{38-32}$$

Astronomical measurements involving light are usually done in wavelengths rather than frequencies, so let us replace f with c/λ and f_0 with c/λ_0, where λ is the measured wavelength and λ_0 is the **proper wavelength.** Also replacing β with v/c in Eq. 38-32, we have

$$\frac{c}{\lambda} = \frac{c}{\lambda_0}\left(1 \pm \frac{v}{c}\right),$$

which leads to
$$v = \pm \frac{\lambda - \lambda_0}{\lambda} c.$$

This is conventionally written as

$$v = \frac{\Delta\lambda}{\lambda} c \quad \text{(radial speed of light source, } v \ll c), \tag{38-33}$$

where $\Delta\lambda$ $(= |\lambda - \lambda_0|)$ is the *wavelength* Doppler shift of the light source. If the source is moving away from us, λ is greater than λ_0 and the Doppler shift is called a *red shift*. (The term does not mean that the detected light is red or even visible; it merely means the wavelength increased.) Similarly, if the source is moving toward us, λ is less than λ_0 and the Doppler shift is called a *blue shift*.

CHECKPOINT 4: The figure shows a source that emits light of proper frequency f_0 while moving directly toward the right with speed $c/4$ as measured from reference frame S. The figure also shows a light detector, which measures a frequency $f > f_0$ for the emitted light. (a) Is the detector moving toward the left or the right? (b) Is the speed of the detector as measured from reference frame S more than $c/4$, less than $c/4$, or equal to $c/4$?

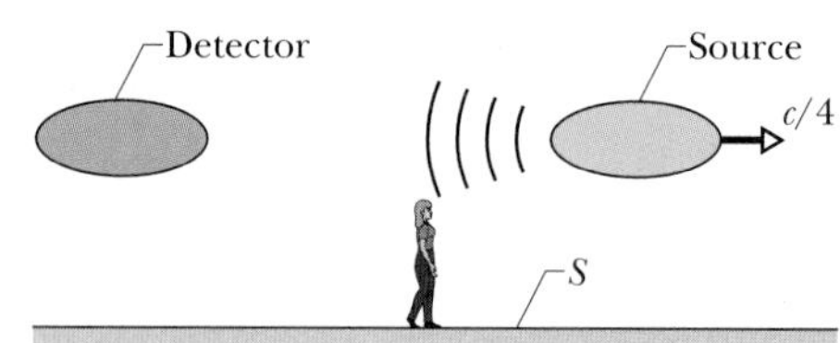

Transverse Doppler Effect

So far, we have discussed the Doppler effect, here and in Chapter 18, only for situations in which the source and the detector move either directly toward or directly away from each other. Figure 38-12 shows a different arrangement, in which a source S moves past a detector D. When S reaches point P, its velocity is perpendicular to the line joining S and D and, at that instant, it is moving neither toward nor away from D. If the source is emitting sound waves of frequency f_0, D detects that frequency (with no Doppler effect) when it intercepts the waves that were emitted at point P. However, if the source is emitting light waves, there is still a Doppler effect, called the **transverse Doppler effect.** In this situation, the detected frequency of the light emitted when the source is at point P is

$$f = f_0\sqrt{1 - \beta^2} \quad \text{(transverse Doppler effect).} \tag{38-34}$$

For low speeds ($\beta \ll 1$), Eq. 38-34 can be expanded in a power series in β and approximated as

$$f = f_0(1 - \tfrac{1}{2}\beta^2) \quad \text{(low speeds).} \tag{38-35}$$

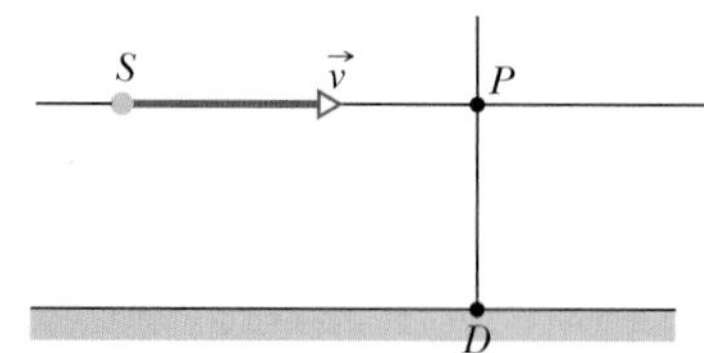

Fig. 38-12 A light source S travels with velocity $\vec{v}$ past a detector at D. The special theory of relativity predicts a transverse Doppler effect as the source passes through point P, where the direction of travel is perpendicular to the line extending through D. Classical theory predicts no such effect.

Here the first term is what we would expect for sound waves and, again, the relativistic effect for low-speed light sources and detectors appears with the β^2 term.

In principle, a police radar unit can determine the speed of a car even when the path of the radar beam is perpendicular (transverse) to the path of the car. However, Eq. 38-35 tells us that because β is small even for a fast car, the relativistic term $\beta^2/2$ in the transverse Doppler effect is extremely small. Thus, $f \approx f_0$ and the radar unit computes a speed of zero. For this reason, police officers always try to direct the radar beam along the car's path to get a Doppler shift that gives the car's actual speed. Any deviation from that alignment works in favor of the motorist, because it reduces the measured speed.

The transverse Doppler effect is really another test of time dilation. If we rewrite Eq. 38-34 in terms of the period T of oscillation of the emitted light wave instead of the frequency, we have, since $T = 1/f$,

$$T = \frac{T_0}{\sqrt{1 - \beta^2}} = \gamma T_0, \tag{38-36}$$

in which T_0 ($= 1/f_0$) is the **proper period** of the source. As comparison with Eq. 38-9 shows, Eq. 38-36 is simply the time dilation formula, since a period is a time interval.

The NAVSTAR Navigation System

Each NAVSTAR satellite continually broadcasts radio signals giving its location, at a set frequency that is controlled by precise atomic clocks. When the signal is sensed by the detector on, say, a commercial aircraft, the frequency has been Doppler-shifted. By detecting the signals from several NAVSTAR satellites simultaneously, the detector can determine the direction to any one of them and the direction of the velocity of that satellite. From the Doppler shift of the signal, the detector then determines the speed of the aircraft.

Let us use some rough numbers to see how well this can be done. The speed of a NAVSTAR satellite relative to the center of Earth is about 1.0×10^4 m/s. The associated β is about 3.0×10^{-5}. Thus, the term $\beta^2/2$ in Eqs. 38-31 and 38-35 (that is, the relativity term) is about 4.5×10^{-10}. In other words, relativity changes the Doppler shift of the detected signal by about 4.5 parts in 10^{10}, which seems hardly worth considering.

However, it is indeed important. The atomic clocks in the satellites are so precise that the variation in the frequency of the satellite signal is only 2 parts in 10^{12}. From Eq. 38-35, we see that β (hence v) depends on the square root of f/f_0. Thus, the clock's frequency variation of 2×10^{-12} causes a variation of

$$\sqrt{2 \times 10^{-12}} = 1.4 \times 10^{-6}$$

in the measured value of the relative speed v between satellite and aircraft.

Since v is due primarily to the satellite's great speed, 1.0×10^4 m/s, this means that v (hence the aircraft's speed) can be determined to an accuracy of about

$$(1.4 \times 10^{-6})(1.0 \times 10^4 \text{ m/s}) = 1.4 \text{ cm/s}.$$

Suppose the aircraft flies for 1 h (3600 s). Knowing the speed to about 1.4 cm/s allows the location at the end of that hour to be predicted to about

$$(0.014 \text{ m/s})(3600 \text{ s}) = 50 \text{ m},$$

which is acceptable in modern navigation.

If relativity effects were not taken into account, the speed of the aircraft could not be known any closer than 21 cm/s, and its location after an hour's flight could not be predicted any better than within 760 m.

Sample Problem 38-5

Figure 38-13*a* shows curves of intensity versus wavelength for light reaching us from interstellar gas on two opposite sides of galaxy M87 (Fig. 38-13*b*). One curve peaks at 499.8 nm; the other at 501.6 nm. The gas orbits the core of the galaxy at a radius r = 100 light-years, apparently moving toward us on one side of the core and moving away from us on the opposite side.

(a) Which curve corresponds to the gas moving toward us? What is the speed of the gas relative to us (and relative to the galaxy's core)?

SOLUTION: The Key Ideas here are these:

1. If the gas were not moving around the galaxy's core, the light from it would be detected at a certain wavelength.
2. The motion of the gas changes the detected wavelength via the Doppler effect, increasing the wavelength for the gas moving away from us and decreasing it for the gas moving toward us.

Thus, the curve peaking at 501.6 nm corresponds to motion away from us, and that peaking at 499.8 nm corresponds to motion toward us.

Let us assume that the increase and the decrease in wavelength due to the motion of the gas are equal in magnitude. Then the unshifted wavelength, which we shall take as the proper wavelength λ_0, must be the average of the two shifted wavelengths:

$$\lambda_0 = \frac{501.6\text{ nm} + 499.8\text{ nm}}{2} = 500.7\text{ nm}.$$

The Doppler shift $\Delta\lambda$ of the light from the gas moving away from us is then

$$\Delta\lambda = |\lambda - \lambda_0| = 501.6\text{ nm} - 500.7\text{ nm} = 0.90\text{ nm}.$$

Substituting this and $\lambda = 501.6$ nm into Eq. 38-33, we find that the speed of the gas is

$$v = \frac{\Delta\lambda}{\lambda}c = \frac{0.90\text{ nm}}{501.6\text{ nm}} 2.998 \times 10^8\text{ m/s} = 5.38 \times 10^5\text{ m/s}. \quad \text{(Answer)}$$

(b) The gas orbits the core of the galaxy because it experiences a gravitational force due to the mass M of the core. What is that mass in multiples of the Sun's mass M_S (= 1.99×10^{30} kg)?

SOLUTION: There are two Key Ideas here:

1. From Eq. 14-1, the magnitude F of the gravitational force on an orbiting gas element of mass m at orbital radius r is

$$F = \frac{GMm}{r^2}.$$

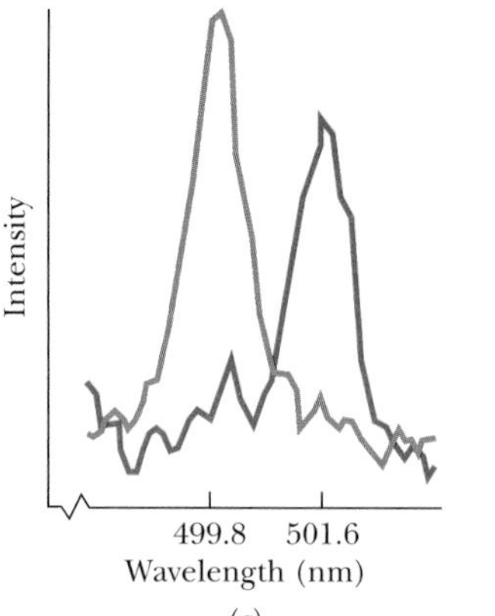

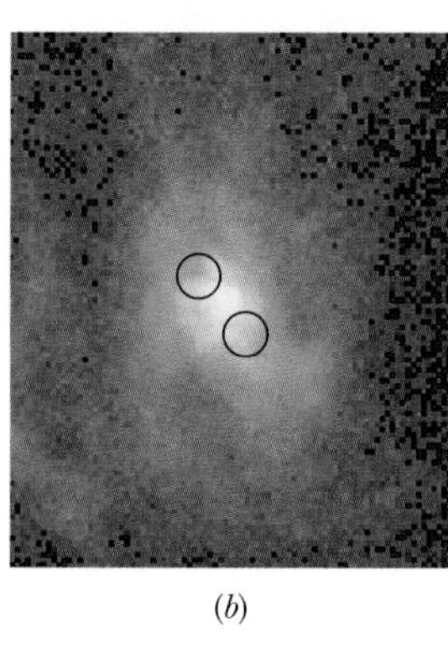

Fig. 38-13 Sample Problem 38-5. (*a*) Plots of intensity versus wavelength for light emitted by gas on opposite sides of galaxy M87 and detected on Earth. (*b*) The central region of M87. The circles indicate the locations of the gas whose intensity is given in (*a*). The core of M87 is halfway between the circles.

2. If the gas element orbits the galaxy core in a circle, then it must have a centripetal acceleration of magnitude $a = v^2/r$, directed toward the core.
3. Newton's second law, written for a radial axis extending from the core to the gas element, tells us that $F = ma$.

Putting these three ideas together, we have

$$\frac{GMm}{r^2} = m\frac{v^2}{r}.$$

Solving this for M and substituting known data, we find

$$M = \frac{v^2 r}{G} = \frac{(5.38 \times 10^5\text{ m/s})^2(100\text{ ly})(9.46 \times 10^{15}\text{ m/ly})}{6.67 \times 10^{-11}\text{ N}\cdot\text{m}^2/\text{kg}^2} = 4.11 \times 10^{39}\text{ kg} = (2.1 \times 10^9)M_S. \quad \text{(Answer)}$$

This result tells us that a mass equivalent to two billion suns has been compacted into the core of the galaxy, strongly suggesting that a "supermassive" black hole occupies the core.

38-11 A New Look at Momentum

Suppose that a number of observers, each in a different inertial reference frame, watch an isolated collision between two particles. In classical mechanics, we have seen that—even though the observers measure different velocities for the colliding particles—they all find that the law of conservation of momentum holds. That is, they find that the total momentum of the system of particles after the collision is the same as it was before the collision.

How is this situation affected by relativity? We find that if we continue to define the momentum $\vec{p}$ of a particle as $m\vec{v}$, the product of its mass and its velocity, total momentum is *not* conserved for the observers in different inertial frames. We have two choices: (1) Give up the law of conservation of momentum or (2) see if we can redefine the momentum of a particle in some new way so that the law of conservation of momentum still holds. The correct choice is the second one.

Consider a particle moving with constant speed v in the positive x direction. Classically, its momentum has magnitude

$$p = mv = m\frac{\Delta x}{\Delta t} \quad \text{(classical momentum)}, \tag{38-37}$$

in which Δx is the distance it travels in time Δt. To find a relativistic expression for momentum, we start with the new definition

$$p = m\frac{\Delta x}{\Delta t_0}.$$

Here, as before, Δx is the distance traveled by a moving particle as viewed by an observer watching that particle. However, Δt_0 is the time required to travel that distance, measured not by the observer watching the moving particle but by an observer moving with the particle. The particle is at rest with respect to this second observer, with the result that the time this observer measures is a proper time.

Using the time dilation formula (Eq. 38-9), we can then write

$$p = m\frac{\Delta x}{\Delta t_0} = m\frac{\Delta x}{\Delta t}\frac{\Delta t}{\Delta t_0} = m\frac{\Delta x}{\Delta t}\gamma.$$

However, since $\Delta x/\Delta t$ is just the particle velocity v,

$$p = \gamma mv \quad \text{(momentum)}. \tag{38-38}$$

Note that this differs from the classical definition of Eq. 38-37 only by the Lorentz factor γ. However, that difference is important: Unlike classical momentum, relativistic momentum approaches an infinite value as v approaches c.

We can generalize the definition of Eq. 38-38 to vector form as

$$\vec{p} = \gamma m\vec{v} \quad \text{(momentum)}. \tag{38-39}$$

This equation gives the correct definition of momentum for all physically possible speeds. For a speed much less than c, it reduces to the classical definition of momentum ($\vec{p} = m\vec{v}$).

38-12 A New Look at Energy

Mass Energy

The science of chemistry was initially developed with the assumption that in chemical reactions, energy and mass are conserved separately. In 1905, Einstein showed that as a consequence of his theory of special relativity, mass can be considered to be another form of energy. Thus, the law of conservation of energy is really the law of conservation of mass–energy.

In a *chemical reaction* (a process in which atoms or molecules interact), the amount of mass that is transferred into other forms of energy (or vice versa) is such a tiny fraction of the total mass involved that there is no hope of measuring the mass change with even the best laboratory balances. Mass and energy truly *seem* to be separately conserved. However, in a *nuclear reaction* (in which nuclei or fundamental particles interact), the energy released is often about a million times greater

TABLE 38-3 The Energy Equivalents of a Few Objects

Object	Mass (kg)	Energy Equivalent	
Electron	9.11×10^{-31}	8.19×10^{-14} J	(= 511 keV)
Proton	1.67×10^{-27}	1.50×10^{-10} J	(= 938 MeV)
Uranium atom	3.95×10^{-25}	3.55×10^{-8} J	(= 225 GeV)
Dust particle	1×10^{-13}	1×10^{4} J	(= 2 kcal)
U.S. penny	3.1×10^{-3}	2.8×10^{14} J	(= 78 GW · h)

than in a chemical reaction, and the change in mass can easily be measured. Taking mass–energy transfers into account in nuclear reactions became routine long ago.

An object's mass m and the energy equivalent E_0 of that mass are related by

$$E_0 = mc^2, \tag{38-40}$$

which, without the subscript 0, is the best-known science equation of all time. This energy that is associated with the mass of an object is called **mass energy** or **rest energy.** The second phrase suggests that E_0 is an energy that the object has even when it is at rest, simply because it has mass. (If you continue your study of physics beyond this book, you will see more refined discussions of the relation between mass and energy. You might even encounter disagreements about just what that relation is and means.)

Table 38-3 shows the mass energy or rest energy of a few objects. The mass energy of, say, a U.S. penny is enormous; the equivalent amount of electric energy would cost well over a million dollars. On the other hand, the entire annual U.S. electric energy production corresponds to a mass of only a few hundred kilograms of matter (stones, burritos, or anything else).

In practice, SI units are rarely used with Eq. 38-40 because they are too large to be convenient. Masses are usually measured in atomic mass units, where

$$1 \text{ u} = 1.66 \times 10^{-27} \text{ kg}, \tag{38-41}$$

and energies are usually measured in electron-volts or multiples of it, where

$$1 \text{ eV} = 1.60 \times 10^{-19} \text{ J}. \tag{38-42}$$

In the units of Eqs. 38-41 and 38-42, the multiplying constant c^2 has the values

$$c^2 = 9.315 \times 10^8 \text{ eV/u} = 9.315 \times 10^5 \text{ keV/u}$$
$$= 931.5 \text{ MeV/u}. \tag{38-43}$$

Total Energy

Equation 38-40 gives an object's mass energy (or rest energy) E_0 that is associated with the object's mass m, regardless of whether the object is at rest or moving. If the object is moving, it has additional energy in the form of kinetic energy K. If we assume that its potential energy is zero, then its total energy E is the sum of its mass energy and its kinetic energy:

$$E = E_0 + K = mc^2 + K. \tag{38-44}$$

Although we shall not prove it, the total energy E can also be written as

$$E = \gamma mc^2, \tag{38-45}$$

where γ is the Lorentz factor for the object's motion.

Since Chapter 7, we have discussed many examples involving changes in the total energy of a particle or a system of particles. However, we did not include mass energy in the discussions because the changes in mass energy were either zero or small enough to be neglected. The law of conservation of total energy still applies even if changes in mass energy are significant. Thus, regardless of what happens to the mass energy, the following statement from Section 8-7 is still true:

➤ The total energy E of an *isolated system* cannot change.

For example, if the total mass energy of two interacting particles in an isolated system decreases, some other type of energy in the system must increase because the total energy cannot change.

In a system undergoing a chemical or nuclear reaction, a change in the total mass energy of the system due to the reaction is often given as a Q value. The Q value for a reaction is obtained from the relation

$$\left(\begin{matrix}\text{system's initial}\\ \text{total mass energy}\end{matrix}\right) = \left(\begin{matrix}\text{system's final}\\ \text{total mass energy}\end{matrix}\right) + Q$$

or

$$E_{0i} = E_{0f} + Q. \tag{38-46}$$

Using Eq. 38-40 ($E_0 = mc^2$), we can rewrite this in terms of the initial *total* mass M_i and the final *total* mass M_f as

$$M_ic^2 = M_fc^2 + Q$$

or

$$Q = M_ic^2 - M_fc^2 = -\Delta M\, c^2, \tag{38-47}$$

where the change in mass due to the reaction is $\Delta M = M_f - M_i$.

If a reaction results in the transfer of energy from mass energy to, say, kinetic energy of the reaction products, the system's total mass energy E_0 (and total mass M) decreases and Q is positive. If, instead, a reaction requires that energy be transferred to mass energy, the system's total mass energy E_0 (and its total mass M) increases and Q is negative.

For example, suppose two hydrogen nuclei undergo a *fusion reaction* in which they join together to form a single nucleus and release two particles. The total mass energy (and total mass) of the resultant single nucleus and two released particles is less than the total mass energy (and total mass) of the initial hydrogen nuclei. Thus, the Q of the fusion reaction is positive, and energy is said to be *released* (transferred from mass energy) by the reaction. This release is important to you, because the fusion of hydrogen nuclei in the Sun is one part of the process that results in sunshine on Earth and makes life here possible.

Kinetic Energy

In Chapter 7 we defined the kinetic energy K of an object of mass m and with a speed v well below c to be

$$K = \tfrac{1}{2}mv^2. \tag{38-48}$$

However, this classical equation is only an approximation that is good enough when the speed is well below the speed of light.

Let us now find an expression for kinetic energy that is correct for *all* physically possible speeds, including speeds close to c. Solving Eq. 38-44 for K and then

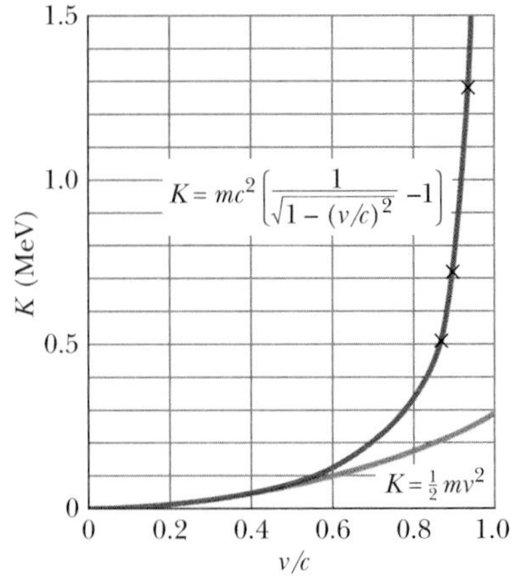

Fig. 38-14 The relativistic (Eq. 38-49) and classical (Eq. 38-48) equations for the kinetic energy of an electron, plotted as a function of v/c, where v is the speed of the electron and c is the speed of light. Note that the two curves blend together at low speeds and diverge widely at high speeds. Experimental data (at the × marks) show that at high speeds the relativistic curve agrees with experiment but the classical curve does not.

substituting for E_0 from Eq. 38-45 lead to

$$K = E - mc^2 = \gamma mc^2 - mc^2$$
$$= mc^2(\gamma - 1) \quad \text{(kinetic energy)}, \tag{38-49}$$

where $\gamma\ (= 1/\sqrt{1 - (v/c)^2})$ is the Lorentz factor for the object's motion.

Figure 38-14 shows plots of the kinetic energy of an electron as calculated with the correct definition (Eq. 38-49) and the classical approximation (Eq. 38-48), both as functions of v/c. Note that on the left side of the graph the two plots coincide; this is the part of the graph—at lower speeds—where we have calculated kinetic energies so far in this book. That part of the graph tells us that we have been justified in calculating kinetic energy with the classical expression of Eq. 38-48. However, on the right side of the graph—at speeds near c—the two plots differ significantly. As v/c approaches 1.0, the plot for the classical definition of kinetic energy increases only moderately while the plot for the correct definition of kinetic energy increases dramatically, approaching an infinite value as v/c approaches 1.0. Thus, when an object's speed v is near c, we *must* use Eq. 38-49 to calculate its kinetic energy.

Figure 38-14 also tells us something about the work we must do on an object to increase the object's speed by, say, 1%. The required work W is equal to the resulting change ΔK in the object's kinetic energy. If the change is to occur on the low-speed left side of Fig. 38-14, the required work might be modest. However, if the change is to occur on the high-speed right side of Fig. 38-14, the required work could be enormous, because the kinetic energy K increases so rapidly there with an increase in speed v. To increase an object's speed to c would require, in principle, an infinite amount of energy; thus, doing so is impossible.

The kinetic energies of electrons, protons, and other particles are often stated with the unit electron-volt or one of its multiples used as an adjective. For example, an electron with a kinetic energy of 20 MeV may be described as a 20 MeV electron.

Momentum and Kinetic Energy

In classical mechanics, the momentum p of a particle is mv and its kinetic energy K is $\frac{1}{2}mv^2$. If we eliminate v between these two expressions, we find a direct relation between momentum and kinetic energy:

$$p^2 = 2Km \quad \text{(classical)}. \tag{38-50}$$

We can find a similar connection in relativity by eliminating v between the relativistic definition of momentum (Eq. 38-38) and the relativistic definition of kinetic energy (Eq. 38-49). Doing so leads, after some algebra, to

$$(pc)^2 = K^2 + 2Kmc^2. \tag{38-51}$$

With the aid of Eq. 38-44, we can transform Eq. 38-51 into a relation between the momentum p and the total energy E of a particle:

$$E^2 = (pc)^2 + (mc^2)^2. \tag{38-52}$$

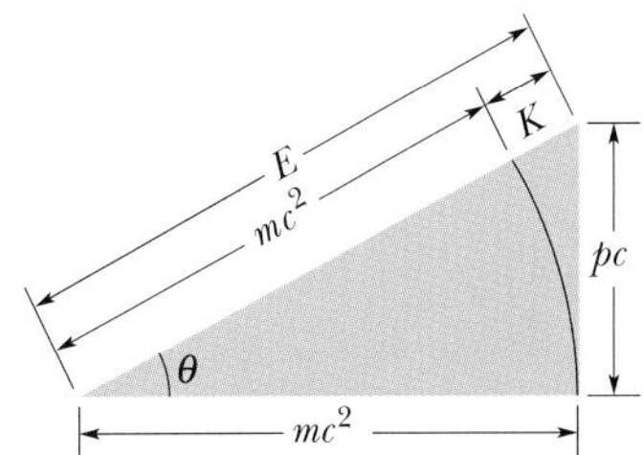

Fig. 38-15 A useful mnemonic device for remembering the relativistic relations among the total energy E, the rest energy or mass energy mc^2, the kinetic energy K, and the momentum p.

The right triangle of Fig. 38-15 can help you keep these useful relations in mind. You can also show that, in that triangle,

$$\sin\theta = \beta \quad \text{and} \quad \cos\theta = 1/\gamma. \tag{38-53}$$

With Eq. 38-52 we can see that the product pc must have the same unit as energy E; thus, we can express the unit of momentum p as an energy unit divided by c. In fact, momentum in fundamental particle physics is often reported in the units MeV/c or GeV/c.

CHECKPOINT 5: Are (a) the kinetic energy and (b) the total energy of a 1 GeV electron more than, less than, or equal to those of a 1 GeV proton?

Sample Problem 38-6

(a) What is the total energy E of a 2.53 MeV electron?

SOLUTION: The **Key Idea** here is that, from Eq. 38-44, the total energy E is the sum of the electron's mass energy (or rest energy) mc^2 and its kinetic energy:

$$E = mc^2 + K. \tag{38-54}$$

The adjective "2.53 MeV" in the problem statement means that the electron's kinetic energy is 2.53 MeV. To evaluate the electron's mass energy mc^2, we substitute the electron's mass m from Appendix B, obtaining

$$mc^2 = (9.109 \times 10^{-31}\ \text{kg})(2.998 \times 10^8\ \text{m/s})^2$$
$$= 8.187 \times 10^{-14}\ \text{J}.$$

Then dividing this result by 1.602×10^{-13} J/MeV gives us 0.511 MeV as the electron's mass energy (confirming the value in Table 38-3). Equation 38-54 then yields

$$E = 0.511\ \text{MeV} + 2.53\ \text{MeV} = 3.04\ \text{MeV}. \quad \text{(Answer)}$$

(b) What is the magnitude p of the electron's momentum, in the unit MeV/c?

SOLUTION: The **Key Idea** here is that we can find p from the total energy E and the mass energy mc^2 via Eq. 38-52,

$$E^2 = (pc)^2 + (mc^2)^2.$$

Solving for pc gives us

$$pc = \sqrt{E^2 - (mc^2)^2}$$
$$= \sqrt{(3.04\ \text{MeV})^2 - (0.511\ \text{MeV})^2} = 3.00\ \text{MeV}.$$

Finally, dividing both sides by c we find

$$p = 3.00\ \text{MeV}/c. \quad \text{(Answer)}$$

Sample Problem 38-7

The most energetic proton ever detected in the cosmic rays coming to Earth from space had an astounding kinetic energy of 3.0×10^{20} eV (enough energy to warm a teaspoon of water by a few degrees).

(a) What were the proton's Lorentz factor γ and speed v (both relative to the ground-based detector)?

SOLUTION: One **Key Idea** here is that the proton's Lorentz factor γ relates its total energy E to its mass energy mc^2 via Eq. 38-45 ($E = \gamma mc^2$). A second **Key Idea** is that the proton's total energy is the sum of its mass energy mc^2 and its (given) kinetic energy K. Putting these ideas together we have

$$\gamma = \frac{E}{mc^2} = \frac{mc^2 + K}{mc^2} = 1 + \frac{K}{mc^2}. \tag{38-55}$$

We can calculate the proton's mass energy mc^2 from its mass given in Appendix B, as we did for the electron in Sample Problem 38-6a. We find that mc^2 is 938 MeV (as listed in Table 38-3). Substituting this and the given kinetic energy into Eq. 38-55, we obtain

$$\gamma = 1 + \frac{3.0 \times 10^{20}\ \text{eV}}{938 \times 10^6\ \text{eV}}$$
$$= 3.198 \times 10^{11} \approx 3.2 \times 10^{11}. \quad \text{(Answer)}$$

This computed value for γ is so large that we cannot use the definition of γ (Eq. 38-8) to find v. Try it; your calculator will tell you that β is effectively equal to 1 and thus that v is effectively equal to c. Actually, v is almost c, but we want a more accurate answer, which we can obtain by first solving Eq. 38-8 for $1 - \beta$. To begin we write

$$\gamma = \frac{1}{\sqrt{1-\beta^2}} = \frac{1}{\sqrt{(1-\beta)(1+\beta)}} \approx \frac{1}{\sqrt{2(1-\beta)}},$$

where we have used the fact that β is so close to unity that $1 + \beta$ is very close to 2. The velocity we seek is contained in the $1 - \beta$

term. Solving for $1 - \beta$ then yields

$$1 - \beta = \frac{1}{2\gamma^2} = \frac{1}{(2)(3.198 \times 10^{11})^2} = 4.9 \times 10^{-24} \approx 5 \times 10^{-24}.$$

Thus,

$$\beta = 1 - 5 \times 10^{-24}$$

and, since $v = \beta c$,

$$v \approx 0.999\,999\,999\,999\,999\,999\,999\,995c. \qquad \text{(Answer)}$$

(b) Suppose that the proton travels along a diameter (9.8×10^4 ly) of the Milky Way galaxy. Approximately how long does the proton take to travel that diameter as measured from the common reference frame of Earth and the galaxy?

SOLUTION: We just saw that this *ultrarelativistic* proton is traveling at a speed barely less than c. Then the Key Idea here is that by the definition of light-year, light takes 1 y to travel 1 ly, so light should take 9.8×10^4 y to travel 9.8×10^4 ly, and this proton should take almost the same time. Thus, from our Earth–Milky Way reference frame, the proton's trip takes

$$\Delta t = 9.8 \times 10^4 \text{ y}. \qquad \text{(Answer)}$$

(c) How long does the trip take as measured in the reference frame of the proton?

SOLUTION: We need four Key Ideas here:

1. This problem involves measurements made from two (inertial) reference frames, one is the Earth–Milky Way frame and the other is attached to the proton.
2. This problem also involves two events: the first is when the proton passes one end of the diameter along the Galaxy, and the second is when it passes the opposite end.
3. The time interval between those two events as measured in the proton's reference frame is the proper time interval Δt_0 because the events occur at the same location in that frame—namely, at the proton itself.
4. We can find the proper time interval Δt_0 from the time interval Δt measured in the Earth–Milky Way frame by using Eq. 38-9 ($\Delta t = \gamma\, \Delta t_0$) for time dilation.

Solving Eq. 38-9 for Δt_0 and substituting γ from (a) and Δt from (b), we find

$$\Delta t_0 = \frac{\Delta t}{\gamma} = \frac{9.8 \times 10^4 \text{ y}}{3.198 \times 10^{11}} = 3.06 \times 10^{-7} \text{ y} = 9.7 \text{ s}. \qquad \text{(Answer)}$$

In our frame, the trip takes 98 000 y. In the proton's frame, it takes 9.7 s! As promised at the start of this chapter, relative motion can alter the rate at which time passes, and we have here an extreme example.

REVIEW & SUMMARY

The Postulates Einstein's **special theory of relativity** is based on two postulates:

1. The laws of physics are the same for observers in all inertial reference frames. No frame is preferred.
2. The speed of light in vacuum has the same value c in all directions and in all inertial reference frames.

The speed of light c in vacuum is an ultimate speed that cannot be exceeded by any entity carrying either energy or information.

Coordinates of an Event Three space coordinates and one time coordinate specify an **event.** One task of special relativity is to relate these coordinates as assigned by two observers who are in uniform motion with respect to each other.

Simultaneous Events If two observers are in relative motion, they will not, in general, agree as to whether two events are simultaneous. If one of the observers finds two events at different locations to be simultaneous, the other will not, and conversely. Simultaneity is *not* an absolute concept but a relative one, depending on the motion of the observer. The relativity of simultaneity is a direct consequence of the finite ultimate speed c.

Time Dilation If two successive events occur at the same place in an inertial reference frame, the time interval Δt_0 between them, measured on a single clock where they occur, is the **proper time** between the events. *Observers in frames moving relative to that frame will measure a larger value for this interval.* For an observer moving with relative speed v, the measured time interval is

$$\Delta t = \frac{\Delta t_0}{\sqrt{1 - (v/c)^2}} = \frac{\Delta t_0}{\sqrt{1 - \beta^2}} = \gamma\, \Delta t_0 \quad \text{(time dilation).} \qquad \text{(38-7 to 38-9)}$$

Here $\beta = v/c$ is the **speed parameter** and $\gamma = 1/\sqrt{1 - \beta^2}$ is the **Lorentz factor.** An important consequence of time dilation is that moving clocks run slow as measured by an observer at rest.

Length Contraction The length L_0 of an object measured by an observer in an inertial reference frame in which the object is at rest is called its **proper length.** *Observers in frames moving relative to that frame and parallel to that length will measure a shorter length.* For an observer moving with relative speed v, the measured length is

$$L = L_0\sqrt{1 - \beta^2} = \frac{L_0}{\gamma} \quad \text{(length contraction).} \qquad \text{(38-13)}$$

The Lorentz Transformation The *Lorentz transformation* equations relate the spacetime coordinates of a single event as seen by observers in two inertial frames, S and S', where S' is moving relative to S with velocity v in the positive x, x' direction. The four coordinates are related by

$$\begin{aligned} x' &= \gamma(x - vt), \\ y' &= y, \\ z' &= z, \\ t' &= \gamma(t - vx/c^2) \end{aligned} \quad \text{(Lorentz transformation equations; valid at all physically possible speeds)} \quad (38\text{-}20)$$

Relativity of Velocities When a particle is moving with speed u' in the positive x' direction in an inertial reference frame S' that itself is moving with speed v parallel to the x direction of a second inertial frame S, the speed u of the particle as measured in S is

$$u = \frac{u' + v}{1 + u'v/c^2} \quad \text{(relativistic velocity).} \quad (38\text{-}28)$$

Relativistic Doppler Effect If a source emitting light waves of frequency f_0 moves directly away from a detector with relative radial speed v (and speed parameter $\beta = v/c$), the frequency f measured by the detector is

$$f = f_0\sqrt{\frac{1-\beta}{1+\beta}}. \quad (38\text{-}30)$$

If the source moves directly toward the detector, the signs in Eq. 38-30 are reversed.

For astronomical observations, the Doppler effect is measured in wavelengths. For speeds much less than c, Eq. 38-30 leads to

$$v = \frac{\Delta\lambda}{\lambda}c, \quad (38\text{-}33)$$

where $\Delta\lambda$ is the *Doppler shift* in wavelength (the magnitude of the change in wavelength) due to the motion.

Transverse Doppler Effect If the relative motion of the light source is perpendicular to a line joining the source and detector, the Doppler frequency formula is

$$f = f_0\sqrt{1-\beta^2}. \quad (38\text{-}34)$$

This **transverse Doppler effect** is due to time dilation.

Momentum and Energy The following definitions of linear momentum $\vec{p}$, kinetic energy K, and total energy E for a particle of mass m are valid at any physically possible speed:

$$\vec{p} = \gamma m\vec{v} \quad \text{(momentum),} \quad (38\text{-}39)$$

$$E = mc^2 + K = \gamma mc^2 \quad \text{(total energy),} \quad (38\text{-}44, 38\text{-}45)$$

$$K = mc^2(\gamma - 1) \quad \text{(kinetic energy).} \quad (38\text{-}49)$$

Here γ is the Lorentz factor for the particle's motion, and mc^2 is the *mass energy* or *rest energy* associated with the mass of the particle. These equations lead to the relationships

$$(pc)^2 = K^2 + 2Kmc^2, \quad (38\text{-}51)$$

and

$$E^2 = (pc)^2 + (mc^2)^2. \quad (38\text{-}52)$$

When a system of particles undergoes a chemical or nuclear reaction, the Q of the reaction is the negative of the change in the system's total mass energy:

$$Q = M_ic^2 - M_fc^2 = -\Delta M\,c^2, \quad (38\text{-}47)$$

where M_i is the system's total mass before the reaction and M_f is its total mass after the reaction.

QUESTIONS

1. In Fig. 38-16, ship A sends a laser pulse to an oncoming ship B, while scout ship C races away. The indicated speeds of the ships are all measured from the same reference frame. Rank the ships according to the speed of the pulse as measured from each ship, greatest first.

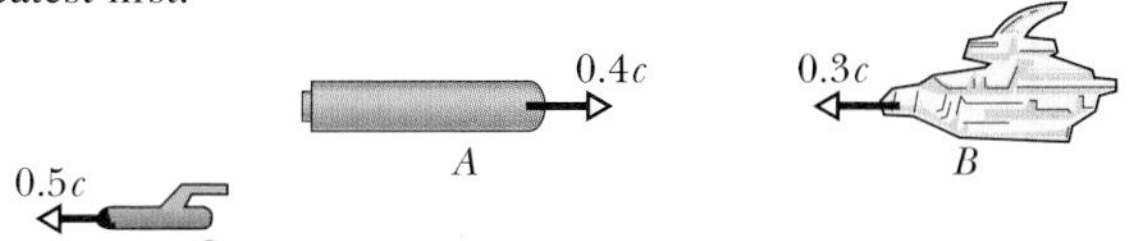

Fig. 38-16 Questions 1 and 7.

2. Figure 38-17a shows two clocks in stationary frame S (they are synchronized in that frame) and one clock in moving frame S'. Clocks C_1 and C_1' read zero when they pass each other. When clocks C_1' and C_2 pass each other, (a) which clock has the smaller reading and (b) which clock measures a proper time?

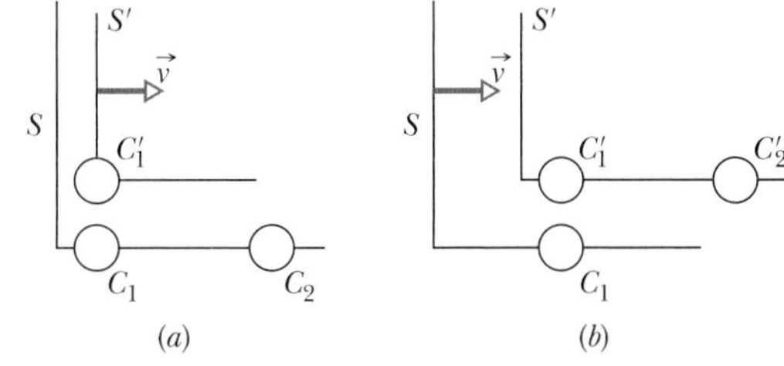

Fig. 38-17 Questions 2 and 3.

3. Figure 38-17b shows two clocks in stationary frame S' (they are synchronized in that frame) and one clock in moving frame S. Clocks C_1 and C_1' read zero when they pass each other. When clocks C_1 and C_2' pass each other, (a) which clock has the smaller reading and (b) which clock measures a proper time?

4. Sam leaves Venus in a spaceship to Mars and passes Sally, who is on Earth, with a relative speed of $0.5c$. (a) Each measures the Venus–Mars voyage time. Who measures a proper time: Sam, Sally, or neither? (b) On the way, Sam sends a pulse of light to Mars. Each measures the travel time of the pulse. Who measures a proper time?

5. Figure 38-18 shows a ship (with on-board reference frame S') passing us (with reference frame S). A proton is fired at nearly the speed of light along the length of the ship, from the front to the rear. (a) Is the spatial separation $\Delta x'$ between the firing of

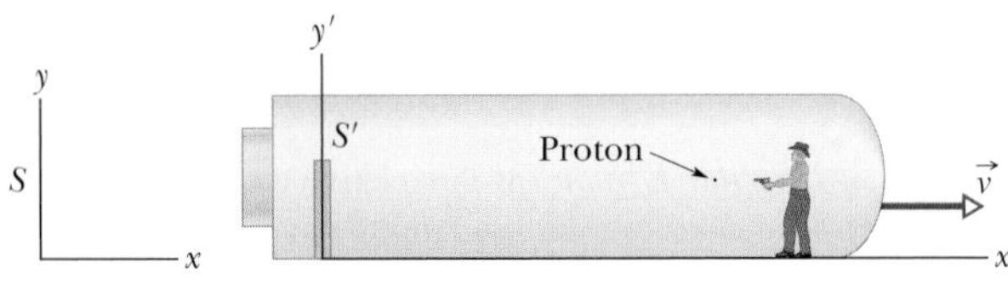

Fig. 38-18 Question 5.

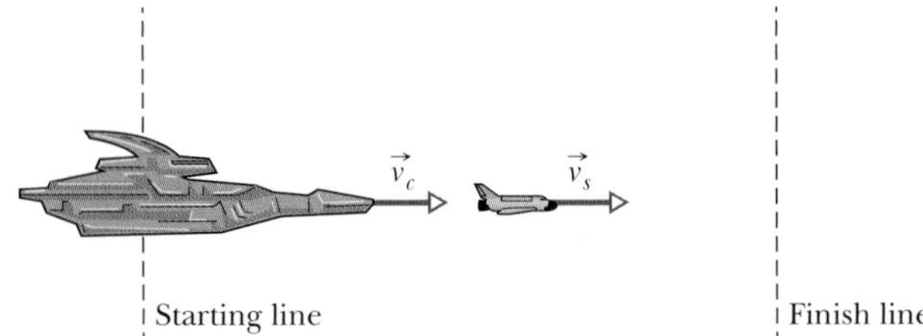

Fig. 38-19 Question 8.

the proton and its impact a positive or negative quantity? (b) Is the temporal separation $\Delta t'$ between those events a positive or negative quantity?

6. (a) In Fig. 38-9, suppose an observer in frame S' measures two events to be at the same location (say, at x') but not at the same time. Can an observer in frame S possibly measure them to be at the same location? (b) If two events occur simultaneously at the same place for one observer, will they be simultaneous for all other observers? (c) Will they occur at the same place for all other observers?

7. Ships A and B in Fig. 38-16 are moving directly toward each other; the velocities indicated are all measured from the same reference frame. Is the speed of ship A relative to ship B more than $0.7c$, less than $0.7c$, or equal to $0.7c$?

8. Figure 38-19 shows one of four star cruisers that are in a race. As each cruiser passes the starting line, a shuttle craft leaves the cruiser and races toward the finish line. You, judging the race, are stationary relative to the starting and finish lines. The speeds v_c of the cruisers relative to you and the speeds v_s of the shuttle craft relative to their starships are, in that order, (1) $0.70c$, $0.40c$; (2) $0.40c$, $0.70c$; (3) $0.20c$, $0.90c$; (4) $0.50c$, $0.60c$. (a) Without written calculation, rank the shuttle craft according to their speeds relative to you, greatest first. (b) Still without written calculation, rank the shuttle craft according to the distances their pilots measure from the starting line to the finish line, greatest first. (c) Each starship sends a signal to its shuttle craft at a certain frequency f_0 as measured on board the starship. Again without written calculation, rank the shuttle craft according to the frequencies they detect, greatest first.

9. While on board a starship, you intercept signals from four shuttle craft that are moving either directly toward or directly away from you. The signals have the same proper frequency f_0. The speed and direction (both relative to you) of the shuttle craft are (a) $0.3c$ toward, (b) $0.6c$ toward, (c) $0.3c$ away, and (d) $0.6c$ away. Rank the shuttle craft according to the frequency you receive, greatest first.

10. The rest energy and total energy, respectively, of three particles, expressed in terms of a basic amount A are (1) A, $2A$; (2) A, $3A$; (3) $3A$, $4A$. Without written calculation, rank the particles according to (a) their mass, (b) their kinetic energy, (c) their Lorentz factor, and (d) their speed, greatest first.

EXERCISES & PROBLEMS

ssm Solution is in the Student Solutions Manual.
www Solution is available on the World Wide Web at: http://www.wiley.com/college/hrw
ilw Solution is available on the Interactive LearningWare.

SEC. 38-2 The Postulates

1E. Quite apart from effects due to Earth's rotational and orbital motions, a laboratory reference frame is not strictly an inertial frame because a particle placed at rest there will not, in general, remain at rest; it will fall. Often, however, events happen so quickly that we can ignore the gravitational acceleration and treat the frame as inertial. Consider, for example, an electron of speed $v = 0.992c$, projected horizontally into a laboratory test chamber and moving through a distance of 20 cm. (a) How long would that journey take, and (b) how far would the electron fall during this interval? What can you conclude about the suitability of the laboratory as an inertial frame in this case? ssm

2E. What fraction of the speed of light does each of the following speeds represent; that is, what is the associated speed parameter β? (a) A typical rate of continental drift (3 cm/y). (b) A highway speed limit of 90 km/h. (c) A supersonic plane flying at Mach 2.5 (1200 km/h). (d) The escape speed of a projectile from the surface of Earth. (e) A typical recession speed of a distant quasar (3.0×10^4 km/s).

SEC. 38-5 The Relativity of Time

3E. The mean lifetime of stationary muons is measured to be 2.2 μs. The mean lifetime of high-speed muons in a burst of cosmic rays observed from Earth is measured to be 16 μs. Find the speed of these cosmic-ray muons relative to Earth. ssm

4E. What must be the speed parameter β if the Lorentz factor γ is (a) 1.01, (b) 10.0, (c) 100, and (d) 1000?

5P. An unstable high-energy particle enters a detector and leaves a track 1.05 mm long before it decays. Its speed relative to the detector was $0.992c$. What is its proper lifetime? That is, how long would the particle have lasted before decay had it been at rest with respect to the detector? ilw

6P. You wish to make a round trip from Earth in a spaceship, traveling at constant speed in a straight line for 6 months and then returning at the same constant speed. You wish further, on your return, to find Earth as it will be 1000 years in the future. (a) How fast must you travel? (b) Does it matter whether you travel in a straight line on your journey? If, for example, you traveled in a circle for 1 year, would you still find that 1000 years had elapsed by Earth clocks when you returned?

SEC. 38-6 The Relativity of Length

7E. A rod lies parallel to the x axis of reference frame S, moving along this axis at a speed of $0.630c$. Its rest length is 1.70 m. What will be its measured length in frame S? ssm

8E. An electron of $\beta = 0.999\ 987$ moves along the axis of an evacuated tube that has a length of 3.00 m as measured by a laboratory observer S at rest relative to the tube. An observer S' at rest relative to the electron, however, would see this tube moving with speed $v\ (= \beta c)$. What length would observer S' measure for the tube?

9E. A meter stick in frame S' makes an angle of 30° with the x' axis. If that frame moves parallel to the x axis of frame S with speed $0.90c$ relative to frame S, what is the length of the stick as measured from S?

10E. The length of a spaceship is measured to be exactly half its rest length. (a) In terms of c, what is the speed of the spaceship relative to the observer's frame? (b) By what factor do the spaceship's clocks run slow, compared to clocks in the observer's frame?

11E. A spaceship of rest length 130 m races past a timing station at a speed of $0.740c$. (a) What is the length of the spaceship as measured by the timing station? (b) What time interval will the station clock record between the passage of the front and back ends of the ship? ssm

12P. (a) Can a person, in principle, travel from Earth to the galactic center (which is about 23 000 ly distant) in a normal lifetime? Explain, using either time-dilation or length-contraction arguments. (b) What constant speed is needed to make the trip in 30 y (proper time)?

13P. A space traveler takes off from Earth and moves at speed $0.99c$ toward the star Vega, which is 26 ly distant. How much time will have elapsed by Earth clocks (a) when the traveler reaches Vega and (b) when Earth observers receive word from the traveler that she has arrived? (c) How much older will Earth observers calculate the traveler to be (measured from her frame) when she reaches Vega than she was when she started the trip? ssm www

SEC. 38-8 Some Consequences of the Lorentz Equations

14E. Observer S reports that an event occurred on the x axis of his reference frame at $x = 3.00 \times 10^8$ m at time $t = 2.50$ s. (a) Observer S' and her frame are moving in the positive direction of x at a speed of $0.400c$. Further, $x = x' = 0$ at $t = t' = 0$. What coordinates does observer S' report for the event? (b) What coordinates would observer S' report if she were moving in the *negative* direction of x at this same speed?

15E. Observer S assigns the spacetime coordinates

$$x = 100\ \text{km} \quad \text{and} \quad t = 200\ \mu\text{s}$$

to an event. What are the coordinates of this event in frame S', which moves in the positive direction of x with speed $0.950c$ relative to S? Assume $x = x' = 0$ at $t = t' = 0$. ssm

16E. Inertial frame S' moves at a speed of $0.60c$ with respect to frame S (Fig. 38-9). Further, $x = x' = 0$ at $t = t' = 0$. Two events are recorded. In frame S, event 1 occurs at the origin at $t = 0$ and event 2 occurs on the x axis at $x = 3.0$ km at $t = 4.0\ \mu$s. What times of occurrence does observer S' record for these same events? Explain the difference in the time order.

17E. An experimenter arranges to trigger two flashbulbs simultaneously, producing a big flash located at the origin of his reference frame and a small flash at $x = 30.0$ km. An observer, moving at a speed of $0.250c$ in the positive direction of x, also views the flashes. (a) What is the time interval between them according to her? (b) Which flash does she say occurs first? ssm www

18P. An observer S sees a big flash of light 1200 m from his position and a small flash of light 720 m closer to him directly in line with the big flash. He determines that the time interval between the flashes is 5.00 μs and the big flash occurs first. (a) What is the relative velocity $\vec{v}$ (give both magnitude and direction) of a second observer S' for whom these flashes occur at the same place in the S' reference frame? (b) From the point of view of S', which flash occurs first? (c) What time interval between them does S' measure?

19P. A clock moves along the x axis at a speed of $0.600c$ and reads zero as it passes the origin. (a) Calculate the clock's Lorentz factor. (b) What time does the clock read as it passes $x = 180$ m? ssm www

20P. In Problem 18, observer S sees the two flashes in the same positions as before, but they now occur closer together in time. How close together in time can they be in the frame of S and still allow the possibility of finding a frame S' in which they occur at the same place?

SEC. 38-9 The Relativity of Velocities

21E. A particle moves along the x' axis of frame S' with a speed of $0.40c$. Frame S' moves with a speed of $0.60c$ with respect to frame S. What is the measured speed of the particle in frame S? ssm

22E. Frame S' moves relative to frame S at $0.62c$ in the positive direction of x. In frame S' a particle is measured to have a velocity of $0.47c$ in the positive direction of x'. (a) What is the velocity of the particle with respect to frame S? (b) What would be the velocity of the particle with respect to S if the particle moved (at $0.47c$) in the *negative* direction of x' in the S' frame? In each case, compare your answers with the predictions of the classical velocity transformation equation.

23E. Galaxy A is reported to be receding from us with a speed of $0.35c$. Galaxy B, located in precisely the opposite direction, is also found to be receding from us at this same speed. What recessional speed would an observer on Galaxy A find (a) for our galaxy and (b) for Galaxy B? ssm

24E. It is concluded from measurements of the red shift of the emitted light that quasar Q_1 is moving away from us at a speed of $0.800c$. Quasar Q_2, which lies in the same direction in space but is closer to us, is moving away from us at a speed $0.400c$. What velocity for Q_2 would be measured by an observer on Q_1?

25P. A spaceship whose rest length is 350 m has a speed of $0.82c$ with respect to a certain reference frame. A micrometeorite, also with a speed of $0.82c$ in this frame, passes the spaceship on an antiparallel track. How long does it take this object to pass the spaceship as measured on the ship? ssm ilw www

26P. An armada of spaceships that is 1.00 ly long (in its rest frame) moves with speed $0.800c$ relative to ground station S. A messenger travels from the rear of the armada to the front with a speed of $0.950c$ relative to S. How long does the trip take as measured (a) in the messenger's rest frame, (b) in the armada's rest frame, and (c) by an observer in frame S?

SEC. 38-10 Doppler Effect for Light

27E. A spaceship, moving away from Earth at a speed of $0.900c$, reports back by transmitting at a frequency (measured in the spaceship frame) of 100 MHz. To what frequency must Earth receivers be tuned to receive the report? ssm

28E. Figure 38-20 is a graph of intensity versus wavelength for light reaching Earth from galaxy NGC 7319, which is about 3×10^8 light-years away. The most intense light is emitted by the oxygen in NGC 7319. In a laboratory that emission is at wavelength $\lambda = 513$ nm, but in the light from NGC 7319 it has been shifted to 525 nm due to the Doppler effect (all the emissions from NGC 7319 have been shifted). (a) What is the radial speed of NGC 7319 relative to Earth? (b) Is the relative motion toward or away from our planet?

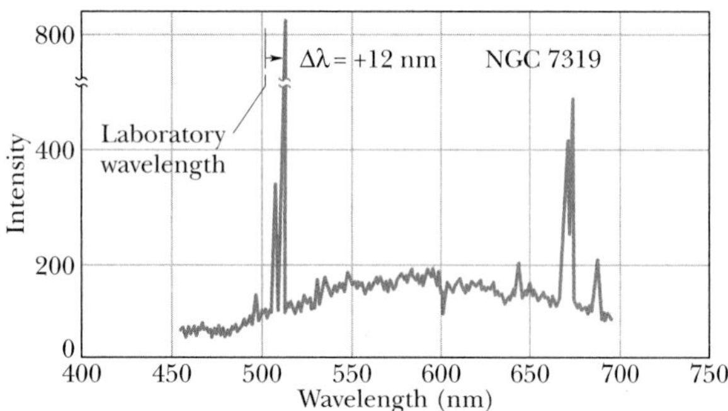

Fig. 38-20 Exercise 28.

29E. Certain wavelengths in the light from a galaxy in the constellation Virgo are observed to be 0.4% longer than the corresponding light from Earth sources. What is the radial speed of this galaxy with respect to Earth? Is it approaching or receding?

30E. Assuming that Eq. 38-33 holds, find how fast you would have to go through a red light to have it appear green. Take 620 nm as the wavelength of red light and 540 nm as the wavelength of green light.

31P. A spaceship is moving away from Earth at a speed of $0.20c$. A light source on the rear of the ship appears blue ($\lambda = 450$ nm) to passengers on the ship. What color would that source appear to an observer on Earth monitoring the receding spaceship? ssm ilw www

SEC. 38-12 A New Look at Energy

32E. How much work must be done to increase the speed of an electron from rest to (a) $0.50c$, (b) $0.990c$, and (c) $0.9990c$?

33E. Find the speed parameter β and Lorentz factor γ for an electron that has a kinetic energy of (a) 1.00 keV, (b) 1.00 MeV, and (c) 1.00 GeV.

34E. Find the speed parameter β and Lorentz factor γ for a particle whose kinetic energy is 10.0 MeV if the particle is (a) an electron, (b) a proton, and (c) an alpha particle.

35E. In terms of c, what is the speed of an electron whose kinetic energy is 100 MeV? ssm

36E. The precise masses in the reaction

$$\text{p} + {}^{19}\text{F} \rightarrow \alpha + {}^{16}\text{O}$$

have been determined to be

$$m(\text{p}) = 1.007825 \text{ u}, \qquad m(\alpha) = 4.002603 \text{ u},$$
$$m(\text{F}) = 18.998405 \text{ u}, \qquad m(\text{O}) = 15.994915 \text{ u}.$$

Calculate the Q of the reaction from these data.

37P. Quasars are thought to be the nuclei of active galaxies in the early stages of their formation. A typical quasar radiates energy at the rate of 10^{41} W. At what rate is the mass of this quasar being reduced to supply this energy? Express your answer in solar mass units per year, where one solar mass unit (1 smu $= 2.0 \times 10^{30}$ kg) is the mass of our Sun. ssm

38P. How much work must be done to increase the speed of an electron from (a) $0.18c$ to $0.19c$ and (b) $0.98c$ to $0.99c$? Note that the speed increase is $0.01c$ in both cases.

39P. A certain particle of mass m has momentum of magnitude mc. What are (a) its speed, (b) its Lorentz factor, and (c) its kinetic energy? ssm

40P. What is the speed of a particle (a) whose kinetic energy is equal to twice its rest energy and (b) whose total energy is equal to twice its rest energy?

41P. What must be the momentum of a particle with mass m so that the total energy of the particle is 3 times its rest energy? ilw

42P. (a) If the kinetic energy K and the momentum p of a particle can be measured, it should be possible to find its mass m and thus identify the particle. Show that

$$m = \frac{(pc)^2 - K^2}{2Kc^2}.$$

(b) Show that this expression reduces to an expected result as $u/c \rightarrow 0$, in which u is the speed of the particle. (c) Find the mass of a particle whose kinetic energy is 55.0 MeV and whose momentum is 121 MeV/c. Express your answer in terms of the mass m_e of the electron.

43P. A 5.00 grain aspirin tablet has a mass of 320 mg. For how many kilometers would the energy equivalent of this mass power an automobile? Assume 12.75 km/L and a heat of combustion of 3.65×10^7 J/L for the gasoline used in the automobile. ssm

44P. The average lifetime of muons at rest is 2.20 μs. A laboratory measurement on muons traveling in a beam emerging from a particle accelerator yields an average muon lifetime of 6.90 μs. What are (a) the speed of these muons in the laboratory, (b) their kinetic energy, and (c) their momentum? The mass of a muon is 207 times that of an electron.

45P. In a high-energy collision between a cosmic-ray particle and a particle near the top of Earth's atmosphere, 120 km above sea level, a pion is created. The pion has a total energy E of 1.35×10^5 MeV and is traveling vertically downward. In the pion's rest frame, the pion decays 35.0 ns after its creation. At what altitude above sea level, as measured from Earth's reference frame, does the decay occur? The rest energy of a pion is 139.6 MeV. ssm www

46P. In Section 29-5 we showed that a particle of charge q and mass m moving with speed v perpendicular to a uniform magnetic field B moves in a circle of radius r given by Eq. 29-16:

$$r = \frac{mv}{qB}.$$

Also, it was demonstrated that the period T of the circular motion is independent of the speed of the particle. These results hold only if $v \ll c$. For particles moving faster, the radius of the circular path must be obtained with

$$r = \frac{p}{qB} = \frac{\gamma mv}{qB} = \frac{mv}{qB\sqrt{1-\beta^2}}.$$

This equation is valid at all speeds. Compute the radius of the path of a 10.0 MeV electron moving perpendicular to a uniform 2.20 T magnetic field, using the (a) classical and (b) relativistic formulas. (c) Calculate the period $T = 2\pi r/v$ of the circular motion using the relativistic formula for r. Is the result independent of the speed of the electron?

47P. Ionization measurements show that a certain low-mass nuclear particle has a charge of $2e$ and is moving with a speed of $0.710c$. The radius of curvature of its path in a magnetic field of 1.00 T is 6.28 m. (The path is a circle whose plane is perpendicular to the magnetic field.) Find the mass of the particle and identify it. [*Hint:* Low-mass nuclear particles are made up of neutrons (which have no charge) and protons (charge $= +e$), in roughly equal numbers. Take the mass of each of these particles to be 1.00 u. Also, see Problem 46.] ssm

48P. A 10 GeV proton in cosmic radiation moves in Earth's magnetic field $\vec{B}$, with its velocity $\vec{v}$ perpendicular to $\vec{B}$, in a region over which the field's average magnitude is 55 μT. What is the radius of the proton's curved path in that region? (See Problem 46.)

49P. A 2.50 MeV electron moves perpendicular to a magnetic field in a path whose radius of curvature is 3.0 cm. What is the magnitude B of the magnetic field? (See Problem 46.)

50P. The proton synchrotron at Fermilab accelerates protons to a kinetic energy of 500 GeV. At such a great energy, relativistic effects are important; in particular, as the speed of the proton increases, the time the proton takes to make a trip around its circular orbit in the synchrotron also increases. In a cyclotron, where the magnetic field magnitude and the oscillator frequency are fixed, this effect of time dilation will put the proton's circling out of synchronization with the oscillator. That eliminates repeated acceleration; hence, the proton will not reach an energy as great as 500 GeV. However, in a synchrotron, both the magnitude of the magnetic field and the oscillation frequency are varied to allow for the change due to time dilation.

At the energy of 500 GeV, calculate (a) the Lorentz factor, (b) the speed parameter, and (c) the magnetic field magnitude at the proton orbit, which has a radius of curvature of 750 m. (See Problem 46; use 938.3 MeV as the proton's rest energy.)

51P*. An alpha particle with kinetic energy 7.70 MeV collides with an ^{14}N nucleus at rest, and the two transform into an ^{17}O nucleus and a proton. The proton is emitted at 90° to the direction of the incident alpha particle and has a kinetic energy of 4.44 MeV. The masses of the various particles are: alpha particle, 4.00260 u; ^{14}N, 14.00307 u; proton, 1.007825 u; and ^{17}O, 16.99914 u. In megaelectron volts, what are (a) the kinetic energy of the oxygen nucleus and (b) the Q of the reaction? (*Hint:* The speeds of the particles are much less than c.)

Additional Problems

52. *The car-in-the-garage problem.* Carman has just purchased the world's longest stretch limo, which has a proper length of $L_c = 30.5$ m. In Fig. 38-21*a*, it is shown parked in front of a garage with a proper length of $L_g = 6.00$ m. The garage has a front door (shown open) and a back door (shown closed). The limo is obviously longer than the garage. Still, Garageman, who owns the garage and knows something about relativistic length contraction, makes a bet with Carman that the limo can fit in the garage with both doors closed. Carman, who dropped the physics course before reaching special relativity, says such a thing, even in principle, is impossible.

To analyze Garageman's scheme, an x_c axis is attached to the limo, with $x_c = 0$ at the rear bumper, and an x_g axis is attached to the garage, with $x_g = 0$ at the (now open) front door. Then Carman is to drive the limo directly toward the front door at a velocity of $0.9980c$ (which is, of course, both technically and financially impossible). Carman is stationary in the x_c reference frame; Garageman is stationary in the x_g reference frame.

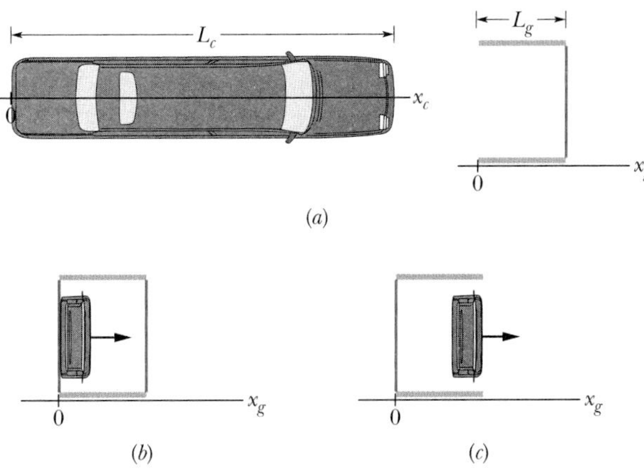

Fig. 38-21 Problem 52.

There are two events to consider. *Event 1:* When the rear bumper clears the front door, the front door is closed. Let the time of this event be zero to both Carman and Garageman: $t_{g1} = t_{c1} = 0$. The event occurs at $x_c = x_g = 0$. Figure 38-21*b* shows event 1 according to the x_g reference frame. *Event 2:* When the front bumper reaches the back door, that door opens. Figure 38-21*c* shows event 2 according to the x_g reference frame.

According to Garageman, (a) what is the length of the limo and (b) what are the spacetime coordinates x_{g2} and t_{g2} of event 2? (c) For how long is the limo temporarily "trapped" inside the garage, with both doors shut?

Now consider the situation from the x_c reference frame, in which the garage comes racing past the limo at a velocity of $-0.9980c$. According to Carman, (d) what is the length of the passing garage, (e) what are the spacetime coordinates x_{c2} and t_{c2} of event 2, (f) is the limo ever in the garage with both doors shut, and (g) which event occurs first? (h) Sketch events 1 and 2 as seen by Carman. (Are the events causally related; that is, does one of them cause the other?) (i) Finally, who wins the bet?

53. *Superluminal jets.* Figure 38-22*a* shows the path taken by a knot in a jet of ionized gas that has been expelled from a galaxy. The knot travels at constant velocity $\vec{v}$ at angle θ from the direction of Earth. The knot occasionally emits a burst of light, which is eventually detected on Earth. Two bursts are indicated in Fig. 38-22*a*, separated by time t as measured in a stationary frame near the bursts. The bursts are shown in Fig. 38-22*b* as if they were photographed on the same piece of film, first when light from burst 1 arrived on Earth and then later when light from burst 2 arrived. The apparent distance D_{app} traveled by the knot between the two bursts is the distance across an Earth-observer's view of the knot's path. The apparent time T_{app} between the bursts is the difference in the arrival times of the light from them. The apparent speed of the knot is then $V_{app} = D_{app}/T_{app}$. In terms of v, t, and θ, what are (a) D_{app} and (b) T_{app}? (c) Evaluate V_{app} for $v = 0.980c$ and $\theta = 30.0°$. When superluminal (faster than light) jets were first observed, they seemed to defy special relativity—at least until the geometry of viewing, such as in Fig. 38-22*a*, was understood.

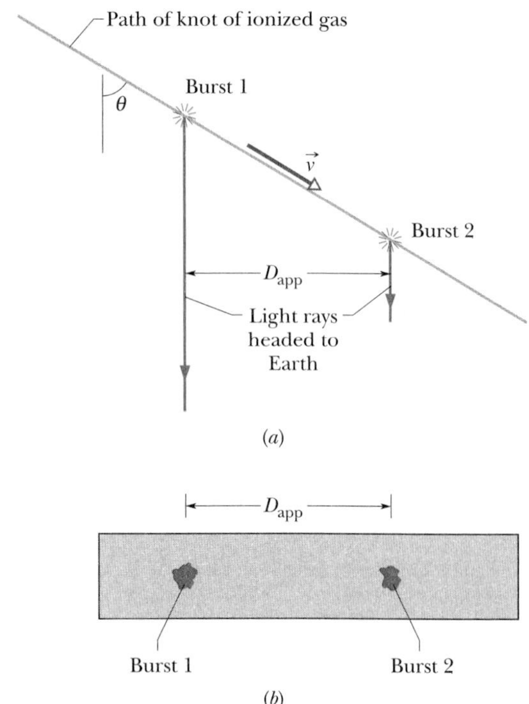

Fig. 38-22 Problem 53.

NEW PROBLEMS

N1. *Relativistic reversal of events.* Figures 38N-1*a* and *b* show the (usual) situation in which a primed reference frame passes an unprimed reference frame, in the common positive direction of the x and x' axes, at a constant relative velocity of magnitude v. We are at rest in the unprimed frame; Bullwinkle, an astute student of relativity in spite of his cartoon upbringing, is at rest in the primed frame. The figures also indicate events A and B that occur at the following spacetime coordinates as measured in our unprimed frame and in Bullwinkle's primed frame:

Event	Unprimed	Primed
A	(x_A, t_A)	(x'_A, t'_A)
B	(x_B, t_B)	(x'_B, t'_B)

In our frame, event A occurs before event B, with temporal separation $\Delta t = t_B - t_A = 1.00\ \mu s$ and spatial separation $\Delta x = x_B - x_A = 400$ m. (a) Let $\Delta t'$ be the temporal separation of the events according to Bullwinkle. Find an expression for $\Delta t'$ in terms of the speed parameter $\beta\ (= v/c)$ and the given data. Graph $\Delta t'$ versus β for the following two ranges of β:

0 to 0.01 (v is low, from 0 to $0.01c$)

0.1 to 1 (v is high, from $0.1c$ to the limit c)

Interpret the two graphs in words. (b) At what value of β is $\Delta t' = 0$?

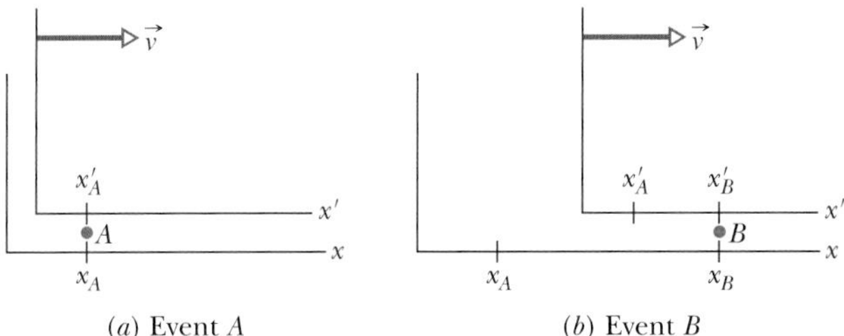

Fig. 38N-1 Problem N1.

N2. Continuation of Problem N1. Let $\Delta x'\ (= x'_B - x'_A)$ be the spatial separation between the two events as determined from the primed frame. (a) Graph $\Delta x'$ versus β from $\beta = 0$ to $\beta = 1$. (b) At what value of β is $\Delta x'$ minimum, and (c) what is that minimum? (d) Interpret the graph in words.

N3. The total energy of a proton passing through a laboratory apparatus is 10.611 nJ. What is its speed? Use the proton mass given in Appendix B under "Best Value," not the commonly remembered rounded number.

N4. *Son of Problem N1.* Reconsider the situation introduced in Problem N1 with this change: The spatial separation according to us is $\Delta x = x_B - x_A = 240$ m (the temporal separation is still $1.00\ \mu s$). Graph (a) $\Delta t' = t'_B - t'_A$ and (b) $\Delta x' = x'_B - x'_A$, each versus β. (c) At what value of β is $\Delta t'$ minimum, and (d) what is that minimum? (e) At what value of β is $\Delta x'$ equal to zero? Interpret, in words, the graphs for (f) the temporal separation and (g) the spatial separation. (h) Physically, why can't the sequence of the two events be reversed for high values of β as occurs in Problem N1?

N5. (a) The energy release in the explosion of 1.00 mol of TNT is 3.40 MJ. The molar mass of TNT is 0.227 kg/mol. What weight of TNT is needed for an explosive release of 1.80×10^{14} J? (b) Can you carry that weight in a backpack or is a truck or train required? (c) Suppose that in an explosion of a fission bomb, 0.080% of the fissionable mass is converted to the released energy. What weight of fissionable material is needed for an explosive release of 1.80×10^{14} J? (d) Can you carry that weight in a backpack or is a truck or train required?

N6. Bullwinkle in reference frame S' passes you in reference frame S along the common directions of the x' and x axes, as in Fig. 38-9. He carries three meter sticks: meter stick 1 is parallel to the x' axis, meter stick 2 is parallel to the y' axis, and meter stick 3 is parallel to the z' axis. On his wristwatch he counts off 15.0 s, which takes 30.0 s according to you. Two events occur during his passage. According to you, event 1 occurs at $x_1 = 33.0$ m and $t_1 = 22.0$ ns, and event 2 occurs at $x_2 = 53.0$ m and $t_2 = 62.0$ ns. According to your measurements, what are the lengths of (a) meter stick 1, (b) meter stick 2, and (c) meter stick 3? According to Bullwinkle, what are (d) the spatial separation and (e) the temporal separation between events 1 and 2, and (f) which event occurs first?

N7. An electron is accelerated from rest through a potential difference of 10.0 MV. In terms of c, what is its speed according to (a) classical (nonrelativistic) physics and (b) relativistic physics?

N8. Reference frame S' passes reference frame S with a certain velocity as in Fig. 38-9. Events 1 and 2 are to have a certain spatial separation $\Delta x'$ according to the S' observer. However, their temporal separation $\Delta t'$ according to that observer has not been set yet. Figure 38N-2 gives their spatial separation Δx according to the S observer as a function of $\Delta t'$ for a range of $\Delta t'$ values. What is $\Delta x'$?

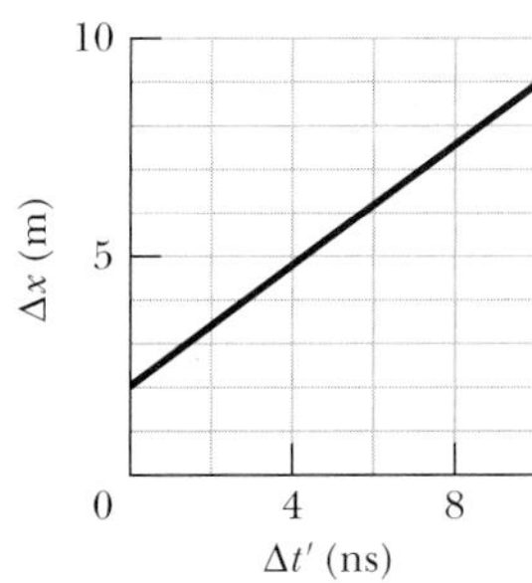

Fig. 38N-2 Problem N8.

N9. Apply the binomial theorem (Appendix E) to the last part of Eq. 38-49 for the kinetic energy of a particle. (a) Retain the first two terms of the expansion to show the kinetic energy in the form

$$K = (\text{first term}) + (\text{second term}).$$

The first term is the classical expression for kinetic energy. The second term is the first-order correction to the classical expression. Assume the particle is an electron. If its speed v is $c/20$, what are the values of (b) the classical expression and (c) the first-order correction? If the electron's speed is $0.80c$, what are the values of (d) the classical expression and (e) the first-order correction? (f) In terms of c, at what speed v does the first-order correction become 10% or greater of the classical expression?

N10. *(Come) back to the future.* Suppose that a father is 20.00 y older than his daughter. He wants to travel outward from Earth for 2.000 y and then back to Earth for another 2.000 y (both intervals as he measures them) such that he is then 20.00 y *younger* than his daughter. In terms of c, what constant speed (relative to Earth) is required for the trip?

N11. *Another approach to velocity transformations.* In Fig. 38N-3, reference frames B and C move past reference frame A in the common direction of their x axes. Represent the x components of the velocities of one frame relative to another with a double subscript. For example, v_{AB} is the x component of the velocity of A relative to B. Similarly, represent the corresponding speed parameters with double subscripts. For example, β_{AB} ($= v_{AB}/c$) is the speed parameter corresponding to v_{AB}. (a) Show that

$$\beta_{AC} = \frac{\beta_{AB} + \beta_{BC}}{1 + \beta_{AB}\beta_{BC}}.$$

Let M_{AB} represent the ratio $(1 - \beta_{AB})/(1 + \beta_{AB})$, and let M_{BC} and M_{AC} represent similar ratios. (b) Show that the relation

$$M_{AC} = M_{AB}M_{BC}$$

is true by deriving the equation of part (a) from it.

Fig. 38N-3 Problem N11.

N12. Continuation of Problem N11. Use the result of part (b) in Problem N11 for the motion along a single axis in the following situation. Frame A is attached to a particle that moves with velocity $+0.500c$ past frame B, which moves past frame C with a velocity of $+0.500c$. What are (a) M_{AC}, (b) β_{AC}, and (c) the velocity of the particle relative to frame C?

N13. Continuation of Problem N11. Let reference frame C move past reference frame D. (a) Show that

$$M_{AD} = M_{AB}M_{BC}M_{CD}.$$

(b) Now put this general result to work: Three particles move parallel to a single axis on which an observer is stationed. Let plus and minus signs indicate the directions of motion along that axis. Particle A moves past particle B at $\beta_{AB} = +0.20$. Particle B moves past particle C at $\beta_{BC} = -0.40$. Particle C moves past observer D at $\beta_{CD} = +0.60$. What is the velocity of particle A relative to observer D? (The solution technique here is *much* faster than using Eq. 38-28.)

N14. As in Fig. 38-9, reference frame S' passes reference frame S with a certain velocity. Events 1 and 2 are to have a certain temporal separation $\Delta t'$ according to the S' observer. However, their spatial separation $\Delta x'$ according to that observer has not been set yet. Figure 38N-4 gives their temporal separation Δt according to the S observer as a function of $\Delta x'$ for a range of $\Delta x'$ values. What is $\Delta t'$?

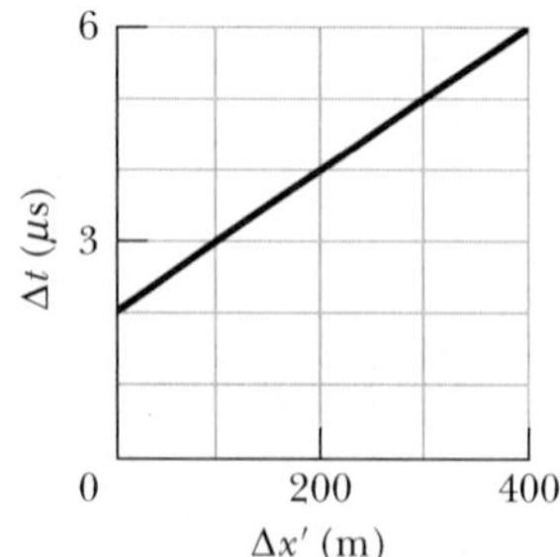

Fig. 38N-4 Problem N14.

N15. In Fig. 38N-5*a*, particle P is to move parallel to the x and x' axes of reference frames S and S', at a certain velocity relative to frame S. Frame S' is to move parallel to the x axis of frame S at velocity v. Figure 38N-5*b* gives the velocity u' of the particle relative to frame S' for a range of values for v. What value will u' have if (a) $v = 0.90c$ and (b) $v \rightarrow c$?

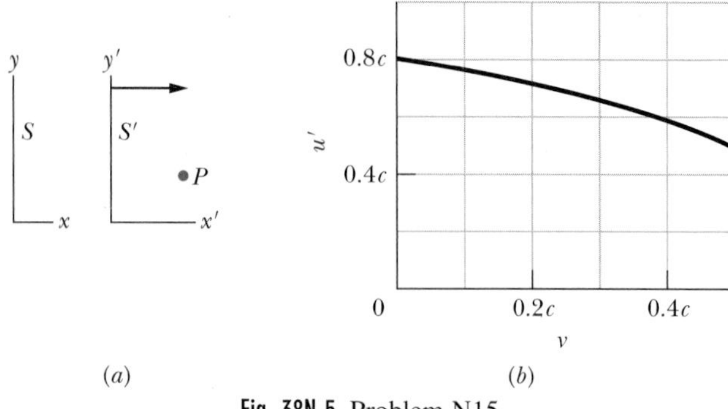

Fig. 38N-5 Problem N15.

N16. A rod is to move at constant speed v along the x axis of reference frame S, with its length parallel to that axis. An observer in frame S is to measure the length L of the rod. Figure 38N-6 gives length L versus speed parameter β for a range of values for β. What is L if $v = 0.95c$?

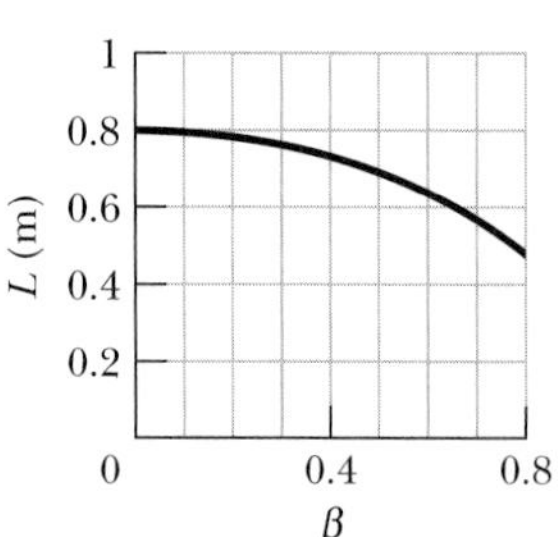

Fig. 38N-6 Problem N16.

N17. Reference frame S' is to pass reference frame S at speed v along the common directions of the x' and x axes, as in Fig. 38-9. An observer who rides along with frame S' is to count off a certain time interval on his wristwatch. The corresponding time interval Δt is to be measured by an observer in frame S. Figure 38N-7 gives Δt versus speed parameter β for a range of values for β. What is interval Δt if $v = 0.98c$?

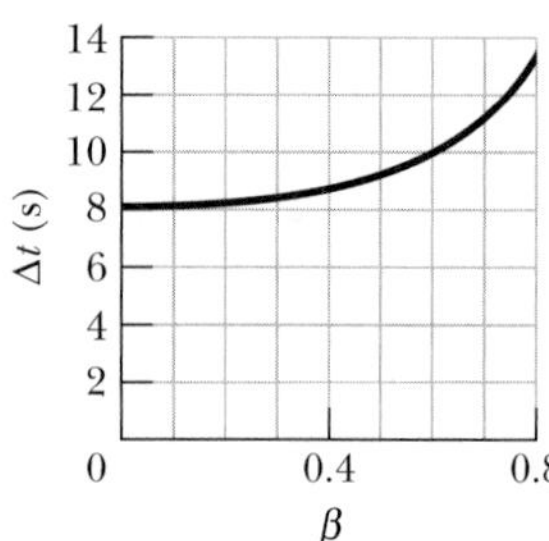

Fig. 38N-7 Problem N17.

N18. In Fig. 38N-8*a*, particle P is to move parallel to the x and x' axes of reference frames S and S', at a certain velocity relative to frame S. Frame S' is to move parallel to the x axis of frame S at velocity v. Figure 38N-8*b* gives the velocity u' of the particle relative to frame S' for a range of values for v. What value will u' have if (a) $v = 0.80c$ and (b) $v \rightarrow c$?

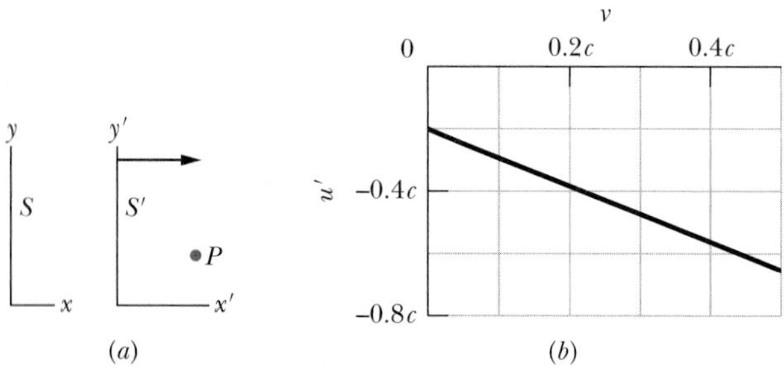

Fig. 38N-8 Problem N18.

39 Photons and Matter Waves

Tracks of tiny vapor bubbles in this bubble-chamber image reveal where electrons (tracks color-coded green) and positrons (red) moved. A gamma ray (which left no track when it entered at the top) kicked an electron out of one of the hydrogen atoms filling the chamber and then converted to an electron–positron pair. Another gamma ray underwent another pair production farther down. These tracks (curved because of a magnetic field) clearly show that electrons and positrons are particles that move along narrow paths. Yet, those particles can also be interpreted in terms of waves.

Can a particle be a wave?

The answer is in this chapter.

39-1 A New Direction

Our discussion of Einstein's theory of relativity took us into a world far beyond that of ordinary experience—the world of objects moving at speeds close to the speed of light. Among other surprises, Einstein's theory predicts that the rate at which a clock runs depends on how fast the clock is moving relative to the observer: the faster the motion, the slower the clock rate. This and other predictions of the theory have passed every experimental test devised thus far, and relativity theory has led us to a deeper and more satisfying view of the nature of space and time.

Now you are about to explore a second world that is outside ordinary experience—the subatomic world. You will encounter a new set of surprises that, though they may sometimes seem bizarre, have led physicists step by step to a deeper view of reality.

Quantum physics, as our new subject is called, answers such questions as: Why do the stars shine? Why do the elements exhibit the order that is so apparent in the periodic table? How do transistors and other microelectronic devices work? Why does copper conduct electricity but glass does not? Because quantum physics accounts for all of chemistry, including biochemistry, we need to understand it if we are to understand life itself.

Some of the predictions of quantum physics seem strange even to the physicists and philosophers who study its foundations. Still, experiment after experiment has proved the theory correct, and many have exposed even stranger aspects of the theory. The quantum world is an amusement park full of wonderful rides that are guaranteed to shake up the commonsense world view you have developed since childhood. We begin our exploration of that quantum park with the photon.

39-2 The Photon, the Quantum of Light

Quantum physics (which is also known as *quantum mechanics* and *quantum theory*) is largely the study of the microscopic world. There many quantities are found only in certain minimum (*elementary*) amounts, or integer multiples of those elementary amounts; they are then said to be *quantized*. The elementary amount that is associated with such a quantity is called the **quantum** of that quantity (*quanta* is the plural).

In a loose sense, U.S. currency is quantized because the coin of least value is the penny, or \$0.01 coin, and the values of all other coins and bills are restricted to integer multiples of that least amount. In other words, the currency quantum is \$0.01, and all greater amounts of currency are of the form n(\$0.01), where n is a positive integer. For example, you cannot hand someone \$0.755 = 75.5(\$0.01).

In 1905, Einstein proposed that electromagnetic radiation (or simply *light*) is quantized and exists in elementary amounts (quanta) that we now call **photons.** This proposal should seem strange to you because we have just spent several chapters discussing the classical idea that light is a sinusoidal wave, with a wavelength λ, a frequency f, and a speed c such that

$$f = \frac{c}{\lambda}. \tag{39-1}$$

Furthermore, in Chapter 34 we discussed the classical light wave as being an interdependent combination of electric and magnetic fields, each oscillating at frequency f. How can this wave of oscillating fields consist of an elementary amount of something—the light quantum? What *is* a photon?

The concept of a light quantum, or a photon, turns out to be far more subtle and mysterious than Einstein imagined. Indeed, it is still very poorly understood. In this book, we shall discuss only some of the basic aspects of the photon concept, somewhat along the lines of Einstein's proposal.

According to that proposal, the quantum of a light wave of frequency f has the energy

$$E = hf \quad \text{(photon energy)}. \tag{39-2}$$

Here h is the **Planck constant,** which has the value

$$h = 6.63 \times 10^{-34}\ \mathrm{J \cdot s} = 4.14 \times 10^{-15}\ \mathrm{eV \cdot s}. \tag{39-3}$$

The least energy a light wave of frequency f can have is hf, the energy of a single photon. If the wave has more energy, its total energy must be an integer multiple of hf, just as the currency in our previous example must be an integer multiple of \$0.01. The light cannot have an energy of $0.6hf$ or $75.5hf$.

Einstein further proposed that when light is absorbed or emitted by an object (matter), the absorption or emission event occurs at the atoms of the object. When light of frequency f is absorbed by an atom, the energy hf of one photon is transferred from the light to the atom. In this *absorption event,* the photon vanishes and the atom is said to absorb it. When light of frequency f is emitted by an atom, an energy hf is transferred from the atom to the light. In this *emission event,* a photon suddenly appears and the atom is said to emit it. Thus, we can have *photon absorption* and *photon emission* by atoms in an object.

For an object consisting of many atoms, there can be many photon absorptions (such as with sunglasses) or photon emissions (such as with lamps). However, each absorption or emission event still involves the transfer of energy equal to that of a single photon of the light.

When we discussed the absorption or emission of light in previous chapters, our examples involved so much light that we had no need of quantum physics, and we got by with classical physics. However, in the late twentieth century, technology became advanced enough that single-photon experiments could be conducted and put to practical use. Since then quantum physics has become part of standard engineering practice, especially in optical engineering.

CHECKPOINT 1: Rank the following radiations according to their associated photon energies, greatest first: (a) yellow light from a sodium vapor lamp, (b) a gamma ray emitted by a radioactive nucleus, (c) a radio wave emitted by the antenna of a commercial radio station, (d) a microwave beam emitted by airport traffic control radar.

Sample Problem 39-1

A sodium vapor lamp is placed at the center of a large sphere that absorbs all the light reaching it. The rate at which the lamp emits energy is 100 W; assume that the emission is entirely at a wavelength of 590 nm. At what rate are photons absorbed by the sphere?

SOLUTION: We assume that all the light emitted by the lamp reaches (and thus is absorbed by) the sphere. Then the **Key Idea** is that the light is emitted and absorbed as photons. The rate R at which photons are absorbed by the sphere is equal to the rate R_{emit} at which photons are emitted by the lamp. That rate is

$$R_{\text{emit}} = \frac{\text{rate of energy emission}}{\text{energy per emitted photon}} = \frac{P_{\text{emit}}}{E}.$$

We then have, from Eq. 39-2 ($E = hf$),

$$R = R_{\text{emit}} = \frac{P_{\text{emit}}}{hf}.$$

Using Eq. 39-1 ($f = c/\lambda$) to substitute for f and then entering known data, we obtain

$$\begin{aligned} R &= \frac{P_{\text{emit}}\lambda}{hc} \\ &= \frac{(100\ \mathrm{W})(590 \times 10^{-9}\ \mathrm{m})}{(6.63 \times 10^{-34}\ \mathrm{J \cdot s})(3.0 \times 10^{8}\ \mathrm{m/s})} \\ &= 2.97 \times 10^{20}\ \text{photons/s}. \end{aligned} \quad \text{(Answer)}$$

39-3 The Photoelectric Effect

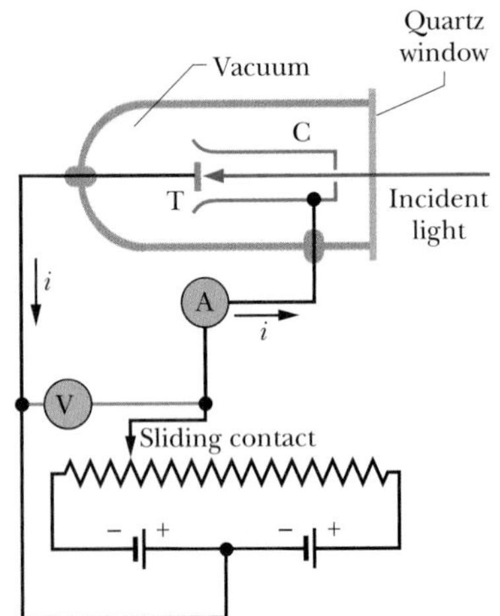

Fig. 39-1 An apparatus used to study the photoelectric effect. The incident light shines on target T, ejecting electrons, which are collected by collector cup C. The electrons move in the circuit in a direction opposite the conventional current arrows. The batteries and the variable resistor are used to produce and adjust the electric potential difference between T and C.

If you direct a beam of light of short enough wavelength onto a clean metal surface, the light will cause electrons to leave that surface (the light will *eject* the electrons from the surface). This **photoelectric effect** is used in many devices, including TV cameras, camcorders, and night vision viewers. Einstein supported his photon concept by using it to explain this effect, which simply cannot be understood without quantum physics.

Let us analyze two basic photoelectric experiments, each using the apparatus of Fig. 39-1 in which light of frequency f is directed onto target T and ejects electrons from it. A potential difference V is maintained between target T and collector cup C to sweep up these electrons, said to be **photoelectrons.** This collection produces a **photoelectric current** i that is measured with meter A.

First Photoelectric Experiment

We adjust the potential difference V by moving the sliding contact in Fig. 39-1 so that collector C is slightly negative with respect to target T. This potential difference acts to slow down the ejected electrons. We then vary V until it reaches a certain value, called the **stopping potential** V_{stop}, at which the reading of meter A has just dropped to zero. When $V = V_{stop}$, the most energetic ejected electrons are turned back just before reaching the collector. Then K_{max}, the kinetic energy of these most energetic electrons, is

$$K_{max} = eV_{stop}, \tag{39-4}$$

where e is the elementary charge.

Measurements show that for light of a given frequency, K_{max} *does not depend on the intensity of the light source.* Whether the source is dazzling bright or so feeble that you can scarcely detect it (or has some intermediate brightness), the maximum kinetic energy of the ejected electrons always has the same value.

This experimental result is a puzzle for classical physics. Classically, the incident light is a sinusoidally oscillating electromagnetic wave. An electron in the target should oscillate sinusoidally due to the oscillating electric force on it from the wave's electric field. If the amplitude of the electron's oscillation is great enough, the electron should break free of the target's surface—that is, be ejected from the target. Thus, if we increase the amplitude of the wave and its oscillating electric field, the electron should get a more energetic "kick" as it is being ejected. *However, that is not what happens.* For a given frequency, intense light beams and feeble light beams give exactly the same maximum kick to ejected electrons.

The actual result follows naturally if we think in terms of photons. Now the energy that can be transferred from the incident light to an electron in the target is that of a single photon. Increasing the light intensity increases the *number* of photons in the light, but the photon energy, given by Eq. 39-2, is unchanged because the frequency is unchanged. Thus, the energy transferred to the kinetic energy of an electron is also unchanged.

Second Photoelectric Experiment

Now we vary the frequency f of the incident light and measure the associated stopping potential V_{stop}. Figure 39-2 is a plot of V_{stop} versus f. Note that the photoelectric effect does not occur if the frequency is below a certain **cutoff frequency** f_0 or, equivalently, if the wavelength is greater than the corresponding **cutoff wavelength** $\lambda_0 = c/f_0$. This is so *no matter how intense the incident light is.*

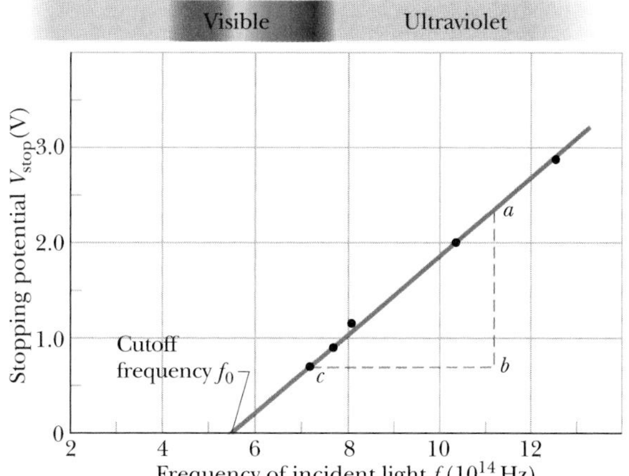

Fig. 39-2 The stopping potential V_{stop} as a function of the frequency f of the incident light for a sodium target T in the apparatus of Fig. 39-1. (Data reported by R. A. Millikan in 1916.)

This is another puzzle for classical physics. If you view light as an electromagnetic wave, you must expect that no matter how low the frequency, electrons can always be ejected by light if you supply them with enough energy—that is, if you use a light source that is bright enough. *That is not what happens.* For light below the cutoff frequency f_0, the photoelectric effect does not occur, no matter how bright the light source.

The existence of a cutoff frequency is, however, just what we should expect if the energy is transferred via photons. The electrons within the target are held there by electric forces. (If they weren't, they would drip out of the target due to the gravitational force on them.) To just escape from the target, an electron must pick up a certain minimum energy Φ, where Φ is a property of the target material called its **work function.** If the energy hf transferred to an electron by a photon exceeds the work function of the material (if $hf > \Phi$), the electron can escape the target. If the energy transferred does not exceed the work function (that is, if $hf < \Phi$), the electron cannot escape. This is what Fig. 39-2 shows.

The Photoelectric Equation

Einstein summed up the results of such photoelectric experiments in the equation

$$hf = K_{max} + \Phi \quad \text{(photoelectric equation).} \tag{39-5}$$

This is a statement of the conservation of energy for a single photon absorption by a target with work function Φ. Energy equal to the photon's energy hf is transferred to a single electron in the material of the target. If the electron is to escape from the target, it must pick up energy at least equal to Φ. Any additional energy ($hf - \Phi$) that the electron acquires from the photon appears as kinetic energy K of the electron. In the most favorable circumstance, the electron can escape through the surface without losing any of this kinetic energy in the process; it then appears outside the target with the maximum possible kinetic energy K_{max}.

Let us rewrite Eq. 39-5 by substituting for K_{max} from Eq. 39-4. After a little rearranging we get

$$V_{stop} = \left(\frac{h}{e}\right)f - \frac{\Phi}{e}. \tag{39-6}$$

The ratios h/e and Φ/e are constants, so we would expect a plot of the measured stopping potential V_{stop} versus the frequency f of the light to be a straight line, as it is in Fig. 39-2. Further, the slope of that straight line should be h/e. As a check, we measure ab and bc in Fig. 39-2 and write

$$\frac{h}{e} = \frac{ab}{bc} = \frac{2.35\text{ V} - 0.72\text{ V}}{(11.2 \times 10^{14} - 7.2 \times 10^{14})\text{ Hz}}$$
$$= 4.1 \times 10^{-15}\text{ V}\cdot\text{s}.$$

Multiplying this result by the elementary charge e, we find

$$h = (4.1 \times 10^{-15}\text{ V}\cdot\text{s})(1.6 \times 10^{-19}\text{ C}) = 6.6 \times 10^{-34}\text{ J}\cdot\text{s},$$

which agrees with values measured by many other methods.

An aside: An explanation of the photoelectric effect certainly requires quantum physics. For many years, Einstein's explanation was also a compelling argument for the existence of photons. However, in 1969 an alternative explanation for the effect was found that used quantum physics but did not need the concept of photons. Light *is* in fact quantized as photons, but Einstein's explanation of the photoelectric effect is not the best argument for that fact.

CHECKPOINT 2: The figure shows data like those of Fig. 39-2 for targets of cesium, potassium, sodium, and lithium. The plots are parallel. (a) Rank the targets according to their work functions, greatest first. (b) Rank the plots according to the value of h they yield, greatest first.

Sample Problem 39-2

A potassium foil is a distance $r = 3.5$ m from an isotropic light source that emits energy at the rate $P = 1.5$ W. The work function Φ of potassium is 2.2 eV. Suppose that the energy transported by the incident light were transferred to the target foil continuously and smoothly (that is, if classical physics prevailed instead of quantum physics). How long would it take for the foil to absorb enough energy to eject an electron? Assume that the foil totally absorbs all the energy reaching it and that the to-be-ejected electron collects energy from a circular patch of the foil whose radius is 5.0×10^{-11} m, about that of a typical atom.

SOLUTION: The **Key Ideas** here are these:

1. The time interval Δt required for the patch to absorb energy ΔE depends on the rate P_{abs} at which the energy is absorbed:
$$\Delta t = \frac{\Delta E}{P_{abs}}.$$
2. If the electron is to be ejected from the foil, the least energy ΔE it must gain from the light is equal to the work function Φ of potassium. Thus,
$$\Delta t = \frac{\Phi}{P_{abs}}.$$
3. Because the patch is totally absorbing, the rate of absorption P_{abs} is equal to the rate P_{arr} at which energy arrives at the patch; that is,
$$\Delta t = \frac{\Phi}{P_{arr}}.$$
4. With Eq. 34-23, we can relate the energy arrival rate P_{arr} to the intensity I of the light at the patch and the area A of the patch:
$$P_{arr} = IA.$$
Then
$$\Delta t = \frac{\Phi}{IA}.$$
5. Because the light source is isotropic, the light intensity I at distance r from the source depends on the rate P_{emit} at which energy is emitted by the source, according to Eq. 34-27:
$$I = \frac{P_{emit}}{4\pi r^2}.$$

Thus, finally, we have

$$\Delta t = \frac{4\pi r^2 \Phi}{P_{emit} A}.$$

The detection area A is $\pi(5.0 \times 10^{-11}\text{ m})^2 = 7.85 \times 10^{-21}\text{ m}^2$, and the work function Φ is 2.2 eV $= 3.5 \times 10^{-19}$ J. Substituting these and other data, we find that

$$\Delta t = \frac{4\pi(3.5\text{ m})^2(3.5 \times 10^{-19}\text{ J})}{(1.5\text{ W})(7.85 \times 10^{-21}\text{ m}^2)}$$
$$= 4580\text{ s} \approx 1.3\text{ h}. \qquad \text{(Answer)}$$

Thus, classical physics tells us that we would have to wait more than an hour after turning on the light source for a photoelectron to be ejected. The actual waiting time is less than 10^{-9} s. Apparently, then, an electron does *not* gradually absorb energy from the light arriving at the patch containing the electron. Rather, either the electron does not absorb any energy at all or it absorbs a quantum of energy instantaneously, by absorbing a photon from the light.

Sample Problem 39-3

Find the work function Φ of sodium from Fig. 39-2.

SOLUTION: The **Key Idea** here is that we can find the work function Φ from the cutoff frequency f_0 (which we can measure on the plot). The reasoning is this: At the cutoff frequency, the kinetic energy K_{max} in Eq. 39-5 is zero. Thus, all the energy hf that is transferred from a photon to an electron goes into the electron's escape, which requires an energy of Φ. Equation 39-5 then gives us, with $f = f_0$,

$$hf_0 = 0 + \Phi = \Phi.$$

In Fig. 39-2, the cutoff frequency f_0 is the frequency at which the plotted line intercepts the horizontal frequency axis, about 5.5×10^{14} Hz. We then have

$$\Phi = hf_0 = (6.63 \times 10^{-34}\ \text{J}\cdot\text{s})(5.5 \times 10^{14}\ \text{Hz})$$
$$= 3.6 \times 10^{-19}\ \text{J} = 2.3\ \text{eV}. \qquad \text{(Answer)}$$

39-4 Photons Have Momentum

In 1916, Einstein extended his concept of light quanta (photons) by proposing that a quantum of light has linear momentum. For a photon with energy hf, the magnitude of that momentum is

$$p = \frac{hf}{c} = \frac{h}{\lambda} \quad \text{(photon momentum)}, \qquad (39\text{-}7)$$

where we have substituted for f from Eq. 39-1 ($f = c/\lambda$). Thus, when a photon interacts with matter, energy *and* momentum are transferred, *as if* there were a collision between the photon and matter in the classical sense (as in Chapter 10).

In 1923, Arthur Compton at Washington University in St. Louis carried out an experiment that supported the view that both momentum and energy are transferred via photons. He arranged for a beam of x rays of wavelength λ to be directed onto a target made of carbon, as shown in Fig. 39-3. An x ray is a form of electromagnetic radiation, at high frequency and thus small wavelength. Compton measured the wavelengths and intensities of the x rays that were scattered in various directions from his carbon target.

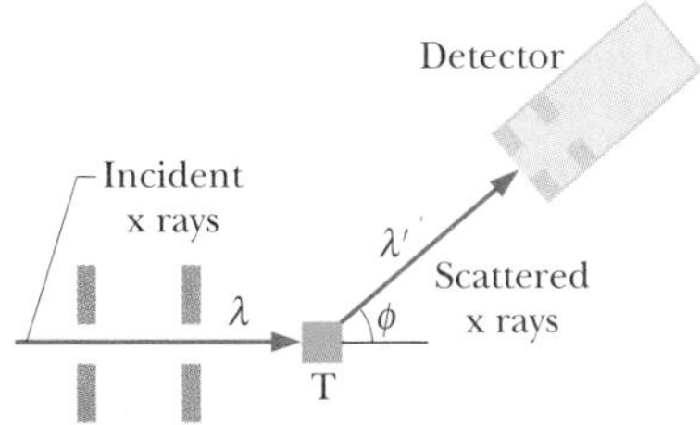

Fig. 39-3 Compton's apparatus. A beam of x rays of wavelength $\lambda = 71.1$ pm is directed onto a carbon target T. The x rays scattered from the target are observed at various angles ϕ to the direction of the incident beam. The detector measures both the intensity of the scattered x rays and their wavelength.

Figure 39-4 shows his results. Although there is only a single wavelength ($\lambda = 71.1$ pm) in the incident x-ray beam, we see that the scattered x rays contain a range

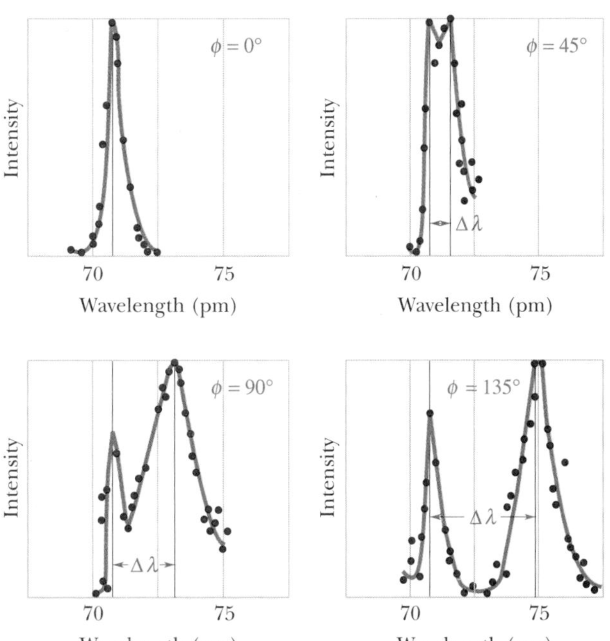

Fig. 39-4 Compton's results for four values of the scattering angle ϕ. Note that the Compton shift $\Delta\lambda$ increases as the scattering angle increases.

of wavelengths with two prominent intensity peaks. One peak is centered about the incident wavelength λ, the other about a wavelength λ' that is longer than λ by an amount $\Delta\lambda$, which is called the **Compton shift.** The value of the Compton shift varies with the angle at which the scattered x rays are detected.

Figure 39-4 is still another puzzle for classical physics. Classically, the incident x-ray beam is a sinusoidally oscillating electromagnetic wave. An electron in the carbon target should oscillate sinusoidally due to the oscillating electric force on it from the wave's electric field. Further, the electron should oscillate at the same frequency as the wave and should send out waves *at this same frequency,* as if it were a tiny transmitting antenna. Thus, the x rays scattered by the electron should have the same frequency, and the same wavelength, as the x rays in the incident beam—but they don't.

Compton interpreted the scattering of x rays from carbon in terms of energy and momentum transfers, via photons, between the incident x-ray beam and loosely bound electrons in the carbon target. Let us see, first conceptually and then quantitatively, how this quantum physics interpretation leads to an understanding of Compton's results.

Suppose a single photon (of energy $E = hf$) is associated with the interaction between the incident x-ray beam and a stationary electron. In general, the direction of travel of the x ray will change (the x ray is scattered) and the electron will recoil, which means that the electron has obtained some kinetic energy. Energy is conserved in this isolated interaction. Thus, the energy of the scattered photon ($E' = hf'$) must be less than that of the incident photon. The scattered x rays must then have a lower frequency f' and thus a longer wavelength λ' than the incident x rays, just as Compton's experimental results in Fig. 39-4 show.

For the quantitative part, we first apply the law of conservation of energy. Figure 39-5 suggests a "collision" between an x ray and an initially stationary free electron in the target. As a result of the collision, an x ray of wavelength λ' moves off at an angle ϕ and the electron moves off at an angle θ, as shown. The conservation of energy then gives us

$$hf = hf' + K,$$

in which hf is the energy of the incident x-ray photon, hf' is the energy of the scattered x-ray photon, and K is the kinetic energy of the recoiling electron. Because the electron may recoil with a speed comparable to that of light, we must use the relativistic expression of Eq. 38-49,

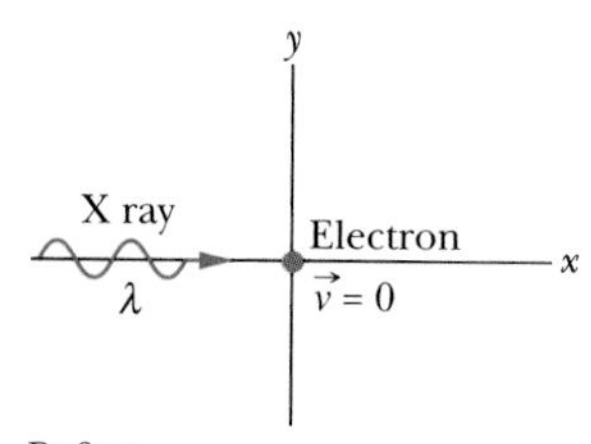

Fig. 39-5 An x ray of wavelength λ interacts with a stationary electron. The x ray is scattered at angle ϕ, with an increased wavelength λ'. The electron moves off with speed v at angle θ.

$$K = mc^2(\gamma - 1),$$

for the electron's kinetic energy. Here m is the electron's mass and γ is the Lorentz factor

$$\gamma = \frac{1}{\sqrt{1 - (v/c)^2}}.$$

Substituting for K in the conservation of energy equation yields

$$hf = hf' + mc^2(\gamma - 1).$$

Substituting c/λ for f and c/λ' for f' then leads to the new energy conservation equation

$$\frac{h}{\lambda} = \frac{h}{\lambda'} + mc(\gamma - 1). \tag{39-8}$$

Next we apply the law of conservation of momentum to the x-ray–electron collision of Fig. 39-5. From Eq. 39-7, the magnitude of the momentum of the incident photon is h/λ, and that of the scattered photon is h/λ'. From Eq. 38-38, the magnitude for the recoiling electron's momentum is γmv. Because we have a two-dimensional situation, we write separate equations for the conservation of momentum along the x and y axes, obtaining

$$\frac{h}{\lambda} = \frac{h}{\lambda'}\cos\phi + \gamma mv\cos\theta \qquad (x \text{ axis}) \tag{39-9}$$

and

$$0 = \frac{h}{\lambda'}\sin\phi - \gamma mv\sin\theta \qquad (y \text{ axis}). \tag{39-10}$$

We want to find $\Delta\lambda$ $(= \lambda' - \lambda)$, the Compton shift of the scattered x rays. Of the five collision variables (λ, λ', v, ϕ, and θ) that appear in Eqs. 39-8, 39-9, and 39-10, we choose to eliminate v and θ, which deal only with the recoiling electron. Carrying out the algebra (it is somewhat complicated) leads to an equation for the Compton shift as a function of the scattering angle ϕ:

$$\Delta\lambda = \frac{h}{mc}(1 - \cos\phi) \qquad \text{(Compton shift)}. \tag{39-11}$$

Equation 39-11 agrees exactly with Compton's experimental results.

The quantity h/mc in Eq. 39-11 is a constant called the **Compton wavelength.** Its value depends on the mass m of the particle from which the x rays scatter. Here that particle is a loosely bound electron, and thus we would substitute the mass of an electron for m to evaluate the *Compton wavelength for Compton scattering from an electron*.

A Loose End

The peak at the incident wavelength λ (= 71.1 pm) in Fig. 39-4 still needs to be explained. This peak arises not from interactions between x rays and the very loosely bound electrons in the target but from interactions between x rays and the electrons that are *tightly* bound to the carbon atoms making up the target. Effectively, each of these latter collisions occurs between an incident x ray and an entire carbon atom. If we substitute for m in Eq. 39-11 the mass of a carbon atom (which is about 22 000 times that of an electron), we see that $\Delta\lambda$ becomes about 22 000 times smaller than the Compton shift for an electron—too small to detect. Thus, the x rays scattered in these collisions have the same wavelength as the incident x rays.

Sample Problem 39-4

X rays of wavelength λ = 22 pm (photon energy = 56 keV) are scattered from a carbon target, and the scattered rays are detected at 85° to the incident beam.

(a) What is the Compton shift of the scattered rays?

SOLUTION: The **Key Idea** here is that the Compton shift is the wavelength change of the x rays due to scattering from loosely bound electrons in a target. Further, that shift depends on the angle at which the scattered x rays are detected, according to Eq. 39-11. Substituting 85° for that angle and 9.11×10^{-31} kg for the electron mass (because the scattering is from electrons) in Eq. 39-11 gives us

$$\Delta\lambda = \frac{h}{mc}(1 - \cos\phi)$$

$$= \frac{(6.63 \times 10^{-34}\ \text{J}\cdot\text{s})(1 - \cos 85^\circ)}{(9.11 \times 10^{-31}\ \text{kg})(3.00 \times 10^{8}\ \text{m/s})}$$

$$= 2.21 \times 10^{-12}\ \text{m} \approx 2.2\ \text{pm}. \qquad \text{(Answer)}$$

(b) What percentage of the initial x-ray photon energy is transferred to an electron in such scattering?

SOLUTION: The **Key Idea** here is to find the *fractional energy loss* (let us call it *frac*) for photons that scatter from the electrons:

$$frac = \frac{\text{energy loss}}{\text{initial energy}} = \frac{E - E'}{E}.$$

From Eq. 39-2 ($E = hf$), we can substitute for the initial energy E and the detected energy E' of the x rays in terms of frequencies. Then, from Eq. 39-1 ($f = c/\lambda$), we can substitute for those frequencies in terms of the wavelengths. We find

$$frac = \frac{hf - hf'}{hf} = \frac{c/\lambda - c/\lambda'}{c/\lambda} = \frac{\lambda' - \lambda}{\lambda'} = \frac{\Delta\lambda}{\lambda + \Delta\lambda}. \quad (39\text{-}12)$$

Substitution of data yields

$$frac = \frac{2.21 \text{ pm}}{22 \text{ pm} + 2.21 \text{ pm}} = 0.091 \quad \text{or} \quad 9.1\%. \quad \text{(Answer)}$$

Although the Compton shift $\Delta\lambda$ is independent of the wavelength λ of the incident x rays (see Eq. 39-11), the *fractional* photon energy loss of the x rays does depend on λ, increasing as the wavelength of the incident radiation decreases, as indicated by Eq. 39-12.

CHECKPOINT 3: Compare Compton scattering for x rays ($\lambda \approx 20$ pm) and visible light ($\lambda \approx 500$ nm) at a particular angle of scattering. Which has the greater (a) Compton shift, (b) fractional wavelength shift, (c) fractional photon energy change, and (d) energy imparted to the electron?

39-5 Light as a Probability Wave

A fundamental mystery in physics is how light can be a wave (which spreads out over a region) in classical physics, whereas it is emitted and absorbed as photons (which originate and vanish at points) in quantum physics. The double-slit experiment of Section 36-4 lies at the heart of this mystery. Let us discuss three versions of that experiment.

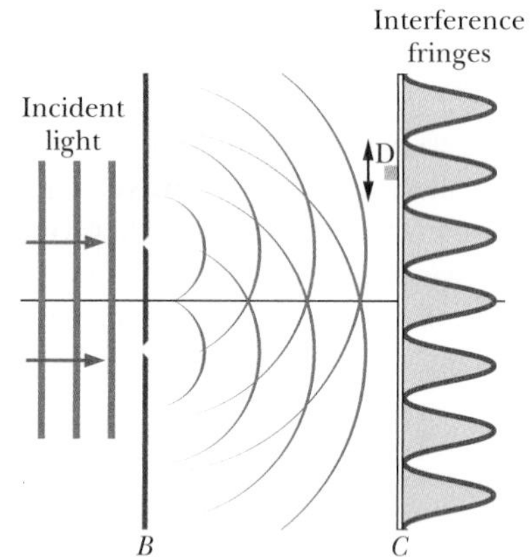

Fig. 39-6 Light is directed onto screen B, which contains two parallel slits. Light emerging from these slits spreads out by diffraction. The two diffracted waves overlap at screen C and form a pattern of interference fringes. A small photon detector D in the plane of screen C generates a sharp click for each photon that it absorbs.

The Standard Version

Figure 39-6 is a sketch of the original experiment carried out by Thomas Young in 1801 (see also Fig. 36-6). Light shines on screen B, which contains two narrow parallel slits. The light waves emerging from the two slits spread out by diffraction and overlap on screen C where, by interference, they form a pattern of alternating intensity maxima and minima. In Section 36-4 we took the existence of these interference fringes as compelling evidence for the wave nature of light.

Let us place a tiny photon detector D at one point in the plane of screen C. Let the detector be a photoelectric device that clicks when it absorbs a photon. We would find that the detector produces a series of clicks, randomly spaced in time, each click signaling the transfer of energy from the light wave to the screen via a photon absorption.

If we moved the detector very slowly up or down as indicated by the black arrow in Fig. 39-6, we would find that the click rate increases and decreases, passing through alternate maxima and minima that correspond exactly to the maxima and minima of the interference fringes.

The point of this thought experiment is as follows. We cannot predict when a photon will be detected at any particular point on screen C; photons are detected at individual points at random times. We can, however, predict that the relative *probability* that a single photon will be detected at a particular point in a specified time interval is proportional to the intensity of the incident light at that point.

We saw in Section 34-4 that the intensity I of a light wave at any point is proportional to the square of E_m, the amplitude of the oscillating electric field vector of the wave at that point. Thus,

> The probability (per unit time interval) that a photon will be detected in any small volume centered on a given point in a light wave is proportional to the square of the amplitude of the wave's electric field vector at that point.

We now have a probabilistic description of a light wave, hence another way to view light. It is not only an electromagnetic wave but it is also a **probability wave.** That is, to every point in a light wave we can attach a numerical probability (per unit time interval) that a photon can be detected in any small volume centered on that point.

The Single-Photon Version

A single-photon version of the double-slit experiment was first carried out by G. I. Taylor in 1909 and has been repeated many times since. It differs from the standard version in that the light source is so extremely feeble that it emits only one photon at a time, at random intervals. Astonishingly, interference fringes still build up on screen C if the experiment runs long enough (several months for Taylor's early experiment).

What explanation can we offer for the result of this single-photon double-slit experiment? Before we can even consider the result, we are compelled to ask questions like these: If the photons move through the apparatus one at a time, through which of the two slits in screen B does a given photon pass? How does a given photon even "know" that there is another slit present so that interference is a possibility? Can a single photon somehow pass through both slits and interfere with itself?

Bear in mind that we can only know when photons interact with matter—we have no way of detecting them without an interaction with matter, such as with a detector or a screen. Thus, in the experiment of Fig. 39-6, we can only know that photons originate at the light source and vanish at the screen. Between source and screen, we cannot know what the photon is or does. However, because an interference pattern eventually builds up on the screen, we can speculate that each photon travels from source to screen *as a wave* that fills up the space between those two objects and then vanishes in a photon absorption, with a transfer of energy and momentum, at some point on the screen.

We *cannot* predict where this transfer will occur (where a photon will be detected) for any given photon originating at the source. However, we *can* predict the probability that a transfer will occur at any given point on the screen. Transfers will tend to occur (and thus photons will tend to be absorbed) in the regions of the bright fringes in the interference pattern that builds up on the screen. Transfers will tend *not* to occur (and thus photons will tend *not* to be absorbed) in the regions of the dark fringes in the built-up pattern. Thus, we can say that the wave traveling from the source to the screen is a *probability wave,* which produces a pattern of "probability fringes" on the screen.

The Single-Photon, Wide-Angle Version

In the past, physicists tried to explain the single-photon double-slit experiment in terms of small packets of classical light waves that are individually sent toward the slits. They would define these small packets as photons. However, modern experiments invalidate this explanation and definition. Figure 39-7 shows the arrangement of one of these experiments, reported in 1992 by Ming Lai and Jean-Claude Diels of the University of New Mexico. Source S contains molecules that emit photons at well separated times. Mirrors M_1 and M_2 are positioned to reflect light that the source emits along two distinct paths, 1 and 2, that are separated by an angle θ, which is close to 180°. This arrangement differs from the standard two-slit experiment, in which the angle between the paths of the light reaching two slits is very small.

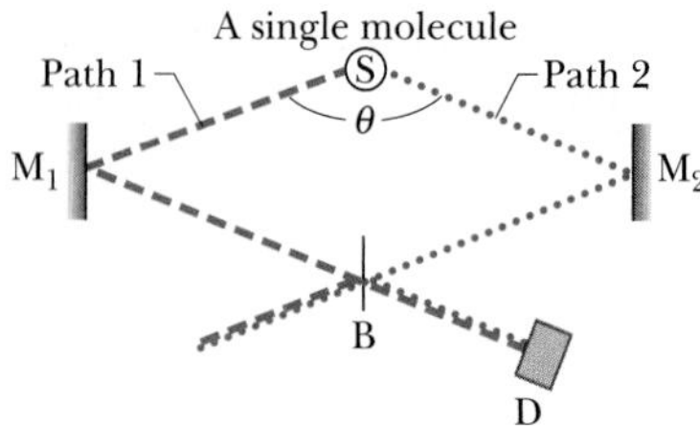

Fig. 39-7 The light from a single photon emission in source S travels over two widely separated paths and interferes with itself at detector D after being recombined by beam splitter B. (After Ming Lai and Jean-Claude Diels, *Journal of the Optical Society of America B*, **9,** 2290–2294, December 1992.)

After reflection from mirrors M_1 and M_2, the light waves traveling along paths 1 and 2 meet at beam splitter B. (A beam splitter is an optical device that transmits half the light incident upon it and reflects the other half.) On the right side of the beam splitter in Fig. 39-7, the light wave traveling along path 2 and reflected by B combines with the light wave traveling along path 1 and transmitted by B. These two waves then interfere with each other as they arrive at detector D (a *photomultiplier tube* that can detect individual photons).

The output of the detector is a randomly spaced series of electronic pulses, one for each detected photon. In the experiment, the beam splitter is moved slowly in a horizontal direction (in the reported experiment, a distance of only about 50 μm maximum), and the detector output is recorded on a chart recorder. Moving the beam splitter changes the lengths of paths 1 and 2, producing a phase shift between the light waves arriving at detector D. Interference maxima and minima appear in the detector's output signal.

This experiment is difficult to understand in traditional terms. For example, when a molecule in the source emits a single photon, does that photon travel along path 1 or path 2 in Fig. 39-7 (or along any other path)? How can it move in both directions at once? To answer, we assume that when a molecule emits a photon, a probability wave radiates in all directions from it. The experiment samples this wave in two of those directions, chosen to be nearly opposite each other.

We see that we can interpret all three versions of the double-slit experiment if we assume that (1) light is generated in the source as photons, (2) light is absorbed in the detector as photons, and (3) light travels between source and detector as a probability wave.

39-6 Electrons and Matter Waves

In 1924 French physicist Louis de Broglie made the following appeal to symmetry: A beam of light is a wave, but it transfers energy and momentum to matter only at points, via photons. Why can't a beam of particles have the same properties? That is, why can't we think of a moving electron—or any other particle, for that matter—as a **matter wave** that transfers energy and momentum to other matter at points?

In particular, de Broglie suggested that Eq. 39-7 ($p = h/\lambda$) might apply not only to photons but also to electrons. We used that equation in Section 39-4 to assign a momentum p to a photon of light with wavelength λ. We now use it, in the form

$$\lambda = \frac{h}{p} \quad \text{(de Broglie wavelength)}, \tag{39-13}$$

to assign a wavelength λ to a particle with momentum of magnitude p. The wavelength calculated from Eq. 39-13 is called the **de Broglie wavelength** of the moving particle. De Broglie's prediction of the existence of matter waves was first verified experimentally in 1927, by C. J. Davisson and L. H. Germer of the Bell Telephone Laboratories and by George P. Thomson of the University of Aberdeen in Scotland.

Figure 39-8 shows photographic proof of matter waves in a more recent experiment. In the experiment, an interference pattern was built up when electrons were sent, *one by one,* through a double-slit apparatus. The apparatus was like the ones we have previously used to demonstrate optical interference, except that the viewing screen was similar to a conventional television screen. When an electron hit the screen, it caused a flash of light whose position was recorded.

The first several electrons (top two photos) revealed nothing interesting and seemingly hit the screen at random points. However, after many thousands of elec-

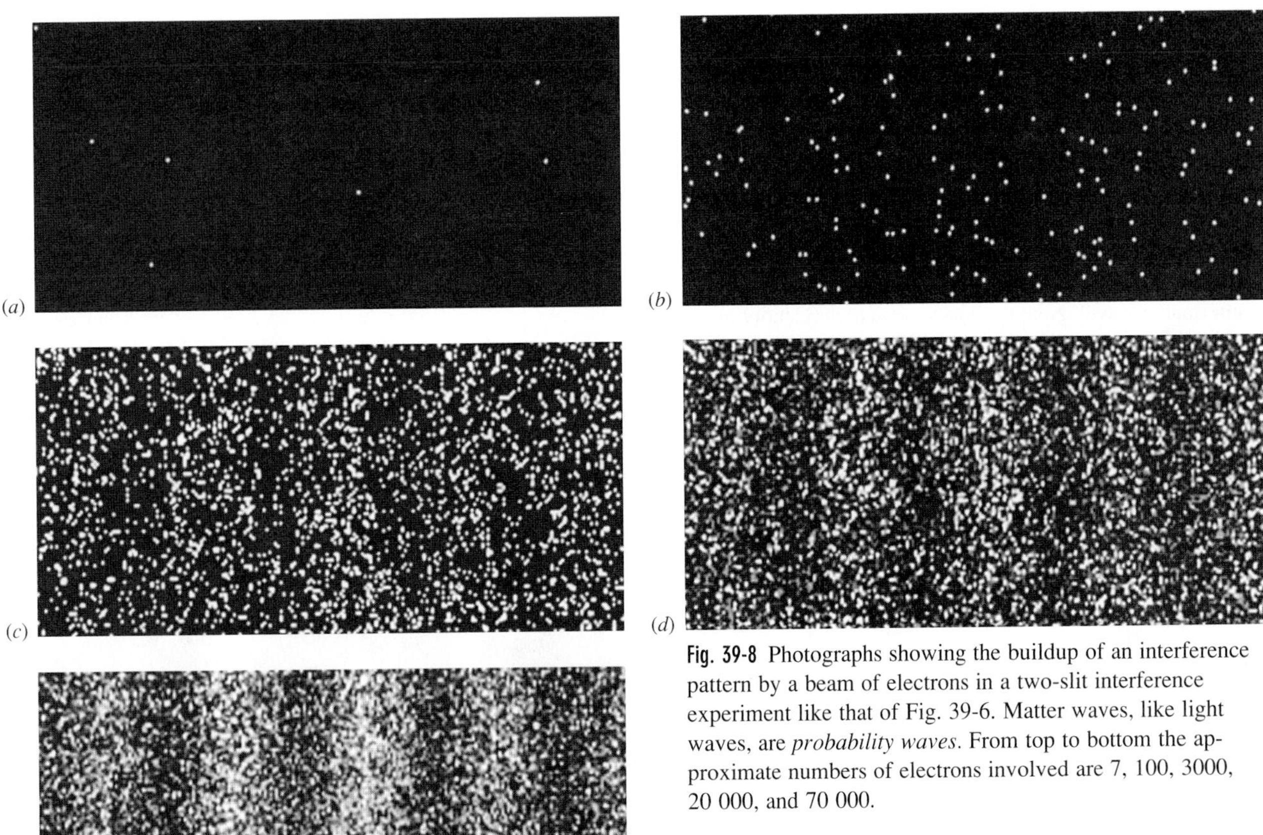

Fig. 39-8 Photographs showing the buildup of an interference pattern by a beam of electrons in a two-slit interference experiment like that of Fig. 39-6. Matter waves, like light waves, are *probability waves*. From top to bottom the approximate numbers of electrons involved are 7, 100, 3000, 20 000, and 70 000.

trons were sent through the apparatus, a pattern appeared on the screen, revealing fringes where many electrons had hit the screen and fringes where few had hit the screen. The pattern is exactly what we would expect for wave interference. Thus, *each* electron passed through the apparatus as a matter wave—the portion that traveled through one slit interfered with the portion that traveled through the other slit. That interference then determined the probability that the electron would materialize at a given point on the screen, hitting the screen there. Many electrons materialized in regions corresponding to bright fringes in optical interference, and few electrons materialized in regions corresponding to dark fringes.

Similar interference has been demonstrated with protons, neutrons, and various atoms. In 1994, it was demonstrated with iodine molecules I_2, which are not only 500 000 times more massive than electrons but far more complex. In 1999, it was demonstrated with the even more complex *fullerenes* (or *buckyballs*) C_{60} and C_{70}. (Fullerenes are soccer-ball-like molecules of carbon atoms, 60 in C_{60} and 70 in C_{70}.) Apparently, such small objects as electrons, protons, atoms, and molecules travel as matter waves. However, as we consider larger and more complex objects, there must come a point at which we are no longer justified in considering the wave nature of an object. At that point, we are back in our familiar nonquantum world, with the physics of earlier chapters of this book. In short, an electron is a matter wave and can undergo interference with itself, but a cat is not a matter wave and cannot undergo interference with itself (which must be a relief to cats).

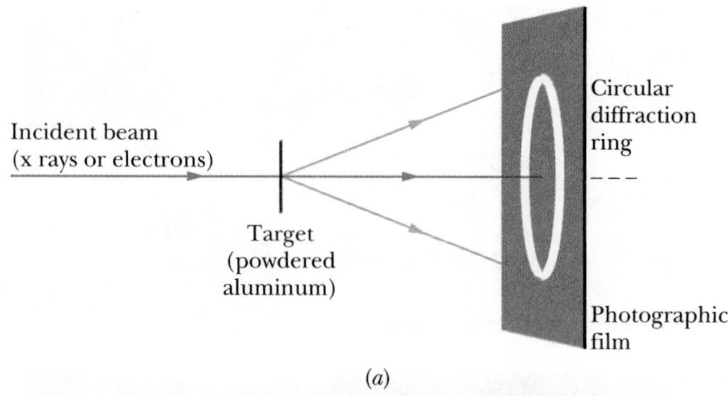

(*a*)

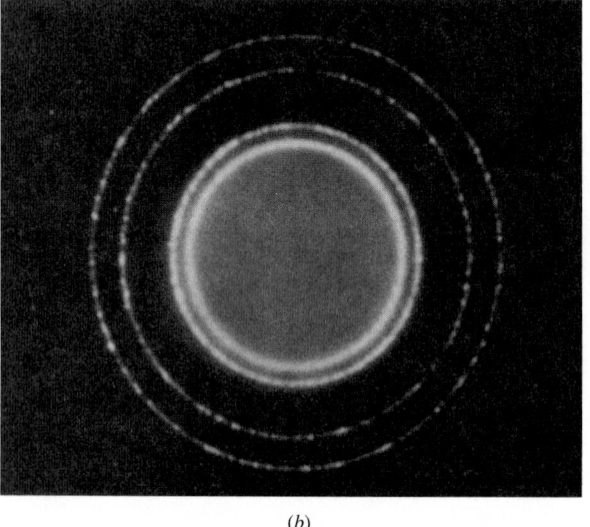

(*b*)

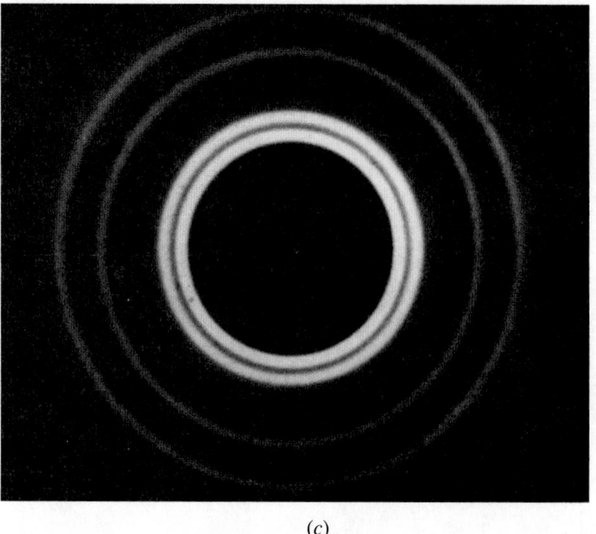

(*c*)

Fig. 39-9 (*a*) An experimental arrangement used to demonstrate, by diffraction techniques, the wavelike character of the incident beam. Photographs of the diffraction patterns when the incident beam is (*b*) an x-ray beam (light wave) and (*c*) an electron beam (matter wave). Note the basic geometrical identity of the patterns.

The wave nature of particles and atoms is now taken for granted in many scientific and engineering fields. For example, electron and neutron diffraction are used to study the atomic structures of solids and liquids, and electron diffraction is used to study the atomic features of surfaces on solids.

Figure 39-9*a* shows an arrangement that can be used to demonstrate the scattering of either x rays or electrons by crystals. A beam of one or the other is directed onto a target consisting of a powder of tiny aluminum crystals. The x rays have a certain wavelength λ. The electrons are given enough energy so that their de Broglie wavelength is the same wavelength λ. The scatter of x rays or electrons by the crystals produces a circular interference pattern on a photographic film. Figure 39-9*b* shows the pattern for the scatter of x rays, whereas Fig. 39-9*c* shows the pattern for the scatter of electrons. The patterns are the same—both x rays and electrons are waves.

Waves and Particles

Figures 39-8 and 39-9 are convincing evidence of the *wave* nature of matter, but we have at least as many experiments that suggest the *particle* nature of matter. Consider the tracks generated by electrons and displayed in the opening photo of this chapter. Surely these tracks—which are strings of bubbles left in the liquid hydrogen that

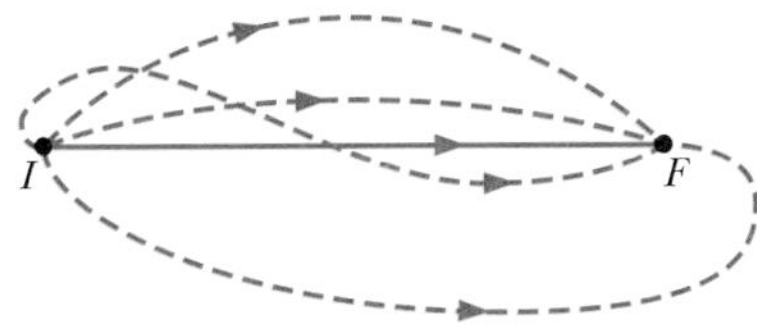

Fig. 39-10 A few of the many paths that connect two particle detection points I and F. Only matter waves that follow paths close to the straight line between these points interfere constructively. For all other paths, the waves following neighboring paths interfere destructively. Thus, a matter wave leaves a straight track.

fills the bubble chamber—strongly suggest the passage of a particle. Where is the wave?

To simplify the situation, let us turn off the magnetic field so that the strings of bubbles will be straight. We can view each bubble as a detection point for the electron. Matter waves traveling between detection points such as I and F in Fig. 39-10 will explore all possible paths, a few of which are shown.

In general, for every path connecting I and F (except the straight-line path), there will be a neighboring path such that matter waves following the two paths cancel each other by interference. This is not true, however, for the straight-line path joining I and F; in this case, matter waves traversing all neighboring paths reinforce the wave following the direct path. You can think of the bubbles that form the track as a series of detection points at which the matter wave undergoes constructive interference.

Sample Problem 39-5

What is the de Broglie wavelength of an electron with a kinetic energy of 120 eV?

SOLUTION: One **Key Idea** here is that we can find the electron's de Broglie wavelength λ from Eq. 39-13 ($\lambda = h/p$) if we first find the magnitude of its momentum p. A second **Key Idea** is that we find p from the given kinetic energy K of the electron. That kinetic energy is much less than the rest energy of an electron (0.511 MeV, from Table 38-3). Thus, we can get by with the classical approximations for momentum p ($= mv$) and kinetic energy K ($= \frac{1}{2}mv^2$).

Eliminating the speed v between these two equations yields

$$\begin{aligned} p &= \sqrt{2mK} \\ &= \sqrt{(2)(9.11 \times 10^{-31}\text{ kg})(120\text{ eV})(1.60 \times 10^{-19}\text{ J/eV})} \\ &= 5.91 \times 10^{-24}\text{ kg}\cdot\text{m/s}. \end{aligned}$$

From Eq. 39-13 then

$$\begin{aligned} \lambda &= \frac{h}{p} \\ &= \frac{6.63 \times 10^{-34}\text{ J}\cdot\text{s}}{5.91 \times 10^{-24}\text{ kg}\cdot\text{m/s}} \\ &= 1.12 \times 10^{-10}\text{ m} = 112\text{ pm}. \end{aligned} \qquad \text{(Answer)}$$

This is about the size of a typical atom. If we increase the kinetic energy, the wavelength becomes even smaller.

✔CHECKPOINT 4: An electron and a proton can have the same (a) kinetic energy, (b) momentum, or (c) speed. In each case, which particle has the shorter de Broglie wavelength?

39-7 Schrödinger's Equation

A simple traveling wave of any kind, be it a wave on a string, a sound wave, or a light wave, is described in terms of some quantity that varies in a wavelike fashion. For light waves, for example, this quantity is $\vec{E}(x, y, z, t)$, the electric field component of the wave. Its observed value at any point depends on the location of that point and on the time at which the observation is made.

What varying quantity should we use to describe a matter wave? We should expect this quantity, which we call the **wave function** $\Psi(x, y, z, t)$, to be more complicated than the corresponding quantity for a light wave because a matter wave, in addition to energy and momentum, transports mass and (often) electric charge. It turns out that Ψ, the uppercase Greek letter psi, usually represents a function that is complex in the mathematical sense; that is, we can always write its values in the form $a + ib$, in which a and b are real numbers and $i^2 = -1$.

In all the situations you will meet here, the space and time variables can be grouped separately and Ψ can be written in the form

$$\Psi(x, y, z, t) = \psi(x, y, z)\, e^{-i\omega t}, \tag{39-14}$$

where ω $(= 2\pi f)$ is the angular frequency of the matter wave. Note that ψ, the lowercase Greek letter psi, represents only the space-dependent part of the complete, time-dependent wave function Ψ. We shall deal almost exclusively with ψ. Two questions arise: What is meant by the wave function, and how do we find it?

What does the wave function mean? It has to do with the fact that a matter wave, like a light wave, is a probability wave. Suppose that a matter wave reaches a particle detector that is small; then the probability that a particle will be detected in a specified time interval is proportional to $|\psi|^2$, where $|\psi|$ is the absolute value of the wave function at the location of the detector. Although ψ is usually a complex quantity, $|\psi|^2$ is always both real and positive. It is, then, $|\psi|^2$, which we call the **probability density,** and not ψ, that has *physical* meaning. Speaking loosely, the meaning is this:

➤ The probability (per unit time) of detecting a particle in a small volume centered on a given point in a matter wave is proportional to the value of $|\psi|^2$ at that point.

Because ψ is usually a complex quantity, we find the square of its absolute value by multiplying ψ by ψ^*, the *complex conjugate* of ψ. (To find ψ^* we replace the imaginary number i in ψ with $-i$, wherever it occurs.)

How do we find the wave function? Sound waves and waves on strings are described by the equations of Newtonian mechanics. Light waves are described by Maxwell's equations. Matter waves are described by **Schrödinger's equation,** advanced in 1926 by Austrian physicist Erwin Schrödinger.

Many of the situations that we shall discuss involve a particle traveling in the x direction through a region in which forces acting on the particle cause it to have a potential energy $U(x)$. In this special case, Schrödinger's equation reduces to

$$\frac{d^2\psi}{dx^2} + \frac{8\pi^2 m}{h^2}[E - U(x)]\psi = 0 \quad \text{(Schrödinger's equation, one-dimensional motion),} \tag{39-15}$$

in which E is the total mechanical energy (potential energy plus kinetic energy) of the moving particle. (We do not consider mass energy in this nonrelativistic equation.) We cannot derive Schrödinger's equation from more basic principles; it *is* the basic principle.

If $U(x)$ in Eq. 39-15 is zero, that equation describes a **free particle**—that is, a moving particle on which no net force acts. The particle's total energy in this case is all kinetic, and thus E in Eq. 39-15 is $\frac{1}{2}mv^2$. That equation then becomes

$$\frac{d^2\psi}{dx^2} + \frac{8\pi^2 m}{h^2}\left(\frac{mv^2}{2}\right)\psi = 0,$$

which we can recast as

$$\frac{d^2\psi}{dx^2} + \left(2\pi\frac{p}{h}\right)^2\psi = 0.$$

To obtain this equation, we replaced mv with the momentum p and regrouped terms.

From Eq. 39-13 we recognize p/h in the equation above as $1/\lambda$, where λ is the de Broglie wavelength of the moving particle. We further recognize $2\pi/\lambda$ as the *angular wave number* k, which we defined in Eq. 17-5. With this substitution, the equation above becomes

$$\frac{d^2\psi}{dx^2} + k^2\psi = 0 \qquad \text{(Schrödinger's equation, free particle).} \tag{39-16}$$

The most general solution of Eq. 39-16 is

$$\psi(x) = Ae^{ikx} + Be^{-ikx}, \tag{39-17}$$

in which A and B are arbitrary constants. You can show that this equation is indeed a solution of Eq. 39-16 by substituting $\psi(x)$ and its second derivative into that equation and noting that an identity results.

If we combine Eqs. 39-14 and 39-17, we find, for the time-dependent wave function Ψ of a free particle traveling in the x direction,

$$\begin{aligned}\Psi(x, t) &= \psi(x)e^{-i\omega t} = (Ae^{ikx} + Be^{-ikx})e^{-i\omega t}\\ &= Ae^{i(kx-\omega t)} + Be^{-i(kx+\omega t)}.\end{aligned} \tag{39-18}$$

Finding the Probability Density $|\psi|^2$

In Section 17-5 we saw that *any function F* of the form $F(kx \pm \omega t)$ represents a traveling wave. This applies to exponential functions like those in Eq. 39-18 as well as to the sinusoidal functions we have used to describe waves on strings. In fact, these two representations of functions are related by

$$e^{i\theta} = \cos\theta + i\sin\theta \quad \text{and} \quad e^{-i\theta} = \cos\theta - i\sin\theta,$$

where θ is any angle.

The first term on the right in Eq. 39-18 thus represents a wave traveling in the direction of increasing x and the second a wave traveling in the negative direction of x. However, we have assumed that the free particle we are considering travels only in the positive direction of x. To reduce the general solution (Eq. 39-18) to our case of interest, we choose the arbitrary constant B in Eqs. 39-18 and 39-17 to be zero. At the same time, we relabel the constant A as ψ_0. Equation 39-17 then becomes

$$\psi(x) = \psi_0\, e^{ikx}. \tag{39-19}$$

To calculate the probability density, we take the square of the absolute value of $\psi(x)$. We get

$$|\psi|^2 = |\psi_0\, e^{ikx}|^2 = (\psi_0^2)\,|e^{ikx}|^2.$$

Now, because

$$|e^{ikx}|^2 = (e^{ikx})(e^{ikx})^* = e^{ikx}\, e^{-ikx} = e^{ikx-ikx} = e^0 = 1,$$

we get

$$|\psi|^2 = (\psi_0^2)(1)^2 = \psi_0^2 \qquad \text{(a constant).}$$

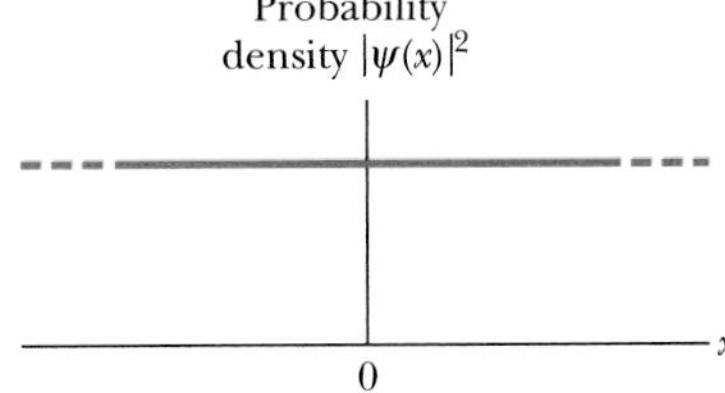

Fig. 39-11 A plot of the probability density $|\psi|^2$ for a free particle moving in the positive x direction. Since $|\psi|^2$ has the same constant value for all values of x, the particle has the same probability of detection at all points along its path.

Figure 39-11 is a plot of the probability density $|\psi|^2$ versus x for a free particle—a straight line parallel to the x axis from $-\infty$ to $+\infty$. We see that the probability density $|\psi|^2$ is the same for all values of x, which means that the particle has equal probabilities of being *anywhere* along the x axis. There is no distinguishing feature by which we can predict a most likely position for the particle. That is, all positions are equally likely.

We'll see what this means in the next section.

39-8 Heisenberg's Uncertainty Principle

Our inability to predict the position of a free particle, as indicated by Fig. 39-11, is our first example of **Heisenberg's uncertainty principle,** proposed in 1927 by German physicist Werner Heisenberg. It states that measured values cannot be assigned to the position $\vec{r}$ and the momentum $\vec{p}$ of a particle simultaneously with unlimited precision.

For the components of $\vec{r}$ and $\vec{p}$, Heisenberg's principle gives the following limits in terms of $\hbar = h/2\pi$ (called "h-bar"):

$$\begin{aligned} \Delta x \cdot \Delta p_x &\geq \hbar \\ \Delta y \cdot \Delta p_y &\geq \hbar \quad \text{(Heisenberg's uncertainty principle).} \\ \Delta z \cdot \Delta p_z &\geq \hbar \end{aligned} \tag{39-20}$$

Here Δx and Δp_x, as examples, represent the intrinsic uncertainties in the measurements of the x components of $\vec{r}$ and $\vec{p}$. Even with the best measuring instruments that technology could ever provide, each product of a position uncertainty and a momentum uncertainty in Eq. 39-20 will be greater than $\hbar$; it can *never* be less.

The particle whose probability density is plotted in Fig. 39-11 is a free particle; that is, no force acts on it, so its momentum $\vec{p}$ must be constant. We implied—without making a point of it—that we can determine $\vec{p}$ with absolute precision; we assumed that $\Delta p_x = \Delta p_y = \Delta p_z = 0$ in Eq. 39-20. That assumption then requires $\Delta x \to \infty$, $\Delta y \to \infty$, and $\Delta z \to \infty$. With such infinitely great uncertainties, the position of the particle is completely unspecified, as Fig. 39-11 shows.

Do not think that the particle *really has* a sharply defined position that is, for some reason, hidden from us. If its momentum can be specified with absolute precision, the words "position of the particle" simply lose all meaning. The particle in Fig. 39-11 can be found *with equal probability* anywhere along the x axis.

Sample Problem 39-6

Assume that an electron is moving along an x axis and that you measure its speed to be 2.05×10^6 m/s, which can be known with a precision of 0.50%. What is the minimum uncertainty (as allowed by the uncertainty principle in quantum theory) with which you can simultaneously measure the position of the electron along the x axis?

SOLUTION: The Key Idea here is that the minimum uncertainty allowed by quantum theory is given by Heisenberg's uncertainty principle in Eq. 39-20. We need only consider components along the x axis because we have motion only along that axis and want the uncertainty Δx in location along that axis. Since we want the minimum allowed uncertainty, we use the equality instead of the inequality in the x-axis part of Eq. 39-20, writing

$$\Delta x \cdot \Delta p_x = \hbar.$$

To evaluate the uncertainty Δp_x in the momentum, we must first evaluate the momentum component p_x. Because the electron's speed v_x is much less than the speed of light c, we can evaluate p_x with the classical expression for momentum instead of using a relativistic expression. We find

$$\begin{aligned} p_x = mv_x &= (9.11 \times 10^{-31}\ \text{kg})(2.05 \times 10^6\ \text{m/s}) \\ &= 1.87 \times 10^{-24}\ \text{kg} \cdot \text{m/s}. \end{aligned}$$

The uncertainty in the speed is given as 0.50% of the measured speed. Because p_x depends directly on speed, the uncertainty Δp_x in the momentum must be 0.50% of the momentum:

$$\begin{aligned} \Delta p_x &= (0.0050)p_x \\ &= (0.0050)(1.87 \times 10^{-24}\ \text{kg} \cdot \text{m/s}) \\ &= 9.35 \times 10^{-27}\ \text{kg} \cdot \text{m/s}. \end{aligned}$$

Then the uncertainty principle gives us

$$\begin{aligned} \Delta x = \frac{\hbar}{\Delta p_x} &= \frac{(6.63 \times 10^{-34}\ \text{J} \cdot \text{s})/2\pi}{9.35 \times 10^{-27}\ \text{kg} \cdot \text{m/s}} \\ &= 1.13 \times 10^{-8}\ \text{m} \approx 11\ \text{nm}, \qquad \text{(Answer)} \end{aligned}$$

which is about 100 atomic diameters. Given your measurement of the electron's speed, it makes no sense to try to pin down the electron's position to any greater precision.

39-9 Barrier Tunneling

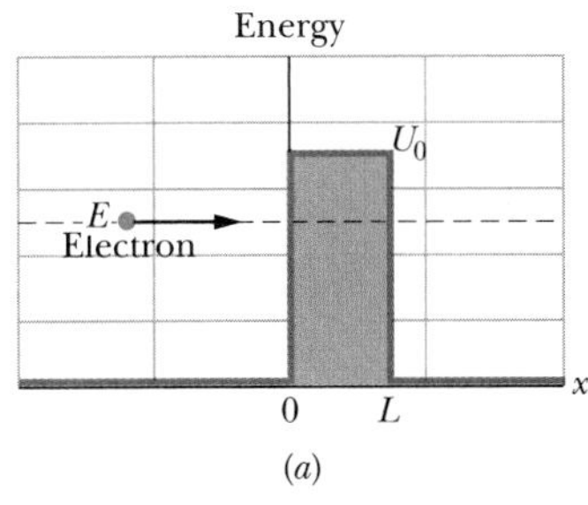

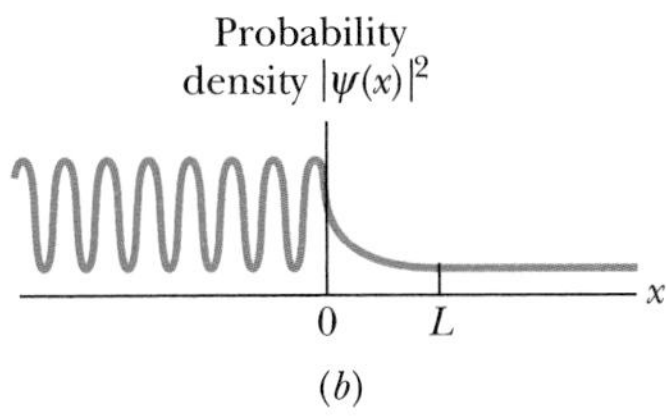

Fig. 39-12 (*a*) An energy diagram showing a potential energy barrier of height U_0 and thickness L. An electron with total energy E approaches the barrier from the left. (*b*) The probability density $|\psi|^2$ of the matter wave representing the electron, showing the tunneling of the electron through the barrier. The curve to the left of the barrier represents a standing matter wave that results from the superposition of the incident and reflected matter waves.

Suppose you repeatedly flip a jelly bean along a tabletop on which a book is positioned somewhere along the jelly bean's path. You would be very surprised to see the jelly bean appear on the other side of the book instead of bouncing back from it. Don't expect this surprising result from jelly beans. However, something very much like it, called **barrier tunneling,** *does* happen for electrons and other particles with small masses.

Figure 39-12*a* shows an electron of total energy E moving parallel to the x axis. Forces act on the electron such that its potential energy is zero except when it is in the region $0 < x < L$, where its potential energy has the constant value U_0. We define this region as a **potential energy barrier** (often loosely called a **potential barrier**) of height U_0 and thickness L.

Classically, because $E < U_0$, an electron approaching the barrier from the left would be reflected from the barrier and would move back in the direction from which it came. In quantum physics, however, the electron is a matter wave and there is a finite chance that it will "leak through" the barrier and appear on the other side. This means that there is a finite probability that the electron will end up on the far side of the barrier, moving to the right.

The wave function $\psi(x)$ describing the electron can be found by solving Schrödinger's equation (Eq. 39-15) separately for the three regions in Fig. 39-12*a*: (1) to the left of the barrier, (2) within the barrier, and (3) to the right of the barrier. The arbitrary constants that appear in the solutions can then be chosen so that the values of $\psi(x)$ and its derivative with respect to x join smoothly (no jumps, no kinks) at $x = 0$ and at $x = L$. Squaring the absolute value of $\psi(x)$ then yields the probability density.

Figure 39-12*b* shows a plot of the result. The oscillating curve to the left of the barrier (for $x < 0$) is a combination of the incident matter wave and the reflected matter wave (which has a smaller amplitude than the incident wave). The oscillations occur because these two waves, traveling in opposite directions, interfere with each other, setting up a standing wave pattern.

Within the barrier (for $0 < x < L$) the probability density decreases exponentially with x. However, provided L is small, the probability density is not quite zero at $x = L$.

To the right of the barrier of Fig. 39-12 (for $x > L$), the probability density plot describes a transmitted (through the barrier) wave with low but constant amplitude. Thus, the electron can be detected in this region but with a relatively small probability. (Compare this part of the figure with Fig. 39-11 for a free particle.)

We can assign a *transmission coefficient* T to the incident matter wave and the barrier in Fig. 39-12*a*. This coefficient gives the probability with which an approaching electron will be transmitted through the barrier—that is, that tunneling will occur. As an example, if $T = 0.020$, then of every 1000 electrons fired at the barrier, 20 (on average) will tunnel through it and 980 will be reflected.

The **transmission coefficient** T is approximately

$$T \approx e^{-2kL}, \tag{39-21}$$

in which

$$k = \sqrt{\frac{8\pi^2 m(U_0 - E)}{h^2}}. \tag{39-22}$$

Because of the exponential form of Eq. 39-21, the value of T is very sensitive to the three variables on which it depends: particle mass m, barrier thickness L, and energy difference $U_0 - E$.

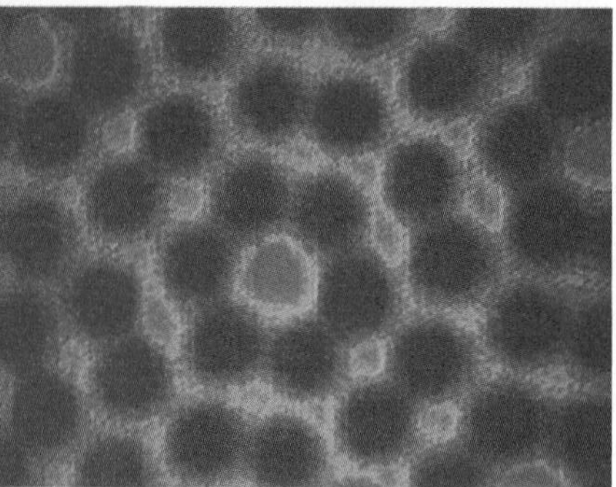

Fig. 39-13 An array of silicon atoms as revealed by a scanning tunneling microscope.

Barrier tunneling finds many applications in technology, among them the tunnel diode, in which a flow of electrons (by tunneling through a device) can be rapidly turned on or off by controlling the barrier height. Because this can be done very quickly (within 5 ps), the device is suitable for applications demanding a high-speed response. The 1973 Nobel prize in physics was shared by three "tunnelers," Leo Esaki (for tunneling in semiconductors), Ivar Giaever (for tunneling in superconductors), and Brian Josephson (for the Josephson junction, a rapid quantum switching device based on tunneling). The 1986 Nobel prize was awarded to Gerd Binnig and Heinrich Rohrer to recognize their development of another useful device based on tunneling, the scanning tunneling microscope.

CHECKPOINT 5: Is the wavelength of the transmitted wave in Fig. 39-12*b* larger than, smaller than, or the same as that of the incident wave?

The Scanning Tunneling Microscope (STM)

A device based on tunneling, the STM allows one to make detailed maps of surfaces, revealing features on the atomic scale with a resolution much greater than can be obtained with an optical or electron microscope. Figure 39-13 shows an example, the individual atoms of the surface being readily apparent.

Figure 39-14 shows the heart of the scanning tunneling microscope. A fine metallic tip, mounted at the intersection of three mutually perpendicular quartz rods, is placed close to the surface to be examined. A small potential difference, perhaps only 10 mV, is applied between tip and surface.

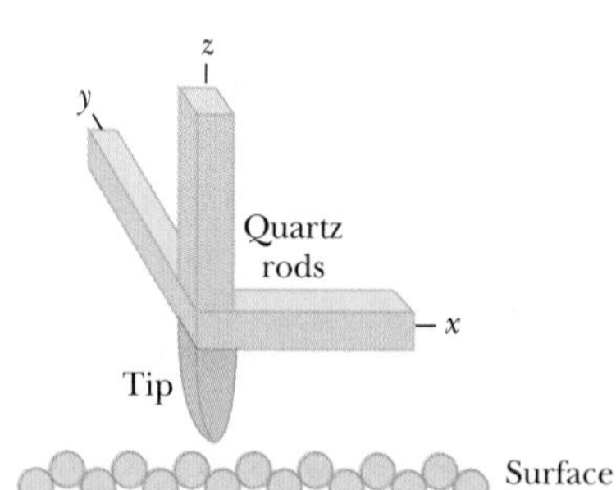

Fig. 39-14 The essence of a scanning tunneling microscope (STM). Three quartz rods are used to scan a sharply pointed conducting tip across the surface of interest and to maintain a constant separation between tip and surface. The tip thus moves up and down to match the contours of the surface, and a record of its movement is a map like that of Fig. 39-13.

Crystalline quartz has an interesting property called *piezoelectricity:* When an electric potential difference is applied across a sample of crystalline quartz, the dimensions of the sample change slightly. This property is used to change the length of each of the three rods in Fig. 39-14, smoothly and by tiny amounts, so that the tip can be scanned back and forth over the surface (in the x and y directions) and also lowered or raised with respect to the surface (in the z direction).

The space between the surface and the tip forms a potential energy barrier, much like that of Fig. 39-12*a*. If the tip is close enough to the surface, electrons from the sample can tunnel through this barrier from the surface to the tip, forming a tunneling current.

In operation, an electronic feedback arrangement adjusts the vertical position of the tip to keep the tunneling current constant as the tip is scanned over the surface. This means that the tip–surface separation also remains constant during the scan. The output of the device—for example, Fig. 39-13—is a video display of the varying vertical position of the tip, hence of the surface contour, as a function of the tip position in the xy plane.

Scanning tunneling microscopes are available commercially and are used in laboratories all over the world.

Sample Problem 39-7

Suppose that the electron in Fig. 39-12*a*, having a total energy E of 5.1 eV, approaches a barrier of height $U_0 = 6.8$ eV and thickness $L = 750$ pm.

(a) What is the approximate probability that the electron will be transmitted through the barrier, to appear (and be detectable) on the other side of the barrier?

SOLUTION: The **Key Idea** here is that the probability we seek is the transmission coefficient T as given by Eq. 39-21 ($T \approx e^{-2kL}$), where k is

$$k = \sqrt{\frac{8\pi^2 m(U_0 - E)}{h^2}}.$$

The numerator of the fraction under the square-root sign is

$$(8\pi^2)(9.11 \times 10^{-31}\text{ kg})(6.8\text{ eV} - 5.1\text{ eV}) \times (1.60 \times 10^{-19}\text{ J/eV}) = 1.956 \times 10^{-47}\text{ J}\cdot\text{kg}.$$

Thus, $$k = \sqrt{\frac{1.956 \times 10^{-47}\text{ J}\cdot\text{kg}}{(6.63 \times 10^{-34}\text{ J}\cdot\text{s})^2}} = 6.67 \times 10^9\text{ m}^{-1}.$$

The (dimensionless) quantity $2kL$ is then

$$2kL = (2)(6.67 \times 10^9\text{ m}^{-1})(750 \times 10^{-12}\text{ m}) = 10.0$$

and, from Eq. 39-21, the transmission coefficient is

$$T \approx e^{-2kL} = e^{-10.0} = 45 \times 10^{-6}. \quad \text{(Answer)}$$

Thus, of every million electrons that strike the barrier, about 45 will tunnel through it.

(b) What is the approximate probability that a proton with the same total energy of 5.1 eV will be transmitted through the barrier, to appear (and be detectable) on the other side of the barrier?

SOLUTION: The **Key Idea** here is that the transmission coefficient T (and thus the probability of transmission) depends on the mass of the particle. Indeed, because mass m is one of the factors in the exponent of e in the equation for T, the probability of transmission is very sensitive to the mass of the particle. This time, the mass is that of a proton (1.67×10^{-27} kg), which is significantly greater than that of the electron in (a). By substituting the proton's mass for the mass in (a) and then continuing as we did there, we find that $T \approx 10^{-186}$. Thus, the probability that the proton will be transmitted is not zero, but barely more than zero. For even more massive particles with the same total energy of 5.1 eV, the probability of transmission is exponentially lower.

REVIEW & SUMMARY

Light Quanta—Photons An electromagnetic wave (light) is quantized, and its quanta are called *photons*. For a light wave of frequency f and wavelength λ, the energy E and momentum magnitude p of a photon are

$$E = hf \quad \text{(photon energy)} \quad (39\text{-}2)$$

and

$$p = \frac{hf}{c} = \frac{h}{\lambda} \quad \text{(photon momentum)}. \quad (39\text{-}7)$$

Photoelectric Effect When light of high enough frequency falls on a clean metal surface, electrons are emitted from the surface by photon–electron interactions within the metal. The governing relation is

$$hf = K_{max} + \Phi, \quad (39\text{-}5)$$

in which hf is the photon energy, K_{max} is the kinetic energy of the most energetic emitted electrons, and Φ is the **work function** of the target material—that is, the minimum energy an electron must have if it is to emerge from the surface of the target. If hf is less than Φ, the photoelectric effect does not occur.

Compton Shift When x rays are scattered by loosely bound electrons in a target, some of the scattered x rays have a longer wavelength than do the incident x rays. This **Compton shift** (in wavelength) is given by

$$\Delta\lambda = \frac{h}{mc}(1 - \cos\phi), \quad (39\text{-}11)$$

in which ϕ is the angle at which the x rays are scattered.

Light Waves and Photons When light interacts with matter, energy and momentum are transferred via photons. When light is in transit, however, we interpret the light wave as a **probability wave,** in which the probability (per unit time) that a photon can be detected is proportional to E_m^2, where E_m is the amplitude of the oscillating electric field of the light wave at the detector.

Matter Waves A moving particle such as an electron or a proton can be described as a **matter wave;** its wavelength (called the **de Broglie wavelength**) is given by $\lambda = h/p$, where p is the momentum of the particle.

The Wave Function A matter wave is described by its **wave function** $\Psi(x, y, z, t)$, which can be separated into a space-dependent part $\psi(x, y, z)$ and a time-dependent part $e^{-i\omega t}$. For a particle of mass m moving in the x direction with constant total energy E through a region in which its potential energy is $U(x)$, $\psi(x)$ can be found by solving the simplified **Schrödinger equation:**

$$\frac{d^2\psi}{dx^2} + \frac{8\pi^2 m}{h^2}[E - U(x)]\psi = 0. \quad (39\text{-}15)$$

A matter wave, like a light wave, is a probability wave in the sense that if a particle detector is inserted into the wave, the probability that the detector will register a particle during any specified time interval is proportional to $|\psi|^2$, a quantity called the **probability density.**

For a free particle—that is, a particle for which $U(x) = 0$—moving in the x direction, $|\psi|^2$ has a constant value for all positions along the x axis.

Heisenberg's Uncertainty Principle The probabilistic nature of quantum physics places an important limitation on detecting a particle's position and momentum. That is, it is not possible to measure the position $\vec{r}$ and the momentum $\vec{p}$ of a particle simultaneously with unlimited precision. The uncertainties in the components of these quantities are given by

$$\Delta x \cdot \Delta p_x \geq \hbar$$
$$\Delta y \cdot \Delta p_y \geq \hbar \quad (39\text{-}20)$$
$$\Delta z \cdot \Delta p_z \geq \hbar.$$

Barrier Tunneling According to classical physics, an incident particle will be reflected from a potential energy barrier whose height is greater than the particle's kinetic energy. According to quantum physics, however, the particle has a finite probability of tunneling through such a barrier. The probability that a given particle of mass m and energy E will tunnel through a barrier of height U_0 and thickness L is given by the transmission coefficient T:

$$T \approx e^{-2kL}, \tag{39-21}$$

where

$$k = \sqrt{\frac{8\pi^2 m(U_0 - E)}{h^2}}. \tag{39-22}$$

QUESTIONS

1. Of the electromagnetic waves generated in a microwave oven and in your dentist's x-ray machine, which has (a) the greater wavelength, (b) the greater frequency, and (c) the greater photon energy?

2. Of the following statements about the photoelectric effect, which are true and which are false? (a) The greater the frequency of the incident light is, the greater is the stopping potential. (b) The greater the intensity of the incident light is, the greater is the cutoff frequency. (c) The greater the work function of the target material is, the greater is the stopping potential. (d) The greater the work function of the target material is, the greater is the cutoff frequency. (e) The greater the frequency of the incident light is, the greater is the maximum kinetic energy of the ejected electrons. (f) The greater the energy of the photons is, the smaller is the stopping potential.

3. According to the figure for Checkpoint 2, is the maximum kinetic energy of the ejected electrons greater for a target made of sodium or of potassium for a given frequency of incident light?

4. In the photoelectric effect (for a given target and a given frequency of the incident light), which of these quantities, if any, depend on the intensity of the incident light beam: (a) the maximum kinetic energy of the electrons, (b) the maximum photoelectric current, (c) the stopping potential, (d) the cutoff frequency?

5. If you shine ultraviolet light on an isolated metal plate, the plate emits electrons for a while. Why does it eventually stop?

6. A metal plate is illuminated with light of a certain frequency. Which of the following determine whether or not electrons are ejected: (a) the intensity of the light, (b) the length of time of exposure to the light, (c) the thermal conductivity of the plate, (d) the area of the plate, (e) the material of the plate?

7. In a Compton-shift experiment, an x-ray photon is scattered in the forward direction, at $\phi = 0$ in Fig. 39-3. How much energy does the electron acquire during this interaction?

8. According to Eq. 39-11 the Compton shift is the same for x rays and for visible light. Why is it that the Compton shift for x rays can be measured readily but that for visible light cannot?

9. Photon A has twice the energy of photon B. (a) Is the momentum of A less than, equal to, or greater than that of B? (b) Is the wavelength of A less than, equal to, or greater than that of B?

10. Photon A is from an ultraviolet tanning lamp and photon B is from a television transmitter. Which has the greater (a) wavelength, (b) energy, (c) frequency, and (d) momentum?

11. The data shown in Fig. 39-4 were taken by directing x rays onto a carbon target. In what essential way, if any, would these data differ if the target were sulfur instead of carbon?

12. An electron and a proton have the same kinetic energy. Which has the greater de Broglie wavelength?

13. (a) If you double the kinetic energy of a nonrelativistic particle, how does its de Broglie wavelength change? (b) What if you double the speed of the particle?

14. The following nonrelativistic particles all have the same kinetic energy. Rank them in order of their de Broglie wavelengths, greatest first: electron, alpha particle, neutron.

15. Figure 39-15 shows four situations in which an electron is moving through a field. It is moving (a) opposite an electric field, (b) in the same direction as an electric field, (c) in the same direction as a magnetic field, (d) perpendicular to a magnetic field. For each situation, is the de Broglie wavelength of the electron increasing, decreasing, or remaining the same?

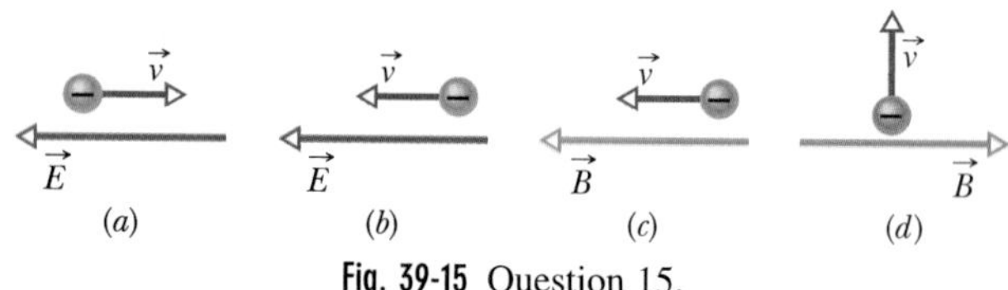

Fig. 39-15 Question 15.

16. A proton and a deuteron, each having a kinetic energy of 3 MeV, approach a potential energy barrier whose height U_0 is 10 MeV. Which particle has the greater chance of tunneling through the barrier? (A deuteron is twice as massive as a proton.)

17. Which has the greater effect on the transmission coefficient T for electron tunneling through a potential energy barrier: (a) raising the barrier height U_0 by 1% or (b) lowering the kinetic energy E of the incident electron by 1%?

18. At the left in Fig. 39-12*b*, why are the minima in the values of $|\psi|^2$ greater than zero?

19. Suppose that the height of the potential energy barrier in Fig. 39-12*a* is infinite. (a) What value would you expect for the transmission coefficient of electrons approaching the barrier? (b) Does Eq. 39-21 predict your expected result?

20. The table gives relative values for three situations for the barrier tunneling experiment of Fig. 39-12. Rank the situations according to the probability of the electron tunneling through the barrier, greatest first.

	Electron Energy	Barrier Height	Barrier Thickness
(a)	E	$5E$	L
(b)	E	$17E$	$L/2$
(c)	E	$2E$	$2L$

EXERCISES & PROBLEMS

ssm Solution is in the Student Solutions Manual.
www Solution is available on the World Wide Web at: http://www.wiley.com/college/hrw
ilw Solution is available on the Interactive LearningWare.

SEC. 39-2 The Photon, the Quantum of Light

1E. Express the Planck constant h in terms of the unit electron-volt-femtoseconds.

2E. Monochromatic light (that is, light of a single wavelength) is to be absorbed by a sheet of photographic film and thus recorded on the film. Photon absorption will occur if the photon energy equals or exceeds the least energy of 0.6 eV needed to dissociate an AgBr molecule in the film. What is the greatest wavelength of light that can be recorded by the film? In what region of the electromagnetic spectrum is this wavelength located?

3E. Show that, for light of wavelength λ in nanometers, the photon energy hf in electron-volts is $1240/\lambda$. ssm

4E. The yellow-colored light from a highway sodium lamp is brightest at a wavelength of 589 nm. What is the photon energy for light at that wavelength?

5E. At what rate does the Sun emit photons? For simplicity, assume that the Sun's entire emission at the rate of 3.9×10^{26} W is at the single wavelength of 550 nm.

6E. A helium–neon laser emits red light at wavelength λ = 633 nm, in a beam of diameter 3.5 mm, and at an energy-emission rate of 5.0 mW. A detector in the beam's path totally absorbs the beam. At what rate per unit area does the detector absorb photons?

7E. A spectral emission line is electromagnetic radiation that is emitted in a wavelength range narrow enough to be taken as a single wavelength. One such emission line that is important in astronomy has a wavelength of 21 cm. What is the photon energy in the electromagnetic wave at that wavelength?

8E. How fast must an electron move to have a kinetic energy equal to the photon energy of sodium light at wavelength 590 nm?

9E. The meter was once defined as 1 650 763.73 wavelengths of the orange light emitted by a source containing krypton-86 atoms. What is the photon energy of that light?

10P. Under ideal conditions, a visual sensation can occur in the human visual system if light of wavelength 550 nm is absorbed by the eye's retina at a rate as low as 100 photons per second. What is the corresponding rate at which energy is absorbed by the retina?

11P. A special kind of lightbulb emits monochromatic light of wavelength 630 nm. Electric energy is supplied to it at the rate of 60 W, and the bulb is 93% efficient at converting that energy to light energy. How many photons are emitted by the bulb during its lifetime of 730 h?

12P. The beam emerging from a 1.5 W argon laser (λ = 515 nm) has a diameter d of 3.0 mm. The beam is focused by a lens system with an effective focal length f_L of 2.5 mm. The focused beam strikes a totally absorbing screen, where it forms a circular diffraction pattern whose central disk has a radius R given by $1.22 f_L \lambda/d$. It can be shown that 84% of the incident energy ends up within this central disk. At what rate are photons absorbed by the screen in the central disk of the diffraction pattern?

13P. An ultraviolet lamp emits light of wavelength 400 nm, at the rate (power) of 400 W. An infrared lamp emits light of wavelength 700 nm, also at the rate of 400 W. (a) Which lamp emits photons at the greater rate and (b) what is that greater rate? ssm

14P. A satellite in Earth orbit maintains a panel of solar cells of area 2.60 m^2 perpendicular to the direction of the Sun's light rays. The intensity of the light at the panel is 1.39 kW/m^2. (a) At what rate does solar energy arrive at the panel? (b) At what rate are solar photons absorbed by the panel? Assume that the solar radiation is monochromatic, with a wavelength of 550 nm, and that all the solar radiation striking the panel is absorbed. (c) How long would it take for a "mole of photons" to be absorbed by the panel?

15P. A 100 W sodium lamp (λ = 589 nm) radiates energy uniformly in all directions. (a) At what rate are photons emitted by the lamp? (b) At what distance from the lamp will a totally absorbing screen absorb photons at the rate of 1.00 $photon/cm^2 \cdot s$? (c) What is the photon flux (photons per unit area per unit time) on a small screen 2.00 m from the lamp? ssm www

SEC. 39-3 The Photoelectric Effect

16E. (a) The least energy needed to eject an electron from metallic sodium is 2.28 eV. Does sodium show a photoelectric effect for red light, with λ = 680 nm? (b) What is the cutoff wavelength for photoelectric emission from sodium? To what color does that correspond?

17E. You wish to pick an element for a photocell that will operate via the photoelectric effect with visible light. Which of the following are suitable (work functions are in parentheses): tantalum (4.2 eV), tungsten (4.5 eV), aluminum (4.2 eV), barium (2.5 eV), lithium (2.3 eV)?

18E. The work functions for potassium and cesium are 2.25 and 2.14 eV, respectively. (a) Will the photoelectric effect occur for either of these elements with incident light of wavelength 565 nm? (b) With light of wavelength 518 nm?

19E. Light strikes a sodium surface, causing photoelectric emission. The stopping potential for the ejected electrons is 5.0 V, and the work function of sodium is 2.2 eV. What is the wavelength of the incident light? ssm

20E. Find the maximum kinetic energy of electrons ejected from a certain material if the material's work function is 2.3 eV and the frequency of the incident radiation is 3.0×10^{15} Hz.

21E. The work function of tungsten is 4.50 eV. Calculate the speed of the fastest electrons ejected from a tungsten surface when light whose photon energy is 5.80 eV shines on the surface.

22P. (a) If the work function for a certain metal is 1.8 eV, what is the stopping potential for electrons ejected from the metal when light of wavelength 400 nm shines on the metal? (b) What is the maximum speed of the ejected electrons?

23P. Light of wavelength 200 nm shines on an aluminum surface. In aluminum, 4.20 eV is required to eject an electron. What is the kinetic energy of (a) the fastest and (b) the slowest ejected electrons? (c) What is the stopping potential for this situation? (d) What is the cutoff wavelength for aluminum? ssm

24P. The wavelength associated with the cutoff frequency for silver is 325 nm. Find the maximum kinetic energy of electrons ejected from a silver surface by ultraviolet light of wavelength 254 nm.

25P. An orbiting satellite can become charged by the photoelectric effect when sunlight ejects electrons from its outer surface. Satellites must be designed to minimize such charging. Suppose a satellite is coated with platinum, a metal with a very large work function ($\Phi = 5.32$ eV). Find the longest wavelength of incident sunlight that can eject an electron from the platinum.

26P. In a photoelectric experiment using a sodium surface, you find a stopping potential of 1.85 V for a wavelength of 300 nm and a stopping potential of 0.820 V for a wavelength of 400 nm. From these data find (a) a value for the Planck constant, (b) the work function Φ for sodium, and (c) the cutoff wavelength λ_0 for sodium.

27P. The stopping potential for electrons emitted from a surface illuminated by light of wavelength 491 nm is 0.710 V. When the incident wavelength is changed to a new value, the stopping potential is 1.43 V. (a) What is this new wavelength? (b) What is the work function for the surface? ssm www

28P. In about 1916, R. A. Millikan found the following stopping-potential data for lithium in his photoelectric experiments:

Wavelength (nm)	433.9	404.7	365.0	312.5	253.5
Stopping potential (V)	0.55	0.73	1.09	1.67	2.57

Use these data to make a plot like Fig. 39-2 (which is for sodium) and then use the plot to find (a) the Planck constant and (b) the work function for lithium.

29P. Suppose the *fractional efficiency* of a cesium surface (with work function 1.80 eV) is 1.0×10^{-16}; that is, on average one electron is ejected for every 10^{16} photons that reach the surface. What would be the current of electrons ejected from such a surface if it were illuminated with 600 nm light from a 2.00 mW laser and all the ejected electrons took part in the charge flow?

30P. X rays with a wavelength of 71 pm are directed onto a gold foil and eject tightly bound electrons from the gold atoms. The ejected electrons then move in circular paths of radius r in a region of uniform magnetic field $\vec{B}$, with $Br = 1.88 \times 10^{-4}$ T · m. Find (a) the maximum kinetic energy of those electrons and (b) the work done in removing them from the gold atoms.

SEC. 39-4 Photons Have Momentum

31E. Light of wavelength 2.4 pm is directed onto a target containing free electrons. (a) Find the wavelength of light scattered at 30° from the incident direction. (b) Do the same for a scattering angle of 120°. ssm

32E. (a) What is the momentum of a photon whose energy equals the rest energy of an electron? What are (b) the wavelength and (c) the frequency of the corresponding radiation?

33E. A certain x-ray beam has a wavelength of 35.0 pm. (a) What is the corresponding frequency? Calculate the corresponding (b) photon energy and (c) photon momentum.

34P. X rays of wavelength 0.010 nm are directed onto a target containing loosely bound electrons. For Compton scattering from one of those electrons, at an angle of 180°, what are (a) the Compton shift, (b) the corresponding change in photon energy, (c) the kinetic energy of the recoiling electron, and (d) the electron's direction of motion?

35P. Show, by analyzing a collision between a photon and a free electron (using relativistic mechanics), that it is impossible for a photon to transfer all its energy to a free electron (and thus for the photon to vanish).

36P. Gamma rays of photon energy 0.511 MeV are directed onto an aluminum target and are scattered in various directions by loosely bound electrons there. (a) What is the wavelength of the incident gamma rays? (b) What is the wavelength of gamma rays scattered at 90.0° to the incident beam? (c) What is the photon energy of the rays scattered in this direction?

37P. Calculate the Compton wavelength for (a) an electron and (b) a proton. What is the photon energy for an electromagnetic wave with a wavelength equal to the Compton wavelength of (c) the electron and (d) the proton? ssm

38P. What is the maximum wavelength shift for a Compton collision between a photon and a free *proton*?

39P. What percentage increase in wavelength leads to a 75% loss of photon energy in a photon–free electron collision? ssm www

40P. Calculate the percentage change in photon energy during a collision like that in Fig. 39-5 for $\phi = 90°$ and for radiation in (a) the microwave range, with $\lambda = 3.0$ cm; (b) the visible range, with $\lambda = 500$ nm; (c) the x-ray range, with $\lambda = 25$ pm; and (d) the gamma-ray range, with a gamma photon energy of 1.0 MeV. (e) What are your conclusions about the feasibility of detecting the Compton shift in these various regions of the electromagnetic spectrum, judging solely by the criterion of energy loss in a single photon–electron encounter?

41P. An electron of mass m and speed v "collides" with a gamma-ray photon of initial energy hf_0, as measured in the laboratory frame. The photon is scattered in the electron's direction of travel. Verify that the energy of the scattered photon, as measured in the laboratory frame, is

$$E = hf_0\left(1 + \frac{2hf_0}{mc^2}\sqrt{\frac{1 + v/c}{1 - v/c}}\right)^{-1}.$$

42P. Show that $\Delta E/E$, the fractional loss of energy of a photon during a collision with a particle of mass m, is given by

$$\frac{\Delta E}{E} = \frac{hf'}{mc^2}(1 - \cos\phi),$$

where E is the energy of the incident photon, f' is the frequency of the scattered photon, and ϕ is defined as in Fig. 39-5.

43P. Consider a collision between an x-ray photon of initial energy 50.0 keV and an electron at rest, in which the photon is scattered backward and the electron is knocked forward. (a) What is the energy of the back-scattered photon? (b) What is the kinetic energy of the electron?

44P. What would be (a) the Compton shift, (b) the fractional Compton shift, and (c) the change in photon energy for light of wavelength 590 nm scattering from a free, initially stationary electron if the scattering is at 90° to the direction of the incident beam? (d) Calculate the same quantities for x rays whose photon energy is 50.0 keV.

45P. What is the maximum kinetic energy of electrons knocked out of a thin copper foil by Compton scattering of an incident beam of 17.5 keV x rays?

46P. Derive Eq. 39-11, the equation for the Compton shift, from Eqs. 39-8, 39-9, and 39-10 by eliminating v and θ.

47P. Through what angle must a 200 keV photon be scattered by a free electron so that the photon loses 10% of its energy?

48P. Show that when a photon of energy E is scattered from a free electron at rest, the maximum kinetic energy of the recoiling electron is given by

$$K_{\text{max}} = \frac{E^2}{E + mc^2/2}.$$

SEC. 39-6 Electrons and Matter Waves

49E. Using the classical equations for momentum and kinetic energy, show that an electron's de Broglie wavelength in nanometers can be written as $\lambda = 1.226/\sqrt{K}$, in which K is the electron's kinetic energy in electron-volts. ssm

50E. A bullet of mass 40 g travels at 1000 m/s. Although the bullet is clearly too large to be treated as a matter wave, determine what Eq. 39-13 predicts for its de Broglie wavelength.

51E. In an ordinary television set, electrons are accelerated through a potential difference of 25.0 kV. What is the de Broglie wavelength of such electrons? (Relativity is not needed.) ssm

52E. Calculate the de Broglie wavelength of (a) a 1.00 keV electron, (b) a 1.00 keV photon, and (c) a 1.00 keV neutron.

53P. The wavelength of the yellow spectral emission line of sodium is 590 nm. At what kinetic energy would an electron have that wavelength as its de Broglie wavelength? ssm

54P. If the de Broglie wavelength of a proton is 100 fm, (a) what is the speed of the proton and (b) through what electric potential would the proton have to be accelerated to acquire this speed?

55P. Neutrons in thermal equilibrium with matter have an average kinetic energy of $(3/2)kT$, where k is the Boltzmann constant and T, which may be taken to be 300 K, is the temperature of the environment of the neutrons. (a) What is the average kinetic energy of such a neutron? (b) What is the corresponding de Broglie wavelength?

56P. An electron and a photon each have a wavelength of 0.20 nm. Calculate (a) their momenta and (b) their energies.

57P. (a) A photon has an energy of 1.00 eV, and an electron has a kinetic energy of that same amount. What are their wavelengths? (b) Repeat for an energy of 1.00 GeV. ssm www

58P. Consider a balloon filled with helium gas at room temperature and pressure. Calculate (a) the average de Broglie wavelength of the helium atoms and (b) the average distance between atoms under these conditions. The average kinetic energy of an atom is equal to $(3/2)kT$, where k is the Boltzmann constant. (c) Can the atoms be treated as particles under these conditions?

59P. Singly charged sodium ions are accelerated through a potential difference of 300 V. (a) What is the momentum acquired by such an ion? (b) What is its de Broglie wavelength? ssm

60P. (a) A photon and an electron both have a wavelength of 1.00 nm. What are the energy of the photon and the kinetic energy of the electron? (b) Repeat for a wavelength of 1.00 fm.

61P. The large electron accelerator at Stanford University provides a beam of electrons with kinetic energies of 50 GeV. Electrons with this energy have small wavelengths, suitable for probing the fine details of nuclear structure via scattering. What de Broglie wavelength does a 50 GeV electron have? How does this wavelength compare with the radius of an average nucleus, taken to be about 5.0 fm? (At this energy you can use the extreme relativistic relationship between momentum and energy, namely, $p = E/c$. This relationship, used for light, is justified when the kinetic energy of a particle is much greater than its rest energy, as in this case.)

62P. The existence of the atomic nucleus was discovered in 1911 by Ernest Rutherford, who properly interpreted some experiments in which a beam of alpha particles was scattered from a metal foil of atoms such as gold. (a) If the alpha particles had a kinetic energy of 7.5 MeV, what was their de Broglie wavelength? (b) Should the wave nature of the incident alpha particles have been taken into account in interpreting these experiments? The mass of an alpha particle is 4.00 u (atomic mass units), and its distance of closest approach to the nuclear center in these experiments was about 30 fm. (The wave nature of matter was not postulated until more than a decade after these crucial experiments were first performed.)

63P. A nonrelativistic particle is moving three times as fast as an electron. The ratio of the de Broglie wavelength of the particle to that of the electron is 1.813×10^{-4}. By calculating its mass, identify the particle. ssm

64P. The highest achievable resolving power of a microscope is limited only by the wavelength used; that is, the smallest item that can be distinguished has dimensions about equal to the wavelength. Suppose one wishes to "see" inside an atom. Assuming the atom to have a diameter of 100 pm, this means that one must be able to resolve a width of, say, 10 pm. (a) If an electron microscope is used, what minimum electron energy is required? (b) If a light microscope is used, what minimum photon energy is required? (c) Which microscope seems more practical? Why?

65P. What accelerating voltage would be required for the electrons of an electron microscope if the microscope is to have the same resolving power as could be obtained using 100 keV gamma rays? (See Problem 64.) ssm

SEC. 39-7 Schrödinger's Equation

66E. (a) Let $n = a + ib$ be a complex number, where a and b are real (positive or negative) numbers. Show that the product nn^* is always a positive real number. (b) Let $m = c + id$ be another complex number. Show that $|nm| = |n|\,|m|$.

67P. Show that Eq. 39-17 is indeed a solution of Eq. 39-16 by substituting $\psi(x)$ and its second derivative into Eq. 39-16 and noting that an identity results.

68P. (a) Write the wave function $\psi(x)$ displayed in Eq. 39-19 in the form $\psi(x) = a + ib$, where a and b are real quantities. (Assume that ψ_0 is real.) (b) Write the time-dependent wave function $\Psi(x, t)$ that corresponds to $\psi(x)$.

69P. Show that the angular wave number k for a nonrelativistic free particle of mass m can be written as

$$k = \frac{2\pi\sqrt{2mK}}{h},$$

in which K is the particle's kinetic energy. ssm

70P. Show that $|\psi|^2 = |\Psi|^2$, with ψ and Ψ related as in Eq. 39-14. That is, show that the probability density does not depend on the time variable.

71P. The function $\psi(x)$ displayed in Eq. 39-19 describes a free particle, for which we assumed that $U(x) = 0$ in Schrödinger's equation (Eq. 39-15). Assume now that $U(x) = U_0 =$ a constant in that equation. Show that Eq. 39-19 is still a solution of Schrödinger's equation, with

$$k = \frac{2\pi}{h}\sqrt{2m(E - U_0)}$$

now giving the angular wave number k of the particle. ssm www

72P. Suppose that we had put $A = 0$ in Eq. 39-17 and relabeled B as ψ_0. What would the resulting wave function then describe? How, if at all, would Fig. 39-11 be altered?

73P. In Eq. 39-18 keep both terms, putting $A = B = \psi_0$. The equation then describes the superposition of two matter waves of equal amplitude, traveling in opposite directions. (Recall that this is the condition for a standing wave.) (a) Show that $|\Psi(x, t)|^2$ is then given by

$$|\Psi(x, t)|^2 = 2\psi_0^2[1 + \cos 2kx].$$

(b) Plot this function, and demonstrate that it describes the square of the amplitude of a standing matter wave. (c) Show that the nodes of this standing wave are located at

$$x = (2n + 1)(\tfrac{1}{4}\lambda), \qquad \text{where } n = 0, 1, 2, 3, \ldots$$

and λ is the de Broglie wavelength of the particle. (d) Write an expression for the most probable locations of the particle.

SEC. 39-8 Heisenberg's Uncertainty Principle

74E. Figure 39-11 shows that because of Heisenberg's uncertainty principle, it is not possible to assign an x coordinate to the position of the electron. (a) Can you assign a y or a z coordinate? (*Hint:* The momentum of the electron has no y or z component.) (b) Describe the extent of the matter wave in three dimensions.

75E. Imagine playing baseball in a universe (not ours!) where the Planck constant is 0.60 J · s. What would be the uncertainty in the position of a 0.50 kg baseball that is moving at 20 m/s along an axis if the uncertainty in the speed is 1.0 m/s? ssm

76E. The uncertainty in the position of an electron is given as 50 pm, which is about equal to the radius of a hydrogen atom. What is the least uncertainty in any simultaneous measurement of the momentum of this electron?

77P. Figure 39-11 shows a case in which the momentum p_x of a particle is fixed so that $\Delta p_x = 0$; then, from Heisenberg's uncertainty principle (Eq. 39-20), the position x of the particle is completely unknown. From the same principle it follows that the opposite is also true; that is, if the position of a particle is exactly known ($\Delta x = 0$), the uncertainty in its momentum is infinite.

Consider an intermediate case, in which the position of a particle is measured, not to infinite precision, but to within a distance of $\lambda/2\pi$, where λ is the particle's de Broglie wavelength. Show that the uncertainty in the (simultaneously measured) momentum is then equal to the momentum itself; that is, $\Delta p_x = p$. Under these circumstances, would a measured momentum of zero surprise you? What about a measured momentum of $0.5p$? Of $2p$? Of $12p$? ssm

78P. You will find in Chapter 40 that electrons cannot move in definite orbits within atoms, like the planets in our solar system. To see why, let us try to "observe" such an orbiting electron by using a light microscope to measure the electron's presumed orbital position with a precision of, say, 10 pm (a typical atom has a radius of about 100 pm). The wavelength of the light used in the microscope must then be about 10 pm. (a) What would be the photon energy of this light? (b) How much energy would such a photon impart to an electron in a head-on collision? (c) What do these results tell you about the possibility of "viewing" an atomic electron at two or more points along its presumed orbital path? (*Hint:* The outer electrons of atoms are bound to the atom by energies of only a few electron-volts.)

SEC. 39-9 Barrier Tunneling

79P. A proton and a deuteron (the latter has the same charge as a proton but twice the mass) strike a potential energy barrier that is 10 fm thick and 10 MeV high. Each particle has a kinetic energy of 3.0 MeV before it strikes the barrier. (a) What is the transmission coefficient for each? (b) What are their respective kinetic energies after they pass through the barrier (assuming that they do so)? (c) What are their respective kinetic energies if they are reflected from the barrier? ssm

80P. Consider a potential energy barrier like that of Fig. 39-12*a* but whose height U_0 is 6.0 eV and whose thickness L is 0.70 nm. What is the energy of an incident electron whose transmission coefficient is 0.0010?

81P. Consider the barrier-tunneling situation in Sample Problem 39-7. What percentage change in the transmission coefficient T occurs for a 1.0% change in (a) the barrier height, (b) the barrier thickness, and (c) the kinetic energy of the incident electron? ssm

82P. (a) Suppose a beam of 5.0 eV protons strikes a potential energy barrier of height 6.0 eV and thickness 0.70 nm, at a rate equivalent to a current of 1000 A. How long would you have to wait—on average—for one proton to be transmitted? (b) How long would you have to wait if the beam consisted of electrons rather than protons?

83P. A 1500 kg car moving at 20 m/s approaches a hill that is 24 m high and 30 m long. Although the car and hill are clearly too large to be treated as matter waves, determine what Eq. 39-21 predicts for the transmission coefficient of the car, as if it could tunnel through the hill as a matter wave. Treat the hill as a potential energy barrier where the potential energy is gravitational.

40 More About Matter Waves

This spectacular computer image was produced in 1993 at IBM's Almaden Research Center in California. The 48 peaks forming the circle mark the positions of individual atoms of iron on a specially prepared copper surface. The circle, which is about 14 nm in diameter, is called a *quantum corral.*

How do these atoms come to be arranged in a circle, and what are the ripples that are trapped within the corral?

The answer is in this chapter.

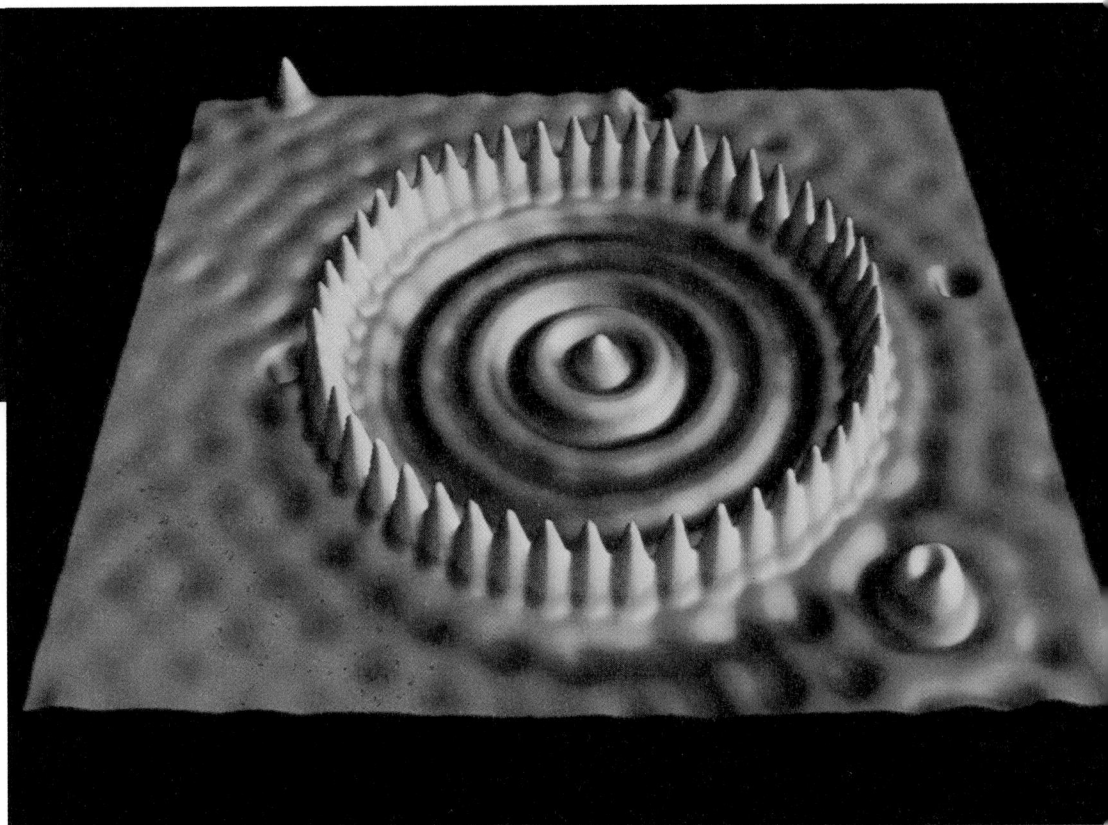

40-1 Atom Building

Early in the twentieth century nobody knew how the electrons in an atom are arranged, what their motions are, how atoms emit or absorb light, or even why atoms are stable. Without this knowledge it is not possible to understand how atoms combine to form molecules or stack up to form solids. As a consequence, the foundations of chemistry—including biochemistry, which underlies the nature of life itself—were more or less a mystery.

In 1926 all these questions and many others were answered with the development of quantum physics. Its basic premise is that moving electrons, protons, and particles of any kind are best viewed as matter waves, whose motions are governed by Schrödinger's equation. Although quantum theory also applies to massive particles, there is no point in treating baseballs, automobiles, planets, and such objects with quantum theory. For such massive, slow-moving objects, Newtonian physics and quantum physics yield the same answers.

Before we can apply quantum physics to the problem of atomic structure, we need to develop some insights by applying quantum ideas in a few simpler situations. These "practice problems" may seem artificial but, as you will see, they provide a firm foundation for understanding a very real problem that we shall analyze in Section 40-8—the structure of the hydrogen atom.

40-2 Waves on Strings and Matter Waves

In Chapter 17 we saw that waves of two kinds can be set up on a stretched string. If the string is so long that we can take it to be infinitely long, we can set up a *traveling wave* of essentially any frequency. However, if the stretched string has only a finite length, perhaps because it is rigidly clamped at both ends, we can set up only *standing waves* on it; further, these standing waves can have only discrete frequencies. In other words, confining the wave to a finite region of space leads to *quantization* of the motion—to the existence of discrete *states* for the wave, each state with a sharply defined frequency.

This observation applies to waves of all kinds, including matter waves. For matter waves, however, it is more convenient to deal with the energy E of the associated particle than with the frequency f of the wave. In all that follows we shall focus on the matter wave associated with an electron, but the results apply to any confined matter wave.

Consider the matter wave associated with an electron moving in the positive x direction and subject to no net force—a so-called *free particle.* The energy of such an electron can have any reasonable value, just as a wave traveling along a stretched string of infinite length can have any reasonable frequency.

Consider next the matter wave associated with an atomic electron, perhaps the *valence* (least tightly bound) electron in a sodium atom. Such an electron—held within the atom by the attractive Coulomb force between it and the positively charged nucleus—is *not* a free particle. It can exist only in a set of discrete states, each having a discrete energy E. This sounds much like the discrete states and quantized frequencies that are available to a stretched string of finite length. For matter waves, then, as for waves of all kinds, we may state a **confinement principle:**

> Confinement of a wave leads to quantization—that is, to the existence of discrete states with discrete energies. The wave can have only those energies.

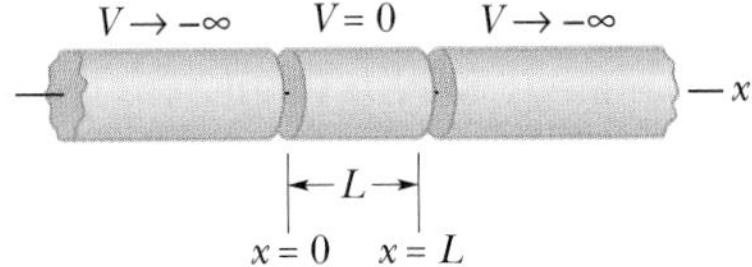

Fig. 40-1 The elements of an idealized "trap" designed to confine an electron to the central cylinder. We take the semi-infinitely long end cylinders to be at an infinitely great negative potential and the central cylinder to be at zero potential.

40-3 Energies of a Trapped Electron

One-Dimensional Traps

Here we examine the matter wave associated with an electron confined to a limited region of space. We do so by analogy with standing waves on a string of finite length, stretched along an x axis and confined between rigid supports. Because the supports are rigid, the two ends of the string are nodes, or points at which the string is always at rest. There may be other nodes along the string, but these two must always be present, as Fig. 17-21 shows.

The states, or discrete standing wave patterns in which the string can oscillate, are those for which the length L of the string is equal to an integer number of half-wavelengths. That is, the string can occupy only states for which

$$L = \frac{n\lambda}{2}, \qquad \text{for } n = 1, 2, 3, \ldots . \tag{40-1}$$

Each value of n identifies a state of the oscillating string; using the language of quantum physics, we can call the integer n a **quantum number.**

For each state of the string permitted by Eq. 40-1, the transverse displacement of the string at any position x along the string is given by

$$y_n(x) = A \sin\left(\frac{n\pi}{L} x\right), \qquad \text{for } n = 1, 2, 3, \ldots , \tag{40-2}$$

in which the quantum number n identifies the oscillation pattern, and A depends on the time at which you inspect the string. (Equation 40-2 is a short version of Eq. 17-47.) We see that for all values of n and for all times, there is a point of zero displacement (a node) at $x = 0$ and at $x = L$, as there must be. Figure 17-20 shows time exposures of such a stretched string for $n = 2$, 3, and 4.

Now let us turn our attention to matter waves. Our first problem is to physically confine an electron that is moving along the x axis so that it remains within a finite segment of that axis. Figure 40-1 shows a conceivable one-dimensional *electron trap*. It consists of two semi-infinitely long cylinders, each of which has an electric potential approaching $-\infty$; between them is a hollow cylinder of length L, which has an electric potential of zero. We put a single electron into this central cylinder to trap it.

The trap of Fig. 40-1 is easy to analyze but is not very practical. Single electrons *can*, however, be trapped in the laboratory with traps that are more complex in design but similar in concept. At the University of Washington, for example, a single electron has been held in a trap for months on end, permitting scientists to make extremely precise measurements of its properties.

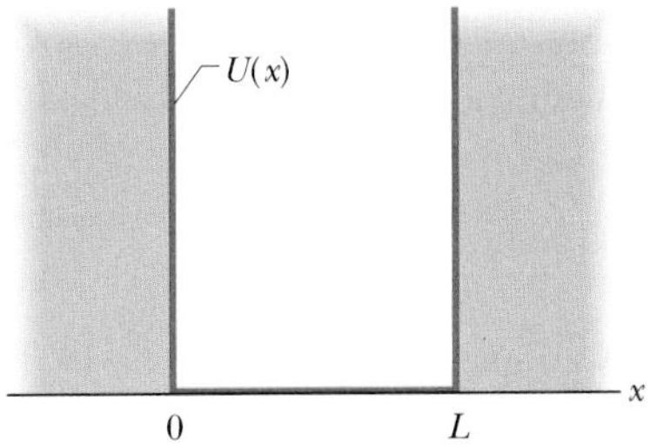

Fig. 40-2 The electric potential energy $U(x)$ of an electron confined to the central cylinder of the idealized trap of Fig. 40-1. We see that $U = 0$ for $0 < x < L$, and $U \to \infty$ for $x < 0$ and $x > L$.

Finding the Quantized Energies

Figure 40-2 shows the potential energy of the electron as a function of its position along the x axis of the idealized trap of Fig. 40-1. When the electron is in the central cylinder, its potential energy U $(= -eV)$ is zero because there the potential V is zero. If the electron could get outside this region, its potential energy would be positive and of infinite magnitude, because there $V \to -\infty$. We call the potential energy pattern of Fig. 40-2 an **infinitely deep potential energy well** or, for short, an *infinite potential well*. It is a "well" because an electron placed in the central cylinder of Fig. 40-1 cannot escape from it. As the electron approaches either end of the cylinder, a force of essentially infinite magnitude reverses the electron's motion, thus trapping it. Because the electron can move along only a single axis, this trap can be called a *one-dimensional infinite potential well.*

Just like the standing wave in a length of stretched string, the matter wave describing the confined electron must have nodes at $x = 0$ and $x = L$. Moreover, Eq. 40-1 applies to such a matter wave if we interpret λ in that equation as the de Broglie wavelength associated with the moving electron.

The de Broglie wavelength λ is defined in Eq. 39-13 as $\lambda = h/p$, where p is the magnitude of the electron's momentum. This magnitude p is related to the electron's kinetic energy K by $p = \sqrt{2mK}$, where m is the mass of the electron. For an electron moving within the central cylinder of Fig. 40-1, where $U = 0$, the total (mechanical) energy E is equal to the kinetic energy. Hence, we can write the de Broglie wavelength of this electron as

$$\lambda = \frac{h}{p} = \frac{h}{\sqrt{2mE}}. \tag{40-3}$$

If we substitute Eq. 40-3 into Eq. 40-1 and solve for the energy E, we find that E depends on n according to

$$E_n = \left(\frac{h^2}{8mL^2}\right)n^2, \quad \text{for } n = 1, 2, 3, \ldots. \tag{40-4}$$

The integer n here is the quantum number of the electron's quantum state in the trap.

Equation 40-4 tells us something important: Because the electron is confined to the trap, it can have only the energies given by the equation. It *cannot* have an energy that is, say, halfway between the values for $n = 1$ and $n = 2$. Why this restriction? Because an electron is a matter wave. Were it, instead, a particle as assumed in classical physics, it could have *any* value of energy while it is confined to the trap.

Figure 40-3 is a graph showing the lowest five allowed energy values for an electron in an infinite well with $L = 100$ pm (about the size of a typical atom). The values are called *energy levels,* and they are drawn in Fig. 40-3 as levels, or steps, on a ladder, in an *energy-level diagram.* Energy is plotted vertically; nothing is plotted horizontally.

The quantum state with the lowest possible energy level E_1 allowed by Eq. 40-4, with quantum number $n = 1$, is called the *ground state* of the electron. The electron tends to be in this lowest energy state. All the quantum states with greater energies (corresponding to quantum numbers $n = 2$ or greater), are called *excited states* of the electron. The state with energy level E_2, for quantum number $n = 2$, is called the *first excited state* because it is the first of the excited states as we move up the energy-level diagram. Similarly, the state with energy level E_3 is called the *second excited state.*

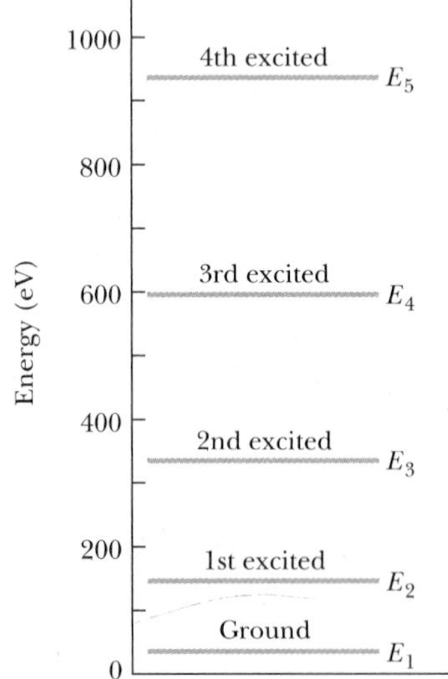

Fig. 40-3 Several of the allowed energies given by Eq. 40-4 for an electron confined to the infinite well of Fig. 40-2. Here width $L = 100$ pm. Such a plot is called an *energy-level diagram.*

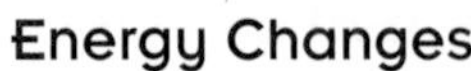

Energy Changes

A trapped electron tends to have the lowest allowed energy, and thus to be in its ground state. It can be changed to an excited state (in which it has greater energy) only if an external source provides the additional energy that is required for the change. Let E_{low} be the initial energy of the electron, and E_{high} be the greater energy in a state that is higher on its energy-level diagram. Then the amount of energy that is required for the electron's change of state is

$$\Delta E = E_{\text{high}} - E_{\text{low}}. \tag{40-5}$$

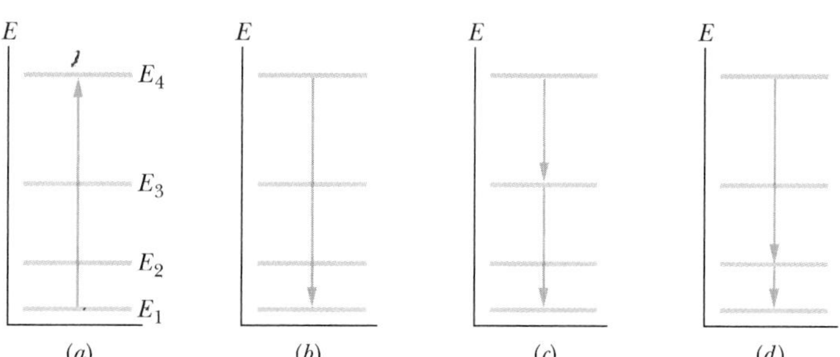

Fig. 40-4 (*a*) Excitation of a trapped electron from the energy level of its ground state to the level of its third excited state. (*b*)–(*d*) Three of four possible ways the electron can de-excite to return to the energy level of its ground state. (Which way is not shown?)

An electron that receives such energy is said to make a *quantum jump* (or *transition*), or to be *excited* from the lower-energy state to the higher-energy state. Figure 40-4*a* represents a quantum jump from the ground state (with energy level E_1) to the third excited state (with energy level E_4). As shown, the jump *must* be from one energy level to another but it can bypass one or more intermediate energy levels.

One way an electron can gain energy to make a quantum jump up to a greater energy level is to absorb a photon. However, this absorption and quantum jump can occur only if the following condition is met:

> If a confined electron is to absorb a photon, the energy hf of the photon must equal the energy difference ΔE between the initial energy level of the electron and a higher level.

Thus, excitation by the absorption of light requires that

$$hf = \Delta E = E_{\text{high}} - E_{\text{low}}. \tag{40-6}$$

When an electron reaches an excited state, it does not stay there but quickly *de-excites* by decreasing its energy. Figures 40-4*b* to *d* represent some of the possible quantum jumps down from the energy level of the third excited state. The electron can reach its ground-state level either with one direct quantum jump (Fig. 40-4*b*) or with shorter jumps via intermediate energy levels (Figs. 40-4*c* and *d*).

One way in which an electron can decrease its energy is by emitting a photon, but only if the following condition is met:

> If a confined electron emits a photon, the energy hf of that photon must equal the energy difference ΔE between the initial energy level of the electron and a lower level.

Thus, Eq. 40-6 applies to both the absorption and the emission of light by a confined electron. That is, the absorbed or emitted light can have only certain values of hf, and thus only certain values of frequency f and wavelength λ.

Aside: Although Eq. 40-6 and what we have discussed about photon absorption and emission can be applied to physical (real) electron traps, they actually cannot be applied to one-dimensional (unreal) electron traps. The reason involves the need to conserve angular momentum in a photon absorption or emission process. In this book, we shall neglect that need and use Eq. 40-6 even for one-dimensional traps.

CHECKPOINT 1: Rank the following pairs of quantum states for an electron confined to an infinite well according to the energy differences between the states, greatest first: (a) $n = 3$ to $n = 1$, (b) $n = 5$ to $n = 4$, (c) $n = 4$ to $n = 3$.

Sample Problem 40-1

An electron is confined to a one-dimensional, infinitely deep potential energy well of width $L = 100$ pm.

(a) What is the least energy the electron can have?

SOLUTION: The **Key Idea** here is that confinement of the electron (a matter wave) to the well leads to quantization of its energy. Because the well is infinitely deep, the allowed energies are given by Eq. 40-4 ($E_n = (h^2/8mL^2)n^2$), with the quantum number n a positive integer. Here, the collection of constants in front of n^2 in Eq. 40-4 is evaluated as

$$\frac{h^2}{8mL^2} = \frac{(6.63 \times 10^{-34}\ \text{J}\cdot\text{s})^2}{(8)(9.11 \times 10^{-31}\ \text{kg})(100 \times 10^{-12}\ \text{m})^2} = 6.031 \times 10^{-18}\ \text{J}. \qquad (40\text{-}7)$$

The least energy of the electron corresponds to the least quantum number, which is $n = 1$ for the ground state of the electron. Thus, Eqs. 40-4 and 40-7 give us

$$E_1 = \left(\frac{h^2}{8mL^2}\right)n^2 = (6.031 \times 10^{-18}\ \text{J})(1^2) \approx 6.03 \times 10^{-18}\ \text{J} = 37.7\ \text{eV}. \qquad \text{(Answer)}$$

(b) How much energy must be transferred to the electron if it is to make a quantum jump from its ground state to its second excited state?

SOLUTION: *First a caution:* Note that, from Fig. 40-3, the *second* excited state corresponds to the *third* energy level, with quantum number $n = 3$. Then one **Key Idea** is that if the electron is to jump from the $n = 1$ level to the $n = 3$ level, the required change in its energy is, from Eq. 40-5,

$$\Delta E_{31} = E_3 - E_1. \qquad (40\text{-}8)$$

A second **Key Idea** is that the energies E_3 and E_1 depend on the quantum number n, according to Eq. 40-4. Therefore, substituting that equation into Eq. 40-8 for energies E_3 and E_1 and using Eq. 40-7 lead to

$$\Delta E_{31} = \left(\frac{h^2}{8mL^2}\right)(3)^2 - \left(\frac{h^2}{8mL^2}\right)(1)^2 = \frac{h^2}{8mL^2}(3^2 - 1^2) = (6.031 \times 10^{-18}\ \text{J})(8) = 4.83 \times 10^{-17}\ \text{J} = 302\ \text{eV}. \qquad \text{(Answer)}$$

(c) If the electron gains the energy for the jump from energy level E_1 to energy level E_3 by absorbing light, what light wavelength is required?

SOLUTION: One **Key Idea** here is that if light is to transfer energy to the electron, the transfer must be by photon absorption. A second **Key Idea** is that the photon's energy must equal the energy difference ΔE between the initial energy level of the electron and a higher level, according to Eq. 40-6 ($hf = \Delta E$). Otherwise, a photon *cannot* be absorbed. Substituting c/λ for f, we can rewrite Eq. 40-6 as

$$\lambda = \frac{hc}{\Delta E}. \qquad (40\text{-}9)$$

For the energy difference ΔE_{31} we found in (b), this equation

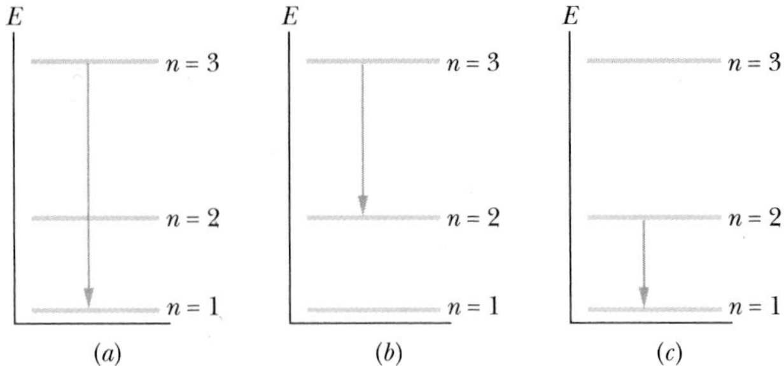

Fig. 40-5 Sample Problem 40-1. De-excitation from the second excited state to the ground state either directly (*a*) or via the first excited state (*b*, *c*).

gives us

$$\lambda = \frac{hc}{\Delta E_{31}} = \frac{(6.63 \times 10^{-34}\ \text{J}\cdot\text{s})(3.0 \times 10^8\ \text{m/s})}{4.83 \times 10^{-17}\ \text{J}} = 4.12 \times 10^{-9}\ \text{m}. \qquad \text{(Answer)}$$

(d) Once the electron has been excited to the second excited state, what wavelengths of light can it emit by de-excitation?

SOLUTION: We have three **Key Ideas** here:

1. The electron tends to de-excite, rather than remain in an excited state, until it reaches the ground state ($n = 1$).
2. If the electron is to de-excite, it must lose just enough energy to jump to a lower energy level.
3. If it is to lose energy by emitting light, then the loss of energy must be by emission of a photon.

Starting in the second excited state (at the $n = 3$ level), the electron can reach the ground state ($n = 1$) by *either* making a quantum jump directly to the ground-state energy level (Fig. 40-5*a*) or by making two *separate* jumps by way of the $n = 2$ level (Figs. 40-5*b* and *c*).

The direct jump involves the same energy difference ΔE_{31} we found in (c). Then the wavelength is the same as we calculated in (c)—except now the wavelength is for light that is emitted, not absorbed. Thus, the electron can jump directly to the ground state by emitting light of wavelength

$$\lambda = 4.12 \times 10^{-9}\ \text{m}. \qquad \text{(Answer)}$$

Following the procedure of part (b), you can show that the energy differences for the jumps of Figs. 40-5*b* and *c* are

$$\Delta E_{32} = 3.016 \times 10^{-17}\ \text{J} \quad \text{and} \quad \Delta E_{21} = 1.809 \times 10^{-17}\ \text{J}.$$

From Eq. 40-9, we then find that the wavelength of the light emitted in the first of these jumps (from $n = 3$ to $n = 2$) is

$$\lambda = 6.60 \times 10^{-9}\ \text{m}, \qquad \text{(Answer)}$$

and the wavelength of the light emitted in the second of these jumps (from $n = 2$ to $n = 1$) is

$$\lambda = 1.10 \times 10^{-8}\ \text{m}. \qquad \text{(Answer)}$$

40-4 Wave Functions of a Trapped Electron

If we solve Schrödinger's equation for an electron trapped in a one-dimensional infinite potential well of width L, we find that the wave functions for the electron are given by

$$\psi_n(x) = A \sin\left(\frac{n\pi}{L}x\right), \qquad \text{for } n = 1, 2, 3, \ldots, \tag{40-10}$$

for $0 \le x \le L$ (the wave function is zero outside that range). We shall soon evaluate the amplitude constant A in this equation.

Note that the wave functions $\psi_n(x)$ have the same form as the displacement functions $y_n(x)$ for a standing wave on a string stretched between rigid supports (see Eq. 40-2). We can picture an electron trapped in a one-dimensional well between infinite-potential walls as being a standing matter wave.

Probability of Detection

The wave function $\psi_n(x)$ cannot be detected or directly measured in any way—we cannot simply look inside the well to see the wave, like we can see a wave in a bathtub of water. All we can do is insert a probe of some kind, to try to detect the electron. At the instant of detection, the electron would materialize at the point of detection, at some position along the x axis within the well.

If we repeated this detection procedure at many positions throughout the well, we would find that the probability of detecting the electron is related to the probe's position x in the well. In fact, they are related by the *probability density* $\psi_n^2(x)$. Recall from Section 39-7 that in general the probability that a particle can be detected in a specified infinitesimal volume centered on a specified point is proportional to $|\psi_n^2|$. Here, with the electron trapped in a one-dimensional well, we are concerned only with detection of the electron along the x axis. Thus, the probability density $\psi_n^2(x)$ here is a probability per unit length along the x axis. (We can omit the absolute value sign here, because $\psi_n(x)$ in Eq. 40-10 is a real quantity, not a complex one.) The probability $p(x)$ that an electron can be detected at position x within the well is

$$\begin{pmatrix}\text{probability } p(x)\\ \text{of detection in width } dx\\ \text{centered on position } x\end{pmatrix} = \begin{pmatrix}\text{probability density } \psi_n^2(x)\\ \text{at position } x\end{pmatrix}(\text{width } dx),$$

or

$$p(x) = \psi_n^2(x)\, dx. \tag{40-11}$$

From Eq. 40-10, we see that the probability density $\psi_n^2(x)$ for the trapped electron is

$$\psi_n^2(x) = A^2 \sin^2\left(\frac{n\pi}{L}x\right), \qquad \text{for } n = 1, 2, 3, \ldots, \tag{40-12}$$

for the range $0 \le x \le L$ (the probability density is zero outside that range). Figure 40-6 shows $\psi_n^2(x)$ for $n = 1, 2, 3$, and 15 for an electron in an infinite well whose width L is 100 pm.

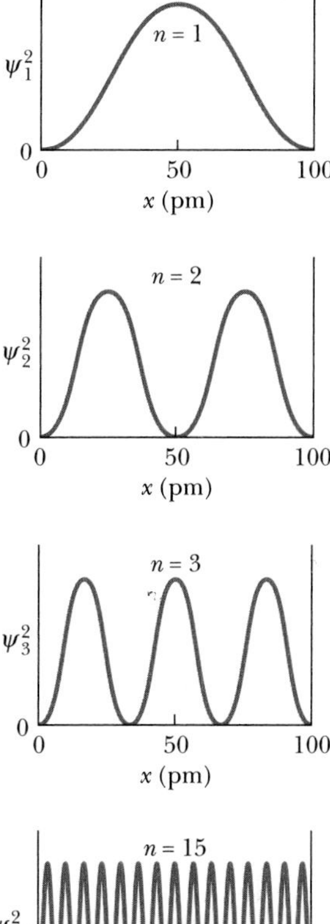

Fig. 40-6 The probability density $\psi_n^2(x)$ for four states of an electron trapped in a one-dimensional infinite well; their quantum numbers are $n = 1, 2, 3$, and 15. The electron is most likely to be found where $\psi_n^2(x)$ is greatest, and least likely to be found where $\psi_n^2(x)$ is least.

To find the probability that the electron can be detected in any finite section of the well—say, between point x_1 and point x_2—we must integrate $p(x)$ between those points. Thus, from Eqs. 40-11 and 40-12,

$$\begin{pmatrix}\text{probability of detection}\\ \text{between } x_1 \text{ and } x_2\end{pmatrix} = \int_{x_1}^{x_2} p(x)$$

$$= \int_{x_1}^{x_2} A^2 \sin^2\left(\frac{n\pi}{L}x\right) dx. \qquad (40\text{-}13)$$

If classical physics prevailed, we would expect the trapped electron to be detectable with equal probabilities in all parts of the well. From Fig. 40-6 we see that it is not. For example, inspection of that figure or of Eq. 40-12 shows that for the state with $n = 2$, the electron is most likely to be detected near $x = 25$ pm and $x = 75$ pm. It can be detected with near-zero probability near $x = 0$, $x = 50$ pm, and $x = 100$ pm.

The case of $n = 15$ in Fig. 40-6 suggests that as n increases, the probability of detection becomes more and more uniform across the well. This result is an instance of a general principle called the **correspondence principle:**

> At large enough quantum numbers, the predictions of quantum physics merge smoothly with those of classical physics.

This principle, first advanced by Danish physicist Niels Bohr, holds for all quantum predictions. It should remind you of a similar principle concerning the theory of relativity—namely, that at low enough particle speeds, the predictions of special relativity merge smoothly with those of classical physics.

CHECKPOINT 2: The figure shows three infinite potential wells of widths L, $2L$, and $3L$; each contains an electron in the state for which $n = 10$. Rank the wells according to (a) the number of maxima for the probability density of the electron and (b) the energy of the electron, greatest first.

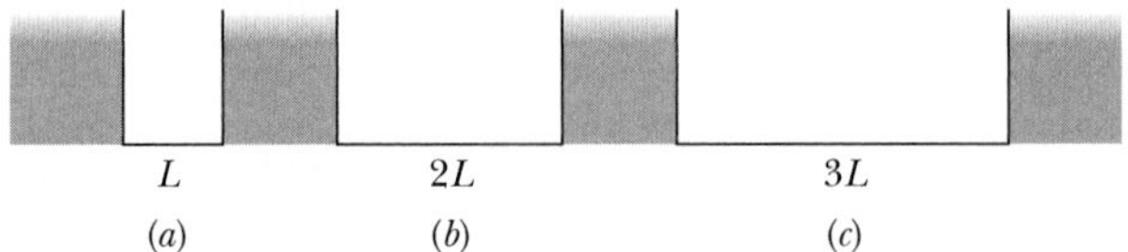

Normalization

The product $\psi_n^2(x)\,dx$ gives the probability that an electron in an infinite well can be detected in the interval of the x axis that lies between x and $x + dx$. We know that the electron must be *somewhere* in the infinite well, so it must be that

$$\int_{-\infty}^{+\infty} \psi_n^2(x)\,dx = 1 \quad \text{(normalization equation)}, \qquad (40\text{-}14)$$

since the probability 1 corresponds to certainty. Although the integral is taken over the entire x axis, only the region from $x = 0$ to $x = L$ makes any contribution to the probability. Graphically, the integral in Eq. 40-14 represents the area under each of the plots of Fig. 40-6.

In Sample Problem 40-2 we shall see that if we substitute $\psi_n^2(x)$ from Eq. 40-12 into Eq. 40-14, it is possible to assign a specific value to the amplitude constant A

that appears in Eq. 40-12; namely, $A = \sqrt{2/L}$. This process of using Eq. 40-14 to evaluate the amplitude of a wave function is called **normalizing** the wave function. The process applies to *all* one-dimensional wave functions.

Zero-Point Energy

Substituting $n = 1$ in Eq. 40-4 defines the state of lowest energy for an electron in an infinite potential well, the ground state. That is the state the confined electron will occupy unless energy is supplied to it to raise it to an excited state.

The question arises: Why can't we include $n = 0$ among the possibilities listed for n in Eq. 40-4? Putting $n = 0$ in this equation would indeed yield a ground-state energy of zero. However, putting $n = 0$ in Eq. 40-12 would also yield $\psi_n^2(x) = 0$ for all x, which we can interpret only to mean that there is no electron in the well. We know that there is, so $n = 0$ is not a possible quantum number.

It is an important conclusion of quantum physics that confined systems cannot exist in states with zero energy. They must always have a certain minimum energy called the **zero-point energy.**

We can make the zero-point energy as small as we like by making the infinite well wider—that is, by increasing L in Eq. 40-4 for $n = 1$. In the limit as $L \to \infty$, the zero-point energy E_1 approaches zero. In this limit, however, with an infinitely wide well, the electron is a free particle, no longer confined in the x direction. Also, because the energy of a free particle is not quantized, that energy can have any value, including zero. Only a confined particle must have a finite zero-point energy and can never be at rest.

✔CHECKPOINT 3: Each of the following particles is confined to an infinite well, and all four wells have the same width: (a) an electron, (b) a proton, (c) a deuteron, and (d) an alpha particle. Rank their zero-point energies, greatest first. The particles are listed in order of increasing mass.

Sample Problem 40-2

Evaluate the amplitude constant A in Eq. 40-10 for an infinite potential well extending from $x = 0$ to $x = L$.

SOLUTION: The Key Idea here is that the wave functions of Eq. 40-10 must satisfy the normalization requirement of Eq. 40-14, which states that the probability that the electron can be detected somewhere along the x axis is 1. Substituting Eq. 40-10 into Eq. 40-14 and taking the constant A outside the integral yield

$$A^2 \int_0^L \sin^2\left(\frac{n\pi}{L}x\right) dx = 1. \qquad (40\text{-}15)$$

We have changed the limits of the integral from $-\infty$ and $+\infty$ to 0 and L because the wave function is zero outside these new limits (so there's no need to integrate out there).

We can simplify the indicated integration by changing the variable from x to the dimensionless variable y, where

$$y = \frac{n\pi}{L}x, \qquad (40\text{-}16)$$

hence

$$dx = \frac{L}{n\pi}dy.$$

When we change the variable, we must also change the integration limits (again). Equation 40-16 tells us that $y = 0$ when $x = 0$ and that $y = n\pi$ when $x = L$, so 0 and $n\pi$ are our new limits. With all these substitutions, Eq. 40-15 becomes

$$A^2 \frac{L}{n\pi} \int_0^{n\pi} (\sin^2 y)\, dy = 1.$$

We can use integral 11 in Appendix E to evaluate the integral, obtaining the equation

$$\frac{A^2L}{n\pi}\left[\frac{y}{2} - \frac{\sin 2y}{4}\right]_0^{n\pi} = 1.$$

Evaluating at the limits yields

$$\frac{A^2L}{n\pi}\frac{n\pi}{2} = 1,$$

so

$$A = \sqrt{\frac{2}{L}}. \qquad \text{(Answer)} \quad (40\text{-}17)$$

This result tells us that the dimension for A^2, and thus for $\psi_n^2(x)$, is an inverse length. This is appropriate because the probability density of Eq. 40-12 is a probability *per unit length*.

Sample Problem 40-3

A ground-state electron is trapped in the one-dimensional infinite potential well of Fig. 40-2, with width $L = 100$ pm.

(a) What is the probability that the electron can be detected in the left one-third of the well (between $x_1 = 0$ and $x_2 = L/3$)?

SOLUTION: One Key Idea here is that if we probe the left one-third of the well, there is no guarantee that we will detect the electron. However, we can calculate the probability of detecting it with the integral of Eq. 40-13. A second Key Idea is that the probability very much depends on which state the electron is in—that is, the value of the electron's quantum number n. Because here the electron is in the ground state, we set $n = 1$ in Eq. 40-13.

We also set the limits of integration as the positions $x_1 = 0$ and $x_2 = L/3$ and, from Sample Problem 40-2, set the amplitude constant A as $\sqrt{2/L}$. We then see that

$$\begin{pmatrix}\text{probability of detection}\\ \text{in left one-third}\end{pmatrix} = \int_0^{L/3} \frac{2}{L}\sin^2\left(\frac{1\pi}{L}x\right) dx.$$

We could find this probability by substituting 100×10^{-12} m for L and then using a graphing calculator or a computer math package to evaluate the integral. Instead, we shall follow the steps of Sample Problem 40-2. From Eq. 40-16, we obtain for the new integration variable y,

$$y = \frac{\pi}{L}x \quad \text{and} \quad dx = \frac{L}{\pi}dy.$$

From the first of these equations, we find the new limits of integration to be $y_1 = 0$ for $x_1 = 0$ and $y_2 = \pi/3$ for $x_2 = L/3$. We then must evaluate

$$\text{probability} = \left(\frac{2}{L}\right)\left(\frac{L}{\pi}\right)\int_0^{\pi/3} (\sin^2 y)\, dy.$$

Using integral 11 in Appendix E, we then find

$$\text{probability} = \frac{2}{\pi}\left(\frac{y}{2} - \frac{\sin 2y}{4}\right)_0^{\pi/3} = 0.20.$$

Thus, we have

$$\begin{pmatrix}\text{probability of detection}\\ \text{in left one-third}\end{pmatrix} = 0.20. \qquad \text{(Answer)}$$

That is, if we repeatedly probe the left one-third of the well, then on average we can detect the electron with 20% of the probes.

(b) What is the probability that the electron can be detected in the middle one-third of the well (between $x_1 = L/3$ and $x_2 = 2L/3$)?

SOLUTION: We now know that the probability of detection in the left one-third of the well is 0.20. A Key Idea here is that by symmetry, the probability of detection in the right one-third of the well is also 0.20. A second Key Idea is that because the electron is certainly in the well, the probability of detection in the entire well is 1.0. Thus, the probability of detection in the middle one-third of the well is

$$\begin{pmatrix}\text{probability of detection}\\ \text{in middle one-third}\end{pmatrix} = 1 - 0.20 - 0.20$$

$$= 0.60. \qquad \text{(Answer)}$$

40-5 An Electron in a Finite Well

A potential energy well of infinite depth is an idealization. Figure 40-7 shows a realizable potential energy well—one in which the potential energy of an electron outside the well is not infinitely great but has a finite positive value U_0, called the **well depth.** The analogy between waves on a stretched string and matter waves fails us for wells of finite depth because we can no longer be sure that matter wave nodes exist at $x = 0$ and at $x = L$. (As we shall see, they don't.)

To find the wave functions describing the quantum states of an electron in the finite well of Fig. 40-7, we *must* resort to Schrödinger's equation, the basic equation of quantum physics. From Section 39-7 recall that, for motion in one dimension, we use Schrödinger's equation in the form of Eq. 39-15:

$$\frac{d^2\psi}{dx^2} + \frac{8\pi^2 m}{h^2}[E - U(x)]\psi = 0. \qquad (40\text{-}18)$$

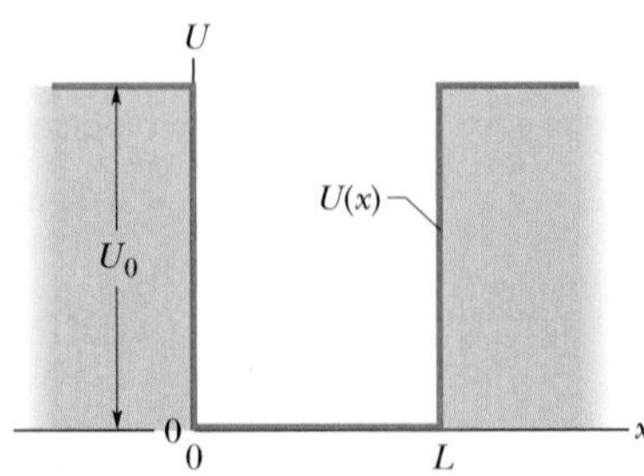

Fig. 40-7 A *finite* potential energy well. The depth of the well is U_0 and its width is L. As in the infinite potential well of Fig. 40-2, the motion of the trapped electron is restricted to the x direction.

Rather than attempting to solve this equation for the finite well, we simply state the results for particular numerical values of U_0 and L. Figure 40-8 shows these results as graphs of $\psi_n^2(x)$, the probability density, for a well with $U_0 = 450$ eV and $L = 100$ pm.

The probability density $\psi_n^2(x)$ for each graph in Fig. 40-8 satisfies Eq. 40-14, the normalization equation, so we know that the areas under all three probability density plots are numerically equal to 1.

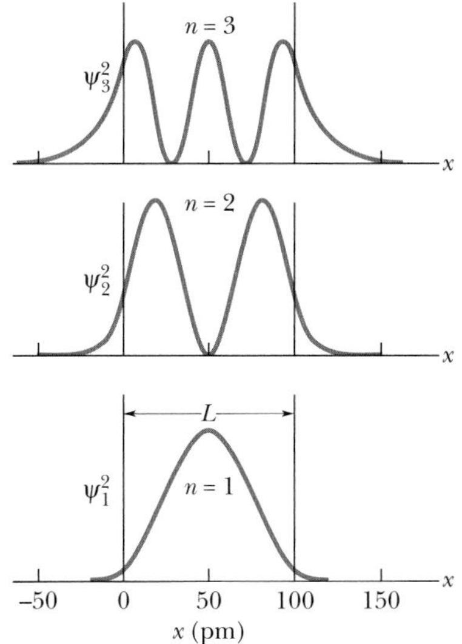

Fig. 40-8 Probability densities $\psi_n^2(x)$ for an electron confined to a finite potential well of depth $U_0 = 450$ eV and width $L = 100$ pm. The only quantum states the electron can have in this well are those that have quantum numbers $n = 1$, 2, and 3.

If you compare Fig. 40-8, for a finite well, with Fig. 40-6, for an infinite well, you will see one striking difference: For a finite well, the electron matter wave penetrates the walls of the well—into a region in which Newtonian mechanics says the electron cannot exist. This penetration should not be surprising, because we saw in Section 39-9 that an electron can tunnel through a potential energy barrier. "Leaking" into the walls of a finite potential energy well is a similar phenomenon. From the plots of ψ^2 in Fig. 40-8, we see that the leakage is greater for greater values of quantum number n.

Because a matter wave *does* leak into the walls of a finite well, the wavelength λ for any given quantum state is greater when the electron is trapped in a finite well than when it is trapped in an infinite well. Equation 40-3 then tells us that the energy E for an electron in any given state is less in the finite well than in the infinite well.

That fact allows us to approximate the energy-level diagram for an electron trapped in a finite well. As an example, we can approximate the diagram for the finite well of Fig. 40-8, which has width $L = 100$ pm and depth $U_0 = 450$ J. The energy-level diagram for an *infinite* well of that width is shown in Fig. 40-3. First we remove the portion of Fig. 40-3 above 450 J. Then we shift the remaining three energy levels down, shifting the level for $n = 3$ the most because the wave leakage into the walls is greatest for $n = 3$. The result is approximately the energy-level diagram for the finite well. The actual diagram is shown in Fig. 40-9.

In that figure, an electron with an energy greater than U_0 (= 450 J) has too much energy to be trapped in the finite well. Thus, it is not confined and its energy is not quantized; that is, its energy is not restricted to certain values. To reach this *nonquantized* portion of the energy-level diagram and thus to be free, a trapped electron must somehow obtain enough energy to have a mechanical energy of 450 J or greater.

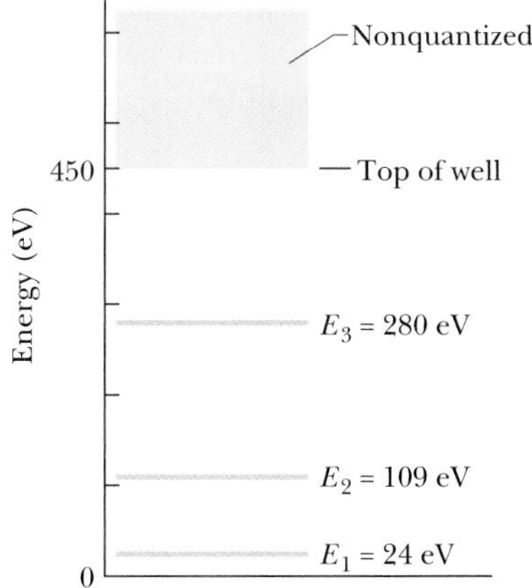

Fig. 40-9 The energy-level diagram corresponding to the probability densities of Fig. 40-8. If an electron is trapped in the finite potential well, it can have only the energies corresponding to $n = 1$, 2, and 3. If it has an energy of 450 eV or greater, it is not trapped and its energy is not quantized.

Sample Problem 40-4

Suppose a finite well with $U_0 = 450$ eV and $L = 100$ pm confines a single electron in its ground state.

(a) What wavelength of light is needed to barely free the electron from the potential well by a single photon absorption?

SOLUTION: One **Key Idea** here is that for the electron to escape from the potential well, it must receive enough energy to put it into the nonquantized energy region of Fig. 40-9. Thus, it must end up with an energy of at least U_0 (= 450 eV). A second **Key Idea** is that the electron is initially in its ground state, with an energy of $E_1 =$ 24 eV. Thus, to barely become free, it must receive an energy of

$$U_0 - E_1 = 450 \text{ eV} - 24 \text{ eV} = 426 \text{ eV}.$$

If it receives this energy from light, then it must absorb a photon with that much energy. From Eq. 40-6, with c/λ substituted for f, we can then write

$$\frac{hc}{\lambda} = U_0 - E_1,$$

from which we find

$$\lambda = \frac{hc}{U_0 - E_1} = \frac{(6.63 \times 10^{-34}\ \text{J}\cdot\text{s})(3.00 \times 10^8\ \text{m/s})}{(426\ \text{eV})(1.60 \times 10^{-19}\ \text{J/eV})} = 2.92 \times 10^{-9}\ \text{m} = 2.92\ \text{nm}. \quad \text{(Answer)}$$

Thus, if the electron absorbs a photon from light of wavelength 2.92 nm, it just barely escapes the potential well.

(b) Can the electron, initially in the ground state, absorb light with a wavelength of 2.00 nm? If so, what then is the electron's energy?

SOLUTION: The **Key Ideas** here are these:

1. In (a) we found that light of 2.92 nm will just barely free the electron from the potential well.
2. We are now considering light with a shorter wavelength of 2.00 nm and thus a greater energy per photon ($hf = hc/\lambda$).
3. Hence, the electron *can* absorb a photon of this light. The energy transfer will not only free the electron but will also provide it with more kinetic energy. Further, because the electron is then no longer trapped, its energy is not quantized and thus there is no restriction on its kinetic energy.

The energy transferred to the electron is the photon energy:

$$hf = h\frac{c}{\lambda} = \frac{(6.63 \times 10^{-34}\ \text{J}\cdot\text{s})(3.00 \times 10^8\ \text{m/s})}{2.00 \times 10^{-9}\ \text{m}} = 9.95 \times 10^{-17}\ \text{J} = 622\ \text{eV}.$$

From (a), the energy required to just barely free the electron from the potential well is $U_0 - E_1$ (= 426 eV). The remainder of the 622 eV goes to kinetic energy. Thus, the kinetic energy of the freed electron is

$$K = hf - (U_0 - E_1) = 622\ \text{eV} - 426\ \text{eV} = 196\ \text{eV}. \quad \text{(Answer)}$$

40-6 More Electron Traps

Here we discuss three types of artificial electron traps.

Nanocrystallites

Perhaps the most direct way to construct a potential energy well in the laboratory is to prepare a sample of a semiconducting material in the form of a powder whose granules are small—in the nanometer range—and of uniform size. Each such granule—each **nanocrystallite**—acts as a potential well for the electrons trapped within it.

Equation 40-4 shows that we can increase the energy of the least energetic quantum state of an electron trapped in an infinite well by reducing the width L of that well. This is also true for the wells formed by individual nanocrystallites. Thus, the smaller the nanocrystallite, the higher its lowest available level—that is, the higher the threshold energy for the photons of light that it can absorb.

If we shine sunlight on a powder of nanocrystallites, the crystallites can absorb all photons with energies above a certain threshold energy E_t (= hf_t). Thus, they can absorb light whose wavelength is *below* a certain threshold λ_t, where

$$\lambda_t = \frac{c}{f_t} = \frac{ch}{E_t}. \tag{40-19}$$

Fig. 40-10 Two samples of powdered cadmium selenide, a semiconductor, differing only in the size of their granules. Each granule serves as an electron trap. The upper sample has the larger granules and consequently the smaller spacing between energy levels and the lower photon energy threshold for the absorption of light. Light not absorbed is scattered, causing the sample to scatter light of greater wavelength and appear red. The lower sample, because of its smaller granules, and consequently its larger level spacing and its larger energy threshold for absorption, appears yellow.

Since light not absorbed is scattered, our powder of nanocrystallites will scatter all wavelengths above λ_t.

We see the powder sample by the light it scatters back to our eyes. Thus, by controlling the size of the nanocrystallites in a sample, we can control the wavelengths of the light scattered by the sample, and hence the sample's color.

Figure 40-10 shows two samples of the semiconductor cadmium selenide, each consisting of a powder of nanocrystallites of uniform size. The upper sample scatters light at the red end of the spectrum. The lower sample differs from the upper sample *only* in that the lower sample is composed of smaller nanocrystallites. For this reason its threshold energy E_t is greater and, from Eq. 40-19, its threshold wavelength λ_t is shorter. The sample takes on a color of shorter wavelength—in this case yellow.

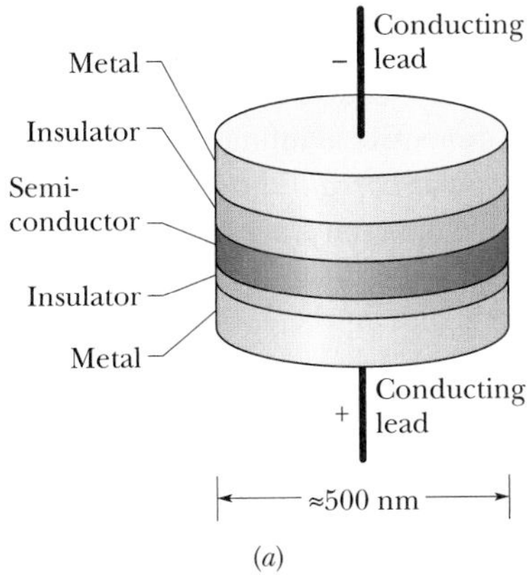

(a)

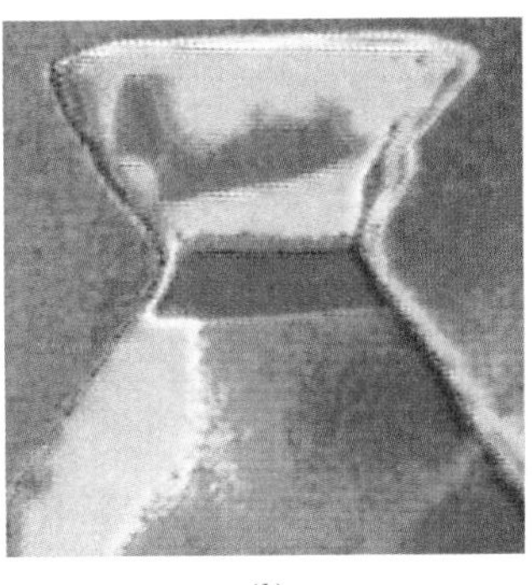
(b)

Fig. 40-11 A quantum dot, or "artificial atom." (*a*) A central semiconducting layer forms a potential energy well in which electrons are trapped. The lower insulating layer is thin enough to allow electrons to be added to or removed from the central layer by barrier tunneling if an appropriate voltage is applied between the leads. (*b*) A photograph of an actual quantum dot. The central purple band is the electron confinement region.

The striking contrast in color between the two samples is compelling evidence of the quantization of the energies of trapped electrons and the dependence of these energies on the size of the electron trap.

Quantum Dots

The highly developed techniques used to fabricate computer chips can be used to construct, atom by atom, individual potential energy wells that behave, in many respects, like artificial atoms. These **quantum dots,** as they are usually called, have promising applications in electron optics and computer technology.

In one such arrangement, a "sandwich" is fabricated in which a thin layer of a semiconducting material, shown in purple in Fig. 40-11*a*, is deposited between two insulating layers, one of which is much thinner than the other. Metal end caps with conducting leads are added at both ends. The materials are chosen to ensure that the potential energy of an electron in the central layer is less than it is in the two insulating layers, causing the central layer to act as a potential energy well. Figure 40-11*b* is a photograph of an actual quantum dot; the well in which individual electrons can be trapped is the purple region.

The lower (but not the upper) insulating layer in Fig. 40-11*a* is thin enough to permit electrons to tunnel through it if an appropriate potential difference is applied between the leads. In this way the number of electrons confined to the well can be controlled. The arrangement does indeed behave like an artificial atom with the property that the number of electrons it contains can be controlled. Quantum dots can be constructed in two-dimensional arrays that could well form the basis for computing systems of great speed and storage capacity.

Quantum Corrals

When a scanning tunneling microscope (described in Section 39-9 and Fig. 39-14) is in operation, its tip exerts a small force on isolated atoms that may be located on an otherwise smooth surface. By careful manipulation of the position of the tip, such isolated atoms can be "dragged" across the surface and deposited at another location. Using this technique, scientists at IBM's Almaden Research Center moved iron atoms across a carefully prepared copper surface, forming the atoms into a circle, which they named a **quantum corral.** The result is shown in the photograph that opens this chapter. Each iron atom in the circle is nestled in a hollow in the copper surface, equidistant from three nearest-neighbor copper atoms. The corral was fabricated at a low temperature (about 4 K) to minimize the tendency of the iron atoms to move randomly about on the surface because of their thermal energies.

The ripples within the corral are due to matter waves associated with electrons that can move over the copper surface but are largely trapped in the potential well of the corral. The dimensions of the ripples are in excellent agreement with the predictions of quantum theory.

40-7 Two- and Three-Dimensional Electron Traps

In the next section, we shall discuss the hydrogen atom as being a three-dimensional finite potential well. As a warm-up for the hydrogen atom, let us extend our discussion of infinite potential wells to two and three dimensions.

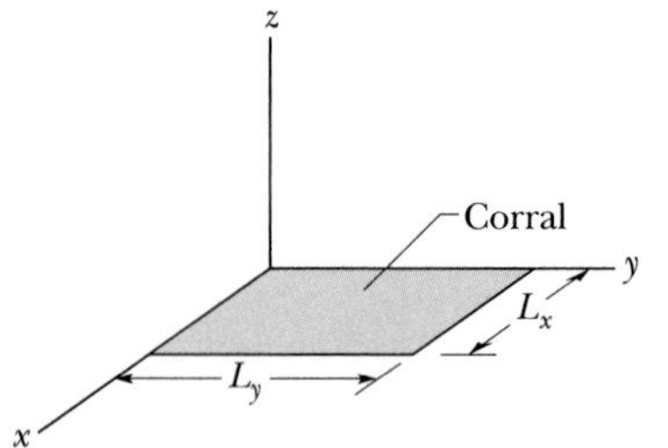

Fig. 40-12 A rectangular corral—a two-dimensional version of the infinite potential well of Fig. 40-2—with widths L_x and L_y.

Rectangular Corral

Figure 40-12 shows the rectangular area to which an electron can be confined by the two-dimensional version of Fig. 40-2—a two-dimensional infinite potential well of widths L_x and L_y. Such a well is called a rectangular *corral*. The corral might be on the surface of a body that somehow prevents the electron from moving parallel to the z axis and thus from leaving the surface. You have to imagine infinite potential energy functions (like $U(x)$ in Fig. 40-2) along each side of the corral, keeping the electron within the corral.

Solution of Schrödinger's equation for the rectangular corral of Fig. 40-12 shows that, for the electron to be trapped, its matter wave must fit into each of the two widths separately, just as the matter wave of a trapped electron must fit into a one-dimensional infinite well. This means the wave is separately quantized in width L_x and in width L_y. Let n_x be the quantum number for which the matter wave fits into width L_x, and let n_y be the quantum number for which the matter wave fits into width L_y. As with a one-dimensional potential well, these quantum numbers can be only positive integers.

The energy of the electron depends on both quantum numbers and is the sum of the energy it would have if it were confined along the x axis alone and the energy it would have if it were confined along the y axis alone. From Eq. 40-4, we can write this sum as

$$E_{nx,ny} = \left(\frac{h^2}{8mL_x^2}\right)n_x^2 + \left(\frac{h^2}{8mL_y^2}\right)n_y^2 = \frac{h^2}{8m}\left(\frac{n_x^2}{L_x^2} + \frac{n_y^2}{L_y^2}\right). \tag{40-20}$$

Excitation of the electron by photon absorption and de-excitation of the electron by photon emission have the same requirements as for one-dimensional traps. The only major difference for the two-dimensional corral is that the energy of any given state depends on two quantum numbers (n_x and n_y) instead of just one (n). In general, different states (with different pairs of values for n_x and n_y) have different energies. However, in some situations, different states can have the same energy. Such states (and their energy levels) are said to be *degenerate*. Degenerate states cannot occur in a one-dimensional well.

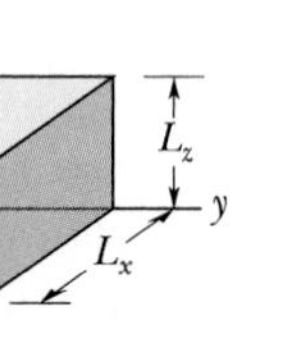

Fig. 40-13 A rectangular box—a three-dimensional version of the infinite potential well of Fig. 40-2—with widths L_x, L_y, and L_z.

Rectangular Box

An electron can also be trapped in a three-dimensional infinite potential well—a *box*. If the box is rectangular as in Fig. 40-13, then Schrödinger's equation shows us that we can write the energy of the electron as

$$E_{nx,ny,nz} = \frac{h^2}{8m}\left(\frac{n_x^2}{L_x^2} + \frac{n_y^2}{L_y^2} + \frac{n_z^2}{L_z^2}\right). \tag{40-21}$$

Here n_z is a third quantum number, for fitting the matter wave into width L_z.

CHECKPOINT 4: In the notation of Eq. 40-20, is $E_{0,0}$, $E_{1,0}$, $E_{0,1}$, or $E_{1,1}$ the ground-state energy of an electron in a rectangular corral?

Sample Problem 40-5

An electron is trapped in a square corral that is a two-dimensional infinite potential well (Fig. 40-12) with widths $L_x = L_y$.

(a) Find the energies of the lowest five energy levels for the electron, and construct an energy-level diagram.

SOLUTION: The **Key Idea** here is that because the electron is trapped in a two-dimensional well that is rectangular, its energy depends on two quantum numbers, n_x and n_y, according to Eq. 40-20. Because the well is square, we can let the widths be $L_x = L_y = L$.

TABLE 40-1 **Energy Levels**

n_x	n_y	Energy*	n_x	n_y	Energy*
1	3	10	2	4	20
3	1	10	4	2	20
2	2	8	3	3	18
1	2	5	1	4	17
2	1	5	4	1	17
1	1	2	2	3	13
			3	2	13

*In multiples of $h^2/8mL^2$

Then Eq. 40-20 simplifies to

$$E_{nx,ny} = \frac{h^2}{8mL^2}(n_x^2 + n_y^2). \tag{40-22}$$

The lowest energy states correspond to low values of the quantum numbers n_x and n_y, which are the positive integers 1, 2, . . . , ∞. Substituting those integers for n_x and n_y in Eq. 40-22, starting with the lowest value 1, we can obtain the energy values as listed in Table 40-1. There we can see that several of the pairs of quantum numbers (n_x, n_y) give the same energy. For example, the (1, 2) and (2, 1) states both have an energy of $5(h^2/8mL^2)$. Each such pair is associated with degenerate energy levels. Note also that, perhaps surprisingly, the (4, 1) and (1, 4) states have less energy than the (3, 3) state.

From Table 40-1 (carefully keeping track of degenerate levels), we can construct the energy-level diagram of Fig. 40-14.

(b) As a multiple of $h^2/8mL^2$, what is the energy difference between the ground state and the third excited state of the electron?

SOLUTION: From Fig. 40-14, we see that the ground state is the (1, 1) state, with an energy of $2(h^2/8mL^2)$. We also see that the third excited state (the third state up from the ground state in the energy-level diagram) is the degenerate (1, 3) and (3, 1) states, with an energy of $10(h^2/8mL^2)$. Thus, the difference ΔE between these two states is

$$\Delta E = 10\left(\frac{h^2}{8mL^2}\right) - 2\left(\frac{h^2}{8mL^2}\right) = 8\left(\frac{h^2}{8mL^2}\right). \quad \text{(Answer)}$$

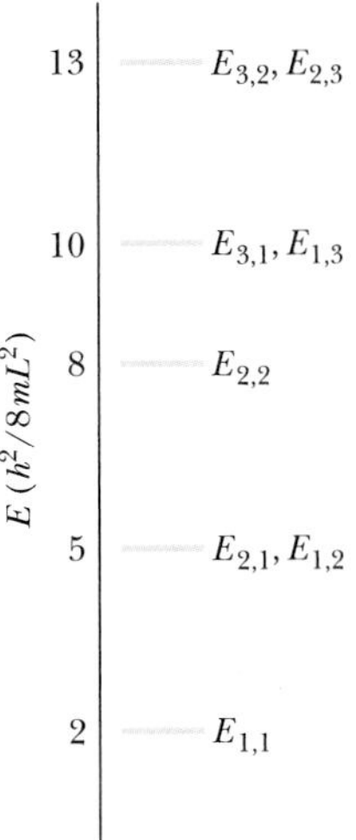

Fig. 40-14 Sample Problem 40-5. Energy-level diagram.

40-8 The Hydrogen Atom

We now move from artificial and fictitious electron traps to natural ones, using the simplest atom—hydrogen—as our example. This atom consists of a single electron (charge $-e$) bound to its central nucleus, a single proton (charge $+e$), by the attractive Coulomb force that acts between them. The hydrogen atom, like all atoms, is an electron trap; it confines its single electron to a region of space. From the confinement principle, we then expect that the electron can exist only in a discrete set of quantum states, each with a certain energy. We wish to identify the energies and the wave functions of these states.

The Energies of the Hydrogen Atom States

In Chapter 25 we wrote Eq. 25-43 for the (electric) potential energy of a two-particle system with charges q_1 and q_2:

$$U = \frac{1}{4\pi\varepsilon_0}\frac{q_1 q_2}{r},$$

where r is the distance between the particles. For the two-particle system of a hydrogen atom, we can write the potential energy as

$$U = \frac{1}{4\pi\varepsilon_0}\frac{(e)(-e)}{r} = -\frac{1}{4\pi\varepsilon_0}\frac{e^2}{r}. \tag{40-23}$$

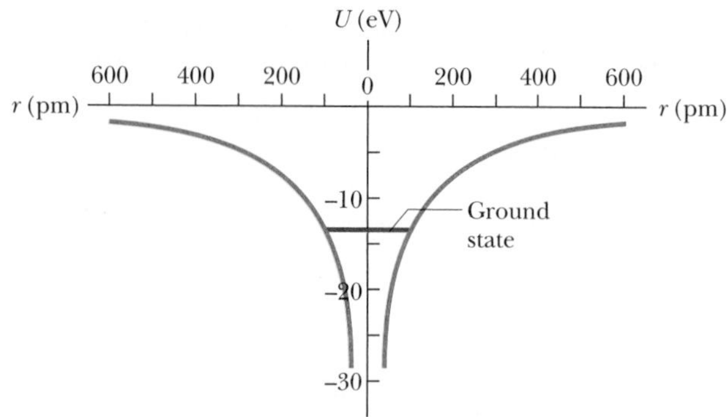

Fig. 40-15 The potential energy U of a hydrogen atom as a function of the separation r between the electron and the central proton. The plot is shown twice (on the left and on the right) to suggest the three-dimensional spherically symmetric trap to which the electron is confined.

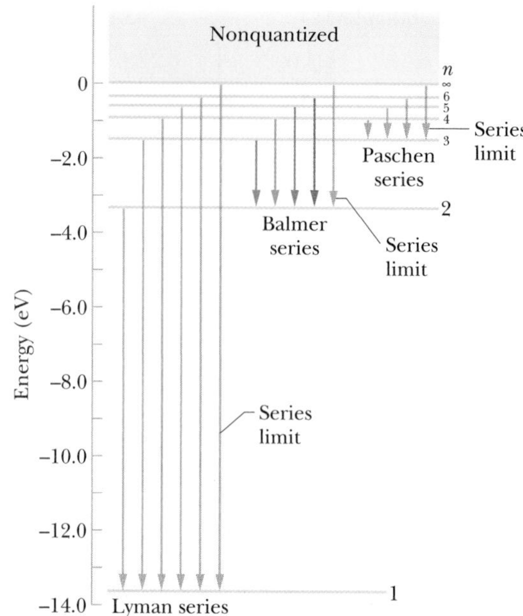

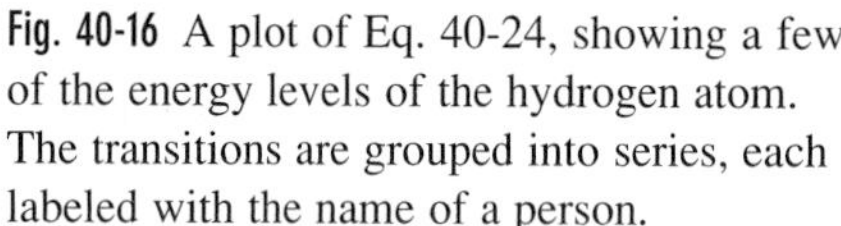
Fig. 40-16 A plot of Eq. 40-24, showing a few of the energy levels of the hydrogen atom. The transitions are grouped into series, each labeled with the name of a person.

The plot of Fig. 40-15 suggests the three-dimensional potential well in which the hydrogen atom's electron is trapped. This well differs from the finite potential well of Fig. 40-7 in that, for the hydrogen atom, U is negative for all values of r because we have (arbitrarily) chosen our zero of potential energy to correspond to $r = \infty$. For the finite well of Fig. 40-7, however, we (equally arbitrarily) chose to assign the zero of potential energy to the region inside the well.

To find the energies of the quantum states of the hydrogen atom, we must solve Schrödinger's equation, with Eq. 40-23 substituted for U in that equation. However, because the electron in the hydrogen atom is trapped in a three-dimensional well, we must use a three-dimensional form of Schrödinger's equation.

Solving that equation reveals that the energies of the electron's quantum states are given by

$$E_n = -\frac{me^4}{8\varepsilon_0^2 h^2}\frac{1}{n^2} = -\frac{13.6 \text{ eV}}{n^2}, \qquad \text{for } n = 1, 2, 3, \ldots, \tag{40-24}$$

where n is a quantum number and m is the mass of an electron. The lowest energy, which is for the ground state with $n = 1$, is indicated in Fig. 40-15. Figure 40-16 shows the energy levels of the ground state and five excited states, each labeled with its quantum number n. It also shows the energy level for the greatest value of n—namely, $n = \infty$—for which $E_n = 0$. For any greater energy, the electron and proton are not bound together (there is no hydrogen atom), and the corresponding region in Fig. 40-16 is like the nonquantized region for the finite well of Fig. 40-9.

The quantized potential energy values given by Eq. 40-24 are actually those of the hydrogen atom—that is, of the *electron* + *proton* system. However, we can usually attribute the energy to the electron alone because its mass is much less than that of the proton. (Similarly, we can attribute the energy of a *ball* + *Earth* system to the ball alone.) Thus, we can say that when an electron is trapped in a hydrogen atom, the *electron* can have only energy values given by Eq. 40-24.

As we have seen with electrons in other potential wells, the electron in a hydrogen atom tends to be in the lowest energy level—that is, in its ground state. It can make a quantum jump up to a higher level, at greater energy, only if it is given the required energy to reach the higher level. One way it can receive that energy is by photon absorption. As we have discussed, this absorption can occur only if the photon's energy hf is equal to the energy difference ΔE between the electron's initial energy level and another level, as given by Eq. 40-6. To decrease its energy, the electron can make a quantum jump down to a lower energy level. If it does so by emitting a photon, the photon's energy hf must again equal the difference ΔE.

Because hf must equal a difference ΔE between the energies of two quantum levels, which can have only certain energy values, a hydrogen atom can emit and absorb light at only certain frequencies f—and thus also at only certain wavelengths λ. Any such wavelength is often called a *line* because of the way it is detected with a spectroscope; thus, a hydrogen atom has *absorption lines* and *emission lines.* A collection of such lines, such as in those in the visible range, is called a **spectrum** of the hydrogen atom.

The lines for hydrogen are said to be grouped into *series,* according to the level that upward jumps start on and downward jumps end on. For example, the emission and absorption lines for all possible jumps up from the $n = 1$ level and down to the $n = 1$ level are said to be in the *Lyman series,* named after the person who first studied those lines. Further, we can say that the Lyman series in the hydrogen spectrum has a *home-base level* of $n = 1$. Similarly, the *Balmer series* has a home-base level of $n = 2$, and the *Paschen series* has a home-base level of $n = 3$.

Some of the downward quantum jumps for these three series are shown in Fig. 40-16. Four lines in the Balmer series are in the visible range, and they are represented in Fig. 40-16 with arrows corresponding to their colors. The shortest of those arrows represents the shortest jump in the series, from the $n = 3$ level to the $n = 2$ level. Thus, that jump involves the least change in the electron's energy and the least emitted photon energy for the series. The emitted light is red. The next jump in the series, from $n = 4$ to $n = 2$, is longer, the photon energy is greater, the wavelength of the emitted light is shorter, and the light is green. The third, fourth, and fifth arrows represent longer jumps and shorter wavelengths. For the fifth jump, the emitted light is in the ultraviolet range and thus is not visible.

The *series limit* of a series is the line produced by the jump between the home-base level and the highest energy level, which is the level with quantum number $n = \infty$. Thus, the series limit is the shortest wavelength in the series. Figure 40-17 is a photograph of the Balmer emission lines taken with a spectroscope (as in Figs. 37-22 and 37-23). The series limit for the series is marked with a small triangle.

Bohr's Theory of the Hydrogen Atom

In 1913, some 13 years before the formulation of Schrödinger's equation, Bohr proposed a model of the hydrogen atom based on a clever combination of classical

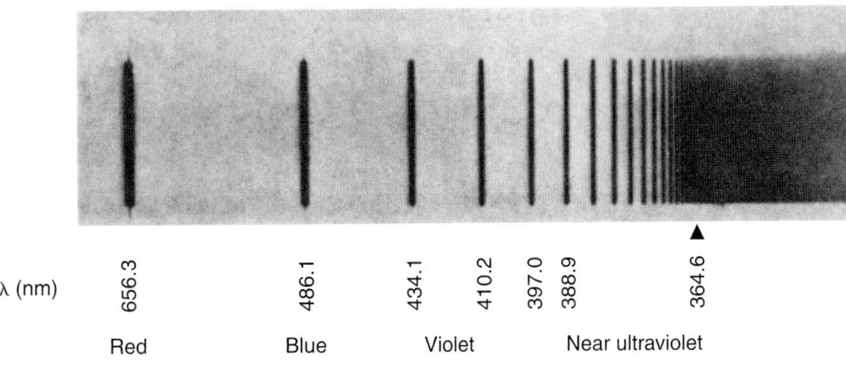

Fig. 40-17 The spectrum of emission lines of the Balmer series of the hydrogen atom. Whereas Fig. 40-16 shows four transitions of this series, along with the series limit, this figure shows about a dozen lines of this series; note how they are progressively closer toward the series limit, which is marked with a triangle.

TABLE 40-2 **Quantum Numbers for the Hydrogen Atom**

Symbol	Name	Allowed Values
n	Principal quantum number	$1, 2, 3, \ldots$
l	Orbital quantum number	$0, 1, 2, \ldots, n-1$
m_l	Orbital magnetic quantum number	$-l, -(l-1), \ldots, +(l-1), +l$

and early quantum concepts. His basic assumption—that atoms exist in discrete quantum states of well-defined energy—was a bold break with classical ideas; it carries over today as an indispensable concept in modern quantum physics. With this assumption, Bohr made skillful use of the correspondence principle (see Section 40-4), not only to derive Eq. 40-24 for the energies of the quantum states of the hydrogen atom but also to derive a numerical value (the *Bohr radius*) for the effective radius of that atom. In spite of its successes, Bohr's specific model of the hydrogen atom, based on the assumption that the electron is a particle that moves in planet-like orbits around the nucleus, was inconsistent with the uncertainty principle and was replaced by the probability density model derived from Schrödinger's work. For Bohr's brilliant achievements, which greatly stimulated progress toward the modern quantum theory, he was awarded the Nobel prize in physics in 1922.

Quantum Numbers for the Hydrogen Atom

Although the energies of the hydrogen atom states can be described by the single quantum number n, the wave functions describing these states require three quantum numbers, corresponding to the three dimensions in which the electron can move. The three quantum numbers, along with their names and the values that they may have, are shown in Table 40-2.

Each set of quantum numbers (n, l, m_l) identifies the wave function of a particular quantum state. The quantum number n, called the **principal quantum number,** appears in Eq. 40-24 for the energy of the state. The **orbital quantum number** l is a measure of the magnitude of the angular momentum associated with the quantum state. The **orbital magnetic quantum number** m_l is related to the orientation in space of this angular momentum vector. The restrictions on the values of the quantum numbers for the hydrogen atom, as listed in Table 40-2, are not arbitrary but come out of the solution to Schrödinger's equation. Note that for the ground state $(n = 1)$, the restrictions require that $l = 0$ and $m_l = 0$. That is, the hydrogen atom in its ground state has zero angular momentum.

CHECKPOINT 5: (a) A group of quantum states of the hydrogen atom has $n = 5$. How many values of l are possible for states within this group? (b) A subgroup of hydrogen atom states within the $n = 5$ group has $l = 3$. How many values of m_l are possible for states within this subgroup?

The Wave Function of the Hydrogen Atom's Ground State

The wave function for the ground state of the hydrogen atom, as obtained by solving the three-dimensional Schrödinger equation and normalizing the result, is

$$\psi(r) = \frac{1}{\sqrt{\pi} a^{3/2}} e^{-r/a} \quad \text{(ground state).} \tag{40-25}$$

Here a is the **Bohr radius,** a constant with the dimension *length.* This radius is loosely taken to be the effective radius of a hydrogen atom and turns out to be a convenient unit of length for other situations involving atomic dimensions. Its value is

$$a = \frac{h^2 \varepsilon_0}{\pi m e^2} = 5.29 \times 10^{-11}\ \text{m} = 52.9\ \text{pm}. \tag{40-26}$$

As with other wave functions, $\psi(r)$ in Eq. 40-25 does not have physical meaning but $\psi^2(r)$ does. It is the probability density—the probability per unit volume—that the electron can be detected. Specifically, $\psi^2(r)\,dV$ is the probability that the electron can be detected in any given (infinitesimal) volume element dV located at radius r from the center of the atom:

$$\begin{pmatrix}\text{probability of detection}\\ \text{in volume } dV\\ \text{at radius } r\end{pmatrix} = \begin{pmatrix}\text{volume probability}\\ \text{density } \psi^2(r)\\ \text{at radius } r\end{pmatrix}(\text{volume } dV). \tag{40-27}$$

Because $\psi^2(r)$ here depends only on r, it makes sense to choose, as a volume element dV, the volume between two concentric spherical shells whose radii are r and $r + dr$. That is, we take the volume element dV to be

$$dV = (4\pi r^2)\,dr, \tag{40-28}$$

in which $4\pi r^2$ is the area of the inner shell and dr is the radial distance between the two shells. Then, combining Eqs. 40-25, 40-27, and 40-28 gives us

$$\begin{pmatrix}\text{probability of detection}\\ \text{in volume } dV\\ \text{at radius } r\end{pmatrix} = \psi^2(r)\,dV = \frac{4}{a^3}\,e^{-2r/a} r^2\,dr. \tag{40-29}$$

Describing the probability of detection is easier if we work with a **radial probability density** $P(r)$ instead of a volume probability density $\psi^2(r)$. This $P(r)$ is a linear probability density such that

$$\begin{pmatrix}\text{radial probability}\\ \text{density } P(r)\\ \text{at radius } r\end{pmatrix}\begin{pmatrix}\text{radial}\\ \text{width } dr\end{pmatrix} = \begin{pmatrix}\text{volume probability}\\ \text{density } \psi^2(r)\\ \text{at radius } r\end{pmatrix}(\text{volume } dV)$$

or

$$P(r)\,dr = \psi^2(r)\,dV. \tag{40-30}$$

Substituting for $\psi^2(r)\,dV$ from Eq. 40-29, we obtain

$$P(r) = \frac{4}{a^3}\,r^2 e^{-2r/a} \quad \text{(radial probability density, hydrogen atom ground state).} \tag{40-31}$$

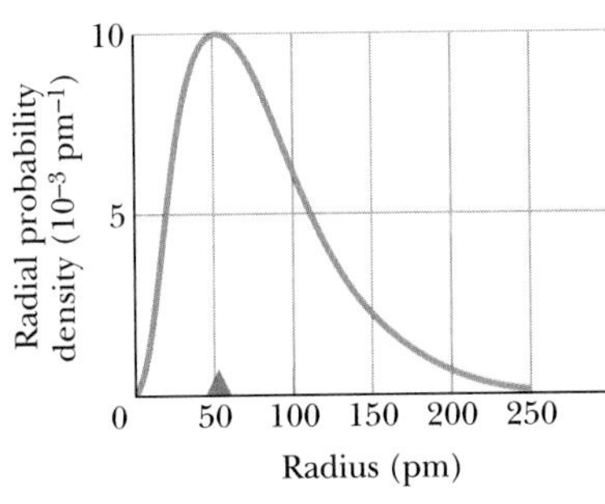

Fig. 40-18 A plot of the radial probability density $P(r)$ for the ground state of the hydrogen atom. The triangular marker is located at one Bohr radius from the origin, and the origin represents the center of the atom.

Figure 40-18 is a plot of Eq. 40-31. The area under the plot is unity; that is,

$$\int_0^\infty P(r)\,dr = 1. \tag{40-32}$$

This equation simply states that in a normal hydrogen atom, the electron must be *somewhere* in the space surrounding the nucleus.

The triangular marker on the horizontal axis of Fig. 40-18 is located one Bohr radius from the origin. The graph tells us that in the ground state of the hydrogen atom, the electron is most likely to be found at about this distance from the center of the atom.

Figure 40-18 conflicts sharply with the popular view that electrons in atoms follow well-defined orbits like planets moving around the Sun. *This popular view, however familiar, is incorrect.* Figure 40-18 shows us all that we can ever know

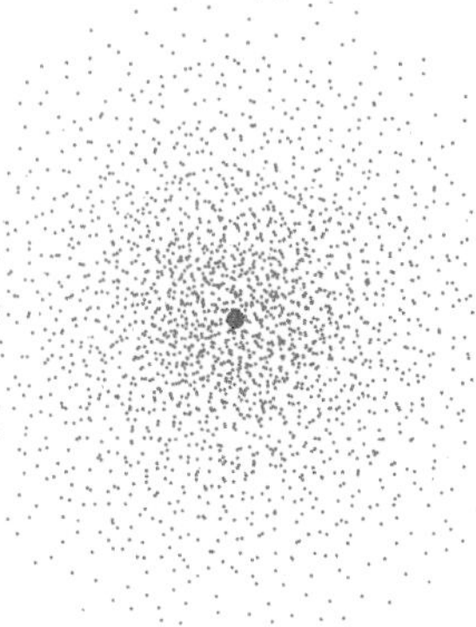

Fig. 40-19 A "dot plot" showing the probability density $\psi^2(r)$—not the *radial* probability density $P(r)$—for the ground state of the hydrogen atom. The density of dots drops exponentially with increasing distance from the nucleus, which is represented here by a red spot. Such dot plots provide a mental image of the "electron cloud" of an atom.

about the location of the electron in the ground state of the hydrogen atom. The appropriate question is not "When will the electron arrive at such-and-such a point?" but "What are the odds that the electron will be detected in a small volume centered on such-and-such a point?" Figure 40-19, which we call a dot plot, suggests the probabilistic nature of the wave function and provides a useful mental model of the hydrogen atom in its ground state. Think of the atom in this state as a fuzzy ball with no sharply defined boundary and no hint of orbits.

It is not easy for a beginner to envision subatomic particles in this probabilistic way. The difficulty is our natural impulse to regard an electron as something like a tiny jelly bean, located at certain places at certain times and following a well-defined path. Electrons and other subatomic particles simply do not behave in this way.

The energy of the ground state, found by putting $n = 1$ in Eq. 40-24, is $E_1 = -13.6$ eV. The wave function of Eq. 40-25 results if you solve Schrödinger's equation with this value of the energy. Actually, you can find a solution of Schrödinger's equation for *any* value of the energy, say $E = -11.6$ eV or -14.3 eV. This may suggest that the energies of the hydrogen atom states are not quantized—but we know that they are.

The puzzle was solved when physicists realized that such solutions of Schrödinger's equation are not physically acceptable because they yield increasingly large values as $r \to \infty$. These "wave functions" tell us that the electron is more likely to be found very far from the nucleus than closer to it, which makes no sense. We get rid of these unwanted solutions by imposing a so-called **boundary condition,** in which we agree to accept only solutions of Schrödinger's equation for which $\psi(r) \to 0$ as $r \to \infty$; that is, we agree to deal only with *confined* electrons. With this restriction, the solutions of Schrödinger's equation form a discrete set, with quantized energies given by Eq. 40-24.

Sample Problem 40-6

(a) What is the wavelength of light for the least energetic photon emitted in the Lyman series of the hydrogen atom spectrum lines?

SOLUTION: One Key Idea here is that for any series, the transition that produces the least energetic photon is the transition between the home-base level that defines the series and the level immediately above it. A second Key Idea is that for the Lyman series, the home-base level is at $n = 1$ (Fig. 40-16). Thus, the transition that produces the least energetic photon is the transition from the $n = 2$ level to the $n = 1$ level. From Eq. 40-24 the energy difference is

$$\Delta E = E_2 - E_1 = -(13.6 \text{ eV})\left(\frac{1}{2^2} - \frac{1}{1^2}\right) = 10.2 \text{ eV}.$$

Then from Eq. 40-6 ($\Delta E = hf$), with c/λ replacing f, we have

$$\lambda = \frac{hc}{\Delta E} = \frac{(6.63 \times 10^{-34} \text{ J}\cdot\text{s})(3.00 \times 10^8 \text{ m/s})}{(10.2 \text{ eV})(1.60 \times 10^{-19} \text{ J/eV})}$$

$$= 1.22 \times 10^{-7} \text{ m} = 122 \text{ nm}. \quad \text{(Answer)}$$

Light with this wavelength is in the ultraviolet range.

(b) What is the wavelength of the series limit for the Lyman series?

SOLUTION: The Key Idea here is that the series limit corresponds to a jump between the home-base level ($n = 1$ for the Lyman series) and the level at the limit $n = \infty$. From Eq. 40-24, the energy difference for this transition is

$$\Delta E = E_\infty - E_1 = -(13.6 \text{ eV})\left(\frac{1}{\infty^2} - \frac{1}{1^2}\right)$$

$$= -(13.6 \text{ eV})(0 - 1) = 13.6 \text{ eV}.$$

The corresponding wavelength is found as in (a) and is

$$\lambda = \frac{hc}{\Delta E}$$

$$= \frac{(6.63 \times 10^{-34} \text{ J}\cdot\text{s})(3.00 \times 10^8 \text{ m/s})}{(13.6 \text{ eV})(1.60 \times 10^{-19} \text{ J/eV})}$$

$$= 9.14 \times 10^{-8} \text{ m} = 91.4 \text{ nm}. \quad \text{(Answer)}$$

Light with this wavelength is also in the ultraviolet range.

Sample Problem 40-7

Show that the radial probability density for the ground state of the hydrogen atom has a maximum at $r = a$.

SOLUTION: One Key Idea here is that the radial probability density for a ground-state hydrogen atom is given by Eq. 40-31,

$$P(r) = \frac{4}{a^3} r^2 e^{-2r/a}.$$

A second Key Idea is that to find the maximum (or minimum) of any function, we must differentiate it and set the result equal to zero. If we differentiate $P(r)$ with respect to r, using derivative 7 of Appendix E and the chain rule for differentiating products, we get

$$\begin{aligned}\frac{dP}{dr} &= \frac{4}{a^3} r^2 \left(\frac{-2}{a}\right) e^{-2r/a} + \frac{4}{a^3} 2r\, e^{-2r/a} \\ &= \frac{8r}{a^3} e^{-2r/a} - \frac{8r^2}{a^4} e^{-2r/a} \\ &= \frac{8}{a^4} r(a - r)e^{-2r/a}.\end{aligned}$$

If we set the right side equal to zero, we obtain an equation that is true if $r = a$. In other words, dP/dr is equal to zero when $r = a$. (Note that we also have $dP/dr = 0$ at $r = 0$ and at $r = \infty$. However, these conditions correspond to a *minimum* in $P(r)$, as you can see in Fig. 40-18.)

Sample Problem 40-8

It can be shown that the probability $p(r)$ that the electron in the ground state of the hydrogen atom will be detected inside a sphere of radius r is given by

$$p(r) = 1 - e^{-2x}(1 + 2x + 2x^2),$$

in which x, a dimensionless quantity, is equal to r/a. Find r for $p(r) = 0.90$.

SOLUTION: The Key Idea here is that there is no guarantee of detecting the electron at any particular radial distance r from the center of the hydrogen atom. However, with the given function, we can calculate the probability the electron will be detected *somewhere* within a sphere of radius r. We seek the radius of a sphere for which $p(r) = 0.90$. Substituting that value in the expression for $p(r)$, we have

$$0.90 = 1 - e^{-2x}(1 + 2x + 2x^2)$$

or

$$10e^{-2x}(1 + 2x + 2x^2) = 1.$$

We must find the value of x that satisfies this equality. It is not possible to solve explicitly for x, but an equation solver on a calculator yields $x = 2.66$. This means that the radius of a sphere such that the electron will be detected inside it 90% of the time is $2.66a$. Mark this position on the horizontal axis of Fig. 40-18—is it a reasonable answer?

Hydrogen Atom States with $n = 2$

According to the requirements of Table 40-2, there are four states of the hydrogen atom with $n = 2$; their quantum numbers are listed in Table 40-3. Consider first the state with $n = 2$ and $l = m_l = 0$; its probability density is represented by the dot plot of Fig. 40-20. Note that this plot, like the plot for the ground state shown in Fig. 40-19, is spherically symmetric. That is, in a spherical coordinate system like that defined in Fig. 40-21, the probability density is a function of the radial coordinate r only and is independent of the angular coordinates θ and ϕ.

TABLE 40-3 **Quantum Numbers for Hydrogen Atom States with $n = 2$**

n	l	m_l
2	0	0
2	1	+1
2	1	0
2	1	−1

It turns out that all quantum states with $l = 0$ have spherically symmetric wave functions. This is reasonable because the quantum number l is a measure of the angular momentum associated with a given state. If $l = 0$, the angular momentum is also zero, which requires that the probability density representing the state have no preferred axis of symmetry.

Dot plots of ψ^2 for the three states with $n = 2$ and $l = 1$ are shown in Fig. 40-22. The probability densities for the states with $m_l = +1$ and $m_l = -1$ are identical. Although these plots are symmetric about the z axis, they are *not* spherically symmetric. That is, the probability densities for these three states are functions of both r and the angular coordinate θ.

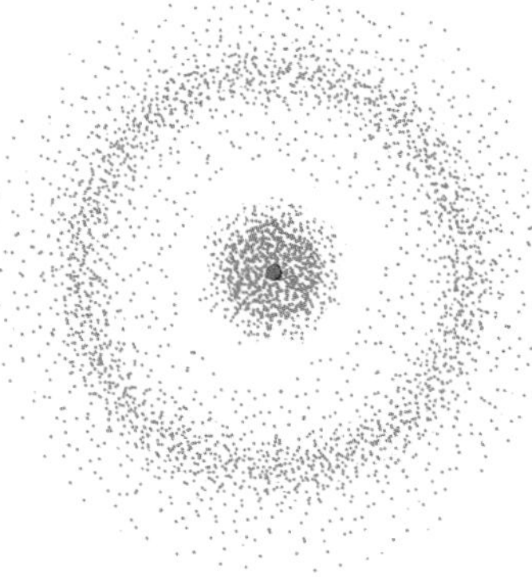

Fig. 40-20 A dot plot showing the probability density $\psi^2(r)$ for the hydrogen atom in the quantum state with $n = 2$, $l = 0$, and $m_l = 0$. The plot has spherical symmetry about the central nucleus. The gap in the dot density pattern marks a spherical surface over which $\psi^2(r) = 0$.

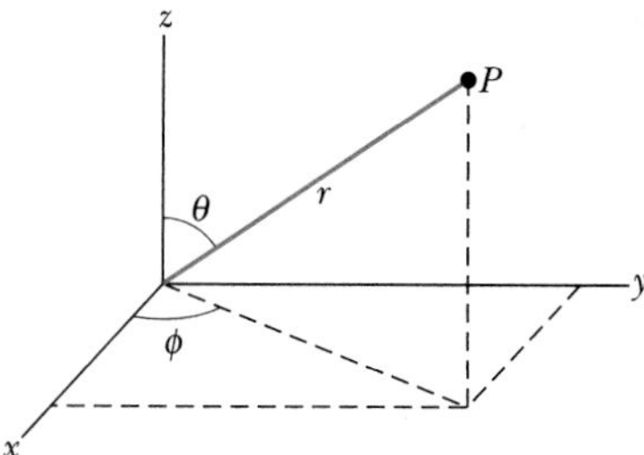

Fig. 40-21 The relationship between the coordinates x, y, and z of the rectangular coordinate system and the coordinates r, θ, and ϕ of the spherical coordinate system. The latter are more appropriate for analyzing situations involving spherical symmetry, such as the hydrogen atom.

Here is a puzzle: What is there about the hydrogen atom that establishes the axis of symmetry that is so obvious in Fig. 40-22? The answer: *absolutely nothing*.

The solution to this puzzle comes about when we realize that all three states shown in Fig. 40-22 have the same energy. Recall that the energy of a state, given by Eq. 40-24, depends only on the principal quantum number n and is independent of l and m_l. In fact, for an *isolated* hydrogen atom there is no way to differentiate experimentally among the three states of Fig. 40-22.

If we add the probability densities for the three states, $n = 2$ and $l = 1$, the combined probability density turns out to be spherically symmetrical, with no unique axis. One can, then, think of the electron as spending one-third of its time in each of the three states of Fig. 40-22, and one can think of the weighted sum of the three independent wave functions as defining a spherically symmetric **subshell,** specified by the quantum numbers $n = 2$, $l = 1$. The individual states will display their separate existence only if we place the hydrogen atom in an external electric or magnetic field. The three states of the $n = 2$, $l = 1$ subshell will then have different energies, and the field direction will establish the necessary symmetry axis.

The $n = 2$, $l = 0$ state, whose probability density is shown in Fig. 40-20, *also* has the same energy as each of the three states of Fig. 40-22. We can view all four states whose quantum numbers are listed in Table 40-3 as forming a spherically symmetric **shell,** specified by the single quantum number n. The importance of shells and subshells will become evident in Chapter 41, where we discuss atoms having more than one electron.

To round out our picture of the hydrogen atom, we display in Fig. 40-23 a dot plot of the radial probability density for a hydrogen atom state with a relatively high quantum number ($n = 45$) and the highest orbital quantum number that the restrictions of Table 40-2 permit ($l = n - 1 = 44$). The probability density forms a ring that is symmetrical about the z axis and lies very close to the xy plane. The mean radius of the ring is n^2a, where a is the Bohr radius. This mean radius is more than 2000 times the effective radius of the hydrogen atom in its ground state.

Figure 40-23 suggests the electron orbit of classical physics. Thus, we have another illustration of Bohr's correspondence principle—namely, that at large quan-

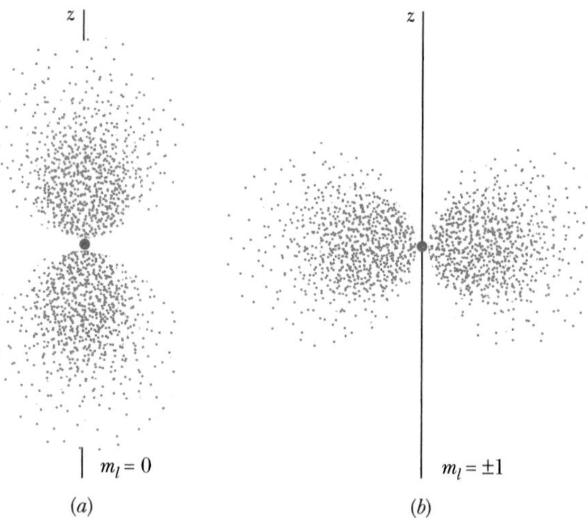

Fig. 40-22 Dot plots of the probability density $\psi^2(r, \theta)$ for the hydrogen atom in states with $n = 2$ and $l = 1$. (*a*) Plot for $m_l = 0$. (*b*) Plot for $m_l = +1$ and $m_l = -1$. Both plots show that the probability density is symmetric about the z axis.

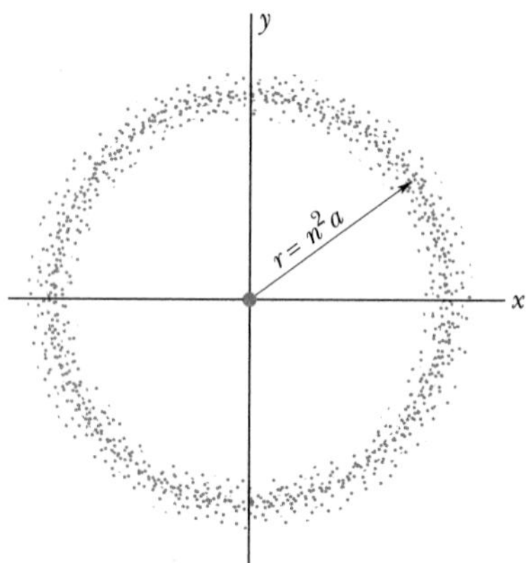

Fig. 40-23 A dot plot of the radial probability density $P(r)$ for the hydrogen atom in a quantum state with a relatively large principal quantum number—namely, $n = 45$—and angular momentum quantum number $l = n - 1 = 44$. The dots lie close to the xy plane, the ring of dots suggesting a classical electron orbit.

tum numbers the predictions of quantum mechanics merge smoothly with those of classical physics. Imagine what a dot plot like that of Figure 40-23 would look like for *really* large values of n and l, say, $n = 1000$ and $l = 999$.

REVIEW & SUMMARY

The Confinement Principle The **confinement principle** applies to waves of all kinds, including waves on a string and the matter waves of quantum physics. It states that confinement leads to quantization—that is, to the existence of discrete states with discrete energies.

An Electron in an Infinite Potential Well An infinite potential well is a device for confining an electron. From the confinement principle we expect that the matter wave representing a trapped electron can exist only in a set of discrete states. For a one-dimensional infinite potential well, the energies associated with these *quantum states* are

$$E_n = \left(\frac{h^2}{8mL^2}\right)n^2, \quad \text{for } n = 1, 2, 3, \ldots, \tag{40-4}$$

in which L is the width of the well and n is a **quantum number.** If the electron is to change from one state to another, its energy must change by the amount

$$\Delta E = E_{\text{high}} - E_{\text{low}}, \tag{40-5}$$

where E_{high} is the higher energy and E_{low} is the lower energy. If the change is done by photon absorption or emission, the energy of the photon must be

$$hf = \Delta E = E_{\text{high}} - E_{\text{low}}. \tag{40-6}$$

The **wave functions** associated with the quantum states are

$$\psi_n(x) = A\sin\left(\frac{n\pi}{L}x\right), \quad \text{for } n = 1, 2, 3, \ldots. \tag{40-10}$$

The **probability density** $\psi_n^2(x)$ for an allowed state has the physical meaning that $\psi_n^2(x)\,dx$ is the probability that the electron will be detected in the interval between x and $x + dx$. For an electron in an infinite well, the probability densities are

$$\psi_n^2(x) = A^2\sin^2\left(\frac{n\pi}{L}x\right), \quad \text{for } n = 1, 2, 3, \ldots. \tag{40-12}$$

At high quantum numbers n, the electron tends toward classical behavior in that it tends to occupy all parts of the well with equal probability. This fact leads to the **correspondence principle:** At large enough quantum numbers, the predictions of quantum physics merge smoothly with those of classical physics.

Normalization and Zero-Point Energy The amplitude A^2 in Eq. 40-12 can be found from the **normalizing equation,**

$$\int_{-\infty}^{+\infty} \psi_n^2(x)\,dx = 1, \tag{40-14}$$

which asserts that the electron must be *somewhere* within the well, because the probability 1 implies certainty.

From Eq. 40-4 we see that the lowest permitted energy for the electron is not zero but the energy that corresponds to $n = 1$. This lowest energy is called the **zero-point energy** of the electron–well system.

An Electron in a Finite Potential Well A finite potential well is one for which the potential energy of an electron inside the well is less than that for one outside the well by a finite amount U_0. The wave function for an electron trapped in such a well extends into the walls of the well.

Two- and Three-Dimensional Electron Traps The quantized energies for an electron trapped in a two-dimensional infinite potential well that forms a rectangular corral are

$$E_{nx,ny} = \frac{h^2}{8m}\left(\frac{n_x^2}{L_x^2} + \frac{n_y^2}{L_y^2}\right), \tag{40-20}$$

where n_x is a quantum number for which the electron's matter wave fits in well width L_x and n_y is a quantum number for which the electron's matter wave fits in well width L_y. Similarly, the energies for an electron trapped in a three-dimensional infinite potential well that forms a rectangular box are

$$E_{nx,ny,nz} = \frac{h^2}{8m}\left(\frac{n_x^2}{L_x^2} + \frac{n_y^2}{L_y^2} + \frac{n_z^2}{L_z^2}\right). \tag{40-21}$$

Here n_z is a third quantum number, for which the matter wave fits in well width L_z.

The Hydrogen Atom The potential energy function for the hydrogen atom is

$$U = -\frac{1}{4\pi\varepsilon_0}\frac{e^2}{r}. \tag{40-23}$$

The energies of the quantum states of the hydrogen atom are found from the three-dimensional form of Schrödinger's equation to be

$$E_n = -\frac{me^4}{8\varepsilon_0^2h^2}\frac{1}{n^2} = -\frac{13.6\text{ eV}}{n^2}, \quad n = 1, 2, 3, \ldots, \tag{40-24}$$

in which n is the **principal quantum number.** The hydrogen atom requires three quantum numbers for its complete description; their names and allowed values are shown in Table 40-2.

The **radial probability density** $P(r)$ for a state of the hydrogen atom is defined so that $P(r)\,dr$ is the probability that the electron will be detected between two concentric shells, centered on the atom's nucleus, whose radii are r and $r + dr$. For the hydrogen atom's ground state,

$$P(r) = \frac{4}{a^3}r^2e^{-2r/a}, \tag{40-31}$$

in which a, the **Bohr radius,** is a length unit whose value is 52.9 pm. Figure 40-18 is a plot of $P(r)$ for the ground state.

Figures 40-20 and 40-22 represent the probability densities (not the *radial* probability densities) for the four hydrogen atom states with $n = 2$. The plot of Fig. 40-20 ($n = 2, l = 0, m_l = 0$) is spherically symmetric. The plots of Fig. 40-22 ($n = 2, l = 1, m_l = 0, +1, -1$) are symmetric about the z axis but, when added together, are also spherically symmetric.

All four states with $n = 2$ have the same energy and may be usefully regarded as a **shell,** identified as the $n = 2$ shell. The three states of Fig. 40-22, taken together, may be regarded as the $n = 2$, $l = 1$ **subshell.** It is not possible to separate the four $n = 2$ states experimentally unless the hydrogen atom is placed in an electric or magnetic field, to permit the establishment of a definite symmetry axis.

QUESTIONS

1. If you double the width of a one-dimensional infinite potential well, (a) is the energy of the ground state of the trapped electron multiplied by 4, 2, $\frac{1}{2}$, $\frac{1}{4}$, or some other number? (b) Are the energies of the higher energy states multiplied by this factor or by some other factor, depending on their quantum number?

2. Three electrons are trapped in three different one-dimensional infinite potential wells of widths (a) 50 pm, (b) 200 pm, and (c) 100 pm. Rank the electrons according to their ground-state energies, greatest first.

3. If you wanted to use the idealized trap of Fig. 40-1 to trap a positron, would you need to change (a) the geometry of the trap, (b) the electric potential of the central cylinder, or (c) the electric potentials of the two semi-infinite end cylinders? (A positron has the same mass as an electron but is positively charged.)

4. An electron is trapped in a one-dimensional infinite potential well in a state with $n = 17$. How many points of (a) zero probability and (b) maximum probability does its matter wave have?

5. Figure 40-24 shows three infinite potential wells, each on an x axis. Without written calculation, determine the wave function ψ for a ground-state electron trapped in each well.

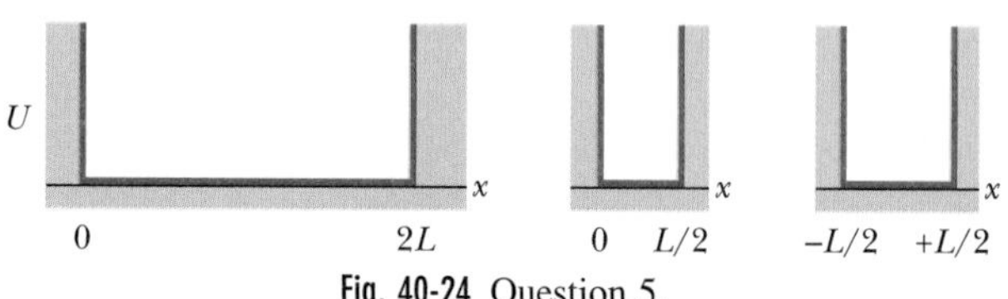

Fig. 40-24 Question 5.

6. Figure 40-25 indicates the lowest energy levels (in electron-volts) for five situations in which an electron is trapped in a one-dimensional infinite potential well. In wells B, C, D, and E, the electron is in the ground state. We shall excite the electron in well A to the fourth excited state (at 25 eV). The electron can then de-excite to the ground state by emitting one or more photons, corresponding to one long jump or several short jumps. What photon *emission* energies of this de-excitation match a photon *absorption* energy (from the ground state) of the other four electrons? Give the corresponding quantum numbers.

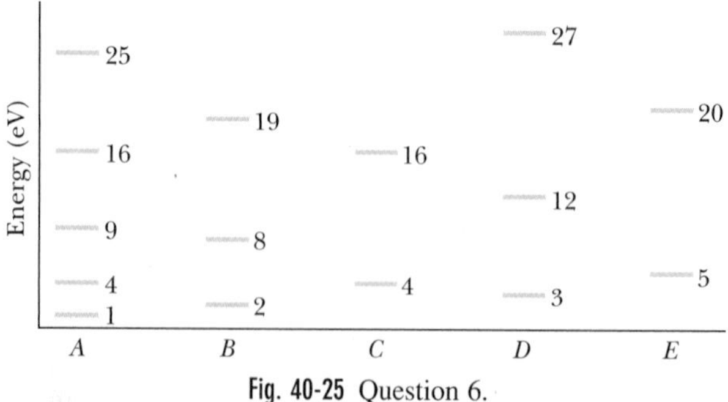

Fig. 40-25 Question 6.

7. Is the ground-state energy of a proton trapped in a one-dimensional infinite potential well greater than, less than, or equal to that of an electron trapped in the same potential well?

8. A proton and an electron are trapped in identical one-dimensional infinite potential wells; each particle is in its ground state. At the center of the wells, is the probability density for the proton greater than, less than, or equal to that of the electron?

9. You want to modify the finite potential well of Fig. 40-7 to allow its trapped electron to exist in more than three quantum states. Could you do so by making the well (a) wider or narrower, (b) deeper or shallower?

10. An electron is trapped in a finite potential well that is deep enough to allow the electron to exist in a state with $n = 4$. How many points of (a) zero probability and (b) probability maximum does its matter wave have within the well?

11. From a visual inspection of Fig. 40-8, rank the quantum numbers of the three quantum states according to the de Broglie wavelength of the electron, greatest first.

12. An electron, trapped in a finite potential energy well such as that of Fig. 40-7, is in its state of lowest energy. Are (a) its de Broglie wavelength, (b) the magnitude of its momentum, and (c) its energy greater than, the same as, or less than they would be if the potential well were infinite, as in Fig. 40-2?

13. The table lists the quantum numbers for five proposed hydrogen atom states. Which of them are not possible?

	n	l	m_l
(a)	3	2	0
(b)	2	3	1
(c)	4	3	−4
(d)	5	5	0
(e)	5	3	−2

14. In 1996 physicists working at an accelerator laboratory suc-

ceeded in producing atoms of antihydrogen. Such atoms consist of a positron moving in the electric field of an antiproton. A positron has the same mass as an electron but the opposite charge. An antiproton has the same mass as a proton but the opposite charge. Would you expect the spectrum of antihydrogen to be the same as that of normal hydrogen or different?

15. (a) From Fig. 40-16, the energy level diagram for the hydrogen atom, you can show that the photon energy of the second spectral line of the Lyman series is equal to the sum of the photon energies of two other lines. What are those lines? (b) The photon energy of the second spectral line of the Lyman series is also equal to the *difference* between the photon energies of two other lines. What are *those* lines?

16. A hydrogen atom is in the third excited state. To what state (give the quantum number n) should it jump to (a) emit light with the longest possible wavelength, (b) emit light with the shortest possible wavelength, and (c) absorb light with the longest possible wavelength?

EXERCISES & PROBLEMS

ssm Solution is in the Student Solutions Manual.
www Solution is available on the World Wide Web at: http://www.wiley.com/college/hrw
ilw Solution is available on the Interactive LearningWare.

SEC. 40-3 Energies of a Trapped Electron

1E. What is the ground-state energy of (a) an electron and (b) a proton if each is trapped in a one-dimensional infinite potential well that is 100 pm wide?

2E. You wish to reduce by one-half the ground-state energy of an electron trapped in a one-dimensional infinite potential well. By what factor must you change the width of the potential well?

3E. Consider an atomic nucleus to be equivalent to a one-dimensional infinite potential well with $L = 1.4 \times 10^{-14}$ m, a typical nuclear diameter. What would be the ground-state energy of an electron if it were trapped in such a potential well? (*Note:* Nuclei do not contain electrons.) ssm

4E. What must be the width of a one-dimensional infinite potential well if an electron trapped in it in the $n = 3$ state is to have an energy of 4.7 eV?

5E. A proton is confined to a one-dimensional infinite potential well 100 pm wide. What is its ground-state energy?

6E. The ground-state energy of an electron trapped in a one-dimensional infinite potential well is 2.6 eV. What will this quantity be if the width of the potential well is doubled?

7E. An electron, trapped in a one-dimensional infinite potential well 250 pm wide, is in its ground state. How much energy must it absorb if it is to jump up to the state with $n = 4$?

8P. An electron is trapped in a one-dimensional infinite potential well. (a) What pair of adjacent energy levels (if any) has an energy difference equal to the energy of the electron in the state with $n = 5$? (b) With $n = 6$?

9P. An electron is trapped in a one-dimensional infinite potential well. Show that the energy difference ΔE between its quantum levels n and $n + 2$ is $(h^2/2mL^2)(n + 1)$.

10P. An electron is trapped in a one-dimensional infinite potential well. (a) What pair of adjacent energy levels (if any) will have three times the energy difference that exists between levels $n = 3$ and $n = 4$? (b) What pair (if any) will have twice that energy difference?

11P. An electron is trapped in a one-dimensional infinite well of width 250 pm and is in its ground state. What are the four longest wavelengths of light that can excite the electron from the ground state via a single photon absorption? ssm www

12P. Suppose that an electron trapped in a one-dimensional infinite well of width 250 pm is excited from its first excited state to its third excited state. (a) In electron-volts, what energy must be transferred to the electron for this quantum jump? If the electron then de-excites by emitting light, (b) what wavelengths can it emit and (c) in which groupings (and orders) can they be emitted? (d) Show the several possible ways the electron can de-excite on an energy-level diagram.

13P. An electron is confined to a narrow evacuated tube of length 3.0 m; the tube functions as a one-dimensional infinite potential well. (a) In electron-volts, what is the energy difference between the electron's ground state and its first excited state? (b) At what quantum number n would the energy difference between adjacent energy levels be 1.0 eV—which is measurable, unlike the result of (a)? At that quantum number, (c) what would be the energy of the electron and (d) would the electron be relativistic?

SEC. 40-4 Wave Functions of a Trapped Electron

14E. An electron that is trapped in a one-dimensional infinite potential well of width L is excited from the ground state to the first excited state. (a) Does that increase, decrease, or have no effect on the probability of detecting the electron in a small length of the x axis (a) at the center of the well and (b) near one of the well walls?

15E. Let ΔE_{adj} be the energy difference between two adjacent energy levels for an electron trapped in a one-dimensional infinite potential well. Let E be the energy of either of the two levels. (a) Show that the ratio $\Delta E_{adj}/E$ approaches the value $2/n$ at large values of the quantum number n. As $n \to \infty$, does (b) ΔE_{adj}, (c) E, or (d) $\Delta E_{adj}/E$ approach zero? (e) What do these results mean in terms of the correspondence principle? ssm

16P. A particle is confined to the one-dimensional infinite potential well of Fig. 40-2. If the particle is in its ground state, what is its probability of detection between (a) $x = 0$ and $x = 0.25L$, (b) $x = 0.75L$ and $x = L$, and (c) $x = 0.25L$ and $x = 0.75L$?

17P. An electron is trapped in a one-dimensional infinite potential well that is 100 pm wide; the electron is in its ground state. What is the probability that you can detect the electron in an interval of width $\Delta x = 5.0$ pm centered at $x =$ (a) 25 pm, (b) 50 pm, and

(c) 90 pm? (*Hint:* The interval Δx is so narrow that you can take the probability density to be constant within it.) ssm

SEC. 40-5 An Electron in a Finite Well

18E. (a) Show that the terms in Schrödinger's equation (Eq. 40-18) have the same dimensions. (b) What is the common SI unit for each of these terms?

19E. An electron in the $n = 2$ state in the finite potential well of Fig. 40-7 absorbs 400 eV of energy from an external source. What is its kinetic energy after this absorption, assuming that the electron moves to a position for which $x > L$? ssm

20E. Figure 40-9 gives the energy levels for an electron trapped in a finite potential energy well 450 eV deep. If the electron is in the $n = 3$ state, what is its kinetic energy?

21P. As Fig. 40-8 suggests, the probability density for the region $x > L$ in the finite potential well of Fig. 40-7 drops off exponentially according to

$$\psi^2(x) = Ce^{-2kx},$$

where C is a constant. (a) Show that the wave function $\psi(x)$ that may be found from this equation is a solution of Schrödinger's equation in its one-dimensional form. (b) What must be the value of k for this to be true?

22P. As Fig. 40-8 suggests, the probability density for an electron in the region $0 < x < L$ for the finite potential well of Fig. 40-7 is sinusoidal, being given by

$$\psi^2(x) = B \sin^2 kx,$$

in which B is a constant. (a) Show that the wave function $\psi(x)$ that may be found from this equation is a solution of Schrödinger's equation in its one-dimensional form. (b) What must be the value of k for this to be true?

23P. Show that for the region $x > L$ in the finite potential well of Fig. 40-7, $\psi(x) = De^{2kx}$ is a solution of Schrödinger's equation in its one-dimensional form, where D is a constant and k is positive. On what basis do we find this mathematically acceptable solution to be physically unacceptable? ssm www

SEC. 40-7 Two- and Three-Dimensional Electron Traps

24E. An electron is contained in the rectangular corral of Fig. 40-12, with widths $L_x = 800$ pm and $L_y = 1600$ pm. What is the electron's ground-state energy in electron-volts?

25E. An electron is contained in the rectangular box of Fig. 40-13, with widths $L_x = 800$ pm, $L_y = 1600$ pm, and $L_z = 400$ pm. What is the electron's ground-state energy in electron-volts?

26P. A rectangular corral of widths $L_x = L$ and $L_y = 2L$ contains an electron. What multiple of $h^2/8mL^2$, where m is the electron's mass, are (a) the energy of the electron's ground state, (b) the energy of its first excited state, (c) the energy of its lowest degenerate states, and (d) the difference between the energies of its second and third excited states?

27P. For Problem 26, at what frequencies can light be absorbed or emitted by the electron for transitions between the lowest five energy levels? Answer in multiples of $h/8mL^2$. ssm www

28P. A cubical box of widths $L_x = L_y = L_z = L$ contains an electron. What multiple of $h^2/8mL^2$, where m is the electron's mass, are (a) the energy of the electron's ground state, (b) the energy of its second excited state, and (c) the difference between the energies of its second and third excited states? How many degenerate states have the energy of (d) the first excited state and (e) the fifth excited state?

29P. For the situation of Problem 28, at what frequencies can light be absorbed or emitted by the electron for transitions between the lowest five energy levels? Answer in multiples of $h/8mL^2$.

SEC. 40-8 The Hydrogen Atom

30E. Verify that the constant appearing in Eq. 40-24 is 13.6 eV.

31E. An atom (not a hydrogen atom) absorbs a photon whose associated frequency is 6.2×10^{14} Hz. By what amount does the energy of the atom increase?

32E. An atom (not a hydrogen atom) absorbs a photon whose associated wavelength is 375 nm and then immediately emits a photon whose associated wavelength is 580 nm. How much net energy is absorbed by the atom in this process?

33E. What is the ratio of the shortest wavelength of the Balmer series to the shortest wavelength of the Lyman series? ssm

34E. (a) What is the energy E of the hydrogen atom electron whose probability density is represented by the dot plot of Fig. 40-20? (b) What minimum energy is needed to remove this electron from the atom?

35E. What are (a) the energy, (b) the magnitude of the momentum, and (c) the wavelength of the photon emitted when a hydrogen atom undergoes a transition from a state with $n = 3$ to a state with $n = 1$? ssm

36E. Repeat Sample Problem 40-6 for the Balmer series of the hydrogen atom.

37E. A neutron, with a kinetic energy of 6.0 eV, collides with a stationary hydrogen atom in its ground state. Explain why the collision must be elastic—that is, why kinetic energy must be conserved. (*Hint:* Show that the hydrogen atom cannot be excited as a result of the collision.) ssm

38E. For the hydrogen atom in its ground state, calculate (a) the probability density $\psi^2(r)$ and (b) the radial probability density $P(r)$ for $r = a$, where a is the Bohr radius.

39E. Calculate the radial probability density $P(r)$ for the hydrogen atom in its ground state at (a) $r = 0$, (b) $r = a$, and (c) $r = 2a$, where a is the Bohr radius.

40E. A hydrogen atom is excited from its ground state to the state with $n = 4$. (a) How much energy must be absorbed by the atom? (b) Calculate and display on an energy-level diagram the different photon energies that may be emitted as the atom returns to its ground state.

41P. How much work must be done to pull apart the electron and the proton that make up the hydrogen atom if the atom is initially in (a) its ground state and (b) the state with $n = 2$? ssm

42P. A hydrogen atom, initially at rest in the $n = 4$ quantum state, undergoes a transition to the ground state, emitting a photon in the process. What is the speed of the recoiling hydrogen atom?

43P. Light of wavelength 486.1 nm is emitted by a hydrogen atom. (a) What transition of the atom is responsible for this radiation? (b) To what series does this transition belong?

44P. What are the widths of the wavelength intervals over which (a) the Lyman series and (b) the Balmer series extend? (Each width begins at the longest wavelength and ends at the series limit.) (c) What are the widths of the corresponding frequency intervals? Express the frequency intervals in terahertz (1 THz $= 10^{12}$ Hz).

45P. In the ground state of the hydrogen atom, the electron has a total energy of -13.6 eV. What are (a) its kinetic energy and (b) its potential energy if the electron is one Bohr radius from the central nucleus? ssm

46P. (a) Find, using the energy-level diagram of Fig. 40-16, the quantum numbers corresponding to a transition in which the wavelength of the emitted radiation is 121.6 nm. (b) To what series does this transmission belong?

47P. A hydrogen atom in a state having a *binding energy* (the energy required to remove an electron) of 0.85 eV makes a transition to a state with an *excitation energy* (the difference between the energy of the state and that of the ground state) of 10.2 eV. (a) What is the energy of the photon emitted as a result of the transition? (b) Identify this transition, using the energy-level diagram of Fig. 40-16.

48P. Verify the wavelengths given in Fig. 40-17 for the visible spectral lines of the Balmer series.

49P. What is the probability that in the ground state of the hydrogen atom, the electron will be found at a radius greater than the Bohr radius? (*Hint:* See Sample Problem 40-8.) ssm www

50P. A hydrogen atom emits light of wavelength 102.6 nm. What are the initial and final quantum numbers for this transition?

51P. Schrödinger's equation for states of the hydrogen atom for which the orbital quantum number l is zero is

$$\frac{1}{r^2}\frac{d}{dr}\left(r^2\frac{d\psi}{dr}\right)+\frac{8\pi^2 m}{h^2}[E-U(r)]\psi=0.$$

Verify that Eq. 40-25, which describes the ground state of the hydrogen atom, is a solution of this equation. ssm

52P. Calculate the probability that the electron in the hydrogen atom, in its ground state, will be found between spherical shells whose radii are a and $2a$, where a is the Bohr radius. (*Hint:* See Sample Problem 40-8.)

53P. Verify that Eq. 40-31, the radial probability density for the ground state of the hydrogen atom, is normalized. That is, verify that

$$\int_0^\infty P(r)\,dr=1$$

is true. ssm

54P. (a) For a given value of the principal quantum number n, how many values of the orbital quantum number l are possible? (b) For a given value of l, how many values of the orbital magnetic quantum number m_l are possible? (c) For a given value of n, how many values of m_l are possible?

55P. What is the probability that an electron in the ground state of the hydrogen atom will be found between two spherical shells whose radii are r and $r+\Delta r$, (a) if $r=0.500a$ and $\Delta r=0.010a$ and (b) if $r=1.00a$ and $\Delta r=0.01a$, where a is the Bohr radius? (*Hint:* Δr is small enough to permit the radial probability density to be taken to be constant between r and $r+\Delta r$.) ssm

56P. For what value of the principal quantum number n would the effective radius, as shown in a probability density dot plot for the hydrogen atom, be 1.0 mm? Assume that l has its maximum value of $n-1$. (*Hint:* Be guided by Fig. 40-23.)

57P*. In Sample Problem 40-7 we showed that the radial probability density for the ground state of the hydrogen atom is a maximum when $r=a$, where a is the Bohr radius. Show that the *average* value of r, defined as

$$r_{avg}=\int P(r)\,r\,dr,$$

has the value $1.5a$. In this expression for r_{avg}, each value of $P(r)$ is weighted with the value of r at which it occurs. Note that the average value of r is greater than the value of r for which $P(r)$ is a maximum. ssm

58P*. The wave function for the hydrogen-atom quantum state with the dot plot shown in Fig. 40-20, which has $n=2$ and $l=m_l=0$, is

$$\psi_{200}(r)=\frac{1}{4\sqrt{2\pi}}a^{-3/2}\left(2-\frac{r}{a}\right)e^{-r/2a},$$

in which a is the Bohr radius and the subscript on $\psi(r)$ gives the values of the quantum numbers n, l, m_l. (a) Plot $\psi^2_{200}(r)$ and show that your plot is consistent with the dot plot of Fig. 40-20. (b) Show analytically that $\psi^2_{200}(r)$ has a maximum at $r=4a$. (c) Find the radial probability density $P_{200}(r)$ for this state. (d) Show that

$$\int_0^\infty P_{200}(r)\,dr=1,$$

and thus that the expression above for the wave function $\psi_{200}(r)$ has been properly normalized.

59P. The wave functions for the three states with the dot plots shown in Fig. 40-22, which have $n=2$, $l=1$, and $m_l=0$, $+1$, and -1, are

$$\psi_{210}(r,\theta)=(1/4\sqrt{2\pi})(a^{-3/2})(r/a)e^{-r/2a}\cos\theta,$$
$$\psi_{21+1}(r,\theta)=(1/8\sqrt{\pi})(a^{-3/2})(r/a)e^{-r/2a}(\sin\theta)e^{+i\phi},$$
$$\psi_{21-1}(r,\theta)=(1/8\sqrt{\pi})(a^{-3/2})(r/a)e^{-r/2a}(\sin\theta)e^{-i\phi},$$

in which the subscripts on $\psi(r,\theta)$ give the values of the quantum numbers n, l, m_l and the angles θ and ϕ are defined in Fig. 40-21. Note that the first wave function is real but the others, which involve the imaginary number i, are complex. (a) Find the probability density for each wave function and show that each is consistent with its dot plot in Fig. 40-22. (b) Add the three probability densities derived in (a) and show that their sum is spherically symmetric, depending only on the radial coordinate r. ssm

41 All About Atoms

Soon after lasers were invented in the 1960s, they became novel sources of light in research laboratories. Today, lasers are ubiquitous and are found in such diverse applications as voice and data transmission, surveying, welding, and grocery-store price scanning. The photograph shows surgery being performed with laser light transmitted via optical fibers. Light from a laser and light from any other source are both due to emissions by atoms.

What, then, is so different about the light from a laser?

The answer is in this chapter.

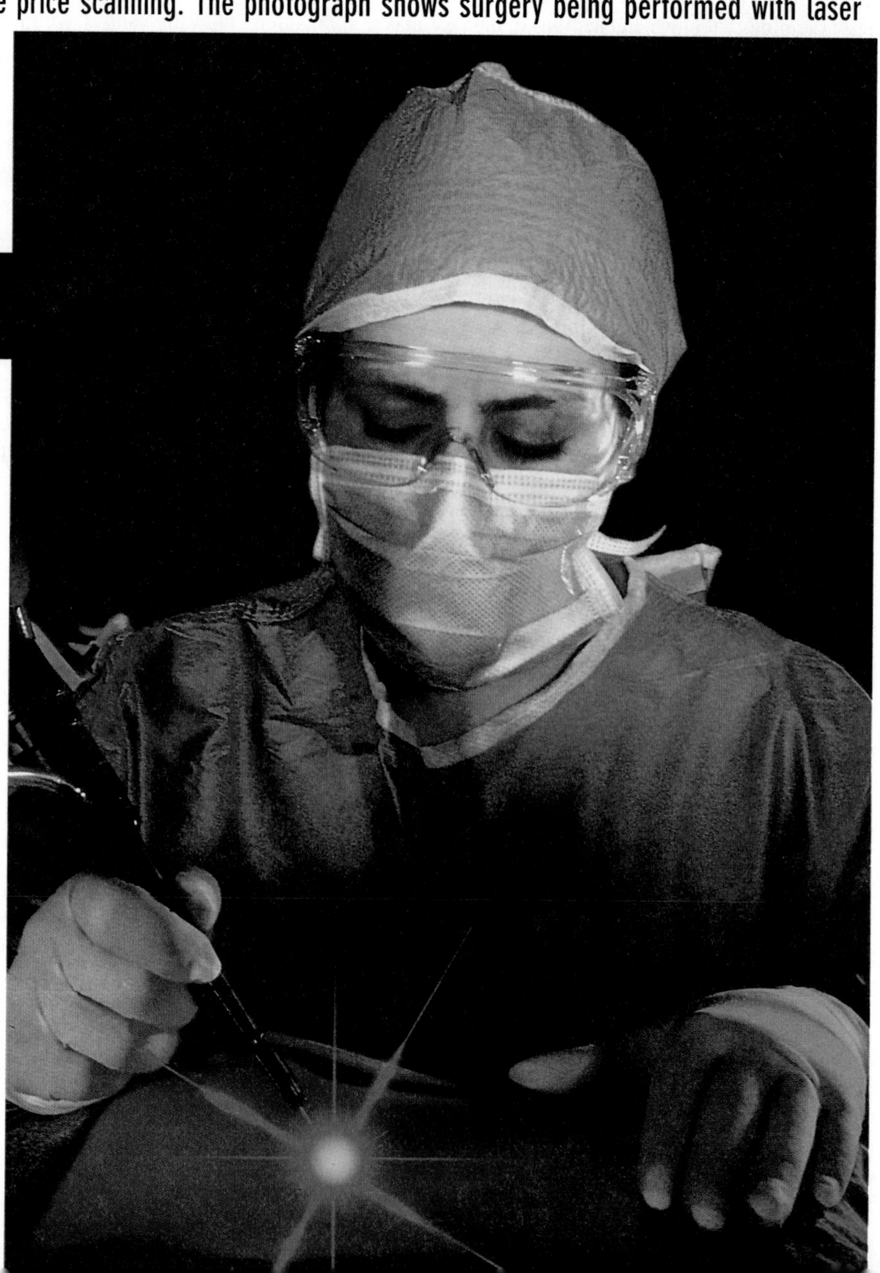

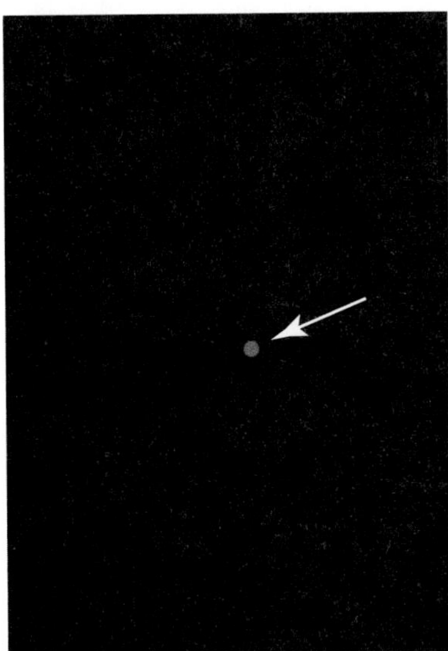

Fig. 41-1 The blue dot is a photograph of the light emitted from a single barium ion held for a long time in a trap at the University of Washington. Special techniques caused the ion to emit light over and over again as it underwent transitions between the same pair of energy levels. The dot represents the cumulative emission of many photons.

41-1 Atoms and the World Around Us

In the early years of the twentieth century many prominent scientists doubted the very existence of atoms. Today, however, every well-informed person believes that atoms exist and are the building blocks of the material world. Today, we can even pick up individual atoms and move them around. That's how the quantum corral on the opening page of Chapter 40 was formed. You can easily count the 48 iron atoms that make up the circle in that image. We can even photograph single atoms by the light they emit. For example, the faint blue dot in Figure 41-1 is due to light emitted by a single barium ion held in a trap at the University of Washington.

41-2 Some Properties of Atoms

You may think the details of atomic physics are remote from your daily life. However, consider how the following properties of atoms—so basic that we rarely think about them—affect the way we live in our world.

Atoms are stable. Essentially all the atoms that form our tangible world have existed without change for billions of years. What would the world be like if atoms continually changed into other forms, perhaps every few weeks or every few years?

Atoms combine with each other. They stick together to form stable molecules and stack up to form rigid solids. An atom is mostly empty space, but you can stand on a floor—made up of atoms—without falling through it.

These basic properties of atoms can be explained by quantum physics, as can the three less apparent properties that follow.

Atoms Are Put Together Systematically

Figure 41-2 shows an example of a repetitive property of the elements as a function of their position in the periodic table (Appendix G). The figure is a plot of the **ionization energy** of the elements; the energy required to remove the most loosely

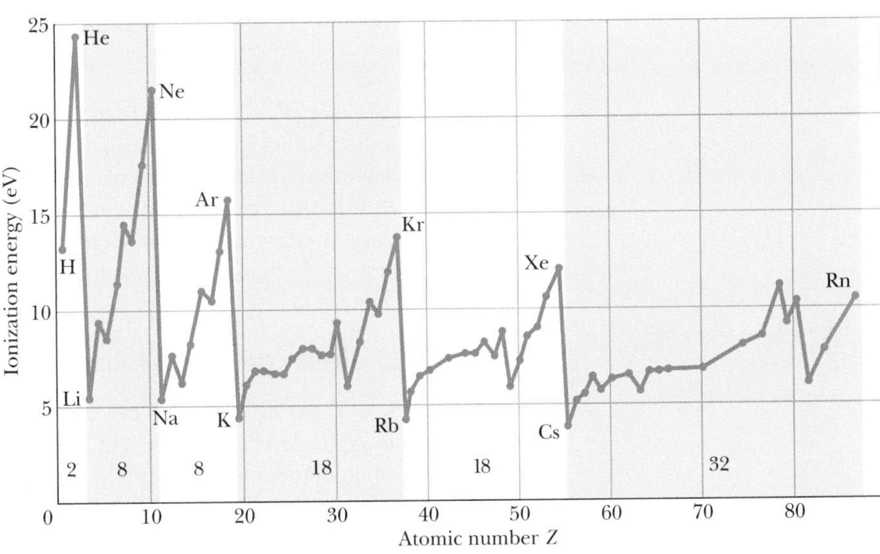

Fig. 41-2 A plot of the ionization energies of the elements as a function of atomic number, showing the periodic repetition of properties through the six complete horizontal periods of the periodic table. The number of elements in each of these periods is indicated.

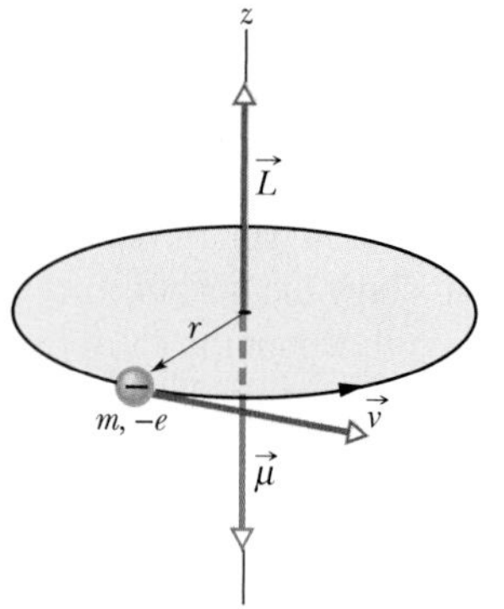

Fig. 41-3 A classical model showing a particle of mass m and charge $-e$ moving with speed v in a circle of radius r. The moving particle has an angular momentum $\vec{L}$ given by $\vec{r} \times \vec{p}$, where $\vec{p}$ is its linear momentum $m\vec{v}$. The particle's motion is equivalent to a current loop that has an associated magnetic momentum $\vec{\mu}$ which is directed opposite $\vec{L}$.

bound electron from a neutral atom is plotted as a function of the position in the periodic table of the element to which the atom belongs. The remarkable similarities in the chemical and physical properties of the elements in each vertical column of the periodic table are evidence enough that the atoms are constructed according to systematic rules.

The elements are arranged in the periodic table in six horizontal **periods;** except for the first, each period starts at the left with a highly reactive alkali metal (lithium, sodium, potassium, and so on) and ends at the right with a chemically inert noble gas (neon, argon, krypton, and so on). Quantum physics accounts for the chemical properties of these elements. The numbers of elements in the six periods are

$$2, 8, 8, 18, 18, \text{ and } 32.$$

Quantum physics predicts these numbers.

Atoms Emit and Absorb Light

We have already seen that atoms can exist only in discrete quantum states, each state having a certain energy. An atom can make a transition from one state to another by emitting light (to jump to a lower energy level E_{low}) or by absorbing light (to jump to a higher energy level E_{high}). As we first discussed in Section 40-3, the light is emitted or absorbed as a photon with energy

$$hf = E_{\text{high}} - E_{\text{low}}. \tag{41-1}$$

Thus, the problem of finding the frequencies of light emitted or absorbed by an atom reduces to the problem of finding the energies of the quantum states of that atom. Quantum physics allows us—in principle at least—to calculate these energies.

Atoms Have Angular Momentum and Magnetism

Figure 41-3 shows a negatively charged particle moving in a circular orbit around a fixed center. As we discussed in Section 32-4, the orbiting particle has both an angular momentum $\vec{L}$ and (since its path is equivalent to a tiny current loop) a magnetic dipole moment $\vec{\mu}$. (Here, for brevity, we drop the subscript orb that we used in Chapter 32.) As Fig. 41-3 shows, vectors $\vec{L}$ and $\vec{\mu}$ are both perpendicular to the plane of the orbit but, because the charge is negative, they point in opposite directions.

The model of Fig. 41-3 is strictly classical and does not accurately represent an electron in an atom. In quantum physics, the rigid orbit model has been replaced by the probability density model, best visualized as a dot plot. In quantum physics, however, it is still true that in general, each quantum state of an electron in an atom involves an angular momentum $\vec{L}$ and a magnetic dipole moment $\vec{\mu}$ that have opposite directions (those vector quantities are said to be *coupled*).

The Einstein–de Haas Experiment

In 1915, well before the discovery of quantum physics, Albert Einstein and Dutch physicist W. J. de Haas carried out a clever experiment designed to show that the angular momentum and magnetic moment of individual atoms are coupled.

Einstein and de Haas suspended an iron cylinder from a thin fiber, as shown in Fig. 41-4*a*. A solenoid was placed around the cylinder but not touching it. Initially,

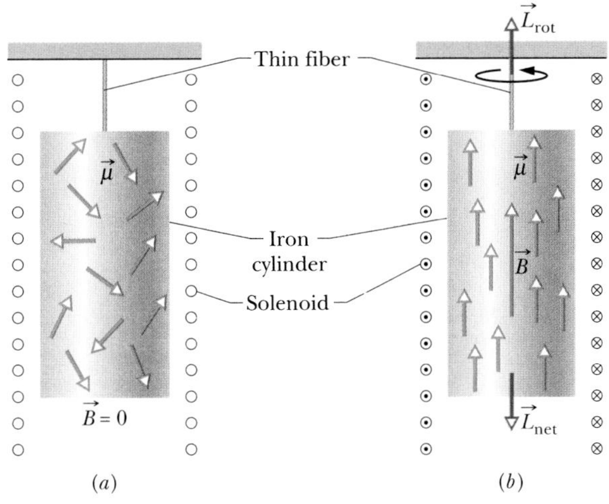

Fig. 41-4 The Einstein–de Haas experimental setup. (*a*) Initially, the magnetic field in the iron cylinder is zero and the magnetic dipole moment vectors $\vec{\mu}$ of its atoms are randomly oriented. The atomic angular momentum vectors (not shown) are directed opposite the magnetic dipole moment vectors and thus are also randomly oriented. (*b*) When a magnetic field $\vec{B}$ is set up along the cylinder's axis, the magnetic dipole moment vectors line up parallel to $\vec{B}$, which means that the angular momentum vectors line up opposite $\vec{B}$. Because the cylinder is initially isolated from external torques, its angular momentum is conserved and the cylinder as a whole must begin to rotate as shown.

the magnetic dipole moments $\vec{\mu}$ of the atoms of the cylinder point in random directions, so their external magnetic effects cancel (Fig. 41-4*a*). However, when a current is switched on in the solenoid (Fig. 41-4*b*) so that a magnetic field $\vec{B}$ is set up parallel to the axis of the cylinder, the magnetic dipole moments of the atoms of the cylinder reorient themselves, lining up with that field. If the angular momentum $\vec{L}$ of each atom is coupled to its magnetic moment $\vec{\mu}$, then this alignment of the atomic magnetic moments must cause an alignment of the atomic angular momenta opposite the magnetic field.

No external torques initially act on the cylinder; thus, its angular momentum must remain at its initial zero value. However, when $\vec{B}$ is turned on and the atomic angular momenta line up antiparallel to $\vec{B}$, they tend to give a net angular momentum $\vec{L}_{net}$ to the cylinder as a whole (directed downward in Fig. 41-4*b*). To maintain zero angular momentum, the cylinder begins to rotate around its central axis to produce an angular momentum $\vec{L}_{rot}$ in the opposite direction (upward in Fig. 41-4*b*).

Were it not for the fiber, the cylinder would continue to rotate for as long as the magnetic field is present. However, the twisting of the fiber quickly produces a torque that momentarily stops the cylinder's rotation and then rotates the cylinder in the opposite direction as the twisting is undone. Thereafter, the fiber will twist and untwist as the cylinder oscillates about its initial orientation in angular simple harmonic motion.

Observation of the cylinder's rotation verified that the angular momentum and the magnetic dipole moment of an atom are coupled in opposite directions. Moreover, it dramatically demonstrated that the angular momenta associated with quantum states of atoms can result in *visible* rotation of an object of everyday size.

41-3 Electron Spin

As we discussed in Section 32-4, whether an electron is *trapped* in an atom or is *free*, it has an intrinsic **spin angular momentum** $\vec{S}$, often called simply **spin.** (Recall that *intrinsic* means that $\vec{S}$ is a basic characteristic of an electron, like its mass and electric charge.) As we shall discuss in the next section, the magnitude of $\vec{S}$ is quantized and depends on a **spin quantum number** s, which is always $\frac{1}{2}$ for electrons (and for protons and neutrons). In addition, the component of $\vec{S}$ measured along any axis is quantized and depends on a **spin magnetic quantum number** m_s, which can have only the value $+\frac{1}{2}$ or $-\frac{1}{2}$.

TABLE 41-1 Electron States for an Atom

Quantum Number	Symbol	Allowed Values	Related to
Principal	n	$1, 2, 3, \ldots$	Distance from the nucleus
Orbital	l	$0, 1, 2, \ldots, (n-1)$	Orbital angular momentum
Orbital magnetic	m_l	$0, \pm 1, \pm 2, \ldots, \pm l$	Orbital angular momentum (z component)
Spin magnetic	m_s	$\pm \frac{1}{2}$	Spin angular momentum (z component)

All states with the same value of n form a **shell.**
There are $2n^2$ states in a shell.

All states with the same values of n and l form a **subshell.**
All states in a subshell have the same energy.
There are $2(2l+1)$ states in a subshell.

The existence of electron spin was postulated on an empirical basis by two Dutch graduate students, George Uhlenbeck and Samuel Goudsmit, from their studies of atomic spectra. The quantum physics basis for electron spin was provided a few years later, by British physicist P. A. M. Dirac, who developed (in 1929) a relativistic quantum theory of the electron.

It is tempting to account for electron spin by thinking of the electron as a tiny sphere spinning about an axis. However, that classical model, like the classical model of orbits, does not hold up. In quantum physics, spin angular momentum is best thought of as a measurable intrinsic property of the electron; you simply can't visualize it with a classical model.

Table 41-1, an extension of Table 40-2, shows the four quantum numbers n, l, m_l, and m_s that completely specify the quantum states of the electron in a hydrogen atom. (Quantum number s is not included because all electrons have the value $s = \frac{1}{2}$.) The same quantum numbers also specify the allowed states of any single electron in a multielectron atom.

41-4 Angular Momenta and Magnetic Dipole Moments

Every quantum state of an electron in an atom has an associated orbital angular momentum and a corresponding orbital magnetic dipole moment. Every electron, whether trapped in an atom or free, has a spin angular momentum and a corresponding spin magnetic dipole moment. We discuss these quantities separately first, and then in combination.

Orbital Angular Momentum and Magnetism

The magnitude L of the **orbital angular momentum** $\vec{L}$ of an electron *in an atom* is quantized; that is, it can have only certain values. These values are

$$L = \sqrt{l(l+1)}\hbar, \tag{41-2}$$

in which l is the orbital quantum number and $\hbar$ is $h/2\pi$. According to Table 41-1, l must be either zero or a positive integer no greater than $n - 1$. For a state with $n = 3$, for example, only $l = 2$, $l = 1$, and $l = 0$ are permitted.

As we discussed in Section 32-4, a magnetic dipole is associated with the orbital angular momentum $\vec{L}$ of an electron in an atom. This magnetic dipole has an **orbital**

magnetic dipole moment $\vec{\mu}_{orb}$, which is related to the angular momentum by Eq. 32-8:

$$\vec{\mu}_{orb} = -\frac{e}{2m}\vec{L}. \tag{41-3}$$

The minus sign in this relation means that $\vec{\mu}_{orb}$ is directed opposite $\vec{L}$. Because the magnitude of $\vec{L}$ is quantized (Eq. 41-2), the magnitude of $\vec{\mu}_{orb}$ must also be quantized and given by

$$\mu_{orb} = \frac{e}{2m}\sqrt{l(l+1)}\hbar. \tag{41-4}$$

Neither $\vec{\mu}_{orb}$ nor $\vec{L}$ can be measured in any way. However, we *can* measure the components of those two vectors along a given axis. Let us imagine that the atom is located in a magnetic field $\vec{B}$; assume that a z axis extends in the direction of the field lines. Then we can measure the z components of $\vec{\mu}_{orb}$ and $\vec{L}$ along that axis.

The components $\mu_{orb,z}$ of the orbital magnetic dipole moment are quantized and given by

$$\mu_{orb,z} = -m_l\mu_B. \tag{41-5}$$

Here m_l is the orbital magnetic quantum number of Table 41-1 and μ_B is the **Bohr magneton**:

$$\mu_B = \frac{eh}{4\pi m} = \frac{e\hbar}{2m} = 9.274 \times 10^{-24}\ \text{J/T} \quad \text{(Bohr magneton)}, \tag{41-6}$$

where m is the electron mass.

The components L_z of the angular momentum are also quantized, and they are given by

$$L_z = m_l\hbar. \tag{41-7}$$

Figure 41-5 shows the five quantized components L_z of the orbital angular momentum for an electron with $l = 2$, as well as the associated orientations of the angular momentum $\vec{L}$. However, *do not take the figure literally* because we cannot detect $\vec{L}$ in any way. Thus, drawing it in a figure like Fig. 41-5 is merely a visual aide. We can extend that visual aide by saying that $\vec{L}$ makes a certain angle θ with the z axis, such that

$$\cos\theta = \frac{L_z}{L}. \tag{41-8}$$

We can call θ the *semi-classical angle* between vector $\vec{L}$ and the z axis, because it is a classical measurement of something that quantum theory tells us cannot be measured.

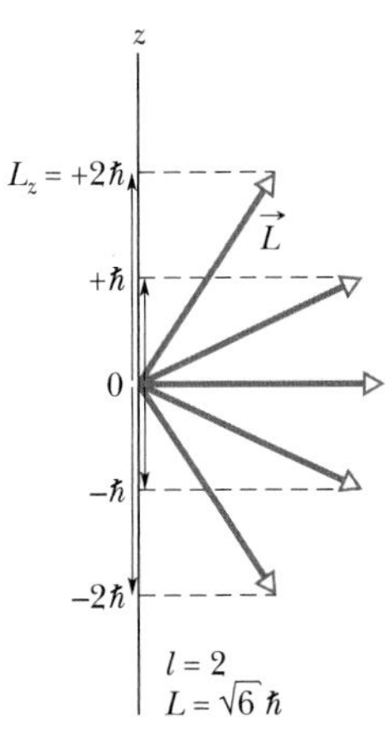

Fig. 41-5 The allowed values of L_z for an electron in a quantum state with $l = 2$. For every orbital angular momentum vector $\vec{L}$ in the figure, there is a vector pointing in the opposite direction, representing the magnitude and direction of the orbital magnetic dipole moment $\vec{\mu}_{orb}$.

Spin Angular Momentum and Spin Magnetic Dipole Moment

The magnitude S of the spin angular momentum $\vec{S}$ of any electron, whether *free or trapped,* has the single value given by

$$S = \sqrt{s(s+1)}\hbar = \sqrt{(\tfrac{1}{2})(\tfrac{1}{2}+1)}\hbar = 0.866\hbar, \tag{41-9}$$

where s $(= \frac{1}{2})$ is the spin quantum number of the electron.

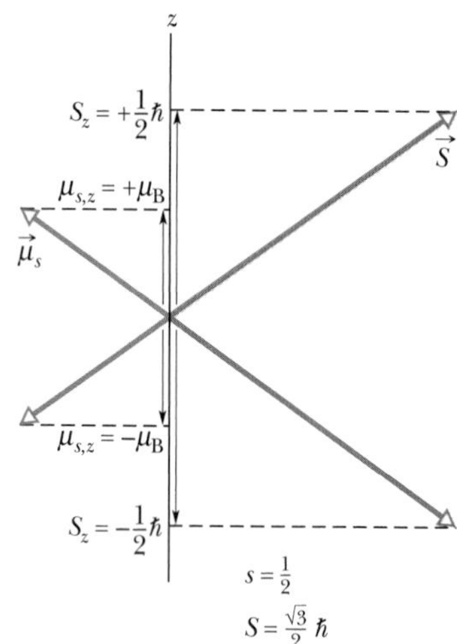

Fig. 41-6 The allowed values of S_z and μ_z for an electron.

As we discussed in Section 32-4, an electron has an intrinsic magnetic dipole that is associated with its spin angular momentum $\vec{S}$, whether the electron is confined to an atom or free. This magnetic dipole has a **spin magnetic dipole moment** $\vec{\mu}_s$, which is related to the spin angular momentum by Eq. 32-2:

$$\vec{\mu}_s = -\frac{e}{m}\vec{S}. \tag{41-10}$$

The minus sign in this relation means that $\vec{\mu}_s$ is directed opposite $\vec{S}$. Because the magnitude of $\vec{S}$ is quantized (Eq. 41-9), the magnitude of $\vec{\mu}_s$ must also be quantized and given by

$$\mu_s = \frac{e}{m}\sqrt{s(s+1)}\hbar. \tag{41-11}$$

Neither $\vec{S}$ nor $\vec{\mu}_s$ can be measured in any way. However, we *can* measure their components along any given axis—call it the z axis. The components S_z of the spin angular momentum are quantized and given by

$$S_z = m_s\hbar, \tag{41-12}$$

where m_s is the spin magnetic quantum number of Table 41-1. That quantum number can have only two values: $m_s = +\frac{1}{2}$ (the electron is said to be *spin up*) and $m_s = -\frac{1}{2}$ (the electron is said to be *spin down*).

The components $\mu_{s,z}$ of the spin magnetic dipole moment are also quantized, and they are given by

$$\mu_{s,z} = -2m_s\mu_B. \tag{41-13}$$

Figure 41-6 shows the two quantized components S_z of the spin angular momentum for an electron and the associated orientations of vector $\vec{S}$. It also shows the quantized components $\mu_{s,z}$ of the spin magnetic dipole moment and the associated orientations of $\vec{\mu}_s$.

Orbital and Spin Angular Momenta Combined

For an atom containing more than one electron, we define a total angular momentum $\vec{J}$, which is the vector sum of the angular momenta of the individual electrons—both their orbital and their spin angular momenta. The number of electrons (and the number of protons) in a neutral atom is the **atomic number** (or **charge number**) Z. Thus, for a neutral atom,

$$\vec{J} = (\vec{L}_1 + \vec{L}_2 + \vec{L}_3 + \cdots + \vec{L}_Z) + (\vec{S}_1 + \vec{S}_2 + \vec{S}_3 + \cdots + \vec{S}_Z). \tag{41-14}$$

Similarly, the total magnetic dipole moment of the multielectron atom is the vector sum of the magnetic dipole moments (both orbital and spin) of its individual electrons. However, because of the factor of 2 in Eq. 41-13, the resultant magnetic dipole moment for the atom does not have the direction of the vector $-\vec{J}$; instead, it makes a certain angle with that vector. The **effective magnetic dipole moment** $\vec{\mu}_{\text{eff}}$ for the atom is the component of the vector sum of the individual magnetic dipole moments in the direction of $-\vec{J}$ (Fig. 41-7).

As you will see in the next section, in typical atoms the orbital angular momenta and the spin angular momenta of most of the electrons sum vectorially to zero. Then $\vec{J}$ and $\vec{\mu}_{\text{eff}}$ of those atoms are due to a relatively small number of electrons, often only a single valence electron.

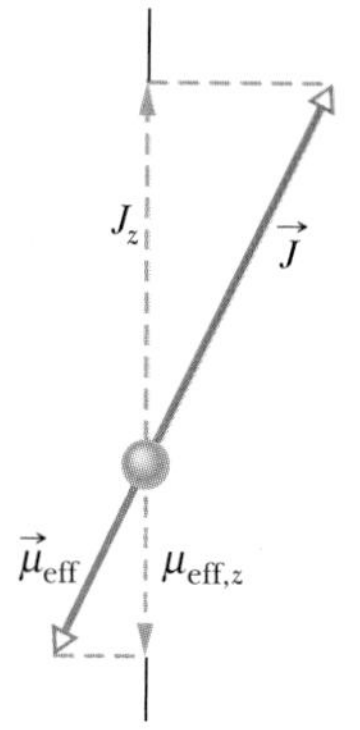

Fig. 41-7 A classical model showing the total angular momentum vector $\vec{J}$ and the effective magnetic moment vector $\vec{\mu}_{\text{eff}}$.

CHECKPOINT 1: An electron is in a quantum state for which the magnitude of the electron's orbital angular momentum $\vec{L}$ is $2\sqrt{3}\hbar$. How many projections of the electron's orbital magnetic dipole moment on a z axis are allowed?

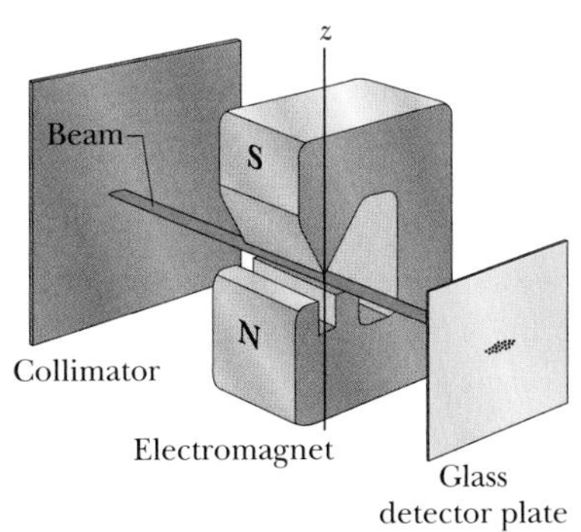

Fig. 41-8 Apparatus used by Stern and Gerlach.

41-5 The Stern–Gerlach Experiment

In 1922, Otto Stern and Walther Gerlach at the University of Hamburg in Germany showed experimentally that the magnetic moment of cesium atoms is quantized. In the Stern–Gerlach experiment, as it is now known, silver is vaporized in an oven, and some of the atoms in that vapor escape through a narrow slit in the oven wall, into an evacuated tube. Some of those escaping atoms then pass through a second narrow slit, to form a narrow beam of atoms (Fig. 41-8). (The atoms are said to be *collimated*—made into a beam—and the second slit is called a *collimator*.) The beam passes between the poles of an electromagnet and then lands on a glass detector plate where it forms a silver deposit.

When the electromagnet is off, the silver deposit is a narrow spot. However, when the electromagnet is turned on, the silver deposit is spread vertically. The spreading occurs because silver atoms are magnetic dipoles, so vertical magnetic forces act on them as they pass through the vertical magnetic field of the electromagnet; these forces deflect them slightly up or down. Thus, by analyzing the silver deposit on the plate, we can determine what deflections the atoms underwent in the magnetic field. When Stern and Gerlach analyzed the pattern of silver on their detector plate, they found a surprise. However, before we discuss that surprise and its quantum implications, let us discuss the magnetic deflecting force acting on the silver atoms.

The Magnetic Deflecting Force on a Silver Atom

We have not previously discussed the type of magnetic force that deflects the silver atoms in a Stern–Gerlach experiment. It is *not* the magnetic deflecting force that acts on a moving charged particle, as given by Eq. 29-2 ($\vec{F} = q\vec{v} \times \vec{B}$). The reason is simple: A silver atom is electrically neutral (its net charge q is zero) and thus this type of magnetic force is also zero.

The type of magnetic force we seek is due to an interaction between the magnetic field $\vec{B}$ of the electromagnet and the magnetic dipole of the individual silver atom. We can derive an expression for the force in this interaction by starting with the potential energy U of the dipole in the magnetic field. Equation 29-38 tells us that

$$U = -\vec{\mu} \cdot \vec{B}, \tag{41-15}$$

where $\vec{\mu}$ is the magnetic dipole moment of a silver atom. In Fig. 41-8, the positive direction of the z axis and the direction of $\vec{B}$ are vertically upward. Thus, we can write Eq. 41-15 in terms of the component μ_z of the atom's magnetic dipole moment along the direction of $\vec{B}$:

$$U = -\mu_z B. \tag{41-16}$$

Then, using Eq. 8-20 ($F = -dU/dx$) for the z axis shown in Fig. 41-8, we obtain

$$F_z = -\frac{dU}{dz} = \mu_z \frac{dB}{dz}. \tag{41-17}$$

This is what we sought—an equation for the magnetic force that deflects a silver atom as the atom passes through a magnetic field.

The term dB/dz in Eq. 41-17 is the *gradient* of the magnetic field along the z axis. If the magnetic field does not change along the z axis (as in a uniform magnetic field or no magnetic field), then $dB/dz = 0$ and a silver atom is not deflected as it moves between the magnet's poles. In the Stern–Gerlach experiment, the poles are designed to maximize the gradient dB/dz, to vertically deflect the silver atoms passing between the poles as much as possible, so that their deflections show up in the deposit on the glass plate.

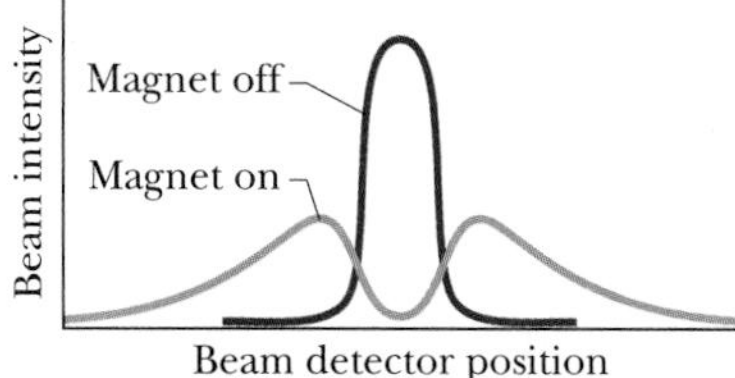

Fig. 41-9 Results of a modern repetition of the Stern–Gerlach experiment. With the electromagnet turned off, there is only a single beam; with the electromagnet turned on, the original beam splits into two subbeams. The two subbeams correspond to parallel and antiparallel alignment of the magnetic moments of cesium atoms with the external magnetic field.

According to classical physics, the components μ_z of silver atoms passing through the magnetic field in Fig. 41-8 should range in value from $-\mu$ (the dipole moment $\vec{\mu}$ is directed straight down the z axis) to $+\mu$ ($\vec{\mu}$ is directed straight up the z axis). Thus, from Eq. 41-17, there should be a range of forces on the atoms, and therefore a range of deflections of the atoms, from a greatest downward deflection to a greatest upward deflection. This means that we should expect the atoms to land along a vertical line on the glass plate. However, this does not happen.

The Experimental Surprise

What Stern and Gerlach found was that the atoms formed two distinct spots on the glass plate, one spot above the point where they would have landed with no deflection and the other spot just as far below that point. This two-spot result can be seen in the plots of Fig. 41-9, which shows the outcome of a more recent version of the Stern–Gerlach experiment. In that version, a beam of cesium atoms (magnetic dipoles like the silver atoms in the original Stern–Gerlach experiment) was sent through a magnetic field with a large vertical gradient dB/dz. The field could be turned on and off, and a detector could be moved up and down through the beam.

When the field was turned off, the beam was, of course, undeflected and the detector recorded the central-peak pattern shown in Fig. 41-9. When the field was turned on, the original beam was split vertically by the magnetic field into two smaller beams, one beam higher than the previously undeflected beam and the other beam lower. As the detector moved vertically up through these two smaller beams, it recorded the two-peak pattern shown in Fig. 41-9.

The Meaning of the Results

In the original Stern–Gerlach experiment, two spots of silver were formed on the glass plate, not a vertical line of silver. This means that the component μ_z along $\vec{B}$ (and z) could not have any value between $-\mu$ and $+\mu$ as classical physics predicts. Instead, μ_z is restricted to only two values, one for each spot on the glass. Thus, the original Stern–Gerlach experiment showed that μ_z is quantized, implying (correctly) that $\vec{\mu}$ is also. Moreover, because the angular momentum $\vec{L}$ of an atom is associated with $\vec{\mu}$, that angular momentum and its component L_z are also quantized.

With modern quantum theory, we can add to the explanation of the two-spot result in the Stern–Gerlach experiment. We now know that a silver atom consists of many electrons, each with a spin magnetic moment and an orbital magnetic moment. We also know that all those moments vectorially cancel out *except* for a single electron, and the orbital dipole moment of that electron is zero. Thus, the combined dipole moment $\vec{\mu}$ of a silver atom is the *spin* magnetic dipole moment of that single electron. According to Eq. 41-13, that means that μ_z can have only two components along the z axis in Fig. 41-8. One component is for quantum number $m_s = +\frac{1}{2}$ (the single electron is spin up), and the other component is for quantum number $m_s = -\frac{1}{2}$ (the single electron is spin down). Substituting into Eq. 41-13 gives us

$$\mu_{s,z} = -2(+\tfrac{1}{2})\mu_B = -\mu_B \quad \text{and} \quad \mu_{s,z} = -2(-\tfrac{1}{2})\mu_B = +\mu_B. \tag{41-18}$$

Then substituting these expressions for μ_z in Eq. 41-17, we find that the force component F_z deflecting the silver atoms as they pass through the magnetic field can have only the two values

$$F_z = -\mu_B\left(\frac{dB}{dz}\right) \quad \text{and} \quad F_z = +\mu_B\left(\frac{dB}{dz}\right), \tag{41-19}$$

which result in the two spots of silver on the glass.

Sample Problem 41-1

In the Stern–Gerlach experiment of Fig. 41-8, a beam of silver atoms passes through a magnetic field gradient dB/dz of magnitude 1.4 T/mm that is set up along the z axis. This region has a length w of 3.5 cm in the direction of the original beam. The speed of the atoms is 750 m/s. By what distance d have the atoms been deflected when they leave the region of the magnetic field gradient? The mass M of a silver atom is 1.8×10^{-25} kg.

SOLUTION: One Key Idea here is that the deflection of a silver atom in the beam is due to an interaction between the magnetic dipole of the atom and the magnetic field, because of the gradient dB/dz. The deflecting force is directed along the field gradient (along the z axis) and is given by Eqs. 41-17. Let us consider only deflection in the positive direction of z; thus, we shall use $F_z = \mu_B(dB/dz)$ from Eqs. 41-19.

A second Key Idea is that we assume the field gradient dB/dz has the same value throughout the region through which the silver atoms travel. Thus, force component F_z is constant in that region, and from Newton's second law, the acceleration a_z of an atom along the z axis due to F_z is also constant and is given by

$$a_z = \frac{F_z}{M} = \frac{\mu_B(dB/dz)}{M}.$$

Because this acceleration is constant, we can use Eq. 2-15 (from Table 2-1) to write the deflection d parallel to the z axis as

$$d = v_{0z}t + \tfrac{1}{2}a_z t^2 = 0t + \tfrac{1}{2}\left(\frac{\mu_B(dB/dz)}{M}\right)t^2. \qquad (41\text{-}20)$$

Because the deflecting force on the atom acts perpendicular to the atom's original direction of travel, the component v of the atom's velocity along the original direction of travel is not changed by the force. Thus, the atom requires time $t = w/v$ to travel through length w in that direction. Substituting w/v for t into Eq. 41-20, we find

$$\begin{aligned} d &= \tfrac{1}{2}\left(\frac{\mu_B(dB/dz)}{M}\right)\left(\frac{w}{v}\right)^2 = \frac{\mu_B(dB/dz)w^2}{2Mv^2} \\ &= (9.27 \times 10^{-24}\ \text{J/T})(1.4 \times 10^3\ \text{T/m}) \\ &\quad \times \frac{(3.5 \times 10^{-2}\ \text{m})^2}{(2)(1.8 \times 10^{-25}\ \text{kg})(750\ \text{m/s})^2} \\ &= 7.85 \times 10^{-5}\ \text{m} \approx 0.08\ \text{mm}. \end{aligned} \qquad \text{(Answer)}$$

The separation between the two subbeams is twice this, or 0.16 mm. This separation is not large but is easily measured.

41-6 Magnetic Resonance

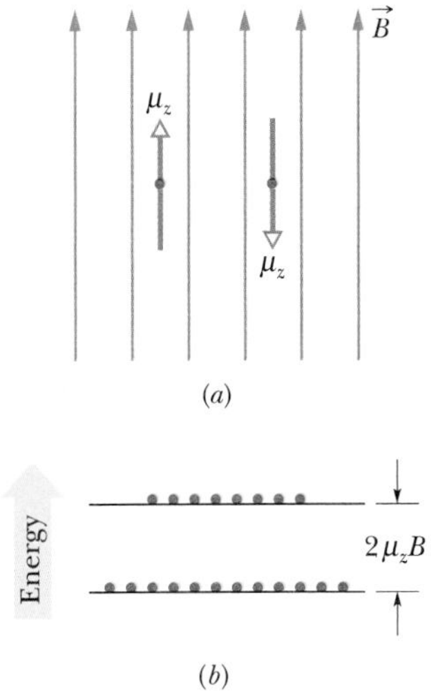

As we discussed briefly in Section 32-4, a proton has an intrinsic spin angular momentum $\vec{S}$ and an associated spin magnetic dipole moment $\vec{\mu}$ that are in the same direction (because the proton is positively charged). If a proton is located in a uniform magnetic field $\vec{B}$ directed along a z axis, the z component μ_z of the spin magnetic dipole moment can have only two quantized orientations: either parallel to $\vec{B}$ or antiparallel to $\vec{B}$, as shown in Fig. 41-10a. From Eq. 29-38, we know that these two orientations differ in energy by $2\mu_z B$, which is the energy involved in reversing a magnetic dipole in a uniform magnetic field. The lower energy state is the one with μ_z parallel to $\vec{B}$, and the higher energy state has μ_z antiparallel to $\vec{B}$.

Let us place a drop of water in a uniform magnetic field $\vec{B}$; then the protons in the hydrogen of the water molecules each have μ_z either parallel or antiparallel to $\vec{B}$. If we next apply to the drop an alternating electromagnetic field of a certain frequency f, the protons in the lower energy state can undergo reversal of their μ_z orientation. This process of reversal is called *spin-flipping* (because the reversal of a proton's magnetic dipole moment requires a reversal of the proton's spin). The frequency f required for the spin-flipping is given by

$$hf = 2\mu_z B, \qquad (41\text{-}21)$$

a condition called **magnetic resonance** (or, as originally, **nuclear magnetic resonance**). In words, if an alternating electromagnetic field is to cause protons to spin-flip in the magnetic field, the photons associated with that field must have an energy hf equal to the energy difference $2\mu_z B$ between the two possible orientations of μ_z (and thus proton spin) in that field.

Once a proton is spin-flipped to the higher energy state, it can drop back to the lower energy state by emitting a photon of the same energy hf given by Eq. 41-21.

Fig. 41-10 (a) A proton (red dot), whose spin component in the direction of an applied magnetic field is $\frac{1}{2}\hbar$, can occupy either of two quantized orientations in an external magnetic field. If Eq. 41-21 is satisfied, the protons in the sample can be induced to flip from one orientation to the other. (b) Normally, there are more protons in the lower energy state than in the higher energy state.

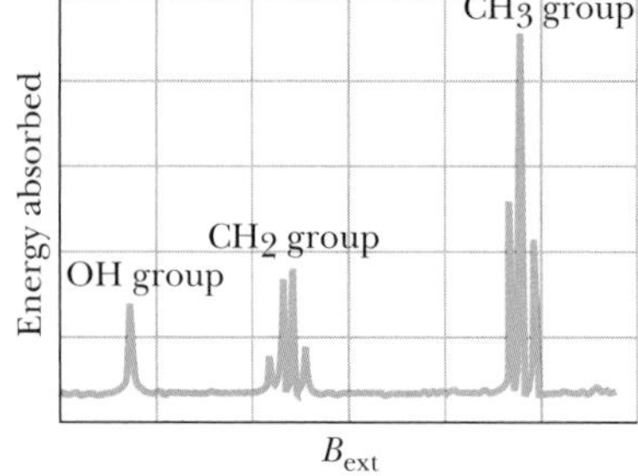

Fig. 41-11 A nuclear magnetic resonance spectrum for ethanol. The spectral lines represent the absorption of energy associated with spin flips of protons. The three groups of lines correspond, as indicated, to protons in the OH group, the CH_2 group, and the CH_3 group of the ethanol molecule. Note that the two protons in the CH_2 group occupy four different local environments. The entire horizontal axis covers less than 10^{-4} T.

Normally more protons are in the lower energy state than in the higher energy state, as Fig. 41-10*b* suggests. This means that there will be a detectable net *absorption* of energy from the alternating electromagnetic field.

The constant field whose magnitude B appears in Eq. 41-21 is actually *not* the imposed external field $\vec{B}_{ext}$ in which the water drop is placed; rather, it is that field as modified by the small, local, internal magnetic field $\vec{B}_{local}$ due to the magnetic moments of the atoms and nuclei near a given proton. Thus, we can rewrite Eq. 41-21 as

$$hf = 2\mu_z(B_{ext} + B_{local}). \tag{41-22}$$

To achieve magnetic resonance, it is customary to leave the frequency f of the electromagnetic oscillations fixed and to vary B_{ext} until Eq. 41-22 is satisfied and an absorption peak is recorded.

Nuclear magnetic resonance is a property that is the basis for a valuable analytical tool, particularly for the identification of unknown compounds. Figure 41-11 shows a **nuclear magnetic resonance spectrum,** as it is called, for ethanol, whose formula we may write as CH_3-CH_2-OH. The various resonance peaks all represent spin flips of protons. They occur at different values of B_{ext}, however, because the local environments of the six protons within the ethanol molecule differ from one another. The spectrum of Fig. 41-11 is a unique signature for ethanol.

Spin technology, called **magnetic resonance imaging** (MRI), has been applied to medical diagnostics with great success. The protons of the various tissues of the human body are situated in many different local magnetic environments. When the body, or part of it, is immersed in a strong external magnetic field, these environmental differences can be detected by spin-flip techniques and translated by computer processing into an image resembling those produced by x rays. Figure 41-12, for example, shows a cross section of a human head imaged by this method.

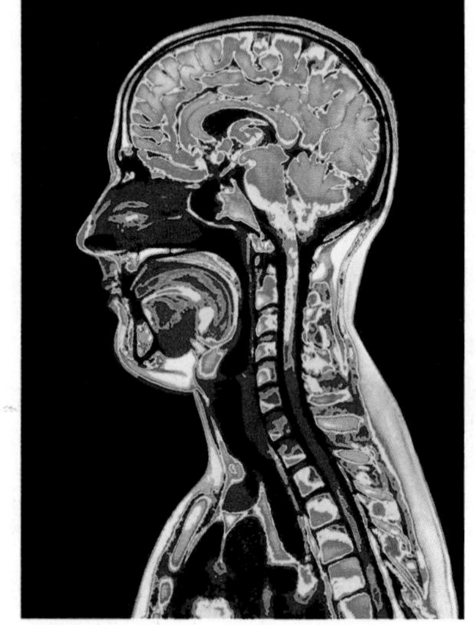

Fig. 41-12 A cross-sectional view of a human head and neck produced by magnetic resonance imaging. Some of the details visible here would not show up on an x-ray image, even with a modern computerized axial tomography scanner (CAT scanner).

Sample Problem 41-2

A drop of water is suspended in a magnetic field $\vec{B}$ of magnitude 1.80 T and an alternating electromagnetic field is applied, its frequency adjusted to produce spin flips of the protons in the water. The component μ_z of the magnetic dipole moment of a proton, measured along the direction of $\vec{B}$, is 1.41×10^{-26} J/T. Assume that the local magnetic fields are negligible compared to $\vec{B}$. What are the frequency f and wavelength λ of the alternating field?

SOLUTION: One **Key Idea** here is that when a proton is located in a magnetic field $\vec{B}$, it has a potential energy because it is a magnetic dipole. A second **Key Idea** is that this potential energy is restricted to two values, with a difference of $2\mu_z B$. The third **Key Idea** is that if the proton is to jump between these two energies (spin-flip), the photon energy hf of the electromagnetic wave must be equal to the energy difference $2\mu_z B$, according to Eq. 41-21. From that equation, we then find

$$f = \frac{2\mu_z B}{h} = \frac{(2)(1.41 \times 10^{-26}\ \text{J/T})(1.80\ \text{T})}{6.63 \times 10^{-34}\ \text{J}\cdot\text{s}}$$

$$= 7.66 \times 10^7\ \text{Hz} = 76.6\ \text{MHz}. \quad \text{(Answer)}$$

The corresponding wavelength is

$$\lambda = \frac{c}{f} = \frac{3.00 \times 10^8\ \text{m/s}}{7.66 \times 10^7\ \text{Hz}} = 3.92\ \text{m}. \quad \text{(Answer)}$$

This frequency and wavelength are in the short radio wave region of the electromagnetic spectrum.

41-7 The Pauli Exclusion Principle

In Chapter 40 we considered a variety of electron traps, from fictional one-dimensional traps to the real three-dimensional trap of a hydrogen atom. In all those examples, we trapped only one electron. However, when we discuss traps containing two or more electrons (as we shall in the next two sections), we must consider a principle that governs any particle whose spin quantum number s is not zero or an integer. This principle applies not only to electrons but also to protons and neutrons, all of which have $s = \frac{1}{2}$. The principle is known as the **Pauli exclusion principle** after Wolfgang Pauli, who formulated it in 1925. For electrons, it states that

> No two electrons confined to the same trap can have the same set of values for its quantum numbers.

As we shall discuss in Section 41-9, this principle means that no two electrons in an atom can have the same four values for the quantum numbers n, l, m_l, and m_s. In other words, the quantum numbers of any two electrons in an atom must differ in at least one quantum number. Were this not true, atoms would collapse, and thus you and the world as you know it could not exist.

41-8 Multiple Electrons in Rectangular Traps

To prepare for our discussion of multiple electrons in atoms, let us discuss two electrons confined to the rectangular traps of Chapter 40. We shall again use the quantum numbers we found for those traps when only one electron was confined. However, here we shall also include the spin angular momenta of the two electrons. To do this, we assume that the traps are located in a uniform magnetic field. Then according to Eq. 41-12, an electron can be either spin up with $m_s = \frac{1}{2}$ or spin down with $m_s = -\frac{1}{2}$. (We shall assume that the magnetic field is very weak so that we can neglect the potential energies of the electrons due to the field.)

As we confine the two electrons to one of the traps, we must keep the Pauli exclusion principle in mind; that is, the electrons cannot have the same set of values for their quantum numbers.

1. *One-dimensional trap.* In the one-dimensional trap of Fig. 40-2, fitting an electron wave to the trap's width L requires the single quantum number n. Therefore, any electron confined to the trap must have a certain value of n, and its quantum number m_s can be either $+\frac{1}{2}$ or $-\frac{1}{2}$. The two electrons could have different values of n, or they could have the same value of n if one of them is spin up and the other is spin down.
2. *Rectangular corral.* In the rectangular corral of Fig. 40-12, fitting an electron wave to the corral's widths L_x and L_y requires the two quantum numbers n_x and n_y. Thus, any electron confined to the trap must have certain values for those two quantum numbers, and its quantum number m_s can be either $+\frac{1}{2}$ or $-\frac{1}{2}$—so now there are three quantum numbers. According to the Pauli exclusion principle, two electrons confined to the trap must have different values for at least one of those three quantum numbers.
3. *Rectangular box.* In the rectangular box of Fig. 40-13, fitting an electron wave to the box's widths L_x, L_y, and L_z requires the three quantum numbers n_x, n_y, and n_z. Thus, any electron confined to the trap must have certain values for these three quantum numbers, and its quantum number m_s can be either $+\frac{1}{2}$ or $-\frac{1}{2}$—so now there are four quantum numbers. According to the Pauli exclusion

principle, two electrons confined to the trap must have different values for at least one of those four quantum numbers.

Suppose we add more than two electrons, one by one, to a rectangular trap in the preceding list. The first electrons naturally go into the lowest possible energy level—they are said to *occupy* that level. However, eventually the Pauli exclusion principle disallows any more electrons from occupying that lowest energy level, and the next electron must occupy the next higher level. When an energy level cannot be occupied by more electrons because of the Pauli exclusion principle, we say that level is **full** or **fully occupied.** In contrast, a level that is not occupied by any electrons is **empty** or **unoccupied.** For intermediate situations, the level is **partially occupied.** The *electron configuration* of a system of trapped electrons is a listing or drawing of the energy levels the electrons occupy, or the set of the quantum numbers of the electrons.

Finding the Total Energy

We shall later want to find the energy of a *system* of two or more electrons confined to a rectangular trap. That is, we shall want to find the total energy for any configuration of the trapped electrons.

For simplicity, we shall assume that the electrons do not electrically interact with one another; that is, we shall neglect the electric potential energies of pairs of electrons. In that case, we can calculate the total energy for any electron configuration by calculating the energy of each electron as we did in Chapter 40, and then summing those energies. (In Sample Problem 41-3 we do so for seven electrons confined to a rectangular corral.)

A good way to organize the energy values of a given system of electrons is with an energy-level diagram *for the system,* just as we did for a single electron in the traps of Chapter 40. The lowest level, with energy E_{gr}, corresponds to the ground state of the system. The next higher level, with energy E_{fe}, corresponds to the first excited state of the system. The next level, with energy E_{se}, corresponds to the second excited state of the system. And so on.

Sample Problem 41-3

Seven electrons are confined to the square corral of Sample Problem 40-5, where the corral is a two-dimensional infinite potential well with widths $L_x = L_y = L$ (Fig. 40-12). Assume that the electrons do not electrically interact with one another.

(a) What is the electron configuration for the ground state of the system of seven electrons?

SOLUTION: We can determine the electron configuration of the system by placing the seven electrons in the corral one by one, to build up the system. One Key Idea here is that because we assume the electrons do not electrically interact with one another, we can use the energy-level diagram for a single trapped electron in order to keep track of how we place the seven electrons in the corral. That *one-electron energy-level diagram* is given in Fig. 40-14 and partially reproduced here as Fig. 41-13*a*. Recall that the levels are labeled as $E_{nx,ny}$ for their associated energy. For example, the lowest level is for energy $E_{1,1}$, where quantum number n_x is 1 and quantum number n_y is 1.

A second Key Idea here is that the trapped electrons must obey the Pauli exclusion principle; that is, no two electrons can have the same set of values for their quantum numbers n_x, n_y, and m_s.

The first electron goes into energy level $E_{1,1}$ and can have $m_s = \frac{1}{2}$ or $m_s = -\frac{1}{2}$. We arbitrarily choose the latter and draw a down arrow (to represent spin down) on the $E_{1,1}$ level in Fig. 41-13*a*. The second electron also goes into the $E_{1,1}$ level but must have $m_s = +\frac{1}{2}$ so that one of its quantum numbers differs from those of the first electron. We represent this second electron with an up arrow (for spin up) on the $E_{1,1}$ level in Fig. 41-13*b*.

Another Key Idea now comes into play: The level for energy $E_{1,1}$ is fully occupied, and thus the third electron cannot have that energy. Therefore, the third electron goes into the next higher level, which is for the equal energies $E_{2,1}$ and $E_{1,2}$ (the level is degenerate). This third electron can have quantum numbers n_x and n_y of either 1 and 2 or 2 and 1, respectively. It can also have a quantum number m_s of either $+\frac{1}{2}$ or $-\frac{1}{2}$. Let us arbitrarily assign it the quantum numbers $n_x = 2$, $n_y = 1$, and $m_s = -\frac{1}{2}$. We then represent it with a down arrow on the level for $E_{1,2}$ and $E_{2,1}$ in Fig. 41-13*c*.

You can show that the next three electrons can also go into

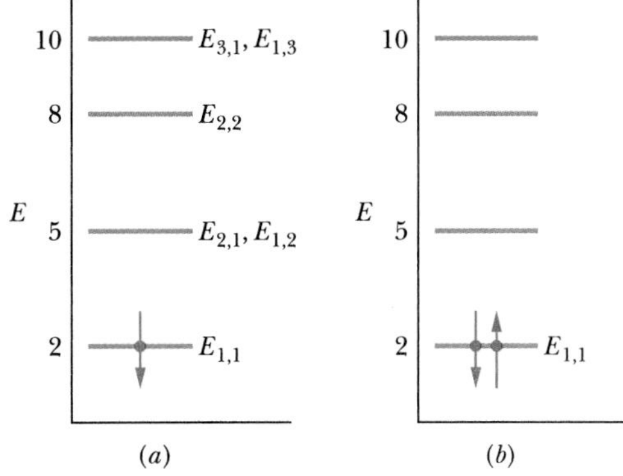

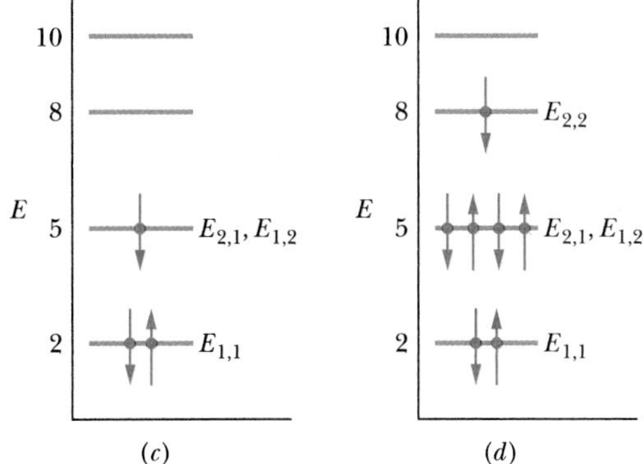

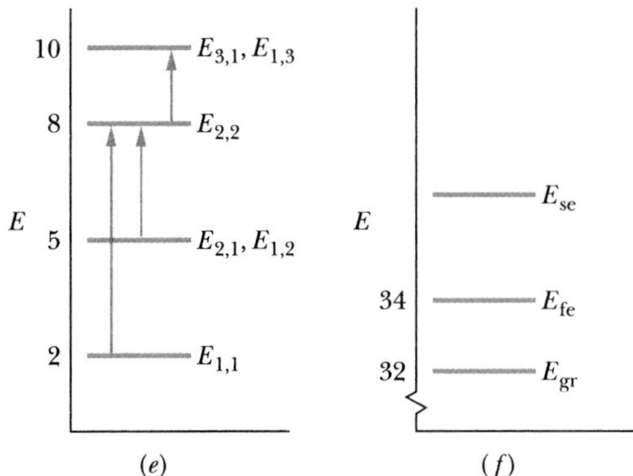

Fig. 41-13 (*a*) Energy-level diagram for one electron in a square corral of widths L. (Energy E is in multiples of $h^2/8mL^2$.) A spin-down electron occupies the lowest level. (*b*) Two electrons (one spin down, the other spin up) occupy the lowest level of the one-electron energy-level diagram. (*c*) A third electron occupies the next energy level. (*d*) The system's ground-state configuration, for all 7 electrons. (*e*) Three transitions to consider as possibly taking the 7-electron system to its first excited state. (*f*) The system's energy-level diagram, for the lowest three total energies of the system (in multiples of $h^2/8mL^2$).

TABLE 41-2 Ground-State Configuration and Energies

n_x	n_y	m_s	Energy*
2	2	$-\frac{1}{2}$	8
2	1	$+\frac{1}{2}$	5
2	1	$-\frac{1}{2}$	5
1	2	$+\frac{1}{2}$	5
1	2	$-\frac{1}{2}$	5
1	1	$+\frac{1}{2}$	2
1	1	$-\frac{1}{2}$	2
		Total	32

*In multiples of $h^2/8mL^2$

the level for energies $E_{2,1}$ and $E_{1,2}$, provided that no set of three quantum numbers is completely duplicated. That level then contains four electrons, with quantum numbers (n_x, n_y, m_s) of

$$(2, 1, -\tfrac{1}{2}), (2, 1, +\tfrac{1}{2}), (1, 2, -\tfrac{1}{2}), (1, 2, +\tfrac{1}{2}),$$

and the level is fully occupied. Thus, the seventh electron goes into the next higher level, which is the $E_{2,2}$ level. Let us arbitrarily assume it is spin down, with $m_s = -\frac{1}{2}$.

Figure 41-13*d* shows all seven electrons on a one-electron energy-level diagram. We now have seven electrons in the corral, and they are in the configuration with the lowest energy that satisfies the Pauli exclusion principle. Thus, the ground-state configuration of the system is that shown in Fig. 41-13*d* and listed in Table 41-2.

(b) What is the total energy of the seven-electron system in its ground state, as a multiple of $h^2/8mL^2$?

SOLUTION: The Key Idea here is that the total energy E_{gr} of the system in its ground state is the sum of the energies of the individual electrons in the system's ground-state configuration. The energy of each electron can be read from Table 40-1, which is partially reproduced in Table 41-2, or from Fig. 41-13*d*. Because there are two electrons in the first (lowest) level, four in the second level, and one in the third level, we have

$$E_{gr} = 2\left(2\frac{h^2}{8mL^2}\right) + 4\left(5\frac{h^2}{8mL^2}\right) + 1\left(8\frac{h^2}{8mL^2}\right) = 32\frac{h^2}{8mL^2}. \quad \text{(Answer)}$$

(c) How much energy must be transferred to the system for it to jump to its first excited state, and what is the energy of that state?

SOLUTION: The Key Ideas here are these:

1. If the system is to be excited, one of the seven electrons must make a quantum jump up the one-electron energy-level diagram of Fig. 41-13*d*.
2. If that jump is to occur, the energy change ΔE of the electron (and thus the system) must be $\Delta E = E_{high} - E_{low}$ (Eq. 40-5),

where E_{low} is the energy of the level where the jump begins and E_{high} is the energy of the level where the jump ends.

3. The Pauli exclusion principle must still apply; in particular, an electron *cannot* jump to a level that is fully occupied.

Let us consider the three jumps shown in Fig. 41-13*e*; all are allowed by the Pauli exclusion principle because they are jumps to empty or partially occupied states. In one of those possible jumps, an electron jumps from the $E_{1,1}$ level to the partially occupied $E_{2,2}$ level. The change in the energy is

$$\Delta E = E_{2,2} - E_{1,1} = 8\frac{h^2}{8mL^2} - 2\frac{h^2}{8mL^2} = 6\frac{h^2}{8mL^2}.$$

(We shall assume that the spin orientation of the electron making the jump can change as needed.)

In another of the possible jumps in Fig. 41-13*e*, an electron jumps from the degenerate level of $E_{2,1}$ and $E_{1,2}$ to the partially occupied $E_{2,2}$ level. The change in the energy is

$$\Delta E = E_{2,2} - E_{2,1} = 8\frac{h^2}{8mL^2} - 5\frac{h^2}{8mL^2} = 3\frac{h^2}{8mL^2}.$$

In the third possible jump in Fig. 41-13*e*, the electron in the $E_{2,2}$ level jumps to the unoccupied, degenerate level of $E_{1,3}$ and $E_{3,1}$. The change in energy is

$$\Delta E = E_{1,3} - E_{2,2} = 10\frac{h^2}{8mL^2} - 8\frac{h^2}{8mL^2} = 2\frac{h^2}{8mL^2}.$$

Of these three possible jumps, the one requiring the least energy change ΔE is the last one. We could consider even more possible jumps, but none would require less energy. Thus, for the system to jump from its ground state to its first excited state, the electron in the $E_{2,2}$ level must jump to the unoccupied, degenerate level of $E_{1,3}$ and $E_{3,1}$, and the required energy is

$$\Delta E = 2\frac{h^2}{8mL^2}. \quad \text{(Answer)}$$

The energy E_{fe} of the first excited state of the system is then

$$E_{\text{fe}} = E_{\text{gr}} + \Delta E$$
$$= 32\frac{h^2}{8mL^2} + 2\frac{h^2}{8mL^2} = 34\frac{h^2}{8mL^2}. \quad \text{(Answer)}$$

We can represent this energy and the energy E_{gr} for the ground state of the system on an energy-level diagram *for the system,* as shown in Fig. 41-13*f*.

41-9 Building the Periodic Table

The four quantum numbers of Table 41-1 identify the quantum states of individual electrons in a multielectron atom. The wave functions for these states, however, are not the same as the wave functions for the corresponding states of the hydrogen atom because, in multielectron atoms, the potential energy associated with a given electron is determined not only by the charge and position of the atom's nucleus but also by the charges and positions of all the other electrons in the atom. Solutions of Schrödinger's equation for multielectron atoms can be carried out numerically—in principle at least—using a computer.

As we discussed in Section 40-8, all states with the same values of the quantum numbers n and l form a subshell. For a given value of l, there are $2l + 1$ possible values of the magnetic quantum number m_l and, for each m_l, there are two possible values for the spin quantum number m_s. Thus, there are $2(2l + 1)$ states in a subshell. It turns out that *all states in a given subshell have the same energy,* its value being determined primarily by the value of n and to a lesser extent by the value of l.

For the purpose of labeling subshells, the values of l are represented by letters:

$$l = 0 \quad 1 \quad 2 \quad 3 \quad 4 \quad 5 \quad \ldots$$

$$s \quad p \quad d \quad f \quad g \quad h \quad \ldots$$

For example, the $n = 3$, $l = 2$ subshell would be labeled the $3d$ subshell.

When we assign electrons to states in a multielectron atom, we must be guided by the Pauli exclusion principle of Section 41-7; that is, no two electrons in an atom can have the same set of the quantum numbers n, l, m_l, and m_s. If this important principle did not hold, *all* the electrons in any atom could jump to the atom's lowest energy level, which would eliminate the chemistry of atoms and molecules, and thus also biochemistry. Let us examine the atoms of a few elements to see how the Pauli exclusion principle operates in the building up of the periodic table.

Neon

The neon atom has 10 electrons. Only two of them fit into the lowest energy subshell, the 1s subshell. These two electrons both have $n = 1$, $l = 0$, and $m_l = 0$, but one has $m_s = +\frac{1}{2}$ and the other has $m_s = -\frac{1}{2}$. The 1s subshell, according to Table 41-1, contains $2(2l + 1) = 2$ states. Because this subshell then contains all the electrons permitted by the Pauli principle, it is said to be **closed.**

Two of the remaining eight electrons fill the next lowest energy subshell, the 2s subshell. The last six electrons just fill the 2p subshell which, with $l = 1$, holds $2(2l + 1) = 6$ states.

In a closed subshell, all allowed z projections of the orbital angular momentum vector $\vec{L}$ are present and, as you can verify from Fig. 41-5, these projections cancel for the subshell as a whole; for every positive projection there is a corresponding negative projection of the same magnitude. Similarly, the z projections of the spin angular momenta also cancel. Thus, a closed subshell has no angular momentum and no magnetic moment of any kind. Furthermore, its probability density is spherically symmetric. Then neon with its three closed subshells (1s, 2s, and 2p) has no "loosely dangling electrons" to encourage chemical interaction with other atoms. Neon, like the other **noble gases** that form the right-hand column of the periodic table, is chemically inert.

Sodium

Next after neon in the periodic table comes sodium, with 11 electrons. Ten of them form a closed neonlike core, which, as we have seen, has zero angular momentum. The remaining electron is largely outside this inert core, in the 3s subshell—the next lowest energy subshell. Because this **valence electron** of sodium is in a state with $l = 0$ (that is, an s state), the sodium atom's angular momentum and magnetic dipole moment must be due entirely to the spin of this single electron.

Sodium readily combines with other atoms that have a "vacancy" into which sodium's loosely bound valence electron can fit. Sodium, like the other **alkali metals** that form the left-hand column of the periodic table, is chemically active.

Chlorine

The chlorine atom, which has 17 electrons, has a closed 10-electron, neonlike core, with 7 electrons left over. Two of them fill the 3s subshell, leaving five to be assigned to the 3p subshell, which is the subshell next lowest in energy. This subshell, which has $l = 1$, can hold $2(2l + 1) = 6$ electrons, so there is a vacancy, or a "hole," in this subshell.

Chlorine is receptive to interacting with other atoms that have a valence electron that might fill this hole. Sodium chloride (NaCl), for example, is a very stable compound. Chlorine, like the other **halogens** that form column VIIA of the periodic table, is chemically active.

Iron

The arrangement of the 26 electrons of the iron atom can be represented as follows:

$$\underbrace{1s^2 \quad 2s^2\ 2p^6 \quad 3s^2\ 3p^6}\,3d^6 \quad 4s^2.$$

The subshells are listed in numerical order and, following convention, a superscript gives the number of electrons in each subshell. From Table 41-1 we can see that an

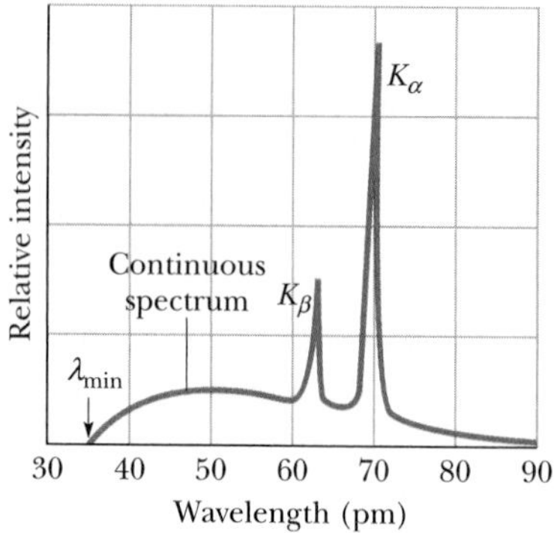

Fig. 41-14 The distribution by wavelength of the x rays produced when 35 keV electrons strike a molybdenum target. The sharp peaks and the continuous spectrum from which they rise are produced by different mechanisms.

s-subshell can hold 2 electrons, a *p*-subshell 6, and a *d*-subshell 10. Thus, iron's first 18 electrons form the five filled subshells that are marked off by the bracket, leaving 8 electrons to be accounted for. Six of the eight go into the 3*d* subshell and the remaining two go into the 4*s* subshell.

The last two electrons do not also go into the 3*d* subshell (which can hold 10 electrons) because the $3d^6\,4s^2$ configuration results in a lower energy state for the atom as a whole than would the $3d^8$ configuration. An iron atom with 8 electrons (rather than 6) in the 3*d* subshell would quickly make a transition to the $3d^6\,4s^2$ configuration, emitting electromagnetic radiation in the process. The lesson here is that except for the simplest elements, the states may not be filled in what we might think of as their "logical" sequence.

41-10 X Rays and the Numbering of the Elements

When a solid target, such as solid copper or tungsten, is bombarded with electrons whose kinetic energies are in the kiloelectron-volt range, electromagnetic radiation called **x rays** is emitted. Our concern here is what these rays—whose medical, dental, and industrial usefulness is so well known and widespread—can teach us about the atoms that absorb or emit them. Figure 41-14 shows the wavelength spectrum of the x rays produced when a beam of 35 keV electrons falls on a molybdenum target. We see a broad, continuous spectrum of radiation on which are superimposed two peaks of sharply defined wavelengths. The continuous spectrum and the peaks arise in different ways, which we next discuss separately.

The Continuous X-Ray Spectrum

Here we examine the continuous x-ray spectrum of Fig. 41-14, ignoring for the time being the two prominent peaks that rise from it. Consider an electron of initial kinetic energy K_0 that collides (interacts) with one of the target atoms, as in Fig. 41-15. The electron may lose an amount of energy ΔK, which will appear as the energy of an x-ray photon that is radiated away from the site of the collision. (Very little energy is transferred to the recoiling atom because of the relatively large mass of the atom; here we neglect that transfer.)

The scattered electron in Fig. 41-15, whose energy is now less than K_0, may have a second collision with a target atom, generating a second photon, whose energy will in general be different from the energy of the photon produced in the first collision. This electron-scattering process can continue until the electron is approximately stationary. All the photons generated by these collisions form part of the continuous x-ray spectrum.

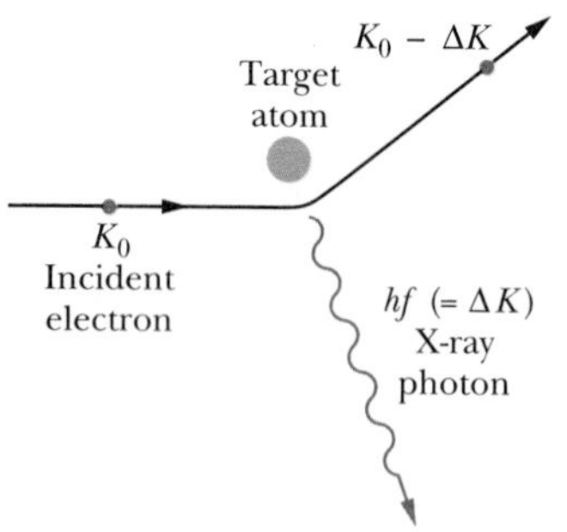

Fig. 41-15 An electron of kinetic energy K_0 passing near an atom in the target may generate an x-ray photon, the electron losing part of its energy in the process. The continuous x-ray spectrum arises in this way.

A prominent feature of that spectrum in Fig. 41-14 is the sharply defined **cutoff wavelength** λ_{min}, below which the continuous spectrum does not exist. This minimum wavelength corresponds to a collision in which an incident electron loses *all* its initial kinetic energy K_0 in a single head-on collision with a target atom. Essentially all this energy appears as the energy of a single photon, whose associated wavelength—the minimum possible x-ray wavelength—is found from

$$K_0 = hf = \frac{hc}{\lambda_{min}},$$

or

$$\lambda_{min} = \frac{hc}{K_0} \quad \text{(cutoff wavelength).} \tag{41-23}$$

The cutoff wavelength is totally independent of the target material. If we were to switch from a molybdenum target to a copper target, for example, all features of the x-ray spectrum of Fig. 41-14 would change *except* the cutoff wavelength.

CHECKPOINT 2: Does the cutoff wavelength λ_{min} of the continuous x-ray spectrum increase, decrease, or remain the same if you (a) increase the kinetic energy of the electrons that strike the x-ray target, (b) allow the electrons to strike a thin foil rather than a thick block of the target material, (c) change the target to an element of higher atomic number?

Sample Problem 41-4

A beam of 35.0 keV electrons strikes a molybdenum target, generating the x rays whose spectrum is shown in Fig. 41-14. What is the cutoff wavelength?

SOLUTION: The **Key Idea** here is that the cutoff wavelength λ_{min} corresponds to an electron transferring (approximately) all of its energy to an x-ray photon, thus producing a photon with the greatest possible frequency and least possible wavelength. From Eq. 41-23, we have

$$\lambda_{min} = \frac{hc}{K_0} = \frac{(4.14 \times 10^{-15}\ \text{eV}\cdot\text{s})(3.00 \times 10^8\ \text{m/s})}{35.0 \times 10^3\ \text{eV}}$$

$$= 3.55 \times 10^{-11}\ \text{m} = 35.5\ \text{pm}. \quad \text{(Answer)}$$

The Characteristic X-Ray Spectrum

We now turn our attention to the two peaks of Fig. 41-14, labeled K_α and K_β. These (and other peaks that appear at wavelengths beyond the wavelength range displayed in Fig. 41-14) form the **characteristic x-ray spectrum** of the target material.

The peaks arise in a two-part process. (1) An energetic electron strikes an atom in the target and, while it is being scattered, the incident electron knocks out one of the atom's deep-lying (low n value) electrons. If the deep-lying electron is in the shell defined by $n = 1$ (called, for historical reasons, the K shell), there remains a vacancy, or *hole,* in this shell. (2) An electron in one of the shells with a higher energy jumps to the K shell, filling the hole in this shell. During this jump, the atom emits a characteristic x-ray photon. If the electron that fills the K-shell vacancy jumps from the shell with $n = 2$ (called the L shell), the emitted radiation is the K_α line of Fig. 41-14; if it jumps from the shell with $n = 3$ (called the M shell), it produces the K_β line, and so on. The hole left in either the L or M shell will be filled by an electron from still farther out in the atom.

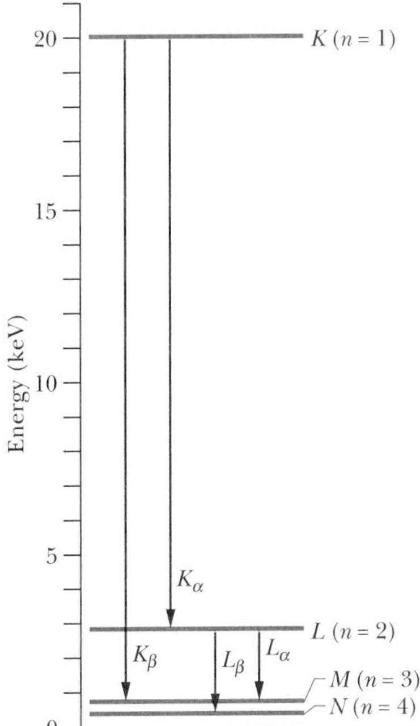

Fig. 41-16 A simplified atomic energy-level diagram for molybdenum, showing the transitions (of holes rather than electrons) that give rise to some of the characteristic x rays of that element. Each horizontal line represents the energy of the atom with a hole (a missing electron) in the shell indicated.

In studying x rays, it is more convenient to keep track of the hole created deep in the atom's "electron cloud" than to record the changes in the quantum state of the electrons that jump to fill that hole. Figure 41-16 does exactly that; it is an energy-level diagram for molybdenum, the element to which Fig. 41-14 refers. The baseline ($E = 0$) represents the neutral atom in its ground state. The level marked K (at $E =$ 20 keV) represents the energy of the molybdenum atom with a hole in its K shell. Similarly, the level marked L (at $E = 2.7$ keV) represents the atom with a hole in its L shell, and so on.

The transitions marked K_α and K_β in Fig. 41-16 are the ones that produce the two x-ray peaks in Fig. 41-14. The K_α spectral line, for example, originates when an electron from the L shell fills a hole in the K shell. In Fig. 41-16, this jump corresponds to a *downward* transition of the hole, from the K level to the L level.

Numbering the Elements

In 1913 British physicist H. G. J. Moseley generated characteristic x rays for as many elements as he could find—he found 38—by using them as targets for electron bombardment in an evacuated tube of his own design. By means of a trolley manipulated by strings, Moseley was able to move the individual targets into the path of an electron beam. He measured the wavelengths of the emitted x rays by the crystal diffraction method described in Section 37-9.

Moseley then sought (and found) regularities in these spectra as he moved from element to element in the periodic table. In particular, he noted that if, for a given spectral line such as K_α, he plotted for each element the square root of the frequency f against the position of the element in the periodic table, a straight line resulted. Figure 41-17 shows a portion of his extensive data. Moseley's conclusion was this:

> We have here a proof that there is in the atom a fundamental quantity, which increases by regular steps as we pass from one element to the next. This quantity can only be the charge on the central nucleus.

Owing to Moseley's work, the characteristic x-ray spectrum became the universally accepted signature of an element, permitting the solution of a number of periodic table puzzles. Prior to that time (1913), the positions of elements in the table were assigned in order of atomic *weight,* although it was necessary to invert this order for several pairs of elements because of compelling chemical evidence; Moseley showed that it is the nuclear charge (that is, the atomic *number* Z) that is the real basis for numbering the elements.

In 1913 the periodic table had several empty squares, and a surprising number of claims for new elements had been advanced. The x-ray spectrum provided a conclusive test of such claims. The lanthanide elements, often called the rare earth elements, had been sorted out only imperfectly because their similar chemical properties made sorting difficult. Once Moseley's work was reported, these elements were properly organized. In more recent times, the identities of some elements beyond uranium were pinned down beyond dispute when the elements became available in quantities large enough to permit a study of their individual x-ray spectra.

It is not hard to see why the characteristic x-ray spectrum shows such impressive regularities from element to element whereas the optical spectrum in the visible and near-visible region does not: The key to the identity of an element is the charge on its nucleus. Gold, for example, is what it is because its atoms have a nuclear charge of $+79e$ (that is, $Z = 79$). An atom with one more elementary charge on its nucleus is mercury; with one fewer, it is platinum. The K electrons, which play such a large role in the production of the x-ray spectrum, lie very close to the nucleus and are thus sensitive probes of its charge. The optical spectrum, on the other hand, involves transitions of the outermost electrons, which are heavily screened from the nucleus by the remaining electrons of the atom and thus are *not* sensitive probes of nuclear charge.

Accounting for the Moseley Plot

Moseley's experimental data, of which the Moseley plot of Fig. 41-17 is but a part, can be used directly to assign the elements to their proper places in the periodic table. This can be done even if no theoretical basis for Moseley's results can be established. However, there is such a basis.

According to Eq. 40-24 the energy of the hydrogen atom is

$$E_n = -\frac{me^4}{8\varepsilon_0^2 h^2}\frac{1}{n^2} = -\frac{13.6\text{ eV}}{n^2}, \qquad \text{for } n = 1, 2, 3, \ldots. \tag{41-24}$$

Consider now one of the two innermost electrons in the K shell of a multielectron atom. Because of the presence of the other K-shell electron, our electron "sees" an effective nuclear charge of approximately $(Z - 1)e$, where e is the elementary charge and Z is the atomic number of the element. The factor e^4 in Eq. 41-24 is the product

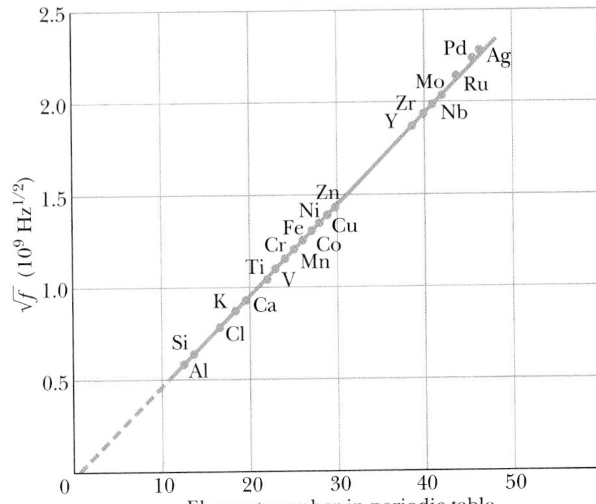

Fig. 41-17 A Moseley plot of the K_α line of the characteristic x-ray spectra of 21 elements. The frequency is calculated from the measured wavelength.

of e^2—the square of hydrogen's nuclear charge—and $(-e)^2$—the square of an electron's charge. For a multielectron atom, we can approximate the effective energy of the atom by replacing the factor e^4 in Eq. 41-24 with $(Z - 1)^2 e^2 \times (-e)^2$, or $e^4(Z - 1)^2$. That gives us

$$E_n = -\frac{(13.6 \text{ eV})(Z - 1)^2}{n^2}. \tag{41-25}$$

We saw that the K_α x-ray photon (of energy hf) arises when an electron makes a transition from the L shell (with $n = 2$ and energy E_2) to the K shell (with $n = 1$ and energy E_1). Thus, using Eq. 41-25, we may write the energy change as

$$\begin{aligned} \Delta E &= E_2 - E_1 \\ &= \frac{-(13.6 \text{ eV})(Z - 1)^2}{2^2} - \frac{-(13.6 \text{ eV})(Z - 1)^2}{1^2} \\ &= (10.2 \text{ eV})(Z - 1)^2. \end{aligned}$$

Then the frequency f of the K_α line is

$$\begin{aligned} f = \frac{\Delta E}{h} &= \frac{(10.2 \text{ eV})(Z - 1)^2}{(4.14 \times 10^{-15} \text{ eV} \cdot \text{s})} \\ &= (2.46 \times 10^{15} \text{ Hz})(Z - 1)^2. \end{aligned} \tag{41-26}$$

Taking the square root of both sides yields

$$\sqrt{f} = CZ - C, \tag{41-27}$$

in which C is a constant ($= 4.96 \times 10^7$ Hz$^{1/2}$). Equation 41-27 is the equation of a straight line. It shows that if we plot the square root of the frequency of the K_α x-ray spectral line against the atomic number Z, we should obtain a straight line. As Fig. 41-17 shows, that is exactly what Moseley found.

CHECKPOINT 3: The K_α x rays arising from a cobalt ($Z = 27$) target have a wavelength of about 179 pm. Is the wavelength of the K_α x rays arising from a nickel ($Z = 28$) target greater than or less than 179 pm?

Sample Problem 41-5

A cobalt target is bombarded with electrons, and the wavelengths of its characteristic x-ray spectrum are measured. There is also a second, fainter characteristic spectrum, which is due to an impurity in the cobalt. The wavelengths of the K_α lines are 178.9 pm (cobalt) and 143.5 pm (impurity), and the proton number for cobalt is $Z_{\text{Co}} = 27$. Determine the impurity using only these data.

SOLUTION: The **Key Idea** here is that the wavelengths of the K_α lines for both the cobalt (Co) and the impurity (X) fall on a K_α Moseley plot, and Eq. 41-27 is the equation for that plot. Substituting c/λ for f in that equation, we obtain

$$\sqrt{\frac{c}{\lambda_{\text{Co}}}} = CZ_{\text{Co}} - C \quad \text{and} \quad \sqrt{\frac{c}{\lambda_{\text{X}}}} = CZ_{\text{X}} - C.$$

Dividing the second equation by the first neatly eliminates C, yielding

$$\sqrt{\frac{\lambda_{\text{Co}}}{\lambda_{\text{X}}}} = \frac{Z_{\text{X}} - 1}{Z_{\text{Co}} - 1}.$$

Substituting the given data yields

$$\sqrt{\frac{178.9 \text{ pm}}{143.5 \text{ pm}}} = \frac{Z_{\text{X}} - 1}{27 - 1}.$$

Solving for the unknown, we find that

$$Z_{\text{X}} = 30.0. \qquad \text{(Answer)}$$

A glance at the periodic table identifies the impurity as zinc.

41-11 Lasers and Laser Light

In the late 1940s and again in the early 1960s, quantum physics made two enormous contributions to technology: the **transistor,** which ushered in the computer revolution, and the **laser.** Laser light, like the light from an ordinary lightbulb, is emitted when atoms make a transition from one quantum state to a quantum state of lower energy. In a laser, however—but not in other light sources—the atoms act together to produce light with several special characteristics:

1. ***Laser light is highly monochromatic.*** Light from an ordinary incandescent lightbulb is spread over a continuous range of wavelengths and is certainly not monochromatic. The radiation from a fluorescent neon sign *is* monochromatic, to about 1 part in about 10^6. However, the sharpness of definition of laser light can be many times greater, as much as 1 part in 10^{15}.
2. ***Laser light is highly coherent.*** Individual long waves (*wave trains*) for laser light can be several hundred kilometers long. When two separated beams that have traveled such distances over separate paths are recombined, they "remember" their common origin and are able to form a pattern of interference fringes. The corresponding *coherence length* for wave trains emitted by a lightbulb is typically less than a meter.
3. ***Laser light is highly directional.*** A laser beam spreads very little; it departs from strict parallelism only because of diffraction at the exit aperture of the laser. For example, a laser pulse used to measure the distance to the Moon generates a spot on the Moon's surface with a diameter of only a few meters. Light from an ordinary bulb can be made into an approximately parallel beam by a lens, but the beam divergence is much greater than for laser light. Each point on a lightbulb's filament forms its own separate beam, and the angular divergence of the overall composite beam is set by the size of the filament.
4. ***Laser light can be sharply focused.*** If two light beams transport the same amount of energy, the beam that can be focused to the smaller spot will have the greater intensity at that spot. For laser light, the focused spot can be so small that an intensity of 10^{17} W/cm^2 is readily obtained. An oxyacetylene flame, by contrast, has an intensity of only about 10^3 W/cm^2.

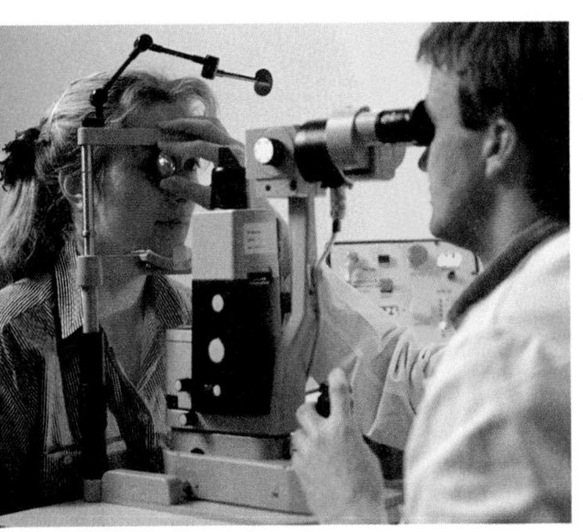

Fig. 41-18 A patient's loose retina is being welded into place by a laser directed into her eye.

Lasers Have Many Uses

The smallest lasers, used for voice and data transmission over optical fibers, have as their active medium a semiconducting crystal about the size of a pinhead. Small as they are, such lasers can generate about 200 mW of power. The largest lasers, used for nuclear fusion research and for astronomical and military applications, fill a large building. The largest such laser can generate brief pulses of laser light with a power level, during the pulse, of about 10^{14} W. This is a few hundred times greater than the total electric power generating capacity of the United States. To avoid a brief national power blackout during a pulse, the energy required for each pulse is stored up at a steady rate during the relatively long interpulse interval.

Among the many uses of lasers are reading bar codes, manufacturing and reading compact discs, performing surgery of many kinds (see the opening photo of this chapter and Fig. 41-18), surveying, cutting cloth in the garment industry (several hundred layers at a time), welding auto bodies, and generating holograms.

41-12 How Lasers Work

The word "laser" is an acronym for "light amplification by the stimulated emission of radiation," so you should not be surprised that **stimulated emission** is the key to laser operation. Einstein introduced this concept in 1917. Although the world had to wait until 1960 to see an operating laser, the groundwork for its development was put in place decades earlier.

Consider an isolated atom that can exist either in its state of lowest energy (its ground state), whose energy is E_0, or in a state of higher energy (an excited state), whose energy is E_x. Here are three processes by which the atom can move from one of these states to the other:

1. ***Absorption.*** Figure 41-19*a* shows the atom initially in its ground state. If the atom is placed in an electromagnetic field that is alternating at frequency f, the atom can absorb an amount of energy hf from that field and move to the higher energy state. From the principle of conservation of energy we have

$$hf = E_x - E_0. \tag{41-28}$$

   We call this process **absorption.**

2. ***Spontaneous emission.*** In Fig. 41-19*b* the atom is in its excited state and no external radiation is present. After a time, the atom will move *of its own accord* to its ground state, emitting a photon of energy hf in the process. We call this process **spontaneous emission**—*spontaneous* because the event was not triggered by any outside influence. The light from the filament of an ordinary lightbulb is generated in this way.

   Normally, the mean life of excited atoms before spontaneous emission occurs is about 10^{-8} s. However, for some excited states, this mean life is perhaps as much as 10^5 times longer. We call such long-lived states **metastable;** they play an important role in laser operation.

3. ***Stimulated emission.*** In Fig. 41-19*c* the atom is again in its excited state, but this time radiation with a frequency given by Eq. 41-28 is present. A photon of energy hf can stimulate the atom to move to its ground state, during which process the atom emits an additional photon, whose energy is also hf. We call this process **stimulated emission**—*stimulated* because the event is triggered by the external photon. The emitted photon is in every way identical to the stimulating photon.

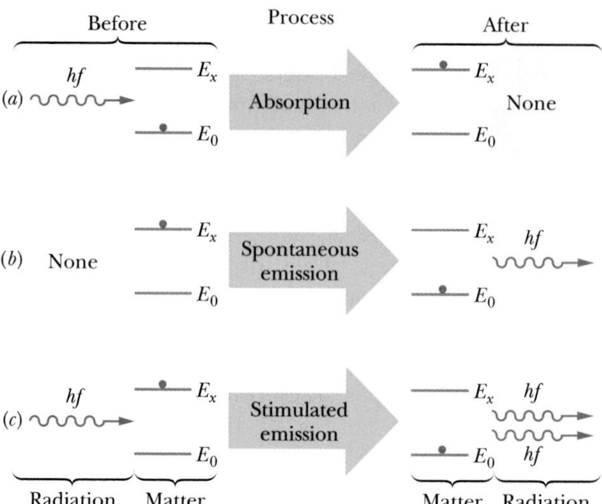

Fig. 41-19 The interaction of radiation and matter in the processes of (*a*) absorption, (*b*) spontaneous emission, and (*c*) stimulated emission. An atom (matter) is represented by the red dot; the atom is in either a lower quantum state with energy E_0 or a higher quantum state with energy E_x. In (*a*) the atom absorbs a photon of energy *hf* from a passing light wave. In (*b*) it emits a light wave by emitting a photon of energy *hf*. In (*c*) a passing light wave with photon energy *hf* causes the atom to emit a photon of the same energy, increasing the energy of the light wave.

Thus, the waves associated with the photons have the same energy, phase, polarization, and direction of travel.

Figure 41-19*c* describes stimulated emission for a single atom. Suppose now that a sample contains a large number of atoms in thermal equilibrium at temperature T. Before any radiation is directed at the sample, a number N_0 of these atoms are in their ground state with energy E_0, and a number N_x are in a state of higher energy E_x. Ludwig Boltzmann showed that N_x is given in terms of N_0 by

$$N_x = N_0 e^{-(E_x - E_0)/kT}, \tag{41-29}$$

in which k is Boltzmann's constant. This equation seems reasonable. The quantity kT is the mean kinetic energy of an atom at temperature T. The higher the temperature, the more atoms—on average—will have been "bumped up" by thermal agitation (that is, by atom–atom collisions) to the higher energy state E_x. Also, because $E_x > E_0$, Eq. 41-29 requires that $N_x < N_0$; that is, there will always be fewer atoms in the excited state than in the ground state. This is what we expect if the level populations N_0 and N_x are determined only by the action of thermal agitation. Figure 41-20*a* illustrates this situation.

If we now flood the atoms of Fig. 41-20*a* with photons of energy $E_x - E_0$, photons will disappear via absorption by ground-state atoms, and photons will be generated largely via stimulated emission of excited-state atoms. Einstein showed that the probabilities per atom for these two processes are identical. Thus, because there are more atoms in the ground state, the *net* effect will be the absorption of photons.

To produce laser light, we must have more photons emitted than absorbed; that is, we must have a situation in which stimulated emission dominates. The direct way to bring this about is to start with more atoms in the excited state than in the ground state, as in Fig. 41-20*b*. However, since such a **population inversion** is not consistent with thermal equilibrium, we must think up clever ways to set up and maintain one.

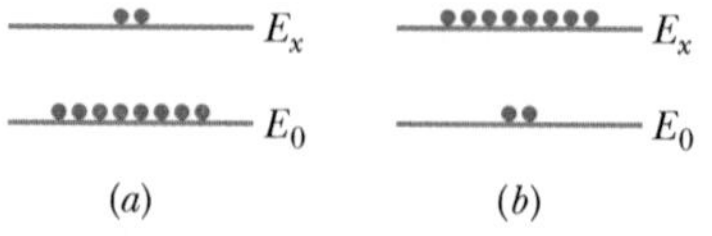

Fig. 41-20 (*a*) The equilibrium distribution of atoms between the ground state E_0 and excited state E_x, accounted for by thermal agitation. (*b*) An inverted population, obtained by special methods. Such an inverted population is essential for laser action.

The Helium–Neon Gas Laser

Figure 41-21 shows a type of laser commonly found in student laboratories. It was developed in 1961 by Ali Javan and his coworkers. The glass discharge tube is filled

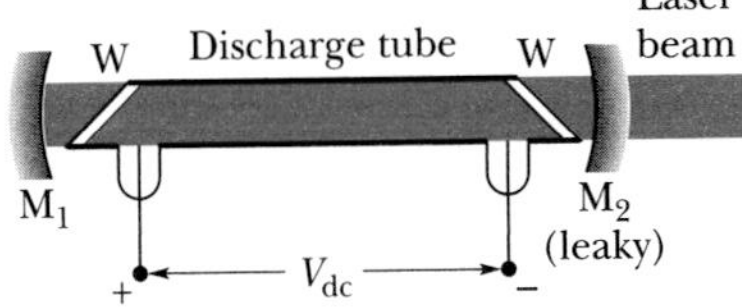

Fig. 41-21 The elements of a helium–neon gas laser. An applied potential V_{dc} sends electrons through a discharge tube containing a mixture of helium gas and neon gas. Electrons collide with helium atoms, which then collide with neon atoms, which emit light along the length of the tube. The light passes through transparent windows W and reflects back and forth through the tube from mirrors M_1 and M_2 to cause more neon atom emissions. Some of the light leaks through mirror M_2 to form the laser beam.

with a 20 : 80 mixture of helium and neon gases, neon being the medium in which laser action occurs.

Figure 41-22 shows simplified energy-level diagrams for the two atoms. An electric current passed through the helium–neon gas mixture serves—through collisions between helium atoms and electrons of the current—to raise many helium atoms to state E_3, which is metastable.

The energy of helium state E_3 (20.61 eV) is very close to the energy of neon state E_2 (20.66 eV). Thus, when a metastable (E_3) helium atom and a ground-state (E_0) neon atom collide, the excitation energy of the helium atom is often transferred to the neon atom, which then moves to state E_2. In this manner, neon level E_2 in Fig. 41-22 can become more heavily populated than neon level E_1.

This population inversion is relatively easy to set up because (1) initially there are essentially no neon atoms in state E_1, (2) the metastability of helium level E_3 ensures a ready supply of neon atoms in level E_2, and (3) atoms in level E_1 decay rapidly (through intermediate levels not shown) to the neon ground state E_0.

Suppose now that a single photon is spontaneously emitted as a neon atom transfers from state E_2 to state E_1. Such a photon can trigger a stimulated emission event, which, in turn, can trigger other stimulated emission events. Through such a chain reaction, a coherent beam of red laser light, moving parallel to the tube axis, can build up rapidly. This light, of wavelength 632.8 nm, moves through the discharge tube many times by successive reflections from mirrors M_1 and M_2 (Fig. 41-21), accumulating additional stimulated emission photons with each passage. M_1 is totally reflecting but M_2 is slightly "leaky" so that a small fraction of the laser light escapes to form a useful external beam.

CHECKPOINT 4: The wavelength of light from laser *A* (a helium–neon gas laser) is 632.8 nm; that from laser *B* (a carbon dioxide gas laser) is 10.6 μm. That from laser *C* (a gallium arsenide semiconductor laser) is 840 nm. Rank these lasers according to the energy interval between the two quantum states responsible for laser action, greatest first.

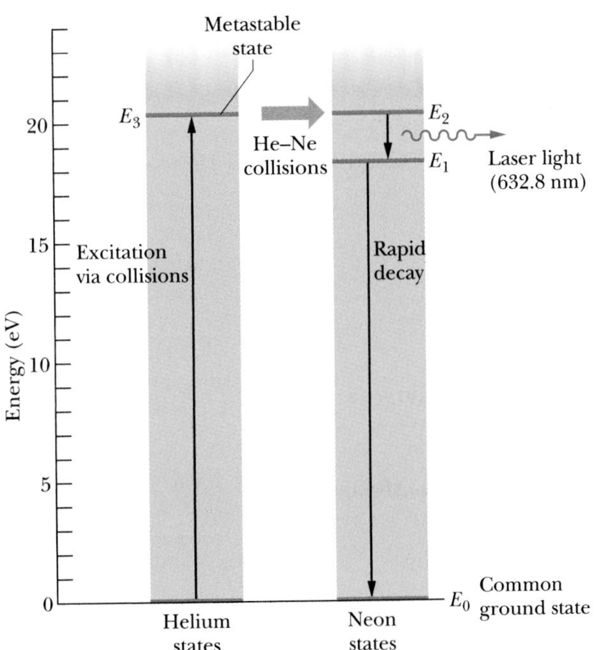

Fig. 41-22 Four essential energy levels for helium and neon atoms in a helium–neon gas laser. Laser action occurs between levels E_2 and E_1 of neon when more atoms are at the E_2 level than at the E_1 level.

Sample Problem 41-6

In the helium–neon laser of Fig. 41-21, laser action occurs between two excited states of the neon atom. However, in many lasers, laser action (*lasing*) occurs between the ground state and an excited state, as suggested in Fig. 41-20.

(a) Consider such a laser that emits at wavelength $\lambda = 550$ nm. If a population inversion is not generated, what is the ratio of the population of atoms in state E_x to the population in the ground state E_0, with the atoms at room temperature?

SOLUTION: One Key Idea here is that the naturally occurring population ratio N_x/N_0 of the two states is due to thermal agitation among the gas atoms, according to Eq. 41-29, which we can write as

$$N_x/N_0 = e^{-(E_x-E_0)/kT}. \qquad (41\text{-}30)$$

To find N_x/N_0 with Eq. 41-30, we need to find the energy separation $E_x - E_0$ between the two states. Here we use another Key Idea: We can obtain $E_x - E_0$ from the given wavelength of 550 nm for the lasing between those two states. We find

$$\begin{aligned} E_x - E_0 &= hf = \frac{hc}{\lambda} \\ &= \frac{(6.63 \times 10^{-34}\ \text{J}\cdot\text{s})(3.00 \times 10^8\ \text{m/s})}{(550 \times 10^{-9}\ \text{m})(1.60 \times 10^{-19}\ \text{J/eV})} \\ &= 2.26\ \text{eV}. \end{aligned}$$

To solve Eq. 41-30, we also need the mean energy of thermal agitation kT for an atom at room temperature (assumed to be 300 K), which is

$$kT = (8.62 \times 10^{-5}\ \text{eV/K})(300\ \text{K}) = 0.0259\ \text{eV}.$$

Substituting the last two results into Eq. 41-30 gives us the population ratio at room temperature:

$$\begin{aligned} N_x/N_0 &= e^{-(2.26\ \text{eV})/(0.0259\ \text{eV})} \\ &\approx 1.3 \times 10^{-38}. \end{aligned} \qquad \text{(Answer)}$$

This is an extremely small number. It is not unreasonable, however. An atom whose mean thermal agitation energy is only 0.0259 eV will not often impart an energy of 2.26 eV to another atom in a collision.

(b) For the conditions of (a), at what temperature would the ratio N_x/N_0 be 1/2?

SOLUTION: The two Key Ideas of (a) apply here but this time we want the temperature T such that thermal agitation has bumped enough neon atoms up to the higher energy state to give $N_x/N_0 = 1/2$. Substituting that ratio into Eq. 41-30, taking the natural logarithm of both sides, and solving for T yield

$$\begin{aligned} T &= \frac{E_x - E_0}{k(\ln 2)} = \frac{2.26\ \text{eV}}{(8.62 \times 10^{-5}\ \text{eV/K})(\ln 2)} \\ &= 38\ 000\ \text{K}. \end{aligned} \qquad \text{(Answer)}$$

This is much hotter than the surface of the Sun. It is clear that if we are to invert the populations of these two levels, some specific mechanism for bringing this about is needed—that is, we must "pump" the atoms. No temperature, however high, will naturally generate a population inversion by thermal agitation.

REVIEW & SUMMARY

Some Properties of Atoms The energies of atoms are quantized; that is, the atoms have only certain specific values of energy associated with different quantum states. Atoms can make transitions between different quantum states by emitting or absorbing a photon; the frequency f associated with that light is given by

$$hf = E_{\text{high}} - E_{\text{low}}, \qquad (41\text{-}1)$$

where E_{high} is the higher energy and E_{low} is the lower energy of the pair of quantum states involved in the transition. Atoms also have quantized angular momenta and magnetic dipole moments.

Angular Momenta and Magnetic Dipole Moments An electron trapped in an atom has an *orbital angular momentum* $\vec{L}$ with a magnitude given by

$$L = \sqrt{l(l+1)}\hbar, \qquad (41\text{-}2)$$

where l is the *orbital quantum number* (which can have the values given by Table 41-1) and where the constant "h-bar" is $\hbar = h/2\pi$. The projection L_z of $\vec{L}$ on an arbitrary z axis is quantized and measurable and can have the values

$$L_z = m_l\hbar, \qquad (41\text{-}7)$$

where m_l is the *orbital magnetic quantum number* (which can have the values given by Table 41-1).

A magnetic dipole is associated with the angular momentum $\vec{L}$ of an electron in an atom. This magnetic dipole has an **orbital magnetic dipole moment** $\vec{\mu}_{\text{orb}}$ that is directed opposite $\vec{L}$:

$$\vec{\mu}_{\text{orb}} = -\frac{e}{2m}\vec{L}, \qquad (41\text{-}3)$$

where the minus sign indicates opposite directions. The projection $\mu_{\text{orb},z}$ of the orbital magnetic dipole moment on the z axis is quantized and measurable and can have the values

$$\mu_{\text{orb},z} = -m_l\mu_B, \qquad (41\text{-}5)$$

where μ_B is the *Bohr magneton:*

$$\mu_B = \frac{eh}{4\pi m} = 9.274 \times 10^{-24}\ \text{J/T}. \qquad (41\text{-}6)$$

An electron, whether trapped or free, has an intrinsic *spin angular momentum* (or just *spin*) $\vec{S}$ with a magnitude given by

$$S = \sqrt{s(s+1)}\hbar, \qquad (41\text{-}9)$$

where s is the *spin quantum number* of the electron, which is always $\frac{1}{2}$. The projection S_z of $\vec{S}$ on an arbitrary z axis is quantized and measurable and can have the values

$$S_z = m_s\hbar, \qquad (41\text{-}12)$$

where m_s is the *spin magnetic quantum number* of the electron, which can be $+\frac{1}{2}$ or $-\frac{1}{2}$.

An electron has an intrinsic magnetic dipole that is associated with its spin angular momentum $\vec{S}$, whether the electron is confined to an atom or free. This magnetic dipole has a **spin magnetic dipole moment** $\vec{\mu}_s$ that is directed opposite $\vec{S}$:

$$\vec{\mu}_s = -\frac{e}{m}\vec{S}. \qquad (41\text{-}10)$$

The projection $\mu_{s,z}$ of the spin magnetic dipole moment $\vec{\mu}_s$ on an arbitrary z axis is quantized and measurable and can have the values

$$\mu_{s,z} = -2m_s\mu_B. \qquad (41\text{-}13)$$

Spin and Magnetic Resonance A proton has an intrinsic spin angular momentum $\vec{S}$ and an associated spin magnetic dipole moment $\vec{\mu}$ that is always in the *same* direction as $\vec{S}$. If a proton is located in an external magnetic field $\vec{B}$, the projection μ_z of $\vec{\mu}$ on an axis z (defined to be along the direction of $\vec{B}$) can have only two quantized orientations: parallel to $\vec{B}$ or antiparallel to $\vec{B}$. The energy difference between these orientations is $2\mu_z B$. The energy required of a photon to *spin-flip* the proton between the two orientations is

$$hf = 2\mu_z(B_{ext} + B_{local}), \qquad (41\text{-}22)$$

where B_{ext} now represents the external field and B_{local} is the local magnetic field set up by the atoms and nuclei surrounding the proton. Detection of such spin flips can lead to *nuclear magnetic resonance spectra* by which specific substances can be identified.

Pauli Exclusion Principle Electrons in atoms and other traps obey the **Pauli exclusion principle,** which requires that *no two electrons in the same atom or any other type of trap can have the same set of quantum numbers.*

Building the Periodic Table The elements are listed in the periodic table in order of increasing atomic number Z; the nuclear charge is Ze, and Z is both the number of protons in the nucleus and the number of electrons in the neutral atom.

States with the same value of n form a **shell,** and those with the same values of both n and l form a **subshell.** In *closed* shells and subshells, which are those that contain the maximum number of electrons, the angular momenta and the magnetic moments of the individual electrons sum to zero.

X Rays and the Numbering of the Elements A **continuous spectrum** of x rays is emitted when high-energy electrons lose some of their energy in a collision with atomic nuclei. The **cutoff wavelength** λ_{min} is the wavelength emitted when such electrons lose *all* their initial energy in a single such encounter and is

$$\lambda_{min} = \frac{hc}{K_0}, \qquad (41\text{-}23)$$

in which K_0 is the initial kinetic energy of the electrons that strike the target.

The **characteristic x ray spectrum** arises when high-energy electrons eject electrons from deep within the atom; when a resulting "hole" is filled by an electron from farther out in the atom, a photon of the characteristic x-ray spectrum is generated.

In 1913 British physicist H. G. J. Moseley measured the frequencies of the characteristic x rays from a number of elements. He noted that when the square root of the frequency is plotted against the position of the element in the periodic table, a straight line results, as in the **Moseley plot** of Fig. 41-17. This allowed Moseley to conclude that the property that determines the position of an element in the periodic table is not its atomic mass but its **atomic number** Z—that is, the number of protons in its nucleus.

Lasers and Laser Light Laser light arises by **stimulated emission.** That is, radiation of a frequency given by

$$hf = E_x - E_0 \qquad (41\text{-}28)$$

can cause an atom to undergo a transition from an upper energy level (of energy E_x) to a lower energy level with a photon of frequency f being emitted. The stimulating photon and the emitted photon are identical in every respect and combine to form laser light.

For the emission process to predominate, there must normally be a **population inversion;** that is, there must be more atoms in the upper energy level than in the lower one.

QUESTIONS

1. An electron in an atom of gold is in a state with $n = 4$. Which of these values of l are possible for it: $-3, 0, 2, 3, 4, 5$?

2. An atom of silver has closed $3d$ and $4d$ subshells. Which subshell has the greater number of electrons, or do they have the same number?

3. An atom of uranium has closed $6p$ and $7s$ subshells. Which subshell has the greater number of electrons?

4. An electron in a mercury atom is in the $3d$ subshell. Which values of m_l are possible for it: $-3, -1, 0, 1, 2$?

5. (a) How many subshells are there in the $n = 2$ shell? How many electron states? (b) Repeat (a) for the $n = 5$ shell.

6. From which atom of each of the following pairs is it easier to remove an electron? (a) Krypton or bromine? (b) Rubidium or cerium? (c) Helium or hydrogen?

7. On what quantum numbers does the energy of an electron depend in (a) a hydrogen atom and (b) a vanadium atom?

8. Label these statements as true or false: (a) One (and only one) of these subshells cannot exist: $2p$, $4f$, $3d$, $1p$. (b) The number of values of m_l that are allowed depends only on l and not on n. (c) There are four subshells with $n = 4$. (d) The least value of n for a given value of l is $l + 1$. (e) All states with $l = 0$ also have $m_l = 0$. (f) There are n subshells for each value of n.

9. Which (if any) of these statements about the Einstein–de Haas experiment or its results are true? (a) Atoms have angular momentum. (b) The angular momentum of atoms is quantized. (c) Atoms have magnetic moments. (d) The magnetic moments of atoms are quantized. (e) The angular momentum of an atom is strongly coupled to its magnetic moment. (f) The experiment relies on the conservation of angular momentum.

10. Consider the elements krypton and rubidium. (a) Which is more suitable for use in a Stern–Gerlach experiment of the kind described in connection with Fig. 41-8? (b) Which, if either, would not work at all?

11. The x-ray spectrum of Fig. 41-14 is for 35.0 keV electrons striking a molybdenum ($Z = 42$) target. If you substitute a silver ($Z = 47$) target for the molybdenum target, will (a) λ_{min}, (b) the wavelength for the K_α line, and (c) the wavelength for the K_β line increase, decrease, or remain unchanged?

12. The K_α x-ray line for any element arises because of a transition between the K shell ($n = 1$) and the L shell ($n = 2$). Figure 41-14 shows this line (for a molybdenum target) occurring at a single wavelength. With higher resolution, however, the line splits into several wavelength components because the L shell does not have a unique energy. (a) How many components does the K_α line have? (b) Similarly, how many components does the K_β line have?

13. Which (if any) of the following is essential for laser action to occur between two energy levels of an atom? (a) There are more atoms in the upper level than in the lower. (b) The upper level is metastable. (c) The lower level is metastable. (d) The lower level is the ground state of the atom. (e) The lasing medium is a gas.

14. Figure 41-22 shows partial energy-level diagrams for the helium and neon atoms that are involved in the operation of a helium–neon laser. It is said that a helium atom in state E_3 can collide with a neon atom in its ground state and raise the neon atom to state E_2. The energy of helium state E_3 (20.61 eV) is close to, but not exactly equal to, the energy of neon state E_2 (20.66 eV). How can the energy transfer take place if these energies are not *exactly* equal?

EXERCISES & PROBLEMS

ssm Solution is in the Student Solutions Manual.
www Solution is available on the World Wide Web at: http://www.wiley.com/college/hrw
ilw Solution is available on the Interactive LearningWare.

SEC. 41-4 Angular Momenta and Magnetic Dipole Moments

1E. Show that $\hbar = 1.06 \times 10^{-34}\ \text{J}\cdot\text{s} = 6.59 \times 10^{-16}\ \text{eV}\cdot\text{s}$.

2E. How many electron states are in these subshells: (a) $n = 4$, $l = 3$; (b) $n = 3$, $l = 1$; (c) $n = 4$, $l = 1$; (d) $n = 2$, $l = 0$?

3E. (a) How many l values are associated with $n = 3$? (b) How many m_l values are associated with $l = 1$? ssm

4E. (a) What is the magnitude of the orbital angular momentum in a state with $l = 3$? (b) What is the magnitude of its largest projection on an imposed z axis?

5E. How many electron states are there in the following shells: (a) $n = 4$, (b) $n = 1$, (c) $n = 3$, (d) $n = 2$?

6E. Write down all the quantum numbers for states that form the subshell with $n = 4$ and $l = 3$.

7E. An electron in a hydrogen atom is in a state with $l = 5$. What is the minimum possible angle between $\vec{L}$ and L_z? ssm

8E. An electron in a multielectron atom has a maximum m_l value of +4. What can you say about the rest of its quantum numbers?

9E. An electron in a multielectron atom is known to have the quantum number $l = 3$. What are its possible n, m_l, and m_s quantum numbers?

10E. How many electron states are there in a shell defined by the quantum number $n = 5$?

11P. An electron is in a state with quantum number $l = 3$. What are the magnitudes of (a) $\vec{L}$ and (b) $\vec{\mu}$? (c) Construct a table of the allowed values of m_l, L_z (in terms of $\hbar$), $\mu_{orb,z}$ (in terms of μ_B), and the semi-classical angle θ between $\vec{L}$ and the positive direction of the z axis. ssm www

12P. An electron is in a state with $n = 3$. What are (a) the number of possible values of l, (b) the number of possible values of m_l, (c) the number of possible values of m_s, (d) the number of states in the $n = 3$ shell, and (e) the number of subshells in the $n = 3$ shell?

13P. If orbital angular momentum $\vec{L}$ is measured along, say, the z axis to obtain a value for L_z, show that

$$(L_x^2 + L_y^2)^{1/2} = [l(l + 1) - m_l^2]^{1/2}\hbar$$

is the most that can be said about the other two components of the orbital angular momentum. ssm

14P. (A correspondence principle problem.) Estimate (a) the quantum number l for the orbital motion of Earth around the Sun and (b) the number of allowed orientations of the plane of Earth's orbit, according to the rules of space quantization. (c) Find θ_{min}, the half-angle of the smallest cone that can be swept out by a perpendicular to Earth's orbit as Earth revolves around the Sun.

SEC. 41-5 The Stern–Gerlach Experiment

15E. Calculate the two possible angles between the electron spin angular momentum vector and the magnetic field in Sample Problem 41-1. Bear in mind that the orbital angular momentum of the valence electron in the silver atom is zero. ssm

16E. Assume that in the Stern–Gerlach experiment as described for neutral silver atoms, the magnetic field $\vec{B}$ has a magnitude of 0.50 T. (a) What is the energy difference between the magnetic moment orientations of the silver atoms in the two subbeams? (b) What is the frequency of the radiation that would induce a transition between these two states? (c) What is its wavelength, and to what part of the electromagnetic spectrum does it belong? The magnetic moment of a neutral silver atom is 1 Bohr magneton.

17E. What is the acceleration of a silver atom as it passes through the deflecting magnet in the Stern–Gerlach experiment of Sample Problem 41-1? ssm

18P. Suppose that a hydrogen atom in its ground state moves 80 cm through and perpendicular to a vertical magnetic field that has a magnetic field gradient $dB/dz = 1.6 \times 10^2$ T/m. (a) What magnitude of force does the field gradient exert on the atom due to the magnetic moment of its electron, which we take to be 1 Bohr magneton? (b) What is the vertical displacement of the atom in the 80 cm of travel if its speed is 1.2×10^5 m/s?

SEC. 41-6 Magnetic Resonance

19E. What is the wavelength associated with a photon that will induce a transition of an electron spin from parallel to antiparallel orientation in a magnetic field of magnitude 0.200 T? Assume that $l = 0$. ssm

20E. The proton, like the electron, has a spin quantum number s of $\frac{1}{2}$. In the hydrogen atom in its ground state ($n = 1$ and $l = 0$), there are two energy levels, depending on whether the electron and proton spins are parallel or antiparallel. If an atom has a spin flip from the state of higher energy to that of lower energy, a photon of wavelength 21 cm is emitted. Radio astronomers observe this 21 cm radiation coming from deep space. What is the effective magnetic field (due to the magnetic dipole moment of the proton) experienced by the electron emitting this radiation?

21E. Excited sodium atoms emit two closely spaced spectrum lines (the *sodium doublet;* see Fig. 41-23) with wavelengths 588.995 nm and 589.592 nm. (a) What is the difference in energy between the two upper energy levels? (b) This energy difference occurs because the electron's spin magnetic moment (= 1 Bohr magneton) can be oriented either parallel or antiparallel to the internal magnetic field associated with the electron's orbital motion. Use your result in (a) to find the strength of this internal magnetic field.

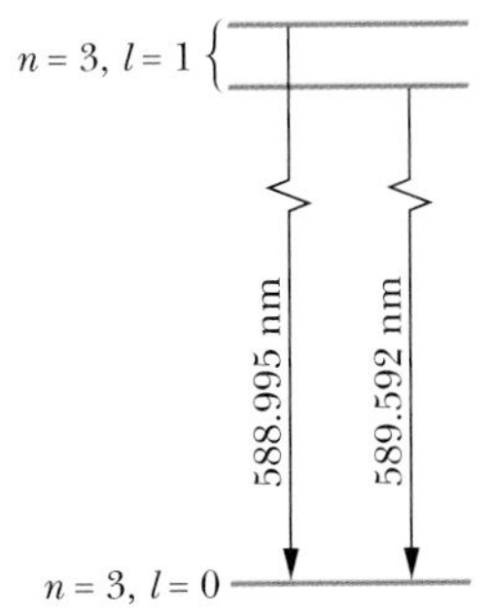

Fig. 41-23 Exercise 21.

22E. An external oscillating magnetic field of frequency 34 MHz is applied to a sample that contains hydrogen atoms. Resonance is observed when the strength of the constant external magnetic field equals 0.78 T. Calculate the strength of the local magnetic field at the site of the protons that are undergoing spin flips, assuming the external and local fields are parallel there. The protons have $\mu_z = 1.41 \times 10^{-26}$ J/T.

SEC. 41-8 Multiple Electrons in Rectangular Traps

23E. Seven electrons are trapped in a one-dimensional infinite potential well of width L. As a multiple of $h^2/8mL^2$, what is the energy of the ground state of the system of seven electrons? Assume that the electrons do not interact with one another, and do not neglect spin.

24E. A rectangular corral of widths $L_x = L$ and $L_y = 2L$ contains seven electrons. As a multiple of $h^2/8mL^2$, what is the energy of the ground state of the system of seven electrons? Assume that the electrons do not interact with one another, and do not neglect spin.

25P. For the situation of Exercise 23, and as multiples of $h^2/8mL^2$, what are the energies of (a) the first excited state, (b) the second excited state, and (c) the third excited state of the system of seven electrons? (d) Construct an energy-level diagram for the lowest four energy levels of the system.

26P. For Exercise 24, and as multiples of $h^2/8mL^2$, what are the energies of (a) the first excited state, (b) the second excited state, and (c) the third excited state of the system of seven electrons? (d) Construct an energy-level diagram for the lowest four energy levels of the system.

27P. A cubical box of widths $L_x = L_y = L_z = L$ contains eight electrons. As a multiple of $h^2/8mL^2$, what is the energy of the ground state of the system of eight electrons? Assume that the electrons do not interact with one another, and do not neglect spin. ssm www

28P. For the situation of Problem 27, and as multiples of $h^2/8mL^2$, what are the energies of (a) the first excited state, (b) the second excited state, and (c) the third excited state of the system of eight electrons? (d) Construct an energy-level diagram for the lowest four energy levels of the system.

SEC. 41-9 Building the Periodic Table

29E. Show that if the 63 electrons in an atom of europium were assigned to shells according to the "logical" sequence of quantum numbers, this element would be chemically similar to sodium.

30E. Consider the elements selenium ($Z = 34$), bromine ($Z = 35$), and krypton ($Z = 36$). In their part of the periodic table, the subshells of the electronic states are filled in the sequence

$$1s \quad 2s \;\; 2p \quad 3s \;\; 3p \;\; 3d \quad 4s \;\; 4p \;\cdots$$

For each of these elements, identify the highest occupied subshell and state how many electrons are in it.

31E. Suppose that the electron had no spin and that the Pauli exclusion principle still held. Which, if any, of the present noble gases would remain in that category?

32E. What are the four quantum numbers for the two electrons of the helium atom in its ground state?

33P. Two electrons in lithium ($Z = 3$) have the quantum numbers $n = 1$, $l = 0$, $m_l = 0$, and $m_s = \pm\frac{1}{2}$. What quantum numbers can the third electron have if the atom is to be in (a) its ground state and (b) its first excited state? ssm

34P. Suppose there are two electrons in the same atom, both of which have $n = 2$ and $l = 1$. (a) If the Pauli exclusion principle did not apply, how many combinations of states would conceivably be possible? (b) How many states does the exclusion principle forbid? Which are they?

35P. Show that the number of states with the same quantum number n is $2n^2$. ssm

SEC. 41-10 X Rays and the Numbering of the Elements

36E. Through what minimum potential difference must an electron in an x-ray tube be accelerated so that it can produce x rays with a wavelength of 0.100 nm?

37E. Knowing that the minimum x-ray wavelength produced by 40.0 keV electrons striking a target is 31.1 pm, determine the Planck constant h.

38E. Show that the cutoff wavelength (in picometers) in the continuous x-ray spectrum from any target is given by $\lambda_{min} = 1240/V$, where V is the potential difference (in kilovolts) through which the electrons are accelerated before they strike the target.

39P. X rays are produced in an x-ray tube by electrons accelerated through an electric potential difference of 50.0 kV. An electron makes three collisions in the target before coming to rest and loses half its remaining kinetic energy in each of the first two collisions. Determine the wavelengths of the resulting photons. (Neglect the recoil of the heavy target atoms.) ssm www

40P. A 20 keV electron is brought to rest by undergoing two successive nuclear encounters such as that of Fig. 41-15, thus transferring its kinetic energy to the energy of two photons. The wavelength associated with the second photon is 130 pm greater than the wavelength associated with the first photon. (a) Find the kinetic energy of the electron after its first encounter. (b) What are the associated wavelengths and energies of the two photons?

41P. Show that a moving electron cannot spontaneously change into an x-ray photon in free space. A third body (atom or nucleus) must be present. Why is it needed? (*Hint:* Examine the conservation of energy and momentum.) ssm

42P. When electrons bombard a molybdenum target, they produce both continuous and characteristic x rays as shown in Fig. 41-14. In that figure the kinetic energy of the incident electrons is 35.0 keV. If the accelerating potential is increased to 50.0 keV, what mean values of (a) λ_{min}, (b) the wavelength of the K_α line, and (c) the wavelength of the K_β line result?

43P. In Fig. 41-14, the x rays shown are produced when 35.0 keV electrons strike a molybdenum ($Z = 42$) target. If the accelerating potential is maintained at this value but a silver ($Z = 47$) target is used instead, what values of (a) λ_{min}, (b) the wavelength of the K_α line, and (c) the wavelength of the K_β line result? The K, L, and M atomic x-ray levels for silver (compare Fig. 41-16) are 25.51, 3.56, and 0.53 keV. ssm

44P. The wavelength of the K_α line from iron is 193 pm. What is the energy difference between the two states of the iron atom that give rise to this transition?

45P. Calculate the ratio of the wavelength of the K_α line for niobium (Nb) to that for gallium (Ga). Take needed data from the periodic table of Appendix G. ssm

46P. From Fig. 41-14, calculate approximately the energy difference $E_L - E_M$ for molybdenum. Compare it with the value that may be obtained from Fig. 41-16.

47P. Here are the K_α wavelengths of a few elements:

Element	λ (pm)	Element	λ (pm)
Ti	275	Co	179
V	250	Ni	166
Cr	229	Cu	154
Mn	210	Zn	143
Fe	193	Ga	134

Make a Moseley plot (like that in Fig. 41-17) from these data and verify that its slope agrees with the value given for C in Section 41-10.

48P. A molybdenum ($Z = 42$) target is bombarded with 35.0 keV electrons and the x-ray spectrum of Fig. 41-14 results. The K_β and K_α wavelengths are 63.0 and 71.0 pm, respectively. (a) What are the corresponding photon energies? (b) It is desired to filter these radiations through a material that will absorb the K_β line much more strongly than it will absorb the K_α line. What substance would you use? The ionization energies of the K electrons in molybdenum and in four neighboring elements are as follows:

	Zr	Nb	Mo	Tc	Ru
Z	40	41	42	43	44
E_K (keV)	18.00	18.99	20.00	21.04	22.12

(*Hint:* A substance will absorb one x radiation more strongly than another if the photons of the first have enough energy to eject a K electron from the atom of the substance but the photons of the second do not.)

49P. A tungsten ($Z = 74$) target is bombarded by electrons in an x-ray tube. (a) What is the minimum value of the accelerating potential that will permit the production of the characteristic K_α and K_β lines of tungsten? (b) For this same accelerating potential, what is λ_{min}? (c) What are the K_α and K_β wavelengths? The K, L, and M energy levels for tungsten (see Fig. 41-16) have the energies 69.5, 11.3, and 2.30 keV, respectively. ssm

50P. The binding energies of K-shell and L-shell electrons in copper are 8.979 and 0.951 keV, respectively. If a K_α x ray from copper is incident on a sodium chloride crystal and gives a first-order Bragg reflection at an angle of 74.1° measured relative to parallel planes of sodium atoms, what is the spacing between these parallel planes?

51P. (a) Using Eq. 41-26, estimate the ratios of photon energies due to K_α transitions in two atoms whose atomic numbers are Z and Z'. (b) What is this ratio for uranium and aluminum? (c) For uranium and lithium?

52P. Determine how close the theoretical K_α x-ray photon energies, as obtained from Eq. 41-27, are to the measured energies of the low-mass elements from lithium to magnesium. To do this, (a) first determine the constant C in Eq. 41-27 to five significant figures by finding C in terms of the fundamental constants in Eq. 41-24 and then using data from Appendix B to evaluate those constants. (b) Next, calculate the percentage deviations of the theoretical from the measured energies. (c) Finally, plot the deviations and comment on the trend. The measured energies (eV) of the K_α photons for these elements are as follows:

Li	54.3	O	524.9
Be	108.5	F	676.8
B	183.3	Ne	848.6
C	277	Na	1041
N	392.4	Mg	1254

(There is actually more than one K_α ray because of the splitting of the L energy level, but that effect is negligible for the elements listed here.)

SEC. 41-12 How Lasers Work

53E. Lasers can be used to generate pulses of light whose durations are as short as 10 fs. (a) How many wavelengths of light (λ = 500 nm) are contained in such a pulse? (b) Supply the missing quantity X (in years):

$$\frac{10 \text{ fs}}{1 \text{ s}} = \frac{1 \text{ s}}{X}.$$

54E. For the conditions of Sample Problem 41-6a, how many moles of neon are needed to put 10 atoms in the excited state E_x?

55E. A hypothetical atom has energy levels uniformly separated by 1.2 eV. At a temperature of 2000 K, what is the ratio of the number of atoms in the 13th excited state to the number in the 11th excited state? ssm

56E. By measuring the go-and-return time for a laser pulse to travel from an Earth-bound observatory to a reflector on the Moon, it is possible to measure the separation between these bodies. (a) What is the predicted value of this time? (b) The separation can be measured to a precision of about 15 cm. To what uncertainty in travel time does this correspond? (c) If the laser beam forms a spot on the Moon 3 km in diameter, what is the angular divergence of the beam?

57E. A hypothetical atom has only two atomic energy levels, separated by 3.2 eV. Suppose that at a certain altitude in the atmosphere of a star there are $6.1 \times 10^{13}/\text{cm}^3$ of these atoms in the higher energy state and $2.5 \times 10^{15}/\text{cm}^3$ in the lower energy state. What is the temperature of the star's atmosphere at that altitude?

58E. A population inversion for two energy levels is often described by assigning a negative Kelvin temperature to the system. What negative temperature would describe a system in which the population of the upper energy level exceeds that of the lower level by 10% and the energy difference between the two levels is 2.1 eV?

59E. A pulsed laser emits light at a wavelength of 694.4 nm. The pulse duration is 12 ps and the energy per pulse is 0.150 J. (a) What is the length of the pulse? (b) How many photons are emitted in each pulse? ssm

60E. A helium–neon laser emits laser light at a wavelength of 632.8 nm and a power of 2.3 mW. At what rate are photons emitted by this device?

61E. A high-powered laser beam (λ = 600 nm) with a beam diameter of 12 cm is aimed at the Moon, 3.8×10^5 km distant. The beam spreads only because of diffraction. The angular location of the edge of the central diffraction disk (see Eq. 37-12) is given by

$$\sin\theta = \frac{1.22\lambda}{d},$$

where d is the diameter of the beam aperture. What is the diameter of the central diffraction disk on the Moon's surface?

62E. Assume that lasers are available whose wavelengths can be precisely "tuned" to anywhere in the visible range—that is, in the range 450 nm $< \lambda <$ 650 nm. If every television channel occupies a bandwidth of 10 MHz, how many channels could be accommodated within this wavelength range?

63E. The active volume of a laser constructed of the semiconductor GaAlAs is only 200 μm^3 (smaller than a grain of sand) and yet the laser can continuously deliver 5.0 mW of power at a wavelength of 0.80 μm. At what rate does it generate photons?

64P. The mirrors in the laser of Fig. 41-21, which are separated by 8.0 cm, form an optical cavity in which standing waves of laser light can be set up. Each standing wave has an integral number n of half wavelengths in the 8.0 cm length, where n is large and the waves differ slightly in wavelength. Near λ = 533 nm, how far apart in wavelength are the standing waves?

65P. The active medium in a particular laser that generates laser light at a wavelength of 694 nm is 6.00 cm long and 1.00 cm in diameter. (a) Treat the medium as an optical resonance cavity analogous to a closed organ pipe. How many standing wave nodes are there along the laser axis? (b) By what amount Δf would the beam frequency have to shift to increase this number by one? (c) Show that Δf is just the inverse of the travel time of laser light for one round trip back and forth along the laser axis. (d) What is the corresponding fractional frequency shift $\Delta f/f$? The appropriate index of refraction of the lasing medium (a ruby crystal) is 1.75. ssm www

66P. A hypothetical atom has two energy levels, with a transition wavelength between them of 580 nm. In a particular sample at 300 K, 4.0×10^{20} such atoms are in the state of lower energy. (a) How many atoms are in the upper state, assuming conditions of thermal equilibrium? (b) Suppose, instead, that 3.0×10^{20} of these atoms are "pumped" into the upper state by an external process, with 1.0×10^{20} atoms remaining in the lower state. What is the maximum energy that could be released by the atoms in a single laser pulse if each atom jumps once between those two states (either via absorption or stimulated emission).

67P. Can an incoming intercontinental ballistic missile be destroyed by an intense laser beam? A beam of intensity 10^8 W/m^2 would probably burn into and destroy a hardened (nonspinning) missile in 1 s. (a) If the laser had 5.0 MW power, 3.0 μm wavelength, and a 4.0 m beam diameter (a very powerful laser indeed), would it destroy a missile at a distance of 3000 km? (b) If the wavelength could be changed, what maximum value would work? Use the equation for the central disk given in Exercise 61. ssm

68P. The beam from an argon laser (of wavelength 515 nm) has a diameter d of 3.00 mm and a continuous energy output rate of 5.00 W. The beam is focused onto a diffuse surface by a lens whose focal length f is 3.50 cm. A diffraction pattern such as that of Fig. 37-9 is formed, the radius of the central disk being given by

$$R = \frac{1.22f\lambda}{d}$$

(see Eq. 37-12 and Sample Problem 37-3). The central disk can be shown to contain 84% of the incident power. (a) What is the radius of the central disk? (b) What is the average intensity (power per unit area) in the incident beam? (c) What is the average intensity in the central disk?

Additional Problems

69. *Martian CO_2 laser*. Where sunlight shines on the atmosphere of Mars, carbon dioxide molecules at an altitude of about 75 km undergo natural laser action. The energy levels involved in the action are shown in Fig. 41-24; population inversion occurs between energy levels E_2 and E_1. (a) What wavelength of sunlight excites the molecules in the lasing action? (b) At what wavelength does lasing occur? (c) In what region of the electromagnetic spectrum do the excitation and lasing wavelengths lie?

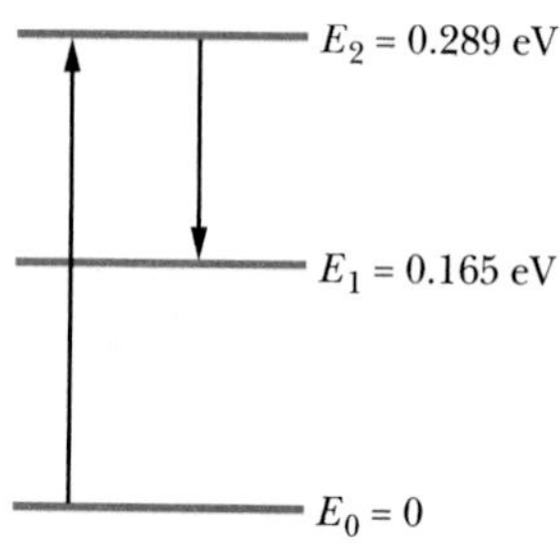

Fig. 41-24 Problem 69.

70. *Comet stimulated emission*. When a comet approaches the Sun, the increased warmth evaporates water from the frozen ice on the surface of the comet nucleus, producing a thin atmosphere of water vapor around the nucleus. Sunlight can then dissociate the water vapor into H and OH. The sunlight can also excite the OH molecules into higher energy levels, two of which are represented in Fig. 41-25.

When the comet is still relatively far from the Sun, the sunlight causes equal excitation to the E_2 and E_1 levels (Fig. 41-25*a*). Hence, there is no population inversion between the two levels. However, as the comet approaches the Sun, the excitation to the E_1 level decreases and population inversion occurs. The reason has to do with one of the many wavelengths—said to be *Fraunhofer lines*—that are missing in sunlight because, as the light travels outward through the Sun's atmosphere, those particular wavelengths are absorbed by the atmosphere.

As a comet approaches the Sun, the Doppler effect due to the comet's speed relative to the Sun shifts the Fraunhofer lines in wavelength, apparently overlapping one of them with the wavelength required for excitation to the E_1 level in OH molecules. Population inversion then occurs in those molecules, and they radiate stimulated emission (Fig. 41-25*b*). For example, as comet Kouhoutek approached the Sun in December 1973 and January 1974, it radiated stimulated emission at about 1666 MHz during mid-January. (a) What was the energy difference $E_2 - E_1$ for that emission? (b) In what region of the electromagnetic spectrum was the emission?

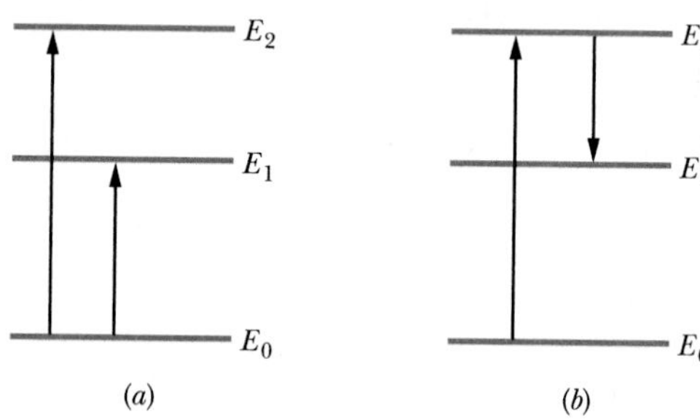

Fig. 41-25 Problem 70.

42 Conduction of Electricity in Solids

A few of the workers at the Fab 11 factory at Rio Rancho, New Mexico. The plant, which represents an investment of $2.5 billion, has a floor area equivalent to that of about two dozen football fields. According to the *New York Times,* the plant "on a high desert mesa in New Mexico is probably the most productive factory in the world, in terms of the value of the goods that it makes."

What do these workers manufacture that requires them to be suited up like astronauts?

The answer is in this chapter.

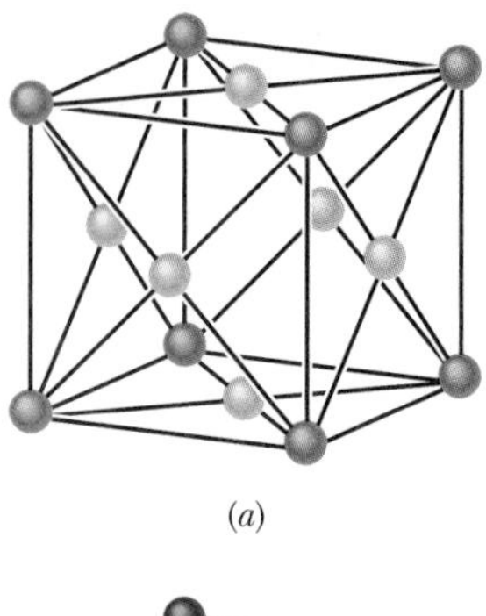

(*a*)

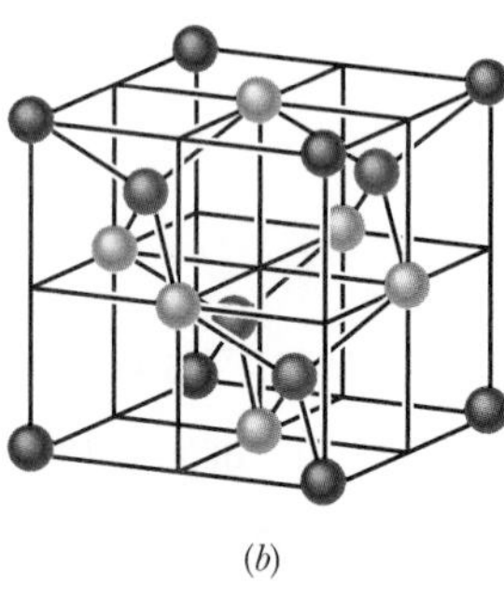

(*b*)

Fig. 42-1 (*a*) The unit cell for copper is a cube. There is one copper atom (darker) at each corner of the cube and one copper atom (lighter) at the center of each face of the cube. The arrangement is called *face-centered cubic.* (*b*) The unit cell for silicon and diamond is also a cube, the atoms being arranged in a so-called *diamond lattice.* There is one atom (darkest) at each corner of the cube and one atom (lightest) at the center of each cube face; in addition, four atoms (medium color) lie within the cube. Every atom is bonded to its four nearest neighbors by a two-electron covalent bond (only the four atoms within the cube show all four *nearest* neighbors).

42-1 Solids

You have seen how well quantum physics works when we apply it to questions involving individual atoms. In this chapter we hope to show, with a single broad example, that this theory works just as well when we apply it to questions involving assemblies of atoms in the form of solids.

Every solid has an enormous range of properties that we can choose to examine. Is it transparent? Can it be hammered out into a thin sheet? At what speeds do sound waves travel through it? Is it magnetic? Is it a good heat conductor? . . . The list goes on and on. However, we choose to focus this entire chapter on a single question: *What are the mechanisms by which a solid conducts, or does not conduct, electricity?* As you will see, quantum physics provides the answer.

42-2 The Electrical Properties of Solids

We shall examine only **crystalline solids**—that is, solids whose atoms are arranged in a repetitive three-dimensional structure called a **lattice.** We shall not consider such solids as wood, plastic, glass, and rubber, whose atoms are not arranged in such repetitive patterns. Figure 42-1 shows the basic repetitive units (the **unit cells**) of the lattice structures of copper, our prototype of a metal, and silicon and diamond, our prototypes of a semiconductor and an insulator, respectively.

We can classify solids electrically according to three basic properties:

1. Their **resistivity** ρ at room temperature, with the SI unit ohm-meter ($\Omega \cdot m$); resistivity is defined in Section 27-4.
2. Their **temperature coefficient of resistivity** α, defined as $\alpha = (1/\rho)(d\rho/dT)$ in Eq. 27-17 and having the SI unit inverse kelvin (K^{-1}). We can evaluate α for any solid by measuring ρ over a range of temperatures.
3. Their **number density of charge carriers** n. This quantity, the number of charge carriers per unit volume, can be found from measurements of the Hall effect, as discussed in Section 29-4, and from other measurements. It has the SI unit inverse cubic meter (m^{-3}).

From measurements of room-temperature resistivity alone, we discover that there are some materials—we call them **insulators**—that for all practical purposes do not conduct electricity at all. These are materials with very high resistivity. Diamond, an excellent example, has a resistivity greater than that of copper by the enormous factor of about 10^{24}.

We can then use measurements of ρ, α, and n to divide most noninsulators, at least at low temperatures, into two major categories: **metals** and **semiconductors.**

Semiconductors have a considerably greater resistivity ρ than metals.

Semiconductors have a temperature coefficient of resistivity α that is both high and negative. That is, the resistivity of a semiconductor *decreases* with temperature, whereas that of a metal *increases.*

Semiconductors have a considerably lower number density of charge carriers n than metals.

Table 42-1 shows values of these quantities for copper, our prototype metal, and silicon, our prototype semiconductor.

Now, with measurements of ρ, α, and n in hand, we have an experimental basis for refining our central question about the conduction of electricity in solids: *What*

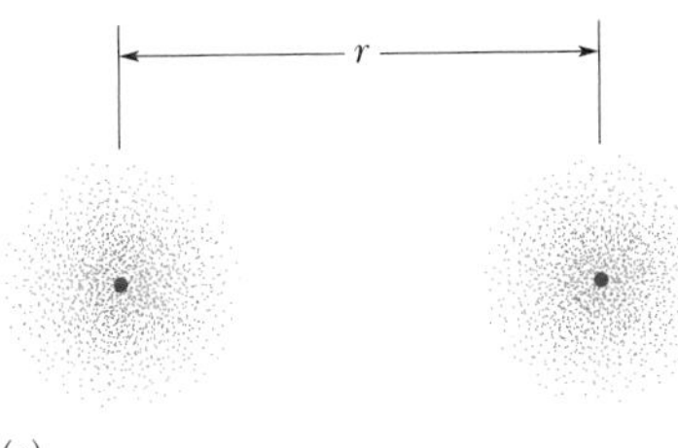

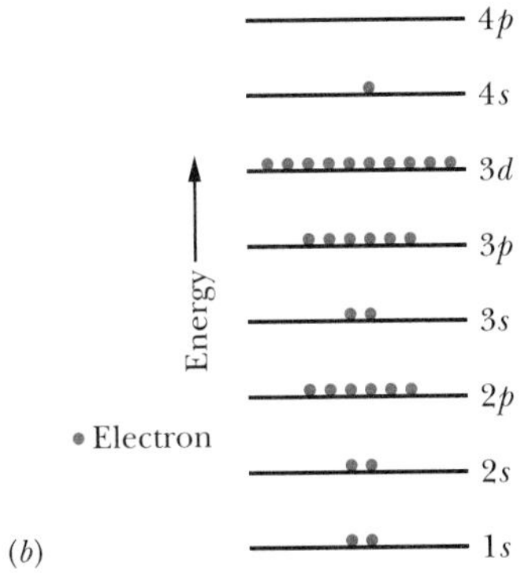

Fig. 42-2 (*a*) Two copper atoms separated by a large distance; their electron distributions are represented by dot plots. (*b*) Each copper atom has 29 electrons distributed among a set of subshells. In the neutral atom in its ground state, all subshells up through the 3*d* level are filled, the 4*s* subshell contains one electron (it can hold two), and higher subshells are empty. For simplicity, the subshells are shown as being evenly spaced in energy.

TABLE 42-1 Some Electrical Properties of Two Materials[a]

		Material	
Property	Unit	Copper	Silicon
Type of conductor		Metal	Semiconductor
Resistivity, ρ	$\Omega \cdot m$	2×10^{-8}	3×10^{3}
Temperature coefficient of resistivity, α	K^{-1}	$+4 \times 10^{-3}$	-70×10^{-3}
Number density of charge carriers, n	m^{-3}	9×10^{28}	1×10^{16}

[a]All values are for room temperature.

features make diamond an insulator, copper a metal, and silicon a semiconductor? Again, quantum physics provides the answers.

42-3 Energy Levels in a Crystalline Solid

The distance between adjacent copper atoms in solid copper is 260 pm. Figure 42-2*a* shows two isolated copper atoms separated by a distance r that is much greater than that. As Fig. 42-2*b* shows, each of these isolated neutral atoms stacks up its 29 electrons in an array of discrete subshells, as follows:

$$1s^2\ 2s^2\ 2p^6\ 3s^2\ 3p^6\ 3d^{10}\ 4s^1.$$

Here we use the shorthand notation of Section 41-9 to identify the subshells. Recall, for example, that the subshell with principal quantum number $n = 3$ and orbital quantum number $l = 1$ is called the 3*p* subshell; it can hold up to $2(2l + 1) = 6$ electrons; the number it actually contains is indicated by a numerical superscript. We see above that the first six subshells in copper are filled, but the (outermost) 4*s* subshell, which can hold 2 electrons, holds only one.

If we bring the atoms of Fig. 42-2*a* closer together, they will—speaking loosely—begin to sense each other's presence. In the language of quantum physics, their wave functions will start to overlap, beginning with those of the outermost electrons.

When the wave functions of the two atoms overlap, we speak not of two independent atoms but of a single two-atom system; here the system contains $2 \times 29 = 58$ electrons. The Pauli exclusion principle also applies to this larger system and requires that each of these 58 electrons occupy a different quantum state. In fact, 58 quantum states are available because each energy level of the isolated atom splits into *two* levels for the two-atom system.

If we bring up more atoms, we gradually assemble a lattice of solid copper. If, say, our lattice contains N atoms, then each level of an isolated copper atom must split into N levels in the solid. Thus, the individual energy levels of the solid form energy **bands,** adjacent bands being separated by an energy **gap,** which represents a range of energies that no electron can possess. A typical band is only a few electron-volts wide. Since N may be of the order of 10^{24}, we see that the individual levels within a band are very close together indeed, and there are a vast number of levels.

Figure 42-3 suggests the band–gap structure of the energy levels in a generalized crystalline solid. Note that bands of lower energy are narrower than those of higher energy. This occurs because electrons that occupy the lower energy bands spend most of their time deep within the atom's electron cloud. The wave functions of these core electrons do not overlap as much as the wave functions of the outer electrons. Hence the splitting of these levels is not as great as it is for the higher energy levels normally occupied by the outer electrons.

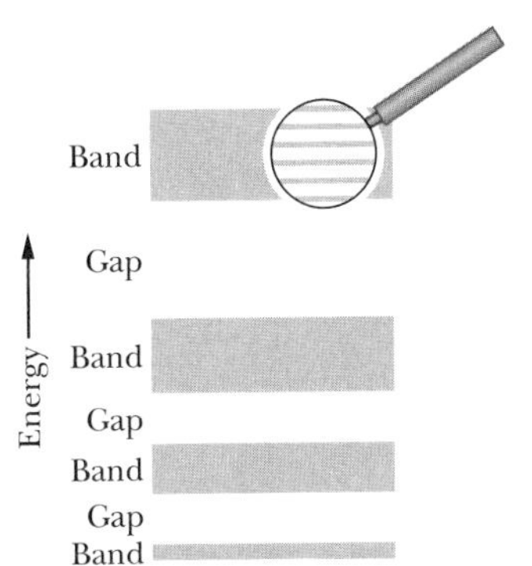

Fig. 42-3 The band–gap pattern of energy levels for an idealized crystalline solid. As the magnified view suggests, each band consists of a very large number of very closely spaced energy levels. (In many solids, adjacent bands may overlap; for clarity, we have not shown this condition.)

42-4 Insulators

A solid is said to be an insulator if no current exists within it when we apply a potential difference across it. For a current to exist, the kinetic energy of the average electron must increase. In other words, some electrons in the solid must move to a higher energy level. However, as Fig. 42-4*a* shows, in an insulator the highest band containing any electrons is fully occupied, and the Pauli exclusion principle keeps electrons from moving to occupied levels.

Thus, the electrons in the filled band of an insulator have no place to go; they are in gridlock. It is as if a child tries to climb a ladder that already has a child standing on each rung; since there are no unoccupied rungs, no one can move.

There are plenty of unoccupied levels (or *vacant levels*) in the band above the filled band in Fig. 42-4*a*. However, if an electron is to occupy one of those levels, it must acquire enough energy to jump across the substantial gap that separates the two bands. In diamond, this gap is so wide (the energy needed to cross it is 5.5 eV, about 140 times the average thermal energy of a free particle at room temperature) that essentially no electron can jump across it. Diamond is thus an insulator, and a very good one.

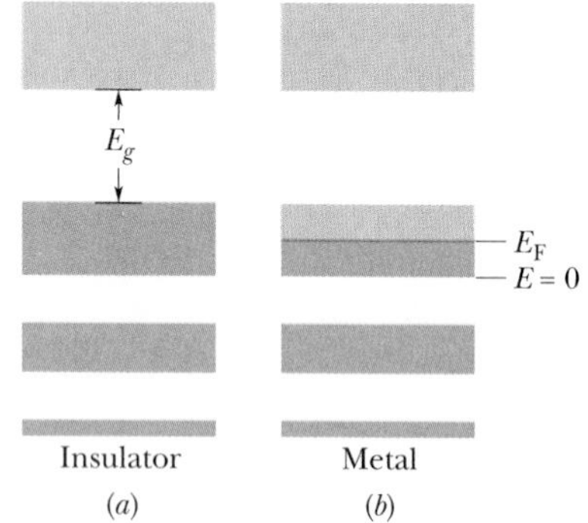

Fig. 42-4 (*a*) The band–gap pattern for an insulator; filled levels are shown in red, and empty levels in blue. Note that the highest filled level lies at the top of a band and the next higher vacant level is separated from it by a relatively large energy gap E_g. (*b*) The band–gap pattern for a metal. The highest filled level, called the Fermi level, lies near the middle of a band. Since vacant levels are available within that band, electrons in the band can easily change levels, and conduction can take place.

Sample Problem 42-1

Approximately what is the probability that, at room temperature (300 K), an electron at the top of the highest filled band in diamond (an insulator) will jump the energy gap E_g in Fig. 42-4*a*. For diamond, E_g is 5.5 eV.

SOLUTION: In Chapter 41 we used Eq. 41-29,

$$\frac{N_x}{N_0} = e^{-(E_x - E_0)/kT}, \tag{42-1}$$

to relate the population N_x of atoms at energy level E_x to the population N_0 at energy level E_0, where the atoms are part of a system at temperature T (measured in kelvins); k is the Boltzmann constant (8.62×10^{-5} eV/K).

A **Key Idea** here is that we can use Eq. 42-1 to *approximate* the probability P that an electron in an insulator will jump the energy gap E_g in Fig. 42-4*a*. To do so, we first set the energy difference $E_x - E_0$ to E_g. Then the probability P of the jump is approximately equal to the ratio N_x/N_0 of the number of electrons just above the energy gap to the number of electrons just below the gap.

For diamond, the exponent in Eq. 42-1 is

$$-\frac{E_g}{kT} = -\frac{5.5 \text{ eV}}{(8.62 \times 10^{-5} \text{ eV/K})(300 \text{ K})} = -213.$$

The required probability is then

$$P = \frac{N_x}{N_0} = e^{-(E_g/kT)} = e^{-213} \approx 3 \times 10^{-93}. \quad \text{(Answer)}$$

This result tells us that approximately 3 electrons out of 10^{93} electrons would jump across the energy gap. Because an actual diamond has less than 10^{23} electrons, we see that the probability of the jump is vanishingly small. No wonder diamond is such a good insulator.

42-5 Metals

The feature that defines a metal is that, as Fig. 42-4*b* shows, the highest occupied energy level falls somewhere near the middle of an energy band. If we apply a potential difference across a metal, a current can exist because there are plenty of vacant levels at nearby higher energies into which electrons (the charge carriers in a metal) can jump. Thus, a metal can conduct electricity because electrons in its highest occupied band can easily move into higher energy levels within that band.

In Section 27-6 we introduced the **free-electron model** of a metal, in which the **conduction electrons** are free to move throughout the volume of the sample like the molecules of a gas in a closed container. We used this model to derive an expression for the resistivity of a metal, assuming that the electrons follow the laws of Newtonian mechanics. Here we use that same model to explain the behavior of the electrons—called the conduction electrons—in the partially filled band of Fig. 42-4*b*. However, we follow the laws of quantum physics by assuming the energies of these electrons to be quantized and the Pauli exclusion principle to hold.

We assume too that the electric potential energy of a conduction electron has the same constant value at all points within the lattice. If we choose this value of the potential energy to be zero, as we are free to do, then the mechanical energy E of the conduction electrons is entirely kinetic.

The level at the bottom of the partially filled band of Fig. 42-4*b* corresponds to $E = 0$. The highest occupied level in this band at absolute zero ($T = 0$ K) is called the **Fermi level,** and the energy corresponding to it is called the **Fermi energy** E_F; for copper, $E_F = 7.0$ eV.

The electron speed corresponding to the Fermi energy is called the **Fermi speed** v_F. For copper the Fermi speed is 1.6×10^6 m/s. This fact should be enough to shatter the popular misconception that all motion ceases at absolute zero; at that temperature—and solely because of the Pauli exclusion principle—the conduction electrons are stacked up in the partially filled band of Fig. 42-4*b* with energies that range from zero to the Fermi energy.

How Many Conduction Electrons Are There?

If we could bring individual atoms together to form a sample of a metal, we would find that the conduction electrons in the metal are the *valence electrons* of the atoms (the electrons in the outer shells of the individual atoms). A *monovalent* atom contributes one such electron to the conduction electrons in a metal; a *bivalent* atom contributes two such electrons. Thus, the total number of conduction electrons

$$\begin{pmatrix}\text{number of conduction}\\ \text{electrons in sample}\end{pmatrix} = \begin{pmatrix}\text{number of atoms}\\ \text{in sample}\end{pmatrix}\begin{pmatrix}\text{number of valence}\\ \text{electrons per atom}\end{pmatrix}. \tag{42-2}$$

(In this chapter, we shall write several equations largely in words because the symbols we have previously used for the quantities in them now represent other quantities.) The *number density* n of conduction electrons in a sample is the number of conduction electrons per unit volume:

$$n = \frac{\text{number of conduction electrons in sample}}{\text{sample volume } V}. \tag{42-3}$$

We can relate the number of atoms in a sample to various other properties of the sample and the material making up the sample with the following equation:

$$\begin{aligned}\begin{pmatrix}\text{number of atoms}\\ \text{in sample}\end{pmatrix} &= \frac{\text{sample mass } M_{\text{sam}}}{\text{atomic mass}} = \frac{\text{sample mass } M_{\text{sam}}}{(\text{molar mass } M)/N_A}\\ &= \frac{(\text{material's density})(\text{sample volume } V)}{(\text{molar mass } M)/N_A},\end{aligned} \tag{42-4}$$

where the molar mass M is the mass of one mole of the material in the sample and N_A is Avogadro's number (6.02×10^{23} mol^{-1}).

Sample Problem 42-2

How many conduction electrons are in a cube of magnesium with a volume of $2.00 \times 10^{-6}\ \text{m}^3$? Magnesium atoms are bivalent.

SOLUTION: The **Key Ideas** here are these:

1. Because magnesium atoms are bivalent, each magnesium atom contributes two conduction electrons.
2. The number of conduction electrons in the cube is related to the number of magnesium atoms in the cube by Eq. 42-2.
3. We can find the number of atoms with Eq. 42-4 and known data about the cube's volume and magnesium's properties.

We can write Eq. 42-4 as

$$\begin{pmatrix}\text{number}\\ \text{of atoms}\\ \text{in sample}\end{pmatrix} = \frac{(\text{material's density})(\text{sample volume } V)N_A}{\text{molar mass } M}.$$

Magnesium has a density of $1.738\ \text{g/cm}^3$ ($= 1.738 \times 10^3\ \text{kg/m}^3$) and a molar mass of 24.312 g/mol ($= 24.312 \times 10^{-3}$ kg/mol) (see Appendix F). The numerator gives us

$$(1.738 \times 10^3\ \text{kg/m}^3)(2.00 \times 10^{-6}\ \text{m}^3)(6.02 \times 10^{23}\ \text{mol}^{-1}) = 2.0926 \times 10^{21}\ \text{kg/mol}.$$

Thus, we have

$$\begin{pmatrix}\text{number of atoms}\\ \text{in sample}\end{pmatrix} = \frac{2.0926 \times 10^{21}\ \text{kg/mol}}{24.312 \times 10^{-3}\ \text{kg/mol}} = 8.61 \times 10^{22}.$$

Using this result and the fact that magnesium atoms are bivalent, we find that Eq. 42-2 yields

$$\begin{pmatrix}\text{number of}\\ \text{conduction electrons}\\ \text{in sample}\end{pmatrix} = (8.61 \times 10^{22}\ \text{atoms})\left(2\,\frac{\text{electrons}}{\text{atom}}\right) = 1.72 \times 10^{23}\ \text{electrons}. \quad \text{(Answer)}$$

Conductivity at $T > 0$

Our practical interest in the conduction of electricity in metals is at temperatures above absolute zero. What happens to the electron distribution of Fig. 42-4*b* at such higher temperatures? As we shall see, surprisingly little.

Of the electrons in the partially filled band of Fig. 42-4*b*, only those that are close to the Fermi energy find unoccupied levels above them, and only those electrons are free to be boosted to these higher levels by thermal agitation. Even at $T = 1000$ K, a temperature at which copper would glow brightly in a dark room, the distribution of electrons among the available levels does not differ much from the distribution at $T = 0$ K.

Let us see why. The quantity kT, where k is the Boltzmann constant, is a convenient measure of the energy that may be given to a conduction electron by the random thermal motions of the lattice. At $T = 1000$ K, we have $kT = 0.086$ eV. No electron can hope to have its energy changed by more than a few times this relatively small amount by thermal agitation alone, so at best, only those few conduction electrons whose energies are close to the Fermi energy are likely to jump to higher energy levels due to thermal agitation. Poetically stated, thermal agitation normally causes only ripples on the surface of the Fermi sea of electrons; the vast depths of that sea lie undisturbed.

How Many Quantum States Are There?

The ability of a metal to conduct electricity depends on how many quantum states are available to its electrons and what the energies of those states are. Thus, a question arises: What are the energies of the individual states in the partially filled band of Fig. 42-4*b*? This question is too difficult to answer because we cannot possibly list the energies of so many states individually. We ask instead: How many states in a unit volume of a sample have energies in the energy range E to $E + dE$? We write this number as $N(E)\,dE$, where $N(E)$ is called the **density of states** at energy E. The conventional unit for $N(E)\,dE$ is states per cubic meter (states/m^3, or simply m^{-3}); the unit for $N(E)$ is states per cubic meter per electron-volt (m^{-3} eV^{-1}).

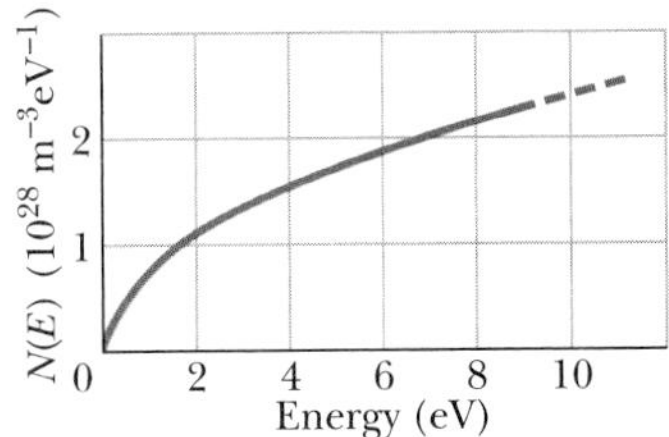

Fig. 42-5 The density of states $N(E)$—that is, the number of electron energy levels per unit energy interval and per unit volume—plotted as a function of electron energy. The density of states function simply counts the available states; it says nothing about whether these states are occupied by electrons.

We can find an expression for the density of states by counting the number of standing electron matter waves that can fit into a box the size of the metal sample we are considering. This is analogous to counting the number of standing waves of sound that can exist in a closed organ pipe. The differences are that our problem is three-dimensional (the organ pipe problem is one-dimensional) and the waves are matter waves (the organ-pipe waves are sound waves). The result of such counting can be shown to be

$$N(E) = \frac{8\sqrt{2}\pi m^{3/2}}{h^3} E^{1/2} \quad \text{(density of states)}, \tag{42-5}$$

where m is the mass of the electron and E is the energy at which $N(E)$ is to be evaluated. Note that nothing in this equation involves the shape of the sample, its temperature, or the material of which it is made. Equation 42-5 is plotted in Fig. 42-5.

CHECKPOINT 1: (a) Is the spacing between adjacent energy levels at $E = 4$ eV in copper larger than, the same as, or smaller than the spacing at $E = 6$ eV? (b) Is the spacing between adjacent energy levels at $E = 4$ eV in copper larger than, the same as, or smaller than the spacing for an identical volume of aluminum at that same energy?

Sample Problem 42-3

(a) Using Fig. 42-5, determine the number of states per electron-volt at 7 eV in a metal sample with a volume V of $2 \times 10^{-9}\ m^3$.

SOLUTION: The **Key Idea** is that we can obtain the number of states per electron-volt at a given energy by using the density of states $N(E)$ at that energy and the sample's volume V. At an energy of 7 eV, this means that

$$\begin{pmatrix}\text{number of states}\\ \text{per eV at 7 eV}\end{pmatrix} = \begin{pmatrix}\text{density of states}\\ N(E) \text{ at 7 eV}\end{pmatrix}\begin{pmatrix}\text{volume } V\\ \text{of sample}\end{pmatrix}.$$

From Fig. 42-5, we see that at an energy E of 7 eV, the density of states is about $2 \times 10^{28}\ m^{-3}\ eV^{-1}$. Thus,

$$\begin{pmatrix}\text{number of states}\\ \text{per eV at 7 eV}\end{pmatrix} = (2 \times 10^{28}\ m^{-3}\ eV^{-1})(2 \times 10^{-9}\ m^3)$$
$$= 4 \times 10^{19}\ eV^{-1}. \quad \text{(Answer)}$$

(b) Next, determine the number of states N in the sample within a *small* energy range ΔE of 0.003 eV, centered at 7 eV.

SOLUTION: From Eq. 42-5 and Fig. 42-5, we know that the density of states is a function of energy E. However, for an energy range ΔE that is small relative to E, we can approximate the density of states (and thus the number of states per electron-volt) to be constant. Thus, at an energy of 7 eV, we find the number of states N in the energy range ΔE of 0.003 eV as

$$\begin{pmatrix}\text{number of states } N\\ \text{in range } \Delta E \text{ at 7 eV}\end{pmatrix} = \begin{pmatrix}\text{number of states}\\ \text{per eV at 7 eV}\end{pmatrix} \times (\text{energy range } \Delta E)$$

or

$$N = (4 \times 10^{19}\ eV^{-1})(0.003\ eV)$$
$$= 1.2 \times 10^{17} \approx 1 \times 10^{17}. \quad \text{(Answer)}$$

The Occupancy Probability $P(E)$

The ability of a metal to conduct electricity depends on the probability that available vacant levels will actually be occupied. Thus, another question arises: If an energy level is available at energy E, what is the probability $P(E)$ that it is actually occupied by an electron? At $T = 0$ K, we know that for all levels with energies below the Fermi energy, $P(E) = 1$, corresponding to a certainty that the level is occupied. We also know that, at $T = 0$ K, for all levels with energies above the Fermi energy, $P(E) = 0$, corresponding to a certainty that the level is *not* occupied. Figure 42-6*a* illustrates this situation.

To find $P(E)$ at temperatures above absolute zero, we must use a set of quantum counting rules called **Fermi–Dirac statistics,** named for the physicists who intro-

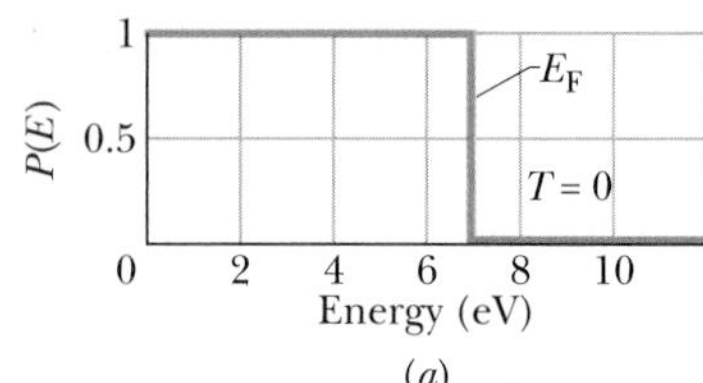

(a)

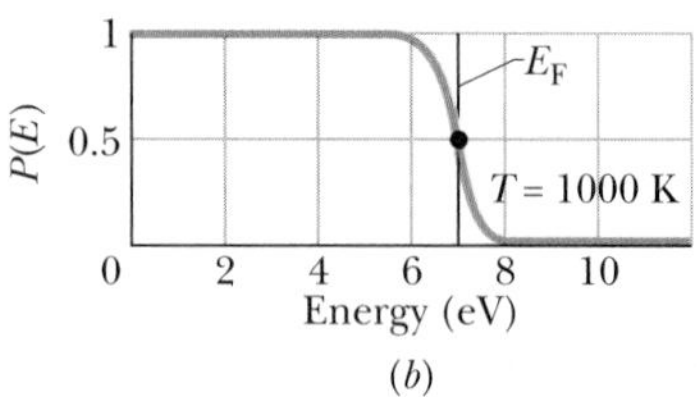

(b)

Fig. 42-6 The occupancy probability $P(E)$ is the probability that an energy level will be occupied by an electron. (*a*) At $T = 0$ K, $P(E)$ is unity for levels with energies E up to the Fermi energy E_F and zero for levels with higher energies. (*b*) At $T = 1000$ K, a few electrons whose energies were slightly less than the Fermi energy at $T = 0$ K move up to states with energies slightly greater than the Fermi energy. The dot on the curve shows that, for $E = E_F$, $P(E) = 0.5$.

duced them. Using these rules, it is possible to show that the **occupancy probability** $P(E)$ is

$$P(E) = \frac{1}{e^{(E - E_F)/kT} + 1} \quad \text{(occupancy probability)}, \tag{42-6}$$

in which E_F is the Fermi energy. Note that $P(E)$ depends not on the energy E of the level but only on the difference $E - E_F$, which may be positive or negative.

To see whether Eq. 42-6 describes Fig. 42-6*a*, we substitute $T = 0$ K in it. Then,

For $E < E_F$, the exponential term in Eq. 42-6 is $e^{-\infty}$, or zero, so $P(E) = 1$, in agreement with Fig. 42-6*a*.

For $E > E_F$, the exponential term is $e^{+\infty}$, so $P(E) = 0$, again in agreement with Fig. 42-6*a*.

Figure 42-6*b* is a plot of $P(E)$ for $T = 1000$ K. It shows that, as stated above, changes in the distribution of electrons among the available states involve only states whose energies are near the Fermi energy E_F. Note that if $E = E_F$ (no matter what the temperature T), the exponential term in Eq. 42-6 is $e^0 = 1$ and $P(E) = 0.5$. This leads us to a more useful definition of the Fermi energy:

> The Fermi energy of a given material is the energy of a quantum state that has the probability 0.5 of being occupied by an electron.

Figures 42-6*a* and *b* are plotted for copper, which has a Fermi energy of 7.0 eV. Thus, for copper both at $T = 0$ K and at $T = 1000$ K, a state at energy $E = 7.0$ eV has a probability of 0.5 of being occupied.

Sample Problem 42-4

(a) What is the probability that a quantum state whose energy is 0.10 eV above the Fermi energy will be occupied? Assume a sample temperature of 800 K.

SOLUTION: The Key Idea here is that the occupancy probability of any state in a metal can be found from Fermi–Dirac statistics according to Eq. 42-6. To apply that equation, let us first calculate its dimensionless exponent:

$$\frac{E - E_F}{kT} = \frac{0.10 \text{ eV}}{(8.62 \times 10^{-5} \text{ eV/K})(800 \text{ K})} = 1.45.$$

Inserting this exponent into Eq. 42-6 yields

$$P(E) = \frac{1}{e^{1.45} + 1} = 0.19 \text{ or } 19\%. \quad \text{(Answer)}$$

(b) What is the probability of occupancy for a state that is 0.10 eV *below* the Fermi energy?

SOLUTION: The Key Idea of part (a) applies here also except that now the state has an energy *below* the Fermi energy. Thus, the exponent in Eq. 42-6 has the same magnitude we found in part (a) but is negative, so Eq. 42-6 now yields

$$P(E) = \frac{1}{e^{-1.45} + 1} = 0.81 \text{ or } 81\%. \quad \text{(Answer)}$$

For states below the Fermi energy, we are often more interested in the probability that the state is *not* occupied. This probability is just $1 - P(E)$, or 19%. Note that it is the same as the probability of occupancy in (a).

How Many *Occupied* States Are There?

Equation 42-5 and Fig. 42-5 tell us how the available states are distributed in energy. The occupancy probability of Eq. 42-6 gives us the probability that any given state will actually be occupied by an electron. To find $N_o(E)$, the density of *occupied* states, we must weight each available state by the appropriate value of the occupancy

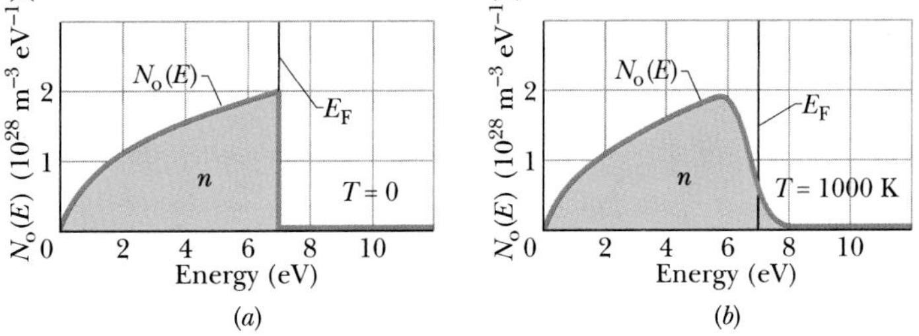

Fig. 42-7 (a) The density of occupied states $N_o(E)$ for copper at absolute zero. The area under the curve is the number density of electrons n. Note that all states with energies up to the Fermi energy $E_F = 7$ eV are occupied, and all those with energies above the Fermi energy are vacant. (b) The same for copper at $T = 1000$ K. Note that only electrons with energies near the Fermi energy have been affected and redistributed.

probability; that is,

$$\begin{pmatrix}\text{density of occupied states}\\ N_o(E)\text{ at energy }E\end{pmatrix} = \begin{pmatrix}\text{density of states}\\ N(E)\text{ at energy }E\end{pmatrix}\begin{pmatrix}\text{occupancy probability}\\ P(E)\text{ at energy }E\end{pmatrix}$$

or

$$N_o(E) = N(E)\,P(E) \quad \text{(density of occupied states).} \tag{42-7}$$

Figure 42-7a is a plot of Eq. 42-7 for copper at $T = 0$ K. It is found by multiplying, at each energy, the value of the density of states function (Fig. 42-5) by the value of the occupancy probability for absolute zero (Fig. 42-6a). Figure 42-7b, calculated similarly, shows the density of occupied states for copper at $T = 1000$ K.

Sample Problem 42-5

If the sample in Sample Problem 42-3 is copper, which has a Fermi energy of 7.0 eV, how many occupied states per electron-volt lie in a narrow energy range around 7.0 eV?

SOLUTION: The **Key Idea** of Sample Problem 42-3a applies here also, except that now we use the density of *occupied* states $N_o(E)$ as given by Eq. 42-7 ($N_o(E) = N(E)\,P(E)$). A second **Key Idea** is that because we want to evaluate quantities for a narrow energy range around 7.0 eV (the Fermi energy for copper), the occupancy probability $P(E)$ is 0.50. From Fig. 42-5, we see that the density of states at 7 eV is $2 \times 10^{18}\ \text{m}^{-3}\,\text{eV}^{-1}$. Thus, Eq. 42-7 tells us that the density of occupied states is

$$N_o(E) = N(E)\,P(E) = (2 \times 10^{28}\ \text{m}^{-3}\,\text{eV}^{-1})(0.50)$$
$$= 1 \times 10^{28}\ \text{m}^{-3}\,\text{eV}^{-1}.$$

Next, we rewrite the equation in Sample Problem 42-3a in terms of occupied states:

$$\begin{pmatrix}\text{number of } occupied\\ \text{states per eV at 7 eV}\end{pmatrix} = \begin{pmatrix}\text{density of } occupied\\ \text{states } N_o(E)\text{ at 7 eV}\end{pmatrix} \times \begin{pmatrix}\text{volume } V\\ \text{of sample}\end{pmatrix}.$$

Substituting our result for $N_o(E)$ and the previously given volume $2 \times 10^{-9}\ \text{m}^3$ for V then gives us

$$\begin{pmatrix}\text{number of occupied}\\ \text{states per eV}\\ \text{at 7 eV}\end{pmatrix} = (1 \times 10^{28}\ \text{m}^{-3}\,\text{eV}^{-1})(2 \times 10^{-9}\ \text{m}^3)$$
$$= 2 \times 10^{19}\ \text{eV}^{-1}. \quad \text{(Answer)}$$

Calculating the Fermi Energy

Suppose we add up (via integration) the number of occupied states per unit volume in Fig. 42-7a at all energies between $E = 0$ and $E = E_F$. The result must equal n, the number of conduction electrons per unit volume for the metal. In equation form, we have

$$n = \int_0^{E_F} N_o(E)\,dE. \tag{42-8}$$

(Graphically, the integral here represents the area under the distribution curve of Fig. 42-7*a*.) Because $P(E) = 1$ for all energies below the Fermi energy, Eq. 42-7 tells us we can replace $N_o(E)$ in Eq. 42-8 with $N(E)$ and then use Eq. 42-8 to find the Fermi energy E_F. If we substitute Eq. 42-5 into Eq. 42-8, we find that

$$n = \frac{8\sqrt{2}\pi m^{3/2}}{h^3} \int_0^{E_F} E^{1/2}\, dE = \frac{8\sqrt{2}\pi m^{3/2}}{h^3} \frac{2E_F^{3/2}}{3}.$$

Solving for E_F now leads to

$$E_F = \left(\frac{3}{16\sqrt{2}\pi}\right)^{2/3} \frac{h^2}{m} n^{2/3} = \frac{0.121h^2}{m} n^{2/3}. \tag{42-9}$$

Thus, when we know n, the number of conduction electrons per unit volume for a metal, we can find the Fermi energy for that metal.

42-6 Semiconductors

If you compare Fig. 42-8*a* with Fig. 42-4*a*, you can see that the band structure of a semiconductor is like that of an insulator. The main difference is that the semiconductor has a much smaller energy gap E_g between the top of the highest filled band (called the **valence band**) and the bottom of the vacant band just above it (called the **conduction band**). Thus, there is no doubt that silicon (E_g = 1.1 eV) is a semiconductor and diamond (E_g = 5.5 eV) is an insulator. In silicon—but not in diamond—there is a real possibility that thermal agitation at room temperature will cause electrons to jump the gap from the valence band to the conduction band.

In Table 42-1 we compared three basic electrical properties of copper, our prototype metallic conductor, and silicon, our prototype semiconductor. Let us look again at that table, one row at a time, to see how a semiconductor differs from a metal.

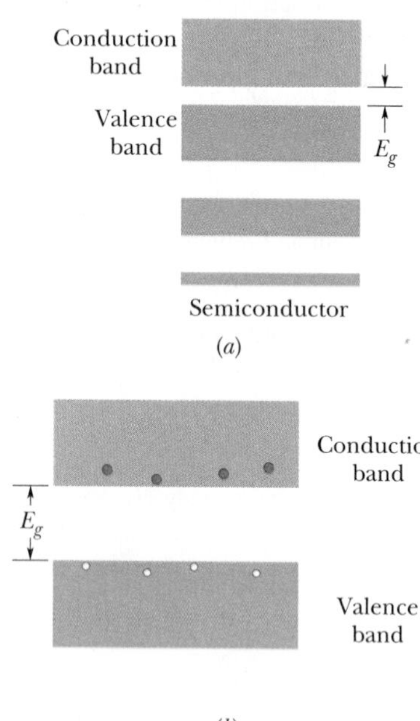

Fig. 42-8 (*a*) The band–gap pattern for a semiconductor. It resembles that of an insulator (see Fig. 42-4*a*) except that here the energy gap E_g is much smaller; thus electrons, because of their thermal agitation, have some reasonable probability of being able to jump the gap. (*b*) Thermal agitation has caused a few electrons to jump the gap from the valence band to the conduction band, leaving an equal number of holes in the valence band.

Number Density of Charge Carriers *n*

The bottom row of Table 42-1 shows that copper has far more charge carriers per unit volume than silicon, by a factor of about 10^{13}. For copper, each atom contributes one electron, its single valence electron, to the conduction process. Charge carriers in silicon arise only because, at thermal equilibrium, thermal agitation causes a certain (very small) number of valence-band electrons to jump the energy gap into the conduction band, leaving an equal number of unoccupied energy states, called **holes,** in the valence band. Figure 42-8*b* shows the situation.

Both the electrons in the conduction band and the holes in the valence band serve as charge carriers. The holes do so by permitting a certain freedom of movement to electrons in the valence band that, in the absence of holes, would be gridlocked. If an electric field $\vec{E}$ is set up in a semiconductor, the electrons in the valence band, being negatively charged, tend to drift in the direction opposite $\vec{E}$. This causes the positions of the holes to drift in the direction of $\vec{E}$. In effect, the holes behave like moving particles of charge $+e$.

It may help to think of a row of cars parked bumper to bumper, with the leading car at one car's length from a barrier. If the leading car moves forward to the barrier, it opens up a car's length space behind it. The second car can then move up to fill that space, allowing the third car to move up, and so on. The motions of the many cars toward the barrier are most simply analyzed by focusing attention on the drift of the single "hole" (parking space) away from the barrier.

In semiconductors, conduction by holes is just as important as conduction by electrons. In thinking about hole conduction, it is well to imagine that all unoccupied states in the valence band are occupied by particles of charge $+e$, and that all electrons in the valence band have been removed, so that these positive charge carriers can move freely throughout the band.

Resistivity ρ

From Chapter 27 recall that the resistivity ρ of a material is $m/e^2n\tau$, where m is the electron mass, e is the fundamental charge, n is the number of charge carriers per unit volume, and τ is the mean time between collisions of the charge carriers. Table 42-1 shows that, at room temperature, the resistivity of silicon is higher than that of copper, by a factor of about 10^{11}. This vast difference can be accounted for by the vast difference in n. Other factors enter, but their effect on the resistivity is swamped by the enormous difference in n.

Temperature Coefficient of Resistivity α

Recall that α (see Eq. 27-17) is the fractional change in resistivity per unit change in temperature:

$$\alpha = \frac{1}{\rho}\frac{d\rho}{dT}. \tag{42-10}$$

The resistivity of copper *increases* with temperature (that is, $d\rho/dT > 0$) because collisions of copper's charge carriers occur more frequently at higher temperatures. Thus, α is *positive* for copper.

The collision frequency also increases with temperature for silicon. However, the resistivity of silicon actually *decreases* with temperature ($d\rho/dT < 0$) because the number of charge carriers n (electrons in the conduction band and holes in the valence band) increases so rapidly with temperature. (More electrons jump the gap from the valence band to the conduction band.) Thus, the fractional change α is *negative* for silicon.

CHECKPOINT 2: The research laboratory of a large corporation developed three new solid materials whose electrical properties are shown here. Anticipating patent applications, the laboratory identified these materials with code names. Classify each material as a metal, an insulator, a semiconductor, or none of the above:

Material (Code Name)	n (m^{-3})	ρ ($\Omega \cdot$ m)	α (K^{-1})
Cleveland	10^{29}	10^{-8}	$+10^{-3}$
Boca Raton	10^{28}	10^{-9}	-10^{-3}
Seattle	10^{15}	10^{3}	-10^{-2}

42-7 Doped Semiconductors

The usefulness of semiconductors in technology can be greatly improved by introducing a small number of suitable replacement atoms (called impurities) into the semiconductor lattice—a process called **doping.** Typically, only about 1 silicon atom in 10^7 is replaced by a dopant atom in the doped semiconductor. Essentially all modern semiconducting devices are based on doped material. Such materials are of two types, called ***n*-type** and ***p*-type**; we discuss each in turn.

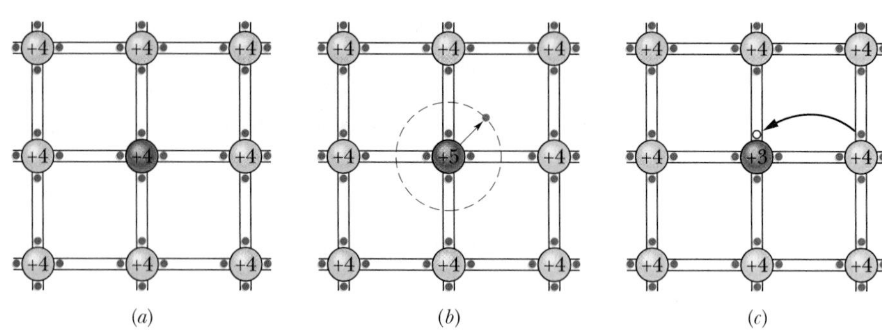

Fig. 42-9 (*a*) A flattened-out representation of the lattice structure of pure silicon. Each silicon ion is coupled to its four nearest neighbors by a two-electron covalent bond (represented by a pair of red dots between two parallel black lines). The electrons belong to the bond—not to the individual atoms—and form the valence band of the sample. (*b*) One silicon atom is replaced by a phosphorus atom (valence = 5). The "extra" electron is only loosely bound to its ion core and may easily be elevated to the conduction band, where it is free to wander through the volume of the lattice. (*c*) One silicon atom is replaced by an aluminum atom (valence = 3). There is now a hole in one of the covalent bonds and thus in the valence band of the sample. The hole can easily migrate through the lattice as electrons from neighboring bonds move in to fill it. Here the hole migrates rightward.

n-Type Semiconductors

The electrons in an isolated silicon atom are arranged in subshells according to the scheme

$$1s^2\ 2s^2\ 2p^6\ 3s^2\ 3p^2,$$

in which, as usual, the superscripts (which add to 14, the atomic number of silicon) represent the numbers of electrons in the specified subshells.

Figure 42-9*a* is a flattened-out representation of a portion of the lattice of pure silicon in which the portion has been projected onto a plane; compare the figure with Fig. 42-1*b*, which represents the unit cell of the lattice in three dimensions. Each silicon atom contributes its pair of 3*s* electrons and its pair of 3*p* electrons to form a rigid two-electron covalent bond with each of its four nearest neighbors. (A covalent bond is a link between two atoms in which the atoms share a pair of electrons.) The four atoms that lie within the unit cell in Fig. 42-1*b* show these four bonds.

The electrons that form the silicon–silicon bonds constitute the valence band of the silicon sample. If an electron is torn from one of these bonds so that it becomes free to wander throughout the lattice, we say that the electron has been raised from the valence band to the conduction band. The minimum energy required to do this is the gap energy E_g.

Because four of its electrons are involved in bonds, each silicon "atom" is actually an ion consisting of an inert neonlike electron cloud (containing 10 electrons) surrounding a nucleus whose charge is $+14e$, where 14 is the atomic number of silicon. The net charge of each of these ions is thus $+4e$, and the ions are said to have a *valence number* of 4.

In Fig. 42-9*b* the central silicon ion has been replaced by an atom of phosphorus (valence = 5). Four of the valence electrons of the phosphorus form bonds with the four surrounding silicon ions. The fifth ("extra") electron is only loosely bound to the phosphorus ion core. On an energy-band diagram, we usually say that such an electron occupies a localized energy state that lies within the energy gap, at an average energy interval E_d below the bottom of the conduction band; this is indicated in Fig. 42-10*a*. Because $E_d \ll E_g$, the energy required to excite electrons from *these* levels into the conduction band is much less than that required to excite silicon valence electrons into the conduction band.

The phosphorus atom is called a **donor** atom because it readily *donates* an electron to the conduction band. In fact, at room temperature virtually *all* the electrons contributed by the donor atoms are in the conduction band. By adding donor atoms, it is possible to increase greatly the number of electrons in the conduction band, by a factor very much larger than Fig. 42-10*a* suggests.

Semiconductors doped with donor atoms are called ***n*-type semiconductors;** the *n* stands for *negative,* to imply that the negative charge carriers introduced into the conduction band greatly outnumber the positive charge carriers, which are the holes

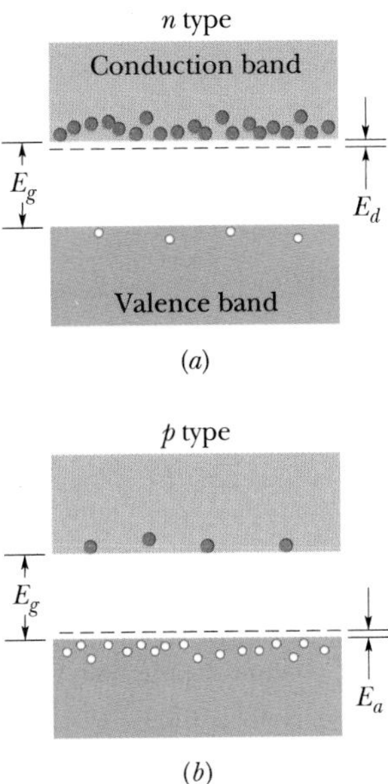

Fig. 42-10 (*a*) In a doped *n*-type semiconductor, the energy levels of donor electrons lie a small interval E_d below the bottom of the conduction band. Because donor electrons can be easily excited to the conduction band, there are now many more electrons in that band. The valence band contains the same small number of holes as before. (*b*) In a doped *p*-type semiconductor, the acceptor levels lie a small energy interval E_a above the top of the valence band. There are now relatively many more holes in the valence band. The conduction band contains the same small number of electrons as before. The ratio of majority carriers to minority carriers in both (*a*) and (*b*) is very much greater than is suggested by these diagrams.

TABLE 42-2 **Properties of Two Doped Semiconductors**

	Type of Semiconductor	
Property	*n*	*p*
Matrix material	Silicon	Silicon
Matrix nuclear charge	$+14e$	$+14e$
Matrix energy gap	1.2 eV	1.2 eV
Dopant	Phosphorus	Aluminum
Type of dopant	Donor	Acceptor
Majority carriers	Electrons	Holes
Minority carriers	Holes	Electrons
Dopant energy gap	0.045 eV	0.067 eV
Dopant valence	5	3
Dopant nuclear charge	$+15e$	$+13e$
Dopant net ion charge	$+e$	$-e$

in the valence band. In *n*-type semiconductors, the electrons are called the **majority carriers,** and the holes the **minority carriers.**

p-Type Semiconductors

Now consider Fig. 42-9*c*, in which one of the silicon atoms (valence = 4) has been replaced by an atom of aluminum (valence = 3). The aluminum atom can bond covalently with only three silicon atoms, so there is now a "missing" electron (a hole) in one aluminum–silicon bond. With a small expenditure of energy, an electron can be torn from a neighboring silicon–silicon bond to fill this hole, thereby creating a hole in *that* bond. Similarly, an electron from some other bond can be moved to fill the second hole. In this way, the hole can migrate through the lattice.

The aluminum atom is called an **acceptor** atom because it readily *accepts* an electron from a neighboring bond—that is, from the valence band of silicon. As Fig. 42-10*b* suggests, this electron occupies a localized acceptor state that lies within the energy gap, at an average energy interval E_a above the top of the valence band. By adding acceptor atoms, it is possible to increase very greatly the number of holes in the valence band, by a factor much larger than Fig. 42-10*b* suggests. In silicon at room temperature, virtually *all* the acceptor levels are occupied by electrons.

Semiconductors doped with acceptor atoms are called ***p*-type semiconductors;** the *p* stands for *positive* to imply that the holes introduced into the valence band, which behave like positive charge carriers, greatly outnumber the electrons in the conduction band. In *p*-type semiconductors, holes are the majority carriers and electrons are the minority carriers.

Table 42-2 summarizes the properties of a typical *n*-type and a typical *p*-type semiconductor. Note particularly that the donor and acceptor ion cores, although they are charged, are not charge *carriers* because at normal temperatures they remain fixed in their lattice sites.

Sample Problem 42-6

The number density n_0 of conduction electrons in pure silicon at room temperature is about $10^{16}\ \text{m}^{-3}$. Assume that, by doping the silicon lattice with phosphorus, we want to increase this number by a factor of a million (10^6). What fraction of silicon atoms must we replace with phosphorus atoms? (Recall that at room temperature, thermal agitation is so effective that essentially every phosphorus atom donates its "extra" electron to the conduction band.)

SOLUTION: One Key Idea here is that, because each phosphorus atom contributes one conduction electron and because we want the total number density of conduction electrons to be $10^6 n_0$, then the number density of phosphorus atoms n_P must be given by

$$10^6 n_0 = n_0 + n_P.$$

Then

$$n_P = 10^6 n_0 - n_0 \approx 10^6 n_0$$
$$= (10^6)(10^{16}\ \text{m}^{-3}) = 10^{22}\ \text{m}^{-3}.$$

This tells us that we must add 10^{22} atoms of phosphorus per cubic meter of silicon.

A second **Key Idea** is that we can find the number density n_{Si} of silicon atoms in pure silicon (before the doping) from Eq. 42-4, which we can write as

$$\begin{pmatrix}\text{number of atoms}\\ \text{in sample}\end{pmatrix} = \frac{(\text{silicon density})(\text{sample volume } V)}{(\text{silicon molar mass } M_{Si})/N_A}.$$

Dividing both sides by the sample volume V to get the number density of silicon atoms n_{Si} on the left, we then have

$$n_{Si} = \frac{(\text{silicon density})N_A}{M_{Si}}.$$

Appendix F tells us that the density of silicon is 2.33 g/cm^3 (= 2330 kg/m^3) and the molar mass of silicon is 28.1 g/mol (= 0.0281 kg/mol). Thus, we have

$$n_{Si} = \frac{(2330\ \text{kg/m}^3)(6.02 \times 10^{23}\ \text{mol}^{-1})}{0.0281\ \text{kg/mol}}$$
$$= 5 \times 10^{28}\ \text{m}^{-3}.$$

The fraction we seek is approximately

$$\frac{n_P}{n_{Si}} = \frac{10^{22}\ \text{m}^{-3}}{5 \times 10^{28}\ \text{m}^{-3}} = \frac{1}{5 \times 10^6}. \quad \text{(Answer)}$$

If we replace only *one silicon atom in five million* with a phosphorus atom, the number of electrons in the conduction band will be increased by a factor of a million.

How can such a tiny admixture of phosphorus have what seems to be such a big effect? The answer is that, although the effect is very significant, it is not "big." The number density of conduction electrons was $10^{16}\ \text{m}^{-3}$ before doping and $10^{22}\ \text{m}^{-3}$ after doping. For copper, however, the conduction-electron number density (given in Table 42-1) is about $10^{29}\ \text{m}^{-3}$. Thus, even after doping, the number density of conduction electrons in silicon remains much less than that of a typical metal, such as copper, by a factor of about 10^7.

42-8 The *p-n* Junction

A ***p-n* junction** (Fig. 42-11*a*) is a single semiconductor crystal that has been selectively doped so that one region is *n*-type material and the adjacent region is *p*-type material. Such junctions are at the heart of essentially all semiconductor devices.

We assume, for simplicity, that the junction has been formed mechanically, by jamming together a bar of *n*-type semiconductor and a bar of *p*-type semiconductor. Thus, the transition from one region to the other is perfectly sharp, occurring at a single **junction plane.**

Let us discuss the motions of electrons and holes just after the *n*-type bar and the *p*-type bar, both electrically neutral, have been jammed together to form the junction. We first examine the majority carriers, which are electrons in the *n*-type material and holes in the *p*-type material.

Motions of the Majority Carriers

If you burst a helium-filled balloon, helium atoms will diffuse (spread) outward into the surrounding air. This happens because there are very few helium atoms in normal air. In more formal language, there is a helium *density gradient* at the balloon–air interface (the number density of helium atoms varies across the interface); the helium atoms move so as to reduce the gradient.

In the same way, electrons on the *n* side of Fig. 42-11*a* that are close to the junction plane tend to diffuse across it (from right to left in the figure) and into the *p* side, where there are very few free electrons. Similarly, holes on the *p* side that are close to the junction plane tend to diffuse across that plane (from left to right) and into the *n* side, where there are very few holes. The motions of both the electrons and the holes contribute to a **diffusion current** I_{diff}, conventionally directed from left to right as indicated in Fig. 42-11*d*.

Recall that the *n*-side is studded throughout with positively charged donor ions,

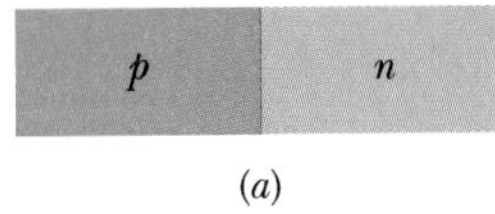

(a)

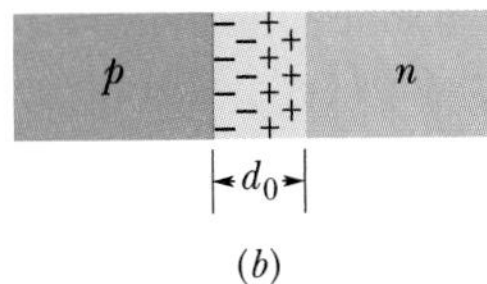

(b)

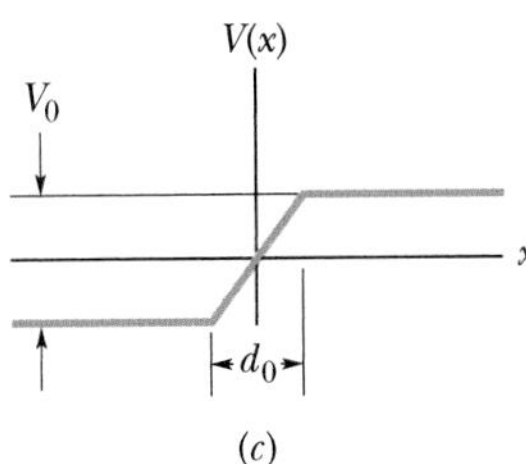

(c)

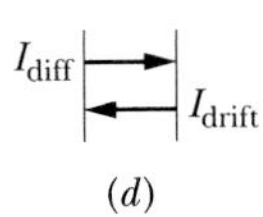

(d)

Fig. 42-11 (*a*) A *p-n* junction. (*b*) Motions of the majority charge carriers across the junction plane uncover a space charge associated with uncompensated donor ions (to the right of the plane) and acceptor ions (to the left). (*c*) Associated with the space charge is a contact potential difference V_0 across d_0. (*d*) The diffusion of majority carriers (both electrons and holes) across the junction plane produces a diffusion current I_{diff}. (In a real *p-n* junction, the boundaries of the depletion zone would not be sharp, as shown here, and the contact potential curve (*c*) would be smooth, with no sharp corners.)

fixed firmly in their lattice sites. Normally, the excess positive charge of each of these ions is compensated electrically by one of the conduction-band electrons. When an *n*-side electron diffuses across the junction plane, however, the diffusion "uncovers" one of these donor ions, thus introducing a fixed positive charge near the junction plane on the *n* side. When the diffusing electron arrives on the *p* side, it quickly combines with an acceptor ion (which lacks one electron), thus introducing a fixed negative charge near the junction plane on the *p* side.

In this way electrons diffusing through the junction plane from right to left in Fig. 42-11*a* result in a buildup of **space charge** on each side of the junction plane, as indicated in Fig. 42-11*b*. Holes diffusing through the junction plane from left to right have exactly the same effect. (Take the time now to convince yourself of that.) The motions of both majority carriers—electrons and holes—contribute to the buildup of these two space charge regions, one positive and one negative. These two regions form a **depletion zone,** so named because it is relatively free of *mobile* charge carriers; its width is shown as d_0 in Fig. 42-11*b*.

The buildup of space charge generates an associated **contact potential difference** V_0 across the depletion zone, as Fig. 42-11*c* shows. This potential difference limits further diffusion of electrons and holes across the junction plane. Negative charges tend to avoid regions of low potential. Thus, an electron approaching the junction plane from the right in Fig. 42-11*b* is moving toward a region of low potential and would tend to turn back into the *n* side. Similarly, a positive charge (a hole) approaching the junction plane from the left is moving toward a region of high potential and would tend to turn back into the *p* side.

Motions of the Minority Carriers

As Fig. 42-10*a* shows, although the majority carriers in *n*-type material are electrons, there are nevertheless a few holes. Likewise in *p*-type material (Fig. 42-10*b*), although the majority carriers are holes, there are also a few electrons. These few holes and electrons are the minority carriers in the corresponding materials.

Although the potential difference V_0 in Fig. 42-11*c* acts as a barrier for the majority carriers, it is a downhill trip for the minority carriers, be they electrons on the *p* side or holes on the *n* side. Positive charges (holes) tend to seek regions of low potential; negative charges (electrons) tend to seek regions of high potential. Thus, both types of carriers are *swept across* the junction plane by the contact potential difference and, together, constitute a **drift current** I_{drift} across the junction plane from right to left, as Fig. 42-11*d* indicates.

Thus, an isolated *p-n* junction is in an equilibrium state in which a contact potential difference V_0 exists between its ends. At equilibrium, the average diffusion current I_{diff} that moves through the junction plane from the *p* side to the *n* side is just balanced by an average drift current I_{drift} that moves in the opposite direction. These two currents cancel because the net current through the junction plane must be zero; otherwise charge would be transferred without limit from one end of the junction to the other.

CHECKPOINT 3: Which of the following five currents across the junction plane of Fig. 42-11*a* must be zero?
(a) the net current due to holes, both majority and minority carriers included
(b) the net current due to electrons, both majority and minority carriers included
(c) the net current due to both holes and electrons, both majority and minority carriers included
(d) the net current due to majority carriers, both holes and electrons included
(e) the net current due to minority carriers, both holes and electrons included

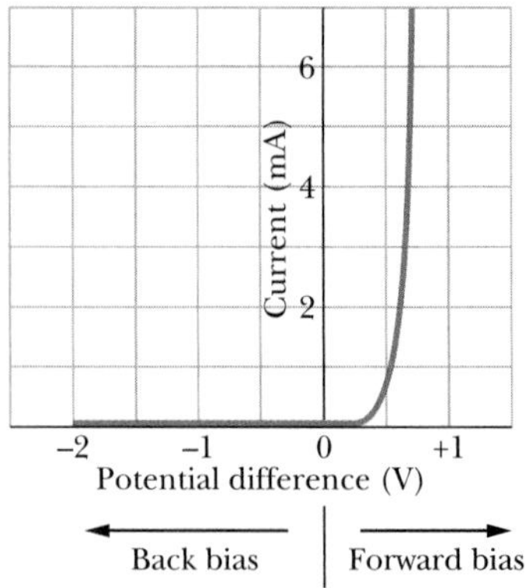

Fig. 42-12 A current–voltage plot for a *p-n* junction, showing that the junction is highly conducting when forward-biased and essentially nonconducting when back-biased.

42-9 The Junction Rectifier

Look now at Fig. 42-12. It shows that, if we place a potential difference across a *p-n* junction in one direction (here labeled + and "Forward bias"), there will be a current through the junction. However, if we reverse the direction of the potential difference, there will be approximately zero current through the junction.

One application of this property is the **junction rectifier,** whose symbol is shown in Fig. 42-13*b*; the arrowhead corresponds to the *p*-type end of the device and points in the allowed direction of conventional current. A sine wave input potential to the device (Fig. 42-13*a*) is transformed to a half-wave output potential (Fig. 42-13*c*) by the junction rectifier; that is, the rectifier acts as essentially a closed switch (zero resistance) for one polarity of the input potential and as essentially an open switch (infinite resistance) for the other.

The average value of the input voltage in Fig. 42-13*a* is zero, but that of the output voltage in Fig. 42-13*c* is not. Thus, a junction rectifier can be used as part of an apparatus to convert an alternating potential difference into a constant potential difference, as for an electronic power supply.

Figure 42-14 shows why a *p-n* junction operates as a junction rectifier. In Fig. 42-14*a*, a battery is connected across the junction with its positive terminal connected at the *p* side. In this **forward-bias connection,** the *p* side becomes more positive than it was before the connection and the *n* side becomes more negative, thus *decreasing* the height of the potential barrier V_0 of Fig. 42-11*c*. More of the majority carriers can now surmount this smaller barrier; hence, the diffusion current I_{diff} increases markedly.

The minority carriers that form the drift current, however, sense no barrier, so the drift current I_{drift} is not affected by the external battery. The nice current balance that existed at zero bias (see Fig. 42-11*d*) is thus upset and, as shown in Fig. 42-14*a*, a large net forward current I_F appears in the circuit.

Another effect of forward bias is to narrow the depletion zone, as a comparison of Figs. 42-11*b* and Fig. 42-14*a* shows. The depletion zone narrows because the reduced potential barrier associated with forward bias must be associated with a smaller space charge. Because the ions producing the space charge are fixed in their lattice sites, a reduction in their number can come about only through a reduction in the width of the depletion zone.

Because the depletion zone normally contains very few charge carriers, it is normally a region of high resistivity. However, when its width is substantially re-

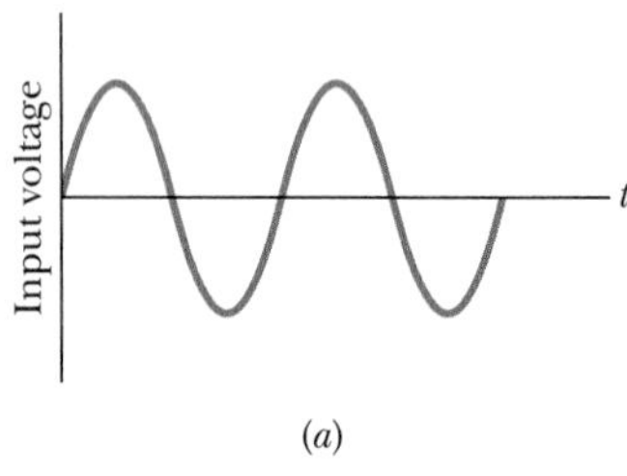

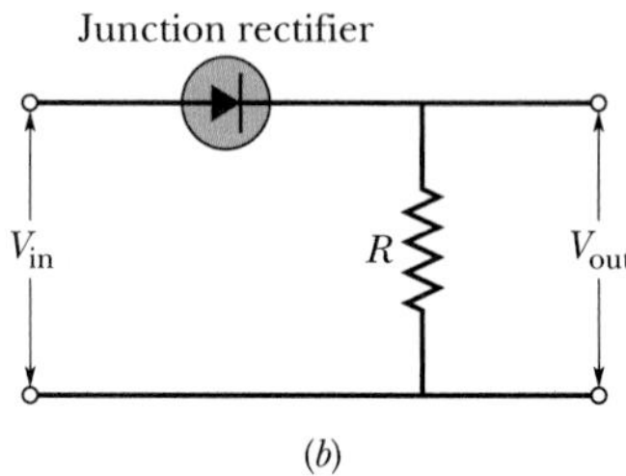

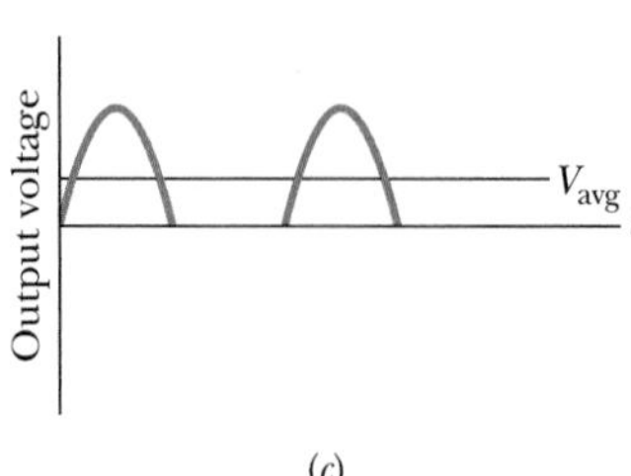

Fig. 42-13 A *p-n* junction connected as a junction rectifier. The action of the circuit in (*b*) is to pass the positive half of the input wave form (*a*) but to suppress the negative half. The average potential of the input wave form is zero; that of the output wave form (*c*) has a positive value V_{avg}.

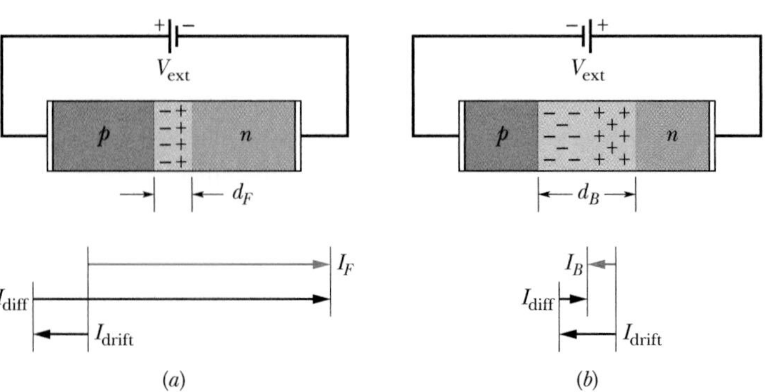

Fig. 42-14 (*a*) The forward-bias connection of a *p-n* junction, showing the narrowed depletion zone and the large forward current I_F. (*b*) The back-bias connection, showing the widened depletion zone and the small back current I_B.

duced by a forward bias, its resistance is also reduced substantially, as is consistent with the large forward current.

Figure 42-14*b* shows the **back-bias** connection, in which the negative terminal of the battery is connected at the *p*-type end of the *p*-*n* junction. Now the applied emf *increases* the contact potential difference, the diffusion current *decreases* substantially while the drift current remains unchanged, and a relatively *small* back current I_B results. The depletion zone *widens,* its *high* resistance being consistent with the *small* back current I_B.

42-10 The Light-Emitting Diode (LED)

Nowadays, we can hardly avoid the brightly colored "electronic" numbers that glow at us from cash registers and gasoline pumps, microwave ovens and alarm clocks, and we cannot seem to do without the invisible infrared beams that control elevator doors and operate television sets via remote control. In nearly all cases this light is emitted from a *p*-*n* junction operating as a **light-emitting diode** (LED). How can a *p*-*n* junction generate light?

Consider first a simple semiconductor. When an electron from the bottom of the conduction band falls into a hole at the top of the valence band, an energy E_g equal to the gap width is released. In silicon, germanium, and many other semiconductors, this energy is largely transformed into thermal energy of the vibrating lattice, and as a result, no light is emitted.

In some semiconductors, however, including gallium arsenide, the energy can be emitted as a photon of energy hf at wavelength

$$\lambda = \frac{c}{f} = \frac{c}{E_g/h} = \frac{hc}{E_g}. \tag{42-11}$$

To emit enough light to be useful as an LED, the material must have a suitably large number of electron–hole transitions. This condition is *not* satisfied by a pure semiconductor because, at room temperature, there are simply not enough electron–hole pairs. As Fig. 42-10 suggests, doping will not help. In doped *n*-type material the number of conduction electrons is greatly increased, but there are not enough holes for them to combine with; in doped *p*-type material there are plenty of holes but not enough electrons to combine with them. Thus, neither a pure semiconductor nor a doped semiconductor can provide enough electron–hole transitions to serve as a practical LED.

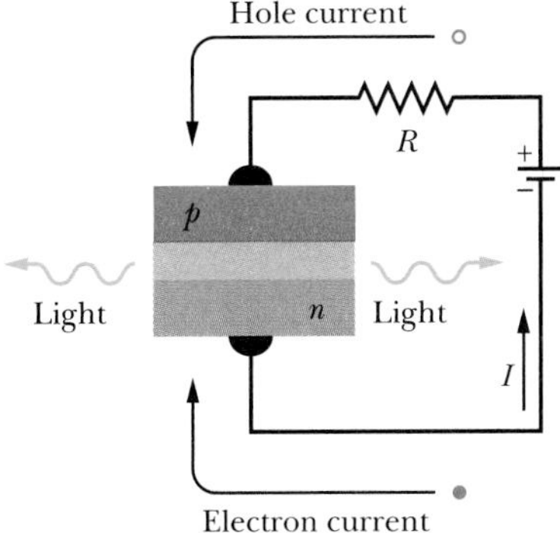

Fig. 42-15 A forward-biased *p*-*n* junction, showing electrons being injected into the *n*-type material and holes into the *p*-type material. (Holes move in the conventional direction of the current I, equivalent to electrons moving in the opposite direction.) Light is emitted from the narrow depletion zone each time an electron and a hole combine across that zone.

What we need is a semiconductor material with a very large number of electrons in the conduction band *and* a correspondingly large number of holes in the valence band. A device with this property can be fabricated by placing a strong forward bias on a heavily doped *p*-*n* junction, as in Fig. 42-15. In such an arrangement the current I through the device serves to inject electrons into the *n*-type material and to inject holes into the *p*-type material. If the doping is heavy enough and the current is great enough, the depletion zone can become very narrow, perhaps only a few micrometers wide. The result is a great number density of electrons in the *n*-type material facing a correspondingly great number density of holes in the *p*-type material, across the narrow depletion zone. With such great number densities so near, many electron–hole combinations occur, causing light to be emitted from that zone. Figure 42-16 shows the construction of an actual LED.

Commercial LEDs designed for the visible region are commonly based on gallium, suitably doped with arsenic and phosphorus atoms. An arrangement in which 60% of the nongallium sites are occupied by arsenic ions and 40% by phosphorus ions results in a gap width E_g of about 1.8 eV, corresponding to red light. Other

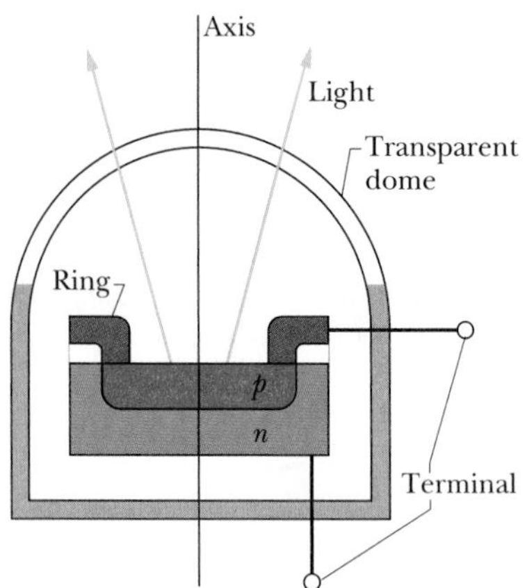

Fig. 42-16 Cross section of an LED (the device has rotational symmetry about the central axis). The p-type material, which is thin enough to transmit light, is in the form of a circular disk. A connection is made to the p-type material through a circular metal ring that touches the disk at its periphery. The depletion zone between the n-type material and the p-type material is not shown.

doping and transition level arrangements make it possible to construct LEDs that emit light in essentially any desired region of the visible and near-visible spectra.

The Photo-Diode

Passing a current through a suitably arranged p-n junction can generate light. The reverse is also true; that is, shining light on a suitably arranged p-n junction can produce a current in a circuit that includes the junction. This is the basis for the **photo-diode.**

When you click your television remote control, an LED in the device sends out a coded sequence of pulses of infrared light. The receiving device in your television set is an elaboration of the simple (two-terminal) photo-diode that not only detects the infrared signals but also amplifies them and transforms them into electrical signals that change the channel or adjust the volume, among other tasks.

The Junction Laser

In the arrangement of Fig. 42-15 there are many electrons in the conduction band of the n-type material and many holes in the valence band of the p-type material. Thus, there is a **population inversion** for the electrons; that is, there are more electrons in higher energy levels than in lower energy levels. As we discussed in Section 41-12, this is normally a necessary—but not a sufficient—condition for laser action.

When a single electron moves from the conduction band to the valence band, it can release its energy as a photon. This photon can stimulate a second electron to fall into the valence band, producing a second photon by stimulated emission. In this way, if the current through the junction is great enough, a chain reaction of stimulated emission events can occur and laser light can be generated. To bring this about, opposite faces of the p-n junction crystal must be flat and parallel, so that light can be reflected back and forth within the crystal. (Recall that in the helium–neon laser of Fig. 41-21, a pair of mirrors served this purpose.) Thus, a p-n junction can act as a **junction laser,** its light output being highly coherent and much more sharply defined in wavelength than light from an LED.

Fig. 42-17 A junction laser developed at the AT&T Bell Laboratories. The cube at the right is a grain of salt.

Junction lasers are built into compact disc (CD) players, where, by detecting reflections from the rotating disc, they are used to translate microscopic pits in the disc into sound. They are also much used in optical communication systems based on optical fibers. Figure 42-17 suggests their tiny scale. Junction lasers are usually designed to operate in the infrared region of the electromagnetic spectrum because optical fibers have two "windows" in that region (at $\lambda = 1.31$ and $1.55\ \mu$m) for which the energy absorption per unit length of the fiber is a minimum.

Sample Problem 42-7

An LED is constructed from a p-n junction based on a certain Ga-As-P semiconducting material whose energy gap is 1.9 eV. What is the wavelength of the emitted light?

SOLUTION: The Key Idea here is to assume that the transitions are from the bottom of the conduction band to the top of the valence band; then Eq. 42-11 holds. From this equation

$$\lambda = \frac{hc}{E_g} = \frac{(6.63 \times 10^{-34}\ \text{J}\cdot\text{s})(3.00 \times 10^8\ \text{m/s})}{(1.9\ \text{eV})(1.60 \times 10^{-19}\ \text{J/eV})}$$

$$= 6.5 \times 10^{-7}\ \text{m} = 650\ \text{nm}. \qquad \text{(Answer)}$$

Light of this wavelength is red.

CHECKPOINT 4: For the LED in this sample problem, is 650 nm (a) the only wavelength that can be emitted, (b) the maximum emitted wavelength, (c) the minimum emitted wavelength, or (d) the average emitted wavelength? (Consider the quantum jump assumed in the solution.)

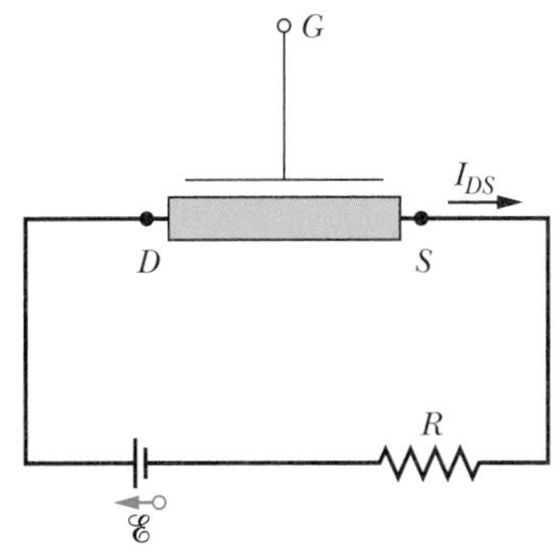

Fig. 42-18 A circuit containing a generalized field effect transistor, in which electrons flow through the device from the source terminal S to the drain terminal D. (The conventional current I_{DS} is in the opposite direction.) The magnitude of I_{DS} is controlled by the electric field set up within the body of the device by a potential applied to G, the gate terminal.

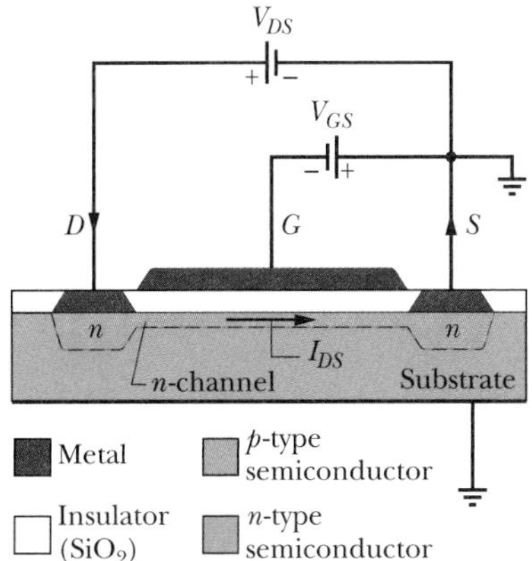

Fig. 42-19 A particular type of field-effect transistor known as a MOSFET. The magnitude of the drain-to-source conventional current through the n channel is controlled by the potential difference V_{GS} applied between the source S and the gate G. A depletion zone that exists between the n-type material and the p-type material is not shown.

42-11 The Transistor

A **transistor** is a three-terminal semiconducting device that can be used to amplify input signals. Figure 42-18 shows a generalized **f**ield-**e**ffect **t**ransistor (FET); in it, the flow of electrons from terminal S (the *source*) to terminal D (the *drain*) can be controlled by an electric field (hence field effect) set up within the device by a suitable electric potential applied to terminal G (the *gate*). Transistors are available in many types; we shall discuss only a particular FET called a MOSFET, or **m**etal-**o**xide-**s**emiconductor-**f**ield-**e**ffect **t**ransistor. The MOSFET has been described as the workhorse of the modern electronics industry.

For many applications the MOSFET is operated in only two states: with the drain-to-source current I_{DS} ON (gate open) or with it OFF (gate closed). The first of these can represent a 1 and the other a 0 in the binary arithmetic on which digital logic is based, and therefore MOSFETs can be used in digital logic circuits. Switching between the ON and OFF states can occur at high speed, so that binary logic data can be moved through MOSFET-based circuits very rapidly. MOSFETs about 500 nm in length—about the same as the wavelength of yellow light—are routinely fabricated for use in electronic devices of all kinds.

Figure 42-19 shows the basic structure of a MOSFET. A single crystal of silicon or other semiconductor is lightly doped to form p-type material. Embedded in this substrate, by heavily "overdoping" with n-type dopants, are two "islands" of n-type material, forming the drain D and the source S. The drain and source are connected by a thin channel of n-type material, called the ***n* channel.** A thin insulating layer of silicon dioxide (hence the O in MOSFET) is deposited on the crystal and penetrated by two metallic terminals (hence the M) at D and S, so that electrical contact can be made with the drain and the source. A thin metallic layer—the gate G—is deposited facing the n channel. Note that the gate makes no electrical contact with the transistor proper, being separated from it by the insulating oxide layer.

Consider first that the source and p-type substrate are grounded (at zero potential) and the gate is "floating"; that is, the gate is not connected to an external source of emf. Let a potential V_{DS} be applied between the drain and the source, such that the drain is positive. Electrons will then flow through the n channel from source to drain, and the conventional current I_{DS}, as shown in Fig. 42-19, will be from drain to source through the n-type material.

Now let a potential V_{GS} be applied to the gate, making it negative with respect to the source. The negative gate sets up within the device an electric field (hence the "field effect") that tends to repel electrons from the n channel into the substrate. This electron movement widens the (naturally occurring) depletion zone between the n channel and the substrate, at the expense of the n channel. The reduced width of the n channel, coupled with a reduction in the number of charge carriers in that channel, increases the resistance of that channel and thus decreases the current I_{DS}. With the proper value of V_{GS}, this current can be shut off completely; hence, by controlling V_{GS}, the MOSFET can be switched between its ON and OFF modes.

Charge carriers do not flow through the *substrate* because the substrate (1) is lightly doped, (2) is not a good conductor, and (3) is separated from the n-channel and the two n-type islands by an insulating depletion zone, not specifically shown in Fig. 42-19. Such a depletion zone always exists at a boundary between n-type material and p-type material, as Fig. 42-11b shows.

Integrated Circuits

Computers and other electronic devices employ thousands (if not millions) of transistors and other electronic components such as capacitors and resistors. These are

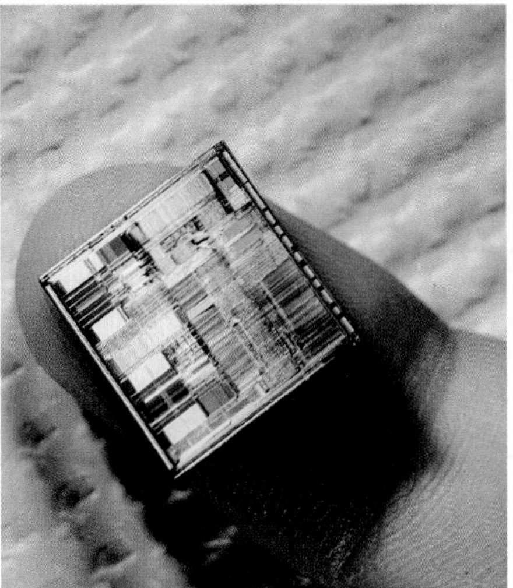

Fig. 42-20 An integrated circuit for the Intel Pentium chip, used mainly in computers. It will be encapsulated in a ceramic coating for installation and use.

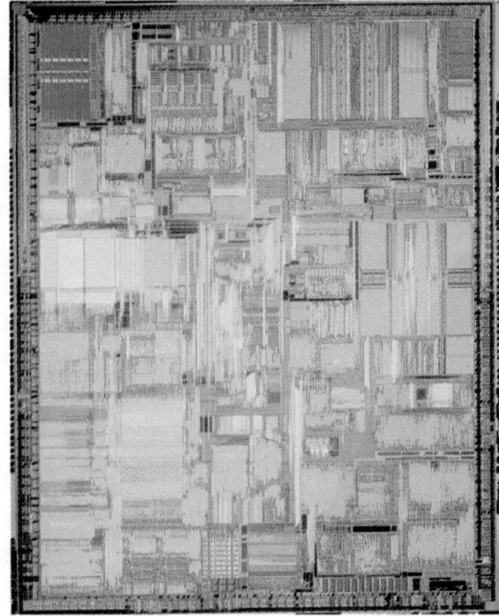

Fig. 42-21 Enlarged photograph of the layout of an Intel chip.

not assembled as separate units but are crafted into a single semiconducting **chip,** forming an **integrated circuit.**

Figure 42-20 shows a Pentium microprocessor chip, manufactured by Intel Corporation. It contains almost 7 million transistors, along with many other electronic components. Figure 42-21 shows a greatly enlarged view of part of the layout of another chip, the different colors identifying different layers of the chip.

At Intel's Rio Rancho plant, chips are fabricated in a 140-step process on 20 cm silicon wafers, each wafer holding about 300 chips. The individual electronic chip components are so small that the tiniest speck of dust can ruin a chip. Precautions are taken to maintain a dust-free atmosphere in the plant's clean rooms, where manufacturing takes place, and which are thousands of times more pristine than a hospital operating room. That is the reason for the workers' protective clothing, shown in the photograph that opens this chapter. As part of the cleanliness program, highly filtered air circulates through the perforated floor at about 30 m/min. There are also air showers and wipedown stations for removing cosmetics from employees.

REVIEW & SUMMARY

Conductors, Semiconductors, and Insulators Three electrical properties that can be used to distinguish among crystalline solids are the **resistivity** ρ, the **temperature coefficient of resistivity** α, and the **number density of charge carriers** n. Solids can be broadly divided into **conductors** (with small ρ) and **insulators** (with large ρ). Conductors can be further divided into **metals** (with small ρ, positive α, large n) and **semiconductors** (with larger ρ, negative α, and smaller n).

Energy Levels and Gaps in a Crystalline Solid An isolated atom can exist in only a discrete set of energy levels. As atoms come together to form a solid, the levels of the individual atoms merge to form the discrete energy **bands** of the solid. These energy bands are separated by energy **gaps,** each of which corresponds to a range of energies that no electron may possess.

Any energy band is made up of an enormous number of very closely spaced levels. The Pauli exclusion principle asserts that only one electron may occupy each of these levels.

Insulators In an insulator, the highest band containing electrons is completely filled and is separated from the vacant band above it by an energy gap so large that electrons can essentially never become thermally agitated enough to jump across the gap.

Metals In a **metal,** the highest band that contains any electrons is only partially filled. The energy of the highest filled level at a temperature of 0 K is called the **Fermi energy** E_F for the metal; for copper, $E_F = 7.0$ eV.

The electrons in the partially filled band are the *conduction electrons* and their number is

$$\begin{pmatrix}\text{number of conduction}\\ \text{electrons in sample}\end{pmatrix} = \begin{pmatrix}\text{number of atoms}\\ \text{in sample}\end{pmatrix} \times \begin{pmatrix}\text{number of valence}\\ \text{electrons per atom}\end{pmatrix}. \quad (42\text{-}2)$$

The number of atoms in a sample is given by

$$\begin{pmatrix}\text{number of atoms}\\ \text{in sample}\end{pmatrix} = \frac{\text{sample mass } M_{\text{sam}}}{\text{atomic mass}} = \frac{\text{sample mass } M_{\text{sam}}}{(\text{molar mass } M)/N_A} = \frac{\begin{pmatrix}\text{material's}\\ \text{density}\end{pmatrix}\begin{pmatrix}\text{sample}\\ \text{volume } V\end{pmatrix}}{(\text{molar mass } M)/N_A}. \quad (42\text{-}4)$$

The number density n of the conduction electrons is

$$n = \frac{\text{number of conduction electrons in sample}}{\text{sample volume } V}. \quad (42\text{-}3)$$

The **density of states** function $N(E)$ is the number of available energy levels per unit volume of the sample and per unit energy interval and is given by

$$N(E) = \frac{8\sqrt{2}\pi m^{3/2}}{h^3} E^{1/2} \quad \text{(density of states)}, \quad (42\text{-}5)$$

where E is the energy at which $N(E)$ is evaluated.

The **occupancy probability** $P(E)$ (the probability that a given available state will be occupied by an electron) is

$$P(E) = \frac{1}{e^{(E - E_F)/kT} + 1} \quad \text{(occupancy probability)}. \quad (42\text{-}6)$$

The **density of occupied states** $N_o(E)$ is given by the product of the two quantities in Eqs. (42-5) and (42-6):

$$N_o(E) = N(E)\,P(E) \quad \text{(density of occupied states)}. \quad (42\text{-}7)$$

The Fermi energy for a metal can be found by integrating $N_o(E)$ for $T = 0$ from $E = 0$ to $E = E_F$. The result is

$$E_F = \left(\frac{3}{16\sqrt{2}\pi}\right)^{2/3} \frac{h^2}{m} n^{2/3} = \frac{0.121h^2}{m} n^{2/3}. \quad (42\text{-}9)$$

Semiconductors The band structure of a **semiconductor** is like that of an insulator except that the gap width E_g is much smaller in the semiconductor. For silicon (a semiconductor) at room temperature, thermal agitation raises a few electrons to the **conduction band,** leaving an equal number of **holes** in the **valence band.** Both electrons and holes serve as charge carriers.

The number of electrons in the conduction band of silicon can be increased greatly by doping with small amounts of phosphorus, thus forming ***n*-type material.** The number of holes in the valence band can be greatly increased by doping with aluminum, thus forming ***p*-type material.**

The p-n Junction A ***p-n* junction** is a single semiconducting crystal with one end doped to form p-type material and the other end doped to form n-type material, the two types meeting at a **junction plane.** At thermal equilibrium, the following occurs at that plane:

The **majority carriers** (electrons on the n side and holes on the p side) diffuse across the junction plane, producing a **diffusion current** I_{diff}.

The **minority carriers** (holes on the n side and electrons on the p side) are swept across the junction plane, forming a **drift current** I_{drift}. These two currents are equal in magnitude, so the net current is zero.

A **depletion zone,** consisting largely of charged donor and acceptor ions, forms across the junction plane.

A **contact potential difference** V_0 develops across the depletion zone.

Applications of the p-n Junction When a potential difference is applied across a p-n junction, the device conducts electricity more readily for one polarity of the applied potential difference than for the other. Thus, a p-n junction can serve as a **junction rectifier.**

When a p-n junction is forward biased, it can emit light, hence can serve as a **light-emitting diode** (LED). The wavelength of the emitted light is

$$\lambda = \frac{c}{f} = \frac{hc}{E_g}. \quad (42\text{-}11)$$

A strongly forward-biased p-n junction with parallel end faces can operate as a **junction laser,** emitting light of a sharply defined wavelength.

MOSFETS In a MOSFET, a type of three-terminal transistor, a potential applied to the **gate** terminal G controls the internal flow of electrons from the **source** terminal S to the **drain** terminal D. Commonly, a MOSFET is operated only in its ON (conducting) or OFF (not conducting) condition. Installed by the thousands and millions on silicon wafers (**chips**) to form **integrated circuits,** MOSFETs form the basis for computer hardware.

QUESTIONS

1. Figure 42-1a shows 14 atoms that represent the unit cell of copper. However, since each of these atoms is shared with one or more adjoining unit cells, only a fraction of each atom belongs to the unit cell shown. What is the number of atoms per unit cell for copper? (To answer, count up the fractional atoms belonging to a single unit cell.)

2. Figure 42-1*b* shows 18 atoms that represent the unit cell of silicon. Fourteen of these atoms, however, are shared with one or more adjoining unit cells. What is the number of atoms per unit cell for silicon? (See Question 1.)

3. Does the interval between adjacent energy levels in the highest occupied band of a metal depend on (a) the material of which the sample is made, (b) the size of the sample, (c) the position of the level in the band, (d) the temperature of the sample, or (e) the Fermi energy of the metal?

4. Compare the drift speed v_d of the conduction electrons in a current-carrying copper wire with the Fermi speed v_F for copper. Is v_d (a) about equal to v_F, (b) much greater than v_F, or (c) much less than v_F?

5. In a silicon lattice, where should you look if you want to find (a) a conduction electron, (b) a valence electron, and (c) an electron associated with the 2*p* subshell of the isolated silicon atom?

6. Which of the following statements, if any, are true? (a) At low enough temperatures, silicon behaves like an insulator. (b) At high enough temperatures, silicon becomes a good conductor. (c) At high enough temperatures, silicon behaves like a metal.

7. The energy gaps E_g for the semiconductors silicon and germanium are, respectively, 1.12 and 0.67 eV. Which of the following statements, if any, are true? (a) Both substances have the same number density of charge carriers at room temperature. (b) At room temperature, germanium has a greater number density of charge carriers than silicon. (c) Both substances have a greater number density of conduction electrons than holes. (d) For each substance, the number density of electrons equals that of holes.

8. An isolated atom of germanium has 32 electrons, arranged in subshells according to this scheme:

$$1s^2\ 2s^2\ 2p^6\ 3s^2\ 3p^6\ 3d^{10}\ 4s^2\ 4p^2.$$

This element has the same crystal structure as silicon and, like silicon, is a semiconductor. Which of these electrons form the valence band of crystalline germanium?

9. Germanium ($Z = 32$) has the same crystal structure and the same bonding pattern as silicon. Is the net charge on a germanium ion within its lattice $+e$, $+2e$, $+4e$, $+28e$, or $+32e$?

10. (a) Of the elements arsenic, indium, tin, gallium, antimony, and boron, which would produce *n*-type material if used as a dopant in silicon? (b) Which would produce *p*-type material? (c) Which would be unsuitable as a dopant? (*Hint*: Consult the periodic table in Appendix G.)

11. A sample of silicon is doped with phosphorus. Which of the following statements, if any, are true? (a) The number of holes in the sample is slightly increased. (b) The sample's resistivity is increased. (c) The sample becomes positively charged. (d) The sample becomes negatively charged. (e) The gap between the valence band and the conduction band decreases slightly.

12. To fabricate an *n*-type semiconductor, would you use (a) silicon doped with arsenic or (b) germanium doped with indium? (*Hint:* Consult the periodic table.)

13. In the biased *p-n* junctions shown in Fig. 42-14, there is an electric field $\vec{E}$ in each of the two depletion zones, associated with the potential difference that exists across that zone. (a) Is $\vec{E}$ directed from left to right or from right to left? (b) Is its magnitude greater for forward bias or for back bias?

14. A certain isolated *p-n* junction develops a contact potential difference V_0 of 0.78 V across its depletion zone. A voltmeter is connected across the terminals of the junction, the positive terminal of the meter being connected to the *p* side of the junction. Will the meter read (a) +0.78 V, (b) −0.78 V, (c) zero, or (d) something else? (*Hint:* Contact potentials appear at the connections between the *p-n* junction and the voltmeter leads.)

15. Which of the following obey Ohm's law: (a) a bar of pure silicon, (b) a bar of *n*-type silicon, (c) a bar of *p*-type silicon, (d) a *p-n* junction?

16. An LED based on a gallium–arsenic–phosphorus semiconducting crystal emits red light. If you look at a white surface through such a crystal, will you see (a) red, (b) blue, (c) nothing, because the crystal is opaque, or (d) white?

EXERCISES & PROBLEMS

ssm Solution is in the Student Solutions Manual.
www Solution is available on the World Wide Web at: http://www.wiley.com/college/hrw
ilw Solution is available on the Interactive LearningWare.

SEC. 42-5 Metals

1E. Copper, a monovalent metal, has molar mass 63.54 g/mol and density 8.96 g/cm^3. What is the number density n of conduction electrons in copper? ssm

2E. Verify the numerical factor 0.121 in Eq. 42-9.

3E. At what pressure, in atmospheres, would the number of molecules per unit volume in an ideal gas be equal to the number density of the conduction electrons in copper, with both gas and copper at temperature $T = 300$ K?

4E. Use Eq. 42-9 to verify 7.0 eV as copper's Fermi energy.

5E. Calculate $d\rho/dT$ at room temperature for (a) copper and (b) silicon, using data from Table 42-1.

6E. What is the number density of conduction electrons in gold, which is a monovalent metal? Use the molar mass and density provided in Appendix F.

7E. (a) Show that Eq. 42-5 can be written as $N(E) = CE^{1/2}$. (b) Evaluate C in terms of meters and electron-volts. (c) Calculate $N(E)$ for $E = 5.00$ eV. ssm

8E. The Fermi energy of copper is 7.0 eV. Verify that the corresponding Fermi speed is 1600 km/s.

9E. What is the probability that a state 0.062 eV above the Fermi energy will be occupied at (a) $T = 0$ K and (b) $T = 320$ K? ssm

10E. Calculate the density of states $N(E)$ for a metal at energy $E =$

8.0 eV and show that your result is consistent with the curve of Fig. 42-5.

11E. Show that Eq. 42-9 can be written as $E_F = An^{2/3}$, where the constant A has the value $3.65 \times 10^{-19}\ \text{m}^2 \cdot \text{eV}$. ssm

12E. Use the result of Exercise 6 to calculate the Fermi energy of gold.

13E. A state 63 meV above the Fermi level has a probability of occupancy of 0.090. What is the probability of occupancy for a state 63 meV *below* the Fermi level?

14P. The Fermi energy for copper is 7.0 eV. For copper at 1000 K, (a) find the energy of the energy level whose probability of being occupied by an electron is 0.90. For this energy, evaluate (b) the density of states $N(E)$ and (c) the density of occupied states $N_o(E)$.

15P. In Eq. 42-6 let $E - E_F = \Delta E = 1.00$ eV. (a) At what temperature does the result of using this equation differ by 1.0% from the result of using the classical Boltzmann equation $P(E) = e^{-\Delta E/kT}$ (which is Eq. 42-1 with two changes in notation)? (b) At what temperature do the results from these two equations differ by 10%? ssm www

16P. Show that $P(E)$, the occupancy probability in Eq. 42-6, is symmetrical about the value of the Fermi energy; that is, show that

$$P(E_F + \Delta E) + P(E_F - \Delta E) = 1.$$

17P. Assume that the total volume of a metal sample is the sum of the volume occupied by the metal ions making up the lattice and the (separate) volume occupied by the conduction electrons. The density and molar mass of sodium (a metal) are 971 kg/m^3 and 23.0 g/mol, respectively; the radius of the Na^+ ion is 98 pm. (a) What percent of the volume of a sample of metallic sodium is occupied by its conduction electrons? (b) Carry out the same calculation for copper, which has density, molar mass, and ionic radius of 8960 kg/m^3, 63.5 g/mol, and 135 pm, respectively. (c) For which of these metals do you think the conduction electrons behave more like a free-electron gas?

18P. Calculate $N_o(E)$, the density of occupied states, for copper at $T = 1000$ K for the energies $E = 4.00$, 6.75, 7.00, 7.25, and 9.00 eV. Compare your results with the graph of Fig. 42-7*b*. The Fermi energy for copper is 7.00 eV.

19P. Calculate the number density (number per unit volume) for (a) molecules of oxygen gas at 0°C and 1.0 atm pressure and (b) conduction electrons in copper. (c) What is the ratio of the latter to the former? (d) What is the average distance between particles in each case? Assume this distance is the edge length of a cube whose volume is equal to the available volume per particle.

20P. What is the probability that an electron will jump across the energy gap E_g (see Fig. 42-4*a*) in a diamond whose mass is equal to the mass of Earth? Use the result of Sample Problem 42-1 and the molar mass of carbon in Appendix F; assume that in diamond there is one valence electron per carbon atom.

21P. The Fermi energy for silver is 5.5 eV. (a) At $T = 0$°C, what are the probabilities that states with the following energies are occupied: 4.4, 5.4, 5.5, 5.6, and 6.4 eV? (b) At what temperature is the probability 0.16 that a state with energy $E = 5.6$ eV is occupied? ssm www

22P. Show that the probability $P(E)$ that an energy level at energy E is not occupied is

$$P(E) = \frac{1}{e^{-\Delta E/kT} + 1},$$

where $\Delta E = E - E_F$.

23P. The Fermi energy of aluminum is 11.6 eV; its density and molar mass are 2.70 g/cm^3 and 27.0 g/mol, respectively. From these data, determine the number of conduction electrons per atom. ssm

24P. At $T = 300$ K, how close to the Fermi energy will we find a state whose probability of occupation by a conduction electron is 0.10?

25P. Silver is a monovalent metal. Calculate (a) the number density of conduction electrons, (b) the Fermi energy, (c) the Fermi speed, and (d) the de Broglie wavelength corresponding to this electron speed. See Appendix F for the needed data on silver. ssm

26P. Zinc is a bivalent metal. Calculate (a) the number density of conduction electrons, (b) the Fermi energy, (c) the Fermi speed, and (d) the de Broglie wavelength corresponding to this electron speed. See Appendix F for the needed data on zinc.

27P. (a) Show that the density of states at the Fermi energy is given by

$$N(E_F) = \frac{(4)(3^{1/3})(\pi^{2/3})mn^{1/3}}{h^2}$$
$$= (4.11 \times 10^{18}\ \text{m}^{-2}\ \text{eV}^{-1})n^{1/3},$$

in which n is the number density of conduction electrons. (b) Calculate $N(E_F)$ for copper using the result of Exercise 1, and verify your calculation with the curve of Fig. 42-5, recalling that $E_F = 7.0$ eV for copper.

28P. (a) Show that the slope dP/dE of Eq. 42-6 at $E = E_F$ is $-1/4kT$. (b) Show that the tangent line to the curve of Fig. 42-6*b* at $E = E_F$ intercepts the horizontal axis at $E = E_F + 2kT$.

29P. Show that, at $T = 0$ K, the average energy E_{avg} of the conduction electrons in a metal is equal to $\frac{3}{5}E_F$. (*Hint:* By definition of average, $E_{avg} = (1/n) \int E\, N_o(E)\, dE$, where n is the number density of charge carriers.) ssm

30P. Use the result of Problem 29 to calculate the total translational kinetic energy of the conduction electrons in 1.0 cm^3 of copper at $T = 0$ K.

31P. (a) Using the result of Problem 29, estimate how much energy would be released by the conduction electrons in a copper coin with mass 3.1 g if we could suddenly turn off the Pauli exclusion principle. (b) For how long would this amount of energy light a 100 W lamp? (*Note:* There is no way to turn off the Pauli principle!)

32P. At 1000 K, the fraction of the conduction electrons in a metal that have energies greater than the Fermi energy is equal to the area under the curve of Fig. 42-7*b* beyond E_F divided by the area under the entire curve. It is difficult to find these areas by direct integration. However, an approximation to this fraction at any temperature T is

$$frac = \frac{3kT}{2E_F}.$$

Note that $frac = 0$ for $T = 0$ K, just as we would expect. What is this fraction for copper at (a) 300 K and (b) 1000 K? For copper,

$E_F = 7.0$ eV. (c) Check your answers by numerical integration using Eq. 42-7.

33P. At what temperature do 1.3% of the conduction electrons in lithium (a metal) have energies greater than the Fermi energy E_F, which is 4.7 eV? (See Problem 32.) ssm

34P. Silver melts at 961°C. At the melting point, what fraction of the conduction electrons are in states with energies greater than the Fermi energy of 5.5 eV? (See Problem 32.)

SEC. 42-6 Semiconductors

35E. (a) What is the maximum wavelength of the light that will excite an electron in the valence band of diamond to the conduction band? The energy gap is 5.5 eV. (b) In what part of the electromagnetic spectrum does this wavelength lie? ssm

36P. The compound gallium arsenide is a commonly used semiconductor, having an energy gap E_g of 1.43 eV. Its crystal structure is like that of silicon, except that half the silicon atoms are replaced by gallium atoms and half by arsenic atoms. Draw a flattened-out sketch of the gallium arsenide lattice, following the pattern of Fig. 42-9*a*. (a) What are the net charges of the gallium and arsenic ion cores? (b) How many electrons per bond are there? (*Hint:* Consult the periodic table in Appendix G.)

37P. (a) Find the angle θ between adjacent nearest-neighbor bonds in the silicon lattice. Recall that each silicon atom is bonded to four of its nearest neighbors. The four neighbors form a regular tetrahedron—a three-sided pyramid whose sides and base are equilateral triangles. (b) Find the bond length, given that the atoms at the corners of the tetrahedron are 388 pm apart.

38P. The occupancy probability function (Eq. 42-6) can be applied to semiconductors as well as to metals. In semiconductors the Fermi energy is close to the midpoint of the gap between the valence band and the conduction band. For germanium, the gap width is 0.67 eV. What is the probability that (a) a state at the bottom of the conduction band is occupied and (b) a state at the top of the valence band is not occupied. Assume that $T = 290$ K. (*Note:* Figure 42-4*b* shows that, in a metal, the Fermi energy lies symmetrically between the population of conduction electrons and the population of holes. To match this scheme in a semiconductor, the Fermi energy must lie near the center of the gap. There need not be an available state at the location of the Fermi energy.)

39P. In a simplified model of an undoped semiconductor, the actual distribution of energy states may be replaced by one in which there are N_v states in the valence band, all these states having the same energy E_v, and N_c states in the conduction band, all these states having the same energy E_c. The number of electrons in the conduction band equals the number of holes in the valence band. (a) Show that this last condition implies that

$$\frac{N_c}{\exp(\Delta E_c/kT) + 1} = \frac{N_v}{\exp(\Delta E_v/kT) + 1},$$

in which

$$\Delta E_c = E_c - E_F \quad \text{and} \quad \Delta E_v = -(E_v - E_F).$$

(*Hint:* See Problem 22.) (b) If the Fermi level is in the gap between the two bands and is far from both bands compared with kT, then the exponentials dominate in the denominators. Under these conditions show that

$$E_F = \frac{(E_c + E_v)}{2} + \frac{kT\ln(N_v/N_c)}{2}$$

and that, if $N_v \approx N_c$, the Fermi level for the undoped semiconductor is close to the gap's center, as stated in Problem 38.

SEC. 42-7 Doped Semiconductors

40P. Pure silicon at room temperature has an electron number density in the conduction band of about $5 \times 10^{15}\ \text{m}^{-3}$ and an equal density of holes in the valence band. Suppose that one of every 10^7 silicon atoms is replaced by a phosphorus atom. (a) Which type will the doped semiconductor be, *n* or *p*? (b) What charge carrier number density will the phosphorus add? (c) What is the ratio of the charge carrier number density (electrons in the conduction band and holes in the valence band) in the doped silicon to that in pure silicon?

41P. What mass of phosphorus is needed to dope 1.0 g of silicon to the extent described in Sample Problem 42-6? ssm

42P. A silicon sample is doped with atoms having donor states 0.110 eV below the bottom of the conduction band. (The energy gap in silicon is 1.11 eV.) (a) If each of these donor states is occupied with a probability of 5.00×10^{-5} at $T = 300$ K, where is the Fermi level with respect to the top of the silicon valence band? (b) What then is the probability that a state at the bottom of the silicon conduction band is occupied?

43P. Doping changes the Fermi energy of a semiconductor. Consider silicon, with a gap of 1.11 eV between the top of the valence band and the bottom of the conduction band. At 300 K the Fermi level of the pure material is nearly at the midpoint of the gap. Suppose that silicon is doped with donor atoms, each of which has a state 0.15 eV below the bottom of the silicon conduction band, and suppose further that doping raises the Fermi level to 0.11 eV below the bottom of that band (Fig. 42-22). (a) For both pure and doped silicon, calculate the probability that a state at the bottom of the silicon conduction band is occupied. (b) Calculate the probability that a donor state in the doped material is occupied. ssm www

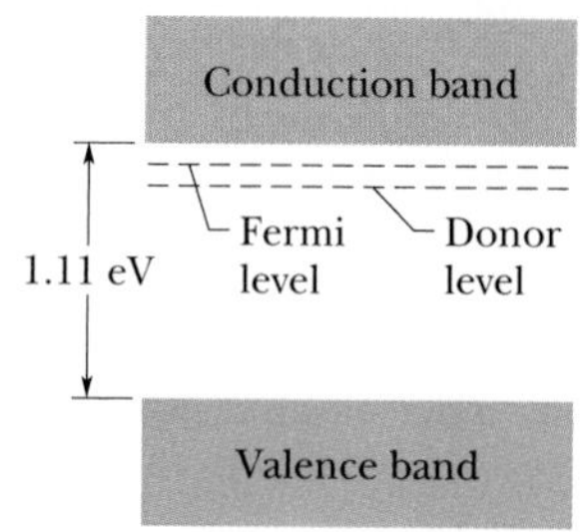

Fig. 42-22 Problem 43.

SEC. 42-9 The Junction Rectifier

44E. For an ideal *p-n* junction rectifier with a sharp boundary between its two semiconducting sides, the current I is related to the potential difference V across the rectifier by

$$I = I_0(e^{eV/kT} - 1),$$

where I_0, which depends on the materials but not on the current or the potential difference, is called the *reverse saturation current*. V is positive if the rectifier is forward-biased and negative if it is back-biased. (a) Verify that this expression predicts the behavior of a junction rectifier by graphing I versus V over the range -0.12 V to $+0.12$ V. Take $T = 300$ K and $I_0 = 5.0$ nA. (b) For the same temperature, calculate the ratio of the current for a 0.50 V forward-bias to the current for a 0.50 V back-bias.

45E. When a photon enters the depletion zone of a *p-n* junction, it can scatter from the valence electrons there, transferring part of its energy to each electron, which then jumps to the conduction band. Thus, the photon creates electron–hole pairs. For this reason, the junctions are often used as light detectors, especially in the x-ray and gamma-ray regions of the electromagnetic spectrum. Suppose a single 662 keV gamma-ray photon transfers its energy to electrons in multiple scattering events inside a semiconductor with an energy gap of 1.1 eV, until all the energy is transferred. Assuming that each of those electrons jumps the gap from the top of the valence band to the bottom of the conduction band, find the number of electron–hole pairs created by the process. ssm

SEC. 42-10 The Light-Emitting Diode (LED)

46P. A potassium chloride crystal has an energy band gap of 7.6 eV above the topmost occupied band, which is full. Is this crystal opaque or transparent to light of wavelength 140 nm?

47P. In a particular crystal, the highest occupied band is full. The crystal is transparent to light of wavelengths longer than 295 nm but opaque at shorter wavelengths. Calculate, in electron-volts, the gap between the highest occupied band and the next higher (empty) band for this material. ssm

SEC. 42-11 The Transistor

48P. A Pentium computer chip, which is about the size of a postage stamp (2.54 cm $\times$ 2.22 cm), contains about 3.5 million transistors. If the transistors are square, what must be their *maximum* dimension? (*Note:* Devices other than transistors are also on the chip, and there must be room for the interconnections among the circuit elements. Transistors smaller than 0.7 μm are now commonly and inexpensively fabricated.)

49P. A silicon-based MOSFET has a square gate 0.50 μm on edge. The insulating silicon oxide layer that separates it from the *p*-type substrate is 0.20 μm thick and has a dielectric constant of 4.5. (a) What is the equivalent gate–substrate capacitance (treating the gate as one plate and the substrate as the other plate)? (b) How many elementary charges e appear in the gate when there is a gate–source potential difference of 1.0 V?

43 Nuclear Physics

Radioactive nuclei that are injected into a patient collect at certain sites within the patient's body, undergo radioactive decay, and emit gamma rays. These gamma rays can be recorded by a detector, and a color-coded image of the patient's body produced on a video monitor. In the images reproduced here (the left one is a front view of a patient and the right one is a back view), you can tell just where the radioactive nuclei have collected (spine, pelvis, and ribs) by the color-coding of brown and orange.

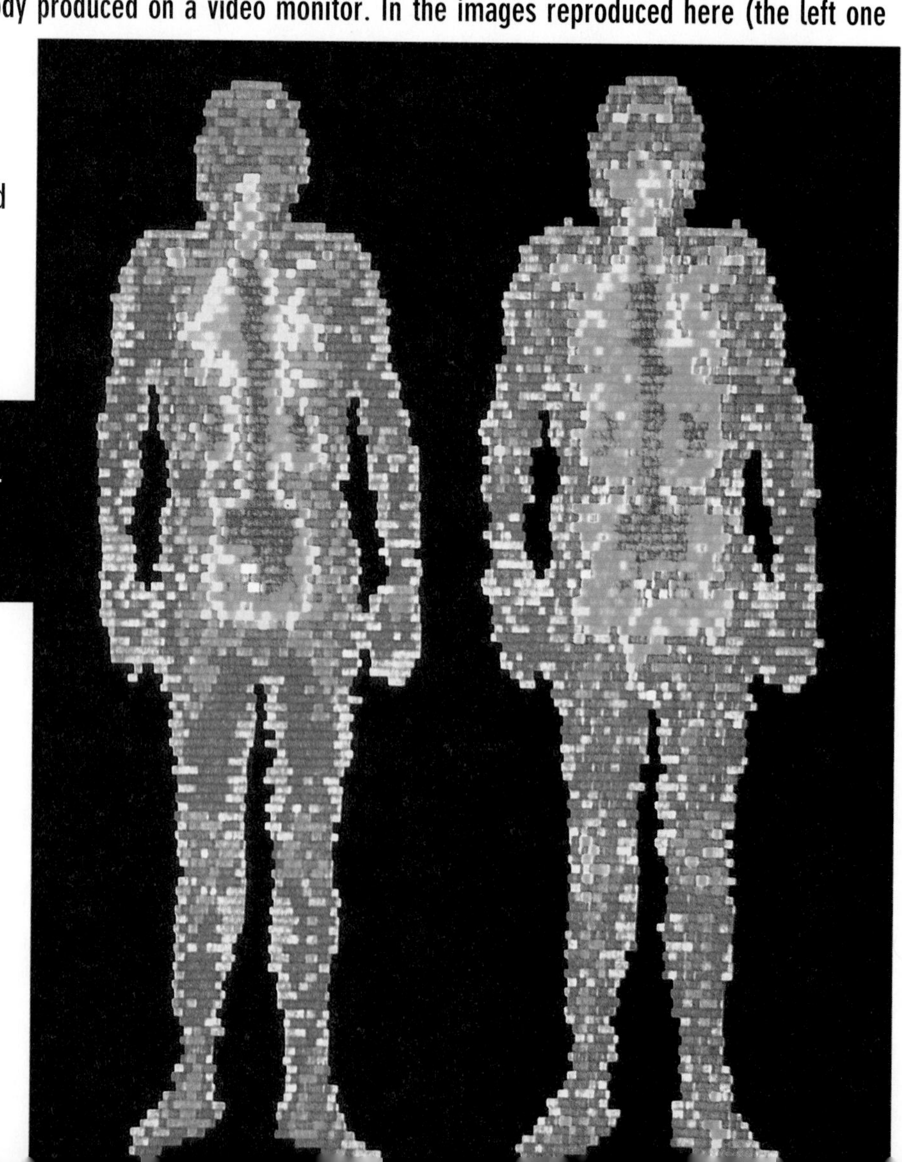

But what actually happens to radioactive nuclei when they undergo decay, and what exactly does "decay" mean?

The answer is in this chapter.

43-1 Discovering the Nucleus

In the first years of the twentieth century not much was known about the structure of atoms beyond the fact that they contain electrons. The electron had been discovered (by J. J. Thomson) in 1897, and its mass was unknown in those early days. Thus, it was not possible even to say how many negatively charged electrons a given atom contained. Atoms were electrically neutral so they must also contain some positive charge, but nobody knew what form this compensating positive charge took.

In 1911 Ernest Rutherford proposed that the positive charge of the atom is densely concentrated at the center of the atom, forming its **nucleus,** and that, furthermore, the nucleus is responsible for most of the mass of the atom. Rutherford's proposal was no mere conjecture but was based firmly on the results of an experiment suggested by him and carried out by his collaborators, Hans Geiger (of Geiger counter fame) and Ernest Marsden, a 20-year-old student who had not yet earned his bachelor's degree.

In Rutherford's day it was known that certain elements, called **radioactive,** transform into other elements spontaneously, emitting particles in the process. One such element is radon, which emits alpha (α) particles with energies of about 5.5 MeV. We now know that these particles are the nuclei of atoms of helium.

Rutherford's idea was to direct energetic alpha particles at a thin target foil and measure the extent to which they were deflected as they passed through the foil. Alpha particles, which are about 7300 times more massive than electrons, have a charge of $+2e$.

Figure 43-1 shows the experimental arrangement of Geiger and Marsden. Their alpha source was a thin-walled glass tube of radon gas. The experiment involves counting the number of alpha particles that are deflected through various scattering angles ϕ.

Figure 43-2 shows their results. Note especially that the vertical scale is logarithmic. We see that most of the particles are scattered through rather small angles

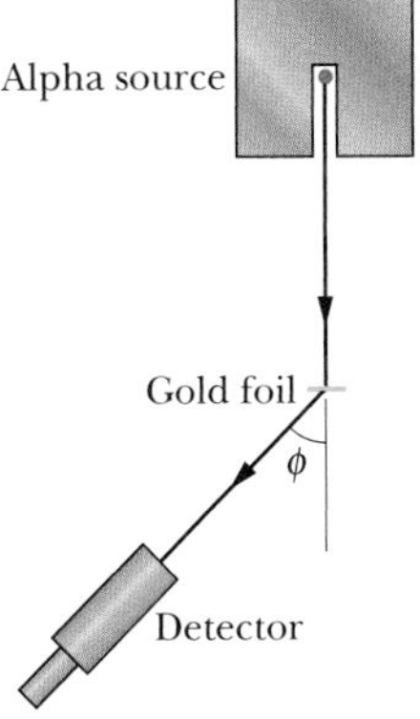

Fig. 43-1 An arrangement (top view) used in Rutherford's laboratory in 1911–1913 to study the scattering of α particles by thin metal foils. The detector can be rotated to various values of the scattering angle ϕ. The alpha source was radon gas, a decay product of radium. With this simple "tabletop" apparatus, the atomic nucleus was discovered.

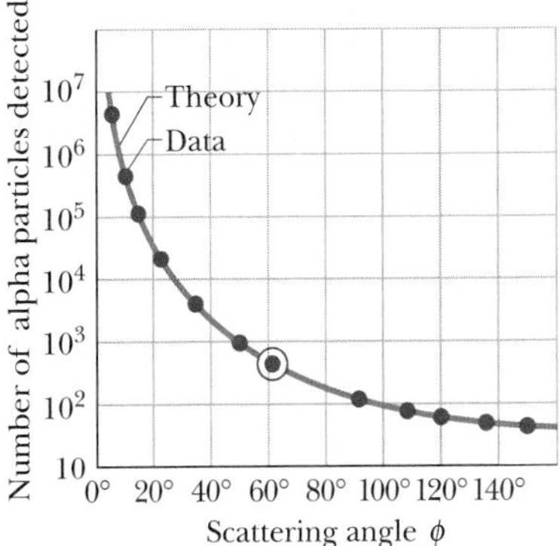

Fig. 43-2 The dots are alpha-particle scattering data for a gold foil, obtained by Geiger and Marsden using the apparatus of Fig. 43-1. The solid curve is the theoretical prediction, based on the assumption that the atom has a small, massive, positively charged nucleus. Note that the vertical scale is logarithmic, covering six orders of magnitude. The data have been adjusted to fit the theoretical curve at the experimental point that is enclosed in a circle.

but—and this was the big surprise—a very small fraction of them are scattered through very large angles, approaching 180°. In Rutherford's words: "It was quite the most incredible event that ever happened to me in my life. It was almost as incredible as if you had fired a 15-inch shell at a piece of tissue paper and it came back and hit you."

Why was Rutherford so surprised? At the time of these experiments, most physicists believed in the so-called plum pudding model of the atom, which had been advanced by J. J. Thomson. In this view the positive charge of the atom was thought to be spread out through the entire volume of the atom. The electrons (the "plums") were thought to vibrate about fixed points within this sphere of positive charge (the "pudding").

The maximum deflecting force that could act on an alpha particle as it passed through such a large positive sphere of charge would be far too small to deflect the alpha particle by even as much as 1°. (The expected deflection has been compared to what you would observe if you fired a bullet through a sack of snowballs.) The electrons in the atom would also have very little effect on the massive, energetic alpha particle. They would, in fact, be themselves strongly deflected, much as a swarm of gnats would be brushed aside by a stone thrown through them.

Rutherford saw that, to deflect the alpha particle backward, there must be a large force; this force could be provided if the positive charge, instead of being spread throughout the atom, were concentrated tightly at its center. Then the incoming alpha particle could get very close to the positive charge without penetrating it; such a close encounter would result in a large deflecting force.

Figure 43-3 shows possible paths taken by typical alpha particles as they pass through the atoms of the target foil. As we see, most are either undeflected or only slightly deflected, but a few (those whose incoming paths pass, by chance, very close to a nucleus) are deflected through large angles. From an analysis of the data, Rutherford concluded that the radius of the nucleus must be smaller than the radius of an atom by a factor of about 10^4. In other words, the atom is mostly empty space.

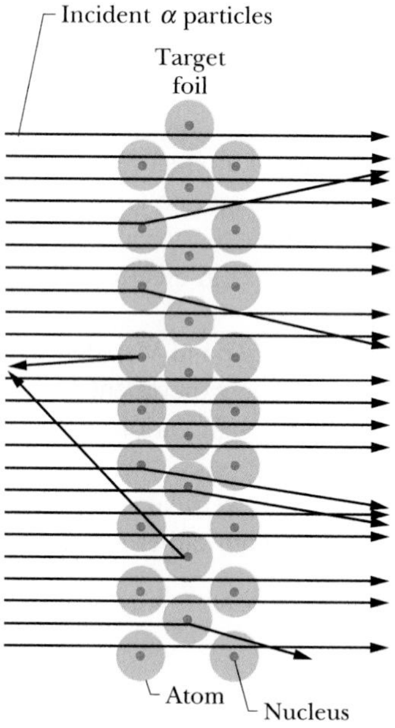

Fig. 43-3 The angle through which an incident alpha particle is scattered depends on how close the particle's path lies to an atomic nucleus. Large deflections result only from very close encounters.

Sample Problem 43-1

A 5.30 MeV alpha particle happens, by chance, to be headed directly toward the nucleus of an atom of gold, which contains 79 protons. How close does the alpha particle get to the center of the nucleus before it comes momentarily to rest and reverses its motion? Neglect the recoil of the relatively massive nucleus.

SOLUTION: The **Key Idea** here is that throughout this process, the total mechanical energy E of the system of alpha particle and gold nucleus is conserved. In particular, the system's initial mechanical energy E_i, before the particle and nucleus interact, is equal to its mechanical energy E_f when the alpha particle momentarily stops. The initial energy E_i is just the kinetic energy K_α of the incoming alpha particle. The final energy E_f is just the electric potential energy U of the system (the kinetic energy is then zero). We can find U with Eq. 25-43 ($U = q_1q_2/4\pi\varepsilon_0 r$).

Let d be the center-to-center distance between the alpha particle and the gold nucleus when the alpha particle is at its stopping point. Then we can write the conservation of energy $E_i = E_f$ as

$$K_\alpha = \frac{1}{4\pi\varepsilon_0}\frac{q_\alpha q_{\text{Au}}}{d},$$

in which q_α (= $2e$) is the charge of the alpha particle (2 protons) and q_{Au} (= $79e$) is the charge of the gold nucleus (79 protons).

Substituting for the charges and solving for d yield

$$d = \frac{(2e)(79e)}{4\pi\varepsilon_0 K_\alpha}$$
$$= \frac{(2 \times 79)(1.60 \times 10^{-19}\ \text{C})^2}{(4\pi)(8.85 \times 10^{-12}\ \text{F/m})(5.30\ \text{MeV})(1.60 \times 10^{-13}\ \text{J/MeV})}$$
$$= 4.29 \times 10^{-14}\ \text{m}. \qquad \text{(Answer)}$$

This is a small distance by atomic standards but not by nuclear standards. It is, in fact, considerably larger than the sum of the radii of the gold nucleus and the alpha particle. Thus, this alpha particle reverses its motion without ever actually "touching" the gold nucleus.

43-2 Some Nuclear Properties

Table 43-1 shows some properties of a few atomic nuclei. When we are interested primarily in their properties as specific nuclear species (rather than as parts of atoms), we call these particles **nuclides.**

Some Nuclear Terminology

Nuclei are made up of protons and neutrons. The number of protons in a nucleus (called the **atomic number** or **proton number** of the nucleus) is represented by the symbol Z; the number of neutrons (the **neutron number**) is represented by the symbol N. The total number of neutrons and protons in a nucleus is called its **mass number** A, so

$$A = Z + N. \tag{43-1}$$

Neutrons and protons, when considered collectively, are called **nucleons.**

We represent nuclides with symbols such as those displayed in the first column of Table 43-1. Consider ^{197}Au, for example. The superscript 197 is the mass number A. The chemical symbol Au tells us that this element is gold, whose atomic number is 79. From Eq. 43-1, the neutron number of this nuclide is 197 − 79, or 118.

Nuclides with the same atomic number Z but different neutron numbers N are called **isotopes** of each other. The element gold has 32 isotopes, ranging from ^{173}Au to ^{204}Au. Only one of them (^{197}Au) is stable; the remaining 31 are radioactive. Such **radionuclides** undergo **decay** (or **disintegration**) by emitting a particle and thereby transforming to a different nuclide.

Organizing the Nuclides

The neutral atoms of all isotopes of an element (all with the same Z) have the same number of electrons and the same chemical properties, and they fit into the same box in the periodic table of the elements. The *nuclear* properties of the isotopes of a given element, however, are very different. Thus, the periodic table is of limited use to the nuclear physicist, the nuclear chemist, or the nuclear engineer.

TABLE 43-1 Some Properties of Selected Nuclides

Nuclide	Z	N	A	Stability[a]	Mass[b] (u)	Spin[c]	Binding Energy (MeV/nucleon)
^{1}H	1	0	1	99.985%	1.007 825	$\frac{1}{2}$	—
^{7}Li	3	4	7	92.5%	7.016 003	$\frac{3}{2}$	5.60
^{31}P	15	16	31	100%	30.973 762	$\frac{1}{2}$	8.48
^{84}Kr	36	48	84	57.0%	83.911 507	0	8.72
^{120}Sn	50	70	120	32.4%	119.902 199	0	8.51
^{157}Gd	64	93	157	15.7%	156.923 956	$\frac{3}{2}$	8.21
^{197}Au	79	118	197	100%	196.966 543	$\frac{3}{2}$	7.91
^{227}Ac	89	138	227	21.8 y	227.027 750	$\frac{3}{2}$	7.65
^{239}Pu	94	145	239	24 100 y	239.052 158	$\frac{1}{2}$	7.56

[a]For stable nuclides, the **isotopic abundance** is given; this is the fraction of atoms of this type found in a typical sample of the element. For radioactive nuclides, the half-life is given.

[b]Following standard practice, the reported mass is that of the neutral atom, not that of the bare nucleus.

[c]Spin angular momentum in units of $\hbar$.

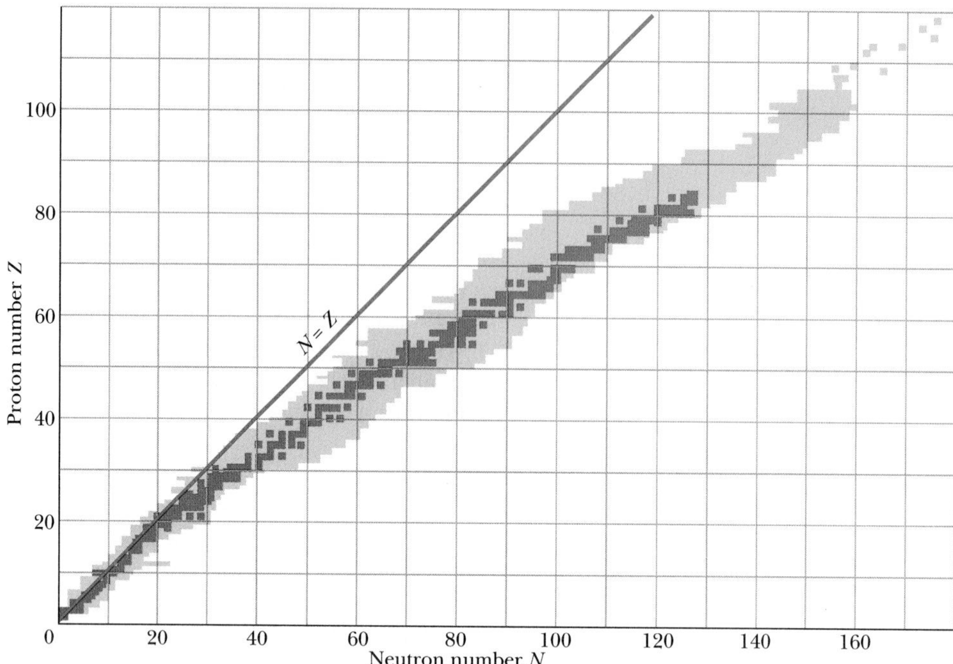

Fig. 43-4 A plot of the known nuclides. The green shading identifies the band of stable nuclides, the beige shading the radionuclides. Low-mass, stable nuclides have essentially equal numbers of neutrons and protons, but more massive nuclides have an increasing excess of neutrons. The figure shows that there are no stable nuclides with $Z > 83$ (bismuth).

We organize the nuclides on a **nuclidic chart** like that in Fig. 43-4, in which a nuclide is represented by plotting its proton number against its neutron number. The stable nuclides in this figure are represented by the green, the radionuclides by the yellow. As you can see, the radionuclides tend to lie on either side of—and at the upper end of—a well-defined band of stable nuclides. Note too that light stable nuclides tend to lie close to the line $N = Z$, which means that they have about the same numbers of neutrons and protons. Heavier nuclides, however, tend to have many more neutrons than protons. As an example, we saw that ^{197}Au has 118 neutrons and only 79 protons, a *neutron excess* of 39.

Nuclidic charts are available as wall charts, in which each small box on the chart is filled with data about the nuclide it represents. Figure 43-5 shows a section of such a chart, centered on ^{197}Au. Relative abundances (usually, as found on Earth) are shown for stable nuclides, and half-lives (a measure of decay rate) are shown for radionuclides. The sloping line points out a line of **isobars**—nuclides of the same mass number, $A = 198$ in this case.

As of early 2000, nuclides with atomic numbers as high as $Z = 118$ ($A = 293$) had been found in laboratory experiments (no elements with Z greater than 92 occur naturally). Although large nuclides generally should be highly unstable and last only a very brief time, certain supermassive nuclides are relatively stable, with fairly long lifetimes. These stable supermassive nuclides form an *island of stability* at high Z and N on a nuclidic chart like Fig. 43-4.

CHECKPOINT 1: Based on Fig. 43-4, which of the following nuclides do you conclude are not likely to be detected: ^{52}Fe ($Z = 26$), ^{90}As ($Z = 33$), ^{158}Nd ($Z = 60$), ^{175}Lu ($Z = 71$), ^{208}Pb ($Z = 82$)?

Nuclear Radii

A convenient unit for measuring distances on the scale of nuclei is the *femtometer* ($= 10^{-15}$ m). This unit is often called the *fermi;* the two names share the same

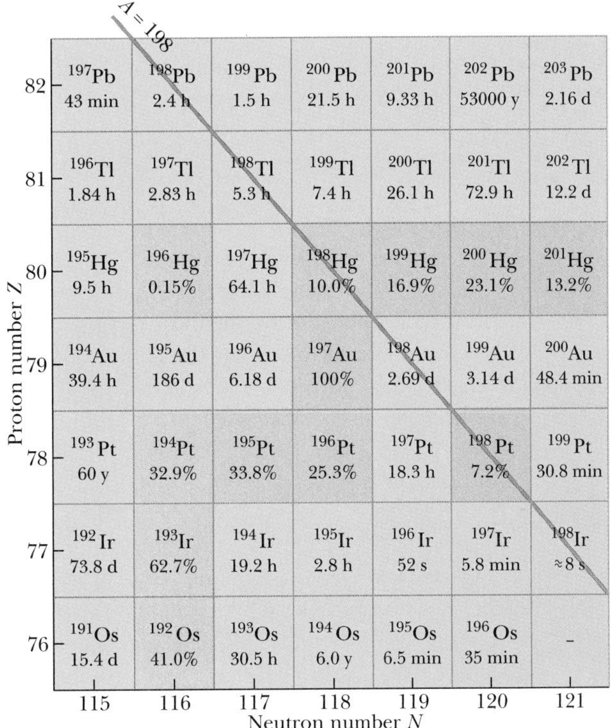

Fig. 43-5 An enlarged and detailed section of the nuclidic chart of Fig. 43-4, centered on ^{197}Au. Green squares represent stable nuclides, for which relative isotopic abundances are given. Beige squares represent radionuclides, for which half-lives are given. Isobaric lines of constant mass number A slope as shown by the example line for $A = 198$.

abbreviation. Thus,

$$1 \text{ femtometer} = 1 \text{ fermi} = 1 \text{ fm} = 10^{-15} \text{ m}. \tag{43-2}$$

We can learn about the size and structure of nuclei by bombarding them with a beam of high-energy electrons and observing how the nuclei deflect the incident electrons. The electrons must be energetic enough (at least 200 MeV) to have de Broglie wavelengths that are smaller than the nuclear structures they are to probe.

The nucleus, like the atom, is not a solid object with a well-defined surface. Furthermore, although most nuclides are spherical, some are notably ellipsoidal. Nevertheless, electron-scattering experiments (as well as experiments of other kinds) allow us to assign to each nuclide an effective radius given by

$$r = r_0 A^{1/3}, \tag{43-3}$$

in which A is the mass number and $r_0 \approx 1.2$ fm. We see that the volume of a nucleus, which is proportional to r^3, is directly proportional to the mass number A and is independent of the separate values of Z and N.

Equation 43-3 does not apply to *halo nuclides,* which are neutron-rich nuclides that were first produced in laboratories in the 1980s. These nuclides are larger than predicted by Eq. 43-3, because some of the neutrons form a *halo* around a spherical core of the protons and the rest of the neutrons. Lithium isotopes give an example. When a neutron is added to ^{8}Li to form ^{9}Li, neither of which are halo nuclides, the effective radius increases by about 4%. However, when two neutrons are added to ^{9}Li to form the neutron-rich isotope ^{11}Li (the largest of the lithium isotopes), they do not join that existing nucleus but instead form a halo around it, increasing the effective radius by about 30%. Apparently this halo configuration involves less energy than a core containing all 11 nucleons. (In this chapter we shall generally assume that Eq. 43-3 applies.)

Nuclear Masses

Atomic masses can be now measured with great precision. Recall from Section 1-6 that such masses are reported in atomic mass units u, chosen so that the atomic mass (not the nuclear mass) of ^{12}C is exactly 12 u. The relation of this unit to the SI mass unit is, approximately,

$$1 \text{ u} = 1.661 \times 10^{-27} \text{ kg}. \tag{43-4}$$

The mass number A of a nuclide is so named because the number represents the mass of the nuclide, expressed in atomic mass units and rounded off to the nearest integer. Thus, the atomic mass of ^{197}Au is 196.966573 u, which we round to 197 u.

In nuclear reactions, the relation $Q = -\Delta m\, c^2$ (Eq. 38-47) is an indispensable workaday tool. As we saw in Section 38-12, Q is the energy released (or absorbed) when the mass of a closed interacting system of particles changes by an amount Δm.

As we also saw in Section 38-12, Einstein's relation $E = mc^2$ tells us that the mass energy of a mass of 1 u is 931.5 MeV. Thus, from Eq. 38-43, we can use

$$c^2 = 931.5 \text{ MeV/u} \tag{43-5}$$

as a convenient conversion between energy measured in millions of electron-volts and mass measured in atomic mass units.

Nuclear Binding Energies

The mass M of a nucleus is *less* than the total mass Σm of its individual protons and neutrons. That means that the mass energy Mc^2 of a nucleus is *less* than the total mass energy $\Sigma(mc^2)$ of its individual protons and neutrons. The difference between these two energies is called the **binding energy** of the nucleus

$$\Delta E_{\text{be}} = \Sigma(mc^2) - Mc^2 \qquad \text{(binding energy).} \tag{43-6}$$

Caution: Binding energy is not an energy that resides in the nucleus. Rather, it is a *difference* in mass energy between a nucleus and its individual nucleons. If we were able to separate a nucleus into its nucleons, we would have to transfer a total energy equal to ΔE_{be} to those particles during the separating process. Although we cannot actually tear apart a nucleus in this way, the nuclear binding energy is still a convenient measure of how well a nucleus is held together.

A better measure is the **binding energy per nucleon** ΔE_{ben}, which is the ratio of the binding energy ΔE_{be} of a nucleus to the number A of nucleons in that nucleus:

$$\Delta E_{\text{ben}} = \frac{\Delta E_{\text{be}}}{A} \qquad \text{(binding energy per nucleon).} \tag{43-7}$$

We can think of the binding energy per nucleon as the average energy needed to separate a nucleus into its individual nucleons.

Figure 43-6 is a plot of the binding energy per nucleon ΔE_{ben} versus mass number A for a large number of nuclei. Those high on the plot are very tightly bound; that is, we would have to supply a great amount of energy per nucleon to break apart one of those nuclei. The nuclei that are lower on the plot, at the left and right sides, are less tightly bound, and less energy per nucleon would be required to break them apart.

These simple statements about Fig. 43-6 have profound consequences. The nucleons in a nucleus on the right side of the plot would be more tightly bound if that nucleus were to split into two nuclei that lie near the top of the plot. Such a process, called **fission,** occurs naturally with large (high mass number A) nuclei such as uranium, which can undergo fission spontaneously (that is, without an external cause

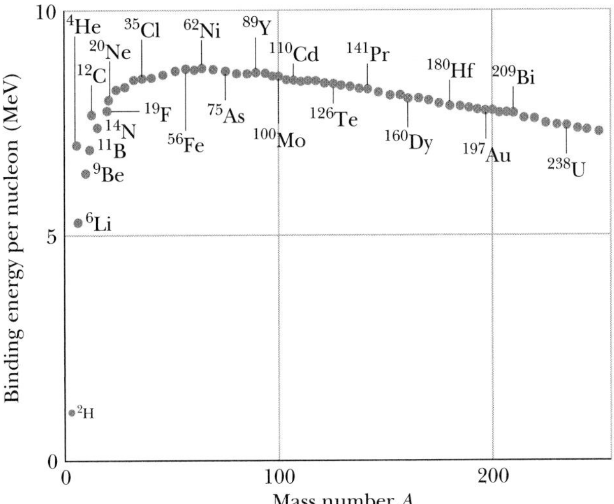

Fig. 43-6 The binding energy per nucleon for some representative nuclides. The nickel nuclide ^{62}Ni has the highest binding energy per nucleon (about 8.794 60 MeV/nucleon) of any known stable nuclide. Note that the alpha particle (^{4}He) has a higher binding energy per nucleon than its neighbors in the periodic table and thus is also particularly stable.

or source of energy). The process can also occur in nuclear weapons in which many uranium or plutonium nuclei are made to fission all at once, to create an explosion.

The nucleons in any pair of nuclei on the left side of the plot would be more tightly bound if the pair were to combine to form a single nucleus that lies near the top of the plot. Such a process, called **fusion,** occurs naturally in stars. Were this not true, the Sun would not shine and thus life could not exist on Earth.

Nuclear Energy Levels

The energy of nuclei, like that of atoms, is quantized. That is, nuclei can exist only in discrete quantum states, each with a well-defined energy. Figure 43-7 shows some of these energy levels for ^{28}Al, a typical low-mass nuclide. Note that the energy scale is in millions of electron-volts, rather than the electron-volts used for atoms. When a nucleus makes a transition from one level to a level of lower energy, the emitted photon is typically in the gamma-ray region of the electromagnetic spectrum.

Nuclear Spin and Magnetism

Many nuclides have an intrinsic *nuclear angular momentum,* or spin, and an associated intrinsic *nuclear magnetic moment.* Although nuclear angular momenta are roughly of the same magnitude as the angular momenta of atomic electrons, nuclear magnetic moments are much smaller than typical atomic magnetic moments.

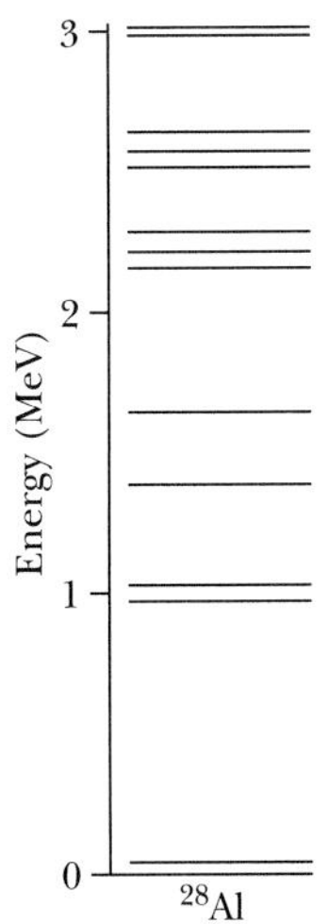

Fig. 43-7 Energy levels for the nuclide ^{28}Al, deduced from nuclear reaction experiments.

The Nuclear Force

The force that controls the motions of atomic electrons is the familiar electromagnetic force. To bind the nucleus together, however, there must be a strong attractive nuclear force of a totally different kind, strong enough to overcome the repulsive force between the (positively charged) nuclear protons and to bind both protons and neutrons into the tiny nuclear volume. The nuclear force must also be of short range because its influence does not extend very far beyond the nuclear "surface."

The present view is that the nuclear force that binds neutrons and protons in the nucleus is not a fundamental force of nature but is a secondary, or "spillover," effect of the **strong force** that binds quarks together to form neutrons and protons. In much the same way, the attractive force between certain neutral molecules is a spillover effect of the Coulomb electric force that acts within each molecule to bind it together.

Sample Problem 43-2

We can think of all nuclides as made up of a neutron-proton mixture that we can call *nuclear matter*. What is the density of nuclear matter?

SOLUTION: One **Key Idea** here is that we can find the (average) density ρ of a nucleus by dividing its total mass by its volume. Let m represent the mass of a nucleon (either a proton or a neutron, because those particles have about the same mass). Then the mass of a nucleus containing A nucleons is Am. Next, we assume the nucleus is spherical with radius r. Then its volume is $\frac{4}{3}\pi r^3$, and we can write the density of the nucleus as

$$\rho = \frac{Am}{\frac{4}{3}\pi r^3}.$$

A second **Key Idea** is that the radius r is given by Eq. 43-3 ($r = r_0 A^{1/3}$), where r_0 is 1.2 fm ($= 1.2 \times 10^{-15}$ m). Substituting for r then leads to

$$\rho = \frac{Am}{\frac{4}{3}\pi r_0^3 A} = \frac{m}{\frac{4}{3}\pi r_0^3}.$$

Note that A has canceled out; thus, this equation for density ρ applies to any nucleus that can be treated as spherical with a radius given by Eq. 43-3. Using 1.67×10^{-27} kg for the mass m of a nucleon, we then have

$$\rho = \frac{1.67 \times 10^{-27}\ \text{kg}}{\frac{4}{3}\pi(1.2 \times 10^{-15}\ \text{m})^3} \approx 2 \times 10^{17}\ \text{kg/m}^3. \quad \text{(Answer)}$$

This is about 2×10^{14} times the density of water.

Sample Problem 43-3

What is the binding energy per nucleon for ^{120}Sn?

SOLUTION: We need two **Key Ideas** here:

1. We can find the binding energy per nucleon ΔE_{ben} if we first find the binding energy ΔE_{be} and then divide by the number of nucleons A in the nucleus, according to Eq. 43-7 ($\Delta E_{ben} = \Delta E_{be}/A$).
2. We can find ΔE_{be} by finding the difference between the mass energy Mc^2 of the nucleus and the total mass energy $\Sigma(mc^2)$ of the individual nucleons that make up the nucleus, according to Eq. 43-6 ($\Delta E_{be} = \Sigma(mc^2) - Mc^2$).

From Table 43-1, we see that a ^{120}Sn nucleus consists of 50 protons ($Z = 50$) and 70 neutrons ($N = A - Z = 120 - 50 = 70$). Thus, we need to imagine a ^{120}Sn nucleus being separated into its 50 protons and 70 neutrons,

$$(^{120}\text{Sn nucleus}) \rightarrow 50\begin{pmatrix}\text{separate}\\ \text{protons}\end{pmatrix} + 70\begin{pmatrix}\text{separate}\\ \text{neutrons}\end{pmatrix}, \quad (43\text{-}8)$$

and then compute the resulting change in mass energy.

For that computation, we need the masses of a ^{120}Sn nucleus, a proton, and a neutron. However, because the mass of a neutral atom (nucleus *plus* electrons) is much easier to measure than the mass of a bare nucleus, calculations of binding energies are traditionally done with atomic masses. Thus, let's modify Eq. 43-8 so that it has a neutral ^{120}Sn atom on the left side. To do that, we include 50 electrons on the left side (to match the 50 protons in the ^{120}Sn nucleus). We must also add 50 electrons on the right side to balance Eq. 43-8. Those 50 electrons can be combined with the 50 protons, to form 50 neutral hydrogen atoms. We then have

$$(^{120}\text{Sn atom}) \rightarrow 50\begin{pmatrix}\text{separate}\\ \text{H atoms}\end{pmatrix} + 70\begin{pmatrix}\text{separate}\\ \text{neutrons}\end{pmatrix}. \quad (43\text{-}9)$$

In Table 43-1, the mass M_{Sn} of a ^{120}Sn atom is 119.902 199 u and the mass m_H of a hydrogen atom is 1.007 825 u; the mass m_n of a neutron is 1.008 665 u. Thus, Eq. 43-6 yields

$$\begin{aligned}\Delta E_{be} &= \Sigma(mc^2) - Mc^2\\ &= 50(m_H c^2) + 70(m_n c^2) - M_{Sn}c^2\\ &= 50(1.007\,825\ \text{u})c^2 + 70(1.008\,665\ \text{u})c^2\\ &\quad - (119.902\,199\ \text{u})c^2\\ &= (1.095\,601\ \text{u})c^2 = (1.095\,601\ \text{u})(931.5\ \text{MeV/u})\\ &= 1020.6\ \text{MeV},\end{aligned}$$

where Eq. 43-5 ($c^2 = 931.5$ MeV/u) provides an easy unit conversion. Note that using atomic masses instead of nuclear masses does not affect the result because the mass of the 50 electrons in the ^{120}Sn atom subtracts out from the mass of the electrons in the 50 hydrogen atoms.

Now Eq. 43-7 gives us the binding energy per nucleon as

$$\begin{aligned}\Delta E_{ben} &= \frac{\Delta E_{be}}{A} = \frac{1020.6\ \text{MeV}}{120}\\ &= 8.51\ \text{MeV/nucleon}. \quad \text{(Answer)}\end{aligned}$$

43-3 Radioactive Decay

As Fig. 43-4 shows, most of the nuclides that have been identified are radioactive. A radioactive nuclide spontaneously emits a particle, transforming itself in the process into a different nuclide, occupying a different square on the nuclidic chart.

Radioactive decay provided the first evidence that the laws that govern the subatomic world are statistical. Consider, for example, a 1 mg sample of uranium metal. It contains 2.5×10^{18} atoms of the very long-lived radionuclide ^{238}U.

The nuclei of these particular atoms have existed without decaying since they were created—well before the formation of our solar system. During any given second only about 12 of the nuclei in our sample will happen to decay by emitting an alpha particle, transforming themselves into nuclei of ^{234}Th. However,

> There is absolutely no way to predict whether any given nucleus in a radioactive sample will be among the small number of nuclei that decay during the next second. All have the same chance.

Although we cannot predict which nuclei in a sample will decay, we can say that if a sample contains N radioactive nuclei, then the rate ($= -dN/dt$) at which nuclei will decay is proportional to N:

$$-\frac{dN}{dt} = \lambda N, \tag{43-10}$$

in which λ, the **disintegration constant** (or **decay constant**) has a characteristic value for every radionuclide. Its SI unit is the inverse second (s^{-1}).

To find N as a function of time t, we first rearrange Eq. 43-10 as

$$\frac{dN}{N} = -\lambda\, dt, \tag{43-11}$$

and then integrate both sides, obtaining

$$\int_{N_0}^{N} \frac{dN}{N} = -\lambda \int_{t_0}^{t} dt,$$

or

$$\ln N - \ln N_0 = -\lambda(t - t_0). \tag{43-12}$$

Here N_0 is the number of radioactive nuclei in the sample at some arbitrary initial time t_0. Setting $t_0 = 0$ and rearranging Eq. 43-12 give us

$$\ln \frac{N}{N_0} = -\lambda t. \tag{43-13}$$

Taking the exponential of both sides (the exponential function is the antifunction of the natural logarithm) leads to

$$\frac{N}{N_0} = e^{-\lambda t}$$

or

$$N = N_0 e^{-\lambda t} \quad \text{(radioactive decay)}, \tag{43-14}$$

in which N_0 is the number of radioactive nuclei in the sample at $t = 0$ and N is the number remaining at any subsequent time t. Note that lightbulbs (for one example) follow no such exponential decay law. If we life-test 1000 bulbs, we expect that they will all "decay" (that is, burn out) at more or less the same time. The decay of radionuclides follows quite a different law.

We are often more interested in the decay rate R ($= -dN/dt$) than in N itself. Differentiating Eq. 43-14, we find

$$R = -\frac{dN}{dt} = \lambda N_0 e^{-\lambda t}$$

or

$$R = R_0 e^{-\lambda t} \quad \text{(radioactive decay)}, \tag{43-15}$$

an alternative form of the law of radioactive decay (Eq. 43-14). Here R_0 is the decay rate at time $t = 0$, and R is the rate at any subsequent time t. We can now rewrite Eq. 43-10 in terms of the decay rate R of the sample as

$$R = \lambda N, \tag{43-16}$$

where R and the number of radioactive nuclei N that have not yet undergone decay must be evaluated at the same instant.

The total decay rate R of a sample of one or more radionuclides is called the **activity** of that sample. The SI unit for activity is the **becquerel,** named for Henri Becquerel, the discoverer of radioactivity:

$$1 \text{ becquerel} = 1 \text{ Bq} = 1 \text{ decay per second.}$$

An older unit, the **curie,** is still in common use:

$$1 \text{ curie} = 1 \text{ Ci} = 3.7 \times 10^{10} \text{ Bq.}$$

Here is an example using these units: "The activity of spent reactor fuel rod #5658 on January 15, 2000, was 3.5×10^{15} Bq ($= 9.5 \times 10^4$ Ci)." Thus, on that day 3.5×10^{15} radioactive nuclei in the rod decayed each second. The identities of the radionuclides in the fuel rod, their disintegration constants λ, and the types of radiation they emit have no bearing on this measure of activity.

Often a radioactive sample will be placed near a detector that, for reasons of geometry or detector inefficiency, does not record all the disintegrations that occur in the sample. The reading of the detector under these circumstances is proportional to (and smaller than) the true activity of the sample. Such proportional activity measurements are reported not in becquerel units but simply in counts per unit time.

There are two common time measures of how long any given type of radionuclides lasts. One measure is the **half-life** $T_{1/2}$ of a radionuclide, which is the time at which both N and R have been reduced to one-half their initial values. The other measure is the **mean life** τ, which is the time at which both N and R have been reduced to e^{-1} of their initial values.

To relate $T_{1/2}$ to the disintegration constant λ, we put $R = \frac{1}{2}R_0$ in Eq. 43-15 and substitute $T_{1/2}$ for t. We obtain

$$\tfrac{1}{2}R_0 = R_0 e^{-\lambda T_{1/2}}.$$

Taking the natural logarithm of both sides and solving for $T_{1/2}$, we find

$$T_{1/2} = \frac{\ln 2}{\lambda}.$$

Similarly, to relate τ to λ, we put $R = e^{-1}R_0$ in Eq. 43-15, substitute τ for t, and solve for τ, finding

$$\tau = \frac{1}{\lambda}.$$

We summarize these results with the following:

$$T_{1/2} = \frac{\ln 2}{\lambda} = \tau \ln 2. \tag{43-17}$$

CHECKPOINT 2: The nuclide ^{131}I is radioactive, with a half-life of 8.04 days. At noon on January 1, the activity of a certain sample is 600 Bq. Using the concept of half-life, without written calculation, determine whether the activity at noon on January 24 will be a little less than 200 Bq, a little more than 200 Bq, a little less than 75 Bq, or a little more than 75 Bq.

Sample Problem 43-4

The table that follows shows some measurements of the decay rate of a sample of ^{128}I, a radionuclide often used medically as a tracer to measure the rate at which iodine is absorbed by the thyroid gland.

Time (min)	R (counts/s)	Time (min)	R (counts/s)
4	392.2	132	10.9
36	161.4	164	4.56
68	65.5	196	1.86
100	26.8	218	1.00

Find the disintegration constant λ and the half-life $T_{1/2}$ for this radionuclide.

SOLUTION: One Key Idea here is that the disintegration constant λ determines the exponential rate at which the decay rate R decreases with time t (as indicated by Eq. 43-15). Therefore, we should be able to determine λ by plotting the measurements of R against the measurement times t.

However, obtaining λ from a plot of R versus t is difficult because R decreases exponentially with t, according to Eq. 43-15. Thus, a second Key Idea is to transform Eq. 43-15 into a linear function of t, so that we can easily find λ. To do so, we take the natural logarithms of both sides of Eq. 43-15. We obtain

$$\ln R = \ln(R_0 e^{-\lambda t}) = \ln R_0 + \ln(e^{-\lambda t}) = \ln R_0 - \lambda t. \tag{43-18}$$

Because Eq. 43-18 is of the form $y = b + mx$, with b and m constants, it is a linear equation giving the quantity $\ln R$ as a function of t. Thus, if we plot $\ln R$ (instead of R) versus t, we should get a straight line. Further, the slope of the line should be equal to $-\lambda$.

Figure 43-8 shows a plot of $\ln R$ versus time t for the given measurements. The slope of the straight line that fits through the plotted points is

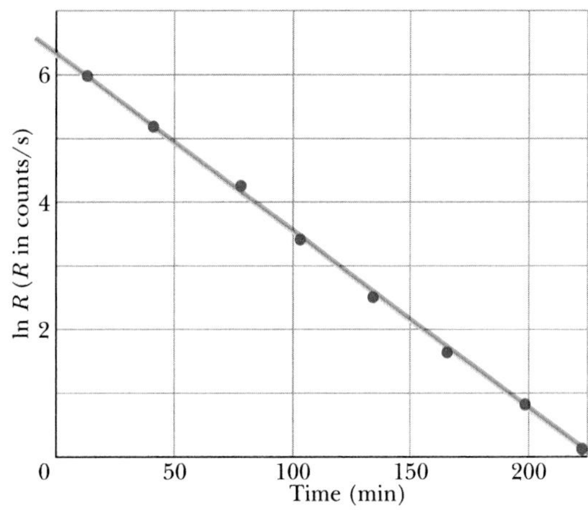

Fig. 43-8 Sample Problem 43-4. A semilogarithmic plot of the decay of a sample of ^{128}I, based on the data in the table.

$$\text{slope} = \frac{0 - 6.2}{225 \text{ min} - 0} = -0.0275 \text{ min}^{-1}.$$

Thus,

$$-\lambda = -0.0275 \text{ min}^{-1}$$

or

$$\lambda = 0.0275 \text{ min}^{-1} \approx 1.7 \text{ h}^{-1}. \quad \text{(Answer)}$$

To find the half-life $T_{1/2}$ of the radionuclide, we use the Key Idea that the time for the decay rate R to decrease by 1/2 is related to the disintegration constant λ via Eq. 43-17 ($T_{1/2} = (\ln 2)/\lambda$). From that equation, we find

$$T_{1/2} = \frac{\ln 2}{\lambda} = \frac{\ln 2}{0.0275 \text{ min}^{-1}} \approx 25 \text{ min}. \quad \text{(Answer)}$$

Sample Problem 43-5

A 2.71 g sample of KCl from the chemistry stockroom is found to be radioactive, and it is decaying at a constant rate of 4490 Bq. The decays are traced to the element potassium and in particular to the isotope ^{40}K, which constitutes 1.17% of normal potassium. Calculate the half-life of this nuclide.

SOLUTION: One Key Idea here is that because the activity R of the sample is apparently constant, we cannot find the half-life $T_{1/2}$ by plotting $\ln R$ versus time t as we did in Sample Problem 43-4. (We would just get a horizontal plot.) However, we can use the following two Key Ideas:

1. We can relate the half-life $T_{1/2}$ to the disintegration constant λ via Eq. 43-17 ($T_{1/2} = (\ln 2)/\lambda$).
2. We can then relate λ to the given activity R of 4490 Bq by means of Eq. 43-16 ($R = \lambda N$), where N is the number of ^{40}K nuclei (and thus atoms) in the sample.

Combining Eqs. 43-17 and 43-16 yields

$$T_{1/2} = \frac{N \ln 2}{R}. \tag{43-19}$$

We know that N in this equation is 1.17% of the total number N_K of potassium atoms in the sample. We also know that N_K must equal the number N_{KCl} of molecules in the sample. We can obtain N_{KCl} from the molar mass M_{KCl} of KCl (the mass of one mole of KCl) and the given mass M_{sam} of the sample by combining Eqs. 20-2 and 20-3 to write

$$N_{KCl} = \left(\begin{matrix}\text{number of moles}\\ \text{in sample}\end{matrix}\right) N_A = \frac{M_{sam}}{M_{KCl}} N_A, \tag{43-20}$$

where N_A is Avogadro's number (6.02×10^{23} mol^{-1}). From Appendix F, we see that the molar mass of potassium is 39.102 g/mol and the molar mass of chlorine is 35.453 g/mol; thus, the molar mass of KCl is 74.555 g/mol. Equation 43-20 then gives us

$$N_{\mathrm{KCl}} = \frac{(2.71\ \mathrm{g})(6.02 \times 10^{23}\ \mathrm{mol^{-1}})}{74.555\ \mathrm{g/mol}} = 2.188 \times 10^{22}$$

as the number of KCl molecules in the sample. Thus, the total number N_K of potassium atoms is also 2.188×10^{22}, and the number of ^{40}K in the sample must be

$$N = 0.0117N_K = (0.0117)(2.188 \times 10^{22}) = 2.560 \times 10^{20}.$$

Substituting this value for N and the given activity of 4490 Bq ($= 4490\ \mathrm{s^{-1}}$) for R into Eq. 43-19 leads to

$$T_{1/2} = \frac{(2.560 \times 10^{20}) \ln 2}{4490\ \mathrm{s^{-1}}} = 3.95 \times 10^{16}\ \mathrm{s} = 1.25 \times 10^{9}\ \mathrm{y}. \quad \text{(Answer)}$$

This half-life of ^{40}K turns out to have the same order of magnitude as the age of the universe. Thus, the activity of ^{40}K in the stockroom sample decreases *very* slowly, too slowly for us to detect during a few days of observation or even an entire lifetime. A portion of the potassium in our bodies consists of this radioisotope, which means that we are all slightly radioactive.

43-4 Alpha Decay

When a nucleus undergoes **alpha decay,** it transforms to a different nuclide by emitting an alpha particle (a helium nucleus, ^{4}He). For example, when uranium ^{238}U undergoes alpha decay, it transforms to thorium ^{234}Th:

$$^{238}\mathrm{U} \rightarrow {}^{234}\mathrm{Th} + {}^{4}\mathrm{He}. \quad (43\text{-}21)$$

This alpha decay of ^{238}U can occur spontaneously (without an external source of energy) because the total mass of the decay products ^{234}Th and ^{4}He is less than the mass of the original ^{238}U. Thus, the total mass energy of the decay products is less than the mass energy of the original nuclide. As defined by Eq. 38-47, in such a process the difference between the initial mass energy and the total final mass energy is called the Q of the process.

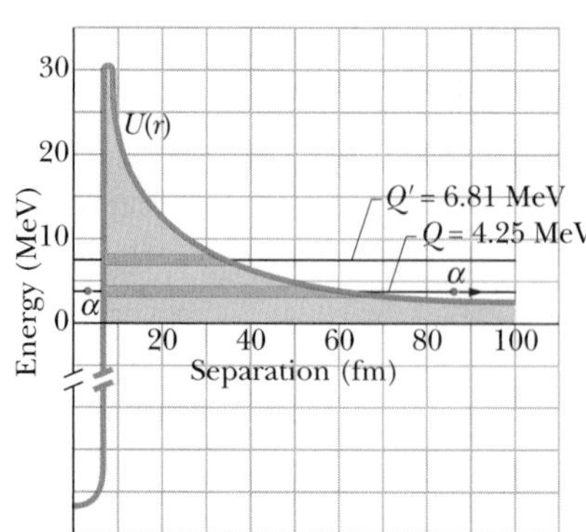

Fig. 43-9 A potential energy function for the emission of an alpha particle by ^{238}U. The horizontal black line marked $Q = 4.25$ MeV shows the disintegration energy for the process. The thick gray portion of this line represents separations r that are classically forbidden to the alpha particle. The alpha particle is represented by a dot, both inside this potential energy barrier (at the left) and outside it (at the right), after the particle has tunneled through. The horizontal black line marked $Q' = 6.81$ MeV shows the disintegration energy for the alpha decay of ^{228}U. (Both isotopes have the same potential energy function because they have the same nuclear charge.)

For a nuclear decay, we say that the difference in mass energy is the decay's *disintegration energy Q*. The Q for the decay in Eq. 43-21 is 4.25 MeV—that amount of energy is said to be released by the alpha decay of ^{238}U, with the energy transferred from mass energy to the kinetic energy of the two products.

The half-life of ^{238}U for this decay process is 4.5×10^9 y. Why so long? If ^{238}U can decay in this way, why doesn't every ^{238}U nuclide in a sample of ^{238}U atoms simply decay at once? To answer the questions, we must examine the process of alpha decay.

We choose a model in which the alpha particle is imagined to exist (already formed) inside the nucleus before it escapes from the nucleus. Figure 43-9 shows the approximate potential energy $U(r)$ of the system consisting of the alpha particle and the residual ^{234}Th nucleus, as a function of their separation r. This energy is a combination of (1) the potential energy associated with the (attractive) strong nuclear force that acts in the nuclear interior and (2) a Coulomb potential associated with the (repulsive) electric force that acts between the two particles before and after the decay has occurred.

The horizontal black line marked $Q = 4.25$ MeV shows the disintegration energy for the process. If we assume that this represents the total energy of the alpha particle during the decay process, then the part of the $U(r)$ curve above this line constitutes a potential energy barrier like that in Fig. 39-12. This barrier cannot be surmounted. If the alpha particle were able to be at some separation r within the barrier, its potential energy U would exceed its total energy E. This would mean, classically, that its kinetic energy K (which equals $E - U$) would be negative, an impossible situation.

We can see now why the alpha particle is not immediately emitted from the ^{238}U nucleus. That nucleus is surrounded by an impressive potential barrier, occupying—if you think of it in three dimensions—the volume lying between two spherical shells (of radii about 8 and 60 fm). This argument is so convincing that we now change our last question and ask: How, since the particle seems permanently trapped inside the nucleus by the barrier, can the ^{238}U nucleus *ever* emit an alpha particle? The answer is that, as you learned in Section 39-9, there is a finite probability that a particle can tunnel through an energy barrier that is classically insurmountable. In fact, alpha decay occurs as a result of barrier tunneling.

Since the half-life of ^{238}U is very long, the barrier is apparently not very "leaky." The alpha particle, presumed to be rattling back and forth within the nucleus, must arrive at the inner surface of the barrier about 10^{38} times before it succeeds in tunneling through the barrier. This is about 10^{21} times per second for about 4×10^9 years (the age of Earth)! We, of course, are waiting on the outside, able to count only the alpha particles that *do* manage to escape.

We can test this explanation of alpha decay by examining other alpha emitters. For an extreme contrast, consider the alpha decay of another uranium isotope, ^{228}U, which has a disintegration energy Q' of 6.81 MeV, about 60% higher than that of ^{238}U. (The value of Q' is also shown as a horizontal black line in Fig. 43-9.) Recall from Section 39-9 that the transmission coefficient of a barrier is very sensitive to small changes in the total energy of the particle seeking to penetrate it. Thus, we expect alpha decay to occur more readily for this nuclide than for ^{238}U. Indeed it does. As Table 43-2 shows, its half-life is only 9.1 min! An increase in Q by a factor of only 1.6 produces a decrease in half-life (that is, in the effectiveness of the barrier) by a factor of 3×10^{14}. This is sensitivity indeed.

TABLE 43-2 Two Alpha Emitters Compared

Radionuclide	Q	Half-Life
^{238}U	4.25 MeV	4.5×10^9 y
^{228}U	6.81 MeV	9.1 min

Sample Problem 43-6

We are given the following atomic masses:

^{238}U	238.050 79 u	^{4}He	4.002 60 u
^{234}Th	234.043 63 u	^{1}H	1.007 83 u
^{237}Pa	237.051 21 u		

Here Pa is the symbol for the element protactinium ($Z = 91$).

(a) Calculate the energy released during the alpha decay of ^{238}U. The decay process is

$$^{238}U \rightarrow {}^{234}Th + {}^{4}He.$$

Note, incidentally, how nuclear charge is conserved in this equation: The atomic numbers of thorium (90) and helium (2) add up to the atomic number of uranium (92). The number of nucleons is also conserved: $238 = 234 + 4$.

SOLUTION: The **Key Idea** here is that the energy released in the decay is the disintegration energy Q, which we can calculate from the change in mass ΔM due to the ^{238}U decay. We use Eq. 38-47,

$$Q = M_i c^2 - M_f c^2, \qquad (43\text{-}22)$$

where the initial mass M_i is that of ^{238}U and the final mass M_f is the sum of the ^{234}Th and ^{4}He masses. As in Sample Problem 43-3, we must do this calculation for neutral atoms—that is, with atomic masses. Using the atomic masses given in the problem statement, Eq. 43-22 becomes

$$\begin{aligned} Q &= (238.050\ 79\ \text{u})c^2 - (234.043\ 63\ \text{u} + 4.002\ 60\ \text{u})c^2 \\ &= (0.004\ 56\ \text{u})c^2 = (0.004\ 56\ \text{u})(931.5\ \text{MeV/u}) \\ &= 4.25\ \text{MeV}. \qquad \text{(Answer)} \end{aligned}$$

Note that using atomic masses instead of nuclear masses does not affect the result because the total mass of the electrons in the products subtracts out from the mass of the nucleons + electrons in the original ^{238}U.

(b) Show that ^{238}U cannot spontaneously emit a proton.

SOLUTION: If this happened, the decay process would be

$$^{238}U \rightarrow {}^{237}Pa + {}^{1}H.$$

(You should verify that both nuclear charge and the number of nucleons are conserved in this process.) Using the same Key Idea as in part (a) and proceeding as we did there, we would find that the mass of the two decay products (= 237.051 21 u + 1.007 83 u) would *exceed* the mass of ^{238}U by $\Delta m = 0.008\ 25$ u, with disintegration energy $Q = -7.68$ MeV. The minus sign indicates that we must *add* 7.68 MeV to a ^{238}U nucleus before it will emit a proton; it will certainly not do so spontaneously.

43-5 Beta Decay

A nucleus that decays spontaneously by emitting an electron or a positron (a positively charged particle with the mass of an electron) is said to undergo **beta decay.** Like alpha decay, this is a spontaneous process, with a definite disintegration energy and half-life. Again like alpha decay, beta decay is a statistical process, governed by Eqs. 43-14 and 43-15. In *beta-minus* (β^-) decay, an electron is emitted by a nucleus, as in the decay

$$^{32}\text{P} \rightarrow {}^{32}\text{S} + \text{e}^- + \nu \quad (T_{1/2} = 14.3 \text{ d}). \tag{43-23}$$

In *beta-plus* (β^+) decay, a positron is emitted by a nucleus, as in the decay

$$^{64}\text{Cu} \rightarrow {}^{64}\text{Ni} + \text{e}^+ + \nu \quad (T_{1/2} = 12.7 \text{ h}). \tag{43-24}$$

The symbol ν represents a **neutrino,** a neutral particle, with very little or no mass, that is emitted from the nucleus along with the electron or positron during the decay process. Neutrinos interact only very weakly with matter and—for that reason—are so extremely difficult to detect that their presence long went unnoticed.*

Both charge and nucleon number are conserved in the above two processes. In the decay of Eq. 43-23, for example, we can write for charge conservation

$$(+15e) = (+16e) + (-e) + (0),$$

because ^{32}P has 15 protons, ^{32}S has 16 protons, and the neutrino ν has zero charge. Similarly, for nucleon conservation, we can write

$$(32) = (32) + (0) + (0),$$

because ^{32}P and ^{32}S each have 32 nucleons and neither the electron nor the neutrino is a nucleon.

It may seem surprising that nuclei can emit electrons, positrons, and neutrinos, since we have said that nuclei are made up of neutrons and protons only. However, we saw earlier that atoms emit photons, and we certainly do not say that atoms "contain" photons. We say that the photons are created during the emission process.

It is the same with the electrons, positrons, and neutrinos emitted from nuclei during beta decay. They are created during the emission process. For beta-minus decay, a neutron transforms into a proton within the nucleus according to

$$\text{n} \rightarrow \text{p} + \text{e}^- + \nu. \tag{43-25}$$

For beta-plus decay, a proton transforms into a neutron via

$$\text{p} \rightarrow \text{n} + \text{e}^+ + \nu. \tag{43-26}$$

Both of these beta-decay processes provide evidence that—as was pointed out—neutrons and protons are not truly fundamental particles. These processes show why the mass number A of a nuclide undergoing beta decay does not change; one of its constituent nucleons simply changes its character according to Eq. 43-25 or 43-26.

*Beta decay also includes *electron capture,* in which a nucleus decays by absorbing one of its atomic electrons, emitting a neutrino in the process. We do not consider that process here. Also, the neutral particle emitted in the decay process of Eq. 43-23 is actually an *antineutrino,* a distinction we shall not make in this introductory treatment.

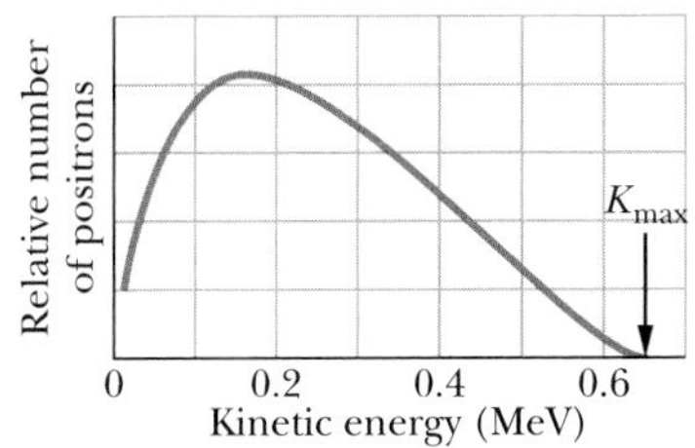

Fig. 43-10 The distribution of the kinetic energies of positrons emitted in the beta decay of ^{64}Cu. The maximum kinetic energy of the distribution (K_{max}) is 0.653 MeV. In all ^{64}Cu decay events, this energy is shared between the positron and the neutrino, in varying proportions. The *most probable* energy for an emitted positron is about 0.15 MeV.

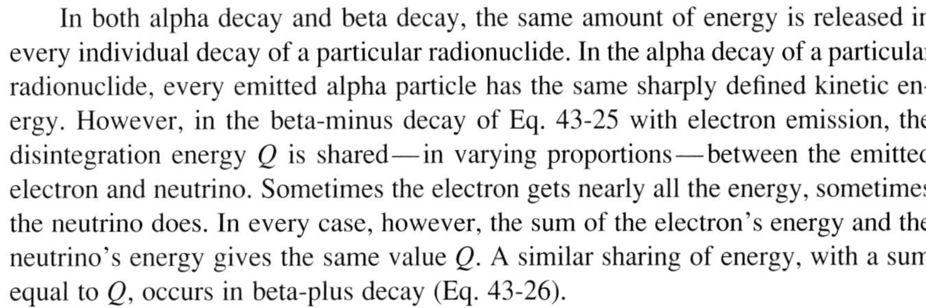

In both alpha decay and beta decay, the same amount of energy is released in every individual decay of a particular radionuclide. In the alpha decay of a particular radionuclide, every emitted alpha particle has the same sharply defined kinetic energy. However, in the beta-minus decay of Eq. 43-25 with electron emission, the disintegration energy Q is shared—in varying proportions—between the emitted electron and neutrino. Sometimes the electron gets nearly all the energy, sometimes the neutrino does. In every case, however, the sum of the electron's energy and the neutrino's energy gives the same value Q. A similar sharing of energy, with a sum equal to Q, occurs in beta-plus decay (Eq. 43-26).

Thus, in beta decay the energy of the emitted electrons or positrons may range from zero up to a certain maximum K_{max}. Figure 43-10 shows the distribution of positron energies for the beta decay of ^{64}Cu (see Eq. 43-24). The maximum positron energy K_{max} must equal the disintegration energy Q because the neutrino carries away approximately zero energy when the positron carries away K_{max}; that is,

$$Q = K_{max}. \tag{43-27}$$

The Neutrino

Wolfgang Pauli first suggested the existence of neutrinos in 1930. His neutrino hypothesis not only permitted an understanding of the energy distribution of electrons or positrons in beta decay but also solved another early beta-decay puzzle involving "missing" angular momentum.

The neutrino is a truly elusive particle; the mean free path of an energetic neutrino in water has been calculated as no less than several thousand light-years. At the same time, neutrinos left over from the big bang that presumably marked the creation of the universe are the most abundant particles of physics. Billions of them pass through our bodies every second, leaving no trace.

In spite of their elusive character, neutrinos have been detected in the laboratory. This was first done in 1953 by F. Reines and C. L. Cowan, using neutrinos generated in a high-power nuclear reactor. (In 1995 Reines, the surviving member of the pair, received a Nobel prize for this work.) In spite of the difficulties of detection, experimental neutrino physics is now a well-developed branch of experimental physics, with avid practitioners at laboratories throughout the world.

The Sun emits neutrinos copiously from the nuclear furnace at its core and, at night, these messengers from the center of the Sun come up at us from below, Earth being almost totally transparent to them. In February 1987, light from an exploding star in the Large Magellanic Cloud (a nearby galaxy) reached Earth after having traveled for 170 000 years. Enormous numbers of neutrinos were generated in this explosion, and about 10 of them were picked up by a sensitive neutrino detector in Japan; Fig. 43-11 shows a record of their passage.

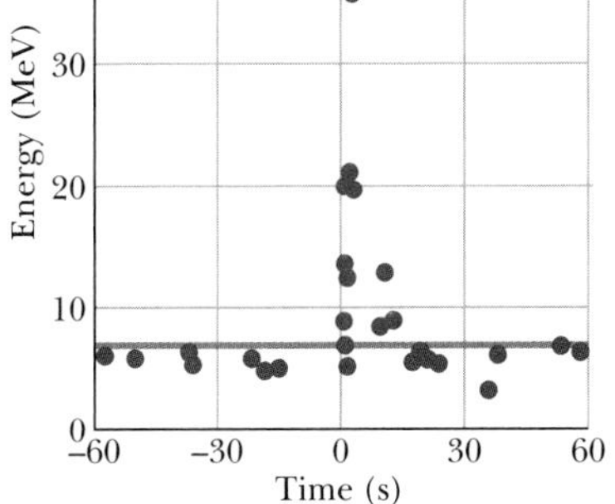

Fig. 43-11 A burst of neutrinos from the supernova SN 1987A, which occurred at (relative) time 0, stands out from the usual *background* of neutrinos. (For neutrinos, 10 is a "burst.") The particles were detected by an elaborate detector housed deep in a mine in Japan. The supernova was visible only in the Southern Hemisphere, so the neutrinos had to penetrate Earth (a trifling barrier for them) to reach the detector.

Radioactivity and the Nuclidic Chart

We can increase the information of the nuclidic chart of Fig. 43-4 by plotting the **mass excess** of each nuclide in a direction perpendicular to the N-Z plane. The mass excess of a nuclide is (in spite of its name) an energy that approximates the nuclide's *total* binding energy. It is defined as $(m - A)c^2$, where m is the atomic mass of the nuclide and A is its mass number, both expressed in atomic mass units, and c^2 is 931.5 MeV/u.

The surface so formed gives a graphic representation of nuclear stability. As Fig. 43-12 shows (for the low mass nuclides), this surface describes a "valley of the

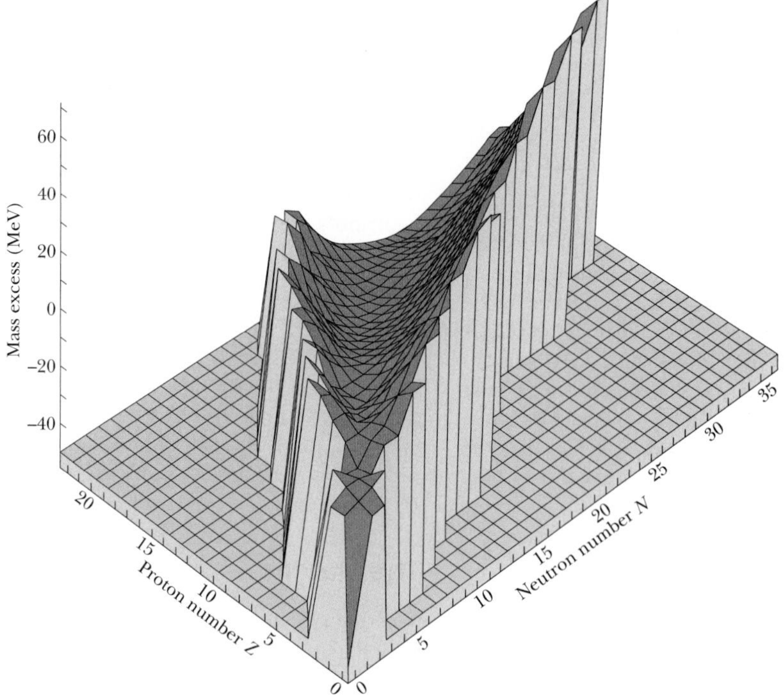

Fig. 43-12 A portion of the valley of the nuclides, showing only the nuclides of low mass. Deuterium, tritium, and helium lie at the nearest end of the plot, with helium at the high point. The valley stretches away from us, with the plot stopping at about $Z = 22$ and $N = 35$. Nuclides with large values of A, which would be plotted much beyond the valley, can decay into the valley by repeated alpha emissions and by fission (splitting of a nuclide).

nuclides," with the stability band of Fig. 43-4 running along its bottom. Nuclides on the proton-rich side of the valley decay into it by emitting positrons, and those on the neutron-rich side do so by emitting electrons.

CHECKPOINT 3: ^{238}U decays into ^{234}Th by the emission of an alpha particle. There follows a chain of further radioactive decays, either by alpha decay or by beta decay. Eventually a stable nuclide is reached and after that, no further radioactive decay is possible. Which of the following stable nuclides is the end product of the ^{238}U radioactive decay chain: ^{206}Pb, ^{207}Pb, ^{208}Pb, or ^{209}Pb? (*Hint:* You can decide by considering the changes in mass number A for the two types of decay.)

Sample Problem 43-7

Calculate the disintegration energy Q for the beta decay of ^{32}P, as described by Eq. 43-23. The needed atomic masses are 31.973 91 u for ^{32}P and 31.972 07 u for ^{32}S.

SOLUTION: The **Key Idea** here is that the disintegration energy Q for the beta decay is the amount by which the mass energy is changed by the decay. Q is given by Eq. 38-47 ($Q = -\Delta M\, c^2$). However, because an individual electron is emitted (and not an electron bound up in an atom), we must be careful to distinguish between nuclear masses (which we do not know) and atomic masses (which we do know). Let the boldface symbols $\mathbf{m}_\text{P}$ and $\mathbf{m}_\text{S}$ represent the nuclear masses of ^{32}P and ^{32}S, and let the italic symbols m_P and m_S represent their atomic masses. Then we can write the change in mass for the decay of Eq. 43-23 as

$$\Delta m = (\mathbf{m}_\text{S} + m_\text{e}) - \mathbf{m}_\text{P},$$

in which m_e is the mass of the electron. If we add and subtract $15m_\text{e}$ on the right side of this equation, we obtain

$$\Delta m = (\mathbf{m}_\text{S} + 16m_\text{e}) - (\mathbf{m}_\text{P} + 15m_\text{e}).$$

The quantities in parentheses are the atomic masses of ^{32}S and ^{32}P, so

$$\Delta m = m_\text{S} - m_\text{P}.$$

We thus see that if we subtract only the atomic masses, the mass of the emitted electron is automatically taken into account. (This procedure will not work for positron emission.)

The disintegration energy for the ^{32}P decay is then

$$\begin{aligned} Q &= -\Delta m\, c^2 \\ &= -(31.972\,07\text{ u} - 31.973\,91\text{ u})(931.5\text{ MeV/u}) \\ &= 1.71\text{ MeV}. \end{aligned} \qquad \text{(Answer)}$$

Experimentally, this calculated quantity proves to be equal to K_{max}, the maximum energy the emitted electrons can have. Although 1.71 MeV is released every time a ^{32}P nucleus decays, in essentially every case the electron carries away less energy than this. The neutrino gets all the rest, carrying it stealthily out of the laboratory.

A fragment of the Dead Sea scrolls and the caves from which the scrolls were recovered.

43-6 Radioactive Dating

If you know the half-life of a given radionuclide, you can in principle use the decay of that radionuclide as a clock to measure time intervals. The decay of very long-lived nuclides, for example, can be used to measure the age of rocks—that is, the time that has elapsed since they were formed. Such measurements for rocks from Earth and the Moon, and for meteorites, yield a consistent maximum age of about 4.5×10^9 y for these bodies.

The radionuclide ^{40}K, for example, decays to ^{40}Ar, a stable isotope of the noble gas argon. The half-life for this decay is 1.25×10^9 y. A measurement of the ratio of ^{40}K to ^{40}Ar, as found in the rock in question, can be used to calculate the age of that rock. Other long-lived decays, such as that of ^{235}U to ^{207}Pb (involving a number of intermediate stages), can be used to verify this calculation.

For measuring shorter time intervals, in the range of historical interest, radiocarbon dating has proved invaluable. The radionuclide ^{14}C (with $T_{1/2} = 5730$ y) is produced at a constant rate in the upper atmosphere as atmospheric nitrogen is bombarded by cosmic rays. This radiocarbon mixes with the carbon that is normally present in the atmosphere (as CO_2) so that there is about one atom of ^{14}C for every 10^{13} atoms of ordinary stable ^{12}C. Through biological activity such as photosynthesis and breathing, the atoms of atmospheric carbon trade places randomly, one atom at a time, with the atoms of carbon in every living thing, including broccoli, mushrooms, penguins, and humans. Eventually an exchange equilibrium is reached at which the carbon atoms of every living thing contain a fixed small fraction of the radioactive nuclide ^{14}C.

This equilibrium persists as long as the organism is alive. When the organism dies, the exchange with the atmosphere stops and the amount of radiocarbon trapped in the organism, since it is no longer being replenished, dwindles away with a half-life of 5730 y. By measuring the amount of radiocarbon per gram of organic matter, it is possible to measure the time that has elapsed since the organism died. Charcoal from ancient campfires, the Dead Sea scrolls, and many prehistoric artifacts have been dated in this way. The age of the scrolls was determined by radiocarbon dating a sample of the cloth used to plug the jars in which the scrolls were sealed.

Sample Problem 43-8

Mass spectrometric analysis of potassium and argon atoms in a Moon rock sample shows that the ratio of the number of (stable) ^{40}Ar atoms present to the number of (radioactive) ^{40}K atoms is 10.3. Assume that all the argon atoms were produced by the decay of potassium atoms, with a half-life of 1.25×10^9 y. How old is the rock?

SOLUTION: The Key Idea here is that if N_0 potassium atoms were present at the time the rock was formed by solidification from a molten form, the number of potassium atoms remaining at the time of analysis is, from Eq. 43-14,

$$N_K = N_0 e^{-\lambda t}, \tag{43-28}$$

in which t is the age of the rock. For every potassium atom that decays, an argon atom is produced. Thus, the number of argon atoms present at the time of the analysis is

$$N_{Ar} = N_0 - N_K. \tag{43-29}$$

We cannot measure N_0, so let's eliminate it from Eqs. 43-28 and 43-29. We find, after some algebra, that

$$\lambda t = \ln\left(1 + \frac{N_{Ar}}{N_K}\right), \tag{43-30}$$

in which N_{Ar}/N_K *can* be measured. Solving for t and using Eq. 43-17 to replace λ with $(\ln 2)/T_{1/2}$ yield

$$t = \frac{T_{1/2} \ln(1 + N_{Ar}/N_K)}{\ln 2}$$

$$= \frac{(1.25 \times 10^9 \text{ y})[\ln(1 + 10.3)]}{\ln 2}$$

$$= 4.37 \times 10^9 \text{ y}. \qquad \text{(Answer)}$$

Lesser ages may be found for other lunar or terrestrial rock samples, but no substantially greater ones. Thus, the solar system must be about 4 billion years old.

43-7 Measuring Radiation Dosage

The effect of radiation such as gamma rays, electrons, and alpha particles on living tissue (particularly our own) is a matter of public interest. Such radiation is found in nature in cosmic rays and arises from radioactive elements in Earth's crust. Radiation associated with some human activities, such as using x rays and radionuclides in medicine and in industry, also contributes.

It is not our task here to explore the various sources of radiation but simply to describe the units in which the properties and effects of such radiations are expressed. We have already discussed the *activity* of a radioactive source. There are two remaining quantities of interest.

1. *Absorbed Dose.* This is a measure of the radiation dose (as energy per unit mass) actually absorbed by a specific object, such as a patient's hand or chest. Its SI unit is the **gray** (Gy). An older unit, the **rad** (from **r**adiation **a**bsorbed **d**ose) is still in common use. The terms are defined and related as follows:

$$1 \text{ Gy} = 1 \text{ J/kg} = 100 \text{ rad}. \tag{43-31}$$

   A typical dose-related statement is: "A whole-body, short-term gamma-ray dose of 3 Gy (= 300 rad) will cause death in 50% of the population exposed to it." Thankfully, our present average absorbed dose per year, from sources of both natural and human origin, is only about 2 mGy (= 0.2 rad).

2. *Dose Equivalent.* Although different types of radiation (gamma rays and neutrons, say) may deliver the same amount of energy to the body, they do not have the same biological effect. The dose equivalent allows us to express the biological effect by multiplying the absorbed dose (in grays or rads) by a numerical **RBE** factor (from **r**elative **b**iological **e**ffectiveness). For x rays and electrons, for example, RBE = 1; for slow neutrons RBE = 5; for alpha particles RBE = 10; and so on. Personnel-monitoring devices such as film badges register the dose equivalent.

   The SI unit of dose equivalent is the **sievert** (Sv). An earlier unit, the **rem,** is still in common use. Their relationship is

$$1 \text{ Sv} = 100 \text{ rem}. \tag{43-32}$$

   An example of the correct use of these terms is: "The recommendation of the National Council on Radiation Protection is that no individual who is (nonoccupationally) exposed to radiation should receive a dose equivalent greater than 5 mSv (= 0.5 rem) in any one year." This includes radiation of all kinds; of course the appropriate RBE factor must be used for each kind.

Sample Problem 43-9

We have seen that a gamma-ray dose of 3 Gy is lethal to half the people exposed to it. If the equivalent energy were absorbed as heat, what rise in body temperature would result?

SOLUTION: One **Key Idea** here is that we can relate an absorbed energy Q and the resulting temperature increase ΔT with Eq. 19-14 ($Q = cm\,\Delta T$). In that equation, m is the mass of the material absorbing the energy and c is the specific heat of that material. Another **Key Idea** is that an absorbed dose of 3 Gy corresponds to an absorbed energy per unit mass of 3 J/kg. Let us assume that c, the specific heat of the human body, is the same as that of water, 4180 J/kg · K. Then we find that

$$\Delta T = \frac{Q/m}{c} = \frac{3 \text{ J/kg}}{4180 \text{ J/kg} \cdot \text{K}} = 7.2 \times 10^{-4} \text{ K} \approx 700\ \mu\text{K}. \quad \text{(Answer)}$$

Obviously the damage done by ionizing radiation has nothing to do with thermal heating. The harmful effects arise because the radiation damages DNA and thus interferes with the normal functioning of tissues in which it is absorbed.

43-8 Nuclear Models

Nuclei are more complicated than atoms. For atoms, the basic force law (Coulomb's law) is simple in form and there is a natural force center, the nucleus. For nuclei the force law is complicated and cannot, in fact, be written down explicitly in full detail. Furthermore, the nucleus—a jumble of protons and neutrons—has no natural force center to simplify the calculations.

In the absence of a comprehensive nuclear *theory,* we turn to the construction of nuclear *models.* A nuclear model is simply a way of looking at the nucleus that gives a physical insight into as wide a range of its properties as possible. The usefulness of a model is tested by its ability to provide predictions that can be verified experimentally in the laboratory.

Two models of the nucleus have proved useful. Although based on assumptions that seem flatly to exclude each other, each accounts very well for a selected group of nuclear properties. After describing them separately, we shall see how these two models may be combined to form a single coherent picture of the atomic nucleus.

The Collective Model

In the *collective model,* formulated by Niels Bohr, the nucleons, moving around within the nucleus at random, are imagined to interact strongly with each other, like the molecules in a drop of liquid. A given nucleon collides frequently with other nucleons in the nuclear interior, its mean free path as it moves about being substantially less than the nuclear radius.

The collective model permits us to correlate many facts about nuclear masses and binding energies; it is useful (as you will see later) in explaining nuclear fission. It is also useful for understanding a large class of nuclear reactions.

Consider, for example, a generalized nuclear reaction of the form

$$X + a \rightarrow C \rightarrow Y + b. \tag{43-33}$$

We imagine that projectile a enters target nucleus X, forming a **compound nucleus** C and conveying to it a certain amount of excitation energy. The projectile, perhaps a neutron, is at once caught up by the random motions that characterize the nuclear interior. It quickly loses its identity—so to speak—and the excitation energy it carried into the nucleus is quickly shared with all the other nucleons in C.

The quasi-stable state represented by C in Eq. 43-33 may have a mean life of 10^{-16} s before it decays to Y and b. By nuclear standards, this is a very long time, being about one million times longer than the time required for a nucleon with a few million electron-volts of energy to travel across a nucleus.

The central feature of this compound-nucleus concept is that the formation of the compound nucleus and its eventual decay are totally independent events. At the time of its decay, the compound nucleus has "forgotten" how it was formed. Hence, its mode of decay is not influenced by its mode of formation. As an example, Fig. 43-13 shows three possible ways in which the compound nucleus ^{20}Ne might be formed and three in which it might decay. Any of the three formation modes can lead to any of the three decay modes.

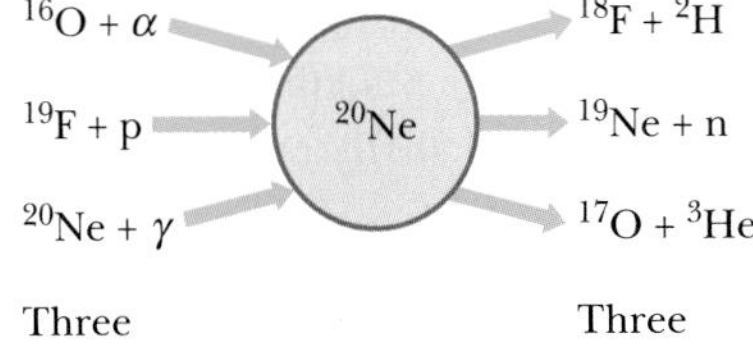

Fig. 43-13 The formation modes and the decay modes of the compound nucleus ^{20}Ne.

The Independent Particle Model

In the collective model, we assume that the nucleons move around at random and bump into each other frequently. The *independent particle model,* however, is based on just the opposite assumption—namely, that each nucleon remains in a well-defined quantum state within the nucleus and hardly makes any collisions at all! The nucleus, unlike the atom, has no fixed center of charge; we assume in this model that each nucleon moves in a potential well that is determined by the smeared-out (time-averaged) motions of all the other nucleons.

A nucleon in a nucleus, like an electron in an atom, has a set of quantum numbers that defines its state of motion. Also, nucleons obey the Pauli exclusion principle, just as electrons do; that is, no two nucleons in a nucleus may occupy the same quantum state at the same time. In this regard, the neutrons and the protons are treated separately, each particle type with its own set of quantum states.

The fact that nucleons obey the Pauli exclusion principle helps us to understand the relative stability of nucleon states. If two nucleons within the nucleus are to collide, the energy of each of them after the collision must correspond to the energy of an *unoccupied* state. If no such state is available, the collision simply cannot occur. Thus, any given nucleon experiencing repeated "frustrated collision opportunities" will maintain its state of motion long enough to give meaning to the statement that it exists in a quantum state with a well-defined energy.

In the atomic realm, the repetitions of physical and chemical properties that we find in the periodic table are associated with a property of atomic electrons—namely, they arrange themselves in shells that have a special stability when fully occupied. We can take the atomic numbers of the noble gases,

$$2, 10, 18, 36, 54, 86, \ldots,$$

as *magic electron numbers* that mark the completion (or closure) of such shells.

Nuclei also show such closed-shell effects, associated with certain **magic nucleon numbers:**

$$2, 8, 20, 28, 50, 82, 126, \ldots.$$

Any nuclide whose proton number Z or neutron number N has one of these values turns out to have a special stability that may be made apparent in a variety of ways.

Examples of "magic" nuclides are ^{18}O ($Z = 8$), ^{40}Ca ($Z = 20$, $N = 20$), ^{92}Mo ($N = 50$), and ^{208}Pb ($Z = 82$, $N = 126$). Both ^{40}Ca and ^{208}Pb are said to be "doubly magic" because they contain both filled shells of protons *and* filled shells of neutrons.

The magic number 2 shows up in the exceptional stability of the alpha particle (^{4}He), which, with $Z = N = 2$, is doubly magic. For example, on the binding energy curve of Fig. 43-6, the binding energy per nucleon for this nuclide stands well above those of its periodic-table neighbors hydrogen, lithium, and berylium. The alpha particle is so tightly bound, in fact, that it is impossible to add another particle to it; there is no stable nuclide with $A = 5$.

The central idea of a closed shell is that a single particle outside a closed shell can be relatively easily removed, but considerably more energy must be expended to remove a particle from the shell itself. The sodium atom, for example, has one (valence) electron outside a closed electron shell. Only about 5 eV is required to strip the valence electron away from a sodium atom; however, to remove a *second* electron (which must be plucked out of a closed shell) requires 22 eV. As a nuclear case, consider ^{121}Sb ($Z = 51$), which contains a single proton outside a closed shell of 50 protons. To remove this lone proton requires 5.8 MeV; to remove a *second* proton, however, requires an energy of 11 MeV. There is much additional experimental evidence that the nucleons in a nucleus form closed shells and that these shells exhibit stable properties.

We have seen that quantum theory can account beautifully for the magic electron numbers—that is, for the populations of the subshells into which atomic electrons are grouped. It turns out that, under certain assumptions, quantum theory can account equally well for the magic nucleon numbers! The 1963 Nobel prize in physics was, in fact, awarded to Maria Mayer and Hans Jensen "for their discoveries concerning nuclear shell structure."

A Combined Model

Consider a nucleus in which a small number of neutrons (or protons) exist outside a core of closed shells that contains magic numbers of neutrons or protons. The outside nucleons occupy quantized states in a potential well established by the central core, thus preserving the central feature of the independent-particle model. These outside nucleons also interact with the core, deforming it and setting up "tidal wave" motions of rotation or vibration within it. These collective motions of the core preserve the central feature of that model. Such a model of nuclear structure thus succeeds in combining the seemingly irreconcilable points of view of the collective and independent-particle models. It has been remarkably successful in explaining observed nuclear properties.

Sample Problem 43-10

Consider the neutron capture reaction

$$^{109}\text{Ag} + \text{n} \rightarrow {}^{110}\text{Ag} \rightarrow {}^{110}\text{Ag} + \gamma, \qquad (43\text{-}34)$$

in which a compound nucleus (^{110}Ag) is formed. Figure 43-14 shows the relative rate at which such events take place, plotted against the energy of the incoming neutron. Find the mean lifetime of this compound nucleus by using the uncertainty principle in the form

$$\Delta E \cdot \Delta t \approx \hbar. \qquad (43\text{-}35)$$

Here ΔE is a measure of the uncertainty with which the energy of a state can be defined. The quantity Δt is a measure of the time available to measure this energy. In fact, here Δt is just t_{avg}, the average life of the compound nucleus before it decays to its ground state.

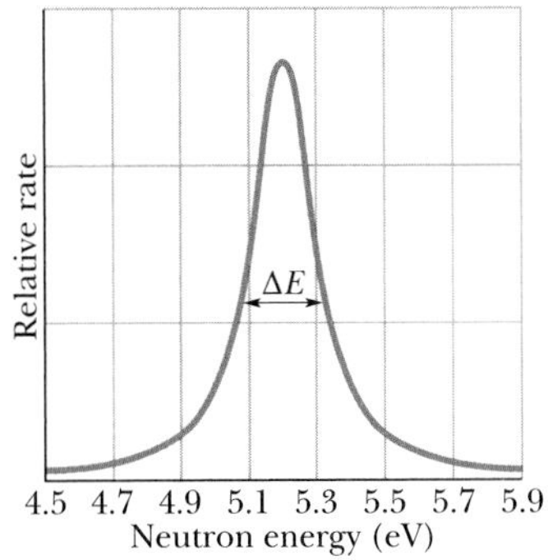

Fig. 43-14 Sample Problem 43-10. A plot of the relative number of reaction events of the type described by Eq. 43-34, as a function of the energy of the incident neutron. The half-width ΔE of the resonance peak is about 0.20 eV.

SOLUTION: We see that the relative reaction rate peaks sharply at a neutron energy of about 5.2 eV. This suggests that we are dealing with a single excited energy level of the compound nucleus ^{110}Ag. When the available energy (of the incoming neutron) just matches the energy of this level above the ^{110}Ag ground state, we have "resonance" and the reaction of Eq. 43-34 really "goes."

However, the resonance peak is not infinitely sharp but has an approximate half-width (ΔE in the figure) of about 0.20 eV. The **Key Idea** here is that we can account for this resonance-peak width by saying that the excited level is not sharply defined in energy, having an energy uncertainty ΔE of about 0.20 eV. Thus, Eq. 43-35 gives us

$$\Delta t = t_{avg} \approx \frac{\hbar}{\Delta E} = \frac{(4.14 \times 10^{-15}\ \text{eV} \cdot \text{s})/2\pi}{0.20\ \text{eV}}$$
$$\approx 3 \times 10^{-15}\ \text{s}. \qquad \text{(Answer)}$$

This is several hundred times greater than the time a 0.20 eV neutron takes to cross the diameter of a ^{109}Ag nucleus. Therefore, the neutron is spending this time of 3×10^{-15} s *as part of* the nucleus.

REVIEW & SUMMARY

The Nuclides Approximately 2000 **nuclides** are known to exist. Each is characterized by an **atomic number** Z (the number of protons), a **neutron number** N, and a **mass number** A (the total number of **nucleons**—protons and neutrons). Thus, $A = Z + N$. Nuclides with the same atomic number but different neutron numbers are **isotopes** of each other. Nuclei have a mean radius r given by

$$r = r_0 A^{1/3}, \qquad (43\text{-}3)$$

where $r_0 \approx 1.2$ fm.

Mass–Energy Exchanges The energy equivalent of one mass unit (u) is 931.5 MeV. The binding energy curve shows that middle-mass nuclides are the most stable and that energy can be released both by fission of high-mass nuclei and by fusion of low-mass nuclei.

The Nuclear Force Nuclei are held together by an attractive force acting among the nucleons. It is thought to be a secondary effect of the **strong force** acting between the quarks that make up the nucleons. Nuclei can exist in a number of discrete energy states, each with a characteristic intrinsic angular momentum and magnetic moment.

Radioactive Decay Most known nuclides are radioactive; they spontaneously decay at a rate R $(= -dN/dt)$ that is proportional to the number N of radioactive atoms present, the proportionality constant being the **disintegration constant** λ. This leads to the law of exponential decay:

$$N = N_0 e^{-\lambda t}, \qquad R = \lambda N = R_0 e^{-\lambda t}$$

(radioactive decay). (43-14, 43-15, 43-16)

The half-life $T_{1/2} = (\ln 2)/\lambda$ of a radioactive nuclide is the time required for the decay rate R (or the number N) in a sample to drop to half its initial value.

Alpha Decay Some nuclides decay by emitting an alpha particle (a helium nucleus, ^{4}He). Such decay is inhibited by a potential energy barrier that cannot be penetrated according to classical physics but is subject to tunneling according to quantum physics. The barrier penetrability, and thus the half-life for alpha decay, is very sensitive to the energy of the emitted alpha particle.

Beta Decay In **beta decay** either an electron or a positron is emitted by a nucleus, along with a neutrino. The emitted particles share the available disintegration energy. The electrons and positrons emitted in beta decay have a continuous spectrum of energies from near zero up to a limit K_{max} $(= Q = -\Delta m\, c^2)$.

Radioactive Dating Naturally occurring radioactive nuclides provide a means for estimating the dates of historic and prehistoric events. For example, the ages of organic materials can often be found by measuring their ^{14}C content; rock samples can be dated using the radioactive isotope ^{40}K.

Radiation Dosage Three units are used to describe exposure to ionizing radiation. The **becquerel** (1 Bq = 1 decay per second) measures the **activity** of a source. The amount of energy actually absorbed is measured in **grays,** with 1 Gy corresponding to 1 J/kg. The estimated biological effect of the absorbed energy is measured in **sieverts;** a dose equivalent of 1 Sv causes the same biological effect regardless of the radiation type by which it was acquired.

Nuclear Models The **collective** model of nuclear structure assumes that nucleons collide constantly and that relatively long-lived **compound nuclei** are formed when a projectile is captured. The formation of a compound nucleus and the eventual decay of that nucleus are totally independent events.

The **independent particle** model of nuclear structure assumes that each nucleon moves, essentially without collisions, in a quantized state within the nucleus. The model predicts nucleon levels and **magic numbers** of nucleons (2, 8, 20, 28, 50, 82, and 126) associated with closed shells of nucleons; nuclides with any of these numbers of neutrons or protons are particularly stable.

The **combined** model, in which extra nucleons occupy quantized states outside a central core of closed shells, is highly successful in predicting many nuclear properties.

QUESTIONS

1. Suppose the alpha particle of Sample Problem 43-1 is replaced with a proton of the same initial kinetic energy and also headed directly toward the nucleus of the gold atom. Will the distance from the center of the nucleus at which the proton stops be greater than, less than, or the same as that of the alpha particle?

2. In your body are there more protons than neutrons, more neutrons than protons, or about the same number of each?

3. The nuclide ^{244}Pu $(Z = 94)$ is an alpha-particle emitter. Into which of the following nuclides does it decay: ^{240}Np $(Z = 93)$, ^{240}U $(Z = 92)$, ^{248}Cm $(Z = 96)$, or ^{244}Am $(Z = 95)$?

4. A certain nuclide is said to be particularly stable. Does its binding energy per nucleon lie slightly above or slightly below the binding energy curve of Fig. 43-6?

5. Is the mass excess of an alpha particle (use a straightedge on Fig. 43-12) greater than or less than the particle's total binding energy (use the binding energy per nucleon from Fig. 43-6)?

6. The radionuclide ^{196}Ir decays by emitting an electron. (a) Into which square in Fig. 43-5 is it transformed? (b) Do further decays then occur?

7. A lead nuclide contains 82 protons. (a) If it also contained 82 neutrons, where would it be located on the plot of Fig. 43-4? (b) If such a nucleus could be formed, would it emit positrons, emit electrons, or be stable? (c) From Fig. 43-4, about how many neutrons do you expect to find in a stable lead nuclide?

8. The nuclide ^{238}U $(Z = 92)$ can fission into two parts that have identical atomic numbers and mass numbers. (a) Is the nuclide ^{238}U above or below the $N = Z$ line of Fig. 43-4? (b) Are the two fragments above or below this line? (c) Are these fragments stable or radioactive?

9. Radionuclides decay exponentially, as in Eq. 43-15. Batteries, stars, and even students also decay, where "decay" stands for "burn out." Do these items decay exponentially?

10. At $t = 0$, a sample of radionuclide A has the same decay rate as a sample of radionuclide B has at $t = 30$ min. The disintegration

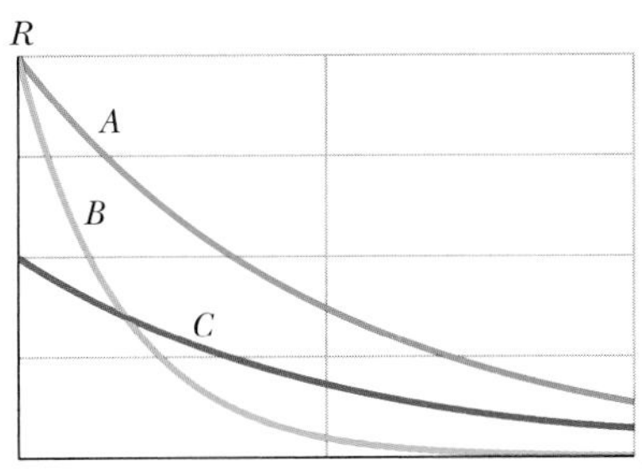

Fig. 43-15 Question 12.

constants are λ_A and λ_B, with $\lambda_A < \lambda_B$. Will the two samples ever have (simultaneously) the same decay rate? (*Hint:* Sketch a graph of their activities.)

11. At $t = 0$, a sample of radionuclide A has twice the decay rate as a sample of radionuclide B. The disintegration constants are λ_A and λ_B, with $\lambda_A > \lambda_B$. Will the two samples ever have (simultaneously) the same decay rate?

12. Figure 43-15 gives the activities of three radioactive samples versus time. Rank the samples according to their (a) half-life and (b) disintegration constant, greatest first. (*Hint:* For (a), use a straightedge on the graph.)

13. If the mass of a radioactive sample is doubled, do (a) the activity of the sample and (b) the disintegration constant of the sample increase, decrease, or remain the same?

14. At $t = 0$ we begin to observe two identical radioactive nuclei with a half-life of 5 min. At $t = 1$ min, one of the nuclei decays. Does that event increase or decrease the chance of the second nucleus decaying in the next 4 min, or is there no effect on the second nucleus?

15. The radionuclide ^{49}Sc has a half-life of 57.0 min. The counting rate of a sample of this nuclide at $t = 0$ is 6000 counts/min above the general background activity, which is 30 counts/min. Without actual computation, determine whether the counting rate of the sample will be about equal to the background rate in about 3 h, 7 h, 10 h, or a time much longer than 10 h.

16. The radionuclides ^{209}At and ^{209}Po emit alpha particles with energies of 5.65 and 4.88 MeV, respectively. Which nuclide has the longer half-life?

17. The magic numbers for nuclei are given in Section 43-8 as 2, 8, 20, 28, 50, 82, and 126. Are nuclides magic (that is, especially stable) when (a) only the mass number A, (b) only the atomic number Z, (c) only the neutron number N, or (d) either Z or N (or both) is equal to one of these numbers? Pick all correct phrases.

18. (a) Which of the following nuclides are magic: ^{122}Sn, ^{132}Sn, ^{98}Cd, ^{198}Au, ^{208}Pb? (b) Which, if any, are doubly magic?

EXERCISES & PROBLEMS

ssm Solution is in the Student Solutions Manual.
www Solution is available on the World Wide Web at: http://www.wiley.com/college/hrw
ilw Solution is available on the Interactive LearningWare.

SEC. 43-1 Discovering the Nucleus

1E. Assume that a gold nucleus has a radius of 6.23 fm and an alpha particle has a radius of 1.80 fm. What energy must an incident alpha particle have in order to "touch" the gold nucleus according to the type of calculation in Sample Problem 43-1?

2E. Calculate the distance of closest approach for a head-on collision between a 5.30 MeV alpha particle and the nucleus of a copper atom.

3P. When an alpha particle collides elastically with a nucleus, the nucleus recoils. Suppose a 5.00 MeV alpha particle has a head-on elastic collision with a gold nucleus that is initially at rest. What is the kinetic energy of (a) the recoiling nucleus and (b) the rebounding alpha particle?

SEC. 43-2 Some Nuclear Properties

4E. The radius of a spherical nucleus is measured, by electron-scattering methods, to be 3.6 fm. What is the likely mass number of the nucleus?

5E. Locate the nuclides displayed in Table 43-1 on the nuclidic chart of Fig. 43-4. Verify that they lie in the stability zone.

6E. A neutron star is a stellar object whose density is about that of nuclear matter, as calculated in Sample Problem 43-2. Suppose that the Sun were to collapse and become such a star without losing any of its present mass. What would be its radius?

7E. The nuclide ^{14}C contains (a) how many protons and (b) how many neutrons?

8E. Using a nuclidic chart, write the symbols for (a) all stable isotopes with $Z = 60$, (b) all radioactive nuclides with $N = 60$, and (c) all nuclides with $A = 60$.

9E. Make a nuclidic chart similar to Fig. 43-5 for the 25 nuclides $^{118-122}$Te, $^{117-121}$Sb, $^{116-120}$Sn, $^{115-119}$In, and $^{114-118}$Cd. Draw in and label (a) all isobaric (constant A) lines and (b) all lines of constant neutron excess, defined as $N - Z$.

10E. The strong neutron excess (defined as $N - Z$) of high-mass nuclei is illustrated by noting that most high-mass nuclides could never fission into two stable nuclei without neutrons being left over. For example, consider the spontaneous fission of a ^{235}U nucleus into two stable *daughter nuclei* with atomic numbers 39 and 53. (a) From Appendix F, what are the daughter elements? From Fig. 43-4, approximately how many neutrons are (b) in the daughter isotopes and (c) left over?

11E. The electric potential energy of a uniform sphere of charge q and radius r is

$$U = \frac{3q^2}{20\pi\varepsilon_0 r}.$$

(a) Find the electric potential energy of the nuclide ^{239}Pu, assumed to be spherical with radius 6.64 fm. (b) For this nuclide, compare the electric potential energy per nucleon, and also per proton, with the binding energy per nucleon of 7.56 MeV. (c) What do you conclude?

12E. Calculate and compare (a) the nuclear mass density ρ_m and (b) the nuclear charge density ρ_q for the fairly low-mass nuclide ^{55}Mn and for the fairly high-mass nuclide ^{209}Bi. (c) Are the differences what you would expect? Explain.

13E. Verify the binding energy per nucleon given in Table 43-1 for ^{239}Pu. The atomic masses you will need are 239.052 16 u (^{239}Pu), 1.007 83 u (^{1}H), and 1.008 67 u (neutron). ssm

14E. (a) Show that an approximate formula for the mass M of an atom is $M = Am_p$, where A is the mass number and m_p is the proton mass. (b) What percent error is committed in using this formula to calculate the masses of the atoms in Table 43-1? The mass of the bare proton is 1.007 276 u. (c) Is this formula accurate enough to be used for calculations of nuclear binding energy?

15E. Nuclear radii may be measured by scattering high-energy electrons from nuclei. (a) What is the de Broglie wavelength for 200 MeV electrons? (b) Are these electrons suitable probes for this purpose? ssm

16E. The characteristic nuclear time is a useful but loosely defined quantity, taken to be the time required for a nucleon with a few million electron-volts of kinetic energy to travel a distance equal to the diameter of a middle-mass nuclide. What is the order of magnitude of this quantity? Consider 5 MeV neutrons traversing a nuclear diameter of ^{197}Au; use Eq. 43-3.

17E. Because a nucleon is confined to a nucleus, we can take the uncertainty in its position to be approximately the nuclear radius r. What does the uncertainty principle say about the kinetic energy of a nucleon in a nucleus with, say, $A = 100$? (*Hint:* Take the uncertainty in momentum Δp to be the actual momentum p.)

18E. The atomic masses of ^{1}H, ^{12}C, and ^{238}U are 1.007 825 u, 12.000 000 u (this one is exact by definition), and 238.050 785 u, respectively. (a) What would these masses be if the mass unit were defined to give the mass of ^{1}H as (exactly) 1.000 000 u? (b) Use your result to suggest why this definition was not made.

19P. (a) Show that the energy associated with the strong force between nucleons in a nucleus is proportional to A, the mass number of the nucleus in question. (b) Show that the energy associated with the Coulomb force between protons in a nucleus is proportional to $Z(Z-1)$. (c) Show that, as we move to larger and larger nuclei (see Fig. 43-4), the importance of the Coulomb force increases more rapidly than does that of the strong force. ssm

20P. You are asked to pick apart an alpha particle (^{4}He) by removing, in sequence, a proton, a neutron, and a proton. Calculate (a) the work required for each step, (b) the total binding energy of the alpha particle, and (c) the binding energy per nucleon. Some needed atomic masses are

^{4}He	4.002 60 u	^{2}H	2.014 10 u
^{3}H	3.016 05 u	^{1}H	1.007 83 u
n	1.008 67 u		

21P. A periodic table might list the average atomic mass of magnesium as being 24.312 u. That average value is the result of *weighting* the atomic masses of the magnesium isotopes according to their natural abundances on Earth. The three isotopes and their masses are ^{24}Mg (23.985 04 u), ^{25}Mg (24.985 84 u), and ^{26}Mg (25.982 59 u). The natural abundance of ^{24}Mg is 78.99% by mass (that is, 78.99% of the mass of a naturally occurring sample of magnesium is due to the presence of ^{24}Mg). Calculate the abundances of the other two isotopes. ssm

22P. To simplify calculations, atomic masses are sometimes tabulated not as the actual atomic mass m but as $(m - A)c^2$, where A is the mass number expressed in atomic mass units. This quantity, usually reported in millions of electron-volts, is called the *mass excess,* represented with symbol Δ. Using data from Sample Problem 43-3, find the mass excesses for (a) ^{1}H, (b) the neutron, and (c) ^{120}Sn.

23P. A penny has a mass of 3.0 g. Calculate the nuclear energy that would be required to separate all the neutrons and protons in this coin from one another. For simplicity assume that the penny is made entirely of ^{63}Cu atoms (of mass 62.929 60 u). The masses of the proton and the neutron are 1.007 83 u and 1.008 67 u, respectively.

24P. Because the neutron has no charge, its mass must be found in some way other than by using a mass spectrometer. When a neutron and a proton meet (assume both to be almost stationary), they combine and form a deuteron, emitting a gamma ray whose energy is 2.2233 MeV. The masses of the proton and the deuteron are 1.007 825 035 u and 2.014 101 9 u, respectively. Find the mass of the neutron from these data, to as many significant figures as the data warrant. (A value of the mass–energy conversion factor c^2 that is more precise than the one presented in the text is 931.502 MeV/u.)

25P. Show that the total binding energy E_{be} of a nuclide is

$$E_{be} = Z\Delta_H + N\Delta_n - \Delta,$$

where Δ_H, Δ_n, and Δ are the appropriate mass excesses (see Problem 22). Using this method, calculate the binding energy per nucleon for ^{197}Au. Compare your result with the value listed in Table 43-1. The needed mass excesses, rounded to three significant figures, are $\Delta_H = +7.29$ MeV, $\Delta_n = +8.07$ MeV, and $\Delta_{197} = -31.2$ MeV. Note the economy of calculation that results when mass excesses are used in place of the actual masses. ssm www

SEC. 43-3 Radioactive Decay

26E. A radioactive nuclide has a half-life of 30 y. What fraction of an initially pure sample of this nuclide will remain undecayed at the end of (a) 60 y and (b) 90 y?

27E. The half-life of a radioactive isotope is 140 d. How many days would it take for the decay rate of a sample of this isotope to fall to one-fourth of its initial value?

28E. The half-life of a particular radioactive isotope is 6.5 h. If there are initially 48×10^{19} atoms of this isotope, how many remain at the end of 26 h?

29E. Consider an initially pure 3.4 g sample of ^{67}Ga, an isotope that has a half-life of 78 h. (a) What is its initial decay rate? (b) What is its decay rate 48 h later? ssm

30E. From data presented in the first few paragraphs of Section 43-3, find (a) the disintegration constant λ and (b) the half-life of ^{238}U.

31E. A radioactive isotope of mercury, ^{197}Hg, decays into gold, ^{197}Au, with a disintegration constant of $0.0108\ h^{-1}$. (a) Calculate its half-life. What fraction of a sample will remain at the end of (b) three half-lives and (c) 10.0 days? ssm

32E. The plutonium isotope ^{239}Pu is produced as a by-product in nuclear reactors and hence is accumulating in our environment. It is radioactive, decaying with a half-life of 2.41×10^4 y. (a) How many nuclei of Pu constitute a chemically lethal dose of 2 mg? (b) What is the decay rate of this amount?

33E. Cancer cells are more vulnerable to x and gamma radiation than are healthy cells. In the past, the standard source for radiation therapy was radioactive ^{60}Co, which decays, with a half-life of 5.27 y, into an excited nuclear state of ^{60}Ni. That nickel isotope then immediately emits two gamma-ray photons, each with an approximate energy of 1.2 MeV. How many radioactive ^{60}Co nuclei are present in a 6000 Ci source of the type used in hospitals? (Energetic particles from linear accelerators are now used in radiation therapy.) ssm

34P. The radionuclide ^{64}Cu has a half-life of 12.7 h. If a sample contains 5.50 g of initially pure ^{64}Cu at $t = 0$, how much of it will decay between $t = 14.0$ h and $t = 16.0$ h?

35P. After long effort, in 1902 Marie and Pierre Curie succeeded in separating from uranium ore the first substantial quantity of radium, one decigram of pure $RaCl_2$. The radium was the radioactive isotope ^{226}Ra, which has a half-life of 1600 y. (a) How many radium nuclei had the Curies isolated? (b) What was the decay rate of their sample, in disintegrations per second? ssm www

36P. The radionuclide ^{32}P ($T_{1/2} = 14.28$ d) is often used as a tracer to follow the course of biochemical reactions involving phosphorus. (a) If the counting rate in a particular experimental setup is initially 3050 counts/s, how much time will the rate take to fall to 170 counts/s? (b) A solution containing ^{32}P is fed to the root system of an experimental tomato plant and the ^{32}P activity in a leaf is measured 3.48 days later. By what factor must this reading be multiplied to correct for the decay that has occurred since the experiment began?

37P. A source contains two phosphorus radionuclides, ^{32}P ($T_{1/2} =$ 14.3 d) and ^{33}P ($T_{1/2} = 25.3$ d). Initially, 10.0% of the decays come from ^{33}P. How long must one wait until 90.0% do so?

38P. Plutonium isotope ^{239}Pu decays by alpha decay with a half-life of 24 100 y. How many milligrams of helium are produced by an initially pure 12.0 g sample of ^{239}Pu at the end of 20 000 y? (Consider only the helium produced directly by the plutonium and not by any by-products of the decay process.)

39P. A 1.00 g sample of samarium emits alpha particles at a rate of 120 particles/s. The responsible isotope is ^{147}Sm, whose natural abundance in bulk samarium is 15.0%. Calculate the half-life for the decay process. ssm

40P. After a brief neutron irradiation of silver, two isotopes are present: ^{108}Ag ($T_{1/2} = 2.42$ min) with an initial decay rate of 3.1×10^5/s, and ^{110}Ag ($T_{1/2} = 24.6$ s) with an initial decay rate of 4.1×10^6/s. Make a semilog plot similar to Fig. 43-8 showing the total combined decay rate of the two isotopes as a function of time from $t = 0$ until $t = 10$ min. We used Fig. 43-8 to illustrate the extraction of the half-life for simple (one isotope) decays. Given only your plot of total decay rate for the two-isotope system here, suggest a way to analyze it in order to find the half-lives of both isotopes.

41P. A certain radionuclide is being manufactured, say, in a cyclotron, at a constant rate R. It is also decaying, with disintegration constant λ. Assume that the production process has been going on for a time that is long compared to the half-life of the radionuclide. Show that the number of radioactive nuclei present after such time remains constant and is given by $N = R/\lambda$. Now show that this result holds no matter how many radioactive nuclei were present initially. The nuclide is said to be in *secular equilibrium* with its source; in this state its decay rate is just equal to its production rate. ssm www

42P. Calculate the mass of a sample of (initially pure) ^{40}K with an initial decay rate of 1.70×10^5 disintegrations/s. The isotope has a half-life of 1.28×10^9 y.

43P. (See Problem 41.) The radionuclide ^{56}Mn has a half-life of 2.58 h and is produced in a cyclotron by bombarding a manganese target with deuterons. The target contains only the stable manganese isotope ^{55}Mn, and the manganese–deuteron reaction that produces ^{56}Mn is

$$^{55}Mn + d \rightarrow {}^{56}Mn + p.$$

After bombardment for a time much longer than 2.58 h, the activity of the target, due to ^{56}Mn, is $8.88 \times 10^{10}\ s^{-1}$. (a) At what constant rate R are ^{56}Mn nuclei being produced in the cyclotron during the bombardment? (b) At what rate are they decaying (also during the bombardment)? (c) How many ^{56}Mn nuclei are present at the end of the bombardment? (d) What is their total mass? ssm

44P. (See Problems 41 and 43.) A radium source contains 1.00 mg of ^{226}Ra, which decays with a half-life of 1600 y to produce ^{222}Rn, a noble gas. This radon isotope in turn decays by alpha emission with a half-life of 3.82 d. (a) What is the rate of disintegration of ^{226}Ra in the source? (b) How long does it take for the radon to come to secular equilibrium with its radium parent? (c) At what rate is the radon then decaying? (d) How much radon is in equilibrium with its radium parent?

45P. One of the dangers of radioactive fallout from a nuclear bomb is its ^{90}Sr, which decays with a 29 year half-life. Because it has chemical properties much like those of calcium, the strontium, if ingested by a cow, becomes concentrated in the cow's milk. Some of the ^{90}Sr ends up in the bones of whoever drinks the milk. The energetic electrons emitted in the beta decay of ^{90}Sr damage the bone marrow and thus impair the production of red blood cells. A 1 megaton bomb produces approximately 400 g of ^{90}Sr. If the fallout spreads uniformly over a 2000 km^2 area, what ground area would hold an amount of radioactivity equal to the "allowed" limit for one person, which is 74 000 counts/s?

SEC. 43-4 Alpha Decay

46E. Consider a ^{238}U nucleus to be made up of an alpha particle (^{4}He) and a residual nucleus (^{234}Th). Plot the electrostatic potential energy $U(r)$, where r is the distance between these particles. Cover the approximate range 10 fm $< r <$ 100 fm and compare your plot with that of Fig. 43-9.

47E. Generally, more massive nuclides tend to be more unstable to alpha decay. For example, the most stable isotope of uranium, ^{238}U, has an alpha decay half-life of 4.5×10^9 y. The most stable isotope of plutonium is ^{244}Pu with an 8.0×10^7 y half-life, and for curium we have ^{248}Cm and 3.4×10^5 y. When half of an original sample of ^{238}U has decayed, what fractions of the original samples of these isotopes of plutonium and curium are left? ssm

48P. Consider that a ^{238}U nucleus emits (a) an alpha particle or (b) a sequence of neutron, proton, neutron, proton. Calculate the energy released in each case. (c) Convince yourself both by reasoned argument and by direct calculation that the difference between these two numbers is just the total binding energy of the alpha particle. Find that binding energy. Some needed atomic and particle masses are

^{238}U	238.050 79 u	^{234}Th	234.043 63 u
^{237}U	237.048 73 u	^{4}He	4.002 60 u
^{236}Pa	236.048 91 u	^{1}H	1.007 83 u
^{235}Pa	235.045 44 u	n	1.008 67 u

49P. A ^{238}U nucleus emits a 4.196 MeV alpha particle. Calculate the disintegration energy Q for this process, taking the recoil energy of the residual ^{234}Th nucleus into account. ssm

50P. Large radionuclides emit an alpha particle rather than other combinations of nucleons because the alpha particle has such a stable, tightly bound structure. To confirm this statement, calculate the disintegration energies for these hypothetical decay processes and discuss the meaning of your findings:

(a) $^{235}\text{U} \rightarrow {}^{232}\text{Th} + {}^{3}\text{He}$,

(b) $^{235}\text{U} \rightarrow {}^{231}\text{Th} + {}^{4}\text{He}$,

(c) $^{235}\text{U} \rightarrow {}^{230}\text{Th} + {}^{5}\text{He}$.

The needed atomic masses are

^{232}Th	232.0381 u	^{3}He	3.0160 u
^{231}Th	231.0363 u	^{4}He	4.0026 u
^{230}Th	230.0331 u	^{5}He	5.0122 u
^{235}U	235.0439 u		

51P. Under certain rare circumstances, a nucleus can decay by emitting a particle more massive than an alpha particle. Consider the decays

$$^{223}\text{Ra} \rightarrow {}^{209}\text{Pb} + {}^{14}\text{C}$$

and

$$^{223}\text{Ra} \rightarrow {}^{219}\text{Rn} + {}^{4}\text{He}.$$

(a) Calculate the Q values for these decays and determine that both are energetically possible. (b) The Coulomb barrier height for alpha-particle emission is 30.0 MeV. What is the barrier height for ^{14}C emission? The needed atomic masses are

^{223}Ra	223.018 50 u	^{14}C	14.003 24 u
^{209}Pb	208.981 07 u	^{4}He	4.002 60 u
^{219}Rn	219.009 48 u		

SEC. 43-5 Beta Decay

52E. Large-mass radionuclides, which may be either alpha or beta emitters, belong to one of four decay chains, depending on whether their mass numbers A are of the form $4n$, $4n + 1$, $4n + 2$, or $4n + 3$, where n is a positive integer. (a) Justify this statement and show that if a nuclide belongs to one of these families, all its decay products belong to the same family. (b) Classify these nuclides as to family: ^{235}U, ^{236}U, ^{238}U, ^{239}Pu, ^{240}Pu, ^{245}Cm, ^{246}Cm, ^{249}Cf, and ^{253}Fm.

53E. A certain stable nuclide, after absorbing a neutron, emits an electron, and the new nuclide splits spontaneously into two alpha particles. Identify the nuclide. ssm

54E. An electron is emitted from a middle-mass nuclide (A = 150, say) with a kinetic energy of 1.0 MeV. (a) What is its de Broglie wavelength? (b) Calculate the radius of the emitting nucleus. (c) Can such an electron be confined as a standing wave in a "box" of such dimensions? (d) Can you use these numbers to disprove the (abandoned) argument that electrons actually exist in nuclei?

55E. The cesium isotope ^{137}Cs is present in the fallout from aboveground detonations of nuclear bombs. Because it decays with a slow (30.2 y) half-life into ^{137}Ba, releasing considerable energy in the process, it is of environmental concern. The atomic masses of the Cs and Ba are 136.9071 and 136.9058 u, respectively; calculate the total energy released in such a decay. ssm

56P. Some radionuclides decay by capturing one of their own atomic electrons, a K-shell electron, say. An example is

$$^{49}\text{V} + \text{e}^- \rightarrow {}^{49}\text{Ti} + \nu, \qquad T_{1/2} = 331 \text{ d}.$$

Show that the disintegration energy Q for this process is given by

$$Q = (m_\text{V} - m_\text{Ti})c^2 - E_K,$$

where m_V and m_Ti are the atomic masses of ^{49}V and ^{49}Ti, respectively, and E_K is the binding energy of the vanadium K-shell electron. (*Hint:* Put $\mathbf{m}_\text{V}$ and $\mathbf{m}_\text{Ti}$ as the corresponding nuclear masses and proceed as in Sample Problem 43-7.)

57P. A free neutron decays according to Eq. 43-26. If the neutron–hydrogen atom mass difference is 840 μu, what is the maximum kinetic energy K_{max} of the electron energy spectrum? ssm

58P. Find the disintegration energy Q for the decay of ^{49}V by K-electron capture, as described in Problem 56. The needed data are m_V = 48.948 52 u, m_Ti = 48.947 87 u, and E_K = 5.47 keV.

59P. The radionuclide ^{11}C decays according to

$$^{11}\text{C} \rightarrow {}^{11}\text{B} + \text{e}^+ + \nu, \qquad T_{1/2} = 20.3 \text{ min}.$$

The maximum energy of the emitted positrons is 0.960 MeV.

(a) Show that the disintegration energy Q for this process is given by

$$Q = (m_C - m_B - 2m_e)c^2,$$

where m_C and m_B are the atomic masses of ^{11}C and ^{11}B, respectively, and m_e is the mass of a positron. (b) Given the mass values $m_C = 11.011\ 434$ u, $m_B = 11.009\ 305$ u, and $m_e = 0.000\ 548\ 6$ u, calculate Q and compare it with the maximum energy of the emitted positron given above. (*Hint:* Let $\mathbf{m}_C$ and $\mathbf{m}_B$ be the nuclear masses and proceed as in Sample Problem 43-7 for beta decay. Note that beta-plus decay is an exception to the general rule that if atomic masses are used in nuclear decay calculations, the mass of the emitted electron is automatically taken care of.)

60P. Two radioactive materials that are unstable with regard to alpha decay, ^{238}U and ^{232}Th, and one that is unstable with regard to beta decay, ^{40}K, are sufficiently abundant in granite to contribute significantly to the heating of Earth through the decay energy produced. The alpha-unstable isotopes give rise to decay chains that stop when stable lead isotopes are formed. The isotope ^{40}K has a single beta decay. Here is the information:

Parent	Decay Mode	Half-Life (y)	Stable End Point	Q (MeV)	f (ppm)
^{238}U	α	4.47×10^9	^{206}Pb	51.7	4
^{232}Th	α	1.41×10^{10}	^{208}Pb	42.7	13
^{40}K	β	1.28×10^9	^{40}Ca	1.31	4

In the table Q is the *total* energy released in the decay of one parent nucleus to the *final* stable end point and f is the abundance of the isotope in kilograms per kilogram of granite; ppm means parts per million. (a) Show that these materials produce energy as heat at the rate of 1.0×10^{-9} W for each kilogram of granite. (b) Assuming that there is 2.7×10^{22} kg of granite in a 20-km-thick spherical shell at the surface of Earth, estimate the power of this decay process over all of Earth. Compare this power with the total solar power intercepted by Earth, 1.7×10^{17} W.

61P*. The radionuclide ^{32}P decays to ^{32}S as described by Eq. 43-23. In a particular decay event, a 1.71 MeV electron is emitted, the maximum possible value. What is the kinetic energy of the recoiling ^{32}S atom in this event? (*Hint:* For the electron it is necessary to use the relativistic expressions for kinetic energy and linear momentum. Newtonian mechanics may safely be used for the relatively slow-moving ^{32}S atom.) ssm www

SEC. 43-6 Radioactive Dating

62E. A 5.00 g charcoal sample from an ancient fire pit has a ^{14}C activity of 63.0 disintegrations/min. A living tree has a ^{14}C activity of 15.3 disintegrations/min per 1.00 g. The half-life of ^{14}C is 5730 y. How old is the charcoal sample?

63E. ^{238}U decays to ^{206}Pb with a half-life of 4.47×10^9 y. Although the decay occurs in many individual steps, the first step has by far the longest half-life; therefore, one can often consider the decay to go directly to lead. That is,

$$^{238}U \rightarrow {}^{206}Pb + \text{various decay products.}$$

A rock is found to contain 4.20 mg of ^{238}U and 2.135 mg of ^{206}Pb. Assume that the rock contained no lead at formation, so all the lead now present arose from the decay of uranium. (a) How many atoms of ^{238}U and ^{206}Pb does the rock now contain? (b) How many atoms of ^{238}U did the rock contain at formation? (c) What is the age of the rock? ssm

64P. A particular rock is thought to be 260 million years old. If it contains 3.70 mg of ^{238}U, how much ^{206}Pb should it contain? See Exercise 63.

65P. A rock, recovered from far underground, is found to contain 0.86 mg of ^{238}U, 0.15 mg of ^{206}Pb, and 1.6 mg of ^{40}Ar. How much ^{40}K will it likely contain? Needed half-lives are listed in Problem 60.

SEC. 43-7 Measuring Radiation Dosage

66E. A Geiger counter records 8700 counts in 1 min. Calculate the activity of the source in becquerels and in curies, assuming that the counter records all decays.

67E. The nuclide ^{198}Au, with a half-life of 2.70 d, is used in cancer therapy. What mass of this nuclide is required to produce an activity of 250 Ci? ssm

68E. An airline pilot spends an average of 20 h per week flying at an altitude of 10 km, at which the dose equivalent due to cosmic radiation is 7.0 μSv/h. What is the annual (52 week) dose equivalent from this source alone? Note that the maximum permitted yearly dose equivalent (from all sources) for the general population is 5 mSv, and for radiation workers it is 50 mSv.

69P. A typical chest x-ray radiation dose is 250 μSv, delivered by x rays with an RBE factor of 0.85. Assuming that the mass of the exposed tissue is one-half the patient's mass of 88 kg, calculate the energy absorbed in joules. ssm

70P. A 75 kg person receives a whole-body radiation dose of 2.4×10^{-4} Gy, delivered by alpha particles for which the RBE factor is 12. Calculate (a) the absorbed energy in joules and (b) the dose equivalent in sieverts and in rem.

71P. An 85 kg worker at a breeder reactor plant accidentally ingests 2.5 mg of ^{239}Pu dust. ^{239}Pu has a half-life of 24 100 y, decaying by alpha decay. The energy of the emitted alpha particles is 5.2 MeV, with an RBE factor of 13. Assume that the plutonium resides in the worker's body for 12 h and that 95% of the emitted alpha particles are stopped within the body. Calculate (a) the number of plutonium atoms ingested, (b) the number that decay during the 12 h, (c) the energy absorbed by the body, (d) the resulting physical dose in grays, and (e) the dose equivalent in sieverts.

SEC. 43-8 Nuclear Models

72E. A typical kinetic energy for a nucleon in a middle-mass nucleus may be taken as 5.00 MeV. To what effective nuclear temperature does this correspond, based on the assumptions of the collective model of nuclear structure?

73E. An intermediate nucleus in a particular nuclear reaction decays within 10^{-22} s of its formation. (a) What is the uncertainty ΔE in our knowledge of this intermediate state? (b) Can this state be called a compound nucleus? (See Sample Problem 43-10.)

74E. In the following list of nuclides, identify (a) those with filled nucleon shells, (b) those with one nucleon outside a filled shell, and (c) those with one vacancy in an otherwise filled shell: ^{13}C, ^{18}O, ^{40}K, ^{49}Ti, ^{60}Ni, ^{91}Zr, ^{92}Mo, ^{121}Sb, ^{143}Nd, ^{144}Sm, ^{205}Tl, and ^{207}Pb.

75P. Consider the three formation processes shown for the compound nucleus ^{20}Ne in Fig. 43-13. Here are some of the masses:

^{20}Ne	19.992 44 u	α	4.002 60 u
^{19}F	18.998 40 u	p	1.007 83 u
^{16}O	15.994 91 u		

What energy must (a) the alpha particle, (b) the proton, and (c) the γ-ray photon have to provide 25.0 MeV of excitation energy to the compound nucleus? ssm www

Additional Problems

76. At the end of World War II, Dutch authorities arrested Dutch artist Hans van Meegeren for treason because, during the war, he sold a masterpiece painting to the infamous Nazi Hermann Goering. The painting, *Christ and His Disciples at Emmaus* by Dutch master Johannes Vermeer (1632–1675), had been discovered in 1937 by van Meegeren, after it had been lost for almost 300 years. Soon after the discovery, art experts proclaimed that *Emmaus* was possibly the best Vermeer ever seen. Selling such a Dutch national treasure to the enemy was unthinkable treason.

However, shortly after being imprisoned, van Meegeren suddenly announced that he, not Vermeer, had painted *Emmaus*. He explained that he had carefully mimicked Vermeer's style, using a 300-year-old canvas and Vermeer's choice of pigments; he had then signed Vermeer's name to the work and baked the painting to give it an authentically old look.

Was van Meegeren lying to avoid a conviction of treason, hoping to be convicted of only the lesser crime of fraud? To art experts, *Emmaus* certainly looked like a Vermeer but, at the time of van Meegeren's trial in 1947, there was no scientific way to answer the question. However, in 1968 Bernard Keisch of Carnegie-Mellon University was able to answer the question with newly developed techniques of radioactive analysis.

Specifically, he analyzed a small sample of white lead-bearing pigment removed from *Emmaus*. This pigment is refined from lead ore, in which the lead is produced by a long radioactive decay series that starts with unstable ^{238}U and ends with stable ^{206}Pb. To follow the spirit of Keisch's analysis, focus on the following abbreviated portion of that decay series, in which intermediate, relatively short-lived radionuclides have been omitted:

$$^{230}\text{Th} \xrightarrow[75.4\text{ ky}]{} {}^{226}\text{Ra} \xrightarrow[1.60\text{ ky}]{} {}^{210}\text{Pb} \xrightarrow[22.6\text{ y}]{} {}^{206}\text{Pb}.$$

The longer and more important half-lives in this portion of the decay series are indicated.

(a) Show that in a unit sample of lead ore, the rate at which the number of ^{210}Pb nuclei changes is given by

$$\frac{dN_{210}}{dt} = \lambda_{226}N_{226} - \lambda_{210}N_{210},$$

where N_{210} and N_{226} are the numbers of ^{210}Pb nuclei and ^{226}Ra nuclei in the unit sample and λ_{210} and λ_{226} are the corresponding disintegration constants.

Because the decay series has been active for billions of years and because the half-life of ^{210}Pb is much less than that of ^{226}Ra, the nuclides ^{226}Ra and ^{210}Pb are in *equilibrium;* that is, their numbers, or concentrations, in the sample do not change. (b) What is the ratio R_{226}/R_{210} of the activities of these nuclides in the unit sample of lead ore? (c) What is the ratio N_{226}/N_{210} of their numbers?

When lead pigment is refined from the ore, most of the ^{226}Ra is eliminated. Assume that only 1.00% remains. Just after the pigment is produced, what are the ratios (d) R_{226}/R_{210} and (e) N_{226}/N_{210}?

Keisch realized that with time the ratio R_{226}/R_{210} of the pigment would gradually change from the value of freshly refined pigment back to the value of the ore as equilibrium between the ^{210}Pb and the remaining ^{226}Ra is established in the pigment. If *Emmaus* were painted by Vermeer and the sample of pigment taken from it were 300 years old when examined in 1968, the ratio would be close to the answer of (b). If *Emmaus* were painted by van Meegeren in the 1930s and the sample were only about 30 years old, the ratio would be close to the answer of (d). Keisch found a ratio of 0.09. (f) Is *Emmaus* a Vermeer?

77. When aboveground nuclear tests were conducted, the explosions shot radioactive dust into the upper atmosphere. Global air circulations then spread the dust worldwide before it settled out on ground and water. One such test was conducted in October 1976. What fraction of the ^{90}Sr produced by that explosion will still exist in October 2006? The half-life of ^{90}Sr is 29 y.

78. A radioactive sample intended for irradiation of a hospital patient is prepared at a nearby laboratory. The sample has a half-life of 83.61 h. What should its initial activity be if its activity is to be 7.4×10^8 Bq when it is used to irradiate the patient 24 h later?

79. The radioactive nuclide ^{99}Tc can be injected into a patient's blood system in order to monitor the blood flow, to measure the blood volume, or to find a tumor, among other goals. The nuclide is produced in a hospital by a "cow" containing ^{99}Mo, a radioactive nuclide that decays to ^{99}Tc with a half-life of 67 h. Once a day, the cow is "milked" for its ^{99}Tc, which is produced in an excited state by the ^{99}Mo; the ^{99}Tc de-excites to its lowest energy state by emitting a gamma-ray photon, which is recorded by detectors placed around the patient. The de-excitation has a half-life of 6.0 h. (a) By what process does ^{99}Mo decay to ^{99}Tc? (b) If a patient is injected with a 8.2×10^7 Bq sample of ^{99}Tc, how many gamma-ray photons are initially produced within the patient each second? (c) If the emission rate of gamma-ray photons from a small tumor that has collected ^{99}Tc is 38 per second at a certain time, how many excited state ^{99}Tc are located in the tumor at that time?

80. Because of the 1986 explosion and fire in a reactor at the Chernobyl nuclear power plant in northern Ukraine, part of the Ukraine is contaminated with ^{137}Ce, which undergoes beta-minus decay with a half-life of 30.2 y. In 1996, the total activity of this contamination over an area of 2.6×10^5 km^2 was estimated to be 1×10^{16} Bq. Assume that the ^{137}Ce is uniformly spread over that area and that the beta-decay electrons travel either directly upward or directly downward. How many beta-decay electrons would you intercept were you to lie on the ground in that area for 1 h (a) in 1996 and (b) today? (You need to estimate your cross-sectional area that intercepts those electrons.)

81. In October 1992, Swiss police arrested two men who were attempting to smuggle osmium out of Eastern Europe for a clandestine sale. However, by error, the smugglers had picked up ^{137}Ce. Reportedly, each smuggler was carrying a 1.0 g sample of ^{137}Ce in a pocket. In bequerels and curies, what was the activity of each sample? ^{137}Ce has a half-life of 30.2 y. (The activities of radioisotopes commonly used in hospitals range up to a few millicuries.)

82. Figure 43-16 shows part of the decay scheme of ^{237}Np on a plot of mass number A versus proton number Z; five lines that represent either alpha decay or beta-minus decay connect dots that represent isotopes. What is the isotope at the end of the five decays (as marked with a question mark in Fig. 43-16)?

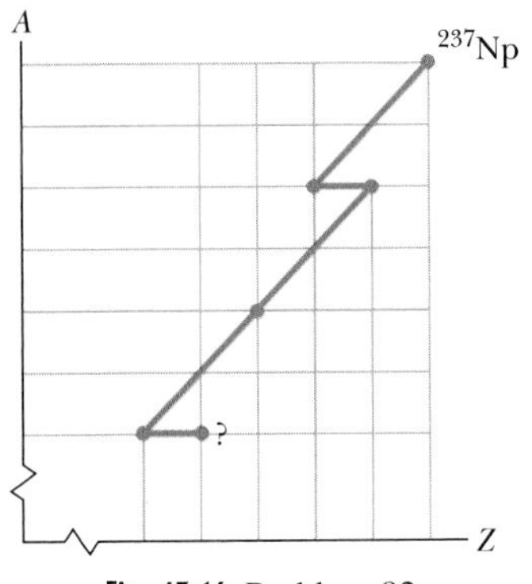

Fig. 43-16 Problem 82.

83. The isotope ^{40}K, with a half-life of 1.26×10^9 y, can decay to either ^{40}Ca or ^{40}Ar. The ratio of the Ca produced to the Ar produced is $8.54/1 = 8.54$. A sample originally had only ^{40}K. It now has equal amounts of ^{40}K and ^{40}Ar; that is, the ratio of K to Ar is $1/1 = 1$. How old is the sample? (*Hint:* Work this like other radioactive-dating problems, except that this decay has two products instead of just one.)

84. The air in some caves has a significant amount of radon gas, which can lead to lung cancer if breathed over a prolonged time. In British caves, the air in the cave with the greatest amount of the gas has an activity per volume of 1.55×10^5 Bq/m^3. Suppose that you spend two full days exploring (and sleeping in) that cave. Approximately how many ^{222}Rn atoms would you take in and out of your lungs during your two-day stay? The radionuclide ^{222}Rn in radon gas has a half-life of 3.82 days. You need to estimate your lung capacity and average breathing rate.

85. A ^{7}Li nucleus with a kinetic energy of 3.00 MeV is sent toward a ^{232}Th nucleus. What is the least center-to-center separation between the two nuclei, assuming that the (more massive) ^{232}Th nucleus does not move?

86. What is the binding energy per nucleon of ^{262}Bh? The mass of the atom is 262.1231 u.

87. How many years are needed to reduce the activity of ^{14}C to 0.020 of its original activity? The half-life of ^{14}C is 5730 y.

88. What is the activity of a 20 ng sample of ^{92}Kr, which has a half-life of 1.84 s?

44 Energy from the Nucleus

This image has transfixed the world since World War II. When Robert Oppenheimer, the head of the scientific team that developed the atomic bomb, witnessed the first atomic explosion, he quoted from a sacred Hindu text: "Now I am become Death, the destroyer of worlds."

What is the physics behind this image that has so horrified the world?

The answer is in this chapter.

44-1 The Atom and Its Nucleus

When we get energy from coal by burning the fuel in a furnace, we are tinkering with atoms of carbon and oxygen, rearranging their outer *electrons* into more stable combinations. When we get energy from uranium in a nuclear reactor, we are again burning a fuel, but then we are tinkering with its nucleus, rearranging its *nucleons* into more stable combinations.

Electrons are held in atoms by the electromagnetic Coulomb force, and it takes only a few electron-volts to pull one of them out. On the other hand, nucleons are held in nuclei by the strong force, and it takes a few *million* electron-volts to pull one of *them* out. This factor of a few million is reflected in the fact that we can extract about that much more energy from a kilogram of uranium than we can from a kilogram of coal.

In both atomic and nuclear burning, the release of energy is accompanied by a decrease in mass, according to the equation $Q = -\Delta m c^2$. The central difference between burning uranium and burning coal is that, in the former case, a much larger fraction of the available mass (again, by a factor of a few million) is consumed.

The different processes that can be used for atomic or nuclear burning provide different levels of power, or rates at which the energy is delivered. In the nuclear case, we can burn a kilogram of uranium explosively in a bomb or slowly in a power reactor. In the atomic case, we might consider exploding a stick of dynamite or digesting a jelly doughnut.

Table 44-1 shows how much energy can be extracted from 1 kg of matter by doing various things to it. Instead of reporting the energy directly, the table shows how long the extracted energy could operate a 100 W lightbulb. Only processes in the first three rows of the table have actually been carried out; the remaining three represent theoretical limits that may not be attainable in practice. The bottom row, the total mutual annihilation of matter and antimatter, is an ultimate energy production goal. In that process, *all* the mass energy is transferred to other forms of energy.

Keep in mind that the comparisons of Table 44-1 are computed on a per-unit-mass basis. Kilogram for kilogram, you get several million times more energy from uranium than you do from coal or from falling water. On the other hand, there is a lot of coal in Earth's crust, and water is easily backed up behind a dam.

44-2 Nuclear Fission: The Basic Process

In 1932 English physicist James Chadwick discovered the neutron. A few years later Enrico Fermi in Rome found that when various elements are bombarded by neutrons, new radioactive elements are produced. Fermi had predicted that the neutron, being uncharged, would be a useful nuclear projectile; unlike the proton or the alpha par-

TABLE 44-1 Energy Released by 1 kg of Matter

Form of Matter	Process	Time[a]
Water	A 50 m waterfall	5 s
Coal	Burning	8 h
Enriched UO_2	Fission in a reactor	690 y
^{235}U	Complete fission	3×10^4 y
Hot deuterium gas	Complete fusion	3×10^4 y
Matter and antimatter	Complete annihilation	3×10^7 y

[a]This column shows the time for which the generated energy could power a 100 W lightbulb.

ticle, it experiences no repulsive Coulomb force when it nears a nuclear surface. Even *thermal neutrons,* which are slowly moving neutrons in thermal equilibrium with the surrounding matter at room temperature, with a mean kinetic energy of only about 0.04 eV, are useful projectiles in nuclear studies.

In the late 1930s physicist Lise Meitner and chemists Otto Hahn and Fritz Strassmann, working in Berlin and following up the work of Fermi and co-workers, bombarded solutions of uranium salts with such thermal neutrons. They found that after the bombardment a number of new radionuclides were present. In 1939 one of the radionuclides produced in this way was positively identified, by repeated tests, as barium. But how, Hahn and Strassmann wondered, could this middle-mass element ($Z = 56$) be produced by bombarding uranium ($Z = 92$) with neutrons?

The puzzle was solved within a few weeks by Meitner and her nephew Otto Frisch. They suggested the mechanism by which a uranium nucleus, having absorbed a thermal neutron, could split, with the release of energy, into two roughly equal parts, one of which might well be barium. Frisch named the process **fission.**

Meitner's central role in the discovery of fission was not fully recognized until recent historical research brought it to light. She did not share in the Nobel prize in chemistry that was awarded to Otto Hahn in 1944. However, both Hahn and Meitner have been honored by having elements named after them: hahnium (symbol Ha, $Z = 105$) and meitnerium (symbol Mt, $Z = 109$).

A Closer Look at Fission

Figure 44-1 shows the distribution by mass number of the fragments produced when ^{235}U is bombarded with thermal neutrons. The most probable mass numbers, occurring in about 7% of the events, are centered around $A \approx 95$ and $A \approx 140$. Curiously, the "double-peaked" character of Fig. 44-1 is still not understood.

In a typical ^{235}U fission event, a ^{235}U nucleus absorbs a thermal neutron, producing a compound nucleus ^{236}U in a highly excited state. It is *this* nucleus that actually undergoes fission, splitting into two fragments. These fragments—between them—rapidly emit two neutrons, leaving (in a typical case) ^{140}Xe ($Z = 54$) and ^{94}Sr ($Z = 38$) as fission fragments. Thus, the overall fission equation for this event is

$$^{235}\text{U} + \text{n} \rightarrow {}^{236}\text{U} \rightarrow {}^{140}\text{Xe} + {}^{94}\text{Sr} + 2\text{n}. \tag{44-1}$$

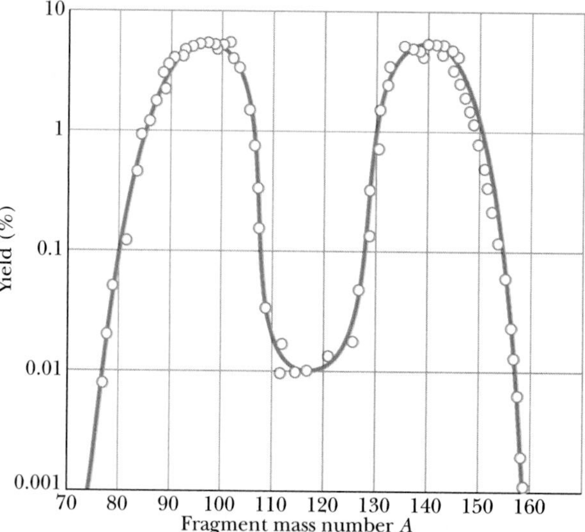

Fig. 44-1 The distribution by mass number of the fragments that are found when many fission events of ^{235}U are examined. Note that the vertical scale is logarithmic.

Note that during the formation and fission of the compound nucleus, there is conservation of the number of protons and of the number of neutrons involved in the process (and thus their total number and the net charge).

In Eq. 44-1, the fragments ^{140}Xe and ^{94}Sr are both highly unstable, undergoing beta decay (with the conversion of a neutron to a proton and the emission of an electron and a neutrino) until each reaches a stable end product. For xenon, the decay chain is

	^{140}Xe →	^{140}Cs →	^{140}Ba →	^{140}La →	^{140}Ce
$T_{1/2}$	14 s	64 s	13 d	40 h	Stable
Z	54	55	56	57	58

(44-2)

For strontium, it is

	^{94}Sr →	^{94}Y →	^{94}Zr
$T_{1/2}$	75 s	19 min	Stable
Z	38	39	40

(44-3)

As we should expect from Section 43-5, the mass numbers (140 and 94) of the fragments remain unchanged during these beta-decay processes, and the atomic numbers (initially 54 and 38) increase by unity at each step.

Inspection of the stability band on the nuclidic chart of Fig. 43-4 shows why the fission fragments are unstable. The nuclide ^{236}U, which is the fissioning nucleus in the reaction of Eq. 44-1, has 92 protons and 236 − 92, or 144, neutrons, for a neutron/proton ratio of about 1.6. The primary fragments formed immediately after the fission reaction have about this same neutron/proton ratio. However, stable nuclides in the middle-mass region have smaller neutron/proton ratios, in the range of 1.3 to 1.4. The primary fragments are thus *neutron rich* (they have too many neutrons) and will eject a few neutrons, two in the case of the reaction of Eq. 44-1. The fragments that remain are still too neutron rich to be stable. Beta decay offers a mechanism for getting rid of the excess neutrons—namely, by changing them into protons within the nucleus.

We can estimate the energy released by the fission of a high-mass nuclide by examining the total binding energy per nucleon ΔE_{ben} before and after the fission. The idea is that fission can occur because the total mass energy will decrease; that is, ΔE_{ben} will *increase* so that the products of the fission are *more* tightly bound. Thus, the energy Q released by the fission is

$$Q = \begin{pmatrix}\text{total final}\\ \text{binding energy}\end{pmatrix} - \begin{pmatrix}\text{initial}\\ \text{binding energy}\end{pmatrix}. \tag{44-4}$$

For our estimate, let us assume that fission transforms an initial high-mass nucleus to two middle-mass nuclei with the same number of nucleons. Then we have

$$Q = \begin{pmatrix}\text{final}\\ \Delta E_{\text{ben}}\end{pmatrix}\begin{pmatrix}\text{final number}\\ \text{of nucleons}\end{pmatrix} - \begin{pmatrix}\text{initial}\\ \Delta E_{\text{ben}}\end{pmatrix}\begin{pmatrix}\text{initial number}\\ \text{of nucleons}\end{pmatrix}. \tag{44-5}$$

From Fig. 43-6, we see that for a high-mass nuclide ($A \approx 240$), the binding energy per nucleon is about 7.6 MeV/nucleon. For middle-mass nuclides ($A \approx 120$), it is about 8.5 MeV/nucleon. Thus, the energy released by fission of a high-mass nuclide to two middle-mass nuclides is

$$Q = \left(8.5\,\frac{\text{MeV}}{\text{nucleon}}\right)(2\text{ nuclei})\left(120\,\frac{\text{nucleons}}{\text{nucleus}}\right) - \left(7.6\,\frac{\text{MeV}}{\text{nucleon}}\right)(240\text{ nucleons}) \approx 200\text{ MeV}. \tag{44-6}$$

CHECKPOINT 1: A generic fission event is

$$^{235}\text{U} + \text{n} \rightarrow X + Y + 2\text{n}.$$

Which of the following pairs *cannot* represent X and Y: (a) ^{141}Xe and ^{93}Sr; (b) ^{139}Cs and ^{95}Rb; (c) ^{156}Nd and ^{79}Ge; (d) ^{121}In and ^{113}Ru?

Sample Problem 44-1

Find the disintegration energy Q for the fission event of Eq. 44-1, taking into account the decay of the fission fragments as displayed in Eqs. 44-2 and 44-3. Some needed atomic and particle masses are

^{235}U	235.0439 u	^{140}Ce	139.9054 u
n	1.008 67 u	^{94}Zr	93.9063 u

SOLUTION: The **Key Ideas** here are (1) that the disintegration energy Q is the energy transferred from mass energy to kinetic energy of the decay products, and (2) that $Q = -\Delta m\, c^2$, where Δm is the change in mass. Because we are to include the decay of the fission fragments, we combine Eqs. 44-1, 44-2, and 44-3 to write the overall transformation as

$$^{235}\text{U} \rightarrow {}^{140}\text{Ce} + {}^{94}\text{Zr} + \text{n}. \tag{44-7}$$

Only the single neutron appears here because the initiating neutron on the left side of Eq. 44-1 cancels one of the two neutrons on the right of that equation. The mass difference for the reaction of Eq. 44-7 is

$$\Delta m = (139.9054\text{ u} + 93.9063\text{ u} + 1.008\,67\text{ u}) - (235.0439\text{ u})$$
$$= -0.223\,53\text{ u},$$

and the corresponding disintegration energy is

$$Q = -\Delta m\, c^2 = (-0.223\,53\text{ u})(931.5\text{ MeV/u})$$
$$= 208\text{ MeV}, \qquad \text{(Answer)}$$

which is in good agreement with our estimate of Eq. 44-6.

If the fission event takes place in a bulk solid, most of this disintegration energy, which first goes into kinetic energy of the decay products, appears eventually as an increase in the internal energy of that body, revealing itself as a rise in temperature. Five or six percent or so of the disintegration energy, however, is associated with neutrinos that are emitted during the beta decay of the primary fission fragments. This energy is carried out of the system and is lost.

44-3 A Model for Nuclear Fission

Soon after the discovery of fission, Niels Bohr and John Wheeler used the collective model of the nucleus (Section 43-8), based on the analogy between a nucleus and a charged liquid drop, to explain its main features. Figure 44-2 suggests how the fission

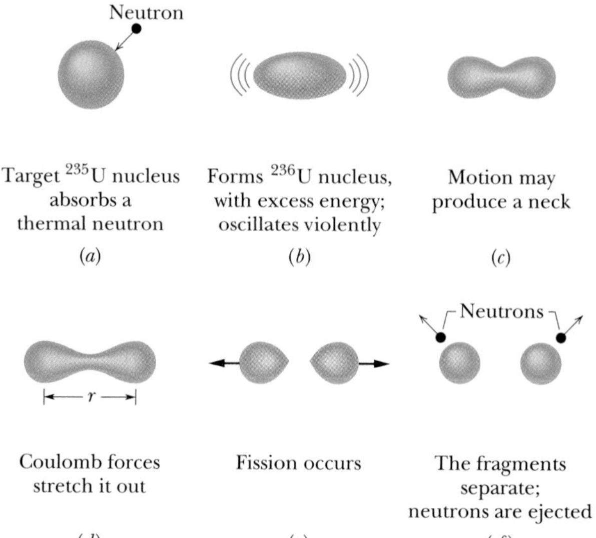

Fig. 44-2 The stages of a typical fission process, according to the collective model of Bohr and Wheeler.

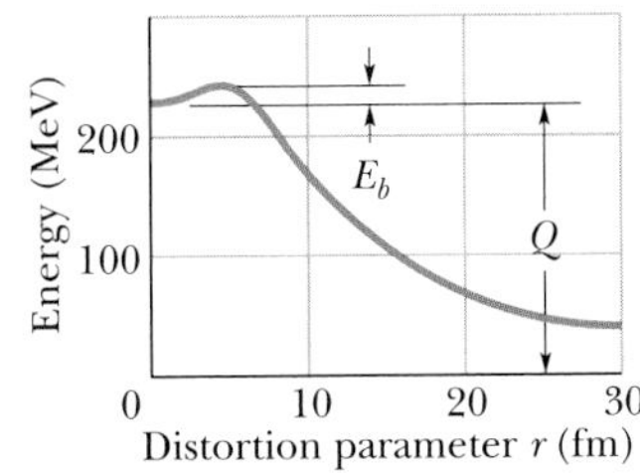

Fig. 44-3 The potential energy at various stages in the fission process, as predicted from the collective model of Bohr and Wheeler. The Q of the reaction (about 200 MeV) and the fission barrier height E_b are both indicated.

process proceeds from this point of view. When a high-mass nucleus—let us say ^{235}U—absorbs a slow (thermal) neutron, as in Fig. 44-2*a*, that neutron falls into the potential well associated with the strong forces that act in the nuclear interior. The neutron's potential energy is then transformed into internal excitation energy of the nucleus, as Fig. 44-2*b* suggests. The amount of excitation energy that a slow neutron carries into a nucleus is equal to the binding energy E_n of the neutron in that nucleus, which is the change in mass energy of the neutron–nucleus system due to the neutron's capture.

Figures 44-2*c* and *d* show that the nucleus, behaving like an energetically oscillating charged liquid drop, will sooner or later develop a short "neck" and will begin to separate into two charged "globs." If the electric repulsion between these two globs forces them far enough apart to break the neck, the two fragments, each still carrying some residual excitation energy, will fly apart (Figs. 44-2*e* and *f*). Fission has occurred.

This model gave a good qualitative picture of the fission process. What remained to be seen, however, was whether it could answer a hard question: Why are some high-mass nuclides (^{235}U and ^{239}Pu, say) readily fissionable by thermal neutrons when other, equally massive nuclides (^{238}U and ^{243}Am, say) are not?

Bohr and Wheeler were able to answer this question. Figure 44-3 shows a graph of the potential energy of the fissioning nucleus at various stages, derived from their model for the fission process. This energy is plotted against the *distortion parameter* r, which is a rough measure of the extent to which the oscillating nucleus departs from a spherical shape. Figure 44-2*d* suggests how this parameter is defined just before fission occurs. When the fragments are far apart, this parameter is simply the distance between their centers.

The energy difference between the initial state and the final state of the fissioning nucleus—that is, the disintegration energy Q—is labeled in Fig. 44-3. The central feature of that figure, however, is that the potential energy curve passes through a maximum at a certain value of r. Thus, there is a *potential barrier* of height E_b that must be surmounted (or tunneled through) before fission can occur. This reminds us of alpha decay (Fig. 43-9), which is also a process that is inhibited by a potential barrier.

We see then that fission will occur only if the absorbed neutron provides an excitation energy E_n great enough to overcome the barrier. This energy E_n need not be *quite* as great as the barrier height E_b because of the possibility of quantum-physics tunneling.

Table 44-2 shows this test of whether capture of a thermal neutron can cause fissioning, for four high-mass nuclides. For each nuclide, the table shows both the barrier height E_b of the nucleus that is formed by the neutron capture and the excitation energy E_n due to the capture. The values of E_b are calculated from the theory of Bohr and Wheeler. The values of E_n are calculated from the change in mass energy due to the neutron capture.

For an example of the calculation of E_n, we can go to the first line in the table,

TABLE 44-2 **Test of the Fissionability of Four Nuclides**

Target Nuclide	Nuclide Being Fissioned	E_n (MeV)	E_b (MeV)	Fission by Thermal Neutrons?
^{235}U	^{236}U	6.5	5.2	Yes
^{238}U	^{239}U	4.8	5.7	No
^{239}Pu	^{240}Pu	6.4	4.8	Yes
^{243}Am	^{244}Am	5.5	5.8	No

which represents the neutron capture process

$$^{235}\text{U} + \text{n} \rightarrow {}^{236}\text{U}.$$

The masses involved are 235.043 923 u for ^{235}U, 1.008 665 u for the neutron, and 236.045 562 u for ^{236}U. It is easy to show that, because of the neutron capture, the mass decreases by 7.026×10^{-3} u. Thus, energy is transferred from mass energy to excitation energy E_n. Multiplying the change in mass by c^2 (= 931.5 MeV/u) gives us $E_n = 6.5$ MeV, which is listed on the first line of the table.

The first and third results in Table 44-2 are historically profound, because they are the reasons why the two atomic bombs used in World War II contained ^{235}U (first bomb) and ^{239}Pu (second bomb). That is, for ^{235}U and ^{239}Pu, $E_n > E_b$. This means that fission by absorption of a thermal neutron is predicted to occur for these nuclides. For the other two nuclides (^{238}U and ^{243}Am), we have $E_n < E_b$; thus, there is not enough energy from a thermal neutron for the excited nucleus to surmount the barrier or to tunnel through it effectively. Instead of fissioning, the nucleus gets rid of its excitation energy by emitting a gamma-ray photon.

^{238}U and ^{243}Am *can* be made to fission, however, if they absorb a substantially energetic (rather than a thermal) neutron. For ^{238}U, for example, the absorbed neutron must have at least 1.3 MeV of energy for this *fast fission* process to have a good chance of occurring.

The two atomic bombs used in World War II depended on the ability of thermal neutrons to cause many high-mass nuclides in the cores of the bombs to fission nearly all at once, so that the fissioning would result in an explosive and devastating output of energy. The first bomb used ^{235}U because enough of it had been refined from uranium ore to make that bomb and a test bomb. (The ore consists mainly of ^{238}U, which, as we have seen, is not caused to fission by thermal neutrons.) The second bomb used ^{239}Pu, based only on theoretical calculations as summarized in Table 44-2, because not enough additional ^{235}U was available when the second bomb was ordered.

44-4 The Nuclear Reactor

For large-scale energy release due to fission, one fission event must trigger others, so that the process spreads throughout the nuclear fuel like flame through a log. The fact that more neutrons are produced in fission than are consumed raises the possibility of just such a **chain reaction,** with each neutron that is produced potentially triggering another fission. The reaction can be either rapid (as in a nuclear bomb) or controlled (as in a nuclear reactor).

Suppose that we wish to design a reactor based on the fission of ^{235}U by thermal neutrons. Natural uranium contains 0.7% of this isotope, the remaining 99.3% being ^{238}U, which is not fissionable by thermal neutrons. Let us give ourselves an edge by artificially *enriching* the uranium fuel so that it contains perhaps 3% ^{235}U. Three difficulties still stand in the way of a working reactor.

1. *The Neutron Leakage Problem.* Some of the neutrons produced by fission will leak out of the reactor and so not be part of the chain reaction. Leakage is a surface effect; its magnitude is proportional to the square of a typical reactor dimension (the surface area of a cube of edge length a is $6a^2$). Neutron production, however, occurs throughout the volume of the fuel and is thus proportional to the cube of a typical dimension (the volume of the same cube is a^3). We can make the fraction of neutrons lost by leakage as small as we wish by making the reactor core large enough, thereby reducing the surface-to-volume ratio (= $6/a$ for a cube).

The scene 20 m from the Chernobyl reactor unit 4 (near Kiev), after it exploded in April 1986. Nearly all the volatile radionuclides inside the reactor were released into the air.

2. *The Neutron Energy Problem.* The neutrons produced by fission are fast, with kinetic energies of about 2 MeV. However, fission is induced most effectively by thermal neutrons. The fast neutrons can be slowed down by mixing the uranium fuel with a substance—called a **moderator**—that has two properties: It is effective in slowing down neutrons via elastic collisions, and it does not remove neutrons from the core by absorbing them so that they do not result in fission. Most power reactors in North America use water as a moderator; the hydrogen nuclei (protons) in the water are the effective component. We saw in Chapter 10 that, if a moving particle has a head-on elastic collision with a stationary particle, the moving particle loses *all* its kinetic energy if the two particles have the same mass. Thus, protons form an effective moderator because they have approximately the same mass as the fast neutrons whose speed we wish to reduce.

3. *The Neutron Capture Problem.* As the fast (2 MeV) neutrons generated by fission are slowed down in the moderator to thermal energies (about 0.04 eV), they must pass through a critical energy interval (from 1 to 100 eV) in which they are particularly susceptible to nonfission capture by ^{238}U nuclei. Such *resonance capture,* which results in the emission of a gamma ray, removes the neutron from the fission chain. To minimize such nonfission capture, the uranium fuel and the moderator are not intimately mixed but are "clumped together," occupying different regions of the reactor volume.

   In a typical reactor, the uranium fuel is in the form of uranium oxide pellets, which are inserted end to end into long hollow metal tubes. The liquid moderator surrounds bundles of these **fuel rods,** forming the reactor **core.** This geometric arrangement increases the probability that a fast neutron, produced in a fuel rod, will find itself in the moderator when it passes through the critical energy interval. Once the neutron has reached thermal energies, it may *still* be captured in ways that do not result in fission (called *thermal capture*). However, it is much more likely that the thermal neutron will wander back into a fuel rod and produce a fission event.

Figure 44-4 shows the neutron balance in a typical power reactor operating at constant power. Let us trace a sample of 1000 thermal neutrons through one complete

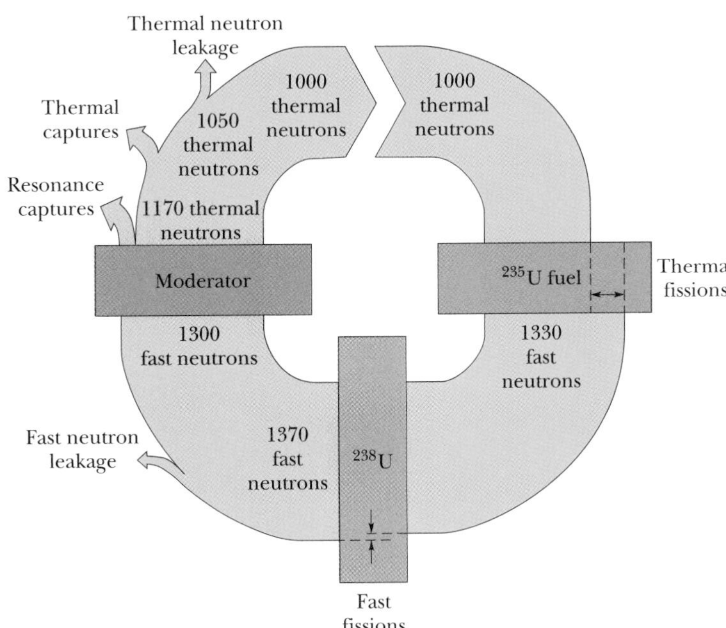

Fig. 44-4 Neutron bookkeeping in a reactor. A generation of 1000 thermal neutrons interacts with the ^{235}U fuel, the ^{238}U matrix, and the moderator. They produce 1370 neutrons by fission; 370 of these are lost by nonfission capture or by leakage; so 1000 thermal neutrons are left to form the next generation. The figure is drawn for a reactor running at a steady power level.

cycle, or *generation,* in the reactor core. They produce 1330 neutrons by fission in the ^{235}U fuel and 40 neutrons by fast fission in ^{238}U, which gives 370 neutrons more than the original 1000, all of them fast. When the reactor is operating at a steady power level, exactly the same number of neutrons (370) is then lost by leakage from the core and by nonfission capture, leaving 1000 thermal neutrons to start the next generation. In this cycle, of course, each of the 370 neutrons produced by fission events represents a deposit of energy in the reactor core, heating up the core.

The *multiplication factor k*—an important reactor parameter—is the ratio of the number of neutrons present at the beginning of a particular generation to the number present at the beginning of the next generation. In Fig. 44-4, the multiplication factor is 1000/1000, or exactly unity. For $k = 1$, the operation of the reactor is said to be exactly *critical,* which is what we wish it to be for steady-power operation. Reactors are actually designed so that they are inherently *supercritical* ($k > 1$); the multiplication factor is then adjusted to critical operation ($k = 1$) by inserting **control rods** into the reactor core. These rods, containing a material such as cadmium that absorbs neutrons readily, can be inserted farther to reduce the operating power level and withdrawn to increase the power level or to compensate for the tendency of reactors to go *subcritical* as (neutron-absorbing) fission products build up in the core during continued operation.

If you pulled out one of the control rods rapidly, how fast would the reactor power level increase? This *response time* is controlled by the fascinating circumstance that a small fraction of the neutrons generated by fission do not escape promptly from the newly formed fission fragments but are emitted from these fragments later, as the fragments decay by beta emission. Of the 370 "new" neutrons produced in Fig. 44-4, for example, perhaps 16 are delayed, being emitted from fragments following beta decays whose half-lives range from 0.2 to 55 s. These delayed neutrons are few in number, but they serve the essential purpose of slowing the reactor response time to match practical mechanical reaction times.

Figure 44-5 shows the broad outlines of an electric power plant based on a *pressurized-water reactor* (PWR), a type in common use in North America. In such

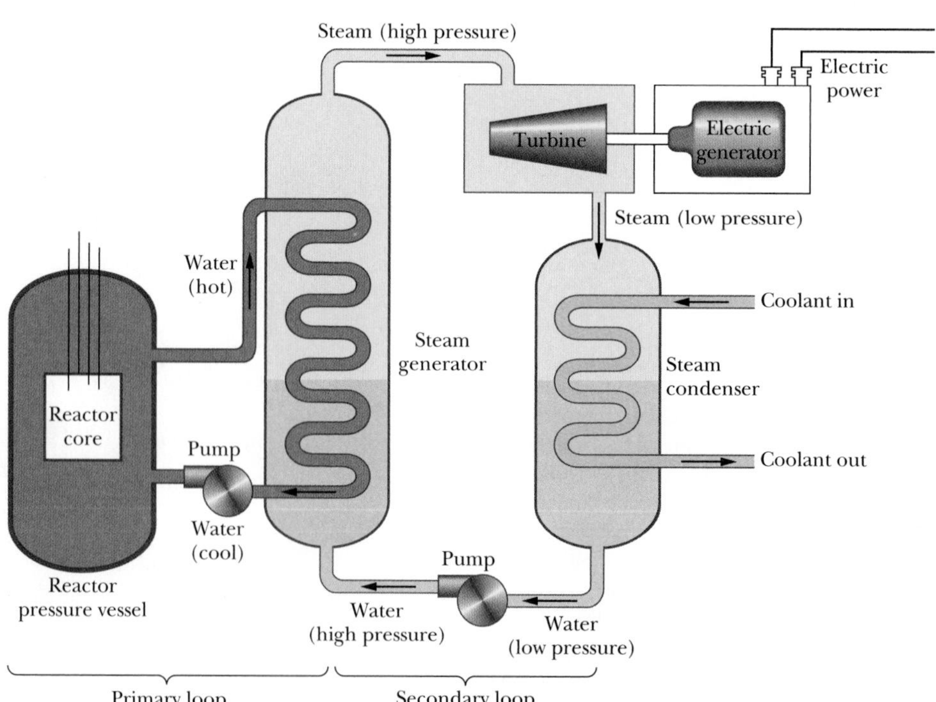

Fig. 44-5 A simplified layout of a nuclear power plant, based on a pressurized-water reactor. Many features are omitted—among them the arrangement for cooling the reactor core in case of an emergency.

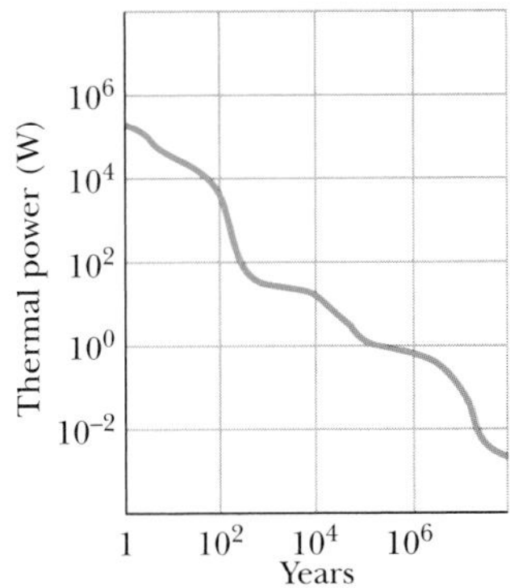

Fig. 44-6 The thermal power released by the radioactive wastes from one year's operation of a typical large nuclear power plant, shown as a function of time. The curve is the superposition of the effects of many radionuclides, with a wide variety of half-lives. Note that both scales are logarithmic.

a reactor, water is used both as the moderator and as the heat transfer medium. In the *primary loop,* water is circulated through the reactor vessel and transfers energy at high temperature and pressure (possibly 600 K and 150 atm) from the hot reactor core to the steam generator, which is part of the *secondary loop*. In the steam generator, evaporation provides high-pressure steam to operate the turbine that drives the electric generator. To complete the *secondary loop,* low-pressure steam from the turbine is cooled and condensed to water and forced back into the steam generator by a pump. To give some idea of scale, a typical reactor vessel for a 1000 MW (electric) plant may be 12 m high and weigh 4 MN. Water flows through the primary loop at a rate of about 1 ML/min.

An unavoidable feature of reactor operation is the accumulation of radioactive wastes, including both fission products and heavy *transuranic* nuclides such as plutonium and americium. One measure of their radioactivity is the rate at which they release energy in thermal form. Figure 44-6 shows the thermal power generated by such wastes from one year's operation of a typical large nuclear plant. Note that both scales are logarithmic. Most "spent" fuel rods from power reactor operation are stored on site, immersed in water; permanent secure storage facilities for reactor waste have yet to be completed.

Much weapons-derived radioactive waste accumulated during World War II and in subsequent years is also still in on-site storage. For example, Fig. 44-7 shows an underground storage tank farm under construction at the Hanford Site in Washington State; each large tank holds 1 ML of highly radioactive liquid waste. There are now 152 such tanks at the site. In addition, much solid waste, both low-level radioactive waste (contaminated clothing, for example) and high-level waste (reactor cores from decommissioned nuclear submarines, for example) is buried in trenches.

Fig. 44-7 An underground tank farm under construction during World War II at the Hanford Site in Washington State. Note the trucks and the workers. Each large tank now holds 1 ML of high-level radioactive waste.

Sample Problem 44-2

A large electric generating station is powered by a pressurized-water nuclear reactor. The thermal power produced in the reactor core is 3400 MW, and 1100 MW of electricity is generated by the station. The fuel charge is 8.60×10^4 kg of uranium, in the form of uranium oxide, distributed among 5.70×10^4 fuel rods. The uranium is enriched to 3.0% ^{235}U.

(a) What is the station's efficiency?

SOLUTION: The **Key Idea** here is the definition of efficiency for this power plant or any other energy device: Efficiency is the ratio of the output power (rate at which useful energy is provided) to the input power (rate at which energy must be supplied). Here the efficiency (eff) is

$$\text{eff} = \frac{\text{useful output}}{\text{energy input}} = \frac{1100\text{ MW (electric)}}{3400\text{ MW (thermal)}}$$
$$= 0.32, \text{ or } 32\%. \qquad \text{(Answer)}$$

The efficiency—as for all power plants—is controlled by the second law of thermodynamics. To run this plant, energy at the rate of 3400 MW − 1100 MW, or 2300 MW, must be discharged as thermal energy to the environment.

(b) At what rate R do fission events occur in the reactor core?

SOLUTION: The Key Ideas here are (1) that the fission events provide the input power P of 3400 MW ($= 3.4 \times 10^9$ J/s) and (2) from Eq. 44-6, the energy Q released by each event is about 200 MeV. Thus, for steady-state operation (P is constant), we find

$$R = \frac{P}{Q} = \left(\frac{3.4 \times 10^9 \text{ J/s}}{200 \text{ MeV/fission}}\right)\left(\frac{1 \text{ MeV}}{1.60 \times 10^{-13} \text{ J}}\right)$$
$$= 1.06 \times 10^{20} \text{ fissions/s}$$
$$\approx 1.1 \times 10^{20} \text{ fissions/s}. \quad \text{(Answer)}$$

(c) At what rate (in kilograms per day) is the ^{235}U fuel disappearing? Assume conditions at start-up.

SOLUTION: Here the Key Idea is that ^{235}U disappears due to two processes: (1) the fission process with the rate calculated in part (b) and (2) the nonfission capture of neutrons at about one-fourth that rate. Thus, the total rate at which ^{235}U disappears is

$$(1 + 0.25)(1.06 \times 10^{20} \text{ atoms/s}) = 1.33 \times 10^{20} \text{ atoms/s}.$$

We next need the mass of each ^{235}U atom. We cannot use the molar mass for uranium listed in Appendix F, because that molar mass is for ^{238}U, the most common uranium isotope. Instead, we shall assume that the mass of each ^{235}U atom in atomic mass units is equal to the mass number A. Thus, the mass of each ^{235}U atom is 235 u ($= 3.90 \times 10^{-25}$ kg). Then the rate at which the ^{235}U fuel disappears is

$$\frac{dM}{dt} = (1.33 \times 10^{20} \text{ atoms/s})(3.90 \times 10^{-25} \text{ kg})$$
$$= 5.19 \times 10^{-5} \text{ kg/s} \approx 4.5 \text{ kg/d}. \quad \text{(Answer)}$$

(d) At this rate of fuel consumption, how long would the fuel supply of ^{235}U last?

SOLUTION: At start-up, we know that the total mass of ^{235}U is 3.0% of the 8.6×10^4 kg of uranium oxide. Then the Key Idea here is that the time T required to consume this total mass of ^{235}U at the steady rate of 4.5 kg/d is

$$T = \frac{(0.030)(8.6 \times 10^4 \text{ kg})}{4.5 \text{ kg/d}} \approx 570 \text{ d}. \quad \text{(Answer)}$$

In practice, the fuel rods must be replaced (usually in batches) before their ^{235}U content is entirely consumed.

(e) At what rate is mass being converted to other forms of energy by the fission of ^{235}U in the reactor core?

SOLUTION: The Key Idea here is that the conversion of mass energy to other forms of energy is linked only to the fissioning that produces the input power (3400 MW), and not to the nonfission capture of neutrons (although both these processes affect the rate at which ^{235}U is consumed). Thus, from Einstein's relation $E = mc^2$, we can write

$$\frac{dm}{dt} = \frac{dE/dt}{c^2} = \frac{3.4 \times 10^9 \text{ W}}{(3.00 \times 10^8 \text{ m/s})^2}$$
$$= 3.8 \times 10^{-8} \text{ kg/s} = 3.3 \text{ g/d}. \quad \text{(Answer)}$$

We see that the mass conversion rate is about the mass of one common coin per day, considerably less than the fuel consumption rate calculated in (c).

✔CHECKPOINT 2: In this sample problem, we saw that the generated power of the nuclear power plant ($P_{gen} = 1100$ MW) was less than the power discharged to the environment ($P_{dis} =$ 2300 MW). Does the second law of thermodynamics: (a) require that P_{gen} always be less than P_{dis}; (b) permit P_{gen} to be greater than P_{dis}; and (c) permit P_{dis} to be zero, assuming optimum reactor design?

44-5 A Natural Nuclear Reactor

On December 2, 1942, when their reactor first became operational (Fig. 44-8), Enrico Fermi and his associates had every right to assume that they had put into operation the first fission reactor that had ever existed on this planet. About 30 years later it was discovered that, if they did in fact think that, they were wrong.

Fig. 44-8 A painting of the first nuclear reactor, assembled during World War II on a squash court at the University of Chicago by a team headed by Enrico Fermi. This reactor, which went critical on December 2, 1942, was built of lumps of uranium embedded in blocks of graphite. It served as a prototype for later reactors whose purpose was to manufacture plutonium for the construction of nuclear weapons.

Some two billion years ago, in a uranium deposit now being mined in Gabon, West Africa, a natural fission reactor apparently went into operation and ran for perhaps several hundred thousand years before shutting down. We can test whether this could actually have happened by considering two questions:

1. *Was There Enough Fuel?* The fuel for a uranium-based fission reactor must be the easily fissionable isotope ^{235}U, which constitutes only 0.72% of natural uranium. This isotopic ratio has been measured for terrestrial samples, in Moon rocks, and in meteorites; in all cases the abundance values are the same. The clue to the discovery in West Africa was that the uranium in that deposit was deficient in ^{235}U, some samples having abundances as low as 0.44%. Investigation led to the speculation that this deficit in ^{235}U could be accounted for if, at some earlier time, the ^{235}U was partially consumed by the operation of a natural fission reactor.

   The serious problem remains that, with an isotopic abundance of only 0.72%, a reactor can be assembled (as Fermi and his team learned) only after thoughtful design and with scrupulous attention to detail. There seems no chance that a nuclear reactor could go critical "naturally."

   However, things were different in the distant past. Both ^{235}U and ^{238}U are radioactive, with half-lives of 7.04×10^8 y and 44.7×10^8 y, respectively. Thus, the half-life of the readily fissionable ^{235}U is about 6.4 times shorter than that of ^{238}U. Because ^{235}U decays faster, there was more of it, relative to ^{238}U, in the past. Two billion years ago, in fact, this abundance was not 0.72%, as it is now, but 3.8%. This abundance happens to be just about the abundance to which natural uranium is artificially enriched to serve as fuel in modern power reactors.

   With this readily fissionable fuel available, the presence of a natural reactor (provided certain other conditions are met) is less surprising. The fuel was there. Two billion years ago, incidentally, the highest order of life-form to have evolved was the blue-green alga.

2. *What Is the Evidence?* The mere depletion of ^{235}U in an ore deposit does not prove the existence of a natural fission reactor. One looks for more convincing evidence.

   If there was a reactor, there must now be fission products. Of the 30 or so elements whose stable isotopes are produced in a reactor, some must still remain. Study of their isotopic abundances could provide the evidence we need.

   Of the several elements investigated, the case of neodymium is spectacularly convincing. Figure 44-9*a* shows the isotopic abundances of the seven stable neodymium isotopes as they are normally found in nature. Figure 44-9*b* shows these abundances as they appear among the ultimate stable fission products of the

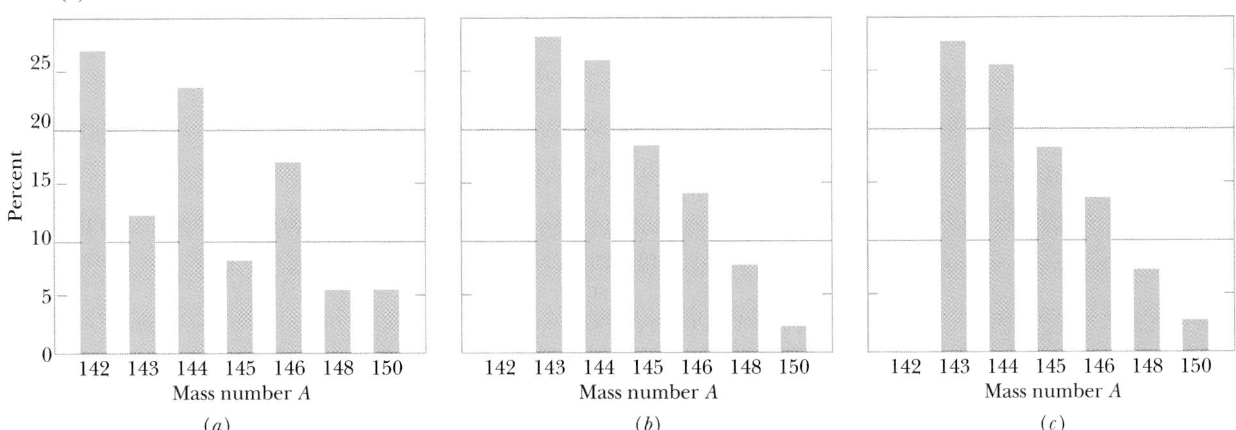

Fig. 44-9 The distribution by mass number of the isotopes of neodymium as they occur in (*a*) natural terrestrial deposits of the ores of this element and (*b*) the spent fuel of a power reactor. (*c*) The distribution (after several corrections) found for neodymium from the uranium mine in Gabon, West Africa. Note that (*b*) and (*c*) are virtually identical and are quite different from (*a*).

fission of ^{235}U. The clear differences are not surprising, considering the totally different origins of the two sets of isotopes. Note particularly that ^{142}Nd, the dominant isotope in the natural element, is absent from the fission products.

The big question is: What do the neodymium isotopes found in the uranium ore body in West Africa look like? If a natural reactor operated there, we would expect to find isotopes from *both* sources (that is, natural isotopes as well as fission-produced isotopes). Figure 44-9*c* shows the abundances after dual-source and other corrections have been made to the data. Comparison of Figs. 44-9*b* and *c* indicates that there was indeed a natural fission reactor at work.

The failure of the fission products of the West African natural reactor to migrate far from their region of production over about 2 billion years may support the idea of long-term storage of radioactive waste in suitably chosen geological environments.

Sample Problem 44-3

The ratio of ^{235}U to ^{238}U in natural uranium deposits today is 0.0072. What was this ratio 2.0×10^9 y ago? The half-lives of the two isotopes are 7.04×10^8 y and 44.7×10^8 y, respectively.

SOLUTION: The **Key Idea** here is that the ratio N_5/N_8 of ^{235}U to ^{238}U at time $t = 0$ was not equal to 0.0072 (the ratio today, at time $t = 2.0 \times 10^9$ y), because those two isotopes have been decaying at different rates. Let $N_5(0)$ and $N_8(0)$ be the numbers of the isotopes in a sample of uranium at $t = 0$, and let $N_5(t)$ and $N_8(t)$ be the numbers of the isotopes at some later time t. Then, for each isotope, we can use Eq. 43-14 to write the number at time t in terms of the number at time $t = 0$:

$$N_5(t) = N_5(0)e^{-\lambda_5 t} \quad \text{and} \quad N_8(t) = N_8(0)e^{-\lambda_8 t},$$

in which λ_5 and λ_8 are the corresponding disintegration constants. Dividing gives

$$\frac{N_5(t)}{N_8(t)} = \frac{N_5(0)}{N_8(0)} e^{-(\lambda_5 - \lambda_8)t}.$$

Because we want the ratio $N_5(0)/N_8(0)$, we rearrange this to be

$$\frac{N_5(0)}{N_8(0)} = \frac{N_5(t)}{N_8(t)} e^{(\lambda_5 - \lambda_8)t}. \tag{44-8}$$

The disintegration constants are related to the half-lives by Eq. 43-17, which yields

$$\lambda_5 = \frac{\ln 2}{T_{1/2,5}} = \frac{\ln 2}{7.04 \times 10^8 \text{ y}} = 9.85 \times 10^{-10} \text{ y}^{-1}$$

and

$$\lambda_8 = \frac{\ln 2}{T_{1/2,8}} = \frac{\ln 2}{44.7 \times 10^8 \text{ y}} = 1.55 \times 10^{-10} \text{ y}^{-1}.$$

The exponent in Eq. 44-8 is then

$$(\lambda_5 - \lambda_8)t = [(9.85 - 1.55) \times 10^{-10} \text{ y}^{-1}](2 \times 10^9 \text{ y}) = 1.66.$$

Equation 44-8 then yields

$$\frac{N_5(0)}{N_8(0)} = \frac{N_5(t)}{N_8(t)} e^{(\lambda_5 - \lambda_8)t} = (0.0072)(e^{1.66}) = 0.0379 \approx 3.8\%. \quad \text{(Answer)}$$

The ratio of ^{235}U to ^{238}U was much greater than this (about 30%) when Earth was formed 4.5 billion years ago.

44-6 Thermonuclear Fusion: The Basic Process

The binding energy curve of Fig. 43-6 shows that energy can be released if two light nuclei combine to form a single larger nucleus, a process called nuclear **fusion.** That process is hindered by the Coulomb repulsion that acts to prevent the two positively charged particles from getting close enough to be within range of their attractive nuclear forces and thus "fusing." The height of this *Coulomb barrier* depends on the charges and the radii of the two interacting nuclei. We show in Sample Problem 44-4 that, for two protons ($Z = 1$), the barrier height is 400 keV. For more highly charged particles, of course, the barrier is correspondingly higher.

To generate useful amounts of energy, nuclear fusion must occur in bulk matter. The best hope for bringing this about is to raise the temperature of the material until the particles have enough energy—due to their thermal motions alone—to penetrate the Coulomb barrier. We call this process **thermonuclear fusion.**

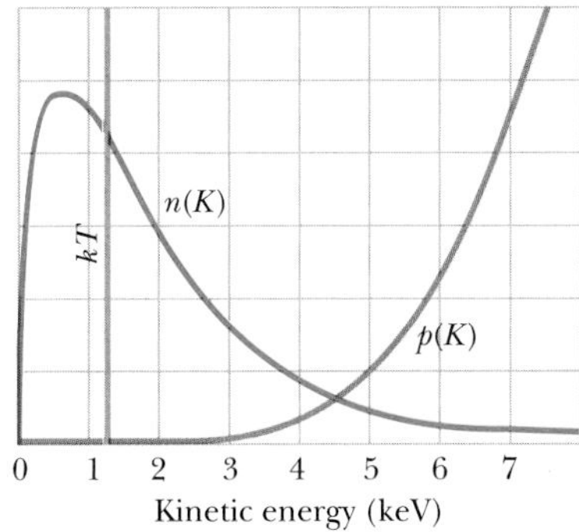

Fig. 44-10 The curve marked $n(K)$ gives the number density per unit energy for protons at the center of the Sun. The curve marked $p(K)$ gives the probability of barrier penetration (and hence fusion) for proton-proton collisions at the Sun's core temperature. The vertical line marks the value of kT at this temperature. Note that the two curves are drawn to (separate) arbitrary vertical scales.

In thermonuclear studies, temperatures are reported in terms of the kinetic energy K of interacting particles via the relation

$$K = kT, \tag{44-9}$$

in which K is the kinetic energy corresponding to the *most probable speed* of the interacting particles, k is the Boltzmann constant, and the temperature T is in kelvins. Thus, rather than saying, "The temperature at the center of the Sun is 1.5×10^7 K," it is more common to say, "The temperature at the center of the Sun is 1.3 keV."

Room temperature corresponds to $K \approx 0.03$ eV; a particle with only this amount of energy could not hope to overcome a barrier as high as, say, 400 keV. Even at the center of the Sun, where $kT = 1.3$ keV, the outlook for thermonuclear fusion does not seem promising at first glance. Yet we know that thermonuclear fusion not only occurs in the core of the Sun but is the dominant feature of that body and of all other stars.

The puzzle is solved when we realize two facts: (1) The energy calculated with Eq. 44-9 is that of the particles with the *most probable* speed, as defined in Section 20-7; there is a long tail of particles with much higher speeds and, correspondingly, much higher energies. (2) The barrier heights that we have calculated represent the *peaks* of the barriers. Barrier tunneling can occur at energies considerably below those peaks, as we saw in the case of alpha decay in Section 43-4.

Figure 44-10 sums things up. The curve marked $n(K)$ in this figure is a Maxwell distribution curve for the protons in the Sun's core, drawn to correspond to the Sun's central temperature. This curve differs from the Maxwell distribution curve given in Fig. 20-7 in that here the curve is drawn in terms of energy and not of speed. Specifically, for any kinetic energy K, the expression $n(K)\,dK$ gives the probability that a proton will have a kinetic energy lying between K and $K + dK$. The value of kT in the core of the Sun is indicated by the vertical line in the figure; note that many of the Sun's core protons have energies greater than this value.

The curve marked $p(K)$ in Fig. 44-10 is the probability of barrier penetration by two colliding protons. The two curves in Fig. 44-10 suggest that there is a particular proton energy at which proton-proton fusion events occur at a maximum rate. At energies much above this value, the barrier is transparent enough but too few protons have these energies, so the fusion reaction cannot be sustained. At energies much below this value, plenty of protons have these energies but the Coulomb barrier is too formidable.

CHECKPOINT 3: Which of these potential fusion reactions will *not* result in the net release of energy: (a) ^{6}Li + ^{6}Li, (b) ^{4}He + ^{4}He, (c) ^{12}C + ^{12}C, (d) ^{20}Ne + ^{20}Ne, (e) ^{35}Cl + ^{35}Cl, and (f) ^{14}N + ^{35}Cl? (*Hint:* Consult the binding energy curve of Fig. 43-6.)

Sample Problem 44-4

Assume a proton is a sphere of radius $R \approx 1$ fm. Two protons are fired at each other with the same kinetic energy K.

(a) What must K be if the particles are brought to rest by their mutual Coulomb repulsion when they are just "touching" each other? We can take this value of K as a representative measure of the height of the Coulomb barrier.

SOLUTION: The **Key Idea** here is that the mechanical energy E of the two-proton system is conserved as the protons move toward each other and momentarily stop. In particular, the initial mechanical energy E_i is equal to the mechanical energy E_f when they stop. The initial energy E_i consists only of the total kinetic energy $2K$ of the two protons. When the protons stop, energy E_f consists only of the electric potential energy U of the system, as given by Eq. 25-43 ($U = q_1q_2/4\pi\varepsilon_0 r$). Here the distance r between the protons when they stop is their center-to-center distance $2R$, and their charges q_1 and q_2 are both e. Then we can write the conservation of energy $E_i = E_f$ as

$$2K = \frac{1}{4\pi\varepsilon_0}\frac{e^2}{2R}.$$

This yields, with known values,

$$K = \frac{e^2}{16\pi\varepsilon_0 R}$$

$$= \frac{(1.60 \times 10^{-19}\ \text{C})^2}{(16\pi)(8.85 \times 10^{-12}\ \text{F/m})(1 \times 10^{-15}\ \text{m})}$$

$$= 5.75 \times 10^{-14}\ \text{J} = 360\ \text{keV} \approx 400\ \text{keV}. \quad \text{(Answer)}$$

(b) At what temperature would a proton in a gas of protons have the average kinetic energy calculated in (a), and thus have energy equal to the height of the Coulomb barrier?

SOLUTION: The Key Idea here is that if we treat the proton gas as an ideal gas, then from Eq. 20-24, the average energy of the protons is $K_{avg} = \frac{3}{2}kT$, where k is the Boltzmann constant. Solving this equation for T and using the result of (a) yield

$$T = \frac{2K_{avg}}{3k} = \frac{(2)(5.75 \times 10^{-14}\ \text{J})}{(3)(1.38 \times 10^{-23}\ \text{J/K})}$$

$$\approx 3 \times 10^9\ \text{K}. \quad \text{(Answer)}$$

The temperature of the core of the Sun is only about 1.5×10^7 K, so it is clear that fusion in the Sun's core must involve protons whose energies are *far* above the average energy.

44-7 Thermonuclear Fusion in the Sun and Other Stars

The Sun radiates energy at the rate of 3.9×10^{26} W and has been doing so for several billion years. Where does all this energy come from? Chemical burning is ruled out; if the Sun had been made of coal and oxygen—in the right proportions for combustion—it would have lasted for only about 1000 y. Another possibility is that the Sun is slowly shrinking, under the action of its own gravitational forces. By transferring gravitational potential energy to thermal energy, the Sun might maintain its temperature and continue to radiate. Calculation shows, however, that this mechanism also fails; it produces a solar lifetime that is too short by a factor of at least 500. That leaves only thermonuclear fusion. The Sun, as you will see, burns not coal but hydrogen, and in a nuclear furnace, not an atomic or chemical one.

The fusion reaction in the Sun is a multistep process in which hydrogen is burned into helium, hydrogen being the "fuel" and helium the "ashes." Figure 44-11 shows the **proton-proton** (p-p) **cycle** by which this occurs.

The p-p cycle starts with the collision of two protons (^{1}H + ^{1}H) to form a deuteron (^{2}H), with the simultaneous creation of a positron (e^+) and a neutrino (ν). The positron very quickly encounters a free electron (e^-) in the Sun and both particles annihilate (see Section 22-6), their mass energy appearing as two gamma-ray photons (γ).

A pair of such events is shown in the top row of Fig. 44-11. These events are actually extremely rare. In fact, only once in about 10^{26} proton-proton collisions is a deuteron formed; in the vast majority of cases, the two protons simply rebound elastically from each other. It is the slowness of this "bottleneck" process that regulates the rate of energy production and keeps the Sun from exploding. In spite of this slowness, there are so very many protons in the huge and dense volume of the Sun's core that deuterium is produced in just this way at the rate of 10^{12} kg/s.

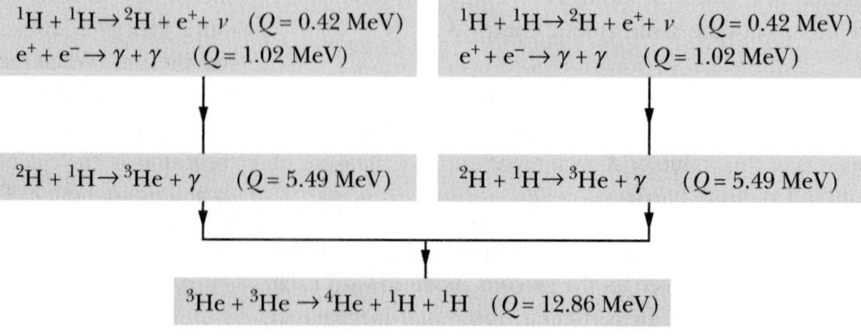

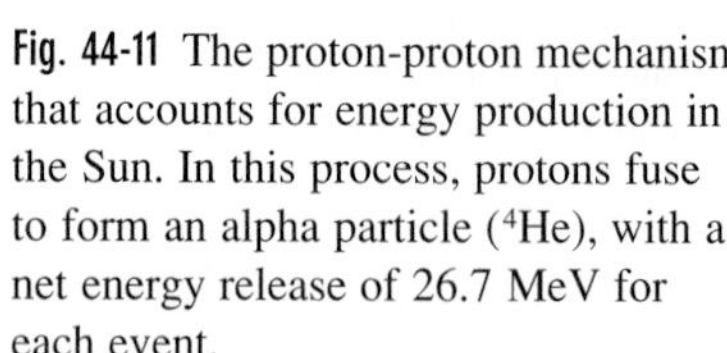

Fig. 44-11 The proton-proton mechanism that accounts for energy production in the Sun. In this process, protons fuse to form an alpha particle (^{4}He), with a net energy release of 26.7 MeV for each event.

Once a deuteron has been produced, it quickly collides with another proton and forms a ^{3}He nucleus, as the middle row of Fig. 44-11 shows. Two such ^{3}He nuclei may eventually (within 10^5 y; there is plenty of time) find each other, forming an alpha particle (^{4}He) and two protons, as the bottom row in the figure shows.

Overall, we see from Fig. 44-11 that the p-p cycle amounts to the combination of four protons and two electrons to form an alpha particle, two neutrinos, and six gamma-ray photons. That is,

$$4\ ^1\text{H} + 2\text{e}^- \rightarrow {}^4\text{He} + 2\nu + 6\gamma. \tag{44-10}$$

Let us now add two electrons to each side of Eq. 44-10, obtaining

$$(4\ ^1\text{H} + 4\text{e}^-) \rightarrow ({}^4\text{He} + 2\text{e}^-) + 2\nu + 6\gamma. \tag{44-11}$$

The quantities in the two sets of parentheses then represent *atoms* (not bare nuclei) of hydrogen and of helium. That allows us to compute the energy release in the overall reaction of Eq. 44-10 (and Eq. 44-11) as

$$\begin{aligned} Q &= -\Delta m\, c^2 \\ &= -[4.002\,603\ \text{u} - (4)(1.007\,825\ \text{u})][931.5\ \text{MeV/u}] \\ &= 26.7\ \text{MeV}, \end{aligned}$$

in which 4.002 603 u is the mass of a helium atom and 1.007 825 u is the mass of a hydrogen atom. Neutrinos have at most a negligibly small mass, and gamma-ray photons have no mass; thus, they do not enter into the calculation of the disintegration energy.

This same value of Q follows (as it must) from adding up the Q values for the separate steps of the proton-proton cycle in Fig. 44-11. Thus,

$$\begin{aligned} Q &= (2)(0.42\ \text{MeV}) + (2)(1.02\ \text{MeV}) + (2)(5.49\ \text{MeV}) + 12.86\ \text{MeV} \\ &= 26.7\ \text{MeV}. \end{aligned}$$

About 0.5 MeV of this energy is carried out of the Sun by the two neutrinos indicated in Eqs. 44-10 and 44-11; the rest (= 26.2 MeV) is deposited in the core of the Sun as thermal energy. That thermal energy is then gradually transported to the Sun's surface where it is radiated away from the Sun as electromagnetic waves, including visible light.

The burning of hydrogen in the Sun's core is alchemy on a grand scale in the sense that one element is turned into another. The medieval alchemists, however, were more interested in changing lead into gold than in changing hydrogen into helium. In a sense, they were on the right track, except that their furnaces were not hot enough. Instead of being at a temperature of, say, 600 K, the ovens should have been at least as hot as 10^8 K.

Hydrogen burning has been going on in the Sun for about 5×10^9 y, and calculations show that there is enough hydrogen left to keep the Sun going for about the same length of time into the future. In 5 billion years, however, the Sun's core, which by that time will be largely helium, will begin to cool and the Sun will start to collapse under its own gravity. This will raise the core temperature and cause the outer envelope to expand, turning the Sun into what is called a *red giant.*

If the core temperature increases to about 10^8 K again, energy can be produced through fusion once more—this time by burning helium to make carbon. As a star evolves further and becomes still hotter, other elements can be formed by other fusion reactions. However, elements more massive than those near the peak of the binding energy curve of Fig. 43-6 cannot be produced by further fusion processes.

Elements with mass numbers beyond the peak of that curve are thought to be formed by neutron capture during cataclysmic stellar explosions that we call *super-*

Fig. 44-12 (*a*) The star known as Sanduleak, as it appeared until 1987. (*b*) We then began to intercept light from the supernova of that star; the explosion was 100 million times brighter than our Sun and could be seen with the unaided eye. The explosion took place 155 000 light-years away and thus actually occurred 155 000 years ago.

novas (Fig. 44-12). In such an event the outer shell of the star is blown outward into space, where it mixes with—and becomes part of—the tenuous medium that fills the space between the stars. It is from this medium, continually enriched by debris from stellar explosions, that new stars form, by condensation under the influence of the gravitational force.

The abundance on Earth of elements heavier than hydrogen and helium suggests that our solar system has condensed out of interstellar material that contained the remnants of such explosions. Thus, all the elements around us—including those in our own bodies—were manufactured in the interiors of stars that no longer exist. As one scientist put it: "In truth, we are the children of the stars."

Sample Problem 44-5

At what rate dm/dt is hydrogen being consumed in the core of the Sun by the p-p cycle of Fig. 44-11?

SOLUTION: The **Key Idea** is that the rate dE/dt at which energy is produced by hydrogen (proton) consumption within the Sun is equal to the rate P at which energy is radiated by the Sun:

$$P = \frac{dE}{dt}.$$

To bring the mass consumption rate dm/dt into this equation, we can rewrite it as

$$P = \frac{dE}{dt} = \frac{dE}{dm}\frac{dm}{dt} \approx \frac{\Delta E}{\Delta m}\frac{dm}{dt}, \tag{44-12}$$

where ΔE is the energy produced when protons of mass Δm are consumed. From our discussion in this section, we know that 26.2 MeV ($= 4.20 \times 10^{-12}$ J) of thermal energy is produced when four protons are consumed. That is, $\Delta E = 4.20 \times 10^{-12}$ J for a mass consumption of $\Delta m = 4(1.67 \times 10^{-27}$ kg). Substituting these data into Eq. 44-12 and using the power P of the Sun given in Appendix C, we find that

$$\frac{dm}{dt} = \frac{\Delta m}{\Delta E}P = \frac{4(1.67 \times 10^{-27}\text{ kg})}{4.20 \times 10^{-12}\text{ J}}(3.90 \times 10^{26}\text{ W})$$
$$= 6.2 \times 10^{11}\text{ kg/s}. \qquad \text{(Answer)}$$

Thus, a huge amount of hydrogen is consumed by the Sun every second. However, you need not worry too much about the Sun running out of hydrogen, because its mass of 2×10^{30} kg will keep it burning for a long, long time.

44-8 Controlled Thermonuclear Fusion

The first thermonuclear reaction on Earth occurred at Eniwetok Atoll on November 1, 1952, when the United States exploded a fusion device, generating an energy release equivalent to 10 million tons of TNT. The high temperatures and densities needed to initiate the reaction were provided by using a fission bomb as a trigger.

A sustained and controllable source of fusion power—a fusion reactor as part of, say, an electric generating plant—is considerably more difficult to achieve. That goal is nonetheless being pursued vigorously in many countries around the world, because many people look to the fusion reactor as the power source of the future, at least for the generation of electricity.

The p-p scheme displayed in Fig. 44-11 is not suitable for an Earth-bound fusion reactor because it is hopelessly slow. The process succeeds in the Sun only because of the enormous density of protons in the center of the Sun. The most attractive reactions for terrestrial use appear to be two deuteron-deuteron (d-d) reactions,

$$^{2}\text{H} + {}^{2}\text{H} \rightarrow {}^{3}\text{He} + \text{n} \qquad (Q = +3.27\text{ MeV}), \qquad (44\text{-}13)$$

$$^{2}\text{H} + {}^{2}\text{H} \rightarrow {}^{3}\text{H} + {}^{1}\text{H} \qquad (Q = +4.03\text{ MeV}), \qquad (44\text{-}14)$$

and the deuteron-triton (d-t) reaction*

$$^{2}\text{H} + {}^{3}\text{H} \rightarrow {}^{4}\text{He} + \text{n} \qquad (Q = +17.59\text{ MeV}). \qquad (44\text{-}15)$$

Deuterium, the source of deuterons for these reactions, has an isotopic abundance of only 1 part in 6700, but is available in unlimited quantities as a component of seawater. Proponents of power from the nucleus have described our ultimate power choice—when we have burned up all our fossil fuels—as either "burning rocks" (fission of uranium extracted from ores) or "burning water" (fusion of deuterium extracted from water).

There are three requirements for a successful thermonuclear reactor:

1. *A High Particle Density n.* The (number) density of interacting particles (the number of, say, deuterons per unit volume) must be great enough to ensure that the d-d collision rate is high enough. At the high temperatures required, the deuterium would be completely ionized, forming a neutral **plasma** (ionized gas) of deuterons and electrons.
2. *A High Plasma Temperature T.* The plasma must be hot. Otherwise the colliding deuterons will not be energetic enough to penetrate the Coulomb barrier that tends to keep them apart. A plasma ion temperature of 35 keV, corresponding to 4×10^8 K, has been achieved in the laboratory. This is about 30 times higher than the Sun's central temperature.
3. *A Long Confinement Time τ.* A major problem is containing the hot plasma long enough to maintain it at a density and a temperature sufficiently high to ensure the fusion of enough of the fuel. It is clear that no solid container can withstand the high temperatures that are necessary, so clever confining techniques are called for; we shall shortly discuss two of them.

It can be shown that, for the successful operation of a thermonuclear reactor using the d-t reaction, it is necessary to have

$$n\tau > 10^{20}\text{ s/m}^3. \qquad (44\text{-}16)$$

This condition, known as **Lawson's criterion,** tells us that we have a choice between confining a lot of particles for a short time or fewer particles for a longer time. Beyond meeting this criterion, it is still necessary that the plasma temperature be high enough.

Two approaches to controlled nuclear power generation are currently under study. Although neither approach has yet been successful, both are being pursued because of their promise and because of the potential importance of controlled fusion to the world's energy problems.

*The nucleus of the hydrogen isotope ^{3}H (tritium) is called the *triton*. It is a radionuclide, with a half-life of 12.3 y.

Fig. 44-13 The Tokamak Fusion Test Reactor at Princeton University.

Magnetic Confinement

In one version of this approach, a suitably shaped magnetic field is used to confine a hot plasma in an evacuated doughnut-shaped chamber called a **tokamak** (the name is an abbreviation consisting of parts of three Russian words). The magnetic forces acting on the charged particles that make up the hot plasma keep the plasma from touching the walls of the chamber. Figure 44-13 shows such a device at the Plasma Physics Laboratory of Princeton University.

The plasma is heated by inducing a current to flow in it and by bombarding the plasma with an externally accelerated beam of particles. The first goal of this approach is to achieve **breakeven,** which occurs when the Lawson criterion is met or exceeded. The ultimate goal is **ignition,** which corresponds to a self-sustaining thermonuclear reaction, with a net generation of energy. As of 2000, ignition has not been achieved, either in tokamaks or in other magnetic confinement devices.

Inertial Confinement

Fig. 44-14 The small spheres on the quarter are deuterium–tritium fuel pellets, designed to be used in a laser fusion chamber.

This approach to confining and heating fusion fuel so that a thermonuclear reaction can occur involves "zapping" a solid fuel pellet from all sides with intense laser beams, evaporating some material from the surface of the pellet. This boiled-off material causes an inward-moving shock wave that compresses the core of the pellet, increasing both its particle density and its temperature. The process is called *inertial confinement* because (a) the fuel is *confined* to the pellet and (b) the particles do not escape from the heated pellet during the very short zapping interval because of their *inertia* (their mass).

Laser fusion, using the inertial confinement approach, is being investigated in many laboratories in the United States and elsewhere. At the Lawrence Livermore Laboratory, for example, deuterium–tritium fuel pellets, each smaller than a grain of sand (Fig. 44-14), are to be zapped by 10 synchronized high-power laser pulses symmetrically arranged around the pellet. The laser pulses are designed to deliver, in total, some 200 kJ of energy to each fuel pellet in less than a nanosecond. This is a delivered power of about 2×10^{14} W during the pulse, which is roughly 100 times the total installed (sustained) electric power generating capacity of the world!

In an operating thermonuclear reactor of the laser-fusion type, fuel pellets are to be exploded—like miniature hydrogen bombs—at a rate of perhaps 10 to 100 per second. The feasibility of laser fusion as the basis of a thermonuclear power reactor has not been demonstrated as of 2000, but research is continuing at a vigorous pace.

Sample Problem 44-6

Suppose a fuel pellet in a laser fusion device contains equal numbers of deuterium and tritium atoms (and no other material). The density $d = 200\ \text{kg/m}^3$ of the pellet is increased by a factor of 10^3 by the action of the laser pulses.

(a) How many particles per unit volume (both deuterons and tritons) does the pellet contain in its compressed state? The molar mass M_d of deuterium atoms is 2.0×10^{-3} kg/mol, and the molar mass M_t of tritium atoms is 3.0×10^{-3} kg/mol.

SOLUTION: The **Key Idea** here is that, for a system consisting of only one type of particle, we can write the (mass) density of the system in terms of the particles' masses and their number density (the number of particles per unit volume):

$$\begin{pmatrix}\text{density,}\\ \text{kg/m}^3\end{pmatrix} = \begin{pmatrix}\text{number density,}\\ \text{m}^{-3}\end{pmatrix}\begin{pmatrix}\text{particle mass,}\\ \text{kg}\end{pmatrix}. \qquad (44\text{-}17)$$

Let n be the total number of particles per unit volume in the compressed pellet. Then the number of deuterium atoms per unit volume is $n/2$, and the number of tritium atoms per unit volume is also $n/2$.

Next, we can extend Eq. 44-17 to the system consisting of the two types of particles by writing the density d^* of the compressed pellet as the sum of the individual densities:

$$d^* = \frac{n}{2}m_d + \frac{n}{2}m_t, \qquad (44\text{-}18)$$

where m_d and m_t are the masses of a deuterium atom and a tritium atom, respectively. We can replace those masses with the given molar masses by substituting

$$m_d = \frac{M_d}{N_A} \quad \text{and} \quad m_t = \frac{M_t}{N_A},$$

where N_A is Avogadro's number. After making those replacements and substituting $1000d$ for the compressed density d^*, we solve Eq. 44-18 for n to obtain

$$n = \frac{2000dN_A}{M_d + M_t},$$

which gives us

$$n = \frac{(2000)(200\ \text{kg/m}^3)(6.02 \times 10^{23}\ \text{mol}^{-1})}{2.0 \times 10^{-3}\ \text{kg/mol} + 3.0 \times 10^{-3}\ \text{kg/mol}}$$
$$= 4.8 \times 10^{31}\ \text{m}^{-3}. \qquad \text{(Answer)}$$

(b) According to Lawson's criterion, how long must the pellet maintain this particle density if breakeven operation is to take place?

SOLUTION: The **Key Idea** here is that, if breakeven operation is to occur, the compressed density must be maintained for a time period τ given by Eq. 44-16 ($n\tau > 10^{20}$ s/m^3). Thus, we have

$$\tau > \frac{10^{20}\ \text{s/m}^3}{4.8 \times 10^{31}\ \text{m}^{-3}} \approx 10^{-12}\ \text{s}. \qquad \text{(Answer)}$$

(The plasma temperature must also be suitably high.)

REVIEW & SUMMARY

Energy from the Nucleus Nuclear processes are about a million times more effective, per unit mass, than chemical processes in transforming mass into other forms of energy.

Nuclear Fission Equation 44-1 shows a **fission** of ^{236}U induced by thermal neutrons bombarding ^{235}U. Equations 44-2 and 44-3 show the beta-decay chains of the primary fragments. The energy released in such a fission event is $Q \approx 200$ MeV.

Fission can be understood in terms of the collective model, in which a nucleus is likened to a charged liquid drop carrying a certain excitation energy. A potential barrier must be tunneled if fission is to occur. Fissionability depends on the relationship between the barrier height E_b and the excitation energy E_n.

The neutrons released during fission make possible a fission **chain reaction.** Figure 44-4 shows the neutron balance for one cycle of a typical reactor. Figure 44-5 suggests the layout of a complete nuclear power plant.

Nuclear Fusion The release of energy by the **fusion** of two light nuclei is inhibited by their mutual Coulomb barrier. Fusion can occur in bulk matter only if the temperature is high enough (that is, if the particle energy is high enough) for appreciable barrier tunneling to occur.

The Sun's energy arises mainly from the thermonuclear burning of hydrogen to form helium by the **proton-proton cycle** outlined in Fig. 44-11. Elements up to $A \approx 56$ (the peak of the binding energy curve) can be built up by other fusion processes once the hydrogen fuel supply of a star has been exhausted.

Controlled Fusion Controlled **thermonuclear fusion** for energy generation has not yet been achieved. The d-d and d-t reactions are the most promising mechanisms. A successful fusion reactor must satisfy **Lawson's criterion,**

$$n\tau > 10^{20}\ \text{s/m}^3, \qquad (44\text{-}16)$$

and must have a suitably high plasma temperature T.

In a **tokamak** the plasma is confined by a magnetic field. In **laser fusion** inertial confinement is used.

QUESTIONS

1. In Table 44-1, does the relation $Q = -\Delta m\, c^2$ apply to (a) all the processes, (b) all the processes except the waterfall, (c) the fission processes only, (d) the fission and fusion processes only?

2. According to Fig. 44-1, fission of ^{235}U by thermal neutrons into two equally massive fragments occurs in about one event in (a) 10 000, (b) 1000, (c) 100, (d) 10.

3. Do the initial fragments formed by fission have (a) more protons than neutrons, (b) more neutrons than protons, or (c) about the same number of each?

4. Consider the fission reaction

$$^{235}\text{U} + \text{n} \rightarrow \text{X} + \text{Y} + 2\text{n}.$$

Rank the following possible nuclides for representation by X (or Y), most likely first: (a) ^{152}Nd, (b) ^{140}I, (c) ^{128}In, (d) ^{115}Pd, (e) ^{105}Mo. (*Hint:* See Fig. 44-1.)

5. Pick the most likely member of each of these pairs to be one of the initial fragments formed by a fission event: (a) ^{93}Sr or ^{93}Ru, (b) ^{140}Gd or ^{140}I, (c) ^{155}Nd or ^{155}Lu. (*Hint:* See Fig. 43-4 and the periodic table.)

6. Suppose that a ^{238}U nucleus "swallows" a neutron and then decays not by fission but by beta decay, emitting an electron and a neutrino. Which nuclide remains after this decay: (a) ^{239}Pu, (b) ^{238}Np, (c) ^{239}Np, or (d) ^{238}Pa?

7. A nuclear reactor is operating at a certain power level, with its multiplication factor k adjusted to unity. If the control rods are used to reduce the power output of the reactor to 25% of its former value, is the multiplication factor now (a) a little less than unity, (b) substantially less than unity, or (c) still equal to unity?

8. A nuclear reactor core should have the smallest possible surface-to-volume ratio. Order the following solids according to their surface-to-volume ratios, greatest first: (a) a cube of edge a, (b) a sphere of radius a, (c) a cone of height a and base radius a, and (d) a cylinder of radius a and height a. (The area of the *curved* surface of the cone is $\sqrt{2}\pi a^2$ and its volume is $\pi a^3/3$.)

9. Figure 44-6 shows how the heat generated by the nuclear waste from one year's operation of a large nuclear power plant decays over time. By what approximate factor has this thermal energy output decreased at the end of 100 years: (a) 20, (b) 200, (c) 2000, (d) more than 2000?

10. Which of these elements is *not* "cooked up" by thermonuclear fusion processes in stellar interiors: (a) carbon, (b) silicon, (c) chromium, (d) bromine?

11. About 2% of the energy generated in the Sun's core by the p-p reaction is carried out of the Sun by neutrinos. Is the energy associated with this neutrino flux (a) equal to, (b) greater than, or (c) less than the energy radiated from the Sun's surface as electromagnetic radiation?

12. Lawson's criterion for the d-t reaction (Eq. 44-16) is $n\tau > 10^{20}$ s/m^3. For the d-d reaction, do you expect the number on the right-hand side to be (a) the same, (b) smaller, or (c) larger?

EXERCISES & PROBLEMS

ssm Solution is in the Student Solutions Manual.
www Solution is available on the World Wide Web at: http://www.wiley.com/college/hrw
ilw Solution is available on the Interactive LearningWare.

SEC. 44-2 Nuclear Fission: The Basic Process

1E. (a) How many atoms are contained in 1.0 kg of pure ^{235}U? (b) How much energy, in joules, is released by the complete fissioning of 1.0 kg of ^{235}U? Assume Q = 200 MeV. (c) For how long would this energy light a 100 W lamp? ssm

2E. Complete the following table, which refers to the generalized fission reaction $^{235}\text{U} + \text{n} \rightarrow \text{X} + \text{Y} + b\text{n}$.

X	Y	b
^{140}Xe	—	1
^{139}I	—	2
—	^{100}Zr	2
^{141}Cs	^{92}Rb	—

3E. At what rate must ^{235}U nuclei undergo fission by neutron bombardment to generate energy at the rate of 1.0 W? Assume that Q = 200 MeV. ssm

4E. The fission properties of the plutonium isotope ^{239}Pu are very similar to those of ^{235}U. The average energy released per fission is 180 MeV. How much energy, in MeV, is released if all the atoms in 1.00 kg of pure ^{239}Pu undergo fission?

5E. Verify that, as stated in Section 44-2, neutrons in equilibrium with matter at room temperature, 300 K, have an average kinetic energy of about 0.04 eV.

6E. Calculate the disintegration energy Q for the fission of ^{98}Mo into two equal parts. The masses you will need are 97.905 41 u for ^{98}Mo and 48.950 02 u for ^{49}Sc. If Q turns out to be positive, discuss why this process does not occur spontaneously.

7E. Calculate the disintegration energy Q for the fission of ^{52}Cr into two equal fragments. The masses you will need are 51.940 51 u for ^{52}Cr and 25.982 59 u for ^{26}Mg. ssm

8E. ^{235}U decays by alpha emission with a half-life of 7.0×10^8 y. It also decays (rarely) by spontaneous fission, and if the alpha decay did not occur, its half-life due to this process alone would be 3.0×10^{17} y. (a) At what rate do spontaneous fission decays occur in 1.0 g of ^{235}U? (b) How many ^{235}U alpha-decay events are there for every spontaneous fission event?

9E. Calculate the energy released in the fission reaction

$$^{235}\text{U} + \text{n} \rightarrow {}^{141}\text{Cs} + {}^{93}\text{Rb} + 2\text{n}.$$

Needed atomic and particle masses are

^{235}U	235.043 92 u	^{93}Rb	92.921 57 u
^{141}Cs	140.919 63 u	n	1.008 67 u

10P. Verify that, as reported in Table 44-1, fissioning of the ^{235}U in 1.0 kg of UO_2 (enriched so that ^{235}U is 3.0% of the total uranium) could keep a 100 W lamp burning for 690 y.

11P. In a particular fission event in which ^{235}U is fissioned by slow neutrons, no neutron is emitted and one of the primary fission fragments is ^{83}Ge. (a) What is the other fragment? (b) How is the disintegration energy Q = 170 MeV split between the two fragments? (c) Calculate the initial speed (just after the fission) of each fragment. ssm www

12P. Consider the fission of ^{238}U by fast neutrons. In one fission event no neutrons are emitted and the final stable end products, after the beta decay of the primary fission fragments, are ^{140}Ce and ^{99}Ru. (a) How many beta-decay events are there in the two beta-decay chains, considered together? (b) Calculate Q for this fission process. The relevant atomic and particle masses are

^{238}U	238.050 79 u	^{140}Ce	139.905 43 u
n	1.008 67 u	^{99}Ru	98.905 94 u

13P. Assume that immediately after the fission of ^{236}U according to Eq. 44-1, the resulting ^{140}Xe and ^{94}Sr nuclei are just touching at their surfaces. (a) Assuming the nuclei to be spherical, calculate the electric potential energy (in MeV) associated with the repulsion between the two fragments. (*Hint:* Use Eq. 43-3 to calculate the radii of the fragments.) (b) Compare this energy with the energy released in a typical fission event. ssm

14P. A ^{236}U nucleus undergoes fission and breaks into two middle-mass fragments, ^{140}Xe and ^{96}Sr. (a) By what percentage does the surface area of the fission products differ from that of the original ^{236}U nucleus? (b) By what percentage does the volume change? (c) By what percentage does the electric potential energy change? The electric potential energy of a uniformly charged sphere of radius r and charge Q is given by

$$U = \frac{3}{5}\left(\frac{Q^2}{4\pi\varepsilon_0 r}\right).$$

SEC. 44-4 The Nuclear Reactor

15E. A 200 MW fission reactor consumes half its fuel in 3.00 y. How much ^{235}U did it contain initially? Assume that all the energy generated arises from the fission of ^{235}U and that this nuclide is consumed only by the fission process. ssm

16E. Repeat Exercise 15 taking into account nonfission neutron capture by the ^{235}U.

17E. ^{238}Np requires 4.2 MeV for fission. To remove a neutron from this nuclide requires an energy expenditure of 5.0 MeV. Is ^{237}Np fissionable by thermal neutrons?

18P. In an atomic bomb, energy release is due to the uncontrolled fission of plutonium ^{239}Pu (or ^{235}U). The bomb's rating is the magnitude of the released energy, specified in terms of the mass of TNT required to produce the same energy release. One megaton (10^6 tons) of TNT releases 2.6×10^{28} MeV of energy. (a) Calculate the rating, in tons of TNT, of an atomic bomb containing 95 kg of ^{239}Pu, of which 2.5 kg actually undergoes fission. (See Exercise 4.) (b) Why is the other 92.5 kg of ^{239}Pu needed if it does not fission?

19P. The thermal energy generated when radiations from radionuclides are absorbed in matter can serve as the basis for a small power source for use in satellites, remote weather stations, and other isolated locations. Such radionuclides are manufactured in abundance in nuclear reactors and may be separated chemically from the spent fuel. One suitable radionuclide is ^{238}Pu ($T_{1/2}$ = 87.7 y), which is an alpha emitter with Q = 5.50 MeV. At what rate is thermal energy generated in 1.00 kg of this material? ssm

20P. (See Problem 19.) Among the many fission products that may be extracted chemically from the spent fuel of a nuclear reactor is ^{90}Sr ($T_{1/2}$ = 29 y). This isotope is produced in typical large reactors at the rate of about 18 kg/y. By its radioactivity it generates thermal energy at the rate of 0.93 W/g. (a) Calculate the effective disintegration energy Q_{eff} associated with the decay of a ^{90}Sr nucleus. (Q_{eff} includes contributions from the decay of the ^{90}Sr daughter products in its decay chain but not from neutrinos, which escape totally from the sample.) (b) It is desired to construct a power source generating 150 W (electric) to use in operating electronic equipment in an underwater acoustic beacon. If the power source is based on the thermal energy generated by ^{90}Sr and if the efficiency of the thermal–electric conversion process is 5.0%, how much ^{90}Sr is needed?

21P. Many fear that helping additional nations develop nuclear power reactor technology will increase the likelihood of nuclear war because reactors can be used not only to produce electrical energy but, as a by-product through neutron capture with inexpensive ^{238}U, to make ^{239}Pu, which is a "fuel" for nuclear bombs. What simple series of reactions involving neutron capture and beta decay would yield this plutonium isotope?

22P. The neutron generation time t_{gen} in a reactor is the average time needed for a fast neutron emitted in one fission to be slowed to thermal energies by the moderator and to initiate another fission. Suppose that the power output of a reactor at time $t = 0$ is P_0. Show that the power output a time t later is $P(t)$, where

$$P(t) = P_0 k^{t/t_{\text{gen}}},$$

and k is the multiplication factor. For constant power output, $k = 1$.

23P. A 66 kiloton atomic bomb (see Problem 18) is fueled with pure ^{235}U (Fig. 44-15), 4.0% of which actually undergoes fission. (a) How much uranium is in the bomb? (b) How many primary fission fragments are produced? (c) How many neutrons generated in the fissions are released to the environment? (On average, each fission produces 2.5 neutrons.) ssm

Fig. 44-15 Problem 23. A "button" of ^{235}U, ready to be recast and machined for a warhead.

24P. The neutron generation time t_{gen} (see Problem 22) in a particular reactor is 1.0 ms. If the reactor is operating at a power level of 500 MW, about how many free neutrons are present in the reactor at any moment?

25P. The neutron generation time (see Problem 22) of a particular reactor is 1.3 ms. The reactor is generating energy at the rate of 1200 MW. To perform certain maintenance checks, the power level must temporarily be reduced to 350 MW. It is desired that the transition to the reduced power level take 2.6 s. To what (constant) value should the multiplication factor be set to effect the transition in the desired time? ssm www

26P. A reactor operates at 400 MW with a neutron generation time (see Problem 22) of 30.0 ms. If its power increases for 5.00 min with a multiplication factor of 1.0003, what is the power output at the end of the 5.00 min?

27P. (a) A neutron of mass m_n and kinetic energy K makes a head-on elastic collision with a stationary atom of mass m. Show that the fractional kinetic energy loss of the neutron is given by

$$\frac{\Delta K}{K} = \frac{4m_n m}{(m + m_n)^2}.$$

(b) Find $\Delta K/K$ for each of the following, acting as the stationary atom: hydrogen, deuterium, carbon, and lead. (c) If $K = 1.00$ MeV initially, how many such head-on collisions would it take to reduce the neutron's kinetic energy to a thermal value (0.025 eV) if the stationary atoms it collides with are deuterium, a commonly used moderator? (In actual moderators, most collisions are not head-on.) ssm

SEC. 44-5 A Natural Nuclear Reactor

28E. How long ago was the ratio $^{235}U/^{238}U$ in natural uranium deposits equal to 0.15?

29E. The natural fission reactor discussed in Section 44-5 is estimated to have generated 15 gigawatt-years of energy during its lifetime. (a) If the reactor lasted for 200 000 y, at what average power level did it operate? (b) How many kilograms of ^{235}U did it consume during its lifetime?

30P. Some uranium samples from the natural reactor site described in Section 44-5 were found to be slightly *enriched* in ^{235}U, rather than depleted. Account for this in terms of neutron absorption by the abundant isotope ^{238}U and the subsequent beta and alpha decay of its products.

31P. Mixed in with ^{238}U, uranium mined today contains 0.72% of fissionable ^{235}U, too little to make reactor fuel for thermal-neutron fission. For this reason, the natural uranium must be enriched with ^{235}U. Both ^{235}U ($T_{1/2} = 7.0 \times 10^8$ y) and ^{238}U ($T_{1/2} = 4.5 \times 10^9$ y) are radioactive. How far back in time would natural uranium have been a practical reactor fuel, with a $^{235}U/^{238}U$ ratio of 3.0%? ssm www

SEC. 44-6 Thermonuclear Fusion: The Basic Process

32E. From information given in the text, collect and write down the approximate heights of the Coulomb barriers for (a) the alpha decay of ^{238}U and (b) the fission of ^{235}U by thermal neutrons.

33E. Calculate the height of the Coulomb barrier for the head-on collision of two deuterons. Take the effective radius of a deuteron to be 2.1 fm. ssm

34E. Verify that the fusion of 1.0 kg of deuterium by the reaction

$$^2H + {^2H} \rightarrow {^3He} + n \qquad (Q = +3.27 \text{ MeV})$$

could keep a 100 W lamp burning for 3×10^4 y.

35E. Methods other than heating the material have been suggested for overcoming the Coulomb barrier for fusion. For example, one might consider particle accelerators. If you were to use two of them to accelerate two beams of deuterons directly toward each other so as to collide head-on, (a) what voltage would each accelerator require for the colliding deuterons to overcome the Coulomb barrier? (b) Why do you suppose this method is not presently used?

36P. Calculate the Coulomb barrier height for two ^7Li nuclei that are fired at each other with the same initial kinetic energy K. (*Hint:* Use Eq. 43-3 to calculate the radii of the nuclei.)

37P. In Fig. 44-10, the equation for $n(K)$, the number density per unit energy for particles, is

$$n(K) = 1.13n \frac{K^{1/2}}{(kT)^{3/2}} e^{-K/kT},$$

where n is the total number density of particles. At the center of the Sun the temperature is 1.50×10^7 K and the average proton energy K_{avg} is 1.94 keV. Find the ratio of the number density of protons at 5.00 keV to that at the mean proton energy.

38P. Expressions for the Maxwell speed distribution for molecules in a gas are given in Chapter 20. (a) Show that the *most probable energy* is given by

$$K_p = \tfrac{1}{2}kT.$$

Verify this result with the energy distribution curve of Fig. 44-10, for which $T = 1.5 \times 10^7$ K. (b) Show that the *most probable speed* is given by

$$v_p = \sqrt{\frac{2kT}{m}}.$$

Find its value for protons at $T = 1.5 \times 10^7$ K. (c) Show that the *energy corresponding to the most probable speed* (which is not the same as the most probable energy) is

$$K_{v,p} = kT.$$

Locate this quantity on the curve of Fig. 44-10.

SEC. 44-7 Thermonuclear Fusion in the Sun and Other Stars

39E. Show that the energy released when three alpha particles fuse to form ^{12}C is 7.27 MeV. The atomic mass of ^4He is 4.0026 u, and that of ^{12}C is 12.0000 u. ssm

40E. We have seen that Q for the overall proton-proton fusion cycle is 26.7 MeV. How can you relate this number to the Q values for the reactions that make up this cycle, as displayed in Fig. 44-11?

41E. At the center of the Sun the density is 1.5×10^5 kg/m^3 and the composition is essentially 35% hydrogen by mass and 65% helium. (a) What is the density of protons at the center of the Sun? (b) How much greater is this than the density of particles in an ideal gas at standard conditions of temperature (0°C) and pressure (1.01×10^5 Pa)?

42P. Verify the three Q values reported for the reactions given in Fig. 44-11. The needed atomic and particle masses are

^1H	1.007 825 u	^4He	4.002 603 u
^2H	2.014 102 u	$e^{\pm}$	0.000 548 6 u
^3He	3.016 029 u		

(*Hint:* Distinguish carefully between atomic and nuclear masses, and take the positrons properly into account.)

43P. The Sun has a mass of 2.0×10^{30} kg and radiates energy at the rate of 3.9×10^{26} W. (a) At what rate does the Sun transfer its mass to other forms of energy? (b) What fraction of its original mass has the Sun lost in this way since it began to burn hydrogen, about 4.5×10^9 y ago? ssm

44P. Calculate and compare the energy released by (a) the fusion of 1.0 kg of hydrogen deep within the Sun and (b) the fission of 1.0 kg of ^{235}U in a fission reactor.

45P. (a) Calculate the rate at which the Sun generates neutrinos. Assume that energy production is entirely by the proton-proton fusion cycle. (b) At what rate do solar neutrinos reach Earth?

46P. In certain stars the *carbon cycle* is more likely than the proton-proton cycle to be effective in generating energy. This cycle is

$$
\begin{aligned}
{}^{12}\text{C} + {}^{1}\text{H} &\rightarrow {}^{13}\text{N} + \gamma, && Q_1 = 1.95 \text{ MeV},\\
{}^{13}\text{N} &\rightarrow {}^{13}\text{C} + e^+ + \nu, && Q_2 = 1.19,\\
{}^{13}\text{C} + {}^{1}\text{H} &\rightarrow {}^{14}\text{N} + \gamma, && Q_3 = 7.55,\\
{}^{14}\text{N} + {}^{1}\text{H} &\rightarrow {}^{15}\text{O} + \gamma, && Q_4 = 7.30,\\
{}^{15}\text{O} &\rightarrow {}^{15}\text{N} + e^+ + \nu, && Q_5 = 1.73,\\
{}^{15}\text{N} + {}^{1}\text{H} &\rightarrow {}^{12}\text{C} + {}^{4}\text{He}, && Q_6 = 4.97.
\end{aligned}
$$

(a) Show that this cycle of reactions is exactly equivalent in its overall effects to the proton-proton cycle of Fig. 44-11. (b) Verify that the two cycles, as expected, have the same Q value.

47P. Coal burns according to the reaction $C + O_2 \rightarrow CO_2$. The heat of combustion is 3.3×10^7 J/kg of atomic carbon consumed. (a) Express this in terms of energy per carbon atom. (b) Express it in terms of energy per kilogram of the initial reactants, carbon and oxygen. (c) Suppose that the Sun (mass = 2.0×10^{30} kg) were made of carbon and oxygen in combustible proportions and that it continued to radiate energy at its present rate of 3.9×10^{26} W. How long would the Sun last? ssm

48P. Assume that the core of the Sun has one-eighth of the Sun's mass and is compressed within a sphere whose radius is one-fourth of the solar radius. Assume further that the composition of the core is 35% hydrogen by mass and that essentially all the Sun's energy is generated there. If the Sun continues to burn hydrogen at the rate calculated in Sample Problem 44-5, how long will it be before the hydrogen is entirely consumed? The Sun's mass is 2.0×10^{30} kg.

49P. A star converts all its hydrogen to helium, achieving a 100% helium composition. Next it converts the helium to carbon via the triple-alpha process,

$$^{4}\text{He} + {}^{4}\text{He} + {}^{4}\text{He} \rightarrow {}^{12}\text{C} + 7.27 \text{ MeV}.$$

The mass of the star is 4.6×10^{32} kg, and it generates energy at the rate of 5.3×10^{30} W. How long will it take to convert all the helium to carbon at this rate? ssm

50P. The effective Q for the proton-proton cycle of Fig. 44-11 is 26.2 MeV. (a) Express this as energy per kilogram of hydrogen consumed. (b) The power of the Sun is 3.9×10^{26} W. If its energy derives from the proton-proton cycle, at what rate is it losing hydrogen? (c) At what rate is it losing mass? Account for the difference in the results for (b) and (c). (d) The mass of the Sun is 2.0×10^{30} kg. If it loses mass at the constant rate calculated in (c), how long will it take to lose 0.10% of its mass?

51P. Figure 44-16 shows an early proposal for a hydrogen bomb. The fusion fuel is deuterium, ^{2}H. The high temperature and particle density needed for fusion are provided by an atomic bomb "trigger," which involves a ^{235}U or ^{239}Pu fission fuel that is arranged to impress an imploding, compressive shock wave on the deuterium. The operative fusion reaction is

$$5\,{}^{2}\text{H} \rightarrow {}^{3}\text{He} + {}^{4}\text{He} + {}^{1}\text{H} + 2\text{n}.$$

(a) Calculate Q for the fusion reaction. For needed atomic masses, see Problem 42. (b) Calculate the rating (see Problem 18) of the fusion part of the bomb if it contains 500 kg of deuterium, 30.0% of which undergoes fusion.

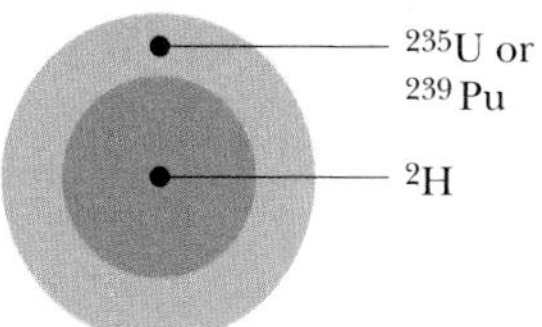

Fig. 44-16 Problem 51.

SEC. 44-8 Controlled Thermonuclear Fusion

52E. Verify the Q values reported in Eqs. 44-13, 44-14, and 44-15. The needed masses are

^{1}H	1.007 825 u	^{4}He	4.002 603 u
^{2}H	2.014 102 u	n	1.008 665 u
^{3}H	3.016 049 u		

53P. Ordinary water consists of roughly 0.0150% by mass of "heavy water," in which one of the two hydrogens is replaced with deuterium, ^{2}H. How much average fusion power could be obtained if we "burned" all the ^{2}H in 1 liter of water in 1 day by somehow causing the deuterium to fuse via the reaction $^2\text{H} + {}^2\text{H} \rightarrow {}^3\text{He} + \text{n}$? ssm

54P. In the deuteron-triton fusion reaction of Eq. 44-15, how is the reaction energy Q shared between the alpha particle and the neutron? Neglect the relatively small kinetic energies of the two combining particles.

45 Quarks, Leptons, and the Big Bang

This color-coded image is effectively a photograph of the universe when it was only 300 000 y old, which was about 15×10^9 y ago. This is what you would have seen then as you looked away in all directions (the view has been condensed to this oval). Patches of light from atoms stretch across the "sky," but galaxies, stars, and planets have not yet formed.

How can such a photograph of the early universe be taken?

The answer is in this chapter.

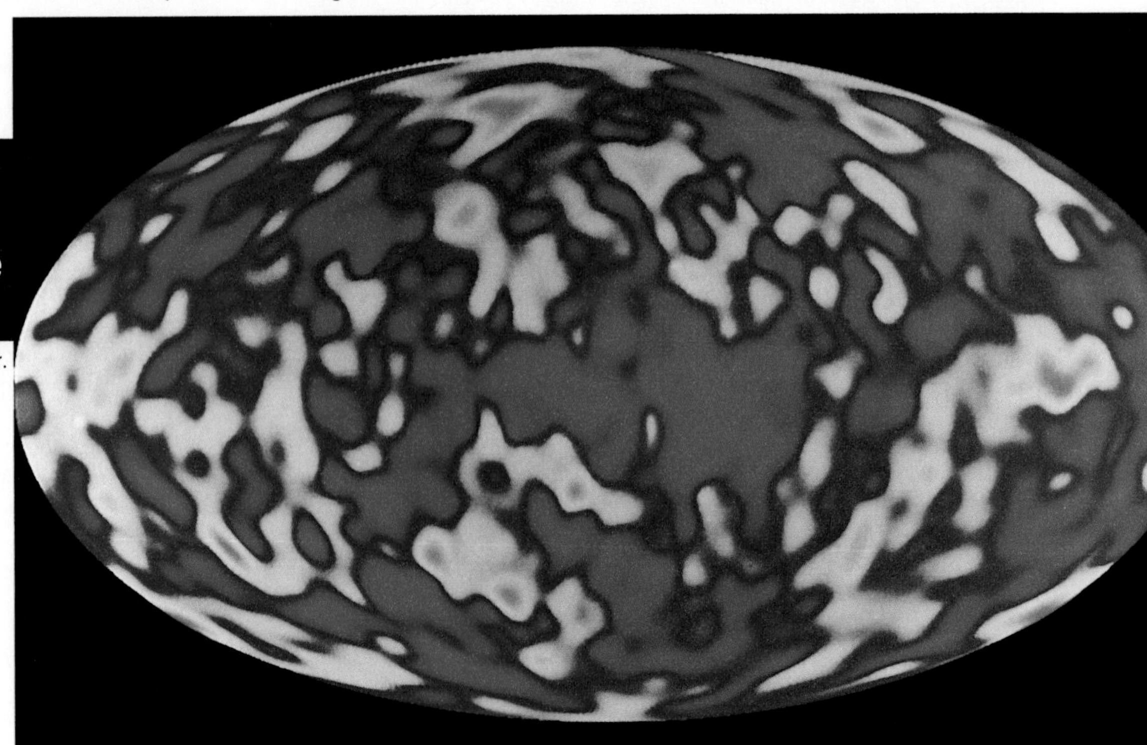

45-1 Life at the Cutting Edge

Physicists often refer to the theories of relativity and quantum physics as "modern physics," to distinguish them from the theories of Newtonian mechanics and Maxwellian electromagnetism, which are lumped together as "classical physics." As the years go by, the word "modern" seems less and less appropriate for theories whose foundations were laid down in the opening years of the twentieth century. Nevertheless, the label hangs on.

In this closing chapter we consider two lines of investigation that are truly "modern" but at the same time have the most ancient of roots. They center around two deceptively simple questions:

What is the universe made of?

How did the universe come to be the way it is?

Progress in answering these questions has been rapid in the last few decades.

Many new insights are based on experiments carried out with large particle accelerators. However, as physicists bang particles together at higher and higher energies using larger and larger accelerators, they come to realize that no conceivable Earth-bound accelerator can generate particles with energies great enough to test their ultimate theories. There has been only one source of particles with these energies, and that was the universe itself within the first millisecond of its existence.

In this chapter you will encounter a host of new terms and a veritable flood of particles with names that you should not try to remember. If you are temporarily bewildered, you are sharing the bewilderment of the physicists who lived through these developments and who at times saw nothing but increasing complexity with little hope of understanding. If you stick with it, however, you will come to share the excitement physicists felt as marvelous new accelerators poured out new results, as the theorists put forth ideas each more daring than the last, and as clarity finally sprang from obscurity.

45-2 Particles, Particles, Particles

In the 1930s, there were many scientists who thought that the problem of the ultimate structure of matter was well on the way to being solved. The atom could be understood in terms of only three particles—the electron, the proton, and the neutron. Quantum physics accounted well for the structure of the atom and for radioactive alpha decay. The neutrino had been postulated and, although not yet observed, had been incorporated by Enrico Fermi into a successful theory of beta decay. There was hope that quantum theory, applied to protons and neutrons, would soon account for the structure of the nucleus. What else was there?

The euphoria did not last. The end of that same decade saw the beginning of a period of discovery of new particles that continues to this day. The new particles have names and symbols such as *muon* (μ), *pion* (π), *kaon* (K), and *sigma* (Σ). All the new particles are unstable; that is, they spontaneously transform into other types of particles according to the same functions of time that apply to unstable nuclei. Thus, if N_0 particles of any one type are present in a sample at time $t = 0$, then the number N of those particles present at some later time t is given by Eq. 43-14,

$$N = N_0 e^{-\lambda t}; \tag{45-1}$$

the rate of decay R, from an initial value of R_0, is given by Eq. 43-15,

$$R = R_0 e^{-\lambda t}; \tag{45-2}$$

Fig. 45-1 The OPAL (omni-purpose apparatus) particle detector at CERN, the European high-energy particle physics laboratory near Geneva, Switzerland. OPAL is designed to measure the energies of particles produced in electron–positron collisions at energies of 50 GeV. Although the detector is huge, it is small compared with the accelerator itself, which is a ring with a circumference of 27 km.

and the half-life $T_{1/2}$, decay constant λ, and mean life τ are related by Eq. 43-17,

$$T_{1/2} = \frac{\ln 2}{\lambda} = \tau \ln 2. \tag{45-3}$$

The half-lives of the new particles range from about 10^{-6} s to 10^{-23} s. Indeed, some of the particles last so briefly that they cannot be detected directly but can only be inferred from indirect evidence.

These new particles are commonly produced in head-on collisions between protons or electrons accelerated to high energies in accelerators at places like Fermilab (near Chicago), CERN (near Geneva), SLAC (at Stanford), and DESY (near Hamburg, Germany). They are discovered with particle detectors that have grown in sophistication until (see Fig. 45-1) they rival the size and complexity of entire accelerators of only a few decades ago.

Today there are several hundred known particles. Naming them has strained the resources of the Greek alphabet, and most are known only by an assigned number in a periodically issued compilation. To make sense of this array of particles, we look for simple physical criteria. We can make a first rough cut among the particles in at least the following three ways.

Fermion or Boson?

All particles have an intrinsic angular momentum called **spin,** as we discussed for electrons, protons, and neutrons in Section 32-4. Generalizing the notation of that section, we can write the component of spin $\vec{S}$ in any direction (assume it to be along a z axis) as

$$S_z = m_s\hbar \qquad \text{for } m_s = s, s-1, \ldots, -s, \tag{45-4}$$

in which $\hbar$ is $h/2\pi$, m_s is the *spin magnetic quantum number,* and s is the *spin quantum number.* The latter can have either positive half-integer values ($\frac{1}{2}, \frac{3}{2}, \ldots$) or nonnegative integer values (0, 1, 2, . . .). For example, an electron has the value $s = \frac{1}{2}$. Hence the spin of an electron (measured along any direction) can have the values

$$S_z = \tfrac{1}{2}\hbar \qquad \text{(spin up)}$$

or

$$S_z = -\tfrac{1}{2}\hbar \qquad \text{(spin down).}$$

Confusingly, the term *spin* is actually used in two ways: It properly means a particle's intrinsic angular momentum $\vec{S}$, but it is often used loosely to mean the particle's spin quantum number s. In the latter case, for example, an electron is said to be a spin-$\frac{1}{2}$ particle.

Particles with half-integer spin quantum numbers (like electrons) are called **fermions,** after Fermi, who (simultaneously with Paul Dirac) discovered the statistical laws that govern their behavior. Like electrons, protons and neutrons also have $s = \frac{1}{2}$ and are fermions.

Particles with zero or integer spin quantum numbers are called **bosons,** after Indian physicist Satyendra Nath Bose, who (simultaneously with Albert Einstein) discovered the governing statistical laws for *those* particles. Photons, which have $s = 1$, are bosons; you will soon meet other particles in this class.

This may seem a trivial way to classify particles, but it is very important for this reason:

Fermions obey the Pauli exclusion principle, which asserts that only a single particle can be assigned to a given quantum state. Bosons *do not* obey this principle. Any number of bosons can occupy a given quantum state.

We saw how important the Pauli exclusion principle is when we "built up" the atoms by assigning (spin-$\frac{1}{2}$) electrons to individual quantum states. Using that principle led to a full accounting of the structure and properties of atoms of different types and of solids such as metals and semiconductors.

Because bosons do *not* obey the Pauli principle, those particles tend to pile up in the quantum state of lowest energy. In 1995 a group in Boulder, Colorado, succeeded in producing a condensate of about 2000 rubidium-87 atoms—they are bosons—in a single quantum state of approximately zero energy.

For this to happen, the rubidium has to be a vapor with a temperature so low and a density so great that the de Broglie wavelengths of the individual atoms are greater than the average separation between the atoms. When this condition is met, the wave functions of the individual atoms overlap and the entire assembly becomes a single quantum system, called a *Bose–Einstein condensate.* Figure 45-2 shows that, as the temperature of the rubidium vapor is lowered to about 1.70×10^{-7} K, this system does indeed "collapse" into a single sharply defined state corresponding to approximately zero speed for its atoms.

Hadron or Lepton?

We can also classify particles in terms of the four fundamental forces that act on them. The *gravitational force* acts on *all* particles, but its effects at the level of subatomic particles are so weak that we need not consider that force (at least not in today's research). The *electromagnetic force* acts on all *electrically charged* particles; its effects are well known and we can take them into account when we need to; we largely ignore this force in this chapter.

We are left with the *strong force,* which is the force that binds nucleons together, and the *weak force,* which is involved in beta decay and similar processes. The weak force acts on all particles, the strong force only on some.

We can, then, roughly classify particles on the basis of whether the strong force acts on them. Particles on which the *strong force* acts are called **hadrons.** Particles

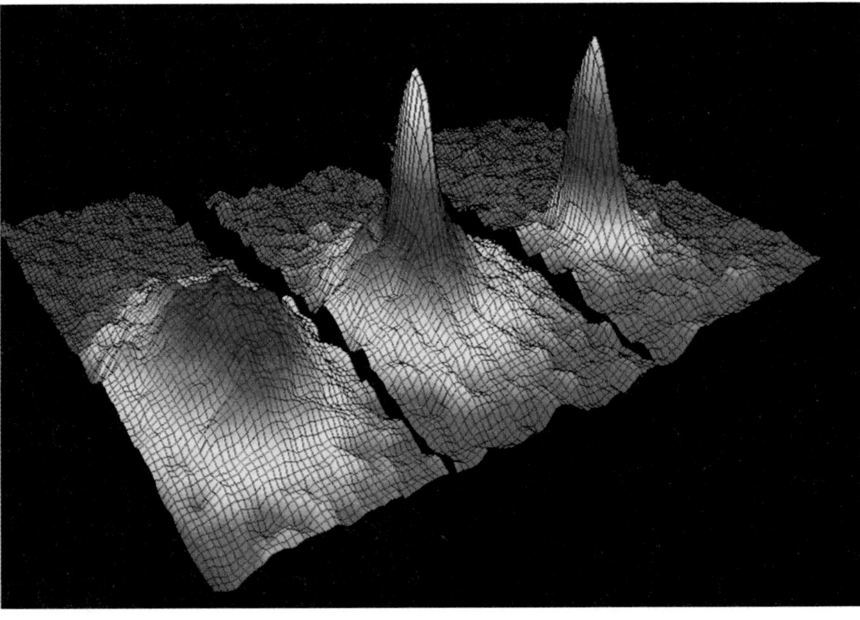

Fig. 45-2 Three plots of the particle speed distribution in a vapor of rubidium-87 atoms. The temperature of the vapor is successively reduced from plot (*a*) to plot (*c*). Plot (*c*) shows a sharp peak centered around zero speed; that is, all the atoms are in the same quantum state. The achievement of such a Bose–Einstein condensate, often called the Holy Grail of atomic physics, was finally recorded in 1995.

on which the strong force does *not* act, leaving the weak force as the dominant force, are called **leptons.** Protons, neutrons, and pions are hadrons; electrons and neutrinos are leptons. You will soon meet other members of these classes.

We can make a further distinction among the hadrons because some of them are bosons (we call them **mesons**); the pion is an example. The other hadrons are fermions (we call them **baryons**); the proton is an example.

Particle or Antiparticle?

In 1928 Dirac predicted that the electron e^- should have a positively charged counterpart of the same mass and spin. The counterpart, the *positron* e^+, was discovered in cosmic radiation in 1932 by Carl Anderson. Physicists then gradually realized that *every* particle has a corresponding **antiparticle.** The members of such pairs have the same mass and spin but opposite signs of electric charge (if they are charged) and opposite signs of quantum numbers that we have not yet discussed.

At first, *particle* was used to refer to the common particles such as electrons, protons, and neutrons, and *antiparticle* referred to their rarely detected counterparts. Later, for the less common particles, the assignment of *particle* and *antiparticle* was made so as to be consistent with certain conservation laws that we shall discuss later in this chapter. (Confusingly, both particles and antiparticles are sometimes called particles when no distinction is needed.) We often, but not always, represent an antiparticle by putting a bar over the symbol for the particle. Thus, p is the symbol for the proton, and $\bar{p}$ (pronounced "p bar") is the symbol for the antiproton.

When a particle meets its antiparticle, the two can *annihilate* each other. That is, the particle and antiparticle disappear and their combined energies reappear in other forms. For an electron annihilating with a positron, this energy reappears as two gamma-ray photons:

$$e^- + e^+ \rightarrow \gamma + \gamma. \tag{45-5}$$

If the electron and positron are stationary when they annihilate, their total energy is their total mass energy, and that energy is then shared equally by the two photons. To conserve momentum and because photons cannot be stationary, the photons fly off in opposite directions.

In 1996 physicists at CERN succeeded in producing, for a few fleeting nanoseconds, a handful of antihydrogen atoms, each consisting of a positron and an antiproton bound together (presumably just as the electron and proton of a hydrogen atom are bound together). Such an assembly of antiparticles is called *antimatter* to distinguish it from an assembly of common particles (*matter*).

One can speculate that there are galaxies of antimatter, complete with atoms, molecules, and even physicists. One can even contemplate the disaster that would occur if, say, an asteroid freed from such a galaxy collided with (hence annihilating) a section of Earth. Thankfully, however, the present view is that not only our galaxy but the universe as a whole consists largely of matter rather than antimatter. (This lack of symmetry is disturbing to a physicist, who normally expects to find symmetry in nature.)

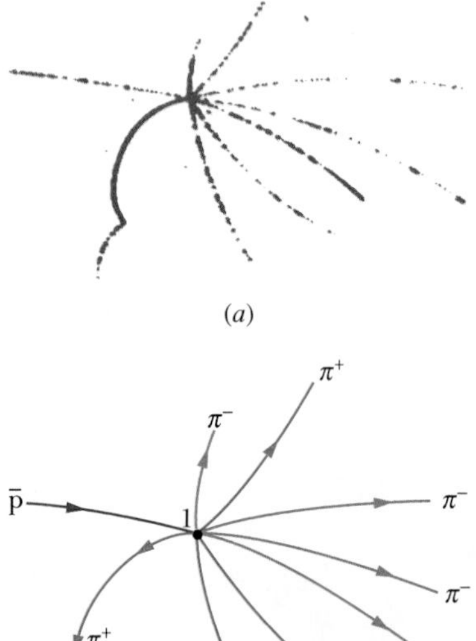

Fig. 45-3 (*a*) A bubble-chamber photograph of a series of events initiated by an antiproton that enters the chamber from the left. (*b*) The tracks redrawn and labeled for clarity. The dots at points 1 and 2 indicate the sites of specific secondary events that are described in the text. The tracks are curved because a magnetic field present in the chamber exerts a deflecting force on each moving charged particle.

45-3 An Interlude

Before pressing on with the task of classifying the particles, let us step aside for a moment and capture some of the spirit of particle research by analyzing a typical particle event—namely, that shown in the bubble-chamber photograph of Fig. 45-3*a*.

The tracks in this figure consist of bubbles formed along the paths of electrically charged particles as they move through a chamber filled with liquid hydrogen. We

TABLE 45-1 **The Particles or Antiparticles Involved in the Event of Fig. 45-3**

Particle	Symbol	Charge q	Mass (MeV/c^2)	Spin s	Identity	Mean Life (s)	Antiparticle
Neutrino	ν	0	0	$\frac{1}{2}$	Lepton	Stable	$\bar{\nu}$
Electron	e^-	-1	0.511	$\frac{1}{2}$	Lepton	Stable	e^+
Muon	μ^-	-1	105.7	$\frac{1}{2}$	Lepton	2.2×10^{-6}	μ^+
Pion	π^+	$+1$	139.6	0	Meson	2.6×10^{-8}	π^-
Proton	p	$+1$	938.3	$\frac{1}{2}$	Baryon	Stable	$\bar{\text{p}}$

can identify the particle that makes a particular track by—among other means—measuring the relative spacing between the bubbles. The chamber lies in a uniform magnetic field that deflects the tracks of positively charged particles counterclockwise and the tracks of negatively charged particles clockwise. By measuring the radius of curvature of a track, we can calculate the momentum of the particle that made it. Table 45-1 shows some properties of the particles and antiparticles that participated in the event of Fig. 45-3*a*, including those that did not make tracks. Following common practice, we express the masses of the particles listed in Table 45-1—and in all other tables in this chapter—in the unit MeV/c^2. The reason for this notation is that the rest energy of a particle is needed more often than its mass. Thus, the mass of a proton is shown in Table 45-1 to be 938.3 MeV/c^2. To find the proton's rest energy, multiply this mass by c^2 to obtain 938.3 MeV.

The general tools used for the analysis of photographs like Fig. 45-3*a* are the laws of conservation of energy, linear momentum, angular momentum, and electric charge, along with other conservation laws that we have not yet discussed. Figure 45-3*a* is actually one of a stereo pair of photographs so that, in practice, these analyses are carried out in three dimensions.

The event of Fig. 45-3*a* is triggered by an energetic antiproton ($\bar{\text{p}}$) that, generated in an accelerator at the Lawrence Berkeley Laboratory, enters the chamber from the left. There are three separate subevents; two occur at points 1 and 2 in Fig. 45-3*b*, and the third occurs out of the frame of the figure. Let's examine each:

1. *Proton-Antiproton Annihilation.* At point 1 in Fig. 45-3*b*, the initiating antiproton (blue track) slams into a proton of the liquid hydrogen in the chamber, and the result is mutual annihilation. We can tell that annihilation occurred while the incoming antiproton was in flight because most of the particles generated in the encounter move in the forward direction—that is, toward the right in Fig. 45-3. From the principle of conservation of linear momentum, the incoming antiproton must have had a forward momentum when it underwent annihilation.

   The total energy involved in the collision of the antiproton and the proton is the sum of the antiproton's kinetic energy and the two (identical) rest energies of those two particles (2×938.3 MeV, or 1876.6 MeV). This is enough energy to create a number of lighter particles and give them kinetic energy. In this case, the annihilation produces four positive pions (red tracks in Fig. 45-3*b*) and four negative pions (green tracks). (For simplicity, we assume that no gamma-ray photons, which would leave no tracks because they lack electric charge, are produced.) Then the annihilation process is

$$\text{p} + \bar{\text{p}} \rightarrow 4\pi^+ + 4\pi^-. \qquad (45\text{-}6)$$

   We see from Table 45-1 that the positive pions (π^+) are *particles* and the negative pions (π^-) are *antiparticles*. The reaction of Eq. 45-6 is a *strong interaction* (it involves the strong force), because all the particles involved are hadrons.

Let us check whether electric charge is conserved in the reaction. To do so, we can write the electric charge of a particle as qe, in which q is a **charge quantum number.** Then determining whether electric charge is conserved in a process amounts to determining whether the initial net charge quantum number is equal to the final net charge quantum number. In the process of Eq. 45-6, the initial net charge number is $1 + (-1)$, or 0, and the final net charge number is $4(1) + 4(-1)$, or 0. Thus, charge *is* conserved.

For the energy balance, note from above that the energy available from the p-$\bar{\text{p}}$ annihilation process is at least the sum of the proton and antiproton rest energies, 1876.6 MeV. The rest energy of a pion is 139.6 MeV, so the rest energies of the eight pions amount to 8×139.6 MeV, or 1116.8 MeV. This leaves at least about 760 MeV to distribute among the eight pions as kinetic energy. Thus, the requirement of energy conservation is easily met.

2. *Pion Decay.* Pions are unstable particles; charged pions decay with a mean life of 2.6×10^{-8} s. At point 2 in Fig. 45-3*b*, one of the positive pions comes to rest in the chamber and decays spontaneously into an antimuon μ^+ (purple track) and a neutrino ν:

$$\pi^+ \rightarrow \mu^+ + \nu. \tag{45-7}$$

The neutrino, being uncharged, leaves no track. Both the antimuon and the neutrino are leptons; that is, they are particles on which the strong force does not act. Thus, the decay process of Eq. 45-7, which is governed by the weak force, is described as a *weak interaction.* The rest energy of an antimuon is 105.7 MeV, so an energy of 139.6 MeV − 105.7 MeV, or 33.9 MeV, is available to share between the antimuon and the neutrino as kinetic energy.

Let us check whether spin angular momentum is conserved by the process of Eq. 45-7. This amounts to determining whether the net component S_z of spin angular momentum along some arbitrary z axis can be conserved by the process. The spin quantum numbers s of the particles in the process are 0 for the pion π^+ and $\frac{1}{2}$ for both the antimuon μ^+ and the neutrino ν. Thus, for π^+, the component S_z must be $0\hbar$, and for μ^+ and ν, it can be either $+\frac{1}{2}\hbar$ or $-\frac{1}{2}\hbar$.

S_z is conserved by the process of Eq. 45-7 if there is *any* way in which the initial S_z ($= 0\hbar$) can be equal to the final net S_z. We see that if one of the products, either μ^+ or ν, has $S_z = +\frac{1}{2}\hbar$ and the other has $S_z = -\frac{1}{2}\hbar$, then their final net value is $0\hbar$. Thus, because S_z can be conserved, the decay process of Eq. 45-7 *can* occur.

From Eq. 45-7, we also see that the net charge is conserved by the process, because before the process the net charge quantum number is $+1$, and after the process it is $+1 + 0 = +1$.

3. *Muon Decay.* Muons (whether μ^- or μ^+) are also unstable, decaying with a mean life of 2.2×10^{-6} s. Although the decay products are not shown in Fig. 45-3*b*, the antimuon produced in the reaction of Eq. 45-7 comes to rest and decays spontaneously according to

$$\mu^+ \rightarrow \text{e}^+ + \nu + \bar{\nu}. \tag{45-8}$$

The rest energy of the antimuon is 105.7 MeV and that of the positron is only 0.511 MeV, leaving 105.2 MeV to be shared as kinetic energy among the three particles produced in the decay process of Eq. 45-8.

You may wonder: Why *two* neutrinos in Eq. 45-8? Why not just one, as in the pion decay in Eq. 45-7? One answer is that the spin quantum numbers of the antimuon, the positron, and the neutrino are each $\frac{1}{2}$; with only one neutrino, the net component S_z of spin angular momentum could not be conserved in the antimuon decay of Eq. 45-8. In Section 45-4 we shall discuss another reason.

Sample Problem 45-1

A stationary positive pion can decay according to

$$\pi^+ \rightarrow \mu^+ + \nu.$$

What is the kinetic energy of the antimuon μ^+? What is the kinetic energy of the neutrino?

SOLUTION: The Key Idea here is that the pion decay process must conserve both total energy and total linear momentum. Let us first write the conservation of total energy (rest energy mc^2 plus kinetic energy K) for the decay process as

$$m_\pi c^2 + K_\pi = m_\mu c^2 + K_\mu + m_\nu c^2 + K_\nu.$$

Because the pion was stationary, its kinetic energy K_π is zero. Then, using the masses listed for m_π, m_μ, and m_ν in Table 45-1, we find

$$\begin{aligned} K_\mu + K_\nu &= m_\pi c^2 - m_\mu c^2 - m_\nu c^2 \\ &= 139.6 \text{ MeV} - 105.7 \text{ MeV} - 0 \\ &= 33.9 \text{ MeV}. \end{aligned} \tag{45-9}$$

We cannot solve Eq. 45-9 for either K_μ or K_ν separately, so let us next apply the principle of conservation of linear momentum to the decay process. Because the pion is stationary when it decays, that principle requires that the muon and neutrino move in opposite directions after the decay. Assume that their motion is along an axis. Then, for components along that axis, we can write the conservation of linear momentum for the decay as

$$p_\pi = p_\mu + p_\nu,$$

which, with $p_\pi = 0$, gives us

$$p_\mu = -p_\nu. \tag{45-10}$$

We want to relate these momenta p_μ and $-p_\nu$ to the kinetic energies K_μ and K_ν so that we can solve for the kinetic energies. Because we have no reason to believe that classical physics can be applied to the motion of the muon and neutrino, we use Eq. 38-51, the momentum–kinetic energy relation from special relativity:

$$(pc)^2 = K^2 + 2Kmc^2. \tag{45-11}$$

From Eq. 45-10, we know that

$$(p_\mu c)^2 = (p_\nu c)^2. \tag{45-12}$$

Substituting from Eq. 45-11 for the left side and then for the right side of Eq. 45-12 yields

$$K_\mu^2 + 2K_\mu m_\mu c^2 = K_\nu^2 + 2K_\nu m_\nu c^2.$$

Setting the neutrino mass $m_\nu = 0$ (as assumed in Table 45-1), substituting $K_\nu = 33.9$ MeV $- K_\mu$ from Eq. 45-9, and then solving for K_μ, we find

$$\begin{aligned} K_\mu &= \frac{(33.9 \text{ MeV})^2}{(2)(33.9 \text{ MeV} + m_\mu c^2)} \\ &= \frac{(33.9 \text{ MeV})^2}{(2)(33.9 \text{ MeV} + 105.7 \text{ MeV})} \\ &= 4.12 \text{ MeV}. \end{aligned} \qquad \text{(Answer)}$$

The kinetic energy of the neutrino is then, from Eq. 45-9,

$$\begin{aligned} K_\nu &= 33.9 \text{ MeV} - K_\mu = 33.9 \text{ MeV} - 4.12 \text{ MeV} \\ &= 29.8 \text{ MeV}. \end{aligned} \qquad \text{(Answer)}$$

We see that, although the magnitudes of the momenta of the two recoiling particles are the same, the neutrino gets the larger share (88%) of the kinetic energy.

Sample Problem 45-2

Stationary protons in a bubble chamber are bombarded by energetic negative pions, and the following reaction occurs:

$$\pi^- + \text{p} \rightarrow \text{K}^- + \Sigma^+.$$

The rest energies of these particles are

π^-	139.6 MeV	K^-	493.7 MeV
p	938.3 MeV	Σ^+	1189.4 MeV

What is the Q of the reaction?

SOLUTION: The Key Idea here is that the Q of a reaction is

$$Q = \begin{pmatrix}\text{initial total}\\ \text{mass energy}\end{pmatrix} - \begin{pmatrix}\text{final total}\\ \text{mass energy}\end{pmatrix}.$$

For the given reaction, we find

$$\begin{aligned} Q &= (m_\pi c^2 + m_\text{p} c^2) - (m_\text{K} c^2 + m_\Sigma c^2) \\ &= (139.6 \text{ MeV} + 938.3 \text{ MeV}) \\ &\quad -(493.7 \text{ MeV} + 1189.4 \text{ MeV}) \\ &= -605 \text{ MeV}. \end{aligned} \qquad \text{(Answer)}$$

The minus sign means that the reaction is *endothermic;* that is, the incoming pion (π^-) must have a kinetic energy greater than a certain threshold value to cause the reaction to occur. The threshold energy is greater than 605 MeV because linear momentum must be conserved, which means that the kaon (K^-) and the sigma (Σ^+) must not only be created but also must be given some kinetic energy. A relativistic calculation whose details are beyond our scope shows that the threshold energy for the reaction is 907 MeV.

45-4 The Leptons

In this and the next section, we discuss some of the particles of one of our classification schemes: lepton or hadron. We begin with the leptons, those particles on which the strong force does *not* act. So far, we have encountered, as leptons, the familiar electron and the neutrino that accompanies it in beta decay. The muon, whose decay is described in Eq. 45-8, is another member of this family. Physicists gradually learned that the neutrino that appears in Eq. 45-7, associated with the production of a muon, is *not the same particle* as the neutrino produced in beta decay, associated with the appearance of an electron. We call the former the **muon neutrino** (symbol ν_μ) and the latter the **electron neutrino** (symbol ν_e) when it is necessary to distinguish between them.

These two types of neutrino are known to be different particles because, if a beam of muon neutrinos (produced from pion decay as in Eq. 45-7) strikes a solid target, *only muons*—and never electrons—are produced. On the other hand, if electron neutrinos (produced by the beta decay of fission products in a nuclear reactor) strike a solid target, *only electrons*—and never muons—are produced.

Another lepton, the **tau,** was discovered at SLAC in 1975; its discoverer, Martin Perl, shared the 1995 Nobel prize in physics. The tau has its own associated neutrino, different still from the other two. Table 45-2 lists all the leptons (both particles and antiparticles); all have a spin quantum number s of $\frac{1}{2}$.

There are reasons for dividing the leptons into three families, each consisting of a particle (electron, muon, or tau), its associated neutrino, and the corresponding antiparticles. Furthermore, there are reasons to believe that there are *only* the three families of leptons shown in Table 45-2. Leptons have no internal structure and no measurable dimensions; they are believed to be truly pointlike fundamental particles when they interact with other particles or with electromagnetic waves.

The Conservation of Lepton Number

According to experiment, particle interactions involving leptons obey a certain conservation law for a quantum number called the **lepton number** L. Each (normal) particle in Table 45-2 is assigned $L = +1$ and each antiparticle is assigned $L = -1$. All other particles, which are not leptons, are assigned $L = 0$. Also according to experiment,

> In all particle interactions, the net lepton number *for each family* is separately conserved.

Thus, there are actually three lepton numbers L_e, L_μ, L_ν, and the net of *each* must

TABLE 45-2 **The Leptons**[a]

Family	Particle	Symbol	Mass (MeV/c^2)	Charge q	Antiparticle
Electron	Electron	e^-	0.511	-1	e^+
	Electron neutrino[b]	ν_e	0	0	$\bar{\nu}_e$
Muon	Muon	μ^-	105.7	-1	μ^+
	Muon neutrino[b]	ν_μ	0	0	$\bar{\nu}_\mu$
Tau	Tau	τ^-	1777	-1	τ^+
	Tau neutrino[b]	ν_τ	0	0	$\bar{\nu}_\tau$

[a]All leptons have a spin quantum number of $\frac{1}{2}$ and are thus fermions.

[b]If the neutrino masses are not zero, they are very small. As of 2000, their values have not been well determined.

remain unchanged during any particle interaction. This experimental fact is called the law of **conservation of lepton number.**

We can illustrate this law by reconsidering the antimuon decay process shown in Eq. 45-8, which we now write more fully as

$$\mu^+ \rightarrow e^+ + \nu_e + \bar{\nu}_\mu. \qquad (45\text{-}13)$$

Consider this first in terms of the muon family of leptons. The μ^+ is an antiparticle (see Table 45-2) and thus has the muon lepton number $L_\mu = -1$. The two particles e^+ and ν_e do not belong to the muon family and thus have $L_\mu = 0$. This leaves $\bar{\nu}_\mu$ on the right which, being an antiparticle, also has the muon lepton number $L_\mu = -1$. Thus, both sides of Eq. 45-13 have the same net muon lepton number—namely, $L_\mu = -1$; if they did not, the μ^+ would not decay by this process.

No members of the electron family appear on the left in Eq. 45-13, so there the net electron lepton number must be $L_e = 0$. On the right side of Eq. 45-13, the positron, being an antiparticle (again see Table 45-2), has the electron lepton number $L_e = -1$. The electron neutrino ν_e, being a particle, has the electron number $L_e = +1$. Thus, the net electron lepton number for these two particles on the right in Eq. 45-13 is also zero; the electron lepton number is also conserved in the process.

No members of the tau family appear on either side of Eq. 45-13, so we must have $L_\tau = 0$ on each side. Thus, each of the lepton quantum numbers L_μ, L_e, and L_τ remains unchanged during the decay process of Eq. 45-13, their constant values being -1, 0, and 0, respectively. This example is but one illustration of the conservation of lepton number; it holds for all particle interactions.

CHECKPOINT 1: (a) The π^+ meson decays by the process $\pi^+ \rightarrow \mu^+ + \nu$. To what lepton family does the neutrino ν belong? (b) Is this neutrino a particle or an antiparticle? (c) What is its lepton number?

45-5 The Hadrons

We are now ready to consider hadrons (baryons and mesons), those particles whose interactions are governed by the strong force. We start by adding another conservation law to our list: conservation of baryon number.

To develop this conservation law, let us consider the proton decay process

$$p \rightarrow e^+ + \nu_e. \qquad (45\text{-}14)$$

This process *never* happens. We should be glad that it does not because otherwise all protons in the universe would gradually change into positrons, with disastrous consequences. Yet this decay process does not violate the conservation laws involving energy, linear momentum, or lepton number.

We account for the apparent stability of the proton—and for the absence of many other processes that might otherwise occur—by introducing a new quantum number, the **baryon number** B, and a new conservation law, the **conservation of baryon number:**

> To every baryon we assign $B = +1$. To every antibaryon we assign $B = -1$. To all particles of other types we assign $B = 0$. A particle process cannot occur if it changes the net baryon number.

In the process of Eq. 45-14, the proton has a baryon number of $B = +1$ and the positron and neutrino both have a baryon number of $B = 0$. Thus, the process does not conserve baryon number and cannot occur.

CHECKPOINT 2: This mode of decay for a neutron is *not* observed:

$$n \rightarrow p + e^-.$$

Which of the following conservation laws does this process violate: (a) energy, (b) angular momentum, (c) linear momentum, (d) charge, (e) lepton number, (f) baryon number? The masses are $m_n = 939.6\text{ MeV}/c^2$, $m_p = 938.3\text{ MeV}/c^2$, and $m_e = 0.511\text{ MeV}/c^2$.

Sample Problem 45-3

Determine whether a stationary proton can decay according to the proposed scheme

$$p \rightarrow \pi^0 + \pi^+.$$

Properties of the proton and the π^+ pion are listed in Table 45-1. The π^0 pion has zero charge, zero spin, and a mass energy of 135.0 MeV.

SOLUTION: The Key Idea here is to see whether the proposed decay violates any of the conservation laws we have discussed. For electric charge, we see that the net charge quantum number is initially +1 and finally is 0 + 1, or +1. Thus, charge is conserved by the decay. Lepton number is also conserved, because none of the three particles is a lepton and thus their lepton numbers are all zero.

Linear momentum can also be conserved: Because the proton is stationary, with zero linear momentum, the two pions must merely move in opposite directions with equal magnitudes of linear momentum (so that their total linear momentum is also zero) to conserve linear momentum. The fact that linear momentum *can* be conserved means that the process does not violate the conservation of linear momentum.

Is there energy for the decay? Because the proton is stationary, that question amounts to asking whether the proton's mass energy is sufficient to produce the mass energies and kinetic energies of the pions. To answer, we evaluate the Q of the decay:

$$\begin{aligned} Q &= \begin{pmatrix}\text{initial total}\\ \text{mass energy}\end{pmatrix} - \begin{pmatrix}\text{final total}\\ \text{mass energy}\end{pmatrix} \\ &= m_p c^2 - (m_0 c^2 + m_+ c^2) \\ &= 938.3\text{ MeV} - (135.0\text{ MeV} + 139.6\text{ MeV}) \\ &= 663.7\text{ MeV}. \end{aligned}$$

The fact that Q is positive indicates that the initial mass energy exceeds the final mass energy. Thus, the proton *does* have enough mass energy to create the pair of pions.

Is spin angular momentum conserved by the decay? This amounts to determining whether the net component S_z of spin angular momentum along some arbitrary z axis can be conserved by the decay. The spin quantum numbers s of the particles in the process are $\frac{1}{2}$ for the proton and 0 for both pions. Thus, for the proton the component S_z can be either $+\frac{1}{2}\hbar$ or $-\frac{1}{2}\hbar$ and for each pion it is $0\hbar$. We see that there is no way that S_z can be conserved. Hence, spin angular momentum is not conserved and the proposed decay of the proton cannot occur.

The decay also violates the conservation of baryon number: The proton has a baryon number of $B = +1$ and both pions have a baryon number of $B = 0$. Thus, nonconservation of baryon number is another reason why the proposed decay cannot occur.

Sample Problem 45-4

A particle called xi-minus and having the symbol Ξ^- decays as follows:

$$\Xi^- \rightarrow \Lambda^0 + \pi^-.$$

The Λ^0 particle (called lambda-zero) and the π^- particle are both unstable. The following decay processes occur in *cascade* until only relatively stable products remain:

$$\Lambda^0 \rightarrow p + \pi^- \qquad \pi^- \rightarrow \mu^- + \bar{\nu}_\mu \qquad \mu^- \rightarrow e^- + \nu_\mu + \bar{\nu}_e.$$

(a) Is the Ξ^- particle a lepton or a hadron? If the latter, is it a baryon or a meson?

SOLUTION: The Key Idea for answering the first question is that only three families of leptons exist (Table 45-2) and none include the Ξ^- particle. Thus, the Ξ^- must be a hadron.

The Key Idea for answering the second question is to determine the baryon number of the Ξ^- particle. If it is +1 or −1, then the Ξ^- is a baryon. If, instead, it is 0, then the Ξ^- is a meson. To see, let us the write the overall decay scheme, from the initial Ξ^- to the final relatively stable products, as

$$\Xi^- \rightarrow p + 2(e^- + \bar{\nu}_e) + 2(\nu_\mu + \bar{\nu}_\mu). \qquad (45\text{-}15)$$

On the right side, the proton has a baryon number of +1 and the electrons and neutrinos have baryon numbers of 0. Thus, the net baryon number of the right side is +1. That must then be the baryon number of the lone Ξ^- particle on the left side. We conclude that the Ξ^- particle is a baryon.

(b) Does the decay process conserve the three lepton numbers?

SOLUTION: The Key Idea here is that any process must separately conserve the net lepton number for each lepton family of Table 45-2. Let us first consider the electron lepton number L_e, which is +1 for the electron e^-, −1 for the anti-electron neutrino $\bar{\nu}_e$, and 0 for the other particles in the overall decay of Eq. 45-15. We see that the net L_e is 0 before the decay and $2[+1 + (-1)] + 2(0 + 0) = 0$ after the decay. Thus, the net electron lepton number *is* conserved. You can similarly show that the net muon lepton number and the net tau lepton number are also conserved.

(c) What can you say about the spin of the Ξ^- particle?

SOLUTION: The Key Idea here is that the overall decay scheme of Eq. 45-15 must conserve the net spin component S_z. Thus, we can determine the spin component S_z of the Ξ^- particle on the left side of Eq. 45-15 by considering the S_z components of the nine particles on the right side. All nine of those particles are spin-$\frac{1}{2}$ particles and thus can have S_z of either $+\frac{1}{2}\hbar$ of $-\frac{1}{2}\hbar$. No matter how we choose between those two possible values of S_z, the net S_z for those nine particles must be a *half-integer* times $\hbar$. Thus, the Ξ^- particle must have S_z of a *half-integer* times $\hbar$, and that means that its spin quantum number s must be a half-integer. (Actually, the quantum number is $\frac{1}{2}$; the Ξ^- particle is listed with other spin-$\frac{1}{2}$ baryons in Table 45-3.)

45-6 Still Another Conservation Law

Particles have intrinsic properties in addition to the ones we have listed so far: mass, charge, spin, lepton number, and baryon number. The first of these additional properties was discovered when researchers observed that certain new particles, such as the kaon (K) and the sigma (Σ), always seemed to be produced in pairs. It seemed impossible to produce only one of them at a time. Thus, if a beam of energetic pions interacts with the protons in a bubble chamber, the reaction

$$\pi^+ + \text{p} \rightarrow \text{K}^+ + \Sigma^+ \tag{45-16}$$

often occurs. The reaction

$$\pi^+ + \text{p} \rightarrow \pi^+ + \Sigma^+, \tag{45-17}$$

which violates no conservation law known in the early days of particle physics, never occurs.

It was eventually proposed (by Murray Gell-Mann in the United States and independently by K. Nishijima in Japan) that certain particles possess a new property, called **strangeness,** with its own quantum number S and its own conservation law. (Be careful not to confuse the symbol S here with spin.) The name *strangeness* arises from the fact that, before the identities of these particles were pinned down, they were known as "strange particles," and the label stuck.

The proton, neutron, and pion have $S = 0$; that is, they are not "strange." It was proposed, however, that the K^+ particle has strangeness $S = +1$ and that Σ^+ has $S = -1$. In the reaction of Eq. 45-16, the net strangeness is initially zero and finally zero; thus, the reaction conserves strangeness. However, in the reaction shown in Eq. 45-17, the final net strangeness is -1; thus, that reaction does not conserve strangeness and cannot occur. Apparently, then, we must add one more conservation law to our list—the **conservation of strangeness:**

> Strangeness is conserved in interactions involving the strong force.

It may seem heavy-handed to invent a new property of particles just to account for a little puzzle like that posed by Eqs. 45-16 and 45-17. However, strangeness and its quantum number soon revealed themselves in many other areas of particle physics, and strangeness is now fully accepted as a legitimate particle attribute, on a par with charge and spin.

Do not be misled by the whimsical character of the name. Strangeness is no more mysterious a property of particles than is charge. Both are properties that particles may (or may not) have; each is described by an appropriate quantum number. Each obeys a conservation law. Still other properties of particles have been discovered and given even more whimsical names, such as *charm* and *bottomness,* but all are perfectly legitimate properties. Let us see, as an example, how the new property of strangeness "earns its keep" by leading us to uncover important regularities in the properties of the particles.

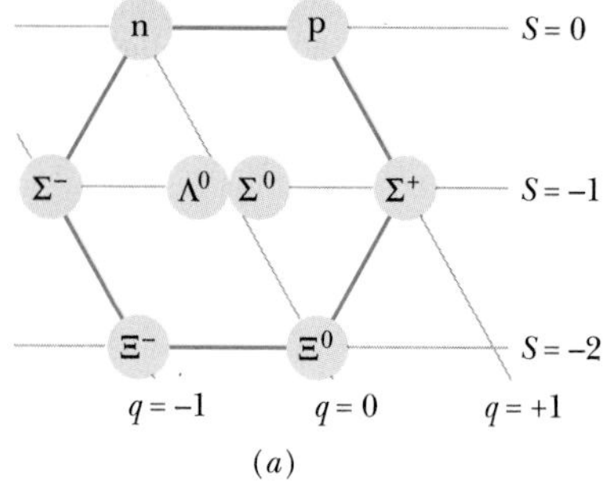

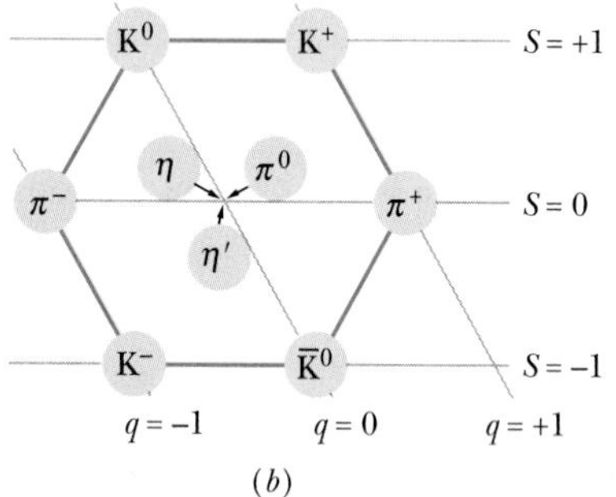

Fig. 45-4 (*a*) The Eightfold Way pattern for the eight spin-$\frac{1}{2}$ baryons listed in Table 45-3. The particles are represented as disks on a strangeness–charge plot, using a sloping axis for the charge quantum number. (*b*) A similar pattern for the nine spin-zero mesons listed in Table 45-4.

45-7 The Eightfold Way

There are eight baryons—the neutron and the proton among them—that have a spin quantum number of $\frac{1}{2}$. Table 45-3 shows some of their other properties. Figure 45-4*a* shows the fascinating pattern that emerges if we plot the strangeness of these baryons against their charge quantum number, using a sloping axis for the charge quantum numbers. Six of the eight form a hexagon with the two remaining baryons at its center.

Let us turn now from the hadrons called baryons to the hadrons called mesons. Nine with a spin of zero are listed in Table 45-4. If we plot them on a sloping strangeness–charge diagram, as in Fig. 45-4*b*, the same fascinating pattern emerges! These and related plots, called the *Eightfold Way* patterns,* were proposed independently in 1961 by Murray Gell-Mann at the California Institute of Technology and by Yuval Ne'eman at Imperial College, London. The two patterns of Fig. 45-4 are representative of a larger number of symmetrical patterns in which groups of baryons and mesons can be displayed.

The symmetry of the Eightfold Way pattern for the spin-$\frac{3}{2}$ baryons (not shown here) calls for *ten* particles arranged in a pattern like that of the tenpins in a bowling alley. However, when the pattern was first proposed, only *nine* such particles were known; the "headpin" was missing. In 1962, guided by theory and the symmetry of the pattern, Gell-Mann made a prediction in which he essentially said:

> There exists a spin-$\frac{3}{2}$ baryon with a charge of -1, a strangeness of -3, and a rest energy of about 1680 MeV. If you look for this *omega minus* particle (as I propose to call it), I think you will find it.

A team of physicists headed by Nicholas Samios of the Brookhaven National Laboratory took up the challenge and found the "missing" particle, confirming all its predicted properties. Nothing beats prompt experimental confirmation for building confidence in a theory!

The Eightfold Way patterns bear the same relationship to particle physics that the periodic table does to chemistry. In each case, there is a pattern of organization

*A borrowing from Eastern mysticism. The "Eight" refers to the eight quantum numbers (only a few of which we have defined here) that are involved in the symmetry-based theory that predicts the existence of the patterns.

TABLE 45-3 **Eight Spin-$\frac{1}{2}$ Baryons**

Particle	Symbol	Mass (MeV/c^2)	Quantum Numbers: Charge q	Quantum Numbers: Strangeness S
Proton	p	938.3	+1	0
Neutron	n	939.6	0	0
Lambda	Λ^0	1115.6	0	−1
Sigma	Σ^+	1189.4	+1	−1
Sigma	Σ^0	1192.5	0	−1
Sigma	Σ^-	1197.3	−1	−1
Xi	Ξ^0	1314.9	0	−2
Xi	Ξ^-	1321.3	−1	−2

TABLE 45-4 **Nine Spin-Zero Mesons**[a]

Particle	Symbol	Mass (MeV/c^2)	Quantum Numbers: Charge q	Quantum Numbers: Strangeness S
Pion	π^0	135.0	0	0
Pion	π^+	139.6	+1	0
Pion	π^-	139.6	−1	0
Kaon	K^+	493.7	+1	+1
Kaon	K^-	493.7	−1	−1
Kaon	K^0	497.7	0	+1
Kaon	$\bar{K}^0$	497.7	0	−1
Eta	η	547.5	0	0
Eta prime	η'	957.8	0	0

[a] All mesons are bosons, having spins of 0, 1, 2, ... The ones listed here all have a spin of 0.

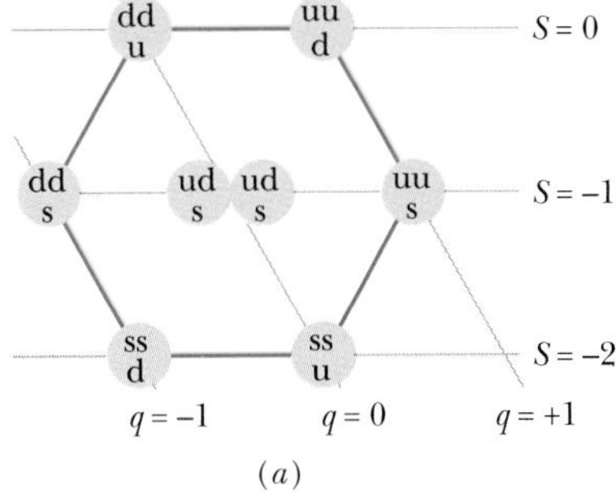

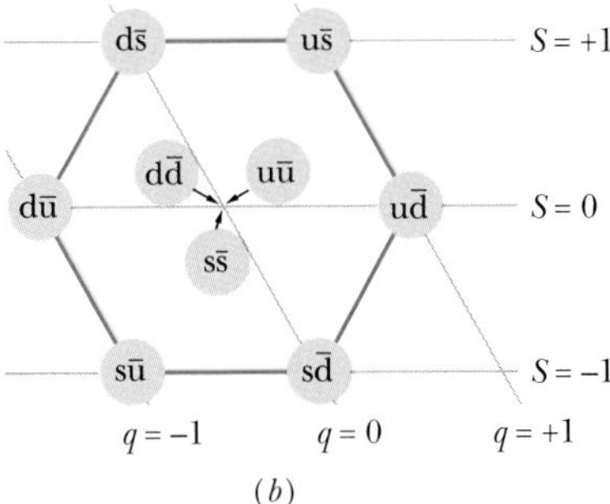

Fig. 45-5 (*a*) The quark compositions of the eight spin-$\frac{1}{2}$ baryons plotted in Fig. 45-4*a*. (Although the two central baryons share the same quark structure, the sigma is an excited state of the lambda, decaying into the lambda by emission of a gamma-ray photon.) (*b*) The quark compositions of the nine spin-zero mesons plotted in Fig. 45-4*b*.

in which vacancies (missing particles or missing elements) stick out like sore thumbs, guiding experimenters in their searches. In the case of the periodic table, its very existence strongly suggests that the atoms of the elements are not fundamental particles but have an underlying structure. Similarly, the Eightfold Way patterns strongly suggest that the mesons and the baryons must have an underlying structure, in terms of which their properties can be understood. That structure can be explained in terms of the *quark model,* which we now discuss.

45-8 The Quark Model

In 1964 Gell-Mann and George Zweig independently pointed out that the Eightfold Way patterns can be understood in a simple way if the mesons and the baryons are built up out of subunits that Gell-Mann called **quarks.** We deal first with three of them, called the *up quark* (symbol u), the *down quark* (symbol d), and the *strange quark* (symbol s), and we assign to them the properties displayed in Table 45-5. (The names of the quarks, along with those assigned to three other quarks that we shall meet later, have no meanings other than as convenient labels. Collectively, these names are called the *quark flavors.* We could just as well call them vanilla, chocolate, and strawberry instead of up, down, and strange.)

The fractional charge quantum numbers of the quarks may jar you a little. However, withhold judgment until you see how neatly these fractional charges account for the observed integral charges of the mesons and the baryons. In all normal situations, whether here on Earth or in any astronomical process, quarks are always bound up together in twos or threes for reasons that are still not well understood. However, in 2000, certain accelerators were finally able to fire atoms into targets with enough energy that the collision freed quarks from one another.

We have seen how we can put atoms together by combining electrons and nuclei. Now let us see how we can put mesons and baryons together by combining quarks. As it turns out,

> There is no known meson or baryon whose properties cannot be understood in terms of an appropriate combination of quarks. Conversely, there is no possible quark combination that does not correspond to an observed meson or baryon.

Let us look first at the baryons of Table 45-3.

Quarks and Baryons

Each baryon is a combination of three quarks; some of the combinations are given in Fig. 45-5*a*. With regard to baryon number, we see that any three quarks (each with $B = +\frac{1}{3}$) yield a proper baryon (with $B = +1$).

Charges also work out, as we can see from three examples. The proton has a quark composition of uud, so its charge quantum number is

$$q(\text{uud}) = \tfrac{2}{3} + \tfrac{2}{3} + (-\tfrac{1}{3}) = +1.$$

The neutron has a quark composition of udd and its charge quantum number is

$$q(\text{udd}) = \tfrac{2}{3} + (-\tfrac{1}{3}) + (-\tfrac{1}{3}) = 0.$$

The Σ^- (sigma-minus) particle has a quark composition of dds and its charge quantum number is

$$q(\text{dds}) = -\tfrac{1}{3} + (-\tfrac{1}{3}) + (-\tfrac{1}{3}) = -1.$$

TABLE 45-5 The Quarks[a]

Particle	Symbol	Mass (MeV/c^2)	Quantum Numbers: Charge q	Strangeness S	Baryon Number B	Antiparticle
Up	u	5	$+\frac{2}{3}$	0	$+\frac{1}{3}$	$\bar{u}$
Down	d	10	$-\frac{1}{3}$	0	$+\frac{1}{3}$	$\bar{d}$
Charm	c	1500	$+\frac{2}{3}$	0	$+\frac{1}{3}$	$\bar{c}$
Strange	s	200	$-\frac{1}{3}$	-1	$+\frac{1}{3}$	$\bar{s}$
Top	t	175 000	$+\frac{2}{3}$	0	$+\frac{1}{3}$	$\bar{t}$
Bottom	b	4300	$-\frac{1}{3}$	0	$+\frac{1}{3}$	$\bar{b}$

[a]All quarks (including antiquarks) have spin $\frac{1}{2}$ and thus are fermions. The quantum numbers, q, S, and B for an antiquark are the negatives of those for the corresponding quark.

The strangeness quantum numbers work out as well. You can check this by using Table 45-3 for the Σ^- and Table 45-5 for the quarks.

Quarks and Mesons

Mesons are quark–antiquark pairs; some of their compositions are given in Fig. 45-5*b*. The quark–antiquark model is consistent with the fact that mesons are not baryons; that is, mesons have a baryon number $B = 0$. The baryon number for a quark is $+\frac{1}{3}$ and for an antiquark is $-\frac{1}{3}$; thus, the combination of baryon numbers in a meson is zero.

Consider the meson π^+, which consists of an up quark u and an antidown quark $\bar{d}$. We see from Table 45-5 that the charge quantum number of the up quark is $+\frac{2}{3}$ and that of the antidown quark is $+\frac{1}{3}$ (the sign is opposite that of the down quark). This adds nicely to a charge quantum number of $+1$ for the π^+ meson; that is,

$$q(\mathrm{u\bar{d}}) = \tfrac{2}{3} + \tfrac{1}{3} = +1.$$

All the charge and strangeness quantum numbers of Fig. 45-5*b* agree with those of Table 45-4 and Fig. 45-4*b*. Convince yourself that all possible up, down, and strange quark–antiquark combinations are used and that all known spin-zero mesons are accounted for. Everything fits.

CHECKPOINT 3: Is a combination of a down quark (d) and an antiup quark ($\bar{u}$) called (a) a π^0 meson, (b) a proton, (c) a π^- meson, (d) a π^+ meson, or (e) a neutron?

A New Look at Beta Decay

Let us see how beta decay appears from the quark point of view. In Eq. 43-23, we presented a typical example of this process:

$$^{32}\mathrm{P} \rightarrow {}^{32}\mathrm{S} + \mathrm{e}^- + \nu.$$

After the neutron was discovered and Fermi had worked out his theory of beta decay, physicists came to view the fundamental beta-decay process as the changing of a neutron into a proton inside the nucleus, according to

$$\mathrm{n} \rightarrow \mathrm{p} + \mathrm{e}^- + \bar{\nu}_\mathrm{e},$$

in which the neutrino is identified more completely. Today we look deeper and see that a neutron (udd) can change into a proton (uud) by changing a down quark into an up quark. We now view the fundamental beta-decay process as

$$\mathrm{d} \rightarrow \mathrm{u} + \mathrm{e}^- + \bar{\nu}_\mathrm{e}.$$

Thus, as we come to know more and more about the fundamental nature of matter, we can examine familiar processes at deeper and deeper levels. We see too that the quark model not only helps us to understand the structure of particles but also clarifies their interactions.

Still More Quarks

There are other particles and other Eightfold Way patterns that we have not discussed. To account for them, it turns out that we need to postulate three more quarks, the *charm quark* c, the *top quark* t, and the *bottom quark* b. Thus, a total of six quarks exist, as listed in Table 45-5.

Note that three quarks are exceptionally massive, the most massive of them (top) being almost 170 times more massive than a proton. To generate particles that contain such quarks, with such large mass energies, we must go to higher and higher energies, which is the reason that these three quarks were not discovered earlier.

The first particle containing a charm quark to be observed was the J/Ψ meson, whose quark structure is $\mathrm{c\bar{c}}$. It was discovered simultaneously and independently in 1974 by groups headed by Samuel Ting at the Brookhaven National Laboratory and Burton Richter at Stanford University.

The top quark defied all efforts to generate it in the laboratory until 1995, when its existence was finally demonstrated in the Tevatron, a large particle accelerator at Fermilab. In this accelerator, protons and antiprotons, each with an energy of 0.9 TeV ($= 9 \times 10^{11}$ eV), are made to collide at the centers of two large particle detectors. In a very few cases, the colliding protons generate a top–antitop ($\mathrm{t\bar{t}}$) quark pair, which *very* quickly decays into particles that can be detected and thus can be used to infer the existence of the top–antitop pair.

Look back for a moment at Table 45-5 (the quark family) and Table 45-2 (the lepton family) and notice the neat symmetry of these two "six-packs" of particles, each dividing naturally into three corresponding two-particle families. In terms of what we know today, the quarks and the leptons seem to be truly fundamental particles with no internal structure.

Sample Problem 45-5

The Ξ^- particle has a spin quantum number s of $\frac{1}{2}$, a charge quantum number q of -1, and a strangeness quantum number S of -2. Also, it is known not to contain a bottom quark. What combination of quarks makes up the Ξ^-?

SOLUTION: From Sample Problem 45-4, we know that the Ξ^- is a baryon. One **Key Idea** then is that it must consist of three quarks (not two as for a meson).

Let us next consider the strangeness $S = -2$ of the Ξ^-. A **Key Idea** here is that only the strange quark s and the antistrange quark $\bar{\mathrm{s}}$ have nonzero values of strangeness (see Table 45-5). Further, because only the strange quark s has a *negative* value of strangeness, Ξ^- must contain that quark. In fact, for Ξ^- to have a strangeness of -2, it must contain two strange quarks.

To determine the third quark, call it x, we can consider the other known properties of Ξ^-. Its charge quantum number q is -1, and the charge quantum number q of each strange quark is $-\frac{1}{3}$. Thus, the third quark x must have a charge quantum number of $-\frac{1}{3}$, so that we can have

$$q(\Xi^-) = q(\mathrm{ssx}) = -\tfrac{1}{3} + (-\tfrac{1}{3}) + (-\tfrac{1}{3}) = -1.$$

Besides quark s, the only quarks with $q = -\frac{1}{3}$ are the down quark d and bottom quark b. Because the problem statement ruled out a bottom quark, the third quark must be a down quark. This conclusion is also consistent with the baryon quantum numbers:

$$B(\Xi^-) = B(\mathrm{ssd}) = \tfrac{1}{3} + \tfrac{1}{3} + \tfrac{1}{3} = +1.$$

Thus, the quark composition of the Ξ^- particle is ssd.

45-9 The Basic Forces and Messenger Particles

We turn now from cataloging the particles to considering the forces between them.

The Electromagnetic Force

At the atomic level, we say that two electrons exert electromagnetic forces on each other according to Coulomb's law. At a deeper level, this interaction is described by a highly successful theory called **quantum electrodynamics** (QED). From this point of view we say that each electron senses the presence of the other by exchanging photons with it.

We cannot detect these photons because they are emitted by one electron and absorbed by the other a very short time later. Because of their undetectable existence, we call them **virtual photons.** Because they communicate between the two interacting charged particles, we sometimes call these photons *messenger particles.*

If a stationary electron emits a photon and remains itself unchanged, energy is not conserved. The principle of conservation of energy is saved, however, by the uncertainty principle, written in the form

$$\Delta E \cdot \Delta t \approx \hbar, \tag{45-18}$$

as discussed in Sample Problem 43-10. Here we interpret this relation to mean that you can "overdraw" an amount of energy ΔE, violating conservation of energy, *provided* you "return" it within an interval Δt given by $\hbar/\Delta E$ so that the violation cannot be detected. The virtual photons do just that. When, say, electron A emits a virtual photon, the overdraw in energy is quickly set right when that electron receives a virtual photon from electron B, and the violation of the principle of conservation of energy for the electron pair is hidden by the inherent uncertainty.

The Weak Force

A theory of the weak force, which acts on all particles, was developed by analogy with the theory of the electromagnetic force. The messenger particles that transmit the weak force between particles, however, are not (massless) photons but massive particles, identified by the symbols W and Z. The theory was so successful that it revealed the electromagnetic force and the weak force as being different aspects of a single **electroweak force.** This accomplishment is a logical extension of the work of Maxwell, who revealed the electric and magnetic forces as being different aspects of a single *electromagnetic* force.

The electroweak theory was specific in predicting the properties of the messenger particles. Their charges and masses, for example, were predicted to be

Particle	Charge	Mass
W	$\pm e$	80.6 GeV/c^2
Z	0	91.2 GeV/c^2

Recall that the proton mass is only 0.938 GeV/c^2; these are massive particles! The 1979 Nobel prize in physics was awarded to Sheldon Glashow, Steven Weinberg, and Abdus Salam for their development of the electroweak theory.

The theory was confirmed in 1983 by Carlo Rubbia and his group at CERN, who experimentally verified both messenger particles and found that their masses agreed with the predicted values. The 1984 Nobel prize in physics went to Rubbia and Simon van der Meer for this brilliant experimental work.

Some notion of the complexity of particle physics in this day and age can be found by looking at an earlier Nobel prize particle physics experiment—the discovery of the neutron. This vitally important discovery was a "tabletop" experiment, employing particles emitted by naturally occurring radioactive materials as projectiles; it was reported in 1932 under the title "Possible Existence of a Neutron," the single author being James Chadwick.

The discovery of the W and Z messenger particles in 1983, by contrast, was carried out at a large particle accelerator, about 7 km in circumference and operating in the range of several hundred billion electron-volts. The principal particle detector alone weighed 20 MN. The experiment employed more than 130 physicists from 12 institutions in 8 countries, along with a large support staff.

The Strong Force

A theory of the strong force—that is, the force that acts between quarks to bind hadrons together—has also been developed. The messenger particles in this case are called **gluons** and, like the photon, they are predicted to be massless. The theory assumes that each "flavor" of quark comes in three varieties that, for convenience, have been labeled *red, yellow,* and *blue.* Thus, there are three up quarks, one of each color, and so on. The antiquarks also come in three colors, which we call *antired, antiyellow,* and *antiblue.* You must not think that quarks are actually colored, like tiny jelly beans. The names are labels of convenience but (for once) they do have a certain formal justification, as you shall see.

The force acting between quarks is called a **color force** and the underlying theory, by analogy with quantum electrodynamics (QED), is called **quantum chromodynamics** (QCD). Apparently, quarks can be assembled only in combinations that are *color-neutral.*

There are two ways to bring about color neutrality. In the theory of actual colors, red + yellow + blue yields white, which is color-neutral; thus, we can assemble three quarks to form a baryon, provided one is a yellow quark, one is a red quark, and one is a blue quark. Antired + antiyellow + antiblue is also white, so that we can assemble three antiquarks (of the proper anticolors) to form an antibaryon. Finally, red + antired, or yellow + antiyellow, or blue + antiblue also yields white. Thus, we can assemble a quark–antiquark combination to form a meson. The color-neutral rule does not permit any other combination of quarks, and none are observed.

The color force not only acts to bind together quarks as baryons and mesons, but it also acts between such particles, in which case it has traditionally been called the strong force. Hence, not only does the color force bind together quarks to form protons and neutrons, but it also binds together the protons and neutrons to form nuclei.

Einstein's Dream

The unification of the fundamental forces of nature into a single force—which occupied Einstein's attention for much of his later life—is very much a current focus of research. We have seen that the weak force has been successfully combined with electromagnetism so that they may be jointly viewed as aspects of a single *electroweak force.* Theories that attempt to add the strong force to this combination—called *grand unification theories* (GUTs)—are being pursued actively. Theories that seek to complete the job by adding gravity—sometimes called *theories of everything* (TOE)—are at an encouraging but speculative stage at this time.

45-10 A Pause for Reflection

Let us put what you have just learned in perspective. If all we are interested in is the structure of the world around us, we can get along nicely with the electron, the neutrino, the neutron, and the proton. As someone has said, we can operate "Spaceship Earth" quite well with just these particles. We can see a few of the more exotic particles by looking for them in the cosmic rays; however, to see most of them, we must build massive accelerators and look for them at great effort and expense.

The reason we must go to such effort is that—measured in energy terms—we live in a world of very low temperatures. Even at the center of the Sun, the value of kT is only about 1 keV. To produce the exotic particles, we must be able to accelerate protons or electrons to energies in the GeV and TeV range and higher.

Once upon a time the temperature everywhere *was* high enough to provide just such energies, and far greater. That time of extremely high temperatures occurred in the **big bang** beginning of the universe, when the universe (and both space and time) first existed. Thus, one reason why scientists study particles at high energies is to understand what the universe was like just after it began.

As we shall discuss shortly, *all* of space within the universe was initially tiny in extent, and the temperature of the particles within that space was incredibly high. With time, however, the universe expanded and cooled to lower temperatures, eventually to the size and temperature we see today.

Actually, the phrase "we see today" is complicated: When we look out into space, we are actually looking back in time, because the light from the stars and galaxies has taken a long time to reach us. The most distant objects that we can detect are **quasars** (*qua*si*s*tell*ar* objects), which are the extremely bright cores of galaxies that are as much as 14×10^9 ly from us. Each such core contains a gigantic black hole; as material (gas and even stars) is pulled into one of those black holes, the material heats up and radiates a tremendous amount of light, enough for us to detect in spite of the huge distance. We "see" a quasar as it once was, when that light began its journey to us billions of years ago.

45-11 The Universe Is Expanding

As we saw in Section 38-10, it is possible to measure the relative speeds at which galaxies are approaching us or receding from us by measuring the shifts in the wavelength of the light they emit. If we look only at distant galaxies, beyond our immediate galactic neighbors, we find an astonishing fact. They are *all* moving away (receding) from us!

In 1929 Edwin P. Hubble established a connection between the apparent speed of recession v of a galaxy and its distance r from us—namely, that they are directly proportional. That is,

$$v = Hr \quad \text{(Hubble's law)}, \tag{45-19}$$

in which H, the proportionality constant, is called the **Hubble constant.** The value of H is usually measured in the unit kilometers per second-megaparsec (km/s · Mpc), where megaparsec is a length unit commonly used in astrophysics and astronomy:

$$1 \text{ Mpc} = 3.084 \times 10^{19} \text{ km} = 3.260 \times 10^{6} \text{ ly}. \tag{45-20}$$

The Hubble constant H has not had the same value since the universe began. Determining its current value is extremely difficult because it involves measurements for very distant galaxies. Recently, one study put the current value of H at 70 ± 7 km/s · Mpc, whereas another study strongly argued for a value of 58 km/s · Mpc.

In this chapter, we shall use a "compromise value":

$$H = 63.0 \text{ km/s} \cdot \text{Mpc} = 19.3 \text{ mm/s} \cdot \text{ly}. \tag{45-21}$$

We interpret the recession of the galaxies to mean that the universe is expanding, much as the raisins in what is to be a loaf of raisin bread grow farther apart as the dough expands. Observers on all other galaxies would find that distant galaxies were rushing away from them also, in accordance with Hubble's law. In keeping with our analogy, we can say that no raisin (galaxy) has a unique or preferred view.

Hubble's law is consistent with the hypothesis that the universe began with the big bang and has been expanding ever since. If we assume that the rate of expansion has been constant (that is, the value of H has been constant), then we can estimate the age T of the universe by using Eq. 45-19. Let us also assume that since the big bang, any given part of the universe (say, a galaxy) has been receding from our location at a speed v given by Eq. 45-19. Then the time required for the given part to recede a distance r is

$$T = \frac{r}{v} = \frac{r}{Hr} = \frac{1}{H} \quad \text{(estimated age of universe).} \tag{45-22}$$

For our compromise value of H in Eq. 45-21, T works out to be 15×10^9 y. Much more sophisticated studies of the expansion of the universe put T between 12×10^9 y and 15×10^9 y.

Sample Problem 45-6

The wavelength shift in the light from a particular quasar indicates that the quasar has a recessional speed of 2.8×10^8 m/s (which is 93% of the speed of light). Approximately how far from us is the quasar?

SOLUTION: The **Key Idea** here is to apply Hubble's law to the given speed v. From Eqs. 45-19 and 45-21, we find

$$r = \frac{v}{H} = \frac{2.8 \times 10^8 \text{ m/s}}{19.3 \text{ mm/s} \cdot \text{ly}}(1000 \text{ mm/m})$$
$$= 14.5 \times 10^9 \text{ ly}. \quad \text{(Answer)}$$

This is only an approximation because the quasar has not always been receding from our location at the same speed v; that is, H has not had its current value throughout the expansion of the universe.

Sample Problem 45-7

A particular emission line, detected in the light from a galaxy, has a wavelength $\lambda_{\text{det}} = 1.1\lambda$, where λ is the proper wavelength of the line. What is the galaxy's distance from us?

SOLUTION: One **Key Idea** here is to assume that Hubble's law ($v = Hr$) applies to the recession of the galaxy. A second **Key Idea** is to assume that the astronomical Doppler shift of Eq. 38-33 ($v = c\,\Delta\lambda/\lambda$) applies to the shift in wavelength due to the recession. We can then set the right side of these two equations equal to each other to write

$$Hr = \frac{c\,\Delta\lambda}{\lambda}, \tag{45-23}$$

which leads us to

$$r = \frac{c\,\Delta\lambda}{H\lambda}. \tag{45-24}$$

In this equation,

$$\Delta\lambda = \lambda_{\text{det}} - \lambda = 1.1\lambda - \lambda = 0.1\lambda.$$

Substituting this into Eq. 45-24 then gives us

$$r = \frac{c(0.1\lambda)}{H\lambda} = \frac{0.1c}{H}$$
$$= \frac{(0.1)(3.0 \times 10^8 \text{ m/s})}{19.3 \text{ mm/s} \cdot \text{ly}}(1000 \text{ mm/m})$$
$$= 1.6 \times 10^9 \text{ ly}. \quad \text{(Answer)}$$

45-12 The Cosmic Background Radiation

In 1965 Arno Penzias and Robert Wilson, of what was then the Bell Telephone Laboratories, were testing a sensitive microwave receiver used for communications research. They discovered a faint background "hiss" that remained unchanged in intensity no matter where their antenna was pointed. It soon became clear that Penzias and Wilson were observing a **cosmic background radiation,** generated in the early universe and filling all space almost uniformly. This radiation, whose maximum intensity occurs at a wavelength of 1.1 mm, has the same distribution in wavelength as does radiation in an enclosure whose walls are held at a temperature of 2.7 K. In this situation the enclosure is the entire universe. Penzias and Wilson were awarded the 1978 Nobel prize in physics for their discovery.

This radiation originated about 300 000 years after the big bang, when the universe suddenly became transparent to electromagnetic waves (the waves were not immediately absorbed by particles). The radiation at that time corresponded to radiation in an enclosure at a temperature of perhaps 10^5 K. As the universe expanded, however, the temperature dropped to its present value of 2.7 K.

45-13 Dark Matter

At the Kitt Peak National Observatory in Arizona, Vera Rubin and her co-worker Kent Ford measured the rotational rates of a number of distant galaxies. They did so by measuring the Doppler shifts of bright clusters of stars located within each galaxy at various distances from the galactic center. As Fig. 45-6 shows, their results were surprising: The orbital speed of stars at the outer visible edge of the galaxy is about the same as that of stars close to the galactic center.

As the solid curve in Fig. 45-6 attests, that is not what we would expect to find if all the mass of the galaxy were represented by visible light. Nor is the pattern found by Rubin and Ford what we find in the solar system. For example, the orbital speed of Pluto (the planet most distant from the Sun) is only about one-tenth that of Mercury (the planet closest to the Sun).

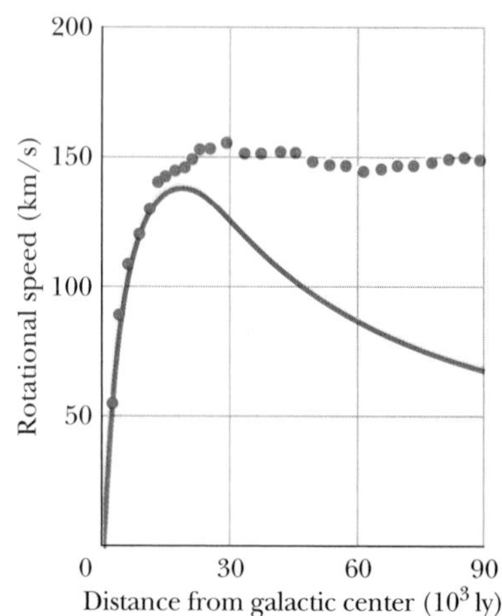

Fig. 45-6 The rotational speed of stars in a typical galaxy as a function of their distance from the galactic center. The theoretical solid curve shows that if a galaxy contained only the mass that is visible, the observed rotational speed would drop off with distance at large distances. The dots are the experimental data, which show that the rotational speed is approximately constant at large distances.

The only explanation for the findings of Rubin and Ford that is consistent with Newtonian mechanics is that a typical galaxy contains much more matter than what we can actually see. In fact, the visible portion of a galaxy represents only about 5 to 10% of the total mass of the galaxy. In addition to these studies of galactic rotation, many other observations lead to the conclusion that the universe abounds in matter that we cannot see.

What then is this **dark matter** that permeates and surrounds a typical galaxy like a gigantic halo whose diameter is perhaps 30 times the diameter of the visible galaxy? Dark matter candidates fall into two classes, whimsically called WIMPs (*w*eakly *i*nteracting *m*assive *p*articles) and MACHOs (*ma*ssive *c*ompact *h*alo *o*bjects). If neutrinos have mass, they are possible WIMP candidates. MACHOs can include objects such as black holes, white dwarf stars, and brown dwarf stars; the latter are Jupiter-size objects that are not massive enough to become actual stars, which shine (and hence are visible) because of fusion.

As of 2000, there is convincing evidence that MACHOs do indeed exist in our own galaxy. Assume that an (invisible) MACHO in our galaxy passes, by chance, between Earth and a star of a nearby galaxy. Einstein, in his general theory of relativity, predicted that light rays passing near any massive object will be deflected by the mass of that object (see Section 14-9). Thus, if star, MACHO, and Earth are aligned, the MACHO will act as a *gravitational lens,* focusing the light rays from the star that pass near it and causing the image of the star to brighten while the MACHO is eclipsing it.

Enough such events have been observed to convince astronomers that MACHOs can account for a substantial fraction (some say 50%) of the dark matter in our own galaxy. Observations are ongoing.

45-14 The Big Bang

In 1985, a physicist remarked at a scientific meeting:

> It is as certain that the universe started with a Big Bang about 15 billion years ago as it is that the Earth goes around the Sun.

This strong statement suggests the level of confidence in which the big bang theory, first advanced by Belgian physicist Georges Lemaître, is held by those who study these matters.

You must not imagine that the big bang was like the explosion of some gigantic firecracker and that, in principle at least, you could have stood to one side and watched. There was no "one side" because the big bang represents the beginning of spacetime itself. From the point of view of our present universe, there is no position in space to which you can point and say, "The big bang happened there." It happened everywhere.

Moreover, there was no "before the big bang," because time *began* with that creation event. In this context, the word "before" loses its meaning. We can, however, conjecture about what went on during succeeding intervals of time after the big bang.

$t \approx 10^{-43}$ **s.** This is the earliest time at which we can say anything meaningful about the development of the universe. It is at this moment that the concepts of space and time come to have their present meanings and the laws of physics as we know them become applicable. At this instant, the entire universe is much smaller than a proton and its temperature is about 10^{32} K.

$t \approx 10^{-34}$ **s.** By this moment the universe has undergone a tremendously rapid inflation, increasing in size by a factor of about 10^{30}. It has become a hot soup of photons, quarks, and leptons, at a temperature of about 10^{27} K.

$t \approx 10^{-4}$ **s.** Quarks can now combine to form protons and neutrons and their antiparticles. The universe has now cooled to such an extent by continued (but much slower) expansion that photons lack the energy needed to break up these new particles. Particles of matter and antimatter collide and annihilate each other. There is a slight excess of matter, which, failing to find annihilation partners, survives to form the world of matter that we know today.

$t \approx$ **1 min.** The universe has now cooled enough so that protons and neutrons, in colliding, can stick together to form the low-mass nuclei ^{2}H, ^{3}He, ^{4}He, and ^{7}Li. The predicted relative abundances of these nuclides are just what we observe in the universe today. There is plenty of radiation present, but light cannot travel far before it interacts with a nucleus. Thus the universe is opaque.

$t \approx$ **300 000 y.** The temperature has now fallen to about 10^4 K, and electrons can stick to bare nuclei when they collide, forming atoms. Because light does not interact appreciably with (uncharged) particles such as neutral atoms, the light is now free to travel great distances. This radiation forms the cosmic *background radiation* discussed in Section 45-12. Atoms of hydrogen and helium, under the influence of gravity, begin to clump together, starting the formation of galaxies and stars.

Early measurements suggested that the cosmic background radiation is uniform in all directions, implying that 300 000 y after the big bang all matter in the universe was uniformly distributed. This finding was most puzzling because matter in the present universe is not uniformly distributed, but instead is collected in galaxies, clusters of galaxies, and superclusters of galactic clusters. There are also vast *voids* in which there is relatively little matter, and there are regions so crowded with matter that they are called *walls*. If the big bang theory of the beginning of the universe is even approximately correct, the seeds for this nonuniform distribution of matter must have been in place before the universe was 300 000 y old and now should show up as a nonuniform distribution of the microwave background radiation.

In 1992, measurements made by NASA's Cosmic Background Explorer (COBE) satellite revealed that the background radiation is, in fact, not perfectly uniform. The image shown on this chapter's opening page was made from those measurements and shows us the universe when it was only 300 000 y old. As you can see, large-scale collecting of matter had already begun; thus, the big bang theory is, in principle, on the right track.

45-15 A Summing Up

Let us, in these closing paragraphs, consider where our rapidly accumulating store of knowledge about the universe is leading us. That it provides satisfaction to a host of curiosity-motivated physicists and astronomers is beyond dispute. However, some view it as a humbling experience in that each increase in knowledge seems to reveal more clearly our own relative insignificance in the grand scheme of things. Thus, in roughly chronological order, we humans have come to realize that

Our Earth is not the center of the solar system.

Our Sun is but one star among many in our galaxy.

Our galaxy is but one of many, and our Sun is an insignificant star in it.

Our Earth has existed for perhaps only a third of the age of the universe and will surely disappear when our Sun burns up its fuel and becomes a red giant.

Our species has inhabited Earth for less than a million years—a blink in cosmological time.

Although our position in the universe may be insignificant, the laws of physics that we have discovered (uncovered?) seem to hold throughout the universe and—as far as we know—since the universe began and for all future time. At least, there is no evidence that other laws hold in other parts of the universe. Thus, until someone complains, we are entitled to stamp the laws of physics "Discovered on Earth." Much remains to be discovered: *"The universe is full of magical things, patiently waiting for our wits to grow sharper."*

REVIEW & SUMMARY

Leptons and Quarks Current research supports the view that all matter is made of six kinds of **leptons** (Table 45-2), six kinds of **quarks** (Table 45-5), and 12 **antiparticles,** one corresponding to each of the leptons and quarks. All these particles have spin quantum numbers equal to $\frac{1}{2}$ and are thus **fermions** (particles with half-integer spin quantum numbers).

The Interactions Particles with electric charge interact through the electromagnetic force by exchanging **virtual photons.** Leptons can interact with each other and with quarks through the **weak force,** via massive W and Z particles as messengers. In addition, quarks interact with each other through the **color force.** The electromagnetic and weak forces are different manifestations of the same force, called the **electroweak force.**

Leptons Three of the leptons (the **electron, muon,** and **tau**) have electric charge equal to $-1e$. There are also three uncharged **neutrinos** (also leptons), one corresponding to each of the charged leptons. The neutrinos have very small, possibly zero, mass. The antiparticles for the charged leptons have positive charge.

Quarks The six quarks (up, down, strange, charm, bottom, and top, in order of increasing mass) each have baryon number $+\frac{1}{3}$ and charge equal to either $+\frac{2}{3}e$ or $-\frac{1}{3}e$. The strange quark has strangeness -1, whereas the others all have strangeness 0. These four algebraic signs are reversed for the antiquarks.

Hadrons: Baryons and Mesons Quarks combine into strongly interacting particles called **hadrons. Baryons** are hadrons with half-integer spin quantum numbers ($\frac{1}{2}$ or $\frac{3}{2}$). **Mesons** are hadrons with integer spin quantum numbers (0 or 1) and thus are **bosons.** Baryons are fermions. Mesons have baryon number equal to zero; baryons have baryon number equal to $+1$ or -1. **Quantum chromodynamics** predicts that the possible combinations of quarks are either a quark with an antiquark, three quarks, or three antiquarks (this prediction is consistent with experiment).

Expansion of the Universe Current evidence strongly suggests that the universe is expanding, with the distant galaxies moving away from us at a rate v given by **Hubble's law:**

$$v = Hr \qquad \text{(Hubble's law).} \tag{45-19}$$

Here we take H, the **Hubble constant,** to have the value

$$H = 63.0 \text{ km/s} \cdot \text{Mpc} = 19.3 \text{ mm/s} \cdot \text{ly}. \tag{45-21}$$

The expansion described by Hubble's law and the presence of ubiquitous background microwave radiation suggest that the universe began in a "big bang" between 12 and 15 billion years ago.

QUESTIONS

1. Not only particles such as electrons and protons but also entire atoms can be classified as fermions or bosons, depending on whether their overall spin quantum numbers are, respectively, half-integral or integral. Consider the helium isotopes ^{3}He and ^{4}He. Which of the following statements is correct? (a) Both are fermions. (b) Both are bosons. (c) ^{4}He is a fermion, and ^{3}He is a boson. (d) ^{3}He is a fermion, and ^{4}He is a boson. (The two helium electrons form a closed shell and play no role in this determination.)

2. Is the direction of the magnetic field in Fig. 45-3*b* out of the page or into the page?

3. Which of the eight pions in Fig. 45-3*b* has the least kinetic energy?

4. An electron cannot decay into two neutrinos. Which of the following conservation laws would be violated if it did: (a) energy, (b) angular momentum, (c) charge, (d) lepton number, (e) linear momentum, (f) baryon number?

5. A proton cannot decay into a neutron and a neutrino. Which of the following conservation laws would be violated if it did: (a) energy (assume the proton is stationary), (b) angular momentum, (c) charge, (d) lepton number, (e) linear momentum, (f) baryon number?

6. A proton has enough mass energy to decay into a shower made up of electrons, neutrinos, and their antiparticles. Which of the following conservation laws would be violated if it did: (a) electron lepton number, (b) angular momentum, (c) charge, (d) muon lepton number, (e) linear momentum, (f) baryon number?

7. As we have seen, the π^- meson has the quark structure $d\bar{u}$. Which of the following conservation laws would be violated if a π^- was formed, instead, from a d quark and a u quark: (a) energy, (b) angular momentum, (c) charge, (d) lepton number, (e) linear momentum, (f) baryon number?

8. A Σ^+ particle has these quantum numbers: strangeness $S = -1$, charge $q = +1$, and spin $s = \frac{1}{2}$. Which of the following quark combinations produce it: (a) dds, (b) $s\bar{s}$, (c) uus, (d) ssu, or (e) $uu\bar{s}$?

9. The left column that follows lists ideas from atomic physics, and the right column from particle physics. Match the items in the two columns.

1. chemistry	a. Eightfold Way patterns
2. electrons	b. missing hadrons
3. periodic table	c. quantum chromodynamics
4. missing elements	d. particle physics
5. quantum mechanics	e. quarks

10. Consider the neutrino whose symbol is $\bar{\nu}_\tau$. (a) Is it a quark, a lepton, a meson, or a baryon? (b) Is it a particle or an antiparticle? (c) Is it a boson or a fermion? (d) Is it stable against spontaneous decay?

11. Match the items in these two columns:

1. tau	a. quark
2. pion	b. lepton
3. proton	c. meson
4. positron	d. baryon
5. charm	e. antiparticle

12. List these particles in order of mass, least first: (a) proton, (b) neutrino, (c) π^+ meson, (d) strange quark, (e) tau, (f) electron, and (g) Σ^-.

13. What are the lepton numbers of these particles: (a) π^-, (b) e^-, (c) μ^+, (d) τ^-, (e) $\bar{\nu}_\mu$?

EXERCISES & PROBLEMS

ssm Solution is in the Student Solutions Manual.
www Solution is available on the World Wide Web at: http://www.wiley.com/college/hrw
ilw Solution is available on the Interactive LearningWare.

SEC. 45-3 An Interlude

1E. Calculate the difference in mass, in kilograms, between the muon and pion of Sample Problem 45-1.

2E. An electron and a positron are separated by a distance r. Find the ratio of the gravitational force to the electric force between them. From the result, what can you conclude concerning the forces acting between particles detected in a bubble chamber?

3E. A neutral pion decays into two gamma rays: $\pi^0 \rightarrow \gamma + \gamma$. Calculate the wavelength of the gamma rays produced by the decay of a neutral pion at rest. ssm

4E. A positively charged pion decays by Eq. 45-7: $\pi^+ \rightarrow \mu^+ + \nu$. What then must be the decay scheme of the negatively charged pion? (*Hint:* The π^- is the antiparticle of the π^+.)

5E. How much energy would be released if Earth were annihilated by collision with an anti-Earth? ssm

6P. Certain theories predict that the proton is unstable, with a half-life of about 10^{32} years. Assuming that this is true, calculate the number of proton decays you would expect to occur in one year in the water of an Olympic-sized swimming pool holding 4.32×10^5 L of water.

7P. Observations of neutrinos emitted by the supernova SN1987a (Fig. 44-12) in the Large Magellanic Cloud place an upper limit of 20 eV on the rest energy of the electron neutrino. Suppose that the rest energy of this neutrino, rather than being zero, is in fact equal to 20 eV. How much slower than the speed of light would be the speed of a 1.5 MeV neutrino that was emitted in a beta decay?

8P. A neutral pion has a rest energy of 135 MeV and a mean life of 8.3×10^{-17} s. If it is produced with an initial kinetic energy of 80 MeV and decays after one mean lifetime, what is the longest possible track this particle could leave in a bubble chamber? Use relativistic time dilation.

9P. The rest energies of many short-lived particles cannot be measured directly but must be inferred from the measured momenta and known rest energies of the decay products. Consider the ρ^0 meson, which decays by the reaction $\rho^0 \rightarrow \pi^+ + \pi^-$. Calculate the rest energy of the ρ^0 meson given that the oppositely directed momenta of the created pions each have magnitude 358.3 MeV/c. See Table 45-4 for the rest energies of the pions. ssm

10P. A positive tau (τ^+, rest energy = 1777 MeV) is moving with 2200 MeV of kinetic energy in a circular path perpendicular to a uniform 1.20 T magnetic field. (a) Calculate the momentum of the tau in kilogram-meters per second. Relativistic effects must be considered. (b) Find the radius of the circular path.

11P. (a) A stationary particle 1 decays into particles 2 and 3, which move off with equal but oppositely directed momenta. Show that the kinetic energy K_2 of particle 2 is given by

$$K_2 = \frac{1}{2E_1}[(E_1 - E_2)^2 - E_3^2],$$

where m_1, m_2, and m_3 are masses and E_1, E_2, and E_3 are the corresponding rest energies. (*Hint:* Follow the arguments of Sample Problem 45-1 except that, in this case, neither of the created particles has zero mass.) (b) Show that the result in (a) yields the kinetic energy of the muon as calculated in Sample Problem 45-1.

SEC. 45-6 Still Another Conservation Law

12E. Verify that the hypothetical proton decay scheme in Eq. 45-14 does not violate the conservation laws of (a) charge, (b) energy, and (c) linear momentum. (d) How about angular momentum?

13E. Which conservation law is violated in each of these proposed decays? Assume that the initial particle is stationary and the decay products have zero orbital angular momentum. (a) $\mu^- \rightarrow e^- + \nu_\mu$; (b) $\mu^- \rightarrow e^+ + \nu_e + \bar{\nu}_\mu$; (c) $\mu^+ \rightarrow \pi^+ + \nu_\mu$. ssm

14P. The A_2^+ particle and its products decay according to the following scheme:

$$A_2^+ \rightarrow \rho^0 + \pi^+, \qquad \mu^+ \rightarrow e^+ + \nu + \bar{\nu},$$
$$\rho^0 \rightarrow \pi^+ + \pi^-, \qquad \pi^- \rightarrow \mu^- + \bar{\nu},$$
$$\pi^+ \rightarrow \mu^+ + \nu, \qquad \mu^- \rightarrow e^- + \nu + \bar{\nu}.$$

(a) What are the final stable decay products? (b) From the evidence, is the A_2^+ particle a fermion or a boson? Is that particle a meson or a baryon? What is its baryon number? (*Hint:* See Sample Problem 45-4.)

SEC. 45-7 The Eightfold Way

15E. The reaction $\pi^+ + p \rightarrow p + p + \bar{n}$ proceeds via the strong interaction. By applying the conservation laws, deduce the charge quantum number, baryon number, and strangeness of the antineutron. ssm

16E. By examining strangeness, determine which of the following decays or reactions proceed via the strong interaction: (a) $K^0 \rightarrow \pi^+ + \pi^-$; (b) $\Lambda^0 + p \rightarrow \Sigma^+ + n$; (c) $\Lambda^0 \rightarrow p + \pi^-$; (d) $K^- + p \rightarrow \Lambda^0 + \pi^0$.

17E. Which conservation law is violated in each of these proposed reactions and decays? (Assume that the products have zero orbital angular momentum.) (a) $\Lambda^0 \rightarrow p + K^-$; (b) $\Omega^- \rightarrow \Sigma^- + \pi^0$ ($S = -3$, $q = -1$, and $m = 1672$ MeV/c^2 for Ω^-); (c) $K^- + p \rightarrow \Lambda^0 + \pi^+$. ssm

18E. Calculate the disintegration energy of the reactions (a) $\pi^+ + p \rightarrow \Sigma^+ + K^+$ and (b) $K^- + p \rightarrow \Lambda^0 + \pi^0$.

19E. A Σ^- particle moving with 220 MeV of kinetic energy decays according to $\Sigma^- \rightarrow \pi^- + n$. Calculate the total kinetic energy of the decay products.

20P. Show that if, instead of plotting strangeness S versus charge q for the spin-$\frac{1}{2}$ baryons in Fig. 45-4*a* and for the spin-zero mesons in Fig. 45-4*b*, the quantity $Y = B + S$ is plotted against the quantity $T_z = q - \frac{1}{2}(B + S)$, then the hexagonal patterns emerge with the use of nonsloping (perpendicular) axes. (The quantity Y is called *hypercharge* and T_z is related to a quantity called *isospin*.)

21P. Use the conservation laws to identify the particle labeled x in each of the following reactions, which proceed by means of the strong interaction: (a) $p + p \rightarrow p + \Lambda^0 + x$; (b) $p + \bar{p} \rightarrow n + x$; (c) $\pi^- + p \rightarrow \Xi^0 + K^0 + x$. ssm www

22P. Consider the decay $\Lambda^0 \rightarrow p + \pi^-$ with the Λ^0 at rest. (a) Calculate the disintegration energy. (b) Find the kinetic energy of the proton. (c) What is the kinetic energy of the pion? (*Hint:* See Problem 11.)

SEC. 45-8 The Quark Model

23E. The quark makeups of the proton and neutron are uud and udd, respectively. What are the quark makeups of (a) the antiproton and (b) the antineutron?

24E. From Tables 45-3 and 45-5, determine the identities of the baryons formed from the following combinations of quarks. Check your answers with the baryon octet shown in Fig. 45-4*a*: (a) ddu; (b) uus; (c) ssd.

25E. Using the up, down, and strange quarks only, construct, if possible, a baryon (a) with $q = +1$ and strangeness $S = -2$ and (b) with $q = +2$ and strangeness $S = 0$. ssm

26E. What quark combinations are needed to form (a) a Λ^0 and (b) a Ξ^0?

27E. There are 10 baryons with spin $\frac{3}{2}$. Their symbols and quantum numbers for charge q and strangeness S are as follows:

	q	S		q	S
Δ^-	-1	0	Σ^{*0}	0	-1
Δ^0	0	0	Σ^{*+}	$+1$	-1
Δ^+	$+1$	0	Ξ^{*-}	-1	-2
Δ^{++}	$+2$	0	Ξ^{*0}	0	-2
Σ^{*-}	-1	-1	Ω^-	-1	-3

Make a charge–strangeness plot for these baryons, using the sloping coordinate system of Fig. 45-4. Compare your plot with this figure.

28P. There is no known meson with charge quantum number $q = +1$ and strangeness $S = -1$ or with $q = -1$ and $S = +1$. Explain why, in terms of the quark model.

29P. The spin-$\frac{3}{2}$ Σ^{*0} baryon (see Exercise 27) has a rest energy of 1385 MeV (with an intrinsic uncertainty ignored here); the spin-$\frac{1}{2}$ Σ^0 baryon has a rest energy of 1192.5 MeV. If each of these particles has a kinetic energy of 1000 MeV, which, if either, is moving faster and by how much?

SEC. 45-11 The Universe Is Expanding

30E. If Hubble's law can be extrapolated to very large distances, at what distance would the apparent recessional speed become equal to the speed of light?

31E. What is the observed wavelength of the 656.3 nm (first Balmer) line of hydrogen emitted by a galaxy at a distance of 2.40×10^8 ly? Assume that the Doppler shift of Eq. 38-33 and Hubble's law apply. ssm www

32E. In the laboratory, one of the lines of sodium is emitted at a wavelength of 590.0 nm. In the light from a particular galaxy, however, this line is seen at a wavelength of 602.0 nm. Calculate the distance to the galaxy, assuming that Hubble's law holds and that the Doppler shift of Eq. 38-33 applies.

33P. Will the universe continue to expand forever? To attack this question, make the (reasonable?) assumption that the recessional speed v of a galaxy a distance r from us is determined only by the matter that lies inside a sphere of radius r centered on us. If the total mass inside this sphere is M, the escape speed v_e from the sphere is $v_e = \sqrt{2GM/r}$ (Eq. 14-27). (a) Show that to prevent unlimited expansion, the average density ρ inside the sphere must be at least equal to

$$\rho = \frac{3H^2}{8\pi G}.$$

(b) Evaluate this "critical density" numerically; express your answer in terms of hydrogen atoms per cubic meter. Measurements of the actual density are difficult and are complicated by the presence of dark matter.

34P. The apparent recessional speeds of galaxies and quasars at great distances are close to the speed of light, so the relativistic Doppler shift formula (Eq. 38-30) must be used. The red shift is reported as fractional red shift $z = \Delta\lambda/\lambda_0$. (a) Show that, in terms of z, the recessional speed parameter $\beta = v/c$ is given by

$$\beta = \frac{z^2 + 2z}{z^2 + 2z + 2}.$$

(b) A quasar detected in 1987 has $z = 4.43$. Calculate its speed parameter. (c) Find the distance to the quasar, assuming that Hubble's law is valid to these distances.

SEC. 45-12 The Cosmic Background Radiation

35P. Due to the presence everywhere of the microwave background radiation, the minimum possible temperature of a gas in interstellar or intergalactic space is not 0 K but 2.7 K. This implies that a significant fraction of the molecules in space that can occupy excited states of low excitation energy may, in fact, be in those excited states. Subsequent de-excitation would lead to the emission of radiation that could be detected. Consider a (hypothetical) molecule with just one excited state. (a) What would the excitation energy have to be for 25% of the molecules to be in the excited state? (*Hint:* See Eq. 41-29.) (b) What would be the wavelength of the photon emitted in a transition back to the ground state?

SEC. 45-13 Dark Matter

36E. What would the mass of the Sun have to be if Pluto (the outermost planet most of the time) were to have the same orbital speed that Mercury (the innermost planet) has now? Use data from Appendix C, and express your answer in terms of the Sun's current mass M. (Assume circular orbits.)

37P. Suppose that the radius of the Sun were increased to 5.90×10^{12} m (the average radius of the orbit of the planet Pluto, the outermost planet), that the density of this expanded Sun were uniform, and that the planets revolved within this tenuous object. (a) Calculate Earth's orbital speed in this new configuration and compare this with its present orbital speed of 29.8 km/s. Assume that the radius of Earth's orbit remains unchanged. (b) What would be the new period of revolution of Earth? (The Sun's mass remains unchanged.) ssm www

38P. Suppose that the matter (stars, gas, dust) of a particular galaxy, of total mass M, is distributed uniformly throughout a sphere of radius R. A star of mass m is revolving about the center of the galaxy in a circular orbit of radius $r < R$. (a) Show that the orbital speed v of the star is given by

$$v = r\sqrt{GM/R^3},$$

and therefore that the period T of revolution is

$$T = 2\pi\sqrt{R^3/GM},$$

independent of r. Ignore any resistive forces. (b) What is the corresponding formula for the orbital period, assuming that the mass of the galaxy is strongly concentrated toward the center of the galaxy, so that essentially all the mass is at distances from the center less than r?

SEC. 45-14 The Big Bang

39E. The wavelength at which a thermal radiator at temperature T radiates electromagnetic waves most intensely is given by Wien's law: $\lambda_{max} = (2898\ \mu m \cdot K)/T$. (a) Show that the energy E of a photon corresponding to that wavelength can be computed from

$$E = (4.28 \times 10^{-10}\ \mathrm{MeV/K})T.$$

(b) At what minimum temperature can this photon create an electron–positron pair (as discussed in Section 22-6)? ssm

40E. Use Wien's law (see Exercise 39) to answer the following questions: (a) The microwave background radiation peaks in intensity at a wavelength of 1.1 mm. To what temperature does this correspond? (b) About 300 000 y after the big bang, the universe became transparent to electromagnetic radiation. Its temperature then was about 10^5 K. What was the wavelength at which the background radiation was then most intense?

Additional Problems

41. Figure 45-7 shows part of the experimental arrangement in which antiprotons were discovered in the 1950s. A beam of 6.2 GeV protons emerged from a particle accelerator and collided with nuclei in a copper target. According to theoretical predictions at the time, collisions with the protons and neutrons in those nuclei should produce antiprotons via the reactions

$$\mathrm{p} + \mathrm{p} \rightarrow \mathrm{p} + \mathrm{p} + \mathrm{p} + \bar{\mathrm{p}}$$

and

$$\mathrm{p} + \mathrm{n} \rightarrow \mathrm{p} + \mathrm{n} + \mathrm{p} + \bar{\mathrm{p}}.$$

However, even if these reactions did occur, they would be rare compared to the reactions

$$\mathrm{p} + \mathrm{p} \rightarrow \mathrm{p} + \mathrm{p} + \pi^+ + \pi^-$$

and

$$\mathrm{p} + \mathrm{n} \rightarrow \mathrm{p} + \mathrm{n} + \pi^+ + \pi^-.$$

Thus, most of the particles produced by the collisions between the 6.2 GeV protons and the copper target were pions.

To prove that antiprotons existed and were also produced by the collisions, particles leaving the target were sent into a series of magnetic fields and detectors as shown in Fig. 45-7. The first magnetic field M1 curved the paths of any charged particles passing through it; moreover, the field was arranged so that the only particles that emerged from it to reach the second magnetic field (Q1) had to be negatively charged (either a $\bar{\mathrm{p}}$ or a π^-) and have a momentum of 1.19 GeV/c. Q1 was a special type of magnetic field (a *quadrapole field*) that focused the particles reaching it into a beam, allowing them to pass through a hole in thick shielding to a *scintillation counter* S1. The travel of a charged particle through such a counter triggers a signal (much like a conventional television screen emits a pulse of light when struck by an electron). Thus, each signal indicated the passage of either a 1.19 GeV/c π^- or (presumably) a 1.19 GeV/c $\bar{\mathrm{p}}$.

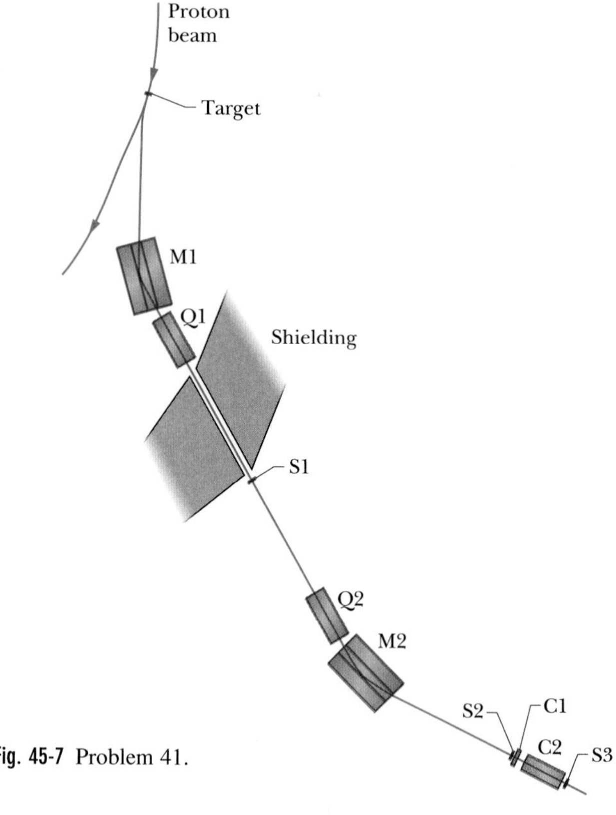

Fig. 45-7 Problem 41.

After being refocused by magnetic field Q2, the particles were directed by magnetic field M2 through a second scintillation counter S2 and then through two *Cerenkov counters* C1 and C2. These latter detectors can be manufactured so that they send a signal only when a particle passes through them with a speed in a certain range. In the experiment, a particle with a speed greater than $0.79c$ would trigger C1 and a particle with a speed between $0.75c$ and $0.78c$ would trigger C2.

There were then two ways to distinguish the predicted rare antiprotons from the abundant negative pions. Both ways involved the fact that the speed of a 1.19 GeV/c $\bar{\mathrm{p}}$ would differ from that of a 1.19 GeV/c π^-: (1) According to calculations, a $\bar{\mathrm{p}}$ would trigger one of the Cerenkov counters and a π^- would trigger the other. (2) Also, the time interval Δt between signals from S1 and S2, which were separated by 12 m, would be one value for a $\bar{\mathrm{p}}$ and another value for a π^-. Thus, if the correct Cerenkov counter were triggered and the time interval Δt had the correct value, the experiment would prove the existence of antiprotons.

What is the speed of (a) an antiproton with a momentum of 1.19 GeV/c and (b) a negative pion with that same momentum? The speed of an antiproton through the Cerenkov detectors would actually be slightly less than calculated here because an antiproton would lose a little energy within the detectors. Which Cerenkov detector was triggered by (c) an antiproton and (d) a negative pion? What time interval Δt indicated the passage of (e) an antiproton and (f) a negative pion? [Problem adapted from O. Chamberlain, E. Segrè, C. Wiegand, and T. Ypsilantis, "Observation of Antiprotons," *Physical Review*, Vol. 100, pp. 947–950 (1955).]

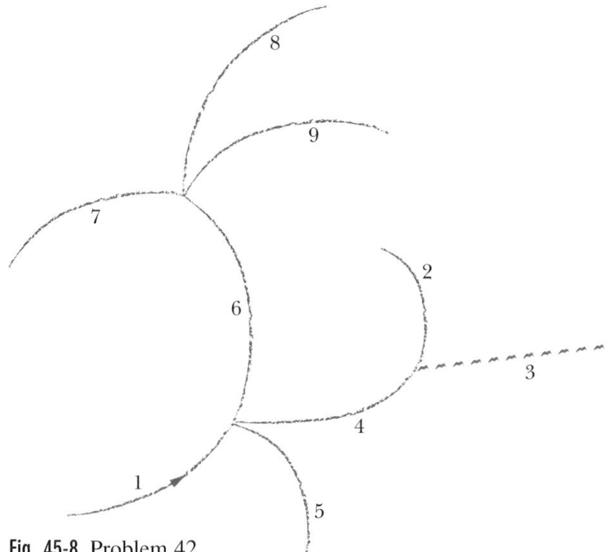

Fig. 45-8 Problem 42.

42. *A particle game.* Figure 45-8 is a sketch of the tracks made by particles in a *fictional* cloud chamber experiment (with a uniform magnetic field directed perpendicular to the page), and the following table gives *fictional* quantum numbers associated with the particles making the tracks. Particle A entered the chamber, leaving track 1 and decaying into three particles. Then the particle creating track 6 decayed into three other particles, and the particle creating track 4 decayed into two other particles, one of which was electrically uncharged—the path of that uncharged particle is represented by the dashed straight line. The particle that created track 8 is known to have a seriousness quantum number of zero.

By conserving the fictional quantum numbers at each decay point and by noting the directions of curvature of the tracks, identify which particle goes with which track. One of the listed particles is not formed; the others appear only once each.

Particle	Charge	Whimsy	Seriousness	Cuteness
A	1	1	−2	−2
B	0	4	3	0
C	1	2	−3	−1
D	−1	−1	0	1
E	−1	0	−4	−2
F	1	0	0	0
G	−1	−1	1	−1
H	3	3	1	0
I	0	6	4	6
J	1	−6	−4	−6

43. *Cosmological red-shift.* The expansion of the universe is often represented with a drawing like Fig. 45-9*a*. In that figure, we are located at the symbol labeled MW (for the Milky Way galaxy), at the origin of an r axis that extends radially away from us in any direction. Other, very distant galaxies are also represented. Superimposed on their symbols are their velocity vectors as inferred from the red-shift of the light reaching us from the galaxies. In accord with Hubble's law, the velocity of each galaxy is proportional to its distance from us. Such drawings can be misleading because they imply (1) that the red-shifts are due to the motions of galaxies relative to us, as they rush away from us through static (stationary) space, and (2) that we are at the center of all this motion.

Actually, the expansion of the universe and the increased separation of the galaxies are due not to an outward rush of the galaxies into pre-existing space but to an expansion of space itself throughout the universe. *Space is dynamic, not static.*

Figures 45-9*b*, *c*, and *d* show a different way of representing the universe and its expansion. Each part of the figure gives part of a one-dimensional section of the universe (along an r axis); the other two spatial dimensions of the universe are not shown. Each of the three parts of the figure show the Milky Way and six other galaxies (represented by dots); the parts are positioned along a time

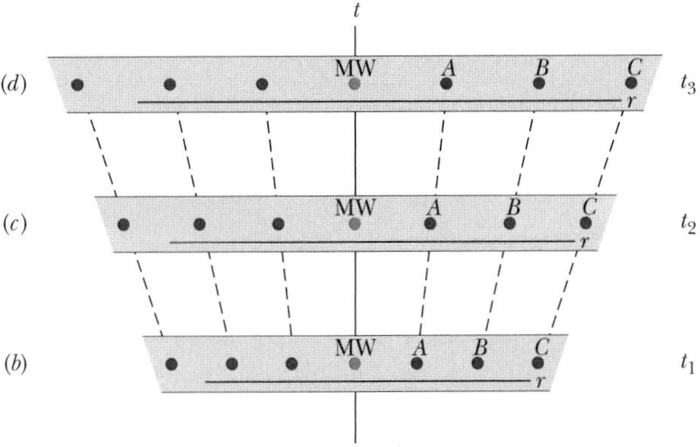

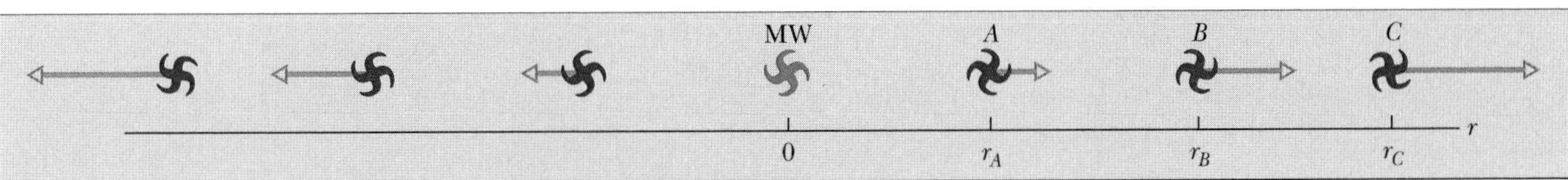

Fig 45-9 Problem 43.

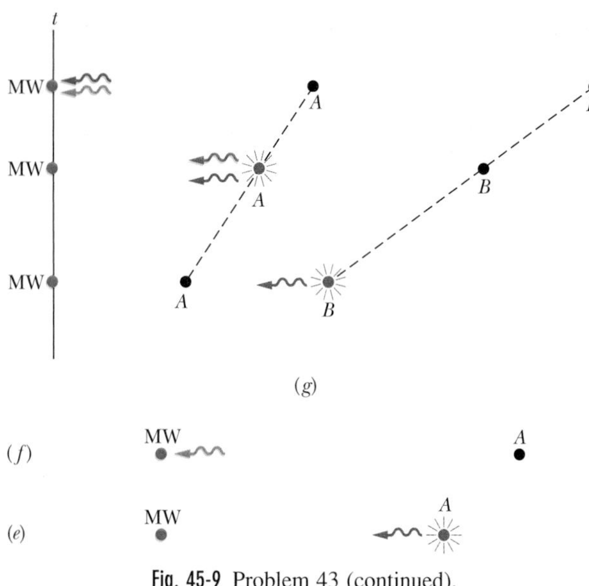

Fig. 45-9 Problem 43 (continued).

axis, with time increasing upward. In part *b*, at the earliest time of the three parts, the Milky Way and the six other galaxies are represented as being relatively close. As time progresses upward in the figures, space expands, causing the galaxies to move apart. Note that the figure parts are drawn relative to the Milky Way, and from that observation point all the other galaxies move away from it because of the expansion. However, there is nothing special about the Milky Way—the galaxies also move away from any other choice of observation point.

Figures 45-9*e* and *f* focus on just the Milky Way galaxy and one of the other galaxies, galaxy *A*, at two particular times during the expansion. In part *e*, galaxy *A* is a distance r from the Milky Way and is emitting a light wave of wavelength λ. In part *f*, after a time interval Δt, that light wave is being detected at Earth. Let us represent the universe's expansion rate per unit length of space with α, which we assume to be constant during time interval Δt. Then during Δt, every unit length of space (say, every meter) expands by $\alpha\,\Delta t$; hence, a distance r expands by $r\alpha\,\Delta t$. The light wave of Figs. 45-9*e* and *f* travels at speed c from galaxy *A* to Earth. (a) Show that

$$\Delta t = \frac{r}{c - r\alpha}.$$

The detected wavelength λ' of the light is greater than the emitted wavelength λ because space expanded during time interval Δt. This increase in wavelength is called the **cosmological red-shift;** it is not a Doppler effect. (b) Show that the change in wavelength $\Delta\lambda$ $(= \lambda' - \lambda)$ is given by

$$\frac{\Delta\lambda}{\lambda} = \frac{r\alpha}{c - r\alpha}.$$

(c) Expand the right side of this equation, using the binomial expansion (given in Appendix E). (d) If you retain only the first term of the expansion, what is the resulting equation for $\Delta\lambda/\lambda$?

If, instead, we assume that Fig. 45-9*a* applies and that $\Delta\lambda$ is due to a Doppler effect, then from Eq. 38-33 we have

$$\frac{\Delta\lambda}{\lambda} = \frac{v}{c},$$

where v is radial velocity of Galaxy *A* relative to Earth. (e) Using Hubble's law, compare this Doppler-effect result with the cosmological-expansion result of (d) and find a value for α. From this analysis you can see that the two results, derived with very different models about the red-shift of the light we detect from distant galaxies, are compatible.

Suppose that the light we detect from Galaxy *A* has a red-shift of $\Delta\lambda/\lambda = 0.050$ and that the expansion rate of the universe has been constant at the current value given in the chapter. (f) Using the result of (b), find the distance between the galaxy and Earth when the light was emitted. Find how long ago the light was emitted by the galaxy (g) by using the result of (a), and (h) by assuming that the red-shift is a Doppler effect. (*Hint:* For (h), the time is just the distance at the time of emission divided by the speed of light, because if the red-shift is just a Doppler effect, the distance does not change during the light's travel to us. Here the two models about the red-shift of the light differ in their results.) (i) At the time of detection, what is the distance between Earth and Galaxy *A*? (We make the assumption that Galaxy *A* still exists; if it ceased to exist, humans would not know about its death until the last light emitted by the galaxy reached Earth.)

Now suppose that the light we detect from Galaxy *B* (Fig. 45-9*g*) has a red-shift of $\Delta\lambda/\lambda = 0.080$. (j) Using the result of (b), find the distance between Galaxy *B* and Earth when the light was emitted. (k) Using the result of (a), find how long ago the light was emitted by Galaxy *B*. (l) When the light that we detect from Galaxy *A* was emitted, what was the distance between Galaxy *A* and Galaxy *B*?

44. Figure 45-10 is a hypothetical plot of the recessional speeds v of galaxies against their distance r from us; the best-fit straight line through the data points is shown. From this plot determine the age of the universe, assuming that Hubble's law holds and that Hubble's constant has had the same value throughout the expansion of the universe.

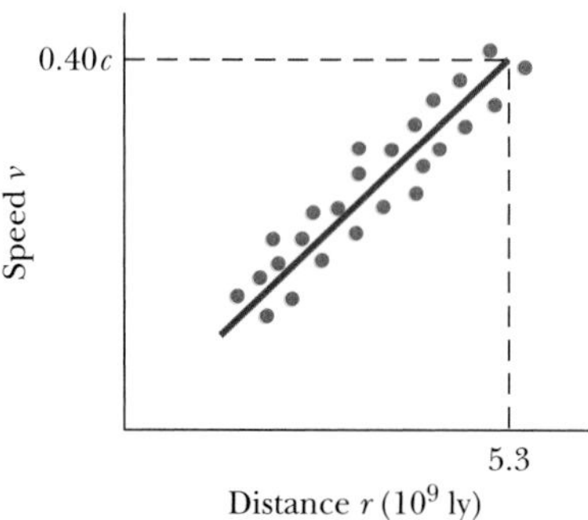

Fig. 45-10 Problem 44.

APPENDIX A
The International System of Units (SI)*

1. The SI Base Units

Quantity	Name	Symbol	Definition
length	meter	m	". . . the length of the path traveled by light in vacuum in 1/299,792,458 of a second." (1983)
mass	kilogram	kg	". . . this prototype [a certain platinum–iridium cylinder] shall henceforth be considered to be the unit of mass." (1889)
time	second	s	". . . the duration of 9,192,631,770 periods of the radiation corresponding to the transition between the two hyperfine levels of the ground state of the cesium-133 atom." (1967)
electric current	ampere	A	". . . that constant current which, if maintained in two straight parallel conductors of infinite length, of negligible circular cross section, and placed 1 meter apart in vacuum, would produce between these conductors a force equal to 2×10^{-7} newton per meter of length." (1946)
thermodynamic temperature	kelvin	K	". . . the fraction 1/273.16 of the thermodynamic temperature of the triple point of water." (1967)
amount of substance	mole	mol	". . . the amount of substance of a system which contains as many elementary entities as there are atoms in 0.012 kilogram of carbon-12." (1971)
luminous intensity	candela	cd	". . . the luminous intensity, in the perpendicular direction, of a surface of 1/600,000 square meter of a blackbody at the temperature of freezing platinum under a pressure of 101.325 newtons per square meter." (1967)

*Adapted from "The International System of Units (SI)," National Bureau of Standards Special Publication 330, 1972 edition. The definitions above were adopted by the General Conference of Weights and Measures, an international body, on the dates shown. In this book we do not use the candela.

2. Some SI Derived Units

Quantity	Name of Unit	Symbol	
area	square meter	m^2	
volume	cubic meter	m^3	
frequency	hertz	Hz	s^{-1}
mass density (density)	kilogram per cubic meter	kg/m^3	
speed, velocity	meter per second	m/s	
angular velocity	radian per second	rad/s	
acceleration	meter per second per second	m/s^2	
angular acceleration	radian per second per second	rad/s^2	
force	newton	N	$kg \cdot m/s^2$
pressure	pascal	Pa	N/m^2
work, energy, quantity of heat	joule	J	$N \cdot m$
power	watt	W	J/s
quantity of electric charge	coulomb	C	$A \cdot s$
potential difference, electromotive force	volt	V	W/A
electric field strength	volt per meter (or newton per coulomb)	V/m	N/C
electric resistance	ohm	Ω	V/A
capacitance	farad	F	$A \cdot s/V$
magnetic flux	weber	Wb	$V \cdot s$
inductance	henry	H	$V \cdot s/A$
magnetic flux density	tesla	T	Wb/m^2
magnetic field strength	ampere per meter	A/m	
entropy	joule per kelvin	J/K	
specific heat	joule per kilogram kelvin	$J/(kg \cdot K)$	
thermal conductivity	watt per meter kelvin	$W/(m \cdot K)$	
radiant intensity	watt per steradian	W/sr	

3. The SI Supplementary Units

Quantity	Name of Unit	Symbol
plane angle	radian	rad
solid angle	steradian	sr

APPENDIX B

Some Fundamental Constants of Physics*

Constant	Symbol	Computational Value	Best (1998) Value	
			Value[a]	Uncertainty[b]
Speed of light in a vacuum	c	3.00×10^{8} m/s	2.997 924 58	exact
Elementary charge	e	1.60×10^{-19} C	1.602 176 462	0.039
Gravitational constant	G	6.67×10^{-11} m^3/s$^2 \cdot$ kg	6.673	1500
Universal gas constant	R	8.31 J/mol $\cdot$ K	8.314 472	1.7
Avogadro constant	N_A	6.02×10^{23} mol^{-1}	6.022 141 99	0.079
Boltzmann constant	k	1.38×10^{-23} J/K	1.380 650 3	1.7
Stefan–Boltzmann constant	σ	5.67×10^{-8} W/m$^2 \cdot$ K^4	5.670 400	7.0
Molar volume of ideal gas at STP[d]	V_m	2.27×10^{-2} m^3/mol	2.271 098 1	1.7
Permittivity constant	ϵ_0	8.85×10^{-12} F/m	8.854 187 817 62	exact
Permeability constant	μ_0	1.26×10^{-6} H/m	1.256 637 061 43	exact
Planck constant	h	6.63×10^{-34} J $\cdot$ s	6.626 068 76	0.078
Electron mass[c]	m_e	9.11×10^{-31} kg	9.109 381 88	0.079
		5.49×10^{-4} u	5.485 799 110	0.0021
Proton mass[c]	m_p	1.67×10^{-27} kg	1.672 621 58	0.079
		1.0073 u	1.007 276 466 88	1.3×10^{-4}
Ratio of proton mass to electron mass	m_p/m_e	1840	1836.152 667 5	0.0021
Electron charge-to-mass ratio	e/m_e	1.76×10^{11} C/kg	1.758 820 174	0.040
Neutron mass[c]	m_n	1.68×10^{-27} kg	1.674 927 16	0.079
		1.0087 u	1.008 664 915 78	5.4×10^{-4}
Hydrogen atom mass[c]	$m_{^1H}$	1.0078 u	1.007 825 031 6	0.0005
Deuterium atom mass[c]	$m_{^2H}$	2.0141 u	2.014 101 777 9	0.0005
Helium atom mass[c]	$m_{^4He}$	4.0026 u	4.002 603 2	0.067
Muon mass	m_μ	1.88×10^{-28} kg	1.883 531 09	0.084
Electron magnetic moment	μ_e	9.28×10^{-24} J/T	9.284 763 62	0.040
Proton magnetic moment	μ_p	1.41×10^{-26} J/T	1.410 606 663	0.041
Bohr magneton	μ_B	9.27×10^{-24} J/T	9.274 008 99	0.040
Nuclear magneton	μ_N	5.05×10^{-27} J/T	5.050 783 17	0.040
Bohr radius	r_B	5.29×10^{-11} m	5.291 772 083	0.0037
Rydberg constant	R	1.10×10^{7} m^{-1}	1.097 373 156 854 8	7.6×10^{-6}
Electron Compton wavelength	λ_C	2.43×10^{-12} m	2.426 310 215	0.0073

[a]Values given in this column should be given the same unit and power of 10 as the computational value.

[b]Parts per million.

[c]Masses given in u are in unified atomic mass units, where 1 u = $1.660\,538\,73 \times 10^{-27}$ kg.

[d]STP means standard temperature and pressure: 0°C and 1.0 atm (0.1 MPa).

*The values in this table were selected from the 1998 CODATA recommended values (www.physics.nist.gov).

APPENDIX C
Some Astronomical Data

Some Distances from Earth

To the Moon*	3.82×10^{8} m	To the center of our galaxy	2.2×10^{20} m
To the Sun*	1.50×10^{11} m	To the Andromeda Galaxy	2.1×10^{22} m
To the nearest star (Proxima Centauri)	4.04×10^{16} m	To the edge of the observable universe	$\sim 10^{26}$ m

*Mean distance.

The Sun, Earth, and the Moon

Property	Unit	Sun		Earth	Moon
Mass	kg	1.99×10^{30}		5.98×10^{24}	7.36×10^{22}
Mean radius	m	6.96×10^{8}		6.37×10^{6}	1.74×10^{6}
Mean density	kg/m^3	1410		5520	3340
Free-fall acceleration at the surface	m/s^2	274		9.81	1.67
Escape velocity	km/s	618		11.2	2.38
Period of rotation[a]	—	37 d at poles[b]	26 d at equator[b]	23 h 56 min	27.3 d
Radiation power[c]	W	3.90×10^{26}			

[a]Measured with respect to the distant stars.
[b]The Sun, a ball of gas, does not rotate as a rigid body.
[c]Just outside Earth's atmosphere solar energy is received, assuming normal incidence, at the rate of 1340 W/m^2.

Some Properties of the Planets

	Mercury	Venus	Earth	Mars	Jupiter	Saturn	Uranus	Neptune	Pluto
Mean distance from Sun, 10^6 km	57.9	108	150	228	778	1430	2870	4500	5900
Period of revolution, y	0.241	0.615	1.00	1.88	11.9	29.5	84.0	165	248
Period of rotation,[a] d	58.7	-243[b]	0.997	1.03	0.409	0.426	-0.451[b]	0.658	6.39
Orbital speed, km/s	47.9	35.0	29.8	24.1	13.1	9.64	6.81	5.43	4.74
Inclination of axis to orbit	$<28°$	$\approx 3°$	23.4°	25.0°	3.08°	26.7°	97.9°	29.6°	57.5°
Inclination of orbit to Earth's orbit	7.00°	3.39°		1.85°	1.30°	2.49°	0.77°	1.77°	17.2°
Eccentricity of orbit	0.206	0.0068	0.0167	0.0934	0.0485	0.0556	0.0472	0.0086	0.250
Equatorial diameter, km	4880	12 100	12 800	6790	143 000	120 000	51 800	49 500	2300
Mass (Earth = 1)	0.0558	0.815	1.000	0.107	318	95.1	14.5	17.2	0.002
Density (water = 1)	5.60	5.20	5.52	3.95	1.31	0.704	1.21	1.67	2.03
Surface value of g,[c] m/s^2	3.78	8.60	9.78	3.72	22.9	9.05	7.77	11.0	0.5
Escape velocity,[c] km/s	4.3	10.3	11.2	5.0	59.5	35.6	21.2	23.6	1.1
Known satellites	0	0	1	2	16 + ring	18 + rings	17 + rings	8 + rings	1

[a]Measured with respect to the distant stars.
[b]Venus and Uranus rotate opposite their orbital motion.
[c]Gravitational acceleration measured at the planet's equator.

APPENDIX D
Conversion Factors

Conversion factors may be read directly from these tables. For example, 1 degree = 2.778×10^{-3} revolutions, so $16.7° = 16.7 \times 2.778 \times 10^{-3}$ rev. The SI units are fully capitalized. Adapted in part from G. Shortley and D. Williams, *Elements of Physics,* 1971, Prentice-Hall, Englewood Cliffs, NJ.

Plane Angle

	°	′	″	RADIAN	rev
1 degree =	1	60	3600	1.745×10^{-2}	2.778×10^{-3}
1 minute =	1.667×10^{-2}	1	60	2.909×10^{-4}	4.630×10^{-5}
1 second =	2.778×10^{-4}	1.667×10^{-2}	1	4.848×10^{-6}	7.716×10^{-7}
1 RADIAN =	57.30	3438	2.063×10^{5}	1	0.1592
1 revolution =	360	2.16×10^{4}	1.296×10^{6}	6.283	1

Solid Angle

1 sphere = 4π steradians = 12.57 steradians

Length

	cm	METER	km	in.	ft	mi
1 centimeter =	1	10^{-2}	10^{-5}	0.3937	3.281×10^{-2}	6.214×10^{-6}
1 METER =	100	1	10^{-3}	39.37	3.281	6.214×10^{-4}
1 kilometer =	10^{5}	1000	1	3.937×10^{4}	3281	0.6214
1 inch =	2.540	2.540×10^{-2}	2.540×10^{-5}	1	8.333×10^{-2}	1.578×10^{-5}
1 foot =	30.48	0.3048	3.048×10^{-4}	12	1	1.894×10^{-4}
1 mile =	1.609×10^{5}	1609	1.609	6.336×10^{4}	5280	1

1 angström = 10^{-10} m
1 nautical mile = 1852 m
= 1.151 miles = 6076 ft
1 fermi = 10^{-15} m
1 light-year = 9.460×10^{12} km
1 parsec = 3.084×10^{13} km
1 fathom = 6 ft
1 Bohr radius = 5.292×10^{-11} m
1 yard = 3 ft
1 rod = 16.5 ft
1 mil = 10^{-3} in.
1 nm = 10^{-9} m

Area

	METER^2	cm^2	ft^2	in.^{2}
1 SQUARE METER =	1	10^{4}	10.76	1550
1 square centimeter =	10^{-4}	1	1.076×10^{-3}	0.1550
1 square foot =	9.290×10^{-2}	929.0	1	144
1 square inch =	6.452×10^{-4}	6.452	6.944×10^{-3}	1

1 square mile = 2.788×10^{7} ft^2 = 640 acres
1 barn = 10^{-28} m^2
1 acre = 43 560 ft^2
1 hectare = 10^{4} m^2 = 2.471 acres

Volume

	METER³	cm³	L	ft³	in.³
1 CUBIC METER =	1	10^6	1000	35.31	6.102×10^4
1 cubic centimeter =	10^{-6}	1	1.000×10^{-3}	3.531×10^{-5}	6.102×10^{-2}
1 liter =	1.000×10^{-3}	1000	1	3.531×10^{-2}	61.02
1 cubic foot =	2.832×10^{-2}	2.832×10^4	28.32	1	1728
1 cubic inch =	1.639×10^{-5}	16.39	1.639×10^{-2}	5.787×10^{-4}	1

1 U.S. fluid gallon = 4 U.S. fluid quarts = 8 U.S. pints = 128 U.S. fluid ounces = 231 in.³

1 British imperial gallon = 277.4 in.³ = 1.201 U.S. fluid gallons

Mass

Quantities in the colored areas are not mass units but are often used as such. When we write, for example, 1 kg "=" 2.205 lb, this means that a kilogram is a *mass* that *weighs* 2.205 pounds at a location where g has the standard value of 9.80665 m/s².

	g	KILOGRAM	slug	u	oz	lb	ton
1 gram =	1	0.001	6.852×10^{-5}	6.022×10^{23}	3.527×10^{-2}	2.205×10^{-3}	1.102×10^{-6}
1 KILOGRAM =	1000	1	6.852×10^{-2}	6.022×10^{26}	35.27	2.205	1.102×10^{-3}
1 slug =	1.459×10^4	14.59	1	8.786×10^{27}	514.8	32.17	1.609×10^{-2}
1 atomic mass unit =	1.661×10^{-24}	1.661×10^{-27}	1.138×10^{-28}	1	5.857×10^{-26}	3.662×10^{-27}	1.830×10^{-30}
1 ounce =	28.35	2.835×10^{-2}	1.943×10^{-3}	1.718×10^{25}	1	6.250×10^{-2}	3.125×10^{-5}
1 pound =	453.6	0.4536	3.108×10^{-2}	2.732×10^{26}	16	1	0.0005
1 ton =	9.072×10^5	907.2	62.16	5.463×10^{29}	3.2×10^4	2000	1

1 metric ton = 1000 kg

Density

Quantities in the colored areas are weight densities and, as such, are dimensionally different from mass densities. See note for mass table.

	slug/ft³	KILOGRAM/ METER³	g/cm³	lb/ft³	lb/in.³
1 slug per foot³ =	1	515.4	0.5154	32.17	1.862×10^{-2}
1 KILOGRAM per METER³ =	1.940×10^{-3}	1	0.001	6.243×10^{-2}	3.613×10^{-5}
1 gram per centimeter³ =	1.940	1000	1	62.43	3.613×10^{-2}
1 pound per foot³ =	3.108×10^{-2}	16.02	16.02×10^{-2}	1	5.787×10^{-4}
1 pound per inch³ =	53.71	2.768×10^4	27.68	1728	1

Time

	y	d	h	min	SECOND
1 year =	1	365.25	8.766×10^3	5.259×10^5	3.156×10^7
1 day =	2.738×10^{-3}	1	24	1440	8.640×10^4
1 hour =	1.141×10^{-4}	4.167×10^{-2}	1	60	3600
1 minute =	1.901×10^{-6}	6.944×10^{-4}	1.667×10^{-2}	1	60
1 SECOND =	3.169×10^{-8}	1.157×10^{-5}	2.778×10^{-4}	1.667×10^{-2}	1

Speed

	ft/s	km/h	METER/SECOND	mi/h	cm/s
1 foot per second =	1	1.097	0.3048	0.6818	30.48
1 kilometer per hour =	0.9113	1	0.2778	0.6214	27.78
1 METER per SECOND =	3.281	3.6	1	2.237	100
1 mile per hour =	1.467	1.609	0.4470	1	44.70
1 centimeter per second =	3.281×10^{-2}	3.6×10^{-2}	0.01	2.237×10^{-2}	1

1 knot = 1 nautical mi/h = 1.688 ft/s   1 mi/min = 88.00 ft/s = 60.00 mi/h

Force

Force units in the colored areas are now little used. To clarify: 1 gram-force (= 1 gf) is the force of gravity that would act on an object whose mass is 1 gram at a location where g has the standard value of 9.80665 m/s^2.

	dyne	NEWTON	lb	pdl	gf	kgf
1 dyne =	1	10^{-5}	2.248×10^{-6}	7.233×10^{-5}	1.020×10^{-3}	1.020×10^{-6}
1 NEWTON =	10^5	1	0.2248	7.233	102.0	0.1020
1 pound =	4.448×10^5	4.448	1	32.17	453.6	0.4536
1 poundal =	1.383×10^4	0.1383	3.108×10^{-2}	1	14.10	1.410×10^{2}
1 gram-force =	980.7	9.807×10^{-3}	2.205×10^{-3}	7.093×10^{-2}	1	0.001
1 kilogram-force =	9.807×10^5	9.807	2.205	70.93	1000	1

1 ton = 2000 lb

Pressure

	atm	$dyne/cm^2$	inch of water	cm Hg	PASCAL	$lb/in.^2$	lb/ft^2
1 atmosphere =	1	1.013×10^6	406.8	76	1.013×10^5	14.70	2116
1 dyne per centimeter^2 =	9.869×10^{-7}	1	4.015×10^{-4}	7.501×10^{-5}	0.1	1.405×10^{-5}	2.089×10^{-3}
1 inch of water[a] at 4°C =	2.458×10^{-3}	2491	1	0.1868	249.1	3.613×10^{-2}	5.202
1 centimeter of mercury[a] at 0°C =	1.316×10^{-2}	1.333×10^4	5.353	1	1333	0.1934	27.85
1 PASCAL =	9.869×10^{-6}	10	4.015×10^{-3}	7.501×10^{-4}	1	1.450×10^{-4}	2.089×10^{-2}
1 pound per inch^2 =	6.805×10^{-2}	6.895×10^4	27.68	5.171	6.895×10^3	1	144
1 pound per foot^2 =	4.725×10^{-4}	478.8	0.1922	3.591×10^{-2}	47.88	6.944×10^{-3}	1

[a]Where the acceleration of gravity has the standard value of 9.80665 m/s^2.

1 bar = 10^6 $dyne/cm^2$ = 0.1 MPa   1 millibar = 10^3 $dyne/cm^2$ = 10^2 Pa   1 torr = 1 mm Hg

Energy, Work, Heat

Quantities in the colored areas are not energy units but are included for convenience. They arise from the relativistic mass–energy equivalence formula $E = mc^2$ and represent the energy released if a kilogram or unified atomic mass unit (u) is completely converted to energy (bottom two rows) or the mass that would be completely converted to one unit of energy (rightmost two columns).

	Btu	erg	ft · lb	hp · h	JOULE	cal	kW · h	eV	MeV	kg	u
1 British thermal unit =	1	1.055×10^{10}	777.9	3.929×10^{-4}	1055	252.0	2.930×10^{-4}	6.585×10^{21}	6.585×10^{15}	1.174×10^{-14}	7.070×10^{12}
1 erg =	9.481×10^{-11}	1	7.376×10^{-8}	3.725×10^{-14}	10^{-7}	2.389×10^{-8}	2.778×10^{-14}	6.242×10^{11}	6.242×10^{5}	1.113×10^{-24}	670.2
1 foot-pound =	1.285×10^{-3}	1.356×10^{7}	1	5.051×10^{-7}	1.356	0.3238	3.766×10^{-7}	8.464×10^{18}	8.464×10^{12}	1.509×10^{-17}	9.037×10^{9}
1 horsepower-hour =	2545	2.685×10^{13}	1.980×10^{6}	1	2.685×10^{6}	6.413×10^{5}	0.7457	1.676×10^{25}	1.676×10^{19}	2.988×10^{-11}	1.799×10^{16}
1 JOULE =	9.481×10^{-4}	10^{7}	0.7376	3.725×10^{-7}	1	0.2389	2.778×10^{-7}	6.242×10^{18}	6.242×10^{12}	1.113×10^{-17}	6.702×10^{9}
1 calorie =	3.969×10^{-3}	4.186×10^{7}	3.088	1.560×10^{-6}	4.186	1	1.163×10^{-6}	2.613×10^{19}	2.613×10^{13}	4.660×10^{-17}	2.806×10^{10}
1 kilowatt-hour =	3413	3.600×10^{13}	2.655×10^{6}	1.341	3.600×10^{6}	8.600×10^{5}	1	2.247×10^{25}	2.247×10^{19}	4.007×10^{-11}	2.413×10^{16}
1 electron-volt =	1.519×10^{-22}	1.602×10^{-12}	1.182×10^{-19}	5.967×10^{-26}	1.602×10^{-19}	3.827×10^{-20}	4.450×10^{-26}	1	10^{-6}	1.783×10^{-36}	1.074×10^{-9}
1 million electron-volts =	1.519×10^{-16}	1.602×10^{-6}	1.182×10^{-13}	5.967×10^{-20}	1.602×10^{-13}	3.827×10^{-14}	4.450×10^{-20}	10^{-6}	1	1.783×10^{-30}	1.074×10^{-3}
1 kilogram =	8.521×10^{13}	8.987×10^{23}	6.629×10^{16}	3.348×10^{10}	8.987×10^{16}	2.146×10^{16}	2.497×10^{10}	5.610×10^{35}	5.610×10^{29}	1	6.022×10^{26}
1 unified atomic mass unit =	1.415×10^{-13}	1.492×10^{-3}	1.101×10^{-10}	5.559×10^{-17}	1.492×10^{-10}	3.564×10^{-11}	4.146×10^{-17}	9.320×10^{8}	932.0	1.661×10^{-27}	1

Power

	Btu/h	ft · lb/s	hp	cal/s	kW	WATT
1 British thermal unit per hour =	1	0.2161	3.929×10^{-4}	6.998×10^{-2}	2.930×10^{-4}	0.2930
1 foot-pound per second =	4.628	1	1.818×10^{-3}	0.3239	1.356×10^{-3}	1.356
1 horsepower =	2545	550	1	178.1	0.7457	745.7
1 calorie per second =	14.29	3.088	5.615×10^{-3}	1	4.186×10^{-3}	4.186
1 kilowatt =	3413	737.6	1.341	238.9	1	1000
1 WATT =	3.413	0.7376	1.341×10^{-3}	0.2389	0.001	1

Magnetic Field

	gauss	TESLA	milligauss
1 gauss =	1	10^{-4}	1000
1 TESLA =	10^{4}	1	10^{7}
1 milligauss =	0.001	10^{-7}	1

1 tesla = 1 weber/meter^2

Magnetic Flux

	maxwell	WEBER
1 maxwell =	1	10^{-8}
1 WEBER =	10^{8}	1

APPENDIX E
Mathematical Formulas

Geometry

Circle of radius r: circumference $= 2\pi r$; area $= \pi r^2$.

Sphere of radius r: area $= 4\pi r^2$; volume $= \frac{4}{3}\pi r^3$.

Right circular cylinder of radius r and height h:
area $= 2\pi r^2 + 2\pi rh$; volume $= \pi r^2 h$.

Triangle of base a and altitude h: area $= \frac{1}{2}ah$.

Quadratic Formula

If $ax^2 + bx + c = 0$, then $x = \dfrac{-b \pm \sqrt{b^2 - 4ac}}{2a}$.

Trigonometric Functions of Angle θ

$\sin\theta = \dfrac{y}{r}$ $\quad\cos\theta = \dfrac{x}{r}$

$\tan\theta = \dfrac{y}{x}$ $\quad\cot\theta = \dfrac{x}{y}$

$\sec\theta = \dfrac{r}{x}$ $\quad\csc\theta = \dfrac{r}{y}$

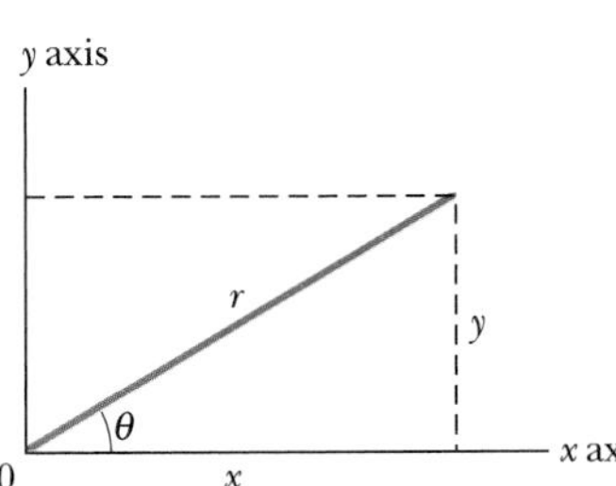

Pythagorean Theorem

In this right triangle,
$a^2 + b^2 = c^2$

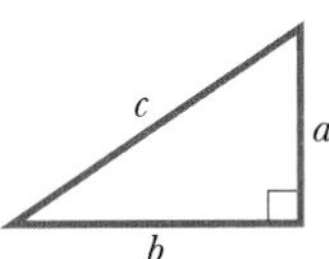

Triangles

Angles are A, B, C

Opposite sides are a, b, c

Angles $A + B + C = 180°$

$\dfrac{\sin A}{a} = \dfrac{\sin B}{b} = \dfrac{\sin C}{c}$

$c^2 = a^2 + b^2 - 2ab\cos C$

Exterior angle $D = A + C$

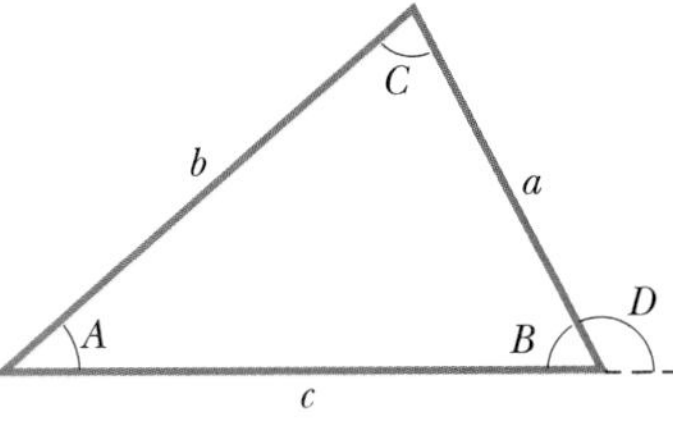

Mathematical Signs and Symbols

$=$ equals

$\approx$ equals approximately

$\sim$ is the order of magnitude of

$\neq$ is not equal to

$\equiv$ is identical to, is defined as

$>$ is greater than ($\gg$ is much greater than)

$<$ is less than ($\ll$ is much less than)

$\geq$ is greater than or equal to (or, is no less than)

$\leq$ is less than or equal to (or, is no more than)

$\pm$ plus or minus

$\propto$ is proportional to

Σ the sum of

x_{avg} the average value of x

Trigonometric Identities

$\sin(90° - \theta) = \cos\theta$

$\cos(90° - \theta) = \sin\theta$

$\sin\theta/\cos\theta = \tan\theta$

$\sin^2\theta + \cos^2\theta = 1$

$\sec^2\theta - \tan^2\theta = 1$

$\csc^2\theta - \cot^2\theta = 1$

$\sin 2\theta = 2\sin\theta\cos\theta$

$\cos 2\theta = \cos^2\theta - \sin^2\theta = 2\cos^2\theta - 1 = 1 - 2\sin^2\theta$

$\sin(\alpha \pm \beta) = \sin\alpha\cos\beta \pm \cos\alpha\sin\beta$

$\cos(\alpha \pm \beta) = \cos\alpha\cos\beta \mp \sin\alpha\sin\beta$

$\tan(\alpha \pm \beta) = \dfrac{\tan\alpha \pm \tan\beta}{1 \mp \tan\alpha\tan\beta}$

$\sin\alpha \pm \sin\beta = 2\sin\frac{1}{2}(\alpha \pm \beta)\cos\frac{1}{2}(\alpha \mp \beta)$

$\cos\alpha + \cos\beta = 2\cos\frac{1}{2}(\alpha + \beta)\cos\frac{1}{2}(\alpha - \beta)$

$\cos\alpha - \cos\beta = -2\sin\frac{1}{2}(\alpha + \beta)\sin\frac{1}{2}(\alpha - \beta)$

Binomial Theorem

$$(1+x)^n = 1 + \frac{nx}{1!} + \frac{n(n-1)x^2}{2!} + \cdots \qquad (x^2 < 1)$$

Exponential Expansion

$$e^x = 1 + x + \frac{x^2}{2!} + \frac{x^3}{3!} + \cdots$$

Logarithmic Expansion

$$\ln(1+x) = x - \tfrac{1}{2}x^2 + \tfrac{1}{3}x^3 - \cdots \qquad (|x| < 1)$$

Trigonometric Expansions (θ in radians)

$$\sin\theta = \theta - \frac{\theta^3}{3!} + \frac{\theta^5}{5!} - \cdots$$

$$\cos\theta = 1 - \frac{\theta^2}{2!} + \frac{\theta^4}{4!} - \cdots$$

$$\tan\theta = \theta + \frac{\theta^3}{3} + \frac{2\theta^5}{15} + \cdots$$

Cramer's Rule

Two simultaneous equations in unknowns x and y,

$$a_1x + b_1y = c_1 \quad \text{and} \quad a_2x + b_2y = c_2,$$

have the solutions

$$x = \frac{\begin{vmatrix} c_1 & b_1 \\ c_2 & b_2 \end{vmatrix}}{\begin{vmatrix} a_1 & b_1 \\ a_2 & b_2 \end{vmatrix}} = \frac{c_1b_2 - c_2b_1}{a_1b_2 - a_2b_1}$$

and

$$y = \frac{\begin{vmatrix} a_1 & c_1 \\ a_2 & c_2 \end{vmatrix}}{\begin{vmatrix} a_1 & b_1 \\ a_2 & b_2 \end{vmatrix}} = \frac{a_1c_2 - a_2c_1}{a_1b_2 - a_2b_1}.$$

Products of Vectors

Let $\hat{\mathrm{i}}$, $\hat{\mathrm{j}}$, and $\hat{\mathrm{k}}$ be unit vectors in the x, y, and z directions. Then

$$\hat{\mathrm{i}}\cdot\hat{\mathrm{i}} = \hat{\mathrm{j}}\cdot\hat{\mathrm{j}} = \hat{\mathrm{k}}\cdot\hat{\mathrm{k}} = 1, \qquad \hat{\mathrm{i}}\cdot\hat{\mathrm{j}} = \hat{\mathrm{j}}\cdot\hat{\mathrm{k}} = \hat{\mathrm{k}}\cdot\hat{\mathrm{i}} = 0,$$

$$\hat{\mathrm{i}}\times\hat{\mathrm{i}} = \hat{\mathrm{j}}\times\hat{\mathrm{j}} = \hat{\mathrm{k}}\times\hat{\mathrm{k}} = 0,$$

$$\hat{\mathrm{i}}\times\hat{\mathrm{j}} = \hat{\mathrm{k}}, \qquad \hat{\mathrm{j}}\times\hat{\mathrm{k}} = \hat{\mathrm{i}}, \qquad \hat{\mathrm{k}}\times\hat{\mathrm{i}} = \hat{\mathrm{j}}.$$

Any vector $\vec{a}$ with components a_x, a_y, and a_z along the x, y, and z axes can be written as

$$\vec{a} = a_x\hat{\mathrm{i}} + a_y\hat{\mathrm{j}} + a_z\hat{\mathrm{k}}.$$

Let $\vec{a}$, $\vec{b}$, and $\vec{c}$ be arbitrary vectors with magnitudes a, b, and c. Then

$$\vec{a}\times(\vec{b}+\vec{c}) = (\vec{a}\times\vec{b}) + (\vec{a}\times\vec{c})$$

$$(s\vec{a})\times\vec{b} = \vec{a}\times(s\vec{b}) = s(\vec{a}\times\vec{b}) \quad (s = \text{a scalar}).$$

Let θ be the smaller of the two angles between $\vec{a}$ and $\vec{b}$. Then

$$\vec{a}\cdot\vec{b} = \vec{b}\cdot\vec{a} = a_xb_x + a_yb_y + a_zb_z = ab\cos\theta$$

$$\vec{a}\times\vec{b} = -\vec{b}\times\vec{a} = \begin{vmatrix} \hat{\mathrm{i}} & \hat{\mathrm{j}} & \hat{\mathrm{k}} \\ a_x & a_y & a_z \\ b_x & b_y & b_z \end{vmatrix}$$

$$= \hat{\mathrm{i}}\begin{vmatrix} a_y & a_z \\ b_y & b_z \end{vmatrix} - \hat{\mathrm{j}}\begin{vmatrix} a_x & a_z \\ b_x & b_z \end{vmatrix} + \hat{\mathrm{k}}\begin{vmatrix} a_x & a_y \\ b_x & b_y \end{vmatrix}$$

$$= (a_yb_z - b_ya_z)\hat{\mathrm{i}} + (a_zb_x - b_za_x)\hat{\mathrm{j}} + (a_xb_y - b_xa_y)\hat{\mathrm{k}}$$

$$|\vec{a}\times\vec{b}| = ab\sin\theta$$

$$\vec{a}\cdot(\vec{b}\times\vec{c}) = \vec{b}\cdot(\vec{c}\times\vec{a}) = \vec{c}\cdot(\vec{a}\times\vec{b})$$

$$\vec{a}\times(\vec{b}\times\vec{c}) = (\vec{a}\cdot\vec{c})\vec{b} - (\vec{a}\cdot\vec{b})\vec{c}$$

Derivatives and Integrals

In what follows, the letters u and v stand for any functions of x, and a and m are constants. To each of the indefinite integrals should be added an arbitrary constant of integration. The *Handbook of Chemistry and Physics* (CRC Press Inc.) gives a more extensive tabulation.

1. $\dfrac{dx}{dx} = 1$
2. $\dfrac{d}{dx}(au) = a\dfrac{du}{dx}$
3. $\dfrac{d}{dx}(u + v) = \dfrac{du}{dx} + \dfrac{dv}{dx}$
4. $\dfrac{d}{dx}x^m = mx^{m-1}$
5. $\dfrac{d}{dx}\ln x = \dfrac{1}{x}$
6. $\dfrac{d}{dx}(uv) = u\dfrac{dv}{dx} + v\dfrac{du}{dx}$
7. $\dfrac{d}{dx}e^x = e^x$
8. $\dfrac{d}{dx}\sin x = \cos x$
9. $\dfrac{d}{dx}\cos x = -\sin x$
10. $\dfrac{d}{dx}\tan x = \sec^2 x$
11. $\dfrac{d}{dx}\cot x = -\csc^2 x$
12. $\dfrac{d}{dx}\sec x = \tan x \sec x$
13. $\dfrac{d}{dx}\csc x = -\cot x \csc x$
14. $\dfrac{d}{dx}e^u = e^u\dfrac{du}{dx}$
15. $\dfrac{d}{dx}\sin u = \cos u\dfrac{du}{dx}$
16. $\dfrac{d}{dx}\cos u = -\sin u\dfrac{du}{dx}$

1. $\displaystyle\int dx = x$
2. $\displaystyle\int au\,dx = a\int u\,dx$
3. $\displaystyle\int (u + v)\,dx = \int u\,dx + \int v\,dx$
4. $\displaystyle\int x^m\,dx = \frac{x^{m+1}}{m + 1} \quad (m \neq -1)$
5. $\displaystyle\int \frac{dx}{x} = \ln|x|$
6. $\displaystyle\int u\frac{dv}{dx}\,dx = uv - \int v\frac{du}{dx}\,dx$
7. $\displaystyle\int e^x\,dx = e^x$
8. $\displaystyle\int \sin x\,dx = -\cos x$
9. $\displaystyle\int \cos x\,dx = \sin x$
10. $\displaystyle\int \tan x\,dx = \ln|\sec x|$
11. $\displaystyle\int \sin^2 x\,dx = \tfrac{1}{2}x - \tfrac{1}{4}\sin 2x$
12. $\displaystyle\int e^{-ax}\,dx = -\frac{1}{a}e^{-ax}$
13. $\displaystyle\int xe^{-ax}\,dx = -\frac{1}{a^2}(ax + 1)e^{-ax}$
14. $\displaystyle\int x^2e^{-ax}\,dx = -\frac{1}{a^3}(a^2x^2 + 2ax + 2)e^{-ax}$
15. $\displaystyle\int_0^\infty x^ne^{-ax}\,dx = \frac{n!}{a^{n+1}}$
16. $\displaystyle\int_0^\infty x^{2n}e^{-ax^2}\,dx = \frac{1\cdot 3\cdot 5\cdots(2n - 1)}{2^{n+1}a^n}\sqrt{\frac{\pi}{a}}$
17. $\displaystyle\int \frac{dx}{\sqrt{x^2 + a^2}} = \ln(x + \sqrt{x^2 + a^2})$
18. $\displaystyle\int \frac{x\,dx}{(x^2 + a^2)^{3/2}} = -\frac{1}{(x^2 + a^2)^{1/2}}$
19. $\displaystyle\int \frac{dx}{(x^2 + a^2)^{3/2}} = \frac{x}{a^2(x^2 + a^2)^{1/2}}$
20. $\displaystyle\int_0^\infty x^{2n+1}\,e^{-ax^2}\,dx = \frac{n!}{2a^{n+1}} \quad (a > 0)$
21. $\displaystyle\int \frac{x\,dx}{x + d} = x - d\ln(x + d)$

APPENDIX F
Properties of the Elements

All physical properties are for a pressure of 1 atm unless otherwise specified.

Element	Symbol	Atomic Number Z	Molar Mass, g/mol	Density, g/cm^3 at 20°C	Melting Point, °C	Boiling Point, °C	Specific Heat, J/(g · °C) at 25°C
Actinium	Ac	89	(227)	10.06	1323	(3473)	0.092
Aluminum	Al	13	26.9815	2.699	660	2450	0.900
Americium	Am	95	(243)	13.67	1541	—	—
Antimony	Sb	51	121.75	6.691	630.5	1380	0.205
Argon	Ar	18	39.948	1.6626×10^{-3}	−189.4	−185.8	0.523
Arsenic	As	33	74.9216	5.78	817 (28 atm)	613	0.331
Astatine	At	85	(210)	—	(302)	—	—
Barium	Ba	56	137.34	3.594	729	1640	0.205
Berkelium	Bk	97	(247)	14.79	—	—	—
Beryllium	Be	4	9.0122	1.848	1287	2770	1.83
Bismuth	Bi	83	208.980	9.747	271.37	1560	0.122
Bohrium	Bh	107	262.12	—	—	—	—
Boron	B	5	10.811	2.34	2030	—	1.11
Bromine	Br	35	79.909	3.12 (liquid)	−7.2	58	0.293
Cadmium	Cd	48	112.40	8.65	321.03	765	0.226
Calcium	Ca	20	40.08	1.55	838	1440	0.624
Californium	Cf	98	(251)	—	—	—	—
Carbon	C	6	12.01115	2.26	3727	4830	0.691
Cerium	Ce	58	140.12	6.768	804	3470	0.188
Cesium	Cs	55	132.905	1.873	28.40	690	0.243
Chlorine	Cl	17	35.453	3.214×10^{-3} (0°C)	−101	−34.7	0.486
Chromium	Cr	24	51.996	7.19	1857	2665	0.448
Cobalt	Co	27	58.9332	8.85	1495	2900	0.423
Copper	Cu	29	63.54	8.96	1083.40	2595	0.385
Curium	Cm	96	(247)	13.3	—	—	—
Dubnium	Db	105	262.114	—	—	—	—
Dysprosium	Dy	66	162.50	8.55	1409	2330	0.172
Einsteinium	Es	99	(254)	—	—	—	—
Erbium	Er	68	167.26	9.15	1522	2630	0.167
Europium	Eu	63	151.96	5.243	817	1490	0.163
Fermium	Fm	100	(237)	—	—	—	—
Fluorine	F	9	18.9984	1.696×10^{-3} (0°C)	−219.6	−188.2	0.753
Francium	Fr	87	(223)	—	(27)	—	—
Gadolinium	Gd	64	157.25	7.90	1312	2730	0.234
Gallium	Ga	31	69.72	5.907	29.75	2237	0.377

Element	Symbol	Atomic Number Z	Molar Mass, g/mol	Density, g/cm^3 at 20°C	Melting Point, °C	Boiling Point, °C	Specific Heat, J/(g · °C) at 25°C
Germanium	Ge	32	72.59	5.323	937.25	2830	0.322
Gold	Au	79	196.967	19.32	1064.43	2970	0.131
Hafnium	Hf	72	178.49	13.31	2227	5400	0.144
Hassium	Hs	108	(265)	—	—	—	—
Helium	He	2	4.0026	0.1664×10^{-3}	−269.7	−268.9	5.23
Holmium	Ho	67	164.930	8.79	1470	2330	0.165
Hydrogen	H	1	1.00797	0.08375×10^{-3}	−259.19	−252.7	14.4
Indium	In	49	114.82	7.31	156.634	2000	0.233
Iodine	I	53	126.9044	4.93	113.7	183	0.218
Iridium	Ir	77	192.2	22.5	2447	(5300)	0.130
Iron	Fe	26	55.847	7.874	1536.5	3000	0.447
Krypton	Kr	36	83.80	3.488×10^{-3}	−157.37	−152	0.247
Lanthanum	La	57	138.91	6.189	920	3470	0.195
Lawrencium	Lr	103	(257)	—	—	—	—
Lead	Pb	82	207.19	11.35	327.45	1725	0.129
Lithium	Li	3	6.939	0.534	180.55	1300	3.58
Lutetium	Lu	71	174.97	9.849	1663	1930	0.155
Magnesium	Mg	12	24.312	1.738	650	1107	1.03
Manganese	Mn	25	54.9380	7.44	1244	2150	0.481
Meitnerium	Mt	109	(266)	—	—	—	—
Mendelevium	Md	101	(256)	—	—	—	—
Mercury	Hg	80	200.59	13.55	−38.87	357	0.138
Molybdenum	Mo	42	95.94	10.22	2617	5560	0.251
Neodymium	Nd	60	144.24	7.007	1016	3180	0.188
Neon	Ne	10	20.183	0.8387×10^{-3}	−248.597	−246.0	1.03
Neptunium	Np	93	(237)	20.25	637	—	1.26
Nickel	Ni	28	58.71	8.902	1453	2730	0.444
Niobium	Nb	41	92.906	8.57	2468	4927	0.264
Nitrogen	N	7	14.0067	1.1649×10^{-3}	−210	−195.8	1.03
Nobelium	No	102	(255)	—	—	—	—
Osmium	Os	76	190.2	22.59	3027	5500	0.130
Oxygen	O	8	15.9994	1.3318×10^{-3}	−218.80	−183.0	0.913
Palladium	Pd	46	106.4	12.02	1552	3980	0.243
Phosphorus	P	15	30.9738	1.83	44.25	280	0.741
Platinum	Pt	78	195.09	21.45	1769	4530	0.134
Plutonium	Pu	94	(244)	19.8	640	3235	0.130
Polonium	Po	84	(210)	9.32	254	—	—
Potassium	K	19	39.102	0.862	63.20	760	0.758
Praseodymium	Pr	59	140.907	6.773	931	3020	0.197
Promethium	Pm	61	(145)	7.22	(1027)	—	—
Protactinium	Pa	91	(231)	15.37 (estimated)	(1230)	—	—
Radium	Ra	88	(226)	5.0	700	—	—
Radon	Rn	86	(222)	9.96×10^{-3} (0°C)	(−71)	−61.8	0.092
Rhenium	Re	75	186.2	21.02	3180	5900	0.134

Element	Symbol	Atomic Number Z	Molar Mass, g/mol	Density, g/cm^3 at 20°C	Melting Point, °C	Boiling Point, °C	Specific Heat, J/(g · °C) at 25°C
Rhodium	Rh	45	102.905	12.41	1963	4500	0.243
Rubidium	Rb	37	85.47	1.532	39.49	688	0.364
Ruthenium	Ru	44	101.107	12.37	2250	4900	0.239
Rutherfordium	Rf	104	261.11	—	—	—	—
Samarium	Sm	62	150.35	7.52	1072	1630	0.197
Scandium	Sc	21	44.956	2.99	1539	2730	0.569
Seaborgium	Sg	106	263.118	—	—	—	—
Selenium	Se	34	78.96	4.79	221	685	0.318
Silicon	Si	14	28.086	2.33	1412	2680	0.712
Silver	Ag	47	107.870	10.49	960.8	2210	0.234
Sodium	Na	11	22.9898	0.9712	97.85	892	1.23
Strontium	Sr	38	87.62	2.54	768	1380	0.737
Sulfur	S	16	32.064	2.07	119.0	444.6	0.707
Tantalum	Ta	73	180.948	16.6	3014	5425	0.138
Technetium	Tc	43	(99)	11.46	2200	—	0.209
Tellurium	Te	52	127.60	6.24	449.5	990	0.201
Terbium	Tb	65	158.924	8.229	1357	2530	0.180
Thallium	Tl	81	204.37	11.85	304	1457	0.130
Thorium	Th	90	(232)	11.72	1755	(3850)	0.117
Thulium	Tm	69	168.934	9.32	1545	1720	0.159
Tin	Sn	50	118.69	7.2984	231.868	2270	0.226
Titanium	Ti	22	47.90	4.54	1670	3260	0.523
Tungsten	W	74	183.85	19.3	3380	5930	0.134
Un-named	Uun	110	(269)	—	—	—	—
Un-named	Uuu	111	(272)	—	—	—	—
Un-named	Uub	112	(264)	—	—	—	—
Un-named	Uut	113	—	—	—	—	—
Un-named	Unq	114	(285)	—	—	—	—
Un-named	Uup	115	—	—	—	—	—
Un-named	Uuh	116	(289)	—	—	—	—
Un-named	Uus	117	—	—	—	—	—
Un-named	Uuo	118	(293)	—	—	—	—
Uranium	U	92	(238)	18.95	1132	3818	0.117
Vanadium	V	23	50.942	6.11	1902	3400	0.490
Xenon	Xe	54	131.30	5.495×10^{-3}	−111.79	−108	0.159
Ytterbium	Yb	70	173.04	6.965	824	1530	0.155
Yttrium	Y	39	88.905	4.469	1526	3030	0.297
Zinc	Zn	30	65.37	7.133	419.58	906	0.389
Zirconium	Zr	40	91.22	6.506	1852	3580	0.276

The values in parentheses in the column of molar masses are the mass numbers of the longest-lived isotopes of those elements that are radioactive. Melting points and boiling points in parentheses are uncertain.

The data for gases are valid only when these are in their usual molecular state, such as H_2, He, O_2, Ne, etc. The specific heats of the gases are the values at constant pressure.

Source: Adapted from J. Emsley, *The Elements,* 3rd ed., 1998, Clarendon Press, Oxford. See also www.webelements.com for the latest values and newest elements.

APPENDIX G

Periodic Table of the Elements

Metals | Metalloids | Nonmetals

THE HORIZONTAL PERIODS	Alkali metals IA	IIA	IIIB	IVB	VB	VIB	VIIB	VIIIB	VIIIB	VIIIB	IB	IIB	IIIA	IVA	VA	VIA	VIIA	Noble gases 0
1	1 H																	2 He
2	3 Li	4 Be											5 B	6 C	7 N	8 O	9 F	10 Ne
3	11 Na	12 Mg											13 Al	14 Si	15 P	16 S	17 Cl	18 Ar
4	19 K	20 Ca	21 Sc	22 Ti	23 V	24 Cr	25 Mn	26 Fe	27 Co	28 Ni	29 Cu	30 Zn	31 Ga	32 Ge	33 As	34 Se	35 Br	36 Kr
5	37 Rb	38 Sr	39 Y	40 Zr	41 Nb	42 Mo	43 Tc	44 Ru	45 Rh	46 Pd	47 Ag	48 Cd	49 In	50 Sn	51 Sb	52 Te	53 I	54 Xe
6	55 Cs	56 Ba	57-71 *	72 Hf	73 Ta	74 W	75 Re	76 Os	77 Ir	78 Pt	79 Au	80 Hg	81 Tl	82 Pb	83 Bi	84 Po	85 At	86 Rn
7	87 Fr	88 Ra	89-103 †	104 Rf	105 Db	106 Sg	107 Bh	108 Hs	109 Mt	110	111	112	113	114	115	116	117	118

Transition metals (groups IIIB–IIB)

Inner transition metals

Lanthanide series *	57 La	58 Ce	59 Pr	60 Nd	61 Pm	62 Sm	63 Eu	64 Gd	65 Tb	66 Dy	67 Ho	68 Er	69 Tm	70 Yb	71 Lu
Actinide series †	89 Ac	90 Th	91 Pa	92 U	93 Np	94 Pu	95 Am	96 Cm	97 Bk	98 Cf	99 Es	100 Fm	101 Md	102 No	103 Lr

The names for elements 104 through 109 (Rutherfordium, Dubnium, Seaborgium, Bohrium, Hassium, and Meitnerium, respectively) were adopted by the International Union of Pure and Applied Chemistry (IUPAC) in 1997. Elements 110, 111, 112, 114, 116, and 118 have been discovered but, as of 2000, have not yet been named. See www.webelements.com for the latest information and newest elements.

ANSWERS

to Checkpoints and Odd-Numbered Questions, Exercises, and Problems

CHAPTER 22

CP **1.** C and D attract; B and D attract **2.** (a) leftward; (b) leftward; (c) leftward **3.** (a) a, c, b; (b) less than **4.** $-15e$ (net charge of $-30e$ is equally shared) **Q** **1.** no, only for charged particles, charged particle-like objects, and spherical shells (including solid spheres) of uniform charge **3.** a and b **5.** $2q^2/4\pi\varepsilon_0 r^2$, up the page **7.** (a) same; (b) less than; (c) cancel; (d) add; (e) the adding components; (f) positive direction of y; (g) negative direction of y; (h) positive direction of x; (i) negative direction of x **9.** (a) possibly; (b) definitely **11.** no (the person and the conductor share the charge) **EP** **1.** 1.38 m **3.** (a) 4.9×10^{-7} kg; (b) 7.1×10^{-11} C **5.** (a) 0.17 N; (b) -0.046 N **7.** either $-1.00\ \mu$C and $+3.00\ \mu$C or $+1.00\ \mu$C and $-3.00\ \mu$C **9.** (a) charge $-4q/9$ must be located on the line joining the two positive charges, a distance $L/3$ from charge $+q$. **11.** (a) 5.7×10^{13} C, no; (b) 6.0×10^5 kg **13.** $q = Q/2$ **15.** (b) $\pm 2.4 \times 10^{-8}$ C **17.** (a) $\frac{L}{2}\left(1 + \frac{1}{4\pi\varepsilon_0}\frac{qQ}{Wh^2}\right)$; (b) $\sqrt{3qQ/4\pi\varepsilon_0 W}$ **19.** -1.32×10^{13} C **21.** (a) 3.2×10^{-19} C; (b) two **23.** 6.3×10^{11} **25.** 122 mA **27.** (a) 0; (b) 1.9×10^{-9} N **29.** (a) ^{9}B; (b) ^{13}N; (c) ^{12}C **N1.** (a) 6.05 cm; (b) 6.05 cm from central bead **N3.** $+13e$ **N5.** (a) positive; (b) $+9$ **N7.** 9.0 kN **N9.** $1.72a$, directly rightward **N11.** $-11.1\ \mu$C **N13.** $q = 0.71Q$ **N15.** (b) $1e$, 0.654 rad; $2e$, 0.889 rad; $3e$, 0.988 rad; $4e$, 1.047 rad; $5e$, 1.088 rad **N17.** (a) positron; (b) electron **N19.** (a) Let $J = qQ/4\pi\varepsilon_0 d^2$. For $\alpha < 0$, $F = -J[\alpha^{-2} + (1 + |\alpha|)^{-2}]$; for $0 < \alpha < 1$, $F = J[\alpha^{-2} - (1 - \alpha)^{-2}]$; for $1 < \alpha$, $F = J[\alpha^{-2} + (\alpha - 1)^{-2}]$

CHAPTER 23

CP **1.** (a) rightward; (b) leftward; (c) leftward; (d) rightward (p and e have same charge magnitude, and p is farther) **2.** all tie **3.** (a) toward positive y; (b) toward positive x; (c) toward negative y **4.** (a) leftward; (b) leftward; (c) decrease **5.** (a) all tie; (b) 1 and 3 tie, then 2 and 4 tie **Q** **1.** (a) toward positive x; (b) downward and to the right; (c) A **3.** two points: one to the left of the particles, the other between the protons **5.** (a) yes; (b) toward; (c) no (the field vectors are not along the same line); (d) cancel; (e) add; (f) adding components; (g) toward negative y **7.** e, b, then a and c tie, then d (zero) **9.** (a) toward the bottom; (b) 2 and 4 toward the bottom, 3 toward the top **11.** (a) 4, 3, 1, 2; (b) 3, then 1 and 4 tie, then 2 **EP** **1.** (a) 6.4×10^{-18} N; (b) 20 N/C **5.** 56 pC **7.** 3.07×10^{21} N/C, radially outward **9.** 50 cm from q_1 and 100 cm from q_2 **11.** 0 **13.** 1.02×10^5 N/C, upward **15.** 6.88×10^{-28} C·m **21.** $q/\pi^2\varepsilon_0 r^2$, vertically downward **23.** (a) $-q/L$; (b) $q/4\pi\varepsilon_0 a(L + a)$ **27.** $R/\sqrt{3}$ **29.** 3.51×10^{15} m/s² **31.** 6.6×10^{-15} N **33.** (a) 1.5×10^3 N/C; (b) 2.4×10^{-16} N, up; (c) 1.6×10^{-26} N; (d) 1.5×10^{10} **35.** (a) 1.92×10^{12} m/s²; (b) 1.96×10^5 m/s **37.** $-5e$ **39.** (a) 2.7×10^6 m/s; (b) 1000 N/C **41.** 27 μm **43.** (a) yes; (b) upper plate, 2.73 cm **45.** (a) 0; (b) 8.5×10^{-22} N·m; (c) 0 **47.** $(1/2\pi)\sqrt{pE/I}$ **N1.** 28% **N3.** (a) -1.72×10^{-15} C/m; (b) -3.82×10^{-14} C/m²; (c) -9.56×10^{-15} C/m²; (d) -1.43×10^{-12} C/m³ **N5.** 2.4×10^{-16} C **N7.** $E = 2k|Q|(\sin\theta/2)/\theta R^2$ **N9.** 217° **N11.** (a) first (top) row: 4, 8, 12; second row: 5, 10, 14; third row; 7, 11, 16; (b) 1.63×10^{-19} C **N13.** (a) $q/8\pi\varepsilon_0 d^2$, to the left; (b) $3q/\pi\varepsilon_0 d^2$, to the right; (c) $7q/16\pi\varepsilon_0 d^2$, to the left **N15.** (a) to the right in the figure; (b) $(2kqQ\cos 60°)/a^2$ **N17.** (a) 47 N/C; (b) 27 N/C **N19.** $4kQ/3d^2$ or $Q/3\pi\varepsilon_0 d^2$ **N21.** (a) 27 km/s; (b) 50 μm **N23.** 5.2 cm **N25.** 1.38×10^{-10} N/C, 180° from $+x$ **N27.** 1.08×10^{-5} N/C, directly rightward **N29.** 1.92×10^{-21} J

CHAPTER 24

CP **1.** (a) $+EA$; (b) $-EA$; (c) 0; (d) 0 **2.** (a) 2; (b) 3; (c) 1 **3.** (a) equal; (b) equal; (c) equal **4.** (a) $+50e$; (b) $-150e$ **5.** 3 and 4 tie, then 2, 1 **Q** **1.** (a) 8 N·m²/C; (b) 0 **3.** (a) all four; (b) neither (they are equal) **5.** (a) S_3, S_2, S_1; (b) all tie; (c) S_3, S_2, S_1; (d) all tie (zero) **7.** 2σ, σ, 3σ; or 3σ, σ, 2σ **9.** (a) all tie ($E = 0$); (b) all tie **EP** **1.** (a) 693 kg/s; (b) 693 kg/s; (c) 347 kg/s; (d) 347 kg/s; (e) 575 kg/s **3.** (a) 0; (b) -3.92 N·m²/C; (c) 0; (d) 0 for each field **5.** 2.0×10^5 N·m²/C **7.** (a) 8.23 N·m²/C; (b) 8.23 N·m²/C; (c) 72.8 pC in each case **9.** 3.54 μC **11.** 0 through each of the three faces meeting at q, $q/24\varepsilon_0$ through each of the other faces **13.** (a) 37 μC; (b) 4.1×10^6 N·m²/C **15.** (a) -3.0×10^{-6} C; (b) $+1.3 \times 10^{-5}$ C **17.** 5.0 μC/m **19.** (a) $E = q/2\pi\varepsilon_0 LR$, radially inward; (b) $-q$ on both inner and outer surfaces; (c) $E = q/2\pi\varepsilon_0 Lr$, radially outward **21.** (a) 2.3×10^6 N/C, radially out; (b) 4.5×10^5 N/C, radially in **23.** 3.6 nC **25.** (b) $\rho R^2/2\varepsilon_0 r$ **27.** (a) 5.3×10^7 N/C; (b) 60 N/C **29.** 5.0 nC/m² **31.** 0.44 mm **33.** (a) $\rho x/\varepsilon_0$; (b) $\rho d/2\varepsilon_0$ **35.** -7.5 nC **39.** -1.04 nC **43.** (a) $E = (q/4\pi\varepsilon_0 a^3)r$; (b) $E = q/4\pi\varepsilon_0 r^2$; (c) 0; (d) 0; (e) inner, $-q$; outer, 0 **45.** $q/2\pi a^2$ **47.** $6K\varepsilon_0 r^3$ **N1.** (a) $+1.8\ \mu$C; (b) $-5.3\ \mu$C; (c) $+8.9\ \mu$C **N3.** (a) $+69.1$ cm and -69.1 cm; (b) $+69.1$ cm **N5.** -5.8 nC/m **N7.** (a) $+2.0$ nC; (b) -1.2 nC; (c) $+1.2$ nC; (d) $+0.80$ nC

N9. (a) 4.2 kN/C; (b) 2.4 kN/C **N11.** −1.70 nC **N13.** (a) $Q/8$; (b) $\frac{1}{2}$ **N15.** 7.1 N·m²/C **N17.** $7E_1/8$ **N19.** 1.125 **N21.** 2.1×10^{17} m/s²

CHAPTER 25

CP **1.** (a) negative; (b) increase **2.** (a) positive; (b) higher **3.** (a) rightward; (b) 1, 2, 3, 5: positive; 4, negative; (c) 3, then 1, 2, and 5 tie, then 4 **4.** all tie **5.** a, c (zero), b **6.** (a) 2, then 1 and 3 tie; (b) 3; (c) accelerate leftward **Q** **1.** (a) higher; (b) positive; (c) negative; (d) all tie **3.** $-4q/4\pi\varepsilon_0 d$ **5.** (a)–(c) $Q/4\pi\varepsilon_0 R$; (d) a, b, c **7.** (a) 2, 4, and then a tie of 1, 3, and 5 (where $E = 0$); (b) negative x direction; (c) positive x direction **9.** (a)–(d) zero **EP** **1.** (a) 3.0×10^5 C; (b) 3.6×10^6 J **3.** (a) 3.0×10^{10} J; (b) 7.7 km/s; (c) 9.0×10^4 kg **5.** 8.8 mm **7.** (a) 136 MV/m; (b) 8.82 kV/m **9.** (b) because $V = 0$ point is chosen differently; (c) $q/(8\pi\varepsilon_0 R)$; (d) potential differences are independent of the choice of $V = 0$ point **11.** (a) $Q/4\pi\varepsilon_0 r$; (b) $\frac{\rho}{3\varepsilon_0}\left(\frac{3}{2}r_2^2 - \frac{1}{2}r^2 - \frac{r_1^3}{r}\right)$, $\rho = \frac{Q}{\frac{4\pi}{3}(r_2^3 - r_1^3)}$; (c) $\frac{\rho}{2\varepsilon_0}(r_2^2 - r_1^2)$, with ρ as in (b); (d) yes **13.** (a) −4.5 kV; (b) −4.5 kV **15.** $x = d/4$ and $x = -d/2$ **17.** (a) 0.54 mm; (b) 790 V **19.** 6.4×10^8 V **21.** $2.5q/4\pi\varepsilon_0 d$ **25.** (a) $-5Q/4\pi\varepsilon_0 R$; (b) $-5Q/4\pi\varepsilon_0(z^2 + R^2)^{1/2}$ **27.** $(\sigma/8\varepsilon_0)[(z^2 + R^2)^{1/2} - z]$ **29.** $(c/4\pi\varepsilon_0)[L - d\ln(1 + L/d)]$ **31.** 17 V/m at 135° counterclockwise from $+x$ **35.** (a) $\frac{Q}{4\pi\varepsilon_0 d(d + L)}$, leftward; (b) 0 **37.** $-0.21q^2/\varepsilon_0 a$ **39.** (a) $+6.0 \times 10^4$ V; (b) -7.8×10^5 V; (c) 2.5 J; (d) increase; (e) same; (f) same **41.** $W = \frac{qQ}{8\pi\varepsilon_0}\left(\frac{1}{r_1} - \frac{1}{r_2}\right)$ **43.** 2.5 km/s **45.** (a) 0.225 J; (b) A, 45.0 m/s²; B, 22.5 m/s²; (c) A, 7.75 m/s; B, 3.87 m/s **47.** 0.32 km/s **49.** 1.6×10^{-9} m **51.** 2.5×10^{-8} C **53.** (a) −180 V; (b) 2700 V, −8900 V **55.** (a) −0.12 V; (b) 1.8×10^{-8} N/C, radially inward **N1.** (a) 3.0 J; (b) $y = -8.5$ m **N3.** $+16q/4\pi\varepsilon_0 a$ **N5.** (a) 3.6 kV; (b) 3.6 kV **N7.** (a) 0.90 J; (b) 4.5 J **N9.** 22 km/s **N11.** (a) $+7.2 \times 10^{-10}$ V; (b) $+2.3 \times 10^{-28}$ J; (c) $+2.4 \times 10^{-29}$ J **N13.** 2.30×10^{-28} J **N15.** 2.3×10^{-22} J **N17.** (a) proton; (b) 65.3 km/s **N19.** (a) 2.72×10^{-14} J; (b) 3.02×10^{-31} kg, about 1/3 of accepted value **N21.** (c) 4.24 V **N23.** 8.8×10^{-14} m **N25.** $(2.9 \times 10^{-2}\ \text{m}^{-3})A$ **N27.** 2.1 d **N29.** (a) −4.8 nm; (b) 8.1 nm; (c) no **N31.** (a) −24 J; (b) 0 **N33.** (a) none; (b) $x = 0.41$ m **N35.** (a) 38 s; (b) 280 days

CHAPTER 26

CP **1.** (a) same; (b) same **2.** (a) decreases; (b) increases; (c) decreases **3.** (a) V, $q/2$; (b) $V/2$, q **4.** (a) $q_0 = q_1 + q_{34}$; (b) equal (C_3 and C_4 are in series) **5.** (a) same; (b)–(d) increase; (e) same (same potential difference across same plate separation) **6.** (a) same; (b) decrease; (c) increase **Q** **1.** a, 2; b, 1; c, 3 **3.** a, series; b, parallel; c, parallel **5.** (a) $C/3$; (b) $3C$; (c) parallel **7.** (a) same; (b) same; (c) more; (d) more **9.** (a) 2; (b) 3; (c) 1 **11.** (a) increases; (b) increases; (c) decreases; (d) decreases; (e) same, increases, increases, increases **EP** **1.** 7.5 pC **3.** 3.0 mC **5.** (a) 140 pF; (b) 17 nC **7.** $5.04\pi\varepsilon_0 R$ **11.** 9090 **13.** 3.16 μF **17.** 43 pF **19.** (a) 50 V; (b) 5.0×10^{-5} C; (c) 1.5×10^{-4} C

21. $$q_1 = \frac{C_1C_2 + C_1C_3}{C_1C_2 + C_1C_3 + C_2C_3}C_1V_0,$$

$$q_2 = q_3 = \frac{C_2C_3}{C_1C_2 + C_1C_3 + C_2C_3}C_1V_0$$

23. 72 F **25.** 0.27 J **27.** (a) 2.0 J **29.** (a) $2V$; (b) $U_i = \varepsilon_0 AV^2/2d$, $U_f = 2U_i$; (c) $\varepsilon_0 AV^2/2d$ **35.** Pyrex **37.** 81 pF/m **39.** 0.63 m² **43.** (a) 10 kV/m; (b) 5.0 nC; (c) 4.1 nC **45.** (a) $C = 4\pi\varepsilon_0\kappa\left(\frac{ab}{b - a}\right)$; (b) $q = 4\pi\varepsilon_0\kappa V\left(\frac{ab}{b - a}\right)$; (c) $q' = q(1 - 1/\kappa)$ **N1.** four **N3.** (a) 60 μC; (b) increase of 60 μC **N5.** 3.6 pC **N7.** 2.0 μF and 6.0 μF **N9.** 1.06 nC **N11.** (a) 24 μC; (b) 6.0 V **N13.** 20 μC **N15.** 45 μC **N17.** (a) 36 μC; (b) 12 μC **N19.** 1.06 nC

CHAPTER 27

CP **1.** 8 A, rightward **2.** (a)–(c) rightward **3.** a and c tie, then b **4.** Device 2 **5.** (a) and (b) tie, then (d), then (c) **Q** **1.** a, b, and c tie, then d (zero) **3.** b, a, c **5.** tie of A, B, and C, then tie of $A + B$ and $B + C$, then $A + B + C$ **7.** (a)–(c) 1 and 2 tie, then 3 **9.** C, A, B **EP** **1.** (a) 1200 C; (b) 7.5×10^{21} **3.** 5.6 ms **5.** (a) 6.4 A/m², north; (b) no, cross-sectional area **7.** 0.38 mm **9.** (a) 2×10^{12}; (b) 5000; (c) 10 MV **11.** 13 min **13.** 2.0×10^{-8} Ω·m **15.** 100 V **17.** 2.4 Ω **19.** 54 Ω **21.** 3.0 **23.** 8.2×10^{-4} Ω·m **25.** 2000 K **27.** (a) 0.43%, 0.0017%, 0.0034% **29.** (a) $R = \rho L/\pi ab$ **31.** 560 W **33.** (a) 1.0 kW; (b) 25¢ **35.** 0.135 W **37.** (a) 10.9 A; (b) 10.6 Ω; (c) 4.5 MJ **39.** 660 W **41.** (a) 3.1×10^{11}; (b) 25 μA; (c) 1300 W, 25 MW **43.** (a) 17 mV/m; (b) 243 J **45.** (a) $J = I/2\pi r^2$; (b) $E = \rho I/2\pi r^2$; (c) $\Delta V = \rho I(1/r - 1/b)/2\pi$; (d) 0.16 A/m²; (e) 16 V/m; (f) 0.16 MV **N1.** (a) 0.43%, 0.0017%, 0.0034% **N3.** (a) 1.3×10^5 A/m²; (b) 94 mV **N5.** 2.3 kW **N7.** (a) 0.67 A; (b) toward the negative terminal **N9.** 3.0×10^6 J/kg **N11.** (a) 17 mV/m; (b) 243 J **N13.** 2.0×10^{-8} Ω·m **N15.** 2.0×10^6 (Ω·m)$^{-1}$ **N17.** 16 **N19.** 13.3 Ω **N21.** 756 kJ **N23.** 146 kJ **N25.** (a) 0.81 mm; (b) 1.0 mm

CHAPTER 28

CP **1.** (a) rightward; (b) all tie; (c) b, then a and c tie; (d) b, then a and c tie **2.** (a) all tie; (b) R_1, R_2, R_3 **3.** (a) less; (b) greater; (c) equal **4.** (a) $V/2$, i; (b) V, $i/2$ **5.** (a) 1, 2, 4, 3; (b) 4, tie of 1 and 2; then 3 **Q** **1.** 3, 4, 1, 2 **3.** (a) no; (b) yes; (c) all tie **5.** parallel, R_2, R_1, series **7.** (a) same; (b) same; (c) less; (d) more **9.** (a) less; (b) less; (c) more **11.** c, b, a **EP** **1.** (a) $320; (b) 4.8 cents **3.** 14 h 24 min **5.** (a) 0.50 A; (b) $P_1 = 1.0$ W, $P_2 = 2.0$ W; (c) $P_1 = 6.0$ W supplied, $P_2 = 3.0$ W absorbed **7.** (a) 14 V; (b) 100 W; (c) 600 W; (d) 10 V, 100 W **9.** (a) 50 V; (b) 48 V; (c) B is connected to the negative terminal **11.** 2.5 V **13.** 8.0 Ω **15.** (a) $r_1 - r_2$; (b) battery with r_1 **19.** 5.56 A **21.** $i_1 =$

50 mA, $i_2 = 60$ mA, $V_{ab} = 9.0$ V **23.** (a) bulb 2; (b) bulb 1 **25.** 3*d* **27.** nine **29.** (a) $R = r/2$; (b) $P_{max} = \mathscr{E}^2/2r$ **31.** (a) 0.346 W; (b) 0.050 W; (c) 0.709 W; (d) 1.26 W; (e) −0.158 W **33.** (a) battery 1, 0.67 A down; battery 2, 0.33 A up; battery 3, 0.33 A up; (b) 3.3 V **35.** (a) Cu: 1.11 A, Al: 0.893 A; (b) 126 m **37.** 0.45 A **39.** −3.0% **45.** 4.6 **47.** (a) 2.41 μs; (b) 161 pF **49.** (a) 0.955 μC/s; (b) 1.08 μW; (c) 2.74 μW; (d) 3.82 μW **51.** (a) 2.17 s; (b) 39.6 mV **53.** (a) 1.0×10^{-3} C; (b) 10^{-3} A; (c) $V_C = 10^3 e^{-t}$ V, $V_R = 10^3 e^{-t}$ V; (d) $P = e^{-2t}$ W **55.** (a) at $t = 0$, $i_1 = 1.1$ mA, $i_2 = i_3 = 0.55$ mA; at $t = \infty$, $i_1 = i_2 = 0.82$ mA, $i_3 = 0$; (c) at $t = 0$, $V_2 = 400$ V; at $t = \infty$, $V_2 = 600$ V; (d) after several time constants ($\tau = 7.1$ s) have elapsed **N1.** (a) 0; (b) 1.25 A, downward **N3.** 10^{-6} **N5.** (a) 12 eV (1.9×10^{-18} J); (b) 6.5 W **N7.** (a) 120 Ω; (b) $i_1 = 51$ mA, $i_2 = i_3 = 19$ mA, $i_4 = 13$ mA **N9.** (a) $i = 0$: current in battery 1 = 0.665 A up, current in battery 2 = 1.26 A up, $V_{AB} = 6.29$ V; $i = 4.0$ A: current in battery 1 = 0.248 A up, current in battery 2 = 3.21 A up, $V_{AB} = -3.93$ V; $i = 8.0$ A: current in battery 1 = 0.169 A down, current in battery 2 = 5.17 A up, $V_{AB} = -14.2$ V; $i = 12$ A: current in battery 1 = 0.587 A down, current in battery 2 = 7.13 A up, $V_{AB} = -24.4$ V; Battery 1 is discharging for $i = 0$ and 4.0 A; it is charging for $i = 6.0$ A and 12 A; Battery 2 is discharging for all values of *i*; (b) emf = 6.29 V, resistance = 2.56 Ω **N11.** (a) $V_T = -ir + \mathscr{E}$; (b) 13.6 V; (c) 0.060 Ω **N13.** (a) 7.50 A, leftward; (b) 10.0 A, leftward; (c) 87.5 W, supplied **N15.** 20 Ω **N17.** (a) 3.0 A; (b) 10 A; (c) 13 A; (d) 1.5 A; (e) 7.5 A **N19.** (a) 3.0 A, downward; (b) 1.6 A, downward; (c) 6.4 W, supplying; (d) 55.2 W, supplying **N21.** (a) 0.33 A, rightward; (b) 720 J **N23.** providing energy, 360 W **N25.** 250 μJ **N27.** (a) 6.4 V; (b) 3.6 W; (c) 17 W; (d) −5.6 W; (e) *a* **N29.** (a) 0; (b) +6.0 V; (c) +22 V

CHAPTER 29

CP **1.** *a*, $+z$; *b*, $-x$; *c*, $\vec{F}_B = 0$ **2.** (a) 2, then tie of 1 and 3 (zero); (b) 4 **3.** (a) $+z$ and $-z$ tie, then $+y$ and $-y$ tie, then $+x$ and $-x$ tie (zero); (b) $+y$ **4.** (a) electron; (b) clockwise **5.** $-y$ **6.** (a) all tie; (b) 1 and 4 tie, then 2 and 3 tie **Q** **1.** (a) no, $\vec{v}$ and $\vec{F}_B$ must be perpendicular; (b) yes; (c) no, $\vec{B}$ and $\vec{F}_B$ must be perpendicular **3.** (a) $\vec{F}_E$; (b) $\vec{F}_B$ **5.** (a) negative; (b) equal; (c) equal; (d) half-circle **7.** (a) $\vec{B}_1$; (b) B_1 into page, B_2 out of page; (c) less **9.** (a) 1, 180°; 2, 270°; 3, 90°; 4, 0°; 5, 315°; 6, 225°; 7, 135°; 8, 45°; (b) 1 and 2 tie, then 3 and 4 tie; (c) 8, then 5 and 6 tie, then 7 **EP** **1.** (a) 6.2×10^{-18} N; (b) 9.5×10^8 m/s²; (c) remains equal to 550 m/s **3.** (a) 400 km/s; (b) 835 eV **5.** (a) east; (b) 6.28×10^{14} m/s²; (c) 2.98 mm **7.** (a) 3.4×10^{-4} T, horizontal and to the left as viewed along $\vec{v}_0$; (b) yes, if its velocity is the same as the electron's velocity **9.** 0.27 mT **11.** 680 kV/m **13.** (b) 2.84×10^{-3} **15.** 21 μT **17.** (a) 2.05×10^7 m/s; (b) 467 μT; (c) 13.1 MHz; (d) 76.3 ns **19.** (a) 0.978 MHz; (b) 96.4 cm **23.** (a) 1.0 MeV; (b) 0.5 MeV **25.** (a) 495 mT; (b) 22.7 mA; (c) 8.17 MJ **27.** (a) 0.36 ns; (b) 0.17 mm; (c) 1.5 mm **29.** (a) $-q$; (b) $\pi m/qB$ **31.** 240 m **33.** 28.2 N, horizontally west **35.** 467 mA, from left to right **37.** 0.10 T, at 31° from the vertical **39.** 4.3×10^{-3} N · m, negative *y* **43.** $2\pi aiB \sin\theta$, normal to the plane of the loop (up) **45.** (a) 540 Ω, connected in series with the galvanometer; (b) 2.52 Ω, connected in parallel **47.** 2.45 A **49.** (a) 12.7 A; (b) 0.0805 N · m **51.** (a) 0.30 J/T; (b) 0.024 N · m **53.** (a) 2.86 A · m²; (b) 1.10 A · m² **55.** (a) $(8.0 \times 10^{-4}\ \text{N}\cdot\text{m})(-1.2\hat{i} - 0.90\hat{j} + 1.0\hat{k})$; (b) -6.0×10^{-4} J **57.** $-(0.10\ \text{V/m})\hat{k}$ **59.** −2.0 T **N1.** (a) −3500 m/s; (b) 7000 m/s **N3.** 4.8×10^{-5} A·m² **N7.** neutron moves tangent to original path, proton moves in a circular orbit of radius 25 cm **N9.** $(0.75\ \text{T})\hat{k}$ **N11.** 1.6×10^{-8} T **N15.** (a) −72 μJ; (b) $(96.0\hat{i} + 48.0\hat{k})$ μN·m **N17.** 8.7 ns **N19.** M/QT

CHAPTER 30

CP **1.** *a*, *c*, *b* **2.** *b*, *c*, *a* **3.** *d*, tie of *a* and *c*, then *b* **4.** *d*, *a*, tie of *b* and *c* (zero) **Q** **1.** *c*, *d*, then *a* and *b* tie **3.** *c*, *a*, *b* **5.** (a) 1, 3, 2; (b) less **7.** *c* and *d* tie, then *b*, *a* **9.** *d*, then tie of *a* and *e*, then *b*, *c* **EP** **1.** (a) 3.3 μT; (b) yes **3.** (a) 16 A; (b) west to east **5.** (a) $\mu_0 qvi/2\pi d$, antiparallel to *i*; (b) same magnitude, parallel to *i* **7.** 2 rad **9.** $\dfrac{\mu_0 i\theta}{4\pi}\left(\dfrac{1}{b} - \dfrac{1}{a}\right)$, out of page **19.** $(\mu_0 i/2\pi w)\ln(1 + w/d)$, up **21.** (a) it is impossible to have other than $B = 0$ midway between them; (b) 30 A **23.** 4.3 A, out of page **25.** 80 μT, up the page **27.** $0.791\mu_0 i^2/\pi a$, 162° counterclockwise from the horizontal **29.** 3.2 mN, toward the wire **31.** (a) $(-2.0\ \text{A})\mu_0$; (b) 0 **35.** $\mu_0 J_0 r^2/3a$ **41.** 0.30 mT **43.** (a) 533 μT; (b) 400 μT **47.** (a) 4.77 cm; (b) 35.5 μT **49.** 0.47 A · m² **51.** (a) 2.4 A · m²; (b) 46 cm **57.** (a) 79 μT; (b) 1.1×10^{-6} N · m **N1.** 14.1*R* **N3.** 3.0 μT **N5.** 1.00 rad **N7.** (a) 30 cm; (b) 2.0 nT; (c) out of page; (d) into page **N9.** 1.28 mm **N11.** 5.0 *μT* **N13.** 256 nT **N15.** 27.5 nT, into the page **N17.** 157 nT **N19.** (a) $(8.0 \times 10^{-3})d$, in teslas, with *d* in meters; (b) 3200 A; (c) $+\hat{k}$ **N21.** 6.0 A, into the page **N23.** (a) *B* from sum: 7.069×10^{-5} T; $\mu_0 in = 5.027 \times 10^{-5}$ T; 40% difference; (b) *B* from sum: 1.043×10^{-4} T; $\mu_0 in = 1.005 \times 10^{-4}$ T; 4% difference; (c) *B* from sum: 2.506×10^{-4} T; $\mu_0 in = 2.513 \times 10^{-4}$ T; 0.3% difference **N25.** 61.3 mA **N27.** (a) $\vec{B} = (\mu_0/2\pi)[i_1/(x - a) + i_2/x]\hat{j}$; (b) $\vec{B} = (\mu_0/2\pi)(i_1/a)(1 + b/2)\hat{j}$

CHAPTER 31

CP **1.** *b*, then *d* and *e* tie, and then *a* and *c* tie (zero) **2.** *a* and *b* tie, then *c* (zero) **3.** *c* and *d* tie, then *a* and *b* tie **4.** *b*, out; *c*, out; *d*, into; *e*, into **5.** d and e **6.** (a) 2, 3, 1 (zero); (b) 2, 3, 1 **7.** *a* and *b* tie, then *c* **Q** **1.** (a) all tie (zero); (b) 2, then tie of 1 and 3 (zero) **3.** (a) into; (b) counterclockwise; (c) larger **5.** *c*, *a*, *b* **7.** *c*, *b*, *a* **9.** (a) more; (b) same; (c) same; (d) same (zero) **EP** **1.** 1.5 mV **3.** (a) 31 mV; (b) right to left **5.** (a) 1.1×10^{-3} Ω; (b) 1.4 T/s **7.** 30 mA **9.** (a) $\mu_0 iR^2\pi r^2/2x^3$; (b) $3\mu_0 i\pi R^2 r^2 v/2x^4$; (c) in the same direction as the current in the large loop **11.** (b) no **13.** 29.5 mC **15.** (a) 21.7 V; (b) counterclockwise **17.** (b) design it so that $Nab = (5/2\pi)$ m² **19.** 5.50 kV **21.** 80 μV, clockwise **23.** (a) 13 μWb/m; (b) 17%; (c) 0 **25.** 3.66 μW **27.** (a) 48.1 mV; (b) 2.67 mA; (c) 0.128 mW **29.** (a) 600 mV, up the page; (b) 1.5 A, clockwise; (c) 0.90 W; (d) 0.18 N; (e) same as (c) **31.** (a) 240 μV; (b) 0.600 mA; (c) 0.144 μW; (d) 2.88×10^{-8} N; (e) same as (c) **33.** (a) 71.5 μV/m; (b) 143 μV/m **37.** 0.10 μWb **41.** let the current change at 5.0 A/s **43.** (b) so that the changing magnetic field of one does not induce current in the other; (c) $L_{eq} = \sum_{j=1}^{N} L_j$ **45.** $6.91\tau_L$ **47.** 46 Ω **49.** (a) 8.45 ns; (b) 7.37 mA **51.** 12.0 A/s **53.** (a) $i_1 = i_2 = 3.33$ A; (b) $i_1 = 4.55$ A, $i_2 = 2.73$ A; (c) $i_1 = 0$, $i_2 = 1.82$ A (reversed); (d) $i_1 = i_2 = 0$ **55.** (a) $i(1 - e^{-Rt/L})$ **57.** 25.6 ms **59.** (a) 97.9 H; (b) 0.196 mJ **63.** (a) 34.2 J/m³; (b) 49.4 mJ **65.** 1.5×10^8 V/m **67.** (a) 1.0 J/m³; (b) 4.8×10^{-15} J/m³ **69.** (a) 1.67 mH; (b) 6.00 mWb **71.** (b) have the turns of the two solenoids wrapped in opposite directions **73.** magnetic field exists only within the cross section of solenoid 1 **75.** (a) $\frac{\mu_0 N l}{2\pi} \ln\left(1 + \frac{b}{a}\right)$; (b) 13 μH

N1. (a) 3.28 ms; (b) 6.45 ms; (c) infinite time; (d) For $R = 6.0\ \Omega$, the current of 2.00 A is the equilibrium current, given by $\mathscr{E}/R = (12\text{ V})/(6.0\ \Omega)$; it takes an infinite time to reach. For $R = 5.00\ \Omega$, the current of 2.00 A is less than the equilibrium current and requires a finite time to reach. (e) 0; (f) 3.00 ms **N3.** 1.2 mΩ **N5.** 18 mV **N7.** 2.9 mV **N9.** 8.0 μA, counterclockwise **N11.** (a) 0.60 mH; (b) 120 **N13.** 12 A/s **N15.** 81.1 μs **N17.** 95.4 Ω **N19.** $B_0(\pi/\tau)r^2 \exp(-t/\tau)$ **N21.** (a) 10 μT, out of the page; (b) 3.3 μT, out of the page

CHAPTER 32

CP **1.** d, b, c, a (zero) **2.** (a) 2; (b) 1 **3.** (a) away; (b) away; (c) less **4.** (a) toward; (b) toward; (c) less **5.** a, c, b, d (zero) **6.** tie of b, c, and d, then a **Q** **1.** supplied **3.** (a) all down; (b) 1 up, 2 down, 3 zero **5.** (a) 1 up, 2 up, 3 down; (b) 1 down, 2 up, 3 zero **7.** (a) 1, up; 2, up; 3, down; (b) and (c) 2, then 1 and 3 tie **9.** (a) rightward; (b) leftward; (c) into **11.** 1, a; 2, b; 3, c and d **EP** **1.** (b) sign is minus; (c) no, there is compensating positive flux through open end near magnet **3.** 47.4 μWb, inward **5.** 55 μT **7.** (a) 31.0 μT, 0°; (b) 55.9 μT, 73.9°; (c) 62.0 μT, 90° **9.** (a) -9.3×10^{-24} J/T; (b) 1.9×10^{-23} J/T **11.** (a) 0; (b) 0; (c) 0; (d) $\pm 3.2 \times 10^{-25}$ J; (e) -3.2×10^{-34} J·s, 2.8×10^{-23} J/T, $+9.7 \times 10^{-25}$ J, $\pm 3.2 \times 10^{-25}$ J **13.** $\Delta\mu = e^2 r^2 B/4m$ **15.** 20.8 mJ/T **17.** yes **19.** (b) K_i/B, opposite to the field; (c) 310 A/m **21.** (a) 3.0 μT; (b) 5.6×10^{-10} eV **23.** 5.15×10^{-24} A·m² **25.** (a) 180 km; (b) 2.3×10^{-5} **27.** 2.4×10^{13} V/m·s **33.** (a) 0.63 μT; (b) 2.3×10^{12} V/m·s **35.** (a) 710 mA; (b) 0; (c) 1.1 A **37.** (a) 2.0 A; (b) 2.3×10^{11} V/m·s; (c) 0.50 A; (d) 0.63 μT·m

N1. (a) 1.1 mWb; (b) inward **N3.** 32.3 mT **N5.** 52 nT·m **N7.** (a) 1.18×10^{-19} T; (b) 1.06×10^{-19} T **N9.** (a) 5.01×10^{-22} T; (b) 4.51×10^{-22} T **N11.** 6.00×10^{-13} T **N13.** (a) 222 μT; (b) 167 μT; (c) 22.7 μT; (d) 1.25 μT; (e) 3.75 μT; (f) 22.7 μT; (g) Because the displacement current in the gap is spread over a larger cross-sectional area, values of B within that area are relatively small. Outside that cross-sectional area, the two values of B are identical. See Fig. 32-22b. **N15.** (a) nine; (b) $4\mu_B = 3.71 \times 10^{-23}$ J/T; (c) $+9.27 \times 10^{-24}$ J; (d) -9.27×10^{-24} J **N17.** (a) 75.4 nT; (b) 67.9 nT **N19.** (a) 27.9 nT; (b) 15.1 nT

CHAPTER 33

CP **1.** (a) $T/2$, (b) T, (c) $T/2$, (d) $T/4$ **2.** (a) 5 V; (b) 150 μJ **3.** (a) remains the same; (b) remains the same **4.** (a) C, B, A; (b) 1, A; 2, B; 3, S; 4, C; (c) A **5.** (a) remains the same; (b) increases **6.** (a) remains the same; (b) decreases **7.** (a) 1, lags; 2, leads; 3, in phase; (b) 3 ($\omega_d = \omega$ when $X_L = X_C$) **8.** (a) increase (circuit is mainly capacitive; increase C to decrease X_C to be closer to resonance for maximum P_{avg}); (b) closer **9.** (a) greater; (b) step-up **Q** **1.** (a) $T/4$; (b) $T/4$; (c) $T/2$ (see Fig. 33-2); (d) $T/2$ (see Eq. 31-37) **3.** b, a, c **5.** (a) 3, 1, 2; (b) 2, tie of 1 and 3 **7.** a, inductor; b, resistor; c, capacitor **9.** (a) leads; (b) capacitive; (c) less **11.** (a) rightward, increase (X_L increases, closer to resonance); (b) rightward, increase (X_C decreases, closer to resonance); (c) rightward, increase (ω_d/ω increases, closer to resonance) **EP** **1.** 9.14 nF **3.** (a) 1.17 μJ; (b) 5.58 mA **5.** with n a positive integer: (a) $t = n(5.00\ \mu\text{s})$; (b) $t = (2n - 1)(2.50\ \mu\text{s})$; (c) $t = (2n - 1)(1.25\ \mu\text{s})$ **7.** (a) 1.25 kg; (b) 372 N/m; (c) 1.75×10^{-4} m; (d) 3.02 mm/s **9.** 7.0×10^{-4} s **11.** (a) 3.0 nC; (b) 1.7 mA; (c) 4.5 nJ **13.** (a) 275 Hz; (b) 364 mA **15.** (a) 6.0 : 1; (b) 36 pF, 0.22 mH **17.** (a) 1.98 μJ; (b) 5.56 μC; (c) 12.6 mA; (d) $-46.9°$; (e) $+46.9°$ **19.** (a) 0.180 mC; (b) $T/8$; (c) 66.7 W **21.** (a) 356 μs; (b) 2.50 mH; (c) 3.20 mJ **23.** Let T_2 (= 0.596 s) be the period of the inductor plus the 900 μF capacitor and let T_1 (= 0.199 s) be the period of the inductor plus the 100 μF capacitor. Close S_2, wait $T_2/4$; quickly close S_1, then open S_2; wait $T_1/4$ and then open S_1. **25.** 8.66 mΩ **27.** (L/R) ln 2 **31.** (a) 0.0955 A; (b) 0.0119 A **33.** (a) 0.65 kHz; (b) 24 Ω **35.** (a) 6.73 ms; (b) 11.2 ms; (c) inductor; (d) 138 mH **37.** (a) $X_C = 0$, $X_L = 86.7\ \Omega$, $Z = 218\ \Omega$, $I = 165$ mA, $\phi = 23.4°$ **39.** (a) $X_C = 37.9\ \Omega$, $X_L = 86.7\ \Omega$, $Z = 206\ \Omega$, $I = 175$ mA, $\phi = 13.7°$ **41.** 1000 V **43.** 89 Ω **45.** (a) 224 rad/s; (b) 6.00 A; (c) 228 rad/s, 219 rad/s; (d) 0.040 **49.** 1.84 A **51.** 141 V **53.** 0, 9.00 W, 2.73 W, 1.82 W **55.** (a) 12.1 Ω; (b) 1.19 kW **57.** (a) 0.743; (b) leads; (c) capacitive; (d) no; (e) yes, no, yes; (f) 33.4 W **59.** (a) 117 μF; (b) 0; (c) 90.0 W, 0; (d) 0°, 90°; (e) 1, 0 **61.** (a) 2.59 A; (b) 38.8 V, 159 V, 224 V, 64.2 V, 75.0 V; (c) 100 W for R, 0 for L and C **63.** (a) 2.4 V; (b) 3.2 mA, 0.16 A **65.** 10 **N1.** 1.59 μF **N3.** (a) 5770 rad/s; (b) 1.09 ms **N5.** (a) 8.84 kHz; (b) 6.00 Ω **N7.** 1.84 kHz **N9.** (a) $X_L = (2\pi)(40\text{ mH})f$; (c) 796 Hz **N11.** (a) 707 Ω; (b) 32.2 mH; (c) 21.9 nF **N19.** (b) 796 Hz; (c) 71 V; (d) 2250 Hz; (e) 94 V; (f) 281 Hz; (g) 94 V **N21.** (b) 61 Hz; (c) 90 Ω; (d) 61 Hz **N23.** (a) 4.00 μF, 5.00 μF, 5.00 μF, 5.00 μF; (b) 1.78 kHz, 1.59 kHz, 1.59 kHz, 1.59 kHz; (c) 12.0 Ω, 12.0 Ω, 6.00 Ω, 4.00 Ω; (d) 19.9 Ω, 22.5 Ω, 19.9 Ω, 19.4 Ω; (e) 0.605 A, 0.535 A, 0.603 A, 0.619 A **N25.** 0.115 A **N27.** (a) 37.0 V; (b) 60.9 V; (c) 113 V; (d) 68.6 W

CHAPTER 34

CP **1.** (a) (Use Fig. 34-5.) On right side of rectangle, $\vec{E}$ is in negative y direction; on left side, $\vec{E} + d\vec{E}$ is greater and in same direction; (b) $\vec{E}$ is downward. On right side, $\vec{B}$ is in negative z direction; on left side, $\vec{B} + d\vec{B}$ is greater and in same direction. **2.** positive direction of x **3.** (a) same; (b) decrease **4.** a, d, b, c (zero) **5.** a **6.** (a) no; (b) yes **Q** **1.** (a) positive direction of z; (b) x **3.** (a) same; (b) increase; (c) decrease **5.** c **7.** a, b, c **9.** none **11.** b **EP** **1.** (a) 0.50 ms; (b) 8.4 min; (c) 2.4 h; (d) 5500 B.C. **3.** (a) 515 nm, 610 nm; (b) 555 nm, 5.41×10^{14} Hz, 1.85×10^{-15} s **5.** it would steadily increase; (b) the summed discrepancies between the apparent time of eclipse and those observed from x; the radius of Earth's orbit **7.** 5.0×10^{-21} H **9.** $B_x = 0$, $B_y = -6.7 \times 10^{-9} \cos[\pi \times 10^{15}(t - x/c)]$, $B_z = 0$ in SI units **11.** 0.10 MJ **13.** 8.88×10^4 m^2 **15.** (a) 16.7 nT; (b) 33.1 mW/m^2 **17.** (a) 6.7 nT; (b) 5.3 mW/m^2; (c) 6.7 W **19.** (a) 87 mV/m; (b) 0.30 nT; (c) 13 kW **21.** 1.0×10^7 Pa **23.** 5.9×10^{-8} Pa **25.** (a) 100 MHz; (b) 1.0 μT along the z axis; (c) 2.1 m^{-1}, 6.3×10^8 rad/s; (d) 120 W/m^2; (e) 8.0×10^{-7} N, 4.0×10^{-7} Pa **29.** 1.9 mm/s **31.** (b) 580 nm **33.** (a) 1.9 V/m; (b) 1.7×10^{-11} Pa **35.** 3.1% **37.** 4.4 W/m^2 **39.** 2/3 **41.** (a) 2 sheets; (b) 5 sheets **43.** 1.48 **45.** 1.26 **47.** 1.07 m **53.** 1.22 **55.** (a) 49°; (b) 29° **57.** (a) cover the center of each face with an opaque disk of radius 4.5 mm; (b) about 0.63 **59.** (a) $\sqrt{1 + \sin^2\theta}$; (b) $\sqrt{2}$; (c) light emerges at the right; (d) no light emerges at the right **61.** 49.0° **63.** (a) 15 m/s; (b) 8.7 m/s; (c) higher; (d) 72° **65.** 1.0 **N1.** 89 cm **N3.** (a) 3.5 μW/m^2; (b) 0.78 μW; (c) 1.5×10^{-17} W/m^2; (d) 110 nV/m; (e) 0.25 fT **N7.** $I(2 - frac)/c$ **N9.** $p_r(\theta) = p_{r\perp} \cos^2\theta$ **N11.** 35° **N13.** 0.21 **N15.** 4.7×10^7 W/m^2 **N17.** 0.031 **N19.** 19.6° or 70.4° (= 90° − 19.6°) **N21.** (a) 26.8°; (b) yes **N23.** (a) 35.1°; (b) 49.9°; (c) 35.1°; (d) 26.1°; (e) 60.7°; (f) 35.3° **N25.** 3.44×10^6 T/s **N27.** (a) 0.33°; (b) 0° **N29.** 0.50 W/m^2 **N31.** (a) z axis; (b) 7.5×10^{14} Hz (c) 1.9 kW/m^2

CHAPTER 35

CP **1.** 0.2d, 1.8d, 2.2d **2.** (a) real; (b) inverted; (c) same **3.** (a) e; (b) virtual, same **4.** virtual, same as object, diverging **Q** **1.** c **3.** (a) a and c; (b) three times; (c) you **5.** convex **7.** (a) decrease; (b) increase; (c) increase **9.** (a) all but variation 2; (b) for 1, 3, and 4: right, inverted; for 5 and 6: left, same **EP** **1.** 40 cm **3.** (a) 3 **7.** new illumination is 10/9 of the old **9.** 10.5 cm **13.** (a) 2.00; (b) none **17.** $i = -12$ cm **19.** 45 mm, 90 mm **23.** 22 cm **27.** same orientation, virtual, 30 cm to the left of the second lens; $m = 1$ **33.** (a) 13.0 cm; (b) 5.23 cm; (c) −3.25; (d) 3.13; (e) −10.2 **35.** (a) 2.35 cm; (b) decrease **37.** (a) 5.3 cm; (b) 3.0 mm **N1.** −2.5 **N3.** +36 cm, −2.0, R, I, same **N5.** −4.4 cm, +0.56, V, NI, opposite **N7.** −72 cm, +3.0, V, NI, opposite **N9.** −14 cm, +0.61, V, NI, opposite **N11.** ±∞, ±∞, either, either, either (ambiguous situation) **N13.** concave, $f = +8.6$ cm, $i = +12$ cm, $m = -0.40$, R, same **N15.** convex, $p = +30$ cm, $m = +0.50$, V, NI, opposite **N17.** concave, $f = +20$ cm, $i = +30$ cm, $m = -0.50$, R, I **N19.** −54 cm, V, NI, same **N21.** +20, R, I, opposite **N23.** +80 cm, R, I, opposite **N25.** +5.3 cm, −0.33, R, I, opposite **N27.** −3.8 cm, +0.38, V, NI, same **N29.** −88 cm, +3.5, V, NI, same **N31.** −8.7 cm, +0.72, V, NI, same **N33.** +84 cm, −1.4, R, I, opposite **N35.** −18 cm, +0.76, V, NI, same **N37.** −30 cm, +0.86, V, NI, same **N39.** +55 cm, −0.74, R, I, opposite **N41.** −2.5 **N43.** converging, $f = +20$ cm, $i = -13$ cm, $m = +1.7$, V, NI, same **N45.** diverging, $f = -5.3$ cm, $i = -4.0$ cm, V, NI, same **N47.** converging, $f = +80$ cm, $i = -20$ cm, V, NI, same **N49.** +24 cm, +6.0, R, NI, opposite **N51.** +3.1 cm, −0.31, R, I, opposite **N53.** −4.6 cm, +0.69, V, NI, same **N55.** −5.5 cm, +0.12, V, NI, same **N57.** $i = (0.5)(2 - n)r/(n - 1)$, to the right of the right side of the sphere **N61.** (a) $\alpha = 0.500$ rad: 7.799 cm; $\alpha = 0.100$ rad: 8.544 cm; $\alpha = 0.0100$ rad: 8.571 cm; mirror equation: 8.571 cm; (b) $\alpha = 0.500$ rad: −13.56 cm; $\alpha = 0.100$ rad: −12.05 cm; $\alpha = 0.0100$ rad: −12.00 cm; mirror equation: −12.00 cm **N63.** +10 cm, +0.75, R, NI, opposite **N65.** −4.0 cm, −1.2, V, I, same **N67.** −5.2 cm, +0.29, V, NI, same

CHAPTER 36

CP **1.** b (least n), c, a **2.** (a) top; (b) bright intermediate illumination (phase difference is 2.1 wavelengths) **3.** (a) 3λ, 3; (b) 2.5λ, 2.5 **4.** a and d tie (amplitude of resultant wave is $4E_0$), then b and c tie (amplitude of resultant wave is $2E_0$) **5.** (a) 1 and 4; (b) 1 and 4 **Q** **1.** a, c, b **3.** (a) 300 nm; (b) exactly out of phase **5.** (a) intermediate closer to maximum, $m = 2$; (b) minimum, $m = 3$; (c) intermediate closer to maximum, $m = 2$; (d) maximum, $m = 1$ **7.** (a)–(c) decrease; (d) blue **9.** (a) maximum; (b) minimum; (c) alternates **11.** (a) 0.5 wavelength; (b) 1 wavelength **EP** **1.** (a) 5.09×10^{14} Hz; (b) 388 nm; (c) 1.97×10^8 m/s **3.** 1.56 **5.** 22°, refraction reduces θ **7.** (a) 3.60 μm; (b) intermediate, closer to fully constructive interference **9.** (a) 0.833; (b) intermediate, closer to fully constructive interference **11.** (a) 0.216 rad; (b) 12.4° **13.** 2.25 mm **15.** 648 nm **17.** 16 **19.** 0.072 mm **21.** 6.64 μm **23.** 2.65 **25.** $y = 27 \sin(\omega t + 8.5°)$ **27.** (a) 1.17 m, 3.00 m, 7.50 m; (b) no **29.** $I = \frac{1}{9}I_m[1 + 8\cos^2(\pi d \sin\theta/\lambda)]$, I_m = intensity of central maximum **31.** fully constructively **33.** 0.117 μm, 0.352 μm **35.** 70.0 nm **37.** 120 nm **39.** (a) 552 nm; (b) 442 nm **43.** 140 **45.** 1.89 μm **47.** 2.4 μm **49.** $\sqrt{(m + \frac{1}{2})\lambda R}$, for $m = 0, 1, 2, \ldots$ **51.** 1.00 m **53.** $x = (D/2a)(m + \frac{1}{2})\lambda$, for $m = 0, 1, 2, \ldots$ **55.** 588 nm **57.** 1.00030 **59.** (a) 0; (b) fully constructive; (c) increase;

(d)

Phase Difference	Position x (μm)	Type
0	≈∞	fc
0.50λ	7.88	fd
1.00λ	3.75	fc
1.50λ	2.29	fd
2.00λ	1.50	fc
2.50λ	0.975	fd

N1. (a) four; (b) three **N5.** the time is longer for the pipeline containing air, by about 1.56 ns **N11.** 131 nm **N13.** 1.89 μm **N15.** 411.4°, 51.4° **N17.** 11 **N19.** (a) 425 nm, 567 nm; (b) longer **N21.** (a) dark; (b) dark; (c) four **N23.** (a) 52.50 nm; (b) 157.5 nm **N25.** 450 nm **N27.** (a) 1.8; (b) 2.2; (c) 1.25 wavelengths **N29.** 0.14 ps **N31.** 3.5 μm

CHAPTER 37

CP **1.** (a) expand; (b) expand **2.** (a) second side maximum; (b) 2.5 **3.** (a) red; (b) violet **4.** diminish **5.** (a) increase; (b) same **6.** (a) left; (b) less **Q** **1.** (a) contract; (b) contract **3.** with megaphone (larger opening, less diffraction) **5.** four **7.** (a) less; (b) greater; (c) greater **9.** (a) decrease; (b) decrease; (c) to the right **11.** (a) increase; (b) first order **EP** **1.** 60.4 μm **3.** (a) $\lambda_a = 2\lambda_b$; (b) coincidences occur when $m_b = 2m_a$ **5.** (a) 70 cm; (b) 1.0 mm **7.** 1.77 mm **11.** (d) 53°, 10°, 5.1° **13.** (b) 0, 4.493 rad, etc.; (c) −0.50, 0.93, etc. **15.** (a) 1.3×10^{-4} rad; (b) 10 km **17.** 50 m **19.** (a) 1.1×10^4 km; (b) 11 km **21.** 27 cm **23.** (a) 0.347°; (b) 0.97° **25.** (a) 8.7×10^{-7} rad; (b) 8.4×10^7 km; (c) 0.025 mm **27.** five **29.** (a) 4; (b) every fourth bright fringe **31.** (a) nine; (b) 0.255 **33.** (a) 3.33 μm; (b) 0, ±10.2°, ±20.7°, ±32.0°, ±45.0°, ±62.2° **35.** three **37.** (a) 6.0 μm; (b) 1.5 μm; (c) m = 0, 1, 2, 3, 5, 6, 7, 9 **39.** 1100 **47.** 3650 **53.** 0.26 nm **55.** 39.8 pm **59.** (a) $a_0/\sqrt{2}$, $a_0/\sqrt{5}$, $a_0/\sqrt{10}$, $a_0/\sqrt{13}$, $a_0/\sqrt{17}$ **61.** 30.6°, 15.3° (clockwise); 3.08°, 37.8° (counterclockwise) **63.** (a) 50 m; (b) no, the width of 10 m is too narrow to resolve; (c) not during daylight, but the light pollution during the night would be a sure sign **N1.** 30.5 μm **N3.** (a) 17.1 m; (b) 1.37×10^{-10} **N5.** 36 cm **N7.** two **N11.** 1.36×10^4 **N19.** 6.8° **N21.** (a) 13; (b) 6 **N23.** 59.5 pm **N27.** 53.4 cm

CHAPTER 38

CP **1.** (a) same (speed of light postulate); (b) no (the start and end of the flight are spatially separated); (c) no (because his measurement is not a proper time) **2.** (a) Sally's; (b) Sally's **3.** a, positive; b, negative; c, positive **4.** (a) right; (b) more **5.** (a) equal; (b) less **Q** **1.** all tie (pulse speed is c) **3.** (a) C_1; (b) C_1 **5.** (a) negative; (b) positive **7.** less **9.** b, a, c, d **EP** **1.** (a) 6.7×10^{-10} s; (b) 2.2×10^{-18} m **3.** 0.99c **5.** 0.445 ps **7.** 1.32 m **9.** 0.63 m **11.** (a) 87.4 m; (b) 394 ns **13.** (a) 26 y; (b) 52 y; (c) 3.7 y **15.** x' = 138 km, t' = −374 μs **17.** (a) 25.8 μs; (b) small flash **19.** (a) 1.25; (b) 0.800 μs **21.** 0.81c **23.** (a) 0.35c; (b) 0.62c **25.** 1.2 μs **27.** 22.9 MHz **29.** 1×10^6 m/s, receding **31.** yellow (550 nm) **33.** (a) 0.0625, 1.00196; (b) 0.941, 2.96; (c) 0.999 999 87, 1960 **35.** 0.999 987c **37.** 18 smu/y **39.** (a) 0.707c; (b) 1.41; (c) 0.414mc^2 **41.** $\sqrt{8}mc$ **43.** 1.01×10^7 km, or 250 Earth circumferences **45.** 110 km **47.** 4.00 u, probably a helium nucleus **49.** 330 mT **51.** (a) 2.08 MeV; (b) −1.18 MeV **53.** (a) $vt \sin\theta$; (b) $t[1 - (v/c)\cos\theta]$; (c) 3.24c **N1.** (b) 0.750 **N3.** 0.999 90c **N5.** (a) 1.2×10^8 N; (b) train; (c) 25 N; (d) backpack **N7.** (a) 6.26c; (b) 0.999c **N9.** (a) $mv^2/2 + 3mv^4/8c^2$; (b) 1.0×10^{-16} J; (c) 1.9×10^{-19} J; (d) 2.6×10^{-14} J; (e) 1.3×10^{-14} J; (f) 0.37c **N13.** (b) +0.44c **N15.** (a) −0.36c; (b) −c **N17.** 40 s

CHAPTER 39

CP **1.** b, a, d, c **2.** (a) lithium, sodium, potassium, cesium; (b) all tie **3.** (a) same; (b)–(d) x rays **4.** (a) proton; (b) same; (c) proton **5.** same **Q** **1.** (a) microwave; (b) x ray; (c) x ray **3.** potassium **5.** positive charge builds up on the plate, inhibiting further electron emission **7.** none **9.** (a) greater; (b) less **11.** no essential change **13.** (a) decreases by a factor of $1/\sqrt{2}$; (b) decreases by a factor of 1/2 **15.** (a) decreasing; (b) increasing; (c) same; (d) same **17.** a **19.** (a) zero; (b) yes **EP** **1.** 4.14 eV · fs **5.** 1.0×10^{45} photons/s **7.** 5.9 μeV **9.** 2.047 eV **11.** 4.7×10^{26} photons **13.** (a) infrared lamp; (b) 1.4×10^{21} photons/s **15.** (a) 2.96×10^{20} photons/s; (b) 48 600 km; (c) 5.89×10^{18} photons/m² · s **17.** barium and lithium **19.** 170 nm **21.** 676 km/s **23.** (a) 2.00 eV; (b) 0; (c) 2.00 eV; (d) 295 nm **25.** 233 nm **27.** (a) 382 nm; (b) 1.82 eV **29.** 9.68×10^{-20} A **31.** (a) 2.7 pm; (b) 6.05 pm **33.** (a) 8.57×10^{18} Hz; (b) 35.4 keV; (c) 1.89×10^{-23} kg · m/s = 35.4 keV/c **37.** (a) 2.43 pm; (b) 1.32 fm; (c) 0.511 MeV; (d) 938 MeV **39.** 300% **43.** (a) 41.8 keV; (b) 8.2 keV **45.** 1.12 keV **47.** 44° **51.** 7.75 pm **53.** 4.3 μeV **55.** (a) 38.8 meV; (b) 146 pm **57.** (a) photon: 1.24 μm; electron: 1.22 nm; (b) 1.24 fm for each **59.** (a) 1.9×10^{-21} kg · m/s; (b) 346 fm **61.** 0.025 fm, about 200 times smaller than a nuclear radius **63.** neutron **65.** 9.70 kV (relativistic calculation), 9.76 kV (classical calculation) **73.** (d) $x = n(\lambda/2)$, where n = 0, 1, 2, 3, . . . **75.** 0.19 m **79.** (a) proton: 9.02×10^{-6}, deuteron: 7.33×10^{-8}; (b) 3.0 MeV for each; (c) 3.0 MeV for each **81.** (a) −20%; (b) −10%; (c) +15% **83.** $T = 10^{-x}$, where $x = 7.2 \times 10^{39}$ (T is very small)

CHAPTER 40

CP **1.** b, a, c **2.** (a) all tie; (b) a, b, c **3.** a, b, c, d **4.** $E_{1,1}$ (neither n_x nor n_y can be zero) **5.** (a) 5; (b) 7 **Q** **1.** (a) 1/4; (b) same factor **3.** c **5.** (a) $(\sqrt{1/L})\sin(\pi/2L)x$; (b) $(\sqrt{4/L})\sin(2\pi/L)x$; (c) $(\sqrt{2/L})\cos(\pi/L)x$ **7.** less **9.** (a) wider; (b) deeper **11.** n = 1, n = 2, n = 3 **13.** b, c, and d **15.** (a) first Lyman plus first Balmer; (b) Lyman series limit minus Paschen series limit **EP** **1.** (a) 37.7 eV; (b) 0.0206 eV **3.** 1900 MeV **5.** 0.020 eV **7.** 90.3 eV **11.** 68.7 nm, 25.8 nm, 13.7 nm, and 8.59 nm **13.** (a) 1.3×10^{-19} eV; (b) about $n = 1.2 \times 10^{19}$; (c) 0.95 J = 5.9×10^{18} eV; (d) yes **15.** (b) no; (c) no; (d) yes **17.** (a) 0.050; (b) 0.10; (c) 0.0095 **19.** 59 eV **21.** (b) $k = (2\pi/h)[2m(U_0 - E)]^{1/2}$ **25.** 3.08 eV **27.** 0.75, 1.00, 1.25, 1.75, 2.00, 2.25, 3.00, 3.75 **29.** 1.00, 2.00, 3.00, 5.00, 6.00, 8.00, 9.00 **31.** 2.6 eV **33.** 4.0 **35.** (a) 12 eV; (b) 6.5×10^{-27} kg · m/s; (c) 103 nm **39.** (a) 0; (b) 10.2 nm⁻¹; (c) 5.54 nm⁻¹ **41.** (a) 13.6 eV; (b) 3.40 eV **43.** (a) n = 4 to n = 2; (b) Balmer series **45.** (a) 13.6 eV; (b) −27.2 eV **47.** (a) 2.6 eV; (b) n = 4 to n = 2 **49.** 0.68 **55.** (a) 0.0037; (b) 0.0054 **59.** (a) $P_{210} = (r^4/8a^5)e^{-r/a}\cos^2\theta$; $P_{21+1} = P_{21-1} = (r^4/16a^5)e^{-r/a}\sin^2\theta$

CHAPTER 41

CP **1.** 7 **2.** (a) decrease; (b)–(c) remain the same

3. less **4.** A, C, B **Q** **1.** 0, 2, and 3 **3.** $6p$ **5.** (a) 2, 8; (b) 5, 50 **7.** (a) n; (b) n and l **9.** a, c, e, f **11.** (a) unchanged; (b) decrease; (c) decrease **13.** a and b **EP** **3.** (a) 3; (b) 3 **5.** (a) 32; (b) 2; (c) 18; (d) 8 **7.** 24.1° **9.** $n > 3$; $m_l = +3, +2, +1, 0, -1, -2, -3$; $m_s = \pm\frac{1}{2}$ **11.** (a) $\sqrt{12}\hbar$; (b) $\sqrt{12}\mu_B$;

(c)

m_l	L_z	$\mu_{orb,z}$	θ
-3	$-3\hbar$	$+3\mu_B$	150°
-2	$-2\hbar$	$+2\mu_B$	125°
-1	$-\hbar$	$+\mu_B$	107°
0	0	0	90°
$+1$	$+\hbar$	$-\mu_B$	73.2°
$+2$	$+2\hbar$	$-2\mu_B$	54.7°
$+3$	$+3\hbar$	$-3\mu_B$	30.0°

15. 54.7° and 125° **17.** 73 km/s² **19.** 5.35 cm **21.** (a) 2.13 meV; (b) 18 T **23.** $44(h^2/8mL^2)$ **25.** (a) $51(h^2/8mL^2)$; (b) $53(h^2/8mL^2)$; (c) $56(h^2/8mL^2)$ **27.** $42(h^2/8mL^2)$ **31.** argon **33.** (a) $(n, l, m_l, m_s) = (2, 0, 0, \pm\frac{1}{2})$; (b) $n = 2, l = 1, m_l = 1, 0$, or -1, $m_s = \pm\frac{1}{2}$ **39.** 49.6 pm, 99.2 pm **43.** (a) 35.4 pm, as for molybdenum; (b) 57 pm; (c) 50 pm **45.** 9/16 **49.** (a) 69.5 kV; (b) 17.9 pm; (c) K_α: 21.4 pm, K_β: 18.5 pm **51.** (a) $(Z - 1)^2/(Z' - 1)^2$; (b) 57.5; (c) 2070 **53.** (a) 6; (b) 3.2×10^6 years **55.** 9.1×10^{-7} **57.** 10 000 K **59.** (a) 3.60 mm; (b) 5.25×10^{17} **61.** 4.7 km **63.** 2.0×10^{16} s^{-1} **65.** (a) 3.03×10^5; (b) 1430 MHz; (d) 3.30×10^{-6} **67.** (a) no; (b) 140 nm **69.** (a) 4.3 μm; (b) 10 μm; (c) infrared

CHAPTER 42

CP **1.** (a) larger; (b) same **2.** Cleveland, metal; Boca Raton, none; Seattle, semiconductor **3.** a, b, and c **4.** b **Q** **1.** 4 **3.** b and c, yes **5.** (a) anywhere in the lattice; (b) in any silicon–silicon bond; (c) in a silicon ion core, at a lattice site **7.** b and d **9.** $+4e$ **11.** none **13.** (a) right to left; (b) back bias **15.** a, b, and c **EP** **1.** 8.49×10^{28} m^{-3} **3.** 3490 atm **5.** (a) $+8.0 \times 10^{-11}$ Ω · m/K; (b) -210 Ω · m/K **7.** (b) 6.81×10^{27} m^{-3} eV$^{-3/2}$; (c) 1.52×10^{28} m^{-3} eV^{-1} **9.** (a) 0; (b) 0.0955 **13.** 0.91 **15.** (a) 2500 K; (b) 5300 K **17.** (a) 90.0%; (b) 12.5%; (c) sodium **19.** (a) 2.7×10^{25} m^{-3}; (b) 8.43×10^{28} m^{-3}; (c) 3100; (d) molecules: 3.3 nm; electrons: 0.228 nm **21.** (a) 1.0, 0.99, 0.50, 0.014, 2.5×10^{-17}; (b) 700 K **23.** 3 **25.** (a) 5.86×10^{28} m^{-3}; (b) 5.52 eV; (c) 1390 km/s; (d) 0.522 nm **27.** (b) 1.80×10^{28} m^{-3} eV^{-1} **31.** (a) 19.8 kJ; (b) 197 s **33.** 200° C **35.** (a) 225 nm; (b) ultraviolet **37.** (a) 109.5°; (b) 235 pm **41.** 0.22 μg **43.** (a) pure: 4.78×10^{-10}; doped: 0.0141; (b) 0.824 **45.** 6.02×10^5 **47.** 4.20 eV **49.** (a) 5.0×10^{-17} F; (b) about $300e$

CHAPTER 43

CP **1.** ^{90}As and ^{158}Nd **2.** a little more than 75 Bq (elapsed time is a little less than three half-lives) **3.** ^{206}Pb **Q** **1.** less **3.** ^{240}U **5.** less **7.** (a) on the $N = Z$ line; (b) positrons; (c) about 120 **9.** no **11.** yes **13.** (a) increases; (b) remains the same **15.** 7 h **17.** d **EP** **1.** 28.3 MeV **3.** (a) 0.390 MeV; (b) 4.61 MeV **7.** (a) six; (b) eight **11.** (a) 1150 MeV; (b) 4.81 MeV/nucleon, 12.2 MeV/proton **15.** (a) 6.2 fm; (b) yes **17.** $K \approx 30$ MeV **21.** ^{25}Mg: 9.303%; ^{26}Mg: 11.71% **23.** 1.6×10^{25} MeV **25.** 7.92 MeV **27.** 280 d **29.** (a) 7.6×10^{16} s^{-1}; (b) 4.9×10^{16} s^{-1} **31.** (a) 64.2 h; (b) 0.125; (c) 0.0749 **33.** 5.3×10^{22} **35.** (a) 2.0×10^{20}; (b) 2.8×10^9 s^{-1} **37.** 209 d **39.** 1.13×10^{11} y **43.** (a) 8.88×10^{10} s^{-1}; (b) 8.88×10^{10} s^{-1}; (c) 1.19×10^{15}; (d) 0.111 μg **45.** 730 cm² **47.** Pu: 1.2×10^{-17}, Cm: $e^{-9173} \approx 0$ **49.** 4.269 MeV **51.** (a) 31.8 MeV, 5.98 MeV; (b) 86 MeV **53.** ^{7}Li **55.** 1.21 MeV **57.** 0.782 MeV **59.** (b) 0.961 MeV **61.** 78.4 eV **63.** (a) U: 1.06×10^{19}, Pb: 0.624×10^{19}; (b) 1.69×10^{19}; (c) 2.98×10^9 y **65.** 1.8 mg **67.** 1.02 mg **69.** 13 mJ **71.** (a) 6.3×10^{18}; (b) 2.5×10^{11}; (c) 0.20 J; (d) 2.3 mGy; (e) 30 mSv **73.** (a) 6.6 MeV; (b) no **75.** (a) 25.4 MeV; (b) 12.8 MeV; (c) 25.0 MeV **77.** 0.49 **79.** (a) beta-minus decay; (b) 8.2×10^7; (c) 1.2×10^6 **81.** 3.2×10^{12} Bq = 86 Ci **83.** 4.28×10^9 y **85.** 1.3×10^{-13} m **87.** 3.2×10^4 y

CHAPTER 44

CP **1.** c and d **2.** (a) no; (b) yes; (c) no **3.** e **Q** **1.** a **3.** b **5.** (a) ^{93}Sr; (b) ^{140}I; (c) ^{155}Nd **7.** c **9.** a **11.** c **EP** **1.** (a) 2.6×10^{24}; (b) 8.2×10^{13} J; (c) 2.6×10^4 y **3.** 3.1×10^{10} s^{-1} **7.** -23.0 MeV **9.** 181 MeV **11.** (a) ^{153}Nd; (b) 110 MeV to ^{83}Ge, 60 MeV to ^{153}Nd; (c) 1.6×10^7 m/s for ^{83}Ge, 8.7×10^6 m/s for ^{153}Nd **13.** (a) 252 MeV; (b) typical fission energy is 200 MeV **15.** 461 kg **17.** yes **19.** 557 W **21.** $^{238}\text{U} + \text{n} \rightarrow {}^{239}\text{U} \rightarrow {}^{239}\text{Np} + \text{e}$, $^{239}\text{Np} \rightarrow {}^{239}\text{Pu} + \text{e}$ **23.** (a) 84 kg; (b) 1.7×10^{25}; (c) 1.3×10^{25} **25.** 0.99938 **27.** (b) 1.0, 0.89, 0.28, 0.019; (c) 8 **29.** (a) 75 kW; (b) 5800 kg **31.** 1.7×10^9 y **33.** 170 keV **35.** (a) 170 kV **37.** 0.151 **41.** (a) 3.1×10^{31} protons/m³; (b) 1.2×10^6 times **43.** (a) 4.3×10^9 kg/s; (b) 3.1×10^{-4} **45.** (a) 1.83×10^{38} s^{-1}; (b) 8.25×10^{28} s^{-1} **47.** (a) 4.1 eV/atom; (b) 9.0 MJ/kg; (c) 1500 y **49.** 1.6×10^8 y **51.** (a) 24.9 MeV; (b) 8.65 megatons **53.** 14.4 kW

CHAPTER 45

CP **1.** (a) the muon family; (b) a particle; (c) $L_\mu = +1$ **2.** b and e **3.** c **Q** **1.** d **3.** the leftmost π^+ pion whose track curves downward **5.** a, b, c, d **7.** c, f **9.** 1d, 2e, 3a, 4b, 5c **11.** 1b, 2c, 3d, 4e, 5a **13.** (a) 0; (b) +1; (c) -1; (d) +1; (e) -1 **EP** **1.** 6.03×10^{-29} kg **3.** 18.4 fm **5.** 1.08×10^{42} J **7.** 2.7 cm/s **9.** 769 MeV **13.** (a) L_e, spin angular momentum; (b) L_μ, charge; (c) energy, L_μ **15.** $q = 0, B = -1, S = 0$ **17.** (a) energy; (b) strangeness; (c) charge **19.** 338 MeV **21.** (a) K^+; (b) $\bar{n}$; (c) K^0 **23.** (a) $\overline{\text{u}}\overline{\text{u}}\overline{\text{d}}$; (b) $\overline{\text{u}}$dd **25.** (a) not possible; (b) uuu **29.** Σ^0, 7530 km/s **31.** 666 nm **33.** (b) 4.5 H-atoms/m³

35. (a) 256 μeV; (b) 4.84 mm **37.** (a) 122 m/s; (b) 246 y **39.** (b) 2.38×10^9 K **41.** (a) $0.785c$; (b) $0.993c$; (c) C2; (d) C1; (e) 51 ns; (f) 40 ns **43.** (c) $r\alpha/c + (r\alpha/c)^2 + (r\alpha/c)^3 + \cdots$; (d) $\Delta\lambda/\lambda = r\alpha/c$; (e) $\alpha = H$; (f) 7.4×10^8 ly; (g) 7.8×10^8 y; (h) 7.4×10^8 y; (i) 7.8×10^8 ly; (j) 1.2×10^9 ly; (k) 1.2×10^9 y; (l) 4.4×10^8 ly

PHOTO CREDITS

CHAPTER 22 Page 505: Michael Watson. Page 506: ©Fundamental Photographs. Page 507: Courtesy Xerox Corporation. Page 508: Johann Gabriel Doppelmayr, *Neuentdeckte Phaenomena von Bewünderswurdigen Würckungen der Natur*, Nuremberg, 1744. Page 516: Courtesy Lawrence Berkeley Laboratory.

CHAPTER 23 Page 520: Tsuyoshi Nishiinoue/Orion Press. Page 534: Russ Kinne/Comstock, Inc.

CHAPTER 24 Page 543: Ralph H. Wetmore II/Tony Stone Images/New York, Inc. Page 554 (left): ©C. Johnny Autery. Page 554 (right): Courtesy E. Philip Krider, Institute for Atmospheric Physics, University of Arizona, Tucson.

CHAPTER 25 Pages 564 and 569: Courtesy NOAA. Page 581: Courtesy Westinghouse Corporation.

CHAPTER 26 Page 588: Bruce Ayres/Tony Stone Images/New York, Inc. Page 589: Paul Silvermann/Fundamental Photographs. Page 600: ©Harold & Ester Edgerton Foundation, 1999, courtesy of Palm Press, Inc. Page 601: Courtesy The Royal Institute, England.

CHAPTER 27 Page 611: ©UPI/Corbis Images. Page 617: The Image Works. Page 625: ©Laurie Rubin. Page 627: Courtesy Shoji Tonaka, International Superconductivity Technology Center, Tokyo, Japan.

CHAPTER 28 Page 633: Hans Reinhard/Bruce Coleman, Inc. Page 634: Courtesy Southern California Edison Company.

CHAPTER 29 Page 658: Johnny Johnson/Tony Stone Images/New York, Inc. Page 659: Ray Pfortner/Peter Arnold, Inc. Page 661: Lawrence Berkeley Laboratory/Photo Researchers. Page 662: Courtesy Dr. Richard Cannon, Southeast Missouri State University, Cape Girardeau. Page 668: Courtesy John Le P. Webb, Sussex University, England. Page 670: Courtesy Dr. L. A. Frank, University of Iowa.

CHAPTER 30 Page 686: Michael Brown/Florida Today/Gamma Liaison. Page 688: Courtesy Education Development Center.

CHAPTER 31 Page 710: Dan McCoy/Black Star. Page 715: Courtesy Fender Musical Instruments Corporation. Page 725: Courtesy The Royal Institute, England.

CHAPTER 32 Page 744: Courtesy A. K. Geim, High Field Magnet Laboratory, University of Nijmegen, The Netherlands. Page 745: Runk/Schoenberger/Grant Heilman Photography. Page 753: Peter Lerman. Page 756: Courtesy Ralph W. DeBlois.

CHAPTER 33 Page 768: Photo by Rick Diaz, provided courtesy Haverfield Helicopter Co. Page 771: Courtesy Agilent Technologies. Page 792: Ted Cowell/Black Star.

CHAPTER 34 Page 801: John Chumack/Photo Researchers. Page 816: Diane Schiumo/Fundamental Photographs. Page 818: *PSSC Physics*, 2nd edition; ©1975 D. C. Heath and Co. with Education Development Center, Newton, MA. Reproduced with permission of Education Development Center. Page 821 (top): Courtesy Bausch & Lomb. Page 821 (bottom): Barbara Filet/Tony Stone Images/New York, Inc. Page 823: Greg Pease/Tony Stone Images/New York, Inc. Page 828: Courtesy Cornell University.

CHAPTER 35 Page 833: Courtesy Courtauld Institute Galleries, London. Page 842: Dr. Paul A. Zahl/Photo Researchers. Page 845: Courtesy Matthew J. Wheeler. Page 856: Piergiorgio Scharandis/Black Star.

CHAPTER 36 Page 861: David Julian/Phototake. Page 866: Runk Schoenberger/Grant Heilman Photography. Page 868: From *Atlas of Optical Phenomena* by M. Cagnet et al., Springer-Verlag, Prentice Hall, 1962. Page 878: Richard Megna/Fundamental Photographs. Page 887: Courtesy Bausch & Lomb.

CHAPTER 37 Page 890: Georges Seurat, *A Sunday on La Grande Jatte*, 1884; oil on canvas (207.5 × 308 cm). Helen Birch Bartlett Memorial Collection, 1926; photograph ©1996, The Art Institute of Chicago. All rights reserved. Page 891: Ken Kay/Fundamental Photographs. Pages 892, 898, and 902: From *Atlas of Optical Phenomena* by Cagnet, Francon, Thierr, Springer-Verlag, Berlin, 1962. Reproduced with permission. Page 900: Warren Rosenberg/BPS/Tony Stone Images/New York, Inc. Page 906: Department of Physics, Imperial College/Science Photo Library/Photo Researchers. Page 907: Kristen Brochmann/Fundamental Photographs. Page 915 (left): Kjell B. Sandved/Bruce Coleman, Inc. Page 915 (right): Pekka Parvianen/Photo Researchers. Page 918: AP/Wide World Photos.

CHAPTER 38 Page 919: Jeffrey Zaruba/Tony Stone Images/New York, Inc. Page 920: Courtesy of the Albert Einstein Archives, The Jewish National & University Library, The Hebrew University of Jerusalem, Israel. Page 940: Courtesy NASA.

CHAPTER 39 Page 953: Lawrence Berkeley Laboratory/Science Photo Library/Photo Researchers. Page 965: Courtesy A. Tonomura, J. Endo, T. Matsuda, and T. Kawasaki/Advanced Research Laboratory, Hitachi, Ltd., Kokubinju, Tokyo; H. Ezawa, Department of Physics, Gakushuin University, Mejiro, Tokyo. Page 966 (left): Courtesy Riber Division of Instruments, Inc. Page 966 (right): From PSSC film "Matter Waves," courtesy Education Development Center, Newton, Massachusetts. Page 972: ©IBMRL/Visuals Unlimited.

CHAPTER 40 Page 979: Courtesy International Business Machines Corporation, Almaden Research Center, CA. Page 990: From *Scientific American*, January 1993, page 122. Reproduced with permission of Michael Steigerwald, Bell Labs–Lucent Technologies. Page 991: From *Scientific American,* September 1995, page 67. Image reproduced with permission of H. Temkin, Texas Tech University. Page 995: From W. Finkelnburg, *Structure of Matter,* Springer-Verlag, 1964. Reproduced with permission.

CHAPTER 41 Page 1006: Larry Mulvehill/Photo Researchers. Page 1007: Courtesy Warren Nagourney. Page 1016: David Job/Tony Stone Images/New York, Inc. Page 1027: Michael Rosenfeld/Tony Stone Images/New York, Inc.

CHAPTER 42 Page 1037: Steve Northrup/New York Times Pictures. Page 1054: Courtesy AT&T. Page 1056: Courtesy Intel.

CHAPTER 43 Page 1062: Elscint/Science Photo Library/Photo Researchers. Page 1079 (top inset): R. Perry/Corbis Sygma. Page 1079 (top): George Rockwin/Bruce Coleman, Inc.

CHAPTER 44 Pages 1092 and 1101: Courtesy U.S. Department of Energy. Page 1099: V. Ivelva/Magnum Photos, Inc. Page 1102: Gary Sheehan, *Birth of the Atomic Age,* 1957. Reproduced courtesy Chicago Historical Society. Page 1108: Courtesy Anglo Australian Telescope Board. Page 1110 (top): Courtesy Princeton University Physics Laboratory. Page 1110 (center): Courtesy Los Alamos National Laboratory, New Mexico. Page 1113: Courtesy Martin Marietta Energy Systems/ U.S. Department of Energy.

CHAPTER 45 Page 1116: Courtesy NASA. Page 1118: David Parker/ Photo Researchers. Page 1119: Courtesy Michael Mathews. Page 1120: Courtesy Lawrence Berkeley Laboratory.

INDEX

Figures are noted by page numbers in *italics;* tables are indicated by t following the page number.

D

F

G

X

Y

Z

Problem Solving Tactics